HUMAN AND
ECOLOGICAL RISK
ASSESSMENT

HUMAN AND ECOLOGICAL RISK ASSESSMENT

Theory and Practice

Edited by

Dennis J. Paustenbach

WILEY-INTERSCIENCE

A JOHN WILEY & SONS, INC., PUBLICATION

For ordering and customer service, call 1-800-CALL-WILEY.

Library of Congress Cataloging-in-Publication Data is available

ISBN 0-471-14747-8

Printed in the United States of America.

10 9 8 7 6

CONTENTS

FOREWORD

EIGHT LESSONS FROM THE REAL WORLD OF RISK ASSESSMENT

"Life is about risk, and it ends badly," Senator Daniel Patrick Moynihan once observed wryly in response to my testimony as U.S. Environmental Protection Agency (EPA) Administrator about risk assessment and the regulatory process. The always thoughtful, often provocative New York senator intended by his observation to remind me to keep things in perspective, not to be defensive about failing to ensure zero risk in environmental regulations. The function of risk assessment, Moynihan believed, is not to drive risks to zero, which would rarely be possible, but to illuminate choices, costs, and priorities. The first discovery in applying risk assessment to the real world is that zero risk is not a prudent objective of policy. Moynihan understood that. But the position is not obvious.

A regulator wishes to provide maximum protection when formulating a new rule. But the very enterprise of honestly assessing, even to the point of quantifying, risks is a confession of limited expectations. Someone still may become ill or even die. Better that the rule protect life and do so unqualifiedly. That, in fact, was the thinking behind the Delaney rule, which prohibited even trace elements in processed foods of substances that had been found, in any quantity, to cause cancer in test animals. As testing became more refined it became more impractical to ensure that no single molecule of such chemicals be present in processed foods. And although Delaney did not apply to fresh fruits and vegetables, moves to eliminate infinitesimally small amounts of carcinogenic chemicals sometimes threatened to raise the costs or reduce the availability of an important source of nutritious foods. In instances where risks can be entirely eliminated, through substituting a different process or chemical for an offending product, the regulator's task is simple. But decisions that reach the EPA Administrator are rarely straightforward and typically involve tradeoffs. "Zero risk" is not ordinarily an option in regulating any more than in other areas of life.

The first of the eight lessons that I learned as administrator was that different federal agencies can come to very different conclusions about risk. It frankly was the biggest and most disturbing of my experiences with risk assessment to learn that the U.S. Food and Drug Administration's regulation of dioxin in the early 1990s was an order of magnitude less conservative than EPA's dioxin standard. The difference had huge economic consequences: if the EPA standard was applied, compliance costs would be billions of dollars more than if the FDA standard were used. When I asked the EPA Science Advisory Board to advise me on which agency's approach was sound, they concluded that both standards had been reached using scientifically accepted methodologies for extrapolating from test animals, one using a method based on the body weight of the rats, the other based on the skin surface of the animals. I found this advice highly disappointing and frustrating. The temptation, if both risk assessments were scientifically valid, was to choose the FDA approach since it entailed the least cost to the economy.

After receiving the scientists' inconclusive judgment about EPA's dioxin standard, and also reading other papers raising serious doubts about our understanding of dioxin, I ordered a full-scale scientific review of dioxin. That review was designed to involve the best governmental and nongovernmental expertise on all aspects of dioxin, and to be fully transparent, peer-reviewed at every stage of the review. It was intended to serve as a model, for all government agencies confronted by expensive and controversial scientific problems and particularly for EPA, which has sometimes been criticized for not opening up its scientific reviews to outside involvement or to public view. That review, only recently completed and released, now stands as perhaps the most extensive and thorough, and longest running, risk assessment ever undertaken. The original impetus for it was my discovery that scientists considered two risk assessment techniques technically valid even though they resulted in widely divergent conclusions. So my second lesson is that differing methodologies result in different conclusions about risk even among experts and in my view make a powerful case for harmonizing protocols in order to retain the confidence of the decisionmakers and the public.

A third lesson, also driven home to me by my encounter with dioxin regulation, was the need to look beyond a simple toxic endpoint such as cancer. One criticism of EPA's early approach to dioxin and other chemicals concerned the agency's focus on cancer in its toxicological characterization. Historically, EPA has given a high priority to cancer in the design of regulations. Some American scientists and many Europeans have criticized what they see as an excessive preoccupation with cancer, in EPA as in the broader society in this country, to the neglect of neurological, fetal, endocrinological, and other impacts. The dioxin study was intended to respond to this criticism by analyzing a broad suite of possible effects on human health.

A regulator must, in fact, balance and weigh various types of risk and illness when making policy. The by-products of chlorination of drinking water expose the public to a lifetime cancer death risk estimated to be one in a hundred thousand, a risk greater than the one-in-a-million range EPA generally prefers. EPA and the broader society accept such a risk because of the much larger offsetting risk posed by cholera, typhus, and various intestinal disorders chlorination protects against. The task for the regulator is to balance the risks and benefits. As assumptions and risk ranges become more transparent, the regulator's decision-making process can seem coolly cerebral, calculating body counts by the numbers. The temptation is to fudge the numbers and not acknowledge the tradeoff that is implicit in a selection of lesser evils. My fourth lesson from my experience is that tradeoffs are unavoidable and that evolving technology and growing transparency will illuminate them more starkly, heightening further the importance of keeping the public's confidence in EPA and other regulatory bodies.

Differing characterizations of risk and varied methodologies for risk assessment are not only found among different agencies. Even within EPA itself, programs take different approaches to risk. In some instances this is inevitable. When regulating for air pollution or pesticide levels in food, for example, the affected public is the entire population. When regulating for Superfund, however, a much smaller population that resides near abandoned hazardous waste dumps is affected. Thus the pesticides' program is concerned with average individual risk, based on assumptions derived from U.S. Agriculture Department studies of how much of different foods Americans actually eat. In contrast, assessments of waste sites might focus on the maximally exposed individual (MEI).

The air program, in setting standards for hazardous air pollutants, bases its decisions on two populations: on aggregate population risk and on maximum individual risk. We used this approach in setting the standard for benzene in 1991, when we set a standard designed to result in no greater than a six-in-ten-thousand risk to the very small percentage of the population that works daily with benzene in coke by-product recovery plants, on benzene storage vessels and the like, and one-in-a-million average lifetime risk to more than 95% of the population. (The Clean Air Act amendments of 1990 required major and expensive changes to coke ovens to make them much safer. I do emphasize that the exposure assumptions are very conservative.) The fifth lesson is that the nature of a problem, its situational reality, the degree to which it affects the larger population or whether it has a more selective impact on certain subsets of the public, justify tailoring the methodology and risk characterization to the specific problem.

A word about exposure assumptions. In 1990, the FDA, implementing a new analytical procedure, was able to detect much smaller concentrations of contaminants than under its previous system. As a result, FDA found residues of the fungicide Procymidon in French, Italian, and a small amount of Spanish wines. Procymidon was widely used in Europe to protect grapes against the fungus botrytis. Botrytis is desirable to concentrate flavor in sauternes, but it's a serious problem for most grape growers. Procymidon had been issued a tolerance or maximum permissible residue level in Europe and Japan, but the manufacturer had never applied for one in the United States, and as long as FDA was not detecting it, there was no concern.

Once we became aware of the presence of an unregistered chemical in imported wine, I ordered a ban on further shipments. The ban caused wine to accumulate in the warehouses and on the docks of European ports. The European Union's (EU's) commissioner for agriculture indicated that more than $400 million of EU exports were being excluded, and there were dark fears that my move was driven by trade and not environmental considerations. Noises were even made about retaliation against U.S. agriculture exports.

EPA pesticides' and toxics' staff determined that based on limited testing that had been done and made public, Procymidon in very large doses had caused cancer in test animals, but that the levels represented in imported wines posed a negligible risk. As soon as we concluded that any risk was negligible, we looked for a means consistent with our laws to allow preexisting stocks of wine to come in.

In the meantime I was visited by the French ambassador who was incredulous that the United States would exclude a half billion dollars' worth of trade when the decision-maker himself and his expert staff considered the threat negligible. The French believe more in a politics of consequence than of process.

The Italian ambassador also paid me a call. He, too, was uncomprehending. "If it's not a risk, why not let it in?" he said. And then he asked about the data from test animals. I told him about the male rats fed high doses of Procymidon who had developed cancer of the testicles. "Of the testicles," he cried. "Italian wine! Nothing could be worse." I had the impression that like a lot of American risk experts he thought we were overdoing it on cancer, but the reproductive organs, that was something much more serious, that got his attention.

Ultimately, we set an interim tolerance for already treated grapes based on available data, and banned imports of all future Procymidon-containing wine pending the full range of tests required to earn registration in the United States. During one meeting with staff on the matter, I ascertained that our initial exposure assumption was based on a consumption over several decades of two liters per day of wine containing Procymidon. I said surely that's unreasonable, no one drinks two liters a day. And one of my staff members quietly replied, "Yes they do; my father does."

There truly is a maximum exposed individual. One may ask fairly whether it is the function of regulators to fashion policies protective of such consumption practices if the cost to the rest of society—to grape growers and wine makers and wine buyers and drinkers—is consequently much higher.

Notice here just how conservative the EPA staff's initial approach was. The two liters containing Procymidon would have had to have come from the 20% of French wine containing Procymidon or the 10% of Italian wine so affected. Letting a little light in on the exposure assumption revealed how unrealistic it was. Nevertheless, the lesson here is that a hugely consequential decision had to be made with partial data. My sixth lession is that much if not most decisionmaking about risk will be tentative and uncertain, valid to the degree it is founded on current science but vulnerable to revision in the light of new research.

Superfund has relied on different exposure assumptions from other EPA programs, though it conducts its risk assessments similarly. The risks it addresses are worst-case, hypothetical present and future risks to the maximum exposed individual, i.e., one who each day consumes two liters of water contaminated by hazardous waste. The program at one time aimed to achieve a risk range in its clean-ups adequate to protect the child who regularly ate 10,000 milligrams of dirt.

And it formerly assumed that all sites, once cleaned up, would be used for residential development, even though many lie within industrial zones. Some of these assumptions have driven clean-up costs to stratospheric levels and, together with liabilities associated with Superfund sites, have resulted in inner-city sites suitable for redevelopment remaining derelict and unproductive. The consequence, in New Jersey and other areas, has been to impose a drag on urban redevelopment in the inner city, and to push new industry to locate in pristine, outlying sites. My seventh lesson is that the resulting loss of property tax revenues, industry, and jobs in many older urbanized areas is itself a kind of pathology our environmental laws should consider, particularly as we have become more sensitive to issues of environmental justice for people of color. Fortunately, during the 1990s, more realistic assumptions about future uses, and consequent changes in cleanup standards, came to be accepted.

An important role of the regulator is to communicate clearly about risks. I once was presented data indicating an unacceptable high residue of the pesticide EBDC in many fruits and vegetables. EBDC at high doses had induced carcinogenic tumors in test animals. I ordered a ban on further applications of this widely used chemical. Scientific staff who briefed me on the available data told me they strongly suspected that further analysis would show that by the time fresh fruits and vegetables passed through supermarkets and reached people's tables there would be negligible or even nondetectable residues of the pesticide. When I conducted the press conference announcing a ban on application of EBDC for more than 40 food products I also said that, if further research indicated that the carcinogenic risk was negligible I would remove the ban. Over the following year additional testing was done and it was reassuring, as EPA scientists had predicted. However, when I then proposed to follow through and remove the ban the same scientists were incredulous. "The press will murder you," one said; "you've admitted that EBDC is a known carcinogen." I went ahead and announced removal of the ban, informing the press that I had promised to follow the science. I invited the press to treat the issue seriously and reminded them that their hyping of an earlier pesticide scare, Alar, had resulted in mothers calling EPA in tears because they had fed their children fresh apples. Press reaction was, in fact, straightforward in reporting on the scientific basis for the decision. Thus, the eighth lesson I learned is that consistent communication about risks, always relying on the science, can go a long way toward warding off public health scares and in shaping a careful and responsible journalistic take on a complex, potentially alarming, problem.

This, then, is the real world of risk assessment. To a regulator it requires a balancing of goods and bads, tradeoffs of apples and oranges. And it rests fundamentally upon sound science, upon samples, tests, studies, comparisons, experimentation. Before there can be specific numerical probabilities of illnesses or deaths for a decisionmaker to ponder, there must have been careful research. Scientific method undergirds the entire regulatory system. This book is an important contribution to the evolving science of risk assessment upon which so much of the integrity and effectiveness of environmental policy rests.

When Dr. Paustenbach's previous text was published in 1989, it filled an important void in the environmental sciences. Before that time, risk assessments were usually conducted by regulatory agencies or those within the regulated community (and their consultants) and were of varying quality. Most of these assessments lacked transparency, that is, few persons knew exactly how the calculations were performed and the basis for the exposure factors and other assumptions. Further, only a few assessments had been published in peer-reviewed journals before 1990 and this tended to inhibit the maturation of the scientific aspects of risk assessment. Thus, his textbook of case studies became a foundation against which others could assess the thoroughness of their work.

This new text comes at a time when the field has passed through its infancy and is now a generally well-respected approach for objectively evaluating environmental issues. Many well-known and respected authors have contributed to this text and have described methods that they have used to evaluate complex environmental questions. Appropriately, an emphasis has been placed on presenting analyses that address topics ranging from risks due to contaminated groundwater, occupational hazards, radionuclide emissions to the community, consumer products, and a variety of risks to wildlife. The overall quality of the text, with the emphasis on providing transparency in the calculations, the quantitative description of uncertainty in the risk estimates, and the importance of proper risk characterization should help ensure that better quality risk assessments are conducted in the coming years. Students and practitioners will benefit significantly from the work of Dr. Paustenbach and his colleagues.

WILLIAM K. REILLY

William K. Reilly is chairman of the Board of the World Wildlife Fund, and was Administrator of the U.S. Environmental Protection Agency under President George H. W. Bush. He headed the U.S. Delegation to the Rio Conference on Environment and Development in 1992.

FOREWORD TO THE FIRST EDITION

Having twice served as Administrator of the Environmental Protection Agency, first in the early 1970s and, most recently, in the mid-1980s, I am convinced that significant differences exist between those two periods of time. In the early 1970s our overriding concern was the gross pollution of our air and our water; this was pollution that we could smell, see, and feel, and that had a significant effect on the environment in which we lived or played. In the mid- to late-1970s, our focus changed and we became more concerned about toxic pollutants—those that affect our health. Cancer and its causes became significant factors in how we feel about environmental contaminants. The concern over cancer coupled with our ability to detect vanishingly small amounts of contaminants dramatically increased the reach and costs of present-day environmental regulations.

The difference in our perception of environmental threats has led us to different approaches in dealing with those threats. It seemed to me in the early 1970s that money alone would solve most of our pollution problems. It soon became obvious that there would never be enough money and that there would always be new environmental problems to solve. The challenge was how to make intelligent judgments about the health risks posed by the myriad of pollutants of concern and which to address first.

When I went back to EPA in 1983, one of my primary goals was to introduce into the EPA decision-making process the concepts of risk assessment and risk management, and to ensure that everybody understood that there was a clear and necessary distinction between the two concepts.

Risk assessment is the scientific evaluation of the human health impacts posed by a particular substance or mixture of substances. Risk management involves a whole host of factors, such as technological feasibility, cost, and public reaction; factors that must be purged from the risk assessment process to the extent possible.

We also tried at the EPA in the mid-1980s to bring some commonality to the risk assessment process for substances that were dealt with by other agencies of the federal government such as the Occupational Safety and Health Administration, the Food and Drug Administration, and the Consumer Product Safety Commission. Our effort in this regard has been modestly successful. It is fair to say that, as a result of the dedication and determination of many in the federal government in recent years, the quantitative approach to analyzing environmental problems, which is the essence of risk assessment, has become generally accepted.

As is clearly demonstrated in this text, many of the ideas which we proposed in 1983 have been implemented in recent assessments. Unlike earlier attempts, scientists have become more comfortable with describing the uncertainties in the assessment process. They also feel more comfortable about stating that sufficient scientific data are not available to reach a firm conclusion.

It is also apparent from the assessments presented in this text that we are more skilled at estimating human exposure and more willing to acknowledge the uncertainties in our estimates of the possible risks associated with exposure to carcinogens and developmental toxicants. Perhaps the most important break-through is that the final decision, the risk management judgment, is no longer confused with the scientific evaluation of the data. This change is important and hopefully permanent.

An area where I felt scientists and risk assessors, in particular, could do a better job was in the communication of risk. We need to describe the hazards posed by suspect substances as clearly as possible, tell people what the known or suspected health problems are, admit our uncertainties, and help the public understand the risk in a larger context. There are a number of examples in this text which do a good job of showing how to present these issues in a comprehensible form.

Scientists should be willing to take a larger role in explaining risks and the risk assessment process to the public. Unfortunately, due to the great pressures on regulatory agencies, the regulated community, and the consultants who serve each of them, scientists have rarely had the opportunity to reduce these often voluminous assessments into papers suitable for publication. Indeed, only a handful of risk assessments addressing specific contaminated sites or chemicals have been published.

For many reasons, Dr. Paustenbach's text is an important and timely contribution to the fields of environmental and occupational health. The breadth of our environmental concerns is clearly illustrated by the diversity of issues discussed here.

He and his colleagues are to be congratulated for having prepared a reference text which presents a large number of rather complex evaluations. This text can serve as an important reference point against which risk assessments of the coming years can be compared. It is my hope that future evaluations will be much improved as a result of the information presented here.

WILLIAM D. RUCKELSHAUS

Former Administrator
United States Environmental Protection Agency
June 1, 1988

PREFACE

Since World War II, most persons living in industrialized nations have enjoyed an amazing improvement in their quality of life and standard of living. For example, mortality at childbirth is no longer considered to be a serious risk, to either the mother or the child. Specific diseases, such as cholera, whopping cough, polio, malaria, diphtheria, measles, and mumps, are now relatively insignificant or have been virtually eliminated in the United States and most other developed countries. Life expectancy continues to increase with each decade and a greater percentage of Americans report that they look forward to living into their seventies, eighties, and beyond. Increased longevity is largely a result of numerous technological advances that have resulted from the synthesis of more than 100,000 different chemicals. Many of these chemicals are pesticides and herbicides which make it possible to feed the world's growing population, as well as life-extending pharmaceuticals.

However, while the existence of such chemicals has improved the quality of life, the improper handling and disposal of many chemicals from about 1900 to 1970 resulted in significant degradation of the environment. It was determined that the presence of industrial chemicals in our food, groundwater, soil, sediment, and ambient air posed some yet-to-be-fully-understood human and ecological hazards. Public concern about these chemicals in the early 1960s, coupled with Rachel Carson's 1962 book entitled *Silent Spring*, essentially launched the first wave of environmentalism in the United States. Since then, virtually everything about the way we handle chemicals—from basic research, through manufacture, to ultimate disposal—has changed. From about 1970–1985 alone, nearly two dozen major pieces of federal legislation and thousands of regulations were promulgated in the United States in an effort to control how chemicals were manufactured, used, distributed, and disposed. However, in spite of implementation of better controls, clean-ups, lesser emissions, and these regulations, the majority of Americans continue to perceive chemicals in the environment to be among the greatest health risks that they face.

During this same 10–15 year period, our analytical methodologies became much more sensitive. Thus, we began to find chemicals in our food, air and water which has previously gone undetected. By 1980, we were able to measure chemical concentration levels in the parts per billion (ppb) and parts per trillion (ppt) range in most environmental media. Due to these incredible technological advancements in analytical chemistry, it was no longer informative to tell the public that "a certain chemical has been detected in a particular media and that at some dose in some animal test that chemical produced some adverse effect." It was obvious that merely detecting even the most acutely toxic substance did not mean that it would pose a significant health hazard. Instead, an approach for making decisions about the significance of environmental sampling results and toxicity data was needed. Thus, the practice of risk assessment drew broad support. Over time, the relatively primitive approaches of the 1970s that were used to characterize risk soon evolved into the current practice of risk assessment. It took only a few years for risk assessments to be considered the primary scientific tool for combining the information from animal toxicity studies, dose–response data, and exposure studies to predict risks to humans and aquatic/avian species. By the mid-1980s, risk assessments were a consideration in virtually all decisionmaking.

Over the past 25 years, the practice of risk assessment has evolved considerably. Initially, the scientific community was quite excited over the possibility of quantitatively predicting risks to humans. Significant sums of money were spent to reduce exposures to certain chemicals when the predicted lifetime cancer risk to an individual exceeded some arbitrary risk criterion, such as 1 in 100,000 or 1 in 1,000,000. Thousands of lawsuits were also filed alleging harm from possible exposure to extremely low concentrations of various substances. Our confidence to precisely predict cancer risks and characterize certain non-cancer hazards, however, eventually eroded as we learned that biology and ecology just aren't that simple. By the late 1980s, it was clear that there were perhaps as many as eight different general modes or mechanisms through which chemicals could cause cancer and that each probably required a different mathematical approach for predicting risk at low doses (something yet to be adequately understood). Scientists learned that not only were toxicity and exposure aspects important, but the persistence of the chemical in humans and the environment also needed to be understood. Thus, the focus on pharmacokinetics and environmental fate/transport. By the 1990s, it became even more clear that our emphasis on cancer effects may have been well intended, but perhaps the non-cancer risks (developmental toxicity, reproductive impairment, endocrine effects, etc.) were even more important than once thought and the cancer risks due to environmental contaminants were relatively inconsequential.

Although the field of risk assessment had matured by the early 1990s, it was often not a transparent process. For example, not all of the bases for exposure calculations were described, the various results from low-dose extrapolation models were not always presented (with confidence limits), the studies which constituted the hazard identification were not always critically evaluated (with weight given to the better studies), the risk characterizations were often one sided or not clearly presented, and the uncertainty in the results was rarely quantitatively described. In large measure, the quality of the assessments stagnated in the 1980s because few, if any, were published in the peer-reviewed literature and there was no compendium or text where persons could see how these assessments were or should be conducted. It was this state of affairs that convinced me that it was time to assemble and share many of the better risk assessments of the period;

an effort that came to fruition in the form of the first edition of this book. I believe that the success of that text, which sold more than 5,000 copies, illustrated that scientists were anxious to learn how to conduct high-quality assessments, and there was an interest in having greater transparency in the process. Hopefully, the significant improvement in the quality of risk assessments conducted over the past ten years is, in part, due to the case studies that were presented in that text.

The practice of risk assessment has significantly improved over the past decade for many reasons. First, more than 200 risk assessments have been published in the peer-reviewed literature. These include comprehensive articles on stationary or mobile sources as well as articles that deal with predicting risks due to exposure to single and multiple chemicals at low doses. Second, at least five peer-reviewed journals now focus on the topic of risk assessment: *Risk Analysis, Regulatory Toxicology and Pharmacology, Journal of Toxicology and Environmental Health, Human and Ecological Risk Assessment*, and *Journal of Environmental Toxicology and Chemistry*. More than a dozen other major journals also occasionally publish articles that focus on some aspect of risk assessment. Third, at least five professional societies place a strong emphasis on risk assessment: Society for Risk Analysis (SRA), International Society for Exposure Assessment (ISEA), International Society for Regulatory Toxicology and Pharmacology (ISRTP), Society of Toxicology (SOT), and Society of Environmental Toxicology and Chemistry (SETAC). Fourth, the U.S. Environmental Protection Agency (EPA) has done an outstanding job at producing documents that have helped standardize and elevate the overall practice of risk assessment by publishing nearly 5,000 pages of general reference and guidance documents. Other nations have also developed a number of publications that have helped bring uniformity to the risk assessment process on an international level.

The purpose of this text, which presents both theory and practice, is to provide the scientific community with an up-to-date single source of information about how to conduct human and ecological risk assessments. The diversity of subjects addressed and the specific cases were intended to share with the reader many of the changes and improvements in the practice of risk assessment that have occurred over the past decade.

The chapters of this book are presented in such a way that the text can be used in graduate level courses, or can serve as a daily reference for practitioners. The first section addresses the basic components of a human health assessment: hazard identification, dose–response assessment, exposure assessment, and risk characterization. The second section deals with the same components, but as they relate to ecological risk assessment. In short, the first six chapters represent the theory portion of the book.

Most of the remaining sections present various case studies that address some of the common environmental and occupational health challenges that scientists have faced over the past 10–15 years. Because it is expected that these same problems (contaminated food, soil, air, water, sediment, and consumer products) will require our attention for at least another 20 years, I believe these case studies will be most helpful to those scientists tasked with characterizing the associated risks. As in the first text, cases involving chemical hazards and exposure to radionuclides are included.

The number of chapters devoted to ecological issues is greater than in the prior text because, in my view, the focus of most of the major risk assessments in the coming years will be driven by hazards to wildlife or contamination of domestic animals and fish that are consumed by humans. Although this field is still evolving, it has made tremendous advances over the past decade and this will undoubtedly continue throughout the next decade. I expect many of the lessons learned in conducting assessments of human health hazards to continue to be experienced by scientists and regulatory agencies in the ecological arena. Due to the myriad of mistakes made and dead ends pursued over the past 20 years on the human health side, it is my expectation that the learning curve in ecological risk assessment should be far less costly.

As before, a section which addresses risk communication and some aspects of risk management is included. Unlike the previous text, a section on evolving issues has been added. Although perhaps surprising, during the early to mid-1990s, it was unclear to me that there were going to be a sufficient number of new and challenging topics to warrant bringing better scientists to the environmental field. I was also concerned that risk assessors had already developed techniques for addressing nearly any question one could raise about human health hazards. At the time, it seemed that most of the exciting improvements were going to be in the field of ecological assessment. However, the introduction of human health concerns about endocrine disruptors, the threat to children's health, chemical mixtures, persistent organic pollutants (POPs), genetically modified foods, subtle non-carcinogenic effects and genomics has convinced me that there is at least another decade of exciting challenges facing risk assessors.

Because risk assessments have definitely earned their place in the decision-making process, it seemed reasonable to add chapters on life-cycle analyses and cost–benefit analysis. After presenting more than 1000 pages illustrating the way to conduct high quality assessments and having illustrated their usefulness and importance, it only made sense to close the text with a discussion of the precautionary principle. For the past five years, it has been speculated by many scientists and regulatory agencies that a scholarly application of the precautionary principle could bring an end to risk assessment and, as such, it seemed an appropriate thought-provoking topic for closing the text.

Risk assessments offer an opportunity for the public to develop an understanding of the critical issues associated with the presence of industrial (and pharmaceutical) chemicals in our environment. It has been, and continues to be, my belief that if emotionalism and subjective claims carry more weight than a thorough and objective analysis, mankind will almost surely compromise its ability to achieve all of the goals of which it is capable. A number of scientists and political scientists have warned us of the hazards of such an approach (see the book *The Demon-Haunted World* by Carl Sagan). If we wish to maintain a standard of living close to that to which we have become accustomed in the developed countries, we need to evaluate the various controllable risks in a uniform and scientifically defensible manner. Such evaluations should help ensure that significant hazards are controlled while insignificant ones are placed much lower in our priorities.

The contributors to this text are among the premier persons in the field. Approximately 60 contributors were drawn from more than a dozen scientific disciplines. They have been responsible for conducting a significant fraction of the important assessments in the United States. Some have helped formulate both domestic and international environmental policies and regulations. It has been an honor to work with them over the three years needed to bring the text to completion. Their qualifications are exceptional and their understanding of the field is validated by the quality of their contributions. I thank each scientist who participated.

Even though I have carefully read and critiqued every chapter, I am unable to endorse uniformly each of the methods used or opinions expressed by the various authors. Since these authors are experts in their respective specialties, it would be presumptuous to have insisted that all of them approach their analysis in exactly the same manner that I might have chosen.

I would especially like to thank my various administrative assistants of the past 4 years for their enthusiasm and support. Specifically, I thank Suzanne Milani for initially saying that "there is at least one more book left in you" and who helped me launch the effort. She was followed by an incredibly disciplined and supportive colleague, Bev Wicker. Lastly, I very much appreciate the support of my current assistant, Neha Patani, who was able to pull together the loose ends and help bring the project to closure.

I also wish to express a special thanks to Bob Esposito of John Wiley & Sons. He is a true professional who values his authors.

It is my hope that you will learn as much from this text as I did in assembling it.

DENNIS J. PAUSTENBACH

Woodside, California

CONTRIBUTORS

BRUCE N. AMES, Department of Molecular and Cell Biology, University of California-Berkeley, Berkeley, California 94720

DEBORAH BENNETT, Harvard School of Public Health, 401 Park Drive, Room 404L, P.O. Box 15677, Boston, Massachusetts 02215

RANJIT BHARVIRKAR, Department of Civil Engineering, North Carolina State University, Raleigh, North Carolina 27695-7908

J. N. BLANCATO, National Exposure Research Laboratory, U.S. EPA, Las Vegas, Nevada 89119

MARGARET A. BRANTON, ARCADIS, 301 East Ocean Boulevard, Suite 1530, Long Beach, California 90802

LARRY BREWER, Springborn Laboratories, Inc., 790 Main Street, Wareham, Massachusetts 02571

TIMOTHY CARROTHERS, Pharsight Corporation, 800 W. El Camino Real, Mountain View, California 94040

GAIL CHARNLEY, HealthRisk Strategies, 826 A Street SE, Washington, DC, 20003

VALERIE A. CRAVEN, Exponent, Inc., 631 First Street, Suite 200, Santa Rosa, California 95404-4716

C. C. DARY, National Exposure Research Laboratory, U.S. EPA, Las Vegas, Nevada 89119

WILLIAM DESVOUSGES, Triangle Economic Research, 2775 Meridian Parkway, Durham, North Carolina 27713

E. DONALD ELLIOTT, Yale Law School, New Haven, Connecticut 06520

SHIH SHING FENG, Applied Biosystems, Inc., 850 Lincoln Centre Drive, Foster City, California 94404

BRENT FINLEY, Exponent, Inc., 631 First Street, Suite 200, Santa Rosa, California 65404

SUSAN M. FLACK, ENSR International, 1544 North Street, Suite 110, Boulder, Colorado 80304

H. CHRISTOPHER FREY, Department of Civil Engineering, North Carolina State University, Raleigh, North Carolina 27695-7908

GEORGE FRIES, Independent Consultant, 2205 Bucknell Terrace, Silver Spring, Maryland 20902

LOIS SWIRSKY-GOLD, Carcinogenic Potency Project, Lawrence Berkeley Laboratory, One Cyclotron Road, Mail Stop: Barker Hall, Berkeley, California 94720

MICHAEL GOODMAN, Exponent, Inc., 310 Montgomery Street, Alexandria, Virginia 22314

JOHN D. GRAHAM, Center for Risk Analysis, Harvard School of Public Health, Boston, Massachusetts 02115

KIRK J. GRIBBEN, Alcoa, Inc., Alcoa Technical Center, 100 Technical Drive, Alcoa Center, Pennsylvania 15069-0001

THOMAS E. GUARDINO, ECONorthwest, Eugene, Oregon 97401

MARK HARRIS, Harris Environmental Risk Management, 1900 Bluffview Court, Flower Mound, Texas 75022

SEAN M. HAYS, Exponent, Inc., 4940 Pearl East Circle, Suite 300, Boulder, Colorado 80301

MIRANDA HENNING, ARCADIS, 24 Preble Street, Suite 100, Portland, Maine 04101

W. E. KASTENBERG, Nuclear Engineering Department, University of California-Berkeley, 4103 Etcheverry, Berkeley, California 94720-1730

BRENT KERGER, HSRI, Inc., 2976 Wellington Circle West, Tallahassee, Florida 32309

JIM KNAAK, University at Buffalo, Department of Pharmacology and Toxicology, 7 Earldom Way, Getzville, New York 14068

LESTER LAVE, Carnegie-Mellon University, GSIA (Tech and Frew St.), Carnegie-Mellon University, Pittsburgh, Pennsylvania 15213

NANCY LA VERDA, Exponent Health Group, 310 Montgomery Street, Alexandria, Virginia 22314

HON WING LEUNG, Independent Consultant, 15 Deer Park Road, Danbury, Connecticut 06811-2717

GEORG LEUBECK, Fred Hutchinson Cancer Research Center, 1100 Fairview Avenue, Seattle, Washington 98109-1024

HEATHER MACLEAN, University of Toronto, Toronto, Ontario, Canada, M51A1

H. SCOTT MATTHEWS, Carnegie-Mellon University, GSIA (Tech and Frew St.), Carnegie-Mellon University, Pittsburgh, Pennsylvania 15213

KRISTY E. MATTHEWS, Triangle Economic Research, 2775 Meridian Parkway, Durham, North Carolina 27713

MONTE A. MAYES, Dow Agro Sciences, 9330 Zionsville Road, Indianapolis, Indiana 46268

T. E. MCKONE, University of California-Berkeley, 140 Warren Hall, #7360, Berkeley, California 94720-7360

HARRY L. MCQUILLEN, U.S. Fish and Wildlife Service, 2800 Cottage Way, Room W-2605 Sacramento, California 95825

SURESH MOOLGAVKAR, Fred Hutchinson Cancer Research Center, 1100 Fairview Avenue, Seattle, Washington 98109-1024

WARNER NORTH, NorthWorks, Inc., 1002 Misty Lane, Belmont, California 94002-3651

ROB PASTOROK, Exponent, Inc., 15375 SE 30th Place, Suite 250, Bellevue, Washington 98007

G. T. PATTERSON, Department of Pesticide Regulation, State of California, Sacramento, California 95814

DENNIS PAUSTENBACH, Exponent, Inc., 149 Commonwealth Drive, Menlo Park, California 94025

BARBARA PETERSEN, Novigen Sciences, Inc., 1730 Rhode Island Ave., NW, Suite 1100, Washington, DC, 20036

PAUL S. PRICE, The LifeLine Group, 129 Oakhurst Road, Cape Elizabeth, Maine 04107

DEBORAH PROCTOR, Exponent, Inc., 320 Goddard Way, Suite 200, Irvine, California 92671

DAVID RABBE, Chemical Land Holdings, Inc., Two Tower Center Boulevard, 10th Floor, East Brunswick, New Jersey 08816

DOUGLAS P. REAGAN, URS Greiner Woodward Clyde, Stanford Place 3, 4582 South Ulster Street, Denver, Colorado 80237

WILLIAM REILLY, Aqua International Partners, L.P., San Francisco, California 94104

FRANK SELKER, Decision Management Associates, LLC, Portland, Oregon 97201

ANNE SERGEANT, U.S. EPA, Office of Research and Development, National Center for Environmental Assessment, 1200 Pennsylvania Avenue, NW, Washington, DC, 20460

JANE SEXTON, Exponent, Inc., 15375 SE 30th Place, Suite 250, Bellevue, Washington 98007

PATRICK SHEEHAN, Exponent, Inc., 1970 Broadway, Suite 250, Oakland, California 94612

WALTER SHIELDS, Exponent, Inc., 15375 SE 30th Place, Suite 250, Bellevue, Washington 98007

THOMAS H. SLOAN, Department of Molecular and Cell Biology, University of California-Berkeley, Berkeley, California 94720

PAUL SLOVIC, Decision Research, 1201 Oak Street, Eugene, Oregon 97401

JESSICA GLICKEN TURNLEY, Galisteo Consulting Group, Inc., 2403 San Mateo Boulevard NE, Suite W-12, Albuquerque, New Mexico 87110

JOHN WARMERDAM, Exponent, Inc., 1970 Broadway, Suite 250, Oakland, California 94612

NADINE M. WEINBERG, ARCADIS, 24 Preble Street, Suite 100, Portland, Maine 04101

TOMAS WIDNER, ENSR International, 1420 Harbor Bay Parkway, Suite 120, Alameda, California 94502

JONATHAN B. WIENER, Center for Environmental Solutions, Duke University, Box 90360, Durham, North Carolina 27708

PAMELA WILLIAMS, Exponent, Inc., 4940 Pearl East Circle, Suite 300, Boulder, Colorado 80301

NATALIE D. WILSON, N.D. Wilson & Associates, LLC, 12948 Victoria Avenue, Huntington Woods, Michigan 48070

ABOUT THE AUTHOR

Dr. Dennis Paustenbach has held numerous leadership positions within two major consulting firms and was the founder of ChemRisk, the largest risk assessment consulting firm in the U.S. during the 1990's. He is board-certified in toxicology, industrial hygiene, and safety and has more than 20 years of experience in risk assessment, environmental engineering, ecotoxicology, and occupational health. Dennis earned a BS in Chemical Engineering, an MS in Industrial Hygiene/Toxicology, and a Ph.D. in Environmental Toxicology. He has directed consulting activities for nearly 700 risk assessments and has published more than 160 peer-reviewed manuscripts in this and related fields. He was also the editor of the most popular book on risk assessment, "The Risk Assessment of Human and Environmental Health Hazards: A Textbook of Case Studies." Dr. Paustenbach has worked on many high-profile environmental projects, including Times Beach, Love Canal, the New Jersey Meadowlands, and the Erin Brockovich case. He has been an adjunct professor at several universities and was a visiting scholar at the Center for Risk Analysis at Harvard. He was recognized as the top risk practitioner by the Society for Risk Analysis in 1998 and received the Arnold J. Lehman Award from the Society of Toxicology in 2002 for his contributions to the fields of Risk Assessment and Toxicology.

SECTION A
Theory: Human Health Assessment

1 Primer on Human and Environmental Risk Assessment

DENNIS J. PAUSTENBACH

Risk Assessment Consultant, 65 Roan Place, Woodside, California

1.1 INTRODUCTION

Since about 1970 the field of risk assessment has received widespread attention within both the scientific and regulatory communities [Starr, 1969; National Academy of Sciences (NAS), 1983; Ruckelshaus, 1984; Commission on Risk Assessment and Risk Management, 1996]. It also attracted the attention of the public (*Wall Street Journal*, 1984a, 1996; Pianin, 2001) and the legal system (Paustenbach, 1989; Breyer, 1993). Beginning in the 1980s, the public began to develop an expectation that risk analysis would help bring order to what appeared to be an unmanageable quantity of scientific and medical data regarding the potential health hazards posed by physical and chemical agents in our environment. It was hoped that risk assessment methodologies would help regulators make better decisions. Such analyses, they hoped, would represent a significant improvement in how decisions are formulated, and be an improvement over the 1950s when scientists tended to give decision makers only the "black-and-white," "yes it is," or "no it isn't" insight obtained from toxicity testing. Risk assessments gained popularity because they offered risk managers, policymakers, and the public a range of options, each having a specific cost and benefit (Graham and Hartwell, 1997).

From 1990 to 2000, the field of risk assessment matured significantly [Graham, 1995; National Research Council (NRC), 1994; Environmental Protection Agency (EPA), 2000a, d]. In the early 1990s, for example, hundreds of risk assessments were conducted of Superfund and Resource Conservation and Recovery Act (RCRA) sites. At the same time, thousands of assessments of incinerators, water discharges, air emissions, contaminated soil, contaminated sediments, consumer products, and other specific events/activities were performed. The quality of these assessments, by and large, were much better than

Human and Ecological Risk Assessment: Theory and Practice, Edited by Dennis J. Paustenbach
ISBN 0-471-14747-8 © 2002 John Wiley & Sons, Inc.

those conducted in the prior 20-year period. In large measure, this was the result of better exposure data, more balanced risk characterizations, the use of Monte Carlo techniques, and the dissemination of information (including the publication of case studies in peer-reviewed journals). The dose extrapolation process was much improved when physiologically based pharmacokinetic (PB-PK) models were applied. All in all, the field progressed nicely since publication of the prior edition of this text in 1989 (hopefully, in part, as a result of it).

Properly conducted risk assessments have received fairly broad acceptance in the United States and most other developed countries. In part, this is because they put into perspective the terms *toxic*, *hazard*, and *risk*. Regrettably, these words have not always been used correctly by either the lay press or those within the scientific community. Toxicity, as noted by Paracelsus in the 1500s, is an inherent property of all substances. All chemical and physical agents can produce adverse health effects, that is, toxicity, at some dose or under specific exposure conditions (concentration plus duration). In short, the dose makes the poison (Paracelsus, 1567). In contrast, exposure to a chemical, for example, benzene, has the *capacity* to produce a particular type of adverse effect, for example, acute myelogenous leukemia (AML), which represents a *hazard*. Risk, however, is the *probability* or likelihood that an adverse outcome will occur in a person or a group that is exposed to a particular concentration or dose of the hazardous agent. For example, characterizing an exposure as being likely to produce one cancer in 100,000 exposed is a risk estimate. Therefore, risk is generally a function of exposure or dose. In this text, the process or procedure used to estimate the likelihood that humans or ecological systems will be affected adversely by a chemical or physical agent under a specific set of conditions is called *health risk assessment*.

The term *risk assessment*, however, has not been used exclusively to describe the likelihood of an adverse response to a chemical or physical agent. In fact, risk assessment has been used to describe the likelihood of a diverse number of unwanted events. These include industrial explosions, workplace injuries, failure of machine parts, a wide variety of natural catastrophes (e.g., earthquakes, tidal waves, hurricanes, volcanic eruptions, tornadoes, blizzards), injury or death due to an array of voluntary activities (e.g., skiing, football, sky diving, flying, hunting), diseases (e.g., cancer, leukemia, developmental toxicity caused by chemical exposure), death due to natural causes (e.g., heart attack, cancer, diabetes), death due to lifestyle (e.g., smoking, alcoholism, diet), and a number of other endpoints (Schwing and Albers, 1980).

The purpose of risk assessments is to provide pertinent information to risk managers, specifically, policymakers and regulators, so that the best possible decisions can be made. As discussed by numerous researchers, factors other than those addressed in a risk assessment can influence decisions about risk (such as social values, severity of the adverse effect, etc.) (Lowrance, 1976, 1984; Wildavsky, 1995; Graham and Hartwell, 1997). However, these considerations usually are outside the scope of classic risk assessments conducted by

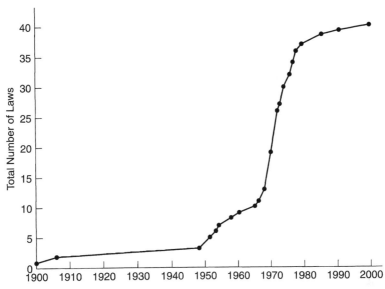

Figure 1.1 Federal legislation dealing with the manufacture, use, transportation, sale, or disposal of hazardous materials.

scientists. In light of the dramatic increase in the number of pieces of legislation that have been passed during the past 30 years (Figure 1.1), it is clear that society has been uncomfortable about the ability of the regulated community (e.g., companies) to "manage" the risks posed by chemical and physical hazards. Starr (1969), in his ground-breaking article on risk assessment, "Social benefits versus technological risk," as well as in his 1985 lecture "Risk management, assessment, and acceptability," discussed some of the conflicts that face those who conduct risk assessments. For example, he noted that "public acceptance of any risk is more dependent on public confidence in risk management than on the quantitative estimates of risk consequences, probabilities, and magnitudes (Starr, 1985)." He was one of the first scientists to draw the distinction between voluntary and involuntary risks; an issue that continues to be a difficult one for risk managers who must often regulate involuntary risks, which would clearly be considered inconsequential when compared with voluntary risks generally accepted by the public (Crouch and Wilson, 1982; Travis et al., 1987; Tengs et al., 1995).

1.1.1 Risk Assessment Versus Risk Management

A health risk assessment is a written document wherein all the pertinent scientific information, regarding toxicology, human experience, environmental fate, and exposure, are assembled, critiqued, and interpreted. The goal of the assessment is to estimate the likelihood of an adverse effect on humans, wildlife, or ecological systems posed by a specific level of exposure to a chemical or physical hazard (Paustenbach, 1989). Risk assessments of environmental hazards, in contrast to the analyses of the likelihood of the occurrence of any undesirable event, for example, earthquakes and machine failure, are thoroughly dependent on the degree of exposure. As a result, assessments of health hazards posed by workplaces, hazardous waste sites, ambient air pollutants, water pollutants, soil contaminants, and food contaminants are dependent on both the potency of the agent and the level of exposure. The same principles apply to assessments of plants, fish, birds, insects, large and small mammals, reptiles, and other wildlife (EPA, 1998c).

The term *risk assessment* has been misused by dozens of scientists whose studies have appeared in the scientific literature during the past 20 years. All too often, risk assessment has incorrectly been characterized as the process used to estimate the low-dose response following exposure to carcinogenic chemicals, often termed *quantitative risk assessment*. Perhaps the best and most widely cited definition of risk assessment is the one suggested by the National Academy of Science (NAS, 1983). In its report, the NAS committee recommended [page 18]:

> *Risk assessment* to mean the characterization of the potential adverse health effects of human exposures to environmental hazards. Risk assessments include several elements: description of the potential adverse health effects based on an evaluation of results of epidemiologic, clinical, toxicologic, and environmental research; extrapolation from those results to predict the type and estimate the extent of health effects in humans under given conditions of exposure; judgments as to the number and characteristics of persons exposed at various intensities and durations; and summary judgments on the existence and overall magnitude of the public-health problem. Risk assessment also includes characterization of the uncertainties inherent in the process of inferring risk.

The NAS committee emphasized that the processes of risk assessment and risk management were to be separate activities. They offered the following definition for risk management (NAS, 1983 pages 18–19):

> The Committee uses the term *risk management* to describe the process of evaluating alternative regulatory actions and selecting among them. Risk management, which is carried out by regulatory agencies under various legislative mandates, is an agency decision-making process that entails consideration of political, social, economic, and engineering information with risk-related information to develop, analyze, and compare regulatory options and to select the appropriate regulatory response to a potential chronic health hazard. The selection process necessarily requires the use of value judgments on such issues as the acceptability of risk and the reasonableness of the costs of control.

One goal of the committee was to encourage scientists, policymakers, and the public to separate clearly the risk assessment from the risk management process. Until this time, many assessments were so laden with value judgments and the subjective views of the risk assessors that the risk manager was unable to separate the scientific data from the wishes of the risk scientist. One significant accomplishment of the NAS committee was the identification of this problem. Since 1983, it has been shown that it is nearly impossible to remove some level of bias from risk assessments.

In 1994, the NAS issued two reports that could be seen as an update on the 1983 document. These were titled "Science and Judgment in Risk Assessment" and "Building Consensus Through Risk Assessment." Later, a third piece entitled "Understanding Risk: Informing Decisions in a Democratic Society" (NRC, 1996) focused on risk characterization and risk communication. These documents, coupled with a series of publications by the EPA Science Advisory Board are an invaluable chronicle of the history of risk assessment (EPA, 1992d, 1995c, 1996a, 2000a).

1.1.2 The Assessment Process

Risk assessment can be divided into four major steps: hazard identification, dose–response assessment, exposure assessment, and risk characterization (NAS, 1983) (Figure 1.2). For some perceived hazards, the risk assessment might stop with the first step, hazard identification, for example, if no adverse effect is identified or if an agency elects to take regulatory action without further analysis. The NAS committee suggested the following definitions for these steps:

Hazard identification is the most easily recognized of the actions of regulatory agencies. It is defined here as the process of determining whether human exposure to an agent could cause an increase in the incidence of a health condition (cancer, birth defect, etc.) or whether exposure by a nonhuman receptor, for example, fish, birds, or other wildlife, might adversely be affected. It involves characterizing the nature and strength of the evidence of causation. Although the question of whether a substance causes cancer or other adverse health effects in humans is theoretically a yes–no question, there are few chemicals or physical agents on which the human data are definitive [National Research Council (NRC), 1983]. Therefore, the question is often restated in terms of effects in laboratory animals or other test systems: Does the agent induce cancer in test animals? Positive answers to such questions are typically taken as evidence that an agent may pose a cancer risk for any exposed human. Information from short-term in vitro tests and structural similarity to known chemical hazards may, in certain circumstances, also be considered as adequate information of identifying a hazard.

Dose–response assessment is the process of characterizing the relation between the dose of an agent administered or received and the incidence of an adverse health effect in exposed populations and estimating the incidence of the effect as a function of exposure to the agent. This process considers such important factors as intensity of exposure, age pattern of exposure, and possibly other variables that might affect response, such as sex, lifestyle, and other mod-

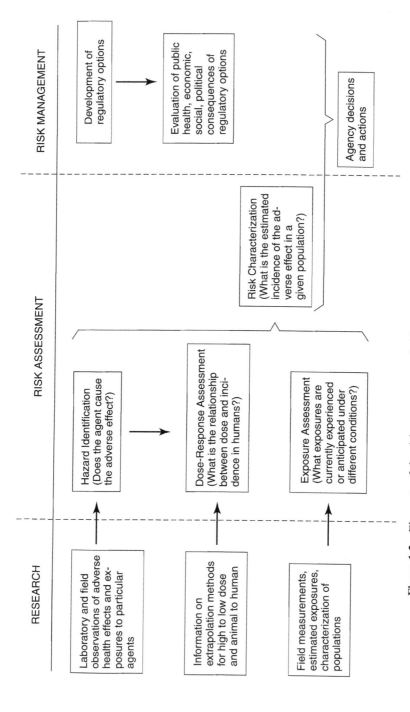

Figure 1.2 Elements of the risk assessment and risk management processes. From NAS (1983).

8

ifying factors. A dose–response assessment usually requires extrapolation from high to low doses and extrapolation from animals to humans, or one laboratory animal species to a species of wildlife. A dose–response assessment should describe and justify the methods of extrapolation used to predict incidence, and it should characterize the statistical and biologic uncertainties in these methods. When possible, the uncertainties should be described numerically rather than qualitatively.

Exposure assessment is the process of measuring or estimating the intensity, frequency, and duration of human or animal exposure to an agent currently present in the environment or of estimating hypothetical exposures that might arise from the release of new chemicals into the environment. In its most complete form, an exposure assessment should describe the magnitude, duration, schedule, and route of exposure; the size, nature, and classes of the human, animal, aquatic, or wildlife populations exposed; and the uncertainties in all estimates. The exposure assessment can often be used to identify feasible prospective control options and to predict the effects of available control technologies for controlling or limiting exposure.

Risk characterization is the process of estimating the incidence of a health effect under the various conditions of human or animal exposure described in the exposure assessment. It is performed by combining the exposure and dose–response assessments. The summary effects of the uncertainties in the preceding steps should be described in this step.

1.2 HEALTH RISK ASSESSMENT: A BRIEF HISTORY

The origins of risk assessment, in its most basic form, can be traced back to that of early humans. Certainly, after humans recognized that the meat of animals represented a source of food when edible plants were not available, they had to weigh the hazard of being mauled by a wild animal, which they hoped to kill and eat, versus the benefit of thwarting starvation. The historical record does not permit us to know how well early humans balanced this risk–benefit relationship; however, it is safe to assume that those who first conducted these assessments fared less well than those who later conducted similar analyses. Almost certainly, the scientists of the latter decades of the 21st century, when studying the health risk analyses conducted in the 1970s to 1990s, will draw a similar analogy to those conducted by primitive humans.

1.2.1 The Earliest Years (ca. 3200 B.C. to A.D. 500)

A thoughtful review of the history of risk analysis has been developed by Covello and Mumpower (1985) and much of the subsequent discussion draws on their work. With respect to the early years, they noted the following:

In the Tigris–Euphrates valley about 3200 B.C. there lived a group called the Asipu. One of their primary functions was to serve as consultants for risky, uncertain, or difficult decisions. If a decision needed to be made concerning a forthcoming risky venture, a proposed marriage arrangement, or a suitable building site, one could consult with a member of the Asipu. The Asipu would identify the important dimensions of the problem, identify alternative actions, and collect data on the likely outcomes (e.g., profit or loss, success or failure) of each alternative. The best available data from their perspective were signs from the gods, which the priestlike Asipu were especially qualified to interpret. The Asipu would then create a ledger with a space for each alternative. If the signs were favorable, they would enter a plus in the space; if not, they would enter a minus. After the analysis was completed, the Asipu would recommend the most favorable alternative. The last step was to issue a final report to the client, etched upon a clay tablet (Oppenheim, 1977).

> According to Grier, the practices of the Asipu mark the first recorded instance of a simplified form of risk analysis. The similarities between the practices and procedures of modern risk analysts and those of their Babylonian forebears underscore the point that people have been dealing with problems of risk for a long time, often in a sophisticated and quantitative way [p. 246].

Covello and Mumpower (1985) implied that religious beliefs played a significant role in the evolution of probability theory and risk analysis.

> An important thread leading to modern quantitative risk analysis can be traced to early religious ideas concerning the probability of an after-life. This should hardly be suprising, considering the salience and seriousness of the risks involved (at least for true believers). Beginning with Plato's *Phaedo* in the 4th century B.C., numerous treatises have been written discussing the risk to one's soul in the afterlife based on how one conducts oneself in the here and now.

> One of the most sophisticated analyses of this issue was carried out by Arnobius the Elder, who lived in the 4th century A.D. in North Africa. Arnobius was a major figure in a pagan church that was competing at the time with the fledgling Christian church. Members of Arnobius'church, who maintained a temple to Venus complete with virgin sacrifices and temple prostitution, led a decadent life in comparison to the austere Christians. Arnobius taunted the Christians for their lives of pointless self abnegation; but, after a revelatory vision, renounced his previous beliefs and attempted to convert to Christianity. The bishop of the Christian church, suspicious of Arnobius' motives and the sincerity of his conversion, refused him the rite of baptism. In an effort to demonstrate the authenticity of his conversion, Arnobius authorized an eight-volume monograph entitled *Against the Pagans*. In this work Arnobius made a number of arguments for Christianity, one of which is particularly relevant to the history of probabilistic risk analysis. After thoroughly discussing the risks and uncertainties associated with decisions affecting one's soul, Arnobius proposed a 2×2 matrix. There are, he argued, two alternatives: "accept Christianity" or "remain a pagan." There are also, he argued, two possible, but uncertain, states of affairs: "God exists" or "God does not exist." If God does not exist there is no difference between the two

alternatives (with the minor exception that Christians may unnecessarily forego some of the pleasures of the flesh enjoyed by pagans). If God exists, however, being a Christian is far better for one's soul then being a pagan [p. 105].

1.2.2   Environmental Risks (ca. A.D. 500 to 1300)

Contamination of the air, water, and land has long been recognized as a potential health problem, but the need to control pollution was not recognized for a rather long period. Covello and Mumpower (1985) in reviewing this topic observed:

> Air pollution (due to dust and smoke from wood and coal fires) has been a ubiquitous problem in congested urban areas since ancient times (Hughes, 1975). The first act of government intervention did not occur until 1285, however, when King Edward I of England responded to a petition from members of the nobility and others concerning the offensive coal smoke in London. Smoke arising from the burning of soft coal had long been a problem in London (White, 1967; Brake, 1975). Edward's response to the petition was one that is now commonly practiced by government risk managers—he established a commission in 1285 to study the problem. In response to the commission's report, several private sector actions were taken, including a voluntary decision by a group of London smiths in 1298 not to "... work at night on account of the unhealthiness of coal and damage to their neighbors" (Hughes, 1975). These voluntary efforts were not sufficient, however, and in 1307 Edward issued a royal proclamation prohibiting the use of soft coal in kilns. Shortly after this, Edward was forced to establish a second commission, the main function of which was to determine why the royal proclamation was not being observed.
>
> Apparently, relatively few advances in the area of health risk assessment, as we would think of it today, were made for nearly 2100 years after the initial observation that the environment could affect human health. The association between malaria and swamps, for example, was established in the 5th century B.C. even though the precise reason for the association remained obscure. Covello and Mumpower (1985) noted that in *Airs, Water, and Places*, thought to have been written by Hippocrates in the 4th or 5th century B.C., an attempt was made to set forth a causal relation between disease and the environment. As early as the 1st century B.C., the Greeks and Romans had observed the adverse effects of exposure to lead (Giltillan, 1965; Nriagu, 1983). Specifically, the Roman Vitruvious (cited in Hughes, 1975) wrote:
>
> "We can take example by the workers in lead who have complexions affected by pallor. For when, in casting, the lead receives the current of air, the fumes from it occupy the members of the body, and burning them thereon, rob the limbs of the virtues of the blood. Therefore it seems that water should not be brought in lead pipes if we desire to have it wholesome." [p. 106]

Our knowledge of the possible role of environmental contamination in effecting human health evolved over a period of several hundred years. Many of the key events are described in Table 1.1.

TABLE 1.1   An Historical Timeline of Events Having Some Significance on the Evolution of the Environmental Movement (based on information in http://www.zoaks.com/ information/envirotimeline/envirotimeline.html)

Ancient Civilization

A.D. 80   The Roman Senate passes a law to protect water stored during dry periods so it can be released for street and sewer cleaning. Aqueducts have to be built because local springs and pools have become polluted.

Middle Ages and the Enlightenment (1300–1700)

1306   Edward I forbids coal burning when English Parliament is in session.

1640   Izaak Walton writes *The Compleat Angler*.

1661   John Evelyn writes "Fumifugium, or the Inconvenience of the Aer and Smoake of London Dissipated" to propose remedies for London's air pollution problem.

1681   William Penn requires Pennsylvania settlers to preserve 1 acre of trees for every 5 acres cleared.

Industrial Revolution (1700–1900)

1739   Benjamin Franklin and neighbors petition Pennsylvania Assembly to stop waste dumping and remove tanneries from Philadelphia's commercial district.

1762–1769   Philadelphia committee led by Benjamin Franklin attempts to regulate waste disposal and water pollution.

1775   English scientist Percival Pott finds that coal is causing an unusually high incidence of cancer among chimney sweeps.

1799   Manhattan Company formed to build water line. Company survives as Chase Manhattan Bank.

1817   U.S. Secretary of Navy authorized to reserve lands producing hardwoods for constructing naval ships.

1832   Arkansas Hot Springs established as a national reservation, setting a precedent for Yellowstone and eventually, a national park system.

1837   Benjamin McCready writes pioneering essay on occupational medicine and conditions of New York City slums.

1842   Edwin Chadwick writes "The Sanitary Condition of the Labouring Population of Great Britain." Report is first scientific inquiry about infectious disease, child mortality, and the link to polluted water supplies and lack of sanitation.

1843   Royal Commission inquiries begin; dreadful working conditions, child labor, public health problems exposed.

1854   John Snow, London doctor, maps spread of cholera in Broad Street neighborhood and traces cases to a contaminated drinking water pump. Snow's epidemiological studies support "contagionist" views, partly supplanting "sanitarian" views about public health.

1863   George Perkins Marsh writes *Man and Nature: The Earth as Modified by Human Action*, with emphasis on forest preservation and soil and water conservation.

1860s–1880s   French scientist Louis Pasteur's germ theory of disease revolutionizes concepts of public health, making it possible to isolate and treat specific diseases.

TABLE 1.1   (continued)

1871	U.S. Fish Commission formed to study decline of coastal fisheries.
1873	London fog kills 1150 people. Similar incidents in 1880, 1882, 1891, and 1892.
1875	British Publish Health Act consolidates authority to deal with pollution, occupational disease, and other problems.
1880s	First U.S. municipal smoke abatement laws aimed at reducing black smoke and ash from factories, railroads, and ships. Regulation under local boards of health.

Progressive Era (1890–1920)

1891	Forest protection bill passes Congress. Thirteen million acres are set aside by 1893.
1892	Sierra Club founded.
1899	Refuse Act prevents some obvious pollution of streams and places Corps of Engineers in charge of permits and regulation.
1900	Automobile is welcomed as bringing relief from pollution. New York City, with 120,000 horses, scrapes up 2.4 million pounds of manure every day.
1905	National Audubon Society organized.
1905	U.S. Forest Service created.
1906	Food and Drug Administration founded.
1907	USDA Animal Health and Plant Health Inspection Service founded.
1908	Swedish chemist Svante Arrhenius argues that the greenhouse effect from coal and petroleum use is warming the globe.
1909	Glasgow, Scotland, winter inversions and smoke accumulations kill over 1000 people.
1909	Bureau of Mines founded to promote safety and welfare of miners. Bureau and the Public Health Service begin studies of lung diseases.
1913	William T. Hornaday, head of New York Zoological Society, writes *Our Vanishing Wildlife, Its Extermination and Preservation.*
1914	Corps of Engineers, Bureau of Mines, and Public Health Service begin pollution surveys of streams and harbors. Reports filed by early 1920s show an accumulation of heavy damage from oil dumping, mine runoff, untreated sewage, and industrial waste.

The Roaring Twenties (1920–1930)

1920	Mineral Leasing Act opens up rich deposits on federal lands for token rental fees.
1921	General Motors researchers discover tetraethyl lead as an antiknock gasoline additive.
1922	Amelia Maggia, first of the "Radium Girls," dies of radiation poisoning. She was a dial painter with U.S. Radium Corporation in Orange, New Jersey.
1924	Oil Pollution Act passed, prohibiting discharge from any vessel within the 3-mile limit, except by accident.
1926	First large-scale survey of air pollution in the United States, in Salt Lake City.

TABLE 1.1   (continued)

1926	First large-scale survey of air pollution in the United States, in Salt Lake City.
1926	Surgeon General's committee of experts reluctantly permit ethyl leaded gasoline back on the fuel market.
1928	Public Health Service begins checking air pollution in eastern U.S. cities, reporting sunlight cut by 20 to 50 percent in New York City.

Depression and World War II (1930–1945)

1930	Meuse River Valley killer smog incident, Belgium, 3-day inversion kills 63 people, with 6000 made ill.
1936	National Wildlife Federation formed.
1936	Alice Hamilton, tireless crusader for worker health, retires from Harvard University faculty.
1939	St. Louis smog episode spurs serious smoke abatement campaign, switch from soft coal to hard coal and fuel oil.

Postwar Era (1945–1960)

1945	Corps of Engineers abandons Potomac River dam after a storm of controversy.
1947	Los Angeles Air Pollution Control District formed.
1948	Twenty people dead, 600 hospitalized in Donora, Pennsylvania, smog attack.
1948	Six hundred deaths in London due to "killer fog." 1948 Aldo Leopold writes *A Sand County Almanac*.
1949	Izaak Walton League writes "Crisis Spots in Conservation," identifying specific water projects to be opposed.
1952	Three to four thousand people die in London "killer fogs."
1953	New York smog incident kills between 170 and 260 people in November.
1955	Congress passes Air Pollution Research Act.
1956	Another killer smog in London; 1000 people die.
1958	First Public Health Service conference on air pollution.
1959	California becomes first to impose automotive emissions standards.
1960	Clean Water Act passes Congress.
1961	International Clean Air Congress held in London.
1961	World Wildlife Fund founded.
1962	Another London smog; 750 people die.

Era of Environmental Reform (1960–1980)

1962	Rachel Carson writes *Silent Spring*.
1963	Congress passes Clean Air Act with $95 million for study and cleanup efforts at local, state, and federal levels.
1963	Nuclear Test Ban Treaty between United States and U.S.S.R. (Russia) stops above-ground tests of nuclear weapons.
1964	Congress passes Wilderness Act, creating National Wilderness Preservation System.
1965	Congress passes Water Quality Act setting standards for states.

TABLE 1.1   (continued)

1965	Weather inversion creates 4-day air pollution incident in New York City; 80 people die.
1967	Environmental Defense Fund formed.
1968	*The Population Bomb* by Paul Erlich published.
1969	Alaska oil fields opened for exploitation.
1970	Dennis Hayes organizes first Earth Day.
1970	Congress establishes Environmental Protection Agency. Also Clean Air Act and National Environmental Policy Act passed.
1972	Congress passes Federal Water Pollution Control Act, Coastal Zone Management Act, and the Ocean Dumping Act.
1973	Eighty nations sign the Convention on International Trade in Endangered Species (CITES).
1973	Arab oil embargo panics U.S. and European consumers; prices quadruple despite the fact that no real shortage exists.
1974	Congress Passes Safe Drinking Water Act.
1975	Atlantic salmon return to Connecticut River after 100-year absence.
1976	National Academy of Science report on CFCs (chlorofluorocarbon) gasses warns of damage to ozone layer.
1976	Congress passes Resource Conservation and Recovery Act (RCRA) to regulate hazardous waste and garbage.
1977	U.S. Department of Energy is created.
1977	Love Canal, New York, evacuated after discovery of hazardous waste near school.
1979	Three Mile Island incident.

Recent Environmental History (1980–2000)

1980	Times Beach incident where waste oil containing dioxin is sprayed on roads
1980	Superfund legislation directs EPA to clean up abandoned toxic waste spills.
1983	Dec. 3, Bhopal disaster. Union Carbide Co. fertilizer plant leaks chemicals that kill 2000 people, and another 8000 die of chronic effects.
1986	April 26. Chernobyl nuclear reactor explodes in Ukraine. Immediate deaths are numbered at 31, midterm deaths are estimated around 4200.
1987	The Montreal Protocol international agreement to phase out ozone-depleting chemicals signed by 24 countries, including the United States, Japan, Canada, and European Economic Community (EEC) nations.
1988	International treaty bans ocean dumping of wastes.
1989	March 24. *Exxon Valdez* oil tanker runs aground in Prince William Sound, Alaska, spilling 11 million gallons.
1990	United Nations warns that global temperature rise might be as much as $2°F$ in 35 years, recommends reducing CO_2 emissions worldwide.
1992	June 3–14 Earth Summit is held in Rio de Janeiro, Brazil.
1997	December 11. Kyoto Protocol adopted by United States and 121 other nations.

1.2.3 Occupational Hazard Assessment (ca. 1300 to 1900)

During the 16th to 18th centuries, the basis for our current approach to health risk assessment, including a sensitivity to the importance of dose–response, became well established. The following advancements were identified by Covello and Mumpower (1985):

- A study by Agricola (1556) linking adverse health effects to various mining and metallurgical practices.
- A study by Evelyn (1661) linking smoke in London to various types of acute and chronic respiratory problems.
- A study by Ramazzini (1700, 1713) indicating that nuns living in Apennine monasteries appeared to have higher frequencies of breast cancer (Ramazzini suggested that this might be due to their celibacy, an observation that is in accord with recent observations that nulliparous women may develop breast cancer more frequently than woman who have had children) (Macmahon and Cole, 1969; Sherman and Korenman, 1974).
- A study by Sir Percival Pott (1775) indicating that juvenile chimney sweeps in England were especially susceptible to scrotal cancer at puberty.
- A study by Hill (1781) linking the use of tobacco snuff with cancer of the nasal passage.
- A study by Ayrton-Paris (1822) as well as by Hutchinson (1887) indicating that occupational and medicinal exposures to arsenic can lead to cancer.
- A study by Chadwick (1842) linking nutrition and sanitary conditions in English slums to various types of ailments.
- A study by Snow (1855) linking cholera outbreaks to contaminated water pumps.
- Studies by Unna (1894) and Dubreuilh (1896) linking sunlight exposure with skin cancer.
- A study by Rehn (1895) linking aromatic amines with bladder cancer.

Covello and Mumpower (1985) observed that:

> despite these studies, progress toward establishing causal links between adverse health effects and different types of hazardous activities was exceedingly slow. It appears that at least two major obstacles impeded progress. The first was the paucity of scientific models of biological, chemical, and physical processes, especially prior to the 17th and 18th centuries. Related to this was the lack of instrumentation and the lack of rigorous observational and experimental techniques for collecting data and testing hypotheses [p. 107].

1.2.4 Occupational Disease Recognition (1900 to 1930)

During the early years of the 20th century, it became clear that the industrial revolution had been responsible for introducing health hazards that were adversely affecting a large number of workers (McCord, 1937; Hamilton, 1929;

Raffle et al., 1987). Dozens of scientific studies appeared in the literature that discussed various unique diseases observed in numerous workplaces. Some of the best accounts were chronicled by Alice Hamilton (Sicherman, 1984). Because of the relatively large numbers of diseases recognized to be associated with exposure to toxicants in the workplace, Harvard University established its industrial hygiene program in 1937. During the ensuing years, numerous other graduate programs in occupational hygiene were established in an effort to train professionals who could recognize, evaluate, and control the causative agents. These included the University of Michigan, Harvard, Wayne State, University of Cincinnati, University of Pittsburgh, University of North Carolina, University of Washington, University of California at Berkeley, and others.

1.2.5   Toxicology Studies and Risk Assessment (1930 to 1940)

According to Friess (1987), beginning in the 1930s the need to protect humans from the adverse effects of chemicals in the workplace, the marketplace, and the environment became a commonly recognized goal in the United States and Europe. The general approach to risk assessment evolved over time, but it was characterized by acceptance of the premise that human health was related to the degree of exposure and the toxicity of the chemical. The setting of permissible exposure limits for the workplace introduced the concept of acceptable levels of exposure to toxic agents (Stokinger, 1981; Paustenbach, 2000).

Friess (1987) has suggested that what we currently call risk assessment began roughly in the 1930s. He noted that assessments took the form of an initial review of the epidemiological data of the worker/user populations for a specific chemical, and the dose–response data collected in tests involving animals. Following the deliberations of a committee of specialists in the health sciences or by an individual, the epidemiologic and animal toxicologic data for the chemical were assessed, and then the dose–response relationship for each serious health effect in the human was estimated (Stokinger, 1970, 1981; Lanier, 1985). The relation could be displayed either as a curve of dosage versus anticipated response or, in an attempt to linearize the relation, as a curve of log dosage versus percentage response. Friess noted that:

> Whatever the display mode, however, the predicted human dose/response curve was then used for two purposes. It could be used to predict human response amplitudes under a specified exposure scenario. Secondly, accepting a 5 percent response amplitude as being essentially a no-effect level within the limits of biological variability in populations, the curve could be used to establish the human No Observable Effect Level (NOEL). This procedure was, and is, a primitive quantitative risk assessment methodology [page 4].

1.2.6   Concern Over Relatively Modest Health Risks (1940 to 1980)

During the 600 years between 1348 and 1948, society's concern for health risks usually focused on those factors that could increase the risk of infectious disease (Table 1.1). This was appropriate since these were the greatest hazard. For example, the 1348 to 1349 epidemic of the Black Death (bubonic plague) killed over one-quarter of the population of Europe, which was approximately 25 million people (Winslow, 1923; Helleiner, 1967; Ziegler, 1969). However, beginning in the late 1940s, after having eliminated many of the truly serious threats to health (often achieved through better understanding, control, and the use of medicinal drugs), our attention began to be diverted to the more subtle and insidious, yet much lower risk, hazards (Eisenbud, 1978). Specifically, society began to focus on the hazards posed by agents found in our environment. (e.g., air, water, soil, food, sediment).

The tremendous increase in the synthesis and manufacture of organic chemicals (Figure 1.3), coupled with the potential problems described in Carson's *Silent Spring* (1962), suggested that pollution of our environment might very well be the next Black Death if attention were not directed to this threat. Interestingly, concurrent with the increased manufacture of industrial chemicals

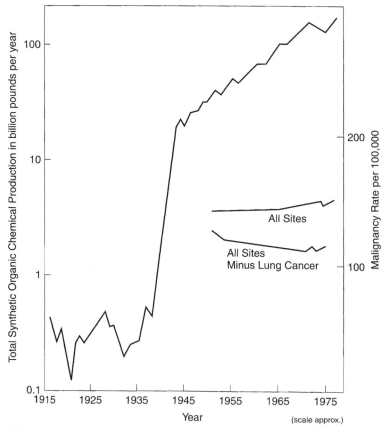

Figure 1.3   In spite of the dramatic increase in the manufacture and presumed increased exposure to synthesized organic chemicals, there has been no apparent increased risk in the cancer rate, even after correction for smoking and other lifestyle factors [American Industrial Health Council (AIHC), 1983].

was the introduction and widespread use of pesticides, including herbicides. As shown in Figure 1.3, there was not a commensurate rise in the cancer rate in the United States as huge quantities of chemicals entered our environment. Concerns about the benefits and risks posed by pesticides have been repeatedly expressed for the the past 50 years since they represented a major contribution to the environmental burden. At about the same time the U.S. Food and Drug Administration (FDA) began to identify and study those chemicals that could hopefully be safely added to foods and drugs (Lehman and Fitzhugh, 1954).

It has been said that society placed a great deal of trust in those firms that directed their efforts to provide for "a better living through chemistry." To a large extent, in fact, this occurred between 1965 and 2000. In retrospect, not enough toxicology studies were conducted to assure their safety given what me know today. However, during this same time period, we recognized that these same chemicals required serious stewardship, for example, they could produce numerous long-term hazards to health and the environment if not maintained, used, and disposed of properly. None of this is surprising since our awareness of the possible hazards had to be obtained, in part through experience.

1.2.7 Setting Acceptable Daily Intakes (1950 to 1970)

The 20 years between 1950 and 1970 *were* important ones in toxicology and risk assessment (Friess, 1987). At least two key reviews in the 1950s serve as milestones; one was by Barnes and Denz (1954) and the other by Lehman and his colleagues at the U.S. Food and Drug Administration (1954) (Friess, 1987). The rationale for and size of the safety factor to be applied to the animal NOEL needed to protect humans evolved from these publications (Dourson and Stara, 1984). For a chemical with a well-defined toxic action at a target tissue, which appears to display a dose threshold and is at least moderately reversible, a safety factor (later called an uncertainty factor) of 100 was proposed. The first factor of 10 was used to extrapolate the NOEL from animals to humans, and the second factor of 10 was to account for the variability in sensitivities within human populations. For more serious irreversible types of effects, even including carcinogenesis, additional safety factors (ranging up to 20) were added to the fundamental 100. For example, at times during the 1960s and 1970s, safety factors of 1000 to 5000 were applied to the apparent NOEL for tumorigenesis observed in chronic animal bioassays in an effort to estimate the "safe dose...." for human populations (Weil, 1972).

1.2.8 The Cancer Hazard (1970 to 1985)

Progress toward the assessment of environmental stressors (e.g., chemical and physical agents) remained at the level of "the dose makes the poison" until about 1974. Some observers, including this author, believe that the vinyl chloride experience may very well have been the critical turning point in the evolution of toxicology and risk assessment. Vinyl chloride (VC) was a chemical that had long been thought so harmless as to be considered sufficiently safe for use as an anesthetic agent for surgery. However, following an epidemiological study of workers involved in its manufacture, it was determined that VC could cause cancer (e.g., angiosarcoma) at levels that were tasteless, odorless, and produced no detectable adverse effects! The vinyl chloride experience sensitized risk scientists and health professionals to the fact that it was quite possible that the background incidence of cancer (believed primarily to be due to diet, smoking, genetic, and other factors) was so high as to possibly mask the adverse effect of a number of carcinogenic agents present in our air, water, and food. As a result, increased emphasis on identifying new approaches to setting acceptable levels of exposure to carcinogens occurred.

The concept of a threshold dose at which no adverse effects would be expected for carcinogens was first challenged in the early 1960s. The classic paper by Mantel and Bryan (1961), which described an approach to estimating virtually safe doses (VSDs), appealed to regulatory agencies who believed that a NOEL for carcinogens might not exist. Soon thereafter, for regulatory purposes, most human exposure to carcinogens was assumed to present a finite incremental cancer risk, irrespective of whether repair processes were operable, after the chemical interacted with DNA (Crump et al., 1976). From this regulatory philosophy evolved the widespread use of mathematical models; those that estimate the excess lifetime cancer risk for humans based on the dose–response curve obtained in animal bioassays. These models had substantial appeal to regulatory agencies; however, it was not clear which model was most likely to provide the best estimate of risk at low doses (Figure 1.4) (Food Safety Council, 1980).

The acceptance of models that predicted the potential upper-bound excess cancer risk of very low doses represented an important turning point in the history of environmental regulation and risk management. Through the use of reasonably standard statistical approaches, the risk estimation approach introduced by Mantel and Bryan in 1961 offered an alternative to the "safety" standard demanded by the Delaney Amendment of 1958. This was later refined by Crump and co-workers (1976) in a very influential paper entitled "Fundamental Carcinogenic Processes and Their Implications for Low Dose Risk Assessment." The approach described by Crump and co-workers became the standard one used by U.S. regulatory agencies from 1976 to the present day for estimating the plausible upper bound of the cancer risk (Anderson et al., 1983; Albert, 1994). Some statisticians believed the confidence bounds on these risk estimates were too great to be a basis for regulatory decision making (Figure 1.5). In 1996, and again in 1999, the EPA proposed cancer guidelines that discussed those situations where use of other low-dose extrapolation models would be considered (EPA, 1996c; 1999d).

1.2.9 Concerns Regarding the Accuracy of Risk Assessments (1980 to 1995)

Friess (1987) predicted that the years between 1985 and 2000 would be a time when public health officials, toxicologists, statisticians, and risk assessors would

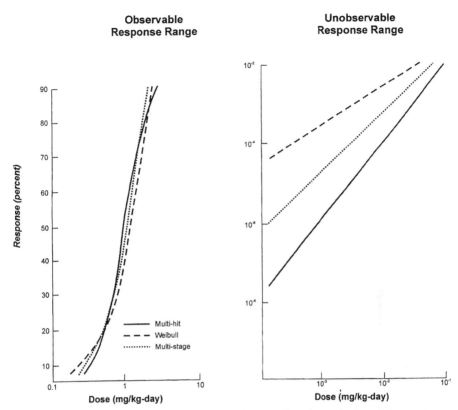

Figure 1.4 Cancer risk assessment relies on the use of mathematical models to extrapolate from the observed experimental range (left plot) to the unobservable environmental exposure range (right plot). All models tend to agree in the high-dose range but may diverge widely in the low-dose range. As a result, the same risk level may be associated with a dose range spanning several orders of magnitude depending on which extrapolation model is used. The U.S. EPA typically relies on the linearized multistage (LMS) model as a default, one of the most conservative of the dose extrapolation models, although alternative models may be used if justified.

begin to question the appropriateness of using so-called cancer models (which are more correctly called dose–response models) to estimate the incidence of tumors in exposed human populations. Indeed, his predictions were accurate. During this 15 to 20-year period, valid scientific arguments were put forth that these models relied on too many assumptions that might apply to initiators of tumorigenesis, but that they were probably not appropriate for promoters (Butterworth and Slaga, 1987; Butterworth et al., 1995; Crump, 1995) and carcinogens that produce tumors through other nongenotoxic mechanisms (Pitot and Dragan, 2001; EPA, 1996a; Powell and Berry, 2000).

Beginning in the mid-1980s, the risk assessment community began to get uneasy about relying on these models in making risk management decisions.

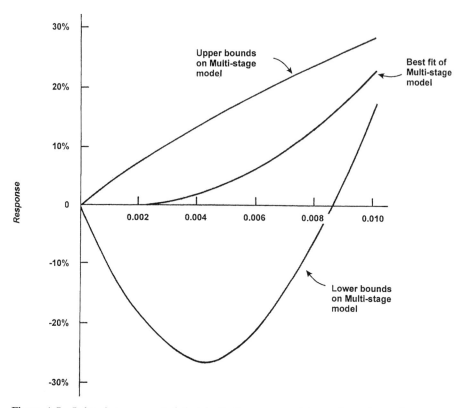

Figure 1.5   It has been suggested that both the upper and lower bounds on the added probability of developing a tumor be presented with each risk assessment. This plot illustrates the marked difference between the upper and lower bounds and the best fit of the multistage model for carcinomas observed in the female rat in the study by Kociba and co-workers using 2,3,7,8-TCDD. From Sielken (1987a).

The questioning of such approaches was illustrated by two pieces that appeared in the *Wall Street Journal*—one in 1984 and the other in 1987. The first addressed the reasonableness of EPA's decision to ban ethylene dibromide (EDB) as a grain fumigant and the second discussed formaldehyde. Havender (1984b) reviewed the ethylene dibromide experience. Dozens of scientific papers and several books were published between 1980 and 1995 about the shortcomings of the risk assessment process:

> Recently, we went through yet another in the parade of hysterias over a carcinogen in the nation's food supply that began with the great cranberry scare of 1959. This time the threat was a grain fumigant, ethylene dibromide, or EDB....

> Does EDB cause cancer in laboratory animals? No knowledgeable person denies this, since all 10 long-term, high dose tests involving male and female rats and mice were unequivocally positive, as was a skin-painting test on mice. Clearly,

EDB must be handled with care. But does this mean that the traces now being found in supermarket food pose a significant hazard to the public and warrant a ban?

According to EPA's estimates, the average person consumes 5 to 10 micrograms of EDB a day—a quantity far too small to be seen with the unaided eye. (By comparison, we typically ingest 140,000 micrograms of pepper—a known carcinogen—each day.) That quantity is less than a quartermillionth of what, on a body weight basis, the rats were given.

In other words, one would have to eat at least 250,000 times as much food every day as we normally do, or some 400 tons daily over a lifetime, to equal the dose that produced cancer in laboratory animals. This huge difference, dwarfing even the 1,000-bottles-of-diet-pop-a-day equivalent human dose that made the saccharin rat tests look ridiculous, is itself enough to justify skepticism about the hazard faced by consumers.

The other reason for not going into a panic over EDB-contaminated foods is that its risk vanishes into insignificance against the background of risks from other, natural carcinogens in food.

Aflatoxin, a mold present in many of the same grain-based foods as EDB (as well as in peanut butter, milk and apple juice) is some 1,000 times more potent than EDB as a carcinogen. Its allowed level in solid food is 20 ppb, which means that consumers are exposed to as much as 20,000 times the carcinogenic hazard from aflatoxin than they would get from EDB present in food at 1 ppb (the "emergency" action level of Florida and some other states).

Against this natural background of dietary carcinogens the risk from 10 micrograms of EDB a day is trivial. Indeed a muffin baked from the most contaminated mix found in California would have, because most of the EDB bakes off, only a fraction of the hazard of a peanut-butter-and-jelly sandwich. If one can eat that sandwich with equanimity, then one should be at least as tranquil about eating corn muffins.

The second commentary, by the editor of the *Wall Street Journal*, described his impression of the wisdom of EPA's position on formaldehyde (*Wall Street Journal*, 1987):

Just presenting the 95%, upper-confidence limit not only grossly distorts the perception of risk, but also unduly alarms the public.

Staffers at the Environmental Protection Agency are calling this widely used medical and industrial chemical [formaldehyde] a "probable human carcinogen," supposedly endangering the health of thousands of Americans. Some EPA officials are therefore urging strict federal regulations on formaldchyde.

To support this claim, the EPA recently produced some seemingly shocking—and highly publicized—statistics on the alleged cancer risks from formaldehyde exposure. The agency purports that three out of every 10,000 garment workers exposed to formaldehyde fumes risk getting cancer. For mobile-home dwellers exposed to formaldehyde-treated pressed wood, the cancer risk over 10 years is said to be two in 10,000. And for residents of conventional homes, the perceived risk is one in 10,000. Given the current widespread fear of cancer. these numbers were certain to scare many people. . . .

EPA's "likely estimate" for cancer risks from formaldehyde exposure among garment workers is actually 4 in 1 *billion*. Similarly, for mobile-home dwellers the "likely" estimate is 2 in 10 billion, and for conventional-home residents 6 in 100 billion . . .

A recent Harvard School of Public Health analysis criticized harshly the federal government's handling of formaldehyde risk assessment. It said that "the true extent of uncertainty about the magnitude of estimated cancer risk is not conveyed to policy makers." As a result, it concluded, "policy guidelines are substituted for scientific judgment in the government risk-assessment process." [p. 5]

The experiences with saccharin, EDB, dioxin, formaldehyde, and methylene chloride proved that the 1980s were difficult years for regulatory agencies that had been mandated to protect the public health, in part, because the human hazard posed by typical levels of exposure in the environment was probably much lower than that predicted by exposure assessments and low-dose extrapolation models used at that time.

1.2.10   Biologically Based Disposition Models (1985 to 2000)

A potentially major advancement in health risk assessment occurred in the early 1980s with the publication of a study by Ramsey and Andersen (1984). This work described a procedure known as physiologically based pharmacokinetic (PB-PK) modeling wherein actual organs and tissue groups were used with weights and blood flows to predict the qualitative time course and distribution of a chemical in the test species. Although their research was an extension of the earlier work of Kety (1951), Mapleson (1963), Riggs (1963), Bischoff (1967), Fiserova-Bergerova (1975), and Davis and Mapleson (1981) and all this was an extension of the seminal work of Haggard (1924), they applied these procedures in a form that was understandable and readily applicable. Moreover, Clewell and Andersen (1985, 1986) showed that with the advent of personal computers, these approaches were within the grasp of virtually all scientists.

The significance of the Ramsey–Andersen work was more widely recognized when the approach was applied to the risk assessment of the bioassay data on methylene chloride (Andersen et al., 1987). Soon after its critical review by several U.S. regulatory agencies, a symposium was jointly sponsored by the National Academy of Science, EPA, FDA, and others, wherein the merit and applications of the PB-PK methodology were discussed. The significance of PB-PK modeling was made evident by the publication of the proceedings of this meeting in Volume 8 of the NAS (1987) series called Drinking Water and Health. At the time, these approaches were hoped to move quantitative risk assessment (and low-dose extrapolation models) to the next level of refinement. This approach is characterized by attempts to incorporate all the available biologic data and our understanding of the mechanisms of cancer into a quantitative estimate of delivered dose and response (Clewell and Andersen, 1985; Krishnan and Andersen, 2001). Although PB-PK models have been developed for nearly 100 chemicals, they have not significantly changed the way the United States regulates chemicals in the workplace or the environment (Table 1.2).

TABLE 1.2 Physiologically Based Pharmacokinetic (PB-PK) Models for Environmental Toxicants

Chemical	
Acetone[a]	2,2′,4,4′5,5′-Hexabromobiphenyl
Acrylonitrile[a]	Hexachlorobenzene[a]
Arsenic[a]	Hexane[a]
Benzene[a]	2-Iodo-3,7,8-trichlorodibenzo-p-dioxin
Benzoic acid	Isoamyl alcohol
Benzo[a]pyrene	Isofenphos
Bromobenzene	Kepone[a]
Bromodichloromethane	Lead[a]
Buta-1,3-diene[a]	Lindane
2-Butoxyethanol[a]	Methanol[a]
Carbon tetrachloride[a]	2-Methoxyethanol
Chlorfenvinphos	Methoxyacetic acid[a]
Chloroalkanes	Methylmercury[a]
Chloroform[a]	Methyl ethyl ketone (MEK)
Chloropentafluorobenzene[a]	Methyl *tert*-butyl ether (MTBE)[a]
Chromium	Nickel
1,2-Dichlorobenzene	Nicotine[a]
1,2-Dichloroethane[a]	Parathion
1,1-Dichloroethylene[a]	Pentachloroethane
1,2-Dichloroethylene	Physostigmine
Dichloromethane[a]	Polychlorinated biphenyls (PCBs)
2,4-Dichlorophenoxyacetic acid[a]	Soman
2,2-Dichloro-1,1,1-trifluoroethane[a]	Styrene[a]
Dieldrin	Toluene[a]
Diisopropylfluorophosphate	2,3,7,8-Tetrabromodibenzo-p-dioxin[a]
5,5′-Dimethyloxazolidine-2,4-dione[a]	2,3,7,8-Tetrachlorodibenzofuran[a]
1,4-Dioxane[a]	2,3,7,8-Tetrachlorodibenzo-p-dioxin[a]
Ethanol	Tetrachloroethylene[a]
Ethyl acrylate	1,1,1-Trichloroethane[a]
Ethyl acetate	Trichloroethylene[a]
Ethylene oxide	1,1,2-Trichloro-1,2,2-trifluoroethane
Fluoride	Vinyl chloride[a]
Fluazifop-butyl[a]	Vinylidene fluoride
Furan[a]	Xylene[a]

Source: Leung (2000).

[a] These chemicals have more than one model.

1.2.11 Biologically Based Cancer Models (1985 to 2000)

Although the development and use of biologically based pharmacokinetic or disposition models represented a significant improvement in our approach to estimating the low-dose response, another potentially important contribution was the development of biologically based cancer models (Moolgavkar, 1986). The first and perhaps most promising of these models was developed by Moolgavkar and Knudson (1981). Despite the diversity of the disease processes categorized under the heading of cancer, their model was compelling even though it was relatively simple. It is a form of the two-stage model initially described by Armitage and Doll in 1954. In this approach, cancer is explained as the end result of two mutagenic events (u1 and u2), corresponding to mutations at a critical gene locus that in the human is duplicated within the genetic material of the cell. As discussed by Andersen (1991), the first event produces an intermediate cell type that may have different growth characteristics than the normal cell but one that is not aggressively malignant. A second irreversible event is necessary to complete the cell transformation process, alter the second locus of the critical gene, and obtain the cancer cell that grows into a tumor by clonal expansion.

Using this model, one can explain how genotoxicants alter mutation rates, how cytotoxicants alter cell death and birth rates of the normal and intermediate cells, and how promoters convey growth advantages on the intermediate cell populations. This approach was first used by Thorslund et al. (1987) and Conolly et al. (1988) and improved upon by Moolgavkar et al. (1986, 1992, 1996). Regrettably, the data upon which to build these models has not been available. As a result, these models have yet to reach their potential and, according to Crump (1995), may never be able to do so.

1.2.12 The Mathematical Modeling and Benchmark Dose Method (1995 to 2000)

Mathematical models may be applied to data on dose–response to reduce the uncertainty in identifying a reliable (i.e., statistically valid) no observed adverse effect level [NOAEL] for determanistic chemicals or, alternatively, the dose at which 10 percent of the study animals are expected to show a response (ED_{10}) (Krewski et al., 1984; Moolgavkar et al., 1999). Examples of relevant curve-fitting models include the probit and Weibull (Park and Snee, 1983; Paustenbach, 1995; Moolgavkar et al., 1999). These models take into account the uncertainty in dose–response curves. Sometimes these models are combined with physiologically based pharmacokinetic (PB-PK) models to predict the response across the doses tested (Reitz et al., 1996).

Since about 1995, EPA and other agencies have begun to use the so-called benchmark dose method to estimate the NOAEL and the ED_{10} (Crump, 1984, 1995; Barnes et al., 1995; EPA, 1995a). As illustrated in Figure 1.6, a statistical fit of a dose–response model to the dose–response data is used to identify an

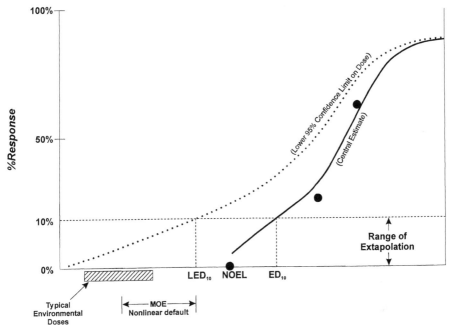

Figure 1.6 Illustration of benchmark dose method for estimating dose corresponding to specified level of increased response; ED_{10} and LED_{10} (the benchmark dose) are central estimate and lower confidence limit of dose corresponding to 10% increase in response, respectively. These are obtained from statistical fitting of dose–response model to dose–response data, and MOE is margin of exposure (safety factor) that can be applied to the benchmark dose.

LED_{10} (the benchmark dose), which is the lower 95 percent confidence limit of the dose, resulting in a response in 10 percent of the study animals. The benchmark dose then is used as the point of departure for establishing allowable exposures to deterministic chemicals, in a manner similar to the approach of determining reference dose (RfDs) from NOAELs by using safety and uncertainty factors.

The rationale supporting use of the LED_{10} as the benchmark dose is that a 10 percent response is at or just below the limit of sensitivity in most animal studies. Use of the lower confidence limit on dose, rather than the best (maximum likelihood) estimate (i.e., the ED_{10}), in determining the benchmark dose accounts for experimental uncertainty; the difference between the lower confidence limit and the best estimate does not provide information on the variability of responses in humans. In risk assessments for deterministic substances, a dose at which significant effects are not observed is not necessarily a dose that results in no effects in any animals, due to the limited sample size. The NOAEL obtained using most study protocols is about the same as an LED_5 or LED_{10}, that is, the lower confidence limit on the dose associated with 5 or 10 percent increase in responses. The ratio of the LED_{10} (or the LED_5) to the estimated dose in humans is called the margin of exposure (MOE) (Figure 1.5).

The benchmark dose method was developed to overcome difficulties with determining the NOAEL based on dose–response data. The potential advantages of the method include the following (Crump, 1984, 1995):

- The benchmark dose method makes use of all the dose–response data by fitting a dose–response model to the data, whereas the determination of a NOAEL generally involves a comparison of responses at discrete doses with responses in control subjects.
- The benchmark dose reflects a sample size more appropriately than the NOAEL because small studies tend to result in smaller benchmark doses, whereas the opposite is the case for NOAELS.
- The NOAEL is constrained to be one of the administered doses, but this is not the case with the benchmark dose method.
- A benchmark dose method can be defined from a data set that does not include a NOAEL.
- The determination of a NOAEL generally involves dose data that are categorized into distinct groups, but this categorization is arbitrary in some studies. Grouping of data into distinct dose categories is not required in the benchmark dose method.
- The benchmark dose approach gained popularity in the late 1990s and was the approach recommended by the EPA Science Advisory Board (SAB) for estimating safe levels of exposure to dioxin (EPA, 2001).

1.2.13   Guidance Documents (1985 to 2000)

In the 1980s, a number of handbooks and guidelines described general approaches to preparing health risk assessments were developed. For example, the U.S. EPA published generic risk assessment guidelines for carcinogens (EPA, 1986a), developmental toxicants (1986b), exposure assessment (1986c), mixtures (1986d), municipal waste combustors (1986e), systemic toxicants (1988a) as well as female (1988b) and male reproductive (1988c) toxicants. The EPA also issued a handbook for conducting endangerment assessments (EPA, 1986f) and risk assessments of Superfund sites (EPA, 1986g). In addition, the State of California published a handbook for conducting assessments of carcinogenic substances (California Department of Health Services, 1985) and guidelines for safe-use determinations for compliance with California's Proposition 65 (California Department of Health Services, 1988).

In the 1990s, many of these guidance documents were updated and yet others were issued. Most of these were published in the United States while some others were developed in other countries. Among the most notable of the more recent guidance documents are the ones for conducting ecological risk assessment (EPA, 1998c), evaluating neurotoxicants (EPA, 1999c), the draft EPA cancer guidelines (1999d), and the exposure factors handbook (EPA, 1997e).

1.3 HELPFUL RESOURCES FOR CONDUCTING RISK ASSESSMENTS

As discussed by the National Academy of Sciences committee convened in 1983, most human or environmental health hazards can be evaluated by dissecting the analysis into four parts. These are hazard identification, dose–response assessment, exposure assessment, and risk characterization. Each is clearly defined in the committee's report *Risk Assessment in* the *Federal Government* (NAS, 1983).

The remainder of this chapter recommends a number of books, book chapters, and scientific papers, including some chapters from this book, which are worth considering when preparing human health or ecological risk assessments. The resources are divided according to the four steps. The works cited are not intended to represent a compilation of all the important papers but rather to illustrate that a wealth of information is available to scientists responsible for conducting these assessments. Nearly 400 additional references are listed at the end of this chapter.

1.3.1 Hazard Identification

The hazard identification step in risk assessment contains a description of a particular chemical or physical agent's capacity to adversely affect, at some dose, the health of biota, fish. wildlife, or humans. It is defined as the process of determining whether exposure to an agent can cause an increased incidence of an adverse health effect (cancer, birth defect, etc.). It involves characterizing the nature and strength of the evidence of causation. Numerous reference texts that describe the toxic effects of about 2000 of the most commonly used chemicals are available. Fewer texts describing the potential adverse effects of xenobiotics on cattle, sheep, fish, birds, and various types of wildlife have been written because of the smaller number of toxicity studies conducted on these nontraditional species.

Discussions about how to conduct toxicology studies, compilations of toxicity test results, or resources listing risk criteria can be found in the following texts:

- *Principles and Methods of Toxicology* (Hayes, 2001)
- *Patty's Toxicology* (Bingham et al., 2001)
- *Documentation of Threshold Limit Values* [American Conference of Governmental Industrial Hygienists (ACGIH), 1999]

- *Sax's Dangerous Properties of Industrial Materials* (Lewis, 2000)
- *General and Applied Toxicology* (Ballantyne et al., 1999)
- *Toxicology* (Marquardt et al., 1999)
- *Immunotoxicology of Environmental and Occupational Metals* (Zelikoff and Thomas, 1998)
- *Encyclopedia of Occupational Health and Safety* (World Health Organization, 1998)
- *Hamilton and Hardy's Industrial Toxicology* (Harbison, 1998)
- *Proctor and Hugh's Chemical Hazards of the Workplace* (Hathaway, Proctor and Hughes, 1996)
- *Registry of Toxic Effects of Chemical Substances* (EPA, 1996)
- *Toxicology: Principles and Applications* (Niesink et al., 1996)
- *Casarett and Doull's Toxicology: The Basic Science of Poisons* (6th Edition) (Klaassen, 2001)
- *Catalog of Teratogenic Agents* (Shepard, 1995)
- *Occupational Diseases: A Guide to Their Recognition* [National Institute for Occupational Safety and Health (NIOSH), 1976]
- *Hazardous Materials Toxicology: Clinical Principles of Environmental Health* (Sullivan and Kreiger, 1992)
- *Chemically Induced Birth Defects* (Schardein, 1992)
- *Air Toxics and Risk Management* (Calabrese and Kenyon, 1991)
- *Handbook of Pesticide Toxicology; 3 volumes* (Hayes and Laws, 1991)
- International Agency for Research on Cancer (IARC) (1985–2000), numerous volumes
- *Handbook of Toxic and Hazardous Chemicals and Carcinogens* (Sittig, 1985)
- *Clinical Toxicology of Commercial Products* (Gosselin et al., 1984)
- *Toxic and Hazardous Industrial Chemicals Safety Manual* (International Technical Information Institute, 1976)
- *Clinical Toxicology* (Thiennes and Haley, 1972)

Each year, new information on the potential adverse effects of xenobiotics on fishes and wildlife becomes available. A few texts have reviewed the overall approach to hazard identification and cataloged some of the adverse effects on fish and other wildlife. The following are useful resources:

- *Development of Methods for Effects-Driven Cumulative Effects Assessment Using Fish Populations: Moose River Project* (Munkittrick et al., 2000)
- *Evaluation of Persistence and Long-Range Transport of Organic Chemicals in the Environment* (Klecka et al., 2000)

- *Evaluating and Communicating Subsistence Seafood Safety in a Cross-Cultural Context: Lessons Learned from Exxon Valdez Oil Spill* (Field et al., 2000)
- *Guidelines for Ecological Risk Assessment* (EPA, 1998c)
- *Ecotoxicology: Ecological Fundamentals of Chemical Exposure and Biological Efforts* (Schurmann and Markell, 1998)
- *Handbook of Ecotoxicology* (Hoffman et al., 1995)
- *Fundamentals of Aquatic Toxicology* (Rand, 1995)
- *Basic Environmental Toxicology* (Cockerham and Shane, 1994)
- *Ecological Assessment of Hazardous Waste Sites* (Maughn, 1993)
- *Ecological Risk Assessment* (Bartell et al., 1992)
- "Risk Assessment for Nitrilotriacetic Acid (NTA)" (Paustenbach, 1989, Chapter 10)
- *The Study of Pollutants in Ecosystems* (Moriarity, 1988)
- *Environmental Hazard Assessment of Effluents* (Bergmann et al., 1986)
- *Handbook of Adverse Effects on the Fathead Minnow* (Brooke et al., 1984)
- *Aquatic Toxicology and Hazard Assessment: Sixth Symposium* (Bishop et al., 1983)
- *Biotransformation and Fate of Chemicals in the Aquatic Environment* (Maki et al., 1980)
- Chapter 8, this volume

Except for the intentional toxic effects posed by pesticides, few general reference texts appear to have been written that review the adverse effects of industrial chemicals on the health or reproductive capacity of birds, large and small wildlife, and the like. General discussions of the hazard identification process for fish and avian species, however, have been published.

- *Multiple Stressors in Ecological Risk and Impact Assessments: Approaches to Risk Estimation* (Ferenc and Foran, 2000)
- *Ecotoxicology and Risk Assessment for Welands* (Lewis et al., 1999)
- *Ecotoxicology: Ecological Fundamentals, Chemical Exposure, and Biological Effects* (Schuurmann and Markett, 1998)
- *Reassessment of Metals Criteria for Aquatic Life Protection* (Bergman et al., 1997)
- *Handbook of Pesticide Toxicology; 3 volumes* (Hayes and Laws, 1991)
- *Handbook of Ecotoxicology* (Hoffmann et al., 1995)
- "An Environmental Risk Assessment of a Pesticide" (Paustenbach, 1989; Chapter 27)

- *Endangerment, Assessment for the Bald Eagle Population Near the Sand Springs Petrochemical Complex Superfund Site* (Paustenbach, 1989, Chapter 29)
- *Examination of Potential Risks from Exposure to Dioxin in Sludge Used to Reclaim Abandoned Strip Mines* (Paustenbach, 1989, Chapter 28)
- *The Risk of Chemicals to Aquatic Environment* (Southworth et al., 1982)
- *Pesticides Studied in Man* (Hayes, 1982)
- *Toxicology of Pesticides* (Hayes, 1975)
- Chapter 24, this volume
- Chapter 22, this volume

Sampling and Statistical Aspects   An important aspect of hazard identification and exposure assessment is a description of the pervasiveness of the hazard. For example, most environmental assessments involving contaminated soil or sediment require knowledge of the concentration of the material in the environment, weighted in some way to account for the geographical magnitude of the contamination, for example, a 1-acre or 300-acre site, a 1000 gal/min or 1,000,000 gal/min stream. All too often, environmental incidents have been described by statements like "concentrations as high as 800 ppm" (parts per million) of a chemical were measured at a 1000-acre waste site. However, following closer examination, one may find that only 1 of 200 samples collected on a 20-acre portion of a 1000-acre site showed this concentration, and that 2 ppm was the mean concentration of the contaminant in the 200 samples.

An appropriate sampling program is critical to the conduct of a health risk assessment. This topic could arguably be part of the exposure assessment, but it has been placed within hazard identification because, if the breadth of contamination is small, no further work may be necessary. Accurate sampling strategies are also important in occupational risk assessment where time-weighted average samples collected over the entire work day, rather than those collected over shorter periods, are known to be the best way to assess the chronic inhalation hazard. When evaluating the potential hazard to wildlife, a thorough sampling plan is no less important since a representative number of fish, birds, rattlesnakes, mice, worms, or other species needs to be collected before the seriousness of the hazard can be characterized.

Schemes or approaches for collecting representative samples of soil, water, air, and other media are perhaps better defined than those used for the collection of birds, fish, mice, or food stuffs. The wealth of information on sampling theory, and the importance of random sampling developed within the statistics community has only recently (e.g., within the past 10 to 15 years) been incorporated into the practice of environmental risk assessment. Scientists involved in these activities should find the following texts useful in designing sampling plans and for interpreting the data:

- "Less Than Obvious: Statistical Treatment of Data Below the Detection Limit" (Helsel, 1990)
- *Lognormal Distribution in Environmental Applications* (EPA, 1997a)
- *Environmental Statistics and Data Analysis* (Ott, 1995a)
- *Supplemental Guidances to RAGS: Calculating the Concentration Term* (EPA, 1992c)
- "Estimating the Mean of Data Sets with Nondetectable Values" (Travis and Land, 1990)
- "Estimation of Averages in Truncated Samples" (Haas, 1990)
- "Evaluation of Statistical Estimation Methods for Lognormally Distributed Variables" (Parkin et al., 1988)
- "A Method for Evaluating the Mean Exposure from a Lognormal Distribution" (Rappaport and Selvin, 1987)
- *Statistical Methods for Environmental Pollution Monitoring* (Gilbert, 1987)
- "Statistical Design and Data Analysis Requirements" (Leidel and Busch, 1985)
- *Handbook for Interpreting Soil Data* (EPA, 1984)

Not only is it important in nearly any risk assessment that samples be collected in a random or representative manner, but the number of samples should be sufficient to conduct a statistically valid analysis. The number needed to ensure statistical validity will be dictated by the variability of the data. The larger the variance, the greater the number of samples needed to define the extent of contamination. These and other important concepts are addressed in the above-mentioned references.

Environmental Fate   When assessing the potential environmental hazard of a chemical in drinking water, air, fly ash, sediment, soil, or groundwater, it is important to understand the behavior of the toxicant in that particular media. For example, the potential groundwater contamination posed by dioxin in soil is entirely different from the hazard posed by the contamination of soil by perchloroethylene (PERC) or other more water-soluble chemicals. Even though dioxin may be orders of magnitude more toxic than PERC, it poses a much lesser hazard to groundwater since it is virtually insoluble in water [less than 2 parts per trillion (ppt)] and because it binds tenaciously to soils and virtually all other solids. A consideration of these types of phenomena is important to the hazard characterization process and has been addressed in the following publications:

- *Fugacity Models* (Mackay, 2000)
- *The Soil Chemistry of Hazardous Materials* (Dragun, 1998)
- *Fundamentals of Ecological Modeling: Developments in Environmental Modeling* (Jorgensed, 1994)

- *Practical Handbook of GroundWater Monitoring* (Nielsen, 1991)
- *Groundwater Models: Scientific and Regulatory Applications* (NRC, 1990)
- *Handbook of Chemical Property Estimation Methods: Environmental Behavior of Organic Compounds* (Lyman and Rosenblat, 1990)
- "Comprehensive Methodology for Assessing the Risks to Humans and Wildlife Posed by Contaminated Soils" (Paustenbach, 1989, Chapter 7)
- *Dynamics, Exposure and Hazard Assessment of Toxic Chemicals* (Haque, 1980)
- *Environmental Risk Analysis for Chemicals in Air, Water, and Soil* (Thibodeaux, 1979)

Chemical Properties   A thorough understanding of the chemical and physical properties of the toxicant is very useful when attempting to understand the severity of the hazard. For example, when chemicals such as the heavy metals, polychlorinated biphenyl (PCB), dichlorodiphenyltrichloroethane (DDT), or the dioxins are not heated, the inhalation hazard is low because of their low volatility. On the other hand, due to their environmental persistence and long biologic half-life, these chemicals may pose a significant hazard if they enter the food chain and accumulate in fish, meat, or milk. In addition, the chemical and physical properties of a chemical will influence its bioavailability when it enters living organisms. The following references should provide insight into understanding the importance of chemical and physical properties in the hazard identification step and, indeed, the entire risk assessment process:

- *Evaluation of Persistence and Long-Range Transport of Organic Chemicals in the Environment* (Klecka et al., 2000)
- *Chemical Transport* (MacKay, 2000)
- *The Handbook of Chemistry and Physics* (Weast, 2000)
- *Principles and Processes for Evaluating Endocrine Disruption in Wildlife* (Kendall et al., 1998)
- *Chemically Induced Alterations in Functional Development and Reproduction of Fishes* (Rolland et al., 1997)
- *Bioavailability* (Hrudey et al., 1996)
- *The Caltox Model* (McKone, 1996)
- *Handbook of Chemical Property Estimation Methods: Environmental Behavior of Organic Compounds* (Lyman and Rosenblat, 1990)
- *Handbook of Organic Industrial Solvents* (American Mutual Insurance Alliance, 1988)
- *Environmental Assessment Technical Handbook* (U.S. Food and Drug Administration, 1984)
- *A Review and Analysis of Parameters for Assessing Transport of Environmentally Released Radionuclides in Agriculture* (Baes et al., 1984)

- Chapter 11, this volume
- "Correlation of Bioconcentration Factors of Chemicals in Aquatic and Terrestrial Organisms with Their Physical and Chemical Properties" (Kenaga, 1980)
- *Understanding Chemicals in the Environment* (Maki et al., 1979)
- "Some Physical Factors in Toxicological Assessment Tests" (Freed et al., 1979)

1.3.2 Dose–Response Assessment

The process of characterizing the relationship between the dose of a substance and the likelihood of an adverse health effect in the exposed population is called the dose response assessment (NAS, 1983). This step should take into account such factors as sex, genetic hypersusceptibility, and other modifying factors, as well as the low-dose extrapolation. For purposes of most risk assessments, the greatest degree of uncertainty will be associated with the extrapolation from the high doses tested in animals to human esposure at the concentrations found in the environment. Typically, environmental exposure is 100- to 10,000-fold below the lowest dose tested in the animal study. It is important in the dose–response assessment that the method used to extrapolate and the justification for selecting that method be described in detail. In the main, although many models are available to estimate risk at very low doses, it is not possible to select the best model based on the fit to the response in observable range. Therefore, either regulatory policy or a weight of biological evidence approach (based on mechanism) is relied upon to make the selection.

Classic Toxicants   Between 1975 and 1995, toxicologists were concerned about identifying the correct approach to extrapolating the results obtained in cancer bioassays to those doses to which most persons were exposed. However, the need to evaluate the risk associated with doses much lower than those tested in experimental animals is equally important when assessing chemicals whose primary adverse effects include noncancer effects (reproductive and developmental toxicity) or where the adverse effect occurs in the liver, kidney, nervous system, respiratory tract, central and peripheral nervous system, the immune system, or other organs (Renwick, 1995; Haber et al., 2001). In the numerous scientific publications that have reviewed the various approaches to extrapolating animal data to identify safe levels of exposure for humans, virtually all have focused on the use of the safety factor or uncertainty factor approach. This is appropriate since all adverse effects, other than cancer and mutation-based developmental effects, are believed to have a threshold dose below which no adverse effects should occur. These approaches have been reviewed by Haber et al. (2001).

The following are excellent references that discuss approaches to setting exposure limits for chemicals that are thought to have thresholds:

- *Noncancer Risk Assessment: Principles and Practice in Environmental and Occupational Settings* (Haber et al., 2001)
- *History and Biological Basis of Occupational Exposure Limits* (Paustenbach, 2000)
- *IRIS: Integrated Risk Information System Database* (EPA, 2000)
- "The Use of an Additional Safety or Uncertainty Factor for Nature of Toxicity in the Estimation of Acceptable Daily Intake and Tolerable Daily Intake Values" (Renwick et al., 1998)
- *Evolution of Science-Based Uncertainty Factors* (Dourson et al., 1996)
- *EPA Guidance for Exposure Risk Assessment* (Barnes et al., 1996)
- "Calculation of Benchmark Doses from Continuous Data" (Crump, 1995)
- "Data-Derived Safety Factors for the Evaluation of Food Additives and Environmental Contaminants" (Renwick, 1993)
- "Safety Factors and Establishment of Acceptable Daily Intake" (Renwick, 1991)
- "Acceptable Daily Intake: Inception, Evaluation and Application" (Lu, 1988)
- "Regulatory History and Experimental Support of Uncertainty (Safety Factors)" (Dourson and Stara, 1983)
- *Methodological Approaches to Deriving Environmental and Occupational Health Standards* (Calabrese, 1978)
- *Drinking Water and Health*, Vols. 1–10 (NAS, 1977–1995)
- "Statistics Versus Safety Factors and Scientific Judgement in the Evaluation of Safety for Man" (Weil, 1972)

Developmental and Reproductive Toxicants   Since 1980, a number of researchers have discussed various approaches to setting acceptable daily intakes or exposure limits for developmental toxicants, as well as reproductive toxicants. At one point in the 1980s, it was suggested that cancer models used for extrapolating bioassay data might be a plausible approach to setting acceptable limits of exposure to these compounds. However, since a threshold is generally believed to exist for developmental and reproductive toxicants, linear, non-thresholds have been considered inappropriate. A number of approaches were proposed in the 1990s for the developmental toxicants and these are recommended for use until better ones are identified.

The following works either present or review the various approaches to setting limits for these categories of toxicants:

- "Proposed Occupational Exposure Limits for Select Glycol Ethers Using PBPK Models and Monte Carlo Simulations" (Sweeney et al., 2001)
- "Issues in Qualitative and Quantitative Risk Analysis for Developmental Toxicity" (Kimmel and Gaylor, 1998)

- "Benchmark Dose Concept Applied to Data from Conventional Development Toxicity Studies" (Kimmel et al., 1995)
- "Dose-Response Assessment for Development Toxicity. II. Comparison of Generic Benchmark Dose Estimates with NOAELs" (Kimmel et al., 1995)
- *Chemically Induced Birth Defects* (Schardein, 1992)
- "A Case Study of Developmental Toxicity Risk Estimation Based on Animal Data" (Paustenbach, 1989, Chapter 21)
- "Cross Species Extrapolations and the Biologic Basis for Safety Factor Determinations in Developmental Toxicology" (Johnson, 1988)
- "Quantification of the Genetic Risk of Environmental Mutagens" (Ehling, 1988)
- *Proposed Guidelines for Assessing Female Reproductive Risk* (EPA, 1988b)
- *Proposed Guidelines for Assessing Male Reproductive Risk* (EPA, 1988c)
- "The Value of Animal Teratogenicity Testing for Predicting Human Risk" (Brown and Fabro, 1988)
- "A Tier System for Developmental Toxicity Evaluations Based on Considerations of Exposure and Effect Relationships" (Johnson, 1987)
- "Evaluation of Developmental Toxicity Data: A Discussion of Some Pertinent Factors, and a Proposal" (Hart et al., 1987)
- *Risk Assessment Guidelines for Suspect Developmental Toxicants* (EPA, 1986)

Carcinogens   The various approaches for predicting the dose–response relationship that have received the most thorough discussion involve those addressing chemical carcinogens. A good deal of the basis for the modeling of chemical carcinogens comes from our experience with human exposure to radiation (BEIR, 1980). Even though experience with exposure to ionizing radiation may give us some insight into the validity of the extrapolation models for some initiators, it remains premature to assume that these models are appropriate for estimating the low-dose response for all chemical carcinogens. In fact, there are many in the scientific community who believe that at least three major categories of carcinogens are likely: cytotoxicants, initiators. and promoters. Each type would require a different approach to assessing the low-dose response (Williams and Weisburger, 1981; Pitot and Dragan, 1995; Powell and Berry, 2000).

Some of the differences between chemical- and radiation-induced carcinogens are that gamma, beta, and alpha radiation are all known to be genotoxic, whereas not all chemical carcinogens act through a genotoxic mechanism. This is an important point since the ability of a chemical to interact directly with DNA is the primary rationale for assuming a linear response at very low doses (Pitot and Dragan, 1995). Low-dose linearity for radiation may be appropriate because it is an initiator and because the dose to which pceple are exposed is linearly related to the internal dose received at the target organ.

However, the internal or delivered dose of a chemical carcinogen to which humans may be exposed will often not be linear at the target organ (Andersen et al., 1989). For example, carcinogens ingested in food must first be diluted in the stomach, transferred into the bloodstream, and transported to other organs, including the liver, where they will be either redistributed as the parent compound or metabolized. Since most environmental chemicals must be metabolized to become reactive, and because the conversion to the metabolite and the likelihood that it will interact with deoxyribonucleic acid (DNA) will often be dose-dependent, this process will frequently be nonlinear (Bus and Gibson, 1995). The parent chemical or metabolite must then be transported through the cell membrane and then through the cytoplasm of the cell before possible interaction with DNA. Pharmacokinetic analyses can be very helpful in estimating the likely target tissue dose based on the administered dose.

The potential strengths and weaknesses of the most commonly accepted approaches to estimating the low-dose response go far beyond the scope of this discussion, and the reader is encouraged to study the following publications, which address this topic:

- *Cancer Risk Assessment* (Park and Hawkins, 2000)
- *Quantitative Estimation of Low Dose Responses* (Ziese et al., 1999)
- *Proposed EPA Guidelines for Carcinogenic Risk Assessment* (EPA, 1999d)
- "Risk Assessment of Nongenotoxic Carcinogens Based upon Cell Proliferation/Death Rates in Rodents" (Gaylor and Zheng, 1996)
- "Predicting Cancer Risk from Vinyl Chloride Exposure with a Physiologically Based Pharmakokinetic Model" (Reitz et al., 1996)
- "The Linearized Multistage Model and the Future of Quantitative Risk Assessment" (Crump, 1996)
- *Quantitative Risk Assessment and Limitations of the Linearized Multistage Model* (Lovell and Thomas, 1996)
- *Low-Dose Extrapolation of Cancer Risks* (International Life Sciences Institute, 1995) (Olin, editor)
- "Use of Mechanistic Models to Estimate Low Dose Cancer Risks" (Crump, 1994a)
- "Carcinogen Risk Assessment in the U.S. Environmental Protection Agency" (Albert, 1994)
- *Quantitative Cancer Modeling and Risk Assessment* (Holland and Sielken, 1993)
- "A Time-to-Response Perspective on Ethylene Oxide Carcinogenicity" (Paustenbach, 1989, Chapter 4)
- *Pharmacokinetics in Low-Dose Extrapolation Using Animal Cancer Data* (Whittemore et al., 1988)
- "Determining Safe Levels of Exposure: Safety Factors or Mathematical Models" (Krewski et al., 1984)

- "The Multistage Model with Time-Dependent Dose Pattern: Applications of Carcinogenic Risk Assessment" (Crump and Howe, 1984)
- "An Improved Procedure for Low-Dose Carcinogenic Risk Assessment from Animal Data" (Crump, 1984)
- "Quantitative Approaches Used to Assess Cancer Risk" (Anderson and the Carcinogen Assessment Group, 1983)
- *Quantitative Risk Assessment* (Food Safety Council, 1980)
- "Fundamental Carcinogenic Processes and Their Implications for Low Dose Risk Assessment" (Crump et al., 1976)
- "Safety Testing of Carcinogenic Agents" (Mantel and Bryan, 1961)

Biologically Based Disposition and Cancer Models   Perhaps one of the most important breakthroughs, which should allow us to conduct much more accurate carcinogenic risk assessments was a mathematical approach used by Bischoff and Brown (1966) and later further refined by Fiserova-Bergerova (1975) and by Ramsey and Andersen (1984). This procedure, known as physiologically based pharmacokinetics (PB-PK) modeling, allows scientists to extrapolate the biologically important dose delivered to target organs more accurately than other approaches (Menzel, 1987). PB-PK models differ from the conventional compartmental models in that they are based to a large extent on the actual physiology of the organism. Instead of using compartments defined by the experimental data themselves, actual organ and tissue groups are used with weights and blood flows from the literature (Bischoff and Brown, 1966; Himmelstein and Lutz, 1979). Instead of composite rate constants determined by fitting the data, actual physicochemical and biochemical constants of the compound are used. The result is a model that predicts the qualitative behavior of the experimental time course *without being based on it*. Refinement of the model to incorporate additional insights gained from comparison with experimental data yields a model that can be used for quantitative extrapolation well beyond the range of experimental conditions (Clewell and Andersen, 1985). Such approaches may also give insight on the reasonableness of low-dose linearity on a chemical-by-chemical basis. These analyses, coupled with pharmacodynamic models, are likely to allow us to make dramatic improvements in the way we assess risks associated with low-level exposures.

The following references are important for understanding PB-PK models or the more newly termed biologically based disposition models:

- "Development of Physiologically-Based Pharmacokinetic Modeling in Toxicology" (Krishnan and Andersen, 2001)
- "Physiologically Based Pharmocokinetic Models" (Leung, 2000)
- "A Physiologically-Based Model for Chromium" (O'Flaherty et al., 2001)
- "Molecular Structure-Based Prediction of the Toxicokinetics of Inhaled Vapors in Humans" (Poulin and Krishnan, 1999)

- "Physiologically Based Modeling of Vinyl Acetate Uptake, Metabolism, and Intracellular pH Changes in the Rat Nasal Cavity" (Plowachalk et al., 1997)
- *Incorporating Monte Carlo Simulation into Physiologically Based Pharmacokinetic Models Using Advanced Continuous Simulation Language (ACSL): A Computational Method* (Thomas et al., 1996)
- "Predicting Cancer Risk from Vinyl Chloride Exposure with a Physiologically Based Pharmacokinetic Model" (Reitz et al., 1996)
- "A Physiologically-Based Model and Risk Assessment for Vinyl Chloride" (Reitz et al., 1994)
- "Physiologically-Based Models" (Leung and Paustenbach, 1995)
- "Physiologically Based Models of Metals" (O'Flaherty, 1998)
- "Physiologically-Based Pharmacokinetics and the Risk Assessment Process for Methylene Chloride" (Andersen et al., 1987)
- "A Physiologically Based Description of the Inhalation Pharmacokinetics of Styrene in Rats and Humans" (Ramsey and Andersen, 1984)

1.3.3 Exposure Assessment

Over the past 15 years, a significant amount of attention and research has been directed to the exposure assessment component of risk assessment (Paustenbach, 2000). This is logical since many of the risk assessments performed during the 1970s tended to make too many conservative assumptions that, as a result, frequently overestimated the actual exposure. Recent interest makes sense because this was the portion of the risk assessment process that, if improved, could increase the accuracy of risk assessments.

Exposure assessment is the process wherein the intensity, frequency, and duration of human exposure to an agent are estimated. The process may also address the exposure of wildlife to chemical agents. An exposure assessment should quantitatively estimate the magnitude (size) of the exposed population, the routes of entry, and the uncertainties in the exposure estimates. Three routes of entry generally need to be considered: inhalation, dermal absorption, and ingestion.

The references listed in the following three sections are useful for conducting exposure assessments and for calculating the uptake of xenobiotics by humans by these three routes of entry.

Inhalation

- Chapter 4, this volume
- "Assessment of Airborne Exposure to Trihalomethanes from Tap Water in Residential Showers and Baths" (Kerger et al., 2000)
- *Methodology for Assessing Health Risks Associated with Multiple Exposure Pathways to Combustor Emissions* (EPA, 1997c)

- *Exposure Factors Handbook* (EPA, 1996d)
- "Assessment of Potential Health Hazards Associated with PCDD and PCDF Emissions from a Municipal Waste Combustor" (Paustenbach, 1989, Chapter 19)
- "Formaldehyde Exposure and Risk in Mobile Homes" (Paustenbach, 1989, Chapter 17)
- *Report of the Task Group on Reference Man* (Snyder, 1975)
- Chapter 11, this volume

Dermal Absorption

- *Bioavailability* (Hrudey et al., 1996)
- *Dermal Exposure Guidance* (EPA, 1992, 2002)
- "Techniques for Estimating the Percutaneous Absorption of Chemicals Due to Environmental and Occupational Exposure" (Leung and Paustenbach, 1994)
- *Contact and Occupational Dermatology* (Marks and Deleo, 1992)
- "Hazard Assessment of 1,1,1-Trichloroethane in Groundwater" (Paustenbach, 1989, Chapter 8)
- *Dermatotoxicology* (Marzulli and Maibach, 1983)
- "Bioavailability of Soil-Bound TCDD: Oral Bioavailability in the Rat" (Shu et al., 1987)
- *The Handbook of Dermaltoxicology* (Maibach, 1975)
- *Exposure Tests for Organic Compounds in Industry* (Piotrowski, 1973)
- Chapter 24, this volume

Ingestion

- *Exposure Factors Handbook*, all three volumes (EPA, 1997e)
- "A Comprehensive Risk Assessment of DEHP as a Component of Baby Pacifiers, Teethers, and Toys" (Paustenbach, 1989, Chapter 26)
- "Risk Analyses of Buried Waste from Electricity Generation" (Paustenbach, 1989, Chapter 15)
- "Methylmercury in Fish: Assessment of Risk for U.S. Consumers" (Paustenbach, 1989, Chapter 25)
- "Infant Exposure Assessment for Breast Milk Dioxins and Furans Derived from Incineration Emissions" (Smith, 1987)
- "Bioavailability of Soil-Borne Polybrominated Biphenyls Ingested by Farm Animals (Fries, 1985)
- Chapter 18, this volume

- Chapter 8, this volume
- Chapter 9, this volume

1.3.4 Risk Characterization

The quantitative estimate of the risk, the size of the exposed population, and the uncertainty surrounding the risk estimates (e.g., the risk characterization) are the topics of principal interest to the regulatory agency or risk manager in arriving at decisions. The risk manager must consider the results of the risk characterization when evaluating the economics, societal aspects, and various benefits of the risk assessment. In general, efforts to improve the risk characterization process have received the least amount of study to date compared with the other three steps in risk assessment. One needs only to review some of the decisions reached in toxic tort cases, decisions by various federal agencies and by industry to understand that improvements continue to be needed.

Certainly, factors such as societal pressure, technical uncertainties, the cost–benefit relationship, the cost–effectiveness relationship, and severity of the potential hazard influence how decision makers respond to the risk assessment. However, better risk characterizations could be developed in risk assessments than have typically been the case. The result of these improvements is that risk managers can make more informed decisions. The following texts and publications are useful sources of information that should be helpful in improving the way risks are characterized:

- Chapter 5, this volume
- *Toward Integrated Environmental Decision-Making* (EPA, 2000a)
- *Science Advisory Board FY 2000 Annual Staff Report: Making Science Real* (EPA, 2000c)
- *Guidance for Risk Characterization* (EPA, 2000e)
- *Understanding Risks: Informed Decisions in a Democratic Society* (NRC, 1996)
- *The Greening of Industry* (Graham and Hartwell, 1996)
- *Beyond the Horizon: Using Foresight to Protect the Environmental Future* (EPA, 1995c)
- *Guidance for Risk Characterizaton* (EPA, 1995a)
- *Science and Judgment in Risk Assessment* (NAS, 1994)
- *Risk and Responsibility* (Leiss and Chociolko, 1994)
- *Worst Things First: The Debate over Risk-Based National Environmental Priorities* (Finkel and Golding, 1994)
- *Risk Assessment Methods* (Covello and Merkhofer, 1993)
- *Readings in Risk* (Glickman and Gough, 1993)
- *Fatal Tradeoffs* (Viscusi, 1992)

- *Safeguarding the Future: Credible Science, Credible Decisions* (EPA, 1992d)
- *Environmental Risk Decisionmaking: A Multidisciplinary Perspective* (Cheshile and Carlisle, 1991)
- *Reducing Risk: Setting Priorities and Strategies for Environmental Protection* (EPA, 1990)
- *Technological Risks* (Lewis, 1990)
- "Legal and Philosophical Aspects of Risk Analysis" (Paustenbach, 1989, Chapter 30)
- *Searching for Safety* (Wildavsky, 1988)
- *Risk Evaluation and Management: Contemporary Issues in Risk Analysis* (Covello et al., 1986)
- "Risk Analysis and Risk Management: An Historical Perspective" (Covello and Mumpower, 1985)
- *Modern Science and Human Values* (Lowrance, 1984)
- *Risk Watch: The Odds of Life* (Urquhart and Heilmann, 1984)
- *The Good News Is The Bad News is Wrong* (Wattenberg, 1984)
- *Risk Assessment in the Federal Government: Managing the Process* (NAS, 1983)
- *Societal Risk Assessment: How Safe Is Safe Enough* (Schwing and Albers, 1980)
- *Of Acceptable Risk* (Lowrance, 1976)
- Chapter 32, this volume

1.3.5 Environmental Risks

The process by which the risks to wildlife and ecosystems are characterized is slightly different than that used for humans. To describe the hazards posed by fairly widespread dissemination of xenobiotics, one needs to understand the physical properties of the compounds, their environmental fate, and their toxic effects, not only to humans but also to aquatic species and other wildlife. Perhaps no other industry has so thoroughly evaluated the potential adverse effects of widespread distribution of xenobiotics in the environment as the detergent industry. This insight is not unexpected, since each day literally millions of pounds of detergents are released into our waterways with potential uptake by a diverse number of organisms, including humans. Some of the more important issues in the evaluation of these hazards have been addressed in the following publications:

- *Guidelines of Ecological Risk Assessment* (EPA, 1998c)
- *Handbook of Ecotoxicology* (Hoffman et al., 1995)
- *Fundamentals of Aquatic Toxicology* (Rand, 1995)
- *Basic Environmental Toxicology* (Cockerham and Shane, 1994)

- *Ecological Assessment of Hazardous Waste Sites* (Maughn, 1993)
- *Ecological Risk Assessment* (Bartell et al., 1992)
- "An Environmental Risk Assessment of a Pesticide" (Paustenbach, 1989, Chapter 27)
- "Detergent Chemicals: A Case Study of a Cationic Surfactant" (Paustenbach, 1989, Chapter 9)
- "Endangerment Assessment for the Bald Eagle Population Near the Sand Springs Pcerochemical Complex Superfund Site" (Paustenbach, 1989, Chapter 29)
- *Environmental Hazard Assessment of Effluents* (Bergmann et al., 1986)
- *Dynamics, Exposure and Hazard Assessment of Toxic Chemicals* (Haque, 1980)
- *Aquatic Toxicology and Hazard Evaluation* (Halter and Johnson, 1977)
- Chapter 6, this volume
- Chapter 24, this volume
- Chapter 22, this volume

1.3.6   Uncertainty Analysis in Health Risk Assessment

Each of the four parts of a risk assessment offers an opportunity for scientists to fall into traps that can dramatically influence the results of the evaluation. During the past 15 years many of the important shortcomings in how to conduct a risk assessment have been identified and, by and large, we have learned that Monte Carlo anlysis can address many of the problems that once plagued the field. The following publications have focused on ways to ensure that the results of risk assessment are presented more fairly, accurately, and transparently:

- *Probabilistic Techniques in Exposure Assessment* (Cullen and Frey, 1999)
- *Risk Assessment Guidance (RAGS3A) for Conducting Probabilistic Risk Assessment* (EPA, 1999d)
- "Using Lognormal Distributions and Lognormal Probability Plots in Probabilistic Risk Assessments" (Burmaster and Huff, 1997)
- "Estimating Exposure Point Concentrations for Surface Soils for Use in Deterministic and Probabilistic Risk Assessments" (Burmaster and Thompson, 1997)
- *Understanding Risks: Informed Decisions in a Democratic Society* (NRC, 1996)
- "Back Calculating Cleanup Targets in Probabilistic Risk Assessments When the Acceptability of Cancer Risk is Defined under Different Risk Mangement Policies (Burmaster and Thompson, 1995)
- "Principles of Good Practice for the Use of Monte Carlo Techniques in Human Health and Ecological Risk Assessment" (Burmaster and Anderson, 1994)

- "The Benefits of Probabilistic Exposure Assessment: Three Case Studies Involving Contaminated Air, Water, and Soil" (Finley and Paustenbach, 1994)
- "Development of a Standard Soil-to-Skin Adherence Probability Density Function for Use in Monte Carlo Analyses of Dermal Exposure" (Finley et al., 1994)
- "Recommended Distributions for Exposure Factors Frequently Used in Health Risk Assessment" (Finley et al., 1994)
- "Using Monte Carlo Simulations in Public Health Risk Assessments: Estimating and Presenting Full Distributions of Risk" (Burmaster and von Stackelberg, 1991)
- "An Assessment and Quantitative Uncertainty Analysis of the Health Risks to Workers Exposed to Chromium Contaminated Soils" (Paustenbach et al., 1991b)
- "Problems Associated with the Use of Conservative Assumptions in Exposure and Risk Analysis" (Paustenbach, 1989, Chapter 14)
- "The Perils of Prudence: How Conservative Risk Assessments Distort Regulation" (Nichols and Zeckhauser, 1988)
- "A Critical Analysis of Risk Assessment of TCDD Contaminated Soils" (Paustenbach et al., 1986)

1.4 PRINCIPLES FOR RISK ANALYSIS

The Society for Risk Analysis (SRA) is the largest society in the world dedicated exclusively to the study of risk analysis. From about 1998–2000, the officers of the society developed some principles that they believed would help guide the practice in the coming years. These were adopted early in 2001 by SRA and are a good set of guidelines (if not a code of conduct) for practitioners. They are as follows:

> Risk analysis applies methods of analysis to matters of risk. Its aim is to increase understanding of the substantive qualities, seriousness, likelihood, and conditions of a hazard or risk and of the options for managing it. Risk analysis is both a profession and an intellectual discipline. These principles are meant to guide both the practice and use of risk analysis.

1. Risk analysis uses observations about what we know to make predictions about what we don't know. Risk analysis is a fundamentally science-based process that strives to reflect the realities of Nature in order to provide useful information for decisions about managing risks. Risk analysis seeks to inform, not to dictate, the complex and difficult choices among possible measures to mitigate risks. Risk analysis enriches fair and transparent deliberative decision-making processes in a democratic society.

2. Risk analysis seeks to integrate knowledge about the fundamental physical, biological, social, cultural, and economic processes that determine human, environmental, and technological responses to a diverse set of circumstances. Because decisions about risks are usually needed when knowledge is incomplete, risk analysts rely on informed judgment and on models reflecting plausible interpretations of the realities of Nature. We do this with a commitment to assess and disclose the basis of our judgments and the uncertainties in our knowledge.

3. Risk analysis relies on both basic and applied research, often integrating information, theories, and analytic tools from a variety of disciplines. As we apply information and tools from diverse disciplines, we seek to give due respect and acknowledgment to the intellectual contributions of those fields while using information standards and criteria appropriate to the policy choices that are at issue.

4. Risk analysts are committed to maintaining and building our professional community as we contribute to advances in our field. We review the work of our peers and help students develop their skills and values. Unless prohibited, we share the data underlying our published analyses in order to facilitate independent reassessment of our own conclusions.

5. The relationship of risk analysts to the sponsors of our efforts is subordinate to our commitment to fairly assess and discuss the risks that are the subjects of our analyses. Risk analysts openly acknowledge our sponsors and our sources of data and support.

1.5   SUMMARY

This brief chapter described many of the key issues that need to be addressed in developing a risk assessment and it identified useful publications for the risk assessor. The art of risk assessment has made significant progress over the past 20 years, especially since the publication of the 1989 edition of this text. The publication of perhaps as many as 15 different related texts and more than 1000 papers which address various aspects of risk assessment over the past 10 years has been the primary reason for these advances. Through the sharing of the experiences and knowledge of the authors who have contributed to this text, it can be anticipated that the overall quality of risk assessment will continue to improve in the coming years.

REFERENCES

Abdel-Rahman, M. S., and Kadry, A. M. 1995. The use of uncertainty factors in deriving RfD's: Overview. *Hum. Ecol. Risk Assess.* 1(5): 614–625.

Abrams, P. A. 1993. Effect of increased productivity on the abundances of trophic levels. *Am. Nat.* 141: 351–371.

Acute Toxicities of Organic Chemicals to Fathead Minnows. Center for Lake Superior Environ. Stud., University of Wisconsin-Superior, Superior, WI, 1984.

Agricola, G. 1950. *De re metallica.* Dover, New York.

Albers, P., Heinz, G. H., and Ohlendorf, H. (Eds.). 2000. *Environmental Contaminants and Terrestrial Vertebrates: Effects on Populations, Communities, and Ecosystems.* Society of Environmental Toxicology and Chemistry, Pensacola, FL.

Albert, R. E. 1994. Carcinogen risk assessment in the Environmental Protection Agency. *Crit. Rev. Toxicol.* 24: 75–85.

Allen, B. C., Kavlock, R. J., Kimmel, C. A., and Faustman, E. M. 1994a. Dose response assessments for developmental toxicity: II. Comparison of generic benchmark dose estimates with NOAEL's. *Fundam. Appl. Toxicol.* 23: 487–495.

Allen, B. C., Kavlock, R. J., Kimmel, C. A., and Faustman, E. M. 1994b. Dose response assessments for developmental toxicity: III. Statistical models. *Fundam. Appl. Toxicol.* 23: 496–509.

Allen, D., Consoli, F. J., Davis, G. A., Fava, J. A., and Warren, J. L. (Eds.). 1997. *Public Policy Applications of Life-Cycle Assessment.* Society of Environmental Toxicology and Chemistry, Pensacola, FL.

Alscher, R. G., and Welburn, A. R. 1994. *Plant Responses to Gaseous Environment.* Chapman and Hall, London.

American Conference of Governmental Industrial Hygienists (ACGIH). 1999. *Documentation of Threshold Limit Values (TLVs),* 6th ed. ACGIH, Cincinnati, OH.

American Conference of Governmental Industrial Hygienists (ACGIH). 2000. *2000 Threshold Limit Values (TLVs)®* for Chemical Substances and Physical Agents and Biological Exposure Indices (BEIs). ACGIH, Cincinnati, OH.

American Industrial Health Council (AIHC). 1983. *Significant Developments Regarding Government Risk Assessment Methodology.* AIHC, Washington, DC.

American Industrial Health Council (AIHC). 1994. *Exposure Factors Sourcebook.* AIHC, Washington, DC.

American Mutual Insurance Alliance (AMIA). 1988. *Handbook of Organic Industrial Solvents.* AMIA, Chicago, IL.

Ames, B. N., Magaw, R., and Gold, L. S. 1987. Ranking possible carcinogenic hazard. *Science* 236: 271–273.

Andersen, M. E. 1991. Quantitiative Risk Assessment and Chemical Carcinogens in Occupational Environments. *Appl. Occup. Env. Hyg* 3: 267–173.

Andersen, M. E. 1981. A physiologically-based toxicokinetic description of the metabolism of inhaled gases and vapors: Analysis at steady-state. *Toxicol. Appl. Pharmacol.* 60: 509–526.

Andersen, M. E. 1995. Development of physiologically based pharmacokinetic and physiologically based pharmacodynamic models for applications in toxicology and risk assessment. *Toxicol Lett* 79(1–3): 35–44.

Andersen, M. E. 1995. Physiologically based pharmacokinetic (PB-PK) models in the study of the disposition and biological effects of xenobiotics and drugs. *Toxicol. Lett.* 82/83: 341–348.

Andersen, M. E., Clewell, H. J., Gargas, M. L., Smith, F. A., and Reitz, R. H. 1987. Physiologically-based pharmacokinetics and the risk assessment process for methylene chloride. *Toxicol. Appl. Pharmacol.* 87: 185–205.

Andersen, M. E., Clewell, H. J., and Krishnan, K. 1995. Tissue dosimetry, pharmacokinetics modeling and interspecies scaling factors. *Risk Anal.* 15: 533–537.

Anderson, E. L., and the Carcinogen Assessment Group of the U.S. Environmental Protection Agency. 1983. Quantitative approaches in use to assess cancer risk. *Risk Anal.* 3: 277–295.

Anderson, J. P. E. et al. (Eds.). 1996. *Pesticides, Soil Microbiology and Soil Quality.* Society of Environmental Toxicology and Chemistry, Pensacola, FL.

Anderson, M. W., Hoel, D. G., and Kaplan, N. L. 1980. A general scheme for the incorporation of pharmacokinetics in low dose risk estimation for chemical carcinogenesis. *Toxicol. Appl. Pharmacol.* 55: 154–161.

Anderson, P. D., and Yuhas, A. L. 1996. Improving risk management by characterizing reality: A benefit of probabilistic risk assessment. *Human Ecol. Risk Assess.* 2: 55–58.

Andrews, J., and Johnson, B. 1996. *Hazardous waste and public health: Health effects of hazardous waste.* Princeton Scientific Publishing Co., Princeton, NJ.

Angell, M. 1996. *Science on Trial.* WW Norton Pub., London, UK.

Angle, C. R., and McIntire, M. S. 1974. Lead in air, dustfall, soil, housedust, milk and water: Correlation with blood lead of urban and suburban school children. *Trace Substances Environ. Health* 6: 23–28.

Armitage, P., and Doll, R. 1954. The age distribution of cancer and a multistage theory of carcinogensis. *Br. J. Cancer* 8: 1–12.

Australian Enironmental Council. 1990. *Guide to Environmental Legislation and Administrative Arrangements in Australia,* 2nd ed. Report No. 18. Commonwealth Government Printer, Canberra, Australia.

Ayrton-Paris, J. 1822. Pharmaecologia.

Baes, C. F., II, Sharp, R. D., Sjoreen, A., and Shor, R. 1984. *A Review and Analysis of Parameters for Assessing Transport of Environmental Released Radionuclides in Agriculture.* ORNL-5786. U.S. Department of Energy, Oak Ridge National Laboratory, Oak Ridge, TN.

Baird, S. J. S., Cohen, J. T., Graham, J. D., Shlyakhter, A. I., and Evans, J. S. 1996. Noncancer risk assessment: A probabilistic alternative to current practice. *Human Ecol. Risk Assess.* 2: 79–102.

Baker, J. E. (Ed.). 1997. *Atmospheric Deposition of Contaminants to the Great Lakes and Coastal Waters.* Society of Environmental Toxicology and Chemistry, Pensacola, FL.

Ballantyne, B., Marrs, T. C., and Syversen, T. 1999. *General and Applied Toxicology,* 2nd ed. MacMillan, London.

Barnard, R. C. 1984. Science, policy and the law: A developing partnership in reducing the risk of cancer. In P. F. Diesler, Jr. (Ed.), *Reducing the Carcinogenic Risks in Industry.* Dekker, New York.

Barnes, D. G., Daston, G. P., Evans, J. S., Jarabek, A. M., Kavlak, R. J., Kimmel, C. A., Park, C., and Spitzer, H. L. 1995. Benchmark dose workshop: Criteria for use of a benchmark dose to estimate reference dose. *Regul. Toxicol. Pharmacol.* 21: 296–306.

Barnes, J., and Denz, F. 1954. Experimental methods used in determining chronic toxicity. *Pharmacol. Rev.* 6: 191–242.

Barnes, D. G., and Dourson, M. 1988. Reference dose (RfD): description and use in health risk assessments. *Regul Toxicol Pharmacol* 8(4): 471–86.

Barnthouse, L., Fava, J., Humphreys, K., Hunt, R., Laibson, S., Noesen, S., Norris, G., Owens, J., Todd, J., Vigon, B., Weitz, K., and Young, J. (Eds.). 1997. *Life-Cycle Impact Assessment: The State of the Art*, 2nd ed. Society of Environmental Toxicology and Chemistry, Pensacola, FL.

Barrett, K. L. et al. (Eds.). *Guidance Document on Regulatory Testing Procedures for Pesticides with Non-Target Arthropods*. Society of Environmental Toxicology and Chemistry, Pensacola, FL.

Bartek, M. J., and LaBudde, J. A. 1975. Percutaneous absorption, in vitro. In H. Maibach (Ed.), *Animal Models in Dermatology*. Churchill-Livingstone, Edinburgh, p. 103.

Bartell, S. M., Gardner, R. H., and O'Neill, R. V. 1992. *Ecological Risk Assessment.* Lewis, Chelsea, MI.

Baumgartner, A., Van Hummelen, P., Lowe, X. R., Adler, I. D., and Wyrobek, A. J. 1999. Numerical and structural chromosomal abnormalities detected in human sperm with a combination of multicolor FISH assays. *Environ. Mol. Mutagen.* 33: 49–58.

Beck, B. D., and Cohen, J. T. 1997. Risk assessment for criteria pollutants versus other non-carcinogens: The difference between implicit and explicit conservatism. *Human Ecol. Risk Assess.* 3: 617–626.

Beck, L. W., Maki, A. W., Artman, N. R., and Wilson, E. R. 1981. Outline and criteria for evaluating the safety of new chemicals. *Regul. Toxicol. Pharmacol.* 1: 19–58.

Beer, T., and Ziolkowski, F. 1995. Environmental Risk Assessment: An Australian Perspective. Commonwealth of Australia.

Begon, M. J., Harper, L., and Townsend, C. R. 1990. *Ecology: Individuals, Populations and Communities.* Blackwell, Boston, Melbourne.

Bell, R. 1992. *Improve Science: Fraud, Compromise, and Political Influence in Scientific Research.* John Wiley and Sons, Inc., New York.

Bender, M. E., and Huggett, R. J. 1984. Fate and effects of kepone in the James River. *Rev. Environ. Toxicol.* 1: 5–50.

Bentkover, J. D., Covello, V. T., and Mumpower, J. 1986. *Benefits Assessment: The State of the Art.* Reidel, Boston, MA.

Bergman, H. L., and Dorward-King, E. J. (Eds.). 1997. *Reassessment of Metals Criteria for Aquatic Life Protection.* Society of Environmental Toxicology and Chemistry, Pensacola, FL.

Bergmann, H. L., Kimerle, R. A., and Maki, A. W. 1986. *Environmental Hazard Assessment of Effluents.* Pergamon, New York.

Berndtson, W. E., and Clegg, E. D. 1992. Developing improved strategies to determine male reproductive risk from environmental toxins. *Theriogenology* 38: 223–237.

Bernstein, P. L. 1996. *Against the Gods: The Remarkable Story of Risk.* John Wiley and Sons, Inc., New York.

Bigham, E., Cohrssen, B., and Powell, C. H. 2001. *Patty's Toxicology*, 5th ed., 9 vols. Wiley, New York.

Binder, S., Sokal, D., and Maughan, D. 1986. Estimating the amount of soil ingested by young children through tracer elements. *Arch. Environ. Health* 41: 341–345.

Biological Exposure to Ionizing Radiation. 1996. A Report. National Council for Radiological Protection (NCRP), Washington, DC.

Bischoff, K. 1967. Applications of a mathematical model for drug distribution in mammals. *Chemical Engineering in Medicine and Biology.* Plenum, New York.

Bischoff, K. B., and Brown, R. G. 1966. Drug distribution in mammals. *Chem. Eng. Prog. Symp. Ser.* 62: 33–45.

Bishop, W. E., Cardwell, R. D., and Heidolph, B. B. 1983. *Aquatic Toxicology and Hazard Assessment: Sixth Symposium.* STP 737. ASTM, Philadelphia, PA.

Bogen, K. T., and Spear, R. C. 1987. Integrating uncertainty and interindividual variability in environmental risk assessment. *Risk Anal.* 7: 427–435.

Bogen, K. T., Keating, G. A., Meissner, S., and Vogel, J. S. 1998. Initial uptake kinetics in human skin exposed to dilute aqueous trichloroethylene. *J. Exp. Anal. Environ. Epidemiol.* 8: 253–271.

Bonazountas, M. 1987. In H. Barth and P. Hermite (Eds.), *Chemical Fate Modeling in Soil Systems: A State of the Art Review in Scientific Basis for Soil Protection in the European Community.* Elsevier, London, pp. 487–566.

Branson, D. R., and Dickson, K. L. 1981. *Aquatic Toxicology and Hazard and Assessment: Fourth Symposium.* STP 737. ASTM, Philadelphia, PA.

Breyer, S. 1993. *Breaking the Vicious Circle: Toward Effective Regulation.* Harvard University Press, Cambridge, MA.

Briand, F., and Cohen, J. E. 1987. Environmental correlates of food chain length. *Science* 238: 956–960.

Briggs, G. G., Bromilow, R. H., and Evans, A. A. 1982. Relationships between lipophilicity and root uptake and translocation of non-ionized chemicals by barley. *Toxicol. Environ. Chem.* 7: 173–189.

Bronmark, C., Klosiewski, S. P., and Stein, R. A. 1992. Indirect effects of predation in a freshwater, benthic food chain. *Ecology* 73: 1662–1674.

Brooke, L. T., Calf, D., Geiger, D. I., and Northcott, C. E. 1984. *Acute Toxicities of Organic Chemicals to Fathead Minnows.* Center for Lake Superior Environmental Studies, University of Wisconsin, Superior, WI.

Browman, M. G., and Chester, S. 1977. The solid-water interface: Transfer of organic pollutants across the solid-water interface. In I. H. Suffet (Ed.), *Fate of Pollutants in the Air and Water Environment,* Part 1. Wiley, New York.

Brown, C. 1978. Statistical aspects of extrapolation of dichotomous dose response data. *J. Natl. Cancer Inst.* 60: 101–108.

Brown, H. S., Guble, R., and Tatelbaum, S. 1988. Methodology for assessing hazards of contaminants to seafood. *Regul. Toxicol. Pharmacol.* 6: 76–100.

Brown, K. G., and Hoel, D. G. 1986. Statistical modeling of animal bioassay data with variable dosing regimens: Example—Vinyl chloride. *Risk Anal.* 6: 155–166.

Brown, N. A., and Fabro, S. 1983. The value of animal teratogenicity testing for predicting human risk. *Clin. Obstet. Gynecol.* 26(2): 467–477.

Bryant, B. 1995. *Environmental Justice.* Island Press, Washington, DC.

Buck, R. J., Hammerstrom, K. A., and Ryan, P. B. 1997. Bias in population estimates of long-term exposure from short-term measurements of individual exposure. *Risk Anal.* 17: 455–465.

Bullard, R. D. 1994. *Unequal Justice.* Sierra Club Books, San Francisco, CA.

Burmaster, D. E. 1998a. A lognormal distribution for time spent showering. *Risk Anal.* 18: 33–36.

Burmaster, D. E. 1998b. Lognormal distributions for skin area as a function of body weight. *Risk Anal.* 18: 27–32.

Burmaster, D. E., and Anderson, P. D. 1994. Principles of good practice for the use of Monte Carlo techniques in human health and ecological risk assessment. *Risk Anal.* 14: 477–491.

Burmaster, D. E., and Huff, D. A. 1997. Using lognormal distributions and lognormal probability plots in probabilistic risk assessments. *Human Ecol. Risk Assess.* 3: 223–234.

Burmaster, D. E., and Thompson, K. M. 1995. Back calculating cleanup targets in probabilistic risk assessments when the acceptability of cancer risk is defined under different risk management policies. *Human Ecol. Risk Assess.* 1: 101–120.

Burmaster, D. E., and Thompson, K. M. 1997. Estimating exposure point concentrations for surface soils for use in deterministic and probabilistic risk assessments. *Human Ecol. Risk Assess.* 3: 363–384.

Burmaster, D. E., and von Stackelberg, K. 1991. Using Monte Carlo simulations in public health risk assessments: Estimating and presenting full distributions of risk. *J. Expos. Anal. Environ. Epidemiol.* 1: 491–521.

Bus, J. S., and Gibson, J. E. 1985. Body defense mechanisms to toxicant exposure. In L. J. Cralley and L. V. Cralley (Eds.), *Patty's Industrial Hygiene and Toxicology*, 2nd ed., Vol. 38. John, New York, pp. 143–174.

Butterworth, B. E., Conolly, R. B., and Morgan, K. T. 1995. A strategy for establishing mode of action of chemical carcinogens as a guide for approaches to risk assessments. *Cancer Lett* 93(1): 129–46.

Butterworth, B. E., and Slaga, T. 1987. *Nongenotoxic Mechanisms in Carcinogenesis: Banbury Report 25.* Cold Spring Harbor Laboratory, Cold Spring Harbor, NY.

Cairns, J., and Cherry, D. S. 1993. Freshwater multi-species test systems. In P. Calow (Ed.), *Handbook of Ecotoxicology.* Blackwell Science, Oxford, pp. 101–116.

Cairns, J., Jr, Dickson, K. L., and Maki, A. W. 1978. *Estimating the Hazards of Chemical Substances to Aquatic Life.* STP 657. ASTM, Philadelphia, PA.

Cairns, J., and Niederlehner, B. R. 1993. Ecological function and resilience: Neglected criteria for environmental impact assessment and ecological risk analysis. *Environ. Professional* 15: 116–124.

Calabrese, E. J. 2001. The frequency of U-shaped dose responses in the toxicological literature. *Toxicol Sciences* 62: 330–338.

Calabrese, E. J., and Baldwin, L. A. 2001. Scientific foundations of hormesis. *Crit. Rev. Toxicol.* 31(4, 5).

Calabrese, E. J., and Baldwin, L. A. 1998. Hormesis as a default parameter in RfD derivation. *Hum. Exp. Toxicol.* 17: 444–447.

Calabrese, E. J., Gilbert, C. E., Kostecki, P. T., Barnes, R., Stanek, E., Veneman, P., Pastides, H., and Edwards, C. 1988. *Epidemiological Study to Estimate How Much Soil Children Eat.* Division of Public Health, University of Massachusetts, Amherst.

Calabrese, E. J., and Kostecki, P. T. 1988. *Soils Contaminated by Petroleum: Environment and Public Health Effects.* Wiley, New York.

Calabrese, E., and Kostecki, P. 1989. *Petroleum Contaminated Soils*, Vols. 1 and 2. Lewis, Chelsea, MI.

Calabrese, E. J., and Kostecki, P. T. 1992. *Risk Assessment and Environmental Fate Methodologies*. Lewis, Ann Arbor, MI.

Calabrese, E. J. 1978. *Methodological Approaches to Deriving Environmental and Occupational Health Standards*. Wiley, New York.

Calabrese, E. J., and Stanek, E. J. 1991. A guide to interpreting soil ingestion studies II: Qualitative and quantitative evidence of soil ingestion. *Regul. Toxicol. Pharmacol.* 13: 278–292.

Calabrese, E. J., Stanek, E. J., and Barnes, R. 1996. Methodology to estimate the amount and particle size of soil ingested by children: Implications for exposure assessment at waste sites. *Regul. Toxicol. Pharmacol.* 24: 264–268.

Calabrese, E. J., Stanek, E. J., Gilbert, C. E., and Barnes, R. M. 1990. Preliminary adult soil ingestion estimates: Results of a pilot study. *Regul. Toxicol. Pharmacol.* 12: 88–95.

California Department of Health Services (CDHS). 1985. *Guidelines for Chemical Carcinogens: Risk Assessments and Their Scientific Rationale*. CDHS, Sacramento, CA.

California Department of Health Services (CDHS). 1988. *Guidelines and Safe Use Determination Procedures for the Safe Drinking Water and Toxic Enforcement Act of 1986*. CDHS, Sacramento, CA.

Carey, J., Cook, P., Giesy, J., Hodson, P., Muir, D., Owens, W., and Solomon, K. (Eds.). 1998. *Ecotoxicological Risk Assessment of the Chlorinated Organic Chemicals*. Society of Environmental Toxicology and Chemistry, Pensacola, FL.

Carnegie Commission. 1993a. *Science, Technology, and Government for a Changing World*. Carnegie Commission, New York.

Carnegie Commission. 1993b. *Risk and the Environment: Improving Regulatory Decisionmaking*. Carnegie Commission, New York.

Carpenter, S. R. 1988. *Complex Interactions in Lake Communities*. Springer, New York.

Carson, R. 1962. *Silent Spring*. Houghton Mifflin, Boston, MA.

Castanho, M. 1987. Methods of soil sampling. *Comments Toxicol.* 1: 221–227.

Center for Risk Analysis. 1994. *Historical Roots of Health Risk Assessment*. Harvard University, School of Public Health, Boston, MA.

Chadwick, E. 1842. *Report of the sanitary condition of the labouring population of Great Britain*. London.

Chapin, R. E., Dutton, S. L., Ross, M. D., and Lamb, J. C. 1985. Effects of ethylene glycol monomethyl ether (EGME) on mating performance and epididymal sperm parameters in F344 rats. *Fundam. Appl. Toxicol.* 5: 182–189.

Chapin, R. E., Gulati, D. K., Fail, P. A., Hope, E., Russell, S. R., Heindel, J. J., George, J. D., Grizzle, T. B., and Teague, J. L. 1993. The effects of feed restriction on reproductive function in Swiss CD-1 mice. *Fundam. Appl. Toxicol.* 20: 15–22.

Chapin, R. E., Harris, M. W., Haseman, J. K., and Wine, R. N. 2000. The control of spermiogenesis in the rat. *J. Androl.* 21(Suppl.): 50.

Chapin, R. E., and Heindel, J. J. 1993. Introduction. In J. J. Heindel and R. E. Chapin (Eds.), *Methods In Toxicology*, Vol. 3B: *Female Reproductive Toxicology*. Academic Press, San Diego, CA, pp. 1–15.

Chapman, P. F., Crane, M., Wiles, J. A., Noppert, F., and McIndoe, E. C. (Eds.). 1996. *Asking the Right Questions: Ecotoxicity and Statistics.* Society of Environmental Toxicology and Chemistry, Pensacola, FL.

Chechile, R. A., and Carlisle, S. 1991. *Environmental Desicionmaking: A Multidisciplinary Perspective.* Van Norstrand, New York.

Chemical Manufacturers Association (CMA). 1984. *Risk Management of Existing Chemicals.* CMA, Washington, DC.

Chen, F., and Schüürmann, G. (Eds.). 1997. *Quantitative Structure-Activity Relationships in Environmental Sciences,* Vol. VII. Society of Environmental Toxicology and Chemistry, Pensacola, FL.

Chiou, C. T. 1981. Partition coefficient and water solubility in environmental chemistry. In J. Saxena and F. Fisher (Eds.), *Hazard Assessment of Chemicals.* Academic, New York, pp. 117–153.

Chiou, C. T., Freed, V. H., Schmedding, D. W., and Kohnert, R. L. 1977. Partition coefficient and bioconcentration of selected organic solvents. *Environ. Sci. Technol.* 11: 475–479.

Christian, M. S. 1983. Assessment of reproductive toxicology: State-of-the-art. In M. S. Christian, W. M. Galbraith, P. Voytek, and M. A. Mehlmann (Eds.), *Assessment of Reproductive and Teratogenic Hazards.* Princeton University Press, Princeton, NJ.

Christian, M. S. 1986. A critical review of multigeneration studies. *J. Am. Coll. Toxicol.* 5: 161–180.

Christian, M. S., and Hoberman, A. M. 1996. Perspectives on the U.S., EEC and Japanese developmental toxicity guidelines. In R. D. Hood (Ed.), *Handbook of Development Toxicology.* CRC Press, Boca Raton, FL, pp. 551–596.

Christian, M. 2001. Test methods for assessing female reproductive and developmental toxicology. In A. W. Hayes (Ed.), *Principles and Methods of Toxicology.* Taylor and Francis, Philadelphia, PA, pp. 1301–1382.

Clausing, O., Brunekreef, A. B., and Van Wijnen, J. H. 1987. A method for estimating soil ingestion by children. *Int. Arch. Occup. Environ. Health* 59: 73–82.

Clayson, D. B., Krewski, D., and Munro, I. 1985. *Toxicological Risk Assessment,* Vols. 1 and 2. CRC Press, Boca Raton, FL.

Clayton, G. D., and Clayton, F. E. 1982. *Patty's Industrial Hygiene and Toxicology,* 3rd rev. ed., Vols. 2A, 2B, and 2C. Wiley, New York.

Clewell, H. J. 1995. The use of physiologically based pharmacokinetic modeling in risk assessment: A case study with methylene chloride. In S. Olin, W. Farland, C. Park, L. Rhomberg, R. Scheuplein, T. Starr, and J. Wilson (Eds.), *Low-Dose Extrapolation of Cancer Risks,* International Life Sciences Institute Press, Washington, DC, pp. 199–222.

Clewell, H. J., and Andersen, M. E. 1985. Risk assessment extrapolations and physiological modeling. *Toxicol. Ind. Health* 1: 111–132.

Clewell, H. J., III, and Jarnot, B. M. 1994. Incorporation of pharmacokinetics in noncancer risk assessment: Example with chloropentafluorobenzene. *Risk Anal.* 14: 265–276.

Clewell, H. J., Gentry, P. R., and Gearhart, J. M. 1997. Investigation of the potential impact of benchmark dose and pharmacokinetic modeling in noncancer risk assessment. *J. Toxicol. Environ. Health* 52: 475–515.

Cochran, W. G. 1977. *Sampling Techniques*, 3rd ed. Wiley, New York.

Cockerham, L. G., and Shane, B. S. 1994. *Basic Environmental Toxicology*. CRC Press, Boca Raton, FL.

Cohen, J. E. 1989. Food webs and community structure. In J. Roughgarden, R. M. May, and S. A. Levin (Eds.), *Perspectives in Ecological Theory*. Princeton University Press, Princeton, NJ.

Colie, C. F. 1993. Male mediated teratogenesis. *Reprod. Toxicol.* 7: 3–9.

Commission on Risk Assessment and Risk Management (CRAM) 1997a. Framework for Environmental Risk Management. Presidential/Congressional Commission on Risk Assessment and Risk Management. Vol. 1. Washington, DC.

Commission on Risk Assessment and Risk Management (CRAM). 1997b. *Risk Assessment and Risk Management in Regulatory Decisionmaking*. Vol. 2. Washington, DC.

Consultancy, J. 1986. Risk assessment for hazardous installations. Commission of the European Communities. Pergamon, Oxford.

Conway, R. A. 1982. *Environmental Risk Analysis for Chemicals*. Van Nostrand-Reinhold, New York.

Cooper, M. 1957. *Pica*. Thomas, Springfield, IL.

Copeland, T. L., Holbrow, A. H., Otani, J. M., Connor, K. T., and Paustenbach, D. J. 1994. Use of probabilistic methods to understand the conservatism in California's approach to assessing health risks posed by air contaminants. *J. Air Waste Manag. Assoc.* 44: 1399–1413.

Copeland, T. L., Paustenbach, D. J., Harris, M. A., and Otani, J. 1993. Comparing the results of a Monte Carlo analysis with EPA's reasonable maximum exposed individual (RMEI): A case study of a former wood treatment site. *Regul. Toxicol. Pharmacol.* 18: 275–312.

Corley, R. A., Markham, D. A., Banks, C., Delorme, P., Masterman, A., and Houle, J. M. 1997. Physiologically based pharmacokinetics and the dermal absorption of 2-butoxyethanol vapor by humans. *Fundam. Appl. Pharmacol.* 39: 120–130.

Cornfield, L. 1977. Carcinogenic risk assessment. *Science* 198: 692–699.

Covello, V., and Merkhofer, M. W. 1993. *Risk Assessment Methods*. Plenum Press, New York.

Covello, V. T., Menkes, J., and Mumpower, J. 1986. *Risk Evaluation and Management. Contemporary Issues in Risk Analysis*, Vol. 1. Plenum, New York.

Covello, V. T., and Mumpower, J. 1985. Risk analysis and risk management: An historical perspective. *Risk Anal.* 5: 103–120.

Covello, V. T., and Lave, L. B. 1987. *Uncertainty in Risk Assessment, Risk Management, and Decision Making, Adv. Risk Anal.*, Vol. 4. Plenum, New York.

Cowan, C., Mackay, D., Feijtel, T. C. J., van de Mcent, D., Di Guardo, A., Davies, J., and Mackay, N. (Eds.). 1995. *The Multi-Media Fate Model: A Vital Tool for Predicting the Fate of Chemicals*. Society of Environmental Toxicology and Chemistry, Pensacola, FL.

Crandall, R. W., and Lave, B. L. 1981. *The Scientific Basis of Risk Assessment*. Brookings Institution, Washington, DC.

Crosby, D., and Wong, A. 1976. Photodegradation of TCDD. *Science.* 195: 1337–1338.

Crouch, E. A. C. 1996. Uncertainty distributions for cancer potency factors: Combining epidemiological studies with laboratory bioassays—the example of acrylonitrile. *Human Ecol. Risk Assess.* 2: 130–149.

Crouch, E. A. C., and Wilson, R. 1982. *Risk/Benefit Analysis.* Ballinger, Cambridge, MA.

Crump, K. 1995. Calculation of benchmark doses from continuous data. *Risk Anal.* 15: 79–85.

Crump, K. S. 1981. An improved procedure for low-dose carcinogenic risk assessment from animal data. *J. Environ. Toxicol.* 5: 339–346.

Crump, K. S. 1985. A new method for determining allowable daily intakes. *Fundam. Appl. Toxicol.* 4: 854–871.

Crump, K. S. 1994a. Use of mechanistic models to estimate low dose cancer risks. *Risk Anal.* 14(6): 1099–1038.

Crump, K. S. 1994b. Limitations of biological models of carcinogenesis for low-dose extrapolation. *Risk Anal.* 14(6): 883–886.

Crump, K. S. 1996. The linearized multistage model and the future of quantitative risk assessment. *Hum. Exper. Toxicol.* 15: 787–798.

Crump, K. S., Clewell, H. J., and Andersen, M. E. 1997. Cancer and non-cancer risk assessment should be harmonized. *Hum. Ecol. Risk Assess.* 3(4): 495–499.

Crump, K. S., Hoel, D. G., Langley, C. H., and Peto, R. 1976. Fundamental carcinogenic processes and their implications for low dose risk assessment. *Cancer Res.* 36: 2913–2979.

Crump, K. S., and Howe, R. B. 1984. The multistage model with time-dependent dose pattern: Applications of carcinogenic risk assessment. *Risk Anal.* 4: 163–176.

Csanady, G. A., Kreuzer, P. E., Baur, C., and Filser, J. G. 1996. A physiological toxicokinetic model for 1,3-butadiene in rodents and man: Blood concentrations of 1,3-butadiene, its metabolically formed epoxides, and of haemoglobin adducts—relevance of glutathione depletion. *Toxicology* 113: 300–305.

Cullen, A. C. 1994. Measures of compounding conservatism in probabilistic risk assessment. *Risk Anal.* 14(4): 389–393.

Cullen, A. C., and Frey, H. C. 1999. *Probabilistic Techniques in Exposure Assessment.* Plenum, New York.

Davis, A., Bloom, N. S., and Que Hee, S. S. 1997. The environmental geochemistry and bioaccessability of mercury in soils and sediments: A review. *Risk Anal.* 17: 557–569.

Davis, N. R., and Mapleson, W. W. 1981. Structure and quantification of a physiological model of the distribution of injected agents and inhaled anaesthetics. *Br J Anaesth* 53(4): 399–405.

DeCaprio, A. P. 1997. Biomarkers: Coming of age for environmental health and risk assessment. *Environ. Sci. Technol.* 31(7): 1837–1848.

deFur, P. L., Crane, M., Ingersoll, C. G., and Tattersfield, L. (Eds.). 1999. *Endocrine Disruption in Invertebrates: Endocrinology, Testing, and Assessment.* Society of Environmental Toxicology and Chemistry, Pensacola, FL.

De Peyster, A., and Day, K. E. (Eds.). 1998. *Ecological Risk Assessment: A Meeting of Policy and Science.* Society of Environmental Toxicology and Chemistry, Pensacola, FL.

DeSesso, J. M. 1997. Comparative embryology. In R. D. Hood (Ed.), *Handbook of Developmental Toxicology*. CRC Press, New York, pp. 111–174.

De Weese, L. R., McEwen, L. C., Hensler, G. L., and Petersen, B. E. 1986. Organo-chlorine contaminants in passeriformes and other avian prey of the peregrine falcon in the western United States. *Environ. Toxicol Chem.* 5: 675–693.

Dickson, K. L., Maki, A. W., and Cairns, J., Jr. 1979. *Analyzing the Hazard Evaluation Process*. American Fisheries Society, Washington, DC.

di Domencio, A., Silano, V., Viviano, G., and Zapponi, G. 1980. Accidental release of 2,3,7,8-tetrachlorodibenzo-*p*-dioxin. Part II. TCDD distribution in the soil surface layer. *Ecotoxicol. Environ. Saf.* 4: 298–320.

Di Giulio, R. T., and Tillitt, D. E. (Eds.). 1999. *Reproductive and Developmental Effects of Contaminants in Oviparous Vertebrates*. Society of Environmental Toxicology and Chemistry, Pensacola, FL.

Dourson, M. 1986. New approaches in the derivation of acceptable daily intakes (ADI). *Comments Toxicol.* 1: 35–48.

Dourson, M. L., Felter, S. P., and Robinson, D. 1998. Evolution of science-based uncertainty factors for noncancer risk assessment. *Regul. Toxicol. Pharmcaol.* 24: 108–120.

Dourson, M. L., and Stara, J. F. 1983. Regulatory history and experimental support of uncertainty (safety factors). *Regul. Toxicol. Pharmacol.* 3: 224–238.

Dourson, M. L., Hertzberg, R., Hartung, R., and Blackburn, K. 1985. Novel methods for the estimation of acceptable daily intake. *Toxicol. Ind. Health.* 1(4): 23–33.

Dragun, J. 1998. *The Soil Chemistry of Hazardous Materials*, 2nd ed. Amherst Scientific, Amherst, MA.

Duan, N. 1991. Stochastic microenvironmental models for air pollution exposure. *J. Exp. Anal. Environ. Epidemiol.* 1(2): 235–257.

Dubreuilh, W. 1896. Des Hyperkeratoses Circonscrites. *Ann. Dermatil. Syphilig* 3: 1158–1204.

Dugard, P. H., Walker, M., Mawdsley, S. J., and Scott, R. C. 1984. Absorption of some glycol ethers through human skin in vitro. *Environ. Health Perspect.* 57: 193–198.

Ebert, E., Harrington, N., Boyle, K., Knight, J., and Keenan, R. 1993. Estimating consumption of freshwater fish among Maine Anglers. *N. Am. J. Fisheries Mgmt.* 13: 737–745.

Ebert, E. S., Price, P. S., and Keenan, R. E. 1994. Selection of fish consumption estimates for use in the regulatory process. *J. Exp. Anal. Environ. Epidemiol.* 4: 373–394.

Ehling, U. H. 1988. Quantification of the genetic risk of environmental mutagens. *Risk Anal.* 8: 45–58.

Eisenbud, M. 1978. *Technology and Health: Human Ecology in Historial Perspective.* New York Univ. Press, New York.

El Sayed, E. I., Graves, J. B., and Bonner, F. L. 1967. Chlorinated hydrocarbon insecticide residues in selected insects and birds found in association with cotton fields. *J. Agric. Food Chem.* 15(6): 1014–1017.

Environmental Protection Agency (EPA). 1979. *Water Related Fate of 129 Priority Pollutants*, Vols. 1 and 2. PB80-204381. EPA, Washington, DC.

Environmental Protection Agency (EPA). 1982a. *Reproductive and Fertility Effects. Pesticide Assessment Guidelines, Subdivision F. Hazard Evaluation: Human and Do-*

mestic Animals. EPA/540/9-82-025. EPA, Office of Pesticides and Toxic Substances, Washington, DC.

Environmental Protection Agency (EPA). 1982b. *Air Quality Criteria for Particulate Matter and Sulfur Oxides.* Vol. 2, EPA-600-8-82-029bF. EPA, Office of Environmental Criteria and Assessment, Research Triangle Park, NC, pp. 5-106–5-112.

Environmental Protection Agency (EPA). 1984. *Handbook for Interpreting Soil Data.* EPA, Research Triangle Park, NC.

Environmental Protection Agency (EPA). 1986a. Guidelines for carcinogen risk assessment. CFR 2984. *Fed. Reg.* 51(185): 33992–34003.

Enviromental Protection Agency (EPA). 1986b. Guidelines for developmental toxicity risk assessment. CFR 2984. *Fed. Reg.* 51(185): 34028–34040.

Environmental Protection Agency (EPA). 1986c. Guidelines for exposure assessment. CFR 2984. *Fed. Reg.* 51(185): 34041–34054.

Environmental Protection Agency (EPA). 1986d. Guidelines for the health assessment of chemical mixtures. CFR 2984. *Fed. Reg.* 51(185): 34014, 34025.

Environmental Protection Agency (EPA). 1986e. *Methodology for the Assessment of Health Risks Associated with Multiple Pathway Exposure to Municipal Waste Combustors.* EPA, Office of Air Quality Planning and Standards, Research Triangle Park, NC.

Environmental Protection Agency (EPA). 1986f. *Superfund Health Assessment Manual.* EPA 540/1-86/060. EPA, Office of Emergency and Remedial Response, Research Triangle Park, NC.

Environmental Protection Agency (EPA). 1986g. *Superfund Exposure Assessment Manual.* OSWFR Directive 9285.5-l. EPA, Research Triangle Park, NC.

Environmental Protection Agency (EPA). 1986h. *Registry of Toxic Effects of Chemical Substances.* EPA, Research Triangle Park, NC.

Environmental Protection Agency (EPA). 1987a. *Handbook for Conducting Endangerment Assessments.* EPA, Research Triangle Park, NC.

Environmental Protection Agency (EPA). 1988a. *OPPTS Harmonized Test Guidelines: 870.3800. Reproduction and Fertility Effects.* EPA, Washington, DC.

Environmental Protection Agency (EPA). 1988b. Proposed guidelines for assessing female reproductive risk. *Fed. Reg.* 53(126): 24834–24847.

Environmental Protection Agency (EPA). 1988c. Proposed guidelines for assessing male reproductive risk. *Fed. Reg.* 53(126): 24850–24869.

Environmental Protection Agency (EPA). 1989a. *Ecological Risk Assessment of Hazardous Waste Site: A Field and Laboratory Reference.* EPA/600/3-89/013. EPA, Corvallis, OR.

Environmental Protection Agency (EPA). 1989b. *Risk Assessment Guidance for Superfund.* Vol. I: *Human Health Evaluation Manual (Part A). Interim Final.* EPA/540/1-89/002. EPA, Office of Emergency and Remedial Response, Washington, DC.

Environmental Protection Agency (EPA). 1990. *Reducing Risk: Setting Priorities and Strategies for Environmental Protection.* EPA/600/R-99/060. EPA, Washington, DC.

Environmental Protection Agency (EPA). 1991a. *Risk Assessment Guidance for Superfund (RAGS).* Vol. I: *Human Health Evaluation Manual (HHEM) (Part B). Development of Risk-Based Preliminary Remediation Goals.* EPA/540/R-92/003. NTIS PB92-963333. EPA, Office of Emergency and Remedial Response, Washington, DC.

Environmental Protection Agency (EPA). 1991b. Guidelines for developmental toxicity risk assessment. Dec. 5, 1991. EPA-Risk Assessment Forum. *Fed. Reg.* 56(234).

Environmental Protection Agency (EPA). 1992a. *Dermal Exposure Assessment: Principles and Applications.* EPA/600/8-91/011. EPA, Office of Health and Environmental Assessment, Office of Research and Development, Washington, DC.

Environmental Protection Agency (EPA). 1992b. Guidelines for exposure assessment. *Fed. Reg.* 57(104): 22888–22938.

Environmental Protection Agency (EPA). 1992c. *Supplemental Guidance to RAGS: Calculating the Concentration Term.* OSWER Directive 9285.7-081. EPA, Office of Solid Waste and Emergency Response, Washington, DC.

Environmental Protection Agency (EPA). 1992d. *Safeguarding the Future: Credible Science, Credible Decisions. Report of the Expert Panel on the Role of Science at EPA.* EPA/600/9-91/050. EPA, Washington, DC.

Environmental Protection Agency (EPA). 1993. Reference dose (RfD): Description and use in health risk assessments (background document 1A), 3/15/93 (updated 5/21/99). Integrated Risk Information System. Available from http://www.epa.gov.ngispgm3/iris/rfd.htm.

Environmental Protection Agency (EPA). 1995a. *Guidance for Risk Characterization.* EPA, Science Policy Council, Washington, DC.

Environmental Protection Agency (EPA). 1995b. *The Use of the Benchmark Dose Approach in Health Risk Assessment.* EPA/630/R-94/007. EPA, Office of Research and Development, Risk Assessment Forum. Washington, DC.

Environmental Protection Agency (EPA). 1995c. *Beyond the Horizon: Using Foresight to Protect the Environmental Future.* EPA-SAB-EC-95-007. EPA, Washington, DC.

Environmental Protection Agency (EPA). 1996a. Draft guidelines for carcinogen risk assessment. *Fed. Reg.* 61(79): 17960–18011.

Environmental Protection Agency (EPA). 1996b. *Exposure Factors Handbook.* Vol. 1: *General Factors-SAB Review Draft.* EPA/600/P-95/002Ba. EPA, Office of Research and Development, Washington, DC.

Environmental Protection Agency (EPA). 1996c. *Exposure Factors Handbook.* Vol. 2: *Food Ingestion Factors-SAB Review Draft.* EPA/600/P-95/002Bb. EPA, Office of Research and Development, Washington, DC.

Environmental Protection Agency (EPA). 1996d. *Exposure Factors Handbook.* Vol. 3: *Activity Factors-SAB Review Draft.* EPA/600/P-95/002P. EPA, Office of Research and Development, Washington, DC.

Environmental Protection Agency (EPA). 1996e. A national agenda to protect children's health from environmental threats (updated 9/11/96). Available from http://occ-env-med.mc.duke.edu/oem/content/epa.htm.

Environmental Protection Agency (EPA). 1996f. *Guidelines for reproductive toxicity risk assessment. Fed. Reg.* 61(212): 56274–56322.

Environmental Protection Agency (EPA). 1996g. *Summary Report for the Workshop on Monte Carlo Analysis.* EPA/630/R-96/010. EPA, Office of Research and Development, Washington, DC.

Environmental Protection Agency (EPA). 1997a. *Lognormal Distribution in Environmental Applications.* EPA/600/R-97/006. EPA, Office of Solid Waste and Emergency Response, Washington, DC.

Environmental Protection Agency (EPA). 1997b. *Guiding Principles for Monte Carlo Analysis.* EPA/630/R-97/001. EPA, Office of Research and Development, Risk Assessment Forum, Washington, DC.

Environmental Protection Agency (EPA). 1997c. *Methodology for Assessing Health Risks Associated with Multiple Exposure Pathways to Combustor Emissions.* NCEA-C-0238. EPA, National Center for Environmental Assessment, Washington, DC.

Environmental Protection Agency (EPA). 1997d. *The Parameter Guidance Document. A Companion to the "Methodology for Assessing Health Risks Associated with Multiple Exposure Pathways to Combustor Emissions."* National Center for Environmental Assessment, Cincinnati, OH.

Environmental Protection Agency (EPA). 1997e. *Exposure Factors Handbook,* Vol. 1: *General Factors.* Office of Research Development, Washington, DC.

Environmental Protection Agency (EPA). 1997f. *Guiding Principles for Monte Carlo Analysis.* Risk Assessment Forum. NTIS PB97-18810 6I.NZ. EPA/630/R-97/001. EPA, Washington, DC.

Environmental Protection Agency (EPA). 1998a. *Health Effects Test Guidelines.* USEPA 712-C-98-189 through USEPA 712-98-351. EPA, Office of Prevention, Pesticides and Toxic Substances, Washington, DC.

Environmental Protection Agency (EPA). 1998b. Guidelines for neurotoxicity risk assessment. *Fed. Reg.* 63(93): 26926–26954.

Environmental Protection Agency (EPA). 1998c. *Guidelines for Ecological Risk Assessment.* Risk Assessment Forum. EPA/630/R-95/002F. EPA, Washington, DC.

Environmental Protection Agency (EPA). 1998d. Framework for addressing key scientific issues presented by the Food Quality Protection Act (FQPA) as developed by the Tolerance Reassessment Advisory Committee (TRAC). *Fed. Reg.* 63(209): 58038–58045.

Environmental Protection Agency (EPA). 1999a. *Sociodemographic Data Used for Identifying Potentially Highly Exposed Populations.* EPA/600/R-99/060. EPA, Washington, DC.

Environmental Protection Agency (EPA). 1999b. Integrated Risk Information System. Available from http://www.epa.gov/iris.

Environmental Protection Agency (EPA). 1999c. *Risk Assessment Guidelines for Dermal Assessment.* EPA, Washington, DC.

Environmental Protection Agency (EPA). 1999d. Guidelines for Carcinogen Risk Assessment (SAB review copy, July 1999). EPA, Washington, DC. http://www.epa.gov/ncea/raf/crasab.htm.

Environmental Protection Agency (EPA). 1999e. *Risk Assessment Guidance (RAGS3A) for Conducting Probabilistic Risk Assessment.* EPA, Washington, DC.

Environmental Protection Agency (EPA). 2000a. *Toward Integrated Environmental Decision-Making.* EPA-SAB-EC-95-007, EPA, Washington, DC.

Environmental Protection Agency (EPA). 2000b. *An Approach to Estimating Exposure to 2,3,7,8 TCDD.* EPA Exposure Assessment Group, Washington, DC.

Environmental Protection Agency (EPA). 2000c. *Science Advisory Panel FY 2000 Annual Staff Report.* EPA-SAB-01-002. EPA, Washington, DC.

Environmental Protection Agency (EPA). 2000d. *Strategic Plan.* EPA-190-R-00-002. EPA, Washington, DC.

Environmental Protection Agency (EPA). 2000e. Risk Characterization Guidance, http://www.epa.gov/ord/spc/rchandbk.pdf., EPA, Washington, DC.

Environmental Protection Agency (EPA). 2001. An SAB Report: Review of the Office of Research and Development's Reassessment of Dioxin. USEPA Science Advisory Board (SAB). Mort Lippman, Chairman. EPA-SAB-EC—01-006. www.epa.gov/sab., EPA, Washington, DC.

Erickson, J. D. 1981. Epidemiology and developmental toxicology. In C. A. Kimmel and J. Beulke-Sam (Eds.), *Developmental Toxicology*. Raven, New York, pp. 289–301.

Ernst, W. 1977. Determination of the bioconcentration potential of marine organisms: A steady-state approach. *Chemosphere* 11: 731–740.

Ernst, W., and Peterson, P. J. 1994. The role of biomarkers in environmental assessment (4). Terrestrial plants. *Ecotoxicology* 3: 180–192.

Eschenroeder, A., Jaeger, R. J., Ospital, J. J., and Doyle, C. 1986. Health risk assessment of human exposure to soil amended with sewage sludge contaminated with polycholorinated dibenzodioxins and dibenzofurans. *Vet. Hum. Toxicol.* 28: 356–442.

Evans, J. S., Graham, J. D., Gray, G. M., and Sielken, R. L., Jr. 1994a. A distributional approach to characterizing low-dose cancer risks. *Risk Anal.* 14: 25–33.

Fabro, S., Schull, G., and Brown, N. A. 1982. The relative teratogenic index and teratogenic potency: Proposed components of developmental toxicity risk assessment. *Teratogen. Carcinogen. Mutagen.* 2: 61–76.

Fairhurst, S. 1995. The uncertainty factor in the setting of occupational exposure standards. *Am. Occup. Hyg.* 39(3): 375–385.

Fan, A., and Chang, L. 1996. Toxicology and Risk Assessment: Principles, Methods and Applications. M. Dekker. New York.

Fava, J. A., Denison, R., Jones, B., Curran, M. A., Vigon, B., Selke, S., and Barnum, J. (Eds.). 1991. *A Technical Framework for Life-Cycle Assessment.* Society of Environmental Toxicology and Chemistry, Pensacola, FL.

Federal Focus. 1996. *Principles of Evaluating Epidemiology Data in Regulatory Risk Assessment.* Federal Focus Publishing, Washington, DC.

Fenske, R. A. 1993. Dermal exposure assessment techniques. *Ann. Occup. Hyg.* 37: 687–706.

Ferenc, S. A., and Foran, J. A. (Eds.). 2000. *Multiple Stressors in Ecological Risk and Impact Assessments: Approaches to Risk Estimation.* Society of Environmental Toxicology and Chemistry, Pensacola, FL.

Field, L. J., Fall, J., Nigbswander, T., Peacock, N., and Varanasi, U. (Eds.). 2000. *Evaluating and Communicating Subsistence Seafood Safety in a Cross-Cultural Context: Lessons Learned from Exxon Valdez Oil Spill.* Society of Environmental Toxicology and Chemistry, Pensacola, FL.

Finkel, A. M., and Evans, J. S. 1987. Evaluating the benefits of uncertainty reduction in environmental health risk management. *J. Air Pollut. Control Assoc.* 37: 1164–1171.

Finkel, A. M., and Golding, D. 1994. *Worst Things First: The Debate over Risk Based National Environmental Priorities.* Resources for Future, Washngton, DC.

Finley, B. L., and Paustenbach, D. J. 1994. The benefits of probabilistic exposure assessment: Three case studies involving contaminated air, water, and soil. *Risk Anal.* 14(1): 53–73.

Finley, B. L., Proctor, D., Scott, P., Harrington, N., Paustenbach, D., and Price, P. 1994. Recommended distributions for exposure factors frequently used in health risk assessment. *Risk Anal.* 14(4): 533–553.

Finley, B. L., Scott, P. K., and Mayhall, D. A. 1994. Development of a standard soil-to-skin adherence probability density function for use in Monte Carlo analyses of dermal exposure. *Risk Anal.* 14: 555–569.

Finley, B. L., Scott, P., and Paustenbach, D. J. 1993. Evaluating the adequacy of maximum contaminant levels as health protective cleanup goals: An analysis based on Monte Carlo techniques. *Reg. Toxicol. Pharmacol.* 18: 438–455.

Fiserova-Bergerova, V. 1975. Mathematical modeling of inhalation exposure. *J. Combust. Toxicol.* 3: 201–210.

Food Safety Council. 1980. Quantitative risk assessment. *Food Safety Assessment.* Ch. 11. Food Safety Council, Washington, DC.

Foran, A. J., and Ferenc, S. A. (Eds.). 1999. *Multiple Stressors in Ecological Risk and Impact Assessment.* Society of Environmental Toxicology and Chemistry, Pensacola, FL.

Francis, B. M., Metcalf, R. L., Lewis, P. A., and Chernoff, N. 1999. Maternal and developmental toxicity of halogenated 4'-nitrodiphenyl ethers in mice. *Teratology* 59: 69–80.

Francis, E. Z., and Kimmel, G. L. 1988. Proceedings of the workshop on one- versus two-generation reproductive effects studies. *J. Am. Coll. Toxicol.* 7: 911–925.

Franzle, O. 1993. *Contaminants in Terrestrial Environments.* Springer, Berlin.

Freed, V. H., Chiou, C. T., Schmeddling, D., and Kohnert, R. 1979. Some physical factors in toxicological assessment tests. *Environ. Health Perspect.* 30: 75–80.

Freeze, R. 2000. *The Environmental Pendulum.* University of California Press, Berkeley, CA.

Frey, H. C., and Rhodes, D. S. 1998. Characterization and simulation of uncertainty frequency distributions: Effects of distribution choice, variability, uncertainty, and parameter dependence. *Hum. Ecol. Risk Assess.* 4: 423–469.

Fries, G. F. 1982. Potential polychlorinated hiphenyl residues in animal products from application of contaminated sewage sludge to land. *J. Environ. Qual.* 11: 14–20.

Fries, G. F. 1985. Bioavailability of soil-borne polybrominated biphenyls ingested by farm animals. *J. Toxicol. Environ. Health* 16: 565–579.

Fries, G. F., and Jacobs, L. W. 1986. Evaluation of residual polybrominated biphenyl contamination present on Michigan farms in 1978. *Mich. Agric. Exp. Stn. Res. Rep.* 477.

Fries, G. F., and Paustenbach, D. J. 1990. Evaluation of potential transmission of 2,3,7,8-tetrachlorodibenzo-*p*-dioxin-contaminated incinerator emissions to humans via foods. *J. Toxicol. Environ. Health* 29: 1–43.

Fries, G. F., Paustenbach, D. J., Mathur, D. B., and Luksemburg, W. J. 1999. A congener specific evaluation of transfer of chlorinated dibenzo-*p*-dioxins and dibenzofurans to milk cows following ingestion of pentachlorophenol-treated wood. *Environ. Sci. Technol.* 33(8): 1165–1170.

Friess, S. 1987. Risk assessment historical perspectives. In *Pharmacokinetics in Risk Assessment: Drinking Water and Health*, Vol. 8. National Academy of Science, Washington, DC, pp. 3–7.

Gargas, M. L., Burgess, R. J., Voisard, D. E., Cason, G. H., and Andersen, M. E. 1989. Partition coefficients of low molecular weight volatile chemicals in various liquids and tissues. *Toxicol. Appl. Pharmacol.* 98: 87–99.

Gaylor, D. 1998. Safety assessment with hormetic effects. *Hum. Exp. Toxicol.* 17: 251–253.

Gaylor, D. W. 1989. Quantitative risk analysis for quantal reproductive and developmental effects. *Environ. Health. Perspect.* 79: 243–246.

Gaylor, D. W. 1992. Incidence of developmental defects at the no observed adverse effect level (NOAEL). *Regul. Toxicol. Pharmacol.* 15: 151–160.

Gaylor, D. W., and Chen, J. J. 1993. Dose–response models for developmental malformations. *Teratology* 47: 291–297.

Gaylor, D. W., and Kodell, R. I. 1980. Linear interpolation algorithm for low dose, risk assessment of toxic substances. *J. Environ. Pathol. Toxicol.* 4: 305–312.

Gaylor, D. W., and Zheng, Q. 1996. Risk assessment of nongenotoxic carcinogens based upon cell proliferation/death rates in rodents. *Risk Anal.* 16: 221–225.

Gibaldi, M., and Perrier, D. 1982. *Pharmacokinetics.* Dekker, New York.

Gilbert, R. O. 1987. *Statistical Methods for Environmental Pollution Monitoring.* Van Nostrand Reinhold, New York.

Glickman, T. S., and Gough, M. 1993. *Readings in Risk.* Resources for Future, Washington, DC.

Goldin, I., and Winters, L. A. 1995. *The Economics of Sustainable Development.* Cambridge University Press, London.

Gosselin, R. E., Gleason, M. N., Hodge, H. C., and Smith, R. P. 1984. *Clinical Toxicology of Commercial Products*, 5th ed. Williams & Wilkins, Baltimore, MD.

Graham, J. D. 1995. Historical perspective on risk assessment in the federal government. *Toxicology* 102: 29–52.

Graham, J. D., Green, L. C., and Roberts, M. J. 1988. In *Search of Safety: Chemicals and Cancer Risks.* Harvard University Press, Cambridge, MA.

Graham, J. D., and Hartwell, J. K. 1997. The Greening of Industry: A Risk Management Approach. Harvard Univ. Press, Cambridge, Mass.

Graham, J. D., and Wiener, J. B. 1995. *Risk vs. Risk: Tradeoffs in Protecting Health and the Environment.* Harvard University Press, Cambridge, MA.

Greenberg, M. S., Burton, G. A., and Fisher, J. W. 1999. Physiologically based pharmacokinetic modeling of inhaled trichloroethylene and its oxidative metabolites in B6C3F1 mice. *Toxicol. Appl. Pharmacol.* 154: 264–278.

Grothe, D. R., Dickson, K. L., and Reed-Judkins, D. K. (Eds.). 1996. *Whole Effluent Toxicity Testing: An Evaluation of Methods and Prediction of Receiving System Impacts.* Society of Environmental Toxicology and Chemistry, Pensacola, FL.

Guidance Document on Testing Procedures for Pesticides in Freshwater Mesocosms. 1991. Society of Environmental Toxicology and Chemistry, Pensacola, FL.

Guzelian, P. S., Henry, C. J., and Olin, S. S. 1992. Similarities and Differences Between Children and Adults: Implications for Risk Assessment. Inter Life Sciences Institute Press, Washington, DC.

Haas, C. N. 1997. Importance of the distributional form in characterizing inputs to Monte Carlo risk assessments. *Risk Anal.* 17: 107–113.

Haber, L., Dollarhide, J., Maier, A., and Dourson, M. L. 2001. Noncancer risk assessment: Principles and practice in environmental and occupational settings. *Patty's Toxicology*, Vol. 9, Ch. 5. E. Bigham, B. Cohrssen, and C. Powell (Eds.). John Wiley and Sons. New York.

Haber, L. T., Allen, B. C., and Kimmel, C. A. 1998. Non-cancer risk assessment for nickel compounds: Issues associated with dose-response modeling of inhalation and oral esposures. *Toxicol. Sci.* 43: 213–229.

Haggard, H. 1924. The absorption, distribution and elimination of ethyl ether. II. Analysis of the mechanism of the absorption and eliminiation of such a gas or vapor as ethyl ether. *J. Biol. Chem.* 59: 753–770.

Hales, B. F., and Robaire, B. 1996. Paternally mediated effects on development. In R. D. Hood (Ed.), *Handbook of Developmental Toxicology*. CRC Press, Boca Raton, FL, pp. 91–107.

Hales, S. G., Feijtel, D., King, J., Fox, L., and Verstraete, M. (Eds.). 1997. *Biodegradation Kinetics: Generation and Use of Data for Regulatory Decision Making*. Society of Environmental Toxicology and Chemistry, Pensacola, FL.

Hallenbeck, W. H., and Cunningham, K. M. 1986. *Quantitative Risk Assessment for Environmental and Occupational Health*. Lewis, Chelsea, MA.

Hamaker, J. W. 1975. The interpretation of soil leaching experiments. In R. Haque and V. H. Freed (Eds.), *Environmental Dynamics of Pesticides*. Plenum, New York, pp. 115–133.

Hamaker, J. W., and Kerlinger, W. O. 1969. Vapor pressure of pesticides. *Adv. Chem. Ser.* 86: 39–54.

Hamaker, J. W., and Thompson, J. M. 1972. Adsorption. In C. A. I. Goring and J. M. Haymaker (Eds.), *Organic Chemicals in the Soil Environment*, Vol. 1. Dekker, New York, pp. 51–122.

Hamilton, A. 1929. *Industrial poisions in the United States*. MacMillan Publishers, New York.

Haque, R. 1980. *Dynamics, Exposure and Hazard Assessment of Toxic Chemicals*. Ann Arbor Science, Ann Arbor, MI.

Harbison, R. D. 1998. *Hamilton and Hardy's Industrial Toxicology*, 5th ed. Mosby, St. Louis, MO.

Harris, M. W., Chapin, R. E., Lockhart, A. C., Jokinen, M. P., Allen, J. D., and Haskins, E. A. 1992. Assessment of a short-term reproductive and developmental toxicity screen. *Fundam. Appl. Toxicol.* 19: 186–196.

Harris, R. L. 2000. *Patty's Industrial Hygiene*, Vols. 1–5, 5th ed. Wiley, New York.

Hart, W. L., Reynolds, R. C., Krasavage, W. J., Kly, T. S., Bell, R. H., and Raleigh, R. L. 1987. Evaluation of developmental toxicity data: A discussion of some pertinent factors, and a proposal. *Risk Anal.* 8: 59–69.

Hartley, H. O., and Sielken, R. L. 1977. Estimation of "safe doses" in carcinogenic experiments. *J. Environ. Pathol.* 1: 241–252.

Hathaway, G., Proctor, N., and Hughes, J. 1996. *Proctor and Hughes Chemical Hazards of the Workplace*, 4th ed. Wiley, New York.

Hattis, D., and Burmaster, D. 1994. Assessment of variability and uncertainty distributions for practical risk analyses. *Risk Anal.* 14: 713–729.

Hattis, D., and Silver, K. 1994. Human interindividual variablity: A major source of uncertainty in assessing risks for noncancer health effect. *Risk Anal.* 14(4): 421–431.

Havender, W. 1984a. Peanut butter sandwich deadlier than muffins containing EDB. *Wall Street Journal*: 1311.

Havender, W. R. 1984. EDB and the marigold option. *Regulation* Jan./Feb.: 13–17.

Hawley, J. 1985. Assessment of health risks from exposure to contaminated soil. *Risk Anal.* 5: 289–302.

Hayes, A. W. 2001. *Principles and Methods of Toxicology*, 4th ed. Taylor and Francis, Philadelphia, PA.

Hayes, W. J., Jr. 1975. *Toxicology of Pesticides.* Waverly, Baltimore, MD.

Hayes, W. J., Jr. 1982. *Pesticides Studied in Man.* Williams & Wilkins, Baltimore, MD.

Hayes, W. J., and Laws, E. 1991. *Handbook of Pesticide Toxicology.* 3 volumes. Academic, San Diego, CA.

Healy, W. B., and Drew, K. R. 1970. Ingestion of soil by hoggets grazing Swedes. *N.A.A. Agric. Res.* 13: 940–944.

Heinonen, O. P., Slone, D., and Shapiro, S. 1977. *Birth Defects and Drugs in Pregnancy.* PSG, Littleton, MA.

Helleiner, O. 1967. The population of Europe from the Black Death of the eve of the vital revolution. *The Cambridge Economic History of Europe.* 4. E. Rich and C. Wilson (Eds.). Cambridge University Press, London.

Helsel, D. R. 1990. Less than obvious: Statistical treatment of data below the detection limit. *Environ. Sci. Technol.* 24: 1766–1774.

Hildrew, A. G. 1992. Food Webs and Species Interactions. In P. Calow and G. E. Petts (Eds.), *The Rivers Handbook. Hydrological and Ecological Principles.* Blackwell, Oxford, pp. 309–330.

Hill, J. 1781. *Cautions Against the Immoderate Use of Snuff.* Baldwin and Jackson, London.

Hill, I. R., Matthiessen, P., and Heimbach, F. (Eds.). 1994. *Guidance Document on Sediment Toxicity Tests and Bioassays for Freshwater and Marine Environments.* Society of Environmental Toxicology and Chemistry, Pensacola, FL.

Himmelstein, K. J., and Lutz, R. J. 1979. A review of the application of physiologically based pharmacokinetic modeling. *J. Pharmacokinet. Biopharm.* 7: 127–137.

Hoffman, D. J., Rattner, B. A., Burton, G. A., Jr., and Cairns, J., Jr. 1995. *Handbook of Ecotoxicology.* CRC Press, Boca Raton, FL.

Hoffman, F. O., and Hammonds, J. S. 1992. *An Introductory Guide to Uncertainty Analysis in Environmental and Health Risk Assessment.* ES/ER/TM-35. Martin Marietta.

Hoffman, F. O., and Hammonds, J. S. 1994. Propagation of uncertainty in risk assessments: The need to distinguish between uncertainty due to lack of knowledge and uncertainty due to variability. *Risk Anal.* 14: 707–711.

Hohenemser, C., and Kasperson, J. X. 1982. *Risk in a Technological Society.* Westview, Denver, CO.

Holmes, K. K., Kissel, J. C., and Richter, K. Y. 1996. Investigation of the influence of oil on soil adherence to skin. *J. Soil Contam.* 5(4): 301–308.

Hood, R. D., and Miller, D. B. 1997. Maternally mediated effects on development. In R. Hood (Ed.), *Handbook of Developmental Toxicology.* CRC Press, Boca Raton, FL, pp. 61–90.

Hrudey, S. E., Chen, W., and Rousseaux, C. 1996. *Bioavailability.* CRC-Lewis, New York.

Hutchinson, J. 1887. Arsenic cancer. *Br. Med. J.* 2: 1280–1284.

Imperato, P., and Mitchell, G. 1985. *Acceptable Risks.* Viking Press, New York.

Implementation Working Group. 1998. *A Science-Based, Workable Framework for Implementing the Food Quality Protection Act. Implementation Working Group's "Road Map" Report.* Jellinek, Schwartz & Connolly, McDermott, Will & Emery and Morgan, Lewis & Bockius, Washington, DC.

Ingersoll, C., Dillon, T., and Biddingen, G. R. (Eds.). 1997. *Ecological Risk Assessments of Contaminated Sediments.* Society of Environmental Toxicology and Chemistry, Pensacola, FL.

Institutes of Medicine. 1994. Veterans and Agent Orange. Health Effects of Herbicides vied in Vietnam, National Academy Press, NY. NY.

International Agency for Research on Cancer. 1999a. IARC Monogragh Summary for polychlorinated dibenzo-*para*-dioxins: 2,3,7,8-tetrachlorodibenzo-*para*-dioxin (Group 1): polychlorinated dibenzo-*para*-dioxins (other than 2,3,7,8-tetrachlorodibenzo-*para*-dioxin): 2,7-DCDD, 1,2,3,6,7,8-/1,2,3,7,8,9-HxCDD, 1,2,3,4,6,7,8-HpCDD (Group 3); dibenzo-*para*-dioxin (Group 3) (updated 5/4/99). Available from http://193.51.164.11/cgi/iHound/chem/iH_chem_frames.html.

International Agency for Research on Cancer. 1999b. Overall evaluations of carcinogenicity to humans. Updated by IARC Jan. 20, 1999. Downloaded Feb. 9, 1999, from http://193.51.164.11/monoeval/crthal.html.

International Agency for Research on Cancer. 1985–2000. Lyon, France.

International Technical Information Institute (ITII). 1976. *Toxic and Hazardous Industrial Chemicals Safety Manual.* ITII, Tokyo.

Jarabek, A. M., Menache, M. G., Overton, J. H. J., Dourson, M. L., and Miller, F. S. 1990. The Environmental Protection Agency's inhalation RfD methodology: Risk assessment for air toxics. *Tox. Ind. Health* 6(5): 279–302.

Jarvinen, A. W., and Ankley, G. (Eds.). 1999. *Linkage of Effects to Tissue Residues: Development of a Comprehensive Database for Aquatic Organisms Exposed to Inorganic and Organic Chemicals.* Society of Environmental Toxicology and Chemistry, Pensacola, FL.

Jarvinen, A. W., and Tyo, R. M. 1978. Toxicity to fathead minnows of endrin in food and water. *Arch. Environ. Chem. Toxicol.* 7(4): 409–421.

Jayjock, M. A. 1997. Uncertainty analysis in the evaluation of exposure. *Am. Ind. Hyg. Assoc. J.* 58(5): 380–382.

Jayjack, M. A., Lynch, J., and Nelson, D. I. (2000). *Risk Assessment Principles for the Industrial Hygienist.* AIHA Press, Fairfax, VA.

Johnson, E. M. 1987. A tier system for developmental toxicity evaluations based on considerations of exposure and effect relationships. *Teratology* 35: 405–427.

Johnson, E. M. 1988. Cross-species extrapolations and the biologic basis for safety factor determinations in developmental toxicology. *Regul. Toxicol. Pharmacol.* 8: 22–36.

Johnson, L., Welsh, T. H., and Wilker, C. E. 1998. Anatomy and physiology of the male reproductive system and potential targets for toxicants. In K. Boekelheide, R. E. Chapin, P. B. Hoyer, and C. Harris (Eds.), *Reproductive and Endocrine Toxicology.* Elsevier, New York, pp. 5–62.

Johnson, P. 1980. The perils of risk avoidance. *Regulation* May/June: 15–19.

Jorgensen, S. E. 1994. *Fundamentals of Ecological Modeling: Developments in Environmental Modeling*, 2nd ed. Elsevier, Amsterdam, The Netherlands.

Jorgensen, S. E. 1997. *An Introduction of Ecosystem Theories: A Pattern*, 2nd ed. Kluer Academic, Dordrecht, The Netherlands.

Jury, W. A., Farmer, W. J., and Spencer, F., IV. 1983. Behavior assessment model for trace organics in soil. Part I. Description of the model. J. Environ. Qual. 12: 558–564.

Jury, W. A., Farmer, W. J., and Spencer, W. F. 1984. Behavior assessment model for trace organics in soil. Part IV. Review of experimental evidence. *J. Environ. Qual.* 13: 580–586.

Kavlok, R. J., Allen, B. C., Fanstman, E. M., and Kimmel, G. A. 1995. Dose response assessments for developmental toxicity: IV. Benchmark doses for fetal weight changes. *Fundam. Appl. Toxicol.* 26: 211–222.

Kenaga, E. 1980. Correlation of bioconcentration factors of chemicals in aquatic and terrestrial organisms with their physical and chemical properties. *Ecotoxicol. Environ. Saf.* 4: 26–38.

Kendall, R. J., Dickerson, R. L., Suk, W. A., and Giesy, J. P. (Eds.). 1998. *Principles and Processes for Evaluating Endocrine Disruption in Wildlife.* Society of Environmental Toxicology and Chemistry, Pensacola, FL.

Kerger, B. D., Schmidt, C. E., and Paustenbach, D. J. 2000. Assessment of airborne exposure to trihalomethanes from tap water in residential showers and baths. *Risk Anal.* 20: 637–651.

Kety, S. 1951. The theory and applications of the exchange of inert gas at the lungs and tissues. *Pharmacol. Rev.* 3: 1–41.

Kezic, S., Mahieu, K., Monster, A. C., and de Wolff, F. A. 1997. Dermal absorption of vaporous and liquid 2-methoxyethanol and 2-ethoxyethanol in volunteers. *Occup. Environ. Med.* 54: 38–43.

Kimbrough, R. 1986. Estimation of amount of soil ingested, inhaled or available for dermal contact. *Comments Toxicol.* 1: 217–221.

Kimbrough, R., Falk, H., Stehr, P., and Fries, G. 1984. Health implications of 2,3,7,8-tetrachlorodibenzo-*p*-dioxin (TCDD) contamination of residential soil. *J. Toxicol. Environ. Health* 14: 47–93.

Kimmel, C. A., and Gaylor, D. W. 1988. Issues in qualitative and quantitative risk analysis for developmental toxicology. *Risk Anal.* 8: 15–20.

Kimmel, C. A., Kavlock, R. J., Allen, B. C., and Faustman, E. M. 1995. Benchmark dose concept applied to data from conventional development toxicity studies. *Toxicol. Lett.* 88: 549–554.

Kimmel, G. L., Clegg, E. D., and Crisp, T. M. 1995. Reproductive toxicity testing: A risk assessment perspective. In R. J. Witorsch (Ed.), *Reproductive Toxicology.* Raven, New York, pp. 75–98.

Kinzelbach, R. 1995. Neozoans in European waters—exemplifying the worldwide process of invasion and species mixing. *Experientia* 51: 526–538.

Kissel, J. C., Richter, K. Y., and Fenske, R. A. 1996a. Factors affecting soil adherence to skin in hand-press trails. *Bull. Environ. Contam. Toxicol.* 56: 722–728.

Kissel, J. C., Richter, K. Y., and Fenske, R. A. 1996b. Field measurement of dermal soil loading attributable to various activities: Implications for exposure assessment. *Risk Anal.* 16(1): 115–125.

Klaassen, C. D. 2001. *Casarett and Doull's Toxicology: The Basic Science of Poisons,* 6th ed. McGraw-Hill, New York.

Klecka, G., Boethling, B., Franklin, J., Graham, D., Grady, L., Howard, P., Kannan, K., Larson, R., Mackay, D., Muir, D., and van de Meent, K. (Eds.). 2000. *Evaluation of Persistence and Long-Range Transport of Organic Chemicals in the Environment.* Society of Environmental Toxicology and Chemistry, Pensacola, FL.

Klinefelter, G. R., and Hess, R. A. 1988. Toxicology of the male excurrent ducts and accessory glands. In K. S. Korach (Ed.), *Reproductive and Developmental Toxicology.* Dekker, New York, pp. 553–591.

Klinefelter, G. R., and Suarez, J. D. 1987. Toxicant-induced acceleration of epididymal sperm transit: Androgen dependent proteins may be involved. *Reprod. Toxicol.* 11: 511–519.

Klinefelter, G. R., and Welch, J. E. 1999. The saga of a male fertility protein (SP22). *Annu. Rev. Biomed. Sci.* 1: 145–184.

Knobil, E., and Neill, J. D. 1998. *Encyclopedia of Reproduction.* Academic, New York.

Knobil, E., Neill, J. D., Greenwald, G. S., Markert, C. L., and Pfaff, D. W. 1994. *The Physiology of Reproduction.* Raven, New York.

Kodell, R. L., Howe, R. B., Chen, J. J., and Gaylor, D. W. 1991. Mathematical modeling of reproductive and developmental toxic effects for quantitative risk assessment. *Risk Anal.* 11: 583–590.

Korschgen, L. J. 1973. Soil-food-chain-pesticide-wildlife relationships in aldrin-treated fields. *J. Wildf. Manag.* 34(1): 186–199.

Krantz, L. 1992. What the odds are. Harper Perennial. NY, NY.

Krewski, D., and Brown, C. 1981. Carcinogenic risk assessment: A guide to the literature. *Biometrics* 37: 353–366.

Krewski, D., Brown, C., and Murdoch, D. 1984. Determining safe levels of exposure: Safety factors or mathematical models. *Fundam. Appl. Toxic.* 4: S383–S394.

Krishnan, K., and Andersen, M. 2001. Physiologically-based pharmacokinetic modeling in toxicology. *Principles and Methods in Toxicology.* Ch. 5. A. W. Hayes (Ed.). Taylor and Francis, Philadelphia, PA.

Krishnan, K., Pelekis, M. L., and Haddad, S. 1995. A simple index for describing the discrepancy between PBPK model simulations and experimental data. *J. Toxicol. Ind. Health* 11: 413–421.

Krzeminski, S. F., Gilbert, J. T., and Ritts, J. A. 1977. A pharmacokinetic model for predicting pesticide residues in fish. *Arch. Environ. Contam. Toxicol.* 5: 157–165.

Kuhn, T. 1977. *The Essential Tension.* The University of Chicago Press, Chicago, IL.

LaGoy, P. 1987. Estimated soil ingestion rates for use in risk assessment. *Risk Anal.* 7: 355–359.

Lamb, J. C., and Foster, P. M. D. 1988. *Physiology and Toxicology of Male Reproduction.* Academic, New York.

Lanier, M. 1985. *The History of the TLV's.* American Conference of Governmental Industrial Hygienists, Cincinnati, OH.

Lave, L. B. 1983. *Quantitative Risk Assessment in Regulation.* Brookings Institution, Washington, DC.

Lehman, A. 1948. The toxicology of the newer agricultural chemicals. *Assn. of Food & Drug Officials of U.S.* 12(3): 82–89.

Lehman, A. J., Vorhes, F. A., et al. 1959. Appraisal of the Safety of Chemicals in Foods, Drugs and Cosmetics. The Association of Food and Drug Officials of the United States. Wash, DC, p. 107.

Lehman, A. J., and Fitzhugh, O. G. 1954. 100-fold margin of safety. *Q. Bull. Assoc. Food Drug Off. U.S.* 18: 33–35.

Leibold, M. A., and Wilbur, H. M. 1992. Interactions between food-web structure and nutrients on pond organisms. *Nature* 360: 341–343.

Leidel, N., and Busch, K. A. 1985. Statistical design and data analysis requirements. In L. J. Cralley and L. V. Cralley (Eds.), *Pattys Industrial Hygiene Toxicology*, 2nd ed. Vol. 3A. Wiley, New York.

Leiss, W., and Chociolko, C. 1994. *Risk and Responsibility.* McGill Queens University Press, London.

Lepow, M. L., Bruckman, L., Robino, R. A., Markowitz, S., Gillette, M., and Kapish, J. 1974. Role of airborne lead in increased body burden of lead in Hartford children. *Environ. Health Perspect.* 6: 99–101.

Leroux, B. G., Leisenring, W., Moolgavkar, S. H., Faustman, E. M. 1996. A biologically-based dose-response model for development toxicology. *Risk Anal.* 16: 449–458.

Leung, H. W. 2000. Physiologically-based pharmacokinetic models. In Ballantyne, Marrs, and Syverson (Eds.), *General and Applied Toxicology. (Second edition).* MacMillan, London, pp. 141–154.

Leung, H. W., and Paustenbach, D. J. 1994. Techniques for estimating the percutaneous absorption of chemicals due to environmental and occupational exposure. *Appl. Environ. Occup. Hyg.* 9(3): 187–197.

Leung, H. W., and Paustenbach, D. J. 1995. Physiologically based pharmacokinetic and pharmacodynamic modeling in health risk assessment and characterization of hazardous substances. *Toxicol. Lett.* 79: 55–65.

Lewis, H. W. 1990. *Technological Risks.* Norton, New York.

Lewis, M. A., Mayer, D., Powell, P., Nelson, M., Klaine, K., Henry, C., and Dickson, K. (Eds.). 1999. *Ecotoxicology and Risk Assessment for Wetlands.* Society of Environmental Toxicology and Chemistry, Pensacola, FL.

Lewis, R. J. 2000. *Sax's Dangerous Properties of Industrial Materials*, 10th ed. Wiley, New York.

Li, L.-H., and Heindel, J. J. 1998. Sertoli cell toxicants. In K. S. Korach (Ed.), *Reproductive and Developmental Toxicology.* Dekker, New York, pp. 655–691.

Lide, D. R. 2000. *Handbook of Chemistry and Physics.* CRC Press, Cleveland, OH.

Lindgreen, B. W. 1980. *Statistical Theory*, 2nd ed. Macmillan, New York.

Linkov, I., Wilson, R., and Gray, G. M. 1998. Anticarcinogenic responses in rodent cancer bioassays are not explained by random effects. *Toxicol. Sci.* 43: 1–9.

Lioy, P. J. 1990. Analysis of total human exposure assessment: A multidisciplinary science for examining human contact with contaminants. *Environ. Sci. Technol.* 24: 938–945.

Lioy, P. J., Wainman, T., and Weisel, C. 1993. A wipe sampler for the quantitative measurement of dust on smooth surfaces: Laboratory performance studies. *J. Exp. Anal. Environ. Epidemiol.* 3: 315–320.

Lioy, P. J., Yiin, L. M., Adgate, J., Weisel, C., and Rhoads, G. G. 1998. The effectiveness of a home cleaning intervention strategy in reducing potential dust and lead exposures. *J. Exp. Anal. Environ. Epidem.* 8(1): 17–35.

Lomborg, B. 2001. *The Skeptical Environmentalist: Measuring the Real State of the World.* Cambridge Univ. Press, Cambridge.

Lovell, D. P., and Thomas, G. 1996. Quantitative risk assessment and the limitations of the linearized multistage model. *Hum. Exp. Toxicol.* 15: 87–104.

Lowrance, W. 1976. *Of Acceptable Risk.* Kaufmann, Los Altos, CA.

Lowrance, W. 1984. *Modern Science and Human Values.* Oxford University Press, London.

Lu, F. C. 1985. Safety assessments of chemicals with threshold effects. *Regul. Toxicol. Pharmacol.* 5: 121–132.

Lu, F. C. 1988. Acceptable daily intake: Inception, evolution, and application. *Regul. Toxicol. Pharmacol.* 8: 45–60.

Lyman, W. J., and Rosenblat, H. 1990. *Handbook of Chemical Property Estimation Methods: Environmental Behavior of Organic Compounds.* American Chemical Society, Washington, DC.

Lynch, M. R. (Ed.). 1995. *Procedures for Assessing the Environmental Fate and Ecotoxicity of Pesticides.* Society of Environmental Toxicology and Chemistry, Pensacola, FL.

Mackay, D. 1991. *Multimedia Environmental Models: The Fugacity Approach.* Lewis, Chelsea, MI.

Mackay, D. 2000. *Chemical Transport.* Lewis Pub., Chelsea, MI.

Mackay, D., and Paterson, S. 1982. Calculating fugacity. *Water Pollut. Res. J.* 16: 59–70.

Maki, A. 1979. An analysis of decision criteria in environmental hazard evaluation programs. In K. L. Dickson, A. W. Maki, and J. Cairns, Jr. (Eds.), *Analyzing the Hazard Evaluation Process.* Am. Fish. Soc., Washington, DC, pp. 83–100.

Maki, A., Dickson, K. L., and Cairns, J., Jr. 1979. *Understanding Chemicals in the Environment.* Am. Soc. Microbiol., Washington, DC.

Maki, A., Dickson, K. L., and Cairns, J., Jr. 1980. *Biotransformation and Fate of Chemicals in the Aquatic Environment.* Am. Soc. Microbiol., Washington, DC.

Mallins, D., and Ostrander, G. 1993. *Aquatic Toxicology.* Lewis Pub., Chelsea, MI.

Manson, J. M., and Kang, Y. J. 1994. Test methods for assessing female reproductive and development toxicology. In A. W. Hayes (Ed.), *Principles and Methods in Toxicology.* Raven, New York, pp. 989–1038.

Manson, J. M., and Wise, L. D. 1991. Teratogens. In M. O. Amdur, J. Doull, and C. D. Klaassen (Eds.), *Casarett and Doull's Toxicology.* Pergamon, New York, pp. 226–254.

Mantel, N., and Bryan, W. R. 1961. "Safety" testing of carcinogenic agents. *J. Natl. Cancer Inst.* 27: 455–460.

Mapleson, W. 1963. An electric analog for uptake and exchange of inert gases and other agents. *J. Appl. Physiol.* 18: 197–204.

Margolis, H. 1996. *Dealing with Risk.* University of Chicago Press, Chicago, IL.

Marquardt, H., Schafer, S. G., McClellan, R. O., and Welsch, F. 1999. *Toxicology.* Academic, San Diego.

Martin, W. E. 1964. Loss of Sr-90, Sr-89 and I-131 from fallout of contaminated plants. *Radiat. Bot.* 4: 275–281.

Marzulli, F. N., and Maibach, I., II. 1983. *Dermatoxicology,* 2nd ed. Hemisphere, Publishing, Washington, DC.

Mattie, D. R., Bates, G. D., Jr., Jepson, G. W., Fisher, J. W., and McDougal, J. N. 1994. Determination of skin: Air partition coefficients for volatile chemicals: Experimental method and applications. *Fund. Appl. Toxicol.* 22: 51–58.

Maughn, J. T. 1993. *Ecological Assessment of Hazardous Waste Sites.* Van Nostrand, New York.

Maxim, L. D., and Harrington, L. 1984. A review of the Food and Drug Administration risk analysis for polychlorinated biphenyls. *Regul. Toxicol. Pharmacol.* 4: 192–199.

Mayo, D., and Hallander, R. 1991. *Acceptable Evidence: Science and Values in Risk Management.* Oxford University Press, New York.

Mazarakis, N. D., Edwards, A. D., and Mehmet, H. 1997. Apoptosis in neural development and disease. *Arch. Dis. Child.* 77: F165–170.

McArthur, B. 1992. Dermal measurement and wipe sampling methods: A review. *Appl. Occup. Environ. Hyg.* 7: 599–606.

MacArthur, R. H. 1995. Fluctuations of animal populations and a measure of community stability. *Ecology* 36: 533–536.

McColl, R. S. 1987. *Environmental Health Risks: Assessment nnd Management.* Institute for Risk Research, Waterloo, Ontario.

McConnell, E., Lucier, G., Rumbaugh, R., Albro, P., Harvan, D., Hass, J., and Harris, M. 1984. Dioxin in soil: Bioavailability after ingestion by rats and guinea pigs. *Science* 223: 1077–1079.

McCord, C. 1937. *A Blind Hog's Acorns.* Bell Publishers, Chicago, IL.

McDougal, J. N. 1996. Physiologically-based pharmacokinetic modeling. In F. N. Marzulli and H. I. Maibach (Eds.), *Dermatoxicology.* Taylor and Francis, Washington, DC.

McDougal, J. N., Jepson, G. W., Clewell, H. J., III, Gargas, M. L., and Andersen, M. E. 1990. Dermal absorption of organic chemical vapors in rats and humans. *Fundam. Appl. Toxicol.* 14: 299–308.

McKone, T. 1995. *The Caltox Model.* Lewis Pub., Chelsea, MI.

McKone, T. E. 1990. Dermal uptake of organic chemicals from a soil matrix. *Risk Anal.* 10: 407–419.

Mehlman, M., and Upton, A. 1994. *The Identification and Control of Environmental and Occupational Diseases: Hazards and Risks of Chemicals in the Oil Refining Industry.* Princeton Scientific Press, Princeton, NJ.

Meistrich, M. L. 1988. Estimation of human reproductive risk from animal studies: Determination of interspecies extrapolation factors for steroid hormone effects on the male. *Risk Anal.* 1: 27–34.

Menzel, D. 1987. Physiological pharmacokinetic modeling. *Environ. Sci. Technol.* 21: 944–950.

Menzer, R. E., and Nelson, J. O. 1986. Water and soil pollutants. In C. D. Klaassen, M. O. Amdur, and J. Doull (Eds.), *Casarett and Doull's Toxicology*, 3rd ed. Macmillan, New York, pp. 825–853.

Michaud, J. M., Huntley, S. L., Sherer, R. A., Gray, M. N., and Paustenbach, D. J. 1994. PCB and dioxin re-entry criteria for building surfaces and air. *J. Exp. Anal. Environ.* Epidemiol. 4(2): 197–227.

Mitsch, W. J., and Gosselink, J. G. 1986. *Wetlands.* Van Nostrand Reinhold, New York.

Moolgavkar, S. H. 1978. The multistage theory of carcinogenesis and the age distribution of cancer in man. *J. Natl. Cancer Inst.* 61: 49–52.

Moolgavkar, S. H. 1986. Carcinogenesis modeling: From molecular biology to epidemiology. *Ann. Rev. Public Health* 7: 151–169.

Moolgavkar, S. H., Day, N. E., and Stevens, R. G. 1980. Two-stage model for carcinogenesis: Epidemiology of breast cancer in females. *J. Natl. Cancer Inst.* 65: 559–569.

Moolgavkar, S. H., and Knudson, A. G. 1981. Mutation and cancer: A model for human carcinogenesis. *J. Natl. Cancer Inst.* 66: 1037–1052.

Moolgavkar, S. H., and Venzon, D. J. 1979. Two event models for carcinogenesis: Incidence curves for childhood and adult tumors. *Math. Biosci.* 47: 55–77.

Morgan, J. N., Berry, M. R., and Graves, R. L. 1997. Effects of commonly used cooking practices on total mercury concentration in fish and their impact on exposure assessments. *J. Exp. Anal. Environ. Epidemiol.* 7: 119–133.

Morgan, M. D., and Henrion, M. 1990. *Uncertainty: A Guide to Dealing with Uncertainty in Quantitative Risk and Policy Analysis.* Cambridge University Press, Cambridge, MA.

Moriarity, B. 1988. *The Study of Pollutants in Ecosystems.* New York, New York.

Morris, J. 2000. *Rethinking Risk and the Precautionary Principle.* Butterworth-Heinemann, Oxford, UK.

Moyle, P. B., and Leidy, R. A. 1992. Loss of biodiversity in aquatic ecosystems: Evidence from fish faunas. In P. S. Fielder and K. J. Subodh (Eds.), *From Conservation Biology: The Theory and Practice of Nature, Conservation, Preservation and Management.* Chapman and Hall, New York, pp. 127–169.

Munkittrick, K., McMaster, M., Van Derkraak, G., Portt, C., Gibbons, W., Farwell, A., and Gray, M. (Eds.). 2000. *Development of Methods for Effects-Driven Cumulative Effects Assessment Using Fish Populations: Moose River Project.* Society of Environmental Toxicology and Chemistry, Pensacola, FL.

Munro, I. C., and Krewski, D. R. 1981. Risk assessment and regulatory decision making. *Food Cosmet. Toxicol.* 19: 549–560.

Mylchreest, E., Sar, M., Cattley, R. C., and Foster, P. M. D. 1999. Disruption of androgen-regulated male reproductive development by di(n-butyl) phthalate during late gestation in rats is different from flutamide. *Toxicol. Appl. Pharmacol.* 156: 81–95.

National Academy of Sciences (NAS). 1975. *Principles for Evaluating Chemicals in the Environment.* NAS, Washington, DC.

National Academy of Sciences (NAS). 1977–1995. *Drinking Water and Health*, Vols. 1–10. NAS, Washington, DC.

National Academy of Sciences (NAS). 1980. *The Effect on Populations of Exposure to Low Levels of Ionizing Radiation*. NAS, Washington, DC.

National Academy of Sciences (NAS). 1999. *Hormonally Active Agents in the Environment*. NAS, Washington, DC.

National Institute for Occupational Safety and Health (NIOSH). 1976. *Occupational Diseases: A Guide to Their Recognition*. NIOSH, Cincinnati, OH.

National Research Council (NRC). 1980a. *Principles of Toxicological Interactions Associated with Multiple Chemical Exposures*. National Academy Press, Washington, DC.

National Research Council (NRC). 1980b. *Lead in the Human Environment*. NRC, Washington, DC.

National Research Council (NRC). 1983. *Risk Assessment in the Federal Government: Managing the Process*. NAS, Washington, DC.

National Research Council (NRC). 1987. *Pharmacokinetics in Risk Assessment: Drinking Water and Health*, Vol. 8. National Academy of Science, Washington, DC.

National Research Coucil (NRC). 1990. *Groundwater Models: Scientific and Regulatory Applications*. NRC, Washington, DC.

National Research Council (NRC). 1994. *Science and Judgment in Risk Assessment*. National Academy Press, Washington, DC.

National Research Council (NRC). 1996. *Linking Science and Technology to Society's Environmental Goals*. NRC, Washington, DC.

National Research Council (NRC). 1996. *Understanding Risk: Informing Decisions in a Democratic Society*. National Academy Press, Washington, DC.

National Reseach Council (NRC). 1997. *Building a Foundation for Sound Environmental Decision making*. NRC, Washington, DC.

National Research Council (NRC). 2000. Waste incinerators and public health. National Academy Press, Washington, DC.

Neely, W. G., Branson, D. R., and Blau, G. E. 1974. The use of the partition coefficient to measure the bioconcentration potential of organic chemicals in fish. *Environ. Sci. Technol.* 8: 1113–1115.

Neumann, D. A., and Kimmel, C. A. 1998. *Human Variability in Response to Chemical Exposure*. ILSI Press, Washington, DC.

Nichols, A. L., and Zeckhauser, R. J. 1988. The perils of prudence: How conservative risk assessments distort regulation. *Regul. Toxicol. Pharmacol.* 8: 61–75.

Nielsen, D. M. 1991. *Practical Handbook of Groundwater Monitoring*. National Water Well Association, Lewis, Ann Arbor, MI.

Niesink, R., deVries, J., and Hollinger, M. 1996. *Toxicology: Principles and Applications*. CRC Press, Boca Raton, FL.

Nohl, J. 1960. *The Black Death*. Cambridge MA, Ballantine Books.

Nordberg, J. 1976. *Effects and Dose Response Relationships of Metals*. Elsevier, Amsterdam, The Netherlands.

Odum, E. P. 1971. *Fundamentals of Ecology*, 3rd ed. Saunders, Philadelphia, PA.

Office of Science and Technology Policy (OSTP). 1985. *Cancer Risk Assessment Guidelines*. OSTP, Washington, DC.

Office of Technology Assessment (OTS). 1981. *Assessment of Technologies for Determining Cancer Risks from the Environment*. OTA, Washington, DC.

O'Flaherty, E. J. 1981. *Toxicant and Drugs: Kinetics and Dynamics*. Wiley, New York.

O'Flaherty, E. J. 1998. Physiologically based models of metal kinetics. *Crit Rev Toxicol* 28(3): 271–317.

O'Flaherty, E. J., Kerger, B. D., Hays, S. M., and Paustenbach, D. J. 2001. A physiologically based model for the ingestion of chromium(III) and chromium(VI) by humans. *Toxicol Sci* 60(2): 196–213.

Olin, S., Farland, W., Park, C., Rhemberg, L., Scheuplein, R., Starr, J., and Wilson, J. 1995. *Low-Dose Extrapolations of Cancer Risks: Issues and Perspectives*. Inter Life Sciences Institute (ILSI) Press, Washington, DC.

Oppenheim, L. 1977. *Ancient Mesopotamia*. University of Chicago Press, Chicago, IL.

Organization for Economic Cooperation and Development (OECD). 1983. *OECD Guidelines for Testing of Chemicals*. Section 4, No. 415: One-Generation Reproduction Toxicity. Adopted May 1983. OECD Brussels.

Organization for Economic Cooperation and Development (OECD). 1996. *OECD Guidelines for Testing of Chemicals*. Section 4, No. 422: Combined Repeated Dose Toxicity Study with the Reproduction/Development Toxicity Screening Test. Adopted March 1999. OECD Brussels.

Organization for Economic Cooperation and Development (OECD). 1998. *The Revised OECD Principles of Good Laboratory Practices*. OECD Brussels.

Ott, W. R. 1995a. *Environmental Statistics and Data Analysis*. CRC Press, Boca Raton, FL.

Ott, W. R. 1995b. Human exposure assessment: The birth of a new science. *J. Exposure Anal. Environ. Epidemiol.* 5(4): 449–472.

Ottoboni, M. A. 1991. *The Dose Makes the Poison*. Second edition. Van Nostrand, New York.

Paracelsus (Theophrastus Ex Hohnenheim Eremita). 1567. Von der Besucht. Dillingen.

Park, C. N., and Hawkins, N. C. 1995. Cancer risk assessment. In L. J. Cralley and J. Bus (Eds.), *Patty's Industrial Hygiene and Toxicology*, Vol. 3B, 3rd ed. Wiley, New York.

Park, C. N., and Snee, R. D. 1983. Quantitative risk assessment: State of the art for carcinogenesis. *Am. Stat.* 37(4): 427–441.

Parks, L. G., Ostby, J. S., Lambright, C. R., Abbott, B. D., Klinefelter, G. R., and Gray, L. E. In press. Perinatal butyl benzyl phthalate (BBP) and bis(2-ethylhexyl) phthalate (DEHP) exposures induce aniandrogenic effects in Sprague-Dawley rats.

Paustenbach, D. J. 1989. A survey of health risk assessment. In D. J. Paustenbach (Ed.), *The Risk Assessment of Environmental and Human Health Hazards*. Wiley, New York.

Paustenbach, D. J. 1995. The practice of health risk assessment in the United States (1975–1995): How the U.S. and other countries can benefit from that experience. *Hum. Ecol. Risk Assess.* 1(1): 29–79.

Paustenbach, D. J. 2000. The history and biological basis of occupational exposure limits for chemical agents. In R. L. Harris (Ed.), *Patty's Industrial Hygiene*, 5th ed., Vol. 3. Wiley, New York, pp. 1903–2000.

Paustenbach, D. J., Clewell, H. J., Gargas, M., and Andersen, M. E. 1988. A physiological pharmacokinetic model for carbon tetrachloride. *Toxicol. Appl. Pharmacol.* 96: 191–211.

Paustenbach, D. J., Bruce, G. M., and Chrostowski, P. 1997a. Current views on the oral bioavailability of inorganic mercury in soil: The impact on health risk assessments. *Risk Anal.* 17: 533–545.

Paustenbach, D. J., Finley, B. L., and Long, T. F. 1997b. The critical role of house dust in understanding the hazards posed by contaminated soils. *Int. J. Toxicol.* 16: 339–362.

Paustenbach, D. J., Leung, H. W., and Rothrock, J. 1999. Health risk assessment. In R. Adams (Ed.), *Occupational skin disease*, 3rd ed. Saunders, Philadelphia, PA, pp. 291–323.

Paustenbach, D. J., Meyer, D. M., Sheehan, P. J., and Lau, V. 1991b. An assessment and quantitative uncertainty analysis of the health risks to workers exposed to chromium contaminated soils. *Toxicol. Ind. Health* 7: 159–196.

Paustenbach, D. J., Rinehart, W. E., and Sheehan, P. J. 1991a. The health hazards posed by chromium-contaminated soils in residential and industrial areas: Conclusions of an expert panel. *Regul. Toxicol. Pharmacol.* 13: 195–222.

Paustenbach, D. J., Shu, H. P., and Murray, F. J. 1986. A critical analysis of risk assessment of TCDD contaminated soils. *Regul. Toxicol. Pharmacol.* 6: 284–307.

Perkins, J. L., Cutter, G. N., and Cleveland, M. S. 1990. Estimating the mean, variance, and confidence limits from censored (limit of detection), lognormally-distributed exposure data. *Am. Ind. Hyg. Assoc. J.* 51: 416–419.

Perreault, S. D. 1997. The mature spermatozoan as a target for reproductive toxicants. In K. Boekelheide, R. E. Chapin, P. B. Hoyer, and C. Harris (Eds.), *Comprehensive Toxicology*. Elsevier Science, New York, pp. 165–179.

Peterson, R. E., Cooke, P. S., Kelce, W. R., and Gray, L. E. 1997. Environmental endocrine disruptors. In K. Boekelheide, R. E. Chapin, P. B. Hoyer, and C. Harris (Eds.), *Comprehensive Toxicology*. Elsevier, New York, pp. 181–192.

Pianin, E. 2001. *Dioxin report by EPA on hold. Washington Post*, April 12, p. A01.

Piotrowski, J. L. 1973. *Exposure Tests for Organic Compounds in Industry*. National Institute of Occupational Safety and Health (NIOSH), Cincinnati, OH.

Pitot, H., and Dragan, Y. 2001. Chemical Carcinogenesis. *Casarett and O'Doull's Toxicology: The Basic Science of Poisons*. C. Klaassen. MacMillan Pub., New York: 241–320.

Plowchalk, D. R., Andersen, M. E., and Boydanffy, M. S. 1997. Physiologically based modeling of vinyl acetate uptake, metabolism, and intracellular pH changes in the rat nasal cavity. *Toxicol Appl Pharmacol* 142(2): 386–400.

Poiger, H., and Schlatter, C. 1980. Influence of solvents and absorbents on dermal and intestinal adsorption of TCDD. *Food Cosmet. Toxicol.* 18: 477–481.

Portier, C. J., and Kaplan, N. L. 1989. Variability of safe estimated when using complicated models of carcinogenic processess. A dose study: Methylene chloride. *Fundam. Appl. Toxicol.* 13: 533–544.

Poulin, P., and Krishnan, K. 1996. Molecular structure-based prediction of the partition coefficients of organic chemicals for physiological pharmacokinetic models. *Toxicol. Meth.* 6: 117–137.

Powell, C. J., and Berry, L. L. 2000. Non-genotoxic and epigenetic carcinogens. In B. Ballantyne, T. Marrs, and T. Syverson (Eds.), *General and Applied Toxicology*, 2nd ed. McMillan, London, pp. 1099–1119.

Powell, M. R. 1999. *Science at EPA: Information in the Regulatory Process*. Resources for the Future, Washington, DC.

Presidential/Congressional Commission on Risk Assessment and Risk Management. 1997. *Risk Assessment and Risk Management in Regulatory Decision-Making*. Final Report, Vol. 2. Washington, DC.

Price, P. S., Su, S. H., and Gray, M. N. 1994. The effect of sampling bias on estimates of angler consumption rates in creel surveys. *J. Exp. Anal. Environ. Epidemiol.* 4: 355–372.

Price, P. S., Su, S. H., Harrington, J. R., and Keenan, R. E. 1996. Uncertainty and variation in indirect exposure assessments: An analysis of exposure to tetrachlorodibenzo-*p*-dioxin from a beef consumption pathway. *Risk Anal.* 16: 263–277.

Proctor, D. M., Zak, M. A., and Finley, B. L. 1997. Resolving uncertainties associated with the construction worker soil ingestion rate: A proposal for risk-based remediation goals. *Hum. Ecol. Risk Assess.* 3: 299–304.

Proctor, N. H., Hughes, J. P., and Fischman, M. 1988. *Chemical Hazards of the Workplace*, 2nd ed. Lippincott, Philadelphia, PA.

Purchase, I. F. H., and Auton, T. R. 1995. Thresholds in chemical carcinogenesis. *Regul. Toxicol. Pharmacol.* 22: 199–205.

Purchase, I. F. H., and Slovic, P. 1999. Quantitative risk assessment breeds fear. *Hum. Ecol. Risk. Assess.* 5(3): 445–453.

Raffle, P., Lee, W., McCallum, P., and Murray, R. 1987. *Hunter's Diseases of Occupations*. Little, Brown, and Company, Boston, MA.

Ramazzini, B. 1700. De Morbia artificium. *Arch. Klin. Chir.* 50: 588.

Ramsey, J. R., and Andersen, M. E. 1984. A physiologically-based description of the inhalation pharmacokinetics of styrene in rats and humans. *Toxicol. Appl. Pharmacol.* 73: 159–175.

Rand, G. M. 1995. *Fundamentals of Aquatic Toxicology*, 2nd ed. Taylor and Francis, Philadelphia, PA.

Rappaport, S. M., and Selvin, J. 1987. A method for evaluating the mean exposure from a lognormal distribution. *Am. Ind. Hyg. Assoc. J.* 48: 374–379.

Rehn, L. 1895. Blasengeschwulste bei Fuchsin-Arbeitern. *Arch. Klin. Chir.* 50: 588–800.

Reinert, K., Bartell, S., and Biddinger, G. (Eds.). 1998. *Ecological Risk Assessment Decision-Support System: A Conceptual Design*. Society of Environmental Toxicology and Chemistry, Pensacola, FL.

Reith, J. P., and Starr, T. B. 1989. Chronic bioassays: Relevance to quantitative risk assessment of carcinogens. *Regul. Toxicol. Pharmacol.* 10: 160–173.

Reitz, R. H., Fox, T. R., and Watanabe, P. G. 1986. The role of pharmacokinetics in risk assessment. *Basic Life Sci.* 38: 499–507.

Reitz, R. H., Gargas, M. L., Andersen, M. E., Provan, W. M., and Green, T. L. 1996. Predicting cancer risk from vinyl chloride exposure with a physiologically based pharmacokinetic model. *Toxicol. Appl. Pharmacol.* 137: 253–267.

Reitz, R. H., Mandrela, A. L., Corley, R. A., Quast, J. F., Gargas, M. L., Andersen, M. E., Staats, D. E., and Conolly, R. B. 1990. Estimating the risk of liver cancer associated with human exposures to chloroform using physiologically-based pharmacokinetic modeling. *Toxicol. Appl. Pharmacol.* 105: 443–459.

Reitz, R. H., Quast, J. F., Stott, W. T., Watanabe, P. G., and Gehring, P. J. 1980. Pharmacokinetics and macromolecular effects of chloroform in rats and mice: Implications for carcinogenic risk estimation. In *Water Chlorination.* Ann Arbor Science Press, Ann Arbor, MI, pp. 983–993.

Renwick, A. G. 1991. Safety factors and establishment of acceptable daily intakes. *Food Addit.* 8(2): 135–150.

Renwick, A. G. 1995. The use of an additional safety or uncertainty factor for nature of toxicity in the estimation of acceptable daily intake and tolerable daily intake values. *Regul. Toxicol. Pharmacol.* 22: 250–261.

Renwick, A. G., and Lazarus, N. R. 1998. Human variability and noncancer risk assessment—an analysis of the default uncertainty factor. *Regul. Toxicol. Pharmacol.* 21(1, Pt. 1): 3–20.

Riggs, D. 1963. *The Mathematical Approach to Physiological Problems.* MIT Press, Cambridge, MA.

Risk Assessment and Environmental Quality Division. 1994. Environmental Quality Objectives in the Netherlands. Ministry of Housing, Spatial Planning, Netherlands.

Robbins, W. A., Rubes, J., Selevan, S. G., and Perreault, S. D. 1999. Air pollution and sperm aneuploidy in healthy young men. *Environ. Epidemiol. Toxicol.* 1: 125–131.

Rodricks, J. V., Brett, S. N., and Wrenn, G. C. 1987. Risk decisions in federal regulatory agencies. *Regul. Toxicol. Pharmacol.* 7: 307–320.

Roes H., Buchet, J. P., and Lauwerys, R. R. 1980. Exposure to lead by the oral and pulmonary routes of children living in the vicinity of a primary lead smelter. *Environment* 22: 81–94.

Rogers, J. M., and Kavlock, R. J. 1996. Developmental toxicology. In C. D. Klaassen (Ed.), *Casarett and Doull's Toxicology: The Basic Science of Poisons,* 5th ed. McGraw-Hill, New York, pp. 301–331.

Rolland, R. M., Gilbertson, M., and Peterson, R. (Eds.). 1997. *Chemically Induced Alterations in Functional Development and Reproduction of Fishes.* Society of Environmental Toxicology and Chemistry, Pensacola, FL.

Romney, E. M., Lindberg, N. G., Hawthorne, H. A., Bystrom, B. B., and Larson, K. H. 1963. Contamination of plant foliage with radioactive nuclides. *Annu. Rev. Plant Physiol.* 14: 271–279.

Rowe, W. D. 1977. *An Anatomy of Risk.* Wiley, New York.

Rubin, C. T. 1994. *The Green Crusade: Retaining the Roots of Environmentalism.* Free Press, New York.

Ruby, M. V., Davis, A., Kempton, J. H., Drexter, J. W., and Bergstrom, P. D. 1992. Lead bioavailability under simulated gastric conditions. *Environ. Sci. Technol.* 26: 1242–1248.

Ruby, M. V., Schoof, R., Brattin, W., Goldade, M., Post, G., Harnois, M., Mosby, D. E., Casteel, S. W., Berti, W., Carpenter, M., Edwards, D., Cragin, D., and Chappell, W. 1999. Advances in evaluating the oral bioavailability of inorganics in soil for use in human health risk assessment. *Environ. Sci. Technol.* 33(21): 3697–3705.

Ruckelhaus, W. 1984b. Vital Speeches of the Day. City News Publ. Co., Southold, NY.

Russell, K. S. 1966. Entry of radioactive materials into plants. In I. S. Russell (Ed.), *Radioactivity and Human Diet*. Pergamon, New York, Chapter 5.

Safe, S. 1996. Endocrine disruptors: An unlikely serious risk. *Wall Street Journal*.

Sagan, C. 1995. *The demon haunted world: Science as a candle in the dark*. Random House, New York.

Savitz, D. A., Sonnefield, N. L., and Olshan, A. F. 1994. Review of epidemiologic studies of paternal occupational exposure and spontaneous abortion. *Am. J. Ind. Med.* 25: 361–383.

Sax, N. I. 1984. *Dangerous Properties of Industrial Materials*, 6th ed. Van Nostrand, New York.

Sayre, J. W., Charney, E., Vestal, J., and Pless, B. 1974. House and hand dust as a potential source of childhood lead exposure. *Am. J. Dis. Child.* 127: 167–170.

Schaeffer, D. J. 1981. Is "No-threshold" a "Non-concept." *Environ. Manag.* 5: 475–481.

Schardein, J. L. 1983. Teratogenic risk assessment. *Issues Rev. Toxicol.* 1: 181–214.

Schardein, J. L. 1992. *Chemically Induced Birth Defects*, 2nd ed. Marcel Dekker, New York.

Schardein, J. L., Schwetz, B. A., and Kopel, M. E. 1985. Species sensitivities and prediction of teratogenic potential. *Environ. Health Perspect.* 61: 55–67.

Schmidt, R. R., and Johnson, E. M. 1997. Principles of teratology. In R. D. Hood (Ed.), *Handbook of Developmental Toxicology*. CRC Press, New York, pp. 3–12.

Schoof, R. A., and Nielsen, J. B. 1997. Evaluation of methods for assessing the oral bioavailability of inorganic mercury in soil. *Risk Anal.* 17: 545–555.

Schrader, S. M. 1997. Male reproductive toxicants. In E. J. Massaro (Ed.), *CRC Handbook of Human Toxicology*. CRC Press, New York, pp. 961–980.

Schrag, S. D., and Dixon, R. L. 1985. Reproductive effects of chemical agents. In R. L. Dixon (Ed.), *Reproductive Toxicology*. Raven, New York, pp. 301–319.

Schultz, J. 1995. *The Ecozones of the World*. Springer, Heidelberg.

Schwing, R. C., and Albers, W. A., Jr. 1980. *Societal Risk Assessment: How Safe Is Safe Enough*. Plenum, New York.

Scott, P. K., Sung, H., Finley, B. L., Schulze, R. H., and Turner, D. B. 1997a. Identification of an accurate soil suspension/dispersion modeling method for use in estimating health-based soil cleanup levels of hexavalent chromium in chromit-ore processing residues. *J. Air Waste Mgmt.* 47(7): 753–765.

Scroti: The Chirurgical Work of Percival Pott. Clark & Collins, London.

Sedman, R., Funk, L. M., and Fountain, R. 1998. Distribution of residence duration in owner occupied housing. *J. Exp. Anal. Environ. Epidemiol.* 8: 51–57.

Segerstrale, U. 2000. *Defenders of the Truth: The Battle for Science in the Sociobiology Debate and Beyond*. Oxford Press, NY, NY.

Seip, H. M., and Heiberg, A. B. 1989. *Risk Management of Chemicals in the Environment*. Plenum, New York.

Sessions, G. 1995. *Deep Ecology for the 21st Century*. Shambhala Pub., Boston.

Shepard, T. H. 1995. *Catalog of Teratogenic Agents*, 8th ed. John Hopkins University Press, Baltimore, MD.

Sheppard, S., Bembridge, J., Holmstrup, M., and Posthuma, L. (Eds.). 1998. *Advances in Earthworm Ecotoxicology*. Society of Environmental Toxicology and Chemistry, Pensacola, FL.

Sherman, B., and Korenman, S. 1974. Inadequare corpus luteum function: A pathophysiological interpretation of human breast cancer epidemiology. *Cancer* 33: 1306–1312.

Shlyakhter, A., Goodman, G., and Wilson, R. 1992. Monte Carlo simulation of rodent carcinogenicity bioassays. *Risk Anal.* 12: 73–82.

Shu, H., Paustenbach, D., Murray, J., Marple, L., Brunck, B., Dei Rossi, D., Webb, A. S., and Tietelbaum, T. 1987. Bioavailability of soil-bound TCDD: Oral bioavailability in the rat. *Fundam. App. J. Toxicol.* 10: 648–654.

Shu, H., Tietelbaum, P., Webb, A. S., Marple, L., Brunck, B., Dei Rossi, D., Murray, J., and Paustenbach, D. 1988. Bioavailability of soil-bound TCDD: Dermal bioavailability in the rat. *Fundam. Appl. Toxicol.* 10: 335–343.

Shuurmann, G., and Markert, B. 1998. *Ecotoxicology: Ecological Fundamenals, Chemical Exposure, and Biological Effects*. Wiley-Interscience, New York.

Sicherman, B. 1984. *Alice Hamilton: A Life in Letters*. Harvard University Press, Cambridge, MA.

Sielken, R. L. 1985. Some issues in the quantitative modeling portion of cancer risk assessment. *Regul. Toxicol. Pharmacol.* 5: 175–181.

Sielken, R. L. 1987a. Quantitative cancer risk assessment for 2,3,7,8-TCDD. *Food Chem. Toxicol.* 25: 257–267.

Sielken, R. L. 1987b. Statistical evaluation reflecting the skewness in the distribution of TCDD levels in human adipose tissue. *Chemosphere* 16: 2135–2140.

Sielken, R. L. 1987c. The capabilities, sensitivity, pitfalls, and future of quantitative risk assessment. In R. S. McColl (Ed.), *Environmental Health Risks: Assessment and Management*. University of Waterloo Press, Waterloo, Ontario, pp. 95–131.

Sielken, Jr., R. L., and Stevenson, D. E. 1997. Opportunities to improve quantitative risk assessment. *Hum. Ecol. Risk Assess.* 3: 479–490.

Sielken, Jr., R. L., and Stevenson, D. E. 1998. Some implications for quantitative risk assessment if hormeseis exists. *Hum. Exp. Toxicol.* 17: 259–262.

Sijm, D., Bruijn, W., de Voogt, D., and de Wolf, J. (Eds.). 1997. *Biotransformation in Environmental Risk Assessment*. Society of Environmental Toxicology and Chemistry, Pensacola, FL.

Silkworth, J., McMartin, D., DeCaprio, A., Rej, R., O'Keefe, P., and Kaminsky, L. 1982. Acute toxicity in guinea pigs and rabbits of soot from a polychlorinated biphenyl-containing transformer fire. *Toxicol. Appl. Pharmacol.* 65: 425–429.

Sittig, M. 1985. *Handbook of Toxic and Hazardous Chemicals and Carcinogens*. Noyes Data Corp., Park Ridge, NJ.

Skrowronski, G. A., Turkall, R. M., and Abdel-Rahman, M. S. 1988. Soil absorption alters bioavailability of benzene in dermally exposed male rats. *Am. Ind. Hyg. Assoc. J.* 49: 506–511.

Slob, W. 1994. Uncertainty analysis in multiplicative models. *Risk Anal.* 14(4): 571–576.

Smith, A. H. 1987. Infant exposure assessment for breast milk dioxins and furans derived from waste incineration emissions. *Risk Anal.* 7: 347–353.

Smith, J. H., Bomberger, D. C., and Haynes, D. L. 1981. Volatilization rates of intermediate and low volatility chemicals from water. *Chemosphere* 10: 281–289.

Smith, R. L. 1994. Use of Monte Carlo simulation for human exposure assessment at a Superfund site. *Risk Anal.* 14(4): 433–439.

Snow, J. 1855. *On the mode of communication of Cholera.* Churchill, London.

Snyder, W. S. 1975. *Report of the Task Group on Reference Man.* ICRP No. 23. Pergamon, New York.

Solbrig, O. T., and Nicolis, G. 1991. *Perspectives in Biological Complexity.* IUBS, Paris.

Southworth, G. R., Parkhust, B. R., Herbes, S. E., and Tsai, S. C. 1982. The risk of chemicals to aquatic environment. In R. A. Conway (Ed.), *Environmental Risk Analysis for Chemicals.* Van Nostrand, New York.

Spacie, A., and Hamelink, J. 1985. Bioaccumulation. In G. M. Rand and S. R. Petrocelli (Eds.), *Fundamentals of Aquatic Toxicology.* McGraw-Hill, New York.

Sparling, D. W., Linder, G., and Bishop, C. (Eds.). 2000. *Ecotoxicology of Amphibians and Reptiles.* Society of Environmental Toxicology and Chemistry, Pensacola, FL.

Spencer, W. F., and Farmer, W. J. 1980. Assessment of the vapor behavior of toxic organic chemicals. In R. Haque (Ed.), *Dynamics, Exposure and Hazard Assessment of Toxic Chemicals.* Ann Arbor Science, Ann Arbor, MI, pp. 1–13 161.

Sprague, J. B. 1973. The ABCs of pollutant bioassay using fish. In J. Cairns, Jr. and K. I. Dickson (Eds.), *Biological Methods for the Assessment of Water Quality.* ASTM, Philadelphia, PA, pp. 6–30.

Stanek, E. J., Calabrese, E. J., and Zorn, M. 2001. Soil ingestion distributions for Monte Carlo risk assessment in children. *Hum. Ecol. Risk Assess.* 7: 357–368.

Stanek, E. J., and Calabrese, E. J. 2000. Daily soil ingestion estimates for children at a Superfund site. *Risk Anal.* 20: 627–635.

Stanek, III, E. J., Calabrese, E. J., and Xu, L. 1998. A caution for Monte Carlo risk assessment of long term exposures based on short term exposure data. *Hum. Ecol. Risk Assess.* 4: 409–422.

Stara, J. F., and Erdreich, L. S. 1985. Advances in Health Risk Assessment for Systemic toxicants and chemical mixtures. Princeton Scientific Publishing, Princeton, NJ.

Starr, C. 1969. Social benetiss versus technological risk. *Science* 165: 1232–1238.

Starr, C. 1985. Risk management, assessment, and acceptability. *Risk Anal.* 5: 97–102.

Stebbing, A. R. D., Travis, C., and Matthiessen, D. (Eds.). 1993. *Environmental Modelling—The Next 10 Years.* Society of Environmental Toxicology and Chemistry, Pensacola, FL.

Stellman, J. M. 1998. *ILO Encyclopedia of Occupational Health and Safety,* 4th ed., 3 vols. Inter Labor Office (ILO), Geneva.

Stern, A. H., Korn, L. R., and Ruppel, B. E. 1996. Estimation of fish consumption and methylmercury intake in the New Jersey population. *J. Exp. Anal. Environ. Epidemiol.* 6: 503–525.

Stickel, L. F. 1973. Chemosphere pesticide reserves in birds and mammals. In C. A. Edwards (Ed.), *Environmental Pollution by Pesticides.* Plenum, New York.

Stickel, W. H., Hayne, D. W., and Stickel, L. F. 1965. Effects of heptachlorocontaminated earthworms on woodcocks. *J. Wild. Manag.* 29(1): 132–146.

Stokinger, H. E. 1981. Threshold limit values. Part I. *Dangerous Properties of Industrial Materials Report.* Van Nostrand-Reinhold, New York, pp. 8–13.

Stokinger, H. 1971. Intended use and application of the TLV's . In *Transactions of the Thirty-third Annual Meeting of the American Conference of Governmental Industrial Hygienists.* 33: 113–116.

Sreit, B. 1995. Energy flow and community structure in freshwater ecosystems. *Experientia* 51: 425–436.

Streit, B., and Stadler, T. 1995. Freshwater invertebrates as model systems in population ecology and genetics. *Experientia* 51: 423–424.

Sublet, V., Zenick, H., and Smith, M. K. 1989. Factors associated with reduced fertility and implantation rates in females mated to acrylamide-treated rats. *Toxicology* 55: 53–67.

Swanson, M., and Socha, A. (Eds.). 1997. *Chemical Ranking and Scoring: Guidelines for Relative Assessments of Chemicals.* Society of Environmental Toxicology and Chemistry, Pensacola, FL.

Sweeney, L. M., Tyler, T. R., Kirman, C. R., Corley, R. A., Reitz, R. H., Paustenbach, D. J., Holson, J. F., Whorton, M. D., Thompson, K. M., and Gargas M. L. 2001. Proposed occupational exposure limits for select ethylene glycol ethers using PBPK models and Monte Carlo simulations. *Toxicol Sci* 62(1): 124–39.

Swenberg, J. A., La, D. K., Scheller, N. A., and Wu, K. Y. 1995. Dose-response relationships for carcinogens. *Toxicol. Lett.* 82: 751–756.

Swindoll, M., Stahl, G., and Ells, D. (Eds.). 2000. *Natural Remediation of Environemtnal Contaminants: Its Role in Ecological Risk Assessment and Risk Management.* Society of Environmental Toxicology and Chemistry, Pensacola, FL.

Tattersfield, L., Matthiessen, D., Campbell, B., Grandy, L., and Länge, J. (Eds.). 1997. *Endocrine Modulators and Wildlife: Assessment and Testing.* Society of Environmental Toxicology and Chemistry, Pensacola, FL.

Tengs, T. O., Adams, M. E., Pliskin, J. S., Safran, D. G., Siegel, J. E., Weinstein, M. C., and Graham, J. D. 1995. Five-hundred life-saving interventions and their cost-effectiveness. *Risk Anal* 15(3): 369–90.

Thibodeaux, L. J. 1979. *Chemodynamics: Environmental Movement of Chemicals in Air, Water and Soil.* Wiley, New York.

Thiennes, C. H., and Haley, T. J. 1972. *Clinical Toxicology*, 5th ed. Lea & Febinger, Philadelphia, PA.

Thomas, J. A. 1991. Toxic responses of the reproductive system. In M. O. Amdur, J. Doull, and C. D. Klaassen (Eds.), *Casarett and Doull's Toxicology.* Pergamon, New York, pp. 484–520.

Thompson, K. M., Burmaster, D. E., and Crouch, E. A. C. 1992. Monte Carlo techniques for quantitative uncertainty analysis in public health risk assessments. *Risk Anal.* 12(1): 53–63.

Thornton, J. 2000. *Pandora's Poison. Chlorine, Health and a New Environmental Strategy.* MIT Press, Cambridge, MA.

Thornton, I., and Abrahams, P. 1981. Soil ingestion as a pathway of metal intake into grazing livestock. In *Proceedings of the International Conference on Heavy Metals in the Environment.* CEP Consultants, Edinburgh, pp. 167–122.

Thorslund, T. W., Brown, C. C., and Charnley, G. 1987. Biologically motivated cancer risk models. *Risk Anal.* 7: 109–119.

Toppari, J., Larsen, J. C., Christiansen, P., Giwercman, A., Granjean, P., Guillette, L. J., Jegou, B., Jensen, T. K., Jouannet, P., Keiding, N., Leffers, H., McLachlan, J. A., Meyer, O., Muller, J., Rajpert-De Meyts, E., Scheike, T., Sharpe, R., Sumpter, J., and Skakkebaek, N. E. 1996. Male reproductive health and environmental xenoestrogens. *Environ. Health Perspect.* 104(Suppl. 4): 741–803.

Travis, C. C., and Land, M. L. 1990. Estimating the mean of data sets with nondetectable values. *Environ. Sci. Technol.* 24: 961–962.

Travis, C. C., White, R. K., and Ward, R. C. 1990. Interspecies extrapolation of pharmacokinetics. *J. Theoret. Biol.* 142: 285–304.

Travis, C. C., Richter, S. A., Crouch, C. E., Wilson, R., and Klema, E. D. 1987. Cancer risk management. *Environ. Sci. Technol.* 21: 415–420.

Tucker, R. K., and Crabtree, D. G. 1970. *Handbook of Toxicity of Pesticides to Wildlife.* Resour. Publ. No. 84. Fish and Wildlife Service, U.S. Department of the Interior, Denver Wildlife Research Center, Denver, CO.

Turnbull, D., and Rodricks, J. 1985. Assessment of possible carcinogenic risk to humans resulting from exposure to di(2-ethylhexyl)phthalate (DEHP). *J. Am. Toxicol.* 4: 111–145.

Tyl, R. W., and Marr, M. C. 1996. Developmental toxicity testing. In R. Hood (Ed.), *Handbook of Developmental Toxicology.* CRC Press, New York, pp. 175–225.

Udo de Haes, H. (Ed.). 1996. *Towards a Methodology for Life-Cycle Assessment.* Society of Environmental Toxicology and Chemistry, Pensacola, FL.

Urquhart, J., and Heilmann, K. 1984. *Risk Watch: The Odds of Life.* Facts on File Pub., New York.

U.S. Food and Drug Administration (USFDA). 1984. *Environmental Assessment Technical Handbook.* Center for Food Safety and Applied Nutrition and Center for Veterinary Medicine, USFDA, Washington, DC.

Van den Berg, M., Ofie, K., and Hutzinger, O. 1984. Uptake and selective retention in rats of orally administered chlorinated dioxins and dibenzofurans from fly-ash and fly-ash extract. *Chemosphere* 12: 537–544.

Van der Eerden, L. J. M. 1992. Fertilizing effects of atmospheric ammonia on seminatural vegetation. Ph.D. Thesis, Vrije Universiteit, Amsterdam.

Van Straalen, N. M., and Denneman, C. A. J. 1989. Ecotoxicological evaluation of soil quality criteria. *Ecotox. Environ. Saf.* 18: 241–251.

Viscusi, K. 1992. *Fatal Tradeoffs.* Oxford University Press, New York.

Vose, D. 1996. *Quantitative Risk Analysis: A Guide to Monte Carlo Simulation Modelling.* Wiley, New York.

Walker, K. 1992. *Australian Environmental Policy: Ten Case Studies.* New South Wales University Press, Kensington, NSW Australia.

Wall Street Journal. 1987. "Scaring the Public." July 7. p. 5.

Wall Street Journal. 1996. Endocrine disruptors: A real threat? July 4.

Wang, G. M., and Schwetz, B. A. 1987. An evaluation system for ranking chemicals with teratogenic potential. *Teratogen. Carcinogen. Mutagen.* 7: 133–139.

Warren-Hicks, W. J., and Moore, D. (Eds.). 1998. *Uncertainty Analysis in Ecological Risk Assessment*. Society of Environmental Toxicology and Chemistry, Pensacola, FL.

Wattenberg, B. J. 1984. *The Good News Is the Bad News Is Wrong*. Simon & Schuster, New York.

Weast, T. 2000. *Handbook of Chemistry and Physics*. CRC Press, Boca Raton, FL.

Weil, C. S. 1970. Selection of number of the valid sampling units and a consideration of their combination in toxicological studies involving reproduction, teratogenesis or carcinogenesis. *Food Cosmet. Toxicol.* 8: 177–182.

Weil, C. S. 1972. Statistics versus safety factors and scientific judgement in the evaluation of safety for man. *Toxicol. Appl. Pharmacol.* 21: 454–463.

Weiss, B. 1988. Neurobehavioral toxicity as a basis for risk assessment. *Trends Pharmacol. Sci.* 9(2): 59–62.

Hathaway, G., Proctor, N., and Hughes, J. 1996. *Proctor and Hughes Chemical Hazards of the Workplace*, 4th ed. Wiley, New York.

Wester, R. C., Bucks, D. A. W., and Maibach, H. I. 1993. Percutaneous absorption of contaminants from soil. In R. G. M. Wang, J. B. Knaak, and H. I. Maibach (Eds.), *Health Risk Assessment: Dermal and Inhalation Exposure and Absorption of Toxicants*. CRC Press, Boca Raton, FL.

Wester, R. C., and Noonan, P. K. 1980. Relevance of animal models for percutaneous absorption. *Int. J. Pharm.* 7: 99–110.

Whipple, C. 1988. *De Minimus Risk*. Macmillan, New York.

Whittemore, A. S., Grosser S. L., and Silvers, A. 1986. Pharmacokinetics in low-dose extrapolation using animal cancer data. *Fundam. Appl. Toxicol.* 7: 183–190.

Wildavsky, A. 1991. *Searching for Safety*. Transaction Pub., New Brunswick, NJ.

Wildavsky, A. 1995. *But is it True? A citizen's guide to environmental health and safety issues*. Harvard University Press, Cambridge, MA.

Williams, G. M., and Weisburger, J. H. 1981. Systematic carcinogen testing through a decision point approval. *Annu. Rev. Pharmacol. Toxicol.* 21: 393–416.

Williams, P. L., James, R. C., and Roberts, S. M. 2000. *Principles of Toxicology: Environmental and Industrial Applications*, 2nd ed. Wiley, New York.

Williams, P. R. D., Scott, P. K., Sheehan, P. J., and Paustenbach, D. J. 2000. A probabilistic assessment of household exposures to MTBE from daily drinking water. *Hum. Ecol. Risk. Assess.* 6: 827–849.

Wilschut, A., and ten Berge, W. F. 1995. Two mathematical skin permeation models for vapours. *Abstracts of Presentations at the Fourth International Prediction of Percutaneous Penetration Conference, Prediction of Percutaneous Penetration*, Vol. 4a. 3M Medica, Italy.

Wilson, E. 1994. *Naturalist*. Island Press, Washington, DC.

Wilson, J. G. 1973. *Environment and Birth Defects*. Academic, New York.

Wilson, J. G., and Warkany, J. 1965. *Teratology: Principles and Techniques*. University of Chicago Press, Chicago, IL.

Wilson, N. D., Shear, N. D., Paustenbach, D. J., and Price, P. S. 1998. The effect of cooking practices on the concentration of DDT and PCB compounds in the edible tissue of fish. *J. Exp. Anal. Environ. Epidemiol.* 8: 423–440.

Winslow, Q. 1923. *The Evolution of and Significance of the Modern Public Health Campaign.* Yale University Press, New Haven, CT.

Wipf, H. K., Homberger, E., Neimer, N., Ranalder, U. B., Vetter, W., and Vuilleumeir, J. P. 1982. TCDD-levels in soil and plant samples from the Seveso area. In O. Hutzinger, R. W. Frei, E. Merian, and P. Pocchiari. (Eds.), *Chlorinated Dioxins and Related Compounds: Impact on the Environment.* Pergamon, New York, pp. 115–126.

Wissmar, R. C., and Swanson, F. J. 1990. In R. J. Naimand and H. Decamps (Eds.), *The Ecology and Management of Aquatic Terrestrial Ecotones.* UNESCO, Parthenon Pub., Paris, pp. 171–198.

Wooster, D. 1994. Predator impacts on stream benthic prey. *Oecologia* 99: 7–15.

Wooton, J. T. 1994. The nature and consequences of indirect effects on ecological communities. *Annu. Rev. Ecol. Syst.* 25: 443–466.

Working, P. K. 1989. *Toxicology of the Male and Female Reproductive Systems.* Hemisphere, New York.

World Health Organization (WHO). 1987. *Occupational Exposure Limits for Airborne Toxic Substances*, 3rd ed. Occup. Saf. Health Ser., No. 37. International Labor Office, WHO, Geneva.

Yang, R. 1994. *Toxicology of Chemical Mixtures: Case Studies, Mechanisms, and Novel Approaches.* Academic Press, New York.

Yang, R. S. H., El-Masri, H. A., Thomas, R. S., Constan, A. A., and Tessari, J. D. 1995. The application of physiologically based pharmacokinetic/pharmacodynamis (PBPK/PD) modeling for exploring risk assessment approaches of chemical mixtures. *Toxicol. Lett.* 79: 193–200.

Zelikoff and Thomas, V. 1998. *Immunotoxicolgy of Environmental and Occupational Metals.*

Zenick, H., Clegg, E. D., Perreault, S. D., Klinefelder, G. R., and Earl Gray, L. 1994. Assessment of male reproductive toxicity: A risk assessment approach. In A. W. Hayes (Ed.), *Principle and Methods of Toxicology.* Raven, New York, pp. 937–988.

Ziegler, P. 1969. *The Black Death.* Penguin Books, Middlesex, England.

Zielhuis, R. L., and van der Kreek, F. W. 1979a. The use of a safety factor in setting health based permissible levels for occupational exposure. Part I. A proposal. *Int. Arch. Occup. Environ. Health* 42: 191–201.

Zielhuis, R. L., and van der Kreek, F. W. 1979b. Calculation of a safety factor in setting health based permissible levels for occupational exposure. Part II. Comparison of extrapolated and published permissible levels. *Int. Arch. Occup. Environ. Health* 42: 203–215.

Zoaks, C. 2000. Envirotimeline. http://www.zoaks/information/envirotimeline/envirotimeline.html.

2 Hazard Identification

DENNIS J. PAUSTENBACH

Risk Assessment Consultant, 65 Roan Place, Woodside, California

2.1 INTRODUCTION

Hazard identification is the determination of whether a particular chemical is or is not causally linked to particular health effects [National Academy of Sciences (NAS), 1983]. Of the four steps in a risk assessment, hazard identification is the most easily recognized in the actions of regulatory agencies. It has been defined as the process of determining whether exposure to an agent can cause an increase in the incidence of adverse health effects (cancer, birth defect, etc.) (NAS, 1983). Although the question of whether a substance causes cancer or other adverse health effects should, theoretically, receive a yes–no answer, it is rarely that simple. In general, the epidemiology or toxicology data are equivocal or the extrapolation of animal data to humans is not straightforward.

The hazard identification portion of a risk assessment was intended to be the first chapter of a four-chapter analysis of a particular chemical. For example, in 1985 it would not be unusual to ask whether 50 ppb of a chemical like trichloroethylene in drinking water posed a significant health hazard. The answer to that question would hopefully be found in a risk assessment of that chemical. In that first chapter of the assessment, the hazard identification step, one would review the toxicology data and the human data. A weight-of-analysis discussion would be offered about which effect seemed to be the most sensitive (the one produced at the lowest dose) and the best study from which one would identify a no-observable-effect level (NOEL). Although both cancer and noncancer effects were discussed, often during the years 1975–1995, the focus was on the carcinogens.

Beginning in the mid-1990s, the hazard identification process often involved a more thorough analyses of the cancer data and discussions of noncancer endpoints (liver toxicity, neurotoxicity, etc) became more sophisticated. The animal data began to be more critically reviewed with respect to its significance to humans (e.g., nitrilotriacetic acid (NTA), saccharin, and dioxin). In recent years,

Human and Ecological Risk Assessment: Theory and Practice, Edited by Dennis J. Paustenbach
ISBN 0-471-14747-8 © 2002 John Wiley & Sons, Inc.

85

especially when conducting risk assessments of hazardous waste sites where dozens of chemicals were evaluated, the hazard identification step was sometimes blended together with the dose–response relationship. Unfortunately, in large measure, most risk assessments conducted since about 1995, which attempt to answer questions about the hazard posed by a chemical in a particular media (such as soil or sediment), are little more than a review of what various regulatory agencies and a few scientists in the field believe are the primary hazard. The classic hazard identification where only one chemical is studied and all possible health effects are evaluated has generally been relegated to organizations that specialize in categorizing chemicals (e.g., carcinogen, neurotoxin) or those who identify safe levels of exposure [e.g., the Environmental Protection Agency (EPA), the International Agency for Research on Cancer (IARC), the Occupational Safety and Hazard Administration (OSHA), and the American Conference of Governmental Industrial Hygienists (ACGIH)].

So, the first question typically considered in the hazard identification of a chemical concerns the types of toxic effects that it can cause. For example, can the chemical damage the liver, the kidney, the lungs, or the reproductive system? Can it cause birth defects, neurotoxic effects, or cancer? The most significant type of hazard information can be obtained through studies of groups of people who happen to have been exposed to the chemical (epidemiologic studies), especially workplace exposure. The other source of information is through controlled laboratory experiments involving various animal species. Several other types of experimental data (e.g., in-vitro or mutagenicity studies) can also be used to assist in characterizing the toxic hazards of a chemical. Some of the basic criteria applied to the hazard identification process are presented in Table 2.1. This chapter will discuss the kinds of data used to identify a chemical as primarily a carcinogenic hazard, as a noncarcinogenic hazard, as an environmental concern, or as a concern to wildlife. Each deserves to be considered within the hazard identification step.

2.2   HAZARD IDENTIFICATION FOR CARCINOGENS

2.2.1   Hazard Identification of Carcinogens: Human Studies

Well-conducted epidemiological studies that show a positive association between an agent and a disease are accepted as the most convincing evidence about human risk (NAS, 1983; NRC, 1994). This evidence is, however, difficult to obtain; often the risk is low, the number of persons exposed is small, the latency period between exposure and disease is long, and the exposures are mixed and multiple. Thus, epidemiologic data require careful interpretation. Even if these problems are solved satisfactorily, the preponderance of chemicals in the environment have not been studied using epidemiologic methods, and the public would not wish to release newly produced substances only to discover years later

TABLE 2.1   Criteria Often Used in Hazard Identification

Epidemiologic Data

- What relative weights should be given to studies with differing results? For example, should positive results outweigh negative results if the studies that yield them are comparable? Should a study be weighted in accord with its statistical power?
- What relative weights should be given to results of different types of epidemiologic studies? For example, should the findings of a prospective study supersede those of a case control study or an ecologic study?
- What statistical significance should be required for results to be considered positive?
- Does a study have special characteristics (such as the questionableness of the control group) that lead one to question the validity of its results?
- What is the significance of a positive finding in a study in which the route of exposure is different from that of a population at potential risk?
- Would evidence on different types of responses be weighted or combined (e.g., data on different tumor sites and data on benign versus malignant tumors)?

Animal-Bioassay Data

- What degree of confirmation of positive results should be necessary? Is a positive result from a single animal study sufficient, or should positive results from two or more animal studies be required? Should negative results be disregarded or given less weight?
- Should a study be weighted according to its quality and statistical power?
- How should evidence of different metabolic pathways or vastly different metabolic rates between animals and humans be factored into a risk assessment?
- How should the occurrence of rare tumors be treated? Should the appearance of rare tumors in a treated group be considered evidence of carcinogenicity even if the finding is not statistically significant?
- How should experimental animal data be used when the exposure routes in experimental animals and humans are different?
- Should a dose-related increase in tumors be discounted when the tumors in question have high or extremely variable spontaneous rates?
- What statistical significance should be required for results to be considered positive?
- Does an experiment have special characteristic (e.g., the presence of carcinogenic contaminants in the test substance) that leads one to question the validity of its results?
- How should findings of tissue damage or other toxic effects be used in the interpretation of tumor data? Should evidence that tumors may have resulted from these effects be taken to mean that they would not be expected to occur at lower doses?
- Should benign and malignant lesions be counted equally?
- Into what categories should tumors be grouped for statistical purposes?
- Should only increases in the numbers of tumors be considered, or should a decrease in the latent period for tumor occurrence also be used as evidence of carcinogenicity?

that they were powerful carcinogenic agents without some kind of assurance about safety. These limitations require reliance on less direct evidence, like animal studies and structure–activity relationships (SAR), to determine whether a health hazard exists.

Epidemiological studies clearly provide the most relevant kind of information for hazard identification, simply because they involve observations of human beings, not laboratory animals. That obvious and substantial advantage is offset to various degrees by the difficulties associated with obtaining and interpreting epidemiologic information. It is often not possible to identify the appropriate populations for study or to obtain the necessary medical information on the health status of individuals in them. Information on the magnitude and duration of chemical exposure, especially that experienced in the distant past, is often available in only qualitative or semiquantitative form (e.g., the number of years worked at low, medium, and high exposure). In recent years, persons have attempted to reconstruct historical exposures using various techniques and to a large extent they have obtained useful results (Paustenbach et al., 1992; Paustenbach, 2000; Widner et al., in press).

Identifying other factors that might influence the health status of a population is often not possible. Epidemiologic studies are not controlled experiments. The investigator identifies an exposure situation and attempts to identify appropriate "control" groups (i.e., unexposed parallel populations), but this is often accomplished with difficulty and any shortcomings make it difficult or impossible to identify cause–effect relationships [OSTP, 1985; International Agency for Research on Cancer (IARC), 1997].

It is rare that convincing causal relationships can be identified with a single epidemiology study (Rothman, 1996). Epidemiologists usually weigh the results from several studies, ideally involving different populations and investigative methods, to determine whether there is a consistent pattern or response among them (Lillienfield, 1995). Some other factors that are often considered are the strength of the statistical association between a particular disease and exposure to the suspect chemical, whether the risk of the disease increases with increasing exposure to the suspect agent, and the degree to which other possible causative factors can be ruled out. Epidemiologists attempt to reach consensus regarding causality by weighing the evidence. Needless to say, different experts will weigh such data differently and consensus typically is not easily achieved (IARC, 1997).

In the case of chemicals suspected of causing cancer in humans, expert groups ("working groups") are regularly convened by the IARC, and other agencies, to consider and evaluate epidemiologic evidence (IARC, 1999). These groups have published their conclusions regarding the "degrees" of strength of the evidence on specific chemicals (sometimes chemical mixtures or even industrial processes when individual causative agents cannot be identified). The highest degree of evidence—sufficient evidence of carcinogenicity—is applied only when a working group agrees that the total body of evidence is convincing with respect to the issue of a cause–effect relationship.

No similar consensus-building organization has been established regarding other forms of toxicity, except perhaps the EPA working group, which identifies reference doses (RfD). There is one group called Toxicology Excellence for Risk Assessment (TERA), which assembles experts and conducts peer reviews, the end result being the categorization of chemicals. Its work has generally been considered of exceptional quality and the results are posted on its web site (TERA, 2001).

2.2.2 Hazard Identification of Carcinogens: Animal Studies

When conducting a hazard identification of carcinogens, by far the most commonly available data are those obtained from animal studies. For example, to date, there have been hundreds of nonpharmaceutical agents (chemicals) evaluated in 2-year chronic bioassays. The assumption that results from animal experiments are applicable to humans is fundamental to toxicology research; this premise underlies much of experimental biology and medicine. Of course, this inference is one form of a "default assumption." In most cases, chemicals that at some dose can cause an increase in cancer in rodents will, at some dose, cause an increase in the incidence of cancer in humans. This assumption is logically extended to both noncarcinogenic and carcinogenic effects. The shortcoming in this assumption, however, is that humans rarely are exposed to the chemical at doses anywhere close to what is administered in animals studies, thus raising the question about the relevance of the results of studies conducted near the maximum tolerance dose (MTD). Although the inference that animal data are applicable to humans usually is valid, there have been a number of chemicals where the animal data have clearly been shown to be of highly uncertain relevance to humans either due to mechanism of action, pharmacokinetic differences, metabolic differences, or of relevance of a target organ (e.g., forestomach tumors in rodents) (Powell and Berry, 2000).

When attempting to identify carcinogens, consistently positive test results in the two sexes and in several strains and species, as well as higher incidences at higher doses constitute the best evidence that the chemical is a true carcinogenic hazard. More often than not, however, such data are either not available or the data are equivocal. Instead, because of the nature of the effect and the limits of detection of animal tests as they are usually conducted, experimental data leading to a positive finding sometimes barely exceed a statistical threshold and may involve tumor types of uncertain relation to human carcinogenesis (NAS, 1983; IARC, 1997; Klaassen, 2001; Hayes, 2001). Interpretation of some, if not most, of the animal bioassay data on most industrial chemicals is difficult for a myriad of reasons. For example, one continually has to deal with thorny issues such as: "Is the tumor incidence significantly greater than controls? Should we consider historical control incidence rates? Should one add tumor types to see if there is a statistical increase, even though most single sites are not statistically increased? Is the site of the tumor increase in an organ not present in humans or not relevant to humans, for example, rodent forestomach? Is a benign tumor of equal concern as malignant tumors? The list of issues to address is quite long. In spite of the different ways that one can evaluate the tumor incidence on the 40 or more sites that are studied, and notwithstanding the uncertainties associated with how to interpret the animal tests, they have, in general, proved to be fairly reliable indicators of the carcinogenic hazard to humans.

But laboratory animals are not human beings, and this obvious fact is one clear disadvantage of animal studies (NAS, 1983; NRC, 1994). Another is the relatively high cost of animal studies containing enough animals to detect an effect of interest; especially carcinogenicity. Like humans, the background incidence in elderly animals is quite high, so it can be difficult to identify whether an increase in the cancer rate is attributable to the agent. The other problem with animal studies evaluating cancer is that at least one dose is going to be very near that which causes frank toxicity; the so-called maximum tolerated dose (MTD). Since humans will never be chronically exposed to doses this great, observations of toxicity in laboratory animals usually requires two acts of extrapolation to predict the human health hazard: interspecies extrapolation and extrapolation from high test doses to lower environmental doses. This is called the dose–response extrapolation (which is addressed elsewhere in this book). There are a number of reasons (based on biologic underpinnings) that animal data can be extrapolated across mammalian species, including *Homo sapiens*, but the scientific basis of such extrapolation is not established with sufficient rigor to allow broad and definitive generalizations to be made (NRC, 1994).

One of the most important reasons for species differences in response to exposure to carcinogens is that toxicity is very often a function of chemical metabolism. Differences among animal species, or even among strains of the same species, with respect to the metabolic handling of a chemical, are not uncommon and can account for toxicity differences (EPA, 1999). For example, we know that some polycyclic aromatic hydrocarbons (PAHs) are converted to a completely different array of metabolites in the mouse, rat, guinea pig, and hamster (Klaassen, 2001). Because in most cases information on a chemical's metabolic profile in humans is lacking (and often unobtainable), identifying the animal species and toxic response most likely to predict the human response accurately is generally not possible. It has become customary to assume, under these circumstances, that in the absence of clear evidence that a particular toxic response is not relevant to human beings, any observation of toxicity in animal species is potentially predictive of response in at least some humans [Environmental Protection Agency (EPA), 1999]. This is not unreasonable, given the significant variation among humans in genetic composition, prior sensitizing events, and concurrent exposure agents. However, when information to the contrary is available, it should be weighed heavily.

In the case of epidemiologic data, IARC expert panels rank evidence of carcinogenicity from animal studies (IARC, 1999). It is generally recognized by experts that evidence of carcinogenicity is most convincing when a chemical

TABLE 2.2 Categorization of Evidence of Carcinogenocity Used by EPA from 1986 to 1996 (EPA, 1986)

Groups	Criteria for Classification
A. Human carcinogen	Sufficient evidence from epidemiologic studies
B. Probable human carcinogen (two subgroups)	Limited evidence from epidemiologic studies and sufficient evidence from animal studies (B1); *or* inadequate evidence from epidemiologic studies (or no data) and sufficient evidence from animal studies (B2)
C. Possible human carcinogens	Limited evidence from animal studies and no human data
D. Not classifiable as to human carcinogenicity	Inadequate human and animal data or no data
E. Evidence of noncarcinogenicity in humans	No evidence of carcinogenicity from adequate human and animal studies

produces excess malignancies in several species and strains of laboratory animals and in both sexes. The observation that a much higher proportion of treated animals than untreated (control) animals develop malignancies adds weight to the evidence of carcinogenicity as a result of exposure. At the other extreme, the observation that a chemical produces only a relatively small increase in the incidence of mostly benign tumors, at a single site of the body, in a single species and sex of test animals does not make a very convincing case for carcinogenicity, although any excess of tumors raises some concern.

The EPA has historically combined human and animal evidence, as shown in Table 2.2, to categorize evidence of carcinogenicity (EPA, 1986, 1996). The agency's evaluations of data on individual carcinogens have been generally similar to those of the IARC (Table 2.3). As discussed later, the 1996 cancer guidelines proposed by EPA, and their 1999 update, are quite different than those issued in 1986. For noncancer health effects, EPA uses categories such as those described in Table 2.4. Animal data on other forms of toxicity are generally evaluated in the same way as carcinogenicity data, although this classification looks at hazard identification (qualitative) and dose–response relationships (quantitative) together. No risk ranking schemes similar to those used for carcinogens have been adopted for the noncarcinogens.

The biologic component of the hazard identification step of a risk assessment generally concludes with a qualitative narrative of the types of toxic responses, if any, that can be caused by the chemical under discussion, the strength of the supporting evidence, the scientific merits of the data, and their value for predicting human toxicity (EPA, 1986c).

TABLE 2.3   IARC Classification of Chemical Carcinogens (IARC, 1999)

- *Group 1: The agent (mixture) is carcinogenic to humans. The exposure circumstance entails exposures that are carcinogenic to humans.*

This category is used when there is sufficient evidence of carcinogenicity to humans. Exceptionally, an agent (mixture) may be placed in this category when evidence of carcinogenicity in humans is less than sufficient but there is sufficient evidence of carcinogenicity in experimental animals and strong evidence in exposed humans that the agent (mixture) acts through a relevant mechanism of carcinogenicity.

- *Group 2*

This category includes agents, mixtures, and exposure circumstance for which, at one extreme, the degree of evidence of carcinogenicity in humans is almost sufficient, as well as those for which, the other extreme, there are no human data but for which there is evidence of carcinogenicity in experimental animals. Agents, mixtures, and exposure circumstances are assigned to either group 2A (probably carcinogenic to humans) or to group 2B (possibly carcinogenic to humans) on the basis of epidemiological and experimental evidence of carcinorgenicity and other relevant data.

- *Group 2A: The agent (mixture) is probably carcinogenic to humans. The exposure circumstance entails exposures that are probably carcinogenic to humans.*

This category is used when there is limited evidence of carcinogenicity in humans and sufficient evidence of carcinogenicity in experimental animals. In some cases, an agent (mixture) may be classified in this category when there is inadequate evidence of carcinogenicity in and sufficient evidence of carcinogenicity in experimental animals and strong evidence that the carcinogenesis is mediated by a mechanism that also operates in humans. Exceptionally, an agent, mixture, or exposure circumstance may be classified in this category solely on the basis of limited evidence of carcinogenicity.

- *Group 2B: The agent (mixture) is possibly carcinogenic to humans. The exposure circumstance entails exposures that are possibly carcinogenic to humans.*

This category is used for agents, mixtures, and exposure circumstances for which there is limited evidence of carcinogenicity in humans and less than sufficient evidence of carcinogenicity in experimental animals. It may also be used when there is inadequate evidence of carcinogenicity in humans but there is sufficient evidence of carcinogenicity in experimental animals. In some instances, an agent, mixture, or exposure circumstance for which there is inadequate evidence of carcinogenicity in humans, but limited evidence of carcinogenicity in experimental animals together with supporting evidence from other relevant data may be placed in this group.

- *Group 3: The agent (mixture or exposure circumstance) is not classifiable as to its carcinogenicity to humans.*

This category is used most commonly for agents, mixtures, and exposure circumstances for which the evidence of carcinogenicity is inadequate in humans and inadequate or limited in experimental animals.

Exceptionally, agents (mixtures) for which the evidence of carcinogenicity is inadequate in humans but sufficient in experimental animals may be placed in this category when there is strong evidence that the mechanism of carcinogenicity in experimental animals does not operate in humans.

Agents, mixtures, and exposure circumstances that do not fall into any other group are also placed in this category.

TABLE 2.3 *(Continued)*

• *Group 4: The agent (mixture) is probably not carcinogenic to humans.*

This category is used for agents or mixtures for which there is evidence suggesting lack of carcinogenicity in humans and in experimental animals. In some instances, agents or mixtures for which there is inadequate evidence of carcinogenicity in humans but evidence suggesting lack of carcinogenicity in experimental animals, consistently and strongly supported by a broad range of other relevant data, may be classified in this group.

TABLE 2.4   Weight of Evidence Classification Methods for Developmental Effects; A Noncancer Effect (EPA, 1986)

Sufficient Evidence

The sufficient evidence category includes data that collectively provide enough information to judge whether a human developmental hazard could exist within the context of dose, duration, timing, and route of exposure. This category includes both human and experimental animal evidence.

Sufficient Human Evidence: This category includes data from epidemiological studies (e.g., case control and cohort studies) that provide convincing evidence for the scientific community to judge that a causal relationship is or is not supported. A case series in conjunction with strong supporting evidence may also be used. Supporting animal data might or might not be available.

Sufficient Experimental Animal Evidence or Limited Human Data: This category includes data from experimental animal studies or limited human data that provide convincing evidence for the scientific community to judge whether the potential for developmental toxicity exists. The minimal evidence necessary to judge that a potential hazard exists generally would be data demonstrating an adverse developmental effect in a single appropriate, well-conducted study in a single experimental species. The minimal evidence needed to judge that a potential hazard does not exist would include data from appropriate, well-conducted laboratory animal studies in several species (at least two) that evaluated a variety of the potential manifestations of developmental toxicity and showed no developmental effects at doses that were minimally toxic to adults.

Insufficient Evidence

This category includes situations for which there is less than the minimal sufficient evidence necessary for assessing the potential for developmental toxicity, such as when no data are available on developmental toxicity, when the available data are from studies in animals or humans that have a limited design (e.g., small numbers, inappropriate dose selection or exposure information, or other uncontrolled factors), when the data are from a single species reported to have no adverse developmental effects, or when data are limited to information on structure–activity relationships, short-term tests, pharmacokinetics, or metabolic precursors.

2.2.3 IARC and Its Role

In 1969, the IARC initiated a program to evaluate the carcinogenic risk of chemicals to humans and to produce monographs on individual chemicals. The monographs program has since been expanded to include consideration of exposures to complex mixtures of chemicals (which occur, e.g., in some occupations and as a result of human habits) and of exposures to other agents, such as radiation and viruses (IARC, 1997).

The objective of the program is to prepare, with the help of international working groups of experts, and to publish in the form of monographs, critical reviews and evaluations of evidence on the carcinogenicity of a wide range of human exposures. The monographs may also indicate where additional research efforts are needed (IARC, 1998). These monographs are true hazard identification documents since no judgment is offered at how to extrapolate the results to very low doses; nor do they offer much advice to risk managers except that a chemical may cause cancer.

The monographs represent the first step in carcinogenic risk assessment, which involves examination of all relevant information to assess the strength of the available evidence that certain exposures could alter the incidence of cancer in humans. The second step is quantitative risk estimation. Detailed, quantitative evaluations of epidemiological data may be made in the monographs, but without extrapolation beyond the range of the data available. That is, quantitative extrapolation from experimental data to the human situation through the use of models such as the Weibull, one hit, or linearized multistage is not undertaken by IARC.

The monographs often assist national and international authorities in classifying whether a chemical is "more likely than not" to fit into a certain classification of carcinogen. For example, many agencies in numerous countries give significant weight to whether IARC classifies a chemical as a "known human carcinogen" or "plausible human carcinogen." Sometimes, the IARC classification will help them formulate decisions concerning preventive measures such as emissions limits for air or water discharges. The evaluations of IARC working groups are scientific, qualitative judgments about the evidence for or against carcinogenicity provided by the available data. These evaluations represent only one part of the body of information on which regulatory measures may be based. Other components of regulatory decisions may vary from one situation to another and from one country to another, responding to different socioeconomic and national priorities. Therefore, no recommendations are offered by IARC with regard to regulation or legislation, which are the responsibility of individual governments and/or other international organizations (IARC, 1998).

The IARC combines both human and animal data in its assessments. Its evaluations of human data address the types of human studies, the quality of the studies, inferences about mechanism, and criteria for causality. When assessing the animal data, the IARC considers the qualitative aspects, quantitative aspects, and statistical analysis of long-term animal studies. In recent years, in vitro studies of genotoxicity and basic mechanism information have also been given weight in the evaluation. The categorization scheme currently used by the IARC for the carcinogens (1999) is presented in Table 2.3.

2.2.4 EPA and Its Role (The 1999 Proposed Cancer Guidelines)

In the late 1980s, the U.S. EPA developed its first set of cancer guidelines (EPA, 1986). At the time they were a useful contribution to the field of hazard identification of carcinogens. Over time, however, due to our increasing understanding of the strengths and weaknesses of animal bioassays, it was clear that these guidelines no longer reflected our best scientific understanding of how carcinogens actually work in the environment. Thus, in 1996, a new set of EPA guidelines was released and a revised draft was issued in 1999. As of the time of this book (i.e., the spring of 2002), the new EPA position has not been finalized, and the EPA is currently evaluating comments from the public.

It is useful to review some of the key issues raised in the current draft (EPA, 1999). One major change is the way hazard evidence is weighed in reaching conclusions about the human carcinogenic potential of agents (Table 2.5). In the 1986 cancer guidelines, tumor findings in animals or humans were the dominant components of decisions. Other information about an agent's properties, its structure–activity relationships to other carcinogenic agents, and its activities in studies of carcinogenic processes was often limited and played only a modulating role as compared with tumor findings. In their 1999 proposal, decisions come from considering all of the evidence. This change recognizes the growing sophistication of research methods, particularly in their ability to reveal the modes of action of carcinogenic agents at cellular and subcellular levels as well as toxicokinetic and metabolic processes. The effect of the change on the assessment of individual agents will depend greatly on the availability of new kinds of data on them in keeping with the state of the art. If these new kinds of data are not forthcoming from public and private research on agents, assessments under these guidelines will not differ significantly from assessments under former guidelines.

Weighing of the evidence includes addressing the likelihood of human carcinogenic effects of the agent and the conditions under which such effects may be expressed, as they are revealed in the toxicological and other biological important features of the agent. (Consideration of actual human exposure and risk implications are done separately: They are not part of the hazard characterization.) In this respect, the guidelines incorporate recommendations of the NRC (1994). In that report, the NRC recommended the expansion of the former concept of hazard identification, which rests on simply a finding of carcinogenic potential, to a concept of characterization that includes dimensions of the expression of this potential. For example, an agent might be observed to be carcinogenic via inhalation exposure and not via oral exposure, or its carcinogenic activity might be secondary to another toxic effect. In addition, the consideration of evidence includes the mode(s) of action of the agent apparent from the available data as a basis for approaching dose–response assessment.

TABLE 2.5   Standard Descriptors (formerly known as categories or classifications) Proposed by EPA in 1999 for Chemicals with Carcinogenic Potential (EPA, 1999)

Carcinogenic to Humans

- This descriptor is appropriate when there is convincing epidemiologic evidence demonstrating causality between human exposure and cancer.
- This descriptor is also appropriate when there is an absence of conclusive epidemiologic evidence to clearly establish a cause and effect relationship between human exposure and cancer, but there is compelling evidence of carcinogenicity in animals and mechanistic information in animals and humans demonstrating similar mode(s) of carcinogenic action. It is used when all of the following conditions are met:
  - There is evidence in human population(s) of association of exposure to the agent with cancer, but not enough to show a causal association, and
  - There is extensive evidence of carcinogenicity, and
  - The mode(s) of carcinogenic action and associated key events have been identified in animals, and
  - The key events that precede the cancer response in animals have been observed in the human population(s) that also shows evidence of an association of exposure to the agent with cancer.

Likely to be Carcinogenic to Humans

This descriptor is appropriate when the available tumor effects and other key data are adequate to demonstrate carcinogenic *potential* to humans. Adequate data are within a spectrum. At one end is evidence for an association between human exposure to the agent and cancer and strong experimental evidence of carcinogenicity in animals; at the other, with no human data, the weight of experimental evidence shows animal carcinogenicity by a mode or modes of action that are relevant or assumed to be relevant to humans.

Suggestive Evidence of Carcinogenicity, but Not Sufficient to Assess Human Carcinogenic Potential

The descriptor is appropriate when the evidence from human or animal data is suggestive of carcinogenicity, which raises a concern for carcinogenic effects but is judged not sufficient for a conclusion as to human carcinogenic potential. Examples of such evidence may include: a marginal increase in tumors that may be exposure-related, or evidence is observed only in a single study, or the only evidence is limited to certain high background tumors in one sex of the species. Dose-response assessment is not indicated for these agents. Further studies would be needed to determine human carcinogenic potential.

Data Are Inadequate for an Assessment of Human Carcinogenic Potential

This descriptor is used when available data are judged inadequate to perform an assessment. This includes a case when there is a lack of pertinent or useful data or when existing evidence is conflicting, e.g., some evidence is suggestive of carcinogenic effects, but other equally pertinent evidence does not confirm a concern.

TABLE 2.5 *(Continued)*

<hr>

Not Likely to Be Carcinogenic to Humans

This descriptor is used when the available data are considered robust for deciding that there is no basis for human hazard concern. The judgment may be based on—

- Extensive human experience that demonstrates lack of carcinogenic effect (e.g., Phenobarbital).
- Animal evidence that demonstrates lack of carcinogenic effect in at least two well-designed and well-conducted studies in two appropriate animal species (in the absence of human data suggesting a potential for cancer effects).
- Extensive experimental evidence showing that the only carcinogenic effects observed in animals are not considered relevant to humans (e.g., showing only effects in the male rat kidney due to accumulation of $\alpha 2\mu$-globulin).
- Evidence that carcinogenic effects are not likely by a particular route of exposure.
- Evidence that carcinogenic effects are not anticipated below a defined dose range.

<hr>

To express the weight of evidence for carcinogenic hazard potential, the 1986 cancer guidelines provided summary rankings for human and animal studies. These summary rankings were integrated to place the overall evidence in classification groups A through E, group A being associated with the greatest probability of human carcinogenicity and group E with evidence of noncarcinogenicity in humans. Data other than tumor findings played a modifying role after initial placement of an agent into a group.

The 1999 proposed guidelines take a different approach, consistent with the change in the basic approach to weighing evidence. No interim classification of tumor findings followed by modifications with other data takes place. Instead, the conclusion reflects the weighing of evidence in one step (Table 2.5). Moreover, standard descriptors of conclusions are employed rather than letter designations, and these are incorporated into a brief narrative description of their informational basis. The narrative with descriptors replaces the previous letter designation.

With respect to low-dose extrapolation, whenever data are sufficient, a biologically based or case-specific dose–response model is developed to relate dose and response data in the range of empirical observation. Otherwise, as a standard, default procedure, a model is used to curve-fit the data. The lower 95 percent confidence limit on dose associated with an estimated 10 percent increased tumor or relevant nontumor response lower effective dose for 10% response $[LED_{10}]$ is identified. This generally serves as the point of departure for extrapolating the relationship to environmental exposure levels of interest when the latter are outside the range of observed data. The environmental exposures of interest may be measured ones or levels of interest to risk managers. Other points of departure may be more appropriate for certain data sets; as described in the guidelines, these may be used instead of the LED_{10}. Additionally, the

LED_{10} is available for comparison with parallel analyses of other carcinogenic agents or of noncancer effects of agents and for gauging and explaining the magnitude of subsequent extrapolation to low-dose levels. The LED_{10} rather than the ED_{10} (the estimate of a 10 percent increased response), is the proposed standard point of departure (see Figure 1.6).

The second step of dose–response assessment is extrapolation to lower dose levels, if needed. This is based on a biologically based or case-specific model if supportable by substantial data. Otherwise, default approaches are applied that accord with the view of mode(s) of action of the agent. These include approaches that assume linearity or nonlinearity of the dose–response relationship or both. The default approach for linearity is to extend a straight line to zero dose, zero response. The default approach for nonlinearity is to use a margin of exposure analysis rather than estimating the probability of effects at low doses. A margin of exposure (MOE) (see Figure 6, Chapter 1) analysis explains the biological considerations for comparing the observed data with the environmental exposure levels of interest and helps in deciding on an acceptable level of exposure in accordance with applicable management factors.

The use of straight-line extrapolation for a linear default is a change from the 1986 guidelines that used the "linearized multistage" (LMS) procedure. This change was made because the former modeling procedure gave an appearance of specific knowledge and sophistication unwarranted for a default. The proposed approach is also more like that employed by the U.S. Food and Drug Administration (USFDA, 1987). The numerical results of the straight-line and LMS procedures are not significantly different (Krewski et al., 1984). The use of a margin of exposure approach was suggested as a new default procedure to accommodate cases in which there is sufficient evidence of a nonlinear dose–response, but not enough evidence to construct a mathematical model for the relationship.

2.3 HAZARD IDENTIFICATION FOR NONCARCINOGENS

When attempting to identify and characterize the noncancer hazard posed by chemicals, the assumed level of confidence in extrapolating the results to humans from animal data has generally been considered higher than for the carcinogens. The only exception might be developmental and reproductive toxicity, due to both species differences and the high doses used in animal studies. One of the primary reasons is that when conducting classic toxicity tests, we need not involve an MTD and, since the effects are thought to have a threshold dose, arguments about the ability to extrapolate results to very low doses (often 1000-fold below the lowest dose tested) are not relevant.

The purpose of toxicological testing is to characterize the potential adverse effects of a chemical on humans through the use of laboratory animals or in vitro systems. The ultimate objective is to identify those substances that might injure humans who might come into contact with them and thus to prevent injury. Four fundamentals are the underpinnings to toxicology (Ottoboni, 1991). First, the magnitude of the biologic response is a function of the concentration of the agent. Second, the concentration at the site of action is related in some predictable and describable manner with the administered dose. Third, the dose and response are causally related. Fourth, for the noncarcinogens, there is a dose at which no adverse effects would be expected. The rest of this section discusses the kinds of tests used to identify noncancer effects in laboratory animals.

2.3.1 Acute Testing

The objectives of acute toxicity testing are to define the intrinsic toxicity of the chemical, to assess the susceptible species, to identify target organs, to provide information for risk assessment after acute exposure to the chemical, and to provide information for the design and selection of dose levels for prolonged studies (Klaassen, 2001; Ballantyne et al., 1999; Hayes, 2001). A battery of acute tests under different conditions and exposure routes should be conducted on chemicals that are likely to be produced in some reasonable quantity or where human exposure cannot be prevented.

By and large, over the past 30–40 years, manufacturers have conducted a basic acute test battery on nearly every chemical to which humans may be routinely or occasionally exposed. This battery usually includes oral, dermal, and inhalation toxicity tests and skin and eye irritation studies. Other tests such as acute preneonatal and neonatal exposure, dermal contact sensitization, and phototoxicity should be considered, depending on the likely degree and type of human exposure. An acute test to estimate the oral lethal dose $(LD)_{50}$ may require as many as 60 animals. Often, 3 or more doses are used and 5 or more animals of each sex are treated. The number of animals used in the dermal and inhalation tests is usually less, while skin and eye irritation tests often require even fewer animals. Standard protocols have been developed, and these should be used if the test results are to be submitted to help support justification of a registration or to meet some other regulatory criteria. A number of texts can be consulted to readily understand how to conduct and interpret standard acute toxicity batteries (Ballantyne et al., 1999; Hayes, 2001), and guidelines have been published by various regulatory agencies.

2.3.2 Subchronic Tests (Mammalian)

Subchronic studies are designed to examine the adverse effects resulting from repeated exposure over a portion of the average life span of an experimental animal. For rodents, these tests are usually 30 to 90 days in duration. Properly designed subchronic studies give valuable information on the cumulative toxicity of a substance, target organs, physiological effects, and metabolic tolerance of a compound following repeated low-dose (relative to acute toxicity testing dose levels) exposure. By monitoring many different parameters of toxicity, including histopathologic evaluation, a wide variety of adverse effects can be detected. The results from such studies can also provide information that will aid in selecting dose levels for chronic, reproductive, and carcinogenicity studies. Subchronic studies are also valuable for establishing dose levels at which no toxicological effects are evident—a critical figure in risk assessment (Klaassen, 2001; Ballantyne et al., 1999; Hayes, 2001). For chemicals where chronic human exposure to low doses is likely, the conduct of subchronic animal studies should be given serious consideration.

The exposure period in subchronic studies may vary, depending on the objective of the study, the species selected for the study, and the route of administration. A generalization that is often made is that subchronic studies do not exceed 10% to 15% of the animal's life span. Like most generalizations, this is not always the case. Oral and inhalation subchronic studies are generally carried out for 3 months in shorter-lived animals (rodents) and about 1 year in longer-lived animals (dogs, monkeys). Subchronic dermal studies are usually performed in 1 month or less. The most common routes of administration used in subchronic toxicity studies are oral, dermal, and respiratory. Wherever feasible, subchronic toxicity studies should expose the animals by the route through which humans are most likely to be exposed . For example, to understand the risk to humans in the workplace to volatile chemicals, inhalation would be the best method for evaluating the risks. As before, a number of texts can be consulted that describe how to conduct and interpret subchronic studies (Klaassen, 2001; Ballantyne et al., 1999; Hayes, 2001).

2.3.3 Chronic Tests (Mammalian)

Long-term toxicity tests are usually defined as studies of longer than 3 months duration, that is, greater than 10 percent of the life span in the laboratory rat. Typically, they have a duration in the vicinity of 18 months to the entire lifetime of the animal. These types of studies can be conducted in all species of laboratory animals, but they are usually conducted in the classic (economically feasible) animals, the mouse or rat. This class of tests encompasses the lifetime toxicity studies, multigeneration reproduction studies, and carcinogenicity studies.

There are two basic reasons for conducting chronic toxicity tests: to produce toxic effect and to define a safe level of exposure (Stevens and Gallo, 1982). The chronic study is defined so as to identify any of the myriad of potential toxic effects of a xenobiotic on structural and functional entities. In contrast to the carcinogenicity studies, which are designed to measure tumor induction, the chronic toxicity study uses a holistic approach to define the etiology of an adverse response to identify the appropriate margin of safety between any proposed use (exposure) levels and those that might produce toxicity.

Fairly large numbers of rodents are used in these tests. The classic chronic toxicity study in rats usually consists of three treatment groups and a control group, all of equal number at the outset, in which the xenobiotic is administered 7 days per week for at least 2 years. Because of the shorter life span of mice, the compound need be given for only 18 months. A second nonrodent species is often required to assure safety, and the choice is usually the purebred beagle. The choice of a larger animal permits more extensive clinical analyses since more blood can be collected from each animal with greater frequency than is possible with rodents. There are several schools of thought on the use of the second species, but the dog should be used with caution since it is a carnivore and often metabolizes compounds differently than an omnivore or herbivore (Stevens and Gallo, 1982).

Perhaps the most intellectually challenging task for the toxicologist responsible for directing a chronic toxicity study is the selection of the dose levels to be tested. There are several suggested approaches. One of these is the approach of the National Cancer Institute's Bioassay Program, which is to conduct a 3-month range-finding study with enough doses to find a level that suppresses body weight gain slightly. This dose is defined as the maximum tolerated dose (MTD) and is selected as the highest dose. Often, 1/10 MTD and 1/100 MTD are selected as the other two test doses.

2.3.4 Developmental Toxicity (Mammalian)

Once known as teratology tests, these toxicological studies have been defined more appropriately as tests for developmental toxicity (EPA, 1986b; Hayes, 2001). There are four primary ways in which altered in utero development can be demonstrated: (i) death of the conceptus, (ii) gross structural abnormality, (iii) in utero growth retardation, or (iv) decrement of anticipated postnatal functional capabilities. These can arise from a variety of causes (Wilson, 1973).

Three terms are often used to describe the results of developmental toxicity tests. For the sake of risk assessment, teratogenic should be used to describe those chemicals that have been shown to produce gross structural abnormalities. Embryotoxic and fetotoxic appear to be the most ill-defined terms. Several studies have used embryotoxic as the sum of all possible toxic actions affecting the embryo, including teratogenic, embryolethal, and other effects. Black and Marks (1986) have proposed that embryotoxicity should describe the loss of an embryo, and the term *fetotoxicity* should be used for less severe effects. Fetotoxicity has also been used to describe the toxic or degenerative effect on fetal tissues and organs after organogenesis (EPA, 1986b; Hayes, 2001). Some authors have suggested that fetotoxicity should be equated with transient effects such that bones and organs would be expected to continue to develop to their normal appearance and function. The EPA guidelines for developmental toxicants (1986b), in contrast, have suggested that embryotoxic and fetotoxic be used to describe a very wide range of adverse effects and that these terms only differentiate the time when the effects are apparent.

2.3.5 Reproductive Toxicity (Mammalian)

In contrast to tests that evaluate developmental toxicity, reproductive tests evaluate chemicals for their ability to affect adversely the fertility of either parent. A number of functional, morphological, and biochemical parameters are available to assess toxic effects on both male and female reproductive function. The functional parameters include reproductive efficiency, cogenesis, and fertilization. Morphological parameters are gross pathology and histopathology. The biochemical parameters include molecular aspects (normal synthesis and metabolism), accessory cell function, and hormonal status. Species survival requires the production and eventual union of the male and female gametes, each with its complement of healthy genes. Of these events and processes, cogenesis, spermatogenesis, and fertilization are studied to evaluate the reproductive process (Hayes, 2001).

These tests typically are conducted for three generations of the test species. Depending on the protocol, the parent may be exposed continuously throughout the key periods prior to and following conception. Thus far, the three-generation protocol has been successful for identifying chemicals that might adversely affect reproduction.

2.4 EVALUATING CHEMICAL AND PHYSICAL PROPERTIES

About 25 years ago it was recognized that both the physical and chemical properties of a substance play a large role in predicting its fate in the environment—an essential part of hazard identification (Cairns et al., 1978; Dickson et al., 1979; Maki et al., 1980; Veith et al., 1980). It was understood then, and now, that if a chemical has a short environmental half-life or if it is not easily transported, it presents a much lesser human or ecological risk than if it has a long half-life or it is easily moved from one media to another, or from one location to another.

In the late 1960s and 1970s, tests to understand these properties began to be conducted on new products (Table 2.6 and 2.7). Together, the toxicity data and the information on fate describe the environmental hazard. Some of the early work involved the study of the relation of physicochemical properties of the organophosphates to their persistence and transport in either biological or environmental systems. It was shown, for example, that the octanol–water partition coefficient gives an indication of the possibility for biological magnification (Kenaga, 1975; MacKay et al., 2000). Similarly, it was shown that the stability of the compound, as evidenced by its resistance to hydrolysis and other degradative reactions, will often account for its persistence—thus allowing the possibility of transport in water or air (Figure 2.1 and Table 2.8).

2.4.1 Water Solubility

Water solubility is an important parameter in understanding environmental fate, which in turn influences the outcome of environmental risk assessments. Together with other physicochemical properties, it can be a useful predictor of the tendency of a chemical to move and distribute between the various environmental compartments (USFDA, 1984; Hoffman et al., 1995). In general, highly water-soluble chemicals are more likely to be transported and distributed by the hydrologic cycle than relatively water-insoluble chemicals. However, many of those chemicals that are known to be significant environmental contaminants,

TABLE 2.6 Candidate Tests for Screening Ecological Impact of New Products

I. Chemical fate (transport, persistence)
  A. Transport
    1. Adsorption isotherm (soil)
    2. Partition coefficient (water–octanol)
    3. Water solubility
    4. Vapor pressure
  B. Other physicochemical properties
    1. Boiling/melting/sublimation points
    2. Density
    3. Dissociation constant
    4. Flammability/explodability
    5. Particle size
    6. pH
    7. Chemical incompatibility
    8. Vapor-phase UV spectrum for halocarbons
    9. UV and visible absorption spectra in aqueous solution
  C. Persistence
    1. Biodegradation
      a. Shake flask procedure following carbon loss
      b. Respirometric method following oxygen (BOD) and/or carbon dioxide
      c. Activated sludge test (simulation of treatment plant)
      d. Methane and CO_2 productions in anaerobic digestion
    2. Chemical degradation
      a. Oxidation (free radical)
      b. Hydrolysis (25°C, pH 5.0 and 9.0)
    3. Photochemical transformation in water

II. Ecological effects
  A. Microbial effects
    1. Cellulose decomposition
    2. Ammonification of urea
    3. Sulfate reduction
  B. Plant effects
    1. Algae inhibition (fresh and seawater, growth, nitrogen fixation)
    2. Duck weed inhibition (increase in fronds or dry weight)
    3. Seed germination and early growth
  C. Animal effects testing
    1. Aquatic invertebrates (*Daphnia*) acute toxicity (first instar)
    2. Fish acute toxicity (96 h)
    3. Quail dietary LC_{50}
    4. Terrestial mammal test
    5. *Daphnia* life-cycle test
    6. *Mysidopsis bahia* life cycle
    7. Fish embryo–juvenile test
    8. Fish bioconcentration test

Source: Toxic substances control; discussion of premanufacture testing policy and technical issues; request for comment, *Federal Register*, **44**, 53, Part IV, pp. 1639–16292, March 16, 1979.

TABLE 2.7   Environmental Processes and Properties That Influence the Degree of Hazard

Process	Key Environmental Property[a]
Physical Transport	
Meteorological transport	Wind velocity
Biouptake	Biomass
Sorption	Organic content of soil or sediments, mass loading of aquatic systems
Volatilization	Turbulence, evaporation rate, reaeration coefficients, soil organic content
Runoff	Precipitation rate
Leaching	Adsorption coefficient
Fallout	Particulate concentration, wind velocity
Chemical Processes	
Photolysis	Solar irradiance, transmissivity of water or air
Oxidation	Concentrations of oxidants and retarders
Hydrolysis	pH, sediment, or soil basicity or acidity
Reduction	Oxygen concentration, ferrous ion concentration, and complexation state
Biological Processes	
Biotransformation	Microorganism population and acclimation level

Source: Mill, T., "Data Needed to Predict the Environmental Fate of Organic Chemicals," presented at Symposium on Environmental Fate and Effects, American Chemical Society, Miami, Florida, September, 1978. Proceedings published by Ann Arbor Science.

[a] At constant temperature.

for example, (DDT) and polychlorinated biphenyls, are those that have very low water solubilities. Their wide distribution is due to their high stability in soil and water and, to a much lesser degree, their vapor transport following evaporation from water. The solubilities of various halogenated hydrocarbons are shown in Table 2.9 and some solubilities for some pesticides are presented in Table 2.10.

2.4.2   Photodegradation (Direct)

Direct photodegradation involves the absorption of light in the ultraviolet–visible (UVV) region by a molecule with a resultant increase in the molecular energy level; the increased energy then chemically transforms the molecule. Determination of the UV absorption spectrum of a substance is a prerequisite to direct photodegradation studies since compounds that do not absorb light in this spectrum will not decompose by direct photodegradation. A molecule that absorbs light in the UVV region, however, does not necessarily undergo chemical transformation since the energy of the molecule may be dissipated in some

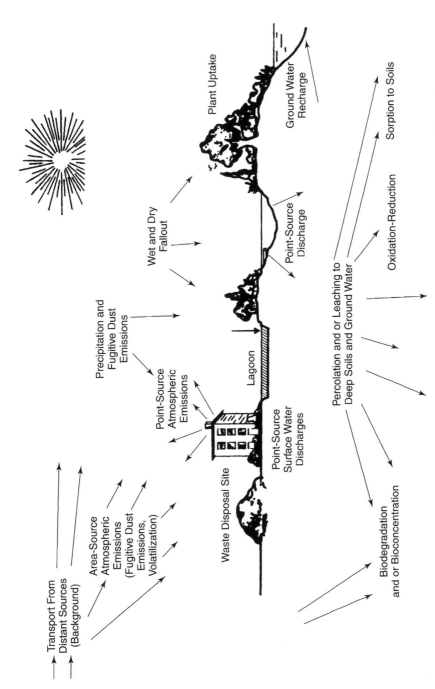

Figure 2.1 Schematic illustrating the transport and fate of atmospheric emissions into various parts of the environment.

105

TABLE 2.8   Solubilities and Partition Coefficients of Various Compounds

Compound	Solubility in Water (ppm)	Log (n-Octanol–Water Partition Coefficient)
Benzene	1,710 (20°C)	2.13
Toluene	470 (16°C)	2.69
Fluorobenzene	1,540 (30°C)	4.27
Chlorobenzene	448 (30°C)	2.84
Bromobenzene	446 (30°C)	2.99
Iodobenzene	340 (30°C)	3.25
p-Dichlorobenzene	79 (25°C)	3.38
Naphthalene	30	3.37
Diphenyl ether	21 (25°C)	4.20
Tetrachloroethylene	400 (25°C)	2.60
Chloroform	7,950 (25°C)	1.97
Carbon tetrachloride	800	2.64
p,p'-DDT	0.0031–0.0034 (25°C)	6.19
p,p'-DDE	0.040 (20°C)	5.69
Benzoic acid	2,700 (18°C)	1.87
Salicylic acid	1,800 (20°C)	2.26
Phenylacetic acid	16,000 (20°C)	1.41
Phenoxyacetic acid	12,000 (10°C)	1.26
2,4-D	890 (25°C)	2.81
2,4,5,2',5'-PCB	0.010 (24°C)	6.11
2,4,5,2',4',5'-PCB	0.00095 (24°C)	6.72
4,4'-PCB	0.062 (20°C)	5.58

Source: Freed et al. (1979).

other manner and the molecule returned to its initial state (USFDA, 1984; Hoffman et al., 1995).

Hydrolysis and biodegradation are generally considered to be the most important degradation pathways for organic chemicals in the aqueous environment, while photodegradation is more important for chemicals in the vapor or gaseous phase (Figure 2.2). Biodegradation or volatilization are generally considered the predominant pathway in the soil environment, although screening tests for determining the degradation of chemicals in soil have not always distinguished between biological, chemical, or photochemical degradation. Photodegradation is a less significant degradation mechanism in soil and water systems because of the limited opportunity for exposure of the substance to sunlight. Therefore, for most chemicals found primarily in water or soil, hydrolysis and biodegradation studies should be performed before other tests. The results from these studies can then be used to decide if additional degradation studies are required. If no degradation pathways have been identified in the environment and if the chemical absorbs in the UV spectrum, or if it has limited volatility, then the direct photodegradability of the substance should be evaluated to determine if the chemical is likely to bioaccumulate in the environment if not well controlled.

TABLE 2.9 Water Solubilities of Various Halogenated Aliphatic Hydrocarbons

Halogenated Aliphatic Hydrocarbon	Solubility (mg/L)
Chloromethane	6,450–7,250 at 20°C
Dichloromethane	13,000–20,000 at 25°C
Trichlormethane (chloroform)	8,200 at 20°C
Tetrachloromethane (carbon tetrachloride)	785 at 20°C
Chloroethane	5,740 at 20°C
1,1-Dichloroethane	5,470 at 20°C
1,2-Dichloroethane	5,500 at 20°C
1,1,1-Trichloroethane	8,690 at 20°C
1,1,2-Trichloroethane	440–4,400 at 20°C
1,1,2-Trichloroethane	4,500 at 20°C
1,1,2,2-Tetrachloroethane	2,900 at 20°C
Hexachloroethane	50 at 22°C
Chloroethane (vinyl chloride)	60 at 10°C
1,1-Dichloroethane	400 at 20°C
1,2-*trans*-Dichloroethane	600 at 20°C
Trichloroethane	1,100 at 20°C

Source: Modified from lecture materials from Tetra Tech, Inc.; distributed at EPA Water Quality Assessment Workshop, June 1981.

2.4.3 Photodegradation (Indirect)

Atmospheric photodegradation involves primarily indirect mechanisms. Indirect photodegradation is the process whereby chemicals react with intermediates that are formed as a result of direct photodegradation. Most frequently in the atmosphere, the indirect process involves free radical formation resulting from the interaction of sunlight with natural constituents, followed by reaction of the free

TABLE 2.10 Water Solubilities of Various Pesticides

Pesticide	Solubility
Acrolein	20.8% at 20°C
Aldrin	17–180 ppb at 25°C
DDD	20–100 ppb at 25°C
DDE	1.2–140 ppb at 20°C
Dieldrin	186–200 ppb at 25°C
Endosulfan	100–260 ppb at 20°C
Endrin	220 ppb at 25°C
Heptachlor	56–180 ppb at 25°C
Heptachlor epoxide	200–350 ppb at 25°C
Hexachlorocyclohexane	0.70–21.3 ppm at 25°C
Lindane	5–12 ppm at 25°C

Source: Modified from lecture materials from Tetra Tech, Inc.; distributed at EPA Water Quality Assessment Workshop, June 1981.

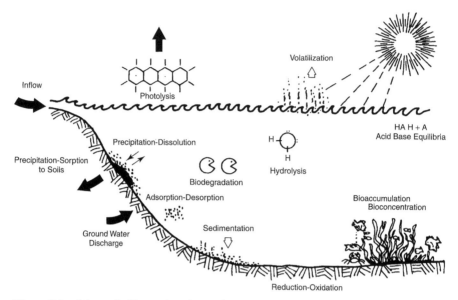

Figure 2.2 Schematic illustrating the various avenues for degradation and movement of xenobiotics in the environment.

radical with the chemical. This free radical reaction can result in the propagation of more reactive species and continued free radical reactions (USFDA, 1984).

Hydroxyl radicals have been implicated as one of the most important reactive species in the photooxidation of organic compounds. A reactivity scale of hydrocarbons based on reactions with hydroxyl radicals has been formulated by Glasson and Tuesday (1970) and others. In general, chemicals containing C–H or C–C bonds will be susceptible to hydroxyl radical attack in the troposphere and will be indirectly photodegraded. Some of the volatile fully halogenated compounds have been alleged to pose a significant hazard to the stratospheric ozone layer. A preliminary estimate of the transfer of halogen from the troposphere to the stratosphere can be determined using various models.

Significant volatility of a chemical should be demonstrated before indirect photodegradation tests of a chemical in its vapor phase are considered. The design of such studies should be based on the reactivity of the substance with hydroxyl radicals. As noted above, biodegradation or hydrolysis (or both) are considered more important degradation processes in soil and water than direct photodegradation. Furthermore, Crosby and Li (1969) predicted long ago that it is both reasonable and probable that in many instances photodegradation will provide the same products as metabolism by plants and microorganisms. They noted that photodegradation is a process that can "open up" a chemical structure ordinarily resistant to metabolism and thus result in an accelerated disappearance of the chemical from the environment. It can also induce chemical transformations entirely separate from those possible by living organisms (USFDA, 1984).

2.4.4 Biodegradation in Soil

Biodegradation of an organic chemical refers to the reduction in molecular complexity owing to the metabolic activity of living organisms, usually microorganisms and particularly bacteria and fungi. When an organic chemical is totally biodegraded in the presence of oxygen (aerobic biodegradation), the end products are inorganic carbon dioxide and water and may also include organic compounds involved in the normal metabolic processes of aerobic microorganisms. Depending on the structure of the chemical, end products may also include other inorganic salts (e.g., nitrates, sulfates). In contrast, when an organic chemical is totally biodegraded in the absence of oxygen (anaerobic biodegradation), the end products theoretically are methane, carbon monoxide, and carbon dioxide and some of the various organic compounds involved in the normal metabolic processes of anaerobic microorganisms (USFDA, 1984).

The carbon dioxide evolution test methods for the biodegradation of xenobiotics usually evaluate the potential biodegradation of an organic chemical to carbon dioxide and water in natural systems, both soil and water. The test is applicable to all chemicals irrespective of water solubility or vapor pressure. The amount of carbon dioxide produced over a given period compared to the amount theoretically possible is used as a measure of biodegradation. Results from this test are considered positive if the actual amount of carbon dioxide produced during the test period in all soils tested is 50 percent or greater than the amount of carbon dioxide that theoretically could be produced from the test chemical. Positive test results indicate that the test chemical will persist indefinitely in soil systems, but reliable biodegradation in the environment may not be assumed (USFDA, 1984).

In contrast, negative test results (less than 50 percent theoretical carbon dioxide production from the test chemical in all soils treated) do not allow distinction among the following: the test chemical may be completely resistant to biodegradation, resistant to biodegradation because of sorption or complexing, biodegraded at a very slow rate, only partially biodegraded (i.e., the identity of the test compound is changed but the compound is not completely changed to carbon dioxide and water), or biodegraded under different biodegradation test conditions (e.g., with lower concentrations of test chemical). In short, this is a screening test. If negative test results are obtained, it may be necessary to rerun the test to analyze for partial biodegradation of the test chemical or to test for biodegradation of the chemical under different conditions. When a negative test result occurs in only one or two of the soils tested, further analysis or testing may also be necessary to determine if the test chemical is actually resistant to biodegradation, degraded at a very slow rate, or only partially degraded in some soil systems (USFDA, 1984).

Biodegradation is the main process by which organic chemicals, following introduction into the environment, are reduced in complexity. Although an organic chemical may be transformed by abiotic mechanisms (e.g., chemical oxidation, hydrolysis, photodegradation) that depend only on light and/or thermal energy, these abiotic mechanisms rarely lead to appreciable changes in chemical structure. Photochemical reactions are much less important for chemicals below the soil surface since the UV light cannot penetrate past about the top 0 to 1 cm (Paustenbach, 1987; Yanders et al., 1985). In contrast, the enzyme-catalyzed metabolic processes of biological systems have the energy, as well as the specificity, to bring about major changes in structure and stability (USFDA, 1984).

2.4.5   Vapor Pressure

Vapor pressure is an important property governing the tendency of a chemical substance to be transported in air and is thus an important parameter in predicting the distribution of chemicals into environmental compartments. For example, vapor pressure data can be used to estimate the losses due to volatilization. This estimate can be used *in* conjunction with values for other parameters of environmental fate (e.g., sorption desorption and degradation) in deciding whether additional tests are necessary for a more complete description of the test chemical's instability and degradation pathways.

Equilibrium vapor pressure can be thought of as the solubility of a chemical substance in air and is dependent on the nature of the chemical and temperature. The vapor pressure of any chemical increases with an increase in temperature. This is because as temperature increases, the kinetic energy or movement of molecules increases, and more high-energy molecules are available to escape into the gaseous state. Vapor pressure values are, therefore, meaningful only if accompanied by the temperature at which they were measured. Vapor pressures of many organic chemicals of environmental interest increase three- to fourfold for each 10°C increase in temperature (USFDA, 1984).

Volatility is the evaporative loss of a substance to the air from the surface of a liquid or solid. Although potential volatility of a chemical is related to its inherent vapor pressure, actual volatization (or vaporization) rates will depend on environmental conditions and on factors that can lessen or enhance the effective vapor pressure or behavior of a chemical at a solid–air or liquid–air interface. For example, some chemicals with very low vapor pressures and low water solubility, such as DDT and polychlorinated biphenyls, because of their low concentrations in the environment, may still be mobilized to significant extent through volatization. The volatization half-lives of various aliphatic hydrocarbons are shown in Table 2.11 and the vapor pressures of select pesticides are presented in Table 2.12.

2.4.6   Dissociation Content

An understanding of the dissociation constant or pK can be useful in the experimental design of tests to measure the environmental fate parameters and ecological effects of a particular chemical. For example, the potential hydrolysis of

TABLE 2.11   Vapor Pressure and Volatilization Half-Life of Various Halogenated Aliphatic Hydrocarbons[a]

Halogenated Aliphatic Hydrocarbon	Vapor Pressure (torr) at 20°C	Volatilization Half-Life[b] (min^2)
Chloromethane	3700	27
Dichloromethane	362	21
Trichloromethane (chloroform)	150	21
Tetrachloromethane (carbon tetrachloride)	90	29
Chloroethane	1000	21
1,1-Dichloroethane	180	22
1,2-Dichloroethane	61	29
1,1,1-Trichloroethane	96	20
1,1,2-Trichloroethane	19	21
1,1,2,2-Tetrachloroethane	5	56
Hexachloroethane	0.4	45
Chloroethene (vinyl chloride)	2660	26
1,1-Dichloroethene	591	22
1,2-*trans*-Dichloroethene	200	22
Trichloroethene	57.9	21
Tetrachloroethene	14	26
1,2-Dichloropropane	42	<50
1,3-Dichloropropene	25	31
Hexachlorobutadiene	0.15	
Hexachlorocyclopentadiene	0081 at 25°C	
Bromomethane	1420	
Bromodichloromethane	50	
Dibromochloromethane	15	
Tribromomethane	10	
Dichlorodifluoromethane	4306	
Trichlorofluoromethane	667	

[a]Modified from lecture materials from Tetra Tech, Inc.; distributed at EPA Water Quality Assessment Workshop, June 1981.

[b]From Dilling (1977). Values were obtained by stirring 1 ppm solutions in an open container at 200 rpm at 25°C; average solution depth was 6.5 cm.

TABLE 2.12   Equilibrium Vapor Pressure and Vapor Density at 30°C

Compound	Vapor Pressure (torr)	Molecular Weight	Vapor Density (g/L)
Lindane	1.28×10^{-4}	291	1.97×10^{-6}
Dieldrin	1.0×10^{-5}	399	2.1×10^{-7}
p,p-DDT	7.16×10^{-7}	354	1.36×10^{-8}
o,p-DDT	5.5×10^{-6}	354	1.03×10^{-7}

Source: Tinsley (1979).

a chemical known to dissociate should be tested at pH values above and below its pK. Knowledge of the pK may be useful in the selection of soils to test for sorption–desorption of a chemical, as only soils with certain pH values might potentially bind it.

The dissociation constant is an equilibrium constant. An equilibrium constant is a measure of the degree to which ionizable chemicals break up into charged constituents owing to the effect of the solvent on the dissolved chemical. By definition, pK is equal in value to the pH at which 50% ionization occurs. Some chemicals have one pK and some chemicals have several.

The distribution of a chemical in the environment is partly a function of the pK of the chemical and the pH of the environment in which the chemical is found. Together, these factors determine the extent to which a substance will exist in the ionized or nonionized form. The extent of ionization of molecules of a chemical will affect the availability of the chemical to enter into physical, chemical, and biological reactions (USFDA, 1984).

Ionic charge can affect a chemical's solubility in water. The ionic charge of a chemical also affects its potential to bind to certain soils and sediments. Anion and cation exchange processes in soils depend on the nature of the soil, pK of the chemical, and pH of the surrounding medium. Most soils are negatively charged. In general, positively charged ions (cations) have a greater potential to bind to negatively charged soil particles than do negatively charged ions (anions) or nonionized species. The ionic charge of a chemical will also affect its potential to partition between lipid or octanol and water, and thus its ability to pass through membranes and its ability to be metabolized (Hayes, 1975). In general, only the nonionized form of an organic substance is capable of entering and passing through lipid membranes (USFDA, 1984).

2.4.7 Ultraviolet–Visible Absorption Spectrum

Absorption spectra give some indication of the wavelengths at which a chemical may be susceptible to direct photodegradation. Before a chemical can undergo a direct photochemical reaction, it must have the ability to absorb energy from wavelengths in the UVV range of the electromagnetic spectrum. Whether photodegradation of a chemical will occur depends on the total energy absorbed in the specific wavelength regions. As an aside, energy absorption is characterized by both the molar extinction coefficient (absorptivity) and the bandwidth. It is worth noting that the absence of measurable absorption does not preclude the possibility of photodegradation through other means, for example, indirect photodegradation of gases [Organization for Economic Cooperation and Development (OECD), 1981, 1997].

The UVV absorption spectrum is a quantitative measure of the ability of a substance to absorb radiation in the electromagnetic spectral region between 200 and 750 nm. It is generally measured with a spectrophotometer and presented as a function of wavelength or wave number. Because of the low cost of this test, it is usually worthwhile to understand when one is first evaluating a chemical that might enter the environment.

2.4.8 Sorption and Desorption

One of the most important factors governing the behavior of chemicals in soil and sediment is the sorption–desorption process. Sorption is a general expression for a process in which a chemical moves from one phase to be accumulated in another, particularly in cases in which the second phase is solid (Weber, 1927b). Sorption of a chemical by soil and sediment can result from adsorption or partitioning. Adsorption is the adhesion of molecules to surfaces of solid bodies with which they are in contact. Ionic species and metal ions exhibit this surface condensation phenomenon. Desorption is the reverse process of absorption. The major forces acting to sorb molecules to the soil include hydrophobic bonding, van der Waals forces, cation exchange, anion exchange, and coordination bonding (Hamaker, 1975; Hamaker and Thompson, 1972; Haque and Schmedding, 1975; Browman and Chesters, 1977; Chiou et al., 1979).

Some of the factors influencing the relative distribution of a test chemical between sorbed and solution phases include physical and chemical parameters of the molecule (e.g., size, shape, water solubility, pK, and polarity), properties of the soil or sediment (e.g., amount and type of clay and organic matter, cation exchange capacity, particle size, and pH, as well as temperature, water content, and salt concentration), and properties of the water in which the test chemical is dissolved (e.g., ionic strength and pH). The organic matter content is probably the most important soil or sediment property determining the sorption of nonionic chemicals (Weed and Weber, 1974; Hamaker, 1975; Stern and Walker, 1978; Karickhoff et al., 1979). Figure 2.3 demonstrates how the percentages of sand, silt, and clay characterize the type of soil, which in turn dictates how it interacts with xenobiotics. The role of specific soil and sediment properties in dictating the sorption of ionic chemicals varies with the type and extent of the charge of the chemical and with the type and extent of the charge of the soil or sediment (USFDA, 1984). Table 2.13 illustrates the relation between water solubility and the soil adsorption coefficient.

2.4.9 Partition Coefficient

The n-octanol–water partition coefficient has often been used to predict the bioaccumulation potential in aquatic and terrestrial organisms and to estimate the amount of sorption to soil and sediment. It is a very useful test to the risk assessor. The processes governed by the partition coefficient are major factors in determining the movement of chemicals in the biosphere. The n-octanol–water partition coefficient ($K_{o/w}$) describes the tendency of a nonionized organic chemical to accumulate in lipid (fatty) tissue and to sorb onto soil particles or onto the surface of organisms or other particulate matter coated with organic material. Although a powerful tool for understanding organic chemicals, it is not a predictor for inorganic chemicals, for metal organic complexes, or for dissociating and ionic organic compounds (USFDA, 1984).

Although numerous systems have been used to measure partition coefficients, such as hexane–water or benzene–water, it has become customary in

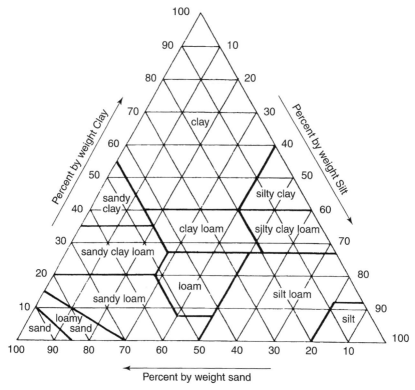

Figure 2.3   Triangle diagram illustrating the various classifications of soil and the criteria by which it is classified.

environmental fate assessment to use the *n*-octanol–water system. *n*-Octanol is considered a good medium for simulating natural fatty substances. Also, the *n*-octanol–water system is widely used as a reference system, and many data using this system have been reported in the literature (Sato and Nakajima, 1979; Tulp and Hutzinger, 1978). The *n*-octanol–water partition coefficients and partition coefficients for a variety of chemicals are presented in Table 2.14 and 2.15, respectively.

If the molecular structure of a relatively simple chemical is known, it is often possible to estimate the *n*-octanol–water partition coefficient. This is because the partition phenomenon exhibits a reasonable additive constitutive property for these molecules. That is, the partition coefficient of a simple molecule can be considered as an additive function of the partition coefficients of component parts of the molecule (Leo et al., 1971), particularly if the components are nonpolar.

2.4.10   Bioconcentration Factor

During the early 1960s, it was hoped that the octanol–water partition coefficient ($K_{o/w}$) would offer environmental toxicologists and risk assessors a simple

TABLE 2.13 Water Solubility, Soil Adsorption Coefficient, and Bioconcentration Factor Data: Experimental and Calculated

Chemical (Use)	Water Solubility (ppm)	Soil Adsorption Coefficient		Bioconcentration Factor Predicted from		Bioconcentration Factor (BCF)
		$K_{o/c}$	$K_{o/c}$	WS	$K_{o/c}$	
Acephate (I)	650,000		3	0.3		
Alachlor (H)	242	190	210	28	9	
Aldicarb (I)	7,800		32	4		
Aldrin (I)	0.013	410	47,600	7,160	22	
Ametryn (H)	185	392	250	33	21	
Aniline	36,600		13	2		
Anthracene	0.073	26,000	18,400	2,700	2,300	
Asulam (H)	5,000	300	40	5	16	
Atrazine (H)	33	149	640	86	7	
Benefin (H)	<1	10,700	>4,400	>620	850	
Bentazon (H)	500	0	140	19	0	
Benzene (H)	1,780	83	71	9	3	
Bifenox (H)	0.35		7,800	1,120		
Biphenyl	7.5		1,400	198		340
Bromacil (H)	8.5	72	1,350	185	3	
Bromobenzene	446	150	150	20		
Butralin (H)	1	8,200	4,400	618	630	
x-sec-Butyl-4-chloro diphenyloxide	0.14		13,000	1,870		298
Captan (F)	<0.5		>6,400	>910		
Carbaryl (I)	40	230	570	77	12	
Carbofuran (I)	415		160	21		
Carbon tetrachloride (IF)	800		110	14		18
Carbophenothion (I)	0.34	45,400	7,900	1,140	4,300	
Chloramben (H)	700	21	120	15	0.8	
Chloramben, methyl ester (H)	120	507	310	41	28	

115

TABLE 2.13 (*Continued*)

Chemical (Use)	Water Solubility (ppm)	Soil Adsorption Coefficient		Bioconcentration Factor Predicted from		Bioconcentration Factor (BCF)
		$K_{o/c}$	$K_{o/c}$	WS	$K_{o/c}$	
Chlorbromuron (H)	50	460	510	68	25	
Chlordane (I)	0.056		21,300	3,140		11,400
Chlorobenzene	448		150	20		12
4-Chlorobiphenyl	1.65		3,300	465		590
4-Chlorodiphenyloxide	3		2,400	330		736
Chloroneb (F)	8	1,200	1,400	190	71	
6-Chloropicolinic acid	3,400	9	50	6.3	0.3	
Chloroxuron (H)	2.7	3,200	2,500	350	220	
Chlorpropham (H)	88	590	370	50	33	
Chlorpyrifos (I)	0.3	13,600	8,500	1,220	1,100	450
Chlorpyrifos methyl (I)	4.0	3,300	2,000	280	230	
Chlorthiamid (H)	950	107	100	13	5	
Crotoxyphos (I)	1,000	170	100	13	8	
Crufomate (I)	200		240	31		
Cyanazine (H)	171	200	260	34	10	
Cycloate (H)	85	345	380	50	18	
2,4-D acid (H)	900	20	100	13	0.8	

Source: Kenaga (1980).

TABLE 2.14   Solubilities and Partition Coefficients of Various Compounds

Compound	Solubility in Water (ppm)	Log (n-Octanol–Water Partition Coefficient)
Benzene	1,710 (20°C)	2.13
Toluene	470 (16°C)	2.69
Fluorobenzene	1,540 (30°C)	4.27
Chlorobenzene	448 (30°C)	2.84
Bromobenzene	446 (30°C)	2.99
Iodobenzene	340 (30°C)	3.25
p-Dichlorobenzene	79 (25°C)	3.38
Naphthalene	30	3.37
Diphenyl ether	21 (25°C)	4.20
Tetrachloroethylene	400 (25°C)	2.60
Chloroform	7,950 (25°C)	1.97
Carbon tetrachloride	800	2.64
p,p'-DDT	0.0031–0.0034 (25°C)	6.19
p,p'-DDE	0.040 (20°C)	5.69
Benzoic acid	2,700 (18°C)	1.87
Salicylic acid	1,800 (20°C)	2.26
Phenylacetic acid	16,000 (20°C)	1.41
Phenoxyacetic acid	12,000 (10°C)	1.26
2,4-D	890 (25°C)	2.81
2,4,5,2',5'-PCB	0.010 (24°C)	6.11
2,4,5,2',4',5'-PCB	0.00095 (24°C)	6.72
4,4'-PCB	0.062 (20°C)	5.58

Source: Freed et al. (1979).

technique for identifying those chemicals that were likely to accumulate in the environment (Table 2.16). The experience with DDT seemed to confirm that the octanol–water partition could identify those chemicals that could bioaccumulate in fish as well as biomagnify in the food chain (microbe to fly to fish to bird). In the late 1970s, it was shown that bioconcentration in the environment could only be partly explained by the octanol–water partition (Table 2.17).

The experience with DDT was an important one. From this, a vocabulary evolved for describing the environmental fate, tansport, and behavior of chemicals (Table 2.18). Although it took many years to characterize, mortalities and reproductive failure in fish and fish-eating birds were linked to unusually high concentrations of DDT or its metabolites in the fat of these animals. Since the top-level carnivores, especially birds, had higher residue concentrations of these chemicals than the food they consumed, it was logical to postulate that accumulation occurred primarily by transfer through the food chain (Spacie and Hamelink, 1985). This idea was supported indirectly by the observation that DDT residues increased in a stepwise fashion from one trophic level to the next. The net efficiency of energy transfer between trophic levels is only about 10%. If the transfer efficiency for a chemical contaminant from food to consumer were greater, say 50 to 100%, and if there were no significant losses from the organism,

TABLE 2.15   Partition Coefficients ($K_{o/w}$) of Various Chemicals

Chemical	Log $K_{o/w}$ (room temperature)
Benzoic acid	1.87
Chloroform	1.97
Benzene	2.13
Salicyclic acid	2.26
Fluorobenzene	2.27
Tetrachloroethylene	2.60
Carbon tetrachloride	2.64
Toluene	2.69
Chlorobenzene	2.84
Malathion	2.89
p-Dichlorobenzene	3.81
Dichlorfenthion	5.14
4,4'-PCB	5.58
p,p-DDE	5.69
2,4,5,2',5'-PCB	6.11
p,p-DDT	6.19
2,4,5,2',4',5'-PCB	6.72

Source: Chiou et al. (1977).

TABLE 2.16   Methods for Estimating the Bioconcentration Factor (BCF) Using Physical Property Data

Octanol–Water Partition Coefficient ($K_{o/w}$)

$$K_{o/w} = \frac{\text{concentration of chemical in octanol phase}}{\text{concentration of chemical in aqueous phase}}$$

For example, $\log \text{BCF} = 0.76 \log K_{o/w} - 23$

Water Solubility (S)

For example, $\log \text{BCF} = 2.791 - 0.564 \log S$

Adsorption Coefficients for Soil and Sediments ($K_{o/c}$)

$$K_{o/c} = \frac{\text{g adsorbed/g organic carbon}}{\text{g/mL solution}}$$

For example, $\log \text{BCF} = 1.119 \log K_{o/c} - 1.579$

TABLE 2.17   Log BCF and Bioconcentration Potential (DDE = 100) for 30 Organic Chemicals as Determined with the Fathead Minnow in 32-day Exposures

Chemical	Mean Exposure C_w (μg/L)	Log BCF	BCF	Bioconcentration Potential[a]
Tris-(2,3-Dibromopropyl phosphate)	47.7	0.44	2.7	<0.1
5-Bromoindole	4.3	1.15	14	<0.1
Hexachlorocyclopentadiene	20.9	1.47	29	<0.1
Diphenylamine	43.7	1.48	30	<0.1
Chlorinated paraffin	49.2	1.69	49	<0.1
Toluene diamine	1.0	1.96	91	0.2
N-Phenyl-2-naphylamine	52.1	2.17	147	0.3
Tricresyl phosphate	31.6	2.22	165	0.3
Lindane	3.4	2.26	180	0.4
Pentachlorophenol	11.1	2.89	770	1.5
2,4,6-Tribromoanisole	4.8	2.94	865	1.7
1,2,4-Trichlorobenzene	1.6	3.32	2,800	5.5
Hexachloronorbornadiene	[b]	3.81	6,400	13
Methoxychlor	3.5	3.92	8,300	16
Heptachlor	3.1	3.98	9,500	19
Heptachloronorbornene	—[b]	4.05	11,100	22
Heptachlorepoxide	1.3	4.16	14,400	28
Mirex	1.2	4.26[c]	18,100	35
Hexabromocyclododecane	6.2	4.26	18,100	35
Hexabromobiphenyl	5.3	4.26	18,100	35
Hexachlorobenzene	2.6	4.27	18,500	36
p,p'-DDT	6.5	4.47[c]	29,400	58
Octachlorostyrene	7.1	4.52	33,000	65
o,p'-DDT	5.1	4.57[c]	37,000	72
Chlordane	5.9	4.58	37,800	74
Aroclor 1016	8.7	4.63[d]	42,500	83
p,p'-DDE	7.3	4.71[e]	51,000	100
Aroclor 1248	4.0	4.85	70,500	138
Aroclor 1254	4.3	5.00[c]	100,000	196
Aroclor 1260	1.0	5.28	194,000	300

Source: Veith et al. (1979).

[a] Bioconcentration potential calculated relative to p,p'-DDE = 100.

[b] Calculated from bioassay with these chemicals.

[c] Geometric mean of two tests.

[d] Geometric mean of five tests.

[e] Geometric mean of three tests.

TABLE 2.18   Terminology Useful in Understanding Behavior of Chemicals in Aquatic Systems[a]

Term	Definition
Uptake	Transfer of a chemical into or onto an aquatic organism. The uptake phase of an accumulation test is the period during which test organisms are exposed to the chemical.
Depuration	Elimination of a chemical from an organism by desorption, diffusion, excretion, egestion, biotransformation, or another route. The depuration phase of a test is the period during which previously exposed organisms are held in uncontaminated water.
Half-life or half-time	Time required for an organism held in clean water to eliminate 50% of the total body burden of tissue concentration of a chemical.
Bioavailability	Term used for the fraction of the total chemical in the surrounding environment which is available for uptake by organisms. The environment may include water, sediment, suspended particles, and food items.
Partitioning	Distribution of a chemical between two immiscible solvents. The partition coefficient (P or $K_{o/w}$) is the ratio of the chemical concentrations in the two solvents at equilibrium. Partition coefficients are commonly measured between n-octanol and water.
Steady-state or dynamic equilibrium	The state at which the competing rates of uptake and elimination of a chemical within an organism or tissue are equal. An apparent steady state is reached when the concentration of a chemical in tissue remains essentially constant during a continuous exposure. Bioconcentration factors are usually measured at steady state.
Compartment	Quantity of chemical that displays uniform rates of uptake and elimination in a biological system and whose kinetics can be distinguished from those of other compartments.
Bioconcentration	The tendency of a chemical to accumulate in a living organism to levels in excess of the concentration in its surrounding environment; e.g., concentration of kepone in fish is hundreds of times higher than the kepone concentration in the water and the concentration of 2,3,7,8-TCDD in a beach mouse can be several hundred times higher than that of the soil in which it lives.

TABLE 2.18 *(Continued)*

Term	Definition
Biomagnification	The process by which the concentration of a chemical in an organism is much greater than in its surrounding environment due not only to bioconcentration but also the uptake of food that has progressively bioconcentrated that chemical in its environment; e.g., flies that bioconcentrate DDT are eaten by frogs, which in turn are eaten by small fish, which are eaten by larger fish, which are eaten by birds.

[a]Based, in part, on Spacie and Hamelink (1985).

then residues would continue to accumulate throughout the life of the consumer. The higher the trophic level, the greater would be the body burden of residues. Although the actual mechanism for such a process was not clear, the concept of *biomagnification* or *transfer up the food chain* became well established (Spacie and Hamelink, 1985).

By the early 1980s it was shown that any one of several simple formulas based primarily on physical properties and especially the octanol–water partition could not consistently predict the concentration of a chemical in various media in the environment. As might have been expected, the processes of biomagnification and bioaccumulation were more complex and therefore not describable by a simple formula or set of formulas. Instead, detailed dynamic models were found to be necessary to distinguish the various routes of uptake by fish and other aquatic species. It is now fairly clear that the degree of accumulation in aquatic organisms depends on the type of food chain, on the availability and persistence of the contaminant in water, metabolism, and especially on the physicochemical properties of the contaminant (Rosenblatt et al., 1996).

The establishment of a bioconcentration factor (BCF) does not mean that aquatic animals in natural environments acquire residues solely from the water. They may take up chemicals both directly from the water and their food. Kepone uptake by estuarine organisms is a good case in point (Table 2.19). The relative contribution from each source depends on the substance, the concentration in the water, length of exposure, nature and degree of contamination in their food, the growth rate of the animal, and metabolism by the animal, but the theoretical limits to an accumulation process are presumed to lie at a value roughly equivalent to the BCF. Thus, by knowing the true BCF for a given chemical, it is possible to make comparisons between different chemicals regarding their accumulation potential.

2.4.11 Hydrolysis

Hydrolysis is one of the most common reactions occurring in the environment, and therefore represents one of the most potentially important pathways for a

TABLE 2.19   Bioconcentration of Kepone from Water by Estuarine Organisms

Species	Concentrations in Water (µg/L)	Duration of Exposure (days)	Body Burden (µg/kg)[a]	BCF[b]
Crassostrea virginica	0.03	30	210	×7,000
(oysters)	0.39	30	2,200	×5,641
M. bahia	0.03	21	120	×4,000
(mysids)	0.41	21	4,800	×11,700
P. pugio	0.02	28	90	×4,500
(grass shrimp)	0.40	28	4,570	×11,425
C. variegatus	0.05	28	370	×7,400
(sheepshead minnow)				
L. xanthurus	0.03	30	90	×3,000
(spot)	0.04	30	940	×2,350

Source: Macek et al. (1979).

[a] Whole-body residue concentrations at the end of exposure.

[b] Ratio of the mean whole-body burden at the end of exposure to the mean measured concentration of kepone in water during the exposure.

chemical. Rates of hydrolysis are independent of many rapidly changing factors that normally affect other degradative processes, such as the amount of sunlight, presence or absence of microbial populations, and extent of oxygen supply. Hydrolysis rates are typically influenced by pH, temperature, and concentration of the chemical, but these properties change seasonally, and slowly, in the aquatic environment (Mabey and Mill, 1978).

Hydrolysis refers to the reaction of an organic chemical (RX) with water, with the resultant net exchange of a group X from the organic chemical for the OH group from the water at the reaction center:

$$RX + HOH \leftrightarrow ROH + HX$$

In aqueous systems, rates of hydrolysis usually show first-order kinetics; that is, the rate of the reaction depends only on the concentration of the organic chemical. This is because water is present in such excess that its concentration does not change during the reaction and thus does not affect reaction rate (USFDA, 1984). The pH of water has a significant effect on the rate of hydrolysis reactions. The pH of natural waters may vary from 5 to 9: The pH of acid precipitation and leachate from mine waste may be as low as 3 to 4. Hydrolysis data are important in the design and interpretation of other environmental fate and effects tests. If a substance is extremely susceptible to hydrolysis, the loss of the compound in a hydrolysis must be taken into account, for example, in aquatic toxicity tests and photodegradation tests. Some examples of rates of hydrolysis are shown in Table 2.20.

TABLE 2.20   Hydrolysis Rates (Half-Life) at pH 7.4

	Half-Life at Temperatures		
Compound	37.5°C	20°C	ΔH (kcal/mol)
Phosmet	1.1 h	7.1 h	19.3
Dialifor	41.8 h	14.0 h	21.2
Malathion	1.3 d	10.5 d	21.6
Methyl chlorpyrifos	2.6 d	12.5 d	16.2
Dicapthon	5.5 d	29.0 d	17.2
Chlorpyrifos	13.4 d	53.0 d	14.2
Parathion	26.8 d	130.0 d	16.3

Source: Freed et al. (1979).

2.4.12   Algae Assay

Algae are simple photosynthetic organisms found in many terrestrial and aquatic habitats. These range from moist soils and surfaces exposed to air, to freshwater ponds, lakes, reservoirs, streams, estuaries, and oceans. In fact, algae may be present wherever there is sufficient moisture and sunlight. Algae are important because they occur symbiotically with fungi in lichens and because they may be associated with, or found in, the cells and tissues of various animal groups (e.g., sponges, Hydra, corals). Algae probably carry out a significant percentage of all photosynthesis that occurs on earth, and they constitute the major component of aquatic ecosystems responsible for the fixation of energy from sunlight (USFDA, 1984). The oxygen generated as a result of this photosynthetic activity is utilized by aquatic animals and contributes in large part to the total reservoir of atmospheric oxygen. Algae are also important because they aid in transforming organic wastes to stable effluents. Because they serve as the foundation of most aquatic food chains and are utilized by many herbivores as a major food source, algae are extremely important in the functioning of aquatic ecosystems (EPA, 1998; Miller and Zepp, 1979). For example, changes in algal growth rate, species composition, maximal standing crop, and photosynthetic rate can have profound effects on other parts of aquatic food chains and other pathways in food webs of aquatic ecosystems.

Many of the problems associated with eutrophication or nutrient enrichment of aquatic ecosystem are due to nutrient uptake and stimulation of algal biomass, Massive algal growths may occur, causing taste, odor, and oxygen depletion problems. The latter may in turn cause fish kills and adversely affect commercial sport fisheries. Some algae release extracellular products, which are toxic to fish, birds, and mammals. Therefore, obtaining an estimate in the laboratory of effects on this significant group is an important part of assessing the potential effects of chemicals on aquatic ecosystems. This is true even though it is acknowledged that it is difficult to extrapolate from data derived in laboratory assay tests to natural aquatic ecosystems (USFDA, 1984).

2.4.13   Cellulose Decomposition

Cellulose is a highly water-insoluble unbranched polysaccharide (complex carbohydrate) and is one of the most abundant organic materials in plants, where it serves as a major supporting material. Dead plant matter consisting largely of cellulose is degraded by a number of species of bacteria and fungi, referred to as decomposers in the carbon cycle. This degradation is accomplished in a stepwise manner going from cellulose to cellobiose to glucose to organic acids and CO_2. Respiratory activity in the producers, herbivores, and carnivores accounts for the return of a considerable amount of biologically fixed carbon as gaseous CO_2 to the atmosphere; however, the most substantial return is accomplished by the respiratory activity of decomposers in their processing of waste materials and dead remains of other trophic (food-web) levels.

The degradation of cellulose is of particular interest to farmers since the fertility of farmland depends in part on the presence of organisms that degrade dead plant matter. The organisms that decompose the various types of cellulose are some of the most important microorganisms contributing to the humification processes in soils. The inhibition of, or interference with, cellulose degradation by toxic chemicals adversely affects the recycling of carbon and soil fertility by retarding the breakdown of the vast amounts of cellulose that enter the soil.

2.4.14   Nitrogen Transformation

Although gaseous nitrogen (N_2) makes up the greatest part of the atmosphere, green plants obtain their nitrogen from the soil solution in the form of ammonia (NH_3) and/or nitrates (NO_2). The main aspects of the biogeochemical cycle for nitrogen are the fixation of gaseous nitrogen, the ammonification of organically found nitrogen, and the processes of nitrification and denitrification. Ammonification is a key initial step in the reintroduction of nitrogen from protein wastes into the soil and is one of the more readily measured reactions of the nitrogen cycle (USFDA, 1984).

The breakdown of proteins and other nitrogen-containing organics in soil and the production of ammonia are the work of widespread and varied microflora. The amino groups are split off to form ammonia in a series of reactions collectively called ammonification. Urea, a waste product found in urine, is also decomposed by numerous microorganisms, resulting in the formation of ammonia. This reaction can serve as a convenient assay method for ammonification activity. There is a strong correlation between an organism's ability to degrade urea and its capacity to degrade protein.

The ammonification test is potentially applicable to all test chemicals except water-insoluble gases. Information from such testing would be used to assess the likelihood that the test chemical interferes with the normal conversion of organically bound nitrogen into ammonia, a critical step in the cycle, which supplies the combined nitrogen required by almost all microorganisms, high plants, and animals (USFDA, 1984).

2.4.15 Seed Germination

Seeds germinate and grow into plants that are able to utilize and convert radiant energy (sunlight) directly, through the process of photosynthesis, into chemical energy that is stored in the form of sugars, starches, and other organic chemicals. These chemicals are in turn used by humans or by other organisms as energy sources. Plants also synthesize other compounds essential to many animals and humans and furnish atmospheric oxygen through photosynthesis. Thus, the maintenance of the biosphere depends on the normal functioning of the plant throughout its life cycle. Angiosperms (dicotyledon and monocotyledon flowering plants) are of particular importance and concern since they are ecologically significant organisms in many ecosystems. They comprise the dominant vegetation in most parts of the United States and are also the source of all major food crops.

Many plant species are especially sensitive to chemicals in the seed germination stage. Certain vital cellular and subcellular processes associated with germination, such as germination cell elongation, cell differentiation, mitosis, and protein and enzyme synthesis, may be affected by xenobiotics (USFDA, 1984). Tests to evaluate seed germination have been developed, and these should be used to evaluate chemicals that can be expected to be widely used in the environment.

2.4.16 Sulfur Transformation

Sulfur is essential to all living organisms as a part of sulfur-containing amino acids. The sulfur cycle is one of the major biogeochemical cycles and involves the release of sulfur as hydrogen sulfide (H_2S) from organic compounds (plant and animal wastes) by anaerobic microbial degradation, enabling the sulfur to be recycled again through living organisms. Anaerobic microbial degradation of organic wastes is carried out by bacteria, which are able to use sulfate rather than oxygen as the acceptor of electrons that are gained upon oxidation of the organic material. These heterotrophic sulfate-reducing bacteria are widely distributed in nature where anoxic conditions exist—as in sewage, sediments, muds, and bovine rumina. Two groups of bacteria are able to reduce sulfate to H_2S (Frobisher et al., 1974). The best known of these are the desulfuration organisms. These organisms have been thoroughly studied because they are responsible for serious odor and corrosion problems associated with sulfate reduction. However, these problems tend to obscure the necessary and beneficial role played by desulfuration in the sulfur cycle (USFDA, 1984).

2.4.17 Microbial Growth Inhibition

The main objective of this test is to determine the lowest concentration of test chemical that will inhibit the growth of tested microbial strains or species. Widespread microbial growth inhibition may result in ecosystem-level effects, which may include, depending on the organisms inhibited, reduction in plant growth or quality through nutritional disturbances (i.e., interruption of nutrient cycling) and interference with the natural degradative functions of microorganisms that play a dominant role in transformations of biotic and xenobiotic wastes. Microorganisms serve many important functions associated with the major biogeochemical cycles, for example, carbon, nitrogen, and sulfur.

2.5   AQUATIC TOXICOLOGY

Aquatic toxicology has been defined as the study of the effects of chemicals and other foreign agents on aquatic organisms with special emphasis on adverse or harmful effects (Rand, 1995). Aquatic toxicity tests are used to evaluate the concentrations of the chemical and the duration of exposure required to produce the criterion effect. The effects of a chemical may be of such minor significance that the aquatic organism is able to carry on its functions in a normal manner and that only under conditions of additional stress (e.g., changes in pH, dissolved oxygen, and temperature) can a chemically induced effect be detected. On the other hand, at sufficiently high concentrations, some chemicals may have the capacity to cause illness and death in some or all aquatic life. A number of species are typically used in these tests (Table 2.21).

Two general approaches may be used to conduct these tests, and each has advantages and limitations. (i) Effects can be studied in controlled laboratory experiments with a limited number of variables. (ii) Effects can be studied in a natural ecosystem (in situ). Until now, the laboratory setting has been favored because of ease and decreased cost compared to that associated with field studies.

2.5.1   Test Methods for Aquatic Species

The principle upon which all toxicity tests are based is that the response of living organisms to the presence (exposure) of toxic agents is dependent upon the dose (exposure level) of the toxic agent. Using this principle, aquatic toxicity tests are designed to describe a concentration–response relationship (referred to as the concentration–response curve when the measured effect is plotted graphically with the concentration). Acute toxicity tests are usually designed to evaluate the concentration–response relationship for survival; whereas chronic studies evaluate sublethal effects such as growth, reproduction, behavior, or biochemical effects and are usually designed to provide an estimate of the concentration that produces no adverse effect.

2.5.2   Acute Toxicity Tests

Acute toxicity tests are short-term tests designed to measure the effects of toxic agents on aquatic species during a short portion of their life span. Acute toxicity

TABLE 2.21 Test Species Commonly Used for Freshwater Toxicity Tests[a,b]

Vertebrates

  Brook trout, *Salvelinus fontinalis*
  Coho salmon, *Oncorhynchus kisutch*
  Chinook salmon, *Oncorhynchus tshawytscha*[a]
  Rainbow trout, *Oncorhynchus mykiss* (formerly *Salmo gairdneri*)
  Goldfish, *Carassius auratus*
  Common carp, *Cyprinus carpio*
  Fathead minnow, *Pimephales promelas*
  White sucker, *Catostomus commersoni*
  Channel catfish, *Ictalurus punctatus*
  Bluegill, *Lepomis macrochirus*
  Green sunfish, *Lepomis cyanellus*
  Northern pike, *Esox lucius*[a]
  Threespine stickleback—*Gasterosteus aculeatus*[a]
  Zebra fish, *Brachydanio rerio*—tropical fish[a]
  Guppy, *Poecilia reticulata*—tropical fish[a]

Invertebrates

  Daphnids, *Daphnia magna, D. pulex, D. pulicaria, Ceriodaphnia dubia*[a]
  Amphipods, *Gammarus lacustris, G. fasciatus, G. pseudolimnaeus*
  Crayfish, *Orconectes* sp., *Cambarus* sp., *Procambarus* sp., *Pacifastacus leniusculus*
  Stoneflies, *Pteronarcys* sp.
  Mayflies, *Baetis* sp., *Ephemerella* sp., *Hexagenia limbata, H. bilineata*
  Midges, *Chironomus tentans, C. riparius*
  Snails, *Physa integra, P. heterostropha, Amnicola limnosa*
  Planaria, *Dugesia tigrina*
  Rotifers, *Brachionus calyciflorus, B. rubens, B. plicatilis* (brackish water)[a]

Source: [a]Rand (1995).
[b]From ASTM, 1992. Copyright ASTM. Reprinted with permission.

tests most often measure effects on survival over a 24- to 96-hour period. The American Society for Testing and Materials (ASTM) has published standard guides on how to perform acute toxicity test for water column and sediment dwelling species for both freshwater and marine invertebrates and fishes. The species most often used in the United States include fathead minnows (*Pimephales promelas*), rainbow trout (*Oncorhynchus mykiss*), bluegill (*Lepomis machrochirus*), Channel catfish (*Ictulurus punctuatus*), sheepshead minnows (*Cyprintodon variegates*), *Daphnia magna, Ceriodaphnia dubia*, amphipod (*Hyalella azteca*), midge (*Chironomous* sp.) duckweed (*Lemna* sp.), green algae (*Selenastrum capricornutum*), marine algae (*Skeletonema costatum*), mayflies (*Hexagenia* sp.), mysid shrimp (*Mysidopsis bahia*), penaid shrimp (*Penaeus* sp.), grass shrimp (*Palaemonetes pugio*), marine amphipod (*Rhepoxinius aboronius*), marine clam (*Ampelisca abdita*), marine worm (*Nereis virens*), and oysters (*Crassotrea virginianicus, Mytilus edulis*, and *Macoma* sp.) (Hoffman et al., 1995).

Acute toxicity tests are usually performed by using five concentrations, a control, and a vehicle control (i.e., solvent control), if a vehicle is used, with 10 to 20 organisms per concentration. Most regulatory guidelines require duplicate exposure levels, although this is not required for pesticide registration. The following general quality control guidelines are usually adhered to: temperature, $\pm 1^\circ C$; pH, 6.5 to 8.5; dissolved oxygen, greater than 60 percent of saturation; hardness, 140 to 160 mg/L as $CaCO_3$; the salinity is controlled to appropriate specified levels for the various test species. All of the above variables, as well as the test concentration, are typically measured at the beginning and end of the study and occasionally more often. This basic design applies for most species (Hoffman et al., 1995).

Endpoints most often measured for acute toxicity tests include a determination of the LC_{50} or EC_{50} (median effective concentration) value, estimate of the acute NOEL, and behavioral observations. The primary endpoint is the LC or EC_{50}. The LC_{50} is a concentration that is estimated to kill 50 percent of a test population. An EC_{50} measures immobilization or an endpoint other than death. LC_{50} and EC_{50} values are measures of central tendency. These endpoints can be determined by a number of statistical approaches. The Litchfield–Wilcoxen approach is most often used, which consists of plotting the survival and test chemical concentration data on log-probability paper, drawing a straight line through the data, checking the goodness of fit of the line with a chi-squared test, and reading the LC or EC_{50} value directly off the graph. Various computer programs are available to perform this calculation. Other common methods include the moving-average and binomial methods. The latter is most often used with data sets where the dose–response curve is steep and no mortality was observed between the concentrations where 0 and 100 percent mortality was observed. A computer program incorporating the probit, moving-average, and binomial methods is available from the EPA, Duluth, Minnesota (EPA, 1995).

2.5.3   Chronic Toxicity Tests for Aquatics

Chronic toxicity tests expose test organisms to a toxicant over a significant portion of the organism's life cycle, typically one tenth or more of the organism's lifetime. Chronic studies usually measure toxicant effects on reproduction, growth, and sublethal effects that can include behavioral, physiological, and biochemical effects. Effects on survival are often determined, but this is not the main purpose of the study. Examples of chronic aquatic toxicity studies include brook trout (*Salvelinus fontinalis*), fathead and sheepshead minnow full life cycle, as well as *Daphnia magna*, *Ceriodaphnia dubia*, zebrafish (*Brachudanio rerio*), and mysid shrimp. Algal tests are typically 4 to 5 days in length and are often reported as acute tests. However, algal species reproduce fast enough that several generations are exposed during a typical study and, therefore, these studies should be classified as chronic studies (Rand, 1995).

Partial life-cycle studies are often referred to as chronic studies; however, only the most sensitive life stage(s) are usually exposed to the toxicant, and these studies are not true chronic studies. They are often referred to as partial chronic or subchronic studies. Common examples of partial life-cycle studies are the fish early life stage studies with the fathead and sheepshead minnows and rainbow trout. These studies expose embryos and young fish (30 to 60 days posthatch) to a toxicant. The early embryonic developmental stages are often the most sensitive and are therefore most often used. Other examples of partial life-cycle studies include tests performed with midges and amphipods. These studies typically evaluate a toxicant's effect on survival, growth, and behavior but not reproduction (Hoffman et al., 1995).

2.6 SEDIMENT TESTS

Sediment toxicity testing has been a rapidly evolving field over the past 10 years. Sediments in natural systems and in test systems are known to reduce the bioavailability of the test chemical (Iannucci and Ludwig, 1999). Bioavailability refers to that fraction of chemical present that is available for uptake by aquatic organisms. It is generally accepted that the unbound, freely available fraction of the chemical present is responsible for toxicity. The extent to which the bioavailability is reduced by sediments is dependent upon the physical-chemical properties of the test chemical and properties of the sediment. Past studies have demonstrated that chemical concentrations that produce biological effects in one sediment type often do not produce effects in other sediments even when the concentration is a factor of 10 higher. The difference is due to the bioavailability of the sediment-sorbed chemical (Hoffman et al., 1995).

The ability to estimate the bioavailability is a key factor in ultimately assessing the hazard of chemicals associated with sediments. Much progress has been made recently in this area. It is now widely recognized that the organic carbon content of the sediment is the factor most responsible for controlling the bioavailability of nonionic (nonpolar) organic chemicals. This concept has been incorporated into an approach termed the *equilibrium partitioning approach* and is being considered by the EPA for use in establishing sediment quality criteria. For some metals (cadmium, nickel, lead), the acid volatile sulfide (AVS) content of the sediments has recently been shown to control the bioavailability of the metal. AVS is a measure of the easily extractable fraction of the total sulfide content associated with sediment mineral surfaces. Metal–sulfide complexes are highly insoluble, which limits the bioavailability of certain metals. When the AVS content of the sediment is exceeded by the metal concentration (on a molar ratio of 1:1), free metal ion toxicity can be expressed. Other approaches are in the developmental stages for additional classes of compounds such as polar ionic chemicals (Hoffman et al., 1995; EPA, 1998).

The recognition that sediments are both a sink and a source for chemicals in natural environments has led to increased interest in sediments and to the development of standard testing techniques for sediment-dwelling organisms. Until recently, most sediment tests were acute studies. Greater emphasis is now being placed on developing sediment toxicity tests with sensitive organisms and sensitive life stages. For example, partial life-cycle test procedures are available for several species of amphipods and the sea urchin. Full life-cycle tests can be performed with the marine worm *Nereis virens* and with freshwater midges (*Chironomus tentans* and *Paratanytarsus disimilis*) and amphipods (*Hyalle azteca*). Partial and full life-cycle tests can be performed with epibenthic species such as *Daphnia magna* and *C. dubia*. These species are tested with sediments present in the test vessels. Pore-water tests offer a potentially sensitive approach to assess sediment safety. The interstitial water is extracted from the sediment, usually by centrifugation, and is used to test a wide variety of test organisms and life stages. This approach allows for the testing of fish early life stages as well as invertebrates (Hoffman et al., 1995).

2.7   BIOCONCENTRATION STUDIES

Bioconcentration is defined as the net accumulation of a material from water into and onto an aquatic organism resulting from simultaneous uptake and depuration. Bioconcentration studies are performed to evaluate the potential for a chemical to accumulate in aquatic organisms, which may subsequently be consumed by higher trophic level organisms, including humans. The extent to which a chemical is concentrated in tissue above the level in water is referred to as the bioconcentration factor (BCF). It is widely recognized that the octanol/water partition coefficient, referred to as $K_{o/w}$ or log P, can be used to estimate the potential for nonionizable organic chemicals to bioconcentrate in aquatic organisms. Octanol is used as a surrogate for tissue lipid in the estimation procedure. Equations used to predict BCFs have been summarized by Lyman et al. (1990).

Methods for conducting bioconcentration studies have been summarized for fishes and saltwater bivalves by ASTM. To date, the scientific community has focused its efforts on developing methods for fishes and bivalves because these species are higher trophic level organisms and are often consumed by humans. In general, the approach for determining the BCF for a given chemical and species is to expose several organisms to the chemical of interest at an exposure level that has environmental relevance and is no more that one tenth of the LC_{50} (lethal concentration) for the species tested. At this exposure level, mortality due to the test chemical can usually be avoided. The test population is sampled repeatedly and tissue residues (usually fillet, viscera, and whole fish) are measured. This is most often done with ^{14}C chemicals to facilitate tissue residue measurements. The study continues until apparent steady state is reached (a plot of tissue chemical concentrations becomes asymptotic with time) or for 28 days. At this point, the remaining fish are placed in clean water and the elimination (depuration) of the chemical from the test species is measured by analyzing tissues at several time intervals. Chronic pharmacokinetic analysis can then be used to predict the bioconcentration of the test chemical in fish or other aquatic species.

2.8 AVIAN TOXICITY TESTS

With the development of literally dozens of pesticides during the past 60 years, the need to evaluate their potential adverse effect on wildlife, and especially birds, is now quite clear. As early as 1979, the EPA under the Federal Insecticide, Fungicide, and Rodenticide Act (FIFRA) established guidelines for assessing the potential effects of pesticides on avian species.

Basic protocols with lethality as the principal endpoint have been used for first-line toxicity testing of birds. These include experiments designed to estimate the acute oral median lethal dosage (LD_{50}), and the 5-day median lethal dietary concentration (LD_{50}), and relevant statistical parameters.

2.8.1 Single-Dose Acute Oral

Reports on single oral dose avian LD_{50}'s contain data for adults of nearly 75 species of birds and more than 1000 chemicals tested. Current full-scale acute oral toxicity tests (tier I testing) required by regulatory guidelines for pesticide registration in the United States require testing with one of two species of birds, usually the mallard (*Anas platyrhnchos*) or the northern bobwhite (*Colinus virginianus*), as well as product analysis and generation of dose–response curves. Acute oral tests have attained general acceptance among environmental toxicologists since these tests are rapid, uncomplicated, and inexpensive, yet statistically reliable within the lethality curve and slope of dosage to death or recovery. These tests often provide insight necessary for further hazard evaluation (Hoffman et al., 1995).

Overnight-fasted birds receive a single dose of test substance at midmorning, usually administered by gavage or capsule, at each of five or six geometrically arranged dosage levels that were predetermined from a preliminary study to span the expected 10 to 90 percent mortality levels. Feed is provided immediately postdosing and observations for signs of intoxication are continued throughout the day. Special attention is given to the length of time to first evidence of toxicity, death, and recovery. Observations are continued twice daily, or more often as indicated, for 2 weeks posttreatment or as long as toxic signs persist. Gross necropsy should be performed on all birds that die and on a subsample of survivors.

Optimal use of the acute text requires statistical estimation of the lethality curve and its midpoint and descriptive information on toxic response. The LD_{50} expressed as milligrams active ingredient per kilogram body mass, its 95 percent confidence interval, and the slope and error of the dose–response curve are derived by probit, logit, or other appropriate analysis. When only the general order of acute toxicity is desired for wide-scale initial comparisons of many species or finished product formulations, then an approximate test of lethality may be used instead permitting conservation of test animals; as few as three groups of three to five birds are tested against a standardized dosage arrangement, and the LD_{50} and its 95 percent confidence interval are calculated from published tables (Hoffman et al., 1995).

2.8.2 Subacute Dietary

The subacute test (LC_{50}), a 5-day feeding trial, is required for two species, including upland game birds and waterfowl, to support registration of pesticides. This test serves as a composite indicator of vulnerability to a contaminated diet, allowing for metabolic changes that occur over time. The test was developed to quantify the toxicity of dietary residues that were considered an important source of exposure of wildlife to environmental contaminants. This test was optimized with young precocial birds, including ducks and quail, but almost any species can be tested provided it can be maintained in captivity in good health and can survive for 5 days without eating. Mortality and signs of intoxication are monitored at least twice daily, and food consumption is measured at 24-hour intervals. After the fifth day, all feed is replaced with untreated feed and the study is continued for at least 3 days or until complete remission of toxic signs. The LC_{50} is expressed as milligram active ingredient per kilogram of feed (or ppm) with its 95 percent confidence interval and slope with error of the dose–response curve as done for acute tests. Results of tests on more than 200 pesticides with young northern bobwhite, coturnix (*Coturnix japonica*), ring-necked pheasants, and mallards have been published. When LC_{50} tests are compared with LD_{50} tests, subacute LC_{50} results often describe relationships among species and chemicals that are quite different from those for LD_{50} results because LC_{50} tests measure ability of birds to cope with toxic feed for a set duration.

Age is another important consideration when evaluating LC_{50} values. There is generally an increase in resistance to chemicals with increasing age during early growth of precocial species. This increase occurs across class of chemical or pesticide and is believed primarily a result of changes in ability to cope with a toxic diet for the exposure duration; older (larger) chicks that eat less proportional to body mass are better able to survive the 5-day trial by reducing food consumption and hence toxic exposure.

2.8.3 Avian Subchronic Dietary (Toxicity)

This test was developed as an extension of the subacute LC_{50} test but with greater emphasis on sublethal indicators of toxicity. It was designed as a precursor to provide a biological indication of the necessity for conducting a full-scale reproductive trial and to provide a possible hazard index based on the ratio of sublethal to lethal toxicity values. The first test of this kind was conducted to compare the effects of organic and inorganic mercury on various physiological parameters, including indicator enzymes and blood chemistries in coturnix through 9 weeks of age, which is full maturity in this species; calculation of periodic EC_{50} values (median effective concentration) for each responding variable was used to develop hazard indices relating the EC_{50} to the oral LD_{50} and 5-day LC_{50}.

2.8.4 Avian Chronic Toxicity Tests: (Reproduction)

One of the primary concerns about the presence of xenobiotics in the environment with respect to the avian hazard is the potential adverse effect on reproduction. Since this can be a rather insidious effect if it were to occur in the field— for example, the population could be markedly diminished before observation begins—it is of critical importance. Testing used to evaluate reproductive toxicity is usually performed on the bobwhite and the mallard. Avian reproduction testing is often required by FIFRA if any of the following conditions apply:

1. Pesticide residues resulting from the proposed use are persistent in the environment to the extent that toxic amounts are expected on avian feed.
2. Pesticide residues are stored or accumulated in plant or animal tissues.
3. Pesticide *is* proposed for use under conditions where birds may be subjected to repeated or continued exposure to the pesticide, especially preceding or during the breeding season.

To satisfy this last requirement, birds are exposed to treated diets beginning not less than 10 weeks before egg laying and extending throughout the laying season. At least two treatment level groups are used. Concentrations for the test substance should be based on residues expected under the proposed use and a multiple, such as 5. Some scientists have suggested that it may be cost effective to consider multiple-level testing, at least three exposure levels, to determine effect and no-effect levels. This may be especially useful for new pesticides that show promise on several crops and where avian exposure may vary or where levels of environmental exposure are not established, or both (Yuch et al., 1994).

Avian reproductive effects studies (tier II) are required by EPA using both waterfowl (mallards) and an upland game species (northern bobwhite) to support the registration of an end-use product that meets one or more of the following criteria: (1) the product in intended for use where birds may be subjected to repeated or continuous exposure to the pesticide or any of its major metabolites or degradation products, especially preceding or during the breeding season; (2) the pesticide, or major metabolites, may persist in the food at toxic levels; (3) persistence and accumulation occurs in plant or animal tissues; and (4) adverse effects occur on mammalian reproduction. The data derived from avian reproductive studies are used to establish the effects of active ingredients of pesticides or other environmental contaminants on reproduction in birds, to determine the stage of the reproductive cycle affected for a given test species, and to compare toxicity information with measured pesticide residues in the field. Other goals include assessment of potential hazard, determination of species differences and defining the need for further testing and field trials. Two distinct protocols have been used for routine screening of contaminant effects on avian reproduction and are described elsewhere.

2.8.5 Pen Field Studies

Simulated testing and actual field testing for mammals and birds are sometimes the best approaches to evaluating safety. These have become more common-place since 1980 and have become more routine for those chemicals that have significant toxicity or where there is persistence. The decision to conduct field tests is usually based on consideration of the physicochemical properties of the pesticide, the proposed use pattern, the likelihood of wildlife exposure to the pesticide under field conditions and at levels expected to be toxic to wildlife, and on review of laboratory data. The major problems with field studies are that they are cumbersome, difficult to analyze, troubled by the unpredictable nature of the wild, and very expensive (Suter, 1993).

2.8.6 Full-Scale Field Studies

Full-scale field tests are often productive if use of the pesticide, or another chemical, is anticipated to have some likelihood of adversely affecting wildlife. Universally acceptable standards for conducting full-scale field tests are not possible because of the variety of ways a pesticide may enter the environment and impact wildlife. In the full-scale field test, the objective is to determine the impact on wildlife populations. Such tests are applicable to use patterns associated with a major wildlife habitat. These would include forests, rangeland, and croplands such as cotton, corn, sorghum, soybeans, rice, and alfalfa. A limited number of full-scale field studies had been conducted through 1985.

2.8.7 Rules of Thumb for Risk Assessment of Avian Hazards

The following guidelines, which can be used in risk assessment calculations, have been offered by Kenaga (1972). Although perhaps dated, these are helpful for conducting assessments where the risk to avian species is a concern:

1. Body weights of birds can be estimated reasonably to within 20% of the actual weight when not known exactly, by use of available literature values.
2. The quantity of food eaten per day by birds can be reasonably estimated when not known exactly. In general, species weighing over 100 g eat less than 20 percent of the body weight in food per day, and those under 10 g may eat as much as 60 percent in dry weight food per day. Based on a linear relationship of body weight to food intake, percent of body weight eaten per day can be estimated within a twofold factor.

3. The upper limits of the amount of residues of a pesticide on various species, shapes, and sizes of plant food and insects immediately after application can be reasonably estimated. In general, the highest residues may be expected on treated food particles that have the highest surface area to weight–volume ratios and greatest exposure to the pesticide application. The upper limits of pesticide residues on plants and their decline with time on various types of crops, as correlated from data on many pesticides, are known.

4. Based upon the information developed in this report on a unit basis, a 1 lb per acre dosage of a noncumulative pesticide could result in a 35-fold difference (7 to 240 ppm) between maximum pesticide residues in different sized natural bird food particles. Also a 20-g bird of nearly any species will likely eat 5 times as much food in milligram per kilogram of body weight per day as a 1000-g bird of nearly any species. Thus, it is possible that a 20-g bird could eat 100 times or more pesticide than a 1000-g bird, in terms of milligram/kilogram-day. Variations in concentrations and dosages of pesticides and variations in good consumption rates of different weights of birds can be estimated to obtain specified milligram/kilogram-day. As has been shown, the maximum residues estimated here rarely occur in nature and therefore mitigating circumstances of stability of the residues, food habits of the birds, and so forth would result in corresponding reduction of this estimate.

5. The estimated or calculated milligram/kilogram-day of intake of pesticides from residues on bird food should be matched with toxicological responses from similar dosages in laboratory or field test results with birds, preferably by use of ad lib dietary feeding tests, rather than acute oral, injection or other less correlatable tests, *not* as suitable for environmental interpretation. Subacute and chronic laboratory toxicity tests are often conducted with diets containing a constant concentration rate of pesticides over a period of days, weeks, or months. Interpretation of such laboratory tests should take into consideration the frequently large and quick rate of decline in pesticide residues that occur on natural bird food, or the dietary intake of birds in such tests should be adjusted accordingly. In addition, the test methods should provide for the study of metabolites of the pesticide or other derived molecules if any occur. Food acceptance of the pesticide in the diet of birds and the relative toxicity of the pesticide to representative species of birds must also be taken into account.

2.9   EVALUATING RISKS TO DOMESTIC ANIMALS

The effects of low-level exposure to xenobiotics unintentionally present in the environment (e.g., polychlorinated biphenyls (PCB), polybrominated biphenyls (PBB), dioxins, metals) have been studied infrequently in domestic animals, especially cows, goats, and sheep. In general, their exposure to chemical contaminants, which have been of widespread concern to humans or aquatic life, has been negligible. More recently, especially because of broad-scale environmental contamination that has been reported at times with PCB, DDT, and heptachlor, a greater number of studies have been conducted (Fries, 1982, 1985; Fries and Jacobs, 1982). Standardized tests for evaluating the effects of xenobiotics on domestic animals have been established for intentional feed additives but not for toxicity testing of unintentional additives. Because of the continuing concern about airborne release of metals and TCDD by incinerators, the evaluation of domestic animals used for food will almost certainly increase in the coming years. Some complex models and field tests have already been developed (Fries and Paustenbach, 1990; Fries et al., 1999).

2.10   EARTHWORMS AND OTHER WILDLIFE

One of the most interesting and challenging areas in environmental assessment is the characterization of the potential hazard to lower wildlife and insects. Heretofore, this area has received little attention, although it could arguably be an important dimension in the evaluation of hazardous waste sites and other hazards. Only a handful of industrial chemicals have been studied thoroughly for their potential adverse effect on lower forms of wildlife such as earthworms, field mice, snakes, ants, or turtles; although a number of pesticides have been studied (Young, 1983; Oliver, 1984; Reineke and Nash, 1984; Young and Cockerham, 1985).

2.11   CLOSING

When a fairly significant area of land or the environment could be accidentally contaminated by a chemical or when it is intentionally applied (e.g., pesticide), it is prudent to have a thorough understanding of the physical and chemical properties of that chemical to appreciate the overall hazard potential. If it is determined that a potentially significant degree of exposure to wildlife might occur, various levels of testing should be conducted. For the pesticides, the EPA has established numerous requirements and has required testing to evaluate the potential adverse affects on aquatic species, large and small animals, lower species, and some plants. The test protocols used to evaluate pesticides are an excellent source of information for developing tests to evaluate various industrial chemicals. Several thoughtful approaches for establishing batteries of tests for new chemicals—including the tier approach—have been proposed, and these should be considered (Dickson et al., 1982; Woltering et al., 1989). One of the debated areas is whether it is necessary to conduct laboratory or field studies to evaluate the hazard to wildlife. In general, laboratory studies are deemed more appropriate as screening tests since the data are easier to collect and interpret, thus making the studies more cost effective (Young and Cockerham, 1985; Rand, 1995).

This chapter represents a brief overview of the various factors to consider when evaluating the potential for a chemical to pose a significant hazard to humans or wildlife (including aquatic species). The problem of hazard identification relies on two central questions. First, what are the specific toxic properties of the chemical to humans or wildlife? Second, is there a genuine potential that the level of exposure could be sufficient under any reasonable condition to pose a health hazard? The topics discussed here and the published materials that are cited should be adequate to assist practitioners in answering these questions.

REFERENCES

Albers, P., Heinz, G. H., and Ohlendorf, H. (Eds.). 2000. *Environmental Contaminants and Terrestrial Vertebrates: Effects on Populations, Communities, and Ecosystems.* Society of Environmental Toxicology and Chemistry, Pensacola, FL.

Allen, D., Consoli, F., Davis, G., Fava, J., and Warren, J. (Eds.). 1997. *Public Policy Applications of Life-Cycle Assessment.* Society of Environmental Toxicology and Chemistry, Pensacola, FL.

American Conference of Governmental Industrial Hygienists (ACGIH). 1999. *Documentation of Threshold Limit Values (TLVs),* 6th ed. ACGIH, Cincinnati, OH.

American Society for Testing and Materials (ASTM). 1980. Conducting acute toxicity tests with fishes, macroinvertebrates, and amphibians. In *Annual Book of ASTM Standards.* E729–780. ASTM, Philadelphia, PA.

Anderson, J. P. E., et al. (Eds.). 1996. *Pesticides, Soil Microbiology and Soil Quality.* Society of Environmental Toxicology and Chemistry, Pensacola, FL.

Baker, J. E. (Ed.). 1997. *Atmospheric Deposition of Contaminants to the Great Lakes and Coastal Waters.* Society of Environmental Toxicology and Chemistry, Pensacola, FL.

Ballantyne, B., Marrs, T., and Syversen, T. 1999. *General and Applied Toxicology.* MacMillan References Limited, London.

Barnes, J. M., and Denz, F. A. 1954. Experimental methods used in determining chronic toxicity. *Pharmacol. Rev.* 6: 191–242.

Barnthouse, L., Fava, J., Humphreys, K., Hunt, R., Laibson, L., Noesen, S., Norris, G., Owens, J., Todd, J., Vigon, B., Weitz, K., and Young, J. (Eds.). 1997. *Life-Cycle Impact Assessment: The State of the Art,* 2nd ed. Society of Environmental Toxicology and Chemistry, Pensacola, FL.

Barnthouse, L. W., and Brown, J. 1994. Issue papper on conceptual model development. In: Ecological risk assessment issue papers. Risk Assessment Forum, U.S. Environmental Protection Agency, Washington, D.C.: 3-1 to 3070.

Barnthouse, L. W., Suter, II, G. W., and Rosen, A. E. 1990. Risks of toxic contaminants to exploited fish populations: influence of life history, data uncertainty, and exploitation intensity. *Environ. Toxicol. Chem.* 9: 297–312.

Barrett, K. L., et al. (Eds.). 1999. *Guidance Document on Regulatory Testing Procedures for Pesticides with Non-Target Arthropods*. Society of Environmental Toxicology and Chemistry, Pensacola, FL.

Bartell, S. M., Gardner, R. H., and O'Neill, R. V. 1992. *Ecological Risk Assessment*. Lewis, Chelsea, MI.

Beck, L. W., Maki, A. W., Artman, N. R., and Wilson, E. R. 1981. Outline and criteria for evaluating the safety of new chemicals. *Regul. Toxicol. Pharmacol.* 1: 19–58.

Bergman, H. L., and Dorward-King, E. J. (Eds.). 1997. *Reassessment of Metals Criteria for Aquatic Life Protection*. Society of Environmental Toxicology and Chemistry, Pensacola, FL.

Bergmann, H. L., Kimerle, R. A., and Maki, A. W. 1986. *Environmental Hazard Assessment of Effluents*. Pergamon, New York.

Bigham, E., Cohrssen, B., and Powell, C. H. 2001. *Patty's Toxicology*, 5th ed., 9 vols. Wiley, New York.

Bishop, W. E., Cardwell, R. D., and Heidolph, B. B. 1983. *Aquatic Toxicology and Hazard Assessment*. Sixth Symposium, STP 737. ASTM, Philadelphia, PA.

Black, D. L., and Marks, T. A. 1986. Inconsistent use of terminology in animal developmental toxicology studies: A discussion. *Teratology* 33: 333–338.

Branson, D. R., and Dickson, K. L. 1981. *Aquatic Toxicology and Hazard Assessment: Fourth Symposium*. STP 737. ASTM, Philadelphia, PA.

Briggs, G. G., Bromilow, R. H., and Evans, A. A. 1982. Relationships between lipophilicity and root uptake and translocation of non-ionized chemicals by barley. *Toxicol. Environ. Chem.* 7: 173–189.

Brody, M. S., Troyer, M. E., et al. 1993. Ecological risk assessment case study: modeling future losses of bottomland forest wetlands and changes in wildlife habitat within a Louisiana basin. In: A review of ecological assessment case studies from a risk assessment perspective. Risk Assessment Forum: U.S. Environmental Protection Agency, Washington, DC: 12-1 to 12-39.

Brooke, L. T., Call, D. J., Geiger, D. L., and Northcott, C. E. (Eds.). 1984. *Acute Toxicities of Organic Chemicals to Fathead Minnows*. Center for Lake Superior Environmental Studies, University of Wisconsin-Superior, Superior.

Browman, M. G., and Chesters, G. 1977. The solid-water interface: Transfer of organic pollutants across the solid-water interface. In I. H. Suffet (Ed.), *Fate of Pollutants in the Air and Water Environment*. Part I. Wiley, New York, pp. 49–101.

Cairns, J. Jr., Dickson, K. L., and Maki, A. W. 1978. *Estimating the hazards of chemical substances to aquatic life STP657*. American Society for Testing and Materials, Philadelphia, PA.

California Department of Health Services (CDHS). 1988. *Guidelines and Safe Use Determination Procedures for the Safe Drinking Water and Toxic Enforcement Act of 1986*. CDHS, Sacramento, CA.

Carey, J., Cook, P., Giesy, J., Hodson, P., Muir, D., Owens, W., and Soloman, K. (Eds.). 1998. *Ecotoxicological Risk Assessment of the Chlorinated Organic Chemicals*. Society of Environmental Toxicology and Chemistry, Pensacola, FL.

Chapin, R. E., and Heindel, J. J. 1993. Introduction. In J. J. Heindel and R. E. Chapin (Ed.), *Methods In Toxicology*, Vol. 3B. *Female Reproductive Toxicology*. Academic, San Diego, CA, pp. 1–15.

Chapman, P. F., Crane, M., Wiles, J. A., Noppert, F., and McIndoe, E. C. (Eds.). 1996. *Asking the Right Questions: Ecotoxicity and Statistics.* Society of Environmental Toxicology and Chemistry, Pensacola, FL.

Chen, F., and Schüürmann, G. (Eds.). 1997. *Quantitative Structure-Activity Relationships in Environmental Sciences,* Vol. VII. Society of Environmental Toxicology and Chemistry, Pensacola, FL.

Chiou, C. T. 1981. Partition coefficient and water solubility in environmental chemistry. In J. Saxena and F. Fisher (Ed.), *Hazard Assessment of Chemicals.* Academic, New York, pp. 117–153.

Christian, M. 2001. Test methods for assessing female reproductive and developmental toxicology. In A. W. Hayes (Ed.), *Principles and Methods of Toxicology.* Taylor and Francis, Philadelphia, PA, pp. 1301–1382.

Clewell, H. J., and Andersen, M. E. 1985. Risk assessment extrapolations and physiological modeling. *Toxicol. Ind. Health* 1: 111–132.

Cockerham, L. G., and Shane, B. S. 1994. *Basic Environmental Toxicology.* CRC Press, Boca Raton, FL.

Commission on Risk Assessment and Risk Management. 1997. *Framework for environmental health risk management. Final Report. Volume 1.* Commission on Risk Assessment and Risk Management, Washington, DC.

Cowan, C., Muckay, D., Feijtel, T. C., van de Meent, D., Di Guardo, A., Davies, J., and Mackay, N. (Eds.). 1995. *The Multi-Media Fate Model: A Vital Tool for Predicting the Fate of Chemicals.* Society of Environmental Toxicology and Chemistry, Pensacola, FL.

Cowan, C. E., Versteg, D. J., Larson, R. J., and Kloepper-Sams, P. J. 1995. Integrated approach for environmental assessment of new and existing substances. *Regul. Toxicol. Pharmacol.* 21: 3–31.

Crosby, D., and Li, M. 1969. Herbicide photodecomposition. *Degradation of Herbicides.* P. Kearney and D. Kaufmann. Dekker, New York: 321–363.

Darnall, K. R., Lloyd, A. C., Winer, A. M., and Pitts, J. N., Jr. 1976. Reactivity scale for atmospheric hydrocarbons based on reaction with hydroxyl radicals. *Environ. Sci. Technol.* 10: 692–696.

DeFur, P. L., Crane, M., Ingersoll, C. G., and Tattersfield, L. (Eds.). 1999. *Endocrine Disruption in Invertebrates: Endocrinology, Testing, and Assessment.* Society of Environmental Toxicology and Chemistry, Pensacola, FL.

De Peyster, A., and Day, K. E. (Eds.). 1998. *Ecological Risk Assessment: A Meeting of Policy and Science.* Society of Environmental Toxicology and Chemistry, Pensacola, FL.

DeSesso, J. M. 1997. Comparative embryology. In R. D. Hood (Ed.), *Handbook of Developmental Toxicology.* CRC Press, New York, pp. 111–174.

Dickson, K. L., Maki, A. W., and Cairns, J., Jr. (Eds.). 1979. *Analyzing the Hazard Evaluation Process.* American Fisheries Society, Washington, DC.

Dickson, K. L., Maki, A. W., and Cairns, J., Jr. (Eds.). 1982. *Modeling the Fate of Chemicals in the Aquatic Environment.* Ann Arbor Science, Ann Arbor, MI.

Di Giulio, R. T., and Tillitt, D. E. (Eds.). 1999. *Reproductive and Developmental Effects of Contaminants in Oviparous Vertebrates.* Society of Environmental Toxicology and Chemistry, Pensacola, FL.

Doyle, G. J., Lloyd, A. C., Darnall, K. R., Winer, A. M., and Pitts, J. N., Jr. 1975. Gas phase kinetic study of relative rates of reaction of selected aromatic compounds with hydroxyl radicals in an environmental chamber. *Environ. Sci. Technol.* 9: 237–241.

Eisenbud, M. 1978. *Environment, Technology, and Health: Human Ecology in Historical Perspective*. New York University Press, New York.

El Beit, I. O. D. 1981. Factors affecting soil residues of dieldrin, endosulfan, Y-HCH, dimethoate, and pyrolan. *Ecotoxicol. Environ. Safety* 5: 135–160.

Environmental Protection Agency (EPA). 1979a. *Water Related Fate of 129 Priority Pollutants*, Vols. 1 and 2. PB80-204381. EPA, Washington, DC.

Environmental Protection Agency (EPA). 1979b. Toxic Substances Control Act: Pre-manufacture testing of new chemical substances. *Fed. Reg.* 44: 16240–16292.

Environmental Protection Agency (EPA). 1986a. Guidelines for exposure assessment. CFR 2984. *Fed. Reg.* 51(185): 34042–34054.

Environmental Protection Agency (EPA). 1986b. Guidelines for health assessment of suspect developmental toxicants. CFR 2984. *Fed. Reg.* 51(185): 34028–34041.

Environmental Protection Agency (EPA). 1986c. *Superfund Health Assessment Manual.* EPA, Office of Emergency and Remedial Response, Washington, DC (produced by ICF Clement, Inc. under EPA contract 68-01-6872).

Environmental Protection Agency (EPA). 1986d. Guidelines for the health risk assessment of chemical mixtures. CFR 2984. *Fed. Reg.* 51: 34014–34027.

Environmental Protection Agency (EPA). 1986e. *Registry of Toxic Effects of Chemical Substances*. EPA, Research Triangle Park, NC.

Environmental Protection Agency (EPA). 1986.

Environmental Protection Agency. 1993a. *Wildlife exposure factors handbook*. Office of Research and Development, U.S. Environmental Protection Agency, Washington, DC.

Environmental Protection Agency. 1994a. *Managing ecological risks at EPA: issues and recommendations for progress.* Center for Environmental Research Information, U.S. Environmental Protection Agency, Washington, DC.

Environmental Protection Agency. 1995b. *EPA risk characterization program. Memorandum to EPA managers from Administrator Carol Browner.*

Environmental Protection Agency (EPA). 1996. *Proposed Guidelines for Carcinogenic Risk Assessment.* EPA/600/P-92/003C. EPA, Office of Reseach and Development, Washington, DC.

Environmental Protection Agency. 1998. *Guidelines for ecological risk assessment*. U.S. Environmental Protection Agency, Washington, DC.

Environmental Protection Agency (EPA). 1999. Updated Draft of Guidelines for Carcinogenic Risk Assessment. Office of Research and Development. EPA, Washington, DC.

Environmental Protection Agency (EPA). 1999. Integrated Risk Information System. Available from *http://www.epa.gov/iris*.

Fava, J. A., Denison, R., Jones, B., Curran, M. A., Vigon, B., Selke, S., and Barnum, J. (Eds.). 1991. *A Technical Framework for Life-Cycle Assessment.* Society of Environmental Toxicology and Chemistry, Pensacola, FL.

Ferenc, S. A., and Foran, J. A. (Eds.). 2000. *Multiple Stressors in Ecological Risk and Impact Assessments: Approaches to Risk Estimation.* Society of Environmental Toxicology and Chemistry, Pensacola, FL.

Field, L. J., Fall, J., Nighswander, T., Peacock, N., and Varanasi, U. (Eds.). 2000. *Evaluating and Communicating Subsistence Seafood Safety in a Cross-Cultural Context: Lessons Learned from* Exxon Valdez *Oil Spill.* Society of Environmental Toxicology and Chemistry, Pensacola, FL.

Foran, A. J., and Ferenc, S. A. (Eds.). 1999. *Multiple Stressors in Ecological Risk and Impact Assessment.* Society of Environmental Toxicology and Chemistry, Pensacola, FL.

Franzle, O. 1993. *Contaminants in Terrestrial Environments.* Springer, Berlin.

Freed, V. H., Chiou, C. T., Schmeddling, D., and Kohnert, R. 1979. Some physical factors in toxicological assessment tests. *Environ. Health Perspect.* 30: 75–80.

Fries, G. 1982. Potential polychlorinated biphenyl residues in animal products from application of contaminated sewage sludge to land. *J. Environ. Qual.* 11: 14–20.

Fries, G. 1985. Bioavailability of soil borne polybrominated biphenyls ingested by farm animals. *J. Toxicol. Environ, Health* 16: 565–579.

Fries, G., and Jacobs, L. 1986. Evaluation of residual polybrominated biphenyl contamination present on Michigan farms in 1978. *Mich. Agric. Exp. Stn., Res. Rep.* 477.

Fries, G. F., and Paustenbach, D. J. 1990. Evaluation of potential transmission of 2,3,7,8-tetrachlorodibenzo-*p*-dioxin-contaminated incinerator emissions to humans via foods. *J. Toxicol. Environ. Health* 29: 1–43.

Fries, G., Paustenbach, D., Mather, D. B., and Luksemberg, W. J. 1999. A congener specific evaluation of transfer of chorinated dibenzo-p-dioxins and dibenzo-p-furans to milk from cows following ingestion of PCP treated wood. *Environ. Sci. Technol.* 33: 1165–1170.

Friess, S. 1987. History of risk assessment. In *Pharmacokinetics in Risk Assessment: Drinking Water and Health*, Vol. 8. National Academy of Science, Washington, DC.

Frobisher, M., Hinsdill, R. D., Crabtree, K. T., and Goodheart, C. R. 1974. *Fundamentals of Microbiology.* Saunders Philadelphia, PA: 674–676.

Gargas, M. L., and Andersen, M. E. 1987. Partition co-efficients for common organic solvents. *Toxicol. Appl. Pharmacol.* 86.

Glasson, W. A., and Tuesday, C. S. 1970. Hydrocarbon reactivities in the atmospheric photooxidation of nitric acid. *Environ. Sci. Technol.* 4: 916–924.

Grothe, D. R., Dickson, K. L., and Reed-Judkns, D. K. (Eds.). 1996. *Whole Effluent Toxicity Testing: An Evaluation of Methods and Prediction of Receiving System Impacts.* Society of Environmental Toxicology and Chemistry, Pensacola, FL.

Guidance Document on Testing Procedures for Pesticides in Freshwater Mesocosms. 1991. Society of Environmental Toxicology and Chemistry, Pensacola, FL.

Hales, S. G., et al. (Eds.). 1997. *Biodegradation Kinetics: Generation and Use of Data for Regulatory Decision Making.* Society of Environmental Toxicology and Chemistry, Pensacola, FL.

Halter, M. T., and Johnson, H. E. 1977. In *Aquatic Toxicology and Hazard Evaluation.* ASTM, Philadelphia, PA, pp. 178–196.

Hamaker, J. W. 1975. The interpretation of soil leaching experiments. In R. Haque and V. H. Freed (Eds.), *Environmental Dynamics of Pesticides*. Plenum, New York, pp. 115–133.

Hamaker, J. W., and Kerlinger, W. O. 1969. Vapor pressure of pesticides. *Adv. Chem. Ser.* 86: 39–54.

Hamaker, J. W., and Thompson, J. M. 1972. Adsorption. In C. A. I. Goring and J. M. Haymaker (Eds.), *Organic Chemicals in the Soil Environment*, Vol. 1. Dekker, New York, pp. 51–122.

Hansch, C., Quinlan, J. E., and Lawrence, G. L. 1986. The linear free energy relationship between partition coefficients and aqueous solubility of organic liquids. *J. Org. Chem.* 33: 347–355.

Haque, R. (Ed.). 1980. *Dynamics, Exposure and Hazard Assessment of Toxic Chemicals*. Ann Arbor Science, Ann Arbor, MI.

Haque, R., and Schmedding, D. 1975. A method of measuring the water solubility of hydrophobic chemcials: Solubility of five polychlorinated biphenyls. *Bull. Environ. Contam. Toxicol.* 14: 13–18.

Harbison, R. D. 1998. *Hamilton and Hardy's Industrial Toxicology*, 5th ed. Mosby Pub. St. Louis, MO.

Hathaway, G., Proctor, N., and Hughes, J. 2000. *Proctor and Hughes Chemical Hazards of the Workplace*, 4th ed. Wiley, New York.

Hayes, A. W. 2001. *Principles and Methods of Toxicology*, 4th ed. Taylor and Francis, Philadelphia, PA.

Hayes, W. J., Jr. 1975. *Toxicology of Pesticides*. Waverly, Baltimore, MD.

Hayes, W. J., Jr. 1982. *Pesticides Studied in Man*. Williams & Wilkins, Baltimore, MD.

Hayes, W. S., and Laws, E. 1991. *Handbook of Pesticide Toxicology*. 3 volumes. Academic, San Diego, CA.

Health Council of the Netherlands (HCN). 1993. Ecotoxicological risk assessment and policy-making in the Netherlands—dealing with uncertainties. *Network* 6(3)/7(1): 8–11.

Hill, I. R., Matthiessen, P., and Heimbach, F. (Eds.). 1994. *Guidance Document on Sediment Toxicity Tests and Bioassays for Freshwater and Marine Environments*. Society of Environmental Toxicology and Chemistry, Pensacola, FL.

Hoffman, D. J., Rattner, B. A., Burton, G. A., and Cairns, J. 1995. *Handbook of Ecotoxicology*, Lewis Publishers Boca Raton, FL.

Ianucci, T., and Ludwig, D. 1999. *Fundamentals of Sediment Toxicology*.

Ikeda, M., Kozumi, A., Kasahara, N., Watanabe, T., Nakatsuka, H., and Sekita, Y. 1987. The statistical approach to the prediction of the possible presence of pollutant chemicals in the environment. *Regul. Toxicol. Pharmacol.* 7: 321–336.

Ingersoll, C., Dillon, T., and Biddinger, G. (Eds.). 1997. *Ecological Risk Assessments of Contaminated Sediments*. Society of Environmental Toxicology and Chemistry, Pensacola, FL.

International Agency for Research on Cancer (IARC). 1985–2000. *The Series of Documents on Carcinogens*. IARC Press, Lyon, France.

International Agency for Research on Cancer (IARC). 1999. Overall evaluations of carcinogenicity to humans. Available from *http://193.51.164.11/monoeval/crthal.html* (Feb. 9, 1999; updated Jan. 20, 1999.).

International Technical Information Institute (ITII). 1976. *Toxic and Hazardous Industrial Chemicals Safety Manual*. ITII, Tokyo.

Jarvinen, A. W., and Ankley, G. (Eds.). 1999. *Linkage of Effects to Tissue Residues: Development of a Comprehensive Database for Aquatic Organisms Exposed to Inorganic and Organic Chemicals*. Society of Environmental Toxicology and Chemistry, Pensacola, FL.

Johnson, L., Welsh, T. H., and Wilker, C. E. 1998. Anatomy and physiology of the male reproductive system and potential targets for toxicants. In K. Boekelheide, R. E. Chapin, P. B. Hoyer, and C. Harris (Eds.), *Reproductive and Endocrine Toxicology*. Elsevier, New York, pp. 5–62.

Jorgensen, S. E. 1994. *Fundamentals of Ecological Modeling, Developments in Environmental Modeling*, 2nd ed. Elsevier, Amsterdam, The Netherlands.

Jury, W. A., Farmer, W. J., and Spencer, W. F. 1984. Behavior assessment model for trace organics in soil. II. Chemical classification and parameter sensitivity. *J. Environ. Qual.* 13: 567–572.

Karickhoff, S. W., Brown, D. S., and Scott, T. A. 1979. Sorption of hydropobic pollutants on natural sediments. *Water Res.* 13: 241–248.

Kenaga, E. E. 1972. Guidelines for environmental study of pesticides: Determination of bioconcentration potential. *Residue Rev.* 44: 73–85.

Kenaga, E. E. 1979. Aquatic test organisms and methods useful for assessment of chronic toxicity of chemicals. In K. L. Dickson, A. W. Maki, and J. Cairns, Jr. (Eds.), *Analyzing the Hazard Evaluation Process*. American Fisheries Society, Washington, DC, pp. 101–111.

Kenaga, E. 1980. Correlation of bioconcentration factors of chemicals in aquatic and terrestrial organisms with their physical and chemical properties. *Ecotoxicol. Environ. Safety* 4: 26–38.

Kenaga, E. E., and Goring, C. A. I. 1980. Relationship between water solubility, soil sorption, octanol water partitioning and concentration of chemicals in biota. *ASTM Spec. Tech. Publ.* STP 707: 78–115.

Kendall, R., Dickerson, R., Giesy, J., and Suk, W. (Eds.). 1998. *Principles and Processes for Evaluating Endocrine Disruption in Wildlife*. Society of Environmental Toxicology and Chemistry, Pensacola, FL.

Klaassen, C. D. 2001. *Casarett and Doull's Toxicology: The Basic Science of Poisons*, 6th ed. McGraw-Hill, New York.

Klecka, G., Bocthlng, B., Franklin, J., Graham, D., Grady, L., Howard, P., Kannan, K., Larson, R., Mackay, D., Muir, D., and van de Meent, K. (Eds.). 2000 *Evaluation of Persistence and Long-Range Transport of Organic Chemicals in the Environment*. Society of Environmental Toxicology and Chemistry, Pensacola, FL.

Knobil, E., and Neill, J. D. 1998. *Encyclopedia of Reproduction*. Academic, New York, NY.

Lee, D. R. 1980. Reference toxicants in quality control of aquatic bioassays. In A. L. Buikema, Jr. and J. Cairns, Jr. (Eds.), *Aquatic Invertebrate Bioassays*. ASTM, Philadelphia, PA, pp. 188–199.

Lehmann, A. J., Vorhes, F. A., et al. 1959. *Appraisal of the Safety of Chemicals in Foods, Drugs and Cosmetics*. Association of Food and Drug Officials of the United States, Washington, DC.

Leo, H., Hansch, C., and Elkins, D. 1971. Partition coefficients and their uses. *Chem. Rev.* 71: 525–616.

Lewis, M. A., et al. (Eds.). 1999. *Ecotoxicology and Risk Assessment for Wetlands*. Society of Environmental Toxicology and Chemistry, Pensacola, FL.

Lewis, R. J., and Irving, N. 2000. *Sax's Dangerous Properties of Industrial Materials*, 10th ed. Wiley, New York.

Li, L.-H., and Heindel, J. J. 1998. Sertoli cell toxicants. In K. S. Korach (Ed.), *Reproductive and Developmental Toxicology*. Marcel Dekker, New York, pp. 655–691.

Lyman and Rosenblatt, D. 1990. *Predicting Chemical Behavior in the Environment*, CRC Press, Boca Raton, FL.

Lynch, D. G., Maceck, G. J., et al. 1994. *Ecological risk assessment case study: assessing the ecological risks of a new chemical under the Toxic Substances Control Act. In: A reivew of ecological assessment case studies from a risk assessment perspective*, Vol. II. Risk Assessment Forum, U.S. Environmental Protection Agency, Washington, DC: 1-1 to 1-35.

Lynch, M. R. (Ed.). 1995. *Procedures for Assessing the Environmental Fate and Ecotoxicity of Pesticides*. Society of Environmental Toxicology and Chemistry, Pensacola, FL.

Mabey, W., and Mill, T. 1978. Critical review of hydrolysis of organic compounds in water under environmental conditions. *J. Phys. Chem. Ref. Data* 7: 383–415.

Macek, K. J., Buxton, K. S., Sauter, S., Gnilka, S., and Dean, J. W. 1976. *Chronic Toxicity of Atrazine to Selected Aquatic Invertebrates and Fishes, Ecol. Res. Ser.* EPA-600/3-76-047. EPA, Washington, DC.

Macek, K. J., Petrocelli, S. R., and Sleight, III, B. H. 1979. Consideration in assessing the potential for, and significance of, biomagnification of chemical residues in aquatic food chains. *ASTM Spec. Tech. Publ.* STP 667: 251–268.

Macek, K. J., and Sleight, III, B. H. 1977. Utility of toxicity tests with embryos and fry of fish in evaluating hazards associated with the chronic toxicity of chemicals to fishes. In F. L. Mayer and J. L. Hamelink (Eds.), *Aquatic Toxicology and Hazard Evoluation*. ASTM, Philadelphia, PA, pp. 137–146.

MacIntosh, D., Suter, II, G. W., and Hoffman, F. O. 1994. Uses of probabilistic exposure models in Ecological risk assessments of contaminated sites. *Risk Anal.* 14(4): 405–419.

Mackay, D., and Paterson, S. 1982. Calculating fugacity. *Environ. Sci. Technol.* 15(9): 1006–1014.

Mackay, D., Paterson, S., and Cheung, B. 1985a. Evaluating the environmental fate of chemicals: The fugacity-level III approach as applied to 2,3,7,8-TCDD. *Chemosphere* 14(6/7): 859–863.

Mackay, D., Paterson, S., Cheung, B., and Neely, W. B. 1985b. Evaluating the environmental behavior of chemicals with a fugacity level III model. *Chemosphere* 14(3/4): 335–374.

Mackay, D., Shiu, W. Y., and Ma, K. C. 2000. *Physical-Chemical Properties and Environmental Fate and Degradation Handbook*. CRCnetBASE 2000. Chapman & Hall, CRCnetBASE, CRC Press, Boca Raton, FL (CD-ROM.).

Maki, A. 1979. An analysis of decision criteria in environmental hazard evaluation programs. In K. L. Dickson, A. W. Maki, and J. Cairns, Jr. (Eds.), *Analyzing the Hazard Evaluation Process*. American Fisheries Society, Washington, DC, pp. 83–100.

Maki, A. W., Dickson, K. L., and Cairns, J., Jr. (Eds.). 1980. *Biotransformation and Fate of Chemicals in the Aquatic Environment.* American Society for Microbiology, Washington, DC.

Manson, J. M., and Kang, Y. J. 1994. Test methods for assessing female reproductive and development toxicology. In A. W. Hayes (Ed.), *Principles and Methods in Toxicology.* Raven Press, New York, pp. 989–1038.

Martin, W. E. 1964. Loss of Sr-90, Sr-89 and I-131 from fallout of contaminated plants. *Radiat. Bot.* 4: 275–281.

Maughn, J. T. 1993. *Ecological Assessment of Hazardous Waste Sites.* Van Nostrand, New York.

Mayer, F. L., and Hamelink, J. L. 1977. *Aquatic Toxicology and Hazard Evaluation.* ASTM, Philadelphia, PA.

McCarty, L., and Mackay, D. 1993. Enhancing ecotoxicological modeling and assessment: body resideus and modes of toxic action. *Environ. Schi. Technol.* 27: 1719–1728.

Miller, G. C., and Zepp, R. G. 1979. Photoreactivity of aquatic pollutants sorbed on suspended sediments. *Environ. Sci. Technol.* 13: 860–863.

Mitsch, W. J., and Gosselink, J. G. 1986. *Wetlands.* Van Nostrand Reinhold, New York.

Moghissi, A. A., Marland, R. E., Congel, F. J., and Eckerman, K. F. 1980. Methodology for environmental human exposure and health risk management. In R. Haque (Ed.), *Dynamics, Exposure, and Hazard Assessment of Toxic Chemicals.* Ann Arbor Science Press, Ann Arbor, MI, Chapter 31.

Motto, H. L., Daines, R. H., Chilko, D. M., and Motto, C. K. 1970. Lead in soils and plants: Its relationship to traffic volumes and proximity to highways. *Environ. Sci. Technol.* 4: 231–237.

Munkittrick, K., McMaster, M., Van der Kraak, G., Portt, C., Gibbons, W., Farwell, A., and Gray, M. (Eds.). 2000. *Development of Methods for Effects-Driven Cumulative Effects Assessment Using Fish Populations: Moose River Project.* Society of Environmental Toxicology and Chemistry, Pensacola, FL.

National Academy of Sciences (NAS). 1983. *Risk Assessment in the Federal Government: Managing the Process.* NAS, Washington, DC.

National Academy of Sciences (NAS). 1999. *Hormonally Active Agents in the Environment.* NAS, Washington, DC.

National Research Council (NRC). 1994. Science and Judgement. NAS Press, Washington, DC.

National Institute for Occupational Safety and Health (NIOSH). 1976. *Occupational Diseases: A Guide to Their Recognition.* NIOSH, Cincinnati, OH.

Neely, W. B. 1977. Material balance analysis of trichlorofluoromethane and carbon tetrachloride in the atmosphere. *Sci. Total Environ.* 8: 267–274.

Neely, W. B., Branson, D. R., and Blau, G. E. 1974. Partition coefficients to measure bioconcentration potential of organic chemicals in fish. *Environ. Sci. Technol.* 8: 1113–1115.

Nielsen, D. M. 1991. *Practical Handbook of Groundwater Monitoring.* National Water Well Assoc. Lewis, Ann Arbor, MI.

Nordberg, T. J. 1976. *Effects and Dose Response Relationships of Metals.* Elsevier, Amsterdam, The Netherlands.

Norberg, T. J., and Mount, D. I. 1985. A new fathead minnow (*Pimephales promelas*) subchronic toxicity test method. *Environ. Toxicol. Chem.* 4(5): 711–718.

Odum, E. P. 1971. *Fundamentals of Ecology*, 3rd ed. Saunders, Philadelphia, PA.

Office of Science and Technology Policy (OSTP). 1985. Chemical carcinogens, a review of the science and its associated principles. *Fed. Regist.* 50(50): 10372–10442.

Oliver, B. G. 1984. Uptake of chlorinated organics from anthropogenically contaminated sediments by obigochaete worms. *Can. J. Fish Aquat. Sci.* 41: 878–883.

Organization for Economic Cooperation and Development (OECD). 1981. *OECD Guidelines for Testing of Chemicals*, Sect. 1: Physical-Chemical Properties. 101. UV-VIS Absorption Spectra. OECD, Paris (available from Publications and Information Center, Washington, DC).

Ottoboni, A. 1991. *The Dose Makes the Poison*. 2nd ed. Van Nostrand, New York.

Parkhurst, B., Warren-Hicks, W., et al. 1995. *Methodology for aquatic ecological risk assessment*. Water Environment Research Foundation, Alexandria, VA.

Pastorak, R. A., Butcher, M., and Nielsen, R. D. 1996. Modeling wildlife exposure to toxic chemicals: trends and recent advances. *Human Ecol. Risk Assess* 2: 444–480.

Pearson, J. G., Foster, R. B., and Bishop, W. E. 1982. *Aquatic Toxicology and Hazard Assessment*. STP 766. ASTM, Philadelphia, PA.

Pfeifer, K. 1984. Bioaccumulation. In *Biomonitoring Applications/Water Quality Management Seminar*. Chemical Manufacturers Association.

Powell, M. R. 1999. *Science at EPA: Information in the Regulatory Process*. Resources for the Future, Washington, DC.

Powell, C. J., and Berry, L. L. 2000. Non-genotoxic and epigenetic carcinogens. In B. Ballantyne, T. Marrs, and T. Syverson (Eds.), *General and Applied Toxicology*, 2nd ed. MacMillan, London, pp. 1099–1119.

Rai, K., and van Ryzin, J. 1979. Risk assessment of toxic environmental substances using a generalized multi-hit dose response model. In N. Breslow and A. Whittemore (Eds.), *Energy and Health*. SIAM, Philadelphia, PA, pp. 99–117.

Rand, G. M. 1995. *Fundamentals of Aquatic Toxicology*, 2nd ed. Taylor and Francis, Philadelphia, PA.

Reinecke, A., and Nash, R. 1984. Toxicology of TCDD and bioavailability by earthworms. *Soil. Biol. Biochem.* 16: 45–49.

Reinert, K., Bartell, S., and Biddnger, G. (Eds.). 1998. *Ecological Risk Assessment Decision-Support System: A Conceptual Design*. Society of Environmental Toxicology and Chemistry, Pensacola, FL.

Riggs, D. S. 1963. *The Mathematical Approach to Physiological Problems*. MIT Press, Cambridge, MA.

Rolland, R. M., Gilbertson, M., and Peterson, R. E. (Eds.). 1997. *Chemically Induced Alterations in Functional Development and Reproduction of Fishes*. Society of Environmental Toxicology and Chemistry, Pensacola, FL.

Rothman, K. 1986. *Modern Epidemiology*. Little, Brown, Boston, MA.

Rupp, E. M., Parzyck, D. C., Walsh, P. J., Booth, R. S., Raridon, R. J., and Whitfield, B. L. 1978. Composite hazard index for assessing limiting exposures to environmental pollutants: Application through a case study. *Environ. Sci. Technol.* 12: 802–807.

Russell, R. S. 1966. Entry of radioactive materials into plants. In R. S. Russell (Ed.), *Radioactivity and Human Diet.* Pergamon, New York, Chapter 5.

Sample, B. E., Opresko, D. M., and Suter, G. W. II. 1996. Toxicological benchmarks for wildlife: 1996 revision. Oak Ridge National Laboratory, Health Sciences Research Division, Oak Ridge, TN.

Sato, A., and Nakajima, T. 1979. Partition coefficients of some aromatic hydrocarbons and ketones in water, blood, and oil. *Brit. J. Ind. Med.* 36: 231–234.

Schrader, S. M. 1997. Male reproductive toxicants. In E. J. Massaro (Ed.), *CRC Handbook of Human Toxicology.* CRC Press, New York, pp. 961–980.

Schrag, S. D., and Dixon, R. L. 1985. Reproductive effects of chemical agents. In R. L. Dixon (Ed.), *Reproductive Toxicology.* Raven, New York, pp. 301–319.

Schultz, J. 1995. *The Ecozones of the World.* Springer Verlag, Heidelberg, Germany.

Shepard, T. H. 1995. *Catalog of Teratogenic Agents*, 8th ed. John Hopkins University Press, Baltimore, MD.

Sheppard, S., Bembridge, J., Holmstrup, M., and Posthuma, L. (Eds.). 1998. *Advances in Earthworm Ecotoxicology.* Society of Environmental Toxicology and Chemistry, Pensacola, FL.

Sijm, D., et al. (Eds.). 1997. *Biotransformation in Environmental Risk Assessment.* Society of Environmental Toxicology and Chemistry, Pensacola, FL.

Snyder, W. S. 1975. *Report of the Task Group on Reference Man.* ICRP No. 23. Pergamon, New York.

Spacie, A., and Hamelink, J. L. 1985. Bioaccumulation. In G. Rand and S. Petrocelli (Eds.), *Fundamentals of Aquatic Toxicology.* Hemisphere, New York, Chapter 17, pp. 495–525.

Sparling, D. W., Linder, G., and Bishop, C. (Eds.). 2000. *Ecotoxicology of Amphibians and Reptiles.* Society of Environmental Toxicology and Chemistry, Pensacola, FL.

Spencer, W. F., and Farmer, W. J. 1980. Assessment of the vapor behavior of toxic organic chemicals. In R. Hague (Ed.), *Dynamics, Exposure and Hazard Assessment of Toxic Chemicals.* Ann Arbor Science, Ann Arbor, MI, pp. 143–161.

Stebbing, A. R. D., Travis C., and Matthiessen, D. (Eds.). 1993. *Environmental Modelling—The Next 10 Years.* Society of Environmental Toxicology and Chemistry, Pensacola, FL.

Stellman, J. M. 1998. *ILO Encyclopedia of Occupational Health and Safety*, 4th ed., 3 vols. Inter Labor Office (ILO), Geneva.

Stern, A. M., and Walker, C. R. 1978. *Hazard Assessment of Toxic substances: Environmental Fate Testing of Organic Chemicals and Ecological Effects Testing.* ASTM Spec. Tech. Publ. STP 657. ASTM, Philadelphia, PA, pp. 81–131.

Stevens and M. Gallo. 1982.

Stickel, L. F. 1973. Chemosphere pesticide reserves in birds and mammals. In C. A. Edwards (Ed.), *Environmental Pollution by Pesticides.* Plenum, New York.

Suter, II, G. W. 1993. *Ecological risk assessment.* Lewis Publishers, Boca Raton, FL.

Suter, II, G. W., Gillett, J., et al. 1994. Issue paper on charecterization of exposure. In: Ecological risk assessment issue papers. Risk Assessment Forum: U.S. Envronmental Protection Agency, Washington, DC: 4-1 to 4-64.

Swanson, M. B., and Socha, A. C. (Eds.). 1997. *Chemical Ranking and Scoring: Guidelines for Relative Assessments of Chemicals.* Society of Environmental Toxicology and Chemistry, Pensacola, FL.

Swindoll, M., Stahl, R. G., and Ellis, S. J. (Eds.). 2000. *Natural Remediation of Environmental Contaminants: Its Role in Ecological Risk Assessment and Risk Management.* Society of Environmental Toxicology and Chemistry, Pensacola, FL.

Tattersfield, L., Matthiessen, P., Campbell, P., Grandy, N., and Large, R. (Eds.). 1997. *Endocrine Modulators and Wildlife: Assessment and Testing.* Society of Environmental Toxicology and Chemistry, Pensacola, FL.

TERA (Toxicology Excellence for Risk Assessment). 2001. www.tera.org. tera@tera.org. Cincinnati, OH.

Thibodeaux, L. J. 1979. *Chemodynamics, Environmental Movement of Chemicals in Air, Water and Soil.* Wiley, New York.

Thiennes, C. H., and Haley, T. J. 1972. *Clinical Toxicology*, 5th ed. Lea & Febinger, Philadelphia, PA.

Thornton, I., and Abrahams, P. 1981. Soil ingestion as a pathway of metal intake into grazing livestock. In *Proceeding of the International Conference on Heavy Metals in the Environment.* CEP Consultants, Edinburgh, pp. 167–172.

Thurston, R. V., Gilfoil, T. A., Meyn, E. L., Zajdel, R. K., Aoki, T. I., and Veith, G. D. 1985. Comparative toxicity of ten organic chemicals to ten common aquatic species. *Water Res.* 19: 1145–1155.

Tinsley, I. J. 1979. *Chemical Concepts in Pollutant Behavior.* Wiley, New York.

Tulp, M. T. M., and Hutzinger, O. 1978. Some thoughts on aqueous solubilities and partition coefficients of PCB, and the mathematical correlation between bioaccumulation and physicochemical properties. *Chemosphere* 10: 849–860.

Udo de Haes, H. (Ed.). 1996. *Towards a Methodology for Life-Cycle Assessment.* Society of Environmental Toxicology and Chemistry, Pensacola, FL.

U. S. Food and Drug Administration (USFDA). 1984. *Environmental Assessment Technical Handbook.* Center for Food Safety and Applied Nutrition and the Center for Veterinary Medicine, USFDA, Washington, DC.

Van den Berg, M., De Vroom, E., Van Greevenbroek, M., Olie, K., and Hutzinger, O. 1985. Bioavailability of PCDDs and PCDFs adsorbed on fly ash in rat, guinea pig and Syrian golden hamster. *Chemosphere* 14: 865–869.

Van Leeuwen, C., Van der Zandt, P., Aldenberg, T., Verhaar, H. J. M., and Hermens, J. L. M. 1992. Application of QSAR. Extrapolation and equilibrium partitioning in aquatic effects assessment. *Environ. Toxicol. Chem.* 11: 267–282.

Veith, G. D., De Foe, D. L., and Bergstedt, B. V. 1979. Measuring and estimating the bioconcentration factor of chemicals in Fish. *J. Fish Res. Board Can.* 36: 1040–1048.

Veith, G. C., Macek, K. J., Petrocelli, S. R., and Carroll, J. 1980. An evaluation of using water partition coefficients and water solubility to estimate bioconcentration factors for organic chemicals in fish. In J. G. Eaton, P. R. Parrish, and A. C. Hendricks (Eds.), *Aquatic Toxicology.* ASTM, Philadelphia, PA, pp. 116–129.

Warren-Hicks, W. J., and Moore, D. R. J. (Eds.). 1998. *Uncertainty Analysis in Ecological Risk Assessment.* Society of Environmental Toxicology and Chemistry, Pensacola, FL.

Weber, J. B. 1972a. Interaction of organic pesticides with particulate matter in aquatic and soil systems. *Adv. Chem. Ser.* 111: 55–120.

Weber, W. J. 1972b. *Physicochemical Processes for Water Quality Control.* Wiley-Interscience, New York.

Weed, S. B., and Weber, J. B. 1974. Pesticide-organic matter interactions. In W. D. Guenzi (Ed.), *Pesticides in Soil and Water.* Soil Science Society of America, Madison, WI, pp. 39–66.

Whitehead, R. G., and Paul, A. A. 1981. Infant growth and human milk requirement. *Lancet* 2: 161–163.

Widner, T. E., Ripple, S. R., and Buddenbaum, J. E. 1996. Identification and screening evaluation of key historical materials and emission sources at the Oak Ridge Reservation. *Health Phys.* 71: 457–469.

Widner, T. E. 2000. *Oak Ridge Dose Reconstruction Project Summary Report.* Prepared for the Tennessee Department of Health. ChemRisk, a Division of McLaren/Hart Evironmental Services. Alameda, CA.

Wiens, J. A., and Parker, K. R. 1995. Analyzing the effects of accidental environmental impacts: approaches and assumptions. *Ecol. Appl.* 5(4): 1069–1083.

Wilson, J. G. 1973. *Environment and Birth Defects.* Academic, New York.

Wissmar, R. C., and Swanson, F. J. 1990. In R. J. Naimand and H. Decamps (Ed.), *The Ecology and Management of Aquatic Terrestrial Ecotones.* UNESCO, Parthenon, Paris, pp. 171–198.

Working, P. K. 1989. *Toxicology of the Male and Female Reproductive Systems.* Hemisphere, New York.

Yanders, A. 1985. Photodegradation of dioxins in soil. *Chemosphere.*

Young, A., and Cockerham, L. 1985. Fate of TCDD in field ecosystem-assessment and significance for Luran exposures. *Dioxins in the Environment.* M. Kamrin and P. Rodgers. Hemisphere, New York pp. 153–171.

Young, A. L. 1983. Long term studies on the persistence and movement of TCDD in a national ecosystem. In R. E. Tucker, A. L. Young, and A. P. Gray (Eds.), *Human and Environmental Risks of Chlorinated Dioxins and Related Compounds.* Plenum, New York, pp. 173–190.

3 Dose–Response Modeling for Cancer Risk Assessment

SURESH H. MOOLGAVKAR and E. GEORG LEUBECK

Fred Hutchinson Cancer Research Center, Seattle, Washington

3.1 INTRODUCTION

Although modern societies constantly strive to balance the health risks of exposure to the products and by-products of industrial processes against the economic benefits that accrue from these activities, the information on which regulatory action is based is often woefully inadequate. Human data on exposure to the agent of interest are often not available and regulation is based on information from experimental studies, which are generally conducted at levels of exposure far higher than those of regulatory interest. The resultant issues of interspecies and low-dose extrapolations are among the most contentious scientific issues of the day.

Whether regulation is based on epidemiologic or toxicologic data, estimation of exposure–response or dose–response relationships is a prerequisite for a rational approach to setting standards for human exposures to potentially toxic substances. In the past couple of decades, a vast biostatistical literature has appeared on exposure–response and dose–response analyses. Summarizing this literature in a single chapter is a formidable task. Although many of the same methods can be used for experimental and epidemiologic studies, we will focus attention in this chapter on statistical methods that have been developed for analyses of epidemiologic studies, and on biologically based mathematical models for analyses of data in which the endpoint of interest is cancer. Moreover, we will not discuss physiologically based pharmacokinetic (PBPK) models, developed to investigate the relationship of exposure to dose by consideration of the uptake, distribution, and disposal of agents of interest. This is not because we consider these models to be unimportant, but because this subject is outside our areas of expertise. Interspecies differences in response to exposure to environmental agents can often be explained, at least partially,

Human and Ecological Risk Assessment: Theory and Practice, Edited by Dennis J. Paustenbach
ISBN 0-471-14747-8 © 2002 John Wiley & Sons, Inc.

in terms of differences in uptake and distribution of the agent. Thus, PBPK models have advanced broadly our understanding of differential species toxicology, and these models are important tools in risk assessment.

Ideally, risk assessments should be based on epidemiologic studies because they offer two obvious major advantages over experimental studies. First, epidemiologic studies are done in the species of ultimate interest, thus finessing the difficult problem of interspecies extrapolation. Second, estimates of risk can be directly obtained for levels of exposure that are close to those typical of "free-living" human populations. Epidemiologic studies are often conducted in cohorts with occupational exposures to the agent of interest, which are typically higher than exposures in the general population. Nevertheless, these exposures are much closer to those in the general population than exposures used in experimental studies. Some of what epidemiologic studies gain in the way of relevance over experimental studies is lost in precision, however. It is generally true that both exposures and disease outcomes are measured with less precision in epidemiologic studies than in experimental studies, leading, possibly, to bias in the estimate of risk. Another, potentially serious, problem arises from the fact that human populations, particularly occupational cohorts, are rarely exposed to single agents. When exposure to multiple agents is involved, the effect of the single agent of interest is often difficult to investigate. Finally, life-style factors such as diet and, in particular, smoking, have a profound effect on human cancer risks. These, potential confounders, are often difficult to control in epidemiologic studies, leading to possible bias in the estimates of risk associated with exposure to the agent of interest.

Epidemiologic studies can be broadly classified into three categories. The cohort study is, at least conceptually, close to the traditional experimental study in that groups of exposed and unexposed individuals are followed in time and the occurrence of disease in the two groups compared. In the case control study, relative risks are estimated from cases of the disease under investigation and suitably chosen controls. In these two types of study, information on exposures and disease is available on an individual basis for all the subjects enrolled in the study. In a third type of study, the ecological study, information is available only on a group basis. Ecological studies have generally been looked upon with disfavor by epidemiologists for reasons that have been extensively discussed elsewhere (Greenland and Morgenstern, 1989; Greenland and Robins, 1994). Nonetheless, they can provide useful information and have played a central role in recent times, particularly in air pollution epidemiology. Because epidemiologic studies are observational (i.e., groups of subjects cannot randomly be assigned to one or another exposure group), careful attention has to be paid to controlling factors that may bias estimates of risk. Thus, controlling for what epidemiologists call "confounding" is of paramount importance both in the design and analyses of epidemiologic studies. The discussion of this fundamental concept is outside the scope of this chapter. The reader is referred to recent excellent texts for details (e.g., Rothman and Greenland, 1998).

When appropriate epidemiologic studies are not available, which is unfortunately the case for most agents of interest, the risk assessor has to rely on experimental data. The twin problems of low-dose and interspecies extrapolation then raise their ugly heads. These are probably the most difficult problems in quantitative risk assessment. While purely statistical methods have been used to address them, the consensus now appears to be moving toward the use of quantitative methods with strong biological underpinnings.

Sophisticated statistical tools have been developed for the analyses of epidemiologic and toxicologic data in the last two decades. Many of these methods fall under the rubric of the so-called relative risk regression models. Additionally, recent research in air pollution epidemiology has exploited regression methods for analyses of time series of counts. Both parametric and semiparametric Poisson regression models have been developed for analyses of these data. Special methods are required when multiple observations are made on the same individual, as is done in panel studies, or in the same geographic location, as is done with Poisson regression analyses of time series of counts. Account must then be taken of serial correlations in the observations. Various statistical methods are used to address this issue. Finally, when the health effect of interest is cancer, stochastic models based on biological considerations can be used for data analyses.

As a complement to the more traditional empirical statistical approaches to analyses of data, the importance of analyses based on ideas of multistage carcinogenesis is threefold. First, the parameters of the model have direct interpretation in biological terms. Thus, analyses using these models often have an external "validity" check based on whether or not the estimates of the parameters are biologically realistic. Second, such analyses can generate biological hypotheses that may be testable either in experimental or epidemiologic studies. For example, recent analyses of epidemiologic (Luebeck et al., 1999) and experimental (Luebeck et al., 1996) data using the TSCE (two-stage clonal expansion, see below) model suggest that the inverse or "paradoxical" dose–rate effect that has been observed following exposure to high LET (linear energy transfer) radiation, such as α particles, is due to an effect on cell proliferation kinetics. Experimental evidence on bystander effects (Iyer and Lehnert, 2000; Barcellos-Hoff and Ravani, 2000) in radiation carcinogenesis from completely independent experiments suggest that this is a plausible explanation of the paradoxical dose rate effect. Third, multistage carcinogenesis models generate a rich and flexible family of hazard (incidence) functions for data analyses. These hazard functions have (qualitative and quantitative) properties that are quite different from the hazard functions in the statistical literature. Furthermore, time and age-related exposures can quite easily be explicitly incorporated into analyses using hazard functions generated by models of multistage carcinogenesis. Thus, for example, quite complicated patterns of exposure to multiple agents typical of many occupational cohort studies can quite easily be explicitly considered in analyses. As a corollary, the risks associated with arbitrary exposure patterns can be explored and used to set guidelines for future exposures.

In this chapter, our main emphasis is on the biologically based approach, and we discuss the empirical statistical approach only briefly.

3.2 MEASURES OF DISEASE FREQUENCY

When discussing dose or exposure response relationships it is important to define clearly what response one is talking about. Often the term dose– or exposure–response is used with no indication of what response means. To define response precisely it is important to have a clear idea of the various commonly used measures of disease frequency. The two fundamental measures of disease frequency used in epidemiology and toxicology are the incidence (or hazard) rate and the probability of disease. The incidence or hazard rate measures the rate (per person per unit time) at which new cases of a disease appear in the population under study. For example, because the incidence rates of many chronic diseases, including cancer, vary strongly with age, a commonly used measure of frequency in epidemiologic studies is the age-specific incidence rate, usually reported in 5-year age categories. For example, the age-specific incidence rate per year in the 5-year age group of 35 to 39 may be estimated as the ratio of the number of new cases of cancer occurring in that age group in a single year and the number of individuals in that age group who are cancer free at the beginning of the year. Strictly speaking, the denominator should not be the total number of individuals who are cancer free at the beginning of the year but the person-years at risk during the year. This is because some individuals contribute less than a full year of experience to the denominator, either because they enter the relevant population after the year has begun (e.g., an individual may reach age 35 sometime during the year) or because they may leave the population before the year is over (e.g., an individual may reach age 40 or die or migrate during the year). Mathematically, the concept of incidence rate is an instantaneous concept and is most precisely defined in terms of the differential calculus. A precise definition of the concept is given in the next section, and the reader is referred to texts on survival analysis (e.g., Kalbfleisch and Prentice, 1980; Cox and Oakes, 1984) for further details.

Another commonly used measure of disease frequency is the probability that an individual will develop disease in a specified period of time. For risk assessment, interest is most often focused on the lifetime probability, often called *lifetime risk* of developing disease. Here lifetime is arbitrarily defined, in the United States usually as 70 years. The incidence (or hazard) rate and probability of developing disease are related by a simple formula. This relationship is expressed by the following equation:

$$P(t) = 1 - \exp\left[-\int_0^t h(s)\, ds\right]$$

where $P(t)$ is the probability of developing the disease of interest by age t, and $h(s)$ is the incidence or hazard rate at age s. Note that although the probability of disease, $P(t)$, is called cumulative incidence in some epidemiology textbooks (e.g., Rothman, 1986), the integral $\int_0^t h(s)\, ds$ is actually the cumulative incidence. When the incidence rate is small, as is true for most chronic diseases, the probability of disease by time t, $P(t)$, is approximately equal to the cumulative incidence:

$$P(t) \approx \int_0^t h(s)\, ds.$$

3.3 EMPIRICAL STATISTICAL METHODS

We will discuss here some of the main statistical tools that have been developed over the last couple of decades for analyses of epidemiologic and toxicologic data. Because most epidemiologic studies are observational, issues of sampling and data analysis are particularly important to assure appropriate interpretation of results in the presence of possible confounding. A discussion of these issues is outside the scope of this chapter.

3.3.1 Statistical Models for Epidemiologic Data

We discuss the use of two broad classes of models, relative regression models, which are widely used for analyses when information is available on each individual in an epidemiologic study, and Poisson regression models, which are used for analyses when data are available on a group level, generally cross-tabulated by exposures of interest.

Relative Risk Regression Models The development here will follow that in the work by Prentice et al. (1986). Although this study was written over a decade ago, it lays out the basic framework for these models.

We introduced the concept of hazard function above as being the appropriate statistical concept that captures the epidemiologic idea of an incidence rate. We begin with a more precise definition of this concept. Consider a large, conceptually infinite, population that is being followed forward in time and about which one wishes to draw inferences regarding the occurrence of some health-related event, generically referred to as a "failure." Typically one is interested in relating the failure to preceding levels of one or more risk factors, such as genetic and life-style factors and exposure to external agents, collectively referred to as covariates. Let $z(t)$ denote the vector of covariates for an individual at time t. Time may be the age of the individual, or, in some settings, it may be more natural to consider other specifications, such as time from a certain calendar date or duration of employment in a specific occupation. Let T denote the time of failure for a subject, and suppose that $Z(t)$ represents the covariate history up to time t. Then the population frequency of failure, which may be thought of as the probability of failure, in a time interval t to $t + \Delta$ with covariate history $Z(t)$, will be denoted by $P[t + \Delta \mid Z(t)]$. The hazard or incidence function (which, if failure refers to death, is often called the force of mortality) is then defined by

$$h[t; Z(t)] = \lim_{\Delta \to 0} P[t + \Delta \mid Z(t); T \geq t]\Delta^{-1} = P'[t \mid Z(t)]/(1 - P[t \mid Z(t)])$$

To simplify notation, we will suppress the dependence of h, P, and so forth on the covariate history $Z(t)$ unless this is not clear from the context. Thus, for example, $h[t; Z(t)]$ will be written as $h(t)$. An intuitive interpretation of the hazard is that it is the rate of failure at time t among those who have not failed up to that time.

Now suppose that one is interested in the incidence of failures among individuals with a specific covariate history, $Z(t)$. For example, one may be interested in the incidence among individuals who are exposed to certain environmental agents thought to be associated with the disease under investigation. Let $Z_0(t)$ represent some standard covariate history; for example, $Z_0(t)$ could be thought of as the covariate history among those not exposed to the agents of interest. One can then write

$$h[t; Z(t)] = h_0(t)RR[t; Z(t)]$$

where $h_0(t) = h[t; Z_0(t)]$ and $RR[t; Z(t)]$ denotes the relative risk of failure at time t associated with covariate history $Z(t)$.

Relative risk regression models attempt to describe risks in populations by focusing on the relative risk function. Various functional forms for RR have been used, the most commonly used being *multiplicative* and *additive* functions of the covariates. The multiplicative model is given by $RR[t; Z(t)] = \exp(\beta_1 z_1 + \beta_2 z_2 + \cdots + \beta_n z_n)$, and the additive model by $RR[t; Z(t)] = 1 + \beta_1 z_1 + \beta_2 z_2 + \cdots + \beta_n z_n$, where z_1 through z_n are the covariates of interest and the β's are parameters to be estimated from the data. Note that the additive model posits that the relative risk is a linear function of the exposures of interest and that the effect of joint exposures is additive. The multiplicative model posits that the logarithm of relative risk is a linear function of the exposures and that the effect of joint exposures is multiplicative. Quite often the relative risk cannot be adequately described by either a multiplicative or an additive model. For example, the relative risk associated with joint exposure to radon and cigarette smoke is greater than additive but less than multiplicative (BEIR IV, 1988). Various mixture models have been proposed (Thomas, 1981; Breslow and Storer, 1985; Guerrero and Johnson, 1982) to address such situations. The use of these models presents special statistical problems (Moolgavkar and Venzon, 1987; Venzon and Moolgavkar, 1988).

There is a vast biostatistical literature on the application of relative risk regression models to the analyses of various study designs encountered in epidemiology, including control of confounding. It is outside the scope of this chapter to review this literature. The interested reader is referred to the appropriate publications (e.g., Breslow and Day, 1980, 1987).

Poisson Regression   Quite often information is available, not on individual members of a study cohort, but on subgroups that are reasonably homogeneous with respect to important characteristics, including exposure, determining disease incidence. As a concrete example, consider the well-known British doctors' study of tobacco smoking and lung cancer (Doll and Peto, 1978). For the cohort of individuals in this study, information on the number of lung cancer deaths is cross-tabulated by daily level of smoking (reported in fairly narrow ranges) and 5-year age categories. Another well-known example is provided by the incidence and mortality data among the cohort of atomic bomb survivors, for which the numbers of cancer cases are reported in cross-tabulated form by (ranges of) age at exposure, total dose received (in narrow ranges) and by 5-year attained age categories. We consider this example in somewhat greater detail below in the context of biologically based models. When data are presented in this way, the method of Poisson regression is often used for analyses. We give a very brief outline of the method here. For more details the reader is referred to the standard text by McCullagh and Nelder (1991).

For Poisson regression, the number of events of the outcome of interest (death or number of cases of disease) in each cell in the cross-tabulated data is assumed to have a Poisson distribution with expectation (mean) that is a function of the covariates of interest. The numbers of events in distinct cells of the cross-tabulated data are assumed to be independent. Suppose that the data are presented in I distinct cross-tabulated cells, and let E_i be the expectation of the number of events in cell i. Suppose that the observed number of events in cell i is O_i. Then under the assumption that the number of events is Poisson distributed, the likelihood of the data is $L = \prod_i \{E_i^{O_i} \exp(-E_i)\}/O_i!$, where the product is taken over all the cells in the cross-tabulated data. The expectations E_i are made functions of the covariates of interest. Generally, $\log(E_i)$ is modeled as a linear function of the covariates. More elaborate functions have been used, however, for example, in the analyses of the atomic bomb survivors data (BEIR V, 1990) and the analyses of lung cancer in cohorts of underground miners (BEIR IV, 1988; Lubin et al., 1994). The expectation has been modeled as well by the hazard function of biologically based carcinogenesis models (e.g., Moolgavkar et al., 1989). One such application is discussed below. Whatever the model form for the expectation, the parameters are estimated by maximizing the likelihood function.

Poisson regression models have played a prominent role in recent analyses of associations between indices of air quality in various urban areas and health outcomes such as mortality (Schwartz and Dockery, 1992; Schwartz, 1993) and hospital admissions (Burnett et al., 1994; Moolgavkar et al., 1997) for specific causes (respiratory disease, heart disease). In these analyses, daily counts of events (deaths or hospital admissions) in a defined geographical area are regressed against levels of air pollution as measured at monitoring stations in that area. Explicitly, the number of events on any given day is assumed to be a Poisson random variable, the expectation of which depends upon indices of air quality and weather on the same or previous days. In this type of study, inferences regarding the association of air pollution with the health events of interest depend upon relating fluctuations in daily counts of events to levels of air pollution on the same or previous days. As indicated above, in the simplest form of Poisson regression the logarithm of the expectation is a linear function of the covariates. This restriction on the shape of the exposure–response function may not be appropriate and, recently, more flexible methods that make no assumptions regarding the shape of this relationship have been introduced for analyses of these data (Health Effects Institute, 1995). An important difference between Poisson regression analyses of air pollution data and the other examples given above (e.g., analysis of the atomic bomb survivor data) is that in the air pollution data information on exposure is available only from central monitors of air quality. It is not possible to form strata of individuals with like exposures within a narrow range. It is not possible, therefore, to investigate the number of deaths or hospital admissions among individuals similarly exposed. This fact makes this type of study of air pollution an ecological study in that exposures and outcomes are known only on the group level, and it is not clear that the number of events is related to the level of exposure.

3.3.2 Statistical Models for Toxicologic Data

Here we focus attention on two classes of exposure– or dose–response models for analyses of response in long-term bioassays. The first is the class of quantal-response models, which are used to investigate the relationship between exposure (or dose) and lifetime probability of response (or tumor if the experiment under consideration is a cancer bioassay). In some bioassays, particularly if serial sacrifices are performed, information may be available not only on whether a particular animal had tumor or not, but also on when (age of the animal) the tumor was observed. In this case, this extra piece of information can be accommodated by using one of the time to response (time to tumor) models.

Quantal-Response Models   There is a whole class of models, the tolerance–distribution models, based on the notion that each animal has its own tolerance to the test agent, and that the toxic response will occur whenever the dose d of the agent exceeds the tolerance. We do not consider these models in this chapter but refer the interested reader to the relevant literature (Prentice, 1976; Krewski and Van Ryzin, 1981; Moolgavkar et al., 1999a).

A quantal-response model currently in wide use by regulatory agencies, the linearized multistage (LMS) model, has its basis in biological considerations.

Suppose $P(d)$ is the lifetime probability of tumor for exposure or dose d. Then, in this model

$$P(d) = 1 - \exp\{-(q_0 + q_1 d + \cdots + q_k d_k)\}$$

This expression for lifetime probability of tumor can be derived from the multistage model after making a number of approximations. These approximations are unlikely to hold, however, in precisely the situations in which the LMS procedure is generally used, that is, with animal data when the probability of tumor is high. Therefore, despite its biological pretensions, this procedure provides an empirical fit to the data. Statistically, the polynomial function of dose used in the equation above permits the description of nonlinear dose–response relationships. A special case of the LMS model is the one-hit model in which $P(d) = 1 - \exp\{-(q_0 + q_1 d)\}$.

The quantal-response models can be fit to data using the maximum-likelihood principle provided that the number of dose groups is larger than the number of unknown parameters. The likelihood of the data, which is binomial, is the product of individual Bernoulli terms contributed by each animal (Krewski and Van Ryzin, 1981; Moolgavkar et al., 1999a).

Time-to-Response Models In some bioassay data, information on the ages at which tumor responses were observed may be available. This information can be incorporated in the analyses by using time-to-tumor models. The fundamental concept required to fit these models to data is that of the hazard function, which we have briefly discussed above. A number of empirical time-to-tumor models were reviewed by Kalbfleisch et al. (1983). Probably the most widely used in the statistical literature is the Weibull model. A time-to-tumor dependent version of the LMS model is also widely used for analyses. A discussion of likelihood construction with these models is outside the scope of this chapter. We point out, however, that the form of the likelihood depends on the amount of information available on tumor onset times, on whether or not the tumor is rapidly lethal, and on the relationship between tumor mortality and causes of death (Krewski et al., 1983; Moolgavkar et al., 1999a). We note that the biologically based models discussed in the next section are time-to-tumor models.

3.4 BIOLOGICALLY BASED DOSE–RESPONSE MODELING

This section comprises the major part of this chapter and discusses dose- or exposure–response analyses within the framework of mathematical models derived from biological considerations. The most commonly used models for this purpose are the Armitage–Doll multistage model and the two-stage clonal expansion (TSCE) model, sometimes also called the MVK (Moolgavkar–

Venzon–Knudson) model. These models can be used for analyses of both time-to-tumor data when they are available and quantal-response data when time-to-tumor information is not available. As we illustrate by examples, these models have been used for analyses of epidemiologic and toxicologic data. As mentioned above, the linearized multistage (LMS) procedure used by regulatory agencies is based on the Armitage–Doll multistage model.

3.4.1   Multistage Carcinogenesis

Current understanding of carcinogenesis as a complex multistage process, modulated by hereditary and environmental influences, is based on observations from histopathological, epidemiologic, and molecular biological studies. Disruption of normal cell proliferation is the *sine qua non* of the malignant state. Conversely, there is accumulating evidence that the kinetics of cell division, cell differentiation (or death), and apoptosis of normal and premalignant cells are important in the carcinogenic process. Increases in cell division rates may lead to increases in the rates of critical mutational events, and an increase in cell division without a compensatory increase in differentiation or apoptosis leads to an increase in the size of critical target cell populations. These observations indicate that carcinogenesis involves successive genomic changes, some of which may result in disruption of normal cellular kinetics and facilitate the acquisition of further mutations, a process that has been compared to Darwinian selection (Cahill et al., 1999). The number of necessary genomic changes required for malignant transformation is not known with certainty for any tumor, although it is thought to be at least two. Intense work in molecular genetics over the last couple of decades has led to some understanding of the critical role of oncogenes and tumor suppressor genes in carcinogenesis (Hanahan and Weinberg, 2000). The concepts of caretaker (guardians of the integrity of the genome) and gatekeeper (regulators of growth) genes have been introduced (Kinzler and Vogelstein, 1997), emphasizing the role both of mutations and cell proliferation in malignant transformation. The role of the microenvironment in which cells find themselves is also increasingly being appreciated (Kinzler and Vogelstein, 1998; Barcellos-Hoff and Ravani, 2000). Nonetheless central questions are: how many rate-limiting steps are involved in carcinogenesis and how are the kinetics of intermediate cell populations different from those of normal cells?

Of particular importance in the risk assessment context are the concepts of initiation and promotion with their implied classification of environmental carcinogens into initiators and promoters. Experimental findings, historically developed from results obtained in the mouse skin carcinogenesis system, led to the operational definitions of the stages of initiation, promotion, and progression. These observations in the mouse skin system have since been extended to other experimental systems, most notably the rodent liver. Initiation is thought to involve (one or more) genomic alterations, promotion is the clonal expansion of initiated cells. Phenotypically altered cell populations in the intermediate stages on the pathway to malignancy have been observed in a variety of epithelial tissues including liver, pancreas, lung, kidney, and urinary bladder in both experimental animals and humans and are generally believed to represent precancerous lesions. Finally, malignant conversion is the process by which one (or a few) of the population of initiated cells becomes fully malignant, probably by acquisition of further genomic alterations.

The observation that both mutation and cell proliferation kinetics play essential roles in carcinogenesis suggests the framework for a mathematical model for carcinogenesis that incorporates the essential biological features of the process. The first multistage models of carcinogenesis were proposed by Nordling (1953) and by Armitage and Doll (1954). About that time it had been observed that the age-specific incidence curves of many common human cancers increased roughly with a power of age. This observation could mathematically be derived from a multistage model for carcinogenesis. These early models ignored cell proliferation, although somewhat later Armitage and Doll (1957) acknowledged the possibility of clonal expansion of intermediate cells in a two-stage model of carcinogenesis and showed that a two-stage model with clonal expansion of intermediate cells could explain the age-specific incidence curves of the common human carcinomas. These ideas were later extended and embellished by Knudson (1971) and by Moolgavkar and Venzon (1979) and Moolgavkar and Knudson (1981).

The main focus of this section is the use of biologically based approaches to model dose–response relationships in cancer. Mathematical models of carcinogenesis have a wider role, however. First, they provide a mathematical framework within which the quantitative aspects of cancer in populations (whether populations of humans in epidemiologic studies or populations of animals in experimental studies) can be viewed and questions about them asked. Thus, for example, models of carcinogenesis have been used for analyses of various epidemiologic and experimental data sets, and biological interpretations of the data have been sought. Second, from the statistical point of view, cancer models provide a rich class of hazard functions for analyses of time-to-tumor data. These models provide a very convenient way to incorporate time- and age-dependent exposure patterns into analyses. Moreover, analyses of epidemiologic data using conventional statistical methods often require rather artificial assumptions, such as multiplicativity of relative risks or additivity of excess risks, which are not required for analyses based on carcinogenesis models. Third, the analyses of intermediate lesions on the pathway to malignancy is greatly facilitated by the use of cancer models.

In the risk assessment context a comprehensive model would incorporate not only the fundamental biological principles underlying carcinogenesis in a pharmacodynamic model, but also the principles underlying intake and disposition of the environmental agent under consideration in a pharmacokinetic model. We are far from developing such a comprehensive model, although some elements of each component of the comprehensive model are in place. In this chapter we intend to cover only pharmacodynamic models. We begin with a brief nontechnical discussion of multistage cancer models and illustrate their use in risk assessment by giving a few epidemiologic examples, as well as one example based on experimental data.

Current multistage models make the following fundamental assumptions: (1) cancers are clonal, that is, malignant tumors arise from a single malignant progenitor cell; (2) each susceptible (stem) cell in a tissue is as likely to become malignant as any other; (3) the process of malignant transformation in a cell is independent of that in any other cell; and (4) once a malignant cell is generated, it gives rise to a detectable tumor with probability 1 after a constant lag time. The last two assumptions, which are made for mathematical convenience, are clearly false. Methods for relaxing these assumptions are currently being investigated.

For more details of the ideas developed here we refer the reader to a recent monograph (Moolgavkar et al., 1999a).

3.4.2   Multistage Models and Incidence Functions

As mentioned earlier, multistage models were first proposed (Nordling, 1953; Armitage and Doll, 1954, 1957) to explain the observation that the age-specific incidence curves of many common adult carcinomas increase roughly with a power of age. In the fitting of these early models to epidemiologic data, a couple of mathematical approximations, which greatly simplified the analyses, were made. The accuracy of these approximations depends crucially on the fact that cancer in human populations is a rare disease. The approximations are quite inaccurate when used for analyses of experimental data in which tumors occur with high probability because the experiments are conducted at high exposure levels of the agent under investigation. Even for epidemiologic data use of these approximations could lead to misinterpretations of the data.

Without going into technical details, we would like to discuss some of the consequences that follow from the multistage nature of carcinogenesis. These consequences can be derived only by a consideration of the exact solution of the underlying mathematical model. The approximate solutions lead to quite different conclusions. Note, the LMS procedure makes even further simplifications and, therefore, should not be considered to be based on biological principles.

The incidence function arising from the concept of multistage carcinogenesis has three important features. These features are not peculiar to a specific multistage model; rather they are true for a broad class of models and thus may be thought of as consequences of multistage carcinogenesis. For technical details the reader is referred to Moolgavkar and Venzon (1979), Moolgavkar and Luebeck (1990), and Heidenreich et al. (1997a).

The first consequence following from a consideration of the exact solution of multistage models is that the incidence function cannot rise indefinitely but must approach a finite asymptote. Recall that the age-specific incidence curves for many cancers increase roughly with a power of age. What the asymptotic behavior of the incidence curve says is that this increase cannot occur indefinitely: After a certain age, the incidence must flatten out. In fact, exactly this behavior has been noted for many of the tumors: The observed incidence at higher ages deviated below the incidence that would have been predicted by a power law relationship. Various explanations were advanced for this observation. For example, it was suggested that cancer cases were being undercounted in the older age groups and that there was something special about the biology of extreme old age that protected against cancer. These explanations may or may not contribute to leveling off of age-specific incidence curves; however, they do not have to be invoked. The exact solution of multistage models predicts exactly this behavior of the incidence function. Thus, the flattening out of the age-specific incidence curve at the older ages is a mathematical consequence of multistage carcinogenesis and could have been predicted if misleading approximations had not been used. The reader is referred to the examples discussed below, and particularly to Figure 3.2a.

The second consequence following from multistage carcinogenesis has to do with the behavior of the age-specific incidence curve after exposure to an environmental carcinogen ceases. Suppose that exposure to an environmental agent increases the rate of one or more of the steps in the carcinogenic pathway, for example, by increasing the rates of specific mutations or of cell divisions. Suppose also that these rates revert to background rates after exposure stops. Then the exact (but not the approximate) solution predicts that the incidence function after exposure ends will eventually approach the incidence function in those individuals who were never exposed. This phenomenon has been observed among ex-smokers, for example. Incidence rates of lung cancer in ex-smokers appear to approach the incidence rates in nonsmokers about 15 years after quitting smoking. A similar phenomenon has been observed among the survivors of the atomic bombings in Hiroshima and Nagasaki. One of the explanations given for this phenomenon is that repair of damaged tissue must have occurred after exposure stopped. This could well be true, but again this explanation need not be invoked. The reversion of the incidence function to background rates is a mathematical consequence of multistage carcinogenesis providing inappropriate approximations are not used. The reader is invited to study the examples shown in Figures 3.2a to 2e.

The third consequence has to do with the incidence of second malignant tumors among individuals who have already had one. A computation using the exact solution shows that the age-specific incidence of second malignant tumors is higher than the age-specific incidence of the first malignant tumor (at the same age). While the incidence of second malignant tumors is difficult to study in human populations because of the treatment intervention after the occurrence of the first tumor, animal experiments appear to show that this is indeed true. The explanation has been advanced that physiological and immunological changes after the first malignancy renders the animal susceptible to a second tumor. This may be true but the higher incidence of second tumors is a logical consequence of multistage carcinogenesis (Dewanji et al., 1991).

3.4.3 Armitage–Doll Model

We gave a precise definition of the concept of incidence (or hazard) because this is the fundamental concept used in fitting multistage models to data. As noted above, the Armitage–Doll model, on which the current LMS procedure for low-dose extrapolation is based, was originally proposed to explain the regularity of the age-specific incidence curves of the common human carcinomas. The age-specific incidence curves of many common adult carcinomas increases roughly with a power of age. Equivalently, a plot of the logarithm of the age-specific incidence curve against the logarithm of age is a straight line, although as pointed out above the observed age-specific incidence rates fall below this line at the older ages. For these, so-called log–log, cancers, Armitage and Doll showed that a multistage model (without consideration of cell proliferation kinetics) predicts this behavior approximately, with the slope of the straight line being one less than the number of stages required for malignant transformation.

The LMS procedure, which is a widely used default procedure for low-dose extrapolation of cancer risks, has its origins in the Armitage–Doll multistage model. To understand the properties of this procedure, it is necessary to understand the Armitage–Doll approximation. The multistage model postulates that a malignant tumor arises in a tissue when a single susceptible cell undergoes malignant transformation via a finite sequence of intermediate stages, the waiting time between any stage and the subsequent one being exponentially distributed. Suppose now that malignancy occurs in n stages and that λ_i is the parameter of the waiting time distribution in stage i. That is, the rate of transition from stage i to stage $i+1$ is λ_i. Then, under the assumptions that malignancy is rare and that each of the transition rates is much smaller than the life span of the species being studied, the multistage model leads to the following expression for the hazard function:

$$h(t) \approx \frac{N}{(n-1)!} \lambda_0 \ldots \lambda_{n-1} t^{n-1}$$

where N is the total number of susceptible cells in the tissue. That is, as mentioned above, the hazard function increases with a power of age one less than the number of stages required for malignant transformation. This form of the multistage model is the basis of the LMS procedure for low-dose extrapolation, which involves a further approximation. It is clear that, with all the approximations made, the LMS procedure is simply an exercise in curve fitting; the hazard function used in the LMS procedure can be substantially different from the exact hazard function generated by the multistage model, particularly when experimental data with high tumor incidence are considered. See the recent International Agency for Research on Cancer (IARC) monograph (Moolgavkar et al., 1999a) for details.

3.4.4 Two-Stage Clonal Expansion (TSCE) Model

The two-stage clonal expansion model posits that malignant transformation of a susceptible cell is the result of two specific, rate-limiting, hereditary (at the level of the cell), and irreversible events. We emphasize that this model should not be interpreted as positing that carcinogenesis results from two specific locus mutations. Rather, this model is best interpreted within the initiation–promotion–progression paradigm of chemical carcinogenesis. Initiation, which confers a growth advantage on the cell, is a rare event and can be thought of as the first rate-limiting step. Mathematically, the process of initiation can reasonably be modeled by a time-dependent Poisson process. Thus, the TSCE model posits that the arrival of cells into the initiated compartment follows a Poisson process, and that promotion consists of the clonal expansion of these cells by a stochastic birth and death process. Finally, one of the initiated cells may be converted into a malignant cell and this conversion may involve one or more mutations. A number of works (Moolgavkar and Venzon, 1979; Moolgavkar and Knudson, 1981; Moolgavkar and Luebeck, 1990; Heidenreich et al., 1997a, b) discuss the biology underlying the model and develop the mathematical and statistical tools required to fit the model to data. When more biological information (on the number of stem cells or on intermediate lesions on the pathway to malignancy, for example) is available, the model can be extended to include this information (Dewanji et al., 1989, 1991).

Because the TSCE model explicitly considers both genomic events and cell proliferation kinetics, it provides a flexible tool for incorporating both genotoxic and nongenotoxic carcinogens in cancer risk assessment. Within the framework of the model, an environmental agent acts by affecting one or more of its parameters. Thus, an environmental agent may affect the rate of one or both stages (initiation and progression) or of cell proliferation kinetics of normal or initiated cells. A purely genotoxic agent would be expected to increase the rates of initiation or progression or both. A pure promoting agent might increase the cell division rate or decrease the rate of apoptosis of initiated cells (and possibly also of normal cells, but this would be expected to have less of an effect on cancer incidence). Compensatory or regenerative cell proliferation occurs at necrogenic doses of many different agents and may explain their promoting effects.

Because the TSCE model explicitly considers initiation and promotion, it can be used for analyses of the intermediate lesions, such as the enzyme-altered foci in rodent hepatocarcinogenesis experiments or the papillomas in mouse skin painting experiments, that arise in many model systems for the study of chemical carcinogenesis. The requisite mathematical expressions required for the analyses of quantitative information on intermediate lesions have been developed and applied to analyses of both altered hepatic foci (Moolgavkar et al.,

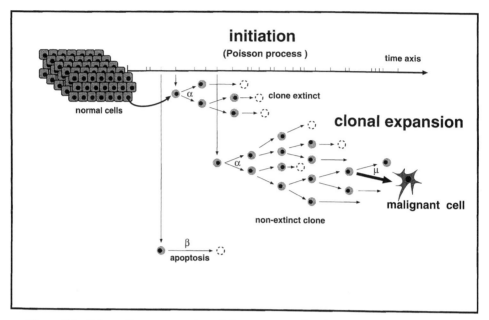

Figure 3.1   Schematic representation of the two-stage clonal expansion (TSCE) model.

1990, 1996; Luebeck et al., 1991, 1995, 2000) and papillomas on the mouse skin (Kopp-Schneider and Portier, 1992). We would like to emphasize that explicit consideration of cell division and apoptosis (rather than just the net proliferation) of initiated cells has important implications. If the rate of apoptosis is greater than zero, then an initiated cell may die without giving rise to a detectable clone. Thus, while initiation is believed to be irreversible on the level of the single cell, it may be reversible on the level of the tissue, that is, initiated cells may die without giving rise to initiated clones (see Figure 3.1). For purposes of quantitative risk assessment analyses of intermediate lesions allows the dose–response curve to be extended to lower doses than would be possible with consideration of malignant lesions alone because intermediate lesions generally occur at doses that are too low for the appearance of malignant lesions with the typical experimental protocol.

3.4.5 Limitations of Biologically Based Models

Although the multistage and the TSCE models are based on biological considerations, they do not begin to capture the complex biological reality of the carcinogenic process. For example, the multistage model ignores cell proliferation and the TSCE model posits only a single intermediate stage with altered cell kinetics. Furthermore, there may be more than a single pathway to malignancy in a tissue. Attempts have been made to develop more general models that accommodate these considerations. The problem with such general models is that there is little quantitative information for the estimation of parameters of the models. Nonetheless, an important questions is, to what extent are conclusions drawn from fitting models to data dependent on the particular model used? While this question is impossible to answer in complete generality, there are some things that can be said. For example, we discussed above some implications of multistage carcinogenesis that hold for a very broad class of models.

In the next section, we give a number of examples of the use of the TSCE model for analyses of real experimental and epidemiologic data sets. The reader should be aware that these conclusions may be model dependent. Some of the conclusions may be amenable to experimental verification. For example, the conclusion, derived from a fit of the TSCE model to radon-induced lung cancer in underground miners, that the inverse dose–rate effect reflects a promotional effect of radon could be experimentally tested. We are currently in the process of developing extensions of the TSCE model to evaluate the robustness of conclusions drawn from fitting the model. The development of such extensions involves a careful balancing act: generalizing the TSCE model while retaining parameter identifiability and estimability.

3.4.6 Some Examples and Applications of the Use of the TSCE Model

We first consider a few hypothetical examples showing the effects of age-dependent increases of initiation and promotion on the age-specific incidence and probability of tumor. These particular examples are instructive in that they show that the temporal behavior of risk following exposure to an environment may not always be intuitive. Next, we consider several examples in which the TSCE model was used for analyses of epidemiologic and experimental data. Our last example describes a recent analysis of data on altered hepatic lesions in rats exposed to TCDD. This study was conducted not for TCDD risk assessment but rather for the elucidation of the effects of chronic TCDD exposure at very low levels on cell kinetics. In the analyses of specific data sets the parameters of the TSCE model, which have direct biological interpretation, are allowed to be functions of the exposure (or dose). Various functional forms are evaluated and a statistical procedure used to determine the optimal functional form. The details are technical and are not provided below. The interested reader is referred to the original works.

Hypothetical Examples As mentioned before, a hallmark of the hazard function of any stochastic multistage model is that it approaches a finite asymptote as a function of age (see, e.g., Moolgavkar and Luebeck, 1990; Little, 1995; Heidenreich et al., 1997b). This observation has important implications for the study of environmental carcinogenesis. For instance, an environmental agent may affect one or more of the model parameters. Within the TSCE model this may be the initiation parameter or a cell kinetic parameter affecting promotion. Now, when exposure to the environmental agent ceases, the affected parameters revert within a short time to background levels. Thus, in many situations, exposure to environmental agents can be modeled by assuming piecewise constant parameters. It was shown by Moolgavkar and Luebeck (1990) that, asymptotically, the hazard function is dominated by its behavior on the last time interval. An immediate consequence of this fact is that, when exposure to an environmental agent ceases, the hazard function approaches the background hazard (hazard in the unexposed) asymptotically. As we will see, the behavior of the hazard function before it approaches its asymptote can be quite complicated, particularly when the environmental agent is a promoter.

To be specific, consider the age-specific incidence and (cumulative) probability of a hypothetical tumor generated by the TSCE model shown in Figure 3.2a. These curves were generated assuming that the total number of stem cells at risk $= 10^7$; the rate of initiation $= 4 \times 10^{-8}$; the rate of progression (rate of the second event) $= 2 \times 10^{-8}$; α, the rate of cell division of initiated cells $= 4$ per cell per year; and β, the rate of apoptosis of initiated cells $= 3.8$ per cell per year. Note that with these assumptions, the TSCE model generates an age-specific incidence curve that approaches an asymptote of 2% per year by age 85, and a (cumulative) probability of tumor of about 0.35 by that age. Figures 3.2b to e show the effects of environmental agents that increase either initiation or promotion $(\alpha - \beta)$ by 35 percent, either between the ages of 20 and 40 or between ages 40 and 60. Of particular interest is the behavior of the hazard function after exposure that affects promotion ceases. For the cases shown, after exposure ceases, the hazard assumes a local maximum, dips, and then approaches the constant baseline asymptote from below. This unintuitive behavior may be explained as follows. The stochastic nature of the two-stage model dictates a distribution of intermediate cells for the individuals in a population of a specific age. After exposure to a carcinogen, individuals who carry a larger number of intermediate cells are at a higher risk to develop a tumor. As a result, with time the surviving individuals (still in the risk set) carry a smaller number of intermediate cells (on average) and are, at least temporarily, at smaller risk for developing the tumor. A similar phenemenon was not observed with exposure to initiators in this example because the distribution of intermediate cells was tighter. The phenomenon observed in this example does not imply, however, that exposure to promoters has a protective effect. Note that the (cumulative) probability of tumor is always higher in the exposed than in the unexposed individuals. Note also that the excess lifetime risk (shown in Figures 3.2c and e), which is a measure often used by regulatory agencies, depends on both the timing of exposure and on the definition of "lifetime."

Emissions from Coke Ovens and Lung Cancer   Emissions from coke ovens contain a complex mixture of polycyclic aromatic hydrocarbons (PAH) and have been associated with lung cancer in occupational cohort studies of coke oven workers. Both the United States Environmental Protection Agency (EPA) and the IARC have determined that coke oven emissions are human carcinogens. In 1984, the EPA used a simple version of the multistage model to analyze the then available epidemiologic data on cohorts of U.S. steel workers

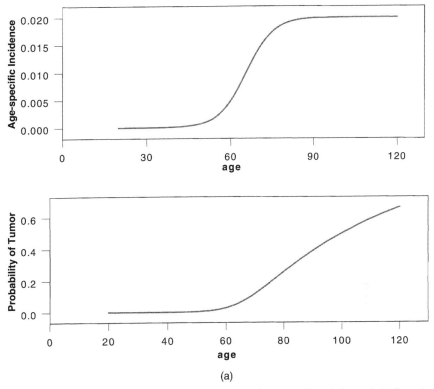

(a)

Figure 3.2 (a) Background age-specific incidence (top panel) and (cumulative) probability (bottom panel) of a hypothetical tumor. These curves were generated using the TSCE model using the following parameters: The number of stem cells, 10^7; the rate of initiation, 4×10^{-8}; the rate of cell division, 4.0 per year; and the rate of cell death, 3.8 per year. (b) Effect of an initiator on age-specific incidence. Two scenarios are shown for exposure to an initiator that increases initiation by 35 percent and leaves all other parameters equal to the background parameters: exposure begins at age 20 and ends at age 40 (dotted); exposure begins at age 40 and ends at age 60 (dash-dotted). Note that the incidence rate among the exposed eventually declines to approach the rate among the unexposed. This is an essential feature of multistage carcinogenesis. (c) Effect of an initiator on cumulative probability. Exposure scenarios as described under (b). Note that the excess risk depends upon both the age at which it is estimated and the period during which exposure occurs. (d) Effect of a promoter on age-specific incidence. Two scenarios are shown for exposure to a promoter that increases $\alpha - \beta$ by 35 percent and leaves all other parameters equal to the background parameters: Exposure begins at age 20 and ends at age 40 (dotted); exposure begins at age 40 and ends at age 60 (dash-dotted). Note that the incidence rate among the exposed eventually approaches the rate among the unexposed. Surprisingly, however, the incidence rate among the exposed actually falls below that in the unexposed population before increasing again. (e) Effect of a promoter on cumulative probability. Exposure scenarios as described under (d). Note that the excess risk depends upon both the age at which it is estimated and the period during which exposure occurs. Note also that although the incidence curve among the exposed falls below that in the unexposed, exposure cannot be considered to be protective because the cumulative probability of tumor is always higher among the exposed.

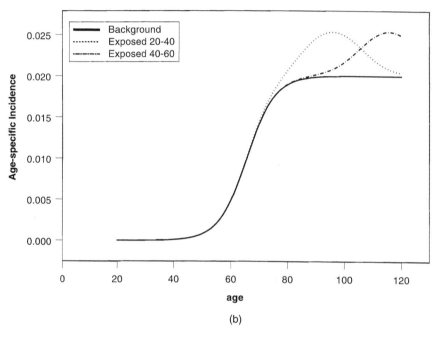

(b)

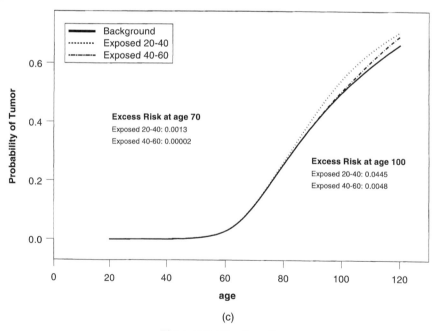

(c)

Figure 3.2 *(Continued)*

EFFECT OF INCREASING PROMOTION

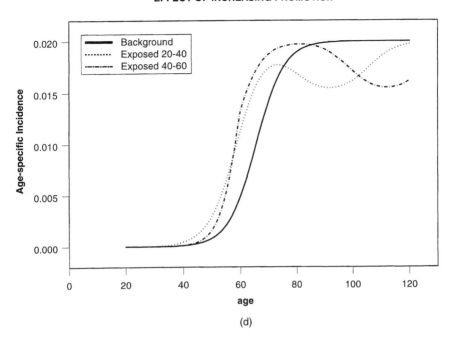

(d)

EFFECT OF INCREASING PROMOTION

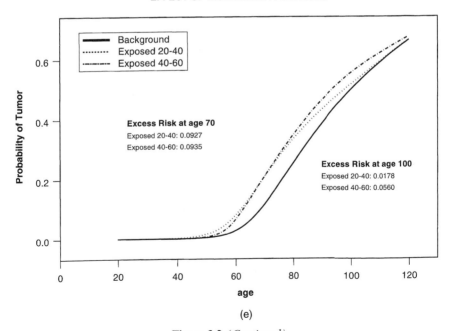

(e)

Figure 3.2 *(Continued)*

and estimated that the unit risk of lung cancer associated with exposure to coke oven emissions was 6.2×10^{-4}. Here unit risk was defined as the excess probability of lung cancer at age 70 associated with continuous exposure to 1 $\mu g/m^3$ from birth. The EPA risk assessment was based on follow-up of the cohorts through 1966. Moreover, the EPA had exposure information only on broad cumulative exposure categories from unpublished tables compiled by Lamd. Since the EPA risk assessment, follow-up of the cohorts has been completed through 1982; there are now four times as many lung cancers in the cohort as were present in 1966. Additionally, detailed exposure histories, reconstructed from job descriptions, are now available for all workers in the cohort.

This updated cohort was used to conduct another risk assessment in 1998 (Moolgavkar et al., 1998). The TSCE model was used to analyze the data and the detailed exposure histories for each worker in the cohort were explicitly considered in the analyses. This ability to explicitly use detailed patterns of exposure is one of the strengths of analyses based on multistage models. A deficiency of the coke oven epidemiologic data is the absence of any information on cigarette smoking, which is potentially a strong confounder of the association between exposure to emissions and lung cancer. An indirect adjustment for smoking could be made, however. It is well known that individuals born at around the same time share many life-style factors, including smoking. Therefore, an indirect way to adjust, at least partially, for smoking is to introduce parameters for birth cohort effects in the models used for analyses of the data. Indeed, birth cohort effects turned out to be highly significant in these data. With the two-stage model applied to the greatly expanded data, the results of analyses indicated that, as expected, emissions from coke oven batteries had significant effects on both the rate of initiation and that of promotion. The exposure–response curve was nonlinear at the higher exposure levels but was close to linear at low levels of exposure. The estimated unit risk, as defined above, was 1.5×10^{-4} with 95 percent confidence interval, derived using Markov chain Monte Carlo methods, equal to 1.2×10^{-4}–1.8×10^{-4}. The interested reader is referred to the recent work by Moolgavkar et al. (1998) for details.

Because the risk assessment for coke oven emissions was based on epidemiologic data, the difficult problem of interspecies extrapolation was completely avoided. There are some concerns, however, about the risk assessment. First, as mentioned above, no smoking information was available and, therefore, a potentially important confounder could not be adequately adjusted in the analyses. Second, although the exposure information was much better than in the original EPA risk assessment, it was still crude compared to the precision with which exposure can be measured in experimental studies. Indeed, it is quite likely that there were substantial exposure measurement errors. How and in which direction these measurement errors biased the estimates of risk is impossible to say. Finally, the unit risk was estimated in a cohort of steel workers. To what extent the estimated risk can be extrapolated to other populations is an open question. Certainly, other factors influencing the risk of lung cancer, such as level of smoking, could modify the effect of coke oven emissions. In our next example, we will illustrate that the risk associated with an agent of interest will depend on background cancer rates in a population, which are determined by exposure to other factors.

Radon and Lung Cancer   This example discusses a recent analysis of the Colorado uranium miners cohort (with follow-up until 1990; Luebeck et al., 1999). Individuals of this cohort were exposed to various levels of α-particle radiation from radon and radon progeny. In total a subcohort of 3238 white male miners of which 354 developed lung cancer were included in the analysis. Lung cancer risks derived from this and other miner cohorts are frequently invoked in determining lung cancer risk from exposure to residential radon [see the BEIR IV (1988) and VI (1998) reports]. In contrast to the previous example, individual level information on tobacco smoking is available here and was included in the analysis. Of interest were also results regarding effects of age and exposure pattern. The TSCE model provides a plausible explanation for the so-called inverse dose rate effect, namely the observation that, for a given total exposure, protraction of that exposure leads to a higher lifetime risk.

Similar to the analysis of the coke oven cohort an identifiable model parameterization was chosen (see, e.g., Heidenreich et al., 1997). The analysis incorporated detailed information on radon/radon progeny exposure from hard rock and uranium mining together with information on cigarette smoking. For details of this analysis we refer the reader to Luebeck et al. (1999). We summarize several important aspects of the analysis:

1. Even though the effect of smoking on lung cancer risk was explicitly modeled, a significant birth cohort effect (BCE) remained. That is, the baseline lung cancer risk showed a linear increase with birth year even after adjusting for smoking information. The lung cancer risks estimated for the subcohort unexposed to radon are comparable to the risks seen in the general white male U.S. population and show the same trend with birth year. See Figure 3.3.

2. The analysis suggested that exposure to radon affected both the rate of initiation of intermediate cells on the pathway to cancer and the rate of cell proliferation of intermediate cells. However, in contrast to the promotional radon effect, which was highly significant, the effect of radon on the rate of initiation was *not* significant. No effect of radon on the rate of malignant conversion was found.

3. The model can be used to study the inverse dose rate effect and to make predictions of dose rate effects at low exposures. The effect was evident for radon exposure levels typical of mines but was attenuated for the more protracted and lower radon exposures in homes (see Figure 3.4).

4. The estimated parameters of the TSCE model can be used to construct the hazard function after exposure ceases. The model predicted an (at-

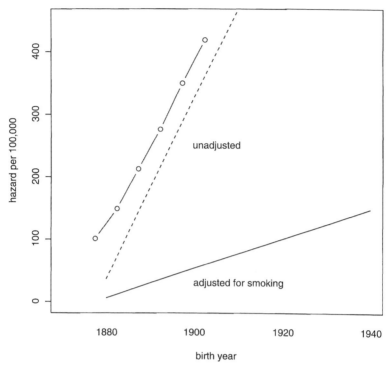

Figure 3.3   Comparison of observed lung cancer mortality rate in the white male U.S. population (age 72.5) as a function of birth cohort with predicted baseline rates in the miners adjusting for birth cohort. Dashed line: risk estimates obtained by ignoring smoking information. Solid line: with smoking information included in the analysis but reporting the risk for nonsmoking miners. Note that smoking explains much of the effect of birth cohort, but that there is a residual effect of birth cohort even after accounting for smoking. The U.S. data are taken from U.S. DHHS (NCI monograph 59) "Cancer Mortality in the U.S.: 1950–1977."

tained) age dependence of lung cancer risk with time since last radon exposure that was in good agreement with observation (Figure 3.5). Note that such a comparison is not possible with the type of relative risk models used in previous analyses.

The inverse dose rate (IDR) effect depends crucially on the interplay between initiation and promotion. Radiation-induced initiation predicts a direct dose rate effect (Moolgavkar, 1997) while radiation-induced promotion predicts an inverse dose rate effect. We believe that for very low exposures and low dose rates the initiation effect predominates, whereas at higher exposure, promotion dominates and is responsible for the IDR effect. Indeed, a TSCE model

exposure centered at age 40 (non-smokers)

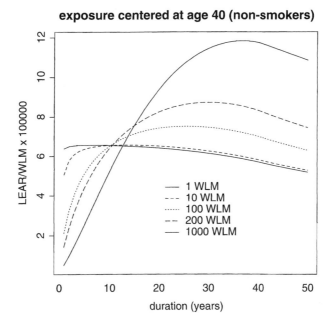

Figure 3.4 Lifetime excess absolute risk per WLM (LEAR/WLM) at age 70 as a function of the duration of exposure (in years) for various total exposures. The exposures were centered at age 40 (nonsmokers).

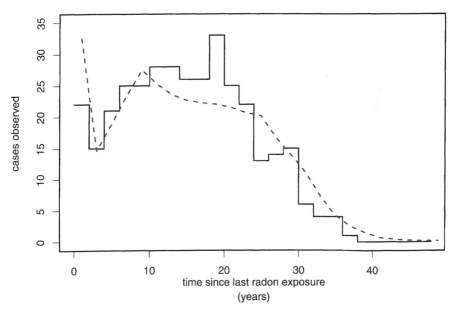

Figure 3.5 Observed (solid line) and expected (dashed line) numbers of lung cancer cases as a function of time since last radon exposure (among whites) and computed on 2-year time intervals.

that does not allow for an increase in net cell proliferation in response to exposure to radon produces no IDR effect.

Whether the predicted promotional response is due to ambient cocarcinogens (uranium ore dust, fossil fuel exhausts) in the mines or to a direct promotional effect in premalignant lesions exposed to α-particles cannot be decided on the basis of these data. However, a reanalysis of the PNL rat study (described in Luebeck et al., 1996) suggests that uranium ore dust may play a promotional role in these lung tumors.

Other explanations for the promotional effect of radon may be given. For instance, it may be possible that α particles trigger "bystander" effects (Little, 1997; Nagasawa, 1999) that could cause cell proliferative responses via changes in gene expression and/or disruption in cytokine signaling between cells. However, these phenomena have been heretofore reported only in in vitro studies involving acute irradiations. Alternatively, we may think of initiated cells as stem cells that have lost the ability to maintain proper homeostasis. In the presence of cytotoxic stress or α-particle-induced cell killing, these cells would be stimulated to divide. Instead of asymmetric cell division with one daughter cell committed to differentiation, the "unstable" initiated cell would, under this scenario, divide with some low probability into two daughter stem cells, leading to expansion of the pool of initiated cells.

Cancer Following Exposure to Low LET Radiation Radon daughters are α emitters, which is high linear energy transfer (LET) radiation. The model has been applied to γ radiation (low LET) as well. Little (1995, 1996) carried out detailed analyses of the A-bomb survivors data using the TSCE model and extensions of it. He concluded that

> without some extra stochastic "stage" appended (such as might be provided by consideration of the process of development of a malignant clone from a single malignant cell) the two-mutations model is perhaps not well able to describe the pattern of excess risk for solid cancers that is often seen after exposure to radiation.

He preferred a three-stage extension of the TSCE model for the A-bomb data. Little's analysis was based on consideration of mortality rather than incidence, however. For cancers that are not rapidly fatal, mortality data are a poor surrogate for incidence. The extra stochastic "stage" that Little deemed necessary for a satisfactory description of the data could be construed to represent the time between occurrence of the malignant tumor and death. In a later work, Little et al. (1996) analyzed the incidence of acute lymphocytic leukemia and of chronic lymphocytic leukemia in England and Wales over the period 1971 to 1988 and concluded that the TSCE model described the incidence of these leukemias well.

Kai et al. (1997) presented analyses of the incidence of three solid cancers—lung, stomach, and colon—among the cohort of A-bomb survivors using the two-mutation model. These analyses showed that the temporal evolution of risk following the (essentially) instantaneous exposure to radiation could be explained entirely by the hypothesis that the exposure resulted in the creation of a (dose-dependent) pool of initiated cells that was added to the pool of spontaneously initiated cells. The dose dependence of initiation was consistent with linearity down to the lowest doses in the cohort. There was no evidence of an age dependence of radiation-induced initiation, suggesting that the high excess relative risk seen in those irradiated as children was not due to an inherently higher susceptibility to radiation. Moolgavkar (1997) discussed some implications of these analyses for assessment of radiation risks in other populations and with protraction of exposure. Heidenreich et al. (1997) analyzed the incidence of all solid cancers combined in the cohort of A-bomb survivors using both exact and approximate solutions of the TSCE model, as well as two empirical models, the "age-at-exposure" model and the "age-attained" model. They concluded that these models, with four parameters estimated for each, described the data well, although the exact TSCE model described some features of the data better than the other models.

In a recent study, Pierce and Mendelsohn (2000) have used the approximate version of the multistage model for analyses of the A-bomb survivors' data. They concluded that the model described the data well and that the pattern of risks was consistent with an equal proportionate increase in mutation rates induced by radiation at each of the stages of the multistage model. Pierce and Mendelsohn (2000) fit their model not to the raw data, however, but to a smoothed version of the data. Moreover, they combined data from many solid cancers, whereas we believe that biologically based analyses should be conducted on a site-specific basis. Figure 3.6 shows the results of analyses of lung cancer among males in the A-bomb cohort. Fits of the Pierce–Mendelsohn, the exact Armitage–Doll, and the TSCE models are shown. For the Pierce–Mendelsohn and the Armitage–Doll models radiation exposure was assumed to cause the same proportionate increase in each of the mutation rates. For the TSCE model, radiation was assumed only to increase the rate of initiation, as in the analysis by Kai et al. (1997) referred to above. Figure 3.6 shows that, at the lower ages, all models describe the data well, but the Pierce–Mendelsohn model, which is based on the Armitage–Doll approximation to the multistage model, fails to fit the downward curvature observed in the data at older ages. The downward curvature is reproduced by the exact solutions of both the multistage and the TSCE models and reflects the approach to an asymptote of the hazard functions of these models.

Occupational Exposure to Refractory Ceramic Fibers and Lung Cancer This example is based on experimental data investigating the carcinogenicity of a class of man-made mineral fibers, the refractory ceramic fibers (RCF). Epidemiologic studies on RCF have so far shown no evidence of increased risk of lung cancer or mesothelioma. The occupational cohorts in which these studies have been conducted are small, however, and the studies, which are continuing,

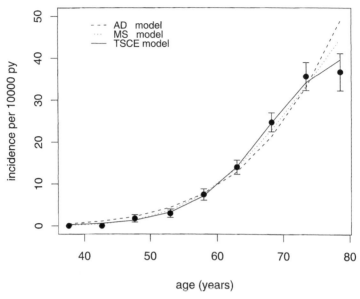

Figure 3.6 Observed (points with error bars) and fitted lung cancer incidence among male A-bomb survivors using the Armitage–Doll (AD) model, the multistage (MS) model, and the two-sage clonal expansion (TSCE) model.

have not been ongoing for a long enough period of time for detection of an effect if there is one. Because of the known carcinogenicity of asbestos, there is concern that the man-made fibers may present a risk of lung cancer and mesothelioma, particularly in the occupational setting. In 1988, IARC evaluated the then available information on RCF and classified these fibers as possible human carcinogens.

We summarize here a recent risk assessment (Moolgavkar et al., 1999b) for RCF based on two long-term oncogenicity studies in male Fischer rats conducted to assess the potential pathogenic effects associated with prolonged inhalation of RCF. Since this risk assessment was based on toxicologic data, both low-dose and interspecies extrapolations had to be addressed. Serial sacrifices were conducted during the study and the fiber burden in the lungs of sacrificed animals were determined by direct counting and reported as number of fibers per milligram of dry lung tissue. With this information, it was possible to determine the temporal pattern of fiber burden in the lung for each animal in the study. This temporal pattern was used explicitly in the TSCE model for analyses of the data. The analogy with the coke oven risk assessment, where detailed patterns of exposure were available for each worker in the cohort, is obvious. The data were consistent with the hypothesis that RCF are initiators of carcinogenesis in the rat lung. A promoting effect could not be ruled out; however, the data set was too small to consider initiation and promotion simultaneously. The likelihood for the initiation model was higher than that for the promotion model. The risk assessment was, therefore, based on treating RCF as an initiator of lung carcinogenesis.

The dose–response relationship appeared to be nonlinear, with an exponential model for dose-related initiation describing the data better than a linear model. The risk of occupational exposure was then estimated in the following steps.

1. The ratio $I(d)/I(0)$ was estimated from analyses of the lung burden data in rats, where $I(d)$ is the rate of initiation at dose (fiber burden) d and $I(0)$ is the rate of background (spontaneous) initiation.

2. The TSCE model was fit to age-specific lung cancer incidence or mortality in the human population to which the extrapolation was to be made, and the parameters of the model estimated, including $I(0)$ for that population. Two populations were used for the extrapolation. First, an occupational cohort of steel workers (not exposed to coke oven emissions) was selected because the requisite data on this population was available (this was a subcohort of the cohort used for the coke oven risk assessment) and because this occupational cohort could be expected to be similar in its smoking and other habits to a cohort of RCF1 workers. A complicating factor is that the workers, although not exposed to coke oven emissions, may have been exposed to other initiators in the workplace. This would tend to increase the estimates of risk based on this cohort. The second population considered was the population of nonsmokers in the American Cancer Society (ACS) cohort. This population could be considered to be representative of a nonsmoking middle-class population in the United States.

3. The estimated ratio $I(d)/I(0)$ from the lung burden studies and the estimated $I(0)$ from the epidemiologic data were used to estimate $I(d)$ in the human population by making the assumption that the ratio is the same in humans and rats. This is the fundamental assumption in this risk assessment. The same assumption was made in the recent CIIT risk assessment of formaldehyde.

4. The deposition clearance model of Yu and colleagues (1997) describing the accumulation of RCF1 in the human lung was used to generate a lung burden profile (fibers per milligram of dry tissue) for exposure scenarios of interest. Various exposure scenarios were considered. Specifically, the EPA default for exposure assumes that human exposure in the occupational setting occurs for 8 hours/day, 5 days/week, 52 weeks/year, starting at age 20 years and continuing until age 50 years.

5. This lung burden profile was used together with the background esti-

mated parameters in the human population and the estimated $I(d)$ in that population to generate incidence curves for lung cancer for various lung burden profiles. Quantities of interest in risk assessment, such as the excess probability of tumor at any age, could then be estimated.

6. Finally, Markov chain Monte Carlo (MCMC) methods were used to generate distributions (and confidence intervals) for the quantities of interest.

For the EPA exposure scenario, the excess probability at age 70 associated with exposure to 1 fiber/cm^3 in the ACS cohort was estimated to be 3.7×10^{-5}, 95 percent confidence interval $= 2.35 \times 10^{-5}$–5.0×10^{-5}. For the steel workers cohort, the risk (excess probability) was estimated to be 1.5×10^{-4}, 95 percent confidence interval $= 0.9 \times 10^{-4}$–2.0×10^{-4}. This example clearly suggests that risk assessments should, to the extent possible, be population specific. A risk assessment in one population may not be applicable to another. For details, the reader is referred to a recent work (Moolgavkar et al., 1998).

Altered Hepatic Lesions in Rats Treated with TCDD   We now describe an analysis of intermediate lesions based on the TSCE model. Here the observations consist of the number and sizes of transections through enzyme-altered lesions on histological liver sections. The fact that the observations are not in three-dimensional space but rather on two-dimensional cross sections is a complication that requires the use of stereological methods. Two avenues can be taken to deal with the stereological problem: First, one might attempt to reconstruct the number and size distribution in three dimensions using an "inverse" stereological transformation and then fit the model to the reconstructed three-dimensional data. Alternatively, one can transform the theoretical model-generated (three-dimensional) number and size distribution into two-dimensional distributions and then fit the data directly in two dimensions. Because the first approach is known to give statistically unstable results, we use the latter approach.

Because the data for this example were reported in terms of the number of visible cell nuclei counted on the individual transections, traditional stereological methods, such as the Wicksell method (Wicksell, 1925) could not be used. Instead, we selected a discrete stereological procedure, which is described schematically in Figure 3.7. To this end, a set of transformation coefficients, a_{nm}, was computed via Monte Carlo simulation. For details of this procedure see Luebeck and de Gunst (2000). Briefly, it was assumed that the observed nuclei were spherically shaped and centered in space at the nodes of a regular cubic lattice. The nuclear diameter was taken to be 10 μm and the lattice spacing (cell dimension) 28 μm (see Jack et al., 1990). Adjustments for the nonzero thickness of the sections were made as well. The derived stereological transformation coefficients a_{nm} are proportional to the probability that an m-cell clone is transected by a random section and shows n-transected nuclei, and the con-

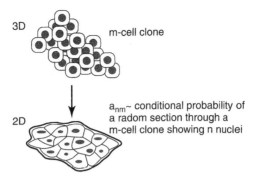

Figure 3.7 Illustration of the stereological problem. The transformation coefficients a_{nm} are proportional to the probability that a m-cell clone is transected by a random section and yields a transection containing n nuclei.

ditional probability that a transection shows exactly $M \geq n \geq n_0$ nuclei is given by

$$q_{n \mid n \geq n_0} = \frac{\sum_{m=n}^{M} \alpha_{nm} p_m}{EZ} \tag{3.1}$$

n_0 and M being the lower and upper size limits, respectively, p_m the model-derived three-dimensional size distribution, and the denominator

$$EZ_{n_0} = \sum_{n=n_0}^{M} \sum_{m=n}^{M} \alpha_{nm} p_m \tag{3.2}$$

represents the expected caliper length, that is, the maximum length perpendicular to the cutting plane of an arbitrary clone.

The data are from an experimental study of enzyme-altered liver foci in female Wistar rats that were treated with diethylnitrosamine (DEN) for 2 weeks at 7 weeks of age. After a resting period of 8 weeks the animals were divided into two treatment groups and treated chronically with either corn oil (vehicle control) or with TCDD. The experimental protocol is shown in Figure 3.8. Groups of 4 to 5 animals were sacrificed 3, 17, 31, 73, and 115 days after start of corn oil/TCDD treatment. The sections taken were also stained immuno-histochemically to reveal the incorporation of bromodeoxyuridine (BrdU) for a determination of the labeling index (LI), which measures cell replication. More experimental details can be found in Stinchcombe et al. (1995).

The TSCE model was adapted for analyses of these data as in previous publications (Moolgavkar et al., 1990; Luebeck et al., 1991, 1995). Here it is

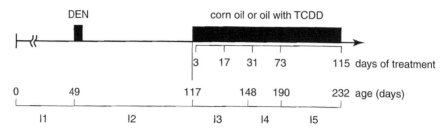

Figure 3.8 Experimental protocol of TCDD experiment schematically. Also shown are the intervals $(I_1, \ldots, I_5)$ used to model the age dependence.

assumed that clones of altered cells arise from a pool of normal susceptible cells (presumably normal hepatocytes) according to a nonhomogeneous Poisson process with rate νX, where X denotes the number of normal cells in 1 mL of liver tissue (see Figure 3.1)

Underlying the clonal expansion of the altered cells is a simple birth-and-death process. Altered cells either divide with cell division rate α or die (and differentiate) with rate β. Malignant transformations of the initiated cells, however, is not considered here.

Mathematical expressions for the size distribution and the expected number of clones as functions of time (or age) can be found in the literature and are not repeated here. For a derivation of the relevant expressions see, for example, Dewanji et al. (1989) and Kopp-Schneider (1992). The analysis of the TCDD data is described in detail in Luebeck et al. (2000). We give a brief discussion of the results here. The measured 3-day LI in foci was roughly constant in all treatment groups and independent of TCDD treatment. Measured values varied between 0.1 and 0.15, or roughly 0.03 to 0.05 per day. No clear time trend was observed. We therefore modeled the cell division rate α as a constant over all time intervals. The estimated cell division rate is 0.06 per day and is in fair agreement with observations.

In the present experiment an attempt was made to also measure the rate of apoptosis. Unfortunately, these measurements are very imprecise because of the short duration of the visible stages of apoptosis (Stinchcombe et al., 1995). In the first 4 weeks after start of TCDD treatment the model predicts the (average) ratio of rates of apoptosis and cell division, β/α, near 0.6 for TCDD and about 1 for controls. In normal adult tissue, where cell death is compensated by cell birth, this ratio is close to 1, reflecting normal homeostatic conditions. In focal tissue homeostasis is compromised, and usually we find that $\beta/\alpha < 1$, which leads to an increase in net cell proliferation.

Thus, the observed proliferative effect of TCDD is accounted for by a reduction of β/α in TCDD-treated animals, although this reduction seems less than what would be expected from direct counts of the number of apoptotic events in GST-P positive foci.

Concerning the process of initiation (of GST-P positive hepatocyctes), the model makes the somewhat startling prediction that acute treatment with DEN leads to protracted initiation, that is, normal hepatocytes appear to acquire the enzyme alteration at substantially higher rates for many weeks after the acute DEN treatment. It is possible that this effect is related to long-lived DNA adducts that are progressively converted into mutations as cells undergo division.

Although we do not find TCDD to increase the total number of de novo initiated GST-P positive cells, we do find an effect of TCDD on the rate of initiation, which can be interpreted as an "acceleration" in the conversion of preinitiated (DNA damaged) cells into cells that express the GST-P positive phenotype.

We conclude that the observed proliferative effect of TCDD is accounted for by the selective outgrowth of GST-P positive clones in response to the inhibitory effects of TCDD on apoptosis. Our analysis also predicts that TCDD interferes with the normal formation of initiated GST-P positive cells after an acute treatment with DEN. We conjecture that in the presence of TCDD the conversion of DNA-damaged cells into initiated cells (that exhibit the GST-P positive phenotype) is advanced by several weeks leading to an apparent initiating effect in the first 30 days or so of TCDD treatment.

3.5 CONCLUDING REMARKS

In this chapter, we have given a brief survey of multistage models of carcinogenesis, with emphasis on their use for quantitative cancer risk assessment. Improvements in risk assessment will depend on the detail and quality of information available from epidemiologic and experimental studies, on better understanding of the biological principles underlying carcinogenesis, and the incorporation of new biological information into cancer models. A good example of a risk assessment based on detailed biological information is provided by a new risk assessment of formaldehyde performed at the Chemical Industry Institute of Toxicology (1999) and currently under review at EPA. This risk assessment uses the TSCE model as the basic pharmacodynamic tool. The risk assessments considered in this chapter make it clear that quantitative assessments of risk should not be reported as single numbers. Clearly, risk will depend on a number of host factors, such as genetic susceptibility, and modifying factors such as age at exposure and exposure to other agents. It is also clear that the uncertainty associated with quantitative assessment of risk is much larger than is suggested by the confidence intervals derived from statistical procedures. Characterization of this uncertainty is one of the biggest challenges facing risk assessment today.

It is unlikely that in the foreseeable future quantitative cancer risk assessment will be based on a complete understanding of the biological processes underlying carcinogenesis. Use of models based on current understanding of carcinogenesis does, however, identify data gaps that need to be filled. The use of such models can also suggest plausible explanations for observed exposure–response relationships. For example, Lutz and Kopp-Schneider (1999) have recently proposed a mechanism for the J-shaped exposure–response relationships reported for some carcinogens. Unfortunately, a biologically based approach to quantitative risk assessment requires more data than are usually available for agents of regulatory interest. Thus, a tiered approach to risk assessment, which would use default procedures when the data are sparse but utilize as much information as possible when data are available, would appear to be the best option. Notwithstanding the fact that such a flexible approach cannot easily be codified into a set of rules, there are encouraging signs that regulatory agencies in the United States are willing to consider it.

REFERENCES

Armitage, P., and Doll, R. 1954. The age distribution of cancer and a multistage theory of carcinogenesis. *Br. J. Cancer* 8: 1–12.

Armitage, P., and Doll, R. 1957. A two-stage theory of carcinogenesis in relation to the age distrobution of human cancer. *Br. J. Cancer* 11: 161–169.

Barcellos-Hoff, M. H., and Ravani, S. A. 2000. Irradiated mammary gland stroma promotes the expression of tumorigenic potential by unirradiated epithelial cells. *Cancer Res.* 60: 1254–1260.

BEIR IV: National Research Council, Committee on the Biological Effects of Ionizing Radiation. 1988. *Health Effects of Radon and Other Internally Deposited Alpha Emitters (BEIR IV)*. National Academy Press, Washington DC.

BEIR V: National Research Council, Committee on the Biological Effects of Ionizing Radiation (BEIR IV). 1990. *Health Effects of Exposure to Low Levels of Ionizing Radiation (BEIR V)*. National Academy Press, Washington DC.

BEIR VI: National Research Council, Committee on the Biological Effects of Ionizing Radiation. 1998. *Health Effects of Exposure to Radon (BEIR VI)*. National Academy Press, Washington DC.

Breslow, N. E., and Day, N. E. 1980. *Statistical Methods in Cancer Research*, Vol. I: *The Analysis of Case-Control Studies*. IARC Scientific Publications No. 32. International Agency for Research on Cancer, Lyon.

Breslow, N. E., and Day, N. E. 1987. *Statistical Methods in Cancer Research*, Vol. II: *The Design and Analysis of Cohort Studies*. IARC Scientific Publications No. 82. International Agency for Research on Cancer, Lyon.

Breslow, N. E., and Storer B. E. 1985. General relative risk functions for case-control studies. *Am. J. Epidemiol.* 122: 149–162.

Burnett, R. T., Dales, R. E., Raizenne, M. E., Krewski, D., Summers, P. W., Roberts, G. R., Raad-Young, M., Dann, T., and Brook, J. 1994. Effects of low ambient levels of ozone and sulfates on the frequency of respiratory admissions to Ontario hospitals. *Environ. Res.* 65: 172–194.

Cahill, D. P., Kinzler, K. W., Vogelstein, B., and Lengauer, C. 1999. Genetic instability and darwinian selection in tumours. *Trends Cell. Biol.* 9: M57–M60.

Chemical Industry Institute of Toxicology. 1999. *Formaldehyde: Hazard Characterization and Dose–Response Assessment for Carcinogenicity by the Route of Inhalation*, Rev. ed. Chemical Industry Institute of Toxicology, Research Triangle Park, NC.

Cox, D. R., and Oakes, D. 1984. *Analysis of Survival Data*. Monographs on Statistics and Applied Probability. Chapman and Hall.

Dewanji, A., Moolgavkar, S. H., and Luebeck, E. G. 1991. Two-mutation model for carcinogenesis: Joint analysis of premalignant and malignant lesions. *Math. Biosci.* 104: 97–109.

Dewanji, A., Venzon, D. J., and Moolgavkar, S. H. 1989. A stochastic two-stage model for cancer risk assessment. II. The number and size of premalignant clones. *Risk Anal.* 9: 179–187.

Doll, R., and Peto, R. 1978. Cigarette smoking and bronchial carcinoma: Dose and time relationships among regular smokers and life-long non-smokers. *J. Epidemiol. Commun. Health* 32: 303–313.

Guerrero, V. M., and Johnson, R. A. 1982. Use of the Box-Cox transformation with binary response models. *Biometrika* 69: 309–341.

Greenland, and Morgenstern 1989.

Greenland, and Robbins 1994.

Hanahan, D., and Weinberg, R. A. 2000. The hallmarks of cancer. *Cell* 100: 57–70.

Health Effects Institute (HEI). (1995). Washington DC.

Heidenreich, W. F., Jacob, P., and Paretzke, H. G. 1997a. Exact solutions of the clonal expansion model and their application to the incidence of solid tumors of atomic bomb survivors. *Radiat. Environ. Biophys.* 36: 45–58.

Heidenreich, W. F., Luebeck, E. G., and Moolgavkar, S. H. 1997b. Some properties of the hazard function of the two-mutation clonal expansion model. *Risk Anal.* 17(3): 391–399.

Iyer, R., and Lehnert, B. E. 2000. Factors underlying the cell growth-related bystander responses to α particles. *Cancer Res.* 60: 1290–1298.

Jack, E. M., Bentley, P., Bieri, F., Muakkassah-Kelly, S. F., Stäubli, W., Suter, J., Waechter, F., and Cruz-Orive, L. M. 1990. Increase in hepatocyte and nuclear volume and decrease in the population of binucleated cells in preneoplastic foci of rat liver: A stereological study using the nucleator method. *Hepatology* 11: 286–297.

Kai, M., Luebeck, E. G., and Moolgavkar, S. H. 1997. Analysis of the incidence of solid cancer among atomic bomb survivors using a two-stage model of carcinogenesis. *Radiat. Res.* 148: 348–358.

Kalbfleisch, J. D., Krewski, D., and Van Ryzin, J. 1983. Dose-response models for time-to-response toxicity data. *Can. J. Stat.* 11: 25–49.

Kalbfleisch, J. D., and Prentice, R. L. 1980. *The Statistical Analysis of Failure Time Data*. Wiley Series in Probability and Mathematical Statistics. Wiley, New York.

Kinzler, K. W., and Vogelstein, B. 1997. Gatekeepers and caretakers. *Science* 386: 761–763.

Kinzler, K. W., and Vogelstein, B. 1998. Landscaping the cancer terrain. *Science* 280: 1036–1037.

Knudson, A. G. 1971. Mutation and cancer: Statistical study of retinoblastoma. *Proc. Nat. Acad. Sci. USA* 68: 820–823.

Kopp-Schneider, A. 1992. Birth-death processes with piecewise constant rates. *Statist. Prob. Lett.* 13: 121–127.

Kopp-Schneider, A., and Portier, C. J. 1992. Birth and death/differentiation rates of papillomas in mouse skin. *Carcinogenesis* 13: 973–978.

Krewski, D., Crump, K. S., Farmer, J., Gaylor, D. W., Howe, R., Portier, C., Salsburg, D., Sielken, R. L., and Van Ryzin, J. 1983. A comparison of statistical methods for low dose extrapolation utilizing time-to-tumor data. *Fundam. Appl. Toxicol.* 3: 140–156.

Krewski, D., and Van Ryzin, J. 1981. Dose-response models for quantal response toxicity data. In M. Csorgo, D. A. Dawson, J. N. K. Rao, and E. Saleh (Eds.), *Statistics and Related Topics*. North-Holland, Amsterdam, pp. 201–231.

Little, J. B. 1997. Commentary: What are the risks of low-level exposure to α-particle radiation from radon? *Proc. Natl. Acad. Sci. USA* 94: 5996–5997.

Little, M. P. 1995. Are two mutations sufficient to cause cancer? Some generalization to the two-mutation model of carcinogenesis of Moolgavkar, Venzon and Knudson, and of the multistage model of Armitage and Doll. *Biometrics* 51: 1278–1291.

Little, M. P. 1996. Generalization of the two-mutation and classical multistage models of carcinogenesis fitted to the Japanese atomic bomb survivor data. *J. Radiol. Prot.* 16: 7–24.

Lubin, J. H., Boice, Jr., J. D., Edling, C., Hornung, R. W., Howe, G., Kunz, E., Kusiak, R. A., Morrison, H. I., Radford, E. P., Samet, J. M., Tirmarche, M., Woodward, A., Yao, S. X., and Pierce, D. A. 1994. *Lung Cancer and Radon: A Joint Analysis of 11 Underground Miners Studies*. Publication No. 94-3644, U.S. National Institutes of Health, Bethesda, MD.

Luebeck, E. G., Buchmann, A., Stinchcombe, S., Moolgavkar, S. H., and Schwarz, M. 2000. Effects of 2,3,7,8-tetrachlorodibenzo-*p*-dioxin (TCDD) on initiation and promotion of GST-P positive foci in rat liver: A quantitative analysis of experimental data using a stochastic model. Submitted.

Luebeck, E. G., Curtis, S. B., Cross, F. T., and Moolgavkar, S. H. 1996. Two-stage model of radon-induced malignant lung tumors in rats: Effects of cell killing. *Radiat. Res.* 145: 163–173.

Luebeck, E. G., Grasl-Kraupp, B., Timmermann-Trosiener, I., Bursch, W., Schulte-Hermann, R., and Moolgavkar, S. H. 1995. Growth kinetics of enzyme altered liver foci in rats treated with phenobarbital or α-hexachlorocyclohexane. *Toxicol. Appl. Pharmacol.* 130: 304–315.

Luebeck, E. G., Heidenreich, W. F., Hazelton, W. D., Paretzke, H. G., and Moolgavkar, S. H. 1999. Biologically-based analysis of the data for the Colorado uranium miners cohort: Age, dose and dose-rate effects. *Radiat. Res.* 152: 339–351.

Luebeck, E. G., Moolgavkar, S. H., Buchmann, A., and Schwarz, M. 1991. Effects of polychlorinated biphenyls in rat liver: Quantitative analysis of enzyme-altered foci. *Toxicol. Appl. Pharmacol.* 111: 469–484.

Lutz, W. K., and Kopp-Schneider, A. 1999. Threshold dose response for tumor induction by genotoxic carcinogens modeled via cell-cycle delay. *Toxicol. Sci.* 49: 110–115.

McCullagh, P., and Nelder, J. A. 1991. *Generalized Linear Models*. Chapman & Hall.

Moolgavkar, S. H. 1997. Stochastic cancer models: Application to analyses of solid cancer incidence in the cohort of A-bomb survivors. *Nucl. Energy* 36(6): 447–451.

Moolgavkar, S. H., Dewanji, A., and Luebeck, G. 1989. Cigarette smoking and lung cancer: Reanalysis of the British doctor's data. *J. Natl. Cancer Inst.* 81: 415–420.

Moolgavkar, S. H., and Knudson, A. G. 1981. Mutation and cancer: A model for human carcinogenesis. *J. Natl. Cancer Inst.* 66: 1037–1052.

Moolgavkar, S. H., Krewski, D., Zeise, L., Cardis, E., and Møller, H. (Eds.). 1999a. *Quantitative Estimation and Prediction of Cancer Risks.* IARC Scientific Publications No. 131. IARC.

Moolgavkar, S. H., and Luebeck, E. G. 1990. Two-event model for carcinogenesis: Biological, mathematical and statistical considerations. *Risk Anal.* 10: 323–341.

Moolgavkar, S. H., Luebeck, E. G., and Anderson, E. L. 1997. Air pollution and hospital admissions for respiratory causes in Minneapolis-St. Paul and Birmingham. *Epidemiology* 8(4): 364–370.

Moolgavkar, S. H., Luebeck, E. G., and Anderson, E. L. 1998. Estimation of unit risk for coke oven emissions. *Risk Anal.* 18(6): 813–825.

Moolgavkar, S. H., Luebeck, E. G., Buchmann, A., and Bock, K. W. 1996. Quantitative analysis of enzyme-altered liver foci in rats initiated with diethylnitrosamine and promoted with 2,3,7,8-tetrachlorodibenzo-*p*-dioxin or 1,2,3,4,6,7,8-heptachlorodibenzo-*p*-dioxin. *Toxicol. Appl. Pharmacol.* 138: 31–42.

Moolgavkar, S. H., Luebeck, E. G., de Gunst, M., Port, R. E., and Schwarz, M. 1990. Quantitative analysis of enzyme-altered foci in rat hepatocarcinogenesis experiments I: Single agent regimen. *Carcinogenesis* 11(8): 1271–1278.

Moolgavkar, S. H., Luebeck, E. G., Turim, J., and Hanna, L. 1999b. Quantitative assessment of the risk of lung cancer associated with occupational exposure to refractory ceramic fibers. *Risk Anal.* 19(4): 599–611.

Moolgavkar, S. H., and Venzon, D. J. 1979. Two-event models for carcinogenesis: Incidence curves for childhood and adult tumors. *Math. Biosci.* 47: 55–77.

Moolgavkar, S. H., and Venzon, D. J. 1987. General relative risk models for epidemiologic studies. *Am. J. Epidemiol.* 126: 949–961.

Nagasawa, H., and Little, J. B. 1999. Unexpected sensitivity to the induction of mutations by very low doses of alpha-particle radiation: Evidence for a Bystander Effect. *Radiat. Res.* 152: 552–557.

Nordling, C. O. 1953. A new theory of the cancer inducing mechanism. *Br. J. Cancer* 7: 68–72.

Pierce, D. A., and Mendelsohn, M. L. 2000. A model for radiation-related cancer suggested by atomic bomb survivor data. *Radiat. Res.* 152: 642–654.

Prentice, R. L. 1976. A generalization of the logit and probit methods for dose response curves. *Biometrics* 32: 761–768.

Prentice, R. L., Moolgavkar, S. H., and Farewell, V. T. 1986. Biostatistical issues and concepts in epidemiologic research. *J. Chronic Dis.* 38: 1169–1183.

Rothman, K. J. 1986. *Modern Epidemiology.* Little, Brown, Boston.

Rothman, K. J., and Greenland, S. 1998. *Moden Epidemiology,* 2nd ed. Lippincott-Raven, New York.

Schwartz, J. 1993. Air pollution and daily mortality in Birmingham, Alabama. *Am. J. Epidemiol.* 137: 1136–1147.

Schwartz, J., and Dockery, D. W. 1992. Increased mortality in Philadelphia associated with daily air pollution concentrations. *Am. Rev. Respir. Dis.* 145: 600–604.

Stinchcombe, S., Buchmann, A., Bock, K. W., and Schwarz, M. 1995. Inhibition of apoptosis during 2,3,7,8-tetrachlorodibenzo-*p*-dioxin-mediated tumour promotion in rat liver. *Carcinogenesis* 16(6): 1271–1275.

Thomas, D. C. 1981. General relative risk models for survival time and matched case-control studies. *Biometrics* 37: 673–686.

Venzon, D. J., and Moolgavkar, S. H. 1988. Origin invariant relative risk functions for case-control and survival studies. *Biometrika* 75: 325–333.

Wicksell, D. S. 1925. The corpuscle problem, Part I. *Biometrika* 17: 87–97.

Yu, C. P., Ding, Y. J., Zhang, L., Oberdorster, G., Mast, R. W., Maxim, L. D., and Utell, M. J. 1997. Retention modeling of refractory ceramic fibers (RCF) in humans. *Reg. Toxicol. Pharmacol.* 25: 18–25.

4 Exposure Assessment*

DENNIS J. PAUSTENBACH

Exponent, Menlo Park, California

4.1 INTRODUCTION

Health risk assessment is the process wherein toxicology data from animal studies and human epidemiology are evaluated, a mathematical formula is applied to predict the response at low doses, and then information about the degree of exposure is used to predict quantitatively the likelihood that a particular adverse response will be seen in a specific human population.[1-3] Regulatory agencies have used the risk assessment process for nearly 50 years, most notably the U.S. Food and Drug Administration (USFDA).[4] However, the difference between assessments performed in the 1950s and 1960s and those performed in the 1980s and 1990s is that dose extrapolation models, quantitative exposure assessments, and quantitative descriptions of uncertainty have been added to the process.[5] Because of increased understanding, the availability of desktop computers, and better quantitative methods for predicting the low-dose response [such as physiologically based pharmacokinetic (PB-PK) models], risk assessments conducted today should provide more accurate risk estimates than in the past.[3,6]

Since 1980, most environmental regulations and some occupational health standards have, at least in part, been based on health risk assessments.[3] They include standards for pesticide residues in crops, drinking water, ambient air, and food additives, as well as exposure limits for contaminants found in indoor air, consumer products, and other media. Risk managers increasingly rely on risk assessment to decide whether a broad array of risks are significant or trivial; an important task since, for example, more than 400 of the about 2000 chemicals routinely used in industry have been labeled carcinogens in various animal studies.[7,8] In theory, the results of risk assessments in the United States should influence virtually all regulatory decisions involving so-called toxic agents.[9]

* Based on: The practice of exposure assessment. In A. W. Hayes (Ed.). *Principles and Methods of Toxicology*, 4th ed. Taylor and Francis, New York, 2001. Reproduced by permission of Routledge, Inc., part of the Taylor and Francis Group.

Human and Ecological Risk Assessment: Theory and Practice, Edited by Dennis J. Paustenbach
ISBN 0-471-14747-8 © 2002 John Wiley & Sons, Inc.

The risk assessment process has four parts: hazard identification, dose–response assessment, exposure assessment, and risk characterization.[7] Although progress has been made over the past 20 years in how to conduct and interpret toxicology and epidemiology studies (e.g., hazard identification), and scientists believe that they are doing a better job of dose–response extrapolation than in the past, most significant advances in the risk assessment process have, over the past ten years, occurred in the field of exposure assessment.[10]

In recent years, an increasing number of environmental scientists have embraced the view that "toxicology data are important, but they do not mean much without quantitative information about human exposure."[11] For this reason, each year since about 1990, the toxicology community has shown increasing interest in understanding the exposure assessment field.[11] Fortunately, a significant amount of research has been conducted to identify better values for many exposure parameters, and improvements have been made in applying these exposure factors to various scenarios.

4.2 BASIC CONCEPTS

4.2.1 Description of Exposure Assessment

Exposure assessment is the step that quantifies the intake of an agent resulting from contact with various environmental media (e.g., air, water, soil, food).[3,12,13] Exposure assessments can address past, current, or future anticipated exposures, although uncertainties can become significant when attempting to anticipate what might have happened or what will happen.[14–19]

Exposure assessment in various forms dates back at least to the early 20th century, and perhaps earlier, particularly in the fields of epidemiology,[20] industrial hygiene,[21,22] and health physics.[23] Epidemiology is the study of disease occurrence and the causes of disease, while the latter fields deal primarily with occupational exposure. Exposure assessment combines elements of all three disciplines and relies upon aspects of statistics, biochemical toxicology, large animal toxicology, atmospheric sciences, analytical chemistry, food sciences, physiology, environmental modeling, and others.[12]

Fundamentally, an exposure assessment describes the nature and size of the various populations exposed to a chemical agent and the magnitude and duration of their exposures.[24,25] They determine the degree of contact a person has with a chemical and estimate the magnitude of the absorbed dose. Several factors need to be considered when estimating the absorbed dose, including exposure duration, exposure route, chemical bioavailability from the contaminated media (e.g., soil), and, sometimes, the unique characteristics of the exposed population (e.g., hairless mice absorb a greater percent of chemical than other mice). By definition, *duration* is the period of time over which the person is exposed. An *acute* exposure generally involves one contact with the chemical; usually for less than a day. An exposure is considered *chronic* when it takes place over a substantial portion of the person's lifetime. Exposures of intermediate duration are called *subchronic*.[12]

Knowledge of the chemical concentration in an environmental medium is essential to determine the magnitude of the absorbed dose. This information is usually obtained by analytical measurements of samples of the contaminated medium (air, water, soil, sediment, food, or dust). Estimates can also be made using mathematical models, such as models relating air concentrations at various distances from a point of release (e.g., a smoke stack) to factors including release rate, weather conditions, distance, and stability of the agent.[26,27]

In general, since about 1995, our ability to perform exposure assessments has matured to a degree that they will usually possess less uncertainty than other steps in the risk assessment. Admittedly, many factors should be considered when estimating exposure; for example, it is a complicated procedure to understand the transport and distribution of a chemical that has been released into the environment. Nonetheless, available data indicate that scientists can now do an adequate job of quantifying chemical concentrations in various media, and the resulting uptake by exposed persons, if they account for all factors that should be considered[28] (Figure 4.1).

The primary routes of human exposure to chemicals in the ambient environment are dust and vapor inhalation, dermal contact with contaminated soils or dusts, and ingestion of contaminated food, water, dust, or soil. In the workplace, the predominant exposure route usually is inhalation, followed by dermal uptake and, to a lesser extent, dust ingestion due to hand-to-mouth contact.[6] Uncertainty in environmental exposure assessment can be greater than in an occupational exposure assessment. However, in many workplaces, there can be large fluctuations in airborne concentrations, a significant difference in work practices of different persons, and there is real difficulty in measuring dermal uptake and incidental ingestion.[22,29,30]

Scientists in the field of radiological health were the first to quantitatively estimate human uptake of environmental contaminants;[31] their work can be a source of valuable information when conducting assessments of chemical contaminants. Since this work, which was conducted after World War II, numerous methodologies for estimating human uptake of environmental contaminants have been described and refined, many over the past decade.[35-37] The availability of information on the degree of exposure associated with various scenarios has increased exponentially, as evidenced by the size of the recent *EPA Exposure Factors Handbook*; a three-volume document containing nearly 1000 pages of information on exposure assessment.[25,38,39,81]

The practice of exposure assessment has changed over time. For example, beginning in the late 1970s, U.S. regulatory policy encouraged or mandated the use of conservative approaches when conducting exposure assessments. This was codified in Environmental Protection Agency's (EPA's) original document entitled *Risk Assessment Guidance for Superfund*.[40] At that time, standardization of exposure assessments used to satisfy regulatory agencies was considered prudent because it guaranteed that risks would not be under-

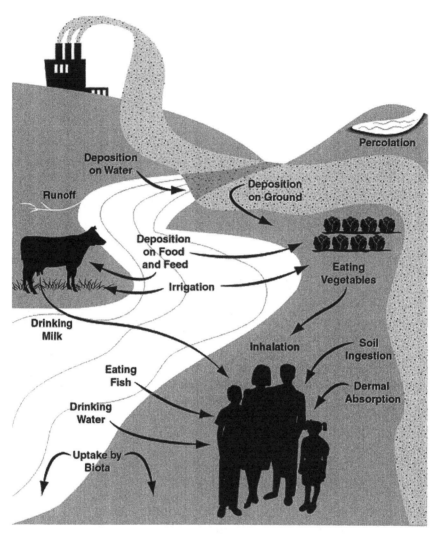

Figure 4.1   Exposure pathways.

estimated to ensure protection of public health. Around 1985, concern evolved that repeated use of conservative exposure factor assumptions was producing unrealistically high estimates of exposure[2,29,41–43] and that the cost of achieving the recommended cleanup levels was becoming unreasonable. Around 1990, risk assessors learned how to apply Monte Carlo techniques to address both typical and highly exposed persons. Application of Monte Carlo techniques to exposure assessment has dramatically improved our understanding of the certainty of our estimates and has decreased the problems associated with the repeated use of conservative assumptions, thereby altering the field permanently.[44–48]

4.2.2   What Is Exposure?

Because there is no agreed-upon definition of where or when exposure takes place, terminology used in published exposure assessment literature is inconsistent. Although there is reasonable agreement that human exposure means contact with the chemical or the agent,[12,40,49] there has not yet been widespread agreement as to whether this means contact with (1) the visible exterior of the person (skin and openings into the body, such as mouth and nostrils), or (2) the so-called exchange boundaries where absorption takes place (skin, lungs, gastrointestinal tract).[12] The differing definitions have led to some ambiguity in the use of terms and units for quantifying exposure.

Some scientists find it helpful to think of the human body as having a hypothetical outer boundary that separates the inside of the body from the outside.[12] The outer boundary of the body consists of the skin and openings into the body, such as the mouth, nostrils, or punctures and lesions in the skin. In most exposure assessments, chemical exposure is defined as contact of the chemical with some part of this boundary. An exposure assessment is the quantitative or qualitative evaluation of that contact. It describes the intensity, frequency, and duration of contact, and often quantifies the rate at which the chemical crosses the boundary (chemical intake or uptake rates), the route of the chemical across the boundary [exposure route (e.g., dermal, oral, or respiratory)], the resulting amount of chemical actually crossing the boundary (dose), and the amount of chemical absorbed (internal dose).[12,50]

Depending on the purpose of the exposure assessment, the numerical output of these analyses may be an estimate of either exposure or dose. If an exposure assessment is being done as part of a risk assessment in support of an epidemiologic study, for example, sometimes only qualitative exposure levels may be calculated. In these situations, categories like low-, medium-, and high-level exposure may be used. In contrast, most risk assessments conducted in recent years attempt to quantitatively predict the absorbed dose and, occasionally, the circulating blood level or the concentration of the toxicant in the target organ.[51,52]

4.2.3   Concepts of Exposure, Intake, Uptake, and Dose

The process of a chemical entering the body can be described in two steps— contact (exposure), followed by actual entry (crossing the boundary). Absorption, either upon crossing the boundary or subsequently, leads to the availability of an amount of chemical to biologically significant sites within the body (target tissue dose). Although the description of contact with the outer boundary is simple conceptually (e.g., mg benzene/cm^2 skin), estimating the degree to which a chemical crosses this boundary is somewhat more complex.

There are two major processes by which a chemical can cross the boundary from "outside to inside" the body (e.g., intake). Intake involves physically moving the chemical in question through an opening in the outer boundary (usually the mouth or nose), typically via inhalation, eating, or drinking. Normally, the chemical is contained in a medium such as air, food, water, or dust/soil. Here, the key question is the mass inhaled or ingested. In this process, mass transfer occurs by bulk flow, and the amount of chemical itself crossing the boundary can be described as a chemical intake rate. The chemical intake rate is the amount of chemical crossing the outer boundary per unit of time, which is the product of the concentration in the media times the ingestion or inhalation rate. Ingestion and inhalation rates are typically expressed as cubic meters of air breathed/hour, kilogram of food ingested/day, liters of water consumed/day, or milligrams of soil/dust ingested per day. Ingestion or inhalation rates typically are not constant over time but will vary within known limits.[12] For example, persons will always breathe between 5 and 25 m^3/day, depending on their degree of physical activity.

Uptake is the second process by which a chemical can cross the boundary from "outside to inside" the body. Uptake, in contrast to intake, involves absorption of the chemical through the skin. Although the chemical is often contained in a carrier medium, the medium itself typically is not absorbed at the same rate as the "contaminant of interest," so estimates of the amount of chemical crossing the boundary cannot be made in the same way as for intake.[12] For example, benzene on the surface of a contaminated soil particle will move quickly through the skin, but benzene in the center of the soil particle may never completely reach the surface and, therefore, it is not bioavailable. Of course, for many inorganic chemicals like arsenic or lead, bioavailability can be very low since the chemical is bound to the interstices. In short, if a chemical cannot be released, it has no bioavailability; therefore, it does not pose a risk.

Dermal absorption is an example of direct uptake across the outer boundary of the body. A chemical uptake rate is the amount of chemical absorbed per unit of time. In this process, mass transfer occurs by diffusion, so uptake will depend on the concentration gradient across the boundary, permeability of the barrier, and other factors.[50,53,54] Chemical uptake rates can be expressed as a function of the exposure concentration, permeability coefficient, and surface area exposed, or as flux.[6]

The conceptual process of contact, entry, and absorption can be used to derive equations for exposure and dose for all exposure routes.

4.2.4 Applied Dose and Potential Dose

Applied dose is the amount of chemical available at the absorption barrier (skin, lung, gastrointestinal tract).[12] It is useful to know the applied dose if a relationship can be established between the applied dose and the internal dose, a relationship that can sometimes be established experimentally. This relationship can be estimated either through modeling or by direct measurement. For example, some researchers have analyzed phenol concentrations in the blood of volunteers over time after placing their hands in a bucket of nitrobenzene in an attempt to quantify the flux rate.[55,56] Usually, it is difficult to measure the applied dose directly, as many of the absorption barriers are internal to the human and not localized in such a way to make measurement easy. An approximation of applied dose can be made, however, using the concept of potential dose.[12] In general, estimating the magnitude of internal dose is easier than external dose.

Potential dose is simply the amount of chemical that is ingested or inhaled, or the amount of chemical contained in material applied to the skin. It is a useful term or concept in those instances when there is a measurable amount of chemical or transport medium, such as eating a certain amount of food or applying a certain amount of material to the skin. The potential dose for ingestion and inhalation is analogous to the administered dose in a dose–response experiment. For example, if plastic pellets are fed to an animal, it reflects the potential dose of humans who eat plastic.[12]

For the dermal route, potential dose is the amount of chemical applied, or the amount of chemical in the medium applied (e.g., as a small amount of soil deposited on the skin). Note that because all of the chemical in the soil particulate is not contacting the skin, this differs from exposure (the concentration in the particulate times the duration of contact) and applied dose (the amount in the layer actually touching the skin).[12]

As previously noted, the amount of chemical that reaches the exchange boundaries of the skin, lungs, or gastrointestinal tract may often be less than the potential dose if the material is only partly bioavailable. For example, less than about 0.1 to 1.0 percent of dioxins or polycyclic aromatic hydrocarbons (PCBs) on fly ash in contact with the skin are likely to penetrate.[57] When bioavailability data are known, adjustments to the potential dose should be made to convert it to applied dose and internal dose.[57,58]

4.2.5 Internal Dose

The amount of chemical that has been absorbed and is available for interaction with biologically significant receptors (e.g., target organs) is called the internal dose. Estimating internal dose is the first objective of a good exposure assessment.[59,60] Transport models are available to assist in this process.[61] Once absorbed, the chemical can be metabolized, stored, excreted, or transported within the body. The amount transported to an individual organ, tissue, or fluid of interest is termed the delivered dose.[62] The delivered dose may be only a small part of the total internal dose. For example, although 1 mg of PCB may be absorbed into the body, at any given time the amount in the liver (the target organ) may only be 0.01 mg. Estimating delivered dose has been among the most exciting areas of exposure assessment research over the past 15 years. Currently, the best approach to estimate delivered dose is to measure blood or to use PB-PK models.[63,64]

The biologically effective dose (BED), or the amount that actually reaches cells, sites, or membranes where adverse effects occur,[65] may represent only a fraction of the delivered dose, but it is obviously the best one for predicting adverse effects. To understand BED is the ultimate goal of exposure assessment. Regrettably, thus far, toxicologists have yet to be able to estimate BED or measure it for most chemicals, except in rare cases.[12]

Currently, most risk assessments dealing with environmental chemicals (as opposed to pharmaceutical assessments) rely upon dose–response relationships based on the potential (administered) dose or the internal dose, because the pharmacokinetic understanding necessary to base relationships on the delivered dose or biologically effective doses is not available for most chemicals.[66] PB-PK models have been developed for about 60 high-volume industrial chemicals.[67] Additional PB-PK models for various chemicals will be developed as more is known about the pharmacokinetics and pharmacology names of environmental chemicals.

In general, it is more convenient in risk assessment to refer to dose rates, or the amount of a chemical dose (applied or internal) per unit time (e.g., mg/day), or as dose rates on a per-unit-body-weight basis (e.g., mg/kg/day). Most exposure data like that found in the various editions of the EPA's *Exposure Factors Handbook* and other guidance documents are presented as dose rates (e.g., grams of fish consumed each day).[25,38,39,68,69]

4.2.6   Exposure and Dose Relationships

Depending on the purpose of the exposure assessment, different estimates of exposure and dose may require calculation. Often, these choices will be made so that the dose metric will be the same as that used in the toxicology study. For example, when trying to understand the hazard posed by an acute toxicant, airborne concentrations as determined during 1- to 30-minute time periods will often be expressed as $\mu g/m^3$, mg/m^3, mg/kg, $\mu g/L$, mg/L, ppb, or ppm.

When risk is a function of time of exposure, exposure or dose profiles can be very useful. In these profiles, the exposure concentration or dose is plotted as a function of time.[70] Concentration and time are used to depict exposure, while amount and time characterize dose.

Such profiles are important for use in risk assessment where the severity of the effect depends on the pattern by which the exposure occurs, rather than on the total (integrated) exposure (Figure 4.2). For example, a developmental toxicant may only produce effects if exposure occurs during a particular stage of development. Similarly, a single acute exposure to very high contaminant levels may induce adverse effects, even if the average exposure is much lower than apparent no-effect levels. To understand the developmental hazard, one must generally consider the pharmacokinetics of the specific chemical. Such profiles will become increasingly important as biologically based dose–response models become available.[63]

Integrated or aggregate exposure is the sum total of exposure to a chemical via all routes of exposure (and all media). With the routine use of personal computers, it is now commonplace to add as many as 6 to 10 different exposure sources per route (e.g., DDT in different fruits and vegetables) and 3 exposure routes (e.g., DDT via food, air, and dermal contact). The units of integrated exposure are concentration times period of time. Integrated exposure is the total "area under the curve" (AUC) of the exposure profile. An exposure profile (a picture of the exposure concentration over time) is particularly useful when

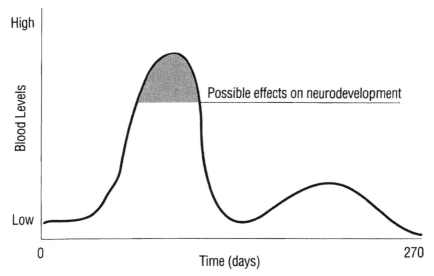

Figure 4.2 Time course of exposure to a developmental toxicant. Note that the shaded portion represents the blood concentration of toxicant which is necessary to offer some probability of an adverse effect on development.

trying to understand occupational exposure because it contains more information than an integrated exposure (a number), including the duration and periodicity of exposure, the peak exposure, and the shape of the area under the time–concentration curve. Such profiles can allow toxicologists to identify those periods of time when risk might be the greatest (e.g., during certain periods of a pregnancy).[12]

The last dose category is the time-weighted average. This is used especially as part of a carcinogen risk assessment. A time-weighted average concentration is the integrated exposure divided by the period when exposure occurs. In cancer risk assessments, the time over which exposure is integrated is usually 70 years.[12] A time-weighted average dose rate is the total dose divided by the time period of dosing, usually expressed in units of mass per unit time, or mass/time normalized to body weight (e.g., mg/kg-day). Time-weighted average dose rates such as the lifetime average daily dose (LADD) are often used in dose–response equations to estimate lifetime risk.

4.2.7 Measures of Dose

For risk assessment purposes, dose estimates should be expressed in a manner that can be compared with available dose–response data from animal or human studies. For example, if data on human exposure is in mg/dL of lead in blood, it would be best to use the blood concentrations in an animal study for the comparison. Frequently, dose–response relationships are based on potential dose (called administered dose in animal studies), although dose–response relationships are sometimes based on internal dose. The measure of dose selected should be based on the mode of action of the adverse effect.[12,60,70–72]

Doses may be expressed in several different ways. Solving Eq. (4.1), for example, gives the dose rate over the time period of interest. The dose-per-unit time is the dose rate, which has units of mass/time. The most common dose measure is average daily dose (ADD), which is used to predict or assess the noncarcinogenic effects of a chemical.

$$\text{ADD} = \frac{C \cdot \text{IR} \cdot B \cdot D}{\text{BW} \cdot \text{AT}} \qquad (4.1)$$

where ADD = potential average daily dose
     BW = body weight
      B = bioavailability
     AT = time period over which the dose is averaged (days)
      C = mean exposure concentration
     IR = ingestion rate
      D = duration

The following example presents a typical calculation.

Example Calculation 1: Determining the Average Daily Dose   A typical American eats a certain amount of lettuce over a lifetime (about 2000 kg). Assume that on any given week, the maximum quantity ingested is 1.5 kg, and the maximum on any one day is 0.4 kg. Assume that the typical aldrin residue is 4 ppm on all lettuce ingested over the person's lifetime. What is the ADD of aldrin for the maximum week? Assume the bioavailability of aldrin in lettuce is 90 percent. Given:

$$C = 4 \text{ mg/kg (aldrin)}$$

$$\text{BW} = 70 \text{ kg}$$

$$\text{AT} = 7 \text{ days}$$

$$\text{IR} = 1.5 \text{ kg}$$

$$B = 0.9$$

Therefore:

$$\text{ADD} = \frac{C \cdot \text{IR} \cdot B}{\text{BW} \cdot \text{AT}}$$

$$\text{ADD} = \frac{(4 \text{ mg/kg})(1.5 \text{ kg})(0.9)}{(70 \text{ kg})(7 \text{ days/week (week)})}$$

$$\text{ADD} = 0.011 \text{ mg/kg-day}$$

When the primary health risk posed by a chemical is cancer, then the biological response is usually described in terms of lifetime probabilities (e.g., the increased risk of developing cancer during a 70-year lifetime is 2 in 100,000). In these circumstances, even though exposure does not occur over the entire lifetime, doses are usually presented as LADDs.[12] The LADD takes the form of Eq. (4.2), with lifetime (LT) replacing the averaging time (AT):

$$\text{LADD}_{\text{pot}} = \frac{C \cdot IR \cdot B \cdot D}{BW \cdot LT} \tag{4.2}$$

Example Calculation 2: Determining the Lifetime Average Daily Dose   What is the LADD in Example 1 involving aldrin in lettuce? Given:

$$C = 4 \text{ mg/kg (aldrin in lettuce)}$$

$$IR = 2000 \text{ kg}$$

$$B = 0.9$$

$$BW = 70 \text{ kg}$$

$$LT = 70 \text{ yr} = 25{,}550 \text{ days}$$

$$D = 70 \text{ yr}$$

where

$$\text{LADD} = \frac{\overline{C} \cdot \overline{IR} \cdot \overline{B}}{BW \cdot LT}$$

Then,

$$\text{LADD} = \frac{(4)\left(\dfrac{2000}{70}\right)(70)(0.9)}{(70)(25{,}550)}$$

$$\text{LADD} = 0.004 \text{ mg/kg-day}$$

Although other measures of chronic dose may be more appropriate for predicting the hazard posed by a chronic toxicant, such as an area-under-the-blood-concentration curve or the peak target tissue concentration, the LADD is the most common dose metric used in carcinogen risk assessment (Figure 4.3).

4.3   CONCEPTUAL APPROACHES TO EXPOSURE ASSESSMENT

4.3.1   Quantifying Exposure

Although exposure assessments are conducted for a variety of reasons, the process of estimating exposure can be approached using one of the following three methods:[12]

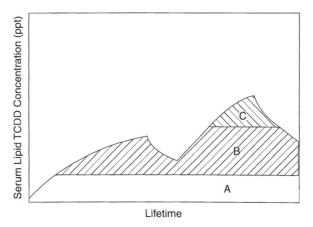

Figure 4.3 Theoretical concentration versus time curve for TCDD illustrating one possible relationship between AUC and response. This figure illustrates the possible combination of AUC and thresholds for production of various responses: area A, no effect; area B, enzyme induction occurs; area C, significant increased cell proliferation. From Aylward et al. (1996).

1. *Direct Measurement Approach.* The exposure can be measured at the point of contact (the outer boundary of the body) while it is taking place, measuring the exposure concentration and time of contact and integrating them (point-of-contact measurement). These assessments are based on direct measurement. An example is the measurement of the amount of contaminated soil on an exposed hand of someone digging a hole to plant a tree. The relevant exposure information would be contaminant concentration in soil ($\mu g/g$), surface area of the hand in contact with the soil (100 m^2), and time of exposure (2 hours).

2. *Exposure Scenario.* The exposure can be hypothetical, which is called an exposure scenario. In these assessments, specific data cannot actually be collected, but the relevant kinds of information can. For example, if an incinerator were built, it cannot be known today how much of each chemical in the airborne emissions would reach the various compartments in the environment (food, soil, sediment, surface water), but one can describe what is likely to occur (a scenario).

3. *Using Biomonitoring Data.* The exposure can be estimated from dose, which in turn can be reconstructed through internal indicators (biomarkers, body burden, excretion levels) after the exposure has taken place (reconstruction). Among the best examples are lead in blood, phenol in urine, volatile hydrocarbons in the breath, and dioxins in blood fat.

These three approaches to exposure quantification (or dose) are independent because each is based on different data. The fact that they are independent measures is useful in verifying or validating the results of the various approaches. Each of the three has strengths and weaknesses; using them in combination can considerably strengthen the credibility of an exposure or risk assessment.[8,12,73] For example, results of the exposure assessment would be validated if one can mathematically predict the absorbed dose per day of a chemical, estimate the resulting blood concentrations, and confirm these estimates by sampling the exposed population.[28]

4.3.2 Estimates Based on Direct Measurement

Point-of-contact or direct exposure assessment evaluates the exposure as it occurs. By measuring chemical concentrations at the interface between the person and the environment as a function of time, this yields an exposure profile. The best known example of point-of-contact measurement is the radiation dosimeter. This small badgelike device measures radiation exposure as it occurs and provides an integrated exposure estimate for the period of time over which the measurement has been taken.[12] Another example is the total exposure assessment methodology (TEAM) studies[74] conducted by EPA. In the TEAM studies, a small pump with a collector and an absorbent was attached to a person's clothing to measure his or her exposure to airborne solvents or other pollutants as it occurred, just as has been done in industrial hygiene studies of the past 60 years.[75] In all of these examples, the measurements are taken at the interface between the person and the environment while exposure is occurring.

The strength of this method is that it directly measures exposure. Providing that the measurement devices are accurate, this method likely gives the most accurate exposure value for the period of time over which the measurement was taken. It is often expensive, however, to use these techniques to evaluate persons in the community, and measurement devices and techniques do not currently exist for all chemicals.

4.3.3 Estimates Based on Exposure Scenarios

Using the exposure scenario approach, the assessor attempts to estimate or predict chemical concentrations in a medium or location and link this information with the time that individuals or populations are in contact with the chemical. An exposure scenario is the set of assumptions describing how this contact takes place. This is, by far, the most common approach to exposure assessment, and such an approach is necessary when trying to predict the impact of events that may occur in the future, such as building a new manufacturing facility or introducing a new pesticide or herbicide.[76–80]

The first step to building a scenario is to determine the concentration of the contaminated media. This is typically accomplished indirectly by measuring, modeling, or using existing data on concentrations in the media of concern, rather than at the point of contact (e.g., pesticide residues on food or metal emissions on residential soils). Assuming that the concentration in the bulk medium is the same as the exposure concentration is a source of potential error and must be discussed in the uncertainty analysis. For example, over the past 20 years, most assessments of contaminated soil hazards were based on soil samples collected in the top 6 inches of soil, even though most persons were exposed routinely to the surface soil (usually the top 1/2 to 2 inches). Arguments can be made in either direction about the appropriateness of this assumption.

The next step is to estimate the contact time, identify who is likely to be exposed, and then develop estimates of the exposure frequency and duration. Like chemical concentration characterization, this is usually done indirectly using demographic data, survey statistics, behavior observation, activity diaries, activity models, or, in the absence of more substantive information, assumptions about behavior.[81,82]

Chemical concentration and population characterizations are ultimately combined in an exposure scenario. One of the major problems in evaluating dose equations is that the limiting assumptions used to derive them (e.g., steady-state assumptions) do not always hold true. Two approaches to this problem are available: (1) to evaluate the exposure or dose equation under conditions when the limiting assumptions do hold true or (2) to build a dynamic model that accounts for both accumulation and degradation. The microenvironment method, which is usually used to evaluate air exposures, is an example of the first approach. This method evaluates segments of time and location when the assumption of constant concentration is approximately true, and then sums the time segments to determine the total exposure for the respiratory route, effectively removing some of the uncertainty. In occupational hygiene, this is done by combining time–motion data with short-term air concentration data. While estimates of exposure concentration and time of contact may be estimated in some situations, the concentration and time-of-contact estimates can be measured for each microenvironment. This avoids much of the error due to summing average values in cases where concentration and time of contact vary widely.

In the second approach, a computer model can efficiently predict dose if enough data are available.[20,26,83] When conducting modeling, there are various tools used to describe uncertainty caused by parameter variation, such as Monte Carlo analysis, and these should be applied in most assessments.

4.3.4 Estimating Exposure Using Biological Monitoring

Exposure can often be estimated after it has taken place. If a total dose is known or can be reconstructed, and information about intake and uptake rates is available, an average past exposure rate can be estimated.[70,84–88] Dose reconstruction relies on measuring biological fluids (blood, urine), hair, nails, or feces after exposure and intake and uptake have already occurred, and using these measurements to back-calculate dose.[70] However, data on body burden levels or biomarkers cannot be used directly unless a relationship can be established between these levels (or biomarker indications) and internal dose. Biological tissue or fluid measurements that reveal the presence of a chemical may directly indicate that an exposure has occurred, provided that the chemical is not a metabolite of other chemicals.[89]

Biological monitoring can be used to evaluate the amount of a chemical in the body by measuring one or more of the following parameters (Table 4.1). In general, if these measurements can be made, then past exposure estimates can be quite accurate. Not all of these can be measured for every chemical:[12]

- The concentration of the chemical itself in biological tissues or sera (blood, urine, breath, hair, adipose tissue, etc.)
- The concentration of the chemical's metabolite(s)
- The biological effect that occurs as a result of human exposure to the chemical (e.g., alkylated hemoglobin or changes in enzyme induction)
- The amount of a chemical or its metabolites bound to target molecules.

The results of biomonitoring can be used to estimate chemical uptake during a specific interval, if background levels do not mask the marker and the relationship between uptake and the selected marker are known.[90] The sampling time for biomarkers is often critical. Establishing a correlation between exposure and measurement of the marker, including pharmacokinetics, is necessary to properly back-calculate historical exposure.[12]

The strengths of this method are that it demonstrates exposure to and absorption of the chemical has actually taken place, and theoretically it can give a good indication of past exposure. The drawbacks are that it will not work for every chemical because of interferences or the reactive nature of the chemical, or because the biological half-life of the agent is too short, it has not been methodologically established for very many chemicals, data relating internal dose to exposure are needed, and it may be expensive.

For those chemicals where biological monitoring can be used to estimate past exposure, the information obtained can be invaluable for conducting retrospective exposure assessments that can be used in epidemiology studies. Some examples of chemicals for which past exposure can reliably be estimated include lead, DDT, mercury, chlordane, dioxin, PBB, PCB, and other persistent organics via sampling of the blood and measurement of mercury in hair.[70]

4.4 INFORMATION UPON WHICH EXPOSURE ASSESSMENTS ARE BASED

Comprehensive exposure assessment of a complex scenario may require several hundred exposure factors to estimate the various chemical concentrations in one of several dozen different media. The most complex exposure assessments are those that address the risks posed by airborne emissions from combus-

TABLE 4.1 Types of Biological Sampling Used to Characterize Exposure-Related Media and Parameters[a]

Type of Measurement (sample)	Usually Attempts to Characterize (whole)	Examples	Typical Information Needed to Characterize Exposure
1. Breath	Total internal dose for individuals or population (usually indicative of relatively recent exposures).	Measurement of volatile organic compounds (VOCs), alcohol (usually limited to volatile compounds).	1. Relationship between individuals and population; exposure history (i.e., steady-state or not) pharmacokinetics (chemical half-life), possible storage reservoirs within the body. 2. Relationship between breath content and body burden.
2. Blood	Total internal dose for individuals or population (may be indicative of either relatively recent exposures to fat-soluble organics or long-term body burden for metals).	Lead studies, pesticides, heavy metals (usually best for soluble compounds, although blood lipid analysis may reveal lipophilic compounds).	1. Same as above 2. Relationship between blood content and body burden.
3. Adipose	Total internal dose for individuals or population (usually indicative of long-term averages for fat-soluble organics).	NHATS, dioxin studies, PCBs (usually limited to lipophilic compounds).	1. Same as above 2. Relationship between adipose content and body burden.

| 4. Nails, hair | Total internal dose for individuals or population (usually indicative of past exposure in weeks to months range; can sometimes be used to evaluate exposure patterns). | Heavy metal studies (usually limited to metals). | 1. Same as above
2. Relationship between nails, hair content and body burden. |
| 5. Urine | Total internal dose for individuals or population (usually indicative of elimination rates); time from exposure to appearance in urine may vary, depending on chemical. | Studies of tetrachloroethylene and trichloroethylene. | 1. Same as above
2. Relationship between urine content and body burden. |

[a]From Ref. 81.

tors.[36,82,92] To estimate the concentration, numerous dispersion models, as well as fate and transport models, may be required. In addition, the assessor may need to search the literature to identify relevant studies from as many as 10 related fields of research (Figure 4.4). Sometimes, hundreds of published papers and government guidance documents need to be evaluated, used, and cited. In short, the exercise can be formidable, especially for assessments involving food chain contamination.

4.4.1   Obtaining Data on Intake and Uptake

The numerous editions of the *Exposure Factors Handbook*[25,38,39] present statistical data on many of the factors used in assessing exposure, including intake rates, and provide citations for primary references. Today, this publication represents the most comprehensive, single source of exposure assessment information. Over the past 15 years, the *Exposure Factors Handbook* has grown to nearly 1000 pages of information. Some of the many intake factors presented include:

- Drinking water consumption rates
- Consumption rates for homegrown fruits, vegetables, beef, and dairy products
- Consumption rates for recreationally caught fish and shellfish
- Incidental soil ingestion rates
- Pulmonary ventilation rates
- Surface area of various parts of the human body.

Table 4.2 presents examples of some of the standard or default exposure factors used in risk assessment.

The *Exposure Factors Handbook* is updated routinely to include additional factors and to include new research data on previously discussed factors. It also provides default parameter values, which can be used when site-specific data are not available. Obviously, general default values should not be used in place of known, valid data that are more relevant to the assessment being conducted. The EPA handbook, though thorough, may not contain all available information on exposure factors or relevant studies, so a supplemental literature search should always be conducted to be sure that all pertinent literature has been identified.

4.4.2   Concentration Measurements in Environmental Media

Measured concentration data can be generated for the exposure assessment by conducting a new field study, or by evaluating data from completed field study results and using them to estimate concentrations. Media measurements taken close to the point of contact are preferable to measurements far removed geo-

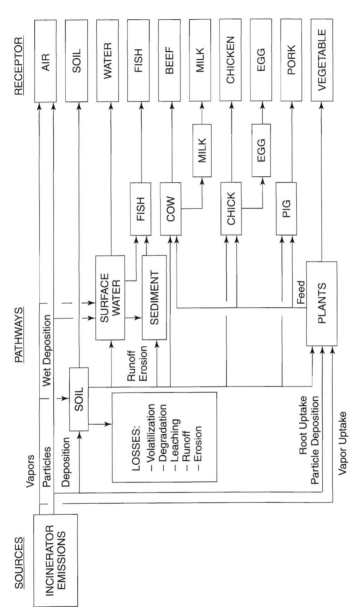

Figure 4.4 EPA's conceptual approach to dealing with direct and indirect exposure pathways as illustrated by assessments of incinerator emissions.

TABLE 4.2   Some Standard Assumptions Used In Regulatory Assessments (EPA, 1996)

Variable	Assumption
Drinking water (ingested)	2 L/day (RME adult)
	1.4 L/day (adult ave.)
	1.0 L/day (child)
	0.1 L/day (incidental ingestion during swimming)
Soil (ingested)	200 mg/day (child ave.)
	800 mg/day (child 90th percentile)
	100 mg/day (adult)
Food	2000 g/day (adult total)
Beef (home grown)	44 g/day (ave.)
100 g/day (all sources)	75 g/day (RME)
Dairy (home grown)	160 g/day (ave.)
400 gm/day (all sources)	300 g/day (RME)
Fruit (home grown)	28 g/day (ave.)
140 g/day (all sources)	42 g/day (RME)
Vegetables (home grown)	50 g/day (ave.)
200 gm/day (all sources)	80 g/day (RME)
Sport fish	30 g/day (ave.)
	140 g/day (RME)
Inhalation	10 m^3/day (ave. 8-h shift.)
	20 m^3/day (adult ave.)
	30 m^3/day (RME)
Body weight	13.2 kg (2–5 yr)
	20.8 kg (6 yr)
	70 kg (adult ave.)
Life span	70 yr
Exposed skin area	0.2 m^2 (adult ave.)
	0.53 m^2 (adult RME)
	1.94 m^2 (male bathing)
	1.69 m^2 (female bathing)
Showering	7 min (ave.)
(5 min. shower uses 40 gals)	12 min. (90th percentile)
Residence time	9 yr (ave.)
	30 yr (RME)

Note: RME, reasonable maximum exposure; ave., average.

graphically or temporally. As the distance from the point of contact increases, the certainty of the data at the point of contact usually decreases, and the obligation for the assessor to show relevance of the data to the assessment at hand becomes greater. For example, an outdoor air measurement, no matter how close it is taken to the point of contact, cannot by itself adequately characterize indoor exposure.[93]

Concentrations often vary considerably from place to place, seasonally and over time due to changing emission and use patterns.[27,29,94] This needs to be considered not only when designing studies to collect new data, but also when evaluating the applicability of existing measurements as estimates of exposure concentrations in a new assessment. It is of particular concern when the measurement data will be used to extrapolate to long time periods, such as lifetimes. Transport and dispersion models are frequently used to help answer these questions.[95]

The exposure assessor is likely to encounter several different types of measurements. One type of measurement used for general indications and concentration trends is outdoor fixed-location monitoring. This measurement is used by EPA and other groups to provide a record of pollutant concentration at one place over time. Nationwide air and water monitoring programs have been established so that baseline values in these environmental media can be documented. Although it is not practical to set up a national monitoring network to gather data for a particular exposure assessment, data from existing networks can be evaluated for relevance to an exposure assessment. These data are often far removed from the point of contact. Adapting data from previous studies usually presents challenges similar to those encountered when using network data. If new data are needed for the assessment, studies can be conducted that measure specific chemicals at specific locations and times.

Indoor air contaminant concentrations can vary as much or more than those in outdoor air.[10] Consequently, indoor exposure is best represented by measurements taken at the point of contact. However, because pollutants such as carbon monoxide can exhibit substantial indoor penetration, indoor exposure estimates should consider potential outdoor, as well as indoor, sources of the contaminant(s) under evaluation.[12]

Food and drinking water measurements can also be made. General characterization of these media, such as market basket studies (where representative diets are characterized), shelf studies (where foodstuffs are taken from store shelves and analyzed), or drinking water quality surveys, are usually far removed from the point of contact for an individual, but may be useful in evaluating exposure concentrations over a large population. Measurements of tap water or foodstuffs in a home and how they are used are closer to the point of contact. In evaluating the relevance of data from previous studies, variation in the distribution systems must be considered, as well as the space–time proximity.[12]

Consumer or industrial product analysis is sometimes done to characterize the chemical concentrations in products. Product formulations can change substantially over time, similar products do not necessarily have similar formulations, and regional differences in product formulation can also occur. These should be considered when determining the relevance of extant data and when setting up sampling plans to gather new data.[12]

Another type of concentration measurement is the microenvironmental measurement. Rather than using measurements to characterize the entire medium, this approach defines specific zones in which the concentration in the medium of interest is thought to be relatively homogenous and then characterizes the concentration in that zone.[12,96] Typical microenvironments include the home or parts of the home, office, automobile, or other indoor settings. Microenvironments can also be divided into time segments (e.g., kitchen during the day, kitchen during the night). This approach can produce measurements that are closely linked with the point of contact, both in location and time, especially when new data are generated for a particular exposure assessment. The more specific the microenvironment, however, the greater the burden on the exposure assessor to establish that the measurements are representative of the population of interest.

The concentration measurement that provides the closest link to the actual point of contact is personal monitoring. In virtually all cases, this information should be the basis of exposure assessments whenever possible (due to data quality requirements).

4.4.3   Models and Their Role

Often the most critical assessment element is estimating pollutant concentrations at exposure points. This is usually carried out by combining field data and modeling results. In the absence of field data, this process often relies on the results of mathematical models such as ISCLT, MODFLOW, etc.[97–101] EPA's Science Advisory Board and others have recommended that modeling ideally should be linked with monitoring data in regulatory assessments, although this is not always possible.

A modeling strategy has several aspects, including setting objectives, model selection, obtaining and installing the code, calibrating and running the computer model, and validation and verification. Many of these aspects are analogous to the quality assurance and quality control measures applied to measurements.

Regardless of whether models are extensively used in an assessment or whether a formal modeling strategy is documented in the exposure assessment plan, when computer simulation models such as fate and transport models and exposure models are used in exposure assessments, the assessor must be aware of the performance characteristics of the model and state how the exposure assessment requirements are satisfied by the model.[83]

The site must be characterized if models are to be used to simulate pollutant behavior at a specific site. Site characterization for any modeling study includes examining all data on the site, such as source characterization, dimensions and topography of the site, location of receptor populations, meteorology, soils, geohydrology, and ranges and distributions of chemical concentrations. For exposure models that simulate both chemical concentration and time of exposure (through behavior patterns), data on these two parameters must be evaluated.[12,16,17]

Criteria are provided by EPA[98] for selecting surface water models and groundwater models, respectively; the reader is referred to these documents for details. Similar selection criteria exist for air dispersion models.[102]

A primary consideration in selecting a model is whether to perform a screening study or a detailed evaluation. A screening study makes a preliminary evaluation of a site or a general comparison between several sites. It may be generic to a type of site (i.e., an industrial segment or a climatic region) or may pertain to a specific site for which sufficient data are not available to properly characterize the site. Screening studies can help direct data collection at the site by, for example, providing an indication of the level of detection and quantification that would be required and the distances and directions from a point of release where chemical concentrations might be expected to be highest.

An example of a screening-level modeling effort would be to estimate the amount of lead deposited by an incinerator onto local crops using a basic air dispersion model, without considering local geographical or weather conditions. The next level of complexity would consider the presence of mountains, their proximity to the stack, the local weather patterns, and the number of atmospheric inversions per year. The next level of analysis could incorporate yet other, more subtle factors.

The value of the screening-level analysis is that it is simple to perform and may indicate that no significant contamination exists. Screening-level models are frequently used to get a first approximation of the concentrations that may be present. Often these models use very conservative assumptions; that is, they tend to overpredict concentrations or exposures. If the results of the conservative screening procedure predict concentrations or exposures at less than a predetermined no-concern level, then more detailed analysis is probably not necessary. If the screening estimates are above that level, refinement of the assumptions or a more sophisticated model are necessary to generate a more realistic estimate.[12]

Screening-level models also help the user conceptualize the physical system, identify important processes, and locate available data. The assumptions used in the preliminary analysis should represent conservative conditions, such that the predicted results overestimate potential conditions, limiting false negatives. If the limited field measurements or screening analyses indicate that a contamination problem may exist, then a detailed modeling study may be useful.

In contrast, the purpose of the detailed evaluation is to use the best data available to make the best estimate of spatial and temporal chemical distributions of a specific site. Detailed studies typically require higher quality data and more sophisticated models.

4.4.4 Accounting for Background Concentrations

Background exposure to xenobiotics, especially environmentally persistent ones, can occur due to natural or anthropogenic sources. In most exposure assessments, background soil concentrations are the focus of attention, but the same issue can be relevant when evaluating sediments, ambient air, groundwater, and vegetation (food stuffs). At some sites, it is important that these so-called background concentrations be accounted for because removing the quantity of toxicant due to humans may, in fact, not appreciably change the concentrations or be sufficient to reduce the risk to acceptable levels. For example, naturally occurring concentrations of lead, arsenic, and cadmium, in some locations, may be higher than cleanup levels established by various regulatory agencies.[102] The exposure assessor should try to determine local background concentrations by gathering data from nearby locations clearly unaffected by the site under investigation or by referring to published works that have assessed this issue.

4.4.5 Description of Background Levels

When assessing soils, background levels can be viewed in at least four different ways:[100]

- *"Pristine" Levels.* Some would like to equate background levels with those associated with the "pristine" state, that is, soils or landscapes unaffected by human activity. This rather idealistic situation probably no longer exists; even in Antarctica, mercury and dioxin concentrations can be detected in some media. Toxic elements mainly associated with the solid phase of some natural material (such as soil dust, plant or volcanic ash, vegetable matter) are relatively mobile in a global sense. For example, Nriagu[103] has suggested that about 40 million tons of heavy metals have been dispersed atmospherically over the many centuries of human activity.

  Increases in pollutant metal concentrations have been measured up to 60 km from smelters, and automotive lead (fine particles) has been measured in soils and rainfall up to about 50 km downwind from major cities. Soil contamination up to 50 to 100 m from highways by automotive lead (coarser particles) is an example of short-range transport, and contrasts with transport of toxic metals on a continental or global scale (e.g., contamination of the Greenland ice sheets from the northern United States, parts of northern Norway from central Germany and the United Kingdom, and the snows of the New Zealand alps by soil dust from inland Australia).[100]

- *"Normal" Levels.* The question could be asked—are soils contaminated at farms in the higher rainfall areas of the Appalachians, which used to receive automotive exhaust particulates from the New York metropolitan area 200 km away. These soils are not pristine, but the chemical concentrations are perfectly safe for growing food, raising farm animals, and residential living. Soils from such areas would have a range of what is often called "normal" background values. To most exposure assessors, this mosaic of normal soils, which is only affected by the minor pollution of everyday activities associated with modern rural and urban life, should be the basis for defining background values. Statistically, this range of normal background values would constitute a single lognormally distributed population. Obviously, one needs to exclude the outliers, or "hot" spots, due to a geochemical anomaly, or localized pollution arising from either industrial emissions, disposal of waste products, or intensive (excessive) use of farm chemicals.[100]

- *Historically Polluted Regions.* Local community and regulatory policies often affect what are defined as background levels. A community with highly developed environmental consciousness may insist on very low, possibly unreasonable, reference values. Some densely populated areas with historically derived pollution, perhaps from former mining activities, may sustain apparently healthy populations who pragmatically must accept higher "background" level values. The cities of Philadelphia, Baltimore, and New York, and parts of Japan, for example, may fit in this category.

- *Geochemical Variation.* Background levels of some potentially toxic elements may vary among geographical regions because of differences in soil type. The resulting concentrations are often called naturally occurring levels. An important factor is the composition of rocks and sediments that weather from soils. Some extreme examples are high concentrations of nickel, cobalt, and chromium in igneous rocks such as basalts that cover extensive areas in western Victoria and Tasmania, Australia; and high concentrations of boron in soils on marine sediments in the Riverina, Mallee, and Wimmera districts of South Australia; and Victoria, Eyre Peninsula, and parts of western Australia.

Some regulatory agencies have provided written guidance describing how to select soil or sediment cleanup values that account for background chemical concentrations. In past years, these have varied significantly, but within the past 5 years, there appears to be some convergence regarding the definition of background, how to measure it, and how it should affect exposure assessment calculations.

4.5 ESTIMATING UPTAKE VIA THE SKIN

When attempting to predict chemical risks in the environmental or occupational setting, the dermal exposure route should nearly always be assessed. In most evaluations of hazardous waste sites and ambient air or water contaminants, this is not a major route of exposure. Although the uptake of chemicals via the skin has generally been overlooked in most workplace exposure assessments, it probably represents a substantial portion of the exposure for many occupations. Even though gloves are more frequently used than in years past and training has increased on the possible hazards of dermal exposure, there is still ample evidence to indicate that, in order to conduct a complete exposure assessment, this route deserves attention.[37,50,105]

In addition to the risks associated with systemic toxicity due to uptake via the skin, it is sometimes necessary to evaluate the allergic contact dermatitis (ACD) hazard. In recent years, techniques have been developed to quantitatively predict the likelihood of illication and induction of ACD.[106] Some regulatory agencies are concerned with ACD and have developed cleanup standards based on this health endpoint.

In the workplace, a worker's skin frequently comes into contact with solvents or chemicals mixed in water (aqueous materials). In most environmental settings where persons can be exposed to contaminated soil or contaminated water, dermal update must be assessed. Fortunately, a good deal of research has been conducted to understand the rate at which chemicals pass through the skin. Percutaneous absorption of neat chemicals (e.g., the pure liquid) had often been studied in humans until the late 1970s.[56,107–113] Because of the potential toxicity of many chemicals and improved laboratory techniques, in vivo human studies have been largely supplanted by experiments with laboratory animals or athymic rodents grafted with human skin.[114] Historical research has shown that, in general, chemical penetration of the human skin is similar to that of a pig or monkey and much slower than that of the rat and rabbit.[115] Thus, for many chemicals, there is some level of confidence that the rate of dermal uptake of a chemical by humans can be inferred from animal data.

Starting in the 1980s, in vitro studies using human skin began to be conducted on a more routine basis. In these studies, a piece of excised skin is attached to a diffusion apparatus with a top chamber to hold the applied chemical and a temperature-controlled bottom chamber containing saline or other fluids (plus a sampling port to withdraw fractions for analysis).[116] Although human forearm skin is optimal, it is difficult to obtain, so abdominal or breast skin is commonly used. Generally, a properly conducted in vitro test can be a reasonably good predictor of the absorption rate in vivo.[27,117] However, due to the fragile nature of the technique, these studies must be carefully interpreted.[118] Often, depending on the conditions of the test, the results are not applicable to humans.

Aside from neat liquids exposure to contaminated water, dermal exposures can also occur through contact with dust or dirt on surfaces, and by way of contact with soil or dust-bound contaminants.[93] Few studies[119–127] have directly estimated soil loading on human skin, and only one of them attempted to measure dermal contact of contaminated equipment by workers.[128] The available studies probably provide sufficient data to generate point estimates of soil adherence and, perhaps, can provide a reasonable probability density function (PDF) for most persons exposed to contaminated soils. The degree of representativeness of the data to the general population is difficult to assess.[129]

Recently, a study measured the adherence of soil to multiple skin surfaces (hands, forearms, lower legs, faces, and feet) under ambient and recreational conditions.[125,126] Dermal loading on the hands was found to vary over five orders of magnitude and to be dependent on the type of activity. Differences between pre- and postactivity adherence demonstrated the episodic nature of dermal contact with soil. However, due to the activity-dependent nature of soil exposure, data from these studies must be interpreted for their relevance to the type of activity, frequency, duration, and otherwise site-specific nature of exposure. The various studies involving contaminated soil are informative for providing an estimate of exposure; however, they are probably a couple orders of magnitude greater than what might be expected in a chemical plant. Nonetheless, this work is a "starting point" for bracketing exposure.

Recently, there has been a reasonable level of research investigating exposure to house dust. The basis for this concern has been increasing evidence that controlling exposure to house dust, especially in homes located near sites with considerable surface soil contamination, is more important for reducing the health hazard than remediating the soil.[93] Of particular interest is the recent work to develop standardized approaches for collecting wipe samples and estimating the amount of dust loading on the palm of the hand.[130,131] Although dermal absorption of toxicants in house dust will almost always pose a relatively low dermal uptake hazard, the uptake due to hand-to-mouth contact can be substantial.

4.5.1   Quantitative Description of Dermal Absorption

For the purposes of risk assessment, percutaneous absorption is defined as transport of externally applied chemicals through cutaneous structures and the extracellular medium to the bloodstream. The simplest way to describe the rate of skin absorption is to apply Fick's first law of diffusion at steady-state:[132,133]

$$J = dQ/dt = Dk\nabla C/e \approx K_p C \qquad (4.3)$$

where $J = dQ/dt$ = chemical flux or rate of chemical absorbed (mg/cm²-h)

D = diffusivity in the stratum corneum (cm²/h)

K = stratum corneum/vehicle partition coefficient of the chemical (unitless)

∇C = concentration gradient (mg/cm³)

e = thickness of the stratum corneum (cm)

K_p = permeability coefficient (cm/h)

C = applied chemical concentration (mg/cm³)

The concentration gradient is equal to the difference between the concentration above and below the stratum corneum. Because the concentration below is small compared to the concentration above, ∇C can be approximated as equal to the applied chemical concentration. From the above equation, it can be seen that the rate of absorption is directly proportional to the applied concentration. The diffusivity represents the rate of migration of the chemical through the stratum corneum. Since the stratum corneum has a non-negligible thickness, there is a period of transient diffusion (lag time), during which the transfer rate rises to reach a steady state. In these studies, the steady state is maintained indefinitely, provided the system remains constant. Depending on the type of chemical, the lag time can range from minutes to days.[50] From an exposure assessment standpoint, if the exposure duration is shorter than the lag time, it is unlikely that there will be any significant systemic absorption.[133,134]

The partition coefficient (K_p) is one of the key parameters that influences the degree to which a chemical penetrates the skin.[133-137] Fatty chemicals tend to accumulate in the stratum corneum. Conversely, the stratum corneum is an effective barrier for hydrophilic substances, which tend to have low skin absorption rates. Because stratum corneum/vehicle partition coefficients are difficult to measure, the three parameters (D, k, and e) are combined to give an overall permeability coefficient (K_p). It is noteworthy that Eq. (4.3) only approximates most in vivo exposure situations in which true steady-state conditions are rarely attained. In spite of its limitations, this equation has yielded satisfactory estimations of the actual absorption rates of chemicals for many situations (Table 4.3).

4.5.2 Pharmacokinetic Models for Estimating the Uptake of Chemicals in Aqueous Solution

Pharmacokinetic models predict the uptake of a chemical through the skin based on fundamental thermodynamics. Several different models have been proposed. For example, a four-compartment pharmacokinetic model was developed in 1982.[135] This model, which uses first-order rate constants, describes chemical movement through the compartments representing the various skin structures. It has been used successfully to predict the chemical disposition in the skin and plasma as a function of their physicochemical properties, and when an input rate constant to the skin surface is added to the model, it can be used to assess vehicle effects. A similar model that treats the barrier membrane as a series of spaces filled with immiscible liquids has also been developed;[136] its advantage is that it allows examination of non-steady-state conditions where Fick's law does not apply.

Under an infinite-dose situation where the amount of a chemical lost by penetration is too small to alter the applied concentration (e.g., where one is swimming), the rate of absorption is essentially linear once steady-state has been reached. In the finite-dose system, however, the chemical solution is applied as a thin film and the concentration decreases as penetration proceeds (e.g., a splash). All other model parameters being the same, penetration is reduced under finite-dose conditions. This is because the chemical concentration is continuously reduced over time, resulting in a decrease in the gradient across the stratum corneum. These modeling results indicate that the mechanism by which fluxes are affected must be considered when extrapolating to non-steady-state conditions.

Although classic pharmacokinetic modeling like that described by Guy et al.[135] can provide a good mathematical description of the disposition of chemicals, it does not depict exactly the biological processes in the intact animal. Fortunately, due to recent improvements in available computer hardware and software, pharmacokinetic methods based on physiological principles are now feasible alternatives for analysis of in vivo skin penetration studies. These so-called PB-PK models realistically describe the disposition of the chemical in the intact animal in terms of rates of blood flow, permeability of membranes, and partitioning of chemicals into tissues.[71,136] Characterizing dermal absorption in terms of actual anatomical, physiological, and biochemical parameters facilitates extrapolations to the real species of interest, humans.

In 1991, a PB-PK model was developed to describe percutaneous absorption

TABLE 4.3 Human Cutaneous Permeability Coefficient Values for Some Industrial Chemicals in Aqueous Medium

	MW	K_{ow}	Observed	Calculated[a]
Organic chemicals				
2-Amino-4-nitrophenol	154.13	21.38	0.00066	0.019
4-Amino-2-nitrophenol	154.13	9.12	0.0028	0.0081
Aniline	93.12	7.94	0.041[b]	0.091
Benzene	78.11	134.90	0.11	0.39
p-Bromophenol	173.02	389.05	0.036	0.25
Butane-2,3-diol	90.12	0.12	<0.00005	0.0009
n-Butanol	74.12	7.59	0.0025	0.024
2-Butanone	72.10	1.94	0.0045	0.007
Carbon disulfide	76.14	100.	0.54[b]	0.3
Chlorocresol	142.58	1258.93	0.055	1.31
S-Chlorophenal	128.56	147.91	0.033	0.19
p-Chlorophenal	128.56	257.04	0.036	0.34
Chloroxylenal	156.61	1621.81	0.059	1.35
m-Cresol	108.13	100.	0.015	0.18
o-Cresol	108.13	100.	0.016	0.18
p-Cresol	108.13	85.11	0.018	0.15
Decanol	158.28	37153.52	0.08	30.11
2,4-Dichlorophenol	163.01	1995.26	0.06	1.5
1,4-Dioxane	88.10	0.38	0.00043	0.0016
Ethanol	46.07	0.49	0.0008	0.0036
2-Ethoxyethanol	90.12	0.29	0.0003	0.0013
Ethylbenzene	106.16	1412.54	1.215[b]	2.65
Ethylether	74.12	6.76	0.016	0.022
p-Ethylphenol	122.17	549.54	0.035	0.79
Heptanol	116.20	257.04	0.038	0.41
Hexanol	102.17	107.15	0.028	0.21
Methanol	32.04	0.17	0.0016	0.0026
Methyl hydroxybenzoate	152.15	91.2	0.0091	0.082
β-Naphthol	144.16	691.83	0.028	0.7
3-Nitrophenol	139.11	100.00	0.0056	0.11
4-Nitrophenol	139.11	81.28	0.0056	0.09
Nitrosodiethanolamine	134.13	0.13	0.0000055	0.0005
Nonanol	144.26	2951.21	0.06	2.99
Octanol	130.22	933.25	0.061	1.19
Pentanol	88.15	36.31	0.006	0.091
Phenol	94.11	32.36	0.0082	0.074
Propanol	60.09	2.00	0.0017	0.0088
Resorcinol	110.11	6.03	0.00024	0.011
Styrene	104.14	891.25	0.635[b]	1.72
Thymol	150.21	1995.26	0.053	1.84
Toluene	92.13	489.78	1.01	1.15
2,4,6-Trichlorophenol	197.46	2344.23	0.059	1.02
3,4-Xylenol	122.16	169.82	0.036	0.25

TABLE 4.3 *(Continued)*

	MW	K_{ow}	Observed	Calculated[a]
Inorganic chemicals				
Cobalt chloride	129.84		0.0004	
Lead acetate	325.29		0.0000042[b]	
Mercuric chloride	271.50		0.00093	
Nickel chloride	129.60		0.001	
Nickel sulfate	154.75		<0.000009	
Silver nitrate	169.87		<0.00035[b]	
Sodium chromate	161.97		0.0021[b]	

Note: From Leung and Paustenbach (1994).
[a] Permeability coefficients calculated using equation presented in Leung and Paustenbach (1994).
[b] All the observed permeability coefficients were obtained by using in vitro techniques except those denoted with superscript b, which were determined in vivo.

of volatile organic contaminants in dilute aqueous solutions.[138] The exposure scenario modeled was either hand or full-body immersion into a vessel of solute-contaminated water. Modeling results suggested that chemical uptake in aqueous solutions is most markedly influenced by epidermal blood flow rates, followed by epidermal thickness and lipid content of the stratum corneum. In general, thicker and fattier skin provides a better barrier to dermal penetration of chemicals. These are precisely the principles under which barrier creams offer their protection for increasing the effective thickness and lipophilicity of the skin. This model also predicted that the dose of some volatile organic chemicals in water absorbed through the skin during a 20-minute bath may be equivalent to the amount inhaled.[138]

Among the most complex and best validated of the various models for dermal uptake of liquids is that developed by McDougal et al.[139] These authors have successfully predicted dermal uptake rates for humans for nearly a dozen chemicals based on animal and physical chemistry data. One advantage of dermal PB-PK models over traditional in vivo methods is their ability to accurately describe nonlinear biochemical and physical processes. For example, describing skin penetration based on blood concentrations or excretion rates as "percent absorbed" assumes that all processes have a simple linear relationship with the exposure concentration. This is often not the case. The kinetics become nonlinear when the absorption, distribution, metabolism, or elimination of a chemical is saturated at high exposure concentrations. This model and models developed since then address this phenomenon in a reasonable manner.

4.5.3 Factors Used to Estimate Dermal Uptake

At least 10 different factors need to be quantitatively accounted for to estimate the likely systemic uptake of a chemical that comes into contact with the skin, either as a liquid or when present in soil or dust.[6,140]

Bioavailability   The typical media of concern for assessing cutaneous contact to environmental chemicals, in contrast with occupational exposure, are house dust, soil, fly ash, and sediment. In the workplace, dermal uptake is due to direct contact with liquids and contact with surfaces contaminated with dirt or liquids. A number of parameters can influence the degree of cutaneous bioavailability of chemicals in complex matrices. These may include aging (time following contamination), soil type (e.g., silt, clay, and sand), type and concentration of co-contaminants (e.g., oil and other organics), and the concentration of the chemical contaminant in the media.[57] The bioavailability of a chemical in soil will usually be affected by its physicochemical properties. Large molecular weight chemicals tend to bind to soil/dust and be less water soluble, while smaller molecules will frequently be water soluble, less tightly bound, and relatively bioavailable.[54,58] The cutaneous bioavailability of perhaps 20 to 30 chemicals in soils has been determined in animals.[57,58,141–145] These studies show that different media and different chemicals can yield dramatically different cutaneous bioavailabilities. The results of these studies, for example, produce values of bioavailability for different chemicals that range from 0.01 to 3% for chemicals in soil.

Skin Surface Area   There is an abundance of information about the surface area of different portions of the body. One simple approach is to use the "rule of nines" for estimating the surface area of the human body:[146] the head and neck are 9 percent, upper limbs are each 9 percent, lower limbs are each 18 percent, and the front or back of the trunk is 18 percent.[2] EPA has estimated an exposed surface area (arms, hands, legs, and feet) of 2900 cm^2 for children 0 to 2 years old, 3400 cm^2 for children 2 to 6 years old, and 2940 cm^2 for adults (an adult is assumed to wear pants, an open-neck short-sleeve shirt, shoes, and no hat or gloves).[12] When assessing chemical exposure in the ambient environment, most of the necessary surface area information can be found in EPA's *Exposure Factors Handbook*.[81] Table 4.4 presents the skin surface areas commonly used when conducting exposure assessments.[146]

Soil Loading on the Skin   A key factor to consider when estimating dermal uptake is the soil-to-skin adherence rate. Values of 0.5 to 0.6 mg/cm^2 and 0.2 to 2.8 mg/cm^2 have been reported for adults and children, respectively.[29,121,147,148] Recent works by Finley et al.,[123] Kissel et al.,[125,126] and Holmes have built on prior studies to show that dermal loading can vary significantly among different activities and different people. Based on data collected in past studies, in 1992 EPA suggested a default soil-to-skin adherence rate of 0.2 mg/cm^2 (median) and 1.0 mg/cm^2 (95th percentile) for an adult. The recent issue of the *Exposure Factors Handbook* gives considerable attention to this topic.[25,39] One approach to improving dermal uptake calculations is to use area-weighted adherence factors as recently suggested by EPA.

TABLE 4.4 Representative Surface Areas of the Human Body (Adult Male)

Body portion	Area (cm^2)
Whole body	18,000
Head and neck	1,620
Head	1,260
Back of head	320
Neck	360
Back of neck	90
Torso	6,480
Back	2,520
Chest	2,520
Sides	1,440
Upper limbs	3,240
Upper arms (elbow–shoulder)	1,440
Lower arms (elbow–wrist)	1,080
Hands	720
Palms	360
Upper arms (back of)	360
Lower arms (back of)	270
Lower limbs	6,480
Thighs	3,240
Lower legs (knee–ankle)	2,160
Feet	1,080
Soles of feet	540
Thighs (back of)	810
Lower legs (back of)	540
Perineum	180

Note: Data adapted from Snyder (1975).

4.5.4 Interpreting Wipe Samples

In some workplaces, wipe sampling has been conducted historically to assess the degree of surface contamination. Hospitals were among the first occupational settings, as long ago as 1940, to rely on this method to determine microbial levels in operating rooms. In pharmaceutical manufacturing, wipe sampling has been used as an indicator of hygienic conditions since the 1960s. The health physics profession has utilized wipe samples extensively as an indicator of the need for better housekeeping and decontamination; this group performed most of the early work in quantifying the relationship of wipe sample concentrations to dermal and oral uptake.

Over the years, few studies have discussed how to collect and interpret wipe samples.[149–152,154,155] When the primary effect of a chemical is skin discoloration, ACD or chloracne wipe sampling was nearly always the preferred approach for assessing the acceptability of the workplace (rather than relying on air samples). Beginning in the 1980s, a substantial number of wipe samples were collected in office buildings contaminated with dioxins and furans after electrical transformer fires to estimate the potential human exposure.[156] The interpretation of these data was often mishandled and, as a result, a number of decisions by risk managers were less than optimal, resulting in significant unnecessary expenses.

Although wipe sampling data have generally been used as an indicator of cleanliness,[151] these data can also be used to estimate systemic uptake of a contaminant if the degree of skin contact with the contaminated surfaces is known. While historical wipe sampling methods were rather imprecise, they were useful for obtaining a rough estimate of the possible exposure, which could be refined later by other means, such as biological monitoring. Again, such approaches have been found useful by those in the nuclear industry.

If one knows that wipe sampling results are representative of what comes into contact with the hands (i.e., actually able to be absorbed), then the procedures for converting wipe sample data to estimates of systemic uptake are straightforward. For example, if one knows the number of times a surface (e.g., valve handle, instrument controller, or drum) is touched, the surface area of the hand touching these items (usually the palm), and the percutaneous chemical absorption rate, then the uptake can be estimated using wipe sample information. One of the most thorough evaluations relying on wipe sample data and time–motion studies was conducted by the National Institute for Occupational Safety and Health (NIOSH) to examine the amount of dioxin absorbed by chemical operators in a 2,4,5-T manufacturing plant. Regrettably, the study was never published.[128] In general, most of the historical wipe sample data within the chemical industry over the past 40 years has not been shared in the literature, nor has it been used in retrospective risk assessments. Perhaps one of the most robust data sets within the chemical industry was collected by Dow Chemical and Monsanto during the 1950s, when they were trying to understand the cause of chloracne, which had been observed in chemical workers.

Until recently, no standardized approaches existed for conducting wipe sampling. Differences in the use of wetting agent (acetone, methylene chloride, water, saline, isopropanol, and ethanol) and sampling media (paper, cotton, and synthetic fibers) produced drastically different results. In some procedures, especially those that used methylene chloride (in which the paint was concurrently stripped by the solvent), the chemical in the paint matrix was assumed to be bioavailable (a completely unreasonable assumption). Clearly, much of the previous work, which measured the amount of chemical released following aggressive scrubbing of the contaminated surfaces with detergent or solvent, did not reflect a realistic exposure scenario. Thus, there has been a need for standard techniques that attempt to mimic the conditions in which a hand comes into contact with a contaminated surface.[152] Virtually all of the best work to date has been conducted by hygienists involved in agricultural exposure assessment.[157–161]

In an attempt to fill this need, fairly sophisticated work to standardize these procedures has been conducted by researchers at Rutgers University. At least two of their wipe sampling procedures and devices have been patented.[130] They have also developed a dry contact sampling device[131] that offers promise for understanding the hazard from surface dusts. The implications of the recent wipe sampling research are that: (1) a minimum number of samples is needed to have statistical confidence; (2) the pressure applied to the cloth during sample collection should be standardized; (3) neat solvent should not be used as a collection media; (4) the size of the sample area needs to be sufficient to collect enough contaminant for quantification; and (5) the technique should be validated by using glove analyses.

Example Calculation 3: Using Wipe Sample Data to Estimate Skin Uptake
Wipe samples have been collected in office buildings where electrical transformers containing PCBs have caught fire and the smoke has been distributed throughout the building in the ventilation system. What dose of dioxin due to skin contact with contaminated building surfaces might be possible if the dioxin concentration in the wipe samples is 10 pg/cm^2 [use Eq. (4.4)]?

$$\text{Uptake} = (C)(A)(r)(B) \tag{4.4}$$

In applying this equation to wipe sampling:

C = concentration of chemical in contaminated surface (mg/cm^2)

A = surface area of one palm (cm^2)

R = removal efficiency of chemical from contaminated surface by skin (unitless)

B = dermal bioavailability (unitless)

The chemical concentration in the wipe samples (C_{wipe}) measures the surface contamination and is related to removal efficiency (R_{wipe}) of the particular wiping procedure (i.e., $C = C_{wipe}/R_{wipe}$). Thus,

$$\text{Uptake} = (C_{wipe})(A)(r)(B)/R_{wipe}$$

where C_{wipe} = 10 pg/cm^2

A = 180 cm^2

r = 10%

B = 1%

R_{wipe} = 50%

Using substitution:

$$\text{Uptake} = (10\ \text{pg/cm}^2)(180\ \text{cm}^2)(0.10)(0.01)/(0.50)$$
$$= 3.6\ \text{pg}$$

4.5.5 Estimating the Dermal Uptake of Chemicals in Soil

One of the most frequently occurring exposure scenarios involving environmental exposures is that of contaminated soil.[29] Unfortunately, dermal uptake of chemicals found on soil has infrequently been evaluated experimentally.[50] A model to estimate the amount of a chemical in soil that crosses the stratum corneum into the underlying tissue layer has been developed.[54] To differentiate this absorptive process from bioavailability, which also includes transport into blood, McKone refers to the percentage of available chemical as an uptake fraction. The approach is based on the fugacity concept, which measures the tendency of a chemical to move from one phase to another. Because the skin has a fat content of about 10 percent and soil has an organic carbon content of 1 to 4 percent, a chemical in soil placed on the skin will move from the soil to the underlying adipose layers of the skin. However, this transfer depends on the period of time between deposition on the skin and removal by evaporative processes. The mass-transfer coefficients of the soil-to-skin layer and the soil-to-air layer define the rate at which these competing processes occur.

Results of this model suggest that the chemical uptake fraction in soil varies with the exposure duration, soil deposition rate, and physical properties of the chemical and is particularly sensitive to the values of the K_{ow}, as well as the mass or depth of soil deposited on the skin. When the amount of soil on the skin is low ($<1\ \text{mg/cm}^2$), a high uptake fraction, approaching unity in some cases, is predicted. With higher soil loading ($20\ \text{mg/cm}^2$), an uptake of only 0.5 percent is predicted. Because of the diverse variations of the uptake fraction with soil loading, results obtained from experiments with a single soil loading should be applied with caution to human soil exposure scenarios.

The dermal uptake of chemicals in soil is a complex process, but its behavior is predictable if the controlling factors are accounted for and quantified.[50,54] In situations involving a relatively thin layer of a chemical on the skin, a few generalizations can be made. First, for chemicals with a high K_{ow} and a low air:water partition coefficient, it is reasonable to assume 100 percent uptake in 12 hours. Second, for chemicals with an air:water partition coefficient greater than 0.01, the uptake fraction is unlikely to exceed 40 percent in 12 hours. Third, for chemicals with an air:water partition coefficient greater than 0.1, one can expect less than 3 percent uptake in 12 hours. In most occupational settings, contaminated soil will rarely be in contact with the skin for greater than 4 hours before it is washed off. Consequently, this should be accounted for when attempting to predict systemic uptake.

4.5.6 Dermal Uptake of Contaminants in Soil

To estimate chemical uptake, one needs to know the percutaneous absorption rate, the exposed skin area, the chemical concentration, and the exposure duration.[104] One scenario would be a thin film of chemical on the skin. For this finite-dose scenario, Eq. (4.5) is useful:

$$\text{Uptake (mg)} = (C)(A)(x)(f)(t) \tag{4.5}$$

where C = concentration of the chemical (mg/cm^2)
   A = skin surface area (cm^2)
   x = thickness of the film layer (cm)
   f = absorption rate (percent per hour)
   t = duration of exposure (hour)

Another scenario would be an excess amount of a chemical on the skin (i.e., infinite dose). In this case, the thickness of the chemical layer is not calculated and steady-state kinetics are assumed. For a chemical in an aqueous or gaseous media:

$$\text{Uptake (mg)} = (C)(A)(K_p)(t)(d) \tag{4.6}$$

where d = distribution factor

For a neat liquid chemical:

$$\text{Uptake (mg)} = (A)(J)(t) \tag{4.7}$$

where K_p = permeability coefficient (cm/h)
   J = flux of chemical $(\text{mg/cm}^2/\text{h})$

EPA has suggested using the following equation for estimating percutaneous absorption of chemicals in soil:[140]

$$\text{Uptake (mg)} = (C)(A)(r)(B) \tag{4.8}$$

where C = concentration of the chemical in soil (mg/g)
   A = skin surface area (cm^2)
   r = soil-to-skin adherence rate (g/cm^2)
   B = cutaneous bioavailability (unitless)

Example Calculation 4: Skin Uptake of a Chemical in Soil A person gardens with soil contaminated on average with 250 ng dioxin per gram of soil (250 ppb). Assuming that the person's hands and lower arms are in contact with the soil, the soil loading is equal to 0.2 mg/cm^2, and the cutaneous bioavailability of dioxin in soil is 1 percent,[57] what is the plausible uptake of dioxin by this person [using Eq. (4.8)]? Assume that the person washes his or her hands every 4 hours and the exposed area of skin is 1800 cm^2.

$$\text{Uptake (ng)} = (C)(A)(r)(B)$$

where $C = 250$ ng/g
$\quad A = 1800$ cm^2
$\quad r = 0.2$ mg/cm^2
$\quad B = 0.01$

By substitution:

$$\text{Uptake} = \left(\frac{250 \text{ ng TCDD}}{1 \text{ g soil}}\right)\left(\frac{0.2 \text{ mg soil}}{\text{cm}^2 \text{ skin}}\right)\left(\frac{1 \text{ g}}{10^3 \text{ mg}}\right)(1800 \text{ cm}^2 \text{ skin})(0.01)$$

$$= 0.9 \text{ ng TCDD}$$

Note: A preferred method for performing this calculation, if data are available, is to use a flux rate (ng/cm^2-h) for the chemical. Assume that rate is 500 ng/cm^2-h:

$$\text{Uptake (ng)} = (C)(J)(A)(t)$$

where $J = 500$ ng/cm^2-h
$\quad t = 4$ h

By substitution:

$$\text{Uptake} = \left(\frac{250 \text{ ng TCDD}}{1 \text{ g soil}}\right)\left(\frac{1 \text{ g}}{10^9 \text{ ng}}\right)(1800 \text{ cm}^2 \text{ skin})(4 \text{ h})\left(\frac{500 \text{ ng}}{\text{cm}^2\text{-h}}\right)$$

$$= 0.9 \text{ ng TCDD}$$

4.5.7 Dermal Uptake of Chemicals in an Aqueous Matrix

Published estimates of dermal uptake of chemicals in water have generally focused on evaluating workplace or environmental exposure. A number of different scenarios has been evaluated.[162–166] If interested in the possible uptake of a chemical present in water, the amount of chlordane absorbed through the skin by a man swimming for 4 hours in water containing 1 ppb chloroform has been estimated.[164] This is useful to compare various approaches. For example, the amount of chloroform absorbed by a boy swimming for 3 hours in water has been calculated.[159] Some have compared the amounts absorbed through the skin during a 10-minute shower versus a 20-minute bath with water containing 1 ppb 1,1,1-trichlorethane.[165]

About 10 years ago, it was recognized that in the indoor environment, dermal exposure to volatile chemicals present in drinking water may not necessarily represent the vast majority of the risk. Specifically, it was found that inhalation exposure due to the release of vapors from liquids to which people were in close contact could be relatively high. For example, comparisons have been made of the chloroform concentration in exhaled breath after a shower to that after an inhalation-only exposure.[164] The concentration after showering was about twice that after the inhalation-only exposure, indicating that the absorbed dose from the skin is approximately equivalent to that from inhalation absorption.

Example Calculation 5: Skin Uptake of a Chemical from Water   A person has filled his swimming pool with shallow well water contaminated with 0.002 mg/mL (2 ppb) toluene. What is the plausible dermal uptake of toluene while swimming in the contaminated water for half an hour? Assume that 18,000 cm^2 of skin are exposed and the K_p is 1.01 cm/h. From Eq. (4.6):

$$Uptake = (C)(A)(K_p)(t)(d)$$

where $C = 0.002$ mg/mL

$\quad\quad A = 18,000$ cm^2

$\quad\quad K_p = 1.01$ cm/h

$\quad\quad t = 0.5$ h

$\quad\quad d =$ distribution factor $(1$ mL of water covers 1 $cm^3)$

By substitution:

$$Uptake = (0.002 \text{ mg/mL})(18,000 \text{ cm}^2)(1.01 \text{ cm/h})(0.5 \text{ h})(1 \text{ mL water}/1 \text{ cm}^3)$$

$$= 18 \text{ mg}$$

4.5.8 Percutaneous Absorption of Liquid Solvents

While the percutaneous absorption of chemical solutes generally proceeds by simple diffusion, the skin uptake of neat chemical liquids is not necessarily exclusively governed by Fick's law. Consequently, the uptake of neat liquid through the skin needs to be estimated using direct in vivo skin contact techniques. Table 4.5 presents the percutaneous absorption rates of some neat (pure liquid) industrial solvents that have been determined in human volunteer studies.

Example Calculation 6: Skin Uptake of a Neat Liquid Chemical   Due to carelessness or a leak, the inside of a glove becomes contaminated with 2-methoxyethanol. How much 2-ME can be absorbed if a worker wears the

TABLE 4.5   Absorption Rates of Some Neat Industrial Liquid Chemicals in Human Skin In Vivo

Chemical	Absorption rate $(mg/cm^2\text{-}h)$
Aniline	0.2–0.7
Benzene	0.24–0.4
2-Butoxyethanol	0.05–0.68
2-(2-Butoxyethoxy)ethanol	0.035
Carbon disulfide	9.7
Dimethylformamide	9.4
Ethylbenzene	22–23
2-Ethoxyethanol	0.796
2-(2-Ethyoxyethoxy)ethanol	0.125
Methanol	11.5
2-Methoxyethanol	2.82
2-(2-Methoxyethoxy)ethanol	0.206
Methyl butyl ketone	0.25–0.48
Nitrobenzene	2
Styrene	9–15
Toluene	14–23
Xylene (mixed)	4.5–9.6
m-Xylene	0.12–0.15

Note: From Leung and Paustenbach (1994).

contaminated glove on one hand for half an hour? Assume the surface area of exposed skin is 360 cm^2 and the flux rate is 2.82 mg/cm^2-h. From Eq. (4.7):

$$\text{Uptake} = (A)(J)(t)$$

where $A = 360$ cm^2
$\quad J = 2.82$ mg/cm^2-h
$\quad t = 0.5$ h

By substitution:

$$\text{Uptake} = (360 \text{ cm}^2)(2.82 \text{ mg/cm}^2\text{-h})(0.5 \text{ h})$$
$$= 508 \text{ mg}$$

To understand the relative hazard from skin exposure versus inhalation exposure, the dose of 2-methoxyethanol absorbed by the same worker via inhalation for 8 hours (10 m^3 of air inhaled), assuming a threshold limit value (TLV) of 16 mg/cm^3, can be estimated and compared to the dose due to inhalation exposure. Assume an 80% inhalation uptake efficiency.

$$\text{Inhalation uptake} = (16 \text{ mg/m}^3)(10 \text{ m}^3)(0.8)$$

$$= 128 \text{ mg}$$

Thus, the uptake of 2-methoxyethanol following 30 minutes of skin exposure of a single hand can be as much as 4 times that from inhalation for 8 hours at the TLV concentration, a presumably safe level of exposure. From this example, it is clear that the cutaneous route of entry can, in some situations, significantly contribute to the total absorbed dose, especially in the occupational setting.

4.5.9   Percutaneous Absorption of Chemicals in the Vapor Phase

Until the 1990s, it was generally assumed that the plausible dose resulting from vapors absorbed through the skin was too low to pose a hazard. Only a few studies have examined this issue.[113,166,167] A few clinical reports have encouraged some limited in vitro research to evaluate the absorption of several chemicals in the gaseous phase through the human skin (Table 4.6). A chamber system to measure the whole-body percutaneous absorption of chemical vapors in rats has been described by McDougal et al.,[53] and this approach has produced some interesting results.[168] In this system, chemical flux across the skin is determined from the chemical concentration in blood during exposure by using a PB-PK model. In most cases, vapor absorption through the skin amounts to less than 10 percent of the total dose received from a combined skin and inhalation

TABLE 4.6   Percutaneous Absorption Rates for Chemical Vapors In Vivo

Chemical	Skin update in combined exposure (%)[a]	Permeability coefficient K_p (cm/h) Rat	Permeability coefficient K_p (cm/h) Human
Styrene	9.4	1.75	0.35–1.42
m-Xylene	3.9	0.72	0.24–0.26
Toluene	3.7	0.72	0.18
Perchloroethylene	3.5	0.67	0.17
Benzene	0.8	0.15	0.08
Halothane	0.2	0.05	
Hexane	0.1	0.03	
Isoflurane	0.1	0.03	
Methylene chloride		0.28	
Dibromomethane		1.32	
Bromochloromethane		0.79	
Phenol			15.74–17.59
Nitrobenzene			11.1
1,1,1-Trichloroethane			0.01

Note: Rat data from McDougal et al. (1990).

[a]In combined exposure, rats are simultaneously absorbing chemical vapors by inhalation and by whole-body absorption of the vapor through the skin.

exposure. While there is good agreement between the rat and human in the relative ranking of the permeability coefficients among the chemicals studied, for an individual chemical the rat skin appears to be two to four times more permeable than the human skin. These observations are consistent with previously reported data.[56,57,111,117,118]

It is generally not necessary to account for the contribution from percutaneous uptake of vapors when the OEL is used as a guideline for acceptable exposure because uptake via this route is usually inherent in the data; that is, the studies of animals or humans from which data were collected were usually exposed via inhalation (whole body) so dermal uptake of the vapor occurred. However, although good work practices and the law require that situations where persons are placed in atmospheres that are life-threatening were it not for a supplied air respirator not occur, sometimes in emergency situations, airline (supplied air) respirators or self-contained breathing apparatus (SCBA) are worn in environments containing chemical concentrations 10-fold to 1000-fold greater than the TLV. In these cases, it is important to account for vapor uptake through either exposed or covered skin.

Although nearly all data on vapor absorption involve bare skin, the role of clothing in preventing skin uptake has occasionally been evaluated. For example, a study of workers wearing denim clothing indicated no decreased uptake of phenol vapors,[111] but found a 20 and 40 percent reduction in nitrobenzene uptake[55] and aniline vapor,[108] respectively. Although standard clothing may slightly decrease the amount of a chemical transferred from air through the skin, it can be a significant source of continuous exposure if the clothing has been contaminated.

Example Calculation 7: Skin Uptake of a Chemical Vapor   Assume that a person needs to repair a leaking pump, so he enters a room wearing an airline respirator. Assume the room contains 500 mg/m^3 nitrobenzene (100 times the current TLV) and it takes 30 minutes to repair the pump. How much nitrobenzene might be absorbed through the skin?

The head, neck, and upper limbs are assumed to be exposed (surface area = 4860 cm^2), and the rest of the body (surface area = 13,140 cm^2) is covered with clothing. Assume the percutaneous K_p of nitrobenzene is 11.1 cm/h, and that the clothing has reduced the skin uptake rate of vapors by about 20 percent.[55]

$$\text{Uptake} = (C)(A)(K_p)(t)$$

$$\text{Uptake through exposed skin} = (500 \text{ mg/m}^3)(4860 \text{ cm}^2)(11.1 \text{ cm/h})$$

$$\times (0.5 \text{ h})(1 \text{ m}^3/10^6 \text{ cm}^3) = 13.5 \text{ mg}$$

$$\text{Uptake through clothing} = (500 \text{ mg/m}^3)(13,140 \text{ cm}^2)(11.1 \text{ cm/h})$$

$$\times (0.8)(0.5 \text{ h})(1 \text{ m}^3/10^6 \text{ cm}^3) = 29 \text{ mg}$$

$$\text{Total uptake} = 13.5 + 29 = 42.5 \text{ mg}$$

From this example, it is clear that if one enters an environment containing a high concentration of an airborne contaminant, even if a supplied-air respirator is worn, the degree of skin uptake of the vapor may be worthy of evaluation to ensure that the worker is protected. In this example, uptake following one day of inhalation exposure at the TLV (5 mg/m^3) results in 50 mg uptake [(10 m^3) (5 mg/m^3)]. These kinds of calculations sometimes have to be conducted in difficult work environments (e.g., submarines, chemical plants during emergency situations, etc.).

4.6 ESTIMATING INTAKE VIA INGESTION

If the appropriate information is available, estimating the intake of various chemicals due to ingestion is a relatively straightforward exercise. In general, one is concerned with the ingestion of the following media: drinking water, other liquids, food, soil, and house dust. Drinking water contamination may occur because of soil contamination from leaking underground storage tanks, landfills, or hazardous waste sites, as well as discharges from contaminated streams or water transport systems. Nearly all foods in Western society contain a number of intentional and unintentional chemicals, including pesticide residues, naturally occurring chemicals, and food additives that serve as preservatives or enhancers of taste or visual appeal. Soils are ingested as a result of eating incompletely washed vegetables, hand-to-mouth contact, and through direct ingestion by children. Soils are also ingested when particles too large to reach the lower respiratory tract are inhaled (and then are swallowed). House dust contaminated with a number of chemicals can be ingested due to contact with foods and hand-to-mouth activities.[93]

4.6.1 Estimating Intake of Chemicals in Drinking Water

Estimating the magnitude of the potential dose of toxics from drinking water requires knowledge of the amount of water ingested, the chemical concentrations in the water, and the chemical bioavailability in the gastrointestinal tract. The amount of water ingested per day varies with each person and is usually related to the amount of physical activity. A good deal of literature has addressed the amount of water ingested by persons engaged in different kinds of activities.[81,99,101]

Currently, EPA suggests that when little is known about the specifics of exposure, a value of 2 L/day for adults and 1 L/day for infants (body weight of less than 10 kg) should be used as the default value. These rates include drinking water consumed in the form of juices and other beverages.

Numerous studies cited in EPA's *Exposure Factors Handbook*[38,39] have generated data on drinking water intake rates. In general, these sources support EPA's use of 2 L/day for adults and 1 L/day for children as upper-percentile tap water intake rates. Many of the studies have reported fluid intake rates for both

TABLE 4.7 Summary of Tap-water Intake by Age

	Intake (mL/d)		Intake (mL/kg/d)	
Age group	Mean	10th–90th Percentiles	Mean	10th–90th Percentiles
Infants (<1 yr)	302	0–649	43.5	0–100
Children (1–10 yr)	736	286–1,294	35.5	12.5–64.4
Teens (11–19 yr)	965	353–1,701	18.2	6.5–32.3
Adults (20–64 yr)	1,366	559–2,268	19.9	8.0–33.7
Adults (65+ yr)	1,459	751–2,287	21.8	10.9–34.7
All ages	1,193	423–2,092	22.6	8.2–39.8

Note: From Ershow and Cantor (1989).

total fluids and tap water. Total fluid intake is defined as consumption of all types of fluids including tap water, milk, soft drinks, alcoholic beverages, and water intrinsic to purchased foods. Total tap water is defined as water consumed directly from the tap as a beverage or used to prepare foods and beverages (i.e., coffee, tea, frozen juices, soups, etc.). Data for both consumption categories are presented in numerous publications. Table 4.7 presents typical information reported from these studies.[12]

All currently available studies on drinking water intake are based on short-term survey data. Although short-term data may be suitable for obtaining mean intake values that are representative of both short- and long-term consumption patterns, upper-percentile values may be different for short-term and long-term data because there is generally more variability in short-term surveys. It should also be noted that most of the currently available drinking water surveys are based on recall. This may be a source of uncertainty in the estimated intake rates because of the subjective nature of this type of survey technique.[12]

To estimate the intake of toxics via direct ingestion of drinking water, the calculation is straightforward:

$$\text{Intake} = (V)(C)(B)$$

where V = volume of water (L/day)

C = concentration of chemical in water (μg/L)

B = bioavailability

One of the more interesting observations of the past 15 years is that ingestion of contaminated drinking water is sometimes not the primary route of exposure to the toxicant in drinking water. Uptake of volatile chemicals via inhalation can be nearly as great in some homes as ingestion, due to the presence of these chemicals in air due to showering, off-gases from the dishwasher, and other opportunities for volatilization of the chemical.[164–168]

4.6.2   Importance of Soil Ingestion When Estimating Human Exposure

Between 1980 and 1995, predicted risks associated with the ingestion of contaminated soil were the primary drivers for remediating many (if not most) hazardous waste sites. As discussed by Paustenbach et al.,[29] there was no better example than the site in Times Beach, Missouri. Billions of dollars can be needed to clean up these kinds of sites to levels that would not pose a significant risk if children actually ate contaminated soil. Because of the expense of clean-ups a good deal of research has been conducted over the past 15 years to attempt to quantitatively understand this route of exposure.

Clearly, the ingestion of soil and house dust is a potential source of human exposure to toxicants. The potential for contaminant exposure via this source is greater for children because they are more likely to ingest greater quantities of soil than adults. Inadvertent soil ingestion among children may occur through the mouthing of objects or hands. Mouthing behavior is considered to be a normal phase of childhood development. Adults may also ingest soil or dust particles that adhere to food, cigarettes, or their hands. Deliberate soil ingestion is defined as pica and is considered to be relatively uncommon. Because normal, inadvertent soil ingestion is more prevalent and data for individuals with pica behavior are limited, the focus of most exposure assessments is on normal levels of soil ingestion that occur as a result of mouthing or unintentional hand-to-mouth activity.[12,29,169,170]

Mouthing activities by children, which are generally accepted as normal and commonplace (e.g., Barltrop[171] estimated that almost 80 percent of all children at age 1 year exhibited mouthing tendencies), are potential exposure routes to trace amounts of soil and/or dust adhering to fingers, hands, and objects placed in the mouth. The available data indicate that soil exposure occurs through several indirect routes:

1. Soil contributes to house dust (e.g., by local dust deposition, mud and dirt carried in by shoes and pets, etc.)
2. House dust (fine particles) adheres to objects and to children's hands
3. Children ingest dust particles when sucking and mouthing objects and fingers

Obviously in some situations, exposure may be direct (a child playing outdoors may eat dirt directly). In other situations, oral exposure may occur via contamination of domestic water supplies or contamination of vegetable produce grown onsite. However, dust in the indoor environment, which may represent the most important source of indirect exposure to soil, needs to be better understood.[93,100]

Many studies have been conducted to estimate the amount of soil ingested by children. Most of the early studies attempted to quantify the amount of soil ingested by measuring the amount of dirt present on children's hands and making generalizations based on behavior. More recently, soil intake studies have been conducted using a methodology that measures trace elements in feces and soil that are believed to be poorly absorbed in the gut. These measurements are used to estimate the amount of soil ingested over a specified period of time.

4.6.3   Studies of Soil Ingestion

In light of the importance of soil ingestion for estimating human exposure to contaminated soil, several literature surveys have been undertaken to identify the typical amount of soil consumed by children and adults.[12,25,29,100,170] Research evaluating lead uptake by children from ingestion of contaminated soil, paint chips, dust, and plaster provides the best source of information. Walter and co-workers[172] estimated that a normal child typically ingests very small quantities of dust or dirt between the ages of 0 and 2 years, the largest quantities between 2 and 7 years, and nearly insignificant amounts thereafter. In the classic text by Cooper,[173] it was noted that the desire of children to eat dirt or place inedible objects in their mouths becomes established in the second year of life and disappears more or less spontaneously by the age of 4 to 5 years. A study by Charney et al.[174] also indicated that mouthing tends to begin at about 18 months and continues through 72 months, depending on several factors such as nutritional and economic status, as well as race. Work by Sayre et al.[175] indicated that ages 2 to 6 years are the important years, but that "intensive mouthing diminishes after 2 to 3 years of age."

An important distinction that is often blurred is the difference between the ingestion of very small quantities of dirt due to mouthing tendencies and the disease known as pica. Children who intentionally eat large quantities of dirt, plaster, or paint chips (1 to 10 g/day), and consequently are at greater risk of developing health problems, can be said to suffer from the disease known as pica. This disease is known as geophasia if the craving is for dirt alone. Geophasia, rather than pica, is generally of greatest concern in areas with contaminated soil.

Duggan and Williams[176] have summarized the literature on the amount of lead ingested through dust and dirt. In their opinion, a quantity of 50 µg of lead was the best estimate for daily ingestion of dust by children. Assuming, on the high side, an average lead concentration of 1000 ppm would indicate a soil and dust ingestion rate of 50 mg/day. Lepow and co-workers[177] estimated an ingestion rate equal to 100 to 250 mg/day (specifically, 10 mg ingested 10 to 25 times a day). Barltrop[171,178] also estimated that the potential uptake of soils and dusts by a toddler is about 100 mg/day. In a Dutch study, the amount of lead on hands ranged from 4 to 12 ng. By assuming maximum lead concentrations of 500 ng/g (concentrations were typically lower) and complete ingestion of the contents adsorbed to a child's hand on 10 separate occasions, the amount of ingested dirt would equal 240 mg. Thus, in order to eat 10,000 mg of soil per day, the rate suggested by the Center for Disease Control, children would have to place their hands into their mouths 410 times a day, a rate that seems improbable.[179,180]

A report by the National Research Council[179] addressing the hazards of lead suggested a rate of 40 mg/day. Day et al.[181] measured the amount of dirt transferred from children's hands (age range from 1 to 3 years) to a sticky sweet, and estimated a daily intake of 2 to 20 sweets would lead to dirt intake of 10 to 1000 mg/day. Bryce-Smith[182] estimated 33 mg/day. In its document addressing lead in air, EPA assumed that children ate 50 mg/day of household dust, 40 mg/day of street dust, and 10 mg/day of dust derived from their parents' clothing (i.e., a total of 100 mg/day).

Lepow et al.[177] measured the dust obtained from children's hands (mean age 4.3 years) using adhesive tape, and hypothesized that up to 10 mg of soil at a time would be ingested when children put their fingers in their mouths. They believed children might do this 10 to 25 times a day, giving a range of soil ingestion from 100 to 250 mg/day.

Kimbrough et al.[170] used a speculative model of soil exposure when estimating the possible risks of contaminated soil at Times Beach, based upon unpublished observations about children's behavior and hand–mouth activity. A few years later, Kimbrough noted that their estimate of up to 10,000 mg/day was clearly a serious overestimate and her current personal estimate would be nearer 50 mg/day.[100]

Hawley[183] developed a complex lifetime exposure model based upon a literature review. For a short time (unspecified) following birth, soil exposure is negligible, then increases linearly to a maximum by age 2.5 years, declines by age 6 years, and rises slightly again in adulthood. The figure for adults (57 mg/day) assumes twice daily ingestion of a quantity of soil corresponding to one-half the inside surface of the fingers and thumbs, and also assumes that adults garden for 8 hours a day for 2 days/week, 5 months of the year. Many of the assumptions in this model could be said to be overly conservative.

La Goy[184] based his soil ingestion estimates upon a review of the literature, in particular using empirical data derived by Binder et al.[185] and Van Wijnen et al.[186] Similarly, Paustenbach[105] based his estimates upon a review of the literature, including the mass-balance quantitative study conducted by Calabrese et al. in 1986.[187]

De Silva[188,189] adopted a different approach that may overcome some of the uncertainties inherent in the assumptions of the above indirect studies. She applied a "slope factor" increase of 0.6 µg/dL in children's blood lead levels for each 1000 ppm increase in soil lead (this factor was developed by Barltrop[178] following his work on blood lead levels in children from villages on old mining sites). De Silva then deduced that an increase of 0.6 µg/dL in blood indicates an extra oral intake of 3.75 µg lead/day, based upon an EPA[97] calculation that an increase of 1.0 µg lead/day in children's diets produces an increase of 0.16 in the blood lead level. With a soil lead value of 1000 ppm, 3.75 mg of soil would contain 3.75 µg of lead, suggesting that 3.75 mg/day (say 4 mg) of soil was ingested by the children. However, the slope factor used here may not be the most appropriate, since mining soil wastes typically have larger sized particles, which tends to decrease lead bioavailability compared with soil contaminated by lead smelter activity, and therefore reduces the slope factor.

A major step forward beyond estimating soil ingestion using indirect measurements was the attempt to study tracer elements found in soil with elements measured in the urine and feces of children. At least six studies have been conducted thus far that have used this approach.[185,187–194] Of these, it appears that the studies by Calabrese and colleagues are the most reliable since prior researchers did not validate their analytical methods to determine whether soil ingestion could accurately be measured at low levels. Other shortcomings have been described elsewhere.

Binder et al.[185] conducted a pilot study involving the analysis of trace elements in children's stool samples. Their data indicated that children 1 to 3 years of age ingest about 180 mg/day of soil (geometric mean), based on the quantity of silicon, aluminum, and titanium found in the feces. However, the limitations of this study and the difficulties encountered in the interpretation of the titanium data appear to be too great for the study to be informative.

In another early tracer study, van Wijnen et al.[186] evaluated the amount of soil eaten by 24 hospitalized and nursery school children. They analyzed the amount of aluminum, titanium, and acid-soluble residue in the feces of children aged 2 to 4 years. The data were normally distributed. They found an average of 105 mg/day of soil in the feces of nursery children, and 49 mg/day in hospitalized children. Even with the limited number of children in the study, the difference between the two groups was significant ($p < 0.01$). If the value for the hospitalized children is assumed to be the background level because these substances are taken in from non-soil sources (e.g., diet and toothpastes), the estimated average amount of soil ingested by the nursery school children would be 56 mg/day. This value is in the lower range of estimates in the literature and supports the use of 100 mg/day as a reasonable daily average uptake of soil by toddlers (ages 2 to 4 years or 1.5 to 3.5 years).

The most thorough and rigorous study, to date, was completed by Calabrese et al.[187,190–194] They quantitatively evaluated six different tracer elements in the stools of 65 school children aged 2 to 4 years. They attempted to evaluate children from diverse socioeconomic backgrounds. This study was more definitive than prior investigations because they analyzed the children's diets, assayed for the presence of tracers in the diapers, assayed house dust and surrounding soil, and corrected for the pharmacokinetics of the tracer materials.

Based on the series of early works by Calabrese et al.[187–193] and others, a few generalizations can be made. The first two studies were difficult to conduct and interpret. Second, only children from a single climate were studied, and it can be expected that rates vary with the amount of time spent indoors and outdoors. Third, only a handful of children have been studied (less than 500), so it is not possible to characterize the percentage of children who might tend to ingest large quantities of soil or house dust. Fourth, the relevant amount of soil or house dust ingested indoors versus outdoors is not known yet. In most cases, the contaminant concentrations in dust can be quite different when found in a carpet versus the yard.[93] Fifth, although there is some degree of uncertainty in the results of the various studies, it appears that a best estimate of soil intake for most children

TABLE 4.8   Values for Childhood and Adult Soil Ingestion Rates That Have Been Used in Health Risk Assessments Conducted Between 1984 and 2000

Author	Age	Soil and dust (mg/d)
Barltrop (1973)	2–6 yr	100
Lepow et al. (1974)	2–6 yr	100–250
Day et al. (1975)	2–6 yr	10–1000
Kimbrough et al. (1984) (CDC)	0–9 mo 9–18 mo 1.5–3.5 yr 3.5–5 yr 5+ yr	0 1000 10,000 1000 100
Hawley (1985)	0–2 yr 2–6 yr 6–18 yr 18–70 yr	Negligible 90 21 57
La Goy (1987)	1–6 yr 1–6 yr	500 (max.) 100 (ave.)
Calabrese et al. (1989)	1–4 yr	27–85 (mean) 9–16 (median)
Paustenbach (1991b)	2–4 yr Adults	25–50 2–5
De Silva (1994)	Children	~4
U.S. EPA (1997b)	Children	200
Calabrese and Stanek (1998)	Children	30–60 (best estimate)
Stanek and Calabrese (2000)	Children	20–40 (best estimate)

resides in the area of 10 to 25 mg/day. It appears that perhaps 1 to 5 percent of the children may ingest much larger amounts during certain days or weeks (e.g., 2000 mg/day), but these tendencies do not occur on a chronic basis.

The issue of how much soil and house dust children eat, as well as the percent of children who are engaged in these activities, remains an active area of research.[186–188] Recent work by Calabrese et al.[194] suggests that prior work yielded reasonable results for purposes of risk assessment. Most of the values discussed here, representing work performed over the past five years, are presented in Table 4.8. Another area of research impacting exposure assessments of contaminated soil, which has been and continues to be actively pursued, is the bioavailability of the contaminant in the soil matrix.[58] As will be discussed later, this factor can be as important as the ingestion rate when estimating the human health hazard.

4.6.4   What Is the Significance of Pica?

There appears to be some confusion in the literature over what constitutes "pica." Pica can be defined as "the habitual ingestion of substances not normally regarded as edible," but some authors have included mouthing and sucking activities in their definitions (e.g., Lourie and Cayman[195]). Others appear to assume that all children with pica necessarily must be habitual soil eaters. In fact, pica behavior may be generalized to the ingestion of many different (nonfood) substances or may be specific to one substance such as paper, soap, or earth. It is likely that repetitive pica behavior specifically for dirt, or habitual geophagia, rarely occurs in the general population in most industrialized countries.[196,197]

Pica should, therefore, be considered a "normal" temporary phenomenon in some children. In the general population, the prevalence of both mouthing and pica, and the range of articles ingested, has been shown to decrease with age.[171] In the 1-year-old age group, 78 percent of children mouthed objects and 35 percent ingested them; this behavior decreased at the age of 4 years, when 33 percent were mouthing and only 6 percent had pica.

It is also relevant to note that in certain circumstances, pica for soil may be culturally determined (such as eating clay, high in silicon and aluminum, for its medical properties in the relief of stomach discomfort and diarrhea by some Aborigines; or the custom of eating earth during pregnancy in certain cultures).[100] For example, it has been suggested that some black women in the southern portions of the United States have a craving for and eat certain clays during pregnancy.

Pica may be associated with physical disorders, including iron deficiency. However, it has been debated whether pica represents a cause or an effect of these deficiencies. Pica can also be associated with mental illness. It has also been reported that 25 percent of institutionalized mentally handicapped adults indulged in pica of one kind or another (including bizarre objects ranging from rags and string to rocks, insects, and feces).[197]

As noted by Taylor,[196] while pica in the form of habitual geophagia would clearly lead (in a very small subset of the population) to a significantly greater risk from exposure to contaminants in soil, the significance of other forms of pica is unclear. Some insight may be gained from the Barltrop et al. study[192] of blood lead levels of children living in various rural villages where soil was both naturally high in lead and contaminated with lead from previous mining activity. This study provided important early evidence that soil and dust are ingested by children and that this can produce a small but significant increase in blood lead levels in relation to soil lead content. Barltrop's study was also important because it showed that lead levels in indoor house dust tended to increase with the soil lead content—house dust levels ranging from about the same at low lead levels (~ 500 ppm), 50 percent at intermediate lead levels (1000 to 10,000 ppm), and about 20 percent in areas with high soil levels ($>10,000$ ppm).[100]

The interesting observation from the point of view of pica was that in areas of low lead contamination (<1000 ppm), the mean blood lead levels of children with pica was only 9 percent higher than children without pica. For areas with high soil lead content (>10,000 ppm), no differences relating to pica were found. Similarly, fecal lead content for children from different areas was not significantly different, irrespective of the pica history of the child. This may well be due to differences in oral bioavailability of lead in the soil. In a similar study, Gallacher et al.[199] failed to demonstrate any relation between children's blood lead levels and reported pica behavior. What did correlate, however, was the amount of lead measured on children's hands using "wet wipes," and this was related more to house dust levels than to outdoor soil lead levels. This clearly suggests that house dust, as well as soil in any contaminated residential site, is a very important factor to consider.[93]

Taken together, these studies suggest that pica may have less impact on most exposure assessments than has been previously reported. Possible explanations include authors' failure to distinguish geophagia from other forms of pica (which might have shown a clearer relationship between blood lead content and geophagia), or that even in the presence of pica, soil ingestion is still low.

4.6.5   Soil Ingestion by Adults

For most persons beyond the ages of 5 to 6 years, the daily uptake of dirt due to intentional ingestion is generally thought to be quite low. With the exception of some lower income persons who eat clays, adults will not usually intentionally ingest dirt or soil. However, there are two other important ways in which adults eat dirt—incidental hand-to-mouth contact and through dust on vegetables. It has been shown that most soil ingested from crops comes from leafy vegetables. Interestingly, investigations at nuclear weapons trials have shown that particles exceeding 45 μm are seldom retained on leaves. Further, superficial contamination by smaller particles is readily lost from leaves, usually by mechanical processes or rain, and certainly by washing. As a result, unless the soil contaminant is absorbed into the plant, superficial contamination of plants by dirt will rarely present a health hazard.[29]

The estimated deposition rate of dust from ambient air in rural environments is about 0.012 μg/cm^2-day, assuming that rural dust contains about 300 μg/g of lead (the substance for which these data were obtained). EPA has estimated that even at relatively high air concentrations (0.45 mg/m^3 total dust), it is unlikely that surface deposition alone can account for more than 0.6 to 1.5 μg lettuce/g dust (2 to 5 μg/g lead) on the surface of lettuce during a 21-day growing period.[29] These data suggest that daily ingestion of dirt and dust by adults due to eating vegetables is unlikely to exceed about 0 to 5 mg/day even if all of the 137 g of leafy and root vegetables, sweet corn, and potatoes consumed by adult males each day were replaced by family garden products.

In its document on lead, EPA uses worst-case assumptions to estimate that persons could take up to 100 μg of lead each day due to unwashed vegetables.

The actual uptake by adults from vegetables should actually be much less, and is probably negligible, because EPA's estimate assumes that all of the suspended dust is contaminated, persons do not wash the vegetables, garden vegetables are eaten throughout the year, rather than only during the growing season, and persons actually replace most vegetables with their own garden products.

With respect to the second route—unwashed vegetables—only a very limited amount of work has been conducted. It has been suggested that the primary route of uptake will be through accidental ingestion of dirt on the hands, which may be of special concern to smokers who tend to have more frequent hand-to-mouth contact. It is true that before the importance of this route of entry was recognized, persons who worked in lead factories between 1890 and 1920 probably received a large portion of their body burden of lead due to poor hygiene; however, such conditions are now rare in the United States and most developed countries.

Some persons have evaluated the exposure experience of agricultural workers who apply or work with pesticide dusts. Due to the frequency and degree of pesticide exposure during its manufacture or application, these data do not appear to be appropriate surrogates for estimating soil uptake from the hands of persons who live on or near sites having contaminated soil. In addition, most of the published studies on pesticides involve liquids such as the organophosphates, rather than "soil-like" particles. Exposure studies of persons who apply granular pesticides might be more useful for defining upper-bound estimates of dermal exposure than estimates based on dusty workplaces.[156,200]

In general, incidental ingestion of contaminated soil due to poor personal hygiene should not constitute a significant hazard; however, this route of entry remains important in industrial settings. For example, Knarr and co-workers[200] showed that the maximum likely uptake of granular and liquid pesticides by applicators via all exposure routes ranges from 2 to 20 mg/day, values that are consistent with the results of other investigations. These estimates would appear to be overestimates of actual conditions at hazardous waste sites since persons involved in remediation will generally wear personal protective equipment when working with the contaminated dirt. For those persons who live offsite, the contribution of contaminated dirt from the site to the overall airborne dust concentration and the resulting deposition of soil or house dust onto skin should not present a hazard. As has been shown in a highway study, dust deposition decreases dramatically with distance from the road. Even having considered the contribution of poor hygiene and soil-contaminated food, the 50-mg/day figure suggested by EPA and other agencies in various guidance documents for estimating soil uptake by adolescents and adults seems a bit too high and, in the view of this writer, a figure of 0 to 10 mg/day seems more reasonable.

At least one study has been conducted to specifically address soil uptake by adults involved in remediating waste sites.[76,201] The results suggest that the amount of soil eaten by these workers is much less than the default value of 100 mg/day suggested by EPA in a number of guidance documents or risk assessments.

4.6.6   Estimating the Intake of Chemicals Via Food

Without question, accurately estimating the ingestion of xenobiotics via foods is one the most complex of all exposure calculations. The hundreds of different possible foods and dozens of different chemicals that can be present as pesticide residue and background concentrations of various chemicals in soil make this a formidable task.

The methodology for estimating uptake via ingestion must account for the quantity of food ingested each day, the concentration of contaminant in the ingested material, and the bioavailability of the contaminant in the media. Over the past 20 years, a significant amount of work has been directed at understanding these exposure factors. Specifically, an entire volume of EPA's *Exposure Factors Handbook* (Vol. II) is devoted to this topic.[81]

The approach to estimating uptake via foods was first applied in the late 1940s by the Food and Drug Administration[202] and had not changed much through 1999. However, because of the passage of the Food Quality Protection Act of 1996 (FQPA), the methodology for estimating uptake from foods will be changing dramatically over the next 5 to 10 years. Specifically, the FQPA requires that all pesticide residues from foods be added together with the goal of understanding the total daily dose of all residual pesticides in the dust. In addition, all pesticide doses that act through the same mechanism or mode of action must be added together. Then, if necessary, the pesticide manufacturers are expected to calculate the necessary residue level so that the total dose does not exceed a fraction of the acceptable daily intake (ADI). Since there are hundreds of foods and dozens of residues, this will be a formidable challenge.

Ingestion of contaminated fruits and vegetables is a potential pathway of human exposure to toxic chemicals. Fruits and vegetables may become contaminated with toxic chemicals by several different pathways. Ambient air pollutants may be deposited on or absorbed by plants or dissolved in rainfall or irrigation waters that contact the plants. Plant roots may also absorb pollutants from contaminated soil and groundwater. The addition of pesticides, soil additives, and fertilizers may also result in food contamination.[81]

The primary information source on consumption rates of fruits and vegetables among the U.S. population is the U.S. Department of Agriculture's (USDA) Nationwide Food Consumption Survey (NFCS) and the USDA Continuing Survey of Food Intakes by Individuals (CSFII). Data from the NFCS have been used in various studies to generate consumer-only and per-capita intake rates for individual fruits and vegetables, as well as total fruits and total vegetables. CSFII data from the 1989 to 1991 survey have been analyzed by EPA to generate per-capita intake rates for various food items and food groups.[81,203,204]

Consumer-only intake is defined as the quantity of fruits and vegetables consumed by individuals who ate these food items during the survey period. Per-capita intake rates are generated by averaging consumer-only intakes over the entire population of users and nonusers. In general, per-capita intake rates are appropriate for use in exposure assessment for which average dose estimates for the general population are of interest because they represent both individuals who ate the foods during the survey period and individuals who may eat the food items at some time, but did not consume them during the survey period. Total fruit intake refers to the sum of all fruits consumed in a day, including canned, dried, frozen, and fresh fruits. Likewise, total vegetable intake refers to the sum of all vegetables consumed in a day, including canned, dried, frozen, and fresh vegetables.

Intake rates may be presented on either an as-consumed or dry-weight basis. As-consumed intake rates (grams/day) are based on the weight of food in the form in which it is consumed. In contrast, dry-weight intake rates are based on the weight of food consumed after the moisture content has been removed. In calculating exposures based on ingestion, the unit of weight used to measure the contaminant concentration in the produce will vary. Intake data from the individual NFCS and CSFII components are based on "as eaten" (i.e., cooked or prepared) forms of the food items or groups. Thus, no corrections are required to account for changes in portion sizes from cooking losses.[205,206]

Estimating source-specific exposures to toxic chemicals in fruits and vegetables may also require information on the amount of fruits and vegetables exposed to or protected from contamination as a result of cultivation practices, the physical nature of the food product itself (i.e., those having protective coverings that are removed before eating would be considered protected), or the amount grown beneath the soil (i.e., most root crops such as potatoes). The percentages of foods grown above and below ground will be useful when the contaminant concentrations in foods are estimated from concentrations in soil, water, and air. For example, vegetables grown below ground would more likely be contaminated by soil pollutants, but leafy above-ground vegetables would more likely be contaminated by deposition of air pollutants on plant surfaces. Some examples of various exposure factors and confidence ratings for liquids and food are presented in Table 4.9.[81]

Individual average daily intake rates calculated from NFCS and CSFII data are based on averages of reported individual intakes over 1 day or 3 consecutive days. Such short-term data are suitable for estimating mean average daily intake rates representative of both short-term and long-term consumption. However, the *distribution* of average daily intake rates generated using short-term data (e.g., 3 day) do not necessarily reflect the long-term *distribution* of average daily intake rates. The distributions generated from short-term and long-term data will differ to the extent that each individual's intake varies from day to day; the distributions will be similar to the extent that individuals' intakes are constant from day to day.[81]

Day-to-day intake variation among individuals will be greatest for food items or groups that are highly seasonal and for items or groups that are eaten year-around but are not typically eaten every day. For these foods, the intake distribution generated from short-term data will not reflect long-term distribution. On the other hand, for broad categories of foods (e.g., vegetables), which are eaten

TABLE 4.9 Summary of Default Exposure Factor Recommendations and Confidence Ratings for Citizens of United States (EPA, 1997)

Exposure factor	Recommendation	Confidence rating
Drinking-water intake rate	21 ml/kg-d or 1.4 L/d (average)	Medium
	34 ml/kg-d or 2.3 L/d (90th percentile)	Medium
	Percentiles and distribution also included	
	Means and percentiles also included for pregnant and lactating women	
Total fruit intake rate	3.4 g/kg-d (per capita average)	Medium
	12.4 g/kg-d (per capita 95th percentile)	Low
	Percentiles also included	
	Means presented for individual fruits	
Total vegetable intake rate	4.3 g/kg-d (per capita average)	Medium
	10 g/kg-d (per capita 95th percentile)	Low
	Percentiles also included	
	Means presented for individual vegetables	
Total meat intake rate	2.1 g/kg-d (per capita average)	Medium
	5.1 g/kg-d (per capita 95th percentile)	Low
	Percentiles also included	
	Percentiles also presented for individual meats	
Total dairy intake rate	8.0 kg-d (per capita average)	Medium
	29.7 g/kg-d (per capita 95th percentile)	Low
	Percentiles also included	
	Means presented for individual dairy products	
Grain intake	4.1 g/kg-d (per capita average)	High
	10.8 g/kg-d (per capita 95th percentile)	Low in long-term upper percentiles
	Percentiles also included	
Breast-milk intake rate	742 ml/d (average)	Medium
	1,033 ml/d (upper percentile)	Medium
Fish intake rate	General population	
	20.1 g/d (total fish) average	High
	14.1 g/d (marine) average	High
	6.0 g/d (freshwater/estuarine) average	High
	63 g/d (total fish) 95th percentile long-term	Medium
	Percentiles also included	
	Serving size	
	129 g (average)	High
	326 g (95th percentile)	High
	Recreational marine anglers	
	2–7 g/d (finfish only)	Medium

TABLE 4.9 *(Continued)*

Exposure factor	Recommendation	Confidence rating
	Recreational freshwater	
	8 g/d (average)	Medium
	25 g/d (95th percentile)	Medium
	Native American subsistence population	
	70 g/d (average)	Medium
	170 g/d (95th percentile)	Low

Note: From U.S. EPA (1997a).

on a daily basis throughout the year with minimal seasonality, the short-term distribution may be a reasonable approximation of the true long-term distribution, although it will show somewhat more variability.

Other relevant fruit and vegetable intake studies include EPA's Dietary Risk Evaluation System (DRES), Office of Pesticide Programs (OPP). The OPP uses the DRES (formerly the Tolerance Assessment System) to assess the dietary risk of pesticide use as part of the pesticide registration process.[205,206] OPP sets tolerances for specific pesticides on raw agricultural commodities based on estimates of dietary risk. These estimates are calculated using pesticide residue data for the food item of concern and relevant consumption data. Intake rates are based primarily on the USDA 1977–1978 NFCS, although intake rates for some food items are based on estimations from production volumes or other data (i.e., some items were assigned an arbitrary value of 0.000001 g/kg-day).[207] OPP has calculated per-capita intake rates of individual fruits and vegetables for 22 subgroups of the population (age, regional, and seasonal) by determining the composition of NFCS food items and disaggregating complex food dishes into their component raw agricultural commodities (RACs).[81,208]

The advantage of using these data is that complex food dishes have been disaggregated to provide intake rates for a very large number of fruits and vegetables. These data are also based on the individual body weights of the respondents. Therefore, using these data to calculate toxic chemical exposure may provide more representative estimates of potential dose per unit body weight. However, because the data are based on the NFCS short-term dietary recall, the same limitations discussed previously for other NFCS data sets also apply here. In addition, consumption patterns may have changed since the data were collected in 1977 to 1978. OPP is in the process of translating consumption information from the USDA CSFII 1989 to 1991 survey to be used in DRES.[81]

USDA has also conducted a study entitled *Food and Nutrient Intakes of Individuals in One Day in the U.S.*[81,203] USDA calculated mean intake rates for total fruits and total vegetables using NFCS data from 1977 to 1978 and 1987 to 1988 and CSFII data from 1994 to 1995.[81,203] Mean per-capita total intake rates are based on intake data for 1 day from the 1977 to 1978 and 1987 to 1988 USDA NFCS, respectively. Data from both surveys are presented in the *Exposure Factors Handbook* to demonstrate that although the 1987 to 1988 survey had fewer respondents, the mean per-capita intake rates for all individuals agree with the earlier survey. Also, slightly different age classifications were used in the two surveys, providing a wider range of age categories from which exposure assessors may select appropriate intake rates. The age groups used in this data set are the same as those used in the 1987 to 1988 NFCS. Information for per-capita intake rates and consumer-only intake rates for various ages of individuals is also available. Intake rates for consumer-only were calculated by dividing the per-capita consumption rate by the fraction of the population using vegetables or fruits in a day.[81]

The advantages of using these data are that they provide intake estimates for all fruits, all vegetables, or all fats combined. Again, these estimates are based on 1-day dietary data that may not reflect usual consumption patterns.[81]

4.6.7   Intake of Fish and Shellfish

Contaminated finfish and shellfish are potential sources of human exposure to toxic chemicals. Pollutants are carried in surface waters but also may be stored and accumulated in sediments as a result of complex physical and chemical processes. Consequently, various aquatic species can be exposed to pollutants and may become sources of contaminated food.[81]

Accurately estimating exposure to various chemicals in a population that consumes fish from a polluted water body requires an estimation of caught-fish intake rates by fishermen and their families. Commercially caught fish are marketed widely, making the prediction of an individual's consumption from a particular commercial source difficult. Because the catch of recreational and subsistence fishermen is generally not diluted in this way, these individuals and their families represent the population that is most vulnerable to exposure by intake of contaminated fish from a specific location.[81]

Over the years, fish consumption survey data have been collected using a number of different approaches that need to be considered when interpreting the survey results. Generally, surveys are either "creel" studies in which fishermen are interviewed while fishing or broader population surveys using mailed questionnaires or phone interviews. Both data types can be useful for exposure assessment purposes, but somewhat different applications and interpretations are needed. In fact, creel study results have often been misinterpreted because of inadequate knowledge of survey principles.[81,209,210]

The typical survey seeks to draw inferences about a larger population from a smaller sample of that population. The larger population from which the survey sample is taken and to which the survey results are generalized denotes the target population of the survey. To generalize from the sample to the target population, the probability of being sampled must be known for each member of the target population. This probability is reflected in weights assigned to each survey respondent, with weights being inversely proportional to sampling probability. When all members of the target population have the same probability of being sampled, all weights can be set to 1 and essentially ignored.[211,212]

In a mail or phone study of licensed anglers, the target population generally involves all licensed anglers in a particular area, and in these studies, the sampling probability is essentially equal for all target population members. In a creel study, the target population is anyone who fishes at the locations being studied; generally in a creel study, the probability of being sampled is not the same for all members of the target population. For instance, if the survey is conducted for 1 day at a site, then it will include all persons who fish there daily, but only about 1/7th of the people who fish there weekly, 1/30th of the people who fish there monthly, and so on. In this example, the probability of being sampled (or inverse weight) is seen to be proportional to the frequency of fishing. However, if the survey involves interviewers who revisit the same site on multiple days, and persons who are only interviewed once for the survey, then the probability of being in the survey is not proportional to frequency; in fact, it increases less proportionally with greater frequency of fishing. If the same site is surveyed every day of the survey period with no reinterviewing, all members of the target population would have the same probability of being sampled, regardless of fishing frequency, implying that the survey weights should all equal 1.[212,213]

On the other hand, if the survey protocol calls for individuals to be interviewed each time an interviewer encounters them (i.e., without regard to whether they were previously interviewed), then the inverse weights will again be proportional to fishing frequency, no matter how many times interviewers revisit the same site. Note that when individuals can be interviewed multiple times, the results of each interview are included as separate records in the database, and the survey weights should be inversely proportional to the expected number of times that an individual's interviews are included in the database.[81,212,213]

Fish and shellfish exposure assessments are among the most complicated of all assessments.[214] A significant portion of the *Exposure Factors Handbook* addresses this topic.[81] Recently, fairly complex Monte Carlo methods have been applied to resolve many of the difficulties estimating exposure of anglers and their families.[48]

4.6.8 Aggregate Exposure and FQPA

Pesticides are regulated under both the Federal Insecticide, Fungicide, and Rodenticide Act (FIFRA) and the Federal Food, Drug, and Cosmetics Act (FFDCA). In 1996, Congress passed the Food Quality Protection Act (FQPA) that amended both FIFRA and FFDCA. These laws mandated the U.S. EPA to register pesticides and set tolerances based on a safety determination, a reasonable certainty that use of a given pesticide or consumption of raw agricultural commodity of processed foods that contain the pesticide and its residues will cause no harm to human health or the environment. The U.S. EPA evaluates risks posed by the use and usage of each pesticide to make a determination of safety. Based upon this determination, the agency regulates pesticides to ensure that use of the chemical is not unsafe.

In the past, the U.S. EPA evaluated the safety of pesticides based on a single-chemical, single-exposure-pathway scenario. However, FQPA requires that the agency consider aggregate exposure in its decision-making process. Section 408(a)(4)(b)(2)(ii) of FFDCA specifies with respect to a tolerance that there must be a determination "that there is a reasonable certainty that no harm will result from aggregate exposure to the pesticide chemical residue, including all anticipated dietary exposures and all other exposures for which there is reliable information." Section (b)(2)(C)(ii)(I) states that "there is a reasonable certainty that no harm will result to infants and children from aggregate exposure to the pesticide chemical residues" *Aggregate dose* is defined as the amount of a single substance available for interaction with metabolic processes or biologically significant receptors from multiple routes of exposure. *Aggregate risk* is defined as the likelihood of the occurrence of an adverse health effect resulting from all routes of exposure to a single substance. Conversely, *cumulative risk* is defined as the likelihood of the occurrence of an adverse health effect resulting from all routes of exposure to a group of substances sharing a common mechanism of toxicity.

As shown in Figure 4.5, the most basic concept underlying all aggregate exposure assessments is that exposure occurs to an individual. The integrity of the data concerning this exposed individual must be maintained throughout the aggregate exposure assessment. In other words, each of the individual "sub-assessments" must be linked back to the same person (394). Because exposures are based on that received by a single individual, aggregate exposure assessments must agree in time, place, and demographic characteristics. Each of these parameters have imbedded attributes that must be matched to create a reasonable assessment. Some of these imbedded attributes include:

- Time (duration, daily, seasonally).
- Place (location and type of home, urbanization, watersheds, region).
- Demographics (age, gender, reproductive status, ethnicity, personal preference).

To develop realistic aggregate exposure and risk assessments requires that the appropriate temporal, spatial, and demographic exposure factors be correctly assigned. Examples of some of these factors include sex- and age-specific body weights, regional specific drinking-water concentrations of the pesticide being considered, seasonally based pesticide residues in food, and frequency of residential pest control representative of housing type. Once an aggregate exposure and risk assessment is completed for one individual, population and subpopulation distributions of exposures and risk may be constructed by probabilistic techniques (161).

An aggregate exposure and risk assessment is distinct from a cumulative risk assessment. Cumulative risk is defined as "the measure or estimate of distribu-

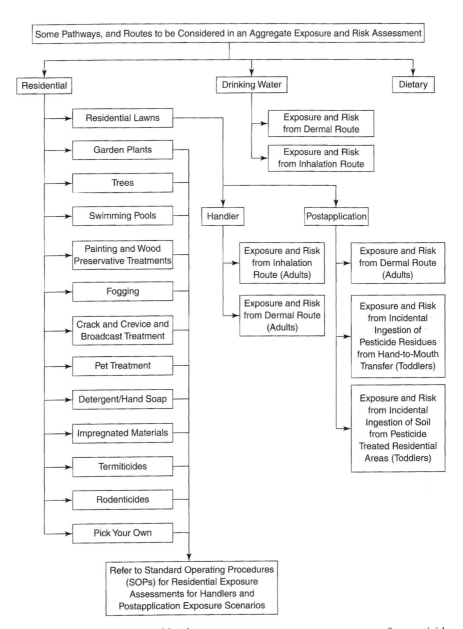

Figure 4.5 Factors to consider in an aggregate exposure assessment of a pesticide. *Source:* U.S. Environmental Protection Agency (1999). *Guidance for performing aggregate exposure and risk assessments* (Draft). Washington, DC.

tions of exposures (doses) for a set of chemicals that act by a common mechanism of toxicity" (390). Cumulative risk assessment evaluates risks from multiple chemicals via all routes and pathways of exposure. The cumulative risk assessment considers the combined toxicological effect of a group of chemicals with a common mechanism of toxicity. The definition of a common mechanism of toxicity is "two or more pesticide chemicals that produce an adverse effect(s) to human health by the same, or essentially the same, sequence of major biochemical events. The underlying basis of the toxicity is the same, or essentially the same, for each chemical" (390). Specific guidance concerning conducting a cumulative risk assessment is currently being developed (161).

4.6.9 Breast Milk

Breast milk is a potential source of exposure to toxic substances for nursing infants. Lipid-soluble chemical compounds accumulate in body fat and may be transferred to breast-fed infants in the lipid portion of breast milk. Because nursing infants obtain most (if not all) of their dietary intake from breast milk, they are especially vulnerable to exposures to these compounds. In fact, the peak body burdens of certain chemicals (like dioxin) can reach their lifetime peak (μg/kg) on the last day of nursing at age 12 to 24 months. Estimating the magnitude of the potential dose to infants from breast milk requires information on the quantity of breast milk consumed per day and the duration (months) over which breastfeeding occurs. Information on the fat content of breast milk is also needed for estimating dose from breast milk residue concentrations that have been indexed to lipid content.[81,215]

Several studies have generated data on breast milk intake.[216–219] Typically, breast milk intake has been measured over a 24-hour period by weighing the infant before and after each feeding without changing its clothing (test weighing). The sum of the difference between the measured weights over the 24-hour period is assumed to be equivalent to the amount of breast milk consumed daily. Intakes measured using this procedure are often corrected for evaporative water losses (insensible water losses) between infant weighings.[218] Neville et al.[219] evaluated the validity of the test-weight approach among bottle-fed infants by comparing the weight of milk taken from bottles with the difference between the before and after feeding weights of infants. Once corrected for insensible water loss, test-weight data were not significantly different from bottle weights. Conversions between weight and volume of breast milk consumed are made using the density of human milk (approximately 1.03 g/mL).[218] Recently, techniques for measuring breast milk intake using stable isotopes have been developed; however, little data based on this new technique have been published.[218]

Studies among nursing mothers in industrialized countries have shown that infant intake averages approximately 750 to 800 g/day (728 to 777 mL/day) during the first 4 to 5 months of life, with a range of 450 to 1200 g/day (437 to 1165 mL/day).[81,218] Similar intakes have also been reported for developing countries.[81,218] Infant birth weight and nursing frequency have been shown to

TABLE 4.10 Default Values for Daily Intakes of Breast Milk

Age	Number of infants surveyed at each time period	Mean intake (ml/d)	Range of daily intake (ml/d)
Completely breast-fed			
1 mo	11	600 ± 159	426–989
3 mo	2	833	645–1000
6 mo	1	682	616–786
Partially breast-fed			
1 mo	4	485 ± 79	398–655
3 mo	11	467 ± 100	242–698
6 mo	6	395 ± 175	147–684
9 mo	3	<554	451–732

Note: From Pao et al. (1982). Data expressed as mean $\pm$ standard deviation.

influence the rate of intake.[81,218] Infants who are larger at birth and/or nurse more frequently have been shown to have higher intake rates. Also, breast milk production among nursing mothers has been reported to be somewhat higher than the amount actually consumed by the infant.[81,212]

Like exposure assessment of fishes, techniques for estimating chemical uptake by children of breast milk continue to evolve. A portion of EPA's *Exposure Factors Handbook* addresses this topic, and a few published papers have offered some novel approaches.[81,216] Some examples of breast milk intake rates and other values are presented in Table 4.10. A distribution for breast milk consumption has been suggested by Copeland et al.

4.7 ESTIMATING UPTAKE VIA INHALATION

Estimating intake via inhalation depends on only a few exposure factors, for example, inhalation rate, airborne chemical concentration, bioavailability, and, if it is a particle, particle size. In general, uncertainty in estimates of intake via inhalation is among the smallest of all exposure calculations.

Inhalation rates are known to vary directly with the amount of physical activity of the persons being evaluated. The default value used by EPA and others is 20 m^3/day. When conducting occupational exposure assessments, it is common to assume that workers inhale about 10 m^3 in an 8-hour workday.[12]

Airborne chemical concentrations are obtained through either direct measurement or modeling. The form of the chemicals in the air will be a gas (includes vapors) or particles (dusts or fumes). Generally, it is assumed that virtually all of the vapors or gases will be absorbed if inhaled.[12,29,161] This may not be the case for volatile chemicals, if the concentration in the blood is approaching steady-state. In those cases, a significant fraction of the inhaled vapors will be present in the exhaled breath and, therefore, not absorbed.[56]

It is usually assumed that if particles enter the lower respiratory tract, they will eventually be absorbed unless the chemical is highly insolvable. Generally, it is assumed that particles less than 100–150 μm are inhalable, but virtually all particles greater than 10 μm (by weight) will be captured in the upper respiratory tract (nose and throat) and then ingested. It has often been assumed that particles less than 10 μm will be captured in the lower respiratory tract and nearly 100 percent of these (by weight) will eventually be absorbed. Notably, for some chemicals, the adverse effect is related to the particle size, so this must be taken into account. For example, it is thought that beryllium particles should be collected in several different size fractions because the adverse effect likely varies according to the particle size.

4.7.1 Various Inhalation Rates

A significant amount of research has been done to correlate various inhalation rates with different tasks and body weights. Most studies on this subject have been summarized in the most recent EPA *Exposure Factors Handbook*.[25] Data are available for dozens of different levels of physical activity and the distributions for several populations are presented.

A number of equations have been proposed for predicting the inhalation rate based on body weight.[81] The *Exposure Factors Handbook* and other sources provide a number of tables that relate physical activity with inhalation rate (see Table 4.11).

TABLE 4.11   Daily Inhalation Rates Estimated From Daily Activities

| Subject | Inhalation rate (IR) | | Daily inhalation rate (DIR)[a] (m³/d) |
	Resting (m³/h)	Light activity (m³/h)	
Adult man	0.45	1.2	22.8
Adult woman	0.36	1.14	21.1
Child (10 yr)	0.29	0.78	14.8
Infant (1 yr)	0.09	0.25	3.76
Newborn	0.03	0.09	0.78

Note: From Snyder (1985). Assumptions made were based on 8 h resting and 16 h light activity for adults and children (10 yr); 14 h resting and 10 h light activity for infants (1 yr); 23 h resting and 1 h light activity for newborns.

$$^a\text{DIR} = \frac{1}{T}\sum_{i=1}^{k} \text{IR}_i t_i$$

where IR_i is corresponding inhalation rate at ith activity, t_i is hours spent during the ith activity, k is number of activity periods, and T is total time of the exposure period (i.e., a day).

4.7.2 Bioavailability of Airborne Chemicals

Because the mass of chemicals inhaled is usually quite small, and because most particles less than 10 μm in diameter are thought to be fairly easily absorbed, it is generally assumed that particles are 100 percent bioavailable after they are trapped in the lower lung. Likewise, it is generally assumed that most vapors and gases are completely absorbed (100 percent bioavailable) if they reach the lower respiratory tract (even though it will often be much less). Both are conservative assumptions that should be reassessed on a case-by-case basis.

4.8 ROLE OF UNCERTAINTY ANALYSIS

Exposure assessment uses a wide array of information sources and techniques. Even when actual exposure-related measurements exist, assumptions or inferences will still be required. Most likely, data will not be available for all aspects of the exposure assessment and these data may be of questionable or unknown quality. In these situations, the exposure assessor will have to rely on a combination of professional judgment, inferences based on analogy with similar chemicals and conditions, estimation techniques, and the like. The net result is that the exposure assessment will be based on a number assumptions with varying degrees of uncertainty.[12]

The decision analysis literature has focused on the importance of explicitly incorporating and quantifying scientific uncertainty in risk assessments.[213,214] Reasons for addressing uncertainties in exposure assessments include:

- Uncertainty information from different sources and of different quality must be combined.
- A decision must be made about whether and how to expend resources to acquire additional information (e.g., production, use, and emissions data; environmental fate information; monitoring data; population data) to reduce the uncertainty.
- So much empirical evidence exists that biases may occur, resulting in so-called best estimates that are not very accurate. Even when all that is needed is a best-estimate answer, the quality of the answer may be improved by incorporating a frank discussion of uncertainty into the analysis.
- Exposure assessment is an iterative process. The search for an adequate and robust methodology to handle the problem at hand may proceed more effectively, and to a more certain conclusion, if the associated uncertainty is explicitly included and it can be used as a guide in the process of refinement.
- A decision is rarely made on the basis of a single piece of analysis. Further, it is rare for there to be one discrete decision; a process of multiple decisions spread over time is the more common occurrence. Chemicals of concern may go through several levels of risk assessment before a final decision is

made. During this process, decisions may be made based on exposure considerations. An exposure analysis that attempts to characterize the associated uncertainty allows the user or decision maker to do a better evaluation in the context of the other factors being considered.

- Exposure assessors have a responsibility to present not just numbers but also a clear and explicit explanation of the implications and limitations of their analyses. Uncertainty characterization helps to achieve this.

Essentially, constructing scientifically sound exposure assessments and analyzing uncertainty go hand in hand. The reward for analyzing uncertainties is knowing that the results have integrity or that significant gaps exist in available information that can make decision making a tenuous process.

4.8.1 Variability Versus Uncertainty

While some authors treat variability as a specific component of uncertainty, EPA[220] and others advise risk assessors (and, by analogy, the exposure assessor) to distinguish between variability and uncertainty.[8] Specifically, uncertainty represents a lack of knowledge about factors affecting exposure or risk, whereas variability arises from true heterogeneity across people, places, or time. In other words, uncertainty can lead to inaccurate or biased estimates, whereas variability can affect the precision of the estimates and the degree to which they can be generalized.

Variability and uncertainty can complement or confound one another. The National Research Council[8] has drawn an instructive analogy based on estimating the distance between the earth and the moon. Prior to fairly recent technological developments, it was difficult to accurately measure this distance, resulting in measurement uncertainty. Because the moon's orbit is elliptical, the distance is a variable quantity. If only a few measurements were taken without knowledge of the elliptical pattern, then either of the following incorrect conclusions might be reached:

- The measurements were faulty, thereby ascribing to uncertainty what was actually caused by variability; or
- The moon's orbit was random, thereby not allowing uncertainty to shed light on seemingly unexplainable differences that are in fact variable and predictable.

A more fundamental error in the above situation might be to incorrectly estimate the true distance and assume that a few observations were sufficient. This latter pitfall—treating a highly variable quantity as if it were invariant or only uncertain—is most relevant to the exposure or risk assessor.[12]

Now consider a situation that relates to exposure, such as estimating the average daily dose by one exposure route—ingestion of contaminated drinking water. Suppose that it is possible to measure an individual's daily water consumption (and concentration of the contaminant) exactly, thereby eliminating uncertainty in the measured daily dose. The daily dose still has an inherent day-to-day variability, however, because of changes in the individual's daily water intake or concentration of the contaminant in the water.[12]

Clearly, it is impractical to measure the individual's dose every day. For this reason, the exposure assessor may estimate the average daily dose (ADD) based on a finite number of measurements, in an attempt to "average out" the day-to-day variability. The individual has a true (but unknown) ADD, which has now been estimated based on a sample of measurements. Because the individual's true average is unknown, it is uncertain how close the estimate is to the true value. Thus, the variability across daily doses has been translated into uncertainty in the ADD. Although the individual's true ADD has no variability, the estimate of the ADD has some uncertainty.[12]

The above discussion pertains to the ADD for one person. Now consider a distribution of ADDs across individuals in a defined population (e.g., the general U.S. population). In this case, variability refers to the range and distribution of ADDs across individuals in the population. By comparison, uncertainty refers to the exposure assessor's state of knowledge about that distribution or about parameters describing the distribution (e.g., mean, standard deviation, general shape, various percentiles).[12]

As noted by the National Research Council,[8] the realms of variability and uncertainty have fundamentally different ramifications for science and judgment. For example, uncertainty may force decision makers to judge how probable it is that exposures have been overestimated or underestimated for every member of the exposed population, whereas variability forces them to cope with the certainty that different individuals are subject to exposures both above and below any of the exposure levels chosen as a reference point.[12]

4.8.2   Types of Variability

Variability in exposure is related to an individual's location, activity, and behavior or preferences at a particular point in time, as well as pollutant emission rates and physical/chemical processes that affect concentrations in various media (e.g., air, soil, food, and water). The variations in pollutant-specific emissions or processes, and in individual locations, activities, or behaviors are not necessarily independent of one another. For example, both personal activities and pollutant concentrations at a specific location might vary in response to weather conditions or between weekdays and weekends.[12]

At a more fundamental level, three types of variability can be distinguished:

- Variability across locations (spatial variability)
- Variability over time (temporal variability)
- Variability among individuals (interindividual variability)

Spatial variability can occur both at regional (macroscale) and local (microscale) levels. For example, fish intake rates can vary depending on the region of the country. Higher consumption may occur among populations located near large bodies of water such as the Great Lakes or coastal areas. As another example, outdoor pollutant levels can be affected at the regional level by industrial activities and at the local level by activities of individuals. In general, higher exposures tend to be associated with closer proximity to the pollutant source, whether it be an industrial plant or related to a personal activity such as showering or gardening. In the context of exposure to airborne pollutants, the concept of a "microenvironment" has been introduced to denote a specific locality (e.g., a residential lot or a room in a specific building) where the airborne concentration can be treated as homogeneous (i.e., invariant) at a particular point in time.

Temporal variability refers to variations over time, whether long or short term. Seasonal fluctuations in weather, pesticide applications, use of wood-burning appliances, and fraction of time spent outdoors are examples of longer-term variability. Examples of shorter-term variability are differences in industrial or personal activities on weekdays versus weekends or at different times of the day.

Interindividual variability can be either of two types: (1) human characteristics such as age or body weight and (2) human behaviors such as location and activity patterns. Each of these variabilities, in turn, may be related to several underlying phenomena that vary. For example, the natural variability in human weight is due to a combination of genetic, nutritional, and other lifestyle or environmental factors. Variability arising from independent factors that combine multiplicatively generally will lead to an approximately log-normal distribution across the population or across spatial/temporal dimensions.

4.8.3 Monte Carlo Analysis

Among the most promising and exciting techniques to emerge in the area of exposure assessment in recent years is the application of Monte Carlo or other probabilistic analyses to environmental health issues.[221–223] Monte Carlo analysis has existed as an engineering analytical tool for many years, but the development of the personal computer and software [e.g., Crystal Ball (Decisioneering, Boulder, CO), @RISK (Palisades Corp., Newfield, NY)] has allowed its application to new areas. As discussed previously, one criticism of many exposure assessments has been a reliance on overly conservative assumptions about exposure, as well as the problem of how to properly account for the highly exposed (but usually small) populations that do exist.[47,217] Monte Carlo techniques offer an approach to addressing this issue.

The probabilistic, or Monte Carlo, model accounts for the uncertainty in select parameters evaluating the range and probability of plausible exposure levels. Instead of specifying input parameters as single values, this model allows for consideration of the probability distributions. The Monte Carlo statistical simulation is a statistical model in which the input parameters to an equation are varied simultaneously. The values are chosen from the parameter distributions, with the frequency of a particular value being equal to the relative frequency of the parameter in the distribution. The simulation involves the following three steps:

1. The probability distribution of each equation parameter (input parameter) is characterized, and the distribution is specified for the Monte Carlo simulation. If the data cannot be fit to a distribution, the data are "bootstrapped" into the simulation, meaning that the input values are randomly selected from the actual data set without a specified distribution.

2. For each iteration of the simulation, one value is randomly selected from each parameter distribution, and the equation is run. Many iterations are performed, such that the random selections for each parameter approximate the distribution of the parameter. Five thousand iterations are typically performed for each dose equation.

3. Each iteration of the equation is evaluated and saved; hence a probability distribution of the equation output (possible doses) is generated.

This technique generates distributions that describe the uncertainty associated with the risk estimate (resultant doses). The predicted dose for every 5th percentile to the 95th percentile of the exposed population and the true mean are calculated. Using these models, the assessor is not forced to rely solely on a single exposure parameter or the repeated use of conservative assumptions to identify the plausible dose and risk estimates. Instead, the full range of possible values and their likelihood of occurrence is incorporated into the analysis to produce the range and probability of expected exposure levels.[69,223-227]

The methodology is illustrated in the following examples. The first example is to understand the time needed to go shopping. Time spent shopping each month (minutes) is estimated by the product of two parameters: the number of trips per month and the total time spent in the store (minutes). Total time spent in the store is the sum of time spent shopping and time spent waiting in line. Using Monte Carlo techniques, a distribution of likely values is associated with each of these parameters. These distributions depend upon the detail of information available to characterize each parameter. For example, the distribution compares all of the information, such as those days when the line at the check-out counter is short, as well as those when the line is long. It is noteworthy that each parameter has a different distribution: log-normal, Gaussian, and square. Total time spent shopping is then calculated repeatedly by combining parameter values that are randomly selected from these distributions. The result is a distribution of likely time spent shopping each month. Using this technique, information concerning each parameter is carried along to the final estimate.

The second example, which directly applies to toxicologists, is to build a distribution that describes the various soil ingestion rates for children. As shown in Figure 4.6, the three pertinent distributions are the basis for constructing the overall exposure distribution.

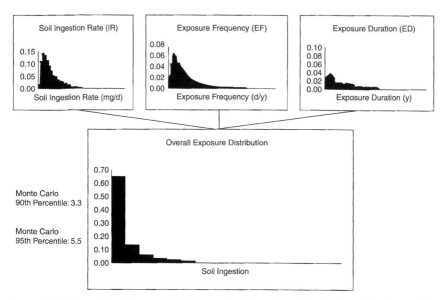

Figure 4.6   Example of how probability density functions (distributions) for three differ-ent related exposure factors are combined to form a distribution for the amount of soil ingested by a population of children. The Monte Carlo technique allows the risk assessor to account for the variability in many exposure parameters within a population and then produce a distribution that characterizes the entire population.

Most of the variables used in an exposure assessment actually exist as ranges, rather than single point values. For instance, the common assumption that adult body weight is 70 kg will be replaced in a Monte Carlo analysis by the appropriate distribution (i.e., normal) of body weights (including maximum, minimum, mean, and standard deviation). Using this approach, virtually every exposure variable, whether physiological, behavioral, environmental, or chro-nological, can be replaced with a probability distribution.[44,69,92,224–238] Since no population (or individual) is exposed to a single concentration, breathes, eats, or drinks at a single rate, or is exposed for the same length of time, it is not appropriate to assess them as such. To be protective, high values are em-ployed, resulting in the problems of compounding conservatisms mentioned previously.[42,43,47,239,240] The probabilistic analysis addresses the main defi-ciencies of the point estimate approach because it imparts more information to risk managers and the public and uses all of the available data.[241,242] The range of values (i.e., the distribution) for all the variables used in an exposure assess-ment is determined (e.g., normal, log-normal, uniform, triangular, etc.) and com-bined into a "distribution of distributions." Because of the extrapolations in-volved and the assumptions made, the area of single greatest uncertainty in risk assessment is associated with the dose–response evaluations.

The Monte Carlo or probabilistic approach deals with the main deficiencies of the point estimate approach because it shares a great deal more information with the risk manager. Instead of presenting a single point estimate of risk, probabilistic analyses characterize a range of potential risks and their likelihood of occurrence. In addition, factors that have the greatest effect on results can be easily identified. For example, in a probabilistic analysis, the risk manager receives the following type of information: The plausible increased cancer risks for the 50th, 95th, and 99th percentiles of the exposed population are 1×10^{-8}, 5×10^{-7}, and 1×10^{-6}, respectively.

It should also be clear that, in addition to exposure variables, data forming the basis of the toxicological criteria [carcinogenic potency factors (CPFs) and reference doses (RfDs)] are also amenable to Monte Carlo–style analysis where a robust database exists.[243–255] As with exposure variables, the advantage to this approach is that it allows all data to be used (and weighted appropriately, where necessary), thus avoiding reliance on a single experiment or endpoint.

Probabilistic analyses have in recent years been recognized in regulatory guidance,[12] and EPA's Risk Assessment Forum has published a document of principles for conducting Monte Carlo analyses[249] (Table 4.12). Like traditional exposure analysis, one challenge to performing a Monte Carlo analysis properly is having appropriate distributions for use in the analysis. Numerous studies on individual variables have been published in the risk assessment literature,[46,241,256–266] and the impact on the distributions employed on the outcome has also been discussed.[267–272] It should be pointed out that these techniques can be combined with other advanced risk assessment methods (i.e., PB-PK modeling) to further reduce uncertainty in risk estimates.[273,274] Recently, two-dimensional Monte Carlo analyses have been developed that take into account both variability and uncertainty. Information appropriate to probabilistic analyses can often be found in published works in fields quite distant from the environmental sciences.

4.8.4 Case Study Using Monte Carlo Technique

An example might be useful. Assume that persons are likely to be exposed to contaminated drinking water at the maximum contaminant limit (MCL). Concern has been raised that these regulatory limits are not sufficiently protective, and that certain federal and state regulatory programs (i.e., Resource Conservation and Recovery Act) are justified in requiring groundwater remediation to levels below that of drinking water standards. To test this supposition, it is necessary to evaluate the possible incremental cancer risk of exposure via tap water ingestion, dermal contact with water while showering, inhalation of indoor vapors, and ingestion of produce irrigated with groundwater, using a probabilistic approach. Probability density functions for each exposure variable (e.g., water ingestion, skin surface area, fraction of exposed skin, showering time, inhalation rate, air exchange and water use rates, exposure time, etc.) are then identified and used in the appropriate exposure equation to calculate dose and risk. A commercially available software package (i.e., @RISK) could be used to conduct the Monte Carlo analysis.[235]

TABLE 4.12   U.S. EPA Guiding Principles for Monte Carlo Analysis

1. Conduct preliminary sensitivity analyses to identify important contributors to the assessment endpoint and its variability and uncertainty.

2. Based on the results of the sensitivity analyses, include probabilistic assessments only for the important pathways and parameters.

3. Use the entire database of information when selecting input distributions.

4. When using surrogate data, identify sources of uncertainty, and whenever possible, validate the use of these data by collecting site/case specific data.

5. If empirical data are collected for use in the assessment, use collection methods that improve the representativeness and quality of these data (especially at the tails of the distribution.)

6. Identify when expert judgment, rather than hard data, is used in the assessment.

7. Separate uncertainty and variability during the analysis.

8. Use appropriate methods to address uncertainty and variability, e.g., two-dimensional Monte Carlo.

9. Discuss the numerical stability of estimates at the tails of the distribution.

10. Identify which sources of uncertainty are addressed by the assessment, and which are not.

11. Provide a detailed description of all models used.

12. Provide a detailed description of the input distributions, including a distinction between variability and uncertainty in these distributions, and a graphical representation of the probability density and cumulative distribution functions.

13. Provide a graphical representation of the probability density and cumulative distribution functions of each output distribution.

14. Consider the potential covariance between important parameters. If the covariance cannot be determined, evaluate the impact of a range of potential covariances on the output distributions.

15. Present point estimates and identify where they fall on the exposure distribution. If there are large differences between point estimates and Monte Carlo estimates, explain if the differences are due to changes in the data or models used.

16. Present results in a tiered approach.

Note: From International Life Science Institute (1998).

Some have suggested that the Latin Hypercube (LHC) approach offers some advantages to traditional approaches for identifying the correct number of iterations. Often, one can reach convergence sooner with LHC than the Monte Carlo option in @RISK/Crystal Ball. In addition, LHC is more reproducible (to the hundredth decimal place). The Monte Carlo option needs more iterations to reach convergence.

The results of such an analysis are presented in Table 4.13.[258] The risk associated with exposure to water at the current MCL level for four different contaminants, as well as the 50th and 95th percentile of exposure as determined by the probabilistic analysis, are shown. At the 50th percentile level ("the best

TABLE 4.13 Risks Calculated for Exposure to Four Halogenated Solvents in Water Using Probabilistic Analysis at the MCL Level and for the 50th and 95th Percentile Exposure

Chemical	50th Percentile risk	95th Percentile risk	MCL risk
Tetrachloroethylene	0	0.000005	0.000007
Chloroform	0.000009	0.00014	0.000017
Bromoform	0.000002	0.000016	0.000023
Vinyl chloride	0.000005	0.000029	0.000054

Note: Adapted from Finley and Paustenbach (1994).

estimate"), the risk ranges from 6×10^{-7} (tetrachloroethylene) to 9×10^{-6} (chloroform), while at the 95th percentile ("the upper-bound risk"), these risks range from 4×10^{-6} (tetrachloroethylene) to 1.5×10^{-4} (chloroform). These values can be compared to the point estimate risks calculated for the MCLs, which range from 7×10^{-6} (tetrachloroethylene) to 5.4×10^{-5} (vinyl chloride). For the 50th percentile (average) person, all calculated risks are within the range of "acceptable" risks adopted by regulatory authorities for Superfund sites (1×10^{-4} to 1×10^{-7}). For the 95th percentile person (upper bound), the risks are still mostly below the 1×10^{-4} benchmark risk level generally used to separate acceptable from unacceptable risks. For tetrachloroethylene, these results are 30 (50th percentile) to 3 (risk at the MCL) times below the reasonable maximum exposure (RME) risk of 2×10^{-5} developed by combining the 95th percentile values for each exposure variable using standard EPA risk assessment methodologies. This point estimate is greater than the 99th percentile of risk and is consistent with statements regarding the conservatism of the RME approach. These results suggest that chemical residues in drinking water at the MCL levels will be health protective and that remedial goals based on de minimis requirements (1×10^{-6}) might be unnecessarily low.[258]

In terms of RME estimates, which often serve as the basis for regulatory decisions, several observations on the utility of probabilistic assessment can be made. First, exposure assessments that incorporate two to three direct exposure pathways usually show that the 95th percentile probabilistic estimates are three to five times below the traditional RME estimates. Second, for multipathway assessments that contain several indirect exposure pathways, the 95th percentile probabilistic estimates can be as much as an order of magnitude below the RME estimates. Third, when the number of distributions used in the exposure assessment is 10 or more, the difference between the 50th and 95th percentile estimates may be between 5 and 10. Finally, in such assessments, the difference between the RME estimates and the 95th percentile probabilistic estimates can be as high as 100. In the probabilistic approach to estimating exposure and risk, the complete range of potential risks can be illustrated along with the likelihood estimates and estimates of uncertainty associated with such risks. While the availability and confidence of distributions for exposure variables differ, risk assessors ought to take advantage of this and similar approaches in their risk assessments to advance and improve the process. Additionally, since the highest degree of uncertainty in risk assessment tends to be the CPFs, attention ought to be directed to applying probabilistic analysis to the development of toxicity criteria in a similar manner.[238,246,247]

4.8.5 Sensitivity Analysis

In addition to establishing exposure and risk distributions, probabilistic analysis can also identify variables with the greatest impact on the estimates and illuminate uncertainties associated with exposure variables through sensitivity analysis.[275–279] This provides some insight into the confidence that resides in exposure and risk estimates and has two important results. First, it identifies the inputs that would benefit most from additional research to reduce uncertainty and improve risk estimates. Second, assuming that a thorough assessment has been conducted, it is possible to phrase the results in more accessible terms, such as "the risk assessment of PCBs in small-mouth bass is based on a large amount of high-quality reliable data, and we have high confidence in the risk estimates derived. The analysis has determined that 90 percent of the increased cancer risk could be eliminated through a ban on carp and catfish, but there is no appreciable reduction in risk from extending such a ban to bass and trout."[252] Such a description provides all stakeholders with considerably more information than a simple point estimate of risk based on a traditional exposure and risk assessment.[3]

If the most "sensitive" exposure variables are based on limited or uncertain data, confidence in these estimates will be poor. Robust data sets, on the other hand, lead to increased confidence in the resulting estimates. In the above example involving small-mouth bass, sensitivity is defined as the ratio of the relative change in risk produced by a unit relative to change in the exposure variables used. A Gaussian approximation (the product of the normalized sensitivity and the standard deviation of the distribution) of intake was used to allow both sensitivity and uncertainty to be gauged. In this case, the true mean of each distribution was chosen as the baseline point value, and the differential value for each variable was calculated by increasing this value by 10%. For each variable, the differential value was substituted, the risks recalculated, and the baseline value replaced.[258] Sensitivity was calculated using the following formula:

$$\text{Sensitivity} = \frac{|\text{Risk}_{\text{baseline}} - \text{Risk}_{10\%}|}{|X_{\text{baseline}} - X_{10\%}|}\,\sigma$$

where $X_{\text{baseline}}, X_{10\%}$ = baseline and differential values for the variable X

$\qquad\qquad \sigma$ = standard deviation for the distribution of variable X

The sensitivity of each variable relative to one another is assessed by summing the unitless sensitivity values and determining the relative percent that each

TABLE 4.14   Results of Sensitivity Analysis for Tetrachloroethylene Exposure in Household Water

Exposure variable	Sensitivity (unitless)	Percentage rank
Shower exposure time	0.000004	55.0%
Exposure duration	0.000001	20.0%
Plant–soil partition factor	0	8.4%
Water ingestion rate	0	4.6%
Surface area of exposed skin	0	4.4%
Body weight	0	3.8%
Dermal permeability constant	0	1.8%
Skin fraction contacting water	0	1.5%

Note: Adapted from Finley and Paustenbach (1994).

variable contributes to the total. Table 4.14 indicates which are the most important variables in the probabilistic analysis for tetrachloroethylene. In this case, the most sensitive exposure variables in household exposure to tap water are exposure time in shower and exposure duration. Relatively small changes in these variables will result in relatively large changes in the risk estimates. Since these estimates are based on actual time-use studies and census information, this suggests a high level of confidence can be placed on this estimate, particularly if site-specific data are being used. If the critical variables (in terms of sensitivity) were not based on robust data, this would suggest that the risk assessment could be improved by additional research on these exposure variables. It is interesting to point out that the form of the distribution chosen for the variables is less important than the validity of the data.[280] When the empirical distribution of the tap water ingestion rate from Erschow and Cantor[281] was substituted with a lognormal distribution developed by Roseberry,[282] the resultant change in the risk estimates was less than 1%.[235]

In this case, the value of the sensitivity analysis is that it allows input variables to be ranked in order of importance and confidence in the output to be established to a higher degree than previously possible. As pointed out by EPA, "[w]here possible, exposure assessors should report variability in exposures as numerical distributions and should characterize uncertainty as probability distributions. They need to identify clearly where they are using point estimates for 'bounding' potential exposure variables or estimates; these point estimates should not be misconstrued to represent, for example, the upper 95th percentile when information on the actual distribution is lacking."[252] As noted by EPA, such explicit presentation of the data reduces the temptation to use the exposure assessment process for veiling policy judgments.[283]

4.9   ISSUES IN EXPOSURE ASSESSMENT

The field of exposure assessment will continue to benefit from ongoing research efforts. The following are some fruitful areas of research.

TABLE 4.15 Effect of Matrix and Aging on the Bioavailability of Lead from Soil

Treatment			
Lead acetate (ppm diet)	Soil lead (ppm)	Tibial lead (standard deviation)	Relative lead absorption
—	—	0.3 (0.3)	—
—	11.3	0	—
50	—	247 (10)	100
50	11.3	130 (30)	53
—	706	40 (6)	16
—	995	108 (26)	44
—	1080	37 (7.3)	15
—	1260	53.6 (7)	22
—	10420	173 (22)	70

Note: Lead acetate in the diet results in an increase in tibial lead, while lead acetate mixed with soil is only 50% as well absorbed. Aged lead from garden soil must reach high levels before significant absorption occurs. Adapted from Chaney et al. (1984).

4.9.1 Bioavailability

Applied research that will improve the practice of exposure assessment is bioavailability, speciation, chemical fate, and the role of biological monitoring. Bioavailability has become an increasingly important aspect of the exposure assessment process.[58,284,285] Alexander[286] has shown that a variety of organic chemicals in soil lose the ability to interact with biological receptors over time, despite the fact that the chemical concentration in soil remains largely the same. The alteration in bioavailability extends across the various routes of exposure as well.[57,141–145,287] Inorganic compounds, even those posing a potentially significant degree of hazard (i.e., cyanide) react similarly.[288–291] These losses in hazard potential are presumably due to irreversible chemical interactions with soil constituents. Table 4.15 indicates that the bioavailability of lead added to soil is immediately halved and that it is further reduced over time.[292] This would suggest that an assumption of 100 percent bioavailability of this compound (and many others) from soil is erroneous. It is also clear that the environmental media in which the compound occurs will also influence its uptake into the body.[287] EPA recognized this fact when it developed two RfDs for manganese depending on whether it occurred in solid matrices (e.g., food, soil, etc.) or water.[293] One simple method to improve bioavailability estimates is to conduct extractions under more biologically relevant conditions.

Bench-scale extraction experiments in simulated gastric fluids or sweat can be used to inexpensively and accurately measure how readily environmental residues can be released from the media in which they occur.[294,295] As with inhalation or ingestion of vapors or solutions, both the release and absorption rates of agents from an environmental matrix (i.e., soil) across biological membranes need to be incorporated into the risk assessment when such data are available, and generated when absent. This need is particularly of issue for assessing dermal exposure. The problem for materials in aqueous solutions is less problematic then from solid matrices.[50] For liquids, permeability constants expressed in terms of agent weight per unit area per time ($mg/cm^2/min$) have been developed for a number of agents, and in vivo and in vitro techniques or mathematical models exist to develop similar flux rates if needed.[296-306] From soil, however, the typical approach in many risk assessments has been to assume a constant percent absorbed from soil adhered to skin as a default. For volatiles, an absorption rate of 25 percent has been used. For semivolatiles and inorganics, absorption rates of 10 and 1 percent have been used, respectively.

Some experimental data for absorption are available for a few agents (e.g., PCBs, DDT, dioxin, benzo[a]pyrene, etc.), suggesting that the simple assumption of a constant percentage absorbed may overestimate or underestimate the dose depending on the agent, co-contaminants, soil type, exposure duration, and similar considerations.[57,144,306,307] The impact of this default approach results in an instantaneous dermal dose being assumed, regardless of whether the soil remains in contact with the skin for one minute or one day. This assumption, together with the questionable route-to-route adjustment of toxicity criteria from oral to dermal previously discussed, results in the dermal absorption of agents from soil, which arguably should present a minor exposure and risk in most cases, being a major driver in the risk assessment of soil-bound contaminants.

4.9.2 Chemical Fate

Risk assessors ought to incorporate information on the fate of chemicals in the environment in their exposure estimates, whenever possible.[289] Many organic compounds tend to degrade over time, and may disappear from exposed surfaces relatively quickly or otherwise change.[308,309] As suggested above, inorganic compounds may also undergo changes in the environment over time that affect their fates.[291,292] Influencing factors include degradation by sunlight, soil and water microbes, evaporation, and chemical interactions. The resultant changes can dramatically alter the outcome of exposure assessments.[285,290] For instance, most criticism of incinerators has focused on the inhalation risk of dioxin emitted from the stacks. As it turns out, the environmental half-life of dioxin (as a vapor) is only 90 minutes because of photolytic degradation. In contrast, the half-life for dioxin in soil or fly ash is 50 to 500 years. The focus of concern is often not the main risk issue when environmental fate is considered because levels and availability change over time.[6] Incorporation of half-life data into risk assessments can have substantial benefits for improving understanding of the potential exposures and risks associated with a specific situation.[309,310] In a similar manner, the risk from persistent contaminants (i.e., DDT) in fish has usually been assessed using results from the analysis of raw fish fillets in combination with assumptions about the size and number of fish meals. The effects of cleaning and cooking on these residues are not typically considered, but have been shown to reduced substantially in many cases (i.e., 50 percent or greater).[311,312] Since many of these risk assessments form the basis of fish advisories or bans with potentially significant economic repercussions, there is obviously an important reason to make these exposure estimates as accurate as possible. Additionally, since there are known health benefits to fish consumption, making recommendations against eating fish based on theoretical risk needs to be rigorously defended.[313]

4.9.3   Biomarkers and Molecular Epidemiology

The past decade has witnessed a dramatic increase in the level of research activity, derivation of theoretical constructs, and development of practical applications for the direct measurement of biological events or responses that result from human exposure to xenobiotics.[90,314] These measurements, conveniently grouped under the descriptor "biological markers" or "biomarkers," reflect molecular and/or cellular alterations that occur along the temporal and mechanistic pathways connecting ambient exposure to a toxicant and eventual disease. As such, an almost limitless array of biomarkers is theoretically available for assessment, and only a minute fraction of these has been recognized and investigated to date.[314-316] Some events that can technically be classified as biomarkers of chemical exposure (e.g., hematological changes following high levels of exposure to lead or benzene, acetylcholinesterase inhibition by organophosphates) have been measured for decades. However, the recent surge of interest in this field has been driven by technical advances in analytical chemistry and molecular genetic techniques and by the recognition that "classical" toxicology and epidemiology may not be able to alone resolve critical questions regarding causation of environmentally induced disease.[90]

Biomarkers are an important component of the emerging discipline of molecular epidemiology, which seeks to expand the capabilities and overcome the limitations of classical epidemiology by incorporating biological measurements collected in exposed humans.[90,317] Early efforts at utilizing biomarkers to make quantitative estimates of exposure and to predict human cancer risk were made by Ehrenberg and Osterman-Golkar.[318] Using ethylene oxide as a model xenobiotic, these investigators explored the use of macromolecular reaction products (i.e., hemoglobin adducts) as internal dosimeters. By employing hemoglobin adduction data, they predicted the level of ambient ethylene oxide that would correspond to a tumorigenic dose of γ radiation, which they termed the "rad-equivalent dose." Seminal work in the area of biomarkers as applied to the molecular epidemiology of cancer was performed by Perera and Weinstein,[319] who proposed the use of such techniques to identify environmental contributors to human cancer incidence. Important early applications of biomarkers to characterize environmental and occupational exposure have also been explored by several other groups in the United States and abroad.

As presented in the original NRC report, biomarkers of internal dose reflect the absorbed fraction of a xenobiotic, that is, the amount of material that has successfully crossed physiological barriers to enter the organism. Consequently, the magnitude of the biomarker accounts for bioavailability and is influenced by numerous parameters such as route of exposure, physiological characteristics of the receptor, and chemical characteristics of the xenobiotic. Generally, simple measurement of xenobiotic levels in biological media (blood, tissue, urine) can provide data on internal dose and this is called biomonitoring.[51] Biomarkers, on the other hand, reflect internal dose (in terms of proximity to downstream events in the sequence) and could include the measurement of a metabolite in selected biological media, particularly if such metabolite is active or critical to the toxic effects seen.[90]

Very useful exposure biomarkers for reactive xenobiotics or their activated (i.e., electrophilic) metabolites are macromolecular reaction products. Substantial research effort has been devoted to the use of protein and deoxyribonucleic acid (DNA) adducts as molecular dosimeters. Ehrenberg and co-workers first proposed using hemoglobin (Hb) adducts to monitor the internal dose of alkenes and epoxides such as ethylene oxide over two decades ago.[318] This methodology has since evolved into a widely used and highly sensitive technique for quantitating N-terminal Hb adducts of a variety of xenobiotic metabolites in human blood. Hb adducts have been employed as internal exposure biomarkers for aromatic amines, nitrosamines, polycyclic aromatic hydrocarbons (PAHs), and other compounds.[90]

Protein and DNA adducts can be considered as biomarkers of either internal or biologically effective dose, depending upon how close their relationship is to actual disease occurrence. The NRC report defined biologically effective dose as *dose at the site of action, dose at the receptor site, or dose to target macromolecules.*[320] This definition is troublesome since, strictly speaking, complete characterization of molecular site and mechanism of action for a given xenobiotic would be necessary to assign a particular measured endpoint as a marker of biologically effective dose. For example, protein adducts cannot be considered as effective dose biomarkers for carcinogens, since they do not satisfy the above criteria. Ambiguity exists even for DNA adducts, since in no reported instance has xenobiotic-induced adduction of a specific base within a particular DNA sequence in a target cell type been unequivocally linked to a specific clinical outcome in people.[90] Despite these uncertainties, adducts in total lymphocyte DNA are considered as appropriate biologically effective dose biomarkers for carcinogens, based upon the postulated mechanism of chemical carcinogenesis and limited experimental data indicating correlations between DNA adducts in lymphocytes and target tissues.[90]

Ideally, a biomarker should be biologically relevant, sensitive, and specific (i.e., valid). In addition, it should be readily accessible, inexpensive, and technically feasible. This combination of requirements is rarely achieved, and some trade-off is inevitable in order to obtain useful biomarker data in a timely manner. A few promising examples are presented in Table 4.16. The validation process for a biomarker involves determining the relationship between the biological parameter measured and both upstream and downstream events in the

TABLE 4.16   Biomarkers Examined for Selected Occupational and Environmental Chemicals

Chemical	Biomarker		
	Exposure	Effect	Susceptibility
PAH	DNA adducts[a]	*hprt* mutation	GST-M1
	Hb adducts	*gpa* mutation	NAT-2
	SA adducts	*fes* oncogene activation	CYP1A1
	Urinary 1-HP[a]	*ras* p21 level	CYP1A2
	Sister chromatid exchange (SCE)	DNA single-strand breaks	
	SCE (high-frequency cells)	Chromosomal aberrations	
		Micronuclei	
1,3-Butadiene	Hb adducts[a]	*hprt* mutation	
	Sister chromatid exchange (SCE)	Chromosomal aberrations	
	Urinary metabolites	Micronuclei	
		ras oncogence activation	
Acrylamide	Hb adducts[a]		
	Urinary metabolites		

Note: From DeCaprio (1997).
[a]Biomarkers for which cumulative data indicate best correlation with ambient exposure.

continuum, that is, the dose–response curve must be characterized.[90] For example, a Hb adduct considered for use as an exposure biomarker for a xenobiotic should exhibit a predictable relationship to ambient exposure level. In addition, if used as a surrogate for DNA adduction, then a reproducible correlation between Hb and DNA adducts must be demonstrated. Biological relevance refers to the nature of the phenomenon being measured and its mechanistic involvement in the pathway from exposure to disease. For biomarkers of exposure, disease relevance is not as critical a requirement as is a predictable exposure–response relationship; the opposite is true for biomarkers of effect.[90]

There are a few terms that are used when discussing biomarkers that help define their usefulness. Sensitivity, for example, reflects the ambient exposure level that can be detected by means of the biomarker. Highly sensitive markers are necessary to quantitate the low ambient levels typical of environmental exposures in industrialized Western nations. Specificity is the probability that the biomarker is indicative of actual exposure to the specific xenobiotic that it is designed to detect. For example, certain macromolecular adducts can be derived from exposure to a number of chemical species and are thus less specific than one unique to a single compound. Biomarkers must also be reasonably accessible. Thus, with the exception of occasional tissue biopsies, samples for use in exposure biomarker studies generally consist of blood, urine, milk, or other readily obtainable biological media. Since these are rarely target tissues for toxicological or carcinogenic effects, it must be inferred how the concentration relates to the incidence of disease. Finally, cost and technical feasibility are important considerations in selection of appropriate biomarkers for applied studies.[90]

4.9.4 Statistical and Analytical Issues

Despite the use of precise and reproducible analytical methods, we often do not have enough data of chemical concentrations to estimate exposure with great certainty. Due to resource availability, over the past 15 to 20 years it has often been the case that a single round of analytical results or samples collected for other purposes[322] serves as input and the surrogate for long-term or lifetime exposure. As noted previously, chemical concentrations vary over both time and space, which makes the task of dose estimation all the more difficult.[311] For instance, to use the (estimated) average dose to predict the typical lifetime dose may seriously overestimate or underestimate the actual dose. Additionally, the average dose may be less important in the biological scheme of things than peak exposures or exposures at specific times (i.e., developmental effects) and ought to be considered as such in the evaluation of exposure.[71] Techniques do exist for estimating long-term exposure from short-term data,[323–325] but the reliability of these estimates is uncertain. Similarly, a variety of mathematical or bench-scale models exist that have been used to estimate exposure in the absence of measurements or long-term monitoring data.[326] As has been noted on several occasions, "all models are wrong, but some are useful," and risk assessors should carefully evaluate mesoscale and microscale models, as well as model outputs, for relevance and accuracy. Often, field measurements can serve as useful and relatively inexpensive "reality checks" on model results.

Equally important in exposure assessment are the statistics used to analyze field data. Environmental data are most often log-normally distributed. Under such conditions, a geometric average is generally assumed to be a better measure of the central tendency of data than the arithmetic mean.[327] Despite this, the arithmetic mean (and the 95 percent upper confidence limit of the arithmetic mean) is typically used to identify environmental concentrations for use in exposure assessment. Since the advances in analytical chemistry have improved our ability to measure trace amounts of chemicals in different media and identify potential sources in some situations, less reliance should be placed on the use of mathematical models to predict the distribution of chemical and physical agents in the environment, and actual field data should be collected.[336,337]

Another important issue in exposure assessment is how the analytical limit of detection (LOD) is handled in calculations. An agent reported as a nondetect may be treated as a numerical zero, or occurring at the LOD or some fraction of the LOD, typically one-half, for purposes of calculating statistics. The manner in which censured data is assessed may affect the outcome of the risk assessment process.[328–335] For instance, analysis of highly contaminated samples or samples containing interfering substances may result in high LODs. Under such conditions and in the absence of additional analysis, assuming that nondetects are present at one-half the LOD could result in the exposure assessment and subsequent risk assessment being driven by compounds that are not truly present in the environmental media. When such an approach is used on a site that may be only 2 to 10 percent contaminated (based on surface area), the predicted severity of the level of contamination will be much higher than what actually occurs.[327]

The practical result of these decisions can be illustrated by considering the following 11 data points resulting from analysis of field samples: Nondetect (ND), ND, ND, ND, ND, 5, 6, 6, 8, 55, and 500 ppm. The results are lognormally distributed as expected. The detection limit is 0.05 ppm, and nondetects are assumed to be present at one-half the detection limit (0.025 ppm). Using these assumptions, the arithmetic mean of the data set is 52.7 ppm, while the geometric mean is 1.3 ppm. The practical consequence of choosing one descriptor over the other may be to misidentify or mischaracterize the dose and ultimately the risk, and will influence regulatory decisions involving remediation and regulation.

4.10 CLOSING THOUGHTS

The field of exposure assessment has evolved significantly over the past 20 yr. We have learned a great deal about where people are exposed to xenobiotics and the relative degree of exposure. Not that long ago, most of our concerns were about industrial chemicals in our water, ambient air, and the soil. Today, we know that indoor exposure to particles, vapors, and gases (influenced by smoking) often represents the predominant source of exposure for most persons. A greater portion of our work in the future will undoubtedly focus on better understanding of both occupational and indoor exposure, rather than environmental exposures.

It is my personal view that of the four portions of a risk assessment, exposure assessment has made the biggest improvement in quality over the 20-yr history of health risk assessment. Usually, exposure assessments will contain less uncertainty than other steps in a risk assessment, especially the dose-response portion. Admittedly, there are a large number of factors to consider when estimating exposure, and it is a complicated procedure to understand the transport and distribution of a chemical that has been released into the environment. Nonetheless, the available data indicate that scientists can do an adequate job of quantifying the concentration of the chemicals in the various media and the resulting uptake by exposed persons if they account for all the factors that should be considered.

There are at least 11 significant lessons we have learned about conducting exposure assessments in recent years. Had we not had to learn through experience, avoiding these lessons could potentially have saved the United States hundreds of millions of dollars and thousands of person-years of work. First, experience has shown that in our attempts to be prudent, we placed too much emphasis on the "so-called" maximally exposed individual (MEI) (117, 242). Often, the results of those analyses were misinterpreted by the public and/or misrepresented by some scientists or lawyers. Often, as a result, poor decisions were made by risk decision makers.

Second, as we have learned how to accurately characterize the risks of exposure for about 95% of the population, more emphasis has been placed on evaluating the various special groups (e.g., Eskimos, subsistence fishermen, dairy farmers). Although the risk for these populations, who can be exposed to particularly high doses (the 95–99.99% group) needs to be understood, the typical levels of exposure for the majority of the population should be the initial focus of the assessment. Perhaps the most significant change in exposure assessment of the past three years has been the national interest in characterizing the risks to children.

The third lesson is a variation of the second: Do not allow the repeated use of conservative assumptions to dictate the results of the assessment. In recent years, many investigators have addressed this issue and have demonstrated its importance. Monte Carlo techniques can generally be successful in addressing this problem.

Fourth, we have learned that risk managers and the public want to understand the statistical confidence in our estimates of risk. Sensitivity analyses can yield important information about the critical exposure variables (132, 144, 290, 356). Further, most risk assessments can benefit from analyses of both variability and uncertainty. Without these, risk managers are not fully informed.

Fifth, we have improved our techniques for statistically handling data; and particularly for samples that have no detectable amount of a contaminant. In the past, regulatory agencies have used the limit of detection (LOD) of the analysis or one-half the LOD in the exposure calculations relying on the premise that the contaminant might be present at that level. We learned that when such an approach is used (without reflection) on a site that may only be 2–10% contaminated (based on surface area), the impact of a few samples on the results could lead us to improper conclusions about the actual level of risk to persons who live there or nearby.

Sixth, we have gained a significant degree of confidence in our ability to estimate historical exposures; so-called dose-reconstruction or retrospective risk assessments, a term that this author coined in 1983. Over the past ten years, for use in epidemiology studies, the likely exposure of workers and/or those in the community nearly forty to fifty years ago have been estimated using chemical usage and emission data, measured data, and models (145, 265, 294, 333, 345, 424).

Seventh, we now understand the need to quantitatively account for indirect pathways of exposure. For example, the uptake of a contaminant in water by humans due to ingestion is obvious (and direct), but the uptake of the same contaminant by garden vegetables due to watering and the uptake via the inhalation of volatile contaminants from the water while showering are indirect pathways that had not always been evaluated in assessments. Perhaps the most important indirect route of exposure, which had not been considered before 1986 when regulating airborne nonvolatile chemicals, was the ingestion of particulate emissions that have deposited onto soil and plants and were subsequently eaten by grazing animals (125, 388). Much additional research in this area will be conducted, and the results will probably change our views about the hazards posed by numerous chemicals.

Eighth, we have learned that children and their exposure patterns are unlike those of adults. As some have said, in more ways than one, children are not miniature adults! Their intake of certain foods, percentage of time spent outdoors, proximity to carpets, and inhalation rates per body weight are all different than adults.

The ninth lesson learned is to use biological monitoring to validate or confirm the predicted degree of human exposure. Over the past 5 to 10 yr, analytical chemists have increased their ability to detect very small quantities of dozens of chemicals in blood, urine, hair, feces, breath, and fat. For many chemicals, these data represent a direct indicator of recent exposure, and in some cases (like PCBs and dioxins), chronic exposure. Validation of our exposure assessments should be one of the major areas of study during the next decade (through both biomonitoring and molecular epidemiology).

Tenth, it has become clear that in most cases, the most significant risks due to exposure to chemicals occur in the workplace. Even though great strides have been made in industrial hygiene over the past 50 yr, the doses to which persons can legally be exposed are much greater (often by a factor of 100) than those to which most persons not in the community will ever be exposed.

Eleventh, and perhaps most important, we have learned that (for most persons) exposures to chemicals and bacteria in the home pose a greater risk than to those in the ambient air or through the ingestion of water. Many fine studies conducted in the 1970s through the current day continue to show that in-home exposures to most chemicals are often about 2–20 times greater than that present in the ambient environment (83, 131, 167, 183, 195, 196, 247, 407). Recently, more than 200 scientists in the fields of epidemiology, exposure assessment, and medicine signed a document called a "consensus statement," which states that future research should focus on personal monitoring; especially of the indoor environment (104).

We have come a long way in a short time. Several professional societies, including the International Society of Exposure Analysis (ISEA), Society for Risk Analysis (SRA), American Industrial Hygiene Association (AIHA), Air and Waste Management Association (AWMA), American Chemical Society (ACS), Society of Toxicology (SOT), International Society for Regulatory Toxicology and Pharmacology (ISRTP), and others, have placed an emphasis on improving the practice of exposure assessment. All indications are that the information we have gained has significantly improved the quality of recent risk assessments, and it can be expected that due to better exposure assessments, future decisions by risk managers will be much better informed.

REFERENCES

1. National Academy of Sciences (NAS). *Risk Assessment in the Federal Government: Managing the Process.* National Academy Press, Washington, DC, 1983.

2. D. J. Paustenbach. *The Risk Assessment of Environmental Hazards: A Textbook of Case Studies.* Wiley, New York, 1989.

3. D. J. Paustenbach. The practice of health risk assessment in the United States (1975–1995): How the U.S. and other countries can benefit from that experience. *Hum. Ecol. Risk Assess.* 1(1): 29–79 (1995).

4. A. J. Lehmann and O. G. Fitzhugh. 100-fold margin of safety. *Q. Bull. Assoc. U.S. Food Drug Admin.* 18: 33 (1954).

5. Center for Risk Analysis. *Historical Roots of Health Risk Assessment.* Harvard University, School of Public Health, Cambridge, MA, 1994.

6. D. J. Paustenbach, H. W. Leung, and J. Rothrock. Risk assessment and the practice of occupational medicine. In R. Adams (Ed.), *Occupational Skin Diseases*, 2nd ed. Taylor and Francis, New York, 1999.

7. National Research Council. *Improving Risk Communication.* National Academy Press, Washington, DC, 1989.

8. National Research Coucil. *Science and Judgment in Risk Assessment.* National Academy Press, Washington, DC, 1994.

9. National Research Council. *Understanding Risk: Informing Decisions in a Democratic Society.* National Academy Press, Washington, DC, 1996.

10. J. W. Roberts, W. T. Budd, J. Chuang, and R. G. Lewis. *Chemical Contaminants in House Dust: Occurrences and Sources.* EPA/600/A-93/215. Environmental Protection Agency, Washington, DC, 1993.

11. L. R. Rhomberg. A survey of methods for chemical risk assessment among federal regulatory agencies. *Hum. Ecol. Risk Assess.* 3: 1029–1196 (1997).

12. U.S. Environmental Protection Agency (EPA). Guidelines for exposure assessment; notice. *Fed. Reg.* 57(104): 22888–22938 (1992).

13. Agency for Toxic Substances and Disease Registry (ATSDR). *Public Health Assessment Guidance Manual.* Lewis, Ann Arbor, MI, 1995.

14. P. A. Stewart and R. F. Herrick. Issues in performing retrospective exposure assessment. *Appl. Occup. Environ. Hyg.* (1991).

15. S. R. Ripple. Looking back: The use of retrospective health risk assessment. *Environ. Sci. Tech.* 26: 1270–1277 (1992).

16. D. J. Paustenbach, P. J. Sheehan, V. Lau, and D. M. Meyer. An assessment and quantitative uncertainty analysis of the health risks to workers exposed to chromium contaminated soils. *Toxicol. Ind. Health* 7: 159–196 (1991).

17. D. J. Paustenbach, R. J. Wenning, V. Lau, N. W. Harrington, D. K. Rennix, and A. H. Parsons. Recent developments on the hazards posed by 2,3,7,8-tetrachlorodibenzo-*p*-dioxin in soil: Implications for setting risk-based cleanup levels at residual and industrial sites. *J. Toxicol. Environ. Health* 36: 103–148 (1992).

18. P. G. Georgopoulos and P. J. Lioy. Conceptual and theoretical aspects of human exposure and dose assessment. *J. Exp. Anal. Environ. Epidemiol.* 4: 253–285 (1994).

19. N. Duan and D. T. Mage. Combination of direct and indirect approaches for exposure assessment. *J. Exp. Anal. Environ. Epidemiol.* 7(4): 439–470 (1997).

20. J. R. Lynch. Measurement of worker exposure. In L. J. Cralley and L. V. Cralley (Eds.), *Patty's Industrial Hygiene and Toxicology*, Vol. 3a: *The Work Environment*, 2nd ed. Wiley-Interscience, New York, 1985, pp. 569–615.

21. C. P. McCord. *Industrial Hygiene for Engineers.* Martin, Chicago, IL, 1943.

22. D. J. Paustenbach. Health risk assessment and the practice of industrial hygiene. *Am. Ind. Hyg. Assoc. J.* 51(7): 339–351 (1990).

23. A. C. Upton. Evolving perspectives on the concept of dose in radiobiology and radiation protection. *Health Phys.* 55(4): 605–614 (1988).

24. U.S. Environmental Protection Agency (EPA). Proposed guidelines for exposure-related measurements. *Fed. Reg.* 53(232): 48830–48853 (1988).

25. U.S. Environmental Protection Agency (EPA). *Exposure Factors Handbook.* Vol. I: *General Factors—SAB Review Draft.* EPA/600/P-95/002Ba. Office of Research and Development, Washington, DC, 1996.

26. P. Zannetti. Particle modeling and its application for simulating air pollution phenomena. In *Environmental Modeling* (with P. Melli). Computational Mechanics Publications, Palo Alto, CA, 1992.

27. P. K. Scott, H. Sung, B. L. Finley, R. H. Schulze, and D. B. Turner. Identification of an accurate soil suspension/dispersion modeling method for use in estimating health-based soil cleanup levels of hexavalent chromium in chromit-ore processing residues. *J. Air Waste Mgmt.* 47(7): 753–765 (1997).

28. D. J. Paustenbach, S. Sururi, and T. Underwood. Comparing the estimated uptake of TCDD using exposure calculations with the actual uptake: A case study of residents of Times Beach, Missouri. *Proc. Int. Dioxin Conf.* Bayreuth, Germany (1997).

29. D. J. Paustenbach, H. P. Shu, and F. J. Murray. A critical examination of assumptions used in risk assessment of dioxin contaminated soil. *Reg. Toxicol. Pharmacol.* 6: 284–307 (1986).

30. M. A. Jaycock, J. R. Lynch, and D. I. Nelson. *Risk Assessment: Principles for the Industrial Hygienist.* American Conference of Governmental Industrial Hygienists, Cincinnati, OH, 2000.

31. E. M. Romney, N. G. Lindberg, H. A. Hawthorne, B. B. Bystrom, and K. H. Larson. Contamination of plant foliage with radioactive nuclides. *Ann. Rev. Plant Physiol.* 14: 271–279 (1963).

32. W. E. Martin. Loss of Sr-90, Sr-89 and I-131 from fallout of contaminated plants. *Radiat. Bot.* 4: 275–281 (1964).

33. R. S. Russell. Entry of radioactive materials into plants. In R. S. Russell (Ed.), *Radioactivity and Human Diet.* Peramon Press, New York, 1966, Chapter 5.

34. C. F. Baes III, R. D. Sharp, A. Sjoreen, and W. R. Shor. *A Review and Analysis of Parameters for Assessing Transport of Environmental Released Radionuclides through Agriculture.* ORNL-5786. U.S. Department of Energy, Oak Ridge National Laboratory, Oak Ridge, TN, 1984.

35. T. E. McKone and K. T. Bogen. Predicting the uncertainties in risk assessment. *Environ. Sci. Technol.* 25: 16–74 (1991).

36. G. F. Fries and D. J. Paustenbach. Evaluation of potential transmission of 2,3,7,8-

tetrachlorodibenzo-*p*-dioxin-contaminated incinerator emissions to humans via foods. *J. Toxicol. Environ. Health.* 29: 1–43 (1990).

37. D. J. Paustenbach, J. Jernigan, R. Bass, R. Kalmes, and P. Scott. A proposed approach to regulating contaminated soil: Identify safe concentrations for seven of the most frequently encountered exposure scenarios. *Regul. Toxicol. Pharm.* 16: 21–56 (1992).

38. U.S. Environmental Protection Agency (EPA). *Exposure Factors Handbook*, Vol. II: *Food Ingestion Factors—SAB Review Draft.* EPA/600/P-95/002Bb. Office of Research and Development, Washington, DC, 1996.

39. U.S. Environmental Protection Agency (EPA). *Exposure Factors Handbook*, Vol. III: *Activity Factors—SAB Review Draft.* EPA/600/P-95/002P. Office of Research and Development, Washington, DC, 1996.

40. U.S. Environmental Protection Agency (EPA). *Risk Assessment Guidance for Superfund*, Vol. I: *Human Health Evaluation Manual (Part A). Interim Final.* Publication 540/1-89/002. Office of Emergency and Remedial Response, Washington, DC, 1989.

41. A. L. Nichols and R. J. Zeckhauser. The perils of prudence: How conventional risk assessments distort regulations. *Regul. Toxicol. Pharmacol.* 8: 61–75 (1988).

42. D. Maxim. Problems associated with the use of conservative assumptions in exposure and risk analysis. In D. J. Paustenbach (Ed.), *The Risk Assessment of Environmental and Human Health Hazards: A Textbook of Case Studies.* Wiley, New York, 1989.

43. A. C. Cullen. Measures of compounding conservatism in probabilistic risk assessment. *Risk Anal.* 14(4): 389–393 (1994).

44. K. M. Thompson and D. E. Burmaster. Parametric distributions for soil ingestion by children. *Risk Anal.* 11: 339–342 (1991).

45. K. M. Thompson, D. E. Burmaster, and E. A. C. Crouch. Monte Carlo techniques for quantitative uncertainty analysis in public health risk assessments. *Risk Anal.* 12(1): 53–63 (1992).

46. B. L. Finley and D. J. Paustenbach. The benefits of probabilistic exposure assessment: Three case studies involving contaminated air, water, and soil. *Risk Anal.* 14(1): 53–73 (1994).

47. D. E. Burmaster and R. H. Harris. The magnitude of compounding conservatisms in Superfund risk assessments. *Risk Anal.* 13: 131–134 (1993).

48. N. D. Wilson, P. Price, and D. J. Paustenbach. An assessment of the risk of DDT and PCB in fish from the Palos Verdes Shelf. In D. J. Paustenbach (Ed.), *Human and Ecological Risk Assessment: Theory and Practice.* Wiley, New York, 2000.

49. M. Allaby. *A Dictionary of the Environment*, 2nd ed. New York University Press, New York, 1983, p. 195.

50. H. W. Leung and D. J. Paustenbach. Techniques for estimating the percutaneous absorption of chemicals due to environmental and occupational exposure. *Appl. Environ. Occup. Hyg.* 9(3): 187–197 (1994).

51. D. J. Paustenbach, H. J. Clewell III, M. L. Gargas, and M. E. Andersen. A physiologically-based pharmacokinetic model for carbon tetrachloride. *Toxicol. Appl. Pharmacol.* 96: 191–211 (1988).

52. R. H. Reitz, M. L. Gargas, M. E. Andersen, W. M. Provan, and T. L. Green. Predicting cancer risk from vinyl chloride exposure with a physiologically based pharmacokinetic model. *Toxicol. Appl. Pharmacol.* 137: 253–267 (1996).

53. J. N. McDougal, G. W. Jepson, H. J. Clewell III, M. L. Gargas, and M. E. Andersen. Dermal absorption of organic chemical vapors in rats and humans. *Fundam. Appl. Toxicol.* 14: 299–308 (1990).

54. T. E. McKone. Dermal uptake of organic chemicals from a soil matrix. *Risk Anal.* 10: 407–419 (1990).

55. J. K. Piotrowski. Further investigations on the evaluation of exposure to nitrobenzene. *Br. J. Ind. Med.* 24: 60–65 (1967).

56. J. Piotrowski. *Exposure Tests for Organic Compounds in Industrial Toxicology.* National Institute for Occupational Safety and Health (NIOSH), Cincinnati, OH, 1977.

57. H. P. Shu, D. J. Paustenbach, F. J. Murray, L. Marple, B. Brunch, D. Dei Rossi, and P. Teitelbaum. Bioavailability of soil-bound TCDD: Dermal bioavailability in the rat. *Fundam. Appl. Toxicol.* 10: 648–654 (1988).

58. S. E. Hrudey, W. Chen, and C. Rousseaux. *Bioavailability.* CRC-Lewis, New York, 1996.

59. J. Ramsey and M. Andersen. A physiologically based description of the inhalation pharmacokinetics of styrene in rats and humans. *Toxicol. Appl. Pharmacol.* 73: 159–175 (1984).

60. U.S. Environmental Protection Agency (EPA). Draft guidelines for carcinogen risk assessment. *Fed. Reg.* 61(79): 17960–18011 (1996).

61. T. E. McKone and K. T. Bogen. Uncertainties in health-risk assessment: An integrated case study based on tetrachloroethylene in California groundwater. *Regul. Toxicol. Pharmacol.* 15: 86–103 (1992).

62. H. J. Clewell. The application of physiologically based pharmacokinetics modeling in human health risk assessment of hazardous substances. *Toxicol. Lett.* 79: 207–217 (1995).

63. M. E. Andersen, H. J. Clewell III, M. L. Gargas, M. G. MacNaughton, R. H. Reitz, R. J. Nolan, and M. J. McKenna. Physiologically based pharmacokinetic modeling with dichloromethane, its metabolite, carbon monoxide, and blood carboxyhemoglobin in rats and humans. *Toxicol. Appl. Pharmacol.* 108: 14–27 (1991). See also M. E. Andersen, H. J. Clewell, and K. Krishnan. Tissue dosimetry, pharmacokinetics modeling and interspecies scaling factors. *Risk Anal.* 15: 533–537 (1995).

64. P. D. Lilly, M. E. Andersen, T. M. Ross, and R. A. Pegram. A physiologically based pharmacokinetic description of the oral uptake, tissue dosimetry, and rates of metabolism of bromodichloromethane in the male rat. *Toxicol. Appl. Pharmacol.* 150(2): 205–217 (1998).

65. National Research Council (NRC). *Human Exposure Assessment for Airborne Pollutants: Advances and Applications.* Committee on Advances in Assessing Human Exposure to Airborne Pollutants. Committee on Geosciences, Environment, and Resources, NRC. National Academy Press, Washington, DC, 1990.

66. C. N. Park and R. D. Snee. Quantitative risk assessment: State-of-the-art for carcinogensis. *Fund. Appl. Toxicol.* 3: 320–333 (1983).

67. H. W. Leung and D. J. Paustenbach. Physiologically based pharmacokinetic and pharmacodynamic modeling in health risk assessment and characterization of hazardous substances. *Toxicol. Lett.* 79: 55–65 (1995).

68. American Industrial Health Council (AIHC). *Exposure Factors Sourcebook.* AIHC, Washington, DC, 1994.

69. B. L. Finley, D. Proctor, P. Scott, N. Harrington, D. Paustenbach, and P. Price. Recommended distributions for exposure factors frequently used in health risk assessment. *Risk Anal.* 14(4): 533–553 (1994).

70. L. L. Aylward, S. M. Hays, N. J. Karch, and D. J. Paustenbach. Relative susceptibility of animals and humans to the cancer hazard posed by 2,3,7,8-tetrachlorodibenzo-*p*-dioxin using internal measures of dose. *Environ. Sci. Technol.* 30(12): 3534 3543 (1996).

71. M. E. Andersen, M. G. MacNaughton, H. J. Clewell, and D. J. Paustenbach. Adjusting exposure limits for long and short exposure periods using a physiological pharmacokinetic model. *Am. Ind. Hyg. Assoc. J.* 48(4): 335–343 (1987).

72. M. E. Andersen and R. B. Conolly. Mechanistic modeling of rodent liver tumor promotion at low levels of exposure: An example related to dose-response relationships for 2,3,7,8-tetrachlorodibenzo-*p*-dioxin. *Hum. Exp. Toxicol.* 17(12): 683–690 (1998).

73. U.S. Office of Science and Technology Policy (DDHS). *Researching Health Risks.* Office of Technology Assessment, Washington, DC, 1993.

74. U.S. Environmental Protection Agency (EPA). *The Total Exposure Assessment Methodology (TEAM) Study. Summary and Analysis,* Vol. 1. EPA/600/6-87/002a. Office of Acid Deposition, Environmental Monitoring and Quality Assurance, Research and Development, 1987.

75. American Conference of Governmental Industrial Hygienists (ACGIH). *Industrial Hygiene Instruments Handbook.* ACGIH, Cincinnati, OH, 1998.

76. D. M. Proctor, M. A. Zak, and B. L. Finley. Resolving uncertainties associated with the construction worker soil ingestion rate: A proposal for risk-based remediation goals. *Hum. Ecol. Risk Assess.* 3: 299–304 (1997).

77. D. J. Paustenbach, W. E. Rinehart, and P. J. Sheehan. The health hazards posed by chromium-contaminated soils in residential and industrial areas: Conclusions of an expert panel. *Regul. Toxicol. Pharmacol.* 13: 195–222 (1991).

78. D. J. Paustenbach, D. M. Meyer, P. J. Sheehan, and V. Lau. An assessment and quantitative uncertainty analysis of the health risks to workers exposed to chromium contaminated soils. *Toxicol. Ind. Health* 7: 159–196 (1991).

79. C. S. Nessel, J. P. Butler, G. B. Post, J. I. Held, M. Gochfeld, and M. A. Gallo. Evaluation of the relative contribution of exposure routes in a health risk assessment of dioxin emissions from a municipal waste incinerator. *J. Exp. Anal. Environ. Epidemiol.* 1: 283–308 (1991).

80. J. B. Knaak, C. C. Dary, G. Patterson, and J. N. Blancato. The worker hazard posed by reentry into pesticide-treated foliage: Reassessment of reentry intervals using foliar residue transfer-percutaneous absorption PB-PK/PD models, with emphasis on isofenphos and parathion. In D. J. Paustenbach (Ed.), *Human and Ecological Risk Assessment: Theory and Practice.* Wiley, New York, 2000.

81. U.S. Environmental Protection Agency (EPA). *Exposure Factors Handbook* (update to the May 1989 ed.). EPA/600/P-95/002Fa. Washington, DC, 1997.

82. U.S. Environmental Protection Agency (EPA). *Methodology for Assessing Health Risks Associated with Multiple Exposure Pathways to Combustor Emissions.* NCEA-C-0238. National Center for Environmental Assessment, Washington, DC, 1997.

83. E. J. Calabrese and P. J. Kostecki. *Risk Assessment and Environmental Fate Methodologies.* Lewis, Ann Arbor, MI, 1993.

84. T. J. Smith, S. K. Hammond, and O. Wong. Health effects of gasoline exposure: I: Exposure assessment for U.S. distribution workers. *Environ. Health Perspect.* 101(6): 13–21 (1993). See also M. F. Hallock, T. J. Smith, S. R. Woskie, and S. K. Hammond. Estimation of historical exposures to machining fluids in the automotive industry. *Am. J. Ind. Med.* 26: 621–634 (1994).

85. N. Plato, S. Krantz, P. Gustavsson, T. J. Smith, and P. Westerholm. A cohort study of Swedish man-made mineral fiber (MMMF) production workers. Part I: Fiber exposure assessment in the rock/slag wool production industry 1938–1990. *Scand. J. Work Environ. Health* 21: 345–352 (1995). See also R. A. Stone, G. M. Marsh, A. O. Youk, T. J. Smith, and M. M. Quinn. Statistical estimation of exposure to fibres in jobs for which no direct measurements are available. *Occup. Hyg.* 3: 91–101 (1996).

86. P. A. Stewart, P. S. J. Lees, and M. Francis. Quantification of historical exposures in occupational cohort studies. *Scand. J. Work Environ. Health* 22: 405–414 (1996).

87. M. Goodman, D. Paustenbach, K. Sipe, C. D. Malloy, P. Chapman, R. Figueroa, K. Zhao, and K. A. Exuzides. Epidemiologic study of pulmonary obstruction in workers occupationally exposed to ethyl- and methyl cyanoacrylate. *J. Toxicol. Environ. Health* (submitted for publication) (2000).

88. N. Sathiakumar, E. Delzell, M. Hovinga, M. Macaluso, J. A. Julian, R. Larson, P. Cole, and D. C. Muir. Mortality from cancer and other causes of death among synthetic rubber workers. *Occup. Environ. Med.* 55(4): 230–235 (1998). See also E. Delzell, N. Sathiakumar, M. Hovinga, M. Macaluso, J. Julian, R. Larson, P. Cole, and D. C. Muir. A follow-up study of synthetic rubber workers. *Toxicology* 113(1–3): 182–189 (1996).

89. A. L. Lynch. *Biological Monitoring.* Wiley, New York, 1994.

90. A. P. DeCaprio. Biomarkers: Coming of age for environmental health and risk assessment. *Environ. Sci. Technol.* 31(7): 1837–1348 (1997).

91. E. D. Pellizzari, R. L. Perritt, and C. A. Clayton. National human exposure assessment survey (NHEXAS): Exploratory survey of exposure among population subgroups in EPA Region V. *J. Expos. Anal. Environ. Epidemiol.* 9: 49–55 (1999).

92. P. S. Price, S. H. Su, J. R. Harrington, and R. E. Keenan. Uncertainty and variation in indirect exposure assessments: An analysis of exposure to tetrachlorodibenzo-*p*-dioxin from a beef consumption pathway. *Risk Anal.* 16: 263–277 (1996).

93. D. J. Paustenbach, B. L. Finley, and T. F. Long. The critical role of house dust in understanding the hazards posed by contaminated soils. *Int. J. Toxicol.* 16: 339–362 (1997).

94. P. K. Scott, M. A. Harris, D. E. Rabbe, and B. L. Finley. Background air concentrations of Cr(VI) in Hudson County, New Jersey: Implications for setting health-based standards for Cr(VI) in soil. *J. Air Waste Mgmt.* 47: 592–600 (1997).

95. E. J. Calabrese and P. T. Kostecki. *Risk Assessment and Environmental Fate Methodologies.* Lewis, Ann Arbor, MI, 1992.

96. P. S. Price, P. K. Scott, N. D. Wilson, and D. J. Paustenbach. An empirical approach for deriving information on total duration of exposure from information on historical exposure. *Risk Anal.* 18: 611–619 (1998).

97. U.S. Environmental Protection Agency (EPA). *Guidelines on Air Quality Models* (rev.). EPA-450/2-78/027R. Office of Air Quality Planning and Standards, Research Triangle Park, NC, 1986.

98. U.S. Environmental Protection Agency (EPA). *Selection Criteria for Mathematical Models Used in Exposure Assessments: Surface Water Models.* EPA-600/8-87/042. NTIS PB88-139928/AS. Office of Health and Environmental Assessment, Office of Research and Development, Washington, DC, 1987.

99. U.S. Environmental Protection Agency (EPA). *Development of Statistical Distributions or Ranges of Standard Factors Used in Exposure Assessments.* EPA 600/8-85-010. Office of Health and Environmental Assessment, Washington, DC, 1985. Available from NTIS, Springfield, VA (PB85-242667).

100. O. El Saadi and A. Langley. *The health risk assessment and management of contaminated sites.* In *Proceedings of a National Workshop on the Health Risk Assessment and Management of Contaminated Sites.* South Australian Health Commission, Adelaide, 1994.

101. U.S. Environmental Protection Agency (EPA). *Methods for Assessing Exposure to Chemical Substances*, Vols. 1–13. EPA-560/5-85/002. NTIS PB86-107067. Office of Toxic Substances, Exposure Evaluation Division, Washington, DC, 1983–1989.

102. J. Dragun. *The Soil Chemistry of Hazardous Materials*, 2nd ed. Amherst Scientific, Amherst, MA, 1998.

103. J. Nriagu. *Heavy Metals in the Environment.* Wiley, New York, 1979.

104. M. Costa, A. Zhitkovich, M. Harris, D. Paustenbach, and M. Gargas. DNA-protein crosslinks produced by various chemicals in cultured human lymphoma cells. *J. Toxicol. Environ. Health.* 30: 101–116 (1997).

105. D. J. Paustenbach. Assessment of the developmental risks resulting from occupational exposure to select glycol ethers within the semiconductor industry. *J. Toxicol. Environ. Heath* 23: 29–75 (1988).

106. J. Nethercott, D. J. Paustenbach, R. Adams, S. Horowitz, B. E. Finley, J. Fowler, J. Marks, C. Morton, and J. Taylor. A study of chromium induced allergic contact dermatitis with 54 volunteers: Implications for environmental risk assessment. *Occup. Environ. Med.* 51(6): 371–380 (1994).

107. R. D. Stewart and H. C. Dodd. Absorption of carbon tetrachloride, trichloroethylene, tetrachloroethylene, methylene chloride, and 1,1,1-trichloroethane through the human skin. *Am. Ind. Hyg. Assoc. J.* 25: 439–446 (1964).

108. T. Dutkiewicz and J. Piotrowski. Experimental investigations on the quantitative estimation of aniline absorption in man. *Pure Appl. Chem.* 3: 319–323 (1961).

109. T. Dutkiewicz and H. Tyras. A study of the skin absorption of ethylbenzene in man. *Br. J. Ind. Med.* 24: 330–332 (1967).

110. R. J. Feldmann and H. I. Maibach. Percutaneous penetration of some pesticides and herbicides in man. *Toxicol. Appl. Pharmacol.* 28: 126–132 (1974).

111. J. Piotrowski. Evaluation of exposure to phenol: Absorption of phenol vapor in the lungs and through the skin and excretion of phenol in urine. *Br. J. Ind. Med.* 28: 172–178 (1971).

112. J. Mraz and M. Nohova. Percutaneous absorption of *N,N*-dimethylformamide in humans. *Int. Arch. Occup. Environ. Health* 64: 79–83 (1992).

113. N. Krivanek, M. McLaughlin, and W. F. Fayweather. Monomethylformamide levels in human urine after repetitive exposure to dimethylformamide. *J. Occup. Med.* 20: 179–187 (1978).

114. G. J. Klain and K. E. Black. Specialized techniques: Congenitally athymic (nude) animal models. In B. W. Kemppainen and W. G. Reifenrath (Eds.), *Methods for Skin Absorption.* CRC Press, Boca Raton, FL, 1990, pp. 165–174.

115. M. J. Bartek, J. A. LaBudde, and H. I. Maibach. Skin permeability in vivo: Comparison in rat, rabbit, pig and man. *J. Invest. Dermatol.* 58: 114–123 (1972).

116. S. W. Frantz. Instrumentation and methodology for in vitro skin diffusion cells. In B. W. Kemppainen and W. G. Reifenrath (Eds.), *Methods for Skin Absorption.* CRC Press, Boca Raton, FL, 1990, pp. 35–59.

117. R. L. Bronaugh, R. F. Stewart, E. R. Congdon, and A. L. Giles, Jr. Methods for in vitro percutaneous absorption studies: I. Comparison with in vitro results. *Toxicol. Appl. Pharmacol.* 62: 474–480 (1982).

118. E. D. Barber, N. M. Teetsel, K. F. Kolberg, and D. Guest. A comparative study of the rates of in vitro percutaneous absorption of eight chemicals using rat and human skin. *Fundam. Appl. Toxicol.* 19: 493–497 (1992).

119. M. L. Lepow, L. Bruckman, M. Gillette, S. Markowitz, R. Robino, and J. Kapish. Investigations into sources of lead in the environment of urban children. *Environ. Res.* 10: 415–426 (1975).

120. H. A. Roels, J. P. Buchet, R. R. Lauwenys, F. Claeys-Thoreau, A. Lafontaine, and G. Verduyn. Exposure to lead by oral and pulmonary routes of children living in the vicinity of a primary lead smelter. *Environ. Res.* 22: 81–94 (1980).

121. S. S. Que Hee, B. Peace, C. S. Scott, C. S. Clark, J. R. Boyle, R. L. Bornschien, and P. B. Hammond. Evolution of efficient methods to sample lead sources, such as house dust and hand dust, in the homes of children. *Environ. Res.* 38: 77–95 (1985).

122. J. H. Driver, J. J. Konz, and G. K. Whitmyre. Soil adherence to human skin. *Bull. Environ. Contam. Toxicol.* 17(9): 1831–1850 (1989).

123. S. C. Sheppard and W. G. Evenden. Contaminant enrichment and properties of soil adhering to skin. *J. Environ. Qual.* 23: 604–613 (1994).

124. B. L. Finley, P. K. Scott, and D. A. Mayhall. Development of a standard soil-to-skin adherence probability density function for use in Monte Carlo analyses of dermal exposure. *Risk Anal.* 14: 555–569 (1994).

125. K. K. Holmes, J. C. Kissel, and K. Y. Richter. Investigation of the influence of oil on soil adherence to skin. *J. Soil Contam.* 5(4): 301–308 (1996).

126. J. C. Kissel, K. Y. Richter, and R. A. Fenske. Field measurement of dermal soil loading attributable to various activities: Implications for exposure assessment. *Risk Anal.* 16(1): 115–125 (1996).

127. J. E. Johnson and J. C. Kissel. Prevalence of dermal pathway dominance in risk assessment of contaminated soils: A survey of Superfund risk assessments, 1989–1992. *Hum. Ecol. Risk Assess.* 2: 356–365 (1996).

128. D. Marlow, M. H. Sweeney, and M. Fingerhut. Estimating the amount of TCDD absorbed by workers who manufactured 2,4,5-T. Paper presented at the Tenth Annual International Dioxin Meeting, Bayreuth, Germany, 1990.

129. D. E. Burmaster and K. M. Thompson. Estimating exposure point concentrations for surface soils for use in deterministic and probabilistic risk assessments. *Hum. Ecol. Risk Assess.* 3: 363–384 (1997).

130. P. J. Lioy, T. Wainman, and C. Weisel. A wipe sampler for the quantitative measurement of dust on smooth surfaces: Laboratory performance studies. *J. Exp. Anal. Environ. Epidemiol.* 3: 315–320 (1993).

131. P. J. Lioy, L. M. Yiin, J. Adgate, C. Weisel, and G. G. Rhoads. The effectiveness of a home cleaning intervention strategy in reducing potential dust and lead exposures. *J. Exp. Anal. Environ. Epidemiol.* (1998).

132. J. Wepierre and J. P. Marty. Percutaneous absorption of drugs. *Trends Pharmacol. Sci.* 1: 23–26 (1979).

133. C. Surber, K. P. Wilhelm, H. I. Maibach, L. I. Hall, and R. H. Guy. Partitioning of chemicals into human stratum corneum: Implications for risk assessment following dermal exposure. *Fundam. Appl. Toxicol.* 15: 99–107 (1990).

134. R. H. Guy, J. Hadgraft, and H. I. Maibach. A pharmacokinetic model for percutaneous absorption. *Int. J. Pharmacol.* 11: 119–129 (1982).

135. M. Gargas, R. J. Burgess, G. E. Voisaro, G. H. Cason, and M. E. Anderson. Partition coefficients of low molecular weight volatile chemicals in various liquids and tissues. *Toxicol. Appl. Pharmacol.* 98: 87–99 (1989).

136. B. D. Anderson, W. I. Higuchiand, and P. V. Raykar. Heterogeneity effects on permeability: Partition coefficient relationships in human stratum corneum. *Pharmacol. Res.* 5: 566–573 (1988).

137. G. L. Flynn. Physicochemical determinants of skin absorption. In T. R. Gerrity and C. J. Henry (Eds.), *Principles of Route-to-Route Extrapolation for Risk Assessment.* Elsevier Science, New York, 1990, pp. 93–127.

138. J. A. Shatkin and H. S. Brown. Pharmacokinetics of the dermal route of exposure to volatile organic chemicals in water. A computer simulation model. *Environ. Res.* 56: 90–108 (1991).

139. J. N. McDougal. Physiologically-based pharmacokinetic modeling. In F. N. Marzulli and H. I. Maibach (Eds.), *Dermatoxicology.* Taylor and Francis, Washington, DC, 1996.

140. U.S. Environmental Protection Agency (EPA). *Dermal Exposure Assessment: Principles and Applications.* EPA-600/8-91/011. Office of Health and Environmental Assessment, Office of Research and Development, Washington, DC, 1992.

141. G. W. Lucier, R. C. Rumbraugh, Z. McCoy, R. Hass, D. Harvon, and P. Albro. Ingestion of soil contaminated with 2,3,7,8-tetrachlorodibenzo-*p*-dioxin (TCDD) alters hepatic enzyme activities in rats. *Fundam. Appl. Toxicol.* 6: 364–371 (1986).

142. T. H. Umbreit, E. J. Hesse, and M. A. Gallo. Acute toxicity of TCDD contaminated soil from an industrial site. *Science* 232: 497–499 (1986).

143. T. H. Umbreit, E. J. Hesse, and M. A. Gallo. Comparative toxicity of TCDD contaminated soil from Times Beach, Missouri and Newark, New Jersey. *Chemosphere* 15: 121–2124 (1986).

144. G. A. Skrowronski, R. M. Turkall, and M. S. Abdel-Rahman. Soil absorption alters bioavailability of benzene in dermally exposed male rats. *Am. Ind. Hyg. Assoc. J.* 49: 506–511 (1988).

145. R. C. Wester, H. I. Maibach, L. Sedik, J. Melendres, M. Wade, and S. Dizio. Percutaneous absorption of pentachlorphenol from soil. *Fundam. Appl. Toxicol.* 20: 68–71 (1993).

146. W. S. Snyder. *Report of the Task Group on Reference Man.* International Commission on Radiological Protection, Pub. No. 23. Pergamon, New York, 1975.

147. California Department of Health Services (CDHS). *Development of Applied Action Levels for Soil Contact: A Scenario for the Exposure of Humans to Soil in a Residential Setting.* CDHS, Sacramento, CA, 1986.

148. J. H. Driver, J. J. Konz, and G. K. Whitmyre. Soil adherence to human skin. *Bull. Environ. Contam. Toxicol.* 17: 1831–1850 (1989).

149. European Center for Ecotoxicology and Toxicology of Chemical (ECETOC). *Strategy for Assigning a "Skin Notation."* Revised ECETOC Document No. 31. ECETOC, Brussels, 1993.

150. European Center for Ecotoxicology and Toxicology of Chemical (ECETOC). *Percutaneous Absorption.* Monograph No. 20. ECETOC, Brussels, 1993.

151. K. Caplan. Wipe sampling criterion. *Am. Ind. Hyg. Assoc. J.* (1993).

152. B. McArthur. Dermal measurement and wipe sampling methods: A review. *Appl. Occup. Environ. Hyg.* 7: 599–606 (1992).

153. D. H. Brouwer and J. J. Van Hemmen. *Elements of a Sampling Strategy for Dermal Exposure Assessment.* [Abs.] International Occupational Hygiene Association. First International Scientific Conference, December 7–10. Brussels, 1992.

154. J. C. Kissel, K. Y. Richter, and R. A. Fenske. Factors affecting soil adherence to skin in hand-press trails. *Bull. Environ. Contam. Toxicol.* 56: 722–728 (1996).

155. R. A. Fenske. Dermal exposure assessment techniques. *Ann. Occup. Hyg.* 37: 687–706 (1993).

156. J. M. Michaud, S. L. Huntley, R. A. Sherer, M. N. Gray, and D. J. Paustenbach. PCB and dioxin re-entry criteria for building surfaces and air. *J. Exp. Anal. Environ. Epidemiol.* 4(2): 197–227 (1994).

157. J. B. Knaak, Y. Iwata, and K. T. Maddy. The worker hazard posed by reentry into pesticide-treated foliage: Development of safe reentry times, with emphasis on chlorhiophos and carbosulfan. In D. J. Paustenbach (Ed.), *The Risk Assessment of Environmental Hazards; A Textbook of Case Studies.* Wiley, New York, 1989, pp. 797–842.

158. T. Lavy, J. Shepard, and D. Bouchard. Field worker exposure and helicopter spray pattern of 2,4,5-T. *Bull. Environ. Contam. Toxicol.* 24(1): 90–96 (1980).

159. W. J. Popendorf and J. T. Leffingwell. Regulating OP pesticide residues for farmworker protection. In: *Residue Review 82.* Springer-Verlag, New York, 1982, pp. 125–201 (1976).

160. T. Lavy, J. Walstad, R. Flynn, and J. Mattice. (2,4-Dichlorophenoxy) acetic acid exposure received by aerial application crews during forest spray operations. *J. Agric. Food Chem.* 30: 375–361 (1982).

161. N. D. Krivanek, M. McLoughlin, and W. E. Fayweather. Monomethylformide levels in human urine after repetitive exposure to dimethylformamide. *J. Occup. Med.* 20: 179–187 (1978).

162. K. Scow, A. E. Wechsler, and J. Stevens. *Identification and Evaluation of Water-borne Routes of Exposure from Other Than Food and Drinking Water.* EPA-440/4-79-016. Environmental Protection Agency, Washington, DC, 1979.

163. J. Byard. Hazard assessment of 1,1,1-trichloroethane in groundwater. In D. J. Paustenbach (Ed.), *The Risk Assessment of Environmental and Human Health Hazards: A Textbook of Case Studies.* Wiley, New York, 1989, pp. 331–334.

164. W. K. Jo, C. P. Weisel, and P. J. Lioy. Routes of chloroform exposure and body burden from showering with chlorinated tap water. *Risk Anal.* 10: 575–580 (1988).

165. B. Kerger and D. Paustenbach. Exposure to 1,1,1 TCE vapors in a home due to contaminated groundwater. *Soc. Toxicol.* (1997).

166. S. Kezic, K. Mahieu, A. C. Monster, and F. A. de Wolff. Dermal absorption of vaporous and liquid 2-methoxyethanol and 2-ethoxyethanol in volunteers. *Occup. Environ. Med.* 54: 38–43 (1997).

167. U.S. Environmental Protection Agency (EPA). *Risk Assessment Guidelines for Dermal Assessment.* EPA, Washington, DC, 1999.

168. D. R. Mattie, G. D. Bates, G. W. Jepson, J. W. Fisher, and J. N. McDougal. Determination of skin: Air partition coefficients for volatile chemicals: Experimental method and applications. *Fundam. Appl. Toxicol.* 22: 51–57 (1994).

169. T. L. Copeland, D. J. Paustenbach, M. A. Harris, and J. Otani. Comparing the results of a Monte Carlo analysis with EPA's reasonable maximum exposed individual (RMEI): A case study of a former wood treatment site. *Regul. Toxicol. Pharmacol.* 18: 275–312 (1993).

170. R. D. Kimbrough, H. Falk, P. Stehr, and G. F. Fries. Health implications of 2,3,7,8-tetrachlorodibenzo-*p*-dioxin (TCDD) contamination of residential soil. *J. Toxicol. Environ. Health* 14: 47–93 (1984).

171. D. Baltrop. The prevalence of pica. *Am. J. Dis. Child.* 112: 116–123 (1966).

172. S. D. Walter, A. J. Yankel, and I. H. von Lindern. Age-specific risk factors for lead absorption in children. *Arch. Environ. Health* 35: 53–58 (1980).

173. M. Cooper. *Pica.* Charles C. Thomas, Springfield, IL, 1957, pp. 60–74.

174. E. Charney, J. Sayre, and M. Coulter. Increased lead absorption in inner city children: Where does the lead come from? *Pediatrics* 65: 226–231 (1980).

175. J. W. Sayre, E. Charney, J. Vostal, and B. Pless. House and hand dust as a potential source of childhood lead exposure. *Am. J. Dis. Child.* 127: 167–170 (1974).

176. M. J. Duggan and S. Williams. Lead-in-dust in city streets. *Sci. Total Environ.* 7: 91–97 (1977).

177. M. L. Lepow, L. Bruckman, R. A. Robino, S. Markowitz, M. Gillette, and J. Kapish. Role of airborne lead in increased body burden of lead in Hartford children. *Environ. Health Perspect.* 6: 99–101 (1974). See also M. L. Lepow, L. Bruckman, M. Gillette, S. Markowitz, R. Robino, and J. Kapish. Investigations into sources of lead in the environment of urban children. *Environ. Res.* 10: 415–426 (1975).

178. D. Baltrop, C. D. Stehlow, I. Thornton, and J. S. Webb. Absorption of lead from dust and soil. *Postgrad. Med. J.* 5: 801–804 (1975).

179. National Research Council (NRC). *Lead in the Environment.* NRC, Washington, DC, 1974.

180. D. J. Paustenbach. Assessing the potential environmental and human health risks of contaminated soil. *Comments Toxicol.* 1: 185–220 (1987).

181. J. P. Day, M. Hart, and M. S. Robinson. Lead in urban street dust. *Nature (Lond.)* 253: 343–345 (1975).

182. D. Bryce-Smith. Lead absorption in children. *Phys. Bull.* 25: 178–181 (1974).

183. J. K. Hawley. Assessment of health risk from exposure to contaminated soil. *Risk Anal.* 5(4): 289–302 (1985).

184. P. K. LaGoy. Estimated soil ingestion rates for use in risk assessment. *Risk Anal.* 7(3): 355–359 (1987).

185. S. Binder, D. Sokal, and D. Maughan. Estimating soil ingestion: The use of tracer elements in estimating the amount of soil ingested by young children. *Arch. Environ. Health* 41: 341–345 (1986).

186. J. H. Van Wijnen, P. Clausing, and B. Brunekreef. Estimated soil ingestion by children. *Environ. Res.* 51: 147–162 (1990).

187. E. J. Calabrese, R. Barnes, E. J. Stanek, H. Pastides, C. E. Gilbert, P. Yeneman, X. Wang, A Laszitty, and P. T. Kostaki. How much soil do young children ingest: An epidemiologic study. *Regul. Toxicol. Pharmacol.* 10: 123–137 (1989).

188. P. E. de Silva. *Assessment of Health Risk to Residents of Contaminated Sites.* AMCOSH Occupational Health Services Report to Gas and Fuel Corporation, Melbourne, Australia, 1991.

189. P. E. de Silva. How much soil do children ingest—A new approach. *App. Occup. Environ. Hyg.* 9: 40–43 (1994).

190. E. J. Calabrese, E. J. Stanek, C. E. Gilbert, and R. M. Barnes. Preliminary adult soil ingestion estimates: Results of a pilot study. *Regul. Toxicol. Pharmacol.* 12: 88–95 (1990).

191. E. J. Stanek and E. J. Calabrese. A guide to interpreting soil ingestion studies I: Development of a model to estimate the soil ingestion detection level of soil ingestion studies. *Regul. Toxicol. Pharmacol.* 13: 263–277 (1991).

192. E. J. Calabrese and E. J. Stanek. A guide to interpreting soil ingestion studies II: Qualitative and quantitative evidence of soil ingestion. *Regul. Toxicol. Pharmacol.* 13: 278–292 (1991).

193. E. J. Stanek and E. J. Calabrese. Daily soil ingestion estimates for children at a Superfund site. *Risk Anal.* 20: 627–635 (2000).

194. E. J. Stanek, E. J. Calabrese, and M. Zorn. Soil ingestion distributions for Monte Carlo risk assessment in children. *Hum. Ecol. Risk Assess.* 7: 357–368 (2001).

195. R. S. Lourie and E. M. Cayman. Why children eat things that are not food. *Children* 10: 143–146 (1963).

196. E. R. Taylor. How much soil do children eat? In E. L. Saadi and A. Langley (Eds.), *The Health Risk Assessment and Management of Contaminated Sites.* South Australian Health Commission, 1983, pp. 72–77.

197. D. C. Danford. Pica and nutrition. *Annu. Rev. Nutrit.* 2: 303–322 (1982).

198. D. Barltrop. Sources and significance of environmental lead for children. In *Proceedings of the International Symposium on Environmental and Health Aspects of Lead.* Commission of European Communities, Center for Information and Documentation, Luxembourg, 1973.

199. J. E. J. Gallacher, P. C. Elwood, K. M. Phillips, B. E. Davies, and D. T. Jones. Relation between pica and blood lead in areas of differing lead exposure. *Arch. Dis. Child.* 59: 40–44 (1984).

200. R. D. Knarr, G. L. Cooper, E. A. Brian, M. G. Kleinschmidt, and D. G. Graham. *Arch. Environ. Contam. Toxicol.* 14: 523 (1985).

201. E. J. Stanek and E. J. Calabrese. Soil ingestion estimates for use in site evaluation based on the best tracer method. *Hum. Ecol. Risk Assess.* 1: 133–156 (1995). See also V. G. Zartarian, A. C. Ferguson, and J. O. Leckie. Quantified mouthing activity data from a four-child pilot field study. *J. Exp. Anal. Environ. Epidemiol.* 8(4): 543–553 (1998).

202. U.S. Department of Agriculture (USDA). Food Consumption: Households in the United States, Seasons and Year 1965–1966. USDA, Washington, DC, 1972.

203. U.S. Department of Agriculture (USDA). *Food and Nutrient Intakes of Individuals in One Day in the United States, Spring 1977.* Nationwide Food Consumption Survey 1977–1978. Preliminary Report No. 2. USDA, Washington, DC, 1980.

204. U.S. Environmental Protection Agency (EPA). *An Estimation of the Daily food Intake Based on Data from the 1977–1978. USDA Nationwide Food Consumption Survey.* EPA-520/1-84-015. Office of Radiation Programs, Washington, DC, 1984.

205. U.S. Department of Agriculture (USDA). *Food and Nutrient Intakes by Individuals in the United States, 1 Day, 1987–88: U.S. Department of Agriculture, Human Nutrition Information Service.* Nationwide Food Consumption Survey 1987–88. NFCS Rpt. No. 87-1-1. USDA, Washington, DC, 1992.

206. E. M. Pao, K. H. Fleming, P. M. Guenther, and S. J. Mickle. *Foods Commonly Eaten by Individuals: Amount per Day and per Eating Occasion.* Home Economics Report No. 44. U.S. Department of Agriculture, Washington, DC, 1982.

207. J. Kariya. Written communication to L. Phillips, Versar, Inc., March 4, 1992.

208. S. B. White, B. Peterson, C. A. Clayton, and D. P. Duncan. *Interim Report Number 1: The Construction of a Raw Agricultural Commodity Consumption Data Base.* Research Triangle Institute for EPA Office of Pesticide Programs, Raleigh, NC, 1983.

209. H. W. Puffer, S. P. Azen, M. J. Duda, and D. R. Young. Consumption rates of potentially hazardous marine fish caught in the metropolitan Los Angeles area. EPA Grant #R807 120010. See also P. S. Price, C. L. Curry, P. E. Goodrum, M. N. Gray, J. I. McCrodden, N. W. Harrington, H. Carlson-Lynch, and R. E. Keenan. Monte Carlo modeling of time-dependent exposures using a microexposure event approach. *Risk Anal.* 16: 339–348 (1996).

209a. P. S. Price, C. L. Curry, P. E. Goodrum, M. N. Gray, J. I. McCrodden, N. W. Harrington, H. Carlson-Lynch, and R. E. Keenan. Monte Carlo modeling of time-dependent exposures using a microexposure event approach. *Risk Anal.* 16: 339–348 (1996).

210. P. S. Price, S. H. Su, and M. N. Gray. The effect of sampling bias on estimates of angler consumption rates in Creel surveys. *J. Exp. Anal. Environ. Epidemiol.* 4: 355–372 (1994).

211. B. Ruffle, D. Burmaster, P. Anderson, and D. Gordon. Lognormal distributions for fish consumption by the general U.S. population. *Risk Anal.* 14(4): 395–404 (1994).

212. E. Ebert, N. Harrington, K. Boyle, J. Knight, and R. Keenan. Estimating consumption of freshwater fish among Maine anglers. *N. Am. J. Fish. Manag.* 13: 737–745 (1993).

213. A. M. Roseberry and D. E. Burmaster. A note: Estimating exposure concentrations of lipophilic organic chemicals to humans via finfish. *J. Exp. Anal. Environ. Epidemiol.* 1: 513–521 (1991).

214. D. M. Murray and D. E. Burmaster. Estimated distributions for average daily consumption of total and self-caught fish for adults in Michigan angler households. *Risk Anal.* 14: 513–520 (1994). See also E. F. Fitzgerald, S.-A. Hwang, K. A. Brix, B. Bush, K. Cook, and P. Worswick. Fish PCB concentrations and consumption patterns among Mohawk women at Akwesasne. *J. Exp. Anal. Environ. Epidemiol.* 5: 1–20 (1995); A. H. Stern, L. R. Korn, and B. E. Ruppel. Estimation of fish consumption and methylmercury intake in the New Jersey population. *J. Exp. Anal. Environ. Epidemiol.* 6: 503–525 (1996); and E. S. Ebert, P. S. Price, and R. E. Keenan. Selection of fish consumption estimates for use in the regulatory process. *J. Exp. Anal. Environ. Epidemiol.* 4: 373–394 (1994).

215. A. H. Smith. Infant exposure assessment for breast milk dioxins and furans derived from waste incineration emissions. *Risk Anal.* 7(3): 347–353 (1987).

216. A. Jones. Risk evaluation of breast milk. *J. Toxicol. Environ. Health* (1994).

217. L. Kohler, G. Meeuwisse, and W. Mortensson. Food intake and growth of infants between six and twenty-six weeks of age on breast milk, cow's milk formula, and soy formula. *Acta Paediatr. Scand.* 73: 40–48 (1984).

218. National Academy of Sciences (NAS). *Nutrition During Lactation.* National Academy Press, Washington, DC, 1991.

219. M. C. Neville, R. Keller, J. Seacat, V. Lutes, M. Neifert, C. Casey, J. Allen, and P. Archer. Studies in human lactation: Milk volumes in lactating women during the onset of lactation and full lactation. *Am. J. Clin. Nutr.* 48: 1375–1386 (1988).

220. C. D. Carrington and P. M. Bolger. Uncertainty and risk assessment. *Hum. Ecol. Risk Assess.* 4: 253–258 (1998).

221. M. D. Morgan and M. Henrion. *Uncertainty: A Guide to Dealing with Uncertainty in Quantitative Risk and Policy Analysis.* Cambridge University Press, Cambridge, MA, 1990.

222. U.S. Environmental Protection Agency (EPA). *Guidance for Risk Characterization.* EPA, Science Policy Council, Washington, DC, 1995.

223. N. Duan and D. T. Mage. Combination of direct and indirect approaches for exposure assessment. *J. Exp. Anal. Environ. Epidemiol.* 7: 439–470 (1997).

224. D. E. Burmaster and K. von Stackelberg. Using Monte Carlo simulations in public health risk assessments: Estimating and presenting full distributions of risk. *J. Exp. Anal. Environ. Epidemiol.* 1: 491–521 (1991).

225. P. D. Anderson and A. L. Yuhas. Improving risk management by characterizing reality: A benefit of probabilistic risk assessment. *Hum. Ecol. Risk Assess.* 2: 55–58 (1996).

226. D. E. Burmaster and P. D. Anderson. Principle of good practice for the use of Monte Carlo techniques in human health and ecological risk assessment. *Risk Anal.* 14: 477–491 (1994).

227. R. L. Smith. Use of Monte Carlo simulation for human exposure assessment at a Superfund site. *Risk Anal.* 14(4): 433–439 (1994).

228. T. S. Glickman. A methodology for estimating time-of-day variations in the size of a population exposed to risk. *Risk Anal.* 6: 317–323 (1986).

229. D. E. Burmaster and N. I. Maxwell. Time and loading—Dependence in the McKone model for dermal uptake of organic chemicals from a soil matrix. *Risk Anal.* 11: 491–497 (1991).

230. M. Israeli and C. B. Nelson. Distribution and expected time of residence for U.S. households. *Risk Anal.* 12: 65–72 (1992).

231. D. M. Murray and D. E. Burmaster. Estimated distributions for total body surface area of men and women in the United States. *J. Exp. Anal. Environ. Epidemiol.* 2: 451–462 (1992).

232. A. C. Taylor, J. S. Evans, and T. E. McKone. The value of animal test information in environmental control decisions. *Risk Anal.* 13: 403–412 (1993).

233. B. Ruffle, D. E. Burmaster, P. D. Anderson, and H. D. Gordon. Lognormal distribution for fish consumption by the general U.S. population. *Risk Anal.* 14: 395–403 (1994).

234. K. M. Thompson, D. E. Burmaster, and E. A. C. Crouch. Monte Carlo techniques for quantitative uncertainty analysis in public health risk assessments. *Risk Anal.* 12: 53–63 (1992).

235. P. R. Trowbridge and D. E. Burmaster. A parametric distribution for the fraction of outdoor soil in indoor dust. *J. Soil Contam.* 6: 161–168 (1997).

236. D. E. Burmaster and K. M. Thompson. Backcalculating cleanup targets in probabilistic risk assessments when the acceptability of cancer risk is defined under different risk management policies. *Hum. Ecol. Risk Assess.* 1: 101–120 (1995).

237. D. E. Burmaster and D. A. Huff. Using lognormal distributions and lognormal probability plots in probabilistic risk assessments. *Hum. Ecol. Risk Assess.* 3: 223–234 (1997).

238. B. Finley, C. Kirman, P. Scott, A. Spivack, T. Bernhardt, J. Warmerdam, and A. Pittignano. A probabilistic risk assessment of a PCB-contaminated waterway: A case study. *J. Soil Contam.* (submitted).

239. B. Allen, R. Gentry, A. Shipp, and C. Van Landingham. Calculation of benchmark doses for reproductive and developmental toxicity observed after exposure to isopropanol. *Regul. Toxicol. Pharmacol.* 28: 38–44 (1998).

240. H. C. Frey and D. S. Rhodes. Characterization and simulation of uncertainty frequency distributions: Effects of distribution choice, variability, uncertainty, and parameter dependence. *Hum. Ecol. Risk Assess.* 4: 423–469 (1998).

241. B. D. Beck and J. T. Cohen. Risk assessment for criteria pollutants versus other non-carcinogens: The difference between implicit and explicit conservatism. *Hum. Ecol. Risk Assess.* 3: 617–626 (1997).

242. C. K. Mertz, P. Slovic, and I. F. H. Purchase. Judgments of chemical risks: Comparisons among senior managers, toxicologists, and the public. *Risk Anal.* 18: 391–403 (1998).

243. R. L. Sielken, Jr. Useful tools for evaluating and presenting more science in quantitative cancer risk assessments. *Toxicol. Subst. J.* 9: 353–404 (1989).

244. A. Shlyakhter, G. Goodman, and R. Wilson. Monte Carlo simulation of rodent carcinogenicity bioassays. *Risk Anal.* 12: 73–82 (1992).

245. J. S. Evans, G. M. Gray, R. L. Sielken, Jr., A. E. Smith, C. Valdez-Flores, and J. D. Graham. Use of probabilistic expert judgment in uncertainty analysis of carcinogenic potency. *Regul. Toxicol. Pharmacol.* 20: 15–36 (1994).

246. J. S. Evans, J. D. Graham, G. M. Gray, and R. L. Sielken, Jr. A distributional approach to characterizing low-dose cancer risks. *Risk Anal.* 14: 25–33 (1994).

247. S. F. Velazquez, P. M. McGinnis, S. T. Vater, W. S. Stiteler, L. A. Knauf, and R. S. Schoeny. Combination of cancer data in quantitative risk assessments: Case study using bromodichloromethane. *Risk Anal.* 14: 285–292 (1994).

248. S. J. S. Baird, J. T. Cohen, J. D. Graham, A. I. Shlyakhter, and J. S. Evans. Non-cancer risk assessment: A probabilistic alternative to current practice. *Hum. Ecol. Risk Assess.* 2: 79–102 (1996).

249. L. A. Cox, Jr. More accurate dose-response estimation using Monte-Carlo uncertainty analysis: The data cube approach. *Hum. Ecol. Risk Assess.* 2: 150–174 (1996).

250. E. A. C. Crouch. Uncertainty distributions for cancer potency factors: Laboratory animal carcinogenicity and interspecies extrapolation. *Hum. Ecol. Risk Assess.* 2: 103–129 (1996).

251. E. A. C. Crouch. Uncertainty distributions for cancer potency factors: Combining epidemiological studies with laboratory bioassays—The example of acrylonitrile. *Hum. Ecol. Risk Assess.* 2: 130–149 (1996).

252. R. L. Sielken, Jr. and C. Valdez-Flores. Comprehensive realism's weight-of-evidence based distributional dose-response characterization. *Hum. Ecol. Risk Assess.* 2: 175–193 (1996).

253. R. L. Sielken, Jr. and D. E. Stevenson. Opportunities to improve quantitative risk assessment. *Hum. Ecol. Risk Assess.* 3: 479–490 (1997).

254. R. A. Hill and S. M. Hoover. Importance of the dose-response model form in probabilistic risk assessment: A case study of health effects from methylmercury in fish. *Hum. Ecol. Risk Assess.* 3: 465–481 (1997).

255. C. P. Boyce. Comparison of approaches for developing distributions for carcinogenic potency factors. *Hum. Ecol. Risk Assess.* 4: 527–578 (1998).

256. T. L. Copeland, D. J. Paustenbach, M. A. Harris, and J. Otani. Comparing the results of a Monte Carlo analysis with EPA's reasonable maximum exposed individual (RMEI): A case study of a former wood treatment site. *Regul. Toxicol. Pharmacol.* 18: 275–312 (1993).

257. T. L. Copeland, A. H. Holbrow, J. M. Otani, K. T. Connor, and D. J. Paustenbach. Use of probabilistic methods to understand the conservatism in California's approach to assessing health risks posed by air contaminants. *J. Air Waste Manag. Assoc.* 44: 1399–1413 (1994).

258. M. L. Gargas, D. J. Paustenbach, B. L. Finley, and T. F. Long. Environmental health risk assessment: Theory and practice. In B. Ballantyne (Ed.), *General and Applied Toxicology.* Turner, London, 1999.

259. B. L. Finley and D. J. Paustenbach. The benefits of probabilistic exposure assessment: Three case studies involving contaminated air, water, and soil. *Risk Anal.* 14: 53–73 (1994).

260. E. J. Stanek and E. J. Calabrese. Improved soil ingestion estimates for use in site evaluations using the best tracer method. *Hum. Ecol. Risk Assess.* 1: 133–157 (1995).

261. E. J. Stanek and E. J. Calabrese. Daily estimates of soil ingestion in children. *Environ. Health Perspect.* 103: 276–285 (1995).

262. U.S. Environmental Protection Agency (EPA). *Guiding Principles for Monte Carlo Analysis.* EPA/630/R-97/001. Risk Assessment Forum, EPA, Office of Research and Development, Washington, DC, 1997.

263. D. E. Burmaster. Lognormal distributions for skin area as a function of body weight. *Risk Anal.* 18: 27–32 (1998).

264. D. E. Burmaster. A lognormal distribution for time spent showering. *Risk Anal.* 18: 33–36 (1998).

265. R. Sedman, L. M. Funk, and R. Fountain. Distribution of residence duration in owner occupied housing. *J. Exp. Anal. Environ. Epidemiol.* 8: 51–57 (1998).

266. A. E. Smith, P. B. Ryan, and J. S. Evans. The effect of neglecting correlations when propagating uncertainty and estimating population distribution of risk. *Risk Anal.* 12: 467–474 (1992).

267. D. Hattis and D. Burmaster. Assessment of variability and uncertainty distributions for practical risk analyses. *Risk Anal.* 14: 713–729 (1994).

268. F. O. Hoffman and J. S. Hammods. Propagation of uncertainty in risk assessments: The need to distinguish between uncertainty due to lack of knowledge and uncertainty due to variability. *Risk Anal.* 14: 707–711 (1994).

269. J. Bukowski, L. Korn, and D. Wartenberg. Correlated inputs in quantitative risk assessment: The effects of distributional shape. *Risk Anal.* 15: 215–219 (1995).

270. J. A. Cooper, S. Ferson, and L. Ginzburg. Hybrid processing of stochastic and subjective uncertainty data. *Risk Anal.* 16: 785–792 (1996).

271. C. N. Haas. Importance of the distributional form in characterizing inputs to Monte Carlo risk assessments. *Risk Anal.* 17: 107–113 (1997).

272. M. M. Hamed and P. B. Bedient. On the effect of probability distributions of input variables in public health risk assessment. *Risk Anal.* 17: 97–105 (1997).

273. W. J. Cronin, E. J. Oswald, M. L. Shelley, J. W. Fisher, and C. D. Fleming. A trichloroethylene risk assessment using a Monte Carlo analysis of parameter uncertainty in conjunction with physiologically-based pharmacokinetic modeling. *Risk Anal.* 15: 555–566 (1995).

274. T. Simon. Combining physiologically based pharmacokinetic modeling with Monte Carlo simulation to derive an acute inhalation guidance value for trichloroethylene. *Regul. Toxicol. Pharmacol.* 26: 257–270 (1997).

275. K. T. Bogen and R. C. Spear. Integrating uncertainty and interindividual variability in environmental risk assessment. *Risk Anal.* 7: 427–435 (1987).

276. R. L. Iman and J. C. Helton. The repeatability of uncertainty and sensitivity analyses for complex probabilistic risk assessments. *Risk Anal.* 11: 591–606 (1991).

277. A. I. Shlyakhter. An improved framework for uncertainty analysis: Accounting for unsuspected errors. *Risk Anal.* 14: 441–447 (1994).

278. R. B. Robinson and B. T. Hurst. Statistical quantification of the sources of variance in uncertainty analysis. *Risk Anal.* 17: 447–454 (1997).

279. S. N. Rai and D. Kreski. Uncertainty and variability analysis in multiplicative risk models. *Risk Anal.* 18: 37–45 (1998).

280. B. L. Finley, P. Scott, and D. J. Paustenbach. Evaluating the adequacy of maximum contaminant levels as health protective cleanup goals: An analysis based on Monte Carlo techniques. *Regul. Toxicol. Pharmacol.* 18: 438–455 (1993).

281. A. G. Ershow and K. P. Cantor. *Total Tapwater Intake in the United States: Population-Based Estimates of Quantities and Sources.* Life Sciences Research Office, Fed. Amer. Soc. Exper. Biol., Bethesda, MD, 1989.

282. A. M. Roseberry and D. E. Burmaster. Lognormal distributions for water intake by children and adults. *Risk Anal.* 12: 99–104 (1992).

283. J. Graham, M. Berry, E. F. Bryan, M. A. Callahan, A. Fan, B. Finley, J. Lynch, T. McKone, H. Ozkaynak, K. Sexton, and K. Walker. The role of exposure databases in risk assessment. *Arch. Environ. Health* 47: 408–420 (1992).

284. R. A. Schoof and J. B. Nielsen. Evaluation of methods for assessing the oral bioavailability of inorganic mercury in soil. *Risk Anal.* 17: 545–555 (1997).

285. D. J. Paustenbach, G. M. Bruce, and P. Chrostowski. Current views on the oral bioavailability of inorganic mercury in soil: The impact on health risk assessments. *Risk Anal.* 17: 533–545 (1997).

286. M. Alexander. How toxic are chemicals in soil? *Environ. Sci. Technol.* 29: 2713–2717 (1995).

287. W. L. Ruoff, G. L. Diamond, S. F. Velazquez, W. M. Stiteler, and D. J. Gefell. Bioavailability of cadmium in food and water: A case study on the derivation of relative bioavailability factors for inorganics and their relevance to the reference dose. *Regul. Toxicol. Pharmacol.* 20: 139–160 (1994).

288. A. Davis, M. V. Ruby, and P. D. Bergstrom. Bioavailability of arsenic and lead from the Butte, Montana, mining district. *Environ. Sci. Technol.* 26: 461–468 (1992).

289. A. Davis, J. W. Drexter, M. V. Ruby, and A. Nicholson. Micromineralogy of mine waste in relation to lead bioavailability, Butte, Montana. *Environ. Sci. Technol.* 27: 1415–1425 (1993).

290. A. Davis, N. S. Bloom, and S. S. Que Hee. The environmental geochemistry and bioaccessability of mercury in soils and sediments: A review. *Risk Anal.* 17: 557–569 (1997).

291. N. S. Shifrin, B. D. Beck, T. D. Gauthier, S. D. Chapnick, and G. Goodman. Chemistry, toxicology, and human health risks of cyanide compounds in soils at former manufactured gas plant sites. *Regul. Toxicol. Pharmacol.* 23: 106–116 (1996).

292. R. L. Chaney, S. B. Sterrett, and H. W. Mielke. The potential for heavy metal exposure from urban gardens and soils. In J. R. Preer (Ed.), *Proceedings of the Symposium on Heavy Metals in Urban Gardens.* Agric. Exper. Sta. Univ. Dist. Columbia, MD. Washington, DC, 1984, pp. 37–44.

293. U.S. Environmental Protection Agency (EPA). *Integrated Risk Information System.* EPA, Washington, DC, 1998.

294. M. V. Ruby, A. Davis, J. H. Kempton, J. W. Drexter, and P. D. Bergstrom. Lead bioavailability under simulated gastric conditions. *Environ. Sci. Technol.* 26: 1242–1248 (1992).

295. S. B. Horowitz and B. L. Finley. Using human sweat to extract chromium from chromite ore processing residue: Applications to setting health-based cleanup levels. *J. Toxicol. Environ. Health* 40: 585–599 (1993).

296. M. J. Bartek, J. A. LaBudde, and H. I. Maibach. Skin permeability in vivo: Comparison in rat, rabbit, pig and man. *J. Invest. Dermatol.* 58: 114–123 (1972).

297. R. C. Wester and P. K. Noonan. Relevance of animal models for percutaneous absorption. *Int. J. Pharmacol.* 7: 99–110 (1980).

298. R. H. Guy, J. Hadgraft, and H. I. Maibach. A pharmacokinetic model for percutaneous absorption. *Int. J. Pharmacol.* 11: 119–129 (1982).

299. C. Surber, K. P. Wilhelm, H. I. Maibach, L. L. Hall, and R. H. Guy. Partitioning of chemicals into human stratum corneum: Implications for risk assessment following dermal exposure. *Fundam. Appl. Toxicol.* 15: 99–107 (1990).

300. S. W. Frantz. Instrumentation and methodology for in vitro skin diffusion cells. In B. W. Kemppainen and W. G. Reifenrath (Eds.), *Methods for Skin Absorption.* CRC Press, Boca Raton, FL, 1990.

301. J. A. Shatkin and H. S. Brown. Pharmacokinetics of the dermal route of exposure to volatile organic chemicals in water. A computer simulation model. *Environ. Res.* 56: 90–108 (1991).

302. E. D. Barber, N. M. Teetsel, K. F. Kolberg, and D. Guest. A comparative study of the rates of in vitro percutaneous absorption of eight chemicals using rat and human skin. *Fundam. Appl. Toxicol.* 19: 493–497 (1992).

303. T. E. McKone. Linking a PB-PK model for chloroform with measured breath concentrations in showers: Implications for dermal exposure models. *J. Exp. Anal. Environ. Epidemiol.* 3: 339–365 (1993).

304. K. T. Bogen. A note on compounded conservatisms. *Risk Anal.* 14: 379–382 (1994).

305. K. T. Bogen, G. A. Keating, S. Meissner, and J. S. Vogel. Initial uptake kinetics in human skin exposed to dilute aqueous trichloroethylene. *J. Exp. Anal. Environ. Epidemiol.* 8: 253–271 (1998).

306. R. C. Wester, D. A. W. Bucks, and H. I. Maibach. Percutaneous absorption of contaminants from soil. In R. G. M. Wang, J. B. Knaak, and H. I. Maibach (Eds.), *Health Risk Assessment: Dermal and Inhalation Exposure and Absorption of Toxicants.* CRC Press, Boca Raton, FL, 1993.

307. R. C. Wester, H. I. Maibach, and L. Sedik. Percutaneous absorption of pentachlorphenol from soil. *Fundam. Appl. Toxicol.* 20: 68–71 (1993).

308. American Chemical Society (ACS). In R. L. Swann and A. Eschenroeder (Eds.), *Fate of Chemicals in the Environment.* ACS Symposium Series 225. ACS, Washington, DC, 1983.

309. D. J. Paustenbach. A survey of environmental risk assessment. In D. J. Paustenbach (Ed.), *The Risk Assessment of Environmental and Human Health Hazards: A Textbook of Case Studies.* Wiley, New York, 1989.

310. S. J. Borgert, S. M. Roberts, R. D. Harbison, and R. C. James. Influence of soil half-life on risk assessment of carcinogens. *Regul. Toxicol. Pharmacol.* 22: 143–151 (1995).

311. J. N. Morgan, M. R. Berry, and R. L. Graves. Effects of commonly used cooking practices on total mercury concentration in fish and their impact on exposure assessments. *J. Exp. Anal. Environ. Epidemiol.* 7: 119–133 (1997).

312. N. D. Wilson, N. D. Shear, D. J. Paustenbach, and P. S. Price. The effect of cooking practices on the concentration of DDT and PCB compounds in the edible tissue of fish. *J. Exp. Anal. Environ. Epidemiol.* 8: 423–440 (1998).

313. K. W. Thomas, L. S. Sheldon, E. D. Pellizzari, R. W. Handy, J. M. Roberds, and M. R. Berry. Testing duplicate diet sample collection methods for measuring personal dietary exposures to chemical contaminants. *J. Exp. Anal. Environ. Epidemiol.* 7: 17–36 (1997).

314. D. B. Hattis. The promise of molecular epidemiology for quantitative risk assessment. *Risk Anal.* 6(2): 181–194 (1986).

315. A. McMillan, A. S. Whittemore, A. Silvers, and Y. DiCiccio. Use of biological markers in risk assessment. *Risk Anal.* 14(5): 807–813 (1994).

316. D. A. Holdway. The role of biomarkers in risk assessment. *Hum. Ecol. Risk Assess.* 2: 263–267 (1996).

317. D. A. Wolfe. Insights on the utility of biomarkers for environmental impact assessment and monitoring. *Hum. Ecol. Risk Assess.* 2: 245–250 (1996).

318. L. Ehrenberg and S. Osterman-Golkar. Alkylation of macromolecules for detecting mutagenic agents. *Teratog. Carcinog. Mutagen.* 1(1): 105–127 (1980).

319. F. P. Perera and I. B. Weinstein. Molecular epidemiology and carcinogen-DNA adduct detection: New approaches to studies of human cancer causation. *J. Chronic Dis.* 35(7): 581–600 (1982).

320. National Research Council. Biological markers in environmental health research. Committee on Biological Markers of the National Research Council. *Env. Health Perspect.* 74: 3–9 (1987).

321. D. J. Hewitt, G. C. Millner, A. C. Nye, M. Webb, and R. G. Huss. Evaluation of residential exposure to arsenic in soil near a superfund site. *Hum. Ecol. Risk Assess.* 1: 323–335 (1995).

322. J. D. Graham, L. Green, and M. J. Roberts. *In Search of Safety: Chemicals and Cancer Risks.* Harvard University Press, Cambridge, MA, 1988, pp. 80–114.

323. W. Slob. A comparison of two statistical approaches to estimate long-term exposure distributions from short-term measurements. *Risk Anal.* 16: 195–200 (1996).

324. R. J. Buck, K. A. Hammerstrom, and P. B. Ryan. Estimating long-term exposures from short-term measurements. *J. Exp. Anal. Environ. Epidemiol.* 5: 359–374 (1995).

325. R. J. Buck, K. A. Hammerstrom, and P. B. Ryan. Bias in population estimates of long-term exposure from short-term measurements of individual exposure. *Risk Anal.* 17: 455–465 (1997).

326. E. J. Stanek, III, E. J. Calabrese, and L. Xu. A caution for Monte Carlo risk assessment of long term exposures based on short term exposure data. *Hum. Ecol. Risk Assess.* 4: 409–422 (1998).

327. K. S. Crump. On summarizing group exposures in risk assessment: Is an arithmetic mean or a geometric mean more appropriate? *Risk Anal.* 18: 293–297 (1998).

328. W. Horwitz. Effects of scientific advances on the decision-making process: Analytical chemistry. *Fund. Appl. Toxicol.* 4: S309–S317 (1984).

329. D. R. Helsel. Less than obvious: Statistical treatment of data below the detection limit. *Environ. Sci. Technol.* 24: 1766–1774 (1990).

330. J. L. Perkins, G. N. Cutter, and M. S. Cleveland. Estimating the mean, variance, and confidence limits from censored (< limit of detection), lognormally-distributed exposure data. *Am. Ind. Hyg. Assoc. J.* 51: 416–419 (1990).

331. C. C. Travis and M. L. Land. Estimating the mean of data sets with nondetectable values. *Environ. Sci. Technol.* 24: 961–962 (1990).

332. R. O. Gilbert. *Statistical Methods for Environmental Pollution Monitoring.* Van Nostrand Reinhold, New York, 1987.

333. C. N. Haas and P. A. Scheff. Estimation of averages in truncated samples. *Environ. Sci. Technol.* 24: 912–919 (1990).

334. S. M. Rappaport and J. Selvin. A method for evaluating the mean exposure from a lognormal distribution. *Am. Ind. Hyg. Assoc. J.* 48: 374–379 (1987).

335. T. B. Parkin, J. J. Melsinger, S. T. Chester, J. L. Starr, and J. A. Robinson. Evaluation of statistical estimation methods for lognormally distributed variables. *Soil Sci. J.* 52: 323–329 (1988).

5 Risk Characterization

PAMELA R. D. WILLIAMS and DENNIS J. PAUSTENBACH
Exponent, Menlo Park, California

5.1 INTRODUCTION

Risk characterization represents the final, and perhaps the most important, step in the risk assessment process. In this step, data on the dose–response relationship of an agent are integrated with estimates of the degree of exposure in a population to characterize the likelihood and severity of health risk [Environmental Protection Agency (EPA), 1995a]. Risk characterizations include both quantitative estimates and qualitative descriptors of risk, as well as discussions about key model assumptions and data uncertainties. The most relevant findings and conclusions about risk are summarized in the risk characterization, which in turn is used to inform risk managers and decision makers [National Research Council (NRC), 1996]. In short, the risk characterization process attempts to make sense of the available data and describe what it means to a broader audience.

Despite the importance of accurate and defensible risk characterizations, this step is often given insufficient attention in health risk evaluations (Paustenbach, 1995). The primary reason for the lack of quality in risk characterizations is that unlike the other three portions of a risk assessment (hazard identification, exposure assessment, and dose–response analysis), it is not possible to rely exclusively on guidance documents and formulas to properly capture the importance of the analysis. A significant command of the nuances of the biology of the chemical, interspecies susceptibility, conflicting animal and human studies, statistics, regulatory history, acceptable risk criteria, and many other factors are needed and these should be weighed and integrated in an attempt to provide the risk manager with all of the relevant information to make an informed decision. Thus, it is not surprising that most risk characterizations fall short of the mark, and ultimately, many well-intended but undefensible decisions have been made over the past 30 years of the environmental revolution (Efron, 1984; Breyer, 1993; Foster et al., 1994; Wildavsky, 1995; Easterbrook, 1995; Freeze, 2000; Milloy, 2001; Lomborg, 2001).

Human and Ecological Risk Assessment: Theory and Practice, Edited by Dennis J. Paustenbach
ISBN 0-471-14747-8 © 2002 John Wiley & Sons, Inc.

Perhaps the single most important shortcoming of historical risk assessments has been the level of, often unjustified, confidence or certainty implied within the analysis or the risk characterization. Scientific analyses, such as those for chemical-specific risks, tend to focus either on quantifying toxic doses (risk criteria such as Reference Dose (RfD), Reference Concentration (RfC), or cancer potency) or the magnitude of public health exposures, with little consideration given to interpreting or summarizing this information in a meaningful way. Even when estimates of risk are provided, these are generally presented as a single numerical value without any discussion of key data uncertainties, model assumptions, or analytical limitations. Users of risk information may therefore misinterpret the findings of a risk assessment or have false impressions about the degree of confidence in reported risk estimates (Nichols and Zeckhauser, 1984; Davies et al., 1987; Graham, 1995a). For example, screening-level analyses, designed to provide a preliminary analysis of the plausible risk or to prioritize potential health hazards, are often interpreted as a quantitative assessment of the actual or likely health risk in a population. In reality, these evaluations tend to be based on very conservative assumptions and/or "worst-case" scenarios and are not representative of exposures or risk in the general population.

Over the past 20 years, regulatory agencies in the United States believe that they have executed their duty to express the uncertainty in their risk estimates by using brief qualitative language that caveats the bound of the analysis. For example, a phrase that was commonly used in EPA's *Health Assessment Documents (HAD)*, including the EPA (1983) report on methylene chloride, states: "the linear extrapolation model used here provides a rough but plausible estimate of the upper limit of risk, and while the true risk is probably not much more than the estimated risk, it could be considerably lower" (page 5-102). This statement, while accurate, provides very little useful information to either the decision-maker or the public. Over the years, the EPA's use of a upper-bound point estimate of risk has been criticized because it does not convey the degree of uncertainty in the estimate, and thus, decision-makers do not know the extent of conservatism (if any) that is provided in the risk estimate (Paustenbach, 1989; NRC, 1994). Similarly, historical exposure evaluations have been based on the maximum exposed individual (MEI) in a population, rather than a typical or even more highly exposed individual. These evaluations, which are typically based on a combination of conservative "worst-case" assumptions, tend to yield scenarios that are not representative of most, if any, of the population. Hence, the results of such evaluations are of little use in societal decision-making (Goldstein, 1989).

It is well accepted that, at a minimum, risk characterizations should contain several types of information. Specifically, the key parameters used in the exposure assessment should be identified and discussed; this includes chemical concentrations at each exposure point, human intake of each chemical on a body-weight basis, route of chemical entry, uptake or absorption for each chemical, and the frequency and duration of exposure (EPA, 1989a). Risk characterizations should also present what is known (and unknown) about the toxicity of each chemical and discuss how the hazard potential may differ based on exposure route, health effect, dose–response model, and intra- or interspecies variability (EPA, 1995b). Finally, all risk characterizations should include a section that discusses data, model, and statistical uncertainties and expresses the level of confidence in exposure and toxicity estimates (EPA, 1989a). Depending on the purpose, risk characterizations may also contain information on the effectiveness of alternative risk management options or the risk of competing or substitute chemicals (NRC, 1996; NRC, 1989; EPA, 1995c; Graham and Hartwell, 1997).

The purpose of this chapter is to provide the reader with a broad, yet comprehensive, overview of the risk characterization process as practiced in the United States (U.S.) and select other countries. By providing criteria and examples for conducting risk characterizations, we anticipate that their quality will improve in the coming years. This review focuses predominantly on chemical-specific risks, although similar procedures and issues may be relevant for other types of risks, such as those associated with exposure to microbial pathogens, radiation exposures and other agents. Specifically, we discuss the following eight topics: (1) objective of risk characterization, (2) guidance documents on risk characterization, (3) key components of risk characterizations, (4) toxicity criteria for evaluating health risks, (5) descriptors used to characterize health risks, (6) methods for quantifying human health risks, (7) key uncertainties in risk characterizations, and (8) the risk decision-making process. As part of this latter section, the practice and role of risk management, cost-benefit analysis, and the precautionary principle are discussed. A brief overview of international aspects of risk characterization is also provided.

5.2 GUIDANCE DOCUMENTS ON RISK CHARACTERIZATION

Interestingly, guidance documents on how to define or conduct risk characterizations have not changed dramatically over the last two decades. In the NRC report published in 1983, *Risk Assessment in the Federal Government: Managing the Process* (commonly referred to as the *Red Book*), risk characterization is broadly defined as "the estimate of the magnitude of the public-health problem" [p. 28]. The report also defined it as "the description of the nature and often the magnitude of human risk, including the attendant uncertainty" (page 3). According to this report, risk characterizations reflect a combination of the exposure and dose–response assessments and serve as the intermediary between risk assessment and risk management (see Figure 5.1). The *Red Book* also specifies that risk characterizations should summarize the key uncertainties in each step of the risk assessment process, particularly those related to statistical and biological uncertainty and the choice of assessment and exposed population. The following, more specific, definition was also provided by NRC (1983):

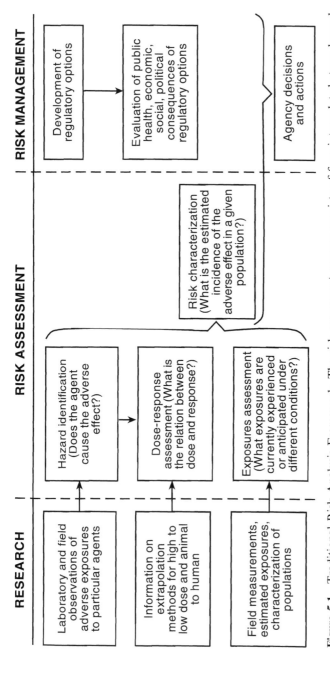

Figure 5.1 Traditional Risk Analysis Framework. The risk assessment process consists of four inter-related steps: hazard identification, dose-response evaluation, exposure assessment, and risk characterization. Risk assessments should be based on the best available science and be as free from value judgments and biases as possible. The results with appropriate description of the uncertainty involved is then provided to risk managers for decisions on actions required to reduce the risks, which include economic, political, and other value judgments. The relationship between basic and applied research, risk assessment, and risk management is also illustrated. The intent is that uncertainty be identified through the risk assessment process and, if critical to understanding the risks, research efforts would be devoted to reducing the uncertainty and improving the estimates of risk. Source: NRC 1983.

Risk characterization is the process of estimating the incidence of a health effect under the various conditions of human exposure described in exposure assessment. It is performed by combining the exposure and dose–response assessments. The summary effects of the uncertainties in the preceding steps are described in this step (page 20). Risk characterization, the estimate of the magnitude of the public-health problem, involves no additional scientific knowledge or concepts. However, the exercise of judgment in the aggregation of population groups with varied sensitivity and different exposure may affect the estimate (page 28).

In a subsequent report published by EPA (1989a), *Risk Assessment Guidance for Superfund* (RAGS), detailed guidance was provided on how to characterize potential cancer and noncancer risks from chemical exposures, particularly at hazardous waste sites. In this widely cited report, which was the primary source of guidance for perhaps hundreds of risk assessments conducted during the 1990s, risk characterization was also defined as the integration of toxicity and exposure assessments into quantitative and qualitative expressions of risk. The RAGS report further stated that risk characterizations should include information on "major assumptions, scientific judgments, and to the extent possible, estimates of uncertainties embodied in the assessment" (page 8-1). Risk characterizations were deemed to be a "key step in the ultimate site decision-making process" and such characterizations were not considered to be complete unless "numerical expressions of risk are accompanied by explanatory text interpreting and qualifying the results" (page 8-1). During the 1980s, there were at least five additional guidance documents published by EPA that discussed how to characterize the uncertainty in exposure assessments and risk assessments for carcinogens, chemical mixtures, and developmental toxicants (EPA, 1985, 1986a–c, 1989b).

The EPA (1992) later developed a special guidance policy on risk characterization for risk assessors and risk managers published in the memorandum, *Guidance on Risk Characterization for Risk Managers and Risk Assessors.* The primary purpose of this guidance was to address problems affecting public perception regarding the reliability of EPA's scientific assessments and related regulatory decisions, and to enhance public confidence in the quality of EPA's scientific work. With regard to risk characterization, the memorandum stated that "numerical risk estimates should always be accompanied by descriptive information carefully selected to ensure an objective and balanced characterization of risk in risk assessment reports and regulatory document" (page 2), and risk evaluations should include a "statement of confidence in the assessment that identifies all major uncertainties along with comment on their influence on the assessment" (page 5). EPA's new policy focused specifically on increasing the clarity, transparency, reasonableness, and consistency of risk characterizations across federal agencies, and emphasized the importance of retaining and communicating critical information from each stage of the risk assessment process. More specific guidance on this latter issue included (page 1):

Specifically, although a great deal of careful analysis and scientific judgment goes into the development of EPA risk assessments, significant information is often omitted as the results of the assessment are passed along in the decision-making process. Often, when risk information is presented to the ultimate decision-maker and to the public, the results have been boiled down to a point estimated of risk. Such "short hand" approaches to risk assessment do not fully convey the range of information considered and used in developing the assessment. In short, informative risk characterization clarifies the scientific basis for EPA decisions, while numbers alone do not give a true picture of the assessment.

The EPA has since updated its 1992 risk characterization policy in response to recommendations made in the NRC (1994) report, *Science and Judgment in Risk Assessment*. Specifically, this report concluded that despite the "fundamentally sound" approach used by EPA to assess health risks, the agency must more clearly establish the scientific and policy basis for risk estimates and better describe the uncertainties in its estimates of risk. This report also recommended that EPA should develop and use an iterative approach to risk assessment, which would not only lead to an improved understanding of the relationship between risk assessment and risk management, but would also allow for improvements in the agency's conservative default-based approach (NRC, 1994). This report also identified six common themes—default options, validation, data needs, uncertainty, variability, and aggregation—that cut across the various stages of risk assessment. According to NRC (1994), if a cross-cutting approach was developed that utilized these themes, many of the problems in risk assessments would be ameliorated, such as the differing opinions in the scientific community, the reluctance to incorporate new information into risk-assessments, and the incompatibility of various inputs to risk characterization.

EPA's (1995a–c) revised guidance materials—*Policy for Risk Characterization, Elements to Consider When Drafting EPA Risk Characterizations*, and *Guidance for Risk Characterization*—more strongly emphasized the need to disclose and describe the uncertainties in risk estimates. In particular, EPA reiterated that risk characterizations should clearly highlight both the confidence and uncertainty in the risk assessment, and that numerical risk estimates should always be accompanied by descriptive and balanced risk information. In the new guidance materials by EPA (1995a), risk characterization was defined as the interface between risk assessment and risk management, that "integrates information from the preceding components of the risk assessment and synthesizes an overall conclusion about risk that is complete, informative and useful for decisionmakers" (page 3). The guidance materials further concluded:

Scientific uncertainty is a fact of life for the risk assessment process, and agency managers almost always must make decisions using assessments that are not as definitive in all important areas as would be desirable. They therefore need to understand the strengths and the limitations of each assessment, and to communicate this information to all participants and the public (page 1). In essence, a risk characterization conveys the assessor's judgment as to the nature and existence (or lack of) human health or ecological risks. Even though a risk characterization describes limitations in an assessment, a balanced discussion of reasonable conclusions and related uncertainties enhances, rather than detracts, from the overall credibility of each assessment (page 3).

More recently, NRC (1996) published a report, *Understanding Risk: Informing Decisions in a Democratic Society*, recommending that risk characterizations should be more than a mere summarization of scientific information. In this report, risk characterizations are defined as an integral part of the entire process of risk decision making that is decision driven, recognizes all significant concerns, and reflects both analysis and deliberation by all interested parties. The NRC (1996) also provided an expanded framework for risk characterizations, in which such characterizations not only describe a potentially hazardous situation, but also enhance practical understanding and illuminate practical choices (see Figure 5.2). The following conveys the "new" definition of risk characterization presented by NRC (1996):

> Risk characterization is a synthesis and summary of information about a potentially hazardous situation that addresses the needs and interests of decision makers and of interested and affected parties. Risk characterization is a prelude to decision making and depends on an iterative, analytic-deliberative process. If the underlying process is unsatisfactory to some or all of the interested and affected parties, the risk characterization will be unsatisfactory as well. A risk characterization can only be as good as the analytic-deliberative process that produces it (page 27).

In their most recent handbook on this issue, EPA (2000a) provides more specific guidance for Agency risk assessors and risk managers on risk characterization principles and practices. This handbook, which implements EPA's 1995 *Risk Characterization Policy*, also includes a number of risk characterization case studies and references. Its primary focus is on increasing the transparency, clarity, consistency and reasonableness of risk characterizations, recognizing that a culture change may be required at EPA in order to meet these objectives. According to EPA (2000a), this handbook provides a "single, centralized body of risk characterization implementation guidance" that can be used to create an "integrated picture" of risk.

The evolving nature of risk characterizations suggests that future efforts to characterize human health risks will be held to a much higher standard than in the past. That is, risk characterizations are expected to provide a much more detailed discussion of key data uncertainties, as well as a description of the range or distribution of plausible risk estimates in a specified population (Paustenbach, 1995; NRC, 1996). Risk characterizations are also expected to provide a fuller discussion of the underlying problem, and in the future, could require more reporting on the costs and benefits of alternate solutions, instead of merely reporting the results of the exposure and dose–response assessment. Various regulatory bodies and private entities will likely continue to seek greater input from risk assessors on how to interpret and communicate scientific findings to a broader audience. To improve risk management decisions in the future, health

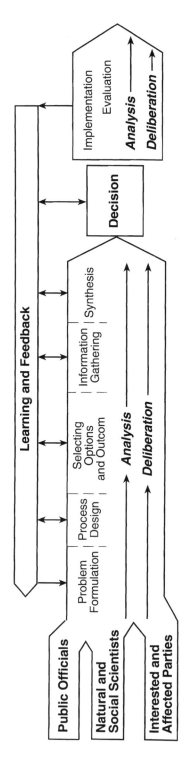

Figure 5.2 Contemporary Risk Analysis Framework. Example of expanded risk analysis framework proposed by NRC (1996), in which it recommended that risk characterizations be more than a mere summarization of scientific information. Specifically, risk characterizations are defined as an integral part of the entire decision-making process that are decision driven, recognize all significant concerns, and reflect both analysis and deliberation by all interested parties.

risk evaluations will also need to consider competing risks or potential risk trade-offs associated with alternative options (Graham and Weiner, 1995; Finkel and Golding, 1994). In short, the risk characterization portion of future risk assessments is going to require a significant amount of transparency and thoughtful discussion about the strengths and weakness of the analysis. The most recent guidance provided by NRC (1996) provides an example of how risk characterizations may evolve in the future (page 16):

> We have concluded that the view of risk characterization as a summary is seriously deficient, and we propose a more robust construction ... If a risk characterization is to fulfill its purpose, it must (1) be decision driven, (2) recognize all significant concerns, (3) reflect both analysis and deliberation, with appropriate input form the interested and affected parties, and (4) be appropriate to the decision.

5.3 KEY COMPONENTS OF RISK CHARACTERIZATION

An important feature of risk characterizations is that they summarize the key findings, assumptions, and/or limitations of the risk assessment. Although the individual components of a risk assessment (hazard identification, dose–response relationship, and exposure assessment) will contain more detailed information about a chemical or agent, the risk characterization should attempt to reiterate the most important issues and interpret this information for a broader audience. For example, risk characterizations would be most effective if they summarized what is known about a chemical's capacity for causing adverse health effects in humans, the key exposure parameters used in the risk assessment, toxicity data related to each chemical of interest, data variability and uncertainty, the level of confidence in exposure and toxicity estimates, and any other risk-related information that might be useful to a decision maker. The following section discusses the key components of risk characterization in greater detail.

5.3.1 Hazard Identification

For any characterization of risk, it is important to know whether the chemical of interest is capable of causing adverse health effects in humans. Information should therefore be presented on the kind and quality of animal or human data, the availability of ancillary information (e.g., structure–activity, genetic toxicity, pharmacokinetics) from other studies, and the weight of evidence from all of these data sources (see Table 5.1).

One of the most important changes in the practice of risk assessment that has occurred over the past 20 years is the recognition that some adverse effects observed in animals will simply not occur in humans. The old mantra "if it happens in animals at some dose, there is a high probability that it could occur

TABLE 5.1 Key Components of Risk Characterization for Hazard Identification

General Components	Specific Components
What is the key toxicological study that provides the basis for health concerns?	• How good is the key study? • Are the data from laboratory or field studies? In single or multiple species? • If the hazard is carcinogenic, were single or multiple tumor sites observed? Were tumors benign or malignant? Were certain tumor types not linked to carcinogenicity? Was the maximum tolerated dose (MTD) used? • If the hazard is not carcinogenic, what endpoints were observed? What is the basis for the critical effect? • Are there other valid studies that support or conflict with this finding?
Besides the health effect observed in the key study, are there other health endpoints of concern?	• What are the significant data gaps?
Are there available epidemiological or clinical data?	• What types of studies were used (e.g., ecologic, case control, cohort)? • Were exposures adequately described? • Were confounding factors adequately accounted for? • Were other causal factors excluded?
How much is known about how (through what biological mechanism) the chemical produces adverse effects?	• What are the relevant studies of mechanisms of action or metabolism? Does this information aid in the interpretation of the toxicity data? • What are the implications for potential health effects?
Are there any nonpositive data in animals or people?	• Were these data considered in the hazard identification?
What are the conclusions of the hazard identification?	• What is the level of confidence in these conclusions? • Are there alternative conclusions that are also supported by the data? • What are the significant data gaps? • What are the major assumptions?

Source: Adapted from EPA (1995b).

in humans at some dose, therefore one needs to regulate accordingly," was rapidly losing favor in the late 1990s. The combination of improved laboratory investigations with animals, which focused on mechanism of action, coupled with a better understanding of human biochemistry and physiology, high-lighted the fact that risk assessments could not be conducted in such a routine manner. For example, it was discovered that the adverse effects observed in rats exposed to high concentrations of airborne formaldehyde would not be ob-served at low doses, and humans in virtually all workplaces were not exposed to concentrations that posed a cancer hazard unless they routinely experienced (probably intolerable) nasal irritation (Starr and Gibson, 1985). It was also discovered that the formation of hyaline droplets in rodents from exposure to various chemicals, including d-limonene and 2,2,4-trimethylpentane (unleaded gasoline), which can lead to rat kidney tumors, would not occur in humans, and therefore these chemicals posed no occupational hazard (Short et al., 1989; Dietrich and Swenberg, 1991). As more was learned about other substances, such as saccharin, vinyl chloride, and methylene chloride, it became increas-ingly clear that the standard assumption that humans are merely "large ro-dents" was not justified.

The complexities of the hazard identification step are well-illustrated by a recent analysis of hexavalent chromium [Cr(VI)], which has been assumed by the State of California to pose a cancer hazard via drinking water. A compre-hensive review of the toxicology and epidemiology literatures on Cr(VI) as part of the hazard identification process, would likely lead most toxicologists to several inferences or conclusions. First, the data indicate that Cr(VI) is an inhalation carcinogen at very high airborne concentrations based on occupa-tional exposure studies during the 1940–1960s (Mancuso, 1997). Second, der-mal exposure to fairly high concentrations of Cr(VI) appears to cause sensitiza-tion in some workers (Proctor et al., 2001a). Third, the literature suggests that ingestion of low concentrations of Cr(VI) in drinking water does not pose a cancer hazard due to the reductive capacity of the stomach (DeFlora, 2000; Proctor et al., 2001a; Last, 2001). Specifically, at nearly any reasonable water concentration, ingested Cr(VI) is reduced in stomach juices to trivalent chro-mium [Cr(III)], a virtually non-toxic chemical. Genotoxicity studies following oral exposures to Cr(VI) have also proven negative, suggesting that Cr(VI) is reduced before systemic absorption in the body (Coogan et al., 1991; Kuy-kendall et al., 1996; Mirsalis et al., 1996). Epidemiological studies of residential and occupational populations, particularly those conducted in the last decade, provide additional evidence that Cr(VI) does not pose an oral cancer risk (Zhang and Li, 1997; Fryzek et al., 2001).

Despite these findings, the approach traditionally used by California's Office of Environmental Health Hazard Assessment (OEHHA) is to label Cr(VI) as a chemical carcinogen by any route, even when the data are only suggestive or are applicable to only one route of exposure, e.g., inhalation. The only excep-tion to this policy choice is when there is an overwhelmingly complete dataset to the contrary (in contrast to a simpler "more likely than not" test). Cr(VI) in drinking water was therefore listed by OEHHA as an oral carcinogen, with a corresponding Public Health Goal of 0.1 ppm (Morey, 1999).

This example illustrates how important it is for risk characterizations to "dig deeper" into the underlying data, and to present data uncertainties and alternative ways of evaluating the biological underpinnings of the hazard identification. Interestingly, a recent panel of experts (convened at the request of the California legislature) concluded that OEHHA did not give adequate weight to all of the scientific evidence, and that EPA's current MCL of 100 ppm for Cr(VI) in drinking water was protective of public health, rather than OEHHA's suggested 0.1 ppm, even if oral exposure to Cr(VI) is assumed to pose a cancer hazard (Last, 2001).

5.3.2 Exposure Parameters

Exposure information needed for risk assessment and characterization include estimated intakes of chemicals by route of entry, important exposure modeling assumptions, and reasonable or likely exposure pathways (EPA, 1989a). For example, a quantiative understanding of the intake of chemicals via different routes of entry is needed to understand the risks posed by chronic, subchronic, or shorter-term exposures. In addition, assumptions used in the exposure modeling may relate to chemical concentration at different exposure points, the frequency and duration of human exposure, and the amount of chemical absorbed into the body via different exposure routes. Multiple pathways, such as inhalation of ambient air, ingestion of drinking water or food, or dermal contact with an agent, may also contribute to individuals' exposures over a specified time period. In essence, characterizing risks require information on the principal pathways, magnitude of human exposure, and the number of people likely to be exposed in a population (see Table 5.2).

For example, the EPA (2000b) recently characterized the various important or dominant exposure pathways for dioxin for the typical American. Specifically, the intake of chlorinated dibenzo dioxins (CDD) and chlorinated dibenzo furans (CDF) and dioxinlike polychlorinated biphenyls (PCBs) were estimated to average about 41 and 24 pg TEQ per day, respectively. The intake of CDD/CDF/PCBs for persons who ingest much more meat and dairy products than the typical citizen (90th percentile) was found to extend to levels at least three times higher than the mean in the general population. Further, EPA estimated that the daily intake could be more than three times higher for a young child as compared to that of an adult, on a body weight basis, due to the dose that could be associated with breast-feeding. The EPA concluded that human intake of CDD/CDF/PCBs is driven almost entirely by the ingestion of foods rather than via inhalation, dermal contact, or the ingestion of water. Only about 30% of the dioxin and PCB congeners with TCDD-like activity were found to account for nearly all of the risk to humans, with virtually all of these being associated with the ingestion of meat, milk (and other dairy products), and fish (EPA, 2000b).

In another study, Williams et al. (2000) estimated the average daily dose

TABLE 5.2 Key Components of Risk Characterization for Exposure Assessment

General Components	Specific Components
What are the most significant sources of environmental exposures?	• Are there data on sources of exposure from different media? • What is the relative contribution of different sources of exposure? • What are the most significant environmental pathways for exposure?
What populations were assessed?	• General population, highly exposed groups, highly susceptible groups? • Number of people likely to be exposed in a population
What was the basis for the exposure assessment?	• Monitoring, modeling, or other analyses of exposure distributions such as Monte Carlo or krieging?
What are the key descriptors of risk?	• What is the range of exposures to average individuals, high-end individuals, general population, high-exposure groups, children, susceptible populations? • How were the central tendency and high-end estimates developed? What factors or methods were used in developing these estimates? • Is there information on highly exposed subgroups? Who are they and what are their levels of exposure? How are they accounted for in the assessment?
Is there reason to be concerned about cumulative or multiple exposures?	• Because of ethnic, racial, or socioeconomic reasons?
What are the conclusions of the exposure assessment?	• What are the results from different approaches (i.e., modeling, monitoring, probability distributions)? • What are the limitations of each approach and the range of most reasonable values? • What is the level of confidence in the results?

Source: Adapted from EPA (1995b).

(ADD) of methyl tert-butyl ether (MTBE) in California from all water-related activities in a household. For the general population, the ADD was estimated to be about 0.17 μg/kg/day and 0.38 μg/kg/day at the 50th and 95th percentiles, respectively. The estimated ADD was approximately the same for more highly exposed households (i.e., homes with contaminated drinking water) at the 50th percentile (0.10 μg/kg-day), but was nearly four times greater at the

95th percentile (1.4 µg/kg-day). For the general population, the consumption of drinking water accounted for the greatest contribution to total MTBE dose at the 50th percentile, but the inhalation of MTBE vapors during showering and other activities accounted for nearly 70 percent of the dose at the 95th percentile. For households with contaminated drinking water, however, ingestion exposures accounted for the greatest contribution to total daily dose at both the 50th and 95th percentiles. Dermal contact with MTBE from water-related activities was found to be an insignificant route of exposure in all households.

These examples illustrate the importance of including exposures via multiple pathways and evaluating the relative contribution of each exposure route to total dose. These examples also highlight how exposures can vary both within a population and between different population groups. Given the significant impact that different exposure parameters can have on final risk estimates, it is essential that risk characterizations provide a clear description of, and rationale for, all exposure assumptions, including how persons are exposed, the exposed population (including sensitive populations), and the exposure frequency and duration.

5.3.3   Toxicity Information

Risk characterizations need to summarize what is known about the toxicity of each chemical, but more importantly, highlight the areas of toxicological uncertainty (EPA, 1989a). For suspected carcinogens, useful information includes the cancer slope factor, type of cancer, and the cancer classification of each chemical. A discussion of the weight of evidence for carcinogenicity should also be provided, and information about the source of concern (either animal or human) needs to be discussed. Ideally, the cancer classifications reported by various agencies in the U.S. and abroad and other scientific bodies (e.g., International Agency for Research on Cancer or IARC) should be discussed, including the reason for any differences among scientific bodies. These latter disparities can often be of significant importance to decision-makers who want to ensure that America does not unnecessarily compromise its international competitiveness. That is, they want to be sure that due to policies in the United States which might insist on conservative approaches to estimate safe levels of exposure, emissions limits are not set that are significantly lower than those in Europe, as this might allow those countries to manufacture products at much lower cost.

For chemicals with noncarcinogenic effects (including carcinogens), key sources of information include established safe levels of exposure and the critical effects associated with varying doses. The toxicity values should be defined on the basis of either an absorbed or administered dose, and these two metrics should be clearly described. Regardless of chemical type, it is important to discuss relevant pharmacokinetic data and uncertainties in any route-to-route extrapolation. In essence, risk assessments and characterizations require information on the biological mechanisms and dose–response relationship underlying observed effects in laboratory or epidemiology studies (see Table 5.3).

TABLE 5.3 Key Components of Risk Characterization for Dose–Response Relationship

General Components	Specific Components
What data were used to develop the dose–response curve?	• Would the results have been significantly different if based on a different data set? • If animal data were used, which species were used? Were data based on the most sensitive species, average of all species, or other? Were any studies excluded and why? • If epidemiological data were used, which studies were used? Were only positive studies used, all studies, or some other combination? Were any studies excluded and why? • Was a meta-analysis performed to combine the epidemiological studies? What approach was used?
What model was used to develop the dose–response curve?	• What rationale supports this choice? Is chemical-specific information available to support this approach? • For noncarcinogenic hazards, how was the RfD/RfC (or acceptable range) calculated? What assumptions and uncertainty factors were used? What is the confidence in the estimates? • For carcinogenic hazards, what dose–response model was used? Linearized multistage (LMS) or other linear-at-low-dose model, a biologically based model based on metabolism data, or data about possible mechanisms of action? What is the basis for selection of the particular dose–response model used? Are there other models that could have been used with equal plausibility and scientific validity?
What is the route and level of exposure observed as compared to expected human exposures?	• Are the available data from the same route of exposure as the expected human exposures? If not, are pharmacokinetic data available to extrapolate across route of exposure? • How far does one need to extrapolate from the observed data to environmental exposures? What is the impact of such an extrapolation?

Source: Adapted from EPA (1995b).

For example, the EPA (2000b) "Reassessment of Dioxin" provides an example of the type of language that is suitable for inclusion in the toxicity assessment portion of the risk characterization. Although critics have observed that EPA's characterization of dioxin presents only one view of how to interpret the data (i.e., it fails to acknowledge that there may be mechanistic reasons why rodents cannot be used to predict the human response), some of the language is nonetheless useful for illustrative purposes here. In this document, EPA (2000b) states:

> There is adequate evidence based on all available information presented in this assessment, as well as discussed elsewhere, to support the inference that humans are likely to respond with a broad spectrum of effects from exposure to dioxin and related compounds. These effects will likely range from biochemical changes at or near background levels of exposure to adverse effects with increasing severity as body burdens increase above background levels. Enzyme induction, changes in hormone levels, and indicators of altered cellular function seen in humans and laboratory animals represent effects of unknown clinical significance but may be early indicators of toxic response … Clearly, adverse effects including, perhaps, cancer may not be detectable until exposures contribute to body burdens that exceed background by one or two orders of magnitude (10–100 times). The mechanistic relationships of biochemical and cellular changes seen at or near background body burden levels to production of adverse effects detectable at higher levels remain uncertain (pages 99–100).

> The deduction that humans are likely to respond with noncancer effects from exposure to dioxin-like compounds is based on the fundamental level at which these compounds impact cellular regulation and the broad range of species that have been demonstrated to respond with adverse effects. For example, because developmental toxicity following exposure to TCDD-like congeners occurs in fish, birds, and mammals, it is likely to occur at some level in humans (page 100).

> Epidemiologic observations of an association between exposures and cancer responses (TCDD); unequivocal positive responses in both sexes, multiple species, multiple sites, and different routes in lifetime bioassays or initiation-promotion protocols or other shorter-term in vivo systems such as transgenic models (TCDD plus numerous PCDDs, PCDFs, and dioxin-like PCBs); and mechanistic or mode-of-action data that are assumed to be relevant to human carcinogenicity, including, for instance, initiation-promotion studies (PCDDs, PCDFs, dioxin-like PCBs) all support the description of complex mixtures of dioxin and related compounds as likely human carcinogens (page 103).

> Background exposures to dioxin and related compounds need to be considered when evaluating both hazard and risk. The term "background" exposure has been used throughout the reassessment to describe exposure of the general population, who are exposed to levels in environmental media (food, air, soil, etc) that have dioxin concentrations within the normal background range (page 108).

The primary strength of this example is that it presents a clear and transparent discussion of EPA's basis for interpretation of the available data. This transparency allows for a constructive dialogue among all parties who may or may not agree with this interpretation. Ultimately, for the risk characterization to serve as a vehicle for informing risk managers, it should present "the other side of the story" if there is substantial legitimate disagreement among the various datasets and experts. In short, dose–response assessments and characterizations should be balanced and address conflicting analyses of equal credibility, and to the extent possible, present a weight-of-evidence discussion about why one approach is superior to another.

For example, there are many epidemiology studies of dioxin and dioxin-like chemicals that have yielded negative or equivocal results, that should have been included in the above risk characterization. In fact, a number of scientists have concluded, based on meta-analyses of the various well-conducted epidemiology studies, that there is no statistical evidence that response markedly changes with increasing dose (Starr, 2001). If true, this would suggest that EPA had to "pick and choose" certain results from these studies in order to support their recommended cancer potency factor. When legitimate disagreements like this exist, the risk characterization would do well to capture them. The 15-year evaluation of the hazards posed by dioxin represents one of the best examples of the dynamic tension among various scientists, regulatory agencies and environmental activists and illustrates that sometimes more data will not necessarily bring clarity to an issue. Indeed, a careful reading of the EPA Science Advisory Board review of dioxin suggests that, even after nearly 5,000 published papers on the subject, there can be a virtual split (50/50) disagreement among experts regarding the hazards posed by a chemical (EPA, 2001a).

5.3.4 Data Variability and Uncertainty

Risk characterizations are typically influenced by both the variability and uncertainty in the exposure and dose–response assessments. Variability refers to true heterogeneity or diversity in a data set or population (NRC, 1994; EPA, 1997b). For example, individuals differ in terms of body weight and breathing rate, and environmental contaminant levels may vary based on temporal and spatial dimensions. For some chemicals, the dose required to cause an adverse health effect can also vary considerably among different segments of the population. For example, persons with deficiencies in alcohol dehydrogenase (ALD) are less tolerant of alcohol, while individuals with glucose-6-phosphate dehydrogenase (G-6-PD) deficiencies are more susceptible to exposures to a number of common industrial chemicals (Klaassen, 2001). Most health risk evaluations require that assumptions be made about the underlying variability in the data, and these should be clearly presented in the risk characterization. Over the last decade, there has been significant progress in identifying how different factors related to health risks vary in a population (Hattis et al., 1999; Finley et al., 1994; EPA, 1997a). In particular, significant attempts have been made to alert the environmental health community that children may be more susceptible than adults to selected chemicals at certain doses, while it is plausible that they will be equally or less susceptible to others (ATSDR, 1997). Incorporating biological variability will be one of the most challenging aspects of risk characterization in the coming decades as we learn more about the genetic basis for differences among individuals (Neumann and Kimmel, 1998).

In contrast with variability, uncertainty refers to a lack of knowledge about specific factors (NRC, 1994; EPA, 1997b). For example, scientific measurements and environmental sampling are subject to some degree of error, and there are many uncertainties associated with the use of different scientific models (EPA, 1995c). Uncertainty also arises from data gaps in the assessment, such as an absence of information on human exposures or the toxicity of a chemical. There are many types of uncertainty in risk characterizations (see Table 5.4). General headings include "parameter uncertainty" (measurement errors, sampling errors, systematic errors) and "model uncertainty" (uncertainty due to necessary simplifications of real-world processes, mis-specification of the model structure, model misuse, use of inappropriate surrogate variables) (Frey, 1992; EPA, 1997b). Another type of uncertainty—"scenario uncertainty" (descriptive errors, aggregation errors, errors in professional judgment, incomplete analysis), has also been reported in the literature (EPA, 1997b).

Recognizing model uncertainty is particularly important, as imperfect mathematical models are typically used to represent scenarios and phenomena of interest in health risk evaluations. For example, the use of different mathematical models to extrapolate from the observed experimental range to the unobservable environmental (human) exposure range in cancer risk evaluations can result in widely divergent results in the low-dose region (see Figure 5.3). In fact, the difference in the best-fit predictions from the most conservative model (i.e., one-hit) to the lease sensitive model (i.e., probit) may range over several orders of magnitude (Gargas et al., 1999). Model uncertainty also refers to the inherent uncertainty in toxicologic tests, which entails judgment about the appropriate animal sex, strain, or species for modeling human exposures.

The key sources of variability and uncertainty in an assessment need to be clearly highlighted in the risk characterization, including a discussion of the relative importance of each of these. Emphasis on this latter point in particular can help prioritize future research needs by identifying those areas of uncertainty in which additional data might have a significant effect on the risk estimate (Cullen and Frey, 1999). Probabilistic modeling techniques, such as Monte Carlo analysis, can be used to incorporate and evaluate sources of variability and uncertainty in risk assessments, and the use of these tools is becoming common practice in most health risk evaluations (EPA, 1997b; Thompson and Graham, 1996; Finley and Paustenbach, 1994).

5.3.5   Level of Confidence

The level of confidence in a risk characterization will depend, in part, on the nature and magnitude of the uncertainties in the assessment. For example, the traditional EPA (1986a) *Cancer Guidelines for Carcinogen Risk Assessment* relies on a weight-of-evidence (WOE) approach for characterizing the extent to which the available data support the hypothesis that an agent causes cancer in humans (see Table 5.5). More recent proposed cancer guidelines by EPA (1996) also utilized a WOE approach, but incorporated three broad narrative descriptions instead of letter designations for cancer classifications (i.e., "known/

TABLE 5.4   Types of Uncertainty in Risk Characterization

Types	Description
	Model Uncertainty
Model structure	Alternative sets of scientific or technical assumptions may be available for developing a model. The uncertainty and implications of these alternative foundations may be evaluated by constructing alternative models and comparing results from each alternative model. In some cases, it may be possible to parameterize alternative model structures into a higher order model, and to evaluate alternative models using traditional sensitivity analysis.
Model detail	Often, models are simplified for purposes of tractability. This can introduce uncertainty in the predictions of simplified models, which can sometimes be gleaned by comparison of their predictions to those of more detailed, inclusive models. In other cases, simple models are developed due to a lack of confidence or knowledge about what the actual model structure should be. Uncertainty about these models may be only qualitatively understood.
Validation	Models for which extensive data are available, and which have been validated for a parameter space of interest, can be evaluated quantitatively in terms of the accuracy and precision of their predictions. However, model uncertainties may exist when which few data are available to test model predictions, requiring more extensive evaluations (e.g., expert judgment).
Extrapolation	Models validated for one portion of a parameter space may be completely inappropriate for making predictions in other regions of the parameter space. This is a key source of model uncertainty.
Model resolution	In numerical models, a spatial and/or temporal grid size must be assumed. The selection of grid size involves a trade-off between computation time and prediction accuracy, thereby introducing uncertainty in the model.
Model boundaries	Any model may have limited boundaries in terms of time, space, number of chemical species, temperature range, types of pathways, etc. The selection of model boundary may be a type of simplification, but within the boundary of the model and parameter space of the problem, the model may be an accurate representation of the real-world phenomenon of interest. However, other overlooked phenomenon not included in the model may introduce uncertainty.

TABLE 5.4 *(Continued)*

Types	Description
Scenario reasonableness	Prior to using a model, an analyst must develop a scenario for the problem of interest. A scenario is a set of assumptions about the nature of the problem to be analyzed and may be constructed to represent an actual environmental problem or may be constructed hypothetically based on policy motivations. To the extent that the scenario fails to consider all factors affecting the key output variable, uncertainty will be introduced.
	Parameter Uncertainty
Empirical quantities	Quantities that are measurable, at least in principle
Random error and statistical variation	Uncertainty associated with imperfections in measurement techniques
Systemic error	Uncertainty that occurs when the mean value of a measured quantity does not converge to the "true" mean value because of biases in measurements and procedures (e.g., imprecise calibration, faulty reading of meters, etc.).
Variability	Uncertainty due to quantities that vary over time, space, or some population of individuals
Inherent randomness or unpredictability	Uncertainty due to quantities that are irreducibly random, even in principle
Lack of empirical bias	Uncertainty due to lack of experience or knowledge of a process or system
Dependence and correlation	Uncertainty that occur when there is more than one uncertain quantity, which may be statistically or functionally dependent
Disagreement	Uncertainty that arises from limited data or alternative theoretical bases for modeling a system or when experts disagree on the interpretation of data or estimates for empirical quantities
Defined constants	Quantities whose values are accepted by convention, but which are actually subject to measurement error
Decision variables	Parameters over which a decision maker exercises control
Value parameters	Preferences or value judgments of a decision maker
Model domain parameters	Parameters associated with a model, but not directly with the phenomenon the model represents

Source: Adapted from Frey (1992).

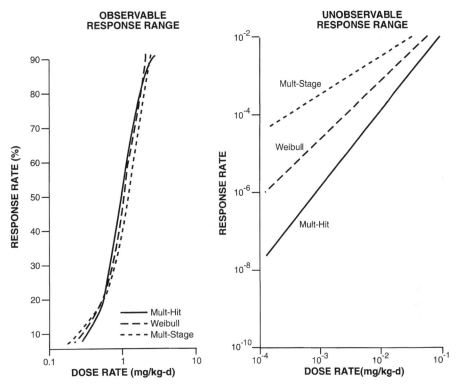

Figure 5.3 Extrapolation from Observed to Unobservable Exposure Ranges in Cancer Risk Assessments Using Alternative Mathematical Models. Cancer risk assessment relies on the use of mathematical models to extrapolate from the observed experimental range (left plot) to the unobservable environmental exposure range (right plot). All models tend to agree in the high-dose range, but may diverge widely in the low-dose range. As a result, the same risk level may be associated with a dose range spanning several orders of magnitude depending on which extrapolation models are used. The EPA typically relies on linearized multistage (LMS) model as a default, one of the most conservative of the dose extrapolation models, although alternative models may be used if justified.

likely," "cannot be determined," and "not likely"). EPA's (1999a) most recent proposed cancer guidelines rely on five narrative descriptions to classify the extent to which an agent causes cancer in humans: "carcinogenic to humans," "likely to be carcinogenic to humans," "suggestive evidence of carcinogenicity, but not sufficient to assess human carcinogenic potential," "not likely to be carcinogenic to humans," and "data are inadequate for an assessment of human cancer potential." In general, EPA's WOE approach considers all scientific information in determining whether and under what conditions an agent may cause cancer in humans

TABLE 5.5   Traditional EPA Classification System for Carcinogens

Human Evidence	Animal Evidence				
	Sufficient	Limited	Inadequate	No Data	No Evidence
Sufficient	A	A	A	A	A
Limited	B1	B1	B1	B1	B1
Inadequate	B2	C	D	D	D
No Data	B2	C	D	D	E
No Evidence	B2	C	D	D	E

Source: EPA (1986a).

Group A	Human carcinogen
Group B	Probable human carcinogen
B1	Limited evidence of carcinogenicity from epidemiology studies
B2	Inadequate human evidence but positive animal evidence
Group C	Possible human carcinogen
Group D	Not classifiable as to human carcinogenicity
Group E	Evidence of noncarcinogenicity for humans

For many chemicals, the EPA also suggests assigning a qualitative descriptor of "high," "medium" or "low" confidence in reported toxicity estimates. Level of confidence discussions are, by their nature, qualitative discussions. The risk assessor, however, is expected to assemble all of the available information and to present a balanced view of what the data show and what might be done with that information. Again, transparency is an important attribute when discussing the level of confidence in a particular opinion or analysis.

For example, the dioxin reassessment by EPA (2000b) states:

> Dioxin and related compounds can produce a wide variety of effects in animals and might produce many of the same effects in humans. Binding of dioxin-like compounds to a cellular protein called the aryl hydrocarbon receptor (AhR) represents the first step in a series of events attributable to exposure to dioxin-like compounds including biochemical, cellular, and tissue-level changes in normal biological processes. A weight-of-the-evidence evaluation suggests that mixtures of dioxin and related compounds are strong cancer promoters and weak direct or indirect initiators, and are likely to present a cancer hazard to humans.

Although such qualitative descriptions can illuminate important caveats and convey the appropriate level of confidence in risk characterizations, the actual quantification of risk will ultimately have the strongest impact on decisionmakers. Numerical risk estimates should therefore be based on the best available data and the appropriate use of exposure and dose–response models to avoid the reliance on poorly characterized risks that have little "confidence" associated with them.

5.3.6 Other Risk-related Information

Depending on the purpose, risk characterizations may contain information on a wide variety of outcomes or consequences. For example, decision makers or interested parties may need information on the economic costs and benefits of alternatives, the secondary effects of hazard events, or the efficacy of alternative regulatory mechanisms (NRC, 1996). Presenting risk information in a broader context can help ensure that risk management decisions will be cost-effective and will not result in unacceptable risk trade-offs from countervailing risks (Graham and Weiner, 1995). For example, eating contaminated fish could plausibly increase an individual's risk of cancer due to the presence of trace concentrations of PCB, DDT, chlordane, or other persistent chemicals. However, the magnitude of risk may not be large enough to warrant a decrease in fish consumption, given the health benefits of substituting beef or other high-fat foods for fish. Indeed, a quantitative risk trade-off analysis by Anderson and Weiner (1995) suggested that a consumer would have to "dread" dying of cancer over ten to hundreds of times more than dying of heart disease in order for the extra cancer risk from eating fish to begin to offset the reduced risk of heart disease from eating fish. Studies by Ponce et al. (2000) and TERA (1999) have reacted similar conclusions in their evaluation of fish advisory risk trade off issues.

Information on the costs of a program or intervention relative to its benefits may be particularly important for risk managers. In a large-scale cost-effectiveness analysis by Tengs et al. (1995), enormous variation was found in the cost of saving one year of life for 587 different life-saving interventions in the United States. For example, the median medical intervention costs about $19,000/life-year, the median injury reduction intervention costs 48,000/life-year, and the median toxin control intervention costs $2,800,000/life-year (see Table 5.6). Such differences in cost-effectiveness data are important because where there are investment inequalities, more lives could be saved by shifting resources (Tengs et al., 1995).

5.4 TOXICITY CRITERIA FOR EVALUATING HEALTH RISKS

The results of a risk characterization are often generated by comparing exposure data to pre-established toxicity criteria to evaluate the potential for adverse effects to occur in a specified population. Such criteria may apply either to the general population or occupational workers, can be based on short-term or longer-term exposures, and may be associated with cancer or noncancer health effects. Established toxicity criteria can also represent enforceable agency standards or may reflect recommended guidance levels or perceived acceptable levels (i.e., background). The following section provides a brief description of common toxicity criteria used in health risk evaluations.

TABLE 5.6   Median of Cost/Life-Year Saved Estimates as a Function of Sector of Society and Type of Intervention

	Type of Intervention			
Sector of Society	Medicine	Fatal Injury Reduction	Toxin Control	All
Health care	$19,000 (n = 310)	N/A	N/A	$19,000 (n = 310)
Residential	N/A	$36,000 (n = 30)	N/A	$36,000 (n = 30)
Transportation	N/A	$56,000 (n = 87)	N/A	$56,000 (n = 87)
Occupational	N/A	$68,000 (n = 16)	$1,400,000 (n = 20)	$350,000 (n = 36)
Environmental	N/A	N/A	$4,200,000 (n = 124)	$4,200,000 (n = 124)
All	$19,000 (n = 310)	$48,000 (n = 133)	$2,800,000 (n = 144)	$42,000 (n = 587)

Tengs et al. (1995).

5.4.1   Reference Doses and Reference Concentrations (Acceptable Doses)

The EPA establishes noncancer criteria for most chemicals, including reference doses (if via ingestion) and reference concentrations (if the chemical is predominantly inhaled). These represent doses to which it is believed that humans can be exposed over a specified period of time without experiencing adverse effects [EPA, 1999b]. Acceptable levels of dermal uptake are usually calculated based on the oral or inhalation criteria. The chronic reference dose (RfD) is defined as "an estimate (with uncertainty spanning perhaps an order of magnitude) of a daily oral exposure to the human population (including sensitive subgroups) that is likely to be without an appreciable risk of deleterious effects during a lifetime" (page 11). The chronic reference concentration (RfC) has the same definition, except it refers to "continuous inhalation exposure" over a lifetime (page 11). Both reference values are usually derived from a NOAEL (no-observable-adverse-effect level), LOAEL (lowest-observable-adverse-effect level), or benchmark dose/concentration, which is normally identified in an animal study, although about 20 percent of these values are based on human epidemiology studies. Uncertainty and/or modifying factors are usually applied to the RfD and the RfC to reflect limitations of the data, such as extrapolation from animal data to humans and intraspecies variability, and to ensure an extra margin of safety for public health (Barnes and Dourson, 1988; Renwick and Lazarus, 1988; Dourson et al., 1996).

The chronic RfDs and RfCs represent agency guidelines that are meant to protect the general population, including susceptible or sensitive individuals.

These reference values are generally considered to be conservative (i.e., especially health protective) because they assume continuous or daily exposures over a lifetime, typically 70 years, and incorporate several uncertainty and/or safety factors. Although technically not enforceable by law, RfDs and RfCs have been used in a wide range of risk assessments or to prioritize risk management goals.

5.4.2 Minimal Risk Levels

The Agency for Toxic Substances and Disease Registry (ATSDR), in conjunction with the EPA, is responsible for establishing a priority list of hazardous substances most commonly found at hazardous waste (Superfund) sites. As part of their responsibilities, ATSDR (2000) prepares toxicological profiles for each substance, and develops minimal risk levels (MRLs) that specify health guidance or acceptable exposure levels. Specifically, MRLs are defined as an "estimate of the daily human exposure to a hazardous substance that is likely to be without appreciable risk of adverse noncancer health effects over a specified duration of exposure" (page 1).

ATSDR's approach is similar to the one used by EPA, in that MRLs are based on noncancer health effects and the NOAEL/uncertainty factor approach. MRLs are also developed separately for oral and inhalation routes of exposure. ATSDR does not derive MRLs for the dermal route of exposure because a suitable method has not yet been identified. The two approaches differ, however, in that MRLs are derived for acute (1 to 14 days), intermediate (15 to 364 days), and chronic (365 days and longer) exposure durations versus the lifetime duration assumed by EPA. MRLs are also based on the most sensitive substance-induced endpoint considered to be of relevance to humans, which may be less severe than those considered under the EPA approach. An understanding of the difference in the rationale between the MRL and the RfD is important since these values may not be quantitatively similar to one another. MRLs are considered to be conservative, and exposures at or above the MRL do not mean that adverse health effects will occur. According to ATSDR (2000), the primary purpose of an MRL is to serve as a "screening level" to identify contaminants and potential health effects that may be of concern at hazardous waste sites; MRLs are not intended to define cleanup goals or action levels.

5.4.3 Dietary Safety Guidelines

Traditional approaches to food safety evaluations have relied on the concept of an acceptable daily intake (ADI), which represents a safe level of exposure to a specific substance in the diet (Rodricks and Taylor, 1983; Winter, 1992). The ADI is defined by EPA (1999b) as "the amount of a chemical a person can be exposed to on a daily basis over an extended period of time (usually a lifetime) without suffering deleterious effects" (page 1). ADI's have been used since 1958 to evaluate the safety of food additives and are generally believed to provide a "reasonable certainty" that "no harm" will result from intended product uses (Rodricks and Taylor, 1983).

The ADI is essentially analogous to the RfD, in that it is derived from a NOAEL (or NOEL) that is divided by an additional safety factor (typically 100) to yield a safe exposure level for noncarcinogenic substances. The primary distinction between the ADI and RfD is that the former approach utilizes management or policy driven "safety factors," while the latter approach uses only data derived or scientific "uncertainty factors." Although the distinction is subtle, it is an important one that can have significant implications on the final recommended or acceptable dose level.

Pesticide "tolerance levels" also represent safety guidelines that are established by EPA (enforced by FDA), and are defined as the maximum legal limit of a pesticide residue that is allowed to remain in or on a raw agricultural commodities, and in some cases, processed foods (NRC, 1987). The tolerance-setting process is required before the use of a pesticide on food crops can be registered, and tolerances are based on a review of toxicity data for each active ingredient and for major impurities or metabolites. For noncarcinogens, acceptable tolerance levels are determined by comparing the theoretical maximum residue contribution (TMRC) for each food form in which the pesticide could occur to the corresponding ADI (NRC, 1987).

More recently, EPA (2000c) developed a population-adjusted dose (PAD) for use under the Food Quality Protection Act (FQPA) that addresses concerns about infant and childhood toxicity. The PAD represents an amount of toxicant to which a person can be safety exposed, and is simply the RfD divided by an additional FQPA safety factor that takes into account potential pre- and post-natal toxicity and completeness of the data for pesticides that exhibit threshold effects.

5.4.4   Cancer Potency

Potency values are established by EPA (or state health agencies) to reflect the relative toxicity of carcinogenic substances. Cancer potencies are typically expressed as a cancer slope factor (CSF) or a unit risk estimate. The EPA (1999b) defines a slope factor as an "upper bound, approximating a 95% confidence limit, on the increased cancer risk from a lifetime exposure to an agent" (page 13). This approach is generally reserved for use in the low-dose region of the dose–response relationship (i.e., exposures corresponding to risks less than 1 in 1000). On the other hand, unit risks are defined as the "upper-bound excess lifetime cancer risk estimated to result from continuous exposure to an agent at a concentration of 1 μg/L in water, or 1 μg/m^3 in air" (page 15). For example, a unit risk estimate of 1.5×10^{-6} μg/L indicates that no more than 1.5 excess tumors could theoretically occur in 1 million persons, if each was exposed daily for a lifetime to 1 μg of the chemical in 1 liter of drinking water.

The characterization of the upper-bound cancer risk estimate, which is calculated in most risk assessments, is one of the most misunderstood and mis-represented of all risk results (Crump, 1996). The major shortcoming in such characterizations is that they generally present only the upper-bound risk and do not present the maximum-likelihood estimate (i.e., "best estimate") of risk. In addition, most risk characterizations do not report lower-bound risk esti-mates (at doses where the upper-bound risk is significant) and many fail to ac-knowledge that the risk may in fact be zero. There are two possible reasons for this latter scenario: either the compound is not a human carcinogen or the dose–response relationship has an effective threshold above the level of expo-sure. It is noteworthy that for mathematical models which do not constrain the results, predicted risks at lower dose levels include both zero and values less than zero (See Figure 5.4) (Sielken, 1987, 1985; Holland and Sielken, 1993).

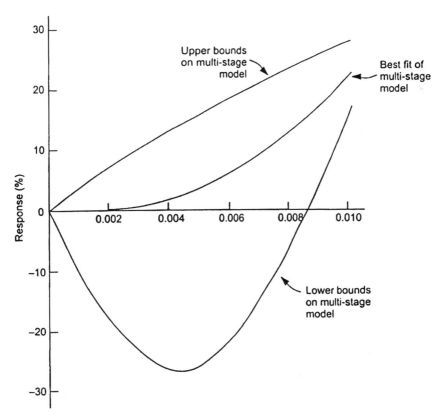

Figure 5.4 Range of Plausible Risks Using Bounding Techniques Inherent in Multi-Stage Model. An illustration of the range of plausible risks using the bounding tech-niques inherent in the multi-stage model. In this case, the multistage model is applied to the results of a Chemical Industry Institute of Toxicology rodent bioassay of formalde-hyde. Source: Adapted from Sielken 1987.

5.4.5 Health Advisories

The EPA (2000d) establishes enforceable and nonenforceable health standards and advisories for drinking water contaminants. Maximum contaminant level goals (MCLGs) are non-enforceable and are defined as "a level at which no known or anticipated adverse effect on the health of persons occur and which allows an adequate margin of safety" (page 4). On the other hand, maximum contaminant levels (MCLs) are enforceable and are defined as "the maximum permissible level of a contaminant in water delivered to users of a public water system" (page 4). In practice, MCLs are set as close to the MCLG as feasible, using the best available treatment technology and taking cost into consideration. Secondary drinking water regulations (SDWRs) are nonenforceable guidelines that relate to cosmetic effects (e.g., tooth or skin discoloration) or aesthetic effects (e.g., taste, odor, or color) of drinking water. Health advisories provide additional information on contaminants that can cause human health effects and are known or anticipated to occur in drinking water. These advisories represent guidance values that are based on noncancer health effects for different durations of exposure (e.g., 1-day, 10-day, and lifetime).

The EPA also establishes health advisories for contaminants found in noncommercial fish and shellfish obtained through sport, recreation, and subsistence activities (EPA, 2000e). These advisories inform the public on which fish to avoid or limit eating due to elevated levels of pollutants and may be targeted to different population groups, such as the general population or sensitive subpopulations (e.g., women and children) or certain species or sizes of fish. These advisories are often based on risk levels that are high relative to those considered acceptable in other situations.

5.4.6 Occupational Limits

Several groups in the United States publish recommended guidelines for occupational exposure to airborne contaminants. The Occupational Safety and Health Administration (OSHA) sets enforceable Permissible Exposure Limits (PELs) to protect workers against the health effects of exposure to hazardous substances. PELs are regulatory limits on the allowable amount or concentration of a substance in the air, which may also contain a skin designation (OSHA, 1999). OSHA PELs are based on an 8-hour time-weighted-average (TWA) exposure. OSHA can begin standard-setting procedures on its own initiative or in response to petitions from other parties such as the National Institute for Occupational Safety and Health (NIOSH). NIOSH conducts research on various occupational safety and health problems, provides technical assistance to OSHA, and often recommends standards for OSHA's adoption.

The American Conference of Governmental Industrial Hygienists (ACGIH) also sets Threshold Limit Values (TLVs) for occupational exposure to airborne contaminants. TLVs are nonenforceable guidelines that represent the average concentration of an airborne contaminant, based on an 8-hour workday and a

40-hour work week, to which nearly all workers may be repeatedly exposed without adverse effect. Short-term exposure levels (STELs) are also nonenforceable guidelines for 15-minute exposures, which are established by ACGIH.

When assessing the hazard posed by occupational exposure to a noncarcinogen, risk assessors and industrial hygienists typically compare the measured airborne concentrations (8-hour TWA) to the occupational exposure limit (PEL or TLV). If the workplace exposure is typically (e.g., 50–95% of the time) below the Occupational Exposure Limit (OEL) it is believed that this is an acceptable level of exposure. In the past, this comparison of exposure to toxicity in occupational settings was termed a "margin of safety" (MOS) approach. Note, however, this term has evolved over time and has a slightly different interpretation in the health (versus occupational) arena. For example, in an evaluation of occupational exposures to 2-methoxyethanol (2-ME), Paustenbach (1988) reported:

> Based on typical exposures in the semiconductor industry and the most appropriate occupational exposure limit, the TLV, the estimated MOS is 53. This MOS suggests that the risk of adverse effects on the offspring of exposed parents is insignificant for those employees who minimize or prevent dermal contact. Assuming that the TLV has been set at a level that was intended to protect the fetus, a margin of safety of one indicates that persons and their offspring can, in general, be exposed to this hazard, yet not be at significant risk of injury.

5.4.7 Background Environmental Levels

There are two types of background levels for a chemical substance. Naturally occurring levels represent concentrations of a substance that are present in the environment without human influence, while anthropogenic levels refer to concentrations of a substance in the environment that are due to man-made sources. Health risks from a particular exposure scenario are often compared an individual's baseline risk from background exposures to the same contaminant. For example, a health risk assessment at a former wood treatment site revealed that the 95th percentile risk for nearby residents was two orders of magnitude lower than the U.S. background cancer risk from the dietary intake of PCDD/PCDFs (Copeland et al., 1993). Specifically, 95 percent of the persons potentially exposed to PCDD/PCDF at this site had an incremental cancer risk less than 1×10^{-7}, while the most likely risk estimate (i.e., 50th percentile) was no greater than 2×10^{-8}. By comparison, background exposures to PCDDs and PCDFs in meat and dairy products were estimated to pose a plausible cancer risk of 1×10^{-5} (Copeland et al., 1993).

In general, the past 25 years of experience has shown that a good understanding of the background concentration of both naturally occurring and anthropogenic chemicals is critical to a proper characterization of risk (Paustenbach, 2000). When assessing soils, for example, background levels can be viewed in at least four different ways (Paustenbach, 2000). The most common approach is to assess "normal levels," which refer to soils affected by only minor pollution of everyday activities associated with modern rural and urban life. In some instances, background soil levels may be defined based on "historically polluted regions" or "geochemical variation" if these locations are associated with higher than normal chemical concentrations. A final approach is to evaluate "pristine levels," or those that are associated with soils or landscapes unaffected by human activity. However, this approach represents idealistic (and unrealistic) conditions for many chemicals, particularly semivolatile persistent chemicals that are globally distributed (e.g., mercury, dioxin, PCBs, and chlordane).

5.5 DESCRIPTORS USED TO CHARACTERIZE HEALTH RISKS

Health risks can be characterized according to several types of risk descriptors. These are based, in part, on the exposure distribution within the population of interest (EPA, 1995c). For example, health risks can be characterized for an individual in a specified population or for the entire population group. Risk characterizations may also be based on average or high-end exposures in a population, and may refer to the general population or a more highly exposed or susceptible population subgroup (EPA, 1995c). Regardless of exposure population, numerical risk estimates may be presented as a single value (i.e., point estimate) or a range of possible values weighted by their likelihood of occurrence (i.e., probability distribution). It should be noted that risk estimates presented as a distribution are typically based on assessments of variability in exposure parameters only, and generally do not include an assessment of uncertainty in biologic parameters. The following section provides a more detailed overview of possible descriptors used to characterize health risks.

5.5.1 Individual/Population Risk

Characterizations of individual risk generally refer to the probability of harm for a random member of the population and are not necessarily representative of a particular individual in the population. The EPA (1995c) has recommended using both "high-end" and "central tendency" descriptors to convey the variability in risk levels for different individuals in the population. High-end descriptors represent plausible estimates of risk for individuals at the upper end of the exposure or risk distribution (e.g., 90th or 95th percentile) but are not meant to represent estimates beyond the true distribution. Central tendency descriptors represent typical or average estimate of risk, based on either the arithmetic mean or the median exposure or risk.

Characterizations of population risk refer to the extent of harm for the population as a whole, and represent the summation of individual risks within a specified group. According to EPA (1995c), population risks can be described as either a probabilistic number of cases or an estimated percentage of the population with risk greater than some level. The former descriptor refers to the probabilistic number of health effect cases estimated in a population over a specified time period and does not represent an actuarial prediction of cases in the population. The latter descriptor, which is most often used for evaluating noncancer effects, is an estimate of the number of people or percentage of the population above a specified risk (or reference) level.

5.5.2 General Population/Highly Exposed or Susceptible

Health risks may be characterized for the general population or for individuals that have higher exposures or susceptibilities to an agent. A "highly exposed" descriptor is useful when there is an identifiable subgroup in a population that is experiencing significantly greater exposure than the larger population (EPA, 1995c). For example, Williams et al. (2000) estimated the cancer and noncancer risk from MTBE drinking water exposures for the general population and a more highly exposed subgroup in California. The latter group was defined as households in California that had contaminated drinking water above the analytical detection limit for MTBE. The increase in lifetime cancer risk was estimated to be about five times greater for the highly exposed group than for the general population but was found to be less than one in a million for both groups at the 95th percentile of exposure (Williams et al., 2000).

A "highly susceptible" descriptor is useful when there is an identifiable subgroup in a population that is more sensitive or susceptible to the toxic effects of an agent than the larger population (EPA, 1995c). Children, pregnant women, elderly people, and individuals with certain illnesses are often considered to be more sensitive than the population as a whole. In such instances, it may be necessary to calculate a separate dose–response relationship or establish unique guidelines for each subgroup. More often than not, there will be very little specific quantitative information about degree of susceptibility. Therefore, it is often better to simply suggest that for a certain population, the margin of exposure should be greater than for the general population.

For example, children are considered to be much more susceptible to the effects of environmental lead than adults (ATSDR, 1999). Children often face greater exposures to lead than do adults because of their unique activities, including breastfeeding and playing with dust, dirt, sand, or paint chips (ATSDR, 1999). The main target for lead toxicity from long-term exposures, either by oral or inhalation routes, is the nervous system. The Centers for Disease Control and Prevention (CDC) has determined that blood lead levels greater than or equal to 10 μg/dL are worthy of further consideration in children ages 1 to 5 years old (ATSDR, 1999). Similarly, persons who are folate-deficient are considered to be more vulnerable to the toxic effects of methanol than healthy individuals. Segments of the population with a high incidence of folate deficiency include pregnant and lactating women, patients with chronic alcoholism, people with a poor diet, and persons with bowel or hematologic disease [Health Effects Institute (HEI), 1999; Osterloh, 1995].

5.5.3 Point Estimate/Risk Distribution

Numerical estimates of risk can be expressed as a point estimate or a distribution of possible risk values. Point estimates refer to the use of a single value for each risk model parameter, resulting in a final estimate of risk that is also a single value. These are often upper-bound values to ensure the inclusion of the maximum plausible risk and are generally not adequate for assessing the actual human risk in a population. Risk distributions, on the other hand, refer to the use of a range of values for selected model parameters weighted by their likelihood of occurrence (EPA, 1997b). Risk estimates are calculated using probability-based techniques, such as Monte Carlo analysis, and can be presented as an entire probability distribution or selected percentiles (e.g., 50th, 90th, and 95th). As mentioned, however, risk distributions may not reflect all sources of variability and uncertainty. In particular, most evaluations do not incorporate the uncertainty in biologic parameters, such as cancer potency, which may be of much greater importance than the variability in exposure parameters.

Although the former risk descriptor may be useful for screening analyses, the latter is considered to be the most optimal method for characterizing health risks (Sielken et al., 1995; Gargas et al., 1999; Thompson and Graham, 1996; Finley et al., 1994). This is because the use of point estimates often results in overly conservative or nonplausible risk values. For example, Copeland et al. (1993) found that the current regulatory approach used to estimate the reasonably maximally exposed (RME) individual can overpredict cancer risks by 10- to 100-fold compared to the plausible risk for even the more highly exposed persons as predicted using Monte Carlo technique (e.g., even at the 95th percentile). Specifically, the incremental lifetime cancer risk from typical exposure to the dioxins and furans in soil for those in the community was estimated to be 2×10^{-8} at the 50th percentile and 1×10^{-7} at the 95th percentile based on the probabilistic assessment. By comparison, the cancer risk for the typical person calculated using the EPA's approach (which was based on default exposure assumptions) was 8×10^{-6}, a 400-fold higher estimate than the 50th percentile risk (Copeland et al., 1993).

Burmaster and Harris (1993) also investigated the magnitude of "compounding conservatisms" in Superfund risk assessments by comparing the results of a point estimate versus probabilistic approach utilizing three exposure factors. In a hypothetical case scenario of workers exposed to chloroform via incidental ingestion of contaminated soil, estimated cancer risks were 8.3×10^{-6} using single "upper bound" values, but were about one-tenth this value at the 95% risk (9.5×10^{-7}) using Monte Carlo simulation (Burmaster and Harris, 1993).

5.6 METHODS FOR QUANTIFYING HUMAN HEALTH RISKS

The process of quantifying health risks differs for carcinogens and noncarcinogens (EPA, 1989a, 1996). For carcinogens, risks are estimated as an incremental or excess individual lifetime cancer based on a specified exposure (absorbed dose). For noncarcinogens, exposure levels over a specified time period are compared to a reference dose derived for a similar exposure period. While the former risk estimate is expressed as the probability of an individual suffering an adverse effect, the latter risk estimate is expressed as the ratio of exposure to toxicity criteria (or vice versa) and is not expressed as a probability. Both cancer and noncancer risk estimates may account for multiple exposure pathways or aggregate risks from multiple substances.

5.6.1 Traditional Regulatory Approach to Characterizing Cancer Risks

Estimates of cancer risk are often based on the simplifying assumption that the dose–response relationship is linear at low doses (EPA, 1989a). Under this assumption, risk is directly related to intake and can be estimated using the following equation, which assumes linearity of response at low doses:

$$\text{Risk} = \text{LADD} \times \text{CSF}$$

where Risk = unitless probability of an individual developing cancer
      LADD = lifetime average daily dose (mg/kg day)
      CSF = cancer slope factor expressed in $(\text{mg/kg day})^{-1}$

The slope factor in this equation usually represents the 95th percent upper confidence limit (UCL) of the probability of response based on experimental animal data used in the multistage model (EPA, 1989a). The slope factor therefore represents an "upper bound" or "plausible upper limit" value. The "true" cancer risk is not expected to exceed this value, and may in fact, be substantially less or may be zero (i.e., because the compound is not a human carcinogen or the dose–response relationship has an effective threshold above the level of exposure). According to EPA (1989a), use of the 95% UCL allows regulators and decision-makers to be "reasonably confident that the 'true risk' will not exceed the risk estimate derived through use of this model and is likely to be less than that predicted" (page 8-6).

For example, the cancer risk from inhalation of hexavalent chromium at a residential site in New Jersey where chromite-ore processing residue was used as fill material was estimated by Sheehan et al. (1991). Specifically, the plausible upper-bound lifetime excess cancer risk from inhalation of airborne soil particles was calculated by multiplying the LADD for a maximally exposed individual (MEI) and a most likely exposed individual (MLEI) by the upper-bound EPA cancer slope factor for hexavalent chromium (see Table 5.7). This example illustrates a rather simplistic approach to estimating cancer risk in which point estimates are used to represent each exposure parameter, including the concentration of hexavelent chromium in ambient air, inhalation rates for different age groups, the fraction of inhaled particles reaching the gut, and the exposure frequency and duration (Sheehan et al., 1991).

TABLE 5.7   Maximum Plausible Excess Cancer Risk for Persons at a Residential Site

Exposure Scenario	LADD (mg/kg-day)	Cancer Potency Factor (mg/kg-day)$^{-1}$	Plausible Cancer Risk
MEI	1.3×10^{-10}	41	5.4×10^{-9}
MLEI	4.4×10^{-10}	41	1.8×10^{-8}

Source: Sheehan et al. (1991).

Paustenbach et al. (1991) conducted a more sophisticated analysis of lifetime cancer risk for workers exposed to hexavalent chromium in ambient air at a partially paved trucking terminal in New Jersey. In this analysis, dose was calculated based a range of possible values to estimate the uptake of inhaled particles, resulting in a distribution of maximum plausible excess cancer risks (see Table 5.8). Note that although the latter analysis utilizes a probabilistic modeling approach to estimate worker exposures, the final cancer risk estimates are still conservative because they are based on EPA's upper-bound cancer slope factor of 41 (mg/kg/day)$^{-1}$ for hexavalent chromium and are not based on a probabilistic analysis of the various results of the low-dose model.

5.6.2   Cancer Risks and Nonlinearity

When there are sufficient data to support an assumption of nonlinearity for carcinogens, a margin of exposure (MOE) analysis is typically used. MOEs are used either when a compound's mode of action leads to a dose response relationship that is nonlinear or when the mode of action may theoretically have a threshold, although the EPA does not generally try to distinguish between these different scenarios (EPA, 1996). The risk in this case is not extrapolated as a probability of an effect at low doses, but rather represents the toxicity point of departure (i.e., the beginning of the extrapolation), divided by the environmental exposure level. Common points of departure include the NOAEL and the LED$_{10}$ (i.e., lower 95% confidence limit on a dose associated with 10% extra risk). In general, the ideal point of departure is the dose where the key events in tumor development would not occur in a heterogeneous human pop-

TABLE 5.8   Maximum Plausible Excess Cancer Risk for Workers Exposed at an Industrial Site

Exposure Level	LADD (mg/kg-day)	Cancer Potency Factor (mg/kg-day)$^{-1}$	Excess Cancer Risk
Median (50th percentile)	2.4×10^{-9}	41	9.8×10^{-8}
Mean	7.8×10^{-9}	41	3.2×10^{-7}
95th percentile	3.3×10^{-8}	41	1.3×10^{-6}

Source: Paustenbach et al. (1991).

ulation, thus representing a "no effect level" (EPA, 1999a). The EPA (1999a) therefore recommends that MOE analyses be based on "precursor responses" rather than tumor incidences, since precursor events can often be detected with greater sensitivity.

Although an RfD or RfC can sometimes be calculated for carcinogens if there is evidence of a biological threshold, in many cases, data on the cancer endpoint may be insufficient to do so. In such instances, a MOE analysis provides useful information on the distance between the exposure of interest and the range of observation where cancer risk is inferred to be sub-linear (EPA, 1999a). Note that is the role of the risk manager to decide whether a given margin of exposure is acceptable under applicable management policy criteria, although the risk characterization provides supporting information to assist the decision-maker in this determination (EPA, 1999a).

As noted in the dioxin reassessment by EPA (2000a) (page 107):

> Generally speaking, when considering either background exposures or incremental exposures plus background, MOEs in the range of 100–1,000 are considered adequate to rule out the likelihood of significant effects occurring in humans based on sensitive animal responses or results from epidemiologic studies. The adequacy of the MOE to be protective of health must take into account the nature of the effect at the "point of departure", the slope of the dose–response curve, the adequacy of the overall database, interindividual variability in the human population, and other factors.

5.6.3 Cancer Risks Based on PBPK Models

Alternative approaches of estimating cancer risk may be appropriate if sufficient information is available on a chemical's mechanism of action or if chemical intakes and risk levels are high (EPA 1989a, 1996). One alternative to the traditional approach to quantitatively estimate the cancer risk is to apply a physiologically based pharmacokinetic (PBPK) model to the toxicology data (Leung, 1991; Leung and Paustenbach, 1995; Clewall, 1995; Clewall et al., 1995; McDougal, 1996; Gargas et al., 1999). In this approach, the PBPK model for the chemical converts the administered dose in the animal study to an equivalent administered dose in humans. For example, if an ingested dose of 10 mg/kg-day of benzene produces a circulating blood level of benzene (or its metabolite phenol) of 0.5 μg/g in the blood of the rat, then a valid PBPK model can estimate the dose that a human needs to ingest to achieve the same circulating blood concentration of benzene (or its metabolite phenol). This approach, therefore, can account for the many metabolic and other pharmacokinetic differences among species so that a better estimate of the risk to humans at various doses can be identified. In general, PBPK modeling involves dividing the animal into relevant compartments and, using chemical and species-specific information, estimating the movement and behavior of the chemical within the body (see Figure 5.5). The model is then tested using experimental results separate from those

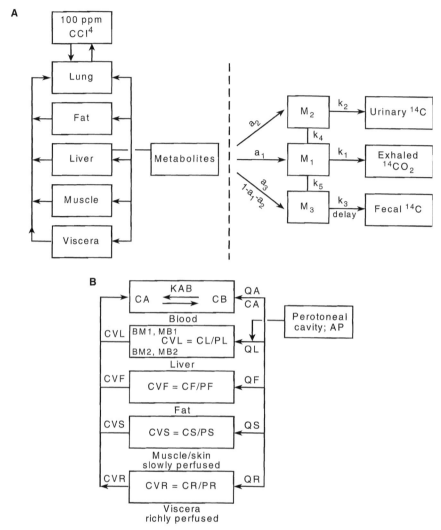

Figure 5.5 Typical Physiologically Based Pharmacokinetic (PBPK) Model. A schematic representation of a typical physiologically based pharmacokinetic (PBPK) model. (a) represents an inhalation model for volatiles (i.e., carbon tetrachloride) and (b) and IP exposure to a non-volatile chemical such as dioxin. The animal is divided into relevant compartments and is validated using experimental data in combination with species-specific information on such items as blood flow, respiration rate, blood-tissue partition coefficients and metabolic rates. Once validated, the model can be used to predict target doses for other species and other routes of exposure.

used to define the parameters in order to validate its predictions (Gargas et al., 1989, McDougal et al., 1990; Mattie et al., 1994; Clewell et al., 1995; Reitz et al., 1996).

There have been several occasions when the PBPK approach has been applied to cancer risk assessment, with two of the better examples being methylene chloride (Andersen et al., 1987) and vinyl chloride (Reitz et al., 1996). In these two examples, the authors were able to explain the basis for the apparent non-linearity and linearity of the carcinogenic response in rodents and humans. In the case of methylene chloride, Andersen et al. (1987) found that one would expect nonlinearity in toxic response (cancer) due to depletion of necessary metaboliz-ing enzymes. This analysis was considered "nearly compelling" from a regula-tory standpoint and was considered in rulemaking for methylene chloride. In the case of vinyl chloride, Reitz et al. (1996) accounted for the metabolic differ-ences between rodents and humans and predicted the difference in the incidence of tumors in these species. Interestingly, this study accurately predicted the inci-dence of the hallmark response, angiosarcoma of the liver, in the worker popu-lation that had historically been exposed to vinyl chloride.

Recently, Kirman et al. (2000) has also suggested an improved approach for characterizing cancer risks that utilizes PBPK models. In this study, the stan-dard EPA approach for evaluating the dose–response relationship of rat brain tumor data for acrylonitrile (AN) was compared to an alternative approach based on analysis of pooled data using PBPK-derived internal dose measures. The cancer potencies predicted for AN by the contrasting assessments were found to be remarkably different—varying by as much as 2 to 4 orders of mag-nitude, with the pooled data assessments yielding lower values (Kirman et al., 2000).

5.6.4 Additivity and Carcinogens

Risks from simultaneous exposure to more than one carcinogenic substance are typically estimated by assuming that the individual risks are additive. This process assumes that intakes of individual substances are relatively small, that there are no synergistic or antagonistic chemical interactions, and that all chem-icals produce the same toxic effect (EPA, 1989a). Note that the summation of several upper-bound risk estimates may result in a total cancer risk estimate that is overly conservative. For example, Cogliano (1997) found that combining the upper-bound cancer risk for several pollutants can lead to an estimate of cancer risk that is about twice as high as the estimated 95 percent upper-bound confi-dence of all the pollutants combined. Because health risks from exposures to chemical mixtures are generally based on a combination of upper-bound risks calculated for individual compounds, these risk assessments tend to be overly conservative (Hwang and Chen, 1999; Gaylor and Chen, 1996; Kodell and Chen, 1994). To reduce some of the uncertainty in risk estimates associated with chemical mixtures, Putzrath (2000) recommends combining the assump-tions concerning the assessment of mixtures of chemicals with the assumptions for evaluating risks of individual chemicals. Other approaches could also be used.

The additive approach also treats all carcinogens as equal, despite potential differences in the underlying database (e.g., animal versus human data) or the weight of evidence for human carcinogenicity. For example, benzene is classified by EPA as a "known human carcinogen" based on evidence presented in numerous occupational epidemiological studies, while 1,3-butadiene is classified as a "probable human carcinogen" based on inadequate human data but sufficient rodent studies (IRIS, 1991, 2000). By comparison, EPA has not formally classified MTBE as an animal or human carcinogen (IRIS, 1993), but an oral slope factor of 1.8×10^{-3} $(mg/kg/day)^{-1}$ has been established as part of California's Public Health Goal for MTBE in drinking water [Office of Environmental Health Hazard Assessment (OEHHA), 1999]. Note that the estimated potency for MTBE is about 8 to 30 times less than the oral slope factor for benzene of 1.5×10^{-2} to 5.5×10^{-2} $(mg/kg/day)^{-1}$. Ideally, risk evaluations based on exposures to multiple gasoline additives or combustion by-products would consider the weight of evidence for each chemical (with respect to its likely ability to be a human cancer hazard) as well the likelihood that the carcinogenic mechanisms act in an additive manner due to a similarity in the mechanism of action.

Cancer risks from multiple exposure pathways are also generally assumed to be additive (EPA, 1989a). This makes sense for a systemic chemical whose toxic effects are similar independent of the route of exposure. For example, ingestion of trichloroethylene probably poses about the same cancer risk when it is inhaled or absorbed through the skin. In contrast, a chemical like Cr(VI) is a known inhalation carcinogen because of its mutagenic effects in the lung, but it poses no carcinogenic risk via ingestion or dermal contact due to the reductive capacity of the stomach and the organics in the skin (DeFlora, 2000; Proctor et al., 2001a).

5.6.5   Hormesis

Hormesis has been defined as a dose–response relationship in which there is a stimulatory response at low-doses, but an inhibitory response at high doses, resulting in a U- or inverted U-shaped dose–response (Calabrese, 2001). This phenomenon, which has been reported hundreds of times in the published literature, indicates that for some chemicals at low doses a lesser incidence of adverse effects than that seen in control animals is observed. To date, regulatory policy in the United States has generally not allowed such observations to be incorporated into rulemaking, or accounted for in the mathematical models used to estimate the cancer risk, because it has been assumed that such observations are an artifact of testing only a limited group of animals and that small doses of "toxic substance" can not be considered advantageous simply as a matter of policy (rather than science).

In recent years, however, the body of scientific evidence supporting the biological basis for this phenomenon has grown, and such considerations should certainly be a part of the risk characterization process (Calabrese and Baldwin, 2001; Calabrese, 2001). Some have argued, for example, that it is intuitive that for some chemicals there will be a hormetic effect because it is well known that there are certain essential metals, which although often carcinogenic in animal studies, must be a part of our diet. A database constructed by the University of Massachusetts has now compiled more than 25,000 published papers that contain nearly 700 dose response relationships. This research group led by Dr. Ed Calabrese has shown that, after the application of rigorous criteria for determining whether a study was of sufficient merit to warrant evaluation, 245 (37% of 668) dose–response relationships from 86 published papers satisfied requirements for evidence of hormesis. A major finding of this research is that when the study design satisfies a set of *a priori* criteria (i.e., a well-defined NOAEL, greater than two doses below the NOAEL, and the end point measured has the capacity to display either stimulatory or inhibitory responses), hormesis is frequently encountered and is broadly represented according to agent, model, and endpoint.

These findings suggest that some of the huge expenditures directed at restricting exposure to exceedingly small concentrations of chemicals in our environment (air, water, food, or soil), may well have yielded no reduction in risk of an adverse effect. Much of the evidence supporting the existence of hormetic effects was recently presented in a special issue of *Critical Reviews in Toxicology* (Calabrese and Baldwin, 2001). The implications of hormesis to risk characterization are potentially profound, and at the least, should be included in these discussions in the coming years.

5.6.6 Traditional Regulatory Approach to Characterizing Noncancer Risks

Estimates of noncancer risk are based on the assumption that there is a level of exposure below which it is unlikely to experience adverse health effects. A common method of evaluating noncancer risks is to generate a "hazard quotient" (HQ), which represents the ratio of exposure to toxicity. Specifically, the HQ is estimated using the following equation (EPA, 1989a):

$$HQ = E/\text{RfD}$$

where HQ = noncancer hazard quotient
E = exposure level or intake (mg/kg/day)
RfD = reference dose (mg/kg/day)

The HQ ratio should not be interpreted as a statistical probability. Instead, estimates less than one (i.e., HQ < 1) indicate that exposures are unlikely to result in any adverse health effect, while estimates greater than one (i.e., HQ > 1) suggest that there may be concern for potential noncancer effects. The level of concern, however, does not increase linearly as the RfD is approached or exceeded. Noncancer risks can be evaluated for different exposure durations, including chronic, subchronic, and shorter-term exposures, as long as the RfD used for comparison is based on the same exposure period (EPA, 1989a).

5.6.7 Additivity and Noncarcinogens

Similar to carcinogens, risks from simultaneous exposure to more than one noncarcinogenic substance or from multiple exposure pathways are generally assumed to be additive by regulatory agencies (based on policies not to possibly underestimate the true risk) Specifically, these effects can be evaluated by summing the individual estimated HQs. The resulting hazard index (HI) assumes that the magnitude of an adverse health effect is directly proportional to the sum of the individual HQ ratios. For example, Williams et al. (2000) estimated potential noncancer risks from oral, inhalation, and dermal exposures to MTBE in drinking water using the HI approach (see Table 5.9). A distribution of hazard quotients for each entry route was first calculated, and then these distributions were combined to yield a distribution of hazard indices for the general population and more highly exposed group, assuming three distinct exposure durations (Williams et al., 2000).

In recent years, there has been an increased level of interest in doing a better job at attempting to understand the kinds of chemicals, which when present in a mixture, are expected to be additive and those that are not. The most widely supported "default" approach was first proposed by toxicologists within the industrial hygiene and toxicology community in the 1950s (Yang, 1994). This approach has been part of the ACGIH guidelines (1999) for more than 40 years and has been generally supported by EPA (1984). The ACGIH TLV Committee approach has been to consider chemicals that act on the same target organ

TABLE 5.9   Estimated Hazard Quotients (HQs) and Hazard Index (HI) for Californians Exposed to MTBE in Drinking Water

	HQ or HI	
Exposure Route	50th Percentile	95th Percentile
General population		
Inhalation	6.5×10^{-5}	3.0×10^{-4}
Ingestion	3.4×10^{-4}	4.8×10^{-4}
Dermal	1.6×10^{-5}	4.6×10^{-5}
Hazard Index (Total all routes)	4.4×10^{-4}	7.1×10^{-4}
Highly exposed population		
Inhalation	4.0×10^{-5}	7.7×10^{-4}
Ingestion	2.0×10^{-4}	2.4×10^{-3}
Dermal	8.7×10^{-5}	1.6×10^{-4}
Hazard Index (total all routes)	2.6×10^{-4}	3.2×10^{-3}

Source: Williams et al. (2000).

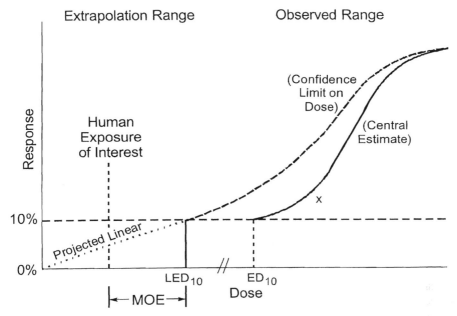

Figure 5.6   Estimating Reference Dose (RfD) Using a Benchmark Dose and Margin of Exposure (MOE) Approach. An alternative approach to the NOAEL method of deriving reference doses is the benchmark dose (BMD) method. In this approach an effect level (usually between a 1 and 10% response rate) is identified from the dose-response curve (solid line) and a bounding estimate on that dose derived (dotted line). A bound on the dose evoking the response of interest is identified and this value is once again divided by appropriate uncertainty factors to derive a reference dose or concentration. In this example, the dose causing a 10% response is chosen (ED_{10}) and a bounding estimate (dotted line) on this response is used to derive the minimum effective dose (MED_{10}). The MED_{10} is then divided by an uncertainty factor to derive the RfD. In this case, the margin of exposure (MOE) is shown indicating where the MED_{10} is found in relation to the typical environmental exposure.

or act through the same mechanism of action as being additive with respect to their hazard. The formula is presented and discussed in the preamble to the annual TLV booklet (ACGIH, 2000).

5.6.8   Margin of Exposure/Margin of Safety

Noncancer risks can also be evaluated using a margin of exposure (MOE) or margin of safety (MOS) approach, which represent alternative ways of comparing toxicity (or point of departure) values to estimated exposure levels (see Figure 5.6). Unlike the HQ approach, which is calculated as the ratio of exposure to toxicity (i.e., RfD), the MOE and MOS are calculated as the ratio of toxicity (i.e., in this case the NOAEL) to exposure. The typical formula for estimating the MOE (or MOS) is as follows (Barnes and Dourson, 1988):

TABLE 5.10   Age-Specific Margin of Safety (MOS) for the Noncarcinogenic Hazard to the Maximally Exposed Individual (MEI) and Most Likely Exposed Individual (MLEI) at a Residential Site

Exposure Scenario	Oral ADD (mg/kg/ day)	RfD Cr(III) (mg/kg/ day)	MOS	Dermal ADD (mg/kg/ day)	RfD Cr(VI) (mg/kg/ day)	MOS
MEI						
Child	6.8×10^{-4}	1.0	1.5×10^3	1.2×10^{-4}	5.0×10^3	4.0×10^1
Youth	3.7×10^{-4}	1.0	2.7×10^3	1.0×10^{-4}	5.0×10^{-3}	5.0×10^1
Adult	2.3×10^{-4}	1.0	4.4×10^3	2.7×10^{-5}	5.0×10^{-3}	1.9×10^2
MLEI						
Child	7.0×10^{-5}	1.0	1.4×10^4	2.0×10^{-6}	5.0×10^{-3}	2.4×10^3
Youth	1.2×10^{-5}	1.0	8.3×10^4	2.0×10^{-6}	5.0×10^{-3}	2.5×10^3
Adult	3.7×10^{-6}	1.0	2.7×10^5	4.5×10^{-7}	5.0×10^{-3}	1.1×10^4

Source: Sheehan et al. (1991).

$$\text{MOE (or MOS)} = \text{NOAEL}/\text{E}$$

where MOE = margin of exposure for noncancer effects;

MOS = margin of safety for noncancer effects;

NOAEL = no observed adverse effect level (mg/kg/day); and

E = exposure level or intake (mg/kg/day).

In general, MOE (or MOS) values that exceed 100 imply an acceptable level of exposure, while values less than 100 have traditionally been used by regulatory agencies as flags for requiring further evaluation (Klaassen, 2001). Sheehan et al. (1991) and Paustenbach et al. (1991) provide examples of estimated MOS values for residents and workers exposed to Cr(VI) vial multiple exposure routes (see Tables 5.10 and 5.11).

Interestingly, there has been little or no guidance documents published that clearly distinguish the MOE from the MOS, or discuss when to use such metrics in favor of the HQ approach. In practice, the difference between MOEs and MOSs appear to be more conceptual than mechanical, in that decisions made based on the latter approach tend to incorporate more management and science-policy judgments, rather than strict scientific principles. These approaches may also be preferred to the more complex and less versatile HQ approach, because they do not require the calculation of a RfD, which can be time consuming and data intensive. As these alternative methods continue to increase in popularity in risk characterizations, more comprehensive guidance on how to interpret and define acceptable MOE/MOS values will become essential.

TABLE 5.11   Distribution of Margin of Safety (MOS) for Noncarcinogenic Hazard for Workers at an Industrial Site

Exposure Route	ADD (mg/kg-day)	RfD Cr(III) (mg/kg-day)	MOS
Soil ingestion			
50th percentile	1.3×10^{-6}	1.0	770,000
Population mean	5.4×10^{-6}	1.0	190,000
95th percentile	2.1×10^{-5}	1.0	48,000
Ingestion of inspired particles			
50th percentile	2.1×10^{-6}	1.0	480,000
Population mean	3.7×10^{-6}	1.0	27,000
95th percentile	1.2×10^{-5}	1.0	83,000
Dermal contact			
50th percentile	2.9×10^{-7}	5.0×10^{-3}	17,000
Population mean	6.5×10^{-6}	5.0×10^{-3}	770
95th percentile	5.5×10^{-5}	5.0×10^{-3}	91

Source: Paustenbach et al. (1991).

5.6.9   Aggregate/Cumulative Risk

As part of the 1996 Food Quality Protection Act (FQPA), the EPA is required to consider aggregate and cumulative exposures and risk when evaluating the safety of pesticides. In the past, EPA evaluated pesticide safety based on a single chemical, single exposure pathway scenario. Aggregate exposure is defined by EPA (1999c) as a "single chemical exposure via the dietary (oral route), drinking water (oral route) and residential (inhalation, dermal and/or oral route) pathways" (page 1). Aggregate exposure analysis therefore considers exposures from all relevant routes and sources for a single chemical, and aggregate risk is defined as "the likelihood of the occurrence of an adverse health effect resulting from all routes of exposure to a single substance" (page 41).

Cumulative risk, on the other hand, is defined by EPA (1999c) as "the likelihood of the occurrence of an adverse health effect resulting from all routes of exposure to a group of substances sharing a common mechanism of toxicity" (page 41). The rationale for this approach is that low-level exposures to multiple substances that cause a common toxic effect by a common mechanism could lead to the same adverse health effect as would a higher level of exposure to any of the chemicals individually (EPA, 1999d). In this context, the EPA (1999d) defines common mechanism of toxicity as "two or more pesticide chemicals or other substances that cause a common toxic effect to human health by the same, or essentially the same, sequence of major biochemical events" (page 4). Under FQPA, aggregate and cumulative exposures and risk apply to both cancer and noncancer endpoints.

Although the greatest focus on how to conduct appropriate aggregate/cumulative risk assessments is currently based in the pesticide arena, such approaches will become increasingly important in other areas and should be considered in future risk characterizations. In fact, a major limitation of traditional noncarcinogenic risk approaches is that they do not account for potential differences in critical toxic effects or mechanism of action, and reference values may have varying levels of confidence associated with them. For example, Wilkinson et al. (2000) found that for chemicals with a "common mechanism of toxicity" but different uncertainty factors (UFs), the choice of noncancer risk method could have a significant impact on the final risk estimate. The major distinction between the five different methods considered—i.e., hazard index (HI), point of departure index (PODI), toxicity equivalence factor (TEF), combined margins of exposure (MOE_T), and cumulative risk index (CRI)— was found to be the point in the process where key uncertainties were considered (Wilkinson et al., 2000).

5.7 KEY UNCERTAINTIES IN RISK CHARACTERIZATIONS

Risk characterization involves integrating information from the first three steps of the risk assessment process—hazard identification, dose–response assessment, and exposure assessment—to derive conclusions about human risk. In all steps, the rationale for selecting or excluding certain data sets as well as a description of key model assumptions and data uncertainties should be clearly presented. The following section highlights the key uncertainties in each step of the risk assessment process.

5.7.1 Uncertainties in Hazard Identification

The primary uncertainty to consider in the hazard identification step is whether a specific chemical or agent is capable of causing in humans the same adverse health effects observed in animal studies. Sometimes this relationship is direct; for example, most agents that cause acute toxicity in the liver of a rodent also cause the same effect in humans (e.g., alcohol and trichloroethylene). In other instances, however, there is uncertainty about the target organ. For example, beryllium does not cause granulomas in rodents following inhalation exposures, but berylliosis (a rare form of lung disease) is observed in humans and is always attributed to exposure to the chemical (Rossman and Jones-William, 1991).

For carcinogens, it is very difficult to predict the relationship between the dose needed to increase the tumor incidence between animals and humans. Sometimes, it is even difficult to predict whether the chemical will be a human carcinogen at any reasonable dose even though it has been shown to be an animal carcinogen. In such instances, one must conduct a careful evaluation of the epidemiology and toxicity data, pharmacokinetic and toxicokinetic data, and information on structure–activity relationship. For example, even though dioxin has increased the tumor incidence rate in virtually every animal study, it remains unclear whether it poses a significant human cancer hazard. Specifically, in at least five different epidemiology studies of humans who have been exposed to dioxin at appreciable doses, there has been no single target organ that has reliably shown an increased tumor incidence (EPA, 2001). Interestingly, even though the cancer rate has sometimes been elevated in a particular tissue, none of these studies have increased the liver cancer rate even though it is consistently the key target in animal studies. Some scientists speculate that perhaps the number and role of AhR in humans is significantly different than in most rodents. Nonetheless, since the mode of action may be similar between animals and humans, and because the overall cancer rate in epidemiology studies has been elevated, the EPA (2001) has recommended that dioxin be categorized as a known human carcinogen.

In similar evaluations of the animal–human relationship, Sielken et al. (1999) found that low exposures to aldrin and dieldren do not significantly increase human cancer risk, despite the much higher risk estimates calculated by EPA based on animal bioassay data (i.e., increased incidence of mouse liver tumors). Green et al. (2000) also found that acetochlor-induced rat nasal tumors do not represent a hazard for humans, due to differences in metabolic conversion and activation pathways between rats and humans.

There are many reasons why a chemical may cause certain kinds of toxic effects in animals but not humans. In particular, there is significant uncertainty in the extrapolation from very high doses administered to test animals to the much lower doses encountered by humans. Difference in uptake rate, mechanisms of action, metabolism, biologic half-life, target tissue susceptibility, ability to repair deoxyribonucleic acid (DNA), and other factors can also account for observed differences in the severity or type of toxic effects between species. There are many examples of clear differences in metabolism and pharmacokinetics. For example, the elimination half-life of dioxin has been found to vary about 10-fold across species—that is, about 2550 days in humans and 30 to 60 days for several rodent species (Olsen, 1994).

Beyond the "internal biologic differences," there are also differences in the efficiency via the three different routes of exposure (ingestion, inhalation, and dermal contact). For example, Ross et al. (2000) found that the dermal absorption of parathion was significantly less in humans compared to several other species (see Table 5.12). In addition, the uptake of lead in the gastrointestinal (GI) tract of different species of animals is also generally much different than that seen in humans. Many chemicals, particularly metals, may pose a carcinogenic hazard for only one exposure route. For example, hexavalent chromium has been found to be carcinogenic if inhaled, but should not pose a cancer hazard if ingested, due to reduction in the stomach. Conversely, arsenic and nickel are both carcinogenic via inhalation, and appear to pose some degree of cancer risk via ingestion. The oral bioavailability of an agent can also be quite different among species based on both chemical and physiological differences. For example, the rat has a forestomach, while humans do not.

On the other hand, chemicals that pose a relatively low toxicity in some laboratory animals may be much more toxic in humans. For example, methanol

TABLE 5.12   **In Vivo Percutaneous Absorption of Chemicals in Humans and Animals**

	Percentage of Dermal Dose Absorbed				
Compound	Man	Monkey	Pig	Rabbit	Rat
Haloprogin	11		20	113	96
N-Acetylcysteine	2		6	2	4
Caffeine	47		32	69	53
Butter yellow	22		42	100	48
Cortisone	3	5	4	30	25
Testosterone	13	18	29	70	47
Parathion	10	30	14	98	95
Lindane	9	16	38	51	31
Malathion	8	19	15	65	
DDT	10	19	43	46	
Propoxur	20				50
Carbaryl	74				96

Source: Ross et al. (2000).

poisoning in humans is generally attributed to metabolic acidosis induced by formic acid (formate) buildup in the body [World Health Organization (WHO), 1997]. Such buildup does not occur in rodents, however, because formate metabolism is about 2.5 times faster in rodents than in humans (Klasassen, 1996). Consequently, animal models are considered to be poor predictors of human health effects from methanol exposures. Over the years, we have failed to identify good animal models for a number of important chemicals including arsenic, beryllium, asbestos, and thalidomide.

5.7.2   Uncertainties in Dose–Response Relationship

There are several key uncertainties to consider when evaluating the underlying biological mechanism or dose–response relationship of a chemical or agent. These include uncertainty in the selection of a particular data set, the reliance on different interspecies scaling factors, and a lack of consensus about a chemical's mechanism of action. For example, there is substantial controversy over dixoin's mode of action and the appropriate way to extrapolate the dose–response curve to low doses. Although many scientists argue that the dose–response relationship is linear at low doses, there is some evidence to suggest a possible threshold effect from dioxin exposure. There is also a lack of scientific consensus on the carcinogenic potency of chloroform due to inconsistencies in the available toxicological database and the belief that chloroform's mode of action may differ between species (Evans et al., 1994). In an effort to reduce these uncertainties, approaches are continuously being developed and refined to quantify the likelihood of carcinogenic risk using decision analytical tools and relying on subjective expert judgments (Fayerweather et al., 1999; Evans et al., 1994).

TABLE 5.13 Formaldehyde Risk Estimate for Rats Based on Alternative Mathematical Models (Excess Lifetime Cancers per 100,000)

Exposure Level (ppm)	Multistage		Probit		Logit		Weibull		Multihit	
	MLE	UCL	MLE	UCL	MLE	UCL	MLE	UCL	MLE	UCL
3	43.4	633	23.9	264	9.9	315	81.5	329	0.8	212
1	7.4	411	3.8	73	16.2	138	27.2	161	0.4	23
0.5	0.4	204	0.4	12	2.9	40	6.5	56	0	0.2
0.1	0	102	0.1	3	0.8	15	2.2	23	0	0

Source: Graham et al. (1988).

Several different types of mathematical models can be used with experimental bioassay data including tolerance distribution models, mechanistic models, and time-to-tumor models. The use of alternative extrapolation models can introduce significant uncertainty in the dose–response assessment and exert a powerful influence on the resulting risk estimate. As illustrated earlier, these models may provide similar estimates for doses in the experimental (observable) range, most models will predict different carcinogenic responses at much lower (unobservable) ranges. For example, significant differences in cancer risk have been observed for formaldehyde exposures using the multistage, probit, logit, Weibull, and multihit models (see Table 5.13). Many other chemicals are considered to have even greater sensitivity to model selection than formaldehyde, particularly when the difference between actual doses and bioassay doses are large (Graham et al., 1988).

Similarly, Holcomb et al. (1999) found significant differences in the predicted frequency of infections from four food-borne pathogens (*Shigella flexneri, Shigella dysenteriae, Campylobacter jejuni,* and *Salmonella typhosa*) in a comparison of six different dose–response models (log-normal, log-logistic, exponential, β-Poisson, and Weibull-gamma). The infectious dose estimated to affect 1 percent of the population ranged from one order of magnitude to as much as nine orders of magnitude, depending on the dose–response model that was used (Holcomb et al., 1999). In an attempt to incorporate such model uncertainties in estimates of microbial risks, Kang et al. (2000) recommends using a "weighted-average" approach based on dose–response estimates from each of the various models.

The applicability of the dose–response data to real-world exposure conditions and potential susceptible populations should also be considered when addressing uncertainty in the risk estimate. For example, Bruckner (2000) found that pesticide toxicity is both age- and compound-dependent and that substantial anatomical, biochemical, and physiological changes during childhood could affect the absorption, distribution, metabolism, and elimination of certain chemicals. However, although age-dependent differences in chemical lethality for some pesticides were found to vary 2- to 3-fold, these differences were generally found to be less than an order of magnitude, suggesting that the existing 10-fold inter-species uncertainty factor for pesticides provides adequate protection of infants and children (Bruckner, 2000; Renwick et al., 2000).

5.7.3 Uncertainties in Exposure Assessment

There are many types of uncertainties to consider in the exposure assessment, including the magnitude of human exposure, the number and type of people likely to be exposed, and all possible exposure pathways (Paustenbach, 2000). Since the mere presence of a chemical in the environment does not in and of itself represent a potential risk to human health, a link must be made between a chemical source and a receptor via an exposure pathway. For example, an assessment of potential health hazards from the construction of a baseball stadium in San Francisco revealed several different types of population groups that might be exposed via multiple routes to chemicals at the site (Brorby et al., 1997). Potential receptors based on current or proposed future site activities included construction and maintenance workers, other on-site workers and visitors, nearby residents or workers, and recreational users and aquatic organisms associated with China Basin Channel or San Francisco Bay. A site conceptual model was developed to identify potential chemical sources (e.g., surface soil, ambient air, indoor air, groundwater, and surface water) and establish links between these and potential receptors (see Figure 5.7).

Additional uncertainties can arise from time–activity patterns in different groups, monitoring studies of chemical concentrations in the environment, modeling of environmental fate and transport of contaminants, and the reliance on unvalidated exposure assumptions. For example, a NRC (1987) report estimated the upper-bound lifetime cancer risk from 28 pesticides in over 200 fresh and processed food items to be 6×10^{-3} based on very conservative exposure assumptions. Specifically, estimates of dietary pesticide exposure were based on EPA's theoretical maximum residue contribution (TMRC) method, which assumes the most severe field application conditions and maximum tolerance values for each pesticide and commodity. In a more recent analysis, Gold et al. (1997) found that estimated dietary cancer risks for individual pesticides were significantly less than that reported by NRC (i.e., by 3 to 6 orders of magnitude) if exposures were calculated based on residue monitoring data collected as part of the Food and Drug Adminstration (FDA) Total Diet Study. Other studies have also noted a significant decrease in calculated pesticide residue risks if worst-case (theoretical) exposure data are substituted with actual exposure data (Archibald and Winter, 1990; McCarthy, 1991).

Major uncertainties have also been associated with the evaluation of aggregate exposures to pesticides under the 1990 Food Quality Protection Act (FQPA). These include limited data on pesticide usage and practices (e.g., label vs. actual), residential environmental factors (e.g., air exchange rate), consumer

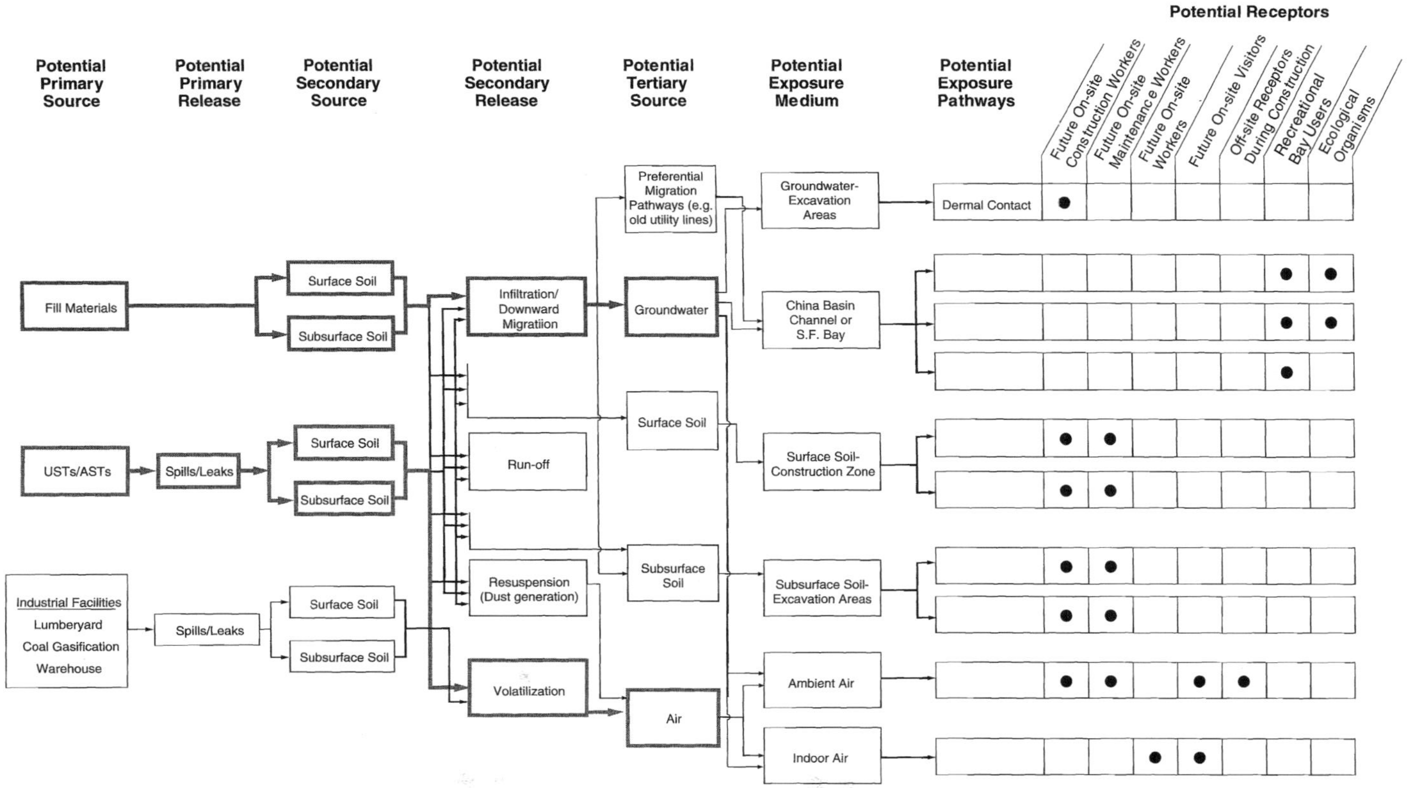

Figure 5.7 Site Conceptual Exposure Model for Baseball Stadium in San Francisco. The site conceptual model identifies potential chemical sources, such as surface soil, ambient air, indoor air, groundwater and surface water, and establish links between these and potential receptors. In this example, potential receptors based on current or proposed future site activities included construction and maintenance workers, other on-site workers and visitors, nearby residents or workers, and recreational users and aquatic organisms associated with China Basin Channel or San Francisco Bay. Primary, secondary, and tertiary sources or release and possible exposure routes are also highlighted. Source: Brorby et al. 1997.

341

nondietary behavior (e.g., time–activity patterns at home), and consumer dietary behavior (e.g., food consumption patterns over time) (Peterson, 2000). To develop realistic exposure and risk assessments requires that the appropriate temporal, spatial, and demographic exposure factors be correctly assigned, and the integrity of the data concerning the exposed individual be maintained through the aggregate exposure assessment (Paustenbach, 2000).

To deal with the uncertainties (and variability) in exposure assessments, the use of Monte Carlo or other probabilistic techniques has increased substantially over the past decade (Morgan & Henrion, 1990; Thompson et al., 1992; Duan and Mage, 1997, Paustenbach, 2000). Such quantitative uncertainty analyses allow distributions (rather than point estimates) to be used for important exposure parameters, which incorporate the range and probability of plausible values (see Figure 5.8).

Virtually every exposure variable, whether physiological, behavioral, environmental, or chronological, can be replaced with a probability distribution (Burmaster and Anderson, 1994; Finley et al., 1994, Price et al., 1996; Burmaster and Thompson, 1997; Frey and Rhodes, 1998; Gargas et al., 1999). The basic steps involved in running a Monte Carlo simulation for an exposure analysis include (Paustenbach, 2000):

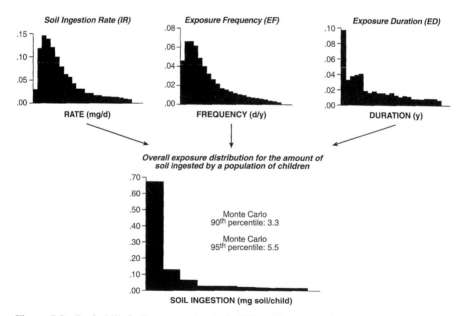

Figure 5.8 Probabilistic Exposure Analysis Using Monte Carlo Techniques. Example of how probability density functions (distributions) for three different related exposure factors are combined to form a distribution for the amount of soil ingested by a population of children. The Monte Carlo technique allows the risk assessor to account for the variability in many exposure parameters within a population and then produce a distribution that characterizes the entire population.

1. The probability distribution of each equation parameter (input parameter) is characterized, and the distribution is specified for the Monte Carlo Simulation. If the data cannot be fit to a distribution, the data are "bootstrapped" into the simulation, meaning that the input values are randomly selected from the actual data set without a specified distribution.

2. For each iteration of the simulation, one value is randomly selected from each parameter distribution, and the equation is run. Many iterations are performed, such that the random selections for each parameter approximate the distribution of the parameter. Five thousand (or more) iterations are typically performed for each dose equation.

3. Each iteration of the equation is evaluated and saved; hence a probability distribution of the equation output (possible doses) is generated.

As mentioned, the probabilistic approach is considered to be optimal for characterizing exposures and health risks, because it does not tend to result in overly conservative or non-plausible risk values (Finley et al., 1994; Sielken et al., 1995; Thompson and Graham, 1996; Gargas et al., 1999). Formal uncertainty analyses can also inform decision-makers and the public about the extent of conservatism that is embedded in default assumptions and risk estimates, and is especially useful in identifying where additional research is likely to resolve major uncertainties (NRC, 1994). The use of Monte Carlo and other techniques will undoubtedly increase in future risk characterizations, and regulatory agencies have already begun to publish guiding principles on how to conduct and interpret such analyses (EPA, 1997b).

5.7.4 Uncertainties in Risk Characterization

The uncertainties associated with risk characterization are generally the result of the combined uncertainties in the site conditions, exposure assumptions, and toxicity criteria. In an attempt to "err on the side of caution" safety factors or other sources of conservatism are often applied during the exposure and dose–response assessments, resulting in a final risk estimate that may be too conservative. For example, in a recent risk characterization of steel industry slag, Proctor et al. (2001b) provided the following discussion related to sources of uncertainty:

> It is recognized that several sources of uncertainty exist in this analysis as in most environmental health risk assessments. The uncertainties of this analysis are compounded because generic applications of slag were assessed, and site-specific information was not available to refine exposure estimates. In addition, one of the most significant uncertainties of this assessment is the assumption that slag presents similar characteristics as soil and that exposure rates for slag and soil are consistent. For example, slag-to-skin adherence is represented by soil-to-skin adherence measurements; soil ingestion rates were used to represent slag ingestion; and particulate suspension was based on soil suspension models. In general, it is

thought that these assumptions overestimate exposure because slag particles are larger and less likely to adhere to the skin or be incidentally ingested as compared to soil (page 25).

Finally, several assumptions were constructed to characterize exposed population activities based on professional judgment regarding human activities during and after slag application in the absence of actual data. In general, it is believed that conservative assumptions were consistently chosen to ensure overestimation of true hazards and risks. For example, for assessing exposures to the farmer scenario, it was assumed that farm fields were entirely slag, without any mixing with soil. This and other conservative assumptions were purposely selected to ensure that the analysis would be considered health-protective by environmental regulators and special interest groups. The stochastic analysis of inhalation exposures to construction workers demonstrated that the selection of multiple upper-bound exposure assumption in the point estimate exposure assessment overestimated the HI by nearly a factor of ten. Therefore, it is reasonable to assume that the selection of upper-bound parameters throughout this assessment has resulted in exaggerated estimates of potential cancer and noncancer risks from exposure to slag (page 26).

One source of uncertainty that is unique to risk characterization, however, is the assumption that the total risk associated with exposure to multiple chemicals is equal to the sum of the individual risks for each chemical (i.e., the risks are additive). Other possible interactions include synergism, where the total risk is higher than the sum of the individual risks, and antagonism, where the total risk is lower than the sum of the individual risks. For example, although there are no data on chemical interactions in humans exposed to chemical mixtures at the dose levels typically observed in environmental exposures, animal studies suggest that synergistic effects will not occur at levels of exposure below their individual effect levels (Seed et al., 1995). As exposure levels approach the individual effect levels, a variety of interactions may occur, including additive, synergistic, and antagonistic interactions (Seed et al., 1995). Animal studies also suggest that interactive effects are unlikely for mixtures of chemicals that affect different target organs (i.e., each chemical acted independently), whereas antagonism may occur for mixtures of chemicals that affect the same target organ, but by different mechanisms.

In a recent risk characterization of air toxics in California, Morello-Frosch et al. (2000) commented on the rationale for and impacts of combining hazardous air pollutant (HAP) exposures to estimate public health risks. Various limitations associated with toxicity information and uncertainties in science policy assumptions were also discussed in this study and include the following (Morello-Frosch et al., 2000):

> For noncancer risk estimates, the total hazard index provides a useful screening-level tool for potential hazards for specific health endpoints or target organ systems, but does not provide an estimate of incidence or probability of effects. Several HAPs in this study have similar health endpoints seen at a variety of experimental conditions. Therefore, while providing a rationale for combing HAP exposures

for a composite hazard index, the results should not be interpreted to mean that the common health effects would necessarily occur at ambient exposure levels. . . . Little is known about how these pollutants interact to fully evaluate the health risks posed by cumulative air toxic exposures. Synergistic or antagonistic interactions among pollutants may mitigate risks in a way that has not been identified in this study (page 284).

Second, health risk estimates are somewhat limited by the availability of toxicity data, much of which is incomplete. Of the 148 pollutants analyzed, only 60% had cancer potency estimates, and 61% had chronic noncancer toxicity values, whereas 28% had no available toxicity information. Regulatory concern traditionally has emphasized cancer, which in part explains the lack of specific hazard information for other chronic outcomes such as developmental, reproductive, and neurological effects (page 284).

More pollutants were associated with potential cancer risks than noncancer effects. Although concentrations for several carcinogenic air toxics were high, other factors may also help explain the predominance of cancer risk estimates. Cancer potencies are derived from occupational studies in humans—typically adult males—when available, and otherwise from toxicological studies in animals . . . because potency estimates derived from human data are generally based on occupational cohorts of healthy males, they may not adequately reflect population variability in susceptibility, including sensitive populations—such as children, who may have increased exposures and are physiologically different from adults in how they metabolize chemicals in the body . . . It is also important to point out that potential overestimation of total cancer risk can result by combining upper-bound cancer potency estimates for several pollutants (page 285).

Finally, it is important to note that many risk characterizations intentionally over-estimate or attempt to generate the "worst case" estimate of risk in order to provide a preliminary screening level analysis. The purpose of such evaluations is simply to rapidly identify those exposure scenarios worthy of further consideration versus those that clearly pose a negligible risk. If estimated risks are acceptable following a screening analysis, a more refined evaluation is generally not warranted, even though there may be significant data gaps or uncertainties. However, if estimated risks are unacceptable based on conservative assumptions, then more data can be used or generated to refine the analysis and develop a better characterization that is less uncertain and prone to overestimation.

5.8 RISK DECISION-MAKING PROCESS

The need to improve risk characterizations has never been greater. Risk characterization serves as the interface between risk assessment and risk management, and hence, is instrumental in conveying the magnitude of a problem and highlighting priority risk areas. Sound characterizations of risk are also important because they allow risk comparisons to be made more accurately, across diverse compounds and settings, and help ensure that risk management actions are targeted appropriately. In addition, a thorough evaluation of key sources of uncertainty and variability in risk characterizations can lead to more productive research efforts and a better use of limited resources in terms of data collection. Finally, a well-prepared and presented risk characterization can be used to communicate risks and the confidence in numerical risk estimates to a broader audience. Consequently, the historical approach used by many risk assessors to present overly conservative or "worst-case" estimates of risk based on limited data will no longer be considered acceptable or necessarily in the best interests of society.

Despite the importance of risk characterizations, many other types of information (other than those considered in classic risk assessment) must be available for decision-making (NRC, 1996). Specifically, the risk decision-making process requires the consideration and weighing of a multitude of factors including economic analyses, legal requirements and acceptable risk criteria, societal values and public perceptions, court decisions, and indirect benefits. For example, recent guidance from the Office of Management and Budget about how to conduct cost-benefit analyses clearly indicates that government bodies want to be presented with the best scientific estimates of the risks and benefits of regulatory decisions as well as quantitative descriptions of the uncertainties in these estimates (OMB, 2001).

Cost-benefit analysis (CBA) involves enumerating all tangible and intangible societal costs and benefits associated with a particular decision or option and these are valued in a common unit, which is typically monetary (Stokey and Zeckhauser, 1978; Weinstein and Fineberg, 1980; Layard and Glaister, 1994; Tietenberg, 1996). An alternative approach that does not involve measuring benefits in economic terms is cost-effectiveness analysis (CEA), in which the benefits of a program are expressed as some unit of output or outcome, such as "number of cases detected" or "number of lives saved" (Weinstein and Fineberg, 1980; Gold et al., 1996; Tietenberg, 1996). Although either approach can be used in risk management evaluations, CBAs have greater flexibility in that they allow for comparisons across diverse programs that may have different health outcomes, and only CBAs can lead to efficient (socially optimal) policy choices. Economic analyses by themselves, however, should not be the primary or the only determinant of decision-making.

Many different legal or statutory requirements may also need to be considered by risk managers, including how to define "negligible" and "acceptable" risks (Mayo and Hollander, 1991). Various criteria exist for prioritizing the different types of risks for management action, including "zero risk" ideals and "de minimis" and "de manifestis" principles (see Figure 5.9). The most commonly cited acceptable risk levels for de minimis risks is one additional cancer case per one million exposed individuals (1×10^{-6}), although in practical applications, an acceptable risk range of 1×10^{-4} to 1×10^{-7} is used to make decisions regarding remedial or regulatory action (Gargas et al., 1999). Acceptable risk levels also tend to be influenced by the number of exposed in-

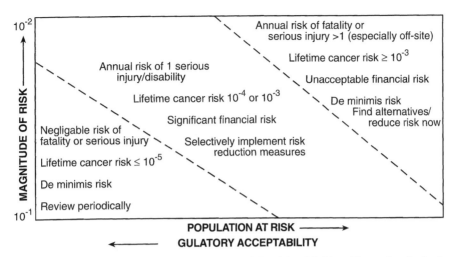

Figure 5.9 Criteria for Risk Prioritization and Decision Making. Example of criteria that exist for prioritizing the different types of risks for management action. These include "de minimis" risks, which are deemed negligible because they result in a lifetime cancer risk less than one in ten thousand, and "de manifestis" risks, which are considered serious because they result in a lifetime cancer risk greater than one in a thousand. The former risk type does not generally warrant any immediate action (although periodic review is recommended), while the latter risk type typically necessitates action leading to an immediate reduction in risk. Lifetime cancer risks that fall between these two ranges may or may not require immediate or longer-term risk reduction strategies, depending on the particular situation. Source: Kolluru 1996.

dividuals. For example, a review of 132 environmental regulations over the last couple of decades reveals that when the number of exposed persons is relatively small, the allowable level of exposure increases for carcinogens, while acceptable risk levels decrease as the number of exposed persons increases (Travis et al., 1987). This concept is illustrated in Figure 5.10. Under the radiation exposure paradigm for stochastic risk management, exposures are deemed acceptable if they correspond to risks less that the maximum allowable risk and they are as low as reasonably achievable (ALAR) (see Figure 5.11).

Societal values are particularly difficult to describe and quantify but often have the greatest influence on the decision-making process. In particular, the "precautionary principle" is based in large part on societal preferences for future environmental protection and safety in situations where very little is known about a chemical or agent. The precautionary principle movement began in Europe in the early 1990s, in which the view was adopted that it would be prudent to work toward the elimination of environmental releases at any measurable quantity of non-naturally occurring substances. Over the past 5–10 years, no single definition for this principle has been accepted by all of the various governmental and non-governmental bodies that are interested in this topic. How-

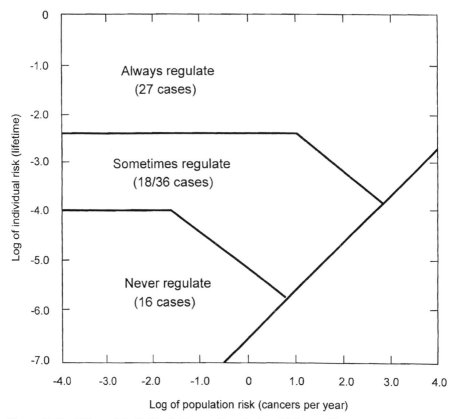

Figure 5.10 Effect of Individual Versus Population Risk on Chemical Carcinogenic Regulation. Represents ranges of individual risk levels inherent in various regulatory decisions as a function of the number of exposed persons. Note that when the number of exposed persons is relatively small, the allowable level of exposure increases. Source: Adapted from Travis et al. 1987.

ever, one of the most widely cited definitions, which embodies the central elements, is as follows (*Rachel's Hazardous Waste News*, 1993):

> The precautionary principle says that to avoid irreparable harm to the environment and to human health, precautionary action should be taken: Where it is acknowledged that a practice (or substance) could cause harm, even without conclusive scientific proof that it has caused harm or does cause harm, the practice (or emissions of the substance) should be prevented and eliminated.

Note that it has been suggested that adoption of this philosophy would spell the end of the practice of risk assessment. Specifically, the precautionary principle contains several views that many risk analysts find troublesome. First, there is no provision for a risk/risk tradeoff. Second, irrespective of the possible benefits

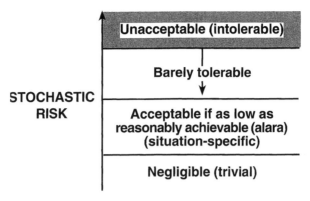

Figure 5.11 Radiation paradigm for stochastic risk management. In this approach, doses are acceptable if they correspond to risks less than the maximum allowable risk and they are as low as reasonably achievable (ALAR). Compliance with the dose limit does not, by itself, determine acceptable risks. This approach differs from other risk oriented paradigms in regards to those risks that are deemed to be "barely tolerable." Under this paradigm such risks reside slightly above those that meet the acceptable ALAR criteria.

to society, no amount of emissions are deemed acceptable. Third, it is inferred that whomever is promoting a technology that might emit some level of toxicants cannot be expected to be objective in their analysis of the possible risks. Critics of the precautionary principle have noted that if adopted, the role of science in the decision making process will be minimized, if not eliminated. This could present a number of dangers since science is the only tool we have to sort out fears of technology versus the real benefits associated with scientific and medical advances. It has also been argued by the developing countries that such an approach places them at a significant disadvantage to the developed nations, since they are willing to bear some degree of degradation of the environment to increase their standard of living to a level that may remotely resemble that of the developed world (Weiner, 2002). In the coming years, many risk managers will therefore be seeking information from the risk characterization, which will assist them in addressing the key concerns raised by the precautionary principal (EEA, 2001). Various court decisions have had a significant impact in risk decision making in the environmental health arena and have been even more pronounced with respect to occupational health. For example, the Supreme Court decision regarding benzene dictated that all future occupational health regulations include a risk assessment component that clearly shows that reduced airborne concentrations of a chemical can be expected to reduce the associated risks (NRC, 1994).

Perhaps the most interesting and unpredictable factor in risk management is the indirect benefits of a particular decision. For example, when EPA decided to aggressively regulate dioxin emissions from incinerators, it became apparent that the air cleaning devices used to remove dioxin would also dramatically reduce metal emissions. Over the years, there have been numerous occasions where the management of a particular chemical resulted in positive benefits with respect to other chemicals. On the other hand, there have been some instances where "less than desirable" or unanticipated events occurred from the management of a particular agent; for example, when one chemical is substituted for another less well-known chemical. As we attempt to improve the risk management process, more emphasis will likely be placed on identifying possible risk trade-offs associated with alternative decisions. Historically, compilation and evaluation of all of these factors has been the role of the risk manager or decision-maker. In fact, most regulatory guidance documents and published papers specify that risk characterizations should remain separate from risk management considerations (NRC, 1983; EPA, 1995a, b). Although it the duty of the risk manager to consider relevant scientific and nonscientific factors, both scientific judgments and policy choices may be involved in the risk assessment process when selecting from among possible inferential bridges. The term "risk assessment policy" has therefore been recommended to differentiate those judgments and choices from the broader social and economic policy issues that are inherent in risk management decisions (NRC, 1994). Indeed, much of the controversy surrounding regulatory decisions has resulted from failing to distinguish risk assessment policy from risk management policy (Bolger et al., 1996). As a guiding principle, persons who generate, review, use, and integrate risk assessments for decision-making should be "sensitive" to distinctions between risk assessment and risk management (EPA, 1995a).

One of the most comprehensive discussions of risk management is presented in the reports by the Commission on Risk Assessment and Risk Management (CRAM) entitled, *Framework for Environmental Health Risk Management* and *Risk Assessment and Risk Management in Regulatory Decision-Making* (CRAM, 1997a, b). The Commission was mandated by Congress in the 1990 Clean Air Act Amendments to investigate and make recommendations about the use of risk assessment and risk management in federal regulatory programs. One of the outcomes of this effort was the development of a framework for making risk management decisions. Risk management was defined in the Commission's reports as comprising the following six stages: formulate the problem in broad context, analyze the risks, define the options, make sound decisions, take actions to implement the decisions, and perform an evaluation of the effectiveness of the actions taken (see Figure 5.12). This framework, along with the CRAM's emphasis on stakeholder involvement, will likely represent the next level of maturation in the field of risk management.

Communicating the results of risk characterizations and risk management decisions will also become an increasingly important aspect of the risk decision-making process. According to NRC (1994), risk communication consists of two parts: communication between the risk assessor and the risk manager and communication between the risk assessment/management team and the public. Risk managers usually base decisions on individual and population risk estimates, presented either as a point estimate (single value) or a range of values (distribu-

Figure 5.12   Risk Decision-Making Framework. Framework for making risk management decisions developed by the Commission on Risk Assessment and Risk Management. Risk management is defined as comprising six stages: formulate the problem in
broad context, analyze the risks, define the options, make sound decisions, take actions
to implement the decisions, and perform an evaluation of the effectiveness of the actions
taken. All stages require significant stakeholder involvement and collaboration. Source:
CRAM 1997a, b.

tion). At a minimum, risk managers will be presented with a qualitative uncertainty analysis, although there has been a growing expectation for quantitative
uncertainty analyses. The general public usually receives much less information
on how risks are characterized than do risk managers (e.g., they may be presented only with the upper-bound value of a point estimate). In most regulatory
situations, the risk management decision and supporting information are published in the *Federal Register* and the public is given the opportunity to comment
on the analysis and resulting decision within a 30- to 60-day period. However,
rising expectations about stakeholder involvement in the risk decision-making
process, along with the greater ease of information transfer using computer technology, will require that a new approach that involves routine interaction with
the public become standard practice in the future.

The importance of risk communication and stakeholder involvement has
been fully recognized within the risk assessment process since the early 1990s.
That is, it became clear that technical documents would have little benefit if
those who would eventually be affected by the outcome did not understand the
risk characterization or were felt left out of the decision-making process (NRC,
1996). A recent EPA Science Advisory Board commentary also affirms that

stakeholder involvement, if properly conducted, can be valuable in supporting high-quality science-based decision-making (EPA, 2001b). The importance of these factors in the field of risk management, however, was understood by many risk scholars and decision makers nearly a decade earlier. For example, in a speech before the National Academy of Sciences, EPA Adminstrator William Ruckelshaus called for a government-wide process for managing risks that involved the public. However, the increased polarization of environmental perspectives may be attributed, in part, to continue imperfect dialog between risk assessors, risk managers, and the public. Consequently, the next era in the history of risk assessment will need to focus on improvements to risk characterization, including better communication of risk findings to decision makers and the public.

5.9 INTERNATIONAL APPROACHES TO RISK CHARACTERIZATION

A cursory review of the risk guidance documents and policies used in some other countries that have been actively involved in practicing health risk assessment over the past 10–15 years indicates that the kind of information presented in other countries is often broader than that contained in most assessments conducted in the U.S. In general, the first three steps—hazard identification, exposure assessment, and estimating an "acceptable" level of exposure—are methodologically very similar to those practiced in the U.S. However, the final step on risk characterization can vary significantly across countries.

In particular, the risk assessment and risk management process appear to be much more intermingled in some other countries than in the U.S. For example, in several policy documents from the United Kingdom (Dept of the Environment, 1991) it is clear that a discussion of cost-benefit and cost-effectiveness analysis, the precautionary principle, and resource evaluation are an integral part of the risk characterization. As in the U.S., a discussion of the uncertainty within each component of the analysis is presented. Although the addition of these other components goes far beyond that which is typically part of a classic risk assessment performed in the U.S., this may simply be a reflection of the separation of responsibilities among government agencies. For example, much of the cost-benefit analysis and cost-effectiveness analyses are conducted by the Office of Management and Budget (OMB) rather than the Environmental Protection Agency (EPA) in the U.S., in an attempt to ensure that the scientific evaluation of the possible health risks not be significantly influenced by the practicalities of economics and feasibility issues.

The Netherlands appear to have an even more integrated approach to the risk assessment and risk characterization process than either the United Kingdom or the U.S. From the onset, consideration is given to various options for dealing with a problem, weighing of the social benefit, cost-benefit analyses, consideration of opportunity costs and other thorny factors that may involve an emphasis on considering "the entire picture" before environmental decisions are made (Ministry of Housing, 1994; Ginjaar, 1996). Due to the size of the country and the importance of capitalizing on their agricultural base, the Netherlands were among the first countries to devise a comprehensive program for evaluating the human and ecological risks posed by contaminated soils (Denneman, 1994). Their risk-based approach was considered among the most respected in the world during the 1990s, and was subsequently adopted by many nations.

Australian regulatory agencies have also relied upon quantitative risk assessments to aid in decision-making beginning in the mid-1980s, and their evaluation process has matured during the 1990s (Walker, 1992; El Saadi, 1992). While their early evaluations focused on risks posed by air pollution, contaminated soils and contaminated sediments, this focus expanded into classic ecological risk assessment in later years. Risk characterizations in Australia appear to have been initially based on regulatory guidance developed in the U.S., but have rapidly evolved to consider cost-benefit and cost-effectiveness evaluations. Berry et al. (1993) provides an example of a complex analysis conducted in Australia that addressed the hazards posed by lead in the environment.

In summary, risk characterizations appear to differ somewhat between the U.S. and some other developed countries. In the U.S., there is a clear desire to keep the risk assessment and risk management processes separate from one another, while some other countries have elected to allow greater intermingling between these processes. It is likely that risk characterizations will continue to evolve worldwide, with perhaps a more uniform approach being adopted.

5.10 FUTURE DIRECTIONS OF RISK CHARACTERIZATION

As mentioned, the evolving nature of risk characterizations in the U.S. suggests that future efforts to characterize human health risks will be held to a much higher standard than in the past. In the future, these characterizations could require more reporting on "non-traditional" aspects for risk assessors, including the costs and benefits or potential risk tradeoffs of alternate solutions. It is also possible that the pressure to bring "the complete picture" to regulators or policy decision-makers will result in a closer alliance between toxicologists, exposure assessors, mathematical modelers, economists, risk communicators, and classic risk assessors. As we have noted, during the early years of risk assessment (circa 1980–1995), it was considered necessary to separate the science issues from the policy issues (e.g., separate risk assessors from risk managers), because it was feared that social pressures and political agendas would influence the outcome of the quantitative analysis. The objective of the separation was to allow the public to clearly understand the science surrounding a specific issue and the associated uncertainties. By maintaining the separation, it was hoped that members of the public would then be able to isolate and understand the other non-scientific factors that were considered by elected officials in reaching a particular decision.

Unfortunately, experience has shown that it has been extremely difficult for personal biases and the policy of various agencies (stated or unstated) to be isolated from the scientific evaluation of the available data. That is, a significant fraction of risk assessors have concluded that it simply is not possible to expect persons to be able to completely remove all biases from their analyses. Therefore, it is likely that there will be a continued effort in the U.S. to discuss within the four corners of a risk assessment, numerous other factors that have been considered by the risk analyst. For example, it has been argued that it is probably better to have both sides of any debate fully shared in a risk characterization (including some policy aspects) so that the public can appreciate the degree of certainty or uncertainty in the scientific basis for a decision, as well as the level of subjectivity in certain decisions by risk managers. The reason for this change in approach is that critics believe that many risk assessments, conducted either by the regulated community or regulatory agencies, have some degree of bias and subjectivity embedded in them, preventing the risk manager from obtaining a completely objective analysis based on the weight of scientific evidence. By incorporating items like cost-benefit or cost-effectiveness analysis, the precautionary principle, regulatory and judicial precedent, and other factors, this may help ensure that the public has access to all relevant information in both the risk assessment and risk management process.

REFERENCES

ACGIH (American Conference of Governmental Industrial Hygienists). 2000. Threshold Limit Values (TLVs) for Chemical Substances and Physical Agents and Biological Exposure Indices (BEIs). ACGIH, 1330 Kemper Meadow Drive, Cincinnati, OH 45240.

Andersen, M. E., Clewell, H. J. III, Gargas, M. L., Smith, F. A. and Reitz, R. H. 1987. Physiologically-based pharmacokinetics and risk assessment process for methylene chloride. Toxicol. Appl. Pharmacol. 87: 185–205.

Anderson, P. D., and Weiner, J. B. 1995. Eating Fish. Risk vs. Risk: Tradeoffs in Protecting Health and the Environment. J. D. Graham and J. B. Weiner (eds), pp. 104–123. Harvard University Press, Cambridge, MA.

Archibald, S. O., Winter, C. K. 1990. Pesticides in Our Food: Assessing the Risk. Chemicals in the Human Food Chain. C. K. Winter, J. N. Seiber, and C. F. Nuckton (eds). Van Nostrand Reinhold. New York, NY.

ATSDR (Agency for Toxic Substances and Disease Registry). 1997. Children's Health Initiative: Acting on the Unique Vulnerability of Children who Dwell near Hazardous Waste Sites. U.S. Dept of Health and Human Services. ATSDR. Atlanta, GA

ATSDR (Agency for Toxic Substances and Disease Registry). 1999. Toxicological Profile for Lead. U.S. Department of Health and Human Services, Public Health Service. July, 1999.

ATSDR (Agency for Toxic Substances and Disease Registry). 2000. Minimal Risk Levels (MRLs) for Hazardous Substances. Last updated on February 17, 2000. URL: *http://www.atsdr.cdc.gov/mrls.html.*

Barnes, D. G. and M. Dourson. 1988. Reference dose (RfD): Description and use in health risk assessments. Reg. Toxicol. Pharmacol., 8: 471–486.

Berry et al. 1993. Reducing Lead Exposure in Australia: Risk Assessment and Analysis of Economics, Social and Environmental Impacts. Final Report. Volumes 1 and 2. Austrialian National Government Publishing Service. Commonwealth of Australia.

Bolger, P. M., C. D. Carrington, and S. H. Henry. 1996. Risk Assessment for Risk Management and Regulatory Decision-Making at the U.S. Food and Drug Administration. Toxicology and Risk Assessment: Principles, Methods and Applications. A. M. Fan and L. W. Chang (eds), pp. 791–798. Marcel Dekker Inc. New York, NY.

Breyer, S. 1993. Breaking the Vicious Circle: Toward Effective Risk Regulation. Harvard University Press. Cambridge, MA.

Brorby, G. P., Spencer, A. L., and Graf, T. E. 1997. Risk-Based Corrective Actions Begin With Risk-Based Investigations – A Case Study. Proceedings of the Petroleum Hydrocarbons & Organic Chemicals in Groundwater Conference: Prevention, Detection, and Remediation November 12–14, 1997.

Bruckner, J. V. 2000. Differences in sensitivity to children and adults to chemical toxicity: the NAS panel report. Toxicol. Pharmacol., 31: 280–285.

Burmaster, D. E., and R. H. Harris. 1993. The magnitude of compounding conservatisms in Superfund risk assessments. Risk Anal., 13: 131–134.

Burmaster, D. E., and Anderson, P. D. 1994. Principles of good practice for the use of Monte Carlo techniques in human health and ecological risk assessment. Risk Anal., 14: 477–491.

Burmaster, D. E., and Thompson, K. M. 1997. Estimating exposure point concentrations for surface soils for use in deterministic and probabilistic risk assessments. Human Ecol. Risk Assess., 3: 363–384.

Calabrese, E. J. 2001. The frequency of U-shaped dose responses in the toxicological literature. Toxicol Sciences, 62: 330–338.

Calabrese, E. J. and L. A. Baldwin. 2001. Scientific foundations of hormesis. Critical Rev. Toxicol., 31: Special issues 4–5.

Clewall, H. J. 1995. The application of physiologically based pharmacokinetics modeling in human health risk assessment of hazardous substances. Toxicol. Lett., 79: 207–217.

Clewell, H. J., Gentry, P. R., Gearhart, J. M., Allen, B. C., and Andersen, M. E. 1995. Considering pharmaxokinetics and mechanistic information in cancer risk assessments for environmental contaminants: examples with vinyl chloride and trichloroethylene. Chemosphere, 31: 2561–2578.

Cogliano, V. J. 1997. Plausible upper bounds: are their sums plausible? Risk Anal., 17: 77–84.

Coogan, T. P., J. Motz, C. A. Snyder, K. S. Squibb, and M. Costa. 1991. Differential DNA-protein crosslinking in lymphocytes and liver following chronic drinking water exposure of rats to potassium chromate. Toxicol. Appl. Pharmacol. 109: 60–72.

Copeland, T. L., Paustenbach, D. J., Harris, M. A., and Otani, J. 1993. Comparing the results of a Monte Carlo analysis with EPA's reasonable maximum exposed individual (RMEI): a case study of a former wood treatment site. Reg. Toxicol. Pharmacol., 18: 275–312.

CRAM (Commission on Risk Assessment and Risk Management). 1997a. Framework for Environmental Risk Management. Presidential/Congressional Commission on Risk Assessment and Risk Management. Final Report: Volume 1. Washington, D.C.

CRAM (Commission on Risk Assessment and Risk Management). 1997b. Risk Assessment and Risk Management in Regulatory Decision-Making. Presidential/Congressional Commission on Risk Assessment and Risk Management. Final Report: Volume 2. Washington, D.C.

Crump, K. S. 1996. The linearized multistage model and the future of quantitative risk assessment. Hum. Exp. Toxicol., 15: 787–798.

Cullen, A. C., and H. C. Frey. 1999. Probabilistic techniques in exposure assessment. Risk Anal., 14: 389–393.

Davies, J. C., V. T. Covello, and F. W. Allen. 1987. Risk Communication. Proceedings of the National Conference on Risk Communication. January 29–31, 1986. The Conservation Foundation. Washington, DC.

Denneman, K. 1995. Guidelines for Soil Remediation. Netherlands.

Dietrich, D. R., and J. A. Swenberg. 1991. The presence of alpha-2u-globulin is necessary for d-Limonene promotion of male rat kidney tumors. Cancer Res., 51: 3512–3521.

Duan, N., and Mage, D. T. 1997. Combination of direct and indirect approaches for exposure assessment. J. Exp. Anal. Environ. Epidemiol., 7: 439–470.

De Flora, S. 2000. Threshold mechanisms and site specificity in chromium(VI) carcinogenesis. Carcinogenesis, 21: 533–541.

Department of the Environment. 1991. Policy Appraisal and the Environment: A Guide for Government Departments. Norwich, United Kingdom.

Dourson, M. L., S. P. Felter, and D. Robinson. 1996. Evolution of science-based UFs in noncancer risk assessment. Reg. Toxicol. Pharmacol., 24: 108–120.

Easterbrook, G. 1995. A Moment On the Earth: The Coming Age of Environmental Optimism. Viking Penguin. New York, NY.

Efron, E. 1984. The Apocalyptics: Cancer and the Big Lie. Simon and Schuster. New York, NY.

El Saadi, O. and Langley, A. 1991. The Health Assessment and Management of Contaminated Sites. South Australian Health Commission. Adelaide, Australia.

EPA (Environmental Protection Agency). 1983. Health Assessment Document for Dichloromethane (Methylene Chloride). Office of Health and Environmental Assessment, Washington, D.C. EPA-600/8-82-0048. December, 1983. External Review Draft.

EPA (Environmental Protection Agency). 1984. Approaches to Risk Assessment of Multiple Chemical Exposures. EPA 600/9-84-008. Washington, DC.

EPA (Environmental Protection Agency). 1985. Methodology for Characterization of Uncertainty in Exposure Assessments. Prepared by Research Triangle Institute. NTIS: PB85-240455.

EPA (Environmental Protection Agency). 1986a. Guidelines for Carcinogen Risk Assessment. U.S. Environmental Protection Agency, 51 Federal Register 33992, September 24, 1986.

EPA (Environmental Protection Agency). 1986b. Guidelines for Health Risk Assessment of Chemical Mixtures. U.S. Environmental Protection Agency, 51 Federal Register 34014, September 24, 1986.

EPA (Environmental Protection Agency). 1986c. Guidelines for the Health Assessment of Suspect Developmental Toxicants. U.S. Environmental Protection Agency, 51 Federal Register 34028, September 24, 1986.

EPA (Environmental Protection Agency). 1989a. Risk Assessment Guidance for Superfund (RAGS). Volume I: Human health evaluation manual (HHEM), Part A, Interim Final, Chapter 8: Risk Characterization. U.S. Environmental Protection Agency, Office of Emergency and Remedial Response, Washington, DC. EPA/540/1-89/002.

EPA (Environmental Protection Agency). 1989b. Proposed Amendments to the Guidelines for the Health Assessment of Suspect Developmental Toxicants. U.S. Environmental Protection Agency, 54 Federal Register 9386, March 6, 1989.

EPA (Environmental Protection Agency). 1992. Guidance on Risk Characterization for Risk Managers and Risk Assessors. United States Environmental Protection Agency, Washington, D.C.

EPA (Environmental Protection Agency). 1995a. Policy for Risk Characterization, United States Environmental Protection Agency, Science Policy Council, Washington, D.C. URL: *http://www.epa.gov/ordntrn/ORD/spc/rcpolicy.htm.*

EPA (Environmental Protection Agency). 1995b. Elements to Consider When Drafting EPA Risk Characterizations, United States Environmental Protection Agency, Science Policy Council, Washington, D.C.. *http://www.epa.gov/ordntrn/ORD/spc/rcelemen.htm.*

EPA (Environmental Protection Agency). 1995c. Guidance for Risk Characterization, United States Environmental Protection Agency, Science Policy Council, Washington, D.C. URL: *http://www.epa.gov/ordntrn/ORD/spc/rcguide.htm.*

EPA (Environmental Protection Agency). 1996. Proposed Guidelines for Carcinogen Risk Assessment. Office of Research and Development. EPA/600/P-92/003C. Washington, DC.

EPA (Environmental Protection Agency). 1997a. Exposure Factors Handbook. U.S. Environmental Protection Agency, Office of Health and Environmental Assessment, Washington, DC.

EPA (Environmental Protection Agency). 1997b. Guiding Principles for Monte Carlo Analysis. U.S. Environmental Protection Agency, Office of Research and Development. EPA/630/R-97/001. Washington, DC.

EPA (Environmental Protection Agency). 1999a. Guidelines for Carcinogen Risk Assessment (SAB review copy, July 1999). Washington, DC. *http://www.epa.gov/ncea/raf/crasab.htm.*

EPA (Environmental Protection Agency). 1999b. EPA Glossary of IRIS Terms. Integrated Risk Information System. Washington, DC. URL: *http://www.epa.gov/ngispgm3/iris/gloss8.htm.*

EPA (Environmental Protection Agency). 1999c. Guidance For Performing Aggregate Exposure and Risk Assessments. Office of Pesticide Programs, Draft. February 1, 1999. Washington, DC.

EPA (Environmental Protection Agency). 1999d. Guidance For Identifying Pesticide Chemicals and Other Substances That Have A Common Mechanism of Toxicity. January 29, 1999. Washington, DC.

EPA (Environmental Protection Agency). 2000a. Science Policy Council Handbook: Risk Characterization: Office of Science Policy, Office of Research and Development. EPA/100/B/00/002. Washington, DC.

EPA (Environmental Protection Agency). 2000b. Exposure and Human Health Reassessment of 2,3,7,8-Tetrachlorodibenzo-p-dioxin (TCDD) and Related Compounds. Part III. Integrated Summary and Risk Characterization for 2,3,7,8-TCDD and Related Compounds. National Center for Environmental Assessment. EPA/600/P-00-001. SAB Review Draft of September. Washington, DC.

EPA (Environmental Protection Agency). 2000c. Available Information on Assessing Exposure from Pesticides in Food: A User's Guide. Office of Pesticide Programs, June 21, 2000. Washington, DC.

EPA (Environmental Protection Agency). 2000d. Drinking Water Health Advisories. Office of Water. Washington, DC. URL: *http://www.epa.gov/ost/drinking/*

EPA (Environmental Protection Agency). 2000e. Consumption Advisories. Office of Water. Washington, DC. URL: *http://www.epa.gov/ost/fish/*

EPA (Environmental Protection Agency). 2001a. An SAB Report: Review of the Office of Research and Development's Reassessment of Dioxin. USEPA Science Advisory Board (SAB). Mort Lippman, Chairman. EPA-SAB-EC—01-006. *www.epa.gov/sab*. Washington, DC.

EPA (Environmental Protection Agency). 2001b. Improved Science-Based Environmental Stakeholder Processes: A Commentary by the EPA Science Advisory Board. EPA-SAB-EC-COM-01-006 Washington, D.C. *http://www.epa.gov/sab/eccm01006.pdf*

EEA (European Environment Agency). 2001. Late Lessons From Early Warnings: The Precautionary Principle 1896–2001 Environmental Issue Report No. 22, Denmark.

Evans, J. S., Gray, G. M., Sielken, R. L., Smith, A. S., Valdez-Flores, C., and Graham, J. D. 1994. Use of probabilistic expert judgment in uncertainty analysis of carcinogenic potency. Reg. Toxicol. Pharmacol., 20: 15–36.

Fayerweather, W. E., Collins, J. J., Schnatter, A. R., Hearne, F. T., Menning, R. A., and Reyner, D. P. 1999. Quantifying uncertainty in a risk assessment using human data. Risk Anal., 19: 1077–1090.

Finkle, A. M. and D. Golding. 1994. Worst Things First. Resources for the Future Publishing. Washington, D.C.

Finley, B. L., and Paustenbach, D. J. 1994. The benefits of probabilistic exposure assessment: three case studies involving contaminated air, water, and soil. Risk Anal., 14: 53–73.

Finley, B., Proctor, D., Scott, P., Harrington, N., Paustenbach, D., and Price, P. 1994. Recommended distributions for exposure factors frequently used in health risk assessment. Risk Anal., 14: 533–553.

Foster, K. R., Bernstein, D. E., and P. W. Huber. 1993. Phantom Risk: Scientific Inference and the Law. MIT Law Press. Cambridge, MA.

Freeze, R. A. 2000. The Environmental Pendulum. University of California Press. Berkeley, CA.

Frey, H. C. 1992. Quantitative Analysis of Uncertainty and Variability in Environmental Policy Making. AAAS/EPA Environmental Science and Engineering Fellow and Research Associated, Center for Energy and Environmental Studies, Carnegie Mellon University, September 1992.

Frey, H. C., and Rhodes, D. S. 1998. Characterization and simulation of uncertainty frequency distributions: effects of distribution choice, variability, uncertainty, and parameter dependence. Human Ecol. Risk Assess., 4: 423–469.

Fryzek, J. P., M. T. Mumma, J. K. McLaughlin, B. E. Henderson, and W. J. Blot. 2001. Cancer mortality in relation to environmental chromium exposure. J. Occ. Env. Med., 43: 635–640.

Gargas, M. L., Finley, B. L., Paustenbach, D. J., and T. F. Long. 1999. Environmental Health Risk Assessment: Theory and Practice. General and Applied Toxicology, Volume 3. B. Ballantyne, T. Marrs, and T. Syversen (eds), 2nd edition, p. 1749–1809. London: Macmillan.

Gaylor, D. W., and Chen, J. J. 1996. A simple upper limit for the sum of the risks of the components in a mixture. Risk Anal., 16: 395–398.

Ginjaar, L. 1996. Risk is More Than a Number: Reflections on the Development of the Environmental Risk Management Approach. Health Council of the Netherlands. Committee on Risk Measures and Risk Assessment. The Ministry of Health, Welfare and Sports. Netherlands.

Gold, M. R., Siegel, J. E., Russsell, L. B., Weinstein, M. C. 1996. Cost-Effectiveness in Health and Medicine. Oxford University Press. New York, NY.

Gold, L. S., Stern, B. R., Slone, T. H., Brown, J. P., Manley, N. B., Ames, B. N. 1997. Pesticide residues in food: investigation of disparities in cancer risk estimates. Cancer Lett., 117: 195–207.

Goldstein, B. D. 1989. The Maximally exposed individual. Environ Forum, Nov–Dec. pp. 13–14.

Graham, J. D. and Hartwell, J. K. 1997. The Greening of Industry: A Risk Management Approach. Harvard University Press. Cambridge, MA.

Graham, J. D. 1995. Historical perspective on risk assessment in the federal government. Toxicology, 102: 29–52.

Graham, J. D., and Weiner, J. B. 1995. Confronting Risk Tradeoffs. Risk vs. Risk: Tradeoffs in Protecting Health and the Environment. J. D. Graham and J. B. Weiner (eds), pp. 1–41. Harvard University Press, Cambridge, MA.

Graham, J. D., Green, L., and Roberts, M. J. 1988. In Search of Safety: Chemicals and Cancer Risks. Harvard University Press, Cambridge, MA.

Green, T., Lee, R., Moore, R. B., Ashby, J., Willis, G. A., Lund, V. J., and Clapp, M. J. L. 2000. Acetochlor-induced rat nasal tumors: further studies on the mode of action and relevance to humans. Reg. Toxicol. Pharmacol., 32: 127–135.

Hattis, D., Banati, P., Goble, R., and Burmaster, D. E. 1999. Human interindividual variability in parameters related to health risks. Risk Anal., 19: 711–724.

HEI (Health Effects Institute). 1999. Reproductive and Offspring Developmental Effects Following Maternal Inhalation Exposure to Methanol in Nonhuman Primates. Research Report Number 89. Cambridge, MA.

Holcomb, D. L., Smith, M. A., Ware, G. O., Hung, Y. C., Brackett, R. E., and Doyle,

M. P. 1999. Comparison of six dose–response models for use with food-borne pathogens. Risk Anal., 19: 1091–1100.

Holland, C. D. and R. L. Sielken. 1993. Quantitative cancer modeling and risk assessment. PTR Prentice Hall. Englewood Cliffs, NJ.

Hwang, J. S., and Chen, J. L. 1999. An evaluation of risk estimation procedures for mixtures of carcinogens. Risk Anal., 19: 1071–1076.

IRIS (Integrated Risk Information System). 1991. 1,3-Butadiene. U.S. Environmental Protection Agency. CASRN 106-99-0. Last Revised: February 1, 1991.

IRIS (Integrated Risk Information System). 1993. Methyl tert-butyl ether (MTBE). U.S. Environmental Protection Agency. CASRN 1634-04-4. Last Revised: March 1, 1993.

IRIS (Integrated Risk Information System). 2000. Benzene. U.S. Environmental Protection Agency. CASRN 71-43-2. Last Revised: January 19, 2000.

Kang, S. H., Kodell, R. L., and Chen, J. L. 2000. Incorporating model uncertainties along with data uncertainties in microbial risk assessment. Reg. Toxicol. Pharmacol., 32: 68–72.

Kirman, C. R., Hays, S. M., Kedderis, G. L., Gargas, M. L., and Strother, D. E. 2000. Improving cancer dose–response characterization by using physiologically based pharmacokinetic modeling: an analysis of pooled data for acrylonitrile-induced brain tumors to assess cancer potency in the rat. Risk Anal., 20: 135–151.

Klaassen, C. D. 1996. Casarett and Doull's Toxicology; The Basic Sciences of Poisons. C. D. Klaassen (ed). McGraw-Hill. New York, NY

Klaassen, C. D. 2001. Casarett and Doull's Toxicology; The Basic Sciences of Poisons. C. D. Klaassen (ed). McGraw-Hill. New York, NY

Kodell, R. L., and Chen, J. J. 1994. Reducing conservatism in risk estimation for mixtures of carcinogens. Risk Anal., 14: 327–332.

Kolluru, R. V. 1996. Risk Assessment and Management: A Unified Approach. Risk Assessment and Management Handbook. R. V. Kolluru, S. M. Bartell, R. M. Pitblado, and R. S. Stricoff (eds), pp. 1.3–1.41. McGraw Hill Inc. New York, NY.

Kuykendall, J. R., B. D. Kerger, E. J. Jarvi, G. E. Corbett, and D. J. Paustenbach. 1996. Measurement of DNA-protein cross-links in human leukocytes following acute ingestion of chromium in drinking water. Carcinogenesis, 17: 1971–1977.

Layard, R., and S. Glaister. 1994. Cost-Benefit Analysis. Cambridge University Press. New York, NY.

Last, G. 2001. Report of the California Expert Panel on Evaluating the Oral Carcinogenicity of Hexavalent Chromium in Drinking Water. Sacramento, CA.

Leung, H. W. 1991. Development and utilization of physiologically based pharmacokinetic models for toxicological applications. J. Toxicol. Environ. Health, 32: 247–267.

Leung, H. W. and Paustenbach, D. J. 1995. Physiologically based pharmacokinetic and pharmacodynamic modeling in health risk assessment and characterization of hazardous substances. Toxicol. Lett., 78: 55–65.

Leung, H. W. (2000). Physiologically based pharmacokinetic modeling. General and Applied Toxicology. B. Ballantyne, T. Marrs, and T. Syversen (eds), pp. 141–154. Volume I. Macmillan Publishing. London, United Kingdom.

Lomberg. 2001. The Skeptical Environmentalist: Measuring the Real State of the World. Cambridge University Press. Cambridge, MA.

Mancuso, T. F. 1997. Chromium as an industrial cracinogen. Part I. Amer. J. Ind. Med. 31: 129–139.

Mattie, D. R., Bates, G. D., Jepson, G. W., Fisher, J. W., and McDougal, J. N. 1994. Determination of skin: air partition coefficients for volatile chemicals: experimental method and applications. Fundam. Appl. Toxicol., 22: 51–57.

Mayo, D. G. and R. D. Hollander. 1991. Acceptable Evidence: Science and Values in Risk Management. Oxford University Press. Oxford, England.

McCarthy, I. F. 1991. Average Residues vs. Tolerances: An Overview of Industry Studies. Pesticide Residues and Food Safety: A Harvest of Viewpoints. Tweedy, Dishburger, Ballantine, McCarthy, and Murphy (eds), pp. 182–191. American Chemical Society. Washington, D.C.

McDougal, J. N. 1996. Physiologically based pharmacokinetic modeling. Dermatoxicology. F. N. Marzulli, and H. I. Maibach (eds), pp. 37–60. Taylor and Francis, London.

McDougal, J. N., Jepson, G. W., Clewell, H. J., Gargas, M. L., and Andersen, M. E. 1990. Dermal absorption of organic chemical vapors in rats and humans. Fundam. Appl. Toxicol., 14: 299–308.

Milloy, S. J. (2001). Junk Science Judo. Cato Institute Press. Washinton, DC.

Ministry of Housing. 1994. Environmental Quality Objectives in the Netherlands. Spatial Planning and Environment. Hague, Netherlands.

Mirsalis, J. C., C. M. Hamilton, K. G. O'Loughlin, D. J. Paustenbach, B. D. Kerger, and S. Patierno. 1996. Chromium (VI) at plausible drinking water concentrations is not genotoxic in the in vivo bone marrow micronucleus or liver unscheduled DNA synthesis assays. Environ. Molec. Mut., 28: 60–63.

Morello-Frosch, R. A., Woodruff, T. J., Axelrad, D. A., and Caldwell, J. C. 2000. Air toxics and health risks in California: the public health implications of outdoor concentrations. Risk Anal., 20: 273–291.

Morey, B. 1999. Public Health Goal for Hexavalent Chromium (Cr VI). Office of Environmental Health Hazard Assessment. Oakland, CA.

Morgan, M. G. and M. Henrion. 1990. Uncertainty: A Guide to Dealing With Uncertainty in Quantitative Risk and Policy Analysis. Cambridge University Press. Cambridge, England.

Neumann, D. A. and C. A. Kimmel. 1998. Human Variability in Response to Chemical Exposure. Inter Life Sciences Institute. Washington, DC.

Nichols, A. L. and R. J. Zeckhauser. 1984. The perils of prudence: how conventional risk assessments distort regulations. Regul Toxicol Pharmacol., 8: 61–71.

NRC (National Research Council). 1996. Understanding Risk: Informing Decisions in a Democratic Society. National Academy Press. Washington, D.C.

NRC (National Research Council). 1996. Linking Science and Technology to Society's Environmental Goals. National Academy Press. Washington, D.C.

NRC (National Research Council). 1994. Science and Judgment in Risk Assessment, National Academy Press. Washington, D.C.

National Research Council. 1989. Improving risk communication. National Academy Press. Washington, D.C.

NRC (National Research Council). 1987. Regulating Pesticides in Food: The Delaney Paradox, National Academy Press. Washington, D.C.

NRC (National Research Council). 1983. Risk Assessment in the Federal Government: Managing the Process, National Academy Press. Washington, D.C.

OEHHA (Office of Environmental Health Hazard Assessment). 1999. Public Health Goal for Methyl Tertiary Butyl Ether (MTBE) in Drinking Water. Office of Environmental Health Hazard Assessment, California Environmental Protection Agency. Sacramento, CA.

OMB (Office of Management and Budget). 2001. Guidelines for Ensuring and Maximizing the Quality, Objectivity, Utility, and Integrity of Information Disseminated by Federal Agencies. Office of Management and Budget, Executive Office of the President. October 1, 2001.

Olsen, J. R. 1994. Pharmacokinetics of Dioxins and Related Chemicals. Dioxins and Health. A. Schecter (ed). Plenum Press. New York, NY.

OSHA (Occupational Safety and Health Administration). 1999. Permissible Exposure Limits. U.S. Department of Labor. URL: *http://www.osha-slc.gov/SLTC/pel/ index.html.*

Osterloh, J. 1995. Study of Neurological Effects of Low Level Methanol in Normal Subjects and Subjects with Susceptibility to Folate Deficiency. Center for Occupational and Environmental Health, University of California, p. 1–64. San Francisco, CA.

Paustenbach, D. J. 1988. Assessment of the developmental risks resulting from occupational exposure to select glycol ethers within the semi-conductor industry. J. Toxicol. Environ. Health, 23: 53–96.

Paustenbach, D. J. 1989. Health risk assessments: opportunities and pitfalls. Columbia J. Environ. Law, 41: 379–410.

Paustenbach, D. J., Meyer, D. M., Sheehan, P. J., Lau, V. 1991. An assessment and quantitative uncertainty analysis of the health risks to workers exposed to chromium contaminated soils. Toxicol. Ind. Health, 7: 159–196.

Paustenbach, D. J. 1995. The practice of health risk assessment in the United States (1975–1995): how the U.S. and other countries can benefit from that experience. Human Ecol. Risk Assess., 1: 29–79.

Paustenbach, D. J. 2000. The practice of exposure assessment: a state-of-the-art review. J. Toxicol. Environ. Health (Part B), 3: 179–291.

Peterson, B. J. 2000. Pesticide residues in food: problems and data needs. Reg. Toxicol. Pharmacol., 31: 297–299.

Ponce, R., Bartell, S. M., Wong, E. Y., LaFlamme, D., Carrington, C., Lee, R. C., Patrick, D. L., Faustman, E. M., and M. Bolger. 2000. Use of quality-adjusted life year weights with dose–response models for public health decisions: a case study of the risks and benefits of fish consumption. Risk Anal., 20: 529–542.

Price, P. S., Curry, C. L., Goodrum, P. E., Gray, M. N., McCrodden, J. I., Harrington, N. W., Carlson-Lynch, H., and Keenan, R. E. 1996. Monte Carlo modeling of time-dependent exposures using a microexposure event approach. Risk Anal., 16: 339–348.

Proctor, D. M., J. M. Otani, B. L. Finley, D. J. Paustenbach, and E. V. Sargent. 2001a . Is hexavalent chromium carcinogenic via ingestion. A state of the science review. J. Toxicol Env Health. (In Press).

Proctor, D. M., Shey, E. C., Fehling, K. A., and Finley, B. L. 2001b. Assessment of human health and ecological risks posed by the environmental uses of steel industry slags. Human Ecol. Risk Assess. (In Press).

Putzrath, R. M. 2000. Reducing uncertainty of risk estimates for mixtures of chemicals within regulatory constraints. Reg. Toxicol. Pharmacol., 31: 44–52.

Rachel's Hazardous Waste News. 1993. The Precautionary Principal. Rachel's Hazardous Waste News.

Reitz, R. H., Garas, M. L., Andersen, M. E., Provan, W. M., and Green, T. L. 1996. Predicting cancer risk from vinyl chloride exposure with a physiologically based pharmacokinetic model. Toxicol. Appl. Pharmacol., 137: 253–267.

Renwick, A. G. and N. R. Lazarus. 1998. Human variability and non-cancer risk assessment: an analysis of default uncertainty factors. Reg. Toxicol. Pharmacol., 27: 3–120.

Renwick, A. G., Dorne, J. L., Walton, K. 2000. An analysis of the need for an additional uncertainty factor for infants and children. Reg. Toxicol. Pharmacol., 31: 286–296.

Rodricks, J. and M. R. Taylor. 1983. Application of risk assessment to food safety decision making. Reg. Toxicol. Pharmacol., 3: 275–307.

Ross, J. H., Dong, M. H., Krieger, R. I. 2000. Conservatism in pesticide exposure assessment. Reg. Toxicol. Pharmacol., 31: 53–58.

Rossman, M. D. and W. Jones-Williams. 1991. Immunopathogenesis of Chronic Beryllium Disease. Beryllium: Biomedical and Environmental Aspects. M. D. Rossman, O. P. Preuss, and M. B. Powers (eds). Williams and Wilkins. Baltimore, MD.

Ruckelshaus, W. D. 1983. Science, risk and public policy. Science, 221: 1026–1028.

Seed, J., R. P. Brown, S. S. Olin, and J. A. Foran. 1995. Chemical mixtures: current risk assessment methodologies and future directions. Reg. Toxicol. Pharmacol., 22: 76–94.

Sheehan, P. J., Meyer, D. M., Sauer, M. M., and Paustenbach D. J. 1991. Assessment of the human health risks posed by exposure to chromium-contaminated soils. J. Toxicol. Environ. Health, 32: 161–201.

Short, B. G., Steinhagen, W. G., and Swenberg, J. A. 1989. Promoting effects of unleaded gasoline and 2,2,4-trimethylpentane on the development of atypical cell foci and renal tubular cell tumors in rats exposed to N-ethyl-N-hydroxyethylnitrosamine. Cancer Res., 49: 6369–6378.

Sielken, R. L. 1985. Some issues in the quantitative modeling portion of cancer risk assessment. Reg. Toxicol. Pharmacol., 5: 175–181.

Sielken, R. L. 1987. Quantitative cancer risk assessments in 2,3,7,8 TCDD. Food Chem. Toxicol., 25: 257–267.

Sielken, R. L., Bretzlaff, R. S., and Stevenson, D. E. 1995. Challenges to default assumptions stimulate comprehensive realism as a new tier in quantitative cancer risk assessment. Regul. Toxicol. Pharmacol., 21: 270–280.

Sielken, R. L., Bretzlaff, R. S., Valdez-Flores, C., Stevenson, D. E., and Jong, G. 1999. Cancer dose–response modeling of epidemiological data on worker exposures to aldrin and dieldrin. Risk Anal., 19: 1101–1111.

Starr, T. B. 2001. Significant shortcomings of the U.S. Environmental Protection Agency's latest draft risk characterization for dioxin-like compounds. Toxicol. Sci., 64: 7–14.

Starr, T. B., and J. E. Gibson. 1985. The mechanistic toxicology of formaldehyde and its implications for quantitative risk estimation. Annu. Rev. Pharmacol. Toxicol., 25: 745–767.

Stokey, E., and R. Zeckhauser. 1978. A Primer for Policy Analysis. W. W. Norton. New York, NY.

Tengs, T. O., Adams, M. E., Pliskin, J. S., Safran, D. G., Siegel, J. E., Weinstein, M. C., and Graham, J. D. 1995. Five-hundred life-saving interventions and their cost-effectiveness. Risk Anal., 13: 369–390.

TERA (1999). Comparative Dietary Risks: Balancing the Risks and Benefits of Fish Consumption. *http://www.tera.org/news/project%20descriptions/diet%20jp.htm.*

Thompson, K. M., Burmaster, D. E., and Crouch, E. A. C. 1992. Monte Carlo techniques for quantitative uncertainty analysis in public health assessments. Risk Anal., 12: 53–63.

Thompson, K. M., and Graham, J. D. 1996. Going beyond the single number: using probabilistic risk assessment to improve risk management. Human Ecol. Risk Assess., 2: 1008–1034.

Tietenberg, T. 1996. Environmental Economics and Policy. Harper Collins. New York, NY.

Travis, C. C., Richter, S. A., Crouch, E. A. C., Wilson, R., and E. D. Klema. 1987. Cancer risk management: a review of 132 federal regulatory decisions. Environ. Sci. Technol., 21: 415–420.

Treuschler, L., Goodman, J., Bus, J., Yang, R., Paustenbach, D., et al. 2001. A report on the importance of mixtures following low level exposure to toxicants. Reg. Toxicol. Pharmacol. (In press).

Walker, K. 1992. Australian Environmental Policy: Ten Case Studies. New South Wales University Press. Kensington, Australia.

Weiner, J. B. 2002. Precaution in a Multi Risk World. Human and Ecological Risk Assessment: Theory and Practice. D. J. Paustenbach (ed). John Wiley and Sons. New York, NY.

Weinstein, M. C., Fineberg, H. V. 1980. Clinical Decision Analysis. W. B. Saunders Company. Philadelphia.

Wildavsky, A. (1995). But Is It True: A Citizen's Guide to Environmental Health and Safety Issues. Harvard University Press. Cambridge, MA.

Wilkinson, C. F., Christoph, G. R., Julien, E., Kelley, J. M., Kronenberg, J., McCarthy, J., and Reiss, R. 2000. Assessing the risks of exposures to multiple chemicals with a common mechanism of toxicity: how to cumulate? Reg. Toxicol. Pharmacol., 31: 30–43.

Williams, P. R. D., Scott, P. K., Sheehan, P. J., and Paustenbach, D. J. 2000. A probabilistic assessment of household exposures to MTBE from drinking water. Human Ecol. Risk Assess., 6: 827–849.

Winter, C. K. 1992. Dietary Pesticide Risk Assessment. Rev. Environ. Contam. Toxicol., 127: 23–67.

WHO (World Health Organization). 1997. Environmental Health Criteria 196; Methanol. Geneva, Switzerland.

Yang, R. S. H. 1994. Toxicology of Chemical Mixtures. Academic Press, NY.

Zhang, J. and S. Li. (1997). Cancer mortality in a Chinese population exposed to hexavalent chromium in water. J. Occup. Environ. Med., 39: 315–319.

SUPPLEMENTARY REFERENCES

Bardos, R. P., E. Damingos, R. Goubier, J. Holst, M. Iterbeke, H. Van Ommen, Y. K. Vlahoyannis, and H. von Deylen. 1994. Survey of EU Member States: Contaminated Land: Definitions, Registers and Priorities of Action. Volume1. AEA Technology. National Environmental Technology Center. Culham, Abingdon, Oxfordshire. OX 14 3DB. United Kingdom.

Bernstein, P. L. (1996). Against the Gods: The Remarkable Story of Risk. John Wiley and Sons. New York, NY.

Chechile, R. A. and S. Carlisle. 1991. Environmental Decision Making: A Multidisciplinary Perspective. Tufts University Center for the study of Decision Making. Van Nostrand. New York, NY.

Department of the Environment. 1995. The Technical Aspects of Controlled Waste Management: Health Effects from Hazardous Waste Landfill Sites. Report CWM/ 057/92. Romney House. London, United Kingdom.

EPA (Environmental Protection Agency). 1986. Explaining Environmental Risk: Some notes on Environmental Risk Communication. P. Sandman (ed). Office of Toxic Substances. Washington, DC 20460.

European Commission of the Health and Consumer Protection Directorate-General. Rue de la Loi 200, B-1049 Bruxelles/Wetstraat 200, B-1049 Brussel. Belgium.

Fumento, M. 1993. Science under siege: Balancing Technology and the Environment. William Morrow and Co. New York, NY.

Graham, J. D. 1995. The Role of Epidemiology in Regulatory Risk Assessment. Elsevier Press. Amsterdam.

Imperato, P. J. and G. Mitchell. 1985. Acceptable Risks: The Daily Risks You and the Government Take With Your Life. Viking Press. New York, NY.

ILSI (International Life Sciences Institute). 1995. Low-dose Extrapolation of Cancer Risks: Issues and Perspectives. S. Olin, W. Farland, C. Park, L. Rhomberg, R. Scheuplein, T. Starr, and J. Wilson (eds). Washington, DC.

JECFA (Joint FAO/WHO Expert Committee on Food Additives). 2001. Opinion of the Scientific Committee on Food on the Risk Assessment of Dioxins and Dioxin-like PCBs in Food. Adopted May 30, 2001.

Moolgavkar, S., D. Krewski, L. Ziese, E. Cardis, and H. Moller. 1999. Quantitative Estimation and Prediction of Human Cancer Risk. IARC Scientific Publication No. 131. Inter Agency for Research on Cancer (IARC). Oxford University Press. Oxford, England.

NRC (National Research Council). 1997. Building a Foundation for Sound Environmental Decisions. National Academy Press. Washington, D.C.

Olin, S. 1998. Aggregage Exposure Assessment. International Life Sciences Institute (ILSI) Washington, DC.

Paustenbach, D. J. 2002. Human and Ecological Risk Assessment: Theory and Practice. John Wiley and Sons. New York, NY.

Powell, M. R. 1999. Science at EPA. Resources for the Future. Washington, D.C.

Rilling, R., H. Sptizer, O. Greene, F. Hucho, and G. Pati. 1992. Challenges: Science and Peace in a Rapidly Changing Environment. Volumes I and II. Symposia held at Technische Universatat in Berlin on Nov. 29–Jan. 12, 1991.

Smith, F. L. 2000. Moving Beyond the Precautionary Principle: The Case for Unbiased Risk Management. Presented to the International Consumers for a Civil Society. Sept. 27, 2000. Text available from the Competitive Enterprise Institute.

Thomas, S. P. and Hrudey, S. E. 1997. Risk in Canada: What We Know and How We Know It. University of Alberta Press. Edmonton, Alberta.

Wernick, I. K. (1995). Community Risk Profiles. The Rockefeller University. New York, NY.

Wildavsky, A. (1988). Searching for Safety. Transaction Publishers. New Brunswick, NJ.

van der Weiden, M. E. J., J. H. M. de Bruijn, and C. J. Van Leeuwen. 1994. Environmental Quality Objectives in the Netherlands: A Review of Environmental Quality Objectives and Their Policy Framework in the Netherlands. M. E. J. van der Weiden (ed). Risk Assessment and Environmental Quality Division. The Hague, Netherlands.

Viscussi, W. Kip. 1992. Fatal Tradeoffs: Public and Private Responsibilities for Risk. Oxford University Press. Oxford, England.

SECTION B
Theory: Ecological Risk Assessment

6 Ecological Risk Assessment: History and Fundamentals

ANNE SERGEANT

U.S. Environmental Protection Agency, Washington, DC

6.1 INTRODUCTION

Ecological risk assessment (ERA) is a process that evaluates the likelihood that adverse ecological effects may occur as a result of exposure to a stressor [Environmental Protection Agency (EPA), 1998]. It is a tool for evaluating information, assumptions, and uncertainties in order to understand the relationships between stressors and ecological effects; its ultimate goal is to inform environmental decision making. ERA asks the question "Is this chemical (or thing, organism, or activity) bad for the ecosystem?"

Let us examine this definition: *Adverse ecological effects* are undesirable changes in an organism or ecosystem component. Examples include species extinction, plant or animal death, and habitat destruction. Of course, the definition of *adverse* depends on the circumstances; one person's crisis may be another's "business as usual." Note that although the principles of ERA can be used to consider beneficial effects, this application is not addressed here. A *stressor* is a chemical, physical, or biological agent (e.g., benzene, excess sediment, or an invasive species) that could cause adverse effects; a risk assessment may consider any or all stressor types. And there can be no risk unless there is *exposure*—the organism or ecosystem component (also known as *receptor*) must contact or otherwise co-occur with the stressor for a link to be made between the stressor and any effects. Finally, the term *likelihood* tells us that we may not be 100 percent sure an effect will occur; a risk assessment examines the strength of the link between stressors and effects. It is not always possible to make a statistical prediction, but risk descriptions do include some statement about how sure we are that the effect will occur or about the connection between an observed effect and a particular stressor. Remember that in ERA (as opposed to, say, the insurance

This chapter is based on information provided in *Guidelines for Ecological Risk Assessment* (published May 14, 1998, Federal Register 63(93): 26846–26924).

Human and Ecological Risk Assessment: Theory and Practice, Edited by Dennis J. Paustenbach
ISBN 0-471-14747-8 © 2002 John Wiley & Sons, Inc.

industry or failure analysis) a risk is something that *could* happen; note also that an observed effect has already happened and therefore has a risk of one.

ERAs have three primary uses: (1) to predict future adverse effects (prospective, e.g., predicting the effects of emissions from a new facility), (2) to evaluate a resource and what might affect it (perhaps for targeting protection efforts), and (3) to determine what might have caused an observed effect (retrospective, e.g., deciding what cleanup action is needed). An individual assessment may combine approaches. For example, a retrospective ERA designed to evaluate why an amphibian population has declined may also be used to choose or prioritize future management actions. Risk assessors may combine retrospective and prospective approaches where ecosystems have a history of previous impacts and there is concern for potential future effects.

6.1.1   Ecological Risk Assessment Process

ERA is built around the characterization of effects and characterization of exposure. These provide the focus for problem formulation, analysis, and risk characterization. The overall process is shown in Figure 6.1. Ecological risk assessment proper is enclosed by the dark solid line. Activities outside this line influence why and how a risk assessment is conducted and how it will be used.

Problem formulation appears at the top. Here, risk managers and risk assessors work together to decide on the assessment's purpose, define the problem, and develop a plan for analyzing and characterizing risk. The basic groundwork for risk communication efforts also begins here. Initial work in problem formulation includes reviewing available information on sources, stressors, effects, and ecosystem and receptor characteristics and then using it to develop assessment endpoints and conceptual models. Either product may be generated first (the order depends on the type of ERA), but both are needed to complete an analysis plan, the final product of problem formulation.

Problem formulation structures *analysis*, shown in the middle box of Figure 6.1. Here data are evaluated to determine how exposure to stressors might occur (exposure characterization) and, given this exposure, what effects can be expected (effects characterization). The first step is to determine the strengths and limitations of data on exposure, effects, and ecosystem and receptor characteristics. The next step is to evaluate potential or actual exposure and expected responses and then generate profiles for exposure and for stressor response.

These profiles are used in *risk characterization*, shown in the third box of Figure 6.1. This final step integrates the exposure and stressor–response profiles. It includes a summary of assumptions, uncertainties, and strengths and limitations of the analyses. Its product is a risk description that places adverse effects in context and describes uncertainty and supporting information.

Although Figure 6.1 shows a sequential process of problem formulation, analysis, and risk characterization, ERAs are often nonlinear, usually iterative, and may involve a lot of interaction between phases. Something learned during analysis or risk characterization can prompt the risk assessor to reevaluate

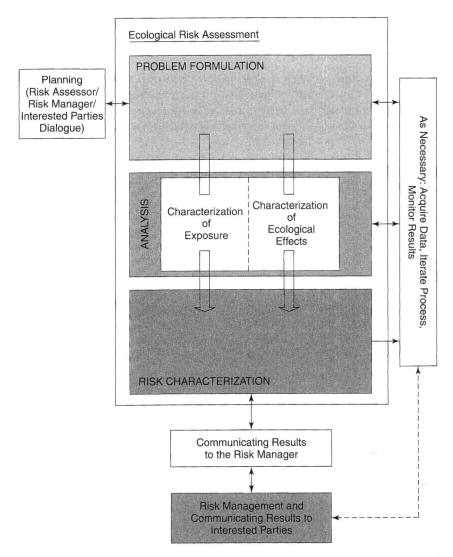

Figure 6.1 The framework for ecological risk assessment (modified from U.S. EPA, 1992a and presented in EPA, 1998).

problem formulation, collect new data, or choose another analytical technique (Table 6.1).

Interactions between risk assessors, risk managers, and, if needed, interested and affected parties (I&APs) are shown in two places. The planning box on the upper left of Figure 6.1 shows where these groups discuss management goals, the ERA's purpose, and the resources available to conduct the work. Note that I&APs are not always part of planning: There is no time for discussion before

TABLE 6.1   Flexibility of the Framework Diagram

The framework process (Figure 6.1) is a general outline for many types of ERAs. It is a flexible process, as illustrated by the examples below:

• Problem formulation usually starts by considering endpoints, stressors, or ecological effects. Problem formulation is generally interactive and iterative.
• In analysis, characterizations of exposure and effects are often intertwined, as when an initial exposure leads to secondary exposures and effects. Analysis strives to illuminate these complex relationships.
• Analysis and risk characterization are shown as separate phases. But some models may combine analysis with data integration during risk characterization.

---

proceeding with an emergency action, and enforcement programs specify what actions must be taken. The box below risk characterization shows where risk assessors formally communicate the results to risk managers. Risk managers or risk communication specialists usually present ERA results to I&APs. These activities are shown outside the ERA process to emphasize that risk assessment and risk management are distinct activities. The former evaluates the likelihood of adverse effects, while the latter responds to an identified risk based on many factors (e.g., social, legal, political, or economic) in addition to risk.

The bar on the right in Figure 6.1 highlights data gathering, iteration, and monitoring. Monitoring data can help identify changes in the ecosystem of interest. They can also be used to evaluate an ERA's predictions. For example, follow-up studies could determine whether mitigation efforts were effective, help verify whether source reduction was effective, or determine the extent of ecosystem recovery. Risk assessors and risk managers can evaluate predictions to gain experience and help improve the risk assessment and risk management processes [Commission on Risk Assessment and Risk Management (CRARM), 1997].

Even though ERA focuses on data analysis and interpretation, the results will not be useful unless the data truly support them. If more data are needed, the analyst may want to suspend the process until they can be collected. The process is more often iterative than linear, since evaluating new information may require revisiting a part of the process or conducting a new assessment (see Table 6.5). The dotted line between the sidebar and the risk management box in Figure 6.1 shows that additional data acquisition, iteration, or monitoring, while important, are not always required.

6.1.2   Ecological Risk Assessment in a Management Context

ERAs are designed and conducted to inform risk management decisions. They provide information about the potential adverse effects of different decisions, including no action. Although managers base their decisions on many factors, ERA provides a scientific process to evaluate ecological risk. Its results are generally used in combination with other information such as engineering or feasibility studies, cost–benefit analyses, legal opinions, and political context.

Ecological Risk Assessment's Contributions to Environmental Decision Making
ERAs may contribute to many types of management actions, including the regulation of hazardous waste sites, industrial chemicals, and pesticides, or the management of watersheds or other ecosystems affected by multiple stressors. ERA has several features that contribute to effective environmental decision making:

- It can accommodate new information as it becomes available.
- It can express changes in effects as a function of changes in exposure to stressors.
- Its focus on specific management questions makes it easy to develop measures of success to determine whether management actions are effective.
- It explicitly evaluates uncertainty.
- It can be used to compare, rank and prioritize risks and can provide data for cost–benefit and cost-effectiveness analyses.
- It considers management goals and objectives as well as scientific issues during problem formulation. This helps ensure that results will be useful to risk managers.

Factors Affecting Value of Ecological Risk Assessment for Environmental Decision Making   Risk managers consider many factors in addition to risk when making their decisions. Legal mandates and political, social, and economic considerations may lead managers to make decisions of varying protectiveness. Reducing risk to the lowest level may be too expensive or technically or societally infeasible. Thus, although ERAs provide critical information to risk managers, they are only part of the environmental decision-making process.

It is usually a good idea to take a broad view during planning. An ERA that focuses too narrowly on one stressor such as a chemical could miss something more important like habitat alteration. However, it may not be possible to expand the scope of an ERA when it is defined by statute (e.g., Superfund ERAs examine stressors associated with a specific site). And not all management alternatives need an ERA. For example, the risks associated with building a hydroelectric dam may be avoided by using alternatives that meet power needs without a new dam.

6.2 PLANNING THE RISK ASSESSMENT

ERAs are conducted to translate environmental data into information risk managers can use to make informed decisions. To make sure ERAs are useful, risk managers and risk assessors (see Tables 6.2 and 6.3) and, where appropriate, interested and affected parties (I&APs) (see Table 6.4), plan the process before beginning problem formulation (see Figure 6.2).

The first step is to decide whether an ERA is really needed. Participants explore what is known about the risk, what can be done to mitigate or prevent it, and whether ERA is the best way to evaluate environmental concerns. An ERA

TABLE 6.2   Who Are Risk Managers?

- Risk managers are people and organizations who have the responsibility or authority to do something about a risk.
- The expression "risk manager" usually means a decision maker in a state or federal regulatory agency who has legal authority to protect or manage a resource. But I&APs who can reduce or mitigate risk may also be risk managers. This is especially true in complex circumstances such as watershed protection activities, where risk is managed by the community.
- Risk management teams may include decision officials in federal, state, local, and tribal governments; commercial, industrial, and private organizations; constituency group leaders; and property owners.

may add little value to the decision process if there are alternatives that circumvent the need for it. Sometimes a back-of-the-envelope calculation using minimal data and a simple model is enough. Or a discussion with I&APs may reveal that there is so much opposition to a particular action that there is no point in considering it, much less going to the trouble of preparing an ERA.

Once an ERA is started, the next step is to ensure that all the right people are involved. Risk management decisions may be made by one person (as is often the case at regulatory agencies) or a group (see Table 6.2). Likewise, the ERA may be prepared by an individual or a team (see Table 6.3). Sometimes I&APs play an important role (Table 6.4). Careful consideration up front about who will participate and how they will do it will ensure successful planning.

The EPA provides perspectives on environmental decision making and risk assessment: ecological concerns in decision making (1994a), ecological entities as candidates for protection (1997a), and an introduction to ERA for risk managers (1995b).

6.2.1   Roles of Risk Managers, Risk Assessors, and Interested and Affected Parties in Planning

Risk managers, risk assessors, and I&APs each bring unique perspectives to planning. The risk manager, usually charged with protecting human health and

TABLE 6.3   Who Are Risk Assessors?

- Risk assessors are specialists in evaluating the potential effects of stressors, tracing the cause of observed effects, or determining what might affect a resources. When a specific process is defined by regulations and guidance, one trained individual may be able to complete the ERA (e.g., premanufacture notice of a chemical). However, for complex assessments, no one person will have the needed breadth of expertise.
- Every risk assessment team should include someone who is knowledgeable and experienced with the ERA process. Other team members such as ecologists, geologists, chemists, toxicologists, and hydrogeologists bring expertise regarding stressors, ecosystems, and scientific issues and techniques.

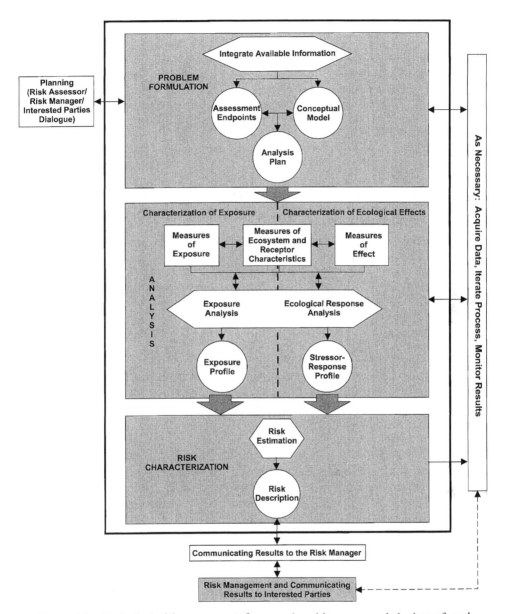

Figure 6.2 Ecological risk assessment framework, with an expanded view of each phase. Within each phase, rectangles designate inputs, hexagons indicate actions, and circles represent outputs.

TABLE 6.4   Who Are Interested and Affected Parties?

- Interested and affected parties (I&APs; sometimes called "stakeholders") may include any level of government, industry, environmental groups, small-business owners, landowners, and other segments of society concerned about an environmental issue or who want to influence risk management decisions.

  Their involvement, particularly during management goal development, may be key to risk reduction or mitigation, since actions are more likely to be implemented when they are backed by consensus.

  Large groups may require trained facilitators and consensus-building techniques such as "dialog" (Martin, 1999; Yalkelovich, 1999) to help them reach agreement.
- In some cases, interested parties may provide important information.

  Local knowledge, particularly in rural communities, and traditional knowledge of native peoples can provide valuable insights about ecological characteristics of a place, past conditions, and current changes.
- The extent of I&AP involvement depends on the individual assessment.

  I&APs may simply provide input to goal development, or they may become risk managers, depending on the degree to which they can take action to manage risk and the regulatory context of the decision. When and how interested parties influence risk assessments and risk management are areas of current discussion (NRC, 1996). See additional information in Table 6.2 and Section 6.2.1.

---

the environment, describes the questions that must be answered before the decision can be made, and what is needed from the risk assessor. To do this, the decision maker considers the *management goal*, which is comprised of an entity of concern, some attribute of interest, and a desired state. For example, this could be "maintain a minnow (entity) population (attribute) that can support the annual Kiddie Fishing Derby (desired state) in Town Pond."

In turn, the risk assessor describes what can be provided to the risk manager, where problems are likely to occur, and where uncertainty may be a problem. They may also offer an ecological perspective on management options and their likelihood of achieving stated goals.

Interested and affected parties may also play an active role in planning, particularly in goal development. The National Research Council (NRC, 1996) describes this as an analytic-deliberative process. Interested and affected parties provide a context for the decision by expressing their concerns about the environment, quality of life, economics, cultural changes, or other values potentially affected by environmental management activities. They may also help develop management options. Finally, where they can do something about risk to ecological values of concern, they may be part of the risk management team.

6.2.2   Products of Planning

Planning generates three outputs: (1) clear management goals, (2) decisions to be made, and (3) the ERA's scope, complexity, and focus. If risk communication is part of the risk management strategy, these topics can be used as an outline for outreach and informational materials.

Management Goals   Management goals contain an *entity*, some *attribute*, and a *desired state*. They reflect the decision maker's overall goals and may address an organism, an ecosystem, or some part or feature of an ecosystem. It is well worth any time spent to explore goals and make sure they are explicit. For example, if the management goal is a general "maintain a sustainable aquatic community," the entity is the aquatic community and the (implicit) attribute is population size of each member species. The desired state is unclear, probably something to the effect of "enough so they'll still be with us next year," and would be much more helpful if it was expressed as something like "that can sustain a sport fishery for organisms x and y and a commercial fishery for organism z." Similarly, "restore a wetland" could be stated more clearly as "restore the prairie pothole wetland at Puddle Prairie State Park (entity) to its areal extent (attribute) before tile drains were installed (desired state)," and "prevent toxicity" could be expressed as "reduce toxic discharges to Lake Serenity (entity; implicit attribute is the aquatic community) to concentrations below ambient water-quality criteria (desired state)." Explicit goals not only guide the risk assessment and subsequent management decisions, they provide a way to tell whether the action taken was successful.

Some ERAs use a *place-based* or *community-based* approach, which generally requires that management goals be developed for each assessment. Management goals for places such as watersheds are based on regulations and on constituency group and public concerns. Facilitated *dialog* is an excellent way to reach consensus on goals. This process is specialized, but briefly its attributes are (1) equality and the absence of coercive influences (everyone has an equal standing and feels free to speak), (2) listening with empathy (everyone is heard and taken seriously), (3) bringing assumptions into the open (people explore and explain what is behind their positions) [see, e.g., Yankelovich (1999) and Martin (1999)]. Goals developed by consensus are usually general, so the planning group should be prepared to spend some time making them specific and measurable enough to use in the ERA.

Early discussions and clear management goals help the risk assessor identify and gather critical information. No matter how they are established, goals that explicitly define what will be protected provide the best foundation for risk assessment objectives and actions to reduce risk.

Management Options   Risk managers must decide how to achieve their management goals. They often start with a list of options, which may range from preventing a new plant's introduction to restoring a degraded ecosystem or reintroducing a native animal. Once options are defined (e.g., leave a contaminated site alone, clean it up, or pave it), ERAs can be used to predict risk across the range of options (and perhaps combined with cost–benefit or other analyses) to aid decision making. When analysts understand the options, they can use them to ensure that the ERA addresses the best range of possibilities.

Risk assessors can also provide a scientific perspective on management options: For example, if the management goal is to reestablish spawning habitat for free-living anadromous salmon in a river impounded by multiple dams, a risk assessor might advise the decision maker that several actions will be needed (dam removal, nutrient input reduction, erosion control, pest management) and that none of them will make any difference unless the dams are removed. If this is not among the management options, the planning group might work together to reconsider the goal, abandon it, or present it as a work in progress with interim goals such as water clarity and quality.

Decision criteria may be used in a tiered framework to determine the best level of detail. Early screening tiers may use predetermined criteria to answer the question "Is there a problem?" In subsequent tiers the question becomes "What, where, and how big is the risk?" Risk managers use the risk characterization results from later tiers with other information to develop a decision, perhaps through formal decision analysis (e.g., Clemen, 1996; Keeney, 1992; Hammond et al., 1999), or managers may request a more detailed ERA to address issues of continuing concern (Table 6.5).

TABLE 6.5   Tiers and Iteration: When Is a Risk Assessment Done?

- Risk assessments range from very simple to complex and time-consuming. How do planners decide the level of effort?

  How many times should the risk assessor revisit data and assumptions?

  How can we tell when the risk assessment is finished?

- Many of these questions can be addressed by using tiered assessments.

  These progress through increasing amounts of data and analysis; generally one only proceeds as far as necessary to feel comfortable with the answer.

  The outcome of a given tier is to either make a management decision, often based on decision criteria, or continue to the next tier.

  Many organizations use this approach (e.g., see Gaudet, 1994; European Community, 1993; Cowan et al., 1995; Baker et al., 1994; Lynch et al., 1994).

- An iteration is a reevaluation that may occur at any time during an ERA.

  It responds to an identified need, new information, or questions raised while conducting an assessment. It may include redoing the ERA with new assumptions and new data.

- Setting up tiered assessments and decision criteria may help avoid iteration.

  Up-front planning and careful development of problem formulation will also avoid having to revisit data, assumptions, and models. However, there is no rule about how many iterations are needed to answer management questions or ensure scientific validity.

  A risk assessment can be considered complete when risk managers have sufficient information and confidence in the results of the risk assessment to make a decision they can defend.

TABLE 6.6   Questions to Ask About Scope and Complexity

- Why are we doing this? Is it mandated, required by a court decision, or meant to provide guidance to a community?
- Will decisions be based on a small area evaluated in depth or a large-scale area but with less detail?
- What are the spatial and temporal boundaries of the problem?
- What information is already available?
- How much more do we need?
- How much time do we have, and what resources are available?
- What limits data collection?
- Is a tiered approach an option?

ERAs for a region or watershed where multiple stressors, ecological values, and political and economic factors influence decision making are more flexible and more complicated. They generally examine the ecological processes most influenced by human actions. ERAs used this way are often based on a general goal and multiple potential decisions. Planning these ERAs naturally takes more time because more management decisions and options are available, and they examine more issues than simpler assessments.

Scope and Complexity   An ERA's purpose determines its scope (e.g., national, regional, site-specific), but its extent, complexity, and confidence level are limited by resources—the data, expertise, time, and money available to do the job. The planning group considers the type of decision (e.g., national policy, local impact), available resources, opportunities for increasing resources (e.g., partnering, new data collection, alternative analytical tools), expertise at hand, and what information is needed to make the decision (see Table 6.6). Detailed analysis is not always needed, and sometimes a simple screening assessment is enough. One way to proceed is by the tiered evaluations described earlier. Where tiers are used, specific management questions and decision criteria should be included in the plan.

Part of the agreement on scope and complexity is based on how sure the decision maker needs to be about risks to decide what to do about them. ERAs completed under legal mandates and likely to be contested or appealed need strict attention to potential sources of uncertainty to ensure that they will stand up in court. The risk manager and risk assessor should be frank with each other about sources of uncertainty, the comfort level needed for the decision, and ways uncertainty can be reduced. For example, successive iterations or tiers of increasing cost and complexity may help reduce uncertainty. Advice on how to address the interplay of management decisions, study boundaries, data needs, uncertainty, and specifying limits on decision errors may be found in EPA's guidance on data quality objectives (EPA, 1994b).

6.2.3 Planning Summary

Planning is complete when the group has decided on (1) management goals, (2) the range of management options the ERA is to address, (3) ERA objectives, including criteria for success, (4) the ERA's focus and scope, and (5) resource availability. Agreements may include the technical approach, spatial scale, temporal scale, and, of course, product deadlines.

6.3 PROBLEM FORMULATION

Problem formulation is a process for developing and evaluating preliminary questions about why ecological effects might occur. This step is crucial—it provides the foundation for the ERA. First participants refine the ERA objectives. Then they evaluate the problem and develop a plan for analyzing data and characterizing risk. Any shortcomings in problem formulation will haunt future work, so it is worth spending time here to refine objectives and plan carefully.

Risk manager and I&AP participation is most important during management option review, assessment endpoint selection, and conceptual model review. The risk manager usually determines how I&APs are involved. Generally, the risk assessor's job is to maintain a scientific approach, the risk manager's task is to describe risk management objectives and the decision to be made, and I&APs provide perspective and context.

6.3.1 Products of Problem Formulation

Problem formulation generates three products: assessment endpoints, conceptual model(s), and an analysis plan. In all but the simplest analyses, the problem formulation process is interactive and iterative and may require several cycles. While problem formulation begins with a review of available information and ends with an analysis plan (Figure 6.3), assessment endpoints and conceptual models are not necessarily developed by a linear process.

6.3.2 Integrating Available Information

To start, the risk assessment team reviews information about stressor sources and characteristics, exposure opportunities, characteristics of the ecosystem(s) potentially at risk, and ecological effects. This is an iterative process that normally occurs throughout problem formulation.

When information is plentiful, problem formulation goes quickly. If not, the ERA may begin with the information at hand, and the problem formulation process helps identify missing data and provides a framework for further data collection. Where data are scant, the limitations of conclusions (i.e., uncertainty) should be acknowledged in the risk characterization.

Despite our limited knowledge of ecosystems and the stressors that affect them, problem formulation offers a systematic approach for organizing and

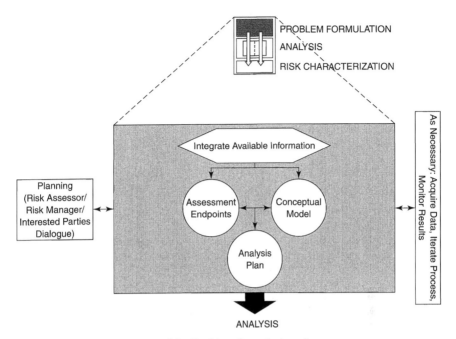

Figure 6.3 Problem formulation phase.

evaluating stressors and possible effects and can function as a preliminary ERA. Table 6.7 presents questions to consider when reviewing preliminary data.

6.3.3 Selecting Assessment Endpoints

Assessment endpoints are explicit expressions of the actual environmental value to be protected, operationally defined by an ecological entity and some attribute. (The entity here is usually an organism or group of organisms.) Note that this is similar to a management goal except for one important difference—it does not include a desired state or direction of change. Assessment endpoints play a critical role in conceptual model development by structuring the process to address management concerns. They should be measurable ecosystem characteristics that represent management goals.

Selection Criteria Ecosystems may be examined at several organizational levels (individual, population, community, ecosystem, landscape) and multiple ecosystem processes. It is rarely clear which of these are most critical to ecosystem function, nor do professionals or the public always agree on which are most valuable. It can be a challenge to choose the characteristics that best address management goals.

TABLE 6.7   Questions to Ask About Source, Stressor, and Exposure Characteristics, Ecosystem Characteristics, and Effects [derived in part from Barnthouse and Brown, (1994)]

Source and Stressor Characteristics

- What is the source? Is it anthropogenic, natural, point source, or nonpoint?
- What type of stressor is it (i.e., chemical, physical, or biological)?
- How intense is it (e.g., the dose or concentration of a chemical, the magnitude or extent of physical disruption, the density or population size of a biological stressor)?
- How does it act on organisms or ecosystem functions?

Exposure Characteristics

- How often does a stressor event occur (e.g., is it isolated, episodic, or continuous; is it subject to natural daily, seasonal, or annual periodicity)?
- How long does it last? How long does it persist (e.g., for chemical, what is its half-life, does it bioaccumulate; for physical, is habitat alteration sufficient to prevent recovery; for biological, will it reproduce and proliferate)?
- What is the timing of exposure? When does it occur in relation to critical organism life cycles or ecosystem events (e.g., reproduction, lake turnover)?
- What is the spatial scale of exposure? Is the extent or influence of the stressor local, regional, global, habitat-specific, or ecosystemwide?
- Where is the stressor? How does it move through the environment (e.g., for chemical, fate and transport; for physical, movement of physical structures; for biological, life history and dispersal characteristics)?

Ecosystems Potentially at Risk

- What are the geographic boundaries? How do they relate to the ecosystem's functional characteristics?
- What key abiotic factors influence the ecosystem (e.g., climatic factors, geology, hydrology, soil type, water quality)?
- Where and how are functional characteristics driving the ecosystem (e.g., energy source and processing, nutrient cycling)?
- What are the ecosystem's structural characteristics (e.g., species number and abundance, trophic relationships)?
- What habitat types are present?
- How do these characteristics influence susceptibility (sensitivity and likelihood of exposure)?
- Are there unique features that are particularly valued (e.g., the last representative of an ecosystem type)?
- What is the landscape context within which the ecosystem occurs?

Ecological Effects

- What type of effects data are available (e.g., field surveys, laboratory tests, or structure–activity relationships)?
- Given the nature of the stressor, what effects are expected?
- Under what circumstances will effects occur?

TABLE 6.8   Salmon and Hydropower: Salmon as the Basis for an Assessment Endpoint

- A hydroelectric dam is proposed for a river. Salmon that spawn in the river are an appropriate choice because they meet the criteria for good assessment endpoints: Both fry and adults are important food sources for other species and prey on aquatic invertebrates (ecological relevance).
- They are sensitive to changes in sedimentation and substrate pebble size, require quality cold-water habitats, and have a hard time climbing fish ladders.
- Dams cause significant, and normally fatal, habitat alteration and physical obstacles to successful salmon breeding and fry survival (susceptibility).
- Salmon support a commercial fishery, some species are endangered, and they have ceremonial importance and are food sources for Native Americans (relevance to management goals).
- Since all three criteria are satisfied, "salmon reproduction and population recruitment" is a good assessment endpoint to evaluate the potential effects of the dam. In addition, actions that protect salmon will likely protect other anadromous fish as well. Note that one assessment endpoint is seldom enough to describe a complex ecosystem.

The criteria used to select assessment endpoints are: (1) ecological relevance (the organism should be a current or historical part of the ecosystem being studied), (2) susceptibility (they should be both exposed to and sensitive to the effects of the stressor), and (3) relevance to management goals (they should be something the manager can do something about). Ecological relevance and susceptibility are essential for scientifically defensible assessment endpoints. However, those that reflect societal values and management goals are more likely to be used in management decisions. Assessment endpoints that meet all three criteria provide the best foundation for an effective ERA (see Table 6.8).

Ecological Relevance   *Ecologically relevant* assessment endpoints reflect important system characteristics and relate functionally to other endpoints (EPA, 1998). The consequences of changes in these endpoints may be quantified (e.g., loss of an influential species leading to changes in community structure) or inferred (survival of individuals is needed to maintain populations). To be relevant, entities must be (or have been) part of the ecosystem of interest.

When screening for ecological relevance, risk assessors usually look for ecosystem components that sustain the natural structure, function, or biodiversity of an ecosystem. They may contribute to the food base (e.g., primary production), provide habitat (for prey or reproduction), promote regeneration of critical resources (decomposition or nutrient cycling), or reflect the structure of the community, ecosystem, or landscape (species diversity or habitat mosaic). In landscape-level ERAs, assessors usually choose several species and an ecosystem process to represent larger functional community or ecosystem processes.

Susceptibility   Ecological resources are considered *susceptible* when they are sensitive to a stressor to which they could be exposed. Susceptibility can usually be identified early in problem formulation.

Sensitivity describes how readily an entity is affected by a particular stressor. It is a function of the stressor's mode of action (e.g., chemical sensitivity is influenced by individual physiology and metabolic pathways) and the entity's life-history characteristics. For example, stream species assemblages that depend on cobble-and-gravel habitat for reproduction are sensitive to fine sediments that fill in spaces between cobbles. Species with long life cycles and low reproductive rates are usually more vulnerable to extinction from increases in mortality than species with short life cycles and high reproductive rates.

Sensitivity may also be a function of life stage; young animals are often more sensitive to stressors than adults. For instance, Pacific salmon eggs and fry are very sensitive to fine-grain sedimentation in river beds because they can be smothered. Energy-intensive events like migration and molting may also increase vulnerability to stressors. Finally, other stressors or natural disturbances may affect sensitivity. For example, the presence of insect pests and disease may increase plants' sensitivity to ozone (Heck, 1993).

Measures of sensitivity may include mortality or adverse reproductive effects from exposure to toxics. Other possible measures include behavioral abnormalities, avoidance of significant food sources and nesting sites, loss of offspring to predation because of the proximity of stressors such as noise, habitat alteration, or loss, community structural changes, or other factors.

Exposure, the other component of susceptibility, can mean co-occurrence, contact, or the absence of contact, depending on the stressor and assessment endpoint. A stressor's origin, movement through environment, and its interaction with the assessment endpoint all influence exposure and how a receptor will respond. To determine which entities are susceptible, the assessor considers their proximity and the timing and intensity of exposure.

Note that it may not be easy to link effects observed at one point to exposure at another. For instance, temperature during egg incubation affects the sex ratio of alligators and marine turtle hatchlings, but population impacts are not observed until years later when the cohort of affected reptiles begins to reproduce. Delayed effects and multiple-stressor exposures complicate susceptibility evaluations (e.g., although toxicity tests may determine receptor sensitivity to one stressor, susceptibility may depend on the co-occurrence of another stressor that significantly alters receptor response) (see example in Table 6.9).

Relevance to Management Goals   Ultimately, an ERA's effectiveness depends on its relevance to management goals—whether it actually contributes to management decisions. Risk managers are more willing to use an ERA when it is based on ecological values that people care about. Candidate assessment endpoints might include endangered species or ecosystems, commercially or recreationally important species, functional attributes that support food sources or flood control (e.g., wetland water sequestration), aesthetic values such as clean air in national parks, or charismatic species such as eagles or whales. However, assessment endpoints based on public perceptions alone could lead to management decisions that do not consider important ecological processes, functions, or interactions.

TABLE 6.9 Sensitivity and Secondary Effects: The Mussels and Fish

- In many streams, native mussels are endangered.
- Management efforts have focused on maintaining suitable habitat for mussels because habitat loss has been considered the greatest threat to this group. However, larval unionid mussels must attach to the gills of a fish host for one month during development.
- Each mussel species attaches to a particular host fish species. So where the fish community has been changed, perhaps due to stressors to which mussels are insensitive, the host fish may no longer be available and mussel larvae will die before reaching maturity.
- No matter how much habitat is restored, mussels will be lost unless the fish community is also restored. In this case, risk is caused by the absence of a critical resource.

The challenge is to find values that meet the selection criteria and are also important to those who make and have to live with the decision—risk managers and the public. Suppose, for example, an assessment is designed to evaluate the risk of applying pesticide around a lake to control insects. At this lake, it turns out that midges are susceptible to the pesticide and form the base of a complex food web that supports native fish popular with local anglers. While both midges and fish are key components of the aquatic community in this scenario, evaluating midges alone would not address both ecological and community concerns (and it would probably be difficult to convince local residents that annoying insects should be protected). Selecting the fishery would allow assessors to characterize the risk to the fishery if the midge population is adversely affected and address both sets of issues. This strategy addresses ecological issues while responding to management concerns. If there is no choice but to use an unpopular assessment endpoint, the risk assessor will need to link it convincingly to values that people do care about.

Practical issues such as what is required by statute (e.g., the Endangered Species Act) or management options may influence assessment endpoint selection. Another concern is whether important variables can be measured directly. Assessment endpoint attributes that can be measured directly are best because they avoid the uncertainty introduced by having to extrapolate or estimate values. Nevertheless, sometimes the only data that are available are those measured indirectly or generated by a model. Note also that data availability and measurement convenience are *not* selection criteria: While established measurement protocols may be familiar and easy to use, convenience is no guarantee of appropriateness.

Defining Assessment Endpoints Two elements are needed to define assessment endpoints: The first is the specific valued *entity*. This can be a species (e.g., eelgrass, piping plover), a functional group (e.g., piscivores), a community (e.g., benthic invertebrates), an ecosystem (e.g., lake), a specific habitat (e.g., wet meadows), a unique place (e.g., a remnant of native prairie), or other entity of concern. The second is the *attribute* or characteristic of that entity that is important to protect and potentially at risk. For piping plovers, it may be nesting and feeding conditions; for a lake, nutrient cycling; a wet meadow, endemic plant community diversity. An assessment endpoint needs both an entity and an attribute to serve as a clear link to management goals and the basis for measurement.

Notice that assessment endpoints are similar to management goals but are distinguished by their neutrality and specificity. Assessment endpoints do not describe a desired achievement (i.e., goal), so they do not contain words like "protect," "maintain," or "restore," or indicate a direction for change such as "loss" or "increase." They are defined by specific entities and their measurable attributes and provide a framework for measuring stress–response relationships. When goals are very broad, it may be difficult to select appropriate assessment endpoints until the goal is broken down into multiple management objectives. Keeney (1992) provides a structured procedure for developing objectives, and Hammond et al. (1999) provide a somewhat simplified approach. Schwartz (1996) provides advice on long-term or "futures" planning. A series of management objectives can clarify the inherent assumptions within the goal and help determine which ecological entities and attributes best represent each objective. From this, multiple assessment endpoints may be selected. See Table 6.10 for examples of management goals and assessment endpoints.

While the assessment endpoint entity influences the scale and character of an ERA, its attribute determines what to measure. When effects can be measured directly, assessment endpoints are the same as their respective measures, and no extrapolations are needed. For example, if the assessment endpoint is "blue jays reproductive success," egg production and fledgling success could be directly measured under different exposure scenarios. In other cases (e.g., toxicity in endangered species), effects cannot be measured directly and surrogate measures of effect must be used. So although assessment endpoints must be defined in terms of measurable attributes, selection does not depend on the ability to measure those attributes directly or on whether methods, models, and data are currently available. For practical reasons, it may be helpful to use assessment endpoints that have well-developed test methods, field measurement techniques, and predictive models (see Suter, 1993a). However, it is not necessary for methods to be standardized protocols; nor should assessment endpoints be selected simply because standardized protocols are readily available.

Clearly defined assessment endpoints provide direction and boundaries for the ERA and can minimize miscommunication and uncertainty. "Ecological integrity" is too vague to be helpful unless it is carefully and specifically defined— by choosing important entities or processes and describing attributes that best represent integrity for that system. Assessment endpoints that are too narrowly defined may not provide data that inform effective risk management. If an assessment focuses only on protecting the habitat of an endangered species, for example, the investigation may overlook other equally important characteristics of the ecosystem and fail to include critical variables (Table 6.9).

TABLE 6.10   Examples of Management Goals and Assessment Endpoints

Case	Regulatory Context/ Management Goal	Assessment endpoint
Assessing Risks of New Chemical Under Toxic Substances Control Act (Lynch et al., 1994)	Protect "the environment" from "an unreasonable risk of injury" (TSCA §2[b][1] and [2]); protect the aquatic environment. Goal was to exceed a concentration of concern on no more than 20 days a year.	Survival, growth, and reproduction of fish, aquatic invertebrates, and algae
Special Review of Granular Carbofuran Based on Adverse Effects on Birds (Houseknecht, 1993)	Prevent ... "unreasonable adverse effects on the environment" (FIFRA §§3[c][5] and 3[c][6]); using cost–benefit considerations. Goal was to have no regularly repeated bird kills.	Individual bird survival
Modeling Future Losses of Bottomland Forest Wetlands (Brody et al., 1993)	National Environmental Policy Act may apply to environmental impact of new levee construction; also Clean Water Act §404.	(1) Forest community structure and habitat value to wildlife species (2) Species composition of wildlife community
Pest Risk Assessment on Importation of Logs From Chile (USDA, 1993)	Assessment was done to help provide a basis for any necessary regulation of the importation of timber and timber products into the United States.	Survival and growth of tree species in the western United States
Baird and McGuire Superfund Site (terrestrial component); (Burmaster et al., 1991; Callahan et al., 1991; Menzie et al., 1992)	Protection of the environment (CERCLA/SARA).	(1) Survival of soil invertebrates (2) Survival and reproduction of song birds
Waquoit Bay Estuary Watershed Risk Assessment (U.S. EPA, 1996b)	Clean Water Act—wetlands protection; water quality criteria— pesticides; endangered species. National Estuarine Research Reserve, Massachusetts, Area of Critical Environmental Concern. Goal was to reestablish and maintain water quality and habitat conditions to support self-sustaining commercial, recreational, and native fish, water-dependent wildlife, and shellfish and to reverse ongoing degradation.	(1) Estuarine eelgrass habitat abundance and distribution (2) Estuarine fish species diversity and abundance (3) Freshwater pond benthic invertebrate species diversity and abundance

Well-selected assessment endpoints are powerful tools. An endpoint that is sensitive to many of the identified stressors, yet responds in different ways to them, will help evaluate the combined effects of multiple stressors while still distinguishing their effects. For example, fish population recruitment may be adversely affected at several life stages, in different habitats, through different ways, and by different stressors. Several measures of effect, exposure, and ecosystem and receptor characteristics could be chosen to evaluate recruitment and provide a basis for distinguishing different stressors, individual effects, and their combined effects.

The assessment endpoint can be used to compare a range of stressors if carefully selected. The National Crop Loss Assessment Network (Heck, 1993) selected crop yields as the assessment endpoint to evaluate the cumulative effects of multiple stressors. Although the primary stressor was ozone, the crop yield endpoint also allowed the risk assessors to consider the effects of sulfur dioxide and soil moisture. Barnthouse et al. (1990) recommended that an endpoint should be selected so that all the effects can be expressed in the same units (e.g., abundance of 1-year-old fish from exposure to toxicity, fishing pressure, and habitat loss). This is especially true when selecting assessment endpoints for multiple stressors. However, most ERAs that examine multiple stressors will need several assessment endpoints.

Final assessment endpoint selection is an important risk manager–risk assessor checkpoint during problem formulation. Risk assessors and risk managers should agree that selected assessment endpoints effectively represent the management goals. In addition, the rationale for their selection should be explicit.

6.3.4 Conceptual Models

Conceptual model development starts with what is known about stressors, potential exposure, and predicted effects to the assessment endpoint. The process also identifies what additional information is needed. Conceptual models have two components: (1) a set of questions that propose predicted relationships between ecological entities and the stressors to which they may be exposed and (2) a diagram of the relationships presented in the risk questions. If possible, they illustrate all relationships being examined; more than one model may be needed. Conceptual models may include ecosystem processes that influence receptor responses or exposure scenarios that qualitatively link land-use activities to stressors. They may describe primary, secondary, and tertiary exposure pathways or co-occurrence between exposure pathways, ecological effects, and ecological receptors. Some of the benefits of conceptual models are featured in Table 6.11.

A carefully prepared conceptual model will provide a helpful reference point throughout the ERA; it can be used to refine models or assumptions, keep the assessment on course, and as an outline when preparing outreach materials.

Risk Questions Risk questions (sometimes called risk hypotheses) are developed to evaluate the relationships between stressors, receptors, and effects. They

TABLE 6.11 Benefits of Conceptual Models

- They help explore the ecosystem of interest.
- They are easily modified as more information becomes available.
- They illustrate what is known and not known and can be used to plan future work.
- They can be used to plan communication efforts.
- They explicitly describe the assumptions being used in the ERA.
- They provide a framework for predicting risks.

are specific questions about potential risk to assessment endpoints and may be based on theory and logic, empirical data, mathematical models, or probability models. They are formulated using a combination of professional judgment and available information on the ecosystem at risk, potential sources of stressors, stressor characteristics, and observed or predicted ecological effects. They may attempt to predict the effects of a stressor before they occur, or explain why observed ecological effects occurred and ultimately what caused the effect. Risk questions may range from very simple (e.g., what is the effect of stressor x on receptor y?) to very complex, as in value-initiated assessments that include prospective and retrospective questions about the effects of multiple stressors on ecological receptors. Although they are sometimes constructed as hypotheses, risk questions are meant to examine relationships in the conceptual model and are not designed for statistically testing null and alternative hypotheses. Suter (1996) discusses how to uses statistical testing in this application.

Early conceptual models are usually quite broad and general in that they identify as many potential relationships as possible. As more information is incorporated, risk assessors sort through potentially large numbers of stressor–effect relationships, and the ecosystem processes that influence them, to identify the questions most appropriate for the analysis phase. The ERA report should document this distillation and refinement process. Examples of risk questions are provided in Table 6.12.

Conceptual Model Diagrams   Conceptual model diagrams illustrate risk questions. They show important pathways clearly and concisely and can be used to generate new questions about relationships. Typically, they are flow diagrams with boxes and arrows to illustrate relationships. It is helpful to use distinct and consistent shapes to distinguish stressors, assessment endpoints, responses, exposure routes, and ecosystem processes, although there is no set configuration or convention. Pictures or cartoons can be very effective (e.g., Bradley and Smith, 1989), and Tufte (1990, 1992, 1997) provides helpful information about visual displays.

Factors to consider when developing conceptual models include the number of relationships, data completeness, how much in known about linkages, and measurement practicability. Several models of varying detail can be more effective than one model that tries to show everything. Flow diagrams that highlight

TABLE 6.12   Examples of Risk Questions

Risk questions include information that sets the problem in perspective.

Stressor-initiated: Chemicals with a high K_{ow} tend to bioaccumulate. Chemical A has a K_{ow} of 5.5 and molecular structure similar to known chemical stressor B.

Question: We know chemical A's K_{ow}, chemical B's mode of action, and the ecosystem's food web. When chemical A is released at a specified rate, will it bioaccumulate enough in 5 years to cause developmental problems in wildlife and fish?

Effects-initiated: Bird kills have repeatedly been noticed on golf courses after the granular form of the highly toxic pesticide carbofuran was applied.

Question: We know that birds die when they consume recently applied granular carbofuran, that the number of dead birds increases as the application level increases, and that exposure can occur when dead and dying birds are consumed by other animals. Will birds of prey and scavenger species will die from eating contaminated birds after granular carbofuran is applied?

Resource-initiated: Waquoit Bay, Massachusetts, supports recreational boating and commercial and recreational shellfishing and is a significant nursery for finfish. Large mats of macroalgae clog the estuary, most of the eelgrass has died, and the scallops are gone.

Question: We know that nutrient loading from septic systems, air pollution, and lawn fertilizers causes eelgrass loss by shading from algal growth and direct toxicity from nitrogen compounds. Will fish and shellfish populations decrease because of loss of eelgrass habitat and periodic hypoxia from excess algal growth and low dissolved oxygen?

data abundance or scarcity can illustrate the risk assessor's confidence in the relationship. They can also show why certain pathways were pursued and others were not.

Uncertainty in Conceptual Models   Conceptual model development may be one of the most important sources of uncertainty in ERA. If important relationships are missed or incorrectly described, the risk characterization may misrepresent actual risks. Uncertainty may arise from ignorance about how the ecosystem functions, failure to identify and interrelate temporal and spatial parameters, omission of stressors, or overlooking secondary effects. Sometimes the assessor may not be sure how a stressor moves through the environment or causes adverse effects. Multiple stressors are a common source of confounding variables, particularly for conceptual models that focus on a single stressor. Although it is impossible to avoid simplification and ignorance, they can be acknowledged by documenting what is known, justifying the model, and ranking model components by uncertainty (see Smith and Shugart, 1994). Conceptual model uncertainty can be explored by considering alternative relationships. If more than one model is plausible, the risk assessor may consider whether it is feasible to compare different models' results or combine them to create a better model.

6.3.5 Analysis Plan

The analysis plan completes problem formulation. Here participants evaluate the risk questions to determine how they can be assessed. The plan delineates the assessment design, data needs, measures, and analytical methods. For some assessments (e.g., for new chemicals), one can use the analysis plan developed when the protocol was established. The more unique and complex the assessment, the more important it is to have a good analysis plan.

The analysis plan includes the pathways and relationships that will be pursued and emphasizes those most likely to contribute to risk. It describes how risk questions were chosen and where more data are needed. It also may compare the confidence desired for the management decision with that expected from alternative analyses to help decide what data are still needed and which analytical approach is best. Final selection is based on the strength of relationships between stressors and effects, exposure pathway completeness, and data quality and availability.

When data are scant and new data cannot be collected, it may be possible to extrapolate from data collected from other organisms or places where similar problems exist. For example, the relationship between nutrient availability and algal growth is well established and consistent and can be acknowledged despite differences in how it is manifested in particular ecosystems. If data must be extrapolated, it is important to identify their source, explain the extrapolation method, and discuss recognized uncertainties.

A phased or tiered approach can improve management decisions where data are less than complete. However, the analysis plan may also include recommendations for new data collection. If it will not be possible to collect new data, this should be acknowledged as a source of uncertainty.

When determining what data to analyze and how to analyze them, consider how these analyses might increase confidence in the conclusions and how they address risk management questions. Risk assessors can ask themselves questions such as: How relevant will the results be to the assessment endpoint(s) and conceptual model(s)? Are there enough good data to be confident in the analyses? How will the analyses help establish cause-and-effect relationships? How will the results address managers' questions? Where are uncertainties likely to become a problem?

Choosing Measures Assessment endpoints and conceptual models help risk assessors identify measurable attributes to quantify and predict change. However, choosing which measures to use is not only challenging, but critical to an ERA's success. There are three categories of measures: *Measures of effect* are measurable changes in an attribute of an assessment endpoint (or a surrogate) in response to a stressor to which it is exposed. *Measures of exposure* describe

TABLE 6.13   Examples of a Management Goal, Assessment Endpoint, and Measures

Goal: Viable, self-sustaining coho salmon population that supports a subsistence and sport fishery.

Assessment Endpoint: Coho salmon breeding success, fry survival, and adult return rates.

Measures of Effects

• Egg and fry response to low dissolved oxygen
• Adult behavior in response to obstacles
• Spawning behavior and egg survival with changes in sedimentation

Measures of Ecosystem and Receptor Characteristics

• Water temperature, water velocity, and physical obstructions
• Abundance and distribution of suitable breeding substrate
• Abundance and distribution of suitable food sources for fry
• Feeding, resting, and breeding behavior
• Natural reproduction, growth, and mortality rates

Measures of Exposure

• Number of hydroelectric dams and associated ease of fish passage
• Toxic chemical concentrations in water, sediment, and fish tissue
• Nutrient and dissolved oxygen levels in ambient waters
• Riparian cover, sediment loading, and water temperature

a stressor's presence and movement in the environment and its contact or co-occurrence with the assessment endpoint. *Measures of ecosystem and receptor characteristics* describe the characteristics that influence the behavior and location of the assessment endpoint, stressor distribution, and assessment endpoint (or surrogate) life-history characteristics that may affect exposure or response to the stressor. Examples of all three are provided in Table 6.13.

Measure selection is particularly complicated when a stressor seems likely to produce a cascade of ecological effects. In these cases, the effect on one entity (i.e., the measure of effect) may become a stressor for other ecological entities (i.e., become a measure of exposure) and may result in impacts on one or more assessment endpoints. For example, if a pesticide reduces earthworm populations, change in earthworm population density could be the direct measure of effect of toxicity and in some cases may be an assessment endpoint. However, the population reduction may then become a secondary stressor to which worm-eating birds become exposed, measured as lowered food supply. This exposure may then result in a secondary measurable effect: starvation. Although "fledging success" may be a directly measurable assessment endpoint, measures of earthworm density, pesticide residue in earthworms and other food sources, availability of alternative foods, nest site quality, and competition for nest sites with other bird species may all be useful. As another example, to assess a cove in Maine for its ability to support lobsters, one might measure lobster age and sex distribution, density per given cove area, habitat extent and quality, and stressors such as predators (including lobstermen).

When assessment endpoint responses cannot be measured directly, surrogates must be used. The selection of what, where, and how to measure surrogate responses determines whether the ERA is still relevant to the risk management decisions. For example, an assessment may evaluate the risk of a pesticide used on seeds to an endangered seed-eating bird to help decide whether to allow that pesticide to be registered for use. The assessment endpoint entity is the endangered species. Example attributes include feeding behavior, survival, growth, and reproduction. While it may be possible to directly collect measures of exposure and life-history characteristics, it would not be appropriate to dose the bird with the pesticide to measure sensitivity. While insectivorous birds may be an adequate surrogate measure of a seedeater's sensitivity to the pesticide, they do not reflect its exposure. So in this case, to evaluate susceptibility, the best surrogate measures would be of other birds with similar life-history characteristics and phylogeny.

The analysis plan is strongest when it is explicit about how measures were selected, what they will be used to evaluate, and the analyses in which they will be used. Associated uncertainties and strategies for addressing them should be included in the plan when possible.

Ensuring That Planned Analyses Meet the Risk Manager's Needs The analysis plan is a risk manager–risk assessor checkpoint. The risk assessor and risk manager review the plan to make sure the analyses will provide information the manager can use when deciding what to do about a risk. These discussions may also identify what can and cannot be done on the basis of a preliminary problem formulation. A review helps establish a balance of decision criteria, data availability, and resource constraints.

The analysis plan outlines analytical methods, risk characterization options, and considerations to be generated (e.g., quotients, narrative discussion, stressor–response curve with probabilities). It describes which analyses will be used and how the risk questions were chosen, as well as potential extrapolations, model characteristics, types of data (including quality), and planned analyses (with specific tests for different types of data) are described. Finally, the plan explains the basis for data selection and how the results will be presented.

The analysis plan summarizes what has been done during problem formulation, shows how the plan relates to management decisions that must be made, and indicates how data and analyses will be used to estimate risks. When the problem is clearly defined and there are enough data to proceed, analysis begins.

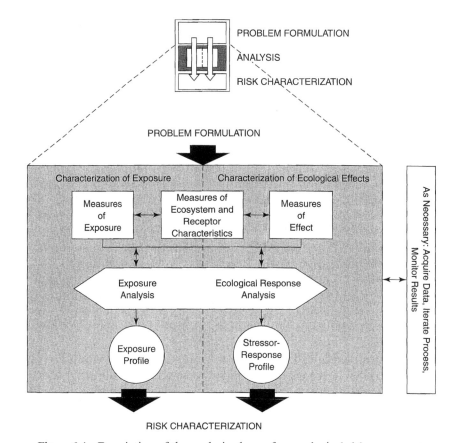

Figure 6.4   Description of the analysis phase of an ecological risk assessment.

6.4   ANALYSIS PHASE

Analysis is a process that examines the two primary components of risk, exposure and effects, and their relationships between each other and ecosystem characteristics. The objective is to provide the ingredients necessary to determine or predict ecological responses to stressors under specific exposure conditions.

The information needs identified during problem formulation should have already been addressed. During analysis (Figure 6.4), the risk assessor:

- Selects the data that will be used (Section 6.4.1)
- Analyzes exposure (Section 6.4.2)
- Analyzes effects (Section 6.4.3)

• Summarizes the conclusions about exposure (Section 6.4.2) and effects (Section 6.4.3). As with problem formulation, the process is not necessarily linear, and more than one pass may be needed.

The analysis phase is flexible, with substantial interaction between the effects and exposure characterizations as illustrated by the dashed line in Figure 6.4. In particular, when secondary stressors and effects are of concern, exposure and effects analyses can become intertwined and difficult to differentiate. In an assessment of bottomland hardwoods, for example (Brody et al., 1993), potential changes in the plant and animal communities under different flooding scenarios were examined. Risk assessors combined the stressor–response and exposure analyses within the FORFLO model for primary effects on the plant community and within the Habitat Suitability Index for secondary effects on the animal community. In addition, the distinction between analysis and risk estimation can become blurred. The model results developed for the bottomland hardwoods assessment were used directly in risk characterization.

The nature of the stressor influences the types of analyses conducted. For chemical stressors, exposure estimates emphasize contact and uptake into the organism, and effects estimations often entail extrapolation from test organisms to the organism of interest. For physical stressors, the initial disturbance may cause primary effects on the assessment endpoint (e.g., loss of wetland acreage). But in many cases secondary effects (e.g., decline of wildlife populations that depend on wetlands) are the main concern; the point of view depends on the assessment endpoints. Because adverse effects can occur even if receptors do not physically contact disturbed habitat, exposure analyses may emphasize co-occurrence with physical stressors rather than contact. For biological stressors, exposure analysis is an evaluation of entry, dispersal, survival, and reproduction (Orr et al., 1993). Because biological stressors can reproduce, interact with other organisms, and evolve over time, exposure and effects cannot always be quantified with confidence, and they are often assessed qualitatively by eliciting expert opinion (Simberloff and Alexander, 1994).

6.4.1 Evaluating Data And Models For Analysis

First, the assessor examines the data and models to ensure that they can be used to evaluate the conceptual model developed in problem formulation.

Strengths and Limitations of Different Types of Data   Data used in the ERA may come from laboratory or field studies or from model outputs. Familiarity with different types of data can help assessors build on strengths and avoid pitfalls. Such a strategy improves confidence in the conclusions.

Because conditions can be controlled in laboratory studies, responses may be less variable and smaller differences easier to detect. However, the controls may limit the range of responses (e.g., animals cannot seek alternative food sources), so they may not reflect responses in the environment. In addition, larger-scale processes are difficult to replicate in the laboratory.

Field surveys measure changes in uncontrolled situations. They often use statistical techniques such as correlation, clustering, or factor analysis to describe an association between a disturbance and an effect such as changes in the physical attributes of watersheds and their associated stream communities. Field surveys are often reported as status-and-trend studies.

Field surveys usually describe exposures and effects (including secondary effects) better than do laboratory studies or theoretical models. They are more important for assessments of multiple stressors or where site-specific factors significantly influence exposure. They are also useful for analyses of larger geographic scales and higher levels of organization. Field survey data are seldom necessary or feasible to collect for screening-level or prospective assessments.

Field surveys should be designed to statistically define one or more of the following:

- Exposure in the system of interest
- Differences in measures of effect between reference sites and study areas
- Lack of differences

Because conditions are not controlled in field studies, variability may be higher and it may be difficult to detect differences. So it is important to verify that studies have sufficient power to detect important differences.

Most data will be reported as measurements for single variables such as a chemical concentration or number of organisms. But sometimes variables are combined and reported as indices such as the rapid bioassessment protocols (EPA, 1989) and the Index of Biotic Integrity, or IBI (Karr, 1981; Karr et al., 1986). Indices can

- Incorporate many attributes of system structure and function
- Evaluate responses from a broad range of anthropogenic stressors
- Minimize the limitations of individual metrics (Barbour et al., 1995)

Indices also have several drawbacks, mostly due to combining heterogeneous variables. Some indices (e.g., the IBI) combine only measures of effects. Differential sensitivity or other factors may make it difficult to attribute causality when many variables are combined. To investigate causality, such indices may need to be separated into their components or analyzed using multivariate methods (Suter, 1993b). It may be difficult to interpret an index that combines measures of exposure and effects because double counting may occur or changes in one variable are linked to changes in another. Professional judgment plays a critical role in developing and applying indices.

Evaluating Measurement or Modeling Studies   The assessor's first task is to determine whether available studies are right for the ERA objectives. Ideally, each

TABLE 6.14   Evaluating a Study for Usefulness in ERA

- Are the study objectives relevant to the risk assessment?
- Are the study variables and conditions comparable to those in the risk assessment?
- Is the study design adequate to meet its objectives?
- Was the study conducted properly?
- How are variability and uncertainty treated and reported?

study should describe the purpose, methods used to collect data, and results. The assessor compares study objectives with those of the risk assessment for consistency. Questions to consider include whether the study met its objectives and whether the data are good enough to be used in the risk assessment. Finally, the risk assessor should identify any areas where existing data do not meet risk assessment needs. Smith and Shugart (1994) and the EPA (1990, 1994b) provide information on how to evaluate data and models. Data are rarely as complete as one would like, but eventually one must decide whether to make do with what is available, generate more with models, or collect new data.

Studies should contain enough information so analysts can reproduce the results, or at least so the details of the author's work can be reviewed. Ideally, the study's complete findings should be available, so the data may be reanalyzed if needed. For models, look for the source code and documentation. A complete report of model results will describe equations, parameter values, any parameter estimation techniques, and results. If a study seems useful but does not provide enough information to decide, consider contacting the principal investigator or other study participants or the model developer. The questions in Table 6.14 may also be helpful.

Assessors can examine the objectives and scope of studies that were designed for purposes other than the risk assessment at hand to identify important uncertainties and ensure that the information is used appropriately. For example, consider studies that measure condition (e.g., stream surveys, population surveys). While the measurements used to evaluate condition may be the same as the measures of effects identified in problem formulation, they must be linked

TABLE 6.15   Considering the Degree of Aggregation in Models

Wiegert and Bartell (1994) suggest the following points for evaluating the proper degree of aggregation or disaggregation:

1. Do not aggregate components with very different flux rates.
2. Do not greatly increase the disaggregation of the structural aspects of the model without similarly increasing the sophistication of the functional relationships and controls.
3. Disaggregate models only as much as required by the goals of the model to facilitate testing.

somehow with stressors to demonstrate causality. In the best case, this means that the stressor was measured at the same time and place as the effect.

What about a model developed for something other than risk assessment? To get an idea of whether it will work, check its intended application, theoretical framework, underlying assumptions, and limiting conditions. For example, a model for chemical transport in the water column alone is of little use for a lipophilic chemical that partitions into sediments.

The variables and conditions examined in a candidate study should be comparable to those identified in problem formulation. A study that examines animal habitat needs in the winter, for example, may miss important breeding-season requirements. Studies that minimize the amount of extrapolation needed are preferred. These are studies that describe:

- Measures of exposure, effects, or ecosystem and receptor characteristics identified in the analysis plan
- Time frame
- Ecosystem and location
- Environmental conditions
- Exposure route.

Study design and implementation can be evaluated to decide whether the study objectives were met and the information is good enough to be used in the risk assessment. Study design provides insight into the sources and magnitude of uncertainty associated with the results. Among the most important design issues of an effects study is whether it has enough statistical power to detect important differences or changes. Because this information is rarely reported (Peterman, 1990), the assessor may need to calculate the magnitude of an effect that could be detected under the study conditions (Rotenberry and Wiens, 1985) or choose another study if one is available.

Part of the exercise examines whether the study was conducted properly:

- For laboratory studies, were test conditions well controlled and control responses within acceptable bounds?
- For field studies, were potentially confounding variables identified and controlled, and were reference sites properly selected?
- For models, do the program's structure and logic make sense? Were algorithms specified correctly? Has the model been validated?

Evaluation is easier if standard methods or quality assurance/quality control (QA/QC) protocols were available and followed. However, it is still a good idea to consider whether the stated precision and accuracy were achieved and whether they are good enough for the risk assessment. For instance, if detection limits for one environmental medium are not achievable in another, it may not be possible to detect concentrations that might cause an effect. Study results can still be useful even if a nonstandard method was used, but it should be clear that the work was conducted properly.

Evaluating Uncertainty   In an ideal world we would be 100 percent sure about our predictions. But uncertainty creeps into all ERAs—through mistakes, assumptions, and imperfect knowledge. Uncertainty analysis can improve credibility because it explicitly describes the size and direction of errors (e.g., we overestimated exposure to be sure we did not miss anything), and it can show where additional data or analytical refinements will do the most good. The objective is to describe and, where possible, quantify what is known and not known about exposure and effects in the system of interest.

Sources of uncertainty include unclear communication about data and how they were manipulated, and errors in the information itself (descriptive errors). These can be examined by reviewing the information and the decisions made when handling it. The study documentation should allow the reader to see whether he or she agrees that the assessor's decisions are valid.

Uncertainties about parameter values include variability, uncertainty about a quantity's true value, and data gaps. The term *variability* is used here to mean true heterogeneity. Examples include variations in soil organic carbon, seasonal differences in animal diets, or differences in chemical sensitivity in different species. Variability is usually addressed in uncertainty analysis, although heterogeneity does not necessarily reflect a lack of knowledge and cannot usually be reduced by collecting more data. It can be described with a distribution or specific percentiles from it (e.g., mean and 95th percentile).

Uncertainty about a quantity's true value may apply to its size, place, or time of occurrence and can usually be reduced by collecting more information. It is described by sampling error (or *variance* in experiments) or measurement error. For biological responses, sampling error can greatly influence a study's ability to detect effects. Properly designed studies use sample sizes large enough to detect important differences, but unfortunately, many studies have sample sizes that are too small to detect anything but gross changes (Smith and Shugart, 1994; Peterman, 1990).

Spatial uncertainties may be described with geographic information systems (GISs). Strategies to reduce uncertainty include ground-truthing and ensuring that the spatial resolution meets the needs of the assessment. A growing literature addresses other issues associated with using spatial data such as colinearity and autocorrelation, boundary and scale effects, lack of true replication (Johnson and Gage, 1997; Fotheringham and Rogerson, 1993; Wiens and Parker, 1995).

Every risk assessor faces situations where data are unavailable or available only for parameters other than the ones really needed. Examples include using laboratory data to estimate a wild animal's response to a stressor or using a bioaccumulation factor measured in a different organism or ecosystem. These data gaps are usually bridged with a combination of scientific analyses, scientific judgment, and perhaps policy decisions. In deriving an ambient water quality criterion, for example, data and analyses are used to construct distributions of species sensitivity for a particular chemical. Policy (science policy in this case) determines how much data are enough and which analyses should be used. Scientific judgment is used to infer that species selected for testing will adequately account for the range of sensitivity of species in the environment. Policy defines the extent to which individual species should be protected (e.g., 90 vs. 95 percent of the species). It is critical to distinguish science from policy and to remember that risk assessment is not the place to make ethical judgments about what is good, bad, acceptable, or unacceptable.

One should make an effort to distinguish variability from uncertainties based on ignorance (e.g., uncertainty about a quantity's true value, or whether a supposed relationship really does exist) when interpreting and communicating results. For instance, in their heron and mink food-web models, MacIntosh et al. (1994) separated expected variability in individual animals' feeding habits from uncertainty about the mean chemical concentration in prey species. They could then place error bounds on the exposure distribution for the animals using the site and estimate the proportion of the animal population that might exceed a toxicity threshold.

When little is known about uncertainty, a useful approach is to compare risks associated with alternative sets of exposure and response assumptions (scenarios). Each scenario is carried through to risk characterization and checked for plausibility (the "laugh test"). Results can be presented as a series of point estimates with different aspects of uncertainty (e.g., exposure, sensitivity, etc.) reflected in each. Classical statistical methods (e.g., confidence limits, percentiles) can readily describe parameter uncertainty. For models, sensitivity analysis can be used to evaluate how the output changes with input variables, and uncertainty can be propagated and analyzed to see how individual parameters can affect overall confidence in the results. Software for Monte Carlo analysis makes is easy to use probabilistic methods; best practices are suggested by the EPA (1996a, 1997b). Other methods such as fuzzy mathematics and Bayesian methods are in the early stages of application to ERA. Note that any method can blur rather than clarify the effect of uncertainty on an assessment's results if used ineptly. A helpful review technique is to ask a colleague who is familiar with the issues but does not expect a particular outcome to check the results.

6.4.2   Characterization of Exposure

Exposure characterization describes potential or actual contact or co-occurrence of stressors with receptors. Two types of data—measures of exposure and measures of ecosystem and receptor characteristics—are used to evaluate stressor sources, receptor behavior, their distribution and movement in the environment, and the extent and pattern of contact or co-occurrence. The objective is to describe how receptors and stressors interact with the environment and each other, and the likelihood of any contact or co-occurrence. Exposure analysis describes

TABLE 6.16   Questions for Source Description

- Where does the stressor originate?
- What environmental media first receive stressors?
- Does the source generate other constituents that influence a stressor's eventual distribution in the environment?
- Are there other sources of the same stressor?
- Are there background sources?
- Is the source still active?
- Does the source produce a distinctive signature that can be seen in the environment or organisms?

exposure in terms of intensity, space, and time in units that can be combined with the effects assessment.

Describing Source(s)   A source can be defined in two general ways: as the place the stressor originates or is released (e.g., a smokestack, historically contaminated sediments) or the management practice or action (e.g., dredging) that produces stressors. In some assessments, the original sources may no longer exist and the source may be defined as the current location of the stressors. For example, contaminated sediments might be considered a source because the industrial plant that produced the chemicals that ended up there no longer operates. A source is the first component of the exposure pathway and influences where and when stressors eventually will be found. Table 6.16 provides some useful questions to consider when describing sources.

Source location and the environmental media that first receive stressors are two areas that deserve particular attention. For chemical stressors, one should also consider whether other constituents emitted by a source influence transport, transformation, or bioavailability of the stressor of interest. The presence of chloride in the feedstock of a coal-fired power plant influences whether mercury is emitted in divalent (e.g., as mercuric chloride) or elemental form (Meij, 1991), for example. In the best case, stressor generation is measured or modeled quantitatively.

Many stressors have natural counterparts or multiple sources, so it may be necessary to characterize these as well. Many chemicals occur naturally (e.g., most metals), are widespread due to other sources (e.g., polycyclic aromatic hydrocarbons in urban ecosystems), or have significant sources outside the boundaries of the current assessment (e.g., atmospheric nitrogen deposited in Chesapeake Bay). Many physical stressors also have natural counterparts. For instance, construction activities may release sediments into a stream in addition to those already coming from a naturally undercut bank. Human activities may also change the magnitude or frequency of natural disturbance cycles. For example, fire suppression may decrease the frequency but increase the severity of fires because fuel accumulates without small fires to consume it.

TABLE 6.17 Additional Questions for Introduction of Biological Stressors

• Is there an opportunity for repeated introduction or escape into the new environment?
• Will the organism be present on a transportable item?
• Are there mitigation requirements or conditions that would kill or impair the organism before entry, during transport, or at the port of entry?

There are several options for evaluating multiple sources:

• Focus on one source and calculate the incremental risks attributable to that source.
• Consider all sources of a stressor and calculate total risks attributable to that stressor. (Relative source attribution can be accomplished as a separate step.)
• Consider all stressors influencing an assessment endpoint and calculate cumulative risks to that endpoint.

Source characterization can be particularly important for introduced biological stressors, since many of the strategies for reducing risks focus on preventing entry in the first place (see Table 6.17). Once the source is identified, the likelihood of entry may be characterized qualitatively. In their risk analysis of Chilean log importation, for example, the assessment team concluded that the beetle *Hylurgus ligniperda* had a high potential for entry into the United States. Their conclusion was based on the beetle's attraction to freshly cut logs and tendency to burrow under the bark, which would provide protection during transport [U.S. Department of Agriculture (USDA), 1993].

Describing Distribution of Stressors or Disturbed Environment The second objective of exposure analysis is to where and when stressors occur in the environment. For physical stressors that directly alter or eliminate portions of the environment, the assessor describes the disturbed environment. Because exposure occurs when receptors co-occur with or contact stressors, this characterization is a prerequisite for estimating exposure. Stressor distribution is examined by evaluating pathways from the source as well as the formation and subsequent distribution of secondary stressors.

People usually think of exposure as actual contact with a stressor, but sometimes a stressor's presence can affect an organism even if there is no contact. For example, staff at a camp noticed a killdeer nest on the sandy beach. In an effort to protect the birds, they inserted sticks into the sand and marked the area with plastic flagging to prevent children from trampling the eggs, which are nearly impossible to see in the sand. Unfortunately, the birds abandoned the nest because they did not have a clear line of sight over the area surrounding the nest. Sticks and plastic tape are clearly not toxic, but the birds found the change to

TABLE 6.18 General Mechanisms of Transport and Dispersal

Physical, Chemical, and Biological Stressors

- By air currents
- In surface water (rivers, lakes, streams)
- Over and/or through the soil surface
- Through groundwater

Primarily Chemical Stressors

- Through the food web

Primarily Biological Stressors

- Splashing or raindrops
- Human activity (boats, campers)
- Passive transmittal by other organisms
- Biological vectors

their habitat intolerable. Many other species require very specific conditions for successful breeding (recall Table 6.9).

Stressors can be transported via many pathways (see Table 6.18). For a chemical stressor, partitioning into various media is estimated using physico-chemical properties such as solubility, vapor pressure, and soil–water distribution coefficient (K_d). For example, chemicals with low water solubility and high K_d tend to be associated with organic carbon—in places like soils, sediments, and biota. From there, the evaluation may examine how the contaminated medium is transported. If a mixture of chemicals is being evaluated, the analysis should consider how its composition may change over time or as it moves through the environment.

The attributes of physical stressors influence where they will go. The size of suspended particles determines where they will eventually deposit in a stream, for example. Physical stressors that destroy or remove ecosystems or portions of them (e.g., fishing or dam construction) often do not need to be modeled— the fish are harvested or the valley is flooded. For these direct disturbances, the challenge is usually to evaluate secondary stressors and effects.

Biological stressors move by diffusion and jump-dispersal (Simberloff and Alexander, 1994). *Diffusion* involves a gradual spread from the establishment site and is primarily a function of reproductive rates and motility. *Jump-dispersal* involves erratic spread, usually by means of a vector. The gypsy moth and zebra mussel have spread this way, the gypsy moth via egg masses on vehicles and the zebra mussel via ballast water. To evaluate dispersion, consider factors such as vector availability, attributes that enhance dispersal (e.g., ability to fly, adhere to objects, feed on multiple species, disperse reproductive units), and habitat or host needs.

Ecosystem characteristics influence all types of stressors; the challenge is to determine which aspects of the ecosystem are most important. Sometimes the factors that influence distribution are known; for example, fine sediments tend to accumulate in low-energy areas such as pools and backwaters. Other cases need more professional judgment. When evaluating whether an introduced insect will become established, for instance, it is useful to know whether the ecosystem is similar to the one where it originated, and whether it feeds on just one or a variety of species in its native habitat.

Secondary stressors may be even more of a concern than the primary stressor. For chemicals, the evaluation usually focuses on metabolites, biodegradation products, or chemicals formed through abiotic processes. As an example, microbial action increases the bioaccumulation of mercury by transforming inorganic forms to organic species. Ecosystem processes may also form secondary stressors: For example, nutrient inputs into an estuary can increase primary production and subsequent decomposition and deplete dissolved oxygen. A combination of factors usually determines bioavailability.

Physical disturbances can also generate secondary stressors, and identifying the one that most affects the assessment endpoint can be a difficult task. The removal of riparian vegetation, for example, may increase nutrient levels, stream temperature, sedimentation, and stream flow extremes. However, temperature change may have the greatest effect on adult salmon mortality in a particular stream.

If stressors have already been released, direct measurement of environmental media or a combination of modeling and measurement is ideal. Models make it possible to investigate the consequences of different management scenarios and may be necessary if the ecosystem cannot be measured directly. They are also useful for quantifying the relationship between sources and stressors: Oberts (1981) related land use to downstream suspended solids concentrations and Novitski (1979) and Johnston et al. (1990) used wetland extent to predict downstream flood peaks.

Describing Contact or Co-occurrence   The third objective is to describe the extent and pattern of co-occurrence or contact between stressors and receptors. This is critical—if there is no exposure, there can be no risk. The analysis may include situations where exposure may occur in the future, where exposure has occurred in the past but is not currently evident (e.g., in some retrospective assessments), and where food or habitat resources may be exposed, resulting in impacts to the assessment endpoint. Exposure can be described in terms of stressor and receptor co-occurrence, actual stressor contact with receptors, or stressor uptake by a receptor. The terms in which exposure is described depend on how the stressor causes adverse effects and how the stressor–response relationship is described. Questions for examining contact or co-occurrence are shown in Table 6.19.

Co-occurrence is particularly useful for evaluating stressors that can cause effects without physically contacting ecological receptors. Whooping cranes

TABLE 6.19   Questions To Ask in Describing Contact or Co-Occurrence

- Must the receptor actually contact the stressor for adverse effects to occur?
- Must the stressor be taken up into a receptor for adverse effects to occur?
- What receptor characteristics will influence contact or co-occurrence?
- Will abiotic ecosystem characteristics influence contact or co-occurrence?
- Will ecosystem processes or other interactions influence contact or co-occurrence?

provide a case in point: They use river sandbars as resting areas, and they prefer sandbars with unobstructed views. Without ever actually contacting the birds, dams can modify the flood regime that maintains the sandbars by scouring and redepositing sand, and obstructions such as bridges can interfere with resting behavior. Co-occurrence is evaluated by comparing stressor and receptor distributions. For instance, stressor location maps may be overlaid with maps of ecological receptors (e.g., bridge placement overlaid on maps showing historical crane resting habitat). Co-occurrence of a biological stressor and receptor may be used, for example, to evaluate exposure when introduced and native species compete for the same resources. GIS is an especially useful tool for evaluating co-occurrence.

Most stressors must contact receptors to cause an effect. Contact is a function of the amount or extent of a stressor in an environmental medium and activity or behavior of the receptors. For biological stressors, risk assessors may not have data and often rely on professional judgment to develop assumptions about when contact occurs.

For chemicals, contact is quantified as the amount of a chemical ingested, inhaled, or in material applied to the skin (potential dose). In its simplest form, it is quantified as an environmental concentration, with the assumption that the chemical is well mixed or that the organism moves randomly through the medium. This approach is commonly used for respired media (water for aquatic organisms, air for terrestrial organisms). Food and soil (ingested media) are usually treated by combining modeled or measured contaminant concentrations with assumptions or parameters describing the contact rate (Table 6.20).

Finally, some stressors must not only be contacted but also reach the target organ. A toxicant that causes liver tumors in fish, for example, must be absorbed and reach the liver to cause the effect. Uptake is evaluated by considering the stressor (e.g., a chemical's form or a pathogen's size), the medium (sorptive properties or presence of solvents), the biological membrane (integrity, permeability), and the organism (sickness, active uptake) (Suter et al., 1994). Because of interactions between these factors, uptake varies on a situation-specific basis. It is usually estimated by modifying a contact estimate with a bioavailability or absorption factor, with a pharmacokinetic model, or by measuring biomarkers or residues in receptors.

Ecosystem and receptor characteristics also influence exposure: For example, naturally anoxic areas above contaminated sediments in an estuary may reduce

TABLE 6.20   Example of an Exposure Equation: Calculating a Potential Dose via Ingestion

$$\text{ADD}_{\text{pot}} = C_k \cdot \text{FR}_k \cdot \text{NIR}_k \cdot n$$

where ADD_{pot} = potential average daily dose (e.g., in mg/kg-day)

C_k = average contaminant concentration in the kth type of food (e.g., in mg/kg wet weight)

FR_k = fraction of intake of the kth food type that is from the contaminated area (unitless)

NIR_k = normalized ingestion rate of the kth food type on a wet-weight basis (e.g., in kg food/kg body-weight-day)

n = number of contaminated food types

Note: A similar equation can be used to calculate uptake by adding an absorption factor that accounts for the fraction of the chemical in the kth food type that is absorbed into the organism.

The choice of potential dose or uptake depends on the form of the stressor–response relationship (EPA, 1993a).

the time bottom-feeding fish spend in contact with sediments and reduce their exposure to contaminants. Biotic interactions can also influence exposure: Competition for resources may force organisms into disturbed areas. The interaction between exposure and receptor behavior can influence both initial and subsequent exposures. Some chemicals may reduce organisms' ability to escape predators and thereby increase predator exposure to the chemical as well as the prey's risk of predation. Or organisms may avoid areas, food, or water with contamination they can detect. While avoidance can reduce exposure to chemicals, it may increase other risks by altering habitat usage or other behavior.

Intensity, temporal, and spatial dimensions should be considered when estimating exposure. Intensity is the most familiar dimension for chemical and biological stressors and may be expressed as the amount of chemical contacted per day or the number of pathogenic organisms per unit area. The temporal dimension comprises duration (the time over which exposure occurs), frequency (how often it occurs), and timing (when it occurs). If exposure is in repeated discrete events of about the same duration (e.g., high-flow events in streams), frequency is the important temporal dimension of exposure. If the repeated events have significant and variable durations, both duration and frequency should be considered. Finally, exposure timing, including the order or sequence of events, can be an important factor. Adirondack Mountain lakes receive high concentrations of hydrogen ions and aluminum during snow melt; this period corresponds to the sensitive life stages of some aquatic organisms.

Spatial extent is another exposure dimension. It is usually expressed as area (e.g., hectares of paved habitat, square meters that exceed a particular chemical threshold). At larger spatial scales, the shape or arrangement of exposure may be an important issue, and area alone may not be enough to describe spatial ex-

TABLE 6.21   Exposure Profile Questions

- How does exposure occur?
- What is exposed?
- How much exposure occurs? When and where does it occur?
- How does exposure vary?
- How uncertain are the exposure estimates?
- What is the likelihood that exposure will occur?

tent for risk assessment. Landscape ecology and GIS provide many options for analyzing and presenting the spatial dimension of exposure (e.g., Pastorok et al., 1996).

Exposure Profile   The final product of exposure analysis is an exposure profile that (1) identifies the receptor and (2) describes the exposure pathways and intensity and spatial and temporal extent of co-occurrence or contact. It also describes the impact of variability and uncertainty on exposure estimates and reaches a conclusion about the likelihood that exposure will occur (Table 6.21).

Exposure should be described in units that can be combined with the effects assessment. For example, exposure should be presented in units of mg/kg-day if the toxicity dose data use the same dose-metric. The objective is to ensure that the information needed for risk characterization has been collected and evaluated. Compiling the exposure profile also provides an opportunity to verify that the important exposure pathways identified in the conceptual model were evaluated. The profile should identify the entity for which the exposure estimates were developed. As an illustration, consider an assessment of risks to grebes feeding in a mercury-contaminated lake. The estimate could describe, say, the local population of grebes feeding there during the summer months.

If exposure can occur through many pathways, it may be useful to rank them, perhaps by their contribution to total exposure. The aforementioned grebes may be exposed to methyl mercury in fish that originated from historically contaminated sediments. They may also be exposed by drinking lake water, but comparing the two exposure pathways may show that the fish pathway contributes the vast majority of exposure to mercury. The assessor should also explain how each of the three exposure dimensions was treated. Grebe exposure might be expressed as the daily potential dose averaged over the summer months and over the extent of the lake. The profile should also describe how exposure can vary depending on receptor attributes or stressor levels: Exposure may be higher for grebes eating a larger proportion of bigger, more contaminated fish.

Variability can be described by using a distribution or by describing where a point estimate is expected to fall on a distribution. Cumulative distribution functions (CDFs) and probability density functions (PDFs) are two common presentation formats. Figures 6.5 and 6.6 show examples of cumulative frequency plots of exposure data. The point estimate/descriptor approach is used when there is

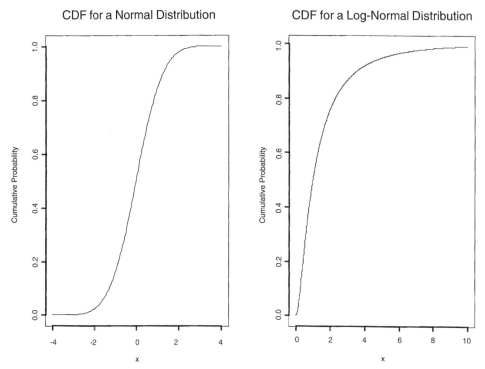

Figure 6.5   Plots of cumulative distribution function (CDF).

not enough information to describe a distribution. Descriptors such as *central tendency* (the mean or median of the distribution), *high end* (estimates that are expected to fall between the 90th and 99.9th percentile of the exposure distribution), and *bounding estimates* (those higher than any actual exposure) may prove helpful.

The exposure profile should summarize important uncertainties. In particular, the assessor should:

- Identify assumptions and describe how they were handled.
- Discuss (and quantify, if possible) the magnitude of sampling and/or measurement error.
- Identify the most sensitive variables influencing exposure.
- Identify which uncertainties can be reduced by collecting more data.

The information above is synthesized to reach a conclusion about the likelihood that exposure will occur. The exposure profile will be combined with the stressor–response profile (the product of the ecological effects characterization discussed in the next section) during risk characterization.

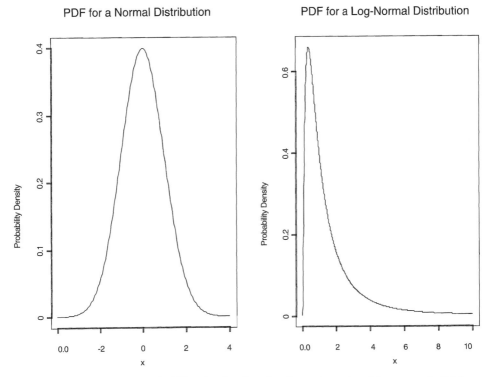

Figure 6.6 Plots for probability density function for a normal and log-normal distribution.

6.4.3 Characterization of Ecological Effects

Effects characterization describes the stressor's effects, links them to the assessment endpoints, and examines how they change with varying stressor levels. The characterization begins by evaluating data to specify the effects elicited, making sure they could actually occur for the chosen assessment endpoints (e.g., for enzyme induction, does the organism have that enzyme system?), and either confirm that assumed conditions are consistent with the conceptual model or modify the conceptual model as needed. Next, the assessor looks at how the stressor causes effects and how effects change with varying stressor levels, and then links the effects with the assessment endpoint. The objective is to describe the relationship between stressor levels and ecological effects and links between measurable ecological effects and assessment endpoints when the latter cannot be directly measured.

Stressor–Response Analysis To evaluate ecological risks, one must examine the relationships between stressors and responses postulated in the analysis plan. For example, an assessor may need point estimates of an effect (such as an LC_{50})

TABLE 6.22   Questions for Stressor–Response Analysis Questions

- Does the assessment require point estimates or stressor–response curves?
- Does the assessment require the establishment of a "no-effect" level?
- Would cumulative effects distributions be useful?
- Will analyses be used as input to a process model?

to compare the effects of several stressors. The shape of the stressor–response curve may be needed to determine the presence or absence of an effects threshold or for evaluating incremental risks, or stressor–response curves may be used as input for effects models. Table 6.22 provides some questions for stressor–response analysis. Assessors usually examine one response (e.g., mortality, mutations) and most quantitative techniques have been developed for univariate analysis. If the response of interest reflects many individual variables (e.g., species abundances in an aquatic community), multivariate techniques may be needed. If it is not possible to quantify stressor–response relationships, they may be described qualitatively (Table 6.23).

Stressor–response relationships are described using intensity, time, or space for consistency with exposure estimates. Intensity is most often used for chemicals (e.g., dose, concentration) and pathogens (e.g., spores per milliliter; propagules per unit of substrate). Point estimates and stressor–response curves can be generated for chemicals and some biological stressors, with the latter response being, for instance, number of organisms infected or actual signs of the pathogen (fruiting bodies, sclerotia, etc.). Exposure duration is also commonly used for chemical stressor–response relationships; for example, median acute effects levels are always associated with a time parameter (e.g., 24 hours). Space or area is usually used for physical stressors, for example, Thomas et al. (1990) related the chance of sighting a spotted owl to the extent of suitable habitat, and Phipps (1979) related tree growth to water table depth.

Risk assessors sometimes interpolate point estimates from a series of responses to which a curve has been fitted to determine particular levels of effect.

TABLE 6.23   Qualitative Stressor–Response Relationships

- The relationship between stressor and response can be described qualitatively, for instance, using categories of high, medium, and low, to describe the intensity of response given exposure to a stressor.
- For example, Pearlstine et al. (1985) assumed that seeds would not germinate if they were inundated with water at the critical time. This stressor–response relationship was described simply as a yes or no.
- In most cases, however, the objective is to describe quantitatively the intensity of response associated with exposure, and in the best case, to describe how intensity of response changes with incremental increases in exposure.

TABLE 6.24   Median Effect Levels

- Median effects are those effects elicited in 50% of the test organisms exposed to a stressor, typically chemical stressors.
- Median effect concentrations can be expressed in terms of lethality or mortality and are known as LC_{50} or LD_{50}, depending on whether concentrations (in the diet or in water) or doses (mg/kg) were used. Median effects other than lethality (e.g., effects on growth) are expressed as EC_{50} or ED_{50}.
- The median effect level is always associated with a time parameter (e.g., 24 or 48 hours). Because these tests seldom exceed 96 hours, their main value lies in evaluating short-term effects of chemicals. Stephan (1977) discusses several statistical methods to estimate the median effect level.

Median effect levels (Table 6.24) are popular because uncertainty is minimized at the midpoint of the regression curve. Table 6.25 shows how statistical hypothesis testing can be used to establish "no-effect" stressor levels based on comparisons between experimental treatments and controls. The risk assessor need not pick a particular effect level of concern; the no-effect level is determined by experimental conditions such as the number of replicates and data variability. This method is only as good as the experiment's power to detect changes, and the no-effect level must be reported.

When comparing sites to reference sites, remember that conditions must be carefully matched to minimize differences other than the stressor, and consider whether confounding factors such covariates should be included in any analysis. One way to match conditions and minimize covariates is to measure effects along a gradient extending from areas where the stressor is present into places where it is not, or from higher to lower concentrations.

If enough data are available, they can be combined to generate multiple-point

TABLE 6.25   No-Effect Levels Derived from Statistical Hypothesis Testing

- Statistical hypothesis tests have typically been used with chronic toxicity tests of chemical stressors that evaluate multiple endpoints.

  For each endpoint, the objective is to determine the highest test level for which effects are not statistically different from the controls (the no-observed-adverse-effect level, NOAEL) and the lowest level at which effects were statistically significant from the control (the lowest-observed-adverse-effect level, LOAEL).

- The range between the NOAEL and the LOAEL is sometimes called the maximum acceptable toxicant concentration, or MATC. The MATC, which can also be reported as the geometric mean of the NOAEL and the LOAEL (i.e., GMATC), provides a useful reference with which to compare toxicities of various chemical stressors.

- Reporting the results of chronic tests in terms of the MATC or GMATC has been widely used within the EPA for evaluating pesticides and industrial chemicals (e.g., Nabholz, 1991).

estimates that can be displayed as cumulative distribution functions. Figure 6.5 shows how this was done for species sensitivity derived from multiple-point estimates (EC_5's) for algae (and one vascular plant) exposed to an herbicide. These distributions can help identify stressor levels that affect a chosen proportion of species. Monte Carlo or other methods can also be used to generate cumulative distribution functions.

To evaluate multiple stressors, the most common approach is to combine stressor–response data for individual stressors. Or the relationship between response and the suite of stressors can be combined in one analysis. It is best to evaluate complex chemical mixtures present in environmental media (e.g., wastewater effluents, contaminated soils) directly, but it is important to consider the relationship between the samples tested and the potential spatial and temporal variability in the mixture. The approach taken depends on whether they can be measured and the assessment needs to project different stressor combinations or the mixture as a whole.

Establishing Cause-and-Effect Relationships (Causality)   Causality is the relationship between cause (a stressor) and effect (response). Without a clear connection between cause and effect, one cannot place much confidence in the ERA's conclusions. Causal relationships are especially important in risk assessments driven by observed effects such as bird or fish kills or a shift in species composition.

Evidence of causality may come from observed (e.g., bird kills are associated with field application of a pesticide) or experimental data (laboratory tests with the pesticides in question show bird kills at levels similar to those found in the field), and associations are even stronger when both types of data are available. But since we cannot always arrange an experiment, scientists have looked for other criteria to support an argument for cause and effect. Table 6.26 presents criteria based on Fox (1991) that are similar to others reviewed by Fox (U.S. Department of Health, Education, and Welfare, 1964; Hill, 1965; Susser, 1986a, b).

The strength of association between stressor and response is often the main reason that adverse effects such as bird kills are linked to specific events or actions. A stronger response to a hypothesized cause is more likely to indicate true causality. A response that follows a change in the hypothesized cause (predictive performance) provides additional evidence.

Demonstrated biological gradients or stressor–response relationships form another important criterion for causality. A stressor–response relationship may be linear, threshold, sigmoidal, or parabolic phenomenon. Biological gradients are generally effects that decrease with distance from a disturbance or toxic discharge.

Repeatedly demonstrated cause-and-effect relationships (consistency of association) also support causality. Consistency may be shown by multiple associations between stressor and response, occurrences in different ecosystems, or associations shown by different methods (Hill, 1965). Fox (1991) adds that in

TABLE 6.26   General Criteria for Causality [Adapted from Fox, (1991)]

Criteria Strongly Affirming Causality

- Strength of association
- Predictive performance
- Demonstration of a stressor–response relationship
- Consistency of association

Criteria Providing a Basis for Rejecting Causality

- Inconsistency in association
- Temporal incompatibility
- Factual implausibility

Other Relevant Criteria

- Specificity of association
- Theoretical and biological plausibility

ecoepidemiology, an association's occurrence in more than one species and population is very strong evidence for causation [e.g., the many bird species killed by carbofuran applications (Houseknecht, 1993)]. Fox also believes that causality is supported if the same incident is observed by different persons under different circumstances and at different times. Sometimes, a stressor may have a distinctive mode of action that suggests its role. Yoder and Rankin (1995) found that patterns of change observed in fish and benthic invertebrate communities could serve as indicators for different types of anthropogenic impact (e.g., nutrient enrichment vs. toxicity). Sometimes, a stressor may have a distinctive mode of action that suggests its role. Yoder and Rankin (1995) found that patterns of change observed in fish and benthic invertebrate communities could serve as indicators for different types of anthropogenic impact (e.g., nutrient enrichment vs. toxicity).

Conversely, inconsistent associations between stressor and response (e.g., the stressor is present without the expected effect, or the effect occurs but the stressor is not found) implies that there is no causal link between them. Temporal (i.e., the presumed cause does not precede the effect) and experimental or observational (factual implausibility) incompatibilities weaken the case for a causal relationship.

For some pathogens, the evaluations proposed by Koch (see Table 6.27) may be useful. For chemicals, ecotoxicologists have slightly modified Koch's postulates to provide evidence of causality (Suter, 1993a). The modifications are:

- The injury, dysfunction, or other putative effect of the toxicant must be regularly associated with exposure to the toxicant and any contributory causal factors.

TABLE 6.27 Koch's Postulates (Pelczar and Reid, 1972)

- A pathogen must be consistently found in association with a given disease.
- The pathogen must be isolated from the host and grown in pure culture.
- When inoculated into test animals, the same disease symptoms must be expressed.
- The pathogen must again be isolated from the test organism.

- Indicators of exposure to the toxicant must be found in the affected organisms.
- The toxic effects must be seen when organisms or communities are exposed to the toxicant under controlled conditions, and any contributory factors should be manifested in the same way during controlled exposures.
- The same indicators of exposure and effects must be identified in the controlled exposures as in the field.

Woodman and Cowling (1987) proposed three rules for establishing the effects of airborne pollutants on the health and productivity of forests: (1) the symptoms seen in individual trees in the forest must be found consistently with the stressor, (2) healthy trees must show the same symptoms when exposed to the stressor under controlled conditions, and (3) natural variation in resistance and susceptibility observed in forest trees must also be seen when clones are exposed to the stressor under controlled conditions.

Linking Measures of Effect to Assessment Endpoints Assessment endpoints cannot always be measured directly. When effects to assessment endpoints cannot be measured directly, explicit links between them are needed; Table 6.28 shows common extrapolations. During analysis, risk assessors should revisit the questions shown in Table 6.29 before proceeding with specific extrapolation approaches in analysis.

TABLE 6.28 Examples of Extrapolations to Link Measures of Effect to Assessment Endpoints

- Every risk assessment has data gaps that should be addressed, but it is not always possible to obtain more information. When there is a lack of time, monetary resources, or a practical means to acquire more data, extrapolations such as those listed below may be the only way to bridge gaps in available data. Extrapolations may be:

    Between taxa (e.g., bluegill to rainbow trout)

    Between responses (e.g., mortality to growth or reproduction)

    From laboratory to field

    Between geographic areas

    Between spatial scales

    From data collected over a short time frame to longer-term effects

TABLE 6.29   Questions Related to Selecting Extrapolation Approaches

- How specific is the assessment endpoint?
- Does the spatial or temporal extent of exposure require more receptors or extrapolation models?
- Are there enough data, and are they good enough for planned extrapolations and models?
- Is the proposed extrapolation technique consistent with ecological information?
- How much uncertainty is acceptable?

Whatever methods are employed to link assessment endpoints with measures of effect, it is important to apply them carefully using enough of the appropriate data. For example, extrapolations between two species may be more credible if factors such as similarities in food preferences, body mass, physiology, and seasonal behavior (e.g., mating and migration habits) are considered (Sample et al., 1996). Extrapolations and models are only as useful as the data on which they are based. Although data are widely available for chemical stressors and aquatic species, they do not exist for all taxa, stressors, or effects. Chemical effects data for wildlife, amphibians, and reptiles are quite limited, and there is even less information on most biological and physical stressors.

Although risks to organisms in the field are best estimated from site-specific studies, these data are seldom available, and risk assessors must extrapolate from laboratory data (see Table 6.30). Factors such as different exposure conditions, predation, competition, or other biotic or abiotic factors not evaluated in the

TABLE 6.30   Questions to Consider When Extrapolating from Effects Observed in the Laboratory to Field Effects of Chemicals

Exposure Factors

How will chemical fate and transport affect exposure in the field?

How comparable are exposure conditions and the timing of exposure? How comparable are the routes of exposure?

How do abiotic factors influence bioavailability and exposure?

What about the organism influences its chances of exposure?

How likely are preference or avoidance behaviors?

Effects Factors

What is known about the biotic and abiotic factors controlling populations of the organisms of concern?

To what degree are critical life-stage data available?

How may exposure to the same or other stressors in the field have altered organism sensitivity?

laboratory may all influence risk. When extrapolating from one geographic area to another (as well as from laboratory tests), consider variations in environmental conditions, spatial scales and heterogeneities, and ecological forcing functions.

Uncertainty factors are empirically derived numbers applied as measure of effects values to produce an estimated stressor level that should not cause adverse effects to the assessment endpoint. They have been developed for chemicals because extensive ecotoxicological data are available, especially for aquatic organisms. They are useful when decisions must be made about stressors in a short time and with little information. Despite their usefulness, uncertainty factors can be misused, especially when applied too conservatively, as when several factors are multiplied together without sufficient justification.

Process models for extrapolation are abstractions of a system or process (Starfield and Bleloch, 1991) that incorporate causal relationships and can generate predictions even when data are not availabile (Wiegert and Bartell, 1994). They make it possible to translate data on individual effects (e.g., mortality, growth, and reproduction) to potential alterations in specific populations, communities, or ecosystems. They can be used to evaluate risk questions about a stressor's effect on an assessment endpoint that cannot easily be tested in the laboratory. Population models describe a group of individuals through time and have been used in ecology and fisheries management and to assess the impacts of power plants and toxicants on specific fish populations (Barnthouse et al., 1987, 1990). See Barnthouse et al. (1986) and Wiegert and Bartell (1994) for reviews of population models. Emlen (1989) has reviewed population models that can be used for terrestrial risk assessment. Community and ecosystem models (e.g., Bartell et al., 1992; O'Neill et al., 1982) are useful when the assessment endpoint involves structural (e.g., community composition) or functional (e.g., primary production) elements. They can also be useful for examining secondary effects. Changes in various community or ecosystem components such as populations, functional types, feeding guilds, or environmental processes can be estimated.

Stressor–Response Profile   The final product of ecological response analysis is a summary of what effects a stressor may elicit. It is also a place to check to be sure that the assessment endpoints and measures of effect identified in the conceptual model were actually evaluated. A useful approach in preparing the stressor–response profile is to imagine that it will be used by someone else to prepare the risk characterization. Table 6.31 describes questions for consideration. The profile should express effects in terms of the assessment endpoint. If it was necessary to extrapolate from measures of effect to the assessment endpoint, or calculate a reference or benchmark dose, describe the extrapolation and its basis, the calculation, and any suspected uncertainties. For additional information on establishing reference concentrations, see Stephan et al. (1985), Van Leeuwen et al. (1992), Wagner and Løkke (1991), and Okkerman et al. (1993). Finally, the assessor should clearly describe major assumptions and default values used in the models.

TABLE 6.31   Questions Addressed by the Stressor–Response Profile

- What ecological entities are affected?
- What is the nature of the effect(s)?
- What is the intensity of the effect(s)?
- Where appropriate, what is the time scale for recovery?
- What causal information links the stressor with any observed effects?
- How do changes in measures of effects relate to changes in assessment endpoints?
- What is the uncertainty associated with the analysis?

6.5   RISK CHARACTERIZATION

Risk characterization (Figure 6.7) is the final phase of ERA, in which the risk assessor (1) clarifies the relationships between stressors, effects, and ecological entities and (2) reaches conclusions about exposure and the adversity of existing or anticipated effects. The assessor uses the results of analysis to estimate risk to the entities included in the assessment endpoints identified in problem formulation. Next, the assessor provides context—the significance of any adverse effects and lines of evidence supporting their likelihood. Finally, the assessor identifies and summarizes the uncertainties, assumptions, and qualifiers in the risk assessment and reports the conclusions.

Conclusions presented in the risk characterization should provide clear information to risk managers in order to be useful for environmental decision making (NRC, 1994). If the risks are not sufficiently defined to inform a management decision, risk managers may elect to repeat one or more phases of the risk assessment process (or, in the worst case, ignore the assessment altogether). Re-evaluating the conceptual model or conducting additional studies may improve the risk estimate. Or a monitoring program may help managers evaluate the consequences of a risk management decision.

6.5.1   Risk Estimation

Risk estimation is the process of integrating exposure and effects data and evaluating any associated uncertainties. The process uses exposure and stressor–response profiles developed according to the analysis plan. Risk estimates can be developed using: (1) field observational studies, (2) categorical rankings, (3) comparisons of single-point exposure and effects estimates, (4) comparisons incorporating the entire stressor–response relationship, (5) incorporation of variability in exposure and/or effects estimates, and (6) process models that rely partially or entirely on theoretical approximations of exposure and effects.

Results of Field Observational Studies   Field observational studies (surveys) provide empirical evidence linking exposure to effects. They measure biological changes in natural settings by collecting actual exposure and effects data for the

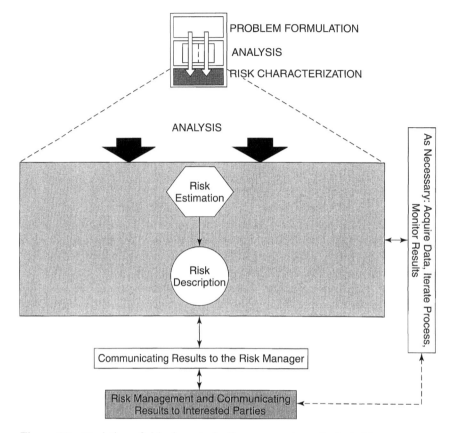

Figure 6.7   Depiction of risk characterization process in ecological risk assessment.

assessment endpoints or their surrogates. Their major advantage is that they can evaluate multiple stressors and complex ecosystem relationships that cannot be replicated in the laboratory. They can be designed to delineate both exposures and effects (including secondary effects) in natural systems, whereas estimates generated from laboratory studies generally delineate either exposures or effects under controlled or prescribed conditions (see Table 6.32).

While field studies best describe reality, as with other kinds of studies they can be limited by (1) a lack of replication, (2) bias in obtaining representative samples, or (3) failure to measure critical components of the system or random variations. Further, effects may be overlooked if the measurements lack the sensitivity to detect them.

If a field survey is used, the assessor should determine whether there is a causal relationship between stressors of interest and observed effects to ensure that they are not attributable to other stressors. Also note that field surveys taken

TABLE 6.32   Example of Field Methods Used for Risk Estimation

- Along with quotients comparing field measures of exposure with laboratory acute toxicity data (see Table 6.34), EPA evaluated the risks of granular carbofuran to birds based on incidents of bird kills following carbofuran applications. More than 40 incidents involving nearly 30 bird species were documented.
- Even though there were many problems with individual field studies (e.g., lack of appropriate control sites, poor data on carcass search efficiencies, no examination of potential synergistic effects of other pesticides, and lack of consideration of other potential receptors such as small mammals), there was so much evidence of mortality associated with carbofuran application that the deficiencies did not alter the assessment's conclusion of high risk (Houseknecht, 1993).

at one point in time cannot usually describe effects other than those associated with past and existing conditions.

Categories and Rankings   In some cases, professional judgment or other qualitative evaluation techniques may be used to rank risks using categories such as low/medium/high or yes/no. This is most often done when data are limited or are not easily expressed in quantitative terms. The U.S. Forest Service risk assessment of pest introduction from importation of logs from Chile used qualitative categories to address limitations in exposure and effects data for the introduced species of concern as well as the resources available for the assessment (see Table 6.33).

Ranking techniques are often used to translate qualitative judgment into mathematical comparisons in comparative risk exercises. For example, Harris et al. (1994) evaluated risk reduction opportunities in Green Bay, Wisconsin, employing an expert panel to compare the relative risk of several stressors against their potential effects. They prepared a mathematical analysis based on fuzzy set theory to rank the risk from each stressor from a number of perspectives, including degree of immediate risk, duration of impacts, and prevention and remediation management.

TABLE 6.33   Using Qualitative Categories to Estimate Risks of an Introduced Species

- The importation of logs from Chile required an assessment of the risks posed by the potential introduction of the bark beetle, *Hylurgus ligniperda* (USDA, 1993).
- Experts judged the potential for colonization and spread of the species, and their opinions were expressed as high, medium, or low as to the likelihood of establishment (exposure) or consequential effects of the beetle. Uncertainties were similarly expressed.
- A ranking scheme was then used to sum the individual elements into an overall estimate of risk (high, medium, or low). Narrative explanations of risk accompanied the overall rankings.

Comparison of point estimates

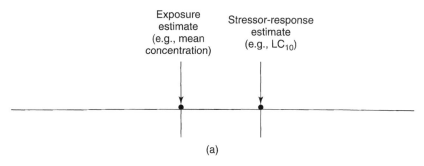

(a)

Comparison of a point estimate of a stressor-response
relationship with uncertainty associated with an exposure
point estimate

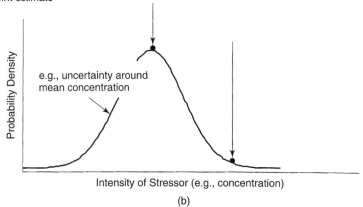

Intensity of Stressor (e.g., concentration)

(b)

Figure 6.8 Risk estimation techniques. a. Comparison of exposure and stressor-
response point estimates. b. Comparison of a point estimate from the stressor-response
relationship with uncertainty associated with an exposure point estimate.

Single-Point Exposure and Effects Comparisons Are there are enough data to
quantify exposure and effects estimates, the simplest comparative approach is
a ratio or quotient (Figure 6.8*a*); this is usually expressed as an exposure concentration divided by an effects concentration. Quotients are most commonly
used for chemical stressors, for which reference or benchmark toxicity values are
widely available (see Table 6.34).

The principal advantages of the quotient method are that (1) it is simple and
quick to use and (2) risk assessors and managers are familiar with its application. It provides an efficient, inexpensive means of identifying high- or low-risk
situations that can allow risk management decisions to be made without the
need for further information.

Quotients may also be used to integrate the risks of multiple chemical stres-

TABLE 6.34   Applying the Quotient Method

- When applying the quotient method to chemical stressors, the effects concentration or dose (e.g., an LC_{50}, LD_{50}, EC_{50}, ED_{50}, NOAEL, or LOAEL) is frequently adjusted by uncertainty factors before division into the exposure number (EPA, 1984; Nabholz, 1991; see Section 6.4.3), although EPA used a slightly different approach in estimating the risks to the survival of birds that forage in agricultural areas where the pesticide granular carbofuran is applied (Houseknecht, 1993).

- In this case, EPA calculated the quotient by dividing the estimated exposure to carbofuran granules in surface soils (expressed as number/ft^2) by the granules/LD_{50} derived from single-dose avian toxicity tests. The calculation yields values in units of LD_{50}/ft^2.

- The assessors assumed that a higher quotient value corresponded to an increased likelihood that a bird would be exposed to lethal levels of granular carbofuran at the soil surface.

- They then estimated minimum and maximum values for LD_{50}/ft^2 for songbirds, upland game birds, and waterfowl that may forage within or near 10 different agricultural crops.

---

sors: First the assessor generates quotients for individual constituents by dividing each exposure level by a corresponding toxicity endpoint (e.g., LC_{50}, EC_{50}, NOAEL) and then sums the results. Note that since this procedure assumes that toxicities are additive or approximately so, this assumption should be explicitly acknowledged in the uncertainty discussion. It is most appropriate when the modes of action of chemicals in a mixture are similar, but there is evidence that additive or near-additive interactions are common even among chemicals with dissimilar modes of action (Könemann, 1981; Broderius, 1991; Broderius et al., 1995; Hermens et al., 1984a, b; McCarty and Mackay, 1993; Sawyer and Safe, 1985).

The quotient method does have its limitations (see Smith and Cairns, 1993; Suter, 1993a): While a quotient can show whether risks are high or low, it probably cannot help a risk manager who needs to reduce risk by some increment. For example, a statement that a mitigation approach will reduce a quotient from 25 to 12 is not helpful because it does not describe what that means for the assessment endpoint. It is also limited by disparities between available data and risk questions. For example, an LC_{50} derived from a 96-hour laboratory test using constant exposure levels is not the best metric for evaluating short-term, pulsed exposures. In addition, the strength of correlation weakens when the quotient method is used, predicting secondary effects such as eutrophication, loss of prey, and opportunities for invasive species.

Finally, the quotient method does not usually consider uncertainty (e.g., extrapolation from the test species). However, some uncertainties can be incorporated into single-point estimates to determine the likelihood that the effect point estimate exceeds the exposure point estimate (Figures 6.8b and 6.9). If exposure

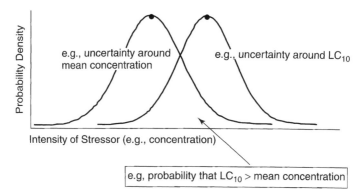

Figure 6.9 Risk estimation techniques: comparison of point estimates with associated uncertainties.

variability is known, then the point estimate of effects can be compared with a cumulative exposure distribution as described in Table 6.35. Despite these limitations, the quotient method remains popular because it is easy to use.

Comparisons Incorporating Entire Stressor–Response Relationship If a curve relating the stressor level to the magnitude of response is available, then one can examine risks associated with many different levels of exposure (Figure 6.10). This approach is particularly useful when the risk assessment outcome is not based on a predetermined decision rule such as a toxicity benchmark level. It also has greater predictive power than the quotient method.

One advantage of such comparisons is that the slope of the effects curve shows the magnitude of change in effects associated with incremental changes

TABLE 6.35   Comparing an Exposure Distribution with a Point Estimate of Effects

- The EPA Office of Pollution Prevention and Toxics uses a Probabilistic Dilution Model (PDM3) to generate a distribution of daily average chemical concentrations based on estimated stream flow variations in a model system.
- The PDM3 compares this exposure distribution with an aquatic toxicity test endpoint to estimate how many days exposure exceeds the endpoint concentration in a 1-year period (Nabholz et al., 1993).
- The frequency of exceedance is based on the duration of the toxicity test used to derive the effects endpoint. Thus, if the endpoint was an acute toxicity level of concern, an exceedance would be identified if the level of concern was exceeded for 4 days or more (not necessarily consecutive).
- The exposure estimates are conservative in that they assume instantaneous mixing of the chemical in the water column and no losses due to physical, chemical, or biodegradation effects.

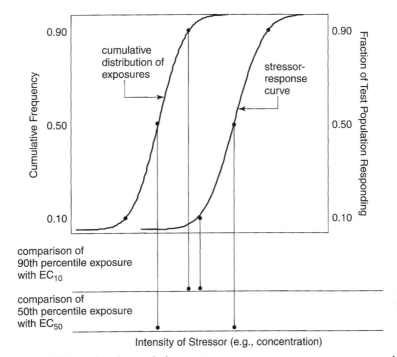

Figure 6.10 Risk estimation techniques: stressor-response curve versus a cumulative distribution of exposures.

in exposure, and changes in the magnitude and likelihood of effects for different exposure scenarios can be used to compare different risk management options. And uncertainty can be incorporated by calculating uncertainty bounds on the stressor–response or exposure estimates.

Like the quotient method, comparisons may be limited by discontinuities with problem formulation and analysis. These limitations include incomplete ability to address secondary effects, the assumption that the exposure pattern used to derive the stressor–response curve matches actual conditions, and failure to consider uncertainties such as extrapolations from tested species.

Comparisons Incorporating Variability in Exposure and Effects If the exposure or stressor–response profiles describe variability, then many different risk estimates can be calculated. Exposure variability can be used to estimate risks to a moderately or highly exposed population segment, while effects variability can be used to estimate risks to average or sensitive groups. This approach can predict changes in the magnitude and likelihood of effects for different exposure scenarios, which makes it especially useful for comparing different risk management options. It also allows one to identify and quantify risks to different segments of the population.

TABLE 6.36   Comparing Cumulative Exposure and Effects Distributions for Chemical Stressors

- Exposure distributions for chemical stressors can be compared with effects distributions derived from point estimates of acute or chronic toxicity values for different species (e.g., Health Council of the Netherlands, 1993; Cardwell et al., 1993; Baker et al., 1994; Solomon et al., 1996). As shown in Figure 5.5 of the EPA ecological risk guidelines, a distribution of exposure concentrations of a herbicide compared with single-species toxicity data for algae (and one vascular plant species) for the same chemical. The degree of overlap of the curves indicates the likelihood that a certain percentage of species may be adversely affected, in that example, the 10th centile of algal species' EC_5 values is exceeded less than 10% of the time.
- The predictive value of this approach is evident. The risk reduction that could be achieved by proposed risk mitigation options can be readily determined by comparing the distribution of expected exposure changes with the effects distribution curve.
- When using effects distributions derived from single-species toxicity data, risk assessors should consider the following questions:

    Do the species for which toxicity test data are available represent those present in the environment?

    Are particularly sensitive (or insensitive) groups of organisms represented in the distribution?

    If a criterion level is selected—for example, protect 95% of species—does the 5% of potentially affected species include organisms of ecological, commercial, or recreational significance?

Limitations include high data requirements and the implicit assumption that the full range of variability in the exposure and effects data is adequately represented. And secondary effects are not readily evaluated with this technique. The assessor may want to corroborate such estimates with field studies or other lines of evidence. Table 6.36 describes the use of cumulative exposure and effects distributions for estimating risk.

Application of Process Models   Process models are mathematical representations of the system under evaluation. They can be useful in both analysis and risk characterization, and are particularly useful for examining "what if" scenarios and for forecasting beyond the limits of observed data. Process models can also consider secondary effects. In addition, some can forecast the combined effects of multiple stressors, such as the effects of multiple chemicals on fish population sustainability (Barnthouse et al., 1990).

Process model outputs may be point estimates, distributions, or correlations; in all cases, careful interpretation is in order; they may imply more certainty than exists and all too often their underlying assumptions accepted on faith. Model outputs seldom express underlying ignorance about species' basic life histories or an ecosystem's structure and function. Since process models are only as good as the assumptions on which they are based, they should be treated as hypothetical descriptions until validated with empirical data. This comparison can show whether our understanding of the system was correct (Johnson, 1995), particularly with respect to the risk questions presented in problem formulation.

6.5.2 Risk Description

After preparing a risk estimate, risk assessors explain the supporting information by evaluating the lines of evidence and the significance of the adverse effects. Ideally, the relationship between the assessment endpoints, measures of effect, and associated lines of evidence will have already been established during analysis. If not, the risk assessor can qualitatively link lines of evidence to the assessment endpoints. Regardless of the technique, the technical narrative supporting a risk estimate is as important as the estimate itself.

Lines of Evidence   The development of lines of evidence provides both a process and a framework for reaching a conclusion regarding confidence in the risk estimate. It is not the kind of proof demanded by experimentalists (Fox, 1991), nor is it a rigorous examination of weights of evidence. The lines-of-evidence approach evaluates all available information, even qualitative evidence. Note that it is very different from the weight-of-evidence approach, which assigns quantitative values to balance opposing factors and reach a conclusion about a "weight." Risk assessors should thoroughly examine all lines of evidence rather than simply reduce them to a system of numeric calculations and results.

Confidence in an ERA's conclusions may be increased by using several lines of evidence to interpret and compare risk estimates. These may be derived from different sources or by different techniques, such as quotient estimates, modeling results, or field observational studies. Risk assessors consider three factors when evaluating lines of evidence: (1) data quality and adequacy, (2) degree and type of uncertainty associated with the evidence, and (3) relationship of the evidence to the risk assessment questions.

Data quality directly influences confidence in a study's results and any conclusions drawn from it. Considerations include whether the experimental design was appropriate for the questions posed in a particular study and whether data quality objectives were clear and attained. One must understand natural variability in assessment endpoint attributes to determine whether there were sufficient data to satisfy the analyses chosen and whether the analyses were sensitive and robust enough to distinguish changes caused by the stressors from other changes.

Each line of evidence has its own set of uncertainties. One major source is extrapolations; the more extrapolations, the more uncertainty introduced into a study. Risk assessors should consider these and any other sources of uncertainty when evaluating the relative importance of particular lines of evidence.

Finally, how directly lines of evidence relate to the risk questions may determine their importance in terms of the assessment endpoint. The strongest lines of evidence are those directly related to the risk questions, and those that establish a cause-and-effect relationship based on a definitive mechanism rather than associations alone.

However, the evaluation process is more that just listing the evidence that supports or refutes the risk estimate. The risk assessor should carefully examine each factor and evaluate its contribution in the risk assessment's context. The importance of lines of evidence is that each and every factor is described and interpreted. Data or study results are often not reported or carried forward in the risk assessment because they are of insufficient quality. Note that such data may still be valuable if they provide information about topics such as how methodologies could be improved and recommendations for further studies.

Sometimes lines of evidence suggest different conclusions. In this case, one should investigate possible reasons for any disagreement rather than ignore inconvenient evidence, perhaps starting by distinguishing between true inconsistencies and due to differences in statistical power. For example, a model may predict adverse effects that were not observed in a field survey. The risk assessor might ask whether the field study design could have detected the predicted difference, and whether model endpoints were comparable with those measured in the study. Conversely, the model may have made unrealistic predictions. While additional iterations and data collection may help resolve uncertainties, resources for this option are not always available.

Lines of evidence that are to be evaluated during risk characterization should be defined early in the risk assessment (during problem formulation) through the development of the conceptual model and selection of assessment endpoints. Further, the analysis plan should incorporate measures that will help interpret the lines of evidence, including methods of reviewing, analyzing, and summarizing the uncertainty in the risk assessment.

Determining Adversity   At this point, the assessment has estimated expected changes in the assessment endpoints and evaluated lines of evidence. The next step is to explain whether these changes are in fact adverse. For our purposes, adverse effects are changes that are undesirable because they alter valued structural or functional attributes of the entities of interest. The risk assessor evaluates the degree of adversity, a challenging task frequently based on professional judgment. The assessor should also recognize that ecological effects will be considered along with economic, legal, or social factors when a risk management decision is made, and be prepared to discuss adverse effects in that context if necessary. The risk manager will use all this information to decide whether a particular adverse effect is acceptable, and may also find it useful when communicating the risk to I&APs.

The criteria for evaluating adverse changes are (1) nature and intensity of effects, (2) spatial and temporal scale, and (3) recovery potential.

TABLE 6.37 What Are Statistically Significant Effects?

- Statistical testing is a "statistical procedure or decision rule that leads to establishing the truth or falsity of a hypothesis ..." (Alder and Roessler, 1972).
- Statistical significance is based on the amount of data, the nature of their distribution, whether intertreatment variance exceeds intratreatment variance in the data, and the a priori significance level (∀).
- The types of statistical tests and the appropriate protocols (e.g., power of test) for these tests should be established as part of the analysis plan.

The nature and intensity of effects help distinguish adverse changes from normal ecosystem variability or those resulting in little or no significant change. For example, for an assessment endpoint involving survival, growth, and reproduction of a species, do predicted effects involve survival and reproduction or only growth? If offspring survival will be affected, by what percentage will it diminish?

Although statistical significance does not necessarily mean ecological significance, risk assessors should consider both ecological and statistical effects when evaluating intensity. For example, a statistically significant 1% decrease in fish growth (see Table 6.37) may not be relevant to an assessment endpoint of fish population viability because other factors such as food supply, disease, and predation are probably more important. And a 10% decline in reproduction may be worse for a population of slowly reproducing trees than for rapidly reproducing planktonic algae.

Natural ecosystem variation can make it difficult to detect stressor-related perturbations. For example, normal intra- and interannual marine fish population variability is several orders of magnitude. Furthermore, cyclic events (e.g., migration, tides) are very important in natural systems and may mask or delay stressor-related effects. It can be very difficult to predict the effects of anthropogenic stressors against this background of variation. Thus, a field study's inability to detect statistically significant effects does not automatically mean that there are none; rather, risk assessors should consider other lines of evidence in reaching their conclusions.

It is also important to consider the location of the effect within the biological hierarchy and the mechanisms that may result in ecological changes. The risk assessor may rely on mechanistic explanations to describe complex ecological interactions and the resulting effects that otherwise may be masked by variability in the ecological components.

The boundaries (global, landscape, ecosystem, organism) of the risk assessment are initially identified in the analysis plan prepared during problem formulation. These spatial and temporal scales are further defined in the analysis phase, where specific exposure and effects scenarios are evaluated. The spatial dimension encompasses both the extent and pattern of effect as well as the context of the effect within the landscape. Factors to consider include the absolute area affected, the extent of critical habitats affected compared with a larger area of interest, and the role or use of the affected area within the landscape.

Adverse effects to assessment endpoints vary with the absolute area of the effect. A larger affected area may be (1) subject to a greater number of other stressors, increasing the complications from stressor interactions, (2) more likely to contain sensitive species or habitats, or (3) more susceptible to landscape-level changes because many ecosystems may be altered by the stressors.

Nevertheless, a smaller area of effect is not always associated with lower risk. The function of an area within the landscape may be more important than the absolute area. Destruction of small but unique areas, such as critical wetlands, may have important effects on local and regional wildlife populations. Also, in river systems, both riffle and pool areas provide important microhabitats that maintain the structure and function of the total river ecosystem. Stressors acting on these microhabitats may result in adverse effects to the entire system.

Spatial factors are important for many species because of the links between ecological landscapes and population dynamics. Links between landscapes can provide refuge for affected populations, and organisms may require corridors between habitat patches for successful migration.

The temporal scale for ecosystems can vary from seconds (photosynthesis, prokaryotic reproduction) to centuries (global climate change). Changes within a forest ecosystem can occur gradually over decades or centuries and may be affected by slowly changing external factors such as climate. When interpreting adversity, risk assessors should recognize that any stressor-induced changes occur within the context of multiple natural time scales. In addition, ecosystem changes may involve intrinsic time lags; so observable responses to a stressor may be delayed. Analysts should make an effort to distinguish a stressor's long-term impacts from its immediately visible effects. For example, visible changes resulting from eutrophication of aquatic systems (turbidity, excessive macrophyte growth, population decline) may not become evident for many years after initial increases in nutrient levels.

Considering the temporal scale of adverse effects leads logically to a consideration of recovery. Recovery is the rate and extent of return of a population or community to some aspect of its condition prior to a stressor's introduction. (While this discussion deals with recovery as a result of natural processes, risk mitigation options may include restoration activities to facilitate or speed up the recovery process.) Because ecosystems are dynamic and, even under natural conditions, constantly changing in response to changes in the physical environment (e.g., weather, natural disturbances) or other factors, it is unrealistic to expect that a system will remain static at some level or return to exactly the same state that it was before it was disturbed (Landis et al., 1993). Therefore, the assessment should be clear about what attributes define "recovered" in a particular system. Examples might include productivity declines in a eutrophic system, reestablishment of a species at a particular density, species recolonization of a damaged habitat, or the restoration of health of diseased organisms.

Recovery can be evaluated in spite of the difficulty in predicting events in ecosystems (e.g., Niemi et al., 1990). For example, it is possible to distinguish changes that are usually reversible (e.g., stream recovery from sewage effluent discharge), frequently irreversible (e.g., establishment of introduced species), and always irreversible (e.g., extinction). Risk assessors should consider the potential irreversibility of significant structural or functional changes in ecosystems or ecosystem components when evaluating adversity. Physical alterations such as deforestation in the coastal hills of Venezuela in recent history and in Britain during the Neolithic period, for example, changed soil structure and seed sources such that forests cannot easily grow again (Fisher and Woodmansee, 1994).

The relative rate of recovery can also be estimated. For instance, fish populations in a stream are likely to recover much faster from exposure to a degradable chemical than from habitat alterations resulting from stream channelization. Risk assessors can use knowledge of factors, such as the temporal scales of organisms' life histories, the availability of adequate stock for recruitment, and the interspecific and trophic dynamics of the populations, in evaluating the relative rates of recovery. A fisheries stock or forest might recover in decades, a benthic invertebrate community in years, and a planktonic community in weeks to months.

Risk assessors should note natural disturbance patterns when evaluating the likelihood of recovery from anthropogenic stressors. Alternatively, if an ecosystem has adapted to a disturbance pattern, it may be affected when the disturbance is removed (e.g., fire-maintained grasslands). The lack of natural analogs makes it difficult to predict recovery from uniquely anthropogenic stressors (e.g., synthetic chemicals).

6.5.3 Reporting Risks

When risk characterization is complete, risk assessors should be able to estimate ecological risks, indicate the overall degree of confidence in the risk estimates, cite lines of evidence supporting the risk estimates, and interpret the adversity of ecological effects. Usually this information is included in a risk assessment report (sometimes called a risk characterization report). The risk assessor should consider the elements listed in Table 6.38 when preparing a risk assessment report.

Like the risk assessment itself, a risk assessment report may be brief or extensive, depending on the nature of and the resources available for the assessment. While it is important to address the elements described in Table 6.38, risk assessors should judge the level of detail required. The report need not be overly complex or lengthy; it is most important that the information required to inform a risk management decision be presented clearly and concisely.

To ensure that readers can understand the ERA and eventually use its conclusions, its results should be *clear* (written in ordinary language accessible to others outside the risk assessment field) and *transparent* (a reader can tell how analyses and conclusions were developed). It also helps if they are *reasonable*

TABLE 6.38   Possible Risk Assessment Report Elements

- Describe planning results.
- Review the conceptual model and the assessment endpoints.
- Describe major data sources and the analytical procedures used.
- Review the stressor–response and exposure profiles.
- Describe risks to the assessment endpoints using risk estimates and adversity evaluations.
- Review and summarize major areas of uncertainty (as well as their direction) and the approaches used to address them.
  - Discuss the degree of scientific consensus in key areas of uncertainty.
  - Identify major data gaps and, if needed, state whether additional data would increase confidence in the assessment results.
  - Discuss science policy judgments or assumptions used to bridge information gaps and the basis for these assumptions.
  - Discuss how the elements of quantitative uncertainty analysis are embedded in the estimate of risk.

(do not contain any unusual or unprecedented conclusions) and *consistent* with other similar analyses. Table 6.39 describes ways to achieve such characteristics.

6.6   RELATING ECOLOGICAL INFORMATION TO RISK MANAGEMENT DECISIONS (COMMUNICATING RISK)

After characterizing risks and preparing a report, risk assessors discuss the results with risk managers. Risk managers use risk assessment results, along with other factors (e.g., economic or legal concerns), in making risk management decisions and as a basis for communicating risks to I&APs and the general public. "Resources" means more than just dollars—it may include personnel, equipment, time, and even political capital.

It is in the assessor's interest to present the results in terms that will be meaningful to the decision maker and other readers: It is very frustrating to have worked hard on a document that is not used to develop the decision it was supposed to inform; so it is a good idea to find out what format the decision maker prefers. If this person only wants the big picture, prepare a summary that tells where supporting information can be found if needed. Or if your reader typically searches for details, by all means provide them—do what it takes to get the assessment read. Although the risk characterization should include enough details somewhere so readers can see clearly how its conclusions were arrived at, eventually the results usually need to be distilled into a smaller package. Depending on the audience, a detailed or "big-picture" extract or summary may be needed. This may seem like a lot of trouble on the assessor's part to produce a small convenience for the reader, but it is well worth the effort if it makes the differ-

TABLE 6.39   Clear, Transparent, Reasonable, and Consistent Risk Characterizations

For Clarity

- Be brief; avoid jargon.
- Make language and organization understandable to risk managers and the informed layperson.
- Fully discuss and explain unusual issues specific to a particular risk assessment.

For Transparency

- Identify the scientific conclusions separately from policy judgments.
- Clearly articulate major differing viewpoints of scientific judgments.
- Define and explain the risk assessment purpose (e.g., regulatory purpose, policy analysis, priority setting).
- Fully explain assumptions and biases (scientific and policy).

For Reasonableness

- Integrate all components into an overall conclusion of risk that is complete, informative, and useful in decision making.
- Acknowledge uncertainties and assumptions in a forthright manner.
- Describe key data as experimental, state-of-the-art, or generally accepted scientific knowledge.
- Identify reasonable alternatives and conclusions that can be derived from the data.
- Define the level of effort (e.g., quick screen, extensive characterization) along with the reason(s) for selecting this level of effort.
- Explain the status of peer review.

For Consistency with Other Risk Characterizations

- Describe how the risks posed by one set of stressors compare with the risks posed by a similar stressor(s) or similar environmental conditions.
- Indicate how the strengths and limitations of the assessment compare with past assessments.

---

ence between making a real contribution to an environmental protection action and being ignored.

Remember also that many people have difficulty with statistics or have developed a mistrust of them. If such readers are part of the audience, straightforward explanations are helpful. And remember that simpler techniques may answer every assessment question. Huff and Geis (1954) have examples of simple, intuitive explanations that may help make statistics more accessible to readers.

If a report will be prepared for an audience with limited literacy or for whom English is a second language, information can be presented by other means. For example, "The State of the Crocodile River" report made its case with map of

TABLE 6.40   Questions Regarding Risk Assessment Results

Questions Principally for Risk Assessors to Ask Risk Managers

- Do you know enough about the risk to make your decision?
- Was the right problem analyzed?
- Was the problem adequately characterized?

Questions Principally for Risk Managers to Ask Risk Assessors

- What effects might occur?
- How bad are they?
- How likely is it that they will occur?
- When and where will they occur?
- How confident are you in your conclusions?
- Do you need more information?
- Do you need to go through the ERA again?
- Will we need to monitor the results of the risk management decision?

Adapted From EPA (1993b).

a river accompanied by pictures of fish, crustaceans, other invertebrates, and plants, with the organisms colored green, yellow, and red to illustrate good, fair, and poor status (South African River Health Programme, 1998).

The questions in Table 6.40 can help risk assessors and risk managers understand each other in discussions about ERA results. Risk managers need to know the major risks to assessment endpoints and get an idea of whether the conclusions are strongly or weakly supported. Insufficient resources, lack of consensus, or other factors may make it impossible to prepare a detailed and well-documented risk characterization. If this is the case, the risk assessor should carefully explain any issues, obstacles, and correctable shortcomings for the risk manager's consideration.

Risk managers also consider social, economic, political, legal, and other issues along with risk assessment results when they make risk management decisions. For example, the risk assessment results may be used in cost–benefit analysis, which may require translating resources (identified through the assessment endpoints) into monetary values. Nontraditional economic considerations such as intergenerational resource values or issues of long-term or irreversible effects cannot be monetized (Costanza et al., 1997); however, they may provide a way to compare the risk assessment results in consistent units such as costs. Risk managers may also consider alternative risk reduction strategies, such as risk mitigation options or substitutions based on relative risk comparisons. For example, mitigation techniques such as buffer strips or lower field application rates can be used to reduce the exposure (and risk) of a pesticide. Finally, risk managers consider public opinion and political demands. Collectively, these other factors may render very high risks acceptable or very low risks unacceptable.

TABLE 6.41   Suggestions for Successful Risk Communication

- Plan carefully and evaluate the success of your communication efforts.
- Coordinate and collaborate with other credible sources.
- Accept and involve the public as a legitimate partner.
- Listen to the public's specific concerns.
- Be honest, frank, and open.
- Speak clearly and with compassion.
- Meet the needs of the media.

From EPA, 1995b.

Risk characterization provides the basis for communicating ecological risks to I&APs and the general public. This task is usually the responsibility of risk managers, but it may be shared with risk assessors. An even better approach is to use communication specialists familiar with common concerns about risk and effective strategies for addressing them. Although the final version of a risk assessment prepared for the government is usually made available to the public, the communication process is best served by tailoring information to a particular audience. This is the time to refer back to the questions posed by I&APs if they were involved in planning the assessment and the conceptual model. Table 6.41 offers suggestions for successful risk communication.

Risk managers typically also decide whether additional follow-up activities are needed. Depending on the importance and visibility of the assessment, confidence in its results, and available resources, it may be advisable to conduct another iteration of the risk assessment in order to inform a final management decision. Another option is to proceed with the decision, implement management action, and develop a monitoring plan to evaluate the results (see Section 6.1). If the decision is to mitigate risks through exposure reduction, for example, monitoring could help determine whether the action taken really did reduce exposure (and effects).

Glossary*

Adverse ecological effects   Changes considered undesirable because they alter valued structural or functional characteristics of ecosystems or their components.

Assessment endpoint   An explicit expression of the environmental value that is to be protected, operationally defined by an ecological entity and its attributes. For example, salmon are valued ecological entities; reproduction and age class structure are some of their important attributes. Together "salmon reproduction and age class structure" form an assessment endpoint.

*Adapted from EPA, 1998.

Attribute   A quality or characteristic of an ecological entity.

Characterization of ecological effects   A portion of the analysis phase of ecological risk assessment that evaluates the ability of a stressor(s) to cause adverse effects under a particular set of circumstances.

Characterization of exposure   A portion of the analysis phase of ecological risk assessment that evaluates the interaction (either by actual contact or co-occurrence) of the stressor with one or more ecological entities.

Community   An assemblage of populations of different species within a specified location in space and time.

Conceptual model   A conceptual model in problem formulation is a written description and visual representation of predicted relationships between ecological entities and the stressors to which they may be exposed.

Cumulative distribution function (CDF)   Cumulative distribution functions are particularly useful for describing the likelihood that a variable will fall within different ranges of x. $F(x)$ (i.e., the value of y at x in a CDF plot) is the probability that a variable will have a value less than or equal to x (Figure 6.5).

Cumulative ecological risk assessment   A process that involves consideration of the aggregate ecological risk to the target entity caused by the accumulation of risk from multiple stressors.

Disturbance   Any event that disrupts ecosystem, community, or population structure and changes resources, substrate availability, or the physical environment [modified from White and Pickett (1985)].

EC$_{50}$ A concentration expected to cause an effect in 50 percent of a group of test organisms.

Ecological entity   A general term that may refer to a species, a group of species, an ecosystem function or characteristic, or a specific habitat.

Ecological relevance   One of the three criteria for assessment endpoint selection. Ecologically relevant endpoints reflect important characteristics of the system and are functionally related to other endpoints.

Ecological risk assessment (ERA)   The process that evaluates the likelihood that adverse ecological effects may occur as a result of exposure to a stressor.

Ecosystem   The biotic community and abiotic environment within a specified location in space and time.

Exposure   The contact or co-occurrence of a stressor with a receptor.

Exposure profile   A summary of the magnitude and spatial and temporal patterns of exposure for the scenarios described in the conceptual model.

Exposure scenario   A set of assumptions concerning how an exposure may take place, including exposure setting, stressor characteristics, and activities that may lead to exposure.

LC$_{50}$ A concentration expected to be lethal to 50 percent of a group of test organisms.

Lines of evidence   Information derived from different sources or by different techniques that can be used to describe and interpret risk estimates. Unlike the term "weight of evidence," it does not necessarily assign quantitative weights to information.

Lowest-observed-adverse-effect level (LOAEL)   The lowest level of a stressor evaluated in a test that causes statistically significant differences from the controls.

Measure of ecosystem and receptor characteristics   Measures that influence the behavior and location of organisms of interest, stressor distribution, and organismal life-history characteristics that may affect exposure or response to the stressor.

Measure of effect   Describes change assessment endpoint (or surrogate) attributes in response to a stressor to which it is exposed. Dose–response data are an example.

Measure of exposure   Describes stressor existence and behavior in the environment and its contact or co-occurrence with the assessment endpoint.

No-observed-adverse-effect level (NOAEL)   The highest level of a stressor evaluated in a test that does not cause statistically significant differences from the controls.

Population   An aggregate of individuals of a species within a specified location in space and time.

Primary effect   An effect where the stressor acts on the ecological component of interest itself, not through effects on other components of the ecosystem (synonymous with direct effect; compare with definition for secondary effect).

Probability density function (PDF)   Probability density functions are particularly useful in describing the relative likelihood that a variable will have different particular values of x. The probability that a variable will have a value within a small interval around x can be approximated by multiplying $f(x)$ (i.e., the value of y at x in a PDF plot) by the width of the interval (Figure 6.6).

Prospective risk assessment   An evaluation of the future risks of a stressor not yet released into the environment or of future conditions resulting from an existing stressor.

Receptor   The ecological entity exposed to the stressor.

Recovery   The rate and extent of return of a population or community to some aspect of its previous condition.

Retrospective risk assessment   An evaluation of the causal linkages between observed ecological effects and a stressor in the environment.

Risk characterization   Integrates exposure and stressor–response to evaluate the likelihood of adverse ecological effects associated with exposure to a stressor.

Secondary effect   An effect where the stressor acts on one component of the ecosystem, which in turn has an effect on the component of interest (synonymous with indirect effects; compare with definition for primary effect).

Source An entity or action that releases a stressor to the environment (or imposes a stressor on the environment).

Stressor Any physical, chemical, or biological entity that can induce an adverse response.

Stressor–response profile A summary of data on the effects of a stressor and the relationship of the data to the assessment endpoint.

Trophic levels A functional classification of taxa within a community that is based on feeding relationships (e.g., aquatic and terrestrial green plants make up the first trophic level and herbivores make up the second).

REFERENCES

Alder, H. L., and Roessler, E. B. 1972. *Introduction to Probability and Statistics.* Freeman, San Francisco, CA.

Baker, J. L., Barefoot, A. C., Beasley, L. E., Burns, L. A., Caulkins, P. P., Clark, J. E., Feulner, R. L., Giesy, J. P., Graney, R. L., Griggs, R. H., Jacoby, H. M., Laskowski, D. A., Maciorowski, A. F., Mihaich, E. M., Nelson, H. P., Jr., Parrish, P. R., Siefert, R. E., Solomon, K. R., and van der Schalie, W. H. (Eds.). 1994. *Aquatic Dialogue Group: Pesticide Risk Assessment and Mitigation.* SETAC Press, Pensacola, FL.

Barbour, M. T., Stribling, J. B., and Karr, J. R. 1995. Multimetric approach for establishing biocriteria and measuring biological condition. In W. S. Davis and T. P. Simon (Eds.), *Biological Assessment and Criteria, Tools for Water Resource Planning and Decision Making.* Lewis, Boca Raton, FL, pp. 63–77.

Barnthouse, L. W., and Brown, J. 1994. Issue paper on conceptual model development. In *Ecological Risk Assessment Issue Papers*, Risk Assessment Forum. EPA/630/R-94/009. U.S. Environmental Protection Agency, Washington, DC, pp. 3-1 to 3-70.

Barnthouse, L. W., O'Neill, R. V., Bartell, S. M., and Suter, G. W., II. 1986. Population and ecosystem theory in ecological risk assessment. In T. M. Poston and R. Purdy (Eds.), *Aquatic Ecology and Hazard Assessment*, 9th symposium. American Society for Testing and Materials, Philadelphia, PA, pp. 82–96.

Barnthouse, L. W., Suter, G. W., II, Rosen, A. E., and Beauchamp, J. J. 1987. Estimating responses of fish populations to toxic contaminants. *Environ. Toxicol. Chem.* 6: 811–824.

Barnthouse, L. W., Suter, G. W., II, and Rosen, A. E. 1990. Risks of toxic contaminants to exploited fish populations: Influence of life history, data uncertainty, and exploitation intensity. *Environ. Toxicol. Chem.* 9: 297–311.

Bartell, S. M., Gardner, R. H., and O'Neill, R. V. 1992. *Ecological Risk Estimation.* Lewis, Boca Raton, FL.

Bradley, C. E., and Smith, D. G. 1989. Plains cottonwood recruitment and survival on a prairie meandering river floodplain. Milk River, Southern Alberta in Northern Canada. *Can. J. Bot.* 64: 1433–1442.

Broderius, S. J. 1991. Modeling the joint toxicity of xenobiotics to aquatic organisms:

Basic concepts and approaches. In M. A. Mayes and M. G. Barron (Eds.), *Aquatic Toxicology and Risk Assessment*, Vol. 14. ASTM STP 1124. American Society for Testing and Materials, Philadelphia, PA, pp. 107–127.

Broderius, S. J., Kahl, M. D., and Hoglund, M. D. 1995. Use of joint toxic response to define the primary mode of toxic action for diverse industrial organic chemicals. *Environ. Toxicol. Chem.* 9: 1591–1605.

Brody, M. S., Troyer, M. E., and Valette, Y. 1993. Ecological risk assessment case study: Modeling future losses of bottomland forest wetlands and changes in wildlife habitat within a Louisiana basin. In *A Review of Ecological Assessment Case Studies from a Risk Assessment Perspective*, Risk Assessment Forum. EPA/630/R-92/005. U.S. Environmental Protection Agency, Washington, DC, pp. 12-1 to 12-39.

Burmaster, D. E., Menzie, C. A., Freshman, J. S., Burris, J. A., Maxwell, N. I., and Drew, S. R. 1991. Assessment of methods for estimating aquatic hazards at Superfund-type sites: A cautionary tale. *Environ. Toxicol. Chem.* 10: 827–842.

Callahan, C. A., Menzie, C. A., Burmaster, D. E., Wilborn, D. C., and Ernst, T. 1991. On-site methods for assessing chemical impacts on the soil environment using earthworms: A case study at the Baird and McGuire Superfund site, Holbrook, Massachusetts. *Environ. Toxicol. Chem.* 10: 817–826.

Cardwell, R. D., Parkhurst, B. R., Warren-Hicks, W., and Volosin, J. S. 1993. Aquatic ecological risk. *Water Environ. Technol.* 5: 47–51.

Clemen, R. T. 1996. *Making Hard Decisions*, 2nd ed. Duxbury, New York.

Commission on Risk Assessment and Risk Management (CRARM). 1997. *Framework for Environmental Health Risk Management. Final Report*, Vol. 1. CRARM, Washington, DC.

Conner, W. H., and Brody, M. 1989. Rising water levels and the future of southeastern Louisiana swamp forests. *Estuaries* 12(4): 318–323.

Costanza, R., d'Arge, R., de Groot, R., Farber, S., Grasso, M., Hannon, B., Limburg, K., Naeem, S., O'Neill, R. V., Paruelo, J., Raskinm, R. G., Sutton, P., and van den Belt, M. 1997. The value of the world's ecosystem services and natural capital. *Nature* 387: 253–260.

Cowan, C. E., Versteeg, D. J., Larson, R. J., and Kloepper-Sams, P. J. 1995. Integrated approach for environmental assessment of new and existing substances. *Regul. Toxicol. Pharmacol.* 21: 3–31.

Emlen, J. M. 1989. Terrestrial population models for ecological risk assessment: A state-of-the-art review. *Environ. Toxicol. Chem.* 8: 831–842.

Environmental Protection Agency (EPA). 1984. *Estimating Concern Levels for Concentrations of Chemical Substances in the Environment*. EPA, Health and Environmental Review Division, Environmental Effects Branch, Washington, DC.

Environmental Protection Agency (EPA). 1990. *Guidance for Data Useability in Risk Assessment*. EPA/540/G-90/008. EPA, Washington, DC.

Environmental Protection Agency (EPA). 1992. *Framework for Ecological Risk Assessment*, Risk Assessment Forum. EPA/630/R-92/001. EPA, Washington, DC.

Environmental Protection Agency (EPA). 1993a. *Wildlife Exposure Factors Handbook*. EPA/600/R-93/187a and 187b. EPA, Office of Research and Development, Washington, DC.

Environmental Protection Agency (EPA). 1993b. *Communicating Risk to Senior EPA Policy Makers: A Focus Group Study*. EPA, Office of Air Quality Planning and Standards, Research Triangle Park, NC.

Environmental Protection Agency (EPA). 1994a. *Managing Ecological Risks at EPA: Issues and Recommendations for Progress*. EPA/600/R-94/183. EPA, Center for Environmental Research Information, Washington, DC.

Environmental Protection Agency (EPA). 1994b. *Guidance for the Data Quality Objectives Process*. EPA QA/G-4. EPA, Quality Assurance Management Staff, Washington, DC.

Environmental Protection Agency (EPA). 1995a. *Ecological Risk: A Primer for Risk Managers*. EPA/734/R-95/001. EPA, Washington, DC.

Environmental Protection Agency (EPA). 1995b. "EPA risk characterization program." Memorandum to EPA managers from Administrator Carol Browner, March 1995.

Environmental Protection Agency (EPA). 1996a. *Summary Report for the Workshop on Monte Carlo Analysis*. EPA/630/R-96/010. EPA, Office of Research and Development, Washington, DC.

Environmental Protection Agency (EPA). 1996b. *Waquoit Bay Watershed. Ecological Risk Assessment Planning and Problem Formulation (Draft)*. Risk Assessment Forum. EPA/630/R-96/004a. EPA, Washington, DC.

Environmental Protection Agency (EPA). 1997a. *Priorities for Ecological Protection: An Initial List and Discussion Document for EPA*. EPA/600/S-97/002. EPA, Office of Research and Development, Washington, DC.

Environmental Protection Agency (EPA). 1997b. *Policy for Use of Probabilistic Analysis in Risk Assessment: Guiding Principles for Monte Carlo Analysis*. EPA/630/R-97/001. EPA, Office of Research and Development, Washington, DC.

Environmental Protection Agency (EPA). 1998. *Guidelines for Ecological Risk Assessment*. Risk Assessment Forum. EPA/630/R-95/002. EPA, Washington, DC.

European Community (EC). 1993. Technical guidance document in support of the risk assessment Commission Directive (93/67/EEC) for new substances notified in accordance with the requirements of Council Directive 67/548/EEC. Brussels, Belgium.

Fisher, S. G., and Woodmansee, R. 1994. Issue paper on ecological recovery. In *Ecological Risk Assessment Issue Papers*, Risk Assessment Forum. EPA/630/R-94/009. U.S. Environmental Protection Agency, Washington, DC, pp. 7-1 to 7-54.

Fotheringham, A. S., and Rogerson, P. A. 1993. GIS and spatial analytical problems. *Int. J. Geograph. Inf. Syst.* 7(1): 3–19.

Fox, G. A. 1991. Practical causal inference for ecoepidemiologists. *J. Toxicol. Environ. Health* 33: 359–373.

Gaudet, C. 1994. *A Framework for Ecological Risk Assessment at Contaminated Sites in Canada: Review and Recommendations*. Environment Canada, Ottawa, Canada.

Hammond, J. S., Keeney, R. J., and Raiffa, H. 1999. *Smart Choices: A Practical Guide to Making Better Decisions*. Harvard Business School Press, Boston, MA.

Harris, H. J., Wenger, R. B., Harris, V. A., and Devault, D. S. 1994. A method for assessing environmental risk: A case study of Green Bay, Lake Michigan, USA. *Environ. Manag.* 18(2): 295–306.

Health Council of the Netherlands. 1993. Ecotoxicological risk assessment and policymaking in the Netherlands—dealing with uncertainties. *Network* 6(3)/7(1): 8–11.

Heck, W. W. 1993. Ecological assessment case study: The National Crop Loss Assessment Network. In *A Review of Ecological Assessment Case Studies from a Risk Assessment Perspective*, Risk Assessment Forum, EPA/630/R-92/005. U.S. Environmental Protection Agency, Washington, DC, pp. 6-1 to 6-32.

Hermens, J., Canton, H., Janssen, P., and De Jong, R. 1984a. Quantitative structure-activity relationships and toxicity studies of mixtures of chemicals with anaesthetic potency: Acute lethal and sublethal toxicity to *Daphnia magna. Aquatic Toxicol.* 5: 143–154.

Hermens, J., Canton, H., Steyger, N., and Wegman, R. 1984b. Joint effects of a mixture of 14 chemicals on mortality and inhibition of reproduction of *Daphnia magna. Aquatic Toxicol.* 5: 315–322.

Hill, A. B. 1965. The environment and disease: Association or causation? *Proc. R. Soc. Med.* 58: 295–300.

Holling, C. S. (Ed.). 1978. *Adaptive Environmental Assessment and Management.* Wiley, Chichester.

Houseknecht, C. R. 1993. Ecological risk assessment case study: Special review of the granular formulations of carbofuran based on adverse effects on birds. In *A Review of Ecological Assessment Case Studies from a Risk Assessment Perspective*, Risk Assessment Forum. EPA/630/R-92/005. U.S. Environmental Protection Agency, Washington, DC, pp. 3-1 to 3-25.

Huff, D., and Geis, I. 1954. *How to Lie with Statistics.* Norton, New York.

Johnson, B. L. 1995. Applying computer simulation models as learning tools in fishery management. *N. Am. J. Fish. Manag.* 15: 736–747.

Johnson, L. B., and Gage, S. H. 1997. Landscape approaches to the analysis of aquatic ecosystems. *Freshwater Biol.* 37: 113–132.

Johnston, C. A., Detenbeck, N. E., and Niemi, G. J. 1990. The cumulative effect of wetlands on stream water quality and quantity: A landscape approach. *Biogeochemistry* 10: 105–141.

Karr, J. R. 1981. Assessment of biotic integrity using fish communities. *Fisheries* 6(6): 21–27.

Karr, J. R., Fausch, K. D., Angermeier, P. L., Yant, P. R., and Schlosser, I. J. 1986. *Assessing Biological Integrity in Running Waters: A Method and Its Rationale.* Illinois Natural History Survey. Special Publication 5. Champaign, IL.

Keeney, R. L. 1992. *Value-Focused Thinking.* Harvard University Press, Cambridge, MA.

Könemann, H. 1981. Fish toxicity tests with mixtures of more than two chemicals: A proposal for a quantitative approach and experimental results. *Aquatic Toxicol.* 19: 229–238.

Landis, W. G., Matthews, R. A., Markiewicz, A. J., and Matthews, G. B. 1993. Multivariate analysis of the impacts of the turbine fuel JP-4 in a microcosm toxicity test with implications for the evaluation of ecosystem dynamics and risk assessment. *Ecotoxicology* 2: 271–300.

Lynch, D. G., Macek, G. J., Nabholz, J. V., Sherlock, S. M., and Wright, R. 1994. Ecological risk assessment case study: Assessing the ecological risks of a new chemical under the Toxic Substances Control Act. In *A Review of Ecological Assessment Case Studies from a Risk Assessment Perspective*, Vol. II, Risk Assessment Forum. EPA/630/R-94/003. U.S. Environmental Protection Agency, Washington, DC, pp. 1-1 to 1-35.

MacIntosh, D. L., Suter, G. W., II, and Hoffman, F. O. 1994. Uses of probabilistic exposure models in ecological risk assessments of contaminated sites. *Risk Anal.* 14(4): 405–419.

Martin, D. 1999. *The Spirit of Dialogue*. International Communities for the Renewal of the Earth, Cross River, NY (available at publication date from Bernard F. Anderson, ICRE, 3248 Middle Ridge Way, Middleton, MD 21769).

McCarty, L. S., and Mackay, D. 1993. Enhancing ecotoxicological modeling and assessment: Body residues and modes of toxic action. *Environ. Sci. Technol.* 27: 1719–1728.

Meij, R. 1991. The fate of mercury in coal-fired power plants and the influence of wet flue-gas desulphurization. *Water Air Soil Pollut.* 56: 21–33.

Menzie, C. A., Burmaster, D. E., Freshman, J. S., and Callahan, C. A. 1992. Assessment of methods for estimating ecological risk in the terrestrial component: A case study at the Baird & McGuire Superfund Site in Holbrook, Massachusetts. *Environ. Toxicol. Chem.* 11: 245–260.

Nabholz, J. V. 1991. Environmental hazard and risk assessment under the United States Toxic Substances Control Act. *Sci. Total Environ.* 109/110: 649–665.

Nabholz, J. V., Miller, P., and Zeeman, M. 1993. Environmental risk assessment of new chemicals under the Toxic Substances Control Act (TSCA) section five. In W. G. Landis, S. G. Hughes, M. Lewis, and J. W. Gorsuch (Eds.), *Environmental toxicology and risk assessment*. ASTM STP 1179. American Society for Testing and Materials, Philadelphia, PA, pp. 40–55.

National Research Council (NRC). 1994. *Science and Judgment in Risk Assessment*. National Academy Press, Washington, DC.

National Research Council (NRC). 1996. *Understanding Risk: Informing Decisions in a Democratic Society*. National Academy Press, Washington, DC.

Niemi, G. J., DeVore, P., Detenbeck, N., Taylor, D., Lima, A., Pastor, J., Yount, J. D., and Naiman, R. J. 1990. Overview of case studies on recovery of aquatic systems from disturbance. *Environ. Manag.* 14: 571–587.

Novitski, R. P. 1979. Hydrologic characteristics of Wisconsin's wetlands and their influence on floods, stream flow, and sediment. In P. E. Greeson, J. R. Clark, and J. E. Clark (Eds.), *Wetland Functions and Values: The State of Our Understanding*. American Water Resources Association, Minneapolis, MN, pp. 377–388.

Oberts, G. L. 1981. Impact of wetlands on watershed water quality. In B. Richardson (Ed.), *Selected Proceedings of the Midwest Conference on Wetland Values and Management*. Freshwater Society, Navarre, MN, pp. 213–226.

Okkerman, P. C., Plassche, E. J. V. D., and Emans, H. J. B. 1993. Validation of some extrapolation methods with toxicity data derived from multispecies experiments. *Ecotoxicol. Environ. Safety* 25: 341–359.

O'Neill, R. V., Gardner, R. H., Barnthouse, L. W., Suter, G. W., II, Hildebrand, S. G., and Gehrs, C. W. 1982. Ecosystem risk analysis: A new methodology. *Environ. Toxicol. Chem.* 1: 167–177.

Orr, R. L., Cohen, S. D., and Griffin, R. L. 1993. *Generic Non-indigenous Pest Risk Assessment Process*. USDA Animal and Plant Health Inspection Service, Beltsville, MD.

Parkhurst, B. R., Warren-Hicks, W., Etchison, T., Butcher, J. B., Cardwell, R. D., and Voloson, J. 1995. *Methodology for Aquatic Ecological Risk Assessment*. RP91-AER-1 1995. Water Environment Research Foundation, Alexandria, VA.

Pastorok, R. A., Butcher, M. K., and Nielsen, R. D. 1996. Modeling wildlife exposure to toxic chemicals: Trends and recent advances. *Hum. Ecol. Risk Assess.* 2: 444–480.

Pearlstine, L., McKellar, H., and Kitchens, W. 1985. Modelling the impacts of a river diversion on bottomland forest communities in the Santee River Floodplain, South Carolina. *Ecol. Model.* 29: 281–302.

Pelczar, M. J., and Reid, R. D. 1972. *Microbiology.* McGraw-Hill, New York.

Peterman, R. M. 1990. The importance of reporting statistical power: The forest decline and acidic deposition example. *Ecology* 71: 2024–2027.

Phipps, R. L. 1979. Simulation of wetlands forest vegetation dynamics. *Ecol. Model.* 7: 257–288.

Rotenberry, J. T., and Wiens, J. A. 1985. Statistical power analysis and community-wide patterns. *Am. Naturalist* 125: 164–168.

Sample, B. E., Opresko, D. M., and Suter, G. W., II. 1996. *Toxicological Benchmarks for Wildlife,* ES/ER/TM-86/R3. Oak Ridge National Laboratory, Health Sciences Research Division, Oak Ridge, TN.

Sawyer, T. W., and Safe, S. 1985. In vitro AHH induction by polychlorinated biphenyl and dibenzofuran mixtures: Additive effects. *Chemosphere* 14: 79–84.

Schindler, D. W. 1987. Detecting ecosystem responses to anthropogenic stress. *Can. J. Fish. Aquat. Sci.* 44(Suppl. 1): 6–25.

Schwartz, P. 1996. *The Art of the Long View: Planning for the Future in an Uncertain World.* Currency/Doubleday, New York.

Simberloff, D., and Alexander, M. 1994. Issue paper on biological stressors. In *Ecological Risk Assessment Issue Papers,* Risk Assessment Forum. EPA/630/R-94/009. U.S. Environmental Protection Agency, Washington, DC, pp. 6-1 to 6-59.

Smith, E. P., and Cairns, J., Jr. 1993. Extrapolation methods for setting ecological standards for water quality: Statistical and ecological concerns. *Ecotoxicology* 2: 203–219.

Smith, E. P., and Shugart, H. H. 1994. Issue paper on uncertainty in ecological risk assessment. In *Ecological Risk Assessment Issue Papers,* Risk Assessment Forum. EPA/630/R-94/009. U.S. Environmental Protection Agency, Washington, DC, pp. 8-1 to 8-53.

Solomon, K. R., Baker, D. B., Richards, R. P., Dixon, K. R., Klaine, S. J., La Point, T. W., Kendall, R. J., Weisskopf, C. P., Giddings, J. M., Geisy, J. P., Hall, L. W., and Williams, W. M. 1996. Ecological risk assessment of atrazine in North American surface waters. *Environ. Toxicol. Chem.* 15(1): 31–76.

South African River Health Programme. 1998. State of the Crocodile River. National Aquatic Ecosystem Biomonitoring Programme. Available from *http://iwqs.pwv.gov.za/michael/biomon/index.htm.*

Starfield, A. M., and Bleloch, A. L. 1991. *Building Models for Conservation and Wildlife Management.* Burgess International Group, Edina, MN.

Stephan, C. E. 1977. Methods for calculating an LC_{50}. In *ASTM Special Technical Publication 634.* American Society for Testing and Materials, Philadelphia, PA, pp. 65–88.

Stephan, C. E., Mount, D. I., Hansen, D. J., Gentile, J. H., Chapman, G. A., and Brungs, W. A. 1985. *Guidelines for Deriving Numerical National Water Quality Criteria for the Protection of Aquatic Organisms and Their Uses.* PB85-227049. Office of Research and Development, U.S. Environmental Protection Agency, Duluth, MN.

Susser, M. 1986a. Rules of inference in epidemiology. *Regul. Toxicol. Pharmacol.* 6: 116–128.

Susser, M. 1986b. The logic of Sir Carl Popper and the practice of epidemiology. *Am. J. Epidemiol.* 124: 711–718.

Suter, G. W., II. 1993a. *Ecological Risk Assessment.* Lewis, Boca Raton, FL.

Suter, G. W., II. 1993b. A critique of ecosystem health concepts and indexes. *Environ. Toxicol. Chem.* 12: 1533–1539.

Suter, G. W., II. 1996. Abuse of hypothesis testing statistics in ecological risk assessment. *Hum. Ecol. Risk Assess.* 2: 331–347.

Suter, G. W., II, Gillett, J. W., and Norton, S. B. 1994. Issue paper on characterization of exposure. In *Ecological Risk Assessment Issue Papers,* Risk Assessment Forum. EPA/630/R-94/009. U.S. Environmental Protection Agency, Washington, DC, pp. 4-1 to 4-64.

Thomas, J. W., Forsman, E. D., Lint, J. B., Meslow, E. C., Noon, B. R., and Verner, J. 1990. *A Conservation Strategy for the Spotted Owl. Interagency Scientific Committee to Address the Conservation of the Northern Spotted Owl.* U.S. Government Printing Office, Washington, DC.

Tufte, E. R. 1990. *Envisioning Information.* Graphics, Cheshire, CT.

Tufte, E. R. 1992. *The Visual Display of Graphical Information.* Graphics, Cheshire, CT.

Tufte, E. R. 1997. *Visual Explanations.* Graphics, Cheshire, CT.

U.S. Department of Agriculture (USDA). 1993. *Pest Risk Assessment of the Importation of* Pinus radiata, Nothofagus dombeyi, *and* Laurelia philippiana *Logs from Chile.* Forest Service Miscellaneous Publication 1517. USDA, Washington, DC.

U.S. Department of Health, Education, and Welfare. 1964. *Smoking and Health. Report of the Advisory Committee to the Surgeon General.* Public Health Service Publication 1103. U.S. Department of Health, Education, and Welfare, Washington, DC.

Van Leeuwen, C. J., Van der Zandt, P. T. J., Aldenberg, T., Verhar, H. J. M., and Hermens, J. L. M. 1992. Extrapolation and equilibrium partitioning in aquatic effects assessment. *Environ. Toxicol. Chem.* 11: 267–282.

Wagner, C., and Løkke, H. 1991. Estimation of ecotoxicological protection levels from NOEC toxicity data. *Water Res.* 25: 1237–1242.

Wiegert, R. G., and Bartell, S. M. 1994. Issue paper on risk integration methods. In *Ecological Risk Assessment Issue Papers.* Risk Assessment Forum. EPA/630/R-94/009. Environmental Protection Agency, Washington, DC, pp. 9-1 to 9-66.

Wiens, J. A., and Parker, K. R. 1995. Analyzing the effects of accidental environmental impacts: Approaches and assumptions. *Ecol. Appl.* 5(4): 1069–1083.

Woodman, J. N., and Cowling, E. B. 1987. Airborne chemicals and forest health. *Environ. Sci. Technol.* 21: 120–126.

Yankelovich, D. 1999. *The Magic of Dialogue: Transforming Conflict into Cooperation.* Simon & Schuster, New York.

Yoder, C. O., and Rankin, E. T. 1995. Biological response signatures and the area of degradation value: New tools for interpreting multi-metric data. In W. S. Davis and T. P. Simon (Eds.), *Biological Assessment and Criteria: Tools for Water Resource Planning and Decision Making.* Lewis, Boca Raton, FL.

SECTION C
Case Studies Involving Contaminated Water

7 Hexavalent Chromium in Groundwater: The Importance of Chemistry and Pharmacokinetics in Quantitating Dose and Risk

BRENT D. KERGER

Health Science Resource Integration, Inc. Tallahassee, FL.

BRENT L. FINLEY and DENNIS J. PAUSTENBACH

ChemRisk®, Alameda, California

7.1 BACKGROUND

In this chapter, we describe a series of research questions aimed to address the possible health risks posed by exposures to hexavalent chromium [Cr(VI)] in groundwater at concentrations up to 4000 parts per billion (ppb). This concentration was considered the plausible upper bound Cr(VI) concentration for long term exposures to residents with impacted groundwater because of the bright yellow coloration of the water at concentrations from 500 to 4000 ppb. We review the available scientific information demonstrating the importance of the chemistry of hexavalent chromium species in determining exposures, absorbed dose, and health risks posed by Cr(VI) in groundwater.

The various forms of hexavalent, trivalent, and metallic chromium that exist as stable species in the environment each exhibit relatively unique chemical properties with respect to their potential transport and dissolution in groundwater. There are many specific hexavalent and trivalent chromium species that exhibit very limited solubility and hence are very unlikely to become important sources of exposure. Similarly, metallic chromium (zero valency) is completely insoluble in water and hence is not likely to become an issue for groundwater

Human and Ecological Risk Assessment: Theory and Practice, Edited by Dennis J. Paustenbach
ISBN 0-471-14747-8   © 2002 John Wiley & Sons, Inc.

exposures. Most of the common trivalent chromium species [Cr(III)] are also very insoluble in water, and fall into the same category of agents unlikely to pose a risk via groundwater. The highly soluble Cr(VI) compounds, most prominently the potassium or sodium salts of chromate or dichromate and chromic acid, are the species most likely to be dissolved in and distributed in groundwater. Although there are certain Cr(III) species that are relatively soluble in water, the much lower toxicity of these species generally precludes these compounds from posing a significant threat to health. Thus, the focus of this chapter is on the chemistry, environmental fate, exposure aspects, pharmacokinetics, and toxicity of the highly soluble hexavalent chromium salts that are found in a limited number of industrial applications.

Potassium chromate and dichromate were used for many years as anti-corrosion and anti-fouling agents in cooling water solutions for certain industrial equipment. As described earlier, cooling towers are one prominent industrial application of highly soluble Cr(VI) salts that can lead to the generation of wastewater that may contain low part per million (ppm) concentrations of hexavalent chromium species dissolved in water. Such industrial wastewater also may include a significant fraction of sediments, rust particles, oils, and other organic and inorganic materials that derive from continuous recirculation of these fluids in industrial cooling towers and similar applications. These anti-fouling agents for use in cooling waters often contained chromium compounds in addition to other anti-fouling chemicals like copper salts. The Cr(VI) concentration maintained in most applications of these cooling tower chemicals ranged from about 5 ppm to 50 ppm.

Understanding the acid-base chemistry of Cr(VI) compounds is central to understanding the fate of dissolved Cr(VI) in groundwater, in the environment generally, and in the human body. The Eh/pH diagram for chromium species presented in Figure 7.1 illustrates the strong influence of acidity in determining the ability of hexavalent species to become rapidly reduced to the less toxic trivalent chromium species. For example, one of the prominent laboratory and industrial uses of soluble chromates has involved the use of highly acidic solutions of chromic acid to thoroughly clean metal and glass surfaces by oxidizing and removing all organic materials from those surfaces. Under low pH conditions, soluble chromates will rapidly attack any organic materials containing nitrogen, oxygen, phosphorus, or sulfur, and in the process are transformed to trivalent chromium species existing as free ions in solution or as covalently liganded sediments or salts that may precipitate out of solution. However, the equilibrium rate constants for Cr(VI) reduction and the types of chromium species normally formed in solution are dramatically altered as the pH approaches neutral to slightly basic.

Neutral to basic pH solutions are associated with much lower rates of Cr(VI) reduction and create a greater potential for persistence of soluble Cr(VI) in groundwater (See Figure 7.1). For example, large volumes of wastewater with a near neutral pH can percolate through soils and enter groundwater if the soil column is relatively inert and contains little organic material (e.g., sandy soils).

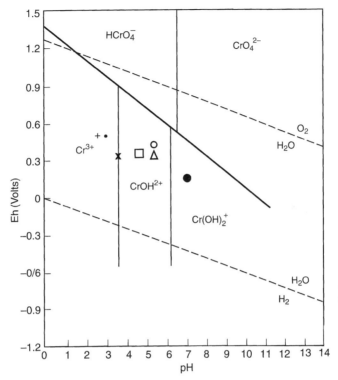

Figure 7.1 Eh-pH predominance diagram for soluble chromium species with location of selected beverages: • = lemonade; + = Kool Aid; ● = tapwater; ○ = tea; △ = drip coffee; □ = percolated coffee; × = orange juice.

Even in instances where the soluble Cr(VI) species are comingled with organic residues and solids that could be sources for reduction to Cr(III), there are certain circumstances (e.g., unlined wastewater storage ponds) where replenished sources of low level Cr(VI) solutions can overwhelm the reduction capacity of these materials. Further, the dilution of Cr(VI) in combination with pH can have an impact on the efficiency of interaction between Cr(VI) species and organic ligands that might augment reduction to Cr(III). Thus, high volumes of relatively low level Cr(VI) concentrations of near neutral pH industrial wastewaters can create a not-so-obvious source for surface release impacts on groundwaters.

Acid-base chemistry of Cr(VI) also has critical importance with respect to the environmental fate of chromium in various media that may come into contact with Cr(VI)-impacted groundwater. Dilution and pH are of similar importance in understanding the nature and extent of indirect Cr(VI) exposures, such as in the use of groundwater for irrigation purposes. For example, various types of fertile soils and/or fertilizing agents are somewhat acidic and contain relatively high amounts of organic ligands that serve as a high capacity source for reduction of Cr(VI) found at relatively low levels in the irrigation water. There are many microenvironments where low to moderate volumes of dilute Cr(VI) solutions can be readily reduced to Cr(III). A household environment example of this is discussed in a later section of this chapter where studies demonstrate rapid Cr(VI) reduction when this water is mixed with common beverages like tea, coffee, lemonade, and orange juice.

In the human body, the acid-base chemistry of Cr(VI) is equally important. For the vast majority of chromium intake related to household uses of water, ingestion is quantitatively more important by orders of magnitude in comparison to potential inhalation of Cr(VI) in aerosols or dermal penetration of Cr(VI). However, the influence of acidic gastric juices on Cr(VI) reduction is known to be relatively rapid and high capacity. Thus, the acid-base chemistry of the gastrointestinal tract, as well as the reductive capacity of the blood, creates a substantial barrier to absorption of Cr(VI) species beyond the stomach. As discussed later, the skin is also a substantial barrier to Cr(VI) uptake that severely limits the likelihood of significant dose or toxicity via dermal contact. If there is some risk associated with groundwater contaminated with concentrations of Cr(VI) that might reasonably be ingested, it might be due to inhalation. This relates to the relatively neutral pH and limited reduction capacity of the mucous and lung fluids, as well as to certain dosimetry issues relating to the inhalation pathway. As discussed further below, the substantial presence of replenishable reducing agents in the blood and internal organs has generally convinced the toxicology and medical community that exposure to Cr(VI) contaminated groundwater will, in the main, rarely be sufficient to pose a substantial health hazard if concentrations are maintained below 100 ppb and probably even much higher concentrations. This view was recently supported by a science advisory board that was convened to assess this issue (Fleegal et al., 2001). Although over the past 5–7 years a number of law suits have been filed that allege an increase above background levels of various illnesses attributed to exposure to Cr(VI) contaminated groundwater, there has been no good scientific evidence to support the breadth of claims shown in Table 7.1.

7.2 MECHANISTIC ASPECTS OF INHALATION AND ORAL TOXICITY OF CR(VI)

The toxicity of Cr(VI) via the inhalation and ingestion routes has been examined in several species, including humans with occupational, accidental, or suicidal exposures [reviewed by the Agency for Toxic Substances and Disease Registry (ATSDR), 1998; DeFlora et al., 1996; International Agency for Research on Cancer (IARC), 1990; DeFlora and Wetterhahn, 1989]. Cr(VI) is well recognized as a portal of entry toxicant in humans and animals, with its pharmacokinetics, dose–response relationships, and mechanism of toxic action all tied to oxidation–reduction reactions in body fluids and tissues. Homeostasis of the human body is highly dependent on keeping any oxidation processes in check,

TABLE 7.1 Partial List of Illnesses or Complaints that Plaintiffs from Various Legal Actions have Claimed to be Possibly Caused by Exposure to Cr(VI)

Non-Hodgkin's Lymphoma	Testicular cancer
Hodgkin's Lymphoma	Non-small-cell carcinoma of the lung
Primary adenocarcinoma of the common duct, involving lymph nodes and gall bladder	Carcinoma of the lung
	Metastatic lesions of the femur
	Pre-cancerous growths on the nose
Metastatic large-cell squamous carcinoma of bronchus and trachea	Lung cancer with metastasis to the brain
	Metastatic renal cell carcinoma
Thrombocythemia (malignancy of bone marrow)	Thyroid cancer
	Ulceration of the colon
Meningothelial meningioma (tumor)	Ulcerative colitis
Prostate cancer	Liver damage
Chronic Lymphocytic Leukemia	Inflammatory bowel disease
Crohn's disease	Sinus problems
Severe Gl distress	Sinus infections
Stomach pain	Bronchitis, leading to pneumonia in one case
Diarrhea	
Asthma	Sarcoidosis
Chronic obstructive lung disease	Severe chronic pulmonary disease
Respiratory distress	Mouth sores
Chronic coughing	Severe dental abscesses and cavities
Breathing difficulties	Cavernous hemangioma in lower lip
Calcified granuioma on lungs	Sore throat
Skin rashes	Dental problems, including crumbling teeth
Scleroderma	
Lupus	Prolapsed uterus
Nose bleeds	Miscarriages
Skin lesions & or irritations	Difficulties with labor
Bleeding gums	Severe/chronic headaches
Ovarian cysts	Joint problems/Pain
Testicular problems	Cavities
Gynecological problems	Weakened immune system
Depression	Cystitis
Chronic fatigue	Difficulties concentrating
Memory loss	Migraine headaches
Loss of attention span	

and the pH, E_h, and other biochemical factors in the blood and tissues are known to have a relatively high and regenerable capacity to effect reduction of oxidizing agents that may enter the body. DeFlora and co-workers (1996) have reviewed the available data relating to the capacity of various body fluids and tissues to reduce and/or sequester Cr(VI), hence reducing its potential for systemic toxicity unless these biochemical defense mechanisms are overwhelmed. As shown in Figure 7.2, the Cr(VI) reduction capacity of fluids and tissues rele-

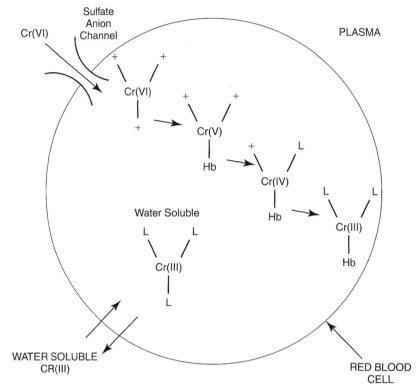

Figure 7.2 Cr(VI) and Cr(III) uptake in red blood cells. This schematic depicts how Cr(VI) readily enters the red blood cell where it is reduced to short-lived reactive intermediates [Cr(V), Cr(IV)] and bound to hemoglobin (Hb) and soluble ligands (L) such as glutathione and amino acids. Essentially complete binding of intracellular Cr(VI) to hemoglobin occurs because >30% of red blood cell mass is hemoglobin. The hemoglobin-bound Cr complexes remain part of the red blood cell for its entire lifespan. Conversely, water-soluble chromium(III) moves across the cell membrane via much slower diffucion and perhaps other processes related to the chemical structure of the attached ligands.

vant to oral exposures is very high, consistent with the observation in humans that suicidal ingestion doses in gram quantities are generally required to cause severe acute toxicity and death (DeFlora et al., 1996). Table 7.2 provides a summary of key animal and human studies regarding no-effect or low-effect levels following chronic oral exposure to Cr(VI) [Gross and Heller, 1946; MacKenzie et al., 1958; Anwar et al., 1961; U.S. Environmental Protection Agency (USEPA), 1996].

In contrast, the inhalation route has relatively low capacity defense mechanisms to protect the lungs from Cr(VI) toxicity, and these defenses can be readily overwhelmed due to uneven dosimetry [i.e., patches of greater

TABLE 7.2   Summary of Key Studies Documenting No Observable Adverse Effect Levels (NOAEL) for Oral Cr(VI) Exposure in Animals and Humans

Animal	Concentration	Duration	Effect Seen
Rat (Gross & Heller, 1946)	134 ppm Cr(VI)	60 days	None
Rat (Mackenzie et al., 1958)	25 ppm Cr(VI)	1 year	None
Dog (Anwar et al., 1961)	11.2 ppm Cr(VI)	4 years	None
Human (USEPA, 1996)	1 ppm Cr(VI)	3 years	None

impaction/deposition of the inhaled Cr(VI) particles in certain regions of the lung] and/or by high dose rate (i.e., high mass inhaled per unit time). This relative sensitivity of the lungs is consistent with the observation of respiratory tract irritation in humans following prolonged exposures to airborne Cr(VI) in the range of 100 to 10,000 $\mu g/m^3$ range (Mancuso, 1951). Such high inhalation exposures in past decades for workers in chromite ore processing and pigment manufacturing occupations have demonstrated clearly elevated rates of lung and nasal sinus cancers attributable to Cr(VI) inhalation (IARC, 1990); however, definitive evidence of clinical disease beyond the portal of entry (i.e., the respiratory tract) has not been revealed in the many occupational studies to date. This absence of systemic effects is consistent with the high capacity of the blood to reduce and/or sequester Cr(VI), thereby preventing any appreciable distribution of this oxidizing agent to other tissues. Table 7.3 provides a summary of key animal and human studies examining no-effect and low-effect levels fol-

TABLE 7.3   Summary of Key Studies Documenting No Effect or Low Effect Levels for Inhalation Cr(VI) Exposures in Animals and Humans

Animal	Concentration	Duration	Effect Seen
Rabbit (Johansson et al., 1986a, 1986b)	900 μg/m^3	6 weeks	None
Rat (Glaser et al., 1985; 1986)	400 μg/m^3	90 days	LDH in BALF
Human (Franchini & Mutti, 1988)	50 μg/m^3	7 yr. avg.	↑ Retinal Binding Protein in Urine
Human (Lindberg & Hedenstierna, 1981)	2 μg/m^3	2.5 yr. avg.	↓ Lung Function
Human (Lindberg & Hedenstierna, 1981)	1 μg/m^3	2.5 yr. avg.	None

lowing chronic inhalation exposure to Cr(VI) (Johansson et al., 1986a, 1986b; Glaser et al., 1985, 1986; Franchini and Mutti, 1988; Lindberg and Hedenstierna, 1981).

The high capacity of red blood cells to rapidly sequester and bind Cr(VI) has been recognized for several decades, and radiolabeled Cr(VI) has even been used as a biomarker to study the life span of red blood cells in humans (Gray and Sterling, 1950). More recently, red blood cell chromium content has been used as a dose biomarker for occupational Cr(VI) exposures (Lewalter et al.,

1985; Korallus, 1986). The chromate anion is taken up by red blood cells via the sulfate anion channel (see Figure 7.2). Once inside the red blood cell, the Cr(VI) is rapidly reduced to unstable intermediates [i.e., short-lived Cr(V) and Cr(IV) species], which become bound to hemoglobin and other intracellular ligands (Lewalter et al., 1985; Ottenwaelder and Wiegand, 1988; Wiegand et al., 1988; Coogan et al., 1991; Miksche et al., 1994; Gray and Sterling, 1950; Weber, 1983; Edel and Sabbioni, 1985). An incorrect paradigm regarding the pharmacokinetics of Cr(VI) and Cr(III) has been recently clarified, that is, the concept that Cr(III), per se, is not taken up by red blood cells or other tissues. The available data on human and animal pharmacokinetics of chromium indicates that when Cr(VI) is reduced to Cr(III) in the body, some fraction of that reduced chromium may be in water-soluble organic-liganded forms that are not prevented from passing through cell membranes. While it is true that the insoluble inorganic forms of Cr(III) [and the Cr(III) ion itself] may not be capable of passing through cell membranes, such is clearly not the case for organic forms of Cr(III) like chromium tripicolinate, chromium trinicotinate, and Cr(VI) that is reduced to Cr(III) in body fluids and/or other organic solutions (Kerger et al., 1997; Anderson, 1986, 1994; Anderson et al., 1997; Mertz, 1969, 1975; Mertz and Roginski, 1971; Gonzales-Vergara et al., 1981; Edel and Sabbioni, 1985; Kortenkamp et al., 1987). This explains the relatively rapid distribution and accumulation in animal tissues (e.g., kidney and liver) of the more bioavailable vitamin forms of Cr(III) (e.g., tripicolinate) as shown by Anderson et al. (1997). The formation of such organic liganded Cr(III) species in body tissues and fluids also explains the tissue distribution and pharmacokinetics of chromium uptake in humans exposed to low to moderate Cr(VI) doses by the oral and dermal routes (Kerger et al., 1996a, 1997; Corbett et al., 1997; Proctor et al., 2002).

7.3 REGULATORY CRITERIA FOR CR(VI) RISK ASSESSMENT AND RISK MANAGEMENT

The U.S. EPA has historically provided toxicity criteria that were developed under the Safe Drinking Water Act and in relation to the human health risk assessment process pertaining to the Superfund program. These include a cancer potency estimate for inhalation exposures (41/mg/kg-day), a chronic reference dose for oral exposures of 0.005 mg/kg-day, and a chronic reference concentration for inhalation exposures of 12 ng/m^3, which was withdrawn in 1992. ATSDR provides a chronic minimal risk level of 20 ng/m^3 for noncancer effects of Cr(VI) inhalation (ATSDR, 1998). The U.S. EPA water standards since 1991 include a maximum contaminant level of 100 ppb for all chromium compounds in drinking water, 1- and 10-day health advisories for young children of 1000 ppb, a longer-term value for children at 200 ppb, a longer-term value for adults at 800 ppb, and a lifetime health advisory for adults at 100 ppb.

The State of California Environmental Protection Agency (Cal-EPA) chose to develop different toxicity criteria for Cr(VI). The drinking water MCL of 50 ppb Cr(VI) remained in effect in California after 1991, when the EPA changed the federal MCL to 100 ppb for all chromium compounds. Cal-EPA also derived its own, more conservative (12-fold higher than the EPA) Cr(VI) cancer slope factor of 510 $(mg/kg\text{-}day)^{-1}$ based on the same epidemiological data for lung cancers, but utilizing different assumptions and risk extrapolation procedures (Cal-EPA, 1994). In addition, whereas U.S. EPA does not consider Cr(VI) to be an oral carcinogen in humans or animals, Cal-EPA has contended that there is sufficient evidence to implicate Cr(VI) as an oral carcinogen in animals, with an estimated oral cancer slope factor of 0.42 $(mg/kg\text{-}day)^{-1}$. Finally, the ambient air exposure limit for Cr(VI) was set at 2 ng/m^3 by Cal-EPA.

7.4   KEY SCIENTIFIC ISSUES AND CONCERNS

The risk assessment issues of concern associated with Cr(VI) contaminated groundwater are varied; so our research was designed to: (1) identify the key sources of Cr(VI) exposures for persons who might have Cr(VI) contaminated water, based on systemic Cr(VI) dose; (2) determine the plausible range of upper-bound residential exposures to Cr(VI) in air and water; (3) evaluate the toxicological significance of the upper-bound Cr(VI) dose by route of exposure; and (4) provide estimates of Cr(VI) exposure and incremental lifetime cancer risks that conservatively evaluate the magnitude of possible hazard or risk to these persons.

The overall approach to address these issues involved performing research that provided substantial demonstrative evidence regarding the likely characteristics of Cr(VI) exposure and uptake via inhalation, ingestion, and dermal contact. This information can then be combined with individual-specific information to calculate likely and upper-bound Cr(VI) doses and incremental lifetime cancer risks that could be placed in perspective by comparison to background doses and adverse effect levels identified in the scientific literature. In essence, the basic principles of toxicology and risk assessment should be applied to evaluate these issues.

7.5   EXPOSURE ASSESSMENT RESEARCH

The discussions below describe three key areas of exposure investigation: (1) observational benchmark studies to determine upper-bound Cr(VI) water concentrations in the community, (2) Cr(VI) reduction capacity studies to examine the fate of Cr(VI) in common beverages made with tap water, and (3) indoor and outdoor studies of airborne Cr(VI) to examine background exposures and specific sources at issue.

7.5.1 Observational Benchmarks

When attempting to identify a plausible upper bound for exposure concentrations for Cr(VI) in drinking water, one approach is to quantify "observational benchmarks", i.e., specific visual observations that help confirm or deny the presence of Cr(VI) in tap water at a certain upper-bound concentration. The simple observation of water color with increasing Cr(VI) or Cr(III) concentration is a very telling observational benchmark. The bright yellow color of Cr(VI) (as potassium dichromate) in water becomes readily apparent at concentrations in the range of 500 to 4000 µg/L, depending on the volume or depth of water examined. For example, in a bathtub filled with a normal depth of Cr(VI)-containing water, an obvious yellow-green tinge is apparent at concentrations of 500 to 1000 µg/L, and an obvious, bright yellow color was observed at Cr(VI) concentrations between 2000 and 4000 µg/L. Similarly, ice cubes made with water containing 2000 to 4000 µg/L of Cr(VI) are clearly discolored, with a more intense yellow color occurring in the center of each cube. Such observations may be useful to confirm or reject the claims regarding potential contamination in water.

7.5.2 Cr(VI) Reduction Capacity of Beverages

There are numerous published studies demonstrating that reducing agents naturally present in body fluids, foods, beverages, and the general environment are capable of reducing hexavalent chromium to trivalent forms (DeFlora et al., 1987, 1989; Capellmann and Bolt, 1992; Donaldson and Barreras, 1966; Korallus et al., 1984; Corbett et al., 1998). This is an important consideration due to the significantly lower toxicity exhibited by Cr(III) compounds compared to Cr(VI) compounds. Accordingly, we developed an experimental protocol to examine the Cr(VI) reduction rate and capacity for common beverages mixed with tap water.

Laboratory studies demonstrated that common beverages such as coffee, tea, and orange juice have a relatively high Cr(VI) reduction capacity [e.g., enough to rapidly reduce >20,000 ppb Cr(VI) in water], probably due to their natural acidity and the presence of such reducing agents as ascorbate and tannic acids in these beverages (Kerger et al., 1996b). Even powdered beverages such as Kool-Aid and lemonade contained sufficient amounts of reducing agents to reduce Cr(VI) in tap water quite readily, but with lower capacity (e.g., 8000 to 10,000 ppb). These data suggest that, even at the expected upper-bound concentrations for drinking water exposures in the affected area (i.e., perhaps up to 4000 µg/L), it was likely that essentially all of the Cr(VI) present in tap water that was consumed as a mixed beverage, such as coffee, tea, and Kool-Aid, was likely to represent intake as Cr(III). Therefore, only the water ingested as unaltered/unmixed tap water was likely to contain Cr(VI) in the affected area. As described later, national surveys of tap water ingestion patterns can be used to allow refinement of the intake rate calculations to account for this likely influence on daily Cr(VI) intake via tap water ingestion.

7.5.3   Indoor and Outdoor Studies of Airborne Cr(VI)

Since Cr(VI) is a nonvolatile metal under ambient conditions and small quantities are readily reduced upon contact with soils and other environmental media, airborne Cr(VI) exposures must involve production of airborne aerosols from the contaminated groundwater. In our field studies, aerosol sources were considered to possibly include: (1) indoor aerosols resulting from showering activities, (2) indoor aerosols resulting from operation of certain air conditioners (the evaporative or "swamp" cooler devices), (3) outdoor aerosols resulting from spray irrigation or washing activities, and (4) outdoor aerosols resulting from cooling tower "drift."

The first two issues were addressed in a series of field studies examining airborne Cr(VI) levels generated from the suspected source activities in a "test house" (discussed below). Outdoor aerosol sources were addressed with source-specific emissions measurements combined with air dispersion modeling using the Industrial Source Complex model. Preliminary modeling of the cooling tower airborne emissions (based on emitted mist or "drift") based on engineering tolerances for the cooling tower drift eliminators (a system of baffles) suggested higher long-term emission rates that were not consistent with the available soil data adjacent to the cooling towers (see Figure 7.3 for cooling tower diagram). Further investigations, including measurement of actual emission rates and particle size distribution for cooling tower drift emissions revealed much lower total drift emissions (about two orders of magnitude less than the engineering tolerance for drift) consistent with the chromium accumulations in soils nearby each cooling tower (Richter et al., 1997). Humidity is an important issue in developing an air dispersion model so that it is consistent with the environmental data.

Dispersion modeling can be applied to examine the magnitude and duration of ambient air Cr(VI) exposure attributable to local irrigation practices (if applicable). If appropriate, historical information on agricultural practices and irrigation type and volume and groundwater levels in an affected area should be considered.

To obtain the measurement sensitivity required to assess background Cr(VI) levels in air (i.e., in the low nanogram per cubic meter range), the triple-impinger sampling method adopted by the American Standards and Testing Materials (ASTM) Method D 5281 was utilized (Sheehan et al., 1992; Finley et al., 1993; Finley and Mayhall, 1994). The test house studies were designed to characterize the Cr(VI) water concentration versus aerosol exposure relationships for showering and operation of the "swamp cooler" desert air conditioner (see Figure 7.4 for swamp cooler diagram). The study conditions and results are described in detail elsewhere (Finley et al., 1997). Briefly, the study showed that no measur-

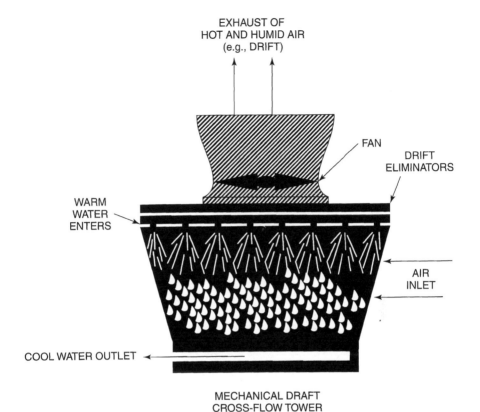

EXHAUST OF
HOT AND HUMID AIR
(e.g., DRIFT)

FAN

DRIFT
ELIMINATORS

WARM
WATER
ENTERS

AIR
INLET

COOL WATER OUTLET

MECHANICAL DRAFT
CROSS-FLOW TOWER

Figure 7.3 How a cooling tower works.

able increase was observed from outdoor to indoor airborne Cr(VI) concentrations as a result of operating a central swamp cooler device with influent Cr(VI) water concentrations of 20,000 µg/L. Indeed, although Cr(VI) concentrations in the recirculation reservoir of the swamp cooler reached a plateau as high as 240,000 µg/L, still there was no increase in the indoor airborne Cr(VI) concentrations. Outdoor and indoor air concentrations in studies conducted during summer and fall seasons were generally in the 1 to 5 ng/m^3 range, as observed in other locations miles away from the facility, and in the Los Angeles basin region (Kerger et al., 1997). It was concluded that swamp coolers did not contribute a significant amount of aerosols to indoor air, explaining the absence of elevated Cr(VI) for indoor versus outdoor air.

Shower air measurements also were made to evaluate the significance of Cr(VI) aerosol exposures during showering activities (Finley et al., 1997). The standard shower/bathtub stall in the test house was outfitted with a mannequin

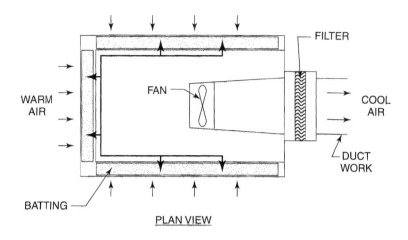

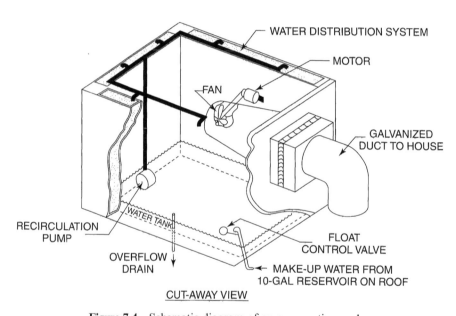

Figure 7.4 Schematic diagram of an evaporative cooler.

simulating an adult standing in the shower and with two ASTM method sampling trains on either side at breathing zone height (see Figure 7.5 for shower configuration). The shower was connected to a recirculating pump such that one of three different Cr(VI) water concentrations (approximately 1000, 5000, and 10,000 μg/L) could be circulated to determine the water concentration versus aerosol exposure concentration relationship. Figure 7.6 provides the results

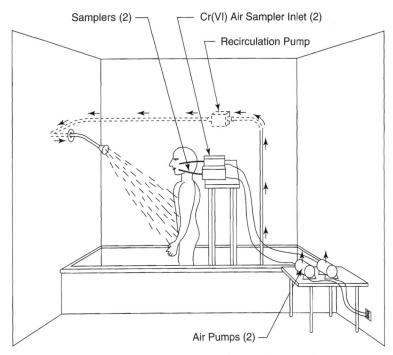

Figure 7.5 Schematic of shower air sampling experiment.

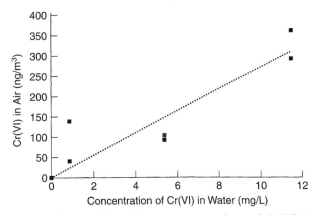

Figure 7.6 A comparison of the airborne concentrations of Cr(VI) in mists during showering at various contaminated water concentrations.

of these experiments, showing that the expected linear relationship was found between Cr(VI) concentrations in the shower water and the shower air. This correlation was used in subsequent exposure calculations to represent personal shower exposures and near-source exposure estimates for irrigation sprinklers fed by wells within the groundwater plume [Shower air Cr(VI) in ng/m^3 = 28 × Cr(VI) water concentration in mg/L).

7.6 RESEARCH ON SYSTEMIC CR(VI) DOSE AND TOXICITY

As mentioned earlier, a number of studies in animals, humans, and in vitro indicate that when relatively low concentrations of Cr(VI) come in contact with biological fluids and tissues, reduction occurs due to the presence of a wide variety of reducing agents that are designed to keep the body in homeostasis (DeFlora and Wetterhahn, 1989; DeFlora et al., 1996, 1997; Kerger et al., 1997). Cr(VI) is known to be rapidly reduced to Cr(III) upon contact with body fluids (DeFlora et al., 1997; see Figure 7.7). If sufficient Cr(VI) is absorbed such that the route-specific barriers to oxidative attack are overwhelmed (i.e., the tissue reduction capacity is exceeded), then Cr(VI) oxidation within the cell may lead to cell damage and/or death (DeFlora and Wetterhahn, 1989). In contrast to the kinetics of Cr(VI), soluble trivalent chromium forms that are liganded to organic molecules such as picolinate and nicotinate are moderately well absorbed and can accumulate in certain tissues (e.g., liver and kidney) without causing toxicity (Anderson et al., 1997). The pharmacokinetic and pharmacodynamic implications of this change in valency are obviously complex and highly route dependent. Given the multiple biological fluids and tissues that serve as barriers to systemic uptake of Cr(VI) and Cr(III) by each route of exposure, it is reasonable to expect that all common routes of exposure will exhibit threshold-dependent dose–response relationships affecting both localized and systemic responses to Cr(VI) exposure (DeFlora et al., 1997; DeFlora, 2000).

There is considerable evidence in the literature pointing to the threshold dependence of both the toxicity and the uptake kinetic studies of ingested Cr(VI) in animals and in humans (MacKenzie et al., 1958; Anwar et al., 1961; Gross and Heller, 1946; Diaz-Mayans, 1986; DeFlora et al., 1997; USEPA, 1996). For example, the studies of Donaldson and Barreras (1966) and DeFlora et al. (1987) provided convincing evidence of the large capacity of acidic gastric juices to convert Cr(VI) to Cr(III) in humans (Figure 7.7). However, these studies did not provide quantitative estimates of the probable threshold for these route-specific defense mechanisms and did not evaluate chronic exposures.

Based on the hypothesis of threshold-dependent systemic uptake and toxicity of Cr(VI), we undertook a series of studies to help further characterize the influence of the natural barriers of the human body to oral and dermal uptake as Cr(VI). Due to the fact that oral exposures would likely represent the greatest mass of Cr(VI) potentially taken into the body as a result of the groundwater contamination, we chose to address this route most thoroughly.

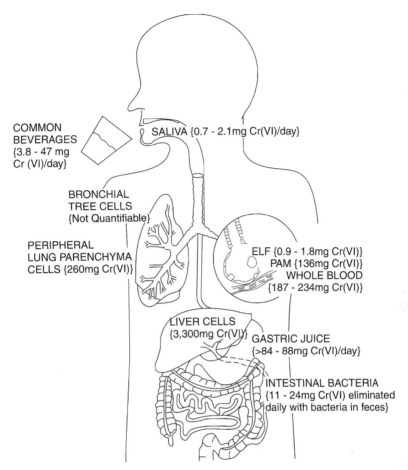

Figure 7.7 Estimates of Cr(VI) sequestration or reduction by organs, cell populations, and fluids in the human body.

Perhaps the most important design element of these studies was assessment of the presence or absence of systemic uptake of chromium in the hexavalent form. To accomplish this, we utilized red blood cell chromium concentrations as a biomarker of Cr(VI) systemic uptake. As discussed earlier, it has been recognized for decades that red blood cells rapidly sequester and reduce/bind Cr(VI) (Lewalter et al., 1985; Korallus, 1986; Gray and Sterling, 1950; Coogan et al., 1991; Miksche et al., 1994). Once Cr(VI) enters the cell, it is apparently reduced to Cr(III) while becoming simultaneously bound to proteins comprising the high content of hemoglobin in these cells. Proteins, amino acids, and other reducing agents present in blood serum also are reported to have some capacity to reduce Cr(VI) (Capellmann and Bolt, 1992; Ottenwaelder and Wiegand, 1988; Wiegand et al., 1988). Consequently, the red blood cells and plasma constituents are probably a formidable barrier to tissue uptake and distribution of chromium in the hexavalent form following uptake of low concentrations (in low ppb) into the bloodstream (DeFlora et al., 1997; Corbett et al., 1998).

7.6.1 Human Studies of Oral Exposure to Cr(VI) in Water

A series of human volunteer studies that were intended to understand the possible threshold-dependent uptake and effects of soluble Cr(VI) concentrations in water up to 10,000 µg/L were conducted. Biomarkers including urine, plasma, and red blood cell chromium concentrations, in addition to clinical screens of blood and urine, were monitored in each of the human volunteer exposure studies to help determine whether or not any appreciable systemic uptake occurred as Cr(VI). Similar methods were utilized to examine potential dermal absorption of Cr(VI) during and after human volunteers soaked in a heated bath containing 22,000 µg/L Cr(VI) as potassium dichromate. Detailed protocols for each study were approved by a Human Use Committee and all volunteers were consenting adults (nearly all toxicologists or physicians).

As summarized in Table 7.4, single-dose and multiple-dose studies were conducted to examine the rate of chromium uptake into the blood and excretion in the urine following ingestion of Cr(VI) at concentrations ranging from 100 to 10,000 µg/L as potassium dichromate (Finley et al., 1997; Kerger et al., 1996a, 1997; Paustenbach et al., 1996; Corbett et al., 1997; Kuykendall et al., 1996). The studies were each designed to examine reasonable upper-bound uptake of chromium in drinking water under a specific schedule of dose ingestion between meals, but no other specific controls on dietary habits. Diet and exercise were recorded by each individual during the study, and each person was also asked to avoid taking vitamin C or chromium supplements. Each urine void during the study was collected separately by the volunteers and analyzed separately for total chromium and creatinine. Blood samples were obtained on a specified schedule and were sent to the laboratory for analysis of total chromium in the plasma fraction and the red blood cell fraction. In certain studies, blood and urine samples were either quick-frozen or analyzed immediately after collection for the presence of Cr(VI). Also, intermittent samples of urine and blood were collected for standard urinalysis, complete blood count, and SMA-24 blood chemistry screens. None of the volunteers experienced any outward adverse health effects as a result of the chromium exposures involved in any of these studies, nor were any of the clinical tests appreciably different when pre- versus postexposure parameters were compared.

The ingestion studies, in sum, demonstrated that exposure to soluble chromates in drinking water at concentrations of 1000 to 5000 ppb are not well absorbed in normal adults and are probably absorbed into the bloodstream almost entirely in trivalent (and probably organic-liganded) forms. Exposures in this range apparently were not toxic following acute (bolus and multiple dose) or subchronic administration to normal adults. Further, it appears that steady-state

TABLE 7.4   Summary of Human Volunteer Studies Involving Ingestion of Water-Soluble Chromium Compounds

Ref.[a]	Subjects and Regimen	Key Observations
1	Five healthy adults per dose group, ages 26 to 63. Single bolus dose of 5000 µg Cr in 0.5 liters of water [10,000 ppb ingested within 2 minutes. Cr administered in separate studies of inorganic Cr(III), organic Cr(III) from Cr(VI) in orange juice], and dichromate in water. Dose taken 2 to 5 hours after last meal.	1. No clinical changes or health impacts. 2. Rapid uptake of Cr observed following all three dosing regimens. 3. Bioavailability and urinary half-life graded response: Cr(III) inorganic < Cr(III) organic < Cr(VI). 4. No sustained elevation of RBC Cr levels, marker for Cr(VI) uptake. 5. No Cr(VI) measurable in plasma or urine at peak of uptake/excretion. 6. No elevation of DNA–protein crosslinks in peripheral lymphocytes.
2	Five healthy adults per dose group; ages 33 to 62. Three daily doses of 0.33 liters Cr(VI) in water for 3 consecutive days. Sequential dosing for 3 days on, 2 days off, at Cr(VI) levels of 100, 500, and 1000 ppb. Doses taken >2 hours after last meal.	1. No clinical changes or health impacts. 2. Cr uptake and excretion were apparent at all dose levels. 3. Bioavailability was a similar, low percentage for all dose levels. 4. No sustained elevation of RBC Cr levels, marker for Cr(VI) uptake.
3	Three healthy adults per dose group; ages 33 to 42. Three daily doses of 0.33 liters Cr(VI) in water for 3 consecutive days, 3 days on, 2 days off, at 5000 and 10,000 ppb. Doses taken >2 hours after last meal.	1. No clinical changes or health impacts. 2. Bioavailability was markedly higher compared to prior study ≤ 1000 ppb. 3. No sustained RBC Cr elevation in 2 of 3 volunteers; one volunteer had higher bioavailability and moderate elevation of plasma and RBC Cr levels, which suggests Cr(III) tissue loading kinetics.
4	One healthy adult, age 42. Ad libitum ingestion of 2 liters per day of 2000 ppb Cr(VI) for 17 consecutive days.	1. No clinical changes or health impacts. 2. Steady-state Cr levels reached in blood and urine within 3 to 7 days. 3. No sustained elevation of RBC Cr levels after cessation of exposure. 4. No elevation of DNA–protein crosslinks in peripheral lymphocytes.

TABLE 7.4 *(Continued)*

Ref.[a]	Subjects and Regimen	Key Observations
5	Four healthy adults, ages 20 to 42. Bathing for 3 hours, submerged to shoulders in heated bath (95°F) with 22,000 ppb Cr(VI).	1. No clinical changes or health impacts. 2. Limited systemic uptake shown by blood and urine Cr levels. 3. No sustained elevation of RBC Cr levels, marker for systemic Cr(VI) uptake. 4. Dermal penetration rate for systemic uptake of total Cr (as cumulative urinary Cr) $\leq 4 \times 10^{-4}$ cm/h.

References: 1 = Kerger et al. (1996a) and Kuykendall et al. (1996); 2 = Finley et al. (1997); 3 = Kerger et al. (1997); 4 = Paustenbach et al. (1996); 5 = Corbett et al. (1997).

conditions are reached rapidly for an individual who drank 2 liters per day of 2000 ppb Cr(VI) water (Paustenbach et al., 1996), which suggests that long-term exposures at these levels are not likely to result in Cr(VI) accumulation in tissues.

7.6.2 In Vitro Studies

In vitro studies were conducted to examine the Cr(VI) reduction capacity of human blood and tissues. In a series of studies using human blood collected using the calcium salt of ethylene diamine tetraacetic acid (EDTA) as an anti-coagulant, we determined that red blood cells have an impressive capacity for Cr(VI) uptake and reduction (Corbett et al., 1998). Even at concentrations of added Cr(VI) of 2000 to 10,000 µg/L in blood, which exceeded the capacity of plasma to reduce all of the Cr(VI), the chromium entering red blood cells was completely converted to Cr(III). These studies also demonstrated that plasma apparently has a variable and substantially lower capacity for Cr(VI) reduction that may be highly dependent on diet and time since meal ingestion, among other individual-specific factors.

Research in conjunction with DeFlora and colleagues (1997) also examined the in vitro Cr(VI) reduction potential of human red blood cells and human liver tissues. As in their previous studies (DeFlora et al., 1987; DeFlora and Wetter-hahn, 1989), the *Salmonella typhimurium* mutagenicity assay was utilized as an indicator for the presence and activity of Cr(VI). These studies further supported our in vitro studies of human blood (Corbett et al., 1998) and illustrated the substantial capacity of gastrointestinal fluids and liver tissues to reduce and/or sequester Cr(VI) (see Figure 7.7). Indeed, these data collectively suggest that the substantial reduction capacity of the blood significantly limits systemic uptake and distribution of Cr(VI) by all routes of exposure. This protective effect of the blood is consistent with the available studies on Cr(VI)-induced carcinogenicity, that is, cancers are only observed at the site of initial absorption (the lungs and nasal sinuses) (IARC, 1990; Costa et al., 1993), where the effective Cr(VI) reduction capacity is low relative to the skin and the stomach (DeFlora et al., 1997; DeFlora, 2000).

Research in conjunction with Kuykendall and colleagues (1996) examined Cr(VI) uptake in cultured cells in humans in vivo using a DNA–protein cross-linking (DPX) assay in peripheral lymphocytes (assay of Zhitkovich and Costa, 1992). Kuykendall et al. (1996) demonstrated that concentrations of Cr(VI) as low as 52 ppb were capable of significantly increasing DPX when added directly to a culture medium of Burkitt's lymphoma cells in vitro. In contrast, DPX measurements of lymphocytes from human volunteers ingesting a 5000-µg bolus of Cr(VI) (0.5 liters at 10,000 ppb) failed to cause any measurable increase in DPX in vivo. These findings further support the appreciable reduction potential influencing Cr(VI) kinetics and toxicity in the human body, especially in regards to oral ingestion.

7.6.3 In Vivo Animal Studies

The in vitro findings regarding Cr(VI) reduction capacity of body fluids and tissues suggested that inconsistent findings reported in various genotoxicity tests of Cr(VI) in vitro and in vivo may be explained by the body's natural barriers to uptake and toxicity by normal routes of exposure. For example, oral exposures to Cr(VI) have not been associated with increased cancer incidence, whereas in vivo genotoxicity assays using intraperitoneal injection as the route of administration were clearly positive (Chorvatovicova et al., 1991, 1993). Research in conjunction with Mirsalis and colleagues (1996) provided genotoxicity dose–response evidence using two standard in vivo genotoxicity assays at Cr(VI) drinking water concentrations of 1000, 5000, and 20,000 µg/L (as potassium dichromate).

They reported that ad libitum and bolus administration at these Cr(VI) concentrations failed to induce positive genotoxicity findings in either the in vivo–in vitro rat hepatocyte unscheduled deoxyribonucleic acid (DNA) synthesis test or the mouse bone marrow micronucleus test. Further details on the methodology and results can be found in Mirsalis et al. (1996). These data further support the hypothesis that plausible upper-bound drinking water concentrations of Cr(VI) may not be associated with significant systemic uptake and distribution of the hexavalent species.

Considered together, the in vivo and in vitro research described above provides several strong lines of evidence indicating that plausible long-term exposures to soluble chromates in drinking water [i.e., likely below 2000 µg/L Cr(VI)] have little or no toxicological significance in normal adults. The weight of evidence indicates that soluble chromates absorbed via ingestion or dermal penetration at these concentrations are rapidly reduced by normal biochemical and physiological barriers specific to each route of exposure (DeFlora, 2000; O'Flaherty et al., 2001).

TABLE 7.5   Possible Sources of Exposure to Contaminated Water

Oral Route Exposures	Inhalation Route Exposures	Dermal Contact Exposures
Tap water ingestion	Shower aerosols	Bathing/washing[b]
Beverages mixed with tap water[a]	Swamp cooler aerosols[a]	Swimming[b]
	Irrigation sprinkler drift	Playing or working in
Ingestion of foods with residues[a]	Cooling tower drift	areas with sprinklers[b]
	Cow wash sprinkler	Contaminated soil contact[b]
Ingestion of local cow's milk[a]	drift	
Ingestion of homegrown vegetables[a]	Splashes while swimming[a]	
	Contaminated soils and dusts[a]	

[a]Excluded as de minimis exposure based on empirical data for Cr(VI) reduction and/or based on screening calculations indicating that relative intake from event would be <1% of route-specific doses that were assessed in detail.

[b]All dermal contact exposures were considered de minimis based on Corbett et al. (1997) showing no appreciable Cr(VI) uptake following human volunteer immersion in heated, chlorinated water containing 22,000 ppb Cr(VI) for 3 h continuously.

7.7   CONDUCTING A RISK ASSESSMENT AND DOSE COMPARISON

To assess risks to those who may be exposed to Cr(VI) contaminated water, one must collect historical data for the purpose of calculating their absorbed dose by the oral and inhalation routes. The exposure events worth considering are summarized in Table 7.5. Dermal uptake can be assumed to be negligible based on the human exposure studies showing no appreciable uptake after spending 3 hours immersed in a heated bath containing 22,000 ppb Cr(VI) (Corbett et al., 1997). Also, several of the ingestion and inhalation exposure events can be inferred to involve either zero or de minimis additional Cr(VI) intake (e.g., less than 1 percent of the assessed dose indicators) and therefore do not need to be included in the exposure calculations. For example, the beverage reduction capacity studies (Kerger et al., 1996b) provided an empirical rationale to exclude the Cr(VI) dose attributable to beverages mixed with tap water since common beverages were capable of rapidly reducing Cr(VI) in tap water up to 8000 ppb. Screening calculations based on the empirical data (e.g., swamp cooler studies) and/or standard risk assessment default assumptions (e.g., soil resuspension and vegetable residues) can be developed to conservatively evaluate and exclude several de minimis intake events (as indicated in Table 7.5).

Table 7.6 provides an overview of the exposure assessment calculations applicable for calculating individual-specific evaluation of Cr(VI) dose and health

TABLE 7.6   Calculation Methods when Conducting Individual-Specific Cr(VI) Assessments

Route	Exposure Calculation
Ingestion	Avrg. daily dose = Avrg. for exposed years: $[AAW \times ATI \times DMF \times EF \times B/ABW]$ Peak daily dose = Maximum for exposed years: $[AAW \times ATI \times DMF \times B/ABW]$
Inhalation	Avrg. daily dose = Avrg. for exposed years: $[AAA \times AIR \times EF \times B/ABW]$ Peak daily dose = Maximum for exposed years: $[AAA \times AIR \times B/ABW]$
Dermal	Assumed negligible based on human studies at 22 ppm × 3 h
Definitions	AAW = annual average Cr(VI) water concentration for location and calendar year ATI = age-specific tap water intake based on Pennington (1983) DMF = desert modifying factor, 2.0 for outside laborers and 1.25 for all others EF = exposure frequency (specific exposure events calculated separately and summed) B = bioavailability (conservatively assumed to be 100% for inhalation and ingestion) ABW = age-specific body weight [based on distributions in Finley et al. (1994)] AIR = age-specific inhalation rate [based on Finley et al. (1994)] AAA = annual average airborne Cr(VI) concentration based on dispersion modeling or empirical data correlated to source-specific water concentrations and calendar year

risks. If conducting a dose-reconstruction, the timing of residence and/or working in the impacted area should be recorded for each person, and age-specific intake and body weight estimates developed. Cr(VI) bioavailability can be assumed to be 100 percent for inhaled aerosols or dusts, and dermal uptake can be assumed to be negligible based on the above human studies. Upper-bound oral doses of Cr(VI) should be calculated only for the estimated fraction of daily tap water ingestion that is not mixed with beverages, e.g., based on national surveys reporting age-, region-, and gender-specific data on tap water ingestion (Pennington, 1983; Ershow and Cantor, 1989). If appropriate, a desert modifying factor can be included to increase the daily water dose for field workers based on desert consumption data by Adolph (1969).

In general, annual average and peak doses should be calculated based on correlation to modeled historical groundwater concentrations. Exposure events should be modeled in accordance with each individual's stated frequency of exposure (e.g., number of showers per day or week, hours per day in the irrigation fields, etc.). Lifetime average daily doses and corresponding cancer risks can be calculated for each person who might have been exposed, and estimates of average and maximum daily doses can be calculated for evaluation of non-cancer endpoints.

Sometimes it is useful to conduct a dose comparison to place into context the calculated average and maximum daily Cr(VI) doses by route. These risks can be placed in perspective by examining the risks of exposure to naturally occurring radionuclides present in the local water supply. For example, using parallel exposure parameters and EPA guidance for assessing radionuclide cancer risks, the incremental lifetime cancer risks attributable to residential exposures to uranium, radium, and radon can often range from 10 to 7,000 per million.

Since this work was performed, additional questions about the cancer hazard posed by the ingestion of chromium(VI) in drinking water have been raised in California. For example, the California EPA published in November of 1998 a draft science policy document entitled "Public Health Goal for Chromium in Drinking Water", which added momentum to the wave of popular concern about chromium(VI) in drinking water. None of the recently published studies on chromium(VI) dose, mechanism, and pharmacokinetics were cited in the original agency document or considered in their evaluation. The authors derived a public health goal of 2.5 ppb for total chromium that would be applicable to all public water purveyors in the State of California (current U.S. EPA MCL is 100 ppb). This criterion was based entirely on the assumption that 7.1% of the total chromium in California sources of public drinking water might be present as chromium(VI), and on protecting against theoretical risks of cancer from ingestion of chromium(VI) at that level (essentially at 7.1% of 2.5 ppb, or 0.18 ppb chromium(VI). Despite all of the available dose-response and mechanistic data concerning chromium(VI) carcinogenicity via ingestion (Proctor et al., 2002a; DeFlora, 2000), the State of California scientific staff stood by their derivation and rationale for the 2.5 ppb Public Health Goal for chromium.

Later, additional concerns were raised in California following reports that chromium(VI) may be present at trace levels (high part per trillion range) in several drinking water sources throughout the state (*Los Angeles Times*, 2001). This reported discovery led the California Legislature to require that a panel of chromium toxicology experts, physicians, and other scientists be convened to assess the potential hazard of these low background levels of chromium(VI) apparently present in public drinking water sources. This expert panel started its work in early 2001 and issued its report later that year. In the main, this panel of primarily full time or former university professors concluded that there was inadequate evidence that ingestion of chromium(VI) at the current U.S. EPA MCL of 100 ppb posed a cancer or non-cancer hazard to humans. Indeed, they concluded that the epidemiological studies and animal toxicity studies upon which the California EPA had based their claims of a cancer hazard were insufficient to reach such a determination (Flegal et al., 2001). In short, their conclusions were similar to those reached by other experts who have performed and/or peer-reviewed the series of studies described in this chapter. A recent epidemiology study adds further support to the panel's views (Fryzek et al., 2001). The California Public Health Goal (PHG) for chromium in water was withdrawn in early 2002.

ACKNOWLEDGMENTS

We gratefully acknowledge the firms who found it appropriate to take on the financial burden and potential litigation risks of conducting extensive research that promoted a better understanding of the potential exposures, systemic doses, and health risks of soluble chromates. We also acknowledge the many scientists that assisted in conducting this research.

REFERENCES

Adolph, E. F. 1969. *Physiology of Man in the Desert.* Hafner, New York.

Agency for Toxic Substances and Disease Registry (ATSDR). 1998. Toxicological profile for chromium, August 1998. U.S. Department of Health and Human Services, Public Health Service, Washington, DC.

Anderson, R. A. 1986. Chromium metabolism and its role in disease processes in man. *Clin. Physiol. Biochem.* 4: 31–41.

Anderson, R. A. 1994. Nutritional and toxicologic aspects of chromium intake: An overview. In W. Mertz et al. (Eds.), *Risk Assessment of Essential Elements.* ILSI Press, Washington, DC, pp. 187–196.

Anderson, R. A., Bryden, N. A., and Polansky, M. M. 1997. Lack of toxicity of chromium chloride and chromium picolinate in rats. *J. Am. Coll. Nutr.* 16(3): 273–279.

Anwar, R. A., Langham, R. F., Hoppert, C. A., et al. 1961. Chronic toxicity studies III. Chronic toxicity of cadmium and chromium in dogs. *Arch. Environ. Health* 3: 456–460.

California Environmental Protection Agency (Cal-EPA). 1994. Memorandum on California cancer potency factors: Update, dated November 1, 1994 from Standards and Criteria Work Group. California Office of Environmental Health Hazard Assessment.

Capellmann, M., and Bolt, H. M. 1992. Chromium(VI) reducing capacity of ascorbic acid and of human plasma in vitro. *Arch. Toxicol.* 66: 45–50.

Chorvatovicova, D., Ginter, E., Kosinova, A., and Zloch, Z. 1991. Effect of vitamins C and E on toxicity and mutagenicity of hexavalent chromium in rat and guinea pig. *Mutat. Res.* 262: 41–46.

Chorvatovicova, D., Kovacikova, Z., Sandula, J., et al. 1993. Protective effect of sulfoethylglucan against hexavalent chromium. *Mutat. Res.* 302: 207–211.

Coogan, T. P., Squibb, K. S., Motz, J., Kinney, P. L., and Costa, M. 1991. Distribution of chromium within cells of the blood. *Toxicol. Appl. Pharmacol.* 108: 157–166.

Corbett, G. E., Dodge, D. G., O'Flaherty, E., Liang, J., Throop, L., Finley, B. L., and Kerger, B. D. 1998. In vitro reduction kinetics of hexavalent chromium in human blood. *Environ. Res.* 7: 7–11.

Corbett, G. E., Finley, B. L., Paustenbach, D. J., and Kerger, B. D. 1997. Systemic uptake of chromium in human volunteers following dermal contact with hexavalent chromium (22 mg/L): Implications for risk assessment. *J. Exposure Anal. Environ. Epidemiol.* 7(2): 179–189.

DeFlora, S., Badolati, G. S., Serra, D., Picciotto, A., Magnolia, M. R., and Savarino, V. 1987. Circadian reduction of chromium in the gastric environment. *Mutat. Res.* 192: 169–174.

DeFlora, S., Camoirano, A., Bagnasco, M., Bennicelli, C., Corbett, G. E., and Kerger, B. D. 1997. Estimates of the chromium(VI) reducing capacity in human body compartments as a mechanism for attenuating its potential toxicity and carcinogenicity. *Carcinogenesis* 18(3): 531–537.

DeFlora, S., Serra, D., Camoirano, A., and Zanacch, P. 1989. Metabolic reduction of chromium as related to its carcinogenic properties. *Biol. Trace Elements Res.* 21: 179–187.

DeFlora, S., and Wetterhahn, K. 1989. Mechanisms of chromium metabolism and genotoxicity. *Life Chem. Rep.* 7: 169–244.

DeFlora, S. 2000. Threshold mechanisms and sites specificity in chromium(VI) carcinogenesis. *Carcinogenesis* 21(4): 533–541.

Diaz-Mayans, J., Laborda, R., and Nunez, A. 1986. Hexavalent chromium effects on motor activity and some metabolic aspects of Wistar albino rats. *Comp. Biochem. Physiol.* 83C: 191–195.

Donaldson, R. M., Jr., and Barreras, R. F. 1966. Intestinal absorption of trace quantities of chromium. *J. Lab. Clin. Med.* 68: 484–493.

Edel, J., and Sabbioni, E. 1985. Pathways of Cr(III) and Cr(VI) in the rat after intratracheal administration. *Hum. Toxicol.* 4: 409–416.

Ershow, A. G., and Cantor, K. P. 1989. Total water and tapwater intake in the United States: Population-based estimates of quantities and sources. National Cancer Institute and Life Sciences Research Office, Federation of American Societies for Experimental Biology, Bethesda, MD.

Finley, B. L., Fehling, K., Falerios M., and Paustenbach, D. J. 1993. Field validation of sampling and analysis of airborne hexavalent chromiuim. *Appl. Occup. Environ. Hyg.* 8: 191–200.

Finley, B. L., Kerger, B. D., Corbett, G. E., Katona, M., Gargas, M., Reitz, R., and Paustenbach, D. J. 1997. Human ingestion of chromium(VI) in drinking water: Pharmacokinetics following repeated exposure. *Toxicol. Appl. Pharmacol.* 142: 151–159.

Finley, B. L., Kerger, B. D., Dodge, D. G., Meyers, S. M., Richter, R. O., and Paustenbach, D. J. 1996. Assessment of airborne hexavalent chromium in the home following use of contaminated tap water. *J. Exposure Anal. Environ. Epidemiol.* 6(2): 229–245.

Finley, F. L., and Mayhall, D. A. 1994. Airborne concentrations of chromium due to contaminated interior building surfaces. *Appl. Occup. Hyg.* 9: 433–441.

Finley, B. L., Proctor, D., Scott, P., Harrington, N., Paustenbach, D., and Price, P. 1994. Recommended distributions for exposure factors frequently used in health risk assessment. *Risk Anal.* 14(4): 533–553.

Flegal, R., Last, J., McConnell, E. E., Schenker, M., and Witschi, H. 2001. Scientific review of toxicological and human health issues related to the development of a public health goal for chromium(VI). Governors Task Force. Sacramento, California.

Franchini, I., and Mutti, A. 1988. Selected toxicological aspects of chromium(VI) compounds. *Sci. Total. Environ.* 71: 379–387.

Fryzek, J. P., Mumma, M. T., McLaughlin, J. K., Henoerson, B. E., and Blot, W. J. 2001. Cancer mortality in relation to environmental chromium exposure. *J. Occ. Env. Med.* 7: 635–640.

Glaser, U., Hochrainer, D., Kloppel, H., and Kuhnen, H. 1985. Low level chromium(VI) inhalation effects on alveolar macrophages and immune functions in Wistar rats. *Arch. Toxicol.* 57: 250–256.

Glaser, U., Hochrainer, D., Kloppel H., and Oldiges, H. 1986. Carcinogenicity of sodium dichromate and chromium(VI/III) oxide aerosols inhaled by male Wistar rats. *Toxicology* 42: 219–232.

Gonzales-Vergara, E., de Gonzales, B. C., Hegenauer, J., and Saltman, P. 1981. Chromium coordination compounds of pyridoxal and nicotinic acid: Synthesis, absorption and metabolism. *Israel J. Chem.* 1: 18–22.

Gray, S. J., and Sterling, K. 1950. The tagging of red cells and plasma proteins with radioactive chromium. *J. Clin. Invest.* 29: 1604–1613.

Gross, W. G., and Heller, V. G. 1946. Chromates in animal nutrition. *J. Ind. Hyg. Toxicol.* 28: 52–56.

International Agency for Research on Cancer (IARC). 1990. *Monographs on the Evaluation of Carcinogenic Risks to Humans*, Volume 49: *Chromium, Nickel and Welding*. World Health Organization/IARC Press, Lyon, France, pp. 49–256.

Johansson, A., Curstedt, T., Rasool, O., Jarstrand, C., and Camner, P. 1986a. Rabbit lung after inhalation of hexavalent and trivalent chromium. *Environ. Res.* 41: 110–119.

Johansson, A., Wiernik, A., Jarstrand, C., et al. 1986b. Rabbit alveolar macrophages after inhalation of hexavalent and trivalent chromium. *Environ. Res.* 39: 372–385.

Kerger, B. D., Finley, B. L., Corbett, G. E., Dodge, D. G., and Paustenbach, D. J. 1997. Ingestion of chromium(VI) in drinking water by human volunteers: Absorption, distribution, and excretion of single and repeated doses. *J. Toxicol. Environ. Health* 50: 67–95.

Kerger, B. D., Paustenbach, D. J., Corbett, G. E., and Finley, B. L. 1996a. Absorption and elimination of trivalent and hexavalent chromium in humans following ingestion of a bolus dose in drinking water. *Toxicol. Appl. Pharmacol.* 141: 145–158.

Kerger, B. D., Richter, R. O., Chute, S. M., Dodge, D. G., Overman, S. K., Liang, J., Finley, B. L., and Paustenbach, D. J. 1996b. Refined exposure assessment for ingestion of tap water contaminated with hexavalent chromium: Considerations of exogenous and endogenous reducing agents. *J. Exposure Anal. Environ. Epidemiol.* 6(2): 163–179.

Korallus, U. 1986. Chromium compounds: Occupational health, toxicological and biological monitoring aspects. *Toxicol. Environ. Chem.* 12: 47–59.

Korallus, U., Harzdorf, C., and Lewalter, J. 1984. Experimental bases for ascorbic acid therapy of poisoning by hexavalent chromium compounds. *Int. Arch. Occup. Environ. Health* 53: 247–256.

Kortenkamp, A., Beyersmann, D., and O'Brien, P. 1987. Uptake of chromium(III) complexes by erythrocytes. *Toxicol. Environ. Chem.* 14: 23–32.

Kuykendall, J. R., Kerger, B. D., Jarvi, E. J., Corbett, G. E., and Paustenbach, D. J. 1996. Measurement of DNA-protein cross-links in human leukocytes following acute ingestion of chromium in drinking water. *Carcinogenesis* 17(9): 1971–1977.

Lewalter, J., Korallus, U., Harzdorf, C., and Weidemann, H. 1985. Chromium bond detection in isolated erythrocytes: A new principle of biological monitoring of exposure to hexavalent chromium. *Int. Arch. Occup. Environ. Health* 55: 305–318.

Lindbergh, E., and Hedenstierna, G. 1981. Chromium plating: Symptoms, findings in the upper airways, and effects on lung function. *Arch. Environ. Health* 38(6): 367–374.

Los Angeles Times. 2001. Bill on chromium VI standard to be proposed, written by Andrew Blankstein. February 23rd. Los Angeles.

MacKenzie, R. D., Byerrum, R. U., Decker, C., Hoppert, C. A., and Langham, F. 1958. Chronic toxicity studies: Hexavalent and trivalent chromium administered in drinking water. *Arch. Ind. Health* 18: 232–234.

Mancuso, T. F. 1951. Occupational cancer and other health hazards in a chromate plant: A medical approach, part II. Clinical and toxicological aspects. *Ind. Med. Surg.* 20(9): 393–403.

Mertz, W. 1969. Chromium occurrence and function in biological systems. *Physiol. Rev.* 49: 163–239.

Mertz, W. 1975. Effects and metabolism of glucose tolerance factor. *Nutr. Rev.* 33: 129–135.

Mertz, W., and Roginski, E. E. 1971. Chromium metabolism: The glucose tolerance factor. In W. Mertz and W. E. Cornatzer (Eds.), *Newer Trace Elements in Nutrition.* Dekker, New York.

Miksche, L., Lewalter, J., and Korallus, U. 1994. Determination of chromium in erythrocytes: A new principle for biological monitoring in chromium(VI) exposed workers. *Proc. DOE Symp. on Biomarkers*, San Diego, CA, February.

Mirsalis, J. C., Hamilton, C. M., O'Loughlin, K. G., Paustenbach, D. J., Kerger, B. D., and Patierno, S. 1996. Chromium(VI) at plausible drinking water concentrations is not genotoxic in the in vivo bone marrow micronucleus or liver UDS assays. *Environ. Mol. Mut.* 28: 60–63.

O'Flaherty, E. J., Kerger, B. D., Hays, S. M., and Paustenbach, D. J. 2001. A physiologically based model for the ingestion of chromium(III) and chromium(VI) by humans. *J. Tox. Sci.* 60: 196–213.

Ottenwaelder, H., and Wiegand, H. J. 1988. Uptake of 51Cr(VI) by human erythrocytes: Evidence for a carrier-mediated transport mechanism. *Sci. Tot. Environ.* 71: 561–566.

Paustenbach, D. J., Hays, S., Brien, B., Dodge, D. G., and Kerger, B. D. 1996. Observation of steady state in blood and urine following human ingestion of hexavalent chromium in drinking water. *J. Toxicol. Environ. Health* 49: 453–461.

Pennington, J. A. 1983. Revision of the total diet study food list and diets. *J. Am. Diet. Assoc. Res.* 82(2): 166–173.

Proctor, D., Paustenbach, D., Harris, M., and Finley, B. 2002a. A state of the art review of the hazards posed by hexavalent chromium. *J. Toxicol. Env. Health* (in press).

Proctor, D., Hays, S. M., Ruby, M. V., Liu, S., Sjong, A., Goodman, M., and Paustenbach, D. 2002b. Rate of hexavalent chromium reduction by human gastric fluid. Presented at the Annual Society of Toxicology Meeting, Nashville, Tennessee. Abstract 1700.

Richter, R. O., Kerger, B. D., and Suder, D. 1997. Key considerations for assessment of chromium(VI) exposures and risks from cooling tower aerosols. *Toxicologist* 36(1): 336. Abstracts Issue of Fundamental and Applied Toxicology, for 1997 set meeting.

Sheehan, P. J., Ricks, R., Ripple, S., and Paustenbach, D. 1992. Field evaluation of a sampling and analytical method for environmental levels of airborne hexavalent chromium. *Am. Ind. Hyg. Assoc. J.* 53: 57–68.

U.S. Environmental Protection Agency (USEPA). 1996. Integrated Risk Information System profile for hexavalent chromium. Office of Health and Environmental Assessment, USEPA, Washington, DC. Downloaded from National Library of Medicine on-line service on January 10, 1996.

Weber, H. 1983. Long-term study of the distribution of soluble chromate-51 in the rat after a single intratracheal administration. *J. Toxicol. Environ. Health* 11: 749–764.

Wiegand, H. J., Ottenwaelder, H., and Bolt, H. M. 1988. Recent advances in biological monitoring of hexavalent chromium compounds. *Sci. Tot. Environ.* 71: 309–315.

Zhitkovich, A., and Costa, M. 1992. A simple, sensitive assay to detect DNA-protein crosslinks in intact cells and in vivo. *Carcinogenesis* 13: 1485–1489.

8 Estimating the Value of Research: Illustrative Calculation for Ingested Inorganic Arsenic

D. WARNER NORTH
NorthWorks, Belmont, California

FRANK SELKER
Decision Management Associates, LLC, Portland, Oregon

THOMAS E. GUARDINO
ECONorthwest, Eugene, Oregon

PREFACE

This work was initially conducted in 1993 as part of a research project on the application of decision analysis concepts to environmental risk assessment and management. The project was supported by the National Science Foundation (NSF) with matching funds from private sector organizations. The text as presented here appeared as a contractor report for NSF. It was presented at the First International Conference on Arsenic, held in New Orleans in August of 1993. It was published in a somewhat abbreviated form in the conference proceedings (*Arsenic Exposure and Health*, Special Issue of *Environmental Geochemistry and Health*, Vol. 16, Northwood, England: Science and Technology Letters, 1994, pp. 1–20.) It is reprinted here with the permission of this publisher.

The purpose of the project was development of methodology. The analysis presented is not a careful elicitation of probabilities from experts but rather a sketch of the problem in quantitative form, using values judged by the authors to be reasonable based on their reading of the literature available in 1993. The

Human and Ecological Risk Assessment: Theory and Practice, Edited by Dennis J. Paustenbach
ISBN 0-471-14747-8 © 2002 John Wiley & Sons, Inc.

intention was to display a useful framework for analysis of a regulatory decision and not a fully developed implementation of the analysis. The illustrative calculations were intended to highlight important uncertainties about the relationship of arsenic exposure to human cancer. These uncertainties included the type of cancer induced by ingested arsenic: whether in addition to skin cancer arsenic exposure causes internal cancers such as bladder and lung. The health benefits from a revised regulatory standard for arsenic in drinking water in reducing cancer incidence depend on the shape of the dose–response relationship at the low doses corresponding to exposure from drinking water in the United States. The conclusion from the illustrative analysis presented in this 1993 study is that value of resolving these uncertainties through additional research is very high, compared to the cost of carrying out such research.

In subsequent international conferences (North et al., 1997), as part of the Arsenic Task Force of the Society for Environmental Geochemistry and Health (Chappell et al., 1997) and in other publications (North, 1998) and meetings with the U.S. Environmental Protection Agency (U.S. EPA) staff (Powell, 1999), the senior author has continued to stress this conclusion about the high value of further research on arsenic. Congress did act in 1996 to defer the deadline for EPA to set an arsenic standard, and it also appropriated money to carry out research on arsenic. However, results from recent research have not yet clarified the extent of the cancer threat posed by arsenic in drinking water at levels below the current US standard of 50 μg/L, or by arsenic in food. The uncertainties as of late 2001 remain substantially the same as in the 1993 illustrative analysis, although more data have become available indicating that arsenic exposure is linked to internal organ cancers.

A crucial issue in the regulatory decision to revise the arsenic drinking water standard involves the burden of proof for departure from EPA's default risk assumptions. The congressionally mandated 1994 National Research Council Report, *Science and Judgment in Risk Assessment*, addressed this broad issue and provided recommendations on the revision of EPA's 1986 cancer risk assessment guidelines, and in particular, recommended that EPA should establish a procedure that permits departures from default attemptions. As of this writing, the EPA cancer guidelines have still not yet been formally revised, although a proposed version has been published (U.S. EPA, 1996). The National Research Council report, *Arsenic in Drinking Water* (National Research Council, 1999) expresses the opinion that: "In light of all the uncertainties on mode of action, the current evidence does not meet EPA's stated criteria (U.S. EPA, 1996) for departure from the default assumption of linearity in this range of extrapolation" (p. 300). The report notes on the same page, ". . . the several modes of action that are considered most plausible would lead to a dose–response curve that exhibits sublinear characteristics at some undetermined region in the low-dose range." The 1999 National Research Council report discusses uncertainty on the shape of the dose–response curve, but essentially uses the default of low-dose linearity as the basis for estimating cancer risk.

During the last year of the Clinton Administration and the first months of the Bush Administration, the arsenic drinking water standard has been the focus of considerable scientific debate and political controversy. EPA issued a proposed standard of 5 µg/L in June of 2000 and then, in January of 2001, a final rule setting a new standard of 10 µg/L. In March of 2001 the new EPA Administrator announced a delay to permit further study, and then in April EPA asked the National Academy of Sciences to carry out an additional study to recommend a standard level between 3 and 20 µg/L, by February of 2002. The National Academy of Sciences report was issued in late 2001, and on October 31 EPA reaffirmed the new standard of 10 µg/L. Implementation of the new standard is still planned for 2006, the compliance date in the Clinton Administration's arsenic rule.

The minutes of the June 2000 meeting of Drinking Water Committee of EPA's Science Advisory Board (U.S. EPA, 2000b) and the subsequent report of this committee (U.S. EPA, 2000a) provide an overview as of 2000 of the scientific uncertainties, analysis, and debate regarding the appropriate level for the federal arsenic drinking water standard. These SAB materials and much other applicable materials on arsenic are available via a website at Harvard University organized by Professor Richard Wilson: http://phys4.harvard.edu/~wilson/arsenic_project_main.html. Comments to EPA comments by Professor Wilson, by Daniel Byrd and Steve Lamm, and the letter from Allan Smith (a key participant in the National Research Council 1999 report on arsenic) available on this site provide good examples of differing viewpoints. Also available on this website are abstracts from the Fourth International Conference on Arsenic Exposure and Health Effects, held in San Diego in the summer of 2000, and information on many other conferences on arsenic. Additional information and references should become available on this website as knowledge and policy on ingested arsenic continue to evolve.

The June 2000 SAB minutes conclude (p. 24–25) with a question from a member of EPA's staff on how EPA can "take the many uncertainties into account quantitatively." John Evans of Harvard, a member of the SAB committee, suggested the use of a " 'Formal Expert Judgment' approach," with the book, *Uncertainty* (Morgan and Henrion 1990) suggested as background reading. Such an approach would be similar to the ideas sketched in our 1993 paper, with careful elicitation of probabilities from appropriate experts. EPA sponsored a conference in the fall of 2000 on the application of such probabilistic methods to dose–response uncertainties (Wilson, 2001).

The value of information concept in decision analysis dates back at least to the mid-1960s. An excellent early reference is Raiffa (1968) and good recent expositions include Clemen (1991) and Morgan and Henrion (1992). An early application to an environmental policy issue involving weather modification is given in Howard et al. (1972). There has been relatively little application of this concept to regulation of toxic chemicals, perhaps in part because of the difficulties involved in placing monetary values on health effects such as human cancer. Reconciling risk assessment and economics is an active area at present within EPA, other U.S. federal agencies, and the professional community.

The report of the Presidential/Congressional Commission on Risk Assessment and Risk Management (1997a, 1997b) and the 1996 National Research

Council report, *Understanding Risk: Informing Decisions in a Democratic Society* both reference this chapter in endorsing the usefulness of the value of information concept.

8.1 INTRODUCTION

When decisions involve uncertainties, more information can be valuable. For example, a weather report can help with planning a weekend outing. Research results on carcinogens can be thought of as information that may have value by leading to improved risk management strategies. In fact, information is the primary product we seek when funding research. We do not usually place an explicit value on information, although actions such as purchasing a newspaper for the weather report or funding research imply values by demonstrating a willingness to pay to acquire information. When evaluating and prioritizing research opportunities, it would be desirable to have an estimate of the value of anticipated results.

Decision analysis offers a method of calculating the value of obtaining new information. The method is based on the observation that information is valuable to the extent that it helps us make better decisions, so we can value information by comparing a decision before and after hypothetical information is available. In essence, methods from decision analysis allow us to ask, "how much better could we do if specific uncertainties were resolved?" For background and further discussion of value-of-information methods, see Raiffa (1968), Howard and Matheson (1983), and Clemen (1991). We illustrate this method first by providing a simple example using a coin-toss game. We then apply the method to a simple research decision regarding a hypothetical carcinogen. Finally, we work through a more realistic, but still illustrative, example: examining research opportunities regarding ingested inorganic arsenic.

In the case of arsenic, we explore how methods for calculating the value of information can be used to help evaluate and prioritize research opportunities. The results are in dollars, allowing us to consider the expected cost effectiveness in evaluating and prioritizing research opportunities. The methods could be similarly applied to research regarding other carcinogens.

8.1.1 Calculating the Value of Information: Coin-Toss Game Example

Imagine that someone offers to pay you $10 if you correctly call a coin toss. Assuming the coin toss is fair, there is a 50 percent probability that your call will be correct. This situation is shown in Figure 8.1. Note that the first, square node of the decision tree is your decision of whether to call heads or tails. After your call, the uncertainty of the toss is resolved, as depicted by the second, circular node. Whether you call heads or tails, you have a 50 percent chance of getting $10 and a 50 percent chance of getting nothing. Thus the "expected value" of this opportunity is $(0.5 \times \$10) + (0.5 \times \$0) = \$5$.

Payoff

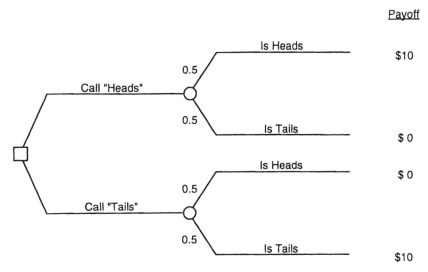

Figure 8.1   Decision tree representing coin-toss game.

Now imagine that you could obtain information revealing whether the toss was a heads or tails before you had to make your call. In this case you could always call the toss correctly, and you would be guaranteed of winning the $10. This situation is shown in Figure 8.2. Notice that we have now reversed the order of the nodes in the decision tree. This is because your information source will now reveal the uncertainty before you need to make the decision.

The coin-toss game is worth $5 without the information and $10 with the information. The value of information is the increase in value, or $5. To summarize, we evaluated a decision with current information and then evaluated the same decision with the opportunity to acquire better information before the decision is made. The difference in expected values is the value of information. This same approach can be used with much more complex, realistic decisions and is the basis for the calculated values of information that we estimate below in the next example and for research pertaining to arsenic.

Notice that the value of information can only be determined in the context of a specific decision. In the coin-toss example the value of information of $5 reflected the particular reward of $10 for a correct call. If the particular reward for a correct call were different, or if the probability of making a correct call with current information were different, then the value of information would also be different.

The $5 value of information in our coin-toss game should be considered an upper bound since we have presumed that the information is correct. The value would be less than $5 if you were provided with information that might be in error. Value-of-information calculations can readily accommodate imperfect information (information that is not definitive and completely reliable), although

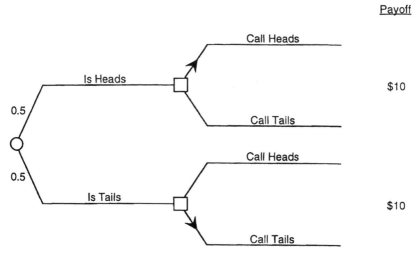

Figure 8.2   Coin-toss decision tree with information regarding the toss outcome (arrows indicate preferred decisions).

the calculations are slightly more involved (Raiffa, 1968; Howard and Matheson, 1983; Clemen, 1991).

In the case of scientific research, however, such calculated values of perfect information are *not* upper bounds. This is because research often yields unexpected results that are helpful with unrelated uncertainties and goals, thereby offering value beyond the initial decision context. The value of unexpected results is not addressed in this chapter. If this value were estimated, it would be added to the values estimated using value-of-information calculations. Thus the actual value of research may be either less than or greater than the calculated value of perfect information. More detailed analysis could go further in the evaluation of both imperfect information and results that may have value in a broader context.

8.1.2   Value of Information in Carcinogen Regulation: A Simple Example

We can extend the general approach illustrated in the coin-toss example to the regulation of carcinogens through another simplified example. Figure 8.3 illustrates a hypothetical regulatory decision to impose new controls on a carcinogen. Imposing the new controls would reduce average exposure in the United States by 50 percent. The decision node (square) at the left of the diagram shows the two possible alternatives in the decision: to adopt or not to adopt the new exposure regulations. The cost of imposing the new regulations is assumed to be one billion dollars per year in this example.

The benefits of such control measures depend critically on the nature of the dose–response relationship for the hypothetical carcinogen. Because the dose–response relationship is uncertain, it is represented in the decision tree diagram

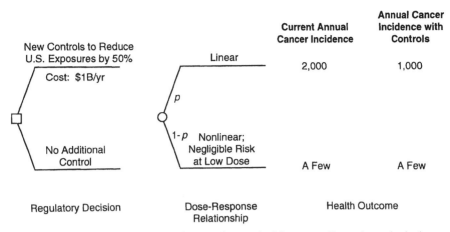

Figure 8.3 Simplified example of a regulatory decision regarding a hypothetical carcinogen.

as an uncertainty node (circle) with two branches. The top branch represents the outcome that the actual dose–response relationship is linear, that is, the risk of cancer is proportional to delivered dose at all levels of exposure. The lower branch represents the outcome that the underlying dose–response relationship is nonlinear, with a negligible risk of health effects at low doses. (In high-dose exposure situations, significant risk may occur, leading to a few cases of cancer annually.) The p next to the upper uncertainty branch represents the probability that the dose–response relationship is linear, while $(1 - p)$ is the probability that the dose–response relationship is nonlinear with very low risks at low exposures.

We assume that if the dose–response relationship is linear, the annual incidence of cancer without controls is anticipated to be 2000 cancer cases. With the proposed controls in place, the number of cancers would be reduced by 50 percent, or 1000 annual cancer cases. However, if the underlying dose–response relationship is nonlinear, the number of annual cancer cases resulting from exposure to the carcinogen would be minimal. (A few annual cancer cases might result from high exposures affecting a small population, such as occupational exposure. If the dose–response is nonlinear, appropriate controls to reduce these high exposures may still be judged appropriate. Such controls targeted only at high exposures might be achieved for much lower cost. We ignore this possibility in our example.)

In setting regulatory policy for the hypothetical carcinogen, a critical trade-off is whether or not the expected benefits of additional controls (i.e., reduced cancer cases) exceeds the cost of implementing the control measures. One way to make this trade-off is to place an explicit assessment on the monetary value of reducing the number of cancers induced by the carcinogen. If we denote the societal value of avoided one cancer case as V, then we can estimate the societal costs and benefits of regulation. Since there are two possible outcomes regarding dose–response, we will need to use a probability-weighted average to estimate expected annual benefits:

$$\text{Expected annual benefit} = p(\text{benefits with linear})$$
$$+ (1 - p)(\text{benefits with nonlinear})$$
$$= p(1000V) + (1 - p)(0)$$
$$= p(1000V)$$

The additional control measures will be cost effective if the expected annual benefit exceeds the control cost of $1 billion per year. The amount by which the expected benefit exceeds the control cost is the expected net benefit of adopting new controls:

$$\text{Expected net benefit} = p(1000V) - \$1 \text{ billion}$$

For example, if the assumed value of an avoided cancer is $1.5 million, and the probability of a linear dose–response relationship is assessed to be 0.75, the expected net benefit of new controls is $0.75 \times (1000 \times \$1.5) - \$1000$, or $125 million. Under these assumptions, adopting new controls for the carcinogen would appear to be good public policy.

As in the coin-toss example, we can use decision analysis techniques to estimate the value of gaining additional information about the nature of the dose–response mechanism for our hypothetical carcinogen. In this example the value of information depends on how the decision to impose new controls would change if the exact form of the dose–response relationship is known before making the decision. Figure 8.4 represents the decision tree in which the uncertainty is resolved before the decision must be made.

In this example, a linear dose–response mechanism will be revealed with a probability of p, while a nonlinear relationship will be revealed with a probability of $(1 - p)$. If we learn that the dose–response relationship is nonlinear, the regulatory decision will be easy: Do not adopt the new control measures because there will be very little reduction in cancer cases. In this case, the net benefit

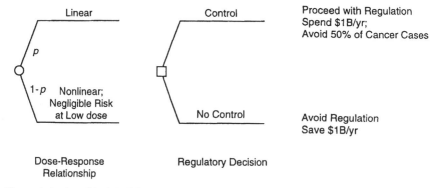

Figure 8.4 Modified decision tree, with resolution of uncertainty placed before the decision, used to estimate the value of information (simplified example with hypothetical carcinogen).

is approximately zero because few if any cancers were avoided and no control costs were incurred.

On the other hand, it we learn that the dose–response mechanism is linear, then our decision will depend on the assumed societal value of an avoided cancer. Specifically, we could decide to impose controls only if the *value* of an avoided cancer is at least as great as the *cost of controls* per avoided cancer. Using the assumptions in this example, the value of an avoided cancer, V, must be at least $1 billion divided by 1000 cancer cases, or $1 million. Thus, for V greater than $1 million, the net benefit of the decision will be positive.

The expected net benefit under perfect information is the probability-weighted average of the two dose–response outcomes, or:

$$\text{Expected net benefit with perfect information}$$
$$= p(1000V - \$1 \text{ billion}) + (1 - p)(0)$$
$$= p(1000V - \$1 \text{ billion})$$

Using our assumed values of $1.5 million and 0.75 for V and p, respectively, the expected net benefit given perfect information about the dose–response mechanism of the carcinogen is $0.75[1000(\$1.5 - \$1)]$, or $375 million. The expected value of perfect information is calculated simply as the increase in expected net benefit given perfect information, that is, $375 million (with perfect information) minus $125 million (without perfect information), or $250 million.

The value of information depends, sometimes in fairly complex ways, on the characteristics of the situation. Figure 8.5 shows how the value of perfect information changes with the value of an avoided cancer (V) and the probability that the dose–response mechanism is linear (p) for this simple example. For all values of V below $1 million the value of information is zero since controls are not cost effective regardless of the dose–response relationship. The diagonal sections of the curves are regions in which the decision with current information is not to control, but the decision would be reversed if it were found that the dose–response relationship is linear. The flat section of the curves are regions in which the decision with current information is to control, but the decision would be reversed if it were found that the dose–response mechanism is nonlinear. The relationship between the value of information and the probabilities and net benefit values may be complex, depending on the details of the decision being analyzed.

As with the coin-toss example, the value of information can only be determined in the context of a specific decision and is not inherent to the uncertain information itself. This value depends closely on two factors: whether or not improved information is likely to change the decision that would be made in the absence of this information and the magnitude of the benefit due to changing the decision.

As illustrated, the results of decision analysis calculations, including those for the value of information, can be sensitive to the probabilities assigned to uncertain outcomes. In the coin-toss example the probabilities were easy to estimate,

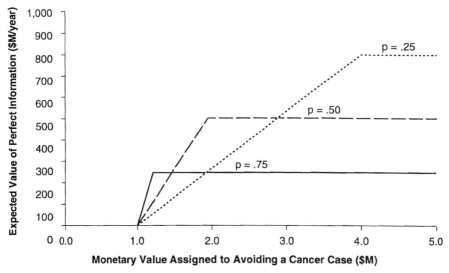

Figure 8.5   Relationship between the value of information, the monetary value placed on avoiding cancer cases, and the likelihood of a linear dose–response relationship (for simplified example using hypothetical carcinogen).

assuming we had a fair coin. However, probabilities in the second example could be difficult to estimate as they often are in real problems. For example, in the context of evaluating research we will need to assign probabilities that various theoretical relationships (e.g., proposed toxicokinetic pathways, toxicodynamic mechanisms of action) are ultimately found to be correct. This may require interviewing a number of experts and assigning probabilities based on their assessments. This dependence on subjective probability assessments can be considered a shortcoming. However, in the absence of definitive information, decisions must be based, implicitly or explicitly, on such subjective judgments. Given this necessity, a method that objectively explores the implications of scientific judgments about alternative hypotheses may be preferable to either ignoring uncertainty or using a less systematic approach.

8.2   EXAMPLES USING INGESTED INORGANIC ARSENIC

Ingested inorganic arsenic has long been associated with increased risk of skin cancer (Hutchinson, 1888). Arsenic is classified by the U.S. Environmental Protection Agency (U.S. EPA) as a known human carcinogen (category A). Risk has primarily been established and quantified through epidemiology studies, particularly the studies carried out among several villages in Taiwan (Tseng et al., 1968; Tseng, 1977). The maximum contaminant level (MCL) standard for drinking water has been set by the U.S. EPA at 50 µg/L (micrograms of arsenic per liter of water). The U.S. EPA and some state agencies are currently considering lowering this drinking water standard to further reduce health risks from ingested inorganic arsenic. However, little is known about arsenic's mechanism of carcinogenicity, and there is no animal species that clearly shows the carcinogenic response shown by humans. A primary challenge of arsenic risk management then is to select cost-effective control levels even though arsenic's potency and mechanism of action remain highly uncertain.

A second challenge is determining what, if any, research is cost effective in helping us improve regulations in the future. This question regarding arsenic has been taken up by several groups, including the U.S. EPA Science Advisory Board (U.S. EPA SAB, 1989, 1992; Fowle, 1992). The 1989 SAB report recommended a revised cancer risk assessment based on biological mechanisms for detoxification. The 1992 SAB report recommends research expected to be possible in 3 to 5 years regarding (1) the form of arsenic responsible for the carcinogenic effect, (2) the variability in and importance of methylation for arsenic detoxification and excretion, and (3) the effects of nutritional and genetic variation on the liver's ability to methylate arsenic. The 1992 SAB report also recommends research that may take longer than 3 to 5 years, including (1) developing an animal model for arsenic's carcinogenic effect leading to a better understanding of arsenic's mechanism of action, (2) differentiating skin cancer caused by arsenic from skin cancer caused by other agents such as sunlight, and (3) investigating the relationship between skin cancer and other skin pathology induced by exposure to arsenic (keratosis). The second challenge, evaluating arsenic research opportunities, is an important part of the risk management response and is the focus of our illustrative example using arsenic.

With the help of a decision tree we can structure the regulatory decision on the arsenic drinking water standard, including uncertainties regarding arsenic's potency and effects. We use this tree to compare *expected effects* of two drinking water standards in terms of costs and human health. We then carry out value-of-information calculations, as illustrated above, to estimate the potential value of research addressing each of the uncertainties considered. Finally, we consider the impacts of varying some of our assumptions in several sensitivity analyses.

We begin our example by constructing a decision tree reflecting the arsenic regulatory decision and important uncertainties (Figure 8.6). The decision tree depicts a regulatory decision and four uncertainties related to the decision.

The first node in Figure 8.6, designated with a square, represents the choice to regulate arsenic at the level of 50 or 20 μg/L. The 50-μg/L level is the current U.S. EPA drinking water standard, and 20 μg/L was selected as one of several possible new standards. Other potential standards could be used equally well as the basis for value-of-information calculations and would yield different results. We explore this further with a sensitivity analysis of a 5-μg/L standard later in this chapter.

The second through fifth nodes, designated with circles, represent four uncertainties regarding risks from arsenic ingestion. Each uncertainty corresponds to an unresolved scientific issue regarding the health risks from ingesting arsenic. Uncertainty at each node is represented by two or three branches, each of which represents a possible outcome when the uncertainty is resolved. This is a simpli-

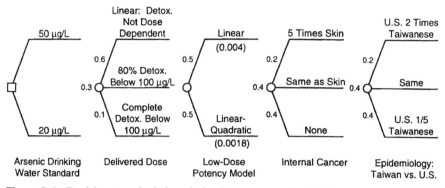

Figure 8.6   Decision tree depicting choice between arsenic drinking water standard of 50 and 20 μg/L with uncertainties regarding arsenic's potency and action. All values are illustrative.

fication in that we represent all possible outcomes at each node with only two or three scenarios. This is analogous to approximating ("discretizing") a continuous probability distribution with a few values. While it is a limited representation of all possible outcomes, it can offer a good approximation while greatly reducing computations. Additional scenarios would be simple to add but were not because: (1) limitations of existing information on the uncertainties and (2) our goal here is to provide an illustrative, rather than definitive, example.

Our assignments of probabilities are consistent with our own reading of the scientific literature and our personal communication with experts. In a more fully implemented analysis a formal process of interviewing experts could be used to assign probabilities to each of the uncertain branches. This analysis should be viewed as illustrative to the extent that results do not reflect an integrated assessment or a consensus among experts. We will now develop our assumptions for each uncertainty node.

8.2.1   Delivered Dose

The first node represents uncertainty regarding the relationship between ingested arsenic and the effective dose delivered to target tissues. We will briefly consider the sources of this uncertainty.

Arsenic is ingested as an inorganic compound ("inorganic arsenic") or as part of an organic molecule ("organic arsenic"). The form of arsenic significantly affects toxicity and the rate of excretion (NAS, 1977). For example, trivalent forms are believed to be more toxic than pentavalent forms, and arsenic in organic compounds generally appears less toxic and more readily excreted than inorganic arsenic (Squibb and Fowler, 1983; U.S. EPA, 1988; Abernathy and Ohanian, 1992). Arsenic usually occurs in water as inorganic arsenate (pentavalent) or as arsenite (trivalent).

After ingestion, inorganic arsenic that is not immediately excreted or absorbed by tissues is progressively methylated to monomethylarsonate (MMA) and then dimethylarsenate (DMA). Methylation is believed to reduce toxicity, although the reduction in toxicity or carcinogenicity does not appear to be complete and has not been fully quantified (Winship, 1984; Marcus and Rispin, 1988; McKinney, 1992; Endo et al., 1992; Harrington-Brock et al., 1993). The mechanisms and kinetics of in vivo methylation are also not well understood, although there is evidence that methylation occurs via a rate-limited enzymatic process that may become saturated or inhibited with very high doses (Valentine et al., 1979; Buchet et al., 1982a, 1982b, 1988).

This understanding of arsenic metabolism has led to speculation that the effective delivered dose will be nonlinear with administered dose: At low doses methylation may more efficiently reduce inorganic arsenic levels and thus partially detoxify administered arsenic, while at higher doses the methylation pathway could be overwhelmed so that tissue levels of inorganic arsenic increase rapidly with dose (U.S. EPA, 1988; Marcus and Rispin, 1988). The existence of such a nonlinearity could imply that low-dose risks are overestimated. However, the issue is not yet resolved since others have found that a roughly constant proportion of ingested inorganic arsenic is methylated across the range of observed doses in past studies (Hopenhayn-Rich et al., 1993; Del Razo, 1994). The potency of inorganic arsenic and its metabolites is further complicated by evidence that potency and carcinogenic effects may depend on exposure to other risk factors (Jan et al., 1986; Li and Rossman, 1991).

It is clear that the metabolism and excretion of arsenicals is complex. Further development of physiologically based pharmacokinetic (PBPK) models will be required to better understand the relation between exposures and the effective delivered dose of dangerous arsenicals. Until such models are developed and validated, the delivered doses of arsenicals remain uncertain.

In the decision tree shown in Figure 8.6, this uncertainty is represented as a node with three branches. Each branch corresponds to an alternative delivered dose relationship that might best represent the actual relationship. The top branch represents the possibility that the relation between exposures and delivered dose is similar at the exposure levels occurring in the United States and exposures that were studied in Taiwan (i.e., the proportion of ingested arsenicals affecting vulnerable tissues is the same). The middle branch represents the possibility that at doses below 100 μg/day, the delivered dose is 20 percent of the doses predicted from higher-exposure epidemiology studies. The bottom branch represents the possibility that detoxification is complete below 100 μg/day. For both the middle and bottom branches, we assume that the detoxification effect disappears when doses reach 200 μg/day. These three relationships between exposure and delivered dose are shown in Figure 8.7.

We assign a probability of 0.6 to the possibility that there is no nonlinearity, a probability of 0.3 to the possibility that there is a nonlinearity with an 80 percent risk reduction below 100 μg/day, and a 0.1 possibility that there is no risk below 100 μg/day. These probabilities are chosen by the authors as illustrative values consistent with our reading of the literature.

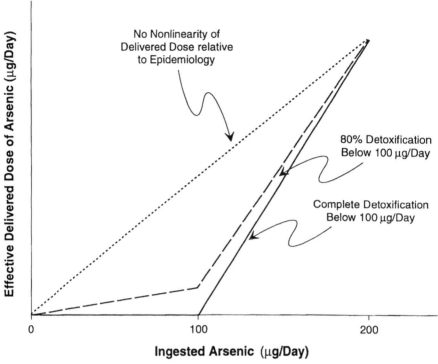

Figure 8.7   Three possible relationships between ingested arsenic and effective dose delivered to vulnerable tissues. Illustrative data.

8.2.2   Low-Dose Model of Potency

The second uncertainty node represents uncertainty regarding the relationship between the dose that actually reaches target tissue and the increased risk of cancer. Dose–response estimates for low doses of arsenic have been based primarily on epidemiological studies of Taiwanese populations (Tseng et al., 1968; Tseng, 1977; U.S. EPA, 1988; Brown et al., 1989). This is better information than that available for many carcinogens because it is from human exposures. Further, the exposures are only about an order of magnitude higher in the studies than exposures of concern in the United States. However, neither these epidemiology data nor available information regarding the mechanism by which arsenic causes cancer is sufficient to reveal with certainty the form of the dose–response relationship (Jacobson-Kram and Montalbano, 1985; U.S. EPA, 1988, Appendix B; Brown et al., 1989; Goldman and Dacre, 1991). When extrapolating existing data to doses that are an order of magnitude lower, minor differences in the shape of the dose–response curve could significantly affect predicted risk.

Arsenic dose–response data have recently been fit to multistage models assuming either low-dose linearity or low-dose quadratic and linear coefficients (U.S. EPA 1988, Appendix B; Brown et al., 1989). While other low-dose forms may prove more accurate, a plausible research goal is to determine whether the

TABLE 8.1 Low-Dose Risks for Arsenic Ingestion

	Linear Low-Dose (10^{-3} Risk at 1 µg/kg/day)	Quadratic Low-Dose (10^{-3} Risk at 1 µg/kg/day)
U.S. EPA (1988), males	5.0	2.3
U.S. EPA (1988), females	3.4	1.0
Brown et al. (1989), males	3.0	1.3
Brown et al. (1989), females	2.1	0.6
Value used (average for males):	4.0	1.8

linear or the combined linear and quadratic low-dose relationship is more accurate. We consider the implications of a pure quadratic (i.e., no linear component) dose–response model later as a sensitivity case.

Table 8.1 summarizes estimates of the risk of getting cancer at some point during a lifetime as a consequence of ingesting arsenic at a rate of 1 µg/kg/day (1 microgram of arsenic per kilogram of body mass every day for 76 years).* The values we use are the average of two of the studies shown: the risk values for males in the U.S. EPA Risk Assessment Forum Report (U.S. EPA, 1988) and for males in Brown et al. (1989). The mean values are 0.004 risk per µg/kg/day assuming a linear low-dose relation and 0.0018 risk per µg/kg/day assuming a linear and quadratic low-dose relation. [Note that we use a single linear-quadratic potency factor across all doses we consider and are thus neglecting the curvature of the linear-quadratic model. This is a good approximation because (1) the quadratic term is very small at the 1 µg/kg/day dose, so the curvature is very small, and (2) the doses we consider are close to 1 µg/kg/day, roughly ranging from 0 to 2 µg/kg/day.] For our illustrative analysis we assume each of the linear and linear-quadratic models has a 50 percent likelihood of more accurately reflecting the true relationship.

8.2.3 Internal Organ Cancers

The third uncertainty node in Figure 8.6 represents uncertainty regarding the risk of internal organ cancers caused by ingested arsenic. While epidemiological studies of arsenic ingestion have focused on skin cancer, several recent studies have suggested that internal organ cancers may also be a concern (Bates et al., 1992; Chen and Chen, 1988; Gibb and Chen, 1989; Chen and Wang, 1990; Smith et al., 1992). Arsenic has been implicated in cancer risk at a variety of internal organs, including the liver, lung, kidney, bladder, nasal cavity, prostate, and gastrointestinal tract.

If arsenic ingestion were found to increase the risk of internal organ cancers, the predicted number and severity of cancers associated with arsenic ingestion could significantly increase. At 50 µg/L of arsenic in drinking water, estimates

---

*We have used 76 years as the median life span in the United States, following Brown et al. (1989).

of risk for internal cancers associated with ingesting arsenic have included zero (implicit in analyses not considering internal cancers), about 7 per 1000 among studied Taiwanese (Chen and Wang, 1990) and 13 per 1000 for one liter of contaminated water per day (Smith et al., 1992). These estimates compare with skin cancer risk estimates (again with consumption of 50 μg/L in water) of 2.6 per thousand using the linear-quadratic low-dose assumption and 5.7 per thousand using the linear low-dose assumption [calculated from Table 8.1: Risk per $1000 = \text{risk } (\mu g/kg/day)^{-1} \times 50 \ (\mu g/L) \times 2 \ (L/day) \times 1/70 \ (kg)^{-1} \times 1000$ people]. Thus the risk of internal cancers associated with ingestion has been suggested to be in the range of zero to about five times the risk of skin cancer (although exposure assumptions have not been entirely comparable). We use a range of 0 to 5 for the risk multipliers for the uncertainty node associated with internal cancers. These multipliers simply scale the total expected risk of cancers. The authors' probabilities are assigned to the branches as shown in Figure 8.6.

Including internal cancer risks requires extending the analysis beyond simply the *number* of cancer cases, because skin cancer associated with arsenic exposure poses a much lower threat to health than internal cancer. If caught early, the skin cancer can be treated inexpensively and effectively with little risk of causing an earlier death. Internal cancers, however, may be difficult to treat and are likely to appreciably shorten life span. For this reason we evaluate skin and internal cancers differently when evaluating results.

8.2.4   Potential Epidemiology Biases

The fourth, and final, uncertainty node shown in Figure 8.6 represents uncertainty regarding how well the results of the Taiwanese studies will predict cancer incidence among the U.S. population. Two major sources of this uncertainty are: (1) the possibility that susceptibility to arsenic-induced cancer may differ between Taiwanese and U.S. populations and (2) aspects of the epidemiology studies that may detract from the accuracy of results. Each is briefly described below.

The major epidemiological studies are based on populations outside of the United States, primarily in Taiwan (Tseng et al., 1968; Tseng, 1977), Mexico (Cebrian et al., 1983), and Germany (Fierz, 1965). There is some concern that differences between populations that have been studied and the U.S. population make extrapolation of results to the United States inaccurate. For example, protein consumption in the United States is significantly higher than among the population studied in Taiwan. Higher protein diets may confer an increased ability to detoxify arsenic, leading to a lower risk associated with a given exposure (Vahter and Marafante, 1987). Further, there may be inherent genetic and physiological differences between populations or combinations of risk factors (Jan et al., 1986; Li and Rossman, 1991) that affect the potency of arsenic as a carcinogen.

There may also be other aspects of the studies that lead to inaccurate risk estimates for those in the United States. For example, the Taiwanese population with high arsenic exposure experienced high mortality from a type of gangrene (blackfoot disease). This may have masked some cancer risk, leading to an underestimate of arsenic cancer risk. Conversely, food consumed by the Taiwanese may have also had elevated levels of arsenic, so doses may have been higher than estimated from water alone. In this case, cancer risk would be overestimated relative to the United States.

Again, each branch of the node representing this uncertainty has an assigned "risk multiplier" that scales risk corresponding to an ultimate resolution of this uncertainty. The values of these multipliers were selected to span the authors' views of uncertainty. The top branch corresponds to the possible outcome that the U.S. population experiences twice the risk suggested by the epidemiology studies. The middle branch corresponds to the possibility that the risk estimates from the epidemiology studies are accurate for the U.S. population, so risks are the same for U.S. citizens as would be predicted by the studies. Finally, the bottom branch corresponds to the U.S. population having risks one fifth of those predicted by the studies. Probabilities are assigned to the branches as shown in Figure 8.6.

The four types of uncertainty illustrated in Figure 8.6 represent much, but not all, of the uncertainty regarding health risks from ingested arsenic.

8.2.5   Exposure Assumptions

The four uncertainty nodes provide a simplified illustrative summary of risks associated with ingesting inorganic arsenic. The remaining information required to estimate risks is a set of projections for arsenic exposure among the U.S. population. Since exposures vary considerably across the population, we will assign fractions of the population to groups with differing estimated exposures. For most people the dominant exposure sources of arsenic are in food and drinking water (U.S. EPA, 1984). We consider each of these pathways.

8.2.6   Drinking Water Exposure

For our analysis we need to estimate exposures to arsenic with the existing 50-μg/L standard and then with a 20-μg/L standard. We begin by briefly reviewing several exposure estimates. Improved exposure estimates are expected to be available in the future.

The 1978 U.S. EPA arsenic regulatory impact analysis suggested that about 150 systems, serving 300,000 people, had arsenic concentrations exceeding 50 μg/L. In 1984 the U.S. EPA estimated that about 110,000 people were estimated to be drinking water over 50 μg/L and roughly 50 million were drinking water with arsenic over 2.5 μg/L (U.S. EPA, 1984). A recent integrated risk assessment (Smith et al., 1992) cited a 1987 U.S. EPA study and estimated that about 350,000 are exposed to water containing over 50 μg/L, about 2.5 million people drink water containing over 25 μg/L, and that average concentration is in the range of 2 to 2.5 μg/L. The U.S. EPA also estimated the population-wide average exposure through drinking water at 2.5 μg/L (U.S. EPA, 1988, Table E-2). U.S. EPA now estimates that there are approximately 1.3 million people

TABLE 8.2   Drinking Water Exposure Assumptions

Percent Population	Number of People (Millions)	Average Arsenic Concentration (µg/L)
Current standard, 50 µg/L		
95.05%	237.6	2.0
4.4%	11.0	10.0
0.55%	1.4	30.0
Average value:		2.51
New standard, 20 µg/L		
95.05%	237.6	2.0
4.95%	12.4	10.0
Average Value:		2.40

drinking water over 20 µg/L, about 12.5 million drinking water over 5 µg/L, and about 36 million are drinking water over 2 µg/L (personal communications with U.S. EPA). Many of the estimates listed above neglect private wells, which may show higher levels of arsenic, especially in the western United States (U.S. EPA, 1988, p. 101).

We primarily rely on personal communications with the U.S. EPA and the Metropolitan Water District in Southern California in selecting our exposure assumptions. In particular, we have assumed three exposure groups, with the population size of each corresponding to the U.S. EPA estimates. We selected plausible estimates of the average exposure within each of these groups, shown in Table 8.2. We assume that for those drinking water over 20 µg/L, the average concentration is 30 µg/L, and for those drinking water between 5 and 20 µg/L, the average concentration is 10 µg/L. Finally, for the 95.05 percent of the population drinking water below 5 µg/L, we assume the average concentration is 2 µg/L. These averages yield a current expected exposure through drinking water of 2.51 µg/L, which is consistent with population average values estimated above. A recent survey of 141 water supplies, serving about one third of the U.S. population produces slightly lower estimates (personal communication with the Metropolitan Water District). In this survey the overall mean level of arsenic in treated water was about 1.4 µg/L (assuming that nondetects were at the detection limit of 0.5 µg/L). The median was 0.6 µg/L. Of the population 98.3 percent had arsenic concentrations less than or equal to 5 µg/L (mean 1.07), 1.2 percent had concentrations of 5 to 20 µg/L (mean 11.4), and one system serving the remaining 0.5 percent of the population had concentrations above 20 (38.6). The samples for this survey were taken primarily from large systems (populations greater than 50,000). Small- and medium-sized systems depend more heavily on groundwater and thus would be expected to have somewhat higher average arsenic levels than those reported in this survey.

The effect of imposing the lower standard is assumed to be that the 1.4 million people now drinking water over 20 µg/L would have their water brought down to an average of 10 µg/L. We selected an average value of less than 20 µg/L because some water supplies may cease using wells with high arsenic or use substitute sources that are well below the 20 μg/L level. Further research regarding compliance with the 1977 50 μg/L standard would verify the degree to which this "compliance overshoot" occurs or is negated by noncompliance at other sites. Note that while the 20-μg/L standard reduces high exposures, it only reduces *average* exposure by about 5 percent, to 2.4 μg/L. This is because most water supplies are already below the 20-μg/L level. This is discussed further with our results.

8.2.7 Food Exposure

For most people the most important exposure route for both organic and inorganic arsenic is food. The 1984 U.S. EPA study of arsenic exposure (U.S. EPA, 1984) uses Food and Drug Administration (FDA) estimates of about 45 μg/day, with regional averages ranging from about 25 to 80 μg/day [see also Pershagen (1986)]. Most of this arsenic was believed to be bound in organic complexes that are readily excreted, although the organic fraction varies widely by food type. The 1988 U.S. EPA Risk Assessment Forum report estimates daily arsenic ingestion in food to be about 50 μg/day, of which 13 μg/day was inorganic arsenic [(U.S. EPA, 1988, pp. 101–104); total intake estimate of 17 to 18 μg/day minus water exposure of 5 μg/day]. U.S. EPA's 1992 Draft Drinking Water Criteria Document on Arsenic estimates adult exposure to arsenic at about 53 μg/day, with 11 to 14 μg/day in inorganic forms (U.S. EPA, 1992). Total arsenic found in FDA's Total Diet Studies have ranged from about 30 to 60 μg/day, with values hovering around 35 μg/day for the past few years. It has been suggested that arsenic in food at a level of 12 μg/day may be a micronutrient essential to good human health (Uthus, 1992).

Regional and individual eating habits can have large impacts on inorganic arsenic exposure, because certain foods are very high in arsenic and because the fraction of total arsenic that is inorganic varies from 0 to 75 percent (U.S. EPA, 1988). This suggests that a wide range of exposures are probably occurring within the U.S. population. However, we are not aware of information characterizing the distribution of exposure. In the absence of detailed information, we arbitrarily selected exposure ranges and population estimates giving about an order of magnitude range in exposures and yielding an average exposure of roughly 13 μg/day (Table 8.3). Better information regarding exposures through food appears to be valuable.

TABLE 8.3 Exposure Assumptions for Inorganic Arsenic in Food

Percent of U.S. Population	Number of People (Millions)	Daily Exposure to Inorganic Arsenic in Food (μg/day)
88%	220	7
10%	25	50
2%	5	100
Average food-borne arsenic exposure:		13.2

8.2.8   Cost of Reducing Drinking Water Standard to 20 μg/L

Based on personal communications with the U.S. EPA and the American Water-works Association (AWWA), we estimate capital costs at $140 million and annual operating costs at $22 million to bring the 600 water systems currently over 20 μg/L to under 20 μg/L. These estimates are preliminary in that the U.S. EPA is expected to provide updated estimates with proposed regulations. To estimate a total annual cost we assume that the capital cost is financed at an interest rate of 5 percent above inflation and amortized (paid off) over 30 years. These assumptions lead to estimated annual capital plus operating costs of $31 million.

Corresponding annual cleanup costs per water system are about $50,000 ($31 million/600 systems) or $22 per person affected ($31 million/1.4 million people). However, costs per system and per household are expected to vary widely depending on the size and type of water system and the water characteristics.

There may be further costs, and potential health and environmental risks, associated with the disposal of water purification by-products. We have not attempted to estimate or include these disposal costs.

8.3   RESULTS AND DISCUSSION

The first step in the value-of-information calculations is analyzing our situation *without* further information. For our example this entails calculating the costs and risks of the current 50 μg/L drinking water standard versus a 20-μg/L standard. The parameters from our decision tree were entered into a computerized spreadsheet to calculate the probabilities and risks of each possible outcome. The risks at each endpoint are calculated using the assumptions regarding exposure, delivered dose, dose–response, and risk multipliers for internal cancer risk and epidemiology uncertainty. The probability at each endpoint is the probability that the information in all branches leading to that point correctly represent the actual underlying risk. This is calculated as the product of probabilities on each branch leading to that endpoint. Once we have the risks and probabilities at each endpoint, we calculate the probability-weighted average, or "expected value," of cancers under each of the two water standards.*

We now go through the calculation needed for one path through the decision tree as an example. First, we assume that risk is being evaluated for the lowest food and water exposure scenarios. The combined consumption of arsenic is 4 μg/day from water and 7 μg/day from food, or 11 μg/day. Assuming an average weight of 70 kg, this corresponds to 0.157 μg/kg/day. We now calculate the risk associated with this dose by working through one scenario of the decision tree, as highlighted in Figure 8.8. Using the middle delivered dose branch reduces this

---

*For the purposes of our illustrative example we neglect concerns about the distribution of risk or societal risk aversion, allowing us to use the expected value as a basis for comparing the alternatives. Issues of risk distribution and aversion could be incorporated into the analysis without changing the overall approach but would add complexity to the evaluation of the decision tree's endpoints.

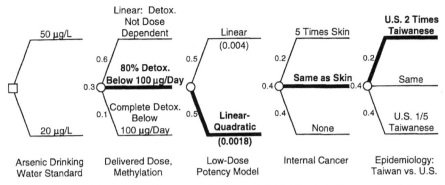

Figure 8.8 Decision tree with one path (scenario) highlighted. Illustrative data and uncertainties.

by 80 percent to 0.031 μg/kg/day. Applying the dose–response risk of 0.004 lifetime cancers per μg/kg/day gives a lifetime risk of 1.26×10^{-4}. Finally, using the risk multipliers of 1 for internal organ cancers and 2 for epidemiology uncertainty gives an estimated risk of 2.5×10^{-4} for skin cancer and the same for internal organ cancers. These are multiplied by 250 million (U.S. population) and divided by a lifetime of 76 years to obtain an annual number of cancer cases of 838 for skin cancer and the same number of internal cancers. Multiplying the probabilities of this particular exposure level (95.05 percent for 4 μg/day in water, 88 percent for 7 μg/day in food) and the four uncertainty branches (30, 50, 40, and 20 percent for the uncertainties selected) yields a total probability of this scenario of 1.00 percent. This scenario's contribution to the overall expected value is the product of the number of cancer cases and probability of the scenario. A similar calculation is carried out for each of the approximately 800 possible combinations of exposure levels and research uncertainty outcomes.

Results of the decision tree calculations are shown in Figure 8.9. The values are expected values, that is, probability-weighted averages across all scenarios. There is a 15 percent chance that the total number of skin and internal cancer cases will exceed the upper error bound shown, and a 15 percent chance that the total will be less than the lower bound shown. The expected number of skin cancer cases associated with arsenic ingestion (lower portion of each bar) at the 50- and 20-μg/L standards are similar to a recent estimate of 1684 cases annually (U.S. EPA, 1988, p. 32). The reduction in cancer cases associated with the reduced drinking water standard is modest: 18 cases of skin cancer and 25 cases of internal organ cancer, for a total of 43, or 1.2 percent of the total of 3472 cancer cases. This is because (1) about 75 percent of exposure to inorganic arsenic is associated with food consumption, and (2) most water systems are already below the 20-μg/L standard. For comparison, Figure 8.9 also shows the risk neglecting all water exposure. This illustrates that in our analysis about three fourths of the total risk from arsenic-induced cancer is due to food-borne arsenic.

Figure 8.10 shows the probability distribution generated by the decision tree for the annual number of cancer cases in the United States associated with arse-

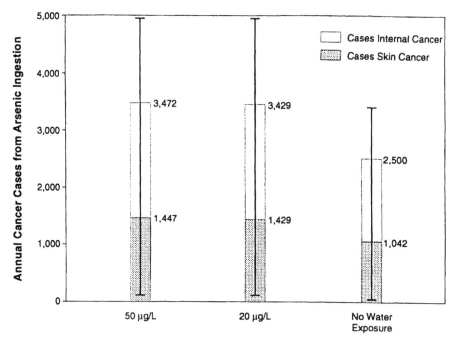

Figure 8.9   Expected values of annual cancer cases for 2 drinking water standards (50 and 20 µg/L) and for the total elimination of drinking water exposure to arsenic. Based on illustrative data. (There is a 70 percent probability that the sum of skin and internal cancer cases will fall between the error bars shown).

nic ingestion with the 50-µg/L standard. These results show a roughly 50 percent chance that fewer than 1000 cases annually are caused by arsenic ingestion, and about a 75 percent chance that annual cases are fewer than 3000. However, there are small probabilities of many more cancer cases, with a 4 percent probability that over 20,000 cases annually are associated with arsenic ingestion. The distribution with the 20-µg/L standard is very similar.

To decide between the 50- and 20-µg/L standards, we must evaluate the benefits of reduced cancer cases versus the costs of implementing the lower standard. To do this we assign monetary values to avoiding a case of skin cancer or an internal organ cancer. Although such valuation is difficult, methods that do not explicitly value health effects must do so implicitly. There has been considerable discussion regarding the use of such valuations in setting public policy (Travis et al., 1987; Miller, 1990). There have also been a number of studies investigating valuations implied by specific regulatory actions. Such valuations are not the purview of scientists and must be made by society and policymakers. For the purposes of our illustrative analysis we will assign values that are somewhat arbitrary, but in our judgment reasonable and consistent with the ranges of values discussed in other regulatory contexts.

For skin cancer the risk of mortality is low with early treatment. However, the risk of mortality is significant with internal cancer. For this reason the value

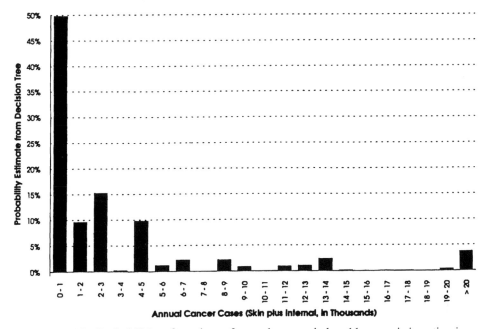

Figure 8.10 Probabilities of numbers of annual cancers induced by arsenic ingestion in the United States under the existing drinking water standard. Based on illustrative data and assumptions, calculated from the decision tree shown in figure 8.6.

assigned to an avoided internal cancers should be considerably higher than the value assigned to an avoided skin cancer. For the purposes of this analysis we initially assign a value of $25,000 to avoiding one case of skin cancer and $1.5 million to avoiding one case of an internal cancer.

Given these assumptions, the costs and benefits of reducing drinking water standards from 50 to 20 µg/L are shown in Table 8.4. As shown, reducing the drinking water standard from 50 to 20 µg/L appears cost effective, offering $7.3 million annually in net benefits. Using a 5 percent real interest rate and a 30-year horizon, this corresponds to a net present value of $112 million in benefits. Of course, this result reflects a number of assumptions and illustrative data, some of which will be revisited below.

8.3.1 Value-of-Information Calculations

The analysis of expected costs and benefits described above is useful for evaluating the best action, but our primary goal is to consider the value of arsenic research options. Using the decision tree and the methods described above, the values of fully resolving the uncertainty at each of the four uncertainty nodes were calculated. The results, as present values over 30 years, are shown in Figure 8.11. The values range from $56 million, for resolving uncertainty regarding the low-dose potency model, to $210 million for resolving uncertainty regarding the importance of internal cancers. Resolving uncertainty regarding either the non-

TABLE 8.4 Annual Costs and Benefits of Reducing the Arsenic Drinking Water Standard from 50 to 20 µg/L[a]

Internal cancer reduction (expected cases):	25.3	
Benefit @ $1.5 million per avoided internal cancer ($M):		$37.9
Skin cancer reduction (expected cases):	18.0	
Benefit @ $25,000 per avoided skin cancer ($M):		$0.4
Total annual benefits ($M):		$38.3
Annual costs ($M):[b]		$31.0
Annual net benefits ($M):		$7.3
Net present value of benefits over 30 years ($M):[c]		$112.0

[a] These results are based on illustrative data and assumptions.
[b] See discussion of cost estimates in text.
[c] Calculated using 5 percent discount rate net of inflation.

linearity in delivered dose or the applicability of epidemiology studies to the United States have values of about $120 to 140 million.

A U.S. EPA spokesman recently estimated the cost of undertaking research on these topics to be about $2.6 million (Fowle, 1991). Although our results are highly approximate, they suggest that the value of research for any one of the four topics considered here are far higher than this cost. Thus it appears that biologically based research regarding health risks associated with arsenic ingestion may be a good public investment.

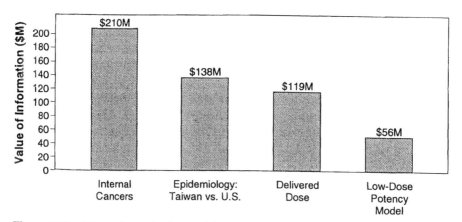

Figure 8.11 Value of completely resolving uncertainty for each of 4 topics. These results are based on illustrative data and assumptions. Values are shown in millions of dollars and are expected net present values over 30-year horizon using 5 percent discount rate net of inflation.

8.3.2 High Stakes and Close Decisions Lead to High Values of Information

To improve our intuition regarding value of information, let us consider two important factors that contributed to the high values of information. First, substantial risks and resources are at stake: the present value of the $31 million annual costs to achieve the 20-µg/L standard over 30 years is $478 million, and the present value of the annual total benefits of $38.3 million is $590 million. Consequently, hundreds of millions of dollars in potential costs and benefits are at stake. It is interesting to note that costs may be about 10 times higher to achieve a 5-µg/L standard and 50 times higher to achieve a 2-µg/L standard (personal communication with U.S. EPA and American Water Works Association). Thus the stakes, and the estimated values of information, would tend to rise considerably for more stringent standards.

Second, costs and benefits are similar and uncertainties are large. In our example, benefits are only about 25 percent greater than costs, and uncertainties affect risks by factors of 2 or greater. As a result, the resolution of an uncertainty can readily change the balance between costs and benefits and can therefore change the decision. In this situation the value of information tends to be high. Conversely, when benefits are far larger than costs, or vice versa, the favored action is clear and the decision tends to be robust. In such one-sided cases, the value of resolving uncertainty is lower, since further information is less likely to change the decision.

We can illustrate both the effect of having higher stakes and having a larger difference in the evaluation of the decision alternative by considering whether to adopt a 5-µg/L standard or keep the 50-µg/L standard. To evaluate this decision we (1) increased the control cost estimate from $31 million annually to $295 annually (capital cost of $1320 million and annual cost of $210 million), again based on recent personal communication with the U.S. EPA and the American Water Works Association, and (2) assumed that under the 5-µg/L standard all water supplies currently exceeding 5 µg/L are brought down to an average of 4 µg/L. The effect of this reduction is to reduce total cancers by an additional 1.5 percent (50 annual cases) below the cancer cases under the 20-µg/L standard, for a total reduction of 2.7 percent (93 annual cases) relative to the 50-µg/L standard.

Valuing avoided cancers as above, the costs of the 5-µg/L standard exceed the benefits by about $154 million annually, or by a present value of $2400 million over 30 years. Thus, with these assumptions, additional regulation to achieve 5 µg/L does not appear cost effective. Further, the decision is not particularly close, with benefits being only about 50 percent of costs. It is not surprising that with the $1.5 million valuation for avoided internal cancer cases the 5-µg/L standard appears less cost effective than the 20-µg/L standard, since the cost is roughly 10 times higher but the number of cancer cases avoided only increases by a factor of 2. If the value of avoided internal cancers is increased to $3 million, then the costs and benefits of the 5-µg/L standard are about equal.

The values of information are also changed in the new decision context of the 5-µg/L standard. Using the $1.5 million value for internal cancer cases, the value of research regarding internal cancers increases by about a factor of 3 to $622 million. This reflects the significantly higher stakes of an action that costs much more but also may avoid more cancer. However, for two of the four uncertainties the value of information is actually reduced to zero. This is because the choice is sufficiently one-sided (costs far greater than benefits) that the decision would not be changed regardless of research outcomes on these two topics. The value of research regarding the applicability of epidemiology results to the United States remains substantial at $75 million but is less than its value of $138 million when the 20-µg/L standard was being considered, again because the choice is now more one-sided. If the value of avoided cancers is raised to $3 million, then all of the uncertainties can easily swing the decision, and the increased stakes lead to very high values of information for all four uncertainties. Present values range from $695 million for uncertainties in the dose–response model to $2145 million for uncertainties regarding internal cancer risk.

8.3.3   Greater Uncertainty Increases the Value of Information

A wider range of possible outcomes regarding an uncertainty tends to increase the value of resolving the uncertainty. Figure 8.12 shows the range of expected cases of internal organ cancers for the resolution of each research uncertainty (with each of the other uncertainties unresolved). For example, the three possible resolutions of uncertainty regarding internal organ cancers result in expected risks ranging from 0 to over 7000 cases of internal organ cancers annually. Notice that the ranges of uncertainty shown in Figure 8.12 for each of the four uncertainties are consistent with the relative values of resolving the uncertainty, as shown in Figure 8.11.

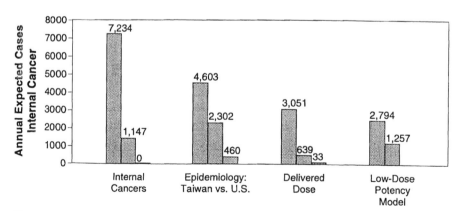

Figure 8.12   Range of expected annual cases of internal organ cancers for resolutions of each uncertainty. (Values shown for each assume other uncertainties are not resolved.) Results are based on illustrative data and assumptions.

To explore the effect of increasing uncertainty, we revisit the range of uncertainty regarding the dose–response model. In the baseline analysis we considered a dose–response model that was purely linear or a model including both linear and quadratic terms. However, models implying either higher or lower risks at low exposures cannot be eliminated using currently available arsenic data. This is because the Taiwanese data have considerable uncertainty in exposure estimates, and exposures for which risks have been quantified are 5 to 50 times higher than exposures of most U.S. citizens. Thus, it is necessary to extrapolate to lower doses, where the nature of the dose–response relationship may be different.

The actual low-dose potency may be found to be either higher or lower than the potency predicted by extrapolation of the linear or linear-quadratic models. For example, some mechanisms of action could suggest decreasing incremental risk as exposure increases. Such a model could predict risk following a curve similar to curve A shown in Figure 8.13. Mechanisms leading to such a relationship could arise, for example, if a metabolic pathway producing a carcinogenic metabolite became saturated at high dose levels (Bailar et al., 1988).

Conversely, mechanisms may be found that suggest very low risks at low doses, with risks increasing more rapidly at moderate and high doses. For example, a mechanism could suggest that the dose–response model is a pure quadratic at low doses, as shown by curve B in Figure 8.13. For example, such a curve may approximate the potency if risk follows second-order kinetics with respect to arsenic concentration. If chromosome alterations at two or more

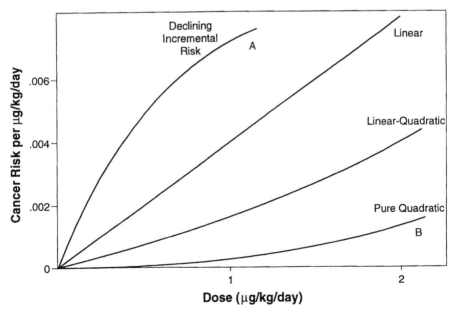

Figure 8.13 Alternative possible dose–response functions for exposure to low doses of inorganic arsenic. Curves are illustrative examples only.

points were required (as in sister chromatid exchanges), such a relationship could hold. If arsenic consumes a cellular resource important to DNA repair, such as methyl donors, cancer risk may remain low until dose reaches a level that begins to exhaust the resource (Mass, 1992), giving a hockey-stick-shaped dose–response.

To demonstrate the effect of increasing the uncertainty of the dose–response model, we keep the linear model as one option, with a lifetime cancer risk of 4.0×10^{-3} for 1 µg/kg/day of arsenic and select a pure quadratic low-dose model as the other alternative. Our least-square fit of a pure quadratic function to data for either 69-year-old Taiwanese males or females yields a risk of 1.1×10^{-4} for 1 µg/kg/day. Fitting a pure quadratic through the lowest dose data for 69-year-old Taiwanese males or females yields risks of 1.8×10^{-4} and 4×10^{-4}, respectively, for 1 µg/kg/day. For our example we select an average value of 2×10^{-4} at 1 µg/kg/day as the risk implied at low doses using a pure quadratic low-dose model. (We emphasize that this very simple derivation is only intended to provide a plausible value for the purpose of illustration.) Thus, we replaced the linear-quadratic low-dose potency of 1.8×10^{-3} per µg/kg/day with our pure-quadratic estimate of 2×10^{-4} at 1 µg/kg/day and recalculated the value of information assuming that the linear and pure quadratic relationships are equally likely. The result is that the value of information for resolving uncertainty regarding the low-dose potency model rises from $56 million to $168 million. With the current standard of 50 µg/L the expected total number of cancer cases also falls by about 27 percent, to 2514. If the pure quadratic dose–response relation were found to be correct, the expected number of cancer cases would fall to 239, or about 10 percent of the expected value with the uncertainty in the dose–response relationship included. The threefold increase in value of information reflects the importance of an uncertainty with such a dramatic impact on estimated risk.

8.3.4   Probabilities of Research Outcomes Affect Value of Information

Probabilities assigned to outcomes also affect the value of information, as was demonstrated in the example of the hypothetical carcinogen. Table 8.5 shows the dependence of the value of information regarding internal cancers from

TABLE 8.5   Effect of Alternative Prior Probabilities on Value of Information[a]

Internal Cancer Risk Scenario	Possible Risk of Internal Cancer			Value of Information ($M)
	$5 \times$ Skin Cancer Risk	Same as Skin Cancer Risk	No Internal Cancer Risk	
Base case	.2	.4	.4	210
Low internal cancer risk	.1	.2	.7	161
High internal cancer risk	.3	.4	.3	165

[a] Based on illustrative data and assumptions.

TABLE 8.6   Effect of Reducing Uncertainty Regarding one Issue on Value of Information from Other Research[a]

Remaining Uncertainty	Value of Information Before Resolution of Internal Cancer Uncertainty ($M)	Value of Information if Internal Cancer Risk Uncertainty Resolved, and Risk Found to Be		
		5 × Skin ($M)	Same as Skin ($M)	None ($M)
Nonlinearity of delivered dose	119	40	84	0
Low-dose potency model	56	0	52	0
Epidemiology applicability to United States	138	1	97	0

[a] Based on illustrative data and assumptions.

ingesting arsenic on the probabilities assigned to internal cancer risk scenarios. Two sets of estimated probabilities are shown in addition to the base case set. One of the new cases has a lower expected risk of internal cancers than the base case and the other has a higher expected risk. The value of information is about 20 percent lower in both cases than it is in the base case. This is because the base case probabilities in the situation where expected benefits and costs of the 20-µg/L standard were closest. Both of the other sets of probabilities lead to decisions in which the comparison of expected costs and benefits is not as close, with a resulting reduction in the value of information. However, in this example (1) the value of information is not very sensitive to probability assignments and (2) the value of information remains high relative to research costs.

8.3.5   Value of Resolving Uncertainty Depends on Other Uncertainties

Resolving one uncertainty usually changes the value of resolving another. For example, Table 8.6 shows how resolving the uncertainty regarding internal cancer will affect the value of information regarding the other uncertainties. The values of resolving each remaining uncertainty is shown for each of the possible resolutions of internal organ cancer risk.

In each case the value of information is reduced by knowledge about the importance of internal cancer. This is consistent with the general (but not universal) trend that as more is known about a problem (i.e., uncertainties are resolved) the value of resolving remaining uncertainties falls. This reflects a sort of declining marginal value of information in that, as we learn more about a situation, the chances are reduced that new information will change our decision making. Since we have not included all possible sources of uncertainty, we will generally underestimate total uncertainty. This result suggests that our values of information may underestimate the true value of resolving uncertainties.

The value of all information falls to zero if internal cancers are not found to be a problem. This is because if internal cancers are not associated with arsenic ingestion, then reducing the standard to 20 μg/L is not cost effective, regardless of the outcome of additional research. The values of information are also lower if internal cancers are found to occur with high frequency relative to skin caner. This is because the reduced standard is likely to be cost effective under these circumstances, regardless of additional findings.

8.4 CONCLUSIONS

Human exposure to carcinogens is a major concern for the public and for environmental policymakers. The primary response is establishing regulations reducing exposures to suspected carcinogens. Second in importance only to this primary regulatory response may be the selection and priority setting of research programs. Over the medium and long term, such research is our principal means for improving the cost effectiveness of risk reduction measures. For this reason, and because research can be expensive and have long lead times, the selection and prioritization of research should be a major component of policies aimed at reducing cancer risks.

Value-of-information calculations offer one means to estimate the potential value of alternative research options. These calculations involve:

1. Identifying the important regulatory decisions
2. Identifying the important uncertainties, and possible resolution of these uncertainties, that affect our ability to regulate cost effectively
3. Obtaining expert assessments of the likelihood that various uncertainties will be resolved in various ways

While numerical results are the primary goal, these steps may also add to the understanding of the decisions and uncertainties in regulating a potential carcinogen.

The coin-toss game showed that the value of information depends on the specific decision being considered. The analysis based on a hypothetical carcinogen reinforced this observation and explored the dependence of results on valuation of health effects and probability assignments.

The arsenic analysis uses illustrative data, although we believe the estimates are realistic based on our review of the literature. The following conclusions and comments should be considered in the context of our assumptions.

First, our calculations suggest that biologically based arsenic research may be highly cost-effective. Under most scenarios we estimate values of resolving uncertainties regarding arsenic potency and mechanism of action to be in the $100 million to $1000 million range. For example, a research program costing $10 million over 5 years is more likely to be an underinvestment than an overinvestment. The values of information shown in this analysis for arsenic research are high because (1) widespread exposure and the potency of arsenic may lead to many cancer cases, (2) costs of reducing exposure through drinking water are high, and (3) uncertainties are great. In one sense these are upper-bound values since they represent the value of results that completely resolve uncertainty regarding various topics. On the other hand, the value of research may exceed these values because research often delivers results and value beyond the initial goals (e.g., results with implications for other carcinogens).

The value of information depends in part on the level of regulation being considered and the values placed on avoided cancer cases. The direction and magnitude of changes in the estimated value of information depend on the particular parameters used in the evaluation and on which uncertainty is being considered. A key factor affecting the value of information is whether or not resolution of a specific uncertainty is likely to change the decision. This depends on whether the decision is close or not, and whether the range of uncertainty is large enough to potentially swing the decision. A second factor affecting the value of information is the magnitude of costs and benefits associated with the decision.

Regarding ingested arsenic, information resolving uncertainty about the risk of internal organ cancers may have the highest value. In the base-case analysis, resolving this uncertainty has a value of over $200 million. If the alternative regulatory level is 5 μg/L or less, and if avoided cases of cancer are assigned a higher value, then (at least for the $3 million/case) the value of information could exceed $1000 million. This information has high value because internal organ cancers are much more dangerous than skin cancer, and there is considerable uncertainty about the magnitude of the risk of internal organ cancer.

Our illustrative analysis suggests that some reduction in drinking water standards may be cost-effective, but it also highlights the limitations of lower standards. First, most people have low exposure to arsenic in drinking water. As a result, reducing the drinking water standard from 50 to 20 μg/L only reduces the population-wide risk associated with drinking water by about 5 percent, and a reduction to 5 μg/L reduces drinking water risk by just another 10 percent. This is because widespread exposure to low doses in water creates most of the population-wide risk. Second, about three-fourths of expected exposure to inorganic arsenic may be from food. This, combined with the widespread low concentration exposure to water-borne arsenic mentioned above, may limit the cancer reductions from lower drinking water standards to only about 1 to 5 percent of total arsenic related cancers. As shown in our example, lowered drinking water standards may still be cost effective, depending on the assumptions used. In addition, there may be important public health issues, and opportunities to reduce risks, associated with arsenic ingested in food.

ACKNOWLEDGMENTS

The authors would like to acknowledge funding for this and related work from the National Science Foundation under Grant No. SES-9022844 with matching funds from the Electric Power Research Institute, the Edison Electric Institute, Monsanto, ARCO, and FMC.

REFERENCES

Abernathy, C. O., and Ohanian, E. V. 1992. Non-carcinogenic effects of inorganic arsenic. *Environ. Geochem. Health* 14: 35–41.

Bailar, J. C. III, Crouch, E. A., Shaikh, R., and Spiegelman, D. 1988. One-hit models of carcinogenesis: Conservative or not? *Risk Anal.* 8(4): 485–497.

Bates, M. N., Smith, A. H., and Hopenhayn-Rich, C. 1992. Arsenic ingestion and internal cancers: A review. *Am. J. Epidemiol.* 135(5): 462–476.

Brown, K. G., Boyle, K. E., Chen, C. W., and Gibb, H. J. 1989. A dose–response analysis of skin cancer from inorganic arsenic in drinking water. *Risk Anal.* 9(4): 519–528.

Buchet, J. P., and Lauwerys, R. 1988. Role of thiols in the in-vitro methylation of inorganic arsenic by rat liver cytosol. *Biochem. Pharmacol.* 37(16): 3149–3153.

Buchet, J. P., Lauwerys, R., Mahieu, P., and Geubel, A. 1982a. Inorganic arsenic metabolism in man. *Arch. Toxicol. Suppl.* 5: 326–327.

Buchet, J. P., Lauwerys, R., and Roels, H. 1982b. Comparison of the urinary excretion of arsenic metabolites after a single oral dose of sodium arsenite, monomethylarsonate or dimethylarsinate. *Int. Arch. Occup. Environ. Health* 48: 71–79.

Cebrian, M. E., Albores, A., Aquilar, M., and Blakely, E. 1983. Chronic arsenic poisoning in the north of Mexico. *Hum. Toxicol.* 2: 121–133.

Chappell, W. R., Beck, B. D., Brown, K. G., Chaney, R., Cothern, R. C. Irgolic, K. J., North, D. W., Thornton, I., and Tsougas, T. A. 1997. Inorganic arsenic: A need and an opportunity to improve risk assessment. *Environ. Health Perspect.* 105(10): 1060–1067.

Chen, C. W., and Chen, C.-J. 1988. Integrated quantitative cancer risk assessment of inorganic arsenic. In *Proceedings of a Symposium on Health Risk Assessment of Environmental, Occupational and Life Style Hazards*, December 20–22, 1988, Taipei, Taiwan.

Chen, C.-J., and Wang, C.-J. 1990. Ecological correlation between arsenic level in well water and age-adjusted mortality from malignant neoplasms. *Cancer Res.* 50: 5470–5474.

Clemen, R. T. 1991. *Making Hard Decisions: An Introduction to Decision Analysis.* PWS-Kent, Boston.

Luz María Del Razo J., José Luis Hernández G., Gonzalo G. Garcia-Vargas, Patricia Ostrosky-Wegman, Christina Cortinas de Nava, and Mariano E. Cebrián. 1994. Urinary excretion of arsenic species in a human population. In W. R. Chappell, C. O. Abernathy, and C. R. Cothern (Eds.), *Arsenic: Exposure and Health.* Special Issue of *Environmental Geochemistry and Health*, Vol. 16. Science and Technology Letters, Northwood, England, pp. 91–100.

Endo, G., Kuroda, K., Okamoto, A., and Horiguchi, S. 1992. Dimethylarsenic acid induces tetraploids in Chinese hamster cells. *Bull. Environ. Contam. Toxicol* 48: 131–137.

Fierz, U. 1965. Catamnestic investigations of the side effects of therapy of skin diseases with inorganic arsenic. *Dermatologica* 131: 41–58.

Fowle, J. R. III. 1991. Health effects of inorganic arsenic in drinking water: research needs—a case study. Presentation to the 25th Conference on Trace Substances in the

Environment and Annual Meeting of the Society for Environmental Geochemistry and Health, Columbia, MO, May 21, 1991.

Fowle, J. R. III. 1992. Health effects of arsenic in drinking water: research needs. *Environ. Geochem. Health* 14: 63–68.

Gibb, H., and Chen, C. 1989. Is inhaled arsenic carcinogenic for sites other than the lung? In U. Mohr (Ed.), *Assessments of Inhalation Hazards*. Springer-Verlag, Berlin, Heidelberg.

Goldman, M. and Dacre, J. C. 1991. Inorganic arsenic compounds: Are they carcinogenic, mutagenic, teratogenic? *Env. Geochem. Health* 13(4): 179–191.

Harrington-Brock, K., Smith, T. W., Doerr, C. L., and Moore, M. M. 1993. Mutagenicity of the human carcinogen arsenic and its methylated metabolites, monomethylarsonic and dimethylarsinic acids in L5178Y TK+/− mouse lymphoma cells. *Environ. Molec. Mutagen.* 21, suppl. 22, Wiley-Liss.

Hopenhayn-Rich, C., Smith, A. H., and Goeden, H. M. 1993. Human studies do not support the methylation threshold hypothesis for the toxicity of inorganic arsenic. *Environ. Res.* 60: 161–177.

Howard, R. A., Matheson, J. E., and North, D. W. 1972. The decision to seed hurricanes. *Science* 176: 1191–1202.

Howard, R. A., and Matheson J. E. (Eds.). 1983. *The Principles and Applications of Decision Analysis*. Strategic Decisions Group, Menlo Park, CA.

Hutchinson, J. 1888. On some examples of arsenic-keratosis of the skin and of arsenic-cancer. *Trans. Pathol. Soc. London* 39: 352–363.

Jacobson-Kram, D., and Montalbano, D. 1985. The reproductive effects assessment group's report on the mutagenicity of inorganic arsenic. *Environ. Mutagen.* 7: 787–804.

Jan, K. Y., Huang, R. Y., and Lee, T. C. 1986. Different modes of action of sodium arsenite, 3-aminobenzamide, and caffeine on the enhancement of ethyl methanesulfonate clastogenicity. *Cyto. Cell Genet.* 41: 202–208.

Li, J.-H., and Rossman, T. G. 1991. Comutagenesis of sodium arsenite with ultraviolet radiation in Chinese hamster V79 cells. *Biol. Metals*, Springer-Verlag.

Marcus, W. L., and Rispin, A. S. 1988. Threshold carcinogenicity using arsenic as an example. In C. R. Cothern, M. A. Mehlman, and W. L. Marcus (Eds.), *Risk Assessment and Risk Management of Industrial and Environmental Chemicals*. Princeton Scientific Publishing, Princeton, NJ.

Mass, M. J. 1992. Human carcinogenesis by arsenic. *Environ. Geochem. Health* 14: 49–54.

McKinney, J. D. 1992. Metabolism and disposition of inorganic arsenic in laboratory animals and humans. *Environ. Geochem. Health* 14: 43–48.

Miller, T. R. 1990. Plausible range for the value of life: Red herring among the mackerel. *J. Forensic Econ.* 3: 17–40.

Morgan, G., and Henrion, M. 1992. *Uncertainty: A Guide to Dealing with Uncertainty in Quantitative Risk and Policy Analysis*. Cambridge University Press, Cambridge.

National Academy of Sciences (NAS). 1977. *Drinking Water and Health*, vol. 1. Prepared by Safe Drinking Water Committee, Advisory Center on Toxicology, National Research Council. National Academy of Sciences, Washington, DC.

National Research Council. 1994. *Science and Judgment in Risk Assessment*. National Academy Press, Washington, DC.

National Research Council. 1996. *Understanding Risk: Informing Decisions in a Democratic Society*. National Academy Press, Washington, DC.

National Research Council. 1999. *Arsenic in Drinking Water*. National Academy Press, Washington, DC.

North, D. W. 1998. Risk assessment using the Taiwan data base: The need for further research. *Hum. Ecol. Risk Assess.* 4(5): 1051–1060.

North, D. W., Selker, F., and Guardino, T. 1993. Estimating the value of research: An illustrative calculation for ingested inorganic arsenic. Decision Focus Incorporated, Mountain View, CA, September.

North, D. W., Selker, F., and Guardino, T. 1994. The value of research on health effects of ingested inorganic arsenic. In W. R. Chappell, C. O. Abernathy, and C. R. Cothern (Eds.), *Arsenic Exposure and Health*. Special Issue of *Environmental Geochemistry and Health*, Vol. 16. *Science and Technology Letters*, Northwood, England, pp. 1–20.

North, D. W., Gibb, H. J., and Abernathy, C. O. 1997. Arsenic: Past, present, and future considerations. In C. O. Abernathy, R. L. Calderon, and W. R. Chappell (Eds.), *Arsenic: Exposure and Health Effects*. Chapman and Hall, London.

Pershagen, G. 1986. Sources of exposure and biological effects of arsenic. In L. Fishbein (Ed.), *Environmental Carcinogens Selected Methods of Analysis, Some Metals: As, Be, Cd, Cr, Ni, Pb, Se, Zn*. International Agency for Research on Cancer (IARC), Lyon, France, pp. 45–61.

Powell, M. R. 1999. *Science at EPA: Information in the Regulatory Process*. Resources for the Future, Washington, DC.

Raiffa, H. 1968. *Decision Analysis: Introductory Lectures on Choices Under Uncertainty*. Addison-Wesley, Reading, MA.

Report of the Presidential/Congressional Commission on Risk Assessment and Risk Management. 1997a. Framework for Environmental Health Risk Management. Final report volume 1. GPO #055-000-00567-2, Washington, DC.

Report of the Presidential/Congressional Commission on Risk Assessment and Risk Management. 1997b. Risk Assessment and Risk Management in Regulatory Decision Making Final report volume 2. GPO #055-000-00567-1, Washington, DC.

Smith, A. H., Hopenhayn-Rich, C., Bates, M. N., Goeden, H. M., Hertz-Picciotto, I., Duggan, H., Wood, R., Kosnett, M., and Smith, M. T. 1992. Cancer risks from arsenic in drinking water. *Environ. Health Perspect.* 97: 259–267.

Squibb, K. S., and Fowler, B. A. 1983. The toxicity of arsenic and its compounds. In B. A. Fowler (Ed.), *Biol. Environ. Effects Arsenic*. Elsevier Science Publishers, Amsterdam, Chapter 3.

Travis, C. C., Crouch, E. A. C., Wilson, R., and Klema, E. D. 1987. Cancer risk management: A review of 133 Federal regulatory decisions. *Env. Sci. Tech.* 21: 415–420.

Tseng, W. P., Chu, H. M., How, S. W., Fong, J. M., Lin, C. S., and Yen, S. 1968. Prevalence of skin cancer in an endemic area of chronic arsenicism in Taiwan. *J. Natl. Cancer Inst.* 40(3): 453–463.

Tseng, W. P. 1977. Effects and dose–response relationships of skin cancer and Blackfoot disease with arsenic. *Environ. Health Perspect.* 19: 109–119.

U.S. Environmental Protection Agency (EPA). September 1984. *Arsenic, Occurrence in Drinking Water, Food, and Air*. Science and Technology Branch, Criteria and Standards Division, Office of Drinking Water. U.S. EPA Contract No. 68-01-6388, Work Assignment 29. U.S. EPA Task Manager, William Coniglio.

U.S. Environmental Protection Agency (EPA). July 1988. *Special Report on Ingested Inorganic Arsenic: Skin Cancer; Nutritional Essentiality*. Risk Assessment Forum. U.S. EPA/625/3-87/013.

U.S. Environmental Protection Agency (EPA). Science Advisory Board. September 1989. *Science Advisory Board's Review of the Issues Relating to Arsenic Contained in the Phase II Proposed Regulations from the Office of Drinking Water*. Letter report to the Administrator.

U.S. Environmental Protection Agency (EPA). Science Advisory Board. May 1992. *Review of Arsenic Research Recommendations: Review by the Drinking Water Committee of the Office of Research and Development's Arsenic Research Recommendations*. EPA-SAB-DWC-92-018.

U.S. Environmental Protection Agency (EPA). Office of Water. 1992. *Second Draft for the Drinking Water Criteria Document on Arsenic*. Prepared Under ICAIR Program No. 1524, for EPA Contract 68-C8-0033, ERG Subcontract No. LSI-8700, ERG Work Assignment No. 2-19, Life Systems, Inc. Work Assignment No. 391524.

U.S. Environmental Protection Agency. 1996. Proposed guidelines for carcinogen risk assessment: Notice. *Fed. Reg.* 61(79): 17959–18011.

U.S. Environmental Protection Agency, 2000a. "Arsenic Proposed Drinking Water Regulation: A Science Advisory Board Review of Certain Elements of the Proposal," A Report by the EPA Science Advisory Board, EPA-SAB-DWC-01-001, December 2000. Internet: http://www.epa.gov/sab/dwc0101.pdf

U.S. Environmental Protection Agency, 2000b. Minutes from the EPA/Science Advisory Board, Drinking Water Subcommittee, June 5–7, 2000. Internet: http://phys4.harvard.edu/~wilson/SABJune_2000_minutes.pdf

Uthus, E. O. 1992. Evidence for arsenic essentiality. *Env. Geochem. Health* 14: 55–58.

Vahter, M., and Marafante, E. 1987. Effects of low dietary intake of methionine, choline, or proteins on the biotransformation of arsenite in the rabbit. *Toxicology Letters* 37(1987): 41–46. Elsevier.

Valentine, J., Kang, H., and Spivey, G. 1979. Arsenic levels in human blood, urine and hair in response to exposure via drinking water. *Environ. Res.* 20: 24–32.

Wilson, J. D., "Advanced Methods of Dose–Response Assessment: Bayesian Approaches—Final Report," Discussion Paper 01-15, Resources for the Future, Washington, D.C., April 2001. Internet: http://www.rff.org.

Winship, W. A. 1984. Toxicity of inorganic arsenic salts. *Adv. Drug React. Act. Poison. Rev.* 3: 129–160.

SECTION D
Case Study Involving Contaminated Soils

9 Risk Assessment of Chromium-Contaminated Soils: Twelve Years of Research to Characterize the Health Hazards

DEBORAH PROCTOR
Exponent, Irvine, California

MARK HARRIS
Harris Environmental Risk Management, Flower Mound, Texas

DAVID RABBE
Chemical Land Holdings, East Brunswick, New Jersey

9.1 INTRODUCTION

The application of health risk assessment to investigate and characterize hazardous waste sites in the United States became a relatively commong practice during the 1980s and 1990s. Indeed, the basic procedures of human health risk assessment were formalized by the U.S. Environmental Protection Agency (USEPA) in a series of guidance documents published over the past 15 years (USEPA, 1986a, 1989, 1991a, 1991b, 1996, 1999a, 1999b). USEPA guidance provides equations, toxicity (i.e., dose–response) criteria, and default exposure assumptions for use by the risk assessor to quantify potential health risks and determine health-protective remedial goals (e.g., cleanup levels). These USEPA default toxicity and exposure assumptions provide conservative input parameters for risk assessment, and their routine use can result in compounding degrees of conservatism and unrealistic exposure and health risk conclusions that can ultimately drive unnecessarily aggressive remedial targets (Nichols and Zechhauser, 1988; Finkel, 1990; Copeland et al., 1994; Finley et al., 1994). Most

Human and Ecological Risk Assessment: Theory and Practice, Edited by Dennis J. Paustenbach
ISBN 0-471-14747-8 © 2002 John Wiley & Sons, Inc.

state environmental agencies use risk assessment paradigms similar to those of the USEPA Superfund guidance for evaluating and conducting risk assessment. Many state programs implement not-so-subtle alterations to the USEPA paradigm, reflecting local legislation and sensitivities, location-specific environmental conditions, and/or the scientific propensities or preferences of the engineers and toxicologists assigned by the state to manage environmental health risks.

At many hazardous waste sites, the repeated use of conservative input parameters and the subsequent overestimation of potential site-related human health risks do not significantly affect the ultimate solution or remedy. Thus, the effort to collect more detailed exposure information and further refine toxicity assessment, to achieve realistic estimations of risk, usually cannot be justified economically. In other situations, the cost of conducting research that could be used to fill data gaps and allow for more precise characterization of the true exposure exceeds the remedial costs, and therefore, the expenditure is not cost effective.

In contrast to these scenarios, at some hazardous waste sites, the identification of scientifically rigorous toxicity criteria and the establishment of precise site-specific exposure factors can dramatically influence the estimation of health risks and the economic costs associated with remedy selection and implementation. Further, an accurate determination of health risk can be important for defending claims of environmental harm or personal injury, as demonstrated by the recent and rapid increase in the value judgments awarded to plaintiffs in environmental cases (e.g., *Anderson et al. v. Pacific Gas and Electric*; *Abel et al. v. Lockheed Martin*). However, the effort involved and resources required to gather information for the purposes of conducting a more scientifically refined exposure and toxicity assessment can be substantial. It is only in the context of large and complex hazardous waste sites, where high-level human exposure is alleged or where remedial and/or potential litigation costs are substantial, that these efforts are justified.

This chapter presents a case study of a group of hazardous waste sites located in northern New Jersey, where a series of unique research efforts were implemented over the past twelve years to improve the accuracy of the risk assessment process. Specifically, this case study documents the use of applied research to advance the characterization of both exposure and toxicity associated with the uncontrolled release of hexavalent and trivalent chromium [Cr(VI) and Cr(III)] in the form of chromite ore processing residue (COPR) in an urban setting. The ultimate goal of the project was to assemble all of the necessary information to identify health-based remedial goals (e.g., soil cleanup standards) using the best science that could be developed. This case study illustrates how research was directed at identifying and filling data gaps to specifically address and replace conservative or default assumptions regarding exposure and health risks identified by the New Jersey Department of Environmental Protection (NJDEP) with refined and site-specific information.

Similar to the approach taken by many state regulatory agencies, NJDEP has modified the USEPA model for human health risk assessment to focus on generating health-based soil cleanup standards, rather than on the calculation of health risk. However, many of the calculations, equations, and assumptions are similar to those used by USEPA. To address the COPR sites in New Jersey, research was applied to most areas of the USEPA risk assessment paradigm (Figure 9.1). To the best of our knowledge, this is the most comprehensive research program ever conducted to minimize the uncertainty in health risk assessment for a group of contaminated sites.

9.2 SITE DESCRIPTION AND HISTORY

9.2.1 History of Chromate Production in Hudson County, New Jersey

Hudson County, New Jersey, is located just west of New York City and is one of the most densely populated counties in the United States, with a density of 12,957 persons/square mile (U.S. Census Bureau, 2002). This area has a rich industrial heritage and at one time was the hub of chromate chemical manufacturing in the United States, with two plants located in Jersey City and one in Kearny (Stapinski, 1993). Key geographic features of the county include the Hackensack River, the New Jersey Meadowlands, and various urban centers such as Jersey City and Kearny. In the past, the county was home to a large industrial base as well as a large residential population. Today, some heavy industrial activities remain, but much of the commercial activity is focused on light manufacturing and trucking/warehousing operations associated with the nearby Port of New York and New Jersey. In the Town of Kearny, light industrial activities are centered in the southern and eastern portions of the town (in and near the Meadowlands) and residential use in the uplands (northern and western portions of the town).

The Martin Dennis Company, Inc. constructed a chromate chemical manufacturing facility in Kearny, New Jersey, in 1916. The plant was located in the south part of the town, which at that time was largely undeveloped due to the presence of the New Jersey Meadowlands. "The meadowlands" is the name of a large tract of wetlands that is located in the counties of Bergen and Hudson [Hackensack Meadowlands Development Commission (HMDC), 2001]. When the Kearny plant was constructed, the wetlands, or meadowlands, contained almost 100,000 acres (Iannuzzi and Ludwig, 2001). However, due to development in this area, only about 8400 acres of wetlands remain today (HMDC, 2001). The Kearny plant was distant to all residential developments but adjoined the meadowlands. A review of aerial photography from 1940 showed that the land adjacent to the plant was either undeveloped or being used in an industrial capacity. During World War II, the Martin Dennis Company and the U.S. government entered into a complex agreement under which the government controlled or owned both the facility and the supply of chromium, while the Martin Dennis Company operated the facility and retained ownership of undeveloped portions of the property. Shortly after the war, the Diamond Alkali Company bought the government's interest in the facility (the actual facility)

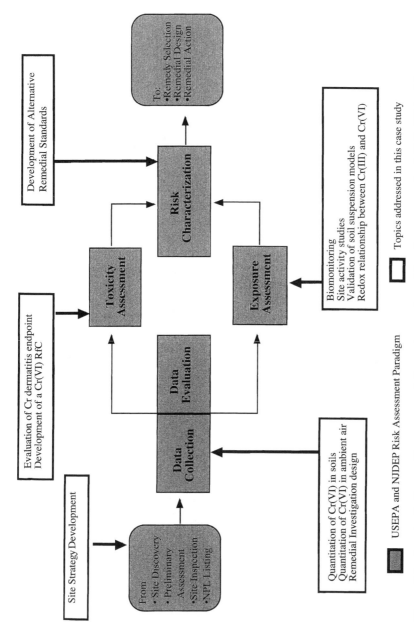

Figure 9.1 Input of site-specific exposure information and toxicity data to standard U.S. EPA risk assessment paradigm.

Site Strategy Development

From
• Site Discovery
• Preliminary Assessment
• Site Inspection
• NPL Listing

Data Collection

Quantitation of Cr(VI) in soils
Quantitation of Cr(VI) in ambient air
Remedial Investigation design

Data Evaluation

Toxicity Assessment

Evaluation of Cr dermatitis endpoint
Development of a Cr(VI) RfC

Exposure Assessment

Biomonitoring
Site activity studies
Validation of soil suspension models
Redox relationship between Cr(III) and Cr(VI)

Risk Characterization

Development of Alternative Remedial Standards

To:
• Remedy Selection
• Remedial Design
• Remedial Action

USEPA and NJDEP Risk Assessment Paradigm

Topics addressed in this case study

and also acquired the Martin Dennis Company, thus gaining control of the entire facility and site property. The facility continued to operate until 1976, when it was closed due to the construction of a more modern plant in North Carolina (Figure 9.2). In 1987, Occidental Chemical Corporation (OCC) purchased the chemicals division of Diamond Shamrock (formerly Diamond Alkali Company). As part of this transaction, Diamond Shamrock indemnified OCC for any environmental claims by government agencies associated with the former chromate chemical manufacturing facility in Kearny. Shortly after the sale of the chemicals company, Diamond Shamrock split into two companies, with one of the newly created companies (Maxus Energy Corporation) assuming the responsibility for addressing environmental claims associated with the OCC indemnification agreement. Maxus Energy Corporation subsequently created a special remedial organization (SRO) to handle the indemnification issues. The characteristics and benefits of SRO's have been discussed previously (Skaggs and Harris, 2001).

There were two additional chromate chemical manufacturing facilities in Jersey City, New Jersey (Paustenbach et al., 1991a, b, 1991; Sheehan et al., 1991). While these plants employed manufacturing operations similar to those of the facility in Kearny, they differed significantly in geographic location. Whereas the facility in Kearny was located in a decidedly nonresidential environment, the two facilities in Jersey City were located in close proximity to residential developments, parks, and the like. The Mutual Chemical Company facility opened in 1905 and closed in 1954 (Honeywell, Inc. is the corporate successor), and the Natural Products facility opened in 1924 and closed in 1964 (PPG Industries, Inc. is the corporate successor for this facility).

9.2.2 Chromate Chemical Production Process and Generation of COPR

The chromate chemical manufacturing process (Figure 9.3) evolved from 1900 to the 1960s, and each manufacturer utilized various modifications of a general process to achieve optimum production levels of sodium dichromate. Of particular significance is the production of a waste product known as chromite ore processing residue, or COPR. In general, about 1 to 1.5 pounds of COPR was generated for each pound of chromate product (hexavalent chromium chemicals) manufactured. The primary product of these three facilities was sodium dichromate, which was used in the production of other chromium-based chemicals such as chromic acid, tanning agents, pigments, and the like. Table 9.1 presents key physical and chemical characteristics of COPR.

9.2.3 Disposal of COPR

Each of the three chromate manufacturing facilities generated large quantities of COPR during the lifetime of the plants. The combined volume of COPR generated by the three plants has been estimated to be at least 2.4 million tons (Figure 9.4).

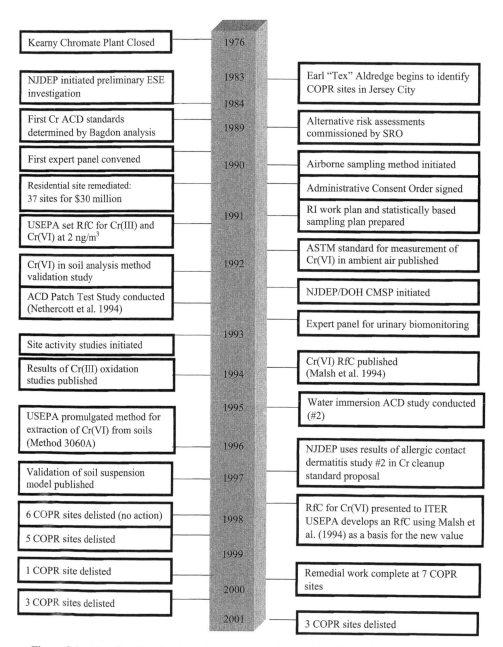

Figure 9.2   Timeline for the chromite ore processing residue (COPR) site investigation, risk assessment and remediation project, 1976 to 2001.

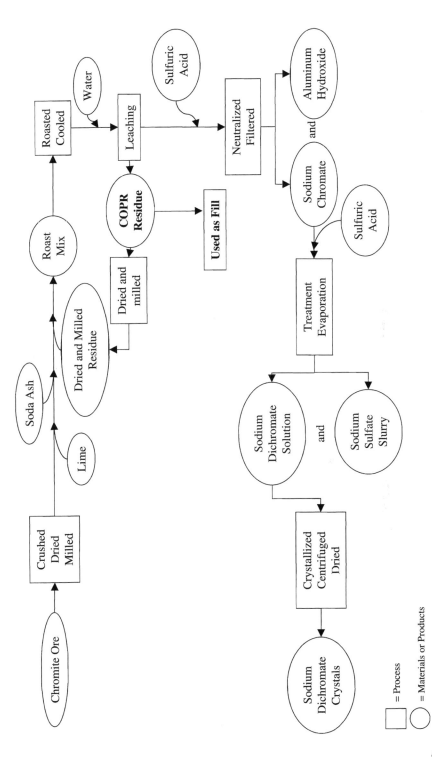

Figure 9.3 Sodium dichromate production process and the generation of chromite ore processing residue (COPR).

519

TABLE 9.1   **Physical and Chemical Characteristics of Chromite Ore Processing Residue (COPR)**

Parameter	Value
Total Cr (mg/kg)	37,000 ± 6,800
Cr(VI) (mg/kg)	9,200 ± 7,100
Iron (mg/kg)	51,000 ± 17,600
Aluminum (mg/kg)	41,000 ± 28,000
Calcium (mg/kg)	238,000 ± 115,000
pH	7–13
E_h (mV)	−150–500
Settled soil bulk density (g/cm^3)	1.2
Packed soil bulk density (g/cm^3)	1.8
Moisture content (%)	5–30

COPR disposal practices in Hudson County, New Jersey, were consistent with typical industrial practices of the time during which these plants operated and with the unique location of the manufacturing facilities. As was common practice prior to the implementation of the first environmental land disposal laws in the 1970s, disposal of industrial wastes was relatively haphazard by current standards. There is little evidence of regulatory oversight of the disposal of COPR prior to 1970. Rather, regulators and industry officials tended to focus their efforts on industrial hygiene practices within the operating facilities [Public Health Service (PHS), 1953]. Indeed, in the case of the Kearny plant, the U.S. government operated the facility during World War II and disposed of COPR in a manner similar to that of other industrial operators before and after the war. Little thought was given to the environmental and human health implications of

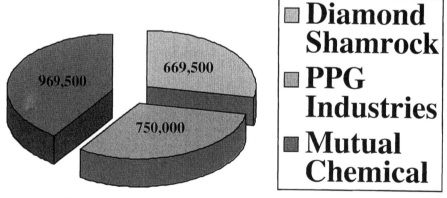

Figure 9.4   Estimated volume of chromite ore processing residue (COPR) produced by the three chromate production plants in hudson county (tons).

allowing the widespread distribution of COPR outside the confines of the plant. Moreover, from an industrial process perspective, the COPR contained relatively low concentrations of hexavalent chromium and was thus thought to be of little health concern based on the knowledge of chromium toxicology at the time.

All three chromate chemical manufacturing facilities were located in or near the New Jersey Meadowlands. As noted by Iannuzzi and Ludwig (2001), during the 20th century, thousands of acres of the New Jersey Meadowlands were filled with soil and refuse to support development. Developers of the lands used all manner of fill, including construction debris, industrial wastes including COPR, chemically-contaminated soils, trash, rocks, and the like. COPR, which was both inexpensive and readily available, was considered an ideal fill material due to its soil-like density following compaction. Known uses of COPR include in construction of buildings (subgrade), for berms and dikes, as backfill during construction of sewer systems, for spot filling in parking lots, and for construction of the New Jersey Turnpike [Environmental Sciences and Engineering (ESE), 1989a; Stapinski, 1993]. In the case of the Diamond Shamrock facility, COPR was initially used as fill for the plant site itself and then for adjacent properties. However, in the 1960s, Diamond Shamrock permitted substantial quantities of COPR to leave the facility in an uncontrolled fashion. This COPR was subsequently used by contractors in the area as fill for various construction projects.

9.2.4   Identification of COPR Sites as a Potential Environmental Hazard

In the 1970s and 1980s, Jersey City was being redeveloped from its industrial heritage into a bedroom community of New York City. As the redevelopment of certain neighborhoods occurred in downtown Jersey City, some areas were found to contain COPR. In 1983, Earl Tex Aldredge, who called himself the "Toxic Avenger" of the Jersey City Hazardous Waste Task Force, identified COPR at a development site where townhouses were being constructed (Stapinski, 1993) (Figure 9.2). Research conducted by the task force and by the NJDEP revealed that construction companies and hauling firms had used or sold the COPR to fill basements of demolished buildings, as base for parking lots and buildings, and/or to fill low-lying areas (ESE, 1989a). As a result of its investigation, the NJDEP estimated that approximately 2 million tons of COPR, containing between 2 and 5 percent total chromium, had been produced by the three chromate production plants (Sheehan et al., 1991). Local residents had become familiar with chromium's signature yellow color on surface soils and in surface water. At neutral pH, Cr(VI) exists as a chromate ion that is bright yellow [International Agency for Research on Cancer (IARC), 1990].

9.2.5   NJDEP Initial Investigations

In 1984, the NJDEP recognized the potential concern for public health and the environment and launched an investigation that consisted of a remedial investigation (RI) of 42 properties potentially affected by COPR; a health risk assessment (HRA) and feasibility study (FS) were also conducted (ESE, 1989a, 1989b, 1989c) (Figure 9.2). The project was funded by the NJDEP's Spill Fund, but costs were ultimately passed on to two of the three companies that produced chromate in Hudson County.

ESE Remedial Investigation   The goal of the ESE remedial investigation was to identify COPR sites and characterize the concentrations of chromium in environmental media, as well as the extent of impact at each site. Most sites were vacant lots that had been residential properties within Jersey City, but industrial and commercial properties, and several recreational areas, including Liberty State Park in Jersey City, were also investigated. A total of 2,138 surface and subsurface soil samples, 31 groundwater samples from 31 monitoring wells located at 7 sites, 33 ambient air samples, 19 surface-water and 11 sediment samples, and 8 building wall wipe samples were collected and analyzed during the RI, which occurred from about 1986 through 1989 (ESE, 1989a). Surface-water and sediment samples were collected from the Upper New York Bay, Penhorn and Caven Creeks, and drainage ditches and puddles of standing water at the sites.

One of the objectives of the RI was to delineate total chromium concentrations in soil in excess of 100 mg/kg. Additionally, sample locations were pre-screened in the field to identify the soil samples that contained the highest concentrations of chromium for subsequent laboratory analysis of Cr(VI). This was performed by screening soil samples in the field for chromium and pH. Only those samples identified in the field as containing higher levels of chromium were analyzed for Cr(VI) in a laboratory, and only the laboratory analytical data were used to assess the potential environmental health risk. Furthermore, the analytical method used to analyze for Cr(VI) was questionable, and at the time of the RI, there was no regulator-approved method to extract Cr(VI) from solid samples. Therefore, the findings of the investigation were biased on the high side, and use of these data in the risk assessment considerably overestimated the true risk.

ESE Risk Assessment   A risk assessment (RA) of four "worst-case" sites and "generic" sites was conducted using the RI data (ESE, 1989c). The RA concluded that the noncancer hazard due to Cr(VI) exposure was significant for worst-case and generic sites. Cancer risks were calculated for potential inhalation of Cr(VI); the theoretical excess cancer risks ranged from one-in-a-million (10^{-6}) for the generic residential properties to one-in-a-thousand (10^{-3}) for industrial sites with heavy truck traffic. The noncancer hazards were assessed by comparing the average daily dose calculated from environmental exposures at each site to the USEPA oral references doses (RfDs) for Cr(VI) and Cr(III). The potential inhalation cancer hazard was evaluated based on a crude soil suspension model, and worst-case input parameters were used consistently. Because ambient Cr(VI) concentrations could not be measured with the available analytical techniques, the model could not be validated. The potential for allergic contact dermatitis (ACD) was not assessed in the ESE risk assessment, although remediation standards were being developed by the NJDEP at that time (1989) to be protective of ACD (Figure 9.2).

ESE Feasibility Study   Based on the RA findings, the state launched a feasibility study (FS) to determine the best method for remediation of the COPR sites (ESE, 1989b). It was concluded that the sites should be cleaned up to a concentration of 100 mg/kg total chromium. The 100 mg/kg total Cr cleanup standard was approximately equal to background levels (Dragun and Chaisson, 1991). The FS investigated dozens of possible alternatives, mostly variations of excavation and disposal options, with cost estimates that ranged up to approximately $650 million per million yards of affected soil. No final remedial approach was selected in the FS; rather, it was concluded that further investigation was needed. The results of the NJDEP's investigation left three responsible parties facing enormous costs for remediation; however, both the site investigation and risk assessment methods were being questioned because they were recognized to severely overestimate the potential exposures and public health risks.

Alternative Risk Assessment   In 1989, the SRO commissioned an alternative health risk assessment of the four representative sites (Paustenbach et al., 1991a, b; Sheehan et al., 1991). The alternative health risk assessment used the data collected by ESE (ESE, 1989a) and assessed the same exposure pathways and scenarios as were evaluated in the NJDEP-commissioned health risk assessment. However, the alternative assessment used more realistic and site-specific exposure assumptions and relied more heavily on exposure data described in the peer-reviewed literature to assess the potential cancer and noncancer health risks. The primary differences in the alternative risk assessment were that they quantitatively accounted for the limited bioavailability of chromium from the COPR matrix, used a simpler but more realistic model of Cr(VI) suspension as dust from soil, and used more realistic estimates of chemical uptake via environmental media (e.g., soil ingestion, vegetable ingestion).

The alternative RA for the residential sites concluded that the potential noncancer health risks posed by oral, dermal, and inhalation exposures to chromium were negligible for workers, residents, and individuals pursuing recreational activities. The potential lung cancer risks from inhalation of Cr(VI) ranged from 1×10^{-12} for workers with average exposure at sites without truck traffic to 2×10^{-7} for the upper bound of potential exposure to workers at sites with heavy truck traffic, and all risks were less than the USEPA de minimus criterion of one in a million (1×10^{-6}). It was recognized, however, that several significant data gaps regarding environmental exposures to chromium in COPR, and the related health risks, remained.

Questionable Conclusions and Data Gaps   As a result of the investigations conducted in the late 1980s, it was clear that there was considerable uncertainty about how to better characterize the potential health risks associated with the

COPR sites. The NJDEP concluded that the sites posed a significant hazard and that COPR must be removed to the assumed background concentration of 100 mg/kg of total chromium. The SRO and two other responsible parties believed that the NJDEP's conclusions about the magnitude of the health risk were flawed because of a strong reliance on overly conservative exposure factors. As a result, in January of 1990, the three responsible parties asked the Industrial Health Foundation to convene an independent expert panel of scientists to review the data, identify data gaps, and offer conclusions regarding the potential hazards posed by the COPR sites (Paustenbach et al., 1991b).

This panel of professors and expert scientists identified the following areas where research was needed if more precise estimates of the risk were to be obtained:

1. *Ambient Air Monitoring.* The analytical limit of detection (LOD) for the Cr(VI) air-sampling method was higher than concentrations in ambient air and higher than levels thought to pose a significant risk for environmental exposures.

2. *Modeling of Soil Suspension to Ambient Air.* No validated model was available to predict the suspension of Cr(VI) from surface soil by wind erosion and vehicle traffic.

3. *Validation of Soil Analytical Method.* No validated method to analyze Cr(VI) in soil and solid media was available.

4. *Threshold for Elicitation of ACD.* There was a very poor understanding of the dose–response relationship for ACD from dermal contact with Cr(VI) in soil, and the exposure model was inaccurate.

5. *Characterization of Chromium Concentrations in Soil.* The data that had been collected were not appropriate for risk assessment because the data sets were biased by focusing on the soils with the highest concentrations of chromium, and thus, long-term average exposures based on these data were overestimated. Because the filling activities at each site varied considerably; it is not feasible to evaluate site-related risks using a generic assessment; rather, site-specific analyses were necessary.

6. *Chromium Chemistry in the Environment.* There was an inadequate understanding of the reduction and oxidation conditions for chromium that could dictate the conversion of chromium between valences in COPR-affected soil.

The panel concluded that if these questions could be answered, then cost-effective remediation strategies could be determined.

9.2.6 Remediation of Residential Sites

Despite the uncertainties in the risk assessment, in 1990, NJDEP ordered one of the responsible parties to conduct remediation of COPR sites in residential areas of Jersey City. The remedial approach involved excavation of 30 relatively small residential properties and removal of the excavated material to a Canadian hazardous waste landfill. NJDEP ordered remediation to 75 mg/kg of total chromium, rather than its original standard of 100 mg/kg. The 75 mg/kg standard for total chromium was set to ensure that Cr(VI) concentrations would be below the NJDEP's new soil standard of 10 mg Cr(VI)/kg soil, which was intended to prevent ACD. The basis for this cleanup level was the assumption that 14 percent of the total Cr was Cr(VI). At the time, no validated method for analyzing Cr(VI) in soil was available, so the cleanup goal had to be set for total chromium, which was measurable. The cost to remediate these 37 small sites was $30 million.

9.2.7 Construction of Interim Remedial Measures

Because of a concern about short-term health hazards, the NJDEP concluded that it would be prudent to require that interim remedial measures (IRMs) be constructed at all sites to "eliminate existing and potential human exposure to chromate chemical production waste" (NJDEP, 1994). The NJDEP required that the 75 ppm total chromium ACD-based soil cleanup standard be used as the determinant of where IRMs should be constructed. Between 1989 and 1994, IRMs were constructed at 37 of the 38 COPR sites in and about the Kearny area by the SRO. Several different types of IRMs were constructed, including (1) geotextile layer overlain by 4 inches of dense graded aggregate underlain by 4 inches of asphalt or (2) plastic liner (30-mm PVC) overlain by 4 inches of dense grade aggregate (Figures 9.5 and 9.6). More than 60 acres of the asphalt IRM

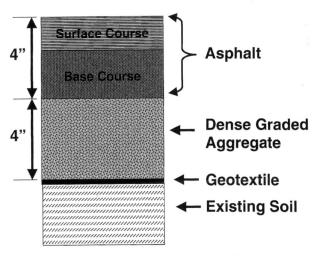

Figure 9.5 Example of interim remedial measure (IRM) paving used at the special remedial organization–managed sites.

Figure 9.6 Photograph of interim remedial measure (IRM) of paving at one of the chromite ore processing residue (COPR) sites managed by the SRO.

and approximately 15 acres of liner-and-stone IRM were installed. The IRMs have been inspected on a routine basis and repaired as needed. The total cost of this program, including development of work plans, construction, final implementation reporting, and maintenance through 2000 has been nearly $20 million.

9.2.8 NJDEP Oversight

As with any large-scale environmental project, the regulatory agency has a significant effect on the characterization of the hazards posed by the site. The regulatory agency can influence the cost of the project, the types of remedies selected, the pace of the remediation, and the public perception of site-related activities. In the case of the NJDEP and the COPR project, initial positions regarding the potential environmental exposures and the associated health risks were developed with the very limited information that was available at the time. As is customary in regulatory risk assessment, data gaps, which in this case were significant, were filled with very conservative, or worst-case, assumptions.

Within the NJDEP, the Site Remediation Program had direct regulatory oversight of the COPR sites. The Administrative Consent Order (ACO), entered into in April 1990 that was applicable to the COPR sites in Kearny, outlined the actions that would be undertaken the SRO. The ACO required the following activities:

1. Construct interim remedial measures (IRMs).
2. Implement a remedial investigation (RI).
3. Conduct a feasibility study (FS).
4. Remediate the COPR sites.

In essence, the ACO required that the sites be secured to prevent future exposure and that the process of assessing the sites and determining remedial alternatives be conducted again. Thus, good science could be used to fill the data gaps identified in the first (ESE) investigations that had been identified by the expert panel and the project team which was being formed to conduct this work.

The Division of Science and Research (DSR) of the NJDEP provided scientific support for various environmental activities. Most activities were aimed at risk assessment issues, and the DSR worked significantly on the COPR project in the early 1990s. Much of DSR's work was aimed specifically at addressing the same data gaps that were identified by the expert panel (Paustenbach et al., 1991b); however, the DSR and the responsible parties generally worked independently on these endeavors, frequently using different methods and reaching different conclusions on the same issues. As a result, from approximately 1989 through 1994, a combative and counterproductive relationship existed, ultimately leading to public outcry and negative media coverage of the involved parties.

The DSR conducted and funded research in many areas, and much of its research was published in the scientific literature to gain support for its positions and counter the positions of the responsible parties, which at that time, were also publishing in the peer-reviewed scientific literature. The DSR investigated the following issues:

1. Soil standards protective of eliciting allergic contact dermatitis (ACD) among pre-sensitized individuals
2. Soil-to-air suspension modeling approaches for the development of Cr(VI) cleanup standards
3. Environmental assessment of indoor Cr dust exposure and the relationship between Cr in dust and urinary Cr in humans in Jersey City and Kearny
4. Reduction/oxidation of Cr in COPR
5. Cr(VI) soil analytical methods
6. Carcinogenicity of COPR in rats [The NJDEP-sponsored study by Dr. Snyder of New York University showed that Cr(VI) in COPR caused no carcinogenic activity in tracheal implantation studies. This study was disregarded by NJDEP when the findings were negative; however, the responsible parties learned of the study after it was complete and encour-

aged the primary researcher, Dr. Snyder, to publish the findings of his investigation (Snyder et al., 1997).]

Had the DSR and responsible parties been able to work cooperatively on these research efforts, fewer public and private resources would have been required to address the data gaps. The relationship between the NJDEP and the SRO improved in the late 1990s, and even the other responsible parties now work more cooperatively with NJDEP to facilitate reasonable risk assessment and management decisions.

9.3   STRATEGIC DIRECTION OF THE SPECIAL REMEDIAL ORGANIZATION

Many large environmental projects are driven down unplanned paths, frequently by regulatory agendas or public opinion, because the managers did not construct and implement a strategy from the beginning of the project to reach a predetermined final endpoint. Environmental project managers frequently modify their strategy in response to the ever-changing policies of regulatory agencies, rather than proceeding in a methodical, stepwise process of investigating and cleaning a site. The development of a strategy not only provides the responsible party with a mechanism for simulating and responding to unplanned scenarios; it also provides guidance for all levels of management when faced with decision making on a real-time basis (e.g., meeting with regulators, local stakeholders, etc.).

At the onset of the project in the late 1980s, one of the responsible parties, Maxus Energy Corporation, established a team of scientists and engineers to address the COPR project. Eventually, they took the additional step of creating the aforementioned SRO to manage the project. The purpose of the group was to develop and implement a strategy for the COPR project that could be followed for the anticipated 10- to 20-year life of the project. Thus, an overall strategy was constructed during the project's infancy and has been implemented over the life of the project.

A specific approach to the human health risk assessment and the identification of health-based chromium cleanup standards were important parts of the overall strategy. An analysis of the problem at project initiation in the late 1980s suggested that cleanup standards would largely dictate the type and size (volume of COPR to be addressed) of the remedial actions that would be required. Furthermore, a review of chromium toxicology suggested that valence differentiation in the environment would be key to correctly calculating health-protective cleanup standards and maximizing the benefit of each dollar committed to remediation.

Key portions of the human health risk assessment strategy for the project were to:

1. Identify and use selected contractors that provide high-quality work while demonstrating the ability to "think outside of the box." No one contractor

would be retained to provide all services. Rather, individuals within various firms with specific relevant expertise were retained. Additionally, it was recognized that for some investigations it would be appropriate to work with the academic community, especially for research-oriented issues and those requiring an independent opinion.

2. Identify and fill data gaps for variables necessary to conduct a human health risk assessment, and develop chromium cleanup standards for both trivalent and hexavalent chromium. The data gaps identified in the NJDEP ESE investigation and those identified by the Industrial Health Foundation expert panel report (Harbison and Rinehart, 1990; Paustenbach et al., 1991b) dictated the research activities that would be conducted.

3. Publish all work performed to address data gaps in the peer-reviewed scientific literature, and utilize the findings to facilitate the decision-making process with the NJDEP.

Implementation of the above strategies contributed greatly to the success of the project.

Selection of staff for the project was key to the overall strategy. Staff were recruited from various environmental consulting firms, with a heavy concentration of young staff managed by a small number of very experienced and well-respected environmental professionals. As noted previously, no single contractor provided the bulk of the staff. Rather, individuals were recruited who possessed specialized skills (risk assessment, analytical methods, site investigation design, etc.). A staff such as this provides exceptional technical capabilities, but the use of more than one firm is likely to be more costly than contracting with any single firm to do all of the work. Academic specialists were engaged to assist on certain complex issues. However, academics generally work at a much slower pace than is typically acceptable for fast-track regulatory efforts, so they generally played a smaller role than the SRO might have preferred.

Publication of the results of most scientific investigations in the peer-reviewed literature was an essential part of the process, and it was critical for gaining acceptance of the responsible parties' findings and recommendations by the regulatory agencies. Without publication and the associated independent peer review, the NJDEP would have been less willing to deviate from traditional and more conservative approaches to risk assessment.

9.4 DATA COLLECTION EFFORTS

The risk assessment strategy called for the collection of data that could be used to characterize the concentrations of Cr(VI) and Cr(III) in environmental media, quantify the potential health risks associated with exposure at the sites under current and future conditions, and if the exposures posed an unacceptable risk, to design and implement a remedy to protect human health and the environment. This section discusses the approaches used to collect site data.

9.4.1 Filling Data Gaps to Reduce Uncertainty in the Site Investigation and Risk Assessment

At project initiation in the late 1980s, there were many significant data gaps that limited characterization of the true health risks posed by the sites and required that conservative assumptions be used for estimating exposure. Because exaggerated risk estimates can lead to unnecessarily restrictive remediation (e.g., remediation to levels that produce no additional reduction of health risk), special efforts were directed at improving data collection techniques.

As there was no sampling method that could measure Cr(VI) in ambient air (much lower concentrations than those found in the workplace), soil suspension models were relied upon to estimate the suspension of Cr(VI) from soil by wind and vehicle traffic (ESE, 1989c). The model-predicted ambient concentrations of Cr(VI) were below the analytical limit of detection of the available method, and as is often the case, the model could not be validated. However, based on empirical analyses of the modeling results, it was believed that the use of conservative assumptions in the model resulted in significantly exaggerated estimates of the airborne Cr(VI) levels. Thus, an air-sampling and analysis method was needed that could measure concentrations of Cr(VI) in ambient air at the 1 ng/m^3 level, approximately 100 times lower than the only other validated method available at the time [e.g., the National Institute for Occupational and Safety Health (NIOSH) method number 7600].

With considerable effort, the project team developed and validated an airborne sampling and analysis method that allowed for the measurement of Cr(VI) in the range of 1 ng/m^3 (Sheehan et al., 1992) (Figure 9.2). With this method, the concentrations of Cr(VI) in ambient air could be measured under typical land-use conditions for all of the Kearny COPR sites before interim remedial measures were implemented. The purpose of this activity was threefold: (1) reduce the use of overly conservative modeling approaches in the assessment of potential health risk by collecting actual airborne sampling data that could be used *in lieu* of the NJDEP modeling approach, (2) validate a new soil suspension modeling approach to allow for the accurate estimation of soil cleanup levels based on acceptable airborne concentrations, and (3) document actual airborne concentrations of Cr(VI) at each site before IRMs were implemented to limit toxic tort and personal injury lawsuits based on inflated estimates of historical airborne exposure levels.

In addition to not having an analytical method for Cr(VI) in air, at the time of the ESE RI and RA (ESE, 1989a, 1989c), there was no validated method to extract Cr(VI) from soil. The method that had been used for the RI (EPA Method 3060) had been rejected because of poor matrix spike recovery in highly organic samples (e.g., anoxic sediments). For this reason, NJDEP had established the 75 mg/kg total chromium remediation goal mentioned previously for soils containing COPR, based on the conservative assumptions that 14 percent of total chromium was Cr(VI), that and a Cr(VI) concentration of 10 mg/kg was protective of allergic contact dermatitis. This "indirect manner" of regulating Cr(VI) in soil was due to the lack of an analytical method of sufficient sensitivity and specificity. The lack of a validated method eliminated the potential use of ideal remediation technologies, which rely upon the reduction of Cr(VI) to Cr(III). Thus, a validated method for measuring Cr(VI) in soils had to be developed and then adopted by the NJDEP and USEPA for use in the RI.

The next challenge was to identify a reasonable approach to characterizing the amount of soil that had been affected. At that time, and to a large extent today, there was no well-established or conventionally used approach for designing investigation programs to achieve a high level of statistical power and confidence in the resulting data set. As mentioned, to reduce laboratory costs and characterize the worst-case risk, the NJDEP used a field screening approach to identify only the most highly affected soil samples in the field, and then only analyzed only those samples for Cr(VI). This approach, as expected, resulted in average concentrations of Cr(VI) that were biased high. To avoid the pitfalls of the ESE investigation approach, it was concluded that a random sampling plan was needed to avoid bias. Further, it was important to collect enough data to be able to predict the average concentrations with a high degree of statistical confidence, power, and precision. Thus, statistical analyses were performed to determine, on a site-by-site basis, how many samples would be necessary for the determination of reliable, site-specific exposure-point concentrations.

9.4.2 Development of Air Sampling and Analysis Methods

In 1987, The Research Triangle Institute (RTI) developed a relatively sensitive sampling and analytical procedure to measure Cr(VI) downwind of industrial point sources for the California Air Resources Board (CARB) (CARB, 1987). It involves collection of airborne Cr(VI) in a series of 500-mL impingers filled with 100 mL of slightly alkaline (pH 8 to 9) buffer solution, which limits the reduction of Cr(VI) to Cr(III). A typical impinger train consists of three impingers, connected in series. The first and second impingers are filled with 0.02 N sodium bicarbonate solution, and the third is empty. The RTI method uses an ion chromatography (IC) column to separate the Cr(VI) from the impinger solution for quantification. Following chromatographic separation, the Cr(VI) is complexed with diphenylcarbazide and measured spectrophotometrically at 520 nm. The limit of detection (LOD) for the RTI method is 0.1 to 1 ng/m^3, depending on whether a preconcentration step is used (CARB, 1987).

To measure the concentrations of Cr(VI) in ambient air at the COPR sites in Kearny, a slightly modified version of the RTI method was used to measure the concentrations of Cr(VI) in ambient air. Modifications included increasing the volume of buffer in each impinger to 200 ml, rather than 100 ml, and the third impinger contained fluid as opposed to being empty. Parameters evaluated during the investigation were collection efficiency, recovery from collection media, stability of the collected analyte, method precision, and method bias (accuracy). Method development was conducted in two phases: (1) initial method assessment and (2) further field evaluation (Sheehan et al., 1992). In phase one, data were collected to assess the effect of various impinger-train variables. Specific objectives were to evaluate the pH stability of the trapping solution, the carry-over of Cr(VI) from one impinger to another, the efficiency of Cr(VI) collection, and the effect of sample storage time on Cr(VI) stability. Phase two consisted of field testing to evaluate method precision, percent total chromium in the hexavalent form, method bias and Cr(VI) stability in ambient and reduced-temperature conditions, conversion of Cr(III) to Cr(VI) during sample collection, and comparison with other Cr(VI) sampling methods (Sheehan et al., 1992).

Method Validation Results   The validation study results are discussed in detail by others (Sheehan et al., 1992; Finley et al., 1993). The results of this exhaustive field study showed that the modified RTI method for collecting airborne Cr(VI) particles met the requirements for precision and efficiency specified by USEPA. As a result, the method was adopted in 1992 as a standard test method for the collection and analysis of Cr(VI) by the American Society for Testing and Materials (ASTM) (Designated method D 5281-92) (Figure 9.2).

Summary of Ambient Air Sampling Results   During 1990 to 1994, airborne concentrations of Cr(VI) were measured at most of the COPR sites managed by the SLO using the validated impinger sampling method. Most of the sites supported industrial or commercial activity, and the others were accessed only occasionally (e.g., railroad track access road). None of the sites supported residential land use, and most were in historically industrialized areas. Some of the sites had bare soil, and others were either partially or entirely paved. Samples were collected prior to the implementation of interim remedial measures to capture the ambient concentrations of Cr(VI) before any remedial action was taken. Most of the sites supported some level of traffic, and many of the sites supported heavy truck traffic. Indoor and outdoor samples for Cr(VI), total chromium, total suspended particulates (TSP), and PM_{10} (particulate matter of less than or equal to 10 μm in diameter) were collected at all sites (except that indoor samples were not collected at sites without buildings). Indoor air samples were also collected at 15 local residential sites as a measure of background Cr(VI) concentrations.

Concentrations of Cr(VI) in total suspended particulates at the industrial sites averaged from 0.39 to 110 ng/m^3 outdoors and from 0.23 to 11 ng/m^3 indoors (Falerios et al., 1992). Concentrations of total chromium were consistently higher, ranging from means of 1.9 to 250 ng/m^3 outdoors and 4.1 to 130 ng/m^3 indoors. The overall ratio of Cr(VI) to total chromium at the industrial sites was 21 percent indoors and 25 percent outdoors. The overall average concentration of Cr(VI) in the homes was 1.2 ng/m^3, with a ratio of Cr(VI) to total chromium of 15 percent (Falerios et al., 1992).

Evaluations of these data revealed that the most significant factor affecting Cr(VI) concentrations in ambient air was the presence of heavy truck traffic on unpaved soil. Although there were limited data to examine the effect of moist soil conditions—because sampling days were targeted for dry weather—the data suggested that moist soil conditions tend to suppress the suspension of Cr(VI) to ambient air. For those sites with sufficient data, statistical analysis indicated significantly lower levels of Cr(VI), Cr(total), and TSP in ambient air were lower during moist soil conditions.

Because the levels of Cr(VI) measured at these sites were nearly always 1000-fold less than the Occupational Safety and Health Administration's (OSHA's) permissible exposure limit (PEL) for Cr(VI) of 52,000 ng/m^3, it was concluded that airborne exposures at these sites did not pose a significant health hazard to onsite or nearby workers.

9.4.3 Development of Cr(VI) Soil Analysis Procedure

Analysis of soil samples for Cr(VI) involves two steps: extraction (or digestion) and analysis. In 1986, the USEPA withdrew a previously approved method for extraction of Cr(VI) from soil samples (Method 3060) (USEPA, 1986). The lack of an approved method greatly jeopardized the project objectives because cleanup standards could only be set for total chromium using an estimate of the fraction of Cr(VI) in total chromium. Further, and perhaps more importantly, treatment techniques that reduce Cr(VI) to essentially nontoxic Cr(III) could not be used because their success could not be measured.

The old USEPA Method 3060 utilized a hot, alkaline (pH 12) solution containing 0.28 M Na_2CO_3 and 0.5 M NaOH to extract Cr(VI) from soil without affecting the valence state. Method 7196A is the analytical method used to determine the concentration of Cr(VI) in an extraction solution. Specifically, once the Cr(VI) is in solution, diphenylcarbazide is added to the sample, and it is acidified to a pH of 2. The diphenylcarbazide reacts selectively with the Cr(VI) in solution to form a red-violet complex. The absorbance of this red-violet complex can then be measured with a spectrophotometer to determine the concentration of Cr(VI). A more detailed description of chromium analytical methods is presented elsewhere (Petura et al., 1998).

The $30 million remediation of the residential sites in Jersey City, performed during 1990, was conducted to meet the 75 mg/kg total chromium cleanup level which is indicative of background concentrations of chromium in the soil (Dragun and Chaisson, 1991). The SRO recognized that the volume of the soils that were removed and the remedy selected (excavation and offsite disposal) resulted from the lack of a validated method for measuring Cr(VI) in soil at that time. For these reasons, it was necessary to develop and seek regulatory approval for a method to analyze Cr(VI) in soils. Consistent with the strategy developed early in the project, the SRO sought and engaged experts in analytical and soil chemistry to address this issue. The experts were given the task of developing and validating a new method for quantifying Cr(VI) in soil.

Method Development and Validation A review of the 1986 USEPA Method 3060 research report indicated that the authors rejected the method because of poor matrix spike recovery in various environmental samples. Furthermore, the authors concluded that in some instances, method-induced oxidation of Cr(III)

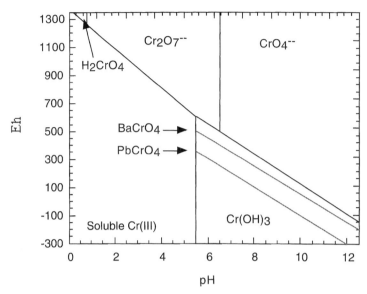

Figure 9.7 E_h/pH phase diagram for chromium.

to Cr(VI) was occurring. However, a careful analysis of this report by the analytical experts generated an alternate hypothesis. Specifically, the existence of Cr(VI) in a solid sample might be precluded if the sample exhibited a reducing environment. This is illustrated in the E_h–pH phase diagram presented in Figure 9.7. For example, environmental samples at a neutral pH and a negative E_h are expected to reduce Cr(VI) to Cr(III). An example of such an environmental matrix would be an anoxic sediment sample. Furthermore, it was hypothesized that the method-induced oxidation of Cr(III) to Cr(VI) was likely to be minimal and could possibly be controlled. Thus, in early 1990, the SRO embarked on a method development and method evaluation study (Figure 9.2). Typically, such studies are conducted at academic institutions or in government-funded laboratories. However, given the urgency for method development and evaluation/approval at some regulatory level for use in RI, it was concluded that it would be necessary to use commercial laboratories for this work to expedite the process.

A method development study was undertaken in 1992 to refine the previously discarded USEPA Method 3060. Over 1500 samples were analyzed, and the method was altered to optimize recovery of Cr(VI). As part of the method development study, changes were also proposed to the analysis method (USEPA SW846 Method 7196A). Table 9.2 summarizes the modifications suggested for these methods. Once method development was completed, a validation study was conducted. The results of these method validation studies have been reported previously (James, 1994; Vitale et al., 1994, 1995). Key conclusions of the study include the following:

TABLE 9.2 Modifications to Extraction and Analysis Methods for Cr(VI) in Solid Matrices

Digestion Item	Method 3060[a]	Modified Method 3060A
1. Sample weight (wet)	100 g	2.5 g
2. Alkaline digest	400 mL	50 mL
3. Final digest volume	1000 mL	100 mL
4. Digestion temperature	Near boiling	90–95°C
5. Digestion time	30–45 min	60 min
6. Nitric acid acidification	pH 7–8	pH 7–8 but if <7, discard digestate and start over

Analysis Item	Method 7196A[b]	Modified Method 7196A
1. Volumes used for analysis	95 mL	45 mL
2. Amount of diphenylcarbazide (DPC) added	2.0 mL	1.0 mL
3. Acidification with H_2SO_4	pH to 1.5–2.5	pH to 1.6–2.2
4. Turbidity	Subtract absorbance observed before DPC addition from absorbance after DPC addition	Same as 7196A, plus filter through 0.45- or 0.1-μm membrane
5. Initial calibration	0.5–5.0 mg/L	0.05–2.0 mg/L

Source: Adapted from Vitale et al. (1994).

[a] As appeared in the 2nd edition of SW-846.
[b] As appeared in the 3rd edition of SW-846.

1. Modified Methods 3060A/7196A for determining total Cr(VI) were satisfactory for use in environmental investigations.

2. Solid samples that exhibit highly reducing conditions cannot support the existence of Cr(VI)—thus, poor matrix spike recoveries are observed.

3. When low Cr(VI) spike recoveries are observed, characterization of the redox environment is important for interpreting the results.

4. Method-induced oxidation of Cr(III) to Cr(VI) occurred only with freshly precipitated $Cr(OH)_3$, which is not likely to exist in environmental samples.

5. Method-induced oxidation could be suppressed by the addition of Mg(II).

6. Method 3060A was more effective at extracting Cr(VI) (both soluble and insoluble Cr(VI) compounds) than other proposed extraction methods that used water or phosphate buffer as an extractant or sample sonication.

The results of these studies supported the use of the Modified Method 3060A/7196A for the determination of Cr(VI) in soil or sediment samples.

Regulatory Approval of Method 3060A/7196A As noted above, the results of the method evaluation studies were positive and were published in the peer-reviewed scientific literature. Armed with the above information, the SRO representatives set out to obtain NJDEP approval for use of this method in the site investigation process. As a result of this work, the NJDEP moved away from the 75 mg/kg soil cleanup level to new soil cleanup criteria of 10 mg of Cr(VI)/kg and 500 mg of total chromium/kg. Moreover, the NJDEP case management team approved the use of these analytical methods for the measurement of Cr(VI) in soil or sediment samples collected for the RI.

At this point, publication in the scientific peer-reviewed literature resulted in an unexpected development. Noting these publications, the USEPA began to consider including the modified methods in the "bible" of USEPA-approved analytical methods, *Manual of Test Methods for Evaluating Solid Wastes* (also called SW846). After independently evaluating the proposed extraction and analysis modifications, the USEPA elected to include the modified methods in the SW846 (USEPA, 1994). On adoption of the modified method (now called SW-846 Method 3060A) and SW-846 Method 7196A by the USEPA, the NJDEP concluded that the Cr(VI) analyses would be considered acceptable in the remedial decision-making process.

9.4.4 Risk-Based Remedial Investigation (RBRI) Design

Because the objective of the overall project plan was to base remediation on risk-based cleanup levels, the data collected during the remedial investigation (RI) had to be sufficient for the purposes of risk assessment. The development of the risk-based site-specific sampling plans for the 38 SRO–managed sites, which occurred in 1990 and 1991 (Figure 9.2), was one of the first such plans developed and implemented in the United States on a large-scale project. Remedial investigation designs now more frequently employ statistical analysis to develop risk-based sampling plans (USEPA, 1996).

The objective of the RI plan was to collect the data needed to provide a high degree of power and confidence in the data used in the risk assessment. The risk assessments of the COPR sites were focused on surface soil, because current and future site users would primarily contact surface soil, and because truck traffic suspends only surface soil. Thus, the sampling plan addressed only sampling of surface soil.

Statistical Design Prior to the implantation of IRMs, samples of surface soil were collected and analyzed for total and hexavalent chromium. However, the Cr(VI) data could not be relied upon, because the majority of the data were collected before Methods 3060A and 7196A were developed and validated for measuring Cr(VI) in soil. Nonetheless, the total chromium data provided a measure of the COPR impacts in surface soil at each site, and they were used as the basis for the site-specific statistically based RI sampling plans. From the onset of sampling and analysis, there was a high degree of heterogeneity in the soil total

Cr and Cr(VI) concentration profiles; some sites had been filled extensively with COPR throughout, and at others, COPR had only been used to fill small areas (e.g., driveways or potholes). Therefore, the sampling plans had to be site-specific to meet the data quality objectives. The project team selected the following data quality objectives for the risk assessment data: 90 percent power, 90 percent confidence, and 10 percent minimum detectable relative difference for each site. These objectives are more stringent than current USEPA data quality objectives for risk assessment (USEPA, 1992).

Because there was expected to be spatial orientation to COPR filling activities, a serpentine sampling plan (USEPA, 1989) was developed for each site to calculate the coefficient of variation (CV) of total chromium concentrations. Serpentine patterns were developed in multiple directions for each site, and the highest calculated CV was selected to ensure that the sites were not undersampled. The number of samples needed to meet data quality objectives with a stratified random sampling plan was calculated. The sites were then each gridded to collect the required number of samples, and the sampling locations were selected using a random number generator from inside each grid. This approach increased the statistical power of analysis because of the random component, while ensuring that the entire site was sampled. In only a few cases, where the filling of COPR had been highly variable, was it necessary to relax the data quality objectives to 2 to 30 percent minimum detectable relative difference, still achieving 90 percent power and 90 percent confidence, to allow for the collection of a reasonable number of samples. Most sites required between 20 and 40 samples to meet the data quality objectives.

Parameters Gathered and Basis

Surface Soil   At each site, surface soils were sampled for total and hexavalent chromium to assess the potential health hazards from chromium-affected soils. Parameters that are used in soil suspension modeling, such as particle size in the top 3 cm and moisture content, were collected at some sites because these parameters are not expected to vary significantly at each site.

Subsurface Soil   Samples representative of Cr(VI) affected soils in either the surface or subsurface were collected for the purpose of measuring the leachability of Cr(VI) and calculating a partitioning coefficient (K_d) to assess the potential for Cr(VI) to concentrate in surface puddles. These data were used with the threshold for elicitation of ACD to set Cr(VI) cleanup levels that would be protective of ACD.

Historical aerial photos were used to assess the extent of filling in the subsurface at each site. These data were used to assess the locations and number of subsurface soil borings. Thus, subsurface soil sampling locations were determined in a biased manner, focused on affected areas, with the objective of identifying the extent of contamination. This is one aspect of the RI work plan that should have been planned differently, and statistical methods should have been used for subsurface soils as well as surface soils. Had the subsurface soil samples been collected using a statistically based sampling plan to achieve risk-based data quality objectives, the NJDEP would have been more likely to allow for the comparison of risk-based soil cleanup levels to site-wide average concentrations [or the 95 percent upper confidence limit (UCL) of the arithmetic mean] in subsurface soils, rather than the maximum detected value, to determine achievement of the cleanup standard.

Groundwater   There was also a concern about groundwater contamination at these sites. Therefore, the sites were subdivided into seven hydrogeologic subregions based on groundwater hydrogeology. Groundwater monitoring wells were located to characterize groundwater on the basis of each hydrogeologic subregion. At sites where there was evidence for potential onsite groundwater contamination, such as the interface of COPR fill with shallow groundwater, monitoring wells were located to determine the highest concentrations of Cr(VI) that could be measured in groundwater. These data were required by NJDEP but were of questionable value from a risk assessment perspective because groundwater in this region is not usable as a drinking water source due to poor quality (saltwater intrusion and high total dissolved solids) and, in some areas, low yield.

9.5 EXPOSURE ASSESSMENT

Exposure assessment is the process of estimating the intensity, frequency, and duration of human exposure (Paustenbach, 2000). The principal elements of exposure assessment for the COPR sites were development of reasonable exposure scenarios for potentially exposed human populations, identification of complete exposure pathways, and estimation of the expected chemical dose (uptake). In 1993 and 1994, an extensive investigation of the most ecologically sensitive area in the vicinity of the COPR sites—the Kearny Marsh (a marshland surrounded on three sides by COPR sites), where elevated levels of total chromium had been measured in sediments—was conducted (Hall and Pulliam, 1995). The ecological study found no biological effects or significant chemical accumulation. Thus, the exposure assessment for the COPR sites focused only on potential human exposures to Cr(VI) and Cr(III) from COPR.

The potentially exposed populations were defined by onsite and surrounding land use. Potentially exposed populations were limited to onsite workers at most sites. At a few sites, offsite residents and/or occasional visitors, such as a utility repairman, also were evaluated. Exposures to chromium in COPR were assumed to occur via inhalation of suspended soil particulates, dermal contact with soil, and incidental soil ingestion. Groundwater was not consumed at these sites, but at some sites, the potential existed for construction or utility workers to contact shallow groundwater during subsurface excavation in areas where the water table is shallow. In these cases, construction workers were assumed to have some dermal contact with the groundwater. Chromium is not volatile, so it is not

TABLE 9.3   Exposure Scenarios and Pathways Typically Considered at Chromite Ore Processing Residue (COPR) Sites in Kearny, NJ

Timeframe[a]	Exposure Medium	Receptor Population	Route of Exposure	Potentially Complete Pathway?
Future	Onsite soil	Adult site workers (industrial/ commercial)	Soil ingestion	Yes
			Dermal contact	Yes
		Adult construction workers	Soil ingestion	Yes
			Dermal contact	Yes
	Onsite air	Adult site workers	Inhalation	Yes
		Adult construction workers	Inhalation	Yes
	Onsite groundwater	Adult site workers	Consumption	No[b]
			Dermal contact	No
		Adult construction workers	Consumption	No[b]
			Dermal contact	Yes
	Offsite air[c]	Offsite adult site workers	Inhalation of particulates	Yes[d]
		Offsite residents (children and adults)[e]	Inhalation of particulates	Yes

[a] Because the interim remedial measures (IRMs) preclude current exposures, the only timeframe assessed is future (baseline conditions), assuming the IRMs deteriorate or are removed in the future.
[b] Groundwater is not used as a potable source in Hudson County, and all sites use the municipal supply.
[c] Offsite air could be affected by wind-blown particulates. If COPR is identified in offsite soil, the offsite property would be identified as a COPR site.
[d] Offsite workers may be exposed via inhalation; however, this scenario is not usually quantified because onsite exposures are assumed to be greater (higher concentration) than offsite exposures.
[e] Inhalation cancer risks to any resident within 400 m of a COPR site are evaluated quantitatively.

necessary or appropriate to assess inhalation of vapors. The exposure scenarios and pathways evaluated for the Kearny industrial COPR sites are presented in Table 9.3.

Several highly specialized site-specific studies were conducted for the COPR sites so that the exposure assessment was as accurate as possible and would not be forced to rely on generic conservative exposure parameters when site-specific data were lacking. The approaches used at the COPR sites discussed in this section are (1) site activity studies conducted at each site, (2) evaluation of the reduction–oxidation (redox) chemistry of chromium in COPR, (3) soil suspension modeling, and (4) biological monitoring. Because the NJDEP uses a risk assessment paradigm wherein cleanup levels are determined based on the "acceptable level of risk" and site-specific exposure conditions, rather than the conventional approach of calculating risk, traditional dose quantification was not performed in the COPR risk assessments.

9.5.1 Site Activity Studies

There are a large number of exposure scenarios and pathways of exposure that could be assessed at a COPR site. However, the exposure scenarios evaluated in the risk assessment were selected based on site-specific conditions, as determined through site activity studies conducted in 1993 through 1996 (Figure 9.2). These studies were conducted at every single site by performing a site visit, reviewing site conditions, and completing a standardized form so that all of the essential information was collected for each site. This was an important element of the initial SRO strategy to ensure that the most accurate information would be used in the risk assessments.

The studies included the collection of data regarding land use on a site and in the surrounding area. Because truck traffic at these sites potentially contributed to significant dust generation and airborne concentrations of Cr(VI) from surface soil, it was important to survey, for each site, the traffic patterns (where cars and trucks drove at the sites), the type of traffic (type of vehicle, number of wheels, vehicle weight and speed), volume of traffic (trips per day), and the time frame (hours per day, month, and year) when traffic would be expected at each site. This information was used as input (1) to model the airborne emission of Cr(VI) from soil to ambient air, assuming that any asphalt or IRM was removed in the future, and (2) to establish health-based cleanup levels, which are also called alternative remediation standards (ARSs), that would be protective of the inhalation pathway of exposure.

Other important site characteristics that were noted and/or obtained from interviews with site representatives included the number of employees and their daily activities, whether the employees worked indoors or outdoors, the presence of yellow or green chromium crystals on building walls or surface soil, the presence and type of IRM, current and historical site operations, and use/disposal of chemicals at the site. The purpose of collecting information about non-COPR-related chemical releases was to allow for the explanation of curious findings, if any, generated during the site investigation. As an example, this information was found to be important at a swimming pool chlorine tablet manufacturing site because the hypocholorite used to produce the swimming pool tablets was ultimately found to bias the Cr(VI) air-sampling results due to analytical interference.

The data from these site surveys were used in site-specific risk assessments to quantify exposures. For sites with no current activity (e.g., abandoned operation or never developed), the site-specific survey data were used to develop distributions of possible exposure conditions for assessing potential future exposures if the site was developed or redeveloped in the future. In doing this, surrounding site usage conditions were incorporated, and worst-case assumptions regarding land use were not necessary.

9.5.2 Reduction–Oxidation Investigation

Early in the project, NJDEP expressed a concern that Cr(III) present in the COPR could be oxidized to Cr(VI) by components of the natural environment, such as manganese oxides. Indeed, some forms of Cr(III) can be oxidized to Cr(VI) under certain conditions (Bartlett and Kimble, 1976; Bartlett and James, 1979; James and Bartlett, 1983a, 1983b; Bartlett, 1991). Based on these studies, the NJDEP Division of Science and Research (DSR) concluded that a low total chromium cleanup standard was necessary.

Drawing upon the advice of experts in the field of Cr soil chemistry, it was believed that this phenomenon was unlikely in COPR, because the forms of Cr(III) in COPR are mostly water insoluble and unavailable for oxidation reactions. The studies conducted by Bartlett and James during the 1970s and 1980s (Bartlett and James, 1979) which first described the oxidation of Cr(III) under environmental conditions, evaluated water-soluble forms of Cr(III) generated in sewage sludge and tanning wastes in the presence of oxidized manganese.

Dr. Bruce James, an expert in chromium soils chemistry, was asked to join the project team in 1992 in an effort to resolve this issue (Figure 9.2). Dr. James, a professor at the University of Maryland, was one of the authors of the studies which initially identified that Cr(III) could be oxidized to Cr(VI) in COPR. To investigate this reaction, COPR samples were subjected to extensive laboratory testing by Dr. James with particular focus on whether Cr(III) in COPR could be oxidized to Cr(VI) in the presence of oxygen or manganese oxides. Based on these tests, he concluded that Cr(III) would not oxidize to Cr(VI) in COPR, even in the presence of manganese oxides or oxygen. Table 9.4 summarizes the results of these experiments. Dr. James published his results in the peer-reviewed scientific literature (James, 1994) and shared his work with NJDEP. The results

TABLE 9.4 Evaluation of Oxidation Potential of Trivalent Chromium Present in Chromite Ore Processing Residue (COPR)

	Mean $\pm$ SEM ($n = 3$)	
	Soil 1	Soil 2
Soil characteristics		
Total Cr (mg/kg)	$1,800 \pm 53$	$10,400 \pm 310$
Total Cr(VI) (mg/kg)	105 ± 2	460 ± 18
Soluble Cr(VI) (mg/kg)	41 ± 0.7	258 ± 3
Valence stability		
Spike + soluble Cr(VI)[a]	107 ± 0.7	323 ± 5
Spike of soluble Cr(III) added (mg/kg)[b]	620	730
Soluble Cr(VI) in Cr(III)-spiked soils (mg/kg)	39 ± 0.2	227 ± 7
Change in Cr(VI) due to added Cr(III) (%)	-4.8	12.0

Source: Adapted from James (1994).

[a] 62 and 73 mg Cr(VI)/kg soil added to soil 1 and 2, respectively.

[b] Added as $Cr(NO_3)_3$. SEM = standard error about the mean.

of these experiments convinced the NJDEP that this issue was not one that warranted further concern.

9.5.3   Soil Suspension Model Validation

Since initiation of the project in the late 1980s, estimating the quantity of soil or road dust that could be suspended to ambient air had been important because the primary health concern was the potential cancer risk associated with inhalation of Cr(VI). Early attempts by the NJDEP to model soil suspension proved to generate incorrect results when compared to actual measured concentrations of Cr(VI) in ambient air (ESE, 1989b). As a result, it was recognized that crude modeling approaches with default exposure parameters would result in overestimation of airborne Cr(VI). Thus, the development and validation of soil suspension (emission) and dispersion modeling approaches for the COPR sites was undertaken. The objective was to establish a method that could be used to set remediation standards for soil based on acceptable levels of Cr(VI) in ambient air.

Soil can be suspended in ambient air by high winds or by mechanical disturbances such as the movement of vehicles over unpaved areas or roads (Cowherd et al. 1985; 1989). Several modeling approaches have been published that estimate the ambient air concentrations of a substance bound to suspended surficial soil particulates. Soils containing COPR have a higher frequency distribution of large particles, with the particle sizes ranging from 0.3 to 17.0 mm and an average particle size of 3.6 mm. Because the particle sizes in COPR-affected soils are quite large, wind erosion has been shown to be a negligible source of particulate emissions for these sites (Scott et al., 1997b).

Particulate Emission Models   At the time of the model validation and continuing to the present, there were two particulate emission models available for characterizing the emissions of soil particulates due to the movement of vehicles over unpaved surfaces: the Rapid Assessment Model (RAM) (Cowherd et al., 1985), and a model presented in the USEPA air pollutant emission factors document (AP-42) (USEPA, 1995a). The same general form that was used for the RAM equation was also used for the AP-42 model, except that the coefficients and exponents of the AP-42 model were estimated using a larger set of data from several additional studies. Because the AP-42 model was based on a more robust data set, it was selected as the particulate emission model for use in the validation study. The following equation was used to express PM_{10} particulate emissions due to vehicle traffic over unpaved roads in terms of a unit (kg) of soil suspended per vehicle kilometer of travel (kg/VKT):

$$E_{PM10} = 1.7k \left(\frac{s}{12}\right) \left(\frac{S}{48}\right) \left(\frac{W}{2.7}\right)^{0.7} \left(\frac{w}{4}\right)^{0.5} \left(\frac{365-p}{365}\right)$$

where E_{PM10} = PM_{10} particulate emission factor; the quantity of emissions from an unpaved road per vehicle kilometer of travel (kg/VKT)

k = particle size multiplier; equal to 0.36 for PM_{10} particulates

s = silt content of the unpaved surface material (%)

S = mean vehicle speed (km/h)

W = mean vehicle weight [megagrams (Mg)]

w = mean number of wheels

p = number of days with at least 0.01 inch of precipitation per year

To convert the emission rate (E_{PM10}) from kg/VKT of PM_{10} particulates to the more standard emission units of g/m^2-sec, the following equation is used for the AP-42 model (USEPA, 1995a):

$$ ER = \frac{E_{PM10} \times TC \times D \times TF \times (1000\ g/kg)}{CF \times A_s} $$

where ER = PM_{10} particulate emission rate (g/m^2-sec)

E_{PM10} = PM_{10} particulate emission factor; the quantity of emissions from an unpaved road per vehicle kilometer of travel (kg/VKT)

TC = daily traffic count for the unpaved area (vehicles/day)

D = average distance a vehicle travels through the unpaved area (km)

CF = conversion factor (3.15×10^7 sec/year)

TF = frequency of traffic (days with traffic/year)

A_s = total source area (m^2)

Air Dispersion Models   Throughout the duration of this project, there have been a variety of air dispersion models available for use, but few were expected to accurately predict dust concentrations at COPR sites. To identify the best one, three criteria were established. The model had to be able to: (1) calculate airborne concentrations from area sources consistent with the emission model output, (2) account for dry particle deposition, (3) calculate annual average concentrations of Cr(VI) for estimating long-term exposures, accounting for the impact of wet weather conditions, and (4) calculate hourly estimates, for comparison with airborne sampling data collected on dry days. Two models were found to meet those requirements: the Industrial Source Complex–Short Term 3 (ISCST3) (USEPA, 1995b) and the Fugitive Dust Model (FDM) (USEPA, 1991d). An analysis of model results demonstrated that both models performed adequately, and that the FDM gave slightly higher (more conservative) estimates of airborne concentrations at the height of the breathing zone (about 6 feet) (Scott et al., 1997a). For this reason, and because the ultimate purpose of the soil suspension modeling effort was to develop health-based cleanup levels, the FDM model was selected.

The AP-42 soil suspension model and FDM air dispersion model were put through a model validation project, which was initiated in 1995 and finalized in 1997 (Figure 9.2). Model performance was judged by modeling airborne concentrations of total suspended particulates (TSP) and Cr(VI) concentrations for traffic and site conditions recorded during specific sampling events, and comparing the results to measured concentrations. The findings of the validation study are as follows:

1. For TSP, the validation study found that the measured data adequately fit the model-predicted values. In all but one case, the model prediction was higher than the measured value.
2. For Cr(VI), the findings also indicated adequate fit of the model-predicted values to those measured. In five of seven cases, the model prediction was higher than the measured value.

Based on this analysis, soil suspension models could be used to predict airborne concentrations of Cr(VI) and TSP. With a validated soil suspension and dispersion modeling approach, health-based cleanup levels for soil that are protective of the theoretical excess cancer risk could be calculated.

9.5.4   Biomonitoring

Chromium Medical Surveillance Program (CMSP)   Urinary biomonitoring for evaluating exposure to chromium has historically been used in occupational settings to monitor for excessive human exposure via inhalation (Anderson et al., 1993). In most cases where urinary monitoring is used to evaluate the inhalation exposure of workers, the form of chromium in the workplace atmosphere is hexavalent chromium, which is more bioavailable than Cr(III) via inhalation, and certainly more bioavailable than either Cr(VI) or Cr(III) by the oral route (Paustenbach et al., 1997). The New Jersey Department of Health (NJDOH) and NJDEP were interested in assessing exposure to environmental chromium among Hudson County residents and workers using biological monitoring. They embarked on a urinary biomonitoring study after a preliminary study of Whitney Young Elementary School students suggested that children who attended the school and lived in areas of COPR-affected soil were more likely to have measurable levels of chromium in their urine than children who did not live in proximity to the COPR sites (NJDOH, 1989). It is important to recognize that the Whitney Young Elementary School study did not find elevated concentrations of chromium in the urine of any of the school children; rather, all the measured concentrations were consistent with background chromium levels from dietary sources (Anderson et al., 1993). The World Health Organization (WHO) has concluded that the use of urinary chromium as a tool for monitoring environmental exposure is not possible due to the very low concentrations found in the environment as compared with uptake in the diet (WHO, 1988).

The circumstances surrounding the NJDOH Chromium Medical Surveillance Program (CMSP) and related studies illustrate how public pressure can push an agency to act even though the scientific bases for the action are weak. It is possible that one of the primary reasons why the NJDOH and NJDEP research staff pursued this project, in the face of an overwhelming preponderance of scientific evidence indicating that only background urinary chromium levels could be measured, was a desire to satisfy the public's concern. Because chromium in the diet will result in urinary chromium levels being measurable in most, if not all, of those who provide samples, the results of the study were potentially problematic. They heightened public concern and could have been used as evidence for toxic tort actions, even if the measured levels were not above "background," because background inter- and intrapersonal variability of spot urine samples can be very high (Bukowski et al., 1992; Paustenbach et al., 1997).

NJDOH and NJDEP Preliminary Urinary Chromium Biomonitoring Studies
A series of smaller studies eventually led to the development and implementation of the CMSP. In April 1989, the NJDOH conducted an evaluation of chromium exposure of children and adults at the Whitney Young Elementary School in Jersey City (NJDOH, 1989). The following month, the NJDOH collected a variety of environmental samples from the school, although the exact type of samples collected (soil, air, dust) has never been specified. Based on these data, the NJDOH closed the school, asserting that it was potentially contaminated with COPR. The children and staff were moved to a nearby school for the remainder of the school year. After the students and teachers were moved, urine samples were collected and analyzed for total chromium. NJDOH then attempted to correlate urinary chromium concentrations with environmental media concentrations at the school or near the homes of the study subjects. NJDOH reported that chromium was detected above the limit of detection (LOD) for 36 percent of children and 16 percent of adults. Moreover, 54 percent of the children who lived on blocks with COPR sites present had urinary chromium levels greater than the LOD. Based on these data, NJDOH concluded that "exposures to chromium-contaminated soils or dusts from the sites are contributing to chromium exposure in children living near the sites" (NJDOH, 1989). During the summer of 1989, an NJDEP contractor sampled the air, walls, and soils at the school for chromium. The contractor was unable to find evidence of significantly elevated chromium levels at the school. However, the NJDOH study was used as the basis to launch the CMSP.

NJDEP subsequently collected household dust samples at homes near COPR sites, and spot urine samples were also collected to determine whether chromium concentrations in dust and in urine could be correlated (Lioy et al., 1992; Stern et al., 1992). The researchers reported that a relationship did exist between chromium concentrations in spot urine samples and in household dust. Specifically, these researchers concluded that "these data show an association between elevated exposure to chromium in household dust and elevated urine levels of chromium, consistent with residential exposure to chromate chemical production waste" (Stern et al., 1992). However, consistent with the findings of the Whitney Young Elementary School study, the measured chromium urinary concentrations were within normal background levels for humans, and the correlation lacked statistical power. The dust concentrations of total chromium were all less than 214 ppm, which is not typical of chromium concentrations in COPR (Table 9.1), and Cr(VI) was not measured in the dust samples. Because the chromium dust concentrations were low, simple kinetic modeling suggested that exposure to chromium in household dust could not significantly alter urinary chromium concentrations.

In contrast to the articles discussed above, a second pair of studies appeared in the scientific literature, co-authored by members of the NJDEP (Bukowksi et al., 1991, 1992). These studies attempted to correlate chromium concentrations in soil and air at a COPR site in Jersey City to urinary and red blood cell (RBC) chromium concentrations of the workers at this location (park rangers). Seventeen workers provided urine and blood samples for chromium analysis. Chromium concentrations in both urine and RBCs were not statistically different from a control group of 35 workers conducting similar activities in another county. Total chromium concentrations in surface soils in Jersey City, proximate to where the 17 employees worked, ranged from 9.6 to 2200 mg/kg. Soils were not analyzed for Cr(VI). Total chromium air concentrations ranged from 0.1 to 0.33 $\mu g/m^3$ as an 8-hour time-weighted average. No relation between chromium concentrations in soil and air at the site and employee urinary chromium concentrations was observed. Based on these findings, the authors concluded: "These results call into question the utility of chromium biomonitoring under environmental exposure conditions" (Bukowski et al., 1992). Yet, even in light of these findings, the NJDOH and NJDEP implemented a broad-scale community biomonitoring program.

Implementation of the CMSP   The NJDEP requested that the responsible parties provide funding for a proposed urinary chromium biomonitoring program to be conducted by the NJDOH. The responsible parties agreed to fund the study if study protocols could be reviewed and agreed upon by outside experts. However, NJDEP and NJDOH would not provide the study protocols prior to the study and would not provide any of the results until the data were published. As a result, the responsible parties chose not to provide funding for this study, and public funds were used. The study was started in 1992 with an anticipated completion date of 1994 (Figure 9.2).

Urinary Biomonitoring Expert Panel   Because of concerns that the CMSP results would be misleading and misinterpreted, a second expert panel was formed, including experts in chromium kinetics, biomonitoring, nutrition, dermatology, and biostatistics (Figure 9.2). The panel reviewed the available information on the study protocol and concluded that the CMSP would not be able to measure environmental chromium exposures in Hudson County. The experts also cautioned that unless the study was carefully designed and implemented, it was highly likely that the researchers could reach spurious conclusions that might lead to unnecessary public concern. The complete panel conclusions are presented in Anderson et al. (1993) and summarized below:

1. The protocols for the study, provided via the Freedom of Information Act (FOIA) process, "do not meet traditional standards of thoroughness for such a costly and important undertaking."
2. Procedures used to collect urine are not sufficiently well controlled to ensure that samples do not inadvertently become contaminated.
3. Most investigations of this type receive the scrutiny of a peer-review committee because the project involves data that are likely to be used in making medical decisions.
4. The use of a single spot urine sample is not an appropriate method for identifying persons exposed to chromium via soil/dust due to a number of shortcomings.
5. The method detection limit for urinary chromium analyses is too high to measure environmental exposures in Hudson County.
6. The method for determining the source of exposure for persons with apparent high levels of urinary chromium is potentially fraught with error.
7. Unwarranted litigation and unnecessary public hysteria could result unless these concerns were addressed.

Understanding the Usefulness of the CMSP Data   In spite of the criticisms and recommendations offered by the expert panel, the study protocol did not change. Thus, additional research, described briefly below, was conducted to allow for proper interpretation of the urinary biomonitoring data (Gargas et al., 1994a, 1994b; Finley et al., 1996).

Ingestion of Chromite Ore Processing Residue   In an effort to understand whether the ingestion of COPR-affected soil or household dust could be measured by urinary biomonitoring, six human volunteers ingested 400 mg of COPR per day for 3 days (low-dose group), and two other human volunteers ingested 2000 mg of COPR per day for 3 days (high-dose group). The COPR ingested by both groups contained 103 ± 20 mg/kg of total chromium and 9.3 ± 3.8 mg/kg of Cr(VI). These exposures were greatly in excess of doses that residents of Jersey City were expected to receive from household dust. The study lasted a total of 6 days, with the first day's urine voids serving as the control data. On the second, third, and fourth days of the study, the volunteers ingested capsules containing the prescribed amounts of COPR. Urine was collected by each volunteer at each voiding over a 6-day period during the study and analyzed for total chromium. The study was unique in that urine from each voiding was analyzed, which addressed the expert panel's concern regarding the intrapersonal variability of spot urine samples. Moreover, each volunteer served as his or her own control. Over 220 samples were analyzed for chromium as part of this study. The study design and results are described in detail elsewhere (Gargas et al., 1994a, 1994b).

The results of the study were as follows:

1. The ingested COPR did not influence the urinary chromium concentrations of the human volunteers.
2. Background concentrations of urinary chromium fluctuated within a large range for the volunteers. Interestingly, the urinary chromium concentrations reported by NJDOH and NJDEP in their preliminary studies were well within the background concentrations reported in these studies.

Table 9.5 presents the mean daily spot urine concentrations for the low- and high-dose groups. It is clear from these results that the COPR had no influence on urinary chromium concentrations. These results provide compelling evidence that the urinary chromium levels measured in Hudson County workers and residents reflected the natural variation of urinary chromium concentrations (Gargas et al., 1994a, 1994b).

Ingestion of Safe Doses of Cr(VI) While the ingestion of COPR provided data that suggested the agencies were simply measuring background variations in urinary chromium, there was concern regarding how to interpret the upper end of the background range of urinary chromium. Indeed, the first series of studies demonstrated that inter- and intrapersonal variability was significant. For example, on the first day of the COPR ingestion studies (prior to COPR

TABLE 9.5 Mean Daily Spot Urine Chromium Concentrations

	Group (µg Cr/g creatinine)	
Day	Low-Dose ($n = 6$) 41 µg Cr/day [3.7 µg Cr(VI)/day]	High-Dose ($n = 2$) 206 µg Cr/day [18.6 µg Cr(VI)/day]
1 (background—no dose)	0.14 ± 0.20 (26)[a]	0.26 ± 0.42 (10)
2 (dose day)	0.18 ± 0.33 (27)	0.19 ± 0.12 (5)
3 (dose day)	0.19 ± 0.35 (31)	0.14 ± 0.09 (12)
4 (dose day)	0.14 ± 0.14 (30)	0.13 ± 0.07 (9)
5	0.11 ± 0.10 (31)	0.11 ± 0.05 (13)

[a] Number of samples collected.
Source: Adapted from Gargas et al. (1994).

ingestion), one individual had a spot urine chromium concentration of 2.9 µg/L. Such a value would likely have been interpreted as evidence of excessive exposure to COPR, when in reality, none had occurred.

To address this concern, additional research was conducted by the responsible parties. In this study, human volunteers ingested "safe" doses of Cr(VI) and Cr(III) in pure form (not in COPR), and urinary chromium was measured. A more complex dosing regimen was employed in this study to address concerns identified during the COPR ingestion study, and to ensure that chromium deficiency in the volunteers was not a problem. A safe dose was defined as the USEPA reference doses (RfDs) of Cr(VI) and Cr(III) [at that time, 0.005 mg/kg-day of Cr(VI) and 1.0 mg/kg-day of Cr(III)]. Potassium chromate was the form of Cr(VI) used in the study, and chromic oxide was used as the Cr(III) compound because these are the forms that are likely to occur in COPR. Figure 9.8 presents the results of this study and provides a comparison to the level of urinary chromium measured by Gargas et al. (1994a) from chromium picolinate ingestion and the levels of urinary chromium measured in the CMSP. Ingestion of the Cr(VI) RfD resulted in urinary chromium concentrations ranging from <0.2 to 97 µg Cr/L. Ingestion of the Cr(III) RfD resulted in urinary chromium

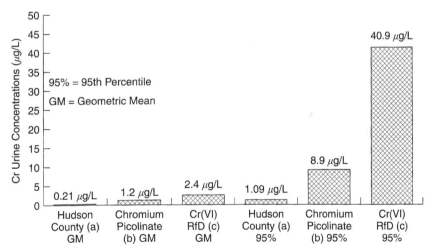

Figure 9.8 Comparison of urinary chromium concentrations in Hudson County residents involved in the NJDOH Chromium Medical Surveillance Project (CMSP) to that in eight human volunteers ingesting chromium picolinate (400 µg/day) to that of six human volunteers ingesting Cr(VI) at the EPA RfD (5 µg/kg-day). Data from Gargas et al. (1994a) and Finley et al. (1996). This plot demonstrates that measurable or elevated levels of chromium in the urine are not necessarily unsafe and that public health studies need to identify the level of urinary chromium that is of biological significance. (*a*) For residents aged 19–40 taken from the CMSP. (*b*) Calculated from pooled data on last day of picolinate dosing in Gargas et al. (1994a). (*c*) Calculated from pooled data during Cr(VI) dosing in Finley et al. (1996).

concentrations ranging from <0.2 to 6.3 µg Cr/L. Clearly, these results suggested that the urinary chromium concentrations measured by NJDOH and NJDEP were well below the USEPA-defined safe oral dose. The results of this study are described more fully elsewhere (Finley et al., 1996).

Release of the CMSP Results   Upon completion of the studies described above, and prior to release of the CMSP, the responsible parties met with senior NJDOH and NJDEP management. The regulatory agencies were surprised that human volunteers would actually ingest COPR. They found the results to be significant. In 1994, after learning of this research, the NJDOH released the results of the CMSP. The researchers had collected over 2500 urine samples from residents and workers in Hudson County (primarily from Jersey City) and reported only limited evidence of Cr exposure. Interestingly, the NJDOH report utilized a novel method for adjusting the urinary chromium concentration of each individual based on diet and other life-style factors, including exercise, smoking, and health status. The adjusted-urinary chromium concentrations were found to be within the range of normal urinary chromium concentrations. However, the CMSP concluded that multiple linear regression models showed that average urinary chromium differences between residents and a comparison group, adjusting for potential confounders, were highest in children under 6 years of age (Fagliano et al., 1997). Although the authors concluded that the CMSP, which was termed a "screening investigation," called for further evaluations of residents and workers, no further work has been conducted, nor is any planned.

Lessons Learned Regarding Urinary Biomonitoring for Environmental Chromium
As a result of the CMSP and the research sponsored by the responsible parties to assess environmental exposure, several limitations of urinary biomonitoring for chromium from environmental exposures were identified and should be considered in any future studies. They are described in detail by Paustenbach et al. (1997) and summarized below:

1. Significant exposure must occur immediately (24 hours) prior to sample collection, and previous exposures from weeks, months, or years prior cannot be measured.
2. Bioaccessibility (fraction of chromium that can be extracted from soil and/or dust) and bioavailability (the fraction that may be systemically absorbed) must be understood.
3. Urinary concentrations from environmental exposures must be higher than the range of background concentrations and the analytical limit of detection.
4. The study sample size must be sufficient to control for inter- and intra-personal variability.

5. The biological significance of measurable levels of chromium in urine must be understood, and elevated levels must be distinguishable from levels of potentially significant harm.

9.6 TOXICITY ASSESSMENT FOR TRIVALENT AND HEXAVALENT CHROMIUM IN COPR

The toxicity assessment for trivalent and hexavalent chromium in COPR mirrors the approach typically used for chromium in toxicity assessments at hazardous waste sites across the United States, and thus this discussion has broad application.

Trivalent chromium is recognized to be an essential micronutrient, and the daily requirement for healthy adults ranges from 50 to 200 µg/day (Anderson and Kozlovsky, 1985). Cr(III) is practically nontoxic, particularly in the water-insoluble forms that exist in COPR, and is not considered to be carcinogenic (Ivankovic and Preussmann, 1975; USEPA, 1998a). By comparison, Cr(VI) is a strong oxidizing agent and has been shown to cause irritation and ulceration of the skin and respiratory tract upon high-level occupational exposure [NIOSH, 1975; Agency for Toxic Substances and Disease Registry (ATSDR), 2000]. Cr(VI) is also recognized to cause hepatic and renal toxicity in humans and animals if ingested (ATSDR, 2000).

Cr(VI) is a contact allergen and can induce type IV allergic reactions among individuals who are sensitized to it (NIOSH, 1975). In industries where workers contact Cr(VI) (e.g., pigment production) and in construction where workers handle wet cement, allergic contact dermatitis (ACD) has been reported (NIOSH, 1975). Cr(VI) is also recognized as a known human carcinogen following inhalation exposure (not dermal contact or ingestion), based on elevated rates of lung cancer among workers from the chromate production, chromate pigment production, chrome plating, and ferrometals industries (IARC, 1990; Cross et al., 1997; USEPA, 1998b).

The manner in which the dose–response relationship for a given chemical is quantified depends on the nature of the adverse health effect. For noncarcinogenic effects, it is usually assumed that a "threshold" level of chemical uptake exists, below which no adverse effects are evident (Paustenbach, 1989a, 1989b; Klassen et al., 1996). In experimental systems such as animal bioassays, the threshold dose is approximated by the dose where no adverse effects are observed—the no-observed-adverse-effect level (NOAEL). The NOAELs are divided by uncertainty factors to estimate the dose that is protective of even sensitive humans, and this dose is termed the reference dose (RfD) for oral exposures, and the reference concentration (RfC) for inhalation exposures. The RfDs and RfCs for Cr(VI) and Cr(III) have been used to assess the noncancer dose–response relationship for chronic exposures to COPR.

For carcinogenic compounds, information from animal bioassays and human epidemiological data (when available) are used to estimate the carcinogenic potency of a particular chemical, which is then used in conjunction with dose estimates to predict the theoretical excess lifetime cancer risk (USEPA, 1989). For Cr(VI), epidemiological data from workers in the chromate production industry (Mancuso, 1975) has been used by USEPA to derive a cancer slope factor (SF), a measure of carcinogenic potency (USEPA, 1984).

In this toxicity assessment, the dose–response relationship for (1) acute noncancer effects (specifically ACD), (2) chronic noncancer effects (as described by RfDs and RfCs), and (3) the increased risk of lung cancer from chronic inhalation exposure to Cr(VI) are quantitatively evaluated. The results of the toxicity assessment are subsequently combined with the exposure assessment to characterize risk and set cleanup levels.

9.6.1   Allergic Contact Dermatitis from Acute Exposure

Chromium-induced ACD has been described as a type IV hypersensitivity that can develop as a result of repeated dermal exposure to soluble forms of Cr(VI) in industrial processes and materials (Adams, 1990). Allergic contact dermatitis resulting from exposure to soluble Cr(VI) has been reported among chromium plating workers (Burrows, 1987), lithographers (Levin et al., 1959), diesel repair shop workers (Winston and Walsh, 1951), leather workers (Morris, 1958), wood preservers and pigment workers (Polak et al., 1973), and cement workers— typically referred to as "cement eczema" (Jaeger and Pelloni, 1950; Anderson, 1960).

No cases of Cr(VI)-induced ACD from contact with COPR in New Jersey (or elsewhere) have been reported. In the CMSP biomonitoring study conducted by the NJDOH of more than 2000 residents and workers who may have contacted COPR environmentally, not a single case of Cr(VI) sensitivity or Cr(VI)-induced ACD could be identified (Fagliano et al., 1997). Nonetheless, the NJDEP has developed environmental standards that protect against Cr(VI)-induced ACD and have required remediation to ACD-based standards for more than 10 years (Stern, 1990; NJDEP, 1995, 1998). The earlier, more stringent, standards have since been abandoned as new information regarding the dose–response relation was developed to specifically address environmental exposures (Nethercott et al., 1994; Fowler et al., 1999). The history of New Jersey ACD standards for Cr(VI) is described in detail because it provides an excellent example of how applied research can resolve uncertainties within toxicity assessment. To date, the NJDEP is the only environmental regulatory agency to have developed chromium standards based on ACD as a health endpoint.

NJDEP 1989 Standard for Cr(VI) in Soil Protective of ACD   In 1989, Dr. Robert Bagdon was funded by the NJDEP to develop standards for chromium in soil that were intended to prevent ACD (Figure 9.2). An extensive review of animal and human literature was conducted (Bagdon and Hazen, 1991). Based on 10 patch-test studies conducted from 1956 to 1980, in which physicians tested patients to identify a threshold for elicitation using the patch test, the concentration necessary to elicit a response in 10 percent of the previously sensitized individuals was estimated. The 10 percent response level is now termed the 10 percent minimum elicitation threshold (or MET).

The concentration of Cr(VI) that elicited a response in 10 percent of the previously sensitized individuals was estimated to be 10 ppm. NJDEP assumed that 10 ppm (mg/L) of Cr(VI) added to a patch equates to 10 mg/kg (ppm) of Cr(VI) in soil. Upon first impression, this assumption may seem correct; however, it is not accurate because the induction of ACD is dependent of the mass of Cr(VI) in contact with the skin, not the concentration of chromium in the solution used to make the patch (Horowitz and Finley, 1994a; Nethercott et al., 1994, 1995).

Bagdon also reviewed the ACD data for Cr(III) and found that these data were far more limited than for Cr(VI). Nonetheless, he identified an equivalent MET for Cr(III) of 500 ppm. NJDEP concluded that the 10 mg/kg standard for Cr(VI) was assumed to be equivalent to 75 mg/kg of total chromium, based on the assumption that 14 percent of total chromium was Cr(VI) in COPR (Bagdon and Hazen, 1991). Because there was no validated analytical method for Cr(VI) in soil in 1990, this was the basis for the 75 mg/kg total chromium cleanup criterion used to remediate the 37 residential sites in Jersey City.

There were many limitations to the NJDEP analysis, of which the most significant was the attempt to convert units from "mg/L" (ppm) of Cr(VI) in solution to an equivalent dose or concentration of soil. A "ppm" in water does not equal a "ppm" in a patch-test material or a "ppm" in soil. The dose of Cr(VI) received through the skin due to contact with water depends on the contact time and the amount of water available. For example, placing your hand in a bathtub of chromium-contaminated water presents an unlimited amount of Cr(VI) that could be transferred across the skin. On the other hand, the amount of chromium available for transport across the skin due to soil containing 1 ppm of Cr(VI) depends on the thickness of the soil layer on the skin surface and the extraction of Cr(VI) from soil by sweat and other liquids on the skin.

The MET necessary to calculate a soil cleanup standard must be in units of the mass of Cr(VI) per surface area of exposed skin (mg/cm^2), not in mg/L in a patch (Horowitz and Finley, 1994b). An additional concern with the NJDEP (1989) analysis was that the studies relied upon on used inconsistent patch-testing methods. For example, Pirila (1954) soaked gauze in Cr(VI) and placed it under plaster on the skin, and Skog and Wahlberg (1969) placed Cr(VI) mixed in petrolatum, to enhance dermal absorption, in a Finn chamber (e.g., approximately a one-half-inch circular plate) applied to the skin. The studies also (1) employed different forms of Cr(VI), (2) included data from tests of materials that were both highly acidic and highly alkaline (which were more likely to induce an irritant response), and (3) used different standards for reading the patch tests. In an attempt to determine whether the available data could shed light on identifying a soil cleanup value, Paustenbach et al. (1992) evaluated the data set, and using only the data for potassium dichromate (the most common testing compound) and different data-fitting techniques, estimated a MET of 150 ppm.

Through the mid-1990s, the NJDEP chose to maintain its position that the historical medical patch-test data could be used to identify soil cleanup levels that would prevent ACD (Bagdon and Hazen, 1991; NJDEP, 1995). Following the conduct of two new studies to evaluate this issue, the NJDEP revised its position in 1998 (NJDEP, 1998) (Figure 9.2).

1994 Human Patch-Testing Studies   In 1993, a group of expert dermatologists was assembled to design a human patch-testing study to specifically provide data that could be used to set Cr(VI) soil standards (Nethercott et al., 1994). In this study, a group of 54 Cr(VI)-sensitized volunteers were patch tested with serial dilutions of Cr(VI) and Cr(III) to determine the cumulative response rate. The physicians used standardized methods for patch testing developed by the North American Contact Dermatitis Group (NACDG, 1989). Rigorous quality control measures were implemented to ensure the accuracy of the exposure and the reliability of the findings. Eligible patients were first prescreened with diagnostic patch tests to ensure that they were allergic to Cr(VI). Only about one-half of the previously identified allergic volunteers reacted to the screening patches; therefore, it is probable that the remaining 54 individuals who participated in the study represented a subset of more highly allergic individuals than the patients used in the older studies, who were not prescreened (Scott and Proctor, 1997).

The results of the human patch-test study indicated that a dose–response relationship for ACD exists in units of mass per area, and that the 10 percent MET based on the cumulative response was 0.089 µg Cr(VI)/cm^2 of skin. Only one of the 54 volunteers may have responded to the Cr(III) challenge at 33 µgCr(III)/m^2 of skin; otherwise, Cr(III) was unable to produce ACD. The data from this study were used to assess the risk of developing ACD from skin contact with COPR in soil. The concentrations of Cr(VI) in soil associated with the MET was calculated as:

$$\text{Soil concentration (mg/kg)} = \frac{\text{MET (mg/cm}^2 \text{ skin)} \times 10^6 \text{ (mg/kg soil)}}{\text{SA (mg soil/cm}^2 \text{ skin) BVA (\%)}}$$

where the soil adherence rate (SA) is the amount of soil that adheres to the skin, and the bioavailability (BVA) is the fraction of Cr(VI) that can be leached from soil by sweat. Using the mass per area dosimetric, and assuming 100 percent bioavailability of Cr(VI) from COPR, Nethercott et al. (1994) concluded that 445 mg/kg of Cr(VI) in soil would not be expected to elicit ACD among 90 percent of Cr(VI)-sensitized individuals. As discussed in the following section, the dermal bioavailability of Cr(VI) from COPR is far less than 100 percent

(Horowitz and Finley, 1994), and thus, the elicitation threshold for Cr(VI) in COPR far exceeds 445 mg/kg Cr(VI) in soil.

Thick and Thin Patches   In a follow-up investigation, Nethercott et al. (1995) tested a subgroup of the cohort with two sets of True Test patches. All of the patches contained 175 ppm of Cr(VI), but half of the patches were sliced to be only one seventh the standard width, to generate "thick" and "thin" patches. The applied dose of the thin patches was 0.13 µg/cm^2, and that for the thick patches was 0.88 µg/cm^3, but both the thick and thin patches contained the same "concentration" of Cr(VI). This study was conducted specifically to address the NJDEP's contention that the concentration (in units of ppm) of Cr(VI) in a patch was the correct dosimetric for assessing dermal exposure to soil. The study found that of the nine volunteers tested, six responded to the thick patch, while none responded to the thin patch. This study clearly demonstrated that the correct dosimetric for assessing exposure to soil is the mass of the allergen per surface area. This principle has been demonstrated for other chemicals as well (Upadhye and Maibach, 1992).

Dermal Bioavailability of Cr(VI) in COPR   Horowitz and Finley (1994a) performed a study to measure the extractable fraction of Cr(VI) from soils containing COPR using real human sweat. Soil samples containing Cr(VI) at 16, 136, and 1,240 mg/kg were sieved to a uniform particle size of <500 µm. Real human sweat was collected from adult male human volunteers who worked at a moderate level in Tyvek suits to facilitate sweat production. Sweat was collected in the safety boots and pored into amber bottles for use in the extraction study. For the soil samples with 136 and 1,240 mg/kg, less than 0.1 percent of the hexavalent chromium leached into the sweat, with a maximum concentration of only 0.133 mg/L of Cr(VI) in sweat. This study suggested that Cr(VI) is tightly bound to COPR, and extraction of Cr(VI) by sweat on the skin is not expected to yield a Cr(VI) concentration that could elicit ACD at concentrations in soil up to 1,240 mg/kg.

Standards for Cr(VI) in Surface Water   After the Nethercott et al. (1994) study findings were released, NJDEP focused on evaluating the ACD hazard for individuals who may come into contact with Cr(VI) in standing water (e.g., puddles). Once again, the NJDEP applied the 10-ppm (mg/L) guideline originally developed by Bagdon and Hazen (1991). Because of the lack of confidence in that criterion in 1995 and 1996, a water immersion study was conducted to evaluate the threshold for ACD among individuals who were known (and proven by patch testing) to be allergic to Cr(VI) (Fowler et al., 1999) (Figure 9.2). In this study, 26 volunteers were exposed to 25 to 29 mg/L of Cr(VI) (as potassium dichromate in solution) by immersing one arm for 30 minutes per day on three consecutive days in a potassium dichromate bath. This concentration range is higher than any measured concentrations of Cr(VI) in puddles at COPR sites; the highest reported concentration is 16 mg/L (ESE, 1989a).

Sixteen of the 21 volunteers demonstrated no response to the challenge. Five volunteers developed a few papules or vesicles (1 to ~15), mild redness, and itching on the Cr(VI)-exposed arm (Fowler et al., 1999). The responses were all very mild. The nature of the responses was evaluated further through collection and histopathological review of skin biopsies. The histopathological diagnosis was focal irritation of the sweat gland. Morphology typical of ACD was not observed. The nature of the mild responses could not be resolved specifically; however, they clearly were not characteristic of ACD or irritant contact dermatitis (ICD), both of which are "eczematous" by definition. The physicians concluded that the response observed was an acute eccrine (sweat gland) reaction. In general, the responses developed rapidly and resolved within a few days, which was a primary consideration for drawing this conclusion.

Based on these results, wherein persons known to be sensitized to chromium were repeatedly exposed to 25 to 29 mg/L of Cr(VI) yet failed to elicit typical ACD or a serious irritant response, it was concluded that casual or occasional exposure to similar concentrations of Cr(VI) in the environment does not pose a dermal hazard. From these findings, in 1997, the NJDEP established the threshold for ACD from contact with water was 25 mg/L (Figure 9.2). Using this standard and water leaching (extraction) studies of COPR and COPR containing soils, site-specific soil cleanup levels that are protective of the 25 mg/L threshold have been developed for each COPR site.

Prevalence of Cr(VI) Sensitivity in the General Population   To understand the cost effectiveness of implementing ACD-based standards, it was necessary to understand the prevalence of Cr(VI) sensitivity in the general population of the United States. In an attempt to address this question, Proctor et al. (1998) reviewed more than 30 published studies from 1950 to 1997. No random survey of the U.S. general population was identified or has been performed to date, but the prevalence of Cr(VI) sensitization among North American clinical cohorts (cohorts of dermatology clinics) was reported to be 1 percent in 1996 [North American Contact Dermatitis Group (NACDG), 1997]. The prevalence of Cr(VI) sensitivity among the general population was calculated by dividing the current U.S. clinical prevalence estimate (1 percent) by a ratio of the Cr(VI) sensitization rate in a clinical population versus a general population in the Netherlands (ratio of 12) to yield an estimated Cr(VI) sensitization rate among the U.S. general population of 0.08 percent (Proctor et al., 1998).

Using this estimate of the prevalence of Cr(VI) sensitization in the general population, a retrospective cost-effectiveness study of the remediation efforts conducted at 37 residential sites in Jersey City was conducted (Proctor et al., 1998). It was determined that the population at risk of developing ACD consisted of 0.139 individuals, and therefore, that the cost to conduct that remediation project, per person at risk of developing ACD at these sites, was $216 million. This cost substantially exceeds the cost per life saved by environmental toxin control at hazardous waste sites of $4.2 million, at the direction of USEPA (Tengs et al., 1995).

9.6.2 Chronic Noncancer Dose–Response Evaluation

The potential noncarcinogenic health effects associated with exposure to Cr(III) and Cr(VI) were evaluated using reference levels established by the USEPA and those published in the peer-reviewed scientific literature (Malsch et al., 1994).

Systemic Noncancer Toxicity Assessment

Trivalent Chromium   An RfD of 1.5 mg/kg-day for insoluble forms of Cr(III) has been determined by the USEPA (1998b). The RfD is based on a study by Ivankovic and Preussmann (1975) in which rats were fed chromic oxide (Cr_2O_3) in baked bread for their lifetime. Animal body weight, organ weight, and food consumption were studied, and histopathology of all major organs was performed. No effects were observed at any of the dose levels, providing an NOAEL of 1468 mg/kg-day (USEPA, 1998b). To this dose, two 10-fold uncertainty factors were applied to account for intraspecies sensitivity and interspecies variability. In addition, the USEPA applied a 10-fold modifying factor. The Ivankovic and Preussmann (1975) study exposed test animals to chromic oxide, the same form of Cr(III) present in COPR. Thus, the USEPA Cr(III) RfD is specifically applicable to the Cr(III) in COPR.

Hexavalent Chromium   The USEPA has developed an RfD for Cr(VI) of 0.003 mg/kg-day based on a rat drinking water study (MacKenzie et al., 1958; U.S. EPA, 1998b). This study involved supplying rats with drinking water containing up to 11 mg/L of Cr(VI) for 1 year, but no effects were observed. Based on the study, reported drinking water consumption rate, and body weight of the rats, an NOAEL of 2.5 mg of Cr(VI)/kg-day was derived. To calculate an RfD, the USEPA applied an uncertainty factor of 100, which accounts for both intraspecies and interspecies variability, a threefold factor to account for the less-than-lifetime exposure and a modifying factor of 3 to account for uncertainties in the toxicological database.

Inhalation Noncancer Toxicity Assessment

Hexavalent Chromium   Occupational exposure via inhalation to low levels of Cr(VI) has been associated with upper respiratory irritation. Nasal irritation is prevalent in workers from the chrome plating, stainless steel, welding, ferrochromium and pigment production, and mining industries (ATSDR, 2000). Effects of irritation reported in the literature include coughing, sneezing, rhinorrhea, epistaxis, and nasal septum perforation/ulceration (Bloomfield and Blum, 1928; Gomes, 1972; Cohen et al., 1974). Additionally, bronchial asthma and increased risk of death due to noncancer respiratory disease (Taylor, 1966; Sorahan et al., 1987; Nemery, 1990; Davies et al., 1991) have been reported from occupational exposures to Cr(VI). Case reports of bronchial asthma (irritant and allergic in nature) associated with occupational exposure have been documented among workers involved in cement production, construction, plating, tanning, welding, and chromate production (PHS, 1953; Park et al., 1994; Wang et al., 1994; Shirakawa and Morimoto, 1996). It should be noted that the majority of the above effects are specific to occupational exposures to chromic acid mists, where information regarding exposure levels, smoking status, and confounding exposures is not available. In a study examining environmental exposures to COPR, the Greater Tokyo Bureau of Hygiene (GTBH) found no differences in pulmonary function between exposed and control populations (Tokyo Bureau of Hygiene, 1989).

The respiratory system in animals is also the primary target for inhalation exposure to Cr(VI). Studies examining Cr(VI) exposures in animals have revealed gross and histological changes to the respiratory tract and altered macrophage function in the lungs (Steffee and Baetler, 1965; Nettesheim et al., 1971; Nettesheim and Szakai, 1972; Glaser et al., 1985; Adachi et al., 1986; Johansson et al., 1986a, 1986b; Adachi, 1987).

Hexavalent Chromium RfCs   The majority of the human inhalation toxicity data for inhaled Cr(VI) is limited to occupational exposures to Cr(VI) in chromic acid mist from chrome-plating operations. However, these data are generally not relevant because chromic acid is not stable in the environment, and environmental Cr(VI) generally exists as a chromate salt. Thus, it is important to set the Cr(VI) RfC for an environmentally relevant form of Cr(VI) (Finley et al., 1992).

In 1991, USEPA developed a reference concentration of 0.002 $\mu g/m^3$ (or 2 ng/m^3) based on a lowest-observed-adverse-effect level (LOAEL) for Cr(VI) of 2 $\mu g/m^3$, reported by Lindberg and Hedenstierna (1983) among chrome platers exposed to Cr(VI) as chromic acid (Figure 9.2). The LOAEL was for nasal irritation, and an NOAEL of 1 $\mu g/m^3$ was reported in the same study. Even though the RfC was developed for Cr(VI), it was also applied to Cr(III) because the USEPA mistakenly thought that chromic acid contains both Cr(VI) and Cr(III). This RfC was never verified or presented in the Integrated Risk Information System (IRIS) database but was published in the Health Effects Assessment Summary Tables (HEAST) (USEPA, 1991a).

In response, the SRO sponsored a review and critique of the USEPA RfCs, which was later published in the peer-reviewed literature (Finley et al., 1992). The evaluation clearly showed that the USEPA RfCs were overly conservative, and USEPA withdrew its 1991 proposal. However, some agencies used the RfCs until alternatives were available. In the 1992 critique, Finley et al. proposed alternative RfCs for both Cr(III) and Cr(VI) as particulates, rather than as acid mists. A Cr(VI) RfC of 1.2 $\mu g/m^3$ was developed from the LOAEL of 3000 $\mu g/m^3$ reported by Steffee and Baetjer (1965) for rats exposed to Cr(VI) as particulates generated from COPR collected from an active chromate production plant. A total uncertainty factor of 300 was applied with a regionally deposited dose ratio (RDDR) to correct for the difference in inhalation kinetics between rats and humans for particles of aerodynamic diameter used in the Steffee and Baetjer (1965) study. As discussed below, Finley et al. (1992) also proposed a value for Cr(III). This RfC is more appropriate for assessing exposures to COPR than the earlier USEPA value because it is based on exposures to Cr(VI) in COPR, rather than in an acid mist.

In 1993, further research was performed by Malsch et al. (1994) to develop an RfC for Cr(VI) particulates using benchmark dose techniques (Malsch et al., 1994) (Figure 9.2). These researchers evaluated the scientific literature to identify studies that provide data that could be used to model the benchmark dose (BD) for Cr(VI). Data from Glaser et al. (1985, 1990) were selected as the most appropriate, and the most sensitive endpoints (e.g., those occurring as at the lowest exposure levels) were symptoms of pulmonary inflammation [e.g., protein in bronchial alveolar lavage fluid (BALF)] (Glaser et al., 1985, 1990). Malsch et al. (1994) modeled the 95 percent lower confidence limit on the 10 percent response level (the benchmark dose, or BD) for several continuous endpoints. BDs ranging from 16 to 67 $\mu g/m^3$ were determined, and to the lowest BD [16 $\mu g/m^3$ for lactate dehydrogenace (LDH) in BALF], an uncertainty factor of 90 was applied with an RDDR of 2.16 to derive an RfC of 0.34 $\mu g/m^3$ (Figure 9.9).

To further validate the Malsch et al. (1994) RfC, the authors presented the value and its derivation to an expert panel of reviewers that was assembled for the International Toxicology Estimates for Risk (ITER). The expert panel reviewed the proposed RfC in three sessions. Once the review was complete and all comments addressed, a documentation package was developed describing the results of the peer review and the panel-approved RfC for Cr(VI) particulates (ITER, 1998). The panel ultimately endorsed an RfC for Cr(VI) of 0.27 $\mu g/m^3$, which was based on an arithmetic average of the BDs calculated for all relevant endpoints in Malsch et al. (1994), divided by an uncertainty factor of 300 and multiplied by an RDDR of 2.16.

In 1998, USEPA followed the lead of the ITER expert review panel and used the published findings of Malsch et al. (1994) to set an RfC for Cr(VI) particulates of 0.1 $\mu g/m^3$ (Figure 9.2). This value is slightly lower than that adopted by the ITER expert panel but was derived in a similar manner. Similar to Malsch et al. (1994), USEPA selected the lowest BD of those calculated for pulmonary irritation symptoms (16 $\mu g/m^3$ for LDH in BALF), rather than a mean of all BDs as selected by the expert panel. Applying an uncertainty factor of 300 and an RDDR of 2.16, USEPA determined an RfC for Cr(VI) particulates of 0.1 $\mu g/m^3$ (USEPA, 1998b). USEPA also set an RfC for Cr(VI) as an acid mist of 0.008 $\mu g/m^3$ based on the results of Lindberg and Hedenstierna (1983).

The initially proposed Cr(VI) RfC of 0.002 $\mu g/m^3$ was 50-fold lower than the final value of 0.1 $\mu g/m^3$, and this change—dramatically, and appropriately—affects the assessment of noncancer effects associated with Cr(VI) in COPR.

Trivalent Chromium In 1998, USEPA reviewed the toxicological literature and concluded that there was inadequate data upon which to base an RfC for Cr(III).

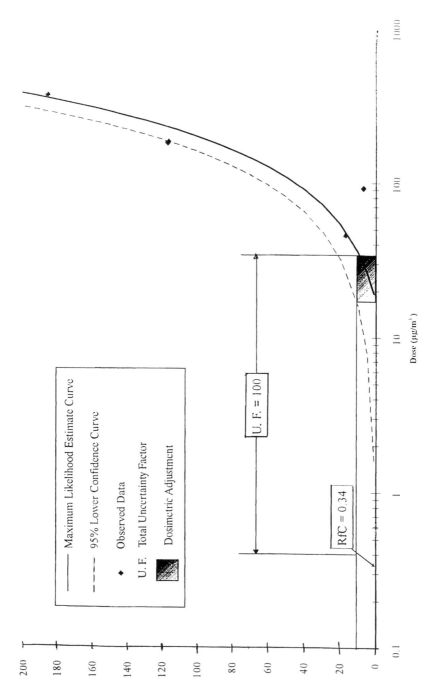

Figure 9.9 Benchmark dose analysis of lactate dehydrogenase in bronchial alveolar lavage fluid (BALF), from Malsch et al. (1994).

As noted previously, Finley et al. (1992) had suggested an RfC of 90 μg/m^3 based on a study by Axelsson et al. (1980) that evaluated pulmonary toxicity among workers at a ferrochrome production plant (Axelsson et al., 1980). This RfC was derived by applying a 10-fold uncertainly factor to account for sensitive subpopulations and an exposure adjustment factor of 2.8 to account for continuous exposure to the reported NOAEL of 2500 μg/m^3. This RfC could be considered applicable to Cr(III) in COPR because the form of Cr(III) in COPR is chromic oxide, which is similar to the form of chromium in ferrochrome alloys produced at the facility.

However, the National Ambient Air Quality Standard for respirable particulates (less than 10 μm in diameter) is 50 μg/m^3 as an annual average, and the proposed standard for particulates less than 2.5 μm in diameter is 15 μg/m^3 (USEPA, 1997). Because these standards are both lower than the Finley et al. (1992) RfC for Cr(III) of 90 μg/m^3, it has been concluded that the inhalation toxicity of Cr(III) is so low as not to warrant quantitative investigation. The findings of another recent rat inhalation toxicity study for Cr(III) as chromic oxide support this conclusion (Derelanko et al., 1999).

9.6.3   Cancer Assessment

Evidence of Carcinogenicity   Epidemiological evidence from the chromate production, chromate pigment production, and chrome-plating industries supports that certain Cr(VI) compounds are carcinogenic to humans upon inhalation (IARC, 1990). It is important to note, however, that the identity and concentrations of specific carcinogenic Cr(VI) compounds, and the influence of confounding variables in affected industries, are not well understood. Historically, these industries experienced Cr(VI) exposures in excess of 1 mg/m^3 (IARC, 1990). Recent improvements in production processes and industrial hygiene practices have reduced Cr(VI) levels in affected industries along with cancer rates (Davies et al., 1991; Korallus et al., 1993; Gibb et al., 2000).

No evidence exists that nonoccupational exposures to Cr(VI) constitute a cancer hazard. Several studies of populations exposed to high levels of Cr(VI) in soil/fill and in drinking water have not reported an increased rate of cancer associated with the exposure (GTBH, 1989; Watt and Eizaguirre, 1991; Zhang and Li, 1997). The Greater Glasgow Health Board (GGHB) conducted a mortality assessment for residents living in a part of Glasgow known to be contaminated with Cr(VI) from COPR. Although the levels of Cr(VI) were not reported, the concentrations of total chromium were reported "to be of the order of 10,000 ppm" (mg/kg). Based on our knowledge of Cr(VI) in COPR in New Jersey, we can estimate that Cr(VI) concentrations were likely in the range of 100 to 1000 mg/kg. The source of the COPR was a chromate production plant that operated in Glasgow from 1930 to 1963. The size of the potentially exposed cohort was a population of approximately 30,000 individuals living in six postal codes. Using the Glasgow Cancer Registry, the GGHB conducted a cohort mortality study of the Cr(VI)-exposed population. The researchers compared mortality rates in the affected areas with those of Glasgow City for the period of 1975 to 1989 and found no increase in the rates of cancer mortality for all cancers, lung cancers, or cancers of any other anatomical site among the exposed population.

In the only two studies of animals exposed to COPR by inhalation and intratracheal injection (Steffee and Baetjer 1965; Snyder et al., 1997), there was no increase in tumor incidence between the exposed and control groups. Snyder et al. (1997) exposed 100 rats to COPR-bearing soil provided by the NJDEP—50 rats to unaffected soil, 50 rats to calcium chromate, and 100 rats to calcium chromate plus COPR-bearing soil. The rats received weekly installations of 1.25 mg/kg of material for the first 6 weeks, and 2.5 mg/kg of material every other week for the next 38 weeks. Primary organs, including the lung, liver, and kidney, were evaluated histopathologically. The study concluded that intratracheal installations with 55 mg/kg of Cr(VI) from COPR-bearing soil delivered over a 44-week period did not cause any lung tumors and did not increase the incidence of other tumors, nor did they increase the incidence of nephropathy (Snyder et al., 1997).

Steffee and Baetjer (1965) administered "chromate residual roast material" from a chromate production facility (assumed to be COPR) to rabbits, guinea pigs, rats, and mice by inhalation at 3 to 4 mg of $Cr(VI)/m^3$ for 5 hours per day, 4 days per week throughout their lifetime. Although severe, this treatment did not result in a statistically significant increase in the incidence of lung tumors among any of the animal groups.

The Snyder et al. (1997) and Steffee and Baetjer (1965) data suggest that COPR is not carcinogenic to animals even under exposure conditions that are far more extreme than those expected in the environment. Further, because the positive control groups used in the Snyder et al. (1997) study (the calcium chromate group and calcium chromate plus COPR-containing soil group) displayed lung tumors, while the group exposed to COPR soil did not, these data suggest that Cr(VI) in COPR-containing soil is not carcinogenic. However, for the purposes of the COPR risk assessments for sites in New Jersey, the potential carcinogenic risk associated with inhaled Cr(VI) in COPR was addressed.

Dose–Response   The accepted regulatory approach for quantitatively evaluating carcinogenicity generally assumes that carcinogenic chemicals should be treated as if they have no threshold of activity (Paustenbach, 1989a). It is assumed by the USEPA that any dose of a carcinogen, no matter how small, presents a cancer risk. However, recent epidemiological and biochemical data suggest that a threshold for carcinogenicity does exist due to extracellular and intracellular reduction of Cr(VI) to the noncarcinogenic Cr(III) form in the respiratory system (Bidstrup, 1989; Davies et al., 1991; Korallus et al., 1993; De Flora, 2000).

Health risks associated with exposure to carcinogens are defined in terms of probabilities. These probabilities identify the likelihood of a carcinogenic response in an individual who receives a given dose of a particular compound. These probabilities are generally estimated using a cancer slope factor (CSF), which is expressed in units of $(mg/kg\text{-}day)^{-1}$. The CSF is multiplied by the lifetime daily human intake of the chemical (dose) (mg/kg-day) to provide an estimate of the theoretical increased cancer risk due to chemical exposure. For inhaled chemicals, including Cr(VI), the CSF is derived from the unit risk value, which is the theoretical excess cancer risk from continuous exposure to 1 $\mu g/m^3$. The USEPA CSF for Cr(VI) via inhalation is 41 $(mg/kg\text{-}day)^{-1}$ (USEPA, 1984, 1998b) and was developed from a unit risk value of 0.012 $(\mu g/m^3)^{-1}$. The CSF has been used to develop inhalation-based cleanup levels (alternative remediation standards) at the COPR sites.

The USEPA CSF for Cr(VI) is likely to be modified in the future based on recently published epidemiological information. The current CSF is based on a combination of the Mancuso (1975) epidemiological study and the Bourne and Yee (1950) industrial hygiene data for a former chromate production plant in Painesville, Ohio. The CSF is based on data with significant and well-recognized limitations (Mundt, 1997; Proctor et al., 1999). These limitations include (1) the airborne exposure data were collected 12 to 18 years after exposures began and likely underestimated actual exposures; (2) the exposure reconstruction is based on a single industrial hygiene study, and therefore, changes in airborne concentrations over time were not determined; (3) the industrial hygiene survey measured only total chromium and total soluble chromium, and therefore, the CSF is not based on speciated Cr(VI) data; and (4) the manner in which the results were reported leaves several data gaps, which require that assumptions about the outcome be made based on professional judgment. More current and improved epidemiological data have been published recently (Gibb et al., 2000), and another study is nearing completion. Thus, it is expected that the USEPA CSF will be reassessed in the near future.

The toxicity criteria used to evaluate Cr(VI) and Cr(III) in COPR are summarized in Table 9.6.

TABLE 9.6   Hexavalent and Trivalent Chromium Toxicity Criteria Used in Chromite Ore Processing Residue (COPR) Risk Assessments[a]

	Inhalation		Oral	Dermal
	Cancer	Noncancer	(Noncancer)	(ACD)
Cr(VI)	CSF of 41 $(mg/kg\text{-}d)^{-1}$	RfC of 0.1 $\mu g/m^3$	RfD of 0.003 mg/kg-d	Minimum elicitation threshold of 25 mg/L
Cr(III)	ND	ND	RfD of 1.5 mg/kg-d	ND

[a]CSF = cancer slope factor, RfC = reference concentration, RfD = reference dose, ACD = allergic contact dermititis, and ND = not determined.

9.7 RISK CHARACTERIZATION

For the COPR sites, risk characterization was performed by comparing the measured concentrations of Cr(VI) and Cr(III) in soil to site-specific alternative remediation standards (ARS) values. The ARS values are determined based on what is deemed to be an acceptable level of exposure, such as a dose equal to the USEPA RfDs and RfCs, the airborne concentration associated with a 1×10^{-6} (one-in-a-million) (10^{-6}) excess cancer risk level (the NJDEP-defined acceptable risk level), or the threshold for elicitation of ACD. This process of risk characterization is essentially the standard USEPA risk characterization process "in reverse," where the toxicity assessment and exposure assessment are used to determine acceptable exposure levels and acceptable soil concentrations (ARS values). The site soil data are then compared to the ARS values to determine the need for remediation. This approach to characterizing risk is used in other states as well. For example, the voluntary remediation programs established in Illinois [Illinois Environmental Protection Agency (IEPA), 1997], Indiana [Indiana Department of Environmental Management (IDEM), 1996], and Michigan [Michigan Department of Environmental Protection (MDNR), 1995] use this approach. It is commonly adopted because the results of the risk assessment can be readily applied to cleanup activities.

9.7.1 Development of Alternative Remediation Standards

For the COPR sites, different ARS values are determined for each pathway of exposure because, unlike risk assessment for most chemicals, there is a different health endpoint for each pathway of exposure. For dermal contact, the most sensitive endpoint is ACD, and thus, there is an ACD ARS for each site, as discussed below. For inhalation, the most sensitive endpoint is the increased risk of lung cancer. For incidental soil ingestion, the most sensitive endpoint is systemic toxicity, such as renal effects, that are not cancerous. Example calculations of how site-specific ARS values are derived for each endpoint and pathway of exposure are described below.

Inhalation Standards Protective of Excess Lung Cancer Risk   The Cr(VI) inhalation ARS for each site is calculated for an acceptable increased cancer risk of 10^{-6}. This acceptable cancer risk level is consistent with New Jersey and USEPA guidance [USEPA, 1990; Industrial Site Recovery Act (ISRA), 1993].

The inhalation ARS values for Cr(VI) at an industrial site where workers are exposed to suspended soil particulates due to heavy truck traffic can be calculated, as an example, using the following equation:

$$\text{ARS}_{\text{inh}} = \frac{\text{Risk} \times \text{BW} \times \text{AT}}{\text{CPF} \times \text{IR} \times \text{EF} \times \text{ED} \times \text{PEF}}$$

where ARS_{inh} = alternative remediation standard for Cr(VI) in soil based on inhalation (mg/kg)

Risk = acceptable risk level (1×10^{-6})

CPF = EPA inhalation cancer potency factor for Cr(VI) [41 (mg/kg-day)$^{-1}$]

IR = inhalation rate (9.1 m^3/workday for an adult)

BW = body weight (70-kg adult)

AT = averaging time (25,550 days in a 70-year lifetime)

ED = exposure duration (assumed upper-bound value of 25 years)

EF = exposure frequency (250 days/year)

PEF = particulate emission factor (kg/m^3) (site-specific, as calculated in Section 9.5.3)

For example, with a PEF of 1×10^{-9} kg/m^3, ARS_{inh}

$$= \frac{10^{-6} \times 70 \text{ kg} \times 25,550 \text{ days}}{41 \text{ (mg/kg-day)}^{-1} \times 9.1 \text{ m}^3/\text{day} \times 250 \text{ days/yr} \times 25 \text{ yr} \times 1 \times 10^{-9} \text{ kg/m}^3}$$

$$= 767 \text{ mg/kg}$$

Inhalation ARS values calculated to date for the COPR sites vary widely, generally ranging from 106 to 7420 mg/kg, with one value as high as 26,300 mg/kg for a remote site with almost no traffic (Table 9.7). Because Cr(III) is not carcinogenic, and there are no established inhalation noncancer standards for Cr(III) because it is relatively innocuous, inhalation ARS values for Cr(III) have not been developed to date.

For some of the industrial sites, the inhalation ARS value is lower than the USEPA soil screening level for Cr(VI) at residential sites—270 mg/kg. This is one rare case where the standard for protection of workers is lower than a standard that would be protective of residential exposures because the potential for exposure is greater.

ARS Values Protective of Allergic Contact Dermatitis ARS values that are protective of ACD are developed on a site-specific basis using the Cr(VI) 25 mg/L (ppm) threshold for elicitation of ACD among presensitized individuals (Fowler et al., 1999), which was developed specifically for these COPR risk assessment and leaching tests designed to simulate the formation of a shallow puddle on surface soil at COPR sites. For these simulations, the target leachate or puddle concentration is 25 ppm of Cr(VI). It should be noted that 25 ppm of Cr(VI) is bright yellow water, and while puddles of yellow water have been observed at the COPR sites, the highest measured concentration of Cr(VI) in a puddle is only 16 ppm (ESE, 1989a).

The NJDEP has estimated that the most shallow puddle has a liquid-to-solid

TABLE 9.7   Alternative Remediation Standards (ARSs) for Special Remedial Organization–Managed Sites

Site Number	Cr(VI) ACD ARS	Cr(VI) ARS for Inhalation	Remedial Action
40	58,800*	235	Deed notice
42 (1 of 1)	228	370	Cap-and-deed notice
42 (1 of 2)	228	760	None required
45	228	410	None required
47	768	265	Excavation and offsite disposal
48	768	NC	Cap-and-deed notice
52	214	205	Excavation and offsite disposal
53	228	1,100	None required
55	225	533	Ex situ treatment with ferrous sulfate
56	265	7,420	Ex situ treatment with ferrous sulfate
59	1180	287	Remedy not complete
62	300	164	Excavation and offsite disposal
110	228	710	None required
145	172	106	Excavation and offsite disposal
148	228	300	None required
167	745	1510	Remedy not complete
169	228	2,200	None required
170	382	159	Excavation and offsite disposal
171	228	3,600	None required
176	546	3930	Remedy not complete
195	99	NC	Excavation and offsite disposal
201	316	26,300	Cap-and-deed notice

*Cr(VI) ACD ARS for alumina hydrate and not COPR
NC: not calculated.

(LSR) ratio of 2:1, and that modification of the ASTM 3987 (ASTM, 1985) water leachate test can be used to simulate the formation of a puddle. The ASTM 3987 test is an 18-hour water extraction test that is typically performed at a liquid-to-solid ratio of 20:1 with constant agitation. Leaching tests cannot specifically assess the soil concentration that would yield a 25 ppm Cr(VI) leachate concentration at a liquid-to-solid ratio of 2:1, because at such a small ratio, there is no free liquid that can be separated from the solid for analysis. The lower the liquid-to-solid ratio, the lower the resulting ARS, because the concentration of Cr(VI) in water that can be generated at a low liquid-to-solid ratio is much higher than at higher ratios (less water available to dilute the Cr(VI) that can be leached out). Thus, the ASTM 3987 tests are run at liquid-to-solid ratio ratios of 4:1, 10:1, 20:1, and 40:1, using the most highly affected Cr(VI) soils from each site. The results are converted to terms of a water-to-soil partitioning coefficient (K_d) (units of L/kg) and are plotted and analyzed by linear regression to extrapolate the measured results that could be achieved at a liquid-to-solid ratio of 2:1.

The ACD ARS values for COPR determined to date have ranged from 99 to 768 mg/kg (Table 9.7). The variability is believed to be related to the efficiency of the production process when the COPR was produced [more Cr(VI) was probably left in the COPR in the earlier years], when the COPR was used as fill, and perhaps most importantly, the environmental degradation conditions at each site. The approach described here for assessing the Cr(VI) concentrations in surface puddles is recognized as being very conservative and far from accurate. Approaches for modeling more realistic liquid-to-solid ratios for puddle formation based on site-specific soil characteristics are currently underway.

Incidental Soil Ingestion-Based Standards   For most of the COPR sites, remediation has been driven by ACD-based standards, and the development of ARS values for the soil ingestion pathway has not been necessary. An ingestion-based ARS value could be calculated according to the following equation:

$$ARS_{ing} = \frac{HI \times RfD \times BW \times AT}{IngR \times EF \times ED}$$

where ARS_{ing} = alternative remediation standard for Cr(VI) in soil based on soil ingestion (mg/kg)

HI = acceptable target hazard index of 1.0

RfD = EPA RfD for Cr(III) or Cr(VI) [1.5 mgCr(III)/kg-day and 0.003 mgCr(VI)/kg-day]

IngR = soil ingestion rate (EPA default values of 0.050 kg soil/day for an adult worker and 0.20 kg soil/day for a child)

BW = body weight (70-kg adult; 15-kg child)

AT = averaging time (9125 days for a 25-year worker exposure duration, and 2190 days for a 6-year residential child exposure duration)

ED = exposure duration (assumed upper-bound value of 25 years for a worker and 6 years for a residential child)

EF = exposure frequency (250 days/year for a worker; 350 days/ year for a resident)

The exposure duration and averaging time cancel out of this equation, and thus, the ARS for ingestion is not sensitive to the period of exposure; although because it is based on the RfD, the period of exposure should be chronic.

For example, the ARS for Cr(VI) at an industrial site is calculated as follows:

$$ARS_{ing} = \frac{1 \times 0.003 \text{ mg/kg-day} \times 70 \text{ kg} \times 9125 \text{ days}}{0.05 \text{ kg/day} \times 250 \text{ days/yr} \times 25 \text{ yr}}$$

$$= 6,132 \text{ mg/kg}$$

The resulting ARS values for an industrial/commercial site for Cr(VI) and Cr(III) are 6132 mg/kg and 3,066,000 mg/kg, respectively, which for Cr(III) exceeds unity and demonstrates that a soil ingestion-based standard is unnecessary at an industrial site where only workers are exposed. ARS values calculated for a residential exposure scenario are 234 mg/kg and 117,300 for Cr(VI) and Cr(III), respectively. This calculation could be refined further to account for the reduction in exposure that occurs during inclement weather conditions, the low oral bioaccessibility of chromium in COPR, and through the use of more refined assumptions regarding the soil ingestion rate and exposure frequency; USEPA default values have been used for these calculations. Because the inhalation- and dermatitis-based ARS values are generally lower than the soil ingestion-based standards, it is not necessary to further refine the soil ingestion-based ARSs.

9.7.2 Qualitative Uncertainty Analysis

There are numerous sources of uncertainty inherent in the risk assessment process. Some level of uncertainty is introduced into the assessment each time an assumption is made. Many assumptions have valid and strong scientific bases, while others are estimates that are usually represented by a range of values. Where there is uncertainty regarding an assumption, a conservative estimate is often chosen to ensure that the assessment will be health protective. This section considers the uncertainties associated with the risk assessment according to each of the major components of the COPR assessments.

Site Characterization   Uncertainties are associated with determination of future land use for these sites and the surrounding area, as well as the exposure scenarios evaluated under these land uses. The uncertainty associated with the future land use of these sites is deemed to be small because these sites have been zoned and used for industrial purposes for at least 50 years. For most sites, high confidence is placed in the assumption that current onsite land use (i.e., industrial) will remain unchanged well into the future. The likelihood that the sites could be rezoned and designated for residential use is considered extremely remote.

Exposure Parameters   Several parameters are incorporated into the exposure assessment that rely on conservative estimates used to define general population behavior. Conservative default values used for exposure parameters (i.e., 25-year exposure duration used for industrial populations, and use of 95th percentile inhalation rates) were chosen for the ARS calculations. The net effect of these conservative exposure assumptions is the overestimation of potential health risks.

Particulate Emission Modeling   There are many uncertainties involved with the particulate emission modeling. Because the particulate emission model is a regression model based on data from several studies, the similarity between traffic conditions at the COPR sites and those on which the model is based will affect the accuracy of the model. In addition, there are uncertainties associated with the FDM for particulate dispersion. A site-specific evaluation of model performance showed that this particulate emission modeling approach produced accurate but consistently conservative results for the COPR sites (Scott et al., 1997a). Therefore, the greatest amount of uncertainty in the particulate emission modeling is associated with future onsite vehicle activity. It was assumed that the type of vehicles that will travel across any of these sites in the future, the frequency of traffic, and the speed at which they will travel in the future, will be similar to current patterns. Any redevelopment of these sites that might occur as part of an area revitalization would result in changes to trucking activities, but the change would more likely result in less truck traffic and the application of additional land cover such as landscaping. Also, for the purpose of estimating worst-case potential future exposures, the presence of existing IRMs was disregarded. For these reasons, it is believed that the current soil suspension modeling approach conservatively estimates the concentrations of Cr(VI) suspended to ambient air from soil at these sites in the future.

Toxicity Criteria The CSF for Cr(VI), and the RfDs and RfCs, are designed by the USEPA to be health protective for sensitive human populations. The RfDs for Cr(III) and Cr(VI) are very conservative because they are both based on studies where the effect level was never determined (the highest dose tested showed no adverse effects). The CSF for Cr(VI) is conservatively applied to Cr(VI) in COPR even though several high-dose animal studies have shown that Cr(VI) in COPR is not carcinogenic (Snyder et al., 1997; Steffee and Baetjer, 1965).

Risk Characterization Risk characterization is performed for these assessments in the manner prescribed by the NJDEP. Rather than assessing risks, ARS values are set to be protective of the level of exposure and risk that the NJDEP deems acceptable. For cancer endpoints, the acceptable risk level is conservatively set at the low end of the USEPA acceptable risk range of 10^{-4} to 10^{-6} (USEPA, 1990). For noncancer endpoints, the hazard index is set at 1, so that the chemical exposure associated with site activities is equal to the RfD or RfC.

9.7.3 Application of ARS Values to Guide Remedial Efforts

Theoretically, the application of standards that are protective of chronic exposures (inhalation and ingestion ARSs) should be based on a comparison of the ARS to the average concentration of Cr(VI) and Cr(III) in the soil that is contacted by the potentially exposed population. The 95% upper confidence limit (UCL) of the arithmetic mean soil concentration for surface soils is the appropriate concentration for comparison with the ARS for both inhalation and ingestion because both standards are protective of long-term average exposures

(USEPA, 1992). Use of the 95% UCL ensures that the estimated mean concentration is not underestimated at the 95% confidence level.

For clean closure, the NJDEP has required that the ARS for inhalation be met based on a comparison of the ARS to the 95 percent UCL for Cr(VI) in all soils (based on the assumption that activities in the future could result in the subsurface soils being excavated and brought to the surface, where they could be contacted). For some sites, the 95 percent UCL concentration in all soils is higher than in just surface soils, and for other sites, the trend is reversed.

The NJDEP has required that the soil ingestion-based ARS be considered a maximum, not-to-be-exceeded value, and that ingestion-based ARS protective of future residential exposure scenarios be considered for each site that does not have a deed restriction precluding residential development. This position is conservative for two reasons: (1) the future land use of most of the current COPR sites can reasonably be assumed to remain industrial based on historical and surrounding land use, and (2) the application of the soil ingestion ARS values to a maximum acceptable concentration makes the assumption that soil ingestion exposure at the site will only occur in one location over the long-term, rather than randomly across the site. These assumptions are unnecessarily conservative and ultimately result in both the implementation of remedial efforts that do not further reduce the potential health risks and increased costs for remedial activities.

The ARS values for ACD are protective of short-term (acute) exposures because, in theory, if a presensitized individual contacts concentrations of Cr(VI) over a threshold for elicitation of ACD, a response (rash) could result. Thus, the ARS values for ACD are also treated as maximum acceptable concentrations, and while conservative, this approach is consistent with the theoretical basis of the ACD toxicity assessment.

9.8   PRACTICAL IMPLICATIONS OF THIS ANALYSIS

9.8.1   Impact of Risk Assessment Activities on Volume of COPR

The effect of the risk assessment activities described in this chapter on the resolution of the public health concerns was significant. Table 9.7 presents a summary of Cr(VI) ARS values for inhalation and ACD endpoints that have been approved to date for various sites addressed thus far. It is important to note that the soil cleanup standard for Cr(III) is numerically so large as to become meaningless in the context of this project (basically, the cleanup value is greater than the concentration found in any sample of soil).

The evolution of cleanup levels for this project is presented in Figure 9.10. As noted earlier in this chapter, the remedial cost of the project was expected to be dictated by the volume of COPR to be addressed. Moreover, it was expected that the volume of COPR to be addressed would decrease as chromium cleanup standards evolved to higher concentrations, particularly when separate standards were assigned to the two different valence states of Cr to reflect their

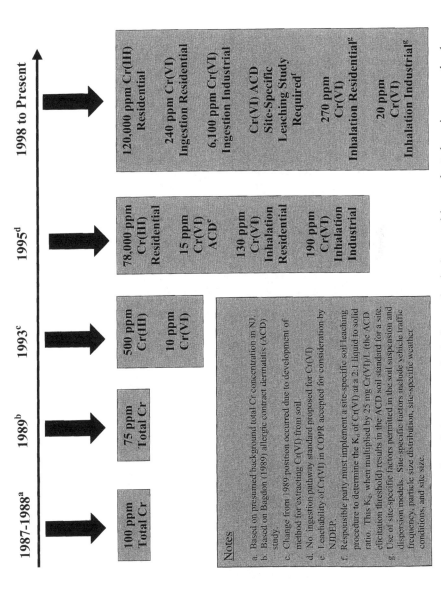

Figure 9.10 Evolution of cleanup standards for total, trivalent, and hexavalent chromium over the duration of the chromite ore processing residue (COPR) project.

571

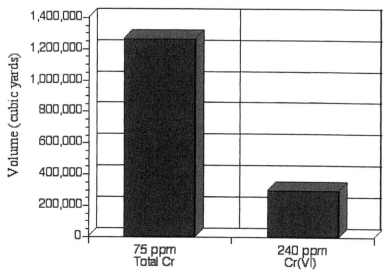

Figure 9.11   Volume of chromite ore processing residue (COPR)-affected soils that must be addressed for the Special Remedial Organization–Managed Sites that are distant from the original plant site at cleanup standards of 75 mg/kg of total chromium and 240 mg/kg of hexavalent chromium.

different toxicities. Indeed, the results of the remedial investigation at the COPR sites verified this hypothesis. Specifically, Figure 9.11 illustrates the impact of increased Cr(VI) cleanup standards on the volume of COPR that must be addressed by the SRO at sites not proximate to the main plant site in Kearny, NJ. A change in the applicable cleanup standard (from 75 mg/kg total chromium to 240 mg/kg Cr(VI); Figure 9.11) resulted in a decrease of approximately 75 percent in the volume of soil to be addressed. This has dramatic implications with regard to remedial action costs, acreage of property affected, and reduced litigation potential.

9.8.2 Impact on Remedy Selection

To date, remedial activities at 21 of 38 COPR sites assigned to the SRO have been completed (or nearly completed). Remedy selection at these 21 sites was radically influenced by the Cr(VI) and Cr(III) cleanup standards obtained from the previous 12 years of research and risk assessment activities. Table 9.7 summarizes the remedial activities (if any) at the 21 sites completed to date. Note that 14 of the 21 sites were closed as clean, with unrestricted use. Of these 14 sites closed as clean, no remedial action was required at seven, and limited hot-spot treatment or removal was required at the other seven. One of the 21 sites was closed as clean for industrial or commercial activities only, and 6 of the 21 sites were closed with a cap-and-contain remedy and deed restrictions for future use. Furthermore, for the sites with the cap-and-contain remedy, the extent of the acreage to be capped was limited to a fraction of the property with Cr(VI) levels above the site-specific standards.

9.8.3   Cost Savings

The risk assessment activities implemented by the SRO clearly resulted in the saving of many millions of dollars and provided remediation activities that ensure the future protection of public health and the environment. At the SRO–managed sites, the volume of COPR to be addressed was reduced by at least 75 percent, or 965,000 cubic yards, for sites not proximate to the former chromate plant site in Kearny (Figure 9.11). By comparison to a remedy consisting of excavation and off-site disposal for soils containing greater than background concentrations of total Cr, the cost savings is estimated to have been $440 million. This estimate assumes: (1) remediation to 75 mg/kg total chromium; (2) cost of $350/ton for design, agency reporting, excavation, transportation, and disposal of COPR soils to a permitted hazardous waste landfill; (3) treatment of contaminated groundwater in the excavation pits; and (4) the purchase of clean fill to replace the COPR soils. Similar cost reductions are expected for the other responsible parties. This estimate does not include any business disruption claims and related litigation from property owners.

In contrast, if all of the sites would have been capped, the impact of the risk assessment activities on remediation costs would have been less significant, as compared to the prior scenario. Clearly, 14 sites would not have been closed as clean for unrestricted use, and liability would have remained with the SRO into the future. Furthermore, cap construction would have occurred over virtually the entire acreage of each property assuming a 75 mg/kg total chromium standard. Under this scenario, remediation cost savings for the SRO-managed sites is estimated to have been a more modest $30 million.

9.9   CONCLUSIONS

This case study illustrates how the incorporation of scientific data from directed research into risk assessment can both improve our understanding of the possible health risks and reduce the final cost of remedial action. It is perhaps the best example to date of the benefits of research to fill the data gaps within an environmental risk assessment.

On several levels, the nearly $10 million in original research that was conducted to characterize the potential health risks posed by COPR serves as a premier case for demonstrating that regulatory agencies and the public can alter their views about "risk" and modify risk management decisions if essential data are gathered. In this case, research was conducted and the results were presented in the peer-reviewed literature. This case study also illustrates, however, that the number of hurdles placed in front of the responsible parties can be substantial and, at times, excessive. Obviously, only large and complex projects such as this one, where remedial obligations can be substantial, warrant the investment in research that was required to characterize the various possible health hazards posed by chromium in COPR. In this case, extrapolating the cost differential for the SRO-managed sites to all the Cr sites in Hudson Cunnly suggests that the responsible parties were able to reduce their potential liabilities by perhaps as much as $1 billion as a result of the investment in research, and in all likelihood this work has been instrumental in minimizing unwarranted personal injury litigation.

This thorough characterization of health risks has also benefited the communities in Hudson County (most notably in the Town of Kearny). Unlike what occurred at Times Beach and Love Canal, residents did not have to be evacuated (with the subsequent problems of disrupted families and the high likelihood of no future recovery of businesses), nor has it suffered any particular stigma associated with the COPR-containing soils. In short, the community has not been significantly affected; property values have been kept intact (or improved), and businesses on affected properties have continued to operate.

The research performed for this project has substantially improved the tools available to characterize chromium in environmental media, and has advanced the level of understanding of the potential health risks associated with environmental exposures to Cr(VI). In total, more than 50 peer-reviewed manuscripts have been published based on this research, new analytical methods for measuring Cr(VI) in both soil and ambient air have been developed, and the toxicity assessment methodologies for dermal and inhalation exposures to Cr(VI) have been refined.

REFERENCES

Adachi, S. 1987. Effects of chromium compounds on the respiratory system. Part 5. Long term inhalation of chromic acid mist in electroplating by C57BL female mice and recapitulation on our experimental studies. *Jpn. J. Ind. Health.* 29: 17–33.

Adachi, S., Yoshimura, H., Katayama, H., and Takemoto, K. 1986. Effects of chromium compounds on the respiratory system. Part 4. Long term inhalation of chromic acid mist in electroplating to ICR female mice. *Jpn. J. Ind. Health.* 28: 283–287.

Adams, R. M. 1990. Allergic contact dermatitis. In R. M. Adams (Ed.), *Occupational Skin Disease*, 2nd ed. W.B. Saunders, Philadephia pp. 26–31.

Agency for Toxic Substances and Disease Registry (ATSDR). 2000. *Toxicological Profile for Chromium*. U.S. Public Health Service, Atlanta, GA.

American Society for Testing and Materials (ASTM). 1985. Method D3987-85 (re-approved 1992).

Anderson, F. E. 1960. Cement and oil dermatitis. The part played by chrome sensitivity. *Br. J. Dermatol.* 72: 108–117.

Anderson, R. A., Colton, T., Doull, J., Marks, J. G., and Smith, R. G. 1993. Designing a biological monitoring program to assess community exposure to chromium: Conclusions of an expert panel. *J. Toxicol. Environ. Health.* 40: 555–583.

Anderson, R. A., and Kozlovsky, A. S. 1985. Chromium intake, absorption and excretion of subjects consuming self-selected diets. *Am. J. Clin. Nutr.* 41: 1177–1183.

Axelsson, G., Rylander, R., and Schmidt, A. 1980. Mortality and incidence of tumors among ferrochromium workers. *Br. J. Indust. Med.* 37: 121–127.

Bagdon, R. E., and Hazen, R. E. 1991. Skin permeation and cutaneous hypersensitivity as a basis for making risk assessments of chromium as a soil contaminant. *Environ. Health. Perspect.* 92: 111–119.

Bartlett, R. J. 1991. Chromium cycling in soils and water: Links, gaps, and methods. *Environ. Health Perspect.* 92: 17–24.

Bartlett, R., and James, B. 1979. Behavior of chromium in soils: III. Oxidation. *J. Environ. Qual.* 8(1): 31–35.

Bartlett, R. J., and Kimble, J. M. 1976. Behavior of chromium in soils: I. Trivalent forms II. hexavalent forms. *J. Environ. Qual.* 5(4): 379–383.

Bidstrup, P. L. 1989. Perspective on safety: Personal opinions. *Am. Ind. Hyg. Assoc. J.* 50(10): 505–509.

Bloomfield, J. J., and Blum, W. 1928. Health hazards in chromium plating. *Pub. Health Prts.* 43: 2330–2347.

Bourne, H. G., and Yee, H. 1950. Occupational cancer in a chromate plant: an environmental appraisal. *Ind. Med. Surgery.* 19(12) 505–567.

Bourne, H. G., and Rushin, W. R. 1950. Atmospheric pollution in the vicinity of a chromate plant. *Ind. Med. Surg.* 19(12): 563–569.

Bukowksi, J. A., Goldstein, M. D., and Johnson, B. B. 1991. Biological markers in chromium exposure assessment: Confounding variables. *Arch. Environ. Health* 46(4): 230–236.

Bukowski, J. A., Goldstein, M. D., Korn, L. R., Rudakewych, M., Shepperly, D., Gates, D., and McLinden, M. 1992. Chromium exposure assessment of outdoor workers in Hudson County, NJ. *Sci. Total Environ.* 122: 291–300.

Burrows, D. 1987. Chromate dermatitis. In H. Maibach (Ed.), *Occupational and Industrial Dermatology*, 2nd ed. Year Book Medical, Chicago, IL, pp. 406–420.

California Air Resources Board (CARB). 1987. Proposed method for the speciation and analysis of hexavalent chromium at ambient atmospheric levels. Memo from CARB to G. Muchison, November 2, 1987.

Cohen, S. R., Davis, D. M., and Kramkowski, R. S. 1974. Clinical manifestations of chromic acid toxicity: Nasal lesions in electroplate workers. *CUTIS* 13: 558–568.

Copeland, T. L., Holbrow, A. M., Otani, J. M., Connor, K. T., and Paustenbach, D. J. 1994. Use of probabilistic methods to understand the conservatism in California's approach to assessing health risks posed by air contaminants. *J. Air Waste Manag. Assoc.* 44: 1399–1413.

Cowherd, C., Englehart, P., Muleski, G., and Kinsey, J. S. 1989. Hazardous waste TSDF fugitive particulate air matter emissions guidance document. U.S. Environmental Protection Agency, Office of Air Quality, Planning and Standards. USEPA 450/3-89-019.

Cowherd, D., Muleski, G. E., Englehart, P. J., and Gillette, D. A. 1985. Rapid assessment of exposure to particulate emissions from surface contamination sites. U.S. Environmental Protection Agency, Office of Air Quality, Planning and Standards. USEPA 600/7-851051.

Cross, H. J., Faux, S. P., Sadhra, S., Sorahan, T., Levy, L. S., Aw, T. C., Braithwaite, R., McRoy, C., Hamilton, L., and Calvert, I. 1997. *Criteria Document for Hexavalent Chromium*. International Chromium Development Association, Adgbaston, Birmingham, UK.

Davies, J. M., Easton, D. F., and Bidstrup, P. L. 1991. Mortality from respiratory cancer and other causes in United Kingdom chromate production workers. *Br. J. Indust. Med.* 48: 299–313.

De Flora, S. 2000. Threshold mechanism and site specificity in chromium(VI) carcinogenesis. *Carcinogenesis* 21(4): 533–541.

Derelanko, M. J., Rinehart, W. E., Hilaski, R. J., Thompson, R. B., and Loser, E. 1999. Thirteen-week subchronic rat inhalation toxicity study with a recovery phase of trivalent chromium compounds, chromic oxide and basic chromium sulfate. *Toxicol. Sci.* 52: 278–288.

Dragun, J. and Chaisson, A. 1991. Elements in North American Soils, page 59 published by Hazardous Materials Control Resources Institute. Greenbelt, MD.

Environmental Sciences and Engineering (ESE). 1989a. Remedial investigation for chromium sites in Hudson County, New Jersey. Prepared for the State of New Jersey Department of Environmental Protection, Trenton, NJ.

Environmental Sciences and Engineering (ESE). 1989b. Risk assessment for chromium sites in Hudson County, New Jersey. Prepared for the State of New Jersey Department of Environmental Protection, Trenton, NJ.

Environmental Sciences and Engineering (ESE). 1989c. Feasibility study for chromium sites in Hudson County, New Jersey. Prepared for the State of New Jersey Department of Environmental Protection, Trenton, NJ.

Fagliano, J., Sarvin, J., Udasin, I., and Gochfeld, M. 1997. Community exposure and medical screening near chromium waste sites in New Jersey. *Reg. Toxicol. Pharmacol.* 26(1): S13–S22.

Falerios, M., Schild, K., Sheehan, P., and Paustenbach, D. J. 1992. Airborne concentrations of trivalent and hexavalent chromium from contaminated soils at unpaved and partially paved commercial/industrial sites. *J. Air Waste Manag. Assoc.* 42: 40–48.

Finkel, A. M. 1990. *Controlling Uncertainty in Risk Management: A Guide for Decision Makers*. Center for Risk Management, Resources for the Future. Washington, DC.

Finley, B., Fehling, K., Falerios, M., and Paustenbach, D. 1993. Field validation for sampling and analysis of airborne hexavalent chromium. *Appl. Occup. Environ. Hyg.* 8(3): 191–200.

Finley, B., Proctor, D., Scott, P., Harrington, N., Paustenbach, D., and Price, P. 1994. Recommended distributions for exposure factors frequently used in health risk assessment. *Risk Anal.* 14(1): 533–553.

Finley, B., Scott, P., Norton, R. L., Gargas, M. L., and Paustenbach, D. 1996. Urinary chromium concentrations in humans following ingestion of safe doses of hexavalent chromium and trivalent chromium: Implications for biomonitoring. *J. Toxicol. Environ. Health* 48: 479–499.

Finley, B. L., Kerger, B. D., Katona, M., Gargas, M. L., Corbett, G. C., and Paustenbach, D. J. 1997. Human ingestion of chromium(VI) in drinking water: Pharmacokinetics following repeated exposure. *Toxicol. Appl. Pharm.* 142: 151–159.

Finley, B. L., Proctor, D. M., and Paustenbach, D. J. 1992. An alternative to the USE-PA's proposed inhalation reference concentrations for hexavalent and trivalent chromium. *Reg. Toxicol. Pharm.* 16: 161–176.

Fowler, J. F., Kauffman, C. L., Marks, J. G., Proctor, D. M., Fredrick, M. M., Otani, J. M., Finley, B. L., Paustenbach, D. J., and Nethercott, J. R. 1999. An environmental hazard assessment of low-level dermal exposure to hexavalent chromium in solution among chromium-sensitized volunteers. *JOEM* 41(3): 150–160.

Gargas, M. L., Norton, R. L., Harris, M. A., Paustenbach, D. J., and Finley, B. L. 1994a. Urinary excretion of chromium following ingestion of chromite-ore processing residues in humans: Implication for biomonitoring. *Risk Anal.* 14(6): 1019–1024.

Gargas, M. L., Norton, R. L., Paustenbach, D. J., and Finley, B. L. 1994b. Urinary excretion of chromium by humans following ingestion of chromium picolinate. *Drug Metab. Disp.* 22(4): 522–529.

Gibb, H., Lees, P. S., and Rooney, B. C. 2000. Lung cancer among workers in chromium chemical production. *Am. J. Ind. Med.* 38(2): 115–126.

Glaser, U., Hochrainer, D., Kloppel, H., and Kuhnen, H. 1985. Low level chromium(VI) inhalation effects on alveolar macrophages and immune functions in Wistar rats. *Arch. Toxicol.* 57: 250–256.

Glaser, U., Hochrainer, D., and Steinhoff, D. 1990. Investigation of irritating properties of inhaled CrVI with possible influence on its carcinogenic action. In N. H. Seemayer and W. Hadnagy (Eds.), *Environmental Hygiene*, Vol. 2. Springer-Verlag, Berlin, Germany pp. 239–245.

Gomes, E. 1972. Incidence of chromium-induced lesions among electroplating workers in Brazil. *Ind. Med.* 41: 21–25.

Greater Tokyo Board of Health (GTBH). 1989. Report concerning the effect of chromium in a health survey (10-year survey). Okayama University Medical School. Tokyo, Japan.

Hackensack Meadowlands Development Commission (HMDC). 2001. *Hackensack Meadowlands Development Commission Mandates: Fact Sheet.* HMDC, Jersey City, NJ.

Hall, W. S., and Pulliam, G. W. 1995. An assessment of metals in an estuarine wetlands ecosystem. *Arch. Environ. Contam. Toxicol.* 29: 164–173.

Harbison, R. D., and Rinehart, W. E. 1990. *Conclusions of the Expert Review Panel on Chromium Contaminated Soil in Hudson County, New Jersey.* Pennsylvania, Industrial Health Foundation, Pittsburgh, PA.

Horowitz, S. B., and Finley, B. L. 1994a. Using human sweat to extract chromium from chromite ore processing residue: Applications to setting health-based cleanup levels. *J. Toxicol. Environ. Health.* 40: 585–599.

Horowitz, S. B., and Finley, B. L. 1994b. Setting health-protective soil concentrations for dermal contact allergens: A proposed methodology. *Reg. Toxicol. Pharm.* 19: 31–47.

Iannuzzi, T., and Ludwig, T. J. 2001. *Ecological and Economic History of an Urban Waterway: The Passaic River.* Amherst Press, Amherst, MA.

Illinois Environmental Protection Agency (IEPA). 1997. *Tiered Approach to Cleanup Objectives.* Illinois Pollution Control Board, Springfield, IL, June 5, 1997.

Indiana Department of Environmental Management (IDEM). 1996. *Resource Guide.* Voluntary Remediation Program, Office of Environmental Management, Indianapolis, IN, July 1996.

Industrial Site Recovery Act (ISRA). 1993. *Section 1-22 of the Public Law-Senate Bill 1070*. Signed June 16, 1993.

International Agency for Research on Cancer (IARC). 1990. IARC monographs on the evaluation of carcinogenic risks to humans, chromium, nickel and welding, World Health Organization. Geneva, Switzerland.

International Toxicology Estimates of Risk (ITER). 1998. *Documentation Package: Chromium VI Inhalation Reference Concentration*. Toxicology Excellence for Risk Assessment (TERA), Cincinnati, OH, February 2, 1998.

Ivankovic, S., and Preussmann, R. 1975. Absence of toxic and carcinogenic effects after administration of high doses of chromic oxide pigment in subacute and long-term feeding experiments in rats. *Food Cosmet. Toxicol.* 13: 347–351.

Jaeger, J., and Pelloni, E. 1950. Test epicutanes aux bichromates postifs danes eczema au ciment. *Dermatologica* 100: 207–216.

James, B. E. 1994. Hexavalent chromium solubility and reduction in alkaline soils enriched with chromite ore processing residue. *J. Environ. Qual.* 23(2): 227–233.

James, B. R., and Bartlett, R. J. 1983a. Behavior of chromium in soils: V. Fate of organically complexed Cr(III) added to soil. *J. Environ. Qual.* 12(2): 169–172.

James, B. R., and Bartlett, R. J. 1983b. Behavior of chromium in soils: VII. Adsorption and reduction of hexavalent forms. *J. Environ. Qual.* 12(2): 177–181.

Johansson, A., Robertson, B., Curstedt, T., and Camner, P. 1986a. Rabbit lung after inhalation of hexa- and trivalent chromium. *Environ. Res.* 41: 110–119.

Johansson, A., Wiernik, A., and Jarstrand, A. C. P. 1986b. Rabbit alveolar macrophages after inhalation of hexa- and trivalent chromium. *Environ. Res.* 39: 372–385.

Klassen, C. D., Amdur, M. A., and Doull, J. 1996. *Cassarett and Doull's Toxicology: The Basic Science of Poisons*. McGraw-Hill, New York.

Korallus, U., Ulm, K., and Steinmann-Steiner-Haldenstaett, W. 1993. Bronchial carcinoma mortality in the German chromate-producing industry: The effects of process modification. *Int. Arch. Occup. Environ. Health* 65: 171–178.

Levin, H. M., Brunner, M. J., and Rattner, H. 1959. Lithographer's dermatitis. *J. Am. Med. Assoc.* 169: 566–569.

Lindberg, E., and Hedenstierna, L. 1983. Chrome plating: Symptoms, findings in the upper airways, and effects on lung function. *Arch. Environ. Health* 38(6): 367–374.

Lindberg, E., and Vesterberg, O. 1983. Urinary excretion of proteins in chrome-platers, exchromeplaters and referents. *Scand. J. Work Environ. Health* 9: 505–510.

Lioy, P. J., Freeman, N. C. G., Wainman, T., Stern, A. H., Boesch, R., Howell, T., and Shupack, S. I. 1992. Microenvironmental analysis of residential exposure to chromium-laden wastes in and around New Jersey homes. *Risk Anal.* 12(1): 287–299.

MacKenzie, R. D., Byerrum, R. U., Decker, C. F., Hoppert, C. A., and Langham, R. F. 1958. Chronic toxicity studies. *AMA Arch. Ind. Health.* 18: 232–234.

Malsch, P. A., Proctor, D. M., and Finley, B. L. 1994. Estimation of a chromium inhalation reference concentration using the benchmark dose method: A case study. *Reg. Toxicol. Pharm.* 20: 58–82.

Mancuso, T. F. 1975. *Consideration of chromium as an industrial carcinogen*. Paper presented at the International Conference on Heavy Metals in the Environment, Toronto, Ontario, Canada.

Michigan Department of Environmental Protection (MDNR). 1995. *Interim Specific Remedial Action Plan.* Environmental Response Division, June 5, 1995. Lansing, MI.

Morris, G. E. 1958. Chrome dermatitis, a study of the chemistry of shoe leather with particular reference to basic chromic sulfate. *AMA Arch. Dermatol.* 78: 612–618.

Mundt, K. 1997. Carcinogenicity of trivalent and hexavalent chromium. *OEM Rept.* 11(11): 95–100.

National Institute for Occupational and Safety Health (NIOSH). 1975. *Occupational Exposure to Chromium (VI).* NIOSH, Cincinnati, OH.

Nemery, B. 1990. Metal toxicity and the respiratory tract. *Eur. Respir. J.* 3: 202–219.

Nethercott, J., Paustenbach, D., Adams, R., Fowler, J., Marks, J., Morton, C., Taylor, J., Horowitz, S., and Finley, B. 1994. A study of chromium induced allergic contact dermatitis with 54 volunteers: Implications for environmental risk assessment. *Occup. Environ. Med.* 51: 371–380.

Nethercott, J., Paustenbach, D., and Finley, B. 1995. Letter to the Editor. A study of chromium induced allergic contact dermatitis with 54 volunteers: Implications for environmental risk assessment. Correspondence *Occup. Environ. Med.* October 1995: 701–702.

Nettesheim, P., Hanna, M. G., Doherty, D. G., Newell, R. F., and Hellman, A. 1971. Effect of calcium chromate dust, influenza virus, and 100 R whole-body X-radiation on lung tumor incidence in mice. *J. Natl. Cancer Inst.* 47: 1129–1144.

Nettesheim, P., and Szakai, K. 1972. Morphogenesis of alveolar bronchiolization. *Lab. Invest.* 26(2): 210–219.

New Jersey Department of Environmental Protection (NJDEP). 1995. *Basis and Background for Soil Cleanup Criteria for Trivalent and Hexavalent Chromium.* NJDEP, Site Remediation Program, Trenton, NJ.

New Jersey Department of Environmental Protection (NJDEP). 1998. *Summary of the Basis and Background of the Soil Cleanup Criteria for Trivalent and Hexavalent Chromium.* NJDEP, Trenton, NJ.

New Jersey Department of Environmental Protection (NJDEP). 1989. *Preliminary Soil Cleanup Standards.* NJDEP, Trenton, NJ.

New Jersey Department of Health (NJDOH). 1989. *Medical Evaluation of Children and Adults of the Whitney Young Jr. School, Jersey City, New Jersey.* NJDOH, Environmental Health Service, Division of Occupational and Environmental Health, Trenton, NJ.

Nichols, A. L., and Zechhauser, R. J. 1988. The perils of prudence: How conventional risk assessments distort regulations. *Reg. Toxicol. Pharmacol.* 8: 61–75.

North American Contact Dermatitis Group (NACDG). 1989. Preliminary studies of the TRUE-test patch test system in the United States. *J. Am. Acad. Dermatol.* 21: 841–843.

North American Contact Dermatitis Group (NACDG). 1997. Prevalence estimates for Cr(VI) allergy 1992–1996. Correspondence from Jim Fowler to Deborah Proctor, Louisville, KY.

Park, H. S., Yu, J. H., and Jung, K. S. 1994. Occupational asthma caused by chromium. *Clin. Exp. Allergy* 24(7): 676–681.

Paustenbach, D. 2000. The practice of exposure assessment: A state-of-art review. *J. Toxicol. Environ. Health Part B* 3: 179–291.

Paustenbach, D. J. 1989a. Important recent advances in the practice of health risk assessment: Implications for the 1990s. *Reg. Toxicol. Pharmacol.* 10: 204–243.

Paustenbach, D. J. 1989b. Health risk assessments: Opportunities and pitfalls. *Col. J. Environ. Law* 14(2): 379–410.

Paustenbach, D. J., Finley, B. L., and Long, T. F. 1997. The critical role of house dust in understanding the hazards posed by contaminated soils. *Int. J. Toxicol.* 16: 339–362.

Paustenbach, D. J., Meyer, D. M., Sheehan, P. J., and Lau, V. 1991a. An assessment and quantitative uncertainty analysis of the health risks to workers exposed to chromium contaminated soils. *Toxicol. Ind. Health* 7(3): 159–196.

Paustenbach, D. J., Rinehart, W. E., and Sheehan, P. J. 1991b. The health hazards posed by chromium-contaminated soils in residential and industrial areas: Conclusions of an expert panel. *Reg. Toxicol. Pharm.* 13: 195–222.

Paustenbach, D. J., Rinehart, W. E., and Sheehan, P. J. 1991c. The health hazards posed by chromium-contaminated soils in residential and industrial areas: Conclusions of an expert panel. *Reg. Toxicol. Pharm.* 13: 195–222.

Paustenbach, D. J., Sheehan, P. J., Paull, J. M., Wisser, L. M., and Finley, B. L. 1992. Review of the allergic contact dermatitis hazard posed by chromium-contaminated soil: Identifying a "safe" concentration. *J. Toxicol. Environ. Health* 37: 177–207.

Petura, J. C., James, B. R., and Vitale, R. J. 1998. Chromium(VI) in soils. In R. A. Meyers (Ed.), *Encyclopedia of Environmental Analysis and Remediation*. Wiley, New York, pp. 1142–1158.

Pirila, V. 1954. On the role of chrome and other trace elements in cement eczema. *Acta Dermato-Venereol* 34: 136–143.

Polak, L., Turk, J. L., and Frey, J. R. 1973. Studies on contact hypersensitivity to chromium compounds. *Progr. Allergy* 17: 145–266.

Proctor, D., Fredrick, M. M., Scott, P. K., Paustenbach, D. J., and Finley, B. L. 1998. The prevalence of chromium allergy in the United States and its implication for setting soil cleanup: A cost-effectiveness case study. *Reg. Toxicol. Pharm.* 28: 27–37.

Proctor, D. M., Panko, J. M., Finley, B. L., Butler, W. J., and Barnhart, R. J. 1999. Commentary—Need for improved science in standard setting for hexavalent chromium. *Reg. Toxicol. Pharm.* 29: 99–101.

Public Health Service (PHS). 1953. *Health of Workers in Chromate Producing Industry. A Study.* U.S. Government Printing Office, Washington, DC.

Scott, P., and Proctor, D. 1997. Evaluation of 10% minimum elicitation threshold for Cr(VI)-induced allergic contact dermatitis using benchmark dose methods. In D. M. Proctor, B. Finley, M. Harris, D. Paustenbach, and D. Rabbe (Eds.), *Chromium in Soil: Perspectives in Chemistry, Health, and Environmental Regulation*, Vol. 6. Lewis, New York, pp. 707–732.

Scott, P. K., Finley, B., Sung, H.-M., Schulze, R. H., and Turner, D. B. 1997a. Identification of an accurate soil suspension/dispersion modeling method for use in estimating health-based soil cleanup levels of hexavalent chromium in chromite ore processing residues. *J. Air Waste Manag. Assoc.* 47: 753–765.

Scott, P. K., Harris, M., Finley, B., and Rabbe, D. 1997b. Background air concentrations of Cr(VI) in Hudson County, New Jersey; implications for setting health-based standards for Cr(VI) in soil. *J. Air Waste Manag. Assoc.* 47: 592–600.

Sheehan, P., Ricks, R., Ripple, S., and Paustenbach, D. 1992. Field evaluation of a sampling and analytical method for environmental levels of airborne hexavalent chromium. *Am. Ind. Hyg. Assoc. J.* 53(1): 57–68.

Sheehan, P. J., Meyer, D. M., Sauer, M. M., and Paustenbach, D. J. 1991. Assessment of the human health risks posed by exposure to chromium-contaminated soils. *J. Toxicol. Environ. Health* 32: 161–201.

Shirakawa, T., and Morimoto, K. 1996. Brief reversible brochospasm resulting form bichromate exposure. *Arch. Environ. Health* 51(3): 221–226.

Skaggs, M. M., and Harris, M. 2001. A management trend: Towards special remedial organizations. Paper presented at the SPE/EPA/DOE Exploration and Production Environmental Conference, Society of Petroleum Engineers, San Antonio, TX.

Skog, E., and Wahlberg, J. E. 1969. Patch testing with potassium dichromate in different vehicles. *Arch. Derm.* 99: 697–700.

Snyder, C. A., Sellakumar, A., and Waterman, S. 1997. An assessment of the tumorigenic properties of a Hudson County soil sample heavily contaminated with hexavalent chromium. *Arch. Environ. Health* 52(3): 220–226.

Sorahan, T., Burges, D. C. L., and Waterhouse, J. A. H. 1987. A mortality study of nickel/chromium platers. *Br. J. Ind. Med.* 44: 250–258.

Stapinski, H. A. 1993. Tainted gold: Chromium contamination in Hudson County. In H. E. Sheehan and R. P. Wedeenpp (Eds.), *Toxic Circles, Environmental Hazards from the Work Place into the Community*. Rutgers University Press. New Brunswick, NJ, pp. 202–230.

Steffee, C. H., and Baetjer, A. M. 1965. Histopathologic effects of chromate chemicals. *Arch. Environ. Health* 11: 66–75.

Stern, A. H. 1990. Soil Cleanup Level Development for Chromite One Processing Residue. Division of Science and Research, NJDEP, Trenton NJ.

Stern, A. H., Freeman, N. C., Pleban, P., Boesch, R. R., Wainman, T., Howell, T., Shupack, S. I., Johnson, B. B., and Lioy, P. 1992. Residential exposure to chromium waste—urine biological monitoring in conjunction with environmental exposure monitoring. *Environ. Res.* 58: 147–162.

Taylor, F. H. 1966. The relationship of mortality and duration of employment as reflected by a cohort of chromate workers. *Am. J. Public Health* 56: 218–229.

Tengs, T. O., Adams, M. E., Piskin, J. S., Satran, B. G., Siegel, J. E., Weinstein, M. C., and Graham, J. D. 1995. Five-hundred life-saving interventions and tier cost-effectiveness. *Risk Anal.* 15(3): 369–390.

Tokyo Bureau of Hygiene 1989. Report concerning the effect of chromium in a health survey (10 year survey). Okayama University Medical School Tokyo, Japan.

Upadhye, M. R., and Maibach, H. I. 1992. Influence of area of application of allergen on sensitization in contact dermatitis. *Contact Dermatitis* 27: 281–286.

U.S. Census Bureau. 2002. Land area, population, and density for states and counties, 1990. State and Counting Quick Facts. http://www.census.gov. Accessed on 1/14/02.

U.S. Environmental Protection Agency (USEPA). 1984. *Health Assessment Document for Chromium*. USEPA, Research Triangle Park, NC.

U.S. Environmental Protection Agency (USEPA). 1986a. *Superfund Public Health Evaluation Manual*. USEPA, Office of Emergency and Remedial Response. USEPA Publication 9285.7-01. Washington, DC.

U.S. Environmental Protection Agency (USEPA). 1986b. *Determination of Stable Valence States of Chromium in Aqueous and Solid Waste Matrices—Experimental Verification of Chemical Behavior*. USEPA, Cincinnati, OH.

U.S. Environmental Protection Agency (USEPA). 1989. *Risk Assessment Guidance for Superfund*, Vol. 1: *Human Health Evaluation Manual (Part A), Interim Final*. Office of Emergency and Remedial Response, USEPA, Washington, DC.

U.S. Environmental Protection Agency (USEPA). 1990. *National Oil and Hazardous Substances Pollution Contingency Plan*. USEPA. 40 CFR Part 300 Appeared on 2-6-90.

U.S. Environmental Protection Agency (USEPA). 1991a. *Health Effects Assessment Summary Tables*. USEPA, Washington, DC.

U.S. Environmental Protection Agency (USEPA). 1991b. *Risk Assessment Guidance for Superfund*, Vol. 1: *Human Health Evaluation Manual (Part B, Development of Risk-based Preliminary Remediation Goals)*. Office of Emergency and Remedial Response, USEPA, Washington, DC.

U.S. Environmental Protection Agency (USEPA). 1991c. *Risk Assessment Guidance for Superfund*, Vol. 1: *Human Health Evaluation Manual (Part C, Risk Evaluation of Remedial Alternatives)*. Office of Emergency and Remedial Response, USEPA, Washington, DC.

U.S. Environmental Protection Agency (USEPA). 1991d. *User's Guide for the Fugitive Dust Model (FDM), User Instructions*. Office of Air Quality Planning and Standards, USEPA, Seattle, WA.

U.S. Environmental Protection Agency (USEPA). 1992. *Guidance for Data Usability in Risk Assessment. Part A*. Office of Emergency and Remedial Response, USEPA, Washington, DC.

U.S. Environmental Protection Agency (USEPA). 1994. *Test Methods for Evaluating Solid Wastes, Physical/Chemical Methods*, 3rd ed. SW-846 2nd update. Office of Solid Waste and Emergency Response, USEPA, Washington, DC.

U.S. Environmental Protection Agency (USEPA). 1995a. Compilation of Air Pollutant Emission Factor. Volume 1: Stationary Point and the Area Source. 4th Edition. PB86-124906. Washington, DC.

U.S. Environmental Protection Agency (USEPA). 1995b. *User's Guide for the Industrial Source Complex (ISC3) Dispersion Modes*. Office of Air Quality Planning and Standards, USEPA, Research Triangle Park, NC.

U.S. Environmental Protection Agency (USEPA). 1996. *Soil Screening Guidance: User's Guide*. Office of Solid Waste and Emergency Response, USEPA, Washington, DC.

U.S. Environmental Protection Agency (USEPA). 1997. 40 CFR 50 Vol. 62(138): 38651–38670.

U.S. Environmental Protection Agency (USEPA). 1998a. *Chromium(III), Insoluble Salts*. USEPA, Washington, DC.

U.S. Environmental Protection Agency (USEPA). 1998b. *Toxicological Review of Hexavalent Chromium, in Support of Summary Information on the Integrated Risk Information System (IRIS)*. USEPA, Washington, DC.

U.S. Environmental Protection Agency (USEPA). 1999a. *Exposure Factors Handbook*. Office of Research and Development, USEPA, Washington, DC.

U.S. Environmental Protection Agency (USEPA). 1999b. *Risk Assessment Guidance for Superfund*, Vol. 3 part A: *Process for Conducting Probabilistic Risk Assessment— Draft*. Office of Emergency and Remedial Response, USEPA, Washington, DC.

Vitale, R. J., Mussoline, G. R., Petura, J. C., and James, B. R. 1994. Hexavalent chromium extraction from soils: Evaluation of an alkaline digestion method. *J. Environ. Qual.* 23: 1249–1256.

Vitale, R. J., Mussoline, G. R., Petura, J. C., and James, B. R. 1995. Hexavalent chromium quantification in soils: An effective and reliable procedure. *Am. Environ. Laborat.* 7(3): 1–10.

Wang, Z. P., Larsson, K., Malmberg, P., Sjogren, B., Hallberg, B. O., and Wrangskog, K. 1994. Asthma, lung function, and bronchial responsiveness in welders. *Am. J. Ind. Med.* 26(6): 741–754.

Watt, G., and Eizaguirre, D. 1991. *Assessment of the Risk to Human Health from Land Contaminated by Chromium Waste*. Department of Public Health, Greater Glasgow Health Board, Glasgow.

Winston, J. R., and Walsh, E. N. 1951. Chromate dermatitis in railroad employees working with diesel locomotives. *J. Am. Med. Assoc.* 147: 1133–1134.

World Health Organization (WHO). 1988. *Chromium*. WHO, Geneva.

Zhang, J., and Li, S. 1997. Cancer mortality in a Chinese population exposed to hexavalent chromium in water. *JOEM* 39(4): 315–319.

SECTION E
Characterizing Exposure to Air Contaminants

10

Quantification of Variability and Uncertainty: Case Study of Power Plant Hazardous Air Pollutant Emissions

H. CHRISTOPHER FREY and RANJIT BHARVIRKAR

North Carolina State University, Raleigh, North Carolina

10.1 INTRODUCTION

Electric power plants are a source of hazardous air pollutant (HAP) emissions, as indicated in a U.S. Environmental Protection Agency (EPA) interim report to Congress on an assessment of human risks due to HAP emissions from power plants (EPA, 1996). The risk assessment includes emissions estimation, air quality modeling, exposure assessment, and dose–response assessment. While electric utilities are currently not subject to the same HAP regulations as other emission sources, the results of the risk analysis may motivate future regulation. Previous analyses, such as by Rubin et al. (1993), have used one-dimensional probabilistic methods to characterize statistical variation in predicted emission rates of air toxics from coal-fired power plants. Frey and Rhodes (1996) and Frey (1998) presented a two-dimensional probabilistic analysis in which both uncertainty and variability in power plant HAP emissions were quantified. In this case study, we focus on refinement of the case studies presented by Frey and Rhodes (1996) and Frey (1998) to illustrate the methods for dealing with variability and uncertainty in environmental models using a two-dimensional approach to probabilistic simulation. This case study focuses on predicting variability and uncertainty in emissions for a single facility for nine HAPs and two different averaging times. The nine HAPs chosen for this case study are arsenic (As), beryllium (Be), cobalt (Co), chromium (Cr), lead (Pb), manganese (Mn), mercury (Hg), nickel (Ni), and selenium (Se). A key methodological aspect of this case study, which differs from previous work, is rigorous analysis of data sets containing nondetected measurements.

Human and Ecological Risk Assessment: Theory and Practice, Edited by Dennis J. Paustenbach
ISBN 0-471-14747-8 © 2002 John Wiley & Sons, Inc.

The case study here is focused upon the emissions of nine HAPs from coal-fired power plants with dry-bottom front-fired wall-fired boilers. This scenario was chosen because (1) this type of power plant comprises a significant portion of installed generating capacity; (2) these nine HAPs are of health and policy concern; and (3) a limited amount of data were readily available for the HAP concentration in coal, the boiler partitioning factor, and the fabric filter partitioning factor for all of the nine HAPs in this type of power plant. This case study illustrates methods for quantifying variability and uncertainty for small data sets (e.g., $n = 3$), for different averaging times (i.e., 3-day averages and annual averages), and for situations involving censored (e.g., nondetected) data.

10.2   COMPILATION OF DATABASE FOR THE CASE STUDY

Many HAPs are present as trace species in coal. Emissions testing of HAPs at power plants has been conducted primarily by the Electric Power Research Institute (EPRI) and the U.S. Department of Energy (DOE). These tests have included characterization of coal properties and inlet and outlet measurements of trace species for each major device of the power plant (e.g., boilers, fabric filters, and others). Figure 10.1 shows a schematic diagram of the power plant considered in this case study. The major devices in this power plant include the boiler and the fabric filter.

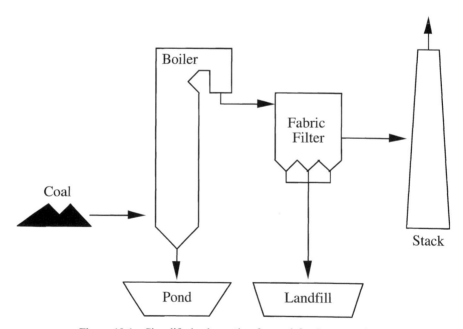

Figure 10.1   Simplified schematic of a coal-fired power plant.

The data for the coal concentration of the HAPs was obtained from the synthesis report prepared by EPRI (EPRI, 1994). The data for the boiler and fabric partitioning factors was obtained from an EPA summary submitted to the Congress as part of an interim report on utility air toxics emissions (EPA, 1996). A power plant emission model, developed by Frey (1998), was used to estimate emissions for a 3-day averaging period.

The model inputs include plant capacity, heat rate (which is inversely related to plant efficiency), coal heating value, 3-day average capacity utilization factor, concentration of the HAPs in coal, and the partitioning factors of the HAPs for the boiler and fabric filters.

$$m = \left(\frac{1000C\ \text{HR}}{\text{HV}}\right) \times 72c_f\left(\frac{f_c}{10^6}\right)f_B f_F \qquad (10.1)$$

where m = mass emission rate of a HAP, lb/3-day averaging period

   c_f = capacity utilization factor (average load over 3 days/maximum possible load)

   C = plant capacity, MW (500 MW is assumed)

   HR = net plant heat rate (Btu/kWh)

   HV = heating value of the coal (Btu/lb)

   f_c = weight concentration of the HAP in the coal (ppm$_w$)

   f_B = partitioning factor for the HAP for the plant boiler (lb in flue gas/lb entering the boiler)

   f_F = partitioning factor for the HAP for the plant fabric filter (lb in flue gas/lb entering the fabric filter system)

The analysis is done with respect to the HAP emissions flow rate. The measurements of air toxic flow rates within the power plant and the emission flow rate from the stack are typically made each day over a 3-day period and averaged. Thus, the reported data represent 3-day averages. The model presented in Eq. (10.1) was used to estimate the emissions of HAPs for a 3-day period. In one year, there are approximately 122 three-day periods. Therefore, variability in emissions over the course of the year can be simulated by a frequency distribution of 122 three-day average values. For each set of 122 three-day periods, an estimate of the annual average emissions can be made.

10.3 DEVELOPMENT OF PROBABILITY DISTRIBUTION MODELS

The case study is developed for a 500-MW plant, with a nominal heat rate of 9780 Btu/kWh. The input assumptions for variability in heat rate, coal heating value, and plant capacity factor from one 3-day period to another are the same as used by Frey and Rhodes (1996). Probability distribution models are developed for the pollutant concentration in coal, boiler partitioning factor, and fabric filter partitioning factor data sets for each of the nine HAP trace species. The case study is based upon a front-fired wall-fired boiler with fabric filter for particulate matter emission control. Parameters for all of the fitted probability distributions were estimated by using the method of maximum-likelihood estimation (MLE).

10.3.1   Parameter Estimation Method

Maximum-likelihood estimation involves the selection of parameter values to specify a parametric probability distribution that is most likely to have yielded the observed random data set (e.g., Cohen and Whitten, 1988). A likelihood function for independent samples is defined as the product of the probability density function of the probability distribution evaluated at each of the random data set sample values and conditioned upon assumed values of the distribution parameters. For a continuous random variable, for which independent samples have been obtained, the likelihood function is

$$L(\theta_1, \theta_2, \ldots, \theta_k) = \prod_{i=1}^{n} f(x_i \mid \theta_1, \theta_2, \ldots, \theta_k) \qquad (10.2)$$

where $\theta_1, \theta_2, \ldots, \theta_k$ = parameters of the parametric probability distribution model

k = number of parameters for the parametric probability distribution model

x_i = values of the random variable, for $i = 1, 2, \ldots, n$

n = number of data points in the data set

f = probability density function

Usually, k is equal to 2 (corresponding to a two-parameter distribution) or 3 (corresponding to a three-parameter distribution). The values of the parameters that maximize the likelihood function are sometimes determined analytically using standard techniques of calculus. In many cases, it is more convenient to work with a log transformation of the likelihood function, referred to as a log-likelihood function. When an analytical solution is not readily available, the maximum-likelihood parameter estimates can be found using numerical techniques such as the Newton–Raphson method or nonlinear programming optimization.

The log-likelihood functions for the normal, log-normal, gamma, Weibull, and beta distributions are shown in Table 10.1. The number of data points is n and each data point is represented as x_i, where i takes the values 1 through n.

TABLE 10.1 Expressions for Log-Likelihood Functions for Data Belonging to Various Probability Distribution Models

Name of Distribution[a]	Log-Likelihood Function
Normal $(\mu = $ mean, $\sigma = $ standard deviation)	$J(\mu,\sigma) = -n \ln \sigma - \dfrac{n}{2} \ln(2\pi) - \displaystyle\sum_{i=1}^{n} \left\{ \dfrac{(x_i - \mu)^2}{2\sigma^2} \right\}$
Log-normal $(\mu = $ mean, $\sigma = $ standard deviation, of log-transformed data)	$J(\mu,\sigma) = -n \ln \sigma - \dfrac{n}{2} \ln(2\pi)$ $- \displaystyle\sum_{i=1}^{n} \left\{ \dfrac{(\ln(x_i) - \mu)^2}{2\sigma^2} \right\}$
Gamma $(\alpha = $ shape, $\beta = $ scale, parameters)	$J(\alpha,\beta) = -n\{\alpha \ln(\beta) + \ln[\Gamma(\alpha)]\}$ $+ \displaystyle\sum_{i=1}^{n} \left\{ (\alpha - 1) \ln(x_i) - \dfrac{x_i}{\beta} \right\}$
Weibull $(\alpha = $ shape, $\beta = $ scale, parameters)	$J(\alpha,\beta) = -n \ln\left(\dfrac{\alpha}{\beta}\right)$ $+ \displaystyle\sum_{i=1}^{n} \left\{ (\alpha - 1) \ln\left(\dfrac{x_i}{\beta}\right) - \left(\dfrac{x_i}{\beta}\right)^{\alpha} \right\}$
Beta $(\alpha = $ shape, $\beta = $ scale, parameters)	$J(\alpha,\beta) = -n \ln\left\{ \dfrac{\Gamma(\alpha)\Gamma(\beta)}{\Gamma(\alpha+\beta)} \right\} + \displaystyle\sum_{i=1}^{n} \{(\alpha - 1) \ln(x_i)$ $- (\beta - 1) \ln(1 - x_i)\}$

[a] Parameter values are different for each type of distribution even though the same symbol may be used to represent parameters of different distributions.

10.3.2 The Need for Improved Methods for Dealing with Non-Detects

A recurring difficulty encountered in investigations of many substances, such as hazardous air pollutant emissions from power plants, is that a substantial portion of the measurements are below the limits of detection for a given sampling and analytical chemistry procedure. Sometimes the limit of detection is referred to as the "sensitivity" of the measurement technique or instrument. Because of random errors inherent in measurements, it is not possible to distinguish a non-detected value from zero. However, this does not mean that the true value of the quantity being measured is equal to zero. This means that the true value of the quantity being measured is unknown but is somewhere between zero and the limit of detection, inclusive.

Measurements below the detection limit (DL) are often reported as "less than detection limit" or non-detects (NDs) rather than as numerical values. It is common in many default procedures for non-detected measurements to be assigned values of zero, the detection limit, or one-half of the detection limit when doing calculations. Each of these three default methods has potentially serious drawbacks.

To illustrate the drawbacks of the default approaches, consider the simple case of attempting to calculate a sample mean from a data set that contains two non-detect values and three values above the limit of detection. Let this data set be $X = \{ND, ND, 1.1, 1.3, 1.8\}$ and let the limit of detection be 1.0. If the non-detects are assigned a value of zero, the mean is estimated to be 0.8. If the non-detects are assigned a value of the limit of detection, the mean is estimated to be 1.2. If the non-detects are assigned a value of one-half of the limit of detection, the estimated mean is 1.0. Which mean is correct?

Because the true value of a non-detect can be anywhere between zero and the limit of detection, it is apparent that assigning a value of zero for all non-detected values will bias the mean on the low side. Conversely, assigning a value of the limit of detection to all non-detected values will bias the mean on the high side. Only under special conditions would one-half of the detection limit produce an unbiased estimate of the mean or of any other statistic estimated from the data set. For example, if the distribution of true values was uniform between zero and the limit of detection, then non-detects would have an average value of one-half of the detection limit. However, if the distribution of values below the detection limit deviates from a uniform distribution, as is often the case, then on average non-detects will not have a value of one-half of the limit of detection. Therefore, the routine use of one-half of the limit of detection will typically produce biased estimates of the mean and of many other statistics.

The default approaches have been used for so many years because they are simple and easy to explain. However, they are always biased if either zero or the detection limit is assigned to non-detects, and they are typically biased if one-half of the detection limit is used. As computing power has increased, it becomes increasingly feasible to employ less biased and more statistically rigorous methods for making inferences from data sets that contain non-detects.

10.3.3 Parameter Estimation for Censored Data Sets

Data sets that contain non-detected values are often referred to as "censored" data sets. A key challenge of particular interest in this work is to fit parametric probability distributions to censored data sets and to make inferences regarding the mean and other statistics taking into account the nondetect data.

There have been extensive investigations of methods for estimating the parameters of probability distributions fitted to a censored data set (David, 1981). Haas and Scheff (1990) have compared various methods described in the environmental science and engineering literature for estimation of means of censored data belonging to normal distributions. In cases where the underlying distribution is known or can be reliably inferred, the MLE estimators have been found to be easier to compute and more accurate as compared to other methods (e.g.,

Newman et al., 1989). Comparative studies have also been conducted by Gilliom and Helsel (1986) with similar findings.

Cohen and Whitten (1988) have developed maximum-likelihood estimators applicable to censored data sets. These estimators were chosen for use in this work. The main reasons for selecting the MLE technique include its applicability to data sampled from various distributions and the ease with which it can be implemented in a computer program. The most general formulation of the likelihood function for a censored data set having multiple detection limits is

$$L(\theta_1, \theta_2, \ldots, \theta_k) = \prod_{i=1}^{n} f(x_i \mid \theta_1, \theta_2, \ldots, \theta_k) \left\{ \prod_{m=1}^{P} \left(\prod_{j=1}^{ND_m} F(DL_m \mid \theta_1, \theta_2, \ldots, \theta_k) \right) \right\}$$

(10.3)

where x_i = detected data point, where, $i = 1, 2, \ldots, n$

$\theta_1, \theta_2, \ldots, \theta_k$ = parameters of the distribution

ND_m = number of nondetects corresponding to detection limit DL_m, where $m = 1, 2, \ldots, P$

P = number of detection limits

f = probability density function

F = cumulative distribution function

Taking the natural logarithm on both sides of Eq. (10.3), the log-likelihood function is obtained as:

$$J(\theta_1, \theta_2, \ldots, \theta_k) = \sum_{i=1}^{n} f(x_i \mid \theta_1, \theta_2, \ldots, \theta_k)$$

$$+ \sum_{m=1}^{P} ND_m \ln\{F(DL_m \mid \theta_1, \theta_2, \ldots, \theta_k)\}$$ (10.4)

The method employed in this work for solving for MLE parameter estimates involves the use of nonlinear optimization to select parameter values that maximum the log-likelihood function.

10.3.4 Data Sets: Pollutant Concentrations in Coal

For three of the pollutants, arsenic, cobalt, and nickel, the data for concentration in coal contained at least one data point reported as a nondetected value. The detection limits for each of the nondetected measurements were also reported. Since the procedures used for measuring the coal concentration of the HAPs were site-specific, the detection limits vary from site to site. For the other six pollutants, there were no nondetected data values.

Two-parameter log-normal, gamma, and Weibull distributions were fitted to the pollutant coal concentration data sets for all pollutants using MLE, except

TABLE 10.2   Summary of Data Sets and Fitted Distributions for Variability in the Concentration of Hazardous Air Pollutant in Bituminous Coal

Coal Concentrations	Number of Detected Values	Number of Non-Detect Values [Detection Limit]	Fitted Distribution	α	β
Arsenic (As)	20	1 (DL = 1.89 ppmw)	Gamma	1.60	6.86
Beryllium (Be)	18	0	Gamma	3.21	0.47
Chromium (Cr)	21	0	Gamma	5.82	2.62
Cobalt (Co)	15	1 (DL = 0.5 ppmw)	Gamma	2.23	2.29
Lead (Pb)	19	0	Gamma	1.74	5.94
Manganese (Mn)	20	0	Gamma	3.06	9.36
Mercury (Hg)	20	0	Gamma	6.29	0.02
Nickel (Ni)	20	2 (DL_1 = 15.0 ppmw, DL_2 = 32.0 ppmw)	Gamma	3.76	4.45
Selenium (Se)	21		Gamma	0.37	7.96

for the nickel coal concentration data set. However, for the nickel coal concentration data set, only the gamma distribution could be fitted. The reason for this is that the MLE method was unable to provide parameter etimates for the log-normal and the Weibull distributions.

After visually comparing the empirical distribution functions and the fitted log-normal, gamma, and Weibull distributions, the two-parameter gamma distribution was selected to represent the coal concentrations of the HAPs. However, it should be noted that the fitted log-normal and Weibull distributions were not substantially different from the fitted gamma distribution. The results for the distributions fitted to the censored data sets are shown graphically. In Table 10.2 the parameters estimated for the gamma distributions fitted to the coal concentration data sets are given.

In Figure 10.2, a comparison of the empirical distribution function of the arsenic coal concentration data set and the fitted gamma distribution is shown.

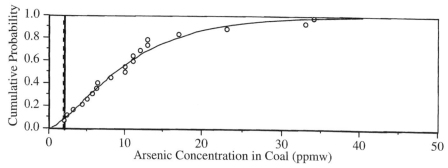

Figure 10.2   Three-day average arsenic concentration in coal fitted to a gamma distribution.

There was one data point below a detection limit of 1.89 ppm$_w$. The data set was plotted using the Hazen "plotting position" formula (Hazen, 1914):

$$F_X(x_i) = \Pr(X < x_i) = \frac{i - 0.5}{n} \qquad \text{for } i = 1, 2, \ldots, n \quad \text{and} \quad x_1 < x_2 < \cdots < x_n$$

$$(10.5)$$

where $\qquad\qquad i =$ rank of the data point when the data set is arranged in an ascending order

$\qquad\qquad\qquad n =$ number of data points

$x_1 < x_2 < \cdots < x_n =$ data points in the rank-ordered data set

$\qquad \Pr(X < x_i) =$ cumulative probability of obtaining a data point whose value is less than x_i

The detection limit associated with the nondetected measurement is less than the values of all of the detected data points in the data set. Therefore, even though the actual value of the nondetect is not known, it is known for certain in this case that it would be the smallest value in the data set and hence would always be assigned the first rank in the ordered data set. Consequently, the detected data points were assigned ranks starting from two. The dotted line shown in Figure 10.2 indicates the detection limit of 1.89 ppm$_w$ associated with the nondetected measurement. From Figure 10.2, the fit appears to be good since all of the data points are situated close to the fitted distribution.

For the cobalt data set shown in Figure 10.3 there was one data point below a detection limit of 0.5 ppm$_w$. Similar to the arsenic coal concentration data set, the detection limit associated with the nondetect is less than the values of all of the detected points in this data set. Therefore the plotting positions are calculated in a similar manner as described in the case of the arsenic coal concentration data set. The vertical dotted line shown in Figure 10.3 indicates the detection limit of 0.5 ppm$_w$ associated with the nondetected measurement. In this case, the fitted distribution tends to underestimate the concentration of cobalt in coal below the median value (equal to 5 ppm$_w$) and to overestimate the con-

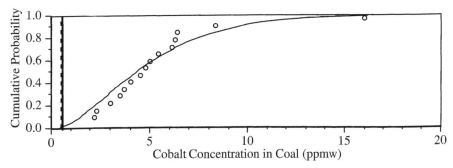

Figure 10.3 Three-day average cobalt concentration in coal fitted to a gamma distribution.

centration above the median value. It appears that the fit is substantially influenced by both the detection limit and the extreme data value of approximately 16 ppm$_w$. However, although the fit is not perfect, it is also not entirely a bad fit. The fit is reasonably good at the upper end of the distribution, and with a relatively small data set such as this, there is typically not a strong quantitative statistical basis for rejecting a fit such as this.

In the case of nickel, there were two data points reported to be below two separate detection limits of 15 and 32 ppm$_w$. However, unlike the arsenic or the cobalt coal concentration data sets, the values of several detected points in the data set were less than the two detection limits. Hence, the ranks of the detected points would vary according to the actual but unknown values of the two nondetects.

For example, consider the smallest detected data point in the data set, which has a value less than the lowest detection limit. Three ranks, corresponding to the following three possibilities, may be assigned to this data point:

1. The values of both of the nondetects are greater than the value of this detected data point, in which case the rank of this detected data point would be one.
2. The value of one nondetect is less than the value of this data point, and the value of the other nondetect is greater than the value of this detected point. In this case, the rank of this detected data point would be two.
3. The values of both of the nondetects are less than the value of this detected data point, in which case the rank of this detected data point would be three.

The three possible ranks (or plotting positions) for all of the detected data points whose values are less than 15.0 ppm$_w$ are calculated in a similar manner.

It is known for certain that the value of the nondetect associated with the detection limit of 15.0 ppm$_w$ will be less than the values of the detected data points that are greater than 15.0 ppm$_w$. In the case of detected data between 15 and 32.0 ppm$_w$, two ranks (or plotting positions) can possibly be assigned. For example, consider the data point having the value 17.0 ppm$_w$ from the nickel coal concentration data set. If the actual value of the nondetect associated with the detection limit of 32.0 ppm$_w$ happened to be less than 17.0 ppm$_w$, the rank of this data point would be 13. If the actual value of the nondetect happened to be greater than 17.0 ppm$_w$, the rank of this data point would be 12 instead of 13. Two possible ranks (or plotting positions) are calculated in a similar manner for all of the detected data points whose values were greater than or equal to 15.0 ppm$_w$ and less than 32.0 ppm$_w$.

For the detected data points whose values are greater than 32.0 ppm$_w$ only one rank (or plotting position) would be possible since it is known for certain that both of the nondetects have values less than 32.0 ppm$_w$.

All possible plotting positions for each of the detected data points in the nickel coal concentration data sets are shown in Figure 10.4. The cumulative

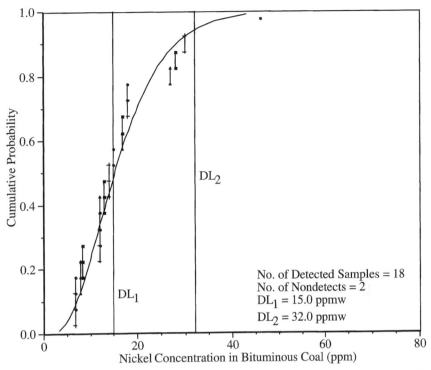

Figure 10.4 Three-day average nickel concentration in coal fitted to a gamma distribution.

distribution function (CDF) for the fitted gamma distribution is also presented in Figure 10.4. The fit generally appears to be good. In the nickel coal concentration data set, there are some data points having the same concentration value. Specifically, there are three data points having a value of 12.0 ppm, two data points having a value of 17.0 ppm, and two data points having a value of 18.0 ppm. Hence, in Figure 10.4, an overlap of plotting positions is seen for the data points having the same concentration value.

For all of the other six pollutant coal concentration data sets, no nondetects were reported. The fitted parametric distributions appear to agree well with the data for beryllium, chromium, lead, and mercury. In the case of manganese the fitted distribution appears to overestimate three data points and underestimate five data points significantly. In the case of selenium, the fitted distribution does not give a good fit. The majority of the data points are overestimated. The shape of the fitted distribution appears to be influenced significantly by a data point having the largest value in the data set. Although it is possible that this data point may be an outlier representing a measurement error, insufficient information is available to make a judgment regarding removal of this extreme value. Therefore, it was deemed more reasonable to include the data point and accept the poor fit rather than to discard the data point simply to obtain a better fit.

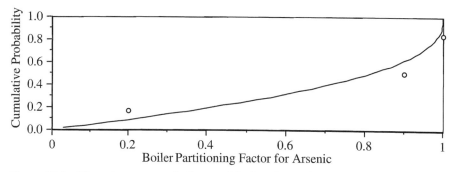

Figure 10.5   Three-day average boiler partitioning factor for arsenic fitted to a beta distribution.

10.3.5   Data Sets: Boiler Partitioning Factors

The number of data points in the boiler partitioning factor data sets was three for all but one of the nine trace species. For the cobalt boiler partitioning factor data set, only two data points were available. The boiler partitioning factors must have numerical values between zero and one. The two-parameter beta distribution was selected to represent all of the boiler partitioning factor data sets.

Figure 10.5 shows the comparison of the empirical distribution and the fitted beta distribution for the 3-day average boiler partitioning factor data set of arsenic. The beta distribution appears to be a good descriptor of the data. Of course, since the beta distribution has two parameters and there are three data points, it is expected to obtain a good fit since the degrees of freedom of the model is similar to the number of data points.

The fitted curve in Figure 10.5 is one possible distribution for variability in the 3-day average boiler partitioning factor for arsenic. If we assume that the data are a random sample of values, then there is a considerable uncertainty regarding the estimate of the population distribution from which the data were drawn because of the random sampling error associated with small data sets. Since there are only three data points, it is possible that other distributions aside from the fitted one would plausibly describe the variability in this partitioning factor. If it is assumed that there are no measurement errors and that the data are a random sample, then a bootstrap simulation can be used to characterize uncertainty associated with random sampling error. Although, in reality there are measurement errors, and the data may not be from a random sample of power plants or a representative sample of actual load conditions, it is still useful to evaluate whether random sampling may be a dominant source of uncertainty in the assessment. The assessment subsequently can be modified, if needed, to characterize additional sources of uncertainty. Section 10.4 describes how uncertainty may be characterized regarding the fitted distribution of variability.

The fits of a beta distribution to the data sets for beryllium, chromium, cobalt, manganese, and mercury were generally very good. The fits for lead, nickel, and

TABLE 10.3 Summary of Data Sets and Fitted Distributions for Variability in the Partitioning of Hazardous Air Pollutant in Dry-Bottom Wall-Fired Boilers

Coal Concentrations	Number of Detected Values	Fitted Distribution	α	β
Arsenic (As)	3	Beta	1.01	0.41
Beryllium (Be)	3	Beta	1.97	0.95
Chromium (Cr)	3	Beta	0.89	0.09
Cobalt (Co)	2	Beta	140.16	4.33
Lead (Pb)	3	Beta	0.38	0.14
Manganese (Mn)	3	Beta	2.99	2.61
Mercury (Hg)	3	Beta	0.74	0.15
Nickel (Ni)	3	Beta	1.35	0.17
Selenium (Se)	3	Beta	0.25	0.13

selenium were not as good because one data point is noticeably misestimated in each case. There are only two data points available for the cobalt boiler partitioning factor data set. The fit will always be good because two parameters were estimated for a data set containing two data points. In Table 10.3 the parameters estimated for the beta distributions fitted to the boiler partitioning factor data sets for all of the nine HAPs are shown.

10.3.6 Data Sets: Fabric Filter Partitioning Factors

The fabric filter partitioning factors can take values between zero and one. The two-parameter beta distribution was selected to represent all of the fabric filter partitioning factor data sets.

As an example, Figure 10.6 shows a comparison of the empirical distribution and the fitted beta distribution for the 3-day average fabric filter partitioning factor data set of arsenic. From Figure 10.6 we can see that the fit appears to be

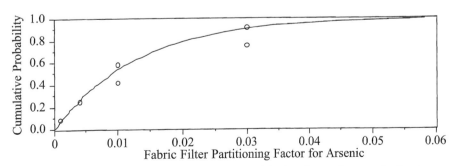

Figure 10.6 Three-day average fabric filter partitioning factor for arsenic fitted to a beta distribution.

good. Specifically, the fitted distribution passes nearly exactly through four of the six data points.

One might ask, what difference does it make whether one uses a parametric distribution fitted to a data set or an empirical distribution of the data themselves as input to a risk assessment model? This question was addressed during a U.S. Environmental Protection Agency sponsored workshop as reported in EPA (1999). In an empirical distribution, each data point is assigned equal probability, no values are interpolated between the data points, and no values are extrapolated beyond the range of the observed data. The workshop participants agreed that whether to use an empirical or parametric distribution is in large part a matter of preference of an analyst. Empirical distributions may be especially useful or appropriate in situations where data sets are large and are not accurately summarized using parametric distributions. Parametric distributions may be more realistic with small data sets in allowing for inferences beyond the range of observed data. With only a small amount of data, it is very unlikely that the true minimum or maximum value of the quantity would be sampled. Therefore, the use of an empirical distribution with a small data set is likely to too narrowly represent the possible range of values. Some adjustments to empirical distributions are sometimes made, such as "mixed empirical-parametric" (MEP) approaches, to allow for extrapolation beyond the range of observed data to characterize lower and upper tails of a distribution. However, as a matter of the analysts' preference in this work, parametric distributions that appear to be consistent with the observed data are used. To the extent that the parametric distributions agree with the observed range of data, one may have some comfort that the extrapolations to the tails of the distribution are at least plausible.

All of the fits for the other eight HAPs also appeared to be good. In Table 10.4 the parameters estimated for the beta distribution fitted to the fabric filter partitioning factor data sets for all of the nine HAPs are given.

TABLE 10.4 Summary of Data Sets and Fitted Distributions for Variability in the Partitioning of Hazardous Air Pollutant in Fabric Filters

Coal Concentrations	Number of Detected Values	Fitted Distribution	α	β
Arsenic (As)	3	Beta	1.01	77.08
Beryllium (Be)	6	Beta	0.62	39.26
Chromium (Cr)	5	Beta	0.16	2.61
Cobalt (Co)	5	Beta	1.28	247.55
Lead (Pb)	6	Beta	1.04	106.68
Manganese (Mn)	6	Beta	0.20	5.13
Mercury (Hg)	5	Beta	0.53	0.12
Nickel (Ni)	6	Beta	0.37	7.96
Selenium (Se)	4	Beta	0.55	1.27

10.4 QUANTIFICATION OF UNCERTAINTY IN THE FREQUENCY DISTRIBUTION FOR MODEL INPUTS

Three of the model inputs—(1) plant heat rate, (2) plant capacity factor, and (3) coal heating value—were assumed to be variable from one 3-day period to another with no uncertainty. This is because plant heat rate, capacity factor, and heating value, which are routinely measured and evaluated, are much better known than are the values of the concentrations of trace species in the coal or of the partitioning factors. Therefore, uncertainty was quantified regarding the fitted distributions for the coal concentrations and the partitioning factors. The method used to estimate uncertainty in these quantities is *bootstrap simulation*, as presented by Efron and Tishirani (1993) and as described by Frey and Rhodes (1996, 1998) and Frey and Burmaster (1999).

In bootstrap simulation, the fitted distribution is treated as an assumed population distribution. Synthetic data sets of the same sample size as the original data set are drawn at random from the fitted distribution, using Monte Carlo simulation. These synthetic data sets are referred to as *bootstrap samples*. The values of any statistic of interest, such as the mean, parameters of a parametric distribution, or percentiles, are calculated based upon the bootstrap sample. The statistics calculated in this manner are referred to as *bootstrap replications* of the statistic. The process is repeated many times to produce hundreds or thousands of bootstrap samples and corresponding bootstrap replications of selected statistics. The distribution of the bootstrap replications for a given statistic can be used to describe a sampling distribution. A sampling distribution is a probability distribution for a statistic. Confidence intervals are estimated based upon sampling distributions. Therefore, one common application of bootstrap simulation is for estimation of confidence intervals for statistics. An advantage of bootstrap simulation over conventional analytical techniques is that bootstrap simulation does not require restrictive assumptions (e.g., normality) as do many analytical techniques.

Bootstrap simulation was used to construct confidence intervals for the fitted parametric distributions. The confidence intervals allow for a comparison of the fitted distribution with the available data that takes into account the uncertainty in the fit associated with random sampling error. If the confidence intervals enclose a reasonable portion of the data, then the analyst has at least partial confirmation that the selected parametric distribution may be at least reasonable.

The number of bootstrap samples simulated was 500. For each bootstrap sample, a parametric distribution was fit, resulting in 500 bootstrap replications of the distribution parameters. Each alternative fitted distribution represents a plausible distribution from which the observed data may have been a random sample. To develop the confidence intervals for the CDF, 500 data points were generated to describe each of the 500 alternative fitted distributions. Hence, a total of 250,000 data points were generated for each variable and uncertain

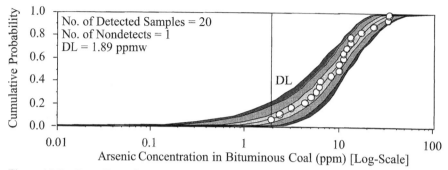

Figure 10.7   Two-dimensional plot showing uncertainty in the frequency distribution of 3-day average arsenic concentration in coal. (*Note: X* axis is logarithmic.)

model input. These values were sorted to yield probability bands for the fitted distribution that in turn enable characterization of confidence intervals for the CDF.

To illustrate the method, the results of the two-dimensional simulation for the pollutant coal concentration, boiler partitioning factor, and fabric filter partitioning factor for three (i.e., arsenic, cobalt, and nickel) of the nine HAP trace species precursors are presented in this section.

In Figure 10.7 for arsenic concentration in coal, 17 of the 20 detected data points lie within the 50 percent confidence interval and all of the detected data points are within the 95 percent confidence interval, indicating that the fitted probability distribution is adequate. The relative uncertainty is largest in the lower percentiles of the distribution because of the nondetected measurements. For example, the 95 percent confidence interval on the 10th percentile of variability is approximately 0.8 to 5 ppm, a factor of approximately 6, compared to a 95 percent confidence interval on the 90th percentile of variability of approximately 15 to 30 ppm, a factor of 2. Thus, the relative range of variability is wider at the lower end of the distribution although the absolute range of uncertainty is higher at the higher end.

In Figure 10.8, the results of the two-dimensional simulation for the arsenic boiler partitioning factor fitted distribution is compared with the empirical distribution. All of the three data points lie within the 90 percent confidence interval. The uncertainty is large because there are only three data points in the data set and because of the large variation in values among the three data points.

For the fabric filter partitioning factor data set for arsenic, shown in Figure 10.9, all except one data point is well within the 50 percent confidence interval range, indicating that the fitted distribution describes the data adequately. The uncertainty in the upper percentiles is relatively large. For example, the 95 percent confidence interval on the 95th percentile of variability ranges from 0.013 to 0.074. The 95 percent confidence interval on the fifth percentile of variability ranges from 6.2×10^{-5} to 0.007. Thus, the uncertainty in the upper percentiles is

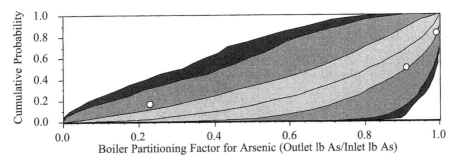

Figure 10.8 Two-dimensional plot showing uncertainty in the frequency distribution of 3-day average boiler partitioning factor for arsenic.

greater than the variability in the data set, which ranges from a minimum data value of 0.055 to a maximum data value of 0.03.

Figure 10.10 shows the two-dimensional plot for the cobalt coal concentration data set. The data points in the central portion of the distribution, within the 25th to 75th percentiles, lie within the 50 percent confidence interval. All of the data points are within the 90 percent confidence interval. Overall, the fit is reasonable but seems to be influenced by the detection limit at the lower end. The relative uncertainty in the lower percentiles is largest because of the non-detect.

In case of the boiler partitioning factor for cobalt, only two data points are available as shown in Figure 10.11. The standard deviation in the data set is very low and therefore the random sampling error is low in this case despite having only two data points. Consequently, the width of the confidence intervals on the frequency distribution is rather narrow compared with the two-dimensional plots for the boiler partitioning factors of other HAPs where three data points are available.

For the cobalt fabric filter partitioning factor data set shown in Figure 10.12, three of the five data points are within the 50 percent confidence interval and

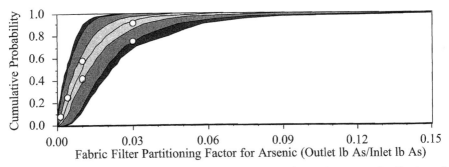

Figure 10.9 Two-dimensional plot showing uncertainty in the frequency distribution of 3-day average fabric filter partitioning factor for arsenic.

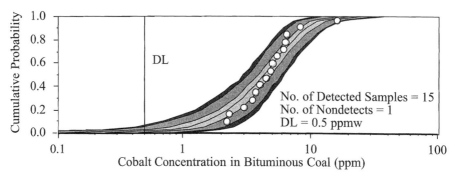

Figure 10.10 Two-dimensional plot showing uncertainty in the frequency distribution of 3-day average cobalt concentration in coal. (*Note:* X axis is logarithmic.)

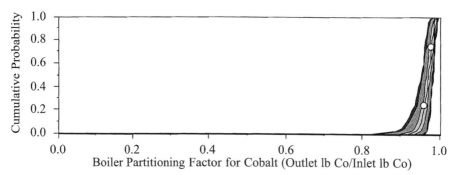

Figure 10.11 Two-dimensional plot showing uncertainty in the frequency distribution of 3-day average boiler partitioning factor for cobalt.

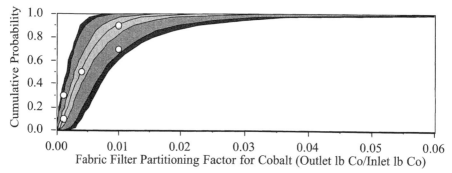

Figure 10.12 Two-dimensional plot showing uncertainty in the frequency distribution of 3-day average fabric filter partitioning factor for cobalt.

all of the points are within the 90 percent confidence interval. Therefore, it can be concluded that the fitted probability distribution describes the data set adequately. As with most positively skewed uncensored data sets, the confidence intervals increase with increasing percentiles of variability.

As previously noted there are two detection limits for the nickel coal concentration data set. The parameters for each bootstrap sample from the fitted distribution were estimated after processing each bootstrap sample using the algorithm shown in Figure 10.13. The processing is done to simulate the multiple nondetects corresponding to the multiple detection limits in the original data set. The assumptions made for developing the algorithm shown in Figure 10.13 include:

- Each detection limit (DL_i) is specific to site i and each data point has its own detection limit.
- Data points in the data set are independent of each other.
- All data points are random samples from the fitted probability distribution model developed in previous section.
- Whether the data point at site i are below DL_i or not is randomly simulated.
- Probability of a nondetect at site i is estimated based on the probability that data in a random sample of total size n are below the value of DL_i.

In the algorithm shown in Figure 10.13, each bootstrap sample of sample size equal to the sum of all detected and nondetected values of the original data set is initially arranged in an ascending order. A cumulative distribution function is developed for the bootstrap sample. For example, we describe the simulation of nondetects in the bootstrap sample for the nickel coal concentration data set. As noted previously, in this data set there are two nondetects corresponding to two detection limits: 15.0 and 32.0 ppm_w and there are 18 detected data points for a total of 20 measurements.

After the 20 bootstrap sample values are arranged in ascending order, the value of each detection limit is compared with all the values in the bootstrap sample. If all of the values in the bootstrap sample are greater than any detection limit, then a nondetect is not simulated for that particular detection limit. Thus, if all the values in the bootstrap sample are greater than 15.0 ppm_w in case of the nickel data set, a nondetect is not simulated for the detection limit of 15.0 ppm_w. If all of the values in the bootstrap sample are not less than 15.0 ppm_w, then the cumulative probability (CP) for 15.0 ppm_w is calculated from the cumulative distribution function of the bootstrap sample. Suppose that the cumulative probability for 15.0 ppm_w turns out to be 0.4. A random number P is generated from a uniform distribution that is bounded by zero and one. Suppose the value of P is 0.3. Therefore CP is greater than P and hence we simulate a nondetect by discarding one randomly selected value, which is less than 15.0 ppm_w, from the bootstrap sample. If the value of P is 0.6, then CP is less than P, and hence we do not simulate a nondetect. This process is repeated for all

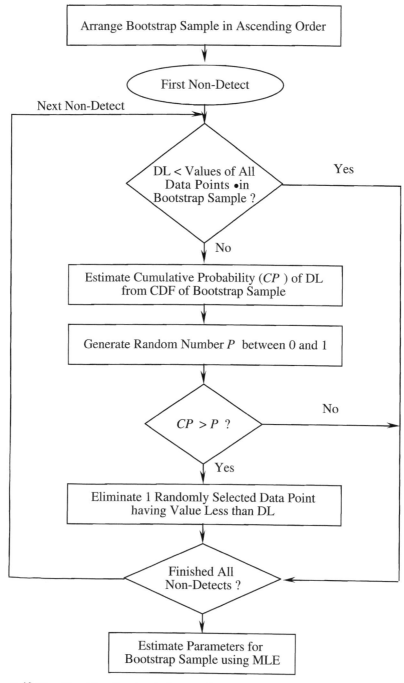

Figure 10.13 Algorithm for simulating nondetects in bootstrap samples generated from a multiply-censored parent population.

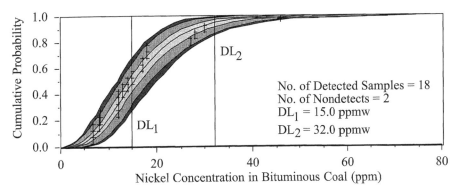

Figure 10.14 Two-dimensional plot showing uncertainty in the frequency distribution of 3-day average nickel concentration in coal.

of the detection limits to randomly simulate the presence of a nondetect in the bootstrap sample.

The results of the two-dimensional simulation for the nickel coal concentration data set are shown in Figure 10.14. The various possible plotting positions for each of the detected data points are also presented in Figure 10.14. All of the detected data points are within the 95 percent confidence intervals regardless of alternative possible plotting positions, indicating that the fitted distribution describes the data adequately.

In Figure 10.15, the results of the two-dimensional simulation for the nickel boiler partitioning factor data set are shown. All of the three data points are within the 95 percent confidence intervals. The 95 percent confidence interval range for uncertainty in the 5th percentile of variability varies from almost zero to one. The 95 percent confidence interval for the median value ranges from 0.75 to 1.0. In this case, the uncertainty in the median is similar to the range of the variability in the data set.

The results for the two-dimensional simulation of the nickel fabric filter partitioning factor data set are presented in Figure 10.16. All of the data points lie

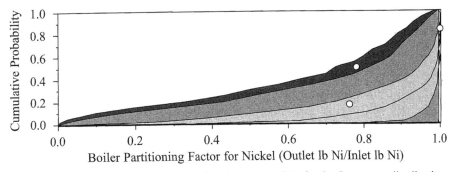

Figure 10.15 Two-dimensional plot showing uncertainty in the frequency distribution of 3-day average boiler partitioning factor for nickel.

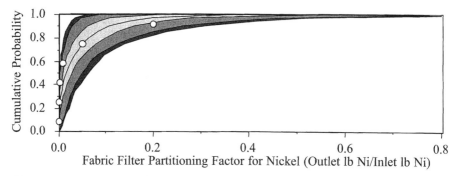

Figure 10.16   Two-dimensional plot showing uncertainty in the frequency distribution of 3-day average fabric filter partitioning factor for nickel.

within the 90 percent confidence interval, indicating that the fitted distribution describes the data adequately. The range of uncertainty in the upper percentiles is comparatively wide. For example, the 95 percent confidence interval for the 95th percentile of variability has a range of 0.04 to 0.4 lb outlet Ni/lb inlet Ni, which is wider than the range of values in the original data set.

As noted in many of the examples presented here, the range of uncertainty in the upper percentiles of the distribution is generally rather large. An important implication of these analyses is that it is typically not possible to make estimates regarding the upper percentiles of pollutant concentrations without a considerable amount of uncertainty.

10.5   ESTIMATING VARIABILITY AND UNCERTAINTY IN A MODEL OUTPUT

Frey (1998) had evaluated the dependencies among all HAPs for each of the three sets of inputs (i.e., coal concentration, boiler and fabric partitioning factors) and had reported that they were not significant. Hence, each input was treated as statistically independent of others.

For each variable and uncertain input, all of the 250,000 data points for each model input were stored in a two-dimensional array (of size 500 × 500) with one dimension representing variability and the other dimension representing uncertainty.

The model inputs for plant heat rate, plant capacity factor, and coal heating value were treated as variable only (i.e., one-dimensional) with distribution assumptions based on those used by Frey and Rhodes (1996). To represent the variability in the plant heat rate, 500 data points were generated based on a normal probability distribution model having a mean of 9780 Btu/kWh and a standard deviation of 580 Btu/kWh. Similarly, 500 data points were generated for the coal heating value based on a normal probability distribution model having a mean of 10,000 Btu/lb and a standard deviation of 180 Btu/lb. For the plant capacity factor, a previously compiled empirical data set containing 500 values was used. The 500 data points were stored in a one-dimensional array (of size 500). Each element of this one-dimensional array has one index associated with it referred to as the variability index (ranging from 1 to 500).

To quantify the variability and uncertainty in the model outputs, the inputs were propagated through the emissions model given in Eq. (10.1). The flow diagram illustrating the propagation of the two-dimensional and one-dimensional inputs through the model to obtain the two-dimensional output is shown in Figure 10.17. The inputs are classified into two categories: (1) both variable and uncertain input and (2) only variable inputs. The propagation of the inputs through the model involves the evaluation of the model [i.e., Eq. (10.1)] 250,000 times. Thus 250,000 values are generated for the model output. These values are stored in a two-dimensional array. The results are then analyzed to calculate confidence intervals for the distribution of variability in the model outputs.

Figure 10.18 shows the results of a two-dimensional simulation of the model output for stack emissions of arsenic. The range of the 95 percent confidence interval for the median value of arsenic emissions for 3-day average emissions varies from approximately 0.2 to 2.5 lb per 3-day period. The range of the 95 percent confidence interval for the 95th percentile of the 3-day periods in a year varies from approximately 2 to 21 lb.

The model output results for predicted cobalt emissions are shown in Figure 10.19. The range of the 95 percent confidence interval for the median value of cobalt emissions for 3-day average emissions varies from approximately 0 to 10 lb. The range of the 95 percent confidence interval for the 95th percentile of the 3-day periods in a year varies from approximately 0 to 350 lb.

The model output results for predicted nickel emissions are shown in Figure 10.20. The range of the 95 percent confidence interval for the median value of nickel emissions for 3-day average emissions varies from approximately 0 to 14 lb. The range of the 95 percent confidence interval for the 95th percentile of the 3-day periods in a year varies from approximately 14 to 145 lb.

The total HAPs emissions are calculated by taking the sum of the individual HAP emissions and the results are shown in Figure 10.21. The range of the 95 percent confidence interval for the median value of total HAP emissions for 3-day average emissions varies from approximately 20 to 80 lb. The range of the 95 percent confidence interval for the 95th percentile for 3-day average emissions varies from approximately 65 to 450 lb.

As noted at the end of the previous section with respect to the uncertainty in the distribution of model inputs, and as observed here with respect to uncertainty in the distribution of model outputs, there is typically large uncertainty in estimates of the upper percentiles of emissions. Similarly, one would expect larger uncertainty in estimates of exposure or risk in the upper percentiles of distributions for variability in these quantities. The fact that there is often large uncertainty in the estimates of the upper tails of distributions implies that is not possible, for example, to make precise statements regarding the 99.9[th] percentile of concentration, exposure, or risk, without also making a statement regarding the level of uncertainty associated with such an estimate.

Figures 10.22–10.25 show results of more direct policy interest. Each of these CDFs represents the uncertainty in the annual average emissions of arsenic, co-

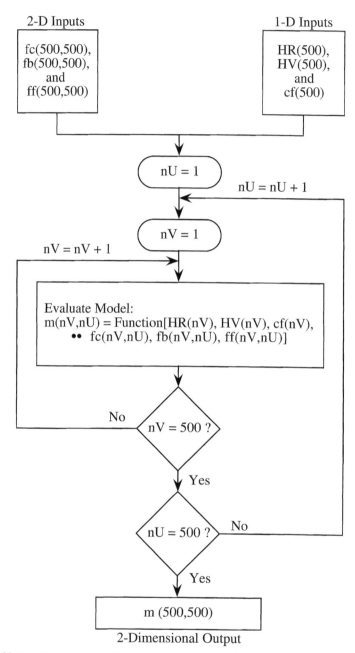

Figure 10.17 Flow diagram illustrating the propagation of two-dimensional (2D) and one-dimensional (1D) inputs through model to obtain two-dimensional output. (*Note:* nU is the uncertainty dimension and nV is the variability dimension.)

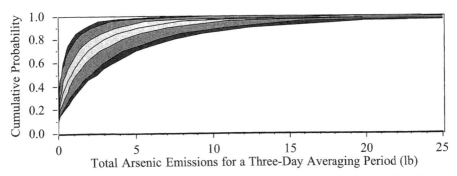

Figure 10.18 Two-dimensional plot illustrating the uncertainty and variability in the total arsenic emissions for a 3-day averaging period.

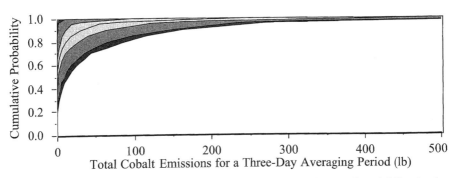

Figure 10.19 Two-dimensional plot illustrating the uncertainty and variability in the total cobalt emissions for a 3-day averaging period.

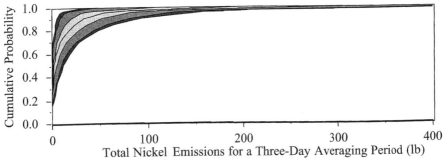

Figure 10.20 Two-dimensional plot illustrating the uncertainty and variability in the total nickel emissions for a 3-day averaging period.

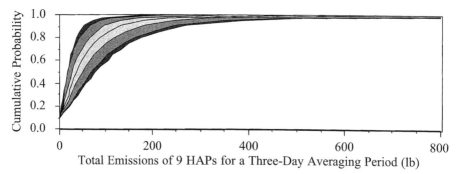

Figure 10.21   Two-dimensional plot illustrating the uncertainty and variability in the total HAPs emissions for a 3-day averaging period.

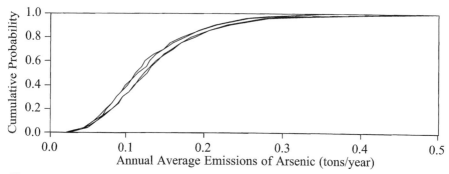

Figure 10.22   Cumulative distribution functions illustrating the year-to-year uncertainty in the average annual arsenic emissions.

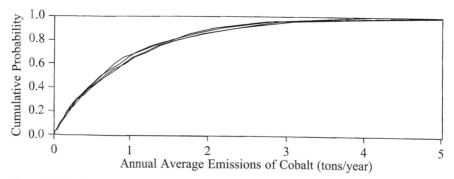

Figure 10.23   Cumulative distribution functions illustrating the year-to-year uncertainty in the average annual cobalt emissions.

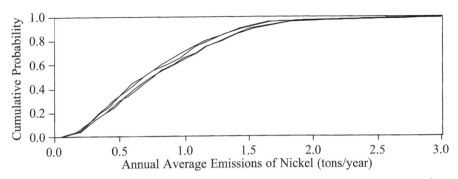

Figure 10.24 Cumulative distribution functions illustrating the year-to-year uncertainty in the average annual nickel emissions.

balt, nickel, and the total of all nine HAPs, respectively. These CDFs were constructed by evaluating the average of 122 three-day periods, which represents approximately one year's worth of emissions. Each CDF represents 500 alternative annual averages. There are a total of four CDFs shown, representing variability in annual averages from 1 year to another over 4 years. The range of uncertainty in annual averages, given by any individual CDF, is much wider than the range of variability in averages from one year to another, given by the relative separation of the CDFs from each other.

The uncertainty in annual average emissions of arsenic is shown in Figure 10.22. There is a 95 percent probability of the power plant emitting less than or equal to 0.3 tons of arsenic in a year.

In the cases of annual cobalt and nickel emissions, shown in Figures 10.23 and 10.24, respectively, there is a 95 percent probability that the emission would be less than or equal to 3.5 and 2 tons, respectively. Figure 10.25 shows the CDFs for annual average emissions of the sum of all the nine HAPs. There is a 95 percent probability that the total HAPs emissions would be less than or equal to 7 tons per year.

10.6 DISCUSSION AND RECOMMENDATIONS

It should be noted that two HAPs, antimony and cadmium, were not included in this analysis due to the difficult nature of the computations involved in these two cases. For both the antimony and cadmium coal concentration data sets, there were several nondetects with different detection limits. In these two cases, the methodology described previously for estimating the parameters of distributions from censored data was not found to be robust. One reason for this failure could be that the log-likelihood function becomes increasingly complicated as the number of detection limits increases and therefore, the particular nonlinear optimization algorithm used here for maximizing the log-likelihood function may not be able to converge on a solution. Furthermore, the implementation of

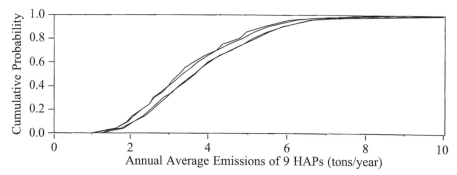

Figure 10.25 Cumulative distribution functions illustrating the year-to-year uncertainty in the average annual total HAPs emissions.

the algorithm for processing the bootstrap samples becomes more complicated as the number of detection limits increases and was beyond the scope of this study. It is recommended that other numerical solution methods be considered in the future.

The results presented here are dependent upon the input assumptions for the model. For example, if log-normal instead of gamma distributions were fitted to the coal concentration data, the amount of uncertainty in the estimated emissions limits might be noticeably different. The sensitivity of two-dimensional HAP emissions model results to the choice between log-normal and gamma distributions fitted to coal data is illustrated in Cullen and Frey (1999).

In this study, only uncertainty due to random sampling error is considered because in small data sets this source of uncertainty may dominate other sources of uncertainty, such as measurement errors. However, it is recommended that uncertainty due to measurement error or other sources should also be considered along with and compared to the uncertainty associated with random sampling error.

The methods demonstrated in this study cannot deal with biases in the data that may result from lack of representativeness. Biases can only be quantified by reference to a datum (or true value), which is unknown in most cases. Alternatively, biases might be quantified based upon encoding expert judgment or reported qualitatively based upon qualitative or semiquantitative methods such as rating factors or rating scores that accompany the probabilistic result. In future, the results of the probabilistic analysis should be reported with either a qualitative rating or a semiquantitative rating that aims to provide an indication of potential biases.

10.7 CONCLUSIONS

The case study presented here illustrates the use of selected techniques for quantifying variability and uncertainty in the inputs to a model, propagating both variability and uncertainty through a model, and making predictions of variability and uncertainty in model outputs. The case study illustrates the importance of averaging time, by considering the averaging times associated with the input data and a different averaging time of more direct policy interest desired in the model output. The case study also illustrates a method for dealing with censored data sets.

10.7.1 Conclusions Regarding the Method and Case Study

The number of available data points for the model inputs ranged between 2 and 21. Uncertainty in the fitted distributions for data sets having few data points (e.g., 3 data points for the arsenic boiler partitioning factor data set) was typically large compared to uncertainty in the fitted distributions for data sets having more data points (e.g., 21 data points in the selenium coal concentration data set).

The range of uncertainty in annual averages was found to be much wider than the range of variability in averages from one year to another. Therefore it can be concluded from this result that the year-to-year variability in the annual average HAPs emissions is negligible as compared to the uncertainty in the annual average emissions. Stated another way, the uncertainty in annual emissions is sufficiently large that, by comparison, variability in emissions from one year to another is less significant.

HAPs emissions from electric utilities are not yet regulated. If an emission limit is selected, the level of confidence associated with the number of utilities that would comply with the selected emission limit can be estimated from the uncertainty estimates calculated in this case study or by applying the method to other types of power plants. For example, if an emissions limit of 2 tons per year is assigned for nickel emissions from power plants, approximately 95 percent of the total number of power plants of the type evaluated would comply with the emission limit.

The approach for fitting distributions to censored data was evaluated using bootstrap simulation. The range of uncertainty in the portion of the distribution extrapolated below the detection limit can be wider, on a relative basis, than the uncertainty associated with the upper tail of the distribution. Whether the additional uncertainty at the lower tail of the distribution is important is context-specific. For example, in the case study presented here, we are typically interested in the upper tail of the distribution of uncertainty in annual emissions. This is not likely to be substantially influenced by the lower tail of the input distributions to the model.

In summary, the case study presented here demonstrates a method for quantification of variability and uncertainty in inputs and outputs of a model, as well as illustrates the policy insights obtained from such an analysis. Although this level of analysis is not required in every instance, it does foster rigorous thinking regarding issues such as the distinction between variability and uncertainty, distinction between averaging times of the available data and averaging times of policy interest, and regarding uncertainty associated with censoring of data. These types of issues are often important in exposure and risk analysis. Therefore, the method and case study presented here serves as a useful conceptual framework for many exposure and risk-related problems.

10.7.2 What Benefits Does This Approach Offer Risk Assessors?

The probabilistic analysis method illustrated here via a detailed case study of HAP emissions offers a number of important benefits to risk assessors and risk managers. These benefits include the following:

- Variability and uncertainty are appropriately distinguished from each other. Variability cannot be reduced, but knowledge of the range of variation can be useful in identifying, for example, high emissions situations, highly exposed individuals, and highly at-risk individudals.
- Knowledge of uncertainty is important in assessing whether reduction in uncertainty would be helpful in clarifying a decision and, if so, in identifying key sources of uncertainty for targeted study.
- The use of parametric distributions to represent small data sets offers the advantage of allowing for reasonable inferences regarding upper and lower tails of the distribution beyond the range of observed data.
- A key challenge in many exposure and risk assessments is that the averaging time of available data and desired model outputs are often not the same. The use of simulation methods such as the one illustrated here allows for reasonable inferences regarding emissions or exposures for one averaging time based upon data collected for a different averaging time.
- Proper treatment of non-detects is important to avoid biases in statistics and distributions estimated from censored data sets. The use of rigorous statistical methods for making inferences from censored data is more credible than the use of default values for non-detects and will improve the acceptance of risk assessments that are based at least in part upon censored data.
- The case study results indicate that uncertainty tends to be large in the upper tail of distributions for variability. From a policy and modeling perspective, this is an important insight. As a practical matter, it is not possible to obtain precise estimates of the upper percentiles of interindividual exposure or risk. However, as shown here, it is possible to make inferences regarding a best estimate of an upper percentile of exposure or risk and regarding a range of uncertainty for the best estimate.

REFERENCES

Cohen, A. C., and Whitten, B. 1988. *Parameter Estimation in Reliability and Life Span Models*. Dekker, New York.

Cullen, A. C., and Frey, H. C. 1999. *Use of Probabilistic Techniques in Exposure Assessment: A Handbook for Dealing with Variability and Uncertainty in Models and Inputs.* Plenum, New York.

David, H. A. 1981. *Order Statistics,* 2nd ed. Wiley, New York.

Efron, B., and Tibshirani, R. J. 1993. *An Intoduction to the Bootstrap, Monographs on Statistics and Applied Probability 57.* Chapman & Hall, New York.

Electric Power Research Institute (EPRI). 1994. *Electric Utility Trace Substances Synthesis Report,* Vol. 2: *Appendices A–N.* TR-104614-V2. EPRI, Palo Alto, CA.

Environmental Protection Agency (EPA). 1996b. *Study of Hazardous Air Pollutant Emissions from Electric Utility Steam Generating Units-Interim Final Report,* Vol. 2: *Appendices A–G,* EPA-453/R-96-013b. EPA, Research Triangle Park, NC.

Frey, H. C. 1998. Methods for quantitative analysis of variability and uncertainty in hazardous air pollutant emissions. Paper No. 98-105B.01. In *Proceedings of the Ninety-First Annual Meeting.* Air and Waste Management Association, Pittsburgh, PA.

Frey, H. C., and Burmaster, D. E. 1999. Methods for characterizing variability and uncertainty: Comparison of bootstrap simulation and likelihood-based approaches. *Risk Anal.* 19(1): 109–130.

Frey, H. C., and Rhodes, D. S. 1996. Characterizing, simulating, and analyzing variability and uncertainty: An illustration of methods using an air toxics emissions example. *Hum. Ecol. Risk Assess.* 2(4): 762–797.

Frey, H. C., and Rhodes, D. S. 1998. Characterization and simulation of uncertain frequency distributions: Effects of distribution choice, variability, uncertainty, and parameter dependence. *Hum. Ecol. Risk Assess.* 4(2): 423–468.

Gilliom, R. J., and Helsel, D. R. 1986. Estimation of distributional parameters for censored trace level water quality data 1. Estimation techniques. *Water Resourc. Res.* 22(2): 135–146.

Haas, C. N., and Scheff, P. A. 1990. Estimation of averages in truncated samples. *Environ. Sci. Technol.* 24(6): 912–919.

Hazen, A. 1914. Storage to be provided in impounding reservoirs for municipal water supply. *Trans. Am. Soc. Civil Eng.* 77: 1539–1640.

Newman, M. C., Dixon, P. M., Looney, B. B., and Pinder, J. E. 1989. Estimating mean and variance for environmental samples with below detection limit observations. *Water Resourc. Bull.* 25(4): 905–915.

Rubin, E. S., Berkenpas, M. P., Frey, H. C., and Toole-O'Neil, B. 1993. Modeling the uncertainty in hazardous air pollutant emissions. In *Proceedings of the Second International Conference on Managing Hazardous Air Pollutants.* TR-104295. Electric Power Research Institute, Palo Alto, CA.

11 Characteristic Time, Characteristic Travel Distance, and Population-Based Potential Dose in a Multimedia Environment: A Case Study

DEBORAH H. BENNETT

Department of Environmental Health, Harvard School of Public Health, Boston, Massachusetts

THOMAS E. MCKONE and WILLIAM E. KASTENBERG

Department of Nuclear Engineering, University of California at Berkeley, Berkeley, California

11.1 INTRODUCTION

The environmental decision-making community is now confronting the potential adverse health and ecological impacts of persistent chemicals such as metals and organic pollutants. The significance of these chemicals is attributed to their potential for health and ecological impacts at low concentrations, coupled with their ability to accumulate and persist in multiple environmental media. Their persistence allows adequate time for environmental interactions and long-range transport to populations far from the source. Persistent organic pollutants (POPs) originate from a broad range of human activities, including combustion for energy production and transportation, industrial processes, and agricultural uses of pesticides.

Environmental impacts have often been classified by the medium to which the pollutants are released, such as air, water, or soil. However, many POPs partition into multiple environmental media. Air emissions can result in con-

Human and Ecological Risk Assessment: Theory and Practice, Edited by Dennis J. Paustenbach
ISBN 0-471-14747-8   © 2002 John Wiley & Sons, Inc.

taminated soil, and contaminated soil can result in air pollution through the mass exchange between the air and soil.

Another limitation for addressing POPs is that pollution is often regulated as a local problem. However, such an approach is not adequate for many POPs because they can be transported long distances in the environment. When a chemical travels long distances, it can cross local regulatory boundaries. In this sense, the scientific or regulatory community has not adequately addressed human exposure to POPs through multiregional, multimedia exposure scenarios.

The objective of this chapter is to present a methodology aimed at answering the following questions:

- What measures can we use to determine whether or not a chemical is a POP with the potential for long-range transport?
- How do we properly quantify the population-based potential dose resulting from a pollutant with the potential for long-range transport?

To answer the first question, we develop a framework to quantify the characteristic time (τ) and characteristic travel distance (CTD) for semivolatile POPs in a multimedia environment. Characteristic time is a measure of temporal persistence; that is, how long a chemical is likely to remain in a multimedia environment after being released to any compartment of that environment. The CTD, on the other hand, is a measure of how far a chemical is likely to travel in the multimedia environment and defines whether a chemical will have a local, regional, or global-scale impact.

A multimedia model is used here to incorporate chemical exchange among air, soil, water, and plants as well as chemical degradation in each compartment. This information is used to quantify the steady-state spread of pollutants between media.

To answer the second question, we present a conceptual model to characterize the population-based potential dose by taking into consideration the CTD and the spatially varying population density. If a chemical travels long distances in the environment, more people are exposed to the chemical, although at lower concentrations. If the dose–respose model is assumed linear, as more people are exposed to the chemical, there is a higher chance of someone experiencing an adverse effect from exposure to the chemical than if only a small number of people are exposed to a chemical. We compare the population-based potential dose calculated with this conceptual model to the potential dose calculated using only the locally exposed population.

The methodologies presented here are appropriate for continuous, large nonpoint atmospheric emissions of organic chemicals, such as the collective emissions from a large urban area including industrial facilities, and combustion emissions for transportation, heating, and electrical generation. The methodologies are appropriate for ubiquitous chemicals with long atmospheric half-lives (several hours or days); a relatively high value for K_{ow}, such that partitioning into vegetation and soil is significant ($K_{ow} > 1 \times 10^6$); and a relatively high vapor pressure (VP) such that there is some partitioning from particles to the gas phase of the atmosphere (VP $> 1 \times 10^{-10}$ Pa). Defining τ and CTD of a chemical will give insight regarding the appropriate scale for regulation. The measure of population-based potential dose can be used by decision makers for a variety of analyses to decide what chemical to use for a process or by policymakers to determine appropriate regulations when evaluating a new chemical as it is being introduced. Potential applications for these measures include risk assessment, pollution prevention assessment, health effects studies, pollutant mass balance studies, life-cycle analyses, sustainability evaluation, and regulatory impact studies. Regulators can then focus on the chemicals with the highest population-based potential dose and corresponding health risk, thereby providing more effective ways to regulate these chemicals.

Two case studies are presented, one for a persistent chemical, 2,3,7,8-tetrachlorodibenzo-*p*-dioxin (TCDD), and one for a nonpersistent chemical, benzo[*a*]pyrene. We will calculate τ, CTD, and population-based potential dose and discuss how the results differ when evaluating a persistent and nonpersistent chemical.

11.1.1 Review of Environmental Modeling

Historically, environmental contamination has been viewed primarily as a local problem, generally affecting one environmental medium and has been regulated accordingly.[1] This approach is changing as an increasing number of examples challenge this view. Traces of several POPs have been found in the Arctic, although there are no sources of organic pollution there.[2,3] This indicates global-scale pollution. Pesticides used in California's central valley have resulted in contamination in the Sierra Nevada Mountain Range,[4,5] an example of both a multimedia problem and long-range transport. The chemicals dichlorodiphenyl-trichloroethane (DDT) and various polychlorinated biphenyls (PCBs), whose use has been either banned or severely restricted in the United States for over 20 years, are currently found in the sediments of the Great Lakes, illustrating persistence.[6,7] The foregoing are all examples of multidimensional problems that are difficult to understand using current models and stimulated the research presented in this chapter.

The EPA Science Advisory Board has recommended that reliable, multimedia models would overcome a significant barrier to risk assessment for evaluating chemicals that partition into multiple environmental media.[8] At present, two types of multimedia transport models are often used: single-region, multimedia models, and simplified global multimedia models. Fugacity-based models that include different compartments, such as air, water, soil (one or more layers), vegetation, and sediment have been developed by Mackay and Paterson[9] and McKone.[10,11] These models are comprehensive with respect to the environmental compartments within the model, yet lack spatial resolution.

Multimedia models that include spatial resolution are also available. For example, the SimpleBox model[12] has a nested set of small-, medium-, and large-scale "unit worlds." Global-scale models have also been developed to determine the distance chemicals are likely to travel in the environment and to look at temperature-dependent trends.[13,14] These models do not include human exposure and thus cannot be used for exposure and risk assessment.

11.1.2 Review of Human Exposure and Dose

Quantifying the potential dose per person has been an essential element in the field of risk assessment since its inception. CalTOX, a fugacity-based multimedia model, calculates the dose from multiple environmental media using 23 exposure pathways.[10] All of these pathways are directly incorporated into the exposure model used in this chapter. Most often, these calculations are carried out for an individual. Sometimes, risk to a population is calculated by multiplying the risk to a representative individual in a given group by the number of individuals in that group. Example calculations of population-based dose can be found in case studies by both Thompson and Evans[15] and Webster and Connett.[16] A study of global chemicals by Travis and Hester[17] calculated the background cancer risk from 11 global pollutants, based on measured background concentrations, demonstrating a need for research quantifying risk from chemicals with a potential for long-range transport.

Starting in 1988, the Environmental Protection Agency (EPA) began requiring companies to report the amount of toxic chemicals released from their facilities through the Toxic Release Inventories program. The simplest method to compare these releases considers the quantity released and the toxicity of the chemical.[18] More advanced methods also include critical factors such as persistence, pollutant fate, or exposure factors.[19,20] It was demonstrated that more advanced methods yield significantly different results than simpler methods. The main drawbacks cited for the advanced methods are that they require more data that are often unavailable[19] and that increasing the complexity also increases the uncertainty.[20]

11.2   METHODS

We first define the multimedia model used to determine the concentrations in the relevant environmental media. The characteristic time is defined and then analytically derived using a multimedia model. We then present a methodology for determining the CTD for airborne semivolatile organic pollutants. A conceptual model for calculating the population-based potential dose is also presented. This requires that we determine the exposure from multiple pathways to the multimedia environment.

11.2.1   Defining the Multimedia Model

We use a model with air, surface water, vegetation, and two soil compartments. A schematic of the model used is shown in Figure 11.1. All phases (i.e., air,

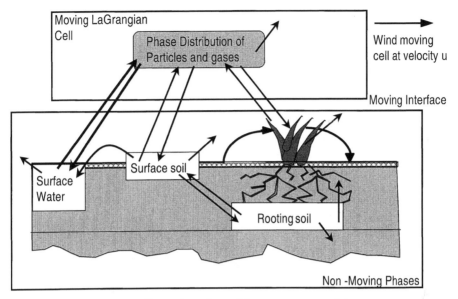

Figure 11.1   Illustration of a multicompartment model system.

water, solids) in an environmental compartment are assumed to be in chemical equilibrium. From the phase composition of each compartment, the fugacity capacity (e.g., the chemical concentration per unit chemical fugacity) can be defined. The fugacity capacity of each compartment can be found in Reference 21.

The steady-state concentration in each environmental compartment is determined from the interactions among the environmental compartments and the decay rate in each compartment. Many of the model compartments and processes needed to define the interactions between compartments have been taken from the CalTOX model.[11,22,23] We use a fugacity-based model, a common approach for describing partitioning in multimedia systems.[9,24]

The chemical exchange between compartments is determined by the transfer coefficients, or T values. These values define the rate of mass transfer per unit mass inventory of the chemical in the compartment from which the chemical is transferred. Table 11.1 lists all of the transport processes used in the model. The equations for the transfer rates between compartments can be found in Reference 21. To calculate the flux from one compartment to another, the mass transfer coefficient is multiplied by the mass in the compartment from which it originates. In the case of diffusion, the net flux is calculated from the gross flux in each direction.

Mass transformation is modeled in each compartment based on pseudo-first-order rate constants taken from experimental or field data. This data is often scarce or highly uncertain and reported values often range over orders of magnitude, especially for vegetation.[25–27] Transformation in air can include reactions with OH radicals and photodegradation. Transformation rates in soil

TABLE 11.1   List of Transfer Pathways in Multimedia Model[a]

	Air to	Surface Soil to	Root-Zone Soil to	Vegetation to	Surface Water to
Air		Particle resuspension, and diffusion		Particle resuspension, diffusion	Diffusion
Surface soil	Wet and dry particle deposition, wet gaseous deposition, and diffusion			Washoff from leaves	
Root-zone soil		Diffusion, advection through rain			
Vegetation	Wet and dry particle deposition and diffusion	Rain splash of particles	Root uptake		
Surface water	Wet and dry particle deposition, wet gaseous deposition, and diffusion	Runoff and erosion			
Flows out of model region	Advection by wind				Advection of surface water

[a] Calculation methods for each pathway can be found in Reference 21.

differ among soil types and include photodegradation on surface soils and degradation by microbial action in deeper soils. Transformation in vegetation can be rapid and includes photodegradation on the leaf surface.[28,29]

Estimates of parameter values can rarely be characterized accurately by a single value due to uncertainty in determining a parameter value, spatial variability, or both. A log-normal probability distribution is assigned to each parameter such that the range conforms to the environmental limits of the selected parameters. The parameter values and associated coefficient of variation (the standard deviation divided by the mean value) used in the case studies are found in Table 11.2.

11.2.2 General Formulation of Characteristic Time

Characteristic time is a measure of temporal persistence; that is, how likely a chemical pollutant is to remain in a multimedia environment after being released to any compartment of that environment. The characteristic time can be determined by finding the overall decay rate of the chemical in a closed, defined landscape system. Because the decay rates in each environmental media can differ significantly for a given chemical, determining the τ in the environment requires knowing both the mass distribution among environmental media and the media-specific half-lives.

The instantaneous mean life or average life expectancy of a molecule in an environmental compartment, τ, is defined as the inverse of the decay rate in the compartment.[30] We will refer to this as τ for that compartment:

$$\tau = \frac{1}{k} \tag{11.1}$$

where k is the decay rate, representing radioactive decay or chemical reactions that irreversibly remove the chemical from the system. In a two-compartment system, such as in Figure 11.2, the effective decay rate is mass averaged between the two compartments, leading to the following instantaneous overall decay rate[31]:

$$k_{\text{overall}} = \frac{M_1 k_1 + M_2 k_2}{M_1 + M_2} \tag{11.2}$$

where M_i is the mass in compartment i (kg).

We prefer steady-state calculation methods to inform decisions regarding chemical impacts in the environment.[31] The steady-state distribution accounts for location of the source and for advective phase-transfer processes while retaining sufficient simplicity to complete calculations in a tractable form, such as a spreadsheet, useful if the output of the analysis is to be utilized as a factor in decision making or subjected to an uncertainty analysis. Methods have been developed using classification trees to determine if a chemical is persistent or

TABLE 11.2   Representative Spatially Averaged Landscape Properties

Landscape Property	Notation	Mean Value	Coefficient of Variation	Reference
Universal gas constant $(Pa\text{-}m^3/mol\text{-}K)$	R	8.31	0	
Ambient environmental temperature (K)	T	288	0.02	57
Yearly average wind speed (m/d)	u	3.46×10^5	0.2	
Relative humidity	rh	0.8	0.1	
Surface area of particles (m^2/m^3)	SA	1.50×10^{-4}	0.1	58
Washout ratio	W_r	5.00×10^4	2	58
Atmospheric dust load (kg/m^3)	ρ_{ba}	5.00×10^{-8}	0.2	58
Dry deposition velocity of air particles (m/d)	v_d	43.2	0.3	58
Boundary layer thickness in air above vegetation (m)	δ_{ap}	0.005	0.2	59
Boundary layer thickness in air above soil (m)	δ_{ag}	0.005	0.2	59
Annual average precipitation (m/d)	$Rain$	2.0×10^{-3}	1	57
Soil runoff rate $(kg/m^2 d)$	$Erosion$	3.0×10^{-3}	0.2	56
Groundwater recharge (m/d)	$Recharge$	1.2×10^{-4}	1	56
Plant–air partition factor, particles $(m^3/kg[FM])$	K_{pa}^{part}	3300	0.1	60
Plant dry-mass fraction	bio_{dm}	0.20	0.2	56
Plant fresh-mass density (kg/m^3)	ρ_p	1.00×10^3	0.2	56
Soil particle density (kg/m^3)	ρ_s	2.60×10^3	0.05	56
Water content in surface soil (%)	β_g	0.17	0.2	56
Air content in the surface soil (%)	α_g	0.40	0.2	56
Water content of root-zone soil (%)	β_s	0.28	0.2	56
Air content of root-zone soil (%)	α_s	0.17	0.2	56
Height of the air compartment (m)	d_a	1000	0.1	
Fraction of area that is surface water	f_w	8.15×10^{-3}	0.2	
Average depth of surface water (m)	d_w	5	1	
Thickness of the ground soil layer (m)	d_g	2.50×10^{-3}	1	
Plant dry mass inventory $(kg[DM]/m^2)$	bio_{inv}	0.40	0.2	
Suspended sediment in surface water (kg/m^3)	ρ_{bw}	0.8	1	
Water runoff rate (m/d)	$Runoff$	2.8×10^{-4}	1	
Organic carbon fraction	f_{oc}	0.03	1	56
Organic carbon fraction in sediments	f_{oc}^{sed}	0.02	1	56
Evaporation rate of surface water (m/d)	$Evaporate$	4.38×10^{-6}	1	56
Sediment particle density (kg/m^3)	ρ_{sd}	2600	0.05	56

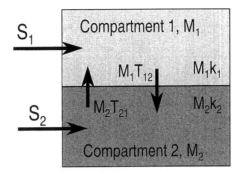

Figure 11.2 Diagram of the two-compartment system used to calculate the characteristic time, τ.

nonpersistent using this measure for characteristic time based on the input properties of the chemical.[31] A similar formulation has been developed by Webster et al.[32]

If more than one environmental compartment influences the overall decay rate in the environment, the effective decay rate equation can be expanded to include any number of environmental compartments. For the multimedia system used in this chapter, we use the following equation:

$$k_{\text{overall}} = \frac{M_a k_a + M_p k_p + M_g k_g + M_s k_s + M_w k_w}{M_a + M_p + M_g + M_s + M_w} \quad (11.3)$$

where the indices a, p, g, s, and w are for the air, plant, ground surface soil, root zone soil, and surface water compartments, respectively.

11.2.3 General Formulation of Characteristic Travel Distance

The CTD is a measure of how far a chemical is likely to travel in the multimedia environment and is derived analytically by following the movement of a particular mass of pollutant in a moving Lagrangian air cell as it interacts with the nonmoving compartments of the environment (i.e., vegetation and soil). We begin with a two-compartment system, a moving (i.e., air) compartment and a nonmoving compartment (i.e., vegetation and soil). This system, illustrated in Figure 11.3, can be thought of as a simple representation of the more complex environmental model shown in Figure 11.1.

We calculate the change in chemical mass in a moving Lagrangian air cell as the cell travels in a one-dimensional band away from the area of release (the source region). The concentration in air is reduced with distance based on degradation in air, transfer to, and subsequent degradation in other media. We chose the CTD as the distance from the source at which the concentration is reduced by 63 percent (i.e., reduced to $1/e$ of the original concentration). By

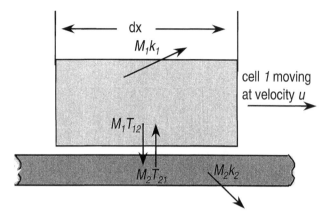

Figure 11.3 Lagrangian system used for determining the characteristic travel distance (CTD).

using a continuous differential Lagrangian structure, the results are obtained in closed analytic form as opposed to a more complex numerical form.

We make the following assumptions: The source term is continuous; the system has reached steady state; there is no lateral air dispersion; the long-term average wind pattern can be represented by an equivalent steady wind rate in one direction; the landscape properties do not vary spatially (or can be spatially averaged); and the atmospheric mixing layer height is constant. We believe these assumptions are justified when considering a large continuous area source, such as the collective sources from a large urban or agricultural region.

The Lagrangian cell represents a small portion of the continuous stream of a pollutant flowing from the source at an average windspeed of u (m/d). At steady state, neither the moving nor the nonmoving phases are accumulating mass with time at a particular location. The mass transferred to the nonmoving phase is equal to the mass decayed in that phase. Additionally, the ratio of the mass in the ground to the mass in the air is spatially independent.

In the Lagrangian model for the airborne pollutant, we balance the time rate of change of mass with both the decay in the moving phase and the net transfer to the nonmoving phase. We define the effective decay rate, $k_{\text{effective}}$ (1/s) as:

$$k_{\text{effective}} = \frac{M_1 k_1 + M_2 k_2}{M_1} \tag{11.4}$$

The effective decay rate is essentially the mass in each compartment multiplied by the decay rate in the corresponding compartment divided by the mass in the moving phase. Solving for the concentration profile as one moves away from the source yields the following result:

$$C_1(x, u) = C_1(0)e^{(-k_{\text{effective}}/u)x} = C_1(0)e^{-x/\text{CTD}} \tag{11.5}$$

We define the CTD to normalize the distance from the source as CTD $=$ $u/k_{\text{effective}}$. The characteristic travel distance is the same for the moving and nonmoving comparments since the ratio between the concentrations is spatially independent. This implies that the concentration decreases at the same rate with distance from the source in both compartments. The complete derivation can be found in Reference 21.

Again, if multiple environmental compartments influence the CTD, it is necessary to include these compartments in the calculation. We expand the definition of CTD to the multicompartment system shown in Figure 11.1 as:

$$k_{\text{effective}} = \frac{M_a k_a + M_p k_p + M_g k_g + M_s k_s + M_w k_w}{M_a} \qquad (11.6)$$

The steady-state mass in all of the compartments is derived in Reference 21.

This method for calculating the CTD has also been applied to other multimedia models.[33] It is interesting to note that the CTD is not necessarily well correlated to τ. A chemical may be persistent but not have a long travel distance if it primarily partitions into the nonmoving phases of the environment.[34]

11.2.4   Population-Based Potential Dose

We develop a conceptual model for calculating the population-based potential dose that incorporates the CTD of a particular chemical and the spatially dependent population density. Humans are exposed to chemicals in the environment through multiple pathways. Exposure is characterized by route of entry as inhalation, ingestion, or dermal uptake. Inhalation exposure includes contact with both indoor and outdoor air. The ingestion pathways include tap water consumption, incidental soil ingestion, and intake of fruits, vegetables, grains, and animal products, such as meat, poultry, eggs, and dairy. The dermal route includes exposure through contaminated water from bathing and recreation, as well as from soil on the skin. The pathways are summarized in Table 11.3. Potential dose is calculated from the contact rate with the exposure media and the chemical concentrations in these exposure media (i.e., tap water, indoor air, etc.).*

For each exposure pathway shown in Table 11.3, the potential dose is calculated from the concentration in the corresponding environmental medium, the relationship between the exposure medium concentration and environmental medium concentration, intake rate, body weight, activity patterns, and ex-

---

*Here we define the potential dose as the amount of chemical that passes into an individual while the actual dose quantifies the amount of chemical that is absorbed into an individual (e.g., the amount of chemical in the air an individual breathes is the potential dose while the actual dose is the portion of that air that passes into the lung tissue). Ideally, risk should be based on the actual dose, but often the potential dose is assumed to equal the dose, an assumption also made in this chapter.

TABLE 11.3   List of Exposure Pathways[a]

Ingestion Pathways

All intake values were correlated per unit body weight, preventing data points based on
  high intake with low body weight
Exposed produce, including grains
Unexposed produce—the concentration in the two types of produce and grains are
  calculated separately. They include exposure to air, soil, and water used for
  irrigation.
Fish—based on surface water concentrations
Meat, milk, and eggs—livestock products are exposed through inhalation, direct
  ingestion of water and soil, and indirect exposure through food contaminated by
  exposure to air and soil.
Soil—both adults and children ingest small amounts of soil through inadvertent hand-
  to-mouth activities
Water—while swimming in surface water
Tap water—concentration linked to concentration in both surface water and
  groundwater, assumed clean in this model

Inhalation Pathways

Breathing rates vary by activity level as do location of activity levels. Breathing rate
  and location need to be linked.
Active outdoors
Resting indoors—the concentration of indoor air includes soil vapors transferred from
  under the house and soil particles transferred to indoor air.
Active indoors
While showering or in the bath—contaminants transferred from tap water

Inhalation Pathways

Showering—from tap water
Swimming—from surface water
Soil—dermal exposure to contaminants in soil can occur during a variety of activities,
  such as construction work, gardening, and recreation outdoors. Children playing
  outdoors also can have rather large soil loading on their skin.

---

[a]Calculation methods for each pathway can be found in the CalTOX manual.[36]

posure duration as[35]:

$$\text{ADD} = C_{\text{env}} \times R \times \frac{\text{CR}}{\text{BW}} \times \frac{\text{ED} \times \text{EF}}{\text{AT}} \tag{11.7}$$

where ADD *is the* average daily dose of chemical via exposure route (mg/kg/
day), C_{env} is the chemical concentration in the environmental medium (mg/kg),
R is the ratio of the environmental concentration and the exposure concentration, CR is the contact rate (kg/day), BW is the body weight (kg), ED is the

exposure duration (years), EF is the exposure frequency (days/year), and AT is the averaging time (days).

The input parameters (e.g., breathing rate, water intake rate, etc.) vary between pathways, thus Equation (11.7) is written differently for each exposure pathway. The risk to an individual due to exposure to a carcinogen is calculated by multiplying the ADD by a cancer potency factor (CPF). In a risk assessment, this equation is most often applied to a site-specific case to determine the risk to an individual. For example, in the case of an air emission, a plume model is used to calculate realistic exposure concentrations for an individual living close to the site. In our case, Eq. (11.7) will be applied to all individuals exposed to the chemical.

Exposure Concentrations Exposure media concentrations may differ from the ambient environmental media concentrations and can be calculated from the ambient air, soil, vegetation, and surface water concentrations. For example, the concentration of a chemical in the indoor air may differ from the concentration in the outdoor air. The indoor air concentration is influenced by the concentration in the outdoor air, the concentration in the soil gas below the house, the concentration in tap water, and the concentration in resuspended particles in the home attributable to soil tracking, due to shoes, clothing, and the fur of pets or particles in the outdoor air that enter the home and are subsequently deposited. Another example is the concentration in meat, which depends on the animal's ingestion of contaminated soil, pasture, and grains, and inhalation of air. Equations relating exposure media concentrations to environmental concentrations were taken directly from the CalTOX model.[36]

Human Activity and Contact The remainder of the terms needed to calculate the ADD in Eq. (11.7) for each exposure pathway relate to various aspects of human activity and contact. Again, the processes used are taken from CalTOX.[36] Exposure duration, the length of time a person is likely to be exposed to a contaminant, is needed to calculate the ADD. However, for population dose, we assume a constant total population and thus calculate the dose as a long-term annual average.

11.2.5 Model of Population Exposure

We use the following equation for the population dose:

$$\text{Population-based potential dose} = \iint P(x, y) \times \text{ADD}(x, y) \, dx \, dy \quad (11.8)$$

where P is the population density (persons/m^2) and ADD is the dose per person (mg/kg-d). In this equation, both the dose per person and the population density can vary spatially. We must, however, determine the appropriate scale to use, the system boundaries, and the population density.

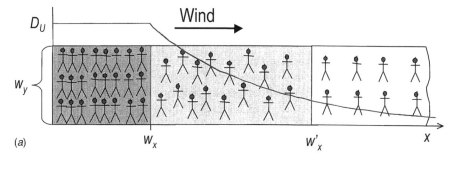

(a)

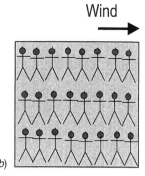

(b)

Figure 11.4 (*a*) Spatial system for calculating population based potential dose. (*b*) Open urban system for calculating population-based potential dose.

When calculating a population dose, we are not concerned with the variability between members of the population. A Monte Carlo simulation that varies uncertainty and variability simultaneously can be used to predict the mean value of the population-based potential dose. If a population risk is determined from the population-based potential dose calculated by Eq. (11.8), a linear dose response curve must be used for the risk measure to be meaningful.

We consider an idealized environmental model with the source term located in the urban region where the population density is highest. We assume a steady wind blowing from the urban region toward the suburban and rural regions, which have lower population densities than the urban region.

Figure 11.4*a* presents a geometry that accounts for the coupling of the higher population density and higher dose per person in the urban region. We call this the spatial model. The concentration is constant in the urban region and decreases exponentially with distance due to decay in the environment as the distance from the source region increases. Population densities are assumed constant in time in the urban, suburban, and rural regions. This geometry uses a wind velocity that always travels in a constant direction with no lateral dispersion. We believe these simplifications are appropriate because the spatial model is designed to compare chemicals, not determine actual risk levels. The

TABLE 11.4   Representative Population Densities and Areas

	Population Density (persons/m^2)	Average Area (m^2)
Urban region	$PD_U = 3.5 \times 10^{-3}$	$A_U = 2 \times 10^8$
Surrounding region	$PD_{SU} = 6 \times 10^{-4}$	$A_{SU} = 5 \times 10^9$
Background	$PD_B = 8 \times 10^{-5}$	

equation for the population-based potential dose for the spatial model is:

$$\text{Population dose} = P_U \times \text{ADD} \times w_y w_x + \int_{w_x}^{w_x'} P_{SU} \times \text{ADD}$$

$$\times e^{-(x-w_x)/\text{CTD}} \times w_y \, dx$$

$$+ \int_{w_x'}^{\infty} P_R \times \text{ADD} \times e^{-(x-w_x)/\text{CTD}} \times w_y \, dx \quad (11.9)$$

the subscripts U, SU, and R refer to urban, suburban, and rural, respectivly; w_x is the width of the urban region in the x direction (m), w_y is the width of the urban region in the y direction (m), and w_x' is the distance at which the population density changes from suburban to rural, $(A_U + A_{SU})/w_y$.

The ADD is calculated from the multimedia, multipathway exposure model. The model region is the size of a representative urban region and we assume an open region (i.e., wind flows out of the region). The size and population densities of the urban, suburban, and rural regions are representative, with the values listed in Table 11.4. The width and length of the urban region, w_x and w_y, are both taken as equal to the square root of the urban area. The CTD is calculated using the methods presented in the previous section.

The spatial model is compared to the urban model shown in Figure 11.4b. In this model, only the urban population is exposed and we use the following equation to calculate the population-based potential dose:

$$\text{Population dose} = P_U \times \text{ADD} \times w_x w_y \quad (11.10)$$

11.3   CASE STUDY

To demonstrate the calculation and evaluation of τ, CTD, and population-based potential dose, we carried out a case study using two chemicals, one that is considered persistent, 2,3,7,8-tetrachlorodibenzo-p-dioxin (TCDD), and one that is not considered persistent, benzo[a]pyrene. TCDD is typically released into the air as a by-product from incineration, combustion of fossil fuels, and industrial processes in urban areas but often contaminates suburban and rural sites as well.[37,38] Airborne 2,3,7,8-TCDD is found in both the gaseous and particulate phases. TCDD is also released to the land and water through industrial sources and landfills. TCDD has limited degradation in soil[39–41] and decays both in air and in vegetation.[29,42–45] In the vapor phase of the atmosphere, reaction with OH radicals is the dominant degradation pathway. TCDD on particles has negligible degradation. The primary degradation process for TCDD in vegetation is the reductive dehalogenation by sunlight, which requires proton donors. Because the lipids in plants are rich sources of proton donors, we expect higher degradation rates in vegetation relative to air.[41,42]

Benzo[a]pyrene is a polyaromatic hydrocarbon that decays rapidly in air and tends to favor the lipid phases of the environment. Benzo[a]pyrene is also a by-product of combustion. This chemical is found at much higher concentrations in urban regions, indicating that it does not travel a long way in the environment.[46] At present, there is no information on the degradation rate of benzo[a]pyrene in vegetation and we assumed a decay rate equal to that in surface soil. The representative values used for all of the chemical properties in calculating the three measures are listed in Table 11.5.

11.3.1   Characteristic Time

The characteristic times for steady dioxin emissions to both air and soil are plotted in Figure 11.5. The characteristic time is over an order of magnitude greater when the pollutant is released directly to the soil compartment. The characteristic time is much less if released to air because a significant portion is decayed in air, consistent with the findings of the dynamic mass balance completed by Eisenberg et al.[47] This demonstrates the importance of determining the location of the source term when calculating τ. Benzo[a]pyrene also has a much longer τ when released to soil than to air. Webster et al.[32] have also developed a method for calculating τ and also found it important to characterize the mode of entry into the environment.

11.3.2   Characteristic Travel Distance

The CTD for benzo[a]pyrene was calculated to be on the order of 30 km. Benzo[a]pyrene has a relatively short CTD due to a relatively rapid degradation rate in air. Consistent with this model prediction, field studies have found benzo[a]pyrene in suburban but not rural regions.[46]

In contrast, the CTD for TCDD was calculated to be on the order of 600 km. Many field studies have found fairly uniform TCDD concentrations in the northern hemisphere[48] while others have found TCDD concentrations in urban regions approximately one order of magnitude greater than concentrations found in rural regions.[49] The calculated CTD is on the same order of magnitude as, or greater than, the distance between urban centers. With a CTD of 600 km, we do expect TCDD to show some reduction in concentration between urban and rural regions. However, we expect less spatial variation if the urban

TABLE 11.5   Representative Chemical Properties Used in the Case Study[26,40,41,43,45,56]

Chemical or Landscape Property	Notation	TCDD Mean Value	TCDD Coefficient of Variation	B[a]P Mean Value	B[a]P Coefficient of Variation
Molecular weight (g/mol)	MW	322	0.01	252	0.01
Octanol–water partition coefficient	K_{ow}	5.70×10^6	1	2.20×10^6	.72
Melting point (K)	T_m	578	0.01	451	0.028
Vapor pressure in (Pa)	VP	1.00×10^{-7}	2	7.13×10^{-7}	.07
Henry's law constant (Pa-m^3/mol)	H	3.75	1.5	0.092	1
Diffusion coefficient in pure air (m^2/s)	D_{air}	4.86×10^{-6}	0.1	5.09×10^{-6}	0.08
Diffusion coefficient; pure water (m^2/s)	D_{water}	5.90×10^{-10}	0.1	6.13×10^{-10}	0.25
Organic carbon partition coefficient	K_{oc}	5.40×10^6	0.1	2.49×10^6	0.9
Biotransfer factor, plant/air (m^3[a]/kg[pFM])	K_{pa}	25000	0.85	5.92×10^5	14
Decay rate in air (1/s)	k_a	8.0×10^{-7}	1.5	1.27×10^{-4}	1
Decay rate is surface water (1/s)				3.47×10^{-6}	1.2
Decay rate in surface soil (1/s)	k_g	2.2×10^{-8}	1.2	3.47×10^{-8}	1.1
Decay rate in root-zone soil (1/s)	k_s	2.1×10^{-10}	1.7	3.47×10^{-8}	1.2
Decay rate in vegetation (1/s)	k_p	1.3×10^{-6}	3.0		

centers are located less than 600 km apart, explaining the fairly uniform concentrations reported in the literature.

To illustrate the possible effect of one urban center on another, we use two large urban centers located 500 km apart as shown in Figure 11.6. We assume an average steady wind from Center A to Center B. The size of the urban centers and their corresponding sources are assumed to be equal. Therefore, each center would have the same concentration if considered independently. Since all of the processes are linear, we can use the principle of superposition and

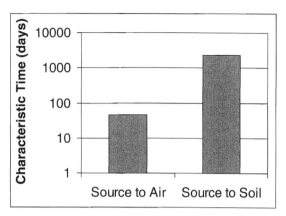

Figure 11.5   Comparison of the characteristic time, τ, for TCDD with a steady source to air and a steady source to soil.

sum the concentrations. The fraction of the initial concentration at each urban center and the cumulative total are plotted with distance in Figure 11.6. Since the centers are slightly less than one CTD apart, approximately one-third of the concentration at Center B results from Center A. If we consider the effect of managing the local sources in Center B relative to input from Center A, we determine that although reducing emissions at Center B will impact concentrations, a more regional approach is necessary for additional reductions. Thus, chemicals with long spatial ranges need to be managed regionally. This knowledge can be used to reduce human exposures through regional strategies that address multiple sources in place of single-source management.

We completed a Monte Carlo uncertainty analysis with 5000 simulations to generate a distribution of plausible CTD values for TCDD with the Crystal Ball software package.[50] The range of values for the CTD of TCDD was from 100 to 1000 km. This range results from both uncertainty in chemical properties and variability in landscape properties.

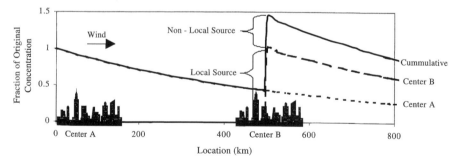

Figure 11.6   For a case study using TCDD, we plot the fraction of the initial concentration with distance from an urban center for Center A, Center B, and the cumulative total.

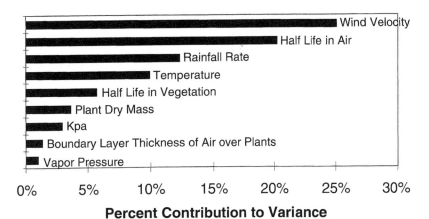

Percent Contribution to Variance

Figure 11.7 Results of sensitivity analysis for the four-compartment steady-state system for the TCDD case study.

To understand which inputs contribute significantly to the uncertainty in the output, a sensitivity analysis was performed to calculate the rank correlation coefficients between CTD and its defining parameters. The rank correlation coefficients are a measure of the strength of the linear relationship between each input and the CTD. This technique considers both the range of uncertainty in the input parameter and the influence of the parameter on the CTD. The rank correlation coefficients are squared and the values normalized to 100 percent to determine the approximate contribution to variance. The most influential parameters are displayed in Figure 11.7. Decreasing the uncertainty and variability in these parameters would have the largest impact for reducing the estimated uncertainty range of CTD. The wind speed, rainfall rate, temperature, and plant biomass are all highly variable and are dependent on site-specific data. The half-life in air, half-life in vegetation, and vegetation–air partition coefficient are all poorly characterized in the literature and thus the uncertainty of CTDs could be reduced as these input values are better defined.

11.3.3 Evaluation of Multimedia Mass Distribution

To gain confidence in the four-compartment model, we used a stationary system with an area of 250,000 km^2, a continuous emission of 1.1 g/d, and a wind speed of 4 m/s through the system.[51,52] These values were selected to represent the conditions in Germany, where background concentrations have been measured.[53] The calculated TCDD concentrations appear to be in agreement with background measurements from Germany, as shown in Table 11.6.[53] Additionally, calculated transfer rates from air to soil are in the range of experimental data.[54] We also compared the model to a vegetation scavenging ratio determined for TCDD, measuring the vegetation concentration resulting from

TABLE 11.6   Fugacity, Inventory, Concentration, and Mass Transformed for Each Compartment for a Fixed Region Using TCDD

Compartment Name	Fugacity (Pa)	Calculated Concentration (gm/m^3)	Measured Concentration[53] (gm/m^3)	Mass Transformed (gm/d)
Air	2.02×10^{-14}	3.50×10^{-15}	3.60×10^{-15}	6.04×10^{-2}
Vegetation	3.85×10^{-15}	1.33×10^{-8}	5.43×10^{-9}	3.73×10^{-1}
Ground surface soil	3.46×10^{-15}	5.41×10^{-8}	5.58×10^{-8}	6.37×10^{-2}
Root-zone soil	3.20×10^{-15}	6.37×10^{-8}	7.09×10^{-8}	5.08×10^{-4}

all processes to the air concentration and found it to be the same order of magnitude.[55]

11.3.4   Population-Based Potential Dose

The uncertainty distributions for the population-based potential dose for both benzo[a]pyrene and TCDD using the spatial and urban models are shown in Figure 11.8. For benzo[a]pyrene, the potential dose is nearly the same using the two calculation methods because very little chemical is carried out of the system by advection. In contrast, for TCDD, the urban model predicts approximately one half the population dose as does the spatial model because much of the chemical is advected out of the model system by wind. This difference is

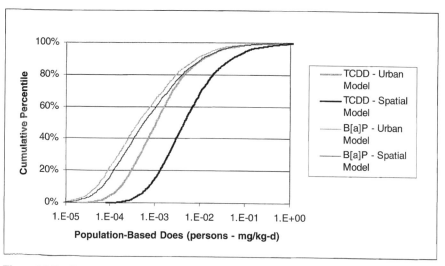

Figure 11.8   Cumulative percentile distribution of population-based potential dose for each calculation method for TCDD (a chemical with a long characteristic travel distance, CTD).

comparable to other sources of uncertainty in the calculation. Thus, for TCDD (and any other chemical with a long CTD such as DDT or hexochloroben-zene), it is important to consider using the spatial model to calculate popula-tion exposure as opposed to the urban model, which only considers the locally exposed population.

11.4 CONCLUSIONS AND RECOMMENDATIONS

In any effort to assess the potential adverse effects of a chemical, τ, CTD, and the population-based potential dose are important components of the analysis. From a policy perspective, these methods need to be transparent and represen-tative of the complex dynamic environment. In this chapter, we demonstrated simple methods for estimating all three measures.

The characteristic time can be used to evaluate whether a chemical is a POP while the CTD determines if a chemical has the potential for long-range trans-port. The duration over which impacts are assessed should be on the same order of magnitude as τ for that chemical. We also showed the importance of using the correct mode of entry into the environment a chemical such as TCDD.

The CTD is essential in any effort to assess the potential adverse effects of a chemical. Often established by political or geographical boundaries (air pollu-tion districts, state boundaries, etc.) or by tradition (i.e., a specified distance from a source), the "regulatory scale" is arbitrary and often not defensible by scientific analysis. In this chapter, we have demonstrated simple methods for making preliminary estimates of the CTD that defines the effective range of impact for a chemical contaminant. The CTD is needed before deciding what scale to use when measuring or modeling the dispersion of an environmental contaminant. This method was evaluated through a case study of TCDD, for which we found a CTD on the order of 600 km. This value is realistic when considered in conjunction with available monitoring data. If a chemical has a CTD on the order of magnitude of, or greater than, the typical distance be-tween urban centers, it is important to regulate the chemical regionally, as op-posed to locally.

When making choices among alternative chemicals for use in a certain pro-cess, evaluating a new chemical upon its introduction to commerce, or deciding if one should further regulate a chemical presently in use, we may want to de-termine the population-based potential dose to that chemical per unit release. If a chemical has a long CTD in the environment, calculating the risk only to in-dividuals near the site may be insufficient, as we must consider the exposure to the population far from the site. We show that as the CTD increases, the differences between calculation methods can be on the same order of magni-tude as contributions from other sources of uncertainty in the calculation (this was the case here for TCDD). It is important to note that the comparison be-tween sources of uncertainty is chemical specific. We recommend that as the

CTD of the chemical pollutant increases, the population-based potential dose should be calculated and the model used for the calculation carefully selected.

In all of the calculations, several assumptions were made, including spatially independent exposure parameters, a linear cancer slope factor at low doses, no dispersion of airborne chemicals, all sources located in urban regions, and a uniform population density in each of the three categories, each assumed to be in a uniform geographical pattern. We have not evaluated the effects of these assumptions on the reliability of the results because this model is intended for screening-level purposes to compare chemicals, not to determine the level of risk.

These measures are useful for determining the potential for exposure prior to introducing a new chemical to the market. The list of possible uses for such measures of persistence include risk assessments, life-cycle impact analyses, development of pollution prevention strategies, evaluation of pollutant mass balances, comparisons between toxic release inventories, and regulatory impact studies.

REFERENCES

1. National Research Council. *Science and Judgment in Risk Assessment.* National Academy Press, Washington, DC, 1994.

2. K. A. Kidd, R. H. Hesslein, B. J. Ross, K. Koczanski, G. R. Stephens, and D. C. G. Muir. Bioaccumulation of organochlorines through a remote freshwater food web in the Canadian Arctic. *Environ. Pollut.* 102: 91–103 (1998).

3. D. C. G. Muir, N. P. Grift, W. L. Lockhart, P. Wilkinson, B. N. Billeck, and G. J. Brunskill. Spatial trends and historical profiles of organochlorine pesticides in arctic lake sediments. *Sci. Tot. Environ.* 161: 447–457 (1995).

4. L. McConnell, J. LeNoir, S. Datta, and J. Seibert. Wet deposition of current-use pesticides in the Sierra Nevada Mountain Range, California, USA. *Environ. Toxicol. Chem.* 17: 1908–1916 (1998).

5. S. Datta, L. McMonnell, J. Baker, J. Lenoir, and J. Seiber. Evidence for atmospheric transport and deposition of polychlorinated biphenyls to the Lake Tahoe basin, California–Nevada. *Environ. Sci. Technol.* 32: 1378–1385 (1998).

6. D. R. Cortes, I. Basu, C. W. Sweet, K. A. Brice, R. M. Hoff, and R. A. Hites. Temporal trends in gas-phase concentrations of chlorinated pesticides measured at the shores of the Great Lakes. *Environ. Sci. Technol.* 32: 1920–1927 (1998).

7. B. R. Hillery, I. Basu, C. W. Sweet, and R. A. Hites. Temporal and spatial trends in a long-term study of gas phase PCB concentrations near the Great Lakes. *Environ. Sci. Technol.* 31: 1811–1816 (1997).

8. Environmental Protection Agency Science Advisor Board (SAB). *A Guide to Risk Ranking, Risk Reduction, and Research Planning.* U.S. Environmental Protection Agency, Washington, DC, 1995.

9. D. Mackay and S. Paterson. Evaluating the multimedia fate of organic chemicals— A level-III fugacity model. *Environ. Sci. Technol.* 25: 427–436 (1991).

10. T. E. McKone. *CalTOX, A Multimedia Total-Exposure Model for Hazardous-Wastes Sites Part I: Executive Summary.* UCRL-CR-111456PTI. Prepared for the Department of Toxic Substances Control, California Environmental Protection Agency, Lawrence Livermore National Laboratory, Livermore, CA, 1993.

11. T. E. McKone. *CalTOX, A Multimedia Total-Exposure Model for Hazardous-Wastes Sites Part II: The Dynamic Multimedia Transport and Transformation Model.* UCRL-CR-111456PTII. Prepared for the Department of Toxic Substances Control, California Environmental Protection Agency, Lawrence Livermore National Laboratory, Livermore, CA, 1993.

12. L. J. Brandes, H. den Hollander, and D. van de Meent. *SimpleBox 2.0: A Nested Multimedia Fate Model for Evaluating the Environmental Fate of Chemicals.* RIVM, The Netherlands, 1996.

13. M. Scheringer. Characterization of the environmental distribution behavior of organic chemicals by means of persistence and spatial range. *Environ. Sci. Technol.* 31: 2891–2897 (1997).

14. F. Wania and D. Mackay. Tracking the distribution of persistent organic pollutants. *Environ. Sci. Technol.* 30: A390–A396 (1996).

15. K. M. Thompson and J. S. Evans. The value of improved national exposure information for perchloroethylene Perc: A case study for dry cleaners. *Risk Anal.* 17: 253–271 (1997).

16. T. Webster and P. Connett. Cumulative impact of incineration on agriculture: A screeing procedure for calculating population risk. *Chemosphere* 19: 597–602 (1989).

17. C. C. Travis and S. Hester. Global chemical pollution. *Environ. Sci. Technol.* 25: 814–819 (1991).

18. A. Horvath, C. Hendrickson, L. Lave, F. McMichael, and T. Wu. Toxic emissions indices for green design and inventory. *Environ. Sci. Technol.* 29: A86–A90 (1995).

19. E. Hertwich, W. Pease, and T. McKone. Evaluating toxic impact assessment methods: What works best? *Environ. Sci. Technol.* 32: A138–A144 (1998).

20. C. Jia, A. Di Guardo, and D. Mackay. Toxic release inventories: Opportunities for improved presentation and interpretation. *Environ. Sci. Technol.* 30: 86A–91A (1996).

21. D. H. Bennett, M. Matthies, T. E. McKone, and W. E. Kastenberg. General formulation of characteristic travel distance for semi-volatile organic chemicals in a multi-media environment. *Environ. Sci. Technol.* 32: 4023–4030 (1998).

22. T. E. McKone. Alternative modeling approaches for contaminant fate in soils: Uncertainty, variability, and reliability. *Reliabil. Eng. Syst. Safety* 54: 165–181 (1996).

23. R. L. Maddalena, T. E. McKone, D. W. Layton, and D. P. H. Hsieh. Comparison of multi-media transport and transformation models: Regional fugacity model vs. CalTOX. *Chemosphere* 30: 869–899 (1995).

24. D. Mackay. *Multimedia Environmental Models, the Fugacity Approach.* Lewis, Chelsea, MI, 1991.

25. D. Komossa and H. Sandermann. Plant metabolic studies of the growth regulator maleic hydrazide. *J. Agric. Food Chem.* 43: 2713–2715 (1995).

26. D. Mackay, W. Y. Shiu, and K. C. Ma. *Illustrated Handbook of Physical-Chemical Properties and Environmental Fate for Organic Chemicals*, Vols. 1–4. Lewis, Boca Raton, FL, 1995.

27. Q. P. Ye, R. K. Puri, S. Kapila, and A. F. Yanders. Studies on the transport and transformation of PCBS in plants. *Chemosphere* 25: 1475–1479 (1992).

28. W. Schwack, W. Andlauer, and W. Armbruster. Photochemistry of parathion in the plant cuticle environment—Model reactions in the presence of 2-propanol and methyl 12-hydroxystearate. *Pesticide Sci.* 40: 279–284 (1994).

29. D. G. Crosby and A. S. Wong. Environmental degradation of 2,3,7,8-tetrachloro-dibenzo-*p*-dioxin (TCDD). *Science* 195: 1337–1338 (1977).

30. J. R. Lamarsh. *Introduction to Nuclear Engineering*, 2nd ed. Addison-Wesley, Reading, MA, 1983.

31. D. H. Bennett, W. E. Kastenberg, and T. E. McKone. General formulation of characteristic time for persistent organic chemicals in a multimedia environment. *Environ. Sci. Technol.* 33: 503–509 (1999).

32. E. Webster, D. Mackay, and F. Wania. Evaluating environmental persistence. *Environ. Toxicol. Chem.* 17: 2148–2158 (1998).

33. A. Beyer, D. Mackay, M. Matthies, F. Wania, and E. Webster. Assessing long-range transport potential of persistent organic pollutants. *Environ. Sci. Technol.* 34: 699–703 (2000).

34. M. Scheringer, D. H. Bennett, T. E. McKone, and K. Hungerbuhler. In R. L. Lipnick, et al. (Eds.), (2002) *Persistent Toxic Bioaccumulative Chemicals: Fate and Exposure*. American Chemical Society, Washington DC (in press).

35. U. S. Environmental Protection Agency (EPA). *Exposure Factors Handbook*. EPA/600/8-89/043. EPA, Office of Health and Environmental Assessment, Washington, DC, 1989.

36. T. E. McKone. *A Multimedia Total-Exposure Model for Hazardous-Wastes Sites Part III: The Multiple-Pathway Exposure Model*. UCRL-CR-111456PtII. Prepared for the State of California, Department Toxic Substances Control, Lawrence Livermore National Laboratory, Livermore, CA, 1993.

37. D. Calamari, E. Bacci, S. Focardi, C. Gaggi, M. Morosini, and M. Vighi. Role of plant biomass in the global environmental partitioning of chlorinated hydrocarbons. *Environ. Sci. Technol.* 25: 1489–1495 (1991).

38. B. D. Eitzer and R. A. Hites. Polychlorinated dibenzo-*para*-dioxins and dibenzofurans in the ambient atmosphere of Bloomington, Indiana. *Environ. Sci. Technol.* 23: 1389–1395 (1989).

39. D. J. Paustenbach, R. J. Wenning, V. Lau, N. W. Harrington, D. K. Rennix, and A. H. Parsons. Recent developments on the hazards posed by 2,3,7,8-tetrachloro-dibenzo-*para*-dioxin in soil—implications for setting risk-based cleanup levels at residential and industrial sites. *J. Toxicol. Environ. Health* 36: 103–149 (1992).

40. A. di Domenico, S. Cerlesi, and S. Ratti. A 2-exponential model to describe the vanishing trend of 2,3,7,8-tetrachlorodibenzodioxin (TCDD) in the soil at Seveso, northern Italy. *Chemosphere* 20: 1559–1566 (1990).

41. M. Arthur and J. Frea. 2,3,7,8-Tetrachlorodibenzo-*p*-dioxin: Aspects of its important properties and its potential biodegradation in soils. *J. Environ. Qual.* 18: 1–11 (1989).

42. F. Schuler, P. Schmid, and C. Schlatter. Photodegradation of polychlorinated dibenzo-*p*-dioxins and dibenzofurans in cuticural waxes of Laural Cherry (*Prunus Laurocerasus*). *Chemosphere* 36: 21–34 (1998).

43. W. W. Brubaker and R. A. Hites. Polychlorinated dibenzo-*p*-dioxins and dibenzo-furans: Gas phase hydroxyl radical reactions and related atmospheric removal. *Environ. Sci. Technol.* 31: 1805–1810 (1997).

44. E. S. C. Kwok, J. Arey, and R. Atkinson. Gas-phase atmospheric chemistry of dibenzo-*p*-dioxin and dibenzofuran. *Environ. Sci. Technol.* 28: 528–533 (1994).

45. J. K. McCrady and S. P. Maggard. Uptake and photodegradation of 2,3,7,8-tetra-chlorodibenzo-*p*-dioxin sorbed to grass foliage. *Environ. Sci. Technol.* 27: 343–350 (1993).

46. D. M. Wagrowski and R. A. Hites. Polycyclic aromatic hydrocarbon accumulation in urban, suburban, and rural vegetation. *Environ. Sci. Technol.* 31: 279–282 (1997).

47. J. N. S. Eisenberg, D. H. Bennett, and T. E. McKone. Chemical dynamics of persistent organic pollutants: A sensitivity analysis relating soil concentration levels to atmospheric emissions. *Environ. Sci. Technol.* 32: 115–123 (1998).

48. C. C. Travis, H. A. Hattemerfrey, and E. Silbergeld. Dioxin, dioxin everywhere. *Environ. Sci. Technol.* 23: 1061–1063 (1989).

49. R. Lohmann and K. C. Jones. Dioxins and furans in air and deposition: A review of levels, behaviour and processes. *Sci. Tot. Environ.* 219: 53–81 (1998).

50. Decisioneering. *Crystal Ball.* Boulder, CO, 1996.

51. H. Fiedler. EPA dioxin-reassessment: Implications for Germany. *Organohalogen Compounds* 22: 209–228 (1995).

52. H. Fiedler and O. Hutzinger. Sources and sinks of dioxins: Germany. *Chemosphere* 25: 1487–1491 (1992).

53. M. S. McLachlan. Bioaccumulation of hydrophoobic chemicals in agricultural food chains. *Environ. Sci. Technol.* 30: 252–259 (1996).

54. C. J. Koester and R. A. Hites. Wet and dry deposition of chlorinated dioxins and furans. *Environ. Sci. Technol.* 26: 1375–1382 (1992).

55. K. Jones and R. Duarte-Davidson. Transfers of airborne PCDD/Fs to bulk deposition collectors and herbage. *Environ. Sci. Technol.* 31: 2937–2943 (1997).

56. CalTOX. Available from *http://www.cwo.com/~herd1/caltox.htm*, 1999.

57. F. Wania and D. Mackay. A global distribution model for persistent organic chemicals. *Sci. Tot. Environ.* 161: 211–232 (1995).

58. T. F. Bidleman. Atmospheric processes. *Environ. Sci. Technol.* 22: 361–367 (1988).

59. L. J. Thibodeaux. *Environmental Chemodynamics: Movement of Chemicals in Air, Water, and Soil.* Wiley, New York, 1996.

60. T. E. McKone and P. B. Ryan. Human exposures to chemicals through food chains—An uncertainty analysis. *Environ. Sci. Technol.* 23: 1154–1163 (1989).

SECTION F
Evaluating Occupational Hazards

12 Methods for Setting Occupational Exposure Limits

HON-WING LEUNG

Independent Consultant, Danbury, Connecticut

12.1 INTRODUCTION

Over the past century there has been an enormous increase in recognition of the potential hazards of exposure to chemical substances in the workplace. Tremendous progress has been made in the prevention of chemically induced occupational illnesses. In the early days, however, investigations of occupational disease were rather crude and relied almost exclusively on anecdotal reports of working conditions. Evaluation of the significance and causality of exposure was hampered by the subjective nature of the observations and by limited analytical capabilities. More recently, advances in the areas of toxicology, industrial hygiene, and analytical chemistry provide the necessary tools for a better understanding of the biologic response and for a more accurate determination of chemical exposure.[1] These technical enhancements, together with the implementation of intensive testing programs, have led to a better assessment of the health risks.[2] Furthermore, considerable resources and effort have been devoted to managing those risks by implementing occupational hygiene programs to control exposures.

Among the programs credited with a substantial improvement in the prevention of occupational illness are those designed to reduce workers' exposure to chemicals. Since exposure to chemicals in the workplace cannot be totally eliminated, it becomes necessary to define some acceptable levels of exposure at which a worker's health will not be jeopardized. These standards are often called occupational exposure limits (OELs), although they are known by various names and acronyms (see below for examples). There are two categories of OELs. One is based on the evaluation of chemical concentrations in environmental media such as air or contaminated surfaces. The other measures the concentration of a chemical or its metabolite in biological media such as urine, blood, or expired air. Many more OELs based on measurements in environ-

Human and Ecological Risk Assessment: Theory and Practice, Edited by Dennis J. Paustenbach
ISBN 0-471-14747-8 © 2002 John Wiley & Sons, Inc.

mental media have been developed because of their relative ease and simplicity of use. At the present time, they enjoy a commanding lead in popularity among hygiene professionals. However, OELs based on biomarkers do offer many unique advantages and undoubtedly will attain wider acceptance when these benefits are recognized.

12.2   HISTORY OF OCCUPATIONAL EXPOSURE LIMITS

The American Conference of Governmental Industrial Hygienists (ACGIH) pioneered in the setting of OELs with the release in 1941 of a first set of 63 standards known as threshold limit values (TLVs).[3] Over the past 50 years, many organizations in numerous countries have also engaged in developing OELs.[4] Presently at the national level in the United States, there are at least six groups that recommend OELs. These include the TLVs of the ACGIH, the recommended exposure limits (RELs) of the National Institute for Occupational Safety and Health (NIOSH), the permissible exposure limits (PELs) of the Occupational Safety and Health Administration (OSHA), the workplace environmental exposure limits (WEELs) of the American Industrial Hygiene Association (AIHA), and the standards for workplace air contaminants of the American National Standards Institute (ANSI). In addition to these national organizations, many local, state, and regional governments have also established OELs.

Outside the United States, as many as 50 other groups have set workplace exposure limits. Prominent among them are the maximum allowable concentrations (MAKs) of the Deutche Forschungsgemeinschaft (DFG) in Germany,[5] and (MACs) of the former Union of Soviet Socialist Republics.[4] Many of these limits are nearly or exactly the same as those developed in the United States. In some cases, such as in Russia, Japan, and the former Soviet bloc countries, the limits are dramatically different. The differences may be due to a number of factors:

1. Difference in the philosophical objective of the limits and the undesirable effects they are meant to minimize or eliminate
2. Difference in the predominant age and sex of the workers
3. The duration of the average workweek
4. The economic state of affairs in that country
5. Consideration of OELs as voluntary guidelines rather than as enforceable standards

In recognition of the problems posed by these differences, the International Occupational Hygiene Association (IOHA) in September of 1997 convened a workshop at the third conference in Crans Montana, Switzerland, on the harmonization of OEL values among various world bodies.

Worldwide, as of 1986, it is estimated that fewer than 1700 OELs had been established.[6] This represents a tiny fraction of the 65,000 chemicals on the Environmental Protection Agency's (EPA's) Toxic Substances Control Act Inventory list. There is clearly a need to establish OELs for more chemicals. To fulfill this need, many chemical and pharmaceutical companies have initiated programs to develop their own OELs.[6] Unfortunately, because of concerns about business confidentiality and litigation, most of the corporate OELs have not been publicized. An effort to make these corporate limits more accessible while overcoming the corporations' concerns is that of the Occupational Toxicology Round Table, which serves as a clearinghouse to compile and disseminate the corporate OELs. The 1998 listing contains corporate OELs for well over 650 new chemicals.

12.3   RISK ASSESSMENT CONSIDERATIONS

OELs are based on the premise that, although all chemical substances are toxic at some concentrations, there exists a level of exposure at which no injurious effect should result (i.e., a threshold).[7] This principle also applies to substances that cause irritation, nuisance, or other forms of stress as the primary biological effects.

It is important to recognize that OELs are meant to protect nearly all workers, but they do not necessarily prevent discomfort or injury for everyone. Because of the wide range in individual susceptibility, a small fraction of workers may experience discomfort at levels at or below the OELs, and some may be affected more seriously by aggravation of a preexisting condition. This tenet of establishing OELs, although less than ideal, is a practical one since controlling exposures to extremely low levels as to protect all hypersusceptible individuals would be infeasible due to either engineering or economic limitations.[8]

OELs are based on the best available information from industrial experience and experimental human and animal studies. The rationale for each of the established values may differ from substance to substance. Protection against impairment of health may be a guiding principle for some, while reasonable freedom from irritation, narcosis, nuisance, or other forms of stress may form the basis for others. The age and completeness of the information available for establishing OELs may also vary; consequently, the precision of each particular OEL is subject to variation. Also, the methods or approach used to set OELs reflect the professional judgment of those developing them. Hence, there is considerable uncertainty inherent in the OEL-setting process.[9]

The criteria used to develop OELs may be classified into four groups: morphologic, functional, biochemical, and miscellaneous (nuisance, cosmetic). Historically most of the OELs have been derived from human data, but in recent years an increasing number of OELs are based on animal data.[10] About half of the existing OELs are set to prevent systemic toxic effects, while 40 percent are based on irritation, and only about 1 to 2 percent are intended to prevent cancer. By setting OELs for cancer-causing agents, the ACGIH and its counterparts throughout the world acknowledge that chemical carcinogens are likely to have a threshold, or at least a practical threshold. To support such a view, they cite biochemical, pharmacokinetic, and toxicological evidence demonstrating inherent, built-in anticarcinogens and other defensive processes in the human bodies. However, there is another equally credible school of thought that there is little or no evidence for the existence of thresholds for chemicals that are genotoxic. To take into account the philosophical postulate that chemical carcinogens do not have a threshold, mathematical modeling approaches for extrapolating the carcinogenic response to low doses have been developed.

12.4   FACTORS TO CONSIDER WHEN SETTING OELS

OELs are essential tools for relating the medical information to potential exposure in the workplace. As such, they must be carefully developed using all available physicochemical, toxicological, and clinical data and reviewed periodically to assure their appropriateness relative to current methods and data.[11] The following describes the types of data that should be considered when setting OELs.

12.4.1   Physicochemical Properties

The physicochemical properties of a substance govern its potential for exposure, mode of entry, and ability to cross biological barriers. For example, the physical state, melting and boiling points, vapor pressure, density, and particle size may indicate whether, under ambient conditions, the chemical is a gas, vapor, mist, or inhalable dust and whether exposure will most likely be by inhalation and/or dermal absorption. The ability of a substance to cross biological barriers and damage tissue is often determined by parameters such as molecular weight, lipid and water solubility, partition coefficient, dissociation constant, pH, and reactivity. Odor threshold is another property to consider, especially for foul-smelling substances.

12.4.2   Nuisance Effects

The kind of nuisance effects usually encountered (objectionable odor or taste, staining of the skin or clothes, etc.) can often be used as warning signs if they occur at levels lower than those where other effects occur. However, to provide a comfortable work environment, OELs need only be set just below nuisance levels.

12.4.3   Pharmacokinetics and Metabolism

Absorption of chemicals occurs when they come in contact with biological barriers they are able to cross. This often depends upon the solubility of the chemical. For example, gases that are highly soluble in water tend to be absorbed from the upper airways of the lung, whereas gases that are less soluble can penetrate to the deeper, air exchange regions of the lung. Dermal absorption, on the other hand, is usually favored by lipophilic compounds. For chemicals that may be dermally absorbed, a "skin" notation is usually added to the numerical OEL. The skin notation specifically identifies those chemicals that dermal exposure can result in absorption and contribute significantly to the total body burden.[12]

The distribution of a chemical depends on its partition coefficient, the perfusion rate to various organs, and the existence of specialized transport systems. The toxicity of a chemical is influenced by the target organ to which it is distributed. Plasma concentrations may serve as an index of tissue burden, but they may not necessarily reflect the concentration at the site of toxic action.[13]

Metabolic pathways are involved in toxifying and detoxifying chemicals. The enzyme systems responsible for these metabolic processes may themselves be either inhibited or activated by chemical exposures. This may lead to a synergistic or antagonistic effect with other chemicals. As a result of different enzyme profiles, human metabolism of an exogenous chemical may be different than that of the species used in a toxicological study. Any health risk assessment must account for such differences. The rate at which chemicals and their toxic metabolites are eliminated from the body is an important consideration when setting OELs. If the biological half-life is short (less than 3 hours), then the compound is considered to be cleared before the next day of exposure.[14] However, all chemicals are not eliminated so rapidly. Highly lipophilic chemicals tend to accumulate in the adipose tissue of the body. Also, repeated exposure to slowly metabolized or eliminated substances may result in a progressive increase in body burden. An adjustment factor has been proposed to account for this in setting OELs.[15]

12.4.4   Toxicity Data

The amount of toxicological data available to conduct a health risk assessment for the purpose of setting OELs varies widely. The nature and quality of the studies, relevance of the experimental routes of exposure, and significance of the observed effects to human health must be evaluated. Only data considered to be of sufficient quality are used, while data of lesser quality are viewed as supportive.

Acute Effects   Acute effects include lethality data, clinical signs, and irritancy data. Acute percutaneous studies can provide an indication of a chemical's potential to be absorbed through the skin. Although acute lethality data are usually not very useful in setting OELs, for some compounds these are the only data available. In such cases, any supplemental information including clinical signs and pharmacotoxic responses observed are especially useful. Irritancy data, especially sensory irritation, are very useful in setting OELs.[16]

Sensitization   Certain chemicals have the potential to induce sensitization in workers. This may involve the skin (allergic dermatitis) and/or the lung (occupational asthma). Although recent research indicates that thresholds for sensitization effects may exist, it is not always practical to develop an OEL to protect individuals who have already been sensitized. It is more appropriate to set an OEL that will prevent an individual from becoming sensitized in the first place.[8] To alert industrial hygienists about chemical sensitizers, the DFG in Germany has pioneered the inclusion of a S notation to its MAKs. Recently, the ACGIH has also begun identifying sensitizers with a SEN notation to its TLVs. Similarly, the AIHA is adding RSEN and DSEN notations to its WEELs for respiratory sensitizer and dermal sensitizer, respectively.

Repeated-Dose Toxicity   Studies involving repeated exposures of animals to toxic chemicals are much more useful for setting OELs than are acute studies. Subacute and subchronic toxicity studies in rodents include exposures for 14, 28, or 90 days or up to 6 months, and chronic toxicity involves exposure duration up to an animal's natural life span. The most useful studies for setting OELs are those for which a no observed adverse effect level (NOAEL) or a lowest observed adverse effect level (LOAEL) has been determined.

Reproductive and Developmental Effects   The endpoints considered most serious in these types of evaluations are teratogenicity and loss of fertility. However, other endpoints that should be considered include embryotoxicity, fetotoxicity, maternal toxicity, and neurobehavioral changes in the offspring. These studies can involve one or more generations.

Genotoxicity Data   Bacterial gene mutation test is often one of the first to be performed on a new chemical substance. Others include both in vitro and in vivo tests for mutagenic and clastogenic effects, such as micronucleus, sister chromatid exchange, chromosome aberration, unscheduled deoxyribonucleic acid (DNA) synthesis, alkaline elusion, dominant/recessive lethal, and transgenic animals. Genotoxicity data should not be used as a substitute for carcinogenicity data. However, the ability of some mutagenicity tests to predict the carcinogenic potential of structurally similar compounds in animals and humans should be considered.

Cancer Data   There are in general two broad types of chemical carcinogens classified according to their mechanism of action: genotoxic and epigenetic. The genotoxic carcinogens are believed to pose a greater concern because they are capable of producing gene mutation or cause injury to DNA. However, recognition that the human body possesses systems to repair the damaged DNA leads many scientists in the OEL-setting organizations to believe that all chemical carcinogens should at least have a "practical threshold." This term simply means that there exists a dose that would not be anticipated to pose a significant cancer risk. Based on this concept, it is therefore possible to set an OEL for any chemical carcinogen.

To alert industrial hygienists about chemical carcinogens, both the ACGIH and DFG have long adopted the use of a cancer notation in their TLVs and MAKs, respectively. Initially chemical carcinogens were put into two categories—human or animal carcinogen—and later the classification has been expanded to account for the various mechanisms of the carcinogenic process.

Biological Relevance of Animal Toxicity Data to Humans Pathological findings in an animal species exposed to a chemical are certainly a cause for alarm. However, in some circumstances, the relevance of the observation in animals to human is uncertain.[17] For example, questions have been raised regarding the relevance of tumors that are unique to certain animal species such as forestomach, Zymbal gland, Harderian gland, and preputial gland, sites that have no human equivalents. Other examples include hyaline droplet nephropathy in male rats related to the deposition of α_{2u}-globulin in the proximal renal tubule.[18] Neither female rats nor any other animal tested, including humans, produce this protein. Similar skepticism applies to chemicals whose rates or pathway of biotransformation (e.g., saturation of metabolic enzymes) or mechanisms of pathogenesis (e.g., peroxisome proliferation) differ significantly between animals and humans.[19] Finally, animals in toxicity testing are routinely exposed to concentrations of chemicals much higher than would be encountered in the occupational setting. While these studies should not be ignored in setting OELs, their relevance needs to be carefully reviewed with respect to other toxicological data using a weight of evidence approach. In most cases these studies should not constitute the major factor driving the OEL.

12.4.5 Hierarchical Ordering of Data

It is difficult to list an absolute hierarchy of toxicological endpoints for setting OELs. Priority should be made on a case-by-case basis using the professional judgment of the toxicologist. However, the following general guidelines should be followed to minimize the uncertainties inherent in data extrapolation:

1. Selection of a NOAEL over a LOAEL will eliminate the need to extrapolate to a dose at which no effect would be expected.
2. Studies performed by inhalation best simulate the primary route of exposure to workplace chemicals. Therefore, selection of an inhalation study would not require the route-to-route extrapolation normally required for studies of oral or parenteral administration.
3. Human data should, in general, take priority over animal data since no interspecies extrapolation would be necessary.
4. Chronic studies, which reflect the long-term repeated exposure in the working environment, are usually more appropriate than acute studies.

5. The most sensitive endpoint that is biologically relevant to humans is preferred.
6. Irreversible effects such as birth anomalies or cancer generally take priority over reversible effects such as irritation or elevated liver enzymes. However, if the LOAEL of a reversible effect is much lower than the NOAEL of an irreversible effect, the former might be selected.

12.4.6 Exposure and Population Parameters

Exposure and population parameters involved in the derivation of OELs include the volume of air inhaled during a work shift, the length of that shift, the duration and type of a handling operation, the number of years of continuous handling of the chemical in question, and certain demographics of the target population. Traditionally, a worker is assumed to be handling a chemical 8 hours a day, 5 days a week, for a 40-year working lifetime. Most OELs are established as the time-weighted average (TWA) concentration for this exposure scenario, to which it is believed that nearly all workers may be repeatedly exposed, day after day, without adverse effect. Excursions above the TWA are permitted provided they are compensated by equivalent excursions below during the workday. The magnitude by which the TWA may be exceeded depends on a number of factors and can vary from substance to substance. As a general rule of thumb, the ACGIH TLV committee recommends that excursions may exceed 3 times the TWA for no more than a total of 30 minutes during a workday, and under no circumstance should they exceed 5 times the TWA.

For certain chemicals that produce acute effects such as irritation and central nervous system depression, a short-term exposure limit (STEL) or ceiling limit may be more appropriate. Both the STEL and ceiling limits denote concentrations that should not be exceeded during any part of the working exposure.

A volume of 10 m^3 of inspired air per 8-hour work shift has been most frequently used for OEL calculations. This figure is derived by assuming that a man engaging in light-duty work has a tidal volume of 1000 cm^3 and a breathing rate of 20 breaths per minute for 8 hours.[20] This inhalation rate is likely an overestimation for many occupations. An extensive survey of the exposure factors in the North American population conducted by the U.S. EPA[21] arrives at a figure of 4.8 m^3 for an average human adult performing light-duty work for 8 hours (Table 12.1). In addition, most jobs, in reality, do not require continuous activity but have many resting periods within the 8-hour workday. The inhalation rate for an average human adult at rest is about 4 m^3. Despite the lower, more realistic estimate of the inhalation rate for a typical worker, the figure of 10 m^3 is entrenched in the OEL-setting process. OELs set with this higher inhalation rate therefore are intrinsically conservative since they may have overestimated the exposure potential by as much as twofold.

TABLE 12.1   Body Weight and Inhalation Rates of Average Human Adult[a]

Activity Level	Body Weight (kg)	Inhalation Rate (m³/8 h)			
		Resting	Light	Moderate	Heavy
Man	78.1	5.6	6.4	20.0	38.4
Woman	65.4	2.4	3.0	12.8	23.2
Man and woman	71.8	4.0	4.8	16.8	31.2

[a] 18 to 75 years of age.

While OELs are established for the typical 8-hour workday, many manufacturing operations employ extended shifts. For these operations the OELs may need to be adjusted to reflect the increased exposure potential.[14,22]

Often OELs are based on an NOAEL identified in an animal toxicity study. To extrapolate the animal NOAEL to the human equivalent, the standard body weight used for a human is 70 kg. OELs are designed to protect a homogeneous population of healthy workers. This differs from ambient air quality standards, which apply to the general population and must consider the most sensitive individuals in the community (e.g., children, the elderly, the infirm, and hypersensitive individuals). However, it is recognized that subpopulations of otherwise healthy workers may be subject to idiosyncratic reactions to certain chemicals. Additional protection or controls may be required to accommodate these unusually sensitive individuals.[8]

12.4.7   Measurement

Measurements of exposures require the availability of sensitive and accurate sampling and analytical methods. While the ability to detect a chemical at the OEL level is important, it should not drive the OEL-setting process. Rather, the OEL should be set based on the scientific principles outlined in the preceding sections, regardless of whether the levels may present a challenge to the analytical chemist. Modern advances in instrumentation and technologies should allow the detection of increasingly lower levels of most chemicals.

Methods developed to support industrial hygiene measurements should be designed according to the length and type of exposure of concern, that is, time-weighted averages or short-term exposures. In addition, where a compound presents an acute hazard, continuous reading methods should be developed to warn of concentrations approaching dangerously high levels.

12.4.8   Documentation

An OEL is only as valid as the data, method, and assumptions used in its development. These should be clearly documented for future reference and verification. In the case of litigation the validity of the data and methods used as well as the qualifications of those involved may be called into question. Appropriate documentation can be used to verify the appropriateness and currency of the data and methods used, and provide a reference point for review when new data or improved scientific methods become available. In addition, the documentation will be used to explain its development to those who must apply it or work within it. Thus, it will assure employees that a thorough evaluation has been performed.

12.4.9 Controlling Exposures

The feasibility or ease of controlling exposures below the OEL should in no way dictate the OEL setting process. OELs represent exposure levels that can be considered acceptable based on an assessment of health risks. The management of the health risks should be handled separately by the industrial hygiene, engineering, and operation management personnel.

OELs can be used in the design and implementation of manufacturing operations, such as in the selection of the appropriate machinery or ventilation. The OEL should be interpreted as an acceptable level of exposure without the use of respirator or other protective equipment. Where engineering control is not sufficient to meet established OELs, personal protective equipment may be used. However, sole reliance on personal protective equipment to meet OELs is strongly discouraged. Implementing administrative controls may effect an additional level of control. These may take the form of restriction of sensitive individuals, such as those previously sensitized to an allergen, or women in the first trimester of pregnancy. Additional administrative controls may include shorter or alternating shifts, job rotation, or enforced breaks.

12.4.10 Medical Surveillance

It may be appropriate to implement a medical surveillance program for those employees involved in the handling of compounds deemed sufficiently hazardous. A medical surveillance program should focus on the most significant and relevant biological endpoint. In general this would be the same endpoint used for the development of the OEL, but may also include medical history, routine examinations, and diagnostic procedures. Medical surveillance programs can yield information specific to individual employees but may also identify trends in a potentially exposed population and may serve to alert health professionals to unknown or unexpected hazards.

12.5 METHODS FOR SETTING OCCUPATIONAL EXPOSURE LIMITS

The following provides a general discussion of the various methods for setting OELs, including their suitability, inherent assumptions, and limitations. It is not possible to develop a step-by-step approach that can be applied to all substances. The choice of any particular method depends greatly on the availability of data. In general, the simplest methods are used for chemicals with no or scanty toxicity data and are the least reliable. More complex methods are more suited for chemicals with a robust database. Generally, the selection of an endpoint should be based on the most sensitive adverse effect. The rationale is that by protecting against the most sensitive effect, all other less sensitive effects will also be automatically prevented. It is important to determine whether the most sensitive endpoint clearly presents an undesirable or injurious effect to the health of the workers. It may be worthwhile to determine and compare OELs based on two different endpoints. However, an OEL based on the most sensitive endpoint and developed by taking into consideration all potential adverse effects should be appropriate for preventing the most serious adverse effects. The ultimate selection will depend on the availability of data and the type of methodology used for the calculation.

Many approaches for deriving OELs from toxicological data have been proposed and put into use over the past half century. Although there are many ways to set OELs, the quantitative relationship between the extent of exposure to a chemical agent and the physiological response of the exposed population (dose–response relationship) is fundamental to the development of the standards. Several common methods and selected examples are presented below to illustrate their application. Figure 12.1 shows the decision logic for the various methods to set OELs.

12.5.1 Structure–Activity Relationships

Analogy   This is the crudest method of setting OELs. It is best used for chemicals lacking toxicity data. In this method, a homologous chemical is assumed to have the same potential to cause a common biological effect as a reference chemical in the same family of compounds. The chemical in question adopts the same OEL value as the reference chemical. One limitation of the analogy method is that it is only applicable to chemicals in a homologous series and isomers. It is the most unreliable method since the OEL of the reference chemical is also frequently established with limited data.

Example 12.1. The TLVs for ethylamine, methylamine, diethylamine, dimethylamine, and trimethylamine are all 5 ppm, based primarily on analogy to prevent irritative effects.

Example 12.2. The TLVs for methyl propyl ketone, methyl isopropyl ketone, and diethyl ketone are all 200 ppm, based primarily on analogy to prevent narcotic effects and significant irritation.

Correlation   This is a variation of the analogy method described above. The correlation method is best suited to a chemical that has no toxicity data but

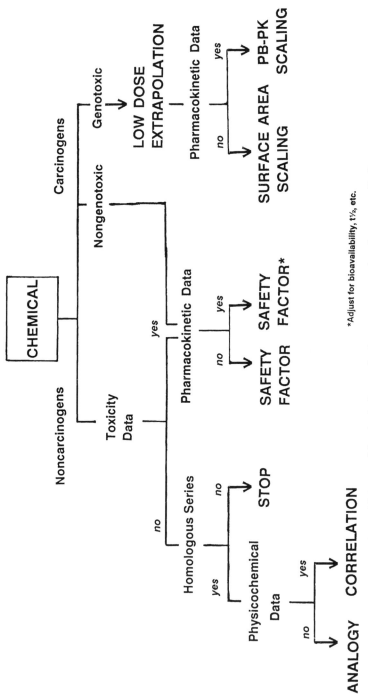

Figure 12.1 Decision logic for methods to set occupational exposure limits.

has some physicochemical data that can be compared with other chemicals in the homologous series. The rationale of this approach is based on the principle that biological effects can be induced by simple physical interactions between a chemical and receptor centers. If the effect of a series of chemicals is purely physical, then the thermodynamic properties can be used to provide a scale to express the potency of each member in a homologous series. The choice of the physicochemical parameter depends on the biological effect on which the OEL is based, for example, vapor pressure, chemical fugacity, acid dissociation constant, partition coefficient, ionization potential, receptor affinity, bonding energy, and intermolecular interaction. Rather than assuming that all chemicals in the same homologous series have identical potency and therefore the same OEL value as in the analogy method, the correlation method sets the OEL value proportional to the magnitude of the physicochemical parameter. The limitation of this method is that the physicochemical parameter must be firmly established as a valid predictor of the biological effect on which the OEL is based.

Example 12.3. Nielson and Alarie[23] determined the median concentration necessary to depress the respiratory rate of mice due to sensory irritation of the upper respiratory tract for a variety of alkylbenzenes. They found that the potency of the alkylbenzenes increased with chain length. However, the ratios of equipotent concentration/saturated vapor concentration for these alkylbenzenes varied very little. Knowing the vapor pressure enabled prediction of the sensory irritation potency and by extension the OELs for these alkylbenzenes. The quantitative relationship for a series of 8 alkylbenzenes is

$$OEL \text{ (ppm)} = 0.0054 \text{ (vapor pressure in ppm)}$$

The predicted values using this relationship were in close agreement with established TLVs for toluene, ethylbenzene, isopropylbenzene, and butyltoluene. OEL values for n-propylbenzene, n-butylbenzene, t-butylbenzene, n-amylbenzene, and n-hexylbenzene, which currently have no established TLVs, were estimated to be 50, 20, 20, 10, and 5 ppm, respectively.

Example 12.4. Leung and Paustenbach[24] observed that there exists an association between the equilibrium proton dissociation constant (pK_a) and the OELs of organic acids that produce irritation as the primary adverse effects. The regression equation for a series of 10 organic acids is $\log OEL \text{ (}\mu mol/m^3\text{)} = 0.43 pK_a + 0.53$. The OEL of a variety of organic acids can be readily predicted using the above relationship.

12.5.2 Uncertainty/Safety Factors

This method establishes OELs by applying an uncertainty and/or safety factor to the NOAEL identified in an animal toxicity study. The rationale of this approach assumes that a chemical has threshold characteristics, and reducing the NOAEL with appropriate factors can derive an acceptable level of exposure for humans. Since nongenotoxic carcinogens are believed to have threshold qualities, this approach is also applicable to these compounds. Uncertainty factors are used to account for the uncertainties in extrapolating toxicity data. These may include extrapolating from a LOAEL to a NOAEL, across one animal species to another, and from different routes of administration. Generally, these factors range from 1 to 10 and are selected based on an evaluation of the appropriateness or scientific validity of the data.[25] The more that is known of the properties of the compound and the better the studies supporting the information, the smaller the uncertainty factors should be. Thus, good science is rewarded since larger uncertainty factors are used when data are incomplete or less relevant. This allows more flexibility to utilize the limited data available. Safety factors are selected based on the level of protection deemed necessary to prevent adverse effects. The size of the safety factor should vary depending on the slope of the dose–response curve. For chemicals that exhibit a very steep dose–response relationship, a larger safety factor should be used. Similarly, a range of values is used to accord a margin of safety commensurate with the severity of the effect:[26]

$$OEL = NOAEL/(UF1)(UF2)(SF)(BR)$$

where UF1, the uncertainty in extrapolation to a chronic exposure NOAEL; UF2, the uncertainty from interspecies extrapolation; SF is the safety factor (the size of this factor depends on the nature and severity of effects); and BR, the breathing rate of a man performing light-duty work in an 8-hour workday ($10 \ m^3$). The composite factor (UF1)(UF2)(SF) typically may range from 1 to 1000. Table 12.2 gives the possible ranges and recommended default values for the adjustment factors used to derive OELs.

Example 12.5. A repeated inhalation study exposed rats, rabbits, and dogs at mean concentrations of 5 and 10 ppm phenyl ether vapor for 7 hours/day, 5

TABLE 12.2   Some Adjustment Factors Occasionally Used to Derive Occupational Exposure Limits

	Possible Range	Recommended Default	Adjustment Basis
Intraspecies variability	1–10	10	3 for pharmacokinetics 3 for pharmacodynamics
Interspecies variability	1–10	Allometric scaling	3 for pharmacokinetics 3 for pharmacodynamics
LOAEL to NOAEL	1–10	3	Chemical specific
Subchronic to chronic	1–10	3	Chemical specific
Route to route	0.01–100	1	Chemical specific
Modifying factor	1–10	1	Chemical specific

days per week for a total of 20 exposures. No signs of toxicity or irritation were observed in animals exposed to 5 ppm. Applying a safety factor of 5 on the NOAEL yields an OEL of 1 ppm.

A variation of the safety factor approach applies to chemicals related to naturally occurring substances produced in the human body. In this method, a level of exposure equivalent to a 1 percent increase in body burden over the daily endogenous production rate is considered not to pose any significant health effect. This is analogous to considering the daily endogenous production rate as the NOAEL and applying a safety factor of 100.

Example 12.6. In human males, the endogenous production of 17β-estradiol is about 50 μg/day. A typical worker is assumed to breathe 10 m^3 of air in an 8-hour workday.

$$\text{OEL for } 17\beta\text{-estradiol} = (50 \ \mu\text{g/day})/(100)(10 \ \text{m}^3/\text{day}) = 0.05 \ \mu\text{g/m}^3$$

12.5.3   Mathematical Models

Low-Dose Extrapolation   This method applies to chemical carcinogens, especially genotoxic carcinogens, which are believed to lack threshold characteristics. Since for these chemicals a response is presumed to occur at any dose, regardless of how small, it becomes necessary to select an arbitrary level that is considered as presenting an insignificant or de minimis level of risk. For workplace standards, such a level is traditionally 1 in a 1000 (10^{-3} or 0.1 percent).[27] In this method the dose–response curve in the observed range of the rodent cancer bioassay or epidemiological study is extrapolated with the aid of a mathematical model downward to yield a dose corresponding to the 10^{-3} response level. The most widely utilized low-dose extrapolation methods are the one-hit, multistage, Weibull, multihit, logit, and probit model,[28] with the linearized multistage model most commonly used by regulatory agencies. There is, however, no compelling scientific basis for the choice of one model over many others.

If the modeling is based on rodent dose–response data, it is necessary to perform an interspecies extrapolation to humans. The extrapolation method used in conjunction with mathematical modeling differs from that for threshold chemicals, which utilizes uncertainty factors. In low-dose extrapolation, interspecies extrapolation is accomplished on the basis of surface area differences between the species. This is equivalent to applying an uncertainty factor of between 6 (rat to human) and 13 (mouse to human). This method of interspecies extrapolation is based on the rationale that tissue burden and hence the sensitivity of a species to a chemical is correlated with the rates of metabolism and clearance of the chemical, which are approximately proportional to the body surface area. Obviously, such a universal correction, which presupposes a chemical's mode of action, is not applicable to all chemicals. For instance, interspecies extrapolation based on surface area differences, which assumes that humans are more susceptible than rodents, is best applied for direct-acting carcinogens.[29] Because physiologically based pharmacokinetic (PBPK) models achieve interspecies extrapolation by taking into account the anatomic, physiological, biochemical, and metabolic differences between species, they may be used to overcome the shortcomings of universal correction.[30,31] However, since PBPK models require a large body of supporting data to construct and validate, such approach is limited to chemicals that have a robust pharmacokinetic and pharmacodynamic database.

The low-dose extrapolation method produces risk estimates that can vary widely with the choice of models and dose–response data and hence has considerable uncertainties. It is also highly conservative since it does not account for biological repair or detoxification processes; consequently, OELs predicted by these models are often too low to be economically or practically feasible.

Example 12.7. Because there is a paucity of human studies available on ethylene oxide, OSHA's risk assessment on ethylene oxide was derived from results of a 2-year inhalation study in the rat.[32] Using the linearized multistage model, OSHA predicted an excess lifetime risk for cancer from exposure to ethylene oxide at 50 ppm to be 63 to 110 per 1000 workers, with a 95 percent upper confidence limit on the excess risk of 101 to 152 deaths per 1000 workers. The risk estimated at 1 ppm was approximately 2.1 to 3.3 excess deaths per 1000. To extrapolate from the animal carcinogenicity data in making risk prediction for humans, OSHA employed a milligram per kilogram body weight per day adjustment to scale the animal doses to equivalent human doses. The total volume of air that a worker was expected to breathe in a normal working day was assumed to be 9.3 m^3, and the typical working lifetime was taken to be 8 hours/day, 5 days/week, 46 weeks/year for 45 years.

Example 12.8. Unlike most chemicals where only animal data are available, there are several human studies on asbestos. OSHA used a mathematical model to describe the relationship between the excess relative risk of lung cancer and asbestos exposure:[33]

$$R = R_e[1 + (K)(f)(d_{t-10})]$$

where R = rate of lung cancer mortality resulting from the asbestos exposure
$\quad R_e$ = rate of expected mortality in the absence of exposure
$\quad K$ = proportionality constant measuring the carcinogenic potency of asbestos exposure
$\quad f$ = intensity of exposure in fibers/cm^3
$\quad d$ = duration of exposure in years
$\quad t$ = time from the onset of asbestos exposure in years (minus 10 years to allow for a minimum latency period)

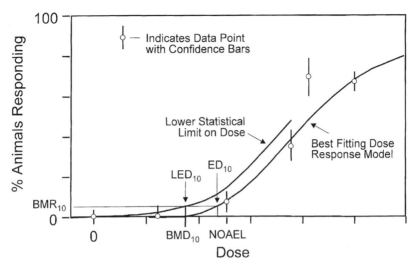

Figure 12.2 Benchmark dose methodology to set occupational exposure limits.

A best estimate of K was obtained from 11 epidemiological studies. Using the equation and a K value of 0.01, OSHA was able to predict a lifetime excess risk of total cancer to various asbestos fiber concentrations. Reduction in the PEL from 2 to 0.2 fibers/cm^3 reduces the calculated risk from 64 to 6.7 per 1000.

Benchmark Dose A recent development in methods to set OELs is the benchmark dose (BMD) approach,[34] which is likened to a hybrid of the mathematical modeling and the uncertainty/safety factor approaches. In this method (Figure 12.2), a mathematical model is used to extrapolate the dose–response data to a level just below that in the observable range, usually 10 percent (ED_{10}) to obtain the BMD.[35] Various mathematical methods may be used, but unlike low–dose extrapolation, the choice of the model is not critical since extrapolation is not taken far beyond the observed dose range for most quantal endpoints. A statistical lower bound on the BMD (LED_{10}) is considered to be the equivalent of an NOAEL. An uncertainty and/or safety factor is then applied to arrive at the OEL:

$$OEL = BMD/(UF)(SF)(BR)$$

The benchmark dose method has several advantages over the uncertainty/safety factor method.[36] For example, while the uncertainty factor method relies on a single point estimate in the dose–response data, the benchmark dose method uses data from the entire dose–response curve and does not require the benchmark dose to be one of the experimental doses. In fact, it does not even require that an NOAEL be identified. The requirement of identifying an

NOAEL has sometimes resulted in costly repetition of studies using additional animals and dose levels. Also since the benchmark dose takes the variability of the data into account, the confidence in a dose estimate increases directly with the quality of the study. Finally, unlike the uncertainty factor method, the benchmark dose approach allows the estimation of risk at given exposure levels.

Example 12.9. Data from a developmental toxicity study for 1,3-butadiene in mice were fit using a log-logistic model.[37] The benchmark dose was estimated to be around 40 ppm. Applying an uncertainty factor of 10 for mouse to human extrapolation and a safety factor of 3 for severity of effects would produce an OEL of about 1.3 ppm.

12.5.4 Performance-Based Occupational Exposure Limits

Many chemical substances possess toxicological or pharmacological properties that make them difficult to set a numerical OEL. For instance, it is often difficult to identify a clear NOAEL for potent sensitizers or carcinogens. This is a particularly acute problem for the pharmaceutical industry where there is an increasing trend toward highly potent drugs. In such situations a performance-based approach toward controlling chemical exposures may be used.[38] The performance-based approach for setting OELs, like the approaches previously described, relies on a health risk assessment of the chemical, but instead of quantifying the risk to arrive at a numerical OEL, places the chemical into a specific exposure control category. The categories dictate the engineering and administrative procedures to maintain risks from the exposures at acceptable levels and are analogous to the biosafety levels 1 to 4 to handle microorganisms of increasing pathogenicity:

1. This category includes chemicals that are relatively nontoxic and produce no systemic effects. The estimated OELs typically range from 1 to 5 mg/m^3.
2. Chemicals in this category generally have little systemic toxicity. Overexposure requires only first-aid or simple medical treatment. The OELs typically range from 0.1 to 1 mg/m^3.
3. Chemicals that have non-life-threatening or incapacitating effects that are normally reversible, but may become irreversible following prolonged exposure. OELs typically range from 1 to 100 $\mu g/m^3$.
4. Chemicals that can produce life-threatening effects. They may also have irreversible effects with disabling consequences. OEL is expected to be below 1 $\mu g/m^3$.
5. This category consists of the most toxic chemicals known, with single doses producing life-threatening effects that require immediate, heroic medical intervention.

Example 12.10. Methyldopa, used as an antihypertensive agent, is an example of a category 1 chemical. The typical therapeutic dose is 0.5 to 2 g/day. It produces a transient, reversible decrease in blood pressure, but no other serious effects. No special containment measures other than good manufacturing practices are required for handling category 1 chemicals.

Example 12.11. Mechlorethamine, a nitrogen mustard, is an example of a category 4 chemical. It has a high order of acute toxicity, is corrosive, and causes toxicity to a variety of tissues including the bone marrow and gastrointestinal tract. It also produces developmental toxicity and malignant tumors at low dosages. To control exposure to a category 4 chemical, closed systems must be used, allowing no open handling.

The advantage of a performance-based approach to control occupational exposure to chemicals is that it categorizes chemicals into common classes that can be easily recognized. It thus helps to put the hazard into perspective. This type of approach is best suited for chemicals that have a low volume of production and a small population potentially exposed. While a performance-based approach offers a practical alternative, this is not a satisfactory substitute for an OEL, since a complete systematic assessment of the potential health effects is not performed, certain adverse effects may be overlooked because they are not specifically sought. Furthermore, the environmental monitoring strategy and time-weighting assumptions are often insufficiently defined.[39] Thus, the performance-based approach should be considered a temporary solution, while awaiting the development of a more extensive health effect database to establish a traditional OEL.

12.5.5   OELs Based on Concentrations in Contaminated Surfaces

While a predominant number of OELs for chemicals is based on the concentration in air, there are some recent attempts to establish surface contamination standards that relate to the air concentration. The health physics profession is the pioneer in this area. The traditional approach health physicists use in setting surface standards is to employ the so-called order of magnitude for their measurement. This order-of-magnitude factor, that is, the ratio of air/surface concentration, is 10^{-6} for routine surveillance of radionuclide contamination. A recent survey, however, found that the reported ratios between air and surface concentrations of industrial chemicals spanned a millionfold.[40] Thus, for industrial hygiene purposes, there appears no simple, quantitative correlation between air and wipe samples. In view of the extremely wide variability, it is recommended that any surface-based OELs should be used only as criteria of cleanliness, but not as criteria of hazard.

In recent years, there has been discussion that a quantitative dermal OEL (DOEL) needs to be developed to protect against excessive uptake of chemicals through the skin.[41] The DOEL refers to the concentration of a chemical deposited on the skin during a work shift. The DOEL is based on the maximum dose of a chemical that can be absorbed through the skin without leading to adverse systemic effects. To derive a DOEL, expressed as the amount of a chemical per unit of skin surface area, one would need to estimate the dermal bioavailability and the area of exposed skin.

$$\text{DOEL} = (\text{Maximum dose})/(\text{Dermal availability})(\text{Exposed skin surface area})$$

Example 12.12. For workers engaged in the manufacturing of the pharmaceutical cyclophosphamide, it is expected that dermal exposure is limited to the hands and lower arms, with a combined total surface area of about 2000 cm^2. Dermal bioavailability is estimated to be about 100 percent. A risk assessment indicates that a reference dose of 7.5 µg is associated with a cancer risk of 4 in 100,000.

$$\text{DOEL} = 7.5 \ \mu g/(1.0)(2000 \ cm^2) = 0.00375 \ \mu g/cm^2 = 4 \ ng/cm^2$$

12.6 BIOMARKERS-BASED OCCUPATIONAL EXPOSURE LIMITS

As discussed above, workplace exposures have historically been evaluated by measuring the chemical concentrations in the air. Air monitoring, however, does not necessarily represent the dose of chemical that has been absorbed into the body since the substance can enter the body by routes other than inhalation, such as through skin absorption. To overcome the shortcomings of air monitoring, biological monitoring techniques, which measure a chemical or its metabolite in specimens of urine, blood, expired air, hair, nails collected from exposed workers, can be used. In 1981, the ACGIH began the development of OELs based on biomarkers known as biological exposure indices (BEIs). The ACGIH defines the BEIs as those representing the levels of determinants that are most likely to be observed in specimens collected from a healthy worker exposed to chemicals to the same extent as a worker with an inhalation exposure to the TLV.[42] Outside the United States, the DFG in Germany has also developed biological monitoring reference values called biological tolerance values (BATs). BATs are defined as the maximum permissible quantity of a chemical, its metabolites, or any deviation from the norm of biological parameters induced by these substances in exposed humans. While air monitoring probably will remain the predominant force in an occupational hygiene program, biological monitoring can be used as a backup to assure the adequacy of control measures or if sampling and analytical methods for the specific air contaminant are not available.

12.7 FACTORS TO CONSIDER WHEN SETTING BEIS

In setting BEIs a good understanding of the relationship between exposure, body burden, and toxic effect is crucial. The pharmacokinetic processes of absorption, distribution, and excretion will determine the burden of a chemical in the body resulting from extrinsic exposures. For airborne exposures, the extent of absorption can also be affected by the pulmonary ventilation and perfusion rates. Thus, depending on their level of physical activity, two individuals working in an identical exposure environment can result in different body burden. Once a chemical has gained entry into the bloodstream, its distribution to other parts of the body depends on its solubility in various tissues and metabolism. Hydrophilic and polar substances will likely be distributed quite uniformly throughout, but lipophilic compounds may accumulate in adipose tissue. Thus, the kinetic behavior of lipid-soluble substances can be very different in workers with different body fat composition. The manifestation of toxic effect following exposure depends on the level and time duration of a chemical in the target tissue (tissue burden). Some effects occur immediately when certain threshold concentration is attained, while some may only be apparent after a period of time.

12.8 METHODS FOR SETTING BEIS

Aside from pharmacokinetic and metabolism data, the most critical information required for setting BEIs are those that can be used to relate the levels of determinants in biological specimens to the airborne exposure levels. These data may come from monitoring workers in field studies or from controlled studies in human volunteers.

Example 12.13. The relationship between inhalation exposure and urinary excretion of methyl ethyl ketone was investigated in four field studies,[43-46] which suggested that an exposure to the TLV level of 200 ppm would result in a urinary concentration of 2.1 to 5.3 mg/L. Based on these data, the ACGIH has recommended a BEI of 2 mg/L for methyl ethyl ketone in urine, collected at the end of the work shift.

Traditionally, BEIs have relied mainly on experimental observations in humans. However, for ethical reasons human exposure studies are seldom conducted. Consequently, there are many chemicals that lack the requisite data for setting BEIs. Recent advances in pharmacokinetic modeling have provided a versatile means of predicting BEIs from ambient OELs.[47] In this approach, a physiologically based pharmacokinetic model is constructed with the physiological parameters set to similute an average human performing light-duty work for 8 hours, followed by a 16-hour resting period before the next shift. Chemical-specific metabolic and kinetic constants are obtained from the literature. The model is then exercised to simulate a chemical input to the system, that is, exposure, which is equal to the OEL level of the corresponding chemical.

Example 12.14. Using a PBPK model,[30] an 8-hour exposure to the TLV level (25 ppm) of 1,4-dioxane is predicted to result in a BEI of 1.7 ppm in end-expired air, 9.5 ppm in mixed-expired air, 11.3 mg/L in blood, and 506.3 mg/L β-hydroxyethoxyacetic acid (the principal metabolite) in the urine at the end of the work shift.

12.9   CONCLUSION

The process of setting OELs is a synthesis of scientific data and professional judgment. The skills and experience that a toxicologist uses in reviewing the scientific evidence of potential harm cannot be replaced by a standard protocol. The toxicologist must make a weight-of-evidence analysis of the data to determine their relevance to the work environment and to evaluate the appropriateness of the different methods available. In the case of the most frequently used method, the uncertainty/safety factor approach, one must also select the appropriate factors based on the credibility or relevance of the data and the severity of the adverse effects. This is an area that is of particular importance. Uncertainty/safety factors have traditionally been set with an eye to conservatism: protecting the employee from adverse effects using huge safety factors multiplied by equally large uncertainty factors to account for biological differences. The value of the composite factor often reached into the thousands and even millions. One must carefully assess the validity of such large and multiplicative factors in light of the scientific evidence. Extremely large safety factors may not be appropriate for effects demonstrating a threshold dose or for compounds metabolized through different pathways in humans and animals. In selecting safety factors, the weight given to genetic toxicology tests must also be carefully considered. Rather than assigning safety factors to specific results, a more appropriate approach may be to select a safety factor on the basis of a thorough review and evaluation of the hazards represented by all the toxicological data.

Since the development of OELs is a data-driven process, the more information there is, the more precise estimates with less uncertainty can be obtained. The OEL-setting process would greatly benefit from additional research in several areas. Most significant would be studies to better relate available oral or parenteral toxicity data to the inhalation route. There is a paucity of inhalation toxicity information for the large majority of chemicals. In an attempt to develop more inhalation toxicity data to support the establishment of ambient air quality standards under the Clean Air Act, the U.S. EPA recently proposed the Hazardous Air Pollutants (HAPs) Test Rule for a first list of 21 chemicals. The required studies, including subchronic toxicity, neurotoxicity, and developmental/reproductive toxicity, will be extremely valuable for setting OELs since they use the route of exposure most relevant to the occupational setting. While it is impractical to conduct long-term inhalation toxicity studies on most chemicals, it is possible to develop pharmacokinetic modeling methods to aid in route-to-route extrapolation and to relate the high-level exposures in animal toxicity studies to the exposures of concern in the occupational environment. Occupational toxicologists would also benefit from additional work on in vitro and in vivo dermal absorption studies and models relating absorption potential to physical parameters. Low-dose extrapolation methods, particularly for effects such as cancer, have been and probably will always be controversial. Resolution of the current dilemmas in this area of risk assessment is unlikely in the near future, although research in the area continues and should receive our support. In the meantime, it is incumbent on toxicologists to evaluate each case and each method on its merits, rather than accepting scientifically questionable risk assessments on the basis of conservatism and habit.

REFERENCES

1. M. Lippmann and B. D. Thurston. Exposure assessment: Input into risk assessment. *J. Occup. Med.* 43: 113–123 (1988).
2. D. J. Paustenbach. Important recent advances in the practice of health risk assessment: Implications for the 1990's. *Regul. Toxicol. Pharmacol.* 10: 204–243 (1989).
3. J. M. Paull. The origin and basis of threshold limit values. *Am. J. Ind. Med.* 5: 227–238 (1984).
4. W. A. Cook. *Occupational Exposure Limits-Worldwide.* American Industrial Hygiene Association, Akron, OH, 1987.
5. D. Henschler. The concept of occupational exposure limits. *Sci. Total Environ.* 19: 9–16 (1991).
6. D. J. Paustenbach and R. Langner. Corporate occupational exposure limits: The current state of affairs. *Am. Ind. Hyg. Assoc. J.* 47: 809–818 (1986).
7. R. L. Zielhaus and W. R. F. Notten. Permissible levels for occupational exposure: Basic concepts. *Int. Arch. Occup. Environ. Health* 42: 269–281 (1979).
8. P. De Silva. TLVs to protect "nearly all workers." *Appl. Ind. Hyg.* 1: 49–53 (1986).
9. R. L. Zielhuis and A. E. Wibowo. Standard setting in occupational health: "Philosophical issues." *Am. J. Ind. Med.* 16: 569–598 (1989).
10. D. J. Paustenbach. The history and biological basis of occupational exposure limits for chemical agents. In R. L. Harris (Ed.), *Patty's Industrial Hygiene and Toxicology*, 5th ed., Vol. 3. Wiley, New York, 2000, pp. 1903–2000.
11. D. M. Galer, H. W. Leung, R. G. Sussman, and R. J. Trzos. Scientific and practical considerations for the development of occupational exposure limits (OELs) for chemical substances. *Regul. Toxicol. Pharmacol.* 15: 291–306 (1992).
12. V. Fiserova-Bergerova, J. T. Pierce, and P. O. Droz. Dermal absorption potential of industrial chemicals criteria for skin notation. *Am. J. Ind. Med.* 17: 617–635 (1990).
13. A. M. Monro. Interspecies comparisons in toxicology: The utility and futility of plasma concentrations of the test substance. *Regul. Toxicol. Pharmacol.* 12: 137–160 (1990).
14. D. J. Paustenbach. Pharmacokinetics and unusual work schedules. In R. L. Harris (Ed.), *Patty's Industrial Hygiene and Toxicology*, 5th ed., Vol. 3. Wiley, New York, 2000, pp. 1787–1901.
15. E. D. Sargent and D. Kirk. Establishing airborne exposure control limits in the pharmaceutical industry. *Am. Ind. Hyg. Assoc. J.* 49: 309–313. (1988).
16. L. E. Kane, C. S. Barrow, and Y. Alarie. A short-term test to predict acceptable

levels of exposures to airborne sensory irritants. *Am. Ind. Hyg. Assoc. J.* 40: 207–229 (1979).

17. B. H. Butterworth. Nongenotoxic carcinogens in the regulatory environment. *Regul. Toxicol. Pharmacol.* 9: 244–256 (1989).

18. J. A. Swenberg. α_{2u}-Globulin nephropathy: Review of the cellular and molecular mechanisms involved and their implications for human risk assessment. *Environ. Health Perspect.* 101(Suppl. 6): 39–44 (1993).

19. D. E. Moody, J. K. Reddy, B. G. Lake, J. A. Popp, and D. H. Reese. Peroxisome proliferation and nongenotoxic carcinogens: Commentary on a symposium. *Fund. Appl. Toxicol.* 16: 233–248 (1991).

20. International Commission on Radiological Protection (ICRP). *Report of the Task Group on Reference Man.* Pergamon, New York.

21. U.S. Environmental Protection Agency (EPA). *Exposure Factors Hanbook.* EPA/600/8-89/043. EPA, Office of Health and Environmental Assessment, Washington, DC, March 1990.

22. D. K. Verma. Adjustment of occupational exposure limits for unusual work schedules. *Am. Ind. Hyg. Assoc. J.* 61: 367–374 (2000).

23. G. D. Nielsen and Y. Alarie. Sensory irritation, pulmonary irritation, and respiratory stimulation by airborne benzene and alkylbenzenes: Prediction of safe industrial exposure levels and correlation with their thermodynamic properties. *Toxicol. Appl. Pharmacol.* 65: 459–477 (1982).

24. H. W. Leung and D. J. Paustenbach. Setting occupational exposure limits for irritant organic acids and bases based on their equilibrium dissociation constants. *Appl. Ind. Hyg.* 3: 115–118 (1988).

25. M. L. Dourson and J. F. Stara. Regulatory history and experimental support and uncertainty (safety) factors. *Regul. Toxicol. Pharmacol.* 3: 224–238 (1983).

26. S. C. Lewis, J. R. Lynch, and A. I. Nikiforov. A new approach to deriving community exposure guidelines from "no-observed-adverse-effect levels." *Regul Toxicol. Pharmacol.* 11: 314–330 (1990).

27. J. V. Rodricks, S. M. Brett, and G. C. Wrenn. Significant risk decisions in federal agencies. *Regul. Toxicol. Pharmacol.* 7: 307–320 (1987).

28. K. S. Crump, D. G. Hoel, C. H. Langley, and R. Peto. Fundamental carcinogenic processes and their implications for low dose risk assessment. *Cancer Res.* 36: 2973–2979 (1976).

29. M. E. Andersen. Tissue dosimetry in risk assessment, or what's the problem anyway? In *Drinking Water and Health, Pharmacokinetics in Risk Assessment*, Vol. 8. National Academy Press, Washington, DC, 1987, pp. 8–23.

30. H. W. Leung. Use of physiologically based pharmacokinetic models to establish biological exposure indexes. *Am. Ind. Hyg. Assoc. J.* 53: 369–374 (1992).

31. H. W. Leung and D. J. Paustenbach. Physiological pharmacokinetic and pharmacodynamic modeling in health risk assessment and characterization of hazardous substances. *Toxicol. Lett.* 79: 55–65 (1995).

32. Occupational Safety and Health Administration. Occupational exposure to ethylene oxide. Final standard. *Fed. Reg.* 49: 46936–47022 (1984).

33. Occupational Safety and Health Administration. Occupational exposure to asbestos. Final Standard. *Fed. Reg.* 51: 22612–22647 (1986).

34. C. A. Kimmel. Quantitative approaches to human risk assessment for noncancer health effects. *Neurotoxicology* 11: 189–198 (1990).

35. K. S. Crump. Calculation of benchmark doses from continuous data. *Risk Anal.* 15: 79–89 (1995).

36. H. J. Clewell, III, P. R. Gentry, and J. M. Gearhart. Investigation of the potential impact of benchmark dose and pharmacokinetic modeling in noncancer risk assessment. *J. Toxicol. Environ. Health* 52: 475–515 (1997).

37. C. A. Kimmel, M. Siegel, T. M. Crisp, and C. W. Chen. Benchmark concentration analysis of 1,3-butadiene reproductive and developmental effects. *Fund. Appl. Toxicol. Suppl.* 30: 146 (1996).

38. B. D. Naumann, E. V. Sargent, B. S. Starkman, W. J. Fraser, G. T. Becker, and G. D. Kirk. Performance-based exposure control limits for pharmaceutical active ingredients. *Am. Ind. Hyg. Assoc. J.* 57: 33–42 (1996).

39. R. Agius. Occupational exposure limits for therapeutic substances. *Ann. Occup. Hyg.* 33: 555–562 (1989).

40. K. J. Caplan. The significance of wipe samples. *Am. Ind. Hyg. Assoc. J.* 54: 70–75 (1993).

41. P. M. J. Bos, D. H. Brouwer, H. Stevenson, P. J. Boogaard, W. L. A. M. de Kort, and J. J. van Hemmen. Proposal for the assessment of quantitative dermal exposure limits in occupational environments: Part 1. Development of a concept to derive a quantitative dermal occupational exposure limit. *Occup. Environ. Med.* 55: 795–804 (1998).

42. V. Fiserova-Bergerova. Development of biological exposure indices (BEIs) and their implementation. *Appl. Ind. Hyg.* 2: 87–92 (1987).

43. M. Miyasaka, M. Kumai, A. Koizumi, T. Watanabe, K. Kurasako, K. Sato, and M. Ikeda. Biological monitoring of occupational exposure to methyl ethyl ketone by means of urinalysis for methyl ethyl ketone itself. *Int. Arch. Occup. Environ. Health* 50: 131–137 (1982).

44. L. Perbellini, F. Brugnone, P. Mozzo, V. Cocheo, and D. Caretta. Methyl ethyl ketone exposure in industrial workers. *Int. Arch. Occup. Environ. Health* 54: 73–81 (1984).

45. S. Ghittori, M. Imbriani, G. Pezzagno, and E. Capodaglio. The urinary concentration of solvents as a biological indicator of exposure. Proposal for biological equivalent exposure limit for nine solvents. *Am. Ind. Hyg. Assoc. J.* 48: 786–790 (1987).

46. C. N. Ong, G. L. Sia, H. Y. Ong, W. H. Phoon, and K. T. Tan. Biological monitoring of occupational exposure to methyl ethyl ketone. *Int. Arch. Occup. Environ. Health* 63: 319–324 (1991).

47. H. W. Leung. Development and utilization of physiologically-based pharmacokinetic models for toxicological applications. *J. Toxicol. Environ. Health* 32: 247–267 (1991).

13 Worker Hazard Posed by Reentry into Pesticide-Treated Foliage: Reassessment of Reentry Levels/Intervals Using Foliar Residue Transfer–Percutaneous Absorption PBPK/PD Models, with Emphasis on Isofenphos and Parathion

J. B. KNAAK

Department of Pharmacology and Toxicology, School of Medicine, and Biomedical Sciences, SUNYAB, Buffalo, New York

C. C. DARY

National Exposure Research Laboratory, U.S. Environmental Protection Agency, Las Vegas, Nevada

G. T. PATTERSON

Department of Pesticide Regulation, State of California, Sacramento, California

J. N. BLANCATO

National Exposure Research Laboratory, U.S. Environmental Protection Agency, Las Vegas, Nevada

13.1 INTRODUCTION

The introduction of organophosphorus (OP) pesticides into modern agriculture has increased production and provided consumers worldwide with high-quality

Human and Ecological Risk Assessment: Theory and Practice, Edited by Dennis J. Paustenbach
ISBN 0-471-14747-8 © 2002 John Wiley & Sons, Inc.

fruits and vegetables. Along with these advances came the early recognition of occupational exposure[1] and the documentation of reentry-related illnesses.[2] The use of these (OP) pesticides, principally ethyl parathion, on citrus in California resulted in a series of serious poisoning incidents among workers reentering treated groves to harvest fruit. Established preharvest intervals (time between the last application of pesticide and harvest) varying from several days to several weeks were originally considered to be adequate to protect the health of workers "entering" or "reentering" a sprayed crop to harvest fruit or vegetables. Workers entering treated crops for activities other than harvesting (e.g., thinning, pruning) were not protected by the preharvest interval. Over the past 25 years California has responded by developing field reentry intervals for toxic organophosphorus pesticides applied to tree fruits and vines and by decreasing the number of dermally toxic OP pesticides registered on tree fruits and vines in the state through its reregistration program.[3] Currently, there are approximately 41 OP pesticides registered by the U.S. Environmental Protection Agency (EPA) with a somewhat smaller number registered for use in California.[4]

This chapter briefly reviews the history of the development of current reentry intervals presented by Knaak et al.[5] with emphasis on the use of foliar residue transport coefficients[6,7] in conjunction with percutaneous absorption physiologically based pharmacokinetic/pharmacodynamic (PBPK/PD) models[8,9] as a mean of developing new field reentry intervals or examining old intervals on dermally toxic pesticides. The authors believe that this approach may also be useful for assessing the risks involved in nonworker exposures (i.e., sensitive individuals such as children) to other toxic environmental residues. PBPK/PD models developed for isofenphos and parathion were used as examples.

13.2 HISTORICAL BACKGROUND

The illnesses among workers who reenter pesticide-treated groves and vineyards were associated with red blood cell cholinesterase depression. An extensive review on the historical background, hazard evaluation, exposure assessment, procedures for calculating a reentry interval, and the mathematics of setting safe levels on foliage was published in the previous volume of this book.[5] Animal or human blood cholinesterase (ChE) dose–response data (e.g., acute, chronic durations of exposure, oral or dermal routes) were used for purposes of estimating safe levels on foliage in conjunction with residue decay data and harvester deposition data.[5,6] This procedure was based in part on the EPA recommendations published in 1984 reentry protection guidelines.[10] EPA's 1997 postapplication guidelines are currently being used for guidance.[11]

The dissipation of OP foliar residues was extensively studied in Florida and the western states.[12-17] Popendorf and Leffingwell[6] and Nigg et al.[7] developed equations involving the transfer of leaf residues in aerosol form (soil dust and water) from leaf surfaces to the skin of workers. The results of dermal dose–ChE response and percutaneous absorption studies involving single percuta-

neous applications of OP and carbamate insecticides were used to estimate toxicity and percutaneous absorption and set field reentry intervals in California.[5]

13.3 RECENT ISSUES

The establishment of reentry intervals in California provided a mechanism for
reducing the hazards involved in reentering pesticide-treated orchards and vineyards. However, the process of setting the intervals did not satisfactorily relate
exposure, absorbed dose and toxicity, reduce the toxicity of the OP pesticides,
or provide information on the relationship between blood acetylcholinesterase
(AChE) inhibition and adverse effects in the nervous system. We believe some
of these issues may be resolved by the use of residue transport coefficients in
conjunction with dermal physiological pharmacokinetic/pharmacodynamic
models.

13.3.1 Measurement of OP Toxicity

Toxicological reviews by EPA on organophosphorus insecticides have raised
new issues over the adverse health effects of OP pesticides. The primary concern is the relationship between the measured plasma and blood cholinesterase
activity and OP neurotoxicity (changes in behavioral and motor activity as well
as brain cholinesterase). According to EPA scientists,[18] animal plasma and red
blood cell cholinesterase inhibition data do not adequately predict effects in the
central nervous system and are biomarkers of exposure and not biomarkers
of adverse effects and cannot be used for determining reference doses (RfDs) or
reference concentrations (RfCs). The EPA also questioned the usefulness of red
blood cell AChE in worker monitoring programs. The EPA "recognizes the
possibility that new analyses may provide additional information on the biological significance of cholinesterase inhibition in red blood cells."[19]

The agency recommends the use of animal no adverse effect levels (NOAELs)
and low adverse effect levels (LOAELs) relevant to adverse neurological effects
in the central nervous system (i.e., brain AChE inhibition, adverse effects on
behavior and motor activity) for defining hazard and setting RfDs [formerly
acceptable daily intakes (ADIs)], for lifetime exposures.[20,21] Clinical effects
in humans and statistically significant inhibition of cholinesterases in humans
were also considered by the agency for setting RfDs. In addition to the NOAEL/
LOAEL approach, a number of other methods are available for characterizing
the dose–response relationship of a chemical. These include: (1) a no-statistical-
significance of trend (NOSTASOT) approach, (2) a benchmark dose (BMD)/
benchmark concentration (BMC) approach, (3) application of Bayesian statistics, and (4) a categorical regression analysis.

Regardless of EPA's scientific position,[18] the Office of Pesticide Programs
Reference Dose Tracking Report indicates that RfDs for chlorpyrifos, diazinon,
ebufos, ethoprop, and fenamiphos are based on plasma ChE inhibition. These
OP pesticides are a sampling of a larger number of OPs listed in the tracking

report. The California Department of Pesticide Regulation (CDPR) recently issued guidelines on the use of cholinesterase inhibition data in risk assessments for OP pesticides that take an intermediate, but scientifically defensible, position in the use of inhibition data.[22]

13.3.2 Worker Monitoring

EPA's deliberations have raised a number of questions in California and Florida concerning the quality of worker monitoring programs, specifically methodologies for determining cholinesterase inhibition and exposure during the work season.

According to Wilson et al.,[23] blood cholinesterase methodologies currently being used in clinical laboratories in California are deficient in a number of quality control procedures. One major problem is the lack of standards for drawing and storage of the samples prior to analysis. The authors recommend that samples be immediately iced to prevent reactivation of enzyme activity. Commercially available kits containing thiocholine require modification (i.e., change in substrate concentrations) before they can be used to simultaneously measure red blood cell (RBC) AChE and plasma BChE. The lack of a reliable ChE assay standard that can be run with each set of samples needs to be addressed.

Krieger[24] believes standardized methods should be used for estimating mixer/loader/applicator (m/l/a) and field worker exposure to pesticides [i.e., pesticide handlers exposure database (PHED) for m/l/a and foliar residues and transfer coefficients for field workers].

Nigg and Knaak[4] recommended a ChE monitoring program in Florida as part of a routine health surveillance program for workers exposed to OP pesticides. Recommendations included but are not limited to (1) obtaining a complete medical history for each employee and a physical examination, (2) obtaining the services of a competent clinical laboratory licensed to run blood cholinesterase tests, (3) measuring a worker's blood cholinesterase activity (at least two analyses) before employment, (4) setting up a monitoring program based on duration and level of exposure to OPs, (5) removing employee from work if a 25 to 30 percent reduction in RBC AChE occurs, (6) determining cause of exposure and making adjustments to prevent repeated exposure, and (7) allowing employee to return to work when RBC AChE activity returns to preexposure values. This recommendation is more restrictive then present practices in California that allow RBC AChE and plasma ChE activity to fall below 70 and 60 percent of baseline, respectively, before the worker must leave the workplace.[23] Workers may not return to work until both enzyme activities return to 80 percent.

13.3.3 Children/Sensitive Subpopulations

The enactment of the Food Quality Protection Act (FQPA, Public Law 104-170) in 1996 altered the established data requirements for pesticide registration and reregistration under the Federal Insecticide Fungicide and Rodenticide Act (FIFRA). The FQPA requires all pesticides to be reevaluated over a 10-year period. The FQPA sets tolerance standards for pesticide residues in raw agricultural commodities and processed food. The tolerance for the pesticide residue in or on food may be considered "safe" when there is reasonable certainty that "aggregate" exposure (i.e., sum total) resulting from the simultaneous dietary ingestion of residues and exposures by "other" pathways and routes results in no harm to human health. Under these conditions, residue concentrations are expected to be measurable in food and water and other media through which aggregate exposure by other pathways and routes might occur. The other pathways might include the inhalation and dermal pathways. The routes of exposure via the inhalation pathway might include respiration of vapors and gases and ingestion of particles. The dermal routes might include percutaneous absorption of the potential dermal dose or nondietary ingestion of residues that have accumulated on the surface of the skin, particularly the hands through hand-to-mouth activity. For aggregate exposures involving chemicals with divergent or poorly understood mechanisms of action, assessment of delivered (internal) dose, and untoward effects would be quite difficult. However, for chemicals such as the OP insecticides with a common mechanism of action as anticholinesterase agents, aggregate and cumulative toxicity might be more easily assessed using a "margin of exposure" (MOE). Multiroute PBPK/PD models are now capable of handling toxicological questions (i.e., internal dose, AChE inhibition, and urinary metabolites) concerning aggregate exposures for individual OP insecticides.

In recent years EPA (National Center for Environmental Research and Quality Assurance [NCERQA, ORD]) has become increasingly concerned with children's vulnerability to toxic substances in the environment (i.e., adverse effects on development including growth, memory, vision, etc.). This concern has resulted in the need to develop dermal/oral exposure, percutaneous absorption, metabolic and health effects data on environmental residues as they pertain to infants, preschool children, and children attending grammar school (kindergarten and through sixth grade) for the purpose of assessing risk and developing cleanup standards.[25] A number of research proposals regarding these needs were recently submitted to the agency and are currently being considered for funding over a 3-year period.

13.4 HAZARD EVALUATION

13.4.1 Properties of Toxic Organophosphorus Pesticides

Seven types of organophosphorus insecticides are produced and used in the United States and are represented by phosphates, phosphorothionates, phosphorothiolates, phosphorodithioates, phosphoroamidates, phosphoramidothioates, and phosphonates. The phosphorothionates and phosphorothiolates are grouped together and designated as phosphorothioates. The phosphorothioates (e.g., parathion) and phosphorodithioates (e.g., azinphosmethyl, methidathion, dimethoate, and phosalone) are oxidized on foliage to produce more toxic products called oxons (phosphates and phosphorothioates).[4]

The organophosphorus pesticides or their oxons inhibit acetylcholinesterase (AChE) in the nervous system and in red blood cells of humans and animals by reacting with the active site of this enzyme. Organophosphorus insecticides vary in their affinity for the enzyme and in their ability to irreversibly phosphorylate the enzyme.[4] Recovery of AChE activity occurs by direct synthesis of new enzyme, dissociation of the enzyme–inhibitor complex, and reactivation of the phosphorylated enzyme. Red blood cell cholinesterase activity measurements in workers and experimental animals are routinely used to estimate exposure and measure indirectly the in vivo effects of these pesticides on nervous system cholinesterase. Hirschberg and Lerman[26] and Lerman et al.[27] reviewed cholinesterase poisoning and the treatment of poisoning in individuals.

Field workers are dermally exposed to residues of organophosphorus insecticides on foliage and soil. EPA regulations[28] currently group pesticide products into three toxicity categories based on the results of dermal LD_{50} studies in the rabbit. Category I materials have a dermal LD_{50} of <200 mg/kg, category II materials a dermal LD_{50} of 200 to 2000 mg/kg, and category III and IV materials a dermal LD_{50} of >2000 mg/kg. The oxons formed from the phosphorothioates and phosphorodithoates have a dermal LD_{50} of <200 mg/kg. Reentry intervals are required by EPA[10,11] and the California Department of Food and Agriculture (CDFA)[29] for pesticides assigned to categories I and II.

Because of the difficulty involved in the quantitative measurement of dermal toxicity in laboratory rats under conditions of acute and chronic exposure, EPA currently recommends the use of oral RfD adjusted for absorption in critical toxicity studies to estimate the risk (hazard index) of a dermally absorbed dose. This is carried out by first calculating a dermally absorbed dose [(DAD) in mg/kg/day]:

$$DAD = \frac{DA \times EF \times ED \times A}{BW \times AT} \qquad (13.1)$$

where DA = absorbed dose in mg/cm^2/h [applied dose in mg/cm$^3 \times K_p$ (cm/h)]

  A = skin surface area available for contact (cm^2)

  EF = exposure frequency (8 hours/day)

  ED = exposure duration (365 days)

  BW = body weight (kg)

  AT = averaging time (days), for noncarcinogenic effects, AT = ED

Then dividing the dermally absorbed dose by the RfD to arrive at a hazard index for the dermal route:[30]

$$\text{Hazard index} = \frac{DAD}{RfD} \qquad (13.2)$$

A DAD for parathion is calculated as follows:

$$DAD = \frac{0.007 \text{ cm/h} \times 0.02 \text{ mg/cm}^3 \times 8 \text{ h/day} \times 365 \text{ days} \times 10^2 \text{ cm}}{0.3 \text{ kg} \times 365 \text{ days}}$$

$$= 0.037 \text{ mg/kg/day} \tag{13.3}$$

Tables 13.1 and 13.2 give the critical studies (NOAELs involving ChE depression) and other studies considered by EPA in establishing reference doses (RfDs) for parathion and isofenphos.[31–38] The K_p value (0.007 cm/h) was taken from the parathion PBPK/PD model described in this chapter. The use of an RfD of 0.05 mg/kg/day for parathion in humans[36] yields a hazard index of 0.74 for a dermally absorbed dose of 0.037 mg/kg/day:

$$\text{Hazard index} = \frac{0.037 \text{ mg/kg/day}}{0.05 \text{ mg/kg/day}} = 0.74 \tag{13.4}$$

A hazard index of 1 or less is considered to be acceptable.

Edson[36] determined the NOAEL of parathion in rats to be 0.02 mg/kg/day when the pesticide was fed over a period of 84 days based on decreases in plasma cholinesterase activity. A minimal effect was found at 0.04 and 0.06 mg/kg/day. Frawley and Fuyat[37] studied the effect of dietary levels of parathion on plasma cholinesterase in the dog. At 1 ppm (0.021 mg/kg/day) a minimal but significant reduction in plasma cholinesterase occurred. At higher dosages (2 ppm, or 0.047 mg/kg/day; 5 ppm, or 0.117 mg/kg/day) plasma cholinesterase was reduced by 60 to 70 percent. EPA selected a 1-year dog feeding study with plasma and RBC ChE inhibition (Lower Effects Level (LEL), 0.01 mg/kg/day) as the toxic endpoint for parathion.

Edson[36] established a NOAEL in humans by administering 0.05 mg/kg/day in the diet for 42 days without a decrease in red-cell cholinesterase. A dose of 0.1 mg/kg/day significantly reduced both plasma and red-cell cholinesterase. In another study, Rider et al.[31] fed parathion to prison volunteers at 3.0, 4.5, 6.0, and 7.5 mg/day. These amounts resulted in the ingestion of 0.043, 0.064, 0.086, and 0.11 mg/kg/day. Plasma cholinesterase was decreased in one subject on day 4 at the highest dosage and in all subjects on day 16. The decrease amounted to 28 percent of the control value. The three lower values resulted in a slight decrease in plasma cholinesterase activity.

The acute oral toxicity of isofenphos was determined to be 40 mg/kg in the rat and approximately 400 mg/kg in the guinea pig. Isofenphos was determined to be a neurotoxicant in the adult hen at single bolus doses greater than its oral LD$_{50}$.[39] The results of a chronic toxicity study used in setting an RfD for isofenphos is given in Table 13.2. A 2-year rat feeding study with RBC ChE inhibition was chosen for isofenphos. The dietary NOEL based on RBC ChE depression was 0.05 mg/kg/day.

TABLE 13.1 Summary of Repeated Dose and Subchronic Toxicity Data for Parathion

Species	Duration of Study	Dosage Levels and No. of Animals per Group	Highest No-Adverse-Effect Level or Lowest-Minimal-Effect Level	Effect Measured	Reference
Human	30-days, single capsule per day	3.0, 4.5, 6.0, 7.5 mg/man; 7 individuals per group, 5 treated and 2 controls/group	NOAEL of 4.0 mg/human (0.057 mg/kg), LOAEL of 6 mg/human (0.0857 mg/kg)	Depression of plasma cholinesterase	Rider et al.[31]
Rat	110 (males) and 120 (females) weeks	0, 0.5, 5.0, and 50 ppm	No NOAEL established	Mortality comparable to controls, no compound-induced oncogenic response	Daley[32]
Mouse	Single dose, virgin, 19th day of pregnancy and of lactation	5 mg/kg; 3 animals/group	No NOAEL established	Plasma and brain ChE Inhibition	Weitman et al., (33).
Rat	Single IP dose	2, 3.5 and 5 mg/kg, 5 rats/dose	Dose related inhibition of brain and plasma ChE 120 min. post injection.	conditioned taste aversion correlated with parathion plasma inhibition (30–40%).	Roney et al., (34)
Rat	16 weeks	5–125 ppm in diet	125 ppm	Mortality	Edson and Noakes[35]
Rat	84 days		0.02 mg/kg/day	No adverse effect	Edson[36]
Dog	24 weeks	1, 2, and 5 ppm (0.021, 0.047, and 0.117 mg/kg per day, respectively)	1 ppm (0.021 mg/kg)	Cholinesterase depression	Frawley and Fuyat[37]
Dog (critical study)	1 year		No NOEL LEL, 0.01 mg/kg/day	RBC and plasma ChE depression	IRIS database[21]

Dog		1–3 mg/kg/day, orally	1 mg/kg/day	Behavioral effects	Hazleton and Holland[38]
Human[a]		90 weeks 30–42 days	NOAEL < 0.078 and > 0.058 mg/kg	Depression of cholinesterase	Edson[36]
Human			0.043 mg/kg/day	Cholinesterase depression	Rider et al.[31]

[a]Study from which the suggested no-adverse-effect level was taken.

TABLE 13.2 Summary of Repeated Dose and Subchronic Toxicity Data for Isofenphos

Species	Duration of Study	Dosage Levels and No. of Animals per Group	Highest No-Adverse-Effect Level or Lowest-Minimal-Effect Level	Effect Measured	Reference
Rat (critical study)	2 years, dietary		NOEL, 0.05 mg/kg/day LEL, 0.5 mg/kg/day	RBC ChE depression	IRIS database[21]

13.4.2 Dermal Dose–ChE Response

A large number of field reentry studies involving farm field workers were conducted in California to determine the relation between foliar pesticide residues in $\mu g/cm^2$ and cholinesterase depression in order to set safe reentry intervals. These studies were costly and often provided little information relating foliar residue data to cholinesterase inhibition because exposure times and foliar residue levels were not sufficient in magnitude to produce a dose-related effect in workers.[5]

Gaines[40] was the first to develop extensive acute dermal toxicity (LD_{50}) data in mg/kg of body weight on organophosphorus pesticides in the rat. As good as the mortality data were, it could not be readily used to relate residue levels or dermal dose in $\mu g/cm^2$ to cholinesterase inhibition. Knaak et al.[41] were the first to develop dermal dose–response curves (ED_{50}) in the rat, relating dose in $\mu g/cm^2$ of skin surface to cholinesterase inhibition for a number of the organophosphates of interest. Figures 13.1 and 13.2 give the dermal dose–response curves obtained by Knaak et al.[41] for paraoxon, parathion, dialifor, phosalone, azinphosmethyl, dimethoate, and methidathion. The dose (ED_{50}) resulting in 50 percent red cell and plasma cholinesterase inhibition after 72 hours of exposure is given along with the slopes of the log-probit regression lines. The 72-hour exposure period simulated a 3-day harvesting period. The results of these studies were used by Knaak et al.[41,42] to establish safe levels on tree foliage (in $\mu g/cm^2$).

Popendorf and Leffingwell[6] used the toxicity data of Grob et al.,[43–46] Gaines,[40] and Knaak et al.[41] to develop an equation for estimating $\Delta AChE$ by an OP pesticide for use in their "uniform field model." Figure 13.3 gives the rat single dermal dose ChE response data developed by Knaak et al.[41] and replotted by Popendorf and Leffingwell.[6] Parathion and the oxons of azinphosmethyl (guthion), chlorthiophos, and parathion were the most toxic compounds studied. Dialifor, methidathion, and chlorthiophos were intermediate in toxicity with phosalone, dimethoate, and azinphosmethyl being the least toxic. In a separate dermal dose ChE–response study involving isofenphos (Fig. 13.4), the dermal dose resulting in 50 percent AChE inhibition in the rat was determined to be 12 mg/kg.[47]

Expressing the applied dose in $\mu g/cm^2$ of skin made it possible to relate the toxicity of the OP pesticide to foliar residues in $\mu g/cm^2$ being transferred to skin of worker on a per hour basis.[6,7] The expression of the dose in $\mu g/cm^2$ of skin was also consistent with the in vivo percutaneous absorption work of Maibach et al.[48] and Maibach and Feldmann.[49]

13.4.3 Percutaneous Absorption

A number of dermatopharmacokinetic studies were conducted using radiolabeled organophosphorus pesticides to determine their fate in animals and humans.[47–53] Fredriksson[52] was among the first to investigate the in vivo and

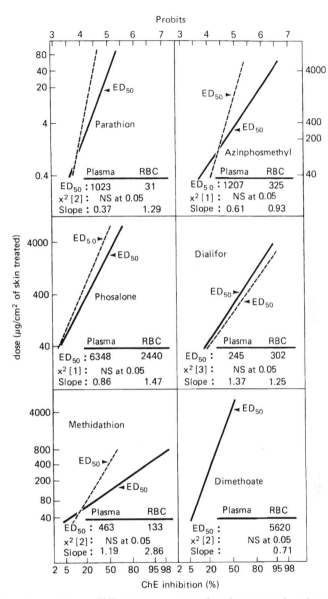

Figure 13.1 Dermal dose–ChE response curves for six organophosphorus pesticides. Male Sprague-Dawley rats weighing 220 to 240 grams were used. A 25-cm^2 area of skin was treated: (-------) plasma ChE, (———) RBC ChE. Blood activity was determined after 72 hours of exposure. (From Knaak et al.[41] Reprinted with permission from Springer-Verlag.)

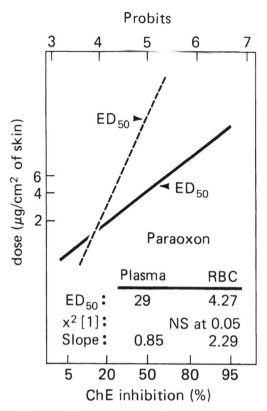

Figure 13.2 Dermal dose–ChE response curves for paraoxon. Male Sprague-Dawley rats weighing 200 to 240 grams were used. A 25-cm^2 area of skin was treated: (--------) plasma ChE, (———) RBC ChE. Blood activity was determined after 72 hours of exposure. (From Knaak et al.[41] Reprinted with permission from Springer-Verlag.)

in vitro percutaneous absorption of ethyl parathion (^{32}P-labeled). The slow rate of absorption of parathion ($0.084\ \mu g\,h^{-1}\,cm^{-2}$) and low anticholinesterase activity suggested this compound was unsuitable as a model for percutaneous absorption studies. Maibach and Feldmann[49] and Feldmann and Maibach[53] studied the percutaneous absorption of [^{14}C]-labeled parathion and malathion in humans. The site of application affected the extent of absorption with absorption varying from 8.6 percent for the forearm to 100 percent for the scrotum.

Percutaneous absorption was reported in percent of the applied dose/hour, per 24 hours or per 5 days, based on intravenous- (IV)-corrected [^{14}C] pesticide recovery data (i.e., urine). Parathion and malathion (4 μg/cm^2) were reportedly absorbed at the rate of 0.22 and 0.41 percent of applied dose/hour, respectively.[53] Wester et al.[51] studied the absorption of [^{14}C] isofenphos in human volunteers. The studies indicated that >90 percent of the isofenphos was lost

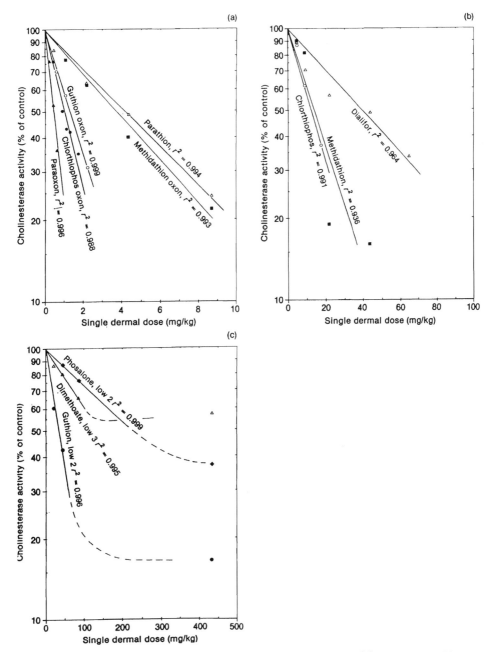

Figure 13.3   (*a*) Dermal dose ChE–response for highly toxic OPs. (*b*) Dermal dose ChE–response for moderately toxic OPs. (*c*) Dermal dose ChE–response for the least toxic OPs. (All data taken from Knaak et al.[41,42])

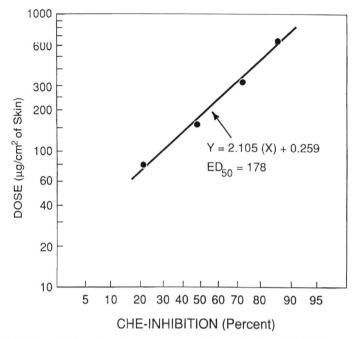

Figure 13.4 Dermal dose–ChE response curve for isofenphos. Male Sprague-Dawley rats weighing 300 grams were used. A 25-cm² area of skin was treated. Blood was taken for analysis 72 hours after topical application of isofenphos. (Data from Knaak et al.[47] Reprinted with permission from IBC Technical Services Ltd.)

by evaporation with ~3.6 percent absorbed over 24 hours. The surface dose was minimal (<1 percent) at the end of 24 hours. Skin stripping showed no residual isofenphos in stratum corneum. In vitro absorption studies utilizing flow-through diffusion methodology with human cadaver skin and human plasma as receptor fluid gave 2.5 percent of the dose absorbed. Skin surface wash at 24 hours recovered 79.7 percent with 6.5 percent remaining in the skin.

The fate, absorption kinetics, and dermal dose–ChE response of topically applied [ring-U-^{14}C] parathion (2.8 mg/kg) and [ring-U-^{14}C] isofenphos (3.2 mg/kg) were individually investigated.[47,50] According to these studies, a topical dose of 40 μg/cm² of parathion was absorbed at the rate of 0.28 μg h^{-1} cm^{-2} (K_p, 0.007 cm/h). The retention time on skin, $t_{1/2}$ of 24.3 to 28.6 hours, was less then its half-life, 28.5 to 39.5 hours, in plasma. Isofenphos was absorbed as readily as parathion (0.74 μg h^{-1} cm^{-2}, K_p, 0.007 cm/h). The retention half-life for a topical dose of 93 μg/cm² of isofenphos was 57.9 hours, while the $t_{1/2}$ for elimination from plasma was 33.9 hours. The time-course recoveries of topically applied ^{14}C-ring-labeled isofenphos and parathion are shown in Figures 13.5 and 13.6. Evaporative losses occurred (total recovery low) with both OP pesticides, with initial losses being greater for parathion.

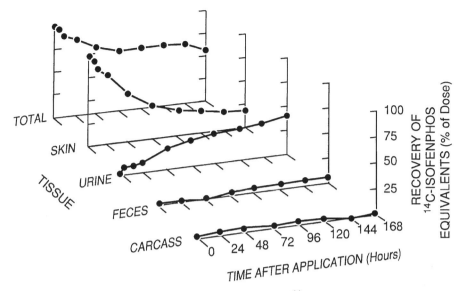

Figure 13.5   Time-course recovery of topically applied [^{14}C-ring] isofenphos equivalents in percentage of applied dose in feces, urine, carcass, and skin (surface and penetrated residues) postapplication. Cumulative percentage plotted for urine and feces. (Data from Knaak et al.[47] Reprinted with permission from IBC Technical Services Ltd.)

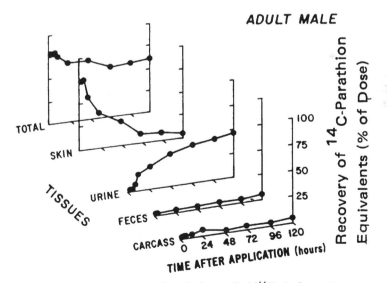

Figure 13.6   Time-course recovery of topically applied [^{14}C-ring] parathion equivalents in percentage of applied dose in feces, urine, carcass, and skin (surface and penetrated residues) postapplication. Cumulative percentage plotted for urine and feces. (Data from Knaak et al.[50] Reprinted with permission from Academic Press.)

The $[^{14}C]$ tissue count data from the published 7-day $[^{14}C\text{-ring}]$ isofenphos and parathion percutaneous studies[47,50] were used to construct and validate the PBPK/PD models using tissue $[^{14}C]$ data, percutaneous permeation constants (K_p, cm/h), and evaporative loss data (Kr, h^{-1}).

13.5   EXPOSURE ASSESSMENT

Exposure assessment is the process by which (1) potentially exposed populations are identified, (2) potential pathways of exposure are identified, and (3) chemical intakes/potential doses are quantified. Exposure is commonly defined as contact of visible external boundaries (i.e., external boundaries such as the mouth, nostrils, and skin) with a chemical agent.[54] As described in the *Guidelines for Exposure Assessment*,[54] exposure is dependent upon the intensity, frequency, and duration of contact. The intensity of contact is typically expressed in terms of the concentration of contaminant per unit mass or volume (i.e., $\mu g/g$, $\mu g/L$, mg/m^3, ppm, etc.) in the media to which humans are exposed.[54]

The development of dislodgeable leaf residue methodology and procedures for measuring the transfer of pesticide foliar residues to workers provided the environmental data for estimating dermal exposure.[6,7,17,54,55]

13.5.1   Dislodgeable Leaf Residue Methodology

Foliar Residues on Tree Fruit Foliage   Preharvest intervals (time between last application, dissipation of residues, and harvest) varying from several days to several weeks were originally considered adequate to protect workers and consumers alike. Gunther et al.[17] were among the first to establish the presence of dislodgeable pesticide residues on citrus foliage. This work led to the development of standardized procedures for the determination of dislodgeable pesticide residues on foliage and on soil surfaces under Western Regional Research Project W-146 "Worker Safety Reentry Intervals for Pesticide-Treated Crops."[55,56] A leaf punch sampler was developed to collect representative leaf samples for analysis of dislodgeable residues. The results of the analysis were expressed in $\mu g/cm^2$ of leaf area for two surface residues (top and bottom surface of leaf). The use of "ppm" was reserved for expressing residues present within the leaf and on/in foliar dust. According to Popendorf and Leffingwell,[6] dust was the vehicle by which OP pesticides were transferred from foliage to workers.

Foliar Residues on Turf Grass   Maddy et al.[57] and Goh et al.[58,59] were among the first to study the dissipation of OP pesticides [chlorpyrifos (0.023 pound A.I./1000 ft^2, $\sim$100 mg/m^2), dichlorvos and diazinon (0.25 pound A.I./1000 ft^2, $\sim$1000 mg/m^2)] on turf to assess potential human exposure and the possible necessity of safety intervals or precautions following application of pesticides to residential and public turf. Most product labels recommend that children and pets be kept off the turf until the spray is dried. Dislodgeable residues were calculated in terms of $\mu g/cm^2$ of grass blade surface area for two-sided residue,[55] $\mu g/cm^2$ of plot sampled, and $\mu g/10$ g of grass sampled. Two methods of leaf surface area measurements were used in the calculation of foliar dislodgeable residues ($\mu g/cm^2$): (1) total leaf surface area (two-sided) by regressing sample of known weight on established regression curve (weight vs. surface area) and (2) leaf area represented by lawn area sampled (100 cm^2). The latter method was considered to more accurately reflect the actual area that could be contacted in the course of human activities on lawn.

Oftanol 5G (U.S. EPA registered product containing 5.23 percent of the active ingredient isofenphos, Mobay Chemical Corporation, Kansas City, MO) was applied in 1983 to 1986 to turf grasses in Orangevale, California, to eradicate Japanese beetle grubs in soil.[60] Application of 5 percent granular formulations of Oftanol with a fertilizer spreader (224 mg/m^2) resulted in the deposition of isofenphos on grass and thatch. Immediately after application, the turf was watered to enable the pesticide to penetrate into the lower thatch and surface of the soil where the grubs feed. Turf was collected at specified time intervals (0, 1, 2, 4, 8, 12, 20, 30, and 40 weeks) after application by inserting a 6.0-cm-diameter stainless steel tube into the turf and soil to a depth of approximately 7.0 cm and removing the plug of turf and soil from the ground soil. The plug was pushed out of the cylinder and the turf cut off with a clean scissors into a glass jar. The thatch was then cut off into a second clean glass jar. Dislodgeable foliar residues of isofenphos and isofenphos oxon were determined by gas chromatography after extraction from leaf surfaces and thatch using a solution of Sur-ten surfactant. The detection limit for dislodgeable isofenphos was 0.5 and 1.0 ppm for oxon. Dislodgeable residues of isofenphos were reported in mg/m^2 of turf for a one-sided residue.

Sears et al.[61] estimated the dislodgeability of isofenphos by scuffling across treated turf (1 m^2 for 1 minute) with cloth pads attached to the bottom of a pair of rubber boots. The dislodgeable fraction of isofenphos was taken just after application and on days 1, 2, 5, 7, and 14 after application (200 mg of isofenphos/m^2) of a liquid formulation. Pads were analyzed for isofenphos by gas chromatography and dislodgeable residues reported in mg/m^2 of turf for a one-sided residue.

13.5.2 Dislodgeable Foliar Residue Dissipation Curves

Parathion Residues on Citrus   Gunther et al.[62] published dissipation curves for foliar applications of parathion. Toxic oxons (P=S to P=O conversion products) were found to be present within the first 3 days postapplication.[62] Parent residues (i.e., parathion) continued to decline while paraoxon residues where more persistent. A dissipation curve for parathion taken from Knaak and Iwata[63] is given in Figure 13.7. The parathion dissipation curve, in conjunction

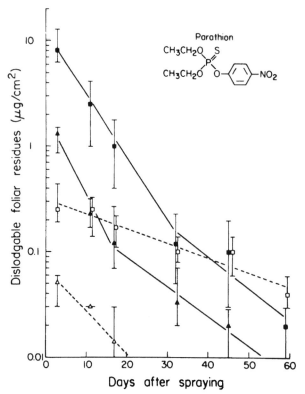

Figure 13.7 Dissipation of parathion (closed symbols) and paraoxon (open symbols) on orange trees by gas chromatography: (■) and (□) 10 pounds A.I. parathion/100 gallons per acre; (▲) and (△) 10 pounds A.I. parathion/1600 gallons per acre. (Figure taken from Knaak and Iwata.[63] Reprinted with permission from the American Chemical Society.)

with those of other OP pesticides, was used in the development of the safe level concept on dislodgeable foliar residues.[41,42]

Isofenphos Residues on Turf   Dislodgeable residues of isofenphos amounting to 14.0 mg/m^2 (ranged from 4.07 to 34.6 mg/m^2, one-sided residue) were found on turf with 22.8 mg/m^2 on thatch immediately after application/watering. These amounts represented only 6 and 10 percent of the amount applied to the area (224 mg of isofenphos/m^2). Dislodgeable isofenphos residues approached the limit of detection 4 weeks after application. The application of two times label rates resulted in 0 time dislodgeable residues of isofenphos as high as 65.7 mg/m^2 of turf (ranged from 5.26 to 65.7 mg/m^2, one-sided residue).[60] In the study by Sears et al.,[61] dislodgeable isofenphos residues amounted to 2.41 percent (4.7 mg/m^2, one-sided residue) of the applied isofenphos directly after application.

13.5.3 Transfer of Foliar and Turf Residues to Workers

Parathion Foliar Residues   Spear et al.,[64] were among the first to examine the nature of dust involved in transporting dislodgeable residues from foliage to the skin of harvesters. Harvesters working in dust-laden foliage, 20 to 30 days after an application, can generate locally high concentrations of pesticides contaminated aerosols. Fall-out of large particles (>50 μm) creates a whole-body dermal dose while smaller particles, although capable of being inhaled, were found to present a minimum respiratory hazard. The rate of transfer of dislodgeable foliar residues to the clothing and skin of workers was investigated using the multilayered gauze pad procedure reported by Durham and Wolfe[65] for liquid aerosol in their work with pesticide applicators. The use of gauze pads to estimate dermal exposures to soil dusts in field studies was later reported by Popendorf,[66] Spear et al.,[64] Popendorf et al.,[67] Popendorf,[68] and Davis.[69] Popendorf,[6,66] Davis,[69] and Nigg et al.[7] described procedures for calculating the rate of exposure from dermally collected and extracted residues. Human dermal surface areas described in Figure 13.8 were used to estimate the dermal dose to the head, neck, arms, hands, and upper and lower body.

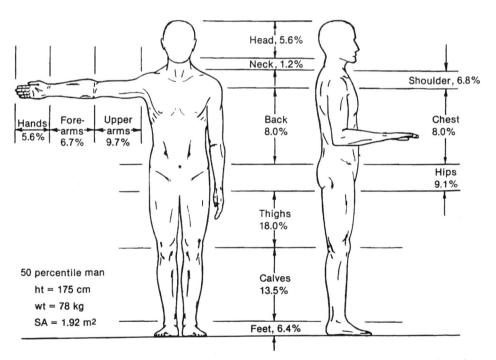

Figure 13.8   Human dermal surface area model derived from mensuration formula and anatomic dimensions. Each percentage corresponds to the proportion of the total surface area (SA) for each location. (From Popendorf and Leffingwell.[6] Reprinted with permission from Springer-Verlag.)

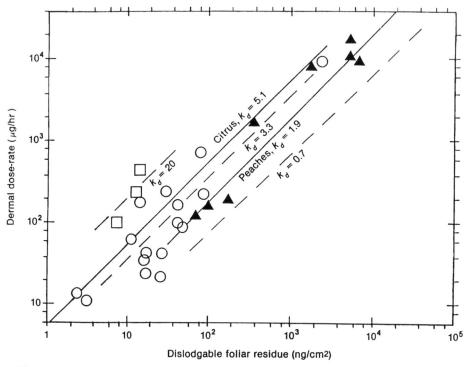

Figure 13.9   Composite dislodgeable residue vs. dermal dose relationships from Spear et al.,[64] Popendorf et al.,[67] and Popendorf.[68] (○) citrus, (□) supraoxon, and (▲) peaches. (From Popendorf and Leffingwell.[6] Reprinted with permission from Springer-Verlag.)

On the basis of field measures in citrus and peaches Popendorf and Leffingwell[6] proposed the following dose rate equation:

$$D' = k_d RT \tag{13.5}$$

where D' = dose in mg, R = the measured residue in ng/cm^2, T = the occupational exposure time in hours, k_d = a crop and/or work practice-specific coefficient. Figure 13.9 gives the relationship between dermal dose–rate (µg/h) and available foliar residue (ng/cm^2) with the slope of the line represented by k_d. In the equation, the k_d value will convert residue in ng/cm^2 on the x axis to dose–rate in µg/h on the y axis based on a one-sided leaf residue. To convert to a two-sided residue, one must divide the residue value by 2. On the basis of this equation a whole-body dermal dose of parathion and paraoxon was calculated for citrus and peach harvesters.

Nigg et al.[7] suggested the following equation for a two-sided residue based on total body exposure of citrus harvesters to chlorobenzilate in Florida: exposure [µg/h = 10^4 times residue (in µg/cm^2)], where 10^4 is the slope of the regression line or transfer coefficient in cm^2/h. Total body exposure was estimated from pads pinned underneath a cotton shirt at the shoulder, chest, back, forearms, and both upper arms. Thigh and shin pads were taped outside regular work clothing. Nigg et al.[7] found that the log-log transformation of data carried out by Popendorf and Leffingwell[6] in California was not necessary for a Florida model. The California data was reanalyzed like the Florida data, yielding smaller probable errors in estimating the dose from the regression line than from transformed data. No statistical differences were observed between the California and the Florida models.

The dislodgeable residue transfer work of Popendorf and Leffingwell[6] and Nigg et al.[7] led to the adoption of standard operating procedures (SOPs) by EPA[70] for estimating exposure from dermal contact. According to an SOP entitled "Postapplication dermal exposure to pesticide residues while harvesting fruit from trees," daily workplace exposures (mg/day) and dose (mg/kg/day) may be estimated using pesticide dislodgeable leaf residue data (DFRs) and crop dislodgeable residue transfer coefficients (TCs).[70] Exposure (mg/day) is calculated in Eq. (13.6) and dose (mg/kg/day) in Eq. (13.7).

$$E_d = \text{DFR} \times \text{CF} \times \text{TC} \times \text{ET} \qquad (13.6)$$

where E_d = exposure on day "d" (mg/day)

    DFR = dislodgeable foliar residue on day "d" ($\mu g/cm^2$)

    CF = conversion factor to convert μg units to mg for daily exposure (0.001 mg/μg)

    TC = transfer coefficient (cm^2/h)

    ET = exposure time (h/day)

Crop and pesticide-specific transfer coefficients were obtained from the California Department of Pesticide Regulations (CDPR). The transfer factors were generated using two-sided residues. Values (cm^2/h) vary from 200 to 10,000. According to EPA the length of the exposure period varies from 8 hours for workers harvesting fruit to approximately 2 hours for children and adults on turf grass. Adult males and females weigh 70 and 60 kg, respectively. Children , 6 years of age weigh 22 kg. Dose is calculated as:

$$D_d = (E_d/\text{BW}) \times A_d \qquad (13.7)$$

where D_d = dermal dose on day "d" (mg/kg/day)

    BW = body weight (kg)

    A_d = dermal absorption in percent (100 percent EPA default value)

The dose is to the total area of exposed skin. The concentration of the pesticide residue per cm^2 of skin and the permeation constant (K_p, cm/h) were not considered in Eq. (13.7). EPA recommends the use of percutaneous absorption data (percent of applied dose) from animal or human studies. Several guide-

TABLE 13.3   Reentry Intervals (REI) for Parathion Applied to Citrus at 10 lb AI/acre

Day[b]	High Exposure Potential[a] TC = 10,000 cm²/h (100 gal/acre)			High Exposure Potential[a] TC = 10,000 cm²/h (1600 gal/acre)		
	DFR (μg/cm²)[c]	Dose (mg/kg/day)[d]	MOE[e]	DFR (μg/cm²)[c]	Dose (mg/kg/day)[d]	MOE[e]
0	10	0.34	0.145	1.5	0.051	0.97
10	3.0	0.10	0.486	0.35	0.012	4.16
20	0.7	0.024	2.08	0.1	0.0034	14.6
30	0.2	0.0069	7.29	0.05	0.0017	29.2
40	0.1	0.0034	14.6	0.025	0.00085	58.3
50	0.04	0.0014	36.5	0.01	0.00034	145.3
60	0.02	0.00069	72.9	0.001	0.000034	1458.0
Safe level	0.09	0.0031	16.2			

[a]Transfer coefficient from Nigg et al.[7]

[b]Days after treatment. Workers wearing long pants, long-sleeved shirts, and no gloves.

[c]Dislodgeable foliar residue (DFR) from parathion dissipation curves, Knaak and Iwata.[63]

[d]Absorbed dose $(mg/kg/day) = [DFR \times TC \times 0.001$ mg/μg unit conversion $\times 0.03$ dermal absorption $\times 8$ h/day]/70 kg body weight.

[e]MOE = oral (human) NOAEL 0.05 mg/kg/day/ absorbed dermal dose in mg/kg/day.

lines are available for measuring percutaneous absorption.[71–74] The EPA uses a 100 percent default dermal absorption factor, while CDPR uses a factor of 50 percent. The CDPR default factor is based on an experimental saturation model with lag time validation against human volunteer dermal absorption studies involving 12 different pesticides.[75] Risk is estimated base on a margin of exposure (MOE) where

$$MOE = \frac{NOEL \ (mg/kg/day)}{Exposure \ (dose \ in \ mg/kg/day \ corrected \ for \ dermal \ absorption)}$$

The "postapplication dermal exposure" SOP was used in Table 13.3 to calculate MOEs for parathion dislodgeable leaf residues. EPA is currently recommending a MOE of 100 or more for OP pesticides. In Table 13.3 a MOE of 16.2 was calculated for the "safe level" of 0.09 μg/cm² used by CDFA for parathion on citrus. This suggests that a MOE of 100 is highly conservative.

Isofenphos Turf Residues   The transfer rate for isofenphos was derived from a study conducted by Harris and Solomon[76] on 2,4-D herbicide. In the study, 10 adult volunteers were exposed for 1.0 hour to residues of 2,4-D on a 2 × 15 m area of turf to which 2,4-D had been applied at the rate of 0.02 lb A.I. per 1000 ft² (4.9 and 10.9 mg of 2,4-D/m², with an average value of 8.45 mg/m², 1 hour after application). During the exposure period, the activity of the volunteers alternated between walking and sitting or lying down on the surface of the turf for intervals of 5 minutes. Five of the volunteers wore long pants, a short-sleeved shirt, socks, and closed footwear. The other half of the volunteers wore shorts, a short-sleeved shirt, and no shoes. The two groups performed the required activities 1 hour after the first application and 1 day following the second application a month later. The herbicide (mean = 227.6 μg) was detected in the urine of three volunteers wearing shorts and exposed to the herbicide 1 hour after application. According to Feldmann and Maibach,[77] humans absorb 5.8 percent of a topically applied dose of 2,4-D. On the basis of this percentage, approximately 4 mg of 2,4-D was transferred to skin/hour per 8.45 mg/m^2 of turf dislodgeables (~480 μg/h per mg/m^2). Using this surrogate transfer rate for isofenphos in granular form, 12 hours after application and watering, approximately 6.7 mg (14 mg/m^2 × 480 μg/h per mg/m^2) would be transferred to skin over the course of 1 hour.

On the basis of the work of Popendorf and Leffingwell,[6] Nigg et al.,[7] and Harris and Solomon[76] a transfer coefficient (k_d) of approximately 5000 may be used for isofenphos on turf for a one-sided residue (i.e., 480 μg/h per 0.10 μg/cm^2 gives a k_d value of 4800 for a one-sided residue and 9800 for a two-sided residue).

13.6 PROCEDURES FOR CALCULATING A REENTRY INTERVAL

In 1971, California established reentry interval regulations specifying the time periods that workers be restricted from activities that involve substantial body contact with foliage in fields of grapes, citrus, peaches, and nectarines after these crops had been treated with any one of 17 OP pesticides or with sulfur.[62] The reentry intervals were selected to take into account the toxicity of the pesticides, foliar residue dissipation rates, human exposure based on cultural practices being employed, pesticide usage patterns, amounts of pesticides used, and the combination patterns of certain pesticide usage.[62] Specific criteria for establishing reasonable safety of a given reentry interval remained to be officially prescribed. Studies carried out by agriculture and public health investigators in the western states, Florida, and Arkansas during the 1970s and 1980s provided technical data for establishing the criteria described in the following sections, setting new reentry intervals, and reassessing old intervals.[5]

13.6.1 U.S. EPA's Recommendations for Setting Reentry Intervals

EPA's 1984 guidelines[10] for reentry protection proposed (1) the interval for residue dissipation to the nondetectable level as the reentry interval and (2) the determination of an allowable human exposure level (AEL, mg/kg/day). According to the guidelines, the reentry level was estimated by plotting foliar residue levels (μg/cm^2) against exposure levels (AEL) in mg/kg/day with the upward sloping line produced by the plot representing the foliar residue transfer coefficient. Knaak et al.[5] estimated reentry intervals for carbosulfan in

TABLE 13.4   Calculation of Reentry Intervals According to U.S. EPA[10] Guidelines with Slight Modification[a]

Days after Spraying	Compound Ratio[b] CS:CF:HCF	NOEL[c] (μg/kg/day)	AEL[d] (μg/kg/day)	Total Dose[e] (μg/h)	Reentry Level[f] (μg/cm^2)
1	94:3.8:1	1224	122	1068	0.21
3	58:3.6:1	1209	121	1059	0.21
7	22.5:2.4:1	1197	120	1050	0.21
10	14.2:1.9:1	1200	120	1050	0.21

[a] Modification involves taking into account all toxic residues present on the foliage and using a total toxic residue curve. Foliar residues: CS = carbosulfan, CF = carbofuran, HCF = 3-hydroxy carbofuran.

[b] Ratios calculated from values used to construct Figure 13.10.

[c] No effect level (NOEL) calculated from data from dermal dose–ChE response curve. NOEL = ED$_{10}$ (25 cm^2)/(0.23 kg/day). Predicted ED$_{10}$ = $[P_1/ED_{10.1} + P_2/ED_{10.2} + \cdots + P_N/ED_{10.N}]^{-1}$, where P = proportion of component in mixture (Finney[82]). ED$_{10}$ values were extrapolated from Figure 13.10. ED$_{10}$ values for CS, CF, and HCF were 11.5, 6.24, and 98.8 μg/cm^2, respectively.

[d] Allowable exposure level (AEL) = NOEL/SF; safety factor (SF) = 10.

[e] Total dose = (AEL)(body weight, 70 kg)/(duration, 8 h/day).

[f] From total dose, reentry level was determined from the graph of whole-body dermal dose (μg/h) versus dislodgeable. Foliar residues (ng/cm^2) were taken from the data of Popendorf and Leffingwell.[6] Total dose was divided by 5.1; the k_d for citrus; derivation was based on the area of only one side of the leaf.

Source: Knaak et al.[5]

Table 13.4 using the AEL method. The NOEL required in the AEL method was calculated from data derived from a rat dermal dose–ChE response curve [NOEL = ED$_{10}$ (μg/25 cm^2)/0.23 kg; AEL = NOEL/SF]. A total allowable dose (μg/h) was determined (AEL × body weight, 70 kg/duration, 8 h/day) and used to relate whole-body dermal dose (μg/h) to dislodgeable residues (μg/cm^2) from the data (k_d of 5100 for a one-sided residue) of Popendorf and Leffingwell.[6] A carbosulfan residue dissipation curve (μg/cm^2 vs. time in days) shown in Figure 13.10 was used to determine the reentry interval. Knaak et al.[5] used dermal dose–ChE response data from animals in place of dermal absorption and oral NOEL data because limited information was available on dermal absorption in animals/humans during the 1970 to 1980s and oral toxicity data for poorly absorbed pesticides overestimated dermal toxicity.

EPA's 1997 guidelines[11] for reentry protection made use of the procedure published in the 1984 guidelines for determining reentry intervals. The guidelines require foliar dissipation and dermal absorption data, foliar residue transfer coefficients, and NOELs from animal or human oral studies to estimate the AEL [reentry dose level (RDL) in the 1997 guidelines]. A reentry interval is selected on the basis of a calculated MOE of 100 or greater. In Table 13.3, a reentry interval between 40 and 50 days was calculated for parathion applied to citrus. The number of days is consistent with CDFA 1989 reentry intervals.

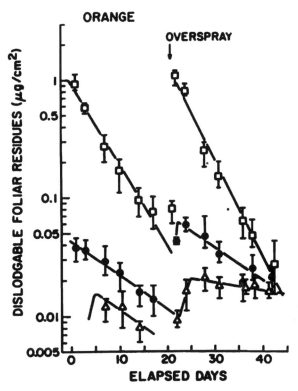

Figure 13.10 Dislodgeable foliar residues of carbosulfan (□), carbofuran (●), and 3-hydroxycarbofuran (△) after treatment of orange trees with Advantage 2.5EC insecticide formulation at 1.5 pounds A.I./200 gallons per acre. Each datum point is the mean value obtained from six replicate field samples, and the vertical lines show the range of values found. (From Knaak et al.[5] Reprinted with permission from American Chemical Society.)

13.6.2 Development of the Safe Level Concept Using Dermal Dose–ChE Response Studies and Field Worker Observations for Setting Reentry Intervals (CDFA)

The California Department of Food and Agriculture (CDFA) used animal dermal dose–ChE response studies in the 1970 to 1980s in conjunction with the results of field worker studies for estimating safe foliar residue levels for a number of organophosphorus pesticides.[5] Table 13.5 gives the dermal dose–ChE response data (ED_{50}) in terms of body weight, total body surface, and the dermal LD_{50} in terms of body weight. On the basis of total body surface, the quantities producing 50 percent red-cell inhibition were 0.33, 2.4, 10.0, 23.0, 25.0, 188.0, and 432.0 µg/cm² of skin, respectively, for paraoxon, parathion, methidathion, dialifor, azinphosmethyl, phosalone, and dimethoate.[5,41,78,79] In Table 13.6, these values and the results of studies conducted by Spear et al.,[80]

TABLE 13.5   Dermal Dose–ChE Response Expressed in Terms of Total Body Surface, Body Weight, and Safety Index

Pesticides	ED$_{50}$ (µg/cm^2 of Body Surface)[a]	ED$_{50}$[b] (mg/kg)	Dermal LD$_{50}$ (mg/kg)	Safety Index LD$_{50}$/ED$_{50}$ (mg/kg)
Paraoxon	0.33 ± 0.2	0.5	2.0[c]	4.0
Parathion	2.4 ± 0.3	3.4	21.0[d]	6.2
Methidathion	10.0 ± 0.3	15.0	150.0[e]	10.0
Dialifor	23.0 ± 0.3	33.0	—	—
Azinphosmethyl	25.0 ± 0.5	35.0	220.0[d]	6.3
Phosalone	188.0 ± 0.4	265.0	1450.0[f]	5.5
Dimethoate	432 ± 2.0	611.0	1420.0[g]	2.3

[a] Pesticides were individually applied in 1.0 mL of acetone to the clipped backs (25 cm^2) of 220–240 grams male rats. Blood was taken 72 hours after application for ChE determination. Response expressed in terms of total body surface (325 cm^2) from dermal dose–ChE response curves in Figures 13.1 and 13.2. Values given with 95 percent confidence limits.

[b] Values determined from dermal dose–ChE response curves.

[c] Estimated.

[d] Gaines.[78]

[e] CIBA-GEIGY Toxicology Bulletin.

[f] Mazuret.[79]

[g] Gaines.[40]

Source: Knaak et al.[5]

Richards et al.,[81] and Popendorf et al.[67] were used to estimate safe levels on foliage.[5,41,42] The field exposure studies involving residue transfer and dermal dose–ChE response in workers established safe levels for azinphosmethyl, azinphos-methyl oxon, phosalone, and paraoxon of 3.1, 0.05, 7.0, and 0.02 µg/cm^2, respectively. These pesticides and their safe foliar levels were used as standards for establishing additional safe levels on foliage for methidathion, methidathion oxon, dialifor, parathion, and dimethoate using their relative toxicities. In practice, the pesticide under investigation and the pesticide standard were grouped according to their slopes, and a safe level for the pesticide under investigation was determined using Eq. 13.8:

Safe level (µg/cm^2) for pesticide under investigation

= Safe level of standard × ED$_{50}$ (µg/cm^2) of pesticide

÷ ED$_{50}$ of standard   (13.8)

The reentry interval in days is obtained from the pesticide dislodgeable foliar residue dissipation curve (µg/cm^2 vs. days) using the established safe level. The

TABLE 13.6 Establishment of Safe Levels on Tree Foliage (in $\mu g/cm^2$) Using the Results of Dermal Dose–ChE Response Studies in Male Rats and Field Reentry Studies

Pesticides[a]	Slopes	ED_{50} ($\mu g/cm^2$ of Body Surface)	Relative Toxicity[b]	Safe Level on Foliage in $\mu g/cm^{2c}$
Paraoxon	2.3	0.33	1.0	0.02[d]
Methidathion	2.9	10.00	30.0	0.60
Azinphosmethyl-Oxon	2.0	0.82	1.0	0.05[e]
Methidathion-oxon	1.8	2.2	3.0	0.15
Dialifor	1.3	23.0	0.12	0.8
Parathion	1.3	2.4	0.013	0.09
Phosalone	1.5	188.0	1.0	7.0[f]
Azinphosmethyl	0.9	25.0	1.0	3.1[g]
Dimethoate	0.7	432.0	17.0	53.0

[a] Pesticide standard is underlined.

[b] ED_{50} of pesticide under investigation divided by ED_{50} of pesticide standard.

[c] Relative toxicity multiplied by safe level of standard.

[d] Spear et al.[80]

[e] Estimated.

[f] Popendorf et al.[67]

[g] Richards et al.[81]

Source: Knaak et al.[5]

dissipation curve in Figure 13.7 shows the relationship between the parathion safe level on citrus (i.e., 0.09 $\mu g/cm^2$) and the reentry interval.

13.6.3 Reentry Intervals for Thions and Oxons (CDFA)

The oxidative conversion of methidathion, azinphosmethyl, and parathion on leaf surfaces to oxons necessitated the development of a procedure for establishing safe levels on foliage for the combined hazard posed by thion and oxon residues. In Table 13.7, this was accomplished for methidathion, azinphosmethyl, and parathion[5,63] by allowing the oxon to be present at a safe level and by reducing the combined residue of oxon and thion to 0.06, 0.02, and 1.6 $\mu g/cm^2$, respectively, for parathion, methidathion, and azinphosmethyl. A safe level for the mixture may also be estimated by determining the toxicity (ED_{50}) of the oxon and thion mixture using the method of Finney[82] in Eq. (13.9):

$$ED_{50} \text{ (mixture, } \mu g/cm^2) = P_1/ED_{50,1} + P_2/ED_{50,2} \ldots P_N/ED_{50,N} \quad (13.9)$$

where P_1 and P_2 are the proportions of oxon and thion, respectively, on foliage after 10, 20, or 30 days. In the case of parathion/paraoxon at 10 days, the ED_{50}

TABLE 13.7 Procedure for Establishing Safe Levels ($\mu g/cm^2$) for Thions + Oxons on Tree Foliage

Application to Citrus	Days Elapsed[a]	Thion[a]	Oxon[a,b]	Thion + Oxon[a,b]	Thion + Oxon × RT[c]	$\dfrac{\text{Thion} + \text{Oxon}}{\text{Thion} + \text{Oxon} \times \text{RT}} \times \text{SL}$[d] for Thion[e]
Parathion 10 lb A.I./1600 gal per acre	10	0.35	0.02	0.37	0.49	0.07
	20	0.09	0.01	0.10	0.16	0.06
Methidathion 5.6 lb A.I./100 gal per acre	10	1.0	0.08	1.08	1.38	0.4
	20	0.25	0.1	0.35	0.73	0.3
	30	0.11	0.08	0.19	0.50	0.2
Azinphosmethyl 6.0 lb A.I./1200 gal per acre	10	1.5	0.05	1.55	2.91	1.7
	20	1.3	0.05	1.35	2.86	1.6
	30	1.1	0.05	1.15	2.52	1.5

[a]Taken from Figure 13.7 and from Figs. 11 and 12 in Knaak et al.[5]

[b]Oxons must be at safe level indicated in Table 13.6. Method assumes oxons will be at a safe level when safe level for thion + oxon is reached.

[c]RT = relative toxicity from Table 13.6 (ED_{50} of thion + ED_{50} of oxon).

[d]SL = safe levels for thions from Table 13.6.

[e]Safe levels for Thion + oxon.

Source: Knaak et al.[5]

of the mixture was 23.4 µg/cm^2, while the safe level for the mixture was 0.067 µg/cm^2 as determined by Eq. (13.9). This value is equivalent to the one given in Table 13.7 for parathion/paraoxon, 10 days after application.

SL, mixture (µg/cm^2)

$$= [ED_{50}, \text{thion} + \text{oxon} \times SL, \text{phosalone}]/ED_{50}, \text{phosalone} \qquad (13.10)$$

The relationship between total parathion/paraoxon residues, their rate of dissipation, and the safe level for total thion and oxon is shown in Figure 13.11. The reentry time in days is obtained from the dislodgeable foliar residue dissipation curve using the safe level of the mixture.

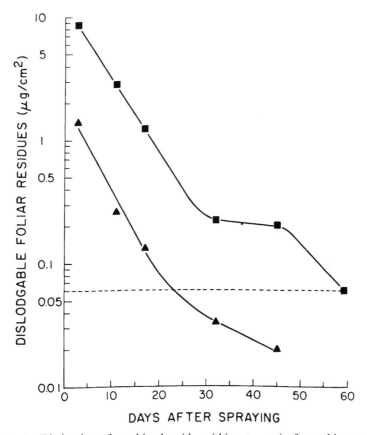

Figure 13.11 Dissipation of combined residues (thion + oxon) of parathion and paraoxon on orange trees: (■) 10 pounds A.I. parathion/100 gallons per acre; (▲) 10 pounds A.I. parathion/1600 gallons per acre. Dashed line is safe level for thion + oxon. Curves are drawn from Figure 13.7. (From Knaak and Iwata.[63] Reprinted with permission from American Chemical Society.)

13.6.4 Probabilistic Modeling to Determine Reentry Intervals (CDPR)

Probabilistic models are currently being used to assess exposure and risk to environment contaminants.[83-88] Guidance for Monte Carlo analysis is provided by a EPA publication on this subject.[89] The CDPR is now using probabilistic modeling to determine reentry intervals when distribution information (range of values) is available for dislodgeable residues (dissipation) and foliar transfers.[90] The probability distribution of the reentry interval is calculated from the full ranges of the individual exposure factors using computer software. According to Ross and Dong[90] relatively few pesticides have the requisite dislodgeable foliar residue and transfer factor distributions to justify using stochastic modeling. If information is available, the procedure reduces the estimated reentry interval at the 95th percentile compared to extreme point estimates. CDPR methodology was used to calculate a 30-day reentry time in Eq. (13.11) for workers harvesting citrus previously sprayed with parathion. The safe level $(0.09 \ \mu g/cm^2)$ on foliage is calculated in Eq. (13.12).

$$T = 30 \ \text{days}$$

$$= \{\ln[(ADD_{SL} \times BW)/(TF \times ABS \times 8 \ h)] - \ln(DFR_0)\}K^{-1} \quad (13.11)$$

$$\ln(DFR_{SL}) = \ln(DFR_0) + KT$$

$$= -2.048 \ (\text{inverse} \ \ln = 0.09 \ \mu g/cm^2) \quad (13.12)$$

where $ADD_{SL} = 5.14 \ \mu g/kg$ BW/day (parathion NOEL/10 for humans)
$\quad \quad$ TF $= 10,000 \ cm^2/h$ transfer factor for a two-sided residue
$\quad \quad$ ABS $= 5$ percent (based on PBPK model)
$\quad \quad DFR_0 = 10 \ \mu g/cm^2$ after the application of 10 pounds A.I. per acre (100 gallons of water), Fig 13.7
$\quad \quad \quad K = -0.157$ ($\mu g/cm^2$ DFR per day, estimated dissipation rate for foliar residue)
$\quad \quad DFR_{SL} =$ safe level, dislodgeable foliar residues

13.6.5 Application of Safe Level Concept to Foliar Residues on Turf to Obtain Reentry Interval

Concern over the safety of children playing on OP-pesticide-treated lawns prompted Maddy et al.[57] and Goh et al.[58,59] to study residues of diazinon, dichorvos, and chlorpyrifos on turf grasses and apply the safe level concept to these residues. Maddy et al.[57] determined a safe level for diazinon by using the ratio of the dermal LD_{50}'s for diazinon and azinphos-methyl and the calculated safe level for azinphos-methyl to derive a safe level of $3.3 \ \mu g/cm^2$ for diazinon.

Goh et al.[58,59] used 0.5 and 0.06 μg/cm^2 as safe levels for chlorpyrifos and dichlorvos. Dislodgeable residues of diazinon, chlorpyrifos, and dichlorvos on turf were less than the established safe levels shortly after application. No reentry interval was established.

A safe level 0.6 μg/cm^2 (6 mg/m^2, two-sided residue) may be derived for isofenphos using the safe level determined for methidathion, as both pesticides are metabolized to oxons, the slopes of their dermal dose–ChE response curves and their ED_{50} values (ChE inhibition) are similar.[41,42] On the basis of the dislodgeable residue studies of Segawa and Powell[60] and Sears at al.,[61] a short reentry interval of 24 hours may be required for isofenphos applied to turf.

13.7 REASSESSMENT OF REENTRY INTERVALS, FOLIAR RESIDUE TRANSFER/PERCUTANEOUS ABSORPTION PBPK/PD MODELS

An increasing number of toxicologists support the development of PBPK/PD models for extrapolating laboratory animal data (i.e., kinetic and toxicity data) from one route of exposure to another (i.e., dermal, oral, inhalation, IV, etc.) and between animals and humans for purposes of conducting risk assessments and establishing RfDs and RfCs. To a much more limited extent these models have been used to estimate dosages received by workers from environmental media (soil, air, water) under largely uncontrolled environmental conditions. PBPK models are capable of handling all the mathematics involved in the determination of an MOE using residue transfer coefficients, skin permeability values (K_p, cm/h), physiological parameters, partition coefficient, metabolic pathways, and metabolic rates constants, while PBPK/PD models provide pharmacodynamic information (i.e., enzyme inhibition).

13.7.1 PBPK Models, Isofenphos, and Parathion

PBPK/PD models are currently available for extrapolating the results of OP studies to humans.[8,9] Knaak et al.[8] developed a multiroute, multianimal PBPK/PD model for isofenphos and presented the model at a symposium sponsored by the American Chemical Society on Biomarkers. The combined metabolic pathway for isofenphos in the rat, guinea pig, and dog is given in Figure 13.12. The isofenphos PBPK/PD model was modified to include passes through the liver for IPS (isopropyl salicylate to salicylic acid) and SA (salicylic acid to 2-hydroxy hippuric acid) (see Fig 13.13) making it possible to use the P-450 catalyzed V_{max}, K_m values develop by Knaak et al.[91] with in vivo [^{14}C-ring] isofenphos rat tissue data.[47] The foliar transfer coefficient ($k_d R$, μg/h) of Nigg et al.[7] was added to the percutaneous absorption mass balance equation for simulating worker exposure to pesticide foliar residues, that is, the mass balance for foliar transfer and percutaneous absorption.

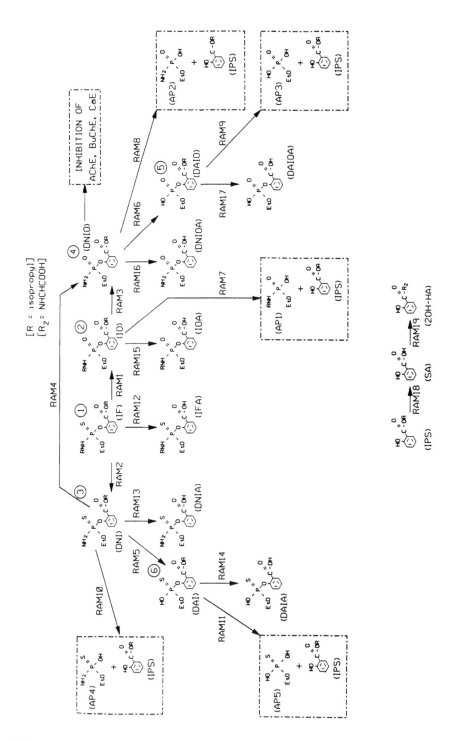

Mass balance for foliar transfer and percutaneous absorption:

$$dA_{surf}/dt = K_p A(C_{sk}/P_{a/sk} - C_{surf}) - K_a A_{surf} + k_d R \quad (\text{pmol h}^{-1})$$

$$A_{surf} = \text{Integ}(dA_{surf}/dt, 0.0)$$

$$C_{surf} = A_{surf}/V_{sk} \quad (\text{pmol cm}^{-3})$$

$$C_{sk} = A_{sk}/V_{sk} \quad (\text{pmol cm}^{-3})$$

$$dA_{air}/dt = K_a * A_{surf} \quad (\text{pmol h}^{-1})$$

where A_{surf} = amount of isofenphos on skin surface, pmol (applied dose)

C_{surf} = concentration on skin surface, pmol cm^{-3}

K_p = permeability constant, cm h^{-1}

$k_d R$ = foliar dose rate to skin, pmol h^{-1}

A_{sk} = amount in skin, pmol

C_{sk} = concentration within skin, pmol cm^{-3}

V_{sk} = volume of skin, pmol cm^{-3}

A_{air} = amount in air, pmol

K_a = rate of loss to air, hour

$P_{a/sk}$ = partition coefficient, air/skin

A = area of treated skin, cm^2

Foliar dose rate $k_d R$ was determined as follows:

Constant $R = 0.1, 1.0, 5.0$ and 10, foliar pesticide concentration in $\mu\text{g cm}^{-2}$

Constant $k_d = 10,000$, slope factor in $\text{cm}^2 \text{h}^{-1}$

$$\text{EXPOS} = k_d * R$$

$$\text{RP} = (\text{EXPOS}/\text{MW})$$

$$k_d R = \text{RP} \times 1.0 \times 10^6, \text{pmol h}^{-1}$$

Figure 13.12 Combined metabolic pathway for isofenphos in the rat, guinea pig, and dog (RAM1) isofenphos (IF) to isofenphos oxon (IO); (RAM2) IF to des N-isopropyl isofenphos (DNI); (RAM3) IO to des N-isopropyl isofenphos oxon (DNIO); (RAM4) DNI to DNIO, (RAM5) DNI to desaminoisofenphos (DAI), (RAM6) DNIO to des aminoisofenphos oxon (DAIO); (RAM7) IO to O-ethylisopropylaminophosphate (AP1) and isopropyl salicylate (IPS); (RAM8) DNIO to O-ethylaminophosphate (AP2) and IPS; (RAM9) DAIO to O-ethylphosphate (AP3) and IPS; (RAM10) DNI to O-ethyl-amino-thiophosphate (AP4) and IPS; (RAM11) DAI to O-ethylthiophosphate (AP5) and IPS; (RAM12) IF to carboxylic acid of isofenphos (IFA); (RAM13) DNI to carboxylic acid of DNI (DNIA); (RAM14) DAI to carboxylic acid of DAI (DAIA); (RAM15) IO to carboxylic acid of IO (IOA); (RAM16) DNIO to carboxylic acid of DNIO (DNIOA); (RAM17) DAIO to carboxylic acid of DAIO (DAIOA); (RAM18) IPS to 2-hydroxy salicylic acid (SA); (RAM19) SA to 2-hydroxy hippuric acid (2-OH-HA). (From Knaak et al.[8] Reprinted with permission from the American Chemical Society.)

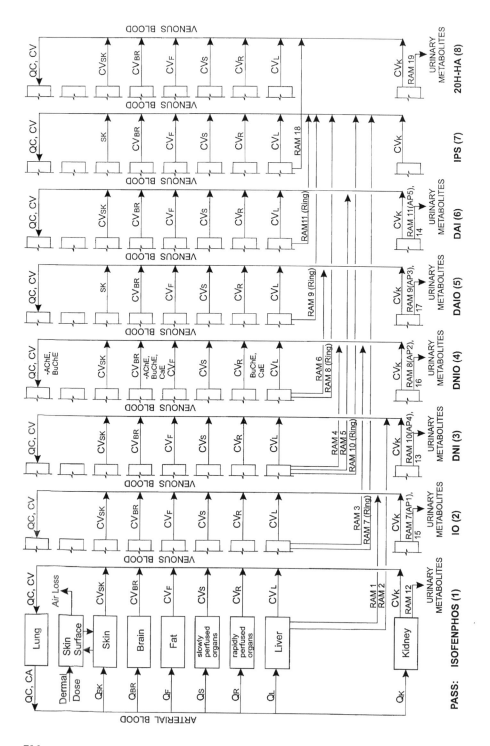

706

In addition to the isofenphos model, a foliar residue transfer/percutaneous absorption parathion PBPK/PD model (Fig 13.14) was developed using the metabolic pathway for parathion in Fig 13.15,[92–97] and the tissue partition coefficients for parathion and paraoxon developed by Jepson et al.[98] and Gearhart et al.[8] V_{max} and K_m values for rat liver P-450 catalyzed desulfuration and oxidative hydrolysis of parathion were obtained from the work of Wallace and Dargan,[99] and paraoxonase activity in rat liver homogenates and microsomes from Gil et al.,[100,101] Gonzalvo et al.,[102] and Wallace and Dargan.[99] Rat serum and brain paraoxonase activity was obtained from Gil et al.[101] and Forsyth and Chambers,[103] respectively. Information on the activity of human serum paraoxonase (PON1) were obtained from Kuo and La Du,[104] Mackness et al.,[105] and Sams and Mason.[106] Table 13.8 gives the physiological parameters taken from ILSI-RSI[107] for the rat and human. Tables 13.9 and 13.10 give the partition coefficients for the parent OPs (isofenphos and parathion) and metabolites in the tissue compartments. Tables 13.11 and 13.12 give the V_{max} and K_m values for rats and humans used in the isofenphos and parathion models, respectively.[8,91,98,99–106,108,109] V_{max} values were scaled to body weight (BW[0.7]) in both models. Human liver P-450 V_{max}, K_m values[91] for the metabolism of isofenphos were used in the human isofenphos and parathion models.

13.7.2 Pharmacodynamics (PD)

In the isofenphos and parathion PBPK/PD models, blood acetylcholinesterase (AChE) and butyrylcholinesterase (BChE), brain [AChE, BChE, and carboxylesterase (CaE)], and liver (CaE and BChE) are inhibited by their oxons, des N-isopropyl isofenphos oxon (DNIO), and paraoxon, respectively. Concentrations of free enzymes in tissues (i.e., C_{AChEB}, etc.) were obtained from Maxwell[110] and are shown in Table 13.13. The bimolecular reaction constants (k_i) describing the overall inhibition were calculated in the two models using literature values for K_a and k_p (or k_{+2}) given in Tables 13.14 and 13.15.[111–114] The equation used in the models for depicting red blood cell AChE inhibition is shown below along with the equation for aging of inhibited AChE:

Figure 13.13 Percutaneous absorption of isofenphos. Physiologically based pharmacokinetic model. Pass 1, IF to DNI, IO, and IFA; pass 2, IO to DNIO; IOA, O-ethylisopropylaminophosphate (AP1), and isopropyl salicylate (IPS); pass 3, DNI to DNIO, DAI, DNIA, O-ethylaminothiophosphate (AP4), and isopropyl salicylate (IPS); pass 4, DNIO to phosphorylate enzymes (AChE, BChE, and CaE), DNIOA, DAIO, aminoethylphosphate (AP2), and isopropyl salicylate; pass 5, DAIO to DAIOA, ethylphosphate (AP3), and isopropyl salicylate (IPS); pass 6, DAI to DAIA, O-ethylthiophosphate (AP5), and isopropyl salicylate (IPS); pass 7, IPS to SA; pass 8, SA to 2OH-HA. (From Knaak et al.[8] Reprinted with permission from the American Chemical Society.)

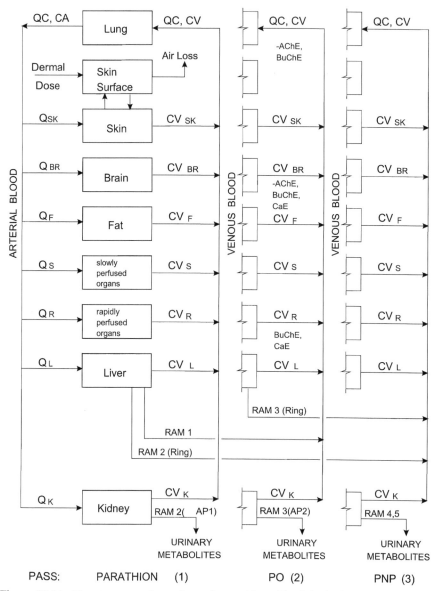

Figure 13.14 Percutaneous absorption of parathion: Physiologically based pharmacokinetic model. Pass 1, P to PO; P to O,O-diethyl phosphorothionate (AP1) and *p*-nitrophenol (PNP); pass 2, PO to phosphorylate enzymes (AChE, BChE, and CaE), O,O-diethyl phosphate (AP2) and PNP; pass 3, PNP to PNP sulfate and PNP glucuronide.

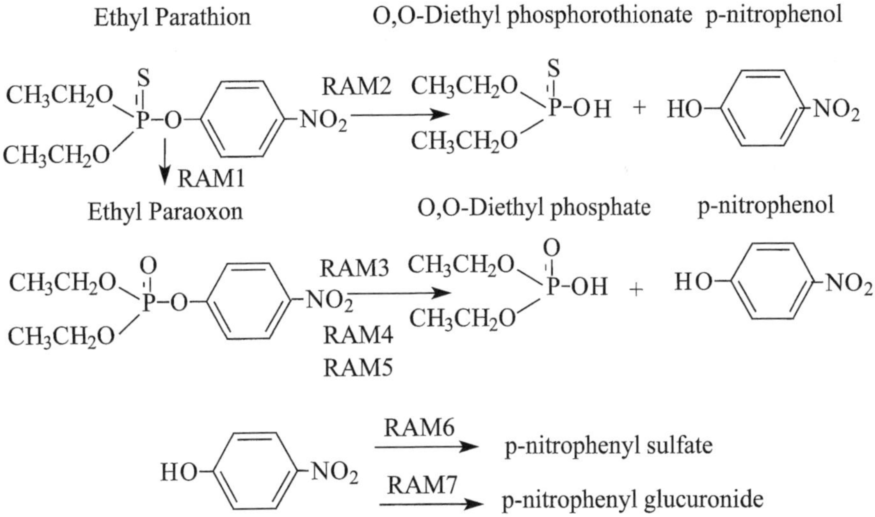

Ethyl Parathion   O,O-Diethyl phosphorothionate p-nitrophenol

Ethyl Paraoxon   O,O-Diethyl phosphate   p-nitrophenol

Figure 13.15 Metabolic pathway for parathion in the rat (Menzie,[92] Neal,[93,94] Nakatsugawa and Dahm,[95] Sultatos and Gagliardi,[96] Zhang and Sultatos[97]). (RAM1, liver) parathion to paraoxon, (RAM2, liver) parathion to O,O-diethyl phosphorothionate and *p*-nitrophenol, (RAM3,4,5, blood, brain and liver) paraoxon to O,O-diethylphosphate and *p*-nitrophenol, (RAM6, liver) *p*-nitrophenol to *p*-nitrophenyl sulfate and (RAM7, liver) *p*-nitrophenol to *p*-nitrophenyl glucuronide.

TABLE 13.8 Parameters Used in the Physiologically Based Pharmacokinetic Models

	Human	Rat
Weights (% of body weight)[a]		
Liver	2.6%	2.7%
Rapidly perfused	4.4%	5.0%
Slowly perfused	58.0%	70.9%
Fat	21.4%	7.0%
Kidney	0.40%	0.76%
Brain	2.00%	0.54%
Flows (allometric constants)[a]		
Alveolar ventilation	15 (L/h)	15 (L/h)
Cardiac output	15 (L/h)	15 (L/h)
Percent of cardiac output[a]		
Liver	23.8%	18.0%
Rapidly perfused	28.3%	30.7%
Slowly perfused	17.0%	27.8%
Fat	5.0%	7.0%
Kidney	13.5%	13.5%
Brain	12.0%	3.0%

[a]Physiological parameters taken from a report prepared by ILSI RSI.[107]

TABLE 13.9   Partition Coefficients Used in the Isofenphos Model[a]

Chemical	L/B	F/B	K/B	Br/B	R/B	S/B
IF	5.42	85	4.0	3.5	3.0	2.5
IO	2.0	5.0	2.0	2.0	2.0	2.0
DNI	2.0	5.0	2.0	1.0	2.0	2.0
DNIO	1.2	2.0	1.2	1.0	1.2	1.2
DAI	1.2	2.0	1.2	1.0	1.2	1.2
DAIO	1.2	2.0	1.2	1.0	1.2	1.2
IPS	1.75	8.0	1.0	1.0	1.0	1.0
SA	1.75	8.0	1.0	1.0	1.0	1.0

[a]Chemicals identified in Figure 13.11. L/B, liver/blood; F/B, fat/blood; K/B, kidney/blood; Br/B, brain/blood; R/B, rapid/blood; S/B, slow/blood.

Equation for the inhibition of blood AChE by oxons:

$$dA_{iAChEB}/dt = (K_{iAChEB} C_{AChEB} C_{DNIOB} V_B), \text{ pmol h}^{-1}$$

where V_B = volume of blood, L

K_{iAChEB} = AChE bimolecular inhibition rate constant, $(\text{pmol L}^{-1})^{-1} \text{h}^{-1}$

C_{AChEB} = concentration of free AChE in blood, pmol L^{-1}

C_{DNIOB} = concentration of DNIO/or paraoxon in the blood, pmol L^{-1}

Equation for aging of inhibited blood AChE:

$$dA_{aiAChEB}/dt = (K_{aiAChEB} A_{iAChEB}) = \text{aging, pmol h}^{-1}$$

where $K_{aiAChEB}$ = rate of aging of phosphorylated AChE, h^{-1}

A_{iAChEB} = amount of phosphorylated (inhibited) AChE, pmol

Similar equations were used in the model for the inhibition of blood plasma (BChE), brain (AChE, BChE, and CaE), and liver (BChE and CaE) enzymes by the oxons. The inhibition equations describe the loss of "oxons" from blood,

TABLE 13.10   Partition Coefficients Used in the Parathion Model[a]

Chemical	L/B	F/B	K/B	Br/B	R/B	S/B
Parathion	5.21	101.2	5.2	4.56	5.21	2.55
Paraoxon	6.62	10.2	6.62	2.31	6.62	3.62
Nitrophenol	1.2	6.0	1.0	1.0	1.0	1.0

[a]Partition coefficients for parathion and paraoxon from Jepson et al.[98] and Gearhart et al.[9] Chemicals identified in Figure 13.14. L/B, liver/blood; F/B, fat/blood; K/B, kidney/blood; Br/B, brain/blood; R/B, rapid/blood; S/B, slow/blood.

TABLE 13.11 V_{max}, K_m **Values Used in Percutaneous Absorption Route, Isofenphos**

Enzymes	Metabolism	Rat[a] V_{max}	Rat[a] K_m	Human[a] V_{max}	Human[a] K_m
P-450[b]					
V_{max1C}	IF to IO	137.2	14.1	1.802	18.4
V_{max2C}	IF to DNI	35.8	9.9	0.901	11.2
V_{max3C}	IO to DNIO	72.6	9.5	5.457	11.6
V_{max4C}	DNI to DNIO	25.1	7.9	0.826	5.8
Deaminases[b]					
V_{max5C}	DNI to DAI	0.0	1.0	0.0	1.0
V_{max6C}	DNIO to DAIO	0.0	1.0	0.0	1.0
OP Hydrolases[c]					
V_{max7C}	IO to Ring/AP1	390	182	390	182
V_{max8C}	DNIO to Ring/AP2	520	182	520	182
V_{max9C}	DAIO to Ring/AP3	0.0	182	0.0	182
V_{max10C}	DNI to Ring/AP4	20	15.0	20	15
V_{max11C}	DAI to Ring/AP5	0.0	182	0.0	182
CaE[d]					
V_{max12C}	IF to IFA	17.6	62	17.6	62
V_{max13C}	DNI to DNIA	270	62	270	62
V_{max14C}	DAI to DAIA	0.0	62	0.0	62
V_{max15C}	IO to IOA	0.0	62	0.0	62
V_{max16C}	DNIO to DNIOA	0.0	62	0.0	62
V_{max17C}	DAIO to DAIOA	0.0	62	0.0	62
V_{max18C}	IPS to SA	25	50	25	50
Glycine conjugation					
V_{max19C}	SA to Hippurate	25	50	25	50

[a] V_{max}, pmoles hr^{-1} kg^{-1} of BW $\times 10^6$; K_m, pmoles L$^{-1} \times 10^6$.
[b] Knaak et al.[8,91]
[c] Wallace and Dargan.[99]
[d] Talcott.[108]

brain, and liver as a result of inhibition (metabolic mass balance involving leaving groups). Aging prevents reactivation and removes the alkyl phosphate from the metabolic mass balance equation. A set of mass balance equations was used to describe changes in B-esterases as a result of reactivation, synthesis, and degradation. Equations are required for maintaining these enzymes at their physiological level over extended period of time. The overall inhibition and recovery of blood AChE was described in the models by the mass balance equations for blood AChE inhibition, reactivation, synthesis and degradation.

Mass balance equation for blood AChE inhibition, reactivation, synthesis and degradation:

$$dA_{\text{AChEB}}/dt = \Sigma(dA_{\text{rAChEB}}/dt + dA_{\text{sAChEB}}/dt - dA_{\text{diAChEB}}/dt - dA_{\text{iAChEB}}/dt)$$

TABLE 13.12 V_{max}, K_m **Values Used in Percutaneous Absorption Route, Parathion**

Enzymes	Metabolism	Rat[a]		Human[a]	
		V_{max}	K_m	V_{max}	K_m
P-450[b]					
V_{max1c}	P to PO	180	10.2	1.35	10.2
V_{max2c}	P to PNP/AP1	231	14.9	2.31	14.9
OP Hydrolases[c]					
V_{max3c} (B)	PO to PNP/AP2	851	2000	851	600
V_{max4c} (Br)	PO to PNP/AP2	21	10	21	10
V_{max5c} (L)	PO to PNP/AP2	1013	290	1013	290
Sulfate conjugation[d]					
V_{max6c}	PNP to sulfate	20	50	20	50
Glucuronide conjugation[d]					
V_{max7c}	PNP to glucuronide	5	50	5	50

[a] V_{max}, pmoles h^{-1} kg^{-1} of BW × 10^6; K_m, pmoles L^{-1} × 10^6.

[b] Wallace and Dargan.[99]

[c] Wallace and Dargan,[99] Gil et al.,[100,101] Gonzalvo et al.,[102] Forsyth and Chambers,[103] Kuo and La Du,[104] Mackness et al.,[105] and Sams and Mason.[106]

[d] Pacifici et al.[109]

P = parathion, PO = paraoxon, AP1 = diethylthiophosphate, AP2 = diethylphosphate, PNP = p-nitrophenol, blood (B), brain (Br), liver (L).

TABLE 13.13   **Tissue AChE, BChE, and ChE**

Enz/Source[a]	Enzyme conc. in tissues in pmol L^{-1}
Blood enzymes	
AChE	1.1×10^3
BChE	4.8×10^3
Brain enzymes	
AChE	37.8×10^3
BChE	12.7×10^3
CaE	549.8×10^3
Liver enzymes	
BChE	8.22×10^3
CaE	455×10^3

[a] Maxwell et al.[110]

TABLE 13.14   Affinity Constants (K_a) and Phosphorylation Constants (k_p) Used to Describe the Inhibition of Tissue AChE, BChE, and CaE by des N-isopropyl Isofenphos Oxon in the Rat and Human

Tissues/Enzymes[a,b]	K_a (pmol L^{-1}) $\times 10^6$	k_p (hr^{-1})	k_i (k_p/K_a) pM$^{-1} \cdot$ hr$^{-1} \times 10^{-6}$
Blood			
AChE	102.0	38.17	0.36
BChE	43.2	21.3	0.49
Brain			
AChE	102.0	38.17	0.36
BChE	43.2	21.3	0.49
CaE	35.0	20.0	0.57
Liver			
BChE	91.0	21.3	0.234
CaE	35.0	20.0	0.60

[a] Wang and Murphy[111] used for AChE.

[b] Values from Cohen et al.,[112] Aldridge and Reiner,[113] Chiu et al.[114] evaluated.

where $dA_{rACHEB}/dt = (K_{rAChEB}A_{iAChEB}) =$ reactivation, pmol h^{-1}

$\qquad K_{rAChEB} =$ rate of reactivation of inhibited AChE (h^{-1})

$\qquad A_{iAChEB} =$ inhibited AChE, pmol

$\qquad dA_{sAChEB}/dt = (K_{sAChEB}C_{AChEB}V_B) =$ synthesis, pmol h^{-1}

$\qquad K_{sAChEB} =$ rate of replacement of red blood cell AChE, h^{-1}

$\qquad C_{AChEB} =$ concentration of free AChE in blood, pmol L^{-1}

$\qquad V_B =$ volume of blood, L

TABLE 13.15   Affinity Constants (K_a) and Phosphorylation Constants (k_p) Used to Describe the Inhibition of Tissue AChE, BChE, and CaE by Paraoxon in the Rat and Human

Tissues/Enzymes[a,b]	K_a (pmol L^{-1}) $\times 10^6$	k_p (hr^{-1})	k_i (k_p/K_a) pM$^{-1} \cdot$ hr$^{-1} \times 10^{-6}$
Blood			
AChE	21.69	38.17	1.76
BChE	9.1	21.3	2.34
Brain			
AChE	21.69	38.17	1.76
BChE	9.1	21.3	2.34
CaE	35.0	20.0	0.57
Liver			
BChE	91	21.3	0.234
CaE	35.0	21.0	0.60

[a] Values from Wang and Murphy[111] used for AChE.

[b] Values from Cohen et al.,[112] Aldridge and Reiner,[113] Chiu et al.[114] evaluated.

$$dA_{diAChEB}/dt = (K_{diAChEB}A_{iAChEB}) + (K_{daiAChEB}A_{aiACHEB})$$
$$= \text{degradation, pmol h}^{-1}$$

$K_{diAChEB}$ = rate of degradation of inhibited AChE, h^{-1}

A_{iAChEB} = amount of inhibited AChE, pmol

$K_{daiAChEB}$ = rate of degradation of aged/inhibited AChE, h^{-1}

$A_{aiACHEB}$ = amount of aged AChE, pmol

The rate of degradation of inhibited enzyme consists of two terms: (1) for degradation of inhibited enzyme and (2) for degradation of aged enzyme.

The inhibition/phosphorylation equations in the two PBPK/PD models were derived from the reaction of OPs (AB) with B-esterases (EH) to form Michaelis–Menten complexes (EHAB), phosphorylated enzymes (EA), and leaving groups (BH) as shown in Eq. (13.13):

$$
\text{EH} + \text{AB} \underset{k_{-1}}{\overset{k_{+1}}{\rightleftharpoons}} \text{EHAB} \xrightarrow{k_{+2}} \text{EA} + \text{BH} \xrightarrow[+\text{H}_2\text{O}]{k_{+3}} \text{EH} + \text{AOH} \qquad (13.13)
$$

with Aged enzyme formed via k_{+4}, and the bimolecular reaction k_i.

The bimolecular reaction, k_i, is the overall reaction involving the formation of the Michaelis–Menten complex (EHAB) and the phosphorylated enzyme (EA), where

$$K_a = \frac{k_{-1}}{k_{+1}} \qquad (13.14)$$

$$k_i = \frac{k_{+2}}{K_a} \qquad (13.15)$$

The rate constants for reactivation of the phosphorylate enzyme and aging are represented in the overall equation by k_{+3} and k_{+4}, respectively. In most cases the reactivation rate, k_{+3}, is small and the phosphorylated enzyme eventually looses an alkoxy group (k_{+4}) as shown in Eq. (13.16) to become an aged enzyme incapable of being hydrolyzed by water or oximes:

$$(13.16)$$

Physiological processes involving enzyme synthesis and degradation (i.e., release and removal of old red cells, synthesis and release of enzymes from the liver) eventually replace phosphorylated and aged enzymes. Total enzyme may be represented by

$$[Et] = [EH] + [EHAB] + [EA] + [EAged] \qquad (13.17)$$

where EH = free enzyme

EHAB = Michaelis complex

EA = phosphorylated enzyme

[EAged] = age enzyme

13.7.3 Output of PBPK/PD Models, Single Topical Dose

Rat Models: Material Balance and Metabolism   Over the time course of 168 hours (Figures 13.5 and 13.6), topically applied [^{14}C-ring]-labeled isofenphos and parathion were either retained on skin (5 to 15.3 percent), lost to air (35 percent), or absorbed through skin (43.2 to 52.0 percent).[47,50] Urinary metabolite data was obtained in the 168-hour ^{14}C-ring-isofenphos study but not in the percutaneous ^{14}C-ring-parathion study. The output from the isofenphos and parathion models (Table 13.16) indicate that practically all the absorbed dosages were eliminated as alkyl phosphates, carboxylic acids derived from isofenphos, and free or conjugated 2-OH benzoic acid and *p*-nitrophenol (Figures 13.12 and 13.15).

The fate of topically applied isofenphos and parathion in rats after 8 hours of exposure (i.e., topical application followed by 8 hours of exposure, washoff, and 16 hours of no exposure) were simulated using the two PBPK/PD models. Eight hours dermal exposures to isofenphos or parathion resulted in the direct absorption of ∼5 percent of the topically applied dose, evaporative losses of ∼5 percent, with ∼90 percent of the applied topical dose being retained on skin.

The toxic oxons, des N-isopropyl isofenphos oxon and paraoxon, formed by the desulfuration of isofenphos and parathion were either hydrolyzed by OP hydrolases (i.e., paraoxonase/A-esterases) in liver and serum to give their respective alkyl phosphates and leaving groups, isopropyl salicylate and *p*-nitrophenol, or they phosphorylated (inhibited) B-esterases in blood (AChE and BChE), brain (AChE, BChE, and CaE), and liver (BChE and CaE).

Inhibition of Rat B-esterases   Dermal dose–response data (Figures 13.1 to 13.4) in the rat provided blood AChE and BChE inhibition data for model validation. The percentage of blood AChE and BChE inhibited by the topical treatments are given in Table 13.17 for isofenphos and parathion. Inhibition of plasma BChE activity followed by no reactivation resulted in BChE inhibition being greater than that obtained for AChE. For purposes of this reentry exercise, no reactivation, aging, synthesis, or degradation was used (i.e., enzyme rate

TABLE 13.16   Fate of ^{14}C-ring Labeled Isofenphos and Parathion Topically Administered to the Rat in Percent of Dose[a]

	Isofenphos[b] (%)			Parathion[c] (%)		
	168 h Study	168 h[d] Model	8 h[d] Model	168 h Study	168 h[e] Model	8 h[e] Model
Loss to air	35.0	44.0	5.07	35.0	38.6	4.18
Retained on skin	15.3	10.0	89.6	5.0	12.3	90.5
Urine and feces	43.2	45.0	4.88	52.0	48.7	4.95
Carcass	6.5	1.0	0.45	8.0	0.4	0.37
Total	100.0	100.0	100.0	100.0	100.0	100.0
Isofenphos metabolites						
Isofenphos (fat, etc.)		0.15	0.014			
Alkyl phosphates	31.5	33.9	3.64			
2-OH benzoic acid	nd[f]	—	—			
Carboxylic acids	11.7	11.8	1.22			
2-OH hippuric acid	31.5	33.9	3.64			
Enz, phosphorylation	nd[f]	0.2	0.013			
Parathion metabolites						
Parathion (fat, etc.)				nd	0.196	0.027
Alkyl phosphates				nd	48.7	4.95
p-Nitrophenol (p-NP)				nd	0.16	0.018
Sulfate of p-NP				nd	38.9	3.94
Glucuronide of p-NP				nd	9.7	0.98
Enz, phosphorylation				nd	00.2	0.018

[a] See metabolic pathways for metabolites, Figures 13.11 and 13.14.
[b] Isofenphos dermal rat dose = 3.2 mg/kg (2.807×10^6 pmol); BW = 0.303 kg, area = 12 cm^2, $K_p = 0.007$, $K_r = 0.0067$; from Knaak et al.[47]
[c] Parathion dermal rat dose = 2.8 mg/kg (2.91×10^6 pmol), BW = 0.303 kg, area = 20 cm^2, $K_p = 0.007$, $K_r = 0.0055$; from Knaak et al.[50]
[d] V_{max}, K_m values, Table 13.11; k_i values, Table 13.14.
[e] V_{max}, K_m values, Table 13.12; k_i values, Table 13.15.
[f] nd, not determined.

constants were set to zero). Equivalent doses of isofenphos (3.2 mg/kg) and parathion (2.8 mg/kg) resulted in less red blood cell AChE inhibition in rats treated with isofenphos (16 percent) than for those rats administered parathion (58 percent). This was expected because parathion (dermal ED$_{50}$ dose) is approximately 3 to 4 times more toxic than isofenphos[47] to the rat. According to Chambers et al.,[115] a 3-mg/kg dose of parathion produces 95 percent CaE inhibition in the liver.

Human Models: Material Balance and Metabolism   Human physiological parameters and metabolic rate data were used to model the fate of topical applications of isofenphos and parathion in workers during 8 and 168 hours of

TABLE 13.17 Percentage of Enzyme Inhibited by Topical Application of Isofenphos or Parathion to the Skin of Rats[a]

Enz/Source	Isofenphos[b] 168 h Study	Isofenphos[b] 168 h[d] Model	Isofenphos[b] 8 h[d] Model	Parathion[c] 168 h Study	Parathion[c] 168 h[e] Model	Parathion[c] 8 h[e] Model
Blood enzymes						
AChE	40–50	16.3	1.83	40–50	58.3	5.9
BChE		22.5	2.43		77.5	7.9
Brain enzymes						
AChE		19.0	1.83		23.7	2.4
BChE		22.4	2.41		31.6	3.2
CaE		25.9	2.8		7.7	0.8
Liver enzymes						
BChE		15.8	1.7		64.9	6.6
CaE		40.6	4.4		100	16.7

[a] See Table 13.16 for recovery and metabolite data; reactivation rates for BChE set to 0.0.

[b] Isofenphos dermal rat dose = 3.2 mg/kg (2.80×10^6 pmol); BW = 0.303 kg, area = 12 cm^2, $K_p = 0.007$, $K_r = 0.0067$; from Knaak et al.[47]

[c] Parathion dermal rat dose = 2.8 mg/kg (2.91×10^6 pmol), BW = 0.303 kg, area = 20 cm^2, $K_p = 0.007$, $K_r = 0.0055$; from Knaak et al.[50]

[d] V_{max}, K_m values, Table 13.11; k_i values, Table 13.14.

[e] V_{max}, K_m values, Table 13.12; k_i values, Table 13.15.

exposure (Table 13.18). Forty-six to 50 percent of topically applied isofenphos and parathion were absorbed over the course of 168 hours with 38 to 44 percent being lost to air and 10 to 12 percent being retained on skin. According to Bronaugh,[116] parathion was absorbed through in vitro human skin to the extent of 39 percent over 160 hours. The use of human P-450, V_{max}, K_m values for isofenphos, resulted in the metabolism of ~53 percent of the topically absorbed dose, with the remaining 48 percent being retained in body fat as isofenphos.[47] Similar percentages were obtained with parathion when human P-450 values for V_{max} and K_m developed for isofenphos were used.[91] Morgan et al.[117] dosed humans orally with methyl and ethyl parathion. Paranitrophenol, a metabolite of each pesticide, was excreted at 37 percent of theoretical while the alkyl phosphates were excreted at more than 50 percent of theoretical in 24 hours.

Published literature on the metabolism of organophosphorus compounds indicate that the rate of biotransformation of OPs (i.e., P=S to P=O conversion) by P-450 enzymes differs between species (i.e., rat and human), but appears to be similar within species. Table 13.18 gives the results when rat P-450 V_{max}, K_m values (scaled to the human) were used to extrapolate rat pharmacokinetic data to humans. The difference was dramatic with practically all the absorbed dose being metabolized leaving only 3 to 4 percent of the parent compound in body fat at the end of 7 days.

TABLE 13.18   Fate of Isofenphos and Parathion Topically Administered to the Human in Percent of Dose (Output from Models)[a]

	Isofenphos[b] (%)			Parathion[c] (%)		
	168 h[d] Model	168 h[e] Model	8 h[e] Model	168 h[f] Model	168 h[g] Model	8 h[g] Model
Loss to air	44.0	44.0	5.1	38.2	38.6	4.2
Retained on skin	10.0	10.0	90.0	12.3	12.3	90.5
Urine and feces	41.7	24.3	1.2	44.3	20.5	0.90
Carcass	4.3	21.7	3.7	5.2	28.6	4.4
Total	100.0	100.0	100.0	100.0	100.0	100.0
Isofenphos metabolites						
Isofenphos (in fat, etc.)	3.5	21.0	2.5			
Alkyl phosphates	22.1	5.8	0.15			
2-OH benzoic acid	—	—	—			
Carboxylic acids	19.9	19.1	1.1			
2-OH hippuric acid	22.1	5.8	0.16			
Inhibited enzymes	0.31	0.04	0.001			
Parathion metabolites						
Parathion (in fat, etc.)				4.0	30.0	2.8
Alkyl phosphate				44.3	18.9	0.92
p-Nitrophenol (p-NP)				2.0	1.4	0.31
Sulfate of p-NP				34.0	13.9	0.43
Glucuronide of p-NP				8.5	3.5	0.12
Inhibited enzymes				0.70	0.26	0.017

[a]See metabolic pathways for metabolites.

[b]Isofenphos dermal human dose = 3.2 mg/kg (6.485×10^8 pmol); BW = 70 kg, area = 1000 cm^2, $K_p = 0.007$, $K_r = 0.0067$; from Knaak et al.[47]

[c]Parathion dermal human dose = 2.8 mg/kg (6.729×10^8 pmol), BW = 70 kg, area = 1000 cm^2, $K_p = 0.007$, $K_r = 0.0055$; from Knaak et al.[50]

[d]V_{max}, K_m values, Table 13.11 (rat values used), k_i values, Table 13.13.

[e]V_{max}, K_m values, Table 13.11 (human values used), k_i values, Table 13.13.

[f]V_{max}, K_m values, Table 13.12 (rat values used), k_i values, Table 13.14.

[g]V_{max}, K_m values, Table 13.12 (human values used), k_i values, Table 13.14.

Inhibition of Human B-esterases by Topical Dose   The simulated effects of topically applied isofenphos and parathion on tissue B-esterases in humans are given in Table 13.19. A topical dose of 3.2 mg/kg of isofenphos produced 6.3 percent red-cell AChE inhibition, while 2.8 mg/kg of parathion resulted in 63.3 percent inhibition of red-cell AChE. Enzyme inhibition was equivalent to the values reported in human studies when P-450 metabolic rate constants obtained from in vitro human liver studies with isofenphos were used.[91] The extrapolation of the isofenphos human liver P-450 values to parathion (PBPK/PD model) also produced results similar to those reported in human studies with parathion.[117] The use of rat P-450 values for isofenphos and parathion resulted

TABLE 13.19 **Percentage of Enzymes Inhibited by Topical Application of Isofenphos or Parathion to the Skin of the Human (Models)**[a]

Enz/Source	Isofenphos[b]			Parathion[c]		
	168 h[d] Model	168 h[e] Model	8 h[e] Model	168 h[f] Model	168 h[g] Model	8 h[g] Model
Blood enzymes						
AChE	45.5	6.3	0.23	100	63.3	2.9
BChE	62.6	8.6	0.31	100	84.2	3.9
Brain enzymes						
AChE	47.0	6.45	0.23	100	72.5	3.3
BChE	61.9	8.5	0.31	100	96.5	4.4
CaE	71.8	9.9	0.36	100	23.5	1.1
Liver enzymes						
BChE	65.9	9.1	0.34	100	89.8	4.3
CaE	100	23.3	0.86	100	100	11.0

[a] See Table 13.19 for recovery and metabolic data.

[b] Isofenphos dermal human dose = 3.2 mg/kg (6.729×10^8 pmol); BW = 70 kg, area = 1000 cm^2, $K_p = 0.007$, $K_r = 0.0067$; from Knaak et al.[47]

[c] Parathion dermal human dose = 2.8 mg/kg (6.729×10^8 pmol); BW = 70 kg, area = 1000 cm^2, $K_p = 0.007$, $K_r = 0.0055$; from Knaak et al.[50]

[d] V_{max}, K_m values, Table 13.11 (rat k_i values used), Table 13.14.

[e] V_{max}, K_m values, Table 13.11 (human k_i values used), Table 13.14.

[f] V_{max}, K_m values, Table 13.12 (rat k_i values used), Table 13.15.

[g] V_{max}, K_m values, Table 13.12 (human k_i values used), Table 13.15.

in >50 and 100 percent inhibition, respectively, of tissue B-esterases (AChE, BChE, and CaE) (Table 13.19). For purposes of this reentry exercise no enzyme reactivation rates were used involving topical or transferred doses.

13.7.4 Output of PBPK/PD Models, Foliar Residues, Metabolism

Parathion: Material Balance The foliar parathion transfer PBPK/PB model was used to simulate 8-hour worker exposures to leaf residues during the harvest of tree crops such as peaches or citrus sprayed with parathion. The transfer coefficient of Nigg et al.[7] was used in the model. Parathion residues of 10 µg/cm^2 occur on citrus in California, 3 days postapplication, when citrus is sprayed at the rate of 10 pounds of active ingredients per acre in 1600 gallons of water (Figure 13.7). These residues decay over the course of 21 days to less than 0.1 µg/cm^2 leaf surface. Worker exposure to parathion foliar residues greater than 0.1 µg/cm^2 of leaf surface was shown by Popendorf and Leffingwell[6] to produce illnesses in field workers harvesting tree fruits. Table 13.20 gives the results of simulated 8-hour worker exposures to foliar residues of parathion on citrus (0.1, 1.0. 5.0, and 10.0 µg/cm^2, two-sided residue). According to the

TABLE 13.20 Percentage of Enzymes Inhibited by Transfer of Parathion Leaf Residues to Skin of Workers[a]

Enz/Source	Parathion Leaf Residues, R ($\mu g/cm^2$)					
	$R = 10$	$R = 5$	$R = 1.0$	$R = 0.1$	$R = 0.09$	
Blood enzymes						
AChE	5.4	2.7	0.55	0.05	0.05	
BChE	7.2	3.6	0.73	0.07	0.07	
Brain enzymes						
AChE	2.7	1.36	0.27	0.027	0.025	
BChE	3.6	1.8	0.36	0.036	0.033	
CaE	0.9	0.44	0.09	0.009	0.008	
Liver enzymes						
BChE	7.9	4.02	0.81	0.08	0.073	
CaE	20.5	10.31	2.08	0.21	0.187	
Dose (pmoles)	2.74×10^9 pmoles	1.3733×10^9 pmoles	2.746×10^8 pmoles	2.746×10^7 pmoles	2.472×10^7 pmoles	
Dose (mg/worker)	800 mg	400 mg	80 mg	8.0 mg	7.2 mg	

[a]Material balance: lost to air, 2.12%, retained on skin, 95.2%, urine and feces, 0.415%, body tissues, 2.7%. Metabolites: parathion in fat, 1.37%, p-nitrophenol, 0.15%, p-nitrophenyl sulfate, 0.21%, p-nitrophenyl glucuronide, 0.05%; BW = 70 kg, area = 1000 cm^2, K_d (TC) = 10,000, 8-h exposure, 24-h metabolism, and enzyme monitoring.

model, foliar residues of 10 µg/cm^2 result in the transfer of 11.42 mg of para-thion (800 mg/70 kg of BW) to the skin of workers over the course of an 8-hour work day. Of the 11.42 mg deposited on skin, ~3.0 percent was absorbed (~50 percent metabolized and ~50 percent retained in body fat as parathion), 2 percent lost to air, and 95 percent remained as a dermal residue at the end of the workday. The parathion present in body fat was small (1.3 percent) at the end of the first day (i.e., 8 hours of exposure followed by a 16-hour period of no exposure).

Inhibition of B-esterases by Foliar Residues of Parathion According to the parathion transfer model (Table 13.20) approximately 1.3 percent of the para-thion residues transferred to skin (11.42 mg) produced 5.4 percent red blood cell AChE inhibition. Parathion in fat was capable of being further metabolized in the liver to produce additional AChE inhibition (5.4 percent) raising the per-cent of inhibition to 11. Additional daily exposures could conceivably results in red-cell AChE inhibitions of more than 30 percent. Under field conditions res-idue levels of 0.1 to 0.09 µg/cm^2 produced little or no red-cell AChE inhibition over the course of several weeks or months of exposure.

Isofenphos, Material Balance, Metabolism The isofenphos foliar transport model utilized the 10,000 k_d factor (two-sided residue) proposed by Nigg et al.[7] and supported by Harris and Solomon[76] for the transport of isofenphos resi-dues on turf to bare feet, ankles, legs, arms, and hands. The material balance data generated by the model is given at the bottom of Table 13.21 for a series of foliar residues (0.09, 0.1, 1.0, 5.0, and 10.0 µg/cm^2) with 0.7 µg/cm^2 (two-sided residue) found on isofenphos-treated turf after application.[60] According to the model, foliar residues of 10 µg/cm^2 resulted in the transfer of 11.42 mg of isofenphos to exposed skin over the course of an 8-hour day. Material bal-ance data indicated that ~95 percent of the transported residue remained on the surface of the skin, ~2.5 percent was lost to air, 0.56 percent was eliminated in urine and feces, and ~2 percent of the dose was retained in tissues. The dose retained in tissues was largely isofenphos present in body fat (~1 percent).

Inhibition of B-esterases by Turf Residues of Isofenphos The transfer of iso-fenphos foliar residues (11.42 mg, 800 mg/70 kg of BW) to skin during the work day resulted in the absorption of ~3 percent of the transferred dose. Ap-proximately 43 percent of the absorbed dose was found in body fat as iso-fenphos with most of the absorbed dose (57 percent) being metabolized to alkyl phosphates, various carboxylic acids and 2-OH hippuric acid. The small amounts of des N-isopropyl isofenphos oxon produced during metabolism re-sulted in ~0.4 percent inhibition of red-cell AChE (Table 13.21). This percent-age will almost double when the isofenphos residing in body fat is metabolized (total inhibition ~1 percent). Less than 0.1 percent red blood cell AChE inhi-bition was obtained when foliar residues were at or below 1.0 µg/cm^2.

TABLE 13.21 Percentage of Enzymes Inhibited by Transfer of Isofenphos Turf Residues to Skin of Workers[a]

Enz/Source	Isofenphos Leaf Residues, R ($\mu g/cm^2$)				
	$R = 10$	$R = 5$	$R = 1.0$	$R = 0.1$	$R = 0.09$
Blood enzymes					
AChE	0.38	0.2	0.03	0.004	0.003
BChE	0.51	0.25	0.05	0.005	0.005
Brain enzymes					
AChE	0.38	0.20	0.04	0.004	0.003
BChE	0.50	0.25	0.05	0.005	0.005
CaE	0.60	0.29	0.06	0.006	0.005
Liver enzymes					
BChE	0.54	0.27	0.05	0.005	0.005
CaE	1.4	0.70	0.14	0.014	0.013
Dose (pmoles)	2.316×10^9 pmoles	1.158×10^9 pmoles	2.316×10^8 pmoles	2.316×10^7 pmoles	2.0845×10^7 pmoles
Dose (mg/worker)	800 mg	400 mg	80 mg	8.0 mg	7.2 mg

[a]Material balance: lost to air, 2.58%, retained on skin, 94.7%, urine and feces, 0.56%, body tissues, 2.16%. Metabolism: Isofenphos in fat, 1.17%, alkyl phosphates, 0.071%, carboxylic acids, 0.504%, 2-OH hippuric acid, 0.071%, 2-OH hippuric acid, 0.071%, carboxylic acids, 0.504%, 2-OH hippuric acid, 0.071%, $BW = 70$ kg, area $= 1000$ cm^2, K_d (TC) $= 10{,}000$, 8-h exposure, 24-h metabolism, and enzyme monitoring.

13.8 DISCUSSION

In the previous edition of this book, we stated that "the large number of variables associated with the reentry problem prevented researchers from coming up with an easy and quick solution to the problem." The reentry problem, however, was resolved by dividing the problem up into three distinct parts: (1) dissipation of the foliar residue, (2) transfer of the residue to the skin and clothing of workers, and (3) percutaneous absorption/dermal dose–ChE response/ChE-NOEL. Since 1989, the emphasis on reentry research has been in the further development of foliar residue transfer coefficients for estimating whole-body exposure (mg/day). The concern of EPA scientists[18] and others regarding the usefulness of cholinesterase inhibition data in risk assessments has largely been resolved by CDPR[22] and by EPA's own use of the data in establishing RfDs, leaving the development and proper use of percutaneous absorption data to be worked on.

The addition of foliar transfer coefficients to PBPK/PD models for parathion and isofenphos made it possible to reexamine the relationship between OP pesticide residues in the workplace, percutaneous absorption and B-esterase inhibition. Evaporative losses from skin were modeled according to rat and human studies.[8,50,51] The models required metabolic pathway data in animals/humans, liver, blood, and brain metabolic rate constants for the major pathways, tissue partition coefficients on parent and metabolites, physiological parameters, and bimolecular rate inhibition constants. The modeling effort suggests that in vitro inhibition data (k_i) may require some in vivo adjustment when used in PBPK/PD models. Equations for the reactivation and aging of inhibited enzymes, degradation of inhibited enzymes, and synthesis of new enzyme were included in the model, but were not used (i.e., rates set to zero) in order to obtain maximum inhibition. The bimolecular inhibition rate constant for paraoxon was used for des N-isopropyl isofenphos oxon (DNIO) because a constant was not available for DNIO.

The human foliar residue transfer–PBPK/PD parathion model (Table 13.20) supports the previously established reentry level of 0.09 μg/cm^2 on citrus. Although a reentry interval has never been established for isofenphos residues on turf, the PBPK/PD isofenphos model (Table 13.21) supports the suggested reentry level of 0.6 μg/cm^2.

EPA's standard methodology (SOPs) for postapplication exposure described in Section 13.5.3 utilizes dislodgeable foliar residue data (μg/cm^2), residue transfer coefficients (cm^2/h), and percutaneous absorption data (default value of 100 percent) to estimate the absorbed dose (mg/kg/day) and MOE. Approximately 3 percent of the transferred foliar residues are absorbed during reentry according to the output from the human foliar residue transfer–PBPK/PD isofenphos and parathion models presented in this chapter. The use of EPA's standard methodology for dislodgeable parathion residues of 0.09 μg/cm^2 on citrus ($\sim$3 percent percutaneous absorption from human PBPK model, 8-hour work day, transfer coefficient of 10,000, and a NOEL of 0.05 mg/kg/day[31]) results in an MOE of 16, while dislodgeable isofenphos residues of 0.6 µg/cm^2 (two-sided residue) on turf found shortly after application (~3 percent percutaneous absorption from human PBPK model, 2-hour exposure, transfer coefficient of 10,000, and a NOEL of 0.05 mg/kg/day[21]) results in an MOE of 10.0. The use of default percutaneous absorption values of 50 to 100 percent resulted in MOEs of less than 1.0 for these exposures. To achieve MOEs of 100, reentry levels of 0.009 and 0.06 µg/cm^2 are needed for parathion and isofenphos, respectively.

According to the dislodgeable residue dissipation curve for parathion taken from Knaak and Iwata (Figure 13.7), a reentry level of 0.09 µg/cm^2 on citrus corresponds to a reentry interval of 21 to 60 days or greater for combined residues (thion + oxon) of parathion and paraoxon. Low gallonage (10 pounds A.I./acre, 100 gallons) resulted in higher residues and longer reentry intervals. In 1989, California reentry intervals for parathion[5] varied between 30 and 60 days and as long as 90 days in certain counties after May 15. Reentry intervals of 90 days were intended to effectively reduce dislodgeable residues to or below 0.09 µg/cm^2 and in some cases perhaps as low as 0.009 µg/cm^2 giving an MOE of >100. The difficulties (i.e., formation of toxic oxons and dissipation of residues) encountered by CDPR in providing field workers reentering parathion-treated crops with a safe working environment made it necessary for the department to cancel its registration and use in California. When possible, CDPR is now using a probabilistic model involving the use of transfer factors and dislodgeable foliar residues to determine reentry intervals for OP and other pesticides. Current CDPR methodology in conjunction with an oral NOEL of 0.05 mg/kg/day for parathion gave a safe dislodgeable foliar level of 0.09 µg/cm^2 for parathion and a reentry time of 30 days consistent with previously published values.

To our knowledge illnesses associated with the use of isofenphos on turf have not been reported. The safe level of 0.6 µg/cm^2 appears to be appropriate for short daily exposures periods (2 hours) to isofenphos on turf grasses. Segawa and Powell[60] and Sears et al.[61] reported dislodgeable isofenphos residues of 0.7 and 0.2 µg/cm^2 of leaf surface (two-sided residue), respectively, after application and 1/10 of this amount after 24 hours. This may be of academic interest because Mobay Chemical petitioned EPA in 1999 to drop the registration of Oftanol due to problems associated with efficacy. On May 26, 1999, EPA published a *Federal Register* notice announcing the final cancellation of pesticide products containing isofenphos.

REFERENCES

1. H. R. Wolf, W. F. Durham, and J. F. Armstrong. Exposure of workers to pesticides. *Arch. Environ. Health* 14: 622–633 (1967).
2. T. R. Milby. Report of the Task Group on Occupational Exposure of Pesticides to the Federal Working Group on Pest Management. U.S. EPA, Washington, DC, Jan. 1974.

3. California Food and Agricultural Code. Article 14, Birth Defect Prevention. Sacramento, CA, 1984.

4. H. N. Nigg and J. B. Knaak. Blood cholinesterases as human biomarkers of organophosphorous pesticide exposure. *Rev. Environ. Contam. Toxicol.* 163: 29–112 (2000).

5. J. B. Knaak, Y. Iwata, and K. T. Maddy. The worker hazard posed by reentry into pesticide-treated foliage: Development of safe reentry times, with emphasis on chlorthiophos and carbosulfan. In D. Paustenbach (Ed.), *The Risk Assessment of Environmental Hazards: A Textbook of Case Studies*. Wiley, New York, 1989, Chapter 24.

6. W. J. Popendorf and J. T. Leffingwell. Regulating OP pesticide residues for farmworker protection. *Residue Rev.* 82: 125–201 (1982).

7. H. N. Nigg, J. H. Stamper, and R. M. Queen. The development and use of a universal model to predict tree crop harvester pesticide exposure. *Am. Ind. Hyg. Assoc. J.* 45: 182–186 (1984).

8. J. B. Knaak, M. A. Al-Bayati, O. G. Raabe, and J. N. Blancato. Use of a multiple pathway and multiroute physiologically based pharmacokinetic model for predicting organophosphorus pesticide toxicity. In J. N. Blancato, R. N. Brown, C. C. Dary, and M. A. Saleh (Eds.), *Biomarkers for Agrochemicals and Toxic Substances*, ACS Symposium Series, 643. American Chemical Society, Washington, DC, 1996.

9. J. M. Gearhart, G. W. Jepson, H. J. Clewell, M. E. Andersen, and R. B. Conolly. Physiologically based pharmacokinetic model for the inhibition of acetylcholinesterase by organophosphate esters. *Environ. Health. Perspect.* 102: 51–60 (1994).

10. U.S. Environmental Protection Agency (USEPA). *Pesticide Assessment Guidelines, Subdivision K, Exposure: Reentry Protection*. USEPA, Washington, DC, 1984.

11. U.S. Environmental Protection Agency (USEPA). *Series 875-Occupational and Residential Exposure Test Guidelines, Group B-Postapplication Exposure Monitoring Test Guidelines*, Version 5.3. USEPA, Office of Pollution Prevention and Toxics, Washington, DC, 1997.

12. H. N. Nigg. Comparison of pesticide particulate recoveries with the vacuum and dislodgeable surface pesticide residue techniques. *Arch. Environ. Contam. Toxicol.* 8: 369–381 (1979).

13. H. N. Nigg and J. C. Allen. A comparison of time and time-weather models for predicting parathion disappearance under California conditions. *Environ. Sci. Technol.* 13: 231–233 (1979).

14. H. N. Nigg, J. C. Allen, and R. W. King. Behavior of parathion residues in the Florida "Valencia" orange agroecosystem. *J. Agric. Food Chem.* 27: 578–582 (1979).

15. J. H. Stamper, H. N. Nigg, and R. M. Queen. Prediction of pesticide dermal exposure and urinary metabolite level of tree crop harvesters from field residues. *Bull. Environ. Contam. Toxicol.* 36: 693–700 (1986).

16. J. H. Stamper, H. N. Nigg, and J. C. Allen. Organophosphate insecticide disappearance from leaf surfaces: An alternative to first-order kinetics. *Environ. Sci. Technol.* 13: 1402–1405 (1979).

17. F. A. Gunther, W. E. Westlake, J. H. Barkley, W. Winterlin, and L. Langbehn. Establishment of dislodgeable residues on leaf surfaces. *Bull. Environ. Contam. Toxicol.* 9: 243–249 (1973).

18. U.S. Environmental Protection Agency (USEPA). *Cholinesterase Inhibition As an Indication of Adverse Toxicological Effect.* Risk Assessment Forum. USEPA, Washington, DC, June 1988.

19. Pesticide and Toxic Chemical News. *Cholinesterase Assay Recommendations May Come by Year End.* Pesticide and Toxic Chemical News. CRC Press, Washington, DC, Dec. 11, 1991, pp. 29–32.

20. D. G. Barnes and M. Dourson. Reference dose (RfD): Description and use in health risk assessments. U.S. Environmental Protection Agency. *Regul. Toxicol. Pharmacol.* 8: 471–486 (1988)

21. Environmental Protection Agency (USEPA). *Integrated Risk Information System Chemical Information Database.* USEPA, Washington, DC, 1997.

22. California Department of Pesticide Regulation. *Use of Cholinesterase Inhibition Data in Risk Assessments for Pesticides.* Sacramento, CA, May 16, 1997.

23. B. W. Wilson, J. R. Sanborn, M. A. O'Malley, J. D. Henderson, and J. R. Billitti. Monitoring the pesticide-exposed worker. In *Occupational Medicine: State of the Art Reviews.* Hanley & Belfus Medical Publishers, Philadelphia, 1998.

24. R. I. Krieger. Pesticide exposure assessment. *Toxicol. Lett.* 82/83: 65–72 (1995).

25. National Center for Environmental Research and Quality Assurance (NCERQA). *1999 Grants for Research Program, Children's Vulnerability to Toxic Substances in the Environment.* U.S. Environmental Protection Agency, Washington, DC.

26. A. Hirschberg and Y. Lerman. Clinical problems in organophosphate insecticide poisoning: The use of a computerized information system. *Fundam. Appl. Toxicol.* 4: PS209–214 (1984).

27. Y. Lerman, A. Hirshberg, and Z. Shteger. Organophosphate and carbamate pesticide poisoning: The usefulness of a computerized clinical information system. *Am. J. Ind. Med.* 6: 17–26 (1984).

28. 40 *Code of Federal Regulations* 162: 10 (July 1, 1983).

29. Title 3, Food and Agriculture, Chapter 6, Pesticides and Control Operations, Group 3, Article 3, Section 6770 of the California Administrative Code, September 16, 1986.

30. U.S. Environmental Protection Agency (USEPA). *Dermal Risk Assessment Guidance Document.* EPA/600/8-91/0118. USEPA, Washington, DC, 1992.

31. J. A. Rider, H. C. Moeller, E. J. Puletti, and J. I. Swader. Toxicity of parathion, systox, octamethyl pyrophosphoramide, and methyl parathion in man. *Toxicol. Appl. Pharmacol.* 14: 603–611 (1969).

32. I. Daly. *Chronic Dietary Ethyl Parathion Study in the Rat.* Biodynamic, East Millstone, NJ, 1977.

33. S. D. Weitman, M. J. Vodicnik, and J. J. Lech. Influence of pregnancy on parathion toxicity and disposition. *Toxicol. Appl. Pharmacol.* 71(2): 215–224 (1983).

34. P. L. Roney, L. G. Costa, and S. D. Murphy. Conditioned taste aversion induced by organophosphate compounds in rats. *Pharmacol. Biochem. Behav.* 24(3): 737–742 (1986).

35. E. F. Edson and D. N. Noakes. The comparative toxicity of six organophosphorus pesticides in the rat. *Toxciol. Appl. Pharmacol.* 2: 523–539 (1960).

36. E. F. Edson. No-effect levels of three organophosphates in the rat, pig, and man. *Food Cosmet. Toxicol.* 2: 311–316 (1964).

37. J. P. Frawley and H. N. Fuyat. Effect of low dietary levels of parathion and systox on blood cholinesterase of dogs. *J. Agric. Food Chem.* 5: 346–348 (1957).

38. L. W. Hazleton and E. G. Holland. Pharmacology and toxicology of parathion. *Adv. Chem. Ser.* 1(31): 31–38 (1950).

39. B. W. Wilson, M. Hooper, E. Chow, R. Higgins, and J. B. Knaak. Antidotes and neuropathic potential of an organophosphate pesticide. *Bull. Environ. Contam. Toxicol.* 33: 386–394 (1984).

40. T. B. Gaines. Acute toxicity of pesticides. *Toxicol. Appl. Pharmacol.* 14: 515–534 (1969).

41. J. B. Knaak, P. Schlocker, C. R. Ackerman, and J. N. Seiber. Reentry research: Establishment of safe pesticide levels on foliage. *Bull. Environ. Contam. Toxicol.* 24: 796–804 (1980).

42. J. B. Knaak. Minimizing occupational exposure to pesticides: Techniques for establishing safe levels of foliar residues. *Residue Rev.* 75: 82–96 (1980).

43. D. Grob, J. L. Lilienthal, Jr., A. M. Harvey, and B. F. Jones. The administration of di-isopropyl fluorophosphate (DFP) to man. I. Effect on plasma and erythrocyte cholinesterase; general systemic effects; use in study of hepatic function and erythropoiesis; and some properties of plasma cholinesterase. *Bull. Johns Hopkins Hosp.* 81: 217–245 (1947).

44. D. Grob, A. M. Harvey, O. R. Langworthy, and J. L. Lilienthal, Jr. The administration of DFP to man. II. Effect of intestinal motility and use in the treatment of abdominal distention. *Bull. Johns Hopkins Hosp.* 81: 245–256 (1947).

45. D. Grob and A. M. Harvey. Observations on the effects of tetraethyl pyrophosphate (TEPP) in man, and on its use in the treatment of myasthenia gravis. *Bull. Johns Hopkins Hosp.* 84: 532–566 (1949).

46. D. Grob and A. M. Harvey. Effects in man of the anti-cholinesterase compound sarin (isopropyl methyl phosphonofluoridate). *J. Clin. Invest.* 37: 350–368 (1958).

47. J. B. Knaak, M. A. Al-Bayati, O. G. Raabe, and J. N. Blancato. In vivo percutaneous absorption studies in the rat: Pharmacokinetics and modeling of isofenphos absorption. In R. C. Scott, R. H. Guy, and J. Hardgraft (Eds.), *Prediction of Percutaneous Penetration: Methods, Measurements, Modeling.* IBC, London, 1990, pp. 1–18.

48. H. I. Maibach, R. J. Feldmann, T. H. Milby, and W. F. Serat. Regional variation in percutaneous penetration in man. *Arch. Environ. Health* 23: 208–211 (1971).

49. H. I. Maibach and R. J. Feldmann. Systemic absorption of pesticides through the skin of man. In *Occupational Exposure to Pesticides. A Report of the Federal Working Group on Pest Management*, Washington, DC, 1974, pp. 120–127.

50. J. B. Knaak, K. Yee, C. R. Ackerman, G. Zweig, D. M. Fry, and B. W. Wilson. Percutaneous absorption and dermal dose-cholinesterase response studies with parathion and carbaryl in the rat. *Toxicol. Appl. Pharmacol.* 76: 252–263 (1984).

51. R. C. Wester, H. I. Maibach, J. Melendres, L. Sedik, J. B. Knaak, and R. Wang. In vivo and in vitro percutaneous absorption and skin evaporation of isofenphos in man. *Fundam. Appl. Toxicol.* 19: 521–526 (1992).

52. T. Fredriksson. Studies on the percutaneous absorption of parathion and paraoxon. III. Rate of absorption of parathion. *Acta. Dermato-Vener.* 41: 353–363 (1961).

53. R. J. Feldmann and H. I. Maibach. Pesticide percutaneous penetration in man (abstract). *J. Invest. Dermatol.* 54: 435 (1970).

54. U.S. Environmental Protection Agency (EPA). *Guidelines for Exposure Assessment. Fed. Reg.* 57: 22888 (1992).

55. Y. Iwata, J. B. Knaak, R. C. Spear, and R. J. Foster. Procedure for the determination of dislodgeable pesticide residues on foliage. *Bull. Environ. Contam. Toxicol.* 18: 649–655 (1977).

56. W. F. Spencer, Y. Iwata, W. W. Kilgore, and J. B. Knaak. Procedure for the determination of pesticide residues on the soil surface. *Bull. Environ. Contam. Toxicol.* 18: 656–662 (1977).

57. K. T. Maddy, W. C. Cusick, and S. Edmiston. *Degradation of Dislodgeable Residues of Chlorpyrifos and Diazinon on Turf: A Preliminary Survey.* Worker Health and Safety Report HS-1196. California Department of Food and Agriculture, Sacramento, CA, 1984.

58. K. S. Goh, S. Edmiston, K. T. Maddy, D. D. Meinders, and S. Margetich. Dissipation of dislodgeable foliar residue of chlorpyrifos and dichlorvos on turf. *Bull. Environ. Contam. Toxicol.* 37: 27–32 (1986).

59. K. S. Goh, S. Edmiston, K. T. Maddy, and S. Margetich. Dissipation of dislodgeable foliar residue for chlorpyrifos and dichlorvos treated lawn: Implication for safe reentry. *Bull. Environ. Contam. Toxicol.* 37: 33–40 (1986).

60. R. T. Segawa and S. J. Powell. *Monitoring the Pesticide Treatments of the Japanese Beetle Project, Sacramento County, California, 1983–1986.* Volume II: *Isofenphos.* California Department of Food and Agriculture, Sacramento, CA, 1989.

61. M. K. Sears, C. Bowhey, H. Braun, and G. R. Stephenson. Dislodgeable residues and persistence of diazinon, chlorpyrifos and isofenphos following their application to turf grass. *Pesticide Sci.* 20: 223–231 (1987).

62. F. A. Gunther, Y. Iwata, G. E. Carman, and C. A. Smith. The citrus reentry problem: Research on its causes and effects, and approaches to its minimization. *Residue Rev.* 67: 1–139 (1977).

63. J. B. Knaak and Y. Iwata. The safe level concept and the rapid field methods: A new approach to solving the reentry problem. In J. R. Plimmer (Ed.), *Pesticide Residue and Exposure*, ACS Symposium Series 182. American Chemical Society, Washington, DC, 1982, pp. 23–39.

64. R. C. Spear, W. J. Popendorf, J. T. Leffingwell, T. H. Milby, J. E. Davies, and W. F. Spencer. Field-workers' response to weathered residues of parathion. *J. Occup. Med.* 19: 406–410 (1977).

65. W. F. Durham and H. R. Wolfe. Measurement of the exposure of workers to pesticides. *Bull. World Health Org.* 26: 75–91 (1962).

66. W. J. Popendorf. An industrial hygiene investigation into the occupational hazard of parathion residues to citrus harvesters. Ph.D. thesis, University of California, Berkeley, 1976.

67. W. J. Popendorf, R. C. Spear, J. T. Leffingwell, J. Yager, and E. Kahn. Harvester exposures to zolone (phosalone) residues in peach orchards. *J. Occup. Med.* 21: 189–194 (1979).

68. W. J. Popendorf. Exploring citrus harvesters' exposure to pesticide contaminated foliar dust. *Am. Ind. Hyg. Assoc. J.* 41: 652–659 (1980).

69. J. E. Davis. Personnel monitoring, minimizing occupational exposure to pesticide residues. *Residue Rev.* 75: 33–50 (1980).

70. U.S. Environmental Protection Agency (EPA). *Human Health Assessment for the Azinphos Methyl Reregistration Eligibility Decision Document (RED).* Case No. 0235. Standard operating procedures for estimating exposure. EPA, Washington, DC, 1999.

71. *OECD Guideline for the Testing of Chemicals, Proposal for a New Guideline: Percutaneous Absorption: In Vivo Method.* Organization for Economic Co-operation and Development Paris, 1994.

72. D. Howes, R. Guy, J. Hadgraft, J. Heylings, U. Hoeck, F. Kemper, H. Maibach, J. P. Marty, H. Merk, J. Parra, D. Rekkas, L. Randelli, H. Schaefer, U. Toucher, and N. Verbiese. Methods for assessing percutaneous absorption. *ECVAM Workshop Report 13, ATLA* 24: 81–106 (1996).

73. U.S. Environmental Protection Agency (EPA). *Health Effects Test Guidelines.* OPPTS 870.7600. *Dermal Penetration.* EPA, Washington, DC, 1998.

74. European Centre for Ecotoxicology and Toxicology of Chemicals (ECETOC). *Percutaneous Absorption.* Monograph 20. ECETOC, Brussels, Belgium, 1993.

75. T. Thongsinthusak, J. H. Ross, S. G. Saiz, and R. Krieger. Estimation of dermal absorption using experimental saturation model. *Reg. Toxicol. Pharmacol.* 29: 37–43 (1999).

76. S. A. Harris and K. R. Solomon. Human exposure to 2,4-D following controlled activities on recently sprayed turf. *J. Environ. Sci. Health, Part B, Pesticides* 27: 9–22 (1992).

77. R. A. Feldmann and H. I. Maibach. Percutaneous penetration of some pesticides and herbicides in man. *Toxicol. Appl. Pharmacol.* 28: 126–132 (1974).

78. T. B. Gaines. The acute toxicity of pesticides to rats. *Toxicol. Appl. Pharmacol.* 2: 88–99 (1960).

79. L. J. Mazuret. Phosalone, methyl-azinphos and parathion, acute percutaneous toxicity in the rat. Unpublished report, 1971.

80. R. C. Spear, W. J. Popendorf, W. F. Spencer, and T. H. Milby. Worker poisonings due to paraoxon residues. *J. Occup. Med.* 19: 411–414 (1977).

81. D. M. Richards, J. F. Kraus, P. Kurtz, N. O. Borhani, R. Mull, W. Winterlin, and W. W. Kilgore. A controlled field trial of physiological responses to organophosphate residues in farm workers. *J. Environ. Pathol. Toxicol.* 2: 493 (1978).

82. D. J. Finney. *Probit Analysis*, 3rd. ed. Cambridge University Press, New York, 1972.

83. T. L. Copeland, D. J. Paustenbach, M. A. Harris, and J. Otani. Comparing the results of a former wood treatment site. *Regul. Toxicol. Pharmacol.* 18: 275–312 (1993).

84. T. L. Copeland, A. H. Holbrow, J. M. Otani, K. T. Connor, and D. J. Paustenbach. Use of probabilistic methods to understand the conservatism in California's approach to assessing health risks posed by air contaminants. *J. Air Waste Manag. Assoc.* 44: 1399–1413 (1994).

85. A. C. Cullen and H. C. Frey. *Probabilistic Techniques in Exposure Assessment.* Plenum, New York, 1999.

86. B. L. Finley and D. J. Paustenbach. The benefits of probabilistic exposure assessment: Three case studies involving contaminated air, water, and soil. *Risk Anal.* 14: 53–73 (1994).

87. B. L. Finley, P. K. Scott, and D. A. Mayhall. Development of a standard soil-to-skin adherence probability density function for use in Monte Carlo analysis of dermal exposure. *Risk Anal.* 14: 555–569 (1994).

88. B. Finley, C. Kirman, P. Scott, A. Spivack, T. Bernhardt, J. Warmerdam, and A. Pittignano. A probabilistic risk assessment of a PCB-contaminated waterway: A case study. *J. Soil Contam.*, in press.

89. U.S. Environmental Protection Agency (EPA). (1997). *Guiding Principles for Monte Carlo Analysis.* Risk Assessment Forum, EPA/630/R-97/001. EPA, Office of Research and Development, Washington, DC.

90. J. H. Ross and M. H. Dong. The use of probabilistic modeling to determine reentry intervals. *Toxicology* 30: 254 (1996).

91. J. B. Knaak, M. A. Al-Bayati, O. G. Raabe, and J. N. Blancato. Development of in vitro V_{max} and K_m values for the metabolism of isofenphos by P-450 enzymes in animals and humans. *Toxicol. Appl. Pharmacol.* 120: 106–113 (1993).

92. C. M. Menzie. *Metabolism of Pesticides.* U.S. Department of the Interior, Fish and Wildlife Service, Special Scientific Report-Wildlife No. 96, 1966.

93. R. A. Neal. Studies on the metabolism of diethyl 4-nitrophenyl phosphorothionate (parathion) in vivo. *Biochem. J.* 103: 183–191 (1967).

94. R. A. Neal. Studies of the enzymic mechanism of the metabolism of diethyl 4-nitrophenyl phosphorothionate (parathion) by rat liver microsomes. *Biochem. J.* 105: 289–297 (1967).

95. T. Nakatsugawa and P. A. Dahm. Microsomal metabolism of parathion. *Biochem. Pharmacol.* 16: 25–38 (1967).

96. L. G. Sultatos and C. L. Gigliardi. Desulfuration of the insecticide parathion by human placenta in vitro. *Biochem. Pharmacol.* 39: 799–801 (1990).

97. H. X. Zhang and L. G. Sultatos. Biotransformation of the organophosphorus insecticides parathion and methyl parathion in male and female rat livers perfused in situ. *Drug Metab. Dispos.* 19: 473–477 (1991).

98. G. W. Jepson, D. K. Hoover, R. K. Black, J. D. McCafferty, D. A. Mahle, and J. M. Gearhart. A partition coefficient determination method for nonvolatile chemicals in biological tissues. *Fundam. Appl. Toxicol.* 22: 519–524 (1994).

99. K. B. Wallace and J. E. Dargan. Intrinsic metabolic clearance of parathion and paraoxon by livers from fish and rodents. *Toxicol. Appl. Pharmacol.* 90: 235–242 (1987).

100. F. Gil, A. Pla, M. C. Gonzalvo, A. F. Hernandez, and E. Villanueva. Partial purification of paraoxonase from rat liver. *Chem.-Biol. Interact.* 87: 69–75 (1993).

101. F. Gil, A. Pla, M. C. Gonzalvo, A. F. Hernandez, and E. Villanueva. Rat liver paraoxonase: Subcellular distribution and characterization. *Chem.-Biol. Interact.* 87: 149–154 (1993).

102. M. C. Gonzalvo, F. Gil, A. F. Hernandez, L. Rodrigo, E. Villanueva, and A. Pla. Human liver paraoxonase (PON1): Subcellular distribution and characterization. *J. Biochem. Mol. Toxicol.* 12(1): 61–69 (1998).

103. C. S. Forsyth and J. E. Chambers. Activation and degradation of the phosphorothionate insecticides parathion and EPN by rat brain. *Biochem. Pharmacol.* 38(10): 1597–1604 (1989).

104. C. L. Kuo and B. N. La Du. Comparison of purified human and rabbit serum paraoxonases. *Drug Metab. Dispos.* 23(9): 935–944 (1995).

105. B. Mackness, M. I. Mackness, S. Arrol, W. Turkie, and P. N. Durrington. Effect of the molecular polymorphisms of human paraoxonase (PON1) on the rate of hydrolysis of paraoxon. *Br. J. Pharmacol.* 122(2): 265–268 (1997).

106. C. Sams and H. J. Mason. Detoxification of organophosphates by A-esterases in human serum. *Hum. Exp. Toxicol.* 18(11): 653–658 (1999).

107. International Life Sciences Institute-Risk Science Institute (ILSI-RSI). *Physiological Parameter Values for PBPK Models.* Report prepared under a cooperative agreement with OHEA. Environmental Protection Agency, Washington, DC, 1994.

108. R. E. Talcott. Hepatic and extrahepatic malathion carboxylesterases. Assay and localization in the rat. *Toxicol. Appl. Pharmacol.* 47: 145–150 (1979).

109. G. M. Pacifici, M. Franchi, C. Bencini, F. Repetti, N. Di Lascio, and G. B. Muraro. Tissue distribution of drug-metabolizing enzymes in humans. *Xenobiotica* 18: 849–856 (1988).

110. D. M. Maxwell, D. E. Lenz, W. A. Groff, A. Kaminski, and H. L. Froehlich. The effects of blood flow and detoxification on in vivo cholinesterase inhibition by Soman in rats. *Toxicol. Appl. Pharmacol.* 88: 66–76 (1987).

111. C. Wang and S. D. Murphy. Kinetic analysis of species differences in acetylcholinesterase sensitivity to organophosphate insecticides. *Toxicol. Appl. Pharmacol.* 66: 409–419 (1982).

112. S. D. Cohen, R. A. Williams, J. E. Killinger, and R. I. Freudenthal. Comparative sensitivity of bovine and rodent acetylcholinesterase to in vitro inhibition by organophosphate insecticides. *Toxicol. Appl. Pharmacol.* 81: 452–459 (1985).

113. W. N. Aldridge and E. Reiner. Enzyme inhibitors as substrates. In A. Neuberger and E. Tatum (Eds.), *North-Holland Research Monographs, Frontiers of Biology*, Vol. 26. North-Holland, London, 1972, p. 236.

114. Y. C. Chiu, A. R. Main, and W. C. Dauterman. Affinity and phosphorylation constants of a series of *O,O*-dialkyl malaoxons and paraoxons with acetylcholinesterase. *Biochem. Pharmacol.* 18: 2171–2177 (1969).

115. J. E. Chambers and R. L. Carr. Inhibition patterns of brain acetylcholinesterase and hepatic and plasma aliesterases following exposures to three phosphorothionate insecticides and their oxons in rats. *Fundam. Appl. Toxicol.* 21: 111–119 (1993).

116. R. L. Bronaugh. In vitro methods for the percutaneous absorption of pesticides. In R. C. Honeycutt, G. Zweig, and N. N. Ragsdale (Eds.), *Dermal Exposure Related to Pesticide Use: Discussion of Risk Assessment.* ACS Symposium Series No. 273. American Chemical Society, Washington, DC, 1985.

117. D. P. Morgan, H. L. Hetzler, E. F. Slach, and L. I. Lin. Urinary excretion of paranitrophenol and alkyl phosphates following ingestion of methyl or ethyl parathion by human subjects. *Arch. Environ. Contam. Toxicol.* 6: 159–173 (1977).

SECTION G
Case Study Involving Exposure to Radionuclides

14 Dose Reconstructions for Radionuclides and Chemicals: Case Study Involving Federal Facilities at Oak Ridge, Tennessee

THOMAS E. WIDNER

ENSR International, Alameda, California

SUSAN M. FLACK

ENSR International, Boulder, Colorado

14.1 INTRODUCTION

Dose reconstruction can be defined as the comprehensive analysis of the exposures that have been received by individuals in the workplace or near facilities that released contaminants to the environment. Dose reconstructions typically focus on retrospective assessment of doses in a realistic manner—in other words, "real doses to real people." Dose reconstructions have been conducted to address public concerns, to provide assurances to state and federal health departments that they have fulfilled their duty to understand public health hazards, and to satisfy concerns of elected officials. Questions that workers and members of the public have asked include:

- Are we exposed?
- Were we exposed in the past, and if so, how much?
- Have we been affected?
- Is our incurred risk of adverse effects increased?
- Is there anything we can do about it?

Human and Ecological Risk Assessment: Theory and Practice, Edited by Dennis J. Paustenbach
ISBN 0-471-14747-8 © 2002 John Wiley & Sons, Inc.

Dose reconstruction provides important details about historical doses and health risks, informing interested parties as to what contaminants people were likely exposed to, who was most likely to have been harmed (sensitive genders, cultural or age groups), where exposures were likely highest and where they were much lower, when exposures were highest, and how exposures varied over time.

Dose reconstruction studies can help in the evaluation of the need for epidemiologic studies and the feasibility of conducting such studies. Epidemiologic studies that have the benefit of dose reconstruction are much more likely to produce meaningful results. Quantitative estimates of past doses are necessary if a study is to assess causal relationships between exposure and outcome or to describe incidences of health effects at varying levels of dose.

While early industrial hygiene activities focused more on recognition of health effects from chemical exposures in the workplace and then their prevention or minimization, the field grew over time to include retrospective estimation of chemical exposures, often in support of occupational epidemiology. Methods for reconstructing occupational exposures based on work history and work environment information include the source–receptor model (Stewart, 1999; Stewart et al., 1996; Smith et al., 1991; Stewart and Herrick, 1991); cumulative exposure indices using air sampling data (Smith et al., 1984); and scaling factors, multipliers, or exposure factors (Seixas et al., 1997; Hornung et al., 1994, 1996; Armstrong et al., 1996; Plato et al., 1995; Kauppinen et al., 1994; Hallock et al., 1994; Smith et al., 1993). Developments in methods for occupational dose reconstruction using incomplete and/or flawed historical information have been described in some recent studies (McLaren/Hart, 1999; Ramachandran and Vincent, 1999; Stewart et al., 1998a, b).

Some of the earliest environmental dose reconstructions involved very persistent organic compounds such as dioxin (Aylward et al., 1996; Scheuplein and Bowers, 1995; Kauppinen et al., 1994). With persistent contaminants, measurements of body tissues and/or body fluids can be used to estimate historical concentrations in the bodies of exposed individuals, provided that basic time histories of exposure and rates and patterns of elimination can be adequately characterized. One of the key issues raised in these studies has been selection of the appropriate measure of lifetime dose for use in evaluating dose–response relationships (such as peak concentrations, average concentrations, or areas under the curve).

One of the areas where environmental dose reconstruction has played an important role, and has experienced significant advancement in terms of methodologies and practices, has been in the evaluation of the potential for off-site health hazards from past operations at sites that were involved in production and testing of U.S. nuclear weapons by the U.S. Department of Energy (DOE) and its predecessor agencies [Miller and Smith, 1996; Office of Technology Assessment (OTA), (1991)]. Dose reconstruction projects at DOE sites typically involve assembly of an independent and comprehensive historical record of operations at a particular facility that goes a long way in pulling away the "cloak of secrecy" that has hidden past operations and releases and led to public distrust. With public involvement, dose reconstruction can be an avenue for providing a thorough public accounting of past practices and releases. As summarized in a National Research Council (NRC) report on radiation dose reconstruction (NRC, 1995), dose reconstructions must meet two criteria; they must withstand scientific scrutiny, and they must satisfy public concern.

The first modern dose reconstruction project performed at a DOE site in the United States involved the Nevada Test Site (NRC, 1995; Voillequé and Gesell, 1990; Thompson and McArthur, 1996; Whicker et al., 1996; Kirchner et al., 1996). Late in the 1970s, it was alleged that leukemias and other cancers had been caused by fallout from aboveground nuclear weapons testing conducted at the Nevada Test Site in the 1950s and 1960s. A dose reconstruction resulted, with assessment of doses from ingestion and inhalation of radionuclides covering four states and parts of five others. Conducted by DOE, this was not an "open" study by current standards.

Public concern about past operations led DOE in 1986 to release documents that detailed past releases from the Hanford Reservation, including the intentional "Green Run" release of iodine-131 in 1949. After an independent panel recommended dose reconstruction at Hanford, DOE directed Battelle Pacific Northwest Laboratories to conduct the Hanford Environmental Dose Reconstruction (HEDR) (NRC, 1995; Shipler et al., 1996; Heeb et al., 1996; Walters et al., 1996; Ramsdell et al., 1996; Anderson et al., 1996; Farris et al., 1996). Independent project direction by a Technical Steering Panel began in 1988, and funding was later transferred to the Centers for Disease Control and Prevention (CDC). HEDR was an open, public study.

The Rocky Flats project was the first major dose reconstruction to address both chemicals and radionuclides (ChemRisk, 1991a, b, 1992b, 1994a–d; Ripple, 1992; Ripple et al., 1996; Mongan et al., 1996a,b; Radiological Assessments Corp. (RAC), 1999). It was initiated shortly after a highly visible 1989 event in which federal agents raided the plant in search of evidence of environmental violations. The Rocky Flats project was conducted by the Colorado Department of Health, under a grant from DOE. States had the option of conducting the studies themselves or having them conducted by CDC.

In 1988, a Historical Dose Evaluation Task Group was chartered within DOE in response to concerns of possible radiological impacts from past operations at the Idaho National Engineering Laboratory (INEL). A 1991 report from the group assessed historical airborne releases (DOE, 1991; Chew et al., 1993). The project was criticized for excluding public participation. CDC is currently conducting dose reconstruction activities at INEL [Sanford Cohen and Associates (SCA) 1993].

Health studies were initiated by the State of Tennessee in 1992, under a grant from DOE. The project initially centered around the Oak Ridge Dose Reconstruction Feasibility Study (ChemRisk, 1993a–e, Widner et al., 1996). The

Oak Ridge Dose Reconstruction followed in 1994 and is the main focus of this chapter (Widner, 2000; Hoffman et al., 1999; Mongan et al., 1999; Price et al., 1999; Gouge et al., 1999; Buddenbaum et al., 1999; Cockroft et al., 1999).

An environmental dose reconstruction was conducted by CDC at the Fernald Feed Materials Production Plant in Ohio (NRC, 1995; Meyer at al., 1996). As a result of past releases from the Fernald facility, payments to local residents were authorized by the courts.

Why don't we just measure contaminants in people today? Measurements of biological markers of past exposures can be useful for some contaminants, and should be considered whenever feasible. For example, information regarding past exposures can be obtained from measurements of:

- Lead in teeth, bone, or blood
- Mercury in hair, blood, or urine
- Polychlorinated biphenyls (PCBs) and some pesticides in adipose tissue (fat)

However, the usefulness of current-day measurements of contaminants in people's bodies can be limited by the following considerations:

- Many contaminants are no longer present in our bodies and offer no measurable "tracers." For example, ^{131}I released from Oak Ridge radioactive lanthanum processing has decayed away, and there are too many other sources of longer-lived ^{129}I from Oak Ridge activities, nuclear weapons fallout, and nuclear processing elsewhere to make it a good indicator of past exposures to ^{131}I from lanthanum processing.
- Without knowledge of an individual's exposure history, it is difficult to say anything about the sources of contaminants that may be found in the body, or to clarify the role of the local facility in contributing to exposure.
- Biological indicators of the exposure a person received are generally not reliable when levels of exposure are comparable to those normally found in the environment.
- A significant degree of interindividual variability in biology and personal behaviors can limit the applicability of results from selected individuals to others.
- Measurements of contaminants in individuals can be invasive and very expensive.

This chapter presents a summary of various investigations performed within the Oak Ridge Health Agreement Studies, in a project that became known as the Oak Ridge Dose Reconstruction. From late 1994 to early 1999, a team of scientists and engineers performed detailed dose reconstruction analyses focused on past releases of radioactive iodine, mercury, and PCBs released from the U.S. government complexes on the Oak Ridge Reservation (the ORR) and radionuclides released from White Oak Creek to the Clinch River. This chapter summarizes the methods and results of these dose reconstructions. In addition, this chapter describes the systematic searching of classified and unclassified historical records that was a vital component of the project and summarizes the less detailed, screening-level assessments that were performed to evaluate the potential off-site health significance of a number of other materials, most notably uranium.

14.2 BACKGROUND

Several characteristics make evaluation of the potential off-site health effects from the ORR very complex and very challenging. In terms of variety and complexity of past operations and materials used (radionuclides and chemicals), the ORR is among the most complex sites in the world. The settings of the three main ORR complexes (the plant sites code-named K-25, X-10, and Y-12), in complex ridge-and-valley terrain, lead to some particularly complex effluent transport patterns and pathways for public exposure. Early photographs of the three main Oak Ridge complexes, as well as the S-50 plant, are shown in Figures 14.1 through 14.4. The potential importance of Oak Ridge releases is

Figure 14.1 Early view of K-25 site, which was also known as Oak Ridge Gaseous Diffusion Plant. K-25 enriched uranium in its U-235 component for approximately 40 years. U.S. DOE photo.

Figure 14.2 Late-1940s view of central portion of X-10 site, now called Oak Ridge National Laboratory. The graphite reactor, then called the Clinton Pile, is the large building at right center. The three-story radioactive lanthanum processing building is to its lower left. U.S. Atomic Energy Commission (AEC) photo.

heightened by the fact that there are communities closer to key production areas than at any other DOE site in the country.

The Oak Ridge Health Agreement Studies used a preliminary screening phase, an iterative assessment approach with predetermined decision guides, and exposure assessment methods that emphasized the use of environmental measurements wherever possible. The investigations of the Oak Ridge Dose Reconstruction used these methods to focus on operations, releases, exposure pathways, locations, and groups of people that are representative of (1) maximum and (2) more typical doses and health risks to people who have lived near the ORR.

14.2.1 Oak Ridge Health Studies Agreement

In 1991, DOE and the State of Tennessee entered into a Health Studies Agreement. One of the goals of the agreement was to assemble a panel to design a study to evaluate the feasibility of doing a dose reconstruction of releases from

Figure 14.3 Early photograph of Oak Ridge Y-12 plant. Initially built to enrich uranium by the electromagnetic process, it later housed operations to enrich lithium and manufacture nuclear weapon components. ORNL photo.

ORR facilities, in effect an independent investigation of the potential for adverse heath effects from past operations. The Oak Ridge Dose Reconstruction Feasibility Study was conducted from 1992 to 1993. In it, investigators took an intense and comprehensive, but relatively quick, "look through the key hole" at past Oak Ridge operations and performed screening evaluations to identify those operations and materials that warranted detailed investigation (Chem-Risk, 1993a–e).

At the close of the feasibility study, the Tennessee Department of Health (TDH) and the Oak Ridge Health Agreement Steering Panel (ORHASP) recommended that a dose reconstruction be conducted for radioactive iodine releases from X-10 radioactive lanthanum processing, mercury releases from Y-12 lithium enrichment, PCBs in the environment near Oak Ridge, and radionuclides released from X-10 to the Clinch River via White Oak Creek. They also called for the study to include systematic searching of historical records, an evaluation of the quality of historical uranium effluent monitoring data, and additional screening of some materials that could not be fully evaluated during the feasibility study.

The Oak Ridge Dose Reconstruction began in late 1994. The methods, accomplishments, and findings of the project are the subjects of this chapter. The

Figure 14.4 Workers disassemble the main process building of the S-50 plant in the 1940s. This facility used the liquid thermal diffusion process for enrichment of uranium and was abandoned after about 1 year of operation. AEC photo.

project was designed to develop estimates of past doses and health risks from the selected contaminants potentially received by people who have lived in off-site areas near the reservation. The dose reconstruction was not designed to include reconstruction of exposures received by workers in the course of their duties at the Oak Ridge plants. And while some projection was done of the potential for health effects to be experienced in the future due to releases up to the current time, this project did not include projection of releases or doses from potential future activities.

14.2.2 Oak Ridge Dose Reconstruction

The Oak Ridge Dose Reconstruction involved seven main components that were conducted concurrently and called project tasks. The first four tasks entailed detailed dose reconstructions for radioactive iodine releases from ra-dioactive lanthanum (RaLa) processing at X-10, mercury releases from lithium enrichment at Y-12, PCBs in the environment near Oak Ridge, and releases of cesium-137 and other waterborne radionuclides from X-10 to the Clinch River

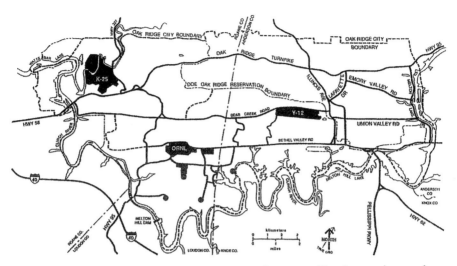

Figure 14.5 Oak Ridge Reservation and surrounding areas. The three main complexes on the reservation are the K-25 site (long known as the Oak Ridge Gaseous Diffusion Plant), the X-10 site (now called Oak Ridge National Laboratory), and the Y-12 plant. The former S-50 plant was within what is now the K-25 site. The reservation is within the city limits of Oak Ridge.

via White Oak Creek. The project also included a systematic search of document repositories (Widner, 2000), an investigation of the quality of airborne and waterborne uranium effluent monitoring at the Oak Ridge complexes (see Appendix A), and performance of additional health risk screening calculations for materials not fully evaluated in the Dose Reconstruction Feasibility Study (see Appendix A).

Although these components were conducted concurrently, the assessments of the individual contaminants were to a certain extent conducted separately. Because people who lived near Oak Ridge may have been exposed to several contaminants at the same time, it is important to look at how the important operations did or did not coincide within the overall time frame of Oak Ridge operations from 1943 through 1994. A map of the Oak Ridge Reservation is presented as Figure 14.5. The identification of the potential time periods and locations of exposure to several (multiple) contaminants is one of the goals of this chapter. Results of the study are discussed in this context in a later section of this chapter, and general time lines of key operations relevant to project tasks are depicted in Figure 14.6.

14.2.3 Decision-Making Processes Used on the Dose Reconstruction Project

Investigators in environmental dose reconstruction projects are being asked to perform increasingly comprehensive investigations of complex sites. In many cases, a wide variety of contaminants, release sources, and exposure pathways

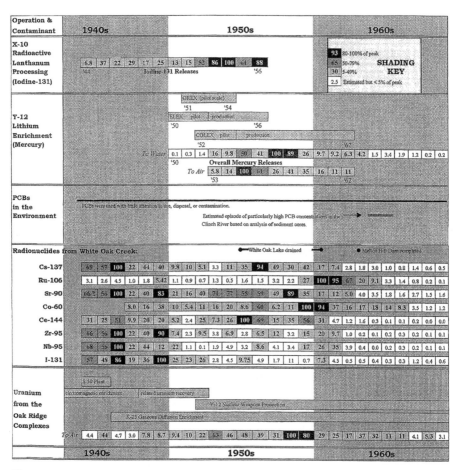

Figure 14.6 Timeline of selected Oak Ridge operations and releases (1944–1969) important to Oak Ridge Dose Reconstruction. Oak Ridge operations continued beyond the 1960s, but peak releases for most contaminants occurred during these first 26 years. The numbers given for each year indicate percentages of the highest annual release of that contaminant.

must be evaluated for their potential significance. The Oak Ridge project was likely the most complex of any conducted to date in this regard. To achieve the goals of the project in light of this complexity, a number of decision-making process were applied to assist in focusing project attention and resources.

Use of Predetermined Decision Guides   More and more commonly, decision guides or "stopping rules" are being considered in the early stages of dose reconstruction projects to guide decisions that will likely be faced later in the study. Decision guides support compromises between the desire for complete investigation of potential hazards of every material that could have been released and the needs to control costs of health studies. They are useful for guiding decisions in screening contaminants or exposure pathways, in setting geographic domains of studies, and in determining when certain elements of an assessment do not warrant further investigation. Possibly useful bases for decision guides include:

- Exposure or dose guidelines
- Epidemiologic detectability of health effects
- Government dose standards
- Minimal risk levels
- Background contaminant concentrations or exposure levels

For this project, a lifetime risk of 1 in 10,000 (0.0001, or 10^{-4}) of developing cancer from cancer-causing agents, and a hazard index of 1 for noncarcinogenic chemicals were adopted as guidelines to support decisions as to whether or not to do dose reconstruction for particular materials. A hazard index is the ratio of the estimated dose that an exposed person has received to the reference dose for the chemical. Published by the U.S. Environmental Protection Agency (EPA), reference doses (RfDs) are estimates of the largest amounts that an individual can take in, per unit body weight, on a daily basis over a lifetime without adverse health effects (even for people who are unusually sensitive). RfDs are generally well below no observed adverse effect levels (NOAELs).

In early phases of investigation, materials or exposure pathways that yielded estimates of risk above these decision guides were given higher priority for further screening or dose reconstruction. These decision guides are not viewed as inflexible because most risk assessments are complex and uncertain, and there may be health effects caused by exposure to multiple contaminants that are not predicted by analyses that address the contaminants one at a time. However, in general a hazard index of 1 is a conservative guideline, since it is well below the NOAEL, which is in turn well below the lowest observed adverse effect level (the LOAEL). We took seriously any risks that approached but did not quite meet guide values.

Screening Calculations In complex investigations, it is wise to focus on those areas that warrant detailed investigation. A summary of the screening methods that we used is presented in Appendix A. An effective method toward achieving appropriate focus in complex investigations is the use of a combination of conservative and nonconservative screening calculations. Conservative screening calculations are designed so that exposures are not underestimated. They typically are based on maximum reported environmental concentrations and the use of models to estimate concentrations where contaminants were not sampled or were sampled but not positively detected. Conservative screening calculations use reasonable maximum parameter values in a deliberate effort to err on the side of safety. Conservative screening calculations yield upper bounds on estimated health risks and are used (with predetermined decision guides) to identify low-priority contaminants.

In contrast, nonconservative screening calculations are designed so that exposures are not overestimated. Nonconservative calculations typically use means or geometric means of concentrations that were measured. In general, margins of safety or conservative biases are removed. In some cases, it is advisable to retain some conservatism in areas with particularly high uncertainty; in these cases, screening is sometimes called refined level I screening. Nonconservative (or reduced conservatism) screening calculations yield estimates of more typical risks and (used with predetermined decision guides) identify high-priority contaminants that warrant immediate consideration. Use of conservative and nonconservative screening allows investigators to separate contaminants into three classes: those of low importance, those warranting immediate attention, and those requiring further investigation to determine their true significance.

Illustrative Example: Quantitative Screening of Materials of Concern Determine if project resources should be spent on a detailed investigation of arsenic releases from burning over 1.2×10^{10} pounds of arsenic-containing coal for power generation at the Oak Ridge Reservation in Tennessee between 1944 and 1962 by evaluating the potential health hazards of historical arsenic releases.

Background The Oak Ridge Dose Reconstruction Feasibility Study used screening calculations to prioritize materials of potential concern so that in the main Dose Reconstruction Study resources could be focused on the most important contaminants, and dilution of resources could be avoided by identifying exposures that were obviously of only minor importance. The method selected to evaluate a given material was dependent on the quantity of the material present on-site, the form and manner in which the material was used, and the availability of environmental monitoring and release data.

Arsenic is a naturally occurring metallic element found in coal. Because the Oak Ridge Reservation facilities were large consumers of electrical power, several coal-fired steam plants were constructed and operated on-site. The storage and use of coal, and the disposal of ash from coal-burning operations, likely resulted in the release of arsenic to air, surface water, and sediments.

Approximately 1.2×10^{10} pounds of coal were burned at the K-25 K-701 power station between 1944 and 1962 (6.6×10^8 lb yr^{-1}). A mean arsenic concentration of 47 mg arsenic kg^{-1} coal was reported in the mid-1970s based on actual measurements of coal samples burned at the Oak Ridge Reservation [Union Carbide Corporation Nuclear Division (UCCND), 1983a]. The caloric value of West Virginia bituminous coal is 14,040 Btu lb^{-1} coal (Avallone and Baumeister, 1996). An emission factor for uncontrolled releases of arsenic from a coal-fired boiler is 1237 lb arsenic per 10^{12} Btu (EPA, 1989). The atmospheric dilution factor (relative concentration, χ/Q) value for releases from K-25 to the nearest community location is 7.4×10^{-7} s m^{-3}. Measured off-site environmental concentrations of arsenic in surface water (0.02 mg L^{-1}) and sediment (55 mg kg^{-1}) were included in the estimates of dose and risk. Biotransfer and bioconcentration factors (BCF) were used to estimate arsenic concentrations in meat, milk, vegetables, and fish (400 BCF $\times$ 0.02 mg L^{-1} = 8 mg kg^{-1}) since no off-site measurement data were located [National Council on Radiation Protection and Measurements (NCRP), 1996b)]. Toxicity criteria used were EPA slope factors for inhalation and ingestion and the oral RfD for arsenic.

Solution   The total mass of arsenic available to be released to the environment per year from direct air releases and/or fly ash disposal is 14,100 kg:

$$M_{As} = (3 \times 10^8 \text{ kg coal}) \times 47 \text{ mg kg}^{-1} \text{ coal} \times \frac{\text{kg}}{1 \times 10^6 \text{ mg}}$$

$$= 14{,}100 \text{ kg}$$

Since no stack release data were located for arsenic, EPA emission factors were used to estimate releases of arsenic to air:

$$\text{Release rate} = \left(\frac{14{,}040 \text{ Btu}}{\text{lb coal}}\right) \times (6.6 \times 10^8 \text{ lb coal yr}^{-1}) \times \left(\frac{1{,}237 \text{ lb As}}{10^{12} \text{ Btu}}\right)$$

$$= 0.16 \text{ g s}^{-1}$$

The predicted air concentration of arsenic in air at the nearest community is 120 ng m^{-3}:

$$C_{air} = (0.16 \text{ g s}^{-1})(7.4 \times 10^{-7} \text{ s m}^{-3})$$

$$= 1.2 \times 10^{-7} \text{ g m}^{-3} = 120 \text{ ng m}^{-3}$$

The level I cancer screening index for the combined inhalation and ingestion of arsenic from coal burned at K-25 is 3.8×10^{-2}. The level II cancer screening index is 8.9×10^{-4}. The level I and level II noncancer screening indices for ingestion of arsenic released from K-25 are 120 and 13, respectively. Arsenic was determined to be a high-priority candidate for further study since both level I and level II screening indices were above the decision guides of a 1×10^{-4} excess lifetime cancer risk and a hazard index of 1.0. However, the NOAEL for arsenic is a factor of three above the oral reference dose. Noncancer screening indices above 3 could indicate that exposures above the NOAEL occurred.

Iterative Assessment Process   Performance of assessments in an iterative fashion can be an effective tool in guiding allocation of resources among different sites, contaminants, exposure routes, or elements of an analysis. In the iterative assessment approach, initial risk calculations are based on information that is readily at hand. The dominant sources of uncertainty in results are identified (typically through Monte Carlo simulation and sensitivity analysis), and further research targets these areas in order to reduce uncertainty. Revised calculations are performed with the new data, with results being more specific and less uncertain. The process is repeated until the uncertainty of results is either acceptable or cannot be further reduced. The iterative assessment process was used in the dose reconstruction to guide allocation of resources between and within the main project tasks that focused on radioiodine, mercury, PCBs, and radionuclides released from X-10 via White Oak Creek.

Uncertainty Analysis and Subjective Confidence Intervals   Although efforts were made to minimize uncertainties in the dose reconstruction, we had imperfect knowledge about most of the inputs to the equations used to reconstruct historical doses. For example, there is uncertainty about the true value of some parameters because exposures occurred in the past, and precise measurements of exposure concentrations or intake rates were not made. In addition, many parameters in the dose equations also exhibit natural variability. For example, within any population, variability in body weights and rates of food and water consumption is expected. Each parameter in the dose equations was characterized by a range of values, called a probability density function (PDF), describing what is known about the uncertainty and variability of that parameter for the population of interest. PDFs were subjectively defined, allowing confidence intervals to be derived within which there is a high probability of including the true but unknown value of the parameter.

When inputs to an equation are defined using distributions, each equation has many possible answers and is solved repeatedly using different values selected from the distributions of input parameters. This process used a computer program and Monte Carlo simulation. Results of the calculations are themselves probability distributions. Dose estimates in this chapter are usually stated as central estimates with 95 percent subjective confidence intervals. The central estimates are median (50th percentile) values and represent a more likely region of doses, and the confidence intervals indicate that the investigators are 95 percent confident that the true values are no lower than the lower confidence limit and no higher than the upper confidence limit.

14.2.4   Technical Highlights of the Oak Ridge Dose Reconstruction

We selected the following aspects of the technical investigations that were undertaken as the highlights of the dose reconstruction project:

1. The estimated upper bound of the total ^{131}I release from Oak Ridge lanthanum processing is about 5 to 6 times the Hanford Green Run release, over double the Windscale release, and about 4000 times the 1979 release from

Three Mile Island. While the average annual airborne ^{131}I release from Hanford was about 8 times the value for X-10, members of the Oak Ridge public lived considerably closer to the release points than at Hanford. Releases from RaLa processing were estimated based on original records that we located that documented irradiation of fuel slugs at X-10 and Hanford and conduct of 731 dissolving batches. These data allowed estimation of iodine inventories in groups of the fuel slugs and quantities of ^{131}I in three physicochemical forms likely released past a caustic scrubber.

2. The modeling of dispersion of airborne [131]I: (a) explicitly included uncertainties in airborne transport of iodine and chemical transformation during transport, (b) accounted for simultaneous dry and wet deposition during wet periods, and (c) accounted for deposition onto trees when leaves were present, distinguishing this from deposition onto pasture grass and adjusting plume depletion accordingly. Data from experiments conducted near Oak Ridge and from the scientific literature were used in specifying feed-to-milk transfer coefficients for ^{131}I, with separate values used for family cows and commercial milk.

3. Calculations of thyroid doses reflected recent ultrasound measurements of thyroid mass, accounting for interindividual variability and uncertainty. The estimation of expected rates of cancer induction included evaluation of (a) age and gender dependency, (b) differences in background rates of thyroid cancer depending on age, gender, and ethnicity, and (c) a biological effectiveness factor for induction of cancer after ^{131}I intake compared to acute, external exposure to X or gamma rays.

4. The number of excess thyroid cancers within 200 km of Oak Ridge was estimated. Estimated health impacts at 41 locations were stated as excess lifetime risks, relative risks, probabilities of causation, and total numbers of cancers. Excess risk of nonneoplastic disease was identified as a potential concern for those who as children drank milk produced near Oak Ridge or were exposed to ^{131}I from both X-10 and fallout from nuclear weapons testing at the Nevada Test Site (NTS). Doses and risks from combined exposure to ^{131}I from X-10 RaLa releases and NTS fallout were estimated for five areas near Oak Ridge.

5. With about 280,000 pounds entering East Fork Poplar Creek (EFPC), the releases of mercury from the Y-12 plant represent a unique case of long-term releases to the environment, with an extensive record of routine effluent measurements. EFPC travels through residential and commercial areas that were fairly well populated. The Y-12 releases were independently estimated based on detailed original records of EFPC concentrations and flow rates that we assembled and analyzed.

6. The 73,000 pounds of mercury estimated released to the air from Y-12 lithium processing is about 1.8 times the estimated total annual release of anthropogenic mercury to the air from noncombustion sources in the United

States in recent years. The Y-12 releases were estimated based on records of indoor air concentrations and ventilation rates of key process buildings. Review of measurements of mercury in tree rings during this project pointed out the apparent importance of evasion of mercury from EFPC to the air as an additional avenue for historical public exposure. This study accounted for releases from Y-12 buildings and from the creek.

7. The assessment of Y-12 mercury releases was one of the first major assessments of public exposures to mercury completed after the EPA issued its Mercury Study Report to Congress. The dose reconstruction for mercury included evaluation of exposures to three species of mercury for adults and children in 17 reference populations, by 16 exposure pathways, for up to 41 individual years of exposure.

8. In the dose reconstruction for PCBs near Oak Ridge, the population threshold for noncancer effects was characterized in an uncertainty analysis of the EPA reference dose and associated safety factors.

9. A two-dimensional analysis of PCB exposures allowed us to determine: (a) the distribution of "true" hazard quotients across the population (doses divided by the threshold dose), (2) the fraction of the population receiving doses above the threshold, and (3) the incremental contribution of Oak Ridge releases to risks from other sources of PCBs.

10. An evaluation of the sustainable fish harvest in EFPC showed that rates of fish consumption from EFPC were very low. PCB and mercury doses from the fish were likely relatively low.

11. Radioactive effluents from X-10 to the Clinch River via White Oak Creek also represent a unique case of long-term releases with extensive monitoring records. Concentration and flow rate measurements were studied in this project, and an assessment of associated uncertainties and biases yielded annual release estimates for eight radionuclides for 1944 through 1991.

12. Concentrations of radionuclides in the Clinch River were estimated using a combination of aquatic transport modeling and environmental measurements, both adjusted for sources of bias and uncertainty, in a single analysis. Site-specific bioconcentration factors for fish were developed using environmental measurements and the scientific literature.

13. Uncertainty in internal and external dosimetry was expressed explicitly in the calculations for radionuclides in the Clinch River, and a dose–response relationship of cancer incidence was expressed for each of 27 organs and for total cancers. Extending to risk accounted for differing radiosensitivity among organs and identified the most important organs.

14. A critical component of this project was the systematic searching of historical records of past Oak Ridge operations and releases. Forty-four document collections at Oak Ridge and at remote locations were targeted in over 15,000 hours of document review. Relevant records were copied, and summary information placed in a database that is now available.

15. Knowledge about past releases of uranium from the Oak Ridge complexes was significantly advanced by the monitoring data review and screening conducted in this project. Airborne releases from Y-12 were independently estimated to have been over seven times those reported by DOE, and releases from K-25/S-50 were 50 percent higher than reported. Screening evaluations were also performed for technetium-99 and neptunium-237, two radionuclides that were contaminants in recycled uranium that was processed at Y-12 and K-25.

16. To evaluate uranium exposures at Scarboro, which likely lies closer to a DOE weapons facility than any other community in the country, monitoring data were used to estimate transport of airborne uranium over Pine Ridge. Ambient airborne uranium measurements in Scarboro from 1986 to 1995 were coupled with estimates of Y-12 uranium releases during the same periods to derive an empirical dispersion factor. This factor was used to estimate airborne uranium levels in Scarboro for all other years of interest.

17. Hundreds of materials were evaluated using (1) qualitative screening, (2) a threshold quantity approach, or (3) a two-level combination of conservative and nonconservative screening. Quantitative screening was conducted for 10 materials or classes of materials and less detailed evaluations completed for 18 others. The two-level screening separated materials into those that warrant immediate attention, those of low importance, and those requiring more investigation to determine their true significance. The screening was supported by use of predetermined "decision guides."

18. Some of the materials subjected to screening were formerly classified at Oak Ridge by their mere presence. All materials can now be named, but special approaches were taken in this public study to evaluate some materials that still have classified aspects of use.

14.2.5 Remainder of This Chapter

The remainder of this chapter contains summaries of the work that was done within the Oak Ridge Dose Reconstruction project. Following the discussions of the detailed dose reconstructions that were performed for I-131, mercury, PCBs, and radionuclides released via White Oak Creek, there is an overall summary of project results across the four detailed studies. Appendix A then describes the less detailed, screening-level assessments that we performed. These focused on uranium and some other materials that were not fully evaluated during the feasibility study.

14.3   IODINE-131 RELEASES FROM X-10 RADIOACTIVE LANTHANUM PROCESSING

14.3.1   Introduction

Oak Ridge National Laboratory (ORNL), originally known by the code name X-10, released radioactive iodine (^{131}I) to the air from 1944 through 1956 as it processed freshly irradiated nuclear reactor fuel (see Figure 14.7). The process recovered RaLa to support weapons development at Los Alamos (see Figure 14.8), for atmospheric radiation tracking, and for radiation warfare experiments [U.S. General Accounting Office (GAO), 1993; Thompson, 1949]. Iodine concentrates in the thyroid gland. Therefore, the health concerns stemming from exposure to ^{131}I include various diseases of the thyroid such as thyroid cancer and non-neoplastic abnormalities such as autoimmune hypothyroidism and Graves disease.

The dose reconstruction for Oak Ridge radioiodine releases investigated the possible risks of thyroid cancer from the releases of ^{131}I between 1944 and 1956

Figure 14.7   Aluminum-clad uranium fuel slugs, such as these being loaded into the Clinton Pile, were used as the source of barium for RaLa production. When irradiated in the reactor and dissolved shortly thereafter by X-10 workers, radionuclides such as iodine-131 were released to the air. U.S. AEC photo.

Figure 14.8 At Los Alamos Laboratory in New Mexico, devices such as this used RaLa to test the implosion process for early atomic weapons. An intense lanthanum radiation source was placed inside the sphere shown in the center of the photo and allowed researchers to characterize how the sphere behaved during implosion caused by conventional high explosives. LANL photo.

at 41 representative locations within 38 km of ORNL, shown in Figure 14.9 (Hoffman et al., 1999). Communities within this region include Oak Ridge, Clinton, Oliver Springs, Kingston, Harriman, Lenoir City, Sweetwater, Maryville, and Knoxville. At each of these locations, the risks of developing thyroid cancer, the relative risk with respect to an unexposed population, and the probability of causation for diagnosed cases of thyroid cancer were determined for individuals of both genders and of various age groups at the time of exposure. The numbers of cancers that could have resulted from exposure to the Oak Ridge ^{131}I releases within 38-, 100-, and 200-km radii of the X-10 facility were also estimated. The overall impact of ^{131}I exposure from the combined contributions of X-10 releases and fallout from atmospheric testing of nuclear weapons at the Nevada Test Site [National Cancer Institute (NCI), 1997] was also evaluated. The results of this study were found to be comparable to those from similar studies conducted at other United States sites. The probability of occurrence of non-neoplastic thyroid disease within 38 km of X-10 was also

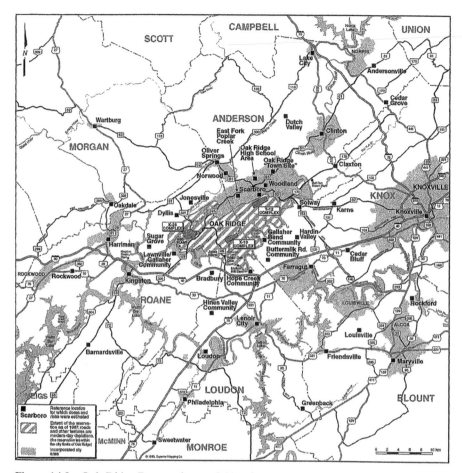

Figure 14.9 Oak Ridge Reservation and 41 reference locations within 38 km at which consequences of iodine-131 releases between 1944 and 1956 were studied. The iodine was released during radioactive lanthanum processing that took place at the X-10 site, located in the center of this map.

evaluated based on new evidence on the doses of radiation required to induce such diseases.

14.3.2 Hazard Identification

It is well established that X and gamma irradiation of the thyroid at absorbed doses approaching 10 centigray (1 cGy = 1 rad) will result in increased incidences of thyroid carcinomas and adenomas in children exposed before the age of 15 (Ron et al., 1995). At higher doses, radiation might also induce nonneoplastic thyroid conditions such as autoimmune hypothyroidism and Graves disease [Institute of Medicine/National Academy of Sciences (IOM/NAS),

1999]. The thyroid gland in children has one of the highest radiogenic risk co-efficients of any organ. Fortunately, fatal thyroid cancers are rare; the 5-year survival rate is 95 percent. The effectiveness of ^{131}I in producing thyroid cancers is a subject that is still under investigation with ongoing epidemiological studies at numerous other locations. The most convincing evidence of the link between ^{131}I exposure and thyroid cancer is still emerging from the follow-up of children exposed to ^{131}I from the 1986 Chernobyl accident (Jacob et al., 1998). Additional supporting evidence exists from animal studies and from epidemiologic investigations of Utah school children exposed to ^{131}I from atmospheric weapons testing at the Nevada Test Site (Kerber et al., 1993).

14.3.3 Dose–Response Evaluation

In this study, a relative risk model was used to estimate the chance of acquiring a thyroid cancer from an absorbed dose of ^{131}I. This model calculates the excess lifetime risk of thyroid cancer per unit dose as the product of the excess relative risk per unit absorbed dose and the background lifetime risk for an unexposed individual, including a series of modifying factors to account for differences in radiosensitivity by gender and age at time of exposure and for the effectiveness of ^{131}I in producing thyroid cancer compared with that of X rays and gamma rays. The background lifetime risk of thyroid cancer was obtained from incidence data for Tennessee, excluding Anderson, Roane, Loudon, and Knox counties, which were affected by X-10 releases (Turri, 1998).

The excess lifetime risk of thyroid cancer per centigray changes markedly depending on the gender of the individual and the age at time of exposure. For females exposed to 1 cGy before the age of 5 years, the excess lifetime risk per centigray ranged from about 5 chances in 100,000 to 16 chances in 10,000, with a central value of 3 chances in 10,000. At the same dose, the risk to males in this age group would be about 4 times less than that for females. Females who were over the age of 20 at time of exposure to 1 cGy would have had risks almost 80 times less, while males over the age of 20 would have had risks about 300 times less than females who were under the age of 5 years.

14.3.4 Exposure Assessment

Over the 13-year period of the RaLa operations at X-10, approximately 30,000 reactor fuel slugs were dissolved in about 731 batches during the process of separating over 19,000 TBq (1 TBq = 10^{12} Bq; approximately 500,000 Ci) of radioactive barium as a source of ^{140}La for shipment to Los Alamos (Hoffman et al., 1999).

Describing Radioiodine Releases   Development of the source term (the amount of ^{131}I released to the atmosphere) involved estimation of quantities of radioiodine released from vents and openings in process buildings and from X-10 stacks during routine and off-normal conditions. For an accident that occurred in April 1954 (Rupp and Witkowski, 1955), releases were estimated for five half-hour periods.

The most important source of radioiodine releases from RaLa processing was exhaust (off gas) from the slug dissolver (Coughlen, 1950). Volatile gases, which included radioiodine, were drawn from the dissolver under negative pressure (see Figure 14.10), through a condenser and a chemical scrubber, and then through piping to reach a 200-foot tall stack. After March 1950, exhausts from the chemical scrubber were routed to a central treatment facility in which contaminated air was passed through an electrostatic precipitator and particulate filters prior to release up a 250-foot brick stack.

The iodine and barium contents of the irradiated fuel slugs from the X-10 Clinton Pile (graphite reactor) and the reactors at Hanford were estimated (Hoffman et al., 1999). For each of the 731 dissolving batches, the potential for "direct" releases of untreated exhaust to the atmosphere through building vents, windows, and other openings was evaluated. Because of the absence of monitoring data, expert opinion was used to quantify the ^{131}I collection efficiency of the condenser and caustic scrubber for routine operations. Expert opinion was also used to quantify the potential degradation of collection efficiencies during the April 1954 accident.

Approximately 8800 to 42,000 Ci (0.3 to 1.6 PBq; 1 PBq $= 10^{15}$ Bq) of ^{131}I was released between 1944 and 1956, of which 6300 to 36,000 Ci (0.23 to 1.3 PBq) was in the elemental (most environmentally reactive) form of iodine; most of the remainder was in the nonreactive volatile organic form. As shown in Table 14.1, the largest releases occurred between 1952 and 1956, when the irradiated uranium fuel slugs came from Hanford reactors. The April 29, 1954, accident released 105 to 500 Ci (3.9 to 21 TBq) over 2.5 hours, accounting for about 6.5 percent of the total releases for 1954. While the period of direct release from the RaLa dissolver appears to have lasted 30 minutes, it took some time for the building ventilation to clear out this airborne contamination. Releases were estimated for five half-hour periods, resulting in a total release duration of 2.5 hours.

Atmospheric Dispersion   After being released into the atmosphere, ^{131}I was transported by the prevailing winds. A fraction of the iodine released in the elemental (reactive) form was chemically transformed during transport to particulate and organic (nonreactive) forms within a few kilometers of the RaLa processing facility (Ludwick, 1964, 1967; Ramsdell et al., 1994; Cambray et al., 1987). The ground-level concentration of ^{131}I in air is affected by several factors including the distance of the location of interest from the RaLa processing facility, the dilution of the concentrations in the air during atmospheric dispersion or mixing, the depletion of iodine from air by the processes of wet and dry deposition (Horst, 1977, 1984), and the chemical form in which iodine is present.

Annual average ground-level concentrations of ^{131}I for routine releases and time-integrated ground-level concentrations of ^{131}I for the 1954 accident were

Figure 14.10 This 1940s photograph shows key components of the Oak Ridge equipment used to separate batches of radioactive barium from irradiated reactor fuel slugs. The cylindrical vessel to the right is the slug dissolver. Above it is a condenser to reduce releases of nitric acid vapors, and to its left is an air mixer. The large tank at the left is a waste neutralizer. The most important path for release of radioactive iodine was through the condenser, the air mixer, a caustic scrubber (not clearly visible) that captured much of the iodine when operating properly, and various stacks that were used over the 13 years of RaLa operations. ORNL photo.

TABLE 14.1 Estimated Annual Iodine-131 Releases from Oak Ridge Radioactive Lanthanum Processing[a]

Year	Elemental Iodine-131 (Ci)			Organic Iodine-131 (Ci)			Particulate Iodine-131 (Ci)		
	2.5%-ile	Central Value	97.5%-ile	2.5%-ile	Central Value	97.5%-ile	2.5%-ile	Central Value	97.5%-ile
				Releases via the Stack[b]					
1944	75	250	660	2.9	12	60	0.015	0.082	0.56
1945	510	1,200	2,400	43	170	810	0.44	1.4	5.6
1946	260	670	1,600	41	160	760	0.0076	0.13	2.0
1947	330	870	2,200	56	220	990	0.0032	0.066	1.0
1948	150	510	1,200	34	130	610	0.22	0.89	4.3
1949	270	790	1,700	41	160	780	0.40	3.1	27
1950	140	400	890	26	100	410	0.025	0.081	0.28
1951	200	500	890	18	71	290	0.0070	0.037	0.18
1952	550	1,600	4,100	110	410	1,700	0.11	0.32	0.83
1953	910	2,700	5,500	150	590	2,400	0.10	0.37	1.6
1954[c]	1,200	2,900	6,500	160	630	2,400	0.082	0.32	1.3
1955	690	1,900	4,400	110	430	1,600	0.055	0.19	0.61
1956	840	2,700	5,700	160	670	2,400	0.10	0.33	1.2
Total[d]	5,600	15,000	36,000	940	3,600	17,000	3.6	6.5	14

Releases from the Building[e]

	0.27	0.52	0.94	0.00064	0.0027	0.015	0.000044	0.00016	0.00061
1944	0.27	0.52	0.94	0.00064	0.0027	0.015	0.000044	0.00016	0.00061
1945	34	62	100	0.084	0.36	1.8	0.006	0.020	0.070
1946	9.2	16	40	0.018	0.10	0.58	0.0015	0.005	0.024
1947	11	19	38	0.028	0.11	0.57	0.0017	0.006	0.026
1948	4.8	16	57	0.015	0.091	0.82	0.00085	0.0048	0.036
1949	5.5	12	27	0.014	0.071	0.43	0.00085	0.0039	0.019
1950	2.2	5.3	13	0.006	0.030	0.15	0.00033	0.0017	0.006
1951	0.75	2.6	5.8	0.0024	0.015	0.066	0.00013	0.00088	0.0032
1952	10	17	25	0.023	0.092	0.37	0.0014	0.0053	0.018
1953	17	40	98	0.051	0.23	1.2	0.0026	0.013	0.057
1954[c]	22	52	99	0.059	0.28	1.3	0.0036	0.016	0.063
1955	12	28	71	0.034	0.18	0.71	0.0020	0.0094	0.038
1956	14	26	43	0.030	0.14	0.56	0.0022	0.0081	0.027
Total[d]	320	500	780	0.70	2.8	14	0.046	0.15	0.54
Total, both sources[d]	6,300	16,000	36,000	940	3,600	17,000	3.8	6.7	14

[a] Values are given for each physicochemical form of iodine and are stated as 95% subjective confidence intervals and central values (50th percentile values).

[b] Via the X-10 stack.

[c] Totals do not include releases from the run 56 accident, but the grand totals do.

[d] Iodine-131 Released, all sources and all forms: 95% confidence interval is 8800–42,000 Ci, with central value of 21,000 Ci.

[e] Directly from the processing building.

estimated using a mathematical model (SORAMI) that accounts for the processes that are important during transport of ^{131}I in the atmosphere. The model was benchmarked using another public-domain model (EPA, 1995a). The model was validated using site-specific release and monitoring data (Bradshaw and Cottrell, 1954; Stanley, 1954), and the validation results indicated that model predictions were within a factor of two of the annual average measurements.

For the analysis of routine releases between 1944 and 1956, observation data in electronic form collected at X-10 from 1987 to 1996 were analyzed statistically to generate surrogate hourly meteorological data. For the April 1954 accident, the dispersion model was supplied with half-hourly meteorological data obtained from records of the specific meteorology prevailing at the time of the accident. Uncertainties in all input parameters to the model were quantified before the concentrations were estimated for each of the 41 locations of interest [NCRP, 1996a; International Atomic Energy Agency (IAEA), 1989]. A plot of upper-bound estimates of time-integrated, ground-level ^{131}I concentrations following the April 1954 accident is shown in Figure 14.11.

Transfer from Air to Vegetation   Consumption of contaminated milk and meat from cattle grazing on pastureland contaminated by the deposition of ^{131}I from air is one of the most important pathways by which ^{131}I enters the human

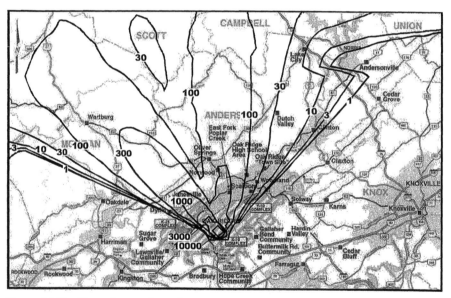

Figure 14.11 Plot of upper bound estimates of 95% subjective confidence interval of time-integrated, ground-level concentrations of total iodine-131 in air following a 1954 accident during RaLa processing ($Bq\,h\,m^{-3}$). Actual concentrations at a given location were likely less than the value for that location represented in this figure. Lower bound estimates do not exceed 30 $Bq\,h\,m^{-3}$ off the reservation.

body (IAEA, 1996a). Once ^{131}I is transported to a given location, it is transferred from the atmosphere to vegetation and the ground surface by precipitation scavenging of the plume and by dry deposition processes [Heinemann and Vogt, 1980; U.S. Nuclear Regulatory Commission; Commission of the European Communities (NRC/CEC), 1994]. The rates of transfer from the atmosphere to vegetation surfaces depend on the chemical form of iodine in air; therefore, the total amount of ^{131}I transferred was estimated by accounting for all three forms of iodine (elemental, particulate, and organic).

For routine releases, annual average concentrations of ^{131}I on vegetation were estimated using a constant rate of deposition of ^{131}I from the air for a given year and the assumption that the annual average concentrations of ^{131}I in vegetation and in air were in equilibrium with each other. For the 1954 accident, deposition of ^{131}I onto vegetation lasted for a period of 2.5 hours, but the ^{131}I remained on the vegetation until the processes of removal from vegetation and natural radioactive decay eliminated it completely. A time-integrated concentration of ^{131}I on the vegetation was estimated to account for the longer-term availability of contaminated feed to cattle (Hoffman et al., 1999).

Transfer from Pasture to Food Products   Once ^{131}I is transferred to the surfaces of vegetation, it is available for ingestion by grazing animals. Once ingested, it is further transferred into milk, meat, cheese, and eggs (NCI, 1997). Estimates of the transfer of ^{131}I from pasture to milk and beef were based on information from the literature (Koranda, 1965; Ng, 1982) and on unpublished measurements of the transfer of ^{131}I into the milk of various breeds of dairy cattle used on farms in East Tennessee during the 1950s and 1960s (Miller, 1973, 1975). No significant difference was observed among the various breeds of cows; however, an inverse relationship was observed between ^{131}I transfer to milk and milk yield for those animals producing less than 10 L d^{-1}. For this reason, a distinction was made between the transfer of ^{131}I into the milk for low-producing "backyard" cows and the higher-producing cows belonging to commercial dairies. Concentrations of ^{131}I were also estimated for goats' milk, human breast milk, cottage cheese, eggs, and beef.

The modeling approach used in this study for estimating concentrations of ^{131}I in milk was validated (Morgan et al., 1963; Morgan and Davis, 1965). Average concentrations of ^{131}I measured in raw milk collected from locations around X-10 in 1962 and 1964 were found to lie well within the 95 percent subjective confidence intervals of average values predicted by the model for eight locations near the sampling sites. Radioiodine releases from ORNL during this period were from reactor operations and radioisotope production.

Concentrations of ^{131}I in animal food products were estimated using mathematical models (Hoffman et al., 1999). The annual average concentration on pasture was related directly to the annual average concentration in animal food products. For the 1954 accident, time-integrated concentrations of ^{131}I in milk and meat were estimated based on the concentrations of ^{131}I in pasture grass.

Distribution of Contaminated Food Products   The distribution of food products from various producers to a potential consumer is a complex process that is difficult to reproduce with high accuracy. Contaminated foodstuffs produced in an affected area may be distributed to other areas not directly exposed by the radioactive plume. Conversely, individuals in the affected area may consume uncontaminated products imported from unaffected areas (Dreicer et al., 1990). In this analysis, food products produced at a given farm were generally assumed to be consumed locally or distributed to the local population for consumption. Milk was often exchanged between counties to cover consumption needs in milk-deficient areas. The effects of the distribution system for a given food type were considered by accounting for radioactive decay between the time of milking or harvest and the time of human consumption of the food product, for the fraction of the food consumed that originated from uncontaminated areas, and for the amount of radioiodine lost during food preparation (IAEA, 1992). Eggs and cottage cheese were assumed to be produced and consumed locally, with the reduction of contamination due to radioactive decay between production and consumption considered.

Intake of ^{131}I via Inhalation and Food Consumption   Human exposure to ^{131}I is dependent on the concentration of ^{131}I in air and in food at a given location and on the rates of inhalation and food consumption. This study included a detailed investigation of inhalation and ingestion rates by gender and age [U.S. Department of Agriculture (USDA), 1965, 1980]. Different rates of inhalation and ingestion were considered for estimating exposures from the routine releases of ^{131}I and from the 1954 accident. Because of the importance of human exposure via the consumption of contaminated milk, this study focused heavily on the estimation of consumption rates of fresh milk for infants and children under the age of 10 (NCI, 1997). These estimates included differences in the consumption of locally produced milk and milk obtained at school (Downen, 1955, 1956). The estimated rates for inhalation of contaminated air for infants, children, teenagers, and adults accounted for the amount of time spent indoors and the differences between the concentration of ^{131}I in air between the indoor and outdoor environments (Snyder et al., 1994).

Consumption of meat, leafy vegetables, eggs, and cheese was also considered in addition to the ingestion of milk. In all, doses and risks were estimated for 11 individual exposure pathways: ingestion of backyard cows' milk, commercial milk, regionally mixed commercial milk, goats' milk, meat, leafy vegetables, eggs, cottage cheese, inhalation of contaminated air, prenatal exposure from ^{131}I ingested by mothers, and ingestion of contaminated mother's milk.

Individuals living near X-10 may have been exposed via more than one exposure pathway at a time. Three special exposure scenarios are designed to match the most likely dietary habits and life-styles in the vicinity of the ORR. The first exposure scenario (called diet 1) refers to individuals living in a rural farm setting. The intake for this exposure scenario is obtained from ingestion of backyard cows' milk, beef, leafy vegetables, eggs, and cheese and from inhalation. The second exposure scenario (diet 2) refers to individuals in a rural area who buy milk from a local dairy farm. They are also exposed to contaminated beef, leafy vegetables, eggs, cottage cheese, and air. The third scenario (diet 3) refers to individuals in a more urban setting, who buy milk and food from the grocery store. The intake for this exposure scenario is obtained from ingestion of regionally averaged commercial milk and from inhalation. Given that the doses and risks from ingestion of goats' milk are substantially larger than the doses and risks from any other exposure pathway, it is addressed separately under the diet 4 scenario.

Estimation of Thyroid Doses from [131]***I Intake***   After it is inhaled or ingested, ^{131}I is absorbed into the bloodstream and then metabolized in a manner that is identical to the absorption and metabolism of stable iodine in the human body. The thyroid gland preferentially absorbs iodine from the extracellular fluid into the thyroid cells and follicles [International Commission on Radiological Protection (ICRP), 1989]. Iodine is then used in the production of hormones essential for human metabolism. In this analysis, the absorbed thyroid dose per unit intake of ^{131}I was estimated as a function of age (Killough and Eckerman, 1986). Since the uncertainty in the dose per unit intake is largely affected by the interindividual variability in the thyroid mass, this investigation employed the most recent information on thyroid volume as determined by ultrasonography. It was found that the mass of the thyroid was somewhat smaller than assumed in past studies (Bier, 1996; Yureiva et al., 1994). However, this finding was offset by the finding of a more rapid biological clearance rate from the thyroid, which resulted in a central estimate of the dose per unit intake that was similar to values recommended by the International Commission on Radiological Protection for children from birth up to age 15.

14.3.5 Risk Characterization

Females born in 1952 who consumed milk from goats (diet 4) that grazed areas adjacent to the ORR received the highest doses and have the highest risks of contacting thyroid cancer during their lifetime. The next highest dose resulted from the consumption of milk from a backyard cow, followed by milk from a local commercial dairy and milk that was regionally mixed. Since the concentration of ^{131}I in regionally mixed retail milk is about the same regardless of location within the 38-km domain, its importance with respect to the consumption of local produce or to inhalation varies from location to location. A sample contour plot of upper bound estimates of excess lifetime risk of developing thyroid cancer for a female born in 1952 on a diet including local produce and backyard cows' milk is shown in Figure 14.12. Other contour plots are presented in the report of the iodine dose reconstruction (Hoffman et al., 1999).

Lower doses are obtained from inhalation or from the consumption of locally produced beef, cottage cheese, mother's milk (with the mother assumed to

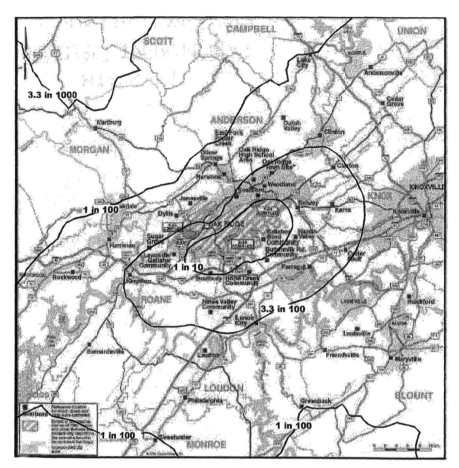

Figure 14.12 Plot of upper bound estimates of 95% subjective confidence interval of excess lifetime risks of developing thyroid cancer for a female born in 1952 on a diet including local produce and backyard cows' milk (diet 1). It is highly unlikely that the actual risk at a given location exceeds the value shown for that location on this figure. Lower bound values are about a factor of 200 lower than the values shown here. Backyard cows were not present in all areas shown on this map.

be on diet 1), or leafy vegetables. The doses from inhalation or from the consumption of one of these food types for a child under the age of 5 at the time of exposure are several hundred to more than 1000 times less important than the dose from the consumption of backyard cows' milk. The thyroid dose from prenatal exposure during the first part of 1952 (assuming the mother to be on diet 1) is about equal to the 5-year total thyroid dose obtained from the consumption of beef or cottage cheese. Risks were estimated specifically for the four diets, each of which consists of a combination of pathways.

Among those 41 reference locations addressed in the dose reconstruction for ^{131}I, the highest doses occurred at Gallaher Bend, located a little more than 6

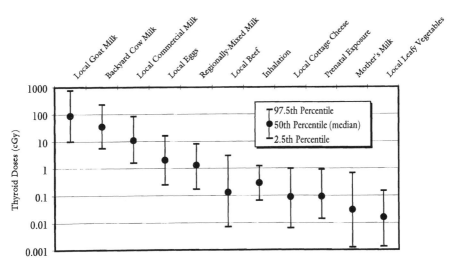

Figure 14.13 Thyroid doses by various exposure pathways estimated in Oak Ridge Dose Reconstruction for females who were born in 1952 and lived near Gallaher Bend. Gallaher Bend is located 6.3 km southeast of the center of the X-10 site, from where iodine-131 was released from 1944 through 1956.

km to the east of X-10, while the lowest doses occurred at Wartburg, located 27 km northwest of X-10. For example, at Gallaher Bend, thyroid doses ranged from about 4 to 250 cGy for individuals of each gender born between 1940 and 1952 who consumed milk from a backyard cow (between one and five 8-ounce glasses each day) and who consumed food products from a local garden or farm. Thyroid doses to a female born in 1952 near Gallaher Bend via the various exposure routes are depicted in Figure 14.13. A similar group of individuals residing in Wartburg would have received doses ranging from about 0.08 to 6 cGy. Doses from the consumption of regionally mixed commercial retail milk ranged from about 0.3 to 10 cGy for individuals born from 1940 to 1952 and did not vary by location.

Detailed summaries of calculated doses and risks at each of the 41 selected locations within 38 km of the X-10 facility are presented in Appendix 11C to the report of the iodine dose reconstruction (Hoffman et al., 1999). Doses were calculated for males and females born in 1920, 1930, 1935, 1940, 1944, 1950, 1952, 1954, and 1956 who lived near X-10 during the period of radioactive lanthanum processing (1944 to 1956). These years of birth were chosen to allow comparison of doses and risks from ^{131}I intake by individuals of different ages at time of exposure. It was assumed that all members of the reference populations received no ^{131}I exposure other than that from the X-10 RaLa releases that occurred from 1944 to 1956. This being the case, it is not important where a person was born or where they lived before 1944 or after 1956. It is very important, however, where in the Oak Ridge area the person lived from 1944 through 1956. Assuming the same dietary sources for ingestion of ^{131}I at a

TABLE 14.2   Excess Lifetime Risks of Incurring Thyroid Cancer for Females Born in Gallaher Bend Area near Oak Ridge, TN, in 1952, Lived There through 1956, and Ate Various Diets That Included Locally Produced Foods

| | 95% Subjective Confidence Interval | | |
Diet Assumptions	Lower Limit	Central Estimate	Upper Limit
Diet 1 = backyard cow milk plus all other local, nonmilk exposure pathways	7.1×10^{-4}	1.1×10^{-2}	1.7×10^{-1}
Diet 2 = local commercial milk plus all other local, nonmilk pathways	3.2×10^{-4}	4.3×10^{-3}	5.9×10^{-2}
Diet 3 = regionally mixed commercial milk plus inhalation	4.4×10^{-5}	4.7×10^{-4}	5.1×10^{-3}
Diet 4 = local goat milk (all other pathways are negligible in comparison)	1.8×10^{-3}	3.1×10^{-2}	4.3×10^{-1}

Note: Values are stated as 95% subjective confidence intervals and central estimates (50th percentile values).

specific location, differences in gender account for only minor differences in the estimation of the thyroid doses. More significant differences are determined by the year of birth, with the lowest doses being for individuals born in 1920, 1930, and 1956. These doses are about one-fourth to one-fifth of the largest doses received by individuals born between the years of 1944 and 1952. Individuals born in 1954 have about the same doses as those born in 1940, which are about 65 percent of the doses for those born between 1944 and 1952.

The highest excess risk of developing thyroid cancer for a female born in 1952 on diet 1 occur at the agricultural communities of Bradbury and Gallaher Bend (Table 14.2). For these locations, based on the assumed quantities of backyard cows' milk, beef, vegetables, eggs, and cheese consumed each day, the risk estimates are confidently above 1 chance in 1000 (1×10^{-3}) but less than 1 chance in 10 (1×10^{-1}). In addition, at these locations, the central estimate of the probability of causation approaches or exceeds 50 percent for females born in 1952 on diets 1, 2, and 4, meaning that a diagnosed thyroid cancer has more than an even chance of being due to exposure to ^{131}I released from X-10. The central estimate of risk for a female born in 1952 on diet 1 is likely to exceed 1 chance in 1000 (1×10^{-3}) up to distances of 35 km to the southwest and more than 38 km to the northeast of X-10. A risk of more than 1 chance in 10,000 (1×10^{-4}) is likely with a subjective confidence of over 50 percent at all locations in the 38-km vicinity of X-10.

Depending on the year of birth, the excess lifetime risk to females is 3 to 4 times larger than the risk to males. At a location such as Bradbury or Gallaher Bend, the lowest risk is for a male born in 1920, who has an excess lifetime risk of thyroid cancer almost 1000 times less than the highest risk for females born in 1952. A female born in 1920 has a risk about 350 times lower than that for a female born in 1952. Individuals of the same gender born in 1944 have about 50 to 60 percent of the risk than those born in 1952 have, while individuals born in 1940 or 1956 have risks about 5 times lower than those for individuals born in 1952. However, a male born in 1940 or 1956 has a risk almost 20 times less than that of a female born in 1952.

The primary locations affected by the April 1954 accident are those to the north and northwest of X-10, such as Jonesville, Norwood, the East Fork Poplar Creek (EFPC) area in Oak Ridge, Oliver Springs, and Wartburg. The excess lifetime risk of thyroid cancer to females on diet 1, exposed in their early childhood at either Jonesville, Norwood, EFPC, or Oliver Springs ranged from 3 or 4 chances in 10 million to nearly 1 chance in 1000. In general, total doses and risks from the 1954 accident are much lower than those from routine emissions.

Within 38 km of X-10, the 95 percent subjective confidence interval of the number of excess thyroid cancers from consumption of cows' milk (commercial and backyard, combined), contaminated by ^{131}I from RaLa processing between 1944 and 1956, range from 6 to 84, with a central estimate of 21. For the consumption of backyard cows' milk alone, the confidence interval ranged from 1 to 33 excess cases, with a central estimate of 7. Commercial cows' milk contributed more to the excess cancers because more individuals were exposed via its consumption (Hoffman et al., 1999).

The 95 percent confidence interval of the number of excess thyroid cancers from drinking of cow's milk contaminated by X-10 ^{131}I releases (commercial and backyard milk, combined) range from 14 to 103 (central estimate of 35) within 100 km of X-10 and from 25 to 149 (central estimate of 58) within 200 km. These cancers are expected to manifest between 1950 and 2020, with most occurring after 1970. These calculations of possible numbers of excess thyroid cancers are based on the linear, no threshold hypothesis. Under that hypothesis, added cancer risk is assumed to be proportional to the collective radiation dose received. This holds whether one has large doses received by a few people or low doses received by a much larger population. Available data cannot exclude the possibility that there is a low-level threshold for causation of cancer. If this is the case, then the values presented above are likely overestimates of true incidence rates.

These estimates were made using a baseline of thyroid cancer diagnoses within the region, with the assumption that the individuals in this population were unexposed. Only about 28 percent of the total thyroid cancers in a population are diagnosed and reported. Therefore, it is likely that the total number of diagnosed and occult cases of thyroid cancer is about 3 to 4 times greater than the total estimates given in this study, which are based on the incidence of thyroid cancer reported only through clinical diagnosis. The clinical significance of an excess incidence of benign nodules is not evaluated in this study, but it is noted that about 9 percent of benign nodules diagnosed through palpation and about 28 percent of those diagnosed through ultrasound will be surgically removed.

14.3.6 Discussion

Most of the uncertainty in the estimates of risk is associated with uncertainty in the estimates of dose (55 percent), followed by uncertainty in the dose–response for groups exposed to external radiation. Uncertainty in the dose estimates is dominated by uncertainty in the concentrations of ^{131}I in milk (45 percent), followed by the uncertainty in the internal dose conversion factor (40 percent). The uncertainty in milk concentrations is dominated by the uncertainty in the transfer of ^{131}I from air to pasture (71 percent). The uncertainty in the internal dose conversion factor is dominated by the uncertainty in the mass of the thyroid in individuals of a given gender and age.

Fallout from atmospheric testing of nuclear weapons at the NTS during 1952, 1953, 1955, and 1957 was a significant contributor to the total ^{131}I exposure for people within 38 km of X-10. Beyond 38 km, ^{131}I from NTS fallout was the dominant source of exposure. For a female born in 1952 who drank backyard cows' milk, central estimates of thyroid dose from exposures from both X-10 releases and NTS fallout within 15 km of X-10 ranged from 25 to 30 cGy, with the 95 percent upper confidence limit exceeding 200 cGy. At all locations within 15 km of X-10, risks to females born in 1952 on rural diet 1 exceeds 1 chance in 1000. The 95 percent upper confidence limit of the excess lifetime risk of thyroid cancer is 6 to 9 chances in 100 up to over 1 chance in 10 at Bradbury and Gallaher Bend. The doses from the combined exposure to ^{131}I from X-10 and NTS fallout are high enough to have possibly manifested excess cases of non-neoplastic disease, namely autoimmune thyroiditis. Available information is not sufficient to support quantitative estimation of the rates of such health effects.

In January 1999, the CDC released a draft final report from the Hanford Thyroid Disease Study, which was conducted by the Fred Hutchinson Cancer Research Center. This 9-year study evaluated whether the occurrence of thyroid disease was associated with radiation dose to the thyroid in a group of 3441 people who were exposed to ^{131}I as children due to releases from the Hanford Nuclear Reservation in the 1940s and 1950s. The draft study results, which are still under review, show that thyroid disease was observed among study participants, but the study was not able to show a relationship between estimated doses to the thyroid from ^{131}I and the level of thyroid disease in that population.

Other recent studies involving people exposed to ^{131}I in the environment have yielded different results. A study of almost 2500 residents of Utah, Nevada, and Arizona who were exposed to NTS fallout as children reported an association between ^{131}I dose to the thyroid gland and the occurrence of thyroid cancer or nodules (Kerber et al., 1993). The recent Institute of Medicine/National Research Council report "Exposure of the American People to Iodine-131 from Nevada Nuclear-Bomb Tests" (IOM/NAS, 1999) states that "there is now strong evidence from Chernobyl that children exposed to I-131 develop thyroid cancer at higher than usual rates." It is important to note that the epidemiologic studies discussed here neither prove nor disprove that a relationship exists between ^{131}I exposure to the thyroid gland and the occurrence of thyroid disease.

14.4 MERCURY RELEASES FROM Y-12 LITHIUM ENRICHMENT

14.4.1 Introduction

Between 1950 and 1963, while lithium was being enriched in its lithium-6 component for use in thermonuclear weapons, many tons of mercury were released to the air and surface waters from the Y-12 plant on the ORR. In December 1981, two brothers, one an employee at ORNL and the other a U.S. Geological Survey (USGS) employee, collected vegetation at Y-12 near East Fork Poplar Creek. They were seeking data to justify a research project. The ORNL employee had become aware of elevated mercury levels in EFPC from a 1978 environmental study by ORNL. The vegetation samples were confiscated by ORNL in April of 1982 and the employee was reprimanded. In discussions between U.S. DOE Oak Ridge Operations and the TDH in 1982, the existence of classified reports describing mercury losses from Y-12 was mentioned. These classified reports were then cited in a newspaper interview by an employee of the State of Tennessee. A local newspaper, the *Appalachian Observer* filed a Freedom of Information Act request in November 1982 for all reports on mercury spills and releases from the ORR. A report (Case, 1977) was released as part of this request and it generated much public and media interest. In May 1983, three days after the release of the 1977 report, a task force was appointed by the Y-12 manager to investigate historical data on mercury operations and releases. Six weeks later, a Congressional subcommittee hearing chaired by Albert Gore, Jr., and Marilyn Lloyd was held in Oak Ridge (LaGrone, 1983).

Preliminary investigations in the Oak Ridge Dose Reconstruction Feasibility Study (ChemRisk, 1993a–e) indicated that mercury releases from operations at the Y-12 plant likely resulted in the highest potential noncancer health risks of any material used in historical activities on the Oak Ridge Reservation. Because of that finding, task 2 of the Oak Ridge Dose Reconstruction was initiated by TDH and ORHASP to bring about a detailed, independent investigation of potential off-site doses and health risks from historical releases of mercury from the Y-12 plant. The objectives of this detailed investigation were to:

- Describe (and independently quantify) past releases of mercury from the reservation.
- Characterize historical environmental concentrations of mercury from those releases.
- Define potential pathways of human exposure to mercury.
- Describe potentially exposed populations.

Figure 14.14   The Y-12 plant received about 24 million pounds of mercury in flasks like these for use in enrichment of lithium in its lithium-6 component from 1950 to 1962 for use in nuclear weapons. About 2 million pounds of this mercury was later reported lost or unaccounted for. U.S. DOE photo.

- Estimate historical human exposures.
- Estimate human health hazards to put the dose estimates in perspective.

The Oak Ridge processes that used the most mercury were the lithium enrichment operations conducted at the Y-12 plant in the 1950s and 1960s. Lithium enrichment operations included three production facilities, requiring over 24 million pounds of mercury (see Figure 14.14). Facilities at Y-12 involved in the Colex (column-based exchange) process for lithium enrichment (see Figure 14.15) released the most significant quantities of mercury [Union Carbide Nuclear Company (UCNC), 1957]. Other processes that used much lower quantities of mercury included facilities built to test or demonstrate other processes for lithium enrichment, production of some nuclear weapon components, processes for chemical recovery or decontamination of nuclear materials, burning of coal in steam plants, and instrumentation.

Our review of lithium enrichment operations and mercury releases included close examination of records assembled by members of a 1983 task force appointed to investigate and quantify mercury releases from Y-12 (UCCND,

Figure 14.15   Workers at the Y-12 plant mercury dumping shed in the 1950s. Pipelines carried the mercury from this open-air facility to the buildings that housed the Colex (column exchange) operations for enrichment of lithium. U.S. DOE photo.

1983b). The investigation included interviews with members of the 1983 task force, review of thousands of task force files and documents archived in the Y-12 records center, and review of classified and unclassified versions of the task force's report. However, our investigation differed from the 1983 Mercury Task Force study in that:

- We identified the references that support the 1983 Mercury Task Force's release estimates and had them made available to the public.
- Our project included a more thorough records review, including an extensive search of boxes of inactive records.
- We took additional steps to verify the data used to estimate historical mercury releases, through review of additional historical drawings and documents.
- Mercury release estimates were revised where more complete information was assembled.
- We estimated that about 62,000 pounds more mercury was released to the air and water than estimated historically.

14.4.2 Hazard Identification

Mercury can exist in the environment in several different forms or species. Mercury used in lithium enrichment operations at Y-12 was elemental mercury, the relatively volatile form of mercury commonly found in thermometers. In the environment, elemental or inorganic mercury can be converted to several different forms. The three primary forms of mercury found in the environment are elemental mercury (the dominant form in air), inorganic mercury (found in soil, water, and food), and organic mercury [commonly found in fish as methylmercury (Bloom, 1992)]. Each of these forms behaves differently in the environment and has been associated with different health effects in people and animals exposed to high concentrations (Schoof and Nielsen, 1997; Paustenbach et al., 1997; Davis et al., 1997). We estimated doses and potential health effects for each form of mercury separately.

Exposure to high doses of inorganic mercury compounds, such as following administration of high doses to laboratory animals or one-time accidental or intentional ingestion of very large doses by humans, has been shown to cause injury to the gastrointestinal (GI) tract and/or to be directly toxic to tubular lining cells in the kidney (Goyer, 1996; Young, 1991; Gerstner and Huff, 1977). The kidneys generally exhibit the highest levels of mercury following exposure to large doses of inorganic mercury compounds (Young, 1991). Toxic effects following low-dose environmental exposures to inorganic mercury compounds have not been documented. The LOAELs and NOAELs used to derive the EPA and the Agency for Toxic Substances and Disease Registry (ATSDR) toxicity criteria for inorganic mercury are all based on observations of adverse effects on the kidney, considered to be the critical effect for inorganic mercury exposure. There are no data on developmental or reproductive toxicity of inorganic mercury in humans following oral exposure, and only very limited data in animals. Studies of reproductive or developmental effects of inorganic mercury in animals suggest NOAELs for these effects are higher than NOAELs for kidney effects. For example, a study of developmental toxicity in hamsters following administration of mercuric chloride identified a NOAEL of 15.7 mg Hg/ kg body weight (Young, 1991; ATSDR, 1997).

Inhaled elemental mercury vapor enters the bloodstream through the thin membranes of the lungs. In its elemental form, mercury readily penetrates the blood-brain barrier and accumulates in the central nervous system (CNS) (Gerstner and Huff, 1977). Acute exposure to very high concentrations of airborne elemental mercury has been associated with injury to the lungs, with symptoms resembling pneumonia, while the critical effect associated with chronic exposures to moderate air concentrations, such as may occur in the workplace, is considered to be CNS effects. Symptoms of elemental mercury toxicity following continued exposure to moderately high concentrations in air range from almost imperceptible disturbance of the CNS, including fine muscle tremors, insomnia, loss of appetite, and effects on the emotional state and memory, to incapacitation (Gerstner and Huff, 1977; ATSDR, 1997). Limited data on developmental or reproductive effects of inhaled elemental mercury in pregnant women or in laboratory animals suggest that reproductive or developmental effects following exposure to high concentrations of elemental mercury vapor may be a concern. For example, Mishinova et al. (1980) reported that the rates of pregnancy and labor complication (reproductive effects) were higher among women exposed to elemental mercury vapors in the workplace than unexposed workers, though dose–response relationships were not established. All of the exposure levels associated with developmental effects were at air concentrations higher than those associated with observations of CNS effects in adult workers $(0.023$ to 0.033 mg m^{-3}). Therefore, toxicity benchmarks based on these lower concentrations are likely to be protective of developmental or reproductive effects.

Studies have shown that methylmercury ingested through fish consumption is readily absorbed through the GI tract (~ 95 percent) and that about 5 percent of absorbed methylmercury partitions to the blood (ATSDR, 1997). Methylmercury can cross the blood-brain barrier and the placenta; thus, women can pass methylmercury to the fetus during pregnancy. The critical effect associated with exposures to methylmercury is assumed to be neurological effects on the developing fetus. Developmental toxicity in infants exposed to methylmercury in utero has been manifested in various levels of brain damage (ranging from cerebral palsy, seizures, and mental retardation to learning disabilities), abnormal muscle tone or reflexes, and delayed onset of walking and talking. Neurological effects associated with postnatal exposures have been observed following short-term exposures to high levels of methylmercury in fish. Effects observed following postnatal exposures included paresthesia (tingling sensation in the extremities), tremors, abnormal reflexes, speech difficulties, and impaired peripheral vision.

14.4.3 Dose–Response Evaluation

To put the Oak Ridge mercury dose estimates in perspective and evaluate the likelihood that the estimated levels of historical exposure caused adverse health effects, we collected, evaluated, and summarized available studies [Marsh et al., 1995; Fawer et al., 1993; National Toxicology Program (NTP), 1993] of the toxicity of different species of mercury through various routes of exposure and established toxicity benchmark values for comparison with the estimated doses. The primary toxicity benchmark values used in this chapter, as summarized in Table 14.3, are EPA RfDs and LOAELs or NOAELs [EPA, 1988; Integrated Risk Information System (IRIS), 1998b; ATSDR, 1997; NTP, 1993]. In the case of inorganic mercury, the associated data were obtained from studies with rats.

14.4.4 Exposure Assessment

Airborne releases of mercury were likely carried by wind within the valley that contains the Y-12 plant or transported across Pine Ridge to nearby residential

TABLE 14.3 Toxicity Benchmarks for Comparison with Results of Oak Ridge Mercury Dose Reconstruction

Exposure Route and Species	No Observed Adverse Effect Level $(mg\,kg^{-1}\,d^{-1})$	EPA Reference Dose[a] $(mg\,kg^{-1}\,d^{-1})$	ATSDR Minimal Risk Level[b] $(mg\,kg^{-1}\,d^{-1})$
Ingestion of inorganic mercury	0.1–0.23 (animal studies)[c]	0.0003	0.002[d]
Inhalation of elemental mercury	0.0029–0.0071 (human studies)[e]	0.000086[f]	0.000057[g]
Ingestion of methylmercury, in utero and child exposure	0.0005 (human studies)[h]	0.0001	0.0005[i]
Ingestion of methylmercury, adult exposure	NA	0.0003[j]	NA

Note: NA = not available.

[a] From IRIS, 1998b.

[b] From ATSDR, 1997.

[c] Minimal data are available. Both data points given are from studies in laboratory animals. The lower limit of the range is based on the intermediate duration LOAEL used as the basis for the EPA RfD divided by an adjustment factor of 3 to extrapolate from a LOAEL to a NOAEL. The upper limit is based on the NOAEL from the NTP intermediate duration study.

[d] For intermediate duration exposures.

[e] Both values given are from studies in humans. The lower limit of the range is based on a NOAEL of 0.010 $mg\,m^{-3}$ (calculated by dividing an average LOAEL of 0.030 $mg\,m^{-3}$ by a LOAEL-to-NOAEL adjustment factor of 3). The upper limit is based on a NOAEL of 0.025 $mg\,m^{-3}$. Milligram-per-kilogram-per-day doses are calculated by multiplying the NOAELs by a breathing rate of 20 $m^3\,d^{-1}$ and dividing by a body weight of 70 kg.

[f] Derived by multiplying EPA's reference concentration (3×10^{-4} $mg\,m^{-3}$) by a breathing rate of 20 $m^3\,d^{-1}$ and dividing by a body weight of 70 kg.

[g] Derived by multiplying ATSDR's minimal risk level for chronic exposure (2×10^{-4} $mg\,m^{-3}$) by a breathing rate of 20 $m^3\,d^{-1}$ and dividing by a body weight of 70 kg.

[h] The data point given is based on studies in humans. Based on ATSDR's estimated 29-g-d^{-1} dose necessary to achieve 5.9 ppm in maternal hair, which was the NOAEL associated with the Seychelles Child Development Study. The 29-g-d^{-1} dose was converted to 0.029 $mg\,d^{-1}$ and then divided by a body weight of 60 kg to yield a NOAEL of 0.0005 mg $kg^{-1}\,d^{-1}$.

[i] For chronic exposure.

[j] From EPA, 1985.

areas. As shown in Figure 14.16, waterborne releases of mercury flowed in EFPC through residential and commercial sections of Oak Ridge.

Releases of Mercury to the Air   Mercury was released from Y-12 to air largely as a result of building ventilation systems installed to lower the concentration of mercury vapor inhaled by workers in the lithium enrichment facilities (see

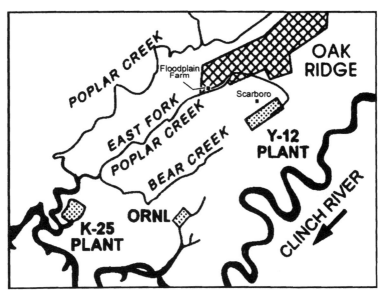

Figure 14.16 Locations and waterways important in the reconstruction of public exposures to releases of mercury from Y-12 plant lithium enrichment operations. Ten years of mercury releases to the air and 31 years of releases to East Fork Poplar Creek were studied in the Oak Ridge Dose Reconstruction.

Figure 14.17). However, airborne mercury in Y-12 exhaust was not routinely monitored. To quantify mercury releases to the air, we located thousands of measurements of mercury in indoor air and used historical engineering drawings to estimate building ventilation rates. The team estimated that about 73,000 more pounds of mercury was released to air from lithium enrichment buildings, a mercury recovery facility, and Y-12 steam plants—about 22,000 pounds more than estimated by the Mercury Task Force (Figure 14.18) (Mongan et al., 1999). This increase was largely due to corrections to calculated average concentrations and estimated building ventilation rates, and inclusion of additional mercury sources.

Releases of Mercury to Surface Waters   Mercury was also released from Y-12 to EFPC. The largest releases to water were the result of an early process in which mercury was washed with nitric acid to remove impurities prior to use for lithium enrichment—the washing process increased the solubility of the mercury and led to an increase in off-site releases as $HgNO_3$. The waters of EFPC near Y-12 have been routinely monitored for mercury since 1953. To quantify waterborne mercury releases, we cross-checked measurements for 1953 through 1993 from several sources and made corrections where mathematical errors had been made. Team members also collected EFPC flow rate measurements

Figure 14.17   The 6-foot-diameter fans shown here were installed in a lithium enrichment building (Colex building 9201-5) in 1956 to increase ventilation and reduce worker exposures. These fans, which could be heard across town, also increased releases of airborne (elemental) mercury to the environment. Airborne releases were estimated in the Oak Ridge Dose Reconstruction based on measured air concentrations and documented building ventilation rates. Y-12 plant photo.

from numerous sources and assembled a more complete data set than the 1983 Mercury Task Force used. We estimated that about 280,000 pounds of mercury was released to EFPC from 1950 to 1993—about 40,000 more than officially reported (Figure 14.18). This increase was largely due to use of more complete mercury concentration and flow rate records. The 280,000 pounds of mercury corresponds to a volume of about 12 cubic yards, as a cubic yard weighs about 22,800 pounds.

Assessment of Off-Site Mercury Exposures   In estimating off-site doses from historical mercury releases, we selected a number of different geographic locations and groups of potentially exposed people northeast of Y-12, along EFPC within the community of Oak Ridge, and farther downstream (DaMassa, 1995). Within these population groups, exposures were characterized for adults and children. In addition, exposures to methylmercury in fish were characterized for in utero exposure because toxicity studies have shown that unborn children may be particularly susceptible to adverse health effects when their mothers consumed contaminated fish during pregnancy.

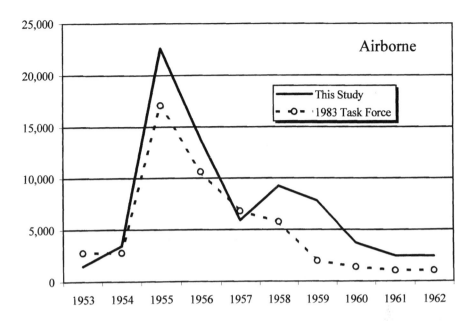

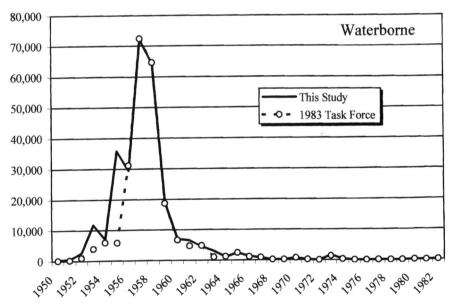

Figure 14.18 Annual airborne (top) and waterborne (bottom) mercury release estimates by Oak Ridge Dose Reconstruction project team and those officially reported by 1983 Mercury Task Force (pounds).

We estimated historical mercury doses for the following populations:

- Wolf Valley residents who lived along the extension of Union Valley on the opposite side of the Clinch River, northeast of the Y-12 plant. Meteorological studies indicate that this is a predominant direction of airflow from Y-12. These individuals could have inhaled airborne mercury, consumed homegrown fruits and vegetables contaminated by airborne mercury, and consumed milk and meat from backyard cattle that grazed on pasture contaminated by airborne mercury. Exposures to this group were evaluated for 1953 to 1962, since this was the period of significant air releases from Y-12.

- Oak Ridge community residents who lived near the EFPC floodplain [Tennessee Valley Authority (TVA), 1959] and may have inhaled mercury volatilized from EFPC and consumed homegrown fruits and vegetables contaminated by mercury that volatilized from EFPC. This population was assumed to have lived within one-half mile of EFPC in the western end of the city of Oak Ridge. Exposures to this group were evaluated for 1950 to 1990.

- Scarboro community residents who lived in Scarboro, historically the closest residential area to the Y-12 facility. These individuals may have inhaled mercury from the Y-12 plant and EFPC, traveled a short distance to fish or play in EFPC, and/or consumed homegrown fruits and vegetables contaminated by airborne mercury. Exposures to this group were evaluated for 1950 to 1990.

- Robertsville School children who attended a junior high school adjacent to the EFPC floodplain. These individuals could have inhaled mercury volatilized from EFPC and came in contact with mercury in floodplain soil. Some children in this age group were also assumed to recreate in EFPC for longer periods and come in contact with EFPC water and sediment. Exposures to this group were evaluated for 1950 to 1990.

- The EFPC floodplain farm family who resided adjacent to the floodplain and consumed homegrown fruits and vegetables contaminated by airborne and soil mercury, consumed milk and meat from cattle that grazed in the floodplain and drank out of the creek, and fished and played in the creek for recreation. Exposures to this group were evaluated for 1950 to 1990. This is not a hypothetical reference population; we interviewed a number of individuals who lived on the floodplain and fit the stated characteristics of EFPC farm family residents.

Inhalation exposures were assumed to be to airborne elemental mercury vapor. Exposures through the remaining pathways, with the exclusion of fish consumption, were assumed to be to inorganic mercury.

Because measurements since 1970 have shown that fish collected downstream of EFPC in Poplar Creek, the Clinch River, and Watts Bar Reservoir contain elevated levels of mercury (Sanders, 1970), we estimated doses to individuals who consumed fish from these waterways. Exposures to mercury in fish, assumed to be methylmercury, were evaluated for three categories of fish consumers based on the number of fish meals consumed per year from Poplar Creek/Clinch River and Watts Bar Reservoir:

- Category 1 Fish Consumers: Assumed to frequently consume fish from these systems (between 52 and 130 meals per year)
- Category 2 Fish Consumers: Assumed to regularly consume fish from these systems (between 17 and 52 meals per year)
- Category 3 Fish Consumers: Assumed to occasionally consume fish from these systems (between 2 and 17 meals per year)

For each population, we estimated annual average daily doses of mercury (stated in terms of daily amounts of mercury taken in per kilogram of body weight) through all applicable exposure pathways. Doses were estimated using equations that take into account the amount of air, water, soil/sediment, food, or fish that was likely inhaled, ingested, or touched and the estimated concentrations of mercury in each medium during a given year. We estimated exposure point concentrations based on historical and current measurements of mercury in different environmental media and/or historical release data and modeling of releases to off-site locations.

We estimated EFPC water concentrations at downstream locations based on concentration and flow rate measurements made near Y-12 between 1953 and 1990 (releases from 1950 through 1952 were estimated rather than measured) and application of factors to account for downstream reduction in concentrations due to mercury loss through volatilization or adherence to sediment and dilution of concentrations by additional inflow to the creek.

We estimated air concentrations at the Wolf Valley location using estimates of annual Y-12 releases to air during 1953 to 1962 and modeling of dispersion off-site. Air concentrations at the Scarboro community due to direct airborne releases from Y-12 were estimated based on the relationship between concentrations of uranium measured in air at Scarboro between 1986 and 1995 and estimated releases of uranium from Y-12 (uranium was also historically released from Y-12 and was evaluated in a screening evaluation undertaken as part of this project; see Appendix A). These concentrations were one of two components of airborne mercury levels at Scarboro, with volatilization of mercury from EFPC contributing the other.

Air concentrations due to volatilization (Lindberg et al., 1991, 1995; Xiao et al., 1991) of mercury from EFPC were estimated based on assumptions about the fraction of the total mercury released from Y-12 that volatilized from EFPC between Y-12 and the junction of EFPC with Poplar Creek. Recent measurements of mercury in rings of red cedars growing in the EFPC floodplain suggest that air concentrations were significantly elevated in the past (Turner and Bloom, 1995). Based on available information (Mongan et al., 1999), we assumed that on average 5 percent (range 1 to 30 percent) of the mercury discharged to EFPC escaped to air between Y-12 and the junction. Air dispersion modeling was performed for releases from each of 403 segments of the creek.

We estimated soil and sediment concentrations in the EFPC floodplain based on sampling conducted during the EFPC floodplain remedial investigation in 1991 and 1992. Historical concentrations were estimated based on statistical analysis of samples collected from areas likely to have been visited by the populations of interest. We estimated soil concentrations at the Scarboro community based on limited soil sampling conducted in Scarboro by Oak Ridge Associated Universities in 1984—these were the only soil samples collected in this area and analyzed for mercury prior to 1998.

Mercury in air and soil can be taken up by fruits or vegetables (Beauford and Barringer, 1977; Stein et al., 1996), which can then be consumed by humans. Mercury in plants, which was taken to have been oxidized from elemental to inorganic mercury, can also transfer into milk and meat when cattle eat the plants. We derived factors to describe these transfers in the environment. We estimated incorporation of airborne mercury into above-ground vegetation, including fruits and vegetables and pasture grass, based on measurements of airborne mercury deposition to vegetation near Oak Ridge in the late 1980s. Transfer of mercury from soil to below-ground vegetables and from soil to pasture grass was estimated based on measurements of mercury in co-located soil and plant samples from Oak Ridge in the mid-1980s and in 1993. Transfer of mercury to milk and meat after consumption of pasture by cattle was estimated based on studies reported in the scientific literature (Ng et al., 1979; Potter et al., 1972; Mullen et al., 1975).

The numbers of fish in EFPC during the years of peak mercury releases were likely low due to poor water quality. However, anecdotal reports suggest that a few people caught and ate fish from EFPC during the 1950s and 1960s. We estimated annual average concentrations of mercury in EFPC fish of the size and type that may have been caught for consumption based on: (1) concentrations measured in fish collected near Oak Ridge after 1970, (2) concentrations measured in fish at other sites with similarly high concentrations in water and/or sediment (EPA, 1997a), (3) information about the maximum possible concentrations in fish of the size likely to have been in EFPC, and (4) evidence of levels of mercury that may be lethal to fish. Historical annual average concentrations in fish from the Clinch River/Poplar Creek and Watts Bar Reservoir that may have been caught for consumption were estimated based on data from sediment cores collected in these systems in the mid-1980s and use of equations describing the relationship between mercury concentrations in sediment and in fish (Wren, 1996). These relationships were established using sediment and fish data from EFPC, Poplar Creek, the Clinch River, and Watts Bar Reservoir.

Illustrative Example: Estimating Transport of Airborne Mercury in Complex Terrain   Without the use of dispersion models, apply an approach for predicting concentrations of airborne mercury in a residential community that is separated from the Y-12 plant by a ridge, with residents as close as 1 km.

Background   In Oak Ridge, Tennessee, there are unique topographical characteristics that make prediction of air concentrations using classical air dispersion models problematic. The major source of historical uranium releases is the Y-12 plant, which is located in Bear Creek Valley. The nearest population is the Scarboro community, which is located 1 km north of Y-12 and is separated from the Y-12 facility by Pine Ridge. Although the predominant wind flows follow the valley, the proximity of the Scarboro community to Y-12 requires the evaluation of potential exposures to historical releases of uranium to air. Based on comparison of total uranium activity measured at Scarboro with remote air-monitoring stations located 20 km from Y-12, it is evident that releases of uranium from Y-12 are transported across Pine Ridge to the Scarboro community.

Pine Ridge represents an elevation change of approximately 200 to 400 feet from Bear Creek Valley. The majority of Y-12 release points and the Scarboro community are both at lower elevations than Pine Ridge. Classical air dispersion model, such as ISCST3 (EPA, 1995a), would overestimate air concentrations at Scarboro since the models do not account for the abrupt change in topography between source and receptor locations.

Solution   An empirical approach based on measured airborne uranium concentrations was developed. Air-monitoring data for uranium were available for the Scarboro community for 1986 to 1995. A relationship between air concentrations of uranium measured at Scarboro, χ (pCi m^{-3}), and the Y-12 uranium release rate estimates, Q (pCi s^{-1}), was derived. This empirical relationship, represented as a relative concentration, χ/Q (s m^{-3}), was used to estimate air concentrations of uranium and mercury in Scarboro for all years of concern, 1944 to 1995:

$$\text{Empirical } \chi/Q(\text{s m}^{-3}) = \frac{\text{Uranium air concentration (pCi m}^{-3})}{\text{Uranium release rate (pCi s}^{-1})}$$

Annual χ/Q values were evaluated for each of the years under consideration (1986 to 1995) for the two uranium isotopes, $^{234/235}$U and ^{238}U. Statistical analyses were performed on the entire 20-value data set to estimate a measure of central tendency. The empirical χ/Q value corresponding to the 95 percent upper confidence limit of the mean was 3.1×10^{-7} s m^{-3}, and the mean value was 2.2×10^{-7} s m^{-3}.

The empirical χ/Q value, stated as a probability density function in the form of a custom distribution incorporating each of the 20 χ/Q values, was applied to estimated annual airborne mercury emissions to estimate airborne mercury concentrations at Scarboro

Air concentration at Scarboro $(\mathrm{mg\,m^{-3}})$

$= $ Annual average release rate $(\mathrm{mg\,s^{-1}}) \times \chi/Q\ (\mathrm{s\,m^{-3}})$

For example, in the year of maximum release of airborne mercury from the Y-12 plant (approximately 22,500 pounds in 1955), using a χ/Q value of 2.2×10^{-7} $\mathrm{s\,m^{-3}}$ that could be sampled from the custom distribution described above, the predicted concentration of mercury in air at the Scarboro community is

$$(320\ \mathrm{mg\,s^{-1}}) \times (2.2 \times 10^{-7}\ \mathrm{s\,m^{-3}}) = 7.0 \times 10^{-5}\ \mathrm{mg\,m^{-3}}$$

Illustrative Example: Methylmercury Dose Calculation for Fish Ingestion   In the actual calculation of methylmercury doses from ingestion of fish, the values for several of the following parameters were randomly selected from distributions of each parameter using Monte Carlo calculation techniques. However, for illustrative purposes, the approximate midpoint of each distribution was used in this example.

A range representing the annual concentration of methylmercury in fish was derived by estimating the 95th percentile confidence intervals about the predicted mean fish concentration associated with "dated" (1946 to 1986) sediment concentrations in cores taken from Watts Bar Reservoir in eastern Tennessee. For each year, a triangular distribution was established using the 5th percentile, mean, and 95th percentile as the minimum, most likely, and maximum fish concentration in the distribution. For 1958, the year with the highest predicted annual average concentration of methylmercury in fish, the minimum value is 0.37 $\mathrm{mg\,kg^{-1}}$, the maximum value is 1.14 $\mathrm{mg\,kg^{-1}}$, and the mean is 0.74 $\mathrm{mg\,kg^{-1}}$.

The oral bioavailability of mercury in food varies from between 60 and 100 percent. A uniform distribution with a minimum of 0.6 and a maximum of 1.0 was used to describe the oral bioavailability of methylmercury in fish. Since the fetus is most sensitive to the toxicity of methylmercury, the dose was calculated using a mean body weight for a woman of childbearing age (62 kg):

$$\mathrm{Dose_{fish\text{-}MeHg}} = \frac{U_f \times \mathrm{FC_{MeHg}} \times B_{\mathrm{oral\text{-}fish}}}{\mathrm{BW}}$$

where $\mathrm{Dose_{fish}} = $ dose $(\mathrm{mg\,kg^{-1}\,d^{-1}})$
$U_f = $ fish consumption rate $(0.030\ \mathrm{kg\,d^{-1}})$
$\mathrm{FC_{MeHg}} = $ concentration of MeHg in fish $(0.74\ \mathrm{mg\,kg^{-1}})$
$B_{\mathrm{oral\text{-}fish}} = $ bioavailability of MeHg in fish $(0.8\ \mathrm{unitless})$
$\mathrm{BW} = $ Body weight $(62\ \mathrm{kg};\ \mathrm{adult\ female})$

and

$$\text{Dose}_{\text{fish-MeHg}} = \frac{0.030 \text{ kg d}^{-1} \times 0.74 \text{ mg kg}^{-1} \times 0.8}{62 \text{ kg}}$$

$$= 2.8 \times 10^{-4} \text{ mg kg}^{-1} \text{ d}^{-1}$$

14.4.5   Risk Characterization

The results of the reconstruction of mercury doses can be characterized as follows (Mongan et al., 1999). For all populations of interest, the highest doses were estimated to have occurred during the mid- to late-1950s. These were the years of highest releases of mercury from Y-12 to air and to EFPC.

Excluding exposures of fish consumers to methylmercury in fish, estimated doses to the EFPC floodplain farm family are the highest of all exposure populations that were evaluated. The estimated total dose to an EFPC floodplain farm family member is dominated by consumption of fruits and vegetables contaminated from airborne mercury and inhalation of airborne mercury that volatilized from EFPC.

Estimated total doses to Wolf Valley ("down valley") residents, resulting from direct air releases of mercury from Y-12, are also dominated by consumption of fruits and vegetables contaminated from airborne mercury. However, the highest doses estimated for this group are about 30 to 40 times lower than the highest doses estimated for the EFPC farm family.

Estimated total doses to Scarboro community residents are dominated by consumption of fruits and vegetables contaminated from airborne mercury, incidental ingestion of waterborne mercury, skin contact with contaminated EFPC water and sediment, and inhalation of airborne mercury due to both direct air releases of mercury from Y-12 and volatilization of mercury from EFPC. The highest estimated inhalation doses (estimated for 1955) are about nine times lower than the highest inhalation doses estimated for the EFPC farm family (estimated for 1957), due largely to the greater distance of Scarboro from EFPC.

Estimated total doses to Robertsville School students are dominated by incidental ingestion of and skin contact with mercury in floodplain soil and in EFPC water.

Estimated doses to community populations 1 and 2, for which exposures from airborne mercury volatilized from EFPC were evaluated, were comprised of inhalation of airborne mercury and consumption of fruits and vegetables contaminated from airborne mercury only.

Estimated methylmercury doses to individuals who consumed fish from Clinch River/Poplar Creek were about fourfold higher than doses estimated for people who consumed the same amount of fish from Watts Bar Reservoir.

For the years that the estimated annual average elemental, total inorganic, or methylmercury doses at the upper bound (97.5th percentile) of the 95 percent subjective confidence interval are less than the corresponding reference dose (RfD), it is not likely that adverse health effects occurred, based on current scientific knowledge. Exceeding the RfD is equivalent to exceeding a hazard index of 1. The following general conclusions can be drawn from our mercury dose reconstruction based on the estimated annual average doses.

Illustrative Example: Dose–Response Relationships for Methylmercury   Calculate two toxicity benchmarks for exposure to methylmercury, the EPA RfD and the ATSDR minimal risk level (MRL).

Background   Studies have shown that methylmercury ingested through fish consumption is readily absorbed through the GI tract ($\sim$ 95 percent) and that about 5 percent of absorbed methylmercury partitions to the blood. Methylmercury can cross the blood-brain barrier and the placenta; therefore, women can pass methylmercury to the fetus during pregnancy. The critical effect associated with exposures to methylmercury is neurological effects on the developing fetus. Developmental toxicity in infants exposed to methylmercury in utero has been manifested in various levels of brain damage ranging from cerebral palsy, seizures and mental retardation to learning disabilities, abnormal reflexes, and delayed onset of walking and talking.

The majority of the dose–response studies for methylmercury establish dose–response relationships based on maternal blood concentrations extrapolated from mercury concentrations measured in hair. Concentrations in blood during pregnancy can be estimated after an exposure has ended because mercury concentrations remain unchanged once incorporated into the hair shaft, and concentrations in newly formed hair are proportional to the simultaneous concentration in the blood. By assuming hair grows 1 cm per month, concentrations in blood in previous months can be calculated.

Both EPA and ATSDR assume a hair:blood concentration ratio of 250:1, an elimination constant of 0.014 d^{-1}, and an average female body weight of 60 kg. EPA assumes a maternal blood volume of 5.4 liters, while ATSDR assumes 4.2 liters. EPA applies an uncertainty factor of 10 and ATSDR assumes an uncertainty factor of 1. EPA assumes a NOAEL of 11 ppm of methylmercury in maternal hair using data from Iraqi exposures in 1971–1972 (Marsh et al., 1987), while ATSDR assumes a NOAEL of 6 ppm of methylmercury in maternal hair using data from the Seychelles Child Development Study (Marsh et al., 1995).

Solution

$$\mathrm{MRL/RfD} = \frac{\left(\dfrac{C_{\mathrm{hair}}}{R_{\mathrm{hair:blood}}}\right) \times b \times V}{A \times f \times \mathrm{BW} \times \mathrm{UF}}$$

where C_{blood} = maternal concentration in blood $(\mathrm{mg\,L}^{-1})$ = $C_{\mathrm{hair}}/R_{\mathrm{hair:blood}}$
 C_{hair} = maternal concentration in hair $(\mathrm{mg\,kg}^{-1})$

$R_{\text{hair:blood}}$ = hair:blood concentration ratio (mg kg^{-1} hair:mg L^{-1} blood)

b = elimination constant (d^{-1})

V = maternal blood volume (L)

A = fraction of mercury in the diet that is absorbed (unitless)

f = fraction of daily intake taken up by blood (unitless)

BW = female body weight (kg)

UF = uncertainty factor (unitless)

and

$$\text{RfD} = \frac{(11/250)(0.014)(5.4)}{(0.95)(0.05)(60)(10)}$$

$$= 1 \times 10^{-4} \ \text{mg kg}^{-1} \, \text{d}^{-1}$$

$$\text{MRL} = \frac{(6/250)(0.014)(4.2)}{(0.95)(0.05)(60)(1)}$$

$$= 5 \times 10^{-4} \ \text{mg kg}^{-1} \, \text{d}^{-1}$$

Inhalation of Airborne (Elemental) Mercury

Comparison to RfDs   The 95 percent UCLs on the estimated inhalation doses of elemental mercury exceeded the RfD at two locations: Scarboro for 1955 and 1957 (child) and the EFPC farm family for 1953 to 1961 (child) and 1955 to 1959 (adult). Central estimates of inhalation dose exceeded the RfD at the farm family location for 1955 and 1957 to 1958.

Comparison to NOAELs   The 95 percent confidence intervals on the estimated annual average elemental mercury doses for all populations and all years were below the NOAEL. The NOAEL was established from studies of workers exposed to airborne mercury vapor for prolonged periods of time—some workers exposed to airborne mercury concentrations above the NOAEL exhibited neurological effects, including hand tremor, increases in memory disturbances, and evidence of dysfunction of the autonomic (involuntary) nervous system. The EPA RfD is about 30 times lower than the NOAEL because it incorporates a conservative safety factor. Health effects in people exposed to elemental mercury below the NOAEL have not been reported.

Populations with Highest Exposures   The highest estimated elemental mercury doses were to children who were members of the EFPC farm family in 1957. The upper bound on the highest estimated annual average elemental mercury inhalation dose is about 13 times higher than the EPA RfD but about one third of the NOAEL (see Figure 14.19).

Estimated doses from inhalation for Scarboro residents during 1953 to 1962

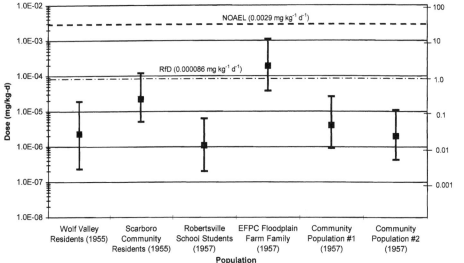

Figure 14.19   Highest estimated elemental mercury doses and hazard indices (for child exposure) for each reference population evaluated in Oak Ridge Dose Reconstruction, with applicable toxicity benchmark values. The year of peak exposure is indicated for each population. Results reflect 95% subjective confidence intervals and central values (medians).

(when air concentrations at this location were assumed to result from both direct airborne mercury releases from Y-12 that were transported over Pine Ridge and volatilization of mercury from EFPC) are about 15 to 40 percent of the inhalation doses estimated for the EFPC farm family during these years. During other years, estimated doses at Scarboro are about 10 percent of doses estimated at the EFPC floodplain farm family location. The higher estimated doses at the EFPC floodplain farm location are due to its closer proximity to EFPC.

Likelihood of Exposures above the RfD, Scarboro Residents   The estimated size of the Scarboro community population was assumed to be between 800 and 1200 individuals per year. Since estimated doses at the 50th percentile for this population were below the RfD for all years, it is likely that doses to most individuals in this population were below the RfD. However, because of the relatively large size of this population, it is likely that inhalation doses to a small number of people in this population during the years of highest mercury releases from Y-12 (1953 to 1962) exceeded the RfD. Deleterious health effects were therefore possible but unlikely.

Likelihood of Exposures above RfD, EFPC Floodplain Farm Family Members
The estimated size of the EFPC floodplain farm family population was very small (a total of between 10 and 50 individuals were assumed in this population

per year). Since estimated doses at the 50th percentile to some members of this population exceeded the RfD during the years of highest mercury releases from Y-12, it is likely that doses to some individuals in this population exceeded the RfD.

Ingestion of and Contact with Inorganic Mercury in Soil, Sediment, Water, Meat, Milk, and Fruits/Vegetables

Comparison to RfDs   The 95 percent UCLs on estimated inorganic mercury doses exceeded the EPA RfD for inorganic mercury for at least one year for all six nonangler populations evaluated in this assessment: Wolf Valley residents (child, 1955), the Scarboro community (child, 1953 to 1962; adult, 1954 to 1959), Robertsville school students (general student, 1955 and 1958; recreator, 1955 to 1958), the EFPC floodplain farm family (child, 1950 to 1970, 1973; adult, 1952 to 1963, 1965), and the two Oak Ridge community populations (community population 1 child, 1955, 1957 to 1958; community population 2 child, 1958). Central estimates of inorganic mercury dose exceeded the RfD for 1955 to 1958 for Scarboro children, 1953 and 1955 to 1959 for EFPC farm family children, and 1955 to 1958 for EFPC farm family adults. The 95 percent lower confidence limit of inorganic mercury dose just reached the RfD for EFPC farm family children in 1958.

Comparison to NOAELs   The 95 percent subjective confidence interval on estimated annual average inorganic mercury doses for all populations and all years were below the NOAEL for inorganic mercury. The NOAEL for inorganic mercury is based on kidney effects observed in rats fed high concentrations of water-soluble mercuric chloride. The EPA RfD is about 3000 times lower than the NOAEL because it incorporates a conservative margin of safety to account for the lack of data on the toxicity of inorganic mercury to humans. Health effects in humans from inorganic mercury at doses at or below the NOAEL have not been reported.

Populations with Highest Exposures   The highest estimated inorganic mercury doses were to children who were members of the EFPC floodplain farm family in 1958. The upper bound on the highest estimated annual average inorganic mercury dose is about 90 times higher than the EPA RfD but about a factor of four below the NOAEL (see Figure 14.20). Doses to these individuals were estimated to be high because they were assumed to live close to EFPC. Inorganic mercury doses to Scarboro community residents during the mid-1950s to early-1960s were also estimated to potentially exceed the RfD because it was assumed that they occasionally recreated in EFPC.

Important Pathways   At five of the six locations where estimated total inorganic mercury doses exceeded the RfD, estimated doses were largely contributed by ingestion of homegrown fruits and vegetables contaminated by airborne

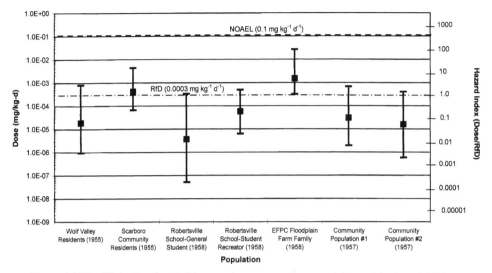

Figure 14.20   Highest estimated inorganic mercury doses and hazard indices (child exposure) for each reference population evaluated in Oak Ridge Dose Reconstruction, with applicable toxicity benchmark values. The year of peak exposure is indicated for each population. Results reflect 95% subjective confidence intervals and central values (medians).

mercury. Contact with contaminated water in EFPC was also an important pathway for Scarboro community residents and EFPC floodplain farm family members.

Likelihood of Exposures above RfD, Wolf Valley Residents   The estimated size of the Wolf Valley population was small (between 30 to 100 people in a given year). For this population, the results of this assessment suggest that only doses to young children may have exceeded the RfD, and only if they consumed very large quantities of homegrown above-ground fruits and vegetables. Because of the small size of this population and the relatively low doses estimated for them, it is likely that the number of individuals in this population who were exposed to inorganic mercury at doses above the RfD was small.

Likelihood of Exposures above RfD, Scarboro Residents   The estimated size of the Scarboro community residents population was relatively large (between 800 and 1200 individuals in a given year). Since estimated doses at the 50th percentile for this population were below the RfD for most years, it is likely that doses to most individuals in this population were below the RfD. However, because of the relatively large size of this population, it is likely that inorganic mercury doses to a moderate number of people in this population during the years of highest mercury releases from Y-12 (1953 to 1962) exceeded the RfD, particularly for those individuals who frequently recreated in EFPC or regularly consumed above-ground fruits/vegetables from backyard gardens. Deleterious health effects were therefore possible but unlikely.

Likelihood of Exposures above RfD, Robertsville School Students The estimated size of the Robertsville School general student population was between 1500 and 2000 students in a given year. Since estimated doses at the 50th percentile for this population were below the RfD for all years, and doses at the 97.5th percentile exceeded the RfD only during a few years in the mid-1950s, it is likely that the number of individuals in this population who were exposed to inorganic mercury at doses above the RfD was small. Doses above the RfD most likely resulted from frequent contact with schoolyard soil and EFPC water and sediment.

Likelihood of Exposures above RfD, EFPC Farm Family Members The estimated size of the EFPC farm family population was very small (10 to 50 individuals in a given year). Because estimated doses at the 50th percentile for this population exceeded the RfD during the years of highest mercury releases from Y-12 (1953 to 1962) and because this population group was assumed to live close to EFPC, it is likely that doses to some individuals in this population exceeded the RfD. Doses above the RfD most likely resulted from frequent contact with floodplain soil and EFPC water and sediment, and eating of "backyard" fruits and vegetables.

Likelihood of Exposures above RfD of Community Populations The estimated size of the community populations was relatively large (1500 to 2000 individuals in a given year). However, results suggest that for these populations, only doses to young children may have exceeded the RfD, and only if they consumed very large quantities of homegrown above-ground fruits and vegetables during the years of highest mercury releases from Y-12 (mid-1950s) and lived closer than one mile from the creek. The number of individuals in these populations exposed to inorganic mercury at doses above the RfD was likely small.

Ingestion of Methylmercury in Fish

Comparison to RfDs (Watts Bar Fish) The 95 percent UCLs on estimated methylmercury doses from consumption of fish exceeded the RfD based on in utero exposures for all years for category 1 fish consumers, 1954 to 1970 and 1974 to 1975 for category 2 fish consumers, and 1957 to 1959 for category 3 fish consumers. Central estimates exceeded the RfD 1950–1981 for category 1 fish consumers and 1956 to 1960 for category 2. During the years of highest mercury releases from Y-12, estimated doses for category 1 fish consumers exceeded the RfD based on in utero exposures even at the lower bound of the distribution (the 2.5th percentile).

Comparison to NOAELs (Watts Bar Fish)   The 95 percent UCL on estimated methylmercury doses exceeded the NOAEL for 1956 to 1960 for category 1 fish consumers, and median doses reached the NOAEL around 1958 to 1959. Estimated doses to category 2 and 3 fish consumers were below the NOAEL. The NOAEL for methylmercury is based on observations of neurological effects in children who were exposed to methylmercury in utero when their mothers consumed methylmercury in fish during pregnancy. Health effects in humans exposed to methylmercury at doses at or below the NOAEL have not been reported.

Exposures to Children (Watts Bar Fish)   Based on our calculations, children who ate as few as three to four meals of fish from Watts Bar Reservoir during the mid- to late-1950s may have been exposed to methylmercury at doses that exceeded the RfD based on in utero exposures. If they ate seven or more meals of fish per year from Watt Bar during these years, it is likely that they were exposed to methylmercury at doses that exceeded the RfD.

Exposures to Adults (Watts Bar Fish)   Based on our calculations, adults who ate nine or more meals of fish from Watts Bar Reservoir during the mid- to late-1950s may have been exposed to methylmercury at doses that exceeded the RfD based on in utero exposures. If they ate about 20 or more meals per year during these years, it is likely that they were exposed to methylmercury at doses that exceeded the RfD. Adults who were not pregnant could have consumed about three times as many fish meals per year as pregnant adult females, without risk of adverse health effects from methylmercury exposure, because it is believed that adults are not as sensitive to adverse health effects from methylmercury exposure as children exposed in utero.

The number of fetuses possibly affected (average doses greater than the NOAEL) was estimated using the average birth rate in the population, the fraction of women of childbearing age, their fish consumption rates, and the fraction of consumers whose doses exceeded the NOAEL for in utero exposure during that year. The estimates were made for Watts Bar Reservoir, Clinch River/Poplar Creek, and EFPC fish consumers and summed over the years of concern. The estimated total number of affected fetuses is uncertain but is nearer to 100 than to 1000.

Comparison to RfDs (Clinch River/Poplar Creek Fish)   As shown in Figure 14.21, the 95 percent UCLs on estimated methylmercury doses from consumption of fish exceeded the RfD based on in utero exposures for all years for category 1 fish consumers, 1950 to 1982 for category 2, and 1950 to 1966 and 1971 to 1972 for category 3. Central estimates exceeded the RfD for all years for category 1, 1950 to 1972 for category 2, and 1956 to 1962 for category 3. Estimated doses exceeded the RfD based on in utero exposures even at the lower confidence limit (the 2.5th percentile) for 1950 to 1975 for category 1 fish consumers and 1950 to 1954 and 1956 to 1964 for category 2 fish consumers.

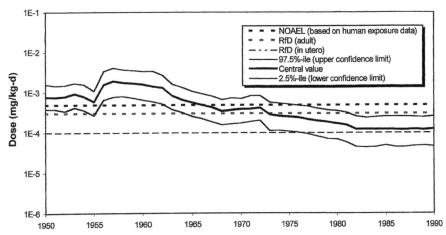

Figure 14.21 Estimated methylmercury doses to adult eaters of fish from Clinch River and Poplar Creek near Oak Ridge Reservation. The solid lines represent subjective confidence bounds and central values (medians) calculated for each year from 1950 through 1990, and the straight horizontal lines allow comparison to toxicologic benchmarks.

Comparison to NOAELs (Clinch River/Poplar Creek Fish) The 95 percent UCL on methylmercury doses exceeded the NOAEL for 1950 to 1975 for category 1 fish consumers, 1950 to 1964 (excepting 1955) for category 2 fish consumers, and 1957 for category 3 fish consumers. Central estimates of dose exceeded the NOAEL for 1950 to 1966 for category 1 and 1956 to 1962 for category 2. For category 1 fish consumers, even the lower confidence limit of methylmercury dose (the 2.5th percentile) exceeded the NOAEL for 1956 to 1962. Health effects in humans exposed to methylmercury at or below the NOAEL have not been reported.

Exposures to Children (Clinch River/Poplar Creek Fish) Children who ate as few as one meal of fish from Clinch River/Poplar Creek during the mid- to late-1950s may have been exposed to methylmercury at doses that exceeded the RfD based on in utero exposures. If they ate two or more meals of fish per year from Clinch River/Poplar Creek during these years, it is likely that they were exposed to methylmercury at doses that exceeded the RfD.

Exposures to Adults (Clinch River/Poplar Creek Fish) Adults who ate two to three or more meals of fish from Clinch River/Poplar Creek during the mid- to late-1950s may have been exposed to methylmercury at doses that exceeded the RfD based on in utero exposures. If they ate five or more meals per year during these years, it is likely that they were exposed to methylmercury at doses that exceeded the RfD. Adults who were not pregnant could have consumed about three times as many fish meals per year as pregnant adults without risk of adverse health effects from methylmercury exposure.

Likelihood of Exposures above RfD (Clinch River/Poplar Creek Fish)   The estimated size of the recreational angler population in Clinch River/Poplar Creek was large (3000 to 10,000 individuals in a given year). Because a large number of people occasionally fished in Clinch River/Poplar Creek and many likely consumed moderate quantities of fish from this system, it is likely that a significant number of people who caught and consumed fish from this system were exposed to methylmercury at doses that exceeded the RfD, particularly if they consumed fish from this system during the mid-1950s and 1960s.

Comparison to RfDs and NOAELs (EFPC Fish)   The 95 percent UCL on estimated methylmercury doses from consumption of EFPC fish by Scarboro residents and EFPC floodplain farm family populations exceeded the RfD for methylmercury (based on in utero exposures) for all years evaluated in this assessment (1950 to 1990). Central estimates exceeded the RfD for 1950 to 1983. However, doses for this population did not exceed the NOAEL.

Interviews with Oak Ridge area residents, including residents of Scarboro and people who lived near EFPC, suggest that the maximum rate of consumption of fish from EFPC was about one fish meal per month. In this assessment, the average consumption rate of fish from EFPC for adults was assumed to be about 2.5 meals per year.

Exposures to Children (EFPC Fish)   Children who ate more than one meal of fish per year from EFPC may have been exposed to methylmercury at doses that exceeded the EPA RfD. If they ate two or more meals of fish per year from EFPC during these years, it is likely that they were exposed to methylmercury at doses that exceeded the RfD.

Exposures to Adults (EFPC Fish)   Adults who ate two to three or more meals of fish per year from EFPC may have been exposed to methylmercury at doses that exceeded the RfD based on in utero exposures. If they ate over five meals per year during these years, it is likely that they were exposed to methylmercury at doses that exceeded the RfD.

14.4.6 Discussion

Based on the results of the mercury dose reconstruction, the following activities may have resulted in exposure to mercury at annual average doses above the reference doses:

- Consumption of any fish from EFPC, the Clinch River, or Poplar Creek
- Consumption of more than three or four meals of fish per year from Watts Bar Reservoir

- Consumption of fruits or vegetables that grow above ground from backyard gardens in Scarboro or within several hundred yards of the EFPC floodplain
- Playing in EFPC more than 10 to 15 hours per year
- Living or attending school within several hundred yards of the EFPC floodplain or in Scarboro (from inhalation of airborne mercury).

The likelihood that these activities resulted in doses above the RfDs was greatest during the mid-1950s to early-1960s.

14.5 POLYCHLORINATED BIPHENYLS (PCBS) IN ENVIRONMENT NEAR OAK RIDGE

14.5.1 Introduction

This section presents the findings of a detailed assessment of the releases of PCBs from the ORR and the potential for adverse effects in the local populations. Beginning with early operations in the 1940s, PCBs were used extensively on the ORR. PCBs were present at Oak Ridge mainly because the ORR was one of the largest users of electrical energy in the United States (TVA, 1988). Because of their good insulating properties and thermal stability, PCBs were present in many electrical components such as transformers and capacitors (see Figure 14.22). In addition, PCBs were used as cutting fluids for lubrication and cooling during certain metal working operations.

Preliminary screening analyses conducted during the Dose Reconstruction Feasibility Study (ChemRisk, 1993a–e), indicated that PCBs potentially represented the most important nonradioactive cancer-causing chemical historically released from the ORR. The study team also found that sources of PCB releases to the environment and the resulting exposures to local populations were poorly characterized. As a result, task 3 of the Oak Ridge Dose Reconstruction was initiated to reconstruct PCB doses to human populations living in and around Oak Ridge.

The objectives of task 3 were to:

- Investigate historical releases of PCBs from the government complexes at Oak Ridge.
- Evaluate PCB levels in environmental media in the ORR area.
- Describe releases of PCBs from other sources in the Oak Ridge area.
- Evaluate potential human exposures and health effects associated with the presence of these contaminants in the environment.

During the first 30 years of operations at the ORR, little or no attention was paid to the use, disposal, or contamination of the environment with PCBs. Few attempts were made to control the release of PCBs to the environment during

Figure 14.22   Because huge quantities of electricity were used by Oak Ridge complexes, PCB-containing components such as transformers and capacitors were plentiful. PCBs were also present in hydraulic fluids, heat-transfer fluids, and cutting oils used in machining. U.S. AEC photo by James E. Westcott.

this period, and minimal effort was made to track or document the amounts of PCBs used, disposed of on site, or released off site. This was because the carcinogenicity of PCBs in laboratory animals was not discovered until the 1970s (Taylor, 1988). In 1977, the manufacture of PCBs was banned in the United States because of evidence that PCBs accumulated in the environment and caused harmful effects (EPA, 1986a).

In the absence of detailed historical records regarding PCB use and disposal at the ORR, it was necessary for us to identify and evaluate all available information regarding processes and disposal practices at the ORR that might have resulted in the release of PCBs to the environment. Data were obtained from a variety of sources. Publicly available documents prepared by ORR contractors, the TVA, and the EPA were obtained and reviewed. Historical records maintained at the ORR were also reviewed to identify relevant processes, accidental spills, and general disposal practices that might have resulted in releases of PCBs (Jordan, 1976). In addition, we obtained information regarding undocumented historical events through interviews with active and retired employees of the ORR and residents of Oak Ridge living adjacent to the facilities. We maintained a detailed system of data management to ensure that the information collected was thoroughly evaluated and that the sources were clearly identified.

Histories of PCBs at the Oak Ridge Sites   In general, it appears that the primary uses of PCBs at the ORR were electrical equipment (i.e., transformers and capacitors), hydraulic fluids, heat-transfer fluids, and cutting oils [Health, Safety, Environment, and Accountability Division (HSEAD), 1992; ORNL, 1991; Jordan, 1976]. In addition, PCBs were present in relatively low levels (less than a few percent) in many products including paints, coatings, adhesives, inks, and gaskets (Versar, 1976). PCB uses, as well as the potential impact of off-site releases, differed among the three main complexes on the ORR. Therefore, separate analyses are presented for each of the facilities in the report of the dose reconstruction for PCBs (Price et al., 1999).

Qualities of PCB Releases   Based on the available information, we determined that developing quantitative estimates of PCB releases from specific release points as a function of time (often called "source terms") would be difficult, if not impossible, for the following reasons. The first problem is the widespread use of PCBs on the ORR and the absence of documentation of releases. From the initial construction of Oak Ridge through the early 1970s, PCBs were viewed as nontoxic, inert substances that offered no particular hazard to workers, the general public, or the environment. As a result, there was no attempt to manage or track the use, release, and disposal of PCBs in any systematic manner. A second problem is that a PCB release that may have happened as a result of a specific event may result in an extended, low-level source of contamination to a body of water (EPA, 1994a). Once PCBs enter a body of water, they may remain localized until a storm event or a change in the fundamental hydrology results in remobilization and additional transport. Rather than basing our risk assessments on quantitative estimates of quantities of PCBs historically released, we estimated past exposures largely based on available environmental measurements of PCBs. Air-related pathways were an exception. These pathways were evaluated using estimates of releases and air dispersion models.

Other Sources of PCBs Although releases to surface water and sediment transport represent the primary transport routes of PCBs to off-site locations, it also necessary to consider other, less significant pathways. For example, there is evidence that burning of PCB-contaminated material associated with both the Y-12 burn yards and the Toxic Substances Control Act (TSCA) incinerator at K-25 (Joyner, 1991) may have resulted in the air releases of PCBs, as well as dioxins and furans produced from the partial incineration of PCBs (Jansson and Sundström, 1982). In addition, there is evidence to suggest that materials containing PCBs, such as used oils or electrical equipment, may have been sold and transported off-site. It has long been recognized that PCBs were used in a large number of facilities throughout the watershed of the Tennessee River and its tributaries, which are shown in Figure 14.23 (TVA, 1988). Because PCBs

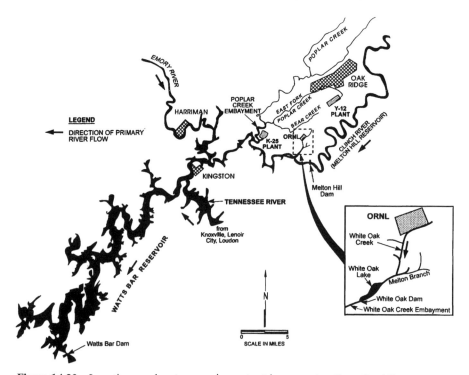

Figure 14.23 Locations and waterways important in reconstruction of public exposures to PCBs in environment near Oak Ridge. Studies indicate that about half of the PCBs in the Clinch River downstream of Oak Ridge came from sources other than the federal complexes, while about 87% of PCBs in Watts Bar Reservoir came from sources on the Tennessee River rather than on the Clinch River.

have been detected in sediment and fish from the Tennessee River above its confluence with the Clinch River and in the Clinch River upstream of the ORR, we collected data on other sources of PCBs entering Watts Bar Reservoir. Available records on PCB use identified more than 22 facilities that managed PCB-containing wastes on portions of the Tennessee River above the Clinch River and on the Clinch River above the ORR.

Two independent approaches were utilized by HydroQual, Inc. in the early 1990s to evaluate the relative fraction of PCBs in Watts Bar fish attributable to Oak Ridge (HydroQual, 1995). One approach involved a straightforward analysis of spatial trends in fish monitoring data. The other approach entailed the development of a sediment transport model (Ziegler and Nisbet, 1995), which was used in conjunction with PCB sediment core data to predict sources and transport of PCBs at various times and locations within the watershed. Based on the results of their analyses, HydroQual concluded that historical releases of PCBs from Oak Ridge were responsible for less than 9 to 13 percent of the currently observed levels in Watts Bar fish and about half of the levels in the Clinch River. HydroQual also concluded that this estimate could be further reduced if sources of PCBs above Melton Hill Dam were considered. In addition, because of the approximate agreement between these two independent measurements, HydroQual concluded that there was strong evidence that the vast majority of PCBs currently detected in fish in the lower Watts Bar occurred as a result of releases to the Tennessee River upstream of the Clinch River. The analyses also indicated that, with the exception of three periods of elevated discharges, PCB releases to the Clinch River from all ORR sources were relatively constant over time, and the total magnitude of annual PCB releases from the 1940s through the 1990s were around 9 kg per year.

14.5.2 Hazard Identification

PCBs are a group of synthetic organic chemicals that contain 209 individual compounds (known as congeners). PCBs are either oily liquids or solids and are colorless to light yellow in color. PCBs enter the environment as mixtures containing a variety of individual components and impurities. Seven types of commercially available PCB mixtures include 35 percent of all the PCBs commercially produced and 98 percent of PCBs sold in the United States since 1970. Some commercial PCB mixtures are known in the United States by their industrial trade name, Aroclor. The name Aroclor 1254, for example, means that the molecule contains 12 carbon atoms (the first two digits) and approximately 54 percent chlorine by weight (second two digits). The more volatile PCBs in water partially evaporate and then return to earth by rainfall, snow, or settling of dust particles. This cycle can be repeated many times. PCBs in water concentrate in fish and can reach levels hundreds of thousand times higher than the levels in water. Extremely small amounts of PCBs can remain in water for years. PCBs bind strongly to soil and sediments and may remain there for several years. PCBs partially evaporate from soil surfaces to air. In general, the breakdown of PCBs in the water and soil occurs over several years or even decades. Sediments containing PCBs at the bottom of a large body of water such as a lake, river, or ocean generally act as a reservoir from which PCBs may be released in small amounts to the water (ATSDR, 1998).

Because of these properties, the majority of PCBs placed in burial grounds or pits have remained in or near these units. Small lakes, ponds, and lagoons have been an integral part of the storm water and wastewater management systems at the ORR. These surface water bodies have served as traps for PCBs, PCB-contaminated oils, and PCB-contaminated sediments and have limited the movement of PCBs off the reservation. Finally, the sediments of Bear Creek, White Oak Creek, and other streams located on the ORR have entrapped a portion of the PCBs released from Y-12 and X-10 and have reduced the amount of PCBs migrating off the reservation (Turner et al., 1988).

PCBs are classified by EPA as a Group B2 carcinogen, indicated by sufficient evidence of cancer in animals and inadequate or no evidence of cancer in humans. EPA has reviewed six studies of the association between occupational exposure to PCBs and cancer mortality and two cases of accidental poisonings, the Yusho incident in Japan and the Yu-Cheng incident in Taiwan. Because of small sample sizes, brief follow-up periods, and confounding exposures to other potential carcinogens, EPA concluded that these studies are inconclusive.

Studies of the general population who were exposed to PCBs by consumption of contaminated food, particularly neurobehavioral evaluations of infants exposed in utero and/or through lactation, have been reported, but the original PCB mixtures, exposure levels, and other details of exposure are not known. Most of the information on health effects of PCB mixtures in humans is available from studies of occupational exposure. Some of these studies examined workers who had some occupational exposure to Aroclor 1254, but sequential or concurrent exposure to other Aroclor mixtures nearly always occurred, exposure involved dermal as well as inhalation routes (relative contribution by each route not known), and monitoring data are lacking or inadequate. Insufficient data are available in these studies to determine possible contributions of Aroclor 1254 alone, extent of direct skin exposure, and possible contaminants. However, it is relevant to note that dermal and ocular effects, including skin irritation, chloracne, hyperpigmentation, and eyelid and conjunctival irritation have been observed in humans occupationally exposed to Aroclor 1254 and other Aroclor formulations (IRIS, 1998c).

Decreased birth weight has also been reported in infants born to women who were occupationally exposed to Aroclor 1016 and other Aroclor formulations, ingested PCB-contaminated fish, and ingested heated Kanechlor PCBs during the Yusho and Yu-Cheng incidents. Due to uncertainties regarding actual sources of PCB exposure, and other confounding factors and study limitations, the decreases in human birth weight cannot be solely attributed to PCBs, particularly specific PCB mixtures. However, due to the consistency with which the effect has been observed, the human data are consistent with the Aroclor-1016-induced decreased birth weight in monkeys reported in the principal studies (IRIS, 1998c).

14.5.3 Dose–Response Evaluation

The toxicity value for the assessment of carcinogenic effects is the cancer slope factor (CSF) (EPA, 1996) and the RfD (IRIS, 1998c) is an estimate of daily exposure that is without appreciable risk of adverse noncarcinogenic effects. The RfD is significantly below the NOAEL and well below the LOAEL (ATSDR, 1998).

In September 1996, the EPA published a reassessment of the carcinogenic potential of PCBs (EPA, 1996). That reassessment considered all available cancer studies of commercial PCB mixtures including a recent oncogenicity feeding study conducted by Brunner et al. (1996). This feeding study is the most comprehensive study of the carcinogenicity of PCBs to date. Conducted in male and female Sprague-Dawley rats using four PCB mixtures with varying degrees of chlorination (Aroclors 1016, 1242, 1254, 1260), the animals were dosed for 7 days a week for 24 months. The study found liver tumors in female rats exposed to Aroclors 1260, 1254, 1242, and 1016 and in male rats exposed to 1260. Earlier studies found high, statistically significant incidences of liver tumors in rats ingesting Aroclor 1260. The CSFs generated from the results of the 1996 study are substantially lower than the previous CSF of 7.7 $(mg/kg-d)^{-1}$. EPA's reassessment specified a new set of CSFs for PCBs depending on the exposure pathway and the degree of chlorination. For PCB mixtures other than those containing minimal amounts of the more highly chlorinated congeners, the central tendency CSFs range from 0.3 to 1 $(mg/kg-d)^{-1}$ and the upper bound CSFs range from 0.4 to 2 $(mg/kg-d)^{-1}$. The lower end of these ranges is to be used for vapor inhalation, dermal exposures, and water ingestion, and the upper end is to be used for soil or sediment ingestion, dust inhalation and food chain exposures. Lower CSFs ranging from 0.04 to 0.07 $(mg/kg-d)^{-1}$ are prescribed for PCB mixtures in which congeners with more than four chlorines comprise less than 0.5 percent of the total PCBs.

The RfD for Aroclor 1254 was used in this assessment to characterize the noncancer effects of PCBs. The Aroclor 1254 RfD is based on a chronic study in which rhesus monkeys were fed capsules containing 0, 5, 20, 40, or 80 μg/kg-d of Aroclor 1254 (Tryphonas, 1989, 1991a; Arnold, 1993a, b). Effects observed in the monkeys and that provide the basis for the RfD, include ocular exudates, inflamed and prominent meibomian glands, distortion in the growth of nails, and decreased immune response to sheep erythrocytes. The dose of 5 μg/kg-d was identified as a LOAEL; however, no NOAEL was identified. In deriving the RfD, one safety of 10 (for sensitive humans) and three factors of 3 (one each for interspecies extrapolation, use of a LOAEL instead of a NOAEL, and subchronic to chronic extrapolation) were applied to the LOAEL, resulting in an RfD of 2×10^{-5} mg/kg-d.

The regulatory processes used in setting the toxicity values are intended to be conservative in the face of uncertainty. As a result, the estimates of the RfD and CSF for PCBs are biased. That is, they are intended to be values that have a high probability of overestimating actual risks. The sources and magnitude of the biases are discussed in the report of our dose reconstruction for PCBs (Price et al., 1999).

14.5.4   Exposure Assessment

We identified potential off-site exposure pathways that were primarily associated with releases to surface water and to air. Releases to surface water are primarily associated with White Oak Creek, Bear Creek, EFPC, Poplar Creek, the Clinch River, and Watts Bar Lake. In general, exposure pathways associated with releases to surface water have included fish consumption, dermal contact with surface water and sediments, and incidental ingestion of surface water and sediments. In addition, based on the available information regarding historical activities in the area, direct contact with floodplain soil as well as pathways associated with bioaccumulation of PCBs in vegetation and animals have been identified as complete exposure pathways for EFPC. Exposure pathways associated with bioaccumulation of PCBs in animals have been identified for Jones Island and the Clinch River. Exposures related to PCB releases in air have included both direct pathways, such as inhalation, and indirect pathways such as bioaccumulation of PCBs in vegetation and animals. These pathways are also likely complete for dioxins and furans that may have been formed during the incineration of PCBs.

We identified potential exposures associated with the historical sale of PCB-containing materials. Waste oils containing less than 500 ppm of PCBs may have been sold by the ORR facilities in the late 1940s (Banic, 1995). Such oil could have been used by local individuals for fuel, dust suppression, or vegetation control. Exposure pathways considered included direct contact with contaminated soil.

An important exposure pathway identified in the report of the dose reconstruction for PCBs (Price et al., 1999) is the accumulation of PCBs in fish and the resulting exposures to the anglers and their families who consume the fish. An important issue in evaluating fish consumption is the frequency and amount of fish that an angler consumes. Fish consumption varies greatly across the local population with some people eating no fish and others obtaining a large amount of their protein needs from the consumption of fish. The types of fish consumers include commercial, recreational, and subsistence anglers.

Price et al. (1999) present an evaluation of the information available on historical fishing activities on the water bodies of interest, identify potentially exposed populations of anglers, and derive estimates of fish consumption rates for the populations that were likely exposed as a result of their fishing activities (Ebert et al., 1993, 1994; Ebert, 1996). Based upon this assessment, eight distinct populations may have received exposure to PCBs through consumption of fish from water bodies in proximity to the ORR. These populations include commercial anglers who fished Watts Bar or Clinch River/Poplar Creek, recreational anglers who fished Watts Bar, Clinch River/Poplar Creek, or EFPC, and subsistence individuals who may have fished any of these water bodies. The report also discusses the data and methodologies used to develop both the point estimates of consumption used in the level I evaluation and the distributions used in the level II and III evaluations.

As described above, potential exposure pathways that were identified included direct exposure to PCBs in water, sediment, floodplain soils, and air, as well as indirect exposure through the ingestion of contaminated food (such as fish, vegetables, beef, and milk). We collected site-specific demographic information (DaMassa, 1996) regarding farming, fishing, and recreational activities through interviews with current and past residents of Oak Ridge. We used this site-specific information, as well as measured levels of PCBs in the various media of concern, to confirm which of the possible exposure pathways actually resulted in exposures to off-site populations. Those that were determined to be "complete" pathways were considered in the level I evaluation to identify the pathways most likely associated with off-site health risks.

Level I Evaluation of PCB Exposures (Screening-Level Assessment)   In the level I evaluation, we characterized the exposure pathways by medium, selected conservative upper-bound exposure parameter values, and developed exposure point concentrations to estimate potential PCB intakes. These intake estimates were then combined with toxicity values to estimate the risks associated with each pathway. These estimates of risk were compared to screening criteria to determine which pathways were most likely to result in risks to off-site populations. The screening criteria or decision guides used were an excess cancer risk of 1 in 10,000 and a nominal hazard quotient (the estimated dose divided by the RfD) equal to 1 for noncancer health effects.

If risk estimates for pathways were below the decision guides, these pathways were set aside from further evaluation. Likely off-site populations were identified for those pathways for which the estimated risks exceeded the decision guides. Because of the conservative exposure and toxicity assumptions used in deriving these estimates of risk, the findings of the level I evaluation cannot be taken as evidence of actual risk. The estimates are best viewed as indications of pathways that warrant additional study.

Exposure point concentrations (EPCs) for the level I evaluation were based on historical data from a variety of sources including TVA and DOE. These data were limited, particularly for years prior to the 1970s. Because this was a retrospective analysis, concentrations reported for any year were treated equally. Soil and sediment samples taken at depth were also considered because these samples may represent historical EPCs. Because of the conservative nature of the level I evaluation, the EPCs for soil, sediment, surface water, drinking water, and aquatic biota were defined as the maximum total PCB concentrations for each medium for each water body. The EPCs for the direct air pathways were modeled using a Gaussian air dispersion model, SCREEN3 (EPA, 1995b). For the indirect air pathways, EPCs were established for vegetables, beef, and milk based on measured or estimated concentrations in various media.

Illustrative Example: PCB Dose Calculation for Fish Ingestion   In the actual calculation of PCB doses from ingestion of fish, the values for several of the following parameters were randomly selected from distributions of each parameter using Monte Carlo calculation techniques. However, for illustrative purposes the approximate midpoint of each distribution was used in this example.

For recreational anglers at Watts Bar Reservoir in Tennessee, a mean consumption rate of 30 g/person-day was derived based on an evaluation of recreational catch statistics reported by the Tennessee Wildlife Resource Agency for the years from 1977 to 1991. This mean was used as the basis for the log-normal distribution of consumption rates for this population. The standard deviation (71 g/person-day) was derived by assuming that the coefficient of variation (2.37) was similar to that reported for other fish consumption studies in Maine and Lake Ontario (2.37 × 30 = 71 g/person-day).

A triangular distribution for PCB concentrations in fish was developed for Watts Bar Reservoir based on the weighted averages calculated for each game species. The minimum, maximum, and mode values were used as the minimum (0.29 $mg\,kg^{-1}$), maximum (1.35 $mg\,kg^{-1}$), and most likely (1.17 $mg\,kg^{-1}$) values for the distribution for all species of game fish.

The fraction of the original PCB concentration remaining after cooking varies from between 40 and 80 percent. A triangular distribution with a minimum of 0.4, a maximum of 0.8, and a most likely value of 0.5 was used to describe the fraction of PCBs remaining after cooking:

$$\text{Dose}_{\text{fish-PCB}} = \frac{U_f \times \text{FC}_{\text{PCB}} \times F_{\text{cook}}}{\text{BW}}$$

where $\text{Dose}_{\text{fish}}$ = dose $(mg\,kg^{-1}\,d^{-1})$
U_f = fish consumption rate $(0.030\ kg\,d^{-1})$
FC_{PCB} = concentration of PCBs in fish $(1.17\ mg\,kg^{-1})$
F_{cook} = fraction of PCBs remaining after cooking (0.5 unitless)
BW = body weight (70 kg)

and

$$\text{Dose}_{\text{fish-PCB}} = \frac{0.030\ kg\,d^{-1} \times 1.17\ mg\,kg^{-1} \times 0.5}{70\ kg}$$

$$= 2.5 \times 10^{-4}\ mg\,kg^{-1}\,d^{-1}$$

Level II Evaluation of PCB Exposures (with Uncertainty Analysis)   After the level I analysis, we focused our efforts to further refine exposures and risks on those pathways that exceeded the decision guides in order to focus project resources on those sources of exposure that had the highest potential for harm. The level I evaluation used a conservative estimate of intake and risk to identify those pathways potentially associated with risks to off-site populations. This analysis was based on a determination of whether there was evidence that an individual in an exposed population could have a risk greater than the decision guides. No determination was made on how likely such a risk would actually occur or the fraction of the population exposed by a pathway would be affected. A proper evaluation of the risks to populations exposed by one or more pathways should focus on the heterogeneity of the population and the uncertainty in the estimates of exposure and risk.

All of the reference populations evaluated had large amounts of variation (heterogeneity) in the doses that specific individuals received. Certain individuals may have received trivial exposures, while others may have been significantly exposed. Therefore, it was critical to determine the fractions of the exposed groups received doses associated with levels of concern.

A second issue is the level of confidence that can be attributed to an estimate of exposure and risk. For certain pathways, there is a high confidence that exposure actually occurred; however, for others the evidence is suggestive but incomplete. PCB exposure from consumption of contaminated fish is well documented. In contrast, exposures to the families on farms that bordered EFPC are quite uncertain because the extent of historical levels of PCB contamination in soils at the farms is largely unknown. Dose–response relationships for PCBs are also highly uncertain (EPA, 1996; Gillis and Price, 1996). Because of these uncertainties, it is important to determine the uncertainty in the estimates of risk made for various portions of the exposed populations.

We characterized the uncertainty and variation in risk to the individuals in the exposed populations. The level II evaluation characterizes the range of doses that plausibly occurred in the exposed populations. Similar to the level I analysis, simple dose models were used for each pathway. Doses were estimated across populations using Monte Carlo analysis.

In the level II evaluation, variability was directly modeled, and estimates of the distribution of doses in the population were directly determined. Uncertainty that occurred due to the lack of knowledge was addressed by using estimates of intake parameters that were biased with respect to uncertainty. Specifically, the parameter values were selected from the upper end of the range of plausible alternative values. The evaluation used the same toxicological criteria (RfD and cancer slope factor) as used in level I. These values overestimate cancer and noncancer risks. Thus, the distribution of cancer and noncancer risks across the population is believed to overestimate the distribution of actual risks in the populations.

The goal of the level II evaluation was to identify those exposed populations where there was some chance that a small fraction (5 percent) of the population received risks in excess of the decision guides. Populations where this was not true were set aside and the remaining populations were subject to additional analyses. The additional analyses included a quantification of the uncertainty in the hazard assessment. In certain populations with risks in excess of the guides, there was insufficient information to allow additional assessments. When this occurred, the lack of information was identified as a critical data gap and was included in recommendations for further work.

The level II assessment evaluated risks to recreational and commercial fish consumers, farm families, and recreational users of surface water bodies near the ORR. Where a population was exposed via multiple pathways, the total intake received from all pathways was estimated. Distributions for those exposure parameters believed to make a significant contribution to the variation in the dose rate were developed, based on the range of available data, by fitting available data to various distribution types (e.g., normal, log-normal) according to accepted methods. Summaries of the parameter values and distributions used for each population, as well as the assumptions and rationale on which they were based, are presented in the report of the dose reconstruction for PCBs (Price et al., 1999).

Level III Evaluation of PCB Exposures (Two-Dimensional Uncertainty Analysis)
The level III evaluation of PCB exposures was performed on recreational anglers fishing Watts Bar and the Clinch River and the children of these anglers. No additional modeling was performed for the commercial angler because their risks in the level II evaluation appeared to be similar or slightly lower than those estimated for recreational anglers. Thus, the level III evaluation for recreational anglers is applicable to commercial anglers as well.

The assessment of noncancer effects in the level II evaluation suggests that the majority of commercial and recreational anglers may have been at risk. The vast majority of anglers had nominal hazard quotients that were greater than one. However, the current noncancer decision guide is based on the assumption that any dose that exceeds the RfD is of some toxicological concern. It is not clear that findings of nominal hazard quotients greater than one (i.e., a dose greater than the RfD) imply the occurrence of adverse effects.

Price et al. (1999) present a characterization of the population threshold for the noncarcinogenic effects of PCBs and uses the characterization to better estimate noncarcinogenic risk. A population threshold is the highest dose that does not cause an adverse effect in an individual who is uniquely sensitive to PCBs. We characterized this threshold using EPA's methodology for setting RfDs (Barnes and Dourson, 1988) and replacing the safety factors (Baird et al., 1996) with distributions (Swartout et al., 1998). The distributions were based on a review of the available literature on generic approaches and by the development of a PCB-specific distribution. The PCB-specific distribution is believed to present the best use of the available toxicological data on PCBs.

As discussed above, the level II evaluation was biased with respect to the uncertainty in the estimate of dose rates and dose response. In the level III evaluation, this uncertainty is quantitatively evaluated along with the information on the variation in dose rates developed in the level II evaluation. Over the past 5 years, techniques have been developed for modeling both the uncertainty and variability of dose rates in exposed populations (Price et al., 1997; Frey and

Rhodes, 1996). One technique, called two-dimensional Monte Carlo, uses a nested loop technique that requires additional model development and computer resources. The result is the distribution of risk estimates across the exposed populations and the uncertainty bounds on those estimates. These estimates take into account the uncertainty in the estimates of dose and of the population threshold of PCB effects.

In the level III assessment, the risks are characterized as the ratio of the fish consumer's dose to the actual population threshold. This ratio is referred to as the "true" hazard quotient. Because the estimates of the population threshold and the dose received are uncertain, the true hazard quotient is described in terms of probability. For example, an angler having a particular estimated exposre may be described as having a 50 percent chance of having a true hazard quotient less than 0.1 and a 95 percent chance of having a true hazard quotient less than 0.5. This uncertainty corresponds to the range of values of the population threshold that is consistent with present toxicology of Aroclor 1254.

The two-dimensional analysis of noncancer risk estimates further characterized risks by: (1) determining the distribution of true hazard quotients across the population, (2) calculating the fraction of the population receiving doses above the threshold of PCB effects, and (3) evaluating the incremental contribution made by ORR releases to the risks from PCBs from other sources.

Illustrative Example: Uncertainty vs. Variability   Characterize the following parameters used to estimate a hazard quotient (HQ) for exposure to PCBs through fish consumption in terms of their uncertainty and variability:

$$HQ = \frac{U_f \times F_{cook} \times FC_{PCB}}{BW \times threshold}$$

where U_f = fish consumption rate $(kg\,d^{-1})$

   FC_{PCB} = concentration of PCBs in fish $(mg\,kg^{-1})$

   F_{cook} = fraction of PCBs remaining after cooking (unitless)

   BW = body weight (kg)

threshold = highest dose that does not cause adverse effects in sensitive individuals $(mg\,kg^{-1}\,d^{-1})$

Background   Much attention has been given to the importance of separating variability and uncertainty in chemical risk assessment (Price et al., 1997; Frey and Rhodes, 1996; Hoffman and Hammonds, 1994). When these factors are treated separately, a risk manager has insight into the level of risk to various members of the exposed population, as well as an understanding of the level of confidence that can be attributed to those estimates.

A parameter can have either an uncertain component, a variable component, or both. Variability results from the fact that no single parameter value represents all individuals in a population. For example, individual anglers can have a wide range of fish consumption rates where no single value represents all of the anglers in a population. In contrast, uncertainty results from insufficient knowledge of either the value of the parameter or the shape of the distribution of the parameter values.

Solution   The fish consumption rate in this example from Watts Bar Reservoir in Tennessee is viewed as having both uncertainty and variability. Uncertainty stems from the fact that studies of fish consumption practices have not been conducted on Watts Bar Reservoir angler populations. The variability in the Watts Bar fish consumption rate was represented by site-specific creel data and information on the coefficient of variation reported in surveys of recreational anglers in areas of the United States other than Tennessee. These data were used to represent the upper bound of the distribution of estimated consumption rates for anglers using Watts Bar. Consumption rates for anglers in Maine (Ebert et al., 1993) were used to represent the lower-bound distribution. The upper-bound distribution likely overestimates intakes due to the conservative assumptions used to derive that distribution. The angler survey from Maine likely underestimates intakes due to the colder weather in Maine as compared to Tennessee (Rupp et al., 1980). It was assumed that the actual distribution of consumption rates fell between the two bounds but was more likely to fall near the upper bound. The distribution for Watts Bar fish consumption rates was represented as triangular with the maximum and mode equal to the upper bound and the minimum equal to the lower bound.

The concentration of PCBs in fish has some uncertainty due to fact that anglers that favor one species of fish over another tend to have different long-term average concentrations of PCBs in the fish they consume. However, consumption of large numbers of fish tends to average out the interfish variation, so that the PCB concentration consumed over time closely resembles the mean concentration. Weighted averages of PCB concentrations in fish were derived for each commonly consumed game species fish for Watts Bar Reservoir by combining relevant data from all available investigations. The variability in the PCB concentration was represented using a triangular distribution with the minimum, mode, and maximum reported values from the data set.

The fraction of PCBs remaining after cooking varies from 40 to 80 percent depending on the cooking method used (Sherer and Price, 1993; ChemRisk, 1992a). The range of PCB concentrations ingested by anglers was defined by the cooking method selected because most anglers will vary their choice of cooking method from meal-to-meal based on the species consumed. This variability was also modeled using a triangular distribution with minimum and maximum from the data set, but since the mode was uncertain, it was allowed to vary with equal probability between the minimum and maximum values.

Body weight is variable and was characterized using distributions provided in the *Exposure Factors Handbook* (EPA, 1997b). Since the data were collected from large numbers of individuals, uncertainty was assumed to be minimal.

Uncertainty in the PCB threshold dose is represented by four safety factors included in the RfD for Aroclor 1254 that describe the true but unknown threshold where the RfD is some point on the lower end of the distribution. Uncertainty is associated with extrapolating from the results of toxicology studies conducted in laboratory animals to draw conclusions about effects in humans. Variability in human response must be addressed to derive an estimate of a dose that is sufficiently low so as to be protective of individuals that are particularly sensitive to a compound.

14.5.5 Risk Characterization

Results of Level I Screening Analysis   Using site-specific information regarding historical activities and exposure pathways, we identified five potentially exposed populations during the screening evaluation:

- Farm families that raised beef, dairy cattle, and vegetables on the floodplain of EFPC
- People who purchased beef and milk from cattle raised in the EFPC floodplain
- Commercial and recreational fish consumers
- Individuals that may have consumed turtles
- Users of surface water for recreational activities

The sizes of these populations vary greatly. The number of anglers using EFPC and the number of farm families are expected to have been small, perhaps less than 20 individuals. In contrast, it is estimated that the number of individuals (anglers and their families) who consumed fish caught in Watts Bar and the Clinch River in the years since the ORR activities began is perhaps 100,000.

Nine exposure pathways exceeded a decision guide in the level I evaluation. They are, for EFPC farm family members, ingestion of sediment, soil, and fish; ingestion of beef, milk, and vegetables that received PCBs from contaminated soil; drinking of milk that received PCBs from contaminated pasture; and contact with contaminated soil. For consumers of fish from Poplar Creek, the Clinch River, or Watts Bar Reservoir, ingestion of fish was the pathway that exceeded the decision guide. These pathways were further evaluated in a second assessment (level II evaluation). In focusing on the pathways that yielded doses above the decision guides, no judgment was made on the acceptability of the risks associated with the pathways that yielded doses below those guide values.

Results of Level II Analysis   The level II evaluation demonstrated that there was considerable variation in both noncarcinogenic and carcinogenic risk estimates for all of the populations evaluated. The risks to adults and children did not differ greatly and, in most cases, the ranges in the risk estimates overlapped. Adults tended to have slightly higher cancer risks because their longer exposure durations resulted in higher lifetime average daily doses. Noncancer risks also tended to be slightly higher. The estimates of risk for the median (50th percentile) and 95th percentile of the cumulative distribution of exposures in each population are presented in Tables 14.4 and 14.5. The majority of the exposures for certain populations (other than those along EFPC) occurred from PCB sources other than the ORR. Estimates of cancer risk from a source of PCBs are directly proportional to the relative contribution of that source. Table 14.4 includes estimates of cancer risks from ORR releases as well as all sources for certain populations. It is more difficult to separate the relative contribution of the ORR releases for estimates of noncancer risks (Table 14.5) because the impact of the additional exposure is a function of the level of exposure from other sources.

Because the sizes of the groups of fish consumers were determined and the distribution of risks was calculated for each population, the analysis also determined the number of excess cases of cancer that would be expected to occur in the populations. Fewer than three excess cases of cancer are expected to occur in the populations of recreational anglers who fished the Clinch River and Watts Bar since the late 1940s. Because the carcinogenic potency of PCBs assumed in this study is thought to overestimate risk (Evans et al., 1994), the actual number of cases is expected to be lower and may be zero. No cases are expected to occur in groups other than the recreational anglers because of the small sizes of these populations.

The dose reconstruction results for each population of interest, stated in terms of the estimates of health risk given in Tables 14.4 and 14.5, can be summarized as follows:

Recreational Fish Consumers   Cancer and noncancer risks for recreational fish consumers at EFPC were lower than for the other bodies of water. Risks for both adults and children were below the decision guides (with the exception of children at the 95th percentile where the nominal hazard quotient was 2 in level II evaluation). The lower fish consumption rates for EFPC, based on its poor quality as a fishery, accounted for the lower risk estimates.

The risks for recreational anglers using Clinch River/Poplar Creek were higher than those at EFPC. The cancer risk for adults at the 95th percentile was 3×10^{-4}. The cancer risk estimate for children at the 95th percentile was less than the decision guide of 1×10^{-4}. The nominal hazard quotient exceeded the noncancer guide of one for both adults and children at the median and 95th percentile in level II evaluation.

Recreational anglers using Watts Bar had the highest cancer and noncancer risks of the three water bodies due to higher levels of PCBs in the fish and greater fish consumption rates. The cancer risk estimate for adults was 6×10^{-4} at the 95th percentile and for children was 1×10^{-4} at the 95th percentile.

TABLE 14.4 Summary of Cancer Risks Evaluated in Level II Analysis of PCBs in Environment near Oak Ridge

Population	Adult			Child		
	East Fork Poplar Creek	Watts Bar Lake	Clinch River/ Poplar Creek	East Fork Poplar Creek	Watts Bar Lake	Clinch River/ Poplar Creek
Recreational fish consumer (from all sources of PCBs)[a]						
95th percentile[b]	c	6×10^{-4}	3×10^{-4}	c	1×10^{-4}	5×10^{-5}
Median	c	4×10^{-5}	2×10^{-5}	c	1×10^{-5}	5×10^{-6}
Recreational fish consumer (from Oak Ridge releases)						
95th percentile	1×10^{-5}	8×10^{-5}	2×10^{-4}	4×10^{-6}	1×10^{-5}	3×10^{-5}
Median	8×10^{-7}	5×10^{-6}	8×10^{-6}	3×10^{-7}	1×10^{-6}	2×10^{-6}
Commercial angler (from all sources of PCBs)						
95th percentile	d	4×10^{-5}	2×10^{-5}	d	9×10^{-5}	2×10^{-5}
Median	d	4×10^{-6}	2×10^{-6}	d	1×10^{-5}	1×10^{-6}
Commercial angler (from Oak Ridge releases)						
95th percentile	d	5×10^{-6}	1×10^{-5}	d	1×10^{-5}	8×10^{-6}
Median	d	5×10^{-7}	8×10^{-7}	d	1×10^{-6}	7×10^{-7}
Farm family						
95th percentile	2×10^{-3}	e	e	9×10^{-4}	e	e
Median	1×10^{-4}	e	e	1×10^{-4}	e	e
Recreational user						
95th percentile	4×10^{-7}	e	e	2×10^{-7}	e	e
Median	3×10^{-8}	e	e	2×10^{-8}	e	e

[a] Estimated risks for all sources include sources of PCBs other than the Oak Ridge complexes.

[b] Numbers are expressed in scientific notation; for example $5 \times 10^{-8} = 0.00000005$.

[c] There are no identified sources of PCB releases to EFPC other than the Y-12 plant.

[d] There has been no commercial fishing of EFPC.

[e] These populations were evaluated only for East Fork Poplar Creek.

TABLE 14.5 Summary of Noncancer Risks (Hazard Quotients) Evaluated in Level II Analysis of PCBs in Environment near Oak Ridge

	Adult			Child		
Population	East Fork Poplar Creek	Watts Bar Lake	Clinch River/ Poplar Creek	East Fork Poplar Creek	Watts Bar Lake	Clinch River/ Poplar Creek
Recreational fish consumer						
95th percentile	1	40	20	2	60	30
Median	0.1	4	2	0.2	5	2
Commercial angler						
95th percentile	[a]	40	6	[a]	50	7
Median	[a]	4	0.6	[a]	5	0.7
Farm family						
95th percentile	100	[b]	[b]	200	[b]	[b]
Median	20	[b]	[b]	40	[b]	[b]
Recreational user						
95th percentile	0.05	[b]	[b]	0.09	[b]	[b]
Median	0.005	[b]	[b]	0.01	[b]	[b]

Note: Hazard quotients are based on estimated doses divided by reference dose for Aroclor 1254 established by the EPA.

[a] There has been no commercial fishing of EFPC.

[b] These populations were evaluated only for East Fork Poplar Creek.

Similar to the Clinch River/Poplar Creek analysis, the nominal hazard quotients for Watts Bar Reservoir in the level II evaluation exceeded the noncancer guide of one for both adults and children at the median and 95th percentile.

As discussed earlier, the contribution of ORR PCB releases likely represented 9 to 13 percent of the total amount of PCBs in Watts Bar. Therefore, at most 13 percent of the total cancer risk for recreational anglers using Watts Bar was attributable to the ORR. Because 13 percent of 6×10^{-4} is less than 1×10^{-4}, ORR releases do not appear to have resulted in cancer risks that exceed the guides.

The percentage of PCBs in the Clinch River that is attributable to the ORR releases is not as well defined. Up to one half of the PCBs in Clinch River fish may have been contributed by other sources on the Clinch River. However, even if half the PCBs were contributed other sources, the ORR releases appear to have resulted in risks above the cancer decision guide.

Commercial Anglers In general, the risks to commercial anglers were similar to, but slightly lower than, risks estimated for recreational fish consumers. At the 95th percentile, all of the cancer risks for these populations were equal to or less than 1×10^{-4}. The nominal hazard quotients for the 95th percentile adult and child were greater than one for the Clinch River/Poplar Creek anglers and

TABLE 14.6 Estimates of True Hazard Quotients for Adult Recreational and Commercial Fish Consumers Evaluated in Level III Analysis of PCBs in Environment Near Oak Ridge

Exposure Level	Watts Bar Reservoir	Clinch River/Poplar Creek
95th percentile	1 (0.2–8)	0.5 (0.08–3)
Median	0.1 (0.02–0.5)	0.05 (0.008–0.3)

Note: The level III analysis included a two-dimensional analysis of uncertainty and variability in noncancer risks to fish consumers. True hazard quotients are equal to the estimated dose divided by the population threshold estimated in this project for noncancer health effects from PCB exposure. These values include sources of PCBs other than the Oak Ridge complexes. Numbers in parentheses are 90% confidence limits.

were higher for the Watts Bar Reservoir anglers in the level II evaluation. As with recreational fish consumers, risks for commercial anglers are affected by other sources of PCBs along the rivers.

EFPC Farm Family The estimates of cancer and noncancer risks were higher for the farm family population than the fish consumer groups. Cancer risk estimates at the 95th percentile for both adults and children exceeded the cancer decision guide. Nominal hazard quotients exceeded the noncancer decision guide for the entire range of the level II uncertainty analysis. It should be noted, however, that the actual concentrations of PCBs in soil at the farms were highly uncertain, and this uncertainty could not be characterized because of a lack of data. Thus, these estimates may not be representative of true hazards associated with those populations.

EFPC Recreational Users Both the noncancer and cancer risks for the recreational users of EFPC were below the decision guides; at the 95th percentile, cancer risks for adults and children were 4×10^{-7} and 2×10^{-7}, respectively. Nominal hazard quotients for both age groups were also less than one.

Results of Level III Evaluation of PCB Exposures Table 14.6 presents the results of the level III evaluation. These estimates of the true hazard quotients provide a more unbiased estimate of noncancer risk than given in Table 14.5. As a result, the values tend to be much lower than the earlier estimates of the nominal hazard quotients.

The confidence intervals of the hazards indices calculated for each of the reference populations addressed in level II or level III assessments of exposures to PCBs are shown in Figure 14.24. These 90 percent confidence intervals are plotted along with the LOAEL value that is used by the EPA as the basis for the RfD for Aroclor 1254 (IRIS, 1998c).

While the risk of noncancer effects can be evaluated in terms of true hazard quotients, the risks can also be evaluated more directly in terms of the fraction

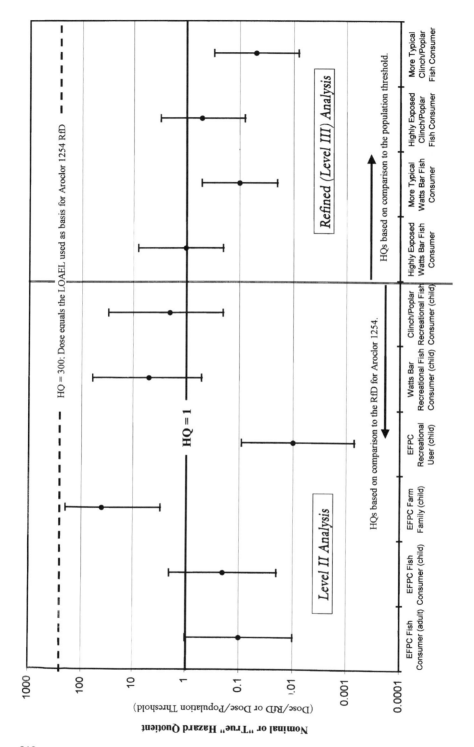

TABLE 14.7 Percentages of Fish-Consuming Populations Receiving a PCB Dose Above the Population Threshold Estimated in the Oak Ridge Dose Reconstruction

Population	Background	Background + Oak Ridge	Change Due to Oak Ridge
Watts Bar fish consumer			
Adult	5.0 (0.61–39)	6.6 (0.82–43)	1.6
Child	7.5 (0.92–44)	8.9 (1.4–48)	1.4
Clinch River/Poplar Creek fish consumer			
Adult	0.55 (0–9.7)	2.2 (0–21)	1.7
Child	0.97 (0–15)	3.8 (0–29)	2.8

Note: "Background" includes risks associated with PCB releases attributable to sources other than the Oak Ridge complexes; 87 and 50% of the total PCB concentrations in Watts Bar and the Clinch River/Poplar Creek, respectively, are assumed to be associated with other sources. Median value are given, with 90% confidence limits given in parentheses.

of the population that have received doses that exceed the population threshold (Table 14.7).

To investigate the impact of the ORR releases of PCBs relative to those from other sources, the exposure model calculations were done with and without the ORR contribution, and incremental changes noted. The analysis indicated that PCB contamination from non-ORR sources resulted in some anglers receiving doses greater than the population threshold. The releases from the ORR resulted in a small increase (1 to 2 percent of fish consumers) in the fraction of individuals receiving doses greater than the population threshold (Table 14.7). Had the ORR releases happened in the absence of the other sources, they would not likely have resulted in adverse effects.

14.5.6 Discussion

Based on the results of the level II evaluation, the following conclusions were reached:

Figure 14.24 Results of analyses of noncancer health hazards from PCBs in environment near Oak Ridge. The level II analysis included an uncertainty analysis, while the level III analysis included a refined, two-dimensional analysis of variability and uncertainty for fish consumers. Values shown are 90% confidence bounds and central values (medians) of hazard quotients, which are calculated by dividing estimated doses by the EPA RfD for Aroclor 1254 (level II analysis) or by the population threshold for noncancer effects estimated in the Oak Ridge Dose Reconstruction (level III analysis). In the refined analysis, the highly exposed fish consumer is represented by the 95th percentile of the population, while the more typically exposed consumer is represented by the 50th percentile.

1. Populations exposed from recreational use of EFPC do not appear to warrant additional assessment. The conservative estimates of cancer and non-cancer risks developed in the level II evaluation did not exceed the decision guides at the 95th percentile.

2. Exposures to PCBs from the consumption of fish in EFPC were also low. Cancer risks were well below the cancer guide. The adult angler was at the non-cancer guide and the child angler was slightly above the guide. Given the limited productivity of the creek and the uncertainty in the estimates of fish consumption, no additional analyses were performed.

3. Cancer and noncancer risks for the farm families greatly exceeded the decision guides. Thus, this population could be at risk from their PCB exposures and should be considered in additional analyses. However, the estimates of PCB exposures used in the assessment are highly uncertain due to the limited data on PCB levels in the farm soils. PCB levels in the sediments and floodplain of EFPC have been characterized to at least some degree in recent remedial investigations; however, the ranges of PCB concentrations historically present in the soils of the fields, pastures, gardens, and areas around the farm houses themselves were not investigated. Because of this data gap, we were unable to further assess risks to the families.

4. Risks to the commercial and recreational anglers using Watts Bar and the Clinch River were similar, suggesting that future assessments need not separate the groups. The carcinogenic risks to each population were above the decision guide. However, when the contribution of the ORR was considered, risks at the 95th percentile were smaller and in the case of Watts Bar were below the decision guide. The Clinch River anglers at the 95th percentile exceeded the guide by a factor of less than two. In addition, the estimate of the total number of cancer cases expected for all populations is small (less than three). Based on these findings, the team did not believe that additional effort to characterize the uncertainty in the carcinogenic risk estimates was warranted. Therefore, no additional work was performed on assessing the carcinogenic effects for these angler populations.

The true hazard quotients from the level III evaluation indicate that the typical (median exposure) fish consumer for either water body is not at risk (true hazard quotient is less than one). The highly exposed (95th percentile) angler may be at risk because the estimates of the true hazard quotients could exceed one. The level III results indicate that portions of the angling populations with very high fish intakes, several percent for Watts Bar and a few percent for the Clinch River, may have received doses in excess of the population threshold. Depending on the unknown value of the population threshold, this finding is likely for Watts Bar and possible for the Clinch River. However, it is not possible to determine the fraction of the population that actually experienced adverse effects. Because of interindividual variation in the tolerance to PCBs, the dose needed to cause an adverse effect in a typical person is higher than the dose that affects sensitive individuals. A small fraction of those who receive doses above the population threshold are likely affected.

14.6 RADIONUCLIDES RELEASED FROM X-10 TO CLINCH RIVER VIA WHITE OAK CREEK

14.6.1 Introduction

In the early days of the Manhattan Project, Clinton Laboratory, also referred to as the X-10 site and now called Oak Ridge National Laboratory, was designed to operate for one year as a pilot plant for the Hanford, Washington, operations for chemical separation of plutonium. All radioactive wastes generated from the X-10 pilot plant were to be stored in large, underground tanks. The original plans changed, and in 1944, the first radioactive effluents from the X-10 site entered White Oak Creek and flowed into White Oak Lake. White Oak Lake served as the final settling basin for contaminants released to White Oak Creek. Radionuclides remaining suspended in the water were released from the X-10 site via the flow over White Oak Dam, which is located 1 km (0.6 mile) upstream from the Clinch River (see Figure 14.25) (Browder, 1949).

Sources of Radioactive Waste During early X-10 operations, the chemical separation pilot plant was the major source of radioactive wastes. Some liquid wastes from the pilot plant were placed in open waste pits dug in the earth; ^{106}Ru began seeping from the pits into White Oak Lake in 1959. Strontium-90 and ^{137}Cs had also been placed in the pits, but these isotopes were retained by nearby soils; however, amounts of ^{106}Ru as high as 7.4×10^{13} Bq (2000 Ci) per year were reportedly released from White Oak Dam from 1959 to 1963 (Lomenick, 1963). From 1944 to 1991, approximately 5.9×10^{15} Bq (160,000 Ci) of radioactivity were released past White Oak Dam to the Clinch; of this amount, 91 percent was tritium (Webster, 1976), and the rest was fission and activation products.

It appears that a secondary source of radionuclide releases to the Clinch River was the scouring of contaminated sediment from White Oak Creek Embayment, the stretch of White Oak Creek between the dam and the Clinch River. After White Oak Lake was drained in 1955, heavy rainfall eroded the exposed bottom sediment, resulting in the relocation of particle-bound radionuclides (primarily ^{137}Cs) to White Oak Creek Embayment (Ohnesorge, 1986). Periodic discharges of water from Melton Hill Dam, which was completed upriver on the Clinch in 1963, resulted in the backflow of water up White Oak Creek Embayment and transport of contaminated sediments into the Clinch River. A coffer cell dam was constructed at the mouth of White Oak Creek in 1990 to prevent the backflow of water up White Oak Creek Embayment. While releases of embayment sediment essentially ceased at that time, releases of waterborne radionuclides continue, albeit at levels much reduced from earlier decades.

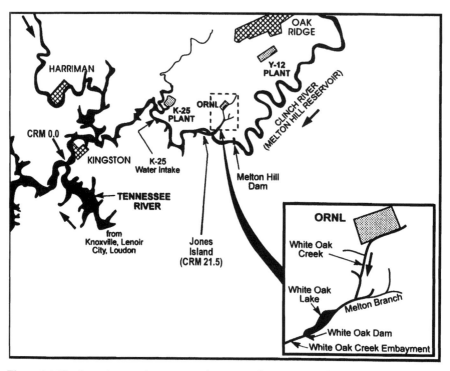

Figure 14.25 Locations and waterways important in reconstruction of public exposures to releases of radionuclides to Clinch River via White Oak Creek on Oak Ridge Reservation. Potential public doses and health risks were studied for people who have lived along the Clinch River from the mouth of White Oak Creek to Kingston, based on 48 years of independently reconstructed releases past White Oak Dam.

The purposes of task 4 of the Oak Ridge Dose Reconstruction were to:

- Estimate the historical releases of radionuclides from the X-10 facility to the Clinch River.
- Evaluate the potential pathways by which members of the public could have been exposed to radioactivity in the Clinch River from 1944 to 1991.
- Calculate radiation doses and health risks to reference individuals who were potentially exposed to radioactivity released to the Clinch River from the X-10 complex.

Direct measurement of the amounts of radionuclides taken up by the organs of specific individuals since 1944 is no longer feasible because most of these radionuclides have short residence times in the human body. Therefore, a dose reconstruction was necessary to estimate past exposures and to interpret the health consequences of these exposures. This dose reconstruction relied upon independent evaluations of the amounts of radionuclides released, reported environmental measurements, and mathematical modeling to estimate past doses and health risks.

14.6.2 Hazard Identification

Twenty-five radionuclides reported to have been released into the Clinch River from the X-10 site from 1944 to 1991 were considered as potential contaminants of concern (Gouge et al., 1999). To focus time and resources on the radionuclides that were most likely to have been important in terms of dose or risk to off-site individuals, a conservative screening evaluation was conducted. The screening analysis identified those radionuclides and pathways for which the estimated human health risk was below a minimum level of concern. Nine exposure pathways and 16 radionuclides, including uranium and tritium, were given low priority for further consideration because conservative screening estimates were at least a factor of 10 below the decision guide of 1 chance in 10,000 for carcinogens. Of the 8 remaining radionuclides, ^{137}Cs, ^{60}Co, ^{106}Ru, and ^{90}Sr were expected to be the most important contributors to radiation dose and subsequent excess health risk.

Internal radiation dosimetry describes the fate of an ingested radionuclide in the body, starting with the passage through the gastrointestinal tract; continuing with absorption to blood, distribution of the radionuclide to various organs in the body, and retention of the radionuclide in these organs; and ending with its elimination by radioactive decay and excretion. During transport through the body, radionuclides emit radiation that results in energy deposition in individual organs and increases the risk of cancer incidence.

The blood transports cesium to all organs and tissues, where it may be absorbed or excreted. Cesium is nearly uniformly distributed throughout the body and is excreted in both urine and feces. Bone and its components are the critical organs for strontium. Data demonstrate the high affinity of mineral bone for strontium, but they also show that the element is not significantly concentrated in any other organ or tissue (Coughtrey and Thorne, 1983). The liver, the kidney, and the urinary bladder tend to accumulate more cobalt within a short time after ingestion or intravenous injection than the other organs. During longer periods of time after administration, cobalt is uniformly distributed in the human body, with the exception of the liver (elevated levels after 1000 days in human subjects) (Leggett, 1997a, b). Only a small fraction of the ingested ruthenium is absorbed, with the rest being rapidly eliminated by fecal excretion. Yamagata et al. (1969) reported that in one male volunteer studied, 95 percent of the administered ruthenium was eliminated within 2 days. Ingested iodine is completely absorbed from the gastrointestinal tract into the bloodstream, where ^{131}I is considered to be metabolized similarly to stable iodine. Usually less than one-fourth of the iodine in the blood stream is taken up by the thyroid gland, while about three-fourths is collected by the kidneys and excreted in the urine.

14.6.3 Dose–Response Evaluation

The estimates of cancer incidence per unit dose that we used were based on cancer incidence data from the Japanese atomic bomb survivors, background incidence rates for cancer in eastern Tennessee, and the use of relative and absolute risk models to transfer epidemiologic findings in the atomic bomb survivors to people exposed to radionuclides released to the Clinch River. Additional details of our approach are given in Section 14.6.5.

14.6.4 Exposure Assessment

Radionuclides Released from White Oak Dam A detailed investigation was performed to identify the methods used for measurements of waterborne radioactivity at White Oak Dam, the methods used for estimation of flow rates at White Oak Dam, and the uncertainties associated with these two types of measurements. Estimates of the quantities of radionuclides historically released from White Oak Dam were based on information from laboratory documents, log books, and interviews with personnel who were responsible for, or involved in, collection and/or analysis of samples of releases from White Oak Dam. Early sampling of water at White Oak Dam is shown in Figure 14.26. Measurements of concentrations of specific radionuclides released from White Oak Dam were available for all years except 1944 to 1948. For these years, releases were calculated based on the estimated fraction that each radionuclide contributed to estimated releases of gross beta-emitting radioactivity. Annual estimates of releases (source terms) were developed for the following radionuclides: cobalt-60, strontium-90, niobium-95, zirconium-95, ruthenium-106, iodine-131, cesium-137, and cerium-144. The uncertainties of the source terms vary over time. We applied factors to account for uncertainty and/or bias introduced by: (1) methods used to back-fit gross beta releases to specific radionuclides, (2) scaling of White Oak Creek flow rates from the Little Chestuee, (3) White Oak Creek flow measurements, (4) flow rate estimation when White Oak Lake was drained, (5) use of nonproportional sampling rates, (6) laboratory processing, (7) detector efficiency, and (8) counting statistics. Peak annual releases for these radionuclides, as well as indications of the lengths of their periods of highest releases, are shown in Table 14.8.

Estimated Concentrations in Water and Sediments Measured concentrations of radionuclides in water are available for many years for several locations downstream from where White Oak Creek enters the Clinch River. These measurements varied somewhat in their locations and methods of measurement, and did not include all the radionuclides of concern. Therefore, mathematical modeling was performed to estimate annual average concentrations of radionuclides in water at specific locations downstream of White Oak Creek. A modified version of the HEC-6 aquatic transport model [U.S. Army Corps of Engineers (ACE), 1993], called HEC-6-R, was used to estimate historical water

Figure 14.26 Water sampling at White Oak Dam on Oak Ridge Reservation in 1949. Starting in 1944, White Oak Creek carried liquid radioactive wastes from X-10 site operations and later from waste disposal pits, trenches, and burial grounds into White Oak Lake. From the lake, water flowed over White Oak Dam and into the Clinch River. ORNL photo.

TABLE 14.8 Summary of Peak Annual Releases for Eight Key Radionuclides Evaluated in Dose Reconstruction for Releases to Clinch River via White Oak Creek

Radionuclide	Peak Annual Release (Ci)			Number of Years $\geq$ 10% of Peak
	Lower Bound	Central Value	Upper Bound	
Cs-137	50	200	510	14
Ru-106	1600	2100	2700	5
Sr-90	68	190	390	18
Co-60	64	85	110	15
Ce-144	70	94	120	13
Zr-95	72	210	440	9
Nb-95	17	200	520	10
I-131	10	68	190	10

Note: Values shown are the highest annual totals estimated over the period 1944–1991, stated as 95% subjective confidence intervals and central values (medians), as well as number of years that releases of each radionuclide were at least 10% of peak year.

concentrations. The estimates of annual releases of specific radionuclides from White Oak Dam were used in the modeling analysis. The uncertainty of the modeled water concentrations was much higher than the uncertainty of water concentrations obtained from measurements; therefore, measurements from specific locations and time periods were used when possible rather than model predictions when there were enough measurements to estimate annual concentrations in water. In particular, the model did not always account well for scouring of sediment after Melton Hill Dam began operation in 1963. Concentrations of radionuclides in river sediment were estimating using the HEC-6-R model to track the sediment inventory in various sections (reaches) of the Clinch River over time. Measurements made in the 1990s were used to calibrate the sediment concentration estimates. Because of the limited availability of measurements, all sediment concentrations used in the risk assessment were based on model estimates.

Estimation of Exposures to Reference Individuals For all locations addressed in this study, the exposure pathways of interest included ingestion of fish, milk, and meat; other exposure pathways of interest varied with location. For the Jones Island area [Clinch River mile (CRM) 21 to 17], the exposure pathways of interest were ingestion of fish, meat, and milk, plus external exposure from standing on or near shoreline sediment. The exposure pathways for the K-25/Grassy Creek area (CRM 17 to CRM 5) included ingestion of fish, meat, milk, and drinking water, plus external exposure to shoreline sediment. For the Kingston Steam Plant area (CRM 5 to CRM 2), the important pathways were ingestion of drinking water, fish, meat, and milk, plus external exposure to shoreline sediment. Exposure pathways for residents of Kingston (CRM 2.0 to CRM 0.0) included ingestion of drinking water, fish, and milk and meat from livestock who drank river water, plus external exposure to shoreline sediment. The areas along the Clinch River that were assessed in the dose reconstruction can be seen Figure 14.25.

For each exposure pathway, reference populations were identified, with varying characteristics and population sizes. There were three categories of fish consumers (Ebert, 1996):

- People in category I were assumed to eat fish regularly, that is, from 52 to 130 meals/year.
- People in category II were assumed to eat from 12 to 52 meals/year.
- People in category III were assumed to eat from 2 to 17 meals/year.

For all categories, it was assumed that 20 to 100 percent of the fish eaten was contaminated, and that 80 to 90 percent of the radioactivity in the fish was retained after processing for consumption.

Two reference individuals, an adult and a child, were considered for the water ingestion pathway. Children were not considered for the K-25/Grassy Creek area or the Kingston Steam Plant area because these are industrial facilities and it is not likely that children would have drunk water from these locations. However, both children and adults were exposed via the Kingston water supply. Multiple reference individuals were considered for the milk ingestion pathway, including children who drank different amounts of home-produced milk depending on whether they were at home or in school. Adults were considered for meat ingestion and external exposure from contaminated sediment.

Estimation of Organ-Specific Radiation Doses The International Commission on Radiological Protection (ICRP) developed a method for calculating radiation doses to people ingesting contaminated food or water (ICRP, 1993). To reflect variability among individuals, ranges of values were developed for the factors that specify the dose per unit intake for given radionuclides. For ^{137}Cs, ^{60}Co, and ^{106}Ru, the ICRP ingestion dose factors were modified by applying several uncertainty factors, the values of which depended on the radionuclide and organ of interest. New dose conversion factors and their uncertainties were calculated for ^{90}Sr and ^{131}I for all internal organs of importance using the ICRP methodology. Each factor was specified as a range of values rather than a point estimate (NCRP, 1996a).

Doses from Fish Ingestion The estimated organ doses to people who ate fish exceeded the dose estimates for all other pathways. The highest doses were for category I fish eaters just below where White Oak Creek enters the Clinch River. Central values of the cumulative doses for 1944 to 1991 for specific organs ranged from 0.31 centisievert (cSv) to the skin to 0.81 cSv to the bone for males and from 0.23 (skin) to 0.60 cSv (bone) for females. One cSv equals 1 rem. The 95 percent subjective confidence intervals ranged from about 0.02 to 8 cSv. Organ doses were generally lower for females than for males, due to the lower ingestion rates assumed for females. For category I fish eaters near Kingston, organ doses are about a factor of 8 to 9 lower than those estimated for the area near Jones Island. Estimated organ doses for categories II and III fish eaters are lower than those for category I in proportion to the lower consumption rates for these categories.

Doses from Other Exposure Pathways   Organ doses from external exposure were about a factor of 1.1 to 3.5 lower than the doses to a category I fish eater near K-25/Grassy Creek, with the largest doses to skin, bone, and thyroid. Adults who spent time along the shoreline but who seldom ate fish probably received the same or higher organ doses from external exposure as from eating fish.

For most organs, doses from drinking water near K-25/Grassy Creek and the Kingston Steam Plant were lower than the doses from external exposure at the same location. However, for the large intestine, bone, and red bone marrow, the doses from drinking water were higher than those from external exposure or eating fish (by category II or III consumers) due to the presence of ^{90}Sr and ^{106}Ru.

Estimated doses from ingestion of meat and milk were lower than those for ingestion of drinking water by factors of about 10 to 1000. The highest doses were to the large intestine, bone, red bone marrow, and (for the ingestion of milk) the thyroid gland.

Illustrative Example   Calculate the radiation dose to the bone surfaces from one year of exposure to ^{137}Cs contamination in shoreline sediment near Clinch River mile 14 in 1947. The following generic approach to estimate the doses for people exposed to contaminated Clinch River shorelines is based on the concentration of the radionuclide in the contaminated sediment multiplied by a dose-rate factor. In the actual calculation of external doses from shoreline exposure, the values for several of the following parameters were repeatedly selected from distributions of each parameter using Monte Carlo simulation techniques. However, for illustrative purposes, single values will be selected from within the appropriate distributions for this example:

$$D = \sum C_s \times \mathrm{DRF}_{\mathrm{ext}} \times \mathrm{EF} \times \Delta t$$

where D = total dose to a given organ (Sv)

$\quad$ C_S = concentration of the radionuclide in the shoreline sediment (Bq kg^{-1})

$\mathrm{DRF}_{\mathrm{ext}}$ = external dose-rate factor (Sv yr^{-1} per Bq kg^{-1}) defined as the dose received during the period of exposure by an individual standing on a shoreline having a unit sediment contamination

$\quad$ EF = exposure frequency (unitless)

$\quad$ Δt = 1 year; summation is performed over the number of exposure years and

$$DRF_{ext} = DRF_{ext,p} \times (z) \times G \times H$$

where $DRF_{ext,p}(z)$ = published dose-rate factors (Eckerman and Leggett, 1996), expressed as a function of z

z = thickness of the contaminated sediment slab (m)

G = geometry adjustment factor (unitless) that accounts for the particular geometry of the shoreline

H = uncertain correction factor (unitless) that accounts for other sources of uncertainty such as the nonhomogeneity of the contamination across the shore and the movement of the individual on the contaminated surface

The solution (based on point estimates selected for example purposes) is as follows:

C_S = 1100 Bq kg^{-1} of ^{137}Cs/^{137}Ba (central estimate for Clinch River mile 14 in 1947)

EF = assume 0.05 (5 percent of the time) for example purposes

Δt = 1 year (to calculate an annual value)

z = 7 cm (selected from a triangular distribution from 2 to 15 cm, with 7 cm most likely)

$DRF_{ext,p}(z)$ = 3.87 × 10^{-6} Sv y^{-1} per Bq kg^{-1} for the bone surface for soil contaminated to a depth of 7 cm

G = 0.58 [derived for 0.661 MeV gamma rays and a shoreline width of 30 m; see Section 10.4.2 of Gouge et al. (1999)]

H = 1.02 [selected from a uniform distribution between 0.95 and 1.05; see Section 10.4.4 of Gouge et al. (1999)]

and thus

$$DRF_{ext} = (3.87 \times 10^{-6} \text{ Sv y}^{-1} \text{ per Bq kg}^{-1}) \times 0.58 \times 1.02$$
$$= 2.29 \times 10^{-6} \text{ Sv y}^{-1} \text{ per Bq kg}^{-1}$$

Therefore

$$D = 1100 \text{ Bq kg}^{-1} \times 2.29 \times 10^{-6} \text{ Sv y}^{-1} \text{ per Bq kg}^{-1} \times 0.05 \times 1 \text{ y}$$
$$= 1.26 \times 10^{-4} \text{ Sv}$$

Thyroid Doses to a Child from Drinking Water and Milk   The 95 percent subjective confidence interval for the doses to a child age 14 or below drinking milk near K-25/Grassy Creek or the Kingston Steam Plant were 0.00058 to 0.054 cSv (0.0062 central value) and 0.00055 to 0.042 cSv (0.0044 central value), respectively. The 95 percent subjective confidence interval for the estimated drinking water dose for a child living in Kingston was 0.000039 to 0.0021 cSv (0.00031 central value) and for the combined pathways (drinking water and milk), 0.00014 to 0.0047 cSv (0.00091 central value). The exposure time for a child drinking water was shorter than that for milk because the Kingston municipal water supply did not become a potential source of contamination until 1955 (Davis, 1997).

14.6.5   Risk Characterization

Organ dose estimates were used to estimate organ-specific and total excess lifetime risks of cancer incidence. Estimates of cancer incidence per unit dose were based on data from the Japanese atomic bomb survivors [United Nations Scientific Committee on the Effects of Atomic Radiation (UNSCEAR), 1994; Thompson et al., 1994], background incidence rates for cancer in eastern Tennessee [Tennessee Cancer Reporting System (TCRS), 1992, 1996], and the use of relative and absolute risk models to transfer epidemiologic findings in the atomic bomb survivors to people exposed to radionuclides released to the Clinch River. The uncertainty due to differences in responses between exposures at high dose rates and low dose rates was considered explicitly in the calculation of risks for each organ. The uncertainty due to differences in responses between exposures at high dose rates and low dose rates was considered explicitly in the calculation of risks for each organ.

Cancer Risks from Fish Ingestion   For any given location, risks of excess lifetime cancer incidence for categories II and III fish eaters are lower than those for category I consumers by factors of about 2 and 8, respectively, in proportion to the lower intake rates assumed for these categories (Figure 14.27). Upper bounds on the total risk from fish consumption for category I fish eaters exceed 1×10^{-3} at Jones Island and 1×10^{-4} at reference locations farther downstream (Figure 14.28). Central estimates exceed 1×10^{-4} at Jones Island for categories I and II fish eaters; they are between 1×10^{-5} and 1×10^{-4} for most other cases addressed, except for category III fish eaters farther downstream than Jones Island, for which they fall between 1×10^{-6} and 1×10^{-5}.

For ingestion of fish from the Jones Island area, the upper bounds on the risk for both males and females were highest (exceeding 1×10^{-4}) for bladder, stomach, lower large intestine, lungs, and red bone marrow (leukemia). For females, the upper bound on the risk estimates for breast also exceeded 1×10^{-4}, as did the liver for males. Although the breast received among the lowest doses, the breast has the highest risk of all the organs examined. For females, the highest risks are for breast and red bone marrow (central values of 5.5×10^{-5}

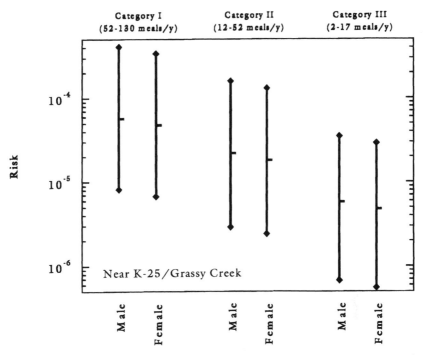

Figure 14.27 Excess lifetime risk of cancer incidence for males and females consuming fish at different rates from Clinch River near K-25 site and Grassy Creek. The vertical lines indicate the 95% subjective confidence intervals of the estimated risks, and central values (medians) are also indicated.

and 2.8×10^{-4}, respectively); for males, the highest risk is for the red bone marrow (central value of 3.7×10^{-5}). The difference between the highest and lowest organ-specific risks at any one location is about a factor of 70 to 80 for females and 40 for males, although the differences in doses were only a factor of 2 to 4. This situation illustrates the great difference in organ sensitivities to radiation-induced cancer and underlines the importance of calculating risks as well as doses in a dose reconstruction study because the organ with the highest dose may not be the organ at highest risk.

For category I fish eaters near Jones Island, the 95 percent subjective confidence interval of the total excess lifetime risk of cancer incidence for all radionuclides and organs was 3.6×10^{-5} to 3.5×10^{-3} (2.8×10^{-3} central value) for males and 2.9×10^{-5} to 2.8×10^{-3} (2.3×10^{-4} central value) for females. The difference in risk between males and females reflects primarily the difference in meal sizes. For both males and females, the largest contribution to the total risk (about 90 percent) is from ^{137}Cs.

For individuals using or living on Watts Bar Reservoir, the exposures, doses, and risks are substantially lower than those for individuals using any segment

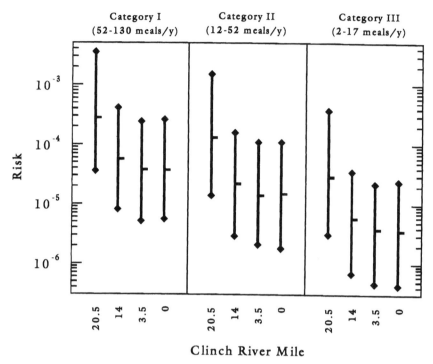

Clinch River Mile

Figure 14.28   Excess lifetime risk of cancer incidence from all exposure pathways for three reference males eating fish at different rates at reference locations along Clinch River (CRM 20.5 = Jones Island area; CRM 14 = K-25/Grassy Creek area; CRM 3.5 = Kingston Steam Plant area; CRM 0 = Kingston area). The vertical lines indicate the 95% subjective confidence intervals of the estimated risks, and central values (medians) are also shown. Risks for females are slightly lower than for males (see Figure 14.27).

of the Clinch River. The best estimate is that exposures from the past consumption of contaminated fish in Watts Bar Reservoir are 4 to 25 times less than for people eating fish from the Clinch River near the K-25/Grassy Creek area, assuming similar ingestion rates.

Cancer Risks from Other Exposure Pathways   Depending on the location, the external radiation dose from shoreline sediments contributes as much as 90 percent of the total risk from all pathways for a category III fish eater; fish ingestion contributes about 10 percent and drinking water from 2 to 30 percent of the total risk of cancer incidence. For category II fish eaters, fish ingestion contributes 30 to 40 percent of the total risk, depending on location, and for category I, about 50 to 60 percent, except for near Jones Island, where the external exposure is low and exposure via drinking water did not occur. For the drinking water or external exposure pathways alone, and combining all pathways for a category III fish eater, upper bounds barely exceed 1×10^{-4}.

Risks of Thyroid Cancer for a Child Who Drank Water and Milk   The highest excess lifetime risk of thyroid cancer for a child who drank water and milk occurred for a girl who drank milk obtained from an area near K-25/Grassy Creek (95 percent subjective confidence interval, 1.1×10^{-7} to 2.5×10^{-5}; central value, 1.8×10^{-6}).

Risk Estimates for Shorter Exposure Periods   In most cases, individuals were not exposed to the various pathways over the entire period from 1944 to 1991. In addition, both the operations at the X-10 site and the releases of radionuclides to the Clinch River changed over time. To account for more realistic exposure times, risks were summarized by decade. The first two decades (1944 to 1953 and 1954 to 1963) produced the highest risks for each pathway and from all pathways combined (Figure 14.29). In the first decade, the ingestion of fish dominated the total risk; however, external exposure to shoreline sediments became increasingly important in later years. Because the ingestion of fish and external exposure to shoreline sediments contributed most of the excess lifetime risk of cancer incidence, ^{137}Cs was the dominant radionuclide in all decades. In addition to risk estimates by decade, estimates of total risk per year at near K-25/Grassy Creek were also made in terms of risk per pound of fish eaten, hour of exposure to shoreline sediment, and liter of water drank.

Results for Some Special Scenarios   Because some people consumed fish bones as well as flesh when eating fish patties (ORNL, 1985), an evaluation was made of the doses and risks from substitution of 8 to 20 percent of a category I fish consumer's intake with fish patties. Doses and risks to bone and red bone marrow were increased about 15 to 25 percent due to increased ingestion of ^{90}Sr that accumulated in the fish bones. However, because ^{90}Sr was a small contributor to total dose and risk from eating fish, the overall risk was not increased by the consumption of fish patties.

Exposures were also evaluated from the consumption of contaminated wildlife (fish, turtles, deer, or waterfowl) from the ORR. Risks per meal (4 to 16 ounces of meat) were estimated for the highest reported contaminant levels in these animals (in the late 1940s for fish, the 1980s for waterfowl, and the early 1990s for turtles and deer) and for more typical levels. For the most contaminated animals, risks were as high as 3×10^{-4} per meal. Risks per meal for more likely values did not exceed 2×10^{-6}. The number of people exposed to contaminated animals from the ORR has not been determined precisely, but it is likely a very small fraction of the total population exposed to contaminated fish, water, or sediment.

14.6.6   Discussion

For all locations and ingestion rates examined, the dominant sources of uncertainty in the risk from fish ingestion are the concentration of ^{137}Cs in fish and the amount of fish eaten. The relative importance of a specific parameter de-

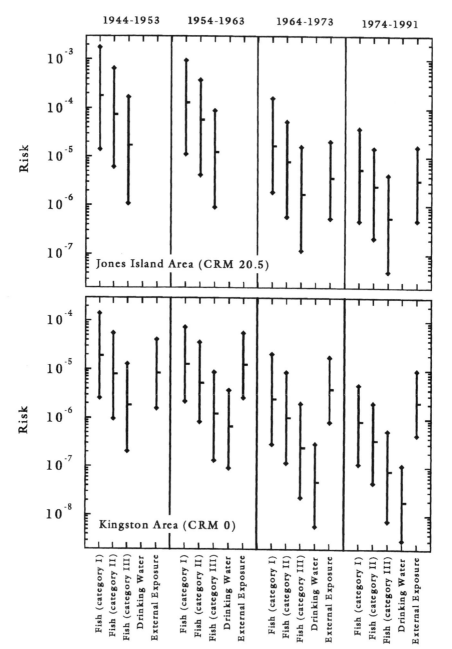

Figure 14.29 Excess lifetime risk of cancer incidence for males from eating fish, drinking water, and external exposure to shoreline sediment near Jones Island (top) and Kingston (bottom) during four time periods. The vertical lines indicate the 95% subjective confidence intervals on the risk estimates, and central values (medians) are also shown. Results for females from fish ingestion are slightly lower than those for males (see Figure 14.27).

pends on the location of exposure and the ingestion rates; in most cases, the bioconcentration factor (Struxness et al., 1967) is the single most important parameter affecting the overall uncertainty. For external exposure, the most important contributors to uncertainty are the concentrations of ^{137}Cs and ^{60}Co in shoreline sediments, followed by the dose-to-risk conversion factors. For internal exposure from drinking water, the most important sources of uncertainty are the amount of the radionuclide consumed, followed by the risk factors and the concentrations of ^{106}Ru and ^{90}Sr in the water. Uncertainty in dosimetry contributes less than 5 percent (internal) or 10 percent (external) of the total uncertainty, while the risk factor (except for internal exposure to ^{137}Cs) contributes 20 to 30 percent. Uncertainties in exposure parameters (such as radionuclide concentrations and amounts of exposure) are dominant for all pathways.

The doses and excess lifetime cancer risks from our dose reconstruction are incremental increases above those from exposure to background sources of radiation. Nevertheless, for the exposure pathways considered in this study, the doses and risks are not large enough for a resulting increase in health effects in the population to be detectable, even by the most thorough of epidemiologic investigations. In most cases, the estimated organ doses are clearly below the limits of epidemiologic detection (1 to 30 cSv) for radiation-induced health outcomes in studies of large groups of people irradiated in utero, as children, or as adults.

Even for category I fish eaters, estimated upper-bound organ doses are below 10 cSv, and central values are below 1 cSv. Lower-bound doses are well below levels that have been considered as limits of epidemiologic detection in studies of other exposed populations (Cowser and Snyder, 1966; Farris et al., 1994). The large uncertainty, combined with the small number of category I fish eaters, diminishes the statistical power available to detect a dose–response relationship through epidemiologic investigation. Therefore, it is unlikely that any trends in incidences of disease in populations that used the Clinch River and Watts Bar Lake after 1943 could be conclusively attributed to releases from the X-10 site, even though increased individual risks have been shown to have resulted.

14.7 OVERALL SUMMARY OF PROJECT RESULTS

Figures 14.30 and 14.31 summarize the maximum estimated health risks for each contaminant addressed with a detailed dose reconstruction as part of this project. Figure 14.30 presents the maximum calculated excess cancer risks for iodine-131, PCBs, and radionuclides released to the Clinch River. The highest values were obtained for iodine-131. For a female who was born in 1952 near Gallaher Bend and drank goat milk (from one to five 8-ounce glasses per day through 1956), we are confident that excess thyroid cancer risk is greater than 1.8 in 1000 but no more than 430 in 1000. The central estimate is 31 in 1000. It

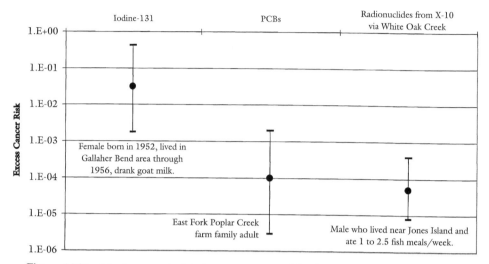

Figure 14.30 Maximum calculated excess cancer risks for carcinogens evaluated in Oak Ridge Dose Reconstruction. Values shown represent subjective confidence intervals and central values (medians).

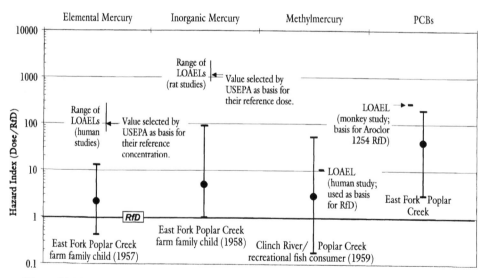

Figure 14.31 Maximum calculated hazard indices for noncarcinogens evaluated in Oak Ridge Dose Reconstruction, with toxicologic benchmark values. Values shown represent subjective confidence intervals and central values (medians).

is important to remember that the cases presented in Figure 14.30 are those that resulted in the highest calculated cancer risks in the assessments of exposures to iodine-131, PCBs, and radionuclides from White Oak Creek. There are other reference populations and other diets for which calculations were performed that are estimated to have received significantly lower doses. In effect, the cases presented in this figure and the figure that follows represent the worst-case exposures addressed in this project.

Figure 14.31 presents the maximum calculated noncancer health risks for mercury (in elemental, inorganic, and methylmercury forms) and for PCBs. The highest hazard index values were obtained for PCB exposures to children in an East Fork Poplar Creek farm family. For a child living on an EFPC farm, we are confident that the screening index is greater than 3 but no more than 200. The central estimate is 40. Again, the cases presented in Figures 14.31 are those that resulted in the highest calculated noncancer hazard indices in the assessments of exposures to mercury and PCBs. There are other reference populations and other diets for which calculations were performed that are estimated to have received significantly lower doses.

The confidence intervals for the maximum calculated hazard indices for exposure to the three forms of mercury and for PCBs presented in Figure 14.31 are graphed along with the values of their EPA reference doses, which in each case correspond to a hazard index of 1, and the published LOAEL value or range of LOAEL values. Except for the case of methylmercury exposures from Clinch River fish, the 95 percent confidence intervals fall below the applicable LOAEL values.

Concurrent Exposures to Several Contaminants   Hundreds of combinations of reference locations and exposure pathways were addressed in our study, for a wide variety of contaminants. Figure 14.6 illustrates how the key operations that were sources of historical releases overlapped in time during a number of periods. While the contaminants studied in this project were to a large extent studied independently, people that have lived in or near Oak Ridge could have been exposed to several contaminants at the same time.

When individuals are exposed to several contaminants at the same time, there are a number of possible ways in which the effects of the individual contaminants can combine. The four most common types of toxicologic interactions are additivity, synergism, potentiation, and antagonism. These forms of interaction can be defined as follows:

- *Additivity* is when the effects of the individual contaminants combine to cause an overall effect that is equal to the sum of their effects as individual contaminants (sometimes depicted $1 + 1 = 2$, using a representation where 0 indicates no effect and increasing numbers indicate increasing adverse effect).
- *Synergism* is when the effects of the individual contaminants combine to cause an overall effect that is greater than the sum of their effects as in-

dividual contaminants $(1 + 1 = 5)$. An example of a synergistic effect is Valium taken with alcohol. For agents to have synergistic toxicity, they must have the same target organ, they must be capable of causing the same adverse response, and they must be present at sufficient concentrations in affected body tissue to cause that response.

- *Potentiation* is when a contaminant would normally have no effect at the exposure level in question, but the presence of the other contaminant activates (or potentiates) the toxicity of the material. The end result is a degree of overall effect that is greater than the sum of the effects of the individual contaminants $(0 + 1 = 3)$.
- *Antagonism* is when toxic materials deactivate or decrease the toxicity of other materials. The end result is a degree of overall effect that is less than the sum of the effects of the individual contaminants, or there is no adverse effect at all $(1 + 1 = 0)$.

Unfortunately, very little is known about the ways in which the effects of the contaminants that were studied in this project combine when people are exposed to environmental concentrations. There are no well-documented studies that demonstrate synergistic or antagonistic effects of the environmental contaminants of interest. Lacking a clear picture of the possible manners in which the contaminants could have interacted toxicologically, the emphasis in this project has been to describe the exposures to each contaminant that could have occurred to a number of reasonable reference populations, and to identify the time periods when exposures to several contaminants could have occurred at the same time.

We prepared eight example exposure conditions (scenarios) to serve as examples of how off-site individuals could have reasonably been exposed to several contaminants during their residence near the ORR (Widner, 2000).

APPENDIX A: SCREENING ASSESSMENT OF HAZARDS FROM OTHER MATERIALS

The purpose of screening in the Oak Ridge Dose Reconstruction was to permit attention and resources to be focused on the most important contaminants and to avoid dilution of resources by identifying situations that are of minor importance (Cockroft et al., 1999). A two-level screening approach was used to:

- Identify those contaminants that produced off-site doses or health risks that are clearly below established minimum levels of concern or decision guides (level I screen)—these materials were assigned a low priority for further study.

- Identify those contaminants that produced off-site doses or health risks that are likely to have been above the established minimum levels of concern or decision guides (level II screen)—these materials were assigned the highest priority for detailed study.

The results of the screening calculations were compared to risk-based decision guides adopted by the ORHASP. As discussed earlier, for radionuclides and carcinogenic chemicals, the decision guide was a lifetime excess cancer incidence of 1 in 10,000 (10^{-4}). For noncarcinogenic chemicals, the decision guide was a hazard index of 1.0.

A.1 LEVEL I SCREENING

The two-level screening approach used different sets of assumptions for releases, environmental transport, exposures, and lifestyles to estimate doses or health risks. Level I screening was designed to estimate the dose or risk to a "maximally exposed" reference individual who should have received the highest exposure and thus would have been most at risk and incorporated conservative parameter values (e.g., intake rates) not expected to lead to an underestimate of risk to any real person. For Level I screening, the screening value was compared to the appropriate decision guide as follows:

- If the screening estimate of risk to the maximally exposed individual was clearly below the decision guide, it was concluded that further study of the contaminant can be deferred until time and resources permit further study, because risks to members of the general population would be even lower. Continued expenditure of time and resources on that contaminant is not justified as long as more important situations warrant study.
- If the screening estimate of risk to the maximally exposed individual was above the decision guide, it was concluded that the contaminant should be further evaluated in the level II screening.

A.2 LEVEL II SCREENING

Level II screening was designed to estimate the dose or risk to a more typical individual in the population of concern. Level II screening incorporated reasonable average or more typical values for the source term and parameter values. It was assumed that the level II screening value underestimated the dose or risk for the most highly exposed individual, although the dose or risk may be overestimated for the general population. For level II screening, the screening value was compared to the appropriate decision guide as follows:

- If the screening value was above the decision guide, it was concluded that the contaminant should be given high priority for detailed study, because it is likely that some individuals received exposures or doses high enough to warrant further study.

- If the screening value was below the decision guide, the contaminant was deferred for further study at a later time, after the highest priority contaminants have been evaluated.

A.3 EQUATIONS AND PARAMETER VALUES USED IN SCREENING

Both level I and II screening calculations used generic equations for calculation of dose and risk. The calculations included all pathways expected to be significant for the contaminant in question, based on the likely behavior patterns of nearby populations, the potential for uptake of the contaminant through food, and the relative toxicity of the contaminant through different exposure routes. Exposure pathways evaluated included inhalation, ground exposure, dermal contact and ingestion of soil or sediment, vegetable ingestion, and ingestion of meat, milk, and/or fish, as appropriate.

Exposure point concentrations used in the level I and II screening were based on available release information (source terms) and/or measured environmental concentrations. For the level I screen, upper bound exposure point concentrations were used, while the level II screening used reasonable average concentrations. For example, in the level I screening, doses and risks were typically calculated using the upper bound (e.g., 95th percentile or maximum) measured or modeled exposure point concentration at the location of the nearest downwind or downstream residence. In the level II screening, doses and risks were typically evaluated using estimates of average measured or modeled exposure point concentrations at the nearest downwind or downstream population center.

Parameter values used to calculate dose incorporated the best available information based on historical knowledge of the Oak Ridge area, literature review, and professional judgment. In some cases, different values were used in the level I and II screening. The parameters that were varied between screening levels were those specific to the target individual, including location and lifestyle factors (such as intake rates, and time spent outdoors), with the level I screening generally incorporating more conservative, or upper bound, estimates of exposure. In most cases, contaminant-specific transfer factors (such as factors describing the transfer of a contaminant to milk or meat or uptake of a contaminant from soil into vegetation) and toxicity values were kept constant for both levels of screening. Toxicity values were established taking into consideration the most sensitive health effects endpoints for the specified target individuals. Care was taken in the level I screening to avoid compounded conservatism leading to unrealistically extreme estimates of the risk posed by a contaminant.

Different exposure durations and averaging times were assumed for radionuclides, carcinogenic chemicals, and noncarcinogenic chemicals. For radionuclides and carcinogenic chemicals, exposure durations of 50 and 10 years were used in the level I and II screening, respectively. For carcinogenic chemicals, the risk was calculated in terms of the total intake averaged over the estimated lifetime, assumed to be 70 years, to give a lifetime average daily intake. For radionuclides, the risk was calculated in terms of the total cumulative dose, and an averaging time was not needed. For noncarcinogenic contaminants, an exposure duration and averaging time of 1 year was used, unless there was evidence that a shorter exposure or averaging time was appropriate for a given contaminant or exposure situation.

The results of the screening analyses were used to identify materials as low, medium, and high priority for further study.

A.4   REVIEW OF HISTORICAL URANIUM EFFLUENT MONITORING AND SCREENING EVALUATION OF POTENTIAL OFF-SITE EXPOSURES

Starting in the early 1940s, large quantities of uranium were processed on the Oak Ridge Reservation to enrich the uranium in its ^{235}U component for nuclear weapon component production and in various research and development projects. One of the earliest devices used to enrich uranium is shown in Figure 14.32. Processes with the potential to release uranium to the environment, including uranium enrichment and fabrication operations and chemical processing of reactor fuel, were conducted in four main facilities at the ORR:

1. Y-12 from 1947 to 1995 (Patton et al., 1963)
2. S-50 from 1944 to 1945 (Fox, 1945)
3. K-25 from 1945–1985 (Rogers, 1985)
4. X-10 from 1944 to 1991 (Feige et al., 1960)

Preliminary investigations in the Oak Ridge Dose Reconstruction Feasibility Study (ChemRisk, 1993a–e) indicated that uranium was not among the list of contaminants that warranted highest priority for detailed investigation of potential off-site health effects. After reviewing the findings of the preliminary feasibility study evaluation of uranium releases, several individuals who had been long-term employees at Oak Ridge uranium facilities and a number of ORHASP members nonetheless recommended that past uranium emissions and potential resulting exposures receive closer examination. These recommendations were based on the following considerations:

Figure 14.32   Calutrons such as this "racetrack" version were used for electromagnetic enrichment of uranium at Oak Ridge Y-12 plant in the 1940s. U.S. DOE photo.

- Available records of past uranium releases were found to be incomplete, and there was knowledge of substantial uranium releases that had gone unmonitored and unreported.
- The different isotopes of uranium had been evaluated separately in the feasibility study.
- The releases from the three ORR complexes (K-25, X-10, and Y-12) had been evaluated separately in the feasibility study.
- There had been no direct evaluation in the feasibility study of the potential combined exposures that members of the public could have received as a result of concurrent releases of all of the uranium isotopes from the three ORR complexes.

When the Oak Ridge Dose Reconstruction was initiated in 1994, it included a screening-level evaluation of Oak Ridge uranium operations and effluent monitoring records to determine if uranium releases from the ORR likely resulted in off-site doses that warrant further study. Our Task 6 report (Buddenbaum et al., 1999) summarizes the methods and results of that evaluation.

Our uranium evaluation followed these basic steps:

- Information that described uranium uses and releases on the ORR was collected.

- Effluent monitoring data were evaluated for quality and for consistency with previous DOE historical uranium release reports.
- Since the airborne effluent monitoring data were found to be incomplete, updated estimates of airborne uranium releases over time were generated using the more complete data available to us.
- Because of the nature of the available data, the screening evaluation of potential off-site exposures to waterborne uranium was based on environmental measurements of uranium in these local surface waters. Waterborne uranium releases from the Oak Ridge complexes were not routinely measured near their individual points of origin like airborne effluents were. Waterborne releases from X-10 were routinely sampled at White Oak Dam, and the uranium isotopes were among those evaluated under the dose reconstruction for releases from White Oak Creek to the Clinch River (Gouge et al., 1999; see Section 14.6). Uranium concentrations were also periodically measured in samples of EFPC water collected just downstream of New Hope Pond on the Y-12 site, and at the confluence of Poplar Creek and the Clinch River near the K-25 site. Early screening of White Oak Creek releases indicated that the uranium isotopes were not among the eight radionuclides that warranted detailed dose reconstruction.
- Air dispersion models were used to estimate uranium air concentrations at selected reference locations near each ORR facility. Due to complexities of the topography surrounding the Y-12 facility, an alternate approach to classical air dispersion modeling was used to estimate uranium air concentrations for the selected reference locations. For each reference location, uranium concentrations in surface water and soil were estimated from environmental measurement data.
- A screening-level evaluation of the potential for health impacts was performed by calculating uranium intakes and associated radiation doses. A two-tiered exposure assessment methodology was employed which provided both upper bound and more typical results. These results are called screening indices. The calculated screening indices were compared to the decision guide established by the ORHASP to assess if releases of a material warrant detailed investigation.

A.4.1 Quantifying Uranium Releases

Independent efforts to reconstruct estimates of past airborne uranium releases focused in most detail on the Y-12 production facility, the K-25 gaseous diffusion plant, and the S-50 liquid thermal diffusion plant. For the Y-12 plant, we quantified releases from operations that were historically monitored based on measurements of indoor uranium concentrations and ventilation exhaust rates or detailed stack sampling and analysis records found on archived computer tapes. For periods in which effluent sampling was not performed or for which

TABLE 14.9   Historical Airborne Uranium Release Totals (kg) for Oak Ridge Y-12 Plant and K-25/S-50 Plants Estimated in Oak Ridge Health Studies and Reported by DOE for Same Facilities

Facility	DOE Reported	This Project	Difference
Y-12 Plant	6,535	50,000	+43,465
K-25 & S-50 Plants	10,713	16,000	+5,287

sampling records could not be found, we estimated air releases using averages of releases for adjacent years or using uranium production data (relative rates of production over time) to scale monitoring results from preceding or subsequent periods for which monitoring data were available.

We developed independent estimates of airborne uranium releases for 1944 to 1988 (Buddenbaum et al., 1999) since the bulk of the releases occurred during this period. DOE release estimates for the period 1989 through 1995 are considered significantly more reliable due to improved monitoring (DOE, 1988).

As shown in Table 14.9 and Figure 14.33, the independent evaluation of past Y-12 airborne total uranium releases yielded results that are over seven times higher than release totals reported by the DOE, with almost 44,000 kg more total uranium released than officially reported (Buddenbaum et al., 1999). The difference between the DOE estimates and ours is largely due to DOE's use of incomplete sets of effluent monitoring data and related documents, together with its use of some annual release estimates that are based on effluent monitoring data that were not adequately corrected to account for sampling biases. Our estimates also include some unmonitored releases that were not included in official release estimates.

The independent evaluation of airborne total uranium releases from K-25 and S-50 was based on analysis of uranium accountability records and incident reports, calculation of purge cascade releases using monitoring data from that system, and use of results of periodic monitoring in three buildings on the K-25 site. The purge cascade was a segment of the gaseous diffusion enrichment equipment that was used to separate and remove light gases that might otherwise have accumulated and blocked flow of the uranium being enriched. A database of over 1200 documented uranium release events was developed using data from over 40 sources, and associated uranium losses were estimated (Buddenbaum et al., 1999).

We reconstructed purge cascade releases for selected time periods. While they were the only airborne releases from K-25 that were monitored on a routine basis (McCall, 1979), purge cascade releases made up a small fraction of total uranium releases from K-25 (e.g., 1.5 percent over 1953 to 1955 and 0.06 percent for 1975). Our uranium screening also included estimates of uranium releases from a series of UF_6 cylinder fire tests conducted in 1965 (Mallett, 1966).

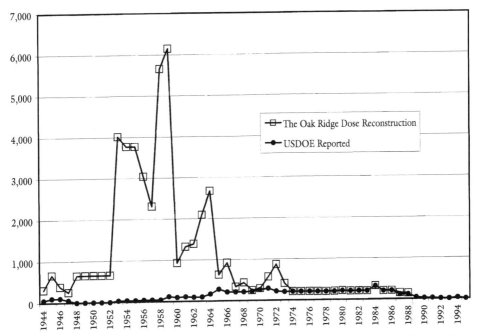

Figure 14.33 Airborne uranium release estimates for Y-12 plant prepared by Oak Ridge Dose Reconstruction project team and published by U.S. DOE (kg).

K-25 airborne releases after 1985 were based on data contained in DOE annual environmental reports. As shown in Table 14.9 and Figure 14.34, the independent evaluation of past K-25/S-50 airborne uranium releases yielded results that are almost 5300 kg greater than the release totals reported by the DOE.

A.4.2   Estimating Uranium Concentrations in the Environment

Once uranium releases had been quantified, various techniques were used to estimate air concentrations at reference locations surrounding the ORR (Buddenbaum et al., 1999). Air dispersion modeling (EPA, 1995a, b) was used to identify the communities surrounding the three facilities that were likely exposed to the highest levels of airborne uranium due to releases from the ORR. Due to the considerable distances between the Y-12, K-25/S-50, and X-10 facilities, three distinct reference locations were used for the exposure assessment. The reference location for each complex was selected based on consideration of housing areas close to the facility, alignment with dominant wind directions, and habitation patterns during the periods of highest releases.

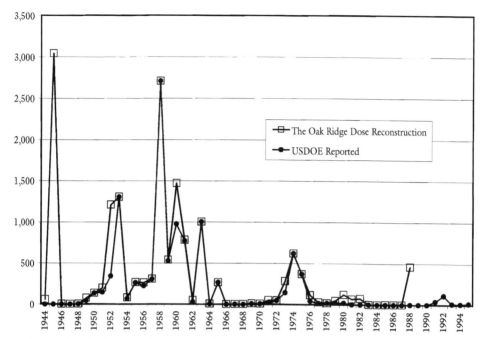

Figure 14.34 Annual airborne uranium release estimates for K-25 and S-50 plants prepared by Oak Ridge Dose Reconstruction project team and published by U.S. DOE (kg).

Three reference locations were selected for use in our uranium screening assessments (see Figure 14.35). Initial screening of exposures at other nearby locations indicated these three reference locations likely received the largest impact due to past releases from the ORR facilities.

Y-12 Reference Location: Scarboro Community   For uranium releases from the Y-12 complex, the Scarboro community was selected as the reference location. The Scarboro community is located approximately 1 km north of Y-12 and is separated from the Y-12 facility by Pine Ridge. The reference location was located at what is currently the Scarboro community center. The proximity of Scarboro to the Y-12 site suggests that screening results would present upper bound values. The closest surface water body to the Scarboro community is EFPC, which runs along the south side of the Y-12 facility, turns toward the north and northwest, and passes about 0.4 mile to the northeast of the populated area of Scarboro at its closest point.

K-25/S-50 Reference Location: Union/Lawnville   For K-25/S-50 releases, the selected reference location was the Union/Lawnville community, which is located approximately 4.5 km south-southwest of the K-25/S-50 complex. Based on the initial air dispersion modeling as well as an assessment of areas around

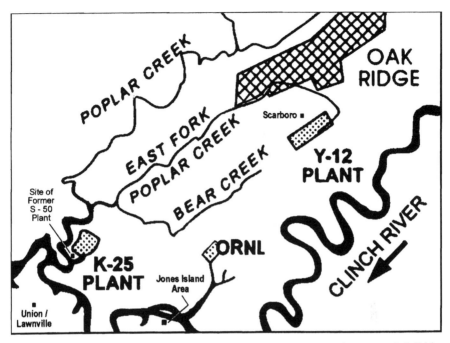

Figure 14.35 Locations of major sources of historical uranium releases on Oak Ridge Reservation. Also shown are the reference locations for which screening-level analyses of potential health risks were performed in the Oak Ridge Dose Reconstruction.

the K-25/S-50 facilities that were inhabited, this community was selected as a suitable reference location for the assessment. The location of the community is defined by the Union Church, which is located on Lawnville Road, approximately 1 km north of Gallaher Road. The primary source of surface water is the Clinch River, which is approximately 1.5 km northeast of Union Church.

X-10 Reference Location: Jones Island (Clinch River)   The selected reference location for X-10 releases was in the area of Jones Island, which is approximately 5 km southwest of the site. This area represents the closest location off reservation from X-10 and is also along a predominant wind direction. Our assessment included evaluation of air exposure pathways from X-10 releases, soil-related pathways based on maximum soil concentrations measured near the reference location, and surface water pathways reflecting consumption of fish from and recreational use of the Clinch River.

A.4.3   Uranium Concentrations in the Air

Due to the complex terrain around the Y-12 facility, any analytical approach to estimating air concentrations at Scarboro that did not reflect the effects of Pine Ridge would lead to overestimation of the fraction of Y-12 releases that were transported to Scarboro. An alternative approach using measured uranium air concentrations at Scarboro was devised for this project. By relating air concentrations measured at Scarboro from 1986 through 1995 [Martin Marietta Energy Systems (MMES), 1987–1995] with Y-12 uranium release estimates for the same years, an empirical relative concentration (χ/Q) relationship was described. This relationship was then applied to all annual release estimates (1944 to 1995) to generate estimates of annual average air concentrations at Scarboro. An air dispersion model was used to estimate concentrations at the reference locations from K-25/S-50 and X-10 releases.

A.4.4   Uranium Concentrations in Surface Waters

The two main surface water bodies addressed in this analysis are the Clinch River and EFPC. Estimates of uranium concentrations in these surface water bodies were derived from available environmental monitoring data (Owings, 1996).

A.4.5   Uranium Concentrations in Soil

Estimates of soil concentrations were based on limited measurements compiled over the years of interest (Hibbits, 1984; Cook et al., 1992). Colocation of soil concentrations and reference locations was not always possible, as sampling locations used for soil measurements were selected based on the monitoring requirements for the facility and were not specific to a community. Therefore, we selected measured soil concentrations from locations closest to each reference location.

Because of the paucity of historical measurements of uranium in the soil near Scarboro and the lack of complete documentation of the methods used for some of the measurements available, some special considerations entered into the assessment of doses to Scarboro residents. The assessment used uranium concentrations measured in surface soil/sediment samples from the EFPC floodplain. The best available measurements were made in studies conducted in the 1980s, and the results were reported as uranium concentrations in units of parts per million (ppm). Detailed information about these data is not available, most significantly the concentrations of the specific uranium isotopes that were present. Evidence of earlier soil sampling in Scarboro was not located during our investigation. We consulted with DOE and current and retired site contractor personnel, who were unable to supply more information regarding the abundance of the uranium isotopes in the soil samples or determine if earlier soil measurements were made in Scarboro.

A.4.6   Screening-Level Evaluation of Public Exposures

Once concentrations of uranium in the applicable environmental media had been quantified, the next step was to evaluate the potential significance of those concentrations. In the case of uranium, which can be chemically toxic as a heavy metal as well as hazardous as a radioactive material, this was done by estimating the radiation doses that could have been received by off-site populations and the total quantities (masses) of uranium that they could have taken into their bodies. Radiation dose estimates were then translated into screening indices, and uranium intakes were used to estimate levels of the metal that might have been present in sensitive body organs, such as the kidneys [National Radiological Protection Board (NRPB), 1996; ICRP, 1995, 1996]. These body burdens were compared to published data that indicate the levels above which uranium, as a toxic heavy metal, can cause adverse health effects.

Our screening assessment evaluated the potential health effects to the individuals that have lived in areas surrounding the ORR. In the more conservative level I assessment, the maximum reported value of 70,000 pCi kg^{-1} of ^{238}U from the EFPC floodplain was used, and the isotopic mixture of natural uranium was assumed in calculating a corresponding $^{234/235}$U concentration of 76,000 pCi kg^{-1}. In the level II assessment, a reported average value of 26 ppm total uranium from the EFPC floodplain was converted to uranium isotope concentrations using similar assumptions. The value of 26 ppm converts to concentrations of 14,000 pCi kg^{-1} of $^{234/235}$U and 12,000 pCi kg^{-1} of ^{238}U. The $^{234/235}$U component of the uranium is most important in terms of doses delivered from uranium exposure, particularly for pathways involving external irradiation of the body.

The second level of screening was considerably less conservative than the level I analysis; less conservative "level II" values were used for various exposure parameters (e.g., consumption rates and fractions of foods contaminated) than were used in the level I screening assessment. The goal in level II assessments is to remove known sources of conservative bias. For soil concentrations, an average value was used in level II compared to a maximum measured value used for the level I assessment. Because of the scarcity of information regarding estimates of uranium concentrations in the environment over the periods of interest, some conservatism was maintained in the uranium concentration estimates used in level II screening to ensure that hazards to a significant portion of the potentially exposed population were not underestimated. Conservatism was probably also introduced by the use of 1980 EFPC floodplain measurements to represent concentrations at Scarboro, which is outside the floodplain. As such, the second level of screening may be more appropriately called a refined level I analysis. The data that are currently available are not sufficient to support a defensible analysis of average or typical exposures to members of the Scarboro community during the years from the community's inception to the present.

A significant factor in the decision to maintain a conservative value of soil concentration in level II screening was the uncertainty concerning the level of ^{235}U enrichment in the soil represented by the value of 26 ppm total uranium. Because of this uncertainty, the concentration corresponding to 14,000 pCi kg^{-1} of $^{234/235}$U (or 26,000 pCi kg^{-1} total uranium) was used, based on the

TABLE 14.10   **Summary of Screening Indices from Oak Ridge Health Studies Evaluation of Past Uranium Releases**

Assessment	Level I Result	Level II Result
Exposures at Scarboro Community due to releases from Y-12 complex	**1.9×10^{-3}**	8.3×10^{-5}
Exposures at Union/Lawnville community due to Releases from K-25 and S-50	**2.7×10^{-4}**	4.0×10^{-5}
Exposures at Jones Island community due to releases from X-10 complex	7.6×10^{-5}	Not performed

Note: Level I screening uses conservative assumptions so that exposures are not underestimated, and level II screening uses reasonable average or typical assumptions so that exposures are not overestimated. Values in boldface exceed project's decision guide of 1×10^{-4} excess lifetime cancer risk.

isotopic composition of natural uranium. To illustrate how the overall results of the assessment would differ if lower concentrations of $^{234/235}$U in soil were assumed, screening indices were also calculated for soil concentrations of 7000 and 2000 pCi kg^{-1} total uranium.

Annual radiation doses from uranium intake and external exposure were calculated for the adult age group for each screening assessment and then converted to screening indices using a dose-to-risk coefficient of 7.3 percent Sv^{-1}(ICRP, 1990). The individual dose conversion factors for ^{234}U, ^{235}U, and ^{238}U were used in estimating internal and external radiation doses from uranium contamination in the environment. Screening indices for the Oak Ridge uranium releases are presented in Table 14.10.

A.4.7   Discussion of Uranium Screening Results

The Scarboro community was associated with the highest total screening index attributable to uranium releases from the Y-12 facility (Buddenbaum et al., 1999). The screening indices were 1.9×10^{-3} for the level I assessment and 8.3×10^{-5} for the level II assessment. These values translate into potential health impacts (excess fatal and nonfatal cancer and severe hereditary effects) of about 2 in 1000 and 8 in 100,000, respectively. While the overall level I screening index for the Scarboro community is above the ORHASP decision guide of 1 in 10,000, the level II value is just barely below that guide value. This indicates that the Y-12 uranium releases are candidates for further study but are not high-priority candidates for further study.

The Y-12 screening indices are most sensitive to $^{234/235}$U and ^{238}U concentrations in soil, $^{234/235}$U concentrations in air, and $^{234/235}$U concentrations in water. The major pathways of concern include ingestion of vegetables grown in contaminated soil, external doses from $^{234/235}$U in soil, inhalation of airborne $^{234/235}$U, and consumption of meat and milk from cattle raised on contaminated pasture. The level II result for the Y-12 assessment in Table 14.10 is based on a $^{234/235}$U soil concentration of 14,000 pCi kg^{-1} (or 26,000 pCi kg^{-1} total uranium). Using a soil value of 7000 pCi kg^{-1} total uranium yields a screening index of 5.8×10^{-5}, a 30 percent reduction from the screening index calculated for the level II assessment. A total uranium soil concentration of 2000 pCi kg^{-1} produces an index of 5.1×10^{-5}, a 40 percent reduction. Note that even though these alternative soil concentrations (7000 and 2000 pCi kg^{-1}) represent 73 and 92 percent reductions in soil concentrations, respectively, the reduction in the screening index for level II is not proportional. The soil pathways represent only 38 percent of the total screening index from $^{234/235}$U and 51 percent from ^{238}U. Since the concentrations in air and water were not changed for the alternative evaluations, a given reduction in soil concentration does not equal a corresponding reduction in the total screening index. Further characterization of the extent of uranium contamination in soils should be a component of any future studies of potential exposures to residents of the Scarboro community.

Air concentrations at the Scarboro community were estimated using the empirical χ/Q approach. This approach used 10 years of measurements of uranium in ambient air at Scarboro with estimates of annual releases from the Y-12 plant to calculate an effective annual dispersion factor that was then used to approximate air concentrations for earlier years. It is important to remember that this approach is reliant upon Scarboro air concentration measurements, which are available only for the period 1986 to 1995, and release estimates for the same years. Differences in operations and release point distributions or characteristics for periods before 1986 could call into question the applicability of the empirical χ/Q value to earlier years. In addition, information was gained late in the project that indicated that Y-12 uranium releases for some of the years used for development of the empirical χ/Q value may have been understated due to omission of some unmonitored release estimates. It was not possible within the scope of this project to evaluate the new data sufficiently to warrant its use in this assessment. If Y-12 uranium releases during years used to develop the empirical χ/Q value applied in this assessment were indeed underreported, that would mean that the associated empirical χ/Q values were overestimated and concentrations at Scarboro that were estimated using that approach were in turn overestimated. It is impossible to gauge the magnitude of any biases potentially introduced by this possible underreporting without closely evaluating the bases of the release estimates during the associated years in the 1980s and 1990s.

For the K-25/S-50 assessment, the total screening index for Union/Lawnville from the level I assessment (3 in 10,000) exceeded the decision guide (Buddenbaum et al., 1999). The less conservative level II screening result did not exceed the guide. This indicates that the K-25/S-50 uranium releases are candidates for further study but are not high-priority candidates for further study. For the level I screening, the air pathways account for approximately 23 percent of the screening index; 76 percent of the total screening index was attributable to the soil pathways. With limited data available to characterize the soil concentrations at Union/Lawnville, these assessments are the best estimates of health impacts within the scope of our screening assessment.

The assessment of releases from X-10 did not yield level I screening indices that exceed the decision guide for level I. The releases from X-10 warrant a lower priority given the pilot-plant nature and relatively short duration of most X-10 uranium operations. Uranium in liquid effluents from X-10's White Oak Creek to the Clinch River was addressed in another component of the Oak Ridge Dose Reconstruction. The preliminary screening analysis for radionuclides in Clinch River water and sediments is described in Section 14.6. In that screening assessment, ^{235}U and ^{238}U are identified as contaminants that were included in the screening analysis. Based on the preliminary screening of White Oak Creek releases, these two uranium isotopes are identified as being among those 16 contaminants that were assigned low priority for further study based on comparison of screening results with the decision guide of 1×10^{-5} excess lifetime cancer risk applied to individual radionuclides within the screening of White Oak Creek releases.

We also used estimates of annual-average intakes of uranium by inhalation and ingestion to evaluate the potential for health effects due to the chemical toxicity of uranium compounds, specifically for damage to the kidneys. Using estimated annual average uranium intake rates via inhalation and ingestion at the Scarboro community, we used biokinetic modeling of uranium retention and excretion in the human body to estimate annual kidney burdens (uranium concentrations in kidney tissue) over the years of interest. Predicted uranium burdens were compared to toxicity thresholds from the scientific literature.

For our conservative screening for chemical toxicity, uranium was assumed to be in its most soluble form (such as uranyl nitrate), and safety factors were included to minimize the potential for underestimation of the potential for toxic effects. As shown in Figure 14.36, estimated kidney burdens resulting from simultaneous intake of uranium by ingestion and inhalation under the Scarboro assessment do not exceed an effects threshold criterion of 1 μg uranium g^{-1} kidney tissue proposed by some scientists but do exceed an effects threshold criterion of 0.02 μg g^{-1} advocated by others who have studied uranium effects in the kidney (Russel et al., 1996; Zhao and Zhao, 1990; Wrenn et al., 1994; Kocher, 1989).

Estimates of annual-average intakes of uranium were also compared to the EPA oral RfD, another method of evaluating the potential effects of ORR uranium exposures. The RfD of 3×10^{-3} mg kg^{-1} d^{-1} is primarily based on animal studies (IRIS, 1998d) and is conservatively set at a level to ensure that there are no adverse effects on renal function. Using estimated annual-average daily uranium intake rates via inhalation and ingestion at the Scarboro community, we determined the annual hazard indices (HIs) shown in Figure 14.37 by dividing the annual-average daily intake rates by the RfD. The average HI is well below unity, which suggests that further study of heavy-metal toxicity from past ORR uranium exposures does not warrant high priority.

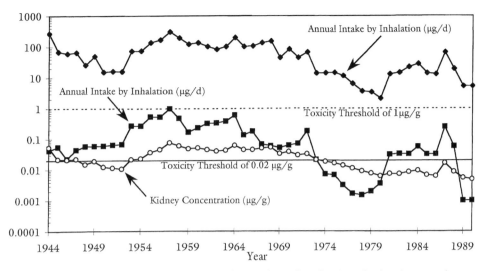

Figure 14.36 Screening-level estimates of annual uranium intakes via simultaneous ingestion and inhalation for Scarboro community resident due to Oak Ridge Y-12 plant uranium releases, with resulting kidney burdens.

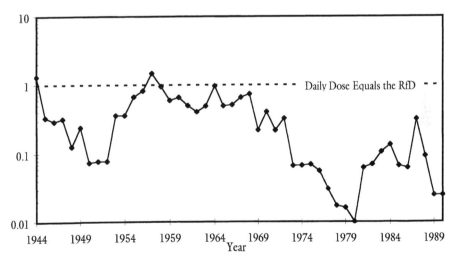

Figure 14.37 Screening-level hazard indices on annual basis for 70-kg person in Scarboro due to uranium releases from Y-12 plant, based on oral reference dose of 3×10^{-3} $\mathrm{mg\,kg^{-1}\,d^{-1}}$.

A.5 SCREENING EVALUATION OF ADDITIONAL MATERIALS OF POTENTIAL CONCERN

This section presents the methods and results of our screening-level evaluations of additional potential materials of concern (Cockroft et al., 1999). Our screening initiatives included quantitative, screening-level evaluations of 10 materials or classes of materials and less detailed evaluations of 18 others. We used three different methods to evaluate the importance of materials in terms of their potential to pose off-site health hazards. The method selected to evaluate a given material was dependent on the quantity of the material present on-site, the form and manner in which the material was used, and the availability of environmental monitoring and release data, as well as whether the material was classified per se (i.e., its mere presence on the ORR remained classified). Fortunately, before this project was completed, the presence of any material on the ORR (at the site level) could be publicly revealed. The methods we used to screen materials are described below.

A.5.1 Qualitative Screening

All materials identified as having been used on the ORR were subject to qualitative screening; for some materials, we determined that based on evaluation of quantities used, forms used, and/or manners of usage, it was unlikely that off-site releases of the material could have been sufficient to pose an off-site health hazard; these materials were not subject to quantitative screening.

Small-quantity materials included chemicals and radionuclides used as calibration standards or check sources for laboratory instruments or analytical methods. Materials used in forms not conducive to off-site release include carbon fibers and glass fibers that were received at the K-25 site as premanufactured filaments wound on spools. These fibers were used in construction of rotors in the centrifuge method of uranium enrichment, in a process by which they were wound on a spool and a plastic binder applied to form the tall, cylindrical rotors. Examples of cases where manners of usage minimized the potential for significant off-site release include liquids, gases, or powders that were kept sealed in cylinders or were processed in containment systems that included multiple barriers against release.

A.5.2 Threshold Quantity Approach

Accurate estimates of inventory quantities of materials used at the Oak Ridge complexes are often not available or in some cases not publicly releasable. It is typically much easier to determine, based on historical records or interviews of active or retired workers, if inventory quantities of a material were below a calculated threshold quantity. For a number of materials, project investigators used conservative assumptions to calculate a "threshold quantity" below which a material was highly unlikely to have posed a risk to human health through off-site releases. Threshold quantities were calculated using the following approach:

1. The maximum allowable air concentration or water concentration of a material was calculated based on the maximum allowable daily dose (assumed to be equivalent to the noncarcinogenic RfD or the dose that would lead to a cancer risk of 1×10^{-6}). To calculate a maximum allowable concentration, the maximum allowable daily dose was multiplied by a typical body weight and divided by a typical breathing or water ingestion rate.

2. The maximum allowable release rate to air or water was then determined by calculating the release rate that would give an air or water concentration equal to the maximum allowable air concentration or water concentration. Release rates were calculated using conservative environmental dispersion or dilution factors.

3. The maximum allowable release rate in grams per second was then converted to a maximum allowable release rate in kilograms per year. This quantity was assumed to be the threshold inventory quantity for the material.

A.5.3 Quantitative Screening

Quantitative screening was performed using a two-level screening approach, each level using a different set of assumptions to calculate potential doses and screening-level risk indices; the goal of this approach is to identify those contaminants that produced doses or health risks to exposed individuals or populations that are clearly below established minimum levels of concern (called a level I screen) and identify those contaminants that produced doses or health risks to exposed individuals or populations that are likely to have been above the established minimum levels of concern (called a level II or refined level I screen).

Both level I and II screening calculations used mathematical equations for calculation of dose and risk through multiple-exposure pathways. These equations relate dose to the exposure point concentration and the magnitude of intake. Pathway equations used in the screening assessment are presented in Appendix B of our screening report (Cockroft et al., 1999). The equations included all pathways potentially significant for the contaminants in question; exposure pathways evaluated included inhalation, ground exposure (for radionuclides), ingestion of soil or sediment, vegetable ingestion, and ingestion of meat, milk, and/or fish.

Parameter values used to calculate dose were selected based on historical knowledge of the Oak Ridge area, literature review, and professional judgment. The parameters that were varied between screening levels included life-style factors such as intake rates and time spent outdoors. Contaminant-specific transfer factors and toxicity values were kept constant for both levels of screening. Different exposure durations and averaging times were assumed for radionuclides, carcinogenic chemicals, and noncarcinogenic chemicals. For radionuclides and carcinogenic chemicals, exposure durations of 50 and 10 years were used in the level I and refined level I screening, respectively. For carcinogenic chemicals, the risk was calculated in terms of the total intake averaged over the estimated lifetime, assumed to be 70 years, to give a lifetime average daily intake. For radionuclides, the risk was calculated in terms of the total cumulative dose, and an averaging time was not needed. For noncarcinogenic materials, an exposure duration and averaging time of one year were used.

The level I screen was designed to estimate the dose or risk to a "maximally exposed" reference individual who should have received the highest exposure and thus would have been most at risk. This level incorporated conservative exposure parameter values (such as intake rates) not expected to lead to an underestimate of risk to any real person in the population of interest. For level I screening, each screening-level risk estimate ("screening index") was compared to the appropriate risk-based decision guide as follows:

- If the screening index for the maximally exposed individual was below the decision guide, it was concluded that further study of the contaminant can be deferred until time and resources permit further study, because risks to members of the general population would be even lower. Continued expenditure of time and resources on that contaminant is not justified as long as there are more important situations to be studied.
- If the screening index for the maximally exposed individual was above the decision guide, it was concluded that the contaminant should be further evaluated in refined level I screening or in level II screening.

Refined level I and level II screens are designed to estimate the dose or risk to a more typical individual in the population of interest than was addressed in level I screening. They incorporated reasonable average or more typical values for the exposure parameter values. It was assumed that the level II screening value underestimated the dose or risk for the most highly exposed individual, although the dose or risk may be overestimated for the general population.

For refined level I or level II screening, each screening index was compared to the appropriate decision guide as follows:

- If the screening index was above the decision guide, it was concluded that the contaminant should be given high priority for detailed study, because it is likely that some individuals received exposures or doses high enough to warrant further investigation.
- If the screening index was below the decision guide, the contaminant was deferred for further study to later phases of the project, after the highest priority contaminants are evaluated.

The refined level I screening evaluations described in this chapter were considerably less conservative than the level I evaluations they followed. For example, many of the exposure parameter values used in the dose and risk calculations are less conservative (more realistic or more typical) than the values of the same parameters used in level I screening. A good example would be the assumed exposure duration for carcinogens, which is 50 years in level I screening and 10 years in level II screening. The refined level I evaluations described in this chapter used the level II exposure parameters.

While a general goal in refined screening is to reduce or eliminate sources of conservative bias, it is not always feasible or advisable to eliminate all conservative bias, and it is often not easy to determine when a sufficient level of realism has been achieved. In the refined screening evaluations described in this chapter, some degree of conservatism was retained, particularly in the estimation of contaminant concentrations in environmental media of interest. One important reason for this is that there were very few measurements of the contaminants of concern made in the environment during the (pre-1970s) periods when levels of many contaminants in the environment were likely the highest. Measurements in process streams or effluents are even more rare. Because of the paucity of information for some vital components of the risk assessment process, some conservatism was retained in the estimation of exposure point concentrations for the refined level I assessments to ensure that exposures were not

TABLE 14.11   Categorization of Potential Materials of Concern Based on Screening Conducted During Oak Ridge Health Studies

Not candidates for further study (conservative screening result less than decision guide)
  From K-25 site, carcinogens: Np-237
  From K-25 site, evaluated qualitatively: carbon fibers, four-ring polyphenyl ether, glass fibers, triplex coating
  From Y-12 plant, carcinogens: Np-237, tritium
  From Y-12 plant, noncarcinogens: beryllium compounds, niobium, tetramethylammoniumborohydride (TMAB) zirconium
  From Y-12 plant, evaluated qualitatively: boron carbide, boron nitride, rubidium nitrate, rubidium bromide, tellurium, titanium boride, yttrium boride, zirconium
Potential candidates for further study (refined screening result less than decision guide)
  From K-25 site, carcinogens: nickel, Tc-99
  From K-25 site, noncarcinogens: copper powder, nickel
  From Y-12 plant, carcinogens: beryllium componds, Tc-99
  From Y-12 plant, noncarcinogens: lithium compounds
  From Oak Ridge Reservation, carcinogens: hexavalent chromium
  From Oak Ridge Reservation, noncarcinogens: hexavalent chromium
High priority for further study (refined screening result greater than decision guide)
  From K-25 site, carcinogens: arsenic
  From K-25 site, noncarcinogens: arsenic
  From Y-12 plant, carcinogens: arsenic
  From Y-12 plant, noncarcinogens: arsenic, lead

underestimated for significant portions of the potentially exposed populations. Because of this, the second-level assessments are called refined level I assessments rather than level II assessments.

A.5.4 Conclusions of the Screening Evaluations

Based on the qualitative and quantitative screening that we performed, it was possible to separate materials into classes based on their apparent importance in terms of potential off-site health hazards. This classification, as summarized in Table 14.11, was to a great degree dependent on the information that is available concerning past uses and releases of the materials of interest. In the course of our work, it was not possible to perform extensive directed searches for records relevant to potential material of concern to the extent that was possible for the operations and contaminants studied in the detailed dose reconstructions. For some materials, very little historical information is available. As a result, it was necessary to make a significant number of conservative assumptions for some materials to ensure that potential doses were not underestimated. If, in the future, more extensive document searching is performed, some of the conclusions reached in the screening evaluations described herein might well change.

ACKNOWLEDGMENTS

We wish to acknowledge and thank the individuals who played important roles in performance of the Oak Ridge Dose Reconstruction. The names of individuals who served as task managers for most of the project are set in italics in the list that follows. Key contributors from McLaren/Hart-ChemRisk included, in alphabetical order, Joanne Avantaggio, Nancy Bonnevie, Gretchen M. Bruce, *John E. (Jack) Buddenbaum*, Brian P. Caldwell, *Jennifer K. Cockroft*, Cynthia Curry, Catherine L. DaMassa, Ellen Ebert, Samer El Sururi, Shih-Shing Feng, Susan M. Flack, Carol Gillis, Patrick Gwinn, Jane Hamblen, Talaat Ijaz, Christopher R. Kirman, James W. Knight, Rajendra Menon, *Thomas R. Mongan, Paul S. Price*, Stephen R. Ripple, Julie Rothrock, Charlie Schmidt, and Thomas E. Widner (project manager). Contributors from SENES Oak Ridge included A. Iulian Apostoaei, Steven Bartell, B. Gordon Blaylock, *Jana H. Gouge, F. Owen Hoffman*, Cathy J. Lewis, Shyam K. Nair, E. Willow Reed, Kathleen M. Thiessen, and Brian A. Thomas. Key contributors from Shonka Research Associates included Timothy E. Bennett, Regan E. Burmeister, Robert E. Burns, Jr., UnYong "Young" Moon, Deborah B. Shonka, and *Joseph J. Shonka*.

We also wish to thank project director Patrick Lipford from the Tennessee Department of Health, Patrick Turri from TDH, and the members of the Oak Ridge Health Agreement Steering Panel, for their untiring support on the project. Under the leadership of Paul Voillequé for most of the project, the steering panel provided high-quality technical oversight and document review during the course of the project, and contributed significantly to the usefulness, understandability, and defensibility of our results. ORHASP members at project completion included James Alexander, Barbara Brooks, Paul Erwin, Joseph Hamilton, Jacqueline Holloway, Patrick Lipford, Norma Morin, Robert Peelle, James Smith, Paul Voillequé, and Nasser Zawia.

REFERENCES

Agency for Toxic Substances and Disease Registry (ATSDR). 1997. *Draft Toxicological Profile for Mercury.* Prepared for the U.S. Department of Health and Human Services. Research Triangle Institute, Atlanta, GA, Aug.

Agency for Toxic Substances and Disease Registry (ATSDR). 1998. *Toxicological Profile for Polychlorinated Biphenyls.* PB/98/101173/AS. U.S. Department of Health and Human Services, Public Health Service, Agency for Toxic Substances and Disease Registry, Atlanta, GA.

American Conference of Governmental Industrial Hygienists (ACGIH). 1996. *Documentation of the Threshold Limit Values*® and Biological Exposure Indices, 9th ed. ACGIH, Cincinnati, OH.

Anderson, D. M., Marsh, T. L., and Deonigi, D. A. 1996. Developing historical food production and consumption data for ^{131}I dose estimates: The Hanford experience. *Health Phys.* 71: 578–587.

Armstrong, T. W., Pearlman, E. D., Schnatter, A. R., Bowes, S. M., Murray, N., and Nicolich, M. J. 1996. Retrospective benzene and total hydrocarbon exposure assessment for a petroleum marketing and distribution worker epidemiology study. *Am. Ind. Hyg. Assoc. J.* 57: 333–343.

Arnold, D. L., Bryce, F., Karpinski, K., Mes, J., Fernie, S., Tryphonas, H., Truelove, J., McGuire, P. F., Burns, D., Tanner, J. R., Stapley, R., Zawidzka, Z. Z., and Basford, D. 1993a. Toxicological consequences of Aroclor 1254 ingestion by female rhesus (*Macaca mulatta*) monkeys, Part 1A: Prebreeding phase—Clinical health findings. *Food Chem. Toxicol.* 31: 799–810.

Arnold, D. L., Bryce, F., Stapley, R., McGuire, P. F., Burns, D., Tanner, J. R., and Karpinski, K. 1993b. Toxicological consequences of Aroclor 1254 ingestion by female rhesus (*Macaca mulatta*) monkeys, Part 1A: Prebreeding phase—Clinical and laboratory findings. *Food Chem. Toxicol.* 31: 811–824.

Avallone, E. A., and Baumeister III, T. (Eds.). 1996. *Marks' Standard Handbook for Mechanical Engineers.* McGraw-Hill, New York.

Aylward, L. L., Hays, S. M., Karch, N. J., and Paustenbach, D. J. 1996. Relative susceptibility of animals and humans to the cancer hazard posed by 2,3,7,8-tetrachlorodibenzo-*p*-dioxin using internal measures of dose. *Environ. Sci. Technol.* 30: 3534–3543.

Baird, S. J. S., Cohen, J. T., Graham, J. D., Shlyakhter, A. I., and Evans, J. S. 1996. Noncancer risk assessment: A probabilistic alternative to current practice. *Hum. Ecol. Risk Assess.* 2(1): 79–102.

Banic, G. 1995. Interviews with George Banic, Oak Ridge, TN, Mar. 15 and May 31.

Barnard, M. G. 1949. 100 Areas Monthly Reports—1949. Hanford Report 45828. Hanford Operations Division. Richland, WA.

Barnes, D. G., and Dourson, M. 1988. Reference dose (RfD): Description and use in health risk assessments. *Regul. Toxicol. Pharmacol.* 8: 471–486.

Baylor, R. 1997. Interviews with R. Baylor, Oak Ridge Reservation Y-12 Classification Office, by Susan Flack, ChemRisk, Alameda, CA, Jan. 17 and July 15.

Beauford, W., and Barringer, A. R. 1977. Uptake and distribution of mercury within higher plants. *Physiologia Pl.* 39: 261–265.

Bier, S. G. 1996. Thyroid mass in children: A comparison of autopsy and ultrasound data. Masters thesis, Nuclear Engineering Department, University of Tennessee, Knoxville, TN.

Bloom, N. S. 1992. On the chemical form of mercury in edible fish and marine invertebrate tissue. *Can. J. Fish. Aquat. Sci.* 49: 1010.

Bradshaw, R. L., and Cottrell, W. D. 1954. *A Study of the Contribution of the RaLa Process to Atmospheric Contamination at ORNL.* ORNL Central Files No. 54-11-86. Health Physics Division, Oak Ridge National Laboratory, Oak Ridge, TN, Nov. 1.

Browder, F. N. 1949. *Liquid Waste Disposal at Oak Ridge National Laboratory.* ORNL-328. Oak Ridge National Laboratory, Oak Ridge, TN.

Brown, J. F. 1994. Unusual congener selection patterns for PCB metabolism and distribution in the rhesus monkey. *Organohalogen Compounds* 21: 29–31.

Brunner, M. J., Sullivan, T. M., and Singer, A. W. 1996. *An Assessment of the Chronic Toxicity and Oncongenicity of Aroclor-1016, Aroclor-1242, Aroclor-1254, and Aroclor-1260 Administered in Diet to Rats.* Study No. SC920192. Chronic toxicity and oncongenicity report. Battelle, Columbus OH.

Buddenbaum, J. E., Burns, Jr., R. E., Cockroft, J. K., Ijaz, T., Shonka, J. J., and Widner, T. E. 1999. *Uranium Releases from the Oak Ridge Reservation—A Review of the Quality of Historical Effluent Monitoring Data and a Screening Evaluation of Potential Off-Site Exposures.* Report of Task 6 of the Oak Ridge Dose Reconstruction. Prepared for the Tennessee Department of Health. ChemRisk, a Service of McLaren/Hart Environmental Services, Alameda, CA, July.

Burger, L. L. 1991. *Fission Product Iodine during Early Hanford-Site Operations: Its Production and Behavior during Fuel Reprocessing, Off-Gas Treatment and Release to the Atmosphere.* Hanford Environmental Dose Reconstruction Project Report, PNL-7210 HEDR. Pacific Northwest Laboratory, Richland, WA, May.

Cambray, R. S., Cawse, P. A., Garland, J. A., Gibson, A. B., Johnson, P., Lewis, G. N. J., Newton, D., Salmon, L., and Wade, B. O. 1987. Observations on radioactivity from the Chernobyl accident. *Nucl. Energy* 26: 77–101.

Case, J. M. 1977. *Unclassified Version of Mercury Elementary at Y-12 Plant 1950 through 1977.* Y/AD-428. Y-12 Plant, Oak Ridge, TN, June 9.

Centers for Disease Control and Prevention (CDC). 1991. *Preventing Lead Poisoning in Young Children.* Centers for Disease Control and Prevention, Atlanta, GA.

ChemRisk. 1991a. *Identification of Chemicals and Radionuclides Used at Rocky Flats.* Report of Task 1 of the Rocky Flats Toxicologic Review and Dose Reconstruction. Prepared for the Colorado Department of Health. ChemRisk, Division of McLaren/Hart, Alameda, CA, Mar.

ChemRisk. 1991b. *Selection of the Chemicals and Radionuclides of Concern.* Report of Task 2 of the Rocky Flats Toxicologic Review and Dose Reconstruction. Prepared for the Colorado Department of Health. ChemRisk, Division of McLaren/Hart, Alameda, CA, June.

ChemRisk. 1992a. *Consumption of Freshwater Fish by Maine Anglers.* ChemRisk, Division of McLaren/Hart, Portland, ME. Revised July 24.

ChemRisk. 1992b. *Reconstruction of Historical Rocky Flats Operations and Identification of Release Points.* Report of Tasks 3 and 4 of the Rocky Flats Toxicologic Review and Dose Reconstruction. Prepared for the Colorado Department of Health. ChemRisk, Division of McLaren/Hart, Alameda, CA, Aug.

ChemRisk. 1993a. *Oak Ridge Health Studies Dose Reconstruction Feasibility Study Report*, Vol. I: *Phase I Overview*. Prepared for the Tennessee Department of Health and the Oak Ridge Health Agreement Steering Panel. ChemRisk, Division of McLaren/Hart, Alameda, CA, Sept.

ChemRisk. 1993b. *Oak Ridge Health Studies Dose Reconstruction Feasibility Study Report*, Vol. II, Part A—Tasks 1 and 2: *A Summary of Historical Activities on the Oak Ridge Reservation with Emphasis on Information Concerning Off-Site Emission of Hazardous Material*. Prepared for the Tennessee Department of Health and the Oak Ridge Health Agreement Steering Panel. ChemRisk, Division of McLaren/Hart, Alameda, CA, Sept.

ChemRisk. 1993c. *Oak Ridge Health Studies Dose Reconstruction Feasibility Study Report*, Vol. II, Part B—Tasks 3 and 4: *Identification of Important Environmental Pathways for Materials Released from the Oak Ridge Reservation*. Prepared for the Tennessee Department of Health and the Oak Ridge Health Agreement Steering Panel. ChemRisk, Division of McLaren/Hart, Alameda, CA, Sept.

ChemRisk. 1993d. *Oak Ridge Health Studies Dose Reconstruction Feasibility Study Report*, Vol. II, Part C—Task 5: *A Summary of Information Concerning Historical Locations and Activities of Populations Potentially Affected by Releases from the Oak Ridge Reservation*. Prepared for the Tennessee Department of Health and the Oak Ridge Health Agreement Steering Panel. ChemRisk, Division of McLaren/Hart, Alameda, CA, Sept.

ChemRisk. 1993e. *Oak Ridge Health Studies Dose Reconstruction Feasibility Study Report*, Vol. II, Part D—Task 6: *Hazard Summaries for Important Materials at the Oak Ridge Reservation*. Prepared for the Tennessee Department of Health and the Oak Ridge Health Agreement Steering Panel. ChemRisk, Division of McLaren/Hart, Alameda, CA, Sept.

ChemRisk. 1994a. *Estimating Historical Emissions from Rocky Flats 1952–1989*. Report of Task 5 of the Rocky Flats Toxicologic Review and Dose Reconstruction. Prepared for the Colorado Department of Public Health and Environment. ChemRisk, Division of McLaren/Hart, Alameda, CA, Mar.

ChemRisk. 1994b. *Demographic and Land Use Reconstruction of the Area Surrounding the Rocky Flats Plant*. Report of Task 7 of the Rocky Flats Toxicologic Review and Dose Reconstruction. Prepared for the Colorado Department of Public Health and Environment. ChemRisk, Division of McLaren/Hart, Alameda, CA, Apr.

ChemRisk. 1994c. *Exposure Pathway Identification and Transport Modeling*. Report of Task 6 of the Rocky Flats Toxicologic Review and Dose Reconstruction. Prepared for the Colorado Department of Public Health and Environment. ChemRisk, Division of McLaren/Hart, Alameda, CA, May.

ChemRisk. 1994d. *Dose Assessment for Historical Contaminant Releases from Rocky Flats*. Report of Task 8 of the Rocky Flats Toxicologic Review and Dose Reconstruction. Prepared for the Colorado Department of Public Health and Environment. ChemRisk, Division of McLaren/Hart, Alameda, CA, Sept.

Chew, E. W., Dickson, R. L., Otis, M. D., Peterson, H. K., and Start, G. E. 1993. Overview of the Idaho National Engineering Laboratory Historical Dose Evaluation. In *Environmental Health Physics*, Proceedings of the Twenty-Sixth Midyear Topical Meeting of the Health Physics Society, Health Physics Society, McLean, VA, Jan. 24–28.

Cockroft, J. K., Bruce, G. M., Buddenbaum, J. E., Burmeister, R. E., Flack, S. M., Ijaz, T., Thiessen, K. M., and Widner, T. E. 1999. *Screening-Level Evaluation of Additional Potential Materials of Concern.* Report of Task 7 of the Oak Ridge Dose Reconstruction. Prepared for the Tennessee Department of Health. ChemRisk, Division of McLaren/Hart Environmental Services, Alameda, CA, July.

Cook, R. B., Adams, S. M., Beauchamp, J. J., Bevelhimer, M. S., Blaylock, B. G., Brandt, C. C., Ford, C. J., Frank, M. L., Gentry, M. J., Holladay, S. K., Hook, L. A., Levine, D. A., Longman, R. C., McGinn, C. W., Skiles, J. L., Suter, G. W., and Williams, L. F. 1992. *Phase 1 Data Summary Report for the Clinch River Remedial Investigation: Health Risk and Ecological Risk Screening Assessment.* ORNL/ER-155. Oak Ridge National Laboratory, Environmental Sciences Division, Oak Ridge, TN, Dec.

Coughlen, C. P. 1950. *Determination of Potential Sources of Area Atmospheric Radioactive Contamination.* ORNL-677. Oak Ridge National Laboratory, Reactor Technology Division, Oak Ridge, TN, June 8.

Coughtrey, P. J., and Thorne, M. C. 1983. *Radionuclide Distribution and Transport in Terrestrial and Aquatic Ecosystems A Critical Review of Data.* Balkema, Rotterdam, pp. 170–210.

Cowser, K. E., and Snyder, W. S. 1966. *Safety Analysis of Radionuclide Release to the Clinch River.* Supplement No. 3 to Status Report No. 5 on Clinch River Study. ORNL-3721. Oak Ridge National Laboratory, Oak Ridge, TN, May.

Cromer, S. 1969. Letter to W. J. Wilcox, Jr., regarding chromate contamination in Y-12 effluent. Y-12 Plant, Oak Ridge, TN, Oct. 1.

DaMassa, C. 1995. Memorandum from C. DaMassa to T. Mongan regarding Oak Ridge area demography, McLaren/Hart Environmental Services, Alameda, CA, Jan. 1.

DaMassa, C. 1996. Preliminary land use data Addressing 4/4/96 memo; response to memo, 6/7/96, farm families—EFPC and Jones Island. Memoranda to J. McCrodden, ChemRisk, Portland, ME, from C. DaMassa, ChemRisk, Alameda, CA.

Davis, A., Bloom, N. S., and Que Hee, S. S. 1997. The environmental geochemistry and bioassessibility of mercury in soils and sediments: A review. *Risk Anal.* 17: 557–570.

Davis, R. 1997. Telephone interview with R. Davis, Kingston Water Treatment Plant employee, by C. Lewis, Mar. 26.

Downen, M. L. 1955. *Milk Consumption in Tennessee Schools, Rural Research Series.* Monograph No. 271. Department of Agricultural Economics and Rural Sociology, Washington, DC, Aug. 1.

Downen, M. L. 1956. *Consumption of Milk in Tennessee Schools, Rural Research Series.* Monograph No. 274. Department of Agricultural Economics and Rural Sociology, Washington, DC, Mar. 31.

Dreicer, M., Bouville, A., and Wachholz, B. W. 1990. Pasture practices, milk distribution, and consumption in the continental U.S. in the 1950s: Analyses and modeling for internal dose estimates. *Health Phys.* 59: 627–636.

Ebert, E. 1996. *Fish Consumption Rate Distributions of Interest for the Dose Reconstruction.* McLaren/Hart, ChemRisk, Portland, ME.

Ebert, E. S., Harrington, N. W., Boyle, K. J., Knight, J. W., and Keenan, R. E. 1993. Estimating consumption of freshwater fish among Maine anglers. *N. Am. J. Fish. Mgt.* 13: 737–745.

Ebert, E. S., Price, P. S., and Keenan, R. E. 1994. Selection of fish consumption estimates for use in the regulatory process. *J. Expos. Anal. Environ. Epidemiol.* 4(3): 373–393.

Eckerman, K. F., and Leggett, R. W. 1996. *DCFPAK: Dose Coefficient Data File Package for Sandia National Laboratory.* ORNL/TM-13347. Oak Ridge National Laboratory, Oak Ridge, TN.

Egli, D. 1985. *The Report of the Joint Task Force on Uranium Recycle Materials Processing.* DOE/OR-859. Oak Ridge Operations, U.S. Department of Energy, Oak Ridge, TN, pp. 13, 40, 67.

Evans, J., Gray, G., Sielken, R., Smith, A., Valdez-Flores, C., and Graham, J. 1994. Use of probabilistic expert judgement in uncertainty analysis of carcinogenic potency. *Regul. Toxicol.* 20: 15–36.

Farris, W. T., Napier, B. A., Ikenberry, T. A., and Shipler, D. B. 1996. Radiation doses from Hanford Site releases to the atmosphere and the Columbia River. *Health Phys.* 71: 588–601.

Farris, W. T., Napier, B. A., Simpson, J. C., Snyder, S. F., and Shipler, D. B. 1994. *Columbia River Pathway Dosimetry Report, 1944–1992. Hanford Environmental Dose Reconstruction Project.* PNWD-2227 HEDR UC-000. Battelle Pacific Northwest Laboratories, Richland, WA.

Fawer, R. F., DeRibaupierre, U., Guillemin, M. P., Berode, M., and Lobe, M. 1983. Measurement of hand tremor inducted by industrial exposure to metallic mercury. *J. Ind. Med.* 40: 204–208.

Feige, Y., Parker, F. L., and Struxness, E. G. 1960. *Analysis of Waste Disposal Practice and Control at Oak Ridge National Laboratory,* Oak Ridge National Laboratory, Oak Ridge, TN, Oct. 4.

Fowlkes, C., Fletcher, M., and Gamble, J. 1959. *Summary of Recirculating Water Treatment Tests for March 1956 through December 1959.* Union Carbide Corporation, Oak Ridge, TN.

Fox, M. C. 1945. *Thermal Diffusion Plant Built Rapidly.* S-50 Plant, Oak Ridge, TN.

Frey, H. C., and Rhodes, D. S. 1996. Characterizing, simulating, and analyzing variability and uncertainty: An illustration of methods using an air toxics emissions example. *Hum. Ecol. Risk Assess.* 2(4): 762–797.

Gerstner, H. B., and Huff, J. E. 1977. Clinical toxicology of mercury. *J. toxicol. Environ. Health* 2: 491–526.

Gillis, C., and Price, P. 1996. Comparison of noncarcinogenic effects and PCB body burdens in rhesus monkeys and humans: Implications for risk assessment (Abstract #748). *Fund. Appl. Toxicol.* 30(1, Pt. 2, *Suppl.*): 146.

Gouge, J. S., Apostoaei, A. I., Blaylock, B. G., Caldwell, B. P., Flack, S. M., Hoffman, F. O., Lewis, C. J., Nair, S. K., Reed, E. W., Thiessen, K. M., Thomas, B. A., and Widner, T. E. 1999. *Radionuclides Released to the Clinch River from White Oak Creek on the Oak Ridge Reservation—An Assessment of Historical Quantities Released, Off-Site Radiation Doses, and Health Risks.* Report of Task 4 of the Oak Ridge Dose Reconstruction. Prepared for the Tennessee Department of Health. ChemRisk, Division of McLaren/Hart Environmental Services, Alameda, CA, July.

Goyer, R. A. 1996. Toxic effects of metals. In C. D. Klaassen (Ed.), *Casarett and Doull's Toxicology, the Basic Science of Poisons,* 5th ed. McGraw-Hill, New York.

Haimes, Y. Y., and Lambert, J. H. 1999. When and how can you specify a probability distribution when you don't know much? Part II. *Risk Anal.* 19: 43–81.

Hallock, M. F., Smith, T. J., Woskie, S. R., and Hammond, S. K. 1994. Estimation of historical exposures to machining fluids in the automotive industry. *Am. J. Ind. Med.* 26: 621–634.

Health, Safety, Environment, and Accountability Division (HSEAD). 1992. *Spill Prevention, Control, and Countermeasures (SPCC) Plan for the Y-12 Plant*, Vol. I. Y/SUB/92-21704/1. Prepared for the U.S. Department of Energy. Health, Safety, Environment, and Accountability Division, Oak Ridge Y-12 Plant, Oak Ridge, TN, Aug.

Heeb, C. M., Gydesen, S. P., Simpson, J. C., and Bates, D. J. 1996. Reconstruction of radionuclide releases from the Hanford Site, 1944–1972. *Health Phys.* 71: 545–555.

Heinemann, K., and Vogt, K. J. 1980. Measurements of the deposition of iodine onto vegetation and of the biological half-life of iodine on vegetation. *Health Phys.* 39: 463–474.

Hibbitts, H. W. 1984. *Transmittal of Environmental Sampling Data for Mercury: February–December*. U.S. DOE, Oak Ridge, TN.

Hoffman, F. O., Apostoaei, A. I., Burns, R. E., Ijaz, T., Lewis, C. J., Nair, S. K., and Widner, T. E. 1999. *Iodine-131 Releases from Radioactive Lanthanum Processing at the X-10 Site in Oak Ridge, Tennessee (1944–1956)—An Assessment of Quantities Released, Off-Site Radiation Doses, and Potential Excess Risks of Thyroid Cancer.* Report of Task 1 of the Oak Ridge Dose Reconstruction. Prepared for the Tennessee Department of Health. ChemRisk, Division of McLaren/Hart Environmental Services, Alameda, CA, July.

Hoffman, F. O., Blaylock, B. G., Travis, C. C., Daniels, K. L., Etnier, E. L., Cowser, K. E., and Weber, C. W. 1984. *Preliminary Screening of Contaminants in Sediments.* ORNL/TM-9370. Oak Ridge National Laboratory, Oak Ridge, TN.

Hoffman, F. O., and Hammonds, J. S. 1994. Propagation of uncertainty in risk assessments: The need to distinguish between uncertainty due to lack of knowledge and uncertainty due to variability. *Risk Anal.* 14: 707–712.

Hornung, R. W., Greife, A. L., Stayner, L. T., Steenland, N. K., Herrick, R. F., Elliott, L. J., Ringenburg, V. L., and Morawetz, J. 1994. Statistical model for prediction of retrospective exposure to ethylene oxide in an occupational mortality study. *Am. J. Ind. Med.* 25: 825–836.

Hornung, R. W., Herrick, R. F., Stewart, P. A., Utterback, D. F., Feigley, C. E., Wall, D. K., Douthit, D. E., and Hayes, R. B. 1996. An experimental design approach to retrospective exposure assessment. *Am. Ind. Hyg. Assoc. J.* 57: 251–256.

Horst, T. W. 1977. A surface depletion model for deposition from a Gaussian plume. *Atmos. Environ.* 11: 41–46.

Horst, T. W. 1984. The modification of plume models to account for dry deposition. *Boundary-Layer Meteorol.* 30: 413–430.

HydroQual. 1995. *PCB and Sediment Fate and Transport Model for the Watts Bar Reservoir*. HydroQual, Mahwah, NJ.

Institute of Medicine/National Academy of Sciences (IOM/NAS). 1999. *Exposure of the American People to Iodine-131 from Nevada Nuclear-Bomb Tests: Review of the National Cancer Institute Report and Public Health Implications.* National Academy Press, Washington, DC.

Integrated Risk Information System (IRIS). 1998a. *Toxicological Summary for Arsenic.* U.S. Environmental Protection Agency, Cincinnati, OH.

Integrated Risk Information System (IRIS). 1998b. *Toxicity Summaries for Mercury (Inorganic), Elemental, and Methylmercury.* U.S. Environmental Protection Agency, Cincinnati, OH.

Integrated Risk Information System (IRIS). 1998c. *Toxicological Summary for Polychlorinated Biphenyls—Aroclor 1016 and Aroclor 1254.* U.S. Environmental Protection Agency, Cincinnati, OH.

Integrated Risk Information System (IRIS). 1998d. *Toxicological Summary for Uranium.* U.S. Environmental Protection Agency, Cincinnati, OH.

Integrated Risk Information System (IRIS). 1999. *Toxicological Summaries for Beryllium, Copper, Chromium, and Nickel.* U.S. Environmental Protection Agency, Cincinnati, OH.

International Atomic Energy Agency (IAEA). 1989. *Evaluating the Reliability of Predictions Made Using Environmental Transfer Models.* Safety Series No. 100. IAEA, Vienna.

International Atomic Energy Agency (IAEA). 1992. *Modelling of Resuspension, Seasonality and Losses during Food Processing.* First report of the VAMP Terrestrial Working Group, IAEA-TECDOC-647. IAEA, Vienna, May.

International Atomic Energy Agency (IAEA). 1996a. *Modelling of Radionuclide Interception and Loss Processes in Vegetation and of Transfer in Semi-natural Ecosystems.* Second report of the VAMP Terrestrial Working Group, IAEA-TECDOC-857. IAEA, Vienna.

International Atomic Energy Agency (IAEA). 1996b. *Draft Revision Annex III: Special Considerations for Assessment of Discharges of Tritium and Carbon-14.* Safety Series No. 57. IAEA, Vienna.

International Commission on Radiological Protection (ICRP). 1989. *Age Dependent Doses to Members of the Public from Intake of Radionuclides: Part 1.* ICRP Publication 56. Pergamon, Oxford.

International Commission on Radiological Protection (ICRP). 1990. *Recommendations of the International Commission on Radiological Protection.* ICRP Publication 60. Pergamon, Oxford. See also. *Ann. ICRP* 21(1–3).

International Commission on Radiological Protection (ICRP). 1993. *Age-Dependent Doses to Members of the Public from Intake of Radionuclides: Part 2 Ingestion Dose Coefficients.* Report No. 67. Pergamon, Oxford, pp. 39–43, 95–107.

International Commission on Radiological Protection (ICRP). 1995. *Age-Dependent Doses to Members of the Public from Intake of Radionuclides: Part 4 Inhalation Dose Coefficients.* ICRP Publication 71, Pergamon, Oxford.

International Commission on Radiological Protection (ICRP). 1996. *Age-Dependent Doses to Members of the Public from Intake of Radionuclides: Part 5 Compilation of Ingestion and Inhalation Dose Coefficients.* ICRP Publication 72. Pergamon, Oxford.

Jacob, P., Goulko, G., Heidenreich, W., Likhtarev, I., Kairo, I., Tronko, N., Bogdanova, T., Kenigsberg, J., Buglova, E., Drozdovich, V., Golovneva, A., Demidchik, E., Balanov, M., Zvonova, I., and Beral, V. 1998. Thyroid cancer risk to children calculated. *Nature* 392: 31–32.

Jansson, B., and Sundström, G. 1982. *Formation of Polychlorinated Dibenzofurans (PCDF) During a Fire Accident in Capacitors Containing Polychlorinated Biphenyls (PCB): Chlorinated Dioxins and Related Compounds*, Vol. 5. Pergamon, New York.

Jordan, R. G. 1976. Memo to W. H. Travis from R. G. Jordan regarding EPA survey of use of PCB's at federal facilities. Report No. 900573. Union Carbide, Oak Ridge, TN, May 6.

Joyner, J. D. 1991. *Final Safety Analysis for TSCA Incinerator*. K-25 Site, Oak Ridge, TN, Mar. 16.

Kauppinen, T. P., Pannett, B., Marlow, D. A., and Kogevinas, M. 1994. Retrospective assessment of exposure through modeling in a study on cancer risks among workers exposed to phenoxy herbicides, chlorophenols and dioxins. *Scand. J. Work Environ. Health* 20: 262–271.

Kerber, R. A., Till, J. E., Simon, S., Lyon, J. L., Thomas, D. C., Preston-Martin, S., Rallison, M. L., Lloyd, R. D., and Stevens, W. A. 1993. A cohort study of thyroid disease in relation to fallout from nuclear weapons testing. *JAMA* 270: 2076–2082.

Killough, G. G., and Eckerman, K. F. 1986. *Age- and Sex-Specific Estimation of Dose to a Normal Thyroid from Clinical Administration of* ^{131}I. ORNL/TM-9800, NUREG/CR-3955. Oak Ridge National Laboratory, Oak Ridge, TN.

Kirchner, T. B., Whicker, F. W., Anspaugh, L. R., and Ng, Y. C. 1996. Estimating internal dose due to ingestion of radionuclides from Nevada Test Site fallout. *Health Phys.* 71: 487–501.

Kocher, D. C. 1989. Relationship between kidney burden and radiation dose from chronic ingestion of U: Implications for radiation standards for the public. *Health Phys.* 57: 9–15.

Koranda, J. J. 1965. *Agricultural Factors Affecting the Daily Intake of Fresh Fallout by Dairy Cows*. UCRL-12470. *Biology and Medicine*. UC-48. TID-4500 (38th ed.). University of California, Lawrence Radiation Laboratory, Livermore, CA, Mar.

LaFrance, L. 1955. Letter from L. LaFrance to D. A. Jennings regarding stack samples at building 9201-4. Y-12 Plant, Oak Ridge, TN, Aug. 19.

LaFrance, L. 1956. Letter from N. E. Bolton to J. L. Williams regarding air analyses in building 9204-2. Y-12 Plant, Oak Ridge, TN, Feb. 9.

LaGrone, J. 1983. Statement of Joe LaGrone, Manager, ORO, USDOE, before the Subcommittees of Energy Research and Production and Oversight and Investigation of the House Committee on Science and Technology [and related testimony]. U.S. DOE, Oak Ridge Operations, Oak Ridge, TN, July.

Leggett, R. W. 1997a. Personal communications with C. Lewis, Oak Ridge, TN, Feb. 2, 1997; Mar. 18, 1997; and May 5, 1997.

Leggett, R. W. 1997b. Dosimetry review comment. Oak Ridge National Laboratory, Oak Ridge, TN, Oct. 1.

Lewis, J. L., Sr. (Ed.) 1989. *Sax's Dangerous Properties of Industrial Materials*, 8th ed. Van Nostrand Reinhold, New York.

Lindberg, S. E., Kim, K. H., Meyers, T. P., and Owens, J. G. 1995. A micrometeorological gradient approach for quantifying air/surface exchange of mercury vapor: Tests over contaminated soils. *Environ. Sci. Technol.* 29: 126–135.

Lindberg, S. E., Turner, R. R., Meyers, T. P., Taylor, Jr., G. E., and Schroeder, W. H. 1991. Atmospheric concentrations and deposition of mercury to a decidious forest at Walker Branch watershed, Tennessee, USA. *Water, Air, Soil Pollut.* 56: 577–594.

Lockheed Martin Energy Systems (LMES). 1997. *Oak Ridge Environmental Information System (OREIS) Environmental Data for Lead Collected during the East Fork Poplar Creek Floodplain and Sewerline Beltway Remedial Investigation.* Oak Ridge National Laboratory, Oak Ridge, TN.

Lomenick, T. F. 1963. Movement of ruthenium in the bed of White Oak Lake. *Health Phys.* 9: 835–845.

Ludwick, J. D. 1964. Investigation of the nature of ^{131}I in the atmosphere. In C. C. Gamertsfelder and J. K. Green (Eds.), *Hanford Radiological Sciences Research and Development Annual Report for 1963.* HW-81746. Hanford Atomic Products Operation, Richland, WA.

Ludwick, J. D. 1967. A Portable Boom-type Air Sampler. In D. W. Pearce and M. R. Compton (Eds.), *Pacific Northwest Laboratory Annual Report for 1966 to the USAEC Division of Biology and Medicine*, Vol. II: *Physical Sciences, Part 1, Atmospheric Sciences.* BNWL-4811. Hanford Atomic Products Operation, Richland, UA, pp. 87–92.

Mallett, A. J. 1966. *ORGDP Container Test and Development Program Fire Tests of UF_6-Filled Cylinders.* ORGDP report K-D-1894. Oak Ridge Gaseous Diffusion Plant, Oak Ridge, TN, Jan. 12.

Marsh, D. O., Clarkson, T. W., Cox, C. C., Myers, G. J., Amin-Zaki, L., and Al-Tikriti, S. 1987. Fetal methylmercury poisoning: Relationship between concentration in single strands of maternal hair and child effects. *Arch. Neurol.* 44(10): 1017–1022.

Marsh, D. O., Clarkson, T. W., Myers, G. J., Davidson, P. W., Cox, C., Cernichiari, E., Tanner, M. A., Lednar, W., Shamlaye, C., Choisy, O., Hoareau, C., and Berlin, M. 1995. The Seychelles study of fetal methylmercury exposure and child development. In: Methylmercury and human health (special issue). *Neurotoxicology* 16(4): 583–596.

Martin Marietta Energy Systems (MMES). 1981. *Clinch River and Poplar Creek Bottom Sediments Data Routine Sampling Program 1975–1981.* K-25 Site, Oak Ridge, TN.

Martin Marietta Energy Systems (MMES). 1986–1995. Oak Ridge Reservation environmental reports—extracted pages. 1987 (ES/ESH-4/V2); 1988 (ES/ESH-8/V2); 1989 (ES/ESH-13/V2); 1990 (ES/ESH-22/V2, errata); 1991 (ES/ESH-22/V2); 1992 (ES/ESH-31/V2); 1993 (ES/ESH-69); 1994 (ES/ESH-69); 1995 (ES/ESH-69). MMES.

Martin Marietta Energy Systems (MMES). 1993. *Oak Ridge Reservation Environmental Report for 1992.* ES/ESH-31. MMES.

McCall, J. S. 1979. *Analytical Support for ORGDP Purge Gas Scrubber, Oak Ridge K-25 Site.* K/TL-842. K-25 Site, Oak Ridge, TN, Feb. 1.

McLaren/Hart. 1999. *Hexavalent Chromium Exposure Reconstruction for Diamond Alkali Company Painesville Ohio Chromate Production Workers from 1940 to 1972.* McLaren/Hart, Inc., Alameda, CA, May 21.

Meyer, K. R., Voillequé, P. G., Schmidt, D. W., Rope, S. K., Killough, G. G., Shleien, B., Moore, R. E., Case, M. J., and Till, J. E. 1996. Overview of the Fernald dosimetry reconstruction project and source term estimates for 1951–1988. *Health Phys.* 71: 425–437.

Miller, C. W., and Smith, J. M. 1996. Why should we do environmental dose reconstruction? *Health Phys.* 71: 420–424.

Miller, J. K., and Swanson, E. W. 1973. Metabolism of ethylenediaminedihydroiodide and sodium or potassium iodide by dairy cows. *J. Dairy Sci.* 56(3): 378–384.

Miller, J. K., Swanson, E. W., and Spalding, G. E. 1975. Iodine absorption, excretion, recycling, and tissue distribution in the dairy cow. *J. Dairy Sci.* 58: 1578–1593.

Mishinova, V. N., Stepanova, P. A., and Zarudin, U. U. 1980. Characteristics of the course of pregnancy and labor in women coming in contact with low concentrations of metallic mercury vapors in manufacturing work places. *Gig. Tr. Prof. Zabol.* 2: 21–23.

Mongan, T. R., Bruce, G. M., Flack, S. M., and Widner, T. E. 1999. *Mercury Releases from Lithium Enrichment at the Oak Ridge Y-12 Plant—A Reconstruction of Historical Releases and Off-Site Doses and Health Risks.* Report of Task 2 of the Oak Ridge Dose Reconstruction. Prepared for the Tennessee Department of Health. ChemRisk, a Division of McLaren/Hart Environmental Services, Alameda, CA, July.

Mongan, T. R., Ripple, S. R., Brorby, G. P., and diTommaso, D. G. 1996a. Plutonium releases from the 1957 fire at Rocky Flats. *Health Phys.* 71: 510–521.

Mongan, T. R., Ripple, S. R., and Winges, K. D. 1996b. Plutonium release from the 903 Pad at Rocky Flats. *Health Phys.* 71: 522–531.

Morgan, K. Z., and Davis, D. M. 1965. *Applied Health Physics Annual Report for 1964.* ORNL-3820. Oak Ridge National Laboratory, Oak Ridge, TN.

Morgan, K. Z., Davis, D. M., and Hart, J. C. 1963. *Applied Health Physics Annual Report for 1962.* ORNL-3490. Oak Ridge National Laboratory, Oak Ridge, TN.

Mullen, A. L., Stanley, R. E., Lloyd, S. R., and Moghissi, A. A. 1975. Absorption, distribution, and milk secretion of radionuclides by the dairy cow. IV. Inorganic radiomercury. *Health Phys.* 28: 685–691.

National Cancer Institute (NCI). 1997. *Estimated Exposures and Thyroid Doses Received by the American People from Iodine-131 in Fallout Following Nevada Atmospheric Nuclear Bomb Tests.* U.S. Department of Health and Human Services, National Institutes of Health, Washington, DC.

National Council on Radiation Protection and Measurements (NCRP). 1996a. *A Guide for the Uncertainty Analysis in Dose and Risk Assessments Related to Environmental Contamination.* NCRP Commentary No. 14. National Council on Radiation Protection and Measurements, Bethesda, MD.

National Council on Radiation Protection and Measurements (NCRP). 1996b. *Screening Models for Releases of Radionuclides to Atmosphere, Surface Water, and Ground.* NCRP Report No. 123. National Council on Radiation Protection and Measurements, Bethesda, MD.

National Radiological Protection Board (NRPB). 1996. *LUDEP 2.0 Personal Computer Program for Calculating Internal Doses Using the ICRP Publication 66 Respiratory Tract Model.* NRPB-SR287. National Radiological Protection Board, Chilton, England.

National Research Council (NRC). 1995. *Radiation Dose Reconstruction for Epidemiologic Uses.* National Academy Press, Washington, DC.

National Toxicology Program (NTP). 1993. *Technical Report on the Toxicology and Carcinogenesis Studies of Mercuric Chloride in F344 Rats and B6C3F1 Mice.* NTP

TR 408. U.S. Department of Health and Human Services, Research Triangle Park, NC, Feb.

Ng, Y. C. 1982. A review of transfer factors for assessing the dose from radionuclides in agricultural products. *Nucl. Safety* 23(1): 57.

Ng, Y. C., Colsher, C. S., Quinn, D. J., and Thompson, S. E. 1977. *Transfer Coefficients for the Prediction of the Dose to Man via the Forage-Cow-Milk Pathway from Radionuclides Released to the Biosphere.* UCRL-51939. Lawrence Livermore Laboratory, Livermore, CA.

Ng, Y. C., Colsher, C. S., and Thompson, S. E. 1979. Transfer factors for assessing the dose from radionuclides in agricultural products. In *Proceedings of an International Symposium on Biological Implications of Radionuclides Released from Nuclear Industries*, Vienna, April 26–March 3, Vol. II, pp. 295–318. IAEA-SM-237/54 [STI/PUB-522 (V.2)]. International Atomic Energy Agency, Vienna.

Oak Ridge Gaseous Diffusion Plant (ORGDP). 1981. *Tabulation of Air Sampling Data for Copper.* K-25 Site, Oak Ridge, TN.

Oak Ridge National Laboratory (ORNL). 1985. *Environmental and Occupational Safety Division Annual Progress Report for 1984.* ORNL-6182. ORNL, Oak Ridge, TN, Dec.

Oak Ridge National Laboratory (ORNL). 1991. *Monitoring Plan for PCBs in the Aquatic Environment of Oak Ridge National Laboratory.* ORNL, Oak Ridge, TN, June 19.

Office of Technology Assessment, U.S. Congress (OTA). 1991. *Complex Cleanup—The Environmental Legacy of Nuclear Weapons Production.* OTA-O-484. U.S. Government Printing Office Washington, DC, Feb.

Ohnesorge, W. F. 1986. *Historical Releases of Radioactivity to the Environment from Oak Ridge National Laboratory.* ORNL/M-135. Oak Ridge National Laboratory, Oak Ridge, TN.

Owings, E. 1995. *Historical Review of Accountable Nuclear Materials at the Y-12 Plant (Unclassified).* Y/DG-256549, rev. 2. Y-12 Plant, Oak Ridge, TN, July 28.

Owings, E. 1996. *Historical Review of Accountable Nuclear Materials at the Y-12 Plant.* Y/EXT-00153/del rev. Y-12 Plant, Oak Ridge, TN, June 3.

Patton, F. S., Googin, J. M., and Griffith, W. L. 1963. *Enriched Uranium Processing.* Pergamon. Oxford.

Paustenbach, D. J., Bruce, G. M., and Chrostowski, P. 1997. Current views on the oral bioavailability of inorganic mercury in soil: Implications for heath risk assessments. *Risk Anal.* 17: 533–544.

Peelle, R. 1999. Personal communication, ORHASP member, Sept.

PEER Consultants. 1993. *Asbestos Inventory and Assessment for the U.S. Department of Energy for the City of Oak Ridge, Tennessee—Final Report.* PEER Consultants, Oak Ridge, TN, Sept.

Pesci, N. 1995. *Tabulation of Lifetime Coal Consumption; Historical Investigation Special Report.* S/R 7, K-701 Power Station/Boiler House. K-25 Site, Oak Ridge, TN.

Physician's Desk Reference (PDR), 49th ed. 1995. Medical Economics Data Production Company, Montvale, NJ.

Plato, N., Krantz, S., Gustavsson, P., Smith, T. J., and Westerholm, P. 1995. Fiber exposure assessment in the Swedish rock wool and slag wool production industry 1938–1990. *Scand. J. Work Environ. Health* 21: 345–352.

Potter, G. D., McIntyre, D. R., and Vattuone, G. M. 1972. Metabolism of Hg-203 administered as HgCl₂ in the dairy cow and calf. *Health Phys.* 22: 13–16.

Price, P. S., Avantaggio, J., Bonnevie, N., Gwinn, P., Hamblen, J., Schmidt, C., and Widner, T. E. 1999. *PCBs in the Environment near the Oak Ridge Reservation—A Reconstruction of Historical Doses and Health Risks.* Report of Task 3 of the Oak Ridge Dose Reconstruction. Prepared for the Tennessee Department of Health. ChemRisk, a Division of McLaren/Hart Environmental Services, Alameda, CA, July.

Price, P. S., Keenan, R. E., Swartout, J. C., Gillis, C. A., Carlson-Lynch, H., and Dourson, M. L. 1997. An approach for modeling noncancer dose responses with an emphasis on uncertainty. *Risk Anal.* 17(4): 427–438.

Radiation Shielding Information Center (RSIC). 1991. *Documentation for CCC-371/ ORIGEN 2.1 Code Package.* Oak Ridge National Laboratory, Oak Ridge, TN.

Radiological Assessments Corporation (RAC). 1999. *Technical Summary Report for the Historical Public Explosures Studies for Rocky Flats Phase II.* RAC, Neeses, SC, September.

Ramachandran, G., and Vincent, J. H. 1999. A Bayesian approach to retrospective exposure assessment. *Appl. Occup. Environ. Hyg.* 14: 547–557.

Ramsdell, J. V., Jr., Simonen, C. A., and Burk, K. W. 1994. *Regional Atmospheric Transport Code for Hanford Emission Tracking (RATCHET).* Hanford Environmental Dose Reconstruction Project. PNWD-2224 HEDR, UC-000. Battelle Pacific Northwest Laboratories, Richland, WA.

Ramsdell, J. V., Jr., Simonen, C. A., Burk, K. W., and Stage, S. A. 1996. Atmospheric dispersion and deposition of ¹³¹I released from the Hanford Site. *Health Phys.* 71: 568–577.

Ripple, S. R. 1992. Looking back—the use of retrospective health risk assessments at nuclear weapons facilities. *Environ. Sci. Technol.* 26: 1270–1277.

Ripple, S. R., Widner, T. E., and Mongan, T. R. 1996. Past radionuclide releases from routine operations at Rocky Flats. *Health Phys.* 71: 502–509.

Rogers, J. 1985. *Oak Ridge Gaseous Diffusion Plant Uranium Discharges.* K-25 Site, Oak Ridge, TN, May 5.

Ron, E., Lubin, J. H., Shore, R. E., Mabuchi, K., Modan, B., Pottern, L. M., Schneider, A. B., Tucker, M. A., and Boice, Jr., J. D. 1995. Thyroid cancer after exposure to external radiation: A pooled analysis of seven studies. *Radiat. Res.* 141: 259–277.

Rupp, A. F., and Witkowski, E. J. 1955. *RaLa Production—1954.* Central Files Number 55-1-211. Oak Ridge National Laboratory, Oak Ridge, TN.

Rupp, E. M., Miller, F. L., and Baes, C. F. 1980. Some results of recent surveys of fish and shellfish consumption by age and region of U.S. residents. *Health Phys.* 39: 165–175.

Russell, J. J., Kathren, R. L., and Dietert, S. E. 1996. A histological kidney study of uranium and non-uranium workers. *Health Phys.* 70: 466–472.

Sanders, M. 1970. *Mercury Analysis of Fish, Water, and Mud Samples Collected in the Oak Ridge Area.* Y/HG-0091/1. Aug. 6.

Sanford Cohen and Associates (SCA). 1993. *Idaho National Engineering Laboratory*

Phase 1 Environmental Dose Reconstruction. Prepared for the Centers for Disease Control and Prevention, Atlanta, GA.

Scheuplein, R. J., and Bowers, J. C. 1995. Dioxin—an analysis of the major human studies: Comparison with animal-based cancer risks. *Risk Anal.* 15: 319–333.

Science Applications International Corporation (SAIC). 1993. *East Fork Poplar Creek—Sewer Line Beltway Remedial Investigation Report,* Vols. I–IV. DOE/OR/ 02-1119&D1. U.S. DOE, Oak Ridge, TN, Apr.

Schoof, R. A., and Nielsen, J. B. 1997. Evaluation of methods for assessing the oral bioavailability of inorganic mercury in soil. *Risk Anal.* 17: 545–556.

Seixas, N. S., Heyer, N J., Welp, E. A. E., and Checkoway, H. 1997. Quantification of historical dust exposures in the diatomaceous earth industry. *Ann. Occup. Hyg.* 41: 591–604.

Sherer, R. A., and Price, P. S. 1993. The effect of cooking processes on PCB levels in edible fish tissue. *Qual. Assuran. Good Prac. Reg. Law* 2(4), 396–407.

Shipler, D. B., Napier, B. A., Farris, W. T., and Freshley, M. D. 1996. Hanford environmental dose reconstruction project—an overview. *Health Phys.* 71: 532–544.

Simon, S. L., and Graham, J. C. 1996. Dose assessment activities in the Republic of the Marshall Islands. *Health Phys.* 71: 438–456.

Smith, T. J. 1992. Occupational exposure and dose over time: Limitations of cumulative exposure. *Am. J. Ind. Med.* 21: 35–51.

Smith, T. J., Hammond, S. K., Hallock, M., and Woskie, S. R. 1991. Exposure assessment for epidemiology: Characteristics of exposure. *Appl. Occup. Environ. Hyg.* 6: 441–447.

Smith, T. J., Hammond, S. K., Laidlaw, F., and Fine, S. 1984. Respiratory exposures associated with silicon carbide production: Estimation of cumulative exposures for an epidemiological study. *Br. J. Ind. Med.* 41: 100–108.

Smith, T. J., Hammond, S. K., and Wong, O. 1993. Health effects of gasoline exposure. I. Exposure assessment for U.S. distribution workers. *Environ. Health Perspect. Suppl.* 101: 13–21.

Snyder, S. F., Farris, W. T., Napier, B. A., Ikenberry, T. A., and Gilbert, R. O. 1994. *Parameters Used in the Environmental Pathways and Radiological Dose Modules (DESCARTES, CIDER, AND CRD Codes) of the Hanford Environmental Dose Reconstruction Integrated Codes (HEDRIC).* PNWD-2033 HEDR-Rev. 1. Battelle Pacific Northwest Laboratories, Richland, WA.

Stanley, W. M., Jr. 1954. Area contamination on April 29, 1954. Oak Ridge National Laboratory, Oak Ridge, TN, intercompany correspondence to C. E. Larson, Apr. 30.

Stein, E. D., Cohen, Y., and Winer, A. M. 1996. Environmental distribution and transformation of mercury compounds. *Crit. Rev. Environ. Sci. Technol.* 26(1): 1–43.

Stewart, P. 1999. Invited articles—Challenges to retrospective exposure assessment. *Scand. J. Work Environ. Health* 25: 505.

Stewart, P. A., and Herrick, R. F. 1991. Issues in performing retrospective exposure assessment. *Appl. Occup. Environ. Hyg.* 6: 421–427.

Stewart, P. A., Lees, P. S., and Francis, M. 1996. Quantification of historical exposures in occupational cohort studies. *Scand. J. Work Environ. Health* 22: 405–414.

Stewart, P. A., Stewart, W. F., Siemiatycki, J., Heineman, E. F., and Dosemeci, M. 1998a. Questionnaires for collecting detailed occupational information for community-based case control studies. *Am. Ind. Hyg. Assoc. J.* 59: 39–44.

Stewart, P. A., Zaebst, D., Zey, J. N., Herrick, R., Dosemeci, M., Hornung, R., Bloom, T., Pottern, L., Miller, B. A., and Blair, A. 1998b. Exposure assessment for a study of workers exposed to acrylonitrile. *Scand. J. Work Environ. Health* 24(Suppl. 2): 42–53.

Struxness, E. G., Carrigan, Jr., P. H., Churchill, M. A., Cowser, K. E., Morton, R. J., Nelson, D. J., and Parker, F. L. 1967. *Comprehensive Report of the Clinch River Study.* ORNL-4035. Oak Ridge National Laboratory, Oak Ridge, TN.

Swartout, J. D., Keenan, R. E., Stickney, J. A., Gillis, C. A., Carlson-Lynch, H. L., Dourson, M. L., Harvey, T., and Price, P. S. 1998. A probabilistic framework for the reference dose. *Fund. Appl. Toxicol.* 36(1, Pt. 2, *Suppl.*): 208.

Taylor, P. R. 1988. *The Health Effects of Polychlorinated Biphenyls.* Harvard School of Public Health, Boston, MA.

Tennessee Cancer Reporting System (TCRS). *Report of Data for 1992.* Tennessee Department of Health, Nashville, TN.

Tennessee Cancer Reporting System (TCRS). *Report of Data for 1996.* Tennessee Department of Health, Nashville, TN.

Tennessee Valley Authority (TVA). 1959. *Floods on Clinch River and East Fork Poplar Creek in Vicinity of Oak Ridge, Tennessee.* Report No. 0-5922. Division of Water Control Planning, Knoxville, TN. Sept.

Tennessee Valley Authority (TVA). 1988. *Distribution and Disposition of Polychlorinated Biphenyl (PCB) Materials at TVA Facilities: Annual Report—Calendar Year 1987.* Tennessee Valley Authority, Division of Water Resources, Chattanooga, TN.

Thompson, C. B., and McArthur, R. D. 1996. Challenges in developing estimates of exposure rate near the Nevada Test Site. *Health Phys.* 71: 470–476.

Thompson, D. E., Mabuchi, K., Ron, E., Soda, M., Tokunaga, M., Ochikubo, S., Sugimoto, S., Ikeda, T., Terasaky, M., Izumi, S., and Preston, D. L. 1994. Cancer incidence in atomic bomb survivors, Part II: Solid tumors, 1958–1987. *Radiat. Res.* 137: S17–S67.

Thompson, W. E., Jr. 1949. *History of the Barium-Lanthanum Process and Production.* ORNL-246 Special (with deletions). Oak Ridge National Laboratory, Oak Ridge, TN, June 22.

Tryphonas, H., Hayward, S., O'Grady, L., Loo, J. C. K., Arnold, D. L., Bryce, F., and Zawidzka, Z. Z. 1989. Immunotoxicity studies of PCB (Aroclor 1254) in the adult rhesus *(Macaca mulatta)* monkey—Preliminary report. *Int. J. Immunoph.* 11(2): 199–206.

Tryphonas, H., Luster, M. I., Schiffman, G., Dawson, L. L., Hodgen, M., Germolec, D., Hayward, S., Bryce, F., Loo, J. C. K., Mandy, F., and Arnold, D. L. 1991a. Effect of chronic exposure of PCB (Aroclor® 1254) on specific and nonspecific immune parameters in the rhesus *(Macaca mulatta)* monkey. *Fund. Appl. Toxicol.* 16: 773–786.

Tryphonas, H., Luster, M. I., White, K. L., Naylor, P. H., Erdos, M. R., Burleson, G. R., Germolec, D., Hodgen, M., Hayward, S., and Arnold, D. L. 1991b. Effects of PCB (Aroclor® 1254) on non-specific immune parameters in rhesus *(Macaca mulatta)* monkeys. *Int. J. Immunoph.* 13(6): 639–648.

Turner, R. R., and Bloom, N. S. 1995. Reconstruction of historical atmospheric mercury releases using analysis of tree rings in red cedar (Juniperus virginia). Poster presentation, Oak Ridge National Laboratory. Oak Ridge, TN.

Turner, R. R., Bogle, M. A., Clapp, R. B., Dearstone, K., Dreier, R. B., Early, T. O., Herbes, S. E., Loar, J. M., Parr, P. D., Southworth, G. R., and Mercier, T. M. 1988. *RCRA Facility Investigation Plan Bear Creek Oak Ridge Y-12 Plant, Oak Ridge, Tennessee.* Report No. Y/TS-417. Prepared for the U.S. Department of Energy. Oak Ridge National Laboratory, Environmental Sciences Division, Oak Ridge, TN.

Turri, P. 1998. Personal communication, Tennessee Department of Health, Nashville, TN.

Union Carbide Corporation (UCC). 1976. *Environmental Monitoring Report, Calendar Year 1975.* Report Y/UB-4. U.S. Energy Research and Development Administration, Oak Ridge Facilities. Oak Ridge, TN.

Union Carbide Corporation (UCC). 1977. *Environmental Monitoring Report, Calendar Year 1976.* Report Y/UB-6. U.S. Energy Research and Development Administration, Oak Ridge Facilities. Oak Ridge, TN.

Union Carbide Corporation Nuclear Division (UCCND). 1950–1970. Oak Ridge Reservation Y-12 plant quarterly reports: Y-1002 through Y-1020, Y-1200 through Y-1220, Y-1421 through Y-1428, Y-1513, Y-1517, Y-1631, Y-1635, Y-1639. Oak Ridge Reservation, Oak Ridge, TN.

Union Carbide Corporation Nuclear Division (UCCND). 1983a. *Assessment of Coal Storage and Steam Management Plans.* Y/EN-4234. Y-12 Plant, Oak Ridge, TN, July 29.

Union Carbide Corporation Nuclear Division (UCCND). 1983b. *The 1983 Mercury Task Force. Mercury at Y-12: A Study of Mercury Use at the Y-12 Plant, Accountability, and Impacts on Y-12 Workers and the Environment—1950–1983.* Y/EX-21/del rev. Y-12 Plant, Oak Ridge, TN, Aug. 18.

Union Carbide Nuclear Company (UCNC). 1957. *Separation of Lithium-6 and Lithium-7.* Y-F40-66/del rev. Y-12 Plant, Oak Ridge, TN, Mar. 29.

United Nations Scientific Committee on the Effects of Atomic Radiation (UNSCEAR). 1994. *Sources and Effects of Ionizing Radiation.* UNSCEAR 1994 Report to the General Assembly with Scientific Annexes, United Nations Publication No. E.94.IX.11. UNSCEAR, Vienna.

U.S. Army Corps of Engineers (ACE). 1993. *HEC-6: Scour and Deposition in Rivers and Reservoirs Users Manual.* ACE, Washington, DC.

U.S. Department of Agriculture (USDA). 1965. *Food and Nutrient Intake of Individuals in the United States. Household Food Consumption Survey 1965–1966.* Report No. 11. Agricultural Research Service, Washington, DC.

U.S. Department of Agriculture (USDA). 1980. *Food and Nutrient Intakes of Individuals in One Day in the United States, Spring 1977. Nationwide Food Consumption Survey 1977–78.* Report No. 2. Agricultural Research Service, Washington, DC.

U.S. Department of Energy (DOE). 1988. *Historical Radionuclide Releases from Current DOE Oak Ridge Operations Office Facilities.* OR-890. DOE Oak Ridge Operations Office, Oak Ridge, TN, May.

U.S. Department of Energy (DOE). 1991. *Idaho National Engineering Laboratory Historical Dose Evaluation*, Vols. 1 and 2. DOE/ID-12119. DOE, Idaho Falls, ID.

U.S. Department of Energy (DOE). 1996. *Remedial Investigation/Feasibility Study of the Clinch River/Poplar Creek Operable Unit*, Vols. 1 and 2. DOE/OR/01-1393. DOE, Oak Ridge, TN.

U.S. Environmental Protection Agency (EPA). 1986a. *Development of Advisory Levels for Polychlorinated Biphenyls (PCBs) Cleanup.* EPA/600-86/002. EPA, Office of Water, Washington, DC.

U.S. Environmental Protection Agency (EPA). 1986b. *Air Quality Criteria for Lead,* Vols. I–IV. EPA 600/8-83-028(a-d). Environmental Criteria and Assessment Office, Office of Research and Development, Research Triangle Park, NC.

U.S. Environmental Protection Agency (EPA). 1988. *Drinking Water Criteria Document for Inorganic Mercury (Final).* PB89-192207. EPA, Cincinnati, OH, July.

U.S. Environmental Protection Agency (EPA). 1989. *Estimating Air Toxics Emissions from Coal and Oil Combustion Sources.* EPA/450/2-89-001. Office of Air Quality Planning and Standards, Washington, DC, Apr.

U.S. Environmental Protection Agency (EPA). 1994a. *Estimating Exposure to Dioxin-Like Compounds,* Vol. 3: *Site-Specific Assessment Procedures.* EPA/600/6-88/005Cc. EPA, Office of Research and Development, Washington, DC, June.

U.S. Environmental Protection Agency (EPA). 1994b. *Guidance Manual for the Integrated Exposure Uptake Biokinetic Model for Lead in Children (Version 0.99D).* EPA, Washington, DC.

U.S. Environmental Protection Agency (EPA). 1994c. *Guidance on Residential Lead-Based Paint, Lead-Contaminated Dust, and Lead-Contaminated Soil.* EPA, Washington, DC.

U.S. Environmental Protection Agency (EPA). 1995a. *User's Guide for the Industrial Source Complex (ISC3) Dispersion Models,* Vols. I and II. ISCST3 version 96113. EPA-454/B-95-003. EPA, Research Triangle Park, NC, Mar.

U.S. Environmental Protection Agency (EPA). 1995b. *Screen3 Model User's Guide.* EPA, Office of Air Quality Planning and Standards, Research Triangle Park, NC, Sept.

U.S. Environmental Protection Agency (EPA). 1996. *PCBs: Cancer Dose-Response Assessment and Application to Environmental Mixtures.* EPA/600/P-96/001. EPA, Office of Research and Development, Washington, DC, Sept.

U.S. Environmental Protection Agency (EPA). 1997a. *Mercury Study Report to Congress,* Vols. 1–5. EPA, Office of Air Quality Planning and Standards and Office of Research and Development, Washington, DC, Dec.

U.S. Environmental Protection Agency (EPA). 1997b. *Exposure Factors Handbook,* Vol. I: *General Factors.* EPA/600/P-95/002Ba. EPA, Office of Research and Development, Washington, DC, Aug.

U.S. Environmental Protection Agency (EPA). 1998. *Region 9 Preliminary Remediation Goals.* EPA, Office of Waste Programs, May 1 update.

U.S. General Accounting Office (GAO). 1993. *Examples of Post World War II Radiation Releases at U.S. Nuclear Sites.* Fact Sheet for the Chairman, Committee on Governmental Affairs, U.S. Senate. GAO/RCED-94-51FS. GAO, Washington, DC, Nov.

U.S. Nuclear Regulatory Commission; Commission of the European Communities (NRC/CEC). 1994. *Probabilistic Accident Consequence Uncertainty Analysis— Dispersion and Deposition Uncertainty Assessment,* Vols. 1, 2. NUREG/CR-6244, EUR 15855EN, SAND94-1453. U.S. Nuclear Regulating Commission, Washington, DC.

Versar. 1976. *PCBs in the United States Industrial Use and Environmental Distribution.* PB-252 012. Prepared for the U.S. Environmental Protection Agency, Office of Toxic Substances. Versar, Springfield, VA, Feb. 25.

Voillequé, P. G., and Gesell, T. F. 1990. Evaluation of environmental radiation exposures from nuclear testing in Nevada: A symposium. *Health Phys.* 59: 501–502.

Walters, W. H., Richmond, M. C., and Gilmore, B. G. 1996. Reconstruction of radioactive contamination in the Columbia River. *Health Phys.* 71: 556–567.

Weber, C. W., and White, J. C. 1977. *Project Progress Report for January 1977— NIOSH Standards Evaluation of Nickel Powder.* K/NI-1, Part 1. Oak Ridge Gaseous Diffusion Plant, Oak Ridge, TN, Feb.

Webster, D. A. 1976. *A Review of Hydrologic and Geologic Conditions Related to the Radioactive Solid Waste Burial Grounds at Oak Ridge National Laboratory, Tennessee.* ORNL/CF/76/358. Oak Ridge National Laboratory, Oak Ridge, TN.

Whicker, F. W., Kirchner, T. B., Anspaugh, L. R., and Ng, Y. C. 1996. Ingestion of Nevada Test Site fallout: Internal dose estimates. *Health Phys.* 71: 477–486.

Widner, T. E. 2000. *Oak Ridge Dose Reconstruction Project Summary Report.* Prepared for the Tennessee Department of Health. ChemRisk, a Division of McLaren/Hart Environmental Services, Alameda, CA, Mar.

Widner, T. E., Ripple, S. R., and Buddenbaum, J. E. 1996. Identification and screening evaluation of key historical materials and emission sources at the Oak Ridge Reservation. *Health Phys.* 71: 457–469.

Wing, J. F. 1980. Letter to H. E. Hodges from J. F. Wing regarding submission of beryllium emission rate data, U.S. Department of Energy, Oak Ridge Reservation, Oak Ridge, TN, Mar. 17.

Wren, C. 1996. *Review of Mercury Levels in Fish, Water, and Sediments.* Progress Report #2 for Oak Ridge Dose Reconstruction Project. Ecological Services for Planning, Guelph, Ontario, Canada, Aug.

Wrenn, M. E., Bertelli, L., Durbin, P. W., Singh, N. P., Kipsztein, J. L., and Eckerman, K. F. 1994. A comprehensive metabolic model for uranium metabolism and dosimetry based on human and animal data. *Radiat. Protect. Dosim.* 53: 255–258.

Xiao, Z. F., Munthe, J., Schroeder, W. H., and Lindqvist, O. 1991. Vertical fluxes of volatile mercury over forest soil and lake surfaces in Sweden. *Tellus* 43B: 267–279.

Yamagata, N., Iwashima, K., Iinuma, T. A., Watari, K., and Nagai, T. 1969. Uptake and retention experiments of radioruthenium in man—I. *Health Phys.* 16: 159–166.

Yarbro, O. 1996. Information for modeling the RaLa process. Memorandum from Orlan Yarbro to Talaat Ijaz, ChemRisk, Cleveland, OH, July 24.

Young, R. A. 1991. *Toxicity Summary for Mercury.* Health and Safety Research Division, Oak Ridge National Laboratory, Oak Ridge, TN, Nov.

Yureiva, N. D., Rafeenko, S. M., Sharipov, V. F., Krupnik, T. A., Dolbeshkin, N. K., and Kovalev, V. M. 1994. Results of the investigation of the health status of children in Mogilev Oblast—a report on the 1994 Chernobyl Sasakawa Project Workshop, Moscow, May 16–17.

Zhao, S. L., and Zhao, F. Y. 1990. Nephrotoxic limit and annual limit on intake for natural U. *Health Phys.* 58: 619–623.

Ziegler, C. K., and Nisbet, B. S. 1995. Long-term simulation of fine-grained sediment transport in large reservoir. *J. Hydraul. Eng.* 121: 173.

SECTION H
Evaluating Risk to Foods

15 Transport of Persistent Organic Pollutants to Animal Products: Fundamental Principles and Application to Health Risk Assessment

GEORGE F. FRIES

Independent Consultant, Silver Spring, Maryland

15.1 INTRODUCTION

Animal products must be evaluated as sources of human exposure to persistent organic pollutants (POPs) because of the importance of these products in most diets. Pathways that include an animal component are often considered the most important sources of human exposure to POPs.[1] Historically, organochlorine insecticides were the primary POPs of concern in regard to the transport of contaminants to foods of animal origin. These insecticides have been restricted or banned in many areas of the world and recent interest has shifted to industrial and contaminant chemicals with dioxinlike activity. The polychlorinated dibenzo-*p*-dioxins (PCDDs), polychlorinated dibenzofurans (PCDFs), and coplanar polychlorinated biphenyls (PCBs) are the most prominent classes of these chemicals.

Low concentrations of POPs occur in diverse areas and environmental media. As a result, body burdens of PCDD/Fs in the general population of the United States average approximately 28 pg/g TCDD toxic equivalents (TEQ) in adipose tissue, and burdens are comparable in other industrialized countries.[1,2] Estimates of human intake of PCDD/F TEQs indicate that more than 90 percent is derived from foods. After excluding fish and other seafoods, it has been concluded that 80 to 95 percent of the food sources are products of animal origin.[3-5] This conclusion, derived for PCDD/Fs, probably also applies to

Human and Ecological Risk Assessment: Theory and Practice, Edited by Dennis J. Paustenbach
ISBN 0-471-14747-8 © 2002 John Wiley & Sons, Inc.

most lipophilic POPs. Thus, it is useful to be able to characterize the transport and fate of POPs in animals and animal production environments when potential human exposure is evaluated.

The fundamental principles of the fate and transport of POPs in animals and animal production systems will be described in this chapter. The application of these principles to risk assessments will be illustrated with a specific example. Emphasis is placed on PCDD/Fs and PCBs in dairy and beef cattle because these animal classes are generally thought to be the most important sources of human exposure and have been the most frequent subject of research. The principles and approaches presented here can be adapted to other POPs and animal species.

15.2   ANIMAL EXPOSURE

15.2.1   General Considerations

Chemical residues in animal products are functions of the levels of contamination in the environment, the exposure of the animal to those contaminants, and the factors that regulate uptake and transport of ingested chemicals in the animal.[6] Knowledge concerning the uptake and transport of PCDD/Fs and PCBs by farm animals has increased substantially in recent years.[7–11] The research on many POPs, however, involves analyses at low concentrations with large relative errors and high costs that limit replication. These factors, together with normal sampling and biological variation, make it difficult to predict the behavior of a specific chemical in an individual animal. This difficulty is not critical because the usual purpose for evaluating uptake and fate of chemicals in animals is to predict or prevent human exposure, which generally involves animal populations at the herd or larger sizes. Uncertainty will decline as animal numbers increase because errors will tend to offset.

The uncertainties associated with identification of sources and measurements of quantities of contaminant ingested are probably greater than the uncertainties associated with quantification of uptake and biotransfer rates of the ingested chemicals. For example, even if aerial deposition on crops were the sole source of a contaminant, feeding and management practices would have important impacts on animal exposure and resulting product residues. These impacts would be related to issues such as the extent of pasture utilization versus hay or silage, and the utilization of nonforage feeds as part of the diet.[6,12,13] Pasture is rarely used for dairy cattle in the United States and the primary forages are hay or silage. On the other extreme, cows may utilize pasture as the sole forage source for 365 days a year in New Zealand.[14] In Europe, pasture may be the primary forage when it is available during the growing season, and stored forages are used for the remainder of the year.[8,11] Numerous other variations in management practices and feed usage will exist depending upon economic and climatic conditions in various regions of the world. Additionally, these factors vary among animal species, and the uncertainties surrounding these factors must be considered in any assessment of human exposure to residues.

15.2.2   Routes of Uptake

Oral ingestion, inhalation, and dermal absorption are potential pathways of contaminant uptake by animals, but only oral ingestion is important. Inhalation contributes less that 1 percent of the intake of PCDD/Fs and PCBs in dairy cattle exposed to normal environmental levels of these compounds.[8,11,15] It is reasonable that this conclusion also applies to other species. Dermal absorption of contaminants such as PCDD/Fs and PCBs in soil or bedding matrices has not been measured in farm animals. Studies in laboratory animals, as well as the usual assumptions made in risk assessments, suggest that contributions to exposure via the dermal pathway are negligible.[16,17]

The water component of oral ingestion also is not an important consideration because normal rate of water consumption and the low solubility of lipophilic compounds would preclude significant intake from this source.[11] Concentrations of PCDD/Fs and PCBs in water were below the detection limits in mass balance studies with dairy cows.[8,15] Thus, oral ingestion of materials other than water is the only pathway of contaminant uptake that requires serious evaluation. This pathway is often thought of in terms of feed, but it may include the deliberate or incidental ingestion of other materials such as bedding, soil, and building materials.

15.2.3   Contamination Sources

Feed contamination occurs by indirect pathways because PCBs never had an important agricultural use, and agricultural chemicals that contained PCDD/F contaminants are no longer used. Deposition of airborne particles and vapors on plant surfaces was originally hypothesized as the primary pathway of feed contamination because PCDD/Fs and PCBs are semivolatile compounds that can be transported aerially over great distances.[1,18-20] Seeds, a major feed source for poultry and swine, were expected to be largely free of residues because of the protection provided by hulls and because lipophilic chemicals are not translocated in plants.[21,22] Cattle, unlike poultry and swine, consume large amounts of herbaceous plant material in the form of pasture, hay, and silage. These considerations led to the conclusion that beef and dairy products would be the primary sources of PCDD/Fs and PCBs in foods.[1,13]

As the database on the occurrence of PCDD/Fs and PCBs in foods increased, several lines of evidence emerged to cast doubt on the primacy of the air–plant–animal exposure pathway of food contamination in the United States. The U.S. Environmental Protection Agency (EPA) surveys of PCDD/Fs and PCBs in the major animal food products did not demonstrate the expected differences in concentrations of PCDD/Fs (TEQ, fat basis) among milk, beef, pork, and poultry samples.[23-26] Differences among species in congener concentration profiles may indicate differing sources or metabolic fates. Different ultimate sources are indicated by the much higher incidence of tetrachlorodibenzo-*p*-dioxin (TCDD) and pentachlorodibenzodioxin (PeCDD) in poultry than in pork or beef, the greater incidence and higher concentrations of 1,2,3,6,7,8-HxCDD in beef and pork than in poultry, and the higher concentrations of coplanar PCBs in beef and poultry than in pork. The high incidence of tetrachlorodibenzofuran (TCDF) in poultry, and the absence of this congener in mammalian products, probably reflects greater metabolic activity in mammals. The concentrations of PCDD/Fs from many sources have distinctive congener profiles.[27] If the diet cannot be measured directly, profiles can be inferred by application of biotransfer coefficients to profiles of the products, and these profiles may be compared to the profiles of potential sources.

Sources of PCDD/Fs that do not involve aerial transport from the original source to feeds have been identified. These include the association of high residue concentrations in beef with pentachlorophenol- (PCP)-treated wood in the animal confinement facilities,[28,29] young chickens with high PCDD concentrations traced to the clay anticaking agent used in soybean meal,[23] and milk contamination in several European countries caused by feeding citrus pulp contaminated during drying.[30] Some sources of PCDD/Fs and PCBs in feeds such as PCBs in fishmeal[11] may only have a tenuous relationship to aerial transport and deposition. The possible importance of recycled animal fats may be inferred from the levels of occurrence of residues in the food surveys.[23-26] As much as 50 percent of the carcass fat may be used for nonfood purposes that include recycling to animal feed. Inappropriate waste disposal as exemplified by the recent Belgium experience in which PCB-containing waste was added to waste fat at a recycling center may also lead to serious residue problems in animal products.[31,32]

Contaminated soil, as a pathway of animal exposure to PCBs and PCDD/Fs, is generally considered less important than direct plant contamination by aerial deposition.[1] However, approximately 85 to 90 percent of the PCDD/Fs deposited on agricultural land reach soil that could be ingested by grazing animals.[13,33] Ingestion of soil by grazing animals is highly variable—abundant herbage and/or the use of supplemental feeds tends to minimize soil ingestion, whereas sparse pasture as the sole source of feed tends to maximize soil ingestion.[6,12,14] Poultry and swine are primarily raised in confinement with little exposure to soil, but these species have the potential for the highest intake as a fraction of total dry matter intake if the animals are in contact with soil.[10,34]

15.3 PHARMACOKINETICS

15.3.1 Dairy Cattle

An explicit or implicit pharmacokinetic model underlies all efforts to quantify residue transport through animals to food products. Models of necessity oversimplify a series of complex environmental and physiological processes. Assumptions that animals are at steady state and in equilibrium with their environment are also an integral part of most models. These assumptions may not always be appropriate because contaminant concentrations and quantities of feed consumed will vary over time; compartment sizes change as weight and body composition changes during growth and lactation; and elimination rates change as lactation and egg production change. These factors, together with the normal variations in analytical determinations, will yield large uncertainties in the measurement of contaminant transport in individual animals.[7,11]

All evidence indicates that active transport is not involved in the uptake and distribution of lipophilic POPs among physiological compartments.[35] Thus, residue distribution among compartments is a function of concentration gradients that will eventually reach an equilibrium based on the fat content of the tissue. The models most frequently used to describe residue transport and disposition in lactating cattle have involved a variant of the two-compartment open system illustrated in Figure 15.1.[35] Although anatomical structures should not be assigned to the compartments, it is generally assumed that the compartment with a slow turnover rate represents the body fat and that the compartment with a rapid turnover rate represents the contents of the gastrointestinal tract. Blood, although treated as a compartment in some models, is a transport medium among the gastrointestinal tract, body fat, and the mammary gland.

The selection of a two-compartment model was based on the biphasic nature of the residue curves in milk following a period of continuous dosing.[37–39] The assumed uniformity of residue concentrations in lipids has received partial

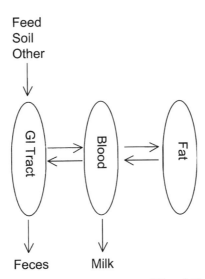

Figure 15.1 Schematic diagram of the transport of lipophilic chemicals in the dairy cow.

confirmation in the analyses of PCDD/Fs and PCBs in the major adipose tissues of cattle.[40-42] Deviations from uniform distribution mainly involve organ tissues or congeners with low toxicity, and thus the consequences for human exposure and risk assessments are minimized. A physiologically based model with added complexity from inclusion of compartments for liver, richly and slowly perfused tissues, and the udder has been proposed.[43] However, the added compartments did not improve concentration predictions significantly. A fugacity model for the behavior of lipophilic compounds in dairy cows successfully described the differences among compounds in transfer rates to milk as a function of the log K_{ow}.[44] Whether log K_{ow} is the controlling factor is uncertain because other factors such as the degree of halogenation in a homologous series are correlated with both log K_{ow} and transfer rates.[7] The other important factor affecting net deposition and transport is metabolism, which in some cases is complete with no residue reaching the food product.

15.3.2 Nonlactating Animals

Models for residue behavior in nonlactating animals are simpler in theory but may be more difficult to apply in practice than models for lactating animals. The simplicity arises because the body burden can be described as a single compartment with concentrations in individual tissues related to lipid content. The difficulties in application arise from the uncertainties in monitoring tissue residue concentrations and the measuring body composition in intact animals. The net fractional uptake of a compound from the gastrointestinal tract is generally treated as a constant for a given matrix. If distribution is uniform in body fat, the lipid-adjusted concentration of a compound in tissues will be a function of lifetime intake, fractional uptake, and the mass of body fat.[44,45]

Growth, body composition, and feed intake functions for various species of farm animals are available, but little has been done to incorporate this information into residue models. An example of insights to be gained from this information is provided in a residue model for growing pigs from birth through attainment of normal slaughter weight.[45] The simulations in Figure 15.2 shows the ratios of accumulated feed intake to the body fat mass under several scenarios. This ratio is equivalent to the bioconcentration factor if absorption from the gastrointestinal tract is complete and there is no elimination. Relatively stable concentrations of residues in body fat are reached when animals are exposed to a constant concentration of contaminant in the diet throughout the postweaning period (Figure 15.2, curve A). This result arises because the relative increase in the mass of body fat is as larger or slightly larger than the increase in accumulated feed intake as the animal ages. Exposures early in life are much less important than exposures at the same dietary concentration nearer slaughter because of dilution through growth (curves B and C). Similar principles apply to other species, but application of the approach to cattle is more complicated because cattle diets are more variable in composition and sources than pig diets during the production cycle. For example, young cattle

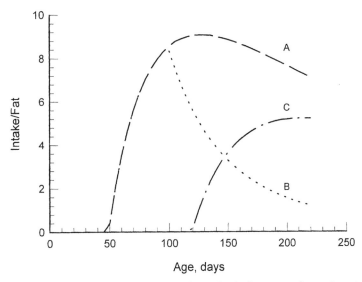

Figure 15.2 Ratio of accumulated feed intake to body fat mass of growing pigs under various exposure scenarios. (A) Contaminated feed from weaning at 50 days to slaughter at 220 days. (B) Contaminated feed from weaning at 50 to 100 days and uncontaminated feed until slaughter at 220 days. (C) Contaminated feed from 120 days until slaughter at 220 days. Adapted from Fries.[43]

may subsist primarily on pasture followed by fattening period on a grain-based diet with potential large differences in residue exposure.[13]

The fate and transport of persistent organics in poultry has received less attention than in other species, but useful analogies can be drawn to processes in mammals. Recent work with PCDD/Fs in laying hens[10,46] confirmed that these compounds were qualitatively similar in behavior to other organohalogen compounds.[47,48] Concentrations of PCDD/Fs in eggs during and following periods of dosing followed curves that were similar to the two-compartment model used for milk. Accumulation of residues in growing birds is expected to follow processes analogous to those discussed for growing pigs.

15.4 TRANSFER COEFFICIENTS

15.4.1 Diet to Milk

Transfer coefficients, which are ratios of the quantities or concentrations of ingested chemicals to the quantities or concentrations in the product, are useful tools for conducting risk assessments and making risk management decisions. The basic assumption in both the measurement and application of transfer coefficients is that the animals are at steady state and that levels of contamination in the environmental are stable.

The three coefficients encountered in the literature to quantitatively characterize the transfer of chemicals from diet to milk or milk fat are bioconcentration factors (BCF), biotransfer factors (BTF), and carryover rates (COR).[11,13,38,45,49] These coefficients are defined by the following equations with the units modified to fit the concentrations usually encountered in work with PCDD/Fs:

$$BCF = C_{MF}/C_{diet} \qquad (15.1)$$

where C_{MF} and C_{diet} are concentrations of POPs in milk fat (pg/g) and in diet dry matter (pg/g).

$$BTF = C_{milk}/I \qquad (15.2)$$

where C_{milk} is the concentration POP in whole milk (pg/kg) and I is the intake of POP (pg/d).

$$COR = 100 \times A_{milk}/I \qquad (15.3)$$

where COR is percent (%), A_{milk} is amount of POP eliminated in milk (pg/d), and I is intake of POP (pg/d).

Five parameters (Table 15.1) provide the only information required to calculate any of the coefficients. Thus, the apparent differences in the coefficients [Eq. (1.3)] are not critical in practice because all can be converted to any of the others if values for the five parameters are available. For example,

$$COR = 100 \times A_{milk}/I = 100(C_{MF} \times M_{fat})/(C_{diet} \times I_{feed})$$
$$= 100(C_{MF}/C_{diet})(M_{fat}/I_{feed})$$

where M_{fat} is the amount of milk fat (kg/d) and I_{feed} is dry matter consumed (kg/d). Since

$$BCF = C_{MF}/C_{diet}$$

TABLE 15.1   Physiological and Residue Measurements Required for Calculation Transfer Coefficients

Measurement	Comments
Feed intake, kg/d	Total on a dry matter basis. Includes soil and extraneous material when relevant. Not required for BCF.
Residue in feed, pg/g	Dry matter basis. Weighted average concentration of a multicomponent diet.
Milk production, kg/d	Not required for BCF.
Milk fat content, %	Not required for BCF.
Residue in fat, pg/g	Alternatively, residue could be measured in whole milk.

it follows that

$$\text{COR} = 100 \times \text{BCF}(M_{\text{fat}}/I_{\text{feed}})$$

The parameters in Table 15.1 are measured with varying degrees of ease and accuracy. The difficulties in obtaining representative samples of large masses of heterogeneous feed, and the variability of analyses carried out near the limits of detection, are major sources of uncertainty. Therefore, coefficients of variation in the range of 40 to 50 percent are not unusual in the results of a single study.

Historically, transfer coefficients were determined by administration of a compound at a fixed daily dose and measurement of concentrations in milk fat when reasonable stability was attained. These conditions are typically reached after 40 to 60 days of dosing, but the animals are not at steady state because concentrations in body fat are expected to increase with longer dosing periods.[38] Improved analytical capability now permits determination of PCDD/F and PCB transfer coefficients in mass balance studies using normally occurring background residues. This method should provide more reliable results if environmental conditions are stable, but the values were not significantly different when coefficients were determined by the two methods sequentially in the same cows.[7]

The BCFs are the easiest of the coefficients to determine and apply because the only measurements required are concentrations of residue in the diet and the product. The BCF approach has been questioned on the grounds that it did not appear to account for multiple pathways of animal exposure.[49] However, this is not a valid criticism because BCFs are based on residue concentrations in the dry matter of the total diet. Thus, if the information is sufficient to calculate the total intake value, which is required to calculate BTFs and CORs, the information also is sufficient to calculate the concentration in the total diet.

A potentially more serious difficulty is that BCFs do not consider quantities of residues ingested or excreted. The fractional absorption of a lipophilic compound from the intestinal tract is generally viewed as a constant for a given compound and matrix.[11,44] Because the absorbed compound must be either stored in fat or eliminated in milk, an animal at steady state will have no net change in body burden, and the fraction of intake eliminated in milk (COR) is expected to be constant regardless of the level of milk fat production. Thus, BCFs may not be constant over a range of production levels. Although there is a positive relationship between dry matter intake and milk production,[50] concentrations will not completely reflect quantities ingested and eliminated because the intake and fat production relationship is not one-to-one. The hypothetical relationship of BCF to production level when COR is constant and cows are at equilibrium is shown in Figure 15.3. In this illustration, BCFs are a reciprocal of fat production with the result that higher values are expected in low producing cows. This relationship was not verified in studies of cows with a wide range in production level.[7] The lack of verification may reflect weight gains and losses that are associated with low and high producing cows,[51] and

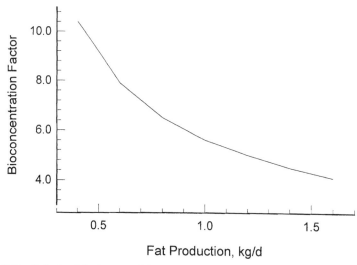

Figure 15.3   Relationship between bioconcentration factors and daily milk fat production of cows if the body burden is at equilibrium and the carryover ratio (COR) is 30 percent.

it illustrates that the assumed steady conditions underlying the transfer coefficients are seldom met in practice.

A comprehensive compilation of published COR values for PCDD/Fs are available.[7] Although derived from diverse studies, the values represented cows with narrow ranges of milk (22 to 28 kg/d) and fat (1.0 to 1.4 kg/d) production that do not differ greatly from the average milk (24 kg/d) and fat (0.9 kg/d) production of cows in the United States. The median COR values from the compilation are listed in Table 15.2 along with BCFs and BTFs derived from the CORs using the production levels of cows in the studies and the estimated dry matter intakes based on those production levels.[50]

The median values listed in Table 15.2 are probably representative of the large herds or populations that would be considered in risk assessment and management decisions under typical conditions in the United States. The ranges of values for individual congeners in single studies were large, with coefficients of variation in the 40 to 50 percent range as cited previously. Residue behavior in individual animals is difficult to predict because of the large errors that can result from sampling and analytical variation. Data on CORs for PCBs are available, but there is considerably less replication than for the PCDD/Fs.[8,11]

15.4.2   Diet to Tissue

Uptake of persistent organics in tissues has not been studied comprehensively because turnover in nonlactating animals is slow and pool sizes change due to growth and fattening. The equations for BTF and BCF can be adapted for tis-

TABLE 15.2 Summary of Median Carryover Ratios (COR), Bioconcentration Factors (BCF), and Biotransfer Factors (BTFs) for PCDD/Fs from Diet to Milk in Dairy Cows[a]

Congener	COR (%)	BTF (d/g)	BCF
2,3,7,8-TCDD	35	17.1	5.7
1,2,3,7,8-PeCDD	28	13.7	4.6
1,2,3,4,7,8-HxCDD	18	8.8	2.9
1,2,3,6,7,8-HxCDD	16	7.8	2.6
1,2,3,7,8,9-HxCDD	12	5.9	2.0
1,2,3,4,6,7,8-HpCDD	2.0	0.98	0.33
1,2,3,4,6,7,8,9-OCDD	0.4	0.20	0.07
2,3,4,7,8-PeCDF	25	12.2	4.1
1,2,3,4,7,8-HxCDF	18	8.8	2.9
1,2,3,6,7,8-HxCDF	16	7.8	2.6
2,3,4,6,7,8-HxCDF	14	6.8	2.3
1,2,3,4,6,7,8-HpCDF	1.9	0.93	0.31
1,2,3,4,7,8,9-HpCDF	4.0	2.0	0.65
1,2,3,4,6,7,8,9-OCDF	0.4	0.20	0.07

[a]Literature sources and derivation of the COR values are cited in Fries et al.[22] BCFs and BTFs were calculated from COR as described in the text. Omitted are 2,3,7,8-TCDF, 1,2,3,7,8-PeCDF, and 1,2,3,7,8,9-HxCDF, which rarely occur in milk.

sues, but COR does not have a tissue analogy unless it is viewed as the body burden divided by lifetime intake. The usefulness of the BTF compilation prepared by Travis and Arms[49] for accumulation of organics in beef is limited. Many of the cited studies were too short for steady state to be established, and some studies involved lactating cattle, which would retain a smaller fraction of the intake than nonlactating cattle. Another difficulty in application of BFTs is that the values are based on a constant unit of daily intake, whereas actual intake is a function of body weight that increases with age.[48]

Review of the BCFs obtained in long-term (> 100 days) beef uptake studies has prompted the conclusion that BCFs for body fat in beef animals and milk fat in dairy cows were approximately equal at steady state.[6,12] Confirmation of this conclusion is provided by simulations with the pig model in which the predicted body concentrations at slaughter weight were are 9- to 10-fold greater than the concentrations in feed fed over a lifetime when uptake of contaminant is 100 percent (Figure 15.2). If the net uptake is in the range of 50 to 60 percent of intake, the BCF would be in the range of 4.5 to 6.0, which is the approximate range of the highest reported BCFs for milk fat (Table 15.2). Thus, BCFs derived for milk fat should also be applicable to beef fat if more reliable information is not available.

15.5 TISSUE AND PRODUCT RELATIONSHIPS

Uniformity, or a predictable relationship, in lipid-adjusted concentrations of residues in tissues is a practical necessity for the conduct of risk assessments and regulatory activities. Without uniformity, analytical resources would be overburdened by the large number of samples that would be required to characterize a carcass. Lipid-adjusted concentrations of organochlorine insecticides and polybrominated biphenyls in depot adipose and muscle tissues of cattle do not differ greatly.[52,53] Exceptions, such as high concentrations in liver and low concentrations in nervous tissue, are not important for risk assessments because of the small contribution of organ meats to total human meat consumption. Distribution data for PCDD/Fs and the dioxinlike PCBs are less extensive, but the available data indicate general agreement among concentrations in subcutaneous, perirenal, and peritoneal fats.[40,42] The relationships of concentrations in muscle to adipose tissue is more complex in that elevated concentrations of heptachlorodibenzodioxins and furans (HpCDD/Fs) and octachlorinated-dibenzodioxins and furans (OCDD/Fs) are often noted in muscle when concentrations of other PCDD/Fs are comparable in muscle and adipose tissue.[40,42] The enhanced concentrations in muscle indicate that properties such as degree of chlorination, molecular weight, or log K_{ow} may affect the ease with which a compound is distributed throughout the body.

A functional relationship also exists between concentrations in body fat and milk fat during steady state and during the depuration-phase animal exposure scenarios. The ratios of concentrations in milk fat to those in body fat is a value between 0.5 and 1.0 for most organohalogen compounds.[38] These ratios in recent work with PCDD/Fs were within one standard deviation of 1:1 for TCDD, the PeCDD/Fs, and HxCDD/Fs, but the ratios were much lower for the HpCDD/Fs and OCDD/Fs.[41]

15.6 EFFECTS OF HALOGENATION AND PHYSICAL PROPERTIES

15.6.1 Metabolism

The number and positions of halogen substitutions have major impacts on uptake and transport of persistent lipophilic compounds. Two or more adjacent nonhalogenated carbons are a typical requirement for metabolism of PCBs.[35] Substitution in the 4, 4′ positions has been suggested to favor persistence in cows.[8] A recent refinement indicates that adjacent unsubstituted meta- and para-positions were more readily metabolized than adjacent unsubstituted meta- and ortho-positions.[11] All PCBs with dioxinlike activity are expected to persist and be present in products if present in the diet. This expectation was confirmed by results of the U.S. food surveys in which PCBs with dioxinlike activity were present in all samples except some pork.[23–25,52]

All PCDD/Fs of toxicological concern have chlorine substitutions in the 2,3,7,8- positions. Of the 2,3,7,8-substituted PCDD/Fs, only the three PCDFs that do not have chlorines in the 4 and 6 positions are metabolized by mammals at a rate sufficient to be of practical importance. Residues of these congeners (2,3,7,8-TCDF, 1,2,3,7,8-PeCDF, and 1,2,3,7,8,9-HxCDF) have only been reported when high doses were administered to experimental animals. The situation with birds is somewhat different in that 2,3,7,8-TCDF was found in nearly all chicken and turkey samples in the U.S. poultry survey.[23] These observations indicate that birds are unable, or have diminished capacity, to metabolize TCDF and the other furans not substituted in the 4 and 6 positions.

15.6.2 Bioaccumulation

The physical property, log K_{ow}, has been suggested as a predictor of bioaccumulation. The initial compilation of the relationship between log K_{ow} and bioaccumulation indicated a positive relationship.[49] A weakness of this formulation was the failure to consider metabolism, which is more prevalent among compounds with low log K_{ow} values. The second weakness was that it did not include compounds with log K_{ow} values greater than 6.5, a value that is lower than the log K_{ow} of all PCDD/Fs and most PCBs that are of toxicological concern.[55,56] The relationship between log K_{ow} and uptake was negative when the log K_{ow} values exceed 6.5.[7,8,57] Equations relating uptake to log K_{ow} have been proposed, but the data bases were small, which may limit general application.[8,57]

An important limitation in the use of log K_{ow} values as predictors of accumulation and transport is that measurements of log K_{ow} by different laboratories and methods may lead to widely differing values.[56] Therefore, it is important that relationships are developed from databases that have been obtained with the same methodology and preferably within the same laboratory. Log K_{ow} values are comparable for compounds with equal degrees of chlorination within a homologous series. The number of halogens may describe accumulation and transport compounds within a homologous series as effectively as the log K_{ow} value.[7]

15.6.3 Toxic Equivalents

The residue concentrations of environmental samples are at times expressed in toxic equivalent (TEQ) concentrations.[58] Although TEQ can be useful short hand for expressing the toxicity of a mixture, the term should not be used for characterizing the transport through environmental and biological systems. The toxicity of the 2,3,7,8-substituted PCDD/Fs generally decreases with increasing chlorination. Transfer coefficients also decrease with increasing chlorination (Table 15.2). Thus, a TEQ in feed of a congener such as TCDD will yield a higher TEQ in the human food product than a TEQ of the more highly chlorinated congeners.

15.7 DEPURATION

Data on elimination rates have been applied in management of incidents involving excessive residues[39] and in pharmacokinetic models to estimate steady-state concentrations from short-term observations.[59,60] Consistent with a two-compartment model (Figure 15.1), concentrations of lipophilic compounds in milk after change from a period of high intake to a period of low intake is described by a two-term exponential equation:

$$C = C_1 e^{-k_1 t} + C_2 e^{-k_2 t} \tag{15.4}$$

where C is concentration in milk, C_1 and C_2 are the initial concentrations of the compartments, k_1 and k_2 are constants, and t is time.[39,41] The first term of Eq. (15.4) is related to clearance of compounds from blood and the gastrointestinal tract, and it contributes little to the concentration in milk after the first 10 days.

The second term reflects elimination from body storage, and half-lives in lactating cows may be as short as 40 to 60 days for chlorinated hydrocarbon insecticides to well over 150 days for the more highly chlorinated PCDD/Fs when measured in midlactation.[38,39,41] The effects of losing and regaining approximately 50 percent of the body fat during lactation[51] on half-lives has not been fully characterized.

Clearance rates, an alternative approach for expressing elimination rates,[36] are the fractions of body burden eliminated per unit time are described by the equation:

$$k_c = E/B \tag{15.5}$$

where k_c is the clearance rate (day^{-1}), E is elimination in milk (pg/day), and B is body burden (pg). The value of k_c is essentially equal to the constant k_2 in Eq. (15.4). The greatest uncertainty is related to the burden value (B) because the term includes estimates rather than direct measurements of body composition.

If the elimination rate constant is known, steady-state concentrations or body burdens can be estimated by rearranging Eq. (15.5). The amounts eliminated daily will equal the amounts of daily uptake at steady state. This approach was used to estimate steady-state concentrations of TCDD from results of a short-term feeding study.[59,60] The resulting BCF was much greater than BCFs in subsequent direct measurements because a one-compartment model was assumed rather than the more appropriate two-compartment system and because the amount of TCDD ingested was used for input rather than the amount absorbed.

15.8   CASE STUDY OF A LANDFILL SITE EVALUATION

A waste disposal company proposed to establish a new landfill equidistant between two metropolitan areas located approximately 80 miles apart. The proposed landfill was of a modern design, and it was proposed to utilize the landfill gas for the production of electricity. An exhaustive site selection process that included an environmental impact assessment was followed. The site selected was in a rural agricultural area where the primary activity was raising beef and dairy cattle. When the local authorities issued a permit for the facility, a group of neighboring farmers appealed the decision. Among the grounds for the appeal was the claim that potential emission of dioxins from the landfill had not been evaluated, and it was alleged that these emissions could be transported aerially and deposited on pastures and other feed crops. The possible increased PCDD/F residues in beef and milk were alleged to not only be a health risk for consumers but also the presence of these residues would harm the reputation of the area as a producer of high-quality animal products.

The waste disposal company carried out an exposure assessment as part of its defense in the appeal hearings. Dioxin emissions in landfill gas had never been considered a problem in the past even though limited literature indicated that facilities of the type proposed could emit dioxins at low concentrations. This literature was reviewed and a conservative value for dioxin emissions was selected. Aerial dispersion and deposition models based on EPA guidelines.[1,61] were applied to the data. The maximum predicted concentrations of PCDD/F congeners in soil and feeds produced at facility boundary are presented in Table 15.3. The predicted concentrations in these matrices decreased by approximately 20 percent for each additional 100-meter distance from the landfill boundary.

TABLE 15.3 Predicted Concentrations of PCDDs and PCDFs in Dry Matter of Soil and Feeds Produced at Property Boundary of Proposed Waste-Handling Facility

Congener	TEF[a]	Predicted Concentration (pg/g)		
		Soil	Forage[b]	Maize Silage
2,3,7,8-TCDD	1	0.34	0.028	0.014
1,2,3,7,8-PeCDD	1	0.28	0.013	0.0063
1,2,3,4,7,8-HxCDD	0.1	0.15	0.0039	0.0019
1,2,3,6,7,8-HxCDD	0.1	0.15	0.0039	0.0019
1,2,3,7,8,9-HxCDD	0.1	0.15	0.0039	0.0019
1,2,3,4,6,7,8-HpCDD	0.01	0.53	0.048	0.024
1,2,3,4,6,7,8,9-OCDD	0.0001	0.53	0.061	0.028
2,3,7,8-TCDF	0.1	0.025	0.0016	0.0008
1,2,3,7,8-PeCDF	0.05	0.040	0.0016	0.0008
2,3,4,7,8-PeCDF	0.5	0.20	0.0055	0.0026
1,2,3,4,7,8-HxCDF	0.1	0.050	0.0009	0.0004
1,2,3,6,7,8-HxCDF	0.1	0.15	0.0027	0.0013
2,3,4,6,7,8-HxCDF	0.1	0.68	0.0089	0.0041
1,2,3,7,8,9-HxCDF	0.1	0.051	0.0007	0.0003
1,2,3,4,6,7,8-HpCDF	0.01	0.47	0.084	0.042
1,2,3,4,7,8,9-HpCDF	0.01	0.53	0.0056	0.0028
1,2,3,4,6,7,8,9-OCDF	0.0001	0.11	0.011	0.0055

[a]Toxic equivalent factors.[58]

[b]Forage includes pasture, hay, and silage produced from grass–legume mixtures.

Several scenarios typical of animal production in the area were evaluated to determine the potential incremental increase in PCDD/F residues due to emissions from the waste-handling facility. It was only necessary to consider the three matrices listed in Table 15.3 when calculating potential animal residues because hulls would protect a locally produced grain crop used for animal feed or the grain would be imported from outside the area of emission deposition. Maize silage is placed in a separate category from the grass–legume forages because maize is considered to have a greater mass per unit surface area than other forages.[1]

The concentration of a congener in the fat of milk or beef is calculated by the equation:

$$C_P = (B \times 0.65 D_{soil} C_{soil}) + (B D_{for} C_{for}) + (B D_{sil} C_{sil}) \qquad (15.6)$$

where C_P is the concentration in the fat of beef or milk, B is the BCF from Table 15.2, 0.65 is a factor for the reduced bioavailability of residues in soil,[1] D_{soil} is fraction of the dry matter intake that is soil, C_{soil} is the congener concentration in soil, D_{for} is fraction of the dry matter intake that is forage, C_{for} is the congener concentration in forage, D_{sil} is fraction of the dry matter intake that is silage, and C_{sil} is the congener concentration in silage. Some of the terms will not apply to specific scenarios.

Calculation of the predicted concentrations for each of scenario is illustrated using TCDD. Similar calculations were made for all of the other congeners with only a change of the specific value of the BCF. The results are summarized in Table 15.4, and the predicted concentrations were converted to TEQs by using the toxic equivalent factors listed in Table 15.3.

The following scenarios were considered;

Beef A. Climatic conditions are such that cattle can graze throughout the year. Cattle are on pasture and receive no other feed. Yearly average soil consumption is assumed to be 4 percent of dry matter intake, which is the default value used in some guidance documents.[1,61] Calculation:

$$C_{TCDD} = (B \times 0.65 D_{soil} C_{soil}) + (B D_{for} C_{for})$$
$$= (5.7 \times 0.65 \times 0.04 \times 0.34) + (5.7 \times 0.96 \times 0.028) = 0.20 \text{ pg/g}$$

Beef B. Same as Beef A except that yearly average soil consumption is assumed to be 6 percent of dry matter intake. This value is more consistent with the stipulated conditions.[13,14] Calculation:

$$C_{TCDD} = (B \times 0.65 D_{soil} C_{soil}) + (B D_{for} C_{for})$$
$$= (5.7 \times 0.65 \times 0.06 \times 0.34) + (5.7 \times 0.94 \times 0.028)$$
$$= 0.22 \text{ pg/g}$$

TABLE 15.4 Predicted Concentrations of PCDDs and PCDFs in Fat of Beef and Milk Produced by Cattle at Property Boundary of Proposed Waste-Handling Facility[a]

Congener	Predicted Concentration (pg/g)				
	Beef A	Beef B	Milk A	Milk B	Milk C
2,3,7,8-TCDD	0.20	0.22	0.13	0.11	0.047
1,2,3,7,8-PeCDD	0.090	0.11	0.057	0.041	0.017
1,2,3,4,7,8-HxCDD	0.022	0.028	0.013	0.0079	0.0033
1,2,3,6,7,8-HxCDD	0.020	0.025	0.012	0.0071	0.0029
1,2,3,7,8,9-HxCDD	0.015	0.019	0.0093	0.0054	0.0022
1,2,3,4,6,7,8-HpCDD	0.020	0.022	0.013	0.011	0.0047
1,2,3,4,6,7,8,9-OCDD	0.005	0.0054	0.0034	0.0030	0.0012
2,3,7,8-TCDF	0.0	0.0	0.0	0.0	0.0
1,2,3,7,8-PeCDF	0.0	0.0	0.0	0.0	0.0
2,3,4,7,8-PeCDF	0.043	0.054	0.026	0.016	0.0065
1,2,3,4,7,8-HxCDF	0.0063	0.0081	0.0037	0.0019	0.0008
1,2,3,6,7,8-HxCDF	0.017	0.022	0.0099	0.0050	0.0020
2,3,4,6,7,8-HxCDF	0.060	0.080	0.034	0.014	0.0058
1,2,3,7,8,9-HxCDF	0.0	0.0	0.0	0.0	0.0
1,2,3,4,6,7,8-HpCDF	0.029	0.030	0.020	0.018	0.0078
1,2,3,4,7,8,9-HpCDF	0.0044	0.0047	0.0029	0.0025	0.0011
1,2,3,4,6,7,8,9-OCDF	0.0009	0.0010	0.0006	0.0005	0.0002
TEQ	0.33	0.37	0.21	0.16	0.069

[a]Exposure scenarios:

Beef A—Cattle on pasture and receive no other feed. Yearly average soil consumption is assumed to be 4% of dry matter intake.

Beef B—Cattle on pasture and receive no other feed. Yearly average soil consumption is assumed to be 6% of dry matter intake.

Milk A—Cattle on pasture. Concentrates consisting of protected seeds and ingredients from outside the area provide 30% of dry matter intake. Yearly average soil consumption is assumed to be 2% of dry matter intake.

Milk B—Cattle are confined and have no access to pasture. Hay produced in the area provides 70% of dry matter intake. Concentrates consisting of protected seeds and ingredients from outside the area provide 30% of dry matter intake.

Milk C—Cattle are confined and have no access to pasture. Hay and silage produced in the area provide 5 and 50% of dry matter intake, respectively. Concentrates consisting of protected seeds and ingredients from outside the area provide 45% of dry matter intake.

Milk A. Cattle on pasture. Concentrates consisting of protected seeds and ingredients from outside the area provide 30 percent of dry matter intake. Yearly average soil consumption is assumed to be 2 percent of dry matter intake, which is the value determined in dairy cows receiving supplemental feed.[14] Calculation:

$$C_{TCDD} = (B \times 0.65 D_{soil} C_{soil}) + (BD_{for} C_{for})$$
$$= (5.7 \times 0.65 \times 0.02 \times 0.34) + (5.7 \times 0.68 \times 0.028)$$
$$= 0.13 \text{ pg/g}$$

Milk B. Cattle are confined and have no access to pasture. Hay produced in the area provides 70 percent of dry matter intake. Concentrates consisting of protected seeds and ingredients from outside the area provide 30 percent of dry matter intake. Calculation:

$$C_{\text{TCDD}} = (BD_{\text{for}} C_{\text{for}})$$
$$= (5.7 \times 0.70 \times 0.028) = 0.11 \text{ pg/g}$$

Milk C. Cattle are confined and have no access to pasture. Hay and silage produced in the area provide 5 and 50 percent of dry matter intake, respectively. Concentrates consisting of protected seeds and ingredients from outside the area provide 45 percent of dry matter intake. Calculation:

$$C_{\text{TCDD}} = (BD_{\text{for}} C_{\text{for}}) + (BD_{\text{sil}} C_{\text{sil}})$$
$$= (5.7 \times 0.05 \times 0.028) + (5.7 \times 0.50 \times 0.014) = 0.047 \text{ pg/g}$$

The highest TEQ values for the scenarios in Table 15.4 were 0.37 pg/g for beef fat and 0.21 pg/g for milk fat. One measure of the significance of these values is a comparison with the average concentrations in U.S. beef and milk. Beef fat in a national survey had average TEQs of 0.35 pg/g (nondetects = 0) and 0.85 pg/g (nondetects = 0.5 × detection limit).[26] Average TEQ concentrations in milk fat were 0.82 pg/g.[25] A high degree of conservatism was incorporated in the projections by using feed and soil concentrations at the property boundary of the proposed facility. It is very unlikely that any animal would consume feed produced exclusively at the property boundary. Finally, the values for all congeners in Table 15.4 are below the normal limits of detection for the standard dioxin analytical methods. For these reasons it was concluded the proposed facility did not pose a significant risk of increased dioxin residues on the neighboring farms.

15.9 CONCLUSIONS

Over the past 20 years our understanding of the movement, uptake, and transport of persistent contaminants to food products obtained from farm animals has matured significantly.[7,10,11,33,44,57] It is now clear that the group of chemicals included in the term "persistent organic pollutants" generally are present in humans as a result of their presence in diets. More than 90 percent of the intake of these 15 to 50 chemicals occurs through ingestion of fish, meat, and dairy products.

This chapter presents the fundamental information necessary for risk assessors to understand the movement of this class of chemicals through the ecosystem and agricultural production systems. If a robust data set of the concentrations of these chemicals in the various media are available, the relatively straightforward calculations illustrated here should be adequate for characterization of the magnitude of the potential human exposure to these chemicals.

REFERENCES

1. U.S. Environmental Protection Agency (EPA). *Estimating Exposure to Dioxin-like Compounds.* EPA/600/P-00/001Bb. Office of Research and Development, EPA, Washington, DC, 2000.

2. J. E. Orban, J. S. Stanley, J. G. Schemberger, and J. C. Remmers. Dioxins and dibenzofurans in adipose tissue of the general US population and selected subpopulations. *Am. J. Public Health* 84: 439–445 (1994).

3. P. Furst, C. Furst, and W. Grobel. Levels of PCDDs and PCDFs in food stuffs from the Federal Republic of Germany. *Chemosphere* 20: 787–792 (1990).

4. R. M. C. Theelen, A. K. D. Leim, W. Slob, and J. H. van Wijnen. Intake of 2,3,7,8 chlorine substituted dioxins, furans, and planar PCBs from food in the Netherlands: Median and distribution. *Chemosphere* 27: 1625–1635 (1993).

5. C. C. Travis and H. A. Hattemer-Frey. Human exposure to dioxin. *Sci. Total Environ.* 104: 97–127 (1991).

6. G. F. Fries. Transport of organic environmental contaminants to animal products. *Rev. Environ. Contam. Toxicol.* 141: 71–109 (1995).

7. G. F. Fries, D. J. Paustenbach, D. B. Mather, and W. J. Luksemburg. A cogener specific evaluation of transfer of chlorinated dibenzo-*p*-dioxin and dibenzofurans in milk of cows followinge ingestion of pentachlorophenol-treated wood. *Environ. Sci. Technol.* 33: 1165–1170 (1999).

8. M. S. McLachlan. Mass balance of polychlorinated biphenyls and other organochlorine compounds in a lactating cow. *J. Agric. Food Chem.* 41: 474–480 (1993).

9. M. McLachlan and W. Richter. Uptake and transfer of PCDD/Fs by cattle fed naturally contaminated feedstuffs and feed contaminated as a result of sewage sludge application. I. Lactating cows. *J. Agric. Food Chem.* 46: 1166–1172 (1998).

10. R. D. Stephens, M. X. Petreas, and D. G. Hayward. Biotransfer and bioaccumulation of dioxins and furans from soil: Chickens as a model for foraging animals. *Sci. Total Environ.* 175: 253–273 (1995).

11. G. O. Thomas, A. J. Sweetman, and K. C. Jones. Input-output balance of polychlorinated biphenyls in a long-term study of lactating dairy cows. *Environ. Sci. Technol.* 33: 104–112 (1999).

12. G. F. Fries. Ingestion of sludge applied organic chemicals by animals. *Sci. Total Environ.* 185: 93–108 (1996).

13. G. F. Fries and D. J. Paustenbach. Evaluation of potential transmission of 2,3,7,8-tetrachlorodibenzo-*p*-dioxin contaminated incinerator emissions to humans via foods. *J. Toxicol. Environ. Health* 29: 1–43 (1990).

14. W. B. Healy. Ingestion of soil by dairy cows. *New Zealand J. Agric. Res.* 11: 487–499.

15. M. S. McLachlan, H. Thoma, M. Reissinger, and O. Hutzinger. PCDD/F in an agricultural food chain. Part 1: PCDD/F mass balance of a lactating cow. *Chemosphere* 20: 1013–1020 (1990).

16. R. D. Kimbrough, H. Falk, P. Stehr, and G. Fries. Health implications of 2,3,7,8-tetrachlorodibenzo-*p*-dioxin (TCDD) contamination of residential soil. *J. Toxicol. Environ. Health* 14: 47–93 (1984).

17. H. Poiger and C. Schlatter. Influence of solvents and absorbents on dermal and intestinal absorption of TCDD. *Food Cosmetic Toxicol.* 18: 477–481 (1980).

18. T. E. Bidleman. Atmospheric processes: Wet and dry deposition of organic compounds are controlled by their vapor-particle partitioning. *Environ. Sci. Technol.* 22: 361–367 (1988).

19. J. M. Czuczwa, B. D. McVeety, and R. A. Hites. Polychlorinated dibenzo-*p*-dioxins and dibenzofurans in sediments from Siskiwit Lake, Isle Royale. *Science* 226: 568–569 (1984).

20. G. O. Thomas, K. E. C. Smith, A. J. Sweetman, and K. C. Jones. Further studies of the air-pasture transfer of polychlorinated biphenyls. *Environ. Pollut.* 102: 119–128 (1998).

21. E. H. Buckley. Accumulation of airborne polychlorinated biphenyls in foliage. *Science* 216: 520–522 (1982).

22. J. K. McCrady, C. MacFarlane, and L. K. Gander. The transport and fate of 2,3,7,8-TCCD in soybean and corn. *Chemosphere* 21: 359–376 (1990).

23. J. Ferrario, C. Byrne, M. Lorber, P. Saunders, W. Leese, A. Dupuy, D. Winters, D. Cleverly, J. Schaum, P. Pinsky, C. Deyrup, R. Ellis, and J. Walcott. A statistical survey of dioxin-like compounds in United States poultry fat. *Organohalogen Compounds* 32: 245–248 (1997).

24. M. Lorber, P. Saunders, J. Ferrario, W. Leese, D. Winters, D. Cleverly, J. Schaum, C. Deyrup, R. Ellis, J. Walcott, A. Dupuy, C. Byrne, and D. McDaniel. A statistical survey of dioxin-like compounds in United States pork. *Organohalogen Compounds* 32: 238–244 (1997).

25. M. N. Lorber, D. L. Winters, J. Griggs, R. Cook, R. Baker, S. Baker, J. Ferrario, C. Byrne, A. Dupuy, and J. Schaum. A national survey of dioxin-like compounds in the United States milk supply. *Organohalogen Compounds* 38: 125–130 (1998).

26. D. Winters, D. Cleverly, K. Maier, A. Dupuy, C. Byrne. C. Deyrup, R. Ellis, J. Ferrario, M. Lorber, D. McDaniel, J. Schaum, and J. Walcott. A statistical survey of dioxin-like compounds in United States beef: A progress report. *Chemosphere* 32: 469–478 (1996).

27. H. Hagenmaier, C. Lindig, and J. She. Correlation of environmental occurrence of polychlorinated dibenzo-*p*-dioxins and dibenzofurans with possible sources. *Chemosphere* 29: 2163–2174 (1994).

28. G. F. Fries, V. J. Feil, and K. L. Davison. The significance of pentachlorophenol-treated wood as a source of dioxin residues in United States beef. *Organohalogen Compounds* 28: 156–159 (1996).

29. G. F. Fries, V. J. Feil, R. G. Zaylskie, K. M. Bialek, and C. P. Rice. Relationship of concentrations of pentachlorophenol and chlorinated dioxins and furans from livestock facilities. *Organohalogen Compounds* 39: 245–248 (1998).

30. R. Malisch. Increase of PCDD/F-contamination of milk and butter in Germany by use of contaminated citrus pulps as component in feed. *Organohalogen Compounds* 38: 65–70 (1998).

31. A. Covaci, J. J. Ryan, and P. Schepens. Patterns of PCBs and PCDFs in chicken

and pork following a Belgian contamination. *Organohalogen Compounds* 47: 349–352 (2000).

32. A. Buekans, K. Schroyens, and D. Liem. PCB in the food chain: The Belgian experience and elements for a risk analysis. *Organohalogen Compounds* 48: 269–272 (2000).

33. K. Welsch-Pausch and M. S. McLachlan. Fate of airborne polychlorinated dibenzo-*p*-dioxins and dibenzofurans in an agricultural ecosystem. *Environ. Pollut.* 102: 129–132 (1998).

34. G. F. Fries, G. S. Marrow, and P. A. Snow. Soil ingestion by swine as a route of contaminant exposure. *Environ. Toxicol. Chem.* 1: 201–204 (1982).

35. H. B. Matthews and R. L. Dedrick. Pharmacokinetics of PCBs. *Ann. Rev. Pharmacol. Toxicol.* 24: 85–103 (1984).

36. M. Gilbaldi. *Biopharmaceutics and Clinical Pharmacokinetics*. Lea & Febiger, Philadelphia, 1977.

37. D. Firestone, M. Clower, A. P. Borsetti, R. H. Teske, and P. E. Long. Polychlorodibenzo-*p*-dioxin and pentachlorophenol in milk and blood of cows fed technical pentachlorophenol. *J. Agric. Food Chem.* 27: 1171–1177 (1979).

38. G. F. Fries. The kinetics of halogenated hydrocarbon retention and elimination in dairy cattle. In G. W. Ivie and H. H. Dorough (Eds.), *Fate of Pesticides in the Large Animal*. Academic, New York, 1977.

39. L. G. M. T. Tuinstra, A. H. Roos, P. L. M. Berender, J. A. van Rhijn, W. A. Traag, and M. J. B. Menglers. Excretion of polychlorinated dibenzo-*p*-dioxins and -furans in milk of cows fed dioxins in the dry period. *J. Agric. Food Chem.* 40: 1772–1776 (1992).

40. M. Lorber, V. Feil, D. Winters, and J. Ferrario. Distribution of dioxins, furans, and coplanar PCBs in different fat matrices in cattle. *Organohalogen Compounds* 32: 327–334 (1998).

41. M. Olling, H. J. G. M. Derks, P. L. M. Berender, A. K. D. Liem, and A. P. J. M. de Jong. Toxicokinetics of eight ^{13}C-labelled polychlorinated dibenzo-*p*-dioxins and -furans in lactating cows. *Chemosphere* 23: 1377–1385 (1991).

42. J. R. Startin, C. Wright, M. Kelly, and N. Harrison. Depletion rates of PCDDs in bull calf tissues. *Organohalogen Compounds* 21: 347–350 (1994).

43. H. J. G. M. Derks, P. L. M. Berender, M. Olling, A. K. D. Liem, and A. P. J. M. de Jong. Pharmacokinetic modeling of polychlorinated dibenzo-*p*-dioxins (PCDDs) and furans (PCDFs) in cows. *Chemosphere* 28: 711–715 (1994).

44. M. S. McLachlan. Model of the fate of hydrophobic contaminants in cows. *Environ. Sci. Technol.* 28: 2407–2414 (1994).

45. G. F. Fries. A model to predict concentrations of lipophilic chemicals in growing pigs. *Chemosphere* 32: 443–451 (1996).

46. M. X. Petreas, L. R. Goldman, D. G. Hayward, R. R. Chang, J. J. Flattery, T. Wiesmuller, R. D. Stephens, D. M. Fry, C. Rappe, S. Bergek, and M. Hjelt. Biotransfer and bioaccumulation of PCDD/PCDFs from soil: Controlled feeding studies of chickens. *Chemosphere* 23: 1731–1741 (1991).

47. G. F. Fries, H. C. Cecil, J. Bitman, and R. J. Lillie. Retention and excretion of polybrominated biphenyls by hens. *Bull. Environ. Contam. Toxicol.* 15: 278–282 (1976).

48. G. F. Fries, R. J. Lillie, H. C. Cecil, and J. Bitman. Retention and excretion of polychlorinated biphenyls by laying hens. *Poultry Sci.* 56: 1275–1280 (1977).

49. C. C. Travis and A. D. Arms. Bioconcentration of organics in beef, milk, and vegetation. *Environ. Sci. Technol.* 22: 271–274 (1988).

50. Subcommittee on Feed Intake. *Predicting Feed Intake of Food-Producing Animals.* National Academy Press, Washington, DC, 1987.

51. S. M. Andrew, D. R. Waldo, and R. A. Erdman. Direct analysis of body composition of dairy cows at three physiological stages. *J. Dairy Sci.* 77: 3022–3033 (1994).

52. G. F. Fries and G. S. Marrow. Distribution of hexachlorobenzene residues in beef steers. *J. Anim Sci.* 45: 1160–1165 (1977).

53. G. F. Fries, G. S. Marrow, and R. M. Cook. Distribution and kinetics of PBB residues in cattle. *Environ. Health Perspect.* 23: 43–50 (1978).

54. D. Winters, D. Cleverly, M. Lorber, K. Maier, A. Dupuy, C. Byrne. C. Deyrup, R. Ellis, J. Ferrario, W. Leese, J. Schaum, and J. Walcott. Coplanar polychlorinated biphenyls (PCBs) in a national sample of beef in the United States: Preliminary results. *Organohalogen Compounds* 28: 350–354 (1996).

55. J. Brodsky and K. Ballschmiter. Reversed phase liquid chromatography of PCBs as a basis for the calculation of water solubility and log K_{ow} for polychlorobiphenyls. *Fresenius Z. Anal. Chem.* 331: 295–301 (1988).

56. W. Y. Shiu, W. Doucette, F. A. P. C. Gobas, A. Andren, and D. MacKay. Physical-chemical properties of chlorinated dibenzo-*p*-dioxins. *Environ. Sci. Technol.* 22: 651–658 (1988).

57. A. J. Sweetman, G. O. Thomas, and K. C. Jones. Modeling the fate and behaviour of lipophilic organic contaminants in lactating dairy cows. *Environ. Pollut.* 104: 261–271 (1999).

58. M. Van den Berg, L. Birnbaum, A. T. C. Bosveld, B. Brunstrom, P. Cook, M. Feely, J. P. Giesy, A. Hanberg, R. Hasegawa, S. W. Kennedy, T. Kubiak, J. C. Larsen, F. X. R. van Leeuwen, A. K. Djiem, C. Nolt, R. E. Peterson, L. Poellinger, S. Safe, D. Schrenk, M. Tillitt, M. Tysklind, M. Younes, F. Waern, and T. Zacharewski. Toxic equivalency factors (TEFs) for PCBs, PCDDs, PCDFs for humans and wildlife. *Environ. Health Perspect.* 106: 775–792 (1998).

59. P. Connett and T. Webster. An estimation of the relative human exposure to 2,3,7,8-TCDD emissions via inhalation and ingestion of cow's milk. *Chemosphere* 16: 2079–2084 (1987).

60. J. B. Stevens and E. N. Gerbec. Dioxin in the agricultural food chain. *Risk Anal.* 8: 329–335 (1988).

61. U.S. Environmental Protection Agency (EPA). *Human Health Risk Assessment Protocol for Hazardous Waste Combustion Facilities.* EPA530-D-98/001a. Office of Solid Waste and Emergency Response, EPA, Washington, DC, 1998.

16 Estimating Dietary Exposure: Methods, Algorithms, and General Considerations

BARBARA PETERSEN

Novigen Sciences, Washington, DC

16.1 INTRODUCTION

Methods for estimating the dietary contribution to exposure will vary depending upon the compound's toxicity profile and the intended purpose for the results. Considerations in defining the approach will include the definition of the chemical to be evaluated and any breakdown products of toxicological significance. Potential biological effects must be carefully considered in planning an exposure assessment. Factors of interest include dose–response relationships, the length of exposure required to produce an adverse effect, sensitive populations, variability, and uncertainty factors. It can be assumed that there will be variability in the intake of any chemical. Therefore, the methodology should describe that variability and wherever possible quantify the uncertainty in the estimates.

In theory it may be possible to use biomarkers to directly estimate dietary exposure. However, such data are virtually never available and it is necessary to indirectly estimate exposure. Indirect exposure is estimated from the concentrations of the chemical in foods along with the frequency of occurrence and the amounts of each food that are consumed by the population being evaluated. It is possible to further refine the estimate by adjusting for the percentage of the chemical that is bioavailable.

The purpose of the assessment plays a critical role in determining the most desirable methodology. The optimum method when the assessment is designed to be conservative (as is often the case for regulatory decision-making applications) is different from the optimum method when the analysis is designed to be as realistic as possible (as in scientific hypothesis testing).

Human and Ecological Risk Assessment: Theory and Practice, Edited by Dennis J. Paustenbach
ISBN 0-471-14747-8 © 2002 John Wiley & Sons, Inc.

Mathematically, exposure is deceptively simple:

Exposure = Concentration in media

$\qquad$ × Consumption of media (quantity and frequency)

$\qquad$ × Proportion absorbed $\hfill$ (16.1)

This equation is modified to incorporate other parameters as necessary. In February, 1997, Food and Agriculture Organization/World Health Organization (FAO/WHO, 1998) convened a joint consultation to develop procedures for estimating dietary exposure at the international level. Two major outcomes of the conference were: (1) mathematical approaches for evaluating exposure that proceed from screening techniques at the international level to refined exposure analyses at the national level and (2) revised procedures for developing regional diets for use in risk assessments for pesticides and contaminants.

In practice, rough, range-finding, or screening-type estimates are often conducted first. Screening estimates are then used to define the actual exposure assessment and often to guide the collection of the appropriate data.

16.2 SCREENING APPROACHES FOR ESTIMATING DIETARY EXPOSURE (RANGE-FINDING METHODS)

Screening or range-finding exposure assessments can be used for the initial evaluation of worst-case exposure. The most widely used methods to determine the order of magnitude of exposure are based on theoretical maximum chemical concentrations and either average or high consumer exposures. These initial calculations usually do not incorporate distributions of either chemical levels or food consumption patterns; similarly, the effects of processing or cooking are not included.

The selection of the most appropriate methodology will depend on: (1) the intended application for the exposure assessment, (2) the biological properties, and (3) the physical and chemical properties of the substance.

The Danish budget method (Hansen, 1979) is widely used in Europe as the first analysis in a tiered approach. The Danish budget method combines estimates of physiological requirements for total food and liquid with a measure of the acceptable daily intake (ADI) to judge whether an intake could possibly be unacceptable. It does not provide an estimate of exposure, but it does provide values that can be used to prioritize resources and to design sampling programs.

Similarly, the U.S. Environmental Protection Agency (EPA) Office of Pesticide Programs (OPP) combines estimates of consumption with extremely high (tolerance) estimates of potential residues in those foods as a Tier 1 screening method for evaluating pesticides (EPA, 1996). Conversely, many programs designed to estimate the intake of essential nutrients would assume the foods contain the lowed possible concentrations—since low exposures are of the greatest concern.

In conclusion, screening methods sacrifice accuracy, in estimating exposure, for speed and simplicity. In the case of the evaluation of toxic effects, screening results that are acceptable indicate that actual exposures will be acceptable since they will be lower than the worst-case estimate. Therefore, it can be assumed that there is no need to expend resources to collect better data or to apply more sophisticated techniques in search of greater accuracy. In contrast, a research project that is attempting to evaluate the cause-and-effect relationship of a chemical and a disease would require more accurate exposure assessments.

16.3 DETAILED EXPOSURE ESTIMATION METHODOLOGY

The basic algorithm [Eq. (16.1)] is modified to allow the analyst to match the available data to the toxicity profile of the chemical. The analyst must decide the period of time to include in the exposure estimate. Exposure can be estimated for an "average" or "median" individual or exposure can be described as the distribution of potential exposures. Examples of common exposure estimates include:

- *Average Exposure.* The most basic models combine data on average food consumption and average concentration levels of the chemical to estimate average exposure. Average chronic exposure is usually estimated on a per-capita consumption basis and is compared to the measurements of biological/toxicological results from lifetime animal feeding studies or other appropriate test results. Single-day or acute exposures may be computed using similar methods, by using food exposure data for a single meal or for the day. Acute exposures are usually compared to results of tests in which the subjects were dosed for a single day.
- *Simple Distribution.* A simple distribution of exposure can be calculated by applying a single, average (or worst-case) estimate of the chemical's concentration level to a distribution of food consumption. Alternatively, a distribution of concentration levels can be applied to an average or worst-case food consumption level.
- *Joint Distribution Probabilistic Assessment.* In creating a joint distribution to estimate exposure, the distribution of food consumption is combined with the distribution of chemical concentration levels. Joint distribution/probabilistic analysis allows the most realistic estimates of exposure.*

In assessing food exposure, it is generally accepted that mean exposure of a population may be reasonably estimated using a one-day recall or diary if the number of subjects is sufficiently large. However, the percentage of the population estimated to be at risk for toxic effects from a chemical will be higher

---

*The statistical tool, Monte Carlo modeling, is the most common method currently used to estimate the joint distribution.

when food exposure is assessed using a one-day recall than with a multiday record or dietary history. This is because extreme levels of exposure (e.g., 90th or 95th percentiles) are invariably higher for a single day than they are for multiple days. In addition, large intraindividual variation associated with one-day surveys may limit the power to detect differences between different population groups (Liu et al., 1978; Beaton et al., 1979; van Staveren et al., 1985).

The U.S. Food and Drug Administration (FDA) has typically estimated the 90th percentile of the population for the U.S. population for food additives; for many pesticides EPA has estimated the average individual (but assumed that the chemical was always present at unusually high levels). The United Kingdom MAFF (UK MAFF, 1995) has routinely estimated the highly exposed consumer as the upper 95th percentile individual. Since 1996, the EPA OPP has used Monte Carlo methods to estimate the exposure to the entire population including up to the upper 99.9th percentile.

16.3.1 Defining Exposure Scenarios

The sources of exposure must be defined along with the determination of magnitude and frequency of exposure. It is usually desirable to know how the chemicals enter the food supply and the timeframes that are involved. The following steps help to define the most appropriate exposure scenario:

- Identify the regulatory standards that apply to the chemical [e.g., WHO, European Union (EU), or U.S. maximum limits].
- Determine the characteristics of intended or accidental additions of the chemical to the food supply, to food contact surfaces, or onto the hands. Determine the impact of processing/cooking and storage of food on the final levels of the chemical.
- Identify and evaluate existing data for both the levels of the chemical and for the appropriate consumption estimates.
- Create or apply draft exposure assessment algorithms that define the parameters to be computed and the models to be used to estimate exposure. If appropriate, develop a tiered methodology that allows a sequence of analyses from range finding to detailed probabilistic exposure assessments.
- Conduct appropriate preliminary exposure assessments including sensitivity analyses to assist in defining the parameters, which need the best available data and/or additional data generation.

Exposure scenarios must address the following topics:

1. *Toxicity Profile Including Toxicity Due to Chronic Exposure versus Acute Exposure.* The length of dosing that is required to elicit a specified biological effect will define the algorithms that are most appropriate.

The relative appropriateness, accuracy, and reliability of the analysis will depend on the source of the chemicals in food, the available data for concentrations of the chemical, and the toxicological characteristics of the chemical. For chemicals in which adverse effects are observed after a single exposure, it is important to estimate exposure by people when they actually ingest food containing the chemical. For chemicals with chronic adverse effects, on the other hand, exposure is usually calculated over days when food is and is not consumed to produce an estimate of average daily exposure.

In summary, the biological effects that are the result of a single or at most few doses will be compared to dietary exposure on a single day. Correspondingly, toxic effects that arise as a result of long-term exposure will be compared to average dietary exposures (usually over a year).

2. *Subpopulations of Interest.* The toxicity of the chemical should be reviewed to determine those subgroups of the population that are of particular interest. Populations can be of interest either because they are unusually susceptible to the effects of the chemical or because they are likely to have unusually high exposures—either because they eat more foods or because those foods contain more residues.

3. *Properties of Chemical as Consumed.* Often when estimating exposure of a food additive or contaminant, it is necessary to define or characterize the chemical in terms of attributes such as structure, volatility, and solubility. For example, does the substance break down during storage, processing, and cooking?

Diets in many countries are highly processed. Therefore, for most assessments, it will be critical to include estimates of the residues in the products as they are consumed (Chin, 1991).

4. *General Information about Presence of Chemical in Foods.* Key information includes sources of the chemical; likely foods that will contain the chemical and the impact of processing, and home preparation on potential residues. In some cases the chemical may only be present in foods from a specific source, for example, imported foods.

16.3.2 Data Used to Refine Exposure Assessments

Estimates of Food Consumption   There have been numerous food consumption surveys conducted to use in estimating the intake of nutrients. Such surveys have been conducted at periodic intervals in many countries. The methods used to conduct dietary surveys are in a continuing state of development and refinement. The survey instruments and procedures to collect and analyze samples are continually being improved, and new compounds are being added to reflect current priorities. The optimal approach for assessing dietary exposures has been debated for years in the United States as well as in other countries [Federation of American Societies of Experimental Biology (FASEB), 1988].

The most appropriate survey to select for use for an exposure assessment must be determined. Considerations can include:

- When were the data collected and are current dietary practices similar enough to be relevant?
- Were appropriate populations and subgroups of the population surveyed?
- Were data collected during all seasons?
- Were the foods of interested included in the survey?
- Was the quantity of each food estimated?

The available information about the survey methodology and the resulting data available from a particular survey is an important determinant of the usefulness of the survey for estimating exposure using the data.

Common Types of Food Consumption Surveys   There are four broad categories of food consumption data: (1) food supply surveys (market disappearance), (2) household or community inventories, (3) household food use, and (4) individual food exposure surveys.

1. *Food Supply Surveys.* Food supply surveys, also called food balance sheets (FBSs) or disappearance data are typically conducted on a countrywide basis each year. These surveys provide data on food availability or disappearance rather than actual food consumption but may be used to estimate indirectly the amounts of foods consumed by the country's population. Food supply data may be useful for setting priorities, analyzing trends, developing policy, and formulating food programs. For some countries, food supply data are the only accessible data representing the country's food consumption. Because similar methods are used around the world, these data may be used to make international comparisons and may also be useful in some epidemiological studies (Sasaki and Kestelhoot, 1992).

FBSs describe a country's food supply during a specified time period. Mean per-capita availability of a food or commodity is calculated by dividing total availability of the food by the total population of the country. FBSs published by the FAO describe the food supply in countries on all continents. European FBSs are also prepared by the Organization for Economic Cooperation and Development (OECD) and the Statistical Office of the European Communities (EUROSTAT). Food supply data in the United States are developed by U.S. Department of Agriculture's (USDA's) Economic Research Service.

WHO GEMS/Food program developed a series of global and regional/cultural diets to use in conducting risk assessments at the international level (WHO, 1988, 1998). These diets are based on FAO FBSs from selected representative countries as well as on expert interpretation. For more than 10 years, the diets provided the food consumption component of the dietary exposure calculations that have been used by WHO for predicting dietary exposure of pesticide residues for the Codex Committee on Pesticide Residues (CCPR) and the Joint Meeting on Pesticide Residues (JMPR).

There are some limitations in the use of the FBSs surveys to estimate exposures. First, waste at the household and individual levels usually is not considered. Therefore, exposure estimates based on food supply data are higher than estimates based on actual food consumption survey data, with the magnitude of the error depending on the quantity of waste produced. Perhaps more importantly for exposure assessment, users of foods cannot be distinguished from nonusers. Therefore, individual variations in exposure cannot be assessed, nor can exposure of potentially sensitive subpopulations be estimated. Finally, food availability is usually reported in terms of raw agricultural commodities. Processed forms of foods are usually not considered, nor is there any way to distinguish use of foods as ingredients. Nonetheless, these data allow assessments to be conducted at the international level using comparable data.

2. *Household Inventories.* Household surveys generally can be categorized as: (1) household or community inventories or (2) household or individual food use. Inventories are accounts of what foods are available in the household. What foods enter the household? Were they purchased, grown, or obtained some other way? What foods are used up by the household? Were they used by household members, guests, and/or tenants? Were they fed to animals?

Inventories vary in precision with which data are collected. Questionnaires may or may not ask about forms of the food (i.e., canned, frozen, and fresh), source (i.e., grown, purchased, or provided through a food program), cost, or preparation. Quantities of foods may be inventoried as purchased, as grown, with inedible parts included or removed, as cooked, or as raw. Such data are available from many countries including Germany, the United Kingdom, Hungary, Poland, Greece, Belgium, Ireland, Luxembourg, Norway, and Spain [Trichopoulou and Lagiou, 1997; Data Food Networking (DAFNE) II, 1995].

3. *Household or Individual Food Use.* Food-use studies, usually conducted at the household or family level, are often used to provide economic data for policy development and planning for feeding programs. Survey methods used include food accounts, inventories, records, and list recalls (Pao et al., 1989; Lee and Nieman, 1993). These methods account for all foods used in the home during the survey period. This includes foods used from what was on hand in the household at the beginning of the survey period and foods brought into the home during the survey period.

Although household food-use data have been used for a variety of purposes, including exposure assessment, serious limitations associated with data from these surveys should be noted. Food waste often is not accounted for. Food purchased and consumed outside the household may or may not be considered. Users of a food within a household cannot be distinguished, and individual variation cannot be determined. Exposures by subpopulations based on age, gender, health status, and other variables for individuals can only be estimated based on standard proportions or equivalents for age/gender categories.

4. *Individual Consumption Studies.* Various survey methods are available for determining consumption of foods and beverages and where these items are consumed (e.g., food frequency interviews, food diaries, 24-hour dietary recalls, multiday weighed food record, and household food purchase records). For example, national dietary surveys were conducted in Australia in 1983 (adults) and 1985 (school children) and the entire population in 1995. Food consumption data are available from food frequency surveys for some states [Australia New Zealand Food Authority (ANZFA), 1997].

Individual exposure studies provide data on food consumption by specific individuals. Methods for assessing food exposures of individuals may be retrospective (e.g., 24-hour or other short-term recalls, food frequencies, and diet histories), prospective (e.g., food diaries, food records, or duplicate portions), or a combination thereof. The most commonly used studies are those using the recall or record method and the food frequency. Each of the methods discussed below may also be applied at the family or household levels.

(a) *Food recall* studies are used to collect information on foods consumed in the past. The unit of observation is the individual or the household. The subject is asked to recall what foods and beverages he or she or the household consumed during a specific period, usually the preceding 24 hours. Since this method depends on memory, foods are quantified retrospectively, often with the aid of pictures, household measures, or two- or three-dimensional food models.

Food recall studies have been used successfully with individuals as young as 6 years of age, and interviewer-administered recalls are usually the method employed for populations with limited literacy. When individuals are not available for an interview or are unable to be interviewed due to age, infirmity, or temporary absence from the household, surrogate respondents are often used (Samet, 1989).

The main disadvantage of the recall method is the potential for error due to faulty memory of respondents. Items that were consumed may be forgotten, or the respondent may recall items consumed that actually were not consumed during the time investigated. To aid recall memories, the interviewer may probe for certain foods or beverages that are frequently forgotten, but this probing has also been shown to introduce potential bias by encouraging reporting of items not actually consumed.

(b) *Food record/diary surveys* collect information about current food exposure by having the subject keep a record of foods and beverages as they are consumed during a specific period. Quantities of foods and beverages consumed are entered in the record usually after weighing, measuring, or recording package sizes. Photographs or other recording devices can be a benefit in improving accuracy.

Data from short-term recalls and from food diaries, which collect detailed information on the kinds and quantities of foods consumed, are generally the most accurate and flexible data to use in assessing exposure of food chemicals. Data from these surveys can be used to estimate either acute or chronic exposure. Average exposure and distributions of exposure can be calculated for subpopulations based on age, gender, ethnic background, socioeconomic status, and other demographic variables, provided that such information is collected for each individual.

In general, data from large surveys using recall/record methods provide the most accurate assessments of exposure to chemicals in food when the chemical under investigation is found in many foods, and when most of these foods are consumed on a regular basis. It is more difficult to capture exposure of infrequently consumed foods using short-term recalls or records.

(c) *Food frequency questionnaire* surveys typically allow qualitative estimates of exposure. A food frequency questionnaire (FFQ) or checklist is used to determine the frequency of consumption of the foods of interest. Usually the food questionnaire is limited to no more than 100 foods. Subjects indicate how many times each day, week, or month they usually consume each food. Occasionally a *semiquantitative FFQ* survey is utilized that will estimate the amounts consumed by having respondents indicate whether their usual portion size is small, medium, or large.

Recommendations to Guide Selection of Food Consumption Data for Most Exposure Assessments For the U.S. population, the data collected in one of two large, national, food consumption surveys, conducted by the U.S. government, will be appropriate for most exposure assessments in that country: (1) the Nationwide Food Consumption Survey conducted by the USDA, beginning in 1935 or (2) the National Health and Nutrition Examination Survey (NHANES) undertaken by the U.S. Department of Health and Human Services, beginning in 1971. Both surveys employ multistage area probability sampling procedures to obtain a sample representative of the population. The surveys are repeated at periodic intervals and are publicly available.

There are some situations that may require more specialized surveys:

1. *If Chemical Is Found in Infrequently Consumed Food.* Some chemicals concentrate in liver and other foods that may be consumed infrequently. However, they may be eaten in significant quantities on those occasions when they are eaten. If an infrequently consumed food, such as liver, is a major source of the chemical under investigation, exposure estimates for that chemical may be low. Exposure assessments for those chemicals will be most accurate when based on surveys in which the frequency of consumption is estimated. In this case, a well-designed FFQ that targets the specific foods of concern should be considered.

2. *Chemicals Found in Only a Few Foods.* If the chemical is found in only a few foods, exposure assessments for that chemical will be most accurate if the data used are from surveys that captured very specific information on foods consumed even when the survey does not capture information about the total diet.

3. *Chemical Is Different in Foods from Different Sources.* If foods from one source, such as institutional foods, have different concentrations of chemicals than those from other sources, it will be appropriate to use food consumption data that permit exclusion of sources that do not contribute. For most foods, the USDA CSFII surveys conducted since 1989 provide extensive detail about the source of each food that each respondent reports consuming.

4. *Foods of Interest Are Associated Only with Specific Population.* Some subpopulations may consume foods that come from a unique source, and special steps are required to accurately assess these groups' exposure. For example, if the chemical is present in fish from the Mississippi river valley and the subpopulation of primary interest is subsistence fisherman who primarily consume fish that were caught from waters in which the chemical being studied is present at higher than typical levels. Vegetarians whose diets are limited, to varying degrees, to nonanimal sources and infants and young children that eat special foods may require more specialized surveys. Commercial infant foods are prepared with special attention to the needs of infants and children, and manufacturers have programs to monitor environmental contaminants. The presence of chemicals in breast milk should be evaluated in assessing nursing infants' exposure of a chemical present in the food supply (Schreiber, 1997), but the national surveys do not contain estimates of breast milk intake.

Estimates of Concentration of Chemical in Foods   The analyst's goal is to characterize the concentration of the chemical in each food. For most analyses, it is important to consider not only the average concentrations but also the variability in concentration.

Existing data can be used to estimate exposure and to guide the collection of future data. A substantial amount of information that is relevant to exposure assessment is available for some chemicals (Graham et al., 1992; Sexton et al., 1992). Such information can be obtained from existing databases on food and drinking water consumption and contaminant residues normally maintained for other purposes (i.e., nutritional exposure or regulatory surveillance). The Total Diet Study (Pennington and Gunderson, 1987) of the FDA, the Nationwide Food Consumption Survey of the USDA, the National Health and Nutrition Examination Surveys (NHANES) of the U.S. National Center for Health Statistics (NCHS, 1982), as well as other national regulatory monitoring programs, can provide information useful for dietary exposure modeling and assessments.

Useful data sources for defining the foods with the highest potential concentration of the chemical include:

- Chemical properties (solubility, heat stability, pH stability, and other properties)
- Metabolites/degradates of potential concern

- Differences in levels of raw versus cooked food
- Levels in home-prepared versus commercially processed foods
- Timeframes for presence of chemical in the food supply or in categories of the food supply—continuously or in periodic intervals
- Impact of processing/home preparation on potential residues

There are three common approaches that are appropriate for collecting data on concentrations of chemicals in food: (1) duplicate diets, (2) market basket or representative sampling, and (3) controlled experimentation. These are discussed below.

The collection of duplicate diets has proven to be a viable, if expensive, sampling approach. Duplicate diet procedures were the method of choice for the U.S. EPA's National Human Exposure Assessment Survey (FASEB, 1993) and the WHO's Global Exposure Monitoring System (WHO, 1983).

Duplicate Diets In duplicate diet sampling, a duplicate portion of all foods and beverages consumed during the monitoring period is collected. Variations of these procedures are used depending on study objectives. Individual food items or food groups and daily food collections may be segregated for independent analysis. Alternatively, it may be necessary to sample foods at the various steps during food preparation to identify the reason for excess contamination. Duplicate diet collections should include all foods consumed, both in and away from the residence (FASEB, 1993), and the monitoring period should be extended to the maximum number of days possible without undue burden on the participant. The researcher must determine the level of information required for various exposure scenarios. Often detailed exposure information can only be obtained at considerable expense and participant burden. For example, detailed dietary histories, segregation of individual food items for analysis, and weighing individual foods may not be necessary when a single measure of total exposure is the goal.

Market Basket and Surveillance Sampling Many countries conduct various types of monitoring or surveillance for chemicals in commodities or in foods (WHO GEMSFOOD, 1997). Monitoring and surveillance studies are conducted to assess compliance with state, national, or international regulations. Samples are collected (often called market basket surveys) to obtain food chemical concentration data that may be used in exposure assessment. A core group of foods that are representative of national dietary patterns is obtained and analyzed to determine the concentrations of the substances of interest. Generally, samples of food are purchased at retail outlets in different regions of the country and prepared as for consumption.

The USDA monitors residue levels in selected fruits and vegetables, meat, and poultry products. The FDA monitors residue levels in all other foods. California, Florida, and a number of other states have monitoring programs. Depending on the specific U.S. monitoring program, foods or commodities may be sampled at the point of entry to the country, at the farm gate, at the food-processing plant, or at the retail level.

Experimental Results Including Controlled Field Studies Controlled experiments are sometimes used to determine the likely levels of chemicals in foods produced under specified conditions and of relevant metabolites. Controlled experiments have also been conducted to identify the source of other chemicals in foods. For example, studies have been conducted to quantify the migration of lead from pottery and glassware into foods and beverages (Bolger et al., 1996). Studies are also conducted to quantiatively determine the impact of processing on residue levels.

Estimation of Variability and Uncertainty in Exposure Assessment Variability and uncertainty arise from variation in individual food consumption practices, differences in the concentrations of the chemical in individual foods, as well as errors in the actual measures. Thus estimating variability and uncertainty will require assessments of the contributions of these different components to the overall estimates.

Food consumption data used in assessing exposure of chemicals in food has a variety of sources of variability and uncertainty. Validation is complicated by the lack of a generally accepted standard. That is, each approach for estimating exposure has its own inherent limitations.

In a few cases, survey methodology has been validated by use of biological markers associated with dietary exposure. Possible sources of biological markers include urine, feces, blood, and tissue samples, but the most easily accessible and therefore most commonly used is urine. Nitrogen content of urine has been used to verify protein exposure. If protein intake calculated from the reported food sources in agreement with protein intake calculated from nitrogen excretion, it is assumed that exposure estimation for other chemicals is reliable (Bingham and Cummings, 1985).

Numerous studies attempting to validate one survey method relative to another have been reported (Bingham et al., 1982; Blake and Durnin, 1963; Block, 1982; Bransby et al., 1948; Fanelli and Stevenhagen, 1986; Grewal et al., 1974; Hussain et al., 1980; Karvetti and Knuts, 1985; Lubbe, 1968; Mahalko et al., 1985; Meredith et al., 1951; Morgan et al., 1987; Morrison et al., 1949; Nettleton et al., 1980; Ramsanen, 1979; Russell-Briefel et al., 1985; Willett et al., 1985; Young et al., 1952). For example, the FFQ, a more recent survey method, has been validated by comparing results of dietary exposure with repeated multiple-day food records, which served as an estimate of usual exposure (Block, 1989; Pietinen et al., 1988a; Willett et al., 1988). Results of validity studies indicated that FFQs could provide useful information about individual nutrient exposures. However, these studies generally have shown better correlations between methods for groups than for individual survey participants.

Confirmation of 24-hour recall, diaries, records, and frequencies has also been reported after comparing estimates of dietary exposure obtained using one of these survey instruments to subjects' actual exposures. Actual exposures may be based on surreptitious observation in cafeterias, institutional dining centers, or other facilities (Baranowski et al., 1986; Gersovitz et al., 1978; Greger and Etnyre, 1978; Madden et al., 1976; Samuelson, 1970; Stunkard and Waxman, 1981).

National consumption surveys and food monitoring programs are not intended to represent the true diets of any individual person. The factors affecting the exposure of an individual or small subset of the population to a specific chemical are not evaluated by using a national average diet and nationwide residue monitoring data (Lioy, 1990). Market-basket surveys reveal little about unique, individual dietary practices that vary from the norm.

Reliability, or reproducibility, is the ability of a method to produce the same or similar estimate on two or more occasions (Block and Hartman, 1989; Pietinen et al., 1988b), whether or not the estimate is accurate. The reliability of food consumption survey data for estimating usual exposure of a population depends somewhat on the number of days of dietary exposure data collected for each individual in the population. The number of days of food consumption data required for reliable estimation of population exposures is related to each subject's day-to-day variation in diet (intraindividual variation) and the degree to which subjects differ from each other in their diets (interindividual variation) (Basiotis et al., 1987; Nelson et al., 1989). When intraindividual variation is small relative to interindividual variation, population exposures can be reliably estimated with consumption data from a smaller number of days than should be obtained when both types of variation are large. Exposure of contaminants can be reliably estimated with fewer days of data when they are present in many foods that are commonly consumed.

Error in individual food consumption surveys may be due to chance or to factors of measurement. Data variability due to chance may be related to the survey sample. Any sample randomly drawn from a population will differ from any other sample, with the degree of difference depending on the size of the sample and the homogeneity of the population from which it was drawn. Error due to chance also arises from data collection at different times of the day, on different days of the week, or in different seasons of the year.

Measurement error may be introduced by the survey instrument, the interviewer, or the respondent. The instrument may bias results when questions are not clear, probes lead the subject to give a desired answer, questions are culture specific, or they do not follow a logical sequence. For self-administered questionnaires, responses will be influenced by the readability level, the use of abbreviations or unfamiliar jargon, clarity of instructions, and amount of space provided for answers. Interviewer bias may be introduced when interviewers make the respondent uncomfortable, are judgmental, or do not use a standard method and/or standard probes.

Respondents may introduce bias when they omit reporting foods they actually ate because they are reluctant to report certain foods or beverages (alcoholic beverages are a good example) or if they are forgetful. Alternatively, they may report the food, but understate the quantity consumed. Foods consumed away from home, particularly on occasions when the focus of attention is on the event rather than on the food, are especially difficult for people to remember. Quantities may be underestimated for similar reasons. Foods and beverages that were not consumed may be reported as consumed because of faulty memories, desire to impress the interviewer, or confusion with similar foods.

Measurement errors also include errors in coding due to unclear handwritten records or erroneous data entry.

16.4 CONCLUSIONS

Exposure to chemicals in foods can be estimated using well-tested statistical methods including Monte Carlo probablistic tools. Food consumption data are available for many populations. The concentrations of some chemicals are available through government monitoring studies. Additional studies will be needed for most chemicals.

The validity of any exposure calculation depends upon the representativeness and reliability of the available data. It also depends upon appropriate selection and use of the algorithms and models. Independent validation and sensitivity analyses to quantify the uncertainty of the estimates are critical components.

REFERENCES

Australia New Zealand Food Authority (ANZFA). 1997. *Dietary Modelling: Principles and Procedures*. Australia New Zealand Food Authority, Canberra, Australia.

Baranowski, T., Dworking, R., Henske, J. C., Clearman, D. R., Dunn, J. K., Nader, P. R., and Hooks, P. C. 1986. The accuracy of children's self-reports of diet: Family health project. *J. Am. Diet. Assoc.* 86: 1380.

Barraj, L., Petersen, B. J., and Moy, G. 1999. Creation of dietary regions using cluster analysis. In press.

Basiotis, P. P., Welsh, S. O., Cronin, J., Kelsay, J. L., and Mertz, W. 1987. Number of days of food intake records required to estimate individual and group nutrient intakes with defined confidence. *J. Nutr.* 117: 1638.

Beaton, G. H., Milner, J., Corey, P., McGuire, V., Cousins, M., Stewart, E., de Ramos, E., Hewitt, D., Grambsch, P. V., Kassim, N., and Little, J. A. 1979. Sources of variance in 24-hour dietary recall data: Implications for nutrition study design and interpretation. *Am. J. Clin. Nutr.* 32: 2456.

Bingham, S., and Cummings, J. H. 1985. Urine nitrogen as an independent validatory measure of dietary intake: A study of nitrogen balance in individuals consuming their normal diet. *Am. J. Clin. Nutr.* 42: 1276.

Bingham, S., Wiggins, H. S., Englyst, H., Seppanen, R., Helms, P., Strand, R., Burton, R., Jorgensen, I. M., Poulsen, L., Paerregaard, A., Bjerrum, L., and James, W. P. 1982. Methods and validity of dietary assessments in four Scandinavian populations. *Nutr. Cancer* 4: 23.

Blake, E. C., and Durnin, J. V. G. A. 1963. Dietary values from a 24-hour recall compared to a 7-day survey on elderly people. *Proc. Nutr. Soc.* 22: 1.

Block, G. 1982. A review of validations of dietary assessment methods. *Am. J. Epidemiol.* 115: 492.

Block, G. 1989. Human dietary assessment: Methods and issues. *Prevent. Med.* 18: 653.

Block, G., and Hartman, A. M. 1989. Issues in reproducibility and validity of dietary studies. *Am. J. Clin. Nutr.* 50: 1133.

Bolger, P. M., Yess, N. J., Gunderson, E. L., Troxell, T. C., and Carrington, C. D. 1996. Identification and reduction of sources of dietary lead in the USA. *Food Addit. Contam.* 13: 53–60.

Bransby, E. R., Daubney, C. G., and King, J. 1948. Comparison of results obtained by different methods of individual dietary survey. *Br. J. Nutr.* 2: 89.

Chin, H. B. 1991. The effect of processing on residues in foods: The food processing industry's residue database. In B. G. Tweedy, H. J. Dishburger, L. G. Ballantine, and J. McCarthy (Eds.), *Pesticide Residues and Food Safety: A Harvest of Viewpoints.* American Chemical Society, Washington, DC.

Codex Alimentarius Commission (CAC). 1998. *Report of the 30th Session of the Codex Committee on Food Additives and Contaminants.* ALINORM 99/12. CAC, Rome.

Dabeka, R. W., McKenzie, A. D., and Lacroix, G. M. A. 1987. Dietary intakes of lead, cadmium arsenic and fluoride by Canadian adults: A 24-hour duplicate-diet study. *Food Addit. Contam.* 4: 89–102.

Data Food Networking (DAFNE) II. 1995. *Network for the Pan-European Food Data Bank Based on Household Budget Surveys.*

Fanelli, M. T., and Stevenhagen, K. J. 1986. Consistency of energy and nutrient intakes of older adults: 24-hour recall vs. 1-day food record. *J. Am. Diet. Assoc.* 86: 664.

Federation of American Societies of Experimental Biology (FASEB). 1988. *Estimation of Exposure to Substances in the Food Supply.* Life Sciences Research Office, Bethesda, MD.

Federation of American Societies of Experimental Biology (FASEB). 1993. *National Human Exposure Assessment Survey Dietary Monitoring Options.* Life Sciences Research Office, Bethesda, MD.

Food and Agriculture Organization/World Health Organization (FAO/WHO). 1995. *Recommendations for the Revision of Guidelines for Predicting Dietary Intakes of Pesticide Residues.* Report of a FAO/WHO Consultation, May 2–6, 1995, York, United Kingdom. WHO/FNU/FOS/95.11. WHO, Geneva, Switzerland.

Food and Agriculture Organization/World Health Organization (FAO/WHO). 1998. *Food Consumption and Exposure Assessment of Chemicals.* Report of a FAO/WHO Consultation, February 10–14, 1997, Geneva. WHO/FSF/FOS/97.5. WHO, Geneva, Switzerland.

Gersovitz, M., Madden, J. P., and Smiciklas-Wright, H. 1978. Validity of the 24-hour dietary recall and seven-day record for group comparisons. *J. Am. Diet. Assoc.* 73: 48.

Graham, J., Walker, K., Berry, M., Bryan, E., Callahan, M., Fan, A., Finley, B., Lynch, J., McKone, T., Ozkaynak, H., and Sexton, K. 1992. The role of exposure data bases in risk assessment. *Arch. Environ. Health* 47: 408–420.

Greger, J. L., and Etnyre, G. M. 1978. Validity of 24-hour recalls by adolescent females. *Am. J. Public Health* 68: 70.

Grewal, T., Gopaldas, T., Gadre, V. J., Shrivastava, S. N., Pranjpe, B. M., Chatterjee, B. N., and Srinivasan, N. 1974. A comparison of weighing and questionnaire dietary survey methods for rural preschool children. *Ind. J. Nutr. Diet.* 11: 224.

Hansen, S. C. 1979. Conditions for use of food additives based on a budget for an acceptable daily intake. *J. Food Protect.* 42: 429.

Hussain, M. A., Abdullah, M., Huda, N., and Ahmad, K. 1980. Studies on dietary survey methodology—a comparison between recall and weighing method in Bangladesh. *Bangladesh Med. Res. Council Bull.* 6: 53.

Karvetti, R. L., and Knuts, L. R. 1985. Validity of the 24-hour recall. *J. Am. Diet. Assoc.* 85: 1437.

Lee, R. D., and Nieman, D. C. 1993. *Nutritional Assessment.* William C. Brown, Dubuque, IA.

Lioy, P. J. 1990. Assessing total human exposure to contaminants. *Environ. Sci. Technol.* 24: 938–945.

Liu, K., Stamler, J., Dyer, A., McKeever, J., and McKeever, P. 1978. Statistical methods to assess and minimize the role of intra-individual variability in obscuring the relationship between dietary lipids and serum cholesterol. *J. Chron. Dis.* 31: 399.

Lubbe, A. M. 1968. A survey of the nutritional status of white school children in Pretoria: Description and comparative study of two dietary survey techniques. *S. Afr. Med. J. Suppl.* 616.

Madden, J. P., Goodman, S. J., and Guthrie, H. A. 1976. Validity of the 24-hr. recall. *J. Am. Diet. Assoc.* 68: 143.

Mahalko, J. P., Johnson, L. K., Gallagher, S. K., and Milne, D. B. 1985. Comparison of dietary histories and seven-day food records in a nutritional assessment of older adults. *Am. J. Clin. Nutr.* 42: 542.

Meredith, A., Matthews, A., Zickefoose, M., Weagley, E., Wayave, M., and Brown, E. G. 1951. How well do school children recall what they have eaten? *J. Am. Diet. Assoc.* 27: 749.

Morgan, K. J., Johnson, S. R., Rizek, R. L., Reese, R., and Stampley, G. L. 1987. Collection of food intake data: An evaluation of methods. *J. Am. Diet. Assoc.* 87: 888.

Morrison, S. D., Russell, F. C., and Stevenson, J. 1949. Estimating food intake by questioning and weighing: A one-day survey of eight subjects. *Proc. Nutr. Soc.* 7: 5.

National Center for Health Statistics (NCHS). 1982. *Second National Health and Nutrition Examination Survey. 1976–80.* Data Tape 5704 (24-Hour Recall, Specific Food Item), NTIS Accession number PB82–142639. National Technical Information Service, Springfield, VA.

Nelson, M., Black, A. E., Morris, J. A., and Cole, T. J. 1989. Between- and within-subject variation in nutrient intake from infancy to old age: Estimating the number of days required to rank dietary intakes with desired precision. *Am. J. Clin. Nutr.* 50: 155.

Nettleton, P., Day, K. C., and Nelson, M. 1980. Dietary survey methods. 2. A comparison of nutrient intakes within families assessed by household measures and the semi-weighed method. *J. Hum. Nutr.* 34: 349.

Pao, E. M., Sykes, K. E., and Cypel, Y. S. 1989. *USDA Methodological Research for Large-Scale Dietary Intake Surveys. 1975–88.* Home Economics Research Report No. 49. U.S. Department of Agriculture, Human Nutrition Information Service, Washington, DC.

Pennington, J. A. T., and Gunderson, E. T. 1987. History of the Food and Drug Administration's Total Diet Study—1961 to 1987. *J. Assoc. Off. Anal. Chem.* 70: 772–782.

Petersen, B., and Barraj, L. 1996. Assessing the intake of contaminants and nutrients: A review of methods. *J. Food Composition Anal.* 9: 243–254.

Petersen, B., and Douglass, J. 1994. Use of food-intake surveys to estimate exposure to nonnutrients. *Am. J. Clin. Nutr.* 59(Suppl.): 2403–2433.

Pietinen, P., Hartman, A. M., Haapa, E., Rasanen, L., Haapakoski, J., Palmgren, J., Albanes, D., Virtamo, J., and Huttenen, J. K. 1988a. Reproducibility and validity of dietary assessment instruments. I. A self-administered food use questionnaire with a portion size booklet. *Am. J. Epidemiol.* 128: 655.

Pietinen, P., Hartman, A. M., Haapa, E., Rasanen, L., Haapakoski, J., Palmgren, J., Albanes, D., Virtamo, J., and Huttenen, J. K. 1988b. Reproducibility and validity of dietary assessment instruments. II. A qualitative food frequency questionnaire. *Am. J. Epidemiol.* 128: 667.

Ramsanen, L. 1979. Nutrition survey of Finnish rural children. VI. Methodological study comparing the 24-hour recall and the dietary history interview. *Am. J. Clin. Nutr.* 32: 2560.

Romesburg, H. C. 1990. *Cluster Analysis for Researchers.* Krieger, Malabar, FL.

Russell-Briefel, R., Caggiula, A. W., and Kuller, L. H. 1985. A comparison of three dietary methods for estimating vitamin A intake. *Am. J. Epidemiol.* 122: 628.

Samet, J. M. 1989. Surrogate measures of dietary intake. *Am. J. Clin. Nutr.* 50: 1139.

Samuelson, G. 1970. An epidemiological study of child health and nutrition in a northern Swedish county. 2. Methodological study of the recall technique. *Nutr. Metab.* 12: 321.

Sasaki, S., and Kestelhoot, H. 1992. Value of Food and Agriculture Organization data on food-balance sheets as a data source for dietary fat intake in epidemiologic studies. *Am. J. Clin. Nutr.* 56: 716.

SCF (Intake and Exposure Working Group). 1994. *Summaries of Food Consumption Databases in the European Union.* CS/Int/Gen2. European Commission, Brussels, Belgium.

Schreiber, J. S. 1997. Transport of organic chemicals to breast milk: Tetrachloroethene case study. In S. Kacew and G. Lambert (Eds.), *Environmental Toxicology and Pharmacology of Human Development.* Taylor and Francis, Washington, DC, pp. 95–143.

Sexton, K., Selevan, S. G., Wagner, D. C., and Lybarger, J. A. 1992. Estimating human exposures to environmental pollutants: Availability and utility of existing databases. *Arch. Environ. Health* 47: 398–407.

Sherlock, J. C., Smart, G. A., Walters, B., Evans, W. H., McWeeny, D. J., and Cassidy, W. 1983. Dietary surveys on a population at Shipham, Somerset, United Kingdom. *Sci. Total Environ.* 29: 121–142.

Stunkard, A. J., and Waxman, M. 1981. Accuracy of self-reports of food intake. *J. Am. Diet. Assoc.* 79: 547.

Trichopoulou, A., and Pagona, L. (Eds.). 1997. *Methodology for the Exploitation of HBS Food Data and Results on Food Availability in 5 European Countries.* DAFNE (Data Food Networking) European Communities.

U.S. Environmental Protection Agency. 1996a. *Exposure factors handbook. Volume II of III: Food ingestion factors—SAB Review Draft.* Washington, DC: U.S. Environmental Protection Agency, Office of Research and Development, EPA/600/P-95/002Bb.

U.S. Environmental Protection Agency. 1996b. *Exposure factors handbook. Volume III of III: Activity factors—SAB Review Draft.* Washington, DC: U.S. Environmental Protection Agency, Office of Research and Development, EPA/600/P-95/002P.

Vahter, M., Berglund, M., Friberg, L., Jorhem, L., Lind, B., Slorach, S., and Akesson, A. 1990. *Dietary Intake of Lead and Cadmium in Sweden.* Var Foda 44(Suppl. 2). National Food Administration, Uppsala, Sweden.

van Staveren, W. A., de Boer, J. O., and Burema, J. 1985. Validity and reproducibility of a dietary history method estimating the usual food intake during one month. *Am. J. Clin. Nutr.* 42: 554.

Willett, W. C., Sampson, L., Stampfer, M. J., Rosner, B., Bain, C., Witschi, J., Hennekens, C. H., and Speizer, F. E. 1985. Reproducibility and validity of a semiquantitative food frequency questionnaire. *Am. J. Epidemiol.* 122: 51.

Willett, W. C., Sampson, L., Browne, M. L., Stampfer, M. J., Rosner, B., Hennekens, C. H., and Speizer, F. E. 1988. The use of a self-administered questionnaire to assess diet four years in the past. *Am. J. Epidemiol.* 127: 188.

World Health Organization (WHO). 1983. *Assessment of Human Exposure to Environmental Pollutants.* EFP/83.52. WHO, Geneva, Switzerland.

World Health Organization (WHO). 1988. *Derived Intervention Levels for Radionuclides in Food.* WHO, Geneva, Switzerland.

World Health Organization (WHO). (1998). *GEMS/Food Regional Diets, Regional per Capita Consumption of Raw and Semi-Processed Agricultural Commodities.* WHO/FSF/FOS 98.3. WHO, Geneva, Switzerland.

Young, C. M., Hagan, G. C., Tucker, R. E., and Foster, W. D. 1952. A comparison of dietary study methods. II. Dietary history vs. seven-day record vs. 24-hour recall. *J. Am. Diet. Assoc.* 28: 218.

17 Analysis of Possible Health Risks to Recreational Fishers Due to Ingesting DDT and PCBs in Fish from Palos Verdes Shelf and Cabrillo Pier

NATALIE D. WILSON and VALERIE A. CRAVEN*
N.D. Wilson & Associates, Huntington Woods, Michigan

PAUL S. PRICE
AMEC Earth & Environmental, Portland, Marine

DENNIS J. PAUSTENBACH*
ChemRisk, Alameda, CA

17.1   INTRODUCTION

The Palos Verdes Shelf is located in the Pacific Ocean off the Palos Verdes Peninsula, west of Los Angeles in the Southern California Bight. Cabrillo Pier is located in the Los Angeles/Long Beach Harbor area (see Figure 17.1). Primarily from the 1950s to 1970s, chlorinated hydrocarbons are believed to have been discharged onto the Palos Verdes Shelf in wastewater from the Los Angeles County Sanitation District's Joint Water Pollution Control Plant located at Whites Point, contributing to elevated concentrations in bottom sediment. Potential risks to human health result from the bioaccumulation of these chemicals in fish that are then caught and consumed by recreational anglers. Therefore, this assessment evaluated the theoretical risks to human health posed by the potential consumption, by recreational anglers, of fish harvested from the

*Currently with Exponent, Inc., Menlo Park, CA 94025.

Human and Ecological Risk Assessment: Theory and Practice, Edited by Dennis J. Paustenbach
ISBN 0-471-14747-8   © 2002 John Wiley & Sons, Inc.

913

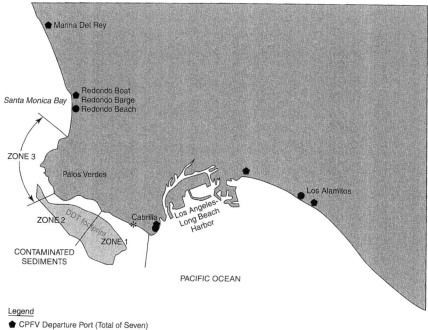

Figure 17.1   Area of interest in the Southern California Bight.

Palos Verdes Shelf and at Cabrillo Pier. The assessment addressed the potential for adverse health effects due to exposure to DDT and its degradation products, DDE and DDD (collectively total DDT or tDDT) and polychlorinated biphenyls (PCBs) in fish consumed by sport fishers. Previous studies [Science Applications International Corporation (SAIC), 1999; U.S. Environmental Protection Agency (EPA), 1996a; Pollock et al., 1990, 1991] have raised concerns about health risks to area anglers due to consumption of white croaker (*Genyonemus lineatus*) and other fish. The EPA is currently contemplating remedial action at the Palos Verdes Shelf to address these concerns. Previous screening and "streamlined" risk assessments have relied upon simple exposure analyses and highly conservative assumptions regarding exposure to estimate upper-bound risk estimates. The methods and results of these prior studies demonstrated the need for a refined risk assessment that used more advanced methods and site-specific information. This risk assessment addressed that need. The purpose of this study was to assess the possible risks of more realistic exposure scenarios at the Palos Verdes Shelf and at Cabrillo Pier and to provide appropriate scientific information for risk managers regarding the need to remediate Palos Verdes Shelf sediments.

17.1.1 Assessment Questions and Endpoints

Our study evaluated the potential cancer risks and noncancer hazards to these anglers of consuming fish containing tDDT and PCBs. Because this risk assessment focused on potential adverse health risks due to eating fish caught only at the Palos Verdes Shelf or at Cabrillo Pier, our results are representative of the incremental risk that tDDT and PCBs in fish consumed from these sites contributes to an angler's total fish consumption risks from all chemicals, sources, and sites. We prepared our analysis to characterize interindividual variability in exposures and risks. Uncertainty in our results was characterized using sensitivity analyses.

17.1.2 Definition of the Population Evaluated

Quantitative risk estimates were calculated for three groups of sport fishers (or anglers). The first group was those anglers who fish at the Palos Verdes Shelf while on commercial passenger fishing vessels (CPFVs), or so-called party-boat anglers. CPFVs leave from several locations within Santa Monica Bay and Los Angeles/Long Beach Harbor, and some of these boats take anglers to the Palos Verdes Shelf for fishing. Some anglers on these trips may harvest and consume fish from the Palos Verdes Shelf. Second, anglers who fish at the Palos Verdes Shelf while on private recreational vessels were evaluated. There are private boat launch sites near enough to the Palos Verdes Shelf that some private boat anglers may harvest and consume fish from the Palos Verdes Shelf. Finally, the third group studied were anglers who catch and consume fish from Cabrillo Pier, a pier/jetty complex located at San Pedro, California, within the Los Angeles/Long Beach Harbor. The Cabrillo Pier anglers were addressed because local surveys suggest that white croaker caught at Cabrillo Pier is consumed more frequently than white croaker caught at the Palos Verdes Shelf or other fishing locations in the area [Southern California Coastal Water Research Project (SCCWRP and MBC), 1994]. This assessment did not quantitatively evaluate health risks for members of the general public who may, despite the ban on commercial fishing for white croaker at the Palos Verdes Shelf, purchase fish at local markets that were harvested from the Palos Verdes Shelf.

17.2 METHODS

For Palos Verdes Shelf and Cabrillo Pier anglers, a microexposure event Monte Carlo analysis was used to characterize potential exposures due to consumption of 13 fish species or species groups caught in the two locations. Consumption of shellfish was not considered because local data indicate that shellfish are only a very minor recreational catch in the Santa Monica Bay region (SCCWRP and MBC, 1994).

Probabilistic techniques such as Monte Carlo analysis have been used to characterize the health risks of populations exposed to various chemicals since about 1990 (Cassin et al., 1998; Bogen et al., 1997; Chan et al., 1997; Carrington et al., 1996; Crouch, 1996a, b; Del Pup et al., 1996; Lipfert et al., 1996; Adams et al., 1994; Finley et al., 1993, 1994; Copeland et al., 1993, 1994; Finley and Paustenbach, 1994; McKone, 1994; Lloyd et al., 1992; McKone and Bogen, 1992; Paustenbach et al., 1991, 1992; Thompson et al., 1992; Whitmyre et al., 1992a, b; Burmaster and Von Stackelberg, 1991; Eschenroeder and Faeder, 1988; Fiering et al., 1984). By 1997, many studies had been published that demonstrated how probabilistic methods represent a significant improvement over deterministic approaches (see, e.g., Goodrum et al., 1996; Richardson and Allan, 1996; Frey and Rhodes, 1996; Cohen et al., 1996; Cullen, 1994; Finley et al., 1993, 1994; Thompson et al., 1992; McKone and Bogen, 1991; Paustenbach et al., 1991). In 1997, the EPA issued policy and guidance (EPA, 1997a) that supported the use of probabilistic techniques in risk assessment. This assessment of the risks to anglers who eat fish from the Palos Verdes Shelf and from Cabrillo Pier was performed in accordance with EPA's *Guiding Principles for Monte Carlo Analysis* (EPA, 1997a) and adheres to EPA's most recent recommended exposure and risk assessment guidance (EPA, 1995, 1997b, c, d, 1999a). In particular, EPA's *Risk Assessment Guidance for Superfund: Volume 3—Part A, Process for Conducting Probabilistic Risk Assessment* (EPA, 1999a) was consulted.

The microexposure event Monte Carlo analysis technique used in this assessment estimates lifetime exposure as the sum of doses received from individual exposure events, each of which is characterized using Monte Carlo methods. For example, this approach considered that the CPFV or private boat anglers may not go to the Palos Verdes Shelf on each trip; that while an angler may catch a fish, he or she may not consume it because that species is not to his or her liking; and that an angler may take more or fewer fishing trips in certain seasons or years than in others. The possible behaviors for a large number of hypothetical anglers are considered to develop an understanding of realistic variations in behaviors, and thus exposures and risks, across the angler population.

In contrast, risk assessments typically rely on assumptions that behavior is constant, rather than variable, over a long period, and one hypothetical combination of behaviors is evaluated to represent the population of anglers. For example, in previous risk assessments for Palos Verdes Shelf or Cabrillo Pier anglers, it was assumed that every angler had behavior that was identical to all other anglers of his or her classification as "typical," "central tendency," "high end," or "maximally exposed." That is, each angler fished at the Palos Verdes Shelf for the same lengthy period of time, regularly caught and ate the same amount of white croaker, and all the white croaker contained the same amounts of tDDT and PCBs. Such an approach does not acceptably characterize risks across the general angler population; in fact, it overestimates the true risk for the vast number of anglers.

17.2.1   Rationale for the Approach

When characterizing the distribution of long-term dose rates that occur as a result of exposures that vary over time, it is important to use a methodology that captures such variation. For example, in the case of fish consumption, lifetime average exposure represent the accumulation of separate chemical doses received from each fish meal, season by season, over the period of years in which the angler fishes. The uptake of chemicals from each fish meal will vary, depending on variables such as season fished, location where the fish is caught, the concentration of chemical in the fish, method of cooking, and the like. Characterizing the impact of this variation presents a challenge to the typical probabilistic and deterministic approaches. As discussed by Price et al. (1996), such analyses may not be appropriate for characterizing the distribution of long-term dose rates that occur as a result of certain kinds of exposure, such as the consumption of fish, that can vary considerably over time.

Deterministic and typical Monte Carlo approaches rely upon the same basic equation to estimate exposure, as depicted in Table 17.1, but these approaches differ in terms of the selection of exposure factor values and the method of calculating lifetime average dose rates. In its most basic form, exposure to chemicals via fish consumption is a function of the intake rate of fish, concentration of chemicals in fish, the period over which consumption of these fishes occurs, body weight, and life span. Deterministic methods rely upon a single-point estimate of the value for each exposure factor to characterize exposure, and these values are typically chosen to be conservative. That is, each value is meant to overestimate the actual value that would occur for most anglers. The result of repeatedly using conservative exposure factor values is that resulting estimates significantly overestimate the dose rate and hence risks for a vast majority of the population. Probabilistic methods such as Monte Carlo analyses use the same mathematical approach to estimate the dose rates as the deterministic approaches, but they use statistical expressions of the range and likelihood of exposure factor values (i.e., "distributions") in place of point estimates. In typical applications of Monte Carlo methodology, these distributions are assumed to be independent of age and to be independent of distributions for other exposure factors.

Distributions depict the range and likelihood of exposure factor values for the population. Figure 17.2 shows the distribution of the number of fishing trips taken in the summer by CPFV anglers and the information that can be conveyed from a distribution. Each bar in Figure 17.2*a* represents the fraction of data associated with the range of the number of trips covered by that bar's width. Accordingly, tall bars represent more popular fishing scenarios, while short bars show less typical situations. For the example in Figure 17.2*a*, it is clear that the fishing frequency reported most often fell in the range from 3 to 9 trips because the tallest bar on the chart covers this range. The full set of bars is referred to as a *histogram*. When a mathematical function is fitted to the histogram, this function is referred to as a *probability density function*, or PDF. The

TABLE 17.1 Comparison of Three Different Methods to Depict Four Different Exposure Assessment Elements

Exposure Assessment Element	Deterministic Approach	Typical Probabilistic Approach	MicroExposure Event Monte Carlo Approach
Method of estimating intake rate	Point estimate (i.e., single value) of intake rate representing long-term average over exposure period	Fixed distribution of intake rates representing long-term average over exposure period for all anglers at all times	Distribution of short-term intake rates for each exposure event
Method of estimating concentration	Point estimate of concentration	Fixed distribution of concentrations for all anglers at all times	Distribution of concentrations for each exposure event
Method of estimating body weight	Point estimate of body weight	Fixed distribution of body weights for all anglers at all times	Distribution of age-specific body weights for each exposure event
Method of calculating lifetime average daily dose (LADD) (mg/kg-day)	LADD^{a} (mg/kg-day) $= \dfrac{\text{IR}(\text{kg/day}) * C(\text{mg/kg}) * \text{EP}(\text{yr})}{\text{BW}(\text{kg}) * L(\text{yr})}$	LADD (mg/kg-day) $= \dfrac{\text{IR}(\text{kg/day}) * C(\text{mg/kg}) * \text{EP}(\text{yr})}{\text{BW}(\text{kg}) * L(\text{yr})}$	LADD^{b} (mg/kg-day) $= \dfrac{\sum\limits_{i=1}^{\text{Nee}} \left(\dfrac{\text{IR}_i(\text{kg}) * C_i(\text{mg/kg})}{\text{BW}_i(\text{kg})} \right)}{L(\text{days})}$

[a] LADD = lifetime average daily dose; IR = intake rate; C = concentration; EP = exposure period; BW = body weight, L = lifespan.

[b] Nee = number of exposure events; i = index for exposure event.

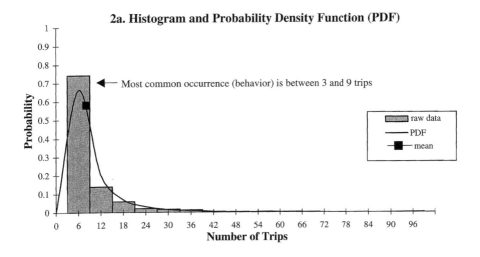

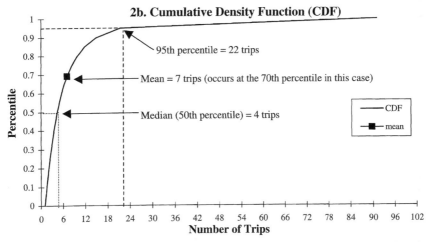

Figure 17.2 Demonstration of the utility of histograms, probability density functions, and cumulative density functions.

line plotted in Figure 17.2*b* is called a *cumulative density function*, or CDF. It represents the running cumulative sum of the height of each bar in Figure 17.2*a*. Points along the line indicate the fraction of fishing frequencies that are less than a certain value. The dotted line in Figure 17.2*b* shows that the median value (i.e., 50th percentile) is four trips. In other words, half of the anglers had fishing frequencies of four trips or fewer, and half fished more than four trips during their lifetime. The mean of seven trips happens, in this case, to correspond to the 70th percentile. The dashed line at the 95th percentile shows that the fishing frequency of 22 trips is exceeded by only 5 percent of anglers. That is, only 5 in 100 persons who are CPFV anglers take more than 22 trips over their lifetime.

In the microexposure event Monte Carlo approach adopted for this assessment, distributions were used to characterize exposure factor values within an analysis structure that takes into account dependencies and correlations among exposure factors. As shown in Table 17.1, these distributions may change from exposure event to exposure event according to whether other factors related to a particular exposure factor value have changed. For example, a dependency that should be accounted for is age. Specifically, if one fishing trip occurs at age 25 and another occurs at age 50, then differences in body weights at those ages can be incorporated by using age-specific body weight distributions. Likewise, certain correlations should be considered. For example, because fish species prefer specific habitats, and fishing trips target specific areas, it is reasonable that species caught from the same location could share a habitat preference such as sandy or rocky bottoms. Similarly, fish that prefer shallow waters, such as surfperch (Embiotocidae spp.) are unlikely to be caught by vessels fishing deep waters.

In this assessment, lifetime average daily dose (LADDs) rates are estimated as the sum of exposures from individual exposure events that occur within the exposure period, divided by a standard lifetime of 70 years. Lifetime doses and risks were calculated repeatedly for the set of hypothetical anglers to reflect the variation across the angler population. Because long-term exposures were constructed from realistic depictions of a series of short-term exposures, our Monte Carlo analyses produced significantly more realistic estimates of long-term dose rates for the population than those produced from deterministic or typical probabilistic methods that rely on extrapolation of short-term data to represent long-term exposures. For example, within the Monte Carlo analysis framework adopted, which is consistent with the event-by-event method for accumulating exposure, fish consumption rate data in units of grams/trip can be incorporated into the calculations. This is important because the *Santa Monica Bay Seafood Consumption Study* (SCCWRP and MBC, 1994) and the Love and Hansen (1999) study directly measured fish consumption rates on a trip-by-trip basis. Appropriately using data in the short-term units in which they were measured promotes realism and accuracy in the long-term exposure assessment. The long-term average fish consumption rate that we report in units of grams/day is a consequence of realistically applying the measured grams/trip rates over the various fishing trips of various anglers. In the deterministic or typical Monte Carlo risk assessment, the long-term average fish consumption rate in grams/day is an input to the analysis that must be determined by some method independent of the risk assessment. The microexposure event Monte Carlo approach applied in this risk assessment employs techniques similar to much of the earliest scientific work performed using Monte Carlo analysis (Ulam, 1991; Eckhardt, 1987; Metropolis, 1987). More recently these techniques have been used to evaluate exposures to pesticides (Price et al., 2001), to lead (Goodrum et al., 1996), to chlorinated solvents in contaminated groundwater (Harrington et al., 1995), and to chlorinated chemicals in fish (Wilson et al., 2001; Simon, 1999; Keenan et al., 1993, 1996). Use of the Monte Carlo approach is particularly appropriate for the evaluation of the CPFV, private boat, and Cabrillo Pier angler populations because of the substantial site-specific information that is available on the fishing and consumption practices of southern California anglers. It is likely that the information on southern California anglers and the fish they consume, which includes surveys of angler behavior, cooking practices, and concentrations of chemicals in fish, makes them one of the most thoroughly studied angler populations.

17.2.2 Sources of Information

The anglers and the fish considered in this study have been the subject of a number of investigations. The results of these studies provide a large database for exposure assessment. Because of the availability of more than 300,000 individual pieces of site-specific information, we considered the database to be adequate to conduct the assessment and hence no new data were collected. The primary sources of site-specific CPFV and Cabrillo Pier angler behavior data (e.g., fishing frequency, consumption rates, species preference, cooking methods, etc.) used in the Monte Carlo analyses were the *Santa Monica Bay Seafood Consumption Study* (SCCWRP and MBC, 1994) and the California Department of Fish and Game (CDFG) surveys (Ally et al., 1991). The *Santa Monica Bay Seafood Consumption Study* (SCCWRP and MBC, 1994) provided information on the fishing behavior of 1243 anglers fishing on pier/jetties, CPFVs, private boats, beaches, and at rocky intertidal sites and their consumption of 73 species at 31 locations in the Santa Monica Bay and Los Angeles Harbor areas. Specifically, 450 CPFV anglers and 198 Cabrillo Pier anglers were interviewed while fishing in this on-site survey conducted between September 1991 and August 1992. Anglers were interviewed over a total of 113 sampling events over 99 days, of which 32 were aboard CPFVs and 8 were at Cabrillo Pier. Interviews with individual anglers were conducted in English, Spanish, Vietnamese, Chinese, and Filipino. These data sources have been used to varying extents in previous risk assessments for Palos Verdes Shelf and Cabrillo Pier anglers (SAIC, 1997a, b, 1999; EPA, 1996a; Pollock et al., 1990, 1991).

The CDFG surveys (Ally et al., 1991) were a fundamental data resource for our analysis for CPFV anglers that was not used in prior risk assessments. These surveys were designed to determine the status of important recreational fisheries, to monitor the fish, and to make management recommendations if necessary. Between 1986 and 1989, CDFG employees took over 1000 randomly selected weekday trips on CPFVs. Once aboard, survey clerks recorded the following information: departure port, boat identification number, date, time of boat departure, trip type, whether the vessel fished in Mexican waters, number of anglers, fish catch by species, number of fish kept by species, fish lengths, time fished in each habitat type and at each specific fishing site (identified by CDFG block number), and bottom depth. Because anglers were not interviewed separately, these data do not provide information on individual anglers' behaviors. However, these data provide information on the frequency with which CPFVs took passengers to fishing sites over the Palos Verdes Shelf. These frequencies were classified by departure port and season. In addition, these data provide information on the types of fish caught at various sites.

The source of site-specific private boat angler behavior data was an intercept survey of private boat anglers returning to boat ramps in the vicinity of the Palos Verdes Shelf conducted under the direction of Drs. Milton Love and Steven Hansen (Love and Hansen, 1999). The target population, sampling sites, questionnaire, and sampling scheme employed by Love and Hansen (1999) together produced survey data more focused for health risk assessment of private boat anglers at the Palos Verdes Shelf than did the *Santa Monica Bay Seafood Consumption Study* (SCCWRP and MBC, 1994).

The Love and Hansen (1999) study was conducted from January 18, 1998, through January 11, 1999. The survey was targeted at private vessels returning to one of three launch sites (i.e., Redondo Beach, Cabrillo Beach, and Los Alamitos Harbor) (see Figure 17.1). These three locations were selected because they are the launch sites closest to the Palos Verdes Shelf; therefore, the authors concluded that they are sites from which private vessels are most likely to fish at the Palos Verdes Shelf. The questionnaire they developed was based on that used in the *Santa Monica Bay Seafood Consumption Study* (SCCWRP and MBC, 1994). One important improvement adopted by Love and Hansen (1999), however, was the collection of information on specifically where the private boat angler caught the fish observed during the interview. Finally, the sampling scheme adapted by Love and Hansen (1999) focused more sampling effort on areas where Palos Verdes Shelf anglers would most likely be encountered than did the *Santa Monica Bay Seafood Consumption Study* (SCCWRP and MBC, 1994). They conducted interviews on a total of 95 days whereas *Santa Monica Bay Seafood Consumption Study* (SCCWRP and MBC, 1994) conducted interviews of CPFV and private boat anglers within the focused geographical range covered by Love and Hansen (1999) on only 36 days. This more focused, intensive sampling effort results in a larger and more applicable data set for purposes of assessing risks to private boat anglers than the *Santa Monica Bay Seafood Consumption Study* (SCCWRP and MBC, 1994).

For the Love and Hansen (1999) study, surveys were conducted on 29 days at Redondo Beach, 31 days at Cabrillo Beach, and 35 days at Los Alamitos Harbor. A total of 737 anglers were identified and assigned a respondent number. Of those 737 anglers, 211 respondents refused to be interviewed, 1 respondent both refused to be interviewed and indicated that he had already been interviewed, and 28 respondents indicated that they had already been interviewed within the period of the survey. The information obtained during the 28 interviews identified as subsequent to an initial interview was removed so the possibility of obtaining information from a respondent more than once during the survey period did not exist. In all, 497 anglers agreed to be interviewed and indicated that they had not been interviewed already over the course of the survey period. Both the total number of private anglers interviewed (526) and the number who provided consumption information (106) are greater in the Love and Hansen (1999) study than the corresponding counts from the *Santa Monica Bay Seafood Consumption Study* (SCCWRP and MBC, 1994).

For the CPFV and Cabrillo Pier angler analyses, chemical concentration data for tDDT and PCBs in fish fillets were compiled from the Los Angeles County Sanitation District (LACSD) (1997), SCCWRP and MBC (1992), Pollock et al. (1991), Risebrough (1987), and Gossett et al., (1983). Specifically, LACSD has collected fish from three arbitrarily defined geographic "zones" of the Palos Verdes Shelf since 1972 (see Figure 17.1). In particular, LACSD measured concentrations of tDDT and PCBs in individual fish of two species consumed by anglers: white croaker and kelp bass (*Paralabrax clathratus*). Because concentrations of tDDT and PCBs in these two species have been declining over the long term, we relied upon data which were collected in August and September of 1996 as the best available, albeit conservative, representation of future concentrations in these fish (LACSD, 1997). For white croaker, the season in which LACSD collected the samples corresponds to the season just prior to spawning, the time of near peak concentrations (Gold et al., 1997; Love et al., 1984). LACSD (1997) reported concentrations of tDDT as both the total detectable DDT and each separate form of DDE, DDD, and DDT. The reported concentrations demonstrate that for both white croaker and kelp bass, DDE constitutes greater than 95 percent of the total DDT reported in each zone of the Palos Verdes Shelf. PCBs are reported as both the total detectable PCBs and separate Aroclors, which include Aroclors 1016, 1221, 1232, 1242, 1248, 1254, and 1260. For both white croaker and kelp bass, Aroclor 1254 consistently constitutes greater than 50 percent of the total detectable PCBs across all zones of the Palos Verdes Shelf.

The Pollock et al. (1991) study reported concentrations of tDDT and PCBs for 16 fish species from 24 locations from Point Dume to Dana Point, including sites on the Palos Verdes Shelf. For each species at each sampled location, concentrations were reported (and only available as) summary statistics (minimum, geometric mean, and maximum) for five composite samples of four fish each. Because the 24 locations represented a variety of habitats and different species prefer different habitats, not all species were collected at each location. When concentrations in a particular fish species were not available for the Palos Verdes Shelf, concentrations in that species at the nearest sampled location were used as surrogates. These samples were collected in the summer of 1987, which is the season just prior to spawning for many fish (Love, 1996); thus these were collected at the time of near peak concentrations. The Pollock et al. (1991) data represent the most complete multispecies data available for the region. They report only tDDT, stating that the total DDT concentration was determined by summing concentrations of the o-p' and p-p' forms of DDT, DDE, and DDD. Likewise, only the total PCBs were reported and these were estimated by summing the concentrations of Aroclors 1254 and 1260.

For the private boat angler analysis, chemical concentration data in fish fillets used to support the analysis were the same as those used for CPFV anglers, with the exception of white croaker data from zone 1 of the Palos Verdes Shelf. Based on the at-sea survey they performed to actually observe fishing from recreational vessels on the Palos Verdes Shelf, it was clear to Love and Hansen (1999) that anglers fishing on the Palos Verdes Shelf caught white croaker only in water shallower than the approximately 50 m depth (i.e., about 165 feet) at which the LACSD collects the samples that are usually used to define the chemical concentrations of white croaker in the Palos Verdes Shelf area. That is, most of the historical data are from areas where fishing is less frequent. Therefore, Love and Hansen (1999) collected white croaker in zone 1 at depths of 20 to 30 m (i.e., about 65 to 100 feet) which corresponded to the depths at which their at-sea survey indicated that anglers typically catch white croaker. Love and Hansen (1999) reported concentrations of tDDT as both the total detectable DDT and as each separate form of DDE, DDD, and DDT. PCBs are reported as both the total detectable PCBs and as separate Aroclors, which include Aroclors 1016, 1221, 1232, 1242, 1248, 1254, and 1260. We used these data to represent the chemical concentrations in white croaker in zone 1 for the private boat angler analysis. The results of using these data in the CPFV analysis are discussed in Section 17.4.

SCCWRP and MBC (1992) provided additional data on concentrations of tDDT and PCBs in white croaker at the Palos Verdes Shelf. Risebrough (1987) and Gossett et al. (1983) were additional data sources for concentrations of tDDT and PCBs in white croaker at Cabrillo Pier. These data did not form the basis of chemical concentration distributions but rather were used as a comparison to the more contemporary data upon which our study ultimately relied.

Table 17.2 presents the 13 principal species and species groups included in the analyses for the CPFV, private boat, and Cabrillo Pier angler populations. These principal species represent those consumed most frequently (i.e., by more than 1 percent of anglers), as reported by the *Santa Monica Bay Seafood Consumption Study* (SCCWRP and MBC, 1994) and the Love and Hansen (1999) study, and those species for which chemical concentration data were available. Species that are without concentration data, but are known to be consumed by some anglers, were included by assigning them concentrations available for species with similar habitat and diet. For example, as shown on Table 17.2, based on what is known about the nonprincipal species yellowtail (*Seriola lalandi*), it was assumed that they contained the same concentration of tDDT and PCBs as the principal species Pacific bonito (*Sarda chiliensis*). Because of

TABLE 17.2 Fish Species and Species Groups Included in the Analyses

Principal Species/Group	Nonprincipal Species in Group due to Habitat and Diet Similarity
Barred sand bass	Triggerfish,[a] octopus,[a] rock crab,[a] sting ray, ocean whitefish,[b] bat ray,[b] spotted sand bass,[b] bass,[b] California skate,[b] shovelnose guitarfish,[b] yellowfin croaker[b]
California halibut	California lizardfish, white sea bass[b]
California scorpionfish	N/A[c]
Chub mackerel	Jack mackerel, Pacific sardine,[a] lingcod
Halfmoon	N/A
Kelp bass	Giant sea bass[b]
Opaleye	N/A
Pacific barracuda	Brown smoothhound,[a] gray smoothhound,[a] smoothound shark,[b] horn shark, leopard shark, shark[a]
Pacific bonito	Yellowtail, dorado,[b] mako,[b] albacore,[b] bluefin tuna[b]
Queenfish	N/A
Rockfishes	Boccacio,[a] treefish, chilipepper,[a] cabezon, California sheephead, olive rockfish,[a] grass rockfish, greenstriped rockfish,[a] copper rockfish,[a] flag rockfish, blue rockfish,[a] squarespot rockfish,[a] Vermillion rockfish,[a] greenspotted rockfish, starry rockfish, gopher rockfish,[a] brown rockfish,[a] red banded rockfish,[a] popeye catalufa[b]
Surfperches	Jacksmelt, blacksmith, rock wrasse,[a] black perch, barred surfperch,[a] shiner perch,[a] white seaperch,[a] walleye surfperch,[a] pile perch,[a] rainbow seaperch[a]
White croaker	Sanddab[b]

[a] Only identified in the *Santa Monica Bay Seafood Consumption Survey* (SCCWRP and MBC, 1994); therefore, only included in the analysis of CPFV and Cabrillo Pier anglers.

[b] Only identified in the Love and Hansen (1999) study; therefore, only included in the analysis for private boat anglers.

[c] N/A indicates that there are no species grouped with the principal species.

the infrequent consumption of the nonprincipal species, any errors introduced by these assumptions would not significantly affect the final estimates of risk.

Information on human body weight distributions for males and females by age was taken from the *Exposure Factors Handbook* (EPA, 1997b). For individuals ages 5 through 19, body weight distributions were available for each year for both males and females. For individuals 20 years old and older, body weight distributions were available for 5-year age groups, but separately for males and females. For individuals 15 years old and older, a minimum body weight of 40 kg (88 lb) and a maximum body weight of 150 kg (330 lb) were established.

17.2.3   Method for Estimating Exposure

As noted previously, the Monte Carlo techniques used in this analysis account for the fact that an individual's exposure to any environmental media is the result of a series of separate exposure events. As such, in these analyses, an exposure event must be clearly defined. In all of the analyses, an exposure event was the consumption of fish harvested on a single fishing trip. The fish may be consumed during a single meal or over multiple meals. This definition allows direct use of trip-specific information available from regional angler intercept surveys to estimate exposures to chemicals in fish. Trip-specific measurements are inherently more accurate than recall-based information.

The CPFV, private boat, and Cabrillo Pier angler analyses used structures similar to each other. Figure 17.3 outlines the structure of the Monte Carlo analyses. Anglers were first assigned lifetime characteristics from site-specific angler data. The analysis then characterized each angler's fishing career (i.e., years fished per lifetime) one year at a time, and within each year, one trip at a time. For example, if fish were caught on a trip and then consumed, the angler's uptake (i.e., absorbed dose) of tDDT and PCBs was calculated and added to the angler's exposure history for the year. The total lifetime dose was defined as the sum of the individual doses from the events that occurred over the years of the angler's fishing career. Using this approach, the average dose received over the angler's lifetime was determined and used in the risk characterization. Lifetime doses and risks were calculated repeatedly for the set of hypothetical anglers to reflect the variation across the angler population.

The calculations for the CPFV, private boat, and Cabrillo Pier anglers were performed using Excel Visual Basic for Applications (Microsoft, 1996). Supporting statistical analyses were conducted using SYSTAT (SYSTAT, 1996), @Risk (Palisade, 1996), Microsoft Access 97 (Microsoft, 1997a), and Microsoft Excel 97 (Microsoft, 1997b). For quality assurance purposes, all data entry and calculations were verified and checked by an individual who did not conduct the original work. The calculations were executed until numerically stable estimates of the mean and the 95th percentile were obtained. Stability was defined as less than a 5 percent average change in these statistics for total cancer risk estimates and total fish consumption rates among analysis runs.

Predicting Exposure for CPFV Anglers at the Palos Verdes Shelf   The CPFV analysis characterized the interindividual variation in long-term rates or uptake of tDDT and PCBs for the CPFV anglers. The structure and organization the analysis were developed to capture the interdependencies and correlations among exposure factors. The sequence of analysis steps allowed consideration of the effects of gender, age, season, fishing avidity, species preference, and cooking preference so that long-term (i.e., multiyear or lifetime) dose rates can be accurately estimated using data collected in short-term (i.e., days, months, or one year) studies.

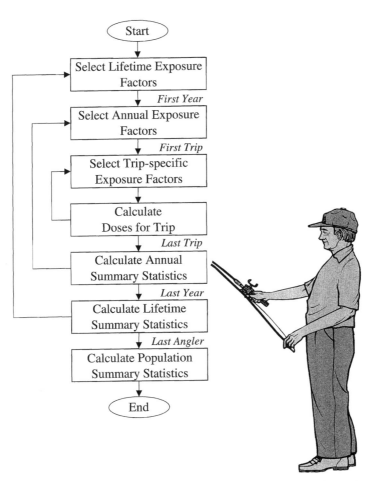

Figure 17.3   Structure and organization of the Monte Carlo analysis for anglers at the Palos Verdes Shelf and Cabrillo Pier.

Simulation of each angler's behavior began by assigning him or her certain personal characteristics that do not change over time. These were the angler's gender, the age at which he or she began fishing, duration of fishing career, the angler's body weight percentile and, as an indicator of avidity, fishing frequency quartile. The angler's gender was chosen at random based on the proportions of male and female anglers in the general angler population covered by the *Santa Monica Bay Seafood Consumption Study* (SCCWRP and MBC, 1994). The distribution of ages at which these anglers began fishing was used as the basis for selecting starting ages for fishing. The numbers of years an angler would fish from CPFVs was developed from *Santa Monica Bay Seafood Consumption Study* (SCCWRP and MBC, 1994) information on number of years fished plus the number of subsequent years estimated using the survey data

(Price et al., 1998). Assignment of a body weight percentile caused body weight to fluctuate with age over time according to the consistent pattern reflected in the *Exposure Factors Handbook* (EPA, 1997b). Assignment of a fishing frequency quartile restricted the number of trips an angler took each season to a particular segment of the distribution. This was done to account for the characteristic that one's general interest in fishing is likely to vary but to be generally consistent over time. For example, an angler being assigned the first frequency quartile resulted in his or her fishing frequency being allowed to vary only up to the 25th percentile; being assigned the second quartile restricted the number of trips per season to between the 26th and 50th percentiles for each season fished, and so on. This approach assumed that a person who was a moderately avid angler at age 17 (e.g., at the 60th percentile of fishing frequency) would, on average, also be a moderately avid angler 10 or 20 years later in life (e.g., at some percentile of fishing frequency within the third quartile).

At this point in the calculations, each CPFV angler was assigned a profile of fish species that would be consumed if caught, and the cooking methods used for preparing the consumed species. These consumption and cooking preferences were assigned once for the angler's entire fishing career in recognition that such preferences are more likely to be consistent from trip to trip than to be random (e.g., always keeping a kelp bass and always frying a kelp bass). A "fishing career" was defined as the total number of years that an angler fishes and served as the exposure duration in the context of the exposure assessment.

The intake of tDDT and PCBs for the first year that the angler fishes was simulated as follows: The first step was to assign the angler a body weight for that year based on the angler's age, gender, and body weight percentile. The seasons fished (i.e., spring, summer, fall, winter, or any combination) were selected as a function of age. The angler's assigned frequency quartile was then used to guide the selection of the number of fishing trips in each season fished. For example, a moderately avid male CPFV angler of typical build might, at 18 years of age, be assigned a median body weight of 71 kg, and be predicted to take 4 trips in the spring and 5 trips in the summer.

The procedure then simulated, trip by trip, the angler's fishing trips for the first season of the first year of fishing. An initial departure port was selected from among the possible CPFV departure ports based on the relative frequency of use by anglers, calculated from locally collected data (SCCWRP and MBC, 1994). The probability that a CPFV trip from that port would visit the Palos Verdes Shelf, which was obtained from the CDFG survey data, was used to determine whether the angler's first trip went to the Palos Verdes Shelf. If the first trip did not go to the Palos Verdes Shelf, the analysis proceeded to the angler's second trip in that season. If the first trip did go to the Palos Verdes Shelf, then one of three zones was selected as the trip's specific destination, again based on the probabilities from the CDFG data. Figure 17.1 shows the departure ports and Palos Verdes Shelf zones used in this analysis; these are the same zones used by the LACSD in its monitoring surveys of fish on the Palos Verdes Shelf. This method properly characterized the frequency of fishing at the Palos Verdes Shelf. Determining the zone of the Palos Verdes Shelf that was fished was important because, for many species, concentrations in different individual fish within the species vary by location within the Palos Verdes Shelf area (LACSD, 1997; Pollock et al., 1991).

The amount of fish consumed was based on three linked events, with the probability that each might occur being based on *Santa Monica Bay Seafood Consumption Study* (SCCWRP and MBC, 1994) information: (1) whether any fish were caught; (2) if fish were caught, the chances of catching a particular combination of one or more species; and (3) consuming the species that were caught. If each of these three events occurred for the current trip, then the concentration of tDDT and PCBs in the fish of each consumed species was estimated based on the distribution of chemical concentration data for fish from the zone where the fish were caught. A consumption rate (grams/trip) for each species was assigned from the distribution of species-specific consumption rates from the *Santa Monica Bay Seafood Consumption Study* (SCCWRP and MBC, 1994). The species-specific cooking method was obtained from the angler's assigned "profile," and a percentage of chemicals lost due to cooking was then selected from the distribution of losses associated with that cooking method (Wilson et al., 1998). The uptake of tDDT and PCBs from consumption of each fish species was calculated from the consumption rates, concentrations of chemicals in fish, and percentages of the chemicals retained after cooking. The uptake of tDDT and PCBs from fish consumed from this trip was then recorded.

In the next step, the CPFV angler's subsequent fishing trips were simulated. The subsequent departure port was chosen based on the likelihood that an angler would use the same or another port given his or her initial departure port. The initially selected departure location influenced the selection of all subsequent departure locations over the angler's fishing career because, for the majority of the anglers, choice of fishing location is based on proximity to one's residence (SCCWRP and MBC, 1994). These subsequent trips were then characterized in the same manner as the first trip.

Once all of the trips for the first season were described, the trips for the remaining seasons of the first year were also accounted for using this Monte Carlo technique. The uptake of tDDT and PCBs by the angler was then summed and recorded for the trips during the first year on which fish from the Palos Verdes Shelf were consumed.

The angler's age was then increased by one year, and the annual setup and cycle of trips were repeated for the next and all subsequent years of the angler's fishing career (i.e., the number of years fished in his or her lifetime). For each CPFV angler, the lifetime dose and risk statistics were calculated from the set of annual doses and recorded, which concluded the simulation for that hypothetical angler.

Predicting Exposure for Private Boat Anglers at the Palos Verdes Shelf   The structure and organization of the private boat analysis is nearly identical to that used in the CPFV analysis. However, as stated above, the source of the information on private boat angler behavior data (e.g., fishing frequency, consumption rates, species preference, cooking methods, etc.) differs from that used in the CPFV analysis. Specifically, the private boat analysis used data collected by the Love and Hansen (1999) study rather than the *Santa Monica Bay Seafood Consumption Study* (SCCWRP and MBC, 1994) because the Love and Hansen (1999) study produced survey data more focused for health risk assessment of private boat anglers at the Palos Verdes Shelf than did the *Santa Monica Bay Seafood Consumption Study* (SCCWRP and MBC, 1994).

The only difference in the structure and organization of the private boat analysis from the structure and organization of the CPFV analysis was the manner in which an angler's likelihood of fishing at the Palos Verdes Shelf was determined. Specifically, the probability that the private boat would visit the Palos Verdes Shelf was obtained directly from information from the Love and Hansen (1999) study. This differs from the structure of the CPFV analysis, which used the probability of the departure location and the probability of fishing at the Palos Verdes Shelf given the departure location based on information from the CDFG surveys to estimate the probability of fishing at the Palos Verdes Shelf because the *Santa Monica Bay Seafood Consumption Study* (SCCWRP and MBC, 1994) did not collect information to support the direct calculation. Love and Hansen's (1999) private boat survey collected information on where the survey participants fished during the trip about which they were interviewed; thus, the probability of fishing at the Palos Verdes Shelf by season was calculated directly without regard to departure location.

Because the Love and Hansen (1999) study collected data on where the private boat survey participants fished during the trip about which they were interviewed, those private boat anglers who fished at the Palos Verdes Shelf were identified. Thus, only those responses from the Love and Hansen (1999) study that correspond to anglers who fished at the Palos Verdes Shelf formed the basis for the probabilities of each of the following three linked events, which determined the amount of fish consumed from the Palos Verdes Shelf: (1) whether any fish were caught from the Palos Verdes Shelf; (2) if fish were caught from the Palos Verdes Shelf, the chances of catching a particular combination of one or more species; and (3) consuming the species that were caught. However, a consumption rate (grams/trip) for each species was assigned from the distribution of species-specific consumption rates derived from the responses of all of the private boat anglers collected in the Love and Hansen (1999) study, not just those anglers who fished at the Palos Verdes Shelf, because of the desire for larger sample sizes for each species.

Predicting Exposure for Cabrillo Pier Anglers   The Cabrillo Pier analysis considered anglers at a fixed location. Therefore, it was simpler in structure than the CPFV or private boat angler analyses because there was no need to include the process of traveling to the Palos Verdes Shelf or visiting a specific zone within the Palos Verdes Shelf. Otherwise, the calculations were the same and were executed in the same manner.

17.2.4 Exposure Factors Included in the Monte Carlo Analyses

Predicting exposures and risks requires defining and specifying numerical values for the inputs, or *exposure factors*, in the exposure equations. Complex analyses like the Monte Carlo analyses for CPFV, private boat, and Cabrillo Pier anglers include more exposure factors than are employed in typical Monte Carlo or deterministic analyses. Through the repeated simulation process, Monte Carlo analyses allow these exposure factors to take on values that may be different for each simulation. The range of values that an exposure factor may take, and the likelihood of the exposure factor taking each value within that range, are described by mathematical expressions drawn from the published literature on that exposure factor; these mathematical expressions are referred to as probability density functions or cumulative density functions. Collectively, these functions are commonly referred to as *distributions* (see Figure 17.2).

Developing exposure factors for the Monte Carlo analyses required identifying a type of distribution and numerical values for the parameters of that distribution. We used standard practices to choose distribution types and parameter values that attained optimal goodness of fit to the underlying data. Parameter values define the range and likelihood of values taken by the exposure factors within the general type of distribution selected. For example, concentrations of chemicals in environmental media such as soil are typically assumed to be log-normally distributed, and the mean and standard deviation of the data are the estimated parameters of the distribution. An empirical cumulative distribution is often used when no standard parametric distribution fits the data; in such a case, percentiles of the data are specified as parameters.

As discussed previously, there is a relative abundance of site-specific data available regarding angler behavior (Love and Hansen, 1999; SCCWRP and MBC, 1994; Ally et al., 1991) and chemical concentrations in fish (Love and Hansen, 1999; LACSD, 1997; SCCWRP and MBC, 1992; Pollock et al., 1991; Risebrough, 1987; Gossett et al., 1983) in southern California waters. We used site-specific data to develop distributions of angler behavior and exposures using discrete probability functions, empirical distribution functions, or standard parametric distributions. All distributions were derived directly from site-specific data or published distributions.

Because the focus of our study was the general angler population at the Palos Verdes Shelf and at the Cabrillo Pier, it was necessary to perform some interim analysis steps. To use data from the *Santa Monica Bay Seafood Consumption Study* (SCCWRP and MBC, 1994) and the Love and Hansen (1999) study, which were on-site intercept studies of angler behavior, we first applied the location sampling weights so that each CPFV or private boat departure location was equally represented.

We also adjusted for avidity bias using the methods of Price et al. (1994). On-site intercept surveys such as the *Santa Monica Bay Seafood Consumption Study* (SCCWRP and MBC, 1994) and the Love and Hansen (1999) study target anglers when and where they are fishing. The on-site intercept survey is the appropriate type of survey to use when targeting a local population of anglers using a particular water body (EPA, 1992b, 1998a). However, on-site intercept surveys are inherently subject to *avidity bias* (i.e., bias toward frequent anglers) because all anglers do not have an equal chance of being included in the survey. In fact, the chance that an angler is included in an on-site intercept survey increases with the number of fishing trips he or she takes. An angler's probability of being included in the survey is equal to the number of days he or she fishes at the survey location during the course of the survey divided by the number of days in the survey period. As an example, for a year-long survey, an angler who fishes once per year has a 1/365 chance of being included on any sampling day, whereas an angler who fishes once per week has a 52/365 chance of being included on any sampling day. As a result, the sample from an on-site intercept survey is not a random sample of anglers; in fact, the survey sample overrepresents frequent anglers. To the extent that fishing frequency is correlated with information of interest in the survey (e.g., fish consumption rates), the raw survey data are biased and hence do not represent the general angler population that is the subject of the exposure assessment.

The concept of avidity bias has been discussed in the statistical literature at least since the 1960s (see, e.g., Robson, 1961), it continues to be a topic of discussion in the fisheries management literature (see, e.g., Pollock et al., 1994), and a correction for avidity bias is used in ongoing fisheries research conducted by the National Marine Fisheries Service (NMFS, 2000). EPA recognized this bias in the *Exposure Factors Handbook* (EPA, 1997c). Avidity bias can be corrected by weighting survey responses according to the inverse of fishing frequency. We applied an avidity bias correction to compensate for the inherent bias in the *Santa Monica Bay Seafood Consumption Study* (SCCWRP and MBC, 1994) and Love and Hansen (1999) data.

Based on the methods presented in Price et al. (1994), the number of days in a typical year and the number of trips that an angler fishes in a year were used to calculate an avidity bias correction factor for each angler's survey responses. This correction factor was calculated as 365 days divided by the number of trips the angler would take during the survey year. While the *Santa Monica Bay Seafood Consumption Study* (SCCWRP and MBC, 1994) and the Love and Hansen (1999) study did not ask how many times an angler had fished in the previous year, anglers were asked how many trips had been taken in the preceding month. To calculate the avidity bias correction factors, we estimated the angler-specific number of trips per year based on the reported number of trips in 4 weeks and the reported fishing seasons. First, for each person, the number of trips in 4 weeks was multiplied by 3 to estimate the number of trips in the season in which the angler was interviewed. The angler's relative avidity was then determined by calculating the angler's trip frequency percentile within the interview season. For the other seasons in which the angler reported having fished, we calculated the number of trips per season based on the seasonal frequency distribution and the angler's frequency percentile. Once the number of trips per season had been determined, they were summed together to estimate the number of trips per year for each angler. The avidity bias correction factors were derived for each angler such that all of the information provided by a single angler was weighted according to the estimated annual fishing frequency.

For each exposure factor, it was our goal to use as our input distribution the least mathematically complex statistical representation that appropriately characterized our knowledge and expectations regarding the likely values that a particular exposure factor might take. Many exposure factors, such as gender or seasons fished, could be characterized simply by assigning discrete probabilities to the available options. Use of standard parametric statistical distributions allows information about the underlying process to be combined with available data to produce more complete representations of interindividual variation (Hattis and Burmaster, 1994; Ruffle et al., 1994; Finley et al., 1994). For example, measured concentrations of chemicals in fish might follow a lognormal distribution. Using a log-normal distribution to represent concentrations in the analysis allows using varying concentration values in the calculations that are consistent with, but not identical to, those actually measured.

Our procedure for developing PDFs was to first attempt to fit standard parametric probability density functions suggested by professional knowledge of the physical process generating the observed values. For example, body weights of individuals at a given age can be expected to vary symmetrically around an average value [International Commission on Radiological Protection (ICRP), 1981]. This suggests that a normal PDF would be appropriate to represent variation in body weights within a physically relevant range. In addition, the total number of years an angler fishes during his or her lifetime can be considered an ongoing process that is typically observed as the number of years that active anglers report having fished at the time of a survey. Therefore, an exponential PDF is an appropriate representation. In the event that parametric PDFs could not be fit successfully using statistical hypothesis testing or visual evaluation of diagnostic graphical representations of the data (e.g., probability plots, histograms, Q–Q plots), or when information was extremely limited, we fitted empirical PDFs to the data. For parametric and empirical PDFs, we established minima and maxima based on physical limitations in input values. For example, concentrations were not allowed to have negative values, and body weights were constrained to biologically appropriate values. This approach is consistent with that recommended by EPA (1997c) and adopted in other studies (see, e.g., Iannuzzi et al., 1996; Hattis and Burmaster, 1994; Finley et al., 1994). For example, we fitted log-normal PDFs to chemical concentration data collected by LACSD (1997) for kelp bass and white croaker at the Palos Verdes Shelf. We truncated the log-normal distributions at the greater of the maximum detected concentration or three standard deviations above the mean. This truncation limit was chosen to allow selection of large but physically plausible concentrations while eliminating the possibility of selecting concentrations that were physically impossible yet within the mathematical range of a standard log-normal distribution. In contrast, for the remaining species, only summary statistics for composite samples from Pollock et al. (1991) were available to represent chemical concentrations. In these cases, we fitted triangular PDFs constructed from the summary statistics.

It was our general practice to make conservative interpretations of the survey data when developing the exposure factor distributions. For example, Love and Hansen (1999) found only one private boat angler who had retained any white croaker from the Palos Verdes Shelf, and that angler was unsure whether it would be consumed. For our analysis, we made the conservative assumption that the white croaker would be eaten in order to develop a nonzero upper-bound probability for consumption of white croaker from the Palos Verdes Shelf by private boat anglers. It would have been equally reasonable to assume that this private boat angler did not eat the fish, which would have resulted in a prediction of a zero probability of consuming white croaker from the Palos Verdes Shelf for private boat anglers. This in turn would have resulted in a prediction of no exposure and thus no risk to private boat anglers from consumption of white croaker at the Palos Verdes Shelf. The overall effect of this and similar conservative interpretations would be to overpredict the probabilities of and magnitude of exposure.

Table 17.3 presents and describes the exposure factor distributions used in the CPFV angler analysis. This table includes: (a) the distribution type selected for each exposure factor, (b) the relevant parameter values for that distribution, and (c) the source or sources of data used to develop the distributions. Table 17.4 describes the exposure factor distributions used in the private boat angler analysis, and Table 17.5 describes those used in the Cabrillo Pier angler analysis.

17.2.5   Approach for Estimating Cancer Risks and Noncancer Hazards

The potential human health risks were characterized using current EPA methods for evaluating carcinogenic risks and noncancer hazards (EPA, 1989a, 1992a, 1995). Upper-bound cancer risks were estimated by multiplying each angler's lifetime average daily dose associated with intake of tDDT and of PCBs, based on a 70-year standard lifetime (EPA, 1997b), by the cancer slope factor for that chemical. The estimated cancer risks were described as "upper bound" due to the use of conservative low-dose extrapolation models to develop the cancer slope factors (Paustenbach, 1995). For each individual, three upper-bound cancer risk estimates were developed: risks associated with tDDT intake, risks associated with PCB intake, and the total risks associated with the intake of both tDDT and PCBs.

Noncancer hazards were estimated using hazard quotients that involve dividing the angler's average daily dose of tDDT or PCBs by the appropriate EPA reference dose. For consistency with the chronic nature of toxicological effects on which reference doses are developed, a minimum averaging time of 2555 days (i.e., 7 years) was used to calculate the average daily doses. Values of hazard quotients less than 1 were taken as evidence that chronic noncancer effects are unlikely to occur.

TABLE 17.3 Distribution Functions for Various Exposure Factors Used in the Commercial Passenger Fishing Vessel (CPFV) Analysis

Exposure Factor (Description)	Distribution Type	Parameter Values						Source
		Min	Mode	Mean	Max	Standard Deviation	Value (Frequency)	
Angler gender (unitless) (Each angler's gender is selected based on the gender distribution.)	Discrete						Male (91%); female (9%)	SCCWRP and MBC (1994)—CPFV and Cabrillo Pier survey data corrected for avidity bias based on self-reported frequency information and adjusted for location sampling bias
Body weight[a] (kg) (Annually, an age- and gender-specific body weight is selected for each angler based on a series of age-specific cumulative body weight distribution functions.)	Empirical	40		78.7	150	13.7		USEPA (1997b)
Age at which fishing begins (years) (The probability of an angler starting to fish at a particular age is described by this distribution.)	Truncated normal	5		26	80	13		SCCWRP and MBC (1994)—CPFV survey data corrected for avidity bias based on self-reported frequency information and adjusted for location sampling bias
Length of fishing career (exposure duration) (years) (The exposure duration distribution describes the probability of an angler fishing for a specific number of years. It is a function of historical years fished, estimates of future years fished, and age at which fishing begins. Duration is limited by the age at which fishing begins; no anglers fish past 90 years of age.)	Exponential	1		7.1	85			Function of historical years fished and estimates of future years fished. Historical data from SCCWRP and MBC (1994)—CPFV survey data corrected for avidity bias based on self-reported frequency information and adjusted for location sampling bias. Future years fished estimated using methods of Price et al. (1998).

(Continued)

TABLE 17.3 (*Continued*)

Exposure Factor (Description)	Distribution Type	Parameter Values						Source
		Min	Mode	Mean	Max	Standard Deviation	Value (Frequency)	
Consuming a particular species given that it is caught and kept (unitless) (This distribution describes the probability of an angler consuming a particular fish species given that he or she catches the fish. Then distributions are used to establish, for each angler's fishing career, which species are eaten if caught.)	Discrete						Chub mackerel (55%); Pacific barracuda (69%); kelp bass (70%); barred sand bass (80%); white croaker (79%); queenfish (26%); halfmoon (95%); California halibut (38%); rockfish (61%); California scorpionfish (76%); surfperch (78%); opaleye (90%); Pacific bonito (66%)	SCCWRP and MBC (1994)—all angler survey data corrected for avidity bias based on self-reported frequency information and adjusted for location sampling bias
Cooking method (unitless) (A cooking method used by the anglers to prepare fish is assigned to an angler for each species eaten and is the same each time that species is consumed.)								
Chub mackerel	Discrete						Soup (4%); raw/smoke/ceviche (9%); bake/boil/steam (9%); broil/BBQ (17%); fry (61%)	SCCWRP and MBC (1994)—all angler survey data corrected for avidity bias based on self-reported frequency information and adjusted for location sampling bias
Pacific barracuda	Discrete						Soup (0.1%); raw/smoke/ceviche (3%); bake/boil/steam (14.7%); broil/BBQ (30.1%); fry (52.1%)	

Kelp bass	Discrete	Soup (3%); raw/smoke/ceviche (5%); bake/boil/steam (16%); broil/**BBQ** (36%); fry (40%)
Barred sand bass	Discrete	Soup (3%); raw/smoke/ceviche (3%); bake/boil/steam (14%); broil/**BBQ** (34%); fry (46%)
White croaker	Discrete	Soup (2%); raw/smoke/ceviche (2%); bake/boil/steam (13%); broil/**BBQ** (16%); fry (67%)
Queenfish	Discrete	Soup (2%); raw/smoke/ceviche (4%); bake/boil/steam (7%); broil/**BBQ** (21%); fry (66%)
Halfmoon	Discrete	Bake/boil/steam (39%); fry (61%)
California halibut	Discrete	Soup (2%); raw/smoke/ceviche (3%); bake/boil/steam (18%); broil/**BBQ** (33%); fry (44%)
Rockfish	Discrete	Soup (3%); raw/smoke/ceviche (5%); bake/boil/steam (15%); broil/**BBQ** (33%); fry (44%)
California scorpionfish	Discrete	Soup (1%); raw/smoke/ceviche (9%); bake/boil/steam (16%); broil/**BBQ** (39%); fry (35%)
Surfperch	Discrete	Soup (0.04%); bake/boil/steam (6%); broil/**BBQ** (26%); fry (68%)

(Continued)

TABLE 17.3 (*Continued*)

Exposure Factor (Description)	Distribution Type	Parameter Values						Source
		Min	Mode	Mean	Max	Standard Deviation	Value (Frequency)	
Opaleye	Discrete						Soup (2%); raw/smoke/ceviche (38%); bake/boil/steam (20%); broil/BBQ (2%); fry (38%)	
Pacific bonito	Discrete						Soup (1%); raw/smoke/ceviche (15%); bake/boil/steam (15%); broil/BBQ (30%); fry (39%)	
Seasons fished[b] (unitless) (This distribution assigns the probability of an angler within a specified age group of fishing in a season or combination of seasons.)								SCCWRP and MBC (1994)—CPFV survey data corrected for avidity bias based on self-reported frequency information and adjusted for location sampling bias
5–18 years old	Discrete						Sp (20%); Su (36%); SpSu (19%); SpFaWi (1%); SpSuFaWi (24%)	
19–64 years old	Discrete						Sp (4%); Su (28%); Fa (4%); Wi (9%); SpSu (11%); SpWi (0.5%); SuFa (3%); FaWi (0.1%); SpSuFa (3%); SpSuFaWi (37%)	
65 years and older	Discrete						Wi (27%); SpSu (27%); SpSuFa (3%); SpSuFaWi (43%)	

Parameter	Distribution	Min	Value	Max	Value	Source / Values
Number of fishing trips per season (unitless) (The number of trips taken by an angler during each season is assigned by this distribution.)						SCCWRP and MBC (1994)— CPFV survey data corrected for avidity bias based on self-reported frequency information and adjusted for location sampling bias
Winter	Truncated log-normal	1	5.0	91	6.7	
Spring	Truncated log-normal	1	5.2	91	7.7	
Summer	Truncated log-normal	1	6.5	91	8.5	
Fall	Truncated log-normal	1	7.6	91	11.2	
Initial departure location (unitless) (Each angler is assigned a departure location for the initial CPFV trip based on this distribution.)	Discrete					LA/Long Beach Harbor Port Complex (18%); Malibu (20%); Marina Del Rey (20%); Redondo Barge (20%); Redondo Boat (18%) — SCCWRP and MBC (1994)— census data
Subsequent departure location given initial departure location (unitless) (For each trip after the first, a subsequent departure location is assigned to an angler based on the initial departure location.)						SCCWRP and MBC (1994)— CPFV survey data on trips in previous 30 days corrected for avidity bias based on self-reported frequency information and normalized for location
Redondo Boat	Discrete					Redondo Boat (54%); Redondo Barge (29%); LA/Long Beach Harbor Port Complex (14%); Malibu (3%)
Malibu	Discrete					Malibu (94%); Marina Del Rey (5%); Redondo Boat (1%)

(Continued)

TABLE 17.3 (*Continued*)

Exposure Factor (Description)	Distribution Type	Parameter Values						Source
		Min	Mode	Mean	Max	Standard Deviation	Value (Frequency)	
Marina Del Rey	Discrete						Marina Del Rey (84%); Redondo Boat (8%); LA/Long Beach Harbor Port Complex (3%); Malibu (5%)	California Department of Fish and Game CPFV survey (Ally et al., 1991) as analyzed by Brooks Marine Consulting
LA/Long Beach Harbor Port Complex	Discrete						LA/Long Beach Harbor Port Complex (85%); Redondo Boat (9%); Marina Del Rey (6%)	
Redondo Barge	Discrete						Redondo Barge (54%); Redondo Boat (29%); LA/Long Beach Harbor Port Complex (14%); Malibu (3%)	
Fishing at Palos Verdes Shelf given departure location (unitless) (This distribution describes the probability of an angler going to the Palos Verdes Shelf given the departure location.)								
LA/Long Beach Harbor Port Complex—all seasons	Discrete						Shelf (15%); no shelf (85%)	
Redondo Boat—winter	Discrete						Shelf (9%); no shelf (91%)	
Redondo Boat—spring	Discrete						Shelf (23%); no shelf (77%)	
Redondo Boat—summer	Discrete						Shelf (29%); no shelf (71%)	
Redondo Boat—fall	Discrete						Shelf (31%); no shelf (69%)	
Redondo Barge, Malibu, Marina Del Rey	Discrete						Shelf (0%); no shelf (100%)	

Parameter	Type	Value	Source
Fishing a particular zone within the Palos Verdes Shelf (unitless) (Based on season, this distribution determines the zone on the Palos Verdes Shelf to which a CPFV goes.)			California Department of Fish and Game CPFV survey (Ally et al, 1991) as analyzed by Brooks Marine Consulting
Winter	Discrete	Zone 1 (51%); zone 2 (33%); zone 3 (16%)	
Spring	Discrete	Zone 1 (39%); zone 2 (25%); zone 3 (36%)	
Summer	Discrete	Zone 1 (31%); zone 2 (20%); zone 3 (49%)	
Fall	Discrete	Zone 1 (40%); zone 2 (26%); zone 3 (34%)	
Catching one or more fish (unitless) (This distribution describes the probability of an angler catching a fish of any species on a trip-by-trip basis.)			SCCWRP and MBC (1994)—CPFV survey data corrected for avidity bias based on self-reported frequency information and adjusted for location sampling bias
Fall and winter	Discrete	Catch (50%); no catch (50%)	
Spring and summer	Discrete	Catch (74%); no catch (26%)	
Catching a particular combination of species (unitless) (This distribution describes the probability of an angler catching a particular combination of fish species.)			SCCWRP and MBC (1994)—CPFV survey data corrected for avidity bias based on self-reported frequency information and adjusted for location sampling bias
Total number of combinations = 66			
Chub mackerel	Discrete	Frequency = 24/66	
Pacific barracuda	Discrete	Frequency = 13/66	

(Continued)

TABLE 17.3 (*Continued*)

Exposure Factor (Description)	Distribution Type	Parameter Values						Source
		Min	Mode	Mean	Max	Standard Deviation	Value (Frequency)	
Kelp bass	Discrete						Frequency = 29/66	SCCWRP and MBC (1994) —survey data on fish consumption rate corrected for avidity bias based on self-reported frequency information and adjusted for location sampling bias
Barred sand bass	Discrete						Frequency = 23/66	
White croaker	Discrete						Frequency = 4/66	
Queenfish	Discrete						Frequency = 0/66	
Halfmoon	Discrete						Frequency = 9/66	
California halibut	Discrete						Frequency = 11/66	
Rockfish	Discrete						Frequency = 14/66	
California scorpionfish	Discrete						Frequency = 15/66	
Surfperch	Discrete						Frequency = 3/66	
Opaleye	Discrete						Frequency = 1/66	
Pacific bonito	Discrete						Frequency = 12/66	
Consumption rate per trip for a particular species (kg/trip) (The mass of a particular fish species consumed on a particular fishing trip is selected on a trip-by-trip basis.)								
Chub mackerel	Empirical	0.001		0.43	3.8	0.66		
Pacific barracuda	Empirical	0.001		0.75	5.3	0.77		
Kelp bass	Empirical	0.001		0.33	3.0	0.46		
Barred sand bass	Empirical	0.001		0.50	7.9	1.00		
White croaker	Empirical	0.001		0.10	1.7	0.22		
Queenfish	Empirical	0.001		0.010	0.020	0.0040		
Halfmoon	Empirical	0.001		0.20	0.64	0.13		
California halibut	Empirical	0.001		0.29	3.1	0.57		
Rockfish	Empirical	0.001		0.22	3.7	0.53		
California scorpionfish	Empirical	0.001		0.20	0.75	0.19		

Surfperch	Empirical	0.001		0.17	2.1	0.37	Pollock et al. (1991)
Opaleye	Empirical	0.001		0.15	0.61	0.090	Pollock et al. (1991)
Pacific bonito	Empirical	0.001		1.20	4.5	0.91	Pollock et al. (1991)
tDDT concentration in fish, Palos Verdes Shelf (ppb) (tDDT concentrations in fish caught in various zones of the Palos Verdes Shelf are selected on a trip by trip basis. The distribution represents the variation of tDDT concentrations across individual fish.)							
Chub mackerel—zone 1	Triangular	5	19	31	70		Pollock et al. (1991)
Chub mackerel—zone 2	Triangular	6	26	38	82		Pollock et al. (1991)
Chub mackerel—zone 3	Triangular	4	14	13	22		Pollock et al. (1991)
Pacific barracuda—all zones	Triangular	9	22	27	51		Pollock et al. (1991)
Kelp bass—zone 1	Truncated log-normal	0		220	1,600	220	LASCD (1997)
Kelp bass—zone 2	Truncated log-normal	0		210	1,600	220	LASCD (1997)
Kelp bass—zone 3	Truncated log-normal	0		150	980	140	LASCD (1997)
Barred sand bass—all zones	Triangular	30	55	56	82		Pollock et al. (1991)
White croaker—zone 1	Truncated log-normal	0		23,000	131,000	18,000	LASCD (1997)
White croaker—zone 2	Truncated log-normal	0		6,000	23,000	2,900	LASCD (1997)
White croaker—zone 3	Truncated log-normal	0		3,300	15,000	1,900	LASCD (1997)
Queenfish—zone 1	Triangular	62	94	99	141		Pollock et al. (1991)
Queenfish—zone 2	Triangular	21	40	44	70		Pollock et al. (1991)

(*Continued*)

943

TABLE 17.3 *(Continued)*

| Exposure Factor (Description) | Distribution Type | \multicolumn{6}{c}{Parameter Values} | | | | | |
		Min	Mode	Mean	Max	Standard Deviation	Value (Frequency)	Source
Queenfish—zone 3	Triangular	31	54	60	95			Pollock et al. (1991)
Halfmoon—all zones	N/A						Concentration = 0	Pollock et al. (1991)
California halibut—all zones	Triangular	2	8	11	23			Pollock et al. (1991)
Rockfish—zone 1	Triangular	26	72	160	383			Pollock et al. (1991)
Rockfish—zone 2	Triangular	37	79	78	118			Pollock et al. (1991)
Rockfish—zone 3	Triangular	40	67	88	158			Pollock et al. (1991)
California scorpionfish—zone 1	Triangular	53	154	190	358			Pollock et al. (1991)
California scorpionfish—zone 2	Triangular	15	41	50	93			Pollock et al. (1991)
California scorpionfish—zone 3	Triangular	16	33	34	54			Pollock et al. (1991)
Surfperch—zone 1	Triangular	17	29	31	48			Pollock et al. (1991)
Surfperch—zone 2	Triangular	23	45	53	92			Pollock et al. (1991)
Surfperch—zone 3	Triangular	22	70	76	137			Pollock et al. (1991)
Opaleye—all zones	N/A						Concentration = 0	Pollock et al. (1991)
Pacific bonito—zone 1	Triangular	7	31	62	147			Pollock et al. (1991)
Pacific bonito—zone 2	Triangular	18	32	38	65			Pollock et al. (1991)
Pacific bonito—zone 3	Triangular	13	18	19	25			Pollock et al. (1991)
PCB concentration in fish, Palos Verdes Shelf (ppb) (PCB concentrations in fish caught in various zones of the Palos Verdes Shelf are selected on a trip by trip basis. The distribution represents the variation of PCB concentrations across individual fish.)								
Chub mackerel—zone 1	Triangular	0	9	20	51			Pollock et al. (1991)
Chub mackerel—zone 2	Triangular	3	13	24	57			Pollock et al. (1991)

Species—zone	Distribution						Reference
Chub mackerel—zone 3	Triangular	2	6	5.7	9		Pollock et al. (1991)
Pacific barracuda—all zones	Triangular	10	16	17	24		Pollock et al. (1991)
Kelp bass—zone 1	Truncated log-normal	0		45	240	29	LACSD (1997)
Kelp bass—zone 2	Truncated log-normal	0		50	260	35	LACSD (1997)
Kelp bass—zone 3	Truncated log-normal	0		48	260	34	LACSD (1997)
Barred sand bass—all zones	Triangular	20	33	31	41		Pollock et al. (1991)
White croaker—zone 1	Truncated log-normal	0		1,700	9,200	1,300	LACSD (1997)
White croaker—zone 2	Truncated log-normal	0		630	2,200	260	LACSD (1997)
White croaker—zone 3	Truncated log-normal	0		420	1,700	220	LACSD (1997)
Queenfish—zone 1	Triangular	15	24	27	41		Pollock et al. (1991)
Queenfish—zone 2	Triangular	10	19	19	27		Pollock et al. (1991)
Queenfish—zone 3	Triangular	8	14	14	19		Pollock et al. (1991)
Halfmoon—all zones	N/A			Concentration = 0			Pollock et al. (1991)
California halibut—all zones	Triangular	0	2	6	16		Pollock et al. (1991)
Rockfish—zone 1	Triangular	0	8	28	76		Pollock et al. (1991)
Rockfish—zone 2	Triangular	7	17	15	22		Pollock et al. (1991)
Rockfish—zone 3	Triangular	7	12	14	22		Pollock et al. (1991)
California scorpionfish—zone 1	Triangular	28	41	40	52		Pollock et al. (1991)
California scorpionfish—zone 2	Triangular	0	2	2.3	5		Pollock et al. (1991)
California scorpionfish—zone 3	Triangular	5	10	10	16		Pollock et al. (1991)
Surfperch—zone 1	Triangular	14	29	30	47		Pollock et al. (1991)
Surfperch—zone 2	Triangular	5	17	21	41		Pollock et al. (1991)
Surfperch—zone 3	Triangular	5	16	16	26		Pollock et al. (1991)
Opaleye—all zones	Triangular	0	1	1.3	3		Pollock et al. (1991)
Pacific bonito—zone 1	Triangular	2	14	42	112		Pollock et al. (1991)
Pacific bonito—zone 2	Triangular	7	13	16	29		Pollock et al. (1991)
Pacific bonito—zone 3	Triangular	4	8	7.7	11		Pollock et al. (1991)

(Continued)

TABLE 17.3 (*Continued*)

Exposure Factor (Description)	Distribution Type	Parameter Values						Source
		Min	Mode	Mean	Max	Standard Deviation	Value (Frequency)	
tDDT cooking loss (unitless) (The tDDT cooking loss distribution describes the percent reduction of tDDT in fish resulting from each cooking method. No loss is associated with making soup or eating fish raw.)								Zabik et al. (1979, 1995a, b, 1996); Smith et al. (1973); Trotter et al. (1988); Skea et al. (1981); Reinert et al. (1972); and Puffer and Gossett (1983), as compiled and analyzed by Wilson et al. (1998)
Bake	Empirical	0		0.26	1	0.25		
Boil	Empirical	0		0.34	1	0.27		
Broil	Empirical	0		0.39	1	0.25		
Fry	Empirical	0		0.41	1	0.23		
Smoke	Empirical	0		0.49	1	0.23		
PCB cooking loss (unitless) (The PCB cooking loss distribution describes the percent reduction of PCBs in fish resulting from each cooking method. No loss is associated with making soup or eating fish raw.)								Zabik et al. (1979, 1995a, b, 1996); Smith et al. (1973); Trotter et al. (1988); Skea et al. (1981); and Puffer and Gossett (1983), as compiled and analyzed by Wilson et al. (1998)
Bake	Empirical	0		0.33	1	0.25		
Boil	Empirical	0		0.52	1	0.30		
Broil	Empirical	0		0.44	1	0.23		
Fry	Empirical	0		0.37	1	0.24		
Smoke	Empirical	0		0.40	1	0.23		

[a] Statistics listed are for a 30-year old male.

[b] Seasons: Sp-spring, Su-summer, Fa-fall, Wi-winter.

946

TABLE 17.4 Distribution Functions for Various Exposure Factors Used in the Private Boat Analysis

Exposure Factor (Description)	Distribution Type	Parameter Values						Source
		Min	Mode	Mean	Maxi	Standard Deviation	Value (Frequency)	
Angler gender (unitless) (Each angler's gender is selected based on the gender distribution.)	Discrete						Male (97%); female (3%)	Love and Hansen (1999)— Private boat survey data corrected for avidity bias based on self-reported frequency information and adjusted for location sampling bias
Body weight[a] (kg) (Annually, an age- and gender-specific body weight is selected for each angler based on a series of age-specific cumulative body weight distribution functions.)	Empirical	40		78.7	150	13.7		USEPA (1997b)
Age at which fishing begins (years) (The probability of an angler starting to fish at a particular age is described by this distribution.)	Truncated normal	5		29	80	13		Love and Hansen (1999)— Private boat survey data corrected for avidity bias based on self-reported frequency information and adjusted for location sampling bias
Length of fishing career (*exposure duration*) (The exposure duration distribution describes the probability of an angler fishing for a specific number of years. It is a function of historical years fished, estimates of future years fished, and age at which fishing begins. Duration is limited by the age at which fishing begins; no anglers fish past 90 years of age.)	Discrete						1 (26.6%); 2 (13.1%); 3 (8.5%); 4 (6.3); 5 (5%); 6 (4.1%); 7 (3.4%); 8 (2.9%); 9 (2.6%); 10 (2.2%) … 80 (0.013%); 81 (0.012%); 82 (0.011%); 83 (0.01%); 84 (0.01%); 85 (0.009%)	Function of historical years fished and estimates of future years fished. Historical data from Love and Hansen (1999)—Private boat survey data corrected for avidity bias based on self-reported frequency information and adjusted for location sampling bias. Future years fished estimated using methods of Price et al. (1998).

(Continued)

947

TABLE 17.4 (*Continued*)

Exposure Factor (Description)	Distribution Type	Parameter Values							Source
		Min	Mode	Mean	Max	Standard Deviation	Value (Frequency)		
Consuming a particular species given that it is caught and kept (unitless) (This distribution describes the probability of an angler consuming a particular fish species given that he or she catches the fish. Then distributions are used to establish, for each angler's fishing career, which species are eaten if caught.)	Discrete						Chub mackerel (6%); Pacific barracuda (77%); kelp bass (68%); barred sand bass (79%); white croaker (8%); queenfish (100%); halfmoon (100%); California halibut (26%); rockfish (80%); California scorpionfish (84%); surfperch (100%); opaleye (100%); Pacific bonito (86%)		Love and Hansen (1999)— Private boat PVS angler[b] survey data corrected for avidity bias based on self-reported frequency information and adjusted for location sampling bias
Cooking method (unitless) (A cooking method used by the anglers to prepare fish is assigned to an angler for each species eaten and is the same each time that species is consumed.)										Love and Hansen (1999)— Private boat survey data corrected for avidity bias based on self-reported frequency information and adjusted for location sampling bias
Chub mackerel	Discrete						Soup/stew (6.1%); smoke/ ceviche (1.8%); bake/boil/ steam (21.6%); broil/BBQ (2.8%); fry (67.7%)		
Pacific barracuda	Discrete						Soup/stew (3%); smoke/ ceviche (15%); bake/boil/ steam (14%); broil/BBQ (59%); fry (9%)		

Kelp bass	Discrete	Soup/stew (0.3%); bake/boil/steam (12%); broil/BBQ (47.7%); fry (40%)
Barred sand bass	Discrete	Soup/stew (4%); smoke/ceviche (3%); bake/boil/steam (6%); broil/BBQ (48%); fry (39%)
White croaker	Discrete	Smoke/ceviche (4%); bake/boil/steam (31%); broil/BBQ (19%); fry (46%)
Queenfish	Discrete	Smoke/ceviche (33%); bake/boil/steam (33%); fry (33%)
Halfmoon	Discrete	Broil/BBQ (4%); fry (96%)
California halibut	Discrete	Smoke/ceviche (1%); bake/boil/steam (8%); broil/BBQ (72%); fry (19%)
Rockfish	Discrete	Bake/boil/steam (9%); broil/BBQ (34%); fry (57%)
California scorpionfish	Discrete	Bake/boil/steam (26%); broil/BBQ (18%); fry (56%)
Surfperch	Discrete	Broil/BBQ (20%); fry (80%)
Opaleye	Discrete	Broil/BBQ (50%); fry (50%)
Pacific bonito	Discrete	Soup/stew (3.7%); smoke/ceviche (9.1%); bake/boil/steam (17.4%); broil/BBQ (60.4%); fry (9.4%)

Seasons fished[c] (unitless) (This distribution assigns the probability of an angler within a specified age group of fishing in a season or combination of seasons.)

Love and Hansen (1999)—Private boat survey data corrected for avidity bias based on self-reported frequency information and adjusted for location sampling bias

(Continued)

TABLE 17.4 (*Continued*)

Exposure Factor (Description)	Distribution Type	Parameter Values						Source
		Min	Mode	Mean	Max	Standard Deviation	Value (Frequency)	
5–18 years old	Discrete						Sp (27%); SpSu (16%); SpSuWi (11%); SpSuFaWi (46%)	Love and Hansen (1999)— Private boat survey data corrected for avidity bias based on self-reported frequency information and adjusted for location sampling bias
19–64 years old	Discrete						Sp (1.5%); Su (26%); Fa (7.3%); SpSu (12%); SpFa (1.3%); SpWi (0.8%); SuFa (2.4%); SuWi (0.7%); FaWi (2%); SpSuFa (3.3%); SpSuWi (3.2%); SpFaWi (1.3%); SuFaWi (5.1%); SpSuFaWi (33.1%)	
65 years and older	Discrete						SpSu (58%); SpSuFa (2.2%); SuFaWi (4.8%); SpSuFaWi (35%)	
Number of fishing trips per season (unitless) (The number of trips taken by an angler during each season is assigned by this distribution.)								
Winter	Truncated log-normal	1		6.0	91	5.4		
Spring	Truncated log-normal	1		6.3	91	4.4		
Summer	Truncated log-normal	1		5.4	91	4.2		
Fall	Truncated log-normal	1		5.9	91	3.6		

Fishing at Palos Verdes Shelf (unitless) (Based on season, this distribution describes the probability of an angler going to the Palos Verdes Shelf.)				Love and Hansen (1999)— Private boat survey data corrected for avidity bias based on self-reported frequency information and adjusted for location sampling bias
	Spring	Discrete	Shelf (42%); no shelf (58%)	
	Summer	Discrete	Shelf (22%); no shelf (78%)	
	Fall	Discrete	Shelf (31%); no shelf (69%)	
	Winter	Discrete	Shelf (27%); no shelf (73%)	
Fishing a particular zone within the Palos Verdes Shelf (unitless) (Based on season, this distribution determines the zone on the Palos Verdes Shelf to which a private boat goes.)				Love and Hansen (1999)— Private boat PVS angler survey data corrected for avidity bias based on self-reported frequency information and adjusted for location sampling bias
	Winter	Discrete	Zone 1 (24%); zone 2 (36%); zone 3 (40%)	
	Spring	Discrete	Zone 1 (27%); zone 2 (35%); zone 3 (38%)	
	Summer	Discrete	Zone 1 (33%); zone 2 (40%); zone 3 (27%)	
	Fall	Discrete	Zone 1 (13%); zone 2 (20%); zone 3 (67%)	
Catching one or more fish (unitless) (This distribution describes the probability of an angler catching a fish of any species on a trip-by-trip basis.)				Love and Hansen (1999)— Private boat PVS angler survey data corrected for avidity bias based on self-reported frequency information and adjusted for location sampling bias

(Continued)

TABLE 17.4 (*Continued*)

Exposure Factor (Description)	Distribution Type	Parameter Values						Source
		Min	Mode	Mean	Max	Standard Deviation	Value (Frequency)	
Spring	Discrete						Catch (58%); no catch (42%)	Love and Hansen (1999)—
Summer	Discrete						Catch (77%); no catch (23%)	Private boat PVS angler
Fall	Discrete						Catch (88%); no catch (12%)	survey data corrected for
Winter	Discrete						Catch (76%); no catch (24%)	avidity bias based on
Catching a particular combination								self-reported frequency
of species (unitless) (This								information and adjusted for
distribution describes the								location sampling bias
probability of an angler								
catching a particular								
combination of fish species.)								
Total number of combinations = 41								
Chub mackerel	Discrete						Probability is included in catch = 0.043	
Pacific barracuda	Discrete						Probability is included in catch = 0.193	
Kelp bass	Discrete						Probability is included in catch = 0.251	
Barred sand bass	Discrete						Probability is included in catch = 0.162	
White croaker	Discrete						Probability is included in catch = 0.026	
Queenfish	Discrete						Probability is included in catch = 0.004	
Halfmoon	Discrete						Probability is included in catch = 0.002	

Species	Distribution	Parameters / Notes			Source
California halibut	Discrete	Probability is included in catch = 0.107			
Rockfish	Discrete	Probability is included in catch = 0.039			
California scorpionfish	Discrete	Probability is included in catch = 0.077			
Surfperch	Discrete	Probability is included in catch = 0.005			
Opaleye	Discrete	Probability is included in catch = 0.004			
Pacific bonito	Discrete	Probability is included in catch = 0.087			Love and Hansen (1999)— Private boat survey data corrected for avidity bias based on self-reported frequency information and adjusted for location sampling bias

Consumption rate per trip for a particular species (kg/trip) (The mass of a particular fish species consumed on a particular fishing trip is selected on a trip-by-trip basis.)

Species	Distribution				
Chub mackerel	Empirical	0.001	1.2	4.9	1.2
Pacific barracuda	Empirical	0.001	1.5	13.5	2.4
Kelp bass	Empirical	0.001	0.60	5.6	0.66
Barred sand bass	Empirical	0.001	0.62	11	1.6
White croaker	Empirical	0.001	0.20	0.8	0.20
Queenfish	Empirical	0.001	0.11	0.11	0
Halfmoon	Empirical	0.001	0.55	0.075	0.0066
California halibut	Empirical	0.001	1.1	3.1	0.67
Rockfish	Empirical	0.001	0.82	5.0	1.4
California scorpionfish	Empirical	0.001	0.18	0.83	0.20
Surfperch	Empirical	0.001	0.61	2.2	0.54
Opaleye	Empirical	0.001	0.15	0.59	0.15
Pacific bonito	Empirical	0.001	2.9	33	3.3

(Continued)

TABLE 17.4 (*Continued*)

tDDT concentration in fish, Palos Verdes Shelf (ppb) (tDDT concentrations in fish caught in various zones of the Palos Verdes Shelf are selected on a trip by trip basis. The distribution represents the variation of tDDT concentrations across individual fish.)

Exposure Factor (Description)	Distribution Type	Parameter Values						Source
		Min	Mode	Mean	Max	Standard Deviation	Value (Frequency)	
Chub mackerel—zone 1	Triangular	5	19	31	70			Pollock et al. (1991)
Chub mackerel—zone 2	Triangular	6	26	38	82			Pollock et al. (1991)
Chub mackerel—zone 3	Triangular	4	14	13	22			Pollock et al. (1991)
Pacific barracuda—all zones	Triangular	9	22	27	51			Pollock et al. (1991)
Kelp bass—zone 1	Truncated log-normal	0		220	1,600	220		LASCD (1997)
Kelp bass—zone 2	Truncated log-normal	0		210	1,600	220		LASCD (1997)
Kelp bass—zone 3	Truncated log-normal	0		150	980	140		LASCD (1997)
Barred sand bass—all zones	Triangular	30	55	56	82			Pollock et al. (1991)
White croaker—zone 1	Truncated log-normal	0		5,900	23,000	5,700		Love and Hansen (1999)
White croaker—zone 2	Truncated log-normal	0		6,000	23,000	2,900		LASCD (1997)
White croaker—zone 3	Truncated log-normal	0		3,300	15,000	1,900		LASCD (1997)
Queenfish—zone 1	Triangular	62	94	99	141			Pollock et al. (1991)
Queenfish—zone 2	Triangular	21	40	44	70			Pollock et al. (1991)

(Continued)

Queenfish—zone 3	Triangular	31	54	60	95		Pollock et al. (1991)
Halfmoon—all zones	N/A					Concentration = 0	Pollock et al. (1991)
California halibut—all zones	Triangular	2	8	11	23		Pollock et al. (1991)
Rockfish—zone 1	Triangular	26	72	160	383		Pollock et al. (1991)
Rockfish—zone 2	Triangular	37	79	78	118		Pollock et al. (1991)
Rockfish—zone 3	Triangular	40	67	88	158		Pollock et al. (1991)
California scorpionfish—zone 1	Triangular	53	154	190	358		Pollock et al. (1991)
California scorpionfish—zone 2	Triangular	15	41	50	93		Pollock et al. (1991)
California scorpionfish—zone 3	Triangular	16	33	34	54		Pollock et al. (1991)
Surfperch—zone 1	Triangular	17	29	31	48		Pollock et al. (1991)
Surfperch—zone 2	Triangular	23	45	53	92		Pollock et al. (1991)
Surfperch—zone 3	Triangular	22	70	76	137		Pollock et al. (1991)
Opaleye—all zones	N/A					Concentration = 0	Pollock et al. (1991)
Pacific bonito—zone 1	Triangular	7	31	62	147		Pollock et al. (1991)
Pacific bonito—zone 2	Triangular	18	32	38	65		Pollock et al. (1991)
Pacific bonito—zone 3	Triangular	13	18	19	25		Pollock et al. (1991)

PCB concentration in fish, Palos Verdes Shelf (ppb) (PCB concentrations in fish caught in various zones of the Palos Verdes Shelf are selected on a trip by trip basis. The distribution represents the variation of PCB concentrations across individual fish.)

Chub mackerel—zone 1	Triangular	0	9	20	51		Pollock et al. (1991)
Chub mackerel—zone 2	Triangular	3	13	24	57		Pollock et al. (1991)
Chub mackerel—zone 3	Triangular	2	6	5.7	9		Pollock et al. (1991)
Pacific barracuda—all zones	Triangular	10	16	17	24		Pollock et al. (1991)
Kelp bass—zone 1	Truncated log-normal	0		45	240	29	LACSD (1997)

TABLE 17.4 (*Continued*)

Exposure Factor (Description)	Distribution Type	Parameter Values						Source
		Min	Mode	Mean	Max	Standard Deviation	Value (Frequency)	
Kelp bass—zone 2	Truncated log-normal	0		50	260	35		LACSD (1997)
Kelp bass—zone 3	Truncated log-normal	0		48	260	34		LACSD (1997)
Barred sand bass—all zones	Triangular	20	33	31	41			Pollock et al. (1991)
White croaker—zone 1	Truncated log-normal	0		600	2,400	500		Love and Hansen (1999)
White croaker—zone 2	Truncated log-normal	0		630	2,200	260		LACSD (1997)
White croaker—zone 3	Truncated log-normal	0		420	1,700	220		LACSD (1997)
Queenfish—zone 1	Triangular	15	24	27	41			Pollock et al. (1991)
Queenfish—zone 2	Triangular	10	19	19	27			Pollock et al. (1991)
Queenfish—zone 3	Triangular	8	14	14	19			Pollock et al. (1991)
Halfmoon—all zones	N/A						Concentration = 0	Pollock et al. (1991)
California halibut—all zones	Triangular	0	2	6	16			Pollock et al. (1991)
Rockfish—zone 1	Triangular	0	8	28	76			Pollock et al. (1991)
Rockfish—zone 2	Triangular	7	17	15	22			Pollock et al. (1991)
Rockfish—zone 3	Triangular	7	12	14	22			Pollock et al. (1991)
California scorpionfish—zone 1	Triangular	28	41	40	52			Pollock et al. (1991)
California scorpionfish—zone 2	Triangular	0	2	2.3	5			Pollock et al. (1991)
California scorpionfish—zone 3	Triangular	5	10	10	16			Pollock et al. (1991)
Surfperch—zone 1	Triangular	14	29	30	47			Pollock et al. (1991)
Surfperch—zone 2	Triangular	5	17	21	41			Pollock et al. (1991)
Surfperch—zone 3	Triangular	5	16	16	26			Pollock et al. (1991)
Opaleye—all zones	Triangular	0	1	1.3	3			Pollock et al. (1991)
Pacific bonito—zone 1	Triangular	2	14	42	112			Pollock et al. (1991)
Pacific bonito—zone 2	Triangular	7	13	16	29			Pollock et al. (1991)
Pacific bonito—zone 3	Triangular	4	8	7.7	11			Pollock et al. (1991)

tDDT cooking loss (unitless) (The tDDT cooking loss distribution describes the percent reduction of tDDT in fish resulting from each cooking method. No loss is associated with making soup or eating fish raw.)

Bake	Empirical	0	0.26	1	0.25	Zabik et al. (1979, 1995a, b, 1996); Smith et al. (1973); Trotter et al. (1988); Skea et al. (1981); Reinert et al. (1972); and Puffer and Gossett (1983), as compiled and analyzed by Wilson et al. (1998)
Boil	Empirical	0	0.34	1	0.27	
Broil	Empirical	0	0.39	1	0.25	
Fry	Empirical	0	0.41	1	0.23	
Smoke	Empirical	0	0.49	1	0.23	

PCB cooking (unitless) (The PCB cooking loss distribution describes the percent reduction of PCBs in fish resulting from each cooking method. No loss is associated with making soup or eating fish raw.)

Bake	Empirical	0	0.33	1	0.25	Zabik et al. (1979, 1995a, b, 1996); Smith et al. (1973); Trotter et al. (1988); Skea et al. (1981); and Puffer and Gossett (1983), as compiled and analyzed by Wilson et al. (1998)
Boil	Empirical	0	0.52	1	0.30	
Broil	Empirical	0	0.44	1	0.23	
Fry	Empirical	0	0.37	1	0.24	
Smoke	Empirical	0	0.40	1	0.23	

[a] Statistics listed are for a 30-year old male.

[b] Private boat PVS anglers are only those anglers who fished at the Palos Verdes Shelf.

[c] Seasons: Sp-spring, Su-summer, Fa-fall, Wi-winter.

TABLE 17.5 Distribution Functions for Various Exposure Factors Used in the Cabrillo Pier Analysis

Exposure Factor (Description)	Distribution Type	Parameter Values						Source
		Min	Mode	Mean	Max	Standard Deviation	Value (Frequency)	
Angler gender (unitless) (Each angler's gender is selected based on the gender distribution.)	Discrete						Male (91%); female (9%)	SCCWRP and MBC (1994)—Cabrillo Pier survey data corrected for avidity bias based on self-reported frequency infor-mation and adjusted for location sampling bias; USEPA (1997b)
Body weight[a] (kg) (Annually, an age- and gender-specific body weight is selected for each angler based on a series of age-specific cumulative body weight distribution functions.)	Empirical	40		78.7	150	13.7		USEPA (1997b)
Age at which fishing begins (years) (The probality of an angler starting to fish at a particular age is described by this distribution.)	Truncated normal	5		33	80	13		SCCWRP and MBC (1994)—Cabrillo Pier survey data corrected for avidity bias based on self-reported frequency information
Length of fishing career (*exposure duration*) (years) (The exposure duration distribution describes the probability of an angler fishing for a specific number of years. It is a function of histo-rical years fished, estimates of future years fished, and age at which fishing begins. Duration is limited by the age at which fishing begins; no anglers fish past 90 years of age.)	Exponential	1		4.1	85			Function of historical years fished and estimates of future years fished. Historical data from SCCWRP and MBC (1994)—Cabrillo Pier survey data corrected for avidity bias based on self-reported frequency information. Future years fished estimated using methods of Price et al. (1998).

Parameter	Type	Values	Source
Consuming a particular species given that it is caught and kept (unitless) (This distribution describes the probability of an angler consuming a particular fish species given that he or she catches the fish. Then distributions are used to establish, for each angler's fishing career, which species are eaten if caught.)	Discrete	Chub mackerel (55%); Pacific barracuda (69%); kelp bass (70%); barred sand bass (80%); white croaker (79%); queenfish (26%); halfmoon (95%); California halibut (38%); rockfish (61%); California scorpionfish (76%); surfperch (78%); opaleye (90%); Pacific bonito (66%)	SCCWRP and MBC (1994)—all angler survey data corrected for avidity bias based on self-reported frequency information and adjusted for location sampling bias
Cooking method (unitless) (A cooking method used by the anglers to prepare fish is assigned to an angler for each species eaten and is the same each time that species is consumed.)			SCCWRP and MBC (1994)—all angler survey data corrected for avidity bias based on self-reported frequency information and adjusted for location sampling bias
Chub mackerel	Discrete	Soup (4%); raw/smoke/ceviche (9%); bake/boil/steam (9%); broil/BBQ (17%); fry (61%)	
Pacific barracuda	Discrete	Soup (0.1%); raw/smoke/ceviche (3%); bake/boil/steam (14.7%); broil/BBQ (30.1%); fry (52.1%)	
Kelp bass	Discrete	Soup (3%); raw/smoke/ceviche (5%); bake/boil/steam (16%); broil/BBQ (36%); fry (40%)	
Barred sand bass	Discrete	Soup (3%); raw/smoke/ceviche (3%); bake/boil/steam (14%); broil/BBQ (34%); fry (46%)	

(Continued)

TABLE 17.5 (*Continued*)

Exposure Factor (Description)	Distribution Type	Parameter Values						Value (Frequency)	Source
		Min	Mode	Mean	Max	Standard Deviation			
White croaker	Discrete							Soup (2%); raw/smoke/ ceviche (2%); bake/boil/ steam (13%); broil/BBQ (16%); fry (67%)	
Queenfish	Discrete							Soup (2%); raw/smoke/ ceviche (4%); bake/boil/ steam (7%); broil/BBQ (21%); fry (66%)	
Halfmoon	Discrete							Bake/boil/steam (39%); fry (61%)	
California halibut	Discrete							Soup (2%); raw/smoke/ ceviche (3%); bake/boil/ steam (18%); broil/BBQ (33%); fry (44%)	
Rockfish	Discrete							Soup (3%); raw/smoke/ ceviche (5%); bake/boil/ steam (15%); broil/BBQ (33%); fry (44%)	
California scorpionfish	Discrete							Soup (1%); raw/smoke/ ceviche (9%); bake/boil/ steam (16%); broil/BBQ (39%); fry (35%)	
Surfperch	Discrete							Soup (0.04%); bake/boil/ steam (6%); broil/BBQ (26%); fry (68%)	
Opaleye	Discrete							Soup (2%); raw/smoke/ ceviche (38%); bake/boil/ steam (20%); broil/BBQ (2%); fry (38%)	

	Distribution						Source
Pacific bonito	Discrete					Soup (1%); raw/smoke/ceviche (15%); bake/boil/steam (15%); broil/BBQ (30%); fry (39%)	SCCWRP and MBC (1994)—Cabrillo Pier survey data corrected for avidity bias based on self-reported frequency information
Seasons fished[b] (unitless) (This distribution assigns the probability of an angler within a specified age group of fishing in a season or combination of seasons.)							
5–18 years old	Discrete					Su (8%); Wi (21%); SpSu (43%); SpSuFa (9%); SpSuFaWi (19%)	
19–64 years old	Discrete					Sp (13%); Su (46%); Fa (7%); Wi (3%); SpSu (10%); SpWi (1%); SuFa (3%); SpSuFa (1%); SpSuWi (1%); SpSuFaWi (15%)	
65 years and older	Discrete					Su (19%); Wi (28%); SpSu (6%); SuFa (3%); SpSuFa (6%); SpSuFaWi (38%)	
Number of fishing trips per season (unitless) (The number of trips taken by an angler during each season is described by this distribution.)							SCCWRP and MBC (1994)—Cabrillo Pier survey data corrected for avidity bias based on self-reported frequency information
Winter	Truncated log-normal	1	5.7	91	4.6		
Spring	Truncated log-normal	1	4.4	91	4.3		
Summer	Truncated log-normal	1	5.1	91	5.1		
Fall	Truncated log-normal	1	5.1	91	4.9		

(Continued)

TABLE 17.5 (*Continued*)

Exposure Factor (Description)	Distribution Type	Parameter Values					Value (Frequency)	Source
		Min	Mode	Mean	Max	Standard Deviation		
Catching one or more fish (unitless) (This distribution describes the probability of an angler catching a fish of any species on a trip-by-trip basis.)	Discrete						Catch (57%); no catch (43%)	SCCWRP and MBC (1994)—Cabrillo Pier survey data corrected for avidity bias based on self-reported frequency information
Catching a particular combination of species (unitless) (This distribution describes the probability of an angler catching a particular combination of fish species.)								SCCWRP and MBC (1994)—Cabrillo Pier survey data corrected for avidity bias based on self-reported frequency information
Total number of combinations = 23								
Chub mackerel	Discrete						Frequency = 10/23	
Pacific barracuda	Discrete						Frequency = 3/23	
Kelp bass	Discrete						Frequency = 1/23	
Barred sand bass	Discrete						Frequency = 1/23	
White croaker	Discrete						Frequency = 10/23	
Queenfish	Discrete						Frequency = 3/23	
Halfmoon	Discrete						Frequency = 0/23	
California halibut	Discrete						Frequency = 7/23	
Rockfish	Discrete						Frequency = 0/23	
California scorpionfish	Discrete						Frequency = 2/23	
Surfperch	Discrete						Frequency = 9/23	
Opaleye	Discrete						Frequency = 1/23	
Pacific bonito	Discrete						Frequency = 1/23	

SCCWRP and MBC (1994) —survey data on fish consumption rate corrected for avidity bias based on self-reported frequency information and adjusted for location sampling bias

Consumption rate per trip for a particular species (kg/trip)
(The mass of a particular fish species consumed on a particular fishing trip is selected on a trip-by-trip basis.)

Species	Distribution					Reference
Chub mackerel	Empirical	0.001	0.43	3.8	0.66	
Pacific barracuda	Empirical	0.001	0.75	5.3	0.77	
Kelp bass	Empirical	0.001	0.33	3.0	0.46	
Barred sand bass	Empirical	0.001	0.50	7.9	1.00	
White croaker	Empirical	0.001	0.10	1.7	0.22	
Queenfish	Empirical	0.001	0.010	0.020	0.0040	
Halfmoon	Empirical	0.001	0.20	0.64	0.13	
California halibut	Empirical	0.001	0.29	3.1	0.57	
Rockfish	Empirical	0.001	0.22	3.7	0.53	
California scorpionfish	Empirical	0.001	0.20	0.75	0.19	
Surfperch	Empirical	0.001	0.17	2.1	0.37	
Opaleye	Empirical	0.001	0.15	0.61	0.090	
Pacific bonito	Empirical	0.001	1.20	4.5	0.91	

tDDT fish concentration, Cabrillo Pier (ppb) (tDDT concentrations in fish caught at Cabrillo Pier are selected on a trip by trip basis. The distribution represents the variation of tDDT concentrations across individual fish.)

Species	Distribution					Reference
Chub mackerel	Triangular	13	29	32	55	Pollock et al. (1991)
Pacific barracuda	Triangular	9	22	27	51	Pollock et al. (1991)
Kelp bass	Triangular	16	45	68	142	Pollock et al. (1991)
Barred sand bass	Triangular	48	79	110	187	Pollock et al. (1991)
White croaker	Triangular	123	914	3,000	8,052	Pollock et al. (1991)

(Continued)

TABLE 17.5 (*Continued*)

| Exposure Factor (Description) | Distribution Type | Parameter Values | | | | | | Source |
		Min	Mode	Mean	Max	Standard Deviation	Value (Frequency)	
Queenfish	Triangular	83	206	230	392			Pollock et al. (1991)
Halfmoon	N/A						Concentration = 0	Pollock et al. (1991)
California halibut	Triangular	2	8	11	23			Pollock et al. (1991)
Rockfish	Triangular	24	42	58	106			Pollock et al. (1991)
California scorpionfish	Triangular	5	14	13	20			Pollock et al. (1991)
Surfperch	Triangular	42	123	150	277			Pollock et al. (1991)
Opaleye	N/A						Concentration = 0	Pollock et al. (1991)
Pacific bonito	Triangular	37	66	76	126			Pollock et al. (1991)
PCB fish concentration, Cabrillo Pier (ppb) (PCB concentrations in fish caught at Cabrillo Pier are selected on a trip by trip basis. The distribution represents the variation of PCB concentrations across individual fish.)								
Chub mackerel	Triangular	14	22	28	49			Pollock et al. (1991)
Pacific barracuda	Triangular	10	16	17	24			Pollock et al. (1991)
Kelp bass	Triangular	6	13	14	22			Pollock et al. (1991)
Barred sand bass	Triangular	13	55	62	117			Pollock et al. (1991)
White croaker	Triangular	13	135	240	589			Pollock et al. (1991)
Queenfish	Triangular	25	82	96	180			Pollock et al. (1991)
Halfmoon	N/A						Concentration = 0	Pollock et al. (1991)
California halibut	Triangular	0	2	6	16			Pollock et al. (1991)
Rockfish	Triangular	3	13	29	70			Pollock et al. (1991)
California scorpionfish	Triangular	0	2	2	4			Pollock et al. (1991)
Surfperch	Triangular	19	61	100	224			Pollock et al. (1991)
Opaleye	Triangular	0	1	1.3	3			Pollock et al. (1991)
Pacific bonito	Triangular	19	35	39	63			Pollock et al. (1991)

tDDT cooking loss (unitless) (The tDDT cooking loss distribution describes the percent reduction of tDDT in fish resulting from each cooking method. No loss is associated with making soup or eating fish raw.)

						Reference
Bake	Empirical	0	0.26	1	0.25	Zabik et al. (1979, 1995a, b, 1996); Smith et al. (1973); Trotter et al. (1988); Skea et al. (1981); Reinert et al. (1972); and Puffer and Gossett (1983), as compiled and analyzed by Wilson et al. (1998)
Boil	Empirical	0	0.34	1	0.27	
Broil	Empirical	0	0.39	1	0.25	
Fry	Empirical	0	0.41	1	0.23	
Smoke	Empirical	0	0.49	1	0.23	

PCB cooking loss (unitless) (The PCB cooking loss distribution describes the percent reduction of PCBs in fish resulting from each cooking method. No loss is associated with making soup or eating fish raw.)

						Reference
Bake	Empirical	0	0.33	1	0.25	Zabik et al. (1979, 1995a, b, 1996); Smith et al. (1973); Trotter et al. (1988); Skea et al. (1981); and Puffer and Gossett (1983), as compiled and analyzed by Wilson et al. (1998)
Boil	Empirical	0	0.52	1	0.30	
Broil	Empirical	0	0.44	1	0.23	
Fry	Empirical	0	0.37	1	0.24	
Smoke	Empirical	0	0.40	1	0.23	

[a] Statistics listed are for a 30-year old male.

[b] Seasons: Sp-spring, Su-summer, Fa-fall, Wi-winter.

We used EPA dose–response criteria for DDT, DDE, and PCBs to estimate carcinogenic risks and noncancer hazards from the distributions of dose rates. For tDDT, the EPA cancer slope factor of 0.34 $(mg/kg\text{-}day)^{-1}$ for DDT and DDE was used to assess cancer risks, and the reference dose of 5×10^{-4} mg/kg-day was used to assess noncancer hazards [Integrated Risk Information System (IRIS), 2002]. Consistent with EPA guidance (IRIS, 2002; EPA, 1996b), the cancer slope factor of 2 $(mg/kg\text{-}day)^{-1}$ was used for PCBs. This value, which is the highest current cancer slope factor for PCBs, was selected because we are estimating risks due to consumption of fish containing PCBs, and we lack information on the distributions of congeners or the average number of chlorines present in the PCBs. The approach we adopted will tend to overestimate, rather than underestimate, the plausible cancer risk. To assess noncancer hazards for PCBs, we selected the reference dose of 2×10^{-5} mg/kg-day developed for Aroclor 1254 (IRIS, 2002). Of the two reference doses available for PCB mixtures, the reference dose for Aroclor 1254 is more relevant for food chain exposures than the reference dose for Aroclor 1016 because PCB congeners that bioaccumulate in fish tend to resemble the more highly chlorinated mixtures such as Aroclor 1254.

17.2.6   Example Calculation

To demonstrate how the Monte Carlo analysis works, we will follow an example angler through his years of fishing from CPFVs on the Palos Verdes Shelf.

To begin, the model characterizes the angler by assigning the lifetime exposure factors, which are listed and described on Table 17.6. The example angler, "Joe", is male. Joe began fishing at the age of 18 and he will fish on party boats until he is 30 years old, which indicates an exposure duration of 12 years. Joe is of typical build, interpreted as a 50th percentile body weight throughout his life. He fishes more frequently than most anglers but is not in the most avid group, which is interpreted via Joe's being assigned the third frequency quartile (i.e., frequency between 50th and 75th percentiles). Accordingly, as Joe's fishing career proceeds, the number of trips he takes each season will be selected from the third quartile of the distribution for that season. A profile of species and cooking preferences is then developed and stored for Joe. Joe consumes all the fish that he catches except California scorpionfish and surfperches. He fries all of the fish that he eats, except for barred sand bass, which he broils.

The analysis then simulates Joe's first fishing year by determining the annual exposure factors (see Table 17.7). Joe will fish in the spring and the summer of the first year. These seasons are chosen from the distribution of combination of seasons fished that applies to 5- to 18-year-olds. The number of trips taken each season is determined based on Joe's avidity quartile. Restricting the trips taken in each season to the third quartile (i.e., between the 50th and 75th percentiles) of each season's distribution yields that four trips will be taken in the spring and five trips in the summer. Joe's body weight is defined to be 70.9 kg based

TABLE 17.6 Lifetime Exposure Factors for Monte Carlo Analysis

Factor	Description	Example Value
Gender	Used to describe the probability of the angler being male or female.	Male
Starting age for fishing	Describes the probability of the angler starting to fish at a particular age.	18
Body weight percentile	Used to select an age- and gender-specific body weight for the angler.	50th
Species preferences Consumed if caught	Describes the probability of an angler eating a particular fish species given that he or she catches the fish.	Eat all species except California scorpionfish and surfperches.
Cooking method if consumed	A cooking method profile is established for each angler for each species that the angler will potentially consume.	Fry all species except barred sand bass, which is broiled.
Fishing frequency quartile (avidity category)	At random, an angler is assigned a quartile for fishing frequency that serves as a representation of his or her avidity throughout his or her fishing career. This quartile is used as a guide for subsequent selection of the fishing frequency for each season fished.	Third quartile (i.e., 50th to 75th percentile frequency).
Duration of fishing career	This distribution describes the probability of an angler fishing for a specified number of years.	12 years

TABLE 17.7   Annual Exposure Factors for Monte Carlo Analysis

Factor	Description	Example Value
Age	Begins with the angler's start age for fishing and is updated each year.	First year, age 18; second year, age 19
Body weight	Based on the angler's gender, current age, and body weight percentile, a body weight is determined.	70.9 kg
Seasons fished	Describes the probability of the angler, within his or her specified age group, fishing in a season or combination of seasons.	Spring and summer
Trips per season	Used to assign the number of trips taken during each season selected.	Four trips in the spring and five trips in the summer

on the body weight distribution for males that are 18 years old and his 50th percentile body weight.

Next, the trip-specific exposure factors (see Table 17.8) are determined beginning with the initial departure location. From the five possible departure locations, Joe's departure location is selected as the Los Angeles/Long Beach Harbor port complex. Based on this departure location, Joe has a 15 percent chance of fishing on the Palos Verdes Shelf and an 85 percent chance of fishing elsewhere. Joe will fish on the Palos Verdes Shelf for this trip and is also determined to fish in zone 1 based on the probability of fishing in each zone of the shelf during the spring season. The odds favor that an angler will catch a fish on spring and summer trips and Joe accomplishes this. From the 66 possible species combinations for CPFV anglers, it is selected that on this trip Joe catches a kelp bass and no other fish. Based on Joe's species and cooking preference profile, he will fry and eat kelp bass. Joe's first fishing trip is displayed in Figure 17.4.

Once Joe has caught and consumed fish, the PCB and tDDT dose is calculated for this trip. The selection from the species-specific distribution indicates that the kelp bass will have an edible mass of 330 grams; therefore, the angler's consumption rate will be 0.33 kg for this trip. The PCB concentration in the kelp bass will be 0.043 mg/kg and the tDDT concentration will be 0.018 mg/kg based on the selections from the distributions for PCBs and tDDT concentrations for kelp bass in zone 1. Because Joe fries the kelp bass, a 33 percent reduction in PCB concentration and a 50 percent reduction in tDDT concentration is selected to result from the cooking process. The dose per trip for each chemical is determined by taking the product of the chemical concentration, the consumption rate, and one minus the cooking loss, then dividing by the body weight. The calculation for the tDDT per trip dose is demonstrated below:

TABLE 17.8 Trip-Specific Exposure Factors for Monte Carlo Analysis

Factor	Description	Example Value
Destination	For CPFV anglers, a distribution is used to describe the probability of an angler departing from a specific location on his or her first trip and another distribution describes the probability of the departure location for each subsequent trip based on the departure location of the first trip. For each trip, a distribution is used to describe the probability of a CPFV angler going to the Palos Verdes Shelf based on the angler's departure location. If an angler goes to the Palos Verdes Shelf, a distribution determines the zone in which the angler fishes. For Cabrillo Pier anglers, the only possible fishing location is the Pier.	Initial departure location is Los Angeles/Long Beach Harbor and fishes on the Palos Verdes Shelf in zone 1.
Catch one or more fish of any species	Describes the probability of the angler catching a fish of any species.	Catch one or more fish.
Species caught (and possible combinations)	Describes the probability of an angler catching a fish or a combination of fish species.	Catch kelp bass only.
Concentrations of chemicals in fish by species	Represents the variation of concentration across individual fish fillets for PCBs and tDDT.	PCB concentration in the kelp bass is 0.043 mg/kg and the tDDT concentration is 0.18 mg/kg.
Consumption rates by species	Used to select the mass of the particular fish species consumed.	Kelp bass has an edible mass of 330 grams; therefore, angler's consumption rate is 0.33 kg for this trip.
Reductions in chemical concentrations due to cooking by species	Chemical cooking loss distribution describes the percent reduction of each chemical resulting from the angler's preferred cooking method.	Frying kelp bass results in a 33% reduction in PCB concentration and a 50 percent reduction in tDDT concentration.

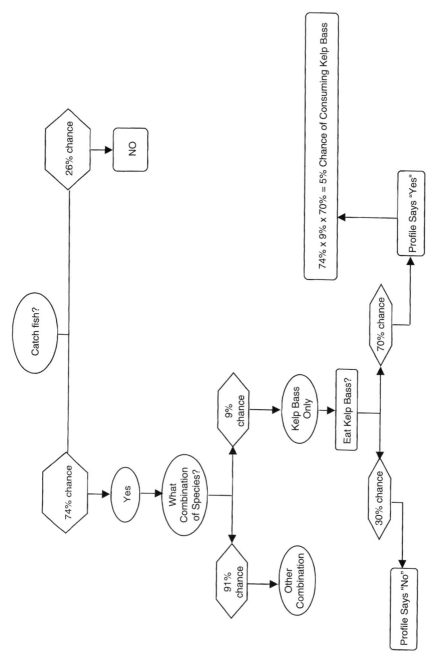

Figure 17.4 Example fishing trip to Palos Verdes Shelf, zone 1.

tDDT per trip dose

$$= \frac{\text{Chemical concentration} \times (1 - \text{fraction lost during cooking}) \times \text{Consumption rate}}{\text{Body weight}}$$

$$0.00042 \text{ mg/kg} = \frac{0.18 \text{ mg/kg} \times (1 - 0.50) \times 0.33 \text{ kg/trip}}{70.9 \text{ kg}}$$

Once the analysis calculates and stores the dose from the first trip, it continues to simulate the remaining trips in the season. The subsequent departure location is chosen with a large probability that the angler will return to the initial departure location. However, the example angler is chosen to depart from Redondo. From this location, a 23 percent chance exists that the CPFV will travel to the Palos Verdes Shelf and a 77 percent chance that the CPFV will not. The CPFV does not travel to the Palos Verdes Shelf; therefore, no dose from Palos Verdes Shelf fish is accrued. The analysis continues to simulate the remaining trips of the spring and then the summer trips. The trip-specific exposure factors are redefined for each trip based on the applicable distributions. Joe consumes fish on only one other trip to the Palos Verdes Shelf during his first fishing year. Once the first year is completed, the trip doses are summed to yield the annual dose for the first year.

Next, the annual exposure factors are determined again, and then the second year is simulated. In Joe's second year, he will fish only in the summer, when he will take six CPFV trips. His age is now 19 years and he weighs 70.9 kg corresponding to his 50th percentile body weight. The analysis continues through the second year and then proceeds through Joe's remaining 10 fishing years. At the end of the exposure period (i.e., after Joe's twelfth year of fishing), the 12 annual doses are summed for PCB and tDDT and the results are lifetime doses for each chemical. From the lifetime dose, the lifetime average daily dose (LADD) and the average daily dose (ADD) are calculated. In Joe's case, his lifetime dose for tDDT is 0.011 mg/kg.

An example calculation for the tDDT LADD follows:

$$\text{tDDT LADD} = \text{Lifetime dose} \times (1/\text{Standard lifetime})$$

$$4.3 \times 10^{-7} \text{ mg/kg-day} = 0.011 \text{ mg/kg} \times (1/25{,}550 \text{ days})$$

Note: 25,550 days = 70 years.

An example calculation for the tDDT ADD follows:

$$\text{tDDT ADD} = \text{Lifetime dose} \times (1/\text{Exposure duration})$$

$$2.5 \times 10^{-6} \text{ mg/kg-day} = 0.011 \text{ mg/kg} \times (1/4380 \text{ days})$$

Note: 4380 days = 12 years.

Using the LADD and the ADD, the upper-bound cancer risk and noncancer hazard quotient can be calculated. An example calculation of the cancer risk due to tDDT follows:

$$\text{Risk due to tDDT} = \text{LADD} \times \text{Cancer slope factor for tDDT}$$

$$1.5 \times 10^{-7} = (4.3 \times 10^{-7} \text{ mg/kg-day}) \times [0.34 \text{ (mg/kg-day)}^{-1}]$$

An example calculation for the tDDT hazard quotient follows:

$$\text{tDDT hazard quotient} = \text{ADD/Noncancer reference dose for tDDT}$$

$$5.0 \times 10^{-3} = 2.5 \times 10^{-6} \text{ mg/kg-day}/5.0 \times 10^{-4} \text{ mg/kg-day}$$

Risks due to tDDT are summed with risks due to PCBs to yield a total upper-bound cancer risk estimate. Hazard quotients are reported separately for tDDT and PCBs.

This example demonstrates the fishing career of only one angler. The analysis actually continues through the fishing careers of other hypothetical CPFV anglers. Separate analyses proceed for hypothetical private boat anglers and Cabrillo Pier anglers. The resulting distributions for risks and hazard quotients are sorted and summarized in the form of percentiles.

17.3   RESULTS

We found that numerical stability of doses and risk estimates was achieved by analyzing 3500 CPFV anglers, 5000 private boat anglers, and 2500 Cabrillo Pier anglers who harvested and consumed at least one fish over their fishing careers from the subject locations. Analysis run times were approximately 20 minutes on a 300-MHz Pentium-based computer. Risk results were rounded to one significant figure in recognition of the limited precision of the underlying data, particularly the EPA's dose–response criteria. The numerical stability of our results was confirmed by approximate 95 percent confidence intervals (based on multiple analysis runs) that were within 10 percent of the values we report.

While we present full distributions of our cancer risk and noncancer hazard results, this discussion focuses on particular statistics of these distributions: the median (i.e., 50th percentile), mean (i.e., average), and 95th percentile. The median is a descriptor of the risk to the central tendency angler. Half of anglers have risks above, and half have risks below, the median value. The mean is presented as a second risk descriptor for the central tendency angler. In this assessment the distribution of risks is right-skewed. That is, most values are clumped at the low end of the distribution with a long tail of relatively few higher values. Under the influence of the relatively few high values, the mean risk typically exceeds the median risk. The mean risks we report fall between the 70th and 80th percentiles of the risk distribution. Hence, the mean risk is typically exceeded by only 20 to 30 percent of the anglers. The 95th percentile risk is the risk exceeded by no more than 5 percent of anglers. We have chosen the 95th percentile risk as a representative risk descriptor for the "high-end" segment of the angler population. According to EPA (1995), "high end descriptors are intended to estimate the exposures that are expected to occur in small, but definable, 'high end' segments of the subject population." Conceptually, according to EPA (1995), "high end exposure means exposure above about the 90th percentile of the population distribution, but not higher than the individual in the population who has the highest exposure." The 95th percentile risk is comparable to the reasonable maximum exposure (RME) or high-end risk in a point estimate risk assessment.

17.3.1   Cancer Risks and Noncancer Hazards for CPFV Anglers

The Monte Carlo analysis predicted that nearly two-thirds (67 percent) of anglers who fish one or more times from CPFVs leaving from Santa Monica and Los Angeles/Long Beach Harbor ports would visit the Palos Verdes Shelf at least once over their fishing careers. Of these, our analysis showed that 91 percent caught and 84 percent consumed at least one Palos Verdes Shelf fish. We also determined that these anglers' fishing careers averaged 9 years in duration and the 95th percentile duration was 25 years. On average, these anglers caught fish on 10 CPFV trips that visited the Palos Verdes Shelf, and fish from 8 of these trips were consumed. Another result of our analysis is that the 95th percentile number of CPFV trips to the Palos Verdes Shelf on which fish were caught was 39, of which 31 yielded fish that were consumed. On average, CPFV anglers catch and consume fish of 4 different species over all of their fishing trips to the Palos Verdes Shelf.

Figure 17.5 presents the distributions of lifetime carcinogenic risk associated with consumption of all species of fish containing tDDT and PCBs for the 57 percent of CPFV anglers who were predicted to consume one or more fish from the Palos Verdes Shelf over their fishing careers. Three distributions are presented: the risks associated with tDDT intake, risks associated with PCB intake, and total risks associated with the intake of both chemicals. The means and 95th percentiles of each distribution are also identified on Figure 17.5. As presented in Table 17.9, the total risk to the median angler was calculated to be 0.05 excess cancers per one million anglers (i.e., 5×10^{-8}). The mean total incremental risk was 2×10^{-7}, and 95 percent of anglers had total risks of 8×10^{-7} or below. These total risks are all below EPA's typical target risk range of 1×10^{-6} to 1×10^{-4} (EPA, 1991). The entire range of risks estimated in this Monte Carlo analysis is less than EPA's most recent estimate (SAIC, 1999) of the high-end or RME risk for the site for white croaker consumers (see Figure 17.5).

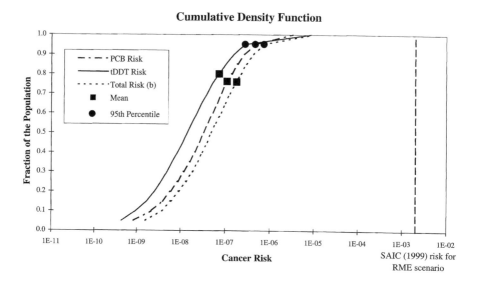

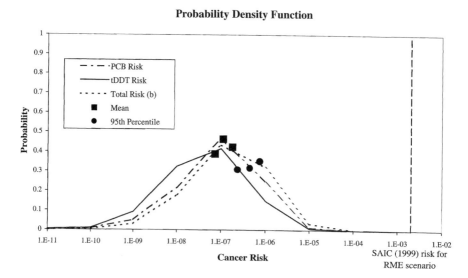

a. These distributions apply only to the 57% of southern California CPFV anglers who consumed fish from the Palos Verdes Shelf; all other CPFV anglers would have no incremental risk.

b. The total risk is the sum of the risks due to ingesting PCBs and tDDT in fish from the Palos Verdes Shelf.

Figure 17.5   Distributions of incremental cancer risks for anglers on commercial passenger fishing vessels (CPFVs) who consume fish containing tDDT and PCBs from the Palos Verdes Shelf.[a]

TABLE 17.9   Results of the Analysis for Commercial Passenger Fishing Vessel (CPFV) Anglers at the Palos Verdes Shelf[a]

	Median	Mean (Average)	95th Percentile
Cancer risks[b]			
Risk due to intake of tDDT	2×10^{-8}	7×10^{-8}	3×10^{-7}
Risk due to intake of PCBs	3×10^{-8}	1×10^{-7}	5×10^{-7}
Total risk (tDDT and PCBs)	5×10^{-8}	2×10^{-7}	8×10^{-7}
Noncancer hazards[c]			
Hazard quotient due to intake of tDDT	0.0006	0.003	0.01
Hazard quotient due to intake of PCBs	0.006	0.02	0.07

[a]Results represent those CPFV anglers who consume fish during their lifetime from the Palos Verdes Shelf (57% of CPFV anglers).

[b]Upper bound of the excess cancer risk.

[c]A hazard quotient less than 1 indicates that no noncancer health effects are expected.

On average, the theoretical cancer risks associated with intake of PCBs in fish exceeded those associated with the intake of tDDT. In fact, for most anglers, PCB risks were more than double the tDDT risks.

Consumption of kelp bass represented the largest average contribution (43 percent) to an angler's total risk due to fish consumption, whereas barred sand bass was consumed by the largest fraction of Palos Verdes Shelf CPFV anglers (57 percent). The mean fish consumption rate for CPFV anglers at the Palos Verdes Shelf was 1.9 g/day, and the 95th percentile consumption rate was 7.5 g/day. These results are very similar to EPA's (1997c) recommended default mean consumption rate for southern California marine anglers of 2.0 g/day and recommended default 95th percentile rate of 5.5 g/day. On a single-species basis, the mean consumption rates for Pacific bonito was highest at 1.1 g/day. The mean consumption rate for white croaker (0.05 g/day) ranked tenth among the 13 species studied.

Figure 17.6 presents the results of the noncancer hazard assessment for the CPFV anglers. Noncancer hazards are expressed in terms of hazard quotients for each chemical. More than 99 percent of the anglers consuming one or more fish from the Palos Verdes Shelf had hazard quotients less than 1. This finding indicates that the intake of tDDT and PCBs in Palos Verdes Shelf fish is highly unlikely to cause noncancer effects among CPFV anglers.

We also considered separately the cancer risks and noncancer hazards to the 9 percent of CPFV anglers whom the Monte Carlo analyses predict would consume one or more white croaker from the Palos Verdes Shelf during their fishing careers so that we could compare our estimates to those prepared by EPA (SAIC, 1999). Our analysis predicted that the median total cancer risk for consumption of white croaker was 8×10^{-8}, the mean risk was 3×10^{-7}, and

Cumulative Density Function

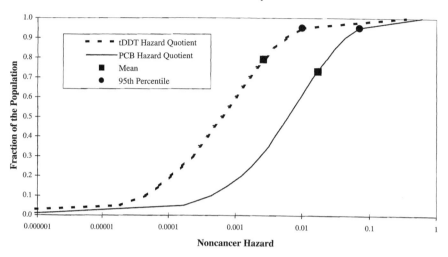

Probability Density Function

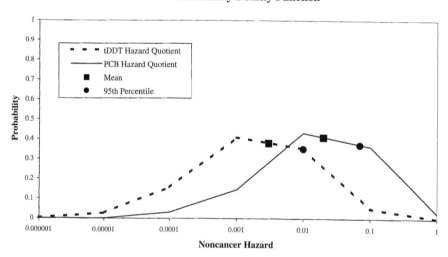

a. These distributions apply only to the 57% of southern California CPFV anglers who consumed fish from the Palos Verdes Shelf; all other CPFV anglers would have no incremental hazard.

Figure 17.6   Distribution of the hazard quotients for noncancer effects for anglers on commercial passenger fishing vessels (CPFVs) who consume fish containing tDDT and PCBs from the Palos Verdes Shelf.[a]

the 95th percentile risk was 1×10^{-6}. On average, for consumption of white croaker only, approximately one-third of the risk was due to intake of PCBs, and two-thirds was due to intake of tDDT in these fish. The mean contribution of white croaker consumption to total cancer risk for consumption of all species was 37 percent. More than 99 percent of the white croaker consumers had noncancer hazard quotients less than 1.

17.3.2 Cancer Risks and Noncancer Hazards for Private Boat Anglers

The Monte Carlo analysis predicted that 96 percent of anglers who fish one or more times from private boats leaving from Redondo Beach, Cabrillo Beach, or Los Alamitos Harbor launch sites would visit the Palos Verdes Shelf at least once over their fishing careers. Of these, our analysis predicted that 98 percent caught and 95 percent consumed at least one Palos Verdes Shelf fish. We also estimated that these anglers' fishing careers averaged 9 years in duration and the 95th percentile duration was 35 years. On average, we estimated that these anglers caught fish on 29 private boat trips that visited the Palos Verdes Shelf and consumed fish from 23 of these trips. Another result of our analysis was that the 95th percentile number of private boat trips to the Palos Verdes Shelf on which fish were caught was 125, of which 97 yielded fish that were consumed. On average, we predicted that private boat anglers caught and consumed fish of 4 different species over all of their fishing trips to the Palos Verdes Shelf.

Figure 17.7 presents the distributions of lifetime carcinogenic risk associated with consumption of all species of fish containing tDDT and PCBs for the 91 percent of private boat anglers who were predicted to consume one or more fish from the Palos Verdes Shelf over their fishing careers. Three distributions are presented: the risks associated with tDDT intake, risks associated with PCB intake, and total risks associated with the intake of both chemicals. The means and 95th percentiles of each distribution are also identified on Figure 17.7. As presented in Table 17.10, for the median private boat angler, we calculated a total incremental cancer risk of 0.3 per one million anglers (i.e., 3×10^{-7}). The mean total incremental risk was 8×10^{-7}, and 95 percent of anglers had total risks of 3×10^{-6} or below. The 95th percentile risk falls within EPA's typical target risk range (EPA, 1991). In fact, more than 99.9 percent of the estimated cancer risks are below or within EPA's target risk range. The entire range of risks estimated in this Monte Carlo analysis is less than EPA's most recent estimate (SAIC, 1999) of the high-end or RME risk for the site for white croaker consumers (see Figure 17.7). On average, the theoretical cancer risks associated with intake of PCBs in fish exceeded those associated with the intake of tDDT. In fact, for most private boat anglers, PCB risks were more than double the tDDT risks.

Consumption of kelp bass represented the largest average contribution (51 percent) to a private boat angler's total risk due to fish consumption, whereas

Cumulative Density Function

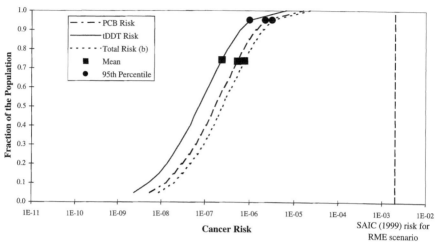

Probability Density Function

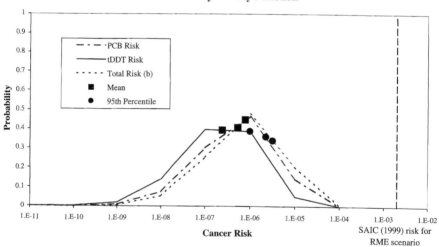

a. These distributions apply only to the 91% of private boat anglers who we predicted consumed fish from the Palos Verdes Shelf; all other private boat anglers would have no incremental risk.

b. The total risk is the sum of the risks due to ingesting PCBs and tDDT in fish from the Palos Verdes Shelf.

Figure 17.7 Distributions of incremental cancer risks for anglers on private boats who consume fish containing tDDT and PCBs from the Palos Verdes Shelf.[a]

TABLE 17.10   Results of the Analysis for Private Boat Anglers at the Palos Verdes Shelf[a]

	Median	Mean (Average)	95th Percentile
Cancer risks[b]			
Risk due to intake of tDDT	7×10^{-8}	2×10^{-7}	1×10^{-6}
Risk due to intake of PCBs	2×10^{-7}	5×10^{-7}	2×10^{-6}
Total risk (tDDT and PCBs)	3×10^{-7}	8×10^{-7}	3×10^{-6}
Noncancer hazards[c]			
Hazard quotient due to intake of tDDT	0.003	0.007	0.03
Hazard quotient due to intake of PCBs	0.03	0.07	0.2

[a] Results represent those private boat anglers who we predicted consumed fish during their lifetime from the Palos Verdes Shelf (91% of private boat anglers).
[b] Upper bound of the excess cancer risk.
[c] A hazard quotient less than 1 indicates that no noncancer health effects are expected.

barred sand bass was consumed by the largest fraction of Palos Verdes Shelf private boat anglers (64 percent). The mean long-term fish consumption rate for private boat anglers at the Palos Verdes Shelf was 11 g/day, and the 95th percentile consumption rate was 36 g/day. On a single-species basis, the mean consumption rate for Pacific barracuda was highest at 5.7 g/day. In contrast, the mean consumption rate for white croaker (0.15 g/day) ranked tenth among the 13 species studied.

Figure 17.8 presents the results of the noncancer hazard assessment for the private boat anglers. More than 99.9 percent of the private boat anglers consuming one or more fish from the Palos Verdes Shelf had hazard quotients less than 1. This finding indicates that the intake of tDDT and PCBs from Palos Verdes Shelf fish is highly unlikely to cause noncancer effects among private boat anglers.

We also considered separately the cancer risks and noncancer hazards to the 4 percent of private boat anglers whom the Monte Carlo analyses predict would consume one or more white croaker from the Palos Verdes Shelf during their fishing careers so that we could compare our estimates to those prepared by EPA (SAIC, 1999). Our analysis predicted that the median total cancer risk for consumption of white croaker was 3×10^{-7}, the mean risk was 4×10^{-7}, and the 95th percentile risk was 2×10^{-6}. On average, for consumption of white croaker only, approximately 40 percent of the risk was due to intake of PCBs, and 60 percent was due to intake of tDDT in these fish. The mean contribution of white croaker consumption to total cancer risk for consumption of all species was 31 percent. More than 99.9 percent of the private boat white croaker consumers had noncancer hazard quotients less than 1.

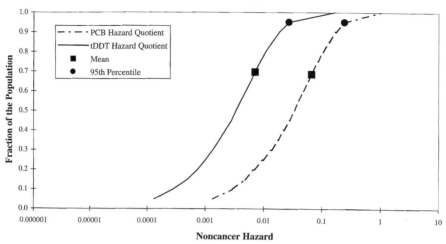

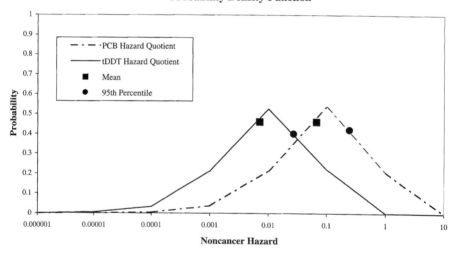

a. These distributions apply only to the 91% of private boat anglers who consumed fish from the Palos Verdes Shelf; all other private boat anglers would have no incremental hazard.

Figure 17.8 Distributions of the hazard quotients for noncancer effects for anglers on private boats who consume fish containing tDDT and PCBs from the Palos Verdes Shelf.[a]

17.3.3 Cancer Risks and Noncancer Hazards for Cabrillo Pier Anglers

Based on our analysis, nearly 99 percent of Cabrillo Pier anglers caught at least one fish and 96 percent consumed a fish during their angling careers. On average, the anglers who consumed fish from Cabrillo Pier caught fish on 23 trips to Cabrillo Pier and consumed fish from 17 trips during a fishing career. Our analysis indicated that the 95th percentile number of trips on which fish were caught during a fishing career was 85, of which 64 yielded fish that were consumed. The average number of different species consumed by Cabrillo Pier anglers was 3. Our analysis indicated that the mean fishing career duration was 5 years, and the 95th percentile duration was 13 years.

Figures 17.9 and 17.10 present cancer risk and noncancer hazard distributions, respectively, for the Cabrillo Pier angler population. For the Cabrillo Pier anglers, as presented in Table 17.11, fifty percent of anglers had total risks for consumption of all species that were less than 2×10^{-7}, and 95 percent of anglers had total risks less than 3×10^{-6}. The mean risk for Cabrillo Pier anglers was 6×10^{-7}. The 95th percentile risk falls within EPA's target risk range (EPA, 1991). In general, about half of the risk was due to intake of PCBs in fish, and about half was due to intake of tDDT. White croaker had the highest average contribution (72 percent) to total risk due to fish consumption and was consumed by the largest fraction of anglers. For the 73 percent of Cabrillo Pier anglers who consumed at least one white croaker, the median risk for consumption of white croaker was 2×10^{-7}, the mean risk was 5×10^{-7}, and the 95th percentile risk was 2×10^{-6}. On average, for consumption of white croaker, approximately one-third of the risk was due to intake of PCBs, and two-thirds was due to intake of tDDT in these fish. The mean fish consumption rate for Cabrillo Pier anglers was 3.4 g/day, and the 95th percentile consumption rate was 12 g/day. On a single-species basis, the mean consumption rate for chub mackerel (*Scomber japonicus*) ranked highest at 2.8 g/day. The mean consumption rate for white croaker ranked fifth at 0.63 g/day.

The hazard quotients for the Cabrillo Pier anglers were less than 1 for more than 99 percent of anglers. These findings clearly show that the probability of noncancer effects due to the intake of tDDT and PCBs in fish is quite low among Cabrillo Pier anglers.

17.4 SENSITIVITY ANALYSIS METHODOLOGY AND RESULTS

To quantitatively describe the uncertainty in the distribution of risk across the angler population, we performed sensitivity analyses for the CPFV and Cabrillo Pier angler analyses to examine the impact of alternative assumptions at various stages in the analyses. In addition, for many exposure factors, we conducted simulations in which we replaced input distributions with default values, which are considered to be conservative. Table 17.12 lists the alternative anal-

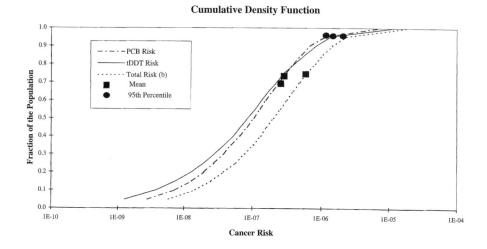

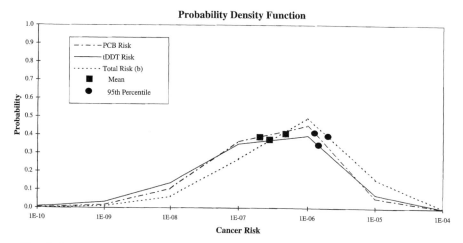

a. These distributions apply only to the 96% of anglers who consumed fish from the Cabrillo Pier; all other Cabrillo Pier anglers would have no incremental risk.

b. The total risk is the sum of the risks due to ingesting PCBs and tDDT in fish from Cabrillo Pier.

Figure 17.9 Distribution of incremental cancer risks for anglers who consume fish containing tDDT and PCBs from Cabrillo Pier.[a]

ysis assumptions and alternative distributions for exposure factors investigated in the sensitivity analyses. This approach to identifying the exposure factors with predominant influence on results is consistent with that suggested in the peer-reviewed literature (Cullen, 1995), by EPA (1995, 1997a), and by the State of California (CalEPA, 1996).

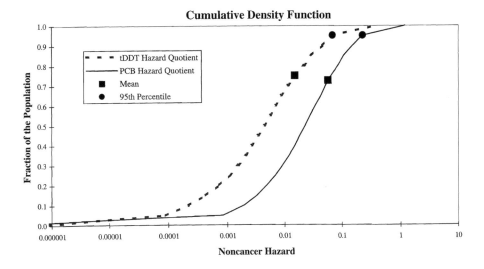

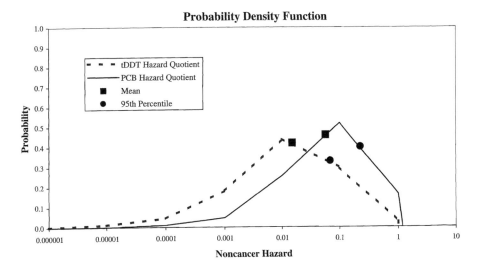

a. These distributions apply only to the 96% of anglers who consumed fish from the Cabrillo Pier; all other
    Cabrillo Pier anglers would have no incremental hazard.

Figure 17.10   Distribution of the hazard quotients for noncancer effects for anglers who
consume fish containing tDDT and PCBs from Cabrillo Pier.[a]

TABLE 17.11   Results of the Analysis for Anglers at Cabrillo Pier

	Median	Mean (Average)	95th Percentile
Cancer risks[b]			
Risk due to intake of tDDT	8×10^{-8}	3×10^{-7}	1×10^{-6}
Risk due to intake of PCBs	1×10^{-7}	3×10^{-7}	1×10^{-6}
Total risk (tDDT and PCBs)	2×10^{-7}	6×10^{-7}	3×10^{-6}
Noncancer hazards[c]			
Hazard quotient due to intake of tDDT	0.005	0.02	0.07
Hazard quotient due to intake of PCBs	0.02	0.06	0.2

[a] Results represent those anglers who consume fish during their lifetime from the Cabrillo Pier (96% of Cabrillo Pier anglers).

[b] Upper bound of the excess cancer risk.

[c] A hazard quotient less than 1 indicates that no noncancer health effects are expected.

Table 17.13 presents summary statistics for alternative chemical concentration distributions developed for tDDT and PCBs in white croaker at Cabrillo Pier. These alternative distributions were based on data collected in 1996 by S. R. Hansen and Associates, nearly 10 years after the data that were used in the Cabrillo Pier baseline analysis were collected.

Table 17.14 presents results of the sensitivity analyses for the CPFV angler analysis in terms of the estimated cancer risks to the median and 95th percentile anglers. Table 17.15 presents results of the sensitivity analyses for noncancer hazards for CPFV anglers. Tables 17.16 and 17.17 present the sensitivity analysis results for Cabrillo Pier anglers for cancer risks and noncancer hazards, respectively. While alternative analysis assumptions and alternative distributions can lower or raise estimates of risk for the exposed populations, the sensitivity analysis showed that nearly any reasonable change in any single exposure factor will have a relatively minor impact on the final estimates of risk for the more highly exposed individuals. All of the alternatives raised or lowered 95th percentile cancer risk and noncancer hazard estimates by a factor of 3 or less.

The largest change in the cancer risk estimates at the 95th percentile for CPFV anglers was observed when the length of an angler's fishing career (i.e., exposure duration) was set to 30 years (or to the exposure duration corresponding to the angler's fishing until 90 years of age, if the angler began fishing after age 60), when each CPFV trip was assigned a 31 percent chance (i.e., the maximum across all departure ports and seasons) of visiting the Palos Verdes Shelf, when the CPFV anglers were assumed to catch and consume fish on every trip, and when the fish consumption rate for each trip was set at the EPA (1997c) default high-end estimate of fish meal size (326 grams). These changes

TABLE 17.12 Description of the Approach for Conducting the Sensitivity Analyses

Distribution and Description of Adjustment	Rationale
CPFV and Cabrillo Pier Analyses	
Exposure duration	
Thirty-year duration	EPA's default value for residential exposure duration (EPA, 1997d); used by SAIC (1999) for RME scenario.
Catch fish	
Catch fish on each trip	Calculate risks given that every angler caught a fish on every trip.
Consume fish	
Consume every fish caught	Determine effect of every angler consuming each fish that was caught.
Catch and consume fish on every trip	Calculate risks given that every angler caught and consumed fish on every trip.
Fish consumption rate	
Consumption equals 326 g/trip for each species caught and consumed	95th percentile for fish serving size from the *Exposure Factors Handbook* (EPA, 1997c).
Cooking loss	
No cooking loss	Determine effect on final results.
PCB and tDDT concentrations in fish	
Correlate PCB and tDDT concentration percentiles	Under certain circumstances, PCB and tDDT concentrations may be correlated such that if one chemical concentration is relatively high in fish, the other is as well.
CPFV Analysis Only	
Probability of going to the Palos Verdes Shelf given departure location	
Regional average probability applied to all trips (0.076)	Use regional average probability of going to the Palos Verdes Shelf.
Maximum probability for region applied to all trips (0.31)	Use highest seasonal probability of going to the Palos Verdes Shelf.
PCB and tDDT concentrations in white croaker from zone 1	
See Table 17.4 for distribution details.	Use data collected by Love and Hansen (1999).
Cabrillo Pier Analysis Only	
PCB and tDDT concentrations in white croaker	
See Table 17.10 for distribution details.	Use data collected by S. R. Hansen & Associates in 1996.

TABLE 17.13 **Alternative Distributions for tDDT and PCB Concentrations in White Croaker at Cabrillo Pier**

Fish Species	Chemical	Distribution Type	Parameter Values				Source
			Minimum (ppm)	Mean (ppm)	Maximum (ppm)	Standard Deviation (ppm)	
White croaker	tDDT	Truncated log-normal	0	1.3	6.8	1.8	S. R. Hansen and Associates (August 1996 data)
White croaker	PCBs	Truncated log-normal	0	0.41	1.3	0.31	S. R. Hansen and Associates (August 1996 data)

TABLE 17.14 Sensitivity Analysis Results for Cancer Risks for Commercial Passenger Fishing Vessel (CPFV) Anglers

Distribution and Description of Adjustment	Cancer Risks[a]			
	Total Risk at Median	Change from Baseline[b] (%)	Total Risk at 95th Percentile	Change from Baseline (%)
Baseline	5×10^{-8}		8×10^{-7}	
Exposure duration				
Thirty-year duration	2×10^{-7}	300	2×10^{-6}	150
Probability of going to the Palos Verdes Shelf given departure location				
Regional average probability applied to all trips (0.076)	5×10^{-8}	0	6×10^{-7}	−25
Maximum probability for region applied to all trips (0.31)	1×10^{-7}	100	2×10^{-6}	150
Catch fish				
Catch fish on each trip	7×10^{-8}	40	1×10^{-6}	25
Consume fish				
Consume every fish caught	7×10^{-8}	40	1×10^{-6}	25
Catch and consume fish on every trip	1×10^{-7}	100	2×10^{-6}	150
Fish consumption rate				
Consumption equals 326 g/trip for each species caught and consumed	5×10^{-8}	0	2×10^{-6}	150
Cooking loss				
No cooking loss	8×10^{-8}	60	1×10^{-6}	25
PCB and tDDT concentrations in fish				
Correlate PCB and tDDT concentration percentiles	5×10^{-8}	0	8×10^{-7}	0
Alternative distributions for tDDT and PCB concentrations in white croaker from zone 1	4×10^{-8c}	-50^{d}	4×10^{-7c}	-60^{d}

[a]Upperbound excess cancer risk due to intake of tDDT and PCBs in fish from Palos Verdes Shelf.
[b]Change from baseline calculated as: [(Total risk after adjustment−Baseline total risk)/(Baseline total risk)] × 100%.
[c]Only results for CPFV anglers who consumed white croaker.
[d]Compared to baseline for CPFV anglers who consumed white croaker. Baseline median and 95th percentile risks for this subpopulation of anglers were 8×10^{-8} and 1×10^{-6}, respectively.

TABLE 17.15 Sensitivity Analysis Results for Noncancer Hazards for Commercial Passenger Fishing Vessel (CPFV) Anglers

Distribution and Description of Adjustment	PCBs				tDDT			
	PCB Hazard[a] at Median	Change from Baseline[b] (%)	PCB Hazard at 95th Percentile	Change from Baseline (%)	tDDT Hazard at Median	Change from Baseline (%)	tDDT Hazard at 95th Percentile	Change from Baseline (%)
Baseline	0.006		0.07		0.0006		0.01	
Exposure duration								
Thirty-year duration	0.007	17	0.08	14	0.0007	17	0.01	0
Probability of going to the Palos Verdes Shelf given departure location								
Regional average probability applied to all trips (0.076)	0.005	−17	0.05	−29	0.0006	0	0.007	−30
Maximum probability for region applied to all trips (0.31)	0.02	233	0.2	186	0.002	233	0.03	200
Catch fish								
Catch fish on each trip	0.009	50	0.1	43	0.0009	50	0.02	100
Consume fish								
Consume every fish caught	0.009	50	0.1	43	0.0009	50	0.01	0
Catch and consume fish on every trip	0.01	67	0.1	43	0.001	67	0.02	100
Fish consumption rate								
Consumption equals 326 g/trip for each species caught and consumed	0.006	0	0.1	43	0.0007	17	0.03	200
Cooking loss								
No cooking loss	0.01	67	0.1	43	0.001	67	0.02	100

PCB and tDDT concentrations in fish

Correlate PCB and tDDT concentration percentiles	0.006	0	0.08	14	0.0007	17	0.01	0
Alternative distributions for tDDT and PCB concentrations in white croaker from zone 1	0.002[c]	−38[d]	0.02[c]	−63[d]	0.0008[c]	−43[e]	0.01[c]	−60[e]

[a] Hazard quotient. A hazard quotient less than 1 indicates that no noncancer health effects are expected.

[b] Change from baseline calculated as: [(Hazard quotient after adjustment—Baseline hazard quotient)/(Baseline hazard quotient)] × 100%.

[c] Only results for CPFV anglers who consumed white croaker.

[d] Compared to baseline CPFV anglers who consumed white croaker. Baseline median and 95th percentile PCB hazards for this subpopulation of anglers were 0.003 and 0.05, respectively.

[e] Compared to baseline CPFV anglers who consumed white croaker. Baseline median and 95th percentile tDDT hazards for this subpopulation of anglers were 0.001 and 0.03, respectively.

TABLE 17.16 Sensitivity Analysis Results for Cancer Risks for Cabrillo Pier Anglers

Distribution and Description of Adjustment	Cancer Risks[a]			
	Total Risk at Median	Change from Baseline[b] (%)	Total Risk at 95th Percentile	Change from Baseline (%)
Baseline	2×10^{-7}		3×10^{-6}	
Exposure duration				
Thirty-year duration	2×10^{-6}	900	1×10^{-5}	233
Catch fish				
Catch fish on each trip	4×10^{-7}	100	5×10^{-6}	67
Consume fish				
Consume every fish caught	3×10^{-7}	50	3×10^{-6}	0
Catch and consume fish on every trip	6×10^{-7}	200	6×10^{-6}	100
Fish consumption rate				
Consumption equals 326 g/trip for each species caught and consumed	6×10^{-7}	200	7×10^{-6}	133
Cooking loss				
No cooking loss	3×10^{-7}	50	5×10^{-6}	67
PCB and tDDT concentrations in fish				
Correlate PCB and tDDT concentration percentiles	2×10^{-7}	0	3×10^{-6}	0
Alternative distributions for tDDT and PCB concentrations in white croaker	1×10^{-7}	−50	1×10^{-6}	−67

[a] Upperbound excess cancer risk due to intake of tDDT and PCBs in fish from Cabrillo Pier.

[b] Change from baseline calculated as: [(Total risk after adjustment—Baseline total risk)/(Baseline total risk)] × 100%

TABLE 17.17 Sensitivity Analysis Results for Noncancer Hazards for Cabrillo Pier Anglers

Distribution and Description of Adjustment	PCBs				tDDT			
	PCB Hazard[a] at Median	Change from Baseline[b] (%)	PCB Hazard at 95th Percentile	Change from Baseline (%)	tDDT Hazard at Median	Change from Baseline (%)	tDDT Hazard at 95th Percentile	Change from Baseline (%)
Baseline	0.02		0.2		0.005		0.07	
Exposure duration								
Thirty-year duration	0.06	200	0.4	100	0.02	300	0.1	43
Catch fish								
Catch fish on each trip	0.05	150	0.4	100	0.009	80	0.1	43
Consume fish								
Consume every fish caught	0.04	100	0.3	50	0.008	60	0.09	29
Catch and consume fish on every trip	0.07	250	0.5	150	0.01	100	0.1	43
Fish consumption rate								
Consumption equals 326 g/trip for each species caught and consumed	0.06	200	0.5	150	0.02	300	0.2	186
Cooking loss								
No cooking loss	0.04	100	0.4	100	0.007	40	0.1	43
PCB and tDDT concentrations in fish								
Correlate PCB and tDDT concentration percentiles	0.03	50	0.3	50	0.005	0	0.08	14
Alternative distributions for tDDT and PCB concentrations in white croaker	0.02	0	0.2	0	0.002	−60	0.03	−57

[a] Hazard quotient. A hazard quotient less than 1 indicates that no noncancer health effects are expected.
[b] Change from baseline calculated as: [(Hazard quotient after adjustment—Baseline hazard quotient)/(Baseline hazard quotient)] × 100%

increased the risk at the 95th percentile threefold over the baseline value to 2×10^{-6}. Specifically, setting the length of an angler's fishing career to 30 years resulted in a 95th percentile exposure duration of 30 years. It is important to note that a 30-year or longer point estimate for length of fishing career is not supported by the available site-specific data (Price et al., 1998; SCCWRP and MBC, 1994). This adjustment was performed for comparison to other risk assessments and should not be considered a "reasonable" alternative assumption for estimating risk. In addition, the alternative when each CPFV trip was assigned a 31 percent chance of visiting the Palos Verdes Shelf is equivalent to the most conservative case, that is, that all CPFV trips leave from Redondo Beach in the fall. Using the Love and Hansen (1999) data for white croaker from zone 1 resulted in a risk estimate for white croaker consumers of 4×10^{-7} at the 95th percentile, a 60 percent reduction as compared to the baseline analysis results for white croaker consumers that used LACSD (1997) data from deeper, less frequented waters. In terms of noncancer hazards to CPFV anglers, the largest change for PCB hazard quotients at the 95th percentile was observed when each CPFV trip was assigned the maximum 31 percent chance of visiting the Palos Verdes Shelf. This alternative, along with when the fish consumption rate for each trip was set at the EPA (1997a) default high-end estimate of fish meal size (326 grams), produced the largest changes for tDDT hazard quotients at the 95th percentile. None of the changes for CPFV anglers resulted in hazard quotients exceeding 1.

For Cabrillo Pier anglers, the largest change in cancer risk estimates at the 95th percentile was also associated with assuming a 30-year fishing career. Using a 30-year exposure duration alternative resulted in a threefold increase over the baseline value at the 95th percentile to 1×10^{-5}. As is the case for CPFV anglers, a 30-year or longer fishing career is not a reasonable assumption for Cabrillo Pier anglers (Price et al., 1998; SCCWRP and MBC, 1994). Using the S. R. Hansen and Associates (1996) data for chemical concentrations in white croaker at Cabrillo Pier (see Table 17.13) resulted in a risk estimate of 1×10^{-6} at the 95th percentile, a 67 percent reduction as compared to the baseline case using Pollock et al. (1991) data. In terms of noncancer hazards to Cabrillo Pier anglers, the largest change for a hazard quotient at the 95th percentile was the nearly threefold change observed when the fish consumption rate for each trip was set at 326 g/trip. None of the changes resulted in hazard quotients exceeding 1.

Varying the assumption regarding meal sizes did not have a large effect on risk or hazard estimates. Hence, we find that although individuals with whom anglers share their fish, that is, nonangler consumers, may have consumption rates that exceed those of anglers, any differences in consumption rates would not significantly affect exposure estimates and the associated risks. Because of this finding and the other factors that would tend to result in nonangler consumers having the same or lower doses and risks than anglers (e.g., dependence on the angler for fish, potentially shorter exposure durations; see Section 17.5.1), we conclude that nonangler consumers in anglers' households would be expected to have a range of doses and risks that are similar to or smaller than the corresponding values for anglers.

Another scenario considered in the sensitivity analysis was that if a fish was harvested and found to have a relatively high concentration of one chemical, concentrations of all chemicals in that fish would be relatively high because of its habitat and environment. Similarly, relatively low concentrations of chemicals might be expected to co-occur. When the concentrations of tDDT and PCBs were correlated such that if an angler consumed a fish with relatively high (or low) tDDT levels, the PCB levels were also relatively high (or low), no significant changes were noted in median or 95th percentile risks for either population.

Overall, these findings suggest that, within our Monte Carlo analysis framework, the adoption of reasonable alternative approaches and exposure factor values does not have a major impact on the overall risk characterizations. Our analyses yielded stable and robust results with respect to reasonable alternative inputs.

17.5 APPLICABILITY OF THE RESULTS TO OTHER POPULATIONS

17.5.1 Nonangler Consumers

Anglers who catch fish at the Palos Verdes Shelf or at Cabrillo Pier may share the fish with other individuals such as family members (e.g., spouses and children) and friends. Thus, individuals other than anglers also may be exposed to tDDT and PCBs in fish. There are only limited data available to describe the exposures to these nonangler consumers. Local data on sport fish consumption rates do not distinguish between anglers and nonangler consumers. The *Santa Monica Bay Seafood Consumption Study* (SCCWRP and MBC, 1994) found that most anglers who kept fish took their fish home and shared them with household members. This suggests that nonangler consumers are generally members of the angler's household and not friends or acquaintances. Unfortunately, only qualitative information was collected on the relative fish consumption frequency and consumption rates of the angler's household members. The lack of data on the consumption rates of nonangler consumers in the household prevents the direct estimation of their exposures and risks. However, it is possible to demonstrate that the results of the analysis of angler exposures and associated risks are a reasonable, if not conservative, measure of the exposures and risk to nonangler consumers.

Nonangler consumers' exposures are largely determined by the exposures to the angler who brings the fish into the household. The frequency of exposure will be the same or less than for the angler. The frequency cannot exceed that of the angler because the nonangler consumer does not consume a fish meal unless the angler has a successful fishing trip. The frequency could, however, be less if for some reason the nonangler does not share in the meal. The species consumed by the other consumers are also determined by the angler (i.e., if the species is not brought home by the angler it cannot be consumed). The number of years a nonangling consumer consumes fish will also be the same or smaller than that of the angler because the nonangler may join or leave the angler's household at some point in an angler's fishing career. As a result, the nonangler may not be present for the full number of years that an angler fishes. Based on these two factors, the long-term exposures of nonangler consumers would be expected to be lower than the anglers' exposures.

Unlike the other factors that determine exposure, the amount of each species consumed by the nonangling consumer could be greater than the angler's consumption rate. The species-specific consumption rates used in this assessment are estimated based on the assumption that all fish brought to a household will be shared equally. Because it is possible that household members may not each consume the same amount of fish, at least some fraction of nonangling consumers may receive doses higher than anglers. The impact of consumption rate will be evaluated via a sensitivity analysis.

17.5.2 Exposures to Children

The limited available data indicate that rates of consumption by children and adults appear to be consistent with each other when normalized to body weight. While there are no reliable data on consumption of sport-caught fish consumption by children, Rupp (1980) reported per-capita rates of consumption for children and adults, based on data provided by the NMFS survey conducted in 1973 and 1974. In his analysis of the NMFS data, Rupp (1980) provided estimates of consumption of freshwater finfish, saltwater finfish, and shellfish for children aged 1 through 18 years as well as adults over the age of 18. We used the reported average consumption rate data provided for each age group and then divided by the average body weight for each age group [EPA, 1989a; Massachusetts Department of Environmental Protection (MADEP), 1995], which yielded estimated rates of consumption in grams per kilogram per day (g/kg-day). Total consumption rates and rates for consumption of each type of fish are very similar when compared on this basis. For example, the national average consumption rates were 0.22 g/kg-day for children aged 1 to 18 years and 0.23 g/kg-day for adults. Similarly, using data provided by Rupp et al. (1980), Pacific region average per-capita consumption rates for both children and adults were 0.24 g/kg-day. Based on these data, it appears that rates of consumption among children and adults, when normalized to body weight, do not differ substantially.

17.5.3 Exposures to Ethnic Subpopulations

This risk assessment relied upon data from the *Santa Monica Bay Seafood Consumption Study* (SCCWRP and MBC, 1994) for CPFV anglers and anglers fishing from Cabrillo Pier and from the Love and Hansen (1999) study for private boat anglers. Both data sources studied the entire angler population including all age, ethnic, and income groups. As discussed below, we did not distinguish among age, ethnic, or income groups in preparing our analysis. Allen et al. (1996) analyzed the *Santa Monica Bay Seafood Consumption Study* (SCCWRP and MBC, 1994) data and observed that while upper decile (i.e., 90th percentile) fish consumption rates based on 4-week recall of fish meals were different across income and ethnic groups, median rates did not vary significantly based on the results of statistical tests. The *Santa Monica Bay Seafood Consumption Study* (Table 12; SCCWRP and MBC, 1994) report indicates that upper decile (i.e., 90th percentile) consumption rates based on 4-week recall for Asian and white anglers exceeded the upper decile for all anglers by 8 and 5 percent, respectively, while the upper decile rates for black and Hispanic anglers were below the overall upper decile by 20 and 40 percent, respectively.

We examined the findings of the *Santa Monica Bay Seafood Consumption Study* to evaluate whether any ethnic groups met the EPA (1995) definition as an "important group," that is, a "highly exposed or highly susceptible group," on the basis of fish consumption rates in order to determine whether such a group should be considered separately in our risk assessment. Specifically, we sought to determine whether any group had fish consumption rates different from the overall rates such that a separate assessment for that group might yield significantly higher exposure and risk results. To make this determination, we evaluated the consumable portion of fish measured for each angler by the survey clerks (kilogram per individual; data from Appendix 12 of SCCWRP and MBC, 1994) because these were the fish consumption rates we intended to use to estimate exposure for CPFV anglers and anglers fishing from Cabrillo Pier.

We compared the distributions of fish consumption rates among Asian (including Chinese, Filipino, Japanese, Korean, and Vietnamese), black, Hispanic, and white anglers. We did not consider the "other" group because it was comprised of multiple ethnic backgrounds and hence is not an ethnically definable group of the population and because its size was far too small to be reliable (i.e., 8 anglers). Table A in Appendix 13 of the *Santa Monica Bay Seafood Consumption Study* (SCCWRP and MBC, 1994) reports that white anglers had average consumption rates that were 8 percent higher than the overall average, while Asian, black, and Hispanic anglers had average consumption rates between 13 and 35 percent less than the overall average. Figure 17.11 demonstrates that white anglers, who formed the majority of anglers in the survey, consistently had higher consumption rates than did the remaining groups, with the exception of the 90th percentile consumption rate for the Asian group, which exceeded the white rate by 6 percent. Based on these findings, we concluded that using all the fish consumption rates over all anglers would not underestimate the intakes of any ethnic subpopulations. We also applied these findings to our analysis of private boat anglers. In short, the exposure estimates and associated risks we report are inclusive of all demographic groups represented by the underlying data.

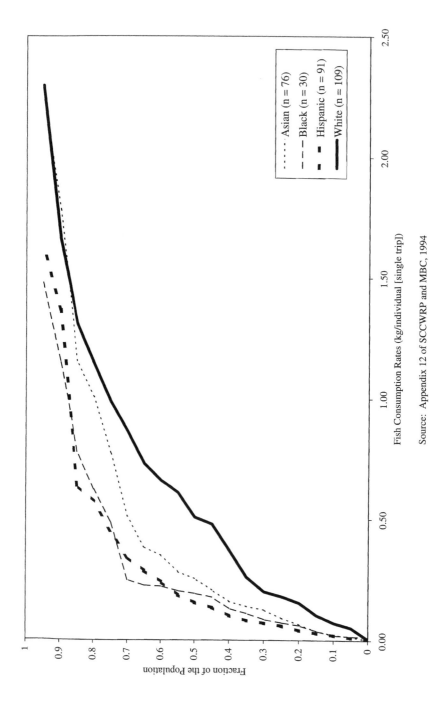

Figure 17.11 Comparison of fish consumption rates from the *Santa Monica Bay Seafood Consumption Study* by ethnic group.

Source: Appendix 12 of SCCWRP and MBC, 1994

17.6 DISCUSSION

17.6.1 Risk Characterization

The fundamental conclusion of our work is that, based on what is known about angler behavior in this geographical area, the levels of tDDT and PCBs in fish at the Palos Verdes Shelf and Cabrillo Pier do not pose a significant risk to human health among recreational fishers or the individuals with whom they share their fish. While no precise estimates of the size of the Palos Verdes Shelf angler population are published, information available from several sources (SCCWRP and MBC, 1994; Love and Hansen, 1999; NMFS, 2000) indicates that the population is small enough that, based on our health risk estimates, no cases of cancer or other health effects would be expected to result from consumption of Palos Verdes Shelf or Cabrillo Pier fish. Our analysis indicates that the majority of the total excess cancer risk due to fish consumption is attributable to PCBs in Palos Verdes Shelf fish. In fact, the analysis suggested that PCB risks were more than double the tDDT risks for most anglers.

This risk assessment is directly applicable to anglers who consume fish from the Palos Verdes Shelf or Cabrillo Pier. Because site-specific data (SCCWRP and MBC, 1994) demonstrate that differences in fish consumption rates among ethnic groups are minor, our findings are applicable to the various ethnic subpopulations within the angler population. Our findings are also applicable to the family members and friends with whom anglers share their fish. As discussed, the estimates of anglers' exposures developed in this assessment are likely to be conservative estimates of exposures to nonangler consumers of fish. Children, despite their lower body weights, are not expected to have higher exposures than adults. Therefore, our findings are also relevant to the evaluation of children's exposures.

Comparison to Background Dietary Intake Residents of the United States have measurable concentrations of tDDT and PCBs in the lipid-rich tissues of the body (e.g., fat, blood serum, etc.). These chemicals are present due to various dietary sources such as meat, milk, fish, and cheese (Lordo et al., 1996). There are no accurate surveys of the amount of tDDT and PCBs to which Americans are currently exposed in their diets. However, surveys of a cross section of the American population have specifically analyzed for persistent chemicals, including tDDT and PCBs, in the adipose tissue and blood. Using pharmacokinetic methods, the daily uptake can be calculated based on concentrations in blood (Leung et al., 1988). The concentrations reported for the American population are approximately 1 to 2 µg (i.e., 1 to 2 millionths of a gram) per kilogram (µg/kg) DDE in blood serum (0.25 mg/kg adipose tissue; Needham, 1997) and 5 µg/kg PCBs in blood serum (0.82 mg/kg in adipose tissue; Condon, 1983). A pertinent comparison for this present risk assessment is to determine the total background intake (i.e., dose) of tDDT and PCBs in foods that must be occurring to produce these background concentrations and then to compare the background intakes to intakes expected in individuals consuming fish from the Palos Verdes Shelf.

To estimate the lifetime average daily dose required to achieve the reported background concentrations of chemicals in blood, it was conservatively assumed that the background concentrations in the published studies represent steady-state concentrations and that elimination of tDDT and PCBs is reasonably described by a first-order process. Using half-lives of 8 years for tDDT (EPA, 1997e) and 7 years for PCBs (EPA, 1996b), the average American is predicted to consume 12 ng (i.e., 12 billionths of a gram) of tDDT per kilogram of body weight per day (ng/kg-day) and 44 ng/kg-day of PCBs to achieve and maintain 1 to 2 μg/kg tDDT and 6 μg/kg PCBs in blood serum, respectively. These values are similar to those presented in Leung et al. (1988). Our Monte Carlo analysis indicates that the mean estimated lifetime average daily doses of tDDT and PCBs by CPFV anglers who consume fish from the Palos Verdes Shelf are 0.2 and 0.05 ng/kg-day, respectively. This contribution represents 2 percent of background for tDDT and 0.1 percent for PCBs. In addition, the analysis for private boat anglers resulted in mean estimated lifetime average daily doses of 0.7 and 0.3 ng/kg-day for tDDT and PCBs, respectively. This contribution represents 6 percent of background for tDDT and 0.7 percent for PCBs. Therefore, from a simple comparative dose standpoint, the contribution of tDDT and PCBs due to eating fish taken from the Palos Verdes Shelf can be considered quite minimal when compared to the amount of tDDT and PCBs that is consumed each day by the typical American.

Lack of Evidence on Human Health Effects   Our risk assessment is based on the assumption that the rodent bioassays for DDT, DDE, DDD, and PCBs indicate that they could pose a cancer hazard to humans. Specifically, all of the calculations rely on EPA guidance that assigns a cancer slope factor to these chemicals (IRIS, 2002). However, there are mixed opinions within the toxicology community about whether any reasonable daily uptake of tDDT or PCBs poses a cancer hazard due to these chemicals' lack of genotoxicity and very low potency in animal studies [Agency for Toxic Substances and Disease Registry (ATSDR), 1994, 1996]. Nonetheless, as discussed previously, even assuming that a cancer hazard is plausible, the risks to anglers and others who consume their fish are expected to be negligible.

The available epidemiologic evidence regarding tDDT also supports our conclusion that health effects are highly unlikely to result from consumption of fish from the Palos Verdes Shelf. The effect of long-term exposure to tDDT in humans was previously evaluated by studies in human volunteers and epidemiological investigations of workers who engaged in the manufacture and formulation of tDDT. For example, human volunteers ingested up to 0.5 mg (i.e., 0.5 thousandths of a gram) per kilogram of body weight per day (mg/kg-day) of tDDT in the form of an oil-filled capsule (Hayes et al., 1956) or an emulsion of milk and peanut oil (Hayes et al., 1971) for up to 21 months. No clinical or laboratory evidence of toxicity was observed in either study. Epidemiological studies that investigated the effects of long-term exposure to high levels of tDDT similarly reported a lack of toxicological consequences of tDDT exposure including liver disease (Ortelee, 1958; Laws et al., 1967, 1973). The maximum daily doses estimated for these workers ranged from 0.26 to 0.6 mg/kg-day for exposure periods of up to 21 years. The average daily intake of tDDT for the general population was estimated to be 4×10^{-4} mg/kg-day during the 1960s (Duggan, 1968), suggesting that nonoccupational exposures to tDDT did not pose a risk to public health during the period of its use as a pesticide. The average daily intake of tDDT estimated for fish consumption from the Palos Verdes Shelf (i.e., 2×10^{-6} mg/kg-day and 4×10^{-6} mg/kg-day as the mean of the average daily dose for CPFV and private boat anglers, respectively) is two orders of magnitude below the historical dietary intake and more than five orders of magnitude below the chronic no-effect exposure levels reported in the human studies. In addition, a 1985 review of the epidemiological data for tDDT and cancer concluded that tDDT has had no significant impact on the incidence of human cancer (Higginson, 1985).

Examination of Profiles of Anglers   Tables 17.18–17.23 present actual profiles of exposure factor values from our analysis results for anglers with total cancer risks corresponding to 50th and 95th percentiles for the CPFV, private boat, and Cabrillo Pier analyses. Examination of these profiles illustrates that many possible combinations of exposure factor values can produce the same risk estimate. The key point is that anglers with 95th percentile risks do not exhibit primarily 95th percentile characteristics. In general, a 95th percentile risk is due to selection of one, or rarely two, exposure factor values near the high end of their distributions. In no case does a 95th percentile angler possess a series of 95th percentile behavior characteristics; in fact, such a case would be inconsistent with the laws of probability. For example, an angler may be very successful at catching fish, but that same angler probably fished for an average number of years and often did not eat all of the fish he or she caught, and the fish he or she did consume had relatively low chemical concentrations. Conversely, another angler may not catch many fish, but, of those caught, most could have been eaten and the chemical concentrations in the fish he or she ate may be near the high end.

Figures 17.12–17.14 compare total risk estimates to values of exposure factors for anglers with risks at or above the 95th percentile. With the exception of the apparent influence of total amount of fish consumed on risks to private boat anglers, no strong relationships between exposure factor values and risk estimates were observed. For CPFV and Cabrillo Pier anglers, neither years fished, total amount of fish consumed, average tDDT concentration in fish, nor the average PCB concentration in fish appear to be particularly influential in predicting whether or not risks to anglers will be in the uppermost 5 percent of the distribution. These analyses corroborate the profiles of anglers with risks at

TABLE 17.18 Profiles of CPFV Anglers with Total Risks at the 50th Percentile (Total Risk = 5×10^{-8})

Years Fished	Number of Successful Palos Verdes Shelf Trips[a]	Mean Fish Consumption Rates		Mean Concentration of Chemicals in Fish Before Cooking (ppb)[c]		Ate White Croaker?
		g/trip	g/day[b]	tDDT	PCBs	
5	7	545	2.09	27	14	No
3	1	766	0.70	188	63	No
14	2	1351	0.53	61	17	No
11	5	527	0.7	49	21	No
18	5	573	0.44	59	17	No
43	3	500	0.10	241	9	No
10	5	815	1.1	29	15	No
5	4	792	1.7	52	13	No
7	5	634	1.2	65	13	No
10	5	796	1.1	16	15	No
4	1	2421	1.66	91	14	No
6	7	501	1.6	27	18	No
5	6	370	1.2	42	28	No
5	2	378	0.41	124	82	No
12	3	467	0.32	131	32	No

[a] Trips that result in catch and consumption of one or more fish species.

[b] Average consumption rate over duration of fishing career. This statistic is an output of the Monte Carlo analysis.

[c] Mean concentration over all fish consumed over duration of fishing career. For demonstration purposes, estimated based on a 40% reduction in chemicals due to cooking and a 70-kg body weight.

the 50th and 95th percentiles (see Tables 17.18, 17.19, 17.22, and 17.23), which also demonstrate that no particular factor drives the risk estimates for CPFV and Cabrillo Pier anglers.

17.6.2 Comparison with Prior Risk Assessments

Tables 17.24 and 17.25 compare our risk estimates with those reported by others for anglers who consume fish from the Palos Verdes Shelf and Cabrillo Pier, respectively. Our results indicate that the exposure scenarios and exposure factor values used in EPA's most recent assessment (SAIC, 1999) and by the State of California (Pollock et al., 1990, 1991) yield risk estimates that are much higher than our estimates for the angler populations studied.

TABLE 17.19 Profiles of CPFV Anglers with Total Risks at the 95th Percentile (Total Risk = 8 × 10⁻⁷)

Years Fished	Number of Successful Palos Verdes Shelf Trips[a]	Mean Fish Consumption Rates		Mean Concentration of Chemicals in Fish Before Cooking (ppb)[c]		Ate White Croaker?
		g/trip	g/day[b]	tDDT	PCBs	
18	51	563	4.4	82	26	Yes
9	34	535	5.5	131	40	Yes
8	8	685	1.9	628	104	No
12	9	553	1.1	790	108	Yes
14	29	834	4.73	107	33	Yes
8	43	852	12.55	46	24	No
29	8	585	0.4	1,280	31	Yes
19	28	1455	6	58	19	Yes
4	14	447	4.3	834	45	Yes
2	20	471	12.9	220	83	No
17	49	859	7	51	18	No
18	48	818	6.0	54	20	Yes
24	30	795	2.7	71	40	Yes
6	8	835	3.1	208	153	Yes
14	14	1430	3.9	106	40	Yes

[a] Trips that result in catch and consumption of one or more fish species.

[b] Average consumption rate over duration of fishing career. This statistic is an output of the Monte Carlo analysis.

[c] Mean concentration over all fish consumed over duration of fishing career. For demonstration purposes, estimated based on a 40% reduction in chemicals due to cooking and a 70-kg body weight.

Our study and the previous risk assessments for these anglers have relied on many of the same data sources. However, the principal reason why our risk estimates are significantly lower than other estimates is that we made more extensive and consistent use of the available data than the previous assessments. In addition, our study was the only one to utilize the CDFG creel survey data that characterized the frequency of CPFV's fishing at the Palos Verdes Shelf and the information from the Love and Hansen (1999) study that characterized the fishing behavior of private boat anglers. Employing Monte Carlo techniques allowed us to use more data from the available sources and to use them in greater depth than have previous assessments conducted by other scientists. Particularly, the structure and organization of our Monte Carlo analysis allowed us to capture the interdependencies and correlations among exposure factors that are present in the data. In constructing the Monte Carlo analyses for

TABLE 17.20   Profiles of Simulated Private Boat Anglers with Total Risks at the 50th Percentile (Total Risk = 3 × 10⁻⁷)

Years Fished	Number of Successful Palos Verdes Shelf Trips[a]	Mean Fish Consumption Rates		Mean Concentration of Chemicals in Fish Before Cooking (ppb)[c]		Ate White Croaker?
		g/trip	g/day[b]	tDDT	PCBs	
4	13	891	7.9	75	28	No
5	24	395	5.2	101	35	No
9	9	985	2.7	148	34	No
11	5	3719	4.6	32	17	No
8	9	2094	6.5	25	16	No
1	4	2794	31	56	33	No
7	7	2131	5.8	42	22	No
1	3	1190	9.8	212	97	No
2	2	5211	14	56	31	No
39	17	974	1.2	75	16	No
12	7	946	1.5	278	18	Yes[d]
4	15	1627	17	29	14	No
1	8	2651	58	62	9	No
2	15	632	13	43	35	No
3	7	1942	12	61	24	No

[a]Trips that result in catch and consumption of one or more fish species.

[b]Average consumption rate over duration of fishing career. This statistic is an output of the Monte Carlo analysis.

[c]Mean concentration over all fish consumed over duration of fishing career. For demonstration purposes, estimated based on a 40% reduction in chemicals due to cooking and a 70-kg body weight.

[d]The result is based on our assumption that the one angler who retained any white croaker from the Palos Verdes Shelf and was unsure whether it would be consumed, actually consumed the fish.

CPFV, private boat, and Cabrillo Pier anglers, we incorporated virtually all of the existing data on the behavior of these local angler populations. We accounted for the major factors necessary to fully describe potential exposures. In particular, we were able to incorporate information on age-related differences in body weight, seasons fished, and trip-to-trip variation in where fishing took place, whether fish were caught, what combination of species was caught, whether any of these fish were consumed, and how they were prepared.

We believe that our approach described a realistic scenario of fishing practices over a lifetime. In contrast, the assessments prepared by EPA and the State of California (SAIC, 1999; Pollock et al., 1990, 1991) relied on combinations of high-end exposure factor values that ignored correlations among the

TABLE 17.21 Profiles of Simulated Private Boat Anglers with Total Risks at the 95th Percentile (Total Risk = 3×10^{-6})

Years Fished	Number of Successful Palos Verdes Shelf Trips[a]	Mean Fish Consumption Rates		Mean Concentration of Chemicals in Fish Before Cooking (ppb)[c]		Ate White Croaker?
		g/trip	g/day[b]	tDDT	PCBs	
16	79	1177	16	69	31	No
40	61	1961	8.2	60	26	No
18	90	1300	18	54	28	No
16	132	1074	24	62	25	No
28	97	1615	15	44	21	No
14	39	1730	13	90	47	No
32	107	1491	14	54	17	No
64	57	1477	4	70	37	Yes[d]
38	81	1756	10	56	20	No
7	80	1775	56	61	23	No
16	95	1464	24	65	23	No
12	105	1615	39	47	16	No
68	152	1588	10	26	14	No
20	68	2103	20	47	21	No
19	81	1064	12	87	39	No

[a] Trips that result in catch and consumption of one or more fish species.

[b] Average consumption rate over duration of fishing career. This statistic is an output of the Monte Carlo analysis.

[c] Mean concentration over all fish consumed over duration of fishing career. For demonstration purposes, estimated based on a 40% reduction in chemicals due to cooking and a 70-kg body weight.

[d] The result is based on our assumption that the one angler who retained any white croaker from the Palos Verdes Shelf and was unsure whether it would be consumed, actually consumed the fish.

exposure factors. These prior assessments based highly uncertain exposure and risk calculations on exposure factor values that, in an effort to not underestimate the true risks, were presumably chosen to be conservative high-end representations. Repeated use of conservative or "reasonable" worst-case values for exposure factors overestimates true exposures and risks (Cullen, 1994; Slob, 1994; McKone and Bogen, 1991), and this overestimation can occur even when each exposure factor value itself is "reasonable" (EPA, 1992a; McKone and Bogen, 1991). By the laws of probability, combining high-end exposure factor values (e.g., 90th or 95th percentile values) yields results that are more extreme than any one of the individual values (i.e., often producing results beyond the 99th or 99.9th percentile values). Many researchers have observed that com-

TABLE 17.22   Profiles of Cabrillo Pier Anglers with Total Risks at the 50th Percentile (Total Risk = 2×10^{-7})

Years Fished	Number of Successful Pier Trips[a]	Mean Fish Consumption Rates		Mean Concentration of Chemicals in Fish Before Cooking (ppb)[c]		Ate White Croaker?
		g/trip	g/day[b]	tDDT	PCBs	
2	8	272	3.0	330	56	Yes
4	3	997	2.0	338	38	Yes
2	8	50	0.55	2024	331	Yes
1	9	275	6.8	426	67	Yes
12	13	148	0.44	617	74	Yes
18	22	317	1.1	49	32	No
3	18	503	8.3	44	29	No
2	5	325	2.2	373	88	Yes
8	17	206	1.2	302	47	Yes
3	13	348	4.1	214	32	Yes
1	5	198	2.7	1224	95	Yes
2	8	81	0.89	1114	189	Yes
4	11	103	0.78	1075	140	Yes
23	9	92	0.099	1457	119	Yes
6	14	165	1.1	397	39	Yes

[a]Trips that result in catch and consumption of one or more fish species.

[b]Average consumption rate over duration of fishing career. This statistic is an output of the Monte Carlo analysis.

[c]Mean concentration over all fish consumed over duration of fishing career. For demonstration purposes, estimated based on a 40% reduction in chemicals due to cooking and a 70-kg body weight.

bining conservative high-end exposure factor values results in unrealistic scenarios with little or no chance of occurring (Cullen, 1994; Hattis and Burmaster, 1994; Lloyd et al., 1992; Whitmyre et al., 1992a). According to EPA (1995), high-end exposures are intended to be plausible estimates of individual exposure expected to occur in small, but definable, high-end segments of the exposed population; conceptually, high-end exposure means exposure above about the 90th percentile of the population, but not beyond the true distribution. However, the combinations of high-end exposure factors relied upon by SAIC (1999) resulted in the unrealistic prediction that high-end anglers will consume over 2350 meals of white croaker from the Palos Verdes Shelf, which would require taking more than 600 CPFV trips or 200 private boat trips to the Palos Verdes Shelf and catching more than a ton of white croaker during their assumed 30-year sport fishing careers.

TABLE 17.23 Profiles of Cabrillo Pier Anglers with Total Risks at the 95th Percentile (Total Risk = 3×10^{-6})

Years Fished	Number of Successful Pier Trips[a]	Mean Fish Consumption Rates		Mean Concentration of Chemicals in Fish Before Cooking (ppb)[c]		Ate White Croaker?
		g/trip	g/day[b]	tDDT	PCBs	
10	71	260	5.1	710	101	Yes
3	97	343	30	361	74	Yes
3	21	537	10	1,660	110	Yes
9	54	337	5.5	549	139	Yes
10	202	451	25	55	37	No
22	62	339	2.6	690	77	Yes
14	97	368	7.0	288	64	Yes
5	52	275	7.8	1,073	123	Yes
5	63	439	15	432	96	Yes
12	85	324	6	469	109	Yes
6	57	241	6.3	1,075	122	Yes
4	61	302	13	1,044	51	Yes
11	72	276	5.0	641	145	Yes
6	11	570	3	3,022	98	Yes
17	35	256	1.4	1,925	99	Yes

[a] Trips that result in catch and consumption of one or more fish species.

[b] Average consumption rate over duration of fishing career. This statistic is an output of the Monte Carlo analysis.

[c] Mean concentration over all fish consumed over duration of fishing career. For demonstration purposes, estimated based on a 40% reduction in chemicals due to cooking and a 70-kg body weight.

The reason that SAIC (1999) predicted such high fish consumption rates is that although consumption rates from the *Santa Monica Bay Seafood Consumption Study* (SCCWRP and MBC, 1994) were based on a 4-week recall period, SAIC extrapolated them to predict a full 30 years (i.e., 360 months) of hypothesized exposure. EPA (1992a) cautions against such extrapolation because it "has a tendency ... to overestimate the exposure levels to the upper end of the distribution." Although the study data clearly indicate that all anglers do not consume self-caught fish each month or season, nor do they all fish in the area for 30 years (Love and Hansen, 1999; Price et al., 1998; SCCWRP and MBC, 1994), SAIC (1999) used the same short-term 90th percentile consumption rate based on a 4-week recall for each and every day throughout their long-term 30-year exposure period. By repeated use of high-end consumption rates, the EPA's exposure scenario exceeded the 99th percentile of exposure suggested by the underlying data.

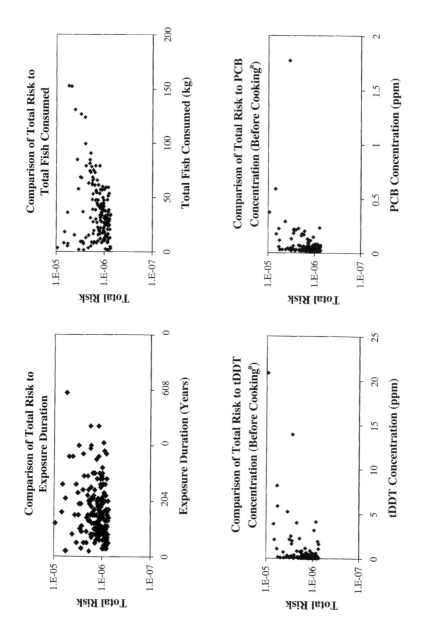

a. Mean percent reduction due to cooking of 40 percent and body weight of 70 kg applied to estimate lifetime average concentrations in fish.

Figure 17.12 Comparison of total risk to various exposure factors for CPFV anglers with risks at or above the 95th percentile.

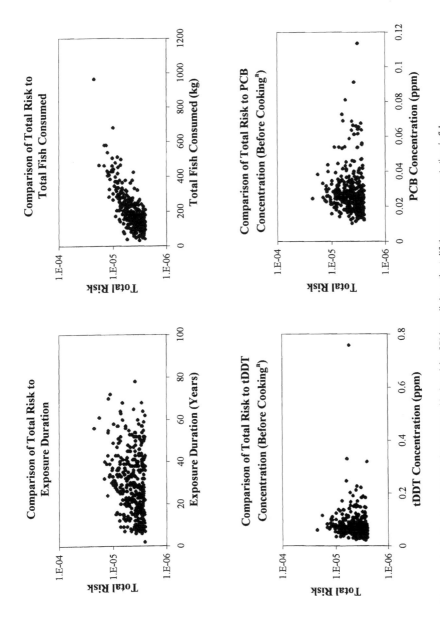

a. Mean percent reduction due to cooking of 40 percent and body weight of 70 kg applied to estimate lifetime average concentrations in fish.

Figure 17.13 Comparison of total risk to various exposure factors for private boat anglers with risks at or above the 95th percentile.

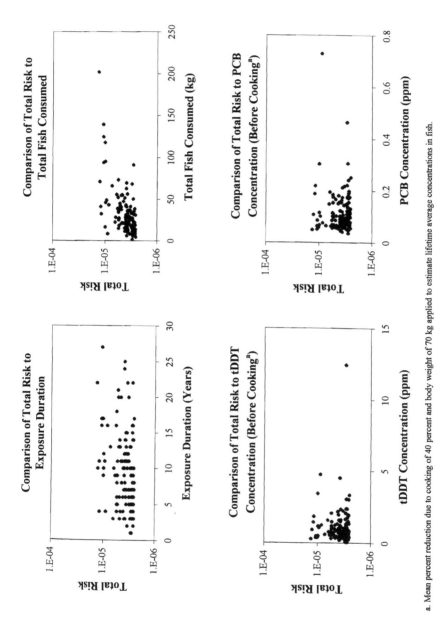

a. Mean percent reduction due to cooking of 40 percent and body weight of 70 kg applied to estimate lifetime average concentrations in fish.

Figure 17.14 Comparison of total risk to various exposure factors for Cabrillo Pier anglers with risks at or above the 95th percentile.

In contrast, our study relied upon the full range of consumption rates measured in the *Santa Monica Bay Seafood Consumption Study* (SCCWRP and MBC, 1994) and the Love and Hansen (1999) study for single fishing trips. In our assessment, when an angler went fishing on the Palos Verdes Shelf and caught fish, he or she had a 10 percent chance of meeting or exceeding the 90th percentile consumption rate for a single trip and a 90 percent chance of having a consumption rate less than the 90th percentile. Because we used a realistic fishing trip-by-fishing trip exposure scenario, our study was able to use the consumption rate information in the way it was measured in the *Santa Monica Bay Seafood Consumption Study* (SCCWRP and MBC, 1994) and the Love and Hansen (1999) study, thus avoiding the overestimation inherent in extrapolating short-term recall information to a long-term exposure scenario. In addition, by considering correlations and dependencies among exposure factors, and by considering variation in behavior consistent with the range and likelihood of observed behavior, the Monte Carlo analysis technique we employed avoided inappropriate combinations of high-end exposure factor values and thus yielded realistic representations of exposures, and particularly of high-end exposures.

Unlike the prior risk assessments, our analysis indicates that the majority of the total incremental cancer risk due to fish consumption is attributable to PCBs in Palos Verdes Shelf fish. In contrast, other studies have reported that tDDT contributes 60 percent or more to the total incremental cancer risk (SAIC, 1999, 1997a, b; EPA, 1996a; Pollock et al., 1990, 1991). This discrepancy is due to the fact that while prior studies have concentrated on theoretical scenarios of white croaker consumption, our study considers consumption of white croaker within the context of a realistic mixed-species diet. While most species of Palos Verdes Shelf fish have higher concentrations of tDDT than of PCBs, only for white croaker is the exceedence of tDDT concentrations over PCB concentrations generally greater in magnitude than a factor of 6, the magnitude of the exceedence of the cancer slope factor used for PCBs over that for tDDT. Thus, focusing exposure scenarios on white croaker consumption distorts the contribution of tDDT and PCBs to an angler's total cancer risk. The analogous argument can be developed for noncancer hazards; the tDDT reference dose exceeds that for PCBs by a factor of 25.

Similarly, our analysis indicates that the risks estimated for Cabrillo Pier anglers are much lower than those previously estimated by the State of California (Pollock et al., 1990, 1991), primarily because the Monte Carlo analysis used information collected directly from Cabrillo Pier anglers in the *Santa Monica Bay Seafood Consumption Study* (SCCWRP and MBC, 1994). In contrast, the exposure assumptions used by the State of California (Pollock et al., 1990, 1991) do not accurately reflect an appropriate range of long-term consumption rates and number of years fished by Cabrillo Pier anglers. For example, in its risk assessments, the State of California used fish consumption rates ranging from 23 g/day (Pollock et al., 1991) based on the unsupported assumption of one fish meal per week to 225 g/day (Pollock et al., 1990) based

TABLE 17.24 Comparison of Risk Estimates and Exposure Factor Values from Risk Assessments for Palos Verdes Shelf Fish Consumers

Study and Risk Estimate[a]	Fish Consumption Rate[b] (g/day)	Years Fished[b]	Number of Successful CPFV Trips to Palos Verdes Shelf	Number of Successful Private Boat Trips to Palos Verdes Shelf	Number of Meals of Palos Verdes Shelf Fish	Amount of Fish Harvested from Palos Verdes Shelf[c] (lbs)
Present study—all fish—CPFV anglers[d]						
Median: 5×10^{-8}	0.93^e	7	3		11^f	11
Mean: 2×10^{-7}	1.9^e	9	8		30^f	29
95th Percentile: 8×10^{-7}	7.5^e	25	31		117^f	111
Present study—all fish—private boat anglers[d]						
Median: 3×10^{-7}	7^e	4		8	88^g	84
Mean: 8×10^{-7}	11^e	9		23	254^g	240
95th Percentile: 3×10^{-6}	36^e	35		97	1070^g	1013
SAIC, 1999—all fish[d]						
Central tendency: 2×10^{-5}	21.4	13.8	222^h	76^i	836^j	791
Present study—white croaker only—CPFV anglers						
Median: 8×10^{-8}	0.02^e	7	1		4^f	4
Mean: 3×10^{-7}	0.05^e	9	1		4^f	4
95th Percentile: 1×10^{-6}	0.22^e	25	4		15^f	14
Present study—white croaker only—private boat anglers						
Median: 3×10^{-7}	0.09^e	4		2	22^g	21
Mean: 4×10^{-7}	0.15^e	9		3	33^g	31
95th Percentile: 2×10^{-6}	0.48^e	35		9	99^g	94
SAIC, 1999—white croaker only						
Central tendency: 9×10^{-6}	0.48	13.8	5^h	2^i	19^j	18
Reasonable maximum exposure: 2×10^{-3}	27.9	30	629^h	215^i	2368^j	2243

[a] Risks represent the total carcinogenic risk estimate (tDDT and PCBs).

[b] Values for SAIC (1999) are the point estimates reported. Values for the present study are statistics from the distributions for consuming anglers.

[c] Based on the assumption that 30% of fish is edible tissue. Amount harvested = [# meals × 129 g/meal × (1 lb/454 g)]/0.3 edible fraction.

[d] Evaluated white croaker and 12 other species.

[e] Fish consumption is used in the analysis on a g/trip basis consistent with the event-by-event approach to estimating exposure. Fish consumption rates in g/day are average over the number of years fished and are computed at the end of each angler's simulated fishing career.

[f] Calculated based on the USEPA (1997c) mean meal size of 129 g (approx. 4.5 oz) and by applying the mean consumption rate per trip for CPFV anglers (486 g/trip) based on the SMBSCS data corrected for avidity bias. # Meals = [# Successful trips × Cons. rate (g/trip)]/129 g/meal.

[g] Calculated based on the USEPA (1997c) mean meal size of 129 g (approx. 4.5 oz) and by applying the mean consumption rate per trip for private boat anglers (1423 g/trip) based on the Love and Hansen (1999) data corrected for avidity bias. # Meals = [# Successful trips × Cons. rate (g/trip)]/129 g/meal.

[h] Estimated based on SAIC's (1999) assumption that all fish consumed come from the Palos Verdes Shelf. Used the mean consumption per trip for CPFV anglers (486 g/trip) based on the SMBSCS data corrected for avidity bias. # Trips = [Cons. rate (g/day) × 365 days/yr × Years fished]/Cons. per trip (g/trip).

[i] Estimated based on SAIC's (1999) assumption that all fish consumed come from the Palos Verdes Shelf. Used the mean consumption per trip for private boat anglers (1423 g/trip) based on the Love and Hansen (1999) data corrected for avidity bias. # Trips = [Cons. rate (g/day) × 365 days/yr × Years fished]/Cons. per trip (g/trip).

[j] Calculated based on the USEPA (1997c) mean meal size of 129 g (approx. 4.5 oz) and using SAIC's (1999) assumption that all fish consumed come from the Palos Verdes Shelf. # Meals = [Cons. rate (g/day) × Years fished × 365 days/yr]/129 g/meal.

1011

TABLE 17.25 Comparison of Risk Estimates and Exposure Factor Values from Risk Assessments for Cabrillo Pier Fish Consumers

Study and Risk Estimate[a]	Fish Consumption Rate[b] (g/day)	Years Fished[b] (Exposure Duration)	Number of Successful Trips to Cabrillo Pier	Number of Meals of Cabrillo Pier Fish	Amount of Fish Harvested from Cabrillo Pier[c] (lb)
Present study[d]					
Median: 2×10^{-7}	1.7[e]	3	8	16[f]	15
Mean: 6×10^{-7}	3.4[e]	5	17	33[f]	31
95th Percentile: 3×10^{-6}	12[e]	13	64	125[f]	118
Pollock et al. (1991)					
Conservative scenario[g]: 9×10^{-4}	23	70	2,341[h]	4,555[i]	4315
Creel scenario[j]: 4×10^{-4}	23	70	2,341[h]	4,555[i]	4315
Pollock et al. (1990)[g]					
7×10^{-4}	23	70	2,341[h]	4,555[i]	4315

[a] Risks represent the total carcinogenic risk estimate (tDDT and PCBs).

[b] Values for Pollock et al. (1990, 1991) are the point estimates reported in those studies. Values for the present study are statistics from the distributions for consuming anglers.

[c] Based on the assumption that 30% of fish is edible tissue. Amount harvested = [# meals × 129 g/meal × (1 lb/454 g)]/0.3 edible fraction.

[d] Evaluated white croaker and 12 other species.

[e] Fish consumption is used in the analysis on a g/trip basis consistent with the event-by-event approach to estimating exposure. Fish consumption rates in g/day are average over the number of years fished and are computed at the end of each angler's simulated fishing career.

[f] Calculated based on the USEPA (1997c) mean meal size of 129 g (approx. 4.5 oz) and by applying the mean consumption rate per trip for Cabrillo Pier anglers (251 g/trip) based on the SMBSCS data corrected for avidity bias. # Meals = [# Successful trips × Cons. rate (g/trip)]/129 g/meal.

[g] Evaluated only white croaker.

[h] Estimated based on the fish consumption rate reported for the study according to the equation: #Trips = [Cons. rate (g/day) × 365 (days/yr) × Years fished (years)]/Cons. per trip (g/trip). Used the mean consumption per trip for Cabrillo Pier anglers (251 g/trip) based on the SMBSCS data corrected for avidity bias.

[i] Calculated based on the USEPA (1997c) mean meal size of 129 g (approx. 4.5 oz) and by applying the mean consumption rate per trip for Cabrillo Pier anglers (251 g/trip). # Meals = [Cons. rate (g/day) × Years fished × 365 days/yr]/129 g/meal.

[j] Evaluated white croaker and 4 other species.

on the 90th percentile fish consumption rate from the Puffer et al. (1981) study, which is known to be overstated due to avidity bias (EPA, 1997c; Price et al., 1994). The average fish consumption rate for Cabrillo Pier anglers predicted by our study was 3.4 g/day and the 95th percentile rate was 12 g/day, estimates which in fact exceed those recently recommended by EPA (1997c) as default rates for southern California marine anglers (i.e., 2.0 g/day at the mean and 5.5 g/day for the 95th percentile). In addition, the State of California (Pollock et al., 1990, 1991) used a 70-year exposure duration in its risk estimates, whereas Price et al.'s (1998) analysis of the available data shows that the mean total number of years fished by Cabrillo Pier anglers is 5 years, and the 95th percentile is 13 years.

For the many reasons that have been discussed here, our results predict that fewer than 1 in 10,000 anglers will experience risks of the magnitude estimated by EPA and the State of California (SAIC, 1997a, b, 1999; EPA, 1996a; Pollock et al., 1990, 1991) due to consumption of tDDT and PCBs in fish from the Palos Verdes Shelf or Cabrillo Pier.

17.6.3 Validation

The net result of our event-by-event methodology is the development of long-term exposure estimates that are consistent with independent general information regarding regional fish consumption rates in general and white croaker catch and consumption in particular.

Our predicted long-term average fish consumption rates for the CPFV anglers at the Palos Verdes Shelf, a small and distinct fishing area, are similar to or exceed those developed by EPA for an angler who fishes throughout southern California marine waters (EPA, 1997c). Because the Palos Verdes Shelf fishing area occupies only a small portion of the fishing area in the Southern California Bight, and because the Palos Verdes Shelf is visited on only 9 to 31 percent of CPFV trips from only two ports within the Southern California Bight, our study may overstate consumption rates for CPFV anglers.

Specifically, our mean and 95th percentile fish consumption rates for CPFV anglers of 1.9 and 7.4 g/day, respectively, are comparable to the mean and 95th percentile rates of 2.0 g/day and 5.5 g/day, respectively developed by EPA from data from 38,000 field interviews conducted by the National Marine Fisheries Service in 1993 (EPA, 1997c). EPA developed these fish consumption rates after first correcting for avidity bias by applying sampling weights proportional to the inverse of the angler's reported fishing frequency, and were based on assumptions that, for those fish that the angler indicated would be kept to eat, 50 percent of a fish's weight was edible, and that the catch would be divided among an average of 2.5 consumers. The total annual fish consumption rate was determined based on the assumption that each trip during the year would yield the same consumption; the annual average daily consumption rate was determined by dividing the total yearly consumption by 365 days.

That our long-term average fish consumption rates for CPFV anglers are more similar to the EPA rates than to those reported in the *Santa Monica Bay Seafood Consumption Study* (SCCWRP and MBC, 1994) for the survey sample is significant for two reasons. First, the concordance confirms that directly extrapolating the *Santa Monica Bay Seafood Consumption Study*'s 28-day consumption rates to a multiyear exposure period without reference to fishing frequency overestimates long-term average consumption rates. Second, the consistency demonstrates the importance of incorporating corrections for avidity bias in order to develop representative consumption rates from intercept survey data.

Our predictions that only 9 percent of CPFV anglers and 4 percent of private boat anglers would eat any white croaker from the Palos Verdes Shelf during their fishing careers, and that the long-term average consumption rate for white croaker ranked tenth in both analyses among the 13 species that we studied, were consistent with several sources. These sources indicate that white croaker, the species upon which prior risk assessments have focused, generally makes up only a small percentage of the total number of fish harvested and consumed by local anglers (Love and Hansen, 1999; SCCWRP and MBC, 1994; Ally et al., 1991; Stull et al., 1987). The *Santa Monica Bay Seafood Consumption Study* found that, of the 108 fish consumed by 232 interviewed CPFV anglers, only 3 were white croaker (SCCWRP and MBC, 1994). Love and Hansen report that, of the 191 fish observed being caught by anglers on CPFVs or private boats during their 1-year at-sea survey conducted in 1997 to 1998, only 12 were white croaker (Love and Hansen, 1999). CDFG data from the period 1986 through 1989 indicate that white croaker catch from CPFVs at the Palos Verdes Shelf area was 0.1 percent of the total fish catch in southern California. Stull et al. (1987) reported that white croaker comprised less than 0.4 percent of the CPFV catch in the Palos Verdes region between 1981 and 1985. Neither the Love and Hansen (1999) data collected during their at-sea survey, the CDFG catch data, nor the Stull et al. (1987) study data were used in our analysis. The consistency of our results with this independent information lends our results credence and supports the overall notion that white croaker is an inappropriate focus for human health risk assessment for anglers at the Palos Verdes Shelf.

17.6.4 Uncertainty

The general concept of uncertainty in the results of a risk assessment arises from uncertainties due to true variability of exposures (commonly referred to as variability or type A uncertainty) and to uncertainties due to lack of knowledge about exposures (commonly referred to as uncertainty or type B uncertainty) (Hoffman and Hammonds, 1994; McKone, 1994). Variation is a natural characteristic of exposures because individuals naturally behave differently from each other. By conducting studies and collecting data, one can learn more about the variability in exposure within a population, but this variability cannot be eliminated (Hattis and Burmaster, 1994; Hoffman and Hammonds, 1994). In contrast, uncertainty refers to our lack of knowledge about an exposure. Uncertainty due to lack of knowledge can be reduced or eliminated as more relevant or reliable data are applied to characterize an exposure (Hattis and Burmaster, 1994; Hoffman and Hammonds, 1994).

Most exposure factors will be subject to varying degrees of variability and uncertainty (Hattis and Burmaster, 1994; Hoffman and Hammonds, 1994; Morgan et al., 1990). Potential sources of uncertainty in the exposure assessment process include uncertainties in the characterization and parameterization of distributions used for exposure factors. These can occur from limitations in the data, the need to use data from surrogate populations, or uncertainty regarding the extent to which the analysis and the studied activity correspond.

Our Monte Carlo analyses were designed to characterize interindividual variation in exposure. This means that our goal was to understand the range of exposures within the population. To limit the uncertainty in our risk estimates, we made use of site-specific data on angling behavior, fish harvesting, and consumption rates that were obtained directly from southern California CPFV, private boat, and Cabrillo Pier anglers. The event-by-event Monte Carlo analysis structure allowed us to use angler behavior data from the *Santa Monica Bay Seafood Consumption Study* (SCCWRP and MBC, 1994) and Love and Hansen (1999) study on the same time scale on which they were collected, thus limiting the potentially significant uncertainties due to extrapolation of short-term data to a long-term phenomenon (EPA, 1992a). This type of extrapolation comprised a major source of uncertainty in previous risk assessments for this site. In developing distributions for exposure factors from the existing data, we attempted to mitigate the effects of the known sources of sampling bias (e.g., avidity bias) in these data. Finally, with the exception of chemical concentration distributions, all of the distributions were developed using a fairly large number of measurements (i.e., greater than 50).

Whenever data or simulations are used for predictive purposes there are uncertainties introduced because the key question of interest—What are the risks to anglers?—cannot be investigated directly. Our risk assessment focused on the behavior of angler populations at the Palos Verdes Shelf and Cabrillo Pier. Our reliance on the *Santa Monica Bay Seafood Consumption Study* (SCCWRP and MBC, 1994) as our primary data source regarding CPFV and Cabrillo Pier angler behavior introduces uncertainty because it was a 1-year study that interviewed 450 CPFV anglers at various regional fishing locations over 32 days, and 198 Cabrillo Pier anglers over 8 days. Similarly, our reliance on the Love and Hansen (1999) study data for characterization of private boat angler behavior introduces uncertainty because it was a 1-year study that interviewed 526 private boat anglers over 95 days. It is almost certainly true that the actual seasons fished, catch success, combinations of species caught, and fish consumption rates of all Palos Verdes Shelf and Cabrillo Pier anglers for the next 1 to 85 years (i.e., the exposure period covered by our study) are not perfectly represented by the survey data collected from CPFV and Cabrillo Pier anglers between September 1991 through August 1992 and from private boat anglers between January 1998 through January 1999. In addition, because we used data from all CPFV anglers interviewed throughout the Santa Monica Bay Region to characterize CPFV anglers fishing at the Palos Verdes Shelf, the range of species caught and consumed, fishing frequencies, and seasons fished are examples of exposure factor distributions that may not be fully representative of the CPFV anglers who fish at the Palos Verdes Shelf. Nonetheless, our use of the data as one-day snapshots of the range of angler behavior interjects less uncertainty into the exposure assessment than would assuming, as EPA did (SAIC, 1999), that all future fishing days for each angler are identical to the interview day.

None of the alternative analytical assumptions or exposure factor characterizations that we evaluated in our sensitivity analyses had an effect on 95th percentile cancer risk or noncancer hazard estimates that was larger than a factor of 3. In our experience, these results illustrate a high degree of stability in the analysis and, as a result, the confidence in our results is quite high. There remain, however, several known but uninvestigated uncertainties that suggest that our Monte Carlo analyses probably overestimate the risks to the subject angler populations.

Some of these uncertainties relate to chemical concentrations in fish. The need due to limited data to rely on chemical concentrations measured in composite samples for species other than kelp bass and white croaker likely underestimated variability in concentrations for anglers who typically eat fewer than four fish of one species from a fishing trip. However, this understatement of variability is not expected to result in a significant underestimate of exposure at the high end, where the lifetime exposure represents accumulated consumption over many fishing trips. In the cases where species were consumed for which concentration data were not available at the Palos Verdes Shelf or at Cabrillo Pier, the use of reasonable surrogate locations or species introduced uncertainty into the exposure calculations. Most of the studies used to determine the concentrations of tDDT and PCBs in fish are based on samples collected just before the spawning seasons for many fish species. Because the lipid content of the fish are elevated just before spawning, concentrations of tDDT and PCBs in fish are higher at these times than in the nonspawning portions of the year (Gold et al., 1997; Larson, 1991; Guillemot et al., 1985). Therefore, these concentration data are biased when used to represent concentrations throughout the year. In addition, the concentrations of chemicals in fish caught on the Palos Verdes Shelf have decreased over time (Stull, 1996). The reasons for this decline are probably twofold: the deposition of new sediments over the more contaminated sediments and the biological degradation of tDDT and PCBs (Quensen et al., 1998). Inasmuch as we are concerned with prospective risks due to fish consumption into the future, when concentrations will likely be lower than they are presently, use of current or historical concentration data creates a conservative bias in our risk estimates.

Our reliance on self-reported information from angler surveys may contribute to an overstatement of risk. Westat (1989) has shown that recall of information regarding participation in activities perceived as pleasurable, including fishing, tends to be overstated by anglers who respond to surveys. For example, it is likely that the 71 percent of anglers who agreed to participate in the *Santa Monica Bay Seafood Consumption Study* (SCCWRP and MBC, 1994) are more enthusiastic and avid anglers than the general angler population. Our correcting for sampling bias due to avidity and our using fish consumption rates derived from observation rather than angler recall was designed to minimize the overstatement of participation and success that is typically found in recall-based angler surveys (Roach et al., 1999; Connelly and Brown, 1995; Westat, 1989; Chase and Harada, 1984; Chosh, 1978), thus minimizing any overstatement of risk due to use of these angler behavior data.

By far the greatest uncertainties in any risk assessment of this nature pertain to the dose–response criteria applied to estimate cancer risk and/or the noncancer hazard from the long-term average dose rates developed in the exposure assessment. The dose–response criteria (i.e., cancer slope factors and reference doses) used in this risk assessment were developed by EPA based on conservative analysis of animal studies where the administered doses far exceeded those encountered by fish-consuming humans. Uncertainties in cancer slope factors result from species-to-species comparison, extrapolation from the high doses administered to test animals to the much lower doses typical of human exposures, data analysis techniques designed to provide upper-bound values, and numerous other biological factors. Similarly, the reference doses used to predict noncancer hazards were developed from animal data to which order-of-magnitude safety factors were applied. The result is reference doses that are almost certainly conservative estimates and thus highly protective.

The uncertainties with respect to dose–response criteria—and the resulting high probability that the risks will be overstated—are particularly large in the case of tDDT. Over the years, many scientists have suggested that EPA's use of the linearized multistage model is not appropriate for evaluation of carcinogenic risks from nongenotoxic chemicals such as tDDT (Butterworth et al., 1995; Cohen, 1995; Cohen and Ellwein, 1992). Specifically, the linearized multistage approach assumes that a chemical poses some carcinogenic hazard at any concentration (Crump, 1981; Crump and Crocket, 1985). While this may be a valid assumption for chemicals that are "initiators" of carcinogenesis, this assumption is almost certainly not valid for a class of compounds known as "promoters," which exhibit no carcinogenic effects below a threshold concentration (Pitot and Dragan, 1991; Williams and Weisburger, 1991; Pitot et al., 1987; Schulte-Hermann, 1985). The carcinogenic effects attributed in the literature to DDT and its metabolites are primarily ascribed to promotion—not to initiation (Williams and Weisburger, 1991; Flodstrom et al., 1990; Warngard et al., 1989; Preat et al., 1986; Shivapurkar et al., 1986; Numoto et al., 1985; Kitagawa et al., 1984; Peraino et al., 1975). The scientific consensus that DDT is, at most, a promoter and not an initiator is evident from the fact that DDT has been used for many years in studies investigating the mechanism of tumor promotion (Ruch et al., 1987, 1994; Leibold and Schwartz, 1993; Klaunig et al., 1990; Klaunig and Ruch, 1987a, b; Ruch and Klaunig, 1986a, b; Williams and Numoto, 1984).

As indicated in EPA's proposed guidelines for cancer risk assessment (EPA, 1996c), compounds that are believed to exhibit a nonlinear dose–response relationship (e.g., promoters) should be evaluated using a margin-of-exposure approach. Such an approach almost always suggests that for a given dose, the cancer risk is much less than that predicted by the linearized multistage model. Evidence that EPA is considering adopting a threshhold approach for some nongenotoxic carcinogens is reflected in its proposal regarding chloroform (Butterworth et al., 1998).

Another approach advocated by EPA is the use of physiologically based pharmacokinetic (PBPK) analyses. As many as 40 chemicals have had validated PBPK models published in the peer-reviewed literature (Leung and Paustenbach, 1995). Recently, a PBPK analysis for tDDT was proposed by Gazi and Conolly (1997) that has the potential to reduce the uncertainty associated with extrapolation of toxicity information from rodents to humans and to thereby aid in the characterization of its toxicity. In spite of the favorable experimental data and the lack of initiator capability, for the sake of this analysis, we have conservatively used EPA's cancer slope factor for DDT and DDE of 0.34 $(mg/kg\text{-}day)^{-1}$, which was developed using the linearized multistage model (IRIS, 2002; EPA, 1986).

As new site-specific data or innovative analysis techniques become available, it will be possible to further reduce—but not eliminate—the uncertainty that is an inherent feature of predictive scientific studies. We view this risk assessment as a significant step in reducing the uncertainty in the screening and "streamlined" risk assessments previously developed for this site. We agree with Fan et al. (1995) who stated: "A concerted attempt to better define the variability and decrease the uncertainty of hazard estimates will result in more efficient protection of the public health and the environment against toxic hazards." As such, this risk assessment provides a more sound basis for decision making regarding tDDT and PCBs in Palos Verdes Shelf fish than the prior studies.

17.7 CONCLUSION

This study of the fish consumption habits of southern California recreational fishers (or anglers) was conducted to determine how much health risk, if any, is posed by tDDT and PCBs in fish from the Palos Verdes Shelf or at Cabrillo Pier. This study found that sport fishers and others who eat fish from those locations are at no significantly increased health risk. The risks were below typical criteria used by regulatory agencies as a basis for concern.

Our study addressed the need for an improvement over the prior risk assessments for Palos Verdes Shelf and Cabrillo Pier anglers conducted by EPA and the State of California. Our study was performed to better estimate the size of health risks to these anglers due to consuming fish. We relied on existing site-specific data. EPA generally favors the use of such site-specific information when possible (see, e.g., EPA, 1989a, b; 1992a, b; 1995; 1997a, b, c, d; 1998b; 1999a, b). We made extensive use of these data in addition to utilizing data that had not been used in previous assessments (i.e., Ally et al., 1991; Love and Hansen, 1999).

Our assessment described the range of risks that may occur across different anglers. The Monte Carlo approach adopted in this assessment was consistent with EPA's (1997a, 1999a) guidelines for use of this methodology. Our Monte Carlo approach represents an important departure from the methods used in the prior assessments because, unlike those assessments, our study made maximum use of the available site-specific data to characterize patterns and changes in fishing behavior that affect risk. Our assessment was also unique because it used a Monte Carlo method that captured the correlations among factors that affect exposure and risk. In particular, our use of Monte Carlo methods to estimate exposure on a fishing trip-by-fishing trip (i.e., event-by-event) basis addressed the need identified by EPA (1997a) to take into consideration any moderate to strong correlations among exposure factors while avoiding the extrapolation of short-term data on angler behavior to long-term scenarios. Our results were validated by their agreement with independent local studies regarding fishing and consumption practices. Finally, we have a high level of confidence in our results because our sensitivity analyses quantitatively investigated the uncertainty associated with our methods and results.

In summary, our risk assessment, built on data collected in five different major local studies, found that there was no significant cancer risk or non-cancer hazard due to consuming sport-caught fish, including white croaker, from either the Palos Verdes Shelf or Cabrillo Pier. The risks to anglers were below the level at which EPA typically requires that action be taken to reduce exposure. The methodology used here should be applicable to characterizing the risks to those who ingest fish from the waterways of most industrialized nations.

REFERENCES

Adams, S. R., Hanna, C. A., Mayernik, J. A., and Mendez, W. M. Jr. 1994. Probabilistic health risk assessment for exposures to estuary sediments and biota contaminated with polychlorinated biphenyls, polychlorinated terphenyls and other toxic substances. *Risk Anal.* 14: 577–594.

Agency for Toxic Substances and Disease Registry (ATSDR). 1994. *Toxicological Profile for 4,4'-DDT, 4-4'-DDE, 4,4'-DDD (Update)*. U.S. Department of Health and Human Service, Public Health Service, Atlanta, GA.

Agency for Toxic Substances and Disease Registry (ATSDR). 1996. *Toxicological Profile for Polychlorinated Biphenyls (Update)*. U.S. Department of Health and Human Service, Public Health Service, Atlanta, GA.

Allen, M. J., Velez, P. V., Diehl, D. W., McFadden, S. E., and Kelsh, M. 1996. Demographic variability in seafood consumption rates among recreational anglers of Santa Monica Bay, California, in 1991–1992. *Fish. Bull.* 94: 597–610.

Ally, J. R. R., Ono, D. S., Read, R. B., and Wallace, M. 1991. *Final Report to the California Department of Fish and Game*. Marine Research Division, Administrative Report 90-2. Long Beach, CA.

Bogen, K. T., Conrado, C. L., and Robison, W. L. 1997. Uncertainty and variability in updated estimates of potential dose and risk at a U.S. nuclear test site—Bikini Atoll. *Health Phys.* 73: 115–126.

Burmaster, D. E., and Von Stackelberg, K. 1991. Using Monte Carlo simulation in public health risk assessments: Estimating and presenting full distributions of risk. *J. Expos. Anal. Environ. Epidemiol.* 1: 491–512.

Butterworth, B. E., Conolly, R. B., and Morgan, K. T. 1995. A strategy for establishing mode of action of chemical carcinogens as a guide for approaches to risk assessments. *Cancer Lett.* 93: 129–146.

Butterworth, B. E., Kedderis, G. L., and Conolly, R. B. 1998. The chloroform cancer risk assessment: A mirror of scientific understanding. *CIIT Activities* 18: 1–10.

California Environmental Protection Agency (CalEPA). 1996. *Air Toxics Hot Spots Program Risk Assessment Guidelines Part IV: Exposure Assessment and Stochastic Analysis*. California Environmental Protection Agency, Sacramento, CA.

Carrington, C. D., Bolger, P. M., and Scheuplein, R. J. 1996. Risk analysis of dietary lead exposure. *Food Addit. Contam.* 13: 61–76.

Cassin, M. H., Lammerding, A. M., Todd, E. C., Ross, W., and McColl, R. S. 1998. Quantitative risk assessment for *Escherichia coli* O157:H7 in ground beef hamburgers. *Int. J. Food Microbiol.* 41: 21–44.

Chan, H. M., Berti, P. R., Receveur, O., and Kuhnlein, H. V. 1997. Evaluation of the population distribution of dietary contaminant exposure in an Arctic population using Monte Carlo statistics. *Environ. Health Perspect.* 105: 316–321.

Chase, D. R., and Harada, M. 1984. Response error in self-reported recreation participation. *J. Leisure. Res.* 15: 322–329.

Chosh, D. N. 1978. Optional recall for discrete events. *Am. Statist. Assoc. 1978 Proc.*, pp. 615–616.

Cohen, S. M. 1995. Human relevance of animal carcinogenicity studies. *Reg. Toxicol. Pharmacol.* 21: 75–80.

Cohen, S. M., and Ellwein, L. B. 1992. Risk assessment based on high dose animal exposure experiments. *Chem. Res. Toxicol.* 5: 742–748.

Condon, S. K. 1996. Personal Communications, August 25 and 18, 1983; Commonwealth of Massachusetts Department of Public Health, Boston, MA, 1983. As cited in Agency for Toxic Substances and Disease Registry (ATSDR), 1996.

Connelly, N. A. and T. L. Brown. 1995. Use of angler diaries to examine biases associated with 12-month recall on mail questionnaires. Transactions of the American Fisheries Society. 124: 413–422.

Copeland, T. L., Paustenbach, D. J., Harris, M. A., and Otani, J. 1993. Comparing the results of a Monte Carlo analysis with EPA's reasonable maximum exposed individual (RMEI): A case study of a former wood treatment site. *Reg. Toxicol. Pharmacol.* 18: 275–312.

Copeland, T. L., Holbrow, A. M., Otani, J. M., Connor, K. T., and Paustenbach, D. J. 1994. Use of probabilistic methods to understand the conservatism in California's approach to assessing health risks posed by air contaminants. *J. Air Waste Manag. Assoc.* 44: 1399–1413.

Crouch, E. A. C. 1996a. Uncertainty distributions for cancer potency factors: Laboratory animal carcinogenicity bioassays and interspecies extrapolation. *Hum. Ecol. Risk Assess.* 2: 103–129.

Crouch, E. A. C. 1996b. Uncertainty distributions for cancer potency factors: Combining epidemiology studies with laboratory bioassays. The example of acrylonitrile. *Hum. Ecol. Risk Assess.* 2: 130–149.

Crump, K. S. 1981. An improved procedure for low-dose carcinogenic risk assessment from animal data. *J. Environ. Pathol. Toxicol.* 5: 675.

Crump, K. S., and Crockett, P. 1985. Improved confidence limits for low-dose carcinogenic risk assessment from animal data. *J. Haz. Mat.* 10: 419–431.

Cullen, A. C. 1994. Measures of compounding conservatism in probabilistic risk assessment. *Risk Anal.* 14: 389–393.

Cullen, A. C. 1995. The sensitivity of probabilistic risk assessment results to alternative model structures: A case study of municipal waste incineration. *J. Air Waste Manag. Assoc.* 45: 538–546.

Del Pup, J., Kmiecik, J., Smith, S., and Reitman, F. 1996. Improvement in human health risk assessment utilizing site- and chemical-specific information: A case study. *Toxicology* 113: 346–350.

Duggan, R. E. 1968. Residues in food and feed. Pesticide residue levels in food in the United States from July 1, 1963 to June 30, 1967. *Pesticide Monit. J.* 2: 2–46.

Eckhardt, R. 1987. Stan Ulam, John von Neumann, and the Monte Carlo method. *Los Alamos Sci.* 15: 131–137.

Eschenroeder, A. Q., and Faeder, E. J. 1988. A Monte Carlo analysis of health risks from PCB-contaminated mineral oil transformer fires. *Risk Anal.* 8: 291–297.

Fan, A., Howd, R., and Davis, B. 1995. Risk assessment of environmental chemicals. *Annu. Rev. Pharmacol. Toxicol.* 35: 341–368.

Fiering, M. B., Wilson, R., Kleiman, E., and Zeise, L. 1984. Statistical distributions of health risks. *Civ. Eng.* 1: 129–138.

Finley, B., and Paustenbach, D. J. 1994. The benefits of probabilistic exposure assessment: Three case studies involving contaminated air, water and soil. *Risk Anal.* 14: 53–73.

Finley, B. L., Proctor, D., Scott, P., Harrington, N. W., Paustenbach, D. J., and Price, P. S. 1994. Recommended distributions for exposure factors frequently used in health risk assessment. *Risk Anal.* 14: 533–553.

Finley, B. L., Scott, P., and Paustenbach, D. J. 1993. Evaluating the adequacy of maximum contaminant levels as health-protective cleanup goals: An analysis based on Monte Carlo techniques. *Reg. Toxicol. Pharmacol.* 18: 438–455.

Flodstrom, S., Hemming, H., Warngard, L., and Ahlborg, U. G. 1990. Promotion of altered hepatic foci development in rat liver, cytochrome P450 enzyme induction and inhibition of cell-cell communication by DDT and some structurally related organohalogen pesticides. *Carcinogenesis* 11: 1413–1417.

Frey, H. C., and Rhodes, D. S. 1996. Characterizing, simulating, and analyzing variability and uncertainty: An illustration of methods using an air toxics emissions approach. *Hum. Ecol. Risk Assess.* 2: 762–797.

Gazi, E., and Conolly, R. B. 1997. Use of DDT/DDE rat and human PBPK models for reducing the uncertainty in human risk assessment. Paper presented at the Annual Meeting of the Society for Risk Analysis, Washington, DC.

Gold, M., Alamillo, J., Fleischli, S., Forrest, J., Gorke, R., Heibshi, L., and Gossett, R. 1997. *Let the Buyer Beware: A Determination of DDT and PCB Concentrations in Commercially Sold White Croaker.* Heal the Bay, CRG Marine Laboratories, and California Wellness Foundation. Samoa Monica.

Goodrum, P. E., Diamond, G. L., Hassett, J. M., and Johnson, D. L. 1996. Monte Carlo modeling of childhood lead exposure: Development of a probabilistic methodology for use with the EPA IEUBK model for lead in children. *Hum. Ecol. Risk Assess.* 2: 681–708.

Gossett, R. W., Puffer, H. W., Arthur, R. H., and Young, D. R. 1983. DDT, PCB and benzo(*a*)pyrene levels in white croaker (*Genyonemus lineatus*) from southern California. *Mar. Poll. Bull.* 14: 60–65.

Guillemot, P. J., Larson, R. J., and Lenarz, W. H. 1995. Seasonal cycles of fat and gonad volume in five species of northern California rockfish (Scorpaenidae: *Sebastes*). *Fish Bull.* 83: 299–311.

Harrington, N. W., Curry, C. L., and Price, P. S. 1995. The Micro Exposure event modeling approach to probabilistic exposure assessment. Paper No. 95-TA42.03. In *Proceedings of the 88th Annual Meeting of the Air and Waste Management Association.* San Antonio, TX. June 18–23, 1995. Air and Waste Management Association. Pittsburgh, PA.

Hattis, D. B., and Burmaster, D. E. 1994. Assessment of variability and uncertainty distributions for practical risk assessments. *Risk Anal.* 14: 713–730.

Hayes, W. J., Dale, W. E., and Pirkle, C. I. 1971. Evidence of safety of long-term, high, oral doses of DDT for man. *Arch. Environ. Health* 22: 119–134.

Hayes, W. J., Durham, W. F., and Cueto, C. 1956. The effect of known repeated oral doses of chlorinophenothane (DDT) in man. *J. Am. Med. Assoc.* 162: 890–897.

Higginson, J. 1985. DDT: Epidemiological evidence. *IARC Sci. Publ.* 65: 107–117.

Hoffman, F. O., and Hammonds, J. S. 1994. Propagation of uncertainty in risk assessments: The need to distinguish between uncertainty due to lack of knowledge and uncertainty due to variability. *Risk Anal.* 14: 707–712.

Iannuzzi, T. J., Harrington, N. W., Shear, N. M., Curry, C. L., Carlson-Lynch, H., Henning, M. H., Su, S. H., and Rabbe, D. E. 1996. Distributions of key exposure factors controlling the uptake of xenobiotic chemicals in an estuarine food web. *Environ. Toxicol. Chem.* 15: 1979–1992.

Integrated Risk Information System (IRIS). 2002. *Integrated Risk Information System.* U.S. Environmental Protection Agency, IRIS, Cincinnati, OH, and the National Library of Medicine, Bethesda, MD.

International Commission on Radiological Protection (ICRP). 1981. *Report of the Task Group on Reference Man.* Pergammon, New York.

Keenan, R. E., Henning, M. H., Goodrum, P. E., Gray, M. N., Sherer, R. A., and Price, P. S. 1993. Using a MicroExposure event modeling approach to probabilistic exposure assessment. In *Proceedings of the 13th International Symposium on Chlorinated Dioxins and Related Compounds*, Vienna, Austria.

Keenan, R. E., Price, P. S., McCrodden, J., and Ebert, E. S. 1996. Using a Micro-Exposure Monte Carlo risk assessment for dioxin in Maine (USA) fish to evaluate the need for fish advisories. *Organohalogen Compounds* 30: 61–65.

Kitagawa, T., Hino, O., Nomura, K., and Sugano, H. 1984. Dose-response studies on promoting and anticarcinogenic effects of phenobarbital and DDT in the rat hepatocarcinogenesis. *Carcinogenesis* 5: 1653–1656.

Klaunig, J. E., and Ruch, R. J. 1987a. Role of cyclic AMP in the inhibition of mouse hepatocyte intercellular communication by liver tumor promoters. *Toxicol. Appl. Pharmacol.* 91: 159–170.

Klaunig, J. E., and Ruch, R. J. 1987b. Strain and species effects on the inhibition of hepatocyte intercellular communication by liver tumor promoters. *Cancer Lett.* 36: 161–168.

Klaunig, J. E., Ruch, R. J., and Weghorst, C. M. 1990. Comparative effects of phenobarbital, DDT, and lindane on mouse hepatocyte gap junctional intercellular communication. *Toxicol. Appl. Pharmacol.* 102: 553–563.

Larson, R. J. 1991. Seasonal cycles of reserves in relation to reproduction in Sebastes. *Environ. Biol. Fishes* 30: 57–70.

Laws, E. R., Curley, A., and Biros, F. J. 1967. Men with extensive occupational exposure to DDT. *Arch. Environ. Health* 15: 766–775.

Laws, E. R., Maddrey, W. C., Curley, A., and Burse, V. W. 1973. Long-term occupational exposure to DDT. *Arch. Environ. Health* 27: 318–321.

Leibold, E., and Schwartz, L. R. 1993. Inhibition of intercellular communication in rat hepatocytes by phenobarbital, 1,1,1-trichloro-2,2-bis(*p*-chlorophenyl)ethane (DDT) and hexachlorocyclohexane (lindane): Modification by antioxidants and inhibitors of cyclo-oxygenase. *Carcinogenesis* 14: 2377–2382.

Leung, H. W., Murray, F. J., and Paustenbach, D. J. 1988. A proposed occupational exposure limit for 2,3,7,8-tetrachlorodibenzo-*p*-dioxin. *Am. Ind. Hyg. Assoc. J.* 49: 466–474.

Leung, H. W., and Paustenbach, D. J. 1995. Physiologically based pharmacokinetic and pharmacodynamic modeling in health risk assessment and characterization of hazardous substances. *Toxicol. Lett.* 79: 55–65.

Lipfert, F. W., Moskowitz, P. D., Fthenakis, V., and Saroff, L. 1996. Probabilistic assessment of health risks of methylmercury from burning coal. *Neurotoxicology* 17: 197–211.

Lloyd, K. J., Thompson, K. M., and Burmaster, D. E. 1992. Probabilistic techniques for backcalculating soil cleanup targets. In K. B. Hoddinott and G. D. Knowles (Eds.), *Superfund Risk Assessment in Soil Contamination Studies*. American Society for Testing and Materials, Philadelphia, PA.

Lordo, R. A., Dinh, K. T., and Schwemberger, J. G. 1996. Semivolatile organic compounds in adipose tissue: Estimated averages for the U.S. population and selected subpopulations. *Am. J. Pub. Health* 86: 1253–1259.

Los Angeles County Sanitation District (LACSD). 1997. *Annual Report 1996*. Los Angeles County Sanitation District, Whittier, CA.

Love, M. 1996. *Probably More Than You Want to Know about the Fishes of the Pacific Coast*, 2nd ed. Really Big Press, Santa Barbara, CA.

Love, M. S., and Hansen, S. 1999. Recreational vessel fishery for white croaker on the Palos Verdes Shelf. Marine Science Institute, University of California, Santa Barbara, CA, and S. R. Hansen and Associates, Occidental, CA.

Love, M. S., McGowen, G. E., Westphal, W., Lavenberg, R. J., and Martin, L. 1984. Aspects of the life history and fishery of the white croaker, *Genyonemus lineatus* (Sciaenidae), off California. *Fish. Bull.* 82: 179–198.

Massachusetts Department of Environmental Protection (MADEP). 1995. *Interim Final Policy*. WSC/ORS-95-141. Massachusetts Department of Environmental Protection, Boston, MA.

McKone, T. E. 1994. Uncertainty and variability in human exposures to soil contaminants through home-grown food: A Monte Carlo assessment. *Risk Anal.* 14: 449–463.

McKone, T. E., and Bogen, K. T. 1991. Predicting the uncertainties in risk assessment. *Environ. Sci. Technol.* 25: 1674–1681.

McKone, T. E., and Bogen, K. T. 1992. Uncertainties in health risk assessment: An integrated case study based on tetrachloroethylene in groundwater. *Reg. Toxicol. Pharmacol.* 15: 86–103.

Metropolis, N. 1987. The beginning of the Monte Carlo method. *Los Alamos Sci.* 15: 125–130.

Microsoft. 1996. *Excel Visual Basic for Applications*, Version 7.0. Microsoft, Palo Alto, CA.

Microsoft. 1997a. *Microsoft Office Suite: Microsoft Access 97*. Microsoft, Palo Alto, CA.

Microsoft. 1997b. *Microsoft Office Suite: Microsoft Excel 97*. Microsoft, Palo Alto, CA.

Morgan, G. M., Henrion, M., and Small, M. 1990. *Uncertainty: A Guide to Dealing with Uncertainty in Quantitative Risk and Policy Analysis*. Cambridge University Press, New York.

National Academy of Sciences (NAS). 1994. *Science and Judgment in Risk Assessment*. National Research Council, National Academy Press, Washington, DC.

National Marine Fisheries Service (NMFS). 2000. Marine Recreational Fisheries Statistics–National Marine Fisheries Service–Marine Recreational Fisheries Statistics Survey. Ed. John F. Witzig. Available from *http://www.st.nmfs.gov/recreational/survey/overview.html*. Retrieved February 5, 2000.

Needham, L. 1997. Personal communication. Centers for Disease Control, Atlanta, GA, October 9.

Numoto, S., Tanaka, T., and Williams, G. M. 1985. Morphologic and cytochemical properties of mouse liver neoplasms induced by diethylnitrosamine and promoted by 4-4'-dichlorodiphenyltrichloroethane, chlordane, or hepatochlor. *Toxicol. Pathol.* 13: 325–333.

Ortelee, M. F. 1958. Study of men with prolonged intensive occupational exposure to DDT. *AMA Arch. Ind. Health* 18: 433–440.

Palisade. 1996. *@Risk Advanced Risk Analysis for* Spreadsheets, Windows Version. Palisade Corporation, Newfield, NY.

Paustenbach, D. J. 1995. The practice of health risk assessment in the United States (1975–1995): How the U.S. and other countries can benefit from that experience. *Hum. Ecol. Risk Assess.* 1: 29–79.

Paustenbach, D. J., Meyer, D. M., Sheehan, P. J., and Lau, V. 1991. An assessment and quantitative uncertainty analysis of the health to workers exposed to chromium contaminated soils. *Toxicol. Ind. Health* 7: 159–196.

Paustenbach, D. J., Wenning, R. J., Lau, V., Harrington, N. W., Rennix, D. K., and Parsons, A. H. 1992. Recent developments on the hazards posed by 2,3,7,8-tetrachlorodibenzo-*p*-dioxin in soil: Implications for setting risk-based cleanup levels at residential and industrial sites. *J. Toxicol. Environ. Health* 36: 103–149.

Peraino, C., Fry, R. J. M., Staffedldt, E., and Christopher, J. P. 1975. Comparative enhancing effects of phenobarbital, amobarbital, diphenyldantoin, and dichlorodiphenyltrichloroethane on 2-acetylaminofluorene-induced hepatic tumorigenesis in the rat. *Cancer Res.* 35: 2884–2890.

Pitot, H. C., and Dragan, Y. P. 1991. Facts and theories concerning the mechanisms of carcinogenesis. *FASEB J.* 5: 2280–2286.

Pitot, H. C., Goldsworthy, T. L., Moran, S., Kennan, W., Glauert, H. P., Maronpot, R. R., and Campbell, H. A. 1987. A method to quantitate the relative initiating and promoting potencies of hepatocarcinogenic agents in their dose-response relationships to altered hepatic foci. *Carcinogenesis* 8: 1491–1499.

Pollock, G. A., Uhaa, I. J., Cook, R. R., Fan, A., Fries, L., and Bounarati, C. 1990. *Evaluation of Health Risks Related to Consumption of Commercial White Croaker (Genyonemus lineatus) from the Fishery on the Palos Verdes Shelf, California.* California Department of Health Services, Sacramento, CA.

Pollock, G. A., Uhaa, I. J., Fan, A. M., Wisniewski, J. A., and Witherell, I. 1991. *Comprehensive Study of Chemical Contamination of Marine Fish from Southern California.* California Environmental Protection Agency, Sacramento, CA.

Pollock, K. H., Jones, C. M., and Brown, T. L. 1994. *Angler Survey Methods and Their Applications in Fisheries Management.* American Fisheries Society, Bethesda, MD.

Preat, V., Gerlache, J. D., Lans, M., Taper, H., and Roberfroid, M. 1986. Comparative analysis of the effect of phenobarbital, dichlorodiphenyltrichloroethane, butylated hydroxytoluene, and nafenopin on rat hepatocarcinogenesis. *Carcinogenesis* 7: 1025–1028.

Price, P. S., Curry, C. L., Goodrum, P. E., Gray, M. N., McCrodden, J. I., Harrington, N. W., Carlson-Lynch, H., and Keenan, R. E. 1996. Monte Carlo modeling of time dependent exposures using a MicroExposure event approach. *Risk Anal.* 16: 339–348.

Price, P. S., Scott, P. K., Wilson, N. D., and Paustenbach, D. J. 1998. An empirical approach for deriving information on total duration of exposure from information on historical exposure. *Risk Anal.* 18: 611–620.

Price, P. S., Su, S. H., and Gray, M. N. 1994. The effect of sampling bias on estimates of angler consumption rates in creel surveys. *J. Exp. Anal. Environ. Epidemiol.* 4: 355–372.

Price, P. S., Young, J. S., and Chaisson, C. F. 2001. Assessing aggregate and cumulative pesticide risks using a probabilistic model. *Ann. Occup. Hyg.* 2001 Apr; 45 Suppl 1: S131–42.

Puffer, H. W., Azen, S. P., Duda, M. J., and Young, D. R. 1981. *Consumption Rates of Potentially Hazardous Marine Fish Caught in the Metropolitan Los Angeles Area.* U.S. Environmental Protection Agency, Corvallis, OR.

Puffer, H. W., and Gossett, R. W. 1983. PCB, DDT, and benzo(*a*)pyrene in raw and pan-fried white croaker (*Genyonemus lineatus*). *Bull. Environ. Contam. Toxicol.* 30: 65–73.

Quensen, J. F., Mueller, S. A., Jain, M. K., and Tiedje, J. M. 1998. Reductive dechlorination of DDE to DDMU in marine sediment microcosms. *Science* 280: 722–724.

Reinert, R. E., Stewart, D., and Seagran, H. L. 1972. Effects of dressing and cooking on DDT concentrations in certain fish from Lake Michigan. *J. Fish. Res. Board Can.* 29: 525–529.

Richardson, G. M., and Allan, M. 1996. A Monte Carlo assessment of mercury exposure and risks from dental amalgam. *Hum. Ecol. Risk Assess.* 2: 709–761.

Risebrough, R. W. 1987. *Distribution of Organic Contaminants in Coastal Areas of Los Angeles and the Southern California Bight.* University of California, Santa Cruz, CA.

Roach, Brian, Joan Trial, and Kevin Boyle. 1999. Comparing 1994 Angler Catch and Harvest Rates from On-site and Mail Surveys on Selected Maine Lakes. *North American Journal of Fisheries Management.* 19(1). February.

Robson, D. S. 1961. On the statistical theory of a roving creel census of fishermen. *Biometrics* Sept.: 415–437.

Ruch, R. J., Bonney, W. J., Sigler, K., Guan, X., Matesic, D., Schafer, L. D., Dupont, E., and Trosko, J. E. 1994. Loss of gap junctions from DDT-treated rat liver epithelial cells. *Carcinogenesis* 15: 301–306.

Ruch, R. J., and Klaunig, J. E. 1986a. Antioxidant prevention of tumor promoter induced inhibition of mouse hepatocyte intercellular communication. *Cancer Lett.* 33: 137–150.

Ruch, R. J., and Klaunig, J. E. 1986b. Effects of tumor promoters, genotoxic carcinogens and hepatocytotoxins on mouse hepatocyte intercellular communication. *Cell. Biol. Toxicol.* 2: 469–483.

Ruch, R. J., Klaunig, J. E., and Pereira, M. A. 1987. Inhibition of intercellular communication between mouse hepatocytes by tumor promoters. *Toxicol. Appl. Pharmacol.* 87: 11–120.

Ruffle, B., Burmaster, D. E., Anderson, P. D., and Gordon, H. D. 1994. Lognormal distributions for fish consumption by the general U.S. population. *Risk Anal.* 14: 395–404.

Rupp, E. M. 1980. Age dependent values of dietary intake for assessing human exposure to environmental pollutants. *Health Phys.* 39: 151–163.

Rupp, E. M., Miller, F. L., and Baes, C. F. 1980. Some results of recent surveys of fish and shellfish consumption by age and region of U.S. residents. *Health Phys.* 39: 165–175.

Science Applications International Corporation (SAIC). 1999. *Human Health Risk Evaluation for Palos Verdes Shelf.* Science Application International Corporation, San Francisco, CA. Prepared for U.S. Environmental Protection Agency, Region IX, San Francisco, CA, April 1999.

Schulte-Hermann, R. 1985. Tumor promotion in the liver. *Arch. Toxicol.* 57: 147–158.

Shivapurkar, N., Hover, K. L., and Poirier, L. A. 1986. Effect of methionine and choline on liver tumor promotion by phenobarbital and DDT in diethylnitrosamine-initiated rats. *Carcinogenesis* 7: 547–550.

Simon, T. W. 1999. Two-dimensional Monte Carlo simulation and beyond: A comparison of several probabilistic risk assessment methods applied to a Superfund site. *Hum. Ecol. Risk Assess.* 5: 823–843.

Skea, J. C., Jackling, S., Symula, J., Simonin, H. A., Harris, E. J., and Colquhoun, J. R. 1981. *Summary of Fish Trimming and Cooking Techniques Used to Reduce Levels of Oil Soluble Contaminants.* Field Toxicant Research Unit, Rome, NY.

Slob, W. 1994. Uncertainty analysis in multiplicative models. *Risk Anal.* 14: 571.

Smith, W. E., Funk, K., and Zabik, M. E. 1973. Effects of cooking on concentrations of PCB and DDT compounds in Chinook (*Oncorhynchus tshawytscha*) and Coho (*O. kisutch*) salmon from Lake Michigan. *J. Fish. Res. Bd. Can.* 30: 702–706.

Southern California Coastal Water Research Project (SCCWRP), MBC Applied Environmental Sciences (MBC), and University of California Santa Cruz Trace Organics Facility. 1992. *Final Report: Santa Monica Bay Seafood Contamination Study.* Prepared for the Santa Monica Bay Restoration Project. Southern California Coastal Water Research Project, Westminster, MBC Applied Environmental Sciences, Costa Mesa, and the University of California, Santa Cruz, CA.

Southern California Coastal Water Research Project (SCCWRP) and MBC. 1994. *Santa Monica Bay Seafood Consumption Study.* Prepared for the Santa Monica Bay Restoration Project. Southern California Coastal Water Research Project, Westminster, and MBC Applied Environmental Sciences, Costa Mesa, CA.

Stull, J. K. 1996. Ocean monitoring off Palos Verdes, southern California. In *Oceans 96, MTS/IEEE Conference Proceedings*, pp. 299–306. Marine Technology Society, Washington, DC.

Stull, J. K., Dryden, K. A., and Gregory, P. A. 1987. *An Historical Review of Fisheries Statistics and Environmental and Societal Influences off the Palos Verdes Peninsula, California.* CalCOFI Rep., Vol. 38. California Cooperative Oceanic Fisheries Investigation, La Jolla, CA.

SYSTAT. 1996. *SYSTAT*, Version 6.0. SYSTAT, Evanston, IL.

Thompson, K. M., Burmaster, D. E., and Crouch, E. A. C. 1992. Monte Carlo techniques for quantitative uncertainty analysis in public health risk assessments. *Risk Anal.* 12: 53–63.

Trotter, W. J., Corneliussen, P. E., Laski, R. R., and Vannelli, J. J. 1988. Levels of polychlorinated biphenyls and pesticides in bluefish before and after cooking. *J. Assoc. Off. Anal. Chem.* 72: 501–503.

Ulam, S. M. 1991. *Adventures of a Mathematician.* University of California Press, Berkeley, CA.

U.S. Environmental Protection Agency (EPA). 1986. *The Assessment of the Carcinogenicity of Dicofol (Kelthane), DDT, DDE, and DDD (TDE).* EPA/600/6-86/001. EPA, Washington, DC.

U.S. Environmental Protection Agency (EPA). 1989a. *Risk Assessment Guidance for Superfund, Vol. I: Human Health Evaluation Manual (Part A)—Interim Final.* EPA/540/1-89-002. EPA, Washington, DC.

U.S. Environmental Protection Agency (EPA). 1989b. *Assessing Human Health Risks from Chemically Contaminated Fish and Shellfish: A Guidance Manual.* EPA-503/8-89-002. EPA, Washington, DC.

U.S. Environmental Protection Agency (EPA). 1991. *Role of the Baseline Risk Assessment in Superfund Remedy Selection Decisions.* EPA, Washington, DC.

U.S. Environmental Protection Agency (EPA). 1992a. *Fed. Reg.* 57(104): 22888–22938.

U.S. Environmental Protection Agency (EPA). 1992b. *Consumption Surveys for Fish and Shellfish: A Review and Analysis of Survey Methods.* EPA/540/1-89-002. EPA, Washington, DC.

U.S. Environmental Protection Agency (EPA). 1995. *Guidance for Risk Characterization.* EPA, Washington, DC.

U.S. Environmental Protection Agency (EPA). 1996a. Memorandum to K. Takata from A. Lincoff and M. Montgomery, Remedial Project Managers, EPA Region IX, Re: Engineering evaluation and cost analysis approval for addressing contaminated marine sediments on the Palos Verdes Shelf. San Francisco, CA.

U.S. Environmental Protection Agency (EPA). 1996b. *PCBs: Cancer Dose-Response Assessment and Application to Environmental Mixtures.* EPA/600/P-96/001. EPA, Washington, DC.

U.S. Environmental Protection Agency (EPA). 1996c. *EPA Proposed Guidelines for Carcinogen Risk Assessment [Signed April 10, 1996].* EPA, Washington, DC.

U.S. Environmental Protection Agency (EPA). 1997a. *Guiding Principles for Monte Carlo Analysis.* EPA/630/R-97/001. EPA, Washington, DC.

U.S. Environmental Protection Agency (EPA). 1997b. *Exposure Factors Handbook,* Vol. I: *General Factors.* EPA/600/P-95/002Fa. EPA, Washington, DC.

U.S. Environmental Protection Agency (EPA). 1997c. *Exposure Factors Handbook,* Vol. II: *Food Ingestion Factors.* EPA/600/P-95/002Fb. EPA, Washington, DC.

U.S. Environmental Protection Agency (EPA). 1997d. *Exposure Factors Handbook,* Vol. III: *Activity Factors.* EPA/600/P-95/002Fa. EPA, Washington, DC.

U.S. Environmental Protection Agency (EPA). 1997e. *Guidance for Assessing Chemical Contaminant Data for Use in Fish Advisories,* Vol. 2: *Risk Assessment and Fish Consumption Limits,* 2nd ed. EPA 823-B-97-009. EPA, Washington, DC.

U.S. Environmental Protection Agency (EPA). 1998a. *Guidance for Conducting Fish and Wildlife Consumption Surveys.* EPA-823-B-98-007. EPA, Washington, DC.

U.S. Environmental Protection Agency (EPA). 1998b. *Ambient Water Quality Criteria Derivation Methodology: Human Health.* Technical support document. Final draft. EPA/822/B-98/005. EPA, Washington, DC.

U.S. Environmental Protection Agency (EPA). 1999a. *Risk Assessment Guidance for Superfund,* Vol. 3, Part A: *Process for Conducting Probabilistic Risk Assessment, Draft.* EPA 000-0-99-000. EPA, Washington, DC.

U.S. Environmental Protection Agency (EPA). 1999b. *Sociodemographic Data Used for Identifying Potentially Highly Exposed Populations.* EPA/600/R-99/060. EPA, Washington, DC.

Warngard, L., Hemming, H., Flodstrom, S., Duddy, S. K., and Kass, G. E. N. 1989. Mechanistic studies on the DDT-induced inhibition of intercellular communication. *Carcinogenesis* 10: 471–476.

Westat. 1989. *Final Report on Investigation of Possible Recall/Reference Period Bias in National Surveys of Fishing, Hunting, and Wildlife—Associated Recreation.* Contract No. 14-16-009-87-008. Westat. Rockville, MD.

Whitmyre, G. K., Driver, J. H., Ginevan, M. E., Tardiff, R. G., and Baker, S. R. 1992a. Human exposure assessment. I: Understanding the uncertainties. *Toxicol. Ind. Health* 8: 297–320.

Whitmyre, G. K., Driver, J. H., Ginevan, M. E., Tardiff, R. G., and Baker, S. R. 1992b. Human exposure assessment. II: Quantifying and reducing the uncertainties. *Toxicol. Ind. Health* 8: 321–342.

Williams, G. M., and Numoto, S. 1984. Promotion of mouse liver neoplasms by the organochlorine pesticides chlordane and heptachlor in comparison to DDT. *Carcinogen. (Lond.)* 5: 1689–1696.

Williams, G. M., and Weisburger, J. H. 1991. Chemical carcinogens. In M. O. Amdur, J. Doull, and C. D. Klaasen (Eds.), *Casarett and Doull's Toxicology*. Pergamon, New York.

Wilson, N. D., Price, P. S., and Paustenbach, D. J. 2001. An event-by-event probabilistic methodology for assessing the health risks of persistent chemicals in fish: a case study at the Palos Verdes Shelf. *J Toxicol Environ Health* 62(8): 595–642.

Wilson, N. D., Shear, N. M., Paustenbach, D. J., and Price, P. S. 1998. The effect of cooking practices on the concentration of DDT and PCB compounds in the edible tissue of fish. *J. Expos. Assess. Environ. Epidemiol.* 8: 423–440.

Zabik, M. E., Hoojjat, P., and Weaver, C. M. 1979. Polychlorinated biphenyls, dieldrin and DDT in lake trout cooked by broiling, roasting, and microwave. *Bull. Environ. Contam. Toxicol.* 21: 136–143.

Zabik, M. E., Booren, A. M., Zabik, M. J., Welch, R., and Humphrey, H. 1996. Pesticide residues, PCBs, and PAHs in baked, charbroiled, salt boiled, and smoked Great Lakes lake trout. *Food Chem.* 55: 231–239.

Zabik, M. E., Zabik, M. J., Booren, A. M., Daubenmire, S., Pascall, M. A., Welch, R., and Humphrey, H. 1995a. Pesticides and total polychlorinated biphenyls residues in raw and cooked walleye and white bass harvested from the Great Lakes. *Bull. Environ. Contam. Toxicol.* 54: 396–402.

Zabik, M. E., Zabik, M. J., Booren, A. M., Nettles, M., Song, J. H., Welch, R., and Humphrey, H. 1995b. Pesticides and total polychlorinated biphenyls in chinook salmon and carp harvested from the Great Lakes: Effects of skin-on and skin-off processing and selected cooking methods. *J. Agric. Food Chem.* 43: 993–1001.

SECTION I
Assessing Risks Associated with Consumer Products

18 Qualitative Health Risk Assessment of Natural Rubber Latex in Consumer Products

SEAN M. HAYS

Exponent, Boulder, Colorado

BRENT L. FINLEY

Exponent, Santa Rosa, California

18.1 INTRODUCTION

In the past 10 years, an increasing number of reports of type I (antibody-mediated) allergic responses to natural rubber latex (NRL) have appeared in the published literature (Landwehr and Boguniewicz, 1996; Turjanmaa et al., 1996). Reported responses range from mild effects such as urticaria and rhinitis to more severe effects such as anaphylactic shock. The first published report of a type I NRL reaction in North America appeared in 1989 (Slater, 1989), and most of the subsequent case reports have come from the health-care industry, where nurses, surgeons, and other health-care staff often wear NRL-containing gloves. Surgical patients, particularly spina bifida patients, constitute another large group of individuals reporting NRL allergic responses (Turjanmaa et al., 1996). Type I NRL allergies have also been recently reported, though far less frequently, in other occupations that use NRL gloves (e.g., food service workers, industrial workers). Conversely, type IV (cell-mediated) allergic responses associated with NRL glove use have been known and reported for several decades. These reactions have typically been attributed to delayed hypersensitivity reactions to various chemicals used in the manufacturing process.

It has been strongly suspected that the causative agents in a type I NRL response are naturally occurring latex proteins. For example, 10 to 15 of the approximately 200 different proteins in raw latex have been identified as being commonly recognized by IgE antibodies in NRL-sensitized patients (Alenius

Human and Ecological Risk Assessment: Theory and Practice, Edited by Dennis J. Paustenbach
ISBN 0-471-14747-8 © 2002 John Wiley & Sons, Inc.

et al., 1994a; Kurup et al., 1995). Several of these proteins are very similar to proteins found in a wide variety of different foods (e.g., bananas, avocados, kiwis) and pollens (e.g., ragweed, timothy grass) that can also trigger type I allergic responses. It has also been suggested that "releasing agents" such as heavily crosslinked cornstarch, which are placed on the interior of NRL gloves to decrease the tackiness and provide for easier donning and removal, may play a role in the development of NRL allergy. Some investigators have suggested that NRL proteins can transfer from the glove interior to the cornstarch particles (Swanson et al., 1994) and that the liberation of these particles into the air serves as the vehicle for inhalation exposure to the NRL proteins.

Because of their ease of use, high degree of tactility, and proven effectiveness as barriers to viruses and other blood-borne disease agents, NRL gloves have been the glove of choice in the health-care industry for several decades. Given the nearly century-long use of latex in the medical field, it is unclear why type I hypersensitivity to latex is only now emerging as a potential health concern. Some researchers have suggested that the increased incidence of case reports in the early 1990s is linked to the Centers for Disease Control and Prevention (CDC) issuance of the Universal Precautions in 1987. This measure recommended that all health-care workers wear gloves to protect themselves (and the patient) from the human immunodeficiency virus (HIV) and other viruses. As a result, the use of NRL gloves in the health-care industry has increased dramatically in the past 10 years. Conversely, it has been suggested that the increase in case reports may be due in part to enhanced physician and patient awareness of NRL allergy. Specifically, it has been noted that there is a potential for diagnostic bias, both with underreporting of a previously unrecognized clinical entity in the past, as well as the misdiagnosis and overreporting of a now well-publicized health issue. But the degree to which such a bias influences the estimates of the incidence and prevalence of NRL allergy cannot be gauged easily (Granady and Slater, 1995).

In this chapter, we examine the weight of evidence linking NRL allergy and exposure to NRL-containing products and conduct a qualitative risk assessment (or perhaps simply a hazard identification). Because most reports of the reported health effects come from health-care workers, this evaluation focuses on the use of NRL gloves in the health-care industry as a case study. Several issues are addressed in this analysis.

First, although NRL glove use has increased significantly over the last decade, it is important to note that hundreds of millions of gloves were used annually in the 1960s through the 1980s. This begs the question: Why were there very few reported cases of NRL allergy during this time frame? Some have suggested that a significant change in the glove manufacturers' process may have occurred in the late 1980s that produced more "allergenic" gloves. Others have suggested that the allergic reactions have always existed, in both health-care workers and the general population, but they were simply not identified as a clinical entity until recently. In this review, we describe the NRL-glove manufacturing process with an emphasis on process changes that occurred in the late 1980s and whether they may have affected the allergenic content of the products. In addition, we examine the weight of evidence supporting a relation between glove use and inhalation responses (such as asthma). Several studies have measured airborne NRL levels in workplaces, but a critical examination of whether these measured concentrations can elicit a response has not been conducted.

Second, we evaluate the epidemiological evidence regarding the incidence of NRL allergy in health-care workers as compared to the general public. Although numerous case reports have been published, only a few studies have actually compared rates of NRL allergy in health-care workers to an "unexposed" population. Some of these studies suggest a higher NRL allergy incidence rate in health-care workers, while others conclude that there is no difference. Nearly all experts in this field agree that, to date, there has not been a rigorous application of epidemiological principles to assess whether a cause-and-effect relation has actually been measured in health-care workers (e.g., Hunt et al., 1995; Yunginger, 1995a).

Third, we assess the degree to which cross-reacting antigens (e.g., food and pollen antigens) might confound the reported observations. While there is no doubt that some individuals may exhibit a type I response upon NRL challenge, the nature of the sensitizing agent is not always clear. For example, an individual who has been sensitized to specific proteins in banana, kiwi, and certain other fruits can test "positive" for latex allergy even though that individual has not been occupationally exposed to latex.

We conclude with a summary of the weight of evidence regarding the hypothesis that glove use is associated with NRL sensitization and offer recommendations for future research that would help reduce uncertainties in understanding this relationship.

18.2 FACTORS THAT INFLUENCE EXPOSURE TO ANTIGENS IN NRL GLOVES

18.2.1 Latex Harvest and Protein Content

The rubber tree (*Heven brasiliensis*), which is tapped to obtain NRL, is grown mostly in the tropical regions of Southeast Asia. Typically, the tree is tapped by making a spiral cut into the wood just below the bark at an angle intended to sever the largest number of latex vessels. This process is continued on a regular basis by reopening the initial cut or making new ones on the tree. As many as 200 different proteins—ranging in molecular weight from less than 5 kilodaltons (kDa) to more than 200 kDa—have been identified in NRL. Some of these proteins are associated with formation of the rubber polymer (i.e., the rubber elongation factor) while others, such as the chitin-binding protein hevein, (lysozyme, aldolase, etc.) function as defense proteins. Similar proteins with similar functions are found widely distributed in nature from plants to molds.

It has been shown that the allergenicity of harvested latex can vary (by up to 25-fold) depending on the season (Yeang, 1997). This potential variance may correspond with changes in plant husbandry practices on rubber plantations as a consequence of the increased demand for latex after the Universal Precautions measure was promulgated. Specifically, more cuts were opened on individual rubber trees, and the trees were tapped more frequently than in the past. It has been demonstrated that the expression of some proteins involved in the biosynthesis of rubber is elevated with increased tapping pressure (Adiwilaga and Kush, 1996), however, it is unclear if allergenic proteins (or the allergenicity of latex) changes with increased tapping pressure.

It has also been proposed that the use of yield stimulants increased around the time that the Universal Precautions were implemented. Yield stimulants such as 2-chloroethanephosphonic acid (ethrel or ethepon) are used to increase the flow of latex (Subramaniam, 1995). It has been shown that the protein content in the common turnip (*Brassica rapa*) was stimulated by salicylic acid or Ethepon (Hanninen et al., 1999). The allergen contents of plants incubated with these agents for various periods of time (up to 192 hours) were tested in an IgE immunoblot assay using sera with antilatex antibodies. Both treatments and time increased the number and intensity of protein bands visible in the immunoblot. The amount of a 25-kDa protein increased more than 10-fold compared to plants serving as untreated controls, and this protein was recognized by the IgG from a rabbit immunized with hevein (a latex-specific protein). Similar studies have shown that treatment of rubber trees with ethylene, ethepon, or salicylic acid induce the expression of hevein (Broekaert et al., 1990; Potter et al., 1993).

In summary, except for Yeang's observation that allergenicity of fresh NRL may change as a function of season (Yeang, 1997), there is little information on the degree to which NRL allergenicity might be influenced by harvest practices. Several studies would seem to suggest that the protein content (and specifically hevein, a protein believed to be one of the more allergenic proteins in latex) increases in latex with increased tapping (wounding) and the use of yield stimulants such as ethepon or ethylene. However, the degree of change, if any, is unknown, and it is therefore not known whether any increased "allergenicity" of harvested NRL occurred as a result of these practices.

18.2.2   Treatment of Natural Rubber Latex

Following the harvest, the NRL is treated with a stabilizer, usually ammonia, to prevent premature coagulation and bacterial growth. Higher concentrations of ammonia (i.e., 0.7 percent) are used when the latex must travel a long distance before treatment. In the late 1980s, the increased consumer demand for NRL gloves resulted in the production facilities being moved closer to the source. As a result, the time interval between harvest and processing into products decreased, which reduced the need for use of high ammonia concentrations as a stabilizing agent.

Several researchers have shown that ammoniation causes degradation of proteins in NRL. Lu et al. (1995) reported that ammonia treatment leads to the breakdown and precipitation of certain latex proteins as both extractable and unextractable antigens. Nonammoniated latex (NAL) was reported to contain 26 distinct proteins, while ammoniated latex (AL) had only 4. When NAL was incubated with ammonia for 7 days, the protein profile was found to resemble that of AL. It is critical to note, however, that the allergenicity of the degraded AL proteins was preserved, as determined by similar inhibition of antilatex rabbit IgG by NAL and AL. Similarly, using antilatex IgE, Alenius et al. (1991) reported identical levels of allergens in NAL and AL extracts, even though the NAL extracts contained much higher levels of protein.

In summary, while it is clear that ammonia does degrade some NRL proteins, it appears that there may not be large differences in the allergenic content of AL and NAL, and therefore, it is unknown whether process changes in the late 1980s resulted in NRL products with greater allergenicity. To truly resolve this question, one might directly compare the allergenicity of glove extracts from gloves manufactured before and after the process changes (i.e., increased use of yield enhancers, decreased time between harvest of latex and manufacturing of gloves, reduced use of ammonia, etc.) (Hunt et al., 1995). To date, such data are not available in the published literature.

18.2.3 Characteristics of NRL Gloves

At present, the Food and Drug Administration (FDA) regulates all NRL gloves marketed in the United States for characteristics such as stress resistance, tensile strength, and sterility. Protein and allergen content are not regulated, but the FDA permits labeling of NRL gloves as "low protein" if the gloves contain less than 50 μg of water-soluble protein per gram of glove material. As noted in several studies that have examined the protein and NRL allergen content of various types of gloves (e.g., Beezhold and Beck, 1992; Yunginger et al., 1993; Alenius et al., 1994a; Yunginger et al., 1994; Yunginger, 1995b; Wrangsjo and Lundberg, 1996; Baur et al., 1997), it is clear that protein and allergen levels can vary by orders of magnitude across different glove lots and glove types. Some investigators have noted that, in general, powdered gloves tend to contain more protein and allergen than powderless gloves, presumably due to the fact that powderless gloves often undergo a chlorination and washing process that powdered gloves do not, which may extract or denature proteins (Yunginger, 1995b; Beezhold et al., 1996). However, because of the aforementioned variability, in several instances, certain brands of powderless gloves have been found to contain higher allergen levels than certain brands of powdered gloves (e.g., see Yunginger et al., 1993; Yunginger, 1995b).

The National Institute of Occupational Safety and Health (NIOSH) recently issued an "alert" that encourages the use of "low-protein" NRL gloves. However, "low-protein" does not necessarily mean "low allergen" content. Indeed, Yunginger's analysis (Yunginger et al., 1993) demonstrated that some gloves that would qualify for low-protein labeling under FDA specifications actually contain higher allergen levels than gloves that would not qualify. Other inconsistencies are apparent; for example, Yunginger et al. (1993) showed that, in general, examination gloves contained less allergen than surgical gloves, while Baur et al. (1997) reported the opposite. It should be noted that all of the above-cited studies were conducted using highly aggressive extraction conditions that are certain to overestimate the amount of allergen that will solubilize or become aerosolized under normal use conditions.

18.2.4   Airborne NRL in the Workplace

In 1988, mineral talc was replaced with cornstarch as the glove powder of choice for examination gloves. The shift to cornstarch occurred 10 years earlier for surgical gloves. The shift in glove powder corresponds with the time period during which more reports of latex sensitization in health-care workers and others exposed to latex gloves started to appear in the peer-reviewed literature. It has been suggested that glove powders can cause respiratory symptoms as a result of one or more latex antigens leaching from the interior surface of the gloves and binding with cornstarch, with the complex being liberated during glove changing. In support of this hypothesis, several studies have shown that NRL allergens can be extracted from cornstarch particles taken from glove interiors. For example, Tomazic et al. (1994) showed that cornstarch from gloves, or cornstarch that was incubated with glove extract, bound a small amount of specific latex IgE antibodies. In short, it appears reasonable to assume that NRL allergens can migrate onto cornstarch particles in measurable concentrations. In addition, airborne levels of cornstarch particles up to 3667 particles/m^3 have been quantified in hospital settings where powdered gloves are used (Newsom, 1997).

Several investigators have attempted to correlate airborne NRL levels in the workplace with glove type and use, and in general the results are consistent with this hypothesis that glove use can result in elevated levels of airborne NRL allergens. For example, Swanson et al. (1994), Tarlo et al. (1994), and Shirakawa et al. (1997) have all reported that airborne NRL allergen levels are measurably higher in areas where high-allergen, powdered latex gloves are in use than where low-allergen, nonpowdered gloves are used. Interestingly, some data suggest that it is the allergen content of the glove, and not the presence or absence of powder, that governs the release of airborne allergen. For example, a prospective study on aeroallergen levels in operating rooms compared the impact of "low" and "high" allergen content gloves in operating rooms on both surgery and nonsurgery days (Heilman et al., 1996). Use of low-allergen gloves resulted in lower aeroallergen levels than did use of high-allergen gloves: mean air levels of allergen were 1.1 ng/m^3 (range = 0.1 to 3.5 ng/m^3) for low-allergen gloves and 13.7 ng/m^3 (range = 2.2 to 56.4 ng/m^3) for high-allergen gloves. However, there was no significant difference between mean aeroallergen levels on non-surgery days (mean $= 0.6$ ng/m^3; range $= 0.1$ to 3.6 ng/m^3) and days when powdered, low-allergen gloves were used. Nonpowdered gloves in general are lower in NRL allergenic content because of additional washing and treatment steps in the manufacturing process.

It is also worth noting that NRL aeroallergens are not ubiquitous throughout the health-care workplace. For example, airborne NRL was found to be mostly absent in the administrative departments of a hospital but measurable in emergency and labor and delivery departments (Page and Esswein, 1999).

In summary, it appears that use of NRL gloves can result in elevated levels of airborne NRL in the workplace. In general, published measurements of airborne NRL in areas where powdered/high-allergen gloves are used vary considerably but are typically on the order of 200 ng/m^3 or lower and appear to be associated with particles larger than 7 μm in mass median aerodynamic diameter (Swanson et al., 1994). It is unclear to what degree the allergenic content of the glove versus the presence/absence of powder is responsible for the presence of the airborne allergens.

18.3 DOSE–RESPONSE ASSESSMENT

It has been shown that direct inhalation challenge with a powdered glove can trigger an allergic response in sensitized individuals (e.g., Lagier et al., 1990; Tarlo et al., 1990). However, there is a paucity of literature as to whether airborne NRL levels in the workplace are sufficient to elicit a response. Some investigators have attempted to correlate glove usage and/or glove type with respiratory symptoms or prevalence of NRL sensitivity. For example, Tarlo et al. (1994) published a case report wherein a laboratory technician was able to tolerate a workplace in which powder-free latex gloves were used, but who suffered asthmatic and anaphylactic attacks when co-workers wore powdered NRL gloves. Similar case reports can be found in the literature [see Hopkins (1995) for a review]. However, the primary shortcoming associated with these studies is that none of them measured airborne NRL, and therefore, it is unknown whether the onset of symptoms was truly a function of NRL exposure. Although it certainly seems plausible, alternative explanations can also be proffered (e.g., the symptoms may have actually been due to cornstarch and not NRL, or the symptoms may have been psychosomatic).

The only published attempt to develop a dose–response relation between airborne NRL levels and respiratory symptoms was conducted by Laoprasert et al. (1998). In that study, respiratory function was monitored while nine latex-sensitive volunteers in a challenge chamber donned and discarded powdered latex, nonpowdered latex, and vinyl gloves for up to one hour. Mean airborne NRL levels during powdered glove use (7600 ng/m^3) were significantly higher than during vinyl glove or nonpowdered latex glove use (mean $= 65$ ng/m^3). Allergy and asthma symptoms were not observed with the nonpowdered latex or vinyl gloves. In four of the volunteers, powdered gloves produced reproducible decreases in respiratory function. However, because the airborne NRL levels in this study (mean $= 7600$ ng/m^3; range $= 93$ to $54,000$ ng/m^3) were much higher than those typically measured in the health-care workplace (usually 200 ng/m^3 or less), the relevance of this study is unclear.

18.4   EPIDEMIOLOGY

Several factors must be considered when evaluating epidemiologic sensitization studies. The first is the method used to diagnose sensitization, as different methods will often yield different prevalence estimates. The prevalence of atopy (defined as exhibiting a positive skin prick test to at least one allergen) in the exposed and control populations must also be understood. Since atopy predetermines susceptibility to development of allergic sensitization, the rates of atopy in the populations studied must either be controlled for or carefully assessed to allow for corrections in the data analysis. Third, the potential for selection bias, particularly in volunteer studies, must be understood and evaluated.

18.4.1   Diagnosis of NRL Allergy

Typically, an NRL type I allergy is diagnosed by relating a positive exposure and symptom history, in conjunction with one or more clinical tests. The most commonly used diagnostic techniques are the in vivo skin prick test (SPT) and in vitro measurement of antilatex IgE antibodies. The SPT involves application of glove extract to the skin (usually forearm), followed by an approximate 1-mm puncture of the skin. A similar procedure with histamine is usually used as a positive control. After an appropriate time interval (15 minutes), the diameter of the wheal and the intensity of the flare are measured. The SPT procedure has been used frequently in Europe and Canada. In the United States, there are no standard commercial latex extracts available for the SPT. The SPT assay is considered (but unproven) by many investigators to be more sensitive and specific than most IgE assays, but it has its limitations. For example, there appears to be a considerable degree of subjectivity and variability in the formulation of the grading criteria and in performing these tests. For example, Porri et al. (1997), Yassin et al. (1994), and Liss et al. (1997) considered a response positive if the wheal was 3 mm larger than the wheal of the negative control and the flare intensity was not graded. Conversely, Palosuo et al. (1998) defined a response as positive if the wheal was half the size of the positive histamine control. Smith used six individual extracts and did not require a positive control or wheal (Smith et al., 1997). Also, allergists typically use extracts of NRL gloves, but these gloves can vary widely in their allergen content (Yunginger, 1995a). Because different investigators use different gloves, different extraction procedures, different extract preparation methods, and different diagnostic criteria, comparison of SPT results from different studies will inherently contain some degree of uncertainty due to difficulty in comparing operator and interpreter variability. In addition, it is very difficult—if not impossible—to prove specificity in a SPT.

The most commonly used in vitro methods are the radioallergosorbent test (RAST) and the enzyme-linked immunometric assays (ELISA and FEIA). In both types of assays, circulating IgE antibodies are assessed, and a grading system is used to determine whether the response is negative, positive, or strongly positive. Some assays are quantitative, and some report levels of specific IgE. Again, there can be some subjectivity as to the grading criteria employed and the minimum response that represents a "positive" effect (e.g., Hunt et al., 1995). Also, as noted by Lebenbom-Mansour et al. (1997), many immunoassays have high specificity (>95 percent) but only fair sensitivity (50 to 70 percent) when compared with history or skin testing. The Ala-STAT test, which is similar in principle to the RAST procedure, has been shown to be highly sensitive (>90 percent), has a specificity of 100 percent in nonatopic individuals, and has been approved by the FDA for diagnosis of latex sensitivity. The drawbacks with all such tests are that they may produce both false positive and false negative results, making identification of the clinical entity and etiologic agent difficult, and that they are simply measures of sensitization and do not necessarily reflect clinical reactivity or sensitivity.

"Challenge" or "provocation" tests are sometimes employed, wherein airway resistance is monitored in participants as they don and remove NRL gloves. Finally, glove "use" tests are conducted occasionally in which the patient wears an NRL glove on one hand while wearing a non-NRL glove on the other. The person is then examined over a period of time (15 minutes) to determine whether symptoms develop. These tests are rarely blinded and are subject to psychological and nonspecific influences. Discussions of these diagnostic procedures can be found in Yunginger et al. (1995b) and Turjanmaa et al. (1995).

Despite the extensive publication record on the topic, there is no standard definition as to what constitutes "latex allergy." Some investigators have diagnosed NRL allergy based simply on the results of a patient interview, while others might require a positive response in two or more clinical tests. Compounding the uncertainty is the fact that some of the diagnostic techniques themselves may contain a significant source of bias (particularly the challenge and use tests, and the history as well). All of these factors make it difficult to evaluate the NRL epidemiology literature as a whole. Still, it should be possible to construct an objective framework of study design criteria by which the "best" studies can be identified for a critical examination.

18.4.2 Desired Characteristics of an Accurate NRL Epidemiology Study

Selection of Test versus Control Groups   To assess whether NRL allergy prevalence in health-care workers (or any other targeted group) is different from that of the general population, it would be preferable to examine individuals selected sequentially from both groups in an unbiased manner. Because volunteers are likely to be more attentive to their health care, and because individuals with symptoms are more likely to volunteer, volunteer bias may yield over-estimates of incidence rates (Grzybowski et al., 1996). Ideally, one would ex-amine *all* employees in a given health-care unit, or obtain close to 100 percent participation of an unselected subset. Volunteer bias and prescreening of par-ticipants should be avoided, and the investigators should be blinded to patient status if possible (atopic, symptomatic, etc.). Identifying an unselected general population is problematic, simply because of the difficulty in encouraging the selected individuals to participate in a blood draw, SPT, or NRL challenge. Evaluation of preoperative surgical patients or any other "patient subpopula-tion" is somewhat biased and may yield an "upper bound" of general NRL prevalence, simply because patients (clinical or hospital) may have had more extensive previous contact with latex-containing devices than the rest of the general population. Although it is not a truly "unselected" general population, some investigators have chosen to evaluate blood samples from blood banks, taking care to exclude samples from health-care workers (e.g., Ownby et al., 1996). A drawback to this unselected population is the inability to perform in vivo assays such as SPT and to obtain a patient history.

Atopy   Atopy is a genetic predisposition toward the development of allergies, and approximately 15 to 20 percent of the general population is atopic (Land-wehr and Boguniewicz, 1996). It has been well established that a large majority of NRL-sensitive individuals are atopic (e.g., Moneret-Vautrin et al., 1993; Yunginger, 1995b; Liebke et al., 1996; Novembre et al., 1997), and many in-vestigators have identified atopy as the primary risk factor in the development of NRL allergy (atopy is the primary risk factor for numerous type I allergies). Hence, it is essential that the percent of atopic participants in the test versus control groups be comparable (or, at least, information on atopic status should be collected during the analysis).

Diagnostic Techniques   Preferably, diagnoses of the "exposed" and "control" populations would be conducted as part of a single study, to minimize the interstudy variability associated with different investigators and different diag-nostic techniques. The challenge and use assays are cumbersome and not par-ticularly sensitive, sufficiently reproducible, quantitative, or specific, and they may contain bias. A diagnosis based on patient recall only (surveys, ques-tionnaires, interviews) is even more subjective and has also been shown to have poor specificity (Lebenbom-Mansour et al., 1997). For example, Cormio et al. (1993) found that, although 49 percent of 77 surgical staff members claimed to have adverse reactions to NRL gloves, only 5.2 percent were found to be al-lergic to NRL (via SPT and subsequent NRL challenge). In general, and as summarized in Table 18.1, the self-reported prevalence rate (approximately 30 to 50 percent) is roughly 10 times the diagnosed rate. It is possible that these self-reported rates include patients who are experiencing type IV reactions. In short, these findings indicate that the results of surveys and questionnaires

TABLE 18.1   Comparison of Reported vs. Actual Incidences of NRL Allergy in Health-Care Workers

Author	Population Size	Self-Reported Incidence (%)	Actual Incidence[a] (%)
Cormio et al., 1993	77	49	5.2
Salkie et al., 1993	203	50	<2
Lagier et al., 1992	248	41	8.40
Turjanmaa et al., 1987	512	24	2.9
Wrangsjo et al., 1994	233	37	3.5
Sussman et al., 1995	50	40	8
Tarlo et al., 1997	203	24	10
Liss et al., 1997	1351	53[b]	12.1
Anellaro et al., 1992	401	35[c]	9.9

[a] Determined as a positive response in an SPT.

[b] Estimated on a subset of 76 participants.

[c] Estimated from percentage of symptomatic patients who were also SPT-positive for latex.

should not be used as accurate indicators of NRL allergy prevalence in the health-care worker population.

Although neither test is completely free of uncertainty, the IgE and SPT assays offer the best measure of sensitivity, accuracy, reproducibility, and objectivity. The commercial Ala-STAT assay (IgE) has the advantage of standardized diagnostic criteria, and therefore, one might be able to statistically compare the results of different studies that used this particular assay. Such a comparison may be invalid for other "nonstandardized" assays (SPT and other IgE assays) that require more subjective grading criteria and diagnoses.

In summary, the ideal study of NRL allergy incidence in health-care workers versus general population would fulfill as many as possible of the following criteria:

- Unbiased (e.g., random or sequential) selection of test and control populations
- Similar percent of atopics, and high participation rate, in both populations
- Retrospective or prospective evaluation of test and control populations (preferably concurrent) using the same diagnostic criteria
- Use of antilatex IgE in serum and/or SPT as the primary diagnostic criterion
- Participants are blinded to study objective
- Diagnosing physicians are blinded to participant status

For the purposes of this review, we emphasize those studies that best fulfill these requirements.

18.4.3   Review of Published Studies

To date, well over 200 individual case reports of NRL allergy have been reported in the medical literature. Most of these reports involve a health-care worker or surgical patient who demonstrates type I symptoms and is diagnosed as NRL-sensitive based on personal history and/or clinical diagnoses. However, while such reports may be useful diagnostic tools and may be useful in identifying potential risk factors, they do not give any particular insight as to the prevalence or etiology of a disease incidence. Therefore, case reports are not considered in this analysis.

Approximately 50 "studies" of NRL allergy prevalence in health-care workers have been published. However, because of the manner in which they were designed, most of these studies are limited in their epidemiological utility. First, many of the studies used volunteers only and/or targeted self-reported or "symptomatic" individuals as their test group (i.e., individuals who were already believed to have NRL allergy because of a pattern of type I symptoms). In addition, almost all studies failed to test a control group, (i.e., a group that does not have occupational exposure to NRL gloves). The lack of other study design elements introduces further uncertainty into the interpretation of these reports. For example, in nearly all the studies, the participants were not blinded as to the purpose of the study, and in most cases, the investigators were aware of the participants' status (atopic, symptomatic, etc.) prior to testing. This allows for the historical analysis to bias the interpretation of the test results.

Table 18.2 summarizes the design elements of approximately 20 studies, which are often cited as "evidence" that health-care workers have elevated rates of NRL sensitization. Each of these studies is lacking in one or more criteria such that, individually or as a whole, they cannot be used to draw rigorous conclusions regarding sensitization rates in the health-care worker population and/or how these rates might compare to that of an unexposed control population. In many of these studies, only symptomatic individuals were tested and/or volunteer bias was present. As noted in Turjanmaa's review of many of these studies (Turjanmaa, 1995), "the prevalence numbers may be biased high because those people with no symptoms related to glove use tend to avoid testing" (Turjanmaa, 1994b). For example, Yassin et al. (1994) tested 224 hospital employees via SPT and reported a 17 percent positive response rate. However, as the authors note, their study may have identified a "falsely high percentage" because participation was strictly voluntary and the study followed an extensive "latex sensitivity awareness" program (100 percent of the positive responders were symptomatic). Further, no control group was used, and specificity could not be proven. Hunt et al. (1995) reported a 30 percent sensitization rate in 342 hospital employees (SPT). The employees were selected consecutively, but only symptomatic individuals were tested and no control group was used. Charous et al. (1994) and Kujala and Reijula (1996) used control groups, but the health-care worker population consisted only of symptomatic individuals. Arellano et al. (1992) evaluated a robust control group (100 con-

TABLE 18.2 Summary of Selected Studies Reporting Latex Allergy Incidence in Health-Care Workers

Author	Population Size	Control Group?	Unbiased Method of Selection?	Diagnostic Methods	Reported Incidence in Health-Care Workers (%)	Reported Incidence in Control Population (%)
Liss et al., 1997	1326	No	No[a,b]	SPT	12.1	—
Gbrzyowski et al., 1996	741	No	Yes	IgE	8.8	—
Yassin et al., 1994	224	No	No[b]	SPT	17.0	—
Hunt et al., 1995	342	No	No[b,c]	SPT	30.0	—
Charous et al., 1994	47	Yes	No[b,c]	IgE	40.0	10.0
Bubak et al., 1992	49	No	No[b,c]	SPT	69.0	—
Lagier et al., 1992	197	No	No[b]	SPT	10.7	—
Arellano et al., 1992	101	Yes	No[b]	SPT	9.9	3[d]
Salkie et al., 1993	203	No	No[b]	IgE	12.0	—
Kujala et al., 1996	25	Yes	No[b,c]	SPT	4.0	0 (0/11)
Sussman et al., 1995	20	No	No[b,c]	SPT	8.0	—
Safadi et al., 1996	41	No	No[b,c]	SPT	10.0	—
Kaczmarek et al., 1996	381	No	No[b]	IgE	5.5	—
Turjanmaa, 1987	512	Yes	UNK	SCT[e]	2.9	0.8
Kibby and Akl, 1997	135	No	No[b]	SPT	8.2	—
Douglas et al., 1997	140	No	No[b]	SPT	22.0	—

[a] Employees documented as being NRL allergic were "encouraged to participate."

[b] Volunteers only.

[c] Symptomatic participants only.

[d] No significant difference ($P = 0.082$) observed between test vs. control groups; controls were atopics only.

[e] Scratch chamber test.

UNK = unknown.

1045

secutive patients without occupational exposure to NRL gloves) and a relatively unselected population of health-care workers. No significant differences were reported in the prevalence of NRL sensitization in the test group (9.9 percent in 101 unselected surgeons, anesthesiologists, and radiologists) versus the control group (3 percent). However, as noted by the authors, the fact that 100 percent of the control group was atopic (versus 38 percent atopy in the test group) biased the results against significant findings. In summary, it might be concluded that in health-care workers who have symptoms of NRL allergy, the reported sensitization rates range from approximately 10 to 30 percent. We refer the reader to Turjanmaa (1994a) for a detailed review of most of these studies.

Most of the published studies are retrospective analyses, although prospective studies have been conducted in a few cases. For example, in a follow-up study of 1326 health-care workers, individuals who were SPT negative were followed prospectively for one year and reevaluated by questionnaire and SPT (Sussman et al., 1998). Four hundred and thirty-five individuals (all of whom worked in hospital settings) participated in the one-year follow-up. Of these, four individuals (seroconversion rate of 0.9 percent) were reported to have become NRL sensitive. Another study (Tarlo et al., 1997) reported that third- and fourth-year dental students had higher incidence rates (as measured via SPT) than first- and second-year students, but because the same group was not examined prospectively, this study is difficult to interpret epidemiologically. One potential shortcoming with the prospective approach is that the "test" group will likely be aware of the study objectives, and this may influence their frequency of glove use. Also, lack of a negative control group makes it difficult to ascertain whether the conversion rate is greater than background. However, a carefully designed study would yield very useful seroconversion rates that would be unattainable otherwise.

Background Prevalence   As summarized in Table 18.3, various researchers have attempted to measure "background" rates of NRL allergy in the "general population" (e.g., Turjanmaa, 1987; Shield and Blaiss, 1992; Lebenbom-Mansour et al., 1997; Ownby et al., 1996). Most of these studies did not actually examine a true representative "general" population but instead evaluated a subpopulation comprising of non-health-care workers (e.g., atopic patients, ambulatory surgical patients, children being screened for food allergies, etc.). None of these studies included a concurrent positive control group. Because different diagnostic techniques were employed in the various studies, it is not surprising that the reported estimates vary considerably. For example, Turjanmaa et al. (1995) reported that 0.12 percent of 804 preoperative adult surgical patients were NRL-sensitive via SPT. Yet in another evaluation of preoperative adult surgical patients, where the investigators took measures to ensure that lab technicians were blinded to the study protocol, 6.7 percent of 996 individuals were found to have antilatex IgE antibodies (Lebenbom-Mansour et al., 1997).

TABLE 18.3 Estimates of Background NRL Allergy Incidence in the General Population

Author	Population Tested	Diagnostic Technique	Reported Incidence (%)	Positive Control Group?
Turjanmaa et al., 1995	804 preop patients	SPT	0.12	No
Lebenbom-Mansour et al., 1997	996 preop patients	IgE	6.7	No
Ownby et al., 1996	1000 blood donors	IgE	6.4	No
Reinheimer and Ownby, 1994	200 allergy patients	IgE	12	No
Shield and Blaiss, 1992	44 atopic children	SPT	6.8	No
Bernardini et al., 1998	1175 schoolchildren	SPT	0.68	No
Porri et al., 1997	258 adults	SPT	6.6	No
Merrett et al., 1995	1436 blood donors in UK	IgE	7.9	No
Porri et al., 1995	195 atopic patients in France	SPT/IgE	5.6/5.8	No
	170 nonatopic patients in France	SPT/IgE	1.2/1.6	No

[a]Visitors to a health-care center for a checkup.

1047

Sensitization rates in children might be expected to yield a more accurate and consistent estimate of background prevalence because children presumably have had less cumulative contact with NRL. On the other hand, it must be noted that infant and child exposure to NRL in baby bottle teats, pacifiers, balloons, and the like may be substantial and could demonstrate a high degree of interindividual variability (Wrangsjo et al., 1992). Indeed, this may partially explain why significant differences in background prevalence estimates have been reported in children. For example, Reinheimer and Ownby (1995) examined 200 consecutive serum samples sent to their laboratory for total IgE measurement (samples sent for latex-specific IgE analysis were excluded). Twelve percent of the samples tested positive using the Ala-STAT IgE assay. A follow-up of the medical charts on the individuals who tested positive showed that most (70.8 percent) of the positive responses were in individuals <18 years of age, and that 91.6 percent of the positive responses indicated no previous exposure to latex. Unfortunately, no inhibitions were performed to demonstrate specificity of the results. Although the test group was not selected randomly, the fact that the follow-up investigation showed that only a small fraction of the positive-results group had any previous latex exposure suggests that this group may have been fairly representative of the general population. On the other hand, Bernardini et al. (1998) reported that only 8 of 1175 (0.7 percent) unselected schoolchildren had a positive SPT response to a commercial latex extract. Once again, it is not clear whether the large differences are due to differences in technique or reflect true heterogeneity in the general population.

Two studies in particular appear to have examined a relatively unselected "general" population. Porri et al. (1997) reported a 6.6 percent sensitization rate (patients who responded to latex with either a positive SPT or a positive RAST) in 258 subjects (of 747 who initially completed a questionnaire) visiting a health-care center for a checkup. The authors stated they felt this rate was "truly representative" of the general population because the sex, age distribution, proportion of subjects wearing latex gloves, reported skin reactions to gloves, and prevalence of symptoms suggestive of atopy were identical in subjects who agreed to participate in the study and in those who did not. More importantly, the authors indicated that subjects who volunteer for a checkup examination in French social security centers do not differ from the general population with regard to age, sex, education, or medical history.

Ownby et al. (1996) examined 1000 blood samples collected from mobile Red Cross donor collection sites. The authors indicated that samples collected at health-care institution sites were excluded, although it is possible that some small fraction of the samples may have come from health-care workers. An NRL allergy prevalence of 6.4 percent was reported using the Ala-STAT analysis for antilatex IgE antibodies. Although the selection process was not completely unbiased (the tested population was a subpopulation of blood donors), this study would appear to be superior to other studies of "background" allergy prevalence, simply because the tested population was relatively unselected (e.g., participants were not "prescreened" for latex use, they were not aware of the study objective, etc.) and the investigators were blinded to participant status.

In summary, a review of the handful of studies that examined NRL sensitization rates in a relatively unselected subset of the adult general population (without concurrent evaluations of health-care workers) suggests that the maximum background prevalence rate of sensitization to NRL in adults is approximately 6 percent.

Selected Studies As shown in Table 18.4, we believe that only a few studies meet a sufficient number of the aforementioned design criteria to permit a comparison of prevalence rates in health-care workers versus the general population. Turjanmaa (1987) reported significantly different ($p < 0.01$, Poisson test) sensitization rates of 2.9 percent in 512 hospital employees and 0.8 percent in 130 concurrent non-health-care participants. The employee population consisted of volunteers from operating and laboratory units of the University Central Hospital in Tampere, Finland, where NRL gloves are used widely. As noted by Grzybowski et al. (1996), Turjanmaa's (1987) worker prevalence rates may have been falsely elevated due to volunteer bias. Also, it should be noted that the employee and control populations were first screened with a "scratch-chamber test," a nonstandard technique that is known to be inaccurate. Only 23 of the 512 employees and 2 of the 130 controls were actually tested via SPT. Hence, although this study included a sizable control group, (1) the "exposed" population likely contained some degree of volunteer bias, (2) the relatively high prevalence of atopy in the health-care workers further biases the findings, and (3) the screening method introduces additional uncertainty in the results.

Gbrzyowski et al. (1996) reported 8.9 percent incidence in 741 registered nurses (of the total 818 eligible employees) using the Ala-STAT IgE assay. The authors were careful to control for volunteer bias and had the highest participation rate (90.6 percent) of any NRL/health-care worker study to date. Although a control group was not formally included as part of this study, the same investigators measured a 6.4 percent NRL seropositive rate in the aforementioned analysis of 1000 blood samples from the general population (Ownby et al., 1996). That study was conducted during the same time frame, using the same laboratory (Henry Ford Hospital), the same technicians, the same diagnostic analysis (Ala-STAT), and the same diagnostic criteria (i.e., 0.35 IU/mL or greater as positive). However, it should be noted that one of the studies employed the tube Ala-STAT assay (Ownby et al., 1996), and the other used the plate version of the Ala-STAT assay (Grzybowski et al., 1996). The prevalence of atopy in each group is unknown. These two studies, if they can be considered comparable, conflict with the findings of Turjanmaa (1987) in that (1) the absolute values of the incidence rates in both groups are much higher, and (2) there is a slight statistical difference ($p = 0.062$) between the health-care worker and control groups. Overall, these two studies would appear to provide a useful comparison of exposed versus control prevalence rates due to the lack of vol-

TABLE 18.4 Summary of Studies That Meet Minimum Design Criteria

Author(s)	Test Population	Control Population	Unbiased Method of Selection?		Participants Blinded to Study Objective?		Investigator Blinded to Participant Status?	Diagnostic Methods	Measured Incidence (%)	Control Group Incidence (%)
			HCW	Control	HCW	Control				
Turjanmaa, 1987	512 hospital employees	130 patients	No	Unknown	No	Unknown	Unknown	SCT[a]/SPT	2.9	0.8
Grzybowski et al., 1996; Ownby et al., 1996 (0–10)	741 nurses	1000 blood donors	Yes[b]	Yes[b]	No	Yes	Yes	IgE	8.9	6.4
DHHS, 1987	146 HCWs	5378 non-HCWs[c]	Yes[b]	Yes[b]	Yes	Yes	Yes	IgE	19.90	18.50
Page and Esswein, 1999	264 HCWs exposed to gloves	255 HCWs with no exposure to NRL gloves	Yes	Yes[b]	Yes	Yes	Unknown	IgE	6.10	6.30

[a] Scratch-chamber test.

[b] All participants were volunteer blood donors.

[c] Worders—classified as working in industries other than health care.

HCW = health-care worker

DHHS = U.S. Department of Health and Human Services

unteer bias in the health-care workers. Still, some uncertainty is introduced by comparing two different studies.

The National Health and Nutritional Examination Survey (NHANES) III (DHHS, 1996), conducted by the CDC, provides another data source for comparison. The purpose of this periodic survey is to determine the health and nutritional status of the U.S. population. In the NHANES III study, blood samples were collected from a cross section of 20,000 volunteers from throughout the United States. The IgE Ala-STAT assay for NRL antibodies was conducted on 5524 adult samples. As summarized in Table 18.4, 146 of the samples were from volunteers who classified themselves as health-care workers. The results indicated that the sensitization rates in health-care workers (19.9 percent) and non-health-care workers (18.5 percent) are roughly the same. Certain elements of the study design are attractive: the use of a standardized assay, the large number of participants, the relatively unselected populations (although some volunteer bias may be present), and the fact that the investigators were blinded to participant status. However, the findings are somewhat curious given the very high prevalence rates reported for both the exposed and control groups. Raising the cut-off for positivity did not influence the rates for different occupations.

A recently published study conducted by NIOSH is arguably the most definitive study to date on the issue of prevalence of NRL allergies in the health-care industry (Page and Esswein, 1999). Reacting to an anonymous complaint by a worker at the Exempla St. Joseph Hospital in Denver, Colorado, NIOSH conducted a health hazard assessment to determine whether latex might be the cause of the worker's complaints. NIOSH selected a worker population exposed to latex within the hospital and an internal control population—hospital administrative personnel—with little to no exposure to latex. NIOSH also took air and dust samples to assess the potential for airborne exposure to latex in the hospital. The participation rate for this study was 80 percent for the control population (comprising workers from 20 nonclinical departments) versus 86 percent for the exposed population (comprising workers from three clinical areas). Therefore, the potential for participation bias seems minimal. NIOSH found a prevalence of sensitization to NRL of 6.1 percent (16/264) for workers exposed to latex, and a prevalence of 6.3 percent (16/255) among workers in the hospital who had no exposure to latex gloves. There was a significant difference between self-reporting of hand dermatitis (23.4 vs. 4.9 percent, $p < 0.01$), rhinoconjunctivitis (16.3 vs. 7.9 percent, $p < 0.01$), and hand urticaria (9.9 vs. 2.1 percent, $p < 0.01$) between those who wore latex gloves and those who did not wear latex gloves, respectively. Latex sensitization was significantly associated with atopy (defined as having a history of allergic rhinitis, asthma, or atopic dermatitis); 81.3 percent of those with latex sensitization were atopic, compared to 59.5 percent of those who were not sensitized. Area air samples for NRL in the nonclinical areas indicated that five of the seven areas had no detectable NRL aeroallergens. Nine of the 16 samples from clinical areas had detectable levels of NRL, ranging from 0.41 to 3.33 ng/m^3. The authors stated: "Levels of airborne, surface and filter dust latex proteins were higher in the work areas of the employees who were not sensitized to latex than those who were sensitized." They concluded: "Neither current nor past occupational latex glove use was a significant risk factor for the development of latex sensitization" (Page and Esswein, 1999).

In summary, there are conflicting findings within the limited number of studies that can be used to compare NRL sensitization rates in health-care workers versus the general population. The weight of evidence might seem to suggest that the prevalence rates are comparable, but some of the studies that support that finding (Grzybowski et al., 1996; DHHS, 1996; Ownby et al., 1996) have shortcomings that prevent a definitive conclusion. The NIOSH study would appear to best satisfy the criteria established for this analysis, and this study suggests that there is no difference in the prevalence of sensitization to NRL between health-care workers and the "unexposed" general population (Page and Esswein, 1999). It seems reasonable to conclude that the overall findings of the more rigorous studies do not support the preponderance of the published claims and assertions on this topic (i.e., that prevalence rates in health-care workers are highly elevated relative to the general population). As mentioned earlier, however, a definitive epidemiological study has yet to be conducted.

18.5   RISK FACTORS AND CONFOUNDERS

Whether or not occupational glove use is responsible for sensitization of health-care workers, the fact remains that approximately 6 percent of the general U.S. population is NRL-sensitized. Clearly, some large fraction of these millions of individuals were sensitized by factors other than glove contact. Investigators have identified a number of known or suspected nonoccupational risk factors for NRL sensitization, and as described below some of these factors may play a role in the increased number of case reports in the past decade.

18.5.1   Atopy

It has long been recognized that atopy is the primary risk factor for developing latex sensitivity. Numerous studies have shown that atopics (or those with a history of atopic disease) represent the majority of latex-sensitized individuals in both occupational and nonoccupational exposures, in both adult and pediatric cases (e.g., Kelly et al., 1994). In a study of atopy as a predictive factor in latex allergy, it was reported that between 5.6 and 7.2 percent of atopics were sensitized to latex, compared to between 1.2 and 2.9 percent of nonatopics (Porri et al., 1995). Similarly, Moneret-Vautrin et al. (1993) reported that nonatopics exposed to latex had a positive SPT of 6.9 percent, compared to 36.4 percent of atopics.

In general, the frequency of latex sensitivity in atopics is about 2 to 4 times greater than in nonatopics. There is little data on the prevalence of atopy in the

U.S. health-care workforce to permit comparisons with the general population. However, reports from other countries suggest that the prevalence of atopy among medical students might be much higher than in the general population. For example, in New Zealand, the prevalence of atopy among 165 medical students was approximately 43 percent (Taylor and Broom, 1981). In Japan, the prevalence of atopy among 100 medical students was 90 percent (Kusunoki et al., 1999). The prevalence of atopy in the general U.S. population is approximately 20 percent (Gergen et al., 1987). Accordingly, if the prevalence of atopy in the U.S. health-care workforce is similar to that measured in other countries, then one would expect the prevalence of NRL sensitization to be approximately 2 to 4 times higher in U.S. health-care workers than in the general U.S. population. If such a difference truly exists, this might partially or entirely explain why health-care workers appear (in some studies) to have a higher rate of NRL sensitization.

One obvious question is: Why would such a difference exist? It has been conjectured that individuals who suffer various health maladies (such as those associated with atopy) during their childhood years are more likely to seek employment in the health-care industry as adults. Alternatively, it has been suggested that single-child families are (1) more likely to have atopic children because the lack of siblings results in a less-developed immune system, and (2) more likely to have children who enter the health-care industry because single-child families tend to be from upper socioeconomic strata, and a large fraction of the health-care industry (surgeons, physicians, registered nurses, etc) come from families in these strata. Clearly, these hypotheses rest on unproven assumptions. It would be useful to understand whether the frequency of atopy in the health-care industry is truly higher than in the general population, and we believe this could be a fruitful area of research.

18.5.2 Environmental and Indoor Pollutants

Numerous environmental and indoor agents have been implicated as potential risk factors for development of sensitivity to latex proteins. For example, exposure to cigarette smoke has been identified in several studies as a risk factor for developing latex sensitivity (Palczynski et al., 1998). Further, latex proteins have shown cross-reactivity with some proteins found in tobacco (Beezhold et al., 1996a; Sowka et al., 1998). Numerous proteins from other plants, including those commonly associated with pollen allergy, also cross-react with latex antigens. These proteins include mugwort, ragweed, blue grass, red mountain cedar, and timothy grass pollen, as well as extracts from decorative *Ficus* and poinsettia plants (Appleyard et al., 1994; Carey et al., 1995; Drouet et al., 1994; Delbourg et al., 1995; Hovanec-Burns et al., 1995; Kurup et al., 1995; Brehler and Theissen, 1996; Miguel et al., 1996; Fuchs et al., 1997; McCullough et al., 1997).

Given the prevalence of decorative plants indoors, this source of allergen may explain, in part, the apparent latex allergy in persons not known to have significant prior exposure to latex. For example, it has been shown that elevated levels of the *Ficus* allergen can be found in the dust collected from the leaf surface and from the floor beneath the plant (Bircher et al., 1995). Brehler and Theissen (1996) reported that nearly 20 percent of latex-sensitive patients were cross-reactive to *Ficus* extracts, and Hovanec-Burns et al. (1995) found that eight out of nine latex-sensitive patients (89 percent) cross-reacted to *Ficus* extracts. The question in any issue involving cross-reactivity is which allergy developed first. In the case of *Ficus*/latex cross-reactivity, Delbourg et al. (1995) reported a latex-sensitive patient who subsequently developed sensitivity to *Ficus*, and a patient in whom the *Ficus* allergy appeared to precede the latex sensitivity. In vitro assays confirmed specific antibodies to both allergens, and cross-reactivity was confirmed by reciprocal inhibition of the two extracts.

Latex allergens are also cross-reactive with fungal allergens, many of which grow indoors and are associated with allergic disease (Li et al., 1995). Consideration of indoor exposure to allergens either as predisposing factors or cross-reactive allergens may prove informative as to the etiology of apparently latex-sensitive patients without any known latex exposure. We suggest that questions regarding potential indoor sources of NRL should be included in future surveys of studies of NRL allergy prevalence. Also, it would be useful to determine whether smoking or exposure to second-hand smoke is associated with latex sensitization.

18.5.3   Hand Eczema

Skin irritation may increase the potential for allergenic proteins to be absorbed systemically and sensitize the immune system (Turjanmaa et al., 1996). As high as 80 percent of NRL-sensitive patients have a history of dermatitis on their hands prior to or coincident with the development of latex sensitivity (Taylor and Praditsuwan, 1996). It has been suggested that this dermatitis may be the result of a preexisting condition or may develop as the result of physical irritation by occlusion or abrasion by glove powders, delayed-contact dermatitis of the type IV variety associated with rubber additives, or type I protein contact dermatitis (Turjanmaa et al., 1996).

18.5.4   Frequency of Exposure

Several conflicting reports exist regarding exposure frequency as a risk factor. For example, Moneret-Vautrin et al. (1993) reported that latex-exposed atopics (health-care workers) were four times as likely to respond with a positive SPT as "nonexposed" atopics. Likewise, nonatopics exposed to latex (health-care workers) were more than 20 times as likely to have a positive SPT as nonatopics who were "unexposed." However, this study suffers from the previously discussed shortcomings of a biased exposed population (sensitized health-care workers were predisposed to participate). This study also leaves unanswered the question of why unexposed atopics or nonatopics should be sensitized to latex at all. Occult exposure to latex in various consumer products, or cross-reactivity to foods containing identical or closely homologous proteins, may explain some, if not all, of this phenomenon. Several studies have found a relation between the number of surgeries and the degree of latex sensitization (Yassin et al., 1992; Ellsworth et al., 1993; Kelly et al., 1994; Shaer et al., 1995; Theissen et al., 1997), while others have not (Golden et al., 1995; Yunginger, 1995a; Lebenbom-Mansour et al., 1997).

18.5.5 Food Allergies

Among the more intriguing observations regarding latex allergy is its relation to food allergens. The observed extensive cross-reactivity has not been explained by any of the theories involving profilins or plant glycoproteins that have been advanced thus far (Kurup et al., 1992; Moneo et al., 1995; Vallier et al., 1995). While profilin from ragweed inhibited latex-specific IgE and is present in other plants (e.g., banana) that are known to cross-react with latex, profilin was barely detectable from glove extracts, and sera from latex-sensitive patients reacted to a 15-kDa protein that was not profilin. Although the ubiquitous nature of profilin in plants may not explain the onset of latex sensitivity, its presence in extracts of other plants or in air samples containing plant material may confound the results of inhibition assays (Vallier et al., 1995). On the other hand, Nieto et al. (1998) recently tested profilin isolated from latex on 17 latex-sensitive patients and found 95 percent with positive SPT, and a third with profilin-specific IgE, suggesting that profilin may be an important allergen, at least in some cases.

As discussed above, atopy is also an important predisposing factor for the development of latex sensitization. Atopic individuals are also at risk for developing other allergies or cross-reactions to allergens from different sources that have certain characteristics in common. In some individuals, a history of food or plant allergies is predictive of latex sensitivity (Lebenbom-Mansour et al., 1997). However, not all of the plant and food allergens cross-react with latex extracts (Turjanmaa et al., 1996), suggesting that only some of the allergens exist in common or that the relative proportions are different or change according to source, season, or other factors. Food allergies may precede the development of latex sensitivity in some cases, while in others, food allergies apparently developed after the latex sensitization occurred (see Blanco et al., 1994; Charous, 1994; Turjanmaa et al., 1996; Hussain et al., 1998). In some individuals with preexisting food allergies, the development of latex sensitivity was accompanied by allergies to new foods (Hussain et al., 1998).

Latex extracts have inhibited the binding of banana and chestnut to IgE, but not vice versa (Rodriguez et al., 1993). Ross et al. (1992) reported that banana-reactive sera had detectable antilatex IgE, but banana IgE was not detected in latex-specific sera. The prevalence of food allergy (as indicated by the presence of IgE specific for one food or more) in latex-sensitive individuals has been reported to be as high as 82 percent (Brehler et al., 1997; Eades et al., 1997).

While a high number of latex-sensitive patients reported symptoms associated with ingestion of various fruits, only one-third of those who were symptomatic had specific IgE toward fruit in their serum. In one study, 27 percent of latex-allergic patients had positive SPT for one or more foods, and roughly one-third of these were symptomatic (Beezhold et al., 1996b). Specific IgE for fruits were found in 100 percent of atopic children displaying symptoms associated with latex, but only in 40 percent of asymptomatic atopic children (Novembre et al., 1997). Only 10 percent of latex-symptomatic children developed clinical symptoms after fruit ingestion. In 137 latex-sensitive patients (confirmed by SPT or IgE antibodies to latex), 21 percent had IgE reactions to various foods (Hussain et al., 1998).

In comparing the occurrence of IgE to various foods from latex-sensitive and non-latex-sensitive individuals, Kim et al. (1998) reported that 47 percent of latex-sensitive patients had IgE to at least one fruit, compared to only 11.4 percent of non-latex-sensitive patients. In terms of specific fruits, 18.8 percent of latex-sensitive individuals responded to avocado extract, compared to only 2.8 percent of non-latex-sensitive individuals. Banana extract elicited responses in 15 percent of latex-sensitive individuals, but only 5.7 percent of non-latex-sensitive individuals responded. Kiwi elicited a response in 13.2 percent of latex-sensitive patients, but only 4.2 percent of non-latex-sensitive individuals responded. No non-latex-sensitive individuals responded to chestnut or peach extracts, while 14.5 percent (chestnut) and 8 percent (peach) of latex-sensitive individuals did respond. In addition to their presence in foods, some cross-reactive proteins (e.g., papain) are used in food additives (e.g., meat tenderizers, beer), pharmaceuticals (e.g., ulcer treatment, cleaning agents, digestive aids), and cosmetic products (Vandenplas et al., 1996). Proteins similar to those found in latex occur in chewing gum and have been reported to provoke oral urticaria in some latex-sensitive patients (Heese et al., 1997).

The preceding discussion suggests that, in some cases, food allergies may have been misdiagnosed as latex allergies, and vice versa, which may explain (in part) why numerous studies have reported both atopic and nonatopic individuals, otherwise unexposed to latex, with apparent latex allergy (Blanco et al., 1994; Ownby, 1995). Proteins from a large number of food plants have been reported as cross-reacting with latex proteins, including avocado, kiwi, papaya, peach, plum, cherry, apricot, loquat, nectarine, orange, peanut, banana, passion fruit, fig, melon, pineapple, mango, coconut, strawberry, apple, chestnut, almond, carrot, tomato, celery, potato, pepper, turnip, wheat, oat, oregano, dill, and sage (M'Raihi et al., 1991; Ross et al., 1992; Anibarro et al., 1993; Rodriguez et al., 1993; Blanco et al., 1994; Charous, 1994; Kurup et al., 1994; Makinen-Kiljunen, 1994; Ahlroth et al., 1995; Baur et al., 1995; Carrillo et al., 1995; Latasa et al., 1995; Lavaud et al., 1995; Alenius et al., 1996; Antico, 1996; Beezhold et al., 1996; Brehler et al., 1997; Gallo et al., 1997; McCullough et al., 1997; Novembre et al., 1997; Wastell, 1997; Hemmer et al., 1998; Hussain et al., 1998; Kagen and Muthiah, 1998).

Recently, an adverse reaction to an herbal tea preparation by a latex-sensitive patient was reported (Pfutzner et al., 1998). This patient was known to be allergic to pollen, in addition to latex. SPT was positive to latex, the herbal tea, various food allergens (unspecified), pollens (i.e., birch, grasses, and mugwort), and condurango bark, one of the tea's ingredients. No other constituents of the tea were positive in this challenge. Inhibition assay with IgE indicated 57 percent inhibition with the condurango bark, compared to 62 percent for NRL extracts. Because condurango bark and similar substances are used in folk remedies, health store preparations, and as flavoring agents without allergenicity testing or quality control, the potential for sensitization of users may be significant.

Shellfish and fish have also been reported to be cross-reactive to latex proteins in some sensitized individuals (Hussain et al., 1998). These findings suggest that cross-reactivity to plant and perhaps some animal proteins can confound the study of latex allergy. Future studies of at-risk populations need to take these factors into account when attempting to elucidate the etiology of latex allergy. For instance, serum tryptase levels are elevated in latex allergic reactions (Volcheck and Li, 1994), but not in reactions resulting from food allergies (Sampson et al., 1992). Ownby (1995) has suggested that serum mast cell tryptase levels may serve as a measure of whether an allergic reaction originates as a food allergy or a latex allergy. Further research is needed to verify the utility of this assay as a marker to determine the etiology of various allergic reactions.

18.6 CONCLUSIONS AND RECOMMENDATIONS

The main findings of this review can be summarized as follows:

- There is no evidence that NRL glove manufacturing process changes yielded gloves with higher allergenic content.
- Much of the epidemiological literature consists of case reports or studies that fail to meet minimum design criteria (particularly lack of control groups).
- Of those studies that permit a valid comparison of latex sensitivity rates in health-care workers versus the general population, the weight of evidence, at the very least, conflicts with the numerous assertions of an "epidemic" and, in fact, is consistent with a conclusion that no difference exists.
- Although NRL glove use in the workplace appears to be correlated with airborne NRL levels, it is unclear whether the measured levels are sufficient to elicit a respiratory response in a sensitized person.

The issues surrounding latex allergy remain confused and controversial. It is important to move beyond the relatively simplistic and essentially anecdotal case reporting that has dominated the medical literature on the subject over the past few years, in order to be more specific about the degree of risk posed by latex allergy to health-care workers and others. To increase our understanding of the latex hypersensitivity issue, it is critical that an accepted case definition of the disease be adopted, ensuring that diagnoses are consistent among researchers and reports. Additionally, an effort should be made to standardize the diagnostic protocols (i.e., in vivo and in vitro tests), including the extracts used in challenges, both to reduce the variability that is obvious in current studies, and to reduce the chance of causing adverse reactions in already allergic patients or inadvertently sensitizing immunologically naive individuals. This standardization, along with the development of validated environmental testing methodologies, would be crucial in developing an understanding of the dose–response and threshold, both for initial sensitization and elicitation of symptoms in allergic patients.

To understand the etiology of the disease and its co-factors and confounders, it would be advisable to develop a detailed patient questionnaire that adds to the known risk factors and investigates other potential sensitization issues that may influence the development of apparent latex sensitization. This information is critical for understanding the actual relative risks associated with various exposures, products, and groups, as well as identifying and controlling for confounders that may skew the results. Toward this end, it would be interesting to undertake a retrospective analysis of patients diagnosed with latex allergy, to try to determine when, where, and how they became sensitized, and to track the incidence and prevalence rates over time. This information could be compared with both a well-designed prospective study of the background occurrence of latex sensitivity, and with studies on the development of latex allergy in naive populations (i.e., medical students) and the prevalence in exposed groups. All these types of studies are critical to gaining a clearer understanding of the issue, as well as developing preventive and intervention strategies to combat the disorder.

REFERENCES

Adiwilaga, K., and Kush, A. 1996. Cloning and characterization of cDNA encoding farnesyl diphosphate synthase from rubber tree (Hevea brasiliensis). *Plant Mol. Biol.* 30(5): 935–946.

Ahlroth, M., Alenius, H., Turjanmaa, K., Makinen-Kiljunen, S., Reunala, T., and Palosuo, T. 1995. Cross-reacting allergens in natural rubber latex and avocado. *J. Allergy Clin. Immunol.* 96(2): 167–173.

Alenius, H., Kurup, V., Kelly, K., Palosuo, T., Turjanmaa, K., and Fink, J. 1994a. Latex allergy: Frequent occurrence of IgE antibodies to a cluster of 11 latex proteins in patients with spina bifida and histories of anaphylaxis. *J. Lab. Clin. Med.* 123(5): 712–720.

Alenius, H., Makinen-Kiljunen, S., Ahlroth, M., Turjanmaa, K., Reunala, T., and Palosuo, T. 1996. Crossreactivity between allergens in natural rubber latex and banana studied by immunoblot inhibition. *Clin. Exper. Allergy* 26(3): 341–348.

Alenius, H., Makinen-Kiljunen, S., Turjanmaa, K., Palosuo, T., and Reunala, T. 1994b. Allergen and protein content of latex gloves. *Ann. Allergy* 73(4): 315–320.

Alenius, H., Turjanmaa, K., Palosuo, T., Makinen-Kiljunen, S., and Reunala, T. 1991. Surgical latex glove allergy: Characterization of rubber protein allergens by immunoblotting. *Int. Arch. Allergy Appl. Immunol.* 96(4): 376–380.

Anibarro, B., Garcia-Ara, M. C., and Pascual, C. 1993. Associated sensitization to latex and chestnut. *Allergy* 48(2): 130–131.

Antico, A. 1996. Oral allergy syndrome induced by chestnut (*Castanea sativa*) [see comments]. *Ann. Allergy Asthma Immunol.* 76(1): 37–40.

Appleyard, J. K., McCullough, J. A., and Ownby, D. R. 1994. Cross-reactivity between latex, ragweed, and blue grass allergens. *J. Allergy Clin. Immunol.* 93: 182.

Anellano, R., Bradley, J., and Sussman, G. 1992. Prevalence of latex sensitization among hospital physicians occupationally exposed to latex gloves. *Anesthesiology* 77(5): 905–908.

Baur, X., Chen, Z., and Allmers, H. 1998. Can a threshold limit value for natural rubber latex airborne allergens be defined? *J. Allergy Clin. Immunol.* 101: 24–27.

Baur, X., Chen, Z., Raulf-Heimsoth, M., and Degens, P. 1997. Protein and allergen content of various natural latex articles. *Allergy* 52(6): 661–664.

Baur, X., Chen, Z., Rozynek, P., Duser, M., and Raulf-Heimsoth, M. 1995. Cross-reacting IgE antibodies recognizing latex allergens, including Hev b 1, as well as papain. *Allergy* 50(7): 604–609.

Beezhold, D., and Beck, W. C. 1992. Surgical glove powders bind latex antigens. *Arch. Surg.* 127(11): 1354–1357.

Beezhold, D., Swanson, M., Zehr, B. D., and Kostyal, D. 1996a. Measurement of natural rubber proteins in latex glove extracts: Comparison of the methods. *Ann. Allergy Asthma Immunol.* 76(6): 520–526.

Beezhold, D. H., Sussman, G. L., Liss, G. M., and Chang, N. S. 1996b. Latex allergy can induce clinical reactions to specific foods. *Clin. Exp. Allergy* 26(4): 416–422.

Bernardini, R., Novembre, E., Ingargiola, A., Veltroni, M., Mugnaini, L., Cianferoni, A., Lombardi, E., and Vierucci, A. 1998. Prevalence and risk factors of latex sensitization in an unselected pediatric population. *J. Allergy Clin. Immunol.* 101: 621–625.

Bircher, A. J., Langauer, S., Levy, F., and Wahl, R. 1995. The allergen of *Ficus benjamina* in house dust [see comments]. *Clin. Exp. Allergy* 25(3): 228–233.

Blanco, C., Carrillo, T., Castillo, R., Quiralte, J., and Cuevas, M. 1994. Latex allergy: Clinical features and cross-reactivity with fruits [see comments]. *Ann. Allergy* 73(4): 309–314.

Brehler, R., and Theissen, U. 1996. *Ficus benjamina* allergie. *Hautarzt* 47: 780–782.

Brehler, R., Theissen, U., Mohr, C., and Luger, T. 1997. "Latex-fruit syndrome": Frequency of cross-reacting IgE antibodies. *Allergy* 52(4): 404–410.

Broekaert, I., Lee, H. I., Kush, A., Chua, N. H., and Raikhel, N. 1990. Wound-induced accumulation of mRNA containing a hevein sequence in laticifers of rubber tree (*Hevea brasiliensis*). *Proc. Natl. Acad. Sci. USA* 87(19): 7633–7637.

Carrillo, T., Blanco, C., Quiralte, J., Castillo, R., Cuevas, M., and de Castro, F. R. 1995. Prevalence of latex allergy among greenhouse workers. *J. Allergy Clin. Immunol.* 96(5): 699–701.

Charous, B. L. 1994. The puzzle of latex allergy: Some answers, still more questions [editorial; comment]. *Ann. Allergy* 73(4): 277–281.

Charous, B. L., Hamilton, R. G., and Yunginger, J. W. 1994. Occupational latex exposure: Characteristics of contact and systemic reactions in 47 workers. *J. Allergy Clin. Immunol.* 94(1): 12–18.

Cormio, L., Turjanmaa, K., Talja, M., Andersson, L. C., and Ruutu, M. 1993. Toxicity and immediate allergenicity of latex gloves. *Clin. Exp. Allergy* 23(7): 618–623.

Delbourg, M. F., Moneret-Vautrin, D. A., Guilloux, L., and Ville, G. 1995. Hypersensitivity to latex and *Ficus benjamina* allergens. *Ann. Allergy Asthma Immunol.* 75(6): 496–500.

Drouet, M., Gros, N., and Sabbah, A. 1994. Latex et pollinose. *Allerg. Immunol. (Paris)* 26(8): 289–292, 295–296.

Eades, J., Keane, B., Cullom, H., and Murray, J. 1997. Prevalence and temporal relationship of food sensitivity in latex allergic patients. *J. Allergy Clin. Immunol.* 99: S503.

Ellsworth, P. I., Merguerian, P. A., Klein, R. B., and Rozycki, A. A. 1993. Evaluation and risk factors of latex allergy in spina bifida patients: Is it preventable? *J. Urol.* 150(2): 691–693.

Fuchs, T., Spitzauer, S., Vente, C., Hevler, J., Kapiotis, S., Rumpold, H., Kraft, D., and Valenta, R. 1997. Natural latex, grass pollen, and weed pollen share IgE epitopes. *J. Allergy Clin. Immunol.* 100(3): 356–364.

Gallo, R., Cozzani, E., and Guarrera, M. 1997. Sensitization to pepper *Capsicum annuum* in a latex-allergic patient. *Contact Derm.* 37(1): 36–37.

Golden, D. B. K., Hamilton, R. G., Birenberg, A., Kreshtool, B., Haglauer, C., and Adkinson, N. F. 1995. Latex sensitization in surgical patients. *J. Allergy Clin. Immunol.* 95(1, Part 2): 157.

Granady, L. C., and Slater, J. E. 1995. The history and diagnosis of latex allergy. *Latex Allergy* 15(1): 21–29.

Grzybowski, M., Ownby, D. R., Peyser, P. A., Johnson, C. C., and Schork, M. A. 1996. The prevalence of anti-latex IgE antibodies among registered nurses. *J. Allergy Clin. Immunol.* 98(3): 535–544.

Hanninen, A. R., Mikkola, J. H., Kalkkinen, N., Turjanmaa, K., Ylitalo, L., Reunala, T., and Palosuo, T. 1999. Increased allergen production in turnip (*Brassica rapa*) by treatments activating defense mechanisms. *J. Allergy Clin. Immunol.* 104(1): 194–201.

Heese, A., Peters, K. P., and Koch, H. U. 1997. Type I allergies to latex and the aeroallergenic problem. *Eur. J. Surg.* Suppl. (579): 19–22.

Heilman, D. K., Jones, R. T., Swanson, M. C., and Yunginger, J. W. 1996. A prospective, controlled study showing that rubber gloves are the major contributor to latex aeroallergen levels in the operating room. *J. Allergy Clin. Immunol.* 98(2): 325–330.

Hemmer, W., Focke, M., Kriechbaumer, N., Gotz, M., and Jarisch, R. 1998. Oilseed rape (*Brassica napus*) pollen allergens cross-react with avocado, banana, and latex. *J. Allergy Clin. Immunol.* 101: S202.

Hopkins, J. 1995. Rubber latex in the air: An occupational and environmental cause of asthma? *Food Chem. Toxic.* 33(10): 895–899.

Hovanec-Burns, D., Ordonez, M., Corrao, M., Enjamun, S., and Unver, E. 1995. Identification of another latex-crossreactive food allergen: Peanut. *J. Allergy Clin. Immunol.* 95(1, Part 2): 150.

Hunt, L. W., Fransway, A. F., Reed, C. E., Miller, L. K., Jones, R. T., Swanson, M. C., and Yunginger, J. W. 1995. An epidemic of occupational allergy to latex involving health care workers. *J. Occup. Environ. Med.* 37(10): 1204–1209.

Hussain, H., Beall, G., and Kim, K. 1998. Prevalence of food allergy in patients with latex allergy. *J. Allergy Clin. Immunol.* 101: S208.

Kagen, S., and Muthiah, R. 1998. Latex and food induced anaphylaxis: Oregano, dill, sage and carrot cross-react with natural rubber latex. *J. Allergy Clin. Immunol.* 101: S208.

Kelly, K. J., Pearson, M. L., Kurup, V. P., et al. 1994. A cluster of anaphylactic reactions in children with spina bifida during general anesthesia: Epidemiologic features, risk factors, and latex hypersensitivity. *J. Allergy Clin. Immunol.* 94(1): 53–61.

Kim, K., Hussain, H., and Beall, G. 1998. Latex allergy and IgE antibodies to foods. *J. Allergy Clin. Immunol.* 101: S208.

Kurup, V. P., Kelly, K. J., Resnick, A., Bansal, N. K., and Fink, J. N. 1992. Characterization of latex antigen and demonstration of latex-specific antibodies by enzyme-linked immunosorbent assay in patients with latex hypersensitivity. *Allergy Proc.* 13(6): 329–334.

Kurup, V. P., Kelly, T., Elms, N., Kelly, K., and Fink, J. 1994. Cross-reactivity of food allergens in latex allergy. *Allergy Proc.* 15(4): 211–216.

Kurup, V. P., Murali, P. S., and Kelly, K. J. 1995. Latex antigens. *Immunol. Allergy Clin. North Am.* 15(1): 45–59.

Kusunoki, T., Hosoi, S., Asai, K., Harazaki, M., and Furusho, K. 1999. Relationships between atopy and lung function: Results from a sample of one hundred medical students in Japan [In Process Citation]. *Ann. Allergy Asthma Immunol.* 83(4): 343–347.

Lagier, F., Badier, M., Charpin, D., Martigny, J., and Vervloet, D. 1990. Latex as aeroallergen. *Lancet* 336: 516–517.

Landwehr, L., and Boguniewicz, M. 1996. Current perspectives on latex allergy. *J. Pediat.* 128(3): 305–312.

Laoprasert, N., Swanson, M., Jones, R., Reed, C., Schroder, D., and Yunginger, J. 1998. Inhalation challenge testing of latex-sensitive health care workers and the effectiveness of laminar flow HEPA-filtered helmets in reducing rhinoconjunctival and asthmatic reactions. *J. Allergy Clin. Immunol.* 102(6): 998–1004.

Latasa, M., Dieguez, I., Sanz, M. L., Parra, A., Pajaron, M. J., and Oehling, A. 1995. Fruit sensitization in patients with allergy to latex. *J. Invest. Allergol. Clin. Immunol.* 5(2): 97–102.

Lavaud, F., Prevost, A., Cossart, C., Guerin, L., Bernard, J., and Kochman, S. 1995. Allergy to latex, avocado pear, and banana: Evidence for a 30 kd antigen in immunoblotting. *J. Allergy Clin. Immunol.* 95(2): 557–564.

Lebenbom-Mansour, M. H., Oesterle, J. R., Ownby, D. R., Jenne, Post, S. K., and Zaglaniczy, K. 1997. The incidence of latex sensitivity in ambulatory surgical patients: A correlation of historical factors with positive serum immunoglobin E levels. *Anesth. Analg.* 85(1): 44–49.

Li, C.-S., Hsu, L.-Y., Chou, C.-C., and Hsieh, K.-H. 1995. Fungus allergens inside and outside the residences of atopic and control children. *Arch. Environ. Health* 50(1): 38–43.

Liebke, C., Niggemann, B., and Wahn, U. 1996. Sensitivity and allergy to latex in atopic and non-atopic children. *Pediatr. Allergy Immunol.* 7(2): 103–107.

Liss, G. M., Sussman, G. L., Deal, K., et al. 1997. Latex allergy: Epidemiological study of 1351 hospital workers. *Occup. Environ. Med.* 54(5): 335–342.

Lu, L. J., Kurup, V. P., Fink, J. N., and Kelly, K. J. 1995. Comparison of latex antigens from surgical gloves, ammoniated and nonammoniated latex: Effect of ammonia treatment on natural rubber latex proteins. *J. Lab. Clin. Med.* 126(ISS 2): P161–168.

Makinen-Kiljunen, S. 1994. Banana allergy in patients with immediate-type hypersensitivity to natural rubber latex: Characterization of cross-reacting antibodies and allergens. *J. Allergy Clin. Immunol.* 93(6): 990–996.

McCullough, J. A., Rau, S. G., White, A. H., and Ownby, D. R. 1997. Evaluation of latex-ragweed cross-reactivity by immunoblot inhibition. *J. Allergy Clin. Immunol.* 99(1, Part 2): S343.

Miguel, A. G., Cass, G. R., Weiss, J., and Glovsky, M. M. 1996. Latex allergens in tire dust and airborne particles. *Environ. Health Perspect.* 104(11): 1180–1186.

Moneo, I., Llamazares, A., Curiel, G., and Martinez, J. 1995. Characterization of latex and chestnut antigens by immunoblotting. *Ann. Allergy Asthma Immunol.* 75(5): 440–444.

Moneret-Vautrin, D. A., Beaudouin, E., Widmer, S., Mouton, C., Kanny, G., Prestat, F., Kohler, C., and Feldmann, L. 1993. Prospective study of risk factors in natural rubber latex hypersensitivity. *J. Allergy Clin. Immunol.* 92(5): 668–677.

M'Raihi, L., Charpin, D., Pons, A., Bongrand, P., and Vervloet, D. 1991. Cross-reactivity between latex and banana. *J. Allergy Clin. Immunol.* 87(1): 129–130.

Nieto, A., Mazon, A., Estornell, F., et al. 1998. Profilin, a relevant allergen in latex allergy. *J. Allergy Clin. Immunol.* 101: S207.

Novembre, E., Bernardini, R., Brizzi, I., Bertini, G., Muganaini, L., Azzari, C., and Vierucci, A. 1997. The prevalence of latex allergy in children seen in a university hospital allergy clinic. *Allergy* 52: 101–105.

Ownby, D. R. 1995. Manifestations of latex allergy. *Immunol. Allergy Clin. North Am.* 15(1): 31–43.

Ownby, D. R., Ownby, H. E., McCullough, J., and Shafer, A. W. 1996. The prevalence of antilatex IgE antibodies in 1000 volunteer blood donors. *J. Allergy Clin. Immunol.* 97(6): 1188–1192.

Page, E. H., and Esswein, E. J. 1999. Health Hazard Evaluation of Exempla St. Joseph Hospital Denver, Colorado. National Institute of Occupational Safety and Health (NIOSH), Cincinnati, OH.

Palczynski, C., Walusiak, J., Ruta, U., Stelmach, I., and Gorski, P. 1998. Associated factors of the latex hypersensitivity in children. *J. Allergy Clin. Immunol.* 101: S205.

Palosuo, T., Makinen-Kiljunen, S., Alenius, H., Reunala, T., Yip, E., and Turjanmaa, K. 1998. Measurement of natural rubber latex allergen levels in medical gloves by allergen-specific IgE-ELISA inhibition, RAST inhibition, and skin prick test. *Allergy* 53: 59–67.

Pfutzner, W., Thomas, P., Rueff, F., and Przybilla, B. 1998. Anaphylactic reaction elicited by condurango bark in a patient allergic to natural rubber latex. *J. Allergy Clin. Immunol.* 101: 281–282.

Porri, F., Lemiere, C., Birnbaum, J., Guilloux, L., Lanteaume, A., Didelot, R., Vervloet, D., and Charpin, D. 1997. Prevalence of latex sensitization in subjects attending health screening: Implications for a perioperative screening. *Clin. Exp. Allergy* 27(4): 413–417.

Potter, S., Uknes, S., Lawton, K., Winter, A. M., Chandler, D., DiMaio, J., Novitzky, R., Ward, E., and Ryals, J. 1993. Regulation of a hevein-like gene in Arabidopsis. *Mol. Plant. Microbe Interact.* 6(6): 680–685.

Reinheimer, G., and Ownby, D. R. 1995. Prevalence of latex-specific IgE antibodies in patients being evaluated for allergy [see comments]. *Ann. Allergy Asthma Immunol.* 74(2): 184–187.

Rodriguez, M., Vega, F., Garcia, M. T., Panizo, C., Laffond, E., Montalvo, A., and Cuevas, M. 1993. Hypersensitivity to latex, chestnut, and banana [see comments]. *Ann. Allergy* 70(1): 31–34.

Ross, B. D., McCullough, J., and Ownby, D. R. 1992. Partial cross-reactivity between latex and banana allergens [see comments]. *J. Allergy Clin. Immunol.* 90(3 Pt 1): 409–410.

Sampson, H. A., Mendelson, L., and Rosen, J. P. 1992. Fatal and near-fatal anaphylactic reactions to food in children and adolescents. *N. Engl. J. Med.* 327–380.

Shaer, C. M., Tosi, L. L., Gonzales, E. C., Mostello, L. A., and Slater, J. E. 1995. Risk factors for development of latex allergy in the spina bifida patient: A review of 96 patients. *J. Allergy Clin. Immunol.* 95(1, Part 2): 156.

Shield, S., and Blaiss, M. 1992. Prevalence of latex sensitivity in children evaluated for inhalant allergy. *Allergy Proc.* 3: 129–131.

Shirakawa, T., Enomoto, T., Shimazu, S.-I., and Hopkin, J. M. 1997. The inverse association between tuberculin responses and atopic disorder. *Science* 275: 77–79.

Slater, J. E. 1989. Rubber anaphylaxis. *N. Engl. J. Med.* 320: 1126–1130.

Smith, C., Garcia, M., and Kim, K. 1997. Diagnostic evaluation of type I latex allergy. *J. Allergy Clin. Immunol.* 99: S495.

Subramaniam, A. 1995. The chemistry of natural rubber latex. *Latex Allergy* 15(1): 1–20.

Sussman, G., Liss, G., Deal, K., et al. 1998. Incidence of latex sensitization among latex glove users. *J. Allergy Clin. Immunol.* 101: 171–180.

Swanson, M. C., Bubak, M. E., Hunt, L. W., Yunginger, J. W., Warner, M. A., and Reed, C. E. 1994. Quantification of occupational latex aeroallergens in a medical center. *J. Allergy Clin. Immunol.* 94(3): 445–451.

Tarlo, S. M., Sussman, G., Contala, A., and Swanson, M. C. 1994. Control of airborne latex by use of powder-free latex gloves. *J. Allergy Clin. Immunol.* 93(6): 985–989.

Tarlo, S. M., Sussman, G. L., and Holness, D. L. 1997. Latex sensitivity in dental students and staff: A cross-sectional study. *J. Allergy Clin. Immunol.* 99(3): 396–401.

Tarlo, S. M., Wong, L., Roos, J., and Booth, N. 1990. Occupational asthma caused by latex in a surgical glove manufacturing plant. *J. Allergy Clin. Immunol.* 85(3): 626–631.

Taylor, B., and Broom, B. C. 1981. Atopy in medical students. *Ann. Allergy* 47(3): 197–199.

Taylor, J. S., and Praditsuwan, P. 1996. Latex allergy. Review of 44 cases including outcome and frequent association with allergic hand eczema. *Arch. Dermatol.* 132(3): 265–271.

Theissen, U., Theissen, J. L., Mertes, N., and Brehler, R. 1997. IgE-mediated hypersensitivity to latex in childhood. *Allergy* 52(6): 665–669.

Tomazic, V. J., Shampaine, E. L., Lamanna, A., Withrow, T. J., Adkinson, N. F., and Hamilton, R. G. 1994. Cornstarch powder on latex products is an allergen carrier. *J. Allergy Clin. Immunol.* 93: 751–758.

Turjanmaa, K. 1987. Incidence of immediate allergy to latex gloves in hospital personnel. *Contact Dermatitis* 17(5): 270–275.

Turjanmaa, K., Alenius, H., Makinen-Kiljunen, S., Reunala, T., and Palosuo, T. 1996. Natural rubber latex allergy. *Allergy* 51(9): 593–602.

Turjanmaa, K., Makinen-Kiljunen, S., Reunala, T., Alenius, H., and Palosuo, T. 1995. Natural rubber latex allergy. The European experience. *Immunol. Allergy Clin. North Am.* 15(1): 71–88.

U.S. Department of Health and Human Services (DHHS). 1996. National Center for Health Statistics. Third National Health and Nutrition Examination Survey, 1988–1994, NHANES III Laboratory Data File (CD-ROM). Public Use Data File Documentation Number 76200. Centers for Disease Control and Prevention, Hyattsville, MD.

Vallier, P., Balland, S., Harf, R., Valenta, R., and Deviller, P. 1995. Identification of profilin as an IgE-binding component in latex from *Hevea brasiliensis*: Clinical implications. *Clin. Exp. Allergy* 25(4): 332–339.

Vandenplas, O. 1995. Occupational asthma caused by natural rubber latex. *Eur. Respir. J.* 8(11): 1957–1965.

Vandenplas, O., Vandezande, L. M., Halloy, J. L., Delwiche, J. P., Jamart, J., and Looze, Y. 1996. Association between sensitization to natural rubber latex and papain. *J. Allergy Clin. Immunol.* 97(6): 1421–1424.

Volcheck, G. W., and Li, J. T. 1994. Elevated serum tryptase level in a case of intraoperative anaphylaxis caused by latex allergy. *Arch. Intern. Med.* 154(19): 2243–2245.

Wastell, C. 1997. Chairman's conclusions. *Eur. J. Surg.* 163(Suppl. 579).

Wrangsjo, K., and Lundberg, M. 1996. Prevention of latex allergy. *Allergy* 51(1): 65–67.

Wrangsjo, K., Montelius, J., and Eriksson, M. 1992. Teats and pacifiers—an allergy risk for infants? *Contact Dermatitis* 27(3): 192–193.

Yassin, M. S., Lierl, M. B., Fischer, T. J., O'Brien, K., Cross, J., and Steinmetz, C. 1994. Latex allergy in hospital employees. *Ann. Allergy* 72(3): 245–249.

Yassin, M. S., Sanyurah, S., Lierl, M. B., Fischer, T. J., Oppenheimer, S., Cross, J., K OB, Steinmetz, C., and Khoury, J. 1992. Evaluation of latex allergy in patients with meningomyelocele. *Ann. Allergy* 69(3): 207–211.

Yeang, H. 1997. Impact of biologic variation on latex allergenicity. *J. Allergy Clin. Immunol.* 101: 145–146.

Yunginger, J. W. 1995a. Natural rubber latex. *Immunol. Allergy Clin. North Am.* 15(3): 583–595.

Yunginger, J. W. 1995b. Variances in antigenicity of latex products. *Immunol. Allergy Clin. North Am.* 15(1): 61–70.

Yunginger, J. W., Jones, R. T., Fransway, A. F., Kelso, J. M., Warner, M. A., and Hunt, L. W. 1994. Extractable latex allergens and proteins in disposable medical gloves and other rubber products. *J. Allergy Clin. Immunol.* 93(5): 836–842.

Yunginger, J. W., Jones, R. T., Fransway, A. F., Kelso, J. M., Warner, M. A., Hunt, L. W., and Reed, C. E. 1993. Latex allergen contents of medical and consumer rubber. Paper presented at the Forty-Ninth Annual Meeting of the American Academy Of Immunology, Chicago, IL, March 91(1, Part 2): 241–262.

SECTION J
Case Studies in Ecological Risk Assessment

19 Determining Values: A Critical Step in Assessing Ecological Risk

DOUGLAS P. REAGAN

URS Corporation, Denver, Colorado

19.1 INTRODUCTION

The current framework of hazardous waste laws, regulations, and guidance documents that provide for the protection of the environment offers only general concepts regarding which aspects of the environment are to be protected. Furthermore, the objectives of environmental protection are often neither clear nor consistent. In some countries "environment" has been defined as the human environment, disregarding ecosystem values. In the United States, the Environmental Protection Agency (EPA, 1989) equates environmental relevance with ecological relevance. This is significant in that it unambiguously recognizes ecological values distinct from human-centered values in the risk assessment process.

In ecological risk assessments, assessment endpoints (AEs) are defined as explicit expressions of the environmental values that are to be protected (Suter, 1989; EPA, 1992). AEs are identified early in an ecological risk assessment to ensure that the assessment focuses on relevant ecological concerns. Proceeding from a toxicological approach, these are often identified by evaluating potential exposure in general terms, then focusing on those organisms most sensitive to the particular environmental stressors (e.g., chemical contaminants), irrespective of their ecological relevance. Such an approach can lead to trivial or even erroneous conclusions (e.g., basing the risk assessment on potential adverse effects on earthworms because of their potential exposure and sensitivity is inappropriate at arid sites where earthworms are not a significant component of regional ecosystems.). Results of such assessments often cleanup criteria that are not based on ecologically relevant endpoints and are, therefore, of little value in making risk management decisions.

Norton (1987) and Harwell et al. (1994) have shown that all values placed

Human and Ecological Risk Assessment: Theory and Practice, Edited by Dennis J. Paustenbach
ISBN 0-471-14747-8 © 2002 John Wiley & Sons, Inc.

on ecological resources are ultimately human values. However, these can be logically subdivided into ecological values and human values. In this context ecological values refer to those that are necessary to maintaining the ecosystem, and human values refer to direct human uses (e.g., hunting, fishing, or timber extraction) and to nonconsumptive considerations (e.g., aesthetic and spiritual values such as wilderness or religious uses by indigenous people). While many of these values are ecosystem specific, some are common to all ecosystems. Identifying all pertinent ecosystem values is a fundamental, but often neglected, step in performing ecological risk assessments.

The EPA (1998) provides three criteria for selecting AEs:

- Ecological relevance
- Relevance to management goals (societal relevance)
- Susceptibility to known or potential stressors

Ecologically relevant endpoints are those that reflect important ecological characteristics and are functionally related to other endpoints. Endpoints based on ecological values that people care about have management relevance (i.e., game species that people value) and are, therefore, more likely to be used in risk management decisions (EPA, 1992, 1998). These two criteria encompass both the relevance to all levels of organization within an ecosystem and to the human values derived from the ecosystem.

The third criterion, susceptibility to environmental stressors, addresses both sensitivity and exposure to stressors. These stressors can be any physical, chemical, or biological entities capable of inducing adverse responses (EPA, 1992). However, most risk assessments focus only on chemical contamination and disregard other related stressors. For example, mine sites may produce environmental stress by the release of chemical contaminants, but waste rock and tailings disposal can produce direct physical stress by burying surface vegetation and depositing sediment in stream channels. The direct physical effects of this discharge can disrupt both terrestrial and aquatic ecosystems, but current regulations are often designed to address chemical stressors and ignore others.

For chemical stressors (i.e., contaminants), properties such as the ability to biomagnify up food chains or affect certain receptor groups (e.g., mammals, fish, invertebrates) are important considerations in an ecological risk assessment. Identifying assessment endpoints based on ecological and management (societal) relevance and subsequently evaluating susceptibility to environmental stressors (e.g., chemical contaminants) can remove the "most sensitive species" bias that in some studies has lead to irrelevant or inappropriate conclusions in ecological risk assessment.

Because all levels of ecological organization may be adversely affected by environmental stressors, the process of determining ecological values can seem daunting. This is particularly true for study sites that support a variety of natural ecosystems. AE selection is all too often an informal process that frequently fails to identify the consequences of stressor effects on ecological values above the population level of organization or focuses on only selected properties (e.g., benthic invertebrate biodiversity). Documentation of the rationale for selecting particular values and not others is generally brief or totally lacking. An inadequate endpoint selection process can result in overlooking pertinent AEs or in not documenting why some potential endpoints were not originally evaluated, thus delaying the completion of the risk assessment. For example, identification of new AEs late in the risk assessment can lead to identification of additional field and/or laboratory investigations, reevaluation of receptors and exposure pathways, and other costly and time-consuming activities. Systematically identifying ecological values in assessing ecological risk eliminates the likelihood of reaching trivial conclusions and incurring costly delays in completing the risk assessment.

19.1.1 Purpose

This chapter describes a systematic and comprehensive process for identifying ecological values to be protected for the ecosystems potentially affected by environmental stressors as defined by the EPA (1992). Examples are provided throughout the text, and a case study is included. These illustrate pertinent aspects of the process of selecting assessment endpoints in ecological risk assessments.

The process of identifying general values, hereafter referred to as general assessment endpoints (GAEs), provides a means of thoroughly documenting the rationale for determining values for each potentially affected ecosystem, irrespective of their potential exposure to environmental stressors. These GAEs encompass ecological and human use values at all levels of ecological organization, thus providing a basis for determining site-specific AEs. Site-specific AEs are then identified by considering which of the general AEs may be susceptible to stressors (i.e., which may have been or are currently being affected by environmental stressors).

The idea for the GAE approach grew out of discussions at the 1993 Pelston Workshop on Sustainable Ecosystem Management, sponsored jointly by the Ecological Society of America (ESA) and the Society for Environmental Toxicology and Chemistry (SETAC). There was no final consensus on the fundamental question: What do we want to sustain? However, there was general agreement that the values to be protected/sustained needed to be identified, and that ecosystem values at all levels of ecological organization would need to be addressed. This approach to determining ecological values was developed further as part of the selection of assessment endpoints for the ecological risk assessment at the Lavaca Bay Superfund site (Parametrix et al., 1996) and presented in conceptual form at the 1996 SETAC Annual Meeting (Reagan et al., 1996). It has since been used successfully in a variety of ecological risk assessments ranging from federal facilities in the United States to the scale of an entire river basin affected by mine waste disposal in Papua New Guinea.

The methodology presented herein has several advantages over less rigorous methods for selecting AEs. First, determining values prior to considering stressor sensitivity avoids the frequent bias of focusing an ecological risk assessment on the "most sensitive species," irrespective of its ecological relevance. Second, because the process can incorporate values identified in other natural resource activities (e.g., ecosystem management, natural resource damage assessment, environmental impact assessment, ISO 14000 compliance), it provides a consistent basis for integration among related processes. Third, the stepwise approach permits a comprehensive and systematic means of addressing all possible endpoints and documenting how each was considered. Fourth, the process provides a basis for stakeholder understanding and involvement in the complex process of identifying ecological and human use values in risk assessment and in subsequent risk management decision making.

19.1.2   Process Overview

Values determination in ecological risk assessments begins by identifying GAEs (the universe of potential ecological values to be protected), then proceeds to identify site-specific AEs. The process of identifying GAEs occurs in two parts. Part one is performed to identify ecological values from a strictly ecological perspective; part two considers the human values associated with the ecological resources of the ecosystem under evaluation.

Identification of ecological values begins with the identification of values common to all ecosystems and progresses to a consideration of values pertinent to the regional ecosystems of interest. This progression is hierarchical and scientifically based, thus providing an objective means of determining which components of the ecosystem are potentially relevant.

The ultimate ecological value to be protected in an ecological risk assessment is a healthy sustainable ecosystem. Recognizing the ecosystem as the appropriate context for evaluating values at all levels of ecological organization is consistent with current principles of ecosystem management and with our current understanding of ecosystems (Kaufmann et al., 1994; Boyce and Haney, 1997; Reichman and Pulliam, 1996). Ecological relevance, as used here, refers to the properties necessary to sustain normal ecosystem structure and function.

The process of identifying ecologically relevant GAEs consists of five steps:

1. Identifying values common to all ecosystems
2. Identifying functional components of regional ecosystems
3. Developing a functional food web of potentially affected ecosystems
4. Determining ecologically relevant attributes of the functional components of these ecosystems
5. Stating the ecologically relevant GAEs

Once these GAEs have been determined, values relevant to management goals (i.e., societal values) are developed. Subsequent sections describe the process of GAE development for a forest ecosystem.

Site-specific AEs are then selected by evaluating the susceptibility of the GAEs to site-related environmental stressors. Potential exposure to environmental stressors can be determined from the conceptual exposure models. Because environmental stressors may differentially affect exposed endpoints (i.e., some chemicals have herbicidal effects while others may affect particular animal groups), GAEs must be examined systematically for each stressor. The result of this process is a list of site-specific assessment endpoints, systematically derived, that provides a firm basis for the ecological risk assessment. Figure 19.1 presents an overview of the value determination process that results in determination of assessment endpoints. Note that the activities for identifying ecological values are sequential and that the activities for determining societal values are not. GAE determination based on ecological relevance is addressed in Section 19.2, Section 19.3 addresses GAE determination based on societal values, and the identification of site-specific assessment endpoints is presented in Section 19.4.

19.2 ECOLOGICALLY RELEVANT VALUES

In broad terms, an ecosystem can be defined as the habitats, both aquatic and terrestrial, of the site or area potentially affected by stressors. Sustaining a healthy ecosystem is the ultimate ecological value to be protected; however, a variety of other ecological values must also be considered. The process of identifying these values begins at the ecosystem level, considers values common to all ecosystems, and progresses to a consideration of values for potentially affected regional ecosystems (Figure 19.2).

19.2.1 Values Common to All Ecosystems

The approach for identifying GAEs proceeds hierarchically from the fundamental value of preserving a healthy and sustainable ecosystem to the identification of valued characteristics common to all ecosystems. These are:

- *Biological Diversity (Biodiversity)*. This property describes ecological structure in terms of components and can include species, community/habitat, and genetic diversity. From a generic perspective, biological diversity is usually considered most relevant in the context of species diversity (i.e., the variety and abundance of species) (Norton, 1987; Magurran, 1988; Kellert, 1996) that is typical of the particular ecosystems under consideration. Communities that are more diverse are not necessarily more relevant than less diverse communities (Paine, 1966), but communities in

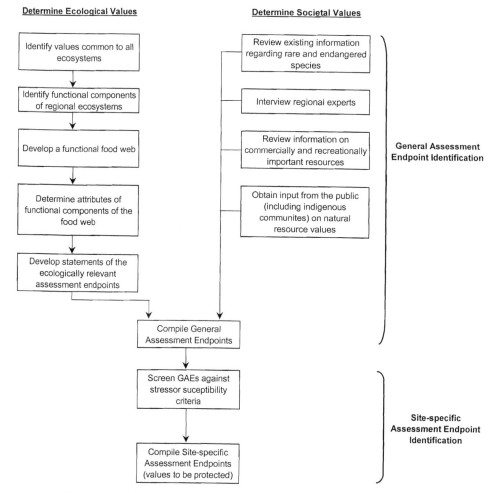

Figure 19.1   Process for determining values in ecological risk assessments.

disturbed ecosystems may be more or less diverse than those in comparable but undisturbed ecosystems.

•   *Functional Integrity.* This property describes ecological function; not only are all of the ecosystem components present, but they interact according to organizational principles typical of that ecosystem. An ecosystem with functional integrity is comprised of living organisms within populations of different taxa, functionally interacting to sustain normal ecosystem interactions. Some organisms modify their surroundings primarily by altering the system's biotic features while others interact to alter ecosystem features such as habitat or food (Reiger, 1993; Jones et al., 1994).

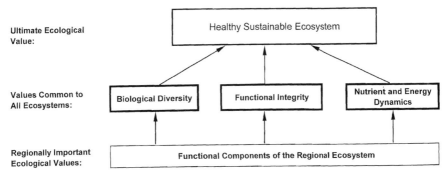

Figure 19.2 Hierarchy of ecological values.

• *Nutrient and Energy Dynamics (Cycling and Transport Processes).* Appropriate nutrient and energy dynamics must be operating within the ecosystem to maintain balanced populations of indigenous species at levels typical of a healthy ecosystem. For an ecosystem to function normally, the flow rates of energy and nutrients (e.g., primary productivity and decomposition) should fall in a range typical of that type of ecosystem. Disruption of flow rates could lead to loss of stability, accumulation of detritus, reduction of energy inputs, or loss of top predators that could alter energy flow patterns and change ecosystem structure (McNaughton, 1978).

These values express the basic considerations of ecosystem structure and function using fundamental ecological concepts expressed in terms that can be evaluated in the risk assessment. Because they are values relevant to all ecosystems (Kormondy, 1969; Brewer, 1979; Odum, 1993), they are appropriate values to be protected in the assessment of ecological risk. Consideration of these properties ensures that these aspects of the ecosystem, each of which can be adversely affected by environmental stressors, addresses the highest levels of ecological organization that can be measured. All other ecologically relevant issues are subsets of these valued characteristics common to all ecosystems.

Values at the ecosystem and community level of ecological organization could be described in other ways (e.g., ecosystem structure and function); however, those defined here generally address composition, organization, and process rates. Additionally, these values are expressed in terms that can be measured directly. For example, stresses on benthic macroinvertebrates might affect benthic diversity, which can be measured directly, and stressors with herbicidal effects could be addressed by looking at rates of vegetation production.

19.2.2 Functional Components of the Regional Ecosystem

Values of regional ecosystems are the next level to be determined. To do this, it is first necessary to identify the principal functional components of the ecosystems. Because food webs provide essential structural organization in ecosystems (Gallopin, 1972) and because all organisms in an ecosystem are part of the food web, the food web concept is used to identify basic functional components of potentially affected ecosystems. Food webs are typically comprised of three basic trophic categories:

- *Producers*—organisms that manufacture food from inorganic compounds by photosynthesis or chemosynthesis (e.g., green plants, chemosynthetic bacteria)
- *Consumers*—organisms that ingest other organisms (e.g., animals that consume plants and/or other animals)
- *Decomposers*—organisms that derive their nourishment from dead organic matter (e.g., fungi and bacteria)

These categories are based on the broad interrelationships among groups of organisms but do not describe the many ways in which these interactions may occur. The three fundamental food web categories need to be categorized into functional groups (components), based on a general knowledge of the species present in regional ecosystems. Consumers are grouped in feeding guilds (e.g., organisms that obtain their food in a functionally similar way). For example, many organisms in a forest eat insects, but those that forage in the canopy perform a different functional role than those that feed in the litter of the forest floor. Therefore, food webs based on feeding guilds facilitate the identification of critical ecosystem functions performed by members of each guild and the interrelationships among guilds that may affect other ecosystem properties.

By using functional rather than taxonomic groups, we are able to determine broad functional aspects of the ecosystem and important interrelationships among components, while avoiding the impossible task of determining the ecological relevance of each species. To illustrate this approach, consider the food web of one of the most complex ecosystems on earth, a tropical rain forest. Functional components were developed on the basis of food, foraging location, and food habits of species present in the forest. Table 19.1 lists the functional components for a tropical rain forest ecosystem in New Guinea.

Based on food type and location, four ecological components have been identified within the herbivorous consumer trophic category of the rain forest food web (Table 19.1). Some feeding guilds have been arbitrarily lumped (e.g., fruit and seed eaters, grazers and browsers) because of the similarity on food type for many species in this particular ecosystem. However, for a savanna grassland, further division into separate components such as seed eaters (granivores) and fruit eaters (frugivores), foliage or leaf eaters (folivores), nectar and pollen feeders (nectarivores), and browsers/grazers would be more appropriate. Similar functional groups have been identified for the other broad trophic categories, based on a detailed knowledge of the organisms present in the ecosystem.

TABLE 19.1   Functional Components of a Tropical Rain Forest Ecosystem in New Guinea

Basic Trophic Category	Functional Component
Producer	Ground layer
	Understory trees
	Canopy trees
	Lianas and epiphytes
	Mycorrhizal fungi (enhancers)
Consumers	Herbivores
	Terrestrial frugivores/granivores
	Arboreal frugivores
	Folivores and browsers
	Nectarivores
	Carnivores and omnivores
	Intermediate and small predators
	Omnivores
	Top terrestrial predators
	Top arboreal predators
	Detritivores and scavengers
Decomposers	Chemical decomposers

These categories could be further subdivided, based on specific knowledge of the importance of some groups. For example, bats and butterflies pollinate different plants, and some relationships are species specific. To the extent that these relationships are known to be important in overall ecosystem function, they can be incorporated into the determination of AEs. The important point is that all species fit into one or more of the functional components. In some instances a species may be classified into more than one of the component categories (e.g., butterflies and moths, which have herbivorous larvae and nectar-feeding adults).

It is important to clearly define the functional components so that the interrelationships among them can be understood. This is particularly true of ecosystems that consist of species that are poorly known and/or have limited distributions, such as those found in New Guinea. The plants and animals of the tropical rain forest of New Guinea have close affinities to Australia but differ from the rest of the world. Tropical rain forests in the Americas, Africa, and Southeast Asia typically have monkeys as prominent arboreal herbivores and cats (e.g., tigers, jaguars) as top predators; however, neither group occurs in New Guinea, which has been isolated from these regions. Tree kangaroos are the ecological equivalent of monkeys, and the largest predators in the forest are pythons (native) and wild dogs (introduced in New Guinea about 2000 years ago).

Information on the species present in the New Guinea rain forest and on their habitats, food habits, and foraging strategies was obtained primarily from

Mackay (1976), Beehler et al. (1986), Flannery (1995). Definitions and representative taxa comprising the functional components of the New Guinea food web listed in Table 19.1 are provided below:

Ground Layer. Herbaceous and woody species usually under 3 m tall (e.g., grasses, ferns, mosses, tree seedlings, and forest herbs)

Understory Trees. Tree species that, when mature, do not emerge into the canopy

Canopy Trees. Tree species that, when mature, reach the canopy

Lianas. Vines and other climbing plants

Epiphytes. Plants growing on other plants

Mycorrhizal Fungi. These fungi comprise a unique functional component of terrestrial ecosystems; they form mutualistic associations with plant roots, helping the plant roots absorb mineral compounds and in return absorbing organic food from the plants. Most tropical plants form such associations, and many plants require mycorrhizal fungi in order to thrive. Consequently, these associations are extremely important in aiding growth and determining plant species composition in tropical rain forests (Bagyaraj, 1989).

Terrestrial Frugivores/Granivores. Animals that consume fruit and seeds but feed at or near ground level (e.g., cassowary, mice, rats, ants)

Arboreal Frugivores. Animals that consume fruit but feed in upper strata of the forest (e.g., pigeons, parrots, cuscus, frugivorous bats)

Folivores and Browsers. Animals that consume leaves and small twigs (e.g., tree kangaroos, ringtail, chewing insects, deer)

Nectarivores. Animals that feed on the pollen and nectar of flowering plants (e.g., bees, beetles, butterflies, honeyeaters, flowerpeckers, nectarivorous bats)

Intermediate Predators. Animals that consume other animals but are also consumed by larger predators (e.g., bandicoot, frogs, insectivorous bats). This category may encompass more that one trophic level; some snakes eat lizards, which eat spiders, which eat predaceous insects, etc.).

Small Predators. This category is optional; it contains spiders, small lizards, and the like but otherwise fits into the same definition as intermediate predators.

Omnivores. Animals that consume both plant and animal matter (e.g., wild boar, berrypeckers)

Top Terrestrial Predators. Large carnivorous animals at the top trophic level of the terrestrial portion of the forest food web (e.g., python, death adder, wild dogs)

Too Arboreal Predators. Large predatory animals at the top of the arboreal portion of the forest food web (e.g., eagle, hawks, owls)

Detritivores and Scavengers. Animals that feed on dead plant or animal material (e.g., termites, black kite)

Chemical Decomposers. Organisms that decompose dead organic matter to simpler chemicals (e.g., bacteria, fungi)

Development of these categories should be performed by an ecologist with at least a general knowledge of the system. Consultation with local experts and local communities should be conducted in addition to reviewing the published literature on ecosystems and components.

Stakeholder involvement is crucial to the successful development of AEs. Ultimately, the effectiveness of an ecological risk assessment depends on how it improves the quality of management decisions, and these decisions are (or at least should be) based on values to be protected. Risk managers are more willing to use a risk assessment as the basis for making remedial decisions if the risk assessment considers ecological values that people care about (EPA, 1998). However, it is not always easy to identify the appropriate suite of stakeholders for a particular site and enlist their involvement. At sites in developed countries, such stakeholders may generally involve local municipalities, national and state agencies, and local communities. At remote sites in developing countries, appropriate stakeholders may additionally include resident indigenous peoples, international and regional nongovernmental organizations (NGOs) such as the Rainforest Action Network or Greenpeace, and development interests (e.g., commercial logging or fishing industries). The discussion of values to be protected among such groups may be contentious and contradictory because some values are clearly in conflict (e.g., protection of tress for timber conflicts with protection of forest wildlife species). Identifying all potential values to be protected and determining which ones will be selected for protection is an extremely valuable means of selecting the appropriate assessment endpoints.

From the standpoint of endpoint selection, it is significant that even a complex ecosystem, such as a tropical rain forest, that is comprised of thousands of plant and potentially millions of animal species can be comprehensively represented in functional terms by less than 20 functional components. It is also noteworthy that despite the greatly different taxonomic components, all tropical rain forests have essentially the same functional groups. This generalization generally holds true for wetlands, freshwater streams, estuaries, and coral reefs; the taxonomic composition may differ, but the functional components and interactions remain much the same.

While exotic plant and animal species are components of many ecosystems, they are generally considered as biological stressors and not as valued components of the ecosystem (Reagan et al., 1999). Therefore, all functional groups (components) should include only native species unless otherwise stated. Some typical exceptions are introduced species that are economically valuable, such as pigs, which have been introduced to many island ecosystems, including New Guinea. In some ecosystems, exotic species have become established and comprise a significant and stable portion of some functional groups (e.g., phragmites in North American wetlands). While these are clearly exotic, they are functionally integrated into the ecosystem and should be appropriately considered. Similar considerations may apply where species have been introduced for recreational purposes (e.g., ring-necked pheasant, rainbow trout in North America) that are now integral components of the ecosystems into which they were introduced (Polis and Winemiller, 1996).

19.2.3   Functional Food Web

The functional components described in the previous section define the general range of feeding preferences and location (strata) in the New Guinea rain forest. A food web based on these functional components is presented in Figure 19.3. The arrows in the food web diagram indicate the direction of flow of energy and nutrients through the food web. Dashed lines indicate the recycling and flux of energy and nutrients as the result of decomposition processes. The diagram shows major pathways; others may be present but are either insufficiently understood or relatively insignificant at this level of conceptualization. Such a diagram is simplistic in that it does not take into account reciprocal predation (food loops) and the division of the food web into day and night compartments, such as occur in tropical forest food webs (Reagan and Waide, 1996). These relationships can be reexamined in the course of selecting site-specific assessment endpoints, where consideration of specific exposure pathways may be important.

19.2.4   Ecologically Relevant Attributes

While feeding relationships are relevant characteristics of each functional component of the terrestrial food web, each component may have additional ecologically relevant attributes that define its overall ecological value. For many functional components, the nontrophic attributes are at least as important as their role in nutrient and energy transfer through the food web.

Relevant attributes of the ecological components of the New Guinea rain forest ecosystem are defined below:

Food. Source of energy and nutrients for ecosystem components

Habitat. Shelter or structural support for other organisms

Energy and Nutrient Fixation. The fundamental process of converting inorganic chemicals to organic compounds that can provide energy and nutrients to other living components of the ecosystem

Decomposition. The breakdown of nonliving organic matter, recycling nutrients and preventing an accumulation of nonliving organic matter that would interrupt energy and nutrient cycling processes

Propagule Dispersal. The distribution of seeds and spores from their origin to other locations. The process is important for recolonization and natural revegetation following disturbance

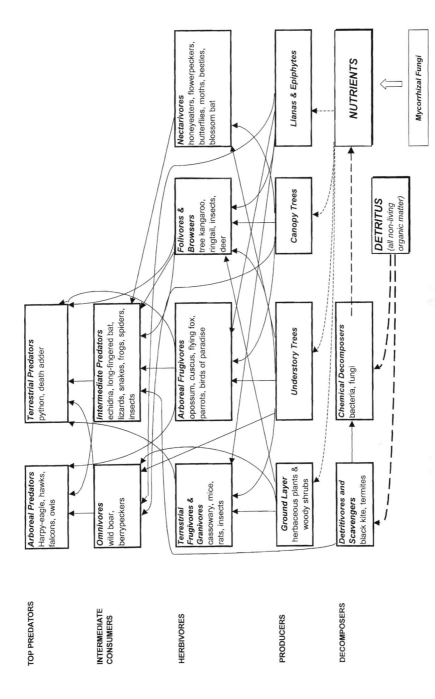

Figure 19.3 Food web based on the functional components of a tropical rain forest in New Guinea.

TOP PREDATORS

INTERMEDIATE
CONSUMERS

HERBIVORES

PRODUCERS

DECOMPOSERS

Arboreal Predators
Harpy-eagle, hawks,
falcons, owls

Terrestrial Predators
python, death adder

Omnivores
wild boar,
berrypeckers

Intermediate Predators
echidna, long-fingered bat,
lizards, snakes, frogs, spiders,
insects

Nectarivores
honeyeaters, flowerpeckers,
butterflies, moths, beetles,
blossom bat

Arboreal Frugivores
opossum, cuscus, flying fox,
parrots, birds of paradise

*Folivores &
Browsers*
tree kangaroo,
ringtail, insects,
deer

*Terrestrial
Frugivores &
Granivores*
cassowary, mice,
rats, insects

Ground Layer
herbaceous plants &
woody shrubs

Understory Trees

Canopy Trees

Lianas & Epiphytes

NUTRIENTS

*Detritivores and
Scavengers*
black kite, termites

Chemical Decomposers
bacteria, fungi

DETRITUS
(all non-living
organic matter)

Mycorrhizal Fungi

1081

Pollination. The cross-fertilization by nectar and pollen feeding animals is the sole means of reproduction in many plant species

Predation. The killing and consumption of other animals by carnivores including top predators, insectivores, and parasites. Predation is a means of exerting control over population dynamics and trophic structure within some ecosystems (Osenberg and Mittelbach, 1996).

Control. Either "top-down" or "bottom-up" effect on the structure (composition and abundance) or function of the ecosystem

The various attributes are then determined for each of the functional components of the ecosystem and arranged in a table displaying the attributes of each functional component (Table 19.2), where each functional component has at least one attribute. While some of these attributes could be considered more important than are others, Table 19.2 provides a means of describing ecological values and subsequently identifying AEs based on these values.

While all organisms are food for other organisms, different functional components provide other services within an ecosystem. Again using the rain forest as an example, trees not only provide food for herbivores, they also provide structural habitat for arboreal species and create shaded microclimates beneath the canopy. Animals that feed on pollen and nectar play a relatively minor role in nutrient and energy transfer through the ecosystem, but they provide the critical function of pollination for many plant species. Seed-eating birds, such as cassowaries and pigeons, not only provide a substantial prey base for predators but frequently bury seeds some distance from the location where they find them, thus playing an important role in seed dispersal and germination (Reagan, 1997). Fruit bats are known to be important seed dispersers for pioneer species in early forest succession. Wunderle (1997) has shown the importance of animal seed dispersers in accelerating the regeneration of tropical forests, and Janzen (1983) believes that seed predators (granivores) have the largest impact on tropical forest structure of any life form.

19.2.5   General Assessment Endpoints Based on Ecological Relevance

The EPA (1992) defines an *endpoint* as "a characteristic of an ecological component ... that may be affected by exposure to a stressor," and defines an *assessment endpoint* as "an explicit expression of the environmental values that is to be protected." Therefore, general assessment endpoints based on functional groups in the New Guinea rain forest ecosystem can be developed directly from Table 19.2, which provides a comprehensive summary of the attributes of all functional components of the ecosystem. The thorough nature of the process ensures that relevant assessment endpoints at all levels of ecological organization will not be overlooked. This is because the ecological components and their attributes (relevant characteristics) were derived using a comprehensive and systematic process based on an understanding of fundamental ecological principles and site-specific information.

TABLE 19.2 Ecologically Relevant Attributes of the Functional Components Identified for a Tropical Rain Forest Ecosystem in New Guinea

Functional Components				Ecological Attributes			
	Food	Habitat	Primary Production	Pollination	Seed Disp.	Decomp.	Control
Ground layer	X	X	X				
Understory trees	X	X	X				
Canopy trees	X	X	X				
Lianas and epiphytes	X	X	X				
Mycorrhizal fungi			X				X
Terrestrial frugivores/granivores	X				X		
Arboreal frugivores	X				X		
Folivores and browsers	X						
Nectarivores	X			X			
Intermediate and small predators	X						
Omnivores	X						
Top terrestrial and arboreal predators							X
Detritivores and scavengers							X
Chemical decomposers							X

The GAEs identified here are considered general endpoints characteristic of a tropical rain forest in New Guinea. Site-specific AEs can be developed that consider exposure pathways, site-specific suites of contaminants, potential receptors and effects, and other factors following the general approach described in Section 19.4.

Globally Relevant Assessment Endpoints   The following endpoints are based on ecological values characteristic of all ecosystems. Detailed descriptions of these endpoints were provided in Section 19.2.1.

- *Biodiversity*—for taxonomic groups, communities, or the entire ecosystem
- *Functional Integrity*—for the entire ecosystem or pertinent subsystems (e.g., soil fauna)
- *Nutrient and energy dynamics*—litter decomposition, productivity, etc.

Regionally Relevant General Assessment Endpoints   Assessment endpoints can be expressed in terms of the significant attributes of the ecological components identified in previous sections. Such statements clearly define the values to be protected and provide a simple means of conveying this information to risk managers and other stakeholders in the ecological risk assessment process.

The following GAE statements are based on the significant attributes of the ecological components of the tropical rain forest ecosystem. Only a few of these statements are provided for illustration; however, in an actual ecological risk assessment all statements would be included for the sake of completeness.

- Canopy trees are valued components of the forest ecosystem because of their role in primary production, providing habitat for arboreal species, and as food for a variety of herbivorous animals.
- Terrestrial granivores and frugivores are valued components of the forest ecosystem because of their importance as a food source to higher level carnivores and their role as seed/propogule dispersers.
- Folivores and browsers are a valued component of the forest ecosystem because of their importance as a food source to higher level carnivores and their role as non-food-chain-based seed/propogule dispersers (e.g., seeds cling to their fur).
- Nectarivores and pollen eaters are valued components of the forest ecosystem because of their importance in plant reproduction (i.e., pollination).
- Mycorrhizae are a valued component of the forest ecosystem because of their importance in nutrient recycling and regeneration of plants.
- Chemical decomposers are a valued component of the forest ecosystem because of their importance in decomposition, nutrient recycling, and as a food source.

19.2.6   General Assessment Endpoints in Major Ecosystem Types

The rain forest example illustrates how an extremely complex and unfamiliar ecosystem could be systematically evaluated to determine relevant ecological values (i.e., GAEs). In practice, most ecosystems evaluated in ecological risk assessments will be less complex.

Many sites will have terrestrial ecosystems composed of a variety of habitat/ vegetation types (e.g., forest, woodland, grassland), freshwater stream ecosystems, or marine/estuarine ecosystems. There is a temptation to regard vegetation types as ecosystems; however, from a food web perspective terrestrial ecosystems usually form a mosaic with many species feeding at the interface between ecosystems or, as in the case of many top predators, foraging throughout all vegetation types.

At the Lavaca Bay Superfund site, GAEs were developed for terrestrial, aquatic, and estuarine ecosystems. The functional components for each of these ecosystems are presented in Table 19.3. Development of the food webs, identification of relevant attributes, and selection of GAEs followed the approach presented earlier in this section. These basic categories form an excellent basis for developing GAEs for any ecosystem or site, as they address the major ecosystems likely to be encountered at a contaminated site or other location subject to an ecological risk assessment.

The GAE approach is flexible and can be adapted to address regionally specific considerations, within or among ecosystems. For example, aquatic and terrestrial species as frequently viewed as components of separate aquatic and terrestrial ecosystems, and ecosystems are frequently developed on the basis of habitats. At sites where the aquatic ecosystems comprise a small but important part of the overall ecosystem, and where there is considerable interaction between terrestrial and aquatic organisms, these two ecosystems can be combined into one integrated ecosystem.

19.3   SOCIETALLY RELEVANT VALUES

Ultimately, the effectiveness of an ecological risk assessment depends on how it improves the quality of risk management decisions. Risk managers are more willing to use a risk assessment as the basis for making remedial decisions if the risk assessment considers ecological values relevant to stakeholders (EPA, 1998). Therefore, to be effective, the ecological risk assessment must consider both the ecological and societal values of ecosystems.

19.3.1   Determining Societal Values

Management goals are inextricably related to the societal values of ecological resources (EPA, 1998). These include formally recognized and protected eco-

TABLE 19.3 Functional Components of the Food Webs for Three General Ecosystem Types

Basci Trophic Category	Functional Components		
	Terrestrial	Freshwater	Estuarine
Producer	Herbaceous vegetation	Phytoplankton	Phytoplankton
	Woody vegetation	Attached algae	Attached algae
	Mycorrhizal fungi (enhancers)	Submergent/emergent vegetation	Submergent/emergent vegetation
		Floating vegetation	
Consumer	Herbivores	Herbivores	Herbivores
	Terrestrial frugivores/granivores	Benthic herbivores	Benthic herbivores
	Arboreal frugivores/granivores	Vertebrate herbivores	Vertebrate herbivores
	Folivores	Zooplankton	Zooplankton
	Browsers		
	Nectarivores		
	Carnivores and omnivores Omnivores	Carnivores and omnivores	Carnivores and omnivores
		Benthic and pelagic omnivores	Benthic omnivores
	Intermediate and small predators	Vertebrate omnivores	Vertebrate omnivores
	Top terrestrial predators	Benthic carnivores	Benthic carnivores
	Top arboreal predators	Nektonic carnivores	Nektonic carnivores
		Semiaquatic carnivores	Semiaquatic carnivores
	Detritivores and scavengers	Detritivores and scavengers	Detritivores and scavengers
Decomposer	Chemical decomposers	Chemical decomposers	Chemical decomposers

logical values such as threatened and endangered species, and recreationally important (e.g., game) species plus regionally valued species and resources. Consistent with the data quality objectives process (EPA, 1993), identification of these values should involve input from risk managers, risk assessors, ecologists, pertinent regulatory authorities, regional experts, and the public. It is particularly important to consult local citizens and understand the values they place on ecological resources. Where the public includes members of indigenous communities (e.g., Indian tribes), the values identified may include a broad range of ecological components and human use values such as food, medicine, and spiritual purposes.

Determining societal values involves a systematic search of all pertinent data sources. These include:

- Reviewing existing information regarding rare and threatened species
- Conducting interviews with regional experts
- Reviewing information on commercially and recreationally important resources
- Obtaining input from the public (including indigenous communities) on natural resource values

These steps are not sequential, but it is important to consult all potential sources. It is also apparent that some values will be in conflict. For example, it may not be possible to fully protect some wildlife species in ecosystems where the primary management goal is for timber extraction.

From an international perspective, societal values can include formally recognized and protected ecological entities such as threatened species. Threatened species are designated by governmental and nongovernmental organizations [e.g., the International Union for the Conservation of Nature and Natural Resources (IUCN)]. In New Guinea, the plumes of many bird species, especially bird-of-paradise, are used in ceremonial costumes, as are the skins of several mammals. Thus local human uses may be in conflict with the global values for the same species.

From a global perspective, many of the species inhabiting New Guinea rain forests are valued because they are threatened with extinction and because of general concerns for loss of tropical biodiversity. The IUCN lists 27 mammal, 31 bird, 10 reptile, 13 fish, and 11 invertebrate species as threatened in New Guinea; most of these inhabit tropical rain forests (Bailie and Groombridge, 1996).

Rare or declining species are frequently accorded legal protection, evidencing society's value for these species. Different levels of protection can be provided by separate governmental entities, complicating the assignment of human values to these species.

This is particularly true in the United States where some of the applicable categories of protection include:

- Endangered (federal, state)
- Threatened (federal, state)
- Species of special concern (federal)
- Species of local concern

Several native wildlife species inhabiting the tropical rain forests of New Guinea are hunted and eaten by local people. These include cuscus, tree kangaroos, cassowaries, and various other birds and mammals. Some species range throughout most of the forest, but others are confined to specific microhabitats. In many developed countries, numerous animal species are valued for recreational hunting (e.g., deer, elk, ducks) or simply for wildlife viewing (e.g., the charismatic megafauna of Africa, birds).

Other ecological values for the ecosystem may be identified, based on a review of the regional management goals and plans for the potentially affected areas. The list of societal values incorporates input from natural resource agencies, NGOs, and other representatives of the public. Such values will be included for consideration as they are identified throughout the ecological risk assessment process.

19.3.2   General Assessment Endpoints Based on Societal Relevance

The following GAEs are based on the societal values of ecological components of the New Guinea forest ecosystem, which were identified in Section 19.2. Only representative GAEs are included for illustrative purposes.

- Game species are valued components of the ecosystem and are to be protected because of their importance for consumptive uses such as hunting and fishing by indigenous communities.
- Threatened and endangered species and their habitats and migratory bird nesting and roosting sites are valued components of the ecosystem to be protected because of their regulatory stature and global value.
- Watershed health (water quality and quantity, erosion control, wetlands, etc.) is a valued component of the ecosystem and is to be protected because of its importance to local communities.
- Certain indigenous plants and animals are valued components of the ecosystem and are to be protected because of their ethnological uses and other consumptive and nonconsumptive uses.

19.4   SELECTING SITE-SPECIFIC ASSESSMENT ENDPOINTS

The GAEs provide a comprehensive summary of the ecological and societal values for the ecosystem. These have been identified to ensure that values at all levels of ecological organization will be appropriately addressed in the subsequent identification of site-specific AEs. Consistent with EPA guidance (EPA, 1992, 1998), the additional criteria applied for the selection of site-specific AEs involve the susceptibility to known or potential environmental stressors.

The process of selecting site-specific assessment endpoints requires an understanding of:

- Which functional components and/or species are potentially affected by environmental stressors (e.g., Is the potentially affected area a grassland, woodland, wetland, river?)
- Which plant and animal groups potentially inhabiting the area of exposure are sensitive to which contaminants
- What exposure pathways exist between contaminant sources and sensitive taxa (e.g., direct exposure, bioaccumulation through the food web)
- Which aspects of ecosystem organization (e.g., biodiversity) could be affected by contamination effects on the potentially exposed functional component(s)
- Which effects are potentially significant enough to require remedial action

Knowing the potentially contaminated habitat(s) is critical to understanding which functional components and taxa, and thus which GAEs, may be affected by contamination. These can be determined by examination of maps showing the distribution of contaminants and habitat types, reference to the comprehensive species list for the potentially affected area, and reconnaissance surveys.

At the Lavaca Bay site, elevated concentrations of mercury were associated with fringing *Spartina* wetlands in the immediate vicinity of the site. Exposure of organisms feeding or residing in these areas was confirmed by tissue sampling, whereas species associated with open water habitats were less exposed. From the suite of GAEs in the estuarine ecosystem, benthos (bottom-dwelling organism) abundance and productivity, zooplankton abundance and productivity, and vertebrate carnivores were retained as AEs for mercury contamination, whereas oyster beds and submergent vegetation were no longer retained (Parametrix, 1996). By using this process, the risk assessment documented that these potential values were considered and clearly indicated why they were not retained.

Sensitivity to known or potential stressors is another pertinent criteria for selecting site-specific AEs. For instance, if a particular bioaccumulative contaminant (e.g., an organochlorine pesticide such as dieldrin) may reach the "top carnivore" functional component of the food web, the taxa comprising that compartment may be differentially sensitive to the contaminant. In the case of dieldrin, raptors would be an appropriate AE representing that GAE because they are more sensitive to dieldrin than are mammalian carnivores. At Rocky Mountain Arsenal, the bald eagle was selected as an endpoint because of its sensitivity to the stressors of concern and its potential exposure through the food web.

Sensitivity of particular taxa to site-specific contaminants is also important in determining which AEs to select. While a number of threatened or endangered species may be present in a contaminated area, the particular taxon (e.g., species) selected as an AE will very much depend on exposure pathways and sensitivity for the contaminants potentially present.

For an endpoint to be affected by a chemical stressor, a potential exposure pathway must exist from the contaminant source to the endpoint. These pathways are generally of two types:

1. Direct pathways in which the receptors/endpoints are in direct contact with physical stressors or with contaminated soil, water, or other physical media.

2. Bioaccumulative pathways in which the contaminant moves from abiotic sources into the food web and into higher trophic levels. In some instances the contaminant may biomagnify (become more highly concentrated) as it progresses to higher levels in food chains.

Because environmental stressors with both types of exposure pathways may be present at a give site, different AEs may be identified for the various contaminants.

Once the relevant taxa for each of the potentially exposed functional components has been determined, the potential effect on higher levels of ecosystem organization can be assessed by examining the functional food web and attributes of each functional component of the web. Thus, a contaminant with fungicidal effects could adversely affect the chemical decomposers and result in direct adverse effects on rates of decomposition and nutrient cycling, in turn adversely affecting primary productivity.

Having proceeded through these steps, it is next appropriate to determine if the scale of the potential adverse effects are sufficient to require remediation. For example, it may be possible to determine that a contaminant is present in abiotic media and that exposure pathways exist to sensitive receptors, but the scale is so small that effects are unlikely to be detected (e.g., pesticides could be present in a few square meters of wetland in which nesting mallards are exposed, but the area is only a fraction of the home range of one mallard pair) and would not likely produce an adverse effect on any GAE.

Scale is also important in deciding how to apply GAEs to the selection of site-specific AEs. For example, contaminated sediments in a spring may have undetectable effects on the total biodiversity of an entire regional ecosystem, but may adversely affect the benthic biodiversity of the spring. These effects of scale on site-specific values should be considered in the process of selecting site-specific AEs.

As indicated above, a number of factors are involved in selecting site-specific assessment endpoints. General assessment endpoints provide a comprehensive context for ensuring that the ecological consequences of contamination are adequately considered in the selection of site-specific assessment endpoints. Identifying GAEs also provides a valuable means of communicating the complex process to the public and documenting the thought processes involved in AE selection.

19.5 CASE STUDY

Throughout this chapter examples have been provided on the application of each step in the identification of GAEs. While these should be sufficient to move through the process for a particular site, an additional study is presented here to illustrate the application of this approach at a large complex site with multiple habitats and a complex history of contamination. The ecological risk assessment at Los Alamos National Laboratory (LANL or the laboratory) was selected for this case study. Reagan et al. (1999) presents detailed description of the assessment endpoint selection process at LANL.

The enormity of the task of identifying assessment endpoints for the LANL ecological risk assessment was apparent by considering the species known to be at the laboratory. There are approximately 500 plant species on or near the laboratory property, 29 mammal species, 200 bird species, 19 reptile species, 8 amphibian species, and many thousands of invertebrate species. The "array of possibilities" for selecting assessment endpoints is very large, although smaller than for a tropical rain forest. A structured process for reaching consensus with regulators and other stakeholders on the array specification was obviously needed to ensure that all relevant valued resources were considered in selecting assessment endpoints, and to provide documentation as to why these resources were selected and others were not. The GAE process was implemented to provide a structured approach for specification of assessment endpoints.

This section provides an overview of the GAE process as it was applied in a pilot implementation of the GAE process at LANL. Those participating in this first attempt at applying the process at the laboratory were members of the New Mexico Environment Department (NMED), New Mexico Game and Fish, U.S. Fish and Wildlife, Department of Energy (DOE), and the laboratory's ER Project, and their consultants. Identification of GAEs is an ongoing process that will incorporate the values of other stakeholders (e.g., pueblos) as the ecological risk assessment process proceeds.

The process was conducted in two parts to identify the ecologically relevant and human values associated with LANL. The Pajarito Plateau ecosystem, which includes LANL, is defined as the habitats, both aquatic and terrestrial, of the Pajarito Plateau, which is situated on the eastern slopes of the Jemez Mountains in northern New Mexico. The laboratory portion of this plateau covers 43 square miles of largely undeveloped land.

19.5.1 Determining Ecological Values

This process involved the five general steps and rationale described in Section 19.2. These steps are described in varying degrees of detail, with more detail provided where the LANL approach varied from the general process description.

Step 1: Identify the Values Common to All Ecosystems. The three principal values common to all ecosystems—biodiversity, functional integrity, and nutrient and energy dynamics—were adopted for LANL. Stakeholders felt that additional definition was necessary to recognize biological diversity at the genetic, species, and ecosystem level. Additional detail was, therefore, provided to define this concept as applied to the ecological risk assessment at LANL.

Step 2: Identify the Functional Components of Regional Ecosystems. Functional components of the Pajarito Pleateau ecosystem were identified from the extensive lists of plants and animals that had been compiled for LANL and the surrounding area over past decades. Knowledge of general food habits and specific information on many species were evaluated in grouping these species into functional components of the ecosystem. Table 19.4 presents a list of functional components and a nonexhaustive list of species comprising each.

Step 3: Develop a Food Web. Because the laboratory encompasses a variety of terrestrial habitats with smaller aquatic habits embedded within them, and because the aquatic habitats are important resources for terrestrial wildlife, one food web was created for both "ecosystems." The food web of the Pajarito Plateau, based on functional components, is presented in Figure 19.4.

Step 4: Determine the Ecologically Relevant Attributes of Each Functional Component of the Ecosystem. Attributes for each of the functional components were similar to those presented in Section 19.2. However, because of debate regarding the actual "control" of an ecosystem by top carnivores, that attribute was replaced by "predation."

Step 5: State the Ecologically Relevant GAEs. These GAEs were stated in general, easily understood terms so that stakeholders, regulators, and subsequent decision makers could understand their meaning. Regionally relevant GAEs can be read off of an attribute table created in step 4. The following three examples illustrate the form of these GAE statements:

- Top carnivores and intermediate carnivores are valued components of the Pajarito Plateau ecosystem because of their role in predation.
- Terrestrial insectivores are a valued component of the Pajarito Plateau ecosystem because of their importance both in predation and as a food source to higher level carnivores.
- Aerial insectivores are a valued component of the Pajarito Plateau ecosystem because of their importance as predators.

The complete list of GAEs for the Pajarito Plateau ecosystem is provided in LANL (1999).

19.5.2 Determining Societal Values of the Regional Ecosystem

The following are the general criteria that the pilot project team used to develop societally relevant GAEs for LANL:

TABLE 19.4 Representative Organisms for Each of the Functional Components of the Pajarito Plateau Ecosystem

Functional Components	Representative Organisms
Producers	Autotrophic organisms
Herbaceous plants	Grasses, forbs, annuals, perennials
Woody shrubs	Chamisa, willow, gambel oak
Conifers	Douglas fir, piñon, spruce, ponderosa pine
Deciduous trees	Aspen, cottonwood, box elder
Submergent, emergent, and floating vascular plants	Cattails, duckweed, watercress
Algae	Green filamentous algae, diatoms
Epiphytes	Lichens, mosses
Mycorrhizae	Mycorrhizal fungi
Consumers	Flesh and plant eaters
Granivores/frugivores (seed and fruit eaters)	Insects (e.g., some ants), rodents, birds
Folivores (leaf eaters)	Insects (e.g., grasshoppers), mammals (e.g., elk)
Browsers	Mammals (e.g., deer, rabbits, and hares)
Nectarivores (nectar and pollen feeders)	Insects (e.g., bees), birds (e.g., hummingbirds), mammals (e.g., some bats)
Fungivores	Insects (e.g., some beetles, flies), mammals [e.g., squirrels and mice (incidental)]
Aquatic herbivores (plant eaters)	Invertebrates (e.g., snails, insects), tadpoles
Parasites	Invertebrates (e.g., ticks, lice, worms)
Terrestrial omnivores	Mammals (e.g., skunk, fox), birds (e.g., robin, raven)
Aquatic omnivores	Invertebrates (e.g., isopods, mollusks)
Aerial insectivores	Mammals (e.g., bats), birds (e.g., flycatchers)
Terrestrial insectivores	Invertebrates (e.g., spiders), mammals (e.g., shrews), reptiles (e.g., lizards)
Intermediate carnivores	Reptiles (e.g., snakes), birds [e.g., kestrel (in part)]
Top carnivores	Mammals (e.g., mountain lion), birds (e.g., red-tailed hawk)
Decomposers	Consumers of dead organic material
Mechanical decomposers	Invertebrates (e.g., earthworms, stoneflies), detritivores (e.g., amphipods), filter feeders (e.g., caddisflies), scavengers (e.g., turkey vultures), shredders (e.g., stoneflies)
Chemical decomposers	Fungi, bacteria

Source: Reagan et al., 1999.

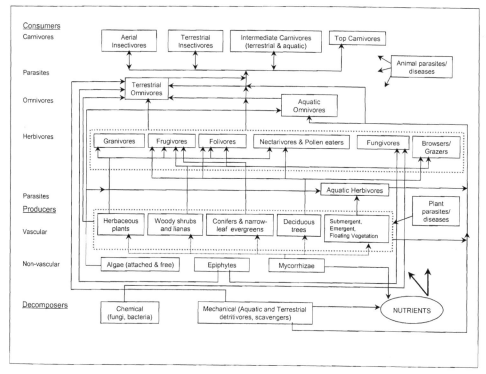

Figure 19.4 Food web of the Pajarito Plateau ecosystem at Los Alamos National Laboratory (from Reagan et al., 1999).

- Identification of societal values should involve input from risk managers, risk assessors, ecologists, appropriate regulatory authorities (e.g., State Department of Game and Fish, U.S. Fish and Wildlife Service), other experts (e.g., anthropologists) tribal representatives and municipalities, and the general public.

- Values include legally protected ecological resources such as threatened and endangered species, as well as recreationally important species (e.g., game and nongame wildlife).

- Other societal values for the ecosystem may be identified based on a review of the management goals and plans for areas potentially affected by laboratory activities. For example, a given area may be under simultaneous management for protection of specific habitat, erosion control, fire suppression, or protection of archeological sites.

- Societal values recognized for the development of GAEs should incorporate concerns for clean water and watershed protection, both of which may fall under the scrutiny of regulatory compliance.

- GAEs should be developed with an eye on neighboring systems of land use and control, as these may impact operations on the area of consideration.

Based on the foregoing criteria, the following societally relevant GAEs were identified during the pilot project:

- Recreationally and commercially important species are valued components of the ecosystem and are to be protected because of their importance for consumptive uses such as hunting and fishing, and for nonconsumptive uses, such as bird watching.
- Threatened and endangered species, their habitats, and migratory bird nesting, roosting, and lighting sites are valued components of the ecosystem to be protected because of their regulatory stature.
- The quality and quantity of water within each watershed are valued components of the ecosystem and require management of point and nonpoint sources of contaminants, consumptive water usage or diversion, erosion and total suspended materials to meet regulatory limits, and total maximum daily loads (TMDLs).
- Certain indigenous plants and animals are valued components of the ecosystem and are to be protected because of their ethnological and other consumptive and nonconsumptive uses.
- The aesthetic quality of the landscape is a valued component of the ecosystem because of its value to society.
- Wetlands within each watershed are valued due to their unique protection by the CWA, as well as their important ecological functions.

This list of GAEs is considered preliminary and subject to change (and become more specific) once additional stakeholders have had an opportunity to participate more fully in the process. The list is also somewhat general but is likely to become more specific as particular locations within LANL are addressed in the risk assessment. Identifying GAEs at LANL is additionally difficult because site contamination potentially affects plant and animal resources used by indigenous peoples of the nearby pueblos of Cochiti, Jemez, San Ildefonso, and Santa Clara. Because specific knowledge of what plants and animals are used by these peoples and for what purposes is secret tribal knowledge, many of the species and their associated values cannot be identified by the GAE process.

Because the ecological risk assessment at LANL is ongoing, it is likely to evolve as this assessment progresses beyond the screening level. Current intentions are to use the GAE approach to integrate the ecological risk assessment with natural resource damage assessment activities at LANL. An integration plan is currently under development that is designed to maximize data usability from the ecological risk assessment for determining and quantifying injury as a basis for developing appropriate types and scales of resource restoration.

19.6 SUMMARY

The success of ecological risk assessments depends to a large extent on appropriate identification of ecological and human values of the potentially affected ecosystems. Applying a comprehensive and systematic approach to identifying these values ensures that all levels of ecological organization and all appropriate species are considered in the selection of AEs. By focusing on identifying values to be protected (i.e., AEs) prior to identifying the endpoints most sensitive to stressors, it is possible to avoid having the "most sensitive species" drive the ecological risk assessment and produce irrelevant results.

Systematically evaluating all species and each level of ecological organization in a functional context makes it is possible to evaluate stressor effects on basic ecosystem process as well as individual taxa. Societal values of the potentially affected ecosystem(s) are simultaneously evaluated by reviewing all pertinent information and through interviews with regional experts and indigenous communities. The methodology permits evaluation of ecosystems at large sites and of complex ecosystems such as tropical rain forests.

The methodology presented herein has several advantages over less rigorous methods for selecting AEs. These include:

- Determining values prior to considering stressor sensitivity
- Providing a consistent basis for integration among related processes
- Permitting a comprehensive and systematic means of addressing all possible endpoints and documenting how each was considered
- Providing a basis for stakeholder understanding and involvement in the complex process of identifying ecological and human use values in risk assessment and in subsequent risk management decision making

Site-specific AEs are identified from the GAEs (all values for an ecosystem) by applying the additional criterion of stressor susceptibility. The overall approach is comprehensive, systematic, and provides a thoroughly documented means of identifying relevant values for ecological risk assessments.

ACKNOWLEDGMENTS

Many groups and individuals have contributed to the development of the approach to selecting assessment endpoints presented in this chapter. The Ecological Society of America and Society for Environmental Toxicology in Chemistry provided the opportunity to formulate some of the original concepts at the 1993 Pelston Conference on Sustainable Ecosystem Management. The Aluminum Company of America (Alcoa) supported an initial version of the general assessment endpoint approach with additional input provided by personnel from Parametrix, Inc., Aquatic Resources Center, Inc., the U.S. Environmental Protection Agency, U.S. Fish and Wildlife Service, NOM, Texas Natural Resources Conservation Commission, Texas General Land Office, and Texas

Parks and Wildlife Department. I especially thank Jim Doenges for providing helpful technical and editorial comments on the original manuscript. Finally, I thank my wife, Julie, for her continued patience, understanding, and support throughout the preparation of this chapter.

REFERENCES

Bagyaraj, D. J. 1989. Mycorrhizas. In H. Lieth and M. J. A. Werger (Eds.), *Ecosystems of the World 14b: Tropical Rain Forest Ecosystems*. Elsevier Scientific Publications, Amsterdam, Chapter 30.

Bailie, J., and Groombridge, B. 1996. *1996 IUCN Red List of Threatened Animals*. International Union for the Conservation of Nature and Natural Resources, Gland, Switzerland.

Beehler, B. M., Pratt, T. K., and Zimmerman, D. A. 1986. *Birds of New Guinea*. Princeton University Press, Princeton, NJ.

Boyce, M. S., and Haney, A. 1997. *Ecosystem Management: Applications for Sustainable Forest and Wildlife Resources*. Yale University Press, New Haven.

Brewer, R. 1979. *Principles of Ecology*. Saunders, Toronto.

Flannery, T. F. 1995. *Mammals of New Guinea*. Cornell University Press, Ithaca, NY.

Gallopin, G. C. 1972. Structural properties of food webs. In B. C. Patten (Ed.), *Systems Analysis and Simulation in Ecology*. Academic, New York, pp. 241–282.

Harwell, M., Gentile, J., Norton, B., and Cooper, W. 1994. Issue paper on ecological significance. In *Ecological Risk Assessment Issue Papers*. Risk Assessment Forum, U.S. Environmental Protection Agency, Washington, DC.

Janzen, D. H. 1983. Food webs: Who eats what, why, how, and with what effect in a tropical forest? In F. B. Golley (Ed.), *Ecosystems of the World 14A: Tropical Rain Forest Ecosystems*. Elsevier Scientific, Amsterdam, Chapter 11.

Jones, C. G., Lawton, J. H., and Shachak, M. 1994. Organisms as ecosystem engineers. *Oikos* 69: 373–386.

Kaufmann, M. R., Graham, R. T., Boyce, Jr., D. A., Moir, W. H., Perry, L., Reynolds, R. T., Bassett, R. L., Mehlhop, P., Edminster, C. B., Block, W. H., and Corn, P. S. 1994. *An Ecological Basis for Ecosystem Management*. General Technical Report RM-246. USDA Forest Service.

Kellert, S. R. 1996. *The Value of Life: Biological Diversity and Human Society*. Island Press/Shearwater Books, Washington, DC.

Kormondy, E. J. 1969. *Concepts of Ecology*. Prentice-Hall, Englewood Cliffs, NJ.

Los Alamos National Laboratory (LANL). 1999. *General Assessment Endpoints for Los Alamos National Laboratory—A Pilot Project*. LA-UR-pending.

Mackay, R. D. 1976. *New Guinea*. Time-Life Books, Amsterdam.

McNaughton, S. J. 1978. Stability and diversity of natural communities. *Nature* 274: 251–253.

Magurran, A. E. 1988. *Ecological Diversity and Its Measurement*. Princeton University Press, Princeton, NJ.

Norton, B. G. 1987. *Why Preserve Natural Variety?* Princeton University Press, Princeton, NJ.

Odum, E. P. 1993. *Ecology and Our Endangered Life Support Systems*, 2nd ed. Sinaur Associates, Sunderland, MA.

Osenberg, C. W., and Mittelbach, G. G. 1996. The relative importance of resource limitation and predator limitation in food chains. In G. A. Polis and K. O. Winemiller (Eds.), *Food Webs: Integration of Patterns and Dynamics*. Chapman and Hall, New York, Chapter 12.

Paine, R. T. 1966. Food web complexity and species diversity. *Am. Nat.* 100: 65–75.

Parametrix. 1996. Ecological risk assessment problem formulation for mercury, Lavaca Bay/Point Comfort Superfund Site. Report to Alcoa.

Parametrix, Woodward-Clyde Consultants, and Aquatic Resources Center. 1996. Identification of general ecologically relevant issues and general assessment endpoints. Report to Alcoa for the Lavaca Bay Superfund Site.

Polis, G. A., and Winemiller, K. O. (Eds.). 1996. *Food Webs: Integration of Patterns and Dynamics*. Chapman and Hall, New York.

Reagan, D. P. 1997. Animal community considerations in the sustainable management of tropical forests. *Trop. Ecol.* 38: 329–332.

Reagan, D., Campbell, T., Gribben, K., Beacham, J., Cardwell, R., Volosin, J., and Kathman, R. 1996. A comprehensive approach to selection endpoints for ecological risk assessments. *Soc. Environ. Toxicol. Chem. Abstr. 17th Annual Meeting*.

Reagan, D. P., Hooten, M. M., Kelly, E. J., Michael, D. I. 1999. General assessment endpoints for ecological risk assessment at Los Alamos National Laboratory. In F. T. Price, K. V. Brix, and N. K. Lane (Eds.), *Environmental Toxicology and Risk Assessment: Recent Achievements in Environmental Fate and Transport*, Vol. 9. ASTM STP 1381. American Scoiety for Testing Materials, West Conshohocken, PA.

Reagan, D. P., and Waide, R. B. (Eds.). 1996. *The Food Web of a Tropical Rain Forest*. University of Chicago Press, Chicago.

Reichman, O. J., and Pulliam, H. R. 1996. The scientific basis for ecosystem management. *Ecol. Applicat.* 6: 694–696.

Reiger, H. A. 1993. The notion of natural and cultural integrity. In S. Woolly, J. Kay, and G. Francis (Eds.), *Ecological Integrity and the Management of Ecosystems*. St. Lucia, Ottawa.

Suter, G. W. II. 1989. Ecological endpoints. In W. Warren-Hicks, B. R. Parkhurst, and S. S. Baker, Jr. (Eds.), *Ecological Assessment of Hazardous Waste Sites: A Field and Laboratory Reference Document*. EPA60013-89/013. U.S. Environmental Protection Agency, Corvallis, OR, pp. 2-1–2-28.

U.S. Environmental Protection Agency (EPA). 1989. *Risk Assessment Guidance for Superfund*, Vol. II: *Environmental Evaluation*. Interim Final EPA 540/1-89/001A. EPA, Washington, DC.

U.S. Environmental Protection Agency (EPA). 1992. *Framework for Ecological Risk Assessment*. EPA/630/R-92/001. Risk Assessment Forum, EPA, Washington, DC.

U.S. Environmental Protection Agency (EPA). 1993. *Data Quality Objectives Process for Superfund: Interim Final Guidance*. EPA540-R-93-071. EPA, Washington, DC.

U.S. Environmental Protection Agency (EPA). 1998. Guidelines for ecological risk assessment. *Fed. Reg.* 63(93): 26846–26924.

Wunderle, J. M., Jr. 1997. The role of animal seed dispersal in accelerating forest regeneration on degraded tropical lands. *Forest Ecol. Manag.* 99: 223–235.

20 Comparison of Aquatic Ecological Risk Assessments at a Former Zinc Smelter and a Former Wood Preservative Site

ROBERT A. PASTOROK, WALTER J. SHIELDS, and
JANE E. SEXTON
Exponent, Bellevue, Washington

20.1 INTRODUCTION

During the last 15 or more years, ecological risk assessment has developed as an important analysis technique to support risk management decisions in environmental regulatory programs [Barnthouse and Suter, 1986; U.S. Environmental Protection Agency (EPA), 1992, 1998; Suter, 1993; Suter et al., 2000; Pastorok et al., 2001]. Risk assessment results aid in evaluating the status of endangered species (e.g., Brook et al., 2000), identifying and ranking areas of concern at hazardous waste sites (e.g., Pastorok et al., 1994; Clifford et al., 1995), and obtaining regulatory approval for new chemical products such as fuel additives or pesticides (e.g., EPA, 1997, 1998). More comparisons of ecological risk assessment case studies are needed to elucidate the commonalities and contrasts in issues as well as assessment approaches at chemically contaminated sites.

In this chapter, we examine two case studies of ecological risk assessments at hazardous waste sites in which the assessment approaches are essentially similar but the industrial activities and primary chemicals of concern are substantially different. We compare the potential risks to aquatic biota posed by metals contamination at an inactive zinc smelter with those posed by creosote and associated organic compounds at an inactive wood preservative site.

The objective in each of these risk assessments was to describe the potential adverse ecological effects associated with contaminants detected at the site in the absence of any action to control or mitigate these releases (i.e., to assess baseline risks under the no-action alternative). The contaminant data and the

Human and Ecological Risk Assessment: Theory and Practice, Edited by Dennis J. Paustenbach
ISBN 0-471-14747-8 © 2002 John Wiley & Sons, Inc.

results of the risk assessment were used to identify and delineate areas of concern at the site and to develop preliminary remediation goals expressed as media-specific chemical concentrations considered to be below thresholds for adverse ecological effects. We focus on the assessment of contaminated sediment at each site because the types of aquatic receptors in each assessment were essentially similar (i.e., aquatic macroinvertebrates and fish) despite some differences in habitat between sites and because toxicity testing was a primary assessment tool in each case.

The original risk assessments were part of separate remedial investigation/feasibility studies (RI/FSs)* designed to determine what, if any, remedial actions were necessary to prevent residual contamination related to site activities from presenting an unacceptable hazard to aquatic macroinvertebrates and fishes. The remedial investigation began in 1994 at the National Zinc Site (NZS) in Bartlesville, Oklahoma, and was completed in 1995 (PTI, 1995). The remedial investigation began in September 1990 at the McCormick & Baxter Creosoting Company Site (MBS) and was completed in March 1992 (PTI, 1992). We first present the elements of each risk assessment, then compare the nature and extent of contamination, the receptors at risk, the magnitude of risk, the spatial distribution of risk, and the resulting remedial actions at the sites.

In accordance with technical guidance from U.S. EPA (1992), both ecological risk assessments followed the four steps outlined below:

- *Problem Formulation.* The physical features, general distribution of chemicals, and ecological receptors (plants and animals) in the study area were described by using existing data. In a preliminary analysis, chemicals of potential concern (CoPCs), physical stressors, ecological receptors, and endpoints to be considered in the assessment were identified.
- *Exposure Assessment.* The ecological receptors likely to contact CoPCs, the means of contact (e.g., ingestion, dermal contact), and the magnitude and frequency of exposure were identified and described.
- *Effects Assessment.* Potential effects of the CoPCs on organisms and toxicity thresholds or concentration–response relationships were assessed.
- *Risk Characterization.* Results of the exposure and effects assessments were combined to evaluate the probability of adverse effects on ecological receptors. The degree of confidence in the risk estimates and the most important sources of uncertainty were also described.

In the following sections, we describe the elements of problem formulation, exposure assessment, effects assessment, and risk estimation for each case study. In the final section, we present a comparative risk characterization and discuss uncertainties in the ecological risk assessments.

---

*Additional information and supporting documentation for these RI/FS projects can be obtained from PTI (1992, 1995) and directly from the authors who were principal investigators for both projects. Both RI/FSs were conducted under the requirements of the U.S. EPA's Superfund program.

20.2 PROBLEM FORMULATION

20.2.1 Site Features and Ecological Receptors

Zinc Smelter Site

Site The NZS consists of areas surrounding a former zinc smelter in Bartlesville, Oklahoma. The ecological study area encompassed approximately 200 acres of grassland area referred to as the South Field, including ponds and intermittent and perennial streams that flow from the site to the Caney River (Figure 20.1). Both terrestrial and aquatic habitats within the South Field were included in the risk assessment. Only the risks to aquatic biota are discussed here because risks to terrestrial receptors that foraged on aquatic biota (e.g., belted kingfisher and great blue heron) were found to be insignificant, and we are comparing this study with the aquatic ecological risk assessment at the MBS where risks to terrestrial receptors were not assessed quantitatively.

Habitats The aquatic habitats (Figure 20.1) considered potentially at risk include two unnamed first-order streams in the South Field (referred to as the North and West tributaries); Eliza Creek, downstream from Highway 123; Sand

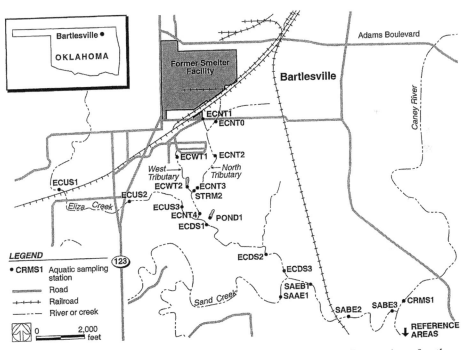

Figure 20.1 National Zinc Site—site location and aquatic sampling stations for the ecological risk assessment.

Creek, from the confluence with Eliza Creek to the confluence with the Caney River; and a small pond located in the South Field east of the confluence of Eliza Creek and the North Tributary.

Receptors   The major aquatic receptors of concern were stream invertebrates and fish. Aquatic vegetation was uncommon in streams of the South Field. Vascular plants were rare in the North and West tributaries and completely absent in Eliza Creek and Sand Creek. Amphibians, reptiles, and avian species that depend on aquatic food resources are described below, but risk assessment results for these species are not considered here (for additional information, see PTI, 1995). Risks to amphibians and reptiles were not assessed quantitatively because of the lack of good toxicity reference values for these groups. Risk to mammalian species was assessed based on empirical surveys of local populations, and risk to birds was assessed based on exposure models. Because the risk to birds and mammals was limited, areas of concern at the site were identified primarily on the basis of risks to aquatic receptors (i.e., benthic macroinvertebrates).

Benthic macroinvertebrate assemblages in the South Field streams were relatively depauperate, possibly as a result of habitat (i.e., soft substrate in these slow-moving streams) and/or contamination. Crayfish (Decopoda) populations occurred only in Eliza Creek. Other taxonomic groups were neither widespread nor abundant; the more common groups found in collections included snails (Gastropoda), dragonfly larvae (Odonata), corixids and belostomatids (Hemiptera), dytiscid beetles (Coleoptera), and aquatic sowbugs (Isopoda).

The most abundant species of fish were bluegill (*Lepomis macrochirus*), mosquito fish (*Gambusia* spp.), red shiner (*Cyprinella lutrensis*), brook silverside (*Labidesthes sicculus*), longear sunfish (*Lepomis megalotis*), sand shiner (*Notropis stramineus*), and green sunfish (*Lepomis cyanellus*). Together, these species accounted for 78 percent of the fishes collected at the site. The most abundant and ubiquitous fishes that would be of a size consumed by great blue herons and belted kingfishers were mosquito fish, various sunfishes (i.e., bluegill, longear sunfish, and green sunfish), and both spotted and largemouth bass. The primary prey items of these receptor species probably include juvenile fishes and invertebrates (crayfish, snails, aquatic insects), taxonomic groups that were potentially exposed to CoPCs.

Amphibians were uncommon in aquatic habitats in all the reference areas and the South Field, including stations along the North and West tributaries and both Eliza and Sand creeks. A recent hatch of several hundred chorus frogs (*Acris crepitans*) was observed in April on the North Tributary in the vicinity of stations ECNT3 and ECWT2, and sporadic observations of frogs were made in July at stations ECNT3 and ECNT4. (Station locations are shown in Figure 20.1.) No amphibians were observed in Eliza or Sand creeks or were caught during seining efforts in July in any reference stream or South Field stream. A few chorus frogs and bullfrogs (*Rana* sp.) were observed during March at reference area TERC, but none was observed during July, August, or September, probably because most of the stream length was dry and the amphibians had moved downstream to the lake habitats.

Reptiles that would be considered part of the aquatic ecosystem include two species of turtles—the snapping turtle (*Chelydra serpentina*) and the red-eared pond slider (*Chrysemys scripta*). Snapping turtles are rarely seen given their cryptic nature and aquatic life history. A total of three snapping turtles were caught during gill netting, one from station POND1 and two from station ECDS1. Red-eared sliders were only caught at station ECDS1 (seven individuals) and station ECUS2 (one individual), although individuals were commonly observed sunning on logs along all sections of Eliza Creek from February through September.

Avian species that depend on fish as a major food source include belted kingfishers (*Megaceryle alcyon*) and great blue herons (*Ardea herodias*). Both species feed at a high trophic level in aquatic food webs and are known to occur in riparian areas of the South Field. Risk assessment results reported by PTI (1995) for these species are not included here because of our focus on comparative risks for aquatic biota only and because risks to piscivores were not significant.

Mammalian species that may forage on fish or aquatic invertebrates include mink (*Mustela vison*), river otter (*Lutra canadensis*), and raccoon (*Procyon lotor*). These species are likely to occur at the NZS. However, a risk assessment was not conducted for these species because the avian receptors were more commonly observed at the site.

Wood Preservative Site

Site   The MBS is located in an industrial area along the Willamette River in Portland, Oregon (Figure 20.2). From 1944 to late 1991, wood products were treated with creosote/oil mixtures, pentachlorophenol/oil mixtures, and a variety of water-based solutions containing arsenic, chromium, copper, and zinc. Releases of creosote into the environment from seeps along the river shoreline were first reported in 1983.

Habitats   The river boundary of the MBS includes an upland beach of sand and cobble (exposed seasonally), an intertidal area of sand and cobble, a shallow (3 to 6 feet) shelf with fine-grained substrate, a shallow backwater on the north side of the site, and a sand bar immediately downstream of the former creosote dock. Several river habitats were defined for analysis of exposure and effects, including the clam, shelf, and fish and crayfish habitat areas (Figure 20.3). The habitat types were based on visual observations and data on grain size composition and organic carbon content of sediments. Human activities have modified most of these habitats. For example, the midchannel of the river adjacent to the site was (and is) dredged for navigation on a regular basis.

Shorelines upstream of the site are highly industrialized, although the downstream area along both sides of the river has relatively dense brush and

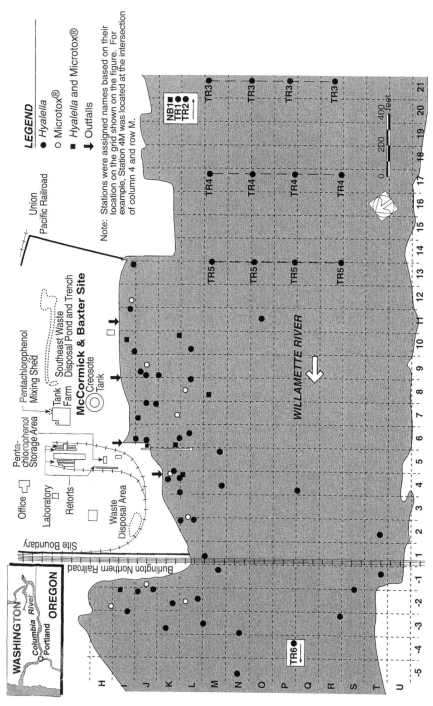

Figure 20.2 McCormick & Baxter Creosoting Company Site—site location and aquatic sampling stations for the ecological risk assessment.

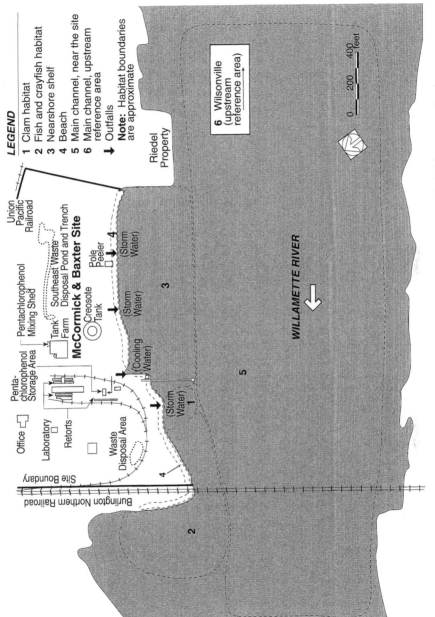

LEGEND

1 Clam habitat
2 Fish and crayfish habitat
3 Nearshore shelf
4 Beach
5 Main channel, near the site
6 Main channel, upstream reference area
→ Outfalls

Note: Habitat boundaries are approximate

6 Wilsonville (upstream reference area)

0   200   400
feet

Union Pacific Railroad

Pentachlorophenol Mixing Shed

Southeast Waste Disposal Pond and Trench

Tank Farm

McCormick & Baxter Site

Creosote Tank

Penta-chlorophenol Storage Area

Office

Laboratory

Retorts

Waste Disposal Area

Site Boundary

Burlington Northern Railroad

Pole Peeler

(Storm Water)

(Storm Water)

(Cooling Water)

(Storm Water)

Riedel Property

WILLAMETTE RIVER

Figure 20.3 McCormick & Baxter Creosoting Company Site—aquatic habitats near the site.

1105

trees for approximately 1.25 miles. Cathedral Park, a maintained grass park, is approximately 1 mile downstream of the site. The Sauvie Island Wildlife Area (SIWA), which is a high-quality wildlife habitat, is located approximately 4 river miles downstream of the site. The species that inhabit the wildlife area [e.g., bald eagle (*Haliaeetus leucocephalus*), sandhill crane (*Grus canadensis*), Canada geese (*Branta canadensis*), and various other waterfowl species] may also migrate through or feed in areas near the MBS. Because most of the shoreline is highly industrialized, however, an environmental management decision was made not to address risks to aquatic birds and mammals during the ecological risk assessment at the MBS.

Receptors The major aquatic receptors of concern were stream invertebrates and fish. Aquatic vegetation is uncommon in this portion of the Willamette River. Amphibians, reptiles, birds, and mammals that depend on aquatic food resources are described below, but the risk assessment did not address these species in a quantitative manner [for additional information, see PTI (1992)].

Comprehensive surveys of benthic macroinvertebrates at the MBS have not been conducted, but the communities were expected to be typical of those of fine-grained sediments in rivers of the Northwest. Macroinvertebrate communities probably include chironomids, oligochaetes, gastropod and bivalve mollusks, isopods, dragonflies, and possibly burrowing mayflies. Crayfish were common near the site.

Fish species found in the area adjacent to the site in relatively high abundance include the northern squawfish (*Ptychocheilus oregonensis*), large-scale sucker (*Catosomus macrocheilus*), chinook salmon (*Oncorhynchus tshawytscha*), coho salmon (*O. kisutch*), and steelhead (*O. mykiss*). Relatively common species include the large-mouth bass (*Micropterus salmoides*), crappie (*Pomoxis* spp.), sculpin (*Cottus* spp.), common carp (*Cyprinus carpio*), white sturgeon (*Acipenser transmontanus*), and yellow perch (*Perca flavescens*). Other species found near the site include the sockeye salmon (*O. nerka*), shad (*Alosa sapidissima*), bluegill (*Lepomis macrochirus*), brown bullhead catfish (*Ictalurus nebulosus*), and channel catfish (*I. punctatus*). Hughes and Gammon (1987) reported that the dominant species in the Willamette River are the northern squawfish, a piscivore, and the large-scale sucker, an omnivore. Ward et al. (1987) also reported that the squawfish is the most abundant piscivore in the lower Columbia basin. The abundance of the northern squawfish does not indicate a high-quality riverine habitat because the species is tolerant of organic pollution, increased temperatures, and sediment loading (Hughes and Gammon, 1987).

Painted turtles (*Chrysemys picts*) have been observed at the site, and amphibian species such as the rough-skinned newt (*Taricha granulosa*), bullfrog (*Rana catesbeiana*), and red-legged frog (*Rana aurora aurora*) are known to be abundant or common at SIWA. Major invertebrate species observed in the river and sediments near the site include crayfish (*Pacifastacus leniusculus*), amphipods (Amphipoda), and two species of clams (Pelecypoda). Ward et al. (1987) collected invertebrates near the MBS and reported large numbers of annelid worms (Oligochaeta) and water fleas (Cladocera). Also found were other arthropod taxa, including copepods (Eucopedpoda) and mysids (Mysidacea), water mites (Hydracarina), dragon- and damselflies (Odonata), and midges (Chironomidae).

Shorebird and wading and diving bird species either observed or expected to be present in the site area include the great blue heron (*Ardea herodias*), double-crested cormorant (*Phalacrocorax auritus*), Canada goose (*Branta canadensis*), mallard (*Anas platyrhynchos*), gulls (*Larus* spp.), green-winged and cinnamon teal (*Anas carolinensis* and *A. cyanoptera*), wood duck (*Aix sponsa*), bufflehead (*Bucephala albeola*), ruddy duck (*Oxyura jamaicensis*), common merganser (*Mergus merganser*), American coot (*Fulica americana*), killdeer (*Charadrius vociferus*), common snipe (*Capella gallinago*), and Western sandpiper (*Ereunetes mauri*).

Mammals that were observed or expected to be present in the site area that may come into contact with riverine sediments and water include the raccoon (*Procyon lotor*), beaver (*Castor canadensis*), nutria (*Myocaster coypu*), black-tailed deer (*Odocoileus hemionus columbianus*), dusky-footed woodrat (*Neotoma fuscipes*), opossum (*Didelphis marsupialis*), vagrant and Trowbridge's shrews (*Sorex vagrans* and *S. trowbridgii*), Townsend's and coast moles (*Scapanus townsendii* and *Neurotrichus gibbsii*), and striped skunk (*Mephitis mephitis*).

20.2.2   Physical and Chemical Stressors

Zinc Smelter Site   Both physical and chemical stressors potentially affected the aquatic ecosystems of the South Field. Intermittent flow in much of the North and West tributaries (Figure 20.1) resulted in a low-quality habitat for stream invertebrates and fishes. The flooding and temperature extremes that occurred during spring and summer of 1994 affected both aquatic and terrestrial systems.

Several bridges on Eliza Creek function as dams, which has converted the upper portion of Eliza Creek near the site into a series of interconnected pools and has thus limited upstream fish movement into the North Tributary.

In the pooled areas of Eliza Creek near the confluence with the North Tributary, silts, clays, and fine organic debris had accumulated on the stream bottom. This siltation resulted in the loss of many benthic macroinvertebrate groups, and those groups that survived are primarily burrowing species. The loss of benthic macroinvertebrates may reduce the food supply for fishes that rely on exposed benthic macroinvertebrates as a resource. The data on sediment grain size distributions at the site indicate that the most depositional habitats were found at the pond in the lower portion of the West Tributary (station ECWT2) and at the pond just north of Eliza Creek (station POND1), where the percentages of fine-grained sediments were high (83 and 93 percent, respectively). Percentages of fine-grained sediments were similar to or less than the mean reference value of 72 percent at all other site stations. Within the North Tributary, depositional habitats were found at stations ECNT0 and

ECNT4, where the percentages of fine-grained sediments (72 and 76 percent, respectively) were similar to the mean reference value.

Chemical stressors include metals from historic runoff from the former smelter, as well as fertilizers and pesticides used in agricultural areas near the South Field and chemicals released from other human activities. Only analytes that exceeded background, reference, or available screening levels were considered CoPCs. For surface water, Oklahoma water quality standards were used as screening criteria.

Arsenic, cadmium, copper, lead, mercury, selenium, silver, and zinc were identified as CoPCs in soils and sediments. Cadmium, lead, selenium, and zinc were identified as CoPCs in surface water. These selected metals occur in materials potentially associated with historical smelter operations, including ore concentrates delivered to the facility; dusts from the transport and storage of ore concentrates and solid waste materials at the facility; atmospheric particulate emissions from roasting, sintering, and smelting processes; and various solid waste materials (e.g., retort and sinter residues, slag, condenser sands). The CoPCs have other anthropogenic sources (e.g., lead and zinc in paints, automobile emissions, automobile tires, and batteries; cadmium and zinc in fertilizers; arsenic and lead in pesticides), and all occur naturally in soils of the Bartlesville area.

Wood Preservative Site Physical stressors on the aquatic ecosystem of the Willamette River near the MBS are primarily either siltation, dredging (in the main channel only), or scouring by ship traffic and floods. The relatively large size of the Willamette River near the site mitigates the effects of extreme climatic events.

Sediment grain size measured near the site was highly variable and, for surface sediments, appears to be seasonally influenced. The central navigation channel was generally composed of intermediate black or brown sand, with 40 to 60 percent fine-grained material, often overlain by green or brown silt or silty sand. Fine-grained material decreases to approximately 20 percent in the mixed sands and gravels toward the site shoreline underneath the railroad bridge and near the shore across from the site. The percent fine-grained material in sediments at upstream and downstream transects and at the upstream reference area was within the range observed midchannel near the site.

Chemical stressors in addition to those associated with the site were diverse [e.g., metals, pesticides, polychlorinated biphenyls (PCBs)] because of the urban-industrial setting of the site, but the primary chemicals of concern were clearly associated with site activities. Releases of wood preservative components to the river probably occurred through various routes, including spillage during delivery by ship at the creosote dock, storm water and cooling water outfalls, drainage from a retort sump, spills and subsequent runoff, and subsurface migration.*

---

*Subsurface flow of nonaqueous phase liquids (NAPLs) resulted in transport of creosote-related contaminants into the nearshore and subsurface sediments in the Willamette River.

Based on the site history and sampling data, the following contaminants were identified at the MBS:

- *Creosote/Polycyclic Aromatic Hydrocarbons (PAHs).* PAHs, as a class of compounds, generally account for 90 percent of the chemical constituents in creosote (Nestler, 1974), which had been used in the wood-preserving processes on-site. PAHs were also the primary CoPCs associated with the diesel oil, medium aromatic treating oils, and other petroleum distillates that were used on-site for fuel and as solvents for creosote and penta-chlorophenol in the wood treatment processes.
- *Pentachlorophenol and Associated Compounds.* Pentachlorophenol was used extensively on-site. Historically, technical-grade pentachlorophenol was purchased in crystalline form, and oil-based pentachlorophenol treat-ing solutions were prepared on-site. Contaminants of technical-grade pentachlorophenol include polychlorinated dibenzo-*p*-dioxins and poly-chlorinated dibenzofurans (PCDDs/PCDFs), hexachlorobenzene, tri- and tetrachlorinated phenols (primarily 2,4,6-tri and 2,4,5,6-tetra isomers), and chlorophenoxyphenols.
- *Metals.* Arsenic, chromium, copper, and zinc are components of various water-based wood preservatives that had been used on-site.

Metals, PAHs, PCDDs, and PCDFs were the CoPCs for the aquatic habitat. Standard site analytes not included in the ecological risk assessment include zinc, carbazole, hexachlorobutadiene, hexachlorobenzene, 2,4-dichlorophenol, and 2,6-dichlorophenol. Of these, hexachlorobutadiene, hexachlorobenzene, and the dichlorophenols were not detected in sediments. Concentrations of zinc were not significantly elevated above local upstream reference values for any station where zinc was detected. Below, we focus on PAHs because they were the primary CoPCs for the ecological risk assessment.

20.2.3 Conceptual Site Model

Zinc Smelter Site   Figure 20.4 shows a conceptual site model for the aquatic ecological risk assessment at the NZS. Possible sources of CoPCs in the South Field included the former smelter, other industrial sources in the local area (various metals), agricultural pesticides (lead and arsenic), fertilizers (cadmium and zinc), and highway runoff (cadmium, lead, and zinc).

Contaminant transport pathways in order of importance for the aquatic re-ceptors at the NZS included:

- Soil erosion and surface runoff from contaminated habitats close to the former smelter facility
- Historical overflows from the facility storm water retention ponds into the North Tributary
- Shallow groundwater transport from beneath the facility

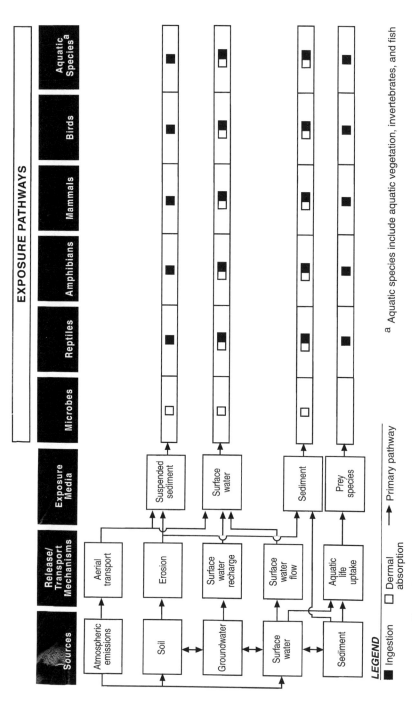

Figure 20.4 Conceptual site model for the aquatic ecological risk assessment at the National Zinc Site.

- Aerial transport of CoPCs from the facility or other industrial/commercial operations adjacent to the facility

Also included in the evaluation of risk was the pond located approximately 500 feet east of the confluence of the North Tributary and Eliza Creek (station POND1). Although this impoundment did not maintain a direct contact with Eliza Creek, periodic seasonal flooding events that inundate portions of the South Field along Eliza Creek would have been a potential source of CoPC loading to the pond. Surface runoff from the adjacent terrestrial habitats of the South Field might also have provided a loading of CoPCs, although the topography was so flat that no discernible drainages connected the South Field to station POND1.

Aquatic biota in the creeks downstream from the former smelter and in the pond were the primary receptors of concern for the surface water and sediment assessments. Risks to piscivorous birds were also estimated based on food-web modeling and are described by PTI (1995).

Wood Preservative Site   Figure 20.5 shows a conceptual site model for the aquatic ecological risk assessment at the MBS. Various areas of the MBS were identified as potential or known sources of contamination. These areas ranged from large areas such as the central process area (e.g., retorts, chemical storage, and underground storage tanks), the tank farm, and the waste disposal area to incidental and minor sources of contamination (e.g., leaking underground storage tanks).

Contaminant transport pathways in order of importance for the aquatic receptors at the MBS included:

- Migration of oil and creosote as NAPLs to river sediments
- Infiltration of contaminants from soil into groundwater and subsequent flow of groundwater to the river
- Storm water runoff, particularly from the central process area
- Aerial transport of CoPCs by wind in a localized area around the site

Aquatic biota in the Willamette River adjacent to the MBS were the primary receptors of concern for the sediment assessment. Risks to piscivorous water birds and mammals and to terrestrial wildlife were considered insignificant because of the low frequency of use of the contaminated areas by these receptors (PTI, 1992).

20.2.4 Endpoints

The assessment endpoints selected for each baseline ecological risk assessment were:

- Fish health

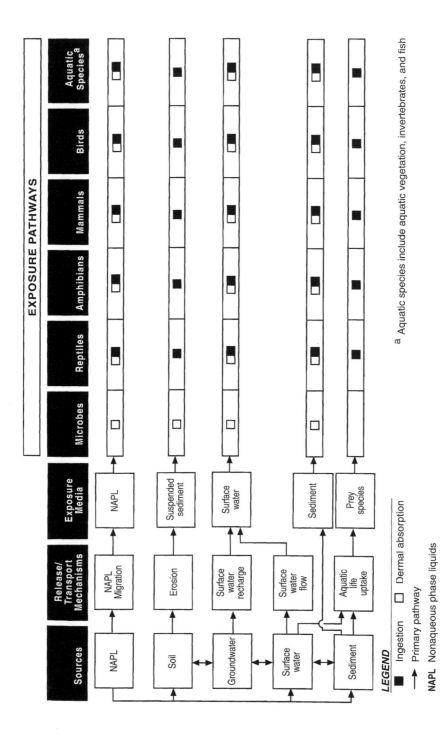

Figure 20.5 Conceptual site model for the aquatic ecological risk assessment at the McCormick & Baxter Creosoting Company site.

- Abundances of key fish species
- Community structure of fishes
- Total abundance of benthic macroinvertebrates
- Species richness of benthic macroinvertebrates

Thus, the objective in each case was to assess the probability of adverse effects on fishes and the potential for CoPC concentrations to cause a decline in their primary food resources (Table 20.1).

Measurement endpoints are characteristics of the ecological system that can be measured quantitatively and related to assessment endpoints. They are generally the results of a chemical measurement, a toxicity test, or a population field survey (EPA, 1992; Suter, 1993). Chemical measurements and sediment toxicity tests were conducted at both sites. Field surveys of biota were limited to a fish community analysis at NZS. Fish community analysis was not conducted at MBS because of the relatively small area of contamination compared with the feeding range of nonmigratory fish in this large, fast-moving river system. At the MBS, fish were sampled for analysis of bioaccumulation as well as gross abnormalities and liver lesions. The latter are morphological effects and therefore are included in the effects assessment, but they may be primarily an indicator of exposure to CoPCs and not necessarily indicative of adverse ecological effects. Benthic macroinvertebrate community analysis was not used in these assessments because physical stressors (intermittent drying, floods and sedimentation at NZS, and dredging and ship traffic at MBS) were so important as confounding variables in assessing the effects of CoPCs.

Both lethal (mortality) and sublethal (e.g., impaired growth or reproduction) endpoints were measured in the toxicity tests conducted at the sites. Measurement endpoints differed somewhat between the two risk assessments and are therefore discussed separately below.

Zinc Smelter Site   The primary measurement endpoints for the NZS assessment were survival, reproduction, and growth of representative test (bioassay) organisms exposed to samples of water or sediment collected at the site. Analyses of fish community structure based on field survey data provided information that supplemented the results of direct testing of water and sediment for toxicity. Measurements of water and sediment chemistry were used to assess exposure of aquatic receptors. Data on bioaccumulation of CoPCs in tissue of algae, benthic macroinvertebrates, crayfish, and fish also provided evidence of exposure. Because these bioaccumulation data were only used for estimating risks to piscivorous birds, they are not reported here [see PTI (1995) for additional information]. Measurement endpoints for assessing risks to aquatic biota included the following:

- *Surface Water.* Toxicity of the surface water to aquatic organisms was evaluated at those stations where concentrations of CoPCs exceeded the Oklahoma water quality standards by a factor of 2 to 10 times. Tests in-

TABLE 20.1 Endpoints and Risk Characterization Approach for Aquatic Ecological Risk Assessment

Study Element	Potential Endpoints		Risk Characterization[a]
	Measurement	Assessment	
Stream water chemistry[b]	CoPC concentrations	Abundance of sensitive aquatic species (i.e., EPA AWQC rationale)	Compare CoPC concentrations with sample-specific Oklahoma water quality standards
Sediment chemistry	CoPC concentrations	Toxicity thresholds for benthic species	Relate significant toxic responses in sediment bioassays to CoPC concentration thresholds
Water toxicity[b]	*Ceriodaphnia* reproduction[c] *Pimephales promelas* survival and growth[c]	Abundance of sensitive aquatic species	Extrapolate qualitatively from measured toxicity to assessment endpoint
Sediment toxicity	*Hyalella azteca* mortality *Chironomus tentans* growth[b] Microtox luminescence[d]	Abundance of benthic species	Extrapolate qualitatively from measured toxicity to assessment endpoint
Bioaccumulation	CoPC concentrations in algae,[b] macroinvertebrates, and fish	Dietary exposure in predators of aquatic species[c]	Correlate tissue concentrations in different taxa and assess potential for dietary exposure and risk in pisoivorous birds[c]
Fish community survey[b]	Abundance of fish species	Abundances of key fish species and community structure	Similarity analyses to compare site with reference area
Fish histopathology[d]	Prevalence of gross abnormalities and liver lesions	Fish health	Compare prevalence of pathological indicators between site and reference area

Note: AWQC, ambient water quality criteria; CoPC, chemical of potential concern.

[a]The probability of adverse effects was assessed either qualitatively or quantitatively, depending on the receptor used and toxicity reference values available in the literature.

[b]National Zinc Site only.

[c]Optional approach to be used only if EPA AWQC are exceeded by 2–10 times (see discussion of approach under Task 2 *Field and Laboratory Investigations* and Task 2.2.1 *Surface Water and Sediment Sampling* of PTI [1994a]).

[d]McCormick & Baxter Site only.

cluded the 96-hour growth and survival bioassay using fathead minnows (*Pimephales promelas*) (EPA, 1994) and the reproductive toxicity test using the daphnid (*Ceriodaphnia dubia*) (EPA, 1994). Additional water samples were collected and analyzed for metals and conventional water quality variables to assist in interpreting the bioassay results.

- *Sediment.* Acute lethal toxicity of sediments to aquatic invertebrates was evaluated by using the amphipod survival test with *Hyalella azteca* [American Society for Testing and Materials (ASTM) (1998)]. Sublethal toxicity of the sediments was evaluated by measuring the growth of midge (*Chironimus tentans*) larvae (ASTM, 1998). Stations were selected for these analyses based on a range of CoPC concentrations. Sediment toxicity samples were also analyzed for total metals, acid-volatile sulfides and simultaneously extracted metals (AVS/SEM), total organic carbon (TOC), and grain size to assist in interpreting toxicity results.

- *Fish Community Structure.* Fish were collected from the NZS and reference areas and enumerated by species to evaluate whether CoPCs were causing apparent changes in fish community structure. The fish community analysis provided an additional line of evidence to compare with the direct testing for surface water toxicity.

The *Ceriodaphnia* and the fathead minnow used in water toxicity tests were selected as representative of indigenous organisms to indicate potential effects on invertebrates and fish in general. The amphipod *H. azteca* and the chironomid *C. tentans* used in sediment toxicity testing were selected as surrogates to indicate potential effects on benthic macroinvertebrates in general. These four species were selected for toxicity testing because they are geographically widespread and protocols for water and sediment testing are well-developed (EPA, 1994; ASTM, 1998).

Wood Preservative Site   The primary measurement endpoints for sediment toxicity were amphipod survival and bacterial activity (as measured by Microtox luminescence testing; Beckman Coulter, Inc., Fullerton, California). Sediment chemistry was used as the primary means of assessing exposure. Data on bioaccumulation of CoPCs in tissue of crayfish and fish also provided evidence of exposure. Because these data were not used directly in risk characterization for aquatic receptors, they are not reported here [see PTI (1992) for additional information]. Fish livers were also collected for histopathological analysis. Water samples were not analyzed for toxicity at the MBS because the primary organic CoPCs were tightly bound to sediments and also because of the continuous dilution from the Willamette River. Measurement endpoints included the following:

Sediment   Sediment samples representing a wide range of CoPC concentrations were analyzed. Acute lethal toxicity was evaluated by using the amphipod survival test with *H. azteca* as at NZS. Sublethal toxicity was evaluated by measuring luminescence in the Microtox test (Beckman Instruments, 1982) instead of by using the midge larval growth assay used at NZS. The use of different measurement endpoints for sublethal toxicity at the two sites was the result of environmental management decisions by the regulatory agencies involved. As at NZS, the sediment samples used for toxicity testing were also analyzed for total metals, TOC, and grain size. AVS and SEM were not measured on sediment samples collected at the MBS because early in the investigation it was apparent that metals were minor CoPCs.

Fish Histopathology   Fish specimens were examined for gross abnormalities and lesions, and a histopathological analysis of the livers of large-scale suckers was performed. This analysis was conducted at the MBS because of reports of an association between PAH contamination of sediments and liver lesions in fishes at other sites (Baumann et al., 1982; Malins et al., 1984; Krahn et al., 1986).

20.2.5   Sampling and Data Analysis Methods

For the NZS study, sampling and analysis details are provided in PTI (1995). Environmental data to support the ecological risk assessment were collected in February and March 1994 and from July to September 1994.

For the MBS assessment, PTI (1992) describes the sampling and laboratory analysis methods. Sediment samples for chemistry and toxicity testing were collected during cruises in 1991 and 1992; fish samples were collected during the 1991 cruise.

Data analysis methods are presented below.

Characterization of Exposure   Exposure concentrations of CoPCs were generally expressed as milligrams/kilogram dry weight chemical concentrations in sediment or µg/L in surface water. For dioxins and related compounds at the MBS, toxicity equivalent concentrations (TECs) were calculated relative to toxicity to fish. To derive "fish TECs," the concentration of each 2,3,7,8-substituted PCDD or PCDF congener was multiplied by a toxicity equivalency factor (TEF) that represents the relative toxicity of that congener compared with that of 2,3,7,8-TCDD. The adjusted concentrations were summed to derive the TEC for that sample. TECs for fish were derived from data for individual PCDD and PCDF congeners by using TEFs for rainbow trout in Walker and Peterson (1991). This approach is analogous to using 2,3,7,8-TCDD toxicity equivalence calculations for human health risk assessment (EPA, 1989a).

Correlation Analysis   The relationships between selected chemical data for environmental media and measurement endpoints were analyzed by using graphical comparisons and linear regression (with and without various transformations of log-normal data). The degree of correlation was typically evaluated by using Spearman's rank correlation coefficient. Spearman rank correlation coefficients indicate the strength of the relationship between two variables based on the rank ordering of the data for each variable. The P value is the probability that the correlation is zero (i.e., complete independence between the variables). The value of the Spearman rank correlation coefficient does not indicate linearity (or nonlinearity) of the relationship, as does a Pearson correlation coefficient. The Spearman rank correlation coefficient was used instead of a Pearson coefficient because of the small sample sizes for some of the data sets and because use of the Spearman coefficient does not require that the data exhibit a bivariate normal distribution. Correlation analysis included comparisons of CoPC concentrations in water and sediments with results of toxicity tests.

Statistical Analyses of Toxicity Tests

Water Toxicity Tests   For the fathead minnow and *Ceriodaphnia* toxicity tests conducted on water samples at the NZS, several kinds of statistical analyses were performed by using Toxstat Version 3.4 (West and Gully, 1994).

For the fathead minnow test, values of percent survival were transformed by using an arcsine transformation (EPA, 1989b). Values of biomass were determined as mean dry weight per surviving fish.

For the *Ceriodaphnia* test, Fisher's exact test was used to test for significant differences in survival between each dilution and the negative control (EPA, 1989b). The reproductive endpoint was evaluated only for dilutions at which survival did not differ significantly from the control value. The number of young was determined per original female. If males were found at test termination, they were not used in the evaluation of the reproductive endpoint because the test focuses on the primary reproductive mode, which is female parthenogenesis.

Values of survival and biomass for the fathead minnow tests and values of mean number of young for the *Ceriodaphnia* tests were compared between each dilution and the negative control by using Dunnett's procedure (for equal sample sizes), the Bonferroni t test (for unequal sample sizes), or Steel's many-one rank test (when the assumptions of normality or homogeneity of variance were not met). Before comparisons with control values were made, data were tested for normality by using the Shapiro–Wilk's test or the chi-square test (depending on the number of test organisms at test termination) and for homogeneity of variance by using Bartlett's test.

For both water toxicity tests at the NZS, results based on the dilution series were used to calculate the no-observable-effect concentration (NOEC), the lowest-observable-effect concentration (LOEC), and the chronic value (ChV; i.e., the geometric mean of the NOEC and LOEC). In addition, the 25-percent inhibition value (IC_{25}) was determined by using EPA-approved software (ICp Calculation Program, Release 2.0, 1993). The IC_{25} is the concentration of test material estimated to cause a 25 percent inhibition of performance of the test organisms. All of the above values were calculated based on both the lethal (survival) and sublethal (growth and reproduction) endpoints, when possible.

In addition, for both water toxicity tests, results for the whole water sample from each site station were compared with the mean reference results by using the t test. Before each comparison, data were tested for homogeneous variance by using the F_{max} test. If the variances were heterogeneous, the data were compared by using the approximate t test.

Sediment Toxicity Tests   For the amphipod and chironomid sediment toxicity tests at the NZS, results for each site station were compared with the mean reference results by using the t test. Before each comparison, data were tested for homogeneous variance by using the F_{max} test. If the variances were heterogeneous, the data were compared by using the approximate t test.

Results of sediment toxicity tests at the MBS were compared with results of tests on local reference area sediments (from transects T3, T4, and T5) and to results from the upstream reference sediment station (Wilsonville) by using the Wilcoxon rank sum test.*

Fish Histopathology   Fish collected from the Willamette River at the MBS were inspected in the field for gross abnormalities or lesions. The livers were sent to the laboratory for slide preparation and histopathological analysis (inspection of tissue slides under a microscope for a variety of abnormalities, including benign and malignant tumors, altered cells, degenerative conditions, and inflammation). The prevalence of liver lesions was compared between the site and the reference area by using chi-square analyses.

Community Analyses   The NZS fish communities were assessed by calculating similarity indices and using cluster analysis. The Simplified Morisita Similarity Index (Krebs, 1989), which accounts for species occurrence and some measure of either relative abundance or biomass, was used because it is generally considered the most robust index of the available similarity indices (Wolda, 1981). A similarity matrix representing all pairwise comparisons of sampling stations was used to create a dendrogram, developed with the unweighted pair group by using an arithmetic averages clustering technique that minimizes within-group variation (SPSS, 1988).

---

*The statistical significance of results for the Microtox test was not based on comparison with results from the upstream reference site (station NB1 at Wilsonville) because of the substantial decrease in luminescence observed in the test of reference site sediments. Station H(−2)a was used as a reference site because, of the remaining Microtox stations, it was located in an area where chemical concentrations in sediments are relatively low and within reference area values. The reason for the significant ($P \leq 0.05$) decrease in Microtox luminescence at reference station NB1 relative to station H(−2)a (PTI, 1992) was not known. However, this decrease may have resulted from a toxic response to contaminants in pore water or another chemical condition at the station that was not reflected in the bulk sediment chemical analyses. However, chemical analyses of sediments from station NB1 showed low concentrations of measured contaminants. Alternatively, the significant ($P \leq 0.05$) response in the Microtox test for reference station NB1 could have been associated with a sample handling error or an aberrant test response that would not have been detected during routine quality assurance processing of the data. Aside from this anomaly, the Microtox results near the site were consistent with a toxic response only at high chemical concentrations.

Preliminary Remediation Goals   Preliminary remediation goals (PRGs) are chemical-specific concentrations in surface water or sediment designed to protect selected receptors from adverse effects associated with high exposure to CoPCs. At the NZS, apparent effects threshold (AET) values were calculated for each sediment CoPC using the results of the chironomids biomass test according to the method of Barrick et al. (1988). The AET is the highest concentration of a chemical (or group of chemicals) in sediment corresponding to a no-effect station. At concentrations above the AET, statistically significant ($P \leq 0.05$) effects on chironomid growth were always observed in the data set. This approach implicitly accounts for any interactive effects of metals and ancillary variables in sediment. The AET approach is especially useful where multiple chemicals of different types (i.e., several metals with different modes of action) may be causing toxicity.

An alternative approach to deriving PRGs was used at the MBS because complete concentration–response data were available for the amphipod test and only PAHs (i.e., individual compounds with similar modes of action) were important in causing sediment toxicity. Threshold chemical concentrations for amphipod survival [i.e., <75 percent survival was the approximate threshold for significant ($P < 0.05$) mortality relative to the response to Wilsonville reference sediments] were derived from logistic concentration–response relationships fit to the data for low-molecular-weight PAHs (LPAHs), high-molecular-weight PAHs (HPAHs), and benzo[a]pyrene. Similar correlation analyses were conducted for the Microtox test results. Standard concentration–response analyses typically included in the Microtox results were also evaluated but are not reported here because the amphipod test was considered a more relevant test for deriving toxicity thresholds to be used as PRGs.

20.3 EXPOSURE ASSESSMENT

20.3.1 Exposure Routes

Despite differences in transport and fate of the CoPCs between the two case study sites, exposure routes are expected to be similar. Exposure routes for benthic macroinvertebrates and fish depend on microhabitat and diet. Aquatic vegetation (ingested by macroinvertebrates and fish) can assimilate CoPCs from surface water or sediments. Benthic macroinvertebrates may be exposed to CoPCs via contact with surface water or sediments, direct or ancillary ingestion of sediment (e.g., direct ingestion by detritus feeders and incidental ingestion by predatory species), ingestion of aquatic vegetation, and ingestion of prey. Fish may be exposed to CoPCs via direct contact with water and sediment and direct ingestion of sediment and food items (i.e., plants, macroinvertebrates, and other fish).

Differences in exposure pathways, including the transport and fate of CoPCs, between the two case study sites are discussed later (see Section 20.6). PTI (1995) and Pastorok et al. (1994) provide detailed scenarios for exposure of ecological receptors at the NZS and MBS, respectively.

20.3.2 Distribution of CoPCs

In the following discussion, sediment concentrations of most organic contaminants are reported in milligrams/kilogram organic carbon. Organic carbon (OC)-normalized concentrations are reported for organic compounds that partition preferentially into organic carbon in sediments because the bioavailability of these contaminants and their transport through the environment is affected by the amount of organic carbon in sediments. Those contaminants that do not partition strongly into organic carbon, such as metals and phenols, are reported in milligrams/kilogram dry weight.

Zinc Smelter Site Concentrations of CoPCs in surface water and sediments in most of the North Tributary and West Tributary and in parts of Eliza Creek were elevated above reference values (Figures 20.6 and 20.7). Spatial distributions of CoPCs in surface water were similar to distributions of CoPCs in sediments. CoPC distributions in surface water and sediment show no influence of the releases from the site on Sand Creek, except for a slight elevation of cadmium concentration in surface water relative to reference area values.

In general, the highest concentrations of most of the eight sediment CoPCs were found in the North Tributary (most frequently at stations ECNT1, ECNT3, and ECNT4), and concentrations declined in downstream areas with increasing distance from the North Tributary. In some cases, concentrations in the lower West Tributary were also substantially elevated (i.e., the elevation above reference value, or EAR, was approximately 85 for cadmium, 4 for copper and lead, and 102 for zinc). In addition, elevated levels of selenium were observed at stations ECDS1 (EAR = 7) and ECDS2 (EAR = 28) downstream from the confluence with the North Tributary. As expected, sediment concentrations of CoPCs were generally highest in areas of sediment accumulation. These patterns indicate relatively local sources of most of the CoPCs in the North Tributary and, in some cases, the lower West Tributary. The dam on Eliza Creek helped restrict further downstream migration of sediment contaminants from the North Tributary.

The potential bioavailability of divalent metals (cadmium, copper, lead, nickel, and zinc) was evaluated by measuring the concentrations of SEMs and AVS in all sediment samples. Because divalent metals tend to bind with AVS and thereby become less bioavailable, their potential for causing toxicity is reduced in the presence of AVS (Ankley et al., 1991). As reported by Ankley et al. (1991), when the molar ratio of the sum of SEM to AVS is less than or equal to 1.0, a sufficient amount of AVS is available to bind with the entire amount of SEM present, and the sediment metals are not expected to be toxic. Conversely, when the SEM/AVS ratio exceeds 1.0, some of the metals are assumed

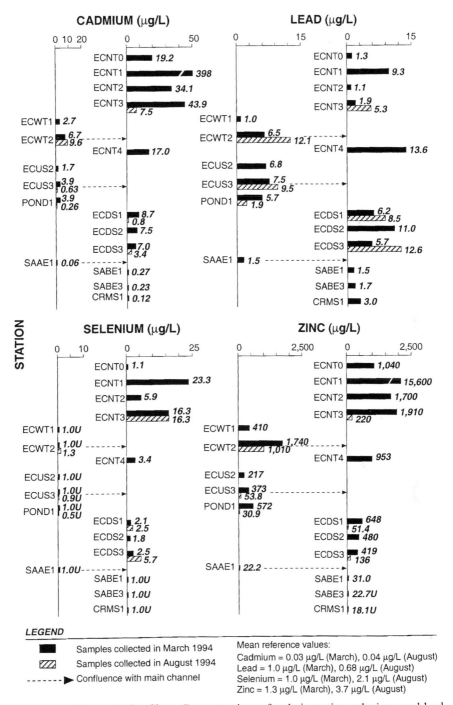

Figure 20.6 National Zinc Site—Concentrations of cadmium, zinc, selenium, and lead in whole water samples.

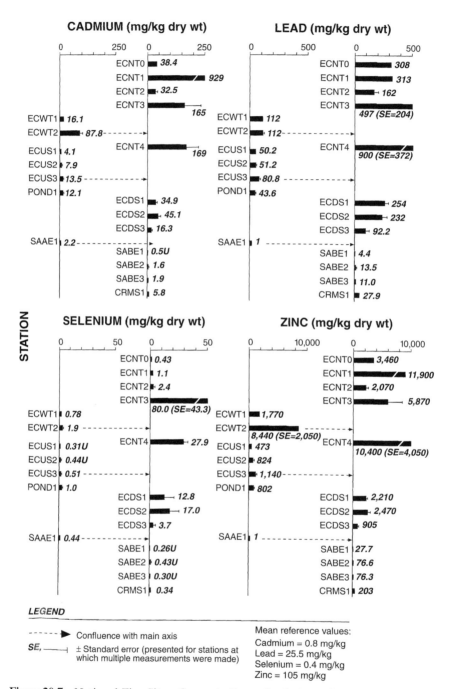

Figure 20.7   National Zinc Site—Concentrations of cadmium, zinc, selenium, and lead in sediment samples.

to be bioavailable and potentially toxic. However, sediments may not exhibit toxicity even when the SEM/AVS ratio exceeds 1.0 because other factors (e.g., complexes with organic material) may reduce the bioavailability or toxicity of the excess amount of metals or the concentration of free SEM may not be above the threshold for some toxic effects.

Except for station ECWT2, the SEM/AVS ratio exceeded 1.0 at all stations, including the reference stations from Kane and Rice creeks (Figure 20.8). The highest SEM/AVS ratio (26.3) was found at station ECNT3 in the North Tributary. The AVS concentrations in Sand Creek and the two reference areas were very low (0.1 to 0.4 mmol/kg), which largely accounted for the high SEM/AVS ratios observed at those relatively uncontaminated areas. The pattern of SEM/AVS ratios found at the study site indicates that a fraction of the divalent metals present in sediments at all stations (except station ECWT2) could be bioavailable in the absence of other limiting factors (e.g., complexation of metals with dissolved organic matter).

Wood Preservative Site   In sediments, the most frequently detected chemicals that also showed substantial elevations above reference values were PAHs, PCDDs, and PCDFs. Because sediment toxicity was mainly correlated with the concentrations of PAHs, only the PAH data are shown here [see PTI (1992) and Pastorok et al. (1994) for data on other CoPCs]. Figure 20.9 presents the median and maximum concentrations in sediments for LPAH and HPAH by habitat type.

The LPAHs and HPAHs showed the widest distribution and greatest range of concentrations throughout sediments near the site (Figure 20.9). The beach habitat area had the highest median and maximum values for both LPAHs (maximum = 400 mg/g OC normalized) and HPAHs (maximum = 140 mg/g OC). The clam, shelf, and fish and crayfish habitat areas had relatively moderate levels of LPAH and HPAH contamination compared with the beach area. The main channel area had the lowest concentrations of LPAHs and HPAHs, which were still approximately 5 to 10 times higher than respective concentrations in the upstream reference area sediments (<0.012 mg/g OC for LPAH and <0.036 mg/g OC for HPAH).

20.4   EFFECTS ASSESSMENT

In this section, we focus on the results of site-specific surveys and toxicity testing as measures of contaminant effects. PTI (1992, 1995) presents summaries of the potential toxicity of CoPCs based on existing literature.

20.4.1   Zinc Smelter Site

For the water toxicity tests, no adverse effects were detected in the fathead minnow survival test for any site stations, and a low level of toxicity was ob-

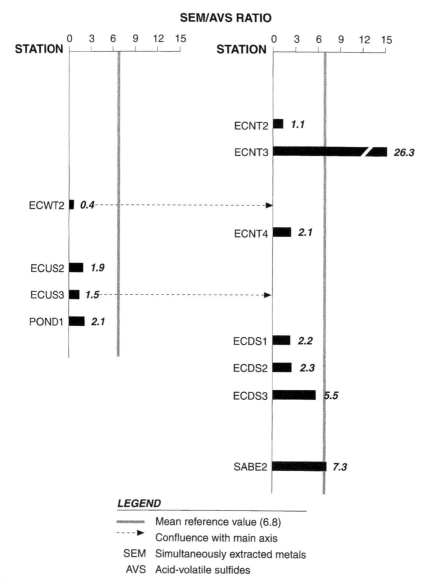

Figure 20.8 National Zinc Site—Simultaneously extracted metals/acid-volatile sulfide ratios for sediments.

served in the cladoceran reproduction test for several site stations (Figure 20.10). Little sediment toxicity was observed at the study area based on the amphipod survival endpoint (Figure 20.11). Sublethal effects expressed as reductions in chironomid growth relative to reference conditions were detected at several stations. Toxicity test results for surface water and sediment are presented in more detail below.

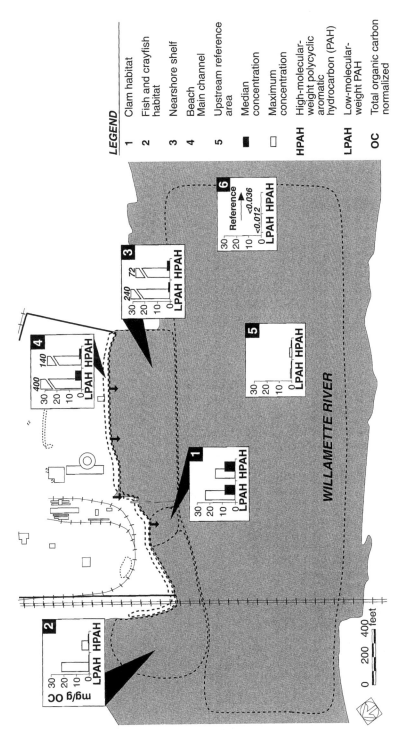

Figure 20.9 McCormick & Baxter Creosoting Company site—low-molecular-weight polycyclic aromatic hydrocarbon and high-molecular-weight polycyclic aromatic hydrocarbon concentrations in surface sediment by habitat area.

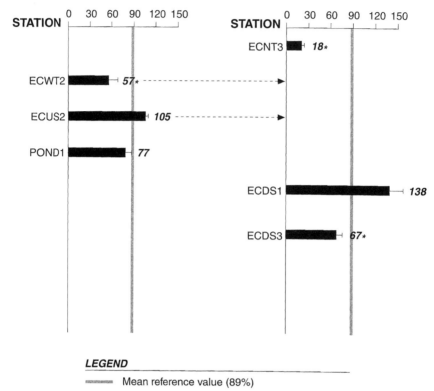

Figure 20.10 National Zinc Site—results of *Ceriodaphnia* test for water, sublethal toxicity.

Water Toxicity Test Results Surface water samples were collected in August for toxicity testing at six site stations (ECNT3, ECDS1, ECDS3, ECWT2, ECUS3, and POND1) and from the Rice Creek reference area (RCREF01). At all site stations, the survival and biomass of fathead minnows did not differ significantly ($P > 0.05$) from the values found at the reference station (RCREF01). The only apparent adverse effect observed in the fathead minnow test was the reduction in growth (as measured by biomass) at the reference station (RCREF01) relative to growth in the control medium (i.e., laboratory water). Although the reason for the reduced biomass at that station was un-

AMPHIPOD SURVIVAL (%)

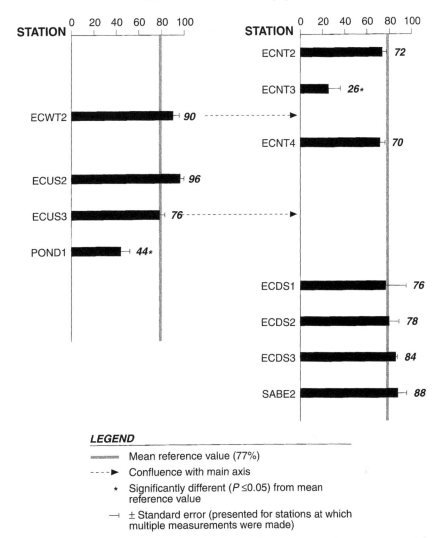

LEGEND

▬▬▬   Mean reference value (77%)

- - -►   Confluence with main axis

*   Significantly different ($P \leq 0.05$) from mean reference value

⊢—   ± Standard error (presented for stations at which multiple measurements were made)

Figure 20.11   National Zinc Site—results of amphipod test for sediment, acute toxicity.

known, metals were unlikely to be responsible given the low concentrations of metals found at the reference site (e.g., Figures 20.6 and 20.7).

By contrast with the fathead minnow test results, a reduction in the number of young produced in the *Ceriodaphnia* test was observed at three site stations (Figure 20.10). One station (ECNT3) was located in the North Tributary, one station (ECDS3) was located downstream in Eliza Creek, and the third station (ECWT2) was located in the lower portion of the West Tributary. However,

TABLE 20.2   Summary of Toxicity Results for *Ceriodaphnia* Test

Station	NOEC[a]	LOEC[a]	ChV[a]	IC$_{25}$[a]	Number of Young[b] (percentage of control)
ECNT3	50[d]	75[d]	61[d]	63[d]	18
ECDS1	100[c]	—	—	—	138
ECDS3	70[d]	100[d]	84[d]	90[d]	67
ECWT2	70[d]	100[d]	84[d]	80[d]	56
ECUS3	100[c]	—	—	—	110
POND1	100[c]	—	—	—	77
RCREF01	100[c]	—	—	—	89

Note: —, values could not be determined because NOEC was 100%; ChV, chronic value; IC$_{25}$, 25% inhibition concentration; LOEC, lowest-observed effect concentration; NOEC, no-observed effect concentration.

[a] Values are expressed as percentages of undiluted field samples.

[b] Based on undiluted field samples (i.e., 100% samples).

[c] Based on both survival and number of young.

[d] Based only on number of young.

the level of observed toxicity was relatively low, and the LOEC in all three cases was 75 to 100 percent (Table 20.2). Based on the calculated chronic values (CVs) and IC$_{25}$ values for the three stations, water from station ECNT3 was more toxic than water from the other two stations, which both exhibited similar levels of toxicity. The numbers of *Ceriodaphnia* young for undiluted field samples are compared between site and reference stations in a spatial context in Figure 20.10. No adverse effects related to survival of the adult test organisms were found.

Sediment Toxicity Test Results   Sediment samples were collected for toxicity testing in August and September from 11 site stations (ECNT2, ECNT3, ECNT4, ECDS1, ECDS2, ECDS3, SABE2, ECWT2, ECUS2, ECUS3, and POND1) and 2 reference areas (KCREF01 and RCREF01). Two toxicity tests were conducted: the 10-day amphipod survival test using *H. azteca* and the 10-day chironomid growth test using the midge *C. tentans*.

In general, few adverse effects were observed in the amphipod test (Figure 20.11). Except for the values of 26 and 44 percent survival observed for stations ECNT3 and POND1, respectively, survival was greater than or equal to 70 percent at all stations. Percentage survival showed a weak relationship with distance downstream from station ECNT3 (Figure 20.11).

By contrast with the amphipod test, adverse sublethal effects on chironomid growth were detected only when results for site stations were compared with the mean reference value for Rice Creek (station RCREF01; Figures 20.11 and 20.12). Mean biomass at the Rice Creek reference area (131 percent relative to control) was more than twice as high as the value observed for the Kane Creek reference area (station KCREF01; 62 percent). This difference was presumed to

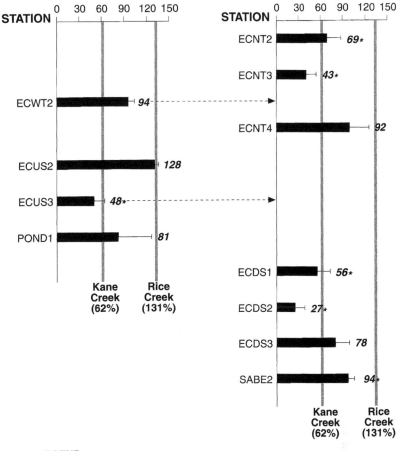

CHIRONOMID BIOMASS
(% of control)

LEGEND

▬▬▬ Mean reference value

- - - ► Confluence with main axis

* Significantly different (*P* ≤0.05) from mean reference value at Rice Creek

⊢ ± Standard error (presented for stations at which multiple measurements were made)

Note: Statistical difference (*P*≤0.05) at Station SABE2 is not considered biologically meaningful because the percent reduction in biomass was less than 10 percent relative to control.

Figure 20.12 National Zinc Site—results of chironomid test for sediment, sublethal toxicity.

represent the natural variability found in reference areas near the NZS. However, because the two mean reference values differed significantly ($P \leq 0.05$; t test), they were not combined for the toxicity evaluation. Instead, test site values were compared with each mean reference value independently.

Chironomid biomass values at 5 of the 11 site stations differed significantly ($P \leq 0.05$) from the reference value of 131 percent found at Rice Creek and showed at least a 10 percent reduction in biomass relative to control. The five stations included the two stations in the upper part of the North Tributary (ECNT2 and ECNT3), one station upstream in Eliza Creek just above the confluence with the North Tributary (ECUS3), and two of the three stations downstream in Eliza Creek (ECDS1 and ECDS2). Although the 6 percent reduction in biomass relative to the control (station SABE2) was statistically significant ($P \leq 0.05$) relative to the Rice Creek reference value, the magnitude of the effect was not biologically meaningful; therefore, it was not considered indicative of toxicity. The statistical significance of the value at station SABE2 was likely an artifact of the low variability in response at that station rather than an indicator of toxicity because the biomass (94 percent of the control value) was relatively high.

The lowest chironomid biomass value (27 percent) was found at station ECDS2 downstream in Eliza Creek. Although low values (43 and 69 percent) were found in the upper part of the North Tributary, a surprisingly high value (92 percent) was found at station ECNT4 in the lower part of that tributary. No consistent pattern of toxicity with distance downstream from the upper part of the North Tributary was found, although the lowest biomass values in upstream and downstream Eliza Creek were found at the stations located closest to the confluence with the North Tributary.

In summary, little sediment toxicity was observed at the site based on the amphipod survival endpoint. No sublethal effects were detected when chironomid biomass at site stations was compared with the mean reference value for Kane Creek. Sublethal toxicity was found in the chironomid test at more than half of the site stations when compared with the mean reference value for Rice Creek alone.

Field Survey of Fish Communities   A total of 1847 individual fish from 32 species were collected at the site, whereas 254 individuals from 12 species were collected at the 2 reference areas (Table 20.3). Although the numbers of individuals and species collected at the site and reference areas were influenced by the number of sampling stations in each area, it is useful to compare the dominant species in the two areas based on their relative abundances in the two assemblages (Table 20.4). In both areas, the two most abundant species were bluegill and mosquito fish. The species accounting for the highest levels of relative abundance at the site were bluegill (28 percent), mosquito fish (17 percent), and red shiner (13 percent). The numerically dominant species at the reference sites were bluegill (31 percent), mosquito fish (23 percent), and longear sunfish (16 percent). Therefore, although sampling intensity was different

TABLE 20.3 Relative Abundances of Fishes Captured at Site and Reference Areas by Species

Family	Species	Common Name	Relative Abundance (%)	
			Onsite	Reference
Lepisosteidae	*Lepisosteus oculatus*	Spotted gar	0.3	—
	Lepisosteus osseus	Longnose gar	<0.1	—
Clupeidae	*Dorosoma cepedianum*	Gizzard shad	1.8	1.2
Cyprinidae	*Campostoma anomalum*	Stoneroller	2.5	—
	Cyprinus carpio	Carp	0.2	—
	Cyprinella lutrensis	Red shiner	13.4	—
	Notropis stramineus	Sand shiner	4.5	—
	Pimephales vigilax	Bullhead minnow	3.1	—
Catostomidae	*Carpiodes carpio*	River carpsucker	0.2	—
	Ictiobus bubalus	Smallmouth buffalo	<0.1	—
	Ictiobus cyprinellus	Bigmouth buffalo	0.2	0.8
	Minytrema melanops	Spotted sucker	0.3	—
Ictaluridae	*Ictalurus natalis*	Yellow bullhead	0.3	1.2
	Ictalurus melas	Black bullhead	0.8	5.1
	Ictalurus punctatus	Channel catfish	0.2	—
	Pylodictis olivaris	Flathead catfish	<0.1	—
Cyprinodontidae	*Fundulus notatus*	Blackstripe topminnow	0.2	—
Poeciliidae	*Gambusia affinis*	Mosquitofish	17.4	22.8
Atherinidae	*Labidesthes sicculus*	Brook silverside	5.1	3.2
Percichthyidae	*Morone chrysops*	White bass	<0.1	—
Centrarchidae	*Lepomis spp.*	Hybrid sunfish	1.4	—
	Lepomis cyanellus	Green sunfish	4.1	9.8
	Lepomis gulosus	Warmouth	0.3	3.2
	Lepomis humilis	Orangespotted sunfish	1.9	1.2
	Lepomis macrochirus	Bluegill	28.2	30.7
	Lepomis megalotis	Longear sunfish	4.9	15.8
	Lepomis microlophus	Redear sunfish	0.6	—
	Micropterus punctulatus	Spotted bass	2.6	—
	Micropterus salmoides	Largemouth bass	2.0	5.1
	Pomoxis annularis	White crappie	3.5	—
Percidae	*Percina caprodes*	Log perch	<0.1	—
	Percina phoxocephala	Slenderhead darter	<0.1	—
Total catch			1,847	254

Note: —, species not captured.

in the two areas, bluegill and mosquito fish accounted for close to half the individuals in both assemblages (46 percent at the study site and 54 percent at the reference sites).

The classification analysis of the fish assemblages at the study and reference sites identified two major groups of stations and a single outlier station based on similarities among their respective assemblages (Figure 20.13). The first

TABLE 20.4 Relative Abundances of Fishes Captured at Site Stations and Reference Areas in Percentages

Family	Common Name	Relative Abundance (%)									
		ECNT2	ECNT3	ECNT4	ECDS1	ECDS3	SABE2	ECUS2	POND1	RCREF01	KCREF01
Lepisosteidae	Spotted gar	—	—	—	3.3	—	—	—	—	—	—
	Longnose gar	—	—	—	—	—	0.3	—	—	—	—
Clupeidae	Gizzard shad	—	—	0.8	3.3	0.9	0.6	3.7	5.6	1.3	—
Cyprinidae	Stoneroller	1.2	8.2	—	—	9.0	1.2	—	—	—	—
	Carp	—	—	—	—	—	—	1.9	—	—	—
	Red shiner	—	—	—	—	1.5	71.2	4.7	—	—	—
	Sand shiner	—	—	—	—	24.1	—	—	—	—	—
	Bullhead minnow	—	—	1.3	15.0	7.3	0.3	1.9	—	—	—
Catostomidae	River carpsucker	—	—	—	1.3	0.3	—	—	—	—	—
	Smallmouth buffalo	—	—	0.3	—	—	—	—	—	—	—
	Bigmouth buffalo	—	—	—	2.0	—	—	—	—	0.9	—
	Spotted sucker	—	—	—	3.3	—	—	—	—	—	—
Ictaluridae	Yellow bullhead	—	—	—	0.7	—	—	—	—	1.3	—
	Black bullhead	—	—	—	4.6	—	—	2.3	1.7	1.8	33.3
	Channel catfish	—	—	—	0.7	0.3	0.3	1.4	—	—	—
	Flathead catfish	—	—	—	—	—	0.3	—	—	—	—

Family	Species									
Cyprinodontidae	Blackstripe topminnow	—	—	0.3	—	—	0.6	—	—	—	—
Poeciliidae	Mosquitofish	69.5	32.0	20.3	2.0	17.4	9.2	2.8	21.7	18.9	55.6
Atherinidae	Brook silverside	—	33.6	1.1	—	0.6	8.3	9.8	—	3.5	—
Percichthyidae	White bass	—	—	—	—	—	0.3	—	—	—	—
Centrarchidae	Hybrid sunfish	—	—	2.4	—	—	—	7.5	—	—	—
	Green sunfish	6.1	3.3	3.5	8.5	2.9	0.6	13.1	—	10.1	7.4
	Warmouth	—	—	0.8	—	0.3	—	—	0.4	3.5	—
	Orangespotted sunfish	—	—	3.7	2.6	0.9	—	6.5	—	1.3	—
	Bluegill	7.3	6.6	59.2	36.0	18.3	—	10.8	62.3	34.4	—
Centrarchidae	Longear sunfish	2.4	8.2	2.4	11.1	6.7	4.6	6.5	—	17.2	3.7
	Redear sunfish	—	—	0.5	—	2.3	—	—	0.4	—	—
	Spotted bass	13.4	8.2	—	1.3	1.7	0.6	0.5	6.9	—	—
	Largemouth bass	—	—	3.2	1.3	4.7	—	2.8	0.4	5.7	—
	White crappie	—	—	0.3	3.3	0.9	0.9	23.8	0.4	—	—
Percidae	Log perch	—	—	—	—	—	0.3	—	—	—	—
	Slenderhead darter	—	—	—	—	—	0.3	—	—	—	—
	Total catch	82	122	375	153	344	326	214	231	227	27

Note: —, species not captured.

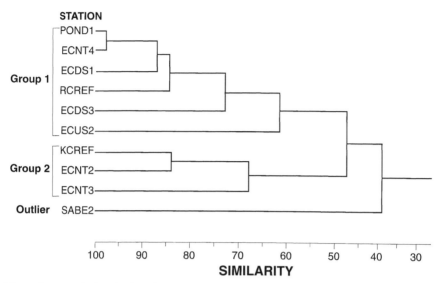

Figure 20.13   National Zinc Site—results of classification analysis of fish assemblages.

group of stations included the pond (station POND1), all three upstream and downstream stations in Eliza Creek (stations ECUS2, ECDS1, and ECDS3), the station in the North Tributary closest to the confluence with Eliza Creek (station ECNT4), and the Rice Creek reference station (station RCREF01). In general, fish assemblages at stations in the first group were dominated by blue-gill. The second group of stations included the two other stations in the North Tributary (stations ECNT2 and ECNT3) and the Kane Creek reference station (station KCREF01). Fish assemblages at stations in the second group generally were dominated by mosquito fish. The single outlier station was station SABE2 in Sand Creek, which was dominated by red shiners (relative abundance, 71.2 percent). This was the only station at which red shiner was the dominant species. The classification results suggest that, aside from the assemblage at station SABE2, the fish assemblages at the study site were similar to those at the reference sites. The relatively unique characteristics of the assemblage at station SABE2 were probably due to the increased flow, different substrate characteristics, and lack of predators (e.g., sunfish and bass) at Sand Creek relative to the other water bodies at which fish assemblages were evaluated.

Ecological Response Analysis   Ecological responses were related to CoPC concentration gradients at the NZS aquatic stations by using the toxicity test results. The fish community analysis was not sensitive to the CoPC concentrations at the site.

Surface Water   The degree of association between the results for the *Ceriodaphnia* test based on number of young and concentrations of each of the four water CoPCs in undiluted samples was evaluated to determine whether concentrations of any of the CoPCs were related to the toxicity. A high degree of association with a CoPC, based on Spearman's rank correlation coefficient (r_s), would indicate a potential causative relationship between the CoPC and the observed toxicity. Because only seven stations were evaluated, any correlation between metal concentration and number of young needed to be relatively high (i.e., $r_s \geq 0.71$ for a positive correlation or $r_s \leq -0.71$ for a negative correlation) to be significant ($P \leq 0.05$).

Concentrations of the four metals in surface water were not significantly correlated with *Ceriodaphnia* reproduction, although total cadmium and zinc showed much higher correlation coefficients [an r_s of -0.64 ($P \leq 0.11$) for cadmium and an r_s of -0.68 ($P \leq 0.06$) for zinc] than did total lead or selenium. These results suggest that elevated cadmium and zinc concentrations may have been partly responsible for the observed toxicity in the *Ceriodaphnia* test. However, because the magnitudes of the correlations of cadmium and zinc with the toxicity results were similar and because these two metals were highly intercorrelated, determining their relative contributions to the observed toxicity was not possible. It does not appear that selenium or lead contributed to the observed toxicity ($r_s < 0.5$ and not significant).

Sediment   The degree of association between the results of the amphipod and chironomid toxicity tests and concentrations of the eight sediment CoPCs was determined by using Spearman's rank correlation coefficient (r_s). Because a moderate number of stations (12) were evaluated, moderate correlations between metal concentrations and toxicity results ($r_s \geq 0.50$ for a positive correlation or $r_s \leq -0.50$ for a negative correlation) would be significant ($P \leq 0.05$).

Concentrations of the eight sediment CoPCs did not correlate significantly ($P > 0.05$) with amphipod survival (a negative correlation of $r_s > -0.40$ was found for all CoPCs). This lack of association was largely the result of the relatively limited range of toxicity observed in the amphipod test.

In contrast with the amphipod test results, chironomid biomass exhibited significant ($P \leq 0.05$) negative correlations with concentrations of several of the eight CoPCs. The CoPCs included cadmium ($r_s = -0.64$), lead ($r_s = -0.59$), selenium ($r_s = -0.55$), and zinc ($r_s = -0.53$), all of which were also water CoPCs. However, based on a parametric correlation coefficient, variation in each of these CoPCs would account for only a minor amount of the variance in toxicity (< 41 percent), which indicates that other factors may have an important effect on the toxicity to chironomids. Chironomid biomass did not correlate significantly ($r_s = -0.13$; $P \geq 0.05$) with the SEM/AVS ratios found for the site sediments (see Section 3.2). The lack of a correlation indicates that AVS did not have a strong influence on the toxicity of the CoPCs or that the proportion of sampled sediments with low SEM relative to that with high metals at this site was a confounding factor. For example, even if the SEM/AVS ratio is greater than 1.0, toxic effects may not be observed if the concentration of SEM is below the effect threshold.

The results of the correlation analysis suggested that cadmium, lead, selenium, and zinc may have been partly responsible for the toxicity observed in the chironomid test when site stations were compared with Rice Creek (no toxicity was detected in comparisons with Kane Creek). However, because the four CoPCs exhibited similar levels of correlation with the chironomid toxicity test results, it was not possible to determine which of the four CoPCs was primarily responsible for the observed toxicity.

Site-Specific Toxicity Thresholds   Sediment AET values derived for the chironomid growth endpoint were:

- 138 mg/kg dry weight for cadmium
- 692 mg/kg dry weight for lead
- 29.2 mg/kg dry weight for selenium
- 12,000 mg/kg dry weight for zinc

These AET values formed the basis for PRGs (i.e., sediment cleanup levels) at the NZS. For surface water, the Oklahoma water quality standards were used as PRGs.

20.4.2 Wood Preservative Site

Sediment Toxicity Test Results   Of the 48 stations tested for toxicity to amphipods, 9 had sediments that were significantly ($P \leq 0.05$) toxic relative to local reference area sediments, and 7 were significantly ($P \leq 0.05$) toxic relative to upstream reference area sediments (Figure 20.14). The lowest survival values were found in the area of the creosote dock, the adjacent shoreline upstream of the creosote dock, and near the railroad bridge.

For the Microtox bioassay, sediments from 8 of 17 stations caused a significant ($P \leq 0.05$) decrease in luminescence relative to sediments from reference station H(-2)* near the MBS (Figure 20.15). The greatest effects (82 to 96 percent decrease in luminescence) were observed within 150 feet of the creosote dock.

All stations with significant ($P \leq 0.05$) effects in the amphipod or the Microtox tests were distributed within 300 feet of the shoreline (Figures 20.14 and 20.15). The limited spatial distribution of sediment toxicity at the MBS suggests that toxicity is mainly related to the high PAH concentrations in sediment resulting from migration of NAPLs and groundwater contaminants.

Fish Histopathology   The data on liver histopathology of large-scale suckers collected during the remedial investigation and available histopathological data

---

*The results from the remote reference site at Wilsonville were not accepted as representative of reference conditions because of significant decreases in luminescence observed in organisms exposed to sediments from station NB1 at Wilsonville.

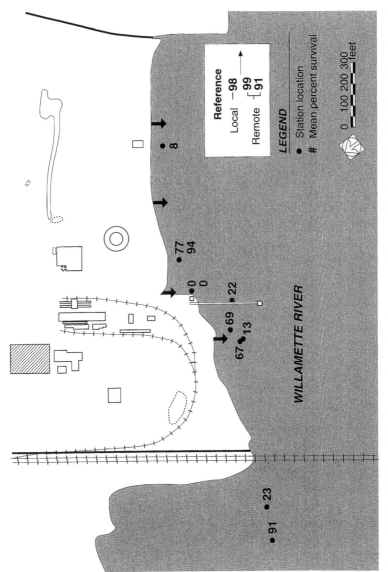

Figure 20.14 McCormick & Baxter Creosoting Company Site—results of amphipod test for sediment, acute toxicity.

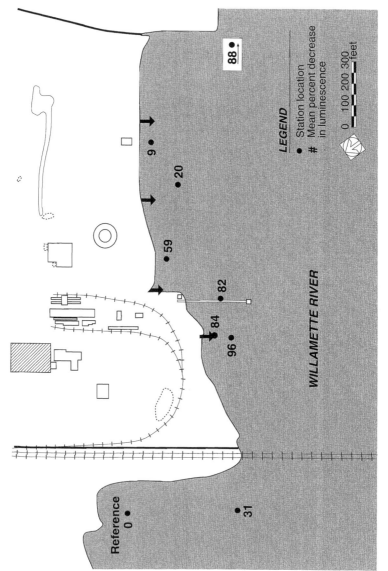

Figure 20.15 McCormick & Baxter Creosoting Company Site—results of Microtox luminescence bioassay on sediments.

on carp collected by Curtis (1991; Curtis et al., 1991) near the MBS suggested that risk to fish populations attributable to chronic toxicity from contamination at the site was low. The most commonly observed abnormal condition in large-scale suckers was the presence of monocellular cell infiltrates indicating mild liver inflammation, which was also observed in the upstream reference area. In general, large-scale suckers that were examined as part of the remedial investigation appeared to be in good health based on their liver condition. No evidence of serious toxic injury (i.e., neoplasia or megalocytic hepatosis) was found in any of the fish livers examined. In addition, Curtis (1991) suggests that only minor abnormalities were found in carp near the site.

PAH contamination of sediments has been shown to be correlated with histopathological abnormalities at a number of sites (Baumann et al., 1982, 1991; Malins et al., 1984; Baumann and Harshbarger, 1985). At MBS, the relatively small area of sediments with high concentrations of PAHs relative to the expected size of the home ranges of large-scale sucker and carp may account for the lack of site-specific histopathological effects. In summary, the histopathology data for fish suggest that risks to populations of mobile fish were low.

Ecological Response Analysis   The sediment toxicity tests provided the best results for quantitative analysis of ecological response to CoPC concentration gradients at the MBS.

Sediment   Mean percentage survival of amphipods was significantly correlated ($P < 0.05$; Spearman rank correlation coefficient) with several organic compounds or groups, including LPAH, HPAH, and benzo[a]pyrene (Figure 20.16). Other contaminants with significant correlations to amphipod survival were frequently undetected in sediments (e.g., hexachlorobenzene and chlorinated phenols), or visual inspection of the data distribution revealed a greater degree of scatter than with PAH compounds (e.g., PCDDs and PCDFs; PTI, 1992). Similar scattered distributions or nonsignificant correlations were found for pentachlorophenol, arsenic, chromium, copper, zinc, and percentage of fine-grained material and TOC in sediments.

Based on the chemical analysis of seven samples of surface sediment and Microtox bioassays of the corresponding pore water samples, the mean percentage decrease in Microtox bioluminescence was negatively correlated ($P < 0.05$; Spearman rank correlation coefficient) with chromium, copper, and TOC. The reasons for these correlations are unknown. However, concentrations of chromium and copper increase in an upstream direction from the site, indicating a source of metals other than the MBS. A response of Microtox to site contaminants in sediments would then result in a negative correlation between Microtox and selected metals. Although Microtox response was apparently positively related to concentrations of LPAH and HPAH (Figure 20.17), correlation coefficients were not significant ($P > 0.05$), possibly because of the small sample size. Pastorok and Becker (1989) found that the Microtox bioassay response was significantly ($P \leq 0.01$) correlated with LPAH and HPAH

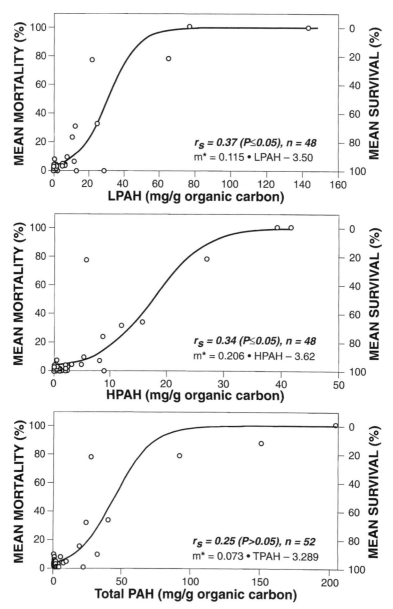

Figure 20.16   McCormick & Baxter Creosoting Company Site—relationship between mean percentage mortality in the amphipod bioassay and polycyclic aromatic hydrocarbon concentrations in sediments.

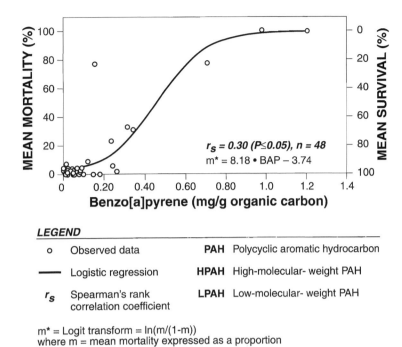

LEGEND

| o | Observed data | **PAH** | Polycyclic aromatic hydrocarbon |

| —— | Logistic regression | **HPAH** | High-molecular- weight PAH |

| r_S | Spearman's rank correlation coefficient | **LPAH** | Low-molecular- weight PAH |

m^* = Logit transform = $\ln(m/(1-m))$
where m = mean mortality expressed as a proportion

Figure 20.16 *(Continued)*

based on pooled data for sediment from three contaminated areas of Puget Sound. In one of the dilution series in which PAH contamination was associated with releases of creosote from a wood treatment facility at Eagle Harbor (Puget Sound), reductions in Microtox bioluminescence results were significantly ($P \leq 0.05$) correlated with PAH concentrations.

Site-Specific Toxicity Thresholds   Threshold chemical concentrations for amphipod survival of <75 percent were derived from logistic concentration–response relationships fit to the data for LPAH, HPAH, and benzo[*a*]pyrene (Figure 20.16). These thresholds are as follows:

- LPAH threshold = 21 mg/kg OC
- HPAH threshold = 12 mg/kg OC
- Benzo[*a*]pyrene threshold = 0.32 mg/kg OC

Logistic concentration–response curves were not fit to data for other chemicals because of small sample size or nonsignificant correlation between amphipod survival and chemical concentration.

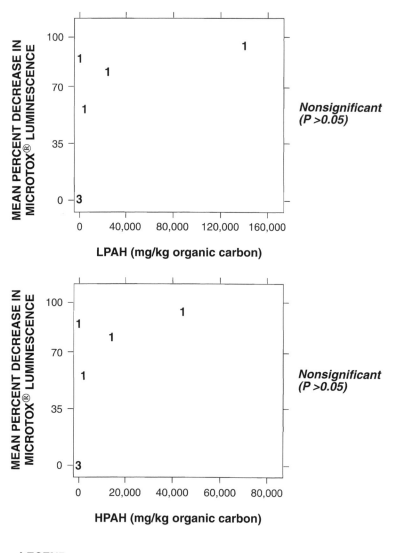

Figure 20.17 Relationship between mean percentage decrease in Microtox bio-luminescence and polycyclic aromatic hydrocarbon concentrations in sediments.

20.5 RISK CHARACTERIZATION

The objective of risk characterization is to evaluate the likelihood of adverse effects on ecological receptors as a result of exposures to CoPC concentrations in abiotic and biotic environmental media. The spatial distribution and ecological significance of risks to aquatic biota are discussed in the following sections.

20.5.1 Toxicity Testing and Biological Indicators

Zinc Smelter Site CoPC concentrations in surface water and sediment were highest in the North and West tributaries and a portion of Eliza Creek downstream of the North Tributary. CoPC distributions in surface water showed no influence of the releases from the site on Sand Creek, except for a slight elevation of cadmium concentration relative to reference area values. Moreover, concentrations of CoPCs in Sand Creek sediments were similar to or lower than mean reference values.

Analyses of the dissolved CoPCs in surface water, the SEM/AVS ratios for sediments, and whole-body samples of aquatic species demonstrated that CoPCs are bioavailable in these areas and are being bioaccumulated to concentrations above reference values (PTI, 1995). Concentrations of CoPCs (except zinc) in crayfish increased along Eliza Creek from upstream of the confluence with the North Tributary (station ECUS2) to downstream stations (PTI, 1995). This trend was particularly apparent for cadmium and selenium. In general, the highest CoPC concentrations in mosquito fish tissue were found in the North Tributary (most frequently at Station ECNT3) for all CoPCs except copper and mercury (PTI, 1995). This pattern was similar to the spatial distribution patterns of most of the water and sediment CoPCs.

Potential ecological risks to aquatic receptors from CoPCs (cadmium, lead, selenium, and zinc) in surface water were confined to the North and lower West tributaries and portions of Eliza Creek (Figure 20.18). The highest risks posed by sediments were associated with the North Tributary (Figure 20.19). Potential ecological risks to aquatic receptors from CoPC concentrations in sediment at the site were evaluated by comparing site-specific toxicity results with spatial patterns of the eight sediment CoPCs. The significant negative correlations found between chironomid growth reduction (based on the toxicity tests) and concentrations of cadmium, lead, selenium, and zinc indicate potential causative relationships for those four metals. Exceedances of AET values for these CoPCs occurred at one station in the North Tributary (station ECNT3) and the two stations in Eliza Creek just upstream and just downstream of the confluence with the North Tributary (Figure 20.19).

The potential ecological significance of the risks posed by surface water and sediment CoPCs (primarily cadmium, lead, selenium, and zinc) was not substantial throughout most of the site. Although toxicity was detected by the *Ceriodaphnia* test, no toxicity was observed in the fathead minnow test results

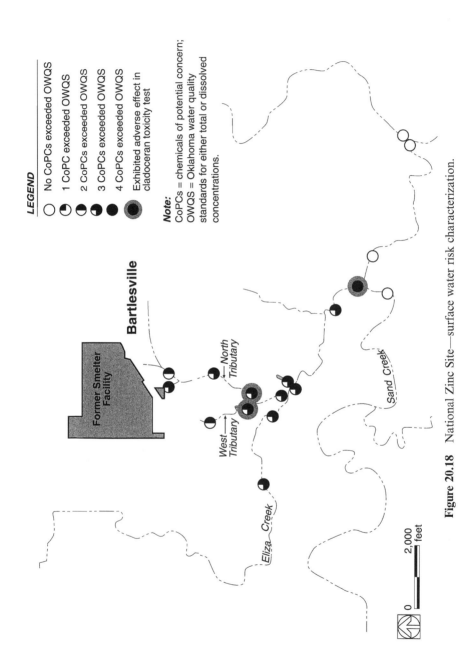

Figure 20.18 National Zinc Site—surface water risk characterization.

1144

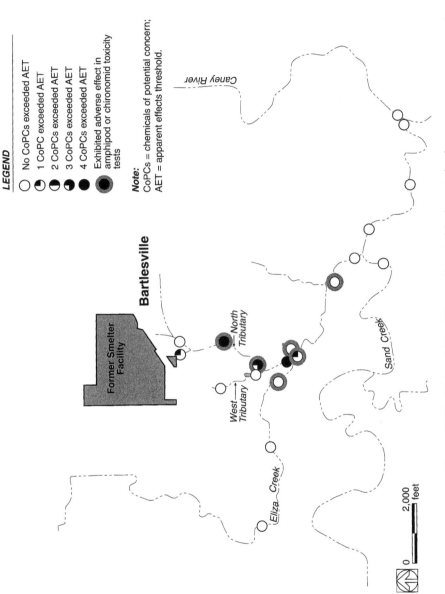

Figure 20.19 National Zinc Site—sediment risk characterization.

for surface water at any of the site stations. Only sublethal toxicity (i.e., reductions in chironomid growth) was found at more than two site stations, and these results may be related to factors other than CoPCs. Lethal toxicity (i.e., amphipod mortality) was found at only two stations (stations ECNT3 and POND1). In addition to CoPCs, high turbidity, siltation, and low concentrations of dissolved oxygen probably stress aquatic species to some degree in most portions of the study area. Thus, the quality of the natural physical habitat was generally low throughout the study area.

Moreover, the fish communities at the NZS were not adversely affected by CoPCs. Aside from the assemblage at station SABE2, the fish assemblages from the study site were similar to those at the reference sites. Dominant species at stations other than SABE2 included mosquito fish and bluegill. The relatively unique characteristics of the assemblage at station SABE2 (dominated by red shiner) were probably due to the increased flow, different substrate characteristics, and lack of predators (e.g., sunfish and bass) at Sand Creek relative to the other water bodies at which fish assemblages were evaluated. The fish community analysis is consistent with the results of the fathead minnow water toxicity test that showed no evidence of adverse effects of the site on fish.

Overall, the areas of concern identified at the NZS were determined based on a finding of significant toxicity in laboratory tests of surface water and sediments from the site, as supported by an evaluation of exceedances of Oklahoma water quality standards or sediment quality values (site-specific AET values).

Wood Preservative Site   Surface water (i.e., storm water) at the site was contaminated with metals, pentachlorophenol, PAHs, and PCDDs/PCDFs because of contamination in surface soils, storm water discharges, and chemical spills. Both surface and subsurface sediments near the site were contaminated with PAHs, in some areas to depths of up to 35 feet below the sediment surface. Limited contamination of sediments with chlorinated phenols, PCDDs/PCDFs, and arsenic had also occurred. Two major areas of sediments were contaminated: an area surrounding the creosote dock that extends along the shoreline and an area around the railroad bridge. These contaminated areas were near the site and can be clearly differentiated from other potential industrial sources of contaminants in the Willamette River.

Tissue samples collected near the site show slight elevations of PAHs (fish) and PCDDs/PCDFs (fish and crayfish) in muscle tissue compared with that of fish and crayfish collected in other industrialized areas of the Willamette River and at upstream and downstream reference areas in the Willamette River (PTI, 1992). Although the bioaccumulation of CoPCs in benthic macroinvertebrates was not measured at the MBS, the finding that PAHs and PCDDs/PCDFs are taken up by crayfish and fish suggests that these compounds are also available to benthic macroinvertebrates.

The results for the *Hyalella* and Microtox bioassays indicate a toxic response to sediment contamination in specific locations near the site and suggest a potential for adverse effects on sedentary benthic species in these locations.

The area of significant ($P \leq 0.05$) toxicity in the bioassays was confined to within approximately 300 feet of the shoreline, and the highly toxic area (i.e., with possibly acute lethal effects) was confined to within approximately 200 feet of the creosote dock (Figures 20.14 and 20.15). Cursory observations indicate that acute toxic effects on crayfish in this area may be associated with creosote releases from shoreline or underwater seeps (PTI, 1995).

Large-scale suckers collected near the site that were examined for histo-pathological lesions showed no adverse effects other than occasional mild inflammation that was also observed in fish collected in reference and other areas of the Willamette River. The data on liver histopathology of large-scale suckers collected as part of the remedial investigation and available histopathological data on carp collected at river mile 7 as part of a separate study (Curtis, 1991; Curtis et al., 1991) suggest that risk to fish populations attributable to chronic toxicity from contamination at the site was low. Acute toxicity to fish was unlikely because of the short-term nature of peak contamination events in the water column (e.g., from creosote seeps) and the possible avoidance by fish of sediments highly contaminated with PAH.

No evidence of adverse biological effects was found throughout most of the main channel of the river. Evaluation of sediment chemistry data for three PAH compounds relative to available EPA-proposed sediment quality criteria supports this conclusion (see below).

20.5.2 Comparison of CoPC Concentrations with Criteria

Comparisons of CoPC concentrations in surface water and sediments with criteria provide some perspective on possible ecological risks at these sites. Such assessments use either generic criteria (e.g., state standards for water quality or EPA-proposed sediment criteria) or site-specific values (e.g., AET or toxicity thresholds derived from concentration–response relationships).

Zinc Smelter Site Exceedances of the Oklahoma water quality standards were found at 14 stations in the study area. The most widespread exceedances were found for cadmium and zinc, whereas exceedances for lead and selenium were relatively localized. For cadmium, selenium, and zinc, the highest ratios to standards were found at station ECNT1, in the upper part of the North Tributary. In addition, exceedances for cadmium, selenium, and zinc were found throughout the North Tributary and showed strong relationships (i.e., declined) with downstream distance from that water body. By contrast, exceedances for lead did not show an apparent relationship with the North Tributary. Toxicity to *Ceriodaphnia* was largely confined to the North and lower West tributaries, although a significant adverse effect was found at station ECDS3 in downstream Eliza Creek. The relatively strong negative correlations found between *Ceriodaphnia* reproduction and concentrations of cadmium and zinc (and, to a lesser extent, selenium) indicate potential causative relationships for those metals.

TABLE 20.5   Comparison of Sediment Quality Values (SQVs) (mg/kg)

	EPA SQVs[a]	NOAA SQVs[b]	Ontario SQVs[c]		Site-Specific SQVs
	Effects Range— Low	Effects Range— Median	Lowest Effects Level	Severe Effects Level	Apparent Effects Threshold
Cadmium	1.2	9	0.6	10	138
Lead	47	110	31	250	692
Selenium	NA	NA	NA	NA	29.2
Zinc	150	270	120	820	12,000

Note: NA, not applicable; NOAA, National Oceanic and Atmospheric Administration.
[a] Long et al. (1995).
[b] Long and Morgan (1991).
[c] Persaud et al. (1992).

CoPCs in sediment exceeded site-specific AET values at five stations (Figure 20.19). Site-specific AET values were much higher than sediment quality values available in the literature (Table 20.5). Use of the generic sediment quality values [i.e., National Oceanic and Atmospheric Administration (NOAA) effects range-low or effects range-median; Ontario sediment quality values] instead of site-specific AETs would have greatly overestimated the risks associated with CoPCs at the NZS. Observed toxicity of sediments was lower than would have been predicted with the generic sediment quality values possibly because the bioavailability of site-related CoPCs was low relative to the bioavailability of corresponding contaminants in sediments used to develop the generic sediment quality values. Alternatively, technical uncertainties in the generic sediment quality values due to problems in the method of their derivation (see, e.g., Sampson et al., 1996a, b) may explain their poor predictive capability in this case.

Wood Preservative Site   Sampling stations where concentrations of three PAH compounds—acenaphthene, phenanthrene, and fluoranthene—in surface sediments exceeded preliminary draft EPA national sediment quality criteria are shown in Figure 20.20.* Exceedances of mean criteria and lower 95 percent confidence limits on the mean criteria are shown. Thirty-one of the 50 stations

---

*EPA sediment quality criteria are currently under development using the equilibrium partitioning method. Preliminary criteria have been proposed for three PAHs found at this site: acenaphthene (120 mg/kg OC), phenanthrene (140 mg/kg OC), and fluoranthene (1020 mg/kg OC). These criteria are consistent with EcoTox thresholds proposed by the EPA (1996) for screening sediment contaminants for potential risk to benthic organisms. However, EPA has not issued final criteria for acenaphthene, fluoranthene, and phenanthrene. Instead, EPA is developing a total PAH sediment criterion (see http://www.epa.gov/ost/cs/manage/strat5.html).

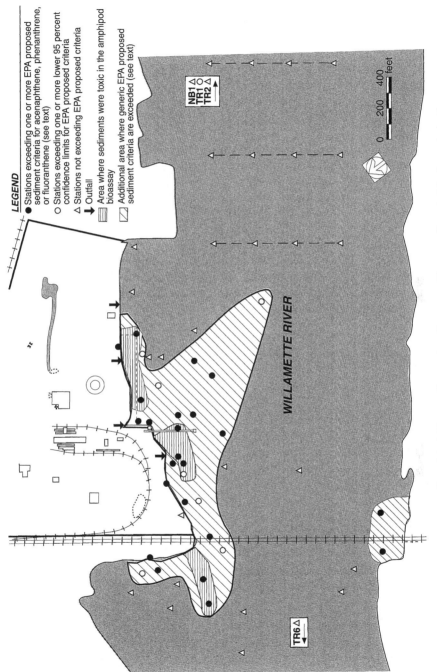

LEGEND

● Stations exceeding one or more EPA proposed sediment criteria for acenaphthene, phenanthrene, or fluoranthene (see text)

○ Stations exceeding one or more lower 95 percent confidence limits for EPA proposed criteria

△ Stations not exceeding EPA proposed criteria

➤ Outfall

▥ Area where sediments were toxic in the amphipod bioassay

▨ Additional area where generic EPA proposed sediment criteria are exceeded (see text)

NB1 △
TR1 ○
TR2 △

0   200   400
feet

WILLAMETTE RIVER

TR6 △

Figure 20.20 McCormick & Baxter Creosoting Company Site— sediment risk characterization.

1149

where PAHs were analyzed in samples of surface sediments showed exceedances of the sediment quality criteria. At 11 stations, the mean sediment quality criteria for all 3 PAH compounds were exceeded. These 11 stations included all of the stations where significant ($P \leq 0.05$) reductions in *Hyalella* survival were observed in sediment bioassays (Figure 20.14). The spatial patterns of sediment criteria exceedances were similar among the 3 PAH compounds (PTI, 1992). Among stations where sediment criteria exceedances were observed, 10 stations showed exceedances of only the lower 95 percent confidence limit of the mean sediment quality criteria for only one of the PAH compounds (usually phenanthrene).

The results of the above comparison of PAH concentrations with preliminary draft EPA sediment quality criteria suggest that benthic organisms within much of the MBS may be exposed to contaminant levels that could cause adverse effects. This conclusion was supported by the results of the Microtox bioassay, at least when relatively sedentary species were considered (i.e., those species that have a small home range relative to the size of the area where sediment criteria were exceeded). Nevertheless, site-specific conditions may mitigate concerns related to exceedances of generic sediment quality criteria (see Section 20.6.6), and results of the *Hyalella* sediment bioassays indicate that at least moderately sensitive species in direct contact with sediments may not experience acute lethal effects over much of the study area.

20.6 COMPARATIVE RISK CHARACTERIZATION

The NZS and the MBS represent contrasting case studies in ecological risk assessment (Table 20.6) because of the different sources of contamination and associated chemicals of concern: metals at the former site and creosote at the latter site. Differences in the chemicals of concern between the sites lead to very different transport pathways, which influence the spatial distribution of contamination and risk. Differences in physical features of the sites lead to some differences in ecological receptors and risk as well. Despite the bioaccumulation of chemicals of concern in invertebrate biota at both sites, bioaccumulation in fish was limited. Direct toxicity to benthic invertebrates was the primary concern at both sites, and risk to fish communities appears minimal.

20.6.1 Site Habitats and Receptors

The size of the lotic systems differs greatly between the sites; relatively small streams flow through the NZS (i.e., channels from a few feet across, including intermittent streams, to 100 feet across), but a major river is adjacent to the MBS (i.e., the tidal portion of the Willamette River, approximately 1800 feet in width near the site). Although crayfish were common at both sites, the benthic macroinvertebrate community was more abundant and diverse at the MBS owing to moderated environmental conditions associated with permanent

TABLE 20.6 Comparison of Elements of Ecological Risk Assessments at Former Zinc Smelter Site and Former Wood Preservative Site

Element	Zinc Smelter Site	Wood Preservative Site
Site location	Bartlesville, Oklahoma	Portland, Oregon
Aquatic habitats	Creeks and streams	Large river
Releases	Stormwater discharges of metals	Creosote spills and waste/ storm discharges
Chemicals of concern	Cd, Pb, Se, Zn	PAHs
Primary transport pathway	Surface runoff/sediment transport	NAPL seepage to sediment
Aquatic receptors (Birds, mammals, amphibians, and reptiles were addressed in the risk assessments but are not included in this comparison)	Benthic macroinvertebrates Fish Bluegill Mosquito fish Red shiner Brook silverside Longear sunfish Sand shiner Green sunfish	Benthic macroinvertebrates Fish Salmonids Largescale sucker Northern squawfish
Assessment endpoints	Fish species abundances and community structure; abundances and species richness of benthic macroinvertebrates	Fish health; abundances and species richness of benthic macroinvertebrates
Measurement endpoints Sediment toxicity Surface water toxicity	*Hyallela* survival *Chironomus* growth *Ceriodaphnia* reproduction *Pimephales* survival and growth	*Hyallela* survival Microtox bioluminescence None (not relevant owing to high flow of Wilamette River)
Fish health	None	Gross lesions and liver histopathology
Fish community	Abundance and community similarity indices	None (area too small in relation to receptor habitat)
Area of toxic sediment	From site downstream for more than 1 mile	Area immediately adjacent to site and within 300 ft of the shore

Note: PAHs, polycyclic aromatic hydrocarbons; NAPL, nonaqueous phase liquid.

stream flow. Because of the small size of the streams at the NZS, physical stressors such as siltation, flooding, and substrate modification probably play a much more important role in controlling the distribution and abundance of macroinvertebrates and fish than they do at the MBS. Nevertheless, the benthic macroinvertebrate community at the MBS may be disturbed by dredging and sediment turbulence caused by ship traffic. Thus, community analyses of macroinvertebrates was omitted from the assessment at both sites. The Willamette

River near the MBS is habitat for major commercial fish, including migratory species such as salmon. Only occasional recreational fishing occurs at the NZS.

20.6.2   Transport of CoPCs and Spatial Distribution of Risk

Cadmium, lead, selenium, and zinc were the chemicals of concern for the aquatic habitat at NZS, whereas PAHs were the main chemicals of concern at the MBS. The differences in the environmental chemistry of these very distinct groups of chemicals leads to important differences in their transport and fate between the two case sites and associated differences in the spatial distribution of risks.

Potential transport pathways for chemicals of concern in the aquatic and riparian habitats at the NZS include soil erosion and runoff during storm events with subsequent deposition into the North Tributary and Eliza Creek, shallow groundwater transport from beneath the NZS facility, historical overflows from the NZS storm water retention ponds into the North Tributary, and aerial transport of chemicals of concern from the NZS facility or other industrial/commercial operations along 14th Street. These transport pathways lead to transport of chemicals of concern approximately 1 mile downstream from the source.

In contrast to the NZS, where surface runoff plays a major role in the distribution of contaminants, subsurface transport of NAPLs and groundwater was the primary mode of transport at the MBS. NAPLs persist in the subsurface, and they can contaminate large volumes of groundwater during long periods of time. Moreover, dense NAPLs can move in a different direction than the groundwater, thereby increasing the complexity of the contaminant plume and the subsequent difficulty and cost of remediation. Subsurface soil contamination had migrated to depths of approximately 80 feet in some areas and from the source areas on-site into subsurface sediments of the Willamette River. Nevertheless, compared with surface water transport, subsurface transport was less extensive spatially in sediments and surface water.

20.6.3   Bioaccumulation

Tissue concentration data (presented in the RI reports) indicated that the bioavailability of the metals of concern at NZS appears to be limited (PTI, 1992). Despite some uptake by lower levels of the aquatic food web (i.e., algae and crayfish), bioaccumulation of cadmium, lead, selenium, and zinc in fish was limited. The high values for site-specific sediment AETs relative to generic sediment quality values (e.g., EPA, 1996; Long et al., 1995) indicate that bioavailability of these metals was low at the NZS. In contrast, PAHs from creosote at the MBS probably had relatively high bioavailability. Despite the ability of crayfish and fish to metabolize PAHs, selected PAH compounds were detected in crayfish and fish tissue at the MBS. Fishes are known to readily metabolize PAHs so bioaccumulation of the parent compounds was limited (Malins et al., 1984; Stein et al., 1984; Meador et al., 1995).

20.6.4   Spatial Extent of Risk

Risk to sediment-dwelling invertebrates was the main ecological concern at these two case study sites. Because of the difference in CoPCs and their main modes of transport at the two sites, the spatial extent of risk varied between the sites. At the NZS, surface water and sediments collected from stations as far as 1 mile downstream of the site were toxic to aquatic invertebrates (*Ceriodaphnia* and *Chironomus*). Because of rapid dilution of PAHs and other CoPCs from the MBS in the Willamette River (which is much larger than any of the creeks at the NZS), surface water toxicity was assumed to be unimportant, and the distribution of PAHs in sediment was the primary factor determining the areas of concern. Moreover, the relatively limited spatial distribution of creosote NAPLs at the MBS resulted in a more limited area of toxic sediments (within 300 feet of the shore adjacent to the site) than at the NZS (1 mile downstream from the source).

Of course, the spatial extent of the distribution of sediment toxicity between dissimilar sites like the NZS and the MBS depends not only on the differences in transport pathways but also on the differences in the concentrations, bioavailability, and inherent toxicity of CoPCs between the sites. Additional comparisons of the spatial extent of risk between sites with different chemicals of concern and different mechanisms of release and transport would provide useful information for designing assessment approaches for future ecological risk assessments.

20.6.5   Risk Assessment Approach

At both sites, a relatively high level of agreement was found between the various chemical and biological indicators and the general distribution of chemicals of concern among major segments of the study areas. Toxicity testing of sediments was a key element of the ecological risk assessment at each site. By using site-specific AET values and sediment toxicity data, significant areas of concern were delineated at the NZS. Use of generic sediment quality values [i.e., EPA (1996) EcoTox values] would have grossly overestimated risks associated with chemicals of concern at the site. Similarly, results of the daphnid and fathead minnow toxicity tests showed that use of Oklahoma state water quality standards would greatly overestimate risk. At the MBS, the amphipod and Microtox toxicity test results indicated that a relatively small area of river sediments was likely to be toxic to benthic biota compared with the area in which generic sediment criteria for PAHs were exceeded. For LPAH, HPAH, and benzo[*a*]pyrene at the MBS, sediment PRGs were expressed as effect thresholds on the respective concentration–response curve rather than as AETs because the PAH mixture was considered responsible for the toxicity and individual compounds have a similar mode of action. In contrast, the potentially different modes of action of the metals that were primary CoPCs at the NZS required that an AET approach be used instead of an approach incorporating an effect threshold on the concentration–response curve.

Toxicity testing of surface water was included in the assessment at the NZS but not at the MBS partly because of the differences in flow between the small creeks at NZS and the large river at the MBS. The Willamette River at the MBS has a much greater capacity to dilute contaminants quickly, and given the transport pathways at the MBS, the potential for surface water toxicity was judged to be of much lower concern than sediment toxicity.

20.6.6 Uncertainties

Despite some uncertainties typical of any ecological risk assessment, the use of multiple indicators of CoPC concentrations (i.e., sediment, water, and tissue concentrations) and adverse biological effects (water toxicity and sediment toxicity) provides a relatively high level of confidence in the results of the aquatic risk assessment for both the NZS and MBS.

The main uncertainties associated with these assessments result from the use of toxicity testing as the primary assessment tool. The amphipod (*Hyalella*) sediment toxicity test is known to have a moderate to high sensitivity relative to a variety of other freshwater sediment bioassays based on results of comparison tests conducted with Great Lakes sediments contaminated by a variety of toxic chemicals, including PAH (Burton, 1991). Because *Hyalella* may not be the most sensitive indicator of the toxicity of the contaminants of concern, the bioassay results should not be interpreted as the worst-case toxicological response. At the NZS, the chironomid growth test provided a more sensitive indicator of ecological effects.

The Microtox pore water test was chosen to supplement the *Hyalella* bioassay at the MBS because it is an extremely sensitive indicator of sublethal effects of contaminants, it provides an environmentally protective estimate of the areal extent of contaminated sediments, and it is relatively economical. Although the relationship of toxicity found by the Microtox test with effects on species found in the lower Willamette River was unknown, strong correlations between Microtox results and those of freshwater bioassays using invertebrates or fish species have been demonstrated in laboratory tests with individual chemicals, mixtures of chemicals, and complex effluents (Bulich et al., 1981; Lebsack et al., 1981; Curtis et al., 1982; Ribo and Kaiser, 1987). Nevertheless, the test may be overly sensitive to some chemicals and may be more suitable as a biomarker of exposure than as a measure of adverse ecological effects.

Finally, the response of individual species to sediment contaminants does not provide a direct measure of the response at the population, community, or ecosystem level. In combination, however, the *Hyalella* and chironomid tests at the NZS or the *Hyalella* and Microtox tests at the MBS provide a sensitive indicator of organism responses to contaminated sediments. The assessments at both sites may have benefited by application of the triad approach (e.g., Chapman et al., 1992) whereby a weight of evidence is considered using sediment toxicity tests, sediment chemistry, and benthic macroinvertebrate community analyses. However, at both of these sites, macroinvertebrate community analysis would have been confounded by physical stressors (e.g., substrate type and physical disturbances).

At the MBS, fish liver histopathological analysis was used as a sensitive indicator of fish health because of the known histopathological responses of fish to PAHs (Malins et al., 1984; Krahn et al., 1986). Fish community structure was not assessed because of the difficulty of conducting quantitative field surveys of fish community structure in a large river and the large spatial range of most fish species relative to the contaminated area of the site. However, liver histopathological conditions may actually represent biomarkers of exposure (Myers et al., 1998) and not adverse effects on individual fish or fish populations. Although links between liver lesions and reproductive effects in English sole have been established in PAH-contaminated areas (Johnson, 2000), the relationship between the prevalence of liver lesions and reproductive effects is probably correlative, not causative

20.7   REMEDIAL ACTIVITIES

At NZS, source controls were implemented at the former smelter site. A 1-mile stretch of the unnamed tributary (i.e., the North Tributary) that flows into Eliza Creek was dewatered and a portion of the sediments were excavated and then hauled to a landfill. The sediments in the West Tributary, as well as Eliza Creek, were allowed to recover naturally.

At MBS, source control activities (i.e., NAPL pumping) have been in place for about 5 years. After NAPL migration ceases, the Oregon Department of Environmental Quality is planning to dredge some of the near-shore sediments with the highest concentrations of PAHs and to place caps over moderately contaminated areas. The agency is using the results of the sediment toxicity tests to help delineate the dredging and capping areas.

20.8   CONCLUSION

These ecological risk assessment case studies demonstrate the use of ecological endpoints in remedial decision making. In both cases, site-specific sediment guidelines were derived from toxicity tests, which, with the chemical concentration data, were used to delineate areas of concern. The use of sediment or water chemistry data combined with bioaccumulation data, toxicity test results, and community data provides a valuable assessment approach (Chapman et al., 1992). At the NZS and the MBS, macroinvertebrate community analysis would have been severely confounded by physical stresses independent of the sites. Fish community analysis was applied at the NZS. Because of the importance of PAHs as CoPCs at the MBS and the small size of the site relative to the ranges of indigenous fishes, fish histopathological analysis was applied instead of community analysis. Ecologically significant adverse effects on fish populations and communities were not found. In each case, bioaccumulation and effects at higher trophic levels were limited.

Comparative risk characterization is a valuable tool for gaining insight into the relative value of assessment tools such as toxicity tests, site-specific criteria, and community surveys, as well as into the relative risks associated with specific chemicals and types of contaminated sites. Comparison of the behavior of chemicals of concern in the environment leads to insights about the spatial scale of potential effects. In general, surface water transport of contaminants yields more extensive areas of contamination than subsurface transport, especially in the case of relatively insoluble materials that form NAPLs (e.g., creosote).

Uncertainties in the results of these two risk assessments relate primarily to extrapolations among species and extrapolations from organism-level effects to population- or community-level effects. The most important uncertainties were related to the extrapolation of laboratory bioassay responses to the population or community responses of benthic macroinvertebrates in the river. In these cases, because the areas of high risk to benthic organisms were well delineated and risks to fish were negligible, decisions could be made based on the direct results of toxicity tests. Nevertheless, in more complex cases, toxicity testing alone may be insufficient. In such cases, toxicity testing provides information for focused surveys of biological communities or for population modeling to assess the ecological significance of risk.

ACKNOWLEDGEMENTS

This chapter is based partly on the ecological risk assessment reports prepared for the National Zinc Site and the McCormick & Baxter Creosoting Company Site. We thank Ron Mellott and Scott Becker for their contributions to the ecological risk assessment at the National Zinc Site, and we thank Rob Barrick, Jennifer Sampson, Greg Linder, and Michael Jacobson for their contribution to the ecological risk assessment at the McCormick & Baxter Creosoting Company Site.

REFERENCES

American Society for Testing and Materials (ASTM). 1998. Standard test methods for measuring the toxicity of sediment-associated contaminants with freshwater invertebrates. E1706-95b. In *Annual Book of Standards*, Vol. 11.05. ASTM, Philadelphia, PA.

Ankley, G. T., Phipps, G. L., Leonard, E. N., Benoit, D. A., Mattson, V. R., Kosian, P. A., Cotter, A. M., Dierkes, J. R., Hansen, D. J., and Mahony, J. D. 1991. Acid-volatile sulfide as a factor mediating cadmium and nickel bioavailability in contaminated sediment. *Environ. Toxicol. Chem.* 10: 1299–1307.

Barnthouse, L. W., and Suter, G. W. (Eds.). 1986. *User's Manual for Ecological Risk Assessment.* Environmental Sciences Division Publication No. 2679. U.S. Environmental Protection Agency, Office of Research and Development, Washington, DC, and Oak Ridge National Laboratory, Oak Ridge, TN.

Barrick, R., Becker, S., Brown, L., Beller, H., and Pastorok, R. 1988. *Sediment Quality Values Refinement: 1988 Update and Evaluation of Puget Sound AET*, Vol. 1: *Final Report.* Prepared for Tetra Tech and U.S. Environmental Protection Agency Region 10, Office of Puget Sound, Seattle, WA. PTI Environmental Services, Bellevue, WA.

Baumann, P. C., and Harshbarger, J. C. 1985. Frequencies of liver neoplasia in a feral fish population and associated carcinogens. *Mar. Environ. Res.* 17: 324–327.

Baumann, P. C., Mac, M. J., Smith, S. B., and Harshbarger, J. C. 1991. Tumor frequencies in walleye (*Stizostedion vitreum*) and brown bullhead (*Ictalurus nebulosus*) and sediment contaminants in tributaries of the Laurentian Great Lakes. *Can. J. Fish. Aquat. Sci.* 48: 1804–1810.

Baumann, P. C., Smith, W. D., and Ribick, M. 1982. Hepatic tumor rates and polynuclear aromatic hydrocarbon levels in two populations of brown bullhead (*Ictalurus nebulosus*). In M. Cooke, A. J. Dennis, and G. L. Fisher (Eds.), *Polynuclear Aromatic Hydrocarbons: Physical and Biological Chemistry.* Sixth International Symposium. Battelle Press, Columbus, OH, and Richland, WA. Springer-Verlag, New York, pp. 93–102.

Beckman Instruments. 1982. *Microtox System Operating Manual.* Beckman Instruments, Carlsbad, CA.

Brook, B. W., O'Grady, J. J., Chapman, A. P., Burgman, M. A., Akçakaya, H. R., and Frankham, R. 2000. Predictive accuracy of population viability analysis in conservation biology. *Nature* 404: 385–387.

Bulich, A. A., Greene, M. W., and Isenberg, D. L. 1981. Reliability of the bacterial luminescence assay for determination of the toxicity of pure compounds and complex effluent. In D. R. Branson and K. L. Dickson (Eds.), *Aquatic Toxicology and Hazard Assessment: Proceedings of the Fourth Annual Symposium.* ASTM STP 737. American Society for Testing and Materials, Philadelphia, PA.

Burton, G. A., Jr. 1991. Assessing the toxicity of freshwater sediments. *Environ. Toxicol. Chem.* 10: 1585–1627.

Chapman, P. M., Power, E. A., and Burton, Jr., G. A. 1992. Integrative assessments in aquatic ecosystems. In G. A. Burton, Jr. (Ed.), *Sediment Toxicity Assessment.* Lewis, Chelsea, MI, pp. 313–340.

Clifford, P. A., Barchers, D. E., Ludwig, D. F., Sielken, R. L., Klingensmith, J. S., Graham, R. V., and Banton, M. I. 1995. An approach to quantifying spatial components of exposure for ecological risk assessment. *Environ. Toxic. Chem.* 14(5): 895–906.

Curtis, C., Lima, A., Lozano, S. J., and Veith, G. D. 1982. Evaluation of a bacterial luminescent bioassay as a method for predicting acute toxicity of organic chemicals to fish. In J. G. Pearson, R. B. Foster, and W. E. Bishop (Eds.), *Aquatic Toxicology and Hazard Assessment: Fifth Conference.* ASTM STP 766. American Society for Testing Materials, Philadelphia, PA, pp. 170–178.

Curtis, L. R. 1991. Update of progress for the Willamette River Toxic Study. January 10, 1991. Oregon State University, Corvallis, OR.

Curtis, L. R., Deinzer, M. L., Williams, D. G., and Hedstrom, O. R. 1991. Toxicity and longitudinal distribution of persistent organochlorines in the Willamette River. Prepared for Oregon Department of Environmental Quality. Oregon State University, Corvallis, OR.

Hughes, R. M., and Gammon, J. R. 1987. Longitudinal changes in fish assemblages and water quality in the Willamette River, Oregon. *Trans. Am. Fish. Soc.* 116: 196–209.

Johnson, L. 2000. *An Analysis in Support of Sediment Quality Thresholds for Polycyclic Aromatic Hydrocarbons (PAHs) to Protect Estuarine Fish.* National Oceanic and Atmospheric Administration, National Marine Fisheries Service, Seattle, WA.

Krahn, M. M., Rhodes, L. D., Myers, M. S., Moore, L. K., MacLeod, Jr., W. D., and Malins, D. C. 1986. Associations between metabolites of aromatic compounds in bile and occurrence of hepatic lesions in English sole (*Parophrys vetulus*) from Puget Sound, Washington. *Arch. Environ. Contam. Toxicol.* 15: 61–67.

Krebs, C. J. 1989. *Ecological Methodology.* Harper & Row, New York.

Lebsack, M. E., Anderson, A. D., DeGraeve, G. M., and Bergman, H. L. 1981. Comparison of bacterial luminescence and fish bioassay results for fossil-fuel process waters and phenolic constituents. In D. R. Branson and K. L. Dickson (Eds.), *Aquatic Toxicology and Hazard Assessment: Fourth Conference.* ASTM STP 737. American Society for Testing and Materials, Philadelphia, PA, pp. 348–356.

Long, E. R., and Morgan, L. G. 1991. *The Potential for Biological Effects of Sediment-Sorbed Contaminants Tested in the National Status and Trends Program.* NOAA Technical Memorandum NOS OMA 52. National Oceanic and Atmospheric Administration, Washington, DC.

Long, E. R., MacDonald, D. D., Smith, S. L., and Calder, F. D. 1995. Incidence of adverse biological effects within ranges of chemical concentrations in marine and estuarine sediments. *Environ. Manag.* 19(1): 81–97.

Malins, D. C., McCain, B. B., Brown, D. W., Chan, S.-L., Myers, M. S., Landahl, J. T., Prohaska, P. G., Friedman, A. J., Rhodes, L. D., Burrows, D. G., Gronlund, W. D., and Hodgins, H. O. 1984. Chemical pollutants in sediments and diseases of bottom-dwelling fish in Puget Sound, Washington. *Environ. Sci. Technol.* 18: 705–713.

Meador, J. P., Stein, J. E., Reichert, W. L., and Varanasi, U. 1995. Bioaccumulation of polycyclic aromatic hydrocarbons by marine organisms. *Rev. Environ. Contamin. Toxicol.* 143: 79–163.

Myers, M. S., Johnson, L. L., Hom, T., Collier, T. K., Stein, J. E., and Varanasi, U. 1998. Toxicopathic hepatic lesions in subadult English sole (*Pleuronectes vetulus*) from Puget Sound, Washington, USA: Relationships with other biomarkers of contaminant exposure. *Marine Environ. Res.* 45: 47–67.

Nestler, F. H. M. 1974. *The Characterization of Wood-Preserving Creosote by Physical and Chemical Methods of Analysis.* FPL 195; U.S. Government Printing Office 1974-754-556/82. U.S. Department of Agriculture Forest Service, Forest Products Laboratory, Madison, WI.

Pastorok, R. A., and Becker, D. S. 1989. *Comparison of Bioassays for Assessing Sediment Toxicity in Puget Sound.* EPA-910/9-89-004. Prepared for U.S. Environmental Protection Agency, Puget Sound Estuary Program, Seattle, WA. PTI Environmental Services, Bellevue, WA.

Pastorok, R. A., Bartell, S. M., Ferson, S., and Ginzburg, L. R. 2001. *Ecological Modeling in Risk Assessment: Chemical Effects on Populations, Ecosystems, and Landscapes.* CRC Press, Lewis, Boca Raton, FL.

Pastorok, R. A., Peek, D. C., Sampson, J. R., and Jacobson, M. A. 1994. Ecological risk assessment for river sediments contaminated by creosote. *Environ. Toxicol. Chem.* 13(12): 1929–1941.

Persaud, D., Jaagumagi, R., and Hayton, A. 1992. *Guidelines for the Protection and Management of Aquatic Sediment in Ontario.* Water Resources Branch, Ontario Ministry of Environment. Queen's Printer for Ontario, Toronto, Ontario, Canada.

PTI. 1992. *McCormick & Baxter Creosoting Company Remedial Investigation Report,* Vol. I. Prepared for Oregon Department of Environmental Quality, Environmental Cleanup Division, Portland, OR. PTI Environmental Services, Bellevue, WA.

PTI. 1995. *National Zinc Site Remedial Investigation and Feasibility Study,* Vol. IV. Ecological Risk Assessment, Operable Unit 2. Prepared for City of Bartlesville, Cyprus Amax Minerals Company and Salomon. PTI Environmental Services, Bellevue, WA.

Ribo, J. M., and Kaiser, K. L. E. 1987. *Photobacterium phosphoreum* toxicity bioassay. I. Test procedures and applications. *Toxicity Assess.* 2: 305–323.

Sampson, J. R., Pastorok, R. A., and Ginn, T. C. 1996a. ER-L and ER-M values should not be used to assess contaminated sediments. *SETAC News* 16(4): 29–31.

Sampson, J. R., Pastorok, R. A., and Ginn, T. C. 1996b. Response to MacDonald et al. (1996) concerning the derivation and use of ERL and ERM values. *SETAC News* 16(6): 19–21.

SPSS. 1988. *SPSS/PC+ Advanced Statistics,* Version 2.0. SPSS, Chicago, IL.

Stein, J. E., Hom, T., and Varanasi, U. 1984. Simultaneous exposure of English sole (*Parophrys vetulus*) to sediment-associated xenobiotics: Part 1—uptake and deposition of 14C-polychlorinated biphenyls and 3H-benzo[*a*]pyrene. *Mar. Environ. Res.* 13: 97–119.

Suter, II, G. W. (Ed.). 1993. *Ecological Risk Assessment.* Lewis, Boca Raton, FL.

Suter, II, G. W., Efroymson, R., Sample, B. E., and Jones, D. S. 2000. *Ecological Risk Assessment for Contaminated Sites.* Lewis, Boca Raton, FL.

U.S. Environmental Protection Agency (EPA). 1989a. *Interim Procedures for Estimating Risks Associated with Exposures to Mixtures of Chlorinated Dibenzo-p-Dioxins and Dibenzofurans (CDDs and CDFs) and 1989 Update.* EPA/625/3-89/016. EPA, Risk Assessment Forum, Washington, DC.

U.S. Environmental Protection Agency (EPA). 1989b. *Short-Term Methods for Estimating the Chronic Toxicity of Effluents and Receiving Waters to Freshwater Organisms,* 2nd ed. EPA 600/4-89/001. EPA, Environmental Monitoring Systems Laboratory, Cincinnati, OH.

U.S. Environmental Protection Agency (EPA). 1992. *Framework for Ecological Risk Assessment.* EPA/630/R-92/001. EPA, Risk Assessment Forum, Washington, DC.

U.S. Environmental Protection Agency (EPA). 1994. *Short-Term Methods for Estimating the Chronic Toxicity of Effluents and Receiving Waters to Freshwater Organisms.* EPA/600/4-91/002. EPA, Environmental Monitoring Systems Laboratory, Cincinnati, OH.

U.S. Environmental Protection Agency (EPA). 1996. *Ecotox thresholds. ECO Update.* EPA 540/F-95/038. *Intermittent Bulletin*, Vol. 3, No. 2. EPA, Office of Emergency and Remedial Response, Washington, DC.

U.S. Environmental Protection Agency (EPA). 1997. *Chemistry Assistance Manual for Premanufacture Notification Submitters.* EPA 744-R-97-003. EPA, Office of Pollution Prevention and Toxics, Washington, DC.

U.S. Environmental Protection Agency (EPA). 1998. Guidelines for ecological risk assessment; notice. *Fed. Reg.* 63(93): 26846–26924.

Walker, M. K., and Peterson, R. E. 1991. Potencies of polychlorinated dibenzo-*p*-dioxin, dibenzofuran, and biphenyl congeners, relative to 2,3,7,8-tetrachlorodibenzo-*p*-dioxin for producing early life stage mortality in rainbow trout (*Oncorhynchus mykiss*). *Aquat. Toxicol.* 21: 219–238.

Ward, D. L., Connolly, P. J., Farr, R. A., and Nigro, A. A. 1987. *Feasibility of Evaluating the Impacts of Waterway Development on Anadromous and Resident Fish in Portland Harbor.* Oregon Department of Fish and Wildlife, Portland, OR.

West and Gulley, D. D. 1994. *Toxstat Version 3.4.* Western Ecosystems Technology, Cheyenne, WY.

Wolda, H. 1981. Similarity indices, sample size and diversity. *Oecologia* 50: 296–302.

21 Integration of Risk Assessment and Natural Resource Damage Assessment: Case Study of Lavaca Bay

KRISTY E. MATHEWS

Triangle Economic Research, Durham, North Carolina

KIRK J. GRIBBEN

Alcoa, Pittsburgh, Pennsylvania

WILLIAM H. DESVOUSGES

Triangle Economic Research, Durham, North Carolina

21.1 INTRODUCTION

In December, 1980, Congress enacted the Comprehensive Environmental Response, Compensation, and Liability Act (CERCLA). This act provides for the identification and cleanup of hazardous waste sites. CERCLA empowers the U.S. Environmental Protection Agency (EPA) to conduct a remedial investigation/feasibility study (RI/FS) of Superfund sites. The remedial investigation (RI) process characterizes and assesses the nature and extent of risk to human health and the environment posed by the hazardous substance release. Much of this book is devoted to implementing human health or ecological risk assessments under CERCLA. Section 107 of CERCLA also authorizes government agencies (hereafter, trustees) to recover natural resource damages that result from hazardous substance releases. These damages include compensation for lost natural resource services, such as habitat services and recreational fishing. The Natural Resource Damage Assessment (NRDA) regulations promulgated

Human and Ecological Risk Assessment: Theory and Practice, Edited by Dennis J. Paustenbach
ISBN 0-471-14747-8 © 2002 John Wiley & Sons, Inc.

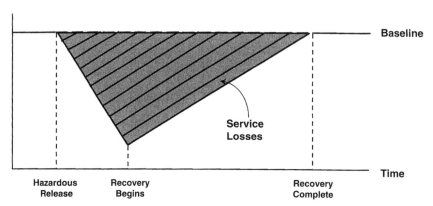

Figure 21.1   Recreational service losses under NRDA.

by the Department of the Interior (DOI) describe a process for estimating natural resource damages [43 Code of Federal Regulations (CFR) 11].*

The goal of this NRDA process is to make the public whole for injuries to natural resources and losses in natural resource services. An injury, as defined by the NRDA regulations, is an observable or measurable adverse change in a natural resource. Natural resource services may be classified into two broad categories. Ecological services are the physical, chemical, and biological functions that one natural resource provides for another resource. Examples include provision of food and nesting habitat. Human-use services refer to the functions that natural resources provide for the public, such as fishing and hunting. The NRDA process measures natural resource services relative to baseline, or the level of services that natural resources would provide in the absence of the injury. If the injury results in a loss of natural resource services, then the public is entitled to compensation for those losses.†

Figure 21.1 illustrates the conceptual basis for NRDA, using recreational fishing as an example. The vertical axis shows the level of fishing, and the horizontal axis shows the passage of time. In this figure, fishing services decline when the injury occurs. For example, a ban on fish consumption is an injury that may result in a decrease in recreational fishing. Over time the fishing ban may become a consumption advisory. Because a consumption advisory is less stringent than a fishing ban, fishing levels may begin to increase, as Figure 21.1 indicates. When fishing services return to baseline levels, the level of services in

---

*The National Oceanic and Atmospheric Administration (NOAA) has also promulgated NRDA regulations for oil spills (15 CFR 990).
†The NRDA regulations for hazardous substance releases (43 CFR 11) and for oil spills (15 CFR 990) describe the NRDA process. For more information on the NRDA process, see References 1 to 3. See Reference 4 for an example of an NRDA for recreational fishing.

the absence of a consumption ban or advisory, then fishing services have been fully restored. The shaded area indicates the service loss that occurred over time. The objectives of the NRDA recreation fishing analysis are to estimate the fishing service losses and determine fair compensation to the public. This compensation can be monetary or in the form of in-kind compensation (restoration projects that enhance or expand fishing services).

At some CERCLA sites, the data needs to support a risk assessment under the RI and injury and damage assessments under the NRDA regulations require similar, but not identical, types of data. Both the RI and the NRDA frameworks permit potentially responsible parties (PRPs) to conduct these assessments, subject to agency and trustee oversight. Because the PRPs ultimately bear the reasonable costs of both assessments, they may find it in their interest to integrate the planning and data collection themselves. If the timing of the two processes is similar, combining the data needs for both the risk assessment and the NRDA may provide opportunities for reducing the total cost of data collection, as well as reducing analyses costs.[5] In addition, PRPs may find management efficiencies from developing an integrated process. The total number of reviews and decisions associated with the two types of assessments studies can be reduced substantially. Finally, PRPs may find that the integrated process is beneficial to reducing the long-term litigation risk associated with NRDA. This chapter presents a case study of combining data needs for a human health risk assessment and an NRDA for recreational fishing losses for the Lavaca Bay National Priorities List (NPL) site in Texas.*

21.2 BACKGROUND ON LAVACA BAY SITE

Lavaca Bay is located on the Texas Gulf Coast between Corpus Christi and Houston. It is part of the larger Matagorda Bay system, which includes Carancahua Bay, Turtle Bay, and Tres Palacios Bay. The Lavaca Bay system consists of Lavaca Bay and several smaller bays such as Cox Bay, Keller Bay, and Chocolate Bay (see Figure 21.2). The Lavaca Bay system covers approximately 64 square miles and offers many fishing opportunities for recreational saltwater anglers. There are fishing access sites in Point Comfort, upper Lavaca Bay, Port Lavaca, Chocolate Bay, Magnolia Beach, and Keller Bay. These sites differ in terms of facilities, type of fishing, and available species of fish.

Local saltwater anglers are the primary users of the Lavaca Bay system. Texas Parks and Wildlife Department (TPWD) creel survey data from 1975 to 1991 show that an annual average of 74 percent of Lavaca Bay anglers originated from the three nearest counties: Calhoun, Jackson, and Victoria.[6]

TPWD creel survey data from 1975 to 1991 also reveal that the top four finfish species caught in the Lavaca Bay system are spotted sea trout (41 per-

---

*The potential ecological service losses evaluated in the NRDA for this site are outside the scope of this chapter and are not discussed here.

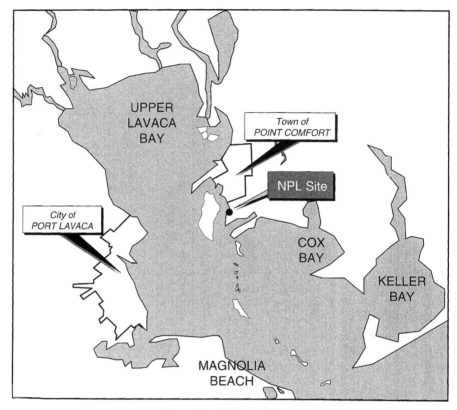

Figure 21.2 Lavaca Bay.

cent), sand sea trout (14 percent), red drum (12 percent), and black drum (11 percent). Combined, these species totaled 78 percent of all catches from 1975 to 1991. Atlantic croaker, southern flounder, gafftopsail catfish, blue catfish, and sheepshead comprise nearly all of the remaining species caught. According to the creel survey data, recreational shellfishing is not popular in Lavaca Bay as crabs, oysters, and shrimp combined for less than 1 percent of the total number of catches.[6,*]

The NPL site is located in Point Comfort, Texas, and encompasses Alcoa's Point Comfort Operations (PCO). PCO covers about 3500 acres of land and is located on the eastern shore of Lavaca Bay, just to the northwest of Cox Bay. Dredge Island is an island in Lavaca Bay, located just west of the facility buildings, that is approximately 375 acres. Historically Dredge Island provided for the disposal of dredge material, gypsum, and wastewater from the chlor-alkali processing unit.

The PCO facility has operated since 1948, when it began operation as an aluminum smelter. The smelting operation utilized alumina as the raw material

*The TPWD creel survey is described in References 7 and 8.

and produced aluminum metal through an electrolytic process. The alumina refining operation, begun in 1959 and continuing today, utilizes bauxite ore to produce alumina. PCO facility operations that have been dismantled and removed include the aluminum smelter, a cryolite plant, a coal tar processing plant, a gas plant, and a chlor-alkali processing unit.

From 1966 to 1970, the PCO facility discharged mercury-containing wastewater into Lavaca Bay from its chlor-alkali processing operations. Alcoa terminated the direct discharge of this wastewater in 1970 after the Texas Water Quality Board notified Alcoa of potential adverse environmental impacts associated with the discharge. Through tissue sampling of aquatic organisms, the Texas Department of Health (TDH) determined that mercury concentrations in finfish and blue crabs near the Alcoa plant were high enough to pose risks to human health. The TDH thus enacted a ban on April 21, 1988, prohibiting the taking of finfish and crabs in waters around the PCO plant. In January, 2000, the TDH reopened the Cox Bay portion of the closure area to recreational fishing.

Figure 21.3 shows the area affected by the 1988 and the 2000 closure orders. The 1988 closure area covers approximately 8 square miles adjacent to the Alcoa plant, and catch-and-release fishing in this area is permitted. Figure 21.3 also shows the different spatial zones, based on sediment concentration of mercury and the relationship to the mercury source, used in the RI analyses. Figure 21.4 depicts the differences in mercury concentrations for the zones within the closure area, as well as the open waters of Lavaca Bay.

The site was placed on the CERCLA NPL in 1994. Alcoa, the State of Texas, and the EPA signed an Administrative Order on Consent under CERCLA in March, 1994, for the conduct of an RI/FS for the site. In 1995, TDH conducted a public health assessment for Lavaca Bay that included a quantitative assessment of potential public health risks from eating contaminated finfish and shellfish from Lavaca Bay.[9] This report recommended conducting a site-specific consumption study of recreational anglers for the site. Remedial activities began in 1995. In 1996, Alcoa and the natural resource trustees began discussions of engaging in a joint NRDA process that would result in a settlement of Alcoa's NRD liability. All parties signed a Memorandum of Agreement in January, 1997.

The timing of the NRDA process coincided with planning for the RI studies. Thus, Alcoa, the EPA, and the trustees agreed to look for opportunities where similar data needs for the RI process and the NRD process would permit cost savings through joint data collections. It was recognized early in the RI process that the most significant human health risk issue is consumption of contaminated fish. A preliminary fish consumption survey[10] identified recreational anglers as a potential subgroup of concern. Moreover, Alcoa and the trustees agreed that the NRDA would include a component to address recreational fishing losses because the TDH closure order constitutes injury under the NRD regulations (43 CFR 11.62). Thus, the specific processes employed at this site permitted consideration of a joint data collection through a fish consumption

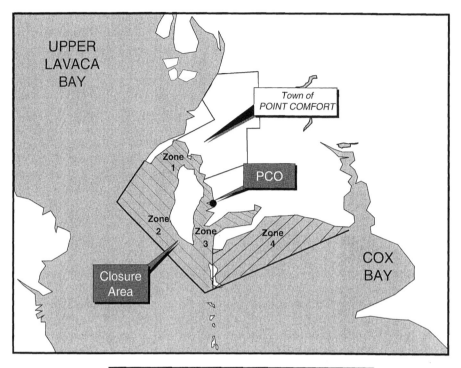

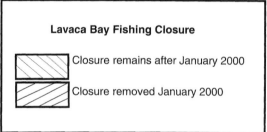

Figure 21.3   Closure area in Lavaca Bay.

study of recreational anglers to provide data for both the human health risk assessment and the NRD recreational fishing study.

21.3 STUDY OBJECTIVES

The principal objective for the fish consumption study was to obtain the necessary information to assess potential risk from ingesting contaminated finfish and shellfish from Lavaca Bay. Chronic exposure to chemicals in contaminated fish and shellfish is typically quantified by an intake or dose with the equation[11]

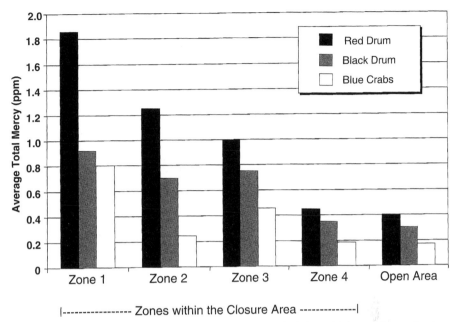

Figure 21.4 Spatial differences of total mercury in finfish and blue crab monitoring data from Lavaca Bay, 1996 to 1999.

$$\text{Intake (mg/kg/day)} = \frac{CF \times IR \times ABS \times FI \times EF \times ED \times AF}{BW \times AT}$$

where CF = concentration of contaminant in fish (mg/kg)

     IR = ingestion rate of fish (g/meal)

  ABS = absorption factor (100%)

     FI = fraction ingested of fish from contaminated source (unitless)

    EF = exposure frequency (days/year)

    ED = exposure duration (years)

    AF = adjustment factor (0.001 kg/g)

  BW = body weight (kg)

   AT = averaging time (days over the exposure period)

The CF can be estimated from previously collected analytical data[12] for various finfish and shellfish species. These data indicate that sport fish (red drum, black drum, spotted seatrout) show higher levels of mercury than species that are commercially harvested in Lavaca Bay (shrimp, crabs, oysters). The finfish and shellfish data also show significant variability in average mercury levels between species (Figure 21.5). Besides the differences in fish tissue concentrations, creel survey data by the TPWD[6] show that anglers target different spe-

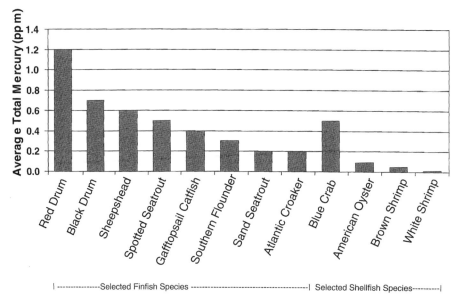

Figure 21.5   Total mercury body burdens for selected species of finfish and shellfish in the closure area of Lavaca Bay, 1996 to 1999.

cies. Moreover, some species (e.g., red drum) have size limits as well as bag limits that do not exist for other sport fish species. All of these factors suggest that fishing pressure and consumption rates differ between species. To understand the mercury concentrations in fish that recreational anglers are exposed to on average (CF), the consumption survey had to determine the species composition that anglers preferentially caught and consumed. Having reliable consumption information at a species level results in a more accurate risk assessment.

Besides an understanding of the mercury levels in fish and species composition, the information considered important to support the risk assessment also included the ingestion rate (IR). Because the objective was to assess risk from chronic exposure, the IR was normalized in units of grams per day. Without a site-specific consumption study, a risk assessment for fish consumption typically relies on published survey data, as summarized by EPA[13] to estimate a default consumption rate. For marine systems, EPA[13] recommends the use of data from a survey conducted by the National Marine Fisheries Service.[14] For the Gulf Coast, EPA[13] recommends a mean intake of 7.2 g/day and a 95th percentile value of 26 g/day. However, at the outset of the RI, the limits of this study were recognized relative to its applicability to recreational anglers in Lavaca Bay. Use of the assumed amount of fish consumption may overestimate or underestimate the actual IR, depending on the actual consumption patterns of Lavaca Bay anglers. At a minimum, however, the use of assumed consumption rates adds substantial uncertainty to estimating the ingestion rate. The objective of the consumption study was to obtain an accurate measure of the grams per day consumed by Lavaca Bay anglers to minimize uncertainty.

Finally, the fraction ingested from the contaminated source (FI) is intended to provide an estimate of the percentage of contaminated fish that are consumed from different locations within Lavaca Bay. Results of the finfish and shellfish sampling reveal that average mercury concentrations in finfish vary according to location (Figure 21.4). For the closure area, finfish mercury concentrations are statistically greater than they are for the remainder of Lavaca Bay. In addition, for the remainder of Lavaca Bay, mercury concentrations for the majority of species are statistically greater than in the background or reference bays.[12] These conclusions imply that understanding exposure and assessing potential human health risks require knowing what fraction of fish are caught and consumed from Lavaca Bay.

A secondary objective for the consumption study was to collect data from recreational anglers for use in the NRD recreational fishing assessment. The technique chosen for this assessment is a random utility model (RUM), which measures the value of a fishing trip in terms of the utility or satisfaction that an angler derives from fishing. RUMs are the preferred assessment technique for measuring potential recreational service losses because they are based on sound economic theory and actual choices of anglers.[4,15-17] The NRDA regulations endorse the use of RUMs as a type of travel cost model (43 CFR 11.83). Thus, the data needs for the recreational fishing assessment correspond to the data requirements for a RUM.

RUMs are based on economic welfare theory and measure angler satisfaction or utility as:

$$V_{ij} = X_{ij}\beta + \varepsilon_{ij}$$

where i = individual angler

j = fishing site

X = matrix of site characteristics

β = vector of the estimated coefficients for X

ε = random error term

Thus, RUMs measure the value of a fishing trip to an angler as a function of the site characteristics of the fishing site. The characteristics of each fishing site, such as fish catch rate, presence of facilities such as a boat ramp or lighted fishing pier, whether or not the fish can be eaten, and distance to the site from the angler's home, distinguish one site from another. The RUM uses anglers' actual fishing site choices to model the factors that influence fishing site selection. By compiling information from anglers on all the fishing sites they visit for each trip and the characteristics of those sites, a RUM explains the relative importance of different site characteristics in the anglers' site choice decisions,

TABLE 21.1 Summary of Data Needs from Recreational Anglers

Type of Data Needed for Each Fishing Trip Taken	Use in Human Health Risk Assessment	Use in NRD RUM for Recreational Fishing
Location of fishing trip	Used to calculate FI	Used to delineate fishing sites and to estimate distance traveled
Number of fish caught and kept, by species	Used to evaluate species composition and to provide check on reported fish consumption	Used to estimate catch and keep rates (site characteristics)
Estimated portion size and meals consumed, by cohort	Used to estimate IR	N/A
Fishing site characteristics (parking, congestion, boat ramp, fishing pier, etc.)	N/A	Used to measure salient site characteristics in the RUM

including the effect of the closure order. An important feature of RUMs is that they evaluate angler utility relative to substitute fishing opportunities.

Table 21.1 summarizes the data needs for both the human health risk assessment and the NRD recreational fishing analysis. To develop a RUM for the Lavaca Bay assessment, the locations of fishing trips made by Lavaca Bay anglers were required. This data need was consistent with the data objectives for the consumption study because the location of the fishing trip is a component of estimating FI. For the consumption study, the location of trip-specific catch was also needed. This data need was consistent with the RUM needs because species caught and catch rates are important site characteristics. The intersection of these data needs provided the impetus for undertaking a joint data collection. Given that the RI fish consumption study plan called for asking saltwater anglers questions about their fishing locations and trip-specific details such as catch, adding site characteristic questions for the RUM was a logical extension of the fish consumption study plan.

21.4 STUDY DESIGN AND IMPLEMENTATION

The Texas Saltwater Fish Survey was designed to collect data for both the consumption analysis and the recreational fishing component of the NRDA for the Lavaca Bay site. This dual role influenced the development of the materials used and the implementation of the study. This section describes the primary study design features. This section also describes the implementation of the Texas Saltwater Fish Survey, which used both telephone and mail modes to collect the data.

21.4.1 Study Design

EPA[18] identifies five survey modes suitable for fish consumption surveys:

- *Recall Telephone.* In this mode, respondents are contacted by phone and asked questions about recent fishing trips and fish consumption. The answers may be recorded on preprinted blank questionnaires or directly into a computerized database designed for the study.
- *Recall Mail.* The recall mail approach is similar to the recall telephone approach in that respondents are asked about recent fishing trips and fish consumption. The primary difference is that respondents receive a written questionnaire in the mail and are asked to complete it and return it.
- *Recall in Person.* In this mode, respondents are asked to recall information about fishing trips and fish consumption while talking face to face with an interviewer. Personal interviews can take place at a variety of locations, including the respondent's home, a centralized interviewing location (such as a fishing club), or at fishing sites.
- *Diary.* The diary approach is a type of mail questionnaire where the respondent is asked to record relevant information concurrently.
- *Creel Census.* Finally, a creel census is the last approach discussed here. These censuses collect on-site harvest information from anglers. Generally, creel census surveys do not ask respondents to recall information on fish caught and consumed, which distinguishes them from the in-person interview described above. However, creel census surveys may ask individuals about their plans for consuming the fish caught.

Each of these modes has advantages and disadvantages, which are discussed in Reference 18. The choice of the specific mode depends on the specific objectives and other characteristics of the study.*

For the Texas Saltwater Fish Survey, the joint data needs dictated that the survey mode could not realistically be a creel census, which is a common choice for consumption studies. Many consumption studies based on on-site data are generally limited to the affected waterbody. This strategy prevents calculating the FI, as the only data collected reflect fish caught at the selected fishing sites. Although on-site data collections can be expanded to include multiple sites, it is difficult to know a priori which other sites anglers who fish in the affected area also use. This inability to perfectly know the substitute sites is a serious limitation of on-site data with respect to RUMs. In a coastal area such as Lavaca Bay, anglers have many nearby choices for fishing access. Including several access points for an on-site data collection would have significantly increased the costs of the data collection. In addition, creel studies collect prospective information on the fate of the fish, which may or may not correspond to actual consumption, particularly for other members of the household. Thus, to calcu-

---

*For additional information on survey modes and designs, see References 19 to 24.

late FI as accurately as possible and to include all relevant substitute fishing sites in the RUM, the survey did not employ a creel census or on-site mode.

Budget and time constraints eliminated other possible collection modes. For example, the budget did not include funding to cover the cost of personal interviews. Moreover, the timing of the study favored a recall approach over a diary approach. Timing constraints suggested a preference for a recall telephone survey over a recall mail survey. The pretesting process revealed that a recall telephone mode was feasible for most respondents. However, for respondents who took many trips, collecting all of the relevant trip information over the telephone would have been inefficient. For these respondents, a mail questionnaire was used. Thus, the study design included a combined mail/telephone mode.

Another principal design feature was the length of the survey period, or the time period for which anglers report their recreational fishing trips. For this study, a one-month period was selected because it was long enough to provide reliable estimates for the exposure period under evaluation. For the risk assessment, the most sensitive endpoint is neurodevelopmental effects in a fetus, with the third trimester of pregnancy the most sensitive period for exposure.[25,26] For the NRDA for recreational fishing, previous experience with recreational angler surveys and the anticipated sample size (see below) suggested that one month of data would provide ample data to estimate fishing losses using a RUM.

The month of November was selected as the study period because the data collection would be timely and November appeared to be a representative fishing month. The temperate climate of coastal Texas permits fishing year-round, with fall one of the peak fishing seasons. November is a "representative" month in that TPWD classifies the first half of November as a "high-use" period and the second half of November as a "low-use" period.[27] In addition, certain sport species of interest, such as flounder and red drum, are prevalent in Lavaca Bay in the fall.[28] Thus, the extrapolation from the survey results to annual representations was reasonably conservative for the baseline risk assessment and NRDA purposes.

Sampling strategy is the final primary design issue. For this study, sampling strategy has three aspects. First, because the study was household-based, the geographic area with the residences of Lavaca Bay anglers had to be identified. The challenge in selecting the relevant geographic area was to capture a sufficiently large area such that the majority of the anglers who use or would consider using Lavaca Bay were included, but not so large a geographic area that Lavaca Bay anglers were rare. If the latter were true, then the data collection would not be cost-effective because identifying that rare angler requires more effort.

Unpublished data from the TPWD provided the basis for selecting the relevant geographic area. TPWD indicate that over 70 percent of the anglers who fish Lavaca Bay come from Calhoun, Jackson, and Victoria counties, the three counties surrounding Lavaca Bay.[6] The TPWD data indicate that residents from any other single county comprise less than 5 percent of Lavaca Bay users. Thus, limiting the geographic area to the three counties resulted in an efficient data collection.

Moreover, the three-county area was appropriate for both the human health risk assessment and the NRDA recreation studies. In terms of the risk assessment, Lavaca Bay anglers from other counties are likely to be infrequent visitors to Lavaca Bay because they have a larger choice of substitute fishing sites, given the distance to Lavaca Bay. Therefore, these more distant anglers are less likely to consume self-caught Lavaca Bay fish. Adopting a larger geographic area for the frame also would have diluted the risk results. For the recreation study, including more distant anglers would not have improved the study.* Distant anglers are more likely to fish from a variety of different locations because as distance from the relevant site increases, so does the number of substitute fishing sites. Including distant anglers would have provided comparatively less information for modeling fishing trips to Lavaca Bay, the primary site of interest. For both studies, limiting the geographic area to the three nearest counties was a reasonable approach.

The next aspect of the sampling strategy was stratification. The sample design included three strata that correspond to the three counties identified above. Calhoun County anglers comprised 50 percent of the sample while Victoria County and Jackson County anglers comprised 30 and 20 percent, respectively. These strata reflect the relative proportions of anglers from the three counties who fish in Lavaca Bay. A stratified design generally reduces sampling error[29] and allows more efficient data collection because anglers who are more likely to regularly fish in Lavaca Bay have a greater probability of inclusion in the study.

The final aspect of the sampling strategy was identifying the target sample size. Given the data objectives for both the consumption and the NRDA studies, this study targeted approximately 2000 angler participants. Because TPWD requires a stamp and license for all saltwater anglers, a computerized listing of saltwater license holders from the three counties provided a census of the target population. This complete listing allowed a fully random sampling process, stratified as described above. Conservative assumptions about respondent cooperation and bad addresses indicated an initial sample size of nearly 3500 anglers across the three-county area (representing 26 percent of the angling population) would provide the desired number of respondents.

21.4.2 Study Implementation

Figure 21.6 provides an overview of the mixed-mode data collection approach. Researchers have found that by combining different approaches and adopting a mixed-mode survey, the disadvantages of any one technique can be over-

---

*The NRDA fishing study included analyses to demonstrate that the exclusion of more distant anglers did not result in an underestimate of losses.

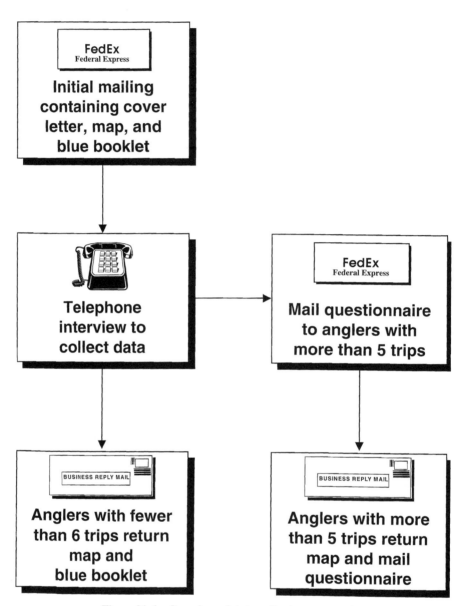

Figure 21.6   Overview of data collection approach.

come.[18,20,22] This combined telephone–mail approach enabled the survey to be completed in a timely manner. This timeliness was important to minimize potential recall bias and to also meet the demands of the risk assessment and NRDA schedules.

The first contact with potential respondents was a mailed package containing an introductory letter explaining the survey, a map of the Matagorda Bay system with drawings of fish meal portion sizes on the back, and a blue booklet. This blue booklet contained instructions for identifying the location of fishing trips on the map, questions related to consumption of self-caught fish, and questions about respondents' opinions on different fishing experiences.* The letter instructed the respondent to identify fishing locations on the map and to answer the questions in the booklet. The letter also informed respondents that a representative of a market research firm would call to collect the information from the blue booklet as well as other information on fishing.

The next contact with respondents was the telephone interview. Near the beginning of the telephone interview, the interviewers asked respondents how many fishing trips they took during November, 1996. For respondents who took five or fewer fishing trips, the telephone interview continued. Telephone interviewers collected the answers to the questions in the blue booklet. In addition, interviewers collected information on fishing trips for the consumption and recreation analyses. The interviewers instructed respondents to return their maps and blue booklets. If respondents took more than five fishing trips, the marketing research firm mailed them a questionnaire containing the same questions asked during the telephone interviews. Mail respondents were asked to complete the questionnaires and return them, along with their maps.

Regardless of whether the respondent participated in the telephone interview or filled out the mail version, the questionnaire itself had three primary types of questions†:

- Questions about the recreational fishing trips during the month of November (fishing site, facilities at the site, distance traveled, and number and species of fish caught). These data are used primarily in the NRDA fishing analysis, although the trip location and the fish caught also play a role in the consumption study.
- Questions about consuming self-caught fish. These questions provide the foundation for the human health risk assessment and fall into two categories. The first is trip-related consumption, where respondents report eating fresh, self-caught fish from their November fishing trips. The second type is preserved consumption, where respondents report consuming previously self-caught fish during November.
- Demographic questions.

One of the first items in the questionnaire asked the location of each fishing trip. The blue booklet sent in the initial mailing provided instructions for respondents to identify the location of their fishing trips by placing numbered stickers on the map. Identifying the location of the trip is crucial to recreation demand modeling because the distance from the respondent's home to the

---

*These opinion questions are used to identify potential restoration options for the NRDA. Because they do not play a role in the consumption analysis, they are not discussed further in this chapter.
†This discussion excludes questions related uniquely to the NRDA for recreational fishing losses.

fishing site is a key determinant of site choice. Then, on a trip-by-trip basis, respondents provided information about the characteristics of each of their fishing trips. The format for collecting these site characteristics was a series of questions that summarized each fishing trip. Respondents named the location of their fishing trips and estimated how long they spent traveling to their fishing sites and how long they spent fishing. Respondents also answered questions about the facilities at the launch or pier site, the congestion at the fishing site, the surroundings or aesthetic view at the site, and the number and type of fish caught and kept. Site characteristics are important in the recreation model because they determine the attractiveness of a particular site and thereby help determine which site an angler will choose.

As part of the individual trip records, respondents provided information about consuming the fish kept from each trip. Collecting consumption data on a per-trip basis offers the advantage of pinpointing where respondents caught their fish. This information allows the estimation of FI from the closure area and from Lavaca Bay. The trip-related consumption information included:

- Number and type of species caught
- Number of each species that were kept and eaten
- Number of meals eaten by the respondent from the fish he/she kept
- Typical portion size for each meal eaten by the respondent
- Number of meals from the fish eaten by the spouse and child, if applicable
- Typical portion size for each meal eaten by the spouse and child, if applicable

Respondents also provided information about their consumption of fish caught in other months but eaten in November. These meals were typically from frozen, smoked, or canned fish. Capturing this information is important because it provides a complete reporting of consumption of self-caught fish for the month of November.

The last section of the questionnaire contained demographic questions that provided information for both the human health risk assessment and the NRD fishing studies. The questions most important for the consumption analysis included the age and sex of the respondents and the number of people in different age groups within their households. These data were used to estimate consumption for the different cohorts or sensitive populations. The demographic questions allow estimating consumption information for women of childbearing age (ages 20 to 39), youths (ages 6 to 19), and small children (less than 6 years old). For the recreation analysis, additional demographic questions such as boat ownership, years of fishing experience, and employment status provide supplemental information on anglers.

Survey administration began on December 16, 1996, when overnight delivery packages were mailed to respondents. Telephone interviewing began a few days later. The study incorporated standard survey protocols to ensure high-

quality data. For example, the study's protocol for the telephone version required the telephone interviewers to attempt to contact the respondent at least
eight times on different days and at different times of the day. For the mail survey, respondents were called three times and asked to return the completed
questionnaires before they were classified as refusals or as unable to contacts
(UTCs). Survey administration concluded on March 2, 1997. In total, 1979 anglers participated in the study, with a response rate of about 83 percent.

21.5 STUDY RESULTS

As described in Section 21.3, this study is designed to measure three facets of
consumption of self-caught fish by recreational anglers. These parameters are:

- Grams per day of self-caught finfish and shellfish consumed by anglers
  from the three-county area (IR)
- The fraction of self-caught fish ingested from Lavaca Bay (FI)
- The species composition of self-caught fish consumed by anglers

In addition to these parameters, the results of the Texas Saltwater Fish Survey
also provide some insight on the issue of subsistence anglers. Finally, this section also describes the use of the survey data in the RUM for the NRDA.

21.5.1 Consumption Rate

Table 21.2 contains the average IR of self-caught fish for this study expressed
as grams per day. This consumption rate incorporates all self-caught finfish and
shellfish consumed from all water locations, as well as any preserved (such
as frozen or smoked) self-caught finfish and shellfish. The statistics provided
include the mean, 95 percent upper confidence limit (UCL) on the mean, and
90th percentile value. The mean and 95 percent UCL provide a descriptor of the
average consumption while the 90th percentile value presents the upper-bound
value. These statistics are presented for each of the cohorts described above,
separately for finfish and shellfish.

  As Table 21.2 indicates, adult men consumed on average 25 g of self-caught
finfish per day. Adult women on average consumed 18 g daily, while women of
childbearing age consumed slightly more per day, 19 g on average. As expected,
children consumed fewer grams than adults do. On average, youths consumed
16 g per day, and small children consumed an average of 11 g per day.

  Shellfish are consumed far less than finfish, with adult men eating less than 2
g per day on average. Adult women and women of childbearing age consumed
an average of about 1 g per day. Youths consumed around 1 g per day on average. Again, small children consumed the least, less than 1 g per day on average. Nearly all of the reported shellfish consumption was from preserved meals,
rather than fresh meals.

TABLE 21.2   Results of Texas Saltwater Fish Survey: Grams per Day of Self-Caught Fish Consumed[a]

Cohort	Mean	95% Upper Confidence Limit on Mean	90th or 95th Percentile of Distribution[b]
Finfish			
Adult men	24.8	27.7	68.1
Adult women	17.9	19.7	47.8
Women of childbearing age	18.8	22.1	45.4
Youths	15.6	17.8	45.4
Small children	11.4	14.2	30.3
Shellfish			
Adult men	1.2	1.6	5.1
Adult women	0.8	1.1	2.4
Women of childbearing age	0.9	1.2	4.0
Youths	0.7	1.0	4.5
Small children	0.4	0.6	2.0

[a]The data in the second column represent the 95 percent UCL of the mean. This statistic means that there is only a 5 percent chance that the mean is greater than the 95 percent UCL. The data in the third column represent the 90th percentile of the distribution. If all observations in this distribution are arranged in ascending order based on their value of the consumption rate, the value of the observation that marks 90 percent of the way through the distribution is the 90th percentile value.

[b]For shellfish, the 95th percentile value is provided in this table because less than 10 percent of the individuals consume shellfish, resulting in a 90th percentile of zero.

For comparison purposes, it is useful to view the results of the Texas Saltwater Fish Survey in the context of other saltwater fish consumption surveys. A review of the literature reveals some studies that measure consumption intake for marine recreational anglers. Table 21.3 contains a summary of these studies. Although each study is unique with regard to its objectives, the target populations and the methods have certain similarities worth mentioning. All of the major studies summarized in Table 21.3 are principally creel surveys or creel surveys supported by telephone interviews. Creel surveys tend to have an inherent avidity bias due to the unequal probabilities of sampling anglers.[13,30] The anglers who go fishing the most have a much higher probability of being interviewed, and therefore the results, left uncorrected, will reflect the resource utilization distribution and are not representative of the target population as a whole.[13] Therefore, as reflected in Table 21.3, the data from creel surveys generally have been reanalyzed by subsequent authors with an attempt to address this avidity bias by correcting with sample weights.

With the exception of the National Marine Fisheries Service (NMFS) study,[14] all target a specific population. Only the NMFS study contains consumption data that are potentially relevant to Lavaca Bay anglers, although the survey itself could not address resource utilization questions (such as species

TABLE 21.3   Summary of Consumption Studies of Marine Recreational Anglers

Author(s)	Location	Year	Data-Collection Approach	Target Population	Grams/Day (90th Percentile)
Puffer et al.[31]	Los Angeles, CA	1980	Creel survey	Sport fishermen	37.0 (225.0)
Price et al.[32] (reanalysis of Ref. 31)	Los Angeles, CA	1994	Creel survey	Sport fishermen	2.9 (35.0)
Pierce et al.[33]	Commencement Bay, WA	1980	Creel survey	Non-commercial fishermen	ND
U.S. EPA[11] (reanalysis of Ref. 33)	Commencement Bay, WA	1989	Creel survey	Non-commercial fishermen	23
Price et al.[32] (reanalysis of Ref. 11)	Commencement Bay, WA	1994	Creel survey	Non-commercial fishermen	1.0 (13.0)
National Marine Fisheries Survey (NMFS)[14]	Nationwide, Gulf states (AL, FL, LA, MS)	1993	Telephone and creel survey	No specific target population	7.2 (26.1)[a]

[a]95th percentile value reported.

composition of catch and consumption and catch location relative to inside or outside of closure area in Lavaca Bay or from outside of Lavaca Bay). It is appropriate to note that EPA[13] selected the NMFS survey as the basis for the recommended default for recreational marine anglers.

The first study[31] is a creel survey of sport anglers in the Los Angeles area. A total of 1059 anglers were interviewed at 12 intercept sites over the course of a year. This study reports median fish consumption rate at 37 g per day, with the 90th percentile rate at 225 g per day. The study also measured different intake rates by age category and by race. The youngest reported age category was under 17 years, with a median intake rate of 27.2 g. Species composition information is also provided but because the prevalent species are not similar to those found in Lavaca Bay, those results are not considered relevant and are not discussed here.

The Puffer et al.[31] study is an intercept study, and the analysis did not weight the results to address the avidity bias. As a consequence, the results are biased upward. In addition, interviews were conducted only with anglers who caught and kept fish. Anglers who did not catch any fish or who released all of the fish they caught were not interviewed. This procedure eliminates the nonconsumers from the database, which overestimates consumption for the target population.[13]

Recognizing the upward bias of the Puffer results, Price et al.[32] attempt to correct for avidity bias in the Puffer et al.[31] results by using inverse fishing frequencies as sampling weights. This study, using approximately the same data, found dramatically lower results with an estimated median at 2.9 g per day and the 90th percentile at 35 g per day. However, EPA[13] concludes that the Price et al.[32] reanalysis, although probably closer to the true consumption rates than the original Puffer et al.[31] study, most likely underestimates consumption. This understatement stems from the violation of equal sampling probabilities relative to fishing frequencies.

The third study[33] is a creel survey of sport anglers in Commencement Bay, Washington. The study included interviews with 508 anglers who caught and kept fish, and it was administered during two different time periods (the summer and the fall). This study did not calculate grams per day, but an EPA analysis[11] of the Pierce et al.[33] data found a median intake rate of 23 g per day. However, the EPA analysis failed to correct for avidity bias and therefore suffers the same upward biases as Puffer et al.[31] Noting this overstatement, a reanalysis was conducted by Price et al.,[32] which corrects for avidity bias in the EPA analysis of the Pierce et al.[33] data by applying sampling weights corresponding to the inverse fishing frequencies. Similar to the experience with the Puffer et al.[31] data, Price et al.[32] found dramatically lower results. Price et al.[32] report a median intake rate of 1.0 g per day with a 90th percentile rate of 13 g per day. However, this reanalysis likely results in an underestimate of consumption because sampling probabilities are less than proportional to fishing frequency.[13]

The last study contained in Table 21.3 is the 1993 NMFS study, on which the EPA[13] bases its default IR. In this combination telephone and creel survey, anglers from all coastal states except Texas and Washington are sampled to estimate the size of the recreational marine finfish catch by location, species, and fishing modes. This study was not designed to collect estimates of consumption for individuals within the survey, which introduces a relatively low level of uncertainty. This analysis assumes there are 2.5 intended consumers for each angler's catch and that half (50 percent) of the weight of the catch is edible. These assumptions are the major limitations of this type of survey. The mean daily intake for anglers in the Gulf region is 7.2 g per day, with the 95th percentile rate of 26.1 g per day.

The results from the Texas Saltwater Fish Survey are based on sound methodologies with the intent to limit sampling bias and problems noted in the major studies reported in Table 21.3. One key difference is that the consumption parameters measured for Lavaca Bay do not depend on the intercept method of data collection. The avidity bias inherent in other studies and subsequent attempts to correct that bias are not present in the Texas Saltwater Fish Survey. Thus, the results from this standpoint are more reliable toward meeting the specific objectives for this survey. However, the reliance on a short-term recall survey method introduces other bias and uncertainty.

The survey method utilized relies on the angler's recall for an estimate of meal frequency and portion. However, experience has demonstrated that there usually is a tendency for recall surveys to overestimate fish consumption.[30,34] There was evidence of this overestimation when the data were validated and approximately 10 percent of anglers reported consuming more fish than they caught and kept. Short-term recall surveys also have other biases. For example, they do not take into account time-varying fishing activities of individuals and do not recognize that some individuals may actually be engaged in more fishing during the survey period while others may fish less. Because the survey was conducted during the fall, which typically shows higher fishing pressure for certain finfish species on the Texas Gulf Coast, this is anticipated to result in a positive bias or more conservative estimate of fish consumption.

Moreover, although some assumptions are made in calculating the consumption parameters, every effort is made to limit the number and types of assumptions made and to make the best assumption possible. For example, in contrast to the NMFS study, the number of consumers in each household is not an assumed number. Moreover, the edible portion of the key species is accounted for on a species-specific basis, rather than an overarching assumption applied to all species. Thus, differences in the results of the Texas Saltwater Fish Survey are the result of improved methodology and assumptions tailored to meet the specific objectives of this study.

21.5.2 Fraction Ingested

The Texas Saltwater Fish Survey collected trip locations (including the location of the water site for boat trips) from anglers. Knowing where anglers caught their fish allows calculation of the FI of self-caught fish from Lavaca Bay. A total of 823 anglers reported taking at least one trip to Lavaca Bay. Of the 2694 trips taken by these anglers, 35 percent occurred in Lavaca Bay and 65 percent occurred in other waters. This calculation suggests a fraction ingested of 35 percent. Thus, the results of the Texas Saltwater Fish Survey indicate that the FI from Lavaca Bay, a much broader geographic area than the closure area, is well below 100 percent for the surveyed population.

21.5.3 Species Composition

The third consumption parameter to be obtained from this study was species composition. Total grams consumed across all households by species were computed. The seven primary species analyzed include:

- Black drum
- Red drum
- Flounder
- Speckled sea trout
- Blue crab
- Shrimp
- Oysters

All other finfish species are grouped into an "other finfish" category. This analysis relied on species preference data from three different sources of fish:

- Fish consumed from Lavaca Bay
- Fish consumed from all waters
- All self-caught finfish and shellfish consumed, including preserved (i.e., frozen or smoked) fish

Table 21.4 contains the results. The percentage of each finfish species consumed does not include the shellfish grams consumed, and vice versa. As the table indicates, red drum comprises the bulk of total finfish grams consumed. Speckled sea trout have the second highest total, and flounder have the third highest. For Lavaca Bay, the gap between red drum and speckled sea trout is wider than it is for all water bodies.

Slight differences in the ranking for black drum and other finfish species can be explained by the differences in the geographic area. Broadening the source to include all waters (and all self-caught fish) introduces additional species. For example, red snapper is a popular fish caught and consumed by many anglers in the study, but it is not typically caught within the boundaries of Lavaca Bay. Thus, the "other finfish species" become more prevalent as the geographic scope is extended. The existence of other species also accounts for the differences in species composition among the sources.

TABLE 21.4 Species Composition of Self-Caught Fish Consumed by Source

Species	Lavaca Bay (%)	All Waters (%)	All Self-Caught Fish[a] (%)
Finfish			
Black drum	8.7	6.4	4.8
Red drum	39.3	35.9	39.1
Flounder	18.9	19.9	17.0
Speckled sea trout	27.3	33.1	33.7
Other finfish species	5.8	4.7	5.4
Shellfish			
Blue crab	0.0	37.1	34.6
Shrimp	100.0	30.9	28.5
Oysters	0.0	32.1	37.0

[a] All self-caught fish include preserved fish (such as smoked or frozen fish) where the location of the catch is not known, as well as fresh fish.

For shellfish, only shrimp are consumed from Lavaca Bay. Although blue crabs and oysters are commercially harvested from Lavaca Bay, the survey did not indicate that individual anglers caught and consumed their own crabs and oysters. This finding is consistent with the TPWD creel data described in Section 21.2. In all waters, shrimp and oysters are equal in terms of total grams consumed. Blue crab, however, is the most popular species for consumption. Out of all self-caught shellfish, oysters account for 37 percent, blue crab for 35 percent, and shrimp for 29 percent. In addition, as described above, the source of the preserved fish meals is not known. Therefore, there is some uncertainty in these estimates.

Because mercury levels varied among species, knowing species composition that anglers caught and consumed is an important component of the baseline risk assessment. Without this information, the default risk assumption would have calculated risk using fish "on average," which according to the finfish data would not have been an accurate predictor of exposure.

21.5.4 Subsistence Angling Population

The results of the Texas Saltwater Fish Survey were reviewed to address the issue of whether there is an indication of any subsistence fishing population. Although there is no agreed-upon approach for identifying subsistence populations, by its very definition, subsistence fishing is motivated by financial reasons and/or ethnic or cultural reasons. Results of the consumption survey's demographic information were viewed to determine whether trends exist in consumption patterns relative to relevant sociodemographic characteristics for anglers surveyed.

First, the average IR (grams per day) for adult men from the survey area was calculated for different income levels and different races. These data are

grams/day

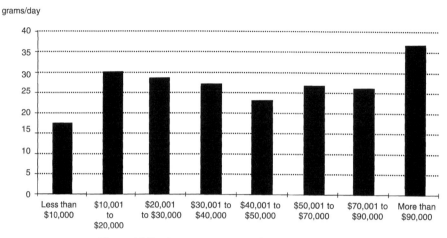

Figure 21.7   Average consumption by income.

presented graphically in Figures 21.7 and 21.8. As Figure 21.7 indicates, the highest consumption rate coincides with the highest household income (i.e., > $90,000/year). Figure 21.8 reveals that white anglers eat more self-caught fish than do anglers from other ethnic groups. Thus, at first glance, the data do not indicate the existence of a subsistence population.

Statistical analyses revealed no relationship between total consumption and relevant socioeconomic characteristics. Correlation analysis, which measures the degree to which differences in one variable move with differences in another variable, indicated no discernible relationship between total grams consumed and the socioeconomic characteristics. Multiple regression analysis, which investigates how combinations of variables explain changes in another variable,

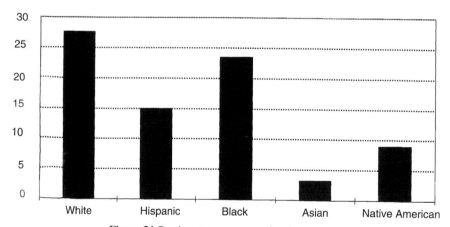

Figure 21.8   Average consumption by race.

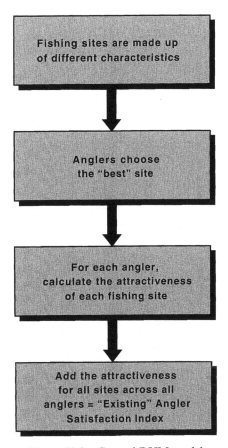

Figure 21.9 General RUM model.

also revealed no discernible statistical relationship. Thus, an analysis of the consumption and socioeconomic information collected in the Texas Saltwater Fish Survey does not indicate the existence of a subsistence population within the recreational fishing population surveyed in the three-county area.

21.5.5 Potential NRD Recreational Fishing Losses

Section 21.3 introduced the RUM and described its professional acceptance for measuring potential recreational fishing losses at NRD sites. Figure 21.9 describes the intuition underlying the RUM, which is based on the premise that site characteristics determine the value of a fishing trip. For this assessment, site characteristics such as fish catch rate, presence of facilities such as a boat ramp or lighted fishing pier, distance to the site from the angler's home, and whether the fish can be eaten are the basis of the model. These characteristics are what distinguish one site from another. Fishing sites are similar to other goods and services in this respect. For example, cars have characteristics that distinguish each from another, such as number of doors, engine size, and air-conditioning. The value of a specific car to a consumer is based on the combination of the specific features. Fishing sites are comparable in that the value of the site is based on the combination of features it offers to anglers.

Based on individual preferences for site characteristics, anglers select the "best" site by choosing the fishing site with the combination of characteristics that gives them the most satisfaction, which is the second box shown in Figure 21.9. The best site may differ for each angler, depending on the distance to the site from his home and other personal preferences. Time and angler income also affect the decision to travel to a specific site. Again, choosing a fishing site is similar to choosing among other goods. When choosing a car, for example, one customer will choose the car with the best gas mileage. Another customer is willing to pay more for a car with leather seats. Anglers have preferences for fishing sites as well. One angler does not want to travel far from home to fish. Another prefers to visit a site where she can launch her boat, even if it is farther from home. Each angler looks at the different choices and chooses the site that provides him or her with the most satisfaction.

The Texas Saltwater Fish Survey data provide the actual fishing-site choices made by anglers from the three-county area. Based on the choices that each angler actually made, the RUM calculates an attractiveness index for each angler for each site, which is the third box shown in Figure 21.9. The model takes into account each angler's preferences about different site characteristics and the different fishing sites available to anglers. Selected sites are more attractive and rate higher on the index.

For example, suppose that Angler A, who does not like to drive far to fish, has two fishing sites within 5 miles of his house. These sites would rate higher on Angler A's attractiveness index than any other sites. Angler B, on the other hand, places a premium on sites where she is likely to catch many sea trout. However, the best sites with sea trout are more than 15 miles from Angler B's house. Even so, on Angler B's attractiveness index, these more distant sites would rate higher than any other closer site.

The fourth box in Figure 21.9 explains that the attractiveness scores for each site for each angler are summed. This sum gives an overall measure of angler satisfaction with the current fishing opportunities. This measure of angler satisfaction reflects the actual choices that anglers make and reveals what characteristics are important to each angler.

Figure 21.10 shows how the general model is used to estimate losses in angler utility associated with the closure area in Lavaca Bay. The first box is the same as the last box on Figure 21.9, which is the existing angler satisfaction index. Because this satisfaction index reflects the current fishing sites that anglers have available to them, it includes the reduced satisfaction from not being able to eat the fish from the closure area. Being able to eat fish caught during fishing trips is important to some anglers, so the consumption ban reduces overall angler satisfaction.

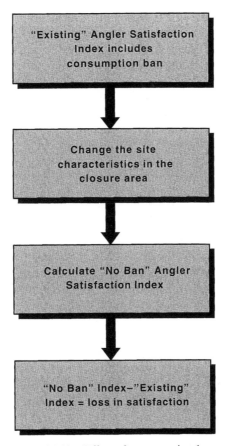

Figure 21.10   Effect of consumption ban.

The second box in Figure 21.10 is the crux of estimating losses in angler utility. To determine how much angler satisfaction has been reduced as a result of the closure order, that site characteristic (i.e., whether or not anglers can eat the fish from that area) is altered in the model by removing the consumption ban. This change permits the measurement of the attractiveness of the closure area without the consumption ban. Thus, this model is like any other model because it predicts how something (in this case, angler satisfaction) would change if the model inputs were different.

As the third box in Figure 21.10 indicates, the model then recalculates the satisfaction index without the consumption ban. As a result, the closure area is more attractive to anglers. This "no-ban" satisfaction index is larger than the existing satisfaction index.

The last box in Figure 21.10 explains that angler utility losses are estimated by subtracting the existing satisfaction index (i.e., the level of angler satisfaction with the consumption ban in place) from the no-ban satisfaction index.

This subtraction reveals how much more satisfaction there would be if the consumption ban were removed. In other words, it reveals how much loss in angler satisfaction has occurred as a result of the consumption ban.

In this specific application for Lavaca Bay, the model relies on a variety of site characteristics to predict fishing site choice selection by anglers. Among others, these site characteristics include catch rates, the type of surroundings, the type of boat launch or other facilities, and travel distance. The effect of the TDH closure order on angler utility is calculated by using a qualitative variable for all trips to the closure area. In a statistical model, a qualitative variable indicates the presence or absence of a certain condition, taking the value of zero when the condition is absent and a value of one when the condition is present. In this model, the closure area variable has a negative influence on site choice selection. It captures the effect of all the unexplained characteristics of the closure area on site selection and, in essence, allows the closure area to be statistically unique from all other fishing sites.

To estimate angler utility in the baseline (i.e., without injury) condition, the qualitative variable for the closure area is given a value of zero. This specification assumes that there is nothing unique about the closure area. To estimate angler utility for the "with-injury" (i.e., with the closure order in place) condition, the closure area variable is given a value of one. This specification assumes that there is something uniquely undesirable about the closure area. The difference between the baseline level of angler utility and the with-injury level of angler utility is the loss in utility attributable to the closure area. This loss in utility provides the basis of the NRDA for recreational fishing losses. Based on the RUMs developed for the recreational fishing assessment, potential fishing losses attributable to the closure order range from 1 to 6 percent of the baseline angler utility levels. This small loss is consistent with the relative size of the closure area and the myriad of nearby, substitute fishing sites.

21.6   DISCUSSION

Results of the Texas Saltwater Fish Survey suggest higher consumption rates on average than other surveys do. This result is expected considering that the target population was recreational anglers in a three-county area along the Texas Gulf Coast. The population is mostly rural with a number of small communities, and the fisheries resources and fishing opportunities are plentiful in this region. A detailed review of the data suggests that this result is not necessarily the result of overall higher consumption by anglers in general, but it appears that there is much higher consumption by few of the more avid angling population. The results of this survey demonstrate that there is a fairly substantial resource available that is utilized by the angling population in general but, most specifically, by a small number of avid anglers. The survey also provided valuable supplemental information and is very important for understanding potential exposures and risks at the site. Specifically:

- Consumption is highly preferential toward the more desirable sport fish species (i.e., red drum, sea trout, etc.).
- Although the TDH has a ban on keeping and consuming finfish and blue crab from the closure area, some fishing and consumption from the closure area occurs.
- Anglers from the three-county area prefer fishing at locations away from the closure area, and actually prefer locations outside of Lavaca Bay.
- Fishing pressure and success is not constant throughout the system but appears to be strongly correlated with some structure, such as fringe marshes, oyster reefs, fishing piers, and the like.

Other advantages of the Texas Saltwater Fish Survey should be considered when contrasted with other surveys. For example, the consumption parameters from this survey measure intake for different cohorts, a characteristic lacking in most other studies. In particular, the intake rates are calculated separately for small children, something no other study has attempted. Besides the site-specific measures that come from the Texas Saltwater Fish Survey, additional advantages stem from coordinating efforts with the NRDA for recreational fishing. For example, the NRD requirements result in a model that predicts the baseline level of fishing, or the level of fishing that would occur in the absence of the closure. This scenario is analogous to a future scenario where meeting the remediation objectives could result in the removal of the closure order.

Section 21.5.5 describes how the RUM is specified. The most important aspect of any RUM designed to address potential NRD fishing losses is adequately capturing the effect of the fishing restrictions on angler behavior. In the Lavaca Bay assessment, the fishing restriction amounts to a consumption ban. At other NRD sites, the fishing restrictions may range from a complete ban on fishing altogether to lesser consumption "advisories" that recommend limited consumption of fish from the contaminated source. RUM specification is flexible enough to account for any of these scenarios. Thus, the RUM developed for an NRDA study can also provide an estimate of the relative change in fishing trips to the affected area in the absence of fishing restrictions. This information translates to an estimate of the relative change in risk for a scenario where fishing restrictions are removed. Because the same anglers who provided data for the consumption analyses also provided data for the RUM, the use of the RUM for this estimate is appropriate.

Another potential benefit of developing a recreational fishing RUM for use in both a human health risk assessment and an NRDA is that RUMs predict angler behavior under alternative scenarios based on actual past behavior under a variety of conditions. Without the benefit of this prediction of how anglers respond to alternative fishing restrictions, predicting risk under alternative scenarios can be difficult. Consumption studies demonstrate that anglers indicate a wide variety of responses that range from changing eating habits to changing target species and from fishing less to fishing at alternative sites.[35-38] These studies reveal that some anglers respond in more than one way, making accurate predictions of risk even more difficult. Because what anglers specifically do determines the risk, having better information about future behaviors based on actual behaviors would provide a more complete picture of actual and future risk.

REFERENCES

1. K. M. Ward and J. W. Duffield. *Natural Resource Damages: Law and Economics.* Wiley, New York, 1992.

2. R. J. Kopp and V. K. Smith. *Valuing Natural Assets: The Economics of Natural Resource Damage Assessment.* Resources for the Future, Washington, DC, 1993.

3. P. D. Boehm, P. D. Galvani, and P. J. O'Donnell. Scientific and legal conundrums in establishing injury and damages: The natural resource assessment regulations. In R. B. Stewart (Ed.), *Natural Resource Damages: A Legal, Economic, and Policy Analysis.* National Legal Center for the Public Interest, Washington, DC, 1995.

4. J. A. Hausman, G. K. Leonard, and D. McFadden. Assessing use value losses caused by natural resource injury. In J. Hausman (Ed.), *Contingent Valuation: A Critical Assessment.* Elsevier Science, North-Holland, 1993.

5. F. E. Sharples, R. W. Dunford, J. J. Bascietto, and G. W. Suter II. Integrating natural resource damage assessment and environmental restoration at federal facilities. *Fed. Facilities Environ. J.* Autumn: 295–317 (1993).

6. Texas Parks and Wildlife Department. Unpublished creel survey data for Lavaca Bay, Austin, TX, undated.

7. L. M. Green, H. R. Osburn, G. C. Matlock, and R. B. Ditton. Development of a social and economic questionnaire for on-site interviews of marine recreational anglers in Texas. *Am. Fish. Soc. Symp.* 12: 406–412 (1991).

8. H. R. Osburn, M. R. Osborn, and H. R. Maddux. Trends in finfish landings by sport-boat fishermen in Texas marine waters, May 1974–May 1987. Management Data Series 150. Texas Parks and Wildlife Department, Coastal Fisheries Branch, Austin, TX, 1988.

9. Texas Department of Health. Preliminary public health assessment for Alcoa (Point Comfort)/Lavaca Bay. Texas Department of Health under cooperative agreement with Agency for Toxic Substances and Disease Registry, Atlanta, GA, 1995.

10. Alcoa. *Volume B7a: Finfish and Shellfish Consumption Study,* Phase 1, General Population Study. Radian International, Austin, TX, April 30, 1996.

11. U.S. Environmental Protection Agency. *Exposure Factors Handbook.* Office of Health and Environmental Assessment, Washington, DC, 1989.

12. Alcoa. *Volume B12a: Finfish and Shellfish Sampling to Support Human Health Baseline Risk Assessment,* data report. Radian International, Austin, TX, May 1997.

13. U.S. Environmental Protection Agency. *Exposure Factors Handbook,* Volume II: *Food Ingestion Factors.* EPA/600/P-95/002Fa. Office of Research and Development, National Center for Environmental Assessment, Washington, DC, Aug. 1997.

14. National Marine Fisheries Service. Marine recreational fisheries statistics survey, Atlantic Coast. U.S. Dept. of Commerce, data tapes provided to U.S. EPA, National Center for Environmental Assessments, 1993.

15. N. E. Bockstael, K. E. McConnell, and I. E. Strand. A random utility model for sportfishing: Some preliminary results for Florida. *Marine Resource Econ.* 6(3): 245–260 (1989).

16. G. R. Parsons and M. S. Needelman. Site aggregation in a random utility model of recreation. *Land Econ.* 68(4): 418–433 (1992).

17. Y. Kaoru, V. K. Smith, and J. L. Liu. Using random utility models to estimate the recreational value of estuarine resources. *Am. J. Agric. Econ.* 77: 141–151 (1995).

18. U.S. Environmental Protection Agency (EPA), *Consumption Surveys for Fish and Shellfish: A Review and Analysis of Survey Methods.* EPA 822/R-92-001. Office of Water, EPA, Washington, DC, 1992.

19. C. H. Backstrom and G. Hursh-Cesar. *Survey Research.* Wiley, New York, 1981.

20. D. A. Dillman. *Mail and Telephone Surveys: The Total Design Method,* Wiley, New York, 1978.

21. P. H. Rossi, J. D. Wright, and A. B. Anderson. *Handbook of Survey Research.* Academic, Orlando, FL, 1983.

22. J. H. Frey. *Survey Research by Telephone.* Sage, Newbury Park, 1989.

23. R. M. Groves, P. B. Biemer, L. E. Lyberg, J. T. Massey, W. L. Nicholls II, and J. Waksberg. *Telephone Survey Methodology.* Wiley, New York, 1988.

24. S. Sudman. *Asking Questions,* Jossey-Bass, San Francisco, CA, 1982.

25. P. M. Rodier, W. S. Webster, and J. Langman. Morphological and behavioral consequences of chemically-induced lesions of the CNS. In H. R. Ellis (Ed.), *Aberrant Development in Infants: Human and Animal Studies.* Lawrence Erlbaum, Hissdale, NJ, 1975, pp. 177–185.

26. J. Langman, W. S. Webster, and P. M. Rodier. Morphological and behavioral abnormalities caused by insults to the CNS in the perinatal period. In C. L. Berry and D. E. Poswillo (Eds.), *Teratology: Trends and Applications.* Springer-Verlag, New York, 1975, pp. 182–200.

27. L. Green. Personal communication with Kristy Mathews of Triangle Economic Research. Texas Parks and Wildlife Department, February 3, 1997.

28. Pasadena Hotspot. Boat Fishing and Wade Fishing Map, Matagorda Bay, Map No. F108. Pasadena Hotspot, Pasadena, TX, undated.

29. R. M. Groves. *Survey Errors and Survey Costs.* Wiley, New York, 1989.

30. E. S. Ebert, P. S. Price, and R. E. Keenan. Selection of fish consumption estimates for use in the regulatory process. *J. Exposure Anal. Environ. Epidemiol.* 4(3): 373–393 (1994).

31. H. W. Puffer, S. P. Azen, M. J. Duda, and D. R. Young. Consumption rates of potentially hazardous marine fish caught in the metropolitan Lost Angeles Area. U.S. Environmental Protection Agency Grant no. R807 120010, 1981.

32. P. Price, S. Su, and M. Gray. *The Effects of Sampling Bias on Estimates of Angler Consumption Rates in Creel Surveys.* ChemRisk, Portland, ME, 1994.

33. R. S. Pierce, D. T. Noviello, and S. H. Rogers. Commencement Bay seafood consumption report. Preliminary report, Tacoma-Pierce County Health Department, Tacoma, WA, 1981.

34. Westat. *Investigation of Possible Recall/Reference Period Bias in National Surveys of Fishing, Hunting, and Wildlife-Associated Recreation.* U.S. Fish and Wildlife Service, Washington, DC, 1989.

35. P. C. West, J. M. Fly, R. Marans, F. Larkins, and D. Rosenblatt. *1991–92 Michigan Sport Anglers Fish Consumption Study*. Final report to the Michigan Great Lakes Protection Fund, Michigan Department of Natural Resources, Ann Arbor, MI, 1993.

36. B. Jones Foire, H. A. Anderson, L. P. Hanrahan, L. J. Olson, and W. C. Sonzogni. Sport fish consumption and body burden levels of chlorinated hydrocarbons: A study of Wisconsin anglers. *Arch. Environ. Health* 44(2): 82–88 (1989).

37. N. A. Connelly, B. A. Knuth, and J. E. Vena. *New York Angler Cohort Study: Health Advisory Knowledge and Related Attitudes and Behavior, with a Focus on Lake Ontario*. Human Dimensions Research Unit, Department of Natural Resources, New York State College of Agriculture and Life Sciences, Series No. 93-9, Sept. 1993.

38. B. A. Knuth, N. A. Connelly, and M. A. Shapiro. *Angler Attitudes and Behavior Associated with Ohio River Health Advisories*. Human Dimensions Research Unit, Department of Natural Resources, New York State College of Agriculture and Life Sciences, Series No. 93-6, July 1993.

SECTION K
Assessing Risks to Birds

22 Using Probabilistic Risk Assessment Methods to Predict Effects of Pesticides on Aquatic Systems and Waterfowl That Use Them

PATRICK J. SHEEHAN and JOHN WARMERDAM

Exponent, Oakland, California

SHIH SHING FENG

PE Biosystems/Celera, Foster City, California

22.1 INTRODUCTION

One of the most comprehensive of the early prospective ecological risk assessments published is the Canadian Wildlife Service study entitled "The Impact of Pesticides on the Ecology of Prairie Nesting Ducks" (Sheehan et al., 1987). This assessment was conducted to evaluate the potential direct toxic effects of the more commonly used organochlorine and organophosphate insecticides and the more recently developed pyrethroid insecticides on aquatic macroinvertebrates in sloughs in the pothole region of Canada and the subsequent indirect impact of invertebrate mortality (loss of food resources) on duckling survival and recruitment. Although this assessment provided a thorough and quantitative analysis of the potential risks to duck populations associated with the aerial spraying of insecticides in the agricultural region of the Canadian Prairie and was used to develop a program to manage those risks, the assessment was incomplete by today's standards in that it did not include a quantitative analysis of uncertainties (SETAC, 1998). A probabilistic approach, where input parameters are characterized as distributions of plausible values, is now employed

Human and Ecological Risk Assessment: Theory and Practice, Edited by Dennis J. Paustenbach
ISBN 0-471-14747-8 © 2002 John Wiley & Sons, Inc.

to represent the variability inherent in diverse populations as well as the uncertainty implicit in the quantification of environmental factors. This case study provides a reanalysis of the Canadian Wildlife Service study data using probabilistic methods to characterize variabilities and uncertainties in estimates of exposure and the direct effects on aquatic macroinvertebrates from pesticides released to prairie pothole sloughs as the result of aerial spraying for agricultural pest control. The assessment also shows how the results of this analysis can be used to relate uncertainties in the magnitude of mortality in the macroinvertebrate community and the food resource for ducklings to associated indirect effects on ducklings and risks to duck populations.

22.2   BACKGROUND

The pothole region of the United States and Canada is the principal grain-growing region of North America. The region is geographically defined as a swath of land crossing the provinces of Alberta, Saskatchewan, and Manitoba in Canada and the states of Minnesota, North and South Dakota, Iowa, and Montana in the United States and is dotted with a vast number of freshwater ponds that are primarily small, shallow, and biologically productive sloughs ringed with aquatic plants. For the purposes of this chapter, *slough* and *pothole* are used interchangeably and refer to any depression holding water. Sloughs are critical habitat for a wide variety of aquatic organisms and wildlife, most notable of which are the estimated 16 million ducks that nest there annually. There is an intimate relationship between the ecology of these sloughs and the essential resources they provide to the waterfowl that inhabit them for part of each year. The overlap between agriculture and waterfowl habitat has been of concern to wildlife agencies for more than a decade. The potential impact of the agricultural use of pesticides in the pothole habitat began to receive attention from the Canadian Wildlife Service in the mid-1980s (Shaw et al., 1984; Mineau et al., 1987; Sheehan et al., 1987; Forsyth, 1989) and the U.S. Fish and Wildlife Service (Grue et al., 1986, 1989).

Assessing the indirect risks to waterfowl posed by pesticide use in the pothole region requires an understanding of the extent of spatial and temporal overlap of pesticide spraying and aquatic macroinvertebrates and ducks in this duck nesting habitat. The prairie pothole region of North America covers about 300,000 mi^2 (770,000 km^2). The average density of sloughs in the pothole region is 12 km^{-2}, with values ranging from 4 to 40 km^{-2} (Smith et al., 1964). Seasonal and annual variability in the number of sloughs is high. The average number of sloughs decreases, due to drying, 20 to 35 percent between May and July; over 80 percent of the ponds lost are less than 0.5 acres (0.2 ha) in size. More than 8 million pairs of ducks breed and rear young in small prairie sloughs each year. Approximately 93 percent of the breeding ducks in the prairie region nest in agricultural land. Based on potential land use in the prairie region, from 60 to 80 percent of the best waterfowl breeding and rearing areas overlap with the best agricultural land in the pothole region (Sheehan et al., 1987). The reproductive process for ducks takes place in these areas between May and September. Typically, this is also the period during which farmers do most of the spraying of grain crops for pest control.

Of the insect pests of prairie crops, grasshoppers are potentially the most important because they occasionally give rise to outbreaks that cover large areas. Grasshopper infestations in the prairies have affected between 3 and 21 million ha annually (Sheehan et al., 1987). Control of these grasshoppers is considered necessary by farmers; thus large quantities of pesticides are applied during outbreaks. The actual areas sprayed with pesticides each year are more difficult to determine. It has been estimated that 3 to 4 million ha are sprayed during years of serious grasshopper outbreaks. This represents about 10 percent of the agriculturally cultivated land in the pothole region. The primary insecticides sprayed during grasshopper infestations are carbofuran, carbaryl, and pyrethroid compounds such as deltamethrin. From 10 to 25 percent of the spraying for grasshoppers is done aerially.

Typically, spraying to control grasshoppers occurs from late May to August. This time frame includes two of the critical periods in the annual reproductive process of all duck species: egg formation and early foraging by newly hatched ducklings. During the reproductive period, adults and ducklings are highly dependent on food from pothole sloughs, particularly aquatic macroinvertebrates. The nutritional demands of hens for egg formation are met by including high quantities of these macroinvertebrates in their diet (Krapu 1974a, b, 1979). All hens consume more than 70 percent animal food during this period (Noyes and Jarvis, 1985). Aquatic insects, crustaceans, and snails are highly selected, presumably because of their protein and calcium content. Although there are less data available on the diet of ducklings than on the diet of laying hens, evidence indicates a likewise high dependence on animal food, primarily aquatic macroinvertebrates, for a period of 1 to 7 weeks posthatch, depending on the species.

There is a substantial spatial and temporal overlap between pesticide spraying and reproducing ducks in the prairie pothole region. This overlap indicates the potential for direct exposure of both adults and ducklings to the pesticides. In addition, due to the importance of macroinvertebrates in the diets of laying hens and ducklings and the widespread aerial application of pesticides for control of grasshoppers, there is also the potential of broad-scale pesticide-induced mortality within the exposed sloughs' macroinvertebrate community and subsequent indirect effects on duck reproductive success.

22.3 PROBLEM FORMULATION

Problem formulation is the first step in an ecological risk assessment. The problem formulation identifies the ecological receptors to be evaluated, the chemicals of interest, the relevant exposure scenarios, measurement and assessment endpoints, the risk assessment approach, and other specifications or limits of the assessment.

22.3.1 Receptors of Interest

Although the risks of pesticide use on duck populations are of ultimate interest, this assessment is focused on estimating the probability and magnitude of the effects of spraying pesticides (aerially) on aquatic macroinvertebrates, a critical food resource for reproducing ducks. Therefore, the aquatic macroinvertebrate community in pothole sloughs is the receptor of primary interest in this assessment.

The high diversity of microhabitats in pothole sloughs supports a diverse invertebrate fauna. Species in 38 orders and families have been reported in these sloughs (Perret, 1962; Sugden, 1964; Swanson et al., 1974). It is the species of classes Gastropoda, Crustacea, and Insecta that are of greatest interest because they are the preferred invertebrate food of ducks. The families of macroinvertebrates from these three classes commonly found in prairie sloughs are given in Table 22.1. The taxa of macroinvertebrates most often cited as dominant in freshwater sloughs are gastropods, amphipods, notonectids, corixids, and the larvae of dipterans, tricopterans, odonates, and ephemeropterans (Joyner, 1980, 1982). These taxa are generally found on submergent vegetation and associated with a variety of substrates such as rock, sand, silt, or detritus. It is clear the dominant macroinvertebrate taxa in these sloughs are also the preferred prey for ducklings (Table 22.2).

The vegetation provides habitat for the greatest number of macroinvertebrates. The abundance of macroinvertebrates associated with aquatic plants has been estimated to range between 1000 and 6000 individuals/m^2 of the vegetated area of slough (Biggs and Malthus, 1982; Scheffer et al., 1984; Keast, 1984). The total invertebrate abundance and/or biomass usually reaches the highest value in late spring or early summer (late May to late June) in pothole sloughs (Swanson et al., 1974; Joyner, 1982).

22.3.2 Chemicals of Interest

The most commonly used insecticides for grasshopper control on cereal grain, oil seed, and forage crops in the prairie pothole region are azinphos-methyl, carbaryl, carbofuran, cypermethrin, deltamethrin, dimethoate, malathion, and methamidophos (Saskatchewan Agriculture, 1984; Manitoba Agriculture, 1984). The three most widely used insecticides in the Canadian prairies for grasshopper control in the late 1980s and 1990s were carbofuran, carbaryl, and deltamethrin. This assessment is focused on two of these compounds: carbaryl and deltamethrin. These compounds represent the widely used organophosphate insecticides and more recently developed synthetic pyrethroid insecticide groups.

22.3.3 Assessment and Measurement Endpoints

Assessment endpoints are the characteristics of the ecological system to be protected. In this case, it is the protection of the macroinvertebrate community

TABLE 22.1 Invertebrate Taxa Commonly Found in Prairie Sloughs

Phylum	Order (Suborder)	Family	Common Name
Protozoa			Protozoans
Porifera	Haploscelerina	Spongillidae	Freshwater sponges
Coelenterata			Hydras
Plathelminthes			Flatworms
Gastroticha			Gastrotrichs
Rotatoria			Rotifers
Nematoda			Nematodes
Annelida			Aquatic earthworms
Hirundinea			Leeches
Gastropoda	Basommatophora	Physidae	Freshwater snails
		Lymnaeidae	(e.g., *Lymnae*, *Physa*, *Helisoma*)
		Planorbidae	
Pelecypoda		Unionidae	Freshwater mussels
		Sphaeriidae	Freshwater clams
Arthropoda	Acarina		
	Trombidiformes		Water mites
Crustacea			
Branchiopoda	Anostraca		Fairy shrimps
	Notostraca		Tadpole shrimps
	Diplostraca		
	Conchostraca		Clam shrimps
	Cladocera		Water fleas
Ostracoda			Seed shrimp
Copepoda			Copepods
Malacostraca	Isopoda		Isopods (aquatic sow bugs)
	Amphipoda		Amphipods (scuds, side swimmers)
Insecta	Collembola		Springtails
	Emphemeroptera		Mayflies
	Odonata		
	Anisoptera		Dragonflies
	Zygoptera		Damselflies
	Hemiptera	Gerridae	Water striders
		Notonectidae	Back swimmers
		Belostomatidae	Giant water bugs
		Corixidae	Water boatmen
	Neuroptera		Dobsonflies
	Trichoptera		Caddisflies
	Coleoptera		Aquatic beetles
		Dystricidae	Predaceous diving beetles
	Diptera		Flies, mosquitoes, midges
		Culicidae	Mosquitos, phantom midges
		Chironomidae	True midges
		Stratiomyidae	Soldier flies

Source: From Perret (1962), Sugden (1964), and Swanson et al. (1974).

TABLE 22.2   **Macroinvertebrate Prey Selected by Flightless Ducklings Foraging in Prairie Pothole Region**

Species	Selected Macroinvertebrate Prey in Approximate Order of Importance	Reference
Dabblers		
Mallard	True fly larvae, pupae (primarily Chironomidae)	Chura, 1961; Perret, 1962
	Caddisfly larvae (primarily Limnophilidae)	
	Aquatic snails (particularly Lymnaeidae)	
	Dragon- and damselfly nymphs (Odonata)	
	Water boatman adults, nymphs (Corixidae)	
	Beetle larvae, adult (Hydrophilidae, Dytiscidae)	
Divers		
Lesser scaup	Scuds (Amphipods)	Bartonek and Hickey,
	Aquatic snails (Lymnaeidae)	1969a, b; Sugden, 1973
	Midge larvae, pupae (Chironomidae)	
	Dragonfly naiads (Zygoptera)	
	Caddisfly larvae (Leptoceridae)	
	Beetle larvae (Haliplidae, Dytiscidae)	

from reductions in survival, growth, and/or reproduction resulting from exposure to the pesticide in water. The measurement endpoint is the comparison of estimated concentrations of the pesticide in water with concentrations determined to be toxic in laboratory toxicity tests with macroinvertebrate species.

22.3.4 Conceptual Assessment Model

The aquatic macroinvertebrate community in pothole sloughs is potentially exposed to pesticides sprayed aerially for grasshopper control when the ponds are directly sprayed and/or when drift from spraying in the upwind direction is deposited on the ponds. The highest concentrations in slough water resulting from spraying occur within the initial 24 hours after spraying as the pesticide mixes into the water column. It is the likelihood and magnitude of invertebrate mortality during the initial high exposure period that are of interest in this assessment. Macroinvertebrates are exposed to the pesticides through gill absorption, dermal uptake, and ingestion of the chemical in water and food. In this case, the primary measure of exposure is the estimated concentration of the pesticide in slough water. Sediment concentrations are not considered in this analysis, as steady-state partitioning between the water column and sediments is unlikely to be reached in the short time frame considered immediately after pesticide application. A distribution of exposure concentration is developed based on the variability and uncertainties in application and slough parameters. Effects on the macroinvertebrate community are characterized using mortality data from laboratory toxicity tests with aquatic insect and crustacean species. Effects distributions representing lethal concentrations for the "community" of test species are developed from test data. The exposure concentration distribution is then compared to the effects distribution to characterize the probability and magnitude of mortality in the exposed aquatic macroinvertebrate community.

22.4 RISK ASSESSMENT APPROACH

To assess the risks of aerial spraying of pesticides on the macroinvertebrate community in the slough ecosystem, a number of exposure assessment, effects assessment, and risk characterization tasks must be completed. These include the following:

Exposure assessment:
  • Estimation of the distribution of concentrations of aerially sprayed pesticides in slough water based on application rates and efficiencies and slough conditions

Effects assessment:
  • Compilation of toxicity test data for aquatic macroinverebrates for the pesticides
  • Development of distributions of effect concentrations for the aquatic macroinvertebrate community

Risk characterization:
  • Comparison of the overlap in the estimated concentration distribution with the lethal effects distributions for the macroinvertebrates
  • Calculation of the probability of the exposure concentration exceeding the effects concentration for macroinvertebrate species
  • Calculation of the probability of the exposure concentration exceeding specific percentiles of the effects concentration

22.5 EXPOSURE ASSESSMENT

Exposure assessment is the process used to characterize the magnitude, frequency, and duration of exposure concentrations or doses of a chemical to the receptors of interest. In this case, exposure is characterized as the estimated distribution of pesticide concentrations in water in oversprayed sloughs. Water column concentrations are estimated based on application rate, application efficiency, slough size and depth, and mixing assumptions.

22.5.1 Considerations

The entry of pesticides into pothole sloughs is expected to occur through direct application (overspray), droplet drift, runoff, or seepage. The likelihood of direct application is highly dependent on the selected mode of application and is greatest for aerial application. Analysis of direct overspray in the pothole region depends on assumptions about applicator ability, specific limitations in aircraft control, the effectiveness of buffer zones, and the size, number, and locations of the sloughs. For small sloughs (less than 0.4 ha) with narrow margins, which are common in cultivated lands, partial overspray is highly probable. These ponds may account for 60 to 80 percent of the sloughs available to ducks in May and June, when pesticide spraying is likely. In the worst case of direct aerial overspray, 100 percent of the applied amount would be deposited on the sloughs and margins. Under a variety of meteorological and application conditions, on-target deposits have been found to vary substantially and rarely approach 100 percent. An average on-target deposit of 50 percent of the application rate has been estimated from a single-swath application. Each additional overflight of a slough can potentially increase the amount of the pesticide deposited into the slough. Overspray deposits of approximately 80 percent of applied quantities are expected from multiswath applications common in grasshopper control (Ware et al., 1970). Therefore, in areas being sprayed aerially, the amount of pesticide reaching slough water is largely a function of the application rate and the number of swaths sprayed.

The second substantial source of pesticide contamination is drift droplet deposition from upwind application. The concentration of pesticide deposited off-target decreases rapidly with distance. Studies show that deposits at 50 m downwind of the direct area of application ranged from 1 to 10 percent of applied quantities (Argauer et al., 1968; Ware et al., 1969; Renne and Wolf, 1979; Currier et al., 1982). In contrast, runoff is expected to be a relatively small contributor in areas of aerial application, perhaps less than 1 percent (Wauchope, 1978). In this case, only deposition from direct overspray due to a single-swath application is considered in estimating pesticide concentrations in slough water.

The other primary factors to be considered in estimating the exposure concentration in slough water are the slough conditions. These include water depth and mixing conditions for incorporating the deposited mass of pesticide and the processes that would remove the pesticide from the water column. Among the assumptions made about these sloughs is that mixing occurs through the entire water column more rapidly than other rate-governed processes (partitioning, degradation) and that the surface film effect and other physical and chemical processes (volatilization, photodegradation) will have negligible influence on water column concentrations in the short term. This is a reasonable assumption because the time frame within which the initial high concentration is achieved in slough water is short, generally less than 24 hours, and is insufficient for rate-governed transfer and removal processes to substantially affect concentrations in the water column.

22.5.2 Exposure Methods

The concentration of the pesticide in slough water shortly after spraying can therefore be estimated as a function of the application rate, the application efficiency, and the depth of the water body that is oversprayed, assuming that aerial applications are generally single swath and the additional contribution from drift is negligible. The application rate is expressed in terms of grams of pesticide applied per square meter; application efficiency as the fraction of the applied pesticide that reaches the water body; and depth of the water body as the average depth, in meters. The concentration in the water column in grams per cubic meter or milligrams per liter due to direct overspray is calculated as follows:

$$\text{Concentration} = \frac{\text{application rate} \times \text{application efficiency}}{\text{depth}}$$

It is expected that water column concentration distributions based on this methodology are representative of operating conditions as drift input, which would increase deposition, is ignored and surface film effects and removal processes, which would limit water column concentrations, are discounted.

22.5.3 Application Rate

Application rates for individual pesticides are recommended by the manufacturer and by governmental agricultural agencies. Rates recommended by insect control agencies in Alberta, Manitoba, and Saskatchewan are 7.5 g/ha for deltamethrin and 1100 g/ha for carbaryl (Sheehan et al., 1987). Variation in application rates is introduced through mechanical, physical, and chemical processes and human error. Applicators introduce error when mixing volumes of pesticides with adjuncts and the spray medium (typically water). Pesticides are metered out through nozzles on the trailing edge of airplane wings. Improperly adjusted or malfunctioning nozzles or pumps affect the application rate. Fluid dynamics (settling, miscibility, and density effects) of the pesticide and mixtures also cause variation in the application rate over the course of a single application. Pilot error with respect to ground speed and application altitude also contributes to variation in application rates. These and other factors combine to produce a range of possible application rates for each pesticide. Little data exist to help define the extent of this type of variability with respect to recommended application rates. We assume that the suggested application rate is the mean value of the distribution of actual application rates and that the standard devi-

TABLE 22.3   Summary Statistics for Application Rates

Pesticide	Mean Value (mg/m^2)	Standard Deviation (mg/m^2)	Minimum Value (mg/m^2)	Maximum Value (mg/m^2)
Carbaryl	110	11	77	143
Deltamethrin	0.75	0.075	0.525	0.975

ation of the application rate is 10 percent of the suggested application rate. We further assume that application rates are normally distributed and that the distribution is truncated on both ends, three standard deviations from the mean. This distribution provides a representation of application rates that includes the variability and uncertainty described previously. Table 22.3 and Figures 22.1 and 22.2 summarize characteristics of the random variables used to describe the application rates for carbaryl and deltamethrin.

22.5.4   Application Efficiency

Application efficiency is a measure of how much of the applied pesticide actually reaches its intended target and is affected primarily by physical and chemical processes. Wind speed and droplet size are the two most important factors in determining application efficiency. Smaller droplets tend to stay airborne longer, leaving them vulnerable to lateral dispersion by wind. The pesticide formulation and the atomization process used for pesticide application determine the droplet size distribution. Pesticide formulations vary in specific characteristics, such as surface tension, density, and viscosity, which affect the droplet size distribution. Other factors that influence application efficiency include application height, relative humidity, ambient temperature, sunlight, and backwash turbulence of the applicator aircraft. Empirical studies show that between 14 and 95 percent of applied pesticides are deposited on-target (Maybank et al.,

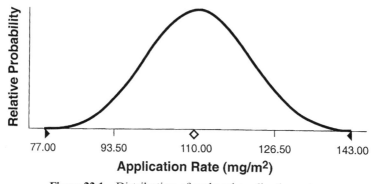

Figure 22.1   Distribution of carbaryl application rates.

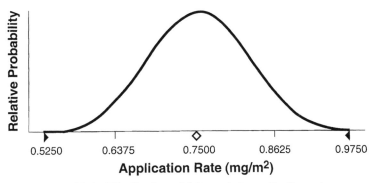

Figure 22.2 Distribution of deltamethrin application rates.

1976, 1978a, b; Renne and Wolf, 1979; Ware et al., 1970, 1984; Hill and Kinniburgh, 1984). None of the data are conclusive in characterizing the shape of the application efficiency distribution. To ensure a conservative concentration estimate, a minimum and maximum application efficiency of 40 and 100 percent are used, and 80 percent is designated as the most likely application efficiency. The distribution of application efficiencies is modeled as a random variable with a triangular distribution, whose characteristics are summarized in Table 22.4 and Figure 22.3.

22.5.5 Pond Depth

The depth of the affected water body is the final variable modeled as a random variable. The four classifications of water bodies used in this study are ephemeral potholes, temporary potholes, semipermanent potholes, and permanent wetlands. The depth of these water bodies varies depending on weather and time of year. The depth and permanency of these water bodies are given in Table 22.5.

Breeding pairs of birds exhibit preferences for different water body types. Kantrud and Stewart (1984) demonstrated that 64 percent of breeding pairs nested in or near semipermanent potholes, 33 percent chose temporary potholes, 2 percent chose permanent wetlands, and 1 percent were found in ephemeral potholes. From this information, a custom distribution is constructed to model water body depth (Table 22.5, Figure 22.4).

TABLE 22.4 Characteristics of Application Efficiency Distribution

Pesticide	Minimum (fraction or target)	Maximum (fraction or target)	Most Likely Value (fraction or target)
Carbaryl	0.4	1.00	0.80
Deltamethrin	0.4	1.00	0.80

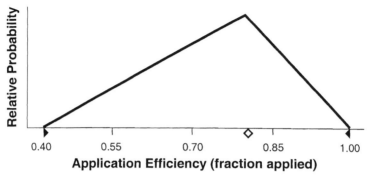

Figure 22.3   Insecticide application efficiency distribution.

TABLE 22.5   Depth and Permanency of Water Body Types

Water Body Type	Depth (m)	Permanency	Probability of Use by Ducklings
Ephemeral potholes	0–0.3	Few days to few weeks	0.01
Seasonal potholes	0.3–0.61	4–12 weeks	0.33
Semipermanent potholes	0.61–1.52	Several years	0.64
Permanent wetlands	1.52–3.0	Indefinite	0.02

22.5.6   Exposure Concentration Distributions

Exposure concentration distributions for carbaryl and deltamethrin were estimated from input parameter distributions using Crystal Ball version 4.0 and 100,000 Monte Carlo simulations. Figures 22.5 and 22.6 present the estimated frequency distributions for carbaryl and deltamethrin concentrations in pothole sloughs. Table 22.6 presents the summary statistics for the estimated pesticide concentrations.

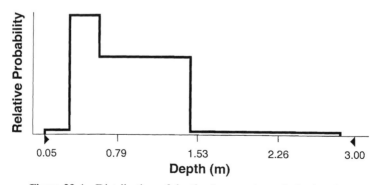

Figure 22.4   Distribution of depth of waters in pothole sloughs.

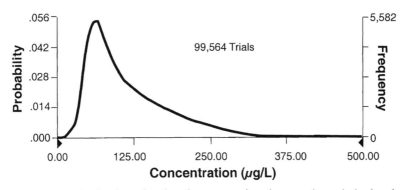

Figure 22.5 Distribution of carbaryl concentrations in water in pothole sloughs.

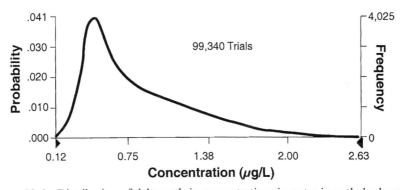

Figure 22.6 Distribution of deltamethrin concentrations in water in pothole sloughs.

A series of standard continuous distributions (including normal, lognormal, and Weibull) were fit to the concentration data for each pesticide. The lognormal distribution provided the best fit of the data, and the concentration distributions are characterized as approximately lognormal. These data are used to characterize the range and distribution of exposure concentrations to which macroinvertebrates might be exposed in oversprayed sloughs.

TABLE 22.6 Summary Statistics for Water Column Concentrations

Pesticide	Mean Value (μg/L)	Standard Deviation (μg/L)	Minimum Value (μg/L)	Maximum Value (μg/L)
Carbaryl	119	84.2	14.5	2080
Deltamethrin	0.811	0.57	0.102	13.9

22.6 EFFECTS ASSESSMENT

The effects assessment is a quantitative description of the relationship between the concentration or dose of the chemical and the type and incidence of adverse effects elicited in exposed receptors. The most commonly reported toxicological benchmark from laboratory toxicity tests is the LC_{50}, that is, the concentration killing 50 percent of the test organisms within a specific exposure period, normally 24, 48, or 96 hours. The LC_{50} data, if reported with confidence limits or the slope of the concentration–response curve, can be used to estimate lethal concentrations for various percentages of the test organisms (e.g., LC_1, LC_{10}, LC_{20}, LC_{30}, LC_{40}, LC_{60}, LC_{70}, LC_{80}, LC_{90}, LC_{99}). These data are then used to develop an effects distribution for each test species and aggregated to develop effects distributions for the "community" of invertebrate test species. It is these lethal concentration distributions that are used to characterize the effects relationship for the macroinvertebrate community.

22.6.1 Toxicity Data

The concentration–response relationships for carbaryl and deltamethrin are characterized based on toxicity test data for aquatic invertebrate species. The effects of acute exposures to carbaryl have been evaluated in tests with 23 aquatic macroinvertebrate species (Table 22.7). These include data for mosquitoes, midges, stoneflies, mayflies, dragonflies, amphipods, and daphnids. The effects of acute exposures to deltamethrin in water have been evaluated for 22 species; however, these represent primarily midges and mosquitoes (Table 22.8). To provide test data for a broader range of macroinvertebrate taxa, data for a toxicologically similar pyrethroid pesticide, cypermethrin, are included in this evaluation (Table 22.8). The 24-hour LC_{50} values for cypermethrin and deltamethrin for the same genera of mosquitoes and midges are generally within a factor of 4 (Sheehan et al., 1987).

22.6.2 Toxicity Model

The response endpoint of interest is the lethal concentration associated with a 24-hour exposure. Since not all invertebrate tests report lethal concentrations for 24-hour exposure periods, the initial step in constructing distributions of lethal concentrations was to extrapolate LC_{50} values for 48- and 96-hour exposure periods to equivalent values for the 24-hour exposure period of interest. This extrapolation was accomplished with an integrated toxicity model based on the premise that lethal toxicity is a function of percent mortality and exposure time. Discounting any lag time or delayed effects in response to exposure, the relationship between the median lethal concentration of the chemical and the duration of exposure can be expressed as $LC_{50} \times$ exposure duration = toxicity constant, which approximates the integrated median lethal exposure. Although the relationship between lethality and time does not hold for all pesticides and aquatic invertebrate species, it has been shown to be a good general model for estimating lethal concentrations from 24- to 96-hour tests (Allison, 1977; Sheehan et al., 1987; French, 1991).

Lethal concentrations are generally estimated from test data using the log-probit model; that is, percent mortality is a lognormal function of exposure concentration [Finney, 1971; U.S. Environmental Protection Agency (EPA), 1975; Stephan, 1977]. Using the log-probit model and the reported or extrapolated 24-hour LC_{50} value and confidence limits or concentration–response slope, the LC_1, LC_{10}, LC_{20}, LC_{30}, LC_{40}, LC_{60}, LC_{70}, LC_{80}, LC_{90}, and LC_{99} were estimated for populations of each invertebrate test species exposed to carbaryl or deltamethrin. For the cases where neither confidence limits nor concentration–response slope were reported, the slope for a taxonomically similar species was used to calculate the lethal concentrations for the selected percentages of response.

22.6.3   Lethal Concentration Profiles

The lethal concentration profiles for populations of invertebrates for which relevant data are available are shown in Tables 22.9 and 22.10 for carbaryl and deltamethrin, respectively. These profiles were used to characterize the lethal concentration distribution for the population of each test species. A similar approach was used by French (1991) to develop the toxicity database that is used in the Natural Resource Damage Assessment Model to estimate mortality of aquatic biota following spills of toxic substances in both freshwater and saltwater environments.

If one assumes that the species tested reasonably represent a cross section of the invertebrates in prairie sloughs, then the lethal concentration profiles for test species can be aggregated to develop effects distributions for the aquatic invertebrate community. That is, probability density distributions can be developed for each lethal concentration percent with the range and shape of the distribution representing the sensitivity of the community to acute exposures to the pesticide. Although it is clear that the test species do not represent the full suite of invertebrate taxa in prairie sloughs (see Table 22.1), they do represent reasonably well the dominant taxa in these freshwater wetlands and the preferred prey of ducklings (see Table 22.2). The taxa of invertebrates most often cited as dominant in prairie sloughs are the gastropods, amphipods, notonectids, corixids, larval dipterns, tricopterns, odonates, and ephemeropterns (Dvorak, 1970; Krull, 1970; Joyner, 1980, 1982; Mittlebach, 1981; Biggs and Malthus, 1982; Dvorak and Best, 1982). Daphnids are also quite common in deeper sloughs (Collias and Collias, 1963). Of the dominant taxa, tests with carbaryl do not include data for gastropods, notonectids, corixids, or tricopterns. For deltamethrin, there are no test data for gastropods, notonectids, tricopterns, or odonates. Perhaps the most important taxa for which no test data are available are the gastropods. While not the most numerous taxa, gastropods often contribute the most to the total invertebrate biomass in prairie sloughs. Available data for

TABLE 22.7 Summary of Toxicity Data for Carbaryl

Species Tested	Stage	Test Type	Toxicity Reported	LC$_{50}$ Concentration, μg/L (95% fiducial limits)	Slope of Concentration–Response Curve	References[a]
Mosquito						
Culex p. pipiens	4th instar	S	24 h LC$_{50}$	75	1.77	Rawash et al., 1975
Aedes cantans	4th instar	S	24 h LC$_{50}$	377	1.86	Rettich, 1977
Aedes vexans	4th instar	S	24 h LC$_{50}$	322	1.63	Rettich, 1977
Aedes excrucians	4th instar	S	24 h LC$_{50}$	145	3.02	Rettich, 1977
Aedes communis	4th instar	S	24 h LC$_{50}$	168	2.59	Rettich, 1977
Aedes punctor	4th instar	S	24 h LC$_{50}$	298	2.10	Rettich, 1977
C. p. pipiens	4th instar	S	24 h LC$_{50}$	333	1.55	Rettich, 1977
Culex p. molestus	4th instar	S	24 h LC$_{50}$	418	1.80	Rettich, 1977
Culiseta annulata	4th instar	S	24 h LC$_{50}$	180	1.62	Rettich, 1977
Aedes aegypti	2nd instar	S	96 h LC$_{50}$	1.9	—	Lakota et al., 1981
Phantom midge, *Chaoborus* sp.	larvae	S	24 h LC$_{50}$	650	1.90	Bluzat and Seuge, 1979
Midge, *Chironomus plumosus*	larvae	S	48 h LC$_{50}$	10	—	Sanders et al., 1983
Stonefly						
Pteronarcys californica	30–35 mm	S	24 h LC$_{50}$	30 (22–40)	—	Sanders and Cope, 1968
	30–35 mm	S	48 h LC$_{50}$	13 (10–16)	—	Sanders and Cope, 1968
	30–35 mm	S	96 h LC$_{50}$	4.8 (3.0–7.7)	—	Sanders and Cope, 1968
Pteronarcella baddia	15–20 mm	S	24 h LC$_{50}$	5.0 (3.6–7.0)	—	Sanders and Cope, 1968
	15–20 mm	S	48 h LC$_{50}$	3.6 (2.9–4.8)	—	Sanders and Cope, 1968
	15–20 mm	S	96 h LC$_{50}$	1.7 (1.4–2.4)	—	Sanders and Cope, 1968
	Nymph		96 h LC$_{50}$	13	—	Woodward and Mauck, 1980
Mayfly						
Cloeon sp.	Nymph	S	48 h LC$_{50}$	480	—	Bluzat and Seuge, 1979
Cloeon dipterum	9.3 mm	S	48 h TLm	370	—	Hashimoto and Nishiuchi, 1981

Species	Size/age		Test	Value	Reference
Dragonfly, *Orthethrum albistylum speciosum*	23 mm	S	48 h TLm	430	Hashimoto and Nishiuchi, 1981
Amphipod				—	
Gammarus lacustris	2 months	S	24 h LC_{50}	40 (32–49)	Sanders, 1969
	2 months	S	48 h LC_{50}	22 (16–30)	Sanders, 1969
	2 months	S	96 h LC_{50}	16 (12–19)	Sanders, 1969
Gammarus fasciatus	—	S	96 h LC_{50}	26 (16–39)	Sanders, 1972
Gammarus pseudolimnaeus	—	S	96 h LC_{50}	4.1–11.9	Woodward and Mauck, 1980
	Mature	S	96 h LC_{50}	16 (12–19)	Sanders et al., 1983
Gammarus pulex	Mature	S	24 h LC_{50}	35 (32–29)	Bluzat and Seuge, 1979
Waterflea					
Daphnia magna	5d	S	48 h LC_{50} immobilization	7.2	Lakota et al., 1981
	1st instar	S	48 h EC_{50} immobilization	5.6 (2.7–12.0)	Sanders et al., 1983
Daphnia pulex	—	S	3 h TLm	30	Hashimoto and Nishiuchi, 1981
	1st instar	S	48 h EC_{50} immobilization	6	Cope, 1966
	—	S	48 h EC_{50} immobilization	6.4 (4.5–8.9)	FWPCA, 1968
Daphnia carinata	2–2.5 mm	S	24 h TLm	100	Santharam et al., 1976
	2–2.5 mm	S	48 h TLm	35	Santharam et al., 1976
Simocephalus serrulatus	1st instar	S	48 h EC_{50} immobilization	8.0	Cope, 1966
	1st instar	S	48 h EC_{50} immobilization	7.6 (6.2–9.3)	Sanders and Cope, 1966

[a]FWPCA, Federal Water Pollution Control Administration.

TABLE 22.8 Summary of Laboratory Studies of Toxicity of Deltamethrin and Cypermethrin

Species Tested	Size/Stage	Test Type	Toxic Effect Reported	Value, µg/L (95% fiducial limits)	Slope of Concentration–Response Curve	References
Deltamethrin						
Mosquito						
Culex p. quinquefasciatus	4 instar	S	24 h LC$_{50}$	0.02	1.68	Mulla et al., 1980
Culex tarsalis	4 instar	S	24 h LC$_{50}$	0.06	1.21	Mulla et al., 1980
Culiseta incidens	4 instar	S	24 h LC$_{50}$	0.3	2.37	Mulla et al., 1980
Aedes nigromaculis	4 instar	S	24 h LC$_{50}$	0.2	2.19	Mulla et al., 1980
Aedes taeniorhynchus	4 instar	S	24 h LC$_{50}$	0.05	1.56	Mulla et al., 1980
Psorophora columbiae	4 instar	S	24 h LC$_{50}$	0.1	2.45	Mulla et al., 1980
Culex pipiens pipiens	4 instar	S	24 h LC$_{50}$	0.19	2.48	Rettich, 1979
Culex pipiens molestus	4 instar	S	24 h LC$_{50}$	0.09	2.05	Rettich, 1979
Culiseta annulata	4 instar	S	24 h LC$_{50}$	0.23	1.84	Rettich, 1979
Aedes cantans	4 instar	S	24 h LC$_{50}$	0.03	2.70	Rettich, 1979
Aedes sticticus	4 instar	S	24 h LC$_{50}$	0.02	2.57	Rettich, 1979
Aedes vexans	4 instar	S	24 h LC$_{50}$	0.09	2.95	Rettich, 1979
Midge						
Tanytarsus spp.	4 instar	S	24 h LC$_{50}$	0.016	2.07	Ali et al., 1978
Procladius spp.	4 instar	S	24 h LC$_{50}$	0.029	2.79	Ali et al., 1978
Chironomus decorus	4 instar	S	24 h LC$_{50}$	0.23	2.54	Ali et al., 1978
Chironomus utahensis	4 instar	S	24 h LC$_{50}$	0.29	2.16	Ali and Mulla, 1978
Crictopus spp.	4 instar	S	24 h LC$_{50}$	0.15	2.16	Ali and Mulla, 1980
Midge, *Dicrotendipea californicus*	4 instar	S	24 h LC$_{50}$	0.14	3.16	Ali and Mulla, 1980
Blackfly, *Simulium virgatum*	Late instar	CF	24 h LC$_{50}$	0.9	1.80	Mohsen and Mulla, 1981
Mayfly, *Baetis parvus*	Nymph, late instar	S	LC$_{50}$ (24 h after 1 h exposure)	0.4	—	Mohsen and Mulla, 1981

Species		Condition	Test	Value	Reference
Caddisfly, *Hydropshche california*	Larva, late instar	S	LC$_{50}$ (24 h after 1 h exposure)	0.4	—	Mohsen and Mulla, 1981
Lobster, *Homarus americanus*	450 g	SR	96 h LC$_{50}$	0.0014	—	Zitko et al., 1979
Waterflea						
Daphnia magna	—	S	24 h LC$_{50}$	8	—	Hoechst Canada, unpublished
	—	S	48 h LC$_{50}$	5	—	Hoechst Canada, unpublished

Cypermethrin

Species		Condition	Test	Value	Reference
Mayfly, *Cloeon dipterum*	Nymph	S	24 h LC$_{50}$	0.6 (0.3–1)	—	Stephenson, 1982
Beetle, *Gyrinua natator*	Adults	S	24 h LC$_{50}$	5	—	Stephenson, 1982
Water boatman, *Corixa punctata*	Adults	S	24 h LC$_{50}$	5	—	Stephenson, 1982
Waterflea, *Daphnia magna*	24 h	S	24 h LC$_{50}$	2 (1.5–3.1)	—	Stephenson, 1982
Amphipod, *Gammarus pulex*	3–8 mm	S	24 h LC$_{50}$	0.1 (0.08–0.2)	—	Stephenson, 1982
Isopod, *Asselus* spp.	3–8 mm	S	24 h LC$_{50}$	0.2 (0.1–0.4)	—	Stephenson, 1982
Mite, *Piona carnea*	adults	S	24 h LC$_{50}$	0.05 (0.03–0.08)	—	Stephenson, 1982
Shrimp, *Crangon septomapinosa*	1.3 g	SR	96 h LC$_{50}$	0.01	—	McLeese et al., 1980
Lobster, *Homarus americanus*	450 g	SR	96 h LC$_{50}$	0.04	—	McLeese et al., 1980

TABLE 22.9 Estimated Lethal Concentrations (μg/L) for Percentages of Aquatic Species Populations Exposed to Carbaryl in Laboratory Toxicity Tests

Species	LC_1	LC_{10}	LC_{20}	LC_{30}	LC_{40}	LC_{50}	LC_{60}	LC_{70}	LC_{80}	LC_{90}	LC_{99}
Mosquito											
Aedes cantans	88.8	170	224	273	323	377	440	521	635	834	1600
Aedes vexans	103.1	172	214	250	285	322	364	415	485	602	1010
Aedes excrucians	11.0	35.2	57.3	81.6	110	145	191	258	367	597	1900
Aedes communis	18.3	49.7	75.5	102	132	168	213	276	374	568	1540
Aedes punctor	52.9	115	160	203	248	298	359	438	556	770	1680
Culex p. pipiens	119.9	190	230	265	298	333	372	418	481	584	925
Culex p. molestus	106.3	197	255	308	361	418	484	567	685	887	1640
Culiseta annulata	58.5	97.1	120	140	160	180	203	231	270	334	554
Aedes aegypti	1.50	3.20	4.30	5.30	6.40	7.60	9.00	10.8	13.5	18.2	37.3
Phantom midge, *Chaoborus* sp.	145.7	286	379	466	554	650	763	908	1110	1480	2900
Midge, *Chironomus plumosus*	4.10	8.30	11.3	14.0	16.9	20.0	23.7	28.5	35.5	47.9	98.2
Stonefly											
Pteronarcys californica	13.4	19.2	22.3	24.9	27.3	30.3	32.4	35.5	39.6	46.0	65.9
Pteronarcella baddia	2.10	3.10	3.60	4.10	4.60	5.00	5.50	6.10	6.90	8.20	12.1
Mayfly, *Cloeon dipterum*	323.8	472	553	620	683	740	817	900	1010	1180	1720
Dragonfly, *Orthethrum abistylum speciosur*	376.3	549	642	721	794	860	950	1050	1170	1370	2000
Amphipod											
Gammarus lacustris	22.5	29.1	32.4	35.0	37.4	40.0	42.2	45.0	48.7	54.2	70.0
Gammarus fasciatus	30.7	52.5	65.7	77.4	88.8	104	115	132	155	194	331
Gammarus pseudolimnae	32.8	43.5	48.9	53.3	57.3	64.0	65.5	70.4	76.7	86.3	114
Gammarus pulex	27.1	30.5	32.0	33.2	34.3	35.0	36.3	37.4	38.8	40.7	45.9
Waterflea											
Daphnia magna	1.559	3.81	5.55	7.28	9.17	11.2	14.0	17.7	23.2	33.7	82.5
Daphnia pulex	5.16	7.96	9.54	10.9	12.2	12.8	15.0	16.7	19.1	22.9	35.3
Simocephalus serrulatus	8.858	11.295	12.5	13.5	14.3	15.2	16.1	17.1	18.4	20.4	26.0

TABLE 22.10 Estimated Concentrations (µg/L) Lethal to Percentage of Population of Aquatic Species Exposed to Deltamethrin in Laboratory Toxicity Tests

Species	LC_1	LC_{10}	LC_{20}	LC_{30}	LC_{40}	LC_{50}	LC_{60}	LC_{70}	LC_{80}	LC_{90}	LC_{99}
Mosquito											
Culex p. quinquefasciatus	0.006	0.010	0.013	0.015	0.018	0.02	0.023	0.026	0.031	0.039	0.067
Culex tarsalis	0.038	0.047	0.051	0.054	0.057	0.06	0.063	0.066	0.070	0.077	0.094
Culiseta incidens	0.040	0.099	0.145	0.192	0.242	0.30	0.372	0.470	0.619	0.905	2.24
Aedes nigromaculis	0.032	0.073	0.104	0.133	0.164	0.20	0.243	0.301	0.386	0.546	1.24
Aedes taeniorhynchus	0.018	0.028	0.034	0.040	0.045	0.05	0.056	0.063	0.073	0.088	0.141
Psorophora columbiae	0.012	0.032	0.047	0.063	0.080	0.10	0.125	0.159	0.212	0.315	0.807
Culex p. pipiens	0.023	0.059	0.089	0.118	0.151	0.19	0.238	0.305	0.407	0.608	1.58
Culex p. molestus	0.017	0.036	0.049	0.062	0.075	0.09	0.108	0.131	0.164	0.226	0.479
Culiseta annulatus	0.056	0.105	0.138	0.168	0.197	0.23	0.268	0.316	0.384	0.502	0.952
Aedes cantans	0.003	0.008	0.013	0.018	0.023	0.03	0.038	0.050	0.069	0.107	0.304
Aedes sticticus	0.002	0.006	0.009	0.012	0.016	0.02	0.025	0.033	0.044	0.067	0.0180
Aedes vixans	0.007	0.023	0.036	0.051	0.069	0.09	0.118	0.158	0.223	0.359	1.12
Midge											
Tanytarsus spp.	0.003	0.006	0.009	0.011	0.013	0.016	0.019	0.023	0.029	0.041	0.087
Procladius spp.	0.003	0.008	0.012	0.017	0.022	0.029	0.037	0.049	0.069	0.108	0.317
Chironomus decorus	0.026	0.070	0.105	0.142	0.182	0.23	0.290	0.373	0.503	0.758	2.02
Chironomus utahansis	0.048	0.108	0.152	0.194	0.239	0.29	0.352	0.433	0.554	0.777	1.75
Crictopus spp.	0.025	0.056	0.079	0.101	0.124	0.15	0.182	0.224	0.286	0.402	0.902
Dicrotendipes californicus	0.010	0.032	0.053	0.077	0.105	0.14	0.187	0.255	0.368	0.611	2.04
Blackfly, *Simulium virgatum*	0.229	0.424	0.549	0.663	0.777	0.90	1.04	1.22	1.48	1.91	3.54
Mayfly, *Cloeon dipterum*	0.112	0.231	0.313	0.391	0.471	0.60	0.666	0.802	1.00	1.36	2.81
Beetle, *Gyrinus natator*	0.764	1.80	2.54	3.28	4.08	5.00	6.10	7.58	9.80	14.0	32.5
Water boatman, *Corixa punctata*	0.764	1.80	2.54	3.28	4.08	5.00	6.10	7.54	9.80	14.0	32.5
Waterflea, *Daphnia magna*	0.253	0.667	1.00	1.35	1.73	2.00	2.74	3.52	4.73	7.11	18.8
Amphipod, *Gammarus pulex*	0.032	0.058	0.074	0.089	0.104	0.10	0.138	0.161	0.193	0.248	0.450
Isopod, *Asellus* spp.	0.032	0.073	0.103	0.133	0.164	0.20	0.244	0.302	0.389	0.551	1.26
Mite, *Piona carnea*	0.013	0.024	0.031	0.037	0.043	0.05	0.057	0.066	0.079	0.101	0.182
Shrimp, *Crangon septomspinosa*	0.008	0.017	0.023	0.029	0.035	0.04	0.051	0.062	0.078	0.107	0.228

other pyrethroid insecticides suggest that the snails are relatively tolerant to these compounds (Sheehan et al., 1987). The aggregated distributions of lethal concentrations are therefore reasonably representative of dominant slough taxa but may overestimate toxicity to the more tolerant invertebrate taxa, such as gastropods.

22.7   RISK CHARACTERIZATION

Risk characterization is the final step in the risk assessment process and is the integration of the exposure and effects analysis to describe the nature and likelihood of adverse effects associated with estimated exposures. To characterize risks to the invertebrate community, the first step is to visually compare the overlap between the estimated exposure concentration distribution for each pesticide and the lethal concentration (effects) distribution.

22.7.1   Overlap of Exposure and Effects Distributions

To graphically demonstrate the results of this analysis, the overlaps of exposure and effects distributions for populations of two test species are shown in Figures 22.7 and 22.8, respectively. The carbaryl exposure distribution is overlapped with the effects distribution for a mosquito (*Culiseta annulata*) and the deltamethrin exposure distribution is overlapped with the effects distribution for a waterflea species (*Daphnia magna*).

To evaluate the likelihood of mortality within the invertebrate community, the distributions of LC_{50} concentrations for all invertebrate test species populations for carbaryl and deltamethrin were overlain on the distributions of slough water concentrations (Figures 22.9 and 22.10, respectively). These data indicate the greater probability of exceeding 50 percent mortality for the variety of species exposed to deltamethrin compared to those exposed to carbaryl.

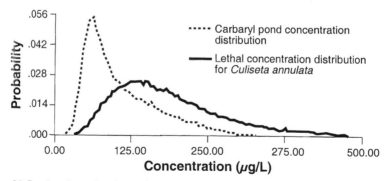

Figure 22.7   Overlap of estimated carbaryl concentration distribution with lethal concentration distribution for *Culiseta annulata*.

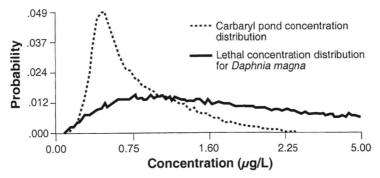

Figure 22.8 Overlap of estimated deltamethrin concentration distribution with lethal concentration distribution for *Daphnia magna*.

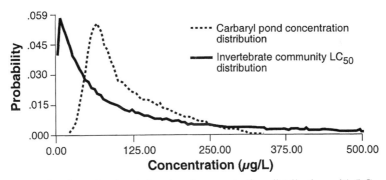

Figure 22.9 Overlap of estimated carbaryl concentration distribution with LC_{50} distribution for exposed aquatic invertebrate community.

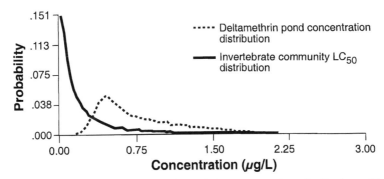

Figure 22.10 Overlap of estimated deltamethrin concentration distribution with LC_{50} distribution for exposed aquatic invertebrate community.

22.7.2 Probability of Exceeding Lethal Concentrations

To quantitatively compare the extent of overlap between exposure and effects distributions requires an application of reliability theory to estimate the "probability of failure." In this case, the probability of failure is defined as the probability that the exposure concentration exceeds the lethal effects concentration. Reliability theory seeks to calculate the probability of occurrence for a particular event or series of events. In its simplest form for engineering applications, two independent variables are used to describe a probable scenario and are defined as resistance R and load Q [Ang and Tang, 1975; Colorado State University (CSU), 1987]. The units of both of these variables are the same and each variable is described by a probability density function. A failure of this system occurs when the load variable exceeds the value of the resistance variable (i.e., $R - Q < 0$). The probability that the load is within the interval defined by $x + dx$ is calculated as the probability of Q at x multiplied by the differential, dx. This is generally expressed for all values of x as $f_Q(x)\,dx$. The probability that failure occurs while Q is within this interval is the probability that Q is within the interval multiplied by the probability that R is less than x, which is the cumulative probability for R at x and is expressed as $F_R(x)$. The probability of failure for all values of Q is then defined as the integral product of these two variables (Ang and Tang, 1975):

$$\mathrm{PF} = \int_{-\infty}^{\infty} F_R(x) f_Q(x)\, dx$$

In most cases, a closed-form solution to this equation is impractical and approximate methods or numerical techniques are relied upon. For problems where both input distributions are normal, an exact solution exists such that a reliability index β is calculated from the mean and variance of both distributions:

$$\beta = \frac{\mu_R - \mu_Q}{\sqrt{\sigma_Q^2 - \sigma_R^2}}$$

where μ is the mean value and σ is the standard deviation.

Beta is evaluated from the standard normal distribution and represents the probability that the load Q exceeds the resistance R for the entire range of values for Q and R. A similar closed-form solution exists when each input variable can be represented by the lognormal distribution (CSU, 1987). In this case, the reliability index is expressed as

$$\beta = \frac{\ln(\mu_R) - (1/2)\ln(1 + V_R^2) - \ln(\mu_Q) + (1/2)\ln(1 + V_Q^2)}{\sqrt{\ln(1 + V_R^2) + \ln(1 + V_Q^2)}}$$

where V is the coefficient of variation.

For this assessment, the random variables considered are all determined, through curve fitting, to be approximately lognormally distributed. The first variable is the concentration of pesticides in prairie potholes following aerial application. This variable is analogous to the load Q described above and is described in units of micrograms per liter. The LC curves for individual invertebrate species are analyzed to derive a resistance curve R, which is also expressed in units of micrograms per liter. The intersection of these two distributions is an expression of reliability and is the estimation of the fraction of each individual prey species that would be expected to die due to pesticide exposure.

The probability that exposure concentrations will exceed lethal concentrations for test species is shown for carbaryl and deltamethrin in Figures 22.11

Figure 22.11   Estimated percent mortality for populations of species for range of estimated distribution of carbaryl exposure concentrations.

Figure 22.12   Estimated percent mortality for populations of species for range of estimated distribution of deltamethrin exposure concentrations.

and 22.12, respectively. Figure 22.11 indicates that greater than 90 percent mortality is expected from the range of exposure concentrations for 9 of the 23 species tested. The more sensitive taxa include one mosquito species, *Aedes aegypti*; the midge, *Chironomus*; two stonefly species; two amphipod species; and the three waterflea species. In contrast, most mosquito species, the phantom midge, mayfly, and dragonfly species populations are expected to show less than 30 percent mortality from the range of estimated carbaryl exposures. Figure 22.12 indicates that greater than 80 percent mortality is expected in populations of 19 of 27 species. Only the whirligig beetle, water boatman, and waterflea are expected to have less than 20 percent mortality when exposed to deltamethrin at estimated concentrations.

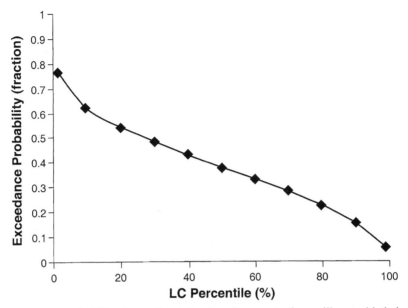

Figure 22.13 Probability that estimated carbaryl concentrations will exceed lethal concentration percentiles for exposed invertebrate community.

This analysis can be extended to estimate the probability of invertebrate community mortality associated with the estimated distribution of exposure concentrations. For clarity, this analysis relies on the assumption that the water column density of all prey species is approximately equal. The analysis can readily be modified to account for actual differences in prey density, mass density, or nutritional value of different prey species. By applying the reliability theory analysis to an aggregation of lethal concentration distributions for the test community, the probability of exceeding a lethal concentration percentile can be estimated. This comparison is shown for estimated carbaryl exposures in Figure 22.13 and for estimated deltamethrin exposures in Figure 22.14. For carbaryl, there is approximately a 50 percent probability of 20 percent mortality in the exposed community and approximately a 20 percent probability that the mortality would reach the 80 percent level. For deltamethrin, there is approximately a 50 percent probability that mortality would reach the 90 percent level in the exposed invertebrate community.

22.7.3 Comparison to Field Studies

One of the major uncertainties in a risk assessment based on laboratory toxicity tests is the correspondence between the level of mortality measured in the laboratory and that observed under field conditions. The response of the invertebrate community in ponds aerially treated with carbaryl has been studied. Hunter et al. (1984) applied carbaryl at 840 g/ha to four woodland ponds (the

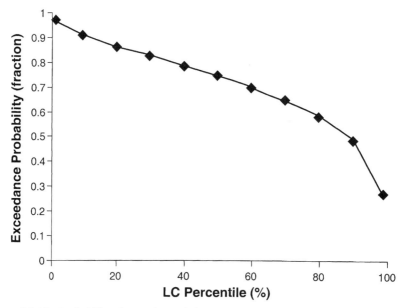

Figure 22.14   Probability that estimated deltamethrin concentrations will exceed lethal concentration percentiles for exposed aquatic invertebrate community.

risk assessment evaluated carbaryl applied at a rate of 1100 g/ha). Amphipod populations were reduced by approximately 99 percent following the carbaryl application and mayfly, and caddisfly populations were reduced by approximately 75 and 66 percent, respectively. However, the numbers of chironomid midges were significantly reduced in only one of the four treated ponds. In the treated ponds, the total numbers of invertebrates were reduced by 30 to 60 percent of the control pond levels, and biomass was reduced by 60 to 75 percent of the pretreatment standing crop. These data agree reasonably well with the risk assessment estimates of approximately a 50 percent probability of 30 percent mortality and approximately a 35 percent probability of 60 percent mortality in the exposed community. The risk assessment also qualitatively predicts the relative high sensitivity of amphipods to carbaryl exposures, but not the relative tolerance of chironomids. Unfortunately, there are no equivalent quantitative field studies of deltamethrin effects on aquatic invertebrates. Tooby et al. (1981) reported that deltamethrin applied to ponds at 50 g/ha severely depleted or eliminated several insect and crustacean families, and Rawn et al. (1983) reported that a single application of 10 g/ha reduced insect emergence. In a field study of cypermethrin, Crossland (1982) reported that the insecticide caused widespread planktonic and benthic invertebrate mortality: 21 of 23 taxa present in pretreatment samples were absent two weeks after application. The risk assessment estimates approximately a 50 percent probability of 90 percent mortality in the invertebrate community following deltamethrin overspray of sloughs.

Clearly, any risk assessment of a complex exposure situation such as evaluated in this analysis will have a relatively high level of uncertainty associated with mortality estimates. However, the probabilistic analysis applied in this assessment provides some quantification of uncertainties in risk estimates, as well as a technique to estimate the probability that exposure concentrations will exceed selected levels of mortality in the exposed invertebrate community. A logical extension to this work is the prediction of the effects of the depletion of the macroinvertebrate community, due to pesticide application, on the recruitment and survival of ducklings living in or near prairie potholes. Prey mass density and duckling energetic requirements could be characterized in a similar fashion to chemical concentrations, and estimates of duckling survival could be calculated. This approach promises to provide methods to estimate both the probability and severity of impacts resulting from exposures, information which has been missing from many of the ecological risk assessments conducted for aquatic systems.

REFERENCES

Ali, A., and Mulla, M. S. 1978. Declining field efficacy of chlopyrifos against chironomid midges and laboratory evaluation of chlorpyrifos against chironomid midges and laboratory evaluation of substitute larvicides. *J. Econ. Entomol.* 71: 778–782.

Ali, A., Mulla, M. S., Pfuntner, A. R., and Luna, L. L. 1978. Pestiferous midges and their control in a shallow residential-recreational lake in southern California. *Mosqu. News* 38(4): 28–535.

Allison, D. T. 1977. *Use of Exposure Units for Estimating Aquatic Toxicity of Organophosphate Pesticides.* EPA600/3-77/077. U.S. Environmental Protection Agency, Washington, DC.

Ang, A. H. S., and Tang, W. H. 1975. *Probability Concepts in Engineering Planning and Design,* Vol. 1: *Basic Principals.* Wiley, New York.

Argauer, R. J., Mason, H. C., Corley, C., Higgins, A. H., Sauls, J. N., and Liljedahl, L. A. 1968. Drift of water-diluted and undiluted formulations of malathion and azinphosmethyl applied by airplane. *J. Agric. Entomol.* 61(4): 1015–1020.

Bartonek, J. C., and Hickey, J. J. 1969a. Food habits of canvasbacks, redheads and lesser scaup in Manitoba. *Condor* 71: 280–290.

Bartonek, J. C., and Hickey, J. J. 1969b. Selective feeding by juvenile diving ducks in summer. *Auk* 86: 443–457.

Biggs, B. J. F., and Malthus, T. J. 1982. Macroinvertebrates associated with various aquatic macrophytes in the backwaters and lakes of the upper Clutha Valley, New Zealand. *New Zealand Mar. Freshwater Res.* 16: 81–88.

Bluzat, R., and Seuge, J. 1979. Etude de la toxicite chronique de deus insectides (carbaryl et lindane) à la generation Fl de *Lymnea stagnalis* L. (Mollusque, Gastéropode, Pulmone). I. Croissance des coquilles. *Hydrobiologia* 65(3): 245–255.

Chura, N. J. 1961. Food availability and preferences of juvenile mallards. *Trans. N. Am. Wildl. Conf.* 26: 121–134.

Collias, N. E., and Collias, E. C. 1963. Selective feeding by wild ducklings of different species. *Wilson Bull.* 75: 6–14.

Colorado State University (CSU). 1987. Reliability based design of transmission line structures. Final report prepared for the Electric Power Research Institute, Colorado State University, Department of Civil Engineering, Fort Collins, CO.

Cope, O. B. 1966. Contamination of the freshwater ecosystem by pesticides. *J. Appl. Ecol. (suppl.)* 3: 33–44.

Crossland, N. O. 1982. Aquatic toxicology of cypermethrin. II. Fate and biological effects in pond experiment. *Aquat. Toxicol.* 2: 205–222.

Currier, W. W., Maccollom, G. B., and Baumann, G. L. 1982. Drift residues of air-applied carbaryl in an orchard environment. *J. Econ. Entomol.* 75: 1065–1068.

Dvorak, J. 1970. A quantitative study on the macrofauna of stands of emergent vegetation in a carp pond of south-west Bohemia. *Rozpr. Cesk. Akad. Ved., Rada Mat. Prir. Ved.* 80: 63–114.

Dvorak, J., and Best, E. P. H. 1982. Macro-invertebrate communities associated with the macrophytes of Lake Vechten: Structural and functional relationships. *Hydrobiologia* 95: 115–126.

Federal Water Pollution Control Administration (FWPCA). 1968. Water quality criteria. Report of the National Tech. Adm. Comm. to Secr. of the Interior FWPCA, U.S. Dept. of the Interior.

Finney, D. J. 1971. *Probit Analysis*, 3rd ed. Cambridge University Press, Cambridge, pp. 1–333.

Forsyth, D. J. 1989. Agricultural chemicals and prairie pothole wetlands: Measuring the needs of the resource and the farmer—Canadian perspective. *Trans. N. Am. Wildl. Nat. Res. Conf.*, 54 pp. 5–66.

French, D. P. 1991. Estimation of exposure and resulting mortality of aquatic biota following spills of toxic substances using a numerical mode. In M. A. Mayes and M. G. Barron (Eds.), *Aquatic Toxicology and Risk Assessment*, Vol. 14. American Society for Testing and Materials, Philadelphia, pp. 35–47.

Grue, C. E., DeWeese, L. R., Mineau, P., Swanson, G. A., Foster, J. R., Arnold, P. M., Huckins, J. N., Sheehan, P. J., Marshall, W. K., and Ludden, A. P. 1986. Potential impacts of agricultural chemicals on waterfowl and other wildlife inhabiting prairie wetlands: An evaluation of research needs and approaches. *Trans. N. Am. Wildl. Nat. Res. Conf.*, 51 pp. 357–383.

Grue, C. E., Tome, M. W., Messmer, T. A., Henry, D. B., Swanson, G. A., and DeWeese, L. R. 1989. Agricultural chemicals and prairie pothole wetlands: Meeting the needs of the resource and the former U.S. perspective. *Trans. N. Am. Wildl. Nat. Res. Conf.*, 54 pp. 43–58.

Hashimoto, Y., and Nishiuchi, Y. 1981. Establishment of bioassay methods for the evaluation of acute toxicity of pesticides to aquatic organisms. *J. Pestic. Sci.* 6(2): 257–263.

Hill, B. D., and Kinniburgh, S. 1984. Aerial deposition of the synthetic pyrethroid deltamethrin. In P. W. Voisey (Ed.), *Proceedings of the Symposium on the Future Role of Aviation in Agriculture*. NRC No. 23504, AFA-TN-17. National Research Council of Canada Associate Committee on Agricultural and Forestry Aviation. Ottawa.

Hunter, M. L. Jr., Witham, J. W., and Dow, H. 1984. Effects of carbaryl-induced depression in invertebrate abundance on the growth and behavior of American black duck and mallard duckling. *Can. J. Zool.* 62: 452–456.

Hoechst Canada. Unpublished data on deltamethrin toxicity tests provided to the Canadian Wildlife Service.

Joyner, D. E. 1980. Influence of invertebrates on pond selection by ducks in Ontario. *J. Wildl. Manag.* 44(3): 700–705.

Joyner, D. E. 1982. Abundance and availability of invertebrates in ponds in relation to dietary requirements of mallards and blue-winged teal at Luther Marsh. *Ont. Field Biol.* 36(1): 19–34.

Kantrud, H. A., and Stewar, R. E. 1984. Ecological distribution and crude density of breeding birds of prairie wetlands. *J. Wildl. Manag.* 48(2): 426–437.

Keast, A. 1984. The introduced aquatic macrophyte *Myriophyllum spicatum* as habitat for fish and their invertebrate prey. *Can. J. Zool.* 62: 1289–1303.

Krapu, G. L. 1974a. Feeding ecology of pintail hens during reproduction. *Auk.* 91: 278–290.

Krapu, G. L. 1974b. Foods of breeding pintails in North Dakota. *J. Wildl. Manag.* 38(3): 408–417.

Krapu, G. L. 1979. Nutrition of female dabbling ducks during reproduction. In T. A. Bookhout (Ed.), *Waterfowl and Wetlands, An Integrated Review*, Proc. Symp. 39th Midwest Fish and Wildlife Conf. Madison, 1977, La Crosse Print Co., La Crosse pp. 59–69.

Krull, J. N. 1970. Aquatic plant-macroinvertebrate associations and waterfowl. *J. Wildl. Manag.* 34(4): 707–718.

Lakota, S., Raszka, A., and Kupczak, I. 1981. Toxic effects of cartap, carbaryl and propoxur on some aquatic organisms. *Acta Hydrobiol.* 23(2): 183–190.

Manitoba Agriculture. 1984. *1984 Manitoba Insect Control Guide.* Manitoba Agriculture, Winnipeg.

Maybank, J., Yoshida, K., Shewchuk, S. R., and Grover, R. 1976. Comparison of swath deposit and drift characteristics of ground-rig and aircraft spray systems; report of the 1975 field trials. *Rep. Saskatchewan Res. Council,* p. 76-1 January. 27 p.

Maybank, J., Yoshida, K., Shewchuk, S. R., and Grover, R. 1978a. Spray drift behavior of aerially-applied herbicide; Report of the 1977 field trials. *Rep. Saskatchewan Res. Council,* March, 78-2. 27 p.

Maybank, J., Yoshida, K., and Grover, R. 1978b. Spray drift from agricultural pesticide applications. *J. Air Pollut. Control Assoc.* 28(10): 1009–1014.

McLeese, D. W., Metcalfe, C. D., and Zitko, V. 1980. Lethality of permethrin, cypermethrin and fenvalerate to salmon, lobster and shrimp. *Bull. Environ. Contam. Toxicol.* 25: 950–955.

Mineau, P., Sheehan, P. J., and Baril, A. 1987. Pesticides and water fowl on the Canadian prairies: A pressing need for research and monitoring. *ICBP Tech. Publ. No. 6,* pp. 133–147.

Mittlebach, G. G. 1981. Patterns of invertebrate size and abundance in aquatic habitats. *Can. J. Fish. Aquat. Sci.* 78: 896–904.

Mohsen, Z. H., and Mulla, M. S. 1981. Toxicity of blackfly larvicidal formulations to some aquatic insects in the laboratory. *Bull. Environ. Contam. Toxicol.* 26: 696–703.

Mulla, M. S., Darwazeh, H. A., and Dhillon, M. S. 1980. New pyrethroids as mosquito larvicides and their effects on non-target organisms. *Mosq. News* 40(1): 6–12.

Noyes, J. H., and Jarvis, R. L. 1985. Diet and nutrition of breeding female redhead and canvasback ducks in Nevada. *J. Wildl. Manag.* 49(1): 203–211.

Perret, N. G. 1962. The spring and summer foods of the common mallard (*Anas platyrhynchos* L.) in southcentral Manitoba. M.Sc. Thesis, University of British Columbia, Vancouver.

Rawash, I. A., Gaaboub, I. A., El-Gayar, F. M., and El-Shazli, A. Y. 1975. Standard curves for nuvacron, malathion, sevin, DDT and kelthane tested against the mosquito *Culex pipiens* L. and the microcrustacean *Daphnia magna* Straus. *Toxicology* 4: 133–144.

Rawn, G. P., Muir, D. C. G., and Webster, G. R. B. 1983. Uptake and persistence of permethrin by fish, vegetation and hydrosoil. In N. K. Kaushik and K. R. Solomon (Eds.), *Proceedings of the Eighth Annual Aquatic Toxicity Workshop*. Can. Tech. Rep. Fish. Aquat. Sci. No. 1151, pp. 195–196.

Renne, D. S., and Wolf, M. A. 1979. Experimental studies of 2,4-D herbicide drift characteristics. *Agric. Meteorol.* 20: 7–24.

Rettich, F. 1977. The susceptibility of mosquito larvae to eighteen insecticides in Czechoslovakia. *Mosq. News* 37(2): 252–257.

Rettich, F. 1979. The toxicity of four synthetic pyrethroids to mosquito larvae and pupae (Diptera, Culicidae) in Czechoslovakia. *Acta Entomol. Bohemoslov.* 76: 395–401.

Sanders, H. O. 1969. *Toxicity of Pesticides to the Crutacean Gammarus lacustrus*. Bureau of Sport Fisheries and Wildlife Technical Paper No. 25. U.S. Fish and Wildlife Service.

Sanders, H. O. 1972. *Toxicity of Some Insecticides to Four Species of Malacostracan Crustaceans*. Technical Paper No. 66. U.S. Fish and Wildlife Service.

Sanders, H. O., and Cope, O. B. 1966. Toxicities of several pesticides to two species of cladocerans. *Trans. Am. Fish. Soc.* 95: 165–169.

Sanders, H. O., and Cope, O. B. 1968. The relative toxicities of several pesticides to naiads of three species of stoneflies. *Limnol. Oceanogr.* 13: 112–117.

Sanders, H. O., Finley, M. T., and Hunn, J. B. 1983. Acute toxicity of six forest insecticides to three aquatic invertebrates and four fishes. Technical Paper No. 110. U.S. Fish and Wildlife Service.

Santharam, K. R., Thayumanavan, B., and Krishnaswamy, S. 1976. Toxicity of some insecticides to *Daphnia carinata* King, an important link in the food chain in the freshwater ecosystems. *Indian J. Ecol.* 3(1): 70–73.

Saskatchewan Agriculture. 1984. Insect control on field crops 1984. Saskatchewan Agriculture, Regina.

Scheffer, M., Achterberg, A. A., and Beltman, B. 1984. Distribution of macroinvertebrates in a ditch in relation to the vegetation. *Freshwater Biol.* 14: 367–370.

SETAC. 1998. *Uncertainty Analysis in Ecological Risk Assessment* (W. J. Warren-Hicks and D. R. J. Moore, Eds.. SETAC, Pensacola, FL.

Shaw, G. C., Smith, D. K., Sheehan, P. J., and Mineau, P. 1984. Environmental concerns. In P. W. Voisey (Ed.), *Proceedings of the Symposium of the Future Role of Aviation in Agriculture*. AFA-TN-17, NR No. 23501. Associate Committee on Agriculture and Forestry Aviation, National Research Council of Canada, Ottawa, pp. 47–63.

Sheehan, P. J., Baril, A., Mineau, P., Smith, D. K., Hartenist, A., and Marshall, W. K. 1987. *The Impact of Pesticides on the Ecology of Prairie Nesting Ducks.* Technical Reports Series No. 19. Canadian Wildlife Service. Ottawa.

Smith, A. G., Stoudt, J. H., and Gollop, J. B. 1964. Prairie potholes and marshes. In J. P. Linduska and A. L. Nelson (Eds.), *Waterfowl Tomorrow.* U.S. Fish and Wildlife Service, Washington, pp. 39–50.

Stephan, C. E. 1977. In F. L. Mayer and J. L. Hamelik (Eds.), *Methods for Calculating an LC_{50} Aquatic Toxicology and Hazard Evaluation.* American Society for Testing and Materials, Philadelphia, PA. pp. 65–84.

Stephenson, R. R. 1982. Aquatic toxicology of cypermethrin. I. Acute toxicity to some freshwater fish and invertebrates in laboratory tests. *Aquat. Toxicol.* 2: 175–185.

Sugden, L. G. 1964. *Food and Food Energy Requirements of Wild Ducklings.* Progress Report No. 1880. Canadian Wildlife Service.

Sugden, L. G. 1973. *Feeding Ecology of Pintail, Gadwall, American Widgeo and Lesser Scaup Ducklings* in Southern Alberta. Report Series No. 24. Canadian Wildlife Service.

Swanson, G. A., Meyer, M. I., and Serie, J. R. 1974. Feeding ecology of breeding blue-winged teals. *J. Wildl. Manag.* 38(3): 396–407.

Tooby, T. E., Thompson, A. N., Rycroft, R. J., Black, I. A., and Hewson, R. T. 1981. *A Pond Study to Investigate the Effects on Fish and Aquatic Invertebrates of Delta-methrin Applied Directly onto Water.* Aquatic Environment Protection 2, Fisheries Laboratory, Burnham-on-Crouch, Essex, England.

U.S. Environmental Protection Agency (EPA). 1975. *Methods for Acute Toxicity with Fish, Macroinvertebrates and Amphibians.* EPA-660/3-75-009. Committee on Methods for Toxicity Tests with Aquatic Organisms, EPA, Corvallis, OR.

Ware, G. W., Buck, N. A., and Estesen, B. J. 1984. Deposit and drift losses from aerial ultra-low-volume and emulsion sprays in Arizona. *J. Econ. Entomol.* 77(2): 298–303.

Ware, G. W., Cahill, W. P., Gerhardt, P. D., and Frost, K. P. 1969. Pesticide drift I. High-clearance vs. aerial application of sprays. *J. Econ. Entomol.* 62(4): 840–843.

Ware, G. W., Cahill, W. P., Gerhardt, P. G., and Witt, J. M. 1970. Pesticide drift IV. On-target deposits from aerial application of sprays. *J. Econ. Entomol.* 63(6): 1982–1983.

Wauchope, R. D. 1978. The pesticide content of surface water draining from agricultural fields: A review. *J. Environ. Qual.* 7(4): 459–472.

Woodward, D. F., and Mauck, W. L. 1980. Toxicity of five forest insecticides to cutthroat trout and two species of aquatic invertebrate. *Bull. Environ. Contam. Toxicol.* 25: 846–853.

Zitko, V., McLeese, D. W., Metcalfe, C. D., and Carson, W. G. 1979. Toxicity of permethrin, decamethrin and related pyrethroids to salmon and lobster. *Bull. Environ. Contam. Toxicol.* 21: 338–343.

23 Methodology for Assessing Risk to Birds Following Application of Sprayable Pesticide to Citrus: Case Study Involving Chlorpyrifos

LARRY W. BREWER* and HARRY L. MCQUILLEN**

Ecotoxicology & Biosystems Associates, Sisters, Oregon

MONTE A. MAYES

Dow AgroSciences, LLC, Indianapolis, Indiana

23.1   INTRODUCTION

Little is left unsaid about the worldwide deleterious effects of organochlorine (OC) insecticides, such as DDT, on wildlife populations, predatory birds in particular. Heavy use of this class of chemicals in North America from the 1940s through the early 1970s is evidenced yet by ubiquitous residues in our environment, even though banned in North America for nearly three decades. Rachel Carson, in her classic work *Silent Spring*, formulated and encapsulated the pesticide-related fears and concerns of biologists and naturalists around the globe.[1] Her book launched an era of new awareness and heightened wariness of the potential toxic effects of pesticides to wildlife. Research has shown that the tremendous environmental persistence of some organochlorine compounds and their ability to bioconcentrate in food chains are the two characteristics that most contributed to their insidious effects on wildlife populations, especially birds.[2]

   The agricultural chemical industry responded to this knowledge by producing new pesticide products using the classes of chemicals known as organophosphorus (OP) and carbamate compounds. While the use of these com-

*Now at Spring Laboratories, Inc.
**Now with U.S. Fish and Wildlife Service.

Human and Ecological Risk Assessment: Theory and Practice, Edited by Dennis J. Paustenbach
ISBN 0-471-14747-8   © 2002 John Wiley & Sons, Inc.

pounds as insecticides occurred during World War II, the use dramatically increased during the 1950s.[2] By the 1980s they had become the standards for invertebrate pest control in nearly all agricultural crops with more than 160 million acre-treatments per year in the United States.[3] These compounds, cholinesterase inhibitors by mode of action, are generally short lived in the environment and generally do not bioconcentrate in food chains. Most are moderately to highly toxic to birds and mammals in laboratory exposure scenarios, and if wildlife exposure in the field is sufficient, acute toxicity and reproductive effects can occur to individual animals.[3,4] However, the numerous OP and carbamate compounds differ in their availability to wildlife in the field based on their physical characteristics (i.e., granular vs. flowable formulations) and environmental behavior. While many cases of OP- and carbamate-related wildlife kills have been reported and the effects to wildlife of exposure to these compounds have been described,[3] there have been few such incidents that have clearly caused population-level effects, as seen with the OC compounds.

Recent observation of large-scale Swainson's hawk die-offs where these birds congregate on their wintering grounds in Argentina, resulting from an OP insecticide used to control grasshoppers, is an example of an eminent threat of serious population effects.[5] The exposure scenario involved large flocks of hawks gorge feeding on sprayed grasshoppers immediately after OP insecticide applications to agricultural crops. Because a large proportion of the total (and relatively small) population of this species was exposed to this scenario, the population was at risk of being diminished at a pace that outstripped their reproductive potential. On the other hand, the use of diazinon on golf courses and turf farms resulted in a large number of reports of die-offs of Canada geese.[6] While it was apparent that many individual Canada geese were being poisoned from grazing on pesticide-treated turf over a number of years, the Canada goose population steadily increased in the areas of concern and throughout the United States during the same period of time. During the 1970s through the early 1990s numerous field studies (mostly not published) were conducted to determine whether and to what degree specific pesticides were likely to cause harm to avian species. Many of these studies resulted in the identification of various levels of avian mortality,[2,7-9] but others clearly demonstrated the absence of negative effects even when exposure was documented.[10-13] It has been this variability in wildlife response to different pesticide products used in different ways in a vast variety of different agricultural crops, rangeland habitats and urban areas that has drawn attention to the need for dependable methods for assessing risk to wildlife posed by pesticide use scenarios.

23.2 RISK ASSESSMENT PROCESS: BRIEF HISTORY

Humans have been making assessments of risk to the environment long before risk assessment terminology was conceived. Early foresters, for example, assessed the possible impacts of logging near streams or on steep hillsides in terms of possible effects on other resources such as soil, water quality, and fish habitat. Wildlife managers have evaluated the population-associated risks of establishing hunting seasons for various species, and farmers have contemplated the risks posed to soil quality by ever-changing cultivation practices. Early risk assessors had no defined approaches or guidelines. Evaluations were based on personal experience or general knowledge and were heavily influenced by cost–benefit considerations. In retrospect, it is clear that financial considerations heavily influenced whether high-risk resource management practices were deemed acceptable. However, as competition for resources increased with decreasing availability and the pollution of our environment increased to the point that it could be seen and smelled in our air and water, greater demands were placed on decision makers to accurately justify resource management and use practices toward the endpoint of keeping the environmental costs and economic benefits more balanced. The National Environmental Policy Act of 1969 established that care for the environment was a national interest, laying down rules and guidelines for regulatory agencies and local governments to evaluate development and resource consumption practices. Requirements for ecological risk assessments in the form of environmental impact statements for such activities began the practice of organized evaluation of environmental risk and cost–benefit analyses. With these requirements and approximately 25 additional pieces of legislation passed between 1970 and 1985, tremendous emphasis was placed on the regulation of the manufacture, transport, and use of hazardous materials, including pesticides.[13] Increased demand for understanding of toxicology, industrial hygiene, analytical chemistry, and environmental safety resulted in remarkable advances in our knowledge and capabilities in these disciplines over the past three decades. One of the many legislative acts was the Federal Insecticide, Fungicide and Rodenticide Act (FIFRA). Under this statute, a pesticide product may be sold or distributed in the United States only if registered or exempted from registration by the U.S. Environmental Protection Agency (EPA). To qualify for registration, a pesticide must not pose unreasonable adverse effects to the environment.[14] The EPA's Office of Pesticide Programs (OPP) was established as the authority responsible for implementing FIFRA. Under the burden of this responsibility, the OPP has progressively refined guidelines for evaluating the risk posed to the environment by pesticides. In 1982 the OPP published *Pesticide Assessment Guidelines, Subdivision E, Hazard Evaluation: Wildlife and Aquatic Organisms* (EPA 6540/9-82-024). Under this guideline, procedures for laboratory tests to evaluate toxic effects of pesticides to wild animals were established to help make assessments of effects more consistent. In 1986, the OPP published *Guidance Document for Conducting Terrestrial Field Studies.*[15] This document provided description of methodology to be used in filed studies to identify and quantify deleterious effects of pesticides to nontarget wildlife, which the OPP required. Until 1992, the combined information generated by laboratory tests and field studies were used to evaluate the effects of pesticides to wildlife. In 1992, the OPP changed its policy in regard to requiring field studies in the pesticide registration process. Under a new paradigm, the OPP published *Framework for Ecological Risk Assessment* (EPA/630/R-92-001, February 1992), in which it provided a step-down process for conducting ecological risk assessment. The new paradigm emphasized the use of laboratory-produced toxicological data, other existing data, and environmental models to assess the risk of chemical-specific pesticide products. During this time the EPA has been conducting initial estimates of risk to avian species via a deterministic method of assessment[17] utilizing laboratory-generated toxicity data and generalized residue estimates (expected environmental concentrations) provided by Hoerger and Kenaga[18] and Kenaga[19] and modified by Fletcher et al.[20] Recently, following recommendations of the FIFRA Scientific Advisory Panel (SAP), the OPP has been striving to develop guidelines for conducting probabilistic assessment of ecological risk to refine its methodology for registering pesticides. The OPP established the Ecological Committee on Formulating Risk Assessment Methodology (ECOFRAM). This committee prepared a report on how to conduct probabilistic risk assessment to evaluate the risk posed by pesticides to nontarget organisms.[21] As a result of these efforts, the OPP is currently formulating a guidance document on this topic.

23.3 CURRENT STATUS OF PESTICIDE RISK ASSESSMENT AT EPA

The EPA is making progress toward developing guidelines for probabilistic risk assessment; however, it is not yet a standard procedure in the agency's pesticide registration process. For the most part, estimations of risk to avian species posed by currently registered or new pesticides have been and are conducted using the deterministic approach mentioned above. In that process environmental pesticide concentration estimates are derived from vegetation residues summarized in the Kenaga nomogram.[19] Since the expected environmental concentration values (for avian food items) used in the deterministic model are based on estimates derived from vegetative residues and were calculated from averaging numerous postapplication residue levels of many pesticides on similar plant groups, the possibility exists for either overestimating or underestimating actual avian exposure potential to specific pesticide products and application circumstances. This is particularly likely given that invertebrates, rather than plant material, make up the major proportion of the avian diet during the typical crop-growing season. The specific residue values from the Kenaga nomogram are assigned to invertebrate food items as follows: Small-insect residues are considered to be the same as residue concentration on forage crops such as alfalfa, and large-insect residues are considered equal to residues on seed-containing pods. Recently, the EPA adopted higher expected residue concentration values based on a review of the Kenega nomogram and the data on which it was based.[17–19] The new values adopted by the EPA estimate large-insect residues to be 15 ppm and small-insect residues to be 135 ppm for each pound of active ingredient applied. While this approach may be useful for general screening and early-tier risk assessment when empirical data are not available, it is generally understood that in assessment of risk posed by chemicals in the environment, particularly when ecological modeling is applied, the use of measured concentrations in specific environmental matrices provides a more accurate assessment than the use of estimated concentrations and reduces the uncertainty of the assessment.[23,24]

The *Final Report of the Avian Effects Dialogue Group, 1988–1993* states: "There is a need to synthesize available information on pesticide residues on insects following pesticide applications to test the assumptions made by Kenaga and the EPA Office of Pesticide Programs. In assessing risk, actual residue data on insects may be more useful than estimated values."[25] When estimating invertebrate residues using the Kenaga nomogram, the EPA often uses the "upper limits" for the residue values as well as the upper limits of toxicity values derived from laboratory toxicity tests such as median lethal concentration and dose (LC_{50}, LD_{50}) tests. Again, this may be justified when empirical data are not available for use in risk assessments in order to assure there is a margin of safety associated with the estimation process. However, if bird food residues are measured in the field, uncertainty is reduced such that the mean residue concentrations are more appropriate for use in the deterministic model. A review of the literature to date indicates that there is insufficient data available on insect residues following pesticide applications. Without such data, it is not possible to thoroughly assess exposure of nontarget wildlife relative to specific pesticide products, crops, application methods, and food types.

23.4 CHLORPYRIFOS USE: PHYSICAL PROPERTIES AND TOXICOLOGY

Chlorpyrifos is used to control a wide variety of insect pests in a large number of agricultural and specialty use patterns throughout the United States. Chlorpyrifos formulations include 15G, a granule containing 15 percent chlorpyrifos by weight; 4E, a 4-pound active ingredient per gallon of emulsifiable concentrate; and 50W, a 50 percent wettable powder. Major crop uses (with percent of total use in parentheses) include corn (55%) peanuts (7%), citrus (6%), alfalfa (6%), cotton (5%), foliar nuts (5%), tobacco (3%), sugar beets (3%), dormant trees (3%), foliar pome fruit (2%), sorghum (1%), and wheat (1%).[26] Important uses of Lorsban 4E insecticide in citurs crops in California include control of aphids, katydids, lepidoptera larvae, scale insects, thrips, and mealybugs.

Chlorpyrifos [O,O-diethyl O-(3,5,6-trichloro-2-pyridinyl) phosphorothioate] is an organophosphorus insecticide characterized by a chlorinated pyridine ring.[26] It is nonpolar and possesses the physical properties detailed in Table 23.1. A detailed discussion of the environmental fate and metabolites of chlor-

TABLE 23.1   Physical and Chemical Properties of Chlorpyrifos

Properties	Reference Value
Physical	
CAS number	2921-88-2
Molecular weight	350.6
Molecular formula	$C_9H_{11}NO_3PSCl_3$
Melting point	41–44°C[a]
Solubility	
Water	1.4 mg/kg
Acetone	650 g/100 g
Vapor pressure	1.8×10^{-5} mm Hg, 2.7 mPa (25°C)
log K_{ow}	5.01
K_{oc}	8500
BCF_{pred}	1546
BCF_{meas} (aquatic)	Average = 1074, $n = 35$
BCF_{meas} (soil)	Insects = <1
	Earthworms = 0.5–2.2
Chemical	
Hydrolysis	$(t_{1/2})$[b]
pH 5	73 days
pH 7	72 days
pH 9	16 days
pH 4.8 + Cu^{2+}	1.2 days
Photolysis: $(t_{1/2})$	
Air	1 day
Water	30 days
Dry soil	Slow
Leaf	Slow
Glass	2 to >50 days
Soil degradation: $(t_{1/2})$	
Wet soil	41 days, $n = 22$[c]
Dry soil	4 days, $n = 4$

[a] Reference 26.

[b] $t_{1/2}$ = half-life.

[c] Average of values for soil incubated at 25°C, 75% 1/3 bar, <11 mg/kg.

Source: Modified from Reference 26.

pyrifos are available in the scientific literature.[26,27] Chlorpyrifos is considered highly toxic to birds and other nontarget organism based on laboratory determinations of LD_{50} and acute dieatary toxicity tests (LC_{50}). Values for these two measures of toxicity in avian species are provided in Tables 23.2 and 23.3, respectively.

TABLE 23.2 Laboratory Acute Oral Toxicity (LD$_{50}$) Values for Chlorpyrifos to Various Bird Species

Test Animal	Age	Purity	LD$_{50}$	95% CI	Reference
Canada goose	Unknown	99.0%	40–80	Unknown	29
Mallard	15–19 days	99.0%	112.0	11.5–1089	29
	2–4 months	99.0%	75.6	35.4–161.0	30
Chukar	3–5 months	99.0%	61.1	47.5–78.6	29
	2–4 months	99.0%	60.7	43.8–83.1	30
Northern bobwhite	10–12 weeks	Technical	32.0	24–43	31
American crow	Unknown	Technical	>32	Unknown	31
Leghorn cockerel	6 weeks	Technical	50–63	Unknown	31
	10–12 days	Technical	25.4	20.8–30.9	32
	Unknown	Technical	31.6	Unknown	33
	15 days	99.9%	34.8	29.3–40.4	34
Beltsville small white turkey	6–7 weeks	Technical	32–63	Unknown	35
Common grackle	Unknown	94.5%	5.62	3.16–10	36
Starling	Unknown	94.5%	75	Unknown	36
Red-winged blackbird	Unknown	94.5%	13.1	7.5–23.7	36
House sparrow	Adult	99.6%	122.0	80–214	37
California quail	5–7 months	94.5%	68.3	40.7–115	29
Rock dove	Unknown	94.5%	10.0	5.62–17.8	35
Japanese quail	2.5 months	94.5%	13.3	7.5–23.7	35

TABLE 23.3 Laboratory Acute Dietary Toxicity (LC$_{50}$) Values for Chlorpyrifos to Various Bird Species

Test Animal	Age	Duration	Purity	LC$_{50}$	95% CI	Reference
Mallard	5–7 days	8 days	Technical	180	150–220	38
	14 days	8 days	Technical	671	322–2170	39
	14 days	8 days	Technical	757	478–1448	39
	14 days	8 days	Technical	900	746–1255	39
	14 days	8 days	Technical	1080	707–2508	39
	Unknown	5–7 days	99.0%	136	84–212	40
Northern	14 days	8 days	Technical	392	293–522	39
bobwhite	14 days	8 days	Technical	421	332–535	39
	14 days	8 days	Technical	353	294–429	39
	14 days	8 days	Technical	397	318–498	39
	Unknown	8 days	99.0%	721	Unknown	41
Pheasant	Unknown	10 days	97.0%	553	421–687	42
Japanese quail	Unknown	14 days	97.0%	299	146–1682	43

23.5 SOME IMPORTANT ASPECTS OF AVIAN FEEDING HABITS

Avian field metabolic rates and food requirements can be accurately estimated based on body weight.[43] The caloric and water content of many bird food items, including seeds and invertebrates, have been determined and published.[44] These data are useful if modeling avian food intake. Most passeriform, galliform, and anseriform species typically present in agricultural environments, even those generally considered granivorous, feed on invertebrate animals during some part of their annual cycle, particularly during the reproductive period.[46,47] Many species feed almost exclusively on insects and other invertebrates, such as earthworms, for the first 5 or 6 weeks of life. Adult birds of many species feed opportunistically on invertebrates throughout their lives. Invertebrates eaten in large quantities by numerous avian species include grasshoppers, crickets, beetles, insect larvae (particularly of the orders Coleoptera and Lepidoptera), and earthworms.[45-48] Examples of passerine species that are often found in and near agricultural crops and feed on such invertebrates include the American robin (*Turdus migratorius*), red-winged blackbird (*Agelaius phoeniceus*), northern mockingbird (*Mimus polyglottos*), brown thrasher (*Toxostoma rufum*), and several sparrow species. Examples of upland bird species that consume this group of invertebrates include the northern bobwhite (*Colinus virginianus*), ring-necked pheasant (*Phasianus colchicus*), and grey partridge (*Perdix perdix*).

Seeds are also common bird food items and are commonly available to birds in and around agricultural crops across the United States but are primarily a fall and winter food item for the majority of species that otherwise concentrate their diet on invertebrates during spring and early summer. Recent literature indicates that bird seed-handling behavior (e.g., seed hulling) significantly reduces exposure to contaminated seeds in many species.[49] In combination with this information, accurate measurements of postapplication residues on seeds is also important in the conduct of avian risk assessments.

23.6 STUDY SCOPE AND DESIGN RATIONALE

The study reported in this chapter was conducted to measure chlorpyrifos residues in potential avian food items for the purpose of estimating potential exposure and subsequent risk of avian species to Lorsban 4E insecticide following its application to citrus fruit. Lorsban applications to citrus typically occur from May to August in the Central Valley of California and other citrus-growing regions. During this time of year, citrus groves are the habitat of numerous insect species that may be utilized by birds as food items. Some of the common insects abundant in citrus groves in Fresno County, California, include crickets (Gryllidae), katydids (Tettigoniidae), cockroaches (Blattidae), grasshoppers (Acrididae), cranefly (Tipulidae) adults and larvae, lacewing (Chrysopidae) adults and larvae, coenosia fly (Anthomyiidae), hover fly (Syrphidae), potato leafhoppers (*Empoasca* sp.), beet armyworms (*Spodoptera exigua*), ground beetles (Carabidae), pill bugs (Isopoda), sow bugs (Isopoda), grey field ants (*Formica cinerea*), anise swallowtail (*Papilio zelicaori*), and others (Devin Carroll, personal communication, Clovis, CA, 1995). Some of these insects are present in the citrus trees and some are present on other vegetation in and around the groves, and numerous species of birds may forage on these insects. Therefore, we evaluated residue concentrations on representative members of the naturally occurring invertebrate populations in the citrus groves to provide the range of residues a foraging bird may encounter in a treated grove. The study design was typical of numerous agricultural field trials conducted annually across the United States. The representative avian food items used during this study consisted of crickets (*Acheta domestica*) enclosed in exposure chambers on the ground, earthworms (*Lumbricus terrestris*) enclosed in subsoil exposure chambers, darkling ground beetle larvae (*Tenebrio molitor*) pinned in place in the citrus trees, wheat (*Triticum* sp.) seeds in seed heads on natural, upright stalks, naturally occurring flying insects, and naturally occurring ground-dwelling insects. This study uses both free-ranging invertebrates and invertebrates enclosed in exposure chambers. The primary reason for using invertebrates in enclosures is to assure that, in the event of rapid invertebrate mortality due to the application of the test substance, adequate samples of invertebrates which were directly exposed to the application are available. When invertebrates die quickly from exposure to the test substance, they are no longer available to be captured in traps or nets and are often difficult or impossible to find for collection by hand. Invertebrates enclosed in exposure chambers are available in adequate numbers for scheduled collections. Additionally, if invertebrates in enclosures do not die after the application of the test substance, they may also provide fresh samples for determining the temporal decline of chemical residues. However, if invertebrates die, their postmortem body weights change rapidly from desiccation and in some cases reabsorption of water as weather conditions change over time. Residue levels of dead invertebrates produce a misleading picture of chemical persistence unless they are normalized to dry weight or fresh weight (weight at time of placement into enclosures). In this study we used the residues on invertebrates enclosed in exposure chambers to measure residue levels 2 hours postapplication. We also collected invertebrates from the enclosures on days 1, 5, and 10 postapplication for the purpose of plotting the estimated natural rate of decline of the test substance on bird food items. For this purpose the measured residue levels of invertebrate samples that were dead prior to collection were normalized to fresh weight. Again, this normalizing corrects for the variability in sample weight resulting primarily from desiccation of the dead invertebrates and the subsequent increase in pesticide concnetration. Based on recent research, it appears that avian species do not readily select desiccated invertebrates as food items.[51] Therefore, we believe the normalized values are most appropriate for avian exposure assessment.

Free-ranging, ground-dwelling invertebrates and free-ranging, flying invertebrates were trapped and collected live at all scheduled collection times. The use of such free-ranging invertebrates is necessary to obtain an accurate estimate of residue levels potentially available to birds on food items. Following application of the test substance, we believe there is a relatively brief period during which dead and/or moribund invertebrates are available to birds as food. Thereafter, the invertebrate population available to birds on a treated field consists of not only invertebrates that were present on the field at the time of application but also individuals that immigrate onto the field or hatch or pupate (e.g., emerged from soil) after the application. The residue levels on such individuals can be highly variable, with the dilution of mean residue levels on the extant insect populations resulting from immigration being greatest in the more mobile species (e.g., flying insects). The residues on these mixed groups (individuals with varying exposure to the test substance) of invertebrates most accurately reflect the residue levels available to birds as they move in and out of a field foraging on the various food types they encounter. The 2-hour samples collected during this study represent those invertebrates that were present on the field during the application of the test substance and includes individuals that received the direct spray. These samples, and in some cases the day 1 samples, represent the maximum residue levels available after application. Subsequent samples of free-ranging invertebrates collected from day 1 through day 10 postapplication provide the best estimate of mean residue levels available to foraging birds. It is important to recognize that after the day 1 sample the samples collected from the enclosures are intended only for use in estimating chemical decline rate because their residues are not representative of residues occurring on living insects, which are subject to predation.

23.7 STUDY AREA

The study site was a portion (12 rows) of a larger 8.1-ha (20-acre) subblock of citrus grove owned by the Harlan Ranch Company. The subblock used had not been treated with chlorpyrifos since May 1996 [applied at 6.0 lb active ingredient (a.i.)/acre]. Harlan Ranch is located approximately 8 km (5 miles) northeast of Clovis, California, in Fresno County (approximately 36°48′ N, 119°45′ W). The site was bordered to the south and east by citrus trees and to the north by state highway 168. A small asphalt access road bordered the site to the west, and dirt maintenance roads defined the limits of treatment to the south and north (Figure 23.1).

Weather in the Central Valley of California during July is typically hot and dry. Rain and thundershowers are rare in summer. Normal average temperatures and rainfall are reported for the Fresno, California, airport from 1948 to 1997 by the National Climatic Data Center (Asheville, NC). Average maximum and minimum July temperatures are 36.8°C (98.2°F) and 18.3°C (65.0°F), respectively. Average total precipitation for July is 0.254 mm (0.01 inch).

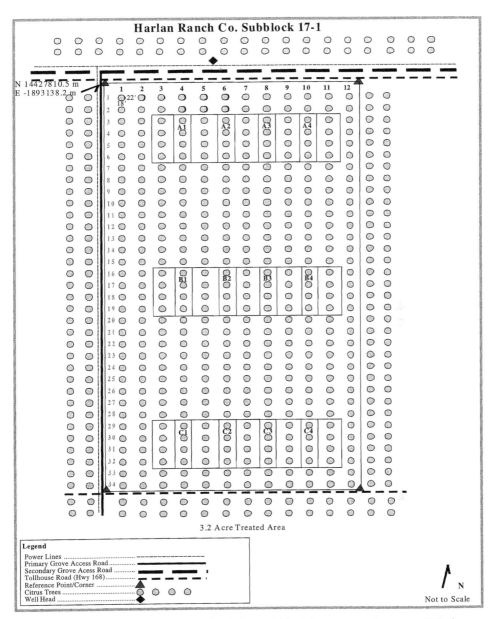

Figure 23.1 Location of main and subplots within 3.2-acre treated area used during EBA study number 079717 conducted at Harlan Ranch, Clovis, CA, July 1997.

23.8 METHODS AND MATERIALS

There were three main plots used during this study (Figure 23.1). These were laid out equidistant from each other within the 1.3-ha (3.2-acre) treated area. Within each main plot, four equally sized subplots were defined for sample collection during different collection periods. The treatment area was determined by counting the number of trees in each row (33), the distance between each tree [5.5 m (18 feet)], and the distance between each row [6.5 m (21 feet)]. To treat 1.3 ha (12,977 m^2), we calculated we needed 12 adjacent tree rows (11 passes × 6.5 m spray swath/pass × 33 tree spaces × 5.5 m/tree space = 12,977 m^2). The outer row on either side of the treated area did not receive treatment.

The location of the first plot was determined by selecting a random number between 3 and 32. The random-number selection was designed to set the plots in from the edge of the treated area by at least one row, east and west, and by at least two trees, north and south. The random number selected was 3. The third tree in the third row from the west edge was selected as the starting point for the first plot. Main plot size was four trees long, north to south, by eight rows wide, east to west. The remaining plots were the same dimensions. Nine trees separated the plots within the treated area, north to south. Plots were designated from north to south as plot A, plot B, and plot C. Within each main plot, we defined four subplots one row wide and four trees long, stretching the entire length of a main plot (Figure 23.2). The subplots were spaced equidistant across the width of each main plot, resulting in a subplot every other tree row. We defined the northernmost tree in each subplot as tree 1. The remaining trees were numbered sequentially from 1 to 4, north to south. Prior to the application of the test substance, we installed two enclosures in each subplot, one for crickets and one for earthworms. The cricket enclosures were buried so the top of the enclosure was nearly even to ground level. Approximately 10 cm of soil and ground litter was added to each cricket enclosure (Figure 23.3). The addition of ground litter to the enclosure was intended to replicate the natural surrounding ground cover. A small water cup filled with small-diameter gravel and potato slices was added to the cricket enclosures as a source of food and water.

The earthworm enclosures were also buried approximately even to ground level. They were then filled with a combination of commercial worm bedding and soil (Figure 23.4). We added approximately 2.5 cm of Brown Bear worm bedding to the bottom of each enclosure and then filled the remainder of the enclosure to within 2.5 cm of the top with soil taken from the hole dug to accommodate the enclosure tub. The soil in the grove was extremely dry so the combination of worm bedding and soil provided nourishment and a moist medium to prevent the earthworms from dying prior to test substance application.

Each enclosure was made from plastic tubs (Rubbermaid or Steralite) measuring 40 × 56 × 15 cm. Drainage holes were cut in the bottom of each tub to prevent flooding during rain or irrigation. The bottom was lined with fiberglass window screen to prevent test species from escaping. The tops were snap-on

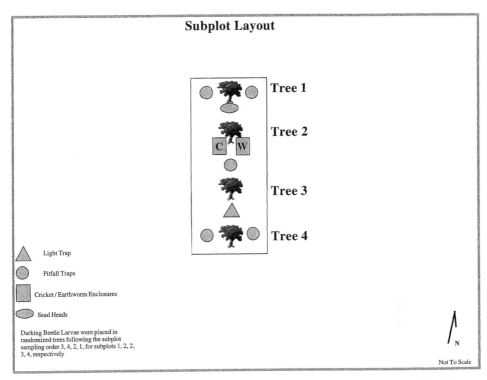

Figure 23.2   Subplot layout used during EBA study number 079717 conducted at Harlan Ranch, Clovis, CA, July 1997.

lids with holes cut in the lid for ventilation. Fiberglass window screen was also secured to the lid to prevent test species escape or depredation of the test species by small mammals or birds. Immediately after test species placement, we attached corrugated cardboard in a tent fashion to the lids of the enclosures. This process allowed air exchange within the enclosures while also shading the enclosures and keeping the interior cooler during periods of direct sunlight. Lids and shade covers were removed immediately prior to test substance application and replaced immediately following application.

Prior to placing the seed heads in the subplots (Figure 23.5), we used 2.5-cm poultry wire to build 30 × 60 × 40 cm wire exclosures. These exclosures were placed over the seed tray to prevent depredation of the seed heads by small mammals and birds. We secured the exclosures in place by weighting the tops with sticks or rocks found in the grove. The exclosures were removed during test substance application and replaced immediately following application. Each seed tray was filled with styrofoam blocks (to hold the seed stalks in an upright fashion) and anchored to the ground by driving 30-cm nail spikes through the tray (and styrofoam) and into the ground. Moth balls were added to each seed tray to aid in preventing depredation of the seed heads. We anchored the seed

Figure 23.3 Cricket enclosure with lid lying adjacent. Litter in enclosure was collected from ground in immediate vicinity.

head stalks to the styrofoam by pushing their ends into the styrofoam and then further securing them using quilting pins through the styrofoam and stalk. This was necessary to prevent the air-blast sprayer from blowing the seed heads to the ground during application.

Samples of earthworms and crickets were placed within each enclosure 24 hours prior to the application. At least 12 earthworms were placed in each earthworm enclosure. The actual number of worms and total weight for each enclosure was recorded immediately prior to placement. At least 50 crickets were placed in each cricket enclosure. Seed heads were also placed in each subplot 24 hours prior to the application. Twelve individual seed heads were anchored into each seed tray and placed on the south side of tree 1 in each subplot, directly under and as close to the tree trunk as possible. Wire enclosures were placed over the seed heads immediately after placement. Seed heads were not weighed prior to placement because the individual seeds are removed from the heads during sample processing.

At least 40 darkling ground beetle larvae were pinned to a single tree within each subplot. Larvae pinning was completed 4 hours prior to the application. Which tree to be used in each subplot was randomized in the order 3, 4, 2, 1. That is, larvae were pinned in tree 3 in all subplots numbered 1 (A1, B1, C1); tree 4 in subplots A2, B2, C2; tree 2 in subplots A3, B3, C3; and tree 1 in sub-

Figure 23.4 Earthworm enclosure with lid adjacent and bucket of worm bedding. Worm bedding was placed in bottom of enclosure to assure adequate nutrition for earthworms. Remainder of enclosure is filled with soil from excavated hole into which it was placed.

plots A4, B4, C4. Four limbs in each tree were selected indiscriminately for pinning, two limbs lowest to the ground and two limbs higher in the tree (up to approximately 2 m). Five larvae were pinned on each limb and five larvae were pinned on the leaves of each selected limb. A piece of white surveyor's flagging was tied to each pinned limb to aid in the recovery process. Larvae were pinned using one "T" pin and one quilting pin, one located superior and one inferior along the longitudinal axis of the body. Placement along the limb was selected to achieve a spiraling effect around the limb, that is, one on top, one on the side, one on the bottom, and so on. Placement on the leaf was alternated between one on top of the leaf and one on the bottom of the leaf. After application, the tree was wrapped using plastic netting to prevent bird depredation of the larvae. The netting was 2.2 m (7.2 feet) tall and was secured to the tree using wire ties.

Flying insect light traps (Figure 23.6) and ground-dwelling insect pitfall traps (Figure 23.7) were placed in each subplot within 24 hours prior to the appropriate sampling period. We initially installed all the pitfall traps prior to the application and put lids on them to prevent contamination. However, the air-blast applicator blew the lids off the traps and the traps filled with irrigation

Figure 23.5 Placing dry wheat stems with seed heads intact on study subplot.

water during irrigation periods. Thus, it was necessary to reinstall the pitfall traps 24 hours prior to each collection period.

Light traps were placed in the appropriate subplot between trees 3 and 4. We attached a 12-V direct-current (dc) battery to the light, inserted a clean, 10-gallon trash bag liner into the trap, and secured it tightly around the edge of the trap. The light was then turned on at approximately sunset on the night prior to the sampling period. The only exception to this was the 2-hour and day 1 sampling periods. The 2-hour and day 1 flying insect and ground-dwelling insect samples were collected the evening of the sampling date. The pitfall traps were constructed using 12-ounce plastic Solo drinking cups, 5 of which were placed in each subplot. To place, we dug holes using a mattock and pick and then buried the cups so they were even to ground level. As much as possible, we prevented contaminated soil from dropping into the cups by removing the lid just prior to the sampling period.

Samples were collected according to the sampling schedule. All earthworms and crickets were removed from their respective enclosures on their respective collection days by sifting through the soil with gloved hands and collecting each individual seen with clean forceps. At no time were test species touched by hand. Once the soil had been sifted thoroughly, the enclosure was removed from its location in the ground and the soil was poured out onto a clean drop cloth. The soil was again hand sifted until all the available individuals (alive or

Figure 23.6 Portable, battery-operated flying insect light trap. Insects approaching light strike plastic deflectors and drop through small opening into bucket where they are unable to find exit hole. This trap can collect hundreds of insects per night.

dead) were collected. After collection, the soil was used to fill the hole. Crickets were placed directly into a sample collection bag, labeled, double bagged, and put immediately on ice in a cooler. Earthworms were placed in a clean, unlabeled "preliminary" bag, and a labeled sample bag was attached securely to that bag. Both bags were then placed in a double bag and placed in the cooler on ice. This procedure allows the earthworms an opportunity to secrete mucus and remove excess soil introduced to their skin through the collection process. Upon returning to the field office, earthworms were transferred to the clean, labeled sample bags, weighed, and frozen.

Seed heads were collected by removing the wire exclosure from the tray, using a clean pair of scissors the stalks were trimmed to approximately 10 cm, and then, using clean gloved hands, the seed heads were bunched together and placed into a clean, unlabeled sample bag. Similar to the earthworm-bagging procedure, a clean, labeled sample bag was attached securely to the preliminary bag and the two were placed in a third bag and put on ice. During sample processing the seeds were removed from the seed heads using clean forceps to pick each seed from the head. The seeds were then placed into the labeled bag, weighed, double bagged, and frozen.

Pinned larvae were removed from their positions in the trees using clean forceps. The pins were pulled gently from the limbs and leaves, and the larvae

Figure 23.7 Preparing pitfall insect trap on study subplot. First, hole is excavated and 16-oz plastic cup is placed into hole so that its top is even with soil surface. Crawling insects fall into cup and cannot climb out because of smooth inner surface.

were grasped with the forceps and dropped into a clean, labeled sample bag. The sample was double bagged and placed on ice in the cooler.

Flying insects were collected from light traps by removing the plastic bag liner and tying the bag closed. A labeled sample bag was attached securely to the liner. The bag was placed immediately on ice in the cooler. During sample processing, the flying insect bag was placed in a freezer for approximately 15 minutes to slow the insects' movement. A small hole was cut in the bottom corner of the bag and the contents poured into the labeled sample bag. The sample was weighed, double bagged, and frozen.

Ground-dwelling insects were collected from the pitfall traps using clean forceps. The contents (insects only) of all five cups was placed in a single labeled sample bag, double bagged, and placed on ice in the cooler.

During sample collection, we took the opportunity to collect any free-ranging insects found within the sampled subplot. These insects were collected using clean forceps, placed in a labeled sample bag, double bagged, and placed on ice in the cooler. All samples were transported to the field office, processed, and placed in the freezer.

Lorsban 4E was applied at the target rate of 2.3 kg a.i./ha (2.0 lb a.i./acre). The application was made to the 1.3-ha area with an air-blast sprayer at a desired rate of 4731 L/ha (500 gal/acre). The application was performed starting at approximately 2200 hours on July 16, 1997, and was completed at approximately 2350 hours on July 16, 1997. A single 500-gallon Air-O-Fan air-blast sprayer drawn by a New Holland 7010 tractor was used to make the application. The tractor was operated by a licensed California state pesticide applicator employed by the Harlan Ranch Company. The sprayer was equipped with D6 and D7 T-Jet nozzles with a similar arrangement on both sides of the sprayer manifold. Spray pressure was set to 150 psi at an average ground speed of 2.044 mph.

The output of the sprayer was calculated by the Harlan Ranch Pest Consultant Advisor (PCA) in a manner typical for that sprayer. We verified the accuracy of the calculations and methodology prior to application. To determine the appropriate application rate, the PCA used the following set of equations:

$$\text{GPM/side} = \frac{\text{GPA} \times \text{MPH} \times \text{row spacing}}{1000}$$

$$\text{Ground speed (ft/sec)} = \text{mph} \times (88/60)$$

where GPM = gallons per minute, GPA = gallons per acre, and mph = miles per hour; 1000 is a constant used in place of unit conversions; 88/60 is a constant for the distance traveled per second at 60 mph; and row spacing is 22 feet. Each tank of test substance was mixed in the sprayer tank using 1.89 L (0.5 gallon) of test substance (4 lb a.i./gal) with 1892.5 L (500 gallons) water.

The sprayer tank was filled three times to treat the entire study area. The application tank solutions were mixed and sampled. Prior to measuring 1.89 L (0.5 gallon) of product for each tank, each Lorsban 4E insecticide container was shaken vigorously. Using a clean, calibrated 2-quart glass Pyrex container, 1.89 L (0.5 gallon) of undiluted product was poured into each tank after it was half to two-thirds full of water. A small (approximately 1 ounce) amount of No Foam was added to each tank to prevent excessive foaming. The mixture was agitated with the agitation auger the entire time of filling.

Weather data were collected at the study site according to the study protocol. A maximum/minimum thermometer and rain gauge were placed in the center of the study site. Readings were taken and recorded daily.

Fortification recoveries in the different sample types ranged from 76.4 to 106.0 percent with a mean of 95.4 percent and a standard deviation of 10.89 percent. The analytical results are summarized in Table 23.4, where the measured and corrected residue values are reported. Measured residue values were corrected for recovery capability of the analytical method using the equation

$$\text{Corrected residue value} = \frac{\text{measured residue}}{\text{fortification \% recovery}}$$

The time of occurrence of peak residues and the temporal pattern of residue decline are often masked by tissue desiccation in those invertebrates that die as

TABLE 23.4   Summary of Samples Collected from Study Plots during EBA Study Number 079717 Conducted at Harlan Ranch in Clovis, CA, July 1997

Sample Number	Collection Time	Plot Number	Sample Type	Sample Weight (g)	Number of Individuals in Sample	Number Dead at Time of Collection
7030	2 h	A3	Pinned larvae	16.9	36	36
7036	2 h	B3	Pinned larvae	19.4	38	0
7040	2 h	C3	Pinned larvae	17.9	35	0
7052	Day 1	A4	Pinned larvae	15.3	36	26
7059	Day 1	B4	Pinned larvae	4.4	10	9
7062	Day 1	C4	Pinned larvae	8.4	18	18
7077	Day 5	A2	Pinned larvae	4.2	15	15
7082	Day 5	B2	Pinned larvae	5.2	17	17
7087	Day 5	C2	Pinned larvae	8.7	27	27
7096	Day 10	A1	Pinned larvae	1	4	4
7105	Day 10	B1	Pinned larvae	3.2	10	10
7111	Day 10	C1	Pinned larvae	3.5	14	14
7033	2 h	A3	Cricket	3.7	14	5
7038	2 h	B3	Cricket	9.9	30	10
7039	2 h	C3	Cricket	8.3	43	33
7054	Day 1	A4	Cricket	6.9	33	32
7058	Day 1	B4	Cricket	8.2	28	28
7060	Day 1	C4	Cricket	10.5	33	33
7072	Day 5	A2	Cricket	1.8	20	20
7083	Day 5	B2	Cricket	4.6	27	27
7091	Day 5	C2	Cricket	0.6	10	10
7099	Day 10	A1	Cricket	3.1	26	26
7107	Day 10	B1	Cricket	2.2	21	21
7115	Day 10	C1	Cricket	3.3	44	44
7034	2 h	A3	Incidental	0.4	12	0
7041	2 h	C3	Incidental	0.5	13	1
7051	Day 1	A4	Incidental	0.3	6	2
7057	Day 1	B4	Incidental	1.2	34	0
7064	Day 1	C4	Incidental	22.6	Unknown	Unknown
7076	Day 5	A2	Incidental	8.5	Unknown	Unknown
7084	Day 5	B2	Incidental	14.3	Unknown	Unknown
7090	Day 5	C2	Incidental	12.5	Unknown	Unknown
7102	Day 10	A1	Incidental	3.5	Unknown	Unknown
7109	Day 10	B1	Incidental	3.3	Unknown	Unknown
7114	Day 10	C1	Incidental	65.2	Unknown	Unknown
7047	2 h	A3	Ground dwelling	0.3	10	0
7050	2 h	B3	Ground dwelling	0.1	10	0
7053	2 h	C3	Ground dwelling	0.4	12	0
7068	Day 1	A4	Ground dwelling	0.7	25	2
7069	Day 1	B4	Ground dwelling	0.3	14	0
7070	Day 1	C4	Ground dwelling	0.1	9	0
7073	Day 5	A2	Ground dwelling	0.1	6	1
7074	Day 5	B2	Ground dwelling	0.5	40	5

TABLE 23.4 *(Continued)*

Sample Number	Collection Time	Plot Number	Sample Type	Sample Weight (g)	Number of Individuals in Sample	Number Dead at Time of Collection
7075	Day 5	C2	Ground dwelling	0.2	7	1
7097	Day 10	A1	Ground dwelling	0.4	11	2
7106	Day 10	B1	Ground dwelling	0.5	15	1
7112	Day 10	C1	Ground dwelling	0.2	Unknown	Unknown
7044	2 h	A3	Flying insect	2	Unknown	Unknown
7045	2 h	B3	Flying insect	2.7	Unknown	Unknown
7046	2 h	C3	Flying insect	1.7	Unknown	Unknown
7065	Day 1	A4	Flying insect	0.6	Unknown	Unknown
7066	Day 1	B4	Flying insect	0.5	Unknown	Unknown
7067	Day 1	C4	Flying insect	1	Unknown	Unknown
7079	Day 5	A2	Flying insect	2.2	Unknown	Unknown
7085	Day 5	B2	Flying insect	0	NA	NA
7086	Day 5	C2	Flying insect	1.8	Unknown	Unknown
7100	Day 10	A1	Flying insect	5.2	Unknown	Unknown
7103	Day 10	B1	Flying insect	4.3	Unknown	Unknown
7104	Day 10	C1	Flying insect	3.1	Unknown	Unknown
7032	2 h	A3	Earthworm	48.4	14	3
7037	2 h	B3	Earthworm	42	13	0
7042	2 h	C3	Earthworm	52.6	14	0
7048	Day 1	A4	Earthworm	60.3	14	0
7056	Day 1	B4	Earthworm	22.7	7	0
7061	Day 1	C4	Earthworm	40.7	11	0
7071	Day 5	A2	Earthworm	39.3	11	0
7080	Day 5	B2	Earthworm	49.8	11	0
7088	Day 5	C2	Earthworm	48.1	13	0
7098	Day 10	A1	Earthworm	19.2	4	0
7108	Day 10	B1	Earthworm	12.1	9	8
7113	Day 10	C1	Earthworm	2.3	1	0
7031	2 h	A3	Seed head	29.3	12	NA
7035	2 h	B3	Seed head	18.1	12	NA
7043	2 h	C3	Seed head	26.2	12	NA
7049	Day 1	A4	Seed head	24.8	12	NA
7055	Day 1	B4	Seed head	21.6	12	NA
7063	Day 1	C4	Seed head	23.1	12	NA
7078	Day 5	A2	Seed head	24.5	12	NA
7081	Day 5	B2	Seed head	21.4	12	NA
7089	Day 5	C2	Seed head	22.2	12	NA
7101	Day 10	A1	Seed head	21.2	12	NA
7110	Day 10	B1	Seed head	20.1	12	NA
7116	Day 10	C1	Seed head	17.5	12	NA
7092	Control	NA	Larvae	113.3	200	0
7093	Control	NA	Seed head	168.9	Unknown	NA
7094	Control	NA	Cricket	209.9	400	Unknown
7095	Control	NA	Earthworm	154.6	36	0

a result of the exposure. To eliminate this effect, residue levels determined for crickets and ground beetle larvae (which were maintained in enclosures and died as a result of direct exposure to the test substance after application) were normalized to their body weights at the time of their placement onto the sub-plots. The equation used to normalize the residue levels is

$$\text{Normalized residue (ppm)} = R_1 - R_1 \frac{W_1 - W_2}{W_1}$$

where R_1 = residue value (corrected ppm from Table 23.4)

W_1 = mean individual sample weight at time of placement (g)

W_2 = mean individual sample weight at time of collection (g)

Again, these calculations apply only to the placed cricket and larvae samples. All other samples, which were alive at collection and therefore experienced no desiccation, were not normalized. Throughout this study, residue values reported for crickets and pinned larvae are normalized values (Table 23.5), and for all other samples the mean corrected residue values (from Table 23.4) are reported.

23.9   RESULTS

The average minimum temperature recorded at the study site was 16.2°C (61.2°F). The average maximum temperature recorded at the study site was 36.7°C (98.1°F). Only a trace of rain, <1 mm, was recorded during the conduct of the study. The average minimum and maximum temperatures and precipitation recorded at the study site were normal for July.

Approximately 3 hours prior to test substance application, 85 to 100 percent of the crickets in each enclosure appeared alive. Following the 2-hour post-application collection period, crickets seen in the enclosures appeared dead, as were the pinned larvae. Table 23.3 presents a summary of all samples collected during the study. This included sample type, collection period, number sampled, and status at sampling. Residues of chlorpyrifos were measured in all sample types collected during this study. Chlorpyrifos concentrations found in the invertebrates and seed samples ranged from <0.02 to 6.77 ppm (Tables 23.4 and 23.5). No measurable residues were found in any control samples (samples collected from the captive invertebrate population prior to application).

Darkling ground beetle larvae pinned in the citrus trees exhibited the highest residues with a geometric mean value of 6.77 ppm in the day 1 samples (Table 23.5). Mean residue concentrations measured in crickets were the second highest detected during this study, at 2.45 ppm in the day 1 samples (Table 23.5). Again, peak mean residue values were well below the 30 ppm predicted by Fletcher et al.[20]

TABLE 23.5 Measured and Corrected Residue Means, Standard Deviations, and Geometric Means for Samples Collected from Study Plots at Harlan Ranch Near Clovis, CA, July 1997

Sample Type	Collection Period	Measured Residues[a] (ppm) Mean	SD	Fortification Recovery (%)	Corrected Residues[b] (ppm) Mean	SD	Geometric Mean (ppm)
Pinned larvae	2 h	6.8	0.35	0.885	7.69	0.4	7.68
	Day 1	8.3	1.3	0.885	9.38	1.47	9.30
	Day 5	7.94	0.61	0.885	8.98	0.69	8.96
	Day 10	6.64	2.89	0.885	7.5	3.26	7.04
Cricket	2 h	4.01	2.78	1.060	3.79	2.62	3.24
	Day 1	4.61	3.41	1.060	4.35	3.22	3.24
	Day 5	7.96	0.53	1.060	7.51	0.5	7.50
	Day 10	6.07	3.3	1.060	5.73	3.11	5.24
Incidental	2 h	1.63	NA	0.935	1.74	NA	1.22
	Day 1	0.48	0.08	0.935	0.51	0.08	0.51
	Day 5	0.81	0.48	0.935	0.86	0.51	0.74
	Day 10	0.49	0.52	0.935	0.53	0.56	0.35
Ground dwelling	2 h	0.37	—	0.764	0.48		NA
	Day 1	0.56	—	0.764	0.74		NA
	Day 5	1.58	—	0.764	2.07		NA
	Day 10	0.12	—	0.764	0.15		NA
Flying insect	2 h	0.3	0.07	1.051	0.29	0.07	0.28
	Day 1	0.06	—	1.051	0.06		NA
	Day 5	0.07	NA	1.051	0.06	0.01	0.06
	Day 10	0.12	0.11	1.051	0.11	0.11	0.08
Earthworms	2 h	0.11	0.10	93.4	0.12	0.11	0.09
	Day 2	0.13	0.16	—	0.14	0.17	0.08
	Day 5	<0.02	<0.02	93.4	NA		NA
	Day 10	0.26	0.37	—	0.28	0.39	0.12
Seeds	2 h	0.17	0.02	104.8	0.16	0.02	0.16
	Day 1	0.18	0.07	104.8	0.17	0.06	0.16
	Day 5	0.09	0.01	104.8	0.08	0.01	0.08
	Day 10	0.05	0.01	104.8	0.05	0.01	0.05

[a] SD = standard deviation.
[b] Corrected residue = measured residue/fortification % recovery.

The mean residue in ground-dwelling insects captured in the pitfall traps was highest on day 5, at 2.07 ppm (Table 23.4), substantially below the 30 ppm predicted by Fletcher et al.[20] The residues decreased to 0.15 ppm by day 10 (Table 23.5). Incidentally collected ground-dwelling insects (free-ranging) exhibited peak residue levels of 1.22 ppm (Table 23.4) in samples collected during the 2-hour sampling period. Following the 2-hour collection, residues dropped substantially for the remainder of the study. The mean residue levels in flying insects taken in light traps were highest during the 2-hour sampling period, at 0.28 ppm (Table 23.4). The residue concentrations for the remaining three collection periods ranged from 0.06 to 0.08 ppm.

Residues measured in earthworms, which were all alive at collection, were low and variable. Residues ranged from undetectable on day 5 to 0.12 on day 10 (Table 23.4). Seeds collected from the standing seed heads contained the lowest measured residues of any of the sample types used during this study. Geometric mean residues ranged from 0.05 to 0.16 ppm (Table 23.4). These values are also considerably lower than the predicted value of 30 ppm based on estimates of Fletcher et al.[20]

23.10   DISCUSSION OF RESEARCH

Getting the test substance applied correctly was critically important in this research process, and the first step to this process was to determine the output of the application machinery. When treating a citrus grove or any orchard crop, the spraying machinery, an air-blast sprayer in this case, travels down the corridor between two rows of trees and sprays both sides. Therefore, a single side of each row of the two rows of trees gets sprayed with each pass. The equation provided in the methods section for calculating output and ground speed provides the data needed to calculate how much chemical to mix into each tank of water to be sprayed out:

$$\mathrm{GPM/side} = \frac{\mathrm{GPA \times MPH \times row\ spacing}}{1000}$$

We desired to spray 500 gal/acre at a ground speed of 2.044 mph, which was determined by testing the tractor over a given distance at a set engine RPM. The tree row spacing was 22 feet and the tree spacing in the rows was 18 feet. Therefore, the calculation for GPM/side was as follows:

$$\mathrm{GPM/side} = \frac{500\ \mathrm{GPA} \times 2.044\ \mathrm{mph} \times 22\ \mathrm{ft}}{1000} = 22.484\ \mathrm{gal/min\ per\ side}$$

The ground speed in feet per second was calculated as

$$\mathrm{Ground\ speed\ (ft/sec)} = 2.044\ \mathrm{mph} \times (88/60)$$

$$= 1.998\ \mathrm{ft/sec,\ or\ 179.87\ ft/min}$$

The sprayer should put out the full 500 gallons in 11.119 minutes [500 gal/ (22.484 GPM/side × 2 sides)]. Given the row spacing of 22 feet (the sprayer swath width) and traveling speed of 179.87 ft/min (3.0 ft/sec × 60 sec/min), this rate would cover 3957.14 ft^2/min. If the tank sprays out in 22.9 minutes, the total coverage per tank is 43,999 ft^2. An acre is 43,560 ft^2, so the application is 100.1 percent of the target application. Simply mixing 2 pounds of active ingredient per tank achieved 2 lb a.i./acre.

The pinned larvae were expected to contain the highest residues because they were placed on the limbs and on both sides of the tree leaves, thus exposing at least half of them directly to the spray. Measured residues increased from the 2-hour sampling to the day 1 sampling and then declined throughout the remainder of the study. Despite direct exposure to the spray, the peak mean residue value on day 1 was considerably lower than the value of 30 ppm (15 ppm/lb a.i. applied) predicted by Fletcher et al.[20]

Reasons for cricket residue concentrations peaking on day 1 rather than day 0 include:

1. Crickets were placed in enclosures on the ground directly under tree canopies. In this position, the enclosures received a direct spray from the airblast sprayer.
2. The ground litter in the enclosures, which mimicked the existing ground litter in the grove, possibly allowed some crickets to be under cover during application, as would be the case for free-ranging insects.
3. Any crickets moving around inside the enclosure following the application (before the spray dried) were likely to be exposed by coming in contact with treated soil and ground litter or ingesting contaminated food. The residues gained from this activity would show up at the next collection period.

Ground-dwelling insects captured in pitfall traps were assumed to have been alive when they wandered into the traps. Thus, desiccation does not play a role in the residue levels. We expected peak residues at the 2-hour and day 1 collections (0.48 and 0.74 ppm, respectively). The reason for the delayed peak residue on day 5 is unknown but may be the result of a sample being inadvertently contaminated with soil particles or inconspicuous vegetation fragments.

The flying insect samples consisted of a combination of large and small insects, but the majority of the sample weight consisted of small insects. The peak mean residue value measured in the 2-hour samples is several orders of magnitude lower than the value of 270 ppm predicted by Fletcher et al.[20] for small insects at this application rate and well below the 30 ppm predicted for large insects. The pattern of residues found in these samples may be explained by the combined effects of resident insects, which survived the application and the influx of nonexposed insects immigrating or pupating into the treated study plots. This is the most mobile group of insects samples, and immigration was expected to be rapid. The residue data support the hypothesis that immigration of new insects occurs quickly after the application.

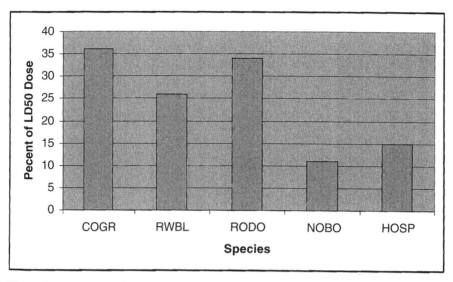

Figure 23.8   Percent of LD_{50} that would be consumed by five bird species if they instantaneously consumed 1 day's total food intake consisting entirely of invertebrates containing highest residue concentration measured during this study (6.77 ppm). It is assumed that birds weighing > 100 g (COGR, RODO, NOBO) consume 50% of their body weight per day and birds weighing < 100 g (RWBL, HOSP) consume 30% of their body weight per day. COGR = common Grackle, RWBL = red-winged blackbird, RODO = rock dove, NOBO = northern Bobwhite, HOPS = house sparrow.

Residue values found in the seeds are lower than all other test species due to the protective nature of the seed heads. The entire seed is not directly exposed to the spray while it is still held in the seed head. By removing the seeds from the sheaths of the seed heads, we measured that portion of the seed head a bird would extract and consume. Hulling of the seed by birds may further reduce their exposure.[50] This is discussed in more detail, as it relates to exposure to birds, in the following section on risk assessment. Seeds that have already dispersed from the seed head and are on the soil surface may exhibit different residue concentrations than those in the seed head during the application.

Figure 23.8 shows the percent of an LD_{50} dose that might be experienced by birds if they consumed a full day's diet consisting entirely of larvae containing the highest concentration measured during this study. The reader should keep in mind this feeding scenario is highly unlikely in the wild. Birds are unlikely to get 100 percent of their feed for an entire day from a single treated field and even more unlikely to consume nothing but exposed food items. The values provided in Figure 23.8 definitely represent an absolute worst-case scenario.

The chemical application to citrus for this study occurred after dark. This is common in this part of California because daytime temperatures make working in protective clothing uncomfortable and dangerous. Also, at night, it is more likely that there is no wind. The insect light traps were turned on immediately after the spraying, so there was a good chance of collecting flying insects that were present during the application. If light traps are to be used when spraying is done in daylight, it is best to make the application in the evening so that the light traps will be effective shortly after spraying. If this cannot be accomplished, other types of insect capture would be necessary.

Because the insect exposure chambers were opened (lids removed) immediately before the pesticide application, the insects within them should have experienced exposure very similar to that of free-ranging insects. If the chambers bias the exposure at all, we believe the effect is in causing slightly elevated exposure.

The methods used in this study worked well. Additionally, by measuring weights of insects prior to placing them into the light traps, we were able to isolate the influence of desiccation on residue levels of dead insects by normalizing the measured residue to their fresh body weight. Whether birds are interested in dead and desiccated insects as a food source is a question not yet thoroughly answered, but recent research on this topic indicates that they are of little interest to birds.[50]

The risk assessment presented below addresses the risk to adult birds in regard to exposure through the ingestion of invertebrates and seed from a pesticide-treated field. Other sources of exposure, such as drinking water, are not addressed. Juvenile birds are addressed to the degree that the LC_{50} values are obtained from tests conducted with birds less than 2 weeks of age. The risk assessment does not address birds in the age range of 1 to 3 days.

23.11 AVIAN RISK ASSESSMENT: DETERMINISTIC APPROACH

Two factors contribute to the estimation of toxicological risk: exposure and toxicity. Exposure can be estimated by determining the amount of a particular type of food a bird might consume and multiplying that amount by the pesticide residue concentration on or in food items. Increased accuracy in the determination of these two values results in reduced uncertainty in the risk assessment. Very small birds, such as swallows, that weigh as little as 15 g, can consume up to 100 percent of their body weight in a day if feeding on food items with high moisture content (e.g., fresh insects).[23,44] However, a bird of similar size consuming food items with low moisture content (e.g., mature millet seeds) can consume only about 50 percent of its body weight.[52,53] Larger birds, such as the American kestrel (*Falco sparverius*) or common grackle (*Quiscalus quiscula*), which weigh approximately 100 g, may consume half their body weight per day when consuming food items with high water content but may consume less than 20 percent of their body weight if feeding on dryer food items.[23]

Tables 23.6 and 23.7 provide a list of laboratory-derived avian toxicity estimates for chlorpyrifos for various bird species and include source references for

TABLE 23.6   Pinned Larvae and Cricket Residues Normalized to Initial Body Weights[a]

Sample Number	Mean Placement Weight per Individual (g)	Mean Collection Weight per Individual (g)	Weight Change (%)	Measured Residue (ppm)	Normalized Residue (ppm)
		Larvae			
2 h					
7030	0.613	0.469	23.5	7.32	5.60
7036	0.605	0.511	15.5	8.11	6.85
7040	0.616	0.511	17.0	7.64	6.34
Mean				7.69	6.27
SD				0.40	0.63
Day 1					
7052	0.615	0.425	30.9	10.92	7.55
7059	0.613	0.440	28.2	9.22	6.62
7062	0.61	0.467	23.4	8.00	6.13
Mean				9.38	6.77
SD				1.47	0.72
Day 5					
7077	0.653	0.280	57.1	9.38	4.02
7082	0.625	0.306	51.0	9.38	4.60
7087	0.653	0.322	50.7	8.18	4.03
Mean				8.98	4.22
SD				0.69	0.33
Day 10					
7096	0.603	0.250	58.5	4.66	1.93
7105	0.658	0.320	51.4	6.77	3.29
7111	0.598	0.250	58.2	11.06	4.62
Mean				7.50	3.28
SD				1.35	
		Crickets			
2 h					
7033	0.26	0.26	0	2.79	2.79
7038	0.38	0.33	13.2	6.76	5.87
7039	0.27	0.19	29.6	1.81	1.28
Mean				3.79	3.31
SD				2.62	2.34
Day 1					
7054	0.28	0.21	25.0	7.40	5.55
7058	0.34	0.29	14.7	4.68	3.99
7060	0.26	0.32	23.1 gain	0.98	0.75
Mean				4.35	3.43
SD				3.22	2.45

TABLE 23.6 *(Continued)*

Sample Number	Mean Placement Weight per Individual (g)	Mean Collection Weight per Individual (g)	Weight Change (%)	Measured Residue (ppm)	Normalized Residue (ppm)
Day 5					
7072	0.27	0.09	66.6	8.09	2.70
7083	0.43	0.17	60.5	7.20	2.84
7091	0.26	0.06	76.9	7.24	1.67
Mean				7.51	2.41
SD				0.50	0.64
Day 10					
7099	0.34	0.12	64.7	3.97	1.40
7107	0.32	0.10	68.8	3.89	1.21
7115	0.42	0.08	81.0	9.32	1.77
Mean				5.73	1.46
SD				3.11	0.28

[a] Normalized to mean individual weight at time of placement on day-1.

each value.[29–43,53–57] Acute oral toxicity (LD_{50}) values range from 5.62 mg/kg body weight for the common grackle to 122 mg/kg body weight for the house sparrow (*Passer domesticus*). Acute dietary toxicity (LC_{50}) values range from 136 ppm for 5- to 7-day-old mallards (*Anas platyrhynchos*) to 721 ppm for 8-day-old northern bobwhite. Avian reproduction studies have shown the no observable effects concentration (NOEC) for mallards and northern bobwhite to be approximately 25 and 125 ppm, respectively.[41–43]

Several factors must be considered when deciding which toxicity value (LC_{50} or LD_{50}) should be used in developing a quotient for risk assessment. The LD_{50} test involves a single bolus of a xenobiotic, which maximizes the toxicokinetics of the xenobiotic. This is demonstrated by the difference between most LD_{50} and LC_{50} values for one species for the same compound. Only in rare situations could a bird in the wild obtain a dose comparable to that obtained in an LD_{50} test. When the pesticide formulation is flowable, the distribution and variability of postapplication residues on food items, under most agricultural circumstances, do not allow a bird the opportunity to quickly consume the amount of residues required to obtain the LD_{50} dose. Therefore, with the exception of treated seed applications, the LC_{50} is a more appropriate model of toxicity and should be used in calculating the risk quotient. Even so, the LC_{50} may also overestimate the toxicity factor when the pesticide product under consideration is characterized by rapid field dissipation. The LC_{50} is generally conducted with constant concentrations of the test substance in feed for 5 days. In this study, for example, the chlorpyrifos residues measured on the bird food items were relatively low to begin with so it is very unlikely that a bird could ingest enough

TABLE 23.7 One-Time Consumption[a] of Food Types Used and Required for Individual to Attain LD$_{50}$ If Consumption Occurred at Time of Peak Residues

	LD$_{50}$[b] (mg/kg)	Body Weight[c] (g)	Percent of Body Weight Required					
			Cricket	Earthworm	Flying Insects	Ground-Dwelling Insects	Seed	Pinned Larvae
House sparrow	122.0	27	3556.9	43571.4	42069.0	5893.7	71764.7	1802.10
Red-winged blackbird	13.1	55	381.9	4678.6	4517.2	632.9	7705.9	193.5
California quail	68.3	184	1991.3	24392.9	23551.7	3299.5	40176.5	1008.9
European starling	75.0	78	2186.6	26785.7	25862.1	3623.2	44117.7	1107.8
Northern bobwhite	32.0	180	932.9	11428.6	11034.5	1545.9	18823.5	472.7
Common grackle	5.62	111	163.8	2007.1	1937.9	271.5	3305.9	83.0
Rock dove	10.0	369	291.5	3571.4	3448.3	483.1	5882.4	147.7
Chukar	61.1	581	1781.3	21821.4	21069.0	2951.7	35941.2	902.5
Residue levels in food types (ppm)			3.43	0.28	0.29	2.07	0.17	6.77

[a]Expressed as percentage of bird's body weight, where

$$\text{Percent body weight required} = \frac{\text{species body weight (kg)} \times LD_{50}\ (mg/kg)/\text{residue concentration (mg/g)}}{\text{species body weight (g)}} \times 100$$

[b]LD$_{50}$ values taken from Table 23.4.

[c]See Reference 13.

TABLE 23.8 **Risk Quotients Derived from Environmental Concentrations on Food Items Measured Divided by Reported LC_{50} Values for Various Bird Species**

	LC_{50}^b (mg/kg)	Cricket	Earthworm	Flying Insects	Ground-Dwelling Insects	Seeds	Pinned Larvae
				Risk Quotients[a]			
Mallard	136	0.03	0.002	0.002	0.01	0.001	0.05
Northern bobwhite	353	0.01	0.0008	0.0008	0.004	0.0005	0.02
Japanese quail	299	0.01	0.0009	0.001	0.005	0.0006	0.02
Pheasant	553	0.006	0.0005	0.0005	0.003	0.0003	0.01
Concentration in food items (ppm)		3.43	0.28	0.29	1.58	0.17	6.77

[a] Risk quotient = concentration (ppm)/LC_{50} (mg/kg).
[b] LC_{50} values taken from Table 23.5.

of a single food item at peak residue levels over 5 days to approach an LC_{50} exposure.

The quotient method used by the EPA for estimating avian risk to a pesticide is calculated as

$$\text{Risk quotient} = \frac{\text{EEC}}{LC_{50}}$$

where EEC is the expected environmental concentration. Using the measured environmental concentration (MEC) in place of the EEC, as derived from the Kenaga nomogram or Fletcher et al.,[18,19] provides a more accurate assessment of risk. To obtain a single value for the MEC estimate from the chlorpyrifos residues measured during this study, we calculated the mean of the residue concentration of the three replicates for each sample collection period and used the mean from the period having the highest average residue value for the risk quotient calculation. Table 23.8 provides the risk quotients calculated using this approach for those bird species for which LC_{50} data are available. The highest quotient calculated (0.07) was for mallards consuming larvae. This value is almost one-seventh of the quotient calculated using residue levels predicted by the Kenaga nomogram (0.48). These results, based on the MEC on typical food items and LC_{50} values, indicate there is very low risk to adult birds posed by the 2.0-lb-a.i./acre application of Lorsban 4E insecticide made during this study.

In some cases the LD_{50} is used rather than the LC_{50} in calculating the risk quotient. The MEC values also reduce uncertainty using this approach. Table 23.8 provides the LD_{50} values and the amount of rapid food consumption (expressed as percent of total body weight) required to provide an LD_{50} for several species. Considering that large birds such as the common grackle are likely to consume slightly less than 50 percent of their body weight per day when feeding on insects such as larvae, it is doubtful that it could ingest a quantity of contaminated larvae in a single feeding episode that would result in harm. For example, a common grackle, the species with the lowest LD_{50}, weighing 111 g, would have to consume, in a single feeding episode, 83 percent of its body weight, or 92 g of larvae (approximately 184 individuals), in order to obtain an LD_{50}. In the case of a 27-g house sparrow, the necessary consumption rate exceeds one thousand times its body weight (Table 23.8). If the common grackle instantaneously consumed 30 percent of its body weight in larvae containing 6.77 ppm chlorpyrifos, it would consume only 36 percent of its LD_{50} (Figure 23.8). Figure 23.8 shows the percent of the LD_{50} ingested by five bird species if a full day's food intake, consisting entirely of larvae containing 6.77 ppm chlorpyrifos, was instantaneously consumed.

The primary question, then, when determining risk to birds, is whether a bird would obtain enough food of the same type quickly enough to obtain a lethal dose. Common grackles are omnivores, which suggests they are likely to consume a variety of food items throughout the day.[12] The likelihood that a bird will consume only one type of food item in the wild seems extremely low, except possibly in the case of spilled grain piles or during the plowing of an agricultural field that has an intense insect larval infestation. Thus, the risk to adult birds exposed to food items treated with chlorpyrifos at 2.0 lb a.i./acre is very low.

The final type of laboratory data that can be included in a risk assessment of oral exposure is avian reproduction study data. The NOEC for chlorpyrifos in both mallards and northern bobwhite is approximately 25 and 125 ppm, respectively when exposed over a long period, usually up to 20 weeks, to a constant dietary concentration. In this study, residue concentrations were measured for only a 10-day period following the application. At no time during this study did residue levels on any of the food items exceed even half of the NOEC for mallards. Thus there appears to be little risk of chronic or reproductive effects.

The delayed exposure observed in earthworms was expected and is believed to be the result of the rate of penetration of the test substance into the soil and the frequency and persistence of earthworm surfacing at night to feed on the organic debris on the surface. Surfacing was believed to be minimal due to relatively dry conditions.

23.12 OVERVIEW

The collection of field residue data and its subsequent use as described in this chapter demonstrates the importance and practicality of using measured environmental factors to refine and improve the dependability of ecological risk assessment. It is clear from these results that the common practice of deriving estimates of invertebrate pesticide residues from generalized plant residue data for the purpose of avian risk estimation results in over estimation of risk by a substantial margin. We have seen that invertebrate species vary in the degree to which they are exposed to pesticides in the field relative to their behavioral patterns. Additionally, it appears that exposed populations of mobile insect species, such as flying insects, are subject to rapid mixing with unexposed populations, resulting in a prey base for birds that does not consist of 100% contaminated prey, even when avian predation occurs directly on or over a recently treated field. The most important observation, however, is that when ecological risk assessment is conducted with environmental, behavioral, or population models that are not validated with appropriate field-derived data, the results of the risk assessment can be misleading. Assumptions associated with uncertainty must be handled with caution and the degree of the uncertainty made clear to the readers of the risk assessment.

The risk assessment process is currently evolving, with substantial interest currently focusing on probabilistic methods.[20] The shift of emphasis from deterministic to probabilistic methods in avian pesticide risk assessment has great potential for increasing our ability to define and understand ecological risk. However, probabilistic models are often more complex, involving numerous assumptions and more uncertainty. Early attempts at probabilistic ecological risk assessments will demonstrate the need for dependable data, and will point out the most obvious ecological and biological knowledge gaps. These will need to be filled by further research if probabilistic methods are to produce a refinement of the dependability and utility of ecological risk assessment.

REFERENCES

1. R. Carson. *Silent Spring*. Houghton Mifflin, Boston, 1962.

2. W. H. Stickel. Some effects of pollutants in terrestrial ecosystems, pp. 25–74, in A. S. McIntyre and C. F. Mills (Eds.). *Ecological Toxicology Research*. Plenum, New York, 1975.

3. M. Eto. *Organophosphorus Pesticides: Organic and Biological Chemistry*. CRC Press, Cleveland, OH, 1974.

4. G. J. Smith. *Toxicity and Pesticide Use Inrlatin to Wildlife: Organophosphorus and Carbamate Compounds*. CRC Press, Boca Raton, FL, 1993.

5. R. DiGiulio and D. Tillitt (Eds). *Reproductive and Developmental Effects of Contaminants in Oviparous Vertebrates*. SETAC special publication series. Setac Press, Pensacol, FL, 447 pp., 1999.

6. B. Woodbridge. Swainson's hawk mortality in Argentina. *Wingspan* 5: 3 (1996).

7. C. P. Stone. Poisoning of wild birds by organophosphate and carbamate pesticides. *N. Y. Fish. Game J.* 26: 37–47 (1979).

8. E. L. Flickenger, K. A. King, W. F. Stout, and M. M. Mohn. Wildlife hazards from Furadan 3G applications to rice in Texaas. *J. Wildlife Manag.* 44: 190–197 (1980).

9. L. W. Brewer, R. J. Kendall, C. J. Driver, C. Zenier, and T. E. Lacher, Jr. Effects of methyl parathion on ducks and duck broods. *Environ. Toxicol. Chem.* 7: 375–379 (1988).

10. R. J. Kendall and L. W. Brewer. American wigeon mortality associated with turf application of Diazinon® AG500. *J. Wildlife Dis.* 28(2): 263–267 (1992).

11. L. W. Brewer, C. J. Driver, R. J. Kendall, T. A. Lacher, and J. Galindo. Avian response to a turf application of Triumph® 4E. *Environ. Toxicol. Chem.* 7: 391–379 (1987).

12. S. L. Tank, L. W. Brewer, M. J. Hooper, G. P. Cobb III, and R. J. Kendall. Survival and pesticide exposure of northern bobwhites (*Colinus virginianus*) and eastern cottontails (*Sylvilagus floridanus*) on agricultural fields treated with Counter® 15G. *Environ. Toxicol. Chem.* 12: 2113–2120 (1992).

13. J. A. Buck, L. W. Brewer, M. J. Hooper, and G. P. Cobb. Monitoring great horned owls for pesticide exposure in south-central Iowa. *J. Wildlife Manag.* 60(2): 321–331 (1996).

14. D. Paustenbach. *The Risk Assessment of Environmental and Human Health Hazards: A Textbook of Case Studies.* Wiley, New York, 1989.

15. E. C. Fite, L. W. Turner, N. J. Cook, and C. S. Stunkard. *Guidance Document for Conducting Terrestrial Field Studies.* Office of Pesticide Programs, U.S. Environmental Protection Agency, Washington, DC, 1988.

16. D. J. Urban and N. J. Cook. *Hazard Evaluation Division Standard Evaluation Procedure: Ecological Risk Assessment.* EPA-540/9-85-001. U.S. Environmental Protection Agency, Washington, DC, 1986.

17. F. Hoerger and E. Kenaga. Pesticide residues in plants: Correlation of representative data as a basis for estimation of their magnitude in the environment. In F. Coulston and F. Korte (Eds.), *Environmental Quality and Safety.* Academic, New York, 1972, pp. 9–28.

18. E. E. Kenaga. Factors to be considered in the evaluation of the toxicity of pesticides to birds and the environment. *Environ. Quality Safety* 2: 166–181 (1973).

19. J. S. Fletcher, J. E. Nellessen, and T. G. Pfleeger. Literature review and evaluation of the EPA food-chain (Kenaga) nomogram, an instrument for estimating pesticide residues on plants. *Environ. Toxicol. Chem.* 13(9): 1383–1391 (1994).

20. ECOFRAM. *Ecological Committee on FIFRA Risk Assessment Methods (ECOFRAM), Final Report, Terrestrial.* U.S. Environmental Protection Agency, Washington, DC, 2000.

21. U.S. Environmental Protection Agency (EPA). *Wildlife Exposure Factors Handbook.* EPA, Washington, DC, 1992.

22. T. W. LaPoint, M. Simini, R. S. Wentsal, K. R. Dixon, and L. W. Brewer. *Standard Approaches to Ecological Risk Assessment on U.S. Army Sites.* U.S. Army Chemical Research, Development and Engineering Center, Aberdeen Proving Grounds, MD, 1994.

23. RESOLVE. *Assessing Pesticides in Birds: Final Report of the Avian Effects Dialogue Group, 1988–1993.* Resolve, Washington, DC, 1994.

24. K. R. Solomon, J. P. Giesy, R. J. Kendall, L. B. Best, J. R. Coats, K. R. Dixon, M. J. Hooper, E. E. Kenaga, and S. T. McMurry. Chlorpyrifos: Ecological risk assessment for birds and mammals in corn angrecosytems. *Hum. Ecol. Risk Assess.*, in press.

25. J. P. Giesy, K. R. Solomon, J. R. Coates, K. R. Dixon, J. M. Giddings, and E. E. Kenaga. Chlorpyrifos: Ecological risk assessment in North American aquatic environments. *Rev. Environ. Contam. Toxicol.* 160: 1–129 (1999).

26. K. D. Racke. Environmental fate of chlorpyrifos. *Rev. Environ. Contam. Toxicol.* 131: 1–154 (1993).

27. K. A. Nagy. Field metabolic rate and food requirement scaling mammals and birds. *Ecol. Monogr.* 57(2): 111–128 (1987).

28. E. F. Hill and M. B. Camardese. Toxicity of anticholinesterase insecticides to birds: Technical grade versus granular formulations. *Ecotoxical. Environ. Safety* 8: 551–563 (1984).

29. E. W. Schafer, Jr. and R. B. Brunton. Chemicals as bird repellents: Two promising agents. *J. Wildlife Manag.* 35: 569–572 (1971).

30. M. Sherman, R. B. Herrick, E. Ross, and M. T. Y. Chang. Further studies on the acute and subacute toxicity of insecticides to chicks. *Toxicol. Appl. Pharmacol.* 11: 49–67 (1967).

31. G. T. Stevenson. A single oral dose LD_{50} toxicity study of Dursban® in broiler type chickens. Research report July 12, 1966. Bioproducts Department, Dow Chemical Company, Midland, MI, 1967.

32. G. T. Stevenson. A single oral dose LD_{50} toxicity study of Dursban® in turkey poults. Research report June 23, 1967. Bioproducts Department, Dow Chemical Company, Midland, MI, 1967.

33. E. W. Schafer, Jr. and R. B. Brunton. Indicator bird species for toxicity determinations: Is the technique usable in test method development? In J. R. Beck (Ed.), *Vertebrate Pest Control and Management Materials.* ASTM STP 680, American Society for Testing and Materials, Philadelphia, PA, 1979, pp. 157–168.

34. T. E. Shellenberger. Toxicology evaluations of DOWCO® 214 with wildlife and DOWCO® 179 with mallard ducks. Letter to E. E. Kenaga. Dow Chemical Company, Midland, MI, 1970.

35. J. D. Gile, J. B. Beavers, and R. Fink. The effect of chemical carriers on avian LC_{50} toxicity tests. *Bull. Environ. Contam. Toxicol.* 31: 195–202 (1983).

36. G. T. Stevenson. A game bird toxicology study—Acute dietary feeding of Dursban® to wild type mallard ducklings. Bioproducts Department Report No. GH-A 122. Dow Chemical Company, Midland, MI, 1965.

37. G. T. Stevenson. The effects of Dursban® *O,O*-diethyl-*O*-3,5,6-trichloro-2pyridyl phosphorothioate on the reproductive capacity of Coturnix quail. Reports GH-A 56 and GH-A 109. Bioproducts Department, Dow Chemical Company, Midland, MI, 1965.

38. E. F. Hill, R. G. Heath, J. W. Spann, and J. D. Williams. *Lethal Dietary Toxicities of Environmental Pollutants to Birds.* Special Scientific Report—Wildlife No. 191. U.S. Fish and Wildlife Service, 61 pp, 1975.

39. R. G. Heath, G. J. W. Spann, E. F. Hill, and J. F. Kreitzer. *Comparative Dietary Toxicities to Birds.* Special Scientific Report—Wildlife No. 152. U.S. Fish and Wildlife Service, 57 pp, 1972.

40. R. Fink. The effect of chlorpyrifos during a one-generation reproduction study on bobwhite. Wildlife International Project No. 103-178. Health and Environmental Sciences Laboratory Report No. GHRC 139. Dow Chemical Company, Midland, MI, 1978.

41. R. Fink. The effect of chlorpyrifos during a one-generation reproduction study on bobwhite. Wildlife International Project No. 103-177. Health and Environmental Sciences Laboratory Report No. GHRC 136. Dow Chemical Company, Midland, MI, 1978.

42. C. B. Schom, U. K. Abbot, and N. Walker. Organophosphorus pesticide effects on domestic and game bird species: Dursban®. *Poult. Sci.* 52: 2083 (1973).

43. J. C. Welty. *The Life of Birds*, 2nd ed. Saunders, Philadelphia, 1975.

44. K. Kaufman. *Lives of North American Birds*. Houghton Mifflin, New York, 1996.

45. A. C. Martin, H. S. Zim, and A. C. Nelson. *American Wildlife and Plants: A Guide to Wildlife Food Habits*. Dover Books, New York, 1961.

46. P. R. Ehrlich, D. S. Dobkin, and D. Wheye. *The Birder's Handbook. A Field Guide to the Natural History of North American Birds*. Simon & Schuster, New York, 1988.

47. J. K. Terres. *Encyclopedia of North American Birds. The Audobon Society*. Knopf, New York, 1980.

48. M. L. Avery, D. L. Fischer, and T. M. Primus. Assessing the hazard to granivorous birds feeding on chemically treated seeds. *Pesticide Sci.* 49: 362–366 (1996).

49. J. Stafford. Avian food selection with application to pesticide risk assessment: Are dead and desiccated insects desirable food. M.S. thesis, Utah State University, 2000.

50. D. J. Forsyth, C. F. Hinks, and N. D. Westcott. Feeding by clay-colored sparrows on grasshoppers and toxicity of carbofuran residues. *Environ. Toxicol. Chem.* 13: 781–788 (1994).

51. U.S. Environmental Protection Agency (EPA). *Pesticide Assessment Guidelines, FIFRA Subdivision E, Hazard Evaluation: Wildlife and Aquatic Organisms. Subsection 70-1d*. EPA, Office of Pesticide Programs, Washington, DC, 1982.

52. R. H. Hudson, R. K. Tucker, and M. A. Haegele. *Handbook of Toxicity of Pesticides to Wildlife*. Resource Publication 153. U.S. Department of the Interior, Fish and Wildlife Service, Washington, DC, 1984.

53. R. K. Tucker and M. A. Haegele. Comparative acute oral toxicity of pesticides to six species of birds. *Toxicol. Appl. Pharmacol.* 20: 57–65 (1971).

54. G. T. Stevenson. An LD_{50} toxicity study of Dursban® in leghorn chickens. Research report July 1, 1963. Bioproducts Department, Dow Chemical Company, Midland, MI, 1963.

55. S. Miyasaki and G. C. Hodgson. Chronic toxicity of Dursban® and its metabolite, 3,5,6-trichloro-2-pyridinol in chickens. *Toxicol. Appl. Pharmacol.* 23: 391–398 (1972).

56. S. P. Gallagher, J. Grimes, J. B. Beavers, and M. J. Jaber. Chlorpyrifos technical: An acute oral toxicity study with the house sparrow. Wildlife International Project No. 103-318a. Health and Environmental Sciences Laboratory Report No. DECO-ES-3133. Dow Chemical Company, Midland, MI, 1996.

24 Using Ecological Risk Assessment to Evaluate Potential Risks Posed by Chemicals to Birds

MIRANDA H. HENNING and NADINE M. WEINBERG
ARCADIS, Portland, Maine

MARGARET A. BRANTON
ARCADIS, Long Beach, California

24.1 INTRODUCTION

Since the U.S. Environmental Protection Agency (USEPA) first released its *Framework for Ecological Risk Assessment* (USEPA, 1992a), the range of tools used for conducting ecological risk assessments (ERAs) and the sophistication of those tools have steadily improved. Because that initial guidance on ERA (USEPA, 1992a) and others issued more recently by various agencies and organizations (e.g., USEPA, 1995, 1997, 1998; California EPA, 1994; Environment Canada, 1994; Massachusetts DEP, 1996; NYSDEC, 1993; WEF, 1993; Wentsel et al., 1996) have generally addressed the overall organizational structure of ERAs, rather than the specific tools that are applied within that structure, risk assessment practitioners have taken the lead in designing, testing, and validating those tools. This chapter provides an overview of the range of tools that have emerged during the past decade for conducting ERAs in which birds are the central receptors of interest.

Birds prove interesting and challenging subjects for evaluation in ERA for a number of reasons. The use of pesticides and other anthropogenic chemicals that cause both acute and chronic effects in avian species gained the attention of the public and the government with the 1962 publication of Rachel Carson's

Human and Ecological Risk Assessment: Theory and Practice, Edited by Dennis J. Paustenbach
ISBN 0-471-14747-8 © 2002 John Wiley & Sons, Inc.

groundbreaking book *Silent Spring* (Carson, 1962). Thereafter, in the 1960s and 1970s, catastrophic declines in a number of bird populations, including brown pelicans, peregrine falcons, and bald eagles (Gress, 1970; Best, 1994; Keith et al., 1971; Nisbet, 1987; Peakall and Kiff, 1988; Postupalsky, 1978; Ratcliffe, 1993), occurred throughout North America and other parts of the world. Studies evaluating these events predated the practice of ERA, and the results of these studies led to the banning of many synthesized chemicals in the early 1970s. Due to their ubiquity, social and ecological value, and susceptibility to many chemicals, birds are routinely considered in ERAs today and, depending on the site, may provide the basis for remediation goals.

The framework for evaluating risks to birds is consistent with the general structure of ERA. Because risks are a function of both exposure and toxicity, potential exposures to receptors of interest must be characterized in order to determine whether adverse effects are expected at that exposure level. The following sections discuss the range of tools available for characterizing exposures and effects in birds potentially exposed to chemicals in the environment, as well as the integration of information on exposures and effects in order to characterize risks.

24.2 TOOLS FOR CHARACTERIZING EXPOSURE TO BIRDS

Although a number of factors complicate the characterization of exposure of birds to site-related chemicals, numerous options are available, ranging from the very simple to the complex. At the most basic levels, exposures may be estimated as a dose (in milligrams per kilogram of body weight per day) based on assumptions reported in the literature regarding a given species' body size and feeding rates and behaviors. Improved accuracy and increased sophistication may be gained by modifying this basic approach in a number of ways. Use of site-specific data on concentrations of chemicals of potential concern (COPCs) in prey, soil, sediment, and surface water is an important first step in moving from a generic assessment to a site-specific ERA. Site specificity may be further increased through the collection of residue data for the species of interest in order to verify exposure and determine site-specific bioaccumulation factors (BAFs). Finally, moving from a deterministic approach to modeling doses probabilistically using Monte Carlo analysis or similar tools provides increased information regarding uncertainty and exposures throughout the population. Following a general discussion of some of the factors that warrant consideration in avian exposure assessment, each of the above tools is discussed.

Several characteristics of wild birds greatly complicate the assessment of their exposure and often introduce uncertainty into the ERA. Avian species are present as resident, breeding, or migratory populations throughout virtually all habitats on the globe. In many instances, avian populations associated with a site only spend a portion of their life cycle at the site; that portion may or may not be a sensitive life stage. The transient nature of many bird species poses a challenge in assessing site-related exposure; if foraging or migratory ranges exceed areas of specific interest, it may be difficult or impossible to allocate exposures from multiple locations.

Exposure to COPCs can be characterized by both duration (acute or chronic) and how direct it is. Indirect and chronic exposure through the food chain is the usual mechanism by which birds are exposed to COPCs. There are exceptions, however, including direct, acute exposure to pesticides (Buckley et al., 1969; Elliott et al., 1996; Hunt, 1964) and lead shot ingestion (Kendall et al., 1996; Longcore et al., 1982; Rocke et al., 1997). Additionally, the foraging behaviors of certain species may provide a direct pathway to COPCs. For example, dabbling ducks stir up and ingest sediment while foraging for benthic invertebrates and vegetation. Although ingestion of sediment can clearly contribute to overall intake of COPCs, this route of exposure generally is secondary to the food chain pathway. In addition, birds of prey may be subject to secondary poisoning through the consumption of accidentally or deliberately poisoned wildlife (Elliott et al., 1997).

Characterizing exposure helps to determine which COPCs have the potential to pose risks to receptors as well as the nature of the potential risk. For instance, effects associated with exposure of different durations or through the food chain versus direct ingestion may determine the endpoints that warrant evaluation. For example, chronic exposure is more likely to contribute to reproductive effects, whereas an extreme acute exposure may result in mortality. In addition, some COPCs may be transformed as they move through the food chain, resulting in either higher or lower toxicity than the parent COPC.

Understanding potential exposure scenarios is also critical in determining which species should be the focus of the ERA (i.e., receptors of interest). The factors that are considered in selecting receptors of interest include feeding guild (i.e., insectivorous, piscivorous, carnivorous), amenability to being studied either in the field or in the laboratory, known sensitivity to the COPCs, and habitat considerations. Avian species frequently evaluated in ERA include tree swallows, great blue herons, and bald eagles. Each represents a distinct feeding guild with unique exposure potential. Although these species do not always have the greatest potential exposure at a given site, they are often used as surrogate species when either the primary species of concern is not responsive to controlled study situations, such as secretive songbird species, or in laboratory studies when a species is threatened or endangered. Although there are uncertainties associated with the use of surrogate species, studies can be designed to approximate the exposure scenario of the species of primary concern.

Regardless of the suite of tools considered and ultimately selected for use in an avian ERA, all of the above factors warrant evaluation during the planning stage. The following sections describe the principal tools available for evaluating avian exposures, including the hazard quotient method, tissue analyses and food chain models, and probabilistic techniques.

24.2.1 Hazard Quotient Method

Perhaps the most simple—and the most frequently used—tool for estimating ecological risks is the hazard quotient (HQ) method [also referred to as the toxicity quotient (TQ) method]. The HQ method refers to the comparison of the potential dose received at a site with a toxicity reference value (TRV) that represents a "safe" dose for that COPC and receptor. (The development TRVs are further discussed under Tools for Evaluating Effects in Birds.) Potential dose is defined as the amount of chemical present in food or water ingested, air inhaled, or material applied to the skin (USEPA, 1992b). As presented by the USEPA (1993a), a general equation for estimating dose for intake processes is

$$D_{pot} = \int_{t_1}^{t_2} C(t) \, IR(t) \, dt$$

where D_{pot} is the total potential dose over time (e.g., total milligrams of chemical intake between times t_1 and t_2), $C(t)$ is the chemical concentration in the environmental medium at time t (e.g., milligrams of chemical per kilogram of medium), and $IR(t)$ is the intake rate of the environmental medium at time t measured as mass ingested or inhaled by an animal per unit time (e.g., kilograms of medium per day). If C and IR are assumed to be constant over time (as they most often are), then the total potential dose can be estimated as

$$D_{pot} = C \times IR \times ED$$

where ED is the exposure duration and equals $t_2 - t_1$. Therefore, if C and IR are constant, the potential average daily dose (ADD_{pot}) for the duration of the exposure, normalized to the animal's body weight (e.g., mg/kg-day), is estimated by dividing total potential dose by ED and by body weight (BW):

$$ADD_{pot} = \frac{C \times IR \times ED}{BW \times ED}$$

or

$$ADD_{pot} = \frac{C \times IR}{BW}$$

If C or IR varies over time, they may be averaged over ED. In addition, a frequency term (FR) or area use factor (AUF) may be used to denote the fraction of the time that an animal is exposed to affected media. This term is often used when the foraging range of a bird species is larger than the area of the site or when the species is migratory. An absorption factor (ABS) is used when an estimate of absorbed dose rather than potential dose is desired (and data are available to support it).

Hence, the dose side of the HQ method requires estimation of a number of exposure factor values, principally concentration of COPC in environmental media, food ingestion rate, water ingestion rate, sediment or soil ingestion rate, food preferences, body weight, absorption factors, arrival and dispersal dates for migratory species, home range size, and foraging range size. Many of these values have been compiled by the USEPA (1993a) for a number of species commonly evaluated, including great blue heron, Canada goose, mallard, lesser scaup, osprey, red-tailed hawk, bald eagle, American kestrel, Northern bob-white, American woodcock, spotted sandpiper, herring gull, belted kingfisher, marsh wren, and American robin. Notably absent from the USEPA's (1993a) compilation are tree swallows, a species that is commonly of interest in ERA because of its preference for consuming emergent insects (which, in turn, may be highly exposed to COPCs in sediment). Hence, literature searches to supplement the USEPA's (1993a) exposure factors may be necessary for completion of the HQ method.

In all cases, exposure factor values should be selected with the objective of most accurately approximating the receptor species' expected behavior at the site. For example, the concentration of COPC in environmental media should reflect the range of concentrations measured throughout the species' foraging range within the site. Hence, measures of central tendency [e.g., median, spatially averaged mean, arithmetic mean, geometric mean, 95 percent upper confidence limit on the mean (95% UCL)] generally are most representative of the range of concentrations to which a subpopulation of a given species would be exposed throughout the period that they are present at the site.

Likewise, measurements of COPCs in prey species should be based on the species and size classes most likely to be consumed by the receptor of interest. For instance, because great blue herons generally consume fish ranging in length from 5 to 30 cm (Henning et al., 1999), fish-sampling efforts should target this size class. Compositing fish samples collected throughout the area in which the receptor species is expected to forage is another option for improving the representativeness of the COPC concentration in prey. When collecting and analyzing earthworms as representative prey of vermivorous species (e.g., American robin, American woodcock), earthworms should not be depurated prior to analysis, as the entire worm (including its gut contents) is consumed by the avian predator. In the USEPA's ongoing ERA for the Housatonic River (Pittsfield, MA), Custer (1998) is collecting prey samples for tree swallows directly from the stomachs of nestlings removed from nest boxes, as well as through ligature, in order to ensure that concentrations of COPCs in prey reflect actual levels of exposure.

With respect to the other exposure factors required, a number of factors warrant consideration in order to ensure that values selected are most representative of the birds for which risks are being evaluated. Because food preferences often vary regionally and depending on prey availability, the literature study upon which food preference assumptions are based should be located in a comparable geographic area and should represent the same season(s) under evaluation. If none of the available studies on prey preferences is directly representative of the site being assessed, the assumed composition of diet may be based on that which maximizes potential exposure to COPCs. For example, the fraction of diet from fish for great blue herons may range from 67.5 to 100 percent (Henning et al., 1999). Because fish may accumulate higher concentrations of COPCs than other prey of great blue herons and because data on concentrations of COPCs in fish are more often available than for other prey, a common assumption employed in HQs for herons is that 100 percent of their diet is composed of fish.

Similarly, home range size and foraging range sizes can be quite variable, depending upon the quality of habitat available, prey density, and receptor species under consideration. As quantitative habitat surveys and surveys of prey density are usually beyond the scope of ERAs, AUFs are most often derived based on conservative assumptions regarding home range size and foraging range size, relative to the size of the habitat within the affected site. Hence, if belted kingfishers have a mean territory size of 1.03 km shoreline along streams in Ohio (Brooks and Davis, 1987) and at least this length of stream is impacted by COPCs, a conservative AUF of 1.0 would generally be incorporated into the HQ.

Screening-level techniques such as the HQ method have been developed to provide a conservative estimate of risk at a relatively low cost. The HQs can be used to eliminate substances from further evaluation when it is clear that there is no potential risk associated with that substance, leaving for further investigation any substances that potentially pose risk (USEPA, 1997). However, ERA guidance (e.g., Massachusetts DEP, 1996; Sample et al., 1996, 1997; USEPA, 1997) stipulates the use of uncertainty factors when extrapolating results across species. This practice adds to the conservatism of the approach such that it may indicate the potential for risk when one does not exist (Moore et al., 1999). Despite these limitations, the HQ approach is most useful as a first phase in ERA because it is relatively fast and inexpensive can be used to refine any future studies to receptors and substances that may pose risk. However, when it has been established that a substance may pose a risk to a receptor and it is carried forward to a subsequent phase of ERA, a limitation of the HQ approach is that risk may be identified when no risk is actually present. This type of error translates into potentially expensive remedial actions, which may carry inherent risks (e.g., dredging, armoring banks).

24.2.2 Tissue Analyses and Food Chain Models

Increased certainty and accuracy can often be realized through collection and analysis of tissue samples from the receptor of interest. Tissue analyses serve several purposes in ERA, including providing direct measures of body burdens in receptors of interest for comparison with data from the literature or laboratory studies. Some COPCs preferentially concentrate in certain organs or body parts. Because bioaccumulative substances (e.g., polychlorinated biphenyls

(PCBs), polychlorinated dibenzo-p-dioxins (PCDDs)) tend to be associated with lipids, tissue concentrations should be lipid normalized rather than wet weight based. Lipid-normalized data can be meaningfully compared among organisms and studies. Tissues that are commonly analyzed include brain, liver, eggs, and whole body. In some instances, the endpoint of interest dictates the tissue that is most relevant. For instance, reproductive studies often analyze COPC concentrations in eggs, as that is the primary source for developing embryos (Fleming et al., 1984; McLane and Hughes, 1980; Tillitt et al., 1992). The choice of tissue analyzed is important when making comparisons with the literature, since it is important to compare like tissues or use conversion factors across tissues where data to support such a conversion are available.

Tissue concentrations are also used in food chain models. Food chain modeling is used to estimate exposure, particularly when a receptor of interest is threatened or endangered, as with the bald eagle, or where it is more efficient to collect tissue data on representative prey of several species so that the data can be used to estimate exposures for multiple receptors. There are significant uncertainties associated with the use of food chain models, however, including the fact that many birds, particularly upper trophic level species, feed opportunistically. In such cases, the food chain model may not be representative of the actual exposure scenario. When direct tissue data are not available on prey, BAFs are derived from soil or sediment concentrations. BAFs that are not site specific contribute uncertainty to the ERA because they are based on specific soil/sediment concentrations as well as on uptake kinetics of particular flora and fauna being evaluated. Furthermore, food chain models are based on species-specific assumptions regarding diet, metabolism, ingestion rates, and foraging. While for some species these exposure factor values are well defined, food chain models frequently rely upon interspecies extrapolations, adding uncertainty and usually conservatism to estimates of exposure, and ultimately risk.

24.2.3 Probabilistic Techniques

Probabilistic exposure assessment differs from traditional "point estimate" (i.e., deterministic) assessment, in that it defines exposure in terms of probability distribution functions (PDFs), rather than single estimates of the reasonable maximum value. Rather than calculating intake just once, thousands of iterations are typically executed, with each iteration selecting input values from the PDFs in a manner that corresponds to the shapes of the various distributions. In comparison to probabilistic techniques, point estimate approaches can be flawed because (a) estimates of intake cannot quantify or even acknowledge uncertainties; (b) the degree of conservatism in the estimate of intake is unknown; (c) conservatisms are applied in many places throughout the evaluation; and (d) HQs cannot evaluate risk reduction associated with remedial actions (Moore et al., 1999).

Distributional data on exposure parameters are available from both the scientific literature and USEPA guidance documents (Henning et al., 1999; Moore et al., 1999; Sample and Suter, 1999; USEPA, 1993a). For example, Henning et al. (1999) identified distributions for body weight, fraction of diet from fish, prey length, distance to foraging site, and feeding territory size for great blue herons. For each exposure parameter, the mean, standard deviation, and PDF type were provided. These distributions were generated using data available in the scientific literature and a combination of statistical techniques and professional judgment. Using this same approach, distributions may be developed using data available in the USEPA's (1993a) *Wildlife Exposure Factors Handbook*. As shown in Moore et al. (1999) and Sample and Suter (1999), distributional data on exposure variables can be used in HQ calculations to develop probabilistic estimates of dose and risk. Specifically, Moore et al. (1999) used PDFs to estimate exposure concentrations of methylmercury and PCBs in media and prey items, assimilation efficiency of constituents, gross energy, metabolic rate, and proportion of fish in diet for kingfishers. These data were then used to estimate total daily intakes. The authors found that, for exposure to methylmercury, the most important PDFs were metabolic rate, gross energy of fish, fish concentration, and proportion of fish in the diet. Finally, data on intake were combined with data on the concentration–response curve for methylmercury and PCBs to determine a risk function for kingfishers exposed to these constituents.

The output of a probabilitic simulation is a PDF, rather than a single-point value. Consequently, a risk management decision must then be made regarding which percentile of the distribution is considered adequately protective. Most often, the 90th, 95th or 99th percentile is selected as a point of comparison to benchmarks of acceptable risk. Reasonable maximum point estimates of risk are usually greater than the 95th percentile of the probabilistic distribution of risk, and midrange point estimates are usually well above the 50th percentile. Risk conclusions yielded by probabilistic analysis may be either greater or less than those yielded by a deterministic approach, depending in large part upon (a) the degree of conservatism built into the deterministic assessment and (b) the percentile of the probabilistic distribution that is selected as a point of comparison to benchmarks of acceptable risk.

Because probabilistic risk assessment offers the very important ability to quantitatively characterize variability and uncertainty, it can prove very useful as a tool for sensitivity analysis even if it is not used to predict risks. By way of example, probabilistic techniques can be used to define the relative contributions of each source of variability and uncertainty to the overall model results. Strategic and informed decisions can then be made regarding whether additional studies should be conducted to reduce the most important sources of uncertainty. In this way, the process of conducting probabilistic analyses can be beneficial even if point estimates are ultimately used to make risk management decisions.

The USEPA (1999a) provides a number of examples in which probabilistic techniques are useful in risk assessment: (a) when screening calculations using conservative point estimates fall above levels of concern; (b) when it is necessary to disclose the degree of bias associated with point estimates of exposure; (c) when it is necessary to rank exposures and/or exposure pathways; (d) when the cost of regulatory or remedial action is high and the exposures are marginal; and (e) when the consequences of simplistic exposure estimates are unacceptable. Because risk assessments conducted using probabilistic techniques provide a much greater range of information on potential risks, they are extremely useful in those situations that indicate that a risk level is exceed based on a conservative point estimate approach.

Although probabilistic techniques are useful in most situations, the principal limitations of employing probabilistic techniques relate to (a) the potential to have that the maximum (or upper percentile) values of the PDF exceed the point estimate prediction of risk; (b) the complexity of risk communication; and (c) the difficulty in achieving consensus when developing PDFs for input variables and when selecting the percentile value that is compared to benchmarks of acceptable risk. However, as more regulatory agencies and programs (e.g., USEPA, 1999a) begin to use and accept probabilistic techniques, difficulties in selecting or identifying distributions should diminish.

24.3 TOOLS FOR CHARACTERIZING EFFECTS IN BIRDS

Several tools currently available to characterize effects in avian ERAs are described below, including development of TRVs, benchmark dose approach, population modeling, laboratory and mesocosm studies, population surveys, and nestbox studies.

24.3.1 Development of TRVs

In the HQ approach, TRVs are the point of comparison (i.e., the denominator) against which potential doses are compared:

$$HQ = \frac{Dose_{pot}}{TRV}$$

All toxicological values chosen for TRV derivation are presented on the basis of milligrams COPC per kilogram body weight per day (mg/kg-day). These units allow comparisons among organisms of different body sizes (Sample et al., 1996).

A variety of approaches are available for deriving TRVs, including regression analyses, toxicity testing, application of extrapolation and uncertainty factors, probabilistic analyses, and others. The most simple (and most common) approach involves application of extrapolation and uncertainty factors (EFs and UFs) to derive TRVs for wildlife receptors from laboratory study results, based on the methodology of Sample et al. (1996). This process involves the determination of a "test species dose" for a critical endpoint from a particular experimental combination of exposure concentration, exposure duration, test species, and COPC. The test species dose from the selected study is then modified to account for the various extrapolations and uncertainties inherent in applying results from a controlled setting to an ecologically relevant settings, as in

$$TRV = \frac{\text{test species dose} \times \text{dose matrix EF}}{\text{duration UF} \times \text{endpoint UF}}$$

Extrapolation and uncertainty factors are based on (a) the dosing matrix used in a laboratory study; (b) the duration of exposure; and (c) the endpoint measured (Calabrese and Baldwin, 1993; Ford et al., 1992; Opresko et al., 1994; Sample et al., 1996; USEPA, 1996a; Watkins and Stelljes, 1993; Wentsel et al., 1994).

The test species dose is a daily dose of a chemical associated with a particular endpoint and effect. In some cases, this dose is explicitly stated within the study; in other cases, only partial or related information is available. For studies that report an effects level as a concentration in food or drinking water but do not report specific body weights or feeding rates of the test species, default weights are used to derive the test species dose:

$$Dose_t = \frac{Cfood_t \times IR_{food}}{BW_t}$$

or

$$Dose_t = \frac{Cwater_t \times IR_{water}}{BW_t}$$

where $Dose_t$ = test species dose of COPEC (mg/kg-day)
  $Cfood_t$ = concentration of COPEC in food (mg/kg)
  IR_{food} = ingestion rate of food by the test species (kg/day)
  BW_t = body weight of the test species (kg)
  $Cwater_t$ = concentration of COPEC in water (mg/L)
  IR_{water} = ingestion rate of water by the test species (L/day)

Because administration of a chemical in a controlled study may use a liquid dosing matrix (e.g., water or corn oil), differential effects due to the dosing matrix may be considered in developing TRVs. If a "dose matrix effect" is indicated, an EF may be incorporated to account for the differential effects of the dosing matrix. To date, however, no comprehensive summaries of dose matrix EFs have been compiled and applicable data are rarely available.

Uncertainty factors may also be included that compensate for differences in exposure durations studied relative to those that are of concern for the ERA. Chronic studies occur over the lifetime or a majority of the life span of the test organism, generally longer than 10 weeks for birds. Additionally, studies in which the test organism is dosed during a critical life stage (e.g., gestation) are grouped with chronic duration studies. Subchronic studies include exposures of 2 weeks to 1 year in duration that do not occur during a critical life stage. Acute studies typically have exposures of less than 2 weeks. No observed adverse effects levels (NOAELs) and lowest observed adverse effects levels (LOAELs) are usually reported from chronic and subchronic studies, with acute studies often reporting frank effect levels [FELs; e.g., median lethal dose (LD$_{50}$) data].

Test species doses from chronic studies should be used preferentially over data from acute and subchronic studies. In cases where chronic data are not available as test species doses, studies involving less than chronic exposures may be used in TRV derivation, with the addition of a duration UF. A number of uncertainty factors have been suggested to account for differences in exposure duration, but most do not have a strong scientific basis or are not well documented. An evaluation of available data on the ratios of acute and subchronic to chronic NOELs (McNamara, 1976; USEPA, 1996a; Weil and McAllister, 1963) indicates that approximately 90 percent of the acute-to-chronic or chronic-to-chronic ratios are less than 8, and the 50th percentile ratio is approximately 3. Based on these data, a UF of 3 is recommended when the TRV is based on subchronic laboratory studies, while a UF of 8 is recommended when only an acute study is available.

Additional UFs may be used to account for uncertainties in extrapolation between effect and no-effect levels. Specifically, a NOAEL$_t$ test species dose maybe estimated from a LOAEL$_t$ value. A UF of 10 is often used with LOAEL$_t$ values to estimate the NOAEL$_t$ (Opresko et al., 1994; Sample et al., 1996); this factor is considered conservative (Sample et al., 1996; USEPA, 1996a). UFs less than 10 may be used if specific information is available that characterizes the dose–response relationship for the observed adverse effect (USEPA, 1996a).

Although body weight scaling factors are often used to account for differences in the body sizes of mammalian test species and mammalian receptors of interest (Sample et al., 1996), scaling factors developed for mammals may not be appropriate for avian interspecies extrapolations. Adjustment factors based on body size for interspecies extrapolation among avian species range from 0.63 to 1.55 (Sample et al., 1996). Given this range, body weight EFs are not recommended in the derivation of avian TRVs. Clearly, one must be careful in extrapolating toxicity results between species because of often extreme differences in sensitivity. Interspecies sensitivity has been the focus of laboratory studies (Brunstrom and Reutergardh, 1986; Brunstrom et al., 1990; Heath et al., 1972; Hill, 1994; Hudson et al., 1984; Peakall and Risebrough, 1989; Schafer 1972; Schafer et al., 1983; Tucker and Haegele, 1971). Among the factors that can affect sensitivity to COPCs are body size, metabolism, and feeding guild. Hill (1994) compiled the results of laboratory toxicity tests on pesticides that determined LD$_{50}$ and median lethal concentration (LC$_{50}$) and determined that neither physiological nor taxonomic similarities can be used to consistently predict the sensitivity of birds to pesticides.

24.3.2   Benchmark Dose Approach

The benchmark dose (BMD) approach offers a clear improvement over the traditional method of developing TRVs (summarized above). This approach has been widely used in establishing toxicity criteria for protection of humans, and it is equally valuable in establishing TRVs for ecological receptors. Bailer and Oris (1993, 1994) applied this approach to ecological receptors a few years before the USEPA first advocated its use for human health risk assessments.

Overall, the BMD approach is a model-derived estimate of the dose at a particular incidence level for the effect reported in a given toxicity bioassay (USEPA, 1996b). The software used to calculate BMDs was developed by the EPA's National Center for Environmental Assessment (NCEA) and is known as Benchmark Dose Software (BMDS). The software is free and can be down-loaded directly from the USEPA's web site (*www.epa.gov*). Output from the model includes information on goodness-of-fit, the BMD, and the lower bound confidence limit on the BMD.

Using information from a given toxicity bioassay on the rates of effects (i.e., incidences) observed in all dose groups tested, BMDS defines the shape of the dose–response curve. Once the shape of the dose–response curve is understood, any incidence rate can be estimated from within or near the range of doses administered. In this way, the effective dose to a certain percent of the population (i.e., ED_{10}) can be estimated and used as a TRV. The ED_{10} represents a more defensible TRV than either the NOAEL or the LOAEL, because the NOAEL and LOAEL both depend upon arbitrarily established doses that may be considerably lower than the ED_{10}. Confidence limits can be used to address data variability to provide conservative estimates of the dose–response curve.

Although application of the BMD approach may be constrained by minimum data requirements, it offers several advantages over the traditional NOAEL/LOAEL approach to deriving TRVs. First, the BMD provides an estimate of dose–response ratio that is within or near the observed effects range. This is especially relevant for ecological data, since the intent is to address population-level effects. For birds, 20 percent incidence levels may be argued for protection of populations (e.g., Suter et al., 1995). Hence, the BMD approach allows the TRV to be set at the incidence judged to be protective of populations.

Second, the TRV is not restricted to predetermined and sometimes arbitrary dose levels, as it is with the NOAEL/LOAEL approach. Consequently, it provides additional flexibility in that a risk assessor is not forced to assume default values for uncertainty factors (e.g., factor of 10) to extrapolate from the LOAEL to the NOAEL. In other words, the BMD approach uses all of the information provided by the multiple dose groups tested, rather than simply focusing on the one dose group in which no effects were observed.

Third, the BMD approach can be used to compare and rank toxicity studies with respect to potency of toxic effects by focusing on the same incidence level (e.g., ED_{10}) across studies. In contrast, comparing NOAELs or LOAELs across studies is not necessarily appropriate, due to the arbitrariness with which dose groups are sometimes defined. Consequently, the BMD approach offers a more reliable method of selecting critical studies and understanding differences in interspecies sensitivity. Finally, the BMD approach may be used to develop either point estimates or a distribution of BMDs representative of sensitive effects in ecological receptors. This flexibility directly supports the conduct of probabilistic ERAs that consider uncertainty and variability in both exposure and toxicity assumptions.

24.3.3 Population Modeling

The USEPA's risk management guidance (1999b) states that populations are the appropriate level of ecological organization for assessment. However, most studies that are conducted in ERAs focus on effects in individuals. The cost and time required to conduct a population-level assessment often preclude the use of field studies that can directly assess population-level impacts. Population modeling provides a tool that can be used to estimate impacts to populations at a fraction of the cost of a population-level field study.

Basic demographic models (e.g., Leslie matrix) require the collection of numerical bookkeeping data, which track changes in population size and structure and incorporate natural history information such as maturation rates, life span, and survivorship (Nunney and Elam, 1994). Submodels are then used to describe specific environmental impacts, including climate, dispersal rates, predator–prey relationships, and anthropogenic impacts (Emlen, 1989). For instance, Dixon et al. (1999) use mathematical models and computer simulations to link potential exposures with mortality and avian reproduction and nestling survival based on acute exposure to pesticides. Spromberg et al. (1998) developed an elaborate simulation model that takes into consideration not only basic demographics and exposure but also varied exposure over time and space to account for both direct exposures and indirect effects on other subpopulations that have never been directly dosed and may continue after the chemical of concern has degraded. While these modeling techniques have not yet been widely applied to field data, the integration of exposure and toxicity data into a basic population model may provide a useful tool in estimating population-level effects in ERA.

Few studies have been published which model avian populations, and these have been focused on extinction (Akcakaya and Atwood, 1997; Maguire et al., 1995) as an endpoint, rather than more local, short-term effects that may be associated with COPCs. Population models have also been used in studies focusing on wildlife management (Brown et al., 1976; Cowardin and Johnson, 1979; Cowardin et al, 1983; Walters et al., 1974; Roseberry, 1979) and effective population sizes (Nunney and Elam, 1994) or the effects of pesticides or other anthropogenic substances on populations over time (Grant et al., 1983; Samuals and Ladino, 1983). Emlen (1989) has explored population modeling for terrestrial species that considered pseudoextinction, which measures the probability of a population decline below a predetermined level or density. This measure may be more useful than extinction in the context of ERA.

Several issues warrant consideration relative to avian population models. Most important in the use of population models is the identification and definition of the population of concern. This categorization should be based on geography, local resources, and other natural history parameters (e.g., migratory patterns, home range, food availability). Once the population is defined, it is then necessary to determine carrying capacities and initial abundances (i.e., what is the starting population) as well as any density-dependent factors such as nest site availability and predation. This information can be used to estimate what happens to the population over time given the reproductive effects seen in test populations. For each age class (i.e., fledgling, subadult, and adult), it is necessary to know survivorship and fecundity (i.e., reproductive success). In addition, depending on the model selected, information on immigration and emigration rates can be used to evaluate dispersal between breeding colonies. Population models can evaluate whether reproductive impairment in some individuals translates into long-term declines in the population.

24.3.4 Laboratory and Mesocosm Studies

Laboratory and mesocosm studies are generally designed to simultaneously collect information regarding both exposures and effects under conditions meant to replicate the actual conditions that are the focus of the ERA. In effect, laboratory and mesocosm studies most often are intended to directly collect information on the assessment endpoint(s), rather than on surrogate measures of exposure and effect. Consequently, such techniques often evaluate the measures of effect and measures of exposure directly, thereby minimizing the need to extrapolate between endpoints, site conditions, species, suites of chemicals, or exposure periods. As would be expected, these more complex studies can be extremely expensive and time consuming; however, they can also yield an improved scientific basis for a remediation decision, significantly reducing the costs of remediation over the long term. Even so, laboratory and mesocosm studies may not be definitive. For example, it may be difficult to obtain sufficient sample sizes to ensure adequate statistical power to discern background exposures from site-related exposures or confounding factors may be associated with laboratory conditions.

24.3.5 Population Surveys

Avian censuses and population surveys have been used only to a limited extent in ERAs. Although the use of population size as a measure of health of a species has been a common tool of biologists for many years (Hutchinson, 1978; Lack, 1954, 1966), several factors limit the use of surveys in ERAs, including (a) the need for extended tracts of undeveloped land made up of uniform habitat type; (b) the difficulty of finding appropriate reference sites of comparable size, habitat type, and quality; (c) time constraints that generally prevent collection of multiple years of data; (d) the challenges of establishing a causal link between observed differences in the composition, density, or diversity of the avian community with exposure to COPCs (rather than natural variability or habitat differences); and (e) limitations in statistical power that would demonstrate that the study design could detect differences in community if they in fact exist. Nonetheless, population surveys can serve as useful tools during problem formulation, specifically for selecting receptors of interest and characterizing the ecological community that is potentially at risk. Such tools are also valuable during risk management, when it is important to understand whether (and the extent to which) remediation alternatives may impact the existing ecological community. Ralph et al. (1993) provides an excellent overview of avian survey techniques; much of the following discussion is excerpted from this source.

Avian population surveys should ideally provide three types of interrelated data: (a) estimates of population sizes and trends for various species of birds; (b) an estimate of demographic parameters (e.g., sex ratios, age distributions, nesting success, survivorship) for at least some of those populations; and (c) habitat data to link the density and demographic parameters to habitat characteristics (e.g., James and Shugart, 1970). Surveys should be replicated over multiple years and study plots to yield the greatest statistical power possible.

The preferred timing of surveys depends on the study design. Demographic monitoring by mist nets or nest searches should span the entire breeding season. In contrast, censuses are generally only conducted during the first half of the breeding season, when birds are most active, paired, on territories, and vocal. Data collected during spring or fall migrations can be confounded by many factors, particularly local weather; they are not generally recommended for ERA purposes because migrants have very limited potential for exposure to COPCs. In contrast, nonbreeding season surveys can be quite valuable because populations are resident and relatively stable in the winter; hence, survivorship and mortality data may be most accurately derived from nonbreeding season surveys (Ralph et al., 1993). Robbins (1972) cautions, however, that winter surveys tend to underestimate population sizes, even under ideal weather conditions.

Several options exist for conducting avian surveys. Stamm et al. (1960) describe methods of mist netting for studying populations of wild birds. Among the advantages of mist netting surveys described by Stamm et al. (1960) are that they provide estimates for a maximum number of species and of individuals, birds of all ages and both sexes, and breeders and nonbreeders. Furthermore, they are applicable to all sorts of habitats and all seasons and allow statistical evaluation of the dependability of the data (Stamm et al., 1960). Mist netting can be used to compare the proportion of juvenile birds captured in mist nets at site and reference plots; this provides one measure of productivity. The sex ratio can be used to assess the species' differential survivorship the previous year and the ability of the population to increase. Capture rate gives an indirect measure of survivorship. Weight, when compared to measures of body size such as wing length, provides a measure of fitness and condition. Mist netting generally yields smaller sample sizes than provided by censusing; therefore, mist netting is of limited use in estimating population sizes (Ralph et al., 1993).

Nest searches provide the most direct measurement of nest productivity and have been used successfully in at least one avian ERA. Henning et al. (1997) monitored approximately 300 natural nests throughout a single breeding season, evaluating reproductive outcome on both an overall and a species-specific basis. Clutch sizes and numbers of young hatched were compared between site and reference groups; no statistically significant differences were observed. Nest success for all species was also well within normal ranges reported in the literature. The authors concluded that there was no evidence that COPCs in the system were adversely affecting reproduction in songbirds at the community level. Because nest finding and subsequent monitoring are labor intensive and require study sites and reference sites with extensive areas, however, this method is unlikely to be more widely adapted for ERA purposes.

Population size is generally measured through censusing, such as point counts, spot mapping methods, strip transect counts, and an area search method. Point counts involve an observer standing in one spot and recording all birds seen or heard at either a fixed distance or unlimited distances. The North American Breeding Bird Survey utilizes point counts to monitor avian populations across the continent. Strip transect surveys [U.S. Department of the Interior, Fish and Wildlife Service (USFWS), 1980] are very similar to point counts, but the observer records all birds seen or heard while traversing each section of a trail. Each section is 100 or 250 m long and is the unit of measurement used throughout the census, ensuring comparability among measures. This method is best used in very open terrain. The area search method consists of a series of three 20-minute point counts in which the observer can move around in a somewhat unrestricted area. Such freedom of movement allows the observer to track down unfamiliar calls and quiet and secretive species. The spot map census method (Williams, 1936; Hall, 1964) is based on the territorial behavior of birds. By marking the locations of observed birds on a detailed map during several (minimum of 8) visits within a breeding season, numbers of territories in an area can be mapped and density calculated (Ralph et al., 1993). Dickson (1978) compared the outcomes of spot mapping and transect surveys, finding that spot mapping resulted in the largest estimates of numbers of birds. For transect methods, Dickson (1978) found that estimates of population sizes increased steadily as the number of counts increased from 6 to 11.

24.3.6 Nest Box Studies

A variation of nest search surveys discussed above is the nest box study. Nest box studies follow reproductive success of certain cavity nesting species (most often tree swallows) in a more controlled setting than allowed by nest search surveys. As discussed below, using nest boxes can substantially decrease the labor required to complete a field reproduction study while controlling for confounding effects of habitat variability.

Nest box studies offer several advantages over other field studies. First, nest box studies allow a large number of receptors to be evaluated within a small area. Although spacing of nest boxes is critical to the study's success, nest boxes can be placed in concentrated areas. As such, it is not necessary to find individual natural nest sites. Second, nest box studies allow evaluation of potential exposures across similar habitats. Because tree swallows easily take up residence in nest box sites that meet a minimum of habitat requirements, the study location can be selected by the researcher to ensure that the habitat at reference sites is comparable to the habitat at the study sites. Third, nest box studies are relatively easy to maintain over several years once the initial boxes have been established. In this way, the same population can be monitored over a long time period.

Several researchers have evaluated reproductive effects in tree swallows using nest box studies (Lombardo, 1994; Rendell and Robertson, 1993; Robertson and Rendell, 1990; Wiggins, 1990; Stutchbury and Robertson, 1988); however, nest box studies have been most recently used to evaluate potential effects of PCBs (McCarty and Secord, 1999; Secord and McCarty, 1997; Froese et al., 1998; Henshel et al., 1999). Although these studies have provided mixed results on potential effects associated with PCB exposures, they have allowed continual refinement of the nest box study technique. As additional data are collected through nest box studies, it is likely that avian ERAs will begin to rely more on the results of field-collected data and less on traditional conservative HQ estimates.

24.4 CHARACTERIZATION OF ECOLOGICAL RISKS TO BIRDS

The final step in conducting ERAs is the integration of information on exposure and effects in order to characterize potential risks. Once a change in an ecological system or function is observed, it is necessary to determine if the effect is caused by the presence of COPCs or a physical disturbance or is part of a natural fluctuation inherent in the species or ecosystem. The question of distinguishing ecological effects attributable to a COPC from natural variability is central to characterizing risks to ecological systems (USEPA, 1993b).

In contrast with human health risks, ecological risk for most avian species is focused on effects at the population or community level. In this respect, risk assessors in government, academia, and consulting recognize that effects to individuals do not matter unless there are some short- or long-term implications for the entire population (as in the case of threatened and endangered species) (USEPA, 1998). The importance of ecological risks at the population level (over the individual level) is clearly recognized; however, most ecological studies used to determine potential adverse effects still focus on effects at the individual level. Although many field-based studies attempt to gather information on population-level effects, the interpretation of results is complicated by several confounding factors (e.g., habitat variables, natural variability). In addition, once an effect is identified, it is necessary to determine if those effects are significant for the population (Gentile et al., 1993; Giesy et al., 1994). It is this issue that is most often debated and without a clear threshold, as ecological systems are too complex to be evaluated based on a single numerical standard.

The ease with which effects are detected depends in part on whether a literature-based study, a laboratory bioassay, or a field survey is conducted. As previously described, there are several advantages and disadvantages associated with each type of study. For extrapolation of effects from laboratory studies to birds in the field, it is essential to examine whether the dose that caused effects in the laboratory also causes the same effects in the field. Although laboratory study results are often not meaningful or relevant for field populations, detecting effects in populations in the field may be difficult, except in extreme exposure cases. The USEPA (1998) has developed criteria for causality based on Fox (1991). Criteria strongly affirming causality include strength of association, predictive performance, demonstration of a stressor–response relationship, and consistency of association. Criteria useful as the basis for rejecting causality include inconsistency in association, temporal incompatibility, and factual implausibility.

When evaluating effects in avian populations, it is important to consider all of these factors in the context of natural variability. For example, most avian populations experience a certain level of annual reproductive failure. In the case of tree swallows, Robertson et al. (1992) reports that nestling success is approximately 79 percent and that fledgling survivorship to subadult varies from 4.3 to 11 percent. Additional data on band recoveries of nestling tree swallows indicate that survivorship in the first year is about 20 percent, increasing to 40 to 60 percent thereafter (Chapman, 1955; Houston and Houston, 1987; Wiggins, 1990). While this amount of loss does not affect the total population over the long term, any reproductive effects associated with exposures to COPCs first must be considered in light of expected background levels of reproductive failures.

As the intensity, predictability, and consistency of an effect increase, so does the likelihood of causality. Although many studies have documented tissue levels of chemicals in birds that died (Fimreite, 1979; Eisler, 1987), such correlations do not prove a cause-and-effect relationship (Burger and Gochfeld, 1997). Controlled laboratory studies will continue to be the predominant source of data on the ecological effects of environmental constituents.

It is an on-going effort to determine what level of effect at an individual or population level is important from both a scientific and a regulatory perspective. Because laboratory studies and even many field studies cannot predict long-term effects to avian populations from exposure to COPCs, extrapolations are often necessary to estimate biological significance. In other cases, effects noted in the field cannot be linked back to a specific cause-and-effect relationship. Halbrook et al. (1999) reported that great blue heron populations near the Oak Ridge National Laboratory (ORNL) were undergoing drastic declines in the 1960s. Although the population had begun to rebound, a field study was undertaken to determine if past releases of COPCs were having any significant effects on the population. The results of the study indicated that concentrations of PCBs and mercury were higher in eggs and chicks in the sites near ORNL compared to reference sites; however, reproductive success measured as number of chicks fledged per nest was similar to reference colonies (Halbrook et al., 1999).

Once an effect is detected, there are several methods by which ecological risks to avian species may be evaluated. This information is then used to determine if the effect noted will lead to significant changes in the population under consideration. The specific method used to evaluate detected effects depends on the type and amount of data available for evaluation. Options include comparing point estimates of exposure levels and effect levels (HQs), using the entire stressor–response relationship, incorporating variability, and using population models (USEPA, 1998). Although HQs are easy to interpret, they have limited use for determining biological significance because they rely heavily on effects at the individual level. In contrast, methods that use field-collected data, incorporate variability, or use population modeling are likely to be better predictors of risk and biological significance, as they provide more detailed data on long-term effects through the incorporation of additional data on species biology and natural history (Barnthouse, 1993).

Many strategies and approaches have been suggested to determine if an effect detected in the laboratory or field is significant on a biological scale. Three approaches are presented here, ranging from qualitative to quantitative. First, on a qualitative level, Duinker and Beanlands (1986) offer a perspective on ecological significance related to a relative ranking of effects. As described in Pascoe et al. (1994), effects can be broken down into one of four categories based on the following definitions:

Major effect: Affects an entire population or species in sufficient magnitude to cause a decline in abundance and/or change in distribution beyond which natural recruitment would not return that population or species, or any population or species dependent upon it, to its natural level.

Moderate effect: Affects a portion of a population and may bring about a change in abundance and/or distribution over one or more generations but does not threaten the integrity of that population or any population dependent upon it.

Minor effect: Affects a specific group of localized individuals within a population over a short period of time but does not affect other trophic levels or the population itself.

Negligible effect: Any effects below the minor category are considered negligible.

Although this approach can provide some useful information, the categorization of effects depends heavily on the type of analysis conducted to determine if an effect is occurring.

Second, to evaluate whether an effect is occurring and whether that effect is significant, the USEPA (1998) recommends using the following criteria: (a) nature of effects and intensity of effects, (b) spatial and temporal scale, and (c) potential for recovery. The nature and intensity of effects should be used to distinguish adverse ecological changes from those within the normal pattern of variability. Although an effect may be statistically significant, it is more important to consider whether that effect also is biologically significant. Those effects that have reproductive implications are generally more important for populations than are other effects. Similarly, the spatial and temporal scale over which effects are noted, as well as the potential for recovery, should be considered when evaluating effects. In this case, extremely localized events that do not have longlasting implications would be considered less significant. Although these criteria are subjective and rely on professional judgment, they provide a useful guide for determining if the effects reported are likely to be significant for a population over the long term, especially if combined with clearly defined assessment and measurement endpoints. If insufficient data are available to evaluate all three criteria, the determination of significance may be more difficult.

In general, better quality and higher quantities of data allow an easier determination of significance both statistically and ecologically (Pascoe et al., 1994). Brunstrom et al. (1990), for example, reported increased ethoxyrisorufin-O-dealkylase (EROD) activity in chicken embryos injected with a polycyclic aromatic hydrocarbon (PAH) mixture. In this case, the effect was significantly different from controls; however, the nature of the effect limits its relevance because changes in EROD activity have not been shown to result in changes in reproductive success or survival of individuals or populations.

Finally, a more quantitative method for determining significance is a weight-of-evidence approach. Using this methodology, data from several laboratory and/or field studies are interpreted together to determine when adverse effects are occurring in an avian population and whether those effects are indeed significant. Formal weight-of-evidence approaches are available (e.g., Menzie et al., 1996) that allow a risk assessor to use a defined process to determine if the reported effects are associated with the COPC or whether these effects are due to natural variability or some other natural stressor. Using these types of approaches ensures that the decision regarding the significance of effects is based more on the results of the science and less on policy implications.

Often environmental agencies use the results of toxicity tests for the most sensitive avian species or life stage to make decisions about other species that may be less sensitive to ecological constituents (USEPA, 1997). In the case of PCBs, for example, NOELs for chickens are as low as 1 mg/kg body weight per day based on hatching success (Scott, 1977). In contrast, studies with mallards report a NOEL of 25 mg/kg body weight per day based on fertility, hatching success, and survival of young (Custer and Heinz, 1980). These results demonstrate that some avian species are likely to be more sensitive to the effects of PCBs. However, the results of studies from a very sensitive species should not simply be used to evaluate effects in other avian species unless other evidence is available to suggest that such criteria are warranted.

When considering the sensitivity of a particular avian population, it is also useful to consider its reproductive strategy. Population ecology has demonstrated that long-lived vertebrates (such as predatory birds) are more sensitive to adult mortality compared to short-lived, highly fecund species such as quail. Conversely, short-lived species are often more vulnerable to catastrophic events that affect critical life stages. As a result, populations in which reproduction and survival are density dependent should generally be less vulnerable than populations with less density dependence. Overall, the response of a population to environmental constituents is influenced by the preexisting pattern of natural environmental variability, the age-specific survival and reproduction of organisms, and the intensity and duration of exposure to the COPCs. Each of these factors can be used to determine if the species of interest may be more sensitive to stressor effects (Barnthouse, 1993).

Often, however, limited data are available on most avian species to determine sensitivity. In these cases, as suggested by Barnthouse (1993), it may be useful to consider the reproductive strategy of the species being evaluated in light of the weight of evidence or other evaluation methods discussed above. For slowly growing populations, risk assessors should consider that there might be less potential for recovery. Similarly, certain populations may have highly sensitive life stages yet effects are only seen at other less sensitive life stages. In this case, exposure may be limited to a localized area and, therefore, have limited intensity and spatial effects.

As described above, population-level effects are most important for determining ecological significance for avian species. The exception to this rule, of course, is for threatened and endangered avian species. In this case, individuals are important because there is usually only a limited number left in the population. A few localized populations also may define the species, and loss of one or more of these could lead to extinction (USEPA, 1999c). For all other avian species, then, should risk assessors consider differences in species sensitivity or increased sensitivities for different life stages? If so, should the threshold by which effects are considered to be significant be lowered to accommodate increased species sensitivities? Given the limited database on most avian receptors, is such an approach even a viable option? Although this chapter cannot answer all of these questions, we have offered some thoughts to consider when evaluating effects in different avian populations.

REFERENCES

Akcakaya, H. R., and Atwood, J. L. 1997. A habitat-based metapopulation model of the California gnatcatcher. *Conserv. Biol.* 11: 422–434.

Bailer, A. J., and Oris, J. T. 1993. Modeling reproductive toxicity in ceriodaphnia tests. *Environ. Toxicol. Chem.* 12: 787–791.

Bailer, A. J., and Oris, J. T. 1994. Assessing toxicity of pollutants in aquatic systems. In *Case Studies in Biometry.* Wiley, New York, pp. 296–306.

Barnthouse, L. J. 1993. Population-level effects. In G. W. Suter II (Ed.), *Ecological Risk Assessment.* Lewis, Ann Arbor.

Best, D., Bowerman, W. W., Kubiak, T. J., Winterstein, S. R., Postupalsky, S., Shield-castle, M. C., and Giesy, J. P., Jr. 1994. Reproductive impairment of bald eagles (*Haliaeetus leucocephalus*) along the Great Lakes shorelines of Michigan and Ohio. In B. U. Meyburg and R. D. Chancellor (Eds.), *Raptor Conservation Today.* Pica Press, East Sussex, U.K. pp. 697–702.

Brooks, R. P., and Davis, W. J. 1987. Habitat selection by breeding belted kingfishers (*Ceryle alcyon*). *Am. Midl. Nat.* 117: 63–70.

Brown, G. M. Jr., Hammack, J., and Tillman, M. F. 1976. Mallard population dynamics and management models. *J. Wildl. Manag.* 40: 542–555.

Brunstrom, B., Bromand, D., and Naf, C. 1990. Embryotoxicity of polycyclic aromatic hydrocarbons (PAHs) in three domestic avian species, and of PAHs and coplanar polychlorinated biphenyls (PCBs) in the common eider. *Environ. Pollut.* 67: 133–143.

Brunstrom, B., and Reutergardh, L. 1986. Differences in sensitivity of some avian species to the embryotoxicity of a PCB, 3,3',4,4'-tetrachlorobiphenyl, injected into the eggs. *Environ. Pollut.* 42: 37–45.

Buckley, J. L., Hickey, J. J., Prestt, I., Stickel, L. F., and Stickel, W. H. 1969. Pesticides as possible factors affecting raptor populations: Summary of a round table discussion and additional comments by the conferees. In J. Hickey (Ed.), *Peregrine Falcon Populations: Their Biology and Decline.* University of Wisconsin Press, Madison, pp. 461–483.

Burger, J., and Gochfeld, M. 1997. Risk, mercury levels, and birds: Relating adverse laboratory effects to field biomonitoring. *Environ. Res.* 75: 160–172.

Calabrese, E. J., and Baldwin, L. A. 1993. *Performing Ecological Risk Assessments.* Lewis, Chelsea, MI.

California EPA. 1994. *Guidance for Ecological Risk Assessment at Hazardous Waste Sites and Permitted Facilities.* California Environmental Protection Agency, Department of Toxic Substances Control, Sacramento, CA.

Carson, R. 1962. *Silent Spring.* Houghton Mifflin, Boston.

Chapman, L. B. 1955. Studies of a tree swallow colony. *Bird-Banding* 25: 45–70.

Cowardin, L. M., and Johnson, D. H. 1979. Mathematics and mallard management. *J. Wildl. Manag.* 43: 18–35.

Cowardin, L. M., Johnson, D. H., Frank, A. M., and Klett, A. T. 1983. Simulating results of management actions on mallard production. *Trans. N. Am. Wildl. Nat. Resour. Conf.* 48: 257–272.

Custer, C. M. 1998. *Research Protocol: Bioaccumulation and Effects of PCBs on Tree Swallows Nesting Along the Housatonic River, Massachusetts.* Study No. WE-98-Cont-08. Upper Mississippi Science Center, Biological Resources Division, U.S. Geological Survey, La Crosse, WI.

Custer, T. W., and Heinz, G. H. 1980. Reproductive success and nest attentiveness of mallard ducks fed Aroclor 1254. *Environ. Pollut. (Ser. A)* 21: 313–318.

Dickson, J. G. 1978. Comparison of breeding bird census techniques. *Am. Birds* 32: 10–13.

Dixon, K. R., Huang, T.-Y., Rummel, K. T., and Sheeler-Gordon, L. L. 1999. An individual-based model for predicting population effects from exposure to agrochemicals. *Aspects Appl. Biol.* 53: 241–251.

Duinker, P. N., and Beanlands, G. E. 1986. The significance of environmental impacts: An exploration of the concept. *Environ. Manag.* 10: 1–10.

Eisler, R. 1987. *Mercury Hazards to Fish, Wildlife, and Invertebrates: A Synoptic Review.* U.S. Fish and Wildlife Service, Laurel, MD, Biol. Rep. 85(1.1).

Elliott, J. E., Langelier, K. M., Mineau, P., and Wilson, L. K. 1996. Poisoning of bald eagles and red-tailed hawks by carbofuran and fensulfothion in the Fraser Delta of British Columbia, Canada. *J. Wildl. Dis.* 32: 486–491.

Elliott, J. E., Wilson, L. K., Langelier, K. M., Mineau, P., and Sinclair, P. H. 1997. Secondary poisoning of birds of prey by the organophosphorous insecticide, phorate. *Ecotoxicology* 6: 219–231.

Emlen, J. M. 1989. Terrestrial population models for ecological risk assessment: A state-of-the-art review. *Environ. Toxicol. Chem* 8: 831–842.

Environment Canada. 1994. *A Framework for Ecological Risk Assessment at Contaminated Sites in Canada: Review and Recommendations.* Scientific Series No. 199. Ecosystem Conservation Directorate, Evaluation and Interpretation Branch, Ottawa, Ontario.

Fimreite, N. 1979. Effect of dietary methyl mercury on ring-necked pheasants. *Can. Wildl. Serv. Occas. Pap.* 9: 1–39.

Fleming, W., Rodgers, J., and Stafford, C. 1984. Contaminants in wood stork eggs and their effects on reproduction, Florida, 1982. *Colonial Waterbirds* 7: 88–93.

Ford, K. L., Applehans, F. M., and Ober, R. 1992. Development of toxicity reference values for terrestrial wildlife. Health and Endangerment, in *HMC/Superfund '92 Proceedings*, December 1–3, Washington, DC. Hazardous Materials Control Resources Institute, Greenbelt, MD, pp. 803–812.

Fox, G. A. 1991. Practical causal inference for ecoepidemiologists. *J. Toxicol. Environ. Health* 33: 359–373.

Froese, K. L., Verbrugge, D. A., Ankley, G. T., Niemi, G. J., Larsen, C. P., and Giesy, J. P. 1998. Bioaccumulation of polychlorinated biphenyls from sediments to aquatic insects and tree swallow eggs and nestlings in Saginaw Bay. Michigan, USA. *Environ. Toxicol. Chem.* 18: 484–492.

Gentile, J. H., Harwell, M. A., van der Schalie, W. H., Norton, S. B., and Rodier, D. J. 1993. Ecological risk assessment: A scientific perspective. *J. Haz. Materials* 35: 241–253.

Giesy, J. P., Ludwig, J. P., and Tillet, D. E. 1994. Deformities in birds of the Great Lakes Region: Assigning causality. *Environ. Sci. Technol.* 28: 128–134.

Grant, W. E., Fraser, S. O., and Isakson, K. G. 1983. Effect of vertebrate pesticides on non-target wildlife populations: Evaluation through modeling. *Ecol. Model* 21: 85–108.

Gress, F. 1970. *Reproductive Status of the California Brown Pelican in 1970 with Notes on Breeding Biology and Natural History.* Wildlife Management Branch Administrative Report No. 70-6. State of California, Department of Fish and Game, Sacramento, CA.

Halbrook, R. S., Brewer, R. L., Jr., and Buehler, D. A. 1999. Ecological risk assessment in a large river-reservoir: 7. Environmental contaminant accumulation and effects in great blue heron. *Environ. Toxicol. Chem.* 18: 641–648.

Hall, G. A. 1964. Breeding-bird censuses—why and how. *Audubon Field Notes* 18: 413–416.

Heath, R. G., Spann, J. W., Kreitzer, J. F., and Vance, C. 1972. Effects of polychlorinated biphenyls on birds. *Proc. XV Int. Ornithol. Congr.*, 15: 475–481.

Henning, M. H., Ebert, E. S., Keenan, R. E., Martin, S. G., and Duncan, J. W. 1997. Assessment of effects of PCB contaminated floodplain soils on reproductive success of insectivorous songbirds. *Chemosphere* 34: 1121–1137.

Henning, M. H., Shear Weinberg, N. M., Wilson, N. D., and Iannuzzi, T. J. 1999. Distributions for key exposure factors controlling the uptake of xenobiotic chemicals by great blue herons (*Ardea herodius*) through ingestion of fish. *Hum. Ecol. Risk Assess.* 5: 125–144.

Henshel, D. S., Sparks, D. W., Mayer, C. A., Lam, Y., Eby, K. B., and Updyke, M. 1999. PCB congener-specific interspecies variations in effects endpoints in passerines. *Proceedings of the Society of Environmental Toxicology and Chemistry 20th Annual Meeting*, Philadelphia, PA, November 14–18. SETAC Press, Pensacola, FL.

Hill, E. F. 1994. Acute and subacute toxicology in evaluation of pesticide hazards to avian wildlife. In R. J. Kendall and T. E. Lacher (Eds.), *Wildlife Toxicology and Population Modeling: Integrated Studies of Agroecosystems*. SETAC Special Publications Series. Lewis Publishers, Boca Raton, FL.

Houston, M. I., and Houston, C. S. 1987. Tree swallow banding near Saskatoon, Saskatchewan. *N. Am. Bird Bander* 12: 103–108.

Hudson, R. H., Tucker, R. K., and Haegele, M. A. 1984. *Handbook of Toxicity of Pesticides to Wildlife*. Resource Publ. 153. U.S. Fish and Wildlife Service, Washington, DC.

Hunt, E. G. 1964. Pesticides as possible factors affecting bird populations: Pesticide residues in fish and wildlife of California. In J. Hickey (Ed.), *Peregrine Falcon Populations: Their Biology and Decline*. University of Wisconsin Press, Madison, pp. 455–460.

Hutchinson, G. E. 1978. *An Introduction to Population Ecology*. Yale University Press, New Haven.

James, F. C., and Shugart, H. H. 1970. A quantitative method of habitat description. *Audubon Field Notes* 24: 727–736.

Keith, J. O., Woods, L. A., Jr., and Hunt, E. G. 1971. Reproductive failure in brown pelicans on the Pacific Coast. *Trans. N. Am. Wildl. Nat. Hist. Conf.* 35: 56–63.

Kendall, R. J., Lacher, T. E., Bunck, C., Daniel, B., Driver, C., Grue, C. E., Leighton, F., Stansley, W., Watanabe, P. G., and Whitworth, M. 1996. *An ecological risk assessment of lead shot exposure in non-waterfowl avian species: Upland game birds and raptors*. *Environ. Toxicol. Chem.* 15: 4–20.

Lack, D. 1954. *The Natural Regulation of Animal Numbers*. Oxford University Press, London.

Lack, D. 1966. *Population Studies of Birds*. Clarendon, Oxford.

Lombardo, M. P. 1994. Nest architecture and reproductive performance in tree swallows (*Tachycineta bicolor*). *Auk* 111: 814–824.

Longcore, J. R., Corr, P. O., and Spencer, J. H. E. 1982. Lead shot incidence in sediments and waterfowl gizzards from Merrymeeting Bay, Maine. *Wildl. Soc. Bull.* 10: 3–10.

Maguire, L. A., Wilhere, G. F., and Dong, Q. 1995. Population variability analysis for red-cockaded woodpeckers in the Georgia Piedmont. *J. Wildl. Manag.* 59: 533–542.

Massachusetts DEP. 1996. *Guidance for Disposal Site Risk Characterization in Support of the Massachusetts Contingency Plan.* BWSC/ORS-95-141. Interim Final Policy, Massachusetts Department of Environmental Protection, Bureau of Waste Site Cleanup and Office of Research and Standards, Boston, MA.

McCarty, J. P., and Secord, A. L. 1999. Nest-building behavior in PCB-contaminated tree swallows. *Auk* 116: 55–63.

McLane, A. R., and Hughes, D. L. 1980. Reproductive success of screech owls fed Aroclor 1248. *Arch. Environm. Contam. Toxicol.* 9: 661–665.

McNamara, B. P. 1976. Concepts in health evaluation of commercial and industrial chemicals. In M. A. Mehlman, R. Shapiro, and H. Blumenthal (Eds.), *Advances in Modern Toxicology*, Vol. 1, Part 1: *New Concepts in Safety Evaluation*. Wiley, New York.

Menzie, C., Henning, M. H., Cura, J., Finkelstein, K., Gentile, J., Maughan, J., Mitchell, D., Petron, S., Potocki, B., Svirsky, S., and Tyler, P. 1996. Special report of the Massachusetts weight-of-evidence workgroup: A weight-of-evidence approach for ecological risks. *Hum. Ecol. Risk Assess.* 2: 277–304.

Moore, D. R. J., Sample, B. E., Suter, G. W., Parkhurst, B. R., and Teed, R. S. 1999. A probabilistic risk assessment of the effects of methylmercury and PCBs on mink and kingfishers along East Fork Poplar Creek, Oak Ridge, Tennessee, USA. *Environ. Toxicol. Chem.* 18: 2941–2953.

New York Department of Environmental Conservation (NYSDEC). 1995. *Technical Guidance for Screening Contaminated Sediments.* NYSDEC, Division of Fish and Wildlife, Albany, NY.

Nisbet, I. C. T. 1989. Organochlorines, reproductive impairment and declines in bald eagle *Haliaeetus leucocephalus* populations: Mechanisms and dose–response relationships. In B.-U. Meyburg, and R. D. Chancellor (Eds.), *Raptors in the Modern World: Proceedings of the Third World Conference on Birds of Prey and Owls,* Israel, March 22–27. WWGBP: Berlin, London, Paris.

Nunney, L., and Elam, D. R. 1994. Estimating the effective population size of conserved populations. *Conserv. Biol.* 8: 175–184.

Opresko, D. M., Sample, B. E., and Suter II, G. W. 1994. *Toxicological Benchmarks for Wildlife*, 1994 rev. EX/ER/TM-86/R1. Oak Ridge National Laboratory, Oak Ridge, TN.

Pascoe, G. A., Blanchet, R. J., Linder, G., Palawski, D., Brumbaugh, W. G., Canfield, T. J., Kemble, N. E., Ingersoll, C. G., Farag, A., and DalSoglio, J. A. 1994. Characterization of ecological risks at the Milltown Reservoir-Clark Fork River sediments superfund site, Montana. *Environ. Toxicol. Chem.* 13: 2043–2058.

Peakall, D. B., and Kiff, L. F. 1988. DDE contamination in peregrines and American kestrels and its effect on reproduction. In T. J. Cade, J. H. Enderson, C. G. Thelander, and C. M. White (Eds.), *Peregrine Falcon Population: Their Management and Recovery*. Peregrine Fund, Boise, pp. 337–350.

Peakall, D., and Risebrough, R. 1989. Comparative toxicology of organochlorine pesticides to avian species. In B. Meyburg and R. Chancellor (Eds.), *Raptor in the Modern World: Proceedings of the Third World Conference on Birds of Prey and Owls,* Israel, March 22–27. WWGBP: Berlin, London, Paris. pp. 461–464.

Postupalsky, S. 1978. *Toxic Chemicals and Cormorant Populations in the Great Lakes.* Report Number 40. Canadian Wildlife Service, Toxicological Division. Madison, WI.

Ralph, C. J., Geupel, G. R., Pyle, P., Martin, T. E., and DeSante, D. F. 1993. *Handbook of Field Methods for Monitoring Landbird.* General Technical Report PSW-GTR-144. U. S. Department of Agriculture, Forest Service, Pacific Southwest Research Station, Albany, CA.

Ratcliffe, D. E. 1993. *The Peregrine Falcon.* T&AD Poyser, London.

Rendell, W. B., and Robertson, R. J. 1993. Cavity size, clutch-size and the breeding ecology of tree swallows *Tachycineta bicolor. Ibis* 135: 305–310.

Robbins, C. S. 1972. An appraisal of the winter bird-population study technique. *Am. Birds* 26.

Robertson, R. J., and Rendell, W. B. 1990. A comparison of the breeding ecology of a secondary cavity nesting bird, the tree swallow (*Tachycineta bicolor*), in nest boxes and natural cavities. *Can. J. Zool.* 68: 1046–1052.

Robertson, R. J., Stutchbury, B. J., and Cohen, R. R. 1992. Tree swallow. In A. Poole, P. Stettenheim, and F. Gill (Eds.), *Birds of North America.* No. 11. The American Ornithologists Union and the Academy of Natural Sciences of Philadelphia, Philadelphia.

Rocke, T. E., Brand, C. J., and Mensik, J. G. 1997. Site-specific lead exposure from lead pellet ingestion in sentinel mallards. *J. Wildl. Manag.* 61: 228–234.

Roseberry, J. L. 1979. Bobwhite population responses to exploitation: Real and simulated. *J. Wildl. Manag.* 43: 285–305.

Sample, B. E., Aplin, M. S., Efroymson, R. A., Suter II, G. W., and Welsh, C. J. E. 1997. *Methods and Tools for Estimation of the Exposure of Terrestrial Wildlife to Contaminants.* ORNL/TM-13391. Oak Ridge National Laboratory, Oak Ridge, TN.

Sample, B. E., Opresko, D. M., and Suter II, G. W. 1996. *Toxicological Benchmarks for Wildlife,* 1996 rev. ES/ER/TM-86/R3. Oak Ridge National Laboratory, Oak Ridge, TN.

Sample, B. E., and Suter II, G. W. 1999. Ecological risk assessment in a large river-reservoir: 4. Piscivorous wildlife. *Environ. Toxicol. Chem.* 18: 610–620.

Samuals, W. B., and Ladino, A. 1983. Calculations of seabird population recovery from potential oil spills in the Mid-Atlantic region of the United States. *Ecol. Model.* 21: 63–84.

Schafer, E. W., Bowles, W. A., Jr., and Hurlbut, J. 1983. The acute oral toxicity, repellency, and hazard potential of 998 chemicals to one or more species of wild and domestic birds. *Arch. Environ. Contam. Toxicol.* 12: 355–382.

Schafer, E. W., Jr. 1972. The acute oral toxicity of 369 pesticidal, pharmaceutical and other chemicals to birds. *Toxicol. Appl. Pharmacol.* 21: 315–330.

Scott, M. L. 1977. Effects of PCBs, DDT and mercury compounds in chickens and Japanese quail. *Fed. Proc.* 36: 1888–1893.

Secord, A. L., and McCarty, J. P. 1977. *Polychlorinated Biphenyl Contamination of Tree Swallows in the Upper Hudson River Valley, New York. Effects on Breeding Biology and Implications for Other Bird Species.* U.S. Fish and Wildlife Service, New York Field Office, Cortland, NY.

Spromberg, J. A., John, B. M., and Landis, W. G. 1998. Metapopulation dynamics: Indirect effects and multiple distinct outcomes in ecological risk assessment. *Environ. Toxicol. Chem.* 17: 1640–1649.

Stamm, D. D., Davis, D. E., and Robbins, C. S. 1960. A method of studying wild bird populations by mist-netting and banding. *Bird Banding* 31: 115–130.

Stutchbury, B. J., and Robertson, R. J. 1998. Within-season and age-related patterns of reproductive performance in female tree swallows (*Tachycineta bicolor*). *Can. J. Zool.* 66: 827–834.

Suter, G. W., Cornaby, B. W., Hadden, C. T., Hull, R. N., Stack, M., and Zafran, F. A. 1995. An approach for balancing health and ecological risks as hazardous waste sites. *Risk Anal.* 15: 221–231.

Tillitt, D. E., Ankley, G. T., Giesy, J. P., Ludwig, J. P., Kurita-Matsuba, H., Weseloh, D. V., Ross, P. S., Bishop, C. A., Sileo, L., Stromborg, K. L., Larson, J., and Kubiak, T. J. 1992. Polychlorinated biphenyl residues and egg mortality in double-crested cormorants from the Great Lakes. *Environ. Toxicol. Chem.* 11: 1281–1288.

Tucker, R. K., and Haegele, M. A. 1971. Comparative acute oral toxicity of pesticides to six species of birds. *Toxicol. Appl. Pharmac.* 20: 57–65.

U.S. Department of the Interior, Fish and Wildlife Service (USFWS). 1980. *Field Guidelines for Using Transects to Sample Nongame Bird Populations.* FWS/OBS-80/58; PB81-235848. USFWS, Biological Services Program, Fort Collins, CO.

U.S. Environmental Protection Agency (USEPA). 1992a. *Framework for Ecological Risk Assessment.* EPA/630/R-92/001. USEPA, Assessment Forum, Washington, DC.

U.S. Environmental Protection Agency (USEPA). 1992b. *Guidelines for Exposure Assessment.* EPA/600/Z-92/001. USEPA, Science Advisory Board, Washington, DC.

U.S. Environmental Protection Agency (USEPA). 1993. *Wildlife Exposure Factors Handbook,* Vol. 1: *Final Report.* EPA/600/R093/187a. USEPA, Office of Research and Development, Washington, DC.

U.S. Environmental Protection Agency (USEPA). 1993b. *Draft Ecological Risk Assessment Issue Papers.* EPA/630/R-94/0004A. USEPA, Office of Research and Development, Washington, DC.

U.S. Environmental Protection Agency (USEPA). 1995. *Supplemental Guidance to RAGS: Region 4 Bulletins. Ecological Risk Assessment.* USEPA, Region 4, Waste Management Division, Atlanta, GA.

U.S. Environmental Protection Agency (USEPA). 1996a. *Review and Analysis of Toxicity Data to Support the Development of Uncertainty Factors for Use in Estimating Risks of Contaminant Stressors to Wildlife.* USEPA, Office of Water, Washington, DC.

U.S. Environmental Protection Agency (USEPA). 1996b. *Benchmark Dose Technical Guidance Document.* EPA/600/P-96/002A. Risk Assessment Forum, USEPA, Washington, DC.

U.S. Environmental Protection Agency (USEPA). 1997. *Ecological Risk Assessment Guidance for Superfund: Process and Design for Conducting Ecological Risk Assess-*

ments. EPA 540-R-97-006. Interim Final. USEPA, Office of Solid Waste and Emergency Response, Edison, NJ.

U.S. Environmental Protection Agency (USEPA). 1998. *Guidelines for Ecological Risk Assessment.* EPA/630/R-95/002F. USEPA, Risk Assessment Forum, Washington, DC.

U.S. Environmental Protection Agency (USEPA). 1999a. *Draft Risk Assessment Guidance for Superfund,* Vol. 3: *Part A, Process for Conducting Probabilistic Risk Assessment.* EPA 000-0-99-000. USEPA, Office of Solid Waste and Emergency Response, Washington, DC.

U.S. Environmental Protection Agency (USEPA). 1999b. *Memorandum: Issuance of Final Guidance: Ecological Risk Assessment and Risk Management Principles for Super fund Sites.* OSWER Directive 9285.7-28P. Stephen Luftig, USEPA, Office of Emergency and Remedial Response, Washington, DC.

U.S. Environmental Protection Agency (USEPA). 1999c. *Ecological Risk Assessment in the Federal Government.* CENR/5-99/001, USEPA, Committee on the Environment and Natural Resources of the National Science and Technology Council.

Walters, C. J., Hilborn, R., Oguss, E., Peterman, R. M., and Stander, J. M. 1974. *Development of a Simulation Model of Mallard Duck Populations. Can. Wild. Serv. Occas. Pap.* 20: 1–35.

Water Environment Federations (WEF). 1993. *Application of Ecological Risk Assessment to Hazardous Waste Site Remediation.* WEF, Alexandria, VA.

Watkins, G. E., and Stelljes, M. E. 1993. A proposed approach to quantitatively assess potential ecological impacts to terrestrial receptors from chemical exposure. In J. W. Gorsuch, F. J. Dwyer, C. G. Ingersoll, and T. W. LaPoint (Eds.), *Environmental Toxicology and Risk Assessment,* Vol. 2 STP 1216. American Society for Testing and Materials, Philadelphia, PA, pp. 442–439.

Weil, C. S., and McAllister, D. D. 1963. Relationship between short- and long-term feeding studies in designing an effective toxicity test. *Agric. Food Chem.* 11: 486–491.

Wentsel, R. S., Checkai, R. T., LaPoint, T. W., Simini, M., Ludwig, D., and Brewer, L. 1994. *Procedural Guidelines for Ecological Risk Assessments at U.S. Army Sites,* Vol. 1. U.S. Army Environmental Center, ERDEC-TR-221. Aberdeen Proving Ground, MD.

Wentsel, R. S., La Point, T. W., Simini, M., Checkai, R. T., Ludwig, D., and Brewer, L. 1996. *Tri-Service Procedural Guidelines for Ecological Risk Assessments.* U.S. Army Edgewood Research, Development and Engineering Center, Aberdeen Proving Ground, MD; Clemson University, Pendleton, SC; Geo-Centers, Inc., Fort Washington, MD; EA Engineering, Science and Technology, Hunt Valley, MD; EBA, Inc., Sisters. Final, June.

Wiggins, D. A. 1990. Clutch size, offspring quality, and female survival in tree swallows—an experiment. *Condor* 92: 534–537.

Williams, A. B. 1936. The composition and dynamics of a beech-maple climax community. *Ecol. Monogr.* 6: 317–408.

SECTION L
Risk Assessment and Life-Cycle Analyses

25 Life-Cycle Impact Analysis: A Challenge for Risk Analysts

H. SCOTT MATTHEWS[1], HEATHER MACLEAN[2], and LESTER LAVE[1]

[1] Carnegie Mellon University, Pittsburgh, Pennsylvania
[2] University of Toronto, Toronto, Ontario, Canada

25.1 INTRODUCTION

Evaluating the environmental and sustainability implications of a product or project requires evaluating these implications from the design stage through manufacture or construction to use, and finally, to the end of life. Life-cycle assessment (LCA) provides a framework for identifying and evaluating these implications. Since the 1960s, considerable progress has been made in developing methods for LCA. However, virtually all analyses stop short of the ultimate goal of LCA, that of assessing the impact of a product, process, or project on human health and the environment.

We discuss the need for LCA, present a brief history and rationale for using LCA, provide an introduction to two life-cycle analytical tools, and present examples of the application of these tools. The applications themselves only attempt to identify and quantify materials and energy use and environmental discharges, the life-cycle inventory analysis stage of LCA, and the life-cycle impact analysis.

25.2 INTRODUCTION TO LIFE-CYCLE ASSESSMENT

The world is filled with controversial environmental problems as well as myriad possible solutions to these problems. Many of these problems are man made—products or services that we consume contribute the lion's share of our environmental problems. However, quantifying the environmental and sustainability

Human and Ecological Risk Assessment: Theory and Practice, Edited by Dennis J. Paustenbach
ISBN 0-471-14747-8 © 2002 John Wiley & Sons, Inc.

implications of a product or process is more difficult than intuition might suggest. Consider the following four recent examples of controversial products:

- Are "throwaway" paper or "durable" foamed plastic or ceramic cups better for the environment? Paper decays in landfills while plastic does not; ceramic cups can provide service for years.
- Are paper grocery bags superior to plastic ones? Are they better than durable canvas or string net bags? Again, the paper decays in a landfill while the plastic does not. The canvas and string bags can last for many uses.
- Are durable cloth diapers superior to disposable paper ones? The cloth diapers are reused while the paper ones go to a landfill after one use.
- Is a battery-powered car with no tailpipe superior to a car with an internal combustion engine? The vehicle emissions from the battery-powered car are zero, in contrast to those for the car with the internal combustion engine.

Intuition can be misleading in each case if attention was focused only on the product, rather than the life cycle of the product. A product's life cycle depends on its design, including the specification of materials, energy efficiency, and so on. Life-cycle discharges start with the extraction of raw materials needed and continue as the product components are manufactured, assembled, and shipped. Then the product is transported to the consumer who uses it and, eventually, throws it away. At that point, it could be landfilled, recycled, or even reused.

One approach to support environmentally conscious decision making is LCA. Life-cycle assessment is a systematic approach linking the product life cycle, from design to disposition, with the environmental impacts generated at each stage. In a U.S. Environmental Protection Agency (EPA) document, Vigon (1993) defines life-cycle assessment as "a concept and a methodology to evaluate the environmental effects of a product or activity holistically, by analyzing the entire life cycle of a particular product, process, or activity. The life cycle assessment consists of three complementary components—inventory, impact, and improvement—and an integrative procedure known as scoping." The life-cycle concept has often been referred to as "cradle-to-grave" assessment. This definition will be clarified below. A life-cycle view of a product is intended to yield environmental improvement by revealing the complete environmental picture of a product, rather than just the emissions generated in the usual course of production by the manufacturer. LCA is useful beyond the scope of a manufacturer as well. Service providers, government agencies, and other interested parties can use these methods to consider the total impact of their global business activities.

The concept of life-cycle analysis was initiated in the 1960s, and the practice of conducting life-cycle studies has existed since the 1970s. Only in the 1990s did researchers attempt to describe a procedure for life-cycle studies that facilitated understanding the overall process, the underlying data, and the inherent assumptions. Vigon (1993) presents the following history of the life cycle inventory stage: "Life-cycle inventory analysis had its beginnings in the 1960s. Concerns over the limitations of raw materials and energy resources sparked interest in finding ways to cumulatively account for energy use and to project future resource supplies and use. In one of the first publications of its kind, Harold Smith reported his calculation of cumulative energy requirements for the production of chemical intermediates and products at the World Energy Conference in 1963. Later in the 1960s, global modeling studies (Meadows, 1972; Goldsmith, 1972) resulted in predictions of the effects of the world's changing population on the demand for finite raw materials and energy resources. The predictions of rapid depletion of fossil fuels and climate changes resulting from excess waste heat stimulated more detailed calculations of energy use and output in industrial processes. During this period, about a dozen studies were performed to estimate costs and environmental implications of alternative sources of energy.

In 1969 researchers initiated a study for the Coca-Cola Company that laid the foundation for the current methods of life-cycle inventory analysis in the United States. This study quantified the raw materials and fuels used and the environmental loadings from the manufacturing processes for alternative containers. Other companies in both the United States and Europe performed similar comparative life-cycle inventory analyses in the early 1970s. At this time, many of the data were derived from publicly available sources such as government documents or technical papers, as specific industrial data were not available. The process of quantifying the resource use and environmental releases of products became known as a resource and environmental profile analysis (REPA), as practiced in the United States. In Europe it was called an "ecobalance." With the formation of public interest groups encouraging industry to ensure the accuracy of information in the public domain and with the oil shortages in the early 1970s, approximately 15 REPAs were performed between 1970 and 1975. Through this period, a protocol or standard research method for conducting these studies was developed. This multistep method requires a number of assumptions. During these years, the assumptions and techniques underwent considerable review by EPA and major industry representatives, resulting in more acceptable methods.

From 1975 through the early 1980s, as interest in these comprehensive studies waned because of the fading influence of the oil crisis, environmental concern shifted to issues of hazardous waste management. However, throughout this time, life-cycle inventory analyses continued to be conducted, and the method was improved through implementing about two studies per year, most of which focused on energy requirements.

During this time, European interest grew with the establishment of an Environment Directorate (DGX1) by the European Commission. European LCA practitioners developed approaches parallel to those being used in the United States. Besides working to standardize pollution regulations throughout Europe, DGX1 issued the Liquid Food Container Directive in 1985, which charged member companies with monitoring the energy and raw materials consumption and solid-waste generation of liquid-food containers. When solid waste became a worldwide issue in 1988, the life-cycle inventory analysis technique again emerged as a tool for analyzing environmental problems. As interest in all areas affecting resources and the environment grows, the method for life-cycle inventory analysis is again being improved. A broad base of consultants and research institutes in North America and Europe have been further refining and expanding the method. With recent emphasis on recycling and composting resources found in the solid-waste stream, approaches for incorporating these waste management options into the life-cycle inventory analysis have been developed. Interest in moving beyond the inventory to analyzing the impacts of environmental resource requirements and emissions brings life-cycle assessment methods to another point of evolution.

Vigon (1993) also reports on the scope of life-cycle inventories:

> Over the past 20 years, most life-cycle inventories have examined different forms of product packaging such as beverage containers, food containers, fast-food packaging, and shipping containers. Many of these inventories have supported efforts to reduce the amount of packaging in the waste stream or to reduce the environmental emissions of producing the packaging. Some studies have looked at actual consumer products, such as diapers and detergents, while others have compared alternative industrial processes for the manufacture of the same product.

The Society for Environmental Toxicology and Chemistry (SETAC) LCA technical framework workshop report published in January 1991 summarized the status of the field at that time and was one of the initial documents that outlined a basis for life cycle studies (SETAC, 1991). The EPA (Vigon, 1993; Curran, 1996) accepted and built on the SETAC framework. The research by the EPA and SETAC led to a four-part approach to LCA that is accepted today:

1. Goal scope and definition
2. Life-cycle inventory (LCI)
3. Life-cycle impact assessment (LCIA)
4. Life-cycle improvement analysis

Step 1 consists of specifically stating the purpose and appropriately identifying the boundaries of the study. SETAC (1991) writes: "The study goal and scope are crucial to managing and coordinating a life-cycle study by bringing together the LCA information needed to make an identified decision and an understanding of the reliability and representativeness of the LCA." It is important that the use for which the study results are intended is clearly stated. Another crucial step in this procedure is defining the "boundary" of the analysis. For example, in the LCA of a paper cup by the manufacturer, the practitioner may decide in the manufacturing phase to consider only the inventory of effects arising from the 10 highest cost items in the production process to save time and effort. The boundary assumption is an important one, as it draws the

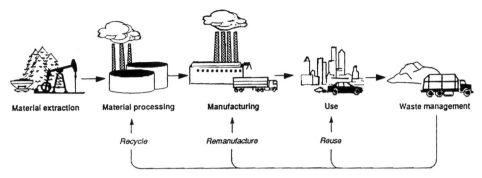

Figure 25.1   Life-cycle assessment involves analysis of the entire supply chain (OTA, 1992).

line around what will be excluded from consideration in the inventory and, inevitably, from the overall assessment. The necessary data collection and interpretation are contingent on a proper understanding of where each stage of a life cycle begins and ends. Any effects that lie outside of the boundary are ignored. This boundary assumption can potentially lead to significant underestimation of the inventory of effects of a product across its life cycle. For a product example, the general scope of each stage is taken from Vigon (1993). Figure 25.1 is a simplified diagram of the life-cycle stages. These stages can be modified for process and project applications.

1. *Raw Materials Acquisition.* This stage of the life cycle of a product includes the removal of raw materials and energy sources from the earth, such as the harvesting of trees or the extraction of crude oil. Transport of the raw materials from the point of acquisition to the point of raw materials processing is considered part of this stage.

2. *Manufacturing.* The manufacturing stage produces the product or package from the raw materials and delivers it to consumers. Three steps are involved in this transformation:

   a. *Materials Manufacture.* This step involves converting a raw material into a form that can be used to fabricate a finished product. For example, several manufacturing activities are required to produce a polyethylene resin from crude oil: The crude oil must be refined; ethylene must be produced in an olefin plant and then polymerized to produce polyethylene; transportation between manufacturing activities and the point of product fabrication is considered part of materials manufacture.

   b. *Product Fabrication.* This step involves processing the manufactured material to create a product ready to be filled or packaged, for example, blow molding a bottle, forming an aluminum can, or producing a cloth diaper.

  c. *Filling/Packaging/Distribution.* This step includes all manufacturing processes and transportation required to fill, package, and distribute a finished product. Energy and environmental wastes caused by transporting the product to retail outlets or to the consumer are accounted for in this step of a product's life cycle.

3. *Use/Reuse/Maintenance.* This is the stage consumers are most familiar with: the actual use, reuse, and maintenance of the product. Energy requirements and environmental wastes associated with product storage and consumption are included in this stage.

4. *Recycle/Waste Management.* Energy requirements and environmental wastes associated with product disposition are included in this stage, as well as post–consumer waste management options such as recycling, composting, and incineration."

The LCI quantifies the inputs and outputs (materials use, energy use, and environmental discharges) associated with each stage of the life cycle. This is typically implemented by either initiating new research to estimate the inventory data or consulting existing databases of inventory information. Data collection is driven by the study's goal. According to the EPA (2000):

> At times site-specific data are needed, such as the data for the manufacture of a certain product being studied. At other times average or commodity data are sufficient, such as when a study is being done at a national level and is not focusing on a particular manufacturer. It is important that the study goal be revisited periodically as data collection progresses to ensure that the goal will be met.

The LCIA further extends the analysis and interprets the results of the inventory in order to assess the impacts of the product or project on human health and the environment. The EPA (2000) reports the following with respect to the LCIA: "Impact indicators are used to measure the potential for the impact to occur rather than directly quantifying actual impacts. This approach works well to simplify the LCA process, making it a more useful tool. A variety of environmental impact indicators and associated indicators have been developed and more continue to be used as the LCA method evolves. The categories for indicators range from a global level, such as contribution to global warming and ozone depletion, to local impacts, such as photochemical smog formation. As an example, a recent study conducted for the EPA defines eight impact categories and indicators for global climate change, stratospheric ozone depletion, acidification, photochemical smog, eutrophication, human toxicity, ecological toxicity, and resource depletion. The International Standardization Organization (ISO) 14042 guidelines for impact assessment describe the need for environmentally relevant indicators and emphasize that the results should be clearly stated in terms of the following criteria:

(a) The ability of the indicator to reflect the consequences of the inventory result on the category endpoint(s), at least qualitatively

(b) The incorporation of environmental data or information in the model, including the environmental condition and the intensity of the category endpoint(s), the spatial extent of projected impacts on category endpoint(s), the temporal aspects, duration, residence time, persistence, timing, and so on, of projected impacts on category endpoints, the reversibility of projected impacts on category endpoints, and the uncertainty of projected impacts with respect to category endpoints

LCA practitioners and developers around the world continue to explore and improve impact assessment methods. Further description of the life-cycle impact assessment method, including discussion on what is and is not LCIA, can be found in the 1997 SETAC report, "Life Cycle Impact Assessment: The State of the Art (Barnthouse, 1997)."

Finally, the improvement analysis is a systematic evaluation of the needs and opportunities to reduce the environmental impacts, energy use, and materials use during the life cycle. This analysis may include both quantitative and qualitative measures of improvements. Introductory material on life-cycle analysis is further documented elsewhere (Barnthouse, 1997; Curran, 1996; Graedel, 1995; Vigon, 1993).

In the four examples cited earlier in this chapter, the total life cycle of each product poses much greater environmental problems than just a single phase. As elaborated below, making paper products require growing, cutting, and transporting trees, making the wood into paper, shipping it to market, and then disposal or recycling. Some steps, such as logging and producing paper, have high environmental damages. Thus, the life cycle of paper has difficulties that are not apparent in thinking of the disposal or recycling of a paper cup or grocery bag. Similarly, canvas or string grocery bags are more resource intensive than plastic or paper ones. Whether the multiuse bag is better depends on how many times it is used. If the string bag is lost or destroyed after a few uses, it is not superior to the disposable bags. If the shopper forgets to use the canvas bags because they are too heavy and clumsy, they are wasted. Cloth diapers require water and energy for washing and drying. Where water is scarce, washing, drying, and water treatment can harm environmental quality more than disposable diapers.

Finally, in comparing the automobiles, their life cycles can be broken down into three basic stages, as shown in Figure 25.2: vehicle manufacture, vehicle use, and end of life. Vehicle use is further broken down into a "fuel" cycle, maintenance, fixed costs (e.g., license fee, insurance), and vehicle operation. The fuel cycle for a gasoline vehicle consists of the processes from crude oil extraction through refining through delivery of the fuel at a retail station. For the electric vehicle, this fuel cycle consists of the steps required to generate the electricity to power the vehicle over its lifetime. Vehicle maintenance consists of the steps required to manufacture replacement parts and to service the vehicle over its lifetime. The vehicle operation consists of the energy used by the vehicle as well as the emissions from the vehicle. In comparing electric and internal combustion engine vehicles, the electric vehicles have large numbers of bat-

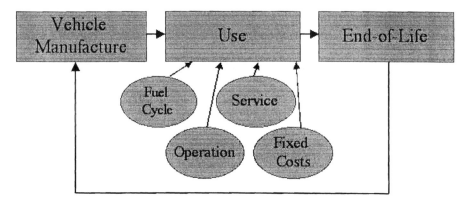

Figure 25.2   Simplified life-cycle inventory diagram for automobiles.

teries. These batteries are made from toxic materials like lead. Mining and smelting the lead and then making and recycling the batteries can generate large amounts of toxic waste, which could be worse for the environment than the air emissions from the internal combustion engine. Additionally, the upstream emissions from electricity generation may be higher than those from refining of gasoline and the tailpipe emissions of an internal combustion engine vehicle. An inevitable requirement in answering any of these comparative questions is some assessment criteria that allow us to weigh impacts of one type versus impacts of another type.

As stated above, a life cycle includes all the steps, from extracting the resources to product disposal. Each life-cycle step has impacts on the environment. For example, raw materials extraction results in depletion of non-renewable resources like petroleum and ores. In the case of ores, mining machinery requires large quantities of energy, generally from burning fossil fuels that release carbon monoxide, nitrogen oxides, and particulate matter into the air. Manufacturing and disposal of a product also require energy and result in discharges. In short, every step of a product's life cycle has both inputs from and discharge to the environment. Over the life cycle, the sum of these inputs and releases can be substantial.

The data burden of a SETAC/EPA LCA can be substantial, due to logistical as well as psychological barriers (e.g., fear of sharing data and therefore revealing confidential corporate information). If the scope of an assessment requires the collection of new inventory data, then existing processes must be measured and analyzed to determine the quantities of inputs and releases for each stage. Alternatively, existing data sources can be used and cited to estimate the effects of particular processes. For example, existing estimates on the environmental impact of electricity generation can be used. But such estimates need to be relevant and specific to the particular application to be of use in such a setting. For example, to yield accurate results, the electricity effects for a particular processing plant need to be relevant to the local mix of electricity purchased. Thus, estimates need to be available that reflect the use of renewable or nonrenewable resources used to make the plant's power. Getting such data could be difficult, as it will rely on fuel and technology assumptions (e.g., data on a power plant burning low-sulfur coal or using flue-gas desulfurization for control of air emissions). At the worst, some form of questionnaire might be used to gather the necessary inventory data. In the end, the success or failure of any LCA will depend greatly on the boundary assumptions, data quality, and level of economic resources available.

Many SETAC-based LCA software tools exist to aid decision makers in performing life-cycle inventories. Most consist of a graphical user interface front end to a database of existing product- or process-specific inventory data. The inventory data may be proprietary, from public data sources, or both. If boundaries are appropriate and care is taken in selecting data sources, meaningful and relevant results are possible. However, the accuracy costs time and money. It is not uncommon for detailed LCA studies to require hundreds of thousands of dollars and 6 months to complete. If time is a primary driver, results may not be available until the next production cycle, which is often too late to make improvements.

The concern over the cost and time required for LCA has resulted in researchers investigating methods to simplify the analysis while still retaining the information needed to satisfy the study goals. One concept that has received attention is streamlining LCA. The EPA (2000) describes this concept as follows:

> A continuing concern over the cost and time required for LCA encouraged some practitioners to investigate the possibility of "streamlining" or simplifying LCA to make it more feasible and more immediately relevant without losing the key features of a life-cycle approach. When the concept of streamlining was first introduced, many LCA practitioners were skeptical, stating that LCA could not be streamlined. Over time, however, there has been growing recognition that "full-scale" LCA and streamlined LCA are not two separate approaches but are, instead, points on a continuum. Most LCA studies will fall somewhere along that continuum, in between the two extremes. As a result, streamlining an LCA becomes part of the scope and goal definition process. For example, as the study team decides what is and is not to be included in the study, they are engaged in streamlining—in addition to determining what will and will not be included, the study team will determine how to best achieve these requirements. The key is to ensure that the streamlining steps are consistent with the study goals and anticipated uses, and that the information produced will meet the users' needs. From this perspective, the scope and goal definition process involves determination of what needs to be included in the study to support the anticipated application and decision.

The concept of streamlined LCA is developed by Graedel (1998), and Bennett (2000) presents an example using the approach.

Despite these advances, this LCA approach still requires setting tight boundaries around the problem to make it tractable. As we show below, the parts of the supply chain that are outside the boundaries are generally important, leading to significant changes in the use of resources and environmental discharges. Thus, while this LCA approach has become easier and cheaper to apply, it still has the inherent difficulty of excluding a significant part of the life cycle.

To summarize, there are two principal difficulties with the SETAC approach to LCA. First, specifying a boundary for the problem is arbitrary. To make the problem tractable, the boundary must be drawn narrowly. However, a narrow boundary excludes much of the life-cycle effects. Second, the process is inherently time consuming and expensive, although the required time and expense can be reduced somewhat by using data from past analyses. However, using past data means that the data are not entirely appropriate or are dated. Even using past data, the analysis is still far from being able to give an answer in real time.

25.3 ECONOMIC INPUT–OUTPUT LIFE-CYCLE ASSESSMENT

The shortcomings discussed above led to the creation of an the economic input–output life-cycle assessment (EIO-LCA) model. This model does not require setting any boundary on the analysis and is essentially free and can be performed in hours, not months. This model is built on the U.S. Department of Commerce's input–output model of the U.S. economy.

Wassily Leontief (1936) developed input–output (I/O) models to relate the production inputs of goods and services in an economy to the production of outputs of the other sectors as well as to the provision of labor and other inputs. The most basic of these I/O models describes an economy only by intersectoral transactions. One of the critical assumptions in an I/O framework is that, generally speaking, firms in each sector purchase some amount of input from firms within the same sector (e.g., an agricultural sector where seeds are purchased to grow vegetables). In more advanced I/O models, households are included. This expansion yields an additional row to represent the provision of the household product labor services and an additional column to show the consumption of goods and services by households. A detailed summary and discussion of I/O theory is included in the Appendix. To simplify computation, the analysis assumes that all inputs and discharges are proportional to output. Thus, if producing $1000 worth of widgets requires 40 gallons of petroleum and produces 400 pounds of CO_2 emissions, then producing $10,000 of widgets would require 400 gallons of petroleum and result in 4000 pounds of CO_2 emissions.

The EIO-LCA model begins with the 485-by-485 1992 U.S. input–output matrix (see the Appendix). These economic data are augmented with corresponding estimates of resource use and environmental flows and emissions from the 485 sectors (see the Appendix for details). The EIO-LCA model can be seen and utilized on the Internet at *http://www.eiolca.net*.

TABLE 25.1 EIO-LCA Supply Chain Effects of Producing $1 Million of Electricity in United States, 1992

Sector	Economic ($, $\times 10^6$)
Electric services (utilities)	1.007134
Coal	0.102573
Other repair and maintenance construction	0.087334
Crude petroleum and natural gas	0.041535
Natural gas distribution	0.037961
Railroads and related services	0.032541
Wholesale trade	0.024300
Petroleum refining	0.023054
Real estate agents, managers, operators, and lessors	0.021044
Banking	0.017472
All other sectors	0.276140
Total	1.671088

Example 25.1: Economywide Impact of Producing Electricity

Data. An average U.S. electric utility generates $1 million of electricity. Use EIO-LCA to estimate economic purchases needed throughout the economy to support this production.

Solution. By entering $1 million as an input to the electric services (utilities) sector of the web model, and choosing to view "economic" results, Table 25.1 shows the 10 most important industry sectors in the supply chain for this additional production as well as a condensed line indicating the amount of purchases coming from the other 475 sectors.

Note that the total amount of purchases from the electric utilities sector is more than $1 million. This is because some amount of electricity is needed to produce electricity. As shown in Table 25.1, the average production of electricity in the United States needs about 10 cents of coal per dollar of electricity, 9 cents of construction (to build the facilities), and petroleum and natural gas. Coal, petroleum, and natural gas are three of the primary sources for generating electricity.

Example 25.2: Economywide Environmental Impact of Producing Electricity

Data. An average U.S. electric utility generates $1 million of electricity. Use EIO-LCA to estimate the conventional pollutants and global warming gases released throughout the economy (i.e., not just those of the electricity-generating faculty) to support this production.

Solution. By entering $1 million as an input to the electric services sector of the web model and choosing conventional pollutant results, Table 25.2 shows the 10 industry sectors that contribute emissions of conventional pollutants to support this additional production as well as a condensed line indicating the amount of purchases coming from the other 475 sectors.

TABLE 25.2 Top Ten Contributing Sectors to Conventional Pollutant and Global Warming Releases Associated with Producing $1 Million of Electricity in United States

	Metric Tons of Releases					
	SO_2	CO	NO_2	VOC	PM_{10}	GWP
Electric services (utilities)	66.5	2.09	30.4	0.238	1.37	11,000
Coal	0.017	0.013	0.018	0.004	0.071	1,100
Crude petroleum and natural gas	0.078	0.134	0.134	0.033	0.001	210
Other repair and maintenance construction	0.000	0.104	0.130	0.001	0.436	4
Natural gas distribution	0.080	0.071	0.039	0.001	0.000	150
Railroads and related services	0.104	0.219	0.785	0.045	0.045	30
Petroleum refining	0.074	0.040	0.049	0.038	0.004	15
Blast furnaces and steel mills	0.025	0.107	0.015	0.005	0.004	20
Water transportation	0.046	0.014	0.041	0.009	0.006	4
Trucking and courier services, except air	0.005	0.198	0.084	0.037	0.000	9
All other sectors	0.137	0.441	0.277	0.128	0.062	43
Total	67.1	3.43	32.0	0.539	2.00	12,600

Note: Totals do not sum due to rounding. VOC = volatile organic compounds; GWP = global warming potential.

When the base economic purchase data are combined with augmented environmental data, many environmental discharges can be estimated. For example, Table 25.2 shows sample EIO-LCA results for emissions of conventional pollutants and global warming. Each of the cell values in Table 25.2 is found by multiplying the economic supply chain amount of purchases from each sector (summarized in Table 25.1) by the relevant conventional pollutant emissions vector (as described in the Appendix).

Note that the sectors summarized in Table 25.2 do not perfectly correspond with those in Table 25.1. This is because some sectors are more polluting than others. For example, while "wholesale trade" is one of the top 10 sectors needed to produce electricity, it is not one of the top 10 sectors in terms of conventional pollutant releases.

25.4 EXTERNAL COST ASSESSMENT

Market incentives motivate businesses to provide the desired goods and services at least cost, even if this means creating a social problem by discharging pollution to the environment. Environment discharges and resultant degradation are not caused by malevolent individuals and companies who desire to harm the environment. Rather, the discharges occur because this is the cheapest way for the plant to deal with unwanted residuals. The resulting environmental pollution is an "externality," an undesired side effect of market behavior. The associated cost is called an external cost. More generally, an externality results whenever private costs (the costs to the firm) are different from social costs (the total costs to society). Economists conclude that the presence of nontrivial externalities requires some sort of government intervention to cure the market failure. Isolating problem commodities and estimating their economywide environmental impacts will help make environmental regulation more effective and efficient.

In the 1970s, Congress created the EPA and passed environmental laws to prevent these externalities. A steel mill might have found that the cheapest way of disposing of waste was to dump it into the environment. To protect the environment, the EPA set up regulations limiting the discharge levels of each pollutant. Relative to what would have happened in the absence of environmental laws and regulations, current discharges are much lower and the environment is much cleaner.

To no one's surprise, the costs of abating air pollution have been substantial. The industries and companies that paid the greatest amount have protested that the regulations were too stringent. Partly as a result, Congress required the EPA to conduct a retrospective benefit–cost analysis of the 1970 Clean Air Act between 1970 and 1990. The EPA (1997) found that the social benefits of abating air pollutants since 1970 exceed costs by a large margin. The EPA analysis does not attempt to justify individual regulations. In particular, the majority of benefits come from eliminating lead from gasoline and reducing the concentration of small particles. Most costs come from reducing the emissions of pollutants causing urban smog (ozone).

25.5 ARE THERE REMAINING AIR POLLUTION PROBLEMS?

While the Clean Air Act has made considerable progress in cleaning the air, current environmental regulation does not eliminate all pollution discharges or even all environmental damage from the discharges. Table 25.3 shows estimates of air emissions of several pollutants. The EPA estimated that 84 million metric tons of total criteria air pollutant emissions and 4.5 billion metric tons of carbon dioxide equivalent global warming potential were emitted as a result of industrial activity in 1992. The majority of emissions are a result of the manufacture and transport of goods. We will discuss the estimation of the external costs later in this chapter.

25.6 TARGETING EXTERNAL ENVIRONMENTAL COSTS

We now focus on current air pollution emissions to ask whether additional abatement is in the public interest and, if so, what is the best approach. The EPA study cited above concludes that the benefits from abating air pollutants have been far greater than the costs. We do not address either the past costs of

TABLE 25.3 Total U.S. Emissions and Estimated External Costs of Various Pollutants, 1992

Species	Emissions (metric tons)	Estimated External Cost ($)
Carbon monoxide (CO)	24 million	13 billion
Nitrogen oxides (NO_x)	17 million	18 billion
Particulate matter (PM 10)	11 million	31 billion
Sulfur dioxide (SO_2)	20 million	36 billion
Volatile organic compounds (VOC)	12 million	17 billion
Global warming potential (in CO_2 equivalent)	4.5 billion	63 billion
Total	—	178 billion

Source: EPA, 1996, 1998a.

Note: The GWP value has been converted from 1.22 billion MMTCE (million metric tons of carbon equivalent) to metric tons of carbon dioxide equivalent, by multiplying by the weight fraction of carbon in carbon dioxide, 44/12.

abatement or the past benefits of abatement in our estimates. We do explore a nonconventional method of giving polluters incentives to abate their emissions. For example, the 1990 Clean Air Act mandated almost a 50 percent reduction in sulfur dioxide emissions but created property rights in emissions and allowed firms to trade emissions permits (Schmalensee, 1998). This novel approach has resulted in greatly reduced abatement costs compared to conventional command and control.

One way to internalize pollution externalities is to levy an effluent fee on emissions. An effluent fee would increase the cost of extracting raw materials. Producers would be motivated to find ways of abating their emissions in order to cut their costs. The high prices of raw materials would lead buyers to seek cheaper materials and otherwise reduce their purchases. The producers converting the raw materials into components would also face high costs due to their emissions. Thus they would seek to lower their costs by finding cheaper materials and by abating their pollution. Throughout the whole supply chain, producers would face higher costs of inputs, higher costs due to effluent fees, and pressure from their customers to lower their costs. By the time commodities were offered for consumption, their costs and prices would reflect their cumulative environmental discharges. At each stage of the supply chain, producers would be motivated to reduce costs. The higher costs would lead to higher prices, causing consumers to look for producers and alternative products that fulfill their demands at lower cost.

Collecting the effluent fees from each producer at each stage of the supply chain would be cumbersome and expensive. It would be far easier to collect the fee at the point of final sale, although, for example, incentives for miners and hazardous waste handlers would not be so clear.

Even if Congress mandated the correct quantity reduction in sulfur dioxide (and other pollutants), that is, the quantity where the marginal social cost of abatement equals the marginal social benefit of abatement, there would still be externalities to be internalized. The producers have played their role by abating to the level where marginal private benefit equals marginal private cost. However, consumers still have to decide how much of each good or service to purchase. Only if consumers make consumption decisions on the basis of prices that reflect all the externalities will social efficiency be attained. The need to charge firms for the marginal social loss from emissions is discussed by Joskow (1992) and Freeman et al. (1992a, b).

Example Problem   One example of the value of external costs is the policy implications of the social values placed on abating ambient air pollution levels. In Table 25.6 below, the median social loss due to emitting a ton of NO_x is $1060 while that for SO_2 is $1800. These estimates could be used by charging polluters $1060 for each ton of NO_x emitted and $1800 for each ton of SO_2 emitted. If so, polluters would be motivated to find all the uses where the cost of abatement are less than $1060 and $1800 per ton, respectively. Charging these effluent fees would lead to substantial revenue collection in most of the United States as well as increased abatement of SO_2 and NO_x. In areas such as Los Angeles, the marginal cost of abating NO_x is estimated to be much larger than the $1060, and so no further abatement is warranted. In contrast, the marginal cost of abating SO_2 is less than $1800, and so additional abatement would be helpful.

25.7   TOTAL COSTS OF MANUFACTURING: DIRECT AND INDIRECT EFFECTS

These estimates can be interpreted either as the amount that would have to be charged if effluent fees were levied at the point of sale for final use or as the cumulative impact of effluent fees charged at each step. In the latter interpretation, the estimate becomes an input into an analysis of the revenues to be raised and the disruption that would be caused by this fee.

Our estimates are for the U.S. economy in 1992. The structure of the I/O model and the nature of our calculation means that we do not allow producers or consumers to react to the effluent fees. Clearly, the response to large fees would be improved controls, switching materials and processes, and changing the demand mix. The estimates do provide an estimate of the external costs from economic activity that U.S. society actually paid in 1992.

However, an I/O model makes no explicit connection between economic purchases of inputs, production, and the external damages that result. For example, employee health effects may result from exposure to a particular substance or process, and companies may incur medical costs as a result. But there is nothing in an I/O model that connects medical costs to the activity that causes them. So I/O models themselves do not capture external costs.

This analysis restricts attention to air emissions from commodity production. A full LCA for a product or a process would include liquid- and solid-waste discharges as well as the environmental discharges associated with use and end of life (Graedel, 1995). For example, assessment of a standard automobile would include petroleum purchases during use. Examples of LCA using the EIO-LCA model appear elsewhere (Hendrickson, 1997; Horvath, 1998; MacLean, 1998).

25.8   EMISSIONS DATA SOURCES

Once economic outputs by sector are estimated, environmental emissions are estimated by multiplying the economic output by the average emission per dollar of output for each of the 498 commodity sectors. For the conventional air pollutants (carbon monoxide, nitrogen oxides, particulate matter, sulfur dioxide, lead, and volatile organic compounds) and carbon dioxide, data on direct emissions for stationary sources are available from the EPA's Aerometric Information Retrieval System (AIRS) database of reported emissions (EPA, 1998b). For sectors not reporting to AIRS (especially those without fixed plants, e.g., trucking), a two-stage estimation is employed. First, fuel consumption is estimated based on the U.S. Department of Commerce (DOC) data on fuel purchases used in the development of the I/O matrices themselves (DOC, 1997). These are typically fossil fuels, and most air emissions come from the combustion of these fuels (e.g., coal, natural gas, petroleum, gasoline). Next, air emissions are estimated based upon standard emission factors for industries and processes, as compiled by the EPA (1995). Emissions of greenhouse gases are translated to equivalent releases of carbon dioxide based on a standard weighting method (Intergovernmental Panel on Climate Change (IPCC), 1995; Wuebbles, 1995). This method determines global warming potential (GWP), which was developed to compare the ability of each greenhouse gas to trap heat in the atmosphere and thus to affect climate change. Note that the conventional pollutants are not part of the GWP method because their specific contribution to climate change is unclear. See the Appendix for details on the construction of the database for pollution emitted by sector as well as a complete presentation of the total sectoral emissions data. Note that lead air emissions have been estimated but are not included in analytical results presented.

Table 25.4 summarizes the total emissions of conventional pollutants for the top 10 sectors [sorted by emissions of particulate matter (PM)].

25.9   DAMAGE VALUATION OF EMISSIONS

So far, we have not discussed how to put a dollar value on the social benefit of abating air pollutants.

TABLE 25.4   Ten Commodity Sectors with Largest PM Emissions

Commodity Sector	Total Air Pollutants Emitted (metric tons)				
	PM	CO	NO$_2$	SO$_2$	VOC
Petroleum and natural gas well drilling	1,090,000	265,000	326,000	12,000	14,000
Other new construction	710,000	168,000	213,000	3,000	2,000
Other repair and maintenance construction	670,000	160,000	200,000	89	830
New residential one-unit structures, nonfarm	461,000	110,000	125,000	1,900	1,500
New office, industrial, and commercial buildings construction	335,000	75,000	100,000	5	1,000
Maintenance and repair of farm and nonfarm residential structures	272,000	66,000	88,000	1,100	770
Electric services (utilities)	233,000	355,000	5,170,000	11,300,000	40,000
Sand and gravel	169,000	95	340	150	160
New residential additions and alterations, nonfarm	168,000	39,000	50,000	660	480
Petroleum, natural gas, and solid mineral exploration	147,000	36,000	38,000	370	2,200

Table 25.5 summarizes some of the studies that have valued the damage caused by SO$_2$ emissions. The second column shows the valuations converted as needed to dollars per metric ton of emission, and the third column shows the value in 1992 dollars using the chain-weighted gross domestic product (GDP) deflator (Federal Reserve Economic Data (FRED), 1998).

As the table reveals, the studies are not in close agreement; the range in 1992 valued estimates, from $760 to $4600 per metric ton, reflects the substantial uncertainties associated with the effects of sulfur dioxide on health, state-to-state variations in exposure, and the valuation of these effects.

Generally, the method of determining these valuations comes from a damage function approach (DFA). The studies translate emissions into ambient air concentrations and then into damage to human health, materials, plants and animals, visibility and aesthetics, and ecology using a dose–response relationship. These human and ecological impacts are translated into economic values, and the externalities are identified (Lee, 1992). In practice, the mortality losses dominate the valuation. The Cifuentes (1993), Zuckerman (1995), and CEC (1993) studies all follow DFAs. Fankhauser does a modified DFA with a stochastic process and discounting that leads to rather high estimates.

Four of the valuation studies were done by state public utility commissions (PUCs). These states wanted to make sure that utilities did not build polluting generators based on inadequate state and federal environmental regulation.

TABLE 25.5 Social Damages per Metric Ton of Sulfur Dioxide Air Emissions

	Valuation	
Estimate	\$/metric ton SO_2	\$/metric ton SO_2
Zuckerman, 1995	4900	4600
CEC/U.S., 1993	3770	3600
BPA, 1987	1950	2350
EPRI, 1990	1730	2080
Nevada PSC, 1990	1700	1800
Elkins et al., 1985	1470	1760
California, 1989	1200	1300
Rowe, 1995	780	940
Repetto, 1990	700	850
Cifuentes, 1993	790	760
Median of studies	1700	1800

Sources: CEC, 1993; Cifuentes, 1993; OTA, 1994; Ottinger, 1992; Zuckerman, 1995.

They sought values so that a utility planning a new-generation unit could account for the external costs of each fuel and technology as well as their private costs. The values were to be independent of the specific location of a generating plant. Using these values, utilities were told to choose the fuel and technology that minimized the sum of private and social costs. Some (e.g., Nevada) were not based on a specific method for internalizing environmental damages and were the result of a political selection among alternatives to meet public goals (Wiel, 1992). These PUC studies are explicitly a public-political valuation of air pollution emissions. While the estimates are informed by the research literature, they represent an interpretation of the valuations by a political body after holding public hearings. Thus, they provide two southwestern and two northeastern state perspectives on the damage associated with emissions of various pollutants, rooted in DFAs. See Matthews (2000) for further description of the use of externality valuation methods and DFAs in environmental analysis.

The use of these varied studies provides a broad range of estimates of the impact of air emissions on health and ecology as well as a range of valuations of the resulting damage. The range reflects real underlying uncertainties, as demonstrated by the EPA's selection of new national ambient air quality standards for small particles and ozone and the storm of criticism it provoked.

Since our I/O data are for 1992, the body of social damage valuations has been converted into 1992 dollars using the chain-weighted GDP deflator (FRED, 1998). The valuations for all six pollutants are shown in Table 25.6.

The EPA's valuations are even more extreme than our range—over \$100,000 per metric ton of PM and roughly \$0 per metric ton of SO_2. This result is a side effect of the EPA decision to allocate all human mortality effects to only PM and lead.

TABLE 25.6   **Unit Social Damage Estimates from Air Emissions of Environmental Externalities**

Species	Number of Studies	External Costs ($/metric ton of air emissions)			
		Minimum	Median	Mean	Maximum
Carbon monoxide (CO)	2	1	520	520	1050
Nitrogen oxides (NO_x)	9	220	1060	2800	9500
Sulfur dioxide (SO_2)	10	770	1800	2000	4700
Particulate matter (PM)	12	950	2800	4300	16200
Volatile organic compounds (VOC)	5	160	1400	1600	4400
Global warming potential (in CO_2 equivalents)	4	2	14	13	23

Sources: CEC, 1993; Frankhauser, 1994; Cifuentes, 1993; Ottinger, 1992; OTA, 1994; Zuckerman, 1995.

In our judgment, the range and median estimates for sulfur dioxide, particles, nitrogen oxides, and volatile organic compounds (VOCs) are the best estimates currently available.*

25.10   EXTERNALITIES FROM COMMODITIES: U.S. ECONOMY IN 1992

The last column of Table 25.3 is the result of multiplying the EPA estimates of total 1992 emissions of each pollutant by the median estimate of dollar damage from Table 25.6. The estimate of total annual environmental damage from the criteria air pollutants is $180 billion in 1992, or 5 percent of 1992 GDP. This is a surprisingly large social damage, given the 28 years of effort resulting from the 1970 Clean Air Act. The estimated annual expenditures on abatement of these pollutants was $20 to $25 billion in 1990 dollars (EPA 1990). In few cases is the marginal cost of abatement as large as the median valuation, and so it is not surprising that the benefit of abatement is much larger than the expenditures on abatement.

Table 25.7 translates the quantity emissions from some example sectors to dollar losses. Shown in the table are the median environmental damage estimates for a $100,000 increase in output from each sector. For example, the median environmental damage from air pollution resulting from producing $100,000 of electricity (using the generation mix in 1992) is $34,000. For electric services, the estimated median air pollution externality is 34 percent of the

---

*The estimates for carbon monoxide are based on fewer studies and are less reliable; estimates for carbon dioxide implied by some estimates of the cost of meeting the Kyoto Agreement are larger than the values we use.

TABLE 25.7 Estimated External Air Pollution Damage from $100,000 of Production (in dollars) from Eight Sample Commodity Sectors

Commodity	Median Species External Costs						Total External Costs		
	CO	GWP	NO_2	SO_2	PM10	VOC	Minimum	Median	Maximum
Electric services	180	17,600	3,400	12,100	560	80	8,600	33,900	94,700
Aluminum	3,000	4,200	1,300	5,600	510	240	3,500	14,900	43,000
Steel	1,400	8,600	910	2,100	440	250	2,500	13,800	34,000
Petroleum refining	330	7,300	1,000	1,400	270	420	2,000	10,700	28,000
Building construction	190	900	300	360	1,100	80	730	2,900	12,100
Automobiles	250	1,400	270	580	100	170	560	2,800	7,900
Aircraft	100	780	140	340	60	70	300	1,500	4,200
Computers	80	640	130	330	50	60	300	1,300	3,700

Note: Totals may not sum due to rounding.

price ($34,000 of $100,000). Producing steel also generates large externalities, 14 percent of the price. In contrast, the direct and indirect external costs from making aircraft and computers are less than 2 percent of the price.

If production was taxed to reflect the environmental damage from emitted air pollution of these six pollutants and no effort was made to abate air pollution emissions, electricity cost would rise by 34 percent and computers would increase 2 percent. Raw materials and intermediate products could be excluded from taxation, resulting in consumer and other final users paying these high taxes. Alternatively, emission taxes could be paid at each stage of production. If emissions per unit of production were unchanged, either method would result in the same tax being paid by consumers. Collecting the tax from consumers would be administratively simpler than a tax at each stage, although the latter would do better at changing information and motivating abatement. However, the incentives to primary and intermediate producers would be less direct.

Benefit–cost studies intended to reflect the full social costs of electricity generation should include the 34 percent external cost amount. A summary of minimum and maximum estimates is also presented in Table 25.7.

Using our method, we have calculated the direct and total social costs of all 500 commodity sectors in the 1992 U.S. I/O table. Table 25.8 shows the 10 commodity sectors in the U.S. economy whose total direct and indirect air pollution emissions have the highest external cost as a percentage of private cost. Also shown in Table 25.8 is the amount of external cost due to production directly in the sector itself, as opposed to the cost for the entire supply chain. This table shows that the carbon black commodity sector causes the highest estimated damage per dollar of output–87 percent. The average external costs for all 500 sectors is also shown.

TABLE 25.8   Ten Commodity Sectors with Highest External Air Pollution Cost Percentages in U.S. Economy, 1992

Commodity Sector	Total (Direct + Indirect) External Cost Percentage	Direct External Cost Percentage
Carbon black	87	82
Electric services (utilities)	34	31
Petroleum and natural gas well drilling	34	31
Petroleum, natural gas, and solid mineral exploration	31	30
Cement, hydraulic	26	19
Lime	22	16
Sand and gravel	20	17
Coal	19	17
Products of petroleum and coal, n.e.c.	18	12
Primary aluminum	15	7
Average over all 500 sectors	4	1

TABLE 25.9 Ten Commodity Sectors with Highest Total External Air Pollution Costs in U.S. Economy, 1992

Commodity Sector	Total External Cost Percentage	1992 Output $(\times 10^9)$	Estimated External Damage $(\times 10^9)$	
			Total	Direct
Electric services (utilities)	34	$171	$58	$53
Petroleum refining	11	$132	$14	$3
Crude petroleum and natural gas	12	$105	$13	$11
Natural gas distribution	15	$77	$11	$6
Trucking and courier services, except air	6	$157	$9	$7
Retail trade, except eating and drinking	2	$523	$9	$2
Wholesale trade	1	$569	$7	$2
Industrial inorganic and organic chemicals	7	$89	$6	$2
Blast furnaces and steel mills	14	$42	$6	$4
Eating and drinking places	2	$281	$6	$0.2
Average over all 500 sectors	4	$22	$0.7	$0.3

Table 25.9 lists the commodity sectors with the largest dollar-valued social damage overall, in contrast to Table 25.8, which focuses on the highest percentage of damages. In Table 25.9, external cost percentages are multiplied by 1992 dollar-valued outputs to determine which sectors cause the most aggregate damage in the economy. The estimated total external damage amounts should not be summed due to double counting: The same external costs may appear for several commodities. For example, some of the total external costs due to electric services come from coal mining. Direct external damages are shown as a final column in Table 25.9.

Electricity services, natural gas distribution, and steel are sectors with relatively high external cost percentages. Electricity leads the list because it also has high output. The other sectors have much smaller percentage externalities but large sectoral outputs. This is particularly true for retail trade and eating and drinking places, where the externality percentages are only 2 percent but the sectoral output is very large. Note that 4 of the top 10 sectors appearing are service related as a result of their large outputs.

Example 25.3: Evaluating the Economic Effect of Restaurants in the Untied States

Problem: What sectors contribute to $100 million of restaurant output? *Hint:* Restaurants are classified by the government as "eating-and-drinking places."

TABLE 25.10 Top Purchases Associated with $100 Million of Demand for Eating-and-Drinking Establishments

Commodity Sector	Economic Effects $(\times 10^6)$
Eating-and-drinking places	$101.89
Wholesale trade	$8.43
Real estate agents, managers, operators, and lessors	$7.22
Advertising	$3.43
Electric services (utilities)	$3.32
Meat animals	$3.21
Meat-packing plants	$3.09
Bread, cake, and related products	$2.84
Trucking and courier services, except air	$2.29
Malt beverages	$2.18
All other sectors	$70.35
Total	$208.24

Solution: Table 25.10 shows a listing of the top direct and indirect purchases for the provision of $100 million of eating-and-drinking establishments (i.e., bars and restaurants) using EIO-LCA. The initial demand of $100 million from the sector actually results in about $102 million of economic activity from the eating-and-drinking sector, as there is some amount of purchases within that sector. However, there is over $106 million of total demand from other sectors. It is generally the demand from these other sectors in the supply chain that leads to environmental damages.

Example 25.4: Real Cost of Restaurants

Problem: Aside from the economic costs from restaurants shown above, what are the external costs from air pollution of providing $100 million of output from restaurants?

Solution: Using EIO-LCA, Table 25.11 shows a detailed listing of the contribution of external costs resulting from direct and indirect air emissions from the production of $100 million worth of eating-and-drinking establishments. The external cost column uses the median valuations from Table 25.6 to value the damages shown in the remaining columns. It is easy to see that direct and indirect purchases of electricity on average result in 12 times more damages than the actual eating-and-drinking service provision.

Congress and the EPA have tended to focus on the industries with the greatest emissions. When an industry has a high ratio of direct emissions costs to output, the EPA and Congress have tended to soften emissions regulations to prevent bankruptcy. We suggest that attention could be focused on the sectors that cause the greatest social damage, especially when that damage is due to suppliers of the sector. For example, eating-and-drinking establishments should

TABLE 25.11   Top Contributing External Cost Sectors Associated with $100 Million of Output from Eating and Drinking Establishments (in millions of dollars.)

Commodity Sector	Median External Costs	CO	GWP	VOC	SO$_2$	PM10	NO$_2$
Electric services (utilities)	1.02	0.00	0.51	0.00	0.39	0.01	0.11
Crude petroleum and natural gas	0.12	0.00	0.11	0.00	0.00	0.00	0.00
Eating and drinking places	0.08	0.01	0.06	0.00	0.01	0.00	0.00
Trucking and courier services, except air	0.08	0.02	0.03	0.01	0.00	0.00	0.02
Natural gas distribution	0.06	0.00	0.05	0.00	0.00	0.00	0.00
Coal	0.06	0.00	0.06	0.00	0.00	0.00	0.00
Paper and paperboard mills	0.04	0.00	0.01	0.00	0.01	0.00	0.00
Other repair and maintenance construction	0.03	0.00	0.00	0.00	0.00	0.03	0.00
Railroads and related services	0.03	0.00	0.01	0.00	0.00	0.00	0.02
Feed grains	0.03	0.01	0.01	0.00	0.00	0.00	0.01
Wholesale trade	0.03	0.01	0.01	0.00	0.00	0.00	0.01
All other sectors	0.47	0.06	0.19	0.04	0.05	0.01	0.02
Total	2.06	0.12	1.05	0.07	0.50	0.07	0.25

Note: Totals may not sum due to rounding.

receive greater attention, particularly in a deregulated era in which they can select their electricity supplier.

The EPA reports that the discharges of sulfur dioxide, particulate matter, nitrogen dioxide, and volatile organic compounds are large, despite decades of regulation. A handful of sectors, such as electricity production, are responsible for the majority of air emissions. For other sectors, their environmental impacts are due to their selection of suppliers among the most polluting sectors. For the most polluting sectors, the key to improving environmental performance is reducing their emissions. For the other sectors, the key to improving their environmental performance is purchasing less of the goods and services of the most polluting sectors.

We caution that our estimates of social cost are uncertain, reflecting the substantial uncertainty in damages and their valuation from air pollutants. These estimates do not include damage from other (toxic) air pollutants or liquid or solid discharges. The estimates show that the social loss from the six air pollutants in 1992 was more than $180 billion (more than 5 percent of GDP). The Council on Environmental Quality (CEQ) reports that annual air pollution control expenditures are on the order of $30 billion. This suggests that the median external costs (totaling almost $200 billion) are significantly higher and there may be room for improvement to bring costs and benefits in line with socially optimal levels. The estimates show that an effluent fee would have increased the price of some commodities substantially.

Another benefit of the use of such valuation is that the EIO-LCA tool becomes more of an assessment mechanism. Investigating only data on the environmental impacts, it is at best a life-cycle inventory tool. With valuation effects, it provides impact information that can be used for assessment purposes.

25.11 CASE STUDY OF COMPARATIVE EIO-LCA EXAMPLE: STEEL VERSUS REINFORCED CONCRETE BRIDGES

Many products and processes can be compared using I/O-based LCA. In this example alternative product designs are compared. There are often important differences in the mix of supply chain purchases necessary to build a product in different ways. This case study uses I/O methods to compare the building of steel and reinforced concrete bridges (see Horvath, 1998).

Background Information on the Bridge Designs The two designs are competitive, with numerous examples of both bridge types in use. The traditional criteria for selecting a particular design have been engineering requirements, initial and life-cycle costs, experience with and availability of a particular material or technology, aesthetics, and ability to erect the structure under local environmental conditions (e.g., climate, topography). In this steel and steel-reinforced concrete comparison case study, it is necessary to consider the economic and environmental implications of the particular bridge material choices and designs. The environmental assessment of the two materials is best performed using LCA.

25.11.1 Design and Construction/Manufacture Economic and Environmental Analysis

Assume for this case study that two alternative designs for a bridge are selected (for details see Spaans, 1997). Using the data for the low bids using steel plate girders and posttensioned concrete girders, Table 25.12 presents the cost figures for the two designs. Assume a concrete deck is designed for both the steel and the concrete girder bridge and the costs of the deck and the cross girders (with the rebar in deck and cross girders) in both cases are comparable. From Table 25.12, the cost of the substructure for the two designs is close, with the concrete bridge being 13 percent less expensive. Thus, only the girders need to be compared since the difference between the two designs is mostly due to the difference in the cost of girder materials. Overall, the concrete alternative has a $1 million lower initial cost (a 37 percent savings).

Use the EIO-LCA method to quantify and assess the economic and environmental effects of the materials extraction, the materials processing, and the manufacturing stages assuming $781,000 worth of steel-reinforced concrete

TABLE 25.12 Low Bids for U.S. 231 over the White
River in Indiana

	Cost
Steel alternative	
Steel girders + bearings	1,756,000 + 93,870
Expansion joints	52,795
Concrete deck and cross girders + rebar	565,628 + 315,748
Substructure	855,039
Bridge, total	3,639,080
Cost per square foot of steel bridge	53.59
Concrete alternative	
Concrete girders + bearings	781,000 + 20,000
Posttensioning + erection	135,000 + 75,000
Expansion joints	34,720
Concrete deck and cross girders + rebar	575,009 + 281,831
Substructure	747,191
Bridge, total	2,649,751
Cost per square foot of concrete bridge	39.02

Source: Spaans, 1997.

girders and $1,756,000 worth of steel girders are purchased (final demand). The two girder materials are represented by sectors in the economic I/O matrix. For the concrete girder, use the sector "concrete products (except block and brick)" [Standard Industrial Classification (SIC) code 3272]. For the steel girder, use the "fabricated structural metal" sector (SIC 3441).

Note: Resource inputs include consumption of electricity, fuels, ores, and fertilizers. Fertilizers are not direct inputs to either steel or concrete manufacturing, but they are part of the long chain of indirect suppliers, that is, upstream suppliers of direct suppliers such as forestry products. Similarly to the resource input requirements, the environmental effects of not only the direct suppliers (such as the cement industry for concrete) but also the indirect suppliers (such as the agricultural sector for concrete) are included in this assessment.

Results of EIO-LCA Model for Bridge Construction Comparison Producing the concrete girders generates an intermediate demand (in input products and services) of $686,000 in the economy (for a total demand of $781,000 + $686,000 = $1,467,000), while producing the steel girders generates an intermediate demand of $2,142,000 (for a total demand of $3,898,000).

Table 25.13 contains a summary of the resource input requirements associated with materials extraction, materials processing, and manufacturing steel

TABLE 25.13   Environmental Effects of Steel-Reinforced Concrete (Concrete + Reinforcing) and Steel Production for an Example Equivalent Bridge Girder Design: Summary of Resource Inputs

Resource Inputs	Unit	Concrete	Steel	Ratio of Concrete to Steel
Electricity	kWh million	0.4	1.4	0.3
Coal and coke				
Anthracite coal	metric ton	0.5	1.4	0.4
Bituminous coal	metric ton	300	800	0.4
Coke	metric ton	8	100	0.1
Total	metric ton	309	901	0.3
Fuels				
Natural gas	metric ton	20	60	0.3
Liquefied natural gas	metric ton	3	5	0.6
Motor gasoline	metric ton	10	20	0.5
Aviation fuel	metric ton	0.1	0.2	0.5
Jet fuel	metric ton	2	7	0.3
Kerosene	metric ton	0.004	0.005	0.8
Light fuel oil	metric ton	30	50	0.6
Heavy fuel oil	metric ton	7	20	0.4
Liquefied petroleum gas	metric ton	6	10	0.6
Total	metric ton	78	172	0.5
Ores				
Iron ore	metric ton	70	900	0.08
Ferrous ore	$	100	700	0.1
Copper ore	metric ton	20	250	0.08
Bauxite	metric ton	2	8	0.3
Gold ore	metric ton	30	250	0.1
Silver ore	metric ton	5	50	0.1
Lead–zinc ore	$	4	300	0.01
Uranium–vanadium ore	$	20	300	0.07
Fertilizers				
Ammonia	metric ton	1	2	0.5
Ammonium nitrate	metric ton	1	0.5	2
Ammonium sulfate	metric ton	0.02	0.04	0.5
Urea	metric ton	0.1	0.1	1
Organic fertilizers	metric ton	0.01	0.01	1
Superphosphate	metric ton	0.3	0.4	0.8
Phosphatic fertilizers	metric ton	0.001	0.001	1
Total	metric ton	2.4	3.0	0.8

and steel-reinforced concrete girders for the highway bridge in our example. With the exception of some fertilizer consumption, all other resource inputs appear to be higher for the steel alternative. Converting fuel usage by type into a common unit, it is roughly 12,000 GJ of energy for the concrete and 33,000 GJ of energy for the steel design.

**TABLE 25.14 Environmental Effects of Steel-Reinforced Concrete
(Concrete + Reinforcing) and Steel Production for Example Equivalent Bridge Girder
Design: Summary of Environmental Outputs**

Environmental outputs	Unit	Concrete	Steel	Ratio of Concrete to Steel
TRI air releases	metric ton	0.1	0.7	0.1
TRI water releases	metric ton	0.02	0.07	0.3
TRI land releases	metric ton	0.06	0.6	0.1
TRI underground releases	metric ton	0.04	0.2	0.2
TRI total releases to the environment	metric ton	0.2	1	0.2
TRI total releases and transfers	metric ton	0.8	7	0.1
CMU-ET for air releases	mt H_2SO_4 equivalent	0.03	0.3	0.1
CMU-ET for water releases	mt H_2SO_4 equivalent	0.01	0.05	0.2
CMU-ET for land releases	mt H_2SO_4 equivalent	0.3	3	0.1
CMU-ET for underground releases	mt H_2SO_4 equivalent	0.02	0.1	0.2
CMU-ET for total releases	mt H_2SO_4 equivalent	0.4	3	0.1
CMU-ET for releases and transfers	mt H_2SO_4 equivalent	2	20	0.1
Ozone depletion potential	mt CFC-11 equivalent	0.002	0.008	0.3
RCRA hazardous waste generated	metric ton	20	70	0.3
RCRA hazardous waste managed	metric ton	20	40	0.5
RCRA hazardous waste shipped	metric ton	3	30	0.1
SO_2	metric ton	10	30	0.3
NO_x	metric ton	4	10	0.4
Methane	metric ton	0.03	0.05	0.6
Volatile organic compounds	metric ton	0.4	0.8	0.5

Note: Numbers may not sum due to rounding. mt = metric ton.

Table 25.14 contains a summary of the outputs associated with the materials extraction, materials processing, and manufacturing stages of steel and concrete bridge girders. Three major groups of environmental impacts are quantified in this assessment: the EPA's Toxics Release Inventory (TRI) chemical emissions, hazardous waste generation, and conventional air pollutant emissions. For the TRI discharges, both the amounts of TRI emissions as reported by facilities to the EPA, without regard to the relative toxicity of the emissions, and the amounts of TRI emissions weighted by relative toxicity using the Carnegie Mellon University-Equivalent Toxicity (CMU-ET) method are assessed. The emissions are higher for the steel girders.

25.11.2 Maintenance Operations during Bridge Lifetime

Maintenance operations during the lifetime of a bridge have environmental consequences. Unfortunately, maintenance costs are difficult to obtain. In this assessment only the painting of the steel structure (perhaps the most important maintenance need for a steel bridge) is considered (Rainer, 1990). Assume that 6,040 m^2 (65,000 ft^2) of steel girder surface needs to be coated for the steel design alternative in the current example and only one coating is required. (Note that the initial painting of the girders and the painting of railings are not assessed.) The R. S. Means (1987) catalog lists a unit material price of $0.10 per ft^2, or $1.08 per m^2 for bridge repainting. Assume that bridge paint is produced by the paints and allied products sector (SIC 285) and it is necessary to purchase $6500 worth of paint for a single job. A maintenance operation such as bridge repainting is repeated several times during the lifetime of a bridge. Rainer (1990) estimates that a typical preventive maintenance plan for a large steel bridge requires a repaint every 8 years. Assume this is the case for ther steel girder bridge in this case study. Assuming the steel bridge will last up to 80 years, it will be repainted 10 times, including the first painting.

It is difficult to assess the relative magnitude of inputs and outputs for the paint without comparing them to another example. Assume the following comparison is valid for this example. Compare the paint figures to the inputs and outputs of steel bridge girders: Table 25.15 compares the resource inputs, and Table 25.16 compares the environmental outputs of the manufacture of the steel girders and the paint for eight repaint jobs for the steel bridge. Note that, as expected, with the exception of conventional air pollutants, all resource requirements and environmental outputs are at least several times higher for the girder production. Of course, if the bridge were repainted more often and in more layers, the differences in the numbers would be smaller. However, Table 25.16 shows a surprising result: Sulfur dioxide, oxides of nitrogen, methane, and volatile organic compound emissions are significantly higher for the paint manufacturing than for the production of all girders for the example highway bridge. Hence the environmental effects of the use phase of products can be very important in LCA. The additional environmental burdens of paint reinforce the conclusion that concrete girders are likely a better environmental choice in this example than steel girders.

25.11.3 End-of-Life Options

Steel Girders Steel bridge girders last a long time. Some steel bridges constructed in the last century still survive with regular maintenance and repair.

TABLE 25.15 Environmental Effects of Paint (for Eight Repaint Jobs) and Steel Manufacturing for a Typical Highway Bridge: Summary of Resource Inputs

Resource inputs	Unit	Paint	Steel	Ratio of Paint to Steel
Electricity	kWh million	0.04	1.4	0.03
Coal and coke				
Anthracite coal	metric ton	0.02	1.4	0.01
Bituminous coal	metric ton	8	800	0.01
Coke	metric ton	0.4	100	0.004
Total	metric ton	8	901	0.009
Fuels				
Natural gas	metric ton	3	60	0.05
Liquefied natural gas	metric ton	0.3	5	0.06
Motor gasoline	metric ton	0.3	20	0.02
Aviation fuel	metric ton	0.004	0.2	0.02
Jet fuel	metric ton	0.2	7	0.03
Kerosene	metric ton	0.0002	0.005	0.04
Light fuel oil	metric ton	2	50	0.04
Heavy fuel oil	metric ton	0.5	20	0.03
Liquefied petroleum gas	metric ton	0.3	10	0.03
Total	metric ton	7	172	0.04
Ores				
Iron ore	metric ton	2	900	0.002
Ferrous ore	$	20	700	0.03
Copper ore	metric ton	3	250	0.01
Bauxite	metric ton	3	8	0.4
Gold ore	metric ton	2	250	0.008
Silver ore	metric ton	0.4	50	0.008
Lead–zinc ore	$	20	300	0.07
Uranium–vanadium ore	$	2	300	0.007
Fertilizers				
Ammonia	metric ton	0.4	2	0.2
Ammonium nitrate	metric ton	0.02	0.5	0.04
Ammonium sulfate	metric ton	0.006	0.04	0.2
Urea	metric ton	0.02	0.1	0.2
Organic fertilizers	metric ton	0.002	0.01	0.2
Superphosphate	metric ton	0.06	0.40	0.2
Phosphatic fertilizers	metric ton	0.0001	0.001	0.1
Total	metric ton	0.5	3	0.2

The decommissioning of steel bridges is often the practice, not because the girders reach the end of their structural life, but because of functional obsolescence: Traffic volumes, loads, or patterns require a wider, stronger, larger, and/ or longer bridge (Lemer, 1996). Often the major traffic routes move away from the bridge. Especially in remote areas or where historic preservation efforts have saved them, many steel bridges have not been deconstructed, but left

TABLE 25.16 Environmental Effects of Paint (for Eight Repaint Jobs) and Steel Manufacturing for Typical Highway Bridge: Summary of Environmental Outputs

Environmental Outputs	Unit	Paint	Steel	Ratio of Paint to Steel
TRI air releases	metric ton	0.06	0.7	0.09
TRI water releases	metric ton	0.008	0.07	0.1
TRI land releases	metric ton	0.008	0.6	0.01
TRI underground releases	metric ton	0.05	0.2	0.3
TRI total releases to the environment	metric ton	0.2	1	0.2
TRI total releases and transfers	metric ton	0.4	7	0.06
CMU-ET for air releases	mt H_2SO_4 equivalent	0.008	0.3	0.03
CMU-ET for water releases	mt H_2SO_4 equivalent	0.003	0.05	0.06
CMU-ET for land releases	mt H_2SO_4 equivalent	0.06	3	0.02
CMU-ET for underground releases	mt H_2SO_4 equivalent	0.03	0.1	0.3
CMU-ET for total releases	mt H_2SO_4 equivalent	0.08	3	0.03
CMU-ET for releases and transfers	mt H_2SO_4 equivalent	0.2	20	0.01
Ozone depletion potential	mt CFC-11 equivalent	0.002	0.008	0.3
RCRA hazardous waste generated	metric ton	8	70	0.1
RCRA hazardous waste managed	metric ton	3	40	0.08
RCRA hazardous waste shipped	metric ton	0.8	30	0.03
SO_2	metric ton	200	30	67
NO_x	metric ton	80	10	8
Methane	metric ton	0.8	0.05	16
Volatile organic compounds	metric ton	20	0.8	25

Note: mt = metric tons.

in place, closed for traffic. In some instances, with smaller bridges, the superstructure has been reused at another location where the old bridge structure was sufficient for the local traffic. This represents a beneficial reuse of steel girders. Since they can be a feedstock for new steel production, it is presumed that, if not left in place or reused, obsolete steel bridge girders are recycled. Comprehensive, national data on steel bridge girder recycling in the United

States are unavailable. Therefore, it is difficult to determine how much steel plate is used in bridge applications on a yearly basis.

Due to the lack of comprehensive data on national steel statistics, data from a joint Federal Highway Administration (FHWA)–EPA study (FHWA–EPA, 1993) is used to estimate the recycling rates for steel girders. This report contains results from a survey of the recycling practices of 29 state highway agencies in the United States conducted at the end of 1992. Data were collected on bridge superstructures, that is, on beams and decks, not solely on steel structural members. Therefore, the results could be skewed if bridge decking reuse or recycling rates had been different from those for the girders. The results of the survey indicate that reuse of obsolete bridge steel superstructures has been practiced by 20 out of 29 states. Rates of reuse range from 1 percent in Maryland and Wyoming to 100 percent in Vermont, with 13 states between 1 and 20 percent and 6 states above 40 percent. However, there is an overlap between reuse and recycling in this report. Recycling of steel superstructures in this context meant cutting, breaking, or modifying the steel superstructure for use in a different highway application or reusing or storing it for subsequent use after straightening, painting, or minor repair. Nine states reported recycling rates between 5 percent (in Arizona) and 100 percent (in Utah), with 7 states achieving more than a 50% recycling rate. Disposal options included sales as scrap, landfilling, giving the superstructure to a contractor or to others, and disposing of unusable or unsuitable items. Therefore, even though the disposal rates were the highest of all three end-of-initial-life options for 21 states out of the 29, it did not automatically mean landfilling. Eleven states explicitly noted that the obsolete steel became property of the contractor who dismantled it, who in turn could have reused, recycled, or eventually landfilled it. One state (Connecticut) noted that only unusable items were landfilled. Three states sold the obsolete steel as scrap. Therefore, the reuse and recycle rates reported by the FHWA–EPA (1993) (17 and 21 percent, respectively) are actually *minimum* rates. For example, Virginia reported that 100 percent of the dismantled steel superstructure was "disposed of" but noted that it was sold as scrap—therefore recycled. For the same state, the recycling rate was reported as zero, when in effect it was 100 percent. Data for five Canadian provinces are also provided in the report, exhibiting a similar trend to the U.S. states.

Concrete Girders Similar to the steel design alternative, the best source of statistics on concrete girder recycling is the joint FHWA–EPA (1993) study. However, the end-of-life options for concrete beams seem to be much more limited than for steel. Of the 27 states responding to the survey in this study, only 6 reported any reuse of old concrete girders, with the rates ranging from 10 to 50 percent. Recycling has only been reported by 4 states, at 10, 50, 70, and 100 percent rates, respectively. Consequently, a large majority of the states reported a 100 percent disposal rate, with the option of the old concrete beams going into landfills or being given to contractors. The contractors may have reused or recycled the old concrete beams, thus raising their reuse and recycling rates, or they may have eventually landfilled them, for lack of a better use. Therefore, as with the steel alternative, the reported reuse and recycling rates may be underestimated. Upon examining the numbers for five Canadian provinces, it is noted that no reuse and only one province's recycling are observed. Of course, data and survey quality issues might make these numbers unreliable or unrepresentative, as with the steel girder option. Unrepresentativeness is possible given the large differences in the reported practices: two states observed a 50 percent reuse rate and one state reported a 100 percent recycling rate when many other states might have landfilled their old concrete beams entirely.

25.11.4 Discussion of Comparative Bridge LCA Example

This case study reports environmental effects associated with the materials extraction, materials processing, and manufacturing stages of steel-reinforced concrete and steel bridge girders for equivalent designs. The concrete design appears to have lower environmental effects overall. Of course, the results might be different for another design as every bridge is unique in its design and material content.

The bridge girders are compared based on summary environmental effects. However, there might be a difference in the expected design life of the two materials. For a more realistic comparison, it is necessary to take into account longevity and annualize environmental effects. However, comprehensive statistics on the expected life of steel and concrete bridges are hard to find. One source (Veshosky and Nickerson, 1993) estimates the life of bridges in Belgium, Japan, Sweden, and Switzerland at 47 to 76 years for steel and 47 to 86 years for reinforced concrete (prestressed concrete bridges last 21 to 86 years); therefore they are comparable. Steel bridges have been constructed for a longer time than concrete bridges. As a result of regular maintenance and extensive repair and modifications, some parts of steel bridges can survive for 100 years or more. The first prestressed concrete bridge was not finished until 1951, and it was the building of the interstate highway system that brought about the prevalence of concrete highway bridges in the United States. Of course, factors other than time may also influence the useful life of bridges: flood, fire, wind, foundation scour, war, and collision. Most importantly, with ever-increasing traffic and changing societal demands, functional obsolescence, not time, might render any type of bridge obsolete long before it fails structurally (Lemer, 1996).

What quantity of resources are embedded in the steel bridges and for what percentage of the national emissions does the manufacturing of steel bridge girders account? Of the currently 580,000 bridges in the United States, it is not known exactly how many have steel versus concrete superstructures. Assume that half of the bridges (290,000) are steel and on average they require about $1.8 million worth of steel girders. Table 25.17 summarizes the percentages of the U.S. national totals of resource inputs into manufacturing steel girders for

TABLE 25.17 Resource Inputs for Manufacturing Steel Girders for 290,000 "Average" Bridges as Percentage of Annual National Totals (U.S.)

Resource Inputs	For 290,000 bridges	Percent of Annual National Total
Electricity, kWh million	406,000	17
Coal and coke, mt		
Anthracite coal	406,000	9
Bituminous coal	232,000,000	39
Coke	29,000,000	239
Natural gas	17,400,000	21
Liquefied natural gas	1,450,000	7
Motor gasoline	5,800,000	2
Aviation fuel	58,000	2
Jet fuel	2,030,000	4
Kerosene	1,450	0.06
Light fuel oil	14,500,000	12
Heavy fuel oil	5,800,000	13
Liquefied petroleum gas	2,900,000	10
Ores		
Iron ore, mt	261,000,000	548
Ferrous ore, $	203,000,000	92
Copper ore, mt	72,500,000	36
Bauxite, mt	2,320,000	287
Gold ore, mt	72,500,000	81
Silver ore, mt	14,500,000	119
Lead–zinc ore, $	87,000,000	33
Uranium–vanadium ore, $	87,000,000	35
Fertilizers, mt		
Ammonia	580,000	6
Ammonium nitrate	145,000	4
Ammonium sulfate	11,600	0.6
Urea	29,000	0.6
Organic fertilizers	2,900	0.2
Superphosphate	116,000	2
Phosphatic fertilizers	290	0.2

Note: Resource inputs and corresponding national totals are based on 1987 data. mt = metric tons.

290,000 "average" bridges. Table 25.18 shows the percentages for environmental outputs. The largest percentage belongs to iron ore. If half of the U.S. bridges were steel, the iron ore consumption would amount to more than 5 years aggregate national demand. Similarly, bauxite consumption would amount to 3 years, and coke consumption would amount to 2 years U.S. demand. Building 290,000 new steel bridges in the United States would raise considerably the environmental outputs as well. Steel bridges in the United States have not been built in 1 year, but over many decades; therefore, the annual input and output totals attributed to bridges are small. However, the

TABLE 25.18   Environmental Outputs for Manufacturing Steel Girders for 290,000 "Average" Bridges as Percentage of Annual National Totals (U.S.)

Environmental Outputs	For 290,000 bridges	Percent of Annual National Total
TRI air releases, mt	203,000	32
TRI water releases, mt	20,300	22
TRI land releases, mt	174,000	140
TRI underground releases, mt	58,000	43
TRI releases to environment, mt	290,000	30
TRI total releases and transfers, mt	2,030,000	83
CMU-ET for air releases, mt H_2SO_4 equiv.	87,000	64
CMU-ET for water releases, mt H_2SO_4 equiv.	14,500	70
CMU-ET for land releases, mt H_2SO_4 equiv.	870,000	109
CMU-ET for underground releases, mt H_2SO_4 equiv.	29,000	70
CMU-ET for total releases, mt H_2SO_4 equiv.	870,000	79
CMU-ET for releases and transfers, mt H_2SO_4 equiv.	5,800,000	66
Ozone depletion potential, mt CFC-11 equiv.	2,320	14
RCRA hazardous waste generated, mt	20,300,000	9
RCRA hazardous waste managed, mt	11,600,000	5
RCRA hazardous waste shipped, mt	8,700,000	55
SO_2, mt	8,700,000	40
NO_x, mt	2,900,000	35
Methane, mt	14,500	40
VOC, mt	232,000	50

Note: Environmental outputs and corresponding national totals are based on 1993 data. mt = metric tons.

comparison to annual national totals is insightful, as we do not regularly think of bridges as sinks of renewable and nonrenewable resources and direct and indirect causes of pollution.

25.11.5   Uncertainties in the Analysis

All the data used in this study are uncertain. For example, concrete bridge girders were estimated by a sector (concrete products) that includes other products, not just girders, with perhaps different environmental implications. Furthermore, the toxic chemical releases data are obtained from the EPA's TRI, collected from manufacturing plants in the United States. Facilities have to report to the TRI, but they do not have to measure their emissions, but rather to estimate. Similar uncertainties exist regarding the other emissions data sources. Therefore, the results obtained in this LCA must be analyzed in light of the uncertainty in the data.

When looking at the initial construction of equivalent bridge designs, steel-reinforced concrete girders appear to have lower overall environmental effects than steel girders. However, steel girders are reusable and recyclable at the end of their useful life. Steel superstructures have had a documented minimum re-use rate of 17 percent and a minimum recycle rate of 21 percent based on a limited U.S. national survey (FHWA–EPA, 1993). For concrete girders, the options have mostly been limited to landfilling in the past. The reuse and recycle rates for steel girders save input resources and presumably reduce environmental pollution compared to using only virgin materials. Recycling rates for steel girders are taken into account in EIO–LCA through the current mix of raw materials for steel mills. In 1987, roughly 20 to 30 percent of steel was recycled. However, Malin (1997) reports that steel plates used to fabricate structural members are increasingly manufactured in minimills which use almost exclusively steel scrap in electric arc furnaces to produce steel. If we assume that the resource inputs and the environmental outputs are lower from minimills than from integrated mills with basic oxygen furnaces, then the results in this study are skewed against steel.

A summary of environmental effects assessed in this case study is given in Table 25.19, but many other environmental burdens in this work have not been included due to the lack of data and, as in the case of visual impacts, the lack of an acceptable metric. For example, the following have not been assessed:

• Dust emissions
• Water usage
• Nonhazardous solid-waste generation and disposal
• Generation and disposal of hazardous waste by type
• Environmental effects of landfills
• Noise and vibration
• Visual impacts

TABLE 25.19   Summary of Environmental Impact of Steel versus Steel-Reinforced Concrete Girders for Typical Highway Bridge

	Manufacturing	Use	End-of-Life Option
Steel	likely higher resource input requirements and environmental outputs	paint, other maintenance	3 reuse and recyclability
Concrete	Better	other maintenance	mostly landfilled

If these (and other) environmental effects were included, the assessment might have yielded different conclusions. Also, the data used in this analysis have large uncertainties associated with them, and they reflect past economic and environmental performance. Therefore, a similar assessment using different designs and baseline years may yield different conclusions. If, however, obsolescence is a main problem, it might not matter if one material lasts longer than the other because a bridge might be decommissioned long before it fails structurally. If indeed this might be the case, bridges should be built from the material that has comparably the lowest environmental burdens. In particular applications, however, engineering, aesthetic, or economic criteria might outweigh the environmental factors.

25.12 INCORPORATION OF RISK ANALYSIS INTO LIFE-CYCLE ASSESSMENT

Now that a straightforward LCA example has been shown, the next step is to consider in a larger sense how LCA inventory data might be used more directly as the input into an assessment method. As stated before, life-cycle inventory analysis is only one of the stages of LCA. Up to this time, inventory analysis has been the focus and endpoint of the vast majority of published studies and life-cycle models. In progressing from inventory to impact analysis, the next life-cycle stage, much more valuable information related to public and ecosystem health and impact must be obtained. However, there is much research to be done in this area and important issues to be addressed and methods to be developed. This section contains a simplified example incorporating risk assessment into LCA. The purpose of this example is to stimulate interest and discussion regarding the incorporation of risk assessment into LCA. Risk assessment is the scientific process that determines the nature of the adverse health effect and the likelihood that humans exposed to a substance will suffer adverse health effects. The following outlines the various steps that can be used to progress from life-cycle inventory information to impact. For this example, the analysis is limited to examining air emissions of a single chemical released from a single facility. Actual inventory information would include the releases of a large number of chemicals from all of the life-cycle stages, raw materials extraction through end of life, including recycling and reuse. However, the example provides a basis for illustrating a method.

One of the few efforts to incorporate risk assessment into LCA is by Seip (2000). The study reports selected results of a life-cycle analysis of paper products from eight Norwegian mills in dilution and exposure/response models and resulting in endpoints related to human and ecosystem health. While the analysis is not reported in detail, the following information is given. The dilution in the water and air of the pollutants was modeled with box models assuming the maximum residence time for the pollutants was 24 hours in air and 1 week for the water pollutants. Factors for the box model for the air pollutants were modified with the use of a Gaussian plume model. For the air pollutants (e.g., sulfur oxides, nitrogen oxides, volatile organic compounds, carbon monoxide, and particulate matter), the endpoint impacts are "excess human exposure" above critical limits and "excess humans at risk." These are both recorded as numbers of people. Seip (page 549) reports the following regarding these categories: "The human exposure attribute tries to capture nuisance conditions, asthma symptoms, excess periods of common colds, etc. The damages are transitory. The excess humans at risk attribute is intended to capture the increased risk of mortality by e.g., angina attacks."

25.13 CASE STUDY INCORPORATING HEALTH RISK ASSESSMENT INTO LCA: BENZENE AIR EMISSIONS FROM AN OIL REFINERY*

25.13.1 Facility and Chemical Background Information

The example facility is an oil refinery in Philadelphia, Pennsylvania, operated by Sun Company. The facility is near a population center (approximately 25,000 people live within 1 mile of the refinery). We focus on benzene emissions and the risk of leukemia. Acute exposure to benzene may cause neurological symptoms like drowsiness, headaches, and unconsciousness. Death may result from exposures to very high levels of the chemical. Chronic noncarcinogenic risks of benzene include damage to the central nervous system, skin, bone marrow, eyes, and respiratory system as well as disorders of the blood. Benzene causes both structural and numerical chromosomal aberrations in humans. Benzene is widely produced in the United States (ranking in the top 20 chemicals for production volume, over 12 million pounds were produced in 1990). Benzene is the Sun Company refinery's top ranked chemical with respect to cancer risk.

25.13.2 Characterization of the Source

Consider annual facility emissions (stack and fugitive) as well as potential emissions from an accident at the facility. The annual facility emissions are from the TRI report of the company and data from the Environmental Defense Fund (EDF) (1997). For this example, refer to the annual stack and fugitive releases from the facility as chronic emissions as they are linked to chronic health effects as distinct from the potential of acute effects due to accidental emissions. It is not necessary to include worker exposure for this example. Typically, life-cycle inventories may not include accident-related emissions; however, this information is valuable for risk assessment and so is included. Assume a 70-year lifetime (25,550 days) for the chronic exposure duration for this example, although this overestimates the risk from this particular source

---

*This case study is adapted from Januschkowetz (1999).

TABLE 25.20   Facility Data: SUN Company, Philadelphia Refinery

Production volume	130,000 bbl/day
Total air emissions, 1996	409,162 lb/yr
Chronic air emissions of benzene, 1996	
Fugitive	39,576 lb/yr
Stack	12,043 lb/yr
Accident emissions of benzene	500 lb
Population around the facility	
0–1 mile zone	25,557 people in 10,032 housing units
0–5 mile zone	787,954 people in 347,958 housing units

Source: EPA, 1997f; EDF, 1997.

since the residents of the area will not spend their entire time in their residence. Assume a 1-day exposure duration for the accident acute emissions cases. Tables 25.20 and 25.21 provide applicable data for this example for the Sun Company facility and atmospheric conditions in the surrounding area.

Assume the following for this example.

- The stack emissions of benzene are continuous and constant at the level of the 1996 TRI data. This allows for steady-state analysis.
- The terrain is relatively flat.
- Average wind velocity and average ambient temperature are constant.
- There are no interactions of the benzene with other pollutants or with atmospheric conditions.
- When the benzene contacts the ground, it is reflected, not absorbed.
- The wind speed is constant with respect to time and elevation.

TABLE 25.21   Facility and Atmospheric Information, Sun Company

Input Parameters	Accident Emission	Chronic Emissions	
		Stack	Fugitive
1. Height of stack above ground (m)	10	12.47	5.0
2. Diameter of opening of stack (m)	2.5	2.07	2.0
3. Velocity of gas emitted from stack (m/s)	12	5	5
4. Temperature of gas as it exits stack (°C)	108	108	108
5. Rate at which pollutant is emitted from stack (g/s)	378	0.1732	0.5693
6. Atmospheric stability	Neutral	Neutral	Neutral
7. Wind velocity (m/s)	4.29	4.29	4.29
8. Distances downwind from facility (km)	0.5, 1.0	0.3, 1.0	0.3, 1.0
9. Ambient temperature (°C)	17.7	17.7	17.7

Note: Assumptions in the table are based on average weather conditions in Philadelphia (LCD, 1997). Stack parameters are taken from Chem (1990).

TABLE 25.22   Threshold Values of Noncarcinogenic and Acute Health Risks

Criteria	Value	Source
15 min short-term exposure limit (STEL)	5 ppm	OSHA (1978), permissible exposure limit (PEL)
8-hour time-weighted average (TWA)	1 ppm	OSHA (1978), PEL
Immediate danger of life or health concentration (IDLH)	3000 ppm	National Research Council (1986)

25.13.3   Threshold Values and Dose–Response Relationship

Threshold values are applicable for acute and noncarcinogenic health risks due to exposure to benzene. These threshold values are primarily derived for occupational exposure to benzene. Use the threshold values for benzene in Table 25.22 for this example.

Researchers have developed different mathematical models to describe the relationship between the level of exposure to a carcinogen and the probability of developing cancer associated with that level (White, 1990). A linear dose–response relationship can be used for this relationship [based on the recommendations of a recent EPA (1998d) report]. Figure 25.3 shows this relationship.

Solution   The following sections outline the steps required to translate the inventory data (emissions in kilograms per year) of benzene from the facility to impact/risk associated with the exposed population. For additional details on the steps required for the risk characterization refer to Naugle (1991).

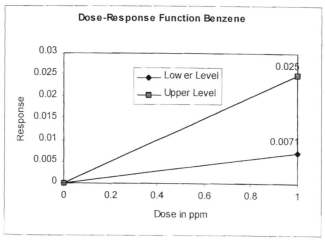

Figure 25.3   Dose–response relationship for human carcinogenic risk of benzene.

25.13.4 Calculation of Benzene Concentration for Exposure Assessment

The inventory data (emissions from the facility) have to be converted to resulting changes in ambient concentrations. The determination of the benzene concentration requires the selection of an appropriate atmospheric model and the application of the model to quantify the pollutant concentrations at various distances from the facility. Ideally a model that considers the emission pattern, meteorology, chemical transformations, and removal processes is necessary for establishing such relationships. There are several models of varying degrees of complexity that enable the expression of a mean concentration of a species emitted from a point source. These include simple atmospheric modeling representations that have analytical solutions such as a box model or the slightly more complex Gaussian plume model as well as sophisticated numerical models. The box model represents a region of the atmosphere as a box that is assumed to have a fixed volume and mass conservation of a pollutant inside the volume. The model assumes uniform mixing within the box. This example employs the more complex Gaussian plume model. This model enables the expression of a mean concentration of a species emitted from a continuous, elevated point source. It is based on the observation that under certain highly idealized conditions the mean concentration has a Gaussian/normal distribution. Dispersion is assumed to be in three directions: downwind, crosswind, and lateral. Although the Gaussian distribution assumption applies only in the case of stationary, homogeneous turbulence, because of its simplicity, it serves as the basis for a large class of atmospheric diffusion formulas in widespread use (Seinfeld, 1998). Seinfeld reports that the justification for these applications is that the dispersion parameters used have been experimentally derived under conditions approximating those of the application.

The mean concentration of benzene downwind of the facility is calculated using the Gaussian plume model. It is necessary to make the simplifying assumptions noted in the example description to account for the limited applicability of the model, uncertainty in actual atmospheric conditions, and chemical release patterns. The inputs to the Gaussian plume model are shown in Tables 25.20–25.22.

The benzene concentrations resulting from the Gaussian plume model are shown in Table 25.23 and in Figure 25.4. The next step is to calculate the lifetime individual risk using these concentrations. The highest benzene concen-

TABLE 25.23 Benzene Concentrations for Accident and Chronic Emissions

		Chronic Emission (ppm)	
Type of Emission	Distance from Facility (km)	Accident (500 lb/accident)	Annual (51,619 lb/year)
Benzene concentration	0.5/0.3	0.97	0.0097
	1	0.46	0.0011

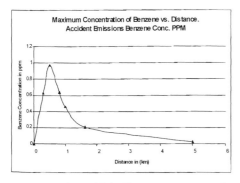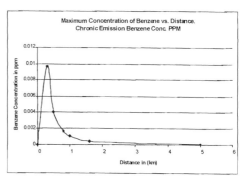

Figure 25.4   Benzene concentration with respect to distance from facility.

tration is at 0.5 km for the accident case and 0.3 km for the chronic case. The benzene concentration decreases after 1 km. These observations result in the specification of the values for parameter 8 in Table 25.21.

Figure 25.5 summarizes the steps used in calculating lifetime pollutant concentration. This range is considered for the minimum and maximum risk values for the calculations. The risk at 1 ppm lifetime exposure ranges from 7.1×10^{-3} to 2.5×10^{-2} (see Figure 25.3). For additional details see Januschkowetz (1999).

25.13.5   Exposure Conditions and Exposed Population

Accident emissions have short time duration, and an exposure duration of 1 day is appropriate in this case. In contrast, chronic emissions are long term. The chronic exposure case is based on the results of the maximum benzene concentrations for the chronic exposure and taking the case of the population within 0.3 km of the facility (maximum chronic benzene concentration) and also within 1.0 km. For the accident exposure, the relevant population is that within 0.5 km of the facility (maximum accident benzene concentration) and at 1.0 km.

25.13.6   Dosimetry Factor

This factor describes the characteristics of contact of the human body with the toxic substance. Since air emissions are being considered, the following factors are important: breathing volume per day, absorption rate, average body weight, average lifetime, and regional surface area of the lung.

25.13.7   Calculation of Lifetime Individual Risk

To derive chronic risk, the lifetime individual risk to the exposed population is calculated from the benzene concentrations and the dose–response relationship.

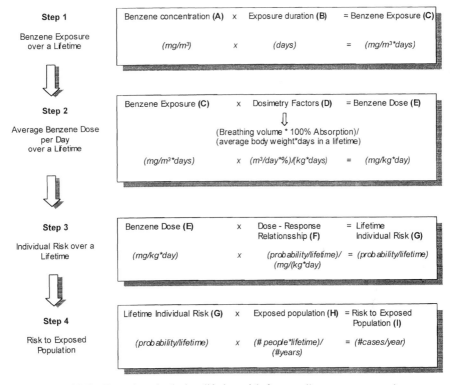

Figure 25.5 Steps in calculating lifetime risk from pollutant concentrations.

The results of the calculation of the accident and chronic lifetime individual risks are shown in Table 25.24. For additional details on the assumptions and calculations refer to Januschkowetz (1999). Interpreting the results from Table 25.24, the accident and chronic lifetime individual risks to the exposed population are small.

TABLE 25.24 Lifetime Individual Risk

	Distance (km)	Emissions (probability/lifetime)	
		Accident	Chronic
Maximum lifetime individual risk (2.5×10^{-2}, 1 ppm)	0.5/0.3	9.51×10^{-7}	2.42×10^{-3}
	1	4.50×10^{-7}	2.72×10^{-5}
Minimum lifetime individual risk (7.1×10^{-3}, 1 ppm)	0.5/0.3	2.07×10^{-7}	6.87×10^{-5}
	1	1.28×10^{-7}	7.72×10^{-6}

25.13.8   Major Uncertainties

This example contains several major sources of uncertainty. These are related to the emissions data, exposure assumptions, dose–response relationship, the model, and its inputs. There are uncertainties in the TRI emissions data associated with the measurements and the reporting processes. Additionally, there are changes in quantities of emissions released over time from facilities. The duration and level of exposure are uncertain. Although benzene has been characterized as a human carcinogen, there is no certain dose–response relationship established. One reason is that only a few human studies are available. Sensitivity analysis is employed to estimate the range of risk. The Gaussian plume model is a representation of the real world and there is much uncertainty related to the model, its inputs, and its outputs. Data were not available on the specific parameters of the facility, and average data from the literature were employed in the example. Additionally average data about the weather conditions in Philadelphia were employed. Finally, when doing the analysis for a particular facility, this model would provide useful information to do an event tree or fault tree analysis to estimate the probability of an accident that would lead to a release of benzene.

25.13.9   Risk Characterization

Incorporating the quantitative results from the example and assessing the methods used and the uncertainties, one is able to characterize the risk. Comparing the concentrations in Table 25.23 with the threshold values (Table 25.22), the health and cancer risks resulting from the exposure of the population to the benzene emissions appear to be extremely small. This population is exposed to many other larger risks in their daily lives (e.g., second-hand smoke). With respect to leukemia risk, there is not a portion of the population that is known to be more susceptible to the risk. Due to the conservative assumptions employed in the analysis, it is concluded that the risks due to the benzene emissions from the facility are small for the exposed population.

25.14   SUMMARY AND CONTRIBUTIONS

Over the past four decades, the environmental implications of our actions received notice and then became a central concern. Analytical tools to quantify the impacts of our environmental discharges have taken longer to develop. A quarter century ago, analysts concluded that it is misleading to focus on discharges into a single medium from a single plant. Rather, improving environmental quality efficiently and effectively requires a life-cycle perspective.

Important progress has been made in life-cycle analysis during the last two decades. A tool that was time consuming and expensive has been transformed into an array of tools that can be low cost and fast. We have presented software that enables the analyst to get a good approximation to the life-cycle environmental discharges, resource use, and energy use for production from each of 485 sectors of the U.S. economy. Current life-cycle analysis tools are able to give a much better characterization of environmental discharges. However, little progress has been made in going from the life-cycle inventory stage to the impact stage. We have presented an illustrative analysis of how to go from environmental discharges to ambient concentrations to health effects. Each of the tools needed to calculate the health impact of these discharges exists, although uncertainty is rampant is actually applying the tools.

Implicitly or explicitly, estimating the desirable level of abatement for each environmental discharge requires more than estimates of the health and environmental effects of each discharge. Policymakers have to weigh the social cost of abating emissions against the social benefit. The cost of abatement is stated in dollar units, rather than units of premature death or disease. To balance the two, a translation is needed of premature death and disease into dollars or of dollars into premature death and disease—two sides of the same coin. Academic researchers and political officials have both attempted the translation, although no one is entirely satisfied with the result. We have presented estimates of the dollar benefit to society of abating each of the pollutants.

Finally, we have provided an alternative to the EPA's traditional command-and-control approach of establishing discharge and ambient concentration standards. The environmental discharges associated with a product or service can be translated into dollar terms and added to the price of the product or service so that the consumer faces the "full social cost" of his or her decisions. This use of market incentives allows each person to make decision as to what is the proper level of discharge abatement.

Despite 30 years of environmental regulation, environmental discharges are still significant. We estimate the remaining air pollution environmental costs to be 5 percent of GDP. Electricity generation has environmental costs that would increase the price by 95 percent if the current costs were internalized.

A great deal of progress has been made in conceptualizing the environmental problems associated with a modern, industrial society. Still greater progress has been made in developing models and tools to enlighten choices of materials, production processes, and consumption.

ACKNOWLEDGMENTS

The authors thank Arpad Horvath of Berkeley, Ruth Reyna-Caamano of Carnegie Mellon, and Antje Januschkowetz of Carnegie Mellon for their contributions of cases and analysis to this chapter. They also thank the support of the Green Design Consortium, the National Science Foundation (Grants EEC-9700568 and DMI 9613405), and the U.S. Environmental Protection Agency (826740-010) for funding support.

APPENDIX

A.1   DESCRIPTION OF INPUT–OUTPUT MODELS

A.1.1   Basis of the EIO-LCA Model

As discussed in the chapter, input–output (I/O) analysis was developed by Leontief in the 1930s. This analysis uses a matrix notation and equations to completely specify the interactions between sectors of the economy when a sector produces output. Specifically, Leontief (1936) proposed an I/O system where the transactions in an economy were represented as a matrix of demands between sectors. When inverted, the model could represent the total output of an economy as a function of the final demands from each of the sectors. Leontief (1970) saw the benefit of linking emissions to the sectoral outputs that result, but his vision included them as additional rows and columns to the model. In essence, "pollution production" became another commodity sector within the economy. Although no pollution was ever "demanded" as in other commodities, its outputs were now part of the model.

One of the side effects of Leontief's method appears to be that this endogenous inclusion of external effects makes the I/O framework more complicated. To solve the model with a range of additional effects required the inversion of a matrix with additional rows and columns. Until even recently, this computation would have been a complicating feature of such a method.

Miller and Blair (1985) proposed that a similar method could exist with external augmentation of the basic I/O model but offered no solution of its equivalence. This external method, economic input–output life cycle assessment (EIO-LCA) has been developed at Carnegie Mellon to specify the expected environmental effects associated with production in the economy (Lave, 1995; Cobas, 1996; Horvath, 1997; Joshi, 1997; Hendrickson, 1998). This model has been made available on the Internet at *http://www.eiolca.net* as a publicly available research tool. Specifically, we link industrial activities to the outputs of conventional pollutants and greenhouse gases. The pollutant emissions considered are carbon monoxide, sulfur dioxide, nitrogen oxides, particulate matter, volatile organic compounds, and global warming potential (in CO_2 mass equivalents).

Figure 25.6 shows the structure of such models, including the standard nomenclature for input-output matrices. In this chapter, matrices and vectors are in boldface type.

Other sectors, representing the GDP categories, are continually added to form more complicated models. Consumption (C), fixed investment (I), government spending (G), and net exports (NX) are examples of these additions. In any I/O model, total commodity output can be represented by the sum across each row's values. A more technical description of an I/O model is shown in Figure 25.7.

Transactions (**A**): Purchases of goods and services between production sectors i: row index j: column index Economically-valued purchases: z_{ij} Technical coefficients: $a_{ij} = z_{ij} / X_j$	Final Demand (**Y**): GDP Components (C+I+G+N X)	Total Commodit y Output (**X**): Output of each sector by summing across each
Value Added: Labor Services, etc.		
Total Industry Output: Column		

Figure 25.6 Layout of advanced input–output models.

Thus for each of the n commodities indexed by i,

$$X_i = z_{i1} + z_{i2} + \cdots + z_{in} + Y_I \tag{A.1}$$

However, I/O models are typically generalized by assuming interindustry flows between sectors can instead be represented as a percentage of sectoral output. This flow is represented by dividing the economically valued flow from sector i to sector j by the total output of sector j:

$$a_{ij} = \frac{z_{ij}}{X_j} \tag{A.2}$$

In such a system, the a_{ij} term is a unitless technical (or I/O) coefficient. For example, if a flow of \$250 of goods goes from sector 3 to sector 4 (z_{34}), and the

	Input to sectors				Intermediate output O	Final demand F	Total output X
Output from sectors	1	2	3	n			
1	X_{11}	X_{12}	X_{13}	X_{1n}	O_1	F_1	X_1
2	X_{21}	X_{22}	X_{23}	X_{2n}	O_2	F_2	X_2
3	X_{31}	X_{32}	X_{33}	X_{3n}	O_3	F_3	X_3
n	X_{n1}	X_{n2}	X_{n3}	X_{nn}	O_n	F_n	X_n
Intermediate input I	I_1	I_2	I_3	I_n			
Value added V	V_1	V_2	V_3	V_n		GDP	
Total input X	X_1	X_2	X_3	X_n			

Figure 25.7 Structure of economic input–output model requirements matrix.

total output of sector 4 (X_4) is \$5000, then $a_{34} = 0.05$. This says that 5 cents worth of inputs from sector 3 is in every dollar's worth of output from sector 4. As a substitution, we can also see from Eq. (A.2) that $z_{ij} = a_{ij}X_j$. This form is more often seen since the system of linear equations corresponding to Eq. (A.1) is typically represented as

$$X_i = a_{i1}X_1 + a_{i2}X_2 + \cdots + a_{in}X_n + Y_i \tag{A.3}$$

It is straightforward to notice that each X_i term on the left has a corresponding term on the right of Eqs. (A.2) and (A.3). Thus all **X** terms are typically moved to the left-hand side of the equation and the whole system written as

$$(1 - a_{11})X_1 - a_{12}X_2 - \cdots - a_{1n}X_n = Y_1$$
$$-a_{21}X_1 + (1 - a_{22})X_2 - \cdots - a_{2n}X_n = Y_2$$
$$\vdots$$
$$-a_{i1}X_1 - a_{i2}X_2 - \cdots + (1 - a_{ii})X_i - \cdots - a_{in}X_n = Y_i \tag{A.4}$$
$$\vdots$$
$$-a_{n1}X_1 - a_{n2}X_2 - \cdots + (1 - a_{nn})X_n = Y_n$$

If we let the matrix **A** contain all of the a_{ij} terms, **X** all the X_i terms, and **Y** the Y_i terms, then system (A.4) can be written more compactly as

$$(\mathbf{I} - \mathbf{A})\mathbf{X} = \mathbf{Y} \tag{A.5}$$

where **I** is the n-by-n identity matrix. This representation takes advantage of the fact that only diagonal entries in the system are $1 - a_{ii}$ terms, and all others are $-a_{ij}$ terms. Finally, we typically want to calculate the total output, **X**, of the economy for various exogenous final demands **Y**, taken as an input to the system. We can take the inverse of **I** − **A** and premultiply it to both sides of Eq. (A.5) to yield the familiar solution

$$\mathbf{X} = (\mathbf{I} - \mathbf{A})^{-1}\mathbf{Y} \tag{A.6}$$

where $(\mathbf{I} - \mathbf{A})^{-1}$ is the Leontief inverse.

A.1.2   Classification of Results in an Input–Output System

Before we proceed, it is important to distinguish between the various results inherent to an I/O model. The results are the total outputs generated from an exogenous final demand. The structure and components of these outputs are worth noting. Due to the intersectoral transactions shown in the matrix **A**, any exogenous final demand will have "ripple" effects throughout the economy expressed in an I/O model. For example, the brief numerical example shown above says that 5 cents of sector 3 product is in sector 4 output. Similarly, probably some amount of sector 4 product is required to make sector 3 output. At face value, this creates a circularity problem in determining results. However, this circularity is completely expressed in the Leontief inverse.

The Leontief inverse can also be expressed as the power-series form

$$\mathbf{X} = \mathbf{IY} + \mathbf{AY} + \mathbf{A}^2\mathbf{Y} + \cdots = [\mathbf{I} + \mathbf{A} + \mathbf{A}^2 + \mathbf{A}^3 + \cdots]\mathbf{Y} \qquad (A.7)$$

The successive terms of the power-series form represent the round-by-round requirements for producing total output $\mathbf{X}$ that are part of the circularity problem.* Equation (A.7) shows that in the first round, $\mathbf{IY}$ is produced (the final demand). In the second round, $\mathbf{AY}$ is needed to produce the $\mathbf{IY}$ of final demand. In the third round, $\mathbf{A}^2\mathbf{Y}$ is needed to produce the $\mathbf{AY}$ from the second round. And so on. The sum of these round-by-round effects, including the final demand, is $\mathbf{X}$.

In this way, the I/O framework shows the "total supply chain effects" of producing goods and services in an economy. We can separate, however, these effects between the "direct" economic effects (i.e., the effects related to the pure production of the final demands) and the "indirect" economic effects, the effects related to producing all goods and services needed to produce the final demand. In our matrix terminology (consistent throughout this chapter), the direct effects are the $[\mathbf{I} + \mathbf{A}]\mathbf{Y}$ term in Eq. (A.7), and the indirect effects are the $[\mathbf{A}^2 + \mathbf{A}^3 + \cdots]\mathbf{Y}$ terms. In all, we call the sum of the direct and indirect effects ($\mathbf{X}$) the "total" economic supply chain effect of production. The magnitude of the direct, indirect, and total effects is completely dependent on the values of the $\mathbf{A}$ matrix. Thus, it is possible that the indirect effects will be larger than the direct effects. When we consider the environmental effects of this production, this same result holds. Since our ultimate goal here is to create a framework for analyzing relative environmental effects of production, the concepts of direct and indirect effects are central.

Input–output tables (providing the matrix $\mathbf{A}$ above) are available for many industrialized countries. For example, the United States, Germany, and Japan all make such tables available.

A.1.3   Incorporation of Environmental Effects into Input–Output Models

Leontief (1970) first suggested the use of I/O models to perform environmental analysis, and it is from this point that two separate methods of incorporating environmental impacts can be generated. The first method is Leontief's pub-

---

*For a more detailed proof and explanation of the power-series form, see Miller and Blair (1985). Due to the computational complexity of generating large inverses, this alternate form is often used since successive terms can be calculated by multiplying the previous term by $\mathbf{A}$.

lished suggestion of endogenizing environmental effects by augmenting the matrix **A** with pollution product rows and columns. An example of this is CO_2 pollution from each sector. The suggestion of endogenizing environmental effects within the I/O model was somewhat problematic at the time. Computational power and capacity were quite expensive, so adding even a single row and column for a particular emission created a problem in solving a larger system of equations.

This is part of the rationale behind the second method, externally augmented I/O models. Externally augmented models keep the consideration of environmental effects external to the I/O model but use the Leontief inverse and output calculations to generate these results. The use of such alternative methods is not itself novel. Miller and Blair (1985) discuss that external methods, if done correctly, would yield computationally equal results, but they offer no proof of their equivalence. See Matthews (1999) for a proof on the equivalency on the two methods.

A.2 DETAILED DESCRIPTION OF THE CONSTRUCTION OF CONVENTIONAL POLLUTANTS DATA

A.2.1 Introduction

Given the method described above, many different vectors of environmental information can be appended to an I/O table to create meaningful analytic results. For example, as shown in Example 25.2, data on releases of conventional pollutants and global warming gases have been appended to the I/O model and can be accessed to estimate the releases generated by each sector needed for production.

In the United States, which is the focus of the EIO-LCA model, the government makes available a significant amount of data at the industry sector level which can be used in the model. Data at the facility level is sometimes limited. Following is a short summary of the effects vectors that have been incorporated into the EIO-LCA model.

A.2.2 Description of Data Sources in the EIO-LCA Model

- Economic purchases from other sectors and of fuels, ores, and fertilizers come from the U.S. Department of Commerce I/O tables and associated work files (DOC, 1997).
- Physical units (kilowatt-hours) of electricity consumption for manufacturing (DOC, 1994a) and mining (DOC, 1994b). Dollar purchases of electricity for all other sectors (DOC, 1997) converted to physical units by average sectoral prices [Department of Energy (DOE), 1996].
- Total energy use (in terrajoules of fuel used) is estimated by taking the purchases of fuels (above) and converting based on the formula

[Fuel consumption (kg) × heat value (MJ/kg)]

+ [30.7% × electricity consumption (kWh) × 3.6 (MJ/kWh)]

Heat values for fuels are taken from Boustead (1979) and Culp (1979). Note that in 1992, 30.7 percent of electricity generation came from hydroelectric, nuclear, wind, geothermal, solar power, and power from biomass (DOE, 1998). This amount of energy was added to energy use from fossil fuel sources to create a more accurate estimate of total energy use.

- Estimates of releases of toxic substances from 1995 TRI releases (EPA, 1997c). These estimates were additionally augmented for mercury releases from electric utilities (which were not reported for 1995) via EPA (1997d).
- Emissions of weighted toxic releases achieved by the CMU-ET method, which converts releases of toxic materials to equivalent releases of sulfuric acid. This is done by using a comparative metric of threshold limit values, (TLVs) from ACGIH (1997). For more information on the CMU-ET method, see Horvath (1995).
- Conventional pollutant releases from the EPA (1998b) Aerometric Information Retrieval System (AIRS). A discussion of the specific adjustments made to create the sectoral estimates of conventional pollutant emissions is given below.
- Similar to mercury emissions above, utility lead emissions incorporated from EPA (1997e).
- Usage and discharge of water for production from DOC (1996).
- Emissions of greenhouse gases estimated by AP-42 emissions factors (EPA, 1995) multiplied by estimate of fuel purchases (referenced above). Global warming potential factors are taken from the Adriaanse (1993) study.
- Generation, management, and shipments of RCRA Subtitle C Hazardous Waste (EPA, 1998c).

A.2.3 Further Detail on Estimation of Conventional Pollutant Emissions

The determination of pollution generated requires a significant amount of data disaggregated by commodity sector. We use a three-tiered set of data to determine the conventional pollutant and greenhouse gas emissions from each of the 485 sectors. The combined sources are:

- Data from the EPA AIRS
- Application of EPA AP-42 emissions factors to sectoral fuel consumption data
- EPA National Air Pollutant Emissions Trends (1900–1996) Report (EPA, 1996)

We first estimate emissions using fuel consumption data from the DOC and EPA AP-42 emission factors. These data derive the expected amounts of each emission resulting from the combustion of fuels given particular technologies in each sector (EPA, 1995). Note that this method is solely used to estimate greenhouse gas emissions.

The second source takes data from the EPA (1998b) AIRS. This web-based system reports emissions data from thousands of facilities that meet threshold reporting requirements for each of the conventional pollutants and provides sufficient detail [including Standard Industrial Classification, (SIC) code] to use them with the I/O model. A mapping from SIC code to I/O sector is provided with the I/O data. Note that this data system, although highly accurate in the data it does contain, represents an underestimation of total emissions of each pollutant since facilities that do not meet the reporting threshold face no reporting requirements. Thus, the fuel consumption data and AIRS data are compared to get a more realistic estimate of emissions.

Finally, data from the EPA (1996) emissions trends report are used as a baseline to check that pollutant data are fairly represented. In general, these data are only needed where sectors lack data in the other two methods. For example, construction and transportation sectors are problematic because they contain many mobile sources and only have fuel consumption data. The emissions trends report provides estimates of emissions from these sectors. The process of combining these sources is described below.

A.2.4   EPA AIRS Database

The EPA's AIRS database is the most extensive air pollution database in the world, containing billions of data points on emissions sources. The system is administered by the EPA Office of Air Quality Planning and Standards (OAQPS), Information Transfer and Program Integration Division (ITPID). The OAQPS establishes national ambient air quality standards for pollutants that are proven detriments to public health, known as criteria pollutants. These criteria air pollutants are carbon monoxide, nitrogen dioxide, sulfur dioxide, particulate matter, lead particulates, and ozone. All but ozone are specifically released by human activity. Volatile organic compounds are a precursor to ozone and are tracked along with the other criteria pollutants as a proxy to actual ozone formation in the atmosphere.

To track air quality standards, one of the requirements of the Clean Air Act is that states maintain air-monitoring stations to assist in the verification of regional ambient air pollution levels. OAQPS in conjunction with state environmental agencies use AIRS to monitor the states' progress in achieving national standards. AIRS has been set up as the sole repository of such information so that states have access to it as their own database of emissions and monitoring data.

Aside from monitoring station data, states also provide an inventory of existing stationary air pollution sources, including an estimate of the quantity of

**TABLE 25.25   AIRS Reporting Threshold
Requirements for Facilities: Criteria Pollutant Emission
(short tons/year)**

Carbon monoxide gas	1000
Nitrogen dioxide gas	100
Sulfur dioxide gas	100
Volatile organic compounds	100
Particulate matter (total)	100
Particulate matter ($<$10)	100
Lead particles	5

Note: VOCs are not criteria pollutants, but they are precursors
of criteria pollutant ozone (smog).

emissions from each source. The source reporting and monitoring data are used
together to determine expected effects from changes in regional emissions. The
AIRS database contains all such information. We have used a part of AIRS
(AIRSData) that is available on the Internet that includes only the reports of
emissions from individual facilities (EPA, 1998b).

To fall under the reporting requirements for the emissions inventories, facilities must first be designated as large sources. The emissions sources data
available in AIRSData give detailed emissions records for each large source in
the United States. Table 25.25 lists criteria pollutants and the threshold
amounts necessary for designation as large sources. As of 1998, the reporting
requirements were are as follows: There are 30,000 total recognized stationary
sources in the United States in 1998. Of these, 8745 facilities report to AIRS.
Examples of the facilities (stationary sources) that report to AIRS are electric
utilities, steel mills, and universities. The AIRS data represents a majority of
the EPA estimated emissions resulting from the provision of goods and services
in the economy. We note that AIRS does not contain any data for mobile or
area sources (e.g., cars and trucks, residences, and dirt roads).

For each reporting source, AIRS includes both a header record of geographical and demographic information, as in the following example:

Facility Name	State	County	City	Year	Industry (SIC)	Facility ID	Region
Bethlehem Steel plant	PA	North-ampton	Bethle-hem	1992	3312: blast furnaces and steel mills	42095-0048	03

Each facility also has an associated pollutant emissions record pertaining to
releases of each of the criteria pollutants:

CO	NO_2	Pb	PM10	PT	SO_2	VOC
36,954	8485		3068	2338	4896	730

Most of the fields in the records have a straightforward meaning and purpose. Following are some notes and specifications of the data available for each facility.

- "Year" notes the last time emissions data were updated for each facility. To reduce paperwork requirements, a facility must only update its reported quantities if they change significantly.
- "SIC" codes and descriptions refer to the Office of Management and Budget's classification system of industries.
- "Facility ID" has the following parts: state code (two digits), county code (three digits), and AIRS facility code (four characters).
- "Region" refers to the EPA region number in which a source is located. There are 10 EPA regions, split geographically across the United States.
- Estimates for each facility are based on the normal operating schedule of a source and include the effects of installed pollution control equipment and regulatory restrictions on operating conditions. The data are thus meant to represent the expected emissions from the source per year.
- Emissions are in short tons (equal to 2000 pounds). When needed, conversions were made assuming 1 ton = 0.909 metric tons

For analytical purposes, all data were aggregated into total emissions for each of the SIC codes and converted from short tons to metric tons. For use within our I/O model, we have mapped SIC codes into their respective detailed I/O sector numbers using the mapping guidance that accompanies the I/O tables (also available at *http://www.eiolca.net/sectors.html*).

A.2.5   AP-42 Emission Factors

Conventional pollutant emissions for each commodity sector were alternatively found based on fuel consumption data (DOC, 1997). The fuel consumption data are part of the 1992 Census of Manufactures data of the U.S. economy and shows the purchases of fuels such as coal, gasoline, and jet fuel on a sector-by-sector basis.

We assumed that purchased fuels are burned for energy purposes and use EPA AP-42 emission factors to assess the expected releases of conventional pollutants and greenhouse gases resulting from combustion (EPA, 1995). Emissions factors are relatively specific; there are many distinct technologies considered for a given sector, and the emission factors provide a fairly good estimate of the emissions that would result from the use of a given amount of fuel. These data are again considered at the SIC level and converted to sectors in our I/O model. The AP-42 results come from Joshi (1997).

One point of note is necessary pertaining to the AP-42 results. The fuel consumption data do not explicitly note the purchases of diesel fuel (or make reference to whether these purchases might be contained within more broadly classified fuels like light and heavy fuel oils). Thus, this method seriously underestimates the emissions of CO and NO_2 that would be expected in certain diesel-intensive sectors (e.g., truck transportation). For these sectors, efforts were made to estimate emissions as described below.

A.2.6 EPA Emission Trends Report

Finally, we consider estimates of sectoral releases of pollutants from the EPA (1996) Emission Trends Report. This report is the result of a substantial effort that includes the submodeling of many industries and activities. Data from AIRS are one of these subcomponents, and a description of the others is beyond the scope of this chapter. However, there is a large amount of proprietary data and detail included in the models that are unavailable publicly or for academic research purposes. In addition, the trends report gives only aggregate estimates, as opposed to the 485 sectors of aggregation needed. Thus, the data from this report are used only to validate or substitute for the other two sources. This was done by comparing summaries of sectoral releases from the emissions trends report with the AP-42 and AIRS data already found. Generally, the only sectors for which emission trends data were needed were sectors with large mobile source emissions, like transportation and construction. Thus, emission trends data were incorporated into the model as an "adjustment" as needed. Table 25.26 summarizes the rationale and method behind all adjustments made using emission trends data. Overall, only 124 data points were adjusted out of the 3395 possible, or about 3 percent.

A.2.7 Combination of Data Sources and Adjustments Made

Using the three data sources, we have constructed an extensive linkage between the 485 commodity sectors and the conventional pollutants released from their activity. Our final data set used a relatively straightforward method to determine its values.

1. If AIRS data existed, AIRS data were used. If no AIRS data were available, the AP-42 data were used.
2. For sectors where the emission trends report value was grossly different than either of the other two values (or where both of these results seemed implausible), the value from the emission trends report was used.

The second point notes the need to replace some of the data due to the diesel issue mentioned above as well as the general problem of AIRS not containing emissions from mobile sources. Table 25.26 details such replacements.

TABLE 25.26 Summary of Adjustments Made Based on Emissions Trends Report Data

Industry	Emission	Rationale and Method
Livestock and agricultural products	CO	Emissions trends (ET) report suggested double the amount of CO emissions from I/O sectors 10100–20600. This is most likely a result of the diesel data issue discussed in text (since the mix of gasoline and diesel vehicles in the industry is roughly 50/50). The distribution of emissions was assumed to hold, and thus doubling each sector's value would incorporate diesel effects.
	NO_2	Due to diesel problem, no NO_2 emissions reported for sectors 10100–20600 (diesel is the primary fuel combustion source of such emissions). Due to predominance of mobile sources, AIRS reported values significantly lower than ET report. Using the percentage distribution of CO emissions across this set of sectors, 1,088,400 metric tons (cited 1,200,000 short tons) of NO_2 from ET were allocated.
Mining	CO	Overruled algorithmic selection of data source for sectors 60200, 80001, 90001. Recall that when both AP-42 and AIRS data are present, AIRS data are selected by default. Inspection of ET data suggested that AIRS was severely underestimating emissions in these sectors; thus AP-42 data were chosen instead.
	PM10	Used AP-42 instead of AIRS data for Sand and Gravel sector (90002).
Construction	CO	Due to diesel problem and mobile sources, AP-42 data still underestimated ET data by a factor of 5 for sectors 110101–120300 (total of 200,000 vs. 1,000,000 metric tons). Assumed distribution was valid and scaled up each sector's value by factor of 5.
	NO_2	Since there is no diesel, no NO_2 AP-42 estimates. Thus, the ET construction sector estimate of 1,252,567 metric tons was distributed across all construction sectors using the percent distribution seen in the CO data.
	PM10	ET releases of PM10 from construction was 4,190,340 metric tons (compared to 0 and 250 for the other sources). Distributed across all construction sectors using CO percentages.
Lumber and Wood	CO	Used AP-42 instead of AIRS data for Logging sector [200100]. (Probably a combination of factors including diesel and mobile source problems)
	NO_2	Used ET estimate for Logging sector since other data underestimated.

TABLE 25.26 *(Continued)*

Industry	Emission	Rationale and Method
Chemicals	Lead	Used AP-42 instead of AIRS data for Industrial inorganic and organic chemicals sector [270100]. Probably a result of threshold reporting levels.
Nonferrous Metal	Lead	Used ET data on Lead releases for Primary nonferrous metals [380501]. Included Primary and secondary lead production.
Transportation (and also Trade)	SO_2	Transportation sector data was closely scrutinized because of supply chain effects. ET report carefully modeled these sectors as well. ET data for SO_2 was substituted for the following sectors: Railroads [650100], Water Trans [650400], Air trans [650500].
	CO	Used AP-42 CO values for Railroads, Local transit [650200], Water trans, Freight forwarders [650701]. For trucking-related sectors (Trucking [650301], Wholesale Trade [690100], and Retail trade [690200]), the ET estimate of emissions from large trucks—1.2 million metric tons—was split across these sectors 60%-20%-20% based on fuel consumption [DOE 97]. DOE [97] also noted that 80% of jet/aviation fuel consumption comes from domestic air carriers, so 80% of ET Air Transportation emissions used here (80% of 820,000).
	NO_2	ET data substituted for Railroads, Water transportation. Used similar method as in CO to extract Air transportation emissions (80% of total). ET report estimated total emissions from large trucks as 2,424,000 tons (or 2,198,568 metric tons). DOE [97] suggested transit is 5% of large truck sources. Local transit [650200] assigned 5% of large truck total (109,928 mt). Remaining 95% of truck emissions distributed amongst Trucking (60%) and Wholesale and Retail trade (20% each).
	VOC	Substituted ET data for VOCs was substituted for Railroads and Water trans. Substituted 80% of ET data on Air transportation emissions (see reference above). Used AP-42 data for Local transit, Trucking, Freight forwarders, and Wholesale and Retail trade.
	PM10	Substituted ET data for Railroads, Water and Air transportation.
	Lead	Substituted 80% of ET data for Air Transportation.
Energy	PM10	Substituted ET data for Electric utilities [680100]. Used AP-42 estimate for Natural gas distribution [680202].

TABLE 25.26 *(Continued)*

Industry	Emission	Rationale and Method
Real Estate	CO	Used AP-42 estimate for Real estate agents and operators [710201]. This number more closely relates to the use of fuel for operations (automobile transportation) that is not represented in AIRS.
Repair Services	CO	Used AP-42 data due to large discrepancy with AIRS for Miscellaneous Repair shops [730101].
Other Services	CO/ VOC	Used AP-42 data due to large discrepancy with AIRS for Other Business Services [730109], Hospitals [770200], Other medical services [770305], Social services [770900], Other state and local government [790300]. For these sectors, we assumed the AP-42 data had been representative of fuel purchases for mobile sources that would not be seen in AIRS data.

REFERENCES

Adriaanse, A. *Environmental Policy Performance Indicators—A Study on the Development of Indicators for Environmental Policy in the Netherlands.* Sdu Uitgeverij Koninginnegracht, May 1993.

American Conference of Governmental Industrial Hygienists. *Guide to Occupational Exposure Values—1997.* Cincinnati, OH: ACGIH, 1997, ISBN 1-882417-20-8.

Barnthouse, L., et al. *Life-Cycle Impact Assessment: The State-of-the-Art,* 2nd edition. The Society of Environmental Toxicology and Chemistry. Pensacola, FL, 1997.

Bennett, E. B., and Graedel, T. E. Conditioned air: Evaluating an environmentally preferable service. *Environmental Science & Technology* 34(4): 541–545 (2000).

Boustead, I., and Hancock, G. F. *Handbook of Industrial Energy Analysis.* Chichester, UK: Ellis Horwood, 1979, ISBN 0-85312-064-1.

Bureau of Economic Analysis, U.S. Department of Commerce. *Documentation for 1992 Benchmark Input–Output Accounts Summary Diskette.* File: SICIO.TXT, Washington, DC, 1997.

Bureau of Labor Statistics. *1992 Consumer Expenditure Survey,* via *ftp://ftp.bls.gov/pub/ special.requests/ce/standard/1992/age.txt*

Cifuentes, L., and Lave, L. B. Economic valuation of air pollution abatement: Benefits from health effects. *Ann. Rev. Energy Environ.* 18: 319–342 (1993).

Commission of the European Communities/United States, 1993. *Externalities of the Fuel Cycle: Externe Project.* Working Documents 1,2,5,9. European Commission, Brussels.

Consoli, F., et al. *Guidelines for Life Cycle Assessment: A Code of Practice.* The Society of Environmental Toxicology and Chemistry. Pensacola, FL, 1993.

Culp, A. W. *Principles of Energy Conversion.* McGraw-Hill Book Company, 1979, ISBN 0-07-014892-9.

Curran, M. A. *Life Cycle Analysis;* Island Press: New York, 1996.

Daly, Herman and John Cobb. *For the Common Good*, Boston, Beacon Press, 1988.

ECO Northwest, et al. *Generic Coal Study: Quantification and Valuation of Environmental Impacts.* Report commissioned by Bonneville Power Administration, January 31, 1987.

Encyclopedia of Chemical Processing and Design, Maketta & Cunningham. Benzene Design problem pp. 209; Table 5. 1990.

Environmental Defense Fund: *Scorecard Report: Sun Co. Inc., Philadelphia.* 1997.

Fankhauser, S. The social costs of greenhouse gas emissions: An expected value approach. *Energy Journal* 15(2): 157.

Fava, J., et al. *A Technical Framework for Life Cycle Assessment.* The Society of Environmental Toxicology and Chemistry. Pensacola, FL, 1994.

Federal Reserve Economic Data, via *http://www.stls.frb.org/fred/data/gdp/gdpctpi*

Flores, E. C. *Life Cycle Assessment Using Input-Output Analysis.* Ph.D. Thesis, Department of Civil and Environmental Engineering, Carnegie Mellon University, April 1996.

Freeman, A. M., Burtraw, D., Harrington, W., and Krupnick, A. Weighing environmental externalities: How to do it right. *Electricity Journal*, August/September 1992, pp. 18–25.

Freeman, A. M., and Krupnick, A. Externality adders: A response to Joskow. *Electricity Journal*, August/September 1992, pp. 61–63.

Goldsmith, E., Allen, R., Allaby, M., Davoll, J., and Lawerence, S. 1972. *A Blueprint for Survival.* 139 pp. Penguin; Hanmondsworth, UK.

Graedel, T. E., and Allenby, B. R. *Industrial Ecology*, Prentice Hall, NJ, 1995.

Graedel, T. E. *Streamlined Life-Cycle Assessment.* Prentice Hall, Upper Saddle River, NJ, 1998.

Hall, B., and Kerr, M. L. *1991–1992 Green Index: A State-by-State Guide to the Nation's Environmental Health*, Washington, DC, Island Press, 1991.

Hartman, R. S., Wheeler, D., and Singh, M. The cost of air pollution abatement. *Applied Economics*, 29: 759–774.

Hendrickson, C. T., Horvath, A., Joshi, S., and Lave, L. B. Use of economic input–output models for environmental life cycle assessment. *Environmental Science & Technology*, April 1998.

Horvath, A., Hendrickson, C., Lave, L. B., McMichael, F. C., and Wu, T. Toxic emissions indices for green design and inventory. *Environmental Science & Technology*, 29(3): 86A–90A (1995).

Horvath, A. *Estimation of the Environmental Implications of Construction Materials and Designs Using Life Cycle Assessment Techniques* Ph.D. Thesis, Department of Civil and Environmental Engineering, Carnegie Mellon University, June 1997.

Intergovernmental Panel on Climate Change, *IPCC Guidelines for National Greenhouse Gas Inventories*, Vol. 1–3, UNEP, OECD, and IPCC, 1995.

Inventory of U.S. Greenhouse Gas Emissions and Sinks: 1990–1996. U.S. EPA (1998 Draft), via *http://www.epa.gov/oppeoee1/globalwarming/inventory/1998-inv.html*

Januschkowetz, A., Reyna-Caamano, R., and Shih, H. H. *Refinery Sun Co. Inc., Philadelphia: Is There a Need of an Environmental Buffer Zone.* Green Design Initiative Technical Report. Carnegie Mellon University, Pittsburgh, PA, June 1999.

Joshi, S. *Comprehensive Product Life Cycle Analysis Using Input–Output Techniques.* Ph.D. Thesis, Heinz School of Public Policy and Management. Carnegie Mellon University, August 1997.

Joskow, P. Weighing environmental externalities: Let's do it right. *Electricity Journal,* May 1992, pp. 53–67.

Koomey, J. *Comparative Analysis of Monetary Estimates of External Environmental Costs Associated with Combustion of Fossil Fuels,* Lawrence Berkeley Laboratory, July 1990.

Lave, L., Cobas, E., Hendrickson, C., and McMichael, F. Using input/output analysis to estimate economy-wide discharges. *Environmental Science & Technology,* Vol. 29, pp. 420A–426A, September 1995.

Lee, R. Estimating the impacts, damages, and benefits of fuel cycles: Insights from an ongoing study. In *Social Costs of Energy,* Proceedings of an International Conference held at Racine, WI, Sept. 8–11, 1992, Hohmeyer and Ottinger, Eds.

Leontief, W. Quantitative Input–Output Relations in the Economic System of the United States. *Review of Economics and Statistics,* 18, no. 3 (August 1936): 105–125.

Leontief, W. Environmental repercussions and the economic structure: An input–output approach. *Review of Economics and Statistics,* Vol. LII, No. 3, August 1970.

Local Climatological Data 1997, Annual Summary with Comparative Data; Philadelphia, PA ISSN 0198-4543.

MacLean, H. L., and Lave, L. B. A life cycle model of an automobile. *Environmental Science and Technology,* July 1999.

Matthews, H. S., and Lave, L. B. Applications of Environmental Valuation for Determining Externality Costs. Environmental Science and Technology, 34(8): 1390–1395 (2000).

Meadows, D. H., et al. *The Limits to Growth: A Report for the Club of Rome's Project on the Predicament of Mankind.* Universe Books, NY, 205.

Miller, R., and Blair, P. *Input–Output Analysis,* Chapter 7, Prentice-Hall, Englewood Cliffs, NJ (USA), 1985.

National Air Pollutant Emission Trends Report, 1900–1996, U.S. EPA, 1997.

Naugle, D. F., and Pierson, T. K. A framework for risk characterization of environmental pollutants. *J. of Air & Waste Management Association* 41(10): 1298–1307 (1991).

Nordhaus, W. D. Optimal greenhouse gas reductions and tax policy in the DICE model. *American Economic Review,* Papers and Proceedings 83(2): 313–317.

Occupational Safety and Health Administration. Occupational exposure to benzene; permanent standard, *Federal Register 43, 5918–5970.* February 10, 1978.

Ottinger, R. L. Pollution taxes—The preferred means of incorporation of environmental externalities. In *Social costs of energy,* Proceedings of an International Conference held at Racine, WI, Sept. 8–11, 1992, Hohmeyer and Ottinger, Eds.

Schmalensee, R., Joskow, P. L., Ellerman, A. D., Montero, J. P., and Bailey, E. M. An Interim Evaluation of Sulfur Dioxide Emissions Trading. *J. of Economic Perspectives,* 12(3): 53–68, Summer 1998.

Seinfeld, J. H., and Pandis, S. N. *Atmospheric Chemistry and Physics: From Air Pollution to Climate Change.* John Wiley & Sons, Inc. New York, NY, 1998.

Seip, K. L., Hallgeir, B., and Johnsen, K. Siting of paper mills: Is a pristine environment an industrial resource? *Environmental Science & Technology*. 34(4): 546–551 (2000).

U.S. Bureau of the Census, *Statistical Abstract of the United States: 1996* (116th edition.), Washington, DC, 1996.

U.S. Congress, Office of Technology Assessment, *Green Products by Design: Choices for a Cleaner Environment*, OTA-E-541 (Washington, DC, U.S. Government Printing Office, October 1992).

U.S. Congress, Office of Technology Assessment, *Studies of the Environmental Costs of Electricity*, OTA ETI-134, Washington, DC, U.S. Government Printing Office, September 1994.

U.S. Department of Commerce. *1992 Census of Manufactures*, Industry Series. Washington, DC: U.S. DOC, 1994.

U.S. Department of Commerce. *1992 Census of Mineral Industries, Subject Series, Fuels and Electric Energy Consumed.* Washington, DC: U.S. DOC, MIC-92-S-2, 1994.

U.S. Department of Commerce. *Water Use in Manufacturing.* 1982 Census of Manufactures, Subject Series, Washington, DC: U.S. Department of Commerce, Economics and Statistics Administration, Bureau of the Census, MC82-S-6, March 1996.

U.S. Department of Commerce, *Input–Output Accounts of the U.S. Economy, 1992 Benchmark.* Computer Diskettes, U.S. DOC, Interindustry Economics Division, Washington, DC, 1997.

U.S. Department of Energy. *Annual Energy Review 1996.* Washington, DC: U.S. DOE, Energy Information Administration, DOE/EIA 0384 (96), 1996.

U.S. Department of Energy. *Transportation Energy Databook*, Edition 17, Center for Transportation Analysis, Oak Ridge, TN, August 1997.

U.S. Department of Energy, Energy Information Administration. *Annual Energy Review 1997.* Washington, DC: DoE, 1998.

U.S. Environmental Protection Agency, Office of Air and Radiation. *The Benefits and Costs of the Clean Air Act: 1970 to 1990.* October 1997.

U.S. Environmental Protection Agency. *Final Regulatory Impact Analysis: Control of Emissions of Air Pollution from Highway Heavy-Duty Engines.* 1997.

U.S. Environmental Protection Agency. *1987–1995 Toxics Release Inventory.* Washington, DC: U.S. EPA, Office of Pollution Prevention and Toxics, CD-ROM, EPA 749-C-97-003, August 1997.

U.S. Environmental Protection Agency. *Mercury Study Report to Congress—Volume II: An Inventory of Anthropogenic Mercury Emissions in the United States.* Washington, DC: U.S. EPA, Office of Air Quality Planning and Standards and Office of Research and Development, EPA-452/R-97-004, December 1997.

U.S. Environmental Protection Agency. *National Air Pollutant Emission Trends, 1900–1996.* Washington, DC: U.S. EPA, Office of Air Quality Planning and Standards, EPA-454/R-97-011, December 1997.

U.S. Environmental Protection Agency. *Envirofacts Report on Sun Co. Inc., Philadelphia*, PA EPA Facility ID: PAD049791098. 1997.

U.S. Environmental Protection Agency. *AIRS Data, http://www.epa.gov/airsweb/sources.htm*

U.S. Environmental Protection Agency. *Solid Waste and Emergency Response. 1995 National Biennial RCRA Hazardous Waste Report.* Washington, DC: U.S. EPA, Obtained through a Freedom of Information Act request, March 1998.

U.S. Environmental Protection Agency. National Center for Environmental Assessment: *Carcinogenic Effects of Benzene: An Update*, April 1998.

U.S. Environmental Protection Agency website. *Life Cycle Assessment Brief. www. epa.gov/ordntrnt/ORD/NRMRL/std/SAB/lca_brief.htm.* Accessed July 9, 2000.

U.S. Environmental Protection Agency. *Environmental Investments: The Cost of a Clean Environment*, Report to the Congress, Office of Policy, Planning, and Evaluation, Washington, DC, 1990.

U.S. Environmental Protection Agency. *Air CHIEF.* CD-ROM, Washington, DC: U.S. EPA, July 1995.

Vigon, B. W., Tolle, D. A., Cornaby, B. W., Latham, H. C., Harrison, C. L., Boguski, T. L., Hunt, R. G., and Sellers, J. D. *Life Cycle Assessment: Inventory Guidelines and Principles*, U.S. Environmental Protection Agency: Washington, DC, February 1993; EPA 600/R-92/245.

White, M. C., Infante, P. F., and Chu, K. C. A Quantitative Estimate of Leukemia Mortality Associated with Occupational Exposure to Benzene. In Glickman, T. S. and M. Gough (Eds.). *Readings in Risk. Resources for the Future*, Washington, DC, (1990).

Wiel, S. Why utilities should incorporate externalities. In *Social Costs of Energy*, Proceedings of an International Conference held at Racine, WI, Sept. 8–11, 1992, Hohmeyer and Ottinger, eds.

Wuebbles, D. J. Weighting functions for ozone depletion and greenhouse gas effects on climate. *Annual Review of Energy and the Environment*, 20: 45–70 (1995).

Zuckerman, B., and Ackerman, F. *The 1994 Update of the Tellus Institute Packaging Study Impact Assessment Method.* SETAC Impact Assessment Working Group Conference, Washington, DC, January 25–26, 1995.

SECTION M
Risk Communication and Risk Management

26 Risk Assessment in Its Social Context

JESSICA GLICKEN TURNLEY

Galisteo Consulting Group, Inc., Albuquerque, New Mexico

26.1 INTRODUCTION

In recent years, there has been an increasing effort to separate risk assessment from risk management (Douglas and Wildavsky, 1982; National Research Council (NRC), 1983; Stern and Fineberg, 1996; Pittinger et al., 1998; Stahl, et al. 2001). As far back as 1979, the president of the National Academy of Sciences said that "the *estimation* of risk is a scientific question—and, therefore, a legitimate activity of scientists in federal agencies, in universities and in the National Research Council. The *acceptability* of a given level of risk, however, is a political question, to be determined in the political arena" (Handler, 1979, in Douglas and Wildavsky, 1982:65). In 1983, the NRC noted that risk assessment means "the characterization of potential adverse health effects of human exposures to environmental hazards," (p. 18) whereas risk management refers to "the process of evaluating alternative regulatory actions and selecting among them" (p. 18). (NRC, 1983).

The NRC's more recent report, *Understanding Risk* (Stern and Fineberg, 1996), characterized these two activities as *analysis* and *deliberation*; where analysis "uses rigorous, replicable methods, evaluated under the agreed protocols of an expert community to arrive at answers to factual questions" (1996:3–4), and deliberation is "any formal or informal process for communication and collective consideration of issues" (1996:4). Risk assessment thus generally refers to the descriptive portion of the activity, where the risk is identified and characterized or described. It is usually conducted by individuals with expertise in science or engineering. Risk management refers to that activity that combines the risk assessment with other information to come to a decision, and thus is prescriptive in nature. Risk managers often are not experts in, or familiar with, the disciplines required to conduct a risk assessment.

Human and Ecological Risk Assessment: Theory and Practice, Edited by Dennis J. Paustenbach
ISBN 0-471-14747-8 © 2002 John Wiley & Sons, Inc.

Under these definitions, it is clear where and how social science can be applied to the deliberative or management portion of the equation. "Acceptable" (as in "acceptable risk") clearly is socially defined, and is time- and space-dependent. The evaluation of alternative actions and the selection among them also is socially conditioned. Social science thus has a clear application to the prescriptive portion of the exercise. However, the application of social science to the descriptive, or risk assessment, activity is less evident. This chapter will identify areas where social science has clear relevance for the risk assessor and introduce some of the disciplines and rigorous methods that can be used in this context.

26.2 USING SOCIAL SCIENCE THEORY AND PRACTICE IN RISK ASSESSMENT PROCESSES

To ensure that a risk assessment yields information that is both relevant and usable by decision makers, risk assessors and risk managers must interact throughout the entire assessment process. The NRC (1996) report emphasized that (Stern and Fineberg, 1996:6)

> the analytic-deliberative process should be *mutual and recursive*. Analysis and deliberation are complementary and must be integrated throughout the process leading to risk characterization: deliberation frames analysis, analysis informs deliberation, and the process benefits from feedback between the two.

A conference of practitioners of both risk management and assessment practitioners, convened by the Society for Environmental Toxicology and Chemistry in 1998 at the end of nearly 2 years of discussion and interaction among these communities, came to a similar conclusion. The framework that emerged from the conference (Pittinger et al., 1998) emphasized continuous and ongoing interaction between the risk management and risk assessment processes and practitioners. The elaboration of this framework in Stahl et al. (2001) continues this theme, emphasizing the importance of effective communication among all parties—risk assessors, risk managers, and other stakeholders—for a socially *and* scientifically acceptable outcome. The risk assessor thus needs to be cognizant of the needs and requirements of the risk manager (the risk manager's "agenda"), the vocabulary in which he both understands and speaks, and the extent and nature of his technical knowledge. This is one area where social science can be very useful to the risk assessor.

The increasing pressure for stakeholder involvement in public decision making in the United States, also characterized as public participation or participation by interested parties, complicates the issue of interaction with decision makers for the risk assessor by adding additional communication dimensions to the process. This pressure toward participative rather than paternalistic decision making* reflects a general decline in confidence of the public in experts and a concurrent increase in the legitimacy of knowledge produced and presented by nonexperts. This is part of a broader social movement characterized as the decline in the power of technocratic government—a "system of governance in which trained experts rule by virtue of their specialized knowledge and position in dominant political and economic institutions" (Fischer, 1990:17). Under the technocratic culture and politics of expertise, begun by McNamera and his "whiz kids" in the 1950s, public policy decisions were made by those who could separate scientific information from input about social and emotional issues and base their decisions on objective, impartial, rational data (Fischer, 1990; McGarity, 1990). However, as a growing body of research is showing, social reaction and response to risk management decisions is influenced as much by public trust in those making and issuing those decisions as it is by confidence in or understanding of the data upon which the decisions are made, that is, the risk assessments upon which they are based (Cvetkovich and Löfstedt, 1999; Renn and Levine, 1991; Leiss, 1986; Nye et al., 1997). Therefore, the risk assessor can no longer assume that the numbers will "speak for themselves" but must pay significant attention to the way in which those numbers (and other scientific outputs from the risk assessment) are communicated to the risk manager and to other stakeholders. This risk assessor thus must also be cognizant of the principles, theory, and approach of participatory processes. Here is another area where a knowledge of social science theory and practice could be useful.

Public participation in any decision-making process is specific to time, site, and issue. The outcome of a particular participatory process serves only to set context for the assessment in question, not to establish precedent for all assessments (Bear, 1994:7). Theoretical treatments of participatory processes in environmental decision making focus primarily on mechanisms for involvement of groups with differing agendas (one of which is the risk assessor and the scientific community he represents) and on negotiation and mediation techniques for bringing those groups to consensus (Bear, 1994; Berry, 1993; Brenneis and M'Gonigle, 1992; Dale and Lane, 1994; Davis and Wurth, 1993; Renn et al., 1995; Rowe and Frewer, 2000; Webler et al., 1995). There also are many case studies of participatory processes in environmental decision making that can be referenced from a lessons-learned standpoint (Beatty, 1991; Finney and Polk, 1995; Hayton, 1993; Kangas, 1994; Renn et al., 1993; Sample, 1993). Appreciating these lessons will require an understanding of how social groups are identified, operate, and communicate—another area addressed by social science.

The remainder of this discussion is divided into three general parts: understanding how stakeholders or interested parties are identified and subsequently involved in the risk assessment process, including issues of communication;

---

* In paternalistic decision making, the government or some other authority entity invites input on terms defined by the authority entity. That entity then makes a unilateral decision (Bear, 1994).

understanding how the problems addressed by risk assessments are defined; and finally, understanding how the results of an assessment are communicated and used. All three of these areas can benefit greatly from the application of social science theory and techniques.

26.3   DEFINING STAKEHOLDERS

Stakeholders have been defined in many ways. Babiuch and Fahar (1994) give 16 definitions, and reference many more. The following statement captures most of the important concepts: "A stakeholder is an individual or group influenced by—and with an ability to significantly impact (either directly or indirectly)—the topical area of interest" (Engi and Glicken, 1995:1). An important aspect of stakeholders is that they are identified and defined *relative to a specific issue*. A review of the literature on folk taxonomies and linguistic classification will provide the risk assessor techniques to identify groups relevant to his or her assessment (Wittgenstein, 1958; de Sassure, 1966; Searle, 1969; Whorf, 1956; Derrida, 1974; Berlin and Kay, 1969; Ellen and Reason, 1979; Geertz, 1973; Bousfield, 1979).

To identify stakeholders in a particular issue is to define the relationship of specific groups to that issue and therefore to each other. It is a systems problem. Social classifications, which are linguistic descriptors of groups of people, only have meaning in terms of their place in the system. Wittgenstein said, "the question, 'what is a word really?' is analogous to 'what is a piece in chess?'" (Wittengstein, 1958:108), that is, one must understand the entire game (language or social system) with all its rules (grammar and semantics, or rules of social interaction) before one can understand what a piece (a word or a social group) is. De Sassure noted that, "concepts are purely differential and defined not by their positive content but negatively by their relations with the other terms of the system. Their most precise characteristic is being what the others are not" (de Sassure, 1966:117). And Derrida added a temporal dimension to the concept of difference, suggesting that, "if words and concepts receive meaning only in sequences of differences, one can justify one's language and one's choice of terms only within a topic [an orientation in space] and an historical strategy. The justification can therefore never be absolute and definitive. It corresponds to a condition of forces and translates an historical calculation" (Derrida, 1974:70). Therefore, to understand these social concepts we call "stakeholders," we must know their spatial and temporal context. To call someone an "environmentalist" in the southwestern United States in the 1990s carries connotations of "antirancher" that are missing in, for example, the deep South and that were absent from the social universe in which, for example, John Muir moved. Note, also, that the name or label applied to the group—environmentalist, scientist, politician—itself defines classes of expected behaviors and generate certain behaviors and attitudes in the speaker. Thus the very process of identifying and including stakeholders helps create and manipulate the social universe. As Geertz said of symbols (and language, including the labels we apply to things, is a symbolic system *par excellence*), they are simultaneously "models of and models for" this social reality (Geertz, 1973).

Studies of group definition and organization could help the risk assessor understand how relevant stakeholder groups are constituted (Radcliffe-Brown, 1952; Knoke, 1990; Davis and Wurth, 1993; Leiden, 1995; Levi-Strauss, 1963; Glicken, 1996). Stakeholder groups can be formal organizations that exist over time (such as the Sierra Club) or informal groups that mobilize only in reference to a particular issue. The assessor must understand the dynamics of these different types of organizations and identify which are at play in reference to the issue that stimulated the risk assessment. This will provide the risk assessor insights as to how these groups might act relative to the information the assessor might present and the scientific process he or she is managing.

Social network analysis examines the structures of power among the components of a social system or social structure (Knoke, 1990; Perrucci and Potter, 1989; J. J. Schensul et al., 1999), and hence can help the risk assessor anticipate various stakeholder groups' capability to mobilize resources and to influence other actors. The increase in both informal expectations and prescriptive laws and regulations about the extent and nature of stakeholder involvement in decision processes may require the risk assessor to engage in interactions with these types of groups as part of a risk management process. The better the assessor's understanding of those groups and how they act, the more likely the assessor is to achieve his or her own goals of appropriate inclusion of scientific information in a risk management decision.

26.4 DEFINING THE RISK PROBLEM

The decision to perform a risk assessment is a decision to allocate resources to a particular problem. This presupposes that a "problem" has been identified, that is, that something of value is seen to be "at risk." That value is defined as a result of social trade-offs that are institutionalized in laws, regulations, and attendant enforcing agencies and social structures.

"Risk," as it is regulated, is a phenomenon of groups. From a statistical standpoint, risk requires a set of occurrences for its calculation. Individual experience can only enter the risk equation in the aggregate, when it becomes a social event rather than an individual experience (Beck, 1995:21). Once risk becomes a social event, it becomes politicized; that is, it comes under the control of the government.

The requirements for the political control of risk are emphasized by the "commons" nature of the risks to health and the environment. One factory will not cause a smog problem, nor will one farmer applying pesticides to his field cause the death of a watershed. But when that one factory is seen as one of a set of factories or pollutant emitters, or that one farmer as one of a set of farmers, their individual contribution to smog or nonpoint pollution can be discussed, understood, and (socially) managed. Government is a formal mechanism of social management. The Western political tradition sees civil society as a mechanism that has emerged in answer to these sorts of commons problems. Government is seen as a social contract that has emerged from the recognition that, as individuals, we do not have the power to preserve "life, liberty, and estate" (Locke, 1690), but that, by uniting into a group through the relinquishment of some individual power to that group, we can accomplish more than we could as individuals (Hobbes, 1651). This assigns legitimacy to the government for control and regulation of perceived threats to society (as distinct from threats to individuals *qua* individuals), which today include environmental damage and anthropogenically induced illnesses (Beck, 1995:21; Ruckelshaus, 1989).

Although risk as it is regulated is a phenomenon of groups, risk as it is perceived is a phenomenon of individuals. Through studies of risk perception and its attendant psychological and sociopsychological dimensions, we can gain understanding of how individuals incorporate considerations of risk into decisions they make. The perception of risk by an individual is a complex interaction of many factors. Actual risk (e.g., mortality or morbidity rates), the catastrophic potential of the event, fear of the unknown (which is a factor of the knowledge base of the target individual), control over the risk, immediacy of the effect, and the relationship of the subject to the risk target (e.g., is it a family member or an unknown individual) all come into play (Bostrom et al., 1994; Douglas and Wildavsky, 1982; Dake, 1991; Fischoff et al., 1978; Lazo et al., 2000; Sjöberj, 2000; Slovic et al., 1979). Slovic's classic 1979 study and Lazo's more recent work (2000) illustrate clearly the differences in the perception of risk between "experts" and lay populations, with some of the rank orders completely inverted by the two populations and associated ratings of the acceptability of the risk very different. Interestingly, we also find a possible sexual bias in risk perception, although results vary, with some studies showing women more likely to perceive risk than men (Bord and O'Connor, 1997; Flynn et al., 1994; Hoefer and Raju, 1991) and others the opposite (Lazo et al., 2000).

We also need to understand the way society is organized through its understanding of risk to protect its way of life (Douglas, 1992). This social organization includes such structures and processes as insurance markets (what are we insuring against and how much are we willing to pay for it?), the trends in the numbers of legal cases having to do with product or professional liability (who bears the financial burden for mistakes—and how do we define mistakes?), and the increasing body of regulations designed to manage social activity to reduce risks of environmental damage or anthropogenically induced illnesses (again— who bears the financial burden, and how do we assign responsibility?). These social perceptions of risk as codified and institutionalized through laws, regulations, and organizations drive the selection of risk assessments to be performed.

Understanding social management structures (such as governments) and the things which they manage (such as risk) requires an understanding of people

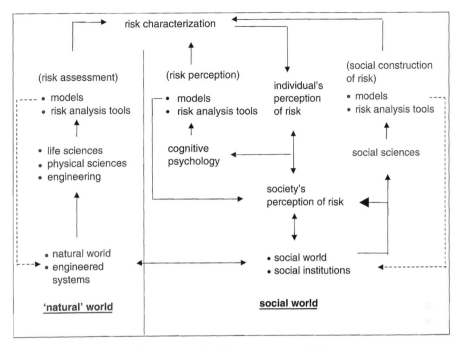

Figure 26.1 Social construction of risk.

acting as individuals and as groups. Characterizing risk requires an understanding of natural phenomena. The relationships among the "natural world" where scientists are employed to characterize risk, the world of individuals where risk perceptions are translated into individual actions, and the social world where individual risk perceptions are translated into collective action as reflected in the development of institutions and social controls, are illustrated in Figure 26.1.

Policy analysis and an understanding of political action and of the construction and evolution of structures of social management can thus help the risk assessor understand the rationale for the decision to perform a particular risk assessment and the universe of possible social actions that might arise as a result of the risk characterization (assessment). This again will help make interaction with risk managers (including stakeholders) more effective.

The (social) decision to perform a risk assessment segues directly into the problem formulation phase of the assessment itself. This phase of the risk assessment requires input from two important areas. Technical information about the environmental resources or aspects of human health in question is, of course, a must. The assessor must also be cognizant of suspected agents of contamination, as well as exposure pathways. The last point segues into the humanistic dimension of the problem formulation stage of a risk assessment. Life-styles and values of impacted groups can have a significant impact on potential exposure rates, as well as on the identification of species of interest to the risk assessment.

This is dramatically illustrated in work with human health and environmental risk assessments at two U.S. Department of Energy (DOE) laboratories (Los Alamos National Laboratory and Pacific Northwest National Laboratory), which are located in remote areas and surrounded by tribal lands with Native Americans living largely traditional life-styles. Risk assessors recognized that exposure scenarios needed to incorporate the ethnobiology of the area (the relationship of culture to the natural environment) to account for non-Western subsistence patterns. The team preparing the environmental impact statement for the Hanford site (see U.S. Department of Energy and Washington State Department of Ecology, 1996) began with CERCLA [Comprehensive environmental response, compensation and liability act] suburban exposure factors and modified them to account for such factors as increased ingestion of locally gathered foods and medicines, and unique exposure pathways such as use of the sweat lodge (*Risk Management Quarterly* 1995). Dermal uptake is of concern for the potters of northern New Mexico around Los Alamos National Laboratory, and the centrality of rituals surrounding big game hunting and ingestion for many of the northern Pueblos may cause additions to lists of species of concern in human health risk assessments.

As the United States becomes much more ethnically diverse, we may find additional disjunctions between standard exposure scenarios and models and the life-styles of urban ethnic communities (such as Asian or Latino) that include diets that are rich in foods or food types not common in Anglo communities. Furthermore, U.S.-generated models and processes are often used as the basis for the development of environmental and ecological risk assessment and management procedures in other countries where dietary and life-style patterns are clearly different than those used to inform the models. Information that will help develop more appropriate exposure scenarios can be gathered through anthropological information elicitation techniques (expert elicitation), which involves the correct identification of appropriate "experts" (note that these may be experts in ritual or in neighborhood activities, rather than in ethnobotany or environmental science per se) and highly developed interviewing skills. Knowledge of local social structure and group values will help guide development of interview protocols. (See LeCompte and Schensul, 1999; Schensul, S. L. et al., 1999; and Bernard, 1988, for basic methodological approaches.)

26.5   COMMUNICATING INFORMATION ABOUT RISK

Communication is key to any applied science effort, but even more to one that incorporates a high level of stakeholder involvement. Risk assessment and risk management are intertwined processes: Risk assessment can be described as a process that begins and ends with social conditions (Glicken and Fairbrother, 1998). "Value information ... is combined with technical information about ecological relationships to frame the questions being asked and to define the conceptual model ... The science ... is embedded in social values" (Glicken and Fairbrother, 1998:784–786). Decision makers identify decisions that need to be made and, through communication with scientists, ascertain that science can provide information critical to that decision. The problem formulation stage of the risk assessment draws heavily on these conversations to ensure that the data collected is relevant to the decision that needs to be made. [The planning step at the beginning of the Environmental Protection Agency [U.S. EPA (1998)] guidelines asks for participation by risk managers.] However, it is important to emphasize that the social context for the science as it helps determine the questions to be asked through the assessment should not impact the integrity of the science performed during the assessment itself. This interaction and separateness between the social context of the risk assessment and the assessment itself are illustrated in the framework shown in Figure 26.2, where the risk assessment step is represented as "data compilation and analysis."

In this model, the risk assessment is embedded in the risk management process. Following the analytic-deliberative process outlined by the NRC (Stern and Fineberg, 1996), the public (stakeholders) would be kept informed of the progress of the assessment, as well as of methodologies and analysis techniques so that the results of the assessment do not come from "black-box science" (Stahl et al., 2001). Risk communication thus must be seen as a process to build trust in the risk assessor, a process that continues throughout the risk assessment and ends (for the assessor) with communication of risk assessment results to the risk manager or decision maker. It should not be perceived as a one-time

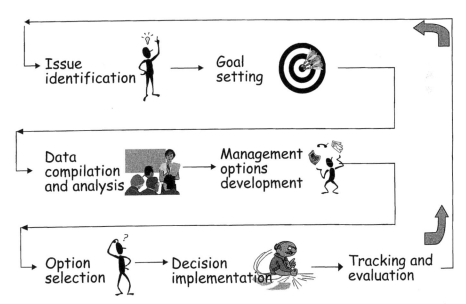

Figure 26.2 Risk management and risk assessment. (Adapted from Pittinger et al., 1998).

event, occurring at the end of the risk assessment and serving as a hand-off of the scientific information to the decision makers. As discussed earlier, this technocratic model is fast loosing credibility in favor of a more participative, trust-based model. Furthermore, if groups with different levels of science education and understanding are involved in the process, the communication process becomes even more complex.

Risk communication is "communication that supplies lay people with the information they need to make informed, independent judgements about risks to health, safety, and the environment." (Bostrom et al., 1994) As with any communication process, risk communication requires effective transmission (i.e., selection of the appropriate media and proper encoding of the message) of the right information to the right audience.

An effective communication event is one that achieves the sender's objectives by sending the desired message through channels and in a format appropriate to the targeted audience. This clearly is a complex process and has a much greater chance of success if that process is a thoughtful and considered one. Figure 26.3 shows the steps involved in any communication event. [See Glicken (1999) for a more complete exposition of the communication process.]

Mistakes are often made at the outset, where the originator of the communication event mistakes a communication event for the process of which it is a part. The success of a public hearing, for example, is the outcome of a whole series of communication events beginning with the communication of the need for such a hearing from a regulatory body to the risk manager; the development and dissemination of the announcement of the hearing; the hearing itself; and the communication of the results of the hearing to appropriate audiences.

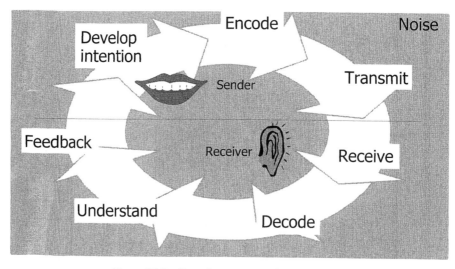

Figure 26.3　Steps in a communication event.

Furthermore, each one of these steps needs to be subdivided into smaller parts. Failure of any one of the parts could lead to failure of the process. Conversely, for the process to succeed, attention must be paid to each part.

Once the communication process is disagreggated into its constituent parts, the communicator must clearly identify his or her objective for each part. The objective of an announcement of a meeting may be to get the right people to the meeting ... or, alternatively, it may be simply to inform the right people that the meeting will happen. A meeting with attendance of zero may be acceptable. The objective of the communication event will help determine the form of the communication and the choice of media, or the communication channel. We will return to communication channels in a moment.

Every communication is an encoded message consisting of a translation of a sender's intent (a cognitive concept) into words or pictures. The encoded message, that is, the words or pictures chosen, are then transmitted to the receiver (the "listener") via some communication vehicle using some communication medium. The receiver runs it through his internal "decoder" to translate the words and pictures back into cognitive concepts. This is all done against a background of considerable "noise," that is, words and pictures that are not directly relevant to the targeted communication.

The encoding process is key to communication success. For a specific communication event to be successful, the audience or receiver (the risk manager or other stakeholders in our case) must be able to understand what is communicated, that is, to transform the words or pictures back into concepts—to "decode" the message. For this to be successful, the communication initiator, the sender, must understand the decoding frame of the receiver and encode his or her message so that it is capable of being decoded.

Sloppy or poor encoding is a primary cause of failure in science communication. Science has a specialized language, a jargon or vocabulary peculiar to it that is not accessible to other parts of the population (Gross, 1996). The term *risk* itself, for example, means very different things to different parts of the population. Everyone is for "environmental quality"—but exactly what this means, again, varies by speaker. For example, at a workshop held in 1993 with stakeholders from a variety of perspectives, including the EPA, public interest groups, academe, and Congress, participants were only able to agree that "environmental quality" meant adherence to the current body of federal, state, and local regulations; the group was not able to come to consensus on a definition independent of law (Glicken and Engi, 1992). Cultural linguistics has developed research and analytical techniques to identify these linguistic discontinuities so that they may be successfully addressed or at least recognized.

Successful encoding thus depends heavily on accurate characterization or description of the audience or communication target or, more accurately, of the decoding framework through which the message must pass before it can be understood by the receiver. Effective audience characterization will help the risk assessor avoid many communication errors and help increase the probability that the scientific information from the risk assessment will be appropriately considered in the risk management decision.

An audience characterization exercise is similar to a characterization of any social group. The characterization exercise draws upon both demographic and psychographic* data, providing the communicator information about culture (values and life-styles), social roles (definitions of expected individual behavior sets), and social organizations (definitions of expected aggregate behaviors). This information is available from a variety of sources, and can be acquired through techniques as diverse as expert elicitation, participant observation (both of which depend heavily on interviewing techniques), surveys, and reviews of census and other published data. These are all basic social science data collection methodologies, each with its own set of protocols and attendant rigor.

An effective audience characterization exercise also will tell the assessor how the target audience best receives information. Some groups respond best to short, concise, oral briefings with only results included; others want extensive information on methodology and are willing to read a great deal of material to get it; still others (particularly neighborhood groups and less well-educated audiences) acquire most of their information from their peers. In these latter cases, identifying a key individual in the community and spending a great deal of time with him/her to ensure that this key individual understands both the content of the risk assessment and its implications for the management decision to be made may be the most effective channel to disseminate information to the full community. If the risk assessor is interested in effective and appropriate use of the results of the risk assessment, the assessor will need to invest time and other resources in managing communication with appropriate parties.

Careless audience characterization can lead to poor choice of communication media. Although the risk assessor may believe he or she has a great deal of important data to communicate to the target audience, if that audience does not have either the time for or the interest in reading, presenting the risk information in a report or document format will be counterproductive. If, however, the risk communication event is recharacterized and an information overview is presented in a short, easy-to-assimilate format designed to stimulate interest in and a request for a longer report, the communicator may have more success. Even if the longer report is not requested, he or she will have communicated an overview and conclusions, and let the target audience know that additional information is available—an important point of credibility and trust for many in the risk arena.

---

*Psychographic data is descriptive data collected about values and life-styles. An audience characterization will lead to descriptions of the decoder frame of the audience. This description will include such dimensions as level of knowledge of the subject to be communicated, perspective on or interest in the topic (i.e., what are they expecting to hear from you), time available to spend on the communication (would they be willing to read a report, or is a one-page flyer more appropriate?), life-style relative to communication media or channels (do they watch a lot of TV? read books? read the newspaper?), and the like.

26.6 SUMMARY

Risk assessment is a scientific exercise and should be conducted in accordance with accepted professional principles and practices. However, the impetus for the conduct of a risk assessment and the use of the output of the risk assessment are socially determined. The embeddedness of the risk assessment in this social process requires that the risk assessor be in constant communication with the society (through its stakeholder representatives) that stimulated the need for the assessment.

We have outlined three general areas where social science can be of value to the risk assessor: stakeholder identification and inclusion, problem identification, and communication.

The requirements for participative processes in risk assessment are part of a larger social trend that recognizes the social embeddedness of science. It does not detract from the objectivity and rationality of the scientific process itself (in our case, the risk assessment), but it does argue that to use science effectively, one must use it in a social context. To that end, the risk assessor often is asked to identify stakeholders relevant to his or her risk assessment. As stakeholder universes are situational, that is, they vary from issue to issue, the risk assessor must be cognizant of, or at least familiar with, stakeholder identification methodologies or experts who can provide such support. Social science can provide the theory as well as the methodology for effective stakeholder identification.

Risk assessments are all motivated by a perceived social problem: something of social value is under stress. Furthermore, the first step in the risk assessment process itself involves identifying assessment endpoints—and these endpoints must be directly related to the problem that prompted the assessment. If the risk assessor can better understand the problem, he or she has a greater chance of identifying endpoints that are relevant to it. Since the problem is socially driven, a better understanding of the complex of social forces that gave rise to that particular problem at that time and place could be an investment with significant positive payoff for the risk assessment.

Communication is a nontrivial part of the risk assessment process. If the results of the risk assessment are poorly understood by those who need to use them, or are understood in a way that was not intended by the risk assessor, the science will not serve society well. If communication is treated as a thoughtful and structured process and given as much attention as the risk assessment itself, the risk assessor can greatly increase the likelihood that the results of the risk assessment will be internalized and seriously considered by the target audience.

Social science has many theoretical constructs, data collection methodologies, and analysis techniques that can serve the risk assessor well. However, it is often treated as a necessary stepchild by scientists conducting risk assessments and given little attention and still fewer resources. The better the assessor understands the society that prompted the risk assessment and understands the principles of effective communication, the more likely it is that the assessment results he or she generates will be correctly interpreted and appropriately used.

REFERENCES

Babiuch, W. M., and Farhar, B. C. 1994. *Stakeholder Analysis Methodologies Resource Book*. NREL/TP461-5857. National Renewable Energy Laboratory, Golden, CO. March.

Bear, D. (Ed.). 1994. *Public Participation in Environmental Decision Making*. American Bar Association, Standing Committee on Environmental Law, Washington DC.

Beatty, K. M. 1991. Public opinion data for environmental decision making: the case of Colorado Springs. *Environ. Impact Assess. Rev.* (11): 29.

Beck, U. 1995. *Ecological Enlightenment: Essays on the Politics of the Risk Society*. (M. A. Ritter, Trans.) Humanities Press, Atlantic Highlands, NJ.

Berlin, B., and Kay, P. 1969. *Basic Color Terms*. California University Press, Berkeley CA.

Bernard, H. R. 1988. *Research Methods in Cultural Anthropology*. Sage Publications, Newbury Park, CA.

Berry, J. M. 1993. Citizen groups and the changing nature of interest group politics in America. *Ann. Am. Acad. Practical Social Sci.* 528: 30–41.

Bord, R. J., and O'Connor, R. E. 1997. The gender gap in environmental attitudes: The case of perceived vulnerability to risk. *Social Sci. Q.* 78: 830–840.

Bostrom, A., Morgan, M. G., Fischoff, B., and Read, D. 1994. What do people know about global climate change? *Risk Anal.* 14: 959–970.

Bousfield, J. 1979. The world seen as a colour chart. In R. F. Ellen and D. Reason (Eds.), *Classifications in Their Social Context*. Academic, New York, pp. 195–220.

Brenneis, K., and M'Gonigle, M. 1992. Public participation: Components of the process. *Environments* 21(3): 5–11.

Cvetkovich, G., and Lofstedt R. (Eds.). 1999. *Social Trust and the Management of Risk*. Earthscan, London.

Dake, K. 1991. Orienting dispositions in the perception of risk. *J. Cross-Cultural Psychol.* 22: 61–82.

Dale, A. P., and Lane, M. B. 1994. Strategic perspectives analysis: A procedure for participatory and political social impact assessment. *Soc. Natural Resourc.* 7: 253–267.

Davis, F. L., and Wurth, A. H., Jr. 1993. American interest group research: Sorting out internal and external perspectives. *Pol. Stud.* 41: 435–542.

Derrida, J. 1974. *Of Grammatology* (G. C. Spivak, Trans.) Johns Hopkins University Press, Baltimore MD.

de Sassure, F. 1996. In C. Bally and A. Sechehaye (Eds.) (W. Baskin, Trans.). *Course in General Linguistics*. McGraw-Hill, New York.

Douglas, M. 1992. *Risk and Blame: Essays in Cultural Theory*. Rutledge, New York.

Douglas, M., and Wildavsky, A. 1982. *Risk and Culture: An Essay on the Selection of Technical and Environmental Dangers*. University of California Press, Berkeley, CA.

Ellen, R. F., and Reason, D. (Eds.). 1979. *Classifications in Their Social Context*. Academic, London.

Engi, D., and Glicken, J. 1995. *The Vital Issues Process: Strategic Planning for a Changing World*. Sandia National Laboratories, SAND95-0845. Albuquerque, NM, 1995.

Finney, C., and Polk, R. E. 1995. Developing stakeholder understanding, technical capability, and responsibiity: The New Bedford harbor superfund forum. *Environ. Impact Assess. Rev.* 15: 517–541.

Fischer, F. 1990. *Technocracy and the Politics of Expertise.* Sage Publications, Newbury Park, CA.

Fischoff, B., Slovic, P., Lichtenstein, S., Read, S., and Combs, B. 1978. How safe is safe enough? A psychometric study of attitudes toward technological risks and benefits. *Policy Sci.* 9: 127–152.

Flynn, J., Slovic, P., and Mertz, C. K. 1994. Gender, race, and perception of environmental health risks. *Risk Anal.* 14(6): 1101–1108.

Geertz, C. 1973. *The Interpretation of Cultures: Selected Essays by Clifford Geertz.* Basic Books, New York.

Glicken, J. 1996. Incorporating stakeholders into risk assessments: Processes and tools to effectively address social concerns. Paper presented at the Society of Environmental Toxicology and Chemistry (SETAC) 17th Annual Meeting, Washington, DC, Pensacola FL.

Glicken, J. 1999. Effective public investment in public decisions. *Sci. Commun.* 2(3): 298–327.

Glicken, J., and Dennis, E. (Eds.). 1992. *Identifying Vital Issues: New Intelligence Strategies for a New World*, Vol. 1: *Vital Issues.* U.S. Department of Energy, Office of Intelligence, June 1992.

Glicken, J., and Fairbrother, A. 1998. Environment and social values. *Hum. Ecol. Risk Assess.* 4(4): 779–786.

Gross, A. G. 1996. *The Rhetoric of Science.* Harvard University Press, Cambridge MA.

Hayton, R. D. 1993. The matter of public participation. *Natural Resourc. J.* 33: 295–281.

Hobbes, T. 1651. *Leviathan* (R. Tuck, Ed.). Cambridge University Press, Cambridge, 1991.

Hoefer, M. P., and Raju, V. S. 1991. Technological risk perception: Similarities and dissimilarities in French and American samples. In B. J. Gamck and W. C. Gekler (Eds.), *The Analysis, Communication and Perception of Risk.* Plenum, New York, pp. 25–30.

Kangas, J. 1994. An approach to public participation in strategic forest management planning. *Forest Ecol. Manag.* 70: 75–88.

Knoke, D. 1990. *Political Networks: The Structural Perspective.* Cambridge University Press, Cambridge.

Lazo, J. K., Kinnell, J. C., and Fisher, A. 2000. Expert and layperson perceptions of ecosystem risk. *Risk Anal.* 20(2): 179–193.

LeCompte, M. D., and Schensul, J. J. 1999. *Ethnographers Toolkit*, Vol. 1: *Designing and Conducting Ethnographic Research*, Alta Mira, Walnut Creek, CA.

Leiden, W. R. 1995. The role of interest groups in policy formulation. *Wash. Law Rev.* 70: 715–724.

Leiss, W. 1986. A value basis for conservation policy. In W. Dunn (Ed.), *Policy Analysis: Perspectives, Concepts and Methods.* Jal, pp. 185–201.

Levi-Strauss, C. 1963. *Structural Anthropology* (C. Jacobson and B. G. Schoepf, Trans.). Basic Books, New York.

Locke, J. 1690. *The Second Treatise of Government* (T. P. Peardon, Ed.). Macmillan, New York, 1952.

McGarity, T. O. 1990. Public participation in risk regulation. *Risk—Issues in Health & Safety.* Spring: 103–130.

National Research Council (NRC). 1983. *Risk Assessment in the Federal Government: Managing the Process.* National Academy Press, Washington, DC.

Nye, J. S., Zelikow, P. D., and King, D. C. 1997. *Why Don't People Trust Government?* Harvard University Press, Cambridge, MA.

Perrucci, R., and Potter, H. R. (Eds.). 1989. *Networks of Power: Organizational Actors at the National, Corporate, and Community Levels.* Aldine de Gruyter, Hawthorne, NY.

Pittinger, C. A., Bachman, R., Barton, A. L., Clark, J. R., deFeur, P. L., Ells, S. J., Slimak, M. W., Stahl, R. G., and Wentsel, R. S. 1998. A multi-stakeholder framework for ecological risk management: Summary of a SETAC technical workshop. Summary of the Society of Environmental Toxicology and Chemistry (SETAC) Workshop on Framework for Ecological Risk Management, June 23–25, 1997, Williamsburg, VA, Pensacola FL, SETAC, 1998.

Radcliffe-Brown, A. R. 1952. *Structure and Function in Primitive Society.* Free Press, New York.

Renn, O., and Levine, D. 1991. Credibility and trust in risk communication. In R. E. Kasperson and P. J. M. Stallen (Eds.), *Communicating Risks to the Public: International Perspectives.* Kluwer Academic, Dordrecht, The Netherlands, pp. 175–218.

Renn, O., Webler, T., Rakel, H., Dienel, P., and Johnson, B. 1993. Public participation in decision making: A three-step procedure. *Policy Sci.* 26: 189–214.

Renn, O., Webler, T., and Wiedmenn, P. 1995. *Fairness and Competence in Citizen Participation: Evaluating Methods of Environmental Discourse.* Kluwer Academic, Dordrecht, The Netherlands.

Risk Management Quarterly. 1995. Lane Environmental and U.S. Department of Energy's Center for Risk Excellence, Richland WA, 5–1:5.

Rowe, G., and Frewer, L. J. 2000. Public participation methods: A framework for evaluation. *Sci. Technol. Hum. Values* 25(1): 3–29.

Ruckelshaus, W. D. 1989. Science, risk, and public policy. In T. Goldfarb (Ed.), *Taking Sides: Clashing Views on Controversial Environmental Issues.* Dushkin, Guildford, CT.

Sample, V. A. 1993. A framework for public participation in natural resource decision making. *J. Forestry* 93: 22–27.

Schensul, J. J., LeCompte, M. D., Trotter II, R. T., Cromley, E., and Singer, M. 1999. *Ethnographers Toolkit, Vol. 4: Mapping Social Networks, Spatial Data, and Hidden Populations.* Alta Mira, Walnut Creek, CA.

Schensul, S. L., Schensul, J. J., and LeCompte, M. D. 1999. *Ethnographers Toolkit,* Vol. 2: *Essential Ethnographic Methods.* Alta Mira, Walnut Creek, CA.

Searle, J. 1969. *Speech Acts.* Cambridge University Press, Cambridge.

Sjöberg, L. 2000. Factors in risk perception. *Risk Anal.* 20(1): 1–12.

Slovic, P., Fischhoff, B., and Lichentenstein, S. 1979. Rating the risks. *Environment* 21(3): 14–20, 36–39.

Stahl, R., Bachman, R., Barton, A., Clark, J., deFur, P., Ells, S., Pittinger, C., Slimak, M., and Wentsel, R. (Eds). 2001. *Risk Management: Ecological Risk-Based Decision-Making.* Society for Environmental Toxicology and Chemistry (SETAC), Pensacola, FL.

Stern, P. C., and Fineberg, H. V. 1996. *Understanding Risk: Informing Decisions in a Democratic Society.* Committee on Risk Characterization. National Academy Press, Washington, DC.

U.S. Department of Energy and Washington State Department of Ecology. 1996. *Final Environmental Impact Statement for the Tank Waste Remediation System* (DOE/EIS-0189), April 1996.

U.S. Environmental Protection Agency (EPA). 1998. *Guidelines for Ecological Risk Assessment.* EPA, Washington, DC.

Webler, T., Kastenholz, H., and Renn, O. 1995. Public participation in impact assessment: A social learning perspective. *Environ. Impact Assess. Rev.* 15(5): 443–463.

Whorf, B. L. 1956. *Language, Thought and Reality.* MIT Press, Boston, MA.

Wittgenstein, L. 1958. *Philosophical Investigations*, 3rd ed. (G. E. M. Anscombe, Trans.). Macmillan, New York.

27 Trust, Emotion, Sex, Politics, and Science: Surveying the Risk-Assessment Battlefield

PAUL SLOVIC

Decision Research, 1201 Oak Street, Eugene, Oregon

27.1 INTRODUCTION

Ironically, as our society and other industrialized nations have expended great effort to make life safer and healthier, many in the public have become more, rather than less, concerned about risk. These individuals see themselves as exposed to more serious risks than were faced by people in the past, and they believe that this situation is getting worse rather than better. Nuclear and chemical technologies (except for medicines) have been stigmatized by being perceived as entailing unnaturally great risks.[1] As a result, it has been difficult, if not impossible, to find host sites for disposing of high-level or low-level radioactive wastes, or for incinerators, landfills, and other chemical facilities.

Public perceptions of risk have been found to determine the priorities and legislative agendas of regulatory bodies such as the Environmental Protection Agency, much to the distress of agency technical experts who argue that other hazards deserve higher priority. The bulk of EPA's budget in recent years has gone to hazardous waste primarily because the public believes that the cleanup of Superfund sites is one of the most serious environmental priorities for the country. Hazards such as indoor air pollution are considered more serious health risks by experts but are not perceived that way by the public.[2]

Great disparities in monetary expenditures designed to prolong life, as shown by Tengs et al.,[3] may also be traced to public perceptions of risk. Such discrepancies are seen as irrational by many harsh critics of public perceptions. These critics draw a sharp dichotomy between the experts and the public. Experts are seen as purveying risk assessments, characterized as objective, analytic, wise, and rational—based on the *real risks*. In contrast, the public is seen to rely on *perceptions of risk* that are subjective, often hypothetical, emotional,

Human and Ecological Risk Assessment: Theory and Practice, Edited by Dennis J. Paustenbach
ISBN 0-471-14747-8   © 2002 John Wiley & Sons, Inc.

foolish, and irrational (see, e.g., Ref. Nos. 4, 5). Weiner[6] defends this dichotomy, arguing that "This separation of reality and perception is pervasive in a technically sophisticated society, and serves to achieve a necessary emotional distance ..." (p. 495).

In sum, polarized views, controversy, and overt conflict have become pervasive within risk assessment and risk management. A desperate search for salvation through risk-communication efforts began in the mid-1980s—yet, despite some localized successes, this effort has not stemmed the major conflicts or reduced much of the dissatisfaction with risk management. This dissatisfaction can be traced, in part, to a failure to appreciate the complex and socially determined nature of the concept "risk." In the remainder of this paper, I shall describe several streams of research that demonstrate this complexity and point toward the need for new definitions of risk and new approaches to risk management.

27.2 THE SUBJECTIVE AND VALUE-LADEN NATURE OF RISK ASSESSMENT

Attempts to manage risk must confront the question: "What is risk?" The dominant conception views risk as "the chance of injury, damage, or loss."[7] The probabilities and consequences of adverse events are assumed to be produced by physical and natural processes in ways that can be objectively quantified by risk assessment. Much social science analysis rejects this notion, arguing instead that risk is inherently subjective.[8-13] In this view, risk does not exist "out there," independent of our minds and cultures, waiting to be measured. Instead, human beings have invented the concept *risk* to help them understand and cope with the dangers and uncertainties of life. Although these dangers are real, there is no such thing as "real risk" or "objective risk." The nuclear engineer's probabilistic risk estimate for a nuclear accident or the toxicologist's quantitative estimate of a chemical's carcinogenic risk are both based on theoretical models, whose structure is subjective and assumption-laden, and whose inputs are dependent on judgment. As we shall see, nonscientists have their own models, assumptions, and subjective assessment techniques (intuitive risk assessments), which are sometimes very different from the scientists' models.

One way in which subjectivity permeates risk assessments is in the dependence of such assessments on judgments at every stage of the process, from the initial structuring of a risk problem to deciding which endpoints or consequences to include in the analysis, identifying and estimating exposures, choosing dose-response relationships, and so on. For example, even the apparently simple task of choosing a risk measure for a well-defined endpoint such as human fatalities is surprisingly complex and judgmental. Table 27.1 shows a few of the many different ways that fatality risks can be measured. How should we decide which measure to use when planning a risk assessment, recognizing that the choice is likely to make a big difference in how the risk is perceived and evaluated?

TABLE 27.1   Some Ways of Expressing Mortality Risks

Deaths per million people in the population
Deaths per million people within x miles of the source of exposure
Deaths per unit of concentration
Deaths per facility
Deaths per ton of air toxic released
Deaths per ton of air toxic absorbed by people
Deaths per ton of chemical produced
Deaths per million dollars of product produced
Loss of life expectancy associated with exposure to the hazard

An example taken from Wilson and Crouch[14] demonstrates how the choice of one measure or another can make a technology look either more or less risky. For example, between 1950 and 1970, coal mines became much less risky in terms of deaths from accidents per ton of coal, but they became marginally riskier in terms of deaths from accidents per employee. Which measure one thinks more appropriate for decision making depends on one's point of view. From a national point of view, given that a certain amount of coal has to be obtained to provide fuel, deaths per million tons of coal is the more appropriate measure of risk, whereas from a labor leader's point of view, deaths per thousand persons employed may be more relevant.

Each way of summarizing deaths embodies its own set of values.[15] For example, "reduction in life expectancy" treats deaths of young people as more important than deaths of older people, who have less life expectancy to lose. Simply counting fatalities treats deaths of the old and young as equivalent; it also treats as equivalent deaths that come immediately after mishaps and deaths that follow painful and debilitating disease. Using "number of deaths" as the summary indicator of risk implies that it is as important to prevent deaths of people who engage in an activity by choice and have been benefiting from that activity as it is to protect those who are exposed to a hazard involuntarily and get no benefit from it. One can easily imagine a range of arguments to justify different kinds of unequal weightings for different kinds of deaths, but to arrive at any selection requires a value judgment concerning which deaths one considers most undesirable. To treat the deaths as equal also involves a value judgment.

27.2.1   The Multidimensionality of Risk

Research has shown that the public has a broad conception of risk, qualitative and complex, that incorporates considerations such as uncertainty, dread, catastrophic potential, controllability, equity, risk to future generations, and so forth, into the risk equation.[16] In contrast, experts' perceptions of risk are not closely related to these dimensions or the characteristics that underlie them. Instead, studies show that experts tend to see riskiness as synonymous with probability of harm or expected mortality, consistent with the ways that risks tend to be characterized in risk assessments (see, for example, Ref. No. 17). As a result of these different perspectives, many conflicts over "risk" may result from experts and laypeople having different definitions of the concept. In this light, it is not surprising that expert recitations of "risk statistics" often do little to change people's attitudes and perceptions.

There are legitimate, value-laden issues underlying the multiple dimensions of public risk perceptions, and these values need to be considered in risk-policy decisions. For example, is risk from cancer (a dreaded disease) worse than risk from auto accidents (not dreaded)? Is a risk imposed on a child more serious than a known risk accepted voluntarily by an adult? Are the deaths of 50 passengers in separate automobile accidents equivalent to the deaths of 50 passengers in one airplane crash? Is the risk from a polluted Superfund site worse if the site is located in a neighborhood that has a number of other hazardous facilities nearby? The difficult questions multiply when outcomes other than human health and safety are considered.

27.2.2   The Risk Game

There are clearly multiple conceptions of risk.[18] Thompson and Dean[19] note that the traditional view of risk characterized by event probabilities and consequences treats the many subjective and contextual factors described above as secondary or accidental dimensions of risk, just as coloration might be thought of as a secondary or accidental dimension of an eye. Accidental dimensions might be extremely influential in the formation of attitudes toward risk, just as having blue or brown coloration may be influential in forming attitudes toward eyes. Furthermore, it may be that all risks possess some accidental dimensions, just as all organs of sight are in some way colored. Nevertheless, accidental dimensions do not serve as criteria for determining whether someone is or is not at risk, just as coloration is irrelevant to whether something is or is not an eye.

I believe that the multidimensional, subjective, value-laden, frame-sensitive nature of risky decisions, as described above, supports a very different view, which Thompson and Dean call "the contextualist conception." This conception places probabilities and consequences on the list of relevant risk attributes along with voluntariness, equity, and other important contextual parameters. On the contextualist view, the concept of risk is more like the concept of a game than the concept of the eye. Games have time limits, rules of play, opponents, criteria for winning or losing, and so on, but none of these attributes is essential to the concept of a game, nor is any of them characteristic of all games. Similarly, a contextualist view of risk assumes that risks are characterized by some combination of attributes such as voluntariness, probability, intentionality, equity, and so on, but that no one of these attributes is essential. The bottom line is that, just as there is no universal set of rules for games, there is no universal set of characteristics for describing risk. The characterization must depend on which risk game is being played.

27.3 SEX, POLITICS, AND EMOTION IN RISK JUDGMENTS

Given the complex and subjective nature of risk, it should not surprise us that many interesting and provocative things occur when people judge risks. Recent studies have shown that factors such as gender, race, political worldviews, affiliation, emotional affect, and trust are strongly correlated with risk judgments. Equally important is that these factors influence the judgments of experts as well as the judgments of laypersons.

27.3.1 Sex

Sex is strongly related to risk judgments and attitudes. Several dozen studies have documented the finding that men tend to judge risks as smaller and less problematic than do women. A number of hypotheses have been put forward to explain these differences in risk perception. One approach has been to focus on biological and social factors. For example, women have been characterized as more concerned about human health and safety because they give birth and are socialized to nurture and maintain life.[20] They have been characterized as physically more vulnerable to violence, such as rape, for example, and this may sensitize them to other risks.[21,22] The combination of biology and social experience has been put forward as the source of a "different voice" that is distinct to women.[23–24]

A lack of knowledge and familiarity with science and technology has also been suggested as a basis for these differences, particularly with regard to nuclear and chemical hazards. Women are discouraged from studying science and there are relatively few women scientists and engineers.[25] However, Barke et al.[26] have found that female physical scientists judge risks from nuclear technologies to be higher than do male physical scientists. Similar results with scientists were obtained by Slovic et al.[27] who found that female members of the British Toxicological Society were far more likely than male toxicologists to judge societal risks as moderate or high. Certainly the female scientists in these studies cannot be accused of lacking knowledge and technological literacy. Something else must be going on.

Hints about the origin of these sex differences come from a study by Flynn, Slovic, and Mertz[28] in which 1,512 Americans were asked, for each of 25 hazard items, to indicate whether the hazard posed (1) little or no risk, (2) slight risk, (3) moderate risk, or (4) high risk to society. The percentage of "high-risk" responses was greater for women on every item. Perhaps the most striking result from this study is shown in Fig. 27.1, which presents the mean risk ratings separately for White males, White females, non-White males, and non-White females. Across the 25 hazards, White males produced risk-perception ratings that were consistently much lower than the means of the other three groups.

Although perceived risk was inversely related to income and educational level, controlling for these differences statistically did not reduce much of the White-male effect on risk perception.

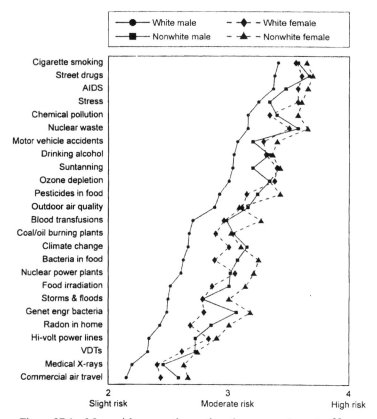

Figure 27.1 Mean risk-perception ratings by race and gender.[28]

When the data underlying Fig. 27.1 were examined more closely, Flynn et al. observed that not all White males perceived risks as low. The "White-male effect" appeared to be caused by about 30% of the White-male sample who judged risks to be extremely low. The remaining White males were not much different from the other subgroups with regard to perceived risk.

What differentiated these White males who were most responsible for the effect from the rest of the sample, including other White males who judged risks as relatively high? When compared to the remainder of the sample, the group of White males with the lowest risk-perception scores were better educated (42.7% college or postgraduate degree vs. 26.3% in the other group), had higher household incomes (32.1% above $50,000 vs. 21.0%), and were politically more conservative (48.0% conservative vs. 33.2%).

Particularly noteworthy is the finding that the low risk-perception subgroup of White males also held very different attitudes from the other respondents. Specifically, they were *more likely* than the others to:

- Agree that future generations can take care of themselves when facing risks imposed on them from today's technologies (64.2% vs. 46.9%).
- Agree that if a risk is very small it is okay for society to impose that risk on individuals without their consent (31.7% vs. 20.8%).
- Agree that science can settle differences of opinion about the risks of nuclear power (61.8% vs. 50.4%).
- Agree that government and industry can be trusted with making the proper decisions to manage the risks from technology (48.0% vs. 31.1%).
- Agree that we can trust the experts and engineers who build, operate, and regulate nuclear power plants (62.6% vs. 39.7%).
- Agree that we have gone too far in pushing equal rights in this country (42.7% vs. 30.9%).
- Agree with the use of capital punishment (88.2% vs. 70.5%).
- Disagree that technological development is destroying nature (56.9% vs. 32.8%).
- Disagree that they have very little control over risks to their health (73.6% vs. 63.1%).
- Disagree that the world needs a more equal distribution of wealth (42.7% vs. 31.3%).
- Disagree that local residents should have the authority to close a nuclear power plant if they think it is not run properly (50.4% vs. 25.1%).
- Disagree that the public should vote to decide on issues such as nuclear power (28.5% vs. 16.7%).

In sum, the subgroup of White males who perceive risks to be quite low can be characterized by trust in institutions and authorities and by anti-egalitarian attitudes, including a disinclination toward giving decision-making power to citizens in areas of risk management.

The results of this study raise new questions. What does it mean for the explanations of gender differences when we see that the sizable differences between White males and White females do not exist for non-White males and non-White females? Why do a substantial percentage of White males see the world as so much less risky than everyone else sees it?

Obviously, the salience of biology is reduced by these data on risk perception and race. Biological factors should apply to non-White men and women as well as to White men and women. The present data thus move us away from biology and toward sociopolitical explanations. Perhaps White males see less risk in the world because they create, manage, control, and benefit from many of the major technologies and activities. Perhaps women and non-White men see the world as more dangerous because in many ways they are more vulnerable, because they benefit less from many of its technologies and institutions, and because they have less power and control over what happens in their communities and their lives. Although the survey conducted by Flynn, Slovic, and

Mertz was not designed to test these alternative explanations, the race and gender differences in perceptions and attitudes point toward the role of power, status, alienation, trust, perceived government responsiveness, and other sociopolitical factors in determining perception and acceptance of risk.

To the extent that these sociopolitical factors shape public perception of risks, we can see why traditional attempts to make people see the world as White males do, by showing them statistics and risk assessments, are often unsuccessful. The problem of risk conflict and controversy goes beyond science. It is deeply rooted in the social and political fabric of our society.

27.3.2   Risk Perception and Worldviews

The influence of social, psychological, and political factors also can be seen in studies examining the impact of worldviews on risk judgments. Worldviews are general social, cultural, and political attitudes that appear to have an influence over people's judgments about complex issues.[29-31] Dake[30] has conceptualized worldviews as "orienting dispositions," because of their role in guiding people's responses. Some of the worldviews identified to date are listed below, along with representative attitude statements:

- Fatalism (e.g., "I feel I have very little control over risks to my health.")
- Hierarchy (e.g., "Decisions about health risks should be left to the experts.")
- Individualism (e.g., "In a fair system, people with more ability should earn more.")
- Egalitarianism (e.g., "If people were treated more equally, we would have fewer problems.")
- Technological Enthusiasm (e.g., "A high-technology society is important for improving our health and social well-being.")

People differ from one another in these views. Fatalists tend to think that what happens in life is preordained. Hierarchists like a society organized such that commands flow down from authorities and obedience flows up the hierarchy. Egalitarians prefer a world in which power and wealth are more evenly distributed. Individualists like to do their own thing, unhindered by government or any other kind of constraints.

Dake,[31,32] Jenkins-Smith[33] and others have measured worldviews with survey techniques and found them to be strongly linked to public perceptions of risk. My colleagues and I have obtained similar results. Peters and Slovic (Ref. No. 34; see also Ref. No. 35), using the same national survey data analyzed for race and gender effects by Flynn et al.,[28] found particularly strong correlations between worldviews and attitudes toward nuclear power. Egalitarians tended to be strongly anti-nuclear; persons endorsing fatalist, hierarchist, and individualistic views tended to be pro-nuclear. Peters and Slovic also showed strong

TABLE 27.2 Percentage of People Who Agreed to Support a New Nuclear Power Plant in Their Community[a]

	Agreement with the worldview question			
	Strongly disagree	Disagree	Agree	Strongly agree
Individualism Worldview				
In a fair system people with more ability should earn more	37.5	37.7	47.2	53.4
Egalitarian Worldview				
What this world needs is a more equal distribution of wealth	73.9	53.7	43.8	33.8

[a]The exact question was: "If your community was faced with a potential shortage of electricity, do you agree or disagree that a new nuclear power plant should be built to supply that electricity?" The cell entries in this table show the percentage of people who agreed with this statement conditioned by whether they agreed or disagreed with questions about individualism and egalitarianism.

correlations between worldviews and perceptions of risk from a wide range of hazards.

Table 27.2 illustrates some of the findings with regard to attitudes toward nuclear power. It shows that people who agreed that "in a fair system people with more ability should earn more" were more likely to support a local nuclear power plant than were people who disagreed with that statement. Similarly, those who agreed with the egalitarian view of equal distribution of wealth were less likely to support a nuclear power plant than were those who disagreed with that view.

27.3.3 Risk Perception, Emotion, and Affect

The studies described in the preceding section illustrate the role of worldviews as orienting mechanisms. Research suggests that emotion is also an orienting mechanism that directs fundamental psychological processes such as attention, memory, and information processing. Emotion and worldviews may thus be functionally similar in that both may help us navigate quickly and efficiently through a complex, uncertain, and sometimes dangerous world.

The discussion in this section is concerned with a subtle form of emotion called affect, defined as a positive (like) or negative (dislike) evaluative feeling toward an external stimulus (e.g., some hazard such as cigarette smoking). Such evaluations occur rapidly and automatically—note how quickly you sense a negative affective feeling toward the stimulus word "hate" or the word "cancer."

Support for the conception of affect as an orienting mechanism comes from a study by Alhakami and Slovic.[36] They observed that, whereas the risks and benefits to society from various activities and technologies (e.g., nuclear power, commercial aviation) tend to be *positively* associated in the world, they are *inversely* correlated in people's minds (higher perceived benefit is associated with lower perceived risk; lower perceived benefit is associated with higher perceived risk). Alhakami and Slovic found that this inverse relationship was linked to people's reliance on general affective evaluations when making risk/benefit judgments. When the affective evaluation was favorable (as with automobiles, for example), the activity or technology being judged was seen as having high benefit and low risk; when the evaluation was unfavorable (e.g., as with pesticides), risks tended to be seen as high and benefits as low. It thus appears that the affective response is primary, and the risk and benefit judgments are derived (at least partly) from it.

Finucane et al.[37] investigated the inverse relationship between risk and benefit judgments under a time-pressure condition designed to limit the use of analytic thought and enhance the reliance on affect. As expected, the inverse relationship was strengthened when time pressure was introduced. A second study tested and confirmed the hypothesis that providing information designed to alter the favorability of one's overall affective evaluation of an item (say nuclear power) would systematically change the risk and benefit judgments for that item. For example, providing information calling people's attention to the benefits provided by nuclear power (as a source of energy) depressed people's perception of the risks of that technology. The same sort of reduction in perceived risk occurred for food preservatives and natural gas, when information about their benefits was provided. Information about risk was also found to alter perception of benefit. A model depicting how reliance upon affect can lead to these observed changes in perception of risk and benefit is shown in Fig. 27.2.

Slovic et al.[38,39] studied the relationship between affect and perceived risk for hazards related to nuclear power. For example, Slovic, Flynn, and Layman asked respondents "What is the first thought or image that comes to mind when you hear the phrase 'nuclear waste repository?'" After providing up to three associations to the repository stimulus, each respondent rated the affective quality of these associations on a five-point scale, ranging from extremely negative to extremely positive.

Although most of the images that people evoke when asked to think about nuclear power or nuclear waste are affectively negative (e.g., death, destruction, war, catastrophe), some are positive (e.g., abundant electricity and the benefits it brings). The affective values of these positive and negative images appear to sum in a way that is predictive of our attitudes, perceptions, and behaviors. If the balance is positive, we respond favorably; if it is negative, we respond unfavorably. For example, the affective quality of a person's associations to a nuclear waste repository was found to be related to whether the person would vote for or against a referendum on a nuclear waste repository and to their judgments regarding the risk of a repository accident. Specifically, more than 90% of those people whose first image was judged very negative said that they would vote against a repository in Nevada; fewer than 50% of those people whose first image was positive said they would vote against the repository.[38]

Using data from the national survey of 1,500 Americans described earlier, Peters and Slovic[34] found that the affective ratings of associations to the stim-

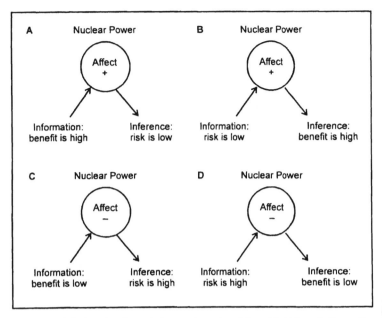

Figure 27.2 Model showing how information about benefit (A) or information about risk (B) could create a more positive affective evaluation of nuclear power and lead to inferences about risk and benefit that are affectively congruent with the information input. Similarly, information could decrease the affective evaluation of nuclear power as in C and D, resulting in inferences that are opposite those in A and B.[37]

ulus "nuclear power" were highly predictive of responses to the question: "If your community was faced with a shortage of electricity, do you agree or disagree that a new nuclear power plant should be built to supply that electricity?" Among the 25% of respondents with the most positive associations to nuclear power, 69% agreed to building a new plant. Among the 25% of respondents with the most negative associations, only 13% agreed.

27.3.4 Worldviews, Affect, and Toxicology

Affect and worldviews seem to influence the risk-related judgments of scientists, as well as laypersons. Evidence for this comes from studies of "intuitive toxicology" that Torbjörn Malmfors, Nancy Neil, Iain Purchase, and I have been conducting in the United States, Canada, and the UK during the past decade. These studies have surveyed both toxicologists and laypersons about a wide range of concepts relating to risks from chemicals. We have examined judgments about the effects of chemical concentration, dose, and exposure on risk. We have also questioned our respondents about the value of animal studies for predicting the effects of chemicals on humans. Before showing how worldviews and affect enter into toxicologists' judgments, a brief description of some basic results will be presented.

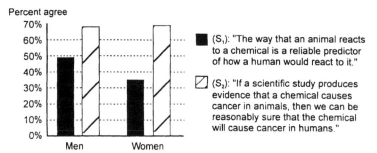

Figure 27.3 Agreement among members of the public in the United States for Statements S_1 and S_2.[40]

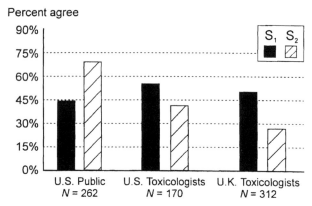

Figure 27.4 Agreement with two statements, S_1 and S_2, regarding the extrapolation of chemical effects in animals to chemical effects in humans.[41]

Consider two survey items that we have studied repeatedly. One is statement S_1: "Would you agree or disagree that the way an animal reacts to a chemical is a reliable predictor of how a human would react to it?" The second statement, S_2, is a little more specific: "If a scientific study produces evidence that a chemical causes cancer in animals, then we can be reasonably sure that the chemical will cause cancer in humans."

When members of the American and Canadian public responded to these items, they showed moderate agreement with S_1; about half the people agreed and half disagreed that animal tests were reliable predictors of human reactions to chemicals. However, in response to S_2, which stated that the animal study found evidence of cancer, there was a jump in agreement to about 70% among both male and female respondents (see Fig. 27.3). The important point about the pattern of response is that agreement was higher on the second item.

What happens if toxicologists are asked about these two statements? Figure 27.4 shows that toxicologists in the United States and toxicologists in the UK

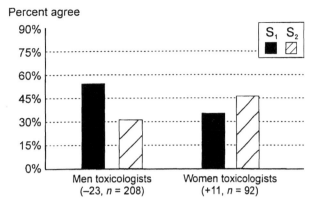

Figure 27.5 Agreement of men and women toxicologists in the United Kingdom with two statements regarding extrapolation of chemical effects in animals to chemical effects in humans.[41]

responded similarly to the public on the first statement but differently on the second. They exhibited the same rather middling level of agreement with the general statement about animal studies as predictors of human health effects.* However, when these studies were said to find evidence of carcinogenicity in animals, then the toxicologists were less likely to agree that the results could be extrapolated to humans. Thus, the same findings which lead toxicologists to be less willing to generalize to humans lead the public to see the chemical as more dangerous for humans.†

Figure 27.5 presents the responses for S_1 and S_2 among men and women toxicologists in the UK (208 men and 92 women). Here we see another interesting finding. The men agree less on the second statement than on the first, but the women agree more, just like the general public. Women toxicologists are more willing than men to say that if a chemical causes cancer in animals, it will likely cause cancer in humans.

We also examined the relative agreement with Statements S_1 and S_2 for each of the British toxicologists in our survey. Greater agreement with S_2 than with S_1 was associated with:

• higher mean perceptions of risk across 25 hazards (the risk-perception index),
• rating pesticides and industrial chemicals as "bad" on a task in which various items were rated on a scale ranging from *good* to *bad,*
• being female,
• being younger,

---

*This is actually a very surprising result, given the heavy reliance on animal studies in toxicology.
† This pattern suggests that animal studies may be scaring the public without informing science.

- agreeing that "I have little control over risks to my health."
- holding an academic position rather than a position in industry,
- disagreeing that "technology is important for social well-being," and
- disagreeing that "economic growth is necessary for good quality of life."

These studies of intuitive toxicology have yielded a number of intriguing findings. One is the low percentage of agreement that animal studies can predict human health effects. Another is that toxicologists show even less confidence in studies that find cancer in animals resulting from chemical exposure. The public, on the other hand, has high confidence in animal studies that find cancer. Disagreements among toxicologists are systematically linked to gender, affiliation (academic vs. other), worldviews, and affect. Thus affective and sociopolitical factors appear to influence scientists' risk evaluations in much the same way as they influence the public's perceptions.*

27.4   THE IMPORTANCE OF TRUST

The research described above has painted a portrait of risk perception influenced by the interplay of psychological, social, and political factors. Members of the public and experts can disagree about risk because they define risk differently, have different worldviews, different affective experiences and reactions, or different social status. Another reason why the public often rejects scientists' risk assessments is lack of trust. Trust in risk management, like risk perception, has been found to correlate with gender, race, worldviews, and affect.

Social relationships of all types, including risk management, rely heavily on trust. Indeed, much of the contentiousness that has been observed in the risk-management arena has been attributed to a climate of distrust that exists between the public, industry, and risk-management professionals (e.g., Ref. Nos. 38 and 42). The limited effectiveness of risk-communication efforts can be attributed to the lack of trust. If you trust the risk manager, communication is relatively easy. If trust is lacking, no form or process of communication will be satisfactory.[43]

27.4.1   How Trust Is Created and Destroyed

One of the most fundamental qualities of trust has been known for ages. Trust is fragile. It is typically created rather slowly, but it can be destroyed in an

---

*Although we have focused only on the relationship between toxicologists' reaction to chemicals and their responses to S_1 and S_2, there were may other links between affect and attitudes in the survey. For example, the very simple bad-good ratings of pesticides correlated significantly $(r = .20)$ with agreement that there is a threshold dose for nongenotoxic carcinogens. The same ratings correlated $-.27$ with the belief that synergistic effects of chemicals cause animal studies of single chemicals to underestimate risk to humans.

instant—by a single mishap or mistake. Thus, once trust is lost, it may take a long time to rebuild it to its former state. In some instances, lost trust may never be regained. Abraham Lincoln understood this quality. In a letter to Alexander McClure, he observed: "If you *once* forfeit the confidence of your fellow citizens, you can *never* regain their respect and esteem" [italics added].

The fact that trust is easier to destroy than to create reflects certain fundamental mechanisms of human psychology called here "the asymmetry principle." When it comes to winning trust, the playing field is not level. It is tilted toward distrust, for each of the following reasons:

1. Negative (trust-destroying) events are more visible or noticeable than positive (trust-building) events. Negative events often take the form of specific, well-defined incidents such as accidents, lies, discoveries of errors, or other mismanagement. Positive events, while sometimes visible, more often are fuzzy or indistinct. For example, how many positive events are represented by the safe operation of a nuclear power plant for one day? Is this one event? Dozens of events? Hundreds? There is no precise answer. When events are invisible or poorly defined, they carry little or no weight in shaping our attitudes and opinions.

2. When events are well-defined and do come to our attention, negative (trust-destroying) events carry much greater weight than positive events.[42]

3. Adding fuel to the fire of asymmetry is yet another idiosyncrasy of human psychology—sources of bad (trust-destroying) news tend to be seen as more credible than sources of good news. The findings reported in Section 3.4 regarding "intuitive toxicology" illustrate this point. In general, confidence in the validity of animal studies is not particularly high. However, when told that a study has found that a chemical is carcinogenic in animals, members of the public express considerable confidence in the validity of this study for predicting health effects in humans.*

4. Another important psychological tendency is that distrust, once initiated, tends to reinforce and perpetuate distrust. Distrust tends to inhibit the kinds of personal contacts and experiences that are necessary to overcome distrust. By avoiding others whose motives or actions we distrust, we never get to see that these people are competent, well-meaning, and trustworthy.

27.4.2 "The System Destroys Trust"

Thus far we have been discussing the psychological tendencies that create and reinforce distrust in situations of risk. Appreciation of those psychological

---

*Further evidence supporting this point comes from a representative sample of the U.S. public surveyed by the author and his colleagues in 1998. Whereas 61% agreed with the general statement about the reliability of animal tests, 72% agreed with S_2, the statement about a test that gave bad news. In response to a new question (S_3) giving good news ("the scientific studies found no evidence that the chemical causes cancer in animals"), only 43% agreed that this enabled us to be reasonably sure the chemical dose no cause cancer in humans.

principles leads us toward a new perspective on risk perception, trust, and conflict. Conflicts and controversies surrounding risk management are not due to public irrationality or ignorance but, instead, can be seen as expected side effects of these psychological tendencies, interacting with a highly participatory democratic system of government and amplified by certain powerful technological and social changes in society. Technological change has given the electronic and print media the capability (effectively utilized) of informing us of news from all over the world—often right as it happens. Moreover, just as individuals give greater weight and attention to negative events, so do the news media. Much of what the media reports is bad (trust-destroying) news.[44,45]

A second important change, a social phenomenon, is the rise of powerful special interest groups, well-funded (by a fearful public) and sophisticated in using their own experts and the media to communicate their concerns and their distrust to the public to influence risk policy debates and decisions.[46] The social problem is compounded by the fact that we tend to manage our risks within an adversarial legal system that pits expert against expert, contradicting each other's risk assessments and further destroying the public trust.

The young science of risk assessment is too fragile, too indirect, to prevail in such a hostile atmosphere. Scientific analysis of risks cannot allay our fears of low-probability catastrophes or delayed cancers unless we trust the system. In the absence of trust, science (and risk assessment) can only feed public concerns, by uncovering more bad news. A single study demonstrating an association between exposure to chemicals or radiation and some adverse health effect cannot easily be offset by numerous studies failing to find such an association. Thus, for example, the more studies that are conducted looking for effects of electric and magnetic fields or other difficult-to-evaluate hazards, the more likely it is that these studies will increase public concerns, even if the majority of these studies fail to find any association with ill health.[47,48] In short, because evidence for lack of risk often carries little weight, risk-assessment studies tend to increase perceived risk.

27.5   RESOLVING RISK CONFLICTS: WHERE DO WE GO FROM HERE?

27.5.1   Technical Solutions to Risk Conflicts

There has been no shortage of high-level attention given to the risk conflicts described above. One prominent proposal by Justice Stephen Breyer[49] attempts to break what he sees as a vicious circle of public perception, congressional overreaction, and conservative regulation that leads to obsessive and costly preoccupation with reducing negligible risks as well as to inconsistent standards among health and safety programs. Breyer sees public misperceptions of risk and low levels of mathematical understanding at the core of excessive regulatory response. His proposed solution is to create a small centralized administrative group charged with creating uniformity and rationality in highly technical areas of risk management. This group would be staffed by civil servants with experience in health and environmental agencies, Congress, and the Office of Management and Budget (OMB). A parallel is drawn between this group and the prestigious Conseil d'Etat in France.

Similar frustration with the costs of meeting public demands led the 104th Congress to introduce numerous bills designed to require all major new regulations to be justified by extensive risk assessments. Proponents of this legislation argued that such measures are necessary to ensure that regulations are based on "sound science" and effectively reduce significant risks at reasonable costs.

The language of this proposed legislation reflects the traditional narrow view of risk and risk assessment based "only on the best reasonably available scientific data and scientific understanding." Agencies are further directed to develop a systematic program for external peer review using "expert bodies" or "other devices comprised of participants selected on the basis of their expertise relevant to the sciences involved" (Ref. No. 50, pp. 57–58). Public participation in this process is advocated, but no mechanisms for this are specified.

The proposals by Breyer and the 104th Congress are typical in their call for more and better technical analysis and expert oversight to rationalize risk management. There is no doubt that technical analysis is vital for making risk decisions better informed, more consistent, and more accountable. However, value conflicts and pervasive distrust in risk management cannot easily be reduced by technical analysis. Trying to address risk controversies primarily with more science is, in fact, likely to exacerbate conflict.

27.5.2   Process-Oriented Solutions

A major objective of this paper has been to demonstrate the complexity of risk and its assessment. To summarize the earlier discussions, danger is real, but risk is socially constructed. Risk assessment is inherently subjective and represents a blending of science and judgment with important psychological, social, cultural, and political factors. Finally, our social and democratic institutions, remarkable as they are in many respects, breed distrust in the risk arena.

Whoever controls the definition of risk controls the rational solution to the problem at hand. If you define risk one way, then one option will rise to the top as the most cost-effective or the safest or the best. If you define it another way, perhaps incorporating qualitative characteristics and other contextual factors, you will likely get a different ordering of your action solutions.[51] Defining risk is thus an exercise in power.

Scientific literacy and public education are important, but they are not central to risk controversies. The public is not irrational. The public is influenced by emotion and affect in a way that is both simple and sophisticated. So are scientists. The public is influenced by worldviews, ideologies, and values. So are scientists, particularly when they are working at the limits of their expertise.

The limitations of risk science, the importance and difficulty of maintaining trust, and the subjective and contextual nature of the risk game point to the need for a new approach—one that focuses on introducing more public participation into both risk assessment and risk decision making to make the decision process more democratic, improve the relevance and quality of technical analysis, and increase the legitimacy and public acceptance of the resulting decisions. Work by scholars and practitioners in Europe and North America has begun to lay the foundations for improved methods of public participation within deliberative decision processes that include negotiation, mediation, oversight committees, and other forms of public involvement.[52-56]

Recognizing interested and affected citizens as legitimate partners in the exercise of risk assessment is no short-term panacea for the problems of risk management. It won't be easy and it isn't guaranteed. But serious attention to participation and process issues may, in the long run, lead to more satisfying and successful ways to manage risk.

ACKNOWLEDGMENTS

Preparation of this paper was supported by the Alfred P. Sloan Foundation, the Electric Power Research Institute, and the National Science Foundation under Grants No. 91-10592 and SBR 94-122754. Reprinted with permission from M. H. Bazerman, D. M. Messick, A. E. Tenbrunsel, & K. A. Wade-Benzoni (Eds.), *Environment, ethics, and behavior* (pp. 277–313). San Francisco, New Lexington, 1997. Copyright © 1997 The New Lexington Press, an imprint of Jossey-Bass Inc., Publishers. All rights reserved.

REFERENCES

1. R. Gregory, J. Flynn, and P. Slovic, "Technological Stigma," *American Scientist* 83, 220–223 (1995).

2. U.S. Environmental Protection Agency (USEPA). Office of Policy Analysis, *Unfinished Business: A Comparative Assessment of Environmental Problems* (Washington, DC, Author, 1987, February).

3. T. O. Tengs, M. E. Adams, J. S. Pliskin, D. G. Safran, J. E. Siegel, M. Weinstein, and J. D. Graham, "Five-Hundred Life-Saving Interventions and Their Cost Effectiveness," *Risk Anal.* 15, 369–390 (1995).

4. R. L. DuPont, *Nuclear Phobia: Phobic Thinking About Nuclear Power* (Washington, DC, The Media Institute, 1980).

5. V. T. Covello, W. G. Flamm, J. V. Rodricks, and R. G. Tardiff, *The Analysis of Actual Versus Perceived Risks* (New York, Plenum, 1983).

6. R. F. Weiner, "Comment on Sheila Jasanoff's Guest Editorial," *Risk Analysis* 13, 495–496 (1993).

7. N. Webster, *Webster's New Twentieth Century Dictionary*, 2nd ed. (New York, Simon & Schuster, 1983).

8. S. O. Funtowicz and J. R. Ravetz, "Three Types of Risk Assessment and the Emergence of Post-Normal Science," in *Social Theories of Risk*, S. Krimsky and D. Golding (eds.) (Westport, CT, Praeger, 1992), pp. 251–273.

9. S. Krimsky and D. Golding (eds.), *Social Theories of Risk* (Westport, CT, Praeger-Greenwood, 1992).

10. H. Otway, "Public Wisdom, Expert Fallibility: Toward a Contextual Theory of Risk," in *Social Theories of Risk*, S. Krimsky and D. Golding (eds.) (Westport, CT, Praeger, 1992), pp. 215–228.

11. N. Pidgeon, C. Hood, D. Jones, B. Turner, and R. Gibson, "Risk Perception," in *Risk: Analysis, Perception and Management*, Royal Society Study Group (ed.) (London, The Royal Society, 1992), pp. 89–134.

12. P. Slovic, "Perception of Risk: Reflections on the Psychometric Paradigm," in *Social Theories of Risk*, S. Krimsky and D. Golding (eds.) (New York, Praeger, 1992), pp. 117–152.

13. B. Wynne, "Risk and Social Learning: Reification to Engagement," in *Social Theories of Risk*, S. Krimsky and D. Golding (eds.) (Westport, CT, Praeger, 1992), pp. 275–300.

14. E. A. C. Crouch and R. Wilson, *Risk/Benefit Analysis* (Cambridge, MA, Ballinger, 1982).

15. National Research Council, *Improving Risk Communication* (Washington, DC, National Academy Press, 1989).

16. P. Slovic, "Perception of Risk," *Science* 236, 280–285 (1987).

17. B. L. Cohen, "Criteria for Technology Acceptability," *Risk Anal.* 5, 1–2 (1985a).

18. K. S. Shrader-Frechette, *Risk and Rationality: Philosophical Foundations for Populist Reforms* (Berkeley, University of California, 1991).

19. P. B. Thompson and W. R. Dean, "Competing Conceptions of Risk," *Risk: Health, Safety Environ.* 7, 361–384 (1996).

20. M. A. E. Steger and S. L. Witt, "Gender Differences in Environmental Orientations: A Comparison of Publics and Activists in Canada and the U.S," *West. Polit. Quart.* 42, 627–649 (1989).

21. T. L. Baumer, "Research on Fear of Crime in the United States," *Victimology* 3, 254–264 (1978).

22. S. Riger, M. T. Gordon, and R. LeBailly, "Women's Fear of Crime: From Blaming to Restricting the Victim," *Victimology* 3, 274–284 (1978).

23. C. Gilligan, *In a Different Voice: Psychological Theory and Women's Development* (Cambridge, MA, Harvard University, 1982).

24. C. Merchant, *The Death of Nature: Women, Ecology, and the Scientific Revolution* (New York, Harper & Row, 1980).

25. J. Alper, "The Pipeline is Leaking Women All the Way Along," *Science* 260, 409–411 (1993).

26. R. Barke, H. Jenkins-Smith, and P. Slovic, "Risk Perceptions of Men and Women Scientists," *Social Sci. Quart.* 78(1), 167–176 (1997).

27. P. Slovic, T. Malmfors, C. K. Mertz, N. Neil, and I. F. H. Purchase, "Evaluating Chemical Risks: Results of a Survey of the British Toxicology Society," *Hum. Exp. Toxicol.* 16, 289–304 (1997).

28. J. Flynn, P. Slovic, and C. K. Mertz, "Gender, Race, and Perception of Environmental Health Risks," *Risk Anal.* 14(6), 1101–1108 (1994).

29. D. M. Buss, K. H. Craik, and K. M. Dake, "Contemporary Worldviews and Perception of the Technological System," in *Risk Evaluation and Management*, V. T. Covello, J. Menkes, and J. L. Mumpower (eds.) (New York, Plenum, 1986), pp. 93–130.

30. K. Dake, "Orienting Dispositions in the Perception of Risk: An Analysis of Contemporary Worldviews and Cultural Biases," *J. Cross-Cult. Psychol.* 22, 61–82 (1991).

31. J. M. Jasper, *Nuclear Politics: Energy and the State in the United States, Sweden, and France* (Princeton, NJ, Princeton University Press, 1990).

32. K. Dake, "Myths of Nature: Culture and the Social Construction of Risk," *J. Social Issues* 48, 21–27 (1992).

33. H. C. Jenkins-Smith, *Nuclear Imagery and Regional Stigma: Testing Hypotheses of Image of Acquisition and Valuation Regarding Nevada*. Technical report Institute for Public Policy, University of New Mexico, (Albuquerque, NM, 1993).

34. E. Peters and P. Slovic, "The Role of Affect and Worldviews as Orienting Dispositions in the Perception and Acceptance of Nuclear Power," *J. Appl. Social Psychol.* 26, 1427–1453 (1996).

35. P. Slovic and E. Peters, "The Importance of Worldviews in Risk Perception," *Risk Dec. Policy* 3(2), 165–170 (1998).

36. A. S. Alhakami and P. Slovic, "A Psychological Study of the Inverse Relationship Between Perceived Risk and Perceived Benefit," *Risk Anal.* 14(6), 1085–1096 (1994).

37. M. L. Finucane, A. Alhakami, P. Slovic, and S. M. Johnson, "The Affect Heuristic in Judgments of Risks and Benefits." *J. Behav. Decision Making* (in press).

38. P. Slovic, J. Flynn, and M. Layman, "Perceived Risk, Trust, and the Politics of Nuclear Waste," *Science* 254, 1603–1607 (1991).

39. P. Slovic, M. Layman, N. Kraus, J. Flynn, J. Chalmers, and G. Gesell, "Perceived Risk, Stigma, and Potential Economic Impacts of a High-Level Nuclear Waste Repository in Nevada," *Risk Anal.* 11, 683–696 (1991).

40. N. N. Kraus, T. Malmfors, and P. Slovic, "Intuitive Toxicology: Expert and Lay Judgments of Chemical Risks," *Risk Anal.* 12, 215–232 (1992).

41. P. Slovic, "Trust, Emotion, Sex, Politics, and Science: Surveying the Risk-Assessment Battlefield," in *Environment, Ethics, and Behavior*, M. H. Bazerman, D. M. Messick, A. E. Tenbrunsel, and K. A. Wade-Benzoni (eds.) (San Francisco, New Lexington, 1997), pp. 277–313.

42. P. Slovic, "Perceived Risk, Trust, and Democracy," *Risk Anal.* 13, 675–682 (1993).

43. J. Fessenden-Raden, J. M. Fitchen, and J. S. Heath, "Providing Risk Information in Communities: Factors Influencing What is Heard and Accepted," *Sci. Technol. Hum. Values* 12, 94–101 (1987).

44. G. Koren and N. Klein, "Bias Against Negative Studies in Newspaper Reports of Medical Research," *J. Am. Med. Assoc.* 266, 1824–1826 (1991).

45. J. Lichtenberg and D. MacLean, "Is Good News No News?" *Geneva Papers Risk Ins.* 17, 362–365 (1992).

46. D. Fenton, "How a PR Firm Executed the Alar Scare," *Wall Street J. A22* (October 3, 1989).

47. D. MacGregor, P. Slovic, and M. G. Morgan, "Perception of Risks from Electromagnetic Fields: A Psychometric Evaluation of a Risk-Communication Approach," *Risk Anal.* 14(5), 815–828 (1994).

48. M. G. Morgan, P. Slovic, I. Nair, D. Geisler, D. MacGregor, B. Fischhoff, D. Lincoln, and K. Florig, "Powerline Frequency Electric and Magnetic Fields: A Pilot Study of Risk Perception," *Risk Anal.* 5, 139–149 (1985).

49. S. Breyer, *Breaking the Vicious Circle: Toward Effective Risk Regulation* (Cambridge, MA, Harvard University, 1993).

50. U.S. Senate, *The Comprehensive Regulatory Reform Act of 1995.* Dole/Johnson discussion draft of S. 5343 (Washington, DC, U.S. Government Printing Office, June 1995).

51. B. Fischhoff, S. Watson, and C. Hope, "Defining Risk," *Policy Sciences* 17, 123–139 (1984).

52. M. R. English, *Siting Low-Level Radioactive Waste Disposal Facilities: The Public Policy Dilemma* (New York, Quorum, 1992).

53. H. Kunreuther, K. Fitzgerald, and T. D. Aarts, "Siting Noxious Facilities: A Test of the Facility Siting Credo," *Risk Anal.* 13, 301–318 (1993).

54. National Research Council. Committee on Risk Characterization, *Understanding Risk: Informing Decisions in a Democratic Society*, P. C. Stern and H. V. Fineberg (eds.) (Washington, DC, National Academy Press, 1996).

55. O. Renn, T. Webler, and B. B. Johnson, "Public Participation in Hazard Management: The Use of Citizen Panels in the U.S.," *Risk—Issues Health Safety* 2(3), 197–226 (Summer, 1991).

56. O. Renn, T. Webler, and P. Wiedemann, *Fairness and Competence in Citizen Participation* (Dordrecht, The Netherlands, Kluwer, 1995).

28 Democratization of Risk Analysis

GAIL CHARNLEY[1] and E. DONALD ELLIOTT[2]
[1]HealthRisk Strategies, Washington, D.C.
[2]Yale Law School, New Haven, Connecticut

28.1 INTRODUCTION

In some ways, assessing and managing risks to our health and our environment have evolved in the past decade toward being less technocratic and more inclusive. Many environmental health regulations and risk management decisions are developed and implemented using transparent processes that include opportunities for consultation and cooperation among regulators, businesses, and communities. This trend reflects the Jeffersonian ideals that people should be involved in their own governance[1] and is a response to a lack of public trust in risk management decisions made by government and industry; expanded public awareness of environmental, health, and safety issues; increased social expectations for improved environmental quality; changes in information technology; and the desire by business and government to demonstrate responsiveness to public concerns.[2] At the same time, it is a natural outgrowth of the interest group pluralism model of administrative action in which regulatory agencies act as brokers for the many relevant interests and perspectives on problems within their jurisdictions.[3]

To a large extent, the body of U.S. laws that seek to establish practices that will ensure safety—or at least mitigate risk—from chemical or other contaminant exposures were established before risk assessment was a well-recognized and codified discipline. Many of those laws were passed in the 1970s in response to public awareness of environmental, health, and safety issues and increased social expectations for improved environmental quality. Most of the methodology of risk assessment was developed in reaction to the calls by those laws to define limits on exposure that will "protect the public health with an adequate margin of safety" according to the Clean Air Act or "protect the public welfare" according to the Clean Water Act. That is, in passing the laws, the U.S. Congress called on the regulatory agencies to develop means to assess risks so as to define exposure levels that would achieve the stated qualitative goals of health protection.[4]

Human and Ecological Risk Assessment: Theory and Practice, Edited by Dennis J. Paustenbach
ISBN 0-471-14747-8   © 2002 John Wiley & Sons, Inc.

Risk assessment has emerged over the last two decades in response to those laws as the dominant paradigm in the United States for including science in regulatory decision making about the best ways to manage threats to health and the environment. Risk assessment is not the only input to decision making in a democracy, of course; social, economic, feasibility, legal, equity, and political considerations also play important roles. The continuing challenge is to maintain a role for risk assessment and to preserve the integrity of science when decision making is influenced by many nontechnical factors.

A number of organizations, court decisions, and scholarly studies have provided momentum for the trend toward more risk-based yet democratic decision making about the best ways to reduce or eliminate health and environmental risks. Others are inconsistent with that trend. This chapter describes some of each.

28.2 RISK ANALYSIS AND THE LAW

Environmental health risk management decision making in the United States was guided by a policy of precaution for many years. Precautionary policies are based on the idea that it is better to be safe than sorry; that is, precaution reflects the need to take action in the face of potentially serious risks without awaiting the results of scientific research that establishes cause-and-effect relationships with full scientific certainty.[5] For example, in the 1950s the Delaney clause required the Food and Drug Administration (FDA) to ban outright food and color additives that had been shown to produce tumors in humans or laboratory animals. In the 1970s, the debate about banning leaded gasoline led to a court decision referred to as the *Ethyl* decision, establishing a legal basis for precautionary decision making. At the time, there was great disagreement about the wisdom of banning leaded gasoline when the benefits of doing so were unclear. But the *Ethyl* decision[6] upheld the decision of the Environmental Protection Agency (EPA) to take a precautionary approach and ban lead anyway, even in the absence of scientific evidence adequate to demonstrate exactly what the benefits of removing it would be.

In 1980, however, the *Benzene* decision overturned the precautionary principle basis of the *Ethyl* decision and substituted an evidence-based principle.[7] The *Benzene* decision struck down a workplace standard for benzene exposure that was based on a policy of trying to reduce concentrations of benzene as far as technologically possible without considering whether existing concentrations posed a significant risk to health. The Supreme Court decided that benzene could be regulated only if it posed a "significant risk of harm." Although the court did not define "significant risk of harm" and did not require that the magnitude of the risk be determined precisely, the decision created a need that quantitative risk assessment filled by providing a means to help decide if a risk is large enough to deserve regulation.

Thus the United States has had a long history of applying the precautionary principle in regulation but has moved away from doing so as we have learned more about risk assessment and its underlying scientific basis. As science-based risk analysis is used to a greater extent in regulation, however, the importance of establishing a factual basis for its application has been highlighted. In the United States, regulatory decisions must be justified by an extensive factual record that is subject to judicial review under the Administrative Procedure Act. For example, only about 10 percent of the manipulation of scientific data done by EPA has been estimated to be necessary to reach a decision; the other 90 percent is required to build the record for court review.[8] On the global scale, precautionary risk management policies are, to a great extent, a response to this U.S. legal tradition that relies heavily on establishing a factual basis for decision making. When Europeans call for decisions based on "the precautionary principle" in international forums, they are challenging the core premise of the American legal culture.

There is a historic ideological commitment in U.S. legal culture that the risks of arbitrary government action are so great that it is better to pay the costs of procedural delay and elaborate legality than to run the risk of unjustified government actions.[9] The cost of elaborate legality can be "paralysis by analysis" or "ossification" of the rulemaking process, however.[10] Opponents of risk analysis justifiably point to the preventable health consequences of delaying regulatory action while scientific analysis is debated. For example, by initially adopting a risk-based approach to setting toxic air pollutant standards, EPA was paralyzed in the 1980s by litigation over the scientific basis of its risk estimates and was able to set standards for only seven chemicals.[11] Adequate control of urban air pollution was thus delayed for at least a decade until technology-based controls were instituted, with associated protracted adverse effects on health. Occupational exposures to many substances are controlled only through voluntary standards at present in the United States because an extensive list of consensus judgment-based standards proposed by the Occupational Safety and Health Administration (OSHA) (and supported by industry and labor) was struck down by the courts due to their supposed inadequate scientific basis.[12]

In Europe and in most of the rest of the industrialized world, including Canada, government regulatory decisions are not subject to judicial challenges in court to nearly the same degree as they are in the United States, and the necessary procedures for marshaling and analyzing scientific evidence before a decision is made are thus nowhere near as great. In a legal culture such as Europe's, which has historically deferred to expert judgment as the basis for regulation, precaution is more easily justified than risk analysis as the basis for decision making. In the United States, however, democratic procedure requires risk analysis to be supported factually despite the potential risks associated with procedural delay itself.

28.3 SCHOLARLY CRITIQUES OF RISK ANALYSIS

The nature, status, and evolution of risk management decision making in the United States has been addressed by a variety of studies. Prominent among those are reports published by the National Academy of Sciences/National Research Council and the Presidential/Congressional Commission on Risk Assessment and Risk Management. Many other studies have either supported or challenged the roles of risk analysis and democracy in decision making. Some of those are described below.

28.3.1 National Academy of Sciences/National Research Council

The National Academy of Sciences/National Research Council (NAS/NRC) 1983 report *Risk Assessment in the Federal Government: Managing the Process* (The Red Book) is given credit for codifying a risk assessment framework and terminology.[13] In that report, the characterization of risk is described somewhat narrowly as "the process of estimating the incidence of a health effect under the various conditions of human exposure" (p. 20). A decade later, *Science and Judgment in Risk Assessment,* which was undertaken specifically to evaluate the practice of risk assessment at EPA, concurred. That report describes the goal of risk characterization as providing an understanding of the type and magnitude of an adverse effect that a particular chemical or emission could cause under particular circumstances.[14] It went a little farther than the Red Book, however, and acknowledged the existence of the risk manager, to whom the results of a risk assessment are to be communicated. The risk manager is then to make risk management decisions on the basis of a risk's public health impact, among other things.

The addition of social scientists to the next NAS/NRC committee that evaluated risk assessment practices in the federal government produced a very different outlook on the characterization of risk and its role in risk management. The committee that wrote *Understanding Risk—Informing Decisions in a Democratic Society*[15] recognized that the Red Book and *Science and Judgment* definitions of risk characterization are the prevailing view at EPA and other agencies, but concluded that the view of risk characterization as a translation or summary is "seriously deficient." The committee emphasized instead that risk characterization should be a *decision-driven activity* directed toward informing choices and solving problems. By acknowledging for the first time that the purpose of risk characterization is to enhance practical understanding and to illuminate practical choices, risk characterization was recognized not as a stand-alone activity whose results are simply to be communicated to a risk manager but as an activity that must be performed as part of a risk management decision-making process.

To reconcile the scientific practice of risk assessment with the needs of democracy, the authors of *Understanding Risk* recommended an *analytic-deliberative process* for risk management decision making. In an analytic-deliberative process, analysis and deliberation are complementary and must be integrated throughout risk characterization: Deliberation frames analysis, analysis informs deliberation, and the process benefits from feedback between the two. Such a process must have an appropriately diverse participation or representation of the spectrum of interested and affected parties, of decision makers, and of specialists in risk analysis, at each step. The report defined *affected parties* as people, groups, or organizations that may experience benefit or harm as a result of a hazard, or of the process leading to risk characterization, or of a decision about risk, noting that such parties need not be aware of the possible harm to be considered affected. *Interested parties* were defined as people, groups, or organizations that decide to become informed about and involved in a risk characterization or decision-making process (and who may or may not be affected parties). A carefully prepared risk assessment will give neither interested nor affected parties the scientific understanding they need if that information is not relevant to the decision to be made, however.

28.3.2 Presidential/Congressional Commission on Risk Assessment and Risk Management

The Presidential/Congressional Commission on Risk Assessment and Risk Management (Risk Commission) also believed that characterization of risk should be performed in the context of a risk management decision-making process. The commission was mandated by Congress when it was last updating the Clean Air Act and could not agree on how risk assessments should be performed after new technology-based standards to control air pollution are in place. The commission's mandate was to investigate and make recommendations about the role of risk assessment and risk management in federal regulatory programs. In volume 1 of its 1997 final report, *Framework for Environmental Health Risk Management*,[16] the commission supported democratic environmental health risk management decision making by promoting a strong role for stakeholders. The commission defined stakeholders as those who are affected by a particular risk or by the decision to be made about managing a risk, concluding that a good risk management decision emerges from a process that elicits the views of those affected by the decision, so that differing technical assessments, public values, knowledge, and perceptions are considered. The commission proposed a six-stage framework to guide risk management decision making that puts stakeholders in the center of the process (Figure 28.1). The Risk Commission believed that stakeholder collaboration is important because there are many conflicting interpretations about the nature and significance of risks, and collaboration provides opportunities to bridge gaps in understanding, language, values, and perceptions. Stakeholders bring to the table important information, knowledge, expertise, and insights for crafting workable solutions and are more likely to accept and implement a risk management decision they have participated in shaping.

28.3.3 Other Studies

Prorisk Assessment The Carnegie Commission on Science, Technology, and Government convened a committee in the early 1990s to review and evaluate

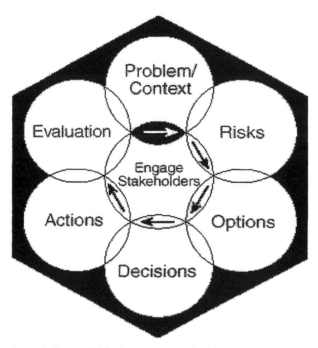

Figure 28.1   Commission on Risk Assessment and Risk Management's Framework for Environmental Health Risk Management Decision-Making.

institutional and administrative issues related to improving environmental regulation. The committee's final report recommended adopting a risk-ranking approach to setting risk management priorities that integrates both good science and societal values.[17] The committee thus recognized that democracy and social choice play a role in decisions about what risks we manage, maintaining that such value choices should not be made covertly by unaccountable "experts"[18] but did not suggest specific mechanisms for effectively integrating science and values.

The Harvard Center for Risk Analysis has produced many studies of the value of risk-based decision making. The center was established to promote reasoned public responses to health, safety, and environmental hazards by taking a "big picture" view of public health. That view compares and ranks a wide range of hazards and develops scientific data that can be used to identify policy choices that are most likely to achieve the greatest health, safety, and environmental benefits with the most efficient use of finite resources. For example, the center's seminal publication *Risk vs. Risk: Tradeoffs in Protecting Health and the Environment* argues for greater reliance on risk analysis and decision science in resource allocation, pointing out that too great a reliance on social choice that is not informed by science leads to misidentification of risk priorities and misallocation of risk management resources.[19] The center's work generally defends greater reliance on scientific knowledge and less emphasis on democratic decision making on the grounds that excessive public participation will leave the risk assessors outside the door and, in the long run, lead to fewer lives saved and less environmental protection.

The decline of science as an important determinant in environmental decision making is also an important concern of Supreme Court Justice Stephen Breyer's book, *Breaking the Vicious Circle—Toward Effective Risk Regulation.*[20] In case after case, the book shows how decision making, particularly in the environmental area, has been politicized while science has been precluded from playing what the author considers to be its rightful role. The author's solution to that problem is to establish a new administrative group of civil servants, explicitly modeled on the French conseil d'état, with interagency jurisdiction and a specified risk-related mission: building an improved, coherent risk-regulating system, setting risk-based priorities, and determining how best to allocate risk management resources. That solution can be considered antithetical to the democratization of risk management decision making.

There are, of course, many other important organizations and publications that have made, and continue to make, valuable contributions to improving risk-based decision making, with varying degrees of support for democratic approaches. The reader is encouraged to consult www.riskworld.com to find links to many of these.

Antirisk Assessment   Despite its important role in regulatory risk management, the practice of risk assessment is not accepted as a useful decision-making tool by all. Risk assessment is considered highly undemocratic by some, who dismiss it as being merely a means to rationalize government decision making and its alleged dominance by industry. The environmental advocacy group Greenpeace claims in its fund-raising videos that the first thing you do when you plan a risk assessment is decide how many people will die. The author of *Pandora's Poison—Chlorine, Health, and a New Environmental Strategy*[21] insists that a "witch's brew of toxic, persistent pollutants has come to blanket the entire planet" (p. vii) because the "apparently sophisticated environmental laws of industrialized countries have failed to halt the tide of contamination" (p. 2). The book assesses health and environmental risks from organochlorine exposure, finds them unacceptably high, and proposes banning them—along with risk assessment—as a risk management solution. In *Protecting Public Health and the Environment—Implementing the Precautionary Principle,*[22] risk assessment is relegated to a "second tier" in the decision-making process. Instead of being used "to quantify an 'acceptable' risk," it should be used to compare alternatives to a risk-producing activity, or to establish priorities, apparently requiring less quantitative analysis and less uncertainty. Decision-making itself should be guided by a policy of precaution that relies on "democratic decision-making structures" (p. 164).

A common theme among risk critics is that risk assessment is undemocratic, morally bankrupt, and ethically impoverished. They contend that it ignores a horde of political and moral questions such as what constitutes "acceptable" risk and who decides what is acceptable. The point such advocates miss is that risk assessment simply characterizes risks; people make political and moral decisions about how much risk is acceptable. Risk assessment is not meant to address ethical issues of safety, people are. Risk assessment is used simply to inform decision making; risk decisions are, ultimately, public policy choices. Risk assessment attempts to bring as much relevant knowledge as possible to participants in a decision, whose job it is to make the value-laden choices. Those who find such decisions unacceptable should not condemn risk assessment as a practice; instead, those who make the decisions and the legal and statutory basis that guides them should be held responsible.

28.4   REGULATORY RISK ANALYSIS

Regulatory risk management decisions are possibly the ultimate embodiment of democratic procedure. They are made in response to the implementation requirements of laws passed by our elected members of Congress; they are required to undergo public review and comment before they are finalized; and they can be subjected to judicial review if suspected as being capricious and arbitrary. Nonetheless, regulatory decisions are often challenged in the court of public opinion as being unfair and inadequate on a variety of grounds held in high esteem by disparate stakeholders. To produce more accepted decisions, many agencies are relying increasingly on stakeholder processes to satisfy the demand for a greater level of democratic participation than is provided by simple public review and comment.

28.4.1   Environmental Protection Agency

EPA has initiated or been a party to a variety of democratic, stakeholder-involvement efforts since the early 1980s. EPA issued an agency-wide Public Participation Policy in 1981 following, appropriately, public discussion, review, and comment. That policy was updated in 2001 and is designed to provide guidance and direction to public officials who manage and conduct EPA programs on reasonable and effective means of involving the public in program decisions. It applies to programs under the Clean Air Act, Resource Conservation and Recovery Act, Toxic Substances Control Act, Federal Insecticide, Fungicide and Rodenticide Act, Safe Drinking Water Act, and the Clean Water Act. Implementation of the Comprehensive Environmental Response, Compensation, and Liability Act, or "Superfund", has also depended heavily on stakeholder involvement at the local level. EPA generally categorizes its stakeholder involvement activities into four major types, defined by the purpose of the interaction and the expectation of the use of the information by EPA and the stakeholders involved: outreach, information exchange, recommendations, and agreements. These four categories are believed to be helpful in clarifying roles and expectations for all involved.

The opportunities for and quality of stakeholder involvement in EPA programs and regions have increased with time and experience, but remain uneven. EPA's new Policy on Public Involvement establishes an agency work group charged with drafting a strategic plan for better integration of agency-wide stakeholder involvement efforts. Its purposes are to strengthen the agency's commitment to public involvement and to establish clear and effective procedures for involving the public in EPA decision making. EPA's Science Advisory Board has expressed concern about increased public involvement in decision making, however, and has made recommendations intended to help the agency preserve a central role for scientific and technical information.

Consumer right-to-know is another democratizing approach to risk management decisions promoted by EPA, through programs such as the Toxic Release Inventory, which requires reporting of chemical emissions, and its new Center for Environmental Information and Statistics. In this case, EPA's role is that of information broker, providing individuals with information about toxic substances that is meant to be useful for individual decisions about controlling and reducing risks. Informational approaches can help reduce risk by informing consumers about risks they can choose to avoid or by generating support for actions to reduce involuntary exposures to risk.

28.4.2 Department of Energy

The Department of Energy (DOE) and the Department of Defense (DOD) environmental programs became democratized when the decisions those agencies were making about when, how, and to what extent remediation should take place at their sites contaminated with chemicals, radioactive materials, biological warfare agents, and munitions, among other things, became the target of quite active public protest. Those agencies were accustomed to technocratic, hierarchical, often top-secret decision making that was not supposed to be questioned. Adapting to the democratic process of stakeholder participation in decision making has been a challenge that those agencies have met with varying degrees of success.

In the case of DOE, the Consortium for Risk Evaluation with Stakeholder Participation (CRESP) has helped facilitate stakeholder involvement in decisions about site remediation. CRESP was established and began operation in 1995 in response to the conclusion by a National Academy of Sciences committee that DOE's Environmental Management Office needed an independent institutional mechanism to develop data and methodology to make risk a key part of its decision making.[23] CRESP's mission is to improve the scientific and technical basis of DOE's environmental management decisions, leading to protective and cost-effective cleanup of the nation's nuclear weapons while enhancing stakeholder understanding of nuclear weapons production facility waste sites. CRESP is organized to provide both guidance to and peer review of the evolving effort to use risk-based methods and evaluations to shape cleanup decisions at DOE sites.

One of the site cleanups that has involved CRESP is underway at DOE's Savannah River site. The Savannah River site was constructed during the early 1950s to produce the basic materials used in the fabrication of nuclear weapons, primarily tritium and plutonium-239. Today, the site both stores and is contaminated by high-level, low-level, and liquid radioactive wastes as well as by radioactive wastes mixed with hazardous chemical wastes. Before CRESP was involved at Savannah River, DOE, EPA, and the states had performed different risk assessments, obtaining conflicting risk estimates due primarily to differences in assumptions about exposure to contaminants through fish consumption. When CRESP became involved, its researchers concluded that the many conflicting assumptions about fish consumption could be overcome by obtaining actual data to replace the assumptions and proceeded to work with local residents to collect the data. Another risk assessment was performed, monitored closely by stakeholders, and a new risk estimate was obtained that was higher than previous estimates. Nonetheless, risks from the approximately 3-mrem radiation exposure occurring through contaminated fish were still considerably lower than risks from background radiation levels of 200 to 400 mrem. The new risk estimate appears to have been credible and accepted by the stakeholders who participated because it directly addressed their concerns and because they had been involved in both research planning and in its actual performance.

28.4.3 State of California

In an effort to streamline risk management decision making by shifting the burden of proof of safety, the voters of California approved the Safe Drinking Water and Toxic Enforcement Act, popularly referred to as Proposition 65, in 1986. Proposition 65 combines a duty-to-warn approach with a shifting of the burden of demonstrating safety from the state regulatory agencies to the proponents of substances found to be carcinogens or reproductive toxins.[24] The basis of the initiative is that no one should knowingly expose another without warning to chemicals known to cause cancer or reproductive toxicity unless the entity responsible for the exposure can demonstrate that the risk is not significant. Acting far more rapidly than any federal agency ever had, California's Health and Welfare Agency issued regulations specifying significant risk levels for hundreds of chemicals. This action caused industries, for example, either to reduce or eliminate toxin exposures or to produce the science and risk assessments necessary to show that exposures were below those associated with significant risk. While Proposition 65 might not be considered a democratic approach to risk management in the purest sense, it effectively put the risk ball in the court of those responsible for toxicant exposures, thereby involving them in their own regulatory decision making whether or not they wished to be. Proposition 65's right-to-know basis provided citizens with information that they could use to make decisions about product use.

28.5 DEMOCRATIC DECISION MAKING: CASE EXAMPLES

The city of Valdez, Alaska offers two examples of stakeholder-based attempts to resolve technically intensive policy disputes related to environmental risks, one successful and one unsuccessful.[25] The first involved competing scientific knowledge claims that led to "dueling scientists" and a stalemate. The second overcame the problems of the first by combining stakeholders' experts to form a single scientific team. The team included stakeholders in defining the scientific issues of concern and in obtaining the data needed to address those concerns.

Large volumes of crude oil are shipped in the Prince William Sound region of Alaska, with oil loaded onto tankers at the port of Valdez at a terminal operated by the Alyeska Pipeline Service Company (Alyeska). Alyeska had supported the establishment of a Regional Citizens' Advisory Council (RCAC) to help oversee environmental management of the marine oil trade there. In the first dispute, Alyeska and the RCAC disagreed about the impact of crude oil vapors emitted by the oil terminal on air quality in the city of Valdez. Alyeska had commissioned a series of air quality studies that examined the levels and sources of airborne volatile organic compounds in Valdez and the RCAC convened a panel of scientists to evaluate the results of the studies. The panel agreed with the findings regarding the levels of ambient airborne benzene but disagreed with the method used to identify the source of the benzene emissions. The two groups of scientists then generated contradictory knowledge claims regarding the sources of benzene, with the RCAC concluding that 90% of it originated at the oil terminal and Alyeska concluding that only 25% originated there. The RCAC asked Alyeska to install vapor control systems and Alyeska refused, unless a significant health risk could be attributed to the terminal. Interviews revealed that the Alyeska scientists questioned the validity of the RCAC models and that RCAC scientists believed the Alyeska results had been manipulated to support the industry's arguments. Mutual suspicions of distorted communication arising from claims of mistaken and manipulated analyses led to an impasse, with neither party accepting the other's interpretation. In the absence of a common foundation of knowledge, further discussion stalled and the Valdez air quality debate remained deadlocked for two years.*

The second dispute involved a debate over the capabilities of the tug vessels used to escort oil tankers in the Sound. The tug vessels' primary purpose was to help correct course errors that might otherwise lead to collisions and oil spills. The RCAC proposed that the oil industry deploy highly maneuverable tractor tug vessels in one region of the Sound and an ocean rescue tug vessel with an enhanced propulsion system in another region of the Sound, on the basis that

---

*Eventually, the debate was superseded by implementation of the 1990 Clean Air Act Amendments and an EPA draft rule requiring a 95% reduction in the emissions of all hazardous air pollutants from the Valdez terminal. Alyeska responded by installing vapor controls. Thus the risk management action taken was in response to impending regulatory requirements, not a result of any determination of potential health effects by a stakeholder process.

doing so would reduce the risk of oil spills. The oil industry opposed the proposal as an unnecessary expense given that existing studies did not demonstrate that those tug vessels would improve safety. The oil industry then proposed to resolve the dispute by performing a comprehensive risk assessment of the oil trade in the Sound. The risk assessment was to be jointly funded and managed through a steering committee comprising RCAC members, oil industry managers, and representatives of the two government regulatory agencies with the appropriate jurisdictions. To avoid "dueling scientists," the steering committee combined the industry's scientific experts with the RCAC's scientific experts to form a single research team. Later interviews found all parties agreeing that if the oil industry had conducted the risk assessment on its own, no one else would have believed the results. By having the participants in the dispute structure and perform the risk assessment jointly, collaborative analysis was used to resolve potentially adversarial technical disagreements.

There were several benefits to using the collaborative model. One benefit was mutual learning among the participants. Frequent meetings led the steering committee to gain a better understanding of the technical dimensions of maritime risk assessment and the research team to better understand the problem at issue and to gather data it would not have otherwise. Steering committee members actually participated in data gathering with the research team. Another benefit resulted from combining resources, making more money available to conduct the work. The results of the risk assessment were accepted as credible by all parties involved in the issue, who agreed that hidden agendas or conspiracies could not influence the collaborative process.

In response to the results of the assessment, the oil industry deployed an ocean rescue tug vessel in the Sound. The risk assessment was not able to determine whether tractor tug vessels would improve the safety provided by the conventional tug vessels already active, however. The governor of Alaska decided the issue by declaring that tractor tug vessels constituted the "best available technology" as required under state law and the oil industry responded with two such vessels on the basis of the policy decision. Thus both science and politics played roles in the outcome.

The two Valdez, Alaska case examples show how involving stakeholders in a democratic attempt to resolve risk-based disputes can either succeed or fail, depending on the role that the participants play in obtaining and in defining the role of technical information.

28.6   CONCLUSION: SCIENCE AND DEMOCRACY

Rachel Carson and her 1962 book *Silent Spring*[26] are often given credit for popularizing awareness of environmental threats and for providing the impetus that led to the drafting of our environmental laws and to the establishment of the EPA. Our environment is undoubtedly better off as a result. Together, our environmental laws and the efforts of governments, businesses, citizens, and municipalities are responsible for giving us the cleanest environment we have had in the United States in 40 years. Countries that lack environmental laws like ours lag way behind us in terms of environmental quality and public health improvements.

The continuing challenge for risk management in a democracy is to maintain a role for risk assessment and to preserve the integrity of science when decision making is influenced by many nontechnical factors. Doing so is particularly challenging when risk management decisions are conducted as collaborative efforts among stakeholders with differing technical knowledge levels, interests, goals, and worldviews. An important concern about increasingly democratic risk management decision making is thus whether stakeholders have the ability to respect and preserve the role that science can play in informing decisions. Some argue that greater stakeholder involvement will marginalize science; others argue that decision making is already tyrannized by science and scientific experts and that involvement of nonscientific, nonexpert stakeholders represents a needed swing of the pendulum back toward an emphasis on social values.

One problem is the rationalization on pseudoscientific grounds of regulatory agency risk management decisions that are actually based on policy or other values, which has been dubbed the "science charade."[27] She has concluded that the deference courts give to agencies in technical areas has produced distorted incentives, providing an open invitation for agencies to make decisions on political grounds while rationalizing them on technical grounds. As Georgetown University Law Professor Steven Goldberg puts it: "Regulatory agencies are regularly accused of being 'captured' by industry, consumer groups, members of Congress or bureaucratic inertia. They are never accused, however, of being captured by scientists."[28] On the other hand, scientists have been accused of failing to place their efforts in an adequate social context, believing that science is separate from social factors or that social factors play minimal roles.[29] Daniel Yankelovich, president of The Public Agenda Foundation, asserts that "in present-day America, a serious gap exists between the point of view of the experts and that of the general public" (p. 91). He believes that in its "eagerness to exalt the truths of science, empiricism has, crudely and blindly, undermined other modes of knowing, including public judgment ... American culture grossly overvalues the importance of information as a form of knowledge and undervalues the importance of cultivating good judgment. It assumes, falsely, that good information automatically leads to good judgment"[30] (p. 10).

Nonetheless, the movement over the last decade toward more inclusive and democratic environmental health risk management decision making reflects an attempt to better integrate scientists' facts with the public's judgment and to put science in its social context. Recent experience shows that the integrity of the science in stakeholder-based decision-making processes can be maintained and its credibility improved when stakeholders are involved in deciding how science will be used to answer their questions and in obtaining the scientific information needed to answer those questions.[31] Implementing the National Academy of Sciences' analytic-deliberative process and following the spirit of the Risk Commission's risk management framework show that science can play an important role in risk management when that role is shaped by stakeholder values to address their concerns.

It is likely, however, that as we struggle to seek the right balance between science and values in order to achieve consensus, risk analysis will play roles of varying importance and influence in risk management decisions. It is important to acknowledge that science may not always be the sole basis for a risk decision and to articulate clearly the extent to which a decision is based on science and on policy. In many cases, science will be one—but not the overriding—consideration. The goal is to strive to maintain the integrity and credibility of science and to define a useful role for scientific information in social decision making.

REFERENCES

1. M. R. English. "Stakeholders" and environmental policymaking. *Center View* 4(2): 1–2 (1996).
2. T. F. Yosie and T. D. Herbst. Managing and communicating stakeholder-based decision making. *Hum. Ecol. Risk Assess.* 4: 643–646 (1998).
3. R. B. Stewart. The reformation of American administrative law. Harvard Law Rev. 88: 1667 (1975).
4. L. R. Rhomberg. *A Survey of Methods for Chemical Health Risk Assessment Among Federal Regulatory Agencies.* Report Prepared for the Commission on Risk Assessment and Risk Management, Washington, DC, 1997.
5. Commission of the European Communities (CEC). *Communication from the Commission on the Precautionary Principle.* COM(2000) 1. CEC, Brussels, Belgium, 2000.
6. *Ethyl Corp. v. EPA*, 541 F.2d 1 (DC Cir.)(en banc), cert. denied, 426 U.S. 941 (1976).
7. *Industrial Union Dept., AFL-CIO v. American Petroleum Inst.*, 448 U.S. 607 (1980); G. Charnley, and E. D. Elliott. Risk versus precaution: A false dichotomy. *Proceedings of the Annual Meeting of the European Section of the Soceity for Risk Analysis.* Edinburgh, Scotland, A. A. Balkema, 2000.
8. US Environmental Protection Agency Office of Solid Waste and Emerency Response Deputy Assistant Administrator Michael Shapiro, personal communication.
9. For a survey of the historic and theoretical justifications of the US tradition of legality and proceduralism in administrative law, see R. J. Pierce, S. A., Shapiro, and P. R. Verkuil, *Administrative Law and Process*, 3rd ed., Chapter 2, pp. 24–41, esp. §2.1.2 ("The role and meaning of separation of powers in democratic theory") and §2.5 ("The administrative process and democratic theory"). Foundation Press, New York, 1999.
10. T. McGarity. Some thoughts on "deossifying" the rulemaking process. *Duke Law J.* 1385 (1992); R. J. Pierce. Seven ways to deossify agency rulemaking. *Admini. Law Rev.* 41: 59 (1995).
11. E. D. Elliott, and E. M. Thomas. Chemicals. In Campbell-Mohn, Futrell, Breen

(Eds.), *Sustainable Environmental Law.* West Publishing Co., St. Paul, 1993, p. 1282.

12. *AFL v. OSHA*, 965 F.2d 962 (11th Cir. 1992).

13. National Academy of Sciences/National Research Council. *Risk Assessment in the Federal Government. Managing the Process.* National Academy Press, Washington, DC, 1983.

14. National Academy of Sciences/National Research Council. *Science and Judgment in Risk Assessment.* National Academy Press, Washington, DC, 1994.

15. National Academy of Sciences/National Research Council. *Understanding Risk. 5Informing Decisions in a Democratic Society.* National Academy Press, Washington, DC, 1996.

16. Presidential/Congressional Commission on Risk Assessment and Risk Management. *Framework for Environmental Health Risk Management.* Final report volume 1. GPO #055-000-00567-2. U.S. Government Printing Office, Washington, DC, 1997.

17. Carnegie Commission on Science, Technology, and Government. *Risk and the Environment. Improving Regulatory Decision Making.* Washington, DC, 1993.

18. For the classical statement of the case against "government by scientists," see D. L. Bazelon, Coping with Technology Through the Legal Process. *Cornell L. Rev.* 62: 817 (1977).

19. J. D. Graham and J. B. Wiener. (Eds.). *Risk vs. Risk: Tradeoffs in Protecting Health and the Environment.* Harvard University Press, Cambridge, MA, 1995.

20. S. Breyer. *Breaking the Vicious Circle. Toward Effective Risk Regulation.* Harvard University Press, Cambridge, MA, 1993.

21. J. Thornton. *Pandora's Poison—Chlorine, Health, and a New Environmental Strategy.* MIT Press, Cambridge, MA, 2000.

22. C. Raffensperger and J. Tickner. (Eds.). *Protecting Public Health and the Environment. Implementing the Precautionary Principle.* Island Press, Washington, DC, 1999.

23. National Academy of Sciences/National Research Council. *Building Consensus through Risk Assessment and Management of the Department of Energy's Environmental Remediation Program.* National Academy Press, Washington, DC, 1994.

24. For a summary of provisions, see Elliott and Thomas (1993), pp. 1296–1297; for an explanation by the principal drafter of the theory behind Proposition 65, see D. Roe, Barking up the right tree: Recent progress in focusing the toxics issue. *Colum. J. Envir. L.* 13: 275 (1988).

25. G. J. Busenberg. Collaborative and adversarial analysis in environmental policy. *Policy Sci.* 32: 1–11 (1999).

26. R. Carson. *Silent Spring.* Houghton-Mifflin, New York, 1962.

27. W. E. Wagner. The science charade in toxic risk regulation. *Columbia Law Rev.* 95: 1613 (1995).

28. S. Goldberg. The reluctant embrace: Law and science in america. *Georgetown Law J.* 75: 1341, 1365 (1987).

29. P. Brown and E. J. Mikkelsen. *No Safe Place.* University of California Press, Berkeley, CA, 1990.

30. D. Yankelovich. *Coming to Public Judgment. Making Democracy Work in a Complex World.* Syracuse University Press, Syracuse, NY, 1991.

31. G. Charnley. *Enhancing the Role of Science in Stakeholder-based Risk Management Decision-Making.* HealthRisk Strategies. Prepared for the American Industrial Health Council and the American Chemistry Council, Washington, DC, 2000. www.riskworld.com

29 Misconceptions About the Causes of Cancer

LOIS SWIRSKY GOLD,[1,2] BRUCE N. AMES,[1] and
THOMAS H. SLONE[1]

[1] Department of Molecular and Cell Biology, University of California
Berkeley, California

[2] Department of Cell and Molecular Biology, Lawrence Berkeley
National Laboratory, Berkeley, California

SUMMARY

The major causes of cancer are: (1) smoking, which accounts for 31 percent of
U.S. cancer deaths and 87 percent of lung cancer deaths, (2) dietary imbalances,
which account for about another third (e.g., lack of sufficient amounts of dietary
fruits and vegetables), (3) chronic infections, mostly in developing countries, and
(4) hormonal factors, which are influenced primarily by life-style. There is no
cancer epidemic except for cancer of the lung due to smoking. Cancer mortality
rates have declined 19 percent since 1950 (excluding lung cancer). Regulatory
policy that focuses on traces of synthetic chemicals is based on misconceptions
about animal cancer tests. Recent research indicates that rodent carcinogens
are not rare. Half of all chemicals tested in standard high-dose animal cancer
tests, whether occurring naturally or produced synthetically, are "carcinogens";
there are high-dose effects in rodent cancer tests that are not relevant to low-
dose human exposures and that contribute to the high proportion of chemicals
that test positive. The focus of regulatory policy is on synthetic chemicals, although 99.9 percent of the chemicals humans ingest are natural. More than 1000
chemicals have been described in coffee: 30 have been tested and 21 are rodent
carcinogens. Plants in the human diet contain thousands of natural "pesticides"
produced by plants to protect themselves from insects and other predators: 71
have been tested and 37 are rodent carcinogens.

There is no convincing evidence that synthetic chemical pollutants are im-

Human and Ecological Risk Assessment: Theory and Practice, Edited by Dennis J. Paustenbach
ISBN 0-471-14747-8 © 2002 John Wiley & Sons, Inc.

portant as a cause of human cancer. Regulations targeted to eliminate low levels of synthetic chemicals are expensive. The Environmental Protection Agency (EPA) has estimated that environmental regulations cost society $140 billion/ year. Others have estimated that the median toxic control program costs 146 times more per hypothetical life-year saved than the median medical intervention. Attempting to reduce tiny hypothetical risks has other costs as well: If reducing synthetic pesticides makes fruits and vegetables more expensive, thereby decreasing consumption, then the cancer rate will increase, especially for the poor. The prevention of cancer will come from knowledge obtained from biomedical research, education of the public, and life-style changes made by individuals. A reexamination of priorities in cancer prevention, both public and private, seems warranted.

In this chapter we highlight nine misconceptions about pollution, pesticides, and the causes of cancer. We briefly present the scientific evidence that undermines each misconception.

MISCONCEPTION 1

Cancer rates are soaring.

Overall cancer death rates in the United States (excluding lung cancer due to smoking) have declined 19 percent since 1950.[1] The types of cancer deaths that have decreased since 1950 are primarily stomach, cervical, uterine, and colorectal. Those that have increased are primarily lung cancer (87 percent is due to smoking, as are 31 percent of all cancer deaths in the United States,[2] melanoma (probably due to sunburns), and non-Hodgkin's lymphoma. If lung cancer is included, mortality rates have increased over time, but recently have declined.[1] For some cancers, mortality rates have begun to decline due in part to early detection, treatment, and improved survival,[2,3] for example, breast cancer in women.[4] The rise in incidence rates in older age groups for some cancers can be explained by known factors such as improved screening. The reason for not focusing on the reported incidence of cancer is that the scope and precision of diagnostic information, practices in screening and early detection, and criteria for reporting cancer have changed so much over time that trends in incidence are not reliable.[4-7] Life expectancy has continued to rise since 1950.[8]

MISCONCEPTION 2

Environmental synthetic chemicals are an important cause of human cancer.

Neither epidemiology nor toxicology supports the idea that exposures to environmental levels of synthetic industrial chemicals are important as a cause of human cancer.[7,9,10] Epidemiological studies have identified several factors that are likely to have a major effect on lowering cancer rates: reduction of smoking, improving diet (e.g., increased consumption of fruits and vegetables), hormonal factors, and control of infections.[10] Although some epidemiological studies find an association between cancer and low levels of industrial pollutants, the associations are usually weak, the results are usually conflicting, and the studies do not correct for potentially large confounding factors such as diet.[10,11] Moreover, exposures to synthetic pollutants are very low and rarely seem toxicologically plausible as a causal factor, particularly when compared to the background of natural chemicals that are rodent carcinogens.[9,12,13] Even assuming that worst-case risk estimates for synthetic pollutants are true risks, the proportion of cancer that the U.S. EPA could prevent by regulation would be tiny.[14] Occupational exposures to some carcinogens cause cancer, though exactly how much has been a controversial issue: A few percent seems a reasonable estimate,[10] much of this from asbestos in smokers. Exposures to substances in the workplace can be much higher than the exposure to chemicals in food, air, or water. Past occupational exposures have sometimes been high, and in risk assessment little quantitative extrapolation may be required from high-dose rodent tests to high-dose occupational exposures. Since occupational cancer is concentrated among small groups with high levels of exposure, there is an opportunity to control or eliminate risks once they are identified; however, current U.S. permissible exposure limits in the workplace are sometimes close to the carcinogenic dose in rodents.[15]

Cancer is due, in part, to normal aging and increases exponentially with age in both rodents and humans.[16] To the extent that the major external risk factors for cancer are diminished, cancer will occur at later ages, and the proportion of cancer caused by normal metabolic processes will increase. Aging and its degenerative diseases appear to be due in part to oxidative damage to deoxyribonucleic acid (DNA) and other macromolecules.[16,17] By-products of normal metabolism—superoxide, hydrogen peroxide, and hydroxyl radical—are the same oxidative mutagens produced by radiation. Mitochondria from old animals leak oxidants[18]: Old rats have about 66,000 oxidative DNA lesions per cell.[19] DNA is oxidized in normal metabolism because antioxidant defenses, though numerous, are not perfect. Antioxidant defenses against oxidative damage include vitamins C and E,[20] most of which come from dietary fruits and vegetables.

Smoking contributes to 31 percent of U.S. cancer, about one-quarter of heart disease, and about 430,000 premature deaths per year in the United States[1,2,10] Tobacco is a known cause of cancer of the lung, mouth, pharynx, larynx, bladder, pancreas, esophagus, and possibly colon. Tobacco causes even more deaths by diseases other than cancer.[21] Smoke contains a wide variety of mutagens and rodent carcinogens. Smoking is also a severe oxidative stress and causes inflammation in the lung. The oxidants in cigarette smoke—mainly nitrogen oxides—deplete the body's antioxidants. Thus, smokers must ingest two to three times more vitamin C than nonsmokers to achieve the same level in blood, but they rarely do. An inadequate concentration of vitamin C in plasma is more common among smokers.[22] Men with inadequate diets or who smoke may damage both their somatic DNA and the DNA of their sperm. When the level of dietary vitamin C is insufficient to keep seminal fluid vitamin C at an adequate level, the oxidative lesions in sperm DNA are increased 2.5 times.[23,24] Male smokers have more oxidative lesions in sperm DNA[25] and more chromosomal abnormalities in sperm[26] than do nonsmokers. It is plausible, therefore, that fathers who smoke may increase the risk of birth defects and childhood cancer in offspring.[23,26] One epidemiological study suggests that the rate of childhood cancers is increased in offspring of male smokers: acute lymphocytic leukemia, lymphoma, and brain tumors were increased three to four times.[27] Risk increased as pack-years of paternal smoking icnreased before conception.[27]

We[10] estimate that unbalanced diets account for about one-third of cancer deaths, in agreement with an earlier estimate of Doll and Peto.[1,2,6] Low intake of fruits and vegetables is an important risk factor for cancer (see Misconception 3). There has been considerable interest in calories (and dietary fat) as a risk factor for cancer, in part because caloric restriction lowers the cancer rate and increases the life span in rodents.[10,28,29]

Chronic inflammation from chronic infection results in the release of oxidative mutagens from phagocytic cells and contributes to cancer.[10,30] White cells and other phagocytic cells of the immune system combat bacteria, parasites, and virus-infected cells by destroying them with potent, mutagenic oxidizing agents. These oxidants protect humans from immediate death from infection, but they also cause oxidative damage to DNA, chronic cell killing with compensatory cell division, and mutation[31,32]; thus they contribute to the carcinogenic process. Antioxidants appear to inhibit some of the pathology of chronic inflammation. Chronic infections are estimated to cause about 21 percent of new cancer cases in developing countries and 9 percent in developed countries.[33]

Endogenous reproductive hormones play a large role in cancer, including that of the breast, prostate, ovary, and endometrium,[34,35] contributing to about 20 percent of all cancer. Many life-style factors such as reproductive history, lack of exercise, obesity, and alcohol influence hormone levels and therefore affect risk.[10,34–36]

Other causal factors in human cancer are excessive alcohol consumption, excessive sun exposure, and viruses. Genetic factors also play a significant role and interact with life-style and other risk factors. Biomedical research is uncovering important genetic variation in humans.

MISCONCEPTION 3

Reducing pesticide residues is an effective way to prevent diet-related cancer.

Reductions in synthetic pesticide use will not effectively prevent diet-related cancer. Fruits and vegetables, which are the source of most pesticide residue exposures to humans, are of major importance for *reducing* cancer; moreover,

TABLE 29.1   Review of Epidemiological Studies on Cancer Showing Protection by Consumption of Fruits and Vegetables

Cancer Site	Fraction of Studies with Significant Cancer Protection	Relative Risk (Median) (low vs. high quartile of consumption)
Epithelial		
Lung	24/25	2.2
Oral	9/9	2.0
Larynx	4/4	2.3
Esophagus	15/16	2.0
Stomach	17/19	2.5
Pancreas	9/11	2.8
Cervix	7/8	2.0
Bladder	3/5	2.1
Colorectal	20/35	1.9
Miscellaneous	6/8	—
Hormone dependent		
Breast	8/14	1.3
Ovary/endometrium	3/4	1.8
Prostate	4/14	1.3
Total	129/172	

Source: Reference 37.

pesticide residues in food are low and frequently not detected (see Misconception 6). Less use of synthetic pesticides would increase costs of fruits and vegetables and thus reduce consumption, especially among people with low incomes, who eat fewer fruits and vegetables and spend a higher percentage of their income on food.

Dietary Fruits and Vegetables and Cancer Prevention   High consumption of fruits and vegetables is associated with a lowered rate of degenerative diseases including cancer, cardiovascular disease, cataracts, and brain dysfunction.[10,16] A review of about 200 epidemiological studies reported a consistent association between low consumption of fruits and vegetables and cancer incidence at many target sites[37–39] (Table 29.1). The quarter of the population with the lowest dietary intake of fruits and vegetables versus the quarter with the highest intake has roughly twice the cancer rate for most types of cancer (lung, larynx, oral cavity, esophagus, stomach, colorectal, bladder, pancreas, cervix, and ovary). Eighty percent of American children and adolescents, and 68 percent of adults[40,41] did not meet the intake recommended by the National Cancer Institute (NCI) and the National Research Council (NRC): five servings of fruits and vegetables per day. Publicity about hundreds of minor hypothetical risks, such as pesticide residues, can result in a loss of perspective on what is important: Half the U.S. population did not name fruit and vegetable consumption as protective against cancer.[42]

Some Micronutrients in Fruits and Vegetables Are Anticarcinogens   Antioxidants such as vitamin C (whose dietary source is fruits and vegetables), vitamin E, and selenium protect against oxidative damage caused by normal metabolism,[19] smoking,[11] and inflammation[16] (see Misconception 2). Micronutrient deficiency can mimic radiation in damaging DNA by causing single- and double-strand breaks, or oxidative lesions, or both.[11] Those micronutrients whose deficiency appears to mimic radiation are folic acid, B_{12}, B_6, niacin, C, E, iron, and zinc, with the laboratory evidence ranging from likely to compelling. The percentage of the population that consumes less than half the recommended daily allowance (RDA) for five of these eight micronutrients is zinc (18 percent), iron (19 percent of menstruating women), C (15 percent), E (20+ percent), and niacin (2 percent). These deficiencies combined with folate, B_{12}, and B_6 (discussed below) may comprise in toto a considerable percentage of the U.S. population.[11]

Folic acid deficiency, one of the most common vitamin deficiencies in the population consuming few dietary fruits and vegetables, causes chromosome breaks in humans.[43] The mechanism of chromosome breaks has been shown to be deficient methylation of uracil to thymine, and subsequent incorporation of uracil into human DNA (4 million/cell).[43] Uracil in DNA is excised by a repair glycosylase with the formation of a transient single-strand break in the DNA; two opposing single-strand breaks cause a double-strand chromosome break, which is difficult to repair. Both high DNA uracil levels and chromosome breaks in humans are reversed by folate administration.[43] Folate supplementation above the RDA minimized chromosome breakage.[44] Folate deficiency has been associated with increased risk of colon cancer[45,46]: in the Nurses' Health Study women who took a multivitamin supplement containing folate for 15 years had a 75 percent lower risk of colon cancer.[47] Folate deficiency also damages human sperm,[48,49] causes neural tube defects in the fetus and an estimated 10 percent of U.S. heart disease.[50] Diets low in fruits and vegetables are commonly low in folate, antioxidants (e.g., vitamin C), and many other micronutrients.[10,37,51] Approximately 10 percent of the U.S. population[52] had a lower folate level than that at which chromosome breaks occur.[43] Nearly 20 years ago, two small studies of low-income (mainly African-American) elderly[53] and adolescents[54] showed that about half the people in both groups studied had folate levels that low; this issue should be reexamined. Recently in the United States, flour, rice, pasta, and cornmeal have been supplemented with folate.[55]

Recent evidence indicates that vitamin B_6 deficiency works by the same mechanism as folate deficiency and causes chromosome breaks (Ingersoll, Shultz, Ames, unpublished). Niacin contributes to the repair of DNA strand breaks by maintaining nicotinamide adenine dinucleotide levels for the poly-ADP-ribose protective response to DNA damage.[56] As a result, dietary insufficiencies of niacin (15 percent of some populations are deficient),[57] folate, and antioxidants may interact synergistically to adversely affect DNA synthesis and repair. Diets deficient in fruits and vegetables are commonly low in folate, antioxidants (e.g., vitamin C), and many other micronutrients, result in DNA damage, and are associated with higher cancer rates.[10,11,37,51]

Micronutrients whose main dietary sources are other than fruits and vegetables are also likely to play a significant role in the prevention and repair of DNA damage and thus are important to the maintenance of long-term health.[11] Deficiency of vitamin B_{12} causes a functional folate deficiency, accumulation of homocysteine (a risk factor for heart disease),[58] and misincorporation of uracil into DNA.[59] Vitamin B_{12} supplementation above the RDA was necessary to minimize chromosome breakage.[44] Strict vegetarians are at increased risk for developing vitamin B_{12} deficiency since the dietary source is animal products.[58]

Optimizing micronutrient intake can have a major effect on health at a low cost.[11] More research in this area, as well as efforts to increase micronutrient intake and improve diets, should be high priorities for public policy.

MISCONCEPTION 4

Human exposures to carcinogens and other potential hazards are primarily to synthetic chemicals.

Contrary to common perception, 99.9 percent of the chemicals humans ingest are natural. The amounts of synthetic pesticide residues in plant foods, for example, are tiny compared to the amount of natural "pesticides" produced by plants themselves.[12,13,60–62] Of all dietary pesticides that humans eat, 99.99 percent are natural: These are chemicals produced by plants to defend themselves against fungi, insects, and other animal predators.[12,60] Each plant produces a different array of such chemicals. On average, Americans ingest roughly 5000 to 10,000 different natural pesticides and their breakdown products. Americans eat about 1500 mg of natural pesticides per person per day, which is about 10,000 times more than they consume of synthetic pesticide residues.[60] Even though only a small proportion of natural pesticides has been tested for carcinogenicity, half of those tested (37/71) are rodent carcinogens; naturally occurring pesticides that are rodent carcinogens are ubiquitous in fruits, vegetables, herbs, and spices[9,13] (Table 29.2). Cooking of foods produces burnt material (about 2000 mg per person per day) that contains many rodent carcinogens.

In contrast, the residues of 200 synthetic chemicals measured by Federal Drug Administration (FDA), including the synthetic pesticides thought to be of greatest importance, average only about 0.09 mg per person per day.[9,12,13] In a single cup of coffee, the natural chemicals that are rodent carcinogens are about equal in weight to an entire year's worth of synthetic pesticide residues that are rodent carcinogens, even though only 3 percent of the natural chemicals in roasted coffee have been adequately tested for carcinogenicity[9] (Table 29.3). This does not mean that coffee or natural pesticides are dangerous, but rather that assumptions about high-dose animal cancer tests for assessing human risk at low doses need reexamination. No diet can be free of natural chemicals that are rodent carcinogens.[13,61,62]

TABLE 29.2   Carcinogenicity Status of Natural Pesticides Tested in Rodents[a]

Carcinogens: $N = 37$	Acetaldehyde methylformylhydrazone, allyl isothiocyanate, arecoline. HCl, benzaldehyde, benzyl acetate, caffeic acid, capsaicin, catechol, clivorine, coumarin, crotonaldehyde, 3,4-dihydrocoumarin, estragole, ethyl acrylate, $N2$-γ-glutamyl-p-hydrazinobenzoic acid, hexanal methylformylhydrazine, p-hydrazinobenzoic acid. HCl, hydroquinone, 1-hydroxy-anthraquinone, lasiocarpine, d-limonene, 3-methoxycatechol, 8-methoxypsoralen, N-methyl-N-formylhydrazine, α-methylbenzyl alcohol, 3-methylbutanal methylformylhydrazone, 4-methylcatechol, methylhydrazine, monocrotaline, pentanal methylformylhydrazone, petasitenine, quercetin, reserpine, safrole, senkirkine, sesamol, symphytine
Noncarcinogens: $N = 34$	Atropine, benzyl alcohol, benzyl isothiocyanate, benzyl thiocyanate, biphenyl, d-carvone, codeine, deserpidine, disodium glycyr-rhizinate, ephedrine sulfate, epigallocatechin, eucalyptol, eugenol, gallic acid, geranyl acetate, β-N-[γ-l(+)-glutamyl]-4-hydroxy-methylphenylhydrazine, glycyrrhetinic acid, p-hydrazinobenzoic acid, isosafrole, kaempferol, dl-menthol, nicotine, norharman, phenethyl isothiocyanate, pilocarpine, piperidine, protocatechuic acid, rotenone, rutin sulfate, sodium benzoate, tannic acid, 1-trans-δ^9-tetrahydrocannabinol, turmeric oleoresin, vinblastine

Note: These rodent carcinogens occur in: absinthe, allspice, anise, apple, apricot, banana, basil, beet, broccoli, Brussels sprouts, cabbage, cantaloupe, caraway, cardamom, carrot, cauliflower, celery, cherries, chili pepper, chocolate, cinnamon, cloves, coffee, collard greens, comfrey herb tea, corn, coriander, currants, dill, eggplant, endive, fennel, garlic, grapefruit, grapes, guava, honey, honeydew melon, horseradish, kale, lemon, lentils, lettuce, licorice, lime, mace, mango, marjoram, mint, mushrooms, mustard, nutmeg, onion, orange, paprika, parsley, parsnip, peach, pear, peas, black pepper, pineapple, plum, potato, radish, raspberries, rhubarb, rosemary, rutabaga, sage, savory, sesame seeds, soybean, star anise, tarragon, tea, thyme, tomato, turmeric, and turnip.
[a]Fungal toxins are not included.
Source: References 61 and 62.

TABLE 29.3   Carcinogenicity in Rodents of Natural Chemicals in Roasted Coffee

Positive: $N = 21$	Acetaldehyde, benzaldehyde, benzene, benzofuran, benzo(a)pyrene, caffeic acid, catechol, 1,2,5,6-dibenzanthracene, ethanol, ethylbenzene, formaldehyde, furan, furfural, hydrogen peroxide, hydroquinone, isoprene, limonene, 4-methylcatechol, styrene, toluene, xylene
Not positive: $N = 8$	Acrolein, biphenyl, choline, eugenol, nicotinamide, nicotinic acid, phenol, piperidine
Uncertain:	Caffeine
Yet to test:	$\sim$1000 chemicals

Source: References 61 and 62.

MISCONCEPTION 5

Cancer risks to humans can be assessed by standard high-dose animal cancer tests.

Approximately half of all chemicals that have been tested in standard animal cancer tests, whether natural or synthetic, are rodent carcinogens (Table 29.4).[61-64] Why such a high positivity rate? In standard cancer tests, rodents are given chronic, near-toxic doses, the maximum tolerated dose (MTD). Evidence is accumulating that cell division caused by the high dose itself, rather than the chemical per se, is increasing the positivity rate. High doses can cause chronic wounding of tissues, cell death, and consequent chronic cell division of neighboring cells, which is a risk factor for cancer.[65] Each time a cell divides, the probability increases that a mutation will occur, thereby increasing the risk for cancer. At the low levels to which humans are usually exposed, such increased cell division does not occur. The process of mutagenesis and carcinogenesis is complicated because many factors are involved: for example, DNA lesions, DNA repair, cell division, clonal instability, apoptosis, and $p53$ (a cell cycle control gene that is mutated in half of human tumors).[66,67] The normal endogenous level of oxidative DNA lesions in somatic cells is appreciable.[19] In addition, tissues injured by high doses of chemicals have an inflammatory immune response involving activation of white cells in response to cell death.[68-75] Activated white cells release mutagenic oxidants (including peroxynitrite, hypochlorite, and H_2O_2). Therefore, the very low levels of chemicals to which

TABLE 29.4   Proportion of Chemicals Evaluated as Carcinogenic

Chemicals tested in both rats and mice[a]	
Chemicals in the Carcinogenic Potency Database (CPDB)	350/590 (59%)
Naturally occurring chemicals in the CPDB	79/139 (57%)
Synthetic chemicals in the CPDB	271/451 (60%)
Chemicals tested in rats and/or mice[a]	
Chemicals in the CPDB	702/1348 (52%)
Natural pesticides in the CPDB	37/71 (52%)
Mold toxins in the CPDB	14/23 (61%)
Chemicals in roasted coffee in the CPDB	21/30 (70%)
Commercial pesticides	79/194 (41%)
Innes negative chemicals retested[a]	17/34 (50%)
Physician's Desk Reference (PDR): drugs with reported cancer tests[b,d]	117/241 (49%)
FDA database of drug submissions[c,d]	125/282 (44%)

[a] From References 61 and 62.

[b] From Davies and Monro.[104]

[c] From Contrera et al.[195]

[d] 140 drugs are in both the FDA and PDR databases.

humans are exposed through water pollution or synthetic pesticide residues may pose no or only minimal cancer risks.

We have discussed[76] the argument that the high positivity rate is due to selecting more suspicious chemicals to test, which is a likely bias since cancer testing is both expensive and time-consuming, making it prudent to test suspicious compounds. One argument against selection bias is the high positivity rate for drugs (Table 29.4), because drug development tends to select chemicals that are not mutagens or expected carcinogens. A second argument against selection bias is that knowledge to predict carcinogenicity in rodent tests is highly imperfect, even now, after decades of testing results have become available on which to base prediction. For example, a prospective prediction exercise was conducted by several experts in 1990 in advance of the 2-year National Toxicology Program (NTP) bioassays. There was wide disagreement among the experts as to which chemicals would be carcinogenic when tested; accuracy varied, thus indicating that predictive knowledge is uncertain.[77] Moreover, if the main basis for selection were suspicion rather than human exposure, then one should select mutagens (80 percent are positive compared to 49 percent of nonmutagens), yet 55 percent of the chemicals tested are nonmutagens.[76]

A 1969 study by Innes et al.[78] has frequently been cited[79] as evidence that the positivity rate is low, because only 9 percent of 119 chemicals tested (primarily pesticides) were positive. However, the Innes tests were only in mice, had only 18 animals per group, and were terminated at 18 months. This protocol lacked the power of modern experiments, in which both rats and mice are tested, with 50 animals per group for 24 months. Of the 34 Innes negative chemicals that have been retested using modern protocols, 17 were positive (Table 29.4).[62,64]

It seems likely that a high proportion of all chemicals, whether synthetic or natural, might be "carcinogens" if run through the standard rodent bioassay at the MTD. For nonmutagens, carcinogenicity would be primarily due to the effects of high doses; for mutagens, it would result from a synergistic effect between cell division at high doses and DNA damage.[80–84] Without additional data on the mechanism of carcinogenesis for each chemical, the interpretation of a positive result in a rodent bioassay is highly uncertain. The carcinogenic effects may be limited to the high dose tested. Analyses of apoptosis and cell proliferation in recent bioassays can help assess the mode of action of a chemical and can be used in risk assessment.[85–87]

Linearity of dose–response seems unlikely in any case due to the inducibility of the numerous defense enzymes that deal with exogenous chemicals as groups (e.g., oxidants and electrophiles) and thus protect us against the natural world of mutagens as well as the small amounts of synthetic chemicals.[60,88–90]

There are validity problems associated with the use of the limited data from animal cancer tests for human risk assessment.[76,91,92] Standard practice in regulatory risk assessment for a given rodent carcinogen has been to extrapolate from the high doses of rodent bioassays to the low doses of most human exposures by multiplying carcinogenic potency in rodents by human exposure. Strikingly, due to the relatively narrow range of doses in 2-year rodent bio-

TABLE 29.5   Cancer Risk Assessment Without Conducting a 2-Year Bioassay[a]

Approach to Risk Assessment	Estimated Regulatory "Safe Dose"
Low-dose linear extrapolation based on the multistage model[b]	
Risk $< 10^{-6}$	$MTD^c/740,000$
Risk $< 10^{-5}$	$MTD/74,000$
Risk $< 10^{-4}$	$MTD/7,400$
Benchmark dose point-of-departure $= LTD_{10}{}^d$ with linear extrapolation	
Risk $< 10^{-6}$	$MTD/700,000$
Risk $< 10^{-5}$	$MTD/70,000$
Risk $< 10^{-4}$	$MTD/7,000$
Reference dose for nonlinear dose–response curve based on uncertainty factors	
$LTD10/1000^e$	$MTD/7,000$
$LTD10/10,000^f$	$MTD/70,000$

[a] From Reference 95 except as noted.

[b] From Gaylor and Gold.[94]

[c] MTD = Maximum tolerated dose (high dose in rodent test).

[d] LTD_{10} = lower confidence limit on dose to produce 10% of rodents with tumors.

[e] Combined uncertainty factors of 10 for animal to human extrapolation, 10 for sensitive humans, and 10 since the LTD_{10} represents a low-observed-adverse-effect level.[196]

[f] Additional uncertainty factor of 10 would be considered to account for possible extra sensitivity of children per the Food Quality Protection Act of 1996 or because of the severity of cancer even from low doses.[197,198]

assays, the small number of animals, and the limited range of tumor incidence rates that could be statistically significant, measures of potency obtained from 2-year bioassays are constrained to a relatively narrow range of values about the MTD (the high dose used in a rodent bioassay). The range of possible values is similarly limited for the EPA potency measure (q_1^*) and the TD_{50} (tumorigenic dose-rate for 50% of test animals). If induced tumors occurred in 100 percent of dosed animals, then the possible values could be more potent, but 100 percent tumor incidence rarely occurs.[64,91,93–95] For example, the dose usually estimated by regulatory agencies to give one cancer in a million can be approximated simply by using the MTD as a surrogate for carcinogenic potency. The "virtually safe dose" (VSD) can be approximated from the MTD. Gaylor and Gold[94] used the ratio MTD/TD_{50} and the relationship between q_1^* and TD_{50} to estimate the VSD. The VSD was approximated by the MTD/740,000 for rodent carcinogens.[94] For 90 percent of the carcinogens, the MTD/740,000 was within a factor of 10 of the VSD (Table 29.5). This is similar to the finding that in near-replicate experiments of the same chemical, potency estimates vary by a factor of 4 around a median value.[63,96,97] Thus, there may be little gain in precision of cancer risk estimates derived from a 2-year bioassay, compared to the estimate based on the MTD from a 90-day study.[98]

Recently, the EPA proposed new carcinogen guidelines[99] that employ a benchmark dose as a point-of-departure (POD) for low-dose risk assessment. If information on the carcinogenic mode of action for a chemical supports a nonlinear dose–response curve below the POD, a margin-of-exposure ratio between the POD and anticipated human exposure would be considered.[87,99] The POD would be divided by uncertainty (safety) factors to arrive at a reference dose that is likely to produce no, or at most, negligible cancer risk for humans. If nonlinearity below the POD is not supported by sufficient evidence, then linear extrapolation from the incidence at the POD to zero would be used for low-dose cancer risk estimation. The carcinogen guidelines suggest that the lower 95 percent confidence limit on the dose estimated to produce an excess of tumors in 10 percent of the animals (LTD_{10}) be used for the POD.

We have shown that, like the TD_{50} or q_1^*, the estimate of the LTD_{10} obtained from 2-year bioassays is constrained to a relatively narrow range of values.[95] Because of this constraint, a simple, quick, and relatively precise determination of the LTD_{10} can be obtained by MTD/7. All that is needed is a 90-day study to establish the MTD. Thus, if the anticipated human exposure were estimated to be small relative to the MTD/7, there may be little value in conducting a chronic 2-year study in rodents because the estimate of cancer risk would be low regardless of the results of a 2-year bioassay. Either linear extrapolation to a risk of less than 1 in 100,000 or use of an uncertainty factor of 10,000 would give the same regulatory "safe dose" (Table 29.5). Linear extrapolation to a virtually safe dose (VSD) associated with a cancer risk estimate of less than one in a million would be 10 times lower than the reference dose based on the LTD_{10}/10,000. Thus, whether the procedure involves a benchmark dose or a linearized model, cancer risk estimation is constrained by the bioassay design.

In regulatory policy, the VSD has been estimated from bioassay results by using a linear model. To the extent that carcinogenicity in rodent bioassays is due to the effects of high doses for the nonmutagens and a synergistic effect of cell division at high doses with DNA damage for the mutagens, then this model is inappropriate and markedly overestimates risk.

MISCONCEPTION 6

The toxicology of synthetic chemicals is different from that of natural chemicals.

It is often assumed that because natural chemicals are part of human evolutionary history, whereas synthetic chemicals are recent, the mechanisms that have evolved in animals to cope with the toxicity of natural chemicals will fail to protect against synthetic chemicals.[79] This assumption is flawed for several reasons[13,60,65]:

Humans have many natural defenses that buffer against normal exposures to toxins[60]; these usually are general rather than tailored to each specific chemical. Thus, the defenses work against both natural and synthetic chemicals. Examples of general defenses include the continuous shedding of cells exposed to toxins—the surface layers of the mouth, esophagus, stomach, intestine, colon, skin, and lungs are discarded every few days; DNA repair enzymes, which repair

DNA that has been damaged from many different sources; and detoxification enzymes of the liver and other organs that generally target classes of toxins rather than individual toxins. That defenses are usually general, rather than specific for each chemical, makes good evolutionary sense. The reason that predators of plants evolved general defenses presumably was to be prepared to counter a diverse and ever-changing array of plant toxins in an evolving world; if a herbivore had defenses against only a set of specific toxins, it would be at a great disadvantage in obtaining new food when favored foods became scarce or evolved new toxins.

Various natural toxins that have been present throughout vertebrate evolutionary history nevertheless cause cancer in vertebrates.[60,62,64,100] Mold toxins, such as aflatoxin, have been shown to cause cancer in rodents and other species, including humans (Table 29.4). Many of the common elements are carcinogenic to humans at high doses (e.g., salts of cadmium, beryllium, nickel, chromium, and arsenic) despite their presence throughout evolution. Furthermore, epidemiological studies from various parts of the world show that certain natural chemicals in food may be carcinogenic risks to humans; for example, the chewing of betel nuts with tobacco is associated with oral cancer.

Humans have not had time to evolve a "toxic harmony" with all of the plants in their diet. The human diet has changed markedly in the last few thousand years. Indeed, very few of the plants that humans eat today (e.g., coffee, cocoa, tea, potatoes, tomatoes, corn, avocados, mangoes, olives, and kiwi fruit) would have been present in a hunter-gatherer's diet. Natural selection works far too slowly for humans to have evolved specific resistance to the food toxins in these relatively newly introduced plants.

Since no plot of land is free from attack by insects, plants need chemical defenses—either natural or synthetic—in order to survive. Thus, there is a trade-off between naturally occurring and synthetic pesticides. One consequence of disproportionate concern about synthetic pesticide residues is that some plant breeders develop plants to be more insect-resistant by making them higher in natural toxins. A recent case illustrates the potential hazards of this approach to pest control: When a major grower introduced a new variety of highly insect-resistant celery into commerce, people who handled the celery developed rashes when they were subsequently exposed to sunlight. Some detective work found that the pest-resistant celery contained 6200 parts per billion (ppb) of carcinogenic (and mutagenic) psoralens instead of the 800 ppb present in common celery.[13,62]

MISCONCEPTION 7

Synthetic chemicals pose greater carcinogenic hazards than natural chemicals.

Gaining a broad perspective about the vast number of chemicals to which humans are exposed is important when assessing relative hazards and setting research and regulatory priorities.[9,10,12,62,79] Rodent bioassays have provided little information about the mechanisms of carcinogenesis that is needed to estimate low-dose risk. The assumption that synthetic chemicals are hazardous, even at the very low levels of human exposure to pollutants in the environment, has led to a bias in testing so that synthetic chemicals account for 76 percent (451/590) of the chemicals tested chronically in both rats and mice even though the vast proportion of human exposures are to naturally occurring chemicals (Table 29.4). The background of natural chemicals has never been systematically tested for carcinogenicity.

One reasonable strategy for setting priorities is to use a rough index to compare and rank possible carcinogenic hazards from a wide variety of chemical exposures at levels that humans typically receive, and then to focus on those that rank highest.[9,62,64] Ranking is a critical first step that can help set priorities when selecting chemicals for chronic bioassay or mechanistic studies, for epidemiological research, and for regulatory policy. Although one cannot say whether the ranked chemical exposures are likely to be of major or minor importance in human cancer, it is not prudent to focus attention on the possible hazards at the bottom of a ranking if, by using the same methodology to identify hazard, there are numerous common human exposures with much greater possible hazards. Our analyses are based on the HERP (Human Exposure/Rodent Potency) index, which indicates what percentage of the rodent carcinogenic potency (TD_{50} in mg/kg/day) a person receives from a given average daily dose for a lifetime exposure (mg/kg/day)[61] (Table 29.6). A ranking based on standard regulatory risk assessment and using the same exposures would be similar.

Overall, our analyses have shown that HERP values for some historically high exposures in the workplace and certain pharmaceuticals rank high, and that there is an enormous background of naturally occurring rodent carcinogens that are present in average consumption or typical portions of common foods, which cast doubt on the relative importance of low-dose exposures to residues of synthetic chemicals such as pesticides.[9,15,62,64] A committee of the NRC/National Academy of Sciences (NAS) recently reached similar conclusions about natural versus synthetic chemicals in the diet and called for further research on natural chemicals.[101]

The HERP ranking in Table 29.6 is for *average* U.S. exposures to all rodent carcinogens in the Carcinogenic Potency Database for which concentration data and average exposure or consumption data were both available, and for which human exposure could be chronic for a lifetime. For pharmaceuticals the doses are recommended doses, and for workplace they are past industry or occupation averages. The 87 exposures in the ranking (Table 29.6) are ordered by possible carcinogenic hazard (HERP), and natural chemicals in the diet are reported in boldface.

Several HERP values make convenient reference points for interpreting Table 29.6. The median HERP value is 0.002 percent, and the background HERP for the average chloroform level in a liter of U.S. tap water is 0.0003 percent. Chloroform is formed as a by-product of chlorination. A HERP of 0.00001 percent

TABLE 29.6 Ranking Possible Carcinogenic Hazards from Average U.S. Exposures to Rodent Carcinogens

Possible Hazard: HERP (%)	Average Daily US Exposure	Human Dose of Rodent Carcinogen	Potency TD$_{50}$ (mg/kg/day)^a		Exposure References
			Rats	Mice	
140	EDB: production workers (high exposure) (before 1977)	Ethylene dibromide, 150 mg	1.52	(7.45)	199, 200
17	Clofibrate	Clofibrate, 2 g	169	ND	201
14	Phenobarbital, 1 sleeping pill	Phenobarbital, 60 mg	(+)	6.09	202
6.8	1,3-Butadiene: rubber industry workers (1978–1986)	1,3-Butadiene, 66.0 mg	(261)	13.9	203
6.2	**Comfrey-pepsin tablets, 9 daily (no longer recommended)**	**Comfrey root, 2.7 g**	626	ND	204, 205
6.1	Tetrachloroethylene: dry cleaners with dry-to-dry units (1980–1990)	Tetrachloroethylene, 433 mg	101	(126)	206
4.0	Formaldehyde: production workers (1979)	Formaldehyde, 6.1 mg	2.19	(43.9)	207
2.4	Acrylonitrile: production workers (1960–1986)	Acrylonitrile, 28.4 mg	16.9	ND	208
2.2	Trichloroethylene: vapor degreasing (before 1977)	Trichloroethylene, 1.02 g	668	(1580)	209
2.1	**Beer, 257 g**	**Ethyl alcohol, 13.1 mL**	9110	(−)	210
1.4	Mobile home air (14 h/day)	Formaldehyde, 2.2 mg	2.19	(43.9)	211
1.3	**Comfrey-pepsin tablets, 9 daily (no longer recommended)**	**Symphytine, 1.8 mg**	1.91	ND	204, 205
0.9	Methylene chloride: workers, industry average (1940s–1980s)	Methylene chloride, 471 mg	724	(1100)	212
0.5	**Wine, 28.0 g**	**Ethyl alcohol, 3.36 mL**	9110	(−)	210

(Continued)

TABLE 29.6 (Continued)

Possible Hazard: HERP (%)	Average Daily US Exposure	Human Dose of Rodent Carcinogen	Potency TD$_{50}$ (mg/kg/day)^a Rats	Potency TD$_{50}$ (mg/kg/day)^a Mice	Exposure References
0.5	Dehydroepiandrosterone (DHEA)	DHEA supplement, 25 mg	68.1	ND	
0.4	Conventional home air (14 h/day)	Formaldehyde, 598 µg	2.19	(43.9)	213
0.2	Fluvastatin	Fluvastatin, 20 mg	125	ND	214
0.1	Coffee, 13.3 g	Caffeic acid, 23.9 mg	297	(4900)	210, 215
0.1	d-Limonene in food	d-Limonene, 15.5 mg	204	(−)	210
0.04	Lettuce, 14.9 g	Caffeic acid, 7.90 mg	297	(4900)	216, 217
0.03	Safrole in spices	Safrole, 1.2 mg	(441)	51.3	218
0.03	Orange juice, 138 g	d-Limonene, 4.28 mg	204	(−)	216, 219
0.03	Comfrey herb tea, 1 cup (1.5 g root) (no longer recommended)	Symphytine, 38 µg	1.91	ND	205
0.03	Tomato, 88.7 g	Caffeic acid, 5.46 mg	297	(4900)	216, 220
0.03	Pepper, black, 446 mg	d-Limonene, 3.57 mg	204	(−)	210, 221
0.02	Coffee, 13.3 g	Catechol, 1.33 mg	88.8	(244)	210, 222, 223
0.02	Furfural in food	Furfural, 2.72 mg	(683)	197	210
0.02	Mushroom (*Agaricus bisporus* 2.55 g)	Mixture of hydrazines,——etc. (whole mushroom)	(−)	20,300	210, 224, 225
0.02	Apple, 32.0 g	Caffeic acid, 3.40 mg	297	(4900)	226, 227
0.02	Coffee, 13.3 g	Furfural, 2.09 mg	(683)	197	210
0.01	BHA: daily U.S. avg. (1975)	BHA, 4.6 mg	606	(5530)	136
0.01	Beer (before 1979), 257 g	Dimethylnitrosamine, 726 ng	0.0959	(0.189)	210, 228, 229
0.008	Aflatoxin: daily U.S. avg. (1984–1989)	Aflatoxin, 18 ng	0.0032	(+)	230
0.007	Cinnamon, 21.9 mg	Coumarin, 65.0 µg	13.9	(103)	231
0.006	Coffee, 13.3 g	Hydroquinone, 333 µg	82.8	(225)	210, 222, 232

0.005	Saccharin: daily U.S. avg. (1977)	Saccharin, 7 mg	2140	(−)	233
0.005	Carrot, 12.1 g	Aniline, 624 µg	194[b]	(−)	216, 234
0.004	Potato, 54.9 g	Caffeic acid, 867 µg	297	(4900)	216, 235
0.004	Celery, 7.95 g	Caffeic acid, 858 µg	297	(4900)	236, 237
0.004	White bread, 67.6 g	Furfural, 500 µg	(683)	197	210
0.003	d-Limonene	Food additive, 475 µg	204	(−)	238
0.003	Nutmeg, 27.4 mg	d-Limonene, 466 µg	204	(−)	210, 239
0.003	Conventional home air (14 h/day)	Benzene, 155 µg	(169)	77.5	213
0.002	Coffee, 13.3 g	4-Methylcatechol, 433 µg	248	ND	210, 232, 240
0.002	Carrot, 12.1 g	Caffeic acid, 374 µg	297	(4900)	216, 237
0.002	Ethylene thiourea: daily U.S. avg. (1990)	Ethylene thiourea, 9.51 µg	7.9	(23.5)	241
0.002	BHA: daily U.S. avg. (1987)	BHA, 700 µg	606	(5530)	136
0.002	DDT: daily U.S. avg. (before 1972 ban)[c]	DDT, 13.8 µg	(84.7)	12.8	242
0.001	Plum, 2.00 g	Caffeic acid, 276 µg	297	(4900)	227, 243
0.001	Pear, 3.29 g	Caffeic acid, 240 µg	297	(4900)	210, 227
0.001	[UDMH: daily U.S. avg. (1988)]	[UDMH, 2.82 µg (from Alar)]	(−)	3.96	226
0.0009	Brown mustard, 68.4 mg	Allyl isothiocyanate, 62.9 µg	96	(−)	210, 244
0.0008	DDE: daily U.S. avg. (before 1972 ban)[d]	DDE, 6.91 µg	(−)	12.5	242
0.0006	Bacon, 11.5 g	Diethylnitrosamine, 11.5 ng	0.0266	(+)	210, 245
0.0006	Mushroom (Agaricus bisporus 2.55 g)	Glutamyl-p-hydrazinobenzoate, 107 µg	ND	277	210, 246
0.0005	Bacon, 11.5 g	Dimethylnitrosamine, 34.5 ng	0.0959	(0.189)	210, 245
0.0004	Bacon, 11.5 g	N-nitrosopyrrolidine, 196 ng	(0.799)	0.679	210, 247
0.0004	EDB: Daily U.S. avg. (before 1984 ban)[d]	EDB, 420 ng	1.52	(7.45)	248
0.0004	Tap water, 1 liter (1987–1992)	Bromodichloromethane, 13 µg	(72.5)	47.7	249
0.0004	TCDD: daily U.S. avg. (1994)	TCDD, 6.0 pg	0.0000235	(0.000156)	155

(Continued)

TABLE 29.6 (*Continued*)

Possible Hazard: HERP (%)	Average Daily US Exposure	Human Dose of Rodent Carcinogen	Potency TD$_{50}$ (mg/kg/day)^a Rats	Mice	Exposure References
0.0003	**Mango, 1.22 g**	**d-Limonene, 48.8 µg**	204	(–)	243, 250
0.0003	**Beer, 257 g**	**Furfural, 39.9 µg**	(683)	197	210
0.0003	Tap water, 1 liter (1987–1992)	Chloroform, 17 µg	(262)	90.3	249
0.0003	Carbaryl: daily U.S. avg. (1990)	Carbaryl, 2.6 µg	14.1	(–)	251
0.0002	**Celery, 7.95 g**	**8-Methoxypsoralen, 4.86 µg**	32.4	(–)	236, 252
0.0002	Toxaphene: daily U.S. avg. (1990)^c	Toxaphene, 595 ng	(–)	5.57	251
0.00009	Mushroom (*Agaricus bisporus*, 2.55 g)	p-Hydrazinobenzoate, 28 µg	ND	454^b	210, 246
0.00008	PCBs: daily U.S. avg. (1984–1986)	PCBs, 98 ng	1.74	(9.58)	153
0.00008	DDE/DDT: daily U.S. avg. (1990)^c	DDE, 659 ng	(–)	12.5	251
0.00007	Parsnip, 54.0 mg	8-Methoxypsoralen, 1.57 µg	32.4	(–)	253, 254
0.00007	Toast, 67.6 g	Urethane, 811 ng	(41.3)	16.9	210, 255
0.00006	**Hamburger, pan fried, 85 g**	**PhIP, 176 ng**	4.22^b	(28.6^b)	216, 256
0.00006	Furfural	Food additive, 7.77 µg	(683)	197	238
0.00005	**Estragole in spices**	**Estragole, 1.99 µg**	ND	51.8	210
0.00005	**Parsley, fresh, 324 mg**	**8-Methoxypsoralen, 1.17 µg**	32.4	(–)	253, 257
0.00005	Estragole	Food additive, 1.78 µg	ND	51.8	238
0.00003	**Hamburger, pan fried, 85 g**	**MeIQx, 38.1 ng**	1.66	(24.3)	216, 256
0.00002	Dicofol: daily U.S. avg. (1990)	Dicofol, 544 ng	(–)	32.9	251
0.00001	**Beer, 257 g**	**Urethane, 115 ng**	(41.3)	16.9	210, 255
0.000006	**Hamburger, pan fried, 85 g**	**IQ, 6.38 ng**	1.65^b	(19.6)	216, 256
0.000005	Hexachlorobenzene: daily U.S. avg. (1990)	Hexachlorobenzene, 14 ng	3.86	(65.1)	251

0.000001	Lindane: daily U.S. avg. (1990)	Lindane, 32 ng	(−)	30.7	251
0.0000004	PCNB: daily U.S. avg. (1990)	PCNB (Quintozene), 19.2 ng	(−)	71.1	251
0.0000001	Chlorobenzilate: daily U.S. avg. (1989)[c]	Chlorobenzilate, 6.4 ng	(−)	93.9	251
0.00000008	Captan: daily U.S. avg. (1990)	Captan, 115 ng	2080	(2110)	251
0.00000001	Folpet: daily U.S. avg. (1990)	Folpet, 12.8 ng	(−)	1550	251
<0.00000001	Chlorothalonil: daily U.S. avg. (1990)	Chlorothalonil, <6.4 ng	828[d]	(−)	251, 258

Note: [Chemicals that occur naturally in foods are in bold.] *Daily human exposure:* Reasonable daily intakes are used to facilitate comparisons. The calculations assume a daily dose for a lifetime. *Possible hazard:* The human dose of rodent carcinogen is divided by 70 kg to give a mg/kg/day of human exposure, and this dose is given as the percentage of the TD_{50} in the rodent (mg/kg/day) to calculate the Human Exposure/Rodent Potency (HERP) index. TD_{50} values used in the HERP calculation are averages calculated by taking the harmonic mean of the TD_{50}'s of the positive tests in that species from the Carcinogenic Potency Database. Average TD_{50} values have been calculated separately for rats and mice, and the more potent value is used for calculating possible hazard.

[a] ND = no data in CPDB; a number in parentheses indicates a TD_{50} value not used in the HERP calculation because TD_{50} is less potent than in the other species. (−) = negative in cancer tests; (+) = positive cancer test(s) not suitable for calculating a TD_{50}.

[b] TD_{50} harmonic mean was estimated for the base chemical from the hydrochloride salt.

[c] No longer contained in any registered pesticide product.[127]

[d] Additional data from the EPA that is not in the CPDB were used to calculate this TD_{50} harmonic mean.

is approximately equal to a regulatory VSD risk of 10^{-6}.[9] Using the benchmark dose approach recommended in the new EPA guidelines with the LTD_{10} as the point of departure (POD), linear extrapolation would produce a similar estimate of risk at 10^{-6} and hence a similar HERP value.[95] If information on the carcinogenic mode of action for a chemical supports a nonlinear dose–response curve, then the EPA guidelines call for a margin of exposure approach with the LTD_{10} as the POD. The reference dose using a safety or uncertainty factor of 1000 (i.e., $LD_{10}/1000$) would be equivalent to a HERP value of 0.001 percent. If the dose–response is judged to be nonlinear, then the cancer risk estimate will depend on the number and magnitude of safety factors used in the assessment.

The HERP ranking maximizes possible hazards to synthetic chemicals because it includes historically high exposure values that are now much lower, for example, DDT, saccharin, and some occupational exposures. Additionally, the values for dietary pesticide residues are averages in the *total diet*, whereas for most natural chemicals the exposure amounts are for concentrations of a chemical in an individual food (i.e., foods for which data are available on concentration and average U.S. consumption).

Table 29.6 indicates that many ordinary foods would not pass the regulatory criteria used for synthetic chemicals. For many natural chemicals the HERP values are in the top half of the table, even though natural chemicals are markedly underrepresented because so few have been tested in rodent bioassays. We discuss several categories of exposure below and indicate that mechanistic data are available for some chemicals, which suggest that the possible hazard may not be relevant to humans or would be low if nonlinearity or a threshold were taken into account in risk assessment.

Occupational Exposures   Occupational and pharmaceutical exposures to some chemicals have been high, and many of the single chemical agents or industrial processes evaluated as human carcinogens have been identified by historically high exposures in the workplace.[102] HERP values rank at the top of Table 29.6 for chemical exposures in some occupations to ethylene dibromide, 1,3-butadiene, tetrachloroethylene, formaldehyde, acrylonitrile, trichloroethylene, and methylene chloride. When exposures are high, the margin of exposure from the carcinogenic dose in rodents is low. The issue of how much human cancer can be attributed to occupational exposure has been controversial, but a few percent seems a reasonable estimate.[10]

In another analysis, we used permissible exposure limits (PELs), recommended in 1989 by the U.S. Occupational Safety and Health Administration (OSHA), as surrogates for actual exposures and compared the permitted daily dose rate for workers with the TD_{50} in rodents (PERP index, Permissible Exposure/Rodent Potency).[15] We found that PELs for 9 chemicals were greater than 10 percent of the rodent carcinogenic dose and for 27 chemicals they were between 1 and 10 percent of the rodent dose. The highest ranking chemicals should be priorities for regulatory scrutiny. In recent years, for two of the top chemicals, 1,3-butadiene and methylene chloride, the PELs have been lowered substantially, and the current PERP values are below 1 percent.

For trichloroethylene (HERP is 2.2 percent for vapor degreasers before 1977), we recently conducted an analysis based on an assumed cytotoxic mechanism of action and using physiologically based pharmacokinetic (PBPK)-effective dose estimates defined as peak concentrations. Our estimates indicate that for occupational respiratory exposures, the PEL for trichloroethylene would produce metabolite concentrations that exceed an acute no-observed-effect level for hepatotoxicity in mice. On this basis the PEL is not expected to be protective. In contrast, the EPA maximum concentration limit (MCL) in drinking water of 5 µg/L based on a linearized multistage model, is more stringent than our MCL based on a 1000-fold safety factor, which is 210 µg/L.[103]

Pharmaceuticals   Some pharmaceuticals that are used chronically are also clustered near the top of the HERP ranking, for example, phenobarbital, clofibrate, and fluvastatin. In Table 29.3 we reported that half the drugs in the Physicians Desk Reference (PDR) with cancer test data are positive in rodent bioassays.[104] Most drugs, however, are used for only short periods, and the HERP values for the rodent carcinogens would not be comparable to the chronic, long-term administration used in HERP. The HERP values for less than chronic administration at typical doses would produce high HERP values, for example, phenacetin (0.3 percent), metronidazole (5.6 percent), and isoniazid (14 percent).

Herbal supplements have recently developed into a large market in the United States based in part on the idea that if it's natural, it's good. They have not been a focus of carcinogenicity testing. The FDA regulatory requirements for safety and efficacy that are applied to pharmaceuticals do not pertain to herbal supplements under the 1994 Dietary Supplement and Health Education Act (DSHEA), and few have been tested for carcinogenicity. Those that are rodent carcinogens tend to rank high in HERP because, like some pharmaceutical drugs, the recommended dose is high relative to the rodent carcinogenic dose. Moreover, under DSHEA the safety criteria that have been used for decades by FDA for food additives that are "generally recognized as safe" (GRAS) are not applicable to dietary supplements,[105] even though supplements are used at higher doses. Comfrey is a medicinal herb whose roots and leaves have been shown to be carcinogenic in rats. The formerly recommended dose of 9 daily comfrey-pepsin tablets has a HERP value of 6.2 percent. Pyrrolizidine alkaloids are unusual constituents of herbal supplements in that many have been tested for carcinogenicity; several are positive in chronic bioassays (lasiocarpine, clivorine, monocrotaline, senkirkine, and riddelliine).[61] Symphytine, a pyrrolizidine alkaloid plant pesticide that is present in comfrey-pepsin tablets and comfrey tea, is a rodent carcinogen; the HERP value for symphytine is 1.3 percent in the pills and 0.03 percent in comfrey herb tea. Recently the FDA issued an advisory to manufacturers of comfrey products to remove them from the market. Comfrey roots and leaves can be bought at health food stores and on the WorldWide Web and can thus be used for tea, although comfrey is recommended for topical use only in the *PDR for Herbal Medicines*.[106] Poisoning epidemics by pyrrolizidine alkaloids have occurred in the developing world. In the United States poisonings, including deaths, have been associated with use of herbal teas containing comfrey.[107]

Dehydroepiandrosterone (DHEA), a natural hormone manufactured as a dietary supplement, has a HERP value of 0.5 percent for the recommended dose of 1 daily capsule containing 25 mg DHEA. DHEA is widely taken in hope of delaying aging and increasing muscle mass, and a 1997 survey reported that it was the fastest-selling product in health food stores.[108] The mechanism of liver carcinogenesis in rats is peroxisome proliferation,[109] like clofibrate.[110] Recent work on the mechanism of peroxisome proliferation in rodents indicates that it is a receptor-mediated response,[111] suggesting a threshold below which tumors are not induced. This mechanism is unlikely to be relevant to humans at any anticipated exposure level.[112,113] Recent analyses of the molecular basis of peroxisome proliferation conclude that there is an apparent lack of a peroxisome proliferative response in humans.[114] A recent review of clinical, experimental, and epidemiological studies concluded that late promotion of breast cancer in postmenopausal women may be stimulated by prolonged intake of DHEA.[115]

Natural Pesticides Natural pesticides, because few have been tested, are markedly underrepresented in our HERP analysis. Importantly, for each plant food listed, there are about 50 additional untested natural pesticides. Although about 10,000 natural pesticides and their breakdown products occur in the human diet,[12] only 71 have been tested adequately in rodent bioassays (Table 29.2). Average exposures to many natural-pesticide rodent carcinogens in common foods rank above or close to the median in the HERP table, ranging up to a HERP of 0.1 percent. These include caffeic acid (in coffee, lettuce, tomato, apple, potato, celery, carrot, plum, and pear), safrole (in spices and formerly in natural root beer before it was banned), allyl isothiocyanate (mustard), *d*-limonene (mango, orange juice, black pepper), coumarin in cinnamon, and hydroquinone, catechol, and 4-methylcatechol in coffee. Some natural pesticides in the commonly eaten mushroom (*Agaricus bisporus*) are rodent carcinogens (glutamyl-*p*-hydrazinobenzoate, *p*-hydrazinobenzoate), and the HERP based on feeding whole mushrooms to mice is 0.02 percent. For *d*-limonene, no human risk is anticipated because tumors are induced only in male rat kidney tubules with involvement of α_{2u}-globulin nephrotoxicity, which does not appear to be relevant for humans.[116–119]

Synthetic Pesticides Synthetic pesticides currently in use that are rodent carcinogens in the CPDB and that are quantitatively detected by the FDA Total Diet Study (TDS) as residues in food, are all included in Table 29.6. Many are at the very bottom of the ranking; however, HERP values are about at the median for ethylene thiourea (ETU), unsymmetrical dimethylhydrazine (UDMH from Alar) before its discontinuance, and DDT before its ban in the United States in 1972. These three synthetic pesticides rank below the HERP values for many naturally occurring chemicals that are common in the diet. The HERP values in Table 29.6 are for residue intake by females 65 and older, since they consume higher amounts of fruits and vegetables than other adult groups, thus maximizing the exposure estimate to pesticide residues. We note that for pesticide residues in the TDS, the consumption estimates for children (mg/kg/day from 1986 to 1991) are within a factor of 3 of the adult consumption (mg/kg/day).[120]

DDT and similar early pesticides have been a concern because of their unusual lipophilicity and persistence, even though there is no convincing epidemiological evidence of a carcinogenic hazard to humans,[121] and although natural pesticides can also bioaccumulate. In a recently completed 24-year study in which DDT was fed to rhesus and cynomolgus monkeys for 11 years, DDT was not evaluated as carcinogenic[122,123] despite doses that were toxic to both liver and central nervous system. However, the protocol used few animals and dosing was discontinued after 11 years, which may have reduced the sensitivity of the study.[62]

Current U.S. exposure to DDT and its metabolites is in foods of animal origin, and the HERP value is low, 0.00008 percent. DDT is often viewed as the typically dangerous synthetic pesticide because it concentrates in adipose tissue and persists for years. DDT was the first synthetic pesticide; it eradicated malaria from many parts of the world, including the United States, and was effective against many vectors of disease such as mosquitoes, tsetse flies, lice, ticks, and fleas. DDT was also lethal to many crop pests and significantly increased the supply and lowered the cost of fresh, nutritious foods, thus making them accessible to more people. DDT was also of low toxicity to humans. A 1970 National Academy of Sciences report concluded: "In little more than two decades DDT has prevented 500 million deaths due to malaria, that would otherwise have been inevitable."[124] There is no convincing epidemiological evidence, nor is there much toxicological plausibility, that the levels of DDT normally found in the environment or in human tissues are likely to be a significant contributor to human cancer.

DDT was unusual with respect to bioconcentration, and because of its chlorine substituents it takes longer to degrade in nature than most chemicals; however, these are properties of relatively few synthetic chemicals. In addition, many thousands of chlorinated chemicals are produced in nature.[125] Natural pesticides can also bioconcentrate if they are fat-soluble. Potatoes, for example, naturally contain the fat-soluble neurotoxins solanine and chaconine,[12,13] which can be detected in the bloodstream of all potato eaters. High levels of these potato neurotoxins have been shown to cause birth defects in rodents.[60]

For ETU the HERP value would be about 10 times lower if the potency value of the EPA were used instead of our TD_{50}; EPA combined rodent results from more than one experiment, including one in which ETU was administered in utero, and obtained a weaker potency.[126] (The CPDB does not include in utero exposures.) Additionally, EPA has recently discontinued some uses of fungicides for which ETU is a breakdown product, and exposure levels are therefore lower.

In 1984 the EPA banned the agricultural use of ethylene dibromide (EDB), the main fumigant in the United States, because of the residue levels found in grain, HERP = 0.0004 percent. This HERP value ranks low, whereas the HERP of 140 percent for the high exposures to EDB that some workers received in the 1970s is at the top of the ranking.[9] Two other pesticides in Table 29.6, toxaphene (HERP = 0.0002 percent) and chlorobenzilate (HERP = 0.0000001 percent), have been canceled.[127,128]

Most residues of synthetic pesticides have HERP values below the median. In descending order of HERP, these are carbaryl, toxaphene, dicofol, lindane, pentachloronitrobenzene (PCNB), chlorobenzilate, captan, folpet, and chlorothalonil. Some of the lowest HERP values in Table 29.6 are for the synthetic pesticides, captan, chlorothalonil, and folpet, which were also evaluated in 1987 by the National Research Council (NRC) and were considered by NRC to have a human cancer risk above 10^{-6}.[129] Why were the EPA risk estimates reported by NRC so high when our HERP values are so low? We have investigated this disparity in cancer risk estimation for pesticide residues in the diet by examining the two components of risk assessment: carcinogenic potency estimates from rodent bioassays and human exposure estimates.[130] We found that potency estimates based on rodent bioassay data are similar whether calculated, as in the NRC report, as the regulatory q_1^* or as the TD_{50} in the CPDB. In contrast, estimates of dietary exposure to residues of synthetic pesticides vary enormously, depending on whether they are based on the theoretical maximum residue contribution (TMRC) calculated by the EPA versus the average dietary residues measured by the FDA in the Total Diet Study (TDS). The EPA's TMRC is the theoretical maximum human exposure anticipated under the most severe field application conditions, which is often a large overestimate compared to the measured residues. For several pesticides, the NRC risk estimate was greater than one in a million whereas the FDA did not detect any residues in the TDS even though the TDS measures residues as low as 1 ppb.[128,130]

Cooking and Preparation of Food Cooking and preparation of food can also produce chemicals that are rodent carcinogens. Alcoholic beverages are a human carcinogen, and the HERP values in Table 29.6 for alcohol in beer (2.1 percent) and wine (0.5 percent) are high in the ranking. Ethyl alcohol is one of the least potent rodent carcinogens in the CPDB, but the HERP is high because of high concentrations in alcoholic beverages and high U.S. consumption. Another fermentation product, urethane (ethyl carbamate), has a HERP value of 0.00001 percent for average beer consumption and 0.00007 percent for average bread consumption (as toast).

Cooking food is plausible as a contributor to cancer. A wide variety of chemicals are formed during cooking. Rodent carcinogens formed include furfural and similar furans, nitrosamines, polycyclic hydrocarbons, and heterocyclic amines. Furfural, a chemical formed naturally when sugars are heated, is a widespread constituent of food flavor. The HERP value for naturally occurring furfural in average consumption of coffee is 0.02 percent and in white bread is 0.004 percent. Furfural is also used as a commercial food additive, and the HERP for total average U.S. consumption as an additive is 0.00006 percent (Table 29.6). Nitrosamines are formed from nitrite or nitrogen oxides (NO_x) and amines in food. In bacon the HERP for diethylnitrosamine is 0.0006 percent, and for dimethylnitrosamine it is 0.0005 percent.

A variety of mutagenic and carcinogenic heterocyclic amines (HA) are formed when meat, chicken, or fish are cooked, particularly when charred. Compared to other rodent carcinogens, there is strong evidence of carcinogenicity for HA in terms of positivity rates and multiplicity of target sites; however, concordance in target sites between rats and mice for these HA is generally restricted to the liver.[131] Under usual cooking conditions, exposures to HA are in the low ppb range, and the HERP values are low: For HA in pan-fried hamburger, the HERP value for 2-amino-1-methyl-6-phenylimidazo[4,5-*b*]-pyridine (PhIP) is 0.00006 percent, for 2-amino-3,8-dimethylimidazo[4,5-*f*]-quinoxaline (MeIQx) 0.00003 percent, and for 2-amino-3-methylimidazo[4,5-*f*]-quinoline (IQ) 0.000006 percent. Carcinogenicity of the 3 HA in the HERP table—IQ, MeIQx, and PhIP—has been investigated in studies in cynomolgus monkeys. IQ administered by gavage rapidly induced a high incidence of hepatocellular carcinoma; if the HERP value were based on the TD_{50} in monkeys, the value would be 0.00002 percent.[132] MeIQx, which induced tumors at multiple sites in rats and mice,[133] did not induce tumors in monkeys.[134] The PhIP study is in progress. Metabolism studies indicate the importance of *N*-hydroxylation in the carcinogenic effect of HA in monkeys.[135] IQ is activated via *N*-hydroxylation and forms DNA adducts; the *N*-hydroxylation of IQ appears to be carried out largely by hepatic CYP3A4 and/or CYP2C9/10, and not by CYP1A2; whereas the poor activation of MeIQx appears to be due to a lack of expression of CYP1A2 and an inability of other cytochromes P450, such as CYP3A4 and CYP2C9/10, to *N*-hydroxylate the quinoxalines. PhIP is activated by *N*-hydroxylation in monkeys and forms DNA adducts, suggesting that it may have a carcinogenic effect.[134,135]

Food Additives Food additives that are rodent carcinogens can be either naturally occurring (e.g., allyl isothiocyanate, furfural, and alcohol) or synthetic [butylated hydroxyanisole (BHA) and saccharin, Table 29.6]. The highest HERP values for average dietary exposures to synthetic rodent carcinogens in Table 29.6 are for exposures in the 1970s to BHA (0.01 percent) and saccharin (0.005 percent). Both are nongenotoxic rodent carcinogens for which data on mechanism of carcinogenesis strongly suggest that there would be no risk to humans at the levels found in food.

BHA is a phenolic antioxidant that is generally regarded as safe (GRAS) by the FDA. By 1987, after BHA was shown to be a rodent carcinogen, its use declined sixfold (HERP = 0.002 percent)[136]; this was due to voluntary replacement by other antioxidants, and to the fact that the use of animal fats and oils, in which BHA is primarily used as an antioxidant, has consistently declined in the United States. The mechanistic and carcinogenicity results on BHA indicate that malignant tumors were induced only at a dose above the MTD at which cell division was increased in the forestomach, which is the only site of tumorigenesis; the proliferation is only at high doses and is dependent on continuous dosing until late in the experiment.[137] Humans do not have a fore-stomach. We note that the dose–response for BHA curves sharply upward, but the potency value used in HERP is based on a linear model; if the California EPA potency value (which is based on a linearized multistage model) were used in HERP instead of TD_{50}, the HERP values for BHA would be 25 times lower.[138]

Saccharin, which has largely been replaced by other sweeteners, has been shown to induce tumors in rodents by a mechanism that is not relevant to humans. Recently, both the National Toxicology Program (NTP) and the International Agency for Research on Cancer (IARC) reevaluated the potential carcinogenic risk of saccharin to humans. NTP delisted saccharin in its *Report on Carcinogens*,[139] and IARC downgraded its evaluation to group 3, "not classifiable as to carcinogenicity to humans."[140] There is convincing evidence that the induction of bladder tumors in rats by sodium saccharin requires a high dose and is related to development of a calcium phosphate-containing precipitate in the urine,[80] which is not relevant to human dietary exposures. In a recently completed 24-year study by NCI, rhesus and cynomolgus monkeys were fed a dose of sodium saccharin that was equivalent to 5 cans of diet soda daily for 11 years.[122] The average daily dose rate of sodium saccharin was about 100 times lower than the dose that was carcinogenic to rats.[62,133] There was no carcinogenic effect in monkeys. There was also no effect on the urine or urothelium and no evidence of increased urothelial cell proliferation or of formation of solid material in the urine.[141] One would not expect to find a carcinogenic effect under the conditions of the monkey study. Additionally, there may be a true species difference because primate urine has a low concentration of protein and is less concentrated (lower osmolality) than rat urine.[141] Human urine is similar to monkey urine in this respect.[80]

For three naturally occurring chemicals that are also produced commercially and used as food additives, average exposure data were available and they are included in Table 29.6. The HERP values are as follows: For furfural the HERP value for the natural occurrence is 0.02 percent compared to 0.00006 percent for the additive; for *d*-limonene the natural occurrence HERP is 0.1 percent compared to 0.003 percent for the additive; and for estragole the HERP is 0.00005 percent for both the natural occurrence and the additive.

Safrole is the principle component (up to 90 percent) of oil of sassafras. It was formerly used as the main flavor ingredient in root beer. It is also present in the oils of basil, nutmeg, and mace.[142] The HERP value for average consumption of naturally occurring safrole in spices is 0.03 percent. In 1960 safrole and safrole-containing sassafras oils were banned from use as food additives in the United States.[143] Before 1960, for a person consuming a glass of sassafras root beer per day for life, the HERP value would have been 0.2 percent.[79] Sassafras root can still be purchased in health food stores and can therefore be used to make tea; the recipe is on the World Wide Web.

Mycotoxins   Of the 23 fungal toxins tested for carcinogenicity, 14 are positive (61 percent) (Table 29.3). The mutagenic mold toxin, aflatoxin, which is found in moldy peanut and corn products, interacts with chronic hepatitis infection in human liver cancer development.[144] There is a synergistic effect in the human liver between aflatoxin (genotoxic effect) and the hepatitis B virus (cell division effect) in the induction of liver cancer.[145] The HERP value for aflatoxin of 0.008 percent is based on the rodent potency. If the lower human potency value calculated by FDA from epidemiological data were used instead, the HERP would be about 10-fold lower.[146] Biomarker measurements of aflatoxin in populations in Africa and China, which have high rates of hepatitis B and C viruses and liver cancer, confirm that those populations are chronically exposed to high levels of aflatoxin.[147,148] Liver cancer is rare in the United States. Hepatitis viruses can account for half of liver cancer cases among non-Asians and even more among Asians in the United States.[149]

Ochratoxin A, a potent rodent carcinogen,[61] has been measured in Europe and Canada in agricultural and meat products. An estimated exposure of 1 ng/kg/day would have a HERP value at about the median of Table 29.6.[150,151]

Synthetic Contaminants Polychlorinated biphenyls (PCBs) and tetrachlorodibenzo-p-dioxin (TCDD), which have been a concern because of their environmental persistence and carcinogenic potency in rodents, are primarily consumed in foods of animal origin. In the United States PCBs are no longer used, but some exposure persists. Consumption in food in the United States declined about 20-fold between 1978 and 1986.[152,153] The HERP value for PCB in Table 29.6 for the most recent reporting in the FDA Total Diet Study (1984 to 1986) is 0.00008 percent, toward the bottom of the ranking, and far below many values for naturally occurring chemicals in common foods. It has been reported that some countries may have higher intakes of PCBs than the United States.[154]

TCDD, the most potent rodent carcinogen, is produced naturally by burning when chloride ion is present, for example, in forest fires or wood burning in homes. EPA[155] proposes that the source of TCDD is primarily from the atmosphere directly from emissions, for example, incinerators, or indirectly by returning dioxin to the atmosphere.[155,156] TCDD bioaccumulates through the food chain because of its lipophilicity, and more than 95 percent of human intake is from animal fats in the diet.[155] Dioxin emissions decreased by 80 percent from 1987 to 1995, which EPA attributes to reduced medical and municipal incineration emissions.[155]

The HERP value of 0.0004 percent for average U.S. intake of TCDD[155] is below the median of the values in Table 29.6. Recently, EPA has reestimated the potency of TCDD based on a body burden dose-metric in humans (rather than intake)[155] and a reevaluation of tumor data in rodents (which determined 2/3 fewer liver tumors).[157] Using this EPA potency for HERP would put TCDD at the median of HERP values in Table 29.6, that is, 0.002 percent.

TCDD exerts many of its harmful effects in experimental animals through binding to the Ah receptor (AhR) and does not have effects in the AhR knockout mouse.[158,159] A wide variety of natural substances also bind to the Ah receptor (e.g., tryptophan oxidation products), and insofar as they have been examined, they have similar properties to TCDD,[60] including inhibition of estrogen-induced effects in rodents.[160] For example, a variety of flavones and other plant substances in the diet, and their metabolites also bind to the Ah receptor, for example, indole-3-carbinol (I3C). I3C is the main breakdown compound of glucobrassicin, a glucosinolate that is present in large amounts in vegetables of the *Brassica* genus, including broccoli, and gives rise to the potent Ah binder, indole carbazole.[161] The binding affinity (greater for TCDD) and amounts consumed (much greater for dietary compounds) both need to be considered in comparing possible harmful effects. Some studies provide evidence of enhancement of carcinogenicity of I3C.[162] Additionally, both I3C and TCDD, when administered to pregnant rats, resulted in reproductive abnormalities in male offspring.[163] Currently, I3C is in clinical trials for prevention of breast cancer[164–166] and also is being tested for carcinogenicity by NTP.[166] I3C is marketed as a dietary supplement at recommended doses about 30 times higher[167] than present in the average Western diet.[166]

TCDD has received enormous scientific and regulatory attention, most recently in an ongoing assessment by the U.S. EPA.[155,156,168,169] Some epidemiologic studies suggest an association with cancer mortality, but the evidence is not sufficient to establish causality. IARC evaluated the epidemiological evidence for carcinogenicity of TCDD in humans as limited.[170] The strongest epidemiological evidence was among highly exposed workers for overall cancer mortality. There is a lack of evidence in humans for any specific target organ. Estimated blood levels of TCDD in studies of those highly exposed workers were similar to blood levels in rats in positive cancer bioassays.[170] In contrast, background levels of TCDD in humans are about 100- to 1000-fold lower than in the rat study. The similarity of worker and rodent blood levels and mechanism of the Ah receptor in both humans and rodents, were considered by IARC when they evaluated TCDD as a group 1 carcinogen in spite of only limited epidemiological evidence. IARC also concluded that "evaluation of the relationship between the magnitude of the exposure in experimental systems and the magnitude of the response (i.e., dose–response relationships) do not permit conclusions to be drawn on the human health risks from background exposures to 2,3,7,8-TCDD." The NTP *Report on Carcinogens* recently evaluated TCDD as "reasonably anticipated to be a human carcinogen," that is, rather than as a known human carcinogen.[171] The EPA draft final report[155] characterized TCDD as a "human carcinogen" but concluded that "there is no clear indication of increased disease in the general population attributable to dioxin-like compounds."[155] Possible limitations of data or scientific tools were given by EPA as possible reasons for the lack of observed effects.

In sum, the HERP ranking in Table 29.6 indicates that when synthetic pesticide residues in the diet are ranked on possible carcinogenic hazard and compared to the ubiquitous exposures to rodent carcinogens, they rank low. Widespread exposures to naturally occurring rodent carcinogens cast doubt on the relevance to human cancer of low-level exposures to synthetic rodent carcinogens. In regulatory efforts to prevent human cancer, the evaluation of low-level exposures to synthetic chemicals has had a high priority. Our results indicate, however, that a high percentage of both natural and synthetic chemicals are rodent carcinogens at the MTD, that tumor incidence data from rodent bioassays are not adequate to assess low-dose risk, and that there is an imbalance in testing of synthetic chemicals compared to natural chemicals. There is an enormous background of natural chemicals in the diet that rank high in possible hazard, even though so few have been tested in rodent bioassays. In Table 29.6, 90 percent of the HERP values are above the level that would approximate a regulatory virtually safe dose of 10^{-6}.

Caution is necessary in drawing conclusions from the occurrence in the diet of natural chemicals that are rodent carcinogens. It is not argued here that these dietary exposures are necessarily of much relevance to human cancer. In fact, epidemiological results indicate that adequate consumption of fruits and vegetables reduces cancer risk at many sites, and that protective factors such as intake of vitamins such as folic acid are important, rather than intake of individual rodent carcinogens (see Misconception 3).

The HERP ranking also indicates the importance of data on mechanism of carcinogenesis for each chemical. For several chemicals, data has recently been generated that indicates that exposures would not be expected to be a cancer risk to humans at the levels consumed in food (e.g., saccharin, BHA, chloroform, d-limonene). Standard practice in regulatory risk assessment for chemicals that induce tumors in high-dose rodent bioassays has been to extrapolate risk to low dose in humans by multiplying potency by human exposure. Without data on mechanism of carcinogenesis, however, the true human risk of cancer at low dose is highly uncertain and could be zero.[9,84,172,173] Adequate risk assessment from animal cancer tests requires more information for a chemical, about pharmacokinetics, mechanism of action, apoptosis, cell division, induction of defense and repair systems, and species differences.

MISCONCEPTION 8

Pesticides and other synthetic chemicals are disrupting hormones.

Synthetic hormone mimics such as organochlorine pesticides have become an environmental issue,[174] which was recently addressed by NAS.[175] We discussed in Misconception 2 that hormone factors are important in human cancer and that life-style factors can markedly change the levels of endogenous hormones. The trace exposures to estrogenic organochlorine residues are tiny compared to the normal dietary intake of naturally occurring endocrine-active chemicals in fruits and vegetables.[176–178] These low levels of human exposure seem toxicologically implausible as a significant cause of cancer or of reproductive abnormalities.[176–179] Synthetic hormone mimics have been proposed as a cause of declining sperm counts, even though it has not been shown that sperm counts are declining.[175,180–184] A recent analysis for the United States examined all available data on sperm counts and found that mean sperm concentrations were higher in New York than all other U.S. cities.[184] When this geographic difference was taken into account, there was no significant change in sperm counts for the past 50 years.[184] Even if sperm counts were declining, there are many more likely causes, such as smoking and diet (Misconception 2).

Some recent studies have compared estrogenic equivalents (EQ) of dietary intake of synthetic chemicals versus phytoestrogens in the normal diet, by considering both the amount humans consume and estrogenic potency. Results support the idea that synthetic residues are orders of magnitude lower in EQ and are generally weaker in potency. One study used a series of in vitro assays and calculated the EQs in extracts from 200 mL of red cabernet wine and the EQs from average intake of organochlorine pesticides.[185] EQs for a single glass of wine ranged from 0.15 to 3.68 µg/day compared to 1.24 ng/day for organochlorine pesticides.[185] Another study[186] compared plasma concentrations of the phytoestrogens genistein and daidzein in infants fed soy-based formula versus cow milk formula or human breast milk. Mean plasma levels were hundreds of times higher for the soy-fed infants than others.

MISCONCEPTION 9

Regulation of low, hypothetical risks is effective in advancing public health.

Since there is no risk-free world and resources are limited, society must set priorities in order to save the greatest number of lives.[187,188] In 1991 the EPA projected that the cost to society of environmental regulations in 1997 would be about $140 billion per year (about 2.6 percent of Gross National Product).[189] Most of this cost would be to the private sector. Several economic analyses have concluded that current expenditures are not cost effective; resources are not being used so as to save the greatest number of lives per dollar. One estimate is that the United States could prevent 60,000 deaths per year by redirecting the same dollar resources to more cost-effective programs.[190] For example, the median toxin control program costs 146 times more per life-year saved than the median medical intervention.[190] This difference is likely to be even greater because cancer risk estimates for toxin control programs are worst-case, hypothetical estimates, and the true risks at low dose are often likely to be zero[9,76,94] (Misconception 5). Some economists have argued that costly regulations intended to save lives may actually increase the number of deaths,[191] in part because they divert resources from important health risks and in part because higher incomes are associated with lower mortality.[192-194] Rules on air and water pollution are necessary (it was a public health benefit to phase lead out of gasoline), and clearly cancer prevention is not the only reason for regulations. However, worst-case assumptions in risk assessment represent a policy decision, not a scientific one, and they confuse attempts to allocate money effectively for risk abatement.

Regulatory efforts to reduce low-level human exposure to synthetic chemicals because they are rodent carcinogens are expensive since they aim to eliminate minuscule concentrations that can now be measured with improved techniques. These efforts distract from the major task of improving public health through increasing scientific understanding about how to prevent cancer (e.g., the role of diet), increasing public understanding of how life-style influences health, and improving our ability to help individuals alter life-style.

ACKNOWLEDGMENTS

This work was supported by a grant from the Office of Energy Research, Office of Health and Environmental Research of the U.S. Department of Energy under Contract DE-AC03-76SF00098 to L.S.G., the National Cancer Institute Outstanding Investigator Grant CA39910 to B.N.A., and by the National Institute of Environmental Health Sciences Center Grant ESO1896.

REFERENCES

1. L. A. G. Ries, M. P. Eisner, C. L. Kosary, B. F. Hankey, B. A. Miller, L. Clegg, and B. K. Edwards (Eds.). *SEER Cancer Statistics Review, 1973–1997*. National Cancer Institute, Bethesda, MD, 2000.

2. American Cancer Society. *Cancer Facts & Figures—2000*. American Cancer Society, Atlanta, GA, 2000.

3. M. S. Linet, L. A. Ries, M. A. Smith, R. E. Tarone, and S. S. Devesa. Cancer surveillance series: Recent trends in childhood cancer incidence and mortality in the United States. *J. Natl. Cancer Inst.* 91: 1051–1058 (1999).

4. R. Peto, J. Boreham, M. Clarke, C. Davies, and V. Beral. UK and USA breast cancer deaths down 25% in year 2000 at ages 20–69 years. *Lancet* 355: 1822 (2000).

5. I. Bailar III and H. L. Gornik. Cancer undefeated. *N. Engl. J. Med.* 336: 1569–1574 (1997).

6. R. Doll and R. Peto. *The Causes of Cancer*. Oxford University Press, New York, 1981.

7. S. S. Devesa, W. J. Blot, B. J. Stone, B. A. Miller, R. E. Tarone, and F. J. Fraumeni, Jr. Recent cancer trends in the United States. *J. Natl. Cancer Inst.* 87: 175–182 (1995).

8. R. N. Anderson. United States life tables, 1997. *Natl. Vital Stat. Rep.* 47(28): 1–37 (1999).

9. L. S. Gold, T. H. Slone, B. R. Stern, N. B. Manley, and B. N. Ames. Rodent carcinogens: Setting priorities. *Science* 258: 261–265 (1992).

10. B. N. Ames, L. S. Gold, and W. C. Willett. The causes and prevention of cancer. *Proc. Natl. Acad. Sci. USA* 92: 5258–5265 (1995).

11. B. N. Ames. Micronutrients prevent cancer and delay aging. *Toxicol. Lett.* 103: 5–18 (1998).

12. B. N. Ames, M. Profet, and L. S. Gold. Dietary pesticides (99.99% all natural). *Proc. Natl. Acad. Sci. USA* 87: 7777–7781 (1990).

13. L. S. Gold, T. H. Slone, and B. N. Ames. Prioritization of possible carcinogenic hazards in food. In D. R. Tennant (Ed.), *Food Chemical Risk Analysis*. Chapman and Hall, London, 1997, pp. 267–295.

14. M. Gough. How much cancer can EPA regulate away? *Risk Anal.* 10: 1–6 (1990).

15. L. S. Gold, G. B. Garfinkel, and T. H. Slone. Setting priorities among possible carcinogenic hazards in the workplace. In C. M. Smith, D. C. Christiani, and K. T. Kelsey (Eds.), *Chemical Risk Assessment and Occupational Health: Current Applications, Limitations, and Future Prospects.* Auburn House, Westport, CT, 1994, pp. 91–103.

16. B. N. Ames, M. K. Shigenaga, and T. M. Hagen. Oxidants, antioxidants, and the degenerative diseases of aging. *Proc. Natl. Acad. Sci. USA* 90: 7915–7922 (1993).

17. K. B. Beckman and B. N. Ames. The free radical theory of aging matures. *Physiol. Rev.* 78: 547–581 (1998).

18. T. M. Hagen, D. L. Yowe, J. C. Bartholomew, C. M. Wehr, K. L. Do, J.-Y. Park, and B. N. Ames. Mitochondrial decay in hepatocytes from old rats: Membrane potential declines, heterogeneity and oxidants increase. *Proc. Natl. Acad. Sci. USA* 94: 3064–3069 (1997).

19. H. J. Helbock, K. B. Beckman, M. K. Shigenaga, P. B. Walter, A. A. Woodall, H. C. Yeo, and B. N. Ames. DNA oxidation matters: The HPLC-electrochemical detection assay of 8-oxo-deoxyguanosine and 8-oxo-guanine. *Proc. Natl. Acad. Sci. USA* 95: 288–293 (1998).

20. C. A. Rice-Evans, J. Sampson, P. M. Bramley, and D. E. Holloway. Why do we expect carotenoids to be antioxidants in vivo? *Free Rad. Res.* 26: 381–398 (1997).

21. Centers for Disease Control and Prevention. Smoking-attributable mortality and years of potential life lost—United States, 1984 [and editorial note—1997]. *MMWR Morb. Mortal. Wkly. Rep.* 46: 444–451 (1997).

22. J. Lykkesfeldt, S. Christen, L. M. Wallock, H. H. Chang, R. A. Jacob, and B. N. Ames. Ascorbate is depleted by smoking and repleted by moderate supplementation: A study in male smokers and nonsmokers with matched dietary and antioxidant intakes. *Am. J. Clin. Nutr.* 71: 530–536 (2000).

23. B. N. Ames, P. A. Motchnik, C. G. Fraga, M. K. Shigenaga, T. M. Hagen, and A. Ohlshan. Antioxidant prevention of birth defects and cancer. In D. R. Mattison (Ed.), *Male-Mediated Developmental Toxicity.* Plenum, New York, 1994, pp. 243–259.

24. C. G. Fraga, P. A. Motchnik, M. K. Shigenaga, H. J. Helbrook, R. A. Jacob, and B. N. Ames. Ascorbic acid protects against endogenous oxidative damage in human sperm. *Proc. Natl. Acad. Sci. USA* 88: 11003–11006 (1991).

25. C. G. Fraga, P. A. Motchnik, A. J. Wyrobek, D. M. Rempel, and B. N. Ames. Smoking and low antioxidant levels increase oxidative damage to sperm DNA. *Mutat. Res.* 351: 199–203 (1996).

26. A. J. Wyrobek, J. Rubes, M. Cassel, D. Moore, S. Perrault, V. Slott, D. Evenson, Z. Zudova, L. Borkovec, S. Selevan, and X. Lowe. Smokers produce more aneuploid sperm than non-smokers. *Am. J. Hum. Genet.* 57(Suppl.) A131: (1995).

27. B.-T. Ji, X.-O. Shu, M. S. Linet, W. Zheng, S. Wacholder, Y.-T. Gao, D.-M. Ying, and F. Jin. Paternal cigarette smoking and the risk of childhood cancer among offspring of nonsmoking mothers. *J. Natl. Cancer Inst.* 89: 238–244 (1997).

28. R. W. Hart, K. Keenan, A. Turturro, K. M. Abdo, J. Leakey, and B. Lyn-Cook. Caloric restriction and toxicity. *Fundam. Appl. Toxicol.* 25: 184–195 (1995).

29. A. Turturro, P. Duffy, R. Hart, and W. T. Allaben. Rationale for the use of

dietary control in toxicity studies-B6C3F$_1$ mouse. *Toxicol. Pathol.* 24: 769–775 (1996).

30. S. Christen, T. M. Hagen, M. K. Shigenaga, and B. N. Ames. Chronic inflammation, mutation, and cancer. In J. Parsonnet (Ed.), *Microbes and Malignancy: Infection as a Cause of Cancer.* Oxford University Press, New York, 1999, pp. 35–88.

31. E. Shacter, E. J. Beecham, J. M. Covey, K. W. Kohn, and M. Potter. Activated neutrophils induce prolonged DNA damage in neighboring cells. Erratum in *Carcinogenesis* 10(3): 628 (1989). *Carcinogenesis* 9: 2297–2304 (1988).

32. K. Yamashina, B. E. Miller, and G. H. Heppner. Macrophage-mediated induction of drug-resistant variants in a mouse mammary tumor cell line. *Cancer Res.* 46: 2396–2401 (1986).

33. P. Pisani, D. M. Parkin, N. Muñoz, and J. Ferlay. Cancer and infection: Estimates of the attributable fraction in 1990. *Cancer Epidemiol. Biomarkers Prev.* 6: 387–400 (1997).

34. B. E. Henderson, R. K. Ross, and M. C. Pike. Towards the primary prevention of cancer. *Science* 254: 1131–1138 (1991).

35. B. E. Henderson and H. S. Feigelson. Hormonal carcinogenesis. *Carcinogenesis* 21: 427–433 (2000).

36. D. J. Hunter and W. C. Willett. Diet, body size, and breast cancer. *Epidemiol. Rev.* 15: 110–132 (1993).

37. G. Block, B. Patterson, and A. Subar. Fruit, vegetables and cancer prevention: A review of the epidemiologic evidence. *Nutr. Cancer* 18: 1–29 (1992).

38. K. A. Steinmetz and J. D. Potter. Vegetables, fruit, and cancer prevention: A review. *J. Am. Diet. Assoc.* 96: 1027–1039 (1996).

39. M. J. Hill, A. Giacosa, and C. P. J. Caygill (Eds.). *Epidemiology of Diet and Cancer.* Ellis Horwood, New York, 1994.

40. S. M. Krebs-Smith, A. Cook, A. F. Subar, L. Cleveland, J. Friday, and L. L. Kahle. Fruit and vegetable intakes of children and adolescents in the United States. *Arch. Pediatr. Adolesc. Med.* 150: 81–86 (1996).

41. S. M. Krebs-Smith, A. Cook, A. F. Subar, L. Cleveland, and J. Friday. US adults' fruit and vegetable intakes, 1989 to 1991: A revised baseline for the Healthy People 2000 objective. *Am. J. Public Health* 85: 1623–1629 (1995).

42. U.S. National Cancer Institute. Why eat five? *J. Natl. Cancer Inst.* 88: 1314 (1996).

43. B. C. Blount, M. M. Mack, C. M. Wehr, J. T. MacGregor, R. A. Hiatt, G. Wang, S. N. Wickramasinghe, R. B. Everson, and B. N. Ames. Folate deficiency causes uracil misincorporation into human DNA and chromosome breakage: Implications for cancer and neuronal damage. *Proc. Natl. Acad. Sci. USA* 94: 3290–3295 (1997).

44. M. Fenech, C. Aitken, and J. Rinaldi. Folate, vitamin B12, homocysteine status and DNA damage in young Australian adults. *Carcinogenesis* 19: 1163–1171 (1998).

45. E. Giovannucci, M. J. Stampfer, G. A. Colditz, E. B. Rimm, D. Trichopoulos, B. A. Rosner, F. E. Speizer, and W. C. Willett. Folate, methionine, and alcohol intake and risk of colorectal adenoma. *J. Natl. Cancer Inst.* 85: 875–884 (1993).

46. J. B. Mason. Folate and colonic carcinogenesis: Searching for a mechanistic understanding. *J. Nutr. Biochem.* 5: 170–175 (1994).

47. E. Giovannucci, M. J. Stampfer, G. A. Colditz, D. J. Hunter, C. Fuchs, B. A. Rosner, F. E. Speizer, and W. C. Willett. Multivitamin use, folate, and colon cancer in women in the Nurses' Health Study. *Ann. Intern. Med.* 129: 517–524 (1998).

48. L. Wallock, R. Jacob, A. Woodall, and B. Ames. Nutritional status and positive relation of plasma folate to fertility indices in nonsmoking men. *FASEB J.* 11: A184 (1997).

49. L. M. Wallock, T. Tamura, C. A. Mayr, K. E. Johnston, B. N. Ames, and R. A. Jacob. Low seminal plasma folate concentrations are associated with low sperm density and count in male smokers and nonsmokers. *Fertil. Steril.* 75: 252–259 (2001).

50. C. J. Boushey, S. A. Beresford, G. S. Omenn, and A. G. Motulsky. A quantitative assessment of plasma homocysteine as a risk factor for vascular disease. Probable benefits of increasing folic acid intakes. *J. Am. Med. Assoc.* 274: 1049–1057 (1995).

51. A. F. Subar, G. Block, and L. D. James. Folate intake and food sources in the US population. *Am. J. Clin. Nutr.* 50: 508–516 (1989).

52. F. R. Senti and S. M. Pilch. Analysis of folate data from the second National Health and Nutrition Examination Survey (NHANES II). *J. Nutr.* 115: 1398–1402 (1985).

53. L. B. Bailey, P. A. Wagner, G. J. Christakis, P. E. Araujo, H. Appledorf, C. G. Davis, J. Masteryanni, and J. S. Dinning. Folacin and iron status and hematological findings in predominately black elderly persons from urban low-income households. *Am. J. Clin. Nutr.* 32: 2346–2353 (1979).

54. L. B. Bailey, P. A. Wagner, G. J. Christakis, C. G. Davis, H. Appledorf, P. E. Araujo, E. Dorsey, and J. S. Dinning. Folacin and iron status and hematological findings in black and Spanish-American adolescents from urban low-income households. *Am. J. Clin. Nutr.* 35: 1023–1032 (1982).

55. P. F. Jacques, J. Selhub, A. G. Bostom, P. W. Wilson, and I. H. Rosenberg. The effect of folic acid fortification on plasma folate and total homocysteine concentrations. *N. Engl. J. Med.* 340: 1449–1454 (1999).

56. J. Z. Zhang, S. M. Henning, and M. E. Swendseid. Poly(ADP-ribose) polymerase activity and DNA strand breaks are affected in tissues of niacin-deficient rats. *J. Nutr.* 123: 1349–1355 (1993).

57. E. L. Jacobson. Niacin deficiency and cancer in women. *J. Am. Coll. Nutr.* 12: 412–416 (1993).

58. V. Herbert and L. J. Filer, Jr. Vitamin B-12. In E. E. Ziegler (Ed.), *Present Knowledge in Nutrition*. ILSI Press, Washington, DC, 1996, pp. 191–205.

59. S. N. Wickramasinghe and S. Fida. Bone marrow cells from vitamin B12- and folate-deficient patients misincorporate uracil into DNA. *Blood* 83: 1656–1661 (1994).

60. B. N. Ames, M. Profet, and L. S. Gold. Nature's chemicals and synthetic chemicals: Comparative toxicology. *Proc. Natl. Acad. Sci. USA* 87: 7782–7786 (1990).

61. L. S. Gold and E. Zeiger (Eds.). *Handbook of Carcinogenic Potency and Genotoxicity Databases*. CRC Press, Boca Raton, FL, 1997.

62. L. S. Gold, N. B. Manley, T. H. Slone, and L. Rohrbach. Supplement to the Carcinogenic Potency Database (CPDB): Results of animal bioassays published in the

general literature in 1993 to 1994 and by the National Toxicology Program in 1995 to 1996. *Environ. Health Perspect.* 107(Suppl. 4): 527–600 (1999).

63. L. S. Gold, L. Bernstein, R. Magaw, and T. H. Slone. Interspecies extrapolation in carcinogenesis: Prediction between rats and mice. *Environ. Health Perspect.* 81: 211–219 (1989).

64. L. S. Gold, T. H. Slone, and B. N. Ames. Overview of analyses of the carcinogenic potency database. In L. S. Gold and E. Zeiger (Eds.), *Handbook of Carcinogenic Potency and Genotoxicity Databases.* CRC Press, Boca Raton, FL, 1997, pp. 661–685.

65. B. N. Ames, L. S. Gold, and M. K. Shigenaga. Cancer prevention, rodent high-dose cancer tests, and risk assessment. *Risk Anal.* 16: 613–617 (1996).

66. J. G. Christensen, T. L. Goldsworthy, and R. C. Cattley. Dysregulation of apoptosis by c-myc in transgenic hepatocytes and effects of growth factors and non-genotoxic carcinogens. *Mol. Carcinog.* 25: 273–284 (1999).

67. L. L. Hill, A. Ouhtit, S. M. Loughlin, M. L. Kripke, H. N. Ananthaswamy, and L. B. Owen-Schaub. Fas ligand: A sensor for DNA damage critical in skin cancer etiology. *Science* 285: 898–900 (1999).

68. D. L. Laskin and K. J. Pendino. Macrophages and inflammatory mediators in tissue injury. *Annu. Rev. Pharmacol. Toxicol.* 35: 655–677 (1995).

69. L. Wei, H. Wei, and K. Frenkel. Sensitivity to tumor promotion of SENCAR and C57Bl/6J mice correlates with oxidative events and DNA damage. *Carcinogenesis* 14: 841–847 (1993).

70. Q. Wei, G. M. Matanoski, E. R. Farmer, M. A. Hedayati, and L. Grossman. DNA repair and aging in basal cell carcinoma: A molecular epidemiology study. *Proc. Natl. Acad. Sci. USA* 90: 1614–1618 (1993). Erratum in *Proc. Natl. Acad. Sci. USA* 90(11): 5378 (1993).

71. D. L. Laskin, F. M. Robertson, A. M. Pilaro, and J. D. Laskin. Activation of liver macrophages following phenobarbital treatment of rats. *Hepatology* 8: 1051–1055 (1988).

72. M. J. Czaja, J. Xu, Y. Ju, E. Alt, and P. Schmiedeberg. Lipopolysaccharide-neutralizing antibody reduces hepatocyte injury from acute hepatotoxin administration. *Hepatology* 19: 1282–1289 (1994).

73. Y. Adachi, L. E. Moore, B. U. Bradford, W. Gao, and R. G. Thurman. Antibiotics prevent liver injury in rats following long-term exposure to ethanol. *Gastroenterology* 108: 218–224 (1995).

74. L. Gunawardhana, S. A. Mobley, and I. G. Sipes. Modulation of 1,2-dichlorobenzene hepatotoxicity in the Fischer-344 rat by a scavenger of superoxide anions and an inhibitor of Kupffer cells. *Toxicol. Appl. Pharmacol.* 119: 205–213 (1993).

75. R. A. Roberts and I. Kimber. Cytokines in non-genotoxic hepatocarcinogenesis. *Carcinogenesis* 20: 1397–1401 (1999).

76. L. S. Gold, T. H. Slone, and B. N. Ames. What do animal cancer tests tell us about human cancer risk? Overview of analyses of the Carcinogenic Potency Database. *Drug Metab. Rev.* 30: 359–404 (1998).

77. G. S. Omenn, S. Stuebbe, and L. B. Lave. Predictions of rodent carcinogenicity testing results: Interpretation in light of the Lave-Omenn value-of-information model. *Mol. Carcinog.* 14: 37–45 (1995).

78. J. R. M. Innes, B. M. Ulland, M. G. Valerio, L. Petrucelli, L. Fishbein, E. R. Hart, A. J. Pallota, R. R. Bates, H. L. Falk, J. J. Gart, M. Klein, I. Mitchell, and J. Peters. Bioassay of pesticides and industrial chemicals for tumorigenicity in mice: A preliminary note. *J. Natl. Cancer Inst.* 42: 1101–1114 (1969).

79. B. N. Ames, R. Magaw, and L. S. Gold. Ranking possible carcinogenic hazards. *Science* 236: 271–280 (1987). Letters: 237: 235 (1987); 237: 1283–1284 (1987); 237: 1399–1400 (1987); 238: 1633–1634 (1987). Technical comment: 240: 1043–1047 (1988).

80. S. M. Cohen. Role of urinary physiology and chemistry in bladder carcinogenesis. *Food Chem. Toxicol.* 33: 715–730 (1995).

81. S. M. Cohen and T. A. Lawson. Rodent bladder tumors do not always predict for humans. *Cancer Lett.* 93: 9–16 (1995).

82. B. E. Butterworth, R. B. Conolly, and K. T. Morgan. A strategy for establishing mode of action of chemical carcinogens as a guide for approaches to risk assessments. *Cancer Lett.* 93: 129–146 (1995).

83. B. N. Ames, M. K. Shigenaga, and L. S. Gold. DNA lesions, inducible DNA repair, and cell division: Three key factors in mutagenesis and carcinogenesis. *Environ. Health Perspect.* 101(Suppl. 5): 35–44 (1993).

84. B. N. Ames and L. S. Gold. Chemical carcinogenesis: Too many rodent carcinogens. *Proc. Natl. Acad. Sci. USA* 87: 7772–7776 (1990).

85. B. E. Butterworth and M. S. Bogdanffy. A comprehensive approach for integration of toxicity and cancer risk assessments. *Regul. Toxicol. Pharmacol.* 29: 23–36 (1999).

86. J. L. Larson, D. C. Wolf, and B. E. Butterworth. Induced cytotoxicity and cell proliferation in the hepatocarcinogenicity of chloroform in female B6C3F₁ mice: Comparison of administration by gavage in corn oil vs. *ad libitum* in drinking water. *Fundam. Appl. Toxicol.* 22: 90–102 (1994).

87. U.S. Environmental Protection Agency (EPA), Office of Science and Technology, Office of Water. *Health Risk Assessment/Characterization of the Drinking Water Disinfection Byproduct Chloroform.* EPA, Washington, DC, 1998.

88. R. Munday and C. M. Munday. Low doses of diallyl disulfide, a compound derived from garlic, increase tissue activities of quinone reductase and glutathione transferase in the gastrointestinal tract of the rat. *Nutr. Cancer* 34: 42–48 (1999).

89. J. E. Trosko. Hierarchical and cybernetic nature of biologic systems and their relevance to homeostatic adaptation to low-level exposures to oxidative stress-inducing agents. *Environ. Health Perspect.* 106(Suppl. 1): 331–339 (1998).

90. T. D. Luckey. Nurture with ionizing radiation: A provocative hypothesis. *Nutr. Cancer* 34: 1–11 (1999).

91. L. Bernstein, L. S. Gold, B. N. Ames, M. C. Pike, and D. G. Hoel. Some tautologous aspects of the comparison of carcinogenic potency in rats and mice. *Fundam. Appl. Toxicol.* 5: 79–86 (1985).

92. D. A. Freedman, L. S. Gold, and T. H. Lin. Concordance between rats and mice in bioassays for carcinogenesis. *Regul. Toxicol. Pharmacol.* 23: 225–232 (1996).

93. D. A. Freedman, L. S. Gold, and T. H. Slone. How tautological are inter-species correlations of carcinogenic potency? *Risk Anal.* 13: 265–272 (1993).

94. D. W. Gaylor and L. S. Gold. Quick estimate of the regulatory virtually safe dose

based on the maximum tolerated dose for rodent bioassays. *Regul. Toxicol. Pharmacol.* 22: 57–63 (1995).

95. D. W. Gaylor and L. S. Gold. Regulatory cancer risk assessment based an a quick estimate of a benchmark dose derived from the maximum tolerated dose. *Regul. Toxicol. Pharmacol.* 28: 222–225 (1998).

96. D. W. Gaylor, J. J. Chen, and D. M. Sheehan. Uncertainty in cancer risk estimates. *Risk Anal.* 13: 149–154 (1993).

97. L. S. Gold, C. Wright, L. Bernstein, and M. de Veciana. Reproducibility of results in "near-replicate" carcinogenesis bioassays. *J. Natl. Cancer Inst.* 78: 1149–1158 (1987).

98. B. N. Ames and L. S. Gold. Perspective: Too many rodent carcinogens: Mitogenesis increases mutagenesis. *Science* 249: 970–971 (1990). Letters: 250: 1498 (1990); 250: 1645–1646 (1990); 251: 12–13 (1991); 251: 607–608 (1991); 252: 902 (1991).

99. U.S. Environmental Protection Agency. Proposed guidelines for carcinogen risk assessment. *Fed. Reg.* 61: 17960–18011 (1996).

100. H. Vainio, J. D. Wilbourn, A. J. Sasco, C. Partensky, N. Gaudin, E. Heseltine, and I. Eragne. Identification des facteurs cancérogènes pour l'homme dans les Monographies du CIRC. *Bull. Cancer* 82: 339–348 (1995).

101. National Research Council. *Carcinogens and Anticarcinogens in the Human Diet: A Comparison of Naturally Occurring and Synthetic Substances.* National Academy Press, Washington, DC, 1996.

102. L. Tomatis and H. Bartsch. The contribution of experimental studies to risk assessment of carcinogenic agents in humans. *Exp. Pathol.* 40: 251–266 (1990).

103. K. T. Bogen and L. S. Gold. Trichloroethylene cancer risk: Simplified calculation of PBPK-based MCLs for cytotoxic endpoints. *Regul. Toxicol. Pharmacol.* 25: 26–42 (1997).

104. T. S. Davies and A. Monro. Marketed human pharmaceuticals reported to be tumorigenic in rodents. *J. Am. Coll. Toxicol.* 14: 90–107 (1995).

105. G. A. Burdock. Dietary supplements and lessons to be learned from GRAS. *Regul. Toxicol. Pharmacol.* 31: 68–76 (2000).

106. J. Gruenwald, T. Brendler, and C. Jaenicke (Eds.). *PDR for Herbal Medicines.* Medical Economics Company, Montvale, NJ, 1998.

107. R. Huxtable. Pyrrolizidine alkaloids: Fascinating plant poisons. *Newslett. Center Toxicol. Southwest Environ. Health Sci. Center* Fall: 1–3 (1995).

108. M. Goldberg. Dehydroepiandrosterone, insulin-like growth factor-I, and prostate cancer. *Ann. Int. Med.* 129: 587–588 (1998).

109. F. Hayashi, H. Tamura, J. Yamada, H. Kasai, and T. Suga. Characteristics of the hepatocarcinogenesis caused by dehydroepiandrosterone, a peroxisome proliferator, in male F-344 rats. *Carcinogenesis* 15: 2215–2219 (1994).

110. J. K. Reddy and S. A. Qureshi. Tumorigenicity of the hypolipidaemic peroxisome proliferator ethyl-α-p-chlorophenoxyisobutyrate (clofibrate) in rats. *Br. J. Cancer* 40: 476–482 (1979).

111. J. M. Ward, J. M. Peters, C. M. Perella, and F. J. Gonzalez. Receptor and nonreceptor-mediated organ-specific toxicity of di(2-ethylhexyl)phthalate (DEHP) in peroxisome proliferator-activated receptor α-null mice. *Toxicol. Pathol.* 26: 240–246 (1998).

112. J. Doull, R. Cattley, C. Elcombe, B. G. Lake, J. Swenberg, C. Wilkinson, G. Williams, and M. van Gemert. A cancer risk assessment of di(2-ethylhexyl)phthalate: Application of the new U.S. EPA Risk Assessment Guidelines. *Regul. Toxicol. Pharmacol.* 29: 327–357 (1999).

113. R. Hertz and J. Bar-Tana. Peroxisome proliferator-activated receptor (PPAR) alpha activation and its consequences in humans. *Toxicol. Lett.* 102/103: 85–90 (1998).

114. N. J. Woodyatt, K. G. Lambe, K. A. Myers, J. D. Tugwood, and R. A. Rovert. The peroxisome proliferator (PP) response element upstream of the human acyl CoA oxidase gene is inactive among a sample human population: Significance for species differences in response to PPs. *Carcinogenesis* 20: 369–372 (1999).

115. B. A. Stoll. Dietary supplements of dehydroepiandrosterone in relation to breast cancer risk. *Eur. J. Clin. Nutr.* 53: 771–775 (1999).

116. G. C. Hard and J. Whysner. Risk assessment of *d*-limonene: An example of male rat-specific renal tumorigens. *Crit. Rev. Toxicol.* 24: 231–254 (1994).

117. U.S. Environmental Protection Agency (EPA). *Report of the EPA Peer Review Workshop on Alpha$_{2u}$-globulin: Association with Renal Toxicity and Neoplasia in the Male Rat.* EPA, Washington, DC, 1991.

118. International Agency for Research on Cancer (IARC). *Some Naturally Occurring Substances: Food Items, Constituents, and Heterocyclic Aromatic Amines and Mycotoxins,* Vol. 56: *IARC Monographs on the Evaluation of Carcinogenic Risk of Chemicals to Humans.* IARC, Lyon, France, 1993.

119. J. M. Rice, R. A. Baan, M. Blettner, C. Genevois-Charmeau, Y. Grosse, D. B. McGregor, C. Partensky, and J. D. Wilbourn. Rodent tumors of urinary bladder, renal cortex, and thyroid gland in IARC Monographs evaluations of carcinogenic risk to humans. *Toxicol. Sci.* 49: 166–171 (1999).

120. U.S. Food and Drug Administration. Food and Drug Administration Pesticide Program: Residue monitoring 1992. *J. Assoc. Off. Anal. Chem.* 76: 127A–148A (1993).

121. T. Key and G. Reeves. Organochlorines in the environment and breast cancer. *Br. Med. J.* 308: 1520–1521 (1994).

122. U. P. Thorgeirsson, D. W. Dalgard, J. Reeves, and R. H. Adamson. Tumor incidence in a chemical carcinogenesis study in nonhuman primates. *Regul. Toxicol. Pharmacol.* 19: 130–151 (1994).

123. S. Takayama, S. M. Sieber, D. W. Dalgard, U. P. Thorgeirsson, and R. H. Adamson. Effects of long-term oral administration of DDT on nonhuman primates. *J. Cancer Res. Clin. Oncol.* 125: 219–225 (1999).

124. National Academy of Sciences. *The Life Sciences: Recent Progress and Application to Human Affairs; the World of Biological Research Requirement for the Future.* Committee on Research in the Life Sciences, Washington, DC, 1970.

125. G. W. Gribble. The diversity of natural organochlorines in living organisms. *Pure Appl. Chem.* 68: 1699–1712 (1996).

126. U.S. Environmental Protection Agency. Ethylene bisdithiocarbamates (EBDCs); Notice of intent to cancel; Conclusion of special review. *Fed. Reg.* 57: 7484–7530 (1992).

127. U.S. Environmental Protection Agency (EPA). *Status of Pesticides in Registration, Reregistration, and Special Review.* EPA, Washington, DC, 1998.

128. L. S. Gold, T. H. Slone, B. N. Ames, and N. B. Manley. Pesticide residues in food and cancer risk: A critical analysis. In R. I. Krieger (Ed.), *Handbook of Pesticide Toxicology.* New York, Academic Press, 2001, pp. 799–843.

129. National Research Council. *Regulating Pesticides in Food: The Delaney Paradox.* National Academy Press, Washington, DC, 1987.

130. L. S. Gold, B. R. Stern, T. H. Slone, J. P. Brown, N. B. Manley, and B. N. Ames. Pesticide residues in food: Investigation of disparities in cancer risk estimates. *Cancer Lett.* 117: 195–207 (1997).

131. L. S. Gold, T. H. Slone, N. B. Manley, and B. N. Ames. Heterocyclic amines formed by cooking food: Comparison of bioassay results with other chemicals in the Carcinogenic Potency Database. *Cancer Lett.* 83: 21–29 (1994).

132. R. H. Adamson, S. Takayama, T. Sugimura, and U. P. Thorgeirsson. Induction of hepatocellular carcinoma in nonhuman primates by the food mutagen 2-amino-3-methylimidazo[4,5-f]quinoline. *Environ. Health Perspect.* 102: 190–193 (1994).

133. L. S. Gold, T. H. Slone, B. N. Ames, N. B. Manley, G. B. Garfinkel, and L. Rohrbach. Carcinogenic Potency Database. In L. S. Gold and E. Zeiger (Eds.), *Handbook of Carcinogenic Potency and Genotoxicity Databases.* CRC Press, Boca Raton, FL, 1997, pp. 1–605.

134. K. Ogawa, H. Tsuda, T. Shirai, T. Ogiso, K. Wakabayashi, D. W. Dalgard, U. P. Thorgeirsson, S. S. Thorgeirsson, R. H. Adamson, and T. Sugimura. Lack of carcinogenicity of 2-amino-3,8-dimethylimidazo[4,5-f]quinoxaline (MeIQx) in cynomolgus monkeys. *Jpn. J. Cancer Res.* 90: 622–628 (1999).

135. E. G. Snyderwine, R. J. Turesky, K. W. Turteltaub, C. D. Davis, N. Sadrieh, H. A. J. Schut, M. Nagao, T. Sugimura, U. P. Thorgeirsson, R. H. Adamson, and S. Snorri. Metabolism of food-derived heterocyclic amines in nonhuman primates. *Mutat. Res.* 376: 203–210 (1997).

136. U.S. Food and Drug Administration (USFDA). *Butylatedhydroxyanisole (BHA) intake: Memo from Food and Additives Color Section to L. Lin.* USFDA, Washington, DC, 1991.

137. D. B. Clayson, F. Iverson, E. A. Nera, and E. Lok. The significance of induced forestomach tumors. *Annu. Rev. Pharmacol. Toxicol.* 30: 441–463 (1990).

138. California Environmental Protection Agency (CalEPA), Standards and Criteria Work Group. *California Cancer Potency Factors: Update.* CalEPA, Sacramento, 1994.

139. National Toxicology Program (NTP). *Ninth Report on Carcinogens.* NTP, Research Triangle Park, NC, 2000.

140. International Agency for Research on Cancer (IARC). *Some Chemicals That Cause Tumours of the Kidney or Urinary Bladder in Rodents and Some Other Substances,* Vol. 73: *IARC Monographs on the Evaluation of Carcinogenic Risk of Chemicals to Humans.* IARC, Lyon, France, 1999.

141. S. Takayama, S. M. Sieber, R. H. Adamson, U. P. Thorgeirsson, D. W. Dalgard, L. L. Arnold, M. Cano, S. Eklund, and S. M. Cohen. Long-term feeding of sodium saccharin to nonhuman primates: Implications for urinary tract cancer. *J. Natl. Cancer Inst.* 90: 19–25 (1998).

142. L. M. Nijssen, C. A. Visscher, H. Maarse, L. C. Willemsens, and M. H. Boelens (Eds.). *Volatile Compounds in Foods. Qualitative and Quantitative Data*, 7th ed. TNO-CIVO Food Analysis Institute, Zeist, The Netherlands, 1996.

143. U.S. Food and Drug Administration. Refusal to extend effective date of statute for certain specified additives in food. *Fed. Reg.* 25: 12412 (1960).

144. G.-S. Qian, R. K. Ross, M. C. Yu, J.-M. Yuan, B. E. Henderson, G. N. Wogan, and J. D. Groopman. A follow-up study of urinary markers of aflatoxin exposure and liver cancer risk in Shanghai, People's Republic of China. *Cancer Epidemiol. Biomarkers Prev.* 3: 3–10 (1994).

145. A. H. Wu-Williams, L. Zeise, and D. Thomas. Risk assessment for aflatoxin B_1: A modeling approach. *Risk Anal.* 12: 559–567 (1992).

146. U.S. Food and Drug Administration (USFDA). *Assessment of Carcinogenic Upper Bound Lifetime Risk from Resulting Aflatoxins in Consumer Peanut and Corn Products. Report of the Quantitative Risk Assessment Committee* USFDA, Washington, DC, 1993.

147. W. A. Pons, Jr. High pressure liquid chromatographic determination of aflatoxins in corn. *J. Assoc. Off. Anal. Chem.* 62: 586–594 (1979).

148. J. D. Groopman, J. Q. Zhu, P. R. Donahue, A. Pikul, L. S. Zhang, J. S. Chen, and G. N. Wogan. Molecular dosimetry of urinary aflatoxin-DNA adducts in people living in Guangxi Autonomous Region, People's Republic of China. *Cancer Res.* 52: 45–52 (1992).

149. M. C. Yu, M. J. Tong, S. Govindarajan, and B. E. Henderson. Nonviral risk factors for hepatocellular carcinoma in a low-risk population, the non-Asians of Los Angeles County, California. *J. Natl. Cancer Inst.* 83: 1820–1826 (1991).

150. T. Kuiper-Goodman and P. M. Scott. Risk assessment of the mycotoxin ochratoxin A. *Biomed. Environ. Sci.* 2: 179–248 (1989).

151. International Life Sciences Institute. Occurrence and significance of ochratoxin A in food. Paper presented at the ILSI Europe Workshop, January 10–12, 1996, Aix-en-Provence, France. *ILSI Europe Newsl.* Feb.: 3 (1996).

152. M. J. Gartrell, J. C. Craun, D. S. Podrebarac, and E. L. Gunderson. Pesticides, selected elements, and other chemicals in adult total diet samples, October 1980–March 1982. *J. Assoc. Off. Anal. Chem.* 69: 146–161 (1986).

153. E. L. Gunderson. Dietary intakes of pesticides, selected elements, and other chemicals: FDA Total Diet Study, June 1984–April 1986. *J. Assoc. Off. Anal. Chem.* 78: 910–921 (1995).

154. World Health Organization (WHO). *Polychlorinated Biphenyls and Terphenyls*, Vol. 140: *Environmental Health Criteria.* WHO, Geneva, 1993.

155. U.S. Environmental Protection Agency (EPA). *Exposure and Human Health Reassessment of 2,3,7,8-Tetrachlorodibenzo-p-Dioxin (TCDD) and Related Compounds. Draft Final.* EPA, Washington, DC, 2000.

156. U.S. Environmental Protection Agency (EPA). *Estimating Exposure to Dioxin-Like Compounds (Review Draft).* EPA, Washington, DC, 1994.

157. D. G. Goodman and R. M. Sauer. Hepatotoxicity and carcinogenicity in female Sprague-Dawley rats treated with 2,3,7,8-tetrachlorodibenzo-*p*-dioxin (TCDD): A pathology working group reevaluation. *Regul. Toxicol. Pharmacol.* 15: 245–252 (1992).

158. L. S. Birnbaum. The mechanism of dioxin toxicity: Relationship to risk assessment. *Environ. Health Perspect.* 102(Suppl. 9): 157–167 (1994).

159. P. M. Fernandez-Salguero, D. M. Hilbert, S. Rudikoff, J. M. Ward, and F. J. Gonzalez. Aryl-hydrocarbon receptor-deficient mice are resistant to 2,3,7,8-tetrachlorodibenzo-*p*-dioxin-induced toxicity. *Toxicol. Appl. Pharmacol.* 140: 173–179 (1996).

160. S. Safe, F. Wang, W. Porter, R. Duan, and A. McDougal. Ah receptor agonists as endocrine disruptors: Antiestrogenic activity and mechanisms. *Toxicol. Lett.* 102/103: 343–347 (1998).

161. C. A. Bradfield and L. F. Bjeldanes. Structure-activity relationships of dietary indoles: A proposed mechanism of action as modifiers of xenobiotic metabolism. *J. Toxicol. Environ. Health* 21: 311–323 (1987).

162. R. H. Dashwood. Indole-3-carbinol: Anticarcinogen or tumor promoter in *Brassica* vegetables? *Chem.-Biol. Interact.* 110: 1–5 (1998).

163. C. Wilker, L. Johnson, and S. Safe. Effects of developmental exposure to indole-3-carbinol or 2,3,7,8-tetrachlorodibenzo-*p*-dioxin on reproductive potential of male rat offspring. *Toxicol. Appl. Pharmacol.* 141: 68–75 (1996).

164. G. J. Kelloff, J. A. Crowell, E. T. Hawk, V. E. Steele, R. A. Lubet, C. W. Boone, J. M. Covey, L. A. Doody, G. S. Omenn, P. Greenwald, W. K. Hong, D. R. Parkinson, D. Bagheri, G. T. Baxter, M. Blunden, M. K. Doeltz, K. M. Eisenhauer, K. Johnson, G. G. Knapp, G. Longfellow, W. F. Malone, S. G. Nayfield, H. E. Seifried, L. M. Swall, and C. C. Sigman. Strategy and planning for chemopreventive drug development: Clinical development plans II. *J. Cell Biochem. Suppl.* 26: 54–71 (1996).

165. G. J. Kelloff, C. W. Boone, J. A. Crowell, V. E. Steele, R. A. Lubet, L. A. Doody, W. F. Malone, E. T. Hawk, and C. C. Sigman. New agents for cancer chemoprevention. *J. Cell Biochem. Suppl.* 26: 1–28 (1996).

166. National Toxicology Program (NTP). *Background Information Indole-3-carbinol (I3C) 700-06-1*. NTP, Research Triangle Park, NC, 2000.

167. Theranaturals. *Theranaturals I3C Caps* (2000) http://www.theranaturals.com.

168. U.S. Environmental Protection Agency (EPA). *Health Assessment Document for 2,3,7,8-Tetrachlorodibenzo-p-Dioxin (TCDD) and Related Compounds*. EPA, Washington, DC, 1994.

169. U.S. Environmental Protection Agency (EPA). *Re-evaluating Dioxin: Science Advisory Board's Review of EPA's Reassessment of Dioxin and Dioxin-like Compounds*. EPA, Washington, DC, 1995.

170. International Agency for Research on Cancer (IARC). *Polychlorinated Dibenzo-para-dioxins and Polychlorinated Dibenzofurans*, Vol. 69: *IARC Monographs on the Evaluation of Carcinogenic Risk of Chemicals to Humans*. IARC, Lyon, France, 1997.

171. U.S. National Toxicology Program. *Ninth Report on Carcinogens*. Department of Health and Human Services, Public Health Service, Research Triangle Park, NC, 2000.

172. D. B. Clayson and F. Iverson. Cancer risk assessment at the crossroads: The need to turn to a biological approach. *Regul. Toxicol. Pharmacol.* 24: 45–59 (1996).

173. J. I. Goodman. A rational approach to risk assessment requires the use of biological information: An analysis of the National Toxicology Program (NTP), final report of the advisory review by the NTP Board of Scientific Counselors. *Regul. Toxicol. Pharmacol.* 19: 51–59 (1994).

174. T. Colborn, D. Dumanoski, and J. P. Myers. *Our Stolen Future: Are We Threatening Our Fertility, Intelligence, and Survival? A Scientific Detective Story.* Dutton, New York, 1996.

175. National Research Council. *Hormonally Active Agents in the Environment.* National Academy Press, Washington, DC, 1999.

176. S. H. Safe. Environmental and dietary estrogens and human health: Is there a problem? *Environ. Health Perspect.* 103: 346–351 (1995).

177. S. H. Safe. Is there an association between exposure to environmental estrogens and breast cancer? *Environ. Health Perspect.* 105(Suppl. 3): 675–578 (1997).

178. S. H. Safe. Endocrine disruptors and human health—Is there a problem? An update. *Environ. Health Perspect.* 108: 487–493 (2000).

179. K. Reinli and G. Block. Phytoestrogen content of foods—a compendium of literature values. *Nutr. Cancer* 26: 123–148 (1996).

180. G. Kolata. Measuring men up, sperm by sperm. *New York Times,* May 5 1996, E4(N), E4(L).

181. S. H. Swan, E. P. Elkin, and L. Fenster. Have sperm densities declined? A reanalysis of global trend data. *Environ. Health Perspect.* 105: 1228–1232 (1997). Letters: 106: A370–371, A420–421.

182. S. Becker and K. Berhane. A meta-analysis of 61 sperm count studies revisited. *Fertil. Steril.* 67: 1103–1108 (1997).

183. J. Gyllenborg, N. E. Skakkebaek, N. C. Nielsen, N. Keiding, and A. Giwercman. Secular and seasonal changes in semen quality among young Danish men: A statistical analysis of semen samples from 1927 donor candidates during 1977–1995. *Int. J. Androl.* 22: 28–36 (1999).

184. J. A. Saidi, D. T. Chang, E. T. Goluboff, E. Bagiella, G. Olsen, and H. Fisch. Declining sperm counts in the United States? A critical review. *J. Urol.* 161: 460–462 (1999).

185. K. Gaido, L. Dohme, F. Wang, I. Chen, B. Blankvoort, K. Ramamoorthy, and S. Safe. Comparative estrogenic activity of wine extracts and organochlorine pesticide residues in food. *Environ. Health Perspect.* 106(Suppl. 6): 1347–1351 (1998).

186. K. D. Setchell, L. Zimmer-Nechemias, J. Cai, and J. E. Heubi. Exposure of infants to phytooestrogens from soy-based infant formula. *Lancet* 350: 23–27 (1997).

187. R. W. Hahn (Ed.). *Risks, Costs, and Lives Saved: Getting Better Results from Regulation.* Oxford University Press, New York, 1996.

188. J. D. Graham and J. B. Wiener (Eds.). *Risk Versus Risk: Tradeoffs in Protecting Health and the Environment.* Harvard University Press, Cambridge, MA, 1995.

189. U.S. Environmental Protection Agency (EPA). *Environmental Investments: The Cost of a Clean Environment.* EPA, Washington, DC, 1991.

190. T. O. Tengs, M. E. Adams, J. S. Pliskin, D. G. Safran, J. E. Siegel, M. C. Weinstein, and J. D. Graham. Five-hundred life-saving interventions and their cost-effectiveness. *Risk Anal.* 15: 369–390 (1995).

191. R. L. Keeney. Mortality risks induced by economic expenditures. *Risk Anal.* 10: 147–159 (1990).

192. A. B. Wildavsky. *Searching for Safety.* Transaction Books, New Brunswick, NJ, 1988.

193. A. B. Wildavsky. *But Is It True? A Citizen's Guide to Environmental Health and Safety Issues.* Harvard University Press, Cambridge, MA, 1995.

194. W. K. Viscusi. *Fatal Tradeoffs: Public and Private Responsibilities for Risk.* Oxford University Press, New York, 1992.

195. J. Contrera, A. Jacobs, and J. DeGeorge. Carcinogenicity testing and the evaluation of regulatory requirements for pharmaceuticals. *Regul. Toxicol. Pharmacol.* 25: 130–145 (1997).

196. D. G. Barnes and M. Dourson. Reference dose (RfD): Description and use in health risk assessments. *Regulatory Toxicol. Pharmacol.* 8: 471–486 (1988).

197. A. G. Renwick. The use of an additional safety or uncertainty factor for nature of toxicity in the estimation of acceptable daily intake and tolerable daily intake values. *Regulatory Toxicol. Pharmacol.* 22: 250–261 (1995).

198. C. S. Schwartz. A semiquantitative method for selection of safety factors in establishing OELs for pharmaceutical compounds. *Hum. Ecol. Risk Assess.* 1: 527–543 (1995).

199. M. G. Ott, H. C. Scharnweber, and R. R. Langner. Mortality experience of 161 employees exposed to ethylene dibromide in two production units. *Br. J. Ind. Med.* 37: 163–168 (1980).

200. J. C. Ramsey, C. N. Park, M. G. Ott, and P. J. Gehring. Carcinogenic risk assessment: Ethylene dibromide. *Toxicol. Appl. Pharmacol.* 47: 411–414 (1978).

201. R. J. Havel and J. P. Kane. Therapy of hyperlipidemic states. *Ann. Rev. Med.* 33: 417 (1982).

202. American Medical Association (AMA), Division of Drugs. *AMA Drug Evaluations.* AMA, Chicago, IL, 1983, pp. 201–202.

203. G. Matanoski, M. Francis, A. Correa-Villaseñor, E. Elliot, C. Santos-Brugoa, and L. Schwartz. Cancer epidemiology among styrene-butadiene rubber workers. *IARC Sci. Pub.* 127: 363–374 (1993).

204. I. Hirono, H. Mori, and M. Haga. Carcinogenic activity of *Symphytum officinale.* *J. Natl. Cancer Inst.* 61: 865–868 (1978).

205. C. C. J. Culvenor, M. Clarke, J. A. Edgar, J. L. Frahn, M. V. Jago, J. E. Peterson, and L. W. Smith. Structure and toxicity of the alkaloids of Russian comfrey (*Symphytum × Uplandicum nyman*), a medicinal herb and item of human diet. *Experientia* 36: 377–379 (1980).

206. J. Andrasik and D. Cloutet. Monitoring solvent vapors in drycleaning plants. *Int. Fabricare Inst. Focus Dry Cleaning* 14(3): 1–8 (1990).

207. D. M. Siegal, V. H. Frankos, and M. Schneiderman. Formaldehyde risk assessment for occupationally exposed workers. *Reg. Toxicol. Pharm.* 3: 355–371 (1983).

208. A. Blair, P. A. Stewart, D. D. Zaebst, L. Pottern, J. N. Zey, T. F. Bloom, B. Miller, E. Ward, and J. Lubin. Mortality of industrial workers exposed to acrylonitrile. *Scand. J. Work Environ. Health* 24(Suppl. 2): 25–41 (1998).

209. N. P. Page and J. L. Arthur. *Special Occupational Hazard Review of Trichloro-*

ethylene. National Institute for Occupational Safety and Health, Rockville, MD, 1978.

210. J. Stofberg and F. Grundschober. Consumption ratio and food predominance of flavoring materials. Second cumulative series. *Perfum. Flavor.* 12: 27–56 (1987).

211. T. H. Connor, J. C. Theiss, H. A. Hanna, D. K. Monteith, and T. S. Matney. Genotoxicity of organic chemicals frequently found in the air of mobile homes. *Toxicol. Lett.* 25: 33–40 (1985).

212. CONSAD Research Corporation. Final report. Economic analysis of OSHA's proposed standards for methylene chloride. OSHA Docket H-71. Oct. 1990.

213. J. McCann, L. Horn, J. Girman, and A. V. Nero. Potential risks from exposure to organic carcinogens in indoor air. In S. S. Sandhu, D. M. deMarini, M. J. Mass, M. M. Moore, and J. L. Mumford (Eds.), *Short-Term Bioassays in the Analysis of Complex Environmental Mixtures*. Plenum, New York, 1987.

214. R. Arky. *Physicians' Desk Reference*, 52nd ed. Medical Economics Company, Montvale, NJ, 1998.

215. R. J. Clarke and R. Macrae (Eds.). *Coffee*, Vols. 1–3. Elsevier, New York, 1988.

216. Technical Assessment Systems (TAS). *Exposure 1 Software Package*. TAS, Washington, DC, 1989. Provided by Barbara Petersen.

217. K. Herrmann. Review on nonessential constituents of vegetables. III. Carrots, celery, parsnips, beets, spinach, lettuce, endives, chicory, rhubarb, and artichokes. *Z. Lebensm. Unters. Forsch.* 167: 262–273 (1978).

218. R. L. Hall, S. H. Henry, R. J. Scheuplein, B. J. Dull, and A. M. Rulis. Comparison of the carcinogenic risks of naturally occurring and adventitious substances in food. In S. L. Taylor and R. A. Scanlan (Eds.), *Food Toxicology: A Perspective on the Relative Risks*. Marcel Dekker, New York, 1989, pp. 205–224.

219. P. Schreier, F. Drawert, and I. Heindze. Über die quantitative Zusammensetzung natürlicher und technologish veränderter pflanzlicher Aromen. *Chem. Mikrobiol. Technol. Lebensm.* 6: 78–83 (1979).

220. H. Schmidtlein and K. Herrmann. Über die Phenolsäuren des Gemüses. II. Hydroxyzimtsäuren und Hydroxybenzoesäuren der Frucht- und Samengemüsearten. *Z. Lebensm. Unters.-Forsch.* 159: 213–218 (1975).

221. T. Hasselstrom, E. J. Hewitt, K. S. Konigsbacher, and J. J. Ritter. Composition of volatile oil of black pepper. *Agric. Food Chem.* 5: 53–55 (1957).

222. R. Tressl, D. Bahri, H. Köppler, and A. Jensen. Diphenole und Caramelkomponenten in Röstkaffees verschiedener Sorten. II. *Z. Lebensm. Unters. Forsch.* 167: 111–114 (1978).

223. W. Rahn and W. A. König. GC/MS investigations of the constituents in a diethyl ether extract of an acidified roast coffee infusion. *J. High Resolut. Chromatogr. Chromatogr. Commun.* 1002: 69–71 (1978).

224. B. Toth and J. Erickson. Cancer induction in mice by feeding of the uncooked cultivated mushroom of commerce *Agaricus bisporus*. *Cancer Res.* 46: 4007–4011 (1986).

225. K. Matsumoto, M. Ito, S. Yagyu, H. Ogino, and I. Hirono. Carcinogenicity examination of *Agaricus bisporus*, edible mushroom, in rats. *Cancer Lett.* 58: 87–90 (1991).

226. U.S. Environmental Protection Agency (EPA), Office of Pesticide Programs.

Daminozide Special Review. Technical Support Document—Preliminary Determination to Cancel the Food Uses of Daminozide. EPA, Washington, DC, 1989.

227. H. D. Mosel and K. Herrmann. The phenolics of fruits. III. The contents of catechins and hydroxycinnamic acids in pome and stone fruits. *Z. Lebensm. Unters. Forsch.* 154: 6–11 (1974).

228. T. Fazio, D. C. Havery, and J. W. Howard. Determination of volatile *N*-nitrosamines in foodstuffs: I. A new clean-up technique for confirmation by GLC-MS. II. A continued survey of foods and beverages. In E. A. Walker, L. Griciute, M. Castegnaro, and M. Borzsonyi (Eds.), *N-Nitroso Compounds: Analysis, Formation and Occurrence.* International Agency for Research on Cancer, Lyon, France, 1980, pp. 419–435.

229. R. Preussmann and G. Eisenbrand. *N*-nitroso carcinogens in the environment. In C. E. Searle (Ed.), *Chemical Carcinogenesis.* ACS Monograph 182. American Chemical Society, Washington, DC, 1984, pp. 829–868.

230. U.S. Food and Drug Administration (USFDA). *Exposure to Aflatoxins.* USFDA, Washington, DC, 1992.

231. S. K. Poole and C. F. Poole. Thin-layer chromatographic method for the determination of the principal polar aromatic flavour compounds of the cinnamons of commerce. *Analyst* 119: 113–120 (1994).

232. L. Heinrich and W. Baltes. Über die Bestimmung von Phenolen im Kaffeegetränk. *Z. Lebensm. Unters. Forsch.* 185: 362–365 (1987).

233. National Research Council. *The 1977 Survey of Industry on the Use of Food Additives.* National Academy Press, Washington, DC, 1979.

234. G. B. Neurath, M. Dünger, F. G. Pein, D. Ambrosius, and O. Schreiber. Primary and secondary amines in the human environment. *Food Cosmet. Toxicol.* 15: 275–282 (1977).

235. H. Schmidtlein and K. Herrmann. Über die Phenolsäuren des Gemüses. IV. Hydroxyzimtsäuren und Hydroxybenzösäuren weiterer Gemüsearten und der Kartoffeln. *Z. Lebensm. Unters. Forsch.* 159: 255–263 (1975).

236. Economic Research Service. *Vegetables and Specialties Situation and Outlook Yearbook.* U.S. Department of Agriculture, Washington, DC, 1994.

237. H. Stöhr and K. Herrmann. Über die Phenolsäuren des Gemüses: III. Hydroxyzimtsäuren und Hydroxybenzoesäuren des Wurzelgemüses. *Z. Lebensm. Unters. Forsch.* 159: 219–224 (1975).

238. F. M. Clydesdale (Ed.). *Food Additives: Toxicology, Regulation, and Properties.* CRC Press, Boca Raton, FL, 1997.

239. E. A. Bejnarowicz and E. R. Kirch. Gas chromatographic analysis of oil of nutmeg. *J. Pharm. Sci.* 52: 988–993 (1963).

240. International Agency for Research on Cancer (IARC). *Coffee, Tea, Mate, Methylxanthines and Methylglyoxal,* Vol. 51: *IARC Monographs on the Evaluation of Carcinogenic Risk of Chemicals to Humans.* IARC, Lyon, France, 1991.

241. U.S. Environmental Protection Agency (EPA). *EBDC/ETU Special Review. DRES Dietary Exposure/Risk Estimates.* EPA, Washington, DC, 1991. memo from R. Griffin to K. Martin.

242. R. E. Duggan and P. E. Corneliussen. Dietary intake of pesticide chemicals in the United States (III), June 1968–April 1970. *Pest. Monit. J.* 5: 331–341 (1972).

243. Economic Research Service. *Fruit and Tree Nuts Situation and Outlook Yearbook.* Department of Agriculture, Washington, DC, 1995.

244. D. G. Carlson, M. E. Daxenbichler, C. H. VanEtten, W. F. Kwolek, and P. H. Williams. Glucosinolates in crucifer vegetables: Broccoli, brussels sprouts, cauliflower, collards, kale, mustard greens, and kohlrabi. *J. Am. Soc. Hort. Sci.* 112: 173–178 (1987).

245. N. P. Sen, S. Seaman, and W. F. Miles. Volatile nitrosamines in various cured meat products: Effect of cooking and recent trends. *J. Agric. Food Chem.* 27: 1354–1357 (1979).

246. Y. Chauhan, D. Nagel, M. Gross, R. Cerny, and B. Toth. Isolation of N_2-[γ-L-(+)-glutamyl]-4-carboxyphenylhydrazine in the cultivated mushroom *Agaricus bisporus. J. Agric. Food Chem.* 33: 817–820 (1985).

247. A. R. Tricker and R. Preussmann. Carcinogenic N-nitrosamines in the diet: Occurrence, formation, mechanisms and carcinogenic potential. *Mutat. Res.* 259: 277–289 (1991).

248. U.S. Environmental Protection Agency (EPA), Office of Pesticide Programs. *Ethylene Dibromide (EDB) Scientific Support and Decision Document for Grain and Grain Milling Fumigation Uses.* EPA, Washington, DC, 1984.

249. American Water Works Association (AWWA), Government Affairs Office. *Disinfectant/Disinfection By-Products Database for the Negotiated Regulation.* AWWA, Washington, DC, 1993.

250. K. H. Engel and R. Tressl. Studies on the volatile components of two mango varieties. *J. Agric. Food Chem.* 31: 796–801 (1983).

251. U.S. Food and Drug Administration. FDA Pesticide Program: Residues in foods 1990. *J. Assoc. Off. Anal. Chem.* 74: 121A–141A (1991).

252. R. C. Beier, G. W. Ivie, E. H. Oertli, and D. L. Holt. HPLC analysis of linear furocoumarins (psoralens) in healthy celery *Apium graveolens. Food Chem. Toxicol.* 21: 163–165 (1983).

253. United Fresh Fruit and Vegetable Association (UFFVA). *Supply Guide: Monthly Availability of Fresh Fruit and Vegetables.* UFFVA, Alexandria, VA, 1989.

254. G. W. Ivie, D. L. Holt, and M. Ivey. Natural toxicants in human foods: Psoralens in raw and cooked parsnip root. *Science* 213: 909–910 (1981).

255. B. J. Canas, D. C. Havery, L. R. Robinson, M. P. Sullivan, F. L. Joe, Jr., and G. W. Diachenko. Chemical contaminants monitoring: Ethyl carbamate levels in selected fermented foods and beverages. *J. Assoc. Off. Anal. Chem.* 72: 873–876 (1989).

256. M. G. Knize, F. A. Dolbeare, K. L. Carroll, D. H. Moore II, and J. S. Felton. Effect of cooking time and temperature on the heterocyclic amine content of fried beef patties. *Food Chem. Toxicol.* 32: 595–603 (1994).

257. S. K. Chaudhary, O. Ceska, C. Tétu, P. J. Warrington, M. J. Ashwood-Smith, and G. A. Poulton. Oxypeucedanin, a major furocoumarin in parsley, *Petroselinum crispum. Planta Med.* 6: 462–464 (1986).

258. U.S. Environmental Protection Agency. *Peer Review of Chlorothalonil.* Office of Pesticides and Toxic Substances, Washington, DC, 1987. Review found in Health Effect Division Document No. 007718.

SECTION N
Evolving Issues

30 Children's Health and Environmental Exposure to Chemicals: Implications for Risk Assessment and Public Health Policy

MICHAEL GOODMAN and NANCY LAVERDA

Exponent, Alexandria, Virginia

30.1 INTRODUCTION

In recent years, the potential health effects of exposure to chemical pollutants among children has become a focus of attention for the general public and the scientific community. The concerns regarding children's exposure to toxic chemicals had existed for some time, however, nearly all of the toxicological and medical literature dealt primarily with acute poisonings due to high-level accidental exposures. At the same time, the preponderance of environmental health research continued to examine the effects of pollutants in the population of maximum convenience and data availability—mostly working-age male adults. The attention started to shift in the mid-1980s following a number of articles in both the scientific and the general press.

A series of publications dealing with chronic exposures of children to lead paint chips and contaminated soil suggested that children demonstrated signs of neurodevelopmental impairment at levels previously considered safe (Needleman, 1983). Furthermore, several authors proposed that children are more sensitive than adults to the toxicological effects of lead at a given dose [Environmental Protection Agency (EPA), 1986].

The concerns about the potential increased vulnerability of children were further elevated by a report of a cluster of childhood leukemia in Woburn, Massachusetts (Lagakos et al., 1986; Cutler et al., 1986). The initial investiga-

Human and Ecological Risk Assessment: Theory and Practice, Edited by Dennis J. Paustenbach
ISBN 0-471-14747-8 © 2002 John Wiley & Sons, Inc.

tion of this cluster suggested that the risk of developing childhood leukemia was associated with consumption of drinking water contaminated with trichloro-ethylene (TCE). Although subsequent reviews of the evidence cast doubt on the validity of the scientific evidence (Kuzmack, 1987; Little, 1999), the Woburn events were highly publicized and provided an inspiration for the popular book and motion picture *A Civil Action*.

In 1989, the Natural Resources Defense Council (NRDC) declared that Alar, a growth regulator used mainly on apples, was a potent cancer-causing agent and that children were at the highest risk of disease. Although the scientific evidence was inconclusive, manufacturers of Alar discontinued its use (Farland et al., 1992).

By the late 1980s, these and other events forced clinicians, toxicologists, risk assessors, and environmental scientists to admit that the scientific community has a poor understanding of children's health issues. The single most important question was: Is there evidence that children's health is at risk following low levels of exposure to chemicals routinely found in the environment (food, air, water, and soil) and generally considered safe? The recognition of knowledge gaps regarding children's health resulted in a marked increase of dialog and research activities in the scientific, regulatory, and public health communities. However, despite multiple general publications and "opinion papers" about the importance of children's health, the data remain sketchy and multiple knowledge gaps are only beginning to close.

30.2   CHRONOLOGY OF REGULATORY EVENTS: 1988–2000

In 1988, the U.S. Congress requested that the National Research Council (NRC) appoint a committee to study the effects of pesticides in the diets of infants and children. The committee was charged with the responsibility to examine the adequacy of existing risk assessment methods with regards to pesticide residues in foods and to identify the issues of greatest concern. As a result of this effort, in 1993 the committee published a report entitled *Pesticides in the Diets of Infants and Children* (NRC, 1993). The committee concluded that qualitative differences between children and adults are the consequence of so-called windows of vulnerability when exposure to a toxicant can permanently alter the function or structure of an organ system. The recommendations of the committee included an extension of the 10-fold uncertainty factor traditionally applied by EPA and the Food and Drug Administration (FDA) to the no-observed-effect level (NOEL) in animal studies involving fetal developmental toxicity. According to the NRC report, the additional uncertainty factor was required "when there is evidence of postnatal developmental toxicity and when data from toxicity testing relative to children is incomplete" (NRC, 1993).

In 1995, Administrator Carol Browner announced EPA's new policy to take into account environmental health risks to infants and children in all risk characterizations and public health standards set for the United States. In the fall of

1996, the EPA released a report entitled *Environmental Health Threats to Children* (EPA, 1996). It outlined the national agenda, which instructed the agency to:

- Ensure that all standards set by EPA are protective of any heightened risks faced by children.
- Develop a scientific research strategy focused on the gaps in knowledge regarding child-specific susceptibility and exposure to environmental pollutants.
- Develop new, comprehensive policies to address cumulative and simultaneous exposures faced by children.
- Expand community right-to-know allowing families to make informed choices concerning environmental exposures to their children.
- Encourage parental responsibility for protecting their children from environmental health threats by providing them with basic information.
- Encourage and expand educational efforts with health-care providers and environmental professionals so they can identify, prevent, and reduce environmental health threats to children.
- Provide the necessary funding to address children's environmental health as a top priority among relative health risks.

Two other important developments in 1996 were the enactments of The Food Quality Protection Act (FQPA) and the Safe Drinking Water Act (SDWA). Both documents contained provisions that support the government's focus on children's environmental health risks. FQPA amended the Federal Insecticide, Fungicide, and Rodenticide Act (FIFRA) and the Federal Food, Drug, and Cosmetic Act (FFDCA). The FQPA fundamentally changed the way EPA regulates pesticides (EPA, 1996) Specifically, EPA was required to apply an additional 10-fold safety factor to regulate pesticide residues in foods previously considered safe. The additional 10-fold uncertainty factor had to be used unless a determination could be made on the basis of reliable data that a lesser margin of safety was protective (EPA, 1996).

In 1997, in agreement with the national agenda (EPA, 1996), EPA established the Office of Children's Health Protection (OCHP). OCHP defined its mission as making the protection of children's health a fundamental goal of public health and environmental protection in the United States with its focus on the following six areas:

- Education and Outreach
- Scientific Data and Methods
- EPA Regulatory Activities
- EPA Children's Health Board
- Children's Health Protection Advisory Committee

- President's Task Force on Environmental Health Risks and Safety Risks to Children

In April 1997, Senator Barbara Boxer (D–CA) and Representative Jim Moran (D–VA) introduced S. 599, the Children's Environmental Protection Act of 1997. This legislation required that all EPA standards be set at levels that protect children with an "adequate margin of safety," and that EPA establish a "safe-for-children product list" that would include only those products or chemicals that minimize potential health risks to children.

In May 1997, Representatives Henry Waxman (D–CA), Jim Saxton (R–NJ), and 92 co-sponsors introduced the Children's Environmental Protection and Right-to-Know Act of 1997 (HR 1636). This legislation called for the disclosure of industrial releases that present a risk to children, reporting of the identity and concentrations of toxic ingredients in consumer products to the Consumer Product Safety Commission, and publication of a list of substances that are carcinogenic, neurotoxic, or toxic to the reproductive system.

In September 1997, the Agency for Toxic Substances and Disease Registry's (ATSDR) Child Health Workgroup of the Board of Scientific Counselors released its final report entitled "Healthy Children, Toxic Environments: Acting on the Unique Vulnerability of Children Who Dwell Near Hazardous Waste Sites" (ATSDR, 1997). The document urged that the following children's health issues had to be addressed in all of ATSDR's public health programs:

- Are children exposed to potentially harmful substances?
- Are any exposure pathways unique to children?
- Do children differ from adults in their weight-adjusted intake of the toxicant?
- Do pharmacokinetic or pharmacodynamic parameters differ between adults and children?
- What are the effects of multiple and cumulative exposures?
- Are latent or delayed effects of early exposure possible?
- At what stage of development is the child exposed?
- Could any developmental processes be altered by the toxicant?
- Are there adequate animal models for childhood exposure after birth?
- What do these models indicate about adverse effects on children who are exposed?
- Are there transgenerational effects?
- Are there ethical and cultural consequences unique to children?
- If children are not included in an agency activity, why are they excluded?

In November 1997, the Natural Resources Defense Council (NRDC) released a new report entitled *Our Children at Risk: The 5 Worst Environmental Threats to Their Health* (Mott et al., 1997). The five threats to children discussed in the report included lead, air pollution, pesticides, environmental tobacco smoke, and drinking water contamination. Lead was called the "single most significant health threat" to children.

In August 1998, EPA and the Department of Health and Human Services (DHHS) announced the establishment of eight Centers of Excellence in Children's Environmental Health Research. The University of Southern California School of Medicine (Los Angeles) received $1.35 million in funding to study the relationship of tobacco smoke, air pollution, and indoor allergens to the development of asthma in inner-city children. The University of Iowa College of Medicine (Iowa City) received $1.21 million to study respiratory illness in children in rural communities. The University of Michigan School of Public Health (Ann Arbor) received $1.3 million to study environmental factors contributing to pediatric asthma in inner-city children. The Johns Hopkins University Children's Center (Baltimore) and Columbia University School of Public Health (New York) received $1.31 and $1.48 million, respectively, to examine the role of particulate matter, environmental tobacco smoke (ETS), and ozone in relation to the rising asthma rates in inner-city children. The University of California at Berkeley School of Public Health (Berkeley) and the University of Washington Department of Environmental Health (Seattle) received $1.18 and $1.35 million, respectively, to evaluate pesticide exposures and pesticide-related health risks in children in agricultural communities. The Mount Sinai School of Medicine (New York) received $1.4 million to ascertain the developmental effects among inner-city children that result from exposure to pollutants in their diet and in their homes.

Following a December 1998 meeting, the FIFRA/FQPA Scientific Advisory Panel (SAP) task force released a draft report entitled "Toxicology Data Requirements for Assessing Risks of Pesticide Exposure to Children's Health" (EPA, 1998a). The task force recommended the following: (1) The core toxicology data set requirements for pesticides under FQPA should include acute and subchronic neurotoxicity studies in rats, immunotoxicity testing, and developmental neurotoxicity testing; (2) guidelines should be developed for direct dosing of neonate animals, as well as criteria for when such studies should be conducted; and (3) guidelines for pharmacokinetic studies should be developed to include consideration of exposure during pregnancy and lactation. The task force also stated that "from the data available ... the default intra-species 10-fold uncertainty factor will be adequate in the majority of cases for protecting children's health when a complete developmental toxicity database is available" (EPA, 1998a). In July 1999, the Environmental Health Committee of the EPA Scientific Advisory Board (SAB) met to discuss EPA's revised cancer risk assessment guidelines. EPA's revised guidelines included special considerations for protecting subpopulations, particularly children, from carcinogens in the environment (EPA, 1999a). The major features of the proposed risk characterization included:

- Weighing of all of the evidence in reaching conclusions about the human carcinogenic potential of agents

- Development of information to understanding the mode of action by which chemical exposures could differentially affect children
- Dose–response assessment for all tumor types, followed by an evaluation of consistency of risk estimates across tumor types, the strength of the mode of action information of each tumor type, and the anticipated relevance of each tumor type in children
- Evaluation of the nature and extent of human variability to the response (e.g., children versus adults, male versus female, etc.)

Public comments on these guidelines were received from the American Industrial Health Council (AIHC), the Chemical Manufacturers Association (CMA), the Chlorine Chemistry Council (CCC), Resources for the Future (RFF), the Natural Resources Defense Council (NRDC), and the National Institute of Environmental Health Sciences (NIEHS).

In September 1999, EPA held a kickoff meeting in Washington, DC, to involve stakeholders in the design and development of a voluntary program to test commercial chemicals to which children may have a high likelihood of exposure. The purpose of the series of meetings, proposed in 1999 and held in 2000 and 2001, was to provide stakeholders with an opportunity to respond to EPA's preliminary thoughts on criteria and considerations that could be used to select chemicals for testing, as well as the test battery to be used in the program.

30.3   DIFFERENCES IN EXPOSURE AND RESPONSE TO TOXIC AGENTS BETWEEN CHILDREN AND ADULTS

A group of individuals loosely characterized as "children" may include a 1-month old infant, a 2-year-old toddler, and a 17-year-old adolescent who, with respect to their responses to various social and environmental influences, are as far apart from each other as they are from adults. These differences explain the lack of agreement regarding what is to be considered "childhood age." According to the FDA's definition, the pediatric age group includes all persons from birth to age 16. However, broader definitions may cover individuals as old as 18 and even those as old as 21 years of age (FDA, 1994).

The effects of chemical, physical, and biological exposures on children's health depend on a number of factors. These include physical location, routes of exposure, absorption, metabolism, and elimination of potential toxic substances. All of these factors may differ in children compared to adults, however, the often-publicized assumption that children are more vulnerable compared to adults appears oversimplified and often not supported by the data.

30.3.1   Differences in Exposure and Absorption

Children generally have different levels of intake of water, air, and certain foods per unit of body weight compared to adults. At different ages, the total intake may be higher, lower, or equal to that of adults. For example, as shown in Figure 30.1, the total weight-adjusted air exchange rate in newborns is lower

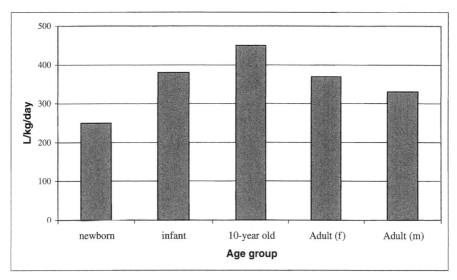

Figure 30.1 Body-weight-adjusted air intake [International Commission on Radiological Protection (IRCP), 1975].

than that in adults. In the subsequent months of life, it rapidly increases, but only slightly exceeds the adult air exchange by the end of infancy. As the child gets older, the air exchange rate continues to increase and reaches its peak around 10-years of age because of relatively low body mass and high level of activity. After that, the air exchange gradually decreases until it reaches adult levels (Plunkett et al., 1992).

The amounts of food and water consumed per kilogram of body weight are higher in infants and gradually decrease with age, approaching those of adults by late adolescence (Figure 30.2). It is also important to consider the composition of children's diets since certain dietary items (e.g., apples) comprise a relatively large proportion of daily intake compared to adults, while other items are hardly present at all (EPA, 1997a).

Surface area per unit of body weight in newborns is approximately three times higher than in adults. The weight-adjusted surface area then rapidly decreases so that by age 3 it becomes roughly 75 percent higher than in adults and reaches its nadir in adolescence. Therefore, dermal exposures to the total body area (e.g., during swimming or bathing) will result in higher doses (per kilogram) in younger children. However, when dermal exposure involves only parts of the body, the resulting weight-adjusted dose in children may be only slightly higher than that in adults (Figure 30.3). For example, the area of the thigh in children under one year of age is half the proportion of the total body surface compared to that in adults (Barkin and Rosen, 1994).

The gastrointestinal (GI) tract of children is known to absorb inorganic substances, particularly heavy metals, more efficiently than that of adults. In the case of lead, for example, the average GI absorption rate in young infants is

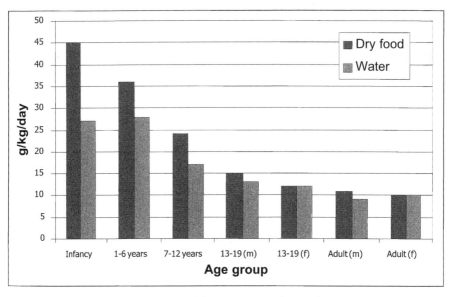

Figure 30.2   Body-weight-adjusted food and water intake (Plunkett et al., 1992).

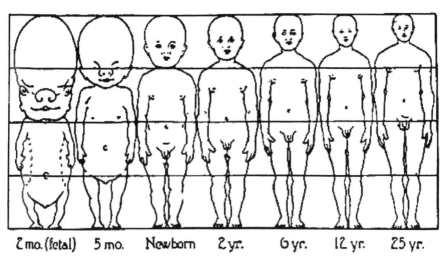

Figure 30.3   Changes in body proportions from second fetal month to adulthood (Robbins et al., 1928).

five times higher than in adults. This is probably also true for other heavy metals, although the data for humans are lacking and most of the available information comes from animal studies (Calabrese, 1986).

The rate of dermal absorption depends heavily on the level of skin keratinization, which appears to reach maturity relatively early in infancy. There is disagreement regarding the role of gestational age. Some studies seem to indicate that premature infants exhibit higher skin permeability than full-term newborns (Greaves et al., 1975), while other studies report that keratinization occurs by 3 to 5 days of life regardless of gestational age (Bearer, 1995).

30.3.2 Differences in Response

Pharmacological data indicate that the differences in responses to different chemicals exhibited by children and adults depend on the substance in question. In some instances, the effects of certain drugs on children are qualitatively different from the effects on adults. For example, the adverse effect of antihistamines on the central nervous system (CNS) of young children often manifests itself as extreme agitation. By contrast, antihistamine toxicity in adults typically presents as CNS depression (Siberry and Iannone, 2000).

Depending on a specific agent's metabolic handling and excretion, infants and children may be more or less sensitive to its toxicity. For instance, chloramphenicol, an antibiotic used for prophylaxis of neonatal sepsis, has been shown to cause severe illness in newborns manifested as pallor and cyanosis. This condition, termed the "gray-baby syndrome," was attributed to the immaturity of neonatal liver function, especially glucuronide conjugation (Evans and Kleiman, 1986).

It is important to note that the relative immaturity of glucuronidation processes in young children as compared to adults may also act as a protective factor. Consider acetaminophen toxicity. Adults eliminate approximately 50 percent of ingested acetaminophen via glucuronide conjugation and 30 percent via sulfate conjugation. By contrast, young children eliminate 45 to 55 percent of acetaminophen as sulfate and only 18 to 30 percent as glucuronide (Miller et al., 1976). In the event of acute acetaminophen overdose, young children are less likely to sustain liver damage than adults (Rodgers and Matyunas, 2000) probably because of the predominance of the nontoxic (i.e., nonglucuronide) pathway (Kauffman, 1992).

The relative sensitivity and resistance to a particular agent may change several times as the child gets older. For example, use of theophylline illustrates the age-specific differences in the P450 cytochrome family of enzymes. P450 cytochrome activity is low in the neonatal period leading to a prolonged half-life and requiring a 12-hour interval between doses. As the P450 cytochrome matures over the next several months, the half-life of theophylline decreases, warranting higher (per kilogram of weight) and more frequent doses. However, in adolescence, the P450 cytochrome activity again decreases and theophylline dosing needs to be adjusted accordingly (Nebert and Gonzalez, 1987).

The kidney excretion of drugs and other chemicals in newborns and young infants is substantially limited by low rates of glomerular filtration and immaturity of the tubular apparatus. As renal function improves, the clearance increases reaching its peak in early adolescence. For example, the half-life of cimetidine is approximately 2.6 hours in newborns, 1.4 hours in children ages 5 through 13, and 2.3 hours in adults (Kacew and Lock, 1988).

Phase I clinical trials of chemotherapeutic compounds directly comparing results between pediatric and adult populations provide probably the best available data on the relative sensitivity of children. One of the objectives of such clinical trials is the determination of the maximally tolerated doses (MTDs) that can be used in phase II trials and, eventually, in clinical practice (Bruckner, 2000). These clinical trials have identified no qualitative and small quantitative (by a factor of ≤ 2) differences between children and adults (Glaubiger et al., 1981; Marsoni et al., 1985; Evans et al., 1989). Importantly, in most instances the MTDs for children were higher than those for adults probably due to more efficient hepatic and renal clearance.

30.4 SELECTED ENVIRONMENTAL CHILDREN'S HEALTH ISSUES

30.4.1 Lead Exposure

The adverse health effects of lead poisoning have been known for years. In the last two decades the environmental levels of lead have decreased due to the introduction of lead-free gasoline, reduced use of lead in manufacturing processes, and discontinued manufacture of household paint containing greater than 0.06 percent lead (EPA, 1990). Changes implemented by the food industry have decreased the amount of lead in food from an estimated 500 µg per person per day in the 1940s to the present level of less than 20 µg/day (EPA, 1986). However, the opportunities for exposure still exist because many buildings still contain lead-based paint and because of the improper removal of lead-based paint (ATSDR, 1999). Moreover, concerns continue to appear regarding previously unaccounted sources of exposure such as lead in candlewicks (Nriagu and Kim, 2000), although the impact of these sources is difficult to assess at this time.

Data from the third National Health and Nutrition Examination Survey (NHANES III), indicated that non-Hispanic white and black children aged 1 to 5 years had mean blood lead levels (BLL) of 3.2 and 5.6 µg/dL, respectively (Pirkle et al., 1994). However, the data also revealed that 8.9 percent of children in this age group still had BLL of 10 µg/dL or higher. More recent data from phase II of NHANES III (1991 to 1994) indicate that 4.4 percent of children aged 1 to 5 have BLLs greater than 10 µg/dL and that the overall mean BLL in this age group is 2.7 µg/dL [Centers for Disease Control and Prevention (CDC), 1997a].

Children have greater potential for exposure to lead due to playing on the ground, hand-to-mouth activity, and, in some children, pica behavior—an appetite for unfit foods or nonfood substances (Juberg et al., 1997; ATSDR, 1999). Although the estimates of pica prevalence may vary, some authors indicate that as many as 11 percent of toddlers have pica (Roberts and Dickey, 1995). Ingestion of household paint dust is probably the main source of lead exposure for inner-city children (ATSDR, 1999). Children absorb about 30 to 40 percent of the ingested lead compared to 5 to 15 percent in adults (Goyer, 1996).

The lead-related health effects include anemia, neurological impairment, colic, and neurobehavioral deficits (ATSDR, 1999). Increased blood pressure can occur in children at a mean BLL of 37.3 µg/dL and decreased hemoglobin has been observed at levels ≥ 40 µg/dL (Factor-Litvak et al., 1996). Renal effects may include aminoaciduria and Fanconi syndrome at levels > 80 µg/dL (Chisolm, 1962; Pueschel et al., 1972). Levels of over 125 µg/dL may result in acute encephalopathy that can be potentially fatal [National Academy of Sciences (NAS), 1972].

Of particular concern are the neurobehavioral effects of lead leading to decreased intelligence, reduced short-term memory, and learning disabilities, with some studies revealing that these effects may persist into adulthood (Juberg et al., 1997). Some studies have found effects on IQ at relatively moderate BLLs in the 30 to 40 µg/dL range (Ruff et al., 1996; ATSDR, 1999).

Meta-analyses using both cross-sectional and prospective studies have concluded that a typical doubling of BLL from 100 to 200 µg/L is associated with an average loss of IQ of 1 to 3 points with no identified threshold (Winneke and Kramer, 1997). However, a causal relation between lead exposure and IQ decrement is hard to confirm or prove. It is possible that lower IQ contributes to elevated BLLs since children who are hyperactive, impulsive, and inattentive are more likely to eat contaminated dirt or paint chips (reverse causality). In addition, many studies failed to control for potential confounders including socioeconomic status, childhood disease, childhood nutrition, parenting skills, child rearing style, parental time spent with the child, skills of other key caretakers, parental IQ levels, and maternal substance abuse or poor nutrition during pregnancy (Juberg et al., 1997).

The CDC provides the following recommendations for screening at 9 to 12 months of age:

<10 µg/dL	Rescreen in 1 year; no additional action necessary
10 to 19 µg/dL	Follow-up testing; family education
20 to 24 µg/dL	Case and clinical management; environmental investigation and control
45 to 69 µg/dL	Same as for 20 to 24 µg/dL except begin within 48 hours
≥70 µg/dL	Same as for 20 to 24 µg/dL but must hospitalize child and begin immediate medical treatment (CDC, 1997b)

At issue is the dose at which lead begins to exert adverse health effects. The term "lead poisoned" may be incorrectly used to define asymptomatic children with BLLs around 10 µg/dL. However, pediatricians have found that symptomatic lead poisoning can occur in children with relatively low BLLs (Piomelli et al., 1984).

30.4.2 Pesticide Exposure

Pesticides pose a risk of acute poisoning if not used or stored properly and not kept out of the reach of children. A recent EPA survey found that 47 percent of all households with children under age 5 had at least one pesticide stored in an unlocked cabinet less than 4 feet off the ground (EPA, 1998b). However, acute pesticide poisonings are rare and their contribution to overall morbidity and mortality is very small. According to the American Association of Poison Control Center's Toxic Exposure Surveillance System (TESS), 90 percent of pesticide exposure cases do not require any medical intervention other than dilution (increased intake of fluids) and observation (Reigart and Roberts, 1999).

A recent analysis of the TESS data showed that there were 103 deaths from pesticide poisonings during a 10-year period from 1985 through 1994. This represents 6 percent of all fatal poisonings due to all nonpharmaceuticals, with a substantial proportion of these deaths due to suicides. For example, data on chlorpyrifos—an organophosphate recently banned because of its alleged unacceptable toxicity—indicate that among eight chlorpyrifos-related deaths that occurred between 1985 and 1995, three were suicidal, three were accidental, and in two cases the circumstances remained unknown (Kingston et al., 1994). Another TESS review indicated that in 1996 there were 10 deaths due to accidental pesticide ingestion and 10 deaths due to ingestion of over-the-counter cold preparations (Litovitz et al., 1997).

In addition to the risk of acute poisoning, a concern exists that low-level long-term exposures may be associated with chronic effects. The potential modes of exposure include pesticide residues in certain foods that constitute a high proportion of children's diets such as apples or presence of pesticide residues in human breast milk (EPA, 1997b). Children's exposure can also occur through playing on floors or lawns or putting things in their mouths.

The hypothesized chronic effects of low-level pesticide exposure include central nervous system damage, cancer, reproductive dysfunction, damage to the endocrine and immune systems, and respiratory illness (EPA, 1998b). Although some authors indicate that children are inherently more sensitive to the potential toxicity of pesticides (Mott et al., 1997), a recent National Academy of Sciences review concluded that this is not necessarily so, and that children may be more or less sensitive depending on their age and on the compound in question (Bruckner, 2000).

It is important to point out that much of the epidemiological research conducted to date relies on interviews and self-administered questionnaires rather than objective biomarker-based measures of exposure. While analytical methods capable of detecting particular pesticides in body fluids are still quite limited, residues of pentachlorophenol, p-dichlorobenzene, carbaryl, dieldrin, chlordane, trans-nonachlor, azinphos-methyl, phosmet, and 2-4-D can be detected in blood and urine (Solomon and Mott, 1998). However, studies using these analytical methods are difficult to carry out on large populations because accurate exposure assessment frequently requires 24-hour urine collection for several days, posing serious logistical difficulties.

A separate source of concern is the use of pesticides in residential settings as reflected in the recent reevaluation of risk associated with the common residential insecticide Dursban (chlorpyrifos). Chlorpyrifos, an organophosphorus insecticide developed in 1962 by Dow Chemical Company, controls a broad spectrum of insects in both agricultural and urban settings. In urban professional products, it is used as a soil-applied termite barrier and is sprayed into cracks and crevices indoors or onto lawns outdoors (Albers et al., 1999).

Recently, two panels of experts conducted comprehensive reviews of the literature on the adverse effects of chlorpyrifos exposure (Clegg and van Gemert, 1999; Albers et al., 1999). The two panels reviewed the evidence on the association between chlorpyrifos and several of the most frequently suggested potential health effects. These included organophosphorus-induced delayed neurotoxicity (OPIDN), neurobehavioral (e.g., mood, attention, and cognitive) disorders, and immunologic disorders.

The first (toxicology and medicine) panel agreed that there was no clear evidence for long-term effects resulting from acute poisoning by organophosphate compounds, other than finding cases of OPIDN from suicidal ingestion. With respect to chronic exposure, the panel agreed that organophosphate compounds do not cause problems in the peripheral or central nervous systems. With respect to neurobehavioral effects, the panel concluded that manifestations are unlikely in the absence of clear cholinesterase inhibition (Clegg and van Gemert, 1999).

The second (epidemiology and medicine) panel expressed some disagreement. The majority of panel members (five members) agreed that the literature reviewed provided little or no scientific evidence that chlorpyrifos exposure causes harm to human health other than its known cholinergic effects associated with acute poisoning. However, those panel members voting in the minority (three members) indicated that the studies reviewed provided inadequate evidence to preclude the possibility of adverse effects to human health from chlorpyrifos exposure at levels associated with its manufacture or professional application. Those voting in the minority suggested further investigation of cohort(s) of workers engaged in either the manufacture or the professional application of chlorpyrifos, or both. The primary health outcomes recommended for study were cognitive and affective disorders, with consideration of the assessment of peripheral neuropathy (Albers et al., 1999).

In June 2000, the EPA announced that it had reached an agreement with Dow Agrosciences, the manufacturer of chlorpyrifos, to phase out its use for nearly all household purposes. The primary reason for the chlorpyrifos ban cited by EPA was the unacceptable risk of chlorpyrifos poisoning. Aware of the EPA's investigations into chlorpyrifos, the United Kingdom Ministry of Agriculture, Fisheries, and Food (MAFF) commissioned its own review of human safety data on chlorpyrifos. The review by its Advisory Committee on Pesticides (ACP) concluded: "The committee is satisfied that there is no evidence for concern about short-term or acute exposures to chlorpyrifos in the UK" [Environmental News Service (ENS), 2000].

It is important to point out that chlopyrifos is only one of several pesticide products that is undergoing EPA reevaluation. Legislative standards referred to as "food tolerances" were originally established by Congress through the Federal Insecticide, Fungicide, and Rodenticide Act (FIFRA) and the Federal Food, Drug, and Cosmetic Act (FFDCA). Food tolerances under these acts published in the Code of Federal Regulations (CFR) included approximately 8350 for residues on raw agricultural commodities (RAC) and about 150 for residues known to concentrate in processed foods (NRC, 1993). In 1996, addressing the issue of children's special exposure patterns, EPA passed the 1996 Food Quality Protection Act (FQPA), which requires EPA to protect child health in the absence of scientific certainty regarding the toxicity of a particular pesticide (EPA, 1998b). By August 2006, EPA plans to complete its review of all tolerances that were in effect in August 1996 when the Food Quality Protection Act was enacted.

30.4.3 Environmental Tobacco Smoke (ETS) Exposure

American children are spending most of their time indoors (Etzel, 1995). Therefore, indoor air quality has become increasingly important. One of the most important preventable sources of indoor air pollution is environmental tobacco smoke (ETS). The term ETS characterizes tobacco combustion products inhaled by nonsmokers in the proximity of burning tobacco. Most potential concerns regarding ETS exposure are associated with sidestream smoke emitted from the burning tip of the cigarette because it contains higher concentrations of toxic compounds (Ericksen et al., 1988).

Several studies have evaluated the association between lower respiratory illness (LRI) and ETS. For example, Margolis et al. (1997) found that infants reportedly exposed to tobacco smoke had a 1.5- to 2-fold increase in incidence of LRI than infants that were not exposed to ETS. However, there was no association between ETS exposure and urinary cotinine, a biomarker for ETS exposure.

Another recent study evaluated whether the risk of bronchial obstruction was increased in ETS-exposed children under 2 years of age. Although there was no clear exposure–response pattern, children exposed to ETS had up to a 50 to 60 percent increase in bronchial obstruction compared to their nonexposed counterparts. Similar effects were present for maternal smoking alone, paternal smoking alone, and both parents smoking (Nafstad et al., 1997). A recent meta-analysis of 38 studies found that children exposed to ETS at home were 57 percent more likely than nonexposed children to contract lower respiratory tract illnesses during the first 3 years of life. The risk was even higher for maternal smoking alone (Strachan and Cook, 1997).

The evidence regarding the association between ETS and middle-ear disease in children appears even stronger [American Council on Science and Health

(ACSH), 1999]. A prospective study carried out in Sweden evaluated several indoor environmental factors. The results indicated no relationship between these factors and middle-ear effusion (MEE), with the exception of parental smoking at home (Iversen et al., 1985). A more recent case control study in Canada also found an association between middle-ear disease and two or more household smokers. These results persisted after adjusting for childcare, infant feeding, socioeconomic status, maternal education, and level of prenatal care (Adair-Bischoff and Sauve, 1998). A day-care center study in the United States evaluated ETS exposure based on serum cotinine and its relationship to otitis media with effusion diagnosed by pneumatic otoscopy. Children with serum cotinine concentrations greater than or equal to 2.5 ng/mL had a 38 percent higher rate of new episodes of otitis media than children with lower or undetectable serum cotinine. Furthermore, the average duration of an otitis media episode was longer among children with elevated cotinine concentrations than in the reference group (Etzel et al., 1992).

Sudden infant death syndrome (SIDS) is the most common cause of death in infants aged 1 month to 1 year (ACSH, 1999). The available evidence suggests that infants of smoking mothers are at increased risk of SIDS independent of other known risk factors, including low birth weight and gestational age (DiFranza and Lew, 1995). For example, a prospective study based on Swedish births between 1983 and 1985 found that maternal smoking doubled the risk of SIDS and a clear dose–response relation by amount smoked was present. Maternal smoking also seemed to influence the time of death, as infants of smokers died at an earlier age (Haglund and Cnattingius, 1990). Malloy et al. (1988) linked birth certificate and infant death certificate data from Missouri for 1979 to 1983 to explore the association of maternal smoking with cause of infant death. Their data included 305,730 singleton white live births, of which 2720 resulted in infant deaths. Mortality associated with passive smoking was particularly high for two causes: respiratory disease, which showed more than a threefold increase in risk, and sudden infant death syndrome, which showed a twofold increase in risk (Malloy et al., 1988). The published literature does not permit a definitive conclusion, however, as to whether the increased risk of SIDS is related to exposure to tobacco smoke during pregnancy (in utero exposure), following birth (postnatal exposure), or both. Thus, at present there is not enough direct evidence to support postnatal exposure to ETS alone as a risk factor for SIDS (ACSH, 1999).

30.5 CHILDHOOD CANCER

Several reports published in the mid-1990s raised concerns that the incidence of childhood cancer, particularly leukemia and brain cancer, might be increasing (Gurney et al., 1996). This prompted speculation that the reported increase was due to environmental causes. However, recent analysis of the SEER data for the years 1975 to 1995 revealed that the rise in leukemia among children was due to an increase confined to 1983 to 1984 and that rates have actually decreased slightly since 1989 (Linet et al., 1999). The rate for brain and other central nervous system cancers rose modestly in a stepwise fashion between 1983 and 1986. The investigators concluded that there has been no substantial change in incidence for the major childhood cancers with rates remaining relatively stable since the mid-1980s. Any observed increases were likely due to diagnostic improvements or changes in reporting methods (Linet et al., 1999).

Another recent examination of the SEER data showed that much of the reported increase in acute lymphocytic leukemia (ALL) was due to changes in diagnostic coding that occurred in the Detroit region. After removal of the Detroit data, the incidence of ALL between 1973 and 1995 remained unchanged (Liberson et al., 2000). Of greater importance is a dramatic decrease in mortality (42 percent) during the last two decades and an increase in the 5-year relative survival rate (almost 70 percent) [National Cancer Institute (NCI), 1996].

Acute lymphocytic leukemia is the most frequent type of childhood cancer in the United States, followed by cancers of the brain and spinal cord combined and non-Hodgkin's lymphoma. Incidence of childhood cancer varies greatly in different parts of the world (NCI, 1996). For example, ALL is responsible for approximately 78 percent of all childhood leukemias in the United States, while acute myelomonocytic leukemia (AMML) is responsible for only 4 percent. But in Turkey, almost half of all childhood leukemias are AMML. In Africa, leukemia is rare but Burkitt's lymphoma accounts for more than half of all childhood cancers (NCI, 1996).

While adult cancers commonly occur in epithelial tissues, cancers of this tissue type rarely occur in children (Chow et al., 1996). Unlike adult cancers, childhood cancers are often composed of embryonal cell types, which are morphologically similar to fetal cells (Chow et al., 1996). Therefore, childhood cancers may arise due to different mechanisms than adult cancers.

Several childhood cancers have known genetic causes. For example, approximately 40 to 50 percent of retinoblastomas are hereditary (Bunin et al., 1989). The incidence of acute leukemia among children with Down's syndrome is 20 to 30 times higher that that of other children (Robison et al., 1984). Many cancers tend to cluster in families, although the precise mechanisms of their development is not known (Little, 1999).

With the exception of in utero exposure to ionizing radiation and leukemia, few associations between environmental risk factors and childhood cancer may be considered established. Intrauterine exposure to diagnostic or therapeutic ionizing radiation is clearly associated with an increased risk of leukemia (Graham et al., 1966). By contrast associations between postnatal exposure to radiation, including therapeutic irradiation, radioactive fallout, and proximity to nuclear plants, showed conflicting results (Little, 1999). Despite extensive study, and wide publicity, the association between electromagnetic field exposure and childhood cancer remains unclear.

There is a substantial body of literature on various environmental, parental occupational, and medical exposures to chemicals and childhood cancer. The hypothesized risk factors include exposure to low-level environmental pollutants, parental smoking, various medications, and parental occupations, especially hydrocarbon-, solvent- and metal-related work (Chow et al., 1996). However, on balance, these studies have shown weak and inconsistent associations and are subject to several methodological problems. These problems include rarity of outcomes, inadequate assessment of the timing of many exposures (preconception, prenatal, postnatal), subject selection and information biases, and failure to control for confounders (Little, 1999). To be considered definitive, future epidemiological studies will have to overcome the problem of heterogeneity of cancer outcomes and problems associated with imprecise and incomplete exposure assessment (Olshan and Daniels, 2000).

A review on childhood cancer would be incomplete without a discussion of clusters, particularly clusters of leukemia. This topic has been the center of heated controversy for the past two decades. A "cluster" is defined as an aggregation of relatively uncommon events or diseases in space and/or time in amounts that are believed or perceived to be greater than could be expected by chance (Last and Abramson, 1988). Putative disease clusters are often perceived to exist on the basis of anecdotal evidence, and much effort is often expended by epidemiologists and biostatisticians in demonstrating whether a true cluster exists [National Conference on Clustering of Health Events (NCCHE), 1990].

The first report documenting cancer clustering appeared in the literature near the turn of the 20th century. In 1905, Arnsperger reported a leukemia cluster in two small towns in Germany (Boyle et al., 1996). Since then, hundreds of childhood cancer clusters have been reported and investigated. There appears to be a consensus among epidemiologists that so far none of these investigations may be considered definitive with respect to determining an underlying cause. For example, the Centers for Disease Control and Prevention in Atlanta investigated 108 clusters over 20 years in 29 U.S. states and five foreign countries. Eighty-one percent of these clusters involved leukemia. No cause was found for any of these clusters (Caldwell, 1990). Many cluster investigations that claimed success in finding a common cause have been subject to criticism due to their inability to overcome one or more of the following limitations:

- Small numbers of cases
- Diverse diagnoses
- Small and changing populations
- Bias and misclassification of exposure

A particular problem with cluster investigations is "boundary shrinkage," that is, bias in defining the boundary of a cluster leading to the overestimation of the disease rate (Olsen et al., 1996). The futility of looking for causes of cancer through cluster analysis is probably explained by the fact that cancers, like other events, are expected to cluster in time and space. For example, a cancer occurrence analysis by county in Minnesota identified 100,000 clusters for 85 cancers; 10,000 of these exceeded the statewide rate at least twofold, and

1500 were statistically significant. A similar analysis for towns, neighborhoods, or school districts would likely produce an infinite number of clusters (Williams, 1998).

In September 1997, the U.S. EPA sponsored the Conference on Preventable Causes of Cancer in Children (Carroquino et al., 1998). Recommendations for future research offered by the participants included more collaborative research integrating epidemiology, molecular biology, toxicology, and risk assessment; the development of better protocols for toxicologic testing of carcinogenicity using young animals; and research focused on specific periods of development during which susceptibility to environmental agents may be enhanced (Carroquino et al., 1998).

30.6   DEVELOPMENTAL DEFECTS

A developmental defect is defined as a structural or functional anomaly that results from an alteration in normal development (NRC, 2000). This definition includes an infinitely large and heterogeneous group of conditions affecting all organs and systems and ranging from severe structural abnormalities incompatible with life to mild functional disorders that may remain undetected.

The leading developmental disorders causing infant mortality are heart defects (31.4 percent), respiratory defects (14.5 percent), nervous system defects (13.1 percent), chromosomal aberrations (13.4 percent), and musculoskeletal anomalies (7.2 percent) (Petrini et al., 1997). More common conditions that are not manifested by overt structural disorders and do not lead to increased mortality are functional deficits. Many of these deficits involve neurobehavioral problems and are not recognized until later in life, most commonly during the early school years.

The causes of developmental disorders are roughly categorized as extrinsic and intrinsic, but the definitions of these two categories are vague and largely overlapping. According to Wilson (1973), 25 percent of congenital anomalies are caused by genetic disorders, 65 to 75 percent are idiopathic, and less than 10 percent are attributable to environmental causes. These observations are in general agreement with more recent reports (Nelson and Holmes, 1989). The most common environmental causes of developmental defects include maternal disease (e.g., diabetes), intrauterine infections (e.g., congenital rubella), and mechanical problems (e.g., amniotic bands) (NRC, 2000).

In all, only about 50 environmental agents are known to cause developmental defects in humans. These include physical exposures, such as ionizing radiation; the so-called life-style chemicals, such as alcohol and illicit drugs; a variety of pharmaceuticals, such as thalidomide and valproic acid; and environmental pollutants, such as lead and methyl mercury. In addition, at least 1200 agents have been shown to cause developmental defects in experimental animals. However, it is not known how many of these 1200 agents actually produce developmental defects in humans (NRC, 2000).

In recent years, a great deal of attention has been placed on the potential neurobehavioral effects of chemical exposures at low (background) levels. There is a growing public concern over pediatric mental health problems, including both overt (e.g., attention-deficit hyperactivity disorder and autism) and subtle (e.g., slightly lower IQ) conditions (Schmidt, 1999).

Attention-deficit hyperactivity disorder (ADHD) is characterized by inappropriate degrees of inattention, impulsiveness, and hyperactivity. Typical symptoms are fidgeting, difficulty remaining seated, distractibility, difficulty following through, impatience, inattentiveness, excessive talking with frequent interruption of others, and engaging in physically dangerous activities (Anderson et al., 1987).

Over the past two decades, multiple community-based studies reported a number of ADHD prevalence estimates ranging from 2 to 17 percent. The dramatic differences in these estimates are due to the choice of informant, methods of sampling and data collection, and the diagnostic definition. Based on a comprehensive review of the literature, Scahill and Schwab-Stone (2000) demonstrated that the best estimate of prevalence is 5 to 10 percent in school-aged children (Scahill and Schwab-Stone, 2000).

Data from the National Ambulatory Medical Care Survey (NAMCS) indicate that the number of office-based visits documenting a diagnosis of ADHD increased from 947,208 in 1990, to 2,357,833 in 1995. After adjustment for population size, these numbers translate into a twofold increase in frequency of office-based visits and a nearly threefold increase in the frequency of prescribed stimulant pharmacotherapy for ADHD (Robison et al., 1999). One needs to interpret these findings with caution as they may possibly reflect improved diagnosis and increased availability of treatment options.

Autism is a childhood disorder characterized by qualitative impairments in social interaction, verbal and nonverbal communication, and a markedly restricted repertoire of activities (Dalton et al., 1992). Despite rhetoric to the contrary [Neuro-Immune Dysfunction Syndromes Research Institute (NIDS), 1999], a recent comprehensive review of the epidemiological literature demonstrated no evidence for a secular increase in the incidence of autism between 1966 and 1998 (Fombonne, 1999).

The current scientific literature contains several reports linking low-level environmental chemical exposures to subclinical adverse developmental outcomes as reflected in neurobehavioral test scores. It is important to realize that the biological meaning of such findings—particularly when the differences between subjects are very small—is not clear. In clinical practice these tests alone are not considered particularly informative because they are viewed as only one part of a comprehensive assessment of neurological function (Matarazzo, 1990).

Despite attempts to standardize neurobehavioral testing for children (Krasnegor et al., 1994), there is little agreement on which specific battery of tests is the most appropriate for environmental research. Several tests may exist for the evaluation of the same neurodevelopmental characteristics within the same age group. For example, the cognitive ability of a 3- to 6-year-old child can be evaluated by using one of the following three instruments: McCarthy Scales of Infant Development, Differential Aptitude Scale, or the Wechsler Primary and Preschool Scale of Intelligence.

Another critical issue is the reliability and validity of these tests. Validity is a measure of how closely the results of a test correspond to the true state of the phenomenon measured. Reliability describes consistency of results when the test is performed more than once on the same individual in different circumstances (Last and Abramson, 1988). The lower the reliability coefficient, the wider the confidence interval for a particular effect, however, if the reliability of a test is known, it is possible to correct for it statistically. The reliability coefficients for different tests are often available from the clinical neuropsychology literature. However, the reliability achieved in a controlled clinical setting may not hold in population research.

Another problem associated with studies of adverse neurodevelopmental effects from environmental exposures stems from the fact that if a study fails to control for potential confounders, many of which are far more powerful predictors of neurodevelopmental problems than the hypothesized environmental exposure, the results of the study are particularly difficult to interpret (Juberg et al., 1997). As noted previously, These potential confounders include socioeconomic status, childhood disease, childhood nutrition, parenting skills, child rearing style, parental time spent with the child, skills of other key caretakers, parental IQ levels, and maternal substance abuse or poor nutrition during pregnancy. In addition, co-exposure to lead, methyl mercury, and other potential neurotoxins may also influence the results.

These methodological problems are exemplified in a recent study, which compared two selected groups of 4- to 5-year-old children residing in the Yaqui Valley of northwestern Mexico (Guilette et al., 1998). The study claimed that these children shared similar genetic backgrounds, diets, water mineral contents, cultural patterns, and social behaviors and that the only major difference was their exposure to pesticides. The authors applied their own Rapid Assessment Tool for Preschool Children (RATPC) and found that the exposed children demonstrated "decreases in stamina, gross and fine eye-hand coordination, 30-minute memory, and the ability to draw a person." Although widely publicized in the general press, the results of this study appear unconvincing for several reasons. First, there is no evidence that the participants (33 "exposed" and 17 "nonexposed") were selected randomly (selection bias). Second, the investigators were aware of the participants' exposure status (interviewer bias). Third, there was no attempt to ascertain pesticide exposure in study participants other than by assuming that the "valley children" were exposed and that the "foothills children" were not exposed. Finally, despite their statements to the contrary, the authors were unable to control for the various confounders that could affect the study results.

A number of studies have evaluated developmental effects of low-level exposure to PCBs and other related contaminants in the United States (Michigan and North Carolina) and in Europe (Netherlands and Faeroe Islands).

For example, a series of studies evaluated the effects of potential prenatal polychlorinated biphenyl (PCB) exposure from maternal Lake Michigan fish consumption on the development of children (Fein et al., 1984; Jacobson et al., 1984, 1985, 1990, 1992; Jacobson and Jacobson, 1996, 1997). Based on their research, the authors of the Michigan studies concluded that in utero exposure to PCBs in concentrations slightly higher than those in the general population could have a long-term impact on intellectual function (Jacobson and Jacobson, 1996).

Critics of the Michigan studies argue that the concentrations of PCBs in the blood of the exposed mothers were only slightly greater than those of mothers who did not eat contaminated fish and were within the range of PCB blood levels reported for the entire North American population. If such trace levels could, indeed, affect intellectual development in infants and children, there should be evidence of widespread intellectual deficits among North American children [American Council on Science and Health (ACSH), 1997]. Moreover, there was no correlation between self-reported fish consumption and umbilical cord serum PCB levels (Schwartz et al., 1983).

Although the levels of PCBs found in breast milk were much higher than those in the maternal blood, there was no association between breast milk PCB concentration and intellectual impairment. One explanation of this observation is that the developing fetal brain is more sensitive to PCB toxicity than is the infant brain. However, it is important to keep in mind that the brain continues to mature and undergoes considerable development long after birth (Needleman, 1999).

In contrast to the Michigan findings, the North Carolina study reported an association between breast milk concentrations of PCBs and a small delay in motor maturation at 24 months, attributed to transplacental exposure to PCBs (Rogan and Gladen, 1991). However, a follow-up study found no association between transplacental or breastfeeding exposure to PCBs and neurobehavioral scores at 3, 4, or 5 years. The authors concluded that the deficits seen at 2 years of age were no longer apparent (Gladen and Rogan, 1991).

Several publications report on a birth cohort of Dutch infants exposed to background PCB levels (Huisman et al., 1995; Koopman-Esseboom et al., 1996; Patandin et al., 1999). The study was carried out in two areas in the Netherlands: Groningen (a semiurban region in the northeast) and Rotterdam (a highly industrialized region in the southwest). The initial study group ($n = 418$) included 104 breast-fed and 107 formula-fed newborns from Groningen and 105 breast-fed and 102 formula-fed newborns from Rotterdam. Exposure to PCBs as well as to dioxins and dibenzofurans involved measuring concentrations in maternal serum and in cord blood and, for those children that were breast-fed, in the breast milk. The study participants underwent neurological examinations between the 10th and 21st day of life, and then at 3 months, 7 months, 18 months, and at 42 months of age.

In the neonatal evaluation, the cord and maternal plasma PCB levels, used as an index of prenatal exposure, were not related to neurological function. However, higher levels of PCBs in breast milk were associated with a higher incidence of hypotonia (Huisman et al., 1995). A follow-up study found that prenatal PCB exposure had a small negative effect on the psychomotor score at 3 months of age. PCB and dioxin exposure through breastfeeding had an adverse effect on the psychomotor outcome at 7 months of age. However, there was no significant influence of the perinatal PCB and dioxin exposure on the mental outcome at 3 and 7 months of age. At 18 months of age, development was unrelated to PCB and dioxin exposure (Koopman-Esseboom et al., 1996).

The most recent follow-up analysis of the Dutch cohort evaluated cognitive development at 42 months of age. This study found that the highest prenatal exposure group (maternal serum concentration of $\Sigma PCB \geq 3$ µg/L) scored four points lower on the Kaufman Assessment Battery for Children compared with the lowest prenatal exposure group ($\Sigma PCB < 1.5$ µg/L). Both lactational exposure and current exposure to PCBs and dioxins were not related to 42-month cognitive performance (Patandin et al., 1999).

The Faeroe Islands study examined 182 singleton term births in families where the typical marine food includes pilot whale (Steuervald et al., 2000). The investigators followed the Dutch study procedures by collecting maternal blood at week 34 of pregnancy and obtaining umbilical cord blood at birth. It is important to note that the PCB levels in the Faeroe Islands cohort were about three times higher than in the Dutch cohort. The neurological examination used the same test battery as was used in the Dutch study and evaluated infants at approximately 2 weeks of age. The results revealed no statistically significant relationship between $\Sigma PCBs$ in both cord blood or breast milk and cognitive or motor outcome. The authors concluded "PCBs had no discernable effect on neurologic function in this study." However, prenatal exposure to methyl mercury showed an association with neurodevelopmental deficit (Steuervald et al., 2000).

The major difficulty in interpreting the results of these neurodevelopmental studies comes from the fact that a multitude of environmental, social, economic, and genetic factors influence intellectual development. It is important to acknowledge that several studies made an attempt to control for some of the potential confounders. However, adequate adjustment for such factors as parental IQ, smoking, alcohol consumption, and illicit drug use is often difficult to achieve. In addition, co-exposure to lead, methyl mercury, and other potential neurotoxins may also influence the results.

30.7   CHILDHOOD ASTHMA

Sufficient evidence exists that asthma prevalence and mortality have increased dramatically over the last several decades. The increase appears to have started in the early 1970s. While in some countries this increase reached epidemic proportions, other countries experienced significantly lower rates. Worldwide data show that asthma is more frequent in the affluent, market economies than in the developing world (Woolcock and Peat, 1997). For example, prevalence of asthma among children in the United States increased by almost 40 percent between 1981 and 1988 (Weitzman et al., 1992). In Australia, asthma prevalence among 8- to 11-year-olds more than doubled within a decade between 1982 and 1992 (Britton et al., 1992; Peat, 1994). By contrast, asthma in developing countries remains relatively rare and the observed increases are less dramatic (Crater and Platts-Mills, 1998).

The International Study of Asthma and Allergies in Childhood found the highest frequencies of asthma in the United Kingdom, Australia, New Zealand, and the Republic of Ireland, followed by North, Central, and South America. The lowest prevalences were reported in several Eastern European countries, Indonesia, Greece, China, Taiwan, Uzbekistan, India, and Ethiopia [International Study of Asthma and Allergies in Childhood Steering Committee (ISAAC), 1992].

One seemingly plausible explanation of the observed increase in asthma is the degraded quality of air due to industrial pollution. However, data show quite the opposite. Increase in asthma has occurred in places where air quality has improved (Lang and Polansky, 1994). At the same time, asthma prevalence is lower in areas with serious outdoor air quality problems (von Mutius, 2000). Although industrial air pollution is most likely not responsible for the increase in asthma prevalence, it is important to draw a distinction between asthma incidence and severity of existing disease. Studies show that children who already have asthma may experience disease exacerbations triggered by poor air quality (Bates, 1995; Schwartz et al., 1993).

Another hypothesis links the increase in asthma incidence with changes in building practices. Between 1950 and 1975 homes became "tighter" causing an increase in indoor allergens such as dust mites, mold, pet dandruff, and cockroach allergen. Research shows that sensitivity to indoor allergens is predominant among children with asthma (Ronmark et al., 1998; Rosenstreich et al., 1997). Therefore, the increase in asthma could be associated with increased exposure to indoor allergens. However, increase in indoor allergens can hardly explain all of the observed increase in asthma. More likely, the increase in sensitivity to indoor allergens is a reflection of the profound life-style changes, which primarily affected economically developed countries. Because children spend most of their time indoors, they are more likely to become sensitive to indoor allergens, termed "antigens of convenience" (Woolcock and Peat, 1997). In addition to antigen exposure, there appears to be an association between asthma and indoor humidity. Increased dampness indoors frequently occurs in houses with tight insulation, particularly in the presence of water intrusion. Dampness in turn can encourage growth of indoor mold. Exposure to molds may be a significant etiologic factor in allergic respiratory disease including asthma (Holt, 1999). Several studies reported associations between damp housing and asthma (Pearce et al., 1998)

A separate type of indoor air pollution affecting childhood asthma is environmental tobacco smoke (ETS). Children exposed to ETS are at increased risk of asthma, particularly when the mother is a smoker (Etzel, 1995). The evidence is the strongest for increases in severity in children who already have asthma, whereas the evidence for asthma incidence is less conclusive (Pearce et al., 1998).

A change in the amount and type of physical activity associated with a mostly indoor lifestyle is another factor, which deserves serious consideration. It is a well-established fact that strenuous exercise may induce an asthma attack. However, research also shows that moderate physical activity has a protective effect (Clark and Cochrane, 1999). Other proposed causes of the increase in asthma incidence include dietary changes, particularly low intake of magnesium, and increased consumption of polyunsaturated fatty acids. Although protective against cardiovascular disease, polyunsaturated fatty acids, particularly of the omega-3 class, competitively inhibit the formation of certain prostaglandins and leukotrienes, which, along with magnesium, may act as natural bronchodilators (Chang et al., 1993).

Despite several proposed explanations, one is forced to conclude that the research conducted to date has failed to identify the cause of the increase in childhood asthma. It is more likely that we are dealing with complex interactions between several endogenous and exogenous factors, the effects of which are extremely difficult to separate.

30.8   CONCLUSIONS

In 1995, children under age 19 made up approximately 28 percent of the U.S. population. This number is projected to increase to 41 percent by 2050 [Public Health Policy Advisory Board (PHPAB), 1999]. Spectacular improvements in childhood and infant mortality rates have been achieved due to the implementation of effective public health measures such as water chlorination, immunizations, universal growth monitoring, and various other health improvements (Last, 1998).

However, only limited information exists regarding the effects of environmental toxic exposures on children's health. Much of the recent scientific discussion has been focused on the proposed additional uncertainty factor to protect children against potentially toxic exposures. The interest in this topic is understandable due to its far-reaching regulatory implications. Generally, the goal of an uncertainty factor is to account for the potential interspecies and interpersonal variability when translating the no-observed adverse effect level (NOAEL) produced by animal studies into exposure limits for human beings. For example, the acceptable daily intake (ADI) for pesticide residues is the NOAEL of subchronic and chronic exposure studies of laboratory animals divided by a safety factor of 100. This safety factor includes two components: a factor of 10 accounting for differences between animals and humans and a factor of 10 accounting for human variability [World Health Organization (WHO), 1987]. A further subdivision of the two safety factor components (interspecies and human variability) has been proposed by Renwick who suggested that each component represents a combination of toxicokinetic and toxicodynamic factors (Renwick et al., 2000).

An additional component of 10 accounting for the differences between children and adults would increase the overall safety factor to 1000. It is important to point out that the concept of the additional component is not new. This approach has been justified and applied in the past when the toxicity of concern was considered serious and irreversible, such as in cases of teratogenicity (Renwick, 1995). The current proposal, however, calls for an additional 10-fold increase in the safety factor in all circumstances. Thus, the discussion regarding the necessity of an additional uncertainty factor needs to focus on one central question: Are children more sensitive than adults beyond the existing 10-fold factor, which currently accounts for the interpersonal variability within human species (Renwick and Lazarus, 1998)?

The National Academy of Sciences Committee on Pesticides in the Diets of Infants and Children addressed this issue by reviewing both animal and human data. The committee found that functional immaturity of children and neonates may confer either greater resistance or greater susceptibility to toxicity depending upon the chemical and the age of the subject. In situations where newborns and children are, indeed, more susceptible, the differences in their susceptibility levels compared to adults are usually no more than two- to threefold and rarely $\geq$10-fold (Bruckner, 2000). Thus, the use of the current safety factors should be sufficient in most cases, and the use of an additional 10-fold factor for infants and children as proposed under FQPA does not appear justifiable. Moreover, an indiscriminant 10-fold decrease in exposure limits may prove detrimental when applied in situations when a particular chemical has known benefits, such as with some pesticides and certain food additives (Renwick et al., 2000).

It is important to note that while concerns regarding the adverse effects of environmental exposures on children are increasing, the mortality indicators are improving. However, as noted in a recent review of childhood mortality published by the PHPAB (1999), this favorable trend contains several important exceptions. Two causes of death have dramatically increased among children during the past two decades: homicide and suicide (PHPAB, 1999).

The leading causes of death among infants are perinatal conditions such as intrauterine growth retardation, respiratory distress syndrome, and maternal complications of pregnancy and perinatal infections, followed by congenital anomalies, SIDS, and injuries. In the 1 to 4 age group, most deaths occur due to injuries followed by congenital anomalies, malignant neoplasms, and homicide. In the 5 to 14 age group, most deaths occur due to injuries followed by malignant neoplasms, homicide, and congenital anomalies. In children 15 years of age or older, the three most common causes of death are injuries, homicide, and suicide (Behrman, 1999).

According to the PHPAB calculations, approximately 10 percent of all deaths in children are attributable to alcohol, either due to alcohol-related congenital anomalies such as fetal alcohol syndrome or because of injuries or violence secondary to alcohol intoxications. Nine percent of deaths occur due to motor vehicle crashes (PHPAB, 1999). This cause of death is rare in the infant age group, at least partially due to the enforcement of infant car-seat use (CDC, 1991). The third, fourth, and fifth in the ranking are illicit drugs, firearms, and tobacco, respectively. It is important to stress that the true impact of tobacco use does not manifest until much later in adulthood. PHPAB calculations also showed that exposure to environmental and occupational toxins accounted for only about 1 percent of deaths in children ages 0 to 19 (PHPAB, 1999). However, as in the case of tobacco, it is not clear whether such exposures in childhood are an important cause of morbidity and mortality later in life.

Unfortunately, the same level of detail is not available for the leading causes of childhood morbidity. The morbidity data are typically less available and less reliable. It appears, however, that the major trends in childhood morbidity are associated with changes in life-style. Recent pediatric literature indicates an alarming trend toward decreased physical activity and increased prevalence of obesity among American children (Gortmaker et al., 1996; Troiano et al., 1995). These changes may be responsible for the dramatic increase in the prevalence of non-insulin-dependent adult-onset diabetes mellitus in children (Fagot-Campagna et al., 2000) and may contribute to the incidence of arterial hypertension in children and adults (CDC, 1997c). The potential contribution of obesity and lack of physical activity to the increasing prevalence of asthma has been discussed earlier.

These data indicate that while addressing potential concerns regarding children's health, one needs to maintain a broad data-driven public health perspective. Reliance on the best available scientific evidence may help identify the most pressing needs and assist in developing the most effective policies to protect children. Yet, the absence of such information may compromise the ability of parents, legislators, health-care providers, educators, and scientists to make informed decisions.

REFERENCES

Adair-Bischoff, C. E., and Sauve, R. S. 1998. Environmental tobacco smoke and middle-ear disease in preschool-age children. *Arch. Pediatr. Adolesc. Med.* 152: 127–133.

Agency for Toxic Substances and Disease Registry (ATSDR). 1997. *Healthy Children— Toxic Environments: Acting on the Unique Vulnerability of Children Who Dwell Near Hazardous Waste Sites.* U.S. Department of Health and Human Services Public Health Service, Atlanta, GA.

Agency for Toxic Substances and Disease Registry (ATSDR). 1999. *Toxicological Profile for Lead (Update).* Agency for Toxic Substances and Disease Registry, Public Health Service, U.S. Department of Health and Human Services, Atlanta, GA.

Albers, J. W., Cole, P., Greenberg, R. S., Mandel, J. S., Monson, R. R., Ross, J. H., Snodgrass, W. R., Spurgeon, A., and van Gemert, M. 1999. Analysis of chlorpyrifos exposure and human health: Expert panel report. *J. Toxicol. Environ. Health. B Crit. Rev.* 2(4): 301–324.

American Council on Science and Health (ACSH). 1997. Position paper of the American Council on Science and Health: Public health concerns about environmental polychlorinated biphenyls (PCBs). *Ecotoxicol. Environ. Safety* 38(2): 71–84.

American Council on Science and Health (ACSH). 1999. *Tobacco Smoke: Health Risk or Health Hype?* American Council on Science and Health, New York, NY.

Anderson, J. C., Williams, S., McGee, R., and Silva, P. A. 1987. DSM-III disorders in pre-adolescent children. *Arch. Gen. Psych.* 44: 69–76.

Barkin, R. M., and Rosen, P. 1994. *Emergency Pediatrics: A Guide to Ambulatory Care,* 4th ed. Mosby, St. Louis, MO.

Bates, D. V. 1995. The effects of air pollution on children. *Environ. Health Perspect.* 103(Suppl. 6): 49–53.

Bearer, C. F. 1995. How are children different from adults? *Environ. Health Perspect.* 103(Suppl. 6): 7–12.

Behrman, R. E. 1999. Overview of pediatrics. In R. E. Behrman et al. (Eds.), *Nelson Textbook of Pediatrics,* 16th ed. Saunders, Philadelphia.

Boyle, P., et al. 1996. Historical aspects of leukemia clusters. In *Methods of Investigating Localized Clustering of Disease.* IARC Scientific, Lyon, France.

Britton, W. J., Woolcock, A. J., Peat, J. K., Sedgwick, C. J., Lloyd, D. M., and Leeder, S. R. 1992. Prevalence of bronchial hyperresponsiveness in children: The relationship between asthma and skin reactivity to allergens in two communities. *Int. J. Epidemiol.* 15(2): 202–209.

Bruckner, J. V. 2000. Differences in sensitivity of children and adults to chemical toxicity: The NAS panel report. *Regul. Toxicol. Pharmacol.* 31(3): 280–285.

Bunin, G. R., Meadows, A. T., Emanuel, B. S., Buckley, J. D., Woods, W. G., and Hammond, G. D. 1989. Pre- and post-conception factors associated with sporadic heritable and nonheritable retinoblastoma. *Cancer Res.* 49(20): 5730–5735.

Calabrese, E. J. 1986. Age and susceptibility to toxic substances. Wiley, New York.

Caldwell, G. G. 1990. Twenty-two years of cancer cluster investigations at the Centers for Disease Control. *Am. J. Epidemiol.* 132: S43–S47.

Carroquino, M. J., Galson, S. K., Licht, J., Amler, R. W., Perera, F. P., Claxton, L. D., and Landrigan, P. J. 1998. The U.S. EPA Conference on Preventable Causes of Cancer in Children: A research agenda. *Environ. Health Perspect.* 106(Suppl. 3): 867–873.

Centers for Disease Control and Prevention (CDC). 1991. Child passenger restraint use and motor-vehicle—related fatalities among children—United States, 1982–1990. *MMWR* 40(34): 600–602.

Centers for Disease Control and Prevention (CDC). 1997a. Adult blood lead epidemiology and surveillance—United States Fourth Quarter 1996. *MMWR* 46(16): 358–359, 367.

Centers for Disease Control and Prevention (CDC). 1997b. Update: Blood lead levels. *MMWR* 46(7): 141–146.

Centers for Disease Control and Prevention (CDC). 1997c. Update: Prevalence of overweight among children, adolescents, and adults—United States, 1988–1994. *MMWR* 46(9): 198–202.

Chang, C. C., Phinney, S. D., Halpern, G. M., and Gershwin, M. E. 1993. Asthma mortality: Another opinion—is it a matter of life and ... bread? *J. Asthma* 30(2): 93–103.

Chisolm, J. J. Jr. 1962. Aminoaciduria as a manifestation of renal tubular injury in lead intoxication and a comparison with patterns of aminoaciduria seen in other diseases. *J. Pediatr.* 60: 1–17.

Chow, W-H., Linet, M. S., Liff, J. M., and Greenberg, R. S. 1996. Cancers in children. In D. Schottenfeld and J. F. Fraumeni (Eds.), *Cancer Epidemiology and Prevention*, 2nd ed. Oxford University Press, New York, pp. 1331–1369.

Clark, C. J., and Cochrane, L. M. 1999. Physical activity and asthma. *Curr. Opin. Pulm. Med.* 5(1): 68–75.

Clegg, D. J., and van Gemert, M. 1999. Expert panel report of human studies on chlorpyrifos and/or other organophosphate exposures. *J. Toxicol. Environ. Health B Crit. Rev.* 2(3): 257–279.

Crater, S. E., and Platts-Mills, T. A. 1998. Searching for the cause of the increase in asthma. *Curr. Opin. Pediatr.* 10(6): 594–599.

Cutler, J. J., Parker, G. S., Rosen, S., Prenney, B., Healey, R., and Caldwell, G. G. 1986. Childhood leukemia in Woburn, Massachusetts. *Public Health Rep.* 101(2): 201–205.

Dalton, R. F., Forman, M. A., and Muller, B. A. 1992. Infantile autism. In R. E. Behrman et al. (Eds.), *Nelson Textbook of Pediatrics*, 14th ed. Saunders, Philadelphia.

DiFranza, J. R., and Lew, R. A. 1995. Effect of maternal cigarette smoking on pregnancy complications and sudden infant death syndrome. *J. Fam. Pract.* 40: 385–394.

Environmental News Service (ENS). 2000. United Kingdom unmoved by U.S. action on Dursban. Available from *http://ens.lycos.com/ens/jul2000*.

Environmental Protection Agency (EPA). 1986. *Air Quality Criteria for Lead.* EPA, Office of Air Quality Planning and Standards, Research Triangle Park, NC.

Environmental Protection Agency (EPA). 1990. *Report on the National Survey of Lead-Based Paint in Housing: Base Report.* EPA/747/R95/003. EPA, Office of Pollution Prevention and Toxics, Washington, DC.

Environmental Protection Agency (EPA). 1996. *Environmental Health Threats to Children.* EPA/175-F-96-001. EPA, Washington, DC.

Environmental Protection Agency (EPA). 1997a. *Exposure Factors Handbook.* EPA/600/P-95/002Fa. EPA, National Center for Environmental Assessment, Washington, DC.

Environmental Protection Agency (EPA). 1997b. Children's exposure to pesticides. STAR Report, Science to Achieve Results, Research in Progress, Office of Research and Development, vol. 1, issue 1.

Environmental Protection Agency (EPA). 1998a. *Toxicology Data Requirements for Assessing Risks of Pesticide Exposure to Children's Health.* EPA, Washington, DC.

Environmental Protection Agency (EPA). 1998b. Health effects of pesticides. In *The EPA Children's Environmental Health Yearbook.* EPA, Washington, DC, Chapter 5.

Environmental Protection Agency (EPA). 1999a. *Guidelines for Carcinogen Risk Assessment.* Risk Assessment Forum, EPA, Washington, DC.

Eriksen, M. P., LeMaistre, C. A., and Newell, G. R. 1988. Health hazards of passive smoking. *Annu. Rev. Public Health.* 9: 47–70.

Etzel, R. A. 1995. Indoor air pollution and childhood asthma: Effective environmental interventions. *Environ. Health Perspect.* 103(Suppl. 6): 55–58.

Etzel, R. A., Pattishall, E. N., Haley, N. J., Fletcher, R. H., and Henderson, F. W. 1992. Passive smoking and middle-ear effusion among children in day care. *Pediatrics.* 90: 228–232.

Evans, L. S., and Kleiman, M. B. 1986. Acidosis as a presenting feature of chloramphenicol toxicity. *J. Pediatr.* 108: 475.

Evans, W. E., Petros, W. P., Relling, M. V., et al. 1989. Clinical pharmacology of cancer chemotherapy in children. *Pediatr. Clin North Am.* 36(5): 1199–1230.

Factor-Litvaki, P., Kline, J. K., Popovac, D., et al. 1996. Blood lead and blood pressure in young children. *Epidemiology,* 7(6): 633–637.

Fagot-Campagna, A., Pettitt, D. J., Engelgau, M. M., Burrows, N. R., Geiss, L. S., Valdez, R., Beckles, G. L., Saaddine, J., Gregg, E. W., Williamson, D. F., and Narayan, K. M. 2000. Type 2 diabetes among North American children and adolescents: An epidemiologic review and a public health perspective. *J Pediatr.* 136(5): 664–672.

Farland, W. H., Fenner-Crisp, P. A., Guzelian, P. S., et al. 1992. Preface to P. S. Guzelian et al. (Eds.), *Similarities and Differences between Children and Adults: Implications for Risk Assessment.* International Life Sciences Institute, Washington, DC.

Fein, G. G., Jacobson, J. L., Jacobson, S. W., et al. 1984. Prenatal exposure to polychlorinated biphenyls: Effects on birth size and gestation age. *J. Pediatr.* 105: 315.

Fombonne, E. 1999. The epidemiology of autism: A review. *Psychol. Med.* 29(4): 769–786.

Food and Drug Administration (FDA). 1994. Specific Requirements on Content and Format of Labeling for Human Prescription Drugs; Revision of "Pediatric Use" Subsection in the Labeling; Final Rule 21, CFR Part 201, Docket No. 92N-0165, Rockville, MD.

Gladen, B. C., and Rogan, W. J. 1991. Effects of perinatal polychlorinated biphenyls and dichlorodiphenyl dichloroethene on later development. *J. Pediatr.* 119(1, Pt. 1): 58–63.

Glaubiger, D. L., von Hoff, D. D., Holcenberg, J. S. et al. 1981. The relative tolerance of children and adults to anticancer drugs. *Front Radiat. Ther. Oncol.* 16: 42–49.

Gortmaker, S. L., Must, A., Sobol, A. M., Peterson, K., Colditz, G. A., and Dietz, W. H. 1996. Television viewing as a cause of increasing obesity among children in the United States, 1986–1990. *Arch. Pediatr. Adolesc. Med.* 150(4): 356–362.

Goyer, R. A. 1996. Toxic effects of metals. In *Casarett and Doull's Toxicology: The Basic Science of Poisons,* 5th ed. McGraw-Hill, New York.

Graham, S., Levin, M. L., Lilienfeld, A. M., Schuman, L. M., Gibson, R., Dowd J. E., and Hempelmann, L. 1966. Preconception, intrauterine, and postnatal irradiation as related to leukemia. *Natl. Cancer Inst. Monogr.* 19: 347–371.

Greaves, S. J., Ferry, D. G., McQueen, E. G., Malcolm, D. S., and Buckfield, P. M. 1975. Serial hexachlorophene blood levels in the premature infant. *N. Z. Med. J.* 81(537): 334–336.

Guillette, E. A., Meza, M. M., Aquilar, M. G., Soto, A. D., and Garcia, I. E. 1998. An anthropological approach to the evaluation of preschool children exposed to pesticides in Mexico. *Environ. Health Perspect.* 106(6): 347–353

Gurney, J. G., Davis, S., Severson, R. K., Fang, J-Y., Ross, J. A., and Robison, L. L. 1996. Trends in cancer incidence among children in the U.S. *Cancer* 78: 532–541.

Haglund, B., and Cnattingius, S. 1990. Cigarette smoking as a risk factor for sudden infant death syndrome: A population-based study. *Am. J. Public Health* 80: 29–32.

Holt, P. G. 1999. Potential role of environmental factors in the etiology and pathogenesis of atopy: A working model. *Environ. Health Perspect.* 107(Suppl. 3): 485–487.

Huisman, M., Koopman-Esseboom, C., Fidler, V., Hadders-Algra, M., van der Paauw, C. G., Tuinstra, L. G. M. Th., Weisglas-Kuperus, N., Sauer, P. J. J., Touwen, B. C. L., and Boersma, E. R. 1995. Perinatal exposure to polychlorinated biphenyls and dioxins and its effect on neonatal neurological development. *Early Hum. Devel.* 41: 111–127.

International Study of Asthma and Allergies in Childhood Steering Committee (ISAAC). 1992. Worldwide variation in prevalence of symptoms of asthma, allergic rhinoconjunctivitis, and atopic eczema. ISAAC. *Lancet* 351(9111): 1225–1232.

Iversen, M., Birch, L., Lundquist, G. R., and Elbrond, O. 1985. Middle-ear effusion in children and the indoor environment: An epidemiological study. *Arch. Environ. Health.* 40: 74–79.

Jacobson, J. L., and Jacobson, S. W. 1996. Intellectual impairment in children exposed to polychlorinated biphenyls in utero. *N. Engl. J. Med.* 335: 783–789.

Jacobson, J. L., and Jacobson, S. W. 1997. Evidence for PCBs as neurodevelopmental toxicants in humans. *Neurotoxicology* 18: 415–424.

Jacobson, J. L., Schwartz, P. M., Fein, G. G., and Dowler, J. K. 1984. Prenatal exposure to an environmental toxin: A test of the multiple effects model. *Devel. Psychobiol.* 20: 523–532.

Jacobson, J. L., Jacobson, S. W., and Humphrey, H. E. B. 1990. Effects of exposure to PCBs and related compounds on growth and activity in children. *Neurotoxicol. Teratol.* 12: 319–326.

Jacobson, J. L., Jacobson, S. W., Padgett, R. J., Brumitt, G. A., and Billings, R. L. 1992. Effects of prenatal PCB exposure on cognitive processing efficiency and sustained attention. *Devel. Psychol.* 28: 297–306.

Jacobson, S. W., Fein, G. G., Jacobson, J. L., Schwartz, P. M., and Dowler, J. K. 1985. The effect of intrauterine PCB exposure on visual recognition memory. *Child Devel.* 56: 853–860.

Juberg, D. R., Kleiman, C. F., and Kwon, S. C. 1997. *Lead and Human Health.* American Council on Science and Health, New York.

Kacew, S., and Lock, S. 1988. *Toxicologic and Pharmacologic Principles in Pediatrics.* Hemisphere Publishing Corp. (HPC), Philadelphia, PA.

Kauffman, R. E. 1992. Acute acetaminophen overdose: An example of reduced toxicity related to developmental differences in drug metabolism. In P. S. Guzelian et al. (Eds.), *Similarities and Differences between Children and Adults: Implications for Risk Assessment.* International Life Sciences Institute, Washington, DC.

Kingston, R. L., Chen, W. L., Borron, S. W., Sioris, L. J., Harris, C. R., and Engebretsen, K. M. 1994. Chlorpyrifos: A ten-year US poison center exposure experience. *Vet. Hum. Toxicol.* 41(2): 87–92.

Koopman-Esseboom, C., Weisglas-Kuperus, N., deRidder, M. A. J., Van der Paauw, C. G., Tuinstra, L. G. M. Th., and Sauer, P. J. J. 1996. Effects of polychlorinated biphenyl/dioxin exposure and feeding type on infants' mental and psychomotor development. *Pediatrics* 97: 700–706.

Krasnegor, N. A., Otto, D. A., Bernstein, J. H., Burke, R., Chappell, W., Eckerman, D. A., Needleman, H. L., Oakley, G., Rogan, W., and Terracciano, G. 1994. Neurobehavioral test strategies for environmental exposures in pediatric populations. *Neurotoxicology* 16(5): 499–509.

Kuzmack, A. M. 1987. Comment on Lagakos, Wessen and Zelen. *J. Am. Stat. Assoc.* 82: 73.

Lagakos, S. W., Wessen, B. J., and Zelen, M. 1986. An analysis of contaminated well water and health effects in Woburn, Massachusetts. *J. Am. Stat. Assoc.* 81: 583–596.

Lang, A. M., Polansky, M. 1994. Patterns of asthma mortality in Philadelphia from 1969 to 1991. *N. Engl. J. Med.* 331(23): 1542–1546.

Last, J. M. 1998. Ethics and public health policy. In R. B. Wallace, B. N. Doebbeling, and J. M. Last (Eds.), *Maxcy-Rosenau-Last Public Health and Preventive Medicine*, 14th ed. Appleton & Lange, Stamford, CT, pp. 35–43.

Last, J. M., and Abramson, J. H. 1988. *A Dictionary of Epidemiology.* International Epidemiological Association, Oxford University Press, New York.

Liberson, G. L., Golden, R. J., Blot, W. J., Fisch, H., and Watson, C. 2000. An examination of sensitivity of reported trends in childhood leukemia incidence rates to geographic location and diagnostic coding (United States). *Cancer Causes Contr.* 11: 413–417.

Linet, M. S., Ries, L. A. G., Smith, M. A., Tarone, R. E., and Devesa, S. S. 1999. Cancer surveillance series: Recent trends in childhood cancer incidence and mortality in the United States. *J. Natl. Cancer Inst.* 91: 1051–1058.

Litovitz, T. L., Smilkstein, M., Felberg, L., et al. 1997. 1996 annual report of the American Association of Poison Control Centers Toxic Exposure Surveillance System. *Am. J. Emerg. Med.* 15(5): 447–500.

Little, J. 1999. *Epidemiology of Childhood Cancer.* IARC, Lyon, France.

Malloy, M. H., Kleinman, J. C., Land, G. H., and Schramm, W. F. 1988. The association of maternal smoking with age and cause of infant death. *Am. J. Epidemiol.* 128: 46–55.

Margolis, P. A., Keyes, L. L., Greenberg, R. A., Bauman, K. E., and LaVange, L. M. 1997. Urinary cotinine and parent history (questionnaire) as indicators of passive smoking and predictors of lower respiratory illness in infants. *Pediatr. Pulmonol.* 23(6): 417–423.

Marsoni, S., Ungerleider, R. S., Hurson, S. B., et al. 1985. Tolerance to antineoplastic agents in children and adults. *Cancer Treat Rep.* 69(11): 1263–1269.

Matarazzo, J. D. 1990. Psychological assessment versus psychological testing. Validation from Binet to the school, clinic, and courtroom. *Am. Psychol.* 45(9): 999–1017.

Miller, R. P., Roberts, R. J., and Fischer, L. J. 1976. Acetaminophen elimination kinetics in neonates, children, and adults. *Clin. Pharmacol. Ther.* 19(3): 284–294.

Mott, L., Fore, D., Curtis, J., and Solomon, G. 1997. *Our Children at Risk: the 5 Worst Environmental Threats to Their Health.* Natural Resources Defense Council, New York.

Nafstad, P., Kongerud, J., Botten, G., Hagen, J. A., and Jaakkola, J. J. 1997. The role of passive smoking in the development of bronchial obstruction during the first 2 years of life. *Epidemiology* 8(3): 293–297.

National Academy of Sciences (NAS). 1972. *Lead: Airborne Lead in Perspective: Biologic Effects of Atmospheric Pollutants.* National Academy of Sciences, Washington, DC, pp. 71–177, 281–313.

National Cancer Institute (NCI). 1996. *Cancer Rates and Risks,* 4th ed. Cancer Statistics Branch, Division of Cancer Prevention and Control, National Cancer Institute, U.S. Department of Health and Human Services, Public Health Service, National Institutes of Health.

National Conference on Clustering of Health Events (NCCHE). 1990. *Am. J. Epidemiol.* 132(1, Suppl.): S1–S202.

National Research Council (NRC). 1993. *Pesticides in the Diets of Infants and Children.* National Academy Press, Washington, DC.

National Research Council (NRC). 2000. *Scientific Frontiers in Developmental Toxicology and Risk Assessment.* Committee on Developmental Toxicology, Board on Environmental Studies and Toxicology, National Research Council, National Academy Press, Washington, DC.

Nebert, D. W., and Gonzalez, F. J. 1987. P450 genes: Structure, evolution, and regulation. *Annu. Rev. Biochem.* 56: 945–993.

Needleman, H. L. 1983. Lead at low dose and the behavior of children. *Neurotoxicology* 4: 121.

Needleman, R. D. 1999. Growth and development. In R. E. Behrman et al. (Eds.), *Nelson Textbook of Pediatrics,* 16th ed. Saunders, Philadelphia.

Nelson, K., and Holmes, L. B. 1989. Malformations due to presumed spontaneous mutations in newborn infants. *N. Engl. J. Med.* 320(1): 19–23.

Neuro-Immune Dysfunction Syndromes Research Institute (NIDS). 1999. Clinical hypothesis—immune "dysfunction/dysregulation"—a reason for childhood neurocognitive dysfunction. Available from *http://nids.net.*

Nriagu, J. O., Kim, M. J. 2000. Emissions of lead and zinc from candles with metal-core wicks. *Sci. Total Environ.* 250(1–3): 37–41.

Olsen, S. F., et al. 1996. Cluster analysis and disease mapping—why, when, and how? A step by step guide. *BMJ* 313: 863–866.

Olshan, A. F., and Daniels, J. L. 2000. Invited commentary: Pesticides and childhood cancer. *Am. J. Epidemiol.* 151(7): 647–649.

Patandin, S., Lanting, C. I., Mulder, P. G. H., Boersma, E. R., Sauer, P. J. J., and Weisglas-Kuperus, N. 1999. Effects of environmental exposure to polychlorinated biphenyls and dioxins on cognitive abilities in Dutch children at 42 months of age. *J. Pediatr.* 134: 33–41.

Pearce, N., Beasley, R., Burgess, C., and Crane, J. 1998. *Asthma Epidemiology: Principles and Methods.* Oxford University Press, New York.

Peat, J. K. 1994. The rising trend in allergic illness: Which environmental factors are important? *Clin. Exp. Allergy* 24(9): 797–800.

Petrini, J., Damus, K., and Johnston, R. B. Jr. 1997. An overview of infant mortality and birth defects in the United States. *Teratology* 56(1–2): 8–10.

Piomelli, S., Rosen, J. F., Chisolm, J. J. Jr., and Graef, J. W. 1984. Management of childhood lead poisoning. *J. Pediatr.* 105: 523.

Pirkle, J. L., Brody, D. J., Gunter, E. W., Kramer, R. A., Paschal, D. C., Flegal, K. M., and Matte, T. D. 1994. The decline in blood lead levels in the United States. The National Health and Nutrition Examination Surveys (NHANES). *JAMA* 272: 284–291.

Plunkett, L. M., Turnbull, D., and Rodricks, J. V. 1992. Differences between adults and children affecting exposure assessment. In P. S. Guzelian et al. (Eds.), *Similarities and Differences between Children and Adults: Implications for Risk Assessment.* International Life Sciences Institute, Washington, DC.

Public Health Policy Advisory Board (PHPAB). 1999. Health and the American child. Part 1: A focus on mortality among children. A report to the Nation from the Public Health Policy Advisory Board (PHPAB), Washington, DC.

Pueschel, S. M., Kopito, L., and Schwachman, H. 1972. Children with an increased lead burden: A screening and follow-up study. *JAMA* 222: 462–466.

Reigart, J. R., and Roberts, J. W. 1999. *Recognition and Management of Pesticide Poisonings.* U.S. Environmental Protection Agency, Washington, DC.

Renwick, A. G. 1995. The use of an additional safety or uncertainty factor for nature of toxicity in the estimation of acceptable daily intake and tolerable daily intake values. *Regul. Toxicol. Pharmacol.* 22: 250–261.

Renwick, A. G., Dorne, J. L., and Walton, K. 2000. An analysis of the need for an additional uncertainty factor for infants and children. *Regul. Toxicol. Pharmacol.* 31: 286–296.

Renwick, A. G., and Lazarus, N. R. 1998. Human variability and non-cancer risk assessment—An analysis of the default uncertainty factor. *Regul. Toxicol. Pharmacol.* 27: 3–20.

Robbins, W. J., Brody, S., Hogan, A. G., et al. 1928. *Growth.* Yale University Press, New Haven, CT.

Roberts, J. W., and Dickey, P. 1995. Exposure of children to pollutants in house dust and indoor air. *Rev. Environ. Contam. Toxicol.* 143: 59–78.

Robison, L. L., Nesbit, M. E., Sather, H. N., Level, C., Shahidi, N., Kennedy, A., and Hammond, D. 1984. Down syndrome and acute leukemia in children: A 10-year retrospective survey from Childrens Cancer Study Group. *J. Pediatr.* 105(2): 235–242.

Robison, L. M., Sclar, D. A., Skaer, T. L., and Galin, R. S. 1999. National trends in the prevalence of attention-deficit/hyperactivity disorder and the prescribing of methylphenidate among school-age children: 1990–1995. *Clin. Pediatr. Phila.* 38(4): 209–217.

Rodgers, G. C., Jr., and Matyunas, N. J. 2000. Poisonings: Drugs, chemicals and plants. In R. E. Behrman et al. (Eds.), *Nelson Textbook of Pediatrics*, 16th ed. Saunders, Philadelphia.

Rogan, W. J., and Gladen, B. C. 1991. PCBs, DDE, and child development at 18 and 24 months. *Ann. Epidemiol.* 1: 407–413.

Ronmark, E., Lundback, B., Jonsson, E., and Platts-Mills, T. 1998. Asthma, type-1 allergy and related conditions in 7- and 8-year-old children in northern Sweden: Prevalence rates and risk factor pattern. *Respir. Med.* 92(2): 316–324.

Rosenstreich, D. L., Eggleston, P., Kattan, M., Baker, D., Slavin, R. G., Gergen, P., Mitchell, H., McNiff-Mortimer, K., Lynn, H., Ownby, D., and Malveaux, F. 1997. The role of cockroach allergy and exposure to cockroach allergen in causing morbidity among inner-city children with asthma. *N. Engl. J. Med.* 336(19): 1356–1363.

Ruff, M. A., Mavkowitz, M. E., Bijur, P. E., et al. 1996. Relationships among blood lead levels, iron deficiency, and cognitive development in two-year-old children. *Environ. Health Perspect.* 104(2): 180–185.

Scahill, L., and Schwab-Stone, M. 2000. Epidemiology of ADHD in school-age children. *Child Adolesc. Psychiatr. Clin. N. Am.* 9(3): 541–555, vii.

Schmidt, C. W. 1999. Poisoning young minds. *Environ. Health Perspect.* 107(6): A302–A307.

Schwartz, J., Slater, D., Larson, T. V., Pierson, W. E., and Koenig, J. Q. 1993. Particulate air pollution and hospital emergency room visits for asthma in Seattle. *Am. Rev. Respir. Dis.* 147(4): 826–831.

Schwartz, P. M., Jacobson, S. W., Fein, G., Jacobson, J. L., Price, H. A. 1983. Lake Michigan fish consumption as a source of PLBs in human cord serum, maternal serum, and milk. *Am. J. Public Health* 73(3): 293–296.

Siberry, G. K., and Iannone, R. 2000. *The Harriet Lane Handbook.* Mosby, Baltimore, MD.

Solomon, G. M., and Mott, L. 1998. *Trouble on the Farm—Growing Up with Pesticides in Agricultural Communities.* Natural Resources Defense Council (NRDC), Washington, DC.

Steuerwald, U., Weihe, P., Jorgensen, P. J., Bjerve, K., Brock, J., Heinzow, B., Budtz-Jorgensen, E., and Grandjean, P. 2000. Maternal seafood diet, methylmercury exposure, and neonatal neurologic function. *J. Pediatr.* 136(5): 599–605.

Strachan, D. P., and Cook, D. G. 1997. Health effects of passive smoking. 1. Parental smoking and lower respiratory illness in infancy and early childhood. *Thorax* 52(10): 905–914.

Troiano, R. P., Flegal, K. M., Kuczmarski, R. J., Campbell, S. M., and Johnson, C. L. 1995. Overweight prevalence and trends for children and adolescents. The National Health and Nutrition Examination Surveys, 1963 to 1991. *Arch. Pediatr. Adolesc. Med.* 149(10): 1085–1091.

von Mutius, E. 2000. The environmental predictors of allergic disease. *J. Allergy Clin. Immunol.* 105(1, Pt. 1): 9–19.

Weitzman, M., Gortmaker, S. L., Sobol, A. M., and Perrin, J. M. 1992. Recent trends in the prevalence and severity of childhood asthma. *JAMA* 268(19): 2673–2677.

Williams, A. N. 1998. Cancer clusters: What role for epidemiology? *Minnesota Med.* 81: 14–17.

Wilson, J. G. 1973. *Environment and Birth Defects.* Academic, New York.

Winneke, G., and Kramer, U. 1997. Neurobehavioral aspects of lead neurotoxicity in children. *Cent. Eur. J. Public Health* 5(2): 65–69.

Woolcock, A. J., and Peat, J. K. 1997. Evidence for the increase in asthma worldwide. In *The Rising Trends of Asthma,* edited by the CIBA foundation. Wiley, New York.

World Health Organization (WHO). 1987. *Principles for the Safety Assessment of Food Additives and Contaminants in Food.* WHO Press, Geneva, Switzerland.

31 Cost–Benefit Analysis

JOHN D. GRAHAM and TIMOTHY J. CARROTHERS

Center for Risk Analysis, Harvard School of Public Health, Boston, Massachusetts

31.1 INTRODUCTION TO COST–BENEFIT ANALYSIS

Cost–benefit analysis is rooted in utilitarian political philosophy and seeks to determine whether a proposed regulation is *efficient*. Efficiency is defined as follows: If those citizens who benefit from a regulation had to pay the entire cost of a regulation, they would nonetheless consider the regulation to be worthwhile (Mishan, 1988). Thus, benefits are defined using "willingness to pay" methods and costs are defined using "willingness to accept" (compensation) methods.

Note that cost–benefit analysis says nothing about who should be forced to pay the costs of a regulation, since that is a *distributional* matter that is considered the province of a politician. The attractive feature of an efficient regulation is that it is possible, at least in theory, to have the winners compensate the losers from the regulation and leave no one worse off and at least some people better off. Even if this compensation is not provided to losers on a regulation-by-regulation basis (an infeasible proposition), it is likely that losers with regard to one regulation will become winners on another regulation. If only efficient regulations are adopted, the average citizen will become better off than if some inefficient regulations are adopted (Leonard and Zeckhauser, 1986).

A key ethical limitation of efficiency is that it does not assure that certain subgroups in society will not be forced to be consistent losers from most or all regulations. These subgroups might be difficult to identify in advance, but they may prove to be the least advantaged members of society such as the poor, the racially oppressed, or members of future generations whose preferences about a current regulation are not yet known and thus cannot be counted. It should also be noted that *efficiency* values nature and ecosystems from a human perspective and thus does not assign any intrinsic rights or value to nonhuman species.

Human and Ecological Risk Assessment: Theory and Practice, Edited by Dennis J. Paustenbach
ISBN 0-471-14747-8 © 2002 John Wiley & Sons, Inc.

These ethical limitations of economic efficiency suggest that the findings of cost–benefit analysis should inform rather than dictate public decisions (Arrow et al., 1996). There will be cases where nonefficiency considerations (notions of justice, rights, and fairness) "trump" the efficiency considerations contained in a cost–benefit analysis. Public debate continues to rage over how trade-offs between efficiency and equity should be resolved (Okun, 1975).

The scope of a cost–benefit analysis is also important. Experience has shown that efforts to reduce *the target risk* can unintentionally increase *countervailing risks*. Thus, a shortsighted analysis may appear efficient, when in truth it is far from so. A proper analysis must carefully explore the possibility of *risk trade-offs* and make changes in its scope to account for reasonably foreseeable *disbenefits* (Graham and Wiener, 1995). In some cases, unforeseen *decreases* in other risks may occur. These risks have been termed *coincident risks*.

As risk-based regulations have grown in number, cost, and complexity, policymakers have expressed increasing interest in obtaining objective information about whether the benefits of these regulations justify their costs. The risk assessment process plays a critical role in any cost–benefit evaluation because the findings of a risk assessment are a necessary input to the benefit side of the ledger (and may also influence the cost side of the ledger if a regulation induces unintended risks to health, safety, and the environment).

This chapter provides the reader a concrete example of how the results of human health risk assessment are used in cost–benefit analysis. For illustrative purposes, a recent U.S. Environmental Protection Agency (EPA) regulation aimed at reducing pollution from passenger cars and light trucks is examined. The regulation is an integrated effort to reduce pollution by changing the sulfur content of fuels and improving the emission control equipment in motor vehicle engines. The ultimate goal is to reduce pollutants that cause smog and particulate matter. EPA's analysis concludes that the estimated benefits of the proposed *integrated* rule ($13.8 to $25.2 billion per year) are likely to exceed the estimated costs of the rule ($5.3 billion per year).

The chapter is organized around the key principles of cost–benefit analysis: societal perspective, analytic time horizons, baseline estimates of pollution-induced risks, decision options, opportunity cost estimation, valuation of human health improvements, discounting of future benefits and costs to their present value, uncertainty analysis, and intangible and nonquantifiable effects. The chapter describes and critiques how EPA analysts handled these issues, and it concludes with some suggestions about how risk assessors could be more helpful to cost–benefit analysts in the future.

31.2 BACKGROUND OF CASE STUDY

Finalized on February 10, 2000, the "Tier 2" motor vehicle emissions regulations are summarized by the following excerpt from the *Federal Register* (EPA, 2000) notice:

Today's action finalizes a major program designed to significantly reduce the emissions from new passenger cars and light trucks, including pickup trucks, vans, minivans, and sport-utility vehicles. These reductions will provide for cleaner air and greater public health protection, primarily by reducing ozone and PM pollution. The program is a comprehensive regulatory initiative that treats vehicles and fuels as a system, combining requirements for much cleaner vehicles with requirements for much lower levels of sulfur in gasoline.

. . .

For cars, light trucks, and larger passenger vehicles, the program will—

• Starting in 2004, through a phase-in, apply for the first time the same set of emission standards ["Tier 2 Standards"] covering passenger cars, light trucks, and large SUVs and passenger vehicles.

. . .

• Apply the same standards to vehicles operated on any fuel.

. . .

• Set more stringent particulate matter standards.
• Set more stringent evaporative emission standards.

For commercial gasoline, the program will—

• Significantly reduce average gasoline sulfur levels nationwide as early as 2000, full phased in in 2006. Refiners will generally add refining equipment to remove sulfur in their refining processes.

. . .

• Enable the new Tier 2 vehicles to meet the emission standards by greatly reducing the degradation of vehicle emission control performance from sulfur in gasoline. Lower sulfur gasoline also appears to be necessary for the introduction of advanced technologies that promise higher fuel economy but are very susceptible to sulfur poisoning (for example, gasoline direct injection engines).

The results of the accompanying cost–benefit analysis are summarized in the following excerpt from the Executive Summary of the Regulatory Impact Analysis (EPA, 1999):

We also made an assessment of the monetary value of the health and general welfare benefits that are expected to result from our standards near full implementation in 2030. We estimate that our Tier 2/gasoline sulfur standards would, in the long term, result in substantial benefits, such as: the yearly avoidance of approximately 4300 premature deaths, approximately 2300 cases of bronchitis, and significant numbers of hospital visits, lost work days, and multiple respiratory ailments (especially those that effect children). Our standards will also produce welfare benefits relating to agricultural crop damage, visibility, and nitrogen deposition in rivers and lakes. Total monetized benefits, however, are driven primarily by the value placed on the reductions in premature deaths.

. . .

The results indicate that using EPAs preferred approach to valuing reductions in premature mortality, total monetary benefits realized after nearly a full turnover of the fleet to Tier 2 vehicles would be approximately $25.2 billion in 2030. . . . Comparing this estimate of the economic benefits with the adjusted cost estimate indicates that the net economic benefit of the Tier 2/gasoline sulfur standards to society are approximately $20 billion in 2030.

31.3 SOCIETAL PERSPECTIVE

One of the first steps in performing an analysis is choosing the perspective of the analysis. If a cost–benefit analysis were done from the perspective of a particular company, the only costs and benefits that would count would be those that influenced the company's profit–loss statement. Yet many health and ecological effects of pollution have no discernible impact on the polluter's profit–loss statement. The *societal perspective* used in cost–benefit analysis is aimed at correcting this problem. In a societal analysis, all adverse and beneficial effects of a regulation (monetary or nonmonetary) are to be counted, regardless of who incurs them. Even effects on future generations and nonhuman species are to be considered, though they are typically valued from the perspective of the current (living) generation of citizens.

In EPA's analysis, the societal perspective was employed on both the benefit and cost side of the ledger. Compliance costs in both the refining and vehicle manufacturing sectors of the economy were counted. On the benefit side, various health benefits to children, middle-aged people, and elderly people were counted, as were benefits of improved recreational visibility and benefits to commercial agriculture.

31.4 ANALYTIC TIME HORIZON

When analyzing a new technology, it is customary to assess all costs and benefits that accrue over the lifetime of the technology (e.g., 10 years for a new passenger car). Yet a longer analytic time horizon may be required when human health and environmental effects are of concern. While a new pollution control device may operate effectively for only 10 years, the benefits of those 10 years of operation may include reduced rates of chronic illness and premature death that will be realized well after the equipment has been retired. For instance, exposures to air pollution as a child may lead to development of chronic bronchitis at an older age. Likewise, it may take many decades to undo the acidification of soil from acid rain caused by sulfur dioxide emissions. Thus, reductions today can lead to benefits far into the future.

The EPA chose to present its analysis as a single year "snapshot" of the yearly benefits and costs expected to be realized once the Tier 2 standards have been fully implemented in 2030. Implementation is defined as the entire fleet of light-duty vehicles completely "turning over," meeting the new standards. The costs used in the estimate are not the actual costs expected in 2030, but instead are adjusted to reflect a long-term average. The EPA states that this lends a more accurate comparison of costs and benefits, as many of the costs may occur "up-front" in implementation. An alternative to the EPA's presentation would be to present a net present value analysis, where the streams of costs and benefits for an analytic time horizon of, say, 75 years are discounted back to the present year using a social discount rate.

With respect to the benefits, the EPA does not adjust for the time lag mentioned in the previous paragraph. Instead, it assumes that all benefits accrue instantaneously. This assumption is not valid for all of the benefit categories. By not discounting for the time "lag" between the reduction in pollution and the health effects reduction associated with it, the EPA introduces an upward bias to its benefit estimate. Because reductions in premature mortality, which may have a latency period anywhere from several years to several decades, comprise over 90 percent of the benefits under EPA's preferred methodology, this bias may be significant. In the face of the large uncertainty regarding the length of the lag, it is disconcerting that EPA fails to present neither a sensitivity analysis nor a valid central estimate of the time lag.

31.5 BASELINE ESTIMATES OF RISK

As a first step toward benefit measurement, it is necessary to quantify the magnitude of health risks under a "do nothing" or "baseline" scenario. One can think of this baseline scenario as the "comparator" or control group that is used to evaluate an intervention of interest.

The EPA performed this task by projecting annualized health risks in the year 2030 from nationwide vehicular emissions in the absence of the Tier 2 rule. Baseline health risks were computed by constructing and linking three types of mathematical models: emissions models, air quality models, and concentration–disease (dose–response) models. Separate modeling was performed for ozone-related health risks and particle-related health risks. Since health-related effects comprised over 90 percent of the total benefits, discussion of visibility and agriculture benefits is omitted here for the sake of brevity.

The EPA's dose–response modeling covered numerous health outcomes, but three assumptions became pivotal to the overall benefit estimate. First, EPA assumed that long-term particulate exposure increases the relative risk of death across the entire population over the age of 30. The magnitude of this chronic mortality effect and the population modeled are based on an epidemiological study by Pope et al. (1995). Second, EPA assumed that long-term exposure to particulate matter increases the risk of chronic bronchitis, an association reported in the literature by Schwartz (1993). Third, EPA assumed that short-term exposure to ozone and particulate matter causes an immediate increase in restricted activity days among the general population, an association reported by Ostro and Rothschild (1989). A detailed discussion of these dose–response assumptions is beyond the scope of this chapter, but it should be noted that the design of the Pope study has led to various concerns regarding residual confounding (Gamble, 1998), and the latter two benefit categories were based on singular results not observed in other published literature.

31.6   DECISION OPTIONS

In a cost–benefit analysis, the analyst compares the baseline, or "business as usual," scenario to one or more alternative decision options. For a decision option to be well defined, it must be feasible to quantify its costs and health risks compared to the baseline option and any other decision options under consideration. Some simple analyses involve a comparison of the baseline scenario to one decision alternative. In more complex analyses, multiple alternatives are compared. When multiple options are compared, it is not sufficient to compare each decision option to the baseline scenario. The analyst must go further and compare the incremental benefits and costs of each option. An option might look attractive when compared to the baseline scenario but have incremental benefits and costs (relative to the next best alternative) that are unattractive.

In EPA's analysis of the proposed integrated rule, the integrated rule was compared to the baseline emission scenario. Another option that was discussed but not analyzed incrementally was a rule that would have required low-sulfur gasoline only in those regions of the country that were in violation of the national ambient air quality standards for ozone and particulate matter ("non-attainment areas"). This targeted rule would have reduced refining costs yet retained the reductions in health risks that were projected to occur in the most heavily polluted regions of the country. EPA did estimate that significant benefits would occur from reducing smog and particulate matter in attainment areas, but these benefits were never compared rigorously to the incremental costs of a national low-sulfur fuel rule.

31.7   OPPORTUNITY COST ESTIMATION

The resources that are consumed by regulatory compliance activities are resources that cannot be dedicated to other productive uses in the economy. The word "resources" refers not to money but to the scarce labor and material inputs that are expended in efforts to comply with a regulation. For example, labor and material resources are expended when a refining process is modified to produce low-sulfur gasoline. The value of these consumed resources in their next best alternative use is the definition of *opportunity cost*.

How do analysts figure out where in the economy these resources would have been used, and how do analysts determine the value of these resources in their alternative uses? Instead of answering these questions on a case-by-case basis, analysts use principles from microeconomics that characterize the operation of a competitive market economy. Under the assumptions of a "perfect" market, the observed market price of any labor or material input is equal to the marginal social value of that input in alternative uses, since suppliers of that input are demanding payment of at least that price for their product. If these suppliers had no alternative uses for the input, the price they could demand would be smaller. Inputs with numerous, high-valued applications in the economy will command a very high price. If the market economy is known to be defective (e.g., due to imperfections such as monopoly or externalities in production of the input), then the observed market price for an input must be replaced by a constructed *shadow price* that accounts for the market imperfection (Mishan, 1988).

The EPA's analysis of the integrated rule employs a traditional microeconomic approach to costing, except for the costing of new refining technologies that are expected to reduce the sulfur content of gasoline with little loss in octane. In costing these innovative technologies, EPA applies a learning factor that reduces the long-term opportunity cost of these technologies compared to what might be predicted by their current price in the market. The learning factor is based on the observation that the costs of previous environmental controls have been consistently overestimated.

31.8 VALUATION OF IMPROVEMENTS IN HUMAN HEALTH

To compare costs (expressed in dollar value) to reductions in human health effects (morbidity or mortality), a common metric must be employed. Although some efforts have been made to convert economic costs into health units, the typical approach is to convert health effects into dollar equivalents. The analytic process of translating human health effects into economic units is called valuation.

In their analysis of the proposed integrated rule, EPA analysts had no direct scientific basis for computing the dollar value of reductions in morbidity and mortality risk induced by pollution. Instead they used the concept of *benefit transfer*, which entails extrapolation of monetized values of health risk from one application in the literature (e.g., workplace safety) to the risks reduced by the integrated rule.

The EPA's range of overall benefit from the rule ($3.2 to $19.5 billion) is based on numerous valuation steps, but two assumptions are particularly important: (1) Each case of premature death from long-term particulate exposure should be valued at $2.73 to $5.89 million, and (2) each case of chronic bronchitis induced by particle pollution should be valued at $74,500 to $319,280. Although each of these assumptions could be discussed in detail, discussion of the first assumption illustrates the uncertainties in benefit transfer methods.

Use of the $5.89 million figure is based on the following benefit transfer assumption: Preventing a statistical fatality from chronic particle exposure in the community is of equal value to preventing a statistical fatality in the workplace from accidents. The $2.73 million figure reflects a downward adjustment for the fact that the average workplace fatality (at age 35) is associated with a larger loss of life expectancy than the average particle-induced fatality from cardio-pulmonary disease. Although EPA uses these two figures to define a "low-end" and "high-end" estimate of benefit, the underlying uncertainty in these figures is greater than this range reflects. The figures are possibly too small if one accounts for the fact that community residents may have stronger health and safety preferences than do workers who have selected and retained blue-collar occupations that involve hazards on the job (Viscusi, 1993). The numbers are possibly too large if EPA analysts have failed to adequately account for the diminished duration and quality of life that is associated with the last third of the human life span (i.e., ages 60 to 90), the time period likely to be impacted most strongly by pollution exposures.

What is needed are validated tools that can be used to generate a substantial body of empirical data on how much money citizens are willing to pay to reduce the risks of pollution. One of the important limitations of existing studies is that respondents to surveys involving willingness to pay do not appear to be appropriately sensitive to the magnitude of risk reduction (Fredrick and Fischoff, 1998; Hammitt and Graham, 1999). A new collaborative program between the National Science Foundation and the EPA is supporting new research projects aimed at addressing these needs. Until such research is completed, the process of valuing health risk reductions will remain highly uncertain.

31.9   DISCOUNTING FUTURE BENEFITS AND COSTS

It is customary to discount future benefits and costs at a rate equivalent to the long-term, inflation-adjusted rate of interest in the economy. Once discounting is performed, the present value of benefits is then compared to the present value of costs. (Alternatively, annualized costs can be compared to annualized benefits for a program with a specified lifetime). Although investment principles provide a clear rationale for discounting future economic consequences to their present value, it is less obvious why future health benefits should be discounted to their present value.

Perhaps the most compelling rationale for discounting future health benefits and costs at the same rate was provided by Keeler and Cretin (1983) of the RAND Corporation. Imagine a program that incurs costs today to avert health impairments in future years. If the health benefits are discounted at a lower rate than economic costs, then it can be shown that society will always be made better off by delaying adoption of the program. This perverse result arises because delaying the program reduces costs by a larger factor than it reduces health benefits (when the discount rate is set lower for health benefits than costs).

It is less clear what precise rate of discount should be selected by analysts.

A recent report by an Expert Panel commissioned by the U.S. Department of Health and Human Services recommended a real discount rate of 3 percent for medical and public health investments but also urged that a sensitivity analysis be conducted over the range from 0 to 5 percent (Gold et al., 1996). The U.S. Office of Management and Budget, in contrast, recommends a rate from 7 to 10 percent depending upon the particular application.

In EPA's analysis of the integrated rule, a real discount rate of 7 percent is used to annualize the capital investments made by refiners and vehicle manufacturers. These annualized costs are then compared to annualized benefits. If a lower rate of discount had been used, the proposed rule would have looked more attractive. However, EPA's main analysis does not take account of the fact that the investments in emission reduction will not produce immediate health benefits due to the lag times that are common for exposure–response relationships involving chronic disease. Although these lag times are likely to be miniscule for the acute changes in hospitalizations and restricted activity days, lag times of years to decades may characterize the exposure–response relationships for chronic bronchitis and the Pope et al. (1995) premature mortality study.

31.10   UNCERTAINTY ANALYSIS

When the inputs to cost–benefit analysis are uncertain, formal tools of uncertainty analysis can be employed to characterize the magnitude and importance of the uncertainties (Morgan and Henrion, 1990; Pate-Cornell, 1996). In the face of uncertainty, EPA chose to present a limited uncertainty analysis. For their primary estimate, they presented the range of benefit calculated using each of the two competing premature mortality valuation methods. For the other uncertainties, they presented the percentage change in benefits, relative to the value-of-statistical-life (VSL)-based central estimate, due to a change in a given assumption. For example, had they used Dockery et al. (1993) instead of Pope et al. (1995) to determine the dose–response function for particulate matter mortality, benefits would have increased by 120 percent.

An important tool of uncertainty analysis that was not employed by EPA was value-of-information (VOI) analysis (Raiffa, 1968; Hirshleifer and Riley, 1992). In a VOI framework, the option of collecting better information prior to making a regulatory decision about the integrated rule would be considered a formal decision option. In this case, the uncertainties are sufficiently large that a formal VOI analysis would be worthwhile.

31.11   INTANGIBLE AND NONQUANTIFIABLE BENEFITS AND COSTS

When cost–benefit analysis is applied to public health and environmental rules, many categories of benefit and cost are difficult to quantify and express in dollar units. When this difficulty arises, it is important to present a qualitative discussion of the nonquantified considerations.

In EPA's analysis of the proposed integrated rule, a serious effort was made to describe the numerous nonquantifiable consequences of the rule, particularly benefits that could not be addressed numerically. Examples of these nonquantified benefits of the proposed rule include premature aging of the lungs from ozone exposure, additional urban ornamental landscaping costs induced by pollution exposure, adverse effects on commercial forests from pollution exposure, and aesthetic injury to forests induced by pollution. The nonquantifiable risks and costs of the rule are discussed in less detail. They include, for example, the risks of octane enhancers that might be added by refiners to compensate for the sulfur reductions and the risks of ultraviolet radiation induced by reductions in ozone concentrations.

31.12 CONCLUSION

Proposed regulations aimed at reducing risk to human health and the environment can be costly and may cause unanticipated risks to human health and the environment. In response to concerns about regulatory costs and unanticipated risks, policymakers are increasingly asking that objective information about a rule's costs and benefits be produced and considered. These cost–benefit analyses can provide useful information for policymakers, but they should not be used to dictate a regulatory decision. Factors other than economic efficiency, such as equity and political feasibility, also play an important role in regulatory decisions.

The case study of EPA's recent proposal to reduce air pollution by regulating fuels and vehicles illustrates the importance of health risk determinations to cost–benefit analysis. The case study also reveals a variety of ways that risk assessors can help cost–benefit analysts in the future. Risk assessments that produce central estimates (e.g., expected values of risk) as well as low-end and high-end estimates of risk will be more useful for cost–benefit analysis than are risk assessments that produce only a single exposure–response coefficient. For risks of premature death, risk assessments could be improved by providing information on both the quantity of life-years that are lost (longevity loss) and the quality of those life years (taking into account the health status of the affected individuals). An important area that requires better collaboration between risk assessors and cost–benefit analysis is the countervailing risks that might be induced by a rule (e.g., the extra ultraviolet radiation caused by a reduction in ground-level ozone concentrations).

Yet the case study also reveals that there are important inputs to cost–benefit analysis that are outside the expertise of risk assessment. These inputs include, most importantly, measurements of citizen preferences for enhanced duration and quality of life. Uncertainty about citizen willingness to pay for improved health status plays a major role in the overall degree of uncertainty about the benefits of EPA's integrated regulation of fuels and vehicles. Moreover, EPA treats the estimated costs of the proposed regulation as a fairly well-known quantity. Yet experience with other rules suggests that the degree of precision in regulatory cost estimates is less than one might think.

Perhaps the single most important lesson from the case study is that it is useful to consider further data collection as an explicit decision option in cost–benefit analysis of rules with major uncertainties. If a formal value-of-information analysis had been conducted of the proposed integrated rule, it is quite possible that the results would have supported further data collection prior to making a decision to regulate fuels and vehicles. The scientific literature already contains several excellent applications of VOI analysis to public health and environmental policy that can be used as models in future cost–benefit analyses (Finkel and Evans, 1987; Taylor et al., 1993).

REFERENCES

Arrow, K. J., Cropper, M. L., Eads, G. C., Hahn, R. W., Lave, L. B., Noll, R. G., Portney, P. R., Russell, M., Schmalensee, R., Smith, U. K., and Stauins, R. N. 1996. Benefit-cost analysis in environmental, health, and safety regulation: A statement of principles. *Science* 272: 221–222.

Dockery, D. W., Pope, C. A., III, Xu, X., Spengler, J. D., Ware, J. H., Fay, M. E., Ferris, B. G., Jr., and Speizer, F. E. 1993. An association between air pollution and mortality in six US cities. *New Engl. J. Med.* 329: 1753–1759.

Finkel, A. M., and Evans, J. S. 1987. Evaluating the benefits of uncertainty reduction in environmental risk management. *J. Air Pollut. Control Assoc.* 37: 1164–1171.

Fredrick, S., and Fischoff, B. 1998. Scope (IN) sensitivity in elicited valuations. *Risk Decision Policy* 3: 109–123.

Gamble, J. F. 1998. PM2.5 and mortality in long-term prospective cohort studies— Cause-effect or statistical associations? *Environ. Health Perspect.* 106: 535–549.

Gold, M. R., Siegel, J. E., Russell, L. B., and Weinstein, M. C. (Eds.). 1996. *Cost-Effectiveness in Health and Medicine.* Oxford University Press, New York.

Graham, J. D., and Wiener, J. B. (Eds.). 1995. *Risk vs. Risk: Tradeoffs in Proteting Health and the Environment.* Harvard University Press, Cambridge, MA.

Hammitt, J., and Graham, J. 1999. Willingness to pay for health protection: Inadequate sensitivity to probability? *J. Risk Uncertainty* 8: 33–62.

Hirshleifer, J., and Riley, J. G. 1992. *The Analytics of Uncertainty and Information.* Cambridge University Press, Cambridge.

Keeler, E. B., and Cretin, S. 1983. Discounting of life-saving and other non-monetary effects. *Manag. Sci.* 29: 300–306.

Leonard, H. B., and Zeckhauser, R. J. 1986. Cost–benefit analysis applied to risks: Its philosophy and legitimacy. In D. MacLean (Ed.), *Values at Risk.* Rowman and Allanheld, Totowa, NJ.

Mishan, E. J. 1988. *Cost Benefit Analysis*, 4th ed. Routledge, New York.

Morgan, M. G., and Henrion, M. 1990. *Uncertainty: A Guide to Dealing with Uncertainty in Risk and Policy Analysis.* Cambridge University Press, Cambridge.

Okun, A. M. 1975. *Equality and Efficiency.* Brookings Institution, Washington, DC.

Ostro, B. D., and Rothschild, S. 1989. Air pollution and acute respiratory morbidity: An observational study of multiple pollutants. *Environ. Res.* 50: 238–247.

Pate-Cornell, M. E. 1996. Uncertainties in risk analysis: Six levels of treatment. *Reliabil. Eng. System Safety* 54: 95–111.

Pope, C. A., Thun, M. J., Namboodiri, M. M., Dockery, D. W., Evans, J. S., Speizer, F. E., and Health, C. W. 1995. Particulate air pollution as a predictor of mortality in a prospective study of US adults. *Am. J. Respir. Critical Care Med.* 151: 669–674.

Raiffa, H. 1968. *Decision Analysis: Introductory Lectures on Choices under Uncertainty.* Random House, New York.

Schwartz, J. 1993. Particulate air pollution and chronic respiratory disease. *Environ. Res.* 62: 7–13.

Taylor, A. C., Evans, J. S., and McKone, T. E. 1993. The value of animal test information in environmental control decisions. *Risk Anal.* 13: 403–412.

U.S. Environmental Protection Agency (EPA). 1999. *Regulatory Impact Analysis—Control of Air Pollution from New Motor Vehicles: Tier 2 Motor Vehicle Emission Standards and Gasoline Sulfur Control Requirements.* EPA-420-R-99-023. EPA, Washington, DC.

U.S. Environmental Protection Agency (EPA). 2000. Control of air pollution from new motor vehicles: Tier 2 motor vehicle emissions standards and gasoline sulfur control requirements. *Feb. Reg.* 65(28): 6698–6870.

Viscusi, W. K. 1993. The value of risks to life and health. *J. Econ. Lit.* 31: 323–344.

32   Precaution in a Multirisk World

JONATHAN B. WIENER

Law School and Nicholas School of the Environment, Duke University, Durham, North Carolina

Heads up: The "precautionary principle" is coming to a law near you. It aspires to answer a timeless question: How should society in general, and regulatory authorities in particular, respond to uncertain risks? This chapter addresses the normative and positive implications of adopting the precautionary principle.

Whereas many analyses have compared precautionary versus reactive regulatory strategies as applied to individual risks taken one at a time, this chapter examines precaution in a world of multiple risks. It distinguishes different versions of the precautionary principle, some of which are more sensible than others. It suggests that precaution against one risk may induce other countervailing risks, so that the ideal is a middle ground of "optimal precaution" rather than maximum precaution. It observes that real-world applications of the precautionary principle, in both the United States and Europe, have often been sensitive to this reality and therefore have moderated the degree of precaution as the legal instrument gains more regulatory teeth and as the countervailing risks rise.

32.1 BACKGROUND

The notion of precautionary regulation has a long history in both Europe and the United States. Prominent endorsements have appeared on both sides of the Atlantic since at least the 1970s (Applegate, 2000; Boehmer-Christiansen, 1994; Cameron & Abouchar, 1991; Sand, 2000). The cognate concept of *vorsorgeprinzip* in German law dates at least to the early 1970s (Boehmer-Christiansen, 1994). In the United States, landmark cases such as *Ethyl Corp. v. EPA* (1976) and *Tennessee Valley Authority v. Hill* (1978) vindicated the notion of precautionary regulation under the Clean Air Act and the Endangered Species Act, re-

Human and Ecological Risk Assessment: Theory and Practice, Edited by Dennis J. Paustenbach
ISBN 0-471-14747-8 © 2002 John Wiley & Sons, Inc.

spectively. Premarket safety review of new drugs under the Federal Food, Drug and Cosmetic Act has an even older pedigree (Applegate, 2000).

In the last decade the ambition of the precautionary principle (PP) has been growing. It is paid homage in several important international agreements, including several treaties on Marine Pollution, the Rio Declaration, the Framework Convention on Climate Change, and the Cartagena Protocol on BioSafety (Bodansky, 1991; Hey, 1992; Sand, 2000; United Nations, 1992b). The treaty that constitutes the European Union (EU) expressly provides that EU policy on the environment "shall be based on the precautionary principle" (EU Treaty, 1993, Article 130R, now Article 174). Proponents have forecast that the PP "could become *the* fundamental principle of environmental protection policy and law" (Cameron and Abouchar, 1991:2). Some assert that the precautionary principle may already be so widely adopted that it is ripening into an enforceable norm of "customary international law," a kind of international legal doctrine from which no nation can dissent (Sand, 2000; Sands, 1995:213). Most recently, the European Commission has formally articulated and endorsed the precautionary principle (European Commission, 2000).

Despite its adoption of a "precautionary preference" in some specific statutes (Applegate, 2000), the United States has not officially adopted the Precautionary Principle as a general basis for regulation. After endorsements of precautionary regulation in cases such as *Ethyl Corp. v. EPA* (1976) and *Tennessee Valley Authority v. Hill* (1978) in the 1970s, the U.S. Supreme Court held in the *Benzene* case (*Industrial Union, AFL-CIO v. American Petroleum Institute*, 1980) that the Occupational Safety and Health Administration (OSHA) cannot simply leap into regulating on the basis of conjecture about uncertain risks. This decision, and a 1983 guidebook from the National Academy of Sciences (NAS), spurred widespread adoption of risk assessment (albeit of ten employing precautionary default assumptions and methods) as the basis for American risk regulation (Jasanoff, 1986, 1995). Along these lines, the United States has resisted blanket statements of the PP in international fora. For example, the United States insisted on qualifying the statement of the Precautionary Principle in the Climate Change Convention (Bodansky, 1993), and the United States responded to the European Commission's recent endorsement of the Precautionary Principle with a long list of skeptical questions (U.S. Department of State, 2000).

Given this history, a common inference today is that Europe endorses the precautionary principle (PP) and seeks proactively to regulate risks, while the United States opposes the precautionary principle and waits more circumspectly for evidence of actual harm before regulating (Lofstedt and Vogel, 2001; Lynch and Vogel, 2000; Richter, 2000). A closely related view is that American regulation is more rooted in the science of formally assessing risks before acting, whereas European regulation is more qualitative and permits action through informal decision making unfettered by science (Jasanoff, 1986, 1998). These juxtapositions partly explain the eagerness among advocates of the PP to have it made part of customary international law—so that the United States, among others, cannot resist it. Across a wide range of examples, observers paint the picture of a civilized, safe Europe confronting a violent, risk-taking America (Daley, 2000; McNeil, 2000).

This chapter suggests that this inference is incorrect, both descriptively and normatively. In a world of multiple risks, the reality is more complicated. First, although precaution can be warranted, it is not universally desirable. Sometimes it is for the best, but sometimes precautionary regulation is too costly, and sometimes—given multiple risks—precaution can even yield a perverse net increase in overall risk. Context is crucial. And different versions of the PP imply different outcomes, some superior to others. Second, when multiple risks are examined, it becomes clear that Europe is not more precautionary than the United States across the board; sometimes Europe does take a more precautionary stance than the United States, but sometimes the roles are reversed and the United States is the more precautionary regulator. Again, context is crucial. Ultimately, the PP offers important insights but remains too simplistic for a complex multirisk world. The PP needs to be refined toward a more mature and sophisticated concept of "optimal precaution" that responds to the reality of a world of multiple risks.

32.2   RISK AND UNCERTAINTY

A brief discussion of risk is warranted to make clear the baseline against which the precautionary principle is being applied. All activities involve risk. Risks beset even the most mundane necessities, such as eating (choking; foodborne disease), breathing (pollution; airborne disease), walking (falling), keeping warm (fire or other energy sources), and sleeping (apnea; bad dreams; oversleeping and missing an appointment). Dealing with risk is an inescapable element of the human condition (Bernstein, 1996). By risk we mean the likelihood (probability) that exposure to a hazard will cause an adverse outcome (harm) to occur, combined with the seriousness of that outcome (e.g. mortality, morbidity, or impaired quality of life). For example, as far as we know today, the probability that exposure to highway driving will cause a fatal accident is significant and clearly greater than zero, whereas the probability that exposure to a cellphone will cause a fatal brain tumor is very low and may well be zero. (Using a cellphone while driving, of course, might raise the probability of the highway accident.)

Some risks are well documented and understood, while others are highly uncertain. That is, our estimates of the probability of an adverse outcome can be more or less confident. For example, we regularly observe that highway accidents kill some motorists, and thus we can be fairly confident in our prediction of the probability of highway fatalities next year; but we are unsure whether cellphones ever cause any brain tumors, and thus we are extremely unsure of our prediction of such fatalities. Even for highway accidents, our prediction of the risk on a given day, or to a given individual, would be highly uncertain. Fundamentally this difference is a matter of degree. All risks are probabilistic and uncertain because we can never know the future with complete certainty (Knight,

1933). Thus, all decisions about the future must be made in the face of uncertainty. We can never be completely certain that something will cause harm; we never have certainty about the risks we incur, or about the opportunities we seek.

A related point is that we can never be completely certain that something is *free* of risk. Any substance or activity could be a hazard that results in harm, if it is experienced in the wrong dose or at the wrong place or time. Even the necessities of life, such as water, salt, oxygen, sunshine, and vitamins, can be harmful or fatal in large quantities (e.g., oxygen poisoning, skin cancer) or in the wrong circumstances (e.g., water in the lungs, salt in the wound). Paracelsus taught that "the dose makes the poison" (Ottoboni, 1984). What is a hazard thus depends not on a classification of intrinsic good versus intrinsic bad, but rather on context.

Moreover, some risks are especially latent: Their adverse impact will only occur a long time (perhaps many years) after the event that set the risk in motion. For example, a highway accident typically causes fatality (if at all) within seconds or minutes after the accident; but if there are any brain tumors caused by cellphone use, it might take many years after the exposure to the cellphone before the tumors become manifest. The longer the latency period between cause and effect, then the earlier (relative to the adverse outcome) measures must be taken if they are to be effective in preventing the outcome. If we wait to observe the latent outcome, it can become too late to take preventive measures.

Meanwhile, the seriousness of the adverse outcome depends on how people evaluate the outcome and its context. A fatality from one cause (e.g., cancer) may be viewed as more serious than a fatality from another cause (e.g., an accident), even controlling for equal probability and latency (Tolley, 1994:323–344). A fatality caused by a familiar, routine hazard (e.g., driving or smoking cigarettes or radon gas in homes) may be viewed as more tolerable than a fatality caused by an unfamiliar, mysterious hazard (e.g., hazardous waste or genetic engineering or radiation from nuclear power plants) (Slovic, 1987). Experts sometimes neglect these differences in valuation and treat a death as a death; other analyses use willingness-to-pay or quality-adjusted-life-year formulations to measure public valuations of risks. Yet basing regulatory policy directly on these attitudes toward risks remains controversial because public risk perceptions and valuations may be based on heuristic errors or on prejudices that good government should discount (Wiener, 1997). If the source of disparate risk valuations is simply factual errors, then educating the public about expert knowledge may help reduce the disparity. On the other hand, if the source of disparate risk valuations is not factual errors but value choices, the question is more difficult. Should democracy reflect all public preferences? One may be tempted to say yes, but the question is not so simple (Sunstein, 1991). Doing so could foster appropriate choices based on public values (Shrader-Frechette, 1991), but it might also elevate mass prejudice to public policy (Cross, 1997). For example, if the public is informed that some risk (say, nuclear power, or transgenic foods, or wolves, or immigrants, or urban youth, or the dark) is not really a significant threat to public well-being, but if the public persists in feeling dread of the unfamiliar (abject fear of the unknown) and therefore presses for regulatory protection, perhaps government should think twice before translating that dread into public policy. Democracy is not simple majority rule; rather, it usually combines majoritarian voting with countermajoritarian constraints such as constitutional rights and checks and balances among branches of government. Likewise, sound regulatory policy entails both responsiveness to public attitudes about risks and enlightened leadership by government officials (Wiener, 1997).

32.3 PRECAUTION AS A PRINCIPLE

In the face of uncertainty about a risk, we often take precautionary measures, such as posting warning labels, driving safely, cooking foods to kill microbes, and saving money for future needs. Yet we never know for sure if these precautionary measures are effective (since, if they are successful, they result in the absence of an adverse outcome that might not have occurred anyway), nor do we know whether they are directed at the most important risks. At the same time, we rarely forego beneficial activities entirely just because they might be risky; we do not forego eating for fear of choking (but we do chew more carefully), nor do we forego crossing the street even though there is an uncertain probability of death (but we do use crosswalks and look both ways). We choose prudent precautions that are proportionate to the expected risk, the cost of sacrifice, and the availability of alternatives.

The precautionary principle seeks to go further; it seeks to impose earlier and more stringent restrictions on potentially risky activities. It invokes common sense adages such as "better safe than sorry" (Margolis, 1996). In particular, it seeks to impose early preventive measures to ward off even those risks for which we have little or no basis on which to predict the future probability of harm.

In the face of probabilistic, uncertain, and latent risks, government has two basic strategies: ex post remedies, ex ante precautions, or both. Ex post remedies include tort law administered by the courts. Ex ante precautions include regulations administered by agencies. Regulations are precautionary measures taken to prevent anticipated future outcomes. Some regulations are more ex ante than others: On the time path over which the risk is forecast to become manifest, some regulations take effect earlier and more stringently than do others. As the reasons accumulate to anticipate a future adverse outcome from a present activity, regulatory agencies may act sooner, and more stringently, to ward off that outcome.

I have spoken thus far of "the" precautionary principle, but the literature reveals no single formulation of the principle. Statements of the precautionary principle are varied and often vague (Applegate, 2000:414–415). One review identified 19 different formulations (Sandin, 1999). Here I will treat the precautionary principle in two ways. First, I will treat it as a statement about the preferred *degree* of precaution. The precautionary principle seeks to advance the *timing* and tighten the *stringency* of ex ante regulation (cf. Applegate, 2000:415–420). On these sliding-scale dimensions, regulation is "more precautionary" when it intervenes earlier and/or more stringently to prevent uncertain future adverse consequences. Second, I will distinguish three particular narrative *versions* of the precautionary principle (cf. Wiener and Rogers, in press). These versions mark three points along the spectrum of degrees of precaution, each a more precautionary exhortation than the last.

32.3.1 Version 1: Uncertainty Does Not Justify Inaction

In its most basic form, the PP permits precautionary regulation in the absence of complete evidence about the particular risk scenario. The most common phrasing is: "Where there are threats of serious or irreversible damage, lack of full scientific certainty shall not be used as a reason for postponing measures to prevent environmental degradation" (Bergen Declaration, 1990). In short, uncertainty does not justify inaction.

This formulation was, for example, the basis for the court's opinion in *Ethyl Corp. v. EPA*. It was deployed to permit the U.S. Environmental Protection Agency (EPA) to regulate lead in gasoline before EPA could demonstrate physical ill effects in exposed human populations. Specifically, it was deployed to counter the argument made by industry that in the face of uncertainty, EPA should not regulate. Similarly, version 1 has been asserted in response to industry's argument that uncertainty about future climate change requires waiting before restricting emissions of greenhouse gases.

But this version of the PP does not go very far. First, it only permits action, rather than compelling it. Second, it only responds to the situation of "lack of full scientific certainty," but there is never "full scientific certainty"; we always face uncertainty, and we must always make decisions under uncertainty. Thus, "lack of full scientific certainty" is not a special difficult case for decision making; it is the general case of all decision making. This version of the PP does not answer the real question, which is *what* action to take in the face of (inevitable) uncertainty. Ban the substance? Require warnings? Investigate options? If we face very high uncertainty about what causes what, it may not even be clear what measures would effectively prevent the anticipated future harm. Even if uncertainty is low, some measures would be disproportionate to the harm to be prevented or would cause other harms. So it is not enough to call for action; the real question is which action.

This version of the PP does, at least, make clear that uncertainty is not a sufficient reason to sit on our hands; it rebuts the oft-heard contention (usually by those about to be regulated) that uncertainty itself precludes regulation. If this contention were valid, ex ante regulation would always be unwarranted (as would all sorts of personal precautions against future risks). If we are ever to act preventively, before deaths occur, then we must act in the face of uncertainty. This first version of the PP states that we may (but it does not say what we should do).

The PP is viewed by some actors as the "antiscience" opposition to "science-based" regulations. But an intelligent version of the precautionary approach can be consistent with scientific principles (Stirling, 1999). Once we get beyond simplistic lists of good and bad substances, we need science to identify which substances and activities pose what degree of risk in what circumstances. One needs to know what to be precautionary about. Because risk assessments inevitably address and involve scientific uncertainty, science remains a necessary element of risk management decisions (and yet science is insufficient to make policy because policy requires forecasting and weighing the consequences of alternative policy options). Thus, version 1 of the PP can be understood as a call for "action" that includes more scientific research to reduce uncertainties and guide the deployment of effective regulatory measures.

32.3.2 Version 2: Uncertain Risk Justifies Action

A second version of the PP is somewhat more aggressive; it says that uncertain risk justifies action. An American conference produced the statement: "When an activity raises threats of harm to human health or the environment, precautionary measures should be taken even if some cause and effect relationships are not fully established" (Wingspread, 1998:353). Similarly, the German Federal Interior Ministry wrote: "The principle of precaution commands that the damages done to the natural world (which surrounds us all) should be avoided *in advance* and in accordance with opportunity and possibility.... it also means acting when conclusively ascertained understanding by science is not yet available" (Boehmer-Christiansen, 1994:37). Or: "According to the precautionary principle, the more uncertain the risk, the more justified is some form of regulatory intervention" (Wagner, 2000:461).

Like the first version of the PP, this version is based on a truism because cause-and-effect relationships are never "fully" or "conclusively" established (even in retrospect); we always deal with uncertain and probabilistic relationships. This version, while calling for proactive precautionary measures, still does not address the real question of *what* measures should be taken. Such measures might include banning the activity, restricting it, requiring warning labels, requiring disclosure of information, or additional research to learn more about the activity, its risks, and its alternatives.

Still, version 2 is more precautionary than version 1, insofar as version 2 impels regulatory intervention rather than just permitting it. Version 2 is not merely a rebuttal to assertions that uncertainty warrants inaction; it is also an affirmative basis for regulation.

32.3.3 Version 3: Shifting the Burden of Proof

A third version of the PP is even more adamant; it insists that uncertainty about risk requires forbidding the potentially risky activity until the proponent of the activity demonstrates that it poses no (or acceptable) risk. "As described in the Wingspread Statement on the Precautionary Principle, the applicant or proponent of an activity or process or chemical needs to demonstrate to the satisfaction of the public and the regulatory community that the environment and public health will be safe. The proof must shift to the party or entity that will benefit from the activity and that is most likely to have the information" (deFur, 1999:345–346).

This version of the PP is more precautionary than versions 1 and 2. It goes beyond permitting or compelling regulation of uncertain risks, by providing an answer to the question that versions 1 and 2 leave unaddressed: what action to take. Version 3 specifies a particular action to take in the face of uncertain risk: Forbid the activity unless a certain standard of proof is met by the proponent. Version 3 has two key components: shifting the *burden* of proof (who must demonstrate risk or safety—must the regulator demonstrate risk, or the regu-latee demonstrate safety?) and setting a *standard* of proof (how risky or how safe—must the regulatee demonstrate no risk, de minimus risk, acceptable risk, or something else?).

Version 3 could invite overregulation, depending on the standard of proof for lifting the ban. What counts as demonstrating that the environment and public health will be "safe"? Some interpretations of version 3 of the PP would impose a standard of proof criterion of "no risk." For example: "One would first compile a list of substances that were known to be toxic. For premarket statutes, firms would not be permitted to introduce substances chemically simi-lar to those on the list without proof beyond a reasonable doubt that they were not toxic. For postmarket statutes, ... [one could] require them to be phased out over time [unless firms] could show beyond a reasonable doubt that any exposures posed no risk of harm or posed no threats of serious damage to health or the environment" (Cranor, 1999:94). Such a measure would represent over-regulation where the risk of the substance is small relative to its benefits. For example, many medicines are toxic (at some dose—as are vitamins and oxygen), and no drug company could offer "proof ... that they were not toxic" or "posed no risk of harm." It is often the very toxicity of medications that enables them to provide a net health benefit (e.g., by killing microorganisms or cancer cells within the body). More generally, because "the dose makes the poison," poten-tially every substance would have to banned if "proof of no toxicity or no harm" were the universal principle.

As a result, real-world applications of version 3 tend to employ a more bal-anced standard of proof. In the *Benzene* case, the Supreme Court held that "safe" does not mean "no risk" but rather means "no significant risk," and then left it up to OSHA to determine which risks are "significant" (*Industrial Union Dept., AFL-CIO*, 1980; cf. Fischhoff et al., 1981). Laws requiring pre-market approval of new products typically require that product proponents "demonstrate acceptable risk" or "show no unreasonable risk." For example, the drug licensing provisions of the U.S. Federal Food Drug & Cosmetic Act grant approval if the proponent demonstrates net benefits to the target patient population. The restrictions on new substances in the Toxic Substances Control Act (TSCA) and the federal pesticides law (Federal Insecticide, Fungicide, and Rodenticide Act, FIFRA) both condition approval on whether the substance poses "unreasonable risk."

Even if a balanced standard of proof is employed, the disincentives to innovation posed by version 3 of the PP may be amplified by the compounding risk aversions of the regulator and the researcher. The regulator will likely want to avoid approving even a net beneficial product that might later prove harmful, while the researcher will want to avoid investing in a beneficial product that is forbidden by the regulator. Hence even with a standard of proof such as "net beneficial," the dual risk aversion of both regulator and researcher may suppress many products that would meet the standard.

On the other hand, version 3 of the PP does contain the useful idea of putting the *burden* of proof (which is different from the *standard* of proof just discussed) on the party best able to generate the information needed to make the decision. That is analogous to the notion of putting the burden of accident avoidance on the least-cost avoider (Calabresi, 1970). It may often make sense to ask industry rather than government to produce much of the data on risk. California's Proposition 65 takes this approach: Substances initially found to be carcinogenic must be accompanied by warning labels (not a prohibition on use), and manufacturers can have the warning requirement lifted if they generate the information needed to demonstrate that the substance falls below a threshold of carcinogenicity.

The stringency of versions 2 and 3 can also be seen by analogy to their use in the area of criminal law. Violence is a problem of public risk and public health. In this context, the precautionary principle would favor earlier and more stringent interventions to prevent the "future dangerousness" of persons who may, with considerable uncertainty, be forecast to commit violence in the future. Version 2 would commit the state to pursuing people who may be future felons but whose latent violence is uncertain. Version 3 would shift the burden of proof from "innocent until proven guilty" to "guilty until proven innocent." On these premises, everyone would be a proper subject of precautionary incarceration or at least monitoring by the state. The real-world example of this strategy is the effort, particularly in the United States, to incarcerate and even execute juvenile offenders to prevent their future dangerousness (Pfeiffer, 1998; Rimer and Boner, 2000; Winterdyk, 1997)—despite scant evidence of any real increase in youth violence (Zimring, 1998). This is just the kind of anticipatory regulatory approach that adherents of the PP advocate for new and potentially risky technologies, but here as applied to young and potentially risky persons. (Similar concerns are raised by precautionary sanctions applied to the mentally ill and to athletes suspected of drug use.) The example is helpful in testing our thinking because it catches most advocates (on both sides) with their shoes on the other feet. Liberals may like precautionary environmental regulation, but recoil at precautionary criminal law; conservatives may have the opposite reaction. Perceptions of dread may attach to synthetic chemicals for some but to youth gangs for others. A more consistent approach to risk and precaution, however, would recognize the commonality of these examples. The youth violence example illustrates the fact that all precautionary regulation can invite false positives, can be counterproductive (here, by inducing recidivism among youths sentenced to adult prisons), and can intrude on personal liberties.

32.4   PRECAUTION AMIDST MULTIPLE RISKS

All of these versions of the PP, and indeed all choices between ex ante and ex post legal systems, confront the trade-off between two kinds of errors: *false negatives* and *false positives*. False negatives occur when an initial finding of no (or acceptable) harm later turns out to have been incorrect. We risk false negatives by presuming "innocent until proven guilty." Waiting before regulating can incur the cost of false negatives. For example, a substance or activity initially deemed unworthy of regulation could later turn out to be harmful; this might be a drug that turns out to pose adverse side effects, or a genetically engineered food (or any food), or an energy source that turns out to emit greenhouse gases or radiation, or a person suspected of a violent crime but not initially prosecuted. By contrast, false positives occur when an initial finding of (unacceptable) harm later turns out to have been incorrect. We risk false positives by presuming "guilty until proven innocent." Precautionary regulations can incur the cost of false positives. For example, a substance or activity banned or restricted could later turn out to be benign; this might be a drug with beneficial therapeutic effects, or a genetically engineered food (or any food), or an energy source that is less hazardous than initially feared, or a person wrongly prosecuted for a crime she did not commit.

Because we must always decide in the face of uncertainty, we always confront the dilemma of whether to try harder to avoid false negatives or false positives. We will not be perfect; we will err, and the question is how to optimize our errors. Some have argued that regulators should be more averse to false negatives (and hence should err on the side of overregulation) on the premise that the cost of underregulating false negatives is health and environmental damage, whereas the cost of overregulating false positives is only money (Page, 1978). That premise is, however, incorrect: The cost of regulating false positives can just as well be health and environmental damage, because the substances and activities restricted offer their own health and environmental benefits (e.g., a therapeutic drug, or genetic engineering that reduces the need for chemical pesticides, or an energy source that reduces emissions of greenhouse gases), or because the substitutes pose risk, or because exaggerated warnings spur panic (and, later, public cynicism), or because overregulating false positives implies restrictions on personal freedom (e.g., prosecuting an innocent suspect of a crime, or restricting consumer choice) (Wiener, 1998).

Put another way, any precautionary intervention can yield unintended side effects. Risk regulation, like medical care, can both heal and hurt (Janicke, 1990;

Moynihan, 1993; Wiener, 1998). Reducing a target risk can increase a counter-vailing risk (Graham and Wiener, 1995). To cite just a few examples: Aspirin treats headaches but causes stomachaches; surgery treats injuries but risks infec-tion; automobile airbags save some adults but kill some children; suppressing forest fires reduces their initial frequency but increases their ultimate severity; regulating pollution into one medium may induce cross-media shifts; hazardous waste cleanups protect future residents but put present workers at risk; banning asbestos reduces cancers but may increase highway accidents from less effective brake linings; regulating outdoor pollution may induce firms to seal the plant and increase indoor pollution; controlling urban ozone may protect lungs but expose skin to more intense ultraviolet radiation; banning drugs may reduce addiction but increase violence; police pursuits may catch fleeing felons but injure bystanders (Graham and Wiener, 1995; Cross, 1996; Sunstein, 1996; Wiener, 1998). These are classic problems; in Joseph Lister's 19th century England, hospital-acquired wound infections killed between 25 and 50 percent of surgical patients; in the United States today the rate of iatrogenic mortality is down to about 0.4 percent, but that rate still implies somewhere near 50,000 to 100,000 deaths a year caused by medical care (Wiener, 1998). Like Odysseus navigating between Scylla and Charybdis, the modern regulator must weigh competing risks.

The basic phenomenon driving these risk–risk trade-offs is the interconnect-edness of multiple risks, derived from the interconnectedness of environmental and social systems. As the pioneering environmentalist John Muir put it, "when we try to pick out anything by itself, we find it hitched to everything else in the universe" (Muir, 1869: 110). Squeezing the balloon (or tugging the web) at one point puts pressure elsewhere. A general shortcoming of the PP is that it ad-dresses risks one at a time as if uncertainty were the crucial issue. But the real-ity is that risks are multiple and trade-offs are the crucial issue. The PP thus neglects interconnectedness and neglects the potential adverse health and envi-ronmental effects of precautionary measures themselves (Graham and Wiener, 1995; Cross, 1996; Margolis, 1996; Wiener, 1998). Ironically, the PP neglects the ecological insight of interconnection.

The PP also speaks as though government regulation were an exogenous remedy for environmental and social ills. It says: Risk warrants regulation. But government regulation is not an exogenous solution to social problems; it is itself an endogenous and fallible human activity, and as such it can create risks—risks that are as real as the risks of market activities. This was the es-sential message of the National Environmental Policy Act (NEPA), the flagship U.S. environmental law that requires environmental impact statements for government projects. It is also the essential problem of iatrogenic injury and of risk–risk trade-offs: sometimes care heals, and sometimes care hurts. That is not to say that regulation can never improve things (Hirschman, 1991); it is just to say that regulatory interventions, like medical care interventions, affect multiple risk variables and generate a portfolio or vector of consequences. Regulatory law can be improved by confronting these trade-offs among targets risks and countervailing risks, and designing regulations to minimize overall risk. The problem is not pollution or market failure per se; the problem is flawed human institutions, including both markets and government. The challenge is to minimize the sum of market risk and regulatory risk.

Optimal regulation in the face of a target risk (TR) and a countervailing risk (CR) would take both seriously and strive to maximize their difference (ΔTR – ΔCR). Uncertainty is not the crucial problem—trade-offs are. Even *certain* risks would not justify regulatory action if ΔCR > ΔTR (such that regulating TR would yield negative net benefits). Meanwhile, standard advice to "muddle through" by "ignoring side effects" (ignore ΔCR) (Lindblom, 1959) is too lax; it will yield more net harm (because of neglected CR) than maximizing (ΔTR – ΔCR) whenever CR is positive and CR can be reduced at less than equal increases in TR. And standard advice to "do no harm" (ensure zero ΔCR), as in the Hippocratic oath, is too stringent; it will yield more net harm (because of neglected TR) than maximizing (ΔTR – ΔCR) whenever CR is positive but squeezing CR to zero will mean larger increases in TR (Wiener, 1998).

Furthermore, version 3 creates an ironic twist: If precautionary measures themselves are human activities that pose some risk, then version 3 of the PP—which forbids any risky "activity" until its proponent demonstrates acceptable risk, or no risk—would forbid many or all precautionary measures themselves. In short, the PP would swallow itself (cf. Cross, 1996). The way out of this paradox is to recognize that all human activities (including regulation) pose risk, and that the real question is not whether to act under uncertainty (we always must), but rather *what action* to take from among the portfolio of alternatives with different risks, costs, and benefits. The astute regulator should optimize across risk–risk trade-offs, minimize overall risk, and seek risk-superior moves that reduce multiple risks in concert (Graham and Wiener, 1995). At the same time, not every countervailing risk deserves extensive attention: The regulator should seek to avoid regulatory side effects up to the point that the benefits of doing so (improved policy outcomes) justify the costs of doing so (chiefly delay) (Wiener, 1998).

In sum, version 1 of the PP is helpful to decision makers as a rebuttal to the mistaken claim that uncertainty warrants inaction. Latent risks mean that inaction under uncertainty may invite future harms. Failing to take action against uncertain risks means incurring the social costs of false negatives in the future. Many cases of latent risks do warrant some precaution: Examples include lead in gasoline, losses of endangered species and biodiversity, resource depletion spurred by the tragedy of open access, and climate change. Simultaneous exposure to multiple stressors may warrant heightened precaution. But PP version 1 leaves unaddressed the serious question, which is what action we should take in the face of uncertain and latent risks. Saying that some precautionary action is warranted does not assist us in deciding which action—whether, say, to reduce greenhouse gas emissions by 5 percent, or 30 percent, or ban all fossil fuel combustion, or let emissions grow unabated while we invest in more resilient adaptation to climate change, or something else. Thus version 1 is helpful but incomplete. On the other hand, versions 2 and 3 of the PP are unhelpful and could be counterproductive. They incur the social costs of unnecessarily regulating false positives. Moreover, version 2 and especially version 3—designed to promote ex ante regulatory action against a single risk—become entangled in the real-world web of multiple risks. By focusing on one risk at a time, they neglect the countervailing risks of regulation; by viewing government intervention as exogenous rather than endogenous, they highlight market failures but neglect government failures. In the real world of multiple risks and imperfect government, aggressive precaution can induce countervailing risks that weaken or even reverse the case for regulation. Precaution itself may be a risky activity; amidst multiple risks, the PP implies that precautions are needed against excessive precaution. Given the reality of multiple interrelated risks, we need a principle of "optimal precaution" rather than of maximum precaution. Precautionary regulation should be followed by continuing surveillance and research to foster learning and adaptive revisions. And we need to seek risk-superior options that reduce multiple risks in concert. Just as Joseph Lister devised antisepsis to reduce wound infection and thereby make surgery safer, we need to devise ways to make precautionary regulation safer (Wiener, 1998).

32.5 PRECAUTION AS APPLIED

The real-world experience of the PP in application illustrates that unbridled precaution could be perverse and that governments therefore apply the PP moderately. Here I offer two perspectives on the real-world application of precaution. First, I argue that the more binding the legal instrument in which the PP is ensconced, the more moderate is its application. Radical versions of the PP are thus more rhetoric than reality. When real regulatory powers (or "teeth") are being exercised, governments have applied the PP more moderately and pragmatically, couching it in balanced criteria for decision making.

Second, I argue that the United States and Europe have not taken simple opposing positions on the PP. In reality, the relative degree of precaution on each side of the Atlantic has varied considerably across topics. The pattern is complex; sometimes Europe is more precautionary, and sometimes the United States is more precautionary. Thus, when the PP is applied in the real world, pragmatic context matters more than ideology.

32.5.1 Precaution and the Teeth of Legal Instruments

Arraying statements of the PP along the sliding scale of "degree of precaution" supports the hypothesis that the PP is stated more strongly in less legally binding texts, and less strongly in more legally binding texts. The strongest statements of the PP—such as the strong form of version 3, forbidding an activity until its proponents demonstrate that it poses no risk—appear in the academic literature (e.g., Cranor, 1999) and in declarations by advocacy groups (Wingspread, 1998).

When the PP is adopted by governments in international fora, it becomes more moderate. The Bergen Declaration (1990)—adopting version 1 (uncertainty does not justify inaction)—was a statement of environment ministers; it bears some governmental imprimatur but has no binding legal effect and represents the views of the arm of government (environment ministries) most likely to favor a strong PP. When the Bergen Declaration's statement was inserted into the Rio Declaration (United Nations, 1992b), it was qualified with the term "cost-effective" added before "measures" (Applegate, 2000:415). Sand (2000:447) reports that this term was inserted by the U.S. lawyer at the fourth session of the UNCED Prepcom in March 1992 over the objections of Europe and Japan. My recollection, however, is that the term "cost-effective" had already been proposed by the United States, and agreed to by other countries, in the preceding negotiations on the Framework Convention on Climate Change; thus the term "cost-effective" appears in the statement of the precautionary principle in Article 3.3 of the Climate Change Convention (United Nations, 1992a). In either case, the point here is that as the PP moved from the declaration of environment ministers at Bergen to the more official (though still nonbinding) declaration signed by presidents and prime ministers at Rio, and the even more potent (signed, ratified, and legally binding) climate change treaty, the PP was moderated. The climate change treaty further moderated the PP by placing it in Article 3 on "principles," the legal effect of which was debated by the parties and doubted by the United States (Bodansky, 1993); the real teeth of the Climate Change Convention were in Article 4, setting emissions limitation obligations, which does not advert to the PP.

When the PP is adopted by governments in national legislation, it becomes more moderate still. Although the EU treaty (1993) states that EU environmental policy shall be "based on the precautionary principle," it does not say what that means. When the organs of European government turned to giving that statement teeth (European Commission, 2000), they translated the PP into a heavily qualified and caveated explanation of what amounts to pragmatic consequentialist decision making. The Commission's statement begins with version 1 (uncertainty does not justify inaction), but then it adds numerous other criteria for ex ante regulation, such as the need for precautionary measures to be proportionate to the risk involved, to maximize net benefits, to take account of the costs and risks of alternatives, to assess the alternative of no action, and to be accompanied by ongoing research to reduce uncertainties and possibly reverse the decision even after the initial precautionary measure has been adopted. Despite having several ambiguities and potential shortcomings, the Commission's communication is in many ways similar to the U.S. guidelines for regulation in Executive Order 12866 (Clinton, 1993) (which requires, among other things, that regulations be based on sound science and maximize net benefits). In both cases, the national governments have articulated sensible, moderate frameworks for risk-based regulation. Of course, the actual regulatory outcomes will depend on how that framework is interpreted and implemented. The point here is that as the European Union has moved from treaty rhetoric to binding implementation, it has moved the PP a long way toward the pragmatic approach to risk regulation already in place in the United States. In particular, the European Union has recognized the problems of false positives and countervailing risks due to excessive precaution and has attempted to write the concomitant qualifications (such as proportionality, net benefits, comparing alternatives, study, and reconsideration after adoption) directly into its conception of the PP.

Member states of the European Union have likewise moderated the PP in actual application. British judicial decisions have so far declined to adopt a strong form of the PP (Sand, 2000:449). France has been inconsistent, sometimes championing the PP (e.g., against imports of British beef and of Swiss bioengineered corn) and sometimes opposing it (e.g., when it was invoked against a French nuclear power plant) (Sand, 2000:448, 450). The French statute (the "Loi Barnier" of 1995) adopting the PP added the qualifications that precautionary actions must be "commensurate" with the risk and "at economically acceptable cost" (Sand, 2000:450). In Germany, where nuclear power plants have the burden of proving safety, the courts have nonetheless held that the PP does not require eliminating all risk and moreover that under the PP the public must tolerate some "residual risks"; the German administrative implementation of the PP requires that precautionary decisions be based on benefit–risk assessments and be commensurate with the risk (Sand, 2000:451). Indeed, German law has a general principle of "proportionality" which constrains precautionary regulation (Emiliou, 1996; Jackson, 1999:604). Sweden takes a more stringent approach, closer to version 3, but even there the strong form of version 3 is only occasionally applied (Sand, 2000:448–449).

Similarly, the application of precaution in the United States has usually been moderate. Although the FFDCA (for new drugs), TSCA (toxic substances), and FIFRA (chemicals) impose premarket screening that shifts the burden of proof to manufacturers, they employ a moderate standard of proof of net benefits or reasonableness (avoidance of unreasonable risk). As to food additives, the Delaney Clause formerly imposed a stringent "no carcinogens" standard of proof (regardless of the triviality of the risk or the countervailing health benefits of the substance), but the 1996 Food Quality Protection Act has moderated that standard (slightly) to a "reasonable certainty of no harm."

The Endangered Species Act (ESA), while prohibiting any federal agency from "jeopardizing" the survival of an endangered species [sec. 7(a)] as a matter of "institutionalized caution" and regardless of the cost (*Tennessee Valley Authority v. Hill*, 1978), nonetheless provides for exemptions that maximize net benefits [sec. 7(h)], and for the determination of a species' "critical habitat" based on economic criteria [sec. 4(b)]. Even though it prohibits persons from "taking" endangered animal species (sec. 9), it allows persons to obtain "incidental take" permits (sec. 10) that immunize such takings if incidental to other valuable economic activities such as agriculture and forestry. It also defines "species" [in secs. 3(16) and 3(8)] in a way that omits microorganisms, perhaps because the drafters viewed the protection of bacteria, fungi, and viruses as not worth the costs; and it allows the government to omit insects that pose an overwhelming risk to humans [sec. 3(6)]. These qualifications of the ESA's otherwise stringent prohibitions on harming endangered species may or may not be desirable. My point here is simply that the initial precautionary edict has been moderated in application.

Another useful example is the phaseout of chlorofluorocarbons (CFCs). The international treaty calling for this phaseout (Montreal Protocol, 1987) cited the PP as part of its rationale. The phaseout regulated different CFCs in proportion to their ozone depletion potential [see 42 USC 7671a(e)]. In implementing this phaseout, national laws allowed for some continued production of CFCs for "essential uses" such as airplane safety and medical devices [42 USC 7671c(d), 7671d(d), 7671i(d)] and for fire suppression if no safe and effective substitutes are available [42 USC 7671c(g)]. In addition, the U.S. statute implementing the phaseout directly confronted the problem that substitutes for CFCs might pose countervailing risks: It instructed EPA to regulate CFC *substitutes* to "reduce overall risks to human health and the environment," including risks of ozone depletion, global warming, toxicity, and other hazards (42 U.S.C. 7671k). Thus, the precautionary effort to protect the stratospheric ozone layer was moderated to treat different CFCs in proportion to their risks, to include caveats for high costs of abatement ("essential uses"), and to deal with the complexities of multiple countervailing risks.

Of course, strict versions of the PP could be enacted into binding law. There are undoubtedly some such examples in place today. The Delaney Clause is one example—though one that has been sharply criticized for being excessive and counterproductive. Another example is the rise of stringent precautionary measures in the criminal law area. There, although the government bears the burden of proof, some stringent precautionary standards have been adopted to incapacitate and deter future dangerousness, such as mandatory minimum sentences and "zero-tolerance" policies on adolescent alcohol use. My point here is simply that, in general, as law becomes more binding, its precautionary tendencies become more moderate. The precautionary rhetoric of crime prevention is arguably even more absolutist than its admittedly absolutist application. And judges may moderate stringent criminal laws in many cases.

In sum: the most aggressive versions of the PP appear in rhetorical and nonbinding declarations. In general (though there are no doubt exceptions), the more binding a legal instrument, the more moderate its application of the PP. This illustrates the move from the absolutist "precautionary principle" toward a more pragmatic "optimal precaution" as precaution must confront the reality of a multirisk world.

32.5.2   Precaution Across the Atlantic

A second illustration of the move toward pragmatic consequentialism in the application of the PP can be seen in the comparison between the United States and Europe. As noted above, the conventional wisdom is that Europe favors precautionary regulation and the United States opposes it. The reality is that the United States and Europe both respond to risks in context. Sometimes Europe is more precautionary; sometimes the United States is more precaution-ary. (For a more detailed comparison on which this section draws, see Wiener and Rogers, in press.)

For example, Europe appears to have been more precautionary than the United States about several types of risks, including the safety of genetically modified foods (Lynch and Vogel, 2000), hormones in beef, including bovine somatatropin (BST) (European Council, 1990, 1999; Vogel, 1995, 1997; Wiener and Rogers, in press), toxic substances, climate change (Sullivan and Jordan, 1997; Wiener, 1999), guns (United Nations, 1998), and antitrust/competition policy (Richter, 2000; Raghavan and Mitchener, 2000).

By contrast, the United States appears to have been more precautionary than Europe about several types of risks, including the licensing of new drugs (e.g., thalidomide, which was licensed in Europe but not in the United States); lead in gasoline (which the United States phased out much earlier and more quickly than did Europe) [compare EPA (1985) with United Kingdom (UK) Department of the Environment (1991)]; depletion of the stratospheric ozone layer (the United States began with a ban on CFCs in spray cans in 1978, almost a decade before Europe acted) (Litfin, 1994:64–67); highway safety (the United States has more stringent speed limits on major highways) (Ibiblio, 2000); nuclear energy (the United States has regulated nuclear power plants more tightly than have France and Germany); bovine spongiform encephalo-pathy (BSE) or "mad cow disease" (the United States banned imports of affected beef years earlier than did Europe, and maintains that ban while Europe has relaxed its restrictions) (Lyall, 2000; European Commission, 1996; U.S. De-partment of Agriculture (USDA) 1991, 2000; UK, 2000; Wiener and Rogers, in press); BSE in blood donations (the United States has banned blood donors who have spent time in Europe) (Hernandez, 2001; Tagliabue, 2001; Wiener and Rogers, in press) choking hazards embedded in food (Harris, 1997); "right to know" requirements [the United States has a suite of provisions such as the Freedom of Information Act (FOIA), Toxics Release Inventory (TRI), Clean Air Act sec. 112(r), Calif. Prop 65, and the OSHA Hazard Communication Standard, while Europeans "can only dream" of this kind of precautionary measure (Sand, 2000:452)]; and legal restraints on dangerous persons such as violent youths (*Stanford v. Kentucky*, 1989), mental health patients (Talbot, 2000), and recovering sex offenders, whose location is publicized in American towns but not in Europe (Hood and Baldwin, 2000).

This is not to say that European or U.S. policy is "better" on any of these examples. As emphasized above, the more precautionary approach is not always the superior approach, or even the more protective approach (given countervailing risks). Rather, this enumeration of conflicting examples only illustrates that neither the European Union nor the United States has a claim to being more precautionary across the board. Relative precaution is context-specific. Thus, this enumeration of transatlantic examples adds further support to the hypothesis that in real-world applications, the precautionary principle is moderated by pragmatic considerations. It does not, however, prove that these real-world applications are converging toward optimal precaution. Nor does it explain what social, political, cultural, economic, or other factors account for the complex observed pattern (see Wiener and Rogers, in press).

32.6 CONCLUSIONS

Some forms of the precautionary principle (PP) could be helpful to policy-makers, who must inescapably act in the face of uncertainty about risk. The PP reminds us that there are real costs of inadequate precaution: false negatives and latent harms that may be left unaddressed or may be exacerbated as regulators watch for signs of greater certainty. But most formulations of the PP seem to have been designed as if regulators addressed one risk at a time, and the only question was how certain they are about this risk. In the real world, regulators are uncertain about all risks, and the real questions are how to deal with multiple risks and alternative actions. In the real world of multiple risks, the PP must be qualified by the recognition that there are real harms from excessive precaution: false positives, cost, inhibited innovation, and the countervailing risks of regulatory interventions.

Of the three versions of the PP examined here, version 1 makes sense but is incomplete: It does not answer the serious question (namely: what action should we take in the face of inevitable uncertainty?). Versions 2 and 3 are flawed insofar as they neglect the countervailing risks of actions to combat target risks. Indeed, if precautionary action itself poses countervailing risks, then a strong version of the PP (such as version 3) swallows itself: It requires precaution about precaution. Hence the PP needs to be qualified to guide intelligent regulation. It must avoid overreacting to false positives, deal with multiple countervailing risks, minimize the sum of market risk and regulatory risk, seek risk-superior moves, and be dynamic and adaptive in the face of changing information.

In practice, the PP has indeed been qualified when actually implemented in binding law. There is no single formulation of the PP. Empirically, the more binding the legal instrument, in general, the less absolutist the version of the PP it contains, and the more it moderates the PP with pragmatic consideration of consequences and alternatives. Across countries and over time, application of the PP has therefore been highly context-specific, variegated, and even inconsistent. This reality sharply undercuts the case for enshrining "the" precautionary principle as a norm of customary international law. If there has been any convergence toward some consensus state practice (which is doubtful), the trend has been toward pragmatic consequentialism, not ideological precaution.

The challenge, then, is to seek risk-superior ways of reducing multiple risks in concert. As with medical care and our daily lives, the goal should be not maximum precaution but an "optimal precaution" that addresses both the risks of inaction and the risks of action.

ACKNOWLEDGMENTS

The author thanks Gail Charnley, David Clarke, Carl Cranor, Don Elliott, John Graham, Ragnar Lofstedt, Sidney Shapiro, Christopher Stone, and participants in workshops at the Toxicology Forum, the AAAS, the SRA, and Cornell Law School for helpful discussion, and Amy Horner for helpful research assistance. I owe special gratitude to Michael Rogers, whose insights and collaboration (Wiener and Rogers, in press) have contributed enormously to this chapter.

REFERENCES

Applegate, J. S. 2000. The precautionary preference: An American perspective on the precautionary principle. *Hum. Ecol. Risk Assess.* 6: 413.

Bergen Declaration. 1990. Bergen ministerial declaration on sustainable development in the ECE Region. UN Doc. A/CONF.151/PC/10. *Yearbook Int. Environ. Law* 1: 429.

Bernstein, P. L. 1996. *Against the Gods: The Remarkable Story of Risk*. Wiley & Sons, New York.

Bodansky, D. 1991. Scientific uncertainty and the precautionary principle. *Environment* September, page 4.

Bodansky, D. 1993. The United Nations Framework Convention on Climate Change: A commentary. *Yale J. Int. Law* 18: 451–504.

Boehmer-Christiansen, S. 1994. The precautionary principle in Germany—Enabling government. In T. O'Riordan and J. Cameron (Eds.), *Interpreting the Precautionary Principle*. Cameron May, London.

Calabresi, G. 1970. *The Cost of Accidents*. Yale University Press, New Haven, CT.

Cameron, J., and Abouchar, J. 1991. The precautionary principle: A fundamental principle of law and policy for the protection of the global environment. *B.C. International & Comparative Law Rev.* 14: 1–27.

Clinton, President W. J. 1993. Executive Order 12866. Regulatory Review. September 30. *Federal Register* 58: 51735.

Cranor, C. F. 1999. Asymmetric information, the precautionary principle and burdens of proof. In C. Raffensperger and J. Tickner (Eds.), *Protecting Public Health and the Environment: Implementing the Precautionary Principle*. Island Press, Washington DC.

Cross, F. B. 1996. Paradoxical perils of the precautionary principle. *Wash. Lee Law Rev.* 53: 851.

Cross, F. B. 1997. The subtle vices behind environmental values. *Duke Environ. L. Policy Forum* 8: 151.

Daley, S. 2000. More and more, Europeans find fault with US: Wide range of events viewed as menacing. *N.Y. Times*, Apr. 9, p. A1.

deFur, P. L. 1999. The precautionary principle: Application to policies regarding endocrine-disrupting chemicals. In C. Raffensperger and J. Tickner (Eds.), *Protecting Public Health and the Environment: Implementing the Precautionary Principle*. Island Press, Washington DC.

Emiliou, N. 1996. *The Principle of Proportionality in European Law*. Kluwer Law International, Cambridge MA.

Ethyl Corp. v. EPA. 1976. 541 F.2d 1 (DC Cir.).

European Commission. 1996. Decision of 27 March 1996 on emergency measures to protect against bovine spongiform encephalopathy. Official Journal of the European Communities, JOL 1996/78-18EN, pp. 47–48.

European Commission. 2000. Communication from the Commission on the Precautionary Principle, COM(2000)1, Brussels, Feb. 2, 2000. Available from *http://europa. eu.int/comm/dgs/health_consumer/library/pub/pub07_en.pdf*.

European Council. 1990. Decision of Apr. 25, 1990, concerning the administration of bovine somatotrophin. 90/218/EEC, JOL 1990/116, p. 27 (and subsequent amendments).

European Council. 1999. Decision of Dec. 17, 1999, concerning the placing on the market and administration of bovine somatotrophin (and repealing Decision 90/218/EEC), 99/879/EEC, JOL.

European Union (EU) Treaty. 1993. Single European Act of 1987, as amended by the Maastricht Treaty on the European Union, 1993. *Int. Legal Materials* 31: 247.

Fischhoff, B., Lichtenstein, S., Slovic, P., Derby, S., and Keeney, R. 1981. *Acceptable Risk*. Cambridge University Press, Cambridge.

Graham, J. D., and Wiener, J. B. (Eds.), 1995. *Risk vs. Risk: Tradeoffs in Protecting Health and the Environment*. Harvard University Press, Cambridge, MA.

Harris, B. 1997. A Comparison of US and EU product safety regulations: A case study. *Risk: Health Safety Environ.* 8: 209.

Hernandez, R. 2001. Citing Med Low, FDA Panel Backs Blood Donor Curbs. NY Times, June 29, p. A23.

Hey, E. 1992. The precautionary concept in environmental policy and law: Institutionalizing caution. *Geo. Int. Environ. Law Rev.* 4: 303.

Hirschman, A. O. 1991. *The Rhetoric of Reaction*. Belknap, Harvard University Press, Cambridge, MA.

Hood, C., and Baldwin, R. 2000. Why is risk regulated in different ways? Beyond the risk society debate. *Risk Regul.* LSE-CARR on-line magazine. Available from *http://www.lse.ac.uk/Depts/carr/nl4.htm*, visited Sept. 25, 2000.

Ibiblio. 2000. Available from *http://www.ibiblio.org/rdu/sl-inter.html*, visited Oct. 5, 2000.

Industrial Union Dept., AFL-CIO v. American Petroleum Institute. 1980. The "Benzene" case. 448 U.S. 607.

Jackson, V. 1999. Ambivalent resistance and comparative constitutionalism: Opening up the conversation on "proportionality," rights and federalism. *Univ. Penna. J. Constitut. Law* 1: 583.

Janicke, M. 1990. *State Failure* (A. Braley, Tr.). Pennsylvania State Univ. Press, University Park PA.

Jasanoff, S. 1986. *Risk Management and Political Culture*. Russell Sage Foundation, New York.

Jasanoff, S. 1995. *Science at the Bar*. Harvard University Press, Cambridge, MA.

Jasanoff, S. 1998. Contingent knowledge: Implications for implementation and compliance. In E. Brown Weiss and H. Jacobson (Eds.), *Engaging Countries: Strengthening Compliance with International Environmental Accords* MIT Press, Cambridge MA.

Knight, F. 1933. *Risk, Uncertainty, and Profit.* Houghton Mifflin, Boston, New York.

Lindblom, C. 1959. The science of "muddling through." *Public Admin. Rev.* 19: 79.

Litfin, K. T. 1994. *Ozone Discourses.* Columbia University Press, New York.

Lofstedt, R., and Vogel, D. 2001. The Changing Character of Regulation: A Comparison of Europe and the United States. *Risk Analysis* 21: 399–405.

Lyall, S. 2000. British wrongly lulled people on "Mad Co," report finds. *N.Y. Times,* Oct. 27, p. A8.

Lynch, D., and Vogel, D. 2000. Apples and oranges: Comparing the regulation of genetically modified food in Europe and the United States. Paper prepared for the American Political Science Association annual meeting, Aug. 31–Sept. 3, Washington DC.

Margolis, H. 1996. *Dealing with Risk.* University of Chicago Press, Chicago IL.

McNeil, D., Jr. 2000. Protests on new genes and seeds grow more passionate in Europe. *N.Y. Times,* Mar. 14, pp. A1, A10.

Montreal Protocol. 1987. Montreal protocol on substances that deplete the stratospheric ozone layer, September 16. *Int. Legal Materials* 26: 1550.

Moynihan, D. P. 1993. Iatrogenic government: Social policy and drug research. *Am. Scholar,* Summer. 351–362.

Muir, J. 1869. *My First Summer in the Sierra* (Sierra Club Books edition, San Francisco, CA, 1988).

Ottoboni, M. A. 1984. *The Dose Makes the Poison: A Plain-Language Guide to Toxicology.* Vincente Books, Berkeley.

Page, T. 1978. A generic view of toxic chemicals and similar risks. *Ecology Law Quarterly* 7: 207.

Pfeiffer, C. 1998. Juvenile crime and violence in europe. *Crime Justice* 23: 255.

Raghavan, A., and Mitchener, B. 2000. EU's antitrust czar isn't afraid to say no; just ask Time Warner. *Wall Street J.,* Oct. 2, p. A1.

Richter, S.-G. 2000. The U.S. consumer's friend, *N.Y. Times,* Sept. 21, p. A31.

Rimer, S., and Boner, R. 2000. Young and condemned: Whether to kill those who killed as youths. *N.Y. Times,* Aug. 22, p. A1.

Sand, P. H. 2000. The precautionary principle: A European perspective. *Hum. Ecol. Risk Assess.* 6: 445.

Sandin, P. 1999. Dimensions of the precautionary principle. *Hum. Ecol. Risk Assess.* 5: 889.

Sands, P. 1995. *Principles of International Environmental Law I.* Manchester University Press, New York.

Shrader-Frechette, K. 1991. *Risk and Rationality: Philosophical Foundations for Populist Reforms.* University of California Press, Berkeley.

Slovic, P. 1987. Perceptions of Risk. *Science* 236: 280.

Stanford v. Kentucky. 1989. 492 U.S. 361, 369.

Stirling, A. 1999. On science and precaution in the management of technological risk. Report prepared for the European Commission Forward Studies Unit, Volume I: *A Synthesis Report of Case Studies*. EUR 19056 EN. European Commission, Joint Research Centre. Ispra, Italy.

Sullivan, K., and Jordan, M. 1997. The challenge: Incorporating many nations' needs into one treaty. *Washington Post*, Nov. 15, p. A1.

Sunstein, C. R. 1991. Preferences and politics. *Philos. Public Affairs* 20: 3.

Sunstein, C. R. 1996. Health-health tradeoffs. In *Free Markets and Social Justice* 298.

Tagliabue, J. 2001. US Plan to Halt Blood Imports Worries Europe. *NY Times*, July 17, p. A1.

Talbot, M. 2000. What's become of the juvenile delinquent? *N.Y. Times Magazine*, Sept. 10. Cover story.

Tennessee Valley Authority v. Hill. 1978. 437 U.S. 153.

Tolley, G. 1994. State of the art health values. In D. Kenkel and R. Fabian (Eds.), *Valuing Health for Policy*. University of Chicago Press, Chicago, IL.

United Kingdom (UK), Department of the Environment (DOE). 1991. *Unleaded Petrol: The UK Position*. UKDOE. London.

United Kingdom (UK). 2000. The Phillips Inquiry. Oct. 26, 2000. Available from *http:// www.bseinquiry.gov.uk/report*.

United Nations. 1992a. Framework convention on climate change, May 29. *Int. Legal Materials* 31: 849.

United Nations. 1992b. Rio declaration on environment and development, June 14. *Int. Legal Materials* 31: 874.

United Nations. 1998. *International Study on Firearm Regulation*. United Nations, New York.

U.S. Department of Agriculture (USDA). 1991. Animal and plant health inspection service, *bovine spongiform encephalopathy*. *Fed. Reg.* 56: 63868.

U.S. Department of Agriculture (USDA). 2000. Animal and plant health inspection service, *bovine spongiform encephalopathy*. Available from *http://www.aphis.usda.gov/ oa/bse/*; Visited on September 27, 2000.

U.S. Department of State. 2000. Questions from the US on the Commission's Communication on the Precautionary Principle, March 2000 (copy on file with author).

U.S. Environmental Protection Agency (EPA). 1985. Phaseout of lead in gasoline. *Fed. Reg.* 50: 9386.

Vogel, D. 1995. *Trading Up: Consumer and Environmental Regulation in a Global Economy*. Harvard University Press, Cambridge.

Vogel, D. 1997. *Barriers or Benefits: Regulation in Transatlantic Trade*. Brookings Institution, Washington, DC.

Wagner, W. E. 2000. The precautionary principle and chemical regulation in the U.S. *Hum. Ecol. Risk Assess.* 6: 459.

Wiener, J. B. 1997. Risk in the republic. *Duke Environ. Law Policy Forum* 8: 1.

Wiener, J. B. 1998. Managing the iatrogenic risks of risk management. *Risk: Environ. Health Safety*. 9: 39–84.

Wiener, J. B. 1999. On the political economy of global environmental regulation. *Georgetown Law J.* 87: 749–775.

Wiener, J. B., and Rogers, M. D. In press. Comparing precaution in the US and Europe. *J. Risk Research* (forthcoming 2002). Available as Working Paper 2001-01. Duke Center for Environmental Solutions, at *http://www.env.duke.edu/solutions/ publications.html.*

Wingspread Statement on the Precautionary Principle. 1998. In C. Raffensperger and J. Tickner (Eds.), *Protecting Public Health and the Environment: Implementing the Precautionary Principle.* Island Press, Washington, DC. (1999).

Winterdyk, J. A. (Ed.). 1997. *Juvenile Justice Systems: International Perspectives.* Canadian Scholars' Press, Toronto.

Zimring, F. E. 1998. The youth violence epidemic: Myth or reality? *Wake Forest Law Rev.* 33: 727.

INDEX